Microwave RF Antennas ar

Ofer Aluf

Microwave RF Antennas and Circuits

Nonlinearity Applications in Engineering

Springer

Ofer Aluf
Netanya
Israel

ISBN 978-3-319-83291-3 ISBN 978-3-319-45427-6 (eBook)
DOI 10.1007/978-3-319-45427-6

Softcover reprint of the hardcover 1st edition 2016

Printed on acid-free paper

This Springer imprint is published by Springer Nature
The registered company is Springer International Publishing AG
The registered company address is: Gewerbestrasse 11, 6330 Cham, Switzerland

Preface

This book on microwave RF circuits: nonlinearity applications in engineering covers and deals with two separate engineering and scientific areas and what between. It gives advance analysis methods for Microwave RF Circuits which represent many applications in engineering. Microwave RF Circuits come in many topological structures and represent many specific implementations which stand the target engineering features. Microwave RF Circuits include RFID antenna systems, microwave elements, microwave semiconductor amplifiers, small-signal (SS) amplifiers and matching networks, power amplifiers, oscillators, filters, antennae systems, and high-power transistor circuit. The basic Microwave RF Circuits can be characterized by some models and the associated equations. The Microwave RF Circuits include RFID ICs and antennas, microstrip, circulators, cylindrical RF network antennas, tunnel diode (TD), bipolar transistors, field-effect transistors, IMPATT amplifiers, small-signal (SS) amplifiers, Bias-T circuits, PIN diode, power amplifiers, LNAs, oscillators, resonators, filters, N-turn antennae, dual spiral coils antennae, Helix antennas, linear dipole and slot array, and hybrid translinear circuit. The Microwave RF Circuits analyze as linear and nonlinear dynamical systems and their dynamics under parameter variations. This book is aimed at newcomers to linear and nonlinear dynamics and chaos Microwave RF Circuits. The presentation stresses analytical and numerical methods, concrete examples, and geometric intuition. The Microwave RF Circuits analysis is developed systematically, starting with first-order differential equations and their bifurcation, followed by phase plane analysis, limit cycles and their bifurcations, chaos, iterated maps, period doubling, renormalization, and strange attractors. Additionally, the book is dealt with delayed Microwave RF Circuits which characterized by overall variables delayed with time. Each variable has specific delay parameter and can be inspected for dynamics. More realistic Microwave RF Circuits models should include some of the past states of Microwave RF Circuits and systems; that is, ideally, a real Microwave RF Circuits should be modeled by differential equations with time delays. The use of delay differential equations (DDEs) in the modeling of Microwave RF Circuits dynamics is currently very active, largely due to progress achieved in the understanding of the dynamics of several classes of delayed differential equations and Microwave RF Circuits and systems. This book is designed for advanced undergraduate or graduate students in electronics, RF and electronic engineering, physics, and mathematics who are interested in Microwave RF Circuits dynamics and innovative analysis methods. It is also addressed to electrical and RF engineers, physics experts and researchers in physics, electronics, engineering and mathematics who use dynamical systems as modeling tools in their studies. Therefore, only a moderate mathematical and electronic semiconductor background in geometry, linear algebra, analysis, and differential equations is required. Each chapter includes various Microwave RF Circuits drawing and their equivalent analyses circuits. Microwave RF Circuits fixed points and stability analysis done by using much estimation. Various bifurcations of Microwave RF Circuits are discussed.

In this book, we try to provide the reader with explicit procedures for application of general Microwave RF Circuits mathematical representations to particular research problems. Special attention is given to numerical implementation of the developed techniques.

Let us briefly characterize the content of each chapter.

Chapter 1. **RFID Antenna Systems Descriptions and Analysis**. In this chapter, RFID antenna systems are described and analyzed. RFID is a dedicated short-range communication (DCRC) technology. RFID system consists of an antenna and a transceiver, which read the radio frequency, and transfers the information to a processing device (reader) and a transponder, or RFID tag. Active RFID tag system includes energy source (battery), and it consumes energy. The active RFID tag system is analyzed as an excitable linear bifurcation system. RFID tag-dimensional parameters are optimized to get the best performances. Under delayed electromagnetic interferences, there are delays in some RFID tag coil variables and we analyze it for stability optimization. There is a unique structure of RFID system, semi-passive RFID tags with double-loop antennae arranged as a shifted gate. The structure is optimized under delayed electromagnetic interferences. RFID tag detector circuit is implemented by using schottky diode, and stability is analyzed for parameter values variation. RFID system burst switch is a very important element, and its behavior in time is inspected. The analysis fills the gap of analytical methods for RFID systems analysis, concrete examples, and geometric examples. One of the crucial RFID system optimization is in electromagnetic environmental which faced RFID system variables delay in time. In some cases, RFID system can be represented as delayed differential equations, which depends on variable parameters and delays. There are practical guidelines that combine graphical information with analytical work to effectively study the local stability of RFID system models involving delay-dependent parameters.

Chapter 2. **Microwave Element Description and Stability Analysis**. In this chapter, microwave element stability is discussed. There are three types of microwave circuits which include microwave elements. The first is a discrete circuit, packaged diodes/transistors mounted in coax and waveguide assemblies. The second is Hybrid MIC (microwave integrated circuit), diodes/transistors and microstrip fabricated separately and then assembled. The third is MMIC (monolithic microwave integrated circuit), diodes, transistors, and microstrip circuits, and other circuit elements, such as lumped capacitors and resistors, which have parasitic effects influenced on overall system stability behavior. Microwave transmission lines are delayed in time and are integral part of power limiter; the stability is inspected for optimization. Reflection-type phase shifter (RTPS) employs a circulator. The RTPS circuit includes microstrip transmission lines with three-port active circulator and analyzes for stability optimization under time delayed. Cylindrical RF network antennas for coupled plasma sources include copper legs. They run as large-volume plasma sources and have stability switching due to system's copper leg parasitic effects. Tunnel diode (TD) is the p-n junction device that exhibits negative resistance. Tunnel diode (TD) can be a microwave oscillator. Transient is in the resonant cavity after turning the bias voltage ON. The resonant circuit with NDR can oscillate. The Tunnel diode (TD) microwave oscillator has parasitic effects in time and delay variables. The stability is optimized when implementing tunnel diode (TD) in microwave oscillator.

Chapter 3. **Microwave Semiconductor Amplifiers Analysis**. In this chapter, microwave semiconductor amplifier circuit analysis is discussed. Microwave semiconductor amplifiers are widely used, and stability analysis is needed. Microwave semiconductors can be bipolar transistors which operate at microwave frequencies, and microwave field-effect transistors (FETs) minimize the adverse effects of transit time and internal capacitance and resistance, IMPATT (impact-ionization avalanche transit time) amplifier which widely used at the high end of the microwave band. Stability of these microwave amplifiers is affected by internal parameter variation and circuit microstrip parasitic effects. IMPATT diodes which are a form of high-power diode are used in high-frequency electronic and microwave devices. FET-combined biasing and matching circuit has many stability issues which must be taken for every RF design, and analysis is done for best performances.

Chapter 4. **Small Signal (SS) Amplifiers and Matching Network Stability Analysis**. In this chapter, small-signal (SS) amplifiers and matching network structures are analyzed for best performances. There are some types of amplifiers. Amplifiers types are zero-frequency amplifiers (DC amplifiers), low-frequency amplifiers (audio amplifiers), and high-frequency amplifiers (RF amplifiers). Amplifiers come in three basic flavors: common base (CB) amplifiers, common collector (CC) amplifiers, and common emitter (CE) amplifiers. It depends whether the base, collector, or emitter is common to both the input and output of the amplifier. When an amplifier's output impedance matches the load impedance, maximum power is transferred to the load and all reflections are eliminated. When an amplifier's output impedance unmatched the load impedance, there are reflections and less than maximum power is transferred to the load. There are instability behaviors in these three types of amplifiers caused by circuit microstrip delays in time parasitic effects. We use RF matching network in our design. There are typical amplifiers matching networks: L matching network, T matching network, and PI matching network. In design of microwave matching network, device parasitic effects of length on RF circuit matching and stability. Bias-T three-port network also suffers from instability under delayed microstrip in time. A PIN diode is suitable for many applications and operates under high level of injection. The PIN diode suffers from instability under parameter variations.

Chapter 5. **Power Amplifier (PA) System Stability Analysis**. In this chapter, power amplifiers (PAs) are analyzed for best performances, and stability was also discussed. Large-signal or power amplifiers (PAs) are used in the output stages of audio amplifier systems to derive a load speaker. There are different types of amplifiers which classified according to their circuit configurations and method of operation. The classification of amplifiers ranged from linear operation with very low efficiency to nonlinear operation but with a much higher efficiency, while others are a compromise between the two. There are two basic amplifier class groups. The first are the classically controlled conduction angle amplifiers forming the more common amplifier classes (A, B, AB, and C). The second set of amplifiers are the newer so-called switching amplifier classes (D, E, F, G, S, T). The most commonly structured amplifier classes are those that are the most common type of amplifier class mainly due to their simple design. We analyze the stability of these amplifiers by inspecting the equivalent circuit differential equations. BJT transistor is replaced by large-signal model in our analysis. The BJT model is known as the Gummel–Poon model. The Ebers–Moll BJT model is a good large signal. We use nonlinear dynamic in our analysis for amplifiers that feed by inputs/outputs exceed certain limits. LNAs are used in many microwave and RF applications. We analyze the stability of wideband low-noise amplifier (LNA) with negative feedback under circuit's parameter variation.

Chapter 6. **Microwave/RF Oscillator Systems Stability Analysis**. In this chapter, our oscillator systems are discussed and their stability behavior is analyzed. Oscillators can be classified into two types: relaxation and harmonic oscillators. A microwave oscillator is an active device to generate power and a resonator to control the frequency of the microwave signal. Important issues in oscillators are frequency stability, frequency tuning, and phase noise. A phase-shift oscillator is a linear electronic oscillator circuit that produces a sine wave output. The feedback network "shifts" the phase of the amplifier output by 180° at the oscillation frequency to give positive feedback, total phase shift of 360°. Phase-shift resonator circuit stability analysis is done by considering BJT small-signal (SS) equivalent circuit model. Closed-loop functioning oscillator can be viewed as feedback system. The oscillation is sustained by feeding back a fraction of the output signal, using an amplifier to gain the signal, and then injecting the energy back into the tank. Closed-loop functioning oscillator stability is inspected and analyze. There are types of transistor oscillators which use feedback and lumped inductance and capacitance resonators. There are three types of transistor LC oscillators, Colpitts, Hartley, and Clapp. In the Hartley oscillator, the feedback is supplied by the inductive divider formed by two inductors. We apply the stability criterion of Liapunov to our system. Colpitts oscillator is the same as Hertley oscillator and instead of using a tapped inductance, Colpitts oscillator uses a tapped capacitance. Colpitts oscillator circuit stability analysis is done by criterion of Liapunov.

Chapter 7. **Filter Systems Stability Analysis**. In this chapter, filter systems in many circuits are inspected for dynamical behavior and stability analysis. The target of analog and RF filtering is to modify the magnitude and phase of signal frequency components. Many analog or radio frequency (RF) circuits perform filtering on the signals passing through them. The analog and RF filter types are defined on the criteria how they modify the magnitude and/or phase of sinusoidal frequency components. Microwave and RF filters pass a range of frequencies and reject other frequencies. A diplexer is a passive device that implements frequency-domain multiplexing. Two ports are multiplexed onto a third port. A diplexer multiplexes two ports onto one port, but more than two parts may be multiplexed. We analyze BPF diplexer circuit stability by using geometric stability switch criteria in delay differential systems. A diplexer filters to pass two bands to separate ports, and stability analysis under parameter variation. The standard local stability analysis about any one of the equilibrium points of dual-band diplexer filter circuit is done. We use crystal in place of LC filter for low-frequency applications. There are lattice crystal filter, half lattice, and cascaded half lattice filters. The standard local stability analysis about any one of the equilibrium point of lattice crystal filter circuit is done. A tunable BPF employing varactor diodes is ideal for many diverse wireless applications. There are two types of tunable BPF employing varactor diodes: top inductively coupled variable BPF and capacitively coupled variable band-pass filter. BPF (varactor diodes) circuit involving N variables and stability behavior is inspected.

Chapter 8. **Antenna System Stability Analysis**. In this chapter, we discussed various antenna systems and behaviors for different conditions for best performances. An antenna is a conductor or group of conductors used for radiating electromagnetic energy into space or collecting electromagnetic energy from space. There are many types of antennas and we discussed those antennas that operate at microwave frequencies. Microwave refer to radio waves with wavelength ranging from as long as one meter to as short as one millimeter with frequencies between 300 MHz and 300 GHz. Another antenna area is for RFID applications. A complete RFID system includes RFID reader and transponder units. N-turn multilayer circular-coil antennas can be integrated with RFID IC for complete RFID tags. We investigate the system stability optimization under delayed electromagnetic interference and parasitic effects. The system is constructed from two antennas: each one N-turn multilayer circular antenna. The standard local stability analysis about any one of the equilibrium points (fixed points) of N-turn multilayer circular-coil antenna RFID system is done. We analyze circuit stability where there is a delay in the first and second RFIDs' N-turn multilayer-coil antenna voltages and antenna voltage derivatives. A double-rectangular spiral antenna is constructed from two antennas, each antenna is a rectangular spiral antenna. Antennas are connected in series with microstrip line and to the RFID IC. The standard local stability analysis about any one of the equilibrium points of RFID tags with double rectangular spiral antenna system is done. A system of single-turn square planar straight thin-film inductor antenna (four segments) is constructed from four straight thin-film inductors which are connected in a single-turn square structure. There are delays in time for the microstrip line parasitic effects, and stability switching is inspected for different values of delay variables. A helical antenna is an antenna consisting of a conducting wire wound in the form of a helix. The helical antennas are mounted over a ground plane. Helical antennas can operate in one of two principal modes: normal mode or axial mode. Helix antenna system stability is inspected under parameter variation.

Chapter 9. **Microwave RF Antennas and Circuits Bifurcation Behavior, Investigation, Comparison and Conclusion**. In this chapter, we summarized the main topics regarding microwave and RF antennas and systems, inspect behavior, dynamics, stability, comparison, and conclusion. Microwave RF antennas are an integral part of every RF or microwave system. An antenna is an electrical device which converts electric power into radio waves, and vice versa. In many wireless applications, antennas are required by radio receiver or transmitter to couple its electrical connection to the electromagnetic field. When we inspect system stability which includes radio waves, we inspect electromagnetic waves which carry signals through the space (or air) at the speed of light with almost no transmission loss. There are mainly two categories of antennas. The first is omnidirectional antenna which receives and/or radiates in all directions. The second is directional antenna which radiates in a particular direction or pattern. Antennas are characterized by a number of parameters, radiation pattern, and the resulting gain. Antenna's gain is dependent on its power in the horizontal directions, and antenna's power gain takes into account the antenna's efficiency (figure of merit). The physical size of an antenna is a practical issue, particularly at lower frequencies. Stability analysis includes a complete RF system with antennas and matching networks.

Netanya, Israel Ofer Aluf

Contents

Introduction

Microwave RF antenna products are currently in a widely use in all aspects of engineering designs. Microwave RF antenna products are transmission lines, coaxial cables, waveguide, strip line and microstrip, microwave semiconductors (PIN diode, RF bipolar transistor, RF FET, varactor, schottky diode, LDMOS, DMOS, GaN devices, etc.), RF combiner and couplers, isolators and circulator, filters, attenuators, switches, phase shifter, detectors, amplifiers, oscillators, tubes, microwave antennas (dipole, slot, horn, spiral, helix, arrays, parabolic dish, phased arrays), low-power communication antennas (ZigBee, RFID/NFC, Bluetooth, Wi-Fi, GPS, etc.). The basic structure of Microwave RF antenna product contains de/multiplexed amplifiers, filters, mixers, etc. The Microwave RF antenna system ports are RF inputs, RF outputs, oscillators, and input control lines). The below figure demonstrates the basic structure of Microwave RF antenna system:

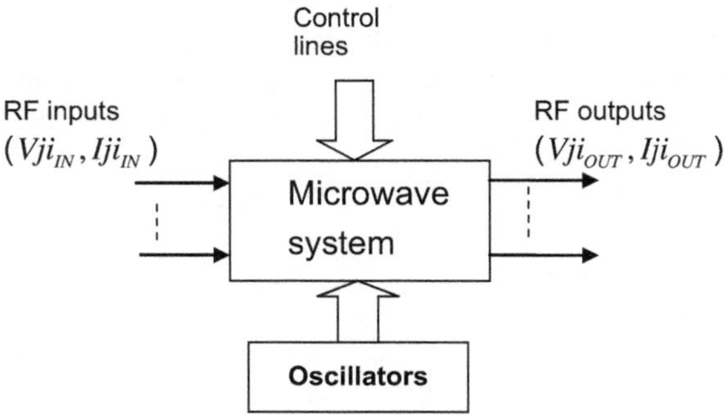

The input control lines can be connected to additional embedded system by many kinds of interfaces (RS232, RS485, UART, SPI, SDIO, etc.). RF inputs can be connected to various antennas and other RF systems. RF outputs can be connected to additional RF systems and devices. The overall Microwave RF antenna system equation can be represented as below:

$$\{Vji_{OUT}, Iji_{OUT}\} = f(Vji, Iji) = \sum_{n}^{i=1} \prod_{i}^{j=1} k_{ji}(Vji_{IN}, Iji_{IN}, \frac{d^n Vji_{IN}}{dt^n}, \frac{d^n Iji_{IN}}{dt^n})$$

The Microwave RF systems can be represented as RF input circuit and RF output circuit. The RF system function contains high-degree derivatives of input and output voltages.

The primary purpose of a Microwave and RF system application functionality on input variables. Many topological Microwave/RF system schematics give a variety of nonlinear behaviors which can be implemented in many engineering areas. Each Microwave/RF system can be represented by a set of differential equations which depend on Microwave/RF system's variable parameters. The investigation of Microwave and RF system's differential equation bifurcation theory, the study of possible changes in the structure of the orbits of a differential equation depending on variable parameters. The book illustrates certain observations and analyzes local bifurcations of an appropriate arbitrary scalar differential equation. Since the implicit function theorem is the main ingredient used in these generalizations, include a precise statement of this theorem. Additional analyze the bifurcations of a Microwave/RF system's differential equation of the circle. The bifurcation behavior of specific differential equations can be encapsulated in certain pictures called bifurcation diagrams. All of that for optimization of Microwave/RF system's parameter optimization—to get the best performance. Dynamics (Chaos, fractals) change with systems that evolute in time. There are two types of dynamical systems: differential equations and iterated map (difference equations). Differential equation has described the evolution of systems in continuous time. Iterated map is arising in problems where the time is discrete. Differential equations can be divided into two main groups: ordinary differential equations and partial differential equations. The differential equation system can be represented as below:

$$\frac{dX_1}{dt} = f_1(x_1, \ldots\ldots, x_n)$$

$$\vdots$$

$$\frac{dX_n}{dt} = f_n(x_1, \ldots\ldots, x_n)$$

$$\dot{X}_i = \frac{dX_i}{dt}$$

Some of the Microwave/RF systems can be represented as an equation in dimension "one." Basic notions of Microwave/RF systems, circuit stability, and bifurcations of vector fields are easily explained for scalar autonomous equations dimension one—because their flows are determined from the equilibrium points. Numerical solutions of such equations lead to scalar maps and show some of the "anomalies" albeit profound and exciting that may arise when numerical approximation is a poor period doubling bifurcation, chaos, etc.

Microwave/RF systems equations can be turned to the dynamics and bifurcations of periodic solutions of no autonomous equations with periodic coefficients' dimension one and one half, where scalar maps reappear naturally as Poincare maps. Microwave/RF system investigates the dynamics of planar autonomous equations—dimension two—where, in addition to equilibria, new dynamical behavior, such as periodic and homoclinic orbits, appears. Microwave/RF system schematic stability of an equilibrium point, subtle topological aspects of linear systems as well as the standard theory of Liapunov functions. Center manifolds and the method of Liapunov–Schmidt to make a reduction to a scalar autonomous equation. Periodic orbit—Poincare—Andronov—Hopf bifurcation—and its analysis can be reduced to that of a nonautonomous periodic equation.

Additionally, we discussed Microwave RF antenna systems with delay elements (parasitic effects, circuit component delays, microstrip delays, etc.). Our Microwave RF antenna system delay differential and delay different model can be analytically used with delay differential equations in dynamically. The need of the incorporation of a time delay is often of the existence of any stage structure. It is often difficult to analytically study models with delay-dependent parameters, even if only a single discrete delay is present. There are practical guidelines that combine graphical information with analytical work to effectively study the local stability of models involving delay-dependent parameters. The stability of a given steady state is simply determined by the graphs of some function of $\tau_1, \ldots, \tau_n$; $n \in \mathbb{N}$ which can be expressed explicitly and thus can be easily depicted by MATLAB and other popular software. We need only look at one such function and locate the zero. This function often has only two zeros, providing thresholds for stability switches. As time delay increases, stability changes from stable to unstable to stable. We emphasize the local stability aspects of some models with delay-dependent parameters. Additionally, there is a general geometric criterion that, theoretically speaking, can be applied to models with many delays, or even distributed delays. The simplest case of a first-order characteristic equation provides more user-friendly geometric and analytic criteria for stability switches. The analytical criteria provided for the first- and second-order cases can be used to obtain some insightful analytical statements and can be helpful for conducting simulations.

Chapter 1
RFID Antennas Systems Descriptions and Analysis

RFID is short for radio frequency identification; RFID is a dedicated short range communication (DSRC) technology. The term RFID is used to describe various technologies that use radio waves to automatically identify people or objects. With RFID, the electromagnetic or electrostatic coupling in the RF (radio frequency) portion of the electromagnetic spectrum is used to transmit signals. RFID system consists of an antenna and a transceiver, which read the radio frequency and transfers the information to a processing device (reader) and a transponder, or RF tag, which contains the RF circuitry and information to be transmitted. The antenna provides the means for the integrated circuit to transmit its information to the reader that converts the radio waves reflected back from the RFID tag into digital information that can then be passed on to computers that can analyze the data. In RFID systems, the tags that hold the data are broken down into two different types. Passive tags use the radio frequency from the reader to transmit their signal and Active tags. Passive tags use the radio frequency from the reader to transmit their signal. Passive tags will generally have their data permanently burned into the tag when it is made, although some can be rewritten. Active tags are much more sophisticated and have an on-board battery for power to transmit their data signal over a greater distance and power random access memory (RAM) giving them the ability to store up to 32,000 bytes of data. RFID systems can use a variety of frequencies to communicate, but because radio waves work and act differently at different frequencies, a frequency for a specific RFID system is often dependent on its application. An RFID system is always made up of two components: transponder, which is located on the object to be identified, detector or reader, which, depending upon design and the technology used, may be a read or write/read device. There is a need to analyzing RFID systems. The analysis is based on nonlinear dynamics and chaos models and shows comprehensive benefits and results. The dynamics of RFID systems provides several ways to use them in a variety of applications covering wide areas. The analysis fills the gap of analytical methods for RFID systems analysis, concrete examples, and geometric examples. The RFID systems analysis is developed systematically, starting with basic passive

O. Aluf, *Microwave RF Antennas and Circuits*,
DOI 10.1007/978-3-319-45427-6_1

and active RFID systems, differential equations and their bifurcations, followed by fixed point analysis, limit cycles and their bifurcations. One of the crucial RFID system optimization is in electromagnetic environmental which faced RFID system variable delay in time. In some cases RFID system can represent as a delayed differential equations which, depending on variable parameters and delays. There are practical guidelines that combine graphical information with analytical work to effectively study the local stability of models involving delay dependent parameters. The stability of a given steady state is determined by the graphs of some function [2–4, 85].

1.1 Active RFID TAGs System Analysis of Energy Consumption as Excitable Linear Bifurcation System

Active RFID Tags have a built in power supply, such as a battery, as well as electronics that perform specialized tasks. By contrast, passive RFID TAGs do not have a power supply and must rely on the power emitted by an RFID Reader to transmit data. Thus, if a reader is not present, the passive TAGs can't communicate a data. Active TAGs can communicate in the absence of a reader. Active RFID Tags system energy consumption can be a function of many variables: q(m), u(m), z (m), t(m), tms(m), when m is the number of TAG IDs which are uniformly distributed in the interval [0,1). It is very important to emphasize that basic Active RFID TAG, equivalent circuit is Capacitor (Cic), Resistor (Ric), L (RFID's Coil inductance as a function of overall Coil's parameters) all in parallel and Voltage generator Vs(t) with serial parasitic resistance. The Voltage generator and serial parasitic resistance are in parallel to all other Active RFID TAG's elements (Cic, Ric, and L (Coil inductance)). The Active RFID TAG equivalent circuit can be represented as a differential equation which depending on variable parameters. The investigation of Active RFID's differential equation based on bifurcation theory, the study of possible changes in the structure of the orbits of a differential equation depending on variable parameters. We first illustrate certain observations and analyze local bifurcations of an appropriate arbitrary scalar differential equation. Finally, investigate Active RFID TAGs system energy for the best performance using an excitable bifurcation diagram. Active RFID Tags have a built in power supply, such as a battery. The major advantages of an active RFID Tags are: It can be read at distances of one hundred feet or more, greatly improving the utility of the device. It may have other sensors that can use electricity for power. The disadvantages of an active RFID Tags are: The TAG cannot function without battery power, which limits the lifetime of the TAG. The TAG is typically more expensive. The TAG is physically larger, which may limit applications. The long term maintenance costs for an active RFID tag can be greater than those of a passive Tag if the batteries are replaced. Battery outages in an active TAGs can result in expensive misreads. Active RFID TAGs may have all or some of the following features: Longest communication range of any TAG, the capability to perform an independent monitoring and control, the capability of initiating communications, the capabilities of performing diagnostics, and the highest data bandwidth. The active RFID TAGs may even be equipped with autonomous networking; the TAGs autonomously determine the best communication path. Mainly active RFID TAGs have a built in power supply, such as battery, as well as electronics that perform specialized tasks. By contrast, passive RFID TAGs do not have a power supply and must rely on the power emitted by an RFID Reader to transmit data. There is an arbitration while reading TAGs (TAGs anti-collision problem). First, identify and then read data stored on RFID Tags [85] (Fig. 1.1).

It is very important to read TAG IDs of all. The Anti-collision protocol based on two methods: ALOHA and its variants and Binary tree search. ALOHA protocol, reducing collisions by separating TAG responds by time (probabilistic and simple). TAG ID may not read for a very long time. The Binary tree search protocol is deterministic in nature. Read all TAGs by successively querying nodes at different levels of the tree with TAG IDs distributed on the tree based on their prefix. Guarantee that all TAGs IDs will be read within a certain time frame. The binary tree search procedure, however, uses up a lot of reader queries and TAG responses by relying on colliding responses of TAGs to determine which sub tree to query next. Higher energy consumption in readers and tags (If they are active TAGs). TAGs can't be assumed to be able to communicate with each other directly. TAGs may not be able of storing states of the arbitration process in their memory. There are three anti-collision protocols: All's include and combine the ideas of a binary tree search protocol with frame slotted ALOHA, deterministic schemes, and energy aware. The first anti-collision protocol is a Multi Slotted (MS) scheme, multiple slots per query to reduce the chances of collision among the TAG responses. The second anti-collision protocol is a Multi Slotted with Selective sleep (MSS) scheme; using sleep commands to put resolved TAGs to sleep during the arbitration process. Both MS and MSS have a probabilistic flavor, TAGs choose a reply slot in a query frame randomly. The third anti-collision protocol is a Multi Slotted with Assigned slots (MAS), assigning tags in each sub tree of the search tree to a specific slot of the query frame. It's a deterministic protocol, including the replay behavior of tags. All three protocols can adjust the frame size used per query. Maximize energy savings at the reader by reducing collisions among TAG responses. The frame size is also chosen based on a specified average time constraint within which all TAGs IDs must be read. The binary search protocols are Binary Tree (BT) and Query Tree

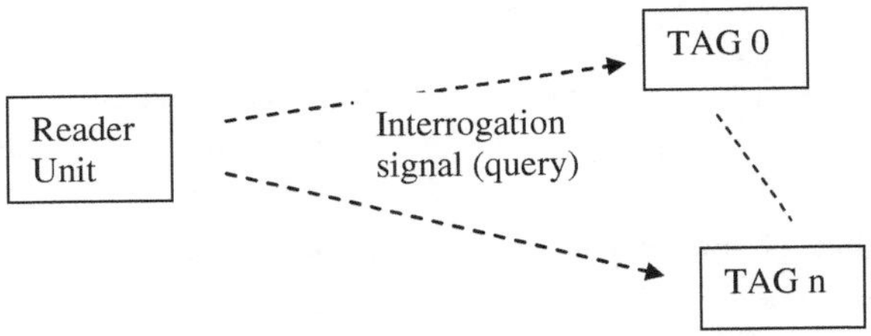

Fig. 1.1 Reader TAG interrogation diagram

(QT). Both work by splitting TAG IDs using queries from the reader until all tags are read. Binary Tree (BT) relies on TAGs remembering the results of previous inquiries by the readers. TAGs susceptible to their power supply. Query Tree (QT) protocol, is a deterministic TAG anti-collision protocol, which is memory loss with TAGs requiring no additional memory except that required to store their ID (Fig. 1.2).

The approach to energy aware anti-collision protocols for RFID systems is to combine the deterministic nature of binary search algorithms along with the simplicity of frame slotted ALOHA to reduce the number of TAG response collisions. The QT protocol relies on colliding responses to queries that are sent to internal nodes of a tree to determine the location of TAG ID. Allow tags to transmit responses within a slotted time frame and thus, try to avoid collisions with responses from other tags. The energy consumption at the reader is a function of the number of queries it sends, and number of slots spent in the receive mode. Energy consumption at an active TAG is a function of the number of queries received by the TAG and the number of responses it sends back. Neglect the energy spent in modes other than transmit and receive for simplicity. Assumption: Time slot in which a reader query or message is sent is equal to the duration as that of a TAG response. The energy model of the reader is based upon a half-duplex operation. Reader transmits energy, and its query for a specific period and then wait in receiving mode with no more energy transmission until the end of the frame. The flow chart for reader query and TAGs: (Fig. 1.3).

Response mechanism is as below: (Fig. 1.4).

Pulse based half duplex operation is termed as sequential (SEQ) operational (Fig. 1.5).

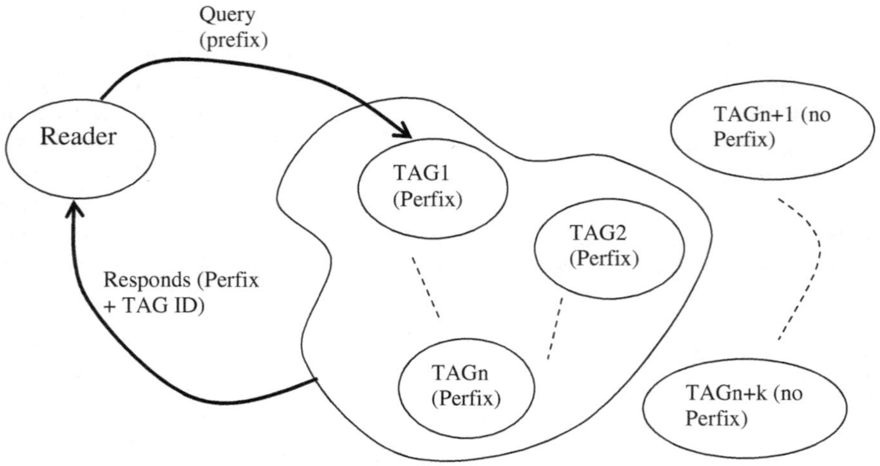

Fig. 1.2 Reader TAGs system query and responds

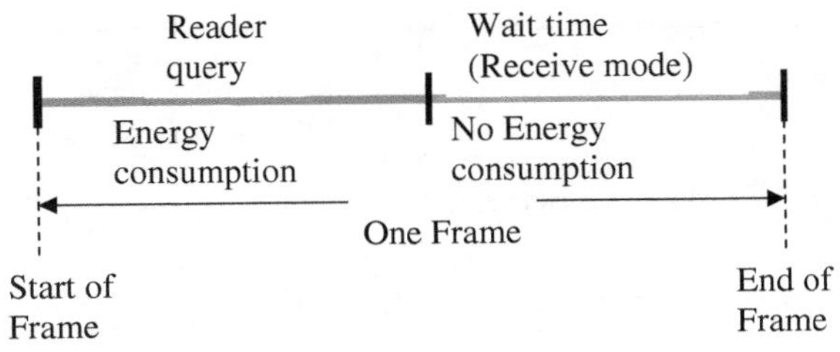

Fig. 1.3 One frame reader query and wait time

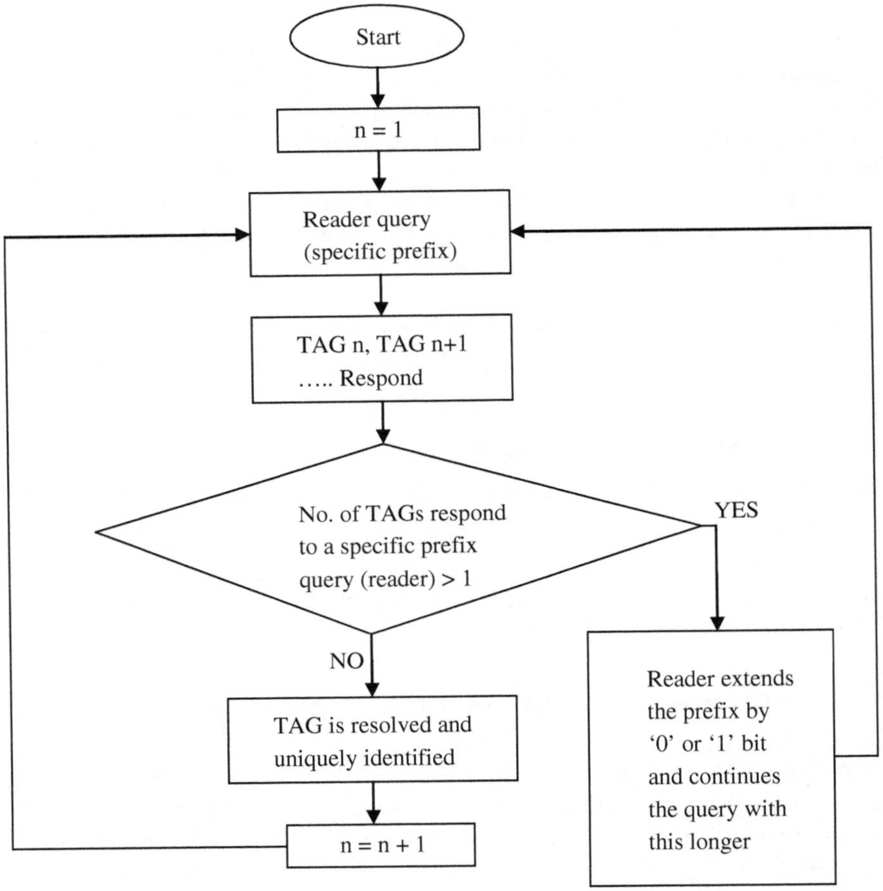

Fig. 1.4 Flow chart for reader query and TAGs

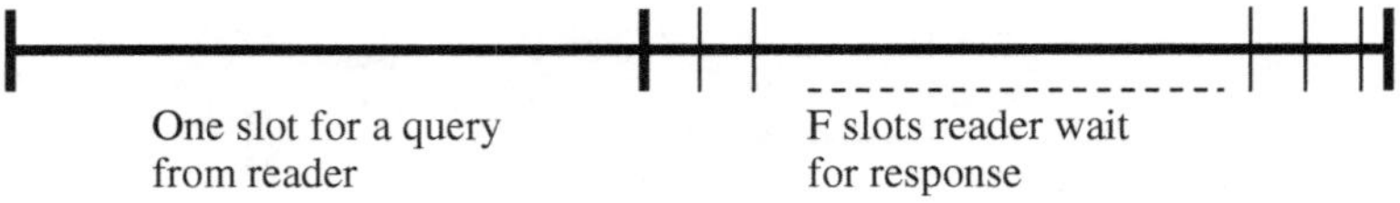

Fig. 1.5 One slot for a query and F slots reader wait for a response

The power required by the reader to transmit	The power required by the reader to receive
PRtx	PRrx

The power required by an active TAG to transmit	The power required by an active TAG to receive
PTtx	PTrx

Reader energy consumption: $q(m) \cdot (PRtx + PRrx \cdot F)$ when q(m) is the number of queries for reading m TAGs. The energy consumption of all active TAGs: $q(m) \cdot PTrx + u(m) \cdot PTtx$ when q(m) is the number of reader queries, u(m) is the number of TAG responses. For MSS scheme (include sleep command) the reader energy consumption is $q(m) \cdot (PRtx + PRrx \cdot F) + z(m) \cdot PRtx$. The total energy consumption for all active TAGs is $q(m) \cdot PTrx + u(m) \cdot PTtx + z(m) \cdot PTrx$, when z(m) is the number of sleep commands issued by the reader. The average analysis of energy consumption:

$$\bar{q}(m) - \textit{average number of reader queires.}$$

$$\bar{u}(m) - \textit{average number of TAG responses.}$$

$$\bar{z}(m) - \textit{average number of sleep commands}$$
$$\textit{issued by the reader (only for MSS Scheme)}$$

$$\bar{t}(m) - \textit{average number of time slots required to read all TAGs.}$$

$$\bar{t}_{MS}(m) - \textit{average number of time slots required to read m TAGs}$$

m TAG IDs are uniformly distributed in the interval [0.1]. We get the expression for one active RFID TAG total energy consumption U(m) = u(m):

$$\text{TAG Power} = \frac{1}{m} \cdot [q(m) \cdot PTrx + U(m) \cdot PTtx + Z(m) \cdot PTrx]$$

Active RFID TAG can represent as a parallel Equivalent Circuit of Capacitor and Resistor in parallel with Supply voltage source (internal resistance) (Fig. 1.6).

The Active RFID TAG Antenna can be represented as Parallel inductor to the basic Active RFID Equivalent Circuit. The simplified complete equivalent circuit of the label is as below: (Fig. 1.7)

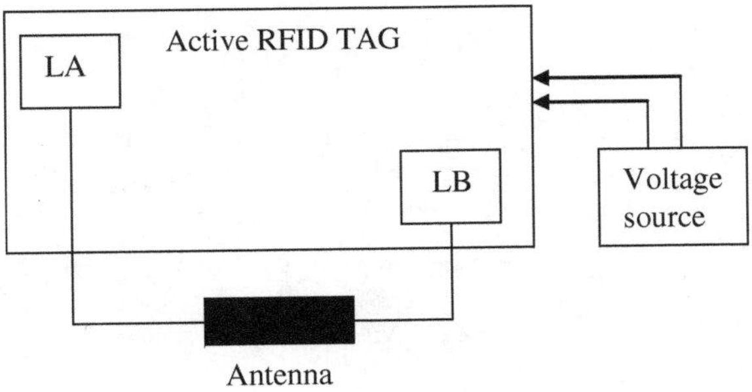

Fig. 1.6 Active RFID TAG system

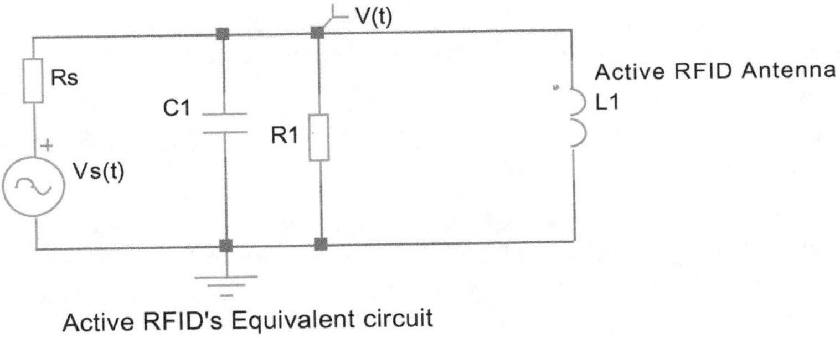

Fig. 1.7 Active RFID TAG's equivalent circuit

$$V_{L_1} = L_1 \cdot \frac{dI}{dt} \Rightarrow I_{L_1} = \frac{1}{L_1} \cdot \int_0^{t_1} V_{L_1} \cdot dt,$$

$$I_{C_1} = C_1 \cdot \frac{dVc_1}{dt}, \sum_{j=1}^{4} I_j = 0$$

$$V = Vc_1 = V_{L_1} = V_{R_1}$$

$$\frac{V}{R1} + C_1 \cdot \frac{dV}{dt} + \frac{1}{L_1} \cdot \int_0^{t_1} V \cdot dt + \frac{V - Vs(t)}{Rs} = 0$$

$$\{\frac{1}{R_1} \cdot \frac{dV}{dt} + C_1 \cdot \frac{d^2V}{dt^2} + \frac{1}{L_1} \cdot V + \frac{dV}{dt} \cdot \frac{1}{Rs}\} \xrightarrow{\frac{dVs(t)}{dt} \to \varepsilon (0 < \varepsilon \ll 1)} 0$$

$$\varepsilon \gg 1 \Rightarrow \ddot{V} \cdot C_1 + (\frac{1}{R_1} + \frac{1}{R_s}) \cdot \dot{V} + \frac{1}{L_1} \cdot \mathrm{V} = \frac{1}{R_s} \cdot \dot{V}_S(\mathrm{t})$$

$$\frac{1}{R_1} \cdot \frac{dV}{dt} + C_1 \cdot \frac{d^2V}{dt^2} + \frac{1}{L_1} \cdot V + [\frac{dV}{dt} - \frac{dVs(t)}{dt}] \cdot \frac{1}{Rs} = 0$$

$$V_2 = \frac{dV_1}{dt} = \frac{dV}{dt}, V_1 = V$$

$$\frac{dV_1}{dt} = V_2, \frac{dV_2}{dt} = -[\frac{1}{C_1 \cdot R_1} + \frac{1}{Rs \cdot C_1}] \cdot V_2 - \frac{1}{C_1 \cdot L_1} \cdot V_1 + \frac{1}{Rs \cdot C_1} \cdot \frac{dVs(t)}{dt}$$

$$\begin{pmatrix} \frac{dV_1}{dt} \\ \frac{dV_2}{dt} \end{pmatrix} = \begin{pmatrix} 0 & 1 \\ -\frac{1}{C_1 \cdot L_1} & -[\frac{1}{C_1 \cdot R_1} + \frac{1}{Rs \cdot C_1}] \end{pmatrix} \cdot \begin{pmatrix} V_1 \\ V_2 \end{pmatrix} + \begin{pmatrix} 0 \\ \frac{1}{Rs \cdot C_1} \cdot \frac{dVs(t)}{dt} \end{pmatrix}$$

$$Lcalc = \left[\frac{\mu_0}{\pi} \cdot [X_1 + X_2 - X_3 + X_4] \cdot Nc^p\right]$$

L1 = Lcalc

$$X_1 = Aavg \cdot \ln\left(\frac{2 \cdot Aavg \cdot Bavg}{d \cdot (Aavg + \sqrt{Aavg^2 + Bavg^2})}\right)$$

$$X_2 = Bavg \cdot \ln\left(\frac{2 \cdot Aavg \cdot Bavg}{d \cdot (Bavg + \sqrt{Aavg^2 + Bavg^2})}\right)$$

$$X_3 = 2 \cdot \left[Aavg + Bavg - \sqrt{[Aavg^2 + Bavg^2]}\right]; X_4 = (Aavg + Bavg)/4$$

The RFID's coil calculation inductance expression is the definition of limits, Estimations: Track thickness t, Al and Cu coils (t > 30 μm). The printed coils as high as possible. Estimation of turn exponent p is needed for inductance calculation.

Coil manufacturing technology	P
Wired	1.8–1.9
Etched	1.75–1.85
Printed	1.7–1.8

Active RFID can be considered as a Van der Pol's system. Van der Pol's equation provides an example of an oscillator with nonlinear damping, energy being dissipated at large amplitudes and generated at low amplitudes. Such systems typically possess limit cycles, sustained oscillations around a state at which energy generation and dissipation balance. The basic Van der Pol's equation can be written in the form: $\ddot{X} + \alpha \cdot \phi(x) \cdot \dot{X} + X = \beta \cdot \rho(t)$.

$$\varepsilon \gg 1 \Rightarrow \ddot{V} \cdot C_1 + (\frac{1}{R_1} + \frac{1}{Rs}) \cdot \dot{V} + \frac{1}{L_1} \cdot V = \frac{1}{Rs} \cdot \dot{V}_S(\mathrm{t})$$

$$\varepsilon \gg 1 \Rightarrow \ddot{V} + \frac{1}{C_1} \cdot (\frac{1}{R_1} + \frac{1}{Rs}) \cdot \dot{V} + \frac{1}{L_1 \cdot C_1} \cdot V = \frac{1}{Rs \cdot C_1} \cdot \dot{V}_S(\mathrm{t})$$

$$X \to V, \alpha \cdot \phi(x) \to \frac{1}{C_1} \cdot (\frac{1}{R_1} + \frac{1}{Rs}); \frac{1}{L_1 \cdot C_1} \to 1, \frac{1}{Rs \cdot C_1} \cdot \dot{V}_S(\mathrm{t}) \to \beta \cdot \rho(t)$$

Let's define:

$$f_s(\mathrm{t}) = \dot{V}_S(\mathrm{t}\ ) \Rightarrow \varepsilon \gg 1 \Rightarrow \ddot{V} \cdot C_1 + (\frac{1}{R_1} + \frac{1}{Rs}) \cdot \dot{V} + \frac{1}{L_1} \cdot V = \frac{1}{Rs} \cdot f_s(\mathrm{t})$$

"f" is a "T" periodic function of the independent variable t, and $\lambda = \frac{1}{Rs}$

The term $\lambda \cdot f_s(\mathrm{t}) = \frac{1}{Rs} \cdot \dot{V}_S(\mathrm{t})$ is called the forcing function $\lambda \to 0 \Rightarrow \frac{1}{Rs} \to 0 \Rightarrow Rs \to \infty$ there is no forcing and the system act as Van Der Pol Oscillator. It is necessary to examine the trajectories (V_1, V_2, and t) of the non-autonomous Active RFID system in $\mathbb{R}^2 x\mathbb{R}$ rather than the orbits in $\mathbb{R}^2$. Equivalently, we may consider the orbits of the Active RFID TAGs three dimensional autonomous systems.

$$\frac{dV_1}{dt} = V_2$$
$$\frac{dV_2}{dt} = -[\frac{1}{C_1 \cdot R_1} + \frac{1}{Rs \cdot C_1}] \cdot V_2 - \frac{1}{C_1 \cdot L_1} \cdot V_1 + \frac{1}{Rs \cdot C_1} \cdot f_s(\mathrm{t}) \ \forall \ f_s(\mathrm{t}) = \dot{V}_S(\mathrm{t})$$
$$\frac{dV_3}{dt} = 1 \ \forall \ (V_3(t) = t)$$

First examine the case of $\lambda = 0 \Rightarrow Rs \cdot C_1 \to \infty, \mathrm{C}_1 = const,\ then\, Rs \to \infty$

The limit cycle, the isolated periodic orbit, of the unforced oscillator of Van Der Pol becomes a cylinder; that is, topologically it is a homomorphism to $S^1 \times \mathbb{R}$. The cylinder is an invariant manifold in the sense that any solution starting on the cylinder remains on it for all positive time. This invariant cylinder attracts all nearby solutions. For $\lambda = 0$, $\lambda \to 0$ $Rs \to \infty$ the Active RFID TAG invariant cylinder is filled with a family of periodic solutions. The cylinder under the projection $\mathbb{R}^2 \times \mathbb{R} \to \mathbb{R}^2$ simply becomes the limit cycle. Actually active RFID TAGs act as periodic forcing with small amplitude, that $|\lambda|$ small. In this case, there is still a cylinder in $\mathbb{R}^2 \times \mathbb{R}$ close to the invariant cylinder of the unforced oscillator. This new cylinder is an invariant manifold of solutions of the forced equation and attracts all nearby solutions. The flow on the invariant cylinder of the forced equation can be quite different from the one of the unforced oscillators. In Active RFID TAG concern to the Van Der Pol's equation, we get the equation:

$$\ddot{X} + \alpha \cdot \phi(x) \cdot \dot{X} + X = \lambda \cdot f_s(t)$$
$$\varepsilon \gg 1 \Rightarrow \ddot{V} + (\frac{1}{R_1} + \frac{1}{Rs}) \cdot \frac{1}{C_1} \cdot \dot{V} + \frac{1}{L_1 \cdot C_1} \cdot V = \frac{1}{Rs \cdot C_1} \cdot f_s(t)$$
$$\varepsilon \gg 1 \Rightarrow \ddot{V} + (\frac{1}{R_1} + \frac{1}{Rs}) \cdot \frac{1}{C_1} \cdot \dot{V} + \frac{1}{L_1 \cdot C_1} \cdot V = \frac{1}{Rs \cdot C_1} \cdot \dot{V}_S(t)$$
$$then\, \phi(x) = 1,\ \alpha = [(\frac{1}{R_1} + \frac{1}{Rs}) \cdot \frac{1}{C_1}],\ \frac{1}{L_1 \cdot C_1} \to 1 (L_1 \cdot C_1 \approx 1)$$

$\phi(x) = 1 > 0 \ \forall \ |\mathrm{t}| > 1\,\mathrm{sec}$, $f_s(\mathrm{t})$ is T periodic and α, β are non negative parameters. $\alpha = (\frac{1}{R_1} + \frac{1}{Rs}) \cdot C_1,\ \beta = \frac{1}{Rs \cdot C_1}$

Unforced investigation: $\lambda = 0 \Rightarrow \frac{1}{Rs} \to 0 \Rightarrow Rs \to \infty$ then we return to Passive RFID TAG since the battery has a very high serial resistance—disconnected status. Active RFID equivalent circuit total TAG power is a summation of all power elements.

$$P_{total} = \sum_{i=1}^{N} P_i = \text{ TAG Power}, \sum_{i=1}^{N} P_i = \frac{1}{m} \cdot [q(m) \cdot P_{Trx} + U(m) \cdot P_{Ttx} + Z(m) \cdot P_{Trx}]$$

$$\sum_{i=1}^{N} P_i = P_{Rs} + P_{C_1} + P_{R_1} + P_{L_1}, \ energy \Rightarrow W(t_0, t) \triangleq \int_{t0}^{t} p(t')dt' = \int_{t0}^{t} v(t') \cdot i(t')dt'$$

$$P(t)_{total} = \frac{dW(t_0, t)}{dt} = \frac{d}{dt}[\sum_{i=1}^{N} w_i], \ energy \Rightarrow w_{inductor} = \frac{1}{2} \cdot L \cdot I^2$$

$$energy \Rightarrow w_{capacitor} = \frac{Q^2}{2 \cdot C}, \ P_{resistor} = I^2 \cdot R, \ P_{R_1} = I_{R_1}^2 \cdot R_1, P_{Rs} = I_{Rs}^2 \cdot Rs$$

$$energy \Rightarrow w_{L_1} = \frac{1}{2} \cdot L_1 \cdot I_{L_1}^2 \Rightarrow P_{L_1} = \frac{d}{dt} w_{L_1} = L \cdot I_{L_1} \cdot \dot{I}_{L_1}$$

$$energy \Rightarrow w_{C_1} = \frac{Q_{C_1}^2}{2 \cdot C_1} \Rightarrow P_{C_1} = \frac{d}{dt} w_{C_1} = \frac{Q_{C_1} \cdot \dot{Q}_{C_1}}{C_1}$$

$$energy \Rightarrow w_{C_1} = \frac{C_1 \cdot V_{C_1}^2}{2} \Rightarrow P_{C_1} = \frac{d}{dt} w_{C_1} = C_1 \cdot V_{C_1} \cdot \dot{V}_{C_1}$$

$$I_{L_1} = \frac{1}{L_1} \cdot \int_{0}^{t} V_{L_1} \cdot dt \Rightarrow \dot{I}_{L_1} = \frac{V_{L_1}}{L_1}; \sum_{i=1}^{N} P_i = I_{R_1}^2 \cdot R_1 + I_{Rs}^2 \cdot Rs + L \cdot I_{L_1} \cdot \dot{I}_{L_1} + \frac{Q_{C_1} \cdot \dot{Q}_{C_1}}{C_1}$$

$$\sum_{i=1}^{N} P_i = \frac{V^2}{R_1} + \frac{[V - Vs(t)]^2}{Rs} + L \cdot I_{L_1} \cdot \dot{I}_{L1} + C_1 \cdot V_{C_1} \cdot \dot{V}_{C_1}$$

$$\sum_{i=1}^{N} P_i = V^2 \cdot [\frac{1}{R_1} + \frac{1}{Rs}] - \frac{2 \cdot V \cdot Vs(t)}{Rs} + \frac{[Vs(t)]^2}{Rs} + \frac{V}{L_1} \cdot \int_{0}^{t} Vdt + C_1 \cdot V \cdot \dot{V}$$

$$\frac{1}{m} \cdot [q(m) \cdot P_{Trx} + U(m) \cdot P_{Ttx} + Z(m) \cdot P_{Trx}] = V^2 \cdot [\frac{1}{R_1} + \frac{1}{Rs}]$$
$$- \frac{2 \cdot V \cdot Vs(t)}{Rs} + \frac{[V_s(t)]^2}{Rs} + \frac{V}{L_1} \cdot \int_{0}^{t} Vdt + C_1 \cdot V \cdot \dot{V}$$

$$\frac{dV_1}{dt} = V_2; \frac{dV_2}{dt} = -[\frac{1}{C_1 \cdot R_1} + \frac{1}{Rs \cdot C_1}] \cdot V_2 - \frac{1}{C_1 \cdot L_1} \cdot V_1 + \frac{1}{Rs \cdot C_1} \cdot \dot{V}_S(\mathrm{t})$$

Now we consider linear system: $\frac{dV_1}{dt} = f(V_1, V_2)$, $\frac{dV_2}{dt} = g(V_1, V_2)$

And suppose that (V_1^*, V_2^*) is a fixed point: $f(V_1^*, V_2^*) = 0, \ g(V_1^*, V_2^*) = 0$

Let $U_1 = V_1 - V_1^*, U_2 = V_2 - V_2^*$ Denote the components of a small disturbance from the fixed point. To see whether the disturbance grows or decays, we need to derive differential equations of U_1 and U_2. Let's do the U1 equation first: $\frac{dU_1}{dt} = \frac{dV_1}{dt}$ Since V_1^* being constant.

$$\frac{dU_1}{dt} = \frac{dV_1}{dt} = f(U_1 + V_1^*,\ U_2 + V_2^*) = f(V_1^*, V_2^*) + U_1 \cdot \frac{\partial f}{\partial V_1} + U_2 \cdot \frac{\partial f}{\partial V_2} + O(U_1^2,\ U_2^2,\ U_1 \cdot U_2)$$

(Taylor series expansion). To simplify the notation, we have written $\frac{\partial f}{\partial V_1}$ and $\frac{\partial f}{\partial V2}$ these partial derivatives are to be evaluated at the fixed point (V_1^*, V_2^*); thus they are numbers, not functions. Also the shorthand notation $O(U_1^2,\ U_2^2,\ U_1 \cdot U_2)$ denotes quadratic termss in U_1 and U_2. Since U_1 and U_2 are small, these quadratic terms are extremely small. Similarly, we find $\frac{dU_2}{dt} = U_1 \cdot \frac{\partial g}{\partial V_1} + U_2 \cdot \frac{\partial g}{\partial V_2} + O(U_1^2,\ U_2^2,\ U_1 \cdot U_2)$, Hence the disturbance (U_1, U_2) evolves according to $\begin{pmatrix} \frac{dU_1}{dt} \\ \frac{dU_2}{dt} \end{pmatrix} = \begin{pmatrix} \frac{\partial f}{\partial V_1} & \frac{\partial f}{\partial V_2} \\ \frac{\partial g}{\partial V_1} & \frac{\partial g}{\partial V_2} \end{pmatrix} \cdot \begin{pmatrix} U_1 \\ U_2 \end{pmatrix}$ + Quadratic terms.

The Matrix $A = \begin{pmatrix} \frac{\partial f}{\partial V_1} & \frac{\partial f}{\partial V_2} \\ \frac{\partial g}{\partial V_1} & \frac{\partial g}{\partial V_2} \end{pmatrix}_{(V_1^*, V_2^*)}$ is called the Jacobian matrix at the fixed point (V_1^*, V_2^*) and the Quadratic terms are tiny, it's tempting to neglect them altogether. If we do that, we obtain the linearized system.

$$\begin{pmatrix} \frac{dU_1}{dt} \\ \frac{dU_2}{dt} \end{pmatrix} = \begin{pmatrix} \frac{\partial f}{\partial V_1} & \frac{\partial f}{\partial V_2} \\ \frac{\partial g}{\partial V_1} & \frac{\partial g}{\partial V_2} \end{pmatrix} \cdot \begin{pmatrix} U_1 \\ U_2 \end{pmatrix}$$

Who's dynamic can be analyzed by the general methods.

$$f(V_1, V_2) = V_2; g(V_1, V_2) = -[\frac{1}{C_1 \cdot R_1} + \frac{1}{Rs \cdot C_1}] \cdot V_2 - \frac{1}{C_1 \cdot L_1} \cdot V_1 + \frac{1}{Rs \cdot C_1} \cdot \dot{V}_S(\mathrm{t})$$

$$\frac{\partial f}{\partial V_1} = 0,\ \frac{\partial f}{\partial V_2} = 1,\ \frac{\partial g}{\partial V_1} = -\frac{1}{C_1 \cdot L_1},\ \frac{\partial g}{\partial V_2} = -(\frac{1}{C_1 \cdot R_1} + \frac{1}{Rs \cdot C_1})$$

$$\begin{pmatrix} \frac{dU_1}{dt} \\ \frac{dU_2}{dt} \end{pmatrix} = \begin{pmatrix} 0 & 1 \\ -\frac{1}{C_1 \cdot L_1} & -[\frac{1}{C_1 \cdot R_1} + \frac{1}{R \cdot C_1}] \end{pmatrix} \cdot \begin{pmatrix} U_1 \\ U_2 \end{pmatrix}$$

The basic Active RFID Forced Van Der Pol's equation

$$\varepsilon \gg 1 \Rightarrow \ddot{V} + (\frac{1}{R_1} + \frac{1}{Rs}) \cdot \frac{1}{C_1} \cdot \dot{V} + \frac{1}{L_1 \cdot C_1} \cdot V = \frac{1}{Rs \cdot C_1} \cdot \dot{V}_S(t)$$

$$then\, \phi(x) = 1,\ \alpha = [(\frac{1}{R_1} + \frac{1}{Rs}) \cdot \frac{1}{C_1}],\ \frac{1}{L_1 \cdot C_1} \to 1(L_1 \cdot C_1 \approx 1); \beta = \frac{1}{Rs \cdot C_1}$$

In our case $\phi(V) = 1$, $\phi(V) > 0$ for $|\mathrm{V}| > 1$ and $\dot{V}_S(\mathrm{t})$ is T periodic and,$(\frac{1}{R_1} + \frac{1}{Rs}) \cdot \frac{1}{C_1}$, $\frac{1}{Rs \cdot C_1}$ is non-negative parameters. It is convenient to rewrite the Active RFID forced Van Der Pol's equation as an autonomous system.

$$\theta = \mathrm{t} \Rightarrow \frac{d\theta}{dt} = 1; \dot{V} = Y - (\frac{1}{R_1} + \frac{1}{Rs}) \cdot \frac{1}{C_1} \cdot \phi(V); \dot{Y} = -V + \frac{1}{R_1 \cdot C_1} \cdot \dot{V}_S(\theta)$$

$$\dot{\theta} = 1; (V, Y, \theta) \in \mathbb{R}^2 x S^1 .$$

$\phi(V) = 1$ remain strictly positive as $|V| \to \infty$ for unforced system, $\frac{1}{R_1 \cdot C_1} \cdot \dot{V}_S(\theta) \to 0$ but $\frac{1}{R_1 \cdot C_1} \neq 0$ then $\dot{V}_S(\theta) = 0$ no energy is supplied to the Active RFID TAG, become Passive RFID TAG. First, we suppose that $\alpha \ll 1((\frac{1}{R_1} + \frac{1}{Rs}) \cdot \frac{1}{C_1} \ll 1)$ is a small parameter, so the autonomous system is a perturbation of linear oscillator. $\dot{V} = Y$, $\dot{Y} = -V$ Has a phase plane filled with circular periodic orbits each period of $2 \cdot \pi$. Using regular perturbation or averaging methods, we can show that precisely one of these orbits is preserved under the perturbation. Selecting the invertible transformation:

$$\begin{pmatrix} \xi 1 \\ \xi 2 \end{pmatrix} = \begin{pmatrix} \cos(t) & -\sin(t) \\ -\sin(t) & -\cos(t) \end{pmatrix} \cdot \begin{pmatrix} V \\ Y \end{pmatrix}$$

Which "freezes" the unperturbed system and

The autonomous system become:

$$\dot{\xi_1} = -(\frac{1}{R_1} + \frac{1}{Rs}) \cdot \frac{1}{C_1} \cdot \cos t \cdot [(\xi_1 \cdot \cos(t) - \xi_2 \cdot \sin(t))^3 /3 - (\xi_1 \cdot \cos(t) - \xi_2 \cdot \sin(t))]$$

$$\dot{\xi_2} = -(\frac{1}{R_1} + \frac{1}{Rs}) \cdot \frac{1}{C_1} \cdot \sin t \cdot [(\xi_1 \cdot \cos(t) - \xi_2 \cdot \sin(t))^3 /3 - (\xi 1 \cdot \cos(t) - \xi 2 \cdot \sin(t))]$$

This transformation is orientation reversing approximations the function $\xi 1$, $\xi 2$ which vary slowly because $\dot{\xi_1}, \dot{\xi_2}$ being small. Integrating each function with respect to time (t) from 0 to $T = 2 \cdot \pi$, holding $\xi 1$, $\xi 2$ fixed we obtain:

$$\dot{\xi_1} = (\frac{1}{R_1} + \frac{1}{Rs}) \cdot \frac{1}{C_1} \cdot \xi_1 \cdot [1 - ({\xi_1}^2 + {\xi_2}^2)/4]/2$$

$$\dot{\xi_2} = (\frac{1}{R_1} + \frac{1}{Rs}) \cdot \frac{1}{C_1} \cdot \xi_2 \cdot [1 - ({\xi_1}^2 + {\xi_2}^2)/4]/2$$

This system is correct at first order, but there is an error of $O([(\frac{1}{R_1}+\frac{1}{Rs})\cdot\frac{1}{C_1}]^2)$. In polar coordinates, we therefore have

$$\dot{r} = \frac{r}{2}\cdot(\frac{1}{R_1}+\frac{1}{Rs})\cdot\frac{1}{C_1}\cdot(1-\frac{r^2}{4}) + O([(\frac{1}{R_1}+\frac{1}{Rs})\cdot\frac{1}{C_1}]^2)$$
$$\dot{\varphi} = 0 + O([(\frac{1}{R_1}+\frac{1}{Rs})\cdot\frac{1}{C_1}]^2)$$

Neglecting the $O([(\frac{1}{R_1}+\frac{1}{Rs})\cdot\frac{1}{C_1}]^2)$ terms this system has an attracting circle of fixed points at r = 2 reflecting the existence of a one parameter family of almost sinusoidal solutions: $V = r(t)\cdot\cos(t+\varphi(t))$ with slowly varying amplitude

$$r(t) = 2 + O([(\frac{1}{R_1}+\frac{1}{Rs})\cdot\frac{1}{C_1}]^2);\ \varphi(t) = \varphi^0 + O([(\frac{1}{R_1}+\frac{1}{Rs})\cdot\frac{1}{C_1}]^2)$$
$$\varphi(t) = \varphi^0 + O([(\frac{1}{R_1}+\frac{1}{Rs})\cdot\frac{1}{C_1}]^2)$$

Constant φ^0 is being determined by initial conditions.

When the value of $(\frac{1}{R_1}+\frac{1}{Rs})\cdot\frac{1}{C_1}$ is not small the averaging procedure no longer works and other methods must be used. The investigation can be done for Active RFID's system forced Van Der Pole. Let's consider $\dot{V}_S(t)\neq 0$ we suppose $(\frac{1}{R_1}+\frac{1}{Rs})\cdot\frac{1}{C_1}, \frac{1}{Rs\cdot C_1} \ll 1$ and use the same transformation as we use in the unforced system $\dot{V}_S(t)\neq 0$. When our interest in the periodic forced response we use the $\frac{2\cdot\pi}{\omega}$ periodic transformation [2–4].

$$\begin{pmatrix}\xi 1\\ \xi 2\end{pmatrix} = \begin{pmatrix}\cos(\omega t) & -\frac{1}{\omega}\cdot\sin(\omega t)\\ -\sin(\omega t) & -\frac{1}{\omega}\cdot\cos(\omega t)\end{pmatrix}\cdot\begin{pmatrix}V\\ Y\end{pmatrix}$$

$$\dot{\xi}_1 = -(\frac{1}{R_1}+\frac{1}{Rs})\cdot\frac{1}{C_1}\cdot\phi(V)\cdot\cos(\omega\cdot t) - (\frac{\omega^2-1}{\omega})\cdot V\cdot\sin(\omega\cdot t) - \frac{1}{Rs\cdot C_1\cdot\omega}\cdot\sin(\omega\cdot t\cdot\dot{V}_S(t))$$
$$\dot{\xi}_2 = (\frac{1}{R_1}+\frac{1}{Rs})\cdot\frac{1}{C_1}\cdot\phi(V)\cdot\sin(\omega\cdot t) - (\frac{\omega^2-1}{\omega})\cdot V\cdot\cos(\omega\cdot t) - \frac{1}{Rs\cdot C_1\cdot\omega}\cdot\cos(\omega\cdot t\cdot\dot{V}_S(t))$$

$\frac{1}{C_1\cdot L_1}\to 1, \phi(V) = 1$ in our case.

$$\dot{\xi}_1 = -(\frac{1}{R_1} + \frac{1}{Rs}) \cdot \frac{1}{C_1} \cdot \cos(\omega \cdot t) - (\frac{\omega^2 - 1}{\omega}) \cdot V \cdot \sin(\omega \cdot t) - \frac{1}{Rs \cdot C_1 \cdot \omega} \cdot \sin(\omega \cdot t \cdot \dot{V}_S(\mathrm{t}))$$

$$\dot{\xi}_2 = (\frac{1}{R_1} + \frac{1}{Rs}) \cdot \frac{1}{C_1} \cdot \sin(\omega \cdot t) - (\frac{\omega^2 - 1}{\omega}) \cdot V \cdot \cos(\omega \cdot t) - \frac{1}{Rs \cdot C_1 \cdot \omega} \cdot \cos(\omega \cdot t \cdot \dot{V}_S(\mathrm{t}))$$

Active RFID TAG system can be represented as Voltage source (internal resistance), Parallel Resistor, Capacitor, and Inductance circuit. Linear bifurcation system explains Active RFID TAG system behavior for any initial condition V(t) and dV(t)/dt . Active RFID's Coil is a very critical element in Active RFID TAG functionality. Optimization can be achieved by Coil's parameters inspection and System bifurcation controlled by them. Spiral, Circles, and other Active RFID phase system behaviors can be optimized for better Active RFID TAG performance and actual functionality. Active RFID TAG losses also controlled for best performance and maximum efficiency.

1.2 RFID TAG's Dimensional Parameters Optimization as Excitable Linear Bifurcation Systems

RFID Equivalent circuits of a Label can be represented as Parallel circuit of Capacitance (Cpl), Resistance (Rpl), and Inductance (Lpc). The Label measurement principle is as follows: Label positioned in defining distance to measurement coil, Low current or voltage source, Measuring of |Z| and Teta of measurement coil, Resonance frequency fro at Teta = 0, Calculation of unloaded quality factor Q0 out of measured bandwidth B0. The Coil design procedure is based on three important steps. The RFID equivalent circuit can be represented as a differential equation which, depending on variable parameter. The investigation of RFID's differential equation based on bifurcation theory [1], the study of possible changes in the structure of the orbits of a differential equation depending on variable parameters. We first illustrate certain observations and analyze local bifurcations of an appropriate arbitrary scalar differential equation [2]. Since the implicit function theorem is the main ingredient used in these generalizations, include a precise statement of this theorem. Additional analyze the bifurcations of a RFID's differential equation on the circle. The bifurcation behavior of specific differential equations can be encapsulated in certain pictures called bifurcation diagrams. Analysis is done for optimization of RFID TAG's dimensional parameters to get the best performance. RFID TAG can be represented as a parallel Equivalent Circuit of Capacitor and Resistor in parallel. For example, see below NXP/PHILIPS ICODE IC, Parallel equivalent circuit and simplified complete equivalent circuit of the label (L1 is the antenna inductance) [7, 85] (Fig. 1.8).

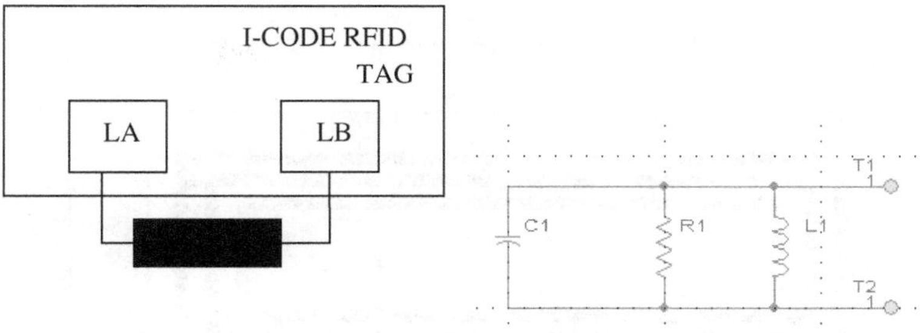

Fig. 1.8 NXP/PHILIPS ICODE IC, Parallel equivalent circuit and simplified complete equivalent circuit of the label (L1 is the antenna inductance)

$$\mathrm{C_1 = Cic + Ccon + Cc,\ R_1 = (Ric \cdot Rpc)/(Ric + Rpc).}$$

$$Vl_1 = L \cdot \frac{dIl}{dt}, Ic_1 = C \cdot \frac{dVc}{dt},\ Il_1 = \frac{1}{L_1} \cdot \int_0^{t_1} Vl_1 \cdot dt$$

$$\sum_{i=1}^{i=3} Ii = 0, \frac{V}{R_1} + C_1 \cdot \frac{dV}{dt} + \frac{1}{L_1} \cdot \int_{t=0}^{t=t_1} V \cdot dt = 0$$

$$\frac{1}{R_1} \cdot \frac{dV}{dt} + C_1 \cdot \frac{d^2V}{dt^2} + \frac{1}{L_1} \cdot V = 0$$

We get differential equation of RFID TAG system which describe the evolution of the system in continues time. V = V(t).

Now I define the following Variable setting definitions:, And get the dynamic equation system: $\frac{dV_1}{dt} = V_2$, $\frac{dV_2}{dt} = -\frac{1}{C_1 \cdot R_1} \cdot V_2 - \frac{1}{C_1 \cdot L_1} \cdot V_1$

The system shape is as nonlinear system equations:

$$\frac{dV_1}{dt} = f_1(V_1, V_2 \ldots, Vn), \frac{dV_2}{dt}$$

The V_1 and V_2 variables are the phase space dimension two. Now Let's Move to three variable system—which the time (t) is the third variable, V3 = t (Fig. 1.9).

$$\frac{dV_1}{dt} = V2, \frac{dV_2}{dt} = -\frac{1}{C_1 \cdot L_1} \cdot V_1 - \frac{1}{C_1 \cdot R_1} \cdot V_2, \frac{dV_3}{dt} = 1$$

$$d = 2 \cdot (t + w)/\pi, Aavg = a_0 - Nc \cdot (g + w), Bavg = b_0 - Nc \cdot (g + w)$$

a_0, b_0—Overall dimensions of the coil. Aavg, Bavg—Average dimensions of the coil. t—Track thickness, w—Track width, g—Gap between tracks. Nc—Number of turns, d—Equivalent diameter of the track. Average coil area; −Ac = Aavg · Bavg. Integrating all those parameters gives the equations for inductance calculation:

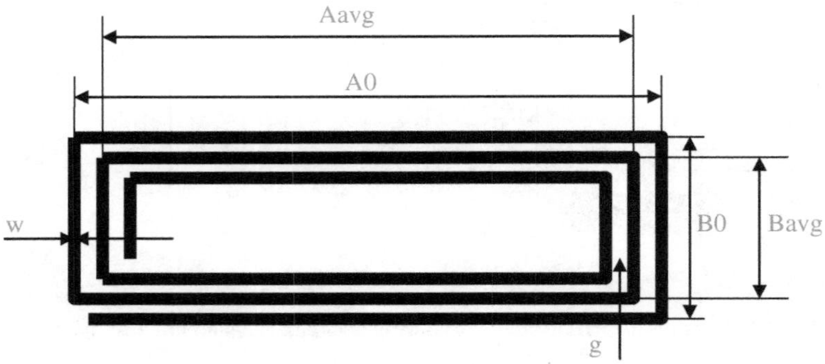

Fig. 1.9 RFID's coil dimensional parameters

$$X_1 = Aavg \cdot \ln\left(\frac{2 \cdot Aavg \cdot Bavg}{d \cdot (Aavg + \sqrt{Aavg^2 + Bavg^2})}\right)$$

$$X_2 = Bavg \cdot \ln\left(\frac{2 \cdot Aavg \cdot Bavg}{d \cdot (Bavg + \sqrt{Aavg^2 + Bavg^2})}\right)$$

$$X_3 = 2 \cdot \left[Aavg + Bavg - \sqrt{[Aavg^2 + Bavg^2]}\right]; X_4 = (Aavg + Bavg)/4$$

The RFID's coil calculation inductance expression is

$$Lcalc = \left[\frac{\mu_0}{\pi} \cdot [X_1 + X_2 - X_3 + X_4] \cdot Nc^p\right], \mathrm{L}_1 = \mathrm{Lcalc}$$

Definition of limits, Estimations: Track thickness t, Al and Cu coils ($t > 30$ µm). The printed coils as high as possible. Estimation of turn exponent p is needed for inductance calculation.

Coil manufacturing technology	P
Wired	1.8–1.9
Etched	1.75–1.85
Printed	1.7–1.8

Now I integrate the Lcalc value inside the differential equations which characterize the RFID system with the Coil inductance.

$$\frac{dV_1}{dt} = 0 \cdot V_1 + 1 \cdot V_2 + 0 \cdot V_3$$
$$\frac{dV_2}{dt} = -\frac{1}{C_1 \cdot \left[\frac{\mu_0}{\pi} \cdot [X_1 + X_2 - X_3 + X_4] \cdot Nc^p\right]} \cdot V_1 - \frac{1}{C_1 \cdot R_1} \cdot V_2 + 0 \cdot V_3$$
$$\frac{dV_3}{dt} = 0 \cdot V_1 + 0 \cdot V_2 + 0 \cdot V_3 + 1$$

The above differential equations can be represented as Matrix formulation:

$$\begin{bmatrix} \frac{dV_1}{dt} \\ \frac{dV_2}{dt} \\ \frac{dV_3}{dt} \end{bmatrix} = \begin{bmatrix} 0 & 1 & 0 \\ \{-\frac{1}{C_1 \cdot \left[\frac{\mu_0}{\pi} \cdot [X_1 + X_2 - X_3 + X_4] \cdot Nc^p\right]}\} & \{-\frac{1}{C_1 \cdot R_1}\} & 0 \\ 0 & 0 & 0 \end{bmatrix} \cdot \begin{bmatrix} V_1 \\ V_2 \\ V_3 \end{bmatrix} + \begin{bmatrix} 0 \\ 0 \\ 1 \end{bmatrix}$$

$$\begin{bmatrix} \frac{dV_1}{dt} \\ \frac{dV_2}{dt} \\ \frac{dV_3}{dt} \end{bmatrix} = \begin{bmatrix} 0 & 1 & 0 \\ \{-\frac{1}{C_1 \cdot \left[\frac{\mu_0}{\pi} \cdot (-X_3 + \sum_{k=1, k \neq 3}^{4} X_k) \cdot Nc^p\right]}\} & \{-\frac{1}{C_1 \cdot R_1}\} & 0 \\ 0 & 0 & 0 \end{bmatrix} \cdot \begin{bmatrix} V_1 \\ V_2 \\ V_3 \end{bmatrix} + \begin{bmatrix} 0 \\ 0 \\ 1 \end{bmatrix}$$

And denote the matrix's elements as functions K_1 and K_2 of Coil overall parameters.

$$K_1 = K_1(a_0, b_0, w, g, d, N_C, t, p, C_1, R_1) = \{-\frac{1}{C_1 \cdot \left[\frac{\mu_0}{\pi} \cdot [X_1 + X_2 - X_3 + X_4] \cdot Nc^p\right]}\}$$
$$K_2 = K_2(a_0, b_0, w, g, d, N_C, t, p, C_1, R_1) = \{-\frac{1}{C_1 \cdot R_1}\}$$

Denote the RFID Matrix systems with those K_1, K_2 parameter function gives:

$$\begin{bmatrix} \frac{dV_1}{dt} \\ \frac{dV_2}{dt} \end{bmatrix} = \begin{bmatrix} 0 & 1 \\ K_1 & K_2 \end{bmatrix} \cdot \begin{bmatrix} V_1 \\ V_2 \end{bmatrix}$$

Now the consideration of trajectories of the form [3]: $V(t) = e^{\lambda \cdot t} \cdot S$, Where S <> 0 is some fixed vector to be determined, and λ is a growth rate, also to be determined. If a such solution exists, they correspond to exponential motion along the line spanned by the vector S. To find the condition on S and λ, we substitute $V(t) = e^{\lambda \cdot t} \cdot S$ into $\dot{V} = A \cdot V; A = \begin{bmatrix} 0 & 1 \\ K_1 & K_2 \end{bmatrix}$ and obtain $\lambda \cdot e^{\lambda \cdot t} \cdot S = e^{\lambda \cdot t} \cdot A \cdot S$ and cancellation the nonzero scalar factor $e^{\lambda \cdot t}$ yields to $\lambda \cdot S = A \cdot S$ which state that the desired straight line solutions exist if S is an eigenvector of A with corresponding eigenvalue λ and the solution is Eigen solution. The eigenvalues of a matrix A are given by the characteristic equation $\det(A - \lambda \cdot I) = 0$ when I is the identity matrix $I = \begin{bmatrix} 1 & 0 \\ 0 & 1 \end{bmatrix}$, we get

$$\Delta = \det(A) = 0 - K_1 = -K_1$$
$$\tau = trace(A) = 0 + K_2 = K_2$$
$$\lambda^2 - \tau \cdot \lambda + \Delta = 0$$
$$\lambda^2 - K_2 \cdot \lambda - K_1 = 0$$
$$\lambda_{1,2} = \frac{1}{2} \cdot K_2 \pm \frac{1}{2} \cdot \sqrt{K_2^2 + 4 \cdot K_1}$$

The above $\lambda_{1,2}$ is a quadratic solution. The typical solution is for the eigenvalues are distinct $\lambda_1 \neq \lambda_2$. In this case, a theorem of linear algebra states that the corresponding eigenvectors S_1 and S_2 are linearly independent, and hence span the entire plane. Any initial condition V_0 can be written as a linear combination of eigenvectors, $V_0 = C_1 \cdot S_1 + C_2 \cdot S_2$. Then the general solution for V(t) it is simply $V(t) = C_1 \cdot e^{\lambda_1 \cdot t} \cdot S_1 + C_2 \cdot e^{\lambda_2 \cdot t} \cdot S_2$. By insertion quadratic solutions into the last V(t) equation we get

$$V(t) = C_1 \cdot e^{\left[\frac{1}{2} \cdot K_2 + \frac{1}{2} \cdot \sqrt{K_2^2 + 4 \cdot K_1}\right] \cdot t} \cdot S_1 + C_2 \cdot e^{\left[\frac{1}{2} \cdot K_2 - \frac{1}{2} \cdot \sqrt{K_2^2 + 4 \cdot K_1}\right] \cdot t} \cdot S_2$$

RFID TAG which gives the best performance is one that his equivalent circuit (Capacitor, Resistor, and Inductance (Antenna) in parallel), and his Voltage/Voltage derivative respect to time phase plane converge (Spiral converge, fixed point respect to the origin, etc.,)

$$\lambda_2 < \lambda_1 < 0$$
$$\frac{1}{2} \cdot K_2 - \frac{1}{2} \cdot \sqrt{K_2^2 + 4 \cdot K_1} < \frac{1}{2} \cdot K_2 + \frac{1}{2} \cdot \sqrt{K_2^2 + 4 \cdot K_1} < 0$$
$$\lambda_1 < 0 \ldots \rightarrow \ldots - K_2 > \sqrt{K_2^2 + 4 \cdot K_1}$$
$$-\{-\frac{1}{C_1 \cdot R_1}\} > \sqrt{\{-\frac{1}{C_1 \cdot R_1}\}^2 + 4 \cdot \{-\frac{1}{C_1 \cdot \left[\frac{\mu_0}{\pi} \cdot [X_1 + X_2 - X_3 + X_4] \cdot Nc^p\right]}\}}$$
$$\frac{1}{C_1 \cdot R_1} > \sqrt{\{\frac{1}{C_1 \cdot R_1}\}^2 - \{\frac{4}{C_1 \cdot \left[\frac{\mu_0}{\pi} \cdot [X_1 + X_2 - X_3 + X_4] \cdot Nc^p\right]}\}}$$
$$\lambda_2 < \lambda_1$$
$$\frac{1}{2} \cdot K_2 - \frac{1}{2} \cdot \sqrt{K_2^2 + 4 \cdot K_1} < \frac{1}{2} \cdot K_2 + \frac{1}{2} \cdot \sqrt{K_2^2 + 4 \cdot K_1}$$
$$0 < \sqrt{K_2^2 + 4 \cdot K_1} \rightarrow K_2^2 + 4 \cdot K_1 > 0 \rightarrow K_2^2 > -4 \cdot K_1$$

$$\{-\frac{1}{C_1 \cdot R_1}\}^2 > -4 \cdot \{-\frac{1}{C_1 \cdot \left[\frac{\mu_0}{\pi} \cdot [X_1 + X_2 - X_3 + X_4] \cdot Nc^p\right]}\}$$
$$\{\frac{1}{C_1 \cdot R_1}\}^2 > \frac{4}{C_1 \cdot \left[\frac{\mu_0}{\pi} \cdot [X_1 + X_2 - X_3 + X_4] \cdot Nc^p\right]}$$

Then both Eigen solutions decay exponentially. The fixed point is a stable node, except eigenvectors are not mutually perpendicular, in general. Trajectories typically approach the origin tangent to the slow Eigen direction, defined as the direction spanned by the eigenvector with the smaller $|\lambda|$. In backward time $t \to \infty$ the trajectories become parallel to the fast Eigen direction [2–4] (Fig. 1.10).

If we reverse all the arrows in the above figure, we obtain a typical phase portrait for an unstable node. Now I investigate the case when eigenvalues are complex number. If the eigenvalues are complex, the fixed point is either a center or a spiral. The origin is surrounded by a family of closed orbits. Note that centers are neutrally stable, since nearby trajectories are neither attracted to nor repelled from the fixed point. A spiral would occur if the RFID system were lightly damped. Then the trajectory would just fail to close, because the RFID system loses a bit of energy on each cycle. To justify these statements, recall that the eigenvalues are $\lambda_{1,2} = \frac{1}{2} \cdot K_2 \pm \frac{1}{2} \cdot \sqrt{K_2^2 + 4 \cdot K_1}; K_2^2 + 4 \cdot K_1 < 0$

To simplify the notation, let's write the eigenvalues as

$$\lambda_{1,2} = \alpha \pm i \cdot \omega\,; \alpha = \frac{1}{2} \cdot K_2\,; \omega = -\frac{1}{2} \cdot \sqrt{K_2^2 + 4 \cdot K_1}$$

$$\omega \neq 0 \,\exists\, V(t) = C_1 \cdot e^{\left[\frac{1}{2} \cdot K_2 + \frac{1}{2} \cdot \sqrt{K_2^2 + 4 \cdot K_1}\right] \cdot t} \cdot S_1 + C_2 \cdot e^{\left[\frac{1}{2} \cdot K_2 - \frac{1}{2} \cdot \sqrt{K_2^2 + 4 \cdot K_1}\right] \cdot t} \cdot S_2$$

$$\text{C's, S's complex, since, } \lambda\text{'s complex}$$

$$V(t) = C_1 \cdot e^{[\alpha + i\omega] \cdot t} \cdot S_1 + C_2 \cdot e^{[\alpha - i\omega] \cdot t} \cdot S_2$$

$$\text{Euler's formula} \to e^{[i \cdot \omega] \cdot t} = \cos[\omega \cdot t] + i \cdot \sin[\omega \cdot t]$$

Hence V(t) is a combination of terms involving

$$e^{\alpha \cdot t} \cdot \cos[\omega \cdot t], e^{\alpha \cdot t} \cdot \sin[\omega \cdot t]$$

Such terms represent exponentially decaying oscillations if $\alpha = \mathrm{Re}(\lambda) < 0$, And growing if $\alpha > 0$. The corresponding fixed points are stable and unstable spirals,

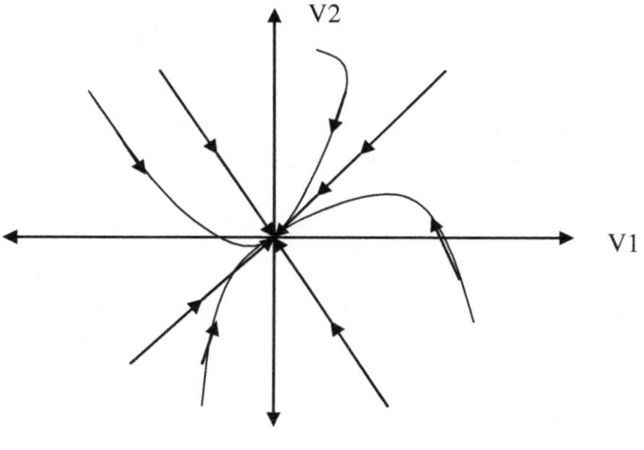

Fig. 1.10 Voltage/Voltage derivative respect to time converge after the reader carrier signal end Case 1

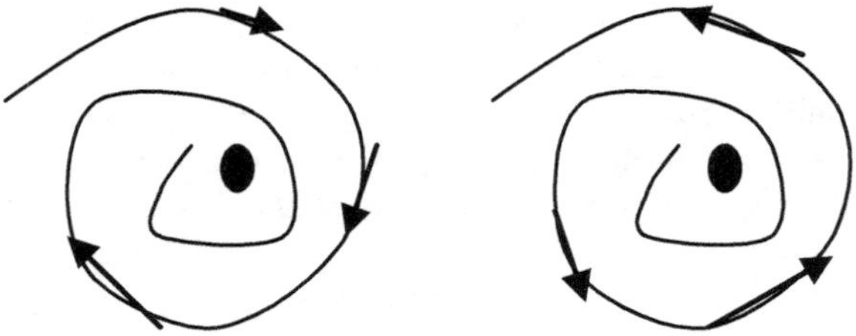

Fig. 1.11 For both centers and spirals, rotation is clockwise or Counter clockwise

respectively. If the eigenvalues are pure imaginary $\alpha = 0$, then all the solutions are periodic with period $T = \frac{2\cdot\pi}{\omega}$. The oscillators have fixed amplitude and the fixed point is centered. For both centers and spirals, it's easy to determine whether the rotation is clockwise or counterclockwise (Fig. 1.11).

$$\alpha = \frac{1}{2}\cdot K_2 = \{-\frac{1}{2\cdot C_1\cdot R_1}\}$$

$$\textit{Decaying oscillators}\ \forall\ \alpha<0 \to \{-\frac{1}{2\cdot C_1\cdot R_1}\}<0 \to \frac{1}{2\cdot C_1\cdot R_1}>0$$

$$\textit{Growing oscillators}\ \forall\ \alpha>0 \to \{-\frac{1}{2\cdot C_1\cdot R_1}\}>0 \to \frac{1}{2\cdot C_1\cdot R_1}<0$$

C_1, $R_1 > 0$ always then only the first behavior, decaying oscillator can exist in our RFID system. In all analysis until now, we have been assuming that the eigenvalues are distinct. What happens if the eigenvalues are equal? Suppose eigenvalues are equal $\lambda_1 = \lambda_2 = \lambda$, then there are two possibilities: either there are two independent eigenvectors corresponding to λ, or there's only one. If there are two independent eigenvectors, then they span the plane and so every vector is an eigenvector with this same eigenvalue λ . To see this, let's write an arbitrary vector X_0 as a linear combination of the two eigenvectors: $X_0 = C_1 \cdot S_1 + C_2 \cdot S_2$.

Then $A\cdot X_0 = A\cdot(C_1\cdot S_1 + C_2\cdot S_2) = C_1\cdot\lambda\cdot S_1 + C_2\cdot\lambda\cdot S_2 = \lambda\cdot X_0$

X0 is also an eigenvector with eigenvalue λ . Since the multiplication by A simply stretches every vector by a factor λ, the matrix must be a multiple of the identity: $A = \begin{bmatrix}\lambda & 0\\ 0 & \lambda\end{bmatrix}$ then if $\lambda \neq 0$, all trajectories are straight lines through the origin $X(t) = e^{\lambda\cdot t}\cdot X_0$ and the fixed point is a star node. On the other hand, if $\lambda = 0$ the whole plane is filled with fixed points. Let's now sketch the above options with RFID Overall parameter restriction. $\lambda_1 = \lambda_2 = \lambda \neq 0$ then (Fig. 1.12)

$$\frac{1}{2}\cdot K_2 + \frac{1}{2}\cdot\sqrt{K_2^2 + 4\cdot K_1} = \frac{1}{2}\cdot K_2 - \frac{1}{2}\cdot\sqrt{K_2^2 + 4\cdot K_1}$$

$$\sqrt{K_2^2 + 4\cdot K_1} = 0 \to K_2^2 + 4\cdot K_1 = 0 \to K_2^2 = -4\cdot K_1$$

$$\frac{\mu_0}{\pi}\cdot[X_1 + X_2 - X_3 + X_4]\cdot Nc^p = C_1\cdot 4\cdot R_1^2$$

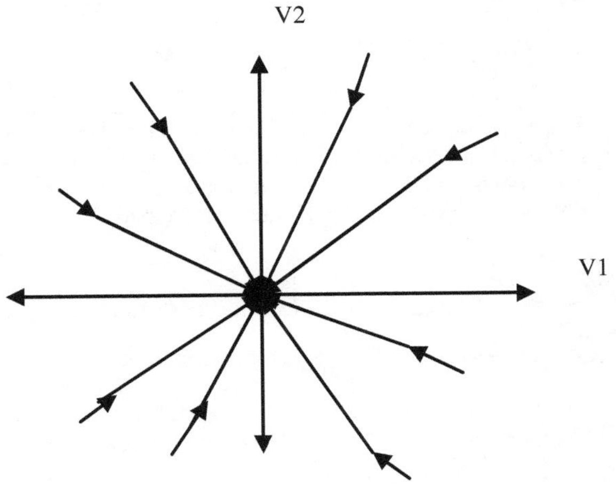

Fig. 1.12 Voltage/Voltage derivative respect to time Converge after the reader Carrier signal end Case 2

Now let's summarize the classification of fixed points in RFID system based on all investigation I did. It is easy to show the type and stability of all the different fixed points on a single diagram [4] (Figs. 1.13 and 1.14).

$$\tau^2 - 4 \cdot \Delta = K_2^2 + 4 \cdot K_1 = 0, \rightarrow K_2 = 2 \cdot \sqrt{-K_1}$$
$$\tau = trace(A) = K_2; \Delta = \det(A) = -K_1$$
$$\lambda^2 - K_2 \cdot \lambda - K_1 = 0; \lambda_{1,2} = \frac{1}{2} \cdot \left[\tau \pm \sqrt{\tau^2 - 4 \cdot \Delta}\right]$$
$$\tau = \lambda_1 + \lambda_2 = K_2; \Delta = \lambda_1 \cdot \lambda_2 = -K_1$$
$$Charecteristic\ equation : (\lambda - \lambda_1) \cdot (\lambda - \lambda_2) = \lambda^2 - \tau \cdot \lambda + \Delta = 0$$

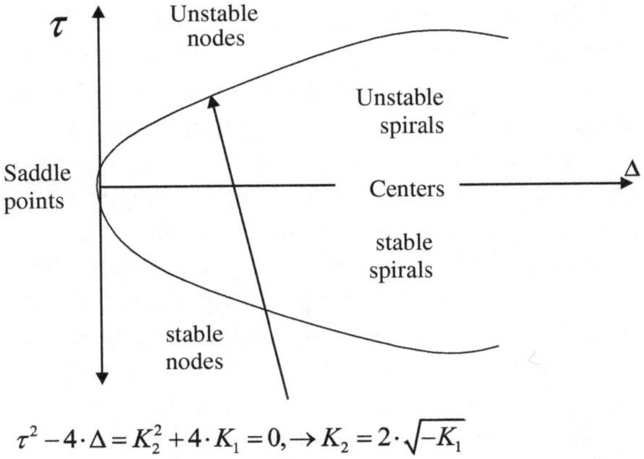

Fig. 1.13 Stable/Unstable diagram

$\Delta < 0....(K_1 > 0)$	$\Delta > 0....(K_1 < 0)$	$\Delta = 0....(K_1 = 0)$
The eigenvalues are real and have opposite sign hence the fixed point is a saddle point,	The eigenvalues are either real with the same sign (nodes), or complex conjugate (spiral & centers).	At least one of the eigenvalues is zero. Then the origin is not an isolated fixed point. There is either a whole line of a fixed point , or a plane of fixed point

Fig. 1.14 Stable Unstable diagram as function of K_1

Nodes satisfy $\tau^2 - 4 \cdot \Delta > 0$ and spirals satisfy $\tau^2 - 4 \cdot \Delta < 0$. The parabola $\tau^2 - 4 \cdot \Delta = 0$ is the borderline between nodes and spirals. Star nodes and degenerate nodes live on this parabola. The stability of the nodes and spirals is determined by τ value. When $\tau < 0$, both eigenvalues have negative real parts, so the fixed point is stable. Unstable spirals and nodes have $\tau > 0$. Neutrally stable centers live on the borderline $\tau = 0$, where eigenvalues are purely imaginary [5].

1.3 RFID TAGs Coil's System Stability Optimization Under Delayed Electromagnetic Interferences

The RFID TAG system has two main variables TAG's voltage and TAG's voltage derivative respect to time. Due to electromagnetic interferences those variables have delays in time domain. We define $\boldsymbol{\tau_1}$ as a time delay respect to TAG's voltage and $\boldsymbol{\tau_2}$ as a time delay respect to TAG's voltage derivative. RFID Equivalent circuits of a Label can be represented as Parallel circuit of Capacitance (Cpl), Resistance (Rpl), and Inductance (Lpc). Our RFID TAG system delay differential and delay different model can be analytically by using delay differential equations in dynamically. The need of the incorporation of a time delay is often of the existence of any stage structure. It is often difficult to analytically study models with delay dependent parameters, even if only a single discrete delay is present. There are practical guidelines that combine graphical information with analytical work to effectively study the local stability of models involving delay dependent parameters. The stability of a given steady state is simply determined by the graphs of some function of τ_1, $\boldsymbol{\tau_2}$ which can be expressed, explicitly and thus can be easily depicted by Matlab and other popular software. We need only look at one such function and locate the zero. This function often has only two zeroes, providing thresholds for stability switches. As time delay increases, stability changes from stable to unstable to stable. We emphasize the local stability aspects of some models with delay dependent parameters. Additionally, there is a general geometric criterion that, theoretically speaking, can be applied to models with many delays, or even distributed delays. The simplest case of a first order characteristic equation, providing more user friendly geometric and analytic criteria for stability switches. The analytical criteria provided for the first and second order cases can be used to obtain some insightful analytical statements and can be helpful for conducting simulations. RFID TAG can be represented as a parallel Equivalent Circuit of Capacitor and Resistor in parallel. For example, see below NXP/PHILIPS ICODE IC, Parallel equivalent circuit and simplified complete equivalent circuit of the label (L1 is the antenna inductance) [6, 8] (Fig. 1.15).

$$\frac{1}{R_1}\cdot\frac{dV}{dt}+C_1\cdot\frac{d^2V}{dt^2}+\frac{1}{L_1}\cdot V=0$$

We get differential equation of a RFID TAG system which describe the evolution of the system in continues time. V = V(t). Now I define the following Variable setting definitions: $V_2=\frac{dV_1}{dt}=\frac{dV}{dt}, V_1=V$. The dynamic equation system: $\frac{dV_1}{dt}=V_2; \frac{dV_2}{dt}=-\frac{1}{C_1\cdot R_1}\cdot V_2-\frac{1}{C_1\cdot L_1}\cdot V_1$ (Fig. 1.16)

$$d=2\cdot(t+w)/\pi; Aavg=a_0-Nc\cdot(g+w); Bavg=b_0-Nc\cdot(g+w)$$

a_0, b_0—Overall dimensions of the coil. Aavg, Bavg—Average dimensions of the coil. t—Track thickness, w—Track width, g—Gap between tracks. Nc—Number of turns, d—Equivalent diameter of the track. Average coil area; $-A_c = A_{avg}\ .\ B_{avg}$. Integrating all those parameters gives the equations for inductance calculation:

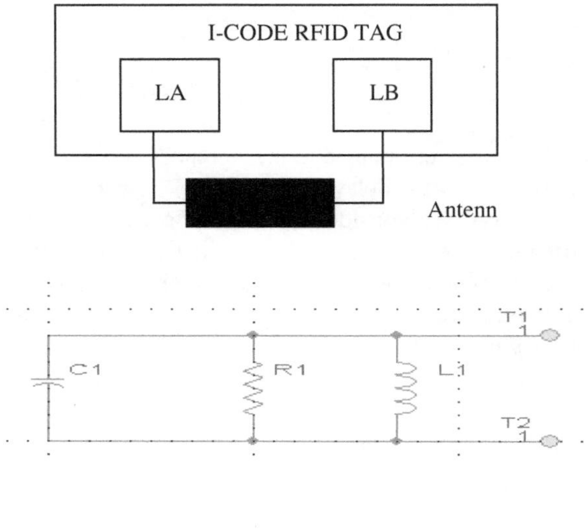

Fig. 1.15 NXP/PHILIPS ICODE IC, Parallel equivalent circuit and simplified complete equivalent circuit of the label (L1 is the antenna inductance)

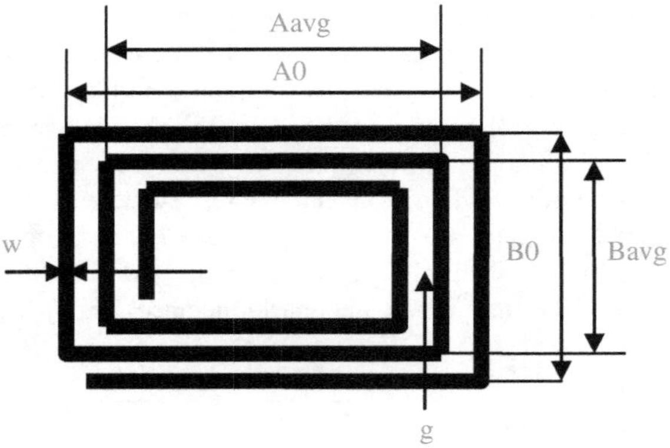

Fig. 1.16 RFID's coil dimensional parameters

$$X_1 = Aavg \cdot \ln\left(\frac{2 \cdot Aavg \cdot Bavg}{d \cdot (Aavg + \sqrt{Aavg^2 + Bavg^2})}\right)$$

$$X_2 = Bavg \cdot \ln\left(\frac{2 \cdot Aavg \cdot Bavg}{d \cdot (Bavg + \sqrt{Aavg^2 + Bavg^2})}\right)$$

$$X_3 = 2 \cdot \left[Aavg + Bavg - \sqrt{\left[Aavg^2 + Bavg^2\right]}\right]; X_4 = (Aavg + Bavg)/4$$

The RFID's coil calculation inductance is $Lcalc = \left[\frac{\mu_0}{\pi} \cdot [X1 + X2 - X3 + X4] \cdot Nc^p\right]$ L1 = Lcalc. Definition of limits, Estimations: Track thickness t, Al and Cu coils (t > 30 μm).

$$\begin{bmatrix} \frac{dV_1}{dt} \\ \frac{dV_2}{dt} \end{bmatrix} = \begin{bmatrix} 0 & 1 \\ \{-\frac{1}{C_1 \cdot \left[\frac{\mu_0}{\pi} \cdot [X_1 + X_2 - X_3 + X_4] \cdot Nc^p\right]}\} & \{-\frac{1}{C_1 \cdot R_1}\} \end{bmatrix} \cdot \begin{bmatrix} V_1 \\ V_2 \end{bmatrix}$$

Due to electromagnetic interferences, we get RFID TAG's voltage and voltage derivative with delays τ_1 and τ_2 respectively $V_1(t) \to V_1(t - \tau_1)$; $V_2(t) \to V_2(t - \tau_1)$. We consider no delay effect on dV_1/dt and dV_2/dt. The RFID TAG's differential equations under electromagnetic interference effects (we consider electromagnetic interferences (delay terms) influence only RFID TAG voltage $V_1(t)$ and voltage derivative $V_2(t)$ respect to time, there is no influence on $dV_1(t)/dt$ and $dV_2(t)/dt$):

$$\frac{dV_1}{dt} = V_2(t - \tau_2)$$
$$\frac{dV_2}{dt} = \{-\frac{1}{C_1 \cdot \left[\frac{\mu_0}{\pi} \cdot [X_1 + X_2 - X_3 + X_4] \cdot Nc^p\right]}\} \cdot V_1(t - \tau_1) - \frac{1}{C_1 \cdot R_1} \cdot V_2(t - \tau_2)$$

To find the Equilibrium points (fixed points) of the RFID TAG system is by

$$\lim_{t \to \infty} V_1(t - \tau_1) = V_1(t); \lim_{t \to \infty} V_2(t - \tau_2) = V_2(t);$$
$$\frac{dV_1(t)}{dt} = 0; \frac{dV_2(t)}{dt} = 0$$
$$\forall\, \mathrm{t} \gg \tau_1; \mathrm{t} \gg \tau_2 \;\exists\; (t - \tau_1) \approx t; (t - \tau_2) \approx t, \mathrm{t} \to \infty$$

We get two equations and the only fixed point is $E^{(0)}(V_1^{(0)}, V_2^{(0)}) = (0,0)$. Stability analysis: The standard local stability analysis about any one of the equilibrium points of RFID TAG system consists in adding to coordinate $[V_1, V_2]$ arbitrarily small increments of exponential form, and retaining the first order terms in v_1, v_2. The system of two homogeneous equations leads to a polynomial characteristic equation in the eigenvalues. The polynomial characteristic equations accept by set the below voltage and voltage derivative respect to time into two RFID TAG system equations. RFID TAG system fixed values with arbitrarily small increments of exponential form $[v_1 \; v_2] \cdot e^{\lambda \cdot t}$ are: i = 0 (first fixed point), i = 1 (second fixed point), i = 2 (third fixed point).

$$V_1(t) = V_1^{(i)} + v_1 \cdot e^{\lambda \cdot t}; V_2(t) = V_2^{(i)} + v_2 \cdot e^{\lambda \cdot t};$$
$$V_1(t - \tau_1) = V_1^{(i)} + v_1 \cdot e^{\lambda \cdot (t - \tau_1)}$$
$$V_2(t - \tau_2) = V_2^{(i)} + v_2 \cdot e^{\lambda \cdot (t - \tau_2)} \;\; \forall \;\; i = 0, 1, 2$$

We choose the above expressions for our $V_1(t)$, $V_2(t)$ as small displacement $[v_1, v_2]$ from the system fixed points in time t = 0.

$$V_1(t = 0) = V_1^{(i)} + v_1; V_2(t = 0) = V_2^{(i)} + v_2$$

for $\lambda < 0$, $t > 0$ the selected fixed point is stable otherwise $\lambda > 0$, $t > 0$ is Unstable. Our system tends to the selected fixed point exponentially for $\lambda < 0$, $t > 0$ otherwise go away from the selected fixed point exponentially. λ Is the eigenvalue parameter which establishes if the fixed point is stable or Unstable, additionally his absolute value ($|\lambda|$) establish the speed of flow toward or away from the selected fixed point [1, 2].

	$\lambda < 0$	$\lambda > 0$
$t = 0$	$V_1(t = 0) = V_1^{(i)} + v_1$ $V_2(t = 0) = V_2^{(i)} + v_2$	$V_1(t = 0) = V_1^{(i)} + v_1$ $V_2(t = 0) = V_2^{(i)} + v_2$
$t > 0$	$V_1(t) = V_1^{(i)} + v_1 e^{-\lvert\lambda\rvert \cdot t}$ $V_2(t) = V_2^{(i)} + v_2 e^{-\lvert\lambda\rvert \cdot t}$	$V_1(t) = V_1^{(i)} + v_1 e^{\lvert\lambda\rvert \cdot t}$ $V_2(t) = V_2^{(i)} + v_2 e^{\lvert\lambda\rvert \cdot t}$
$t < 0$	$V_1(t \to \infty) = V_1^{(i)}$ $V_2(t \to \infty) = V_2^{(i)}$	$V_1(t \to \infty, \lambda > 0) \sim v_1 e^{\lvert\lambda\rvert \cdot t}$ $V_2(t \to \infty, \lambda > 0) \sim v_2 e^{\lvert\lambda\rvert \cdot t}$

The speeds of flow toward or away from the selected fixed point for RFID TAG system voltage and voltage derivative respect to time are

$$\frac{dV_1(t)}{dt} = \lim_{\Delta t \to 0} \frac{V_1(t + \Delta t) - V_1(t)}{\Delta t} = \lim_{\Delta t \to 0} \frac{V_1^{(i)} + v_1 \cdot e^{\lambda \cdot (t + \Delta t)} - [V_1^{(i)} + v_1 \cdot e^{\lambda \cdot t}]}{\Delta t}$$

$$= \lim_{\Delta t \to 0} \frac{v_1 \cdot e^{\lambda \cdot t} \cdot [e^{\lambda \cdot \Delta t} - 1]}{\Delta t} \overset{e^{\lambda \cdot \Delta t} \approx 1 + \lambda \cdot \Delta t}{\longrightarrow} \lim_{\Delta t \to 0} \frac{v_1 \cdot e^{\lambda \cdot t} \cdot [1 + \lambda \cdot \Delta t - 1]}{\Delta t} = \lambda \cdot v_1 \cdot e^{\lambda \cdot t}$$

$$\frac{dV_2(t)}{dt} = \lim_{\Delta t \to 0} \frac{V_2(t + \Delta t) - V_2(t)}{\Delta t} = \lim_{\Delta t \to 0} \frac{V_2^{(i)} + v_2 \cdot e^{\lambda \cdot (t + \Delta t)} - [V_2^{(i)} + v_2 \cdot e^{\lambda \cdot t}]}{\Delta t}$$

$$= \lim_{\Delta t \to 0} \frac{v_2 \cdot e^{\lambda \cdot t} \cdot [e^{\lambda \cdot \Delta t} - 1]}{\Delta t} \overset{e^{\lambda \cdot \Delta t} \approx 1 + \lambda \cdot \Delta t}{\longrightarrow} \lim_{\Delta t \to 0} \frac{v_2 \cdot e^{\lambda \cdot t} \cdot [1 + \lambda \cdot \Delta t - 1]}{\Delta t} = \lambda \cdot v_2 \cdot e^{\lambda \cdot t}$$

And the time derivative of the above equations:

$$\frac{dV_1(t)}{dt} = v_1 \cdot \lambda \cdot e^{\lambda \cdot t}; \frac{dV_2(t)}{dt} = v_2 \cdot \lambda \cdot e^{\lambda \cdot t};$$

$$\frac{dV_1(t - \tau_1)}{dt} = v_1 \cdot \lambda \cdot e^{\lambda \cdot (t - \tau_1)} = v_1 \cdot \lambda \cdot e^{\lambda \cdot t} \cdot e^{-\tau_1 \cdot \lambda}$$

$$\frac{dV_2(t - \tau_2)}{dt} = v_2 \cdot \lambda \cdot e^{\lambda \cdot (t - \tau_2)} = v_2 \cdot \lambda \cdot e^{\lambda \cdot t} \cdot e^{-\tau_2 \cdot \lambda}$$

First, we take the RFID TAG's voltage (V_1) differential equation: $\frac{dV_1}{dt} = V_2$ and adding to its coordinates $[V_1 V_2]$. Arbitrarily small increments of exponential form $[v_1 v_2] \cdot e^{\lambda \cdot t}$ and retaining the first order terms in v_1, v_2.

$$\lambda \cdot v_1 \cdot e^{\lambda \cdot t} = V_2^{(i)} + v_2 \cdot e^{\lambda \cdot t} \; ; \; V_2^{(i=0)} = 0 \; ; \; -\lambda \cdot v_1 + v_2 = 0$$

Second, we take the RFID TAG's voltage (V_2) differential equation: $\frac{dV_2}{dt} = \{-\frac{1}{C_1 \cdot \left[\frac{\mu_0}{\pi} \cdot [X_1 + X_2 - X_3 + X_4] \cdot N_c^p\right]}\} \cdot V_1(t) - \frac{1}{C_1 \cdot R_1} \cdot V_2(t)$ and adding to its coordinates $[V_1 V_2]$ arbitrarily small increments of exponential form $[v_1\ v_2] \cdot e^{\lambda \cdot t}$ and retaining the first order terms in v_1, v_2.

$$\frac{dV_2}{dt} = \{-\frac{1}{C_1 \cdot \left[\frac{\mu_0}{\pi} \cdot [X_1 + X_2 - X_3 + X_4] \cdot N_c^p\right]}\} \cdot V_1(t) - \frac{1}{C_1 \cdot R_1} \cdot V_2(t)$$

$$\lambda \cdot v_2 \cdot e^{\lambda \cdot t} = \{-\frac{1}{C_1 \cdot \left[\frac{\mu_0}{\pi} \cdot [X_1 + X_2 - X_3 + X_4] \cdot N_c^p\right]}\} \cdot (V_1^{(i)} + v_1 \cdot e^{\lambda \cdot t})$$

$$- \frac{1}{C_1 \cdot R_1} \cdot (V_2^{(i)} + v_2 \cdot e^{\lambda \cdot t})$$

$$V_1^{(i)} = 0;\ V_2^{(i)} = 0$$

$$-\lambda \cdot v_2 - \{\frac{1}{C_1 \cdot [\frac{\mu_0}{\pi} \cdot [X_1 + X_2 - X_3 + X_4] \cdot N_c^p]}\} \cdot v_1 - \frac{1}{C_1 \cdot R_1} \cdot v_2 = 0$$

We can summery our system eigenvalues equations: $-\lambda \cdot v_1 + v_2 = 0$

$$-\{\frac{1}{C_1 \cdot [\frac{\mu_0}{\pi} \cdot [X_1 + X_2 - X_3 + X_4] \cdot N_c^p]}\} \cdot v_1 - \lambda \cdot v_2 - \frac{1}{C_1 \cdot R_1} \cdot v_2 = 0$$

$$\begin{pmatrix} -\lambda & 1 \\ -\{\frac{1}{C_1 \cdot [\frac{\mu_0}{\pi} \cdot [X_1 + X_2 - X_3 + X_4] \cdot N_c^p]}\} & -\lambda - \frac{1}{C_1 \cdot R_1} \end{pmatrix} \cdot \begin{pmatrix} v_1 \\ v_2 \end{pmatrix} = 0$$

$$A - \lambda \cdot I = \begin{pmatrix} -\lambda & 1 \\ -\{\frac{1}{C_1 \cdot [\frac{\mu_0}{\pi} \cdot [X_1 + X_2 - X_3 + X_4] \cdot N_c^p]}\} & -\lambda - \frac{1}{C_1 \cdot R_1} \end{pmatrix};$$

$$\det(A - \lambda \cdot I) = 0$$

$$\lambda \cdot (\lambda + \frac{1}{C_1 \cdot R_1}) + \frac{1}{C_1 \cdot [\frac{\mu_0}{\pi} \cdot [X_1 + X_2 - X_3 + X_4] \cdot N_c^p]} = 0$$

We get two eigenvalues: λ_1, λ_2. If $\lambda_1 < 0$, $\lambda_2 < 0$ then we have stable node.

If $\lambda_1 > 0$, $\lambda_2 > 0$ then we have unstable node. If $\lambda_1 \cdot \lambda_2 < 0$ then we have saddle point.

If $\lambda_1 = \lambda_2 < 0$ then we have attracting focus. If we have $\lambda_1 < \lambda_2 = 0$ then we have attracting line. If we have $\lambda_1 = 0 < \lambda_2$ then we have repelling line. If we have

$0 < \lambda_1 = \lambda_2$ then we have repelling focus. If λ_1, λ_2 are complex conjugate and the real part is negative than we have attracting spiral otherwise (positive real part) repelling spiral. If the real part is zero then we have a center. We define

$$V_1(t-\tau_1) = V_1^{(i)} + v_1 \cdot e^{\lambda\cdot(t-\tau_1)}; V_2(t-\tau_2) = V_2^{(i)} + v_2 \cdot e^{\lambda\cdot(t-\tau_2)}.$$

Then we get two delayed differential equations respect to adding to its coordinates $[V_1 V_2]$ arbitrarily small increments of exponential form$[v_1\ v_2] \cdot e^{\lambda \cdot t}$.

$$v_1 \cdot \lambda \cdot e^{\lambda\cdot t} = V_2^{(i)} + v_2 \cdot e^{\lambda\cdot(t-\tau_2)}; V_2^{(i=0)} = 0 \Rightarrow v_1 \cdot \lambda \cdot e^{\lambda\cdot t} = v_2 \cdot e^{\lambda\cdot(t-\tau_2)}$$

$$\begin{aligned}\lambda \cdot v_2 \cdot e^{\lambda\cdot t} = \{ & -\frac{1}{C_1 \cdot \left[\frac{\mu_0}{\pi} \cdot [X_1 + X_2 - X_3 + X_4] \cdot Nc^p\right]}\} \cdot V_1^{(i)} - \frac{1}{C_1 \cdot R_1} \cdot V_2^{(i)} \\ & + \{-\frac{1}{C_1 \cdot \left[\frac{\mu_0}{\pi} \cdot [X_1 + X_2 - X_3 + X_4] \cdot Nc^p\right]}\} \cdot v_1 \cdot e^{\lambda\cdot(t-\tau_1)} \\ & - \frac{1}{C_1 \cdot R_1} \cdot v_2 \cdot e^{\lambda\cdot(t-\tau_2)}\end{aligned}$$

In the equilibrium fixed point $V_1^{(i=0)} = V_2^{(i=0)} = 0$ and in the equilibrium fixed point

$$V_1^{(i=0)} = V_2^{(i=0)} = 0; \{-\frac{1}{C_1 \cdot \left[\frac{\mu_0}{\pi} \cdot [X_1 + X_2 - X_3 + X_4] \cdot Nc^p\right]}\} \cdot V_1^{(i)} - \frac{1}{C_1 \cdot R_1} \cdot V_2^{(i)} = 0$$

$$\lambda \cdot v_2 \cdot e^{\lambda\cdot t} = \{-\frac{1}{C_1 \cdot \left[\frac{\mu_0}{\pi} \cdot [X_1 + X_2 - X_3 + X_4] \cdot Nc^p\right]}\} \cdot v_1 \cdot e^{\lambda\cdot(t-\tau_1)} - \frac{1}{C_1 \cdot R_1} \cdot v_2 \cdot e^{\lambda\cdot(t-\tau_2)}$$

We define $f_{\#}(X_1, X_2, etc....) = \left[\frac{\mu_0}{\pi} \cdot [X_1 + X_2 - X_3 + X_4] \cdot Nc^p\right]$. The small increments Jacobian of our RFID TAG system:

$$\begin{bmatrix} -\lambda & e^{-\lambda\cdot\tau_2} \\ -\frac{1}{C_1 \cdot f_{\#}} \cdot e^{-\lambda\cdot\tau_1} & -\frac{1}{C_1 \cdot R_1} \cdot e^{-\lambda\cdot\tau_2} - \lambda \end{bmatrix} \cdot \begin{pmatrix} v_1 \\ v_2 \end{pmatrix} = 0;$$

$$A - \lambda \cdot I = \begin{bmatrix} -\lambda & e^{-\lambda\cdot\tau_2} \\ -\frac{1}{C_1 \cdot f_{\#}} \cdot e^{-\lambda\cdot\tau_1} & -\frac{1}{C_1 \cdot R_1} \cdot e^{-\lambda\cdot\tau_2} - \lambda \end{bmatrix}$$

$$\det|A - \lambda \cdot I| = 0; D(\lambda, \tau_1, \tau_2) = \lambda^2 + \lambda \cdot \frac{1}{C_1 \cdot R_1} \cdot e^{-\lambda\cdot\tau_2} + \frac{1}{C_1 \cdot f_{\#}} \cdot e^{-\lambda\cdot(\tau_1 + \tau_2)}$$

We have three stability analysis cases: $\tau_1 = \tau; \tau_2 = 0$ or $\tau_2 = \tau; \tau_1 = 0$ or $\tau_1 = \tau_2 = \tau$ otherwise $\tau_1 \neq \tau_2$. We need to get characteristics equations as all above stability analysis cases. We study the occurrence of any possible stability switching, resulting from the increase of the value of the time delay τ for the general characteristic equation $D(\lambda,\tau)$. $D(\lambda,\tau) = P_n(\lambda,\tau) + Q_m(\lambda,\tau)\cdot e^{-\lambda\tau}$

The expression for $P_n(\lambda,\tau)$ is $P_n(\lambda,\tau) = \sum_{k=0}^{n} P_k(\tau)\cdot\lambda^k = P_0(\tau) + P_1(\tau)\cdot\lambda + P_2(\tau)\cdot\lambda^2 + P_3(\tau)\cdot\lambda^3 + \ldots$.

The expression for $Q_m(\lambda,\tau)$ is $Q_m(\lambda,\tau) = \sum_{k=0}^{m} q_k(\tau)\cdot\lambda^k = q_0(\tau) + q_1(\tau)\cdot\lambda + q_2(\tau)\cdot\lambda^2 + \ldots$.

First, we analyze RFID Tag system second order characteristic equation for. The first case we analyze is when there is a delay in RFID Label voltage and no delay in voltage time derivative [4, 5].

$$D(\lambda,\tau_1=\tau,\tau_2=0) = \lambda^2 + \lambda\cdot\frac{1}{C_1\cdot R_1} + \frac{1}{C_1\cdot f_{\#}}\cdot e^{-\lambda\cdot\tau_1};$$

$$D(\lambda,\tau) = P_n(\lambda,\tau) + Q_m(\lambda,\tau)\cdot e^{-\lambda\tau}$$

The expression for $P_n(\lambda,\tau)$:

$$P_n(\lambda,\tau) = \sum_{k=0}^{n} P_k(\tau)\cdot\lambda^k = P_0(\tau) + P_1(\tau)\cdot\lambda + P_2(\tau)\cdot\lambda^2 = \lambda^2 + \lambda\cdot\frac{1}{C_1\cdot R_1};$$

$$P_2(\tau) = 1; P_1(\tau) = \frac{1}{C_1\cdot R_1}; P_0(\tau) = 0$$

The expression for $Q_m(\lambda,\tau)$: $Q_m(\lambda,\tau) = \sum_{k=0}^{m} q_k(\tau)\cdot\lambda^k = q_0(\tau) = \frac{1}{C_1\cdot f_{\#}}$. Our RFID system second order characteristic equation: $D(\lambda,\tau) = \lambda^2 + a(\tau)\cdot\lambda + b(\tau)\cdot\lambda\cdot e^{-\lambda\cdot\tau} + c(\tau) + d(\tau)\cdot e^{-\lambda\cdot\tau}$

Then $a(\tau) = \frac{1}{C_1\cdot R_1}; \mathrm{b}(\tau) = 0; \mathrm{c}(\tau) = 0; \mathrm{d}(\tau) = \frac{1}{C_1\cdot f_{\#}}\tau \in R_{+0}$ and $a(\tau), \mathrm{b}(\tau), \mathrm{c}(\tau), \mathrm{d}(\tau) : R_{+0} \to R$ are differentiable functions of the class $C^1(R_{+0})$, such that $\mathrm{c}(\tau) + \mathrm{d}(\tau) = \frac{1}{C_1\cdot f_{\#}} \neq 0$ for all $\tau \in R_{+0}$ and for any $\tau, \mathrm{b}(\tau), \mathrm{d}(\tau)$ are not simultaneously zero. We have

$$P(\lambda,\tau) = P_n(\lambda,\tau) = \lambda^2 + a(\tau)\cdot\lambda + c(\lambda) = \lambda^2 + \frac{1}{C_1\cdot R_1}\cdot\lambda$$

$$Q(\lambda,\tau) = Q_m(\lambda,\tau) = b(\tau)\cdot\lambda + d(\tau) = \frac{1}{C_1\cdot f_{\#}}$$

We assume that $P_n(\lambda,\tau) = P_n(\lambda)$ and $Q_m(\lambda,\tau) = Q_m(\lambda)$ can't have common imaginary roots. That is, for any real number; $\frac{1}{C_1\cdot f_{\#}} - \omega^2 + i\cdot\omega\cdot\frac{1}{C_1\cdot R_1} \neq 0$

$$F(\omega,\tau) = |P(i\cdot\omega,\tau)|^2 - |Q(i\cdot\omega,\tau)|^2 = (c-\omega^2)^2 + \omega^2\cdot a^2 - (\omega^2\cdot b^2 + d^2)$$

$F(\omega,\tau) = \omega^4 + \omega^2\cdot\frac{1}{(C_1\cdot R_1)^2} - \frac{1}{(C_1\cdot f_{\#})^2}$; Hence $F(\omega,\tau) = 0$ implies $\omega^4 + \omega^2\cdot\frac{1}{(C_1\cdot R_1)^2} - \frac{1}{(C_1\cdot f_{\#})^2} = 0$ and its roots are given by

$$\omega_+^2 = \frac{1}{2}\cdot\{(b^2 + 2\cdot c - a^2) + \sqrt{\Delta}\} = \frac{1}{2}\cdot\{\sqrt{\Delta} - \frac{1}{(C_1\cdot R_1)^2}\};$$

$$\omega_-^2 = \frac{1}{2}\cdot\{(b^2 + 2\cdot c - a^2) - \sqrt{\Delta}\} = -\frac{1}{2}\cdot\{\sqrt{\Delta} + \frac{1}{(C_1\cdot R_1)^2}\}$$

$$\omega_-^2 = \frac{1}{2}\cdot\{(b^2 + 2\cdot c - a^2) - \sqrt{\Delta}\} = -\frac{1}{2}\cdot\{\sqrt{\Delta} + \frac{1}{(C_1\cdot R_1)^2}\};$$

$$\Delta = (b^2 + 2\cdot c - a^2) - 4\cdot(c^2 - d^2) = \frac{1}{C_1^2}\cdot[(\frac{2}{f_{\#}})^2 - \frac{1}{R_1^2}]$$

Therefore the following holds: $2\cdot\omega_{+/-}^2 - (b^2 + 2\cdot c - a^2) = \pm\sqrt{\Delta}$; $2\cdot\omega_{+/-}^2 + \frac{1}{(C_1\cdot R_1)^2} = \pm\sqrt{\Delta}$

Furthermore

$$P_R(i\cdot\omega,\tau) = c(\tau) - \omega^2(\tau) = -\omega^2(\tau); P_I(i\cdot\omega,\tau) = \omega(\tau)\cdot a(\tau) = \omega(\tau)\cdot\frac{1}{C_1\cdot R_1}$$

$$Q_R(i\cdot\omega,\tau) = d(\tau) = \frac{1}{C_1\cdot f_{\#}}; Q_I(i\cdot\omega,\tau) = \omega(\tau)\cdot b(\tau) = 0$$

Hence

$$\sin\theta(\tau) = \frac{-P_R(i\cdot\omega,\tau)\cdot Q_I(i\cdot\omega,\tau) + P_I(i\cdot\omega,\tau)\cdot Q_R(i\cdot\omega,\tau)}{|Q(i\cdot\omega,\tau)|^2}$$
$$\cos\theta(\tau) = -\frac{P_R(i\cdot\omega,\tau)\cdot Q_R(i\cdot\omega,\tau) + P_I(i\cdot\omega,\tau)\cdot Q_I(i\cdot\omega,\tau)}{|Q(i\cdot\omega,\tau)|^2}$$
$$\sin\theta(\tau) = \frac{-(c-\omega^2)\cdot\omega\cdot b + \omega\cdot a\cdot d}{\omega^2\cdot b^2 + d^2} = \omega\cdot\frac{f_{\#}}{R_1};$$
$$\cos\theta(\tau) = -\frac{(c-\omega^2)\cdot d + \omega^2\cdot a\cdot b}{\omega^2\cdot b^2 + d^2} = \omega^2\cdot C_1\cdot f_{\#}$$

Which jointly with $\omega^4 + \omega^2\cdot\frac{1}{(C_1\cdot R_1)^2} - \frac{1}{(C_1\cdot f_{\#})^2} = 0$ Defines the maps $S_n(\tau) = \tau - \tau_n(\tau); \tau \in \mathrm{I}, \mathrm{n} \in \mathbb{N}_0$, that are continuous and differentiable in τ based on Lemma 1.1. Hence we use theorem 1.2. This proves the theorem 1.3 and theorem 1.4. Remark: a, b, c, d parameters are independent of delay parameter τ even we use $a(\tau), \mathrm{b}(\tau), \mathrm{c}(\tau), \mathrm{d}(\tau)$. Second, we analyze RFID Tag system second order characteristic equation for $\tau_1 = 0; \tau_2 = \tau$. The second case we analyze is when there is no delay in RFID Label voltage and there is a delay in voltage time derivative.

$$D(\lambda, \tau_1 = 0, \tau_2 = \tau) = \lambda^2 + \lambda\cdot\frac{1}{C_1\cdot R_1}\cdot e^{-\lambda\cdot\tau_2} + \frac{1}{C_1\cdot f_{\#}}\cdot e^{-\lambda\cdot\tau_2}$$
$$D(\lambda, \tau_1 = 0, \tau_2 = \tau) = \lambda^2 + (\lambda\cdot\frac{1}{C_1\cdot R_1} + \frac{1}{C_1\cdot f_{\#}})\cdot e^{-\lambda\cdot\tau};$$
$$D(\lambda,\tau) = P_n(\lambda,\tau) + Q_m(\lambda,\tau)\cdot e^{-\lambda\tau}$$

The expression for $P_n(\lambda,\tau)$ is $P_n(\lambda,\tau) = \sum_{k=0}^{n} P_k(\tau)\cdot\lambda^k = P_0(\tau) + P_1(\tau)\cdot\lambda + P_2(\tau)\cdot\lambda^2 = \lambda^2$

$P_2(\tau) = 1; P_1(\tau) = 0; P_0(\tau) = 0$. The expression for $Q_m(\lambda,\tau)$ is

$$Q_m(\lambda,\tau) = \sum_{k=0}^{m} q_k(\tau)\cdot\lambda^k = \lambda\cdot\frac{1}{C_1\cdot R_1} + \frac{1}{C_1\cdot f_{\#}};$$
$$q_0(\tau) = \frac{1}{C_1\cdot f_{\#}}; q_1(\tau) = \frac{1}{C_1\cdot R_1}; q_2(\tau) = 0$$

Our RFID system second order characteristic equation:

$$D(\lambda,\tau) = \lambda^2 + a(\tau)\cdot\lambda + b(\tau)\cdot\lambda\cdot e^{-\lambda\cdot\tau} + c(\tau) + d(\tau)\cdot e^{-\lambda\cdot\tau}$$

$$a(\tau) = 0; b(\tau) = \frac{1}{C_1\cdot R_1}; c(\tau) = 0; d(\tau) = \frac{1}{C_1\cdot f_{\#}}$$

And in the same manner like our previous case analysis:

$$P(\lambda,\tau) = P_n(\lambda,\tau) = \lambda^2 + a(\tau)\cdot\lambda + c(\lambda) = \lambda^2;$$

$$Q(\lambda,\tau) = Q_m(\lambda,\tau) = b(\tau)\cdot\lambda + d(\tau) = \lambda\cdot\frac{1}{C_1\cdot R_1} + \frac{1}{C_1\cdot f_{\#}}$$

We assume that $P_n(\lambda,\tau) = P_n(\lambda)$ and $Q_m(\lambda,\tau) = Q_m(\lambda)$ can't have common imaginary roots. That is, for any real number ω; $p_n(\lambda = i\cdot\omega,\tau) + Q_m(\lambda = i\cdot\omega,\tau) \neq 0$

$$\frac{1}{C_1\cdot f_{\#}} - \omega^2 + i\cdot\omega\cdot\frac{1}{C_1\cdot R_1} \neq 0;$$

$$F(\omega,\tau) = |P(i\cdot\omega,\tau)|^2 - |Q(i\cdot\omega,\tau)|^2 = (c-\omega^2)^2 + \omega^2\cdot a^2 - (\omega^2\cdot b^2 + d^2)$$

$F(\omega,\tau) = \omega^4 - \omega^2\cdot\frac{1}{(C_1\cdot R_1)^2} - \frac{1}{(C_1\cdot f_{\#})^2}$

Hence $F(\omega,\tau) = 0$ implies $\omega^4 - \omega^2\cdot\frac{1}{(C_1\cdot R_1)^2} - \frac{1}{(C_1\cdot f_{\#})^2} = 0$ And its roots are given by

$$\omega_+^2 = \frac{1}{2}\cdot\{(b^2 + 2\cdot c - a^2) + \sqrt{\Delta}\} = \frac{1}{2}\cdot\{\sqrt{\Delta} + \frac{1}{(C_1\cdot R_1)^2}\};$$

$$\omega_-^2 = \frac{1}{2}\cdot\{(b^2 + 2\cdot c - a^2) - \sqrt{\Delta}\} = \frac{1}{2}\cdot\{-\sqrt{\Delta} + \frac{1}{(C_1\cdot R_1)^2}\}$$

$$\Delta = (b^2 + 2\cdot c - a^2) - 4\cdot(c^2 - d^2) = \frac{1}{C_1^2}\cdot[(\frac{2}{f_{\#}})^2 + \frac{1}{R_1^2}]$$

Therefore the following holds:

$$2\cdot\omega_{+/-}^2 - (b^2 + 2\cdot c - a^2) = \pm\sqrt{\Delta}; 2\cdot\omega_{+/-}^2 + \frac{1}{(C_1\cdot R_1)^2} = \pm\sqrt{\Delta}$$

Furthermore

$$P_R(i\cdot\omega,\tau) = c(\tau) - \omega^2(\tau) = -\omega^2(\tau);$$

$$P_I(i\cdot\omega,\tau) = \omega(\tau)\cdot a(\tau) = 0; Q_R(i\cdot\omega,\tau) = d(\tau) = \frac{1}{C_1\cdot f_{\#}}$$

$$Q_I(i\cdot\omega,\tau) = \omega(\tau)\cdot b(\tau) = \omega(\tau)\cdot\frac{1}{C_1\cdot R_1};$$

$$\sin\theta(\tau) = \frac{-P_R(i\cdot\omega,\tau)\cdot Q_I(i\cdot\omega,\tau) + P_I(i\cdot\omega,\tau)\cdot Q_R(i\cdot\omega,\tau)}{|Q(i\cdot\omega,\tau)|^2}$$

$$\cos\theta(\tau) = -\frac{P_R(i\cdot\omega,\tau)\cdot Q_R(i\cdot\omega,\tau) + P_I(i\cdot\omega,\tau)\cdot Q_I(i\cdot\omega,\tau)}{|Q(i\cdot\omega,\tau)|^2};$$

$$\sin\theta(\tau) = \frac{-(c-\omega^2)\cdot\omega\cdot b + \omega\cdot a\cdot d}{\omega^2\cdot b^2 + d^2} = \frac{\omega^3\cdot C_1\cdot R_1}{\omega^2 + (\frac{R_1}{f_{\#}})^2}$$

$$\cos\theta(\tau) = -\frac{(c-\omega^2)\cdot d + \omega^2\cdot a\cdot b}{\omega^2\cdot b^2 + d^2} = \frac{\omega^2\cdot C_1\cdot\frac{R_1}{f_{\#}}}{\omega^2 + (\frac{R_1}{f_{\#}})^2}$$

Which jointly with

$$\omega^4 - \omega^2\cdot\frac{1}{(C_1\cdot R_1)^2} - \frac{1}{(C_1\cdot f_{\#})^2} = 0$$

Defines the maps $S_n(\tau) = \tau - \tau_n(\tau); \tau \in \mathrm{I}, \mathrm{n} \in \mathbb{N}_0$

Defines the maps $S_n(\tau) = \tau - \tau_n(\tau); \tau \in \mathrm{I}, \mathrm{n} \in \mathbb{N}_0$ are continuous and differentiable in τ based on Lemma 1.1. Hence we use theorem 1.2. This proves the theorem 1.3 and theorem 1.4. Remark: a, b, c, d parameters are independent of delay parameter τ even we use $a(\tau), \mathrm{b}(\tau), \mathrm{c}(\tau), \mathrm{d}(\tau)$ [4, 5].

Third, we analyze RFID Tag system second order characteristic equation for $\tau_1{=}\tau; \tau_2 = \tau$. The third case we analyze is when there is delay both in RFID Label voltage and voltage time derivative [4, 5].

$$D(\lambda,\tau_1=\tau,\tau_2=\tau) = \lambda^2 + \lambda\cdot\frac{1}{C_1\cdot R_1}\cdot e^{-\lambda\cdot\tau} + \frac{1}{C_1\cdot f_{\#}}\cdot e^{-\lambda\cdot\tau\cdot 2};$$

$$D(\lambda,\tau_1=\tau_2=\tau) = \lambda^2 + (\lambda\cdot\frac{1}{C_1\cdot R_1} + \frac{1}{C_1\cdot f_{\#}}\cdot e^{-\lambda\cdot\tau})\cdot e^{-\lambda\cdot\tau}$$

$$D(\lambda,\tau) = P_n(\lambda,\tau) + Q_m(\lambda,\tau)\cdot e^{-\lambda\tau}$$

The expression for $P_n(\lambda,\tau)$ is

$$P_n(\lambda,\tau) = \sum_{k=0}^{n} P_k(\tau)\cdot\lambda^k = P_0(\tau) + P_1(\tau)\cdot\lambda + P_2(\tau)\cdot\lambda^2 = \lambda^2; P_2(\tau) = 1;$$

$$P_1(\tau) = 0; P_0(\tau) = 0.$$

The expression for $Q_m(\lambda,\tau)$ is $Q_m(\lambda,\tau) = \sum_{k=0}^{m} q_k(\tau)\cdot\lambda^k = \lambda\cdot\frac{1}{C_1\cdot R_1} + \frac{1}{C_1\cdot f_{\#}}\cdot e^{-\lambda\tau}$

Taylor expansion: $e^{-\lambda\tau} \approx 1 - \lambda\cdot\tau + \frac{\lambda^2\cdot\tau^2}{2}$ since we need n > m [BK] analysis we choose $e^{-\lambda\tau} \approx 1 - \lambda\cdot\tau$.

$$Q_m(\lambda,\tau) = \sum_{k=0}^{m} q_k(\tau)\cdot\lambda^k = \lambda\cdot\frac{1}{C_1}\left(\frac{1}{R_1} - \frac{\tau}{f_{\#}}\right) + \frac{1}{C_1\cdot f_{\#}};$$
$$q_0(\tau,\lambda) = \frac{1}{C_1\cdot f_{\#}};\, q_1(\tau) = \frac{1}{C_1}\cdot\left(\frac{1}{R_1} - \frac{\tau}{f_{\#}}\right);\, q_2(\tau) = 0$$

Our RFID system second order characteristic equation:

$$D(\lambda,\tau) = \lambda^2 + a(\tau)\cdot\lambda + b(\tau)\cdot\lambda\cdot e^{-\lambda\cdot\tau} + c(\tau) + d(\tau)\cdot e^{-\lambda\cdot\tau}$$
$$a(\tau) = 0;\,\mathrm{b}(\tau) = \frac{1}{C_1}\cdot\left(\frac{1}{R_1} - \frac{\tau}{f_{\#}}\right);\,\mathrm{c}(\tau) = 0;\,\mathrm{d}(\tau) = \frac{1}{C_1\cdot f_{\#}}$$

And in the same manner like our previous case analysis: $P(\lambda,\tau) = P_n(\lambda,\tau) = \lambda^2$; $Q(\lambda,\tau) = Q_m(\lambda,\tau) = \lambda\cdot\frac{1}{C_1}\left(\frac{1}{R_1} - \frac{\tau}{f_{\#}}\right) + \frac{1}{C_1\cdot f_{\#}}$

We assume that $P_n(\lambda,\tau) = P_n(\lambda)$ and $Q_m(\lambda,\tau)$ can't have common imaginary roots. That is, for any real number ω; $p_n(\lambda = i\cdot\omega,\tau) + Q_m(\lambda = i\cdot\omega,\tau) \neq 0$

$$-\omega^2 + i\cdot\omega\cdot\frac{1}{C_1}\left(\frac{1}{R_1} - \frac{\tau}{f_{\#}}\right) + \frac{1}{C_1\cdot f_{\#}} \neq 0;$$
$$F(\omega,\tau) = |P(i\cdot\omega,\tau)|^2 - |Q(i\cdot\omega,\tau)|^2;\, P(i\cdot\omega,\tau) = -\omega^2$$
$$P_R(i\cdot\omega,\tau) = -\omega^2;\, P_I(i\cdot\omega,\tau) = 0;$$
$$Q(\lambda = i\cdot\omega,\tau) = i\cdot\omega\cdot\frac{1}{C_1}\left(\frac{1}{R_1} - \frac{\tau}{f_{\#}}\right) + \frac{1}{C_1\cdot f_{\#}}$$
$$Q_I(\lambda = i\cdot\omega,\tau) = \omega\cdot\frac{1}{C_1}\left(\frac{1}{R_1} - \frac{\tau}{f_{\#}}\right);\, Q_R(\lambda = i\cdot\omega,\tau) = \frac{1}{C_1\cdot f_{\#}}$$
$$|P(i\cdot\omega,\tau)|^2 = P_I^2 + P_R^2;\, |Q(i\cdot\omega,\tau)|^2 = Q_I^2 + Q_R^2;$$
$$|P(i\cdot\omega,\tau)|^2 = P_I^2 + P_R^2 = \omega^4$$
$$|Q(i\cdot\omega,\tau)|^2 = \omega^2\cdot\frac{1}{C_1^2}\left(\frac{1}{R_1} - \frac{\tau}{f_{\#}}\right)^2 + \frac{1}{(C_1\cdot f_{\#})^2};$$
$$F(\omega,\tau) = \omega^4 - \omega^2\cdot\frac{1}{C_1^2}\left(\frac{1}{R_1} - \frac{\tau}{f_{\#}}\right)^2 - \frac{1}{(C_1\cdot f_{\#})^2}$$

Hence $F(\omega,\tau) = 0$ implies $\omega^4 - \omega^2\cdot\frac{1}{C_1^2}\left(\frac{1}{R_1} - \frac{\tau}{f_{\#}}\right)^2 - \frac{1}{(C_1\cdot f_{\#})^2} = 0$

$$F_{\omega} = 4 \cdot \omega^3 - 2 \cdot \omega \cdot \frac{1}{C_1^2} (\frac{1}{R_1} - \frac{\tau}{f_{\#}})^2 = 2 \cdot \omega \cdot [2 \cdot \omega^2 - \frac{1}{C_1^2} (\frac{1}{R_1} - \frac{\tau}{f_{\#}})^2];$$
$$F_{\tau} = \frac{2 \cdot \omega^2}{C_1^2 \cdot f_{\#}} \cdot (\frac{1}{R_1} - \frac{\tau}{f_{\#}})$$
$$P_{I\omega} = 0; P_{R\omega} = -2 \cdot \omega;$$
$$Q_{I\omega} = \frac{1}{C_1} \cdot [\frac{1}{R_1} - \frac{\tau}{f_{\#}}]; Q_{R\omega} = 0; P_{I\tau} = 0; P_{R\tau} = 0$$
$$Q_{I\tau} = -\frac{\omega}{C_1 \cdot f_{\#}}; Q_{R\tau} = 0$$

The expressions for U, V can be derived easily [BK]:

$$U = (P_R \cdot P_{I\omega} - P_I \cdot P_{R\omega}) - (Q_R \cdot Q_{I\omega} - Q_I \cdot Q_{R\omega});$$
$$V = (P_R \cdot P_{Ix} - P_I \cdot P_{Rx}) - (Q_R \cdot Q_{Ix} - Q_I \cdot Q_{Rx})$$
$$V = \frac{\omega}{C_1^2 \cdot f_{\#}^2}; \mathrm{U} = \frac{1}{C_1^2 \cdot f_{\#}} \cdot [\frac{\tau}{f_{\#}} - \frac{1}{R_1}]; \omega_{\tau} = -\frac{F_{\tau}}{F_{\omega}}$$

And we get the expression: $\omega_{\tau} = -\frac{\frac{\omega}{C_1^2 \cdot f_{\#}} \cdot (\frac{1}{R_1} - \frac{\tau}{f_{\#}})}{2 \cdot \omega^2 - \frac{1}{C_1^2} (\frac{1}{R_1} - \frac{\tau}{f_{\#}})^2}$

Defines the maps $S_n(\tau) = \tau - \tau_n(\tau); \tau \in \mathrm{I}, \mathrm{n} \in \mathbb{N}_0$

Defines the maps $S_n(\tau) = \tau - \tau_n(\tau); \tau \in \mathrm{I}, \mathrm{n} \in \mathbb{N}_0$ that are continuous and differentiable in τ based on Lemma 1.1 (see Appendix A). Hence we use theorem 1.2. This proves the theorem 1.3 and theorem 1.4 (see Appendix D).

Remark Taylor approximation for $e^{-\lambda\tau} \approx 1 - \lambda \cdot \tau$ giving us a good stability analysis, approximation only for a restricted delay time interval.

Now we discuss RFID TAG system stability analysis under delayed variables in time. Our RFID homogeneous system for v_1, v_2 leads to a characteristic equation for the eigenvalue λ having the form $P(\lambda) + Q(\lambda) \cdot e^{-\lambda \cdot \tau} = 0$; first case $\tau_1 = \tau; \tau_2 = 0$. $D(\lambda, \tau_1 = \tau, \tau_2 = 0) = \lambda^2 + \lambda \cdot \frac{1}{C_1 \cdot R_1} + \frac{1}{C_1 \cdot f_{\#}} \cdot e^{-\lambda \cdot \tau_1}$. We use different parameters terminology from our last characteristics parameters definition:

$$k \to j; p_k(\tau) \to a_j; q_k(\tau) \to c_j; n = 2; m = 0$$

Additionally $P_n(\lambda, \tau) \to P(\lambda); Q_m(\lambda, \tau) \to Q(\lambda)$ then $P(\lambda) = \sum_{j=0}^{2} a_j \cdot \lambda^j$ and $Q(\lambda) = \sum_{j=0}^{0} c_j \cdot \lambda^j$.

$$P(\lambda) = \lambda^2 + \lambda \cdot \frac{1}{C_1 \cdot R_1}; Q(\lambda) = \frac{1}{C_1 \cdot f_{\#}} \quad n, m \in \mathbb{N}_0, \ n > m$$

And $a_j, c_j : \mathrm{R}_{+0} \to R$ are continuous and differentiable function of τ such that $a_0 + c_0 \neq 0$. In the following "–" denotes complex and conjugate. $P(\lambda), Q(\lambda)$

Are analytic functions in λ and differentiable in τ. And the coefficients:

$$\{a_j(C_1, R_1), c_j(C_1, \text{antenna parametrs})\} \in \mathbb{R}$$

Depend on RFID C_1, R_1 values and antenna parameters but not on τ. $a_0 = 0, a_1 = \frac{1}{C_1 \cdot R_1}, a_2 = 1, a_3 = 0; c_0 = \frac{1}{C_1 \cdot f_{\#}}, c_1 = c_2 = 0$

Unless strictly necessary, the designation of the varied arguments $(R_1, C_1, \text{antenna parametrs})$ will subsequently be omitted from P, Q, a_j, c_j. The coefficients a_j, c_j are continuous, and differentiable functions of their arguments, and direct substitution shows that $a_0 + c_0 \neq 0; \frac{1}{C_1 \cdot f_{\#}} \neq 0$

$\forall C_1$, antenna parameters $\in \mathbb{R}_+$ I.e. $\lambda = 0$ is not a root of the characteristic equation. Furthermore $P(\lambda)$, $Q(\lambda)$ are analytic functions of λ for which the following requirements of the analysis (see Kuang 1993, Sect. 3.4) can also be verified in the present case [4, 5].

(a) If $\lambda = i \cdot \omega$, $\omega \in \mathbb{R}$ then $P(i \cdot \omega) + Q(i \cdot \omega) \neq 0$, i.e. P and Q have no common imaginary roots. This condition was verified numerically in the entire $(R_1, C_1, \text{antenna parametrs})$ domain of interest.

(b) $|Q(\lambda)/P(\lambda)|$ Is bounded for $|\lambda| \to \infty$, $\mathrm{Re}\lambda \geq 0$. No roots bifurcation from ∞. Indeed, in the limit $|Q(\lambda)/P(\lambda)| = |\frac{1}{C_1 \cdot f_{\#} \cdot (\lambda^2 + \lambda \cdot \frac{1}{C_1 \cdot R_1})}|$

(c)
$$F(\omega) = |P(i \cdot \omega)|^2 - |Q(i \cdot \omega)|^2; F(\omega, \tau) = \omega^4 + \omega^2 \cdot \frac{1}{(C_1 \cdot R_1)^2} - \frac{1}{(C_1 \cdot f_{\#})^2}$$

Has at most a finite number of zeros. Indeed, this is a bi-cubic polynomial in ω (second degree in ω^2).

(d) Each positive root $\omega(C_1, R_1, \text{antenna parametrs})$ of $\mathrm{F}(\omega) = 0$ is continuous and differentiable with respect to C_1, R_1, antenna parametrs.
The condition can only be assessed numerically.

In addition, since the coefficients in P and Q are real, we have, and $\overline{Q(-i \cdot \omega)} = Q(i \cdot \omega)$ thus, $\omega > 0$ may be an eigenvalue of the characteristic equation. The analysis consists in identifying the roots of the characteristic equation situated on the imaginary axis of the complex λ-plane, whereby increasing the parameters C_1, R_1, antenna parametrs and delay τ, Reλ may, at the crossing, Change its sign from (−) to (+), i.e. from stable focus $E^{(0)}(V_1^{(0)}, V_2^{(0)}) = (0, 0)$ to an unstable one, or vice versa.

This feature may be further assessed by examining the sign of the partial derivatives with respect to C_1, R_1 and antenna parameters.

$$\wedge^{-1}(C_1) = (\frac{\partial \mathrm{Re}\lambda}{\partial C_1})_{\lambda = i\cdot\omega}, \mathrm{R}_1, \text{antenna parameters} = const$$

$$\wedge^{-1}(\mathrm{R}_1) = (\frac{\partial \mathrm{Re}\lambda}{\partial \mathrm{R}_1})_{\lambda = i\cdot\omega}, C_1, \text{antenna parameters} = const$$

$$\wedge^{-1}(f_{\#}) = (\frac{\partial \mathrm{Re}\lambda}{\partial f_{\#}})_{\lambda = i\cdot\omega}, C_1, \mathrm{R}_1 = const; \wedge^{-1}(\tau) = (\frac{\partial \mathrm{Re}\lambda}{\partial \tau})_{\lambda = i\cdot\omega}, C_1, \mathrm{R}_1,$$

$$\text{antenna parameters} = const; \text{ where } \omega \in \mathbb{R}_+.$$

In the first case $\tau_1 = \tau; \tau_2 = 0$ we get the following results

$$P_R(i\cdot\omega) = -a_2\cdot\omega^2 + a_0 = -\omega^2; P_I(i\cdot\omega) = -a_3\cdot\omega^3 + a_1\cdot\omega = \omega\cdot\frac{1}{C_1\cdot R_1}$$

$$Q_R(i\cdot\omega) = -c_2\cdot\omega^2 + c_0 = \frac{1}{C_1\cdot f_{\#}}; Q_I(i\cdot\omega) = c_1\cdot\omega = 0; F(\omega) = 0$$

$$\omega = \pm\sqrt{\frac{1}{2\cdot(C_1\cdot R_1)^2} \pm \frac{1}{2}\cdot\sqrt{\frac{1}{(C_1\cdot R_1)^4} + 4\cdot\frac{1}{[C_1\cdot f_{\#}]^2}}}; \frac{1}{(C_1\cdot R_1)^4} + 4\cdot\frac{1}{[C_1\cdot f_{\#}]^2} > 0$$

Always and additional for $\omega \in R$; $\frac{1}{2\cdot(C_1\cdot R_1)^2} \pm \frac{1}{2}\cdot\sqrt{\frac{1}{(C_1\cdot R_1)^4} + 4\cdot\frac{1}{[C_1\cdot f_{\#}]^2}} > 0$

And there are two options: first always exist $\frac{1}{2\cdot(C_1\cdot R_1)^2} + \frac{1}{2}\cdot\sqrt{\frac{1}{(C_1\cdot R_1)^4} + 4\cdot\frac{1}{[C_1\cdot f_{\#}]^2}} > 0$ Second $\frac{1}{2\cdot(C_1\cdot R_1)^2} - \frac{1}{2}\cdot\sqrt{\frac{1}{(C_1\cdot R_1)^4} + 4\cdot\frac{1}{[C_1\cdot f_{\#}]^2}} > 0$

Not exist and always negative for any RFID TAG overall parameter values.

Writing $P(\lambda) = P_R(\lambda) + i\cdot P_I(\lambda)$ and $Q(\lambda) = Q_R(\lambda) + i\cdot Q_I(\lambda)$, and inserting $\lambda = i\cdot\omega$

Into the RFID characteristic equation, ω must satisfy the following:

$$\sin\omega\cdot\tau = g(\omega) = \frac{-P_R(i\cdot\omega)\cdot Q_I(i\cdot\omega) + P_I(i\cdot\omega)\cdot Q_R(i\cdot\omega)}{|Q(i\cdot\omega)|^2}$$

$$\cos\omega\cdot\tau = h(\omega) = -\frac{P_R(i\cdot\omega)\cdot Q_R(i\cdot\omega) + P_I(i\cdot\omega)\cdot Q_I(i\cdot\omega)}{|Q(i\cdot\omega)|^2}$$

Where $|Q(i\cdot\omega)|^2 \neq 0$ in view of requirement (a) above, and $(g,h) \in R$. Furthermore, it follows above $\sin\omega\cdot\tau$ and $\cos\omega\cdot\tau$ equations that, by squaring and adding the sides, ω must be a positive root of $F(\omega) = |P(i\cdot\omega)|^2 - |Q(i\cdot\omega)|^2 = 0$.

Note that $F(\omega)$ is independent of τ. Now it is important to notice that if $\tau \notin I$ (assume that $I \subseteq R_{+0}$ is the set where $\omega(\tau)$ is a positive root of $F(\omega)$ and for $\tau \notin I$, $\omega(\tau)$ is not defined. Then for all τ in I $\omega(\tau)$ is satisfied that $F(\omega,\tau) = 0$)

Then there are no positive $\omega(\tau)$ solutions of $F(\omega,\tau)=0$, and we cannot have stability switches. For any $\tau \in I$, where $\omega(\tau)$ is a positive solution of $F(\omega,\tau)=0$ We can define the angle $\theta(\tau) \in [0, 2\cdot\pi]$ as the solution of

$$\sin\theta(\tau) = \frac{-P_R(i\cdot\omega)\cdot Q_I(i\cdot\omega) + P_I(i\cdot\omega)\cdot Q_R(i\cdot\omega)}{|Q(i\cdot\omega)|^2}$$

$$\cos\theta(\tau) = -\frac{P_R(i\cdot\omega)\cdot Q_R(i\cdot\omega) + P_I(i\cdot\omega)\cdot Q_I(i\cdot\omega)}{|Q(i\cdot\omega)|^2}$$

And the relation between the argument $\theta(\tau)$ and $\omega(\tau)\cdot\tau$ for $\tau\in I$ must be $\omega(\tau)\cdot\tau = \theta(\tau) + n\cdot 2\cdot\pi \; \forall\, n \in \mathbb{N}_0$. Hence we can define the maps $\tau_n : I \to R_{+0}$ given by $\tau_n(\tau) = \frac{\theta(\tau)+n\cdot 2\cdot\pi}{\omega(\tau)}; n\in\mathbb{N}_0, \tau\in I$. Let us introduce the functions $I \to R$; $S_n(\tau) = \tau - \tau_n(\tau)$, $\tau \in I$, $n\in\mathbb{N}_0$ that is continuous and differentiable in τ. In the following, the subscripts λ, ω, R_1, C_1 and RFID TAG antenna parameters $(w, g, B_0, A_0, A_{avg}, B_{avg}, etc.,)$ indicate the corresponding partial derivatives. Let us first concentrate on $\wedge(x)$, remember in $\lambda(R_1, C_1, w, g, B_0, A_0, A_{avg}, B_{avg}, etc.,)$ and $\omega(R_1, C_1, w, g, B_0, A_0, A_{avg}, B_{avg}, etc.,)$, and keeping all parameters except one (x) and τ. The derivation closely follows that in reference [BK]. Differentiating RFID characteristic equation $P(\lambda)+Q(\lambda)\cdot e^{-\lambda\cdot\tau}=0$ with respect to specific parameter (x), and inverting the derivative, for convenience, one calculates:

Remark $x = R_1, C_1, w, g, B_0, A_0, A_{avg}, B_{avg}, etc.,$

$$\left(\frac{\partial\lambda}{\partial x}\right)^{-1} = \frac{-P_\lambda(\lambda,x)\cdot Q(\lambda,x) + Q_\lambda(\lambda,x)\cdot P(\lambda,x) - \tau\cdot P(\lambda,x)\cdot Q(\lambda,x)}{P_x(\lambda,x)\cdot Q(\lambda,x) - Q_x(\lambda,x)\cdot P(\lambda,x)}$$

Where $P_\lambda = \frac{\partial P}{\partial \lambda}, \ldots$ etc., substituting $\lambda = i\cdot\omega$, and bearing i $\overline{P(-i\cdot\omega)} = P(i\cdot\omega)$, $\overline{Q(-i\cdot\omega)} = Q(i\cdot\omega)$ then $i\cdot P_\lambda(i\cdot\omega) = P_\omega(i\cdot\omega)$ and $i\cdot Q_\lambda(i\cdot\omega) = Q_\omega(i\cdot\omega)$ and that on the surface $|P(i\cdot\omega)|^2 = |Q(i\cdot\omega)|^2$, one obtains:

$$\left(\frac{\partial\lambda}{\partial x}\right)^{-1}\Big|_{\lambda=i\cdot\omega} = \left(\frac{i\cdot P_\omega(i\cdot\omega,x)\cdot\overline{P(i\cdot\omega,x)} + i\cdot Q_\lambda(i\cdot\omega,x)\cdot\overline{Q(\lambda,x)} - \tau\cdot|P(i\cdot\omega,x)|^2}{P_x(i\cdot\omega,x)\cdot\overline{P(i\cdot\omega,x)} - Q_x(i\cdot\omega,x)\cdot\overline{Q(i\cdot\omega,x)}}\right)$$

Upon separating into real and imaginary parts, with $P = P_R + i\cdot P_I$; $Q = Q_R + i\cdot Q_I$; $P_\omega = P_{R\omega} + i\cdot P_{I\omega}$;$Q_\omega = Q_{R\omega} + i\cdot Q_{I\omega}$; $P_x = P_{Rx} + i\cdot P_{Ix}$; $Q_x = Q_{Rx} + i\cdot Q_{Ix} P^2 = P_R^2 + P_I^2$. When (x) can be any RFID TAG parameters R_1, C_1, and time delay τ etc., Where for convenience, we have dropped the arguments $(i\cdot\omega, x)$, and where

$$F_\omega = 2\cdot[(P_{R\omega}\cdot P_R + P_{I\omega}\cdot P_I) - (Q_{R\omega}\cdot Q_R + Q_{I\omega}\cdot Q_I)];$$
$$F_x = 2\cdot[(P_{Rx}\cdot P_R + P_{Ix}\cdot P_I) - (Q_{Rx}\cdot Q_R + Q_{Ix}\cdot Q_I)]$$
$$\omega_x = -F_x/F_\omega$$

We define U and V:

$$U = (P_R \cdot P_{I\omega} - P_I \cdot P_{R\omega}) - (Q_R \cdot Q_{I\omega} - Q_I \cdot Q_{R\omega})$$
$$V = (P_R \cdot P_{Ix} - P_I \cdot P_{Rx}) - (Q_R \cdot Q_{Ix} - Q_I \cdot Q_{Rx})$$

We choose our specific parameter as time delay x = τ. $P_{I\tau} = 0; P_{R\tau} = 0; Q_{I\tau} = 0; Q_{R\tau} = 0 \Rightarrow V = 0$

$$U = \frac{\omega^2}{C_1 \cdot R_1}; P^2 = \omega^4 + \omega^2 \cdot \frac{1}{(C_1 \cdot R_1)^2};$$
$$F_\tau = 0; \frac{\partial F}{\partial \omega} = F_\omega = 2 \cdot [2 \cdot \omega^3 + \omega \cdot \frac{1}{(C_1 \cdot R_1)^2}]$$
$$F(\omega, \tau) = 0$$

And differentiating with respect to τ and we get

$$F_\omega \cdot \frac{\partial \omega}{\partial \tau} + F_\tau = 0; \tau \in I \Rightarrow \frac{\partial \omega}{\partial \tau} = -\frac{F_\tau}{F_\omega}; \wedge^{-1}(\tau) = (\frac{\partial \mathrm{Re}\lambda}{\partial \tau})_{\lambda = i \cdot \omega}$$
$$\wedge^{-1}(\tau) = \mathrm{Re}\{\frac{-2 \cdot \left[U + \tau \cdot |P|^2\right] + i \cdot F_\omega}{F_\tau + i \cdot 2 \cdot \left[V + \omega \cdot |P|^2\right]}\} - \frac{2 \cdot \omega^2 + \frac{1}{(C_1 \cdot R_1)^2}}{\omega^4 + \omega^2 \cdot \frac{1}{(C_1 \cdot R_1)^2}};$$
$$sign\{\wedge^{-1}(\tau)\} = sign\{(\frac{\partial \mathrm{Re}\lambda}{\partial \tau})_{\lambda = i \cdot \omega}\}$$
$$sign\{\wedge^{-1}(\tau)\} = sign\{F_\omega\} \cdot sign\{\tau \cdot \frac{\partial \omega}{\partial \tau} + \omega + \frac{U \cdot \frac{\partial \omega}{\partial \tau} + V}{|P|^2}\};$$
$$\frac{\partial \omega}{\partial \tau} = \omega_\tau = -\frac{F_\tau}{F_\omega}; F_\tau = 0 \Rightarrow \frac{\partial \omega}{\partial \tau} = 0$$

Then we get $sign\{\wedge^{-1}(\tau)\} = sign\{2 \cdot \omega \cdot [2 \cdot \omega^2 + \frac{1}{(C_1 \cdot R_1)^2}]\} \cdot sign\{\omega\}$

Result: $\wedge^{-1}(\tau) > 0$ for all ω, R_1, C_1 values. The sign of $\wedge^{-1}(\tau)$ is independent of τ values, then in the first case $\tau_1 = \tau; \tau_2 = 0$ there is no stability switch for different values τ. We now inspect the third interesting case when $\tau_1 = \tau; \tau_2 = \tau$. The third case we analyze is when there are delays both in RFID Label voltage and voltage time derivative [4, 5].

$$D(\lambda, \tau_1 = \tau, \tau_2 = \tau) = \lambda^2 + \lambda \cdot \frac{1}{C_1 \cdot R_1} \cdot e^{-\lambda \cdot \tau} + \frac{1}{C_1 \cdot f_\#} \cdot e^{-\lambda \cdot \tau \cdot 2}$$

Taylor expansion:

$$e^{-\lambda \tau} \approx 1 - \lambda \cdot \tau + \frac{\lambda^2 \cdot \tau^2}{2}$$

Since we need n > m [BK] analysis, we choose $e^{-\lambda\tau} \approx 1 - \lambda \cdot \tau$ then we get our RFID system second order characteristic equation:

$$D(\lambda,\tau) = \lambda^2 + a(\tau)\cdot\lambda + b(\tau)\cdot\lambda\cdot e^{-\lambda\cdot\tau} + c(\tau) + d(\tau)\cdot e^{-\lambda\cdot\tau}$$

$$a(\tau) = 0; \mathrm{b}(\tau) = \frac{1}{C_1}\cdot(\frac{1}{R_1} - \frac{\tau}{f_{\#}}); \mathrm{c}(\tau) = 0; \mathrm{d}(\tau) = \frac{1}{C_1\cdot f_{\#}}$$

$$F(\omega,\tau) = |P(i\cdot\omega,\tau)|^2 - |Q(i\cdot\omega,\tau)|^2 = (c-\omega^2)^2 + \omega^2\cdot a^2 - (\omega^2\cdot b^2 + d^2)$$

$$F(\omega,\tau) = \omega^4 - \omega^2\cdot\frac{1}{(C_1\cdot R_1)^2} - \frac{1}{(C_1\cdot f_{\#})^2}$$

Hence $F(\omega,\tau) = 0$ implies $\omega^4 - \omega^2\cdot\frac{1}{C_1^2}(\frac{1}{R_1} - \frac{\tau}{f_{\#}})^2 - \frac{1}{(C_1\cdot f_{\#})^2} = 0$ and its roots are given by

$$\omega_+^2 = \frac{1}{2}\cdot\{(b^2 + 2\cdot c - a^2) + \sqrt{\Delta}\} = \frac{1}{2}\cdot\{\sqrt{\Delta} + \frac{1}{C_1^2}(\frac{1}{R_1} - \frac{\tau}{f_{\#}})^2\}$$

$$\omega_-^2 = \frac{1}{2}\cdot\{(b^2 + 2\cdot c - a^2) - \sqrt{\Delta}\} = \frac{1}{2}\cdot\{-\sqrt{\Delta} + \frac{1}{C_1^2}(\frac{1}{R_1} - \frac{\tau}{f_{\#}})^2\}$$

$$\Delta = (b^2 + 2\cdot c - a^2) - 4\cdot(c^2 - d^2) = \frac{1}{C_1^2}\cdot(\frac{1}{R_1} - \frac{\tau}{f_{\#}})^2 + \frac{4}{(C_1\cdot f_{\#})^2}$$

Therefore the following holds: $2\cdot\omega_{+/-}^2 - (b^2 + 2\cdot c - a^2) = \pm\sqrt{\Delta}$

$$\sin\theta(\tau) = \frac{-P_R(i\cdot\omega,\tau)\cdot Q_I(i\cdot\omega,\tau) + P_I(i\cdot\omega,\tau)\cdot Q_R(i\cdot\omega,\tau)}{|Q(i\cdot\omega,\tau)|^2}$$

$$\cos\theta(\tau) = -\frac{P_R(i\cdot\omega,\tau)\cdot Q_R(i\cdot\omega,\tau) + P_I(i\cdot\omega,\tau)\cdot Q_I(i\cdot\omega,\tau)}{|Q(i\cdot\omega,\tau)|^2}$$

$$\sin\theta(\tau) = \frac{-(c-\omega^2)\cdot\omega\cdot b + \omega\cdot a\cdot d}{\omega^2\cdot b^2 + d^2} = \frac{\omega^3\cdot\frac{1}{C_1}\cdot(\frac{1}{R_1} - \frac{\tau}{f_{\#}})}{\omega^2\cdot\frac{1}{C_1^2}(\frac{1}{R_1} - \frac{\tau}{f_{\#}})^2 + \frac{1}{(C_1\cdot f_{\#})^2}}$$

$$\cos\theta(\tau) = -\frac{(c-\omega^2)\cdot d + \omega^2\cdot a\cdot b}{\omega^2\cdot b^2 + d^2} = \frac{\omega^2\cdot\frac{1}{C_1\cdot f_{\#}}}{\omega^2\cdot\frac{1}{C_1^2}(\frac{1}{R_1} - \frac{\tau}{f_{\#}})^2 + \frac{1}{(C_1\cdot f_{\#})^2}}$$

For our stability switching analysis, we choose typical RFID parameter values:

$$C_1 = 23\,\mathrm{pF}; R_1 = 100\,k\Omega = 10^5; \mathrm{L}_{calc} = f_{\#} = 2.65\,\mathrm{mH}$$

Then

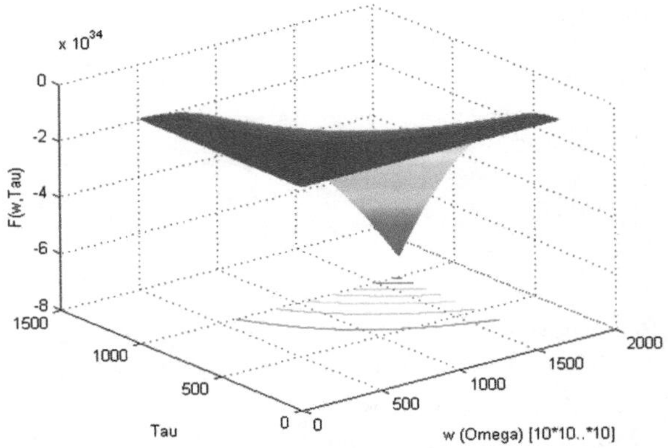

Fig. 1.17 RFID TAG F (ω, τ) function for $\tau_1 = \tau_2 = \tau$

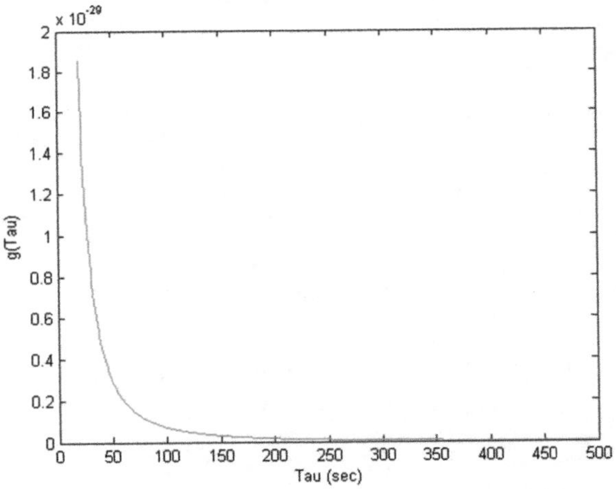

Fig. 1.18 RFID TAG stability switch diagram based on different delay values of our RFID TAG system

$$\frac{1}{C_1^2} = 1.89 \times 10^{21}; \frac{1}{C_1^2 \cdot f_{\#}^2} = 2.69 \times 10^{26}$$

We find those ω, τ values which fulfill $F(\omega, \tau) = 0$. We ignore negative, complex, and imaginary values of ω for specific τ values. The table gives the list. $\tau \in [0.001..10]$. And can be expressed by straight line ($\omega = \tau \cdot 1.64 \times 10^{13}$) (Fig. 1.17).

τ	ω
0.001	1.64×10^{10}
0.01	1.64×10^{11}
0.05	8.2×10^{11}
0.1	1.64×10^{12}

(continued)

(continued)

τ	ω
0.2	3.28×10^{12}
1	1.64×10^{13}
5	8.2×10^{13}
10	1.64×10^{14}

Remark In the above figure ω variable is 10^{10} units.

<u>MATLAB</u>: [w,t] = meshgrid(1:1:1640,0:0.01:10);

f = w. * w. * w. * w − w. * w. * 1.89 * 10^21. * (10^−5 − (t./(2.65 * 10^−3))).^2−2.69 * 10^26; meshc(f); $\%\omega \to w, \tau \to t$

We plot the stability switch diagram based on different delay values of our RFID TAG system (Fig. 1.18).

$$\wedge^{-1}(\tau) = (\frac{\partial \mathrm{Re}\lambda}{\partial \tau})_{\lambda = i \cdot \omega} = \mathrm{Re}\{\frac{-2 \cdot [U + \tau \cdot |P|^2] + i \cdot F_\omega}{F_\tau + i \cdot 2 \cdot [V + \omega \cdot |P|^2]}\}$$

$$\wedge^{-1}(\tau) = (\frac{\partial \mathrm{Re}\lambda}{\partial \tau})_{\lambda = i \cdot \omega} = \frac{2 \cdot \{F_\omega \cdot (V + \omega \cdot P^2) - F_\tau \cdot (U + \tau \cdot P^2)\}}{F_\tau^2 + 4 \cdot (V + \omega \cdot P^2)^2}$$

$$g(Tau) = \wedge^{-1}(\tau) = (\frac{\partial \mathrm{Re}\lambda}{\partial \tau})_{\lambda = i \cdot \omega}$$

The stability switch occurs only on those delay values (τ) which fit the equation: $\tau = \frac{\theta_+(\tau)}{\omega_+(\tau)}$ and $\theta_+(\tau)$ is the solution of

$$\sin\theta(\tau) = \frac{\omega^3 \cdot \frac{1}{C_1} \cdot (\frac{1}{R_1} - \frac{\tau}{f_\#})}{\omega^2 \cdot \frac{1}{C_1^2}(\frac{1}{R_1} - \frac{\tau}{f_\#})^2 + \frac{1}{(C_1 \cdot f_\#)^2}}; \cos\theta(\tau) = \frac{\omega^2 \cdot \frac{1}{C_1 \cdot f_\#}}{\omega^2 \cdot \frac{1}{C_1^2}(\frac{1}{R_1} - \frac{\tau}{f_\#})^2 + \frac{1}{(C_1 \cdot f_\#)^2}}$$

When $\omega = \omega_+(\tau)$ if only ω_+ is feasible. Additionally When all RFID TAG parameters are known and the stability switch due to various time delay values τ is described in the below expression:

$$\begin{aligned} sign\{\wedge^{-1}(\tau)\} &= sign\{F_\omega(\omega(\tau), \tau)\} \cdot sign\{\tau \cdot \omega_\tau(\omega(\tau)) \\ &+ \omega(\tau) + \frac{U(\omega(\tau)) \cdot \omega_\tau(\omega(\tau)) + V(\omega(\tau))}{|P(\omega(\tau))|^2}\} \end{aligned}$$

Remark we know $F(\omega, \tau) = 0$ implies its roots $\omega_i(\tau)$ and finding those delays values τ which ω_i is feasible. There are τ values which ω_i is a complex or imaginary number, then unable to analyze stability [4, 5].

RFID TAGs environment is characterized by electromagnetic interferences which can influence the RFID TAGs stability in time. There are two main RFID

TAGs variables which are affected by electromagnetic interferences, the voltage developed on the RFID Label and his voltage time derivative respectively. Each RFID Label variable under electromagnetic interferences is characterized by time delay respectively. The two time delays are not the same, but can be categorized to some sub cases due to interferences behavior. The first case is when there is RFID Label voltage time delay, but no voltage derivative time delay. The second case is when there is no RFID Label voltage time delay, but there is a voltage derivative time delay. The third case is when both RFID Label voltage time delay and voltage derivative time delay exist. For simplicity of our analysis we consider the third case, two delays are the same (there is a difference but it is neglected in our analysis). In each case we derive the related characteristic equation. The characteristic equation is dependent on RFID Label overall parameters and interferences time delay. Upon mathematics manipulation and [BK] theorems and definitions we derive the expression which gives us a clear picture on RFID Label stability map. The stability map gives all possible options for stability segments, each segment belongs to different time delay value segment. RFID Label stability analysis can be influenced either by TAG overall parameter values. We left this analysis and do not discuss it in the current chapter.

Lemma 1.1 *Assume that $\omega(\tau)$ is a positive and real root of $F(\omega,\tau)=0$*

Defined for $\tau \in I$, this is continuous and differentiable. Assume further that if $\lambda = i\cdot\omega$, $\omega \in R$, then $P_n(i\cdot\omega,\tau)+Q_n(i\cdot\omega,\tau)\neq 0$, $\tau \in \mathrm{R}$ hold true. Then the functions $S_n(\tau)$, $n \in N_0$, are continuous and differentiable on I.

Theorem 1.2 *Assume that $\omega(\tau)$ is a positive real root of $F(\omega,\tau)=0$ defined for $\tau \in I$, $I \subseteq R_{+0}$, and at some $\tau^* \in I$, $S_n(\tau^*)=0$ for some $n \in N_0$ then a pair of simple conjugate pure imaginary roots $\lambda_+(\tau^*)=i\cdot\omega(\tau^*)$, $\lambda_-(\tau^*)=-i\cdot\omega(\tau^*)$ of $D(\lambda,\tau)=0$ exist at $\tau=\tau^*$ which crosses the imaginary axis from left to right if $\delta(\tau^*)>0$ and cross the imaginary axis from right to left if $\delta(\tau^*)<0$ where*

$$\delta(\tau^*) = sign\{\frac{d\mathrm{Re}\lambda}{d\tau}|_{\lambda=i\omega(\tau^*)}\} = sign\{F_\omega(\omega(\tau^*),\tau^*)\}\cdot sign\{\frac{dS_n(\tau)}{d\tau}|_{\tau=\tau^*}\}$$

The theorem becomes $sign\{\frac{d\mathrm{Re}\lambda}{d\tau}|_{\lambda=i\omega\pm}\} = sign\{\pm\Delta^{1/2}\}\cdot sign\{\frac{dS_n(\tau)}{d\tau}|_{\tau=\tau^*}\}$

Theorem 1.3 *The characteristic equation: $\tau_1=\tau, \tau_2=0$; $\tau_1=0, \tau_2=\tau$*

$$D(\lambda,\tau) = \lambda^2 + a(\tau)\cdot\lambda + b(\tau)\cdot\lambda\cdot e^{-\lambda\cdot\tau} + c(\tau) + d(\tau)\cdot e^{-\lambda\cdot\tau};$$

$$D(\lambda,\tau_1,\tau_2) = \lambda^2 + \lambda\cdot\frac{1}{C_1\cdot R_1}\cdot e^{-\lambda\cdot\tau_2} + \frac{1}{C_1\cdot f_\#}\cdot e^{-\lambda\cdot(\tau_1+\tau_2)}$$

Has a pair of simple and conjugate pure imaginary roots $\lambda=\pm\omega(\tau^*)$, $\omega(\tau^*)$ Real at $\tau^*\in I$ if $S_n(\tau^*)=\tau^*-\tau_n(\tau^*)=0$ for some $n\in N_0$. If $\omega(\tau^*)=\omega_+(\tau^*)$, this pair of simple conjugate pure imaginary roots crosses the imaginary axis from left to right if $\delta_+(\tau^*)>0$ and crosses the imaginary axis from right to left if $\delta_+(\tau^*)<0$ where $\delta_+(\tau^*)=sign\{\frac{d\mathrm{Re}\lambda}{d\tau}|_{\lambda=i\omega_+(\tau^*)}\}=sign\{\frac{dS_n(\tau)}{d\tau}|_{\tau=\tau^*}\}$. If

$\omega(\tau^*) = \omega_-(\tau^*)$, this pair of simple conjugate pure imaginary roots cross the imaginary axis from left to right if $\delta_-(\tau^*) > 0$ and crosses the imaginary axis from right to left If $\delta_-(\tau^*) < 0$ where $\delta_-(\tau^*) = sign\{\frac{d\mathrm{Re}\lambda}{d\tau}|_{\lambda=i\omega_-(\tau^*)}\} = -sign\{\frac{dS_n(\tau)}{d\tau}|_{\tau=\tau^*}\}$. If $\omega_+(\tau^*) = \omega_-(\tau^*) = \omega(\tau^*)$ then $\delta(\tau^*) = 0$ and $sign\{\frac{d\mathrm{Re}\lambda}{d\tau}|_{\lambda=i\omega(\tau^*)}\} = 0$, the same is true when $S_n'(\tau^*) = 0$. The following result can be useful in identifying values of τ where the stability switches happened.

Theorem 1.4 *Assume that for all $\tau \in I$, $\omega(\tau)$ is defined as a solution of $F(\omega, \tau) = 0$ then $\delta_\pm(\tau) = sign\{\pm\Delta^{1/2}(\tau)\} \cdot signD_\pm(\tau)$*

$$D_\pm(\tau) = \omega_\pm^2 \cdot [(\omega_\pm^2 \cdot b^2 + d^2) + a' \cdot (c - \omega_\pm^2) + b \cdot d' - b' \cdot d - a \cdot c'] + \omega_\pm \cdot \omega_\pm' \cdot [\tau \cdot (\omega_\pm^2 \cdot b^2 + d^2) - b \cdot d + a \cdot (c - \omega_\pm^2) + 2 \cdot \omega_\pm^2 \cdot a]$$

$$a' = \frac{da(\tau)}{d\tau}; \mathrm{b}' = \frac{db(\tau)}{d\tau}; \mathrm{c}' = \frac{dc(\tau)}{d\tau}; \mathrm{d}' = \frac{dd(\tau)}{d\tau}$$

1.4 Semi-Passive RFID Tags with Double Loop Antennas Arranged as a Shifted Gate System for Stability Optimization Under Delayed Electromagnetic Interferences

A semi-passive tags operate similarly to passive RFID tags. However, they contain a battery that enables long reading distance and also enables the tag to operate independently of the reader. Semi-Passive TAGs with double loop antennas arranged as a shifted gate system influence by electromagnetic interferences which effect there stability behavior. Semi-Passive RFID TAGs system with a battery is like a Reader unit and aimed to improve the communication performance by using double loop antennas in walk-through gate arrangement in various TAGs orientations of the RFID system operating mainly in the LF band. The below figure describes the double loop antennas as a shifted gate in x-direction [8] (Fig. 1.19).

The antenna gate is shifted to avoid cancellation of magnetic fields between two TAGs, and to improve the magnetic-field distribution. The RFID system at Low Frequency (LF) band has been widely adopted. The RFID tags for this application have usually installed in applications such that the orientation of tag id difficult to fix for transferring data with RFID reader. Most of the LF-RFID reader antennas are rectangular or circular loops, but these antennas cannot generate sufficient field

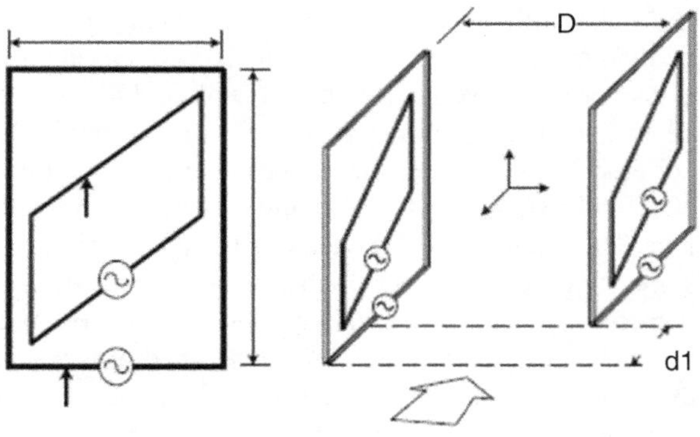

Fig. 1.19 Double loop antennas arranged as a shifted gate in x-direction

strengths in some locations and/or tag antenna orientations. The double loop antennas arranged as a shifted gate improves magnetic-field distribution in a region of interest suitable for communication with various tag orientations, and enhance the communication distance [1]. The antenna gate is shifted to avoid cancellation of magnetic fields between two gate antennas. The gate antenna consists of two sides of rectangular loops with two types of excitations; i.e., in phase and 180° out of phase. When two antennas are excited in phase, the directions of the currents flowing in two loops are in the same direction, resulting in the cancellation of magnetic fields in the x-direction in the middle region of the gate. When two antennas are excited 180° out of phase, the direction of the currents flowing in two antennas is in the opposite direction, resulting in the cancellation of magnetic fields in the y-direction in the middle region of the gates. Thus, the gate antennas are arranged as a shifted gate to maintain magnetic fields in the middle region. The double loop antenna is employed due to the fact that this antenna consists of two parallel loops (primary and secondary loops). The shape of the primary loop is rectangular for generating the magnetic field in the y-direction. The secondary loop is always within the primary loop, and is optimized such that the magnetic fields in x- and z-directions are strongly generated. D is the separation distance between gate antennas, and d_1 is the shifted distance in the x-direction. Due to electromagnetic interferences there are differences in time delays with respect to gate antenna's first and second loop voltages and voltages derivatives. The delayed voltages are $V_{i1}(t - \tau_1)$ and $V_{i2}(t - \tau_2)$ respectively ($\tau_1 \neq \tau_2$) and delayed voltages derivatives are $dV_{i1}(t - \Delta_1)/dt$, $dV_{i2}(t - \Delta_2)/dt$ respectively ($\Delta_1 \neq \Delta_2; \tau_1 \geq 0; \tau_2 \geq 0; \Delta_1, \Delta_2 \geq 0$). The Semi-Passive RFID TAG with double loop antennas equivalent circuit can represent as a delayed differential equations which depending on variable parameters and delays. Our Semi-Passive RFID TAG system delay differential and delay different model can be analytically by using delay differential equations in dynamically. The need of the incorporation of a time delay is often of the existence of any stage structure. It is often difficult to analytically study models with delay dependent parameters, even if only a single discrete delay is present. There are practical guidelines that combine graphical information with analytical work to effectively study the local stability of models involving delay dependent parameters. The stability of a given steady state is simply determined by the graphs of some function of τ_1, τ_2 which can be expressed, explicitly and thus can be easily depicted by Matlab and other popular software. We need only look at one such function and locate the zeros. This function often has only two zeros, providing thresholds for stability switches. As time delay increases, stability changes from stable to unstable to stable. We emphasize the local stability aspects of some models with delay dependent parameters. Additionally, there is a general geometric criterion that, theoretically speaking, can be applied to models with many delays, or even distributed delays. The simplest case of a first order characteristic equation, providing more user friendly geometric and analytic criteria for stability switches. The analytical criteria provided for the first and second order cases can be used to obtain some insightful analytical statements and can be helpful for conducting simulations [5, 6]. Semi-Passive RFID TAG with double loop antenna can be represented as a two inductors in series (L_{11} and L_{12} for the first double loop gate antenna) with parasitic resistance r_{P1}. The double loop antennas in series are connected in parallel to Semi-Passive RFID TAG. The Equivalent Circuit of Semi-Passive RFID TAG is Capacitor (C_1) and Resistor (R_1) in parallel with voltage generator $V_{s1}(t)$ and parasitic resistance r_{S1}. In case we have Passive RFID TAG switch S_1 is OFF otherwise is ON (Reader/Active RFID system) and long distance is achievable. The second double loop gate antenna is defined as two inductors in series L_{21} and L_{22} with series parasitic resistor r_{P2}. $V_{s2}(t)$ and parasitic resistance r_{S2} are belong to the second gate antenna system with another Semi-Passive RFID TAG [1].

L_{11} and L_{12} are mostly formed by traces on planar PCB. $2 \cdot L_m$ element represents the mutual inductance between L_{11} and L_{12}. We consider that the double loop antennas parameter values ($L_{a1}, L_{a2}, L_{b1}, L_{b2}, a_1, a_2$) are the same in the first and second gates. Since two inductors (L_{11}, L_{12}) are in series and there is a mutual inductance between L_{11} and L_{12}, the total antenna inductance L_T: $L_T = L_{11} + L_{12} + 2 \cdot L_m$ and $L_m = K \cdot \sqrt{L_{11} \cdot L_{12}}$. L_m is the mutual inductance between L_{11} and L_{12}. K is the coupling coefficient of two inductors $0 \le K \le 1$. We start with the case of passive RFID TAG which switch S1 is OFF. I(t) is the current that flow through a double loop antenna. V_{11} and V_{12} are the voltages on L_{11} and L_{12} respectively. V_m is the voltage on double loop antenna mutual inductance element.

$$V_{11} = L_{11} \cdot \frac{dI}{dt}; V_{12} = L_{12} \cdot \frac{dI}{dt}; V_{CD} = I \cdot r_{p1}; V_m = 2 \cdot L_m \cdot \frac{dI}{dt};$$
$$V_{AB} = V_{R_1} = V_{C_1} = V_{11} + V_{12} + V_{CD} + V_m; I_{C_1} = C_1 \cdot \frac{dV_{C_1}}{dt}$$

$$V_{AB} = V_{R_1} = V_{C_1} = V_{11} + V_{12} + V_{CD} + V_m; I_{C_1} = C_1 \cdot \frac{dV_{C_1}}{dt};$$
$$I_{C_1} + I_{R_1} + I = 0 \Rightarrow C_1 \cdot \frac{dV_{C_1}}{dt} + \frac{V_{C_1}}{R_1} + I = 0; L_{11} \neq L_{12}$$

$$\frac{dV_{C_1}}{dt} = \frac{dV_{11}}{dt} + \frac{dV_{12}}{dt} + \frac{dV_{CD}}{dt} + \frac{dV_m}{dt};$$
$$I = \frac{1}{L_{11}} \cdot \int V_{11} \cdot dt = \frac{1}{L_{12}} \cdot \int V_{12} \cdot dt;$$
$$V_{CD} = I \cdot r_{p1} = \frac{r_{p1}}{L_{11}} \cdot \int V_{11} \cdot dt = \frac{r_{p1}}{L_{12}} \cdot \int V_{12} \cdot dt$$

$$\frac{dV_{CD}}{dt} = \frac{r_{p1}}{L_{11}} \cdot V_{11} + \frac{r_{p1}}{L_{12}} \cdot V_{12}; V_{11} = \frac{L_{11}}{L_{12}} \cdot V_{12}; V_{12} = \frac{L_{12}}{L_{11}} \cdot V_{11};$$
$$I = \frac{1}{L_{11}} \cdot \int V_{11} \cdot dt = \frac{1}{L_{12}} \cdot \int V_{12} \cdot dt \Rightarrow \frac{dI}{dt} = \frac{1}{L_{11}} \cdot V_{11} = \frac{1}{L_{12}} \cdot V_{12}$$

$$V_m = 2 \cdot L_m \cdot \frac{dI}{dt} = 2 \cdot K \cdot \sqrt{L_{11} \cdot L_{12}} \cdot \frac{1}{L_{11}} \cdot V_{11} = 2 \cdot K \cdot \sqrt{\frac{L_{12}}{L_{11}}} \cdot V_{11};$$
$$\frac{dV_m}{dt} = 2 \cdot K \cdot \sqrt{\frac{L_{12}}{L_{11}}} \cdot \frac{dV_{11}}{dt}$$

We get the following differential equation respect to $V_{11}(t)$ variable, η_1, η_2, η_3 are global parameters.

$$\frac{d^2V_{11}}{dt^2} \cdot \eta_1 + \frac{dV_{11}}{dt} \cdot \eta_2 + V_{11} \cdot \eta_3 = 0$$
$$\eta_1 = C_1 \cdot (1 + \frac{L_{12}}{L_{11}} + 2 \cdot K \cdot \sqrt{\frac{L_{12}}{L_{11}}});$$
$$\eta_2 = \frac{C_1 \cdot r_{p1}}{L_{11}} + \frac{1}{R_1} \cdot (1 + \frac{L_{12}}{L_{11}} + 2 \cdot K \cdot \sqrt{\frac{L_{12}}{L_{11}}});$$
$$\eta_2 = \frac{C_1 \cdot r_{p1}}{L_{11}} + \frac{1}{R_1} \cdot \frac{\eta_1}{C_1}; \eta_3 = \frac{1}{L_{11}} \cdot (1 + \frac{r_{p1}}{R_1})$$
$$\eta_1 = \eta_1(C_1, L_{12}, L_{11}, K); \eta_2 = \eta_2(C_1, r_{p1}, L_{12}, L_{11}, K, R_1);$$
$$\eta_3 = \eta_3(L_{11}, r_{p1}, R_1); V'_{11} = \frac{dV_{11}}{dt}; \frac{dV'_{11}}{dt} = \frac{d^2V_{11}}{dt^2}$$
$$\frac{dV'_{11}}{dt} = -V'_{11} \cdot \frac{\eta_2}{\eta_1} - V_{11} \cdot \frac{\eta_3}{\eta_1}; \frac{dV_{11}}{dt} = V'_{11}.$$

In the same manner we find our V_{12} differential equation. We get the following differential equation respect to $V_{12}(t)$ variable, ξ_1, ξ_2, ξ_3 are global parameters.

$$\frac{d^2V_{12}}{dt^2}\cdot\xi_1+\frac{dV_{12}}{dt}\cdot\xi_2+V_{12}\cdot\xi_3=0;\xi_1=C_1\cdot(1+\frac{L_{11}}{L_{12}}+2\cdot K\cdot\sqrt{\frac{L_{11}}{L_{12}}});$$
$$\xi_2=\frac{C_1\cdot r_{p1}}{L_{12}}+\frac{1}{R_1}\cdot(1+\frac{L_{11}}{L_{12}}+2\cdot K\cdot\sqrt{\frac{L_{11}}{L_{12}}});\xi_3=\frac{1}{L_{12}}\cdot(1+\frac{r_{p1}}{R_1})$$
$$\xi_2=\frac{C_1\cdot r_{p1}}{L_{12}}+\frac{1}{R_1}\cdot\frac{\xi_1}{C_1};V'_{12}=\frac{dV_{12}}{dt};\frac{dV'_{12}}{dt}=\frac{d^2V_{12}}{dt^2};$$
$$\xi_1=\xi_1(C_1,L_{12},L_{11},K);\xi_2=\xi_2(C_1,r_{p1},L_{12},L_{11},K,R_1)$$
$$\xi_3=\xi_3(L_{12},r_{p1},R_1);\frac{dV'_{12}}{dt}=-V'_{12}\cdot\frac{\xi_2}{\xi_1}-V_{12}\cdot\frac{\xi_3}{\xi_1};\frac{dV_{12}}{dt}=V'_{12}.$$

Summary: We get our RFID double loop antennas system's four differential equations.

$$\frac{dV'_{11}}{dt}=-V'_{11}\cdot\frac{\eta_2}{\eta_1}-V_{11}\cdot\frac{\eta_3}{\eta_1};\frac{dV_{11}}{dt}=V'_{11};\frac{dV'_{12}}{dt}=-V'_{12}\cdot\frac{\xi_2}{\xi_1}-V_{12}\cdot\frac{\xi_3}{\xi_1};\frac{dV_{12}}{dt}=V'_{12}$$

$$\begin{pmatrix}\frac{dV'_{11}}{dt}\\ \frac{dV_{11}}{dt}\\ \frac{dV'_{12}}{dt}\\ \frac{dV_{12}}{dt}\end{pmatrix}=\begin{pmatrix}\Gamma_{11} & \cdots & \Gamma_{14}\\ \vdots & \ddots & \vdots\\ \Gamma_{41} & \cdots & \Gamma_{44}\end{pmatrix}\cdot\begin{pmatrix}V'_{11}\\ V_{11}\\ V'_{12}\\ V_{12}\end{pmatrix};\Gamma_{11}=-\frac{\eta_2}{\eta_1};\Gamma_{12}=-\frac{\eta_3}{\eta_1};$$
$$\Gamma_{33}=-\frac{\xi_2}{\xi_1};\Gamma_{34}=-\frac{\xi_3}{\xi_1};\Gamma_{21}=\Gamma_{43}=1$$

$$\Gamma_{13}=\Gamma_{14}=\Gamma_{22}=\Gamma_{23}=\Gamma_{24}=\Gamma_{31}=\Gamma_{32}=\Gamma_{41}=\Gamma_{42}=\Gamma_{44}=0$$

The RFID double loop antennas system's primary and secondary loops are composed of a thin wire or a thin plate element (Fig. 1.20). Units are all in cm, and a_1, a_2 are radiuses of the primary and secondary wires in cm. There inductances can be calculated by the following formulas (Fig. 1.21):

$$L_{11}=4\cdot\{L_{b1}\cdot\ln[\frac{2\cdot A_1}{a_1\cdot(L_{b1}+l_{c_1})}]+L_{a1}\cdot\ln[\frac{2\cdot A_1}{a_1\cdot(L_{b1}+l_{c_1})}]+2\cdot[a_1+l_{c_1}-(L_{a1}+L_{b1})]\}$$

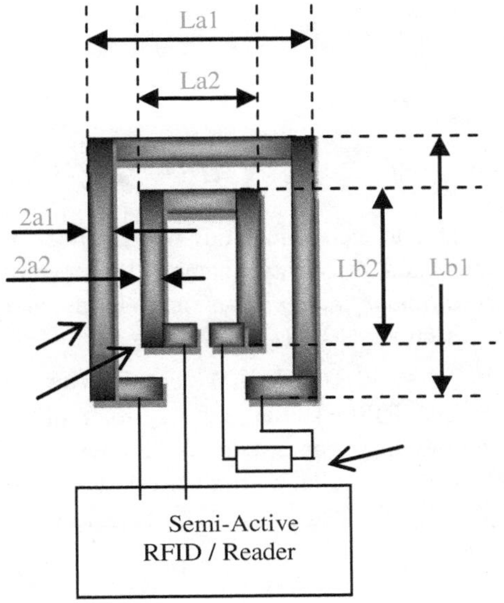

Fig. 1.20 Double loop antennas in series with parasitic resistance and Semi-Passive RFID TAG

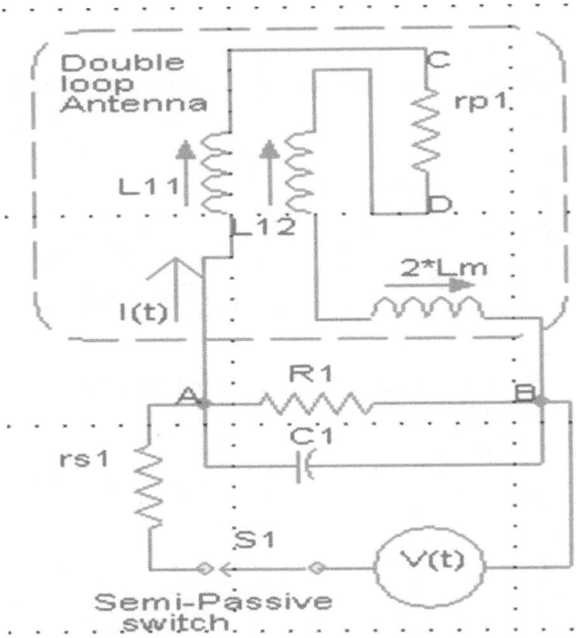

Fig. 1.21 Equivalent circuit of double loop antennas in series with Semi-Passive RFID TAG

$$L_{12} = 4 \cdot \{L_{b2} \cdot \ln[\frac{2 \cdot A_2}{a_2 \cdot (L_{b2} + l_{c_2})}] + L_{a2} \cdot \ln[\frac{2 \cdot A_2}{a_2 \cdot (L_{b2} + l_{c_2})}] + 2 \cdot [a_2 + l_{c_2} - (L_{a2} + L_{b2})]\}$$

$$l_{c_1} = \sqrt{L_{a1}^2 + L_{b1}^2}; A_1 = L_{a1} \cdot L_{b1}; l_{c_2} = \sqrt{L_{a2}^2 + L_{b2}^2}; A_2 = L_{a2} \cdot L_{b2}$$

Due to electromagnetic interferences, we get a shifted gate RFID system's primary and secondary antennas loops voltages with delays τ_1 and τ_2 respectively. Additionally, we get antennas loop voltages derivatives with delays Δ_1 and Δ_2 respectively. $V_{11}(t) \to V_{11}(t - \tau_1)$; $V_{12}(t) \to V_{12}(t - \tau_2)$; $V'_{11}(t) \to V'_{11}(t - \Delta_1)$ $V'_{12}(t) \to V'_{12}(t - \Delta_2)$. We consider no delay effect on $\frac{dV_{11}}{dt}; \frac{dV_{12}}{dt}; \frac{dV'_{11}}{dt}; \frac{dV'_{12}}{dt}$.

The RFID shifted gate system, differential equations under electromagnetic interferences (delays terms) influence only RFID double loop voltages $V_{11}(t)$, $V_{12}(t)$ and voltages derivatives $V'_{11}(t)$ and $V'_{12}(t)$ respect to time, there is no influence on $\frac{dV_{11}(t)}{dt}; \frac{dV_{12}(t)}{dt}; \frac{dV'_{11}(t)}{dt}; \frac{dV'_{12}(t)}{dt}$.

$$\begin{pmatrix} \frac{dV'_{11}}{dt} \\ \frac{dV_{11}}{dt} \\ \frac{dV'_{12}}{dt} \\ \frac{dV_{12}}{dt} \end{pmatrix} = \begin{pmatrix} \Gamma_{11} & \dots & \Gamma_{14} \\ \vdots & \ddots & \vdots \\ \Gamma_{41} & \dots & \Gamma_{44} \end{pmatrix} \cdot \begin{pmatrix} V'_{11}(t - \Delta_1) \\ V_{11}(t - \tau_1) \\ V'_{12}(t - \Delta_2) \\ V_{12}(t - \tau_2) \end{pmatrix}$$

To find equilibrium points (fixed points) of the RFID shifted gate system is by

$$\lim_{t \to \infty} V_{11}(t - \tau_1) = V_{11}(t), \lim_{t \to \infty} V_{12}(t - \tau_2) = V_{12}(t),$$
$$\lim_{t \to \infty} V'_{11}(t - \Delta_1) = V'_{11}(t), \lim_{t \to \infty} V'_{12}(t - \Delta_2) = V'_{12}(t)$$
$$\frac{dV_{11}(t)}{dt} = 0; \frac{dV_{12}(t)}{dt} = 0; \frac{dV'_{11}(t)}{dt} = 0; \frac{dV'_{12}(t)}{dt} = 0. \ \forall\, t \gg \tau_1; t \gg \tau_2; t \gg \Delta_1; t \gg \Delta_2$$
$$\exists\, (t - \tau_1) \approx t; (t - \tau_2) \approx t; (t - \Delta_1) \approx t; (t - \Delta_2) \approx t, \ t \to \infty.$$

We get four equations and the only fixed point is $E^{(0)}(V_{11}^{\prime(0)}, V_{11}^{(0)}, V_{12}^{\prime(0)}, V_{12}^{(0)}) = (0, 0, 0, 0)$ since

$$\eta_3 \neq 0 \;\&\; \eta_1 \neq 0 \Rightarrow \Gamma_{12} \neq 0; \xi_3 \neq 0 \;\&\; \xi_1 \neq 0 \Rightarrow \Gamma_{34} \neq 0$$

Stability analysis: The standard local stability analysis about any one of the equilibrium points of RFID shifted gate system consists in adding to coordinate $[V'_{11}\ V_{11}\ V'_{12}\ V_{12}]$ arbitrarily small increments of exponential form$[v'_{11}\ v_{11}\ v'_{12}\ v_{12}]\cdot e^{\lambda \cdot t}$, and retaining the first order terms in $V'_{11}\ V_{11}\ V'_{12}\ V_{12}$. The system of four homogeneous equations leads to a polynomial characteristic equation in the eigenvalues λ. The polynomial characteristic equations accept by set the below voltages and voltages derivative respect to time into two RFID shifted gate system equations.

RFID shifted gate system fixed values with arbitrarily small increments of exponential form $[v'_{11}\ v_{11}\ v'_{12}\ v_{12}]\cdot e^{\lambda \cdot t}$ are: i = 0 (first fixed point), i = 1 (second fixed point), i = 2 (third fixed point), etc.,

$$V'_{11}(t) = V'^{(i)}_{11} + v'_{11}\cdot e^{\lambda \cdot t};\ V_{11}(t) = V^{(i)}_{11} + v_{11}\cdot e^{\lambda \cdot t}$$
$$V'_{12}(t) = V'^{(i)}_{12} + v'_{12}\cdot e^{\lambda \cdot t};\ V_{12}(t) = V^{(i)}_{12} + v_{12}\cdot e^{\lambda \cdot t}$$

We choose the above expressions for our $V'_{11}(t)$, $V_{11}(t)$ and $V'_{12}(t)$, $V_{12}(t)$ as small displacement $[v'_{11}\ v_{11}\ v'_{12}\ v_{12}]$ from the system fixed points in time t = 0.

$$V'_{11}(t=0) = V'^{(i)}_{11} + v'_{11};\ V_{11}(t=0) = V^{(i)}_{11} + v_{11}$$
$$V'_{12}(t=0) = V'^{(i)}_{12} + v'_{12};\ V_{12}(t=0) = V^{(i)}_{12} + v_{12}$$

For $\lambda < 0$, $t > 0$ the selected fixed point is stable otherwise $\lambda > 0$, $t > 0$ is Unstable. Our system tends to the selected fixed point exponentially to $\lambda < 0$, $t > 0$ otherwise go away from the selected fixed point exponentially. λ is the eigenvalue parameter which establishes if the fixed point is stable or Unstable, additionally his absolute value ($|\lambda|$) establish the speed of flow toward or away from the selected fixed point [2, 3].

	λ < 0	λ > 0
t = 0	$V'_{11}(t=0) = V'^{(i)}_{11} + v'_{11}$ $V_{11}(t=0) = V^{(i)}_{11} + v_{11}$ $V'_{12}(t=0) = V'^{(i)}_{12} + v'_{12}$ $V_{12}(t=0) = V^{(i)}_{12} + v_{12}$	$V'_{11}(t=0) = V'^{(i)}_{11} + v'_{11}$ $V_{11}(t=0) = V^{(i)}_{11} + v_{11}$ $V'_{12}(t=0) = V'^{(i)}_{12} + v'_{12}$ $V_{12}(t=0) = V^{(i)}_{12} + v_{12}$
t > 0	$V'_{11}(t) = V'^{(i)}_{11} + v'_{11}\cdot e^{-\lvert\lambda\rvert \cdot t}$ $V_{11}(t) = V^{(i)}_{11} + v_{11}\cdot e^{-\lvert\lambda\rvert \cdot t}$ $V'_{12}(t) = V'^{(i)}_{12} + v'_{12}\cdot e^{-\lvert\lambda\rvert \cdot t}$ $V_{12}(t) = V^{(i)}_{12} + v_{12}\cdot e^{-\lvert\lambda\rvert \cdot t}$	$V'_{11}(t) = V'^{(i)}_{11} + v'_{11}\cdot e^{\lvert\lambda\rvert \cdot t}$ $V_{11}(t) = V^{(i)}_{11} + v_{11}\cdot e^{\lvert\lambda\rvert \cdot t}$ $V'_{12}(t) = V'^{(i)}_{12} + v'_{12}\cdot e^{\lvert\lambda\rvert \cdot t}$ $V_{12}(t) = V^{(i)}_{12} + v_{12}\cdot e^{\lvert\lambda\rvert \cdot t}$

(continued)

(continued)

	$\lambda < 0$	$\lambda > 0$
t > 0 $t \to \infty$	$V'_{11}(t \to \infty) = V'^{(i)}_{11}$ $V_{11}(t \to \infty) = V^{(i)}_{11}$ $V'_{12}(t \to \infty) = V'^{(i)}_{12}$ $V_{12}(t \to \infty) = V^{(i)}_{12}$	$V'_{11}(t \to \infty, \lambda > 0) \approx v'_{11} \cdot e^{\lvert\lambda\rvert \cdot t}$ $V_{11}(t \to \infty, \lambda > 0) \approx v_{11} \cdot e^{\lvert\lambda\rvert \cdot t}$ $V'_{12}(t \to \infty, \lambda > 0) \approx v'_{12} \cdot e^{\lvert\lambda\rvert \cdot t}$ $V_{12}(t \to \infty, \lambda > 0) \approx v_{11} \cdot e^{\lvert\lambda\rvert \cdot t}$

The speeds of flow toward or away from the selected fixed point for RFID shifted gate system voltages and voltages derivatives respect to time are

$$\frac{dV'_{11}(t)}{dt} = \lim_{\Delta t \to 0} \frac{V'_{11}(t+\Delta t) - V'_{11}(t)}{\Delta t} = \lim_{\Delta t \to 0} \frac{V'^{(i)}_{11} + v'_{11} \cdot e^{\lambda \cdot (t+\Delta t)} - [V'^{(i)}_{11} + v'_{11} \cdot e^{\lambda \cdot t}]}{\Delta t}$$
$$= \lim_{\Delta t \to 0} \frac{v'_{11} \cdot e^{\lambda \cdot t} \cdot [e^{\lambda \cdot \Delta t} - 1]}{\Delta t} \longrightarrow e^{\lambda \cdot \Delta t} \approx 1 + \lambda \cdot \Delta t \lambda \cdot v'_{11} \cdot e^{\lambda \cdot t}$$
$$\frac{dV_{11}(t)}{dt} = \lambda \cdot v_{11} \cdot e^{\lambda \cdot t}; \frac{dV_{12}(t)}{dt} = \lambda \cdot v_{12} \cdot e^{\lambda \cdot t};$$
$$\frac{dV'_{12}(t)}{dt} = \lambda \cdot v'_{12} \cdot e^{\lambda \cdot t}; \frac{dV'_{11}(t-\Delta_1)}{dt} = \lambda \cdot v'_{11} \cdot e^{\lambda \cdot t} \cdot e^{-\lambda \cdot \Delta_1}$$

$$\frac{dV'_{11}(t-\Delta_1)}{dt} = \lambda \cdot v'_{11} \cdot e^{\lambda \cdot t} \cdot e^{-\lambda \cdot \Delta_1}; \frac{dV_{11}(t-\tau_1)}{dt} = \lambda \cdot v_{11} \cdot e^{\lambda \cdot t} \cdot e^{-\lambda \cdot \tau_1}$$
$$\frac{dV_{12}(t-\tau_2)}{dt} = \lambda \cdot v_{12} \cdot e^{\lambda \cdot t} \cdot e^{-\lambda \cdot \tau_2}; \frac{dV'_{12}(t-\Delta_2)}{dt} = \lambda \cdot v'_{12} \cdot e^{\lambda \cdot t} \cdot e^{-\lambda \cdot \Delta_2}$$

First, we take the RFID shifted gate voltages V_{11}, V_{12} differential equations: $\frac{dV_{11}}{dt} = V'_{11}; \frac{dV_{12}}{dt} = V'_{12}$ and adding coordinates $[V'_{11}\ V_{11}\ V'_{12}\ V_{12}]$ arbitrarily small increments of exponential terms $[v'_{11}\ v_{11}\ v'_{12}\ v_{12}] \cdot e^{\lambda \cdot t}$ and retaining the first order terms in $v'_{11}\ v_{11}\ v'_{12}\ v_{12}$.

$$\lambda \cdot v_{11} \cdot e^{\lambda \cdot t} = V'^{(i)}_{11} + v'_{11} \cdot e^{\lambda \cdot t}; V'^{(i)}_{11} = 0; \lambda \cdot v_{11} = v'_{11} \Rightarrow -\lambda \cdot v_{11} + v'_{11} = 0$$
$$\lambda \cdot v_{12} \cdot e^{\lambda \cdot t} = V'^{(i)}_{12} + v'_{12} \cdot e^{\lambda \cdot t}; V'^{(i)}_{12} = 0; \lambda \cdot v_{12} = v'_{12} \Rightarrow -\lambda \cdot v_{12} + v'_{12} = 0$$

Second, we take the RFID shifted gate's voltages derivative V'_{11}, V'_{12} differential equations:

$$\frac{dV'_{11}}{dt} = \Gamma_{11} \cdot V'_{11} + \Gamma_{12} \cdot V_{11}; \frac{dV'_{12}}{dt} = \Gamma_{33} \cdot V'_{12} + \Gamma_{34} \cdot V_{12}$$

And adding coordinates $[V'_{11}\ V_{11}\ V'_{12}\ V_{12}]$ arbitrarily small increments of exponential terms $[v'_{11}\ v_{11}\ v'_{12}\ v_{12}] \cdot e^{\lambda \cdot t}$ and retaining the first order terms in $v'_{11}\ v_{11}\ v'_{12}\ v_{12}$.

$$\lambda \cdot v_{11}^{'} \cdot e^{\lambda \cdot t} = \Gamma_{11} \cdot [V_{11}^{'(i)} + v_{11}^{'} \cdot e^{\lambda \cdot t}] + \Gamma_{12} \cdot [V_{11}^{(i)} + v_{11} \cdot e^{\lambda \cdot t}] \ ; \ V_{11}^{'(i=0)} = 0 \ ; \ V_{11}^{(i=0)} = 0$$

$$\lambda \cdot v_{11}^{'} = \Gamma_{11} \cdot v_{11}^{'} + \Gamma_{12} \cdot v_{11} \Rightarrow -\lambda \cdot v_{11}^{'} + \Gamma_{11} \cdot v_{11}^{'} + \Gamma_{12} \cdot v_{11} = 0$$

$$\lambda \cdot v_{12}^{'} \cdot e^{\lambda \cdot t} = \Gamma_{33} \cdot [V_{12}^{'(i)} + v_{12}^{'} \cdot e^{\lambda \cdot t}] + \Gamma_{34} \cdot [V_{12}^{(i)} + v_{12} \cdot e^{\lambda \cdot t}] \ ; \ V_{12}^{'(i=0)} = 0 \ ; \ V_{12}^{(i=0)} = 0$$

$$\lambda \cdot v_{12}^{'} = \Gamma_{33} \cdot v_{12}^{'} + \Gamma_{34} \cdot v_{12} \Rightarrow -\lambda \cdot v_{12}^{'} + \Gamma_{33} \cdot v_{12}^{'} + \Gamma_{34} \cdot v_{12} = 0$$

We can summery our eigenvalues equations: $(-\lambda + \Gamma_{11}) \cdot v_{11}^{'} + \Gamma_{12} \cdot v_{11} = 0$

$$v_{11}^{'} - \lambda \cdot v_{11} = 0 \ ; \ (-\lambda + \Gamma_{33}) \cdot v_{12}^{'} + \Gamma_{34} \cdot v_{12} = 0 \ ; \ v_{12}^{'} - \lambda \cdot v_{12} = 0$$

$$\begin{pmatrix} \Omega_{11} & \dots & \Omega_{14} \\ \vdots & \ddots & \vdots \\ \Omega_{41} & \cdots & \Omega_{44} \end{pmatrix} \cdot \begin{pmatrix} v_{11}^{'} \\ v_{11} \\ v_{12}^{'} \\ v_{12} \end{pmatrix} = 0 \ ; \ \Omega_{11} = -\lambda + \Gamma_{11} \ ; \ \Omega_{12} = \Gamma_{12} \ ; \ \Omega_{13} = 0 \ ; \ \Omega_{14} = 0$$

$$\Omega_{21} = 1 \ ; \ \Omega_{22} = -\lambda \ ; \ \Omega_{23} = 0 \ ; \ \Omega_{24} = 0 \ ; \ \Omega_{31} = 0 \ ; \ \Omega_{32} = 0 \ ; \ \Omega_{33} = -\lambda + \Gamma_{33} \ ; \ \Omega_{34} = \Gamma_{34}$$

$$\Omega_{41} = 0 \ ; \ \Omega_{42} = 0 \ ; \ \Omega_{43} = 1 \ ; \ \Omega_{44} = -\lambda$$

$$A - \lambda \cdot I = \begin{pmatrix} \Omega_{11} & \dots & \Omega_{14} \\ \vdots & \ddots & \vdots \\ \Omega_{41} & \cdots & \Omega_{44} \end{pmatrix} \ ; \ \det(A - \lambda \cdot I) = 0$$

$$\det(A - \lambda \cdot I) = (-\lambda + \Gamma_{11}) \cdot \det \begin{pmatrix} -\lambda & 0 & 0 \\ 0 & -\lambda + \Gamma_{33} & \Gamma_{34} \\ 0 & 1 & -\lambda \end{pmatrix} - \Gamma_{12} \cdot \det \begin{pmatrix} 1 & 0 & 0 \\ 0 & -\lambda + \Gamma_{33} & \Gamma_{34} \\ 0 & 1 & -\lambda \end{pmatrix}$$

$$\det(A - \lambda \cdot I) = (-\lambda + \Gamma_{11}) \cdot (-\lambda) \cdot [(-\lambda + \Gamma_{33}) \cdot (-\lambda) - \Gamma_{34}] - \Gamma_{12} \cdot [(-\lambda + \Gamma_{33}) \cdot (-\lambda) - \Gamma_{34}]$$

$$\lambda^4 - \lambda^3 \cdot (\Gamma_{33} + \Gamma_{11}) + \lambda^2 \cdot (\Gamma_{11} \cdot \Gamma_{33} - \Gamma_{34} - \Gamma_{12}) + \lambda \cdot (\Gamma_{11} \cdot \Gamma_{34} + \Gamma_{12} \cdot \Gamma_{33}) + \Gamma_{12} \cdot \Gamma_{34} = 0$$

Eigenvalues stability discussion: Our Semi-passive RFID tags with double loop antenna system involving N variables ($N > 2, N = 4$), the characteristic equation is of degree $N = 4$ and must often be solved numerically. Expect in some particular cases, such an equation has ($N = 4$) distinct roots that can be real or complex. These values are the eigenvalues of the 4×4 Jacobian matrix (A). The general rule is that the Steady State (SS) is stable if there is no eigenvalue with positive real part. It is sufficient that one eigenvalue is positive for the steady state to be unstable. Our 4-variables $(V'_{11}, V_{11}, V'_{12}, V_{12})$ system has four eigenvalues. The type of behavior can be characterized as a function of the position of these eigenvalues in the Re/Im plane. Five non-degenerated cases can be distinguished: (1) the four eigenvalues are real and negative (stable steady state), (2) the four eigenvalues are real, three of them are negative (unstable steady state), (3) and (4) two eigenvalues are complex conjugates with a negative real part and the other eigenvalues are real negative (stable steady state), two cases can be distinguished depending on the relative value of the real part of the complex eigenvalues and of the real one, (5) two eigenvalues are complex conjugates with a negative real part and other eigenvalues real are positive (unstable steady state) [2–4].

We define

$$V'_{11}(t-\Delta_1) = V'^{(i)}_{11} + v'_{11} \cdot e^{\lambda\cdot(t-\Delta_1)}; V_{11}(t-\tau_1) = V^{(i)}_{11} + v_{11} \cdot e^{\lambda\cdot(t-\tau_1)}$$
$$V'_{12}(t-\Delta_2) = V'^{(i)}_{12} + v'_{12} \cdot e^{\lambda\cdot(t-\Delta_2)}; V_{12}(t-\tau_2) = V^{(i)}_{12} + v_{12} \cdot e^{\lambda\cdot(t-\tau_2)}$$

Then we get four delayed differential equations with respect to coordinates $[V'_{11}\ V_{11}\ V'_{12}\ V_{12}]$ arbitrarily small increments of exponential $[v'_{11}\ v_{11}\ v'_{12}\ v_{12}] \cdot e^{\lambda\cdot t}$.

$$\lambda \cdot e^{\lambda\cdot t} \cdot v'_{11} = \Gamma_{11} \cdot e^{\lambda\cdot(t-\Delta_1)} \cdot v'_{11} + \Gamma_{12} \cdot e^{\lambda\cdot(t-\tau_1)} \cdot v_{11}; \lambda \cdot e^{\lambda\cdot t} \cdot v_{11} = e^{\lambda\cdot(t-\Delta_1)} \cdot v'_{11}$$
$$\lambda \cdot e^{\lambda\cdot t} \cdot v'_{12} = \Gamma_{33} \cdot e^{\lambda\cdot(t-\Delta_2)} \cdot v'_{12} + \Gamma_{34} \cdot e^{\lambda\cdot(t-\tau_2)} \cdot v_{12}; \lambda \cdot e^{\lambda\cdot t} \cdot v_{12} = e^{\lambda\cdot(t-\Delta_2)} \cdot v'_{12}$$

In the equilibrium fixed point $V'^{(i=0)}_{11} = 0$, $V^{(i=0)}_{11} = 0$, $V'^{(i=0)}_{12} = 0$, $V^{(i=0)}_{12} = 0$. The small increments Jacobian of our RFID shifted gate system is as bellow:

$$\Upsilon_{11} = -\lambda + \Gamma_{11} \cdot e^{-\lambda\cdot\Delta_1}; \Upsilon_{12} = \Gamma_{12} \cdot e^{-\lambda\cdot\tau_1}; \Upsilon_{13} = 0; \Upsilon_{14} = 0;$$
$$\Upsilon_{21} = e^{-\lambda\cdot\Delta_1}; \Upsilon_{22} = -\lambda; \Upsilon_{23} = 0; \Upsilon_{24} = 0; \Upsilon_{31} = 0; \Upsilon_{32} = 0$$

$$\Upsilon_{33} = -\lambda + \Gamma_{33} \cdot e^{-\lambda\cdot\Delta_2}; \Upsilon_{34} = \Gamma_{34} \cdot e^{-\lambda\cdot\tau_2}; \Upsilon_{41} = 0; \Upsilon_{42} = 0;$$

$$\begin{pmatrix} \Upsilon_{11} & \cdots & \Upsilon_{14} \\ \vdots & \ddots & \vdots \\ \Upsilon_{41} & \cdots & \Upsilon_{44} \end{pmatrix} \cdot \begin{pmatrix} v'_{11} \\ v_{11} \\ v'_{12} \\ v_{12} \end{pmatrix} = 0; \Upsilon_{43} = e^{-\lambda\cdot\Delta_2}; \Upsilon_{44} = -\lambda$$

$$A - \lambda \cdot I = \begin{pmatrix} \Upsilon_{11} & \cdots & \Upsilon_{14} \\ \vdots & \ddots & \vdots \\ \Upsilon_{41} & \cdots & \Upsilon_{44} \end{pmatrix}; \det |A - \lambda \cdot I| = 0$$

$$D(\lambda,\tau_1,\tau_2,\Delta_1,\Delta_2) = \lambda^4 + \Gamma_{12}\cdot\Gamma_{34}\cdot e^{-\lambda\cdot[\sum_{i=1}^{2}\tau_i + \sum_{j=1}^{2}\Delta_j]}$$
$$+\lambda\cdot\{\Gamma_{11}\cdot\Gamma_{34}\cdot e^{-\lambda\cdot[\tau_2+\sum_{j=1}^{2}\Delta_j]} + \Gamma_{33}\cdot\Gamma_{12}\cdot e^{-\lambda\cdot[\tau_1+\sum_{j=1}^{2}\Delta_j]}\}$$
$$+\lambda^2\cdot\{-\Gamma_{34}\cdot e^{-\lambda\cdot(\Delta_2+\tau_2)} - \Gamma_{12}\cdot e^{-\lambda\cdot(\Delta_1+\tau_1)} + \Gamma_{11}\cdot\Gamma_{33}\cdot e^{-\lambda\cdot\sum_{j=1}^{2}\Delta_j}\}$$
$$-\lambda^3\cdot\{\Gamma_{33}\cdot e^{-\lambda\cdot\Delta_2} + \Gamma_{11}\cdot e^{-\lambda\cdot\Delta_1}\}$$

We have three stability cases: $\tau_1 = \tau_2 = \tau$ & $\Delta_1 = \Delta_2 = 0$ or $\tau_1 = \tau_2 = 0$ & $\Delta_1 = \Delta_2 = \Delta$ or $\tau_1 = \tau_2 = \Delta_1 = \Delta_2 = \tau_\Delta$ otherwise $\tau_1 \neq \tau_2$ & $\Delta_1 \neq \Delta_2$ and they are positive parameters. There are other possible simple stability cases: $\tau_1 = \tau; \tau_2 = 0; \Delta_1 = \Delta_2 = 0$ or $\tau_1 = 0; \tau_2 = \tau; \Delta_1 = \Delta_2 = 0$. $\tau_1 = \tau_2 = 0; \Delta_1 = \Delta; \Delta_2 = 0$ or $\tau_1 = \tau_2 = 0; \Delta_1 = 0; \Delta_2 = \Delta$. We need to get characteristics equations for all above stability analysis cases. We study the occurrence of any possible stability switching, resulting from the increase the value of the time delays τ, Δ, τ_Δ for the general characteristic equation $D(\lambda, \tau/\Delta/\tau_\Delta)$. If we choose τ parameter, then $D(\lambda,\tau) = P_n(\lambda,\tau) + Q_m(\lambda,\tau)\cdot e^{-\lambda\tau}$. The expression for $P_n(\lambda,\tau)$; $P_n(\lambda,\tau) = \sum_{k=0}^{n} P_k(\tau)\cdot\lambda^k = P_0(\tau) + P_1(\tau)\cdot\lambda + P_2(\tau)\cdot\lambda^2 + P_3(\tau)\cdot\lambda^3 + \ldots$.

The expression for $Q_m(\lambda,\tau)$ is $Q_m(\lambda,\tau) = \sum_{k=0}^{m} q_k(\tau)\cdot\lambda^k = q_0(\tau) + q_1(\tau)\cdot\lambda + q_2(\tau)\cdot\lambda^2 + \ldots$.

First, we discuss RFID shifted gate system fourth order characteristic equation for $\tau_1 = \tau; \tau_2 = 0; \Delta_1 = \Delta_2 = 0$. The first case we analyze is when there is a delay in RFID first gate's primary loop antenna voltage and no delay in secondary loop antenna voltage. Additionally, there is no delay in the gate's primary and secondary loop antennas voltages derivatives [5, 6]. The general characteristic equation D(λ, τ) is ad follow:

$$D(\lambda,\tau) = \lambda\cdot\Gamma_{11}\cdot\Gamma_{34} + \lambda^2\cdot(\Gamma_{11}\cdot\Gamma_{33} - \Gamma_{34}) - \lambda^3\cdot(\Gamma_{33}+\Gamma_{11})$$
$$+\lambda^4 + \{\Gamma_{12}\cdot\Gamma_{34} + \lambda\cdot\Gamma_{33}\cdot\Gamma_{12} - \lambda^2\cdot\Gamma_{12}\}\cdot e^{-\lambda\cdot\tau}$$
$$D(\lambda,\tau) = P_n(\lambda,\tau) + Q_m(\lambda,\tau)\cdot e^{-\lambda\tau}; n = 4; m = 2; n > m.$$

The expression for $P_n(\lambda,\tau)$ is

$$P_n(\lambda,\tau) = \sum_{k=0}^{n} P_k(\tau)\cdot\lambda^k = P_0(\tau) + P_1(\tau)\cdot\lambda + P_2(\tau)\cdot\lambda^2 + P_3(\tau)\cdot\lambda^3$$
$$+ P_4(\tau)\cdot\lambda^4 = \lambda\cdot\Gamma_{11}\cdot\Gamma_{34} + \lambda^2\cdot(\Gamma_{11}\cdot\Gamma_{33} - \Gamma_{34}) - \lambda^3\cdot(\Gamma_{33}+\Gamma_{11}) + \lambda^4$$
$$P_0(\tau) = 0; P_1(\tau) = \Gamma_{11}\cdot\Gamma_{34}; P_2(\tau) = \Gamma_{11}\cdot\Gamma_{33} - \Gamma_{34}; P_3(\tau) = -(\Gamma_{33}+\Gamma_{11}); P_4(\tau) = 1$$

The expression for $Q_m(\lambda,\tau)$ is

$$Q_m(\lambda,\tau) = \sum_{k=0}^{m} q_k(\tau)\cdot\lambda^k = q_0(\tau) + q_1(\tau)\cdot\lambda + q_2(\tau)\cdot\lambda^2 = \Gamma_{12}\cdot\Gamma_{34} + \lambda\cdot\Gamma_{33}\cdot\Gamma_{12} - \lambda^2\cdot\Gamma_{12}$$

$$q_0(\tau) = \Gamma_{12}\cdot\Gamma_{34}; q_1(\tau) = \Gamma_{33}\cdot\Gamma_{12}; q_2(\tau) = -\Gamma_{12}$$

The homogeneous system for $V'_{11}\ V_{11}\ V'_{12}\ V_{12}$ leads to a characteristic equation for the eigenvalue λ having the form $P(\lambda)+Q(\lambda)\cdot e^{-\lambda\cdot\tau}=0; P(\lambda)=\sum_{j=0}^{4} a_j\cdot\lambda^j; Q(\lambda)=\sum_{j=0}^{2} c_j\cdot\lambda^j$

And the coefficients $\{a_j(q_i,q_k),\ c_j(q_i,q_k)\}\in\mathbb{R}$ depend on q_i, q_k, but not on τ. q_i, q_k are any two shifted gate system's parameters, other parameters keep as a constant. $a_0=0; a_1=\Gamma_{11}\cdot\Gamma_{34}; a_2=\Gamma_{11}\cdot\Gamma_{33}-\Gamma_{34}; a_3=-(\Gamma_{33}+\Gamma_{11}); a_4=1$ $c_0=\Gamma_{12}\cdot\Gamma_{34}; c_1=\Gamma_{33}\cdot\Gamma_{12}; c_2=-\Gamma_{12}$. Unless strictly necessary, the designation of the varied arguments (q_i,q_k) will subsequently be omitted from P, Q, a_j, c_j. The coefficients a_j, c_j are continuous, and differentiable functions of their arguments, and direct substitution shows that $a_0 + c_0 \neq 0$ for $\forall q_i, q_k \in \mathbb{R}_+$, i.e.

λ = 0 is not a of $P(\lambda)+Q(\lambda)\cdot e^{-\lambda\cdot\tau}=0$. Furthermore, P(λ), Q(λ) are analytic functions of λ, for which the following requirements of the analysis [BK] can also be verified in the present case:

(a) If $\lambda = i\cdot\omega$, $\omega\in\mathbb{R}$, then $P(i\cdot\omega)+Q(i\cdot\omega)\neq 0$.
(b) $|Q(\lambda)/P(\lambda)|$ is bounded for $|\lambda|\to\infty$, $\mathrm{Re}\lambda\geq 0$. No roots bifurcation from ∞.
(c) $F(\omega)=|P(i\cdot\omega)|^2-|Q(i\cdot\omega)|^2$ Has a finite number of zeros. Indeed, this is a polynomial in ω.
(d) Each positive root $\omega(q_i,q_k)$ of F(ω) = 0 is continuous and differentiable respect to q_i, q_k.

We assume that $P_n(\lambda,\tau)=P_n(\lambda)$ and $Q_m(\lambda,\tau)=Q_m(\lambda)$. It can't have common imaginary roots. That is, for any real number ω;

$$p_n(\lambda=i\cdot\omega,\tau)+Q_m(\lambda=i\cdot\omega,\tau)\neq 0;$$
$$p_n(\lambda=i\cdot\omega,\tau)=i\cdot\omega\cdot\Gamma_{11}\cdot\Gamma_{34}+i\cdot\omega^3\cdot(\Gamma_{33}+\Gamma_{11})-\omega^2\cdot(\Gamma_{11}\cdot\Gamma_{33}-\Gamma_{34})+\omega^4$$
$$Q_m(\lambda=i\cdot\omega,\tau)=i\cdot\omega\cdot\Gamma_{33}\cdot\Gamma_{12}+\Gamma_{12}\cdot\Gamma_{34}+\omega^2\cdot\Gamma_{12}$$
$$p_n(\lambda=i\cdot\omega,\tau)+Q_m(\lambda=i\cdot\omega,\tau)=\Gamma_{12}\cdot\Gamma_{34}+\omega^2\cdot[\Gamma_{12}-\Gamma_{11}\cdot\Gamma_{33}+\Gamma_{34}]+\omega^4$$
$$+i\cdot\omega\cdot[\Gamma_{33}\cdot\Gamma_{12}+\Gamma_{11}\cdot\Gamma_{34}]+i\cdot\omega^3\cdot[\Gamma_{33}+\Gamma_{11}]\neq 0$$

$$|P(i\cdot\omega,\tau)|^2=\omega^2\cdot\Gamma_{11}^2\cdot\Gamma_{34}^2+\omega^4\cdot\{2\cdot\Gamma_{11}\cdot\Gamma_{34}\cdot(\Gamma_{33}+\Gamma_{11})+(\Gamma_{11}\cdot\Gamma_{33}-\Gamma_{34})^2\}$$
$$+\omega^6\cdot\{(\Gamma_{33}+\Gamma_{11})^2-2\cdot(\Gamma_{11}\cdot\Gamma_{33}-\Gamma_{34})\}+\omega^8$$
$$|Q(i\cdot\omega,\tau)|^2=\Gamma_{12}^2\cdot\Gamma_{34}^2+\omega^2\cdot\Gamma_{12}^2\cdot(2\cdot\Gamma_{34}+\Gamma_{33}^2)+\omega^4\cdot\Gamma_{12}^2$$
$$F(\omega,\tau)=|P(i\cdot\omega,\tau)|^2-|Q(i\cdot\omega,\tau)|^2=-\Gamma_{12}^2\cdot\Gamma_{34}^2+\omega^2\cdot\{\Gamma_{11}^2\cdot\Gamma_{34}^2-\Gamma_{12}^2\cdot(2\cdot\Gamma_{34}+\Gamma_{33}^2)\}$$
$$+\omega^4\cdot\{2\cdot\Gamma_{11}\cdot\Gamma_{34}\cdot(\Gamma_{33}+\Gamma_{11})+(\Gamma_{11}\cdot\Gamma_{33}-\Gamma_{34})^2-\Gamma_{12}^2\}$$
$$+\omega^6\cdot\{(\Gamma_{33}+\Gamma_{11})^2-2\cdot(\Gamma_{11}\cdot\Gamma_{33}-\Gamma_{34})\}+\omega^8$$

We define the following parameters for simplicity:

$$\begin{aligned}
&\Xi_0 = -\Gamma_{12}^2 \cdot \Gamma_{34}^2; \Xi_2 = \Gamma_{11}^2 \cdot \Gamma_{34}^2 - \Gamma_{12}^2 \cdot (2 \cdot \Gamma_{34} + \Gamma_{33}^2); \\
&\Xi_4 = 2 \cdot \Gamma_{11} \cdot \Gamma_{34} \cdot (\Gamma_{33} + \Gamma_{11}) + (\Gamma_{11} \cdot \Gamma_{33} - \Gamma_{34})^2 - \Gamma_{12}^2 \\
&\Xi_6 = (\Gamma_{33} + \Gamma_{11})^2 - 2 \cdot (\Gamma_{11} \cdot \Gamma_{33} - \Gamma_{34}); \Xi_8 = 1 \\
F(\omega, \tau) &= |P(i \cdot \omega, \tau)|^2 - |Q(i \cdot \omega, \tau)|^2 = \Xi_0 + \Xi_2 \cdot \omega^2 \\
&+ \Xi_4 \cdot \omega^4 + \Xi_6 \cdot \omega^6 + \Xi_8 \cdot \omega^8 = \sum_{k=0}^{4} \Xi_{2 \cdot k} \cdot \omega^{2 \cdot k}
\end{aligned}$$

Hence $F(\omega, \tau) = 0$ implies $\sum_{k=0}^{4} \Xi_{2 \cdot k} \cdot \omega^{2 \cdot k} = 0$ and its roots are given by solving the above polynomial. Furthermore $P_R(i \cdot \omega, \tau) = -\omega^2 \cdot (\Gamma_{11} \cdot \Gamma_{33} - \Gamma_{34}) + \omega^4$

$$\begin{aligned}
&P_R(i \cdot \omega, \tau) = -\omega^2 \cdot (\Gamma_{11} \cdot \Gamma_{33} - \Gamma_{34}) + \omega^4 \\
&P_I(i \cdot \omega, \tau) = \omega \cdot \{\Gamma_{11} \cdot \Gamma_{34} + \omega^2 \cdot (\Gamma_{33} + \Gamma_{11})\}; \\
&Q_R(i \cdot \omega, \tau) = \Gamma_{12} \cdot \Gamma_{34} + \omega^2 \cdot \Gamma_{12}; Q_I(i \cdot \omega, \tau) = \omega \cdot \Gamma_{33} \cdot \Gamma_{12}
\end{aligned}$$

Hence

$$\sin \theta(\tau) = \frac{-P_R(i \cdot \omega, \tau) \cdot Q_I(i \cdot \omega, \tau) + P_I(i \cdot \omega, \tau) \cdot Q_R(i \cdot \omega, \tau)}{|Q(i \cdot \omega, \tau)|^2}$$

$$\cos \theta(\tau) = -\frac{P_R(i \cdot \omega, \tau) \cdot Q_R(i \cdot \omega, \tau) + P_I(i \cdot \omega, \tau) \cdot Q_I(i \cdot \omega, \tau)}{|Q(i \cdot \omega, \tau)|^2}$$

$$\sin \theta(\tau) = \frac{\begin{array}{c}\{\Gamma_{11} \cdot \Gamma_{33} - \Gamma_{34} - \omega^2\} \cdot \omega^3 \cdot \Gamma_{33} \cdot \Gamma_{12} \\ + \omega \cdot \{\Gamma_{11} \cdot \Gamma_{34} + \omega^2 \cdot (\Gamma_{33} + \Gamma_{11})\} \cdot \{\Gamma_{12} \cdot \Gamma_{34} + \omega^2 \cdot \Gamma_{12}\}\end{array}}{\Gamma_{12}^2 \cdot \Gamma_{34}^2 + \omega^2 \cdot \Gamma_{12}^2 \cdot (2 \cdot \Gamma_{34} + \Gamma_{33}^2) + \omega^4 \cdot \Gamma_{12}^2}$$

$$\cos \theta(\tau) = -\frac{\begin{array}{c}\omega^2 \cdot \{\Gamma_{34} - \Gamma_{11} \cdot \Gamma_{33} + \omega^2\} \cdot \{\Gamma_{12} \cdot \Gamma_{34} + \omega^2 \cdot \Gamma_{12}\} \\ + \omega^2 \cdot \{\Gamma_{11} \cdot \Gamma_{34} + \omega^2 \cdot (\Gamma_{33} + \Gamma_{11})\} \cdot \Gamma_{33} \cdot \Gamma_{12}\end{array}}{\Gamma_{12}^2 \cdot \Gamma_{34}^2 + \omega^2 \cdot \Gamma_{12}^2 \cdot (2 \cdot \Gamma_{34} + \Gamma_{33}^2) + \omega^4 \cdot \Gamma_{12}^2}$$

Which jointly with $F(\omega, \tau) = 0 \Rightarrow \sum_{k=0}^{4} \Xi_{2 \cdot k} \cdot \omega^{2 \cdot k} = 0$ that is continuous and differentiable in τ based on Lemma 1.1. Hence we use theorem 1.2. This proves the theorem 1.3. Remark: RFID shifted gate system parameters are independent of the delay parameter τ.

Second, we discuss RFID shifted gate system fourth order characteristic equation for $\tau_1 = \tau_2 = \tau$ & $\Delta_1 = \Delta_2 = 0$. The second case we analyze is when there is a delay in RFID gate's primary and secondary loop antenna voltages ($\tau_1 = \tau_2 = \tau$) and no delay in the gate's primary and secondary loop antennas voltages derivatives [5, 6]. The general characteristic equation D(λ, τ) is ad follow:

$$D(\lambda,\tau) = \lambda^4 - \lambda^3 \cdot (\Gamma_{33} + \Gamma_{11}) + \lambda^2 \cdot \Gamma_{11} \cdot \Gamma_{33} + \{\Gamma_{12} \cdot \Gamma_{34} \cdot e^{-\lambda\cdot\tau} + \lambda \cdot (\Gamma_{11} \cdot \Gamma_{34} + \Gamma_{12} \cdot \Gamma_{33}) - \lambda^2 \cdot (\Gamma_{34} + \Gamma_{12})\} \cdot e^{-\lambda\cdot\tau}$$

Under Taylor series approximation: $e^{-\lambda\cdot\tau} \approx 1 - \lambda \cdot \tau + \frac{1}{2} \cdot \lambda^2 \cdot \tau^2$. The Maclaurin series is a Taylor series expansion of a $e^{-\lambda\cdot\tau}$ function about zero (0). We get the following general characteristic equation D(λ, τ) under Taylor series approximation:

$$e^{-\lambda\cdot\tau} \approx 1 - \lambda \cdot \tau + \frac{1}{2} \cdot \lambda^2 \cdot \tau^2.$$

$$D(\lambda,\tau) = \lambda^4 - \lambda^3 \cdot [\Gamma_{33} + \Gamma_{11}] + \lambda^2 \cdot \Gamma_{11} \cdot \Gamma_{33} + \{\Gamma_{12} \cdot \Gamma_{34} + \lambda \cdot [\Gamma_{11} \cdot \Gamma_{34} + \Gamma_{12} \cdot \Gamma_{33} - \Gamma_{12} \cdot \Gamma_{34} \cdot \tau] + \lambda^2 \cdot [\frac{1}{2} \cdot \Gamma_{12} \cdot \Gamma_{34} \cdot \tau^2 - \Gamma_{34} - \Gamma_{12}]\} \cdot e^{-\lambda\cdot\tau}$$

$$D(\lambda,\tau) = P_n(\lambda,\tau) + Q_m(\lambda,\tau) \cdot e^{-\lambda\tau}; n = 4; m = 2; n > m.$$

The expression for $P_n(\lambda,\tau)$ is

$$P_n(\lambda,\tau) = \sum_{k=0}^{n} P_k(\tau) \cdot \lambda^k = P_0(\tau) + P_1(\tau) \cdot \lambda + P_2(\tau) \cdot \lambda^2 + P_3(\tau) \cdot \lambda^3 + P_4(\tau) \cdot \lambda^4 = \lambda^4 - \lambda^3 \cdot [\Gamma_{33} + \Gamma_{11}] + \lambda^2 \cdot \Gamma_{11} \cdot \Gamma_{33}$$

$$P_0(\tau) = 0; P_1(\tau) = 0; P_2(\tau) = \Gamma_{11} \cdot \Gamma_{33}; P_3(\tau) = -[\Gamma_{33} + \Gamma_{11}]; P_4(\tau) = 1.$$

The expression for $Q_m(\lambda,\tau)$ is $Q_m(\lambda,\tau) = \sum_{k=0}^{m} q_k(\tau) \cdot \lambda^k = q_0(\tau) + q_1(\tau) \cdot \lambda + q_2(\tau) \cdot \lambda^2$

$$Q_m(\lambda,\tau) = \sum_{k=0}^{m} q_k(\tau) \cdot \lambda^k = q_0(\tau) + q_1(\tau) \cdot \lambda + q_2(\tau) \cdot \lambda^2$$

$$Q_m(\lambda,\tau) = \sum_{k=0}^{m} q_k(\tau) \cdot \lambda^k = \Gamma_{12} \cdot \Gamma_{34} + \lambda \cdot [\Gamma_{11} \cdot \Gamma_{34} + \Gamma_{12} \cdot \Gamma_{33} - \Gamma_{12} \cdot \Gamma_{34} \cdot \tau] + \lambda^2 \cdot [\frac{1}{2} \cdot \Gamma_{12} \cdot \Gamma_{34} \cdot \tau^2 - \Gamma_{34} - \Gamma_{12}]; q_0(\tau) = \Gamma_{12} \cdot \Gamma_{34}$$

$$q_1(\tau) = \Gamma_{11} \cdot \Gamma_{34} + \Gamma_{12} \cdot \Gamma_{33} - \Gamma_{12} \cdot \Gamma_{34} \cdot \tau; q_2(\tau) = \frac{1}{2} \cdot \Gamma_{12} \cdot \Gamma_{34} \cdot \tau^2 - \Gamma_{34} - \Gamma_{12}$$

The homogeneous system for $V'_{11}\ V_{11}\ V'_{12}\ V_{12}$ leads to a characteristic equation for the eigenvalue λ having the form

$$P(\lambda,\tau)+Q(\lambda,\tau)\cdot e^{-\lambda\cdot\tau}=0;P(\lambda)=\sum_{j=0}^{4}a_j\cdot\lambda^j;Q(\lambda)=\sum_{j=0}^{2}c_j\cdot\lambda^j$$

And the coefficients $\{a_j(q_i,q_k,\tau),\ c_j(q_i,q_k,\tau)\}\in\mathbb{R}$ depend on q_i,q_k and delay τ. q_i,q_k are any two shifted gate system's parameters, other parameters keep as a constant.

$$a_0=0;a_1=0;a_2=\Gamma_{11}\cdot\Gamma_{33};a_3=-[\Gamma_{33}+\Gamma_{11}];a_4=1$$
$$c_0=\Gamma_{12}\cdot\Gamma_{34};c_1=\Gamma_{11}\cdot\Gamma_{34}+\Gamma_{12}\cdot\Gamma_{33}-\Gamma_{12}\cdot\Gamma_{34}\cdot\tau;$$
$$c_2=\frac{1}{2}\cdot\Gamma_{12}\cdot\Gamma_{34}\cdot\tau^2-\Gamma_{34}-\Gamma_{12}$$

Unless strictly necessary, the designation of the varied arguments (q_i,q_k) will subsequently be omitted from P, Q, a_j, c_j. The coefficients a_j, c_j are continuous, and differentiable functions of their arguments, and direct substitution shows that $a_0+c_0\neq 0$ for $\forall q_i,q_k\in\mathbb{R}_+$, i.e. λ = 0 is not a of $P(\lambda,\tau)+Q(\lambda,\tau)\cdot e^{-\lambda\cdot\tau}=0$. We assume that $P_n(\lambda,\tau)$ and $Q_m(\lambda,\tau)$ can't have common imaginary roots. That is, for any real number ω:

$$p_n(\lambda=i\cdot\omega,\tau)+Q_m(\lambda=i\cdot\omega,\tau)\neq 0;p_n(\lambda=i\cdot\omega,\tau)$$
$$=\omega^4+i\cdot\omega^3\cdot(\Gamma_{33}+\Gamma_{11})-\omega^2\cdot\Gamma_{11}\cdot\Gamma_{33}$$

$$Q_m(\lambda=i\cdot\omega,\tau)=\Gamma_{12}\cdot\Gamma_{34}+i\cdot\omega\cdot[\Gamma_{11}\cdot\Gamma_{34}+\Gamma_{12}\cdot\Gamma_{33}-\Gamma_{12}\cdot\Gamma_{34}\cdot\tau]$$
$$-\omega^2\cdot[\frac{1}{2}\cdot\Gamma_{12}\cdot\Gamma_{34}\cdot\tau^2-\Gamma_{34}-\Gamma_{12}]$$
$$p_n(\lambda=i\cdot\omega,\tau)+Q_m(\lambda=i\cdot\omega,\tau)=\omega^4-\omega^2\cdot[\frac{1}{2}\cdot\Gamma_{12}\cdot\Gamma_{34}\cdot\tau^2-\Gamma_{34}-\Gamma_{12}+\Gamma_{11}\cdot\Gamma_{33}]$$
$$+\Gamma_{12}\cdot\Gamma_{34}+i\cdot\omega^3\cdot(\Gamma_{33}+\Gamma_{11})+i\cdot\omega\cdot[\Gamma_{11}\cdot\Gamma_{34}$$
$$+\Gamma_{12}\cdot\Gamma_{33}-\Gamma_{12}\cdot\Gamma_{34}\cdot\tau]\neq 0;|P(i\cdot\omega,\tau)|^2$$
$$=\omega^8+\omega^6\cdot\{(\Gamma_{33}+\Gamma_{11})^2-2\cdot\Gamma_{11}\cdot\Gamma_{33}\}+\omega^4\cdot\Gamma_{11}^2\cdot\Gamma_{33}^2$$

$$|Q(i\cdot\omega,\tau)|^2=\Gamma_{12}^2\cdot\Gamma_{34}^2+\omega^2\cdot\{[\Gamma_{11}\cdot\Gamma_{34}+\Gamma_{12}\cdot\Gamma_{33}-\Gamma_{12}\cdot\Gamma_{34}\cdot\tau]^2$$
$$-2\cdot\Gamma_{12}\cdot\Gamma_{34}\cdot[\frac{1}{2}\cdot\Gamma_{12}\cdot\Gamma_{34}\cdot\tau^2-\Gamma_{34}-\Gamma_{12}]\}$$
$$+\omega^4\cdot[\frac{1}{2}\cdot\Gamma_{12}\cdot\Gamma_{34}\cdot\tau^2-\Gamma_{34}-\Gamma_{12}]^2$$

We need to find the expression for $F(\omega,\tau) = |P(i\cdot\omega,\tau)|^2 - |Q(i\cdot\omega,\tau)|^2$

$$\begin{aligned}F(\omega,\tau) &= |P(i\cdot\omega,\tau)|^2 - |Q(i\cdot\omega,\tau)|^2 = \omega^8 + \omega^6\cdot\{(\Gamma_{33}+\Gamma_{11})^2 - 2\cdot\Gamma_{11}\cdot\Gamma_{33}\}\\ &+\omega^4\cdot\{\Gamma_{11}^2\cdot\Gamma_{33}^2 - [\frac{1}{2}\cdot\Gamma_{12}\cdot\Gamma_{34}\cdot\tau^2 - \Gamma_{34} - \Gamma_{12}]^2\}\\ &-\omega^2\cdot\{[\Gamma_{11}\cdot\Gamma_{34} + \Gamma_{12}\cdot\Gamma_{33} - \Gamma_{12}\cdot\Gamma_{34}\cdot\tau]^2\\ &-2\cdot\Gamma_{12}\cdot\Gamma_{34}\cdot[\frac{1}{2}\cdot\Gamma_{12}\cdot\Gamma_{34}\cdot\tau^2 - \Gamma_{34} - \Gamma_{12}]\} - \Gamma_{12}^2\cdot\Gamma_{34}^2\end{aligned}$$

We define the following parameters for simplicity:

$$\begin{aligned}\Xi_0 &= -\Gamma_{12}^2\cdot\Gamma_{34}^2;\ \Xi_2 = -[\Gamma_{11}\cdot\Gamma_{34} + \Gamma_{12}\cdot\Gamma_{33} - \Gamma_{12}\cdot\Gamma_{34}\cdot\tau]^2\\ &\quad +2\cdot\Gamma_{12}\cdot\Gamma_{34}\cdot[\frac{1}{2}\cdot\Gamma_{12}\cdot\Gamma_{34}\cdot\tau^2 - \Gamma_{34} - \Gamma_{12}]\\ \Xi_4 &= \Gamma_{11}^2\cdot\Gamma_{33}^2 - [\frac{1}{2}\cdot\Gamma_{12}\cdot\Gamma_{34}\cdot\tau^2 - \Gamma_{34} - \Gamma_{12}]^2;\\ \Xi_6 &= (\Gamma_{33}+\Gamma_{11})^2 - 2\cdot\Gamma_{11}\cdot\Gamma_{33};\ \Xi_8 = 1\\ F(\omega,\tau) &= |P(i\cdot\omega,\tau)|^2 - |Q(i\cdot\omega,\tau)|^2\\ &= \Xi_0 + \Xi_2\cdot\omega^2 + \Xi_4\cdot\omega^4 + \Xi_6\cdot\omega^6 + \Xi_8\cdot\omega^8 = \sum_{k=0}^{4}\Xi_{2\cdot k}\cdot\omega^{2\cdot k}\end{aligned}$$

Hence $F(\omega,\tau) = 0$ implies $\sum_{k=0}^{4}\Xi_{2\cdot k}\cdot\omega^{2\cdot k} = 0$ and its roots are given by solving the above polynomial. Furthermore $P_R(i\cdot\omega,\tau) = \omega^4 - \omega^2\cdot\Gamma_{11}\cdot\Gamma_{33}$

$$\begin{aligned}P_R(i\cdot\omega,\tau) &= \omega^4 - \omega^2\cdot\Gamma_{11}\cdot\Gamma_{33}\\ P_I(i\cdot\omega,\tau) &= \omega^3\cdot(\Gamma_{33}+\Gamma_{11})\\ Q_R(i\cdot\omega,\tau) &= \Gamma_{12}\cdot\Gamma_{34} - \omega^2\cdot[\frac{1}{2}\cdot\Gamma_{12}\cdot\Gamma_{34}\cdot\tau^2 - \Gamma_{34} - \Gamma_{12}];\\ Q_I(i\cdot\omega,\tau) &= \omega\cdot[\Gamma_{11}\cdot\Gamma_{34} + \Gamma_{12}\cdot\Gamma_{33} - \Gamma_{12}\cdot\Gamma_{34}\cdot\tau]\end{aligned}$$

Hence

$$\sin\theta(\tau) = \frac{-P_R(i\cdot\omega,\tau)\cdot Q_I(i\cdot\omega,\tau) + P_I(i\cdot\omega,\tau)\cdot Q_R(i\cdot\omega,\tau)}{|Q(i\cdot\omega,\tau)|^2}$$

$$\cos\theta(\tau) = -\frac{P_R(i\cdot\omega,\tau)\cdot Q_R(i\cdot\omega,\tau) + P_I(i\cdot\omega,\tau)\cdot Q_I(i\cdot\omega,\tau)}{|Q(i\cdot\omega,\tau)|^2}$$

$$\sin\theta(\tau) = \frac{\begin{array}{l} -\{\omega^4 - \omega^2\cdot\Gamma_{11}\cdot\Gamma_{33}\}\cdot\omega\cdot[\Gamma_{11}\cdot\Gamma_{34}+\Gamma_{12}\cdot\Gamma_{33}-\Gamma_{12}\cdot\Gamma_{34}\cdot\tau] \\ +\omega^3\cdot(\Gamma_{33}+\Gamma_{11})\cdot\{\Gamma_{12}\cdot\Gamma_{34}-\omega^2\cdot[\frac{1}{2}\cdot\Gamma_{12}\cdot\Gamma_{34}\cdot\tau^2-\Gamma_{34}-\Gamma_{12}]\}\end{array}}{\begin{array}{l}\Gamma_{12}^2\cdot\Gamma_{34}^2+\omega^2\cdot\{[\Gamma_{11}\cdot\Gamma_{34}+\Gamma_{12}\cdot\Gamma_{33}-\Gamma_{12}\cdot\Gamma_{34}\cdot\tau]^2 \\ -2\cdot\Gamma_{12}\cdot\Gamma_{34}\cdot[\frac{1}{2}\cdot\Gamma_{12}\cdot\Gamma_{34}\cdot\tau^2-\Gamma_{34}-\Gamma_{12}]\} \\ +\omega^4\cdot[\frac{1}{2}\cdot\Gamma_{12}\cdot\Gamma_{34}\cdot\tau^2-\Gamma_{34}-\Gamma_{12}]^2\end{array}}$$

$$\cos\theta(\tau) = -\frac{\begin{array}{l}\{\omega^4-\omega^2\cdot\Gamma_{11}\cdot\Gamma_{33}\}\cdot\{\Gamma_{12}\cdot\Gamma_{34}-\omega^2\cdot[\frac{1}{2}\cdot\Gamma_{12}\cdot\Gamma_{34}\cdot\tau^2 \\ -\Gamma_{34}-\Gamma_{12}]\}+\omega^4\cdot(\Gamma_{33}+\Gamma_{11})\cdot[\Gamma_{11}\cdot\Gamma_{34}+\Gamma_{12}\cdot\Gamma_{33} \\ -\Gamma_{12}\cdot\Gamma_{34}\cdot\tau]\end{array}}{\begin{array}{l}\Gamma_{12}^2\cdot\Gamma_{34}^2+\omega^2\cdot\{[\Gamma_{11}\cdot\Gamma_{34}+\Gamma_{12}\cdot\Gamma_{33}-\Gamma_{12}\cdot\Gamma_{34}\cdot\tau]^2 \\ -2\cdot\Gamma_{12}\cdot\Gamma_{34}\cdot[\frac{1}{2}\cdot\Gamma_{12}\cdot\Gamma_{34}\cdot\tau^2-\Gamma_{34}-\Gamma_{12}]\} \\ +\omega^4\cdot[\frac{1}{2}\cdot\Gamma_{12}\cdot\Gamma_{34}\cdot\tau^2-\Gamma_{34}-\Gamma_{12}]^2\end{array}}$$

That is a continuous and differentiable in τ based on Lemma 1.1. Hence we use theorem 1.2. This proves the theorem 1.3. Third, we discuss RFID shifted gate system fourth order characteristic equation for $\tau_1 = \tau_2 = \Delta_1 = \Delta_2 = \tau_\Delta$. The third case we analyze is when there is a delay in RFID gate's primary and secondary loop antenna voltages ($\tau_1 = \tau_2 = \Delta_1 = \Delta_2 = \tau_\Delta$) and delay in the gate's primary and secondary loop antennas voltages derivatives [5, 6]. The general characteristic equation $D(\lambda, \tau)$ is as follows:

$$\begin{aligned} D(\lambda,\tau_\Delta) &= \lambda^4 + \{\Gamma_{12}\cdot\Gamma_{34}\cdot e^{-\lambda\cdot 3\cdot\tau_\Delta} + \lambda\cdot(\Gamma_{11}\cdot\Gamma_{34}+\Gamma_{12}\cdot\Gamma_{33})\cdot e^{-\lambda\cdot 2\cdot\tau_\Delta} \\ &\quad + \lambda^2\cdot(-\Gamma_{34}+\Gamma_{11}\cdot\Gamma_{33}-\Gamma_{12})\cdot e^{-\lambda\cdot\tau_\Delta} - \lambda^3\cdot(\Gamma_{33}+\Gamma_{11})\}\cdot e^{-\lambda\cdot\tau_\Delta} \end{aligned}$$

The Maclaurin series is a Taylor series expansion of $e^{-\lambda\cdot\tau}; e^{-2\cdot\lambda\cdot\tau}; e^{-3\cdot\lambda\cdot\tau}$ functions about zero (0). We get the following general characteristic equation $D(\lambda, \tau)$ under Taylor series approximation:

$$\begin{aligned} &e^{-\lambda\cdot\tau} \approx 1-\lambda\cdot\tau; e^{-\lambda\cdot 2\cdot\tau} \approx 1-\lambda\cdot 2\cdot\tau \\ &e^{-\lambda\cdot 3\cdot\tau} \approx 1-\lambda\cdot 3\cdot\tau \end{aligned}$$

.

$$D(\lambda, \tau_\Delta) = \lambda^4 + \{\Gamma_{12} \cdot \Gamma_{34} + \lambda \cdot (\Gamma_{11} \cdot \Gamma_{34} + \Gamma_{12} \cdot \Gamma_{33} - \Gamma_{12} \cdot \Gamma_{34} \cdot 3 \cdot \tau_\Delta)$$
$$+ \lambda^2 \cdot (\Gamma_{11} \cdot \Gamma_{33} - \Gamma_{12} - \Gamma_{34} - [\Gamma_{11} \cdot \Gamma_{34} + \Gamma_{12} \cdot \Gamma_{33}] \cdot 2 \cdot \tau_\Delta)$$
$$+ \lambda^3 \cdot ([\Gamma_{34} - \Gamma_{11} \cdot \Gamma_{33} + \Gamma_{12}] \cdot \tau_\Delta - \Gamma_{33} - \Gamma_{11})\} \cdot e^{-\lambda \cdot \tau_\Delta}$$

$$D(\lambda, \tau) = P_n(\lambda, \tau) + Q_m(\lambda, \tau) \cdot e^{-\lambda\tau}; n = 4; m = 3; n > m.$$

The expression for $P_n(\lambda, \tau)$ being

$$P_n(\lambda, \tau) = \sum_{k=0}^{n} P_k(\tau) \cdot \lambda^k = P_0(\tau) + P_1(\tau) \cdot \lambda + P_2(\tau) \cdot \lambda^2 + P_3(\tau) \cdot \lambda^3$$
$$+ P_4(\tau) \cdot \lambda^4 = \lambda^4$$
$$P_0(\tau) = 0; P_1(\tau) = 0; P_2(\tau) = 0; P_3(\tau) = 0; P_4(\tau) = 1$$

The expression for $Q_m(\lambda, \tau)$ being

$$Q_m(\lambda, \tau) = \sum_{k=0}^{m} q_k(\tau) \cdot \lambda^k = q_0(\tau) + q_1(\tau) \cdot \lambda + q_2(\tau) \cdot \lambda^2 + q_3(\tau) \cdot \lambda^3$$
$$Q_m(\lambda, \tau) = \Gamma_{12} \cdot \Gamma_{34} + \lambda \cdot (\Gamma_{11} \cdot \Gamma_{34} + \Gamma_{12} \cdot \Gamma_{33} - \Gamma_{12} \cdot \Gamma_{34} \cdot 3 \cdot \tau_\Delta)$$
$$+ \lambda^2 \cdot (\Gamma_{11} \cdot \Gamma_{33} - \Gamma_{12} - \Gamma_{34} - [\Gamma_{11} \cdot \Gamma_{34} + \Gamma_{12} \cdot \Gamma_{33}] \cdot 2 \cdot \tau_\Delta)$$
$$+ \lambda^3 \cdot ([\Gamma_{34} - \Gamma_{11} \cdot \Gamma_{33} + \Gamma_{12}] \cdot \tau_\Delta - \Gamma_{33} - \Gamma_{11})$$
$$q_0(\tau) = \Gamma_{12} \cdot \Gamma_{34}; q_1(\tau) = \Gamma_{11} \cdot \Gamma_{34} + \Gamma_{12} \cdot \Gamma_{33} - \Gamma_{12} \cdot \Gamma_{34} \cdot 3 \cdot \tau_\Delta;$$
$$q_2(\tau) = \Gamma_{11} \cdot \Gamma_{33} - \Gamma_{12} - \Gamma_{34} - [\Gamma_{11} \cdot \Gamma_{34} + \Gamma_{12} \cdot \Gamma_{33}] \cdot 2 \cdot \tau_\Delta$$
$$q_3(\tau) = [\Gamma_{34} - \Gamma_{11} \cdot \Gamma_{33} + \Gamma_{12}] \cdot \tau_\Delta - \Gamma_{33} - \Gamma_{11}$$

A homogeneous system for V'_{11} V_{11} V'_{12} V_{12} leads to a characteristic equation for the eigenvalue λ having the form $P(\lambda, \tau) + Q(\lambda, \tau) \cdot e^{-\lambda \cdot \tau} = 0; P(\lambda) = \sum_{j=0}^{4} a_j \cdot \lambda^j; Q(\lambda) = \sum_{j=0}^{3} c_j \cdot \lambda^j$ and the coefficients $\{a_j(q_i, q_k, \tau), c_j(q_i, q_k, \tau)\} \in \mathbb{R}$ depend on q_i, q_k and delay τ. q_i, q_k are any two shifted gate system's parameters, other parameters kept as a constant.

$$a_0 = 0; a_1 = 0; a_2 = 0; a_3 = 0; a_4 = 1; c_0 = \Gamma_{12} \cdot \Gamma_{34};$$
$$c_1 = \Gamma_{11} \cdot \Gamma_{34} + \Gamma_{12} \cdot \Gamma_{33} - \Gamma_{12} \cdot \Gamma_{34} \cdot 3 \cdot \tau_\Delta$$
$$c_0 = \Gamma_{12} \cdot \Gamma_{34}; c_1 = \Gamma_{11} \cdot \Gamma_{34} + \Gamma_{12} \cdot \Gamma_{33} - \Gamma_{12} \cdot \Gamma_{34} \cdot 3 \cdot \tau_\Delta;$$
$$c_2 = \Gamma_{11} \cdot \Gamma_{33} - \Gamma_{12} - \Gamma_{34} - [\Gamma_{11} \cdot \Gamma_{34} + \Gamma_{12} \cdot \Gamma_{33}] \cdot 2 \cdot \tau_\Delta$$
$$c_3 = [\Gamma_{34} - \Gamma_{11} \cdot \Gamma_{33} + \Gamma_{12}] \cdot \tau_\Delta - \Gamma_{33} - \Gamma_{11}.$$

Unless strictly necessary, the designation of the varied arguments (q_i, q_k) will subsequently be omitted from P, Q, a_j, c_j. The coefficients a_j, c_j are continuous, and differentiable functions of their arguments, and direct substitution shows that $a_0 + c_0 \neq 0$ for $\forall q_i, q_k \in \mathbb{R}_+$, i.e.

$\lambda = 0$ is not a $P(\lambda, \tau) + Q(\lambda, \tau) \cdot e^{-\lambda \cdot \tau} = 0$. We assume that $P_n(\lambda, \tau)$ and $Q_m(\lambda, \tau)$ can't have common imaginary roots. That is, for any real number ω:

$$p_n(\lambda = i \cdot \omega, \tau) + Q_m(\lambda = i \cdot \omega, \tau) \neq 0$$

$$\begin{aligned} p_n(\lambda = i \cdot \omega, \tau) &= \omega^4; Q_m(\lambda = i \cdot \omega, \tau) = \\ &\Gamma_{12} \cdot \Gamma_{34} - \omega^2 \cdot (\Gamma_{11} \cdot \Gamma_{33} - \Gamma_{12} - \Gamma_{34} - [\Gamma_{11} \cdot \Gamma_{34} + \Gamma_{12} \cdot \Gamma_{33}] \cdot 2 \cdot \tau_\Delta) \\ &+ i \cdot \omega \cdot (\Gamma_{11} \cdot \Gamma_{34} + \Gamma_{12} \cdot \Gamma_{33} - \Gamma_{12} \cdot \Gamma_{34} \cdot 3 \cdot \tau_\Delta) \\ &- i \cdot \omega^3 \cdot ([\Gamma_{34} - \Gamma_{11} \cdot \Gamma_{33} + \Gamma_{12}] \cdot \tau_\Delta - \Gamma_{33} - \Gamma_{11}) \\ p_n(\lambda = i \cdot \omega, \tau) &+ Q_m(\lambda = i \cdot \omega, \tau) = \\ &\Gamma_{12} \cdot \Gamma_{34} - \omega^2 \cdot (\Gamma_{11} \cdot \Gamma_{33} - \Gamma_{12} - \Gamma_{34} - [\Gamma_{11} \cdot \Gamma_{34} + \Gamma_{12} \cdot \Gamma_{33}] \cdot 2 \cdot \tau_\Delta) \\ &+ \omega^4 + i \cdot \omega \cdot (\Gamma_{11} \cdot \Gamma_{34} + \Gamma_{12} \cdot \Gamma_{33} - \Gamma_{12} \cdot \Gamma_{34} \cdot 3 \cdot \tau_\Delta) \\ &- i \cdot \omega^3 \cdot ([\Gamma_{34} - \Gamma_{11} \cdot \Gamma_{33} + \Gamma_{12}] \cdot \tau_\Delta - \Gamma_{33} - \Gamma_{11}) \neq 0; |P(i \cdot \omega, \tau)|^2 = \omega^8 \end{aligned}$$

$$\begin{aligned} |Q_m(\lambda = i \cdot \omega, \tau)|^2 = \Gamma_{12}^2 \cdot \Gamma_{34}^2 &+ \omega^2 \cdot \{(\Gamma_{11} \cdot \Gamma_{34} + \Gamma_{12} \cdot \Gamma_{33} - \Gamma_{12} \cdot \Gamma_{34} \cdot 3 \cdot \tau_\Delta)^2 \\ &- 2 \cdot \Gamma_{12} \cdot \Gamma_{34} \cdot (\Gamma_{11} \cdot \Gamma_{33} - \Gamma_{12} - \Gamma_{34} - [\Gamma_{11} \cdot \Gamma_{34} + \Gamma_{12} \cdot \Gamma_{33}] \cdot 2 \cdot \tau_\Delta)\} \\ &+ \omega^4 \cdot \{(\Gamma_{11} \cdot \Gamma_{33} - \Gamma_{12} - \Gamma_{34} - [\Gamma_{11} \cdot \Gamma_{34} + \Gamma_{12} \cdot \Gamma_{33}] \cdot 2 \cdot \tau_\Delta)^2 \\ &- 2 \cdot (\Gamma_{11} \cdot \Gamma_{34} + \Gamma_{12} \cdot \Gamma_{33} - \Gamma_{12} \cdot \Gamma_{34} \cdot 3 \cdot \tau_\Delta) \cdot ([\Gamma_{34} - \Gamma_{11} \cdot \Gamma_{33} \\ &\quad + \Gamma_{12}] \cdot \tau_\Delta - \Gamma_{33} - \Gamma_{11})\} \\ &+ \omega^6 \cdot ([\Gamma_{34} - \Gamma_{11} \cdot \Gamma_{33} + \Gamma_{12}] \cdot \tau_\Delta - \Gamma_{33} - \Gamma_{11})^2 \end{aligned}$$

We need to find the expression for $F(\omega, \tau) = |P(i \cdot \omega, \tau)|^2 - |Q(i \cdot \omega, \tau)|^2$

$$\begin{aligned} F(\omega, \tau) = |P(i \cdot \omega, \tau)|^2 - |Q(i \cdot \omega, \tau)|^2 &= \omega^8 - \omega^6 \cdot ([\Gamma_{34} - \Gamma_{11} \cdot \Gamma_{33} + \Gamma_{12}] \cdot \tau_\Delta - \Gamma_{33} - \Gamma_{11})^2 \\ &- \omega^4 \cdot \{(\Gamma_{11} \cdot \Gamma_{33} - \Gamma_{12} - \Gamma_{34} - [\Gamma_{11} \cdot \Gamma_{34} + \Gamma_{12} \cdot \Gamma_{33}] \cdot 2 \cdot \tau_\Delta)^2 \\ &- 2 \cdot (\Gamma_{11} \cdot \Gamma_{34} + \Gamma_{12} \cdot \Gamma_{33} - \Gamma_{12} \cdot \Gamma_{34} \cdot 3 \cdot \tau_\Delta) \cdot ([\Gamma_{34} - \Gamma_{11} \cdot \Gamma_{33} + \Gamma_{12}] \cdot \tau_\Delta - \Gamma_{33} - \Gamma_{11})\} \\ &- \omega^2 \cdot \{(\Gamma_{11} \cdot \Gamma_{34} + \Gamma_{12} \cdot \Gamma_{33} - \Gamma_{12} \cdot \Gamma_{34} \cdot 3 \cdot \tau_\Delta)^2 \\ &- 2 \cdot \Gamma_{12} \cdot \Gamma_{34} \cdot (\Gamma_{11} \cdot \Gamma_{33} - \Gamma_{12} - \Gamma_{34} - [\Gamma_{11} \cdot \Gamma_{34} + \Gamma_{12} \cdot \Gamma_{33}] \cdot 2 \cdot \tau_\Delta)\} - \Gamma_{12}^2 \cdot \Gamma_{34}^2 \end{aligned}$$

We define the following parameters for simplicity:

$$\Xi_0 = -\Gamma_{12}^2 \cdot \Gamma_{34}^2; \Xi_2 = -\{(\Gamma_{11} \cdot \Gamma_{34} + \Gamma_{12} \cdot \Gamma_{33} - \Gamma_{12} \cdot \Gamma_{34} \cdot 3 \cdot \tau_\Delta)^2 \\ - 2 \cdot \Gamma_{12} \cdot \Gamma_{34} \cdot (\Gamma_{11} \cdot \Gamma_{33} - \Gamma_{12} - \Gamma_{34} - [\Gamma_{11} \cdot \Gamma_{34} + \Gamma_{12} \cdot \Gamma_{33}] \cdot 2 \cdot \tau_\Delta)\}$$

$$\Xi_4 = -\{(\Gamma_{11} \cdot \Gamma_{33} - \Gamma_{12} - \Gamma_{34} - [\Gamma_{11} \cdot \Gamma_{34} + \Gamma_{12} \cdot \Gamma_{33}] \cdot 2 \cdot \tau_\Delta)^2 \\ - 2 \cdot (\Gamma_{11} \cdot \Gamma_{34} + \Gamma_{12} \cdot \Gamma_{33} - \Gamma_{12} \cdot \Gamma_{34} \cdot 3 \cdot \tau_\Delta) \cdot ([\Gamma_{34} - \Gamma_{11} \cdot \Gamma_{33} + \Gamma_{12}] \cdot \tau_\Delta - \Gamma_{33} - \Gamma_{11})\}$$

$$\Xi_6 = -([\Gamma_{34} - \Gamma_{11} \cdot \Gamma_{33} + \Gamma_{12}] \cdot \tau_\Delta - \Gamma_{33} - \Gamma_{11})^2; \Xi_8 = 1$$

$$F(\omega, \tau) = |P(i \cdot \omega, \tau)|^2 - |Q(i \cdot \omega, \tau)|^2 = \Xi_0 + \Xi_2 \cdot \omega^2 + \Xi_4 \cdot \omega^4 + \Xi_6 \cdot \omega^6 + \Xi_8 \cdot \omega^8 = \sum_{k=0}^{4} \Xi_{2\cdot k} \cdot \omega^{2\cdot k}$$

Hence $F(\omega, \tau) = 0$ implies $\sum_{k=0}^{4} \Xi_{2\cdot k} \cdot \omega^{2\cdot k} = 0$ and its roots are given by solving the above polynomial. Furthermore

$$P_R(i \cdot \omega, \tau) = \omega^4; P_I(i \cdot \omega, \tau) = 0$$

$$Q_R(i \cdot \omega, \tau) = \Gamma_{12} \cdot \Gamma_{34} - \omega^2 \cdot (\Gamma_{11} \cdot \Gamma_{33} - \Gamma_{12} - \Gamma_{34} - [\Gamma_{11} \cdot \Gamma_{34} + \Gamma_{12} \cdot \Gamma_{33}] \cdot 2 \cdot \tau_\Delta)$$

$$Q_I(i \cdot \omega, \tau) = \omega \cdot \{(\Gamma_{11} \cdot \Gamma_{34} + \Gamma_{12} \cdot \Gamma_{33} - \Gamma_{12} \cdot \Gamma_{34} \cdot 3 \cdot \tau_\Delta) \\ - \omega^2 \cdot ([\Gamma_{34} - \Gamma_{11} \cdot \Gamma_{33} + \Gamma_{12}] \cdot \tau_\Delta - \Gamma_{33} - \Gamma_{11})\}$$

Hence

$$\sin\theta(\tau) = \frac{-P_R(i \cdot \omega, \tau) \cdot Q_I(i \cdot \omega, \tau) + P_I(i \cdot \omega, \tau) \cdot Q_R(i \cdot \omega, \tau)}{|Q(i \cdot \omega, \tau)|^2}$$

$$\cos\theta(\tau) = -\frac{P_R(i \cdot \omega, \tau) \cdot Q_R(i \cdot \omega, \tau) + P_I(i \cdot \omega, \tau) \cdot Q_I(i \cdot \omega, \tau)}{|Q(i \cdot \omega, \tau)|^2}$$

$$\sin\theta(\tau) = \frac{-\omega^5 \cdot \{(\Gamma_{11} \cdot \Gamma_{34} + \Gamma_{12} \cdot \Gamma_{33} - \Gamma_{12} \cdot \Gamma_{34} \cdot 3 \cdot \tau_\Delta) - \omega^2 \cdot ([\Gamma_{34} - \Gamma_{11} \cdot \Gamma_{33} + \Gamma_{12}] \cdot \tau_\Delta - \Gamma_{33} - \Gamma_{11})\}}{\begin{array}{l}\Gamma_{12}^2 \cdot \Gamma_{34}^2 + \omega^2 \cdot \{(\Gamma_{11} \cdot \Gamma_{34} + \Gamma_{12} \cdot \Gamma_{33} - \Gamma_{12} \cdot \Gamma_{34} \cdot 3 \cdot \tau_\Delta)^2 \\ - 2 \cdot \Gamma_{12} \cdot \Gamma_{34} \cdot (\Gamma_{11} \cdot \Gamma_{33} - \Gamma_{12} - \Gamma_{34} - [\Gamma_{11} \cdot \Gamma_{34} + \Gamma_{12} \cdot \Gamma_{33}] \\ \cdot 2 \cdot \tau_\Delta)\} + \omega^4 \cdot \{(\Gamma_{11} \cdot \Gamma_{33} - \Gamma_{12} - \Gamma_{34} - [\Gamma_{11} \cdot \Gamma_{34} \\ + \Gamma_{12} \cdot \Gamma_{33}] \cdot 2 \cdot \tau_\Delta)^2 - 2 \cdot (\Gamma_{11} \cdot \Gamma_{34} + \Gamma_{12} \cdot \Gamma_{33} - \Gamma_{12} \cdot \Gamma_{34} \cdot 3 \cdot \tau_\Delta) \\ \cdot ([\Gamma_{34} - \Gamma_{11} \cdot \Gamma_{33} + \Gamma_{12}] \cdot \tau_\Delta - \Gamma_{33} - \Gamma_{11})\} + \omega^6 \cdot ([\Gamma_{34} - \Gamma_{11} \cdot \Gamma_{33} \\ + \Gamma_{12}] \cdot \tau_\Delta - \Gamma_{33} - \Gamma_{11})^2\end{array}}$$

$$\cos\theta(\tau) = -\frac{\omega^4 \cdot \{\Gamma_{12} \cdot \Gamma_{34} - \omega^2 \cdot (\Gamma_{11} \cdot \Gamma_{33} - \Gamma_{12} - \Gamma_{34} - [\Gamma_{11} \cdot \Gamma_{34} + \Gamma_{12} \cdot \Gamma_{33}] \cdot 2 \cdot \tau_\Delta)\}}{\begin{array}{l}\Gamma_{12}^2 \cdot \Gamma_{34}^2 + \omega^2 \cdot \{(\Gamma_{11} \cdot \Gamma_{34} + \Gamma_{12} \cdot \Gamma_{33} - \Gamma_{12} \cdot \Gamma_{34} \cdot 3 \cdot \tau_\Delta)^2 \\ - 2 \cdot \Gamma_{12} \cdot \Gamma_{34} \cdot (\Gamma_{11} \cdot \Gamma_{33} - \Gamma_{12} - \Gamma_{34} - [\Gamma_{11} \cdot \Gamma_{34} + \Gamma_{12} \cdot \Gamma_{33}] \cdot 2 \cdot \tau_\Delta)\} \\ + \omega^4 \cdot \{(\Gamma_{11} \cdot \Gamma_{33} - \Gamma_{12} - \Gamma_{34} - [\Gamma_{11} \cdot \Gamma_{34} + \Gamma_{12} \cdot \Gamma_{33}] \cdot 2 \cdot \tau_\Delta)^2 - 2 \cdot (\Gamma_{11} \cdot \Gamma_{34} \\ + \Gamma_{12} \cdot \Gamma_{33} - \Gamma_{12} \cdot \Gamma_{34} \cdot 3 \cdot \tau_\Delta) \cdot ([\Gamma_{34} - \Gamma_{11} \cdot \Gamma_{33} + \Gamma_{12}] \cdot \tau_\Delta - \Gamma_{33} - \Gamma_{11})\} \\ + \omega^6 \cdot ([\Gamma_{34} - \Gamma_{11} \cdot \Gamma_{33} + \Gamma_{12}] \cdot \tau_\Delta - \Gamma_{33} - \Gamma_{11})^2\end{array}}$$

It is continuous and differentiable in τ_Δ based on Lemma 1.1. Hence we use theorem 1.2. This proves the theorem 1.3. Next we analyze RFID shifted gate system stability analysis under delayed variables in time. Our RFID shifted gate homogeneous system for $v'_{11}\ v_{11}\ v'_{12}\ v_{12}$ leads to a characteristic equation for the eigenvalue λ having the form $P(\lambda) + Q(\lambda) \cdot e^{-\lambda \cdot \tau} = 0$; Second case $\tau_1 = \tau_2 = \tau$; $\Delta_1 = \Delta_2 = 0$.

$$\begin{aligned} D(\lambda, \tau_1 = \tau_2 = \tau, \Delta_1 = \Delta_2 = 0) &= \lambda^4 - \lambda^3 \cdot (\Gamma_{33} + \Gamma_{11}) + \lambda^2 \cdot \Gamma_{11} \cdot \Gamma_{33} \\ &+ \{\Gamma_{12} \cdot \Gamma_{34} \cdot e^{-\lambda \cdot \tau} + \lambda \cdot (\Gamma_{11} \cdot \Gamma_{34} + \Gamma_{12} \cdot \Gamma_{33}) - \lambda^2 \cdot (\Gamma_{34} + \Gamma_{12})\} \cdot e^{-\lambda \cdot \tau} \end{aligned}$$

Under Taylor series approximation: $e^{-\lambda \cdot \tau} \approx 1 - \lambda \cdot \tau + \frac{1}{2} \cdot \lambda^2 \cdot \tau^2$. The Maclaurin series is a Taylor series expansion of a $e^{-\lambda \cdot \tau}$ function about zero (0). We get the following general characteristic equation D(λ, τ) under Taylor series approximation: $e^{-\lambda \cdot \tau} \approx 1 - \lambda \cdot \tau + \frac{1}{2} \cdot \lambda^2 \cdot \tau^2$.

$$\begin{aligned} D(\lambda, \tau) &= \lambda^4 - \lambda^3 \cdot [\Gamma_{33} + \Gamma_{11}] + \lambda^2 \cdot \Gamma_{11} \cdot \Gamma_{33} + \{\Gamma_{12} \cdot \Gamma_{34} \\ &+ \lambda \cdot [\Gamma_{11} \cdot \Gamma_{34} + \Gamma_{12} \cdot \Gamma_{33} - \Gamma_{12} \cdot \Gamma_{34} \cdot \tau] + \lambda^2 \cdot [\frac{1}{2} \cdot \Gamma_{12} \cdot \Gamma_{34} \cdot \tau^2 - \Gamma_{34} - \Gamma_{12}]\} \cdot e^{-\lambda \cdot \tau} \end{aligned}$$

We use different parameters terminology from our last characteristics parameters definition: $k \to j; p_k(\tau) \to a_j; q_k(\tau) \to c_j; n = 4; m = 2; n > m$

Additionally $P_n(\lambda, \tau) \to P(\lambda); Q_m(\lambda, \tau) \to Q(\lambda)$ then $P(\lambda) = \sum_{j=0}^{4} a_j \cdot \lambda^j$; $Q(\lambda) = \sum_{j=0}^{2} c_j \cdot \lambda^j$

$$\begin{aligned} P_\lambda &= \lambda^4 - \lambda^3 \cdot [\Gamma_{33} + \Gamma_{11}] + \lambda^2 \cdot \Gamma_{11} \cdot \Gamma_{33}; \\ Q_\lambda &= \Gamma_{12} \cdot \Gamma_{34} + \lambda \cdot [\Gamma_{11} \cdot \Gamma_{34} + \Gamma_{12} \cdot \Gamma_{33} - \Gamma_{12} \cdot \Gamma_{34} \cdot \tau] \\ &+ \lambda^2 \cdot [\frac{1}{2} \cdot \Gamma_{12} \cdot \Gamma_{34} \cdot \tau^2 - \Gamma_{34} - \Gamma_{12}] \end{aligned}$$

$n, m \in \mathbb{N}_0$, $n > m$ and $a_j, c_j : \mathrm{R}_{+0} \to R$ are continuous and differentiable function of τ such that $a_0 + c_0 \neq 0$. In the following "—" denotes complex and conjugate. $P(\lambda), Q(\lambda)$ are analytic functions in λ and differentiable in τ. The coefficients

$\{a_j(C_1, R_1, \text{gate antenna parametrs}) \text{ and } c_j(C_1, R_1, \tau, \text{gate antenna parametrs })\} \in \mathbb{R}$ depend on RFID shifted gate system's C_1, R_1,τ values and antenna parameters. $a_0 = 0; a_1 = 0; a_2 = \Gamma_{11} \cdot \Gamma_{33}; a_3 = -[\Gamma_{33} + \Gamma_{11}]; a_4 = 1$
$c_0 = \Gamma_{12} \cdot \Gamma_{34}; c_1 = \Gamma_{11} \cdot \Gamma_{34} + \Gamma_{12} \cdot \Gamma_{33} - \Gamma_{12} \cdot \Gamma_{34} \cdot \tau;$
$c_2 = \frac{1}{2} \cdot \Gamma_{12} \cdot \Gamma_{34} \cdot \tau^2 - \Gamma_{34} - \Gamma_{12}$

Unless strictly necessary, the designation of the varied arguments $(R_1, C_1, \tau, \text{gate antenna parametrs})$ will subsequently be omitted from P, Q, a_j, c_j. The coefficients a_j, c_j are continuous, and differentiable functions of their arguments, and direct substitution shows that $a_0 + c_0 \neq 0; \Gamma_{12} \cdot \Gamma_{34} \neq 0$.

$$\frac{\eta_3 \cdot \xi_3}{\eta_1 \cdot \xi_1} = \frac{\frac{1}{L_{11} \cdot L_{12}} \cdot (1 + \frac{r_{p1}}{R_1})^2}{C_1^2 \cdot (1 + \frac{L_{12}}{L_{11}} + 2 \cdot K \cdot \sqrt{\frac{L_{12}}{L_{11}}}) \cdot (1 + \frac{L_{11}}{L_{12}} + 2 \cdot K \cdot \sqrt{\frac{L_{11}}{L_{12}}})} \neq 0$$
$$\forall C_1, \text{ gate antenna parameters} \in \mathbb{R}_+$$

i.e. $\lambda = 0$ is not a root of the characteristic equation. Furthermore $P(\lambda)$, $Q(\lambda)$ are analytic functions of λ for which the following requirements of the analysis (see Kuang 1993, Sect. 3.4) can also be verified in the present case [5, 6].

(a) If $\lambda = i \cdot \omega, \omega \in \mathbb{R}$ then $P(i \cdot \omega) + Q(i \cdot \omega) \neq 0$, i.e. P and Q have no common imaginary roots. This condition was verified numerically in the entire $(R_1, C_1, \text{antenna parametrs})$ domain of interest.
(b) $|Q(\lambda)/P(\lambda)|$ is bounded for $|\lambda| \to \infty$, $\text{Re}\lambda \geq 0$. No roots bifurcation from ∞. Indeed, in the limit.

$$|\frac{Q(\lambda)}{P(\lambda)}| = |\frac{\{\Gamma_{12} \cdot \Gamma_{34} + \lambda \cdot [\Gamma_{11} \cdot \Gamma_{34} + \Gamma_{12} \cdot \Gamma_{33} - \Gamma_{12} \cdot \Gamma_{34} \cdot \tau] + \lambda^2 \cdot [\frac{1}{2} \cdot \Gamma_{12} \cdot \Gamma_{34} \cdot \tau^2 - \Gamma_{34} - \Gamma_{12}]\}}{\lambda^4 - \lambda^3 \cdot [\Gamma_{33} + \Gamma_{11}] + \lambda^2 \cdot \Gamma_{11} \cdot \Gamma_{33}}|$$

(c) $$F(\omega) = |P(i \cdot \omega)|^2 - |Q(i \cdot \omega)|^2$$
$$\begin{aligned} F(\omega, \tau) = |P(i \cdot \omega, \tau)|^2 - |Q(i \cdot \omega, \tau)|^2 &= \omega^8 + \omega^6 \cdot \{(\Gamma_{33} + \Gamma_{11})^2 - 2 \cdot \Gamma_{11} \cdot \Gamma_{33}\} \\ &+ \omega^4 \cdot \{\Gamma_{11}^2 \cdot \Gamma_{33}^2 - [\frac{1}{2} \cdot \Gamma_{12} \cdot \Gamma_{34} \cdot \tau^2 - \Gamma_{34} - \Gamma_{12}]^2\} \\ &- \omega^2 \cdot \{[\Gamma_{11} \cdot \Gamma_{34} + \Gamma_{12} \cdot \Gamma_{33} - \Gamma_{12} \cdot \Gamma_{34} \cdot \tau]^2 \\ &- 2 \cdot \Gamma_{12} \cdot \Gamma_{34} \cdot [\frac{1}{2} \cdot \Gamma_{12} \cdot \Gamma_{34} \cdot \tau^2 - \Gamma_{34} - \Gamma_{12}]\} - \Gamma_{12}^2 \cdot \Gamma_{34}^2 \end{aligned}$$

Has at most a finite number of zeros. Indeed, this is a polynomial in ω (Degree in ω^8).
(d) Each positive root $\omega(R_1, C_1, \tau, \text{gate antenna parametrs})$ of $F(\omega) = 0$ being continuous and differentiable with respect to R_1, C_1, τ, gate antenna parametrs. This condition can only be assessed numerically.

In addition, since the coefficients in P and Q are real, we have $\overline{P(-i \cdot \omega)} = P(i \cdot \omega)$, and $\overline{Q(-i \cdot \omega)} = Q(i \cdot \omega)$ thus $\lambda = i \cdot \omega$, $\omega > 0$ maybe on eigenvalue of characteristic equations. The analysis consists in identifying the roots of the characteristic equation situated on the imaginary axis of the complex λ-plane, whereby increasing the parameters R_1, C_1, τ, gate antenna parametrs, Reλ may, at the crossing, Change its sign from (−) to (+), i.e. from a stable focus $E^{(0)}(V_{11}'^{(0)}, V_{11}^{(0)}, V_{12}'^{(0)}, V_{12}^{(0)}) = (0,0,0,0)$ to an unstable one, or vice versa. This feature may be further assessed by examining the sign of the partial derivatives with respect to C_1, R_1, τ and gate antenna parameters.

$$\wedge^{-1}(C_1) = (\frac{\partial \mathrm{Re}\lambda}{\partial C_1})_{\lambda = i \cdot \omega}, \mathrm{R}_1, \tau, \text{gate antenna parametrs} = const$$

$$\wedge^{-1}(\mathrm{R}_1) = (\frac{\partial \mathrm{Re}\lambda}{\partial \mathrm{R}_1})_{\lambda = i \cdot \omega}, C_1, \tau, \text{gate antenna parametrs} = const$$

$$\wedge^{-1}(L_{11}) = (\frac{\partial \mathrm{Re}\lambda}{\partial L_{11}})_{\lambda = i \cdot \omega}, C_1, \mathrm{R}_1, \tau = const; \wedge^{-1}(L_{12}) = (\frac{\partial \mathrm{Re}\lambda}{\partial L_{12}})_{\lambda = i \cdot \omega}, C_1, \mathrm{R}_1, \tau = const$$

$$\wedge^{-1}(\tau) = (\frac{\partial \mathrm{Re}\lambda}{\partial \tau})_{\lambda = i \cdot \omega}, C_1, \mathrm{R}_1, \text{gate antenna parametrs} = const; \text{where } \omega \in \mathbb{R}_+.$$

In the case $\tau_1 = \tau_2 = \tau; \Delta_1 = \Delta_2 = 0$ we get the following results:

$$P_R(i \cdot \omega, \tau) = \omega^4 - \omega^2 \cdot \Gamma_{11} \cdot \Gamma_{33}; P_I(i \cdot \omega, \tau) = \omega^3 \cdot (\Gamma_{33} + \Gamma_{11});$$
$$Q_R(i \cdot \omega, \tau) = \Gamma_{12} \cdot \Gamma_{34} - \omega^2 \cdot [\frac{1}{2} \cdot \Gamma_{12} \cdot \Gamma_{34} \cdot \tau^2 - \Gamma_{34} - \Gamma_{12}]$$
$$Q_I(i \cdot \omega, \tau) = \omega \cdot [\Gamma_{11} \cdot \Gamma_{34} + \Gamma_{12} \cdot \Gamma_{33} - \Gamma_{12} \cdot \Gamma_{34} \cdot \tau]$$

$$\Xi_0 = -\Gamma_{12}^2 \cdot \Gamma_{34}^2; \Xi_2 = -[\Gamma_{11} \cdot \Gamma_{34} + \Gamma_{12} \cdot \Gamma_{33} - \Gamma_{12} \cdot \Gamma_{34} \cdot \tau]^2 + 2 \cdot \Gamma_{12} \cdot \Gamma_{34} \cdot [\frac{1}{2} \cdot \Gamma_{12} \cdot \Gamma_{34} \cdot \tau^2 - \Gamma_{34} - \Gamma_{12}]$$
$$\Xi_4 = \Gamma_{11}^2 \cdot \Gamma_{33}^2 - [\frac{1}{2} \cdot \Gamma_{12} \cdot \Gamma_{34} \cdot \tau^2 - \Gamma_{34} - \Gamma_{12}]^2;$$
$$\Xi_6 = (\Gamma_{33} + \Gamma_{11})^2 - 2 \cdot \Gamma_{11} \cdot \Gamma_{33}; \Xi_8 = 1$$

$$F(\omega, \tau) = |P(i \cdot \omega, \tau)|^2 - |Q(i \cdot \omega, \tau)|^2 = \Xi_0 + \Xi_2 \cdot \omega^2 + \Xi_4 \cdot \omega^4 + \Xi_6 \cdot \omega^6 + \Xi_8 \cdot \omega^8 = \sum_{k=0}^{4} \Xi_{2 \cdot k} \cdot \omega^{2 \cdot k}$$

Hence $F(\omega, \tau) = 0$ implies $\sum_{k=0}^{4} \Xi_{2 \cdot k} \cdot \omega^{2 \cdot k} = 0$ When writing $P(\lambda) = P_R(\lambda) + i \cdot P_I(\lambda)$ and $Q(\lambda) = Q_R(\lambda) + i \cdot Q_I(\lambda)$, and inserting $\lambda = i \cdot \omega$ Into RFID Gate system's characteristic equation, ω must satisfy the following:

$$\sin\omega\cdot\tau = g(\omega) = \frac{-P_R(i\cdot\omega)\cdot Q_I(i\cdot\omega) + P_I(i\cdot\omega)\cdot Q_R(i\cdot\omega)}{|Q(i\cdot\omega)|^2}$$

$$\cos\omega\cdot\tau = h(\omega) = -\frac{P_R(i\cdot\omega)\cdot Q_R(i\cdot\omega) + P_I(i\cdot\omega)\cdot Q_I(i\cdot\omega)}{|Q(i\cdot\omega)|^2}$$

Where $|Q(i\cdot\omega)|^2 \neq 0$ in view of requirement (a) above, and $(g,h)\in R$. Furthermore, it follows above $\sin\omega\cdot\tau$ and $\cos\omega\cdot\tau$ equations that, by squaring and adding the sides, ω must be a positive root of $F(\omega) = |P(i\cdot\omega)|^2 - |Q(i\cdot\omega)|^2 = 0$. Note that $F(\omega)$ is dependent of τ. Now it is important to notice that if $\tau \notin I$ (assume that $I \subseteq R_{+0}$ is the set where $\omega(\tau)$ is a positive root of $F(\omega)$ and for $\tau \notin I$,$\omega(\tau)$ is not defined. Then for all τ in I $\omega(\tau)$ is satisfied that $F(\omega,\tau)=0$). Then there are no positive $\omega(\tau)$ solutions for $F(\omega,\tau)=0$, and we cannot have stability switches. For any $\tau\in I$ where $\omega(\tau)$ is a positive solution of $F(\omega,\tau)=0$, we can define the angle $\theta(\tau)\in[0,2\cdot\pi]$ as the solution of

$$\sin\theta(\tau) = \frac{-P_R(i\cdot\omega)\cdot Q_I(i\cdot\omega) + P_I(i\cdot\omega)\cdot Q_R(i\cdot\omega)}{|Q(i\cdot\omega)|^2}$$

$$\cos\theta(\tau) = -\frac{P_R(i\cdot\omega)\cdot Q_R(i\cdot\omega) + P_I(i\cdot\omega)\cdot Q_I(i\cdot\omega)}{|Q(i\cdot\omega)|^2}$$

And the relation between the argument $\theta(\tau)$ and $\omega(\tau)\cdot\tau$ for $\tau\in I$ must be $\omega(\tau)\cdot\tau = \theta(\tau) + n\cdot 2\cdot\pi \;\forall\, n\in\mathbb{N}_0$. Hence we can define the maps $\tau_n : I \to R_{+0}$ given by

$$\tau_n(\tau) = \frac{\theta(\tau) + n\cdot 2\cdot\pi}{\omega(\tau)};\, n\in\mathbb{N}_0, \tau\in I$$

Let us introduce the functions $I\to R$; $S_n(\tau) = \tau - \tau_n(\tau)$, $\tau\in I$, $n\in\mathbb{N}_0$ (187)

That is a continuous and differentiable in τ. In the following, the subscripts λ, ω, R_1, C_1 and RFID Gate antenna parameters $(L_{a1}, L_{a2}, L_{b1}, L_{b2}, a_1, a_2)$ indicate the corresponding partial derivatives. Let us first concentrate on $\wedge(x)$ remembering in $\lambda(L_{a1}, L_{a2}, L_{b1}, L_{b2}, a_1, a_2)$ and $\omega(L_{a1}, L_{a2}, L_{b1}, L_{b2}, a_1, a_2)$, and keeping all parameters except one (x) and τ. The derivation closely follows that in reference [BK]. Differentiating RFID characteristic equation $P(\lambda) + Q(\lambda)\cdot e^{-\lambda\cdot\tau} = 0$ with respect to specific parameter (x), and inverting the derivative, for convenience, one calculates:

Remark

$$x = R_1, C_1, L_{a1}, L_{a2}, L_{b1}, L_{b2}, a_1, a_2, etc.,$$

$$\left(\frac{\partial\lambda}{\partial x}\right)^{-1} = \frac{-P_\lambda(\lambda,x)\cdot Q(\lambda,x) + Q_\lambda(\lambda,x)\cdot P(\lambda,x) - \tau\cdot P(\lambda,x)\cdot Q(\lambda,x)}{P_x(\lambda,x)\cdot Q(\lambda,x) - Q_x(\lambda,x)\cdot P(\lambda,x)}$$

Where $P_\lambda = \frac{\partial P}{\partial \lambda}, \ldots$ etc., Substituting $\lambda = i \cdot \omega$, and bearing i $\overline{P(-i \cdot \omega)} = P(i \cdot \omega)$, $\overline{Q(-i \cdot \omega)} = Q(i \cdot \omega)$ then $i \cdot P_\lambda(i \cdot \omega) = P_\omega(i \cdot \omega)$ and $i \cdot Q_\lambda(i \cdot \omega) = Q_\omega(i \cdot \omega)$. That on the surface $|P(i \cdot \omega)|^2 = |Q(i \cdot \omega)|^2$, one obtains

$$(\frac{\partial \lambda}{\partial x})^{-1}|_{\lambda = i \cdot \omega} = (\frac{i \cdot P_\omega(i \cdot \omega, x) \cdot \overline{P(i \cdot \omega, x)} + i \cdot Q_\lambda(i \cdot \omega, x) \cdot \overline{Q(\lambda, x)} - \tau \cdot |P(i \cdot \omega, x)|^2}{P_x(i \cdot \omega, x) \cdot \overline{P(i \cdot \omega, x)} - Q_x(i \cdot \omega, x) \cdot \overline{Q(i \cdot \omega, x)}})$$

Upon separating into real and imaginary parts, with

$$P = P_R + i \cdot P_I; Q = Q_R + i \cdot Q_I; P_\omega = P_{R\omega} + i \cdot P_{I\omega}$$
$$Q_\omega = Q_{R\omega} + i \cdot Q_{I\omega}; P_x = P_{Rx} + i \cdot P_{Ix}$$
$$Q_x = Q_{Rx} + i \cdot Q_{Ix}; P^2 = P_R^2 + P_I^2$$

When (x) can be any RFID Gate parameters R_1, C_1, And time delay τ etc. Where for convenience, we have dropped the arguments $(i \cdot \omega, x)$, and where

$$F_\omega = 2 \cdot [(P_{R\omega} \cdot P_R + P_{I\omega} \cdot P_I) - (Q_{R\omega} \cdot Q_R + Q_{I\omega} \cdot Q_I)]$$
$$F_x = 2 \cdot [(P_{Rx} \cdot P_R + P_{Ix} \cdot P_I) - (Q_{Rx} \cdot Q_R + Q_{Ix} \cdot Q_I)]; \omega_x = -F_x / F_\omega.$$

We define U and V:

$$U = (P_R \cdot P_{I\omega} - P_I \cdot P_{R\omega}) - (Q_R \cdot Q_{I\omega} - Q_I \cdot Q_{R\omega});$$
$$V = (P_R \cdot P_{Ix} - P_I \cdot P_{Rx}) - (Q_R \cdot Q_{Ix} - Q_I \cdot Q_{Rx})$$

We choose our specific parameter as time delay x = τ.

$$P_{R\omega} = 2 \cdot \omega \cdot [2 \cdot \omega^2 - \Gamma_{11} \cdot \Gamma_{33}]; P_{I\omega} = 3 \cdot \omega^2 \cdot (\Gamma_{33} + \Gamma_{11});$$
$$P_{R\tau} = 0; P_{I\tau} = 0; Q_{R\tau} = -\omega^2 \cdot \Gamma_{12} \cdot \Gamma_{34} \cdot \tau; Q_{I\tau} = -\omega \cdot \Gamma_{12} \cdot \Gamma_{34}$$
$$P_{R\tau} = 0; P_{I\tau} = 0; Q_{R\tau} = -\omega^2 \cdot \Gamma_{12} \cdot \Gamma_{34} \cdot \tau; Q_{I\tau} = -\omega \cdot \Gamma_{12} \cdot \Gamma_{34};$$
$$P_{R\omega} \cdot P_R = 2 \cdot \omega^3 \cdot [2 \cdot \omega^4 - 3 \cdot \omega^2 \cdot \Gamma_{11} \cdot \Gamma_{33} + \Gamma_{11}^2 \cdot \Gamma_{33}^2]$$
$$P_{I\omega} \cdot P_I = 3 \cdot \omega^5 \cdot (\Gamma_{33} + \Gamma_{11})^2; \omega_\tau = -F_\tau / F_\omega;$$
$$Q_{R\omega} = -2 \cdot \omega \cdot [\frac{1}{2} \cdot \Gamma_{12} \cdot \Gamma_{34} \cdot \tau^2 - \Gamma_{34} - \Gamma_{12}]$$
$$Q_{I\omega} = \Gamma_{11} \cdot \Gamma_{34} + \Gamma_{12} \cdot \Gamma_{33} - \Gamma_{12} \cdot \Gamma_{34} \cdot \tau$$
$$Q_{R\omega} \cdot Q_R = -2 \cdot \omega \cdot [\frac{1}{2} \cdot \Gamma_{12} \cdot \Gamma_{34} \cdot \tau^2 - \Gamma_{34} - \Gamma_{12}] \cdot [\Gamma_{12} \cdot \Gamma_{34} - \omega^2 \cdot (\frac{1}{2} \cdot \Gamma_{12} \cdot \Gamma_{34} \cdot \tau^2 - \Gamma_{34} - \Gamma_{12})]$$
$$Q_{I\omega} \cdot Q_I = \omega \cdot [\Gamma_{11} \cdot \Gamma_{34} + \Gamma_{12} \cdot \Gamma_{33} - \Gamma_{12} \cdot \Gamma_{34} \cdot \tau]^2;$$
$$F_\tau = 2 \cdot [(P_{R\tau} \cdot P_R + P_{I\tau} \cdot P_I) - (Q_{R\tau} \cdot Q_R + Q_{I\tau} \cdot Q_I)]$$

$$F_\tau = 2 \cdot \omega^2 \cdot \Gamma_{12} \cdot \Gamma_{34} \cdot [\Gamma_{11} \cdot \Gamma_{34} + \Gamma_{12} \cdot \Gamma_{33} - \tau \cdot \omega^2 \cdot (\frac{1}{2} \cdot \Gamma_{12} \cdot \Gamma_{34} \cdot \tau^2 - \Gamma_{34} - \Gamma_{12})]$$
$$P_R \cdot P_{I\omega} = 3 \cdot \omega^4 \cdot (\omega^2 - \Gamma_{11} \cdot \Gamma_{33}) \cdot (\Gamma_{33} + \Gamma_{11});$$
$$P_I \cdot P_{R\omega} = 2 \cdot \omega^4 \cdot (\Gamma_{33} + \Gamma_{11}) \cdot (2 \cdot \omega^2 - \Gamma_{11} \cdot \Gamma_{33})$$
$$Q_R \cdot Q_{I\omega} = [\Gamma_{12} \cdot \Gamma_{34} - \omega^2 \cdot (\frac{1}{2} \cdot \Gamma_{12} \cdot \Gamma_{34} \cdot \tau^2 - \Gamma_{34} - \Gamma_{12})] \cdot [\Gamma_{11} \cdot \Gamma_{34} + \Gamma_{12} \cdot \Gamma_{33} - \Gamma_{12} \cdot \Gamma_{34} \cdot \tau]$$
$$Q_I \cdot Q_{R\omega} = -2 \cdot \omega^2 \cdot (\Gamma_{11} \cdot \Gamma_{34} + \Gamma_{12} \cdot \Gamma_{33} - \Gamma_{12} \cdot \Gamma_{34} \cdot \tau) \cdot (\frac{1}{2} \cdot \Gamma_{12} \cdot \Gamma_{34} \cdot \tau^2 - \Gamma_{34} - \Gamma_{12})$$
$$V = (P_R \cdot P_{I\tau} - P_I \cdot P_{R\tau}) - (Q_R \cdot Q_{I\tau} - Q_I \cdot Q_{R\tau});$$
$$P_R \cdot P_{I\tau} = 0; P_I \cdot P_{R\tau} = 0$$
$$Q_R \cdot Q_{I\tau} = -\omega \cdot \Gamma_{12} \cdot \Gamma_{34} \cdot [\Gamma_{12} \cdot \Gamma_{34} - \omega^2 \cdot (\frac{1}{2} \cdot \Gamma_{12} \cdot \Gamma_{34} \cdot \tau^2 - \Gamma_{34} - \Gamma_{12})];$$
$$Q_I \cdot Q_{R\tau} = -\omega^3 \cdot \Gamma_{12} \cdot \Gamma_{34} \cdot \tau \cdot [\Gamma_{11} \cdot \Gamma_{34} + \Gamma_{12} \cdot \Gamma_{33} - \Gamma_{12} \cdot \Gamma_{34} \cdot \tau]$$
$$F(\omega, \tau) = 0.$$

Differentiating with respect to τ and we get

$$F_\omega \cdot \frac{\partial \omega}{\partial \tau} + F_\tau = 0; \tau \in I \Rightarrow \frac{\partial \omega}{\partial \tau} = -\frac{F_\tau}{F_\omega}; \wedge^{-1}(\tau) = (\frac{\partial \mathrm{Re}\lambda}{\partial \tau})_{\lambda = i \cdot \omega}$$
$$\wedge^{-1}(\tau) = \mathrm{Re}\{\frac{-2 \cdot [U + \tau \cdot |P|^2] + i \cdot F_\omega}{F_\tau + i \cdot 2 \cdot [V + \omega \cdot |P|^2]}\}; \frac{\partial \omega}{\partial \tau} = \omega_\tau = -\frac{F_\tau}{F_\omega}$$
$$sign\{\wedge^{-1}(\tau)\} = sign\{(\frac{\partial \mathrm{Re}\lambda}{\partial \tau})_{\lambda = i \cdot \omega}\};$$
$$sign\{\wedge^{-1}(\tau)\} = sign\{F_\omega\} \cdot sign\{\tau \cdot \frac{\partial \omega}{\partial \tau} + \omega + \frac{U \cdot \frac{\partial \omega}{\partial \tau} + V}{|P|^2}\}$$

We shall presently examine the possibility of stability transitions (bifurcations) in a shifted gate, double loop RFID system, about the equilibrium point $E^{(0)}(V_{11}'^{(0)}, V_{11}^{(0)}, V_{12}'^{(0)}, V_{12}^{(0)})$ as a result of a variation of delay parameter τ. The analysis consists in identifying the roots of our system characteristic equation situated on the imaginary axis of the complex λ-plane, Whereby increasing the delay parameter τ, Re λ may at the crossing, changes its sign from − to +, i.e. from a stable focus $E^{(*)}$ to an unstable one, or vice versa. This feature may be further assessed by examining the sign of the partial derivatives with respect to τ, $\wedge^{-1}(\tau) = (\frac{\partial \mathrm{Re}\lambda}{\partial \tau})_{\lambda = i \cdot \omega}$

$$\wedge^{-1}(\tau) = (\frac{\partial \mathrm{Re}\lambda}{\partial \tau})_{\lambda = i \cdot \omega}, C_1, \mathrm{R}_1, \text{gate antenna parametrs} = const \text{ where } \omega \in \mathbb{R}_+.$$

For our stability switching analysis, we choose typical RFID shifted gate parameters values: L_{11} = 4.5 mH, L_{12} = 2.5 mH, C_1 = 23 pF, R_1 = 100 kΩ = 10^5, r_{p1} = 100 Ω, K = 0.6, 2·L_m = 0.004 ($2 \cdot L_m = 2 \cdot K \cdot \sqrt{L_{11} \cdot L_{12}}$). $\eta_1 = 56.22 \times 10^{-12}$

$$\eta_2 = 2.49 \times 10^{-5}; \eta_3 = 222.42; \xi_1 = 101.2 \times 10^{-12}; \xi_2 = 4.492 \times 10^{-5}$$

$$\xi_3 = 400.4; \Gamma_{11} = -\frac{\eta_2}{\eta_1} = -4.42 \times 10^5; \Gamma_{12} = -\frac{\eta_3}{\eta_1} = -3.95 \times 10^{12};$$

$$\Gamma_{33} = -\frac{\xi_2}{\xi_1} = -4.43 \times 10^5; \Gamma_{34} = -\frac{\xi_3}{\xi_1} = -3.95 \times 10^{12}.$$

$$\Gamma_{21} = \Gamma_{43} = 1; \Gamma_{13} = \Gamma_{14} = \Gamma_{22} = \Gamma_{23} = \Gamma_{24} = 0;$$

$$\Gamma_{31} = \Gamma_{32} = \Gamma_{41} = \Gamma_{42} = \Gamma_{44} = 0$$

Then we get the expression $F(\omega, \tau)$ for a typical RFID shifted gate parameters values.

$$\begin{aligned} F(\omega, \tau) &= |P(i \cdot \omega, \tau)|^2 - |Q(i \cdot \omega, \tau)|^2 = \omega^8 + \omega^6 \cdot 39.16 \times 10^{10} \\ &+ \omega^4 \cdot \{383.17 \times 10^{20} - [7.8 \times 10^{24} \cdot \tau^2 + 7.9 \times 10^{12}]^2\} \\ &- \omega^2 \cdot \{[34.94 \times 10^{17} - 15.6 \times 10^{24} \cdot \tau]^2 \\ &- 31.2 \times 10^{24} \cdot [7.8 \times 10^{24} \cdot \tau^2 + 7.9 \times 10^{12}]\} - 243.39 \times 10^{48} \end{aligned}$$

We find those ω, τ values which fulfill $F(\omega, \tau) = 0$. We ignore negative, complex, and imaginary values of ω for specific τ values. $\tau \in [0.001..10]$

And we can be express by 3D function $F(\omega, \tau) = 0$. Since it is a very complex function, we recommend to solve it numerically rather than analytic.

We plot the stability switch diagram based on different delay values of our RFID double gate system. Since it is a very complex function, we recommend to solve it numerically rather than analytic.

$$\Lambda^{-1}(\tau) = \left(\frac{\partial \mathrm{Re}\lambda}{\partial \tau}\right)_{\lambda = i \cdot \omega} = \mathrm{Re}\left\{\frac{-2 \cdot [U + \tau \cdot |P|^2] + i \cdot F_\omega}{F_\tau + i \cdot 2 \cdot [V + \omega \cdot |P|^2]}\right\};$$

$$\Lambda^{-1}(\tau) = \left(\frac{\partial \mathrm{Re}\lambda}{\partial \tau}\right)_{\lambda = i \cdot \omega} = \frac{2 \cdot \{F_\omega \cdot (V + \omega \cdot P^2) - F_\tau \cdot (U + \tau \cdot P^2)\}}{F_\tau^2 + 4 \cdot (V + \omega \cdot P^2)^2}$$

The stability switch occurs only on those delay values (τ) which fit the equation: $\tau = \frac{\theta_+(\tau)}{\omega_+(\tau)}$ and $\theta_+(\tau)$ is the solution of $\sin\theta(\tau) = \ldots; \cos\theta(\tau) = \ldots$ when $\omega = \omega_+(\tau)$ if only ω_+ is feasible. Additionally, when all RFID double gate system's parameters are known and the stability switch due to various time delay values τ is described in the following expression:

$$sign\{\wedge^{-1}(\tau)\} = sign\{F_{\omega}(\omega(\tau),\tau)\} \cdot sign\{\tau \cdot \omega_{\tau}(\omega(\tau)) + \omega(\tau) + \frac{U(\omega(\tau)) \cdot \omega_{\tau}(\omega(\tau)) + V(\omega(\tau))}{|P(\omega(\tau))|^2}\}$$

Remark: we know $F(\omega,\tau) = 0$ implies its roots $\omega_i(\tau)$ and finding those delays values τ which ω_i is feasible. There are τ values which are ω_i complex or imaginary numbered, then unable to analyze stability [5, 6].

Semi-passive RFID Tags with the double loop antennas environment is characterized by electromagnetic interferences which can influence the shifted gate system stability in time. There are four main RFID double loop antenna variables which are affected by electromagnetic interferences, first and second loop antenna voltages and voltages derivatives respectively. Each loop antennas voltage variable under electromagnetic interferences are characterized by time delay respectively. The two time delays are not the same, but can be categorized to some sub cases due to interferences behavior. The first case we analyze is when there is a delay in RFID first gate's primary loop antenna voltage and no delay in secondary loop antenna voltage. The second case we analyze is when there is a delay in RFID gate's primary and secondary loop antenna voltages ($\tau_1 = \tau_2 = \tau$) and no delay in the gate's primary and secondary loop antennas voltages derivatives [5, 6]. The third case we analyze is when there is a delay in RFID gate's primary and secondary loop antenna voltages ($\tau_1 = \tau_2 = \Delta_1 = \Delta_2 = \tau_\Delta$) and delay in the gate's primary and secondary loop antennas voltages derivatives [4, 5]. For simplicity of our analysis we consider in the third case all delays are the same (there is a difference but it is neglected in our analysis). In each case we derive the related characteristic equation. The characteristic equation is dependent on double loop antennas overall parameters and interferences time delay. Upon mathematics manipulation and [BK] theorems and definitions we derive the expression which gives us a clear picture on double loop antennas stability map. The stability map gives all possible options for stability segments, each segment belongs to different time delay value segment. Double loop antennas arranged as a shifted gate's stability analysis can be influenced either by system overall parameter values [5, 6].

1.5 RFID TAGs Detectors Stability Analysis Under Delayed Schottky Diode's Internal Elements in Time

The RFID market is growing and several cost, size and DC power constraints in the TAG itself have forced designers to abandon super heterodyne receivers for older and simpler crystal video receiver. Consisting of a simple detector circuit and a printed antenna, this receiver can face a stability issues due to delay elements in time. The Schottky diode detector demodulates the signal and sends the data on to the digital circuit of the TAG; this is the so-called "wake up" signal. A simple RFID TAG receiver block diagram includes input antenna signal with series resistance, inductor (choke), Schottky diode, and output capacitor. At a small signal (RF Input) levels, the Schottky diode can be represented by a linear equivalent circuit. Due to Schottky parasitic delayed in time, there is a stability issue by analyzing the detector operation. We include two parasitic delay elements in the Schottky equivalent circuit. We define τ_1, τ_2 as delays in time, respectively, for the Schottky equivalent circuit. We consider first those two delays in time are not equal $\tau_1 \neq \tau_2$ then another three cases $\tau_1 = \tau$ & $\tau_2 = 0$, $\tau_2 = \tau$ & $\tau_1 = 0$, $\tau_1 = \tau_2 = \tau$. The RFID receiver detector delayed in time equivalent circuit can represent as a delayed differential equations which depending on variable parameters and delays. The investigation of our RFID receiver detector system, differential equation based on bifurcation theory [1], the study of possible changes in the structure of the orbits of a delayed differential equation depending on variable parameters. We first illustrate certain observations and analyze local bifurcations of an appropriate arbitrary scalar delayed differential equation [2]. RFID receiver detector stability analysis is done under different time delays respect to currents and currents derivative. All of that for optimization of RFID receiver detector equivalent circuit parameter analysis to get the best performance. RFID system, the reader or interrogator sends a modulated RF signal which is received by the TAG. The Schottky diode detector demodulates the signal and sends the data on to the digital circuits of the TAG. The reader stops sending modulated data and illuminates the TAG with continuous wave (CW) or un-modulated signal. The TAG's FSK encoder and switch driver switch the load placed on the TAG's antenna from one state to another, causing the radar cross section of the TAG to be changed. The weak signal reflected from the TAG is modulated; this signal is then detected by the reader's receiver. In this way the reader and TAG can communicate using RF generated only in the reader. The key performance parameter for RFID TAG detector diode is operating in the square law region in voltage sensitivity. For incoming RF small signal from the RFID reader to the TAG, we can use Schottky diode which represented by a linear equivalent circuit. Rj is the junction resistance (Rv or video resistance) of the diode, where RF power is converted into video voltage output. For maximum output, all the incoming RF voltages should ideally appear across Rj. Cj is the junction capacitance of the diode chip itself. It is a parasitic element which shorts out the junction resistance, shunting the RF energy to the series resistance Rs. Rs is a parasitic resistance representing losses in the diode's bond wire, the bulk silicon at the base of the chip and other loss mechanisms. The RF voltage appearing across Rs results in power lost as heat. Lp and Cp are package parasitic inductance and capacitance, respectively. Unlike the two chips parasitic, they can easily be tuned out with an external impedance matching network. The package parasitic inductance Lp has a parasitic delay element in time (τ_1). The resistance losses in the diode's bond wire have a parasitic delay element in time (τ_2). V(t) represents the RFID tag antenna voltage in time, the incoming RF small signal from the RFID reader. We consider ideal delay lines (TAU1, TAU2), $V_{\tau_1} \to \varepsilon_1 V_{\tau_1} \to \varepsilon_2; \varepsilon_1, \varepsilon_2 \ll \varepsilon > 0$ [85] (Fig. 1.22).

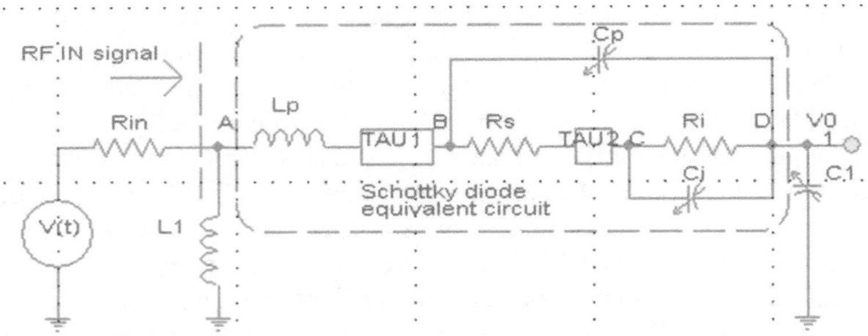

Fig. 1.22 RFID TAG receiver detector equivalent circuit

$$\frac{V(t)-V_A}{R_{in}} = I_{R_{in}}; I_{R_{in}} = I_{L_1} + I_{L_P}; V_{\tau_1} \to \varepsilon_1; V_{\tau_1} \to \varepsilon_2; \varepsilon_1, \varepsilon_2 \ll \varepsilon > 0; V_A - V_B = L_P \cdot \frac{dI_{L_P}}{dt}$$

$$I_{L_P} = I_{C_P} + I_{R_S}; I_{R_S} = \frac{V_B - V_C}{R_S}; V_A = L_1 \cdot \frac{dI_{L_1}}{dt}; I_{C_P} = C_P \cdot \frac{d(V_B - V_D)}{dt}; I_{R_j} = \frac{V_C - V_D}{R_j}$$

$$I_{C_j} = C_j \cdot \frac{d(V_C - V_D)}{dt}; I_{R_S} = I_{R_j} + I_{C_j}; I_{C_1} = C_1 \cdot \frac{dV_D}{dt}; I_{C_1} = I_{C_P} + I_{R_j} + I_{C_j}$$

$$\frac{dV_D}{dt} = \frac{I_{C_1}}{C_1} \Rightarrow I_{C_j} = C_j \cdot \frac{d(V_C - V_D)}{dt} = C_j \cdot [\frac{dV_C}{dt} - \frac{dV_D}{dt}] \Rightarrow I_{C_j} = C_j \cdot [\frac{dV_C}{dt} - \frac{dV_D}{dt}]$$

$$I_{C_P} = C_P \cdot \frac{d(V_B - V_D)}{dt} = C_P \cdot [\frac{dV_B}{dt} - \frac{dV_D}{dt}] = C_P \cdot [\frac{dV_B}{dt} - \frac{I_{C_1}}{C_1}]; \frac{V(t) - V_A}{R_{in}} = I_{R_{in}} = I_{L_1} + I_{L_P}$$

$$\frac{V(t)}{R_{in}} - \frac{L_1}{R_{in}} \cdot \frac{dI_{L_1}}{dt} = I_{L_1} + I_{L_P}; I_{R_{in}} = I_{L_1} + I_{L_P} \Rightarrow I_{L_1} = I_{R_{in}} - I_{L_P} = \frac{V(t) - V_A}{R_{in}} - I_{L_P}$$

$$\frac{V(t)}{R_{in}} - \frac{L_1}{R_{in}} \cdot \frac{d}{dt}[\frac{V(t) - V_A}{R_{in}} - I_{L_P}] = I_{L_1} + I_{L_P} = I_{R_{in}}; I_{C_P} = C_P \cdot [\frac{dV_B}{dt} - \frac{I_{L_P}}{C_1}]$$

$$I_{R_j} = \frac{V_C - V_D}{R_j} \Rightarrow I_{R_j} \cdot R_j = V_C - V_D; I_{C_j} = C_j \cdot \frac{d(V_C - V_D)}{dt} = C_j \cdot \frac{d(I_{R_j} \cdot R_j)}{dt} = C_j \cdot R_j \cdot \frac{dI_{R_j}}{dt}$$

$$I_{L_P} = I_{C_P} + I_{R_S} \Rightarrow I_{C_P} = I_{L_P} - I_{R_S}; I_{C_1} = I_{C_P} + I_{R_j} + I_{C_j} = I_{L_P} - I_{R_S} + I_{R_j} + I_{C_j}$$

$$I_{R_S} = I_{R_j} + I_{C_j} \Rightarrow I_{C_1} = I_{L_P} - I_{R_S} + I_{R_j} + I_{C_j} = I_{L_P} - (I_{R_j} + I_{C_j}) + I_{R_j} + I_{C_j} = I_{L_P}$$

$$I_{R_{in}} = \frac{V(t) - V_A}{R_{in}} = \frac{V(t)}{R_{in}} - \frac{1}{R_{in}} \cdot L_1 \cdot \frac{dI_{L_1}}{dt} = \frac{1}{R_{in}} \cdot [V(t) - L_1 \cdot \frac{dI_{L_1}}{dt}]; I_{L_1} = I_{R_{in}} - I_{L_P}; I_{C_P} = I_{L_P} - I_{R_S}$$

$$I_{C_1} = I_{L_P}; I_{R_S} = I_{R_j} + I_{C_j}; I_{C_j} = C_j \cdot [\frac{dV_C}{dt} - \frac{I_{C_1}}{C_1}] = C_j \cdot \frac{d}{dt}[I_{R_j} \cdot R_j] = C_j \cdot R_j \cdot \frac{dI_{R_j}}{dt}$$

$$I_{C_P} = C_P \cdot [\frac{dV_B}{dt} - \frac{I_{C_1}}{C_1}]; V_A - V_B = L_P \cdot \frac{dI_{L_P}}{dt} \Rightarrow L_1 \cdot \frac{dI_{L_1}}{dt} - V_B = L_P \cdot \frac{dI_{L_P}}{dt} \Rightarrow V_B = L_1 \cdot \frac{dI_{L_1}}{dt} - L_P \cdot \frac{dI_{L_P}}{dt}$$

$$\frac{dV_B}{dt} = L_1 \cdot \frac{d^2 I_{L_1}}{dt^2} - L_P \cdot \frac{d^2 I_{L_P}}{dt^2}; I_{C_P} = C_P \cdot [\frac{dV_B}{dt} - \frac{I_{C_1}}{C_1}] = C_P \cdot [L_1 \cdot \frac{d^2 I_{L_1}}{dt^2} - L_P \cdot \frac{d^2 I_{L_P}}{dt^2} - \frac{I_{C_1}}{C_1}]$$

$$I_{L_1} = I_{R_{in}} - I_{L_P} = \frac{V(t) - V_A}{R_{in}} - I_{L_P} = \frac{V(t)}{R_{in}} - \frac{L_1}{R_{in}} \cdot \frac{dI_{L_1}}{dt} - I_{L_P}; I_{C_P} = I_{L_P} - I_{R_S}; I_{C_1} = I_{L_P}$$

$$I_{L_1} = \frac{V(t)}{R_{in}} - \frac{L_1}{R_{in}} \cdot \frac{dI_{L_1}}{dt} - I_{L_P} \Rightarrow \frac{dI_{L_1}}{dt} = \frac{1}{R_{in}} \cdot \frac{dV(t)}{dt} - \frac{L_1}{R_{in}} \cdot \frac{d^2 I_{L_1}}{dt^2} - \frac{dI_{L_P}}{dt}$$

$$\frac{L_1}{R_{in}} \cdot \frac{d^2 I_{L_1}}{dt^2} = \frac{1}{R_{in}} \cdot \frac{dV(t)}{dt} - \frac{dI_{L_P}}{dt} - \frac{dI_{L_1}}{dt} \Rightarrow \frac{d^2 I_{L_1}}{dt^2} = \frac{1}{L_1} \cdot \frac{dV(t)}{dt} - \frac{R_{in}}{L_1} \cdot \frac{dI_{L_P}}{dt} - \frac{R_{in}}{L_1} \cdot \frac{dI_{L_1}}{dt}$$

$$I_{R_S} = I_{R_j} + I_{C_j} \Rightarrow \frac{V_B - V_C}{R_S} = I_{R_j} + I_{C_j};$$

$$I_{C_j} = I_{R_S} - I_{R_j}; I_{C_j} = C_j \cdot R_j \cdot \frac{dI_{R_j}}{dt} \Rightarrow I_{R_S} - I_{R_j} = C_j \cdot R_j \cdot \frac{dI_{R_j}}{dt}$$

$$I_{C_P} = C_P \cdot [\frac{dV(t)}{dt} - R_{in} \cdot \frac{dI_{L_P}}{dt} - R_{in} \cdot \frac{dI_{L_1}}{dt} - L_P \cdot \frac{d^2 I_{L_P}}{dt^2} - \frac{I_{C_1}}{C_1}]$$

$$I_{L_P} - I_{R_S} = C_P \cdot [\frac{dV(t)}{dt} - R_{in} \cdot \frac{dI_{L_P}}{dt} - R_{in} \cdot \frac{dI_{L_1}}{dt} - L_P \cdot \frac{d^2 I_{L_P}}{dt^2} - \frac{I_{C_1}}{C_1}]$$

$$I_{C_1} = I_{L_P} \Rightarrow I_{L_P} - I_{R_S} = C_P \cdot [\frac{dV(t)}{dt} - R_{in} \cdot \frac{dI_{L_P}}{dt} - R_{in} \cdot \frac{dI_{L_1}}{dt} - L_P \cdot \frac{d^2 I_{L_P}}{dt^2} - \frac{I_{L_P}}{C_1}]$$

$$I_{R_S} = \frac{V_B - V_C}{R_S} \Rightarrow V_B - V_C = I_{R_S} \cdot R_S;$$

$$I_{C_P} = C_P \cdot \frac{d(V_B - V_D)}{dt} \Rightarrow \frac{I_{C_P}}{C_P} = \frac{d}{dt}(V_B - V_D) \Rightarrow V_B - V_D = \frac{1}{C_P} \cdot \int I_{C_P} \cdot dt$$

$$I_{C_j} = C_j \cdot \frac{d(V_C - V_D)}{dt} \Rightarrow \frac{I_{C_j}}{C_j} = \frac{d(V_C - V_D)}{dt} \Rightarrow V_C - V_D = \frac{1}{C_j} \cdot \int I_{C_j} \cdot dt$$

$$(*) V_B - V_D = \frac{1}{C_P} \cdot \int I_{C_P} \cdot dt; (**) V_C - V_D = \frac{1}{C_j} \cdot \int I_{C_j} \cdot dt$$

$$(*) - (**) \to V_B - V_C = \frac{1}{C_P} \cdot \int I_{C_P} \cdot dt - \frac{1}{C_j} \cdot \int I_{C_j} \cdot dt \Rightarrow I_{R_S} \cdot R_S = \frac{1}{C_P} \cdot \int I_{C_P} \cdot dt - \frac{1}{C_j} \cdot \int I_{C_j} \cdot dt$$

$$I_{R_S} \cdot R_S = \frac{1}{C_P} \cdot \int I_{C_P} \cdot dt - \frac{1}{C_j} \cdot \int I_{C_j} \cdot dt \Rightarrow R_S \cdot \frac{dI_{R_S}}{dt} = \frac{1}{C_P} \cdot I_{C_P} - \frac{1}{C_j} \cdot I_{C_j}$$

$$R_S \cdot \frac{dI_{R_S}}{dt} = \frac{1}{C_P} \cdot I_{C_P} - \frac{1}{C_j} \cdot I_{C_j} \Rightarrow R_S \cdot \frac{dI_{R_S}}{dt} = \frac{1}{C_P} \cdot (I_{L_P} - I_{R_S}) - \frac{1}{C_j} \cdot (I_{R_S} - I_{R_j})$$

$$R_S \cdot \frac{dI_{R_S}}{dt} = \frac{1}{C_P} \cdot (I_{L_P} - I_{R_S}) - \frac{1}{C_j} \cdot (I_{R_S} - I_{R_j}) = \frac{1}{C_P} \cdot I_{L_P} + \frac{1}{C_j} \cdot I_{R_j} - I_{R_S} \cdot (\frac{1}{C_P} + \frac{1}{C_j})$$

$$R_S \cdot \frac{dI_{R_S}}{dt} = \frac{1}{C_P} \cdot I_{L_P} + \frac{1}{C_j} \cdot I_{R_j} - I_{R_S} \cdot (\frac{1}{C_P} + \frac{1}{C_j}) \Rightarrow \frac{dI_{R_S}}{dt} = \frac{1}{R_S \cdot C_P} \cdot I_{L_P} + \frac{1}{R_S \cdot C_j} \cdot I_{R_j} - I_{R_S} \cdot (\frac{1}{R_S \cdot C_P} + \frac{1}{R_S \cdot C_j})$$

We define $Y = I_{L_P} \Rightarrow \frac{dI_{R_S}}{dt} = \frac{1}{R_S \cdot C_P} \cdot Y + \frac{1}{R_S \cdot C_j} \cdot I_{R_j} - I_{R_S} \cdot (\frac{1}{R_S \cdot C_P} + \frac{1}{R_S \cdot C_j})$

$$Y = I_{L_P} \Rightarrow \frac{dI_{R_S}}{dt} = \frac{1}{R_S \cdot C_P} \cdot Y + \frac{1}{R_S \cdot C_j} \cdot I_{R_j} - I_{R_S} \cdot (\frac{1}{R_S \cdot C_P} + \frac{1}{R_S \cdot C_j})$$

$$\frac{V(t)}{R_{in}} - \frac{L_1}{R_{in}} \cdot \frac{dI_{L_1}}{dt} = I_{L_1} + I_{L_P} \Rightarrow \frac{V(t)}{R_{in}} - I_{L_1} - I_{L_P}$$

$$= \frac{L_1}{R_{in}} \cdot \frac{dI_{L_1}}{dt} \Rightarrow \frac{V(t)}{L_1} - \frac{R_{in}}{L_1} \cdot I_{L_1} - \frac{R_{in}}{L_1} \cdot I_{L_P} = \frac{dI_{L_1}}{dt}$$

$$I_{L_P} - I_{R_S} = C_P \cdot [\frac{dV(t)}{dt} - R_{in} \cdot \frac{dI_{L_P}}{dt} - R_{in} \cdot \frac{dI_{L_1}}{dt} - L_P \cdot \frac{d^2 I_{L_P}}{dt^2} - \frac{I_{L_P}}{C_1}]$$

$$I_{L_P} - I_{R_S} = C_P \cdot [\frac{dV(t)}{dt} - R_{in} \cdot \frac{dI_{L_P}}{dt} - R_{in} \cdot (\frac{V(t)}{L_1} - \frac{R_{in}}{L_1} \cdot I_{L_1} - \frac{R_{in}}{L_1} \cdot I_{L_P}) - L_P \cdot \frac{d^2 I_{L_P}}{dt^2} - \frac{I_{L_P}}{C_1}]$$

$$I_{L_P} - I_{R_S} = C_P \cdot [\frac{dV(t)}{dt} - R_{in} \cdot \frac{dI_{L_P}}{dt} - \frac{R_{in} \cdot V(t)}{L_1} + \frac{R_{in}^2}{L_1} \cdot I_{L_1} + \frac{R_{in}^2}{L_1} \cdot I_{L_P} - L_P \cdot \frac{d^2 I_{L_P}}{dt^2}$$
$$- \frac{I_{L_P}}{C_1}]$$

$$- I_{L_P} + I_{R_S} + C_P \cdot \frac{dV(t)}{dt} - C_P \cdot R_{in} \cdot \frac{dI_{L_P}}{dt} - \frac{C_P \cdot R_{in} \cdot V(t)}{L_1}$$
$$+ C_P \cdot \frac{R_{in}^2}{L_1} \cdot I_{L_1} + C_P \cdot \frac{R_{in}^2}{L_1} \cdot I_{L_P} - C_P \cdot L_P \cdot \frac{d^2 I_{L_P}}{dt^2} - \frac{C_P \cdot I_{L_P}}{C_1} = 0$$

$$- C_P \cdot L_P \cdot \frac{d^2 I_{L_P}}{dt^2} - C_P \cdot R_{in} \cdot \frac{dI_{L_P}}{dt} + I_{L_P} \cdot [C_P \cdot \frac{R_{in}^2}{L_1} - \frac{C_P}{C_1} - 1]$$
$$+ I_{R_S} + C_P \cdot \frac{R_{in}^2}{L_1} \cdot I_{L_1} - \frac{C_P \cdot R_{in} \cdot V(t)}{L_1} + C_P \cdot \frac{dV(t)}{dt} = 0$$

We define: $Y = I_{L_P}; X = \frac{dI_{L_P}}{dt}; \frac{dX}{dt} = \frac{d^2 I_{L_P}}{dt^2}; \frac{dY}{dt} = \frac{dI_{L_P}}{dt} = X$ then we get the expression:

$$- C_P \cdot L_P \cdot \frac{dX}{dt} - C_P \cdot R_{in} \cdot X + Y \cdot [C_P \cdot \frac{R_{in}^2}{L_1} - \frac{C_P}{C_1} - 1]$$
$$+ I_{R_S} + C_P \cdot \frac{R_{in}^2}{L_1} \cdot I_{L_1} - \frac{C_P \cdot R_{in} \cdot V(t)}{L_1} + C_P \cdot \frac{dV(t)}{dt} = 0$$

$$C_P \cdot L_P \cdot \frac{dX}{dt} = -C_P \cdot R_{in} \cdot X + Y \cdot [C_P \cdot \frac{R_{in}^2}{L_1} - \frac{C_P}{C_1} - 1]$$
$$+ I_{R_S} + C_P \cdot \frac{R_{in}^2}{L_1} \cdot I_{L_1} - \frac{C_P \cdot R_{in} \cdot V(t)}{L_1} + C_P \cdot \frac{dV(t)}{dt} = 0$$

$$\frac{dX}{dt} = -\frac{R_{in}}{L_P} \cdot X + Y \cdot [\frac{R_{in}^2}{L_1 \cdot L_P} - \frac{1}{C_1 \cdot L_P} - \frac{1}{C_P \cdot L_P}]$$
$$+ I_{R_S} \cdot \frac{1}{C_P \cdot L_P} + \frac{R_{in}^2}{L_1 \cdot L_P} \cdot I_{L_1} - \frac{R_{in} \cdot V(t)}{L_1 \cdot L_P} + \frac{1}{L_P} \cdot \frac{dV(t)}{dt}$$

$$\frac{dY}{dt} = X; \frac{dI_{L_1}}{dt} = \frac{V(t)}{L_1} - \frac{R_{in}}{L_1} \cdot I_{L_1} - \frac{R_{in}}{L_1} \cdot Y; \frac{dI_{R_j}}{dt} = \frac{1}{C_j \cdot R_j} \cdot I_{R_S} - \frac{1}{C_j \cdot R_j} \cdot I_{R_j}$$
$$\frac{dI_{R_S}}{dt} = \frac{1}{R_S \cdot C_P} \cdot Y + \frac{1}{R_S \cdot C_j} \cdot I_{R_j} - I_{R_S} \cdot \frac{1}{R_S} \cdot (\frac{1}{C_P} + \frac{1}{C_j})$$

We have five variables in our system: X, Y, I_{L_1}, I_{R_j}, I_{R_S} and we can represent our system as the following set of differential equations matrix representation.

$$\begin{pmatrix} \frac{dX}{dt} \\ \frac{dY}{dt} \\ \frac{dI_{L_1}}{dt} \\ \frac{dI_{R_j}}{dt} \\ \frac{dI_{R_S}}{dt} \end{pmatrix} = \begin{pmatrix} \Xi_{11} & \cdots & \Xi_{1n} \\ \vdots & \ddots & \vdots \\ \Xi_{m1} & \cdots & \Xi_{mn} \end{pmatrix}_{n=m=5} \cdot \begin{pmatrix} X \\ Y \\ I_{L_1} \\ I_{R_j} \\ I_{R_S} \end{pmatrix} + \begin{pmatrix} -\frac{R_{in}}{L_1 \cdot L_P} \\ 0 \\ \frac{1}{L_1} \\ 0 \\ 0 \end{pmatrix} \cdot V(t) + \begin{pmatrix} \frac{1}{L_P} \\ 0 \\ 0 \\ 0 \\ 0 \end{pmatrix} \cdot \frac{dV(t)}{dt}$$

$$\Xi_{11} = -\frac{R_{in}}{L_P}; \Xi_{12} = \frac{R_{in}^2}{L_1 \cdot L_P} - \frac{1}{C_1 \cdot L_P} - \frac{1}{C_P \cdot L_P} = \frac{1}{L_P} \cdot (\frac{R_{in}^2}{L_1} - \frac{1}{C_1} - \frac{1}{C_P}); \Xi_{13} = \frac{R_{in}^2}{L_1 \cdot L_P}; \Xi_{14} = 0$$
$$\Xi_{15} = \frac{1}{C_P \cdot L_P}; \Xi_{21} = 1; \Xi_{22} = \Xi_{23} = \Xi_{24} = \Xi_{25} = 0; \Xi_{31} = 0; \Xi_{32} = -\frac{R_{in}}{L_1}; \Xi_{33} = -\frac{R_{in}}{L_1}$$
$$\Xi_{34} = \Xi_{35} = 0; \Xi_{41} = \Xi_{42} = \Xi_{43} = 0; \Xi_{44} = -\frac{1}{C_j \cdot R_j}; \Xi_{45} = \frac{1}{C_j \cdot R_j}; \Xi_{51} = 0; \Xi_{52} = \frac{1}{R_S \cdot C_P}$$
$$\Xi_{53} = 0; \Xi_{54} = \frac{1}{R_S \cdot C_j}; \Xi_{55} = -\frac{1}{R_S} \cdot (\frac{1}{C_P} + \frac{1}{C_j}).$$

We consider RF in signal $V(t) = A_0 + f(t); |f(t)| < 1 \,\&\, A_0 \gg |f(t)|$ then $V(t)|_{A_0 \gg |f(t)|}$ $V(t)|_{A_0 \gg |f(t)|} = A_0 + f(t) \approx A_0$; $\frac{dV(t)}{dt}|_{A_0 \gg |f(t)|} = \frac{df(t)}{dt} \to \varepsilon$. We can our matrix representation: $\varepsilon \to 0$. Due to parasitic delay elements in Schottky equivalent circuit, τ_1 for the current flow through Schottky diode's package parasitic inductance (L_P) and τ_2 for the current flow through Schottky diode's parasitic resistance (R_S).

$$Y(t) = I_{L_P}(t) \rightarrow Y(t-\tau_1) = I_{L_P}(t-\tau_1); I_{R_S}(t) \rightarrow I_{R_S}(t-\tau_2).$$

And $X(t) = \frac{dI_{L_P}(t)}{dt}; I_{L_1}(t); I_{R_j}(t)$. We consider no delay effects on $\frac{dY}{dt} = \frac{dI_{L_P}}{dt}; \frac{dI_{R_S}}{dt}$. To find equilibrium points (fixed points) of the RFID tag detector, we define

$$\lim_{t\to\infty} Y(t-\tau_1) = Y(t); \lim_{t\to\infty} I_{L_P}(t-\tau_1) = I_{L_P}(t); \lim_{t\to\infty} I_{R_S}(t-\tau_2) = I_{R_S}(t)$$

$$\begin{pmatrix} \frac{dX}{dt} \\ \frac{dY}{dt} \\ \frac{dI_{L_1}}{dt} \\ \frac{dI_{R_j}}{dt} \\ \frac{dI_{R_S}}{dt} \end{pmatrix} = \begin{pmatrix} \Xi_{11} & \cdots & \Xi_{1n} \\ \vdots & \ddots & \vdots \\ \Xi_{m1} & \cdots & \Xi_{mn} \end{pmatrix}_{n=m=5} \cdot \begin{pmatrix} X \\ Y \\ I_{L_1} \\ I_{R_j} \\ I_{R_S} \end{pmatrix} + \begin{pmatrix} -\frac{R_{in}}{L_1 \cdot L_P} \\ 0 \\ \frac{1}{L_1} \\ 0 \\ 0 \end{pmatrix} \cdot A_0 + \varepsilon$$

In equilibrium points (fixed points)

$$\frac{dY}{dt} = \frac{dI_{L_P}}{dt} = 0; \frac{dI_{R_S}}{dt} = 0 \,\forall t \gg \tau_1,\, t \gg \tau_2$$
$$\exists\, (t-\tau_1) \approx t,\, (t-\tau_2) \approx t,\, t \rightarrow \infty$$

We get five equations:

$$-\frac{R_{in}}{L_P} \cdot X^* + Y^* \cdot \left[\frac{R_{in}^2}{L_1 \cdot L_P} - \frac{1}{C_1 \cdot L_P} - \frac{1}{C_P \cdot L_P}\right] + I_{R_S}^* \cdot \frac{1}{C_P \cdot L_P}$$
$$+ \frac{R_{in}^2}{L_1 \cdot L_P} \cdot I_{L_1}^* - \frac{R_{in} \cdot V(t)}{L_1 \cdot L_P} + \frac{1}{L_P} \cdot \frac{dV(t)}{dt} = 0$$

$$X^* = 0; \frac{V(t)}{L_1} - \frac{R_{in}}{L_1} \cdot I_{L_1}^* - \frac{R_{in}}{L_1} \cdot Y^* = 0; \frac{1}{C_j \cdot R_j} \cdot I_{R_S}^* - \frac{1}{C_j \cdot R_j} \cdot I_{R_j}^* = 0$$
$$\frac{1}{R_S \cdot C_P} \cdot Y^* + \frac{1}{R_S \cdot C_j} \cdot I_{R_j}^* - I_{R_S}^* \cdot \frac{1}{R_S} \cdot \left(\frac{1}{C_P} + \frac{1}{C_j}\right) = 0$$

Since $X^* = 0$ then

$$Y^* \cdot \left[\frac{R_{in}^2}{L_1 \cdot L_P} - \frac{1}{C_1 \cdot L_P} - \frac{1}{C_P \cdot L_P}\right] + I_{R_S}^* \cdot \frac{1}{C_P \cdot L_P}$$
$$+ \frac{R_{in}^2}{L_1 \cdot L_P} \cdot I_{L_1}^* - \frac{R_{in} \cdot V(t)}{L_1 \cdot L_P} + \frac{1}{L_P} \cdot \frac{dV(t)}{dt} = 0.$$
$$\frac{V(t)}{L_1} - \frac{R_{in}}{L_1} \cdot I_{L_1}^* - \frac{R_{in}}{L_1} \cdot Y^* = 0 \Rightarrow Y^* = \frac{V(t)}{R_{in}} - I_{L_1}^*.$$

Then

$$\frac{1}{R_S \cdot C_P} \cdot \left(\frac{V(t)}{R_{in}} - I_{L_1}^*\right) + \frac{1}{R_S \cdot C_j} \cdot I_{R_j}^* - I_{R_S}^* \cdot \frac{1}{R_S} \cdot \left(\frac{1}{C_P} + \frac{1}{C_j}\right) = 0$$

$$\left(\frac{V(t)}{R_{in}} - I_{L_1}^*\right) \cdot \left[\frac{R_{in}^2}{L_1 \cdot L_P} - \frac{1}{C_1 \cdot L_P} - \frac{1}{C_P \cdot L_P}\right] + I_{R_S}^* \cdot \frac{1}{C_P \cdot L_P}$$
$$+ \frac{R_{in}^2}{L_1 \cdot L_P} \cdot I_{L_1}^* - \frac{R_{in} \cdot V(t)}{L_1 \cdot L_P} + \frac{1}{L_P} \cdot \frac{dV(t)}{dt} = 0$$

We get three equations: $\frac{1}{C_j \cdot R_j} \cdot I_{R_S}^* - \frac{1}{C_j \cdot R_j} \cdot I_{R_j}^* = 0$

$$\frac{1}{C_j \cdot R_j} \cdot I_{R_S}^* - \frac{1}{C_j \cdot R_j} \cdot I_{R_j}^* = 0$$

$$\frac{1}{R_S \cdot C_P} \cdot \left(\frac{V(t)}{R_{in}} - I_{L_1}^*\right) + \frac{1}{R_S \cdot C_j} \cdot I_{R_j}^* - I_{R_S}^* \cdot \frac{1}{R_S} \cdot \left(\frac{1}{C_P} + \frac{1}{C_j}\right) = 0$$

$$\left(\frac{V(t)}{R_{in}} - I_{L_1}^*\right) \cdot \left[\frac{R_{in}^2}{L_1 \cdot L_P} - \frac{1}{C_1 \cdot L_P} - \frac{1}{C_P \cdot L_P}\right] + I_{R_S}^* \cdot \frac{1}{C_P \cdot L_P}$$
$$+ \frac{R_{in}^2}{L_1 \cdot L_P} \cdot I_{L_1}^* - \frac{R_{in} \cdot V(t)}{L_1 \cdot L_P} + \frac{1}{L_P} \cdot \frac{dV(t)}{dt} = 0$$

$$\frac{1}{C_j \cdot R_j} \cdot I_{R_S}^* - \frac{1}{C_j \cdot R_j} \cdot I_{R_j}^* = 0 \Rightarrow I_{R_j}^* = I_{R_S}^*$$

We get two equations:

$$\frac{1}{R_S \cdot C_P} \cdot \left(\frac{V(t)}{R_{in}} - I_{L_1}^*\right) + \frac{1}{R_S \cdot C_j} \cdot I_{R_S}^* - I_{R_S}^* \cdot \frac{1}{R_S} \cdot \left(\frac{1}{C_P} + \frac{1}{C_j}\right) = 0$$

$$\left(\frac{V(t)}{R_{in}} - I_{L_1}^*\right) \cdot \left[\frac{R_{in}^2}{L_1 \cdot L_P} - \frac{1}{C_1 \cdot L_P} - \frac{1}{C_P \cdot L_P}\right] + I_{R_S}^* \cdot \frac{1}{C_P \cdot L_P}$$
$$+ \frac{R_{in}^2}{L_1 \cdot L_P} \cdot I_{L_1}^* - \frac{R_{in} \cdot V(t)}{L_1 \cdot L_P} + \frac{1}{L_P} \cdot \frac{dV(t)}{dt} = 0$$

By mathematic manipulation, we get the following two equations:

$$\frac{V(t)}{R_{in}} - I_{L_1}^* - I_{R_S}^* = 0 \Rightarrow I_{R_S}^* = \frac{V(t)}{R_{in}} - I_{L_1}^*$$

$$I_{L_1}^* \cdot \left(\frac{1}{C_1} + \frac{1}{C_P}\right) + I_{R_S}^* \cdot \frac{1}{C_P}$$
$$+ V(t) \cdot \left\{\frac{1}{R_{in}} \cdot \left[\frac{R_{in}^2}{L_1} - \frac{1}{C_1} - \frac{1}{C_P}\right] - \frac{R_{in}}{L_1}\right\} + \frac{dV(t)}{dt} = 0$$

We define for simplicity:

$$\Omega = \frac{1}{R_{in}} \cdot [\frac{R_{in}^2}{L_1} - \frac{1}{C_1} - \frac{1}{C_P}] - \frac{R_{in}}{L_1}$$

$$I_{R_S}^* = \frac{V(t)}{R_{in}} - I_{L_1}^*; I_{L_1}^* \cdot (\frac{1}{C_1} + \frac{1}{C_P}) + I_{R_S}^* \cdot \frac{1}{C_P} + V(t) \cdot \Omega + \frac{dV(t)}{dt} = 0$$

$$I_{L_1}^* \cdot (\frac{1}{C_1} + \frac{1}{C_P}) + (\frac{V(t)}{R_{in}} - I_{L_1}^*) \cdot \frac{1}{C_P} + V(t) \cdot \Omega + \frac{dV(t)}{dt} = 0$$

$$\Rightarrow I_{L_1}^* \cdot \frac{1}{C_1} + V(t) \cdot [\frac{1}{R_{in} \cdot C_P} + \Omega] + \frac{dV(t)}{dt} = 0$$

$$I_{L_1}^* \cdot \frac{1}{C_1} + V(t) \cdot [\frac{1}{R_{in} \cdot C_P} + \Omega] + \frac{dV(t)}{dt} = 0$$

$$\Rightarrow I_{L_1}^* = -C_1 \cdot \{V(t) \cdot [\frac{1}{R_{in} \cdot C_P} + \Omega] + \frac{dV(t)}{dt}\}$$

$$I_{R_S}^* = \frac{V(t)}{R_{in}} + C_1 \cdot \{V(t) \cdot [\frac{1}{R_{in} \cdot C_P} + \Omega] + \frac{dV(t)}{dt}\}$$

$$= V(t) \cdot \{\frac{1}{R_{in}} + C_1 \cdot [\frac{1}{R_{in} \cdot C_P} + \Omega]\} + C_1 \cdot \frac{dV(t)}{dt}$$

We define:

$$\Omega_1 = \frac{1}{R_{in}} + C_1 \cdot [\frac{1}{R_{in} \cdot C_P} + \Omega]; I_{R_S}^* = V(t) \cdot \Omega_1 + C_1 \cdot \frac{dV(t)}{dt}$$

$$I_{R_j}^* = I_{R_S}^* \Rightarrow I_{R_j}^* = V(t) \cdot \Omega_1 + C_1 \cdot \frac{dV(t)}{dt}; X^* = 0$$

$$Y^* = \frac{V(t)}{R_{in}} - I_{L_1}^* = V(t) \cdot \{\frac{1}{R_{in}} + C_1 \cdot [\frac{1}{R_{in} \cdot C_P} + \Omega]\} + C_1 \cdot \frac{dV(t)}{dt}$$

We can summery our system fixed points in the next table:

Fixed point coordinates $E^*(X^*, Y^*, I_{L_1}^*, I_{R_j}^*, I_{R_S}^*)$	Fixed points expression $V(t) = A_0 + f(t)$ $\lvert f(t)\rvert < 1 \,\&\, A_0 \gg \lvert f(t)\rvert$	$V(t)\rvert_{A_0 \gg \lvert f(t)\rvert} = A_0 + f(t) \approx A_0$ $\frac{dV(t)}{dt}\rvert_{A_0 \gg \lvert f(t)\rvert} = \frac{df(t)}{dt} \to \varepsilon$
X^*	0	0
Y^*	$V(t) \cdot \{\frac{1}{R_{in}} + C_1 \cdot [\frac{1}{R_{in} \cdot C_P} + \Omega]\} + C_1 \cdot \frac{dV(t)}{dt}$	$A_0 \cdot \{\frac{1}{R_{in}} + C_1 \cdot [\frac{1}{R_{in} \cdot C_P} + \Omega]\}$
$I_{L_1}^*$	$-C_1 \cdot \{V(t) \cdot [\frac{1}{R_{in} \cdot C_P} + \Omega] + \frac{dV(t)}{dt}\}$	$-C_1 \cdot A_0 \cdot [\frac{1}{R_{in} \cdot C_P} + \Omega]$
$I_{R_j}^*$	$V(t) \cdot \Omega_1 + C_1 \cdot \frac{dV(t)}{dt}$	$A_0 \cdot \Omega_1$
$I_{R_S}^*$	$V(t) \cdot \{\frac{1}{R_{in}} + C_1 \cdot [\frac{1}{R_{in} \cdot C_P} + \Omega]\} + C_1 \cdot \frac{dV(t)}{dt}$	$A_0 \cdot \{\frac{1}{R_{in}} + C_1 \cdot [\frac{1}{R_{in} \cdot C_P} + \Omega]\}$

Stability analysis: The standard local stability analysis about any one of the equilibrium points of the RFID tag detector system consists in adding to coordinate $[X, Y, I_{L_1}, I_{R_j}, I_{R_s}]$ arbitrarily small increments of exponential form $[x, y, i_{L_1}, i_{R_j}, i_{R_s}] \cdot e^{\lambda \cdot t}$ and retaining the first order terms in $X, Y, I_{L_1}, I_{R_j}, I_{R_s}$. The system of five homogeneous equations leads to a polynomial characteristic equation in the eigenvalues. The polynomial characteristic equations accept by set the below currents and currents derivative with respect to time into RFID tag detector system equations. RFID tag detector system fixed values with arbitrarily small increments of exponential form $[x, y, i_{L_1}, i_{R_j}, i_{R_s}] \cdot e^{\lambda \cdot t}$ are: j = 0 (first fixed point), j = 1 (second fixed point), j = 2 (third fixed point), etc.

$$X(t) = X^{(j)} + x \cdot e^{\lambda \cdot t}; Y(t) = Y^{(j)} + y \cdot e^{\lambda \cdot t};$$
$$Y(t - \tau_1) = Y^{(j)} + y \cdot e^{\lambda \cdot (t - \tau_1)}; I_{L_1}(t) = I_{L_1}^{(j)} + i_{L_1} \cdot e^{\lambda \cdot t}$$
$$I_{R_j}(t) = I_{R_j}^{(j)} + i_{R_j} \cdot e^{\lambda \cdot t}; I_{R_S}(t) = I_{R_S}^{(j)} + i_{R_S} \cdot e^{\lambda \cdot t};$$
$$I_{R_S}(t - \tau_2) = I_{R_S}^{(j)} + i_{R_S} \cdot e^{\lambda \cdot (t - \tau_2)}.$$

We choose these expressions for ourselves $X(t), Y(t), I_{L_1}(t)$ and $I_{R_j}(t), I_{R_S}(t)$ as a small displacement $[x, y, i_{L_1}, i_{R_j}, i_{R_s}]$ from the RFID tag detector system fixed points in time t = 0.

$$X(t = 0) = X^{(j)} + x; Y(t = 0) = Y^{(j)} + y;$$
$$I_{L_1}(t = 0) = I_{L_1}^{(j)} + i_{L_1}; I_{R_j}(t = 0) = I_{R_j}^{(j)} + i_{R_j}$$
$$I_{R_S}(t = 0) = I_{R_S}^{(j)} + i_{R_S}$$

For $\lambda < 0$, $t > 0$, the selected fixed point is stable otherwise $\lambda > 0$, $t > 0$ is unstable. Our system tends to the selected fixed point exponentially for $\lambda < 0$, $t > 0$ otherwise go away from the selected fixed point exponentially. λ Is the eigenvalue parameter which is established if the fixed point is stable or unstable; additionally, his absolute value ($|\lambda|$) establishes the speed of flow toward or away from the selected fixed point (Yuri 1995; Jack and Huseyin 1991). The speeds of flow toward or away from the selected fixed point for Schottky detector system currents and currents derivatives with respect to time are

$$\frac{dX(t)}{dt} = \lim_{\Delta t \to \infty} \frac{X(t + \Delta t) - X(t)}{\Delta t} = \lim_{\Delta t \to \infty} \frac{X^{(j)} + x \cdot e^{\lambda \cdot (t + \Delta t)} - [X^{(j)} + x \cdot e^{\lambda \cdot t}]}{\Delta t} \overset{e^{\lambda \cdot \Delta t} \approx 1 + \lambda \cdot \Delta t}{=} \lambda \cdot x \cdot e^{\lambda \cdot t}$$
$$\frac{dY(t)}{dt} = \lim_{\Delta t \to \infty} \frac{Y(t + \Delta t) - Y(t)}{\Delta t} = \lim_{\Delta t \to \infty} \frac{Y^{(j)} + y \cdot e^{\lambda \cdot (t + \Delta t)} - [Y^{(j)} + y \cdot e^{\lambda \cdot t}]}{\Delta t} \overset{e^{\lambda \cdot \Delta t} \approx 1 + \lambda \cdot \Delta t}{=} \lambda \cdot y \cdot e^{\lambda \cdot t}$$
$$\frac{dI_{L_1}(t)}{dt} = \lambda \cdot i_{L_1} \cdot e^{\lambda \cdot t}; \frac{dI_{R_j}(t)}{dt} = \lambda \cdot i_{R_j} \cdot e^{\lambda \cdot t}; \frac{dI_{R_S}(t)}{dt} = \lambda \cdot i_{R_S} \cdot e^{\lambda \cdot t}; \frac{dY(t - \tau_1)}{dt} = \lambda \cdot y \cdot e^{\lambda \cdot t} \cdot e^{-\lambda \cdot \tau_1}$$
$$\frac{dI_{R_S}(t - \tau_2)}{dt} = \lambda \cdot i_{R_S} \cdot e^{\lambda \cdot t} \cdot e^{-\lambda \cdot \tau_2}$$

First, we take Schottky detector variable $X, Y, I_{L_1}, I_{R_j}, I_{R_s}$ differential equations and adding to coordinate $[X, Y, I_{L_1}, I_{R_j}, I_{R_s}]$ arbitrarily small increments of exponential terms $[x, y, i_{L_1}, i_{R_j}, i_{R_s}] \cdot e^{\lambda \cdot t}$ and retaining the first order terms in $x, y, i_{L_1}, i_{R_j}, i_{R_s} (V(t) \to \varepsilon; \frac{dV(t)}{dt} \to \varepsilon)$ then

$$E^*(X^*, Y^*, I^*_{L_1}, I^*_{R_j}, I^*_{R_S}) = (0,0,0,0,0);$$
$$X^{(j=0)} = 0, Y^{(j=0)} = 0, I^{(j=0)}_{L_1} = 0, I^{(j=0)}_{R_j} = 0, I^{(j=0)}_{R_S} = 0.$$

We can see that our fixed point is a saddle node. We define $Y(t-\tau_1) = Y^{(j)} + y \cdot e^{\lambda \cdot (t-\tau_1)}$ and $I_{R_S}(t-\tau_2) = I^{(j)}_{R_S} + i_{R_S} \cdot e^{\lambda \cdot (t-\tau_2)}$. Then we get five delayed differential equations with respect to coordinates $[X, Y, I_{L_1}, I_{R_j}, I_{R_s}]$ arbitrarily small increments of exponential $[x, y, i_{L_1}, i_{R_j}, i_{R_s}] \cdot e^{\lambda \cdot t}$. We consider no delay effects on $\frac{dY(t)}{dt}$ and $\frac{dI_{R_S}(t)}{dt}$. We get the following equations:

	$\lambda < 0$	$\lambda > 0$
t = 0	$X(t=0) = X^{(j)} + x$	$X(t=0) = X^{(j)} + x$
	$Y(t=0) = Y^{(j)} + y$	$Y(t=0) = Y^{(j)} + y$
	$I_{L_1}(t=0) = I^{(j)}_{L_1} + i_{L_1}$	$I_{L_1}(t=0) = I^{(j)}_{L_1} + i_{L_1}$
	$I_{R_j}(t) = I^{(j)}_{R_j} + i_{R_j}$	$I_{R_j}(t) = I^{(j)}_{R_j} + i_{R_j}$
	$I_{R_S}(t=0) = I^{(j)}_{R_S} + i_{R_S}$	$I_{R_S}(t=0) = I^{(j)}_{R_S} + i_{R_S}$
t > 0	$X(t) = X^{(j)} + x \cdot e^{-\lvert\lambda\rvert \cdot t}$	$X(t) = X^{(j)} + x \cdot e^{\lvert\lambda\rvert \cdot t}$
	$Y(t) = Y^{(j)} + y \cdot e^{-\lvert\lambda\rvert \cdot t}$	$Y(t) = Y^{(j)} + y \cdot e^{\lvert\lambda\rvert \cdot t}$
	$I_{L_1}(t) = I^{(j)}_{L_1} + i_{L_1} \cdot e^{-\lvert\lambda\rvert \cdot t}$	$I_{L_1}(t) = I^{(j)}_{L_1} + i_{L_1} \cdot e^{\lvert\lambda\rvert \cdot t}$
	$I_{R_j}(t) = I^{(j)}_{R_j} + i_{R_j} \cdot e^{-\lvert\lambda\rvert \cdot t}$	$I_{R_j}(t) = I^{(j)}_{R_j} + i_{R_j} \cdot e^{\lvert\lambda\rvert \cdot t}$
	$I_{R_S}(t) = I^{(j)}_{R_S} + i_{R_S} \cdot e^{-\lvert\lambda\rvert \cdot t}$	$I_{R_S}(t) = I^{(j)}_{R_S} + i_{R_S} \cdot e^{\lvert\lambda\rvert \cdot t}$
t > 0; t → ∞	$X(t \to \infty) = X^{(j)}$	$X(t \to \infty, \lambda > 0) = x \cdot e^{\lvert\lambda\rvert \cdot t}$
	$Y(t \to \infty) = Y^{(j)}$	$Y(t \to \infty, \lambda) = y \cdot e^{\lvert\lambda\rvert \cdot t}$
	$I_{L_1}(t \to \infty) = I^{(j)}_{L_1}$	$I_{L_1}(t \to \infty, \lambda > 0) = i_{L_1} \cdot e^{\lvert\lambda\rvert \cdot t}$
	$I_{R_j}(t \to \infty) = I^{(j)}_{R_j}$	$I_{R_j}(t \to \infty, \lambda > 0) = i_{R_j} \cdot e^{\lvert\lambda\rvert \cdot t}$
	$I_{R_S}(t \to \infty) = I^{(j)}_{R_S}$	$I_{R_S}(t \to \infty, \lambda > 0) = i_{R_S} \cdot e^{\lvert\lambda\rvert \cdot t}$

$$\begin{aligned}\lambda \cdot x \cdot e^{\lambda \cdot t} = & -\frac{R_{in}}{L_P} \cdot [X^{(j)} + x \cdot e^{\lambda \cdot t}] + [Y^{(j)} + y \cdot e^{\lambda \cdot (t-\tau_1)}] \cdot [\frac{R^2_{in}}{L_1 \cdot L_P} - \frac{1}{C_1 \cdot L_P} - \frac{1}{C_P \cdot L_P}] \\ & + [I^{(j)}_{R_S} + i_{R_S} \cdot e^{\lambda \cdot (t-\tau_2)}] \cdot \frac{1}{C_P \cdot L_P} + \frac{R^2_{in}}{L_1 \cdot L_P} \cdot [I^{(j)}_{L_1} + i_{L_1} \cdot e^{\lambda \cdot t}] \\ & - \frac{R_{in} \cdot V(t)}{L_1 \cdot L_P} + \frac{1}{L_P} \cdot \frac{dV(t)}{dt}; V(t), \frac{dV(t)}{dt} \to \varepsilon\end{aligned}$$

$$\lambda \cdot x \cdot e^{\lambda \cdot t} = -\frac{R_{in}}{L_P} \cdot X^{(j)} - \frac{R_{in}}{L_P} \cdot x \cdot e^{\lambda \cdot t} + Y^{(j)} \cdot [\frac{R_{in}^2}{L_1 \cdot L_P} - \frac{1}{C_1 \cdot L_P} - \frac{1}{C_P \cdot L_P}]$$
$$+ y \cdot [\frac{R_{in}^2}{L_1 \cdot L_P} - \frac{1}{C_1 \cdot L_P} - \frac{1}{C_P \cdot L_P}] \cdot e^{\lambda \cdot (t-\tau_1)} I_{R_S}^{(j)} \cdot \frac{1}{C_P \cdot L_P}$$
$$+ i_{R_S} \cdot \frac{1}{C_P \cdot L_P} \cdot e^{\lambda \cdot (t-\tau_2)} + \frac{R_{in}^2}{L_1 \cdot L_P} \cdot I_{L_1}^{(j)} + \frac{R_{in}^2}{L_1 \cdot L_P} \cdot i_{L_1} \cdot e^{\lambda \cdot t}$$

$$\lambda \cdot x \cdot e^{\lambda \cdot t} = -\frac{R_{in}}{L_P} \cdot X^{(j)} + Y^{(j)} \cdot [\frac{R_{in}^2}{L_1 \cdot L_P} - \frac{1}{C_1 \cdot L_P} - \frac{1}{C_P \cdot L_P}]$$
$$+ I_{R_S}^{(j)} \cdot \frac{1}{C_P \cdot L_P} + \frac{R_{in}^2}{L_1 \cdot L_P} \cdot I_{L_1}^{(j)} - \frac{R_{in}}{L_P} \cdot x \cdot e^{\lambda \cdot t}$$
$$+ y \cdot [\frac{R_{in}^2}{L_1 \cdot L_P} - \frac{1}{C_1 \cdot L_P} - \frac{1}{C_P \cdot L_P}] \cdot e^{\lambda \cdot (t-\tau_1)}$$
$$+ i_{R_S} \cdot \frac{1}{C_P \cdot L_P} \cdot e^{\lambda \cdot (t-\tau_2)} + \frac{R_{in}^2}{L_1 \cdot L_P} \cdot i_{L_1} \cdot e^{\lambda \cdot t}$$

At fixed point:

$$-\frac{R_{in}}{L_P} \cdot X^{(j)} + Y^{(j)} \cdot [\frac{R_{in}^2}{L_1 \cdot L_P} - \frac{1}{C_1 \cdot L_P} - \frac{1}{C_P \cdot L_P}] + I_{R_S}^{(j)} \cdot \frac{1}{C_P \cdot L_P} + \frac{R_{in}^2}{L_1 \cdot L_P} \cdot I_{L_1}^{(j)} = 0$$

Then

$$-x \cdot e^{\lambda \cdot t} \cdot [\lambda + \frac{R_{in}}{L_P}] + y \cdot [\frac{R_{in}^2}{L_1 \cdot L_P} - \frac{1}{C_1 \cdot L_P} - \frac{1}{C_P \cdot L_P}] \cdot e^{\lambda \cdot (t-\tau_1)}$$
$$+ i_{R_S} \cdot \frac{1}{C_P \cdot L_P} \cdot e^{\lambda \cdot (t-\tau_2)} + \frac{R_{in}^2}{L_1 \cdot L_P} \cdot i_{L_1} \cdot e^{\lambda \cdot t} = 0$$
$$\frac{dY}{dt} = X \Rightarrow \lambda \cdot y \cdot e^{\lambda \cdot t} = X^{(j)} + x \cdot e^{\lambda \cdot t}.$$

At fixed point $X^{(j)} = 0 \Rightarrow -x + \lambda \cdot y = 0$

$$\lambda \cdot i_{L_1} \cdot e^{\lambda \cdot t} = \frac{V(t)}{L_1} - \frac{R_{in}}{L_1} \cdot [I_{L_1}^{(j)} + i_{L_1} \cdot e^{\lambda \cdot t}] - \frac{R_{in}}{L_1} \cdot [Y^{(j)} + y \cdot e^{\lambda \cdot (t-\tau_1)}]; V(t) \to \varepsilon$$
$$\lambda \cdot i_{L_1} \cdot e^{\lambda \cdot t} = -\frac{R_{in}}{L_1} \cdot I_{L_1}^{(j)} - \frac{R_{in}}{L_1} \cdot Y^{(j)} - i_{L_1} \cdot \frac{R_{in}}{L_1} \cdot e^{\lambda \cdot t} - y \cdot \frac{R_{in}}{L_1} \cdot e^{\lambda \cdot (t-\tau_1)}.$$

At fixed point $-\frac{R_{in}}{L_1} \cdot I_{L_1}^{(j)} - \frac{R_{in}}{L_1} \cdot Y^{(j)} = 0$ then

$$- \lambda \cdot i_{L_1} \cdot e^{\lambda \cdot t} - i_{L_1} \cdot \frac{R_{in}}{L_1} \cdot e^{\lambda \cdot t} - y \cdot \frac{R_{in}}{L_1} \cdot e^{\lambda \cdot (t - \tau_1)} = 0$$

$$\lambda \cdot i_{R_j} \cdot e^{\lambda \cdot t} = \frac{1}{C_j \cdot R_j} \cdot [I_{R_S}^{(j)} + i_{R_S} \cdot e^{\lambda \cdot (t - \tau_2)}] - \frac{1}{C_j \cdot R_j} \cdot [I_{R_j}^{(j)} + i_{R_j} \cdot e^{\lambda \cdot t}]$$

$$- \lambda \cdot i_{R_j} \cdot e^{\lambda \cdot t} - i_{R_j} \cdot \frac{1}{C_j \cdot R_j} \cdot e^{\lambda \cdot t} + i_{R_S} \cdot \frac{1}{C_j \cdot R_j} \cdot e^{\lambda \cdot (t - \tau_2)}$$
$$+ \frac{1}{C_j \cdot R_j} \cdot I_{R_S}^{(j)} - \frac{1}{C_j \cdot R_j} \cdot I_{R_j}^{(j)} = 0$$

At fixed point $\frac{1}{C_j \cdot R_j} \cdot I_{R_S}^{(j)} - \frac{1}{C_j \cdot R_j} \cdot I_{R_j}^{(j)} = 0$ then $-i_{R_j} \cdot e^{\lambda \cdot t} \cdot [\lambda + \frac{1}{C_j \cdot R_j}] + i_{R_S} \cdot \frac{1}{C_j \cdot R_j} \cdot e^{\lambda \cdot (t - \tau_2)} = 0$

$$\lambda \cdot i_{R_S} \cdot e^{\lambda \cdot t} = \frac{1}{R_S \cdot C_P} \cdot [Y^{(j)} + y \cdot e^{\lambda \cdot (t - \tau_1)}] + \frac{1}{R_S \cdot C_j} \cdot [I_{R_j}^{(j)} + i_{R_j} \cdot e^{\lambda \cdot t}]$$
$$- [I_{R_S}^{(j)} + i_{R_S} \cdot e^{\lambda \cdot (t - \tau_2)}] \cdot \frac{1}{R_S} \cdot (\frac{1}{C_P} + \frac{1}{C_j})$$

$$\lambda \cdot i_{R_S} \cdot e^{\lambda \cdot t} = \frac{1}{R_S \cdot C_P} \cdot Y^{(j)} + y \cdot \frac{1}{R_S \cdot C_P} \cdot e^{\lambda \cdot (t - \tau_1)} + \frac{1}{R_S \cdot C_j} \cdot I_{R_j}^{(j)}$$
$$+ i_{R_j} \cdot \frac{1}{R_S \cdot C_j} \cdot e^{\lambda \cdot t} - I_{R_S}^{(j)} \cdot \frac{1}{R_S} \cdot (\frac{1}{C_P} + \frac{1}{C_j})$$
$$- i_{R_S} \cdot \frac{1}{R_S} \cdot (\frac{1}{C_P} + \frac{1}{C_j}) \cdot e^{\lambda \cdot (t - \tau_2)}$$

$$\lambda \cdot i_{R_S} \cdot e^{\lambda \cdot t} = \frac{1}{R_S \cdot C_P} \cdot Y^{(j)} + \frac{1}{R_S \cdot C_j} \cdot I_{R_j}^{(j)} - I_{R_S}^{(j)} \cdot \frac{1}{R_S} \cdot (\frac{1}{C_P} + \frac{1}{C_j})$$
$$+ y \cdot \frac{1}{R_S \cdot C_P} \cdot e^{\lambda \cdot (t - \tau_1)} + i_{R_j} \cdot \frac{1}{R_S \cdot C_j} \cdot e^{\lambda \cdot t}$$
$$- i_{R_S} \cdot \frac{1}{R_S} \cdot (\frac{1}{C_P} + \frac{1}{C_j}) \cdot e^{\lambda \cdot (t - \tau_2)}$$

At fixed point $\frac{1}{R_S \cdot C_P} \cdot Y^{(j)} + \frac{1}{R_S \cdot C_j} \cdot I_{R_j}^{(j)} - I_{R_S}^{(j)} \cdot \frac{1}{R_S} \cdot (\frac{1}{C_P} + \frac{1}{C_j}) = 0$ then

$$- i_{R_S} \cdot e^{\lambda \cdot t} \cdot [\lambda + \frac{1}{R_S} \cdot (\frac{1}{C_P} + \frac{1}{C_j}) \cdot e^{-\lambda \cdot \tau_2}]$$
$$+ y \cdot \frac{1}{R_S \cdot C_P} \cdot e^{\lambda \cdot (t - \tau_1)} + i_{R_j} \cdot \frac{1}{R_S \cdot C_j} \cdot e^{\lambda \cdot t} = 0$$

We can summarize our last results:

$$-x\cdot[\lambda+\frac{R_{in}}{L_P}]+y\cdot[\frac{R_{in}^2}{L_1\cdot L_P}-\frac{1}{C_1\cdot L_P}-\frac{1}{C_P\cdot L_P}]\cdot e^{-\lambda\cdot\tau_1}$$
$$+\frac{R_{in}^2}{L_1\cdot L_P}\cdot i_{L_1}+i_{R_S}\cdot\frac{1}{C_P\cdot L_P}\cdot e^{-\lambda\cdot\tau_2}=0$$
$$x-\lambda\cdot y=0$$
$$-y\cdot\frac{R_{in}}{L_1}\cdot e^{-\lambda\cdot\tau_1}-i_{L_1}\cdot[\frac{R_{in}}{L_1}+\lambda]=0$$
$$-i_{R_j}\cdot[\lambda+\frac{1}{C_j\cdot R_j}]+i_{R_S}\cdot\frac{1}{C_j\cdot R_j}\cdot e^{-\lambda\cdot\tau_2}=0$$
$$y\cdot\frac{1}{R_S\cdot C_P}\cdot e^{-\lambda\cdot\tau_1}+i_{R_j}\cdot\frac{1}{R_S\cdot C_j}-i_{R_S}\cdot[\lambda+\frac{1}{R_S}\cdot(\frac{1}{C_P}+\frac{1}{C_j})\cdot e^{-\lambda\cdot\tau_2}]=0$$

The small increments Jacobian of our RFID Schotky detector system is as follows:

$$\begin{pmatrix}\Upsilon_{11} & \dots & \Upsilon_{15}\\ \vdots & \ddots & \vdots\\ \Upsilon_{51} & \dots & \Upsilon_{55}\end{pmatrix}\cdot\begin{pmatrix}x\\ y\\ i_{L_1}\\ i_{R_j}\\ i_{R_S}\end{pmatrix}=0;\ \Upsilon_{11}=-\frac{R_{in}}{L_P}-\lambda;$$

$$\Upsilon_{12}=[\frac{R_{in}^2}{L_1\cdot L_P}-\frac{1}{C_1\cdot L_P}-\frac{1}{C_P\cdot L_P}]\cdot e^{-\lambda\cdot\tau_1}$$
$$\Upsilon_{13}=\frac{R_{in}^2}{L_1\cdot L_P};\ \Upsilon_{14}=0;\ \Upsilon_{15}=\frac{1}{C_P\cdot L_P}\cdot e^{-\lambda\cdot\tau_2};$$
$$\Upsilon_{21}=1;\ \Upsilon_{22}=-\lambda;\ \Upsilon_{23}=\Upsilon_{24}=\Upsilon_{25}=0$$
$$\Upsilon_{31}=0;\ \Upsilon_{32}=-\frac{R_{in}}{L_1}\cdot e^{-\lambda\cdot\tau_1};\ \Upsilon_{33}=-\frac{R_{in}}{L_1}-\lambda;$$
$$\Upsilon_{34}=0;\ \Upsilon_{35}=0;\ \Upsilon_{41}=\Upsilon_{42}=\Upsilon_{43}=0$$
$$\Upsilon_{44}=-\frac{1}{C_j\cdot R_j}-\lambda;\ \Upsilon_{45}=\frac{1}{C_j\cdot R_j}\cdot e^{-\lambda\cdot\tau_2};$$
$$\Upsilon_{51}=0;\ \Upsilon_{52}=\frac{1}{R_S\cdot C_P}\cdot e^{-\lambda\cdot\tau_1};\ \Upsilon_{53}=0$$
$$\Upsilon_{54}=\frac{1}{R_S\cdot C_j};\ \Upsilon_{55}=-\frac{1}{R_S}\cdot(\frac{1}{C_P}+\frac{1}{C_j})\cdot e^{-\lambda\cdot\tau_2}-\lambda$$

$$|A-\lambda\cdot I|=\begin{pmatrix}\Upsilon_{11} & \dots & \Upsilon_{15}\\ \vdots & \ddots & \vdots\\ \Upsilon_{51} & \dots & \Upsilon_{55}\end{pmatrix};\ \det|A-\lambda\cdot I|=0$$

We define for simplicity the following parameters:

$$\sigma_1 = -\frac{R_{in}}{L_P}; \sigma_2 = \frac{R_{in}^2}{L_1 \cdot L_P} - \frac{1}{C_1 \cdot L_P} - \frac{1}{C_P \cdot L_P};$$
$$\sigma_3 = \frac{R_{in}^2}{L_1 \cdot L_P}; \sigma_4 = \frac{1}{C_P \cdot L_P}; \sigma_5 = -\frac{R_{in}}{L_1}$$
$$\sigma_6 = \frac{1}{C_j \cdot R_j}; \sigma_7 = \frac{1}{R_S \cdot C_P}; \sigma_8 = \frac{1}{R_S \cdot C_j}; \sigma_9 = -\frac{1}{R_S} \cdot (\frac{1}{C_P} + \frac{1}{C_j})$$

$$\Upsilon_{11} = \sigma_1 - \lambda; \Upsilon_{12} = \sigma_2 \cdot e^{-\lambda \cdot \tau_1}; \Upsilon_{13} = \sigma_3; \Upsilon_{14} = 0;$$
$$\Upsilon_{15} = \sigma_4 \cdot e^{-\lambda \cdot \tau_2}; \Upsilon_{21} = 1; \Upsilon_{22} = -\lambda; \Upsilon_{23} = \Upsilon_{24} = \Upsilon_{25} = 0$$
$$\Upsilon_{31} = 0; \Upsilon_{32} = \sigma_5 \cdot e^{-\lambda \cdot \tau_1}; \Upsilon_{33} = \sigma_5 - \lambda;$$
$$\Upsilon_{34} = 0; \Upsilon_{35} = 0; \Upsilon_{41} = \Upsilon_{42} = \Upsilon_{43} = 0$$
$$\Upsilon_{44} = -\sigma_6 - \lambda; \Upsilon_{45} = \sigma_6 \cdot e^{-\lambda \cdot \tau_2}; \Upsilon_{51} = 0;$$
$$\Upsilon_{52} = \sigma_7 \cdot e^{-\lambda \cdot \tau_1}; \Upsilon_{53} = 0; \Upsilon_{54} = \sigma_8; \Upsilon_{55} = \sigma_9 \cdot e^{-\lambda \cdot \tau_2} - \lambda$$

We need to find $D(\tau_1, \tau_2)$ for the following cases: (A) $\tau_1 = \tau; \tau_2 = 0$ (B) $\tau_1 = 0; \tau_2 = \tau$ (C) $\tau_1 = \tau_2 = \tau$. We need to get characteristics equations for all above stability analysis cases. We study the occurrence of any possible stability switching, resulting from the increase of the value of the time delays τ_1, τ_2 for the general characteristic equation $D(\tau_1, \tau_2)$. If we choose τ as a parameter, then the expression: $D(\lambda, \tau) = P_n(\lambda, \tau) + Q_m(\lambda, \tau) \cdot e^{-\lambda \cdot \tau}; n, m \in \mathbb{N}_0; n > m$ [12, 18, 19].

$$\det\begin{pmatrix} \Upsilon_{11} & \dots & \Upsilon_{15} \\ \vdots & \ddots & \vdots \\ \Upsilon_{51} & \dots & \Upsilon_{55} \end{pmatrix} = (\sigma_1 - \lambda) \cdot (-\lambda) \cdot \det\begin{pmatrix} \sigma_5 - \lambda & 0 & 0 \\ 0 & -(\sigma_6 + \lambda) & \sigma_6 \cdot e^{-\lambda \cdot \tau_2} \\ 0 & \sigma_8 & (\sigma_9 \cdot e^{-\lambda \cdot \tau_2} - \lambda) \end{pmatrix}$$
$$- \sigma_2 \cdot e^{-\lambda \cdot \tau_1} \cdot \det\begin{pmatrix} \sigma_5 - \lambda & 0 & 0 \\ 0 & -(\sigma_6 + \lambda) & \sigma_6 \cdot e^{-\lambda \cdot \tau_2} \\ 0 & \sigma_8 & (\sigma_9 \cdot e^{-\lambda \cdot \tau_2} - \lambda) \end{pmatrix}$$
$$+ \sigma_3 \cdot \{\det\begin{pmatrix} \sigma_5 \cdot e^{-\lambda \cdot \tau_1} & 0 & 0 \\ 0 & -(\sigma_6 + \lambda) & \sigma_6 \cdot e^{-\lambda \cdot \tau_2} \\ \sigma_7 \cdot e^{-\lambda \cdot \tau_1} & \sigma_8 & (\sigma_9 \cdot e^{-\lambda \cdot \tau_2} - \lambda) \end{pmatrix}$$
$$+ \lambda \cdot \det\begin{pmatrix} 0 & 0 & 0 \\ 0 & -(\sigma_6 + \lambda) & \sigma_6 \cdot e^{-\lambda \cdot \tau_2} \\ 0 & \sigma_8 & (\sigma_9 \cdot e^{-\lambda \cdot \tau_2} - \lambda) \end{pmatrix}\}$$
$$+ \sigma_4 \cdot e^{-\lambda \cdot \tau_2} \cdot \{\det\begin{pmatrix} \sigma_5 \cdot e^{-\lambda \cdot \tau_1} & \sigma_5 - \lambda & 0 \\ 0 & 0 & -(\sigma_6 + \lambda) \\ \sigma_7 \cdot e^{-\lambda \cdot \tau_1} & 0 & \sigma_8 \end{pmatrix}$$
$$+ \lambda \cdot \det\begin{pmatrix} 0 & \sigma_5 - \lambda & 0 \\ 0 & 0 & -(\sigma_6 + \lambda) \\ 0 & 0 & \sigma_8 \end{pmatrix}\}$$

$$\det\begin{pmatrix} 0 & 0 & 0 \\ 0 & -(\sigma_6+\lambda) & \sigma_6\cdot e^{-\lambda\cdot\tau_2} \\ 0 & \sigma_8 & (\sigma_9\cdot e^{-\lambda\cdot\tau_2}-\lambda) \end{pmatrix} = 0;\ \det\begin{pmatrix} 0 & \sigma_5-\lambda & 0 \\ 0 & 0 & -(\sigma_6+\lambda) \\ 0 & 0 & \sigma_8 \end{pmatrix} = 0$$

We get the following expression:

$$\det\begin{pmatrix} \Upsilon_{11} & \cdots & \Upsilon_{15} \\ \vdots & \ddots & \vdots \\ \Upsilon_{51} & \cdots & \Upsilon_{55} \end{pmatrix} = (\sigma_1-\lambda)\cdot(-\lambda)\cdot\det\begin{pmatrix} \sigma_5-\lambda & 0 & 0 \\ 0 & -(\sigma_6+\lambda) & \sigma_6\cdot e^{-\lambda\cdot\tau_2} \\ 0 & \sigma_8 & (\sigma_9\cdot e^{-\lambda\cdot\tau_2}-\lambda) \end{pmatrix}$$

$$-\sigma_2\cdot e^{-\lambda\cdot\tau_1}\cdot\det\begin{pmatrix} \sigma_5-\lambda & 0 & 0 \\ 0 & -(\sigma_6+\lambda) & \sigma_6\cdot e^{-\lambda\cdot\tau_2} \\ 0 & \sigma_8 & (\sigma_9\cdot e^{-\lambda\cdot\tau_2}-\lambda) \end{pmatrix}$$

$$+\sigma_3\cdot\det\begin{pmatrix} \sigma_5\cdot e^{-\lambda\cdot\tau_1} & 0 & 0 \\ 0 & -(\sigma_6+\lambda) & \sigma_6\cdot e^{-\lambda\cdot\tau_2} \\ \sigma_7\cdot e^{-\lambda\cdot\tau_1} & \sigma_8 & (\sigma_9\cdot e^{-\lambda\cdot\tau_2}-\lambda) \end{pmatrix}$$

$$+\sigma_4\cdot e^{-\lambda\cdot\tau_2}\cdot\det\begin{pmatrix} \sigma_5\cdot e^{-\lambda\cdot\tau_1} & \sigma_5-\lambda & 0 \\ 0 & 0 & -(\sigma_6+\lambda) \\ \sigma_7\cdot e^{-\lambda\cdot\tau_1} & 0 & \sigma_8 \end{pmatrix}$$

<u>First expression</u>:

$$\det\begin{pmatrix} \sigma_5-\lambda & 0 & 0 \\ 0 & -(\sigma_6+\lambda) & \sigma_6\cdot e^{-\lambda\cdot\tau_2} \\ 0 & \sigma_8 & (\sigma_9\cdot e^{-\lambda\cdot\tau_2}-\lambda) \end{pmatrix} = (\sigma_5-\lambda)\cdot\det\begin{pmatrix} -(\sigma_6+\lambda) & \sigma_6\cdot e^{-\lambda\cdot\tau_2} \\ \sigma_8 & (\sigma_9\cdot e^{-\lambda\cdot\tau_2}-\lambda) \end{pmatrix}$$

$$= (\sigma_5-\lambda)\cdot\{-(\sigma_6+\lambda)\cdot(\sigma_9\cdot e^{-\lambda\cdot\tau_2}-\lambda)-\sigma_8\cdot\sigma_6\cdot e^{-\lambda\cdot\tau_2}\}$$

$$= (\sigma_5-\lambda)\cdot\{-\sigma_6\cdot\sigma_9\cdot e^{-\lambda\cdot\tau_2}+\sigma_6\cdot\lambda-\lambda\cdot\sigma_9\cdot e^{-\lambda\cdot\tau_2}+\lambda^2-\sigma_8\cdot\sigma_6\cdot e^{-\lambda\cdot\tau_2}\}$$

$$= (\sigma_5-\lambda)\cdot\{\sigma_6\cdot\lambda+\lambda^2-[\sigma_6\cdot\sigma_9+\sigma_8\cdot\sigma_6+\lambda\cdot\sigma_9]\cdot e^{-\lambda\cdot\tau_2}\}$$

$$\det\begin{pmatrix} \sigma_5-\lambda & 0 & 0 \\ 0 & -(\sigma_6+\lambda) & \sigma_6\cdot e^{-\lambda\cdot\tau_2} \\ 0 & \sigma_8 & (\sigma_9\cdot e^{-\lambda\cdot\tau_2}-\lambda) \end{pmatrix}$$

$$= (\sigma_5-\lambda)\cdot\det\begin{pmatrix} -(\sigma_6+\lambda) & \sigma_6\cdot e^{-\lambda\cdot\tau_2} \\ \sigma_8 & (\sigma_9\cdot e^{-\lambda\cdot\tau_2}-\lambda) \end{pmatrix}$$

$$= (\sigma_5-\lambda)\cdot\{-(\sigma_6+\lambda)\cdot(\sigma_9\cdot e^{-\lambda\cdot\tau_2}-\lambda)-\sigma_8\cdot\sigma_6\cdot e^{-\lambda\cdot\tau_2}\}$$

$$= (\sigma_5-\lambda)\cdot\{-\sigma_6\cdot\sigma_9\cdot e^{-\lambda\cdot\tau_2}+\sigma_6\cdot\lambda-\lambda\cdot\sigma_9\cdot e^{-\lambda\cdot\tau_2}+\lambda^2-\sigma_8\cdot\sigma_6\cdot e^{-\lambda\cdot\tau_2}\}$$

$$= (\sigma_5-\lambda)\cdot\{\sigma_6\cdot\lambda+\lambda^2-[\sigma_6\cdot\sigma_9+\sigma_8\cdot\sigma_6+\lambda\cdot\sigma_9]\cdot e^{-\lambda\cdot\tau_2}\}$$

$$= \sigma_5 \cdot \sigma_6 \cdot \lambda + \sigma_5 \cdot \lambda^2 - \sigma_5 \cdot [\sigma_6 \cdot \sigma_9 + \sigma_8 \cdot \sigma_6 + \lambda \cdot \sigma_9] \cdot e^{-\lambda \cdot \tau_2}$$
$$- \sigma_6 \cdot \lambda^2 - \lambda^3 + \lambda \cdot [\sigma_6 \cdot \sigma_9 + \sigma_8 \cdot \sigma_6 + \lambda \cdot \sigma_9] \cdot e^{-\lambda \cdot \tau_2}$$
$$= \sigma_5 \cdot \sigma_6 \cdot \lambda + \sigma_5 \cdot \lambda^2 - [\sigma_5 \cdot \sigma_6 \cdot \sigma_9 + \sigma_5 \cdot \sigma_8 \cdot \sigma_6 + \lambda \cdot \sigma_5 \cdot \sigma_9] \cdot e^{-\lambda \cdot \tau_2}$$
$$- \sigma_6 \cdot \lambda^2 - \lambda^3 + [\lambda \cdot (\sigma_6 \cdot \sigma_9 + \sigma_8 \cdot \sigma_6) + \lambda^2 \cdot \sigma_9] \cdot e^{-\lambda \cdot \tau_2}$$
$$= \sigma_5 \cdot \sigma_6 \cdot \lambda + (\sigma_5 - \sigma_6) \cdot \lambda^2 - \lambda^3$$
$$+ \{-\sigma_5 \cdot \sigma_6 \cdot (\sigma_9 + \sigma_8) + \lambda \cdot (\sigma_6 \cdot \sigma_9 + \sigma_8 \cdot \sigma_6 - \sigma_5 \cdot \sigma_9) + \lambda^2 \cdot \sigma_9\} \cdot e^{-\lambda \cdot \tau_2}$$

We define for simplicity:

$$\psi_1 = \sigma_5 \cdot \sigma_6; \psi_2 = \sigma_5 - \sigma_6;$$
$$\psi_3 = -\sigma_5 \cdot \sigma_6 \cdot (\sigma_9 + \sigma_8)$$
$$\psi_4 = \sigma_6 \cdot \sigma_9 + \sigma_8 \cdot \sigma_6 - \sigma_5 \cdot \sigma_9$$

Then we define

$$\det \begin{pmatrix} \sigma_5 - \lambda & 0 & 0 \\ 0 & -(\sigma_6 + \lambda) & \sigma_6 \cdot e^{-\lambda \cdot \tau_2} \\ 0 & \sigma_8 & (\sigma_9 \cdot e^{-\lambda \cdot \tau_2} - \lambda) \end{pmatrix}$$
$$= \psi_1 \cdot \lambda + \psi_2 \cdot \lambda^2 - \lambda^3 + \{\psi_3 + \lambda \cdot \psi_4 + \lambda^2 \cdot \sigma_9\} \cdot e^{-\lambda \cdot \tau_2}$$

<u>Second expression</u>:

$$\det \begin{pmatrix} \sigma_5 \cdot e^{-\lambda \cdot \tau_1} & 0 & 0 \\ 0 & -(\sigma_6 + \lambda) & \sigma_6 \cdot e^{-\lambda \cdot \tau_2} \\ \sigma_7 \cdot e^{-\lambda \cdot \tau_1} & \sigma_8 & (\sigma_9 \cdot e^{-\lambda \cdot \tau_2} - \lambda) \end{pmatrix}$$
$$= \sigma_5 \cdot e^{-\lambda \cdot \tau_1} \cdot \det \begin{pmatrix} -(\sigma_6 + \lambda) & \sigma_6 \cdot e^{-\lambda \cdot \tau_2} \\ \sigma_8 & (\sigma_9 \cdot e^{-\lambda \cdot \tau_2} - \lambda) \end{pmatrix}$$
$$= \sigma_5 \cdot e^{-\lambda \cdot \tau_1} \cdot \{-(\sigma_6 + \lambda) \cdot (\sigma_9 \cdot e^{-\lambda \cdot \tau_2} - \lambda) - \sigma_8 \cdot \sigma_6 \cdot e^{-\lambda \cdot \tau_2}\}$$
$$= \sigma_5 \cdot e^{-\lambda \cdot \tau_1} \cdot \{-\sigma_6 \cdot \sigma_9 \cdot e^{-\lambda \cdot \tau_2} + \sigma_6 \cdot \lambda - \lambda \cdot \sigma_9 \cdot e^{-\lambda \cdot \tau_2} + \lambda^2 - \sigma_8 \cdot \sigma_6 \cdot e^{-\lambda \cdot \tau_2}\}$$
$$= \sigma_5 \cdot e^{-\lambda \cdot \tau_1} \cdot \{\sigma_6 \cdot \lambda + \lambda^2 - [\sigma_6 \cdot \sigma_9 + \sigma_8 \cdot \sigma_6 + \lambda \cdot \sigma_9] \cdot e^{-\lambda \cdot \tau_2}\}$$
$$= (\sigma_6 \cdot \lambda + \lambda^2) \cdot \sigma_5 \cdot e^{-\lambda \cdot \tau_1} - \sigma_5 \cdot [\sigma_6 \cdot \sigma_9 + \sigma_8 \cdot \sigma_6 + \lambda \cdot \sigma_9] \cdot e^{-\lambda \cdot (\tau_2 + \tau_1)};$$

$\psi_5 = \sigma_6 \cdot \sigma_9 + \sigma_8 \cdot \sigma_6$

$$\det\begin{pmatrix} \sigma_5 \cdot e^{-\lambda\cdot\tau_1} & 0 & 0 \\ 0 & -(\sigma_6+\lambda) & \sigma_6 \cdot e^{-\lambda\cdot\tau_2} \\ \sigma_7 \cdot e^{-\lambda\cdot\tau_1} & \sigma_8 & (\sigma_9 \cdot e^{-\lambda\cdot\tau_2} - \lambda) \end{pmatrix}$$
$$= (\sigma_6 \cdot \lambda + \lambda^2) \cdot \sigma_5 \cdot e^{-\lambda\cdot\tau_1} - \sigma_5 \cdot [\psi_5 + \lambda \cdot \sigma_9] \cdot e^{-\lambda\cdot(\tau_2+\tau_1)}$$

Third expression:

$$\det\begin{pmatrix} \sigma_5 \cdot e^{-\lambda\cdot\tau_1} & (\sigma_5 - \lambda) & 0 \\ 0 & 0 & -(\sigma_6+\lambda) \\ \sigma_7 \cdot e^{-\lambda\cdot\tau_1} & 0 & \sigma_8 \end{pmatrix}$$
$$= \sigma_5 \cdot e^{-\lambda\cdot\tau_1} \cdot \det\begin{pmatrix} 0 & -(\sigma_6+\lambda) \\ 0 & \sigma_8 \end{pmatrix}$$
$$- (\sigma_5 - \lambda) \cdot \det\begin{pmatrix} 0 & -(\sigma_6+\lambda) \\ \sigma_7 \cdot e^{-\lambda\cdot\tau_1} & \sigma_8 \end{pmatrix}$$
$$= -(\sigma_5 - \lambda) \cdot \det\begin{pmatrix} 0 & -(\sigma_6+\lambda) \\ \sigma_7 \cdot e^{-\lambda\cdot\tau_1} & \sigma_8 \end{pmatrix} = -(\sigma_5 - \lambda) \cdot \sigma_7 \cdot e^{-\lambda\cdot\tau_1} \cdot (\sigma_6 + \lambda)$$
$$= -(\sigma_5 - \lambda) \cdot \sigma_7 \cdot (\sigma_6 + \lambda) \cdot e^{-\lambda\cdot\tau_1} = \sigma_7 \cdot (-\sigma_5 \cdot \sigma_6 - \sigma_5 \cdot \lambda + \lambda \cdot \sigma_6 + \lambda^2) \cdot e^{-\lambda\cdot\tau_1}$$
$$= \sigma_7 \cdot (-\sigma_5 \cdot \sigma_6 + \lambda \cdot [\sigma_6 - \sigma_5] + \lambda^2) \cdot e^{-\lambda\cdot\tau_1}$$

$$\psi_1 = \sigma_5 \cdot \sigma_6; \psi_2 = \sigma_5 - \sigma_6 \Rightarrow -\psi_2 = \sigma_6 - \sigma_5$$

$$\det\begin{pmatrix} \sigma_5 \cdot e^{-\lambda\cdot\tau_1} & (\sigma_5 - \lambda) & 0 \\ 0 & 0 & -(\sigma_6+\lambda) \\ \sigma_7 \cdot e^{-\lambda\cdot\tau_1} & 0 & \sigma_8 \end{pmatrix} = \sigma_7 \cdot (-\psi_1 - \lambda \cdot \psi_2 + \lambda^2) \cdot e^{-\lambda\cdot\tau_1}$$

We integrate our expression in below $D(\tau_1, \tau_2)$ expression.

$$\det\begin{pmatrix} \Upsilon_{11} & \cdots & \Upsilon_{15} \\ \vdots & \ddots & \vdots \\ \Upsilon_{51} & \cdots & \Upsilon_{55} \end{pmatrix} = (\sigma_1 - \lambda) \cdot (-\lambda) \cdot \det\begin{pmatrix} \sigma_5 - \lambda & 0 & 0 \\ 0 & -(\sigma_6+\lambda) & \sigma_6 \cdot e^{-\lambda\cdot\tau_2} \\ 0 & \sigma_8 & (\sigma_9 \cdot e^{-\lambda\cdot\tau_2} - \lambda) \end{pmatrix}$$
$$- \sigma_2 \cdot e^{-\lambda\cdot\tau_1} \cdot \det\begin{pmatrix} \sigma_5 - \lambda & 0 & 0 \\ 0 & -(\sigma_6+\lambda) & \sigma_6 \cdot e^{-\lambda\cdot\tau_2} \\ 0 & \sigma_8 & (\sigma_9 \cdot e^{-\lambda\cdot\tau_2} - \lambda) \end{pmatrix}$$
$$+ \sigma_3 \cdot \det\begin{pmatrix} \sigma_5 \cdot e^{-\lambda\cdot\tau_1} & 0 & 0 \\ 0 & -(\sigma_6+\lambda) & \sigma_6 \cdot e^{-\lambda\cdot\tau_2} \\ \sigma_7 \cdot e^{-\lambda\cdot\tau_1} & \sigma_8 & (\sigma_9 \cdot e^{-\lambda\cdot\tau_2} - \lambda) \end{pmatrix}$$
$$+ \sigma_4 \cdot e^{-\lambda\cdot\tau_2} \cdot \det\begin{pmatrix} \sigma_5 \cdot e^{-\lambda\cdot\tau_1} & \sigma_5 - \lambda & 0 \\ 0 & 0 & -(\sigma_6+\lambda) \\ \sigma_7 \cdot e^{-\lambda\cdot\tau_1} & 0 & \sigma_8 \end{pmatrix}$$

$$\det\begin{pmatrix} \Upsilon_{11} & \cdots & \Upsilon_{15} \\ \vdots & \ddots & \vdots \\ \Upsilon_{51} & \cdots & \Upsilon_{55} \end{pmatrix} = (\sigma_1 - \lambda) \cdot (-\lambda) \cdot [\psi_1 \cdot \lambda + \psi_2 \cdot \lambda^2 - \lambda^3$$

$$+ \{\psi_3 + \lambda \cdot \psi_4 + \lambda^2 \cdot \sigma_9\} \cdot e^{-\lambda \cdot \tau_2}]$$
$$- \sigma_2 \cdot e^{-\lambda \cdot \tau_1} \cdot [\psi_1 \cdot \lambda + \psi_2 \cdot \lambda^2 - \lambda^3$$
$$+ \{\psi_3 + \lambda \cdot \psi_4 + \lambda^2 \cdot \sigma_9\} \cdot e^{-\lambda \cdot \tau_2}]$$
$$+ \sigma_3 \cdot [(\sigma_6 \cdot \lambda + \lambda^2) \cdot \sigma_5 \cdot e^{-\lambda \cdot \tau_1}$$
$$- \sigma_5 \cdot [\psi_5 + \lambda \cdot \sigma_9] \cdot e^{-\lambda \cdot (\tau_2 + \tau_1)}]$$
$$+ \sigma_4 \cdot e^{-\lambda \cdot \tau_2} \cdot [\sigma_7 \cdot (-\psi_1 - \lambda \cdot \psi_2 + \lambda^2) \cdot e^{-\lambda \cdot \tau_1}]$$

$$\det\begin{pmatrix} \Upsilon_{11} & \cdots & \Upsilon_{15} \\ \vdots & \ddots & \vdots \\ \Upsilon_{51} & \cdots & \Upsilon_{55} \end{pmatrix} = (\lambda^2 - \sigma_1 \lambda) \cdot [\psi_1 \cdot \lambda + \psi_2 \cdot \lambda^2 - \lambda^3$$

$$+ \{\psi_3 + \lambda \cdot \psi_4 + \lambda^2 \cdot \sigma_9\} \cdot e^{-\lambda \cdot \tau_2}]$$
$$- [(\psi_1 \cdot \lambda + \psi_2 \cdot \lambda^2 - \lambda^3) \cdot \sigma_2 \cdot e^{-\lambda \cdot \tau_1}$$
$$+ \sigma_2 \cdot \{\psi_3 + \lambda \cdot \psi_4 + \lambda^2 \cdot \sigma_9\} \cdot e^{-\lambda \cdot (\tau_1 + \tau_2)}]$$
$$+ \sigma_3 \cdot (\sigma_6 \cdot \lambda + \lambda^2) \cdot \sigma_5 \cdot e^{-\lambda \cdot \tau_1}$$
$$- \sigma_3 \cdot \sigma_5 \cdot [\psi_5 + \lambda \cdot \sigma_9] \cdot e^{-\lambda \cdot (\tau_2 + \tau_1)}$$
$$+ (-\psi_1 \cdot \sigma_4 \cdot \sigma_7 - \lambda \cdot \psi_2 \cdot \sigma_4 \cdot \sigma_7 + \lambda^2 \cdot \sigma_4 \cdot \sigma_7) \cdot e^{-\lambda \cdot (\tau_1 + \tau_2)}$$

$$\det\begin{pmatrix} \Upsilon_{11} & \cdots & \Upsilon_{15} \\ \vdots & \ddots & \vdots \\ \Upsilon_{51} & \cdots & \Upsilon_{55} \end{pmatrix} = \psi_1 \cdot \lambda^3 + \psi_2 \cdot \lambda^4 - \lambda^5$$

$$+ \{\psi_3 \cdot \lambda^2 + \lambda^3 \cdot \psi_4 + \lambda^4 \cdot \sigma_9\} \cdot e^{-\lambda \cdot \tau_2}$$
$$- \sigma_1 \cdot \psi_1 \cdot \lambda^2 - \sigma_1 \cdot \psi_2 \cdot \lambda^3 + \sigma_1 \cdot \lambda^4$$
$$+ \{-\sigma_1 \cdot \psi_3 \cdot \lambda - \sigma_1 \cdot \psi_4 \cdot \lambda^2 - \sigma_1 \cdot \sigma_9 \cdot \lambda^3\} \cdot e^{-\lambda \cdot \tau_2}$$
$$- (\psi_1 \cdot \lambda + \psi_2 \cdot \lambda^2 - \lambda^3) \cdot \sigma_2 \cdot e^{-\lambda \cdot \tau_1}$$
$$- \sigma_2 \cdot \{\psi_3 + \lambda \cdot \psi_4 + \lambda^2 \cdot \sigma_9\} \cdot e^{-\lambda \cdot (\tau_1 + \tau_2)}$$
$$+ (\sigma_3 \cdot \sigma_6 \cdot \lambda + \sigma_3 \cdot \lambda^2) \cdot \sigma_5 \cdot e^{-\lambda \cdot \tau_1}$$
$$- \sigma_3 \cdot \sigma_5 \cdot [\psi_5 + \lambda \cdot \sigma_9] \cdot e^{-\lambda \cdot (\tau_2 + \tau_1)}$$
$$+ (-\psi_1 \cdot \sigma_4 \cdot \sigma_7 - \lambda \cdot \psi_2 \cdot \sigma_4 \cdot \sigma_7 + \lambda^2 \cdot \sigma_4 \cdot \sigma_7) \cdot e^{-\lambda \cdot (\tau_1 + \tau_2)}$$

$$\det\begin{pmatrix} \Upsilon_{11} & \dots & \Upsilon_{15} \\ \vdots & \ddots & \vdots \\ \Upsilon_{51} & \cdots & \Upsilon_{55} \end{pmatrix} = -\sigma_1 \cdot \psi_1 \cdot \lambda^2 + (\psi_1 - \sigma_1 \cdot \psi_2) \cdot \lambda^3 + (\psi_2 + \sigma_1) \cdot \lambda^4 - \lambda^5$$

$$\begin{aligned} &- (\psi_1 \cdot \lambda + \psi_2 \cdot \lambda^2 - \lambda^3) \cdot \sigma_2 \cdot e^{-\lambda \cdot \tau_1} \\ &+ (\sigma_3 \cdot \sigma_6 \cdot \lambda + \sigma_3 \cdot \lambda^2) \cdot \sigma_5 \cdot e^{-\lambda \cdot \tau_1} \\ &+ \{\psi_3 \cdot \lambda^2 + \lambda^3 \cdot \psi_4 + \lambda^4 \cdot \sigma_9\} \cdot e^{-\lambda \cdot \tau_2} \\ &+ \{-\sigma_1 \cdot \psi_3 \cdot \lambda - \sigma_1 \cdot \psi_4 \cdot \lambda^2 - \sigma_1 \cdot \sigma_9 \cdot \lambda^3\} \cdot e^{-\lambda \cdot \tau_2} \\ &- \sigma_2 \cdot \{\psi_3 + \lambda \cdot \psi_4 + \lambda^2 \cdot \sigma_9\} \cdot e^{-\lambda \cdot (\tau_1 + \tau_2)} \\ &- \sigma_3 \cdot \sigma_5 \cdot [\psi_5 + \lambda \cdot \sigma_9] \cdot e^{-\lambda \cdot (\tau_2 + \tau_1)} \\ &+ (-\psi_1 \cdot \sigma_4 \cdot \sigma_7 - \lambda \cdot \psi_2 \cdot \sigma_4 \cdot \sigma_7 + \lambda^2 \cdot \sigma_4 \cdot \sigma_7) \cdot e^{-\lambda \cdot (\tau_1 + \tau_2)} \end{aligned}$$

$$\det\begin{pmatrix} \Upsilon_{11} & \dots & \Upsilon_{15} \\ \vdots & \ddots & \vdots \\ \Upsilon_{51} & \cdots & \Upsilon_{55} \end{pmatrix} = -\sigma_1 \cdot \psi_1 \cdot \lambda^2 + (\psi_1 - \sigma_1 \cdot \psi_2) \cdot \lambda^3 + (\psi_2 + \sigma_1) \cdot \lambda^4 - \lambda^5$$

$$\begin{aligned} &+ (-\psi_1 \cdot \sigma_2 \cdot \lambda - \psi_2 \cdot \sigma_2 \cdot \lambda^2 + \sigma_2 \cdot \lambda^3) \cdot e^{-\lambda \cdot \tau_1} \\ &+ (\sigma_3 \cdot \sigma_6 \cdot \sigma_5 \cdot \lambda + \sigma_3 \cdot \sigma_5 \cdot \lambda^2) \cdot e^{-\lambda \cdot \tau_1} \\ &+ \{\psi_3 \cdot \lambda^2 + \lambda^3 \cdot \psi_4 + \lambda^4 \cdot \sigma_9\} \cdot e^{-\lambda \cdot \tau_2} \\ &+ \{-\sigma_1 \cdot \psi_3 \cdot \lambda - \sigma_1 \cdot \psi_4 \cdot \lambda^2 - \sigma_1 \cdot \sigma_9 \cdot \lambda^3\} \cdot e^{-\lambda \cdot \tau_2} \\ &+ \{-\sigma_2 \cdot \psi_3 - \lambda \cdot \sigma_2 \cdot \psi_4 - \lambda^2 \cdot \sigma_2 \cdot \sigma_9\} \cdot e^{-\lambda \cdot (\tau_1 + \tau_2)} \\ &+ [-\sigma_3 \cdot \sigma_5 \cdot \psi_5 - \lambda \cdot \sigma_3 \cdot \sigma_5 \cdot \sigma_9] \cdot e^{-\lambda \cdot (\tau_2 + \tau_1)} \\ &+ (-\psi_1 \cdot \sigma_4 \cdot \sigma_7 - \lambda \cdot \psi_2 \cdot \sigma_4 \cdot \sigma_7 + \lambda^2 \cdot \sigma_4 \cdot \sigma_7) \cdot e^{-\lambda \cdot (\tau_1 + \tau_2)} \end{aligned}$$

$$\det\begin{pmatrix} \Upsilon_{11} & \dots & \Upsilon_{15} \\ \vdots & \ddots & \vdots \\ \Upsilon_{51} & \cdots & \Upsilon_{55} \end{pmatrix} = -\sigma_1 \cdot \psi_1 \cdot \lambda^2 + (\psi_1 - \sigma_1 \cdot \psi_2) \cdot \lambda^3 + (\psi_2 + \sigma_1) \cdot \lambda^4 - \lambda^5$$

$$\begin{aligned} &+ \{(\sigma_3 \cdot \sigma_6 \cdot \sigma_5 - \psi_1 \cdot \sigma_2) \cdot \lambda \\ &+ (\sigma_3 \cdot \sigma_5 - \psi_2 \cdot \sigma_2) \cdot \lambda^2 + \sigma_2 \cdot \lambda^3\} \cdot e^{-\lambda \cdot \tau_1} \\ &+ \{-\sigma_1 \cdot \psi_3 \cdot \lambda + (\psi_3 - \sigma_1 \cdot \psi_4) \cdot \lambda^2 \\ &+ (\psi_4 - \sigma_1 \cdot \sigma_9) \cdot \lambda^3 + \lambda^4 \cdot \sigma_9\} \cdot e^{-\lambda \cdot \tau_2} \\ &+ \{-\sigma_2 \cdot \psi_3 - \sigma_3 \cdot \sigma_5 \cdot \psi_5 - \psi_1 \cdot \sigma_4 \cdot \sigma_7 \\ &- (\psi_2 \cdot \sigma_4 \cdot \sigma_7 + \sigma_2 \cdot \psi_4 + \sigma_3 \cdot \sigma_5 \cdot \sigma_9) \cdot \lambda \\ &+ (\sigma_4 \cdot \sigma_7 - \sigma_2 \cdot \sigma_9) \cdot \lambda^2\} \cdot e^{-\lambda \cdot (\tau_1 + \tau_2)} \end{aligned}$$

We define for simplicity the following parameters:

$$\Theta_2 = -\sigma_1 \cdot \psi_1; \Theta_3 = \psi_1 - \sigma_1 \cdot \psi_2; \Theta_4 = \psi_2 + \sigma_1; \Theta_5 = -1$$
$$\mathrm{A}_1 = \sigma_3 \cdot \sigma_6 \cdot \sigma_5 - \psi_1 \cdot \sigma_2; \mathrm{A}_2 = \sigma_3 \cdot \sigma_5 - \psi_2 \cdot \sigma_2; \mathrm{A}_3 = \sigma_2$$
$$\mathrm{B}_1 = -\sigma_1 \cdot \psi_3; \mathrm{B}_2 = \psi_3 - \sigma_1 \cdot \psi_4; \mathrm{B}_3 = \psi_4 - \sigma_1 \cdot \sigma_9; \mathrm{B}_4 = \sigma_9$$
$$C_0 = -\sigma_2 \cdot \psi_3 - \sigma_3 \cdot \sigma_5 \cdot \psi_5 - \psi_1 \cdot \sigma_4 \cdot \sigma_7;$$
$$C_1 = -(\psi_2 \cdot \sigma_4 \cdot \sigma_7 + \sigma_2 \cdot \psi_4 + \sigma_3 \cdot \sigma_5 \cdot \sigma_9)$$
$$C_2 = \sigma_4 \cdot \sigma_7 - \sigma_2 \cdot \sigma_9$$

$$\det\begin{pmatrix} \Upsilon_{11} & \dots & \Upsilon_{15} \\ \vdots & \ddots & \vdots \\ \Upsilon_{51} & \cdots & \Upsilon_{55} \end{pmatrix} = \sum_{l=2}^{5} \Theta_l \cdot \lambda^l + [\sum_{k=1}^{3} \mathrm{A}_k \cdot \lambda^k] \cdot e^{-\lambda \cdot \tau_1} + [\sum_{k=1}^{4} \mathrm{B}_k \cdot \lambda^k] \cdot e^{-\lambda \cdot \tau_2} + [\sum_{k=0}^{2} C_k \cdot \lambda^k] \cdot e^{-\lambda \cdot (\tau_1 + \tau_2)}$$

$$D(\tau_1, \tau_2) = \sum_{l=2}^{5} \Theta_l \cdot \lambda^l + [\sum_{k=1}^{3} \mathrm{A}_k \cdot \lambda^k] \cdot e^{-\lambda \cdot \tau_1} + [\sum_{k=1}^{4} \mathrm{B}_k \cdot \lambda^k] \cdot e^{-\lambda \cdot \tau_2} + [\sum_{k=0}^{2} C_k \cdot \lambda^k] \cdot e^{-\lambda \cdot (\tau_1 + \tau_2)}$$

Three cases: (A) $\tau_1 = \tau; \tau_2 = 0$ (B)$\tau_1 = 0; \tau_2 = \tau$ (C) $\tau_1 = \tau_2 = \tau$.

(A)
$$\tau_1 = \tau; \tau_2 = 0; D(\tau) = \sum_{l=2}^{5} \Theta_l \cdot \lambda^l + [\sum_{k=1}^{4} \mathrm{B}_k \cdot \lambda^k] + [\sum_{k=1}^{3} \mathrm{A}_k \cdot \lambda^k] \cdot e^{-\lambda \cdot \tau} + [\sum_{k=0}^{2} C_k \cdot \lambda^k] \cdot e^{-\lambda \cdot \tau}$$

$$D(\tau_1 = \tau; \tau_2 = 0) = \sum_{l=2}^{5} \Theta_l \cdot \lambda^l + [\sum_{k=1}^{4} \mathrm{B}_k \cdot \lambda^k]$$

$$+ [\sum_{k=1}^{3} \mathrm{A}_k \cdot \lambda^k] \cdot e^{-\lambda \cdot \tau} + [\sum_{k=0}^{2} C_k \cdot \lambda^k] \cdot e^{-\lambda \cdot \tau}$$

$$D(\tau_1 = \tau; \tau_2 = 0) = \mathrm{B}_1 \cdot \lambda + \sum_{l=2}^{4} (\Theta_l + \mathrm{B}_l) \cdot \lambda^l$$

$$+ \Theta_5 \cdot \lambda^5 + [C_0 + \sum_{l=1}^{2} (\mathrm{A}_l + C_l) \cdot \lambda^l + \mathrm{A}_3 \cdot \lambda^3] \cdot e^{-\lambda \cdot \tau}$$

$$D(\lambda, \tau) = P_n(\lambda, \tau) + Q_m(\lambda, \tau) \cdot e^{-\lambda \cdot \tau}; n, m \in \mathbb{N}_0; n > m$$

$$P_n(\lambda, \tau) = \mathrm{B}_1 \cdot \lambda + \sum_{l=2}^{4} (\Theta_l + \mathrm{B}_l) \cdot \lambda^l + \Theta_5 \cdot \lambda^5; n = 5;$$

$$Q_m(\lambda, \tau) = [C_0 + \sum_{l=1}^{2} (\mathrm{A}_l + C_l) \cdot \lambda^l + \mathrm{A}_3 \cdot \lambda^3]; m = 3$$

$$P_n(\lambda, \tau) = \sum_{k=0}^{n} P_k(\tau) \cdot \lambda^k = P_0(\tau) + P_1(\tau) \cdot \lambda + P_2(\tau) \cdot \lambda^2 + P_3(\tau) \cdot \lambda^3 + \ldots;$$

$$Q_m(\lambda, \tau) = \sum_{k=0}^{m} q_k(\tau) \cdot \lambda^k = q_0(\tau) + q_1(\tau) \cdot \lambda + q_2(\tau) \cdot \lambda^2 + \ldots$$

$$D(\lambda, \tau) = P_n(\lambda, \tau) + Q_m(\lambda, \tau) \cdot e^{-\lambda \tau}; n = 5; m = 3; n > m$$

$$P_n(\lambda, \tau) = \sum_{k=0}^{n} P_k(\tau) \cdot \lambda^k = P_0(\tau) + P_1(\tau) \cdot \lambda + P_2(\tau) \cdot \lambda^2$$

$$+ P_3(\tau) \cdot \lambda^3 + P_4(\tau) \cdot \lambda^4 + P_5(\tau) \cdot \lambda^5$$

$$P_0 = 0; P_1 = \mathrm{B}_1; P_2 = \Theta_2 + \mathrm{B}_2; P_3 = \Theta_3 + \mathrm{B}_3; P_4 = \Theta_4 + \mathrm{B}_4; P_5 = \Theta_5$$

$$Q_m(\lambda, \tau) = \sum_{k=0}^{m} q_k(\tau) \cdot \lambda^k = q_0(\tau) + q_1(\tau) \cdot \lambda + q_2(\tau) \cdot \lambda^2 + q_3(\tau) \cdot \lambda^3;$$

$$q_0(\tau) = C_0; q_1(\tau) = \mathrm{A}_1 + C_1; q_2(\tau) = \mathrm{A}_2 + C_2$$

$$q_3(\tau) = \mathrm{A}_3.$$

The homogeneous system for $X, Y, I_{L_1}, I_{R_j}, I_{R_S}$ leads to a characteristic equation for the eigenvalue λ having the form $P(\lambda, \tau) + Q(\lambda, \tau) \cdot e^{-\lambda \cdot \tau} = 0$; $P(\lambda) = \sum_{j=0}^{5} a_j \cdot \lambda^j; Q(\lambda) = \sum_{j=0}^{3} c_j \cdot \lambda^j$. The coefficients $\{a_j(q_i, q_k, \tau), c_j(q_i, q_k, \tau)\} \in \mathbb{R}$ depend on

q_i, q_k and delay τ. q_i, q_k are any Schottky detector's global parameters, other parameters kept as a constant.

$$a_0 = 0; a_1 = \mathrm{B}_1; a_2 = \Theta_2 + \mathrm{B}_2; a_3 = \Theta_3 + \mathrm{B}_3$$
$$a_4 = \Theta_4 + \mathrm{B}_4; a_5 = \Theta_5; c_0(\tau) = C_0;$$
$$c_1(\tau) = \mathrm{A}_1 + C_1; c_2(\tau) = \mathrm{A}_2 + C_2; c_3(\tau) = \mathrm{A}_3.$$

Unless strictly necessary, the designation of the varied arguments (q_i, q_k) will subsequently be omitted from P, Q, a_j, and c_j. The coefficients a_j, c_j are continuous, and differentiable functions of their arguments, and direct substitution shows that $a_0 + c_0 \neq 0$ for $\forall q_i, q_k \in \mathbb{R}_+$; that is, $\lambda = 0$ is not of $P(\lambda) + Q(\lambda) \cdot e^{-\lambda \cdot \tau} = 0$. Furthermore, $P(\lambda)$, $Q(\lambda)$ are analytic functions of λ, for which the following requirements of the analysis (Kuang J and Cong Y 2005; Kuang Y 1993) can also be verified in the present case:

(a) If $\lambda = i \cdot \omega$, $\omega \in \mathbb{R}$, then $P(i \cdot \omega) + Q(i \cdot \omega) \neq 0$.
(b) $|Q(\lambda)/P(\lambda)|$ is bounded for $|\lambda| \to \infty$, $\mathrm{Re}\lambda \geq 0$. No roots bifurcation from ∞.
(c) $F(\omega) = |P(i \cdot \omega)|^2 - |Q(i \cdot \omega)|^2$ has a finite number of zeros. Indeed, this is a polynomial in ω.
(d) Each positive root $\omega(q_i, q_k)$ of $F(\omega) = 0$ is continuous and differentiable with respect to q_i, q_k.

We assume that $P_n(\lambda, \tau)$ and $Q_m(\lambda, \tau)$ cannot have common imaginary roots. That is for any real number ω;

$$p_n(\lambda = i \cdot \omega, \tau) + Q_m(\lambda = i \cdot \omega, \tau) \neq 0.$$

$$p_n(\lambda = i \cdot \omega, \tau) = \mathrm{B}_1 \cdot i \cdot \omega + \sum_{l=2}^{4} (\Theta_l + \mathrm{B}_l) \cdot (i \cdot \omega)^l + \Theta_5 \cdot (i \cdot \omega)^5$$
$$= i \cdot \omega \cdot \mathrm{B}_1 + \sum_{l=2}^{4} (\Theta_l + \mathrm{B}_l) \cdot i^l \cdot \omega^l + i \cdot \Theta_5 \cdot \omega^5$$

$$\sum_{l=2}^{4} (\Theta_l + \mathrm{B}_l) \cdot i^l \cdot \omega^l = -(\Theta_2 + \mathrm{B}_2) \cdot \omega^2 + (\Theta_2 + \mathrm{B}_2) \cdot \omega^4 - (\Theta_2 + \mathrm{B}_2) \cdot \omega^3 \cdot i$$

$$p_n(\lambda = i \cdot \omega, \tau) = -(\Theta_2 + \mathrm{B}_2) \cdot \omega^2 + (\Theta_2 + \mathrm{B}_2) \cdot \omega^4 + i \cdot [\omega \cdot \mathrm{B}_1 - (\Theta_2 + \mathrm{B}_2) \cdot \omega^3 + \Theta_5 \cdot \omega^5]$$

$$Q_m(\lambda = i \cdot \omega, \tau) = C_0 + \sum_{l=1}^{2} (\mathrm{A}_l + C_l) \cdot (i \cdot \omega)^l - i \cdot \mathrm{A}_3 \cdot \omega^3;$$

$$\sum_{l=1}^{2} (\mathrm{A}_l + C_l) \cdot (i \cdot \omega)^l = i \cdot \omega \cdot (\mathrm{A}_1 + C_1) - (\mathrm{A}_2 + C_2) \cdot \omega^2$$

$$Q_m(\lambda = i \cdot \omega, \tau) = C_0 - (\mathrm{A}_2 + C_2) \cdot \omega^2 + i \cdot [\omega \cdot (\mathrm{A}_1 + C_1) - \mathrm{A}_3 \cdot \omega^3]$$

$$
\begin{aligned}
& p_n(\lambda = i \cdot \omega, \tau) + Q_m(\lambda = i \cdot \omega, \tau) \\
& = C_0 - (\Theta_2 + \mathrm{B}_2) \cdot \omega^2 - (\mathrm{A}_2 + C_2) \cdot \omega^2 \\
& \quad + (\Theta_2 + \mathrm{B}_2) \cdot \omega^4 + i \cdot [\omega \cdot \mathrm{B}_1 + \omega \cdot (\mathrm{A}_1 + C_1) \\
& \quad - (\Theta_2 + \mathrm{B}_2) \cdot \omega^3 - \mathrm{A}_3 \cdot \omega^3 + \Theta_5 \cdot \omega^5] \neq 0
\end{aligned}
$$

$$
\begin{aligned}
& p_n(\lambda = i \cdot \omega, \tau) + Q_m(\lambda = i \cdot \omega, \tau) \\
& = C_0 - (\Theta_2 + \mathrm{B}_2 + \mathrm{A}_2 + C_2) \cdot \omega^2 \\
& \quad + (\Theta_2 + \mathrm{B}_2) \cdot \omega^4 + i \cdot [\omega \cdot (\mathrm{A}_1 + C_1 + \mathrm{B}_1) \\
& \quad - (\Theta_2 + \mathrm{B}_2 + \mathrm{A}_3) \cdot \omega^3 + \Theta_5 \cdot \omega^5] \neq 0
\end{aligned}
$$

$$
\begin{aligned}
|P(i \cdot \omega, \tau)|^2 &= [-(\Theta_2 + \mathrm{B}_2) \cdot \omega^2 + (\Theta_2 + \mathrm{B}_2) \cdot \omega^4]^2 \\
& \quad + [\omega \cdot \mathrm{B}_1 - (\Theta_2 + \mathrm{B}_2) \cdot \omega^3 + \Theta_5 \cdot \omega^5]^2 \\
&= (\Theta_2 + \mathrm{B}_2)^2 \cdot \omega^4 + (\Theta_2 + \mathrm{B}_2)^2 \cdot \omega^8 \\
& \quad - 2 \cdot (\Theta_2 + \mathrm{B}_2)^2 \cdot \omega^6 + \omega^2 \cdot \mathrm{B}_1^2 - \mathrm{B}_1 \cdot (\Theta_2 + \mathrm{B}_2) \cdot \omega^4 \\
& \quad + \mathrm{B}_1 \cdot \Theta_5 \cdot \omega^6 - (\Theta_2 + \mathrm{B}_2) \cdot \mathrm{B}_1 \cdot \omega^4 \\
& \quad + (\Theta_2 + \mathrm{B}_2)^2 \cdot \omega^6 - (\Theta_2 + \mathrm{B}_2) \cdot \Theta_5 \cdot \omega^8 + \Theta_5 \cdot \mathrm{B}_1 \cdot \omega^6 \\
& \quad - \Theta_5 \cdot (\Theta_2 + \mathrm{B}_2) \cdot \omega^8 + \Theta_5^2 \cdot \omega^{10}
\end{aligned}
$$

$$
\begin{aligned}
|P(i \cdot \omega, \tau)|^2 &= \omega^2 \cdot \mathrm{B}_1^2 + [(\Theta_2 + \mathrm{B}_2) - 2 \cdot \mathrm{B}_1] \cdot (\Theta_2 + \mathrm{B}_2) \cdot \omega^4 \\
& \quad + [2 \cdot \mathrm{B}_1 \cdot \Theta_5 - (\Theta_2 + \mathrm{B}_2)^2] \cdot \omega^6 + [(\Theta_2 + \mathrm{B}_2) - 2 \cdot \Theta_5] \cdot (\Theta_2 + \mathrm{B}_2) \cdot \omega^8 + \Theta_5^2 \cdot \omega^{10}
\end{aligned}
$$

$$
\begin{aligned}
|Q(i \cdot \omega, \tau)|^2 &= [C_0 - (\mathrm{A}_2 + C_2) \cdot \omega^2]^2 + [\omega \cdot (\mathrm{A}_1 + C_1) - \mathrm{A}_3 \cdot \omega^3]^2 = C_0^2 + (\mathrm{A}_2 + C_2)^2 \cdot \omega^4 \\
& \quad - 2 \cdot C_0 \cdot (\mathrm{A}_2 + C_2) \cdot \omega^2 + \omega^2 \cdot (\mathrm{A}_1 + C_1)^2 + \mathrm{A}_3^2 \cdot \omega^6 - 2 \cdot (\mathrm{A}_1 + C_1) \cdot \mathrm{A}_3 \cdot \omega^4
\end{aligned}
$$

$$
\begin{aligned}
|Q(i \cdot \omega, \tau)|^2 &= C_0^2 + [(\mathrm{A}_1 + C_1)^2 - 2 \cdot C_0 \cdot (\mathrm{A}_2 + C_2)] \cdot \omega^2 \\
& \quad + [(\mathrm{A}_2 + C_2)^2 - 2 \cdot (\mathrm{A}_1 + C_1) \cdot \mathrm{A}_3] \cdot \omega^4 + \mathrm{A}_3^2 \cdot \omega^6
\end{aligned}
$$

$$
\begin{aligned}
F(\omega, \tau) &= |P(i \cdot \omega, \tau)|^2 - |Q(i \cdot \omega, \tau)|^2 = \omega^2 \cdot \mathrm{B}_1^2 \\
& \quad + [(\Theta_2 + \mathrm{B}_2) - 2 \cdot \mathrm{B}_1] \cdot (\Theta_2 + \mathrm{B}_2) \cdot \omega^4 \\
& \quad + [2 \cdot \mathrm{B}_1 \cdot \Theta_5 - (\Theta_2 + \mathrm{B}_2)^2] \cdot \omega^6 + [(\Theta_2 + \mathrm{B}_2) \\
& \quad - 2 \cdot \Theta_5] \cdot (\Theta_2 + \mathrm{B}_2) \cdot \omega^8 + \Theta_5^2 \cdot \omega^{10} \\
& \quad - \{C_0^2 + [(\mathrm{A}_1 + C_1)^2 - 2 \cdot C_0 \cdot (\mathrm{A}_2 + C_2)] \cdot \omega^2 \\
& \quad + [(\mathrm{A}_2 + C_2)^2 - 2 \cdot (\mathrm{A}_1 + C_1) \cdot \mathrm{A}_3] \cdot \omega^4 + \mathrm{A}_3^2 \cdot \omega^6\}
\end{aligned}
$$

$$
\begin{aligned}
F(\omega,\tau) = |P(i\cdot\omega,\tau)|^2 - |Q(i\cdot\omega,\tau)|^2 = -C_0^2 \\
&+ \{\mathrm{B}_1^2 - [(\mathrm{A}_1 + C_1)^2 - 2\cdot C_0\cdot(\mathrm{A}_2 + C_2)]\}\cdot\omega^2 \\
&+ \{[(\Theta_2 + \mathrm{B}_2) - 2\cdot\mathrm{B}_1]\cdot(\Theta_2 + \mathrm{B}_2) \\
&- [(\mathrm{A}_2 + C_2)^2 - 2\cdot(\mathrm{A}_1 + C_1)\cdot\mathrm{A}_3]\}\cdot\omega^4 \\
&+ \{[2\cdot\mathrm{B}_1\cdot\Theta_5 - (\Theta_2 + \mathrm{B}_2)^2] - \mathrm{A}_3^2\}\cdot\omega^6 \\
&+ [(\Theta_2 + \mathrm{B}_2) - 2\cdot\Theta_5]\cdot(\Theta_2 + \mathrm{B}_2)\cdot\omega^8 + \Theta_5^2\cdot\omega^{10}
\end{aligned}
$$

We define the following parameters for simplicity: $\Pi_0, \Pi_2, \Pi_4, \Pi_6, \Pi_8, \Pi_{10}$

$$
\begin{aligned}
&\Pi_0 = -C_0^2; \Pi_2 = \mathrm{B}_1^2 - [(\mathrm{A}_1 + C_1)^2 - 2\cdot C_0\cdot(\mathrm{A}_2 + C_2)] \\
&\Pi_4 = [(\Theta_2 + \mathrm{B}_2) - 2\cdot\mathrm{B}_1]\cdot(\Theta_2 + \mathrm{B}_2) - [(\mathrm{A}_2 + C_2)^2 - 2\cdot(\mathrm{A}_1 + C_1)\cdot\mathrm{A}_3] \\
&\Pi_6 = [2\cdot\mathrm{B}_1\cdot\Theta_5 - (\Theta_2 + \mathrm{B}_2)^2] - \mathrm{A}_3^2; \Pi_8 \\
&\quad = [(\Theta_2 + \mathrm{B}_2) - 2\cdot\Theta_5]\cdot(\Theta_2 + \mathrm{B}_2); \Pi_{10} = \Theta_5^2
\end{aligned}
$$

Hence $F(\omega,\tau) = 0$ implies $\sum_{k=0}^{5} \Pi_{2\cdot k}\cdot\omega^{2\cdot k} = 0$ and its roots are given by solving the above polynomial.

$$
\begin{aligned}
P_R(i\cdot\omega,\tau) &= -(\Theta_2 + \mathrm{B}_2)\cdot\omega^2 + (\Theta_2 + \mathrm{B}_2)\cdot\omega^4 \\
P_I(i\cdot\omega,\tau) &= \omega\cdot\mathrm{B}_1 - (\Theta_2 + \mathrm{B}_2)\cdot\omega^3 + \Theta_5\cdot\omega^5; \\
Q_R(i\cdot\omega,\tau) &= C_0 - (\mathrm{A}_2 + C_2)\cdot\omega^2 \\
Q_I(i\cdot\omega,\tau) &= \omega\cdot(\mathrm{A}_1 + C_1) - \mathrm{A}_3\cdot\omega^3
\end{aligned}
$$

Hence

$$
\sin\theta(\tau) = \frac{-P_R(i\cdot\omega,\tau)\cdot Q_I(i\cdot\omega,\tau) + P_I(i\cdot\omega,\tau)\cdot Q_R(i\cdot\omega,\tau)}{|Q(i\cdot\omega,\tau)|^2}
$$

$$
\cos\theta(\tau) = -\frac{P_R(i\cdot\omega,\tau)\cdot Q_R(i\cdot\omega,\tau) + P_I(i\cdot\omega,\tau)\cdot Q_I(i\cdot\omega,\tau)}{|Q(i\cdot\omega,\tau)|^2}
$$

We use different parameters terminology from our last characteristics parameters definition: $k \to j; p_k(\tau) \to a_j; q_k(\tau) \to c_j; n = 5; m = 3; n > m$

Additionally $P_n(\lambda,\tau) \to P(\lambda); Q_m(\lambda,\tau) \to Q(\lambda)$ Then $P(\lambda) = \sum_{j=0}^{5} a_j\cdot\lambda^j; Q(\lambda) = \sum_{j=0}^{3} c_j\cdot\lambda^j$

$$P_\lambda = a_0 + a_1 \cdot \lambda + a_2 \cdot \lambda^2 + a_3 \cdot \lambda^3 + a_4 \cdot \lambda^4 + a_5 \cdot \lambda^5;$$
$$Q_\lambda = c_0 + c_1 \cdot \lambda + c_2 \cdot \lambda^2 + c_3 \cdot \lambda^3$$

$n, m \in \mathbb{N}_0$, $n > m$ and $a_j, c_j : \mathrm{R}_{+0} \to R$ are continuous and differentiable function of τ such that $a_0 + c_0 \neq 0$. In the following "—" denotes complex and conjugate. $P(\lambda), Q(\lambda)$ are analytic functions in λ and differentiable in τ.

The coefficients $\{a_j(L_P, L_1, C_f, R_{in}, R_s, C_P, R_j, \tau, \ldots)$ and $c_j(L_P, L_1, C_f, R_{in}, R_s, C_P, R_j, \tau, \ldots)\} \in \mathbb{R}$ depend on RFID TAG detector system's $L_P, L_1, C_f, R_{in}, R_s, C_P, R_j, \tau, \ldots$ values. Unless strictly necessary, the designation of the varied arguments: $(L_P, L_1, C_f, R_{in}, R_s, C_P, R_j, \tau, \ldots)$ will subsequently be omitted from P, Q, a_j, c_j. The coefficients a_j, c_j are continuous, and differentiable functions of their arguments, and direct substitution shows that $a_0 + c_0 \neq 0$; $a_0 = 0; c_0(\tau) = C_0$

$$\begin{aligned} C_0 &= -\sigma_2 \cdot \psi_3 - \sigma_3 \cdot \sigma_5 \cdot \psi_5 - \psi_1 \cdot \sigma_4 \cdot \sigma_7 \\ &\to -\sigma_2 \cdot \psi_3 - \sigma_3 \cdot \sigma_5 \cdot \psi_5 - \psi_1 \cdot \sigma_4 \cdot \sigma_7 \neq 0. \\ &- [\frac{R_{in}^2}{L_1 \cdot L_P} - \frac{1}{C_1 \cdot L_P} - \frac{1}{C_P \cdot L_P}] \cdot \psi_3 \\ &+ \frac{R_{in}^2}{L_1 \cdot L_P} \cdot \frac{R_{in}}{L_1} \cdot \psi_5 - \psi_1 \cdot \frac{1}{C_P \cdot L_P} \cdot \frac{1}{R_S \cdot C_P} \neq 0 \\ &\forall L_P, L_1, C_f, R_{in}, R_s, C_P, R_j, \tau, \ldots \in \mathbb{R}_+ \end{aligned}$$

i.e. $\lambda = 0$ is not a root of the characteristic equation. Furthermore $P(\lambda)$, $Q(\lambda)$ are analytic functions of λ for which the following requirements of the analysis (see Kuang 1993, Sect. 3.4) can also be verified in the present case [6, 7].

(a) If $\lambda = i \cdot \omega$, $\omega \in \mathbb{R}$ then $P(i \cdot \omega) + Q(i \cdot \omega) \neq 0$, i.e. P and Q have no common imaginary roots. This condition was verified numerically in the entire $(L_P, L_1, C_f, R_{in}, R_s, C_P, R_j, \tau, \ldots)$ domain of interest.

(b) $|Q(\lambda)/P(\lambda)|$ is bounded for $|\lambda| \to \infty$, $\mathrm{Re}\lambda \geq 0$. No roots bifurcation from ∞. Indeed, in the limit:

$$|\frac{Q(\lambda)}{P(\lambda)}| = |\frac{c_0 + c_1 \cdot \lambda + c_2 \cdot \lambda^2 + c_3 \cdot \lambda^3}{a_0 + a_1 \cdot \lambda + a_2 \cdot \lambda^2 + a_3 \cdot \lambda^3 + a_4 \cdot \lambda^4 + a_5 \cdot \lambda^5}|$$

(c)
$$F(\omega) = |P(i \cdot \omega)|^2 - |Q(i \cdot \omega)|^2;$$
$$F(\omega, \tau) = |P(i \cdot \omega, \tau)|^2 - |Q(i \cdot \omega, \tau)|^2 = \sum_{k=0}^{5} \Pi_{2 \cdot k} \cdot \omega^{2 \cdot k}$$

Has at most a finite number of zeros. Indeed, this is a polynomial in ω (Degree in ω^{10}).

(d) Each positive root $\omega(L_P, L_1, C_f, R_{in}, R_s, C_P, R_j, \tau, \ldots)$ of $F(\omega) = 0$ is continuous and differentiable with respect to $L_P, L_1, C_f, R_{in}, R_s, C_P, R_j, \tau, \ldots$.

The condition can only be assessed numerically.

In addition, since the coefficients in P and Q are real, we have $\overline{P(-i \cdot \omega)} = P(i \cdot \omega)$, and $\overline{Q(-i \cdot \omega)} = Q(i \cdot \omega)$ thus, $\omega > 0$ maybe on eigenvalue of characteristic equations. The analysis consists in identifying the roots of the characteristic equation situated on the imaginary axis of the complex λ-plane, whereby increasing the parameters $L_P, L_1, C_f, R_{in}, R_s, C_P, R_j, \tau, \ldots$, Reλ may, at the crossing, change its sign from (−) to (+), i.e. from a stable focus $E^{(0)}(X^{(0)}, Y^{(0)}, I_{L_1}^{(0)}, I_{R_j}^{(0)}, I_{R_S}^{(0)})|$ $V(t)|_{A_0 \gg |f(t)|} = A_0 + f(t) \approx A_0$ $= (0,0,0,0,0)$ to an unstable
$\frac{dV(t)}{dt}|_{A_0 \gg |f(t)|} = \frac{df(t)}{dt} \to \varepsilon$
$A_0 \to \varepsilon$

one, or vice versa. This feature may be further assessed by examining the sign of the partial derivatives with respect to $L_P, L_1, C_f, R_{in}, R_s, C_P, R_j, \tau, \ldots$ and gate antenna parameters.

$$\wedge^{-1}(L_P) = \left(\frac{\partial \mathrm{Re}\lambda}{\partial L_P}\right)_{\lambda = i\cdot\omega}, L_1, C_f, R_{in}, R_s, C_P, R_j, \tau, \ldots = const;$$

$$\wedge^{-1}(L_1) = \left(\frac{\partial \mathrm{Re}\lambda}{\partial L_1}\right)_{\lambda = i\cdot\omega}, L_P, C_f, R_{in}, R_s, C_P, R_j, \tau, \ldots = const$$

$$\wedge^{-1}(C_f) = \left(\frac{\partial \mathrm{Re}\lambda}{\partial C_f}\right)_{\lambda = i\cdot\omega}, L_P, L_1, R_{in}, R_s, C_P, R_j, \tau, \ldots = const;$$

$$\wedge^{-1}(R_{in}) = \left(\frac{\partial \mathrm{Re}\lambda}{\partial R_{in}}\right)_{\lambda = i\cdot\omega}, L_P, L_1, C_f, R_s, C_P, R_j, \tau, \ldots = const$$

$$\wedge^{-1}(R_{in}) = \left(\frac{\partial \mathrm{Re}\lambda}{\partial R_{in}}\right)_{\lambda = i\cdot\omega}, L_P, L_1, C_f, R_s, C_P, R_j, \tau, \ldots = const;$$

$$\wedge^{-1}(R_s) = \left(\frac{\partial \mathrm{Re}\lambda}{\partial R_s}\right)_{\lambda = i\cdot\omega}, L_P, L_1, C_f, R_{in}, C_P, R_j, \tau, \ldots = const$$

$$\wedge^{-1}(C_P) = \left(\frac{\partial \mathrm{Re}\lambda}{\partial C_P}\right)_{\lambda = i\cdot\omega}, L_P, L_1, C_f, R_{in}, R_s, R_j, \tau, \ldots = const;$$

$$\wedge^{-1}(R_j) = \left(\frac{\partial \mathrm{Re}\lambda}{\partial R_j}\right)_{\lambda = i\cdot\omega}, L_P, L_1, C_f, R_{in}, R_s, C_P, \tau, \ldots = const$$

$$\wedge^{-1}(\tau) = \left(\frac{\partial \mathrm{Re}\lambda}{\partial \tau}\right)_{\lambda = i\cdot\omega}, L_P, L_1, C_f, R_{in}, R_s, C_P, R_j, \ldots = const; \omega \in \mathbb{R}_+.$$

When writing $P(\lambda) = P_R(\lambda) + i \cdot P_I(\lambda)$ and $Q(\lambda) = Q_R(\lambda) + i \cdot Q_I(\lambda)$, and inserting $\lambda = i \cdot \omega$ into active RFID Schottky detector system's characteristic equation ω must satisfy the following:

$$\sin(\omega \cdot \tau) = g(\omega) = \frac{-P_R(i\cdot\omega)\cdot Q_I(i\cdot\omega) + P_I(i\cdot\omega)\cdot Q_R(i\cdot\omega)}{|Q(i\cdot\omega)|^2}$$

$$\cos(\omega \cdot \tau) = h(\omega) = -\frac{P_R(i\cdot\omega)\cdot Q_R(i\cdot\omega) + P_I(i\cdot\omega)\cdot Q_I(i\cdot\omega)}{|Q(i\cdot\omega)|^2}$$

Where $|Q(i \cdot \omega)|^2 \neq 0$ in view of requirement (a) above, and $(g, h) \in R$. Furthermore, it follows above $\sin \omega \cdot \tau$ and $\cos \omega \cdot \tau$ equations that, by squaring and adding the sides, ω must be a positive root of $F(\omega) = |P(i \cdot \omega)|^2 - |Q(i \cdot \omega)|^2 = 0$. Note: $F(\omega)$ is dependent on τ. Now it is important to notice that if $\tau \notin I$ (assume that $I \subseteq R_{+0}$ is the set where $\omega(\tau)$ is a positive root of $F(\omega)$ and for, $\tau \notin I$, $\omega(\tau)$ is not defined. Then for all τ in I $\omega(\tau)$ is satisfied that $F(\omega, \tau) = 0$). Then there are no positive $\omega(\tau)$ solutions for $F(\omega, \tau) = 0$, and we cannot have stability switches. For $\tau \in I$ where $\omega(\tau)$ is a positive solution of $F(\omega, \tau) = 0$, we can define the angle $\theta(\tau) \in [0, 2 \cdot \pi]$ as the solution of

$$\sin \theta(\tau) = \ldots; \cos \theta(\tau) = \ldots;$$

$$\sin \theta(\tau) = \frac{-P_R(i \cdot \omega) \cdot Q_I(i \cdot \omega) + P_I(i \cdot \omega) \cdot Q_R(i \cdot \omega)}{|Q(i \cdot \omega)|^2};$$

$$\cos \theta(\tau) = -\frac{P_R(i \cdot \omega) \cdot Q_R(i \cdot \omega) + P_I(i \cdot \omega) \cdot Q_I(i \cdot \omega)}{|Q(i \cdot \omega)|^2}.$$

And the relation between the argument $\theta(\tau)$ and $\omega(\tau) \cdot \tau$ for $\tau \in I$ must be $\omega(\tau) \cdot \tau = \theta(\tau) + n \cdot 2 \cdot \pi \; \forall \, n \in \mathbb{N}_0$. Hence we can define the maps $\tau_n : I \to R_{+0}$ given by $\tau_n(\tau) = \frac{\theta(\tau) + n \cdot 2 \cdot \pi}{\omega(\tau)}; n \in \mathbb{N}_0, \tau \in I$. Let us introduce the functions $I \to R$; $S_n(\tau) = \tau - \tau_n(\tau)$, $\tau \in I$, $n \in \mathbb{N}_0$ that is continuous and differentiable in τ. In the following, the subscripts $\lambda, \omega, L_P, L_1, C_f, R_{in}, R_s, C_P, R_j, \ldots$ indicate the corresponding partial derivatives. Let us first concentrate on $\wedge(x)$, remember in $\lambda(L_P, L_1, C_f, R_{in}, R_s, C_P, R_j, \ldots)$ and $\omega(L_P, L_1, C_f, R_{in}, R_s, C_P, R_j, \ldots)$, and keeping all parameters except one (x) and τ. The derivation closely follows that in reference [BK]. Differentiating RFID TAG detector system characteristic equation $P(\lambda) + Q(\lambda) \cdot e^{-\lambda \cdot \tau} = 0$ with respect to specific parameter (x), and inverting the derivative, for convenience, one calculates:

Remark

$$x = L_P, L_1, C_f, R_{in}, R_s, C_P, R_j, \ldots, etc.,$$

$$\left(\frac{\partial \lambda}{\partial x}\right)^{-1} = \frac{-P_\lambda(\lambda, x) \cdot Q(\lambda, x) + Q_\lambda(\lambda, x) \cdot P(\lambda, x) - \tau \cdot P(\lambda, x) \cdot Q(\lambda, x)}{P_x(\lambda, x) \cdot Q(\lambda, x) - Q_x(\lambda, x) \cdot P(\lambda, x)}$$

Where $P_\lambda = \frac{\partial P}{\partial \lambda}, \ldots$ etc., substituting $\lambda = i \cdot \omega$ and bearing $\overline{P(-i \cdot \omega)} = P(i \cdot \omega)$; $\overline{Q(-i \cdot \omega)} = Q(i \cdot \omega)$ then $i \cdot P_\lambda(i \cdot \omega) = P_\omega(i \cdot \omega)$; $i \cdot Q_\lambda(i \cdot \omega) = Q_\omega(i \cdot \omega)$ and that on the surface $|P(i \cdot \omega)|^2 = |Q(i \cdot \omega)|^2$, one obtains:

$$\left(\frac{\partial \lambda}{\partial x}\right)^{-1}\Big|_{\lambda = i \cdot \omega} = \left(\frac{i \cdot P_\omega(i \cdot \omega, x) \cdot \overline{P(i \cdot \omega, x)} + i \cdot Q_\lambda(i \cdot \omega, x) \cdot \overline{Q(\lambda, x)} - \tau \cdot |P(i \cdot \omega, x)|^2}{P_x(i \cdot \omega, x) \cdot \overline{P(i \cdot \omega, x)} - Q_x(i \cdot \omega, x) \cdot \overline{Q(i \cdot \omega, x)}}\right)$$

Upon separating into real and imaginary parts, with $P = P_R + i \cdot P_I$; $Q = Q_R + i \cdot Q_I$

$$P_\omega = P_{R\omega} + i \cdot P_{I\omega}; Q_\omega = Q_{R\omega} + i \cdot Q_{I\omega}; P_x = P_{Rx} + i \cdot P_{Ix};$$
$$Q_x = Q_{Rx} + i \cdot Q_{Ix}; P^2 = P_R^2 + P_I^2.$$

When (x) can be any RFID Schottky detector parameter's $L_P, L_1, C_f, R_{in}, \ldots$ and time delay τ etc. Where for convenience, we have dropped the arguments $(i \cdot \omega, x)$, and where

$$F_\omega = 2 \cdot [(P_{R\omega} \cdot P_R + P_{I\omega} \cdot P_I) - (Q_{R\omega} \cdot Q_R + Q_{I\omega} \cdot Q_I)]$$
$$F_x = 2 \cdot [(P_{Rx} \cdot P_R + P_{Ix} \cdot P_I) - (Q_{Rx} \cdot Q_R + Q_{Ix} \cdot Q_I)]\omega_x = -F_x/F_\omega.$$

We define U and V:

$$U = (P_R \cdot P_{I\omega} - P_I \cdot P_{R\omega}) - (Q_R \cdot Q_{I\omega} - Q_I \cdot Q_{R\omega});$$
$$V = (P_R \cdot P_{Ix} - P_I \cdot P_{Rx}) - (Q_R \cdot Q_{Ix} - Q_I \cdot Q_{Rx})$$

We choose our specific parameter as time delay x = τ.

$$Q_I = \omega \cdot (A_1 + C_1) - A_3 \cdot \omega^3$$
$$P_R = -(\Theta_2 + B_2) \cdot \omega^2 + (\Theta_2 + B_2) \cdot \omega^4; P_I = \omega \cdot B_1 - (\Theta_2 + B_2) \cdot \omega^3 + \Theta_5 \cdot \omega^5;$$
$$Q_R = C_0 - (A_2 + C_2) \cdot \omega^2$$
$$P_{R\omega} = 4 \cdot (\Theta_2 + B_2) \cdot \omega^3 - 2 \cdot (\Theta_2 + B_2) \cdot \omega = 2 \cdot (\Theta_2 + B_2) \cdot \omega \cdot (2 \cdot \omega^2 - 1)$$
$$P_{I\omega} = B_1 - 3 \cdot (\Theta_2 + B_2) \cdot \omega^2 + 5 \cdot \Theta_5 \cdot \omega^4;$$
$$Q_{R\omega} = -2 \cdot (A_2 + C_2) \cdot \omega; Q_{I\omega} = (A_1 + C_1) - 3 \cdot A_3 \cdot \omega^2$$
$$P_{R\tau} = 0; P_{I\tau} = 0; Q_{R\tau} = 0; Q_{I\tau} = 0; \omega_\tau = -F_\tau/F_\omega$$
$$\begin{aligned} P_{R\omega} \cdot P_R &= 2 \cdot (\Theta_2 + B_2) \cdot \omega \cdot (2 \cdot \omega^2 - 1) \cdot [(\Theta_2 + B_2) \cdot \omega^4 - (\Theta_2 + B_2) \cdot \omega^2] \\ &= 2 \cdot (\Theta_2 + B_2) \cdot \omega \cdot (2 \cdot \omega^2 - 1) \cdot (\Theta_2 + B_2) \cdot \omega^2 \cdot [\omega^2 - 1] \\ &= 2 \cdot (\Theta_2 + B_2)^2 \cdot \omega^3 \cdot (2 \cdot \omega^2 - 1) \cdot [\omega^2 - 1] \end{aligned}$$
$$P_{R\omega} \cdot P_R = 2 \cdot (\Theta_2 + B_2)^2 \cdot \omega^3 \cdot (2 \cdot \omega^2 - 1) \cdot [\omega^2 - 1];$$
$$Q_{R\omega} \cdot Q_R = -2 \cdot (A_2 + C_2) \cdot \omega \cdot [C_0 - (A_2 + C_2) \cdot \omega^2]$$
$$F_\tau = 2 \cdot [(P_{R\tau} \cdot P_R + P_{I\tau} \cdot P_I) - (Q_{R\tau} \cdot Q_R + Q_{I\tau} \cdot Q_I)] = 0;$$
$$P_R \cdot P_{I\omega} = (\Theta_2 + B_2) \cdot \omega^2 \cdot (\omega^2 - 1) \cdot [B_1 - 3 \cdot (\Theta_2 + B_2) \cdot \omega^2 + 5 \cdot \Theta_5 \cdot \omega^4]$$
$$P_I \cdot P_{R\omega} = 2 \cdot \omega^2 \cdot [B_1 - (\Theta_2 + B_2) \cdot \omega^2 + \Theta_5 \cdot \omega^4] \cdot (\Theta_2 + B_2) \cdot (2 \cdot \omega^2 - 1).$$
$$Q_R \cdot Q_{I\omega} = [C_0 - (A_2 + C_2) \cdot \omega^2] \cdot [(A_1 + C_1) - 3 \cdot A_3 \cdot \omega^2];$$
$$Q_I \cdot Q_{R\omega} = -2 \cdot \omega^2 \cdot [(A_1 + C_1) - A_3 \cdot \omega^2] \cdot (A_2 + C_2)$$
$$V = (P_R \cdot P_{I\tau} - P_I \cdot P_{R\tau}) - (Q_R \cdot Q_{I\tau} - Q_I \cdot Q_{R\tau}) = 0. F(\omega, \tau) = 0$$

Differentiating with respect to τ and we get

$$F_\omega \cdot \frac{\partial\omega}{\partial\tau} + F_\tau = 0; \tau \in I \Rightarrow \frac{\partial\omega}{\partial\tau} = -\frac{F_\tau}{F_\omega}; \wedge^{-1}(\tau) = (\frac{\partial \mathrm{Re}\lambda}{\partial\tau})_{\lambda=i\cdot\omega}; \frac{\partial\omega}{\partial\tau} = \omega_\tau = -\frac{F_\tau}{F_\omega}$$

$$\wedge^{-1}(\tau) = \mathrm{Re}\{\frac{-2\cdot[U+\tau\cdot|P|^2]+i\cdot F_\omega}{F_\tau + i\cdot 2\cdot[V+\omega\cdot|P|^2]}\}; sign\{\wedge^{-1}(\tau)\} = sign\{(\frac{\partial \mathrm{Re}\lambda}{\partial\tau})_{\lambda=i\cdot\omega}\}$$

$$sign\{\wedge^{-1}(\tau)\} = sign\{F_\omega\}\cdot sign\{\tau\cdot\frac{\partial\omega}{\partial\tau} + \omega + \frac{U\cdot\frac{\partial\omega}{\partial\tau}+V}{|P|^2}\}.$$

We shall presently examine the possibility of stability transitions (bifurcations) RFID TAG detector system, about the equilibrium point $E^{(0)}(X^{(0)}, Y^{(0)}, I_{L_1}^{(0)}, I_{R_j}^{(0)}, I_{R_S}^{(0)}) = (0,0,0,0,0)$ as a result of a variation of delay parameter τ. The analysis consists in identifying the roots of our system characteristic equation situated on the imaginary axis of the complex λ-plane.

Where by increasing the delay parameter τ, Re λ may at the crossing, changes its sign from − to +, i.e. from a stable focus $E^{(*)}$ to an unstable one, or vice versa. This feature may be further assessed by examining the sign of the partial derivatives with respect to τ,

$$\wedge^{-1}(\tau) = (\frac{\partial \mathrm{Re}\lambda}{\partial\tau})_{\lambda=i\cdot\omega}$$

$$\wedge^{-1}(\tau) = (\frac{\partial \mathrm{Re}\lambda}{\partial\tau})_{\lambda=i\cdot\omega}, L_P, L_1, C_f, R_{in}, R_s, C_P, R_j, \ldots = const; \omega \in \mathbb{R}_+.$$

$$\begin{aligned} U &= (P_R\cdot P_{I\omega} - P_I\cdot P_{R\omega}) - (Q_R\cdot Q_{I\omega} - Q_I\cdot Q_{R\omega}) \\ &= (\Theta_2 + \mathrm{B}_2)\cdot\omega^2\cdot(\omega^2-1)\cdot[\mathrm{B}_1 - 3\cdot(\Theta_2+\mathrm{B}_2)\cdot\omega^2 + 5\cdot\Theta_5\cdot\omega^4] \\ &\quad - 2\cdot\omega^2\cdot[\mathrm{B}_1 - (\Theta_2+\mathrm{B}_2)\cdot\omega^2 + \Theta_5\cdot\omega^4]\cdot(\Theta_2+\mathrm{B}_2)\cdot(2\cdot\omega^2-1) \\ &\quad - [C_0 - (\mathrm{A}_2 + C_2)\cdot\omega^2]\cdot[(\mathrm{A}_1+C_1) - 3\cdot\mathrm{A}_3\cdot\omega^2] \\ &\quad - 2\cdot\omega^2\cdot[(\mathrm{A}_1+C_1) - \mathrm{A}_3\cdot\omega^2]\cdot(\mathrm{A}_2+C_2) \end{aligned}$$

The single diode detector, R_L is the video load resistance which not seen in RFID TAG receiver detector equivalent circuit. L_1, the shunt inductance, provides a current return path for the diode, and is chosen to be large compared to diode impedance at the input or RF frequency. C_1, the bypass capacitance, is chosen to be sufficiently large that is capacitive reactance is small compared to the diode impedance, but small enough to avoid having it resistance load the video circuit. P_{in} is the RF input power applied to the detector circuit and V_O is the output voltage appearing across R_L. L_P is packaged parasitic inductance (Schottky linear equivalent circuit). C_P is package parasitic capacitance. R_S is the diode's parasitic series resistance. C_j is junction parasitic capacitance, and R_j is the diode's junction resistance. L_P, C_P, and R_L are constants. R_S has some small variation with temperature, but that variation is not a significant parameter in this analysis. C_j is a function of both temperature and DC bias, but this analysis concerns itself with the zero bias detectors and the variation with temperature is not significant. R_j is a key element in equivalent circuit—its behavior clearly will affect the performance of the detector circuit. For our stability switching analysis, we choose typical Schottky detector parameter values: $L_P = 2$ nH, $R_S = 1.5\ \Omega$, $C_P = 0.08$ pF, $C_j = 0.2$ pF, $R_j = 500\ \Omega$, $R_L = 100\ K\Omega$, $R_{in} = 1\ K\Omega$, $L_1 = 1$ mH, $C_1 = 1\ \mu F$

$$\sigma_1 = -5 \times 10^{11}; \sigma_2 = -6.2492 \times 10^{21}; \sigma_3 = 5 \times 10^{17};$$
$$\sigma_4 = 6.25 \times 10^{21}; \sigma_5 = -10^6; \sigma_6 = 10^{10}$$
$$\sigma_7 = 8.33 \times 10^{12}; \sigma_8 = 3.33 \times 10^{12}; \sigma_9 = -1.155 \times 10^{13};$$
$$\psi_1 = -10^{16}; \psi_2 = -1.0001 \times 10^{10}$$
$$\psi_3 = -8.22 \times 10^{28}; \psi_4 = -8.2212 \times 10^{22}; \psi_5 = -8.22 \times 10^{22};$$
$$\Theta_2 = -5 \times 10^{27}; \Theta_3 = -5.0005 \times 10^{21}$$
$$\Theta_4 = -5.1 \times 10^{11}; \Theta_5 = -1; A_1 = -6.2497 \times 10^{37};$$
$$A_2 = -6.2498 \times 10^{31}; A_3 = -6.2492 \times 10^{21}$$
$$B_1 = -4.11 \times 10^{40}; B_2 = -4.1106 \times 10^{34};$$
$$B_3 = -5.8572 \times 10^{24}; B_4 = -1.155 \times 10^{13}$$
$$C_0 = 6.8997 \times 10^{48}; C_1 = 6.9178 \times 10^{42};$$
$$C_2 = -2.0116 \times 10^{34}; \Pi_0 = -4.7606 \times 10^{97}$$
$$\Pi_2 = -4.8132 \times 10^{85}; \Pi_4 = -3.3789 \times 10^{75};$$
$$\Pi_6 = -1.6897 \times 10^{69}; \Pi_8 = 1.6897 \times 10^{69}; \Pi_{10} = 1$$

Then we get the expression for $F(\omega,\tau)$ Schottky diode detector parameter values. We find those ω, τ values which fulfill $F(\omega,\tau) = 0$. We ignore negative, complex, and imaginary values of ω for specific τ values. $\tau \in [0.001\ldots10]$, we can be express by 3D function $F(\omega,\tau) = 0$. We plot the stability switch diagram based on different delay values of our Schottky diode detector.

$$\wedge^{-1}(\tau) = \left(\frac{\partial \mathrm{Re}\lambda}{\partial \tau}\right)_{\lambda=i\cdot\omega} = \mathrm{Re}\left\{\frac{-2\cdot[U+\tau\cdot|P|^2]+i\cdot F_\omega}{F_\tau + i\cdot 2\cdot[V+\omega\cdot|P|^2]}\right\}$$
$$\wedge^{-1}(\tau) = \left(\frac{\partial \mathrm{Re}\lambda}{\partial \tau}\right)_{\lambda=i\cdot\omega} = \frac{2\cdot\{F_\omega\cdot(V+\omega\cdot P^2) - F_\tau\cdot(U+\tau\cdot P^2)\}}{F_\tau^2 + 4\cdot(V+\omega\cdot P^2)^2}$$

The stability switch occurs only on those delay values (τ) which fit the equation: $\tau = \frac{\theta_+(\tau)}{\omega_+(\tau)}$ and $\theta_+(\tau)$ is the solution of $\sin\theta(\tau) = \ldots; \cos\theta(\tau) = \ldots$ when $\omega = \omega_+(\tau)$ if only ω_+ is feasible. Additionally, when all Schottky diode detector's parameters are known and the stability switch due to various time delay values τ is described in the following expression:

$$sign\{\wedge^{-1}(\tau)\} = sign\{F_{\omega}(\omega(\tau),\tau)\} \cdot sign\{\tau \cdot \omega_{\tau}(\omega(\tau)) + \omega(\tau) + \frac{U(\omega(\tau)) \cdot \omega_{\tau}(\omega(\tau)) + V(\omega(\tau))}{|P(\omega(\tau))|^2}\}$$

Remark we know $F(\omega,\tau) = 0$ implies its roots $\omega_i(\tau)$ and finding those delays values τ which ω_i is feasible. There are τ values which give complex ω_i or imaginary number, then unable to analyze stability [6, 7]. F function is independent on τ the parameter $F(\omega) = 0$.

The results: We find those ω, τ values which fulfill $F(\omega,\tau) = 0$. We ignore negative, complex, and imaginary values of ω. We define new MATLAB script parameters: $\pi_{2k} \to G_{2k}$ (k = 0…5). Running a MATLAB script to find ω values, gives the following results:

$$F(\omega) = 0 \Rightarrow \omega_1 = 1.0\mathrm{e}+034*; \omega_2 = 0 + 4.1106\mathrm{i}; \omega_3 = 0 - 4.1106\mathrm{i}; \omega_4,\ldots,\omega_{11} = 0$$

MATLAB script: G0 = −4.7606 * 1e97; G2 = −4.8132 * 1e85; G4 = -3.3789 * 1e75; G6 = −1.6897 * 1e69; G8 = 1.6897 * 1e69; G10 = 1; p = [G10 0 G8 0 G6 0 G4 0 G2 0 G0]; r = roots(p).

Next is to find those ω, τ values which fulfil $\sin\theta(\tau) = \ldots$; $\sin(\omega \cdot \tau) = \frac{-P_R \cdot Q_I + P_I \cdot Q_R}{|Q|^2}$ and $\cos\theta(\tau) = \ldots$; $\cos(\omega \cdot \tau) = -\frac{(P_R \cdot Q_R + P_I \cdot Q_I)}{|Q|^2}$; $|Q|^2 = Q_R^2 + Q_I^2$. Finally, we plot the stability switch diagram

$$g(\tau) = \wedge^{-1}(\tau) = (\frac{\partial \mathrm{Re}\lambda}{\partial \tau})_{\lambda = i \cdot \omega}$$

$$g(\tau) = \wedge^{-1}(\tau) = (\frac{\partial \mathrm{Re}\lambda}{\partial \tau})_{\lambda = i \cdot \omega} = \frac{2 \cdot \{F_{\omega} \cdot (V + \omega \cdot P^2) - F_{\tau} \cdot (U + \tau \cdot P^2)\}}{F_{\tau}^2 + 4 \cdot (V + \omega \cdot P^2)^2}$$

$$sign[g(\tau)] = sign[\wedge^{-1}(\tau)] = sign[(\frac{\partial \mathrm{Re}\lambda}{\partial \tau})_{\lambda = i \cdot \omega}] = sign[\frac{2 \cdot \{F_{\omega} \cdot (V + \omega \cdot P^2) - F_{\tau} \cdot (U + \tau \cdot P^2)\}}{F_{\tau}^2 + 4 \cdot (V + \omega \cdot P^2)^2}]$$

Since $F_{\tau}^2 + 4 \cdot (V + \omega \cdot P^2)^2 > 0$ then

$$sign[\wedge^{-1}(\tau)] = sign\{F_\omega \cdot (V + \omega \cdot P^2) - F_\tau \cdot (U + \tau \cdot P^2)\}$$

$$sign[\wedge^{-1}(\tau)] = sign\{[F_\omega] \cdot [(V + \omega \cdot P^2) - \frac{F_\tau}{F_\omega} \cdot (U + \tau \cdot P^2)]\};$$

$$\omega_\tau = -\frac{F_\tau}{F_\omega}; \omega_\tau = (\frac{\partial\omega}{\partial\tau})^{-1} = -\frac{\partial F/\partial\omega}{\partial F/\partial\tau}$$

$$sign[\wedge^{-1}(\tau)] = sign\{[F_\omega] \cdot [V + \omega_\tau \cdot U + \omega \cdot P^2 + \omega_\tau \cdot \tau \cdot P^2]\};$$

$$sign[\wedge^{-1}(\tau)] = sign\{[F_\omega] \cdot [\frac{1}{P^2}] \cdot [\frac{V + \omega_\tau \cdot U}{P^2} + \omega + \omega_\tau \cdot \tau]\}$$

$$sign[\frac{1}{P^2}] > 0 \Rightarrow sign[\wedge^{-1}(\tau)] = sign\{[F_\omega] \cdot [\frac{V + \omega_\tau \cdot U}{P^2} + \omega + \omega_\tau \cdot \tau]\}$$

$$sign[\wedge^{-1}(\tau)] = sign[F_\omega] \cdot sign[\frac{V + \omega_\tau \cdot U}{P^2} + \omega + \omega_\tau \cdot \tau];$$

$$F_\omega = 2 \cdot [(P_{R\omega} \cdot P_R + P_{I\omega} \cdot P_I) - (Q_{R\omega} \cdot Q_R + Q_{I\omega} \cdot Q_I)]$$

We check the sign of $\wedge^{-1}(\tau)$ according the following rule:

$sign[F_\omega]$	$sign[\frac{V+\omega_\tau \cdot U}{P^2} + \omega + \omega_\tau \cdot \tau]$	$sign[\wedge^{-1}(\tau)]$
$\pm$	$\pm$	$+$
$\pm$	$\mp$	$-$

If $sign[\Lambda^{-1}(\tau)] > 0$ then the crossing proceeds from (−) to (+) respectively (stable to unstable). If $sign[\Lambda^{-1}(\tau)] < 0$ then the crossing proceeds from (+) to (−) respectively (unstable to stable). Anyway the stability switching can occur only for $\omega = 1.0e + 034$ or $\omega = 0$ [30, 32].

1.6 RFID System Burst Switch Stability Analysis Under Delayed Internal Diode Circuitry Parasitic Effects in Time

There are systems which converting Radio Frequency (RF) energy into a Direct Current (DC). In other areas, the circuit has been used to provide DC power to operate remote autonomous devices that have no on-board power supply. In the case of the part, a battery controlled by the burst switch is used to power the device. CMOS (silicon) devices are equipped with a form of sleep circuitry with a current draw at a minimum during sleep. An external input signal is used to wake-up the device. The use of the switch requires considerably more design and analysis to avoid false wake-up states and to ensure functionality under adverse conditions. A simple generic burst switch is constructed from input RFID rectangular spiral antenna, matching network, voltage doubler and load. The voltage doubler unit is constructed from two diodes D_1 and D_2 with parasitic effects, delay in time. One of

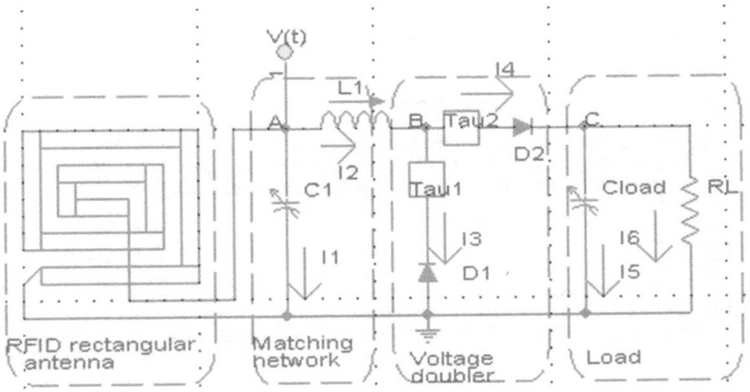

Fig. 1.23 A simple generic RFID burst switch diagram

the difficulties with the simple RF wake-up circuit is that spurious RF energy (noise) could potentially waken the sleeping device. Thus, it may be necessary to interface a low power or passive circuit (essentially a filter) between the RF switch and the higher power consuming receiver. Spurious RF energy is presented in our system as delay RFID antenna voltage and voltage derivative in time. The low power circuit (filter) could be any low-power device that can be turned on for a short period of time, increment a counter(s) and go back to sleep. In effect, this device acts like a receiver. A watchdog timer may be used to reset the device after extended noisy periods or after long intervals of inactivity. V_A is the voltage on the RFID rectangular spiral antenna.

$$V(t) = V_A(t); \frac{dV(t)}{dt} = \frac{dV_A(t)}{dt}$$

We define: $V_1(t) = V(t); V_2(t) = \frac{dV(t)}{dt} = \frac{dV_1(t)}{dt}$. Tau1 ($\tau_1$) and Tau2 ($\tau_2$) delay lines represent our diodes D1 and D1 parasitic effect delay In time, $V_{\tau_1}, V_{\tau_2} \to \varepsilon$ (Fig. 1.23)

$$I_{D_1}(t) \to I_{D_1}(t - \tau_1); I_{D_2}(t) \to I_{D_2}(t - \tau_2).$$

RFID burst switch matching network design: The matching network match between RFID rectangular spiral antenna impedance to our load impedance. First, we need to calculate our matching network input impedance Zin [85] (Fig. 1.24).

Rectangular spiral RFID antenna length calculation and Inductance/resistance

We have the following rectangular spiral RFID antenna and first we need to calculate the total length.

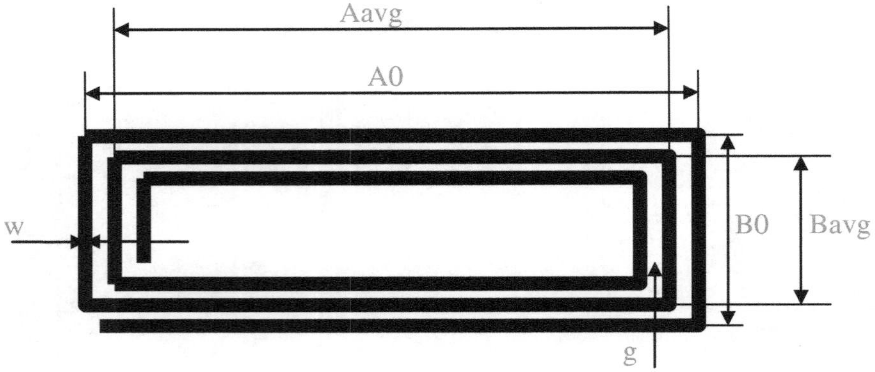

A_0, B_0—Overall dimensions of the coil. Aavg, Bavg—Average dimensions of the coil. w—Track width, g—Gap between tracks, t—Track thickness, Average coil

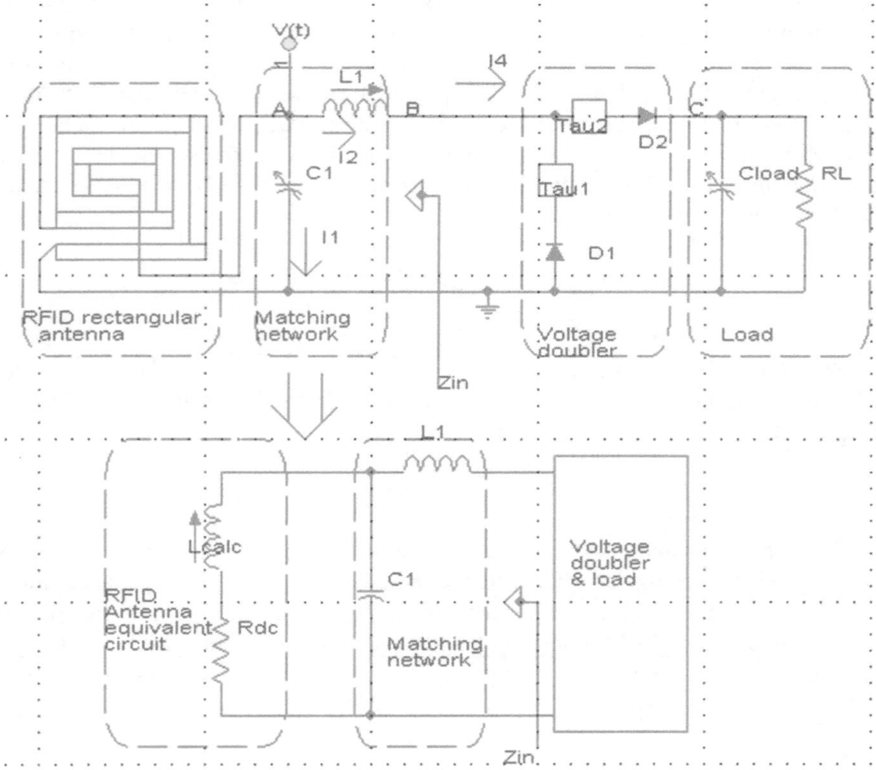

Fig. 1.24 A simple generic RFID burst switch equivalent circuit

area: Ac = Aavg•Bavg. Nc—Number of turns. L_0 is the length of the first turn $l_0 = 2 \cdot (A_0 + B_0) - (w + g)$. l_k is the length of turn k + 1 [7, 8].

$$k = 1 \Rightarrow L_1 = A_0 - (w+g) + B_0 - 2 \cdot (w+g) + A_0 - 2 \cdot (w+g) + B_0 - 3 \cdot (w+g)$$
$$k = 2 \Rightarrow L_2 = A_0 - 3 \cdot (w+g) + B_0 - 4 \cdot (w+g) + A_0 - 4 \cdot (w+g) + B_0 - 5 \cdot (w+g)$$
$$k = 3 \Rightarrow L_3 = A_0 - 5 \cdot (w+g) + B_0 - 6 \cdot (w+g) + A_0 - 6 \cdot (w+g) + B_0 - 7 \cdot (w+g)$$

$$\begin{aligned} L_T = L_0 + \sum_{k=1}^{N_c-1} \{&A_0 - [1 + (k-1) \cdot 2] \cdot (w+g) + B_0 - [2 + (k-1) \cdot 2] \cdot (w+g) \\ &+ A_0 - [2 + (k-1) \cdot 2] \cdot (w+g) + B_0 - [3 + (k-1) \cdot 2] \cdot (w+g)\} \end{aligned}$$

$$\begin{aligned} &\sum_{k=1}^{N_c-1} \{A_0 - [1 + (k-1) \cdot 2] \cdot (w+g) + B_0 - [2 + (k-1) \cdot 2] \cdot (w+g) \\ &\quad + A_0 - [2 + (k-1) \cdot 2] \cdot (w+g) + B_0 - [3 + (k-1) \cdot 2] \cdot (w+g)\} \\ &= \sum_{k=1}^{N_C-1} \{2 \cdot (A_0 + B_0) - 8 \cdot k \cdot (w+g)\} \\ &= 2 \cdot \sum_{k=1}^{N_C-1} \{(A_0 + B_0) - 4 \cdot k \cdot (w+g)\} \\ &= 2 \cdot (A_0 + B_0) \cdot (N_C - 1) - 2 \cdot \sum_{k=1}^{N_C-1} [4 \cdot k \cdot (w+g)] \\ &= 2 \cdot (A_0 + B_0) \cdot (N_C - 1) - 8 \cdot (w+g) \cdot \sum_{k=1}^{N_C-1} k; \sum_{k=1}^{N_C-1} k = N_C - 1 \end{aligned}$$

$$\begin{aligned} &\sum_{k=1}^{N_c-1} \{A_0 - [1 + (k-1) \cdot 2] \cdot (w+g) + B_0 - [2 + (k-1) \cdot 2] \cdot (w+g) \\ &\quad + A_0 - [2 + (k-1) \cdot 2] \cdot (w+g) + B_0 - [3 + (k-1) \cdot 2] \cdot (w+g)\} \\ &= 2 \cdot (A_0 + B_0) \cdot (N_C - 1) - 8 \cdot (w+g) \cdot (N_C - 1) \\ &= 2 \cdot (N_C - 1) \cdot [A_0 + B_0 - 4 \cdot (w+g)] \end{aligned}$$

$$\begin{aligned} L_T &= L_0 + 2 \cdot (N_C - 1) \cdot [A_0 + B_0 - 4 \cdot (w+g)] \\ &\quad \cdot [A_0 + B_0 - 4 \cdot (w+g)] = 2 \cdot (A_0 + B_0) - (w+g) \\ &= L_0 + 2 \cdot (A_0 + B_0) \cdot (1 + N_C - 1) - (w+g) \cdot [1 + 8 \cdot (N_C - 1)] \\ &= L_0 + 2 \cdot (A_0 + B_0) \cdot N_C - (w+g) \cdot (8 \cdot N_C - 7) \end{aligned}$$

Final result: $L_T = L_0 + 2 \cdot (A_0 + B_0) \cdot N_C - (w+g) \cdot (8 \cdot N_C - 7)$

$$L_T = 2 \cdot (A_0 + B_0) - (w+g) + 2 \cdot (A_0 + B_0) \cdot N_C - (w+g) \cdot (8 \cdot N_C - 7)$$
$$L_T = 2 \cdot (A_0 + B_0) \cdot (1 + N_C) - (w+g) \cdot [1 + 8 \cdot N_C - 7]$$
$$L_T = 2 \cdot (A_0 + B_0) \cdot (1 + N_C) - (w+g) \cdot [8 \cdot N_C - 6]$$
$$L_T = 2 \cdot (A_0 + B_0) \cdot (1 + N_C) - 2 \cdot (w+g) \cdot [4 \cdot N_C - 3]$$
$$L_T = 2 \cdot \{(A_0 + B_0) \cdot (1 + N_C) - (w+g) \cdot [4 \cdot N_C - 3]\}$$

The DC resistance of rectangular spiral RFID antenna:

$$R_{DC} = \frac{L_T}{\sigma \cdot S} = \frac{L_T}{\sigma \cdot \pi \cdot a^2}$$

L_T—total length of the wire. σ—conductivity of the wire (mohm/m). S—Cross section area $\pi \cdot a^2$. a—radius of the wire.

$$R_{DC} = \frac{L_T}{\sigma \cdot S} = \frac{L_T}{\sigma \cdot \pi \cdot a^2} = \frac{2 \cdot \{(A_0 + B_0) \cdot (1 + N_C) - (w+g) \cdot [4 \cdot N_C - 3]\}}{\sigma \cdot \pi \cdot a^2}$$
$$R_{DC}|_{S=w \cdot t} = \frac{L_T}{\sigma \cdot S} = \frac{L_T}{\sigma \cdot \pi \cdot w \cdot t} = \frac{2 \cdot \{(A_0 + B_0) \cdot (1 + N_C) - (w+g) \cdot [4 \cdot N_C - 3]\}}{\sigma \cdot \pi \cdot w \cdot t}$$

$$L_{calc} = \frac{\mu_0}{\pi} \cdot (\sum_{k=1, k \neq 3}^{4} X_k - X_3) \cdot N_c^P; X_1 = A_{avg} \cdot \ln(\frac{2 \cdot A_{avg} \cdot B_{avg}}{d \cdot (A_{avg} + \sqrt{A_{avg}^2 + B_{avg}^2})})$$

$$X_2 = B_{avg} \cdot \ln(\frac{2 \cdot A_{avg} \cdot B_{avg}}{d \cdot (B_{avg} + \sqrt{A_{avg}^2 + B_{avg}^2})});$$

$$X_3 = 2 \cdot (A_{avg} + B_{avg} - \sqrt{A_{avg}^2 + B_{avg}^2}); X_4 = \frac{(A_{avg} + B_{avg})}{4}$$
$$d = \frac{2 \cdot (t+w)}{\pi}; A_{avg} = A_0 - N_c \cdot (g+w); B_{avg} = B_0 - N_c \cdot (g+w)$$

ω—Angular frequency.

$$Z_{ant} = R_{DC} + j \cdot \omega \cdot L_{calc}; Z_{in} = Z_{ant} \parallel Z_{c_1} + j \cdot \omega \cdot L_1$$
$$= \frac{Z_{ant} \cdot Z_{c_1}}{Z_{ant} + Z_{c_1}} + j \cdot \omega \cdot L_1; Z_{c_1} = \frac{1}{j \cdot \omega \cdot C_1}$$
$$Z_{in} = \frac{(R_{DC} + j \cdot \omega \cdot L_{calc}) \cdot \frac{1}{j \cdot \omega \cdot C_1}}{R_{DC} + j \cdot \omega \cdot L_{calc} + \frac{1}{j \cdot \omega \cdot C_1}} + j \cdot \omega \cdot L_1$$
$$= \frac{R_{DC} + j \cdot \omega \cdot L_{calc}}{[1 - \omega^2 \cdot L_{calc} \cdot C_1] + j \cdot \omega \cdot C_1 \cdot R_{DC}} + j \cdot \omega \cdot L_1$$
$$Z_{in} = \frac{R_{DC} + j \cdot \omega \cdot L_{calc}}{[1 - \omega^2 \cdot L_{calc} \cdot C_1] + j \cdot \omega \cdot C_1 \cdot R_{DC}} \cdot \frac{[1 - \omega^2 \cdot L_{calc} \cdot C_1] - j \cdot \omega \cdot C_1 \cdot R_{DC}}{[1 - \omega^2 \cdot L_{calc} \cdot C_1] - j \cdot \omega \cdot C_1 \cdot R_{DC}} + j \cdot \omega \cdot L_1$$

$$Z_{in} = \frac{R_{DC} \cdot [1 - \omega^2 \cdot L_{calc} \cdot C_1] - j \cdot \omega \cdot C_1 \cdot R_{DC}^2 + j \cdot \omega \cdot L_{calc} \cdot [1 - \omega^2 \cdot L_{calc} \cdot C_1] + \omega^2 \cdot L_{calc} \cdot C_1 \cdot R_{DC}}{[1 - \omega^2 \cdot L_{calc} \cdot C_1]^2 + \omega^2 \cdot C_1^2 \cdot R_{DC}^2} + j \cdot \omega \cdot L_1$$

$$Z_{in} = \frac{R_{DC} \cdot [1 - \omega^2 \cdot L_{calc} \cdot C_1] + \omega^2 \cdot L_{calc} \cdot C_1 \cdot R_{DC} + j \cdot \omega \cdot \{L_{calc} \cdot [1 - \omega^2 \cdot L_{calc} \cdot C_1] - C_1 \cdot R_{DC}^2\}}{[1 - \omega^2 \cdot L_{calc} \cdot C_1]^2 + \omega^2 \cdot C_1^2 \cdot R_{DC}^2} + j \cdot \omega \cdot L_1$$

$$Z_{in} = \frac{R_{DC} \cdot [1 - \omega^2 \cdot L_{calc} \cdot C_1] + \omega^2 \cdot L_{calc} \cdot C_1 \cdot R_{DC}}{[1 - \omega^2 \cdot L_{calc} \cdot C_1]^2 + \omega^2 \cdot C_1^2 \cdot R_{DC}^2} + j \cdot \omega \cdot \frac{\{L_{calc} \cdot [1 - \omega^2 \cdot L_{calc} \cdot C_1] - C_1 \cdot R_{DC}^2\}}{[1 - \omega^2 \cdot L_{calc} \cdot C_1]^2 + \omega^2 \cdot C_1^2 \cdot R_{DC}^2} + j \cdot \omega \cdot L_1$$

$$Z_{in} = \frac{R_{DC} \cdot [1 - \omega^2 \cdot L_{calc} \cdot C_1] + \omega^2 \cdot L_{calc} \cdot C_1 \cdot R_{DC}}{[1 - \omega^2 \cdot L_{calc} \cdot C_1]^2 + \omega^2 \cdot C_1^2 \cdot R_{DC}^2} + j \cdot \omega \cdot [\frac{\{L_{calc} \cdot [1 - \omega^2 \cdot L_{calc} \cdot C_1] - C_1 \cdot R_{DC}^2\}}{[1 - \omega^2 \cdot L_{calc} \cdot C_1]^2 + \omega^2 \cdot C_1^2 \cdot R_{DC}^2} + L_1]$$

$$A_1 = \frac{R_{DC} \cdot [1 - \omega^2 \cdot L_{calc} \cdot C_1] + \omega^2 \cdot L_{calc} \cdot C_1 \cdot R_{DC}}{[1 - \omega^2 \cdot L_{calc} \cdot C_1]^2 + \omega^2 \cdot C_1^2 \cdot R_{DC}^2};$$

$$B_1 = \omega \cdot [\frac{\{L_{calc} \cdot [1 - \omega^2 \cdot L_{calc} \cdot C_1] - C_1 \cdot R_{DC}^2\}}{[1 - \omega^2 \cdot L_{calc} \cdot C_1]^2 + \omega^2 \cdot C_1^2 \cdot R_{DC}^2} + L_1]$$

$$Z_{in} = A_1 + j \cdot B_1$$

If we neglect voltage doubler unit's parasitic elements (Inductance, capacitance, and resistances) then it is transparent and the load is connected directly to a matching unit.

$$Z_{load} = \frac{\frac{1}{j \cdot \omega \cdot C_{load}} \cdot R_L}{\frac{1}{j \cdot \omega \cdot C_{load}} + R_L} = \frac{R_L}{1 + j \cdot \omega \cdot C_{load} \cdot R_L} = \frac{R_L}{1 + j \cdot \omega \cdot C_{load} \cdot R_L} \cdot \frac{(1 - j \cdot \omega \cdot C_{load} \cdot R_L)}{(1 - j \cdot \omega \cdot C_{load} \cdot R_L)}$$

$$Z_{load} = \frac{R_L}{1 + j \cdot \omega \cdot C_{load} \cdot R_L} \cdot \frac{(1 - j \cdot \omega \cdot C_{load} \cdot R_L)}{(1 - j \cdot \omega \cdot C_{load} \cdot R_L)} = \frac{R_L - j \cdot \omega \cdot C_{load} \cdot R_L \cdot R_L}{1 + \omega^2 \cdot C_{load}^2 \cdot R_L^2}$$

$$Z_{load} = \frac{R_L}{1 + \omega^2 \cdot C_{load}^2 \cdot R_L^2} - j \cdot \frac{\omega \cdot C_{load} \cdot R_L \cdot R_L}{1 + \omega^2 \cdot C_{load}^2 \cdot R_L^2}$$

$$A_2 = \frac{R_L}{1 + \omega^2 \cdot C_{load}^2 \cdot R_L^2}; B_2 = \frac{\omega \cdot C_{load} \cdot R_L \cdot R_L}{1 + \omega^2 \cdot C_{load}^2 \cdot R_L^2}; Z_{load} = A_2 - j \cdot B_2$$

If $Z_{in} = Z_{load}^*$ (complex conjugate) then maximum power is transferred from the RFID rectangular spiral antenna to the load (no power reflections). For perfect match $A_1 = A_2$ and $B_1 = B_2$.

$$A_1 = A_2 \Rightarrow \frac{R_{DC} \cdot [1 - \omega^2 \cdot L_{calc} \cdot C_1] + \omega^2 \cdot L_{calc} \cdot C_1 \cdot R_{DC}}{[1 - \omega^2 \cdot L_{calc} \cdot C_1]^2 + \omega^2 \cdot C_1^2 \cdot R_{DC}^2} = \frac{R_L}{1 + \omega^2 \cdot C_{load}^2 \cdot R_L^2}$$

$$B_1 = B_2 \Rightarrow \omega \cdot [\frac{\{L_{calc} \cdot [1 - \omega^2 \cdot L_{calc} \cdot C_1] - C_1 \cdot R_{DC}^2\}}{[1 - \omega^2 \cdot L_{calc} \cdot C_1]^2 + \omega^2 \cdot C_1^2 \cdot R_{DC}^2} + L_1] = \frac{\omega \cdot C_{load} \cdot R_L \cdot R_L}{1 + \omega^2 \cdot C_{load}^2 \cdot R_L^2}$$

Remark our matching is dependent on the angular frequency value, $\omega = 2 \cdot \pi \cdot f$.

Stability analysis: we need to write our system, differential equations and analyze our system behavior under parameter variation.

$$I_{R_{DC}} = I_{L_{calc}}; I_{L_{calc}} = I_1 + I_2; I_1 = I_{C_1}; I_2 = I_{L_1};$$
$$I_2 = I_3 + I_4; I_4 = I_{D_2}; -I_3 = I_{D_1}; I_4 = I_5 + I_6$$
$$I_5 = I_{C_{load}}; I_6 = I_{R_L} = \frac{V_C}{R_L}; V_{C_1} = V(t); V_{L_1} = V_A - V_B;$$
$$V_{D_2} = V_B - V_C; V_B = -V_{D_1}; V_C = V_{C_{load}} = V_{R_L}$$
$$V_A = V_{C_1}; V_A = V(t) = V_{L_{calc}} + V_{R_{DC}}; I_{C_1} = C_1 \cdot \frac{dV(t)}{dt};$$
$$I_{C_{load}} = C_{load} \cdot \frac{dV_C}{dt}; V_{L_1} = L_1 \cdot \frac{dI_{L_1}}{dt}$$
$$V_{L_{calc}} = L_{calc} \cdot \frac{dI_{L_{calc}}}{dt}; V_{R_{DC}} = R_{DC} \cdot I_{R_{DC}};$$
$$V_{D_1} = V_t \cdot \ln[\frac{I_{D_1}}{I_0} + 1]; V_{D_2} = V_t \cdot \ln[\frac{I_{D_2}}{I_0} + 1]$$
$$I_{L_{calc}} = I_1 + I_2; I_{L_{calc}} = C_1 \cdot \frac{dV(t)}{dt} + I_{L_1}; V_{L_1} = L_1 \cdot \frac{dI_{L_1}}{dt};$$
$$V_{L_1} = V_A - V_B = V_{L_{calc}} + I_{L_{calc}} \cdot R_{DC} + V_{D_1}$$
$$V_{L_1} = V_{L_{calc}} + I_{L_{calc}} \cdot R_{DC} + V_t \cdot \ln[\frac{I_{D_1}}{I_0} + 1]$$
$$\Rightarrow L_1 \cdot \frac{dI_{L_1}}{dt} = L_{calc} \cdot \frac{dI_{L_{calc}}}{dt} + I_{L_{calc}} \cdot R_{DC} + V_t \cdot \ln[\frac{I_{D_1}}{I_0} + 1]$$
$$L_1 \cdot \frac{dI_{L_1}}{dt} = L_{calc} \cdot \frac{d}{dt}[C_1 \cdot \frac{dV(t)}{dt} + I_{L_1}] + [C_1 \cdot \frac{dV(t)}{dt} + I_{L_1}] \cdot R_{DC} + V_t \cdot \ln[\frac{I_{D_1}}{I_0} + 1]$$
$$L_1 \cdot \frac{dI_{L_1}}{dt} = L_{calc} \cdot C_1 \cdot \frac{d^2V(t)}{dt^2} + L_{calc} \cdot \frac{dI_{L_1}}{dt} + C_1 \cdot R_{DC} \cdot \frac{dV(t)}{dt} + I_{L_1} \cdot R_{DC} + V_t \cdot \ln[\frac{I_{D_1}}{I_0} + 1]$$
$$L_1 \cdot \frac{dI_{L_1}}{dt} = L_{calc} \cdot C_1 \cdot \frac{d^2V(t)}{dt^2} + L_{calc} \cdot \frac{dI_{L_1}}{dt} + C_1 \cdot R_{DC} \cdot \frac{dV(t)}{dt} + I_{L_1} \cdot R_{DC} + V_t \cdot \ln[\frac{I_{D_1}}{I_0} + 1]$$
$$(*)[L_1 - L_{calc}] \cdot \frac{dI_{L_1}}{dt} = L_{calc} \cdot C_1 \cdot \frac{d^2V(t)}{dt^2} + C_1 \cdot R_{DC} \cdot \frac{dV(t)}{dt} + I_{L_1} \cdot R_{DC} + V_t \cdot \ln[\frac{I_{D_1}}{I_0} + 1]$$

$$I_2 = I_3 + I_4 \Rightarrow I_{L_1} = -I_{D_1} + I_{D_2}; V_{D_2} = V_B - V_C = -V_{D_1} - V_{C_{load}};$$

$$V_{C_{load}} = V_C; V_t \cdot \ln[\frac{I_{D_2}}{I_0} + 1] = -V_t \cdot \ln[\frac{I_{D_1}}{I_0} + 1] - V_{C_{load}}$$

$$I_4 = I_5 + I_6 \Rightarrow I_{D_2} = C_{load} \cdot \frac{dV_C}{dt} + \frac{V_C}{R_L};$$

$$I_{L_1} = -I_{D_1} + I_{D_2} \Rightarrow \frac{dI_{L_1}}{dt} = -\frac{dI_{D_1}}{dt} + \frac{dI_{D_2}}{dt}$$

$$V_t \cdot \ln[\frac{I_{D_2}}{I_0} + 1] = -V_t \cdot \ln[\frac{I_{D_1}}{I_0} + 1] - V_{C_{load}} \Rightarrow V_t \cdot \ln[\frac{I_{D_2}}{I_0} + 1] = -V_t \cdot \ln[\frac{I_{D_1}}{I_0} + 1] - V_C$$

$$V_t \cdot \ln[\frac{I_{D_2}}{I_0} + 1] = -V_t \cdot \ln[\frac{I_{D_1}}{I_0} + 1] - V_C \Rightarrow V_C = -V_t \cdot \ln[\frac{I_{D_1}}{I_0} + 1] - V_t \cdot \ln[\frac{I_{D_2}}{I_0} + 1]$$

$$\frac{dV_C}{dt} = -V_t \cdot \frac{1}{[\frac{I_{D_1}}{I_0} + 1]} \cdot \frac{1}{I_0} \cdot \frac{dI_{D_1}}{dt} - V_t \cdot \frac{1}{[\frac{I_{D_2}}{I_0} + 1]} \cdot \frac{1}{I_0} \cdot \frac{dI_{D_2}}{dt}$$
$$\Rightarrow \frac{dV_C}{dt} = -V_t \cdot \frac{1}{[I_{D_1} + I_0]} \cdot \frac{dI_{D_1}}{dt} - V_t \cdot \frac{1}{[I_{D_2} + I_0]} \cdot \frac{dI_{D_2}}{dt}$$

$$(**) I_{D_2} = C_{load} \cdot \{-V_t \cdot \frac{1}{[I_{D_1} + I_0]} \cdot \frac{dI_{D_1}}{dt} - V_t \cdot \frac{1}{[I_{D_2} + I_0]} \cdot \frac{dI_{D_2}}{dt}\}$$
$$+ \frac{1}{R_L} \cdot \{-V_t \cdot \ln[\frac{I_{D_1}}{I_0} + 1] - V_t \cdot \ln[\frac{I_{D_2}}{I_0} + 1]\}$$

$$(*)[L_1 - L_{calc}] \cdot \frac{dI_{L_1}}{dt} = L_{calc} \cdot C_1 \cdot \frac{d^2V(t)}{dt^2} + C_1 \cdot R_{DC} \cdot \frac{dV(t)}{dt}$$
$$+ I_{L_1} \cdot R_{DC} + V_t \cdot \ln[\frac{I_{D_1}}{I_0} + 1]$$

$$[L_1 - L_{calc}] \cdot [-\frac{dI_{D_1}}{dt} + \frac{dI_{D_2}}{dt}] = L_{calc} \cdot C_1 \cdot \frac{d^2V(t)}{dt^2} + C_1 \cdot R_{DC} \cdot \frac{dV(t)}{dt}$$
$$+ [-I_{D_1} + I_{D_2}] \cdot R_{DC} + V_t \cdot \ln[\frac{I_{D_1}}{I_0} + 1]$$

We define the following new variables:

$$X = \frac{dI_{D_2}}{dt}; Y = \frac{dI_{D_1}}{dt}; Z = \frac{dV(t)}{dt}; \frac{dZ}{dt} = \frac{d^2V(t)}{dt^2}$$
$$[L_1 - L_{calc}] \cdot [-Y + X] = L_{calc} \cdot C_1 \cdot \frac{dZ}{dt} + C_1 \cdot R_{DC} \cdot Z$$
$$+ [-I_{D_1} + I_{D_2}] \cdot R_{DC} + V_t \cdot \ln[\frac{I_{D_1}}{I_0} + 1]$$

$$(**) I_{D_2} = C_{load} \cdot \{-V_t \cdot \frac{1}{[I_{D_1} + I_0]} \cdot Y - V_t \cdot \frac{1}{[I_{D_2} + I_0]} \cdot X\} + \frac{1}{R_L} \cdot \{-V_t \cdot \ln[\frac{I_{D_1}}{I_0} + 1] - V_t \cdot \ln[\frac{I_{D_2}}{I_0} + 1]\}$$

$$I_{D_2} = -\frac{C_{load} \cdot V_t}{[I_{D_1} + I_0]} \cdot Y - \frac{C_{load} \cdot V_t}{[I_{D_2} + I_0]} \cdot X - \frac{V_t}{R_L} \cdot \{\ln[\frac{I_{D_1}}{I_0} + 1] + \ln[\frac{I_{D_2}}{I_0} + 1]\}$$

$$I_{D_2} = -\frac{C_{load} \cdot V_t}{[I_{D_1} + I_0]} \cdot Y - \frac{C_{load} \cdot V_t}{[I_{D_2} + I_0]} \cdot X - \frac{V_t}{R_L} \cdot \ln\{[\frac{I_{D_1}}{I_0} + 1] \cdot [\frac{I_{D_2}}{I_0} + 1]\}$$

$$\frac{C_{load} \cdot V_t}{[I_{D_2} + I_0]} \cdot X = -I_{D_2} - \frac{C_{load} \cdot V_t}{[I_{D_1} + I_0]} \cdot Y - \frac{V_t}{R_L} \cdot \ln\{[\frac{I_{D_1}}{I_0} + 1] \cdot [\frac{I_{D_2}}{I_0} + 1]\}$$

$$X = -\frac{I_{D_2} \cdot [I_{D_2} + I_0]}{C_{load} \cdot V_t} - \frac{[I_{D_2} + I_0]}{[I_{D_1} + I_0]} \cdot Y - \frac{[I_{D_2} + I_0]}{R_L \cdot C_{load}} \cdot \ln\{[\frac{I_{D_1}}{I_0} + 1] \cdot [\frac{I_{D_2}}{I_0} + 1]\}$$

$$X = \frac{dI_{D_2}}{dt} \Rightarrow \frac{dI_{D_2}}{dt} = -\frac{I_{D_2} \cdot [I_{D_2} + I_0]}{C_{load} \cdot V_t} - \frac{[I_{D_2} + I_0]}{[I_{D_1} + I_0]} \cdot Y - \frac{[I_{D_2} + I_0]}{R_L \cdot C_{load}} \cdot \ln\{[\frac{I_{D_1}}{I_0} + 1] \cdot [\frac{I_{D_2}}{I_0} + 1]\}$$

$$(*) [L_1 - L_{calc}] \cdot [-Y + X] = L_{calc} \cdot C_1 \cdot \frac{dZ}{dt} + C_1 \cdot R_{DC} \cdot Z + [-I_{D_1} + I_{D_2}] \cdot R_{DC} + V_t \cdot \ln[\frac{I_{D_1}}{I_0} + 1]$$

$$L_{calc} \cdot C_1 \cdot \frac{dZ}{dt} = [L_1 - L_{calc}] \cdot [-Y + X] - C_1 \cdot R_{DC} \cdot Z - [-I_{D_1} + I_{D_2}] \cdot R_{DC} - V_t \cdot \ln[\frac{I_{D_1}}{I_0} + 1]$$

$$\frac{dZ}{dt} = \frac{[L_1 - L_{calc}]}{L_{calc} \cdot C_1} \cdot [-Y + X] - \frac{R_{DC}}{L_{calc}} \cdot Z - [-I_{D_1} + I_{D_2}] \cdot \frac{R_{DC}}{L_{calc} \cdot C_1} - \frac{V_t}{L_{calc} \cdot C_1} \cdot \ln[\frac{I_{D_1}}{I_0} + 1]$$

We can summerize our RFID burst switch system, differential equations:

$$\frac{dZ}{dt} = \frac{[L_1 - L_{calc}]}{L_{calc} \cdot C_1} \cdot [-Y + X] - \frac{R_{DC}}{L_{calc}} \cdot Z - [-I_{D_1} + I_{D_2}] \cdot \frac{R_{DC}}{L_{calc} \cdot C_1} - \frac{V_t}{L_{calc} \cdot C_1} \cdot \ln[\frac{I_{D_1}}{I_0} + 1]$$

$$\frac{dZ}{dt} = \frac{[L_1 - L_{calc}]}{L_{calc} \cdot C_1} \cdot [-Y - \frac{I_{D_2} \cdot [I_{D_2} + I_0]}{C_{load} \cdot V_t} - \frac{[I_{D_2} + I_0]}{[I_{D_1} + I_0]} \cdot Y - \frac{[I_{D_2} + I_0]}{R_L \cdot C_{load}} \cdot \ln\{[\frac{I_{D_1}}{I_0} + 1] \cdot [\frac{I_{D_2}}{I_0} + 1]\}] - \frac{R_{DC}}{L_{calc}} \cdot Z - [-I_{D_1} + I_{D_2}] \cdot \frac{R_{DC}}{L_{calc} \cdot C_1} - \frac{V_t}{L_{calc} \cdot C_1} \cdot \ln[\frac{I_{D_1}}{I_0} + 1]$$

$$\frac{dI_{D_2}}{dt} = -\frac{I_{D_2} \cdot [I_{D_2} + I_0]}{C_{load} \cdot V_t} - \frac{[I_{D_2} + I_0]}{[I_{D_1} + I_0]} \cdot Y - \frac{[I_{D_2} + I_0]}{R_L \cdot C_{load}} \cdot \ln\{[\frac{I_{D_1}}{I_0} + 1] \cdot [\frac{I_{D_2}}{I_0} + 1]\}; \frac{dI_{D_1}}{dt} = Y$$

We have four variables in our system: Z, I_{D_1}, I_{D_2}, Y and we can represent our system as the following set of differential equations: $\frac{dZ}{dt} = \xi_1(Z, I_{D_1}, I_{D_2}, Y)$

$$\frac{dI_{D_2}}{dt} = \xi_2(Z, I_{D_1}, I_{D_2}, Y); \frac{dI_{D_1}}{dt} = \xi_3(Z, I_{D_1}, I_{D_2}, Y);$$

$$\xi_1 = \xi_1(Z, I_{D_1}, I_{D_2}, Y); \xi_2 = \xi_2(Z, I_{D_1}, I_{D_2}, Y)$$

$$\xi_3 = \xi_3(Z, I_{D_1}, I_{D_2}, Y)$$

$$\xi_1 = \frac{[L_1 - L_{calc}]}{L_{calc} \cdot C_1} \cdot [-Y - \frac{I_{D_2} \cdot [I_{D_2} + I_0]}{C_{load} \cdot V_t} - \frac{[I_{D_2} + I_0]}{[I_{D_1} + I_0]} \cdot Y - \frac{[I_{D_2} + I_0]}{R_L \cdot C_{load}} \cdot \ln\{[\frac{I_{D_1}}{I_0} + 1] \cdot [\frac{I_{D_2}}{I_0} + 1]\}] - \frac{R_{DC}}{L_{calc}} \cdot Z - [-I_{D_1} + I_{D_2}] \cdot \frac{R_{DC}}{L_{calc} \cdot C_1} - \frac{V_t}{L_{calc} \cdot C_1} \cdot \ln[\frac{I_{D_1}}{I_0} + 1]$$

$$\xi_2 = -\frac{I_{D_2} \cdot [I_{D_2} + I_0]}{C_{load} \cdot V_t} - \frac{[I_{D_2} + I_0]}{[I_{D_1} + I_0]} \cdot Y - \frac{[I_{D_2} + I_0]}{R_L \cdot C_{load}} \cdot \ln\{[\frac{I_{D_1}}{I_0} + 1] \cdot [\frac{I_{D_2}}{I_0} + 1]\}; \xi_3 = Y$$

RFID system burst switch's voltage doubler unit is constructed from two diodes D_1 and D_2 with parasitic effects, delay in time. D_1 current delay in time $I_{D_1}(t) \to I_{D_1}(t - \tau_1)$ and D_2 current delay in time $I_{D_2}(t) \to I_{D_2}(t - \tau_2)$. Spurious RF energy is presented in our system as delay RFID antenna voltage (V (t)) and voltage derivative (dV(t)/dt) in time. We neglect voltage delayed in time and consider only voltage derivative delay in time (Δ).

$$Z(t) = \frac{dV(t)}{dt}; Z(t) \to Z(t-\Delta)$$

We consider no delay effects on $\frac{dZ}{dt} = \frac{dI_{D_1}}{dt}; \frac{dI_{D_2}}{dt}$. To find equilibrium points (fixed points) of the RFID system burst switches, we define $\lim_{t\to\infty} I_{D_1}(t-\tau_1) = I_{D_1}(t)$

$$\lim_{t\to\infty} I_{D_2}(t-\tau_2) = I_{D_2}(t); \lim_{t\to\infty} Z(t-\Delta) = Z(t)$$

In equilibrium points (fixed points)

$$\frac{dI_{D_1}}{dt} = 0; \frac{dI_{D_2}}{dt} = 0; \frac{dZ}{dt} = 0 \,\forall t \gg \tau_1,\, t \gg \tau_2,$$
$$t \gg \Delta \,\exists\, (t-\tau_1) \approx t,\, (t-\tau_2) \approx t, (t-\Delta) \approx t,\, t \to \infty.$$
$$\frac{dI_{D_1}}{dt} = 0 \Rightarrow Y^* = 0; \frac{dI_{D_2}}{dt} = 0 \Rightarrow -\frac{I^*_{D_2} \cdot [I^*_{D_2} + I_0]}{C_{load} \cdot V_t}$$
$$-\frac{[I^*_{D_2} + I_0]}{R_L \cdot C_{load}} \cdot \ln\{[\frac{I^*_{D_1}}{I_0} + 1] \cdot [\frac{I^*_{D_2}}{I_0} + 1]\} = 0$$
$$-\frac{I^*_{D_2} \cdot [I^*_{D_2} + I_0]}{C_{load} \cdot V_t} - \frac{[I^*_{D_2} + I_0]}{R_L \cdot C_{load}} \cdot \ln\{[\frac{I^*_{D_1}}{I_0} + 1] \cdot [\frac{I^*_{D_2}}{I_0} + 1]\} = 0$$
$$\Rightarrow -\frac{[I^*_{D_2} + I_0]}{C_{load}} \cdot (\frac{I^*_{D_2}}{V_t} + \frac{1}{R_L} \cdot \ln\{[\frac{I^*_{D_1}}{I_0} + 1] \cdot [\frac{I^*_{D_2}}{I_0} + 1]\}) = 0$$

Case I:

$$-\frac{[I^*_{D_2} + I_0]}{C_{load}} = 0 \Rightarrow I^*_{D_2} = -I_0 \Rightarrow \frac{dZ}{dt} = 0$$
$$\Rightarrow [I^*_{D_1} + I_0] \cdot \frac{R_{DC}}{C_1} - R_{DC} \cdot Z^* - \frac{V_t}{C_1} \cdot \ln[\frac{I^*_{D_1}}{I_0} + 1] = 0$$

Case II:

$$\frac{I^*_{D_2}}{V_t} + \frac{1}{R_L} \cdot \ln\{[\frac{I^*_{D_1}}{I_0} + 1] \cdot [\frac{I^*_{D_2}}{I_0} + 1]\} = 0 \Rightarrow \frac{R_L}{V_t} \cdot I^*_{D_2} + \ln\{[\frac{I^*_{D_1}}{I_0} + 1] \cdot [\frac{I^*_{D_2}}{I_0} + 1]\} = 0$$
$$\ln\{[\frac{I^*_{D_1}}{I_0} + 1] \cdot [\frac{I^*_{D_2}}{I_0} + 1]\} = -\frac{R_L}{V_t} \cdot I^*_{D_2} \Rightarrow e^{-\frac{R_L}{V_t} \cdot I^*_{D_2}} = [\frac{I^*_{D_1}}{I_0} + 1] \cdot [\frac{I^*_{D_2}}{I_0} + 1].$$

Let us define

$$\phi = -\frac{R_L}{V_t}\cdot I^*_{D_2}; e^{\phi} = \sum_{n=0}^{\infty}\frac{\phi^n}{n!} = 1+\phi+\frac{\phi^2}{2!}+\frac{\phi^3}{3!}+\cdots$$

$$\Rightarrow e^{-\frac{R_L}{V_t}\cdot I^*_{D_2}} = \sum_{n=0}^{\infty}\frac{(-\frac{R_L}{V_t}\cdot I^*_{D_2})^n}{n!}$$

$$e^{-\frac{R_L}{V_t}\cdot I^*_{D_2}} = \sum_{n=0}^{\infty}\frac{(-\frac{R_L}{V_t}\cdot I^*_{D_2})^n}{n!} = \sum_{n=0}^{\infty}\frac{(-1)^n\cdot(\frac{R_L}{V_t}\cdot I^*_{D_2})^n}{n!}$$

$$= 1-\frac{R_L}{V_t}\cdot I^*_{D_2}+\frac{(\frac{R_L}{V_t}\cdot I^*_{D_2})^2}{2!}-\frac{(\frac{R_L}{V_t}\cdot I^*_{D_2})^3}{3!}+\cdots$$

$$e^{-\frac{R_L}{V_t}\cdot I^*_{D_2}} = 1-\frac{R_L}{V_t}\cdot I^*_{D_2}+\frac{1}{2}\cdot(\frac{R_L}{V_t})^2\cdot(I^*_{D_2})^2-\frac{1}{6}\cdot(\frac{R_L}{V_t})^3\cdot(I^*_{D_2})^3+\cdots$$

For easy investigation, we take $e^{-\frac{R_L}{V_t}\cdot I^*_{D_2}} = 1-\frac{R_L}{V_t}\cdot I^*_{D_2} \Rightarrow 1-\frac{R_L}{V_t}\cdot I^*_{D_2} = [\frac{I^*_{D_1}}{I_0}+1]\cdot[\frac{I^*_{D_2}}{I_0}+1]$

$$e^{-\frac{R_L}{V_t}\cdot I^*_{D_2}} = 1-\frac{R_L}{V_t}\cdot I^*_{D_2} \Rightarrow 1-\frac{R_L}{V_t}\cdot I^*_{D_2} = [\frac{I^*_{D_1}}{I_0}+1]\cdot[\frac{I^*_{D_2}}{I_0}+1]$$

$$1-\frac{R_L}{V_t}\cdot I^*_{D_2} = [\frac{I^*_{D_1}}{I_0}+1]\cdot[\frac{I^*_{D_2}}{I_0}+1] \Rightarrow \frac{I^*_{D_1}\cdot I^*_{D_2}}{I_0^2}+\frac{I^*_{D_1}}{I_0}+\frac{I^*_{D_2}}{I_0}+\frac{R_L}{V_t}\cdot I^*_{D_2} = 0$$

$$\frac{dZ}{dt} = 0 \Rightarrow \frac{[L_1-L_{calc}]}{L_{calc}\cdot C_1}\cdot[-\frac{I^*_{D_2}\cdot[I_{D_2}+I_0]}{C_{load}\cdot V_t}$$
$$-\frac{[I^*_{D_2}+I_0]}{R_L\cdot C_{load}}\cdot\ln\{[\frac{I^*_{D_1}}{I_0}+1]\cdot[\frac{I^*_{D_2}}{I_0}+1]\}]-\frac{R_{DC}}{L_{calc}}\cdot Z^*$$
$$-[-I^*_{D_1}+I^*_{D_2}]\cdot\frac{R_{DC}}{L_{calc}\cdot C_1}-\frac{V_t}{L_{calc}\cdot C_1}\cdot\ln[\frac{I^*_{D_1}}{I_0}+1] = 0$$

Remark Our system, equilibrium points (fixed points) can be calculated numerically rather than analytically (Case I and Case II). For both cases Y* = 0, Y* = dI_{D1}/dt = 0. At equilibrium no current is flowing through D_1, D_1 is in OFF state ($V_B > 0$).

The standard local stability analysis about any one of the equilibrium points of the RFID system burst switch consists in adding to coordinate $[Z, I_{D_1}, I_{D_2}, Y]$ arbitrarily small increments of exponential form $[z, i_{D_1}, i_{D_2}, y]\cdot e^{\lambda\cdot t}$ and retaining the first order terms in Z, I_{D_1}, I_{D_2}, Y. The system of three homogeneous equations leads to a polynomial characteristic equation in the eigenvalues. The polynomial characteristic equations accept by set the below currents and currents derivative with respect to time into RFID system burst switch equations. RFID system burst switches fixed values with arbitrarily small increments of exponential form

$[z, i_{D_1}, i_{D_2}, y] \cdot e^{\lambda \cdot t}$ are: j = 0 (first fixed point), j = 1 (second fixed point), j = 2 (third fixed point), etc.

$$Z(t) = Z^{(j)} + z \cdot e^{\lambda \cdot t}$$
$$I_{D_1}(t) = I_{D_1}^{(j)} + i_{D_1} \cdot e^{\lambda \cdot t}; I_{D_2}(t) = I_{D_2}^{(j)} + i_{D_2} \cdot e^{\lambda \cdot t};$$
$$Y(t) = Y^{(j)} + y \cdot e^{\lambda \cdot t}; I_{D_1}(t - \tau_1) = I_{D_1}^{(j)} + i_{D_1} \cdot e^{\lambda \cdot (t - \tau_1)}$$
$$I_{D_2}(t - \tau_2) = I_{D_2}^{(j)} + i_{D_2} \cdot e^{\lambda \cdot (t - \tau_2)}; Z(t - \Delta) = Z^{(j)} + z \cdot e^{\lambda \cdot (t - \Delta)}.$$

We choose these expressions for our $Z(t), I_{D_1}(t), I_{D_2}(t), Y(t)$ as a small displacement $[z, i_{D_1}, i_{D_2}, y]$ from the RFID system burst switch fixed points in time t = 0.

$$I_{D_1}(t = 0) = I_{D_1}^{(j)} + i_{D_1}; I_{D_2}(t = 0) = I_{D_2}^{(j)} + i_{D_2}; Y(t = 0) = Y^{(j)} + y; Z(t = 0) = Z^{(j)} + z.$$

For $\lambda < 0$, $t > 0$, the selected fixed point is stable otherwise $\lambda > 0$, $t > 0$ is unstable. Our system tends to the selected fixed point exponentially for $\lambda < 0$, $t > 0$ otherwise go away from the selected fixed point exponentially. λ Is the eigenvalue parameter which is established if the fixed point is stable or unstable; additionally, his absolute value ($|\lambda|$) establishes the speed of flow toward or away from the selected fixed point (Yuri 1995; Jack and Huseyin 1991). The speeds of flow toward or away from the selected fixed point for RFID system burst switch diodes (D_1 and D_2) currents and antenna voltage derivative with respect to time are

$$\frac{dZ(t)}{dt} = \lim_{\Delta t \to \infty} \frac{Z(t + \Delta t) - Z(t)}{\Delta t} = \lim_{\Delta t \to \infty} \frac{Z^{(j)} + z \cdot e^{\lambda \cdot (t + \Delta t)} - [Z^{(j)} + z \cdot e^{\lambda \cdot t}]}{\Delta t}$$
$$= \stackrel{e^{\lambda \cdot \Delta t} \approx 1 + \lambda \cdot \Delta t}{\longrightarrow} \lambda \cdot z \cdot e^{\lambda \cdot t}$$

$$\frac{dI_{D_1}(t)}{dt} = \lim_{\Delta t \to \infty} \frac{I_{D_1}(t + \Delta t) - I_{D_1}(t)}{\Delta t} = \lim_{\Delta t \to \infty} \frac{I_{D_1}^{(j)} + i_{D_1} \cdot e^{\lambda \cdot (t + \Delta t)} - [I_{D_1}^{(j)} + i_{D_1} \cdot e^{\lambda \cdot t}]}{\Delta t}$$
$$= \stackrel{e^{\lambda \cdot \Delta t} \approx 1 + \lambda \cdot \Delta t}{\longrightarrow} \lambda \cdot i_{D_1} \cdot e^{\lambda \cdot t}$$

$$\frac{dI_{D_2}(t)}{dt} = \lim_{\Delta t \to \infty} \frac{I_{D_2}(t + \Delta t) - I_{D_2}(t)}{\Delta t} = \lim_{\Delta t \to \infty} \frac{I_{D_2}^{(j)} + i_{D_2} \cdot e^{\lambda \cdot (t + \Delta t)} - [I_{D_2}^{(j)} + i_{D_2} \cdot e^{\lambda \cdot t}]}{\Delta t}$$
$$= \stackrel{e^{\lambda \cdot \Delta t} \approx 1 + \lambda \cdot \Delta t}{\longrightarrow} \lambda \cdot i_{D_2} \cdot e^{\lambda \cdot t}$$

$$\frac{dY(t)}{dt} = \lim_{\Delta t \to \infty} \frac{Y(t + \Delta t) - Y(t)}{\Delta t} = \lim_{\Delta t \to \infty} \frac{Y^{(j)} + y \cdot e^{\lambda \cdot (t + \Delta t)} - [Y^{(j)} + y \cdot e^{\lambda \cdot t}]}{\Delta t}$$
$$= \stackrel{e^{\lambda \cdot \Delta t} \approx 1 + \lambda \cdot \Delta t}{\longrightarrow} \lambda \cdot y \cdot e^{\lambda \cdot t}$$

$$\frac{dI_{D_1}(t - \tau_1)}{dt} = \lambda \cdot i_{D_1} \cdot e^{\lambda \cdot t} \cdot e^{-\lambda \cdot \tau_1}; \quad \frac{dI_{D_2}(t - \tau_2)}{dt} = \lambda \cdot i_{D_2} \cdot e^{\lambda \cdot t} \cdot e^{-\lambda \cdot \tau_2};$$
$$\frac{dZ(t - \Delta)}{dt} = \lambda \cdot z \cdot e^{\lambda \cdot t} \cdot e^{-\lambda \cdot \Delta}$$

First, we take RFID system burst switch variable Z, I_{D_1}, I_{D_2}, Y differential equations and adding to coordinate $[Z, I_{D_1}, I_{D_2}, Y]$ arbitrarily small increments of exponential terms $[z, i_{D_1}, i_{D_2}, y] \cdot e^{\lambda \cdot t}$ and retaining the first order terms in z, i_{D_1}, i_{D_2}, y.

$$Z(t) = Z^{(j)} + z \cdot e^{\lambda \cdot t} \Rightarrow \frac{dZ(t)}{dt} = z \cdot \lambda \cdot e^{\lambda \cdot t}$$

$$\begin{aligned}\frac{dZ}{dt} = & \frac{[L_1 - L_{calc}]}{L_{calc} \cdot C_1} \cdot [-Y - \frac{I_{D_2} \cdot [I_{D_2} + I_0]}{C_{load} \cdot V_t} - \frac{[I_{D_2} + I_0]}{[I_{D_1} + I_0]} \cdot Y \\ & - \frac{[I_{D_2} + I_0]}{R_L \cdot C_{load}} \cdot \ln\{[\frac{I_{D_1}}{I_0} + 1] \cdot [\frac{I_{D_2}}{I_0} + 1]\}] - \frac{R_{DC}}{L_{calc}} \cdot Z \\ & - [-I_{D_1} + I_{D_2}] \cdot \frac{R_{DC}}{L_{calc \cdot C_1}} - \frac{V_t}{L_{calc} \cdot C_1} \cdot \ln[\frac{I_{D_1}}{I_0} + 1]\end{aligned}$$

$$\begin{aligned}z \cdot \lambda \cdot e^{\lambda \cdot t} = & \frac{[L_1 - L_{calc}]}{L_{calc} \cdot C_1} \cdot [-(Y^{(j)} + y \cdot e^{\lambda \cdot t}) - \frac{(I_{D_2}^{(j)} + i_{D_2} \cdot e^{\lambda \cdot t}) \cdot [I_{D_2}^{(j)} + i_{D_2} \cdot e^{\lambda \cdot t} + I_0]}{C_{load} \cdot V_t} \\ & - \frac{[I_{D_2}^{(j)} + i_{D_2} \cdot e^{\lambda \cdot t} + I_0]}{[I_{D_1}^{(j)} + i_{D_1} \cdot e^{\lambda \cdot t} + I_0]} \cdot (Y^{(j)} + y \cdot e^{\lambda \cdot t}) \\ & - \frac{[I_{D_2}^{(j)} + i_{D_2} \cdot e^{\lambda \cdot t} + I_0]}{R_L \cdot C_{load}} \cdot \ln\{[\frac{I_{D_1}^{(j)} + i_{D_1} \cdot e^{\lambda \cdot t}}{I_0} + 1] \cdot [\frac{I_{D_2}^{(j)} + i_{D_2} \cdot e^{\lambda \cdot t}}{I_0} + 1]\}] \\ & - \frac{R_{DC}}{L_{calc}} \cdot (Z^{(j)} + z \cdot e^{\lambda \cdot t}) \\ & - [-(I_{D_1}^{(j)} + i_{D_1} \cdot e^{\lambda \cdot t}) + I_{D_2}^{(j)} + i_{D_2} \cdot e^{\lambda \cdot t}] \cdot \frac{R_{DC}}{L_{calc} \cdot C_1} \\ & - \frac{V_t}{L_{calc} \cdot C_1} \cdot \ln[\frac{I_{D_1}^{(j)} + i_{D_1} \cdot e^{\lambda \cdot t}}{I_0} + 1]\end{aligned}$$

$$\begin{aligned}z \cdot \lambda \cdot e^{\lambda \cdot t} = & \frac{[L_1 - L_{calc}]}{L_{calc} \cdot C_1} \cdot [-(Y^{(j)} + y \cdot e^{\lambda \cdot t}) \\ & - \frac{I_{D_2}^{(j)} \cdot (I_{D_2}^{(j)} + I_0) + I_{D_2}^{(j)} \cdot i_{D_2} \cdot e^{\lambda \cdot t} + (I_{D_2}^{(j)} + I_0) \cdot i_{D_2} \cdot e^{\lambda \cdot t} + i_{D_2}^2 \cdot e^{2 \cdot \lambda \cdot t}}{C_{load} \cdot V_t} \\ & - \frac{[(I_{D_2}^{(j)} + I_0) + i_{D_2} \cdot e^{\lambda \cdot t}]}{[(I_{D_1}^{(j)} + I_0) + i_{D_1} \cdot e^{\lambda \cdot t}]} \cdot \frac{[(I_{D_1}^{(j)} + I_0) - i_{D_1} \cdot e^{\lambda \cdot t}]}{[(I_{D_1}^{(j)} + I_0) - i_{D_1} \cdot e^{\lambda \cdot t}]} \cdot (Y^{(j)} + y \cdot e^{\lambda \cdot t}) \\ & - \frac{[I_{D_2}^{(j)} + i_{D_2} \cdot e^{\lambda \cdot t} + I_0]}{R_L \cdot C_{load}} \cdot \ln\{[\frac{I_{D_1}^{(j)} + i_{D_1} \cdot e^{\lambda \cdot t}}{I_0} + 1] \cdot [\frac{I_{D_2}^{(j)} + i_{D_2} \cdot e^{\lambda \cdot t}}{I_0} + 1]\}] \\ & - \frac{R_{DC}}{L_{calc}} \cdot (Z^{(j)} + z \cdot e^{\lambda \cdot t}) \\ & - [-I_{D_1}^{(j)} - i_{D_1} \cdot e^{\lambda \cdot t} + I_{D_2}^{(j)} + i_{D_2} \cdot e^{\lambda \cdot t}] \cdot \frac{R_{DC}}{L_{calc} \cdot C_1} \\ & - \frac{V_t}{L_{calc} \cdot C_1} \cdot \ln[\frac{I_{D_1}^{(j)} + i_{D_1} \cdot e^{\lambda \cdot t}}{I_0} + 0]\end{aligned}$$

$$z \cdot \lambda \cdot e^{\lambda \cdot t} = \frac{[L_1 - L_{calc}]}{L_{calc} \cdot C_1} \cdot [-(Y^{(j)} + y \cdot e^{\lambda \cdot t})$$
$$- \frac{I_{D_2}^{(j)} \cdot (I_{D_2}^{(j)} + I_0) + I_{D_2}^{(j)} \cdot i_{D_2} \cdot e^{\lambda \cdot t} + (I_{D_2}^{(j)} + I_0) \cdot i_{D_2} \cdot e^{\lambda \cdot t} + i_{D_2}^2 \cdot e^{2 \cdot \lambda \cdot t}}{C_{load} \cdot V_t}$$
$$- \{\frac{(I_{D_2}^{(j)} + I_0) \cdot (I_{D_1}^{(j)} + I_0) - (I_{D_2}^{(j)} + I_0) \cdot i_{D_1} \cdot e^{\lambda \cdot t} + (I_{D_1}^{(j)} + I_0) \cdot i_{D_2} \cdot e^{\lambda \cdot t} - i_{D_2} \cdot i_{D_1} \cdot e^{2 \cdot \lambda \cdot t}}{(I_{D_1}^{(j)} + I_0)^2 - i_{D_1}^2 \cdot e^{2 \cdot \lambda \cdot t}}\}$$
$$\cdot (Y^{(j)} + y \cdot e^{\lambda \cdot t})$$
$$- \frac{[I_{D_2}^{(j)} + i_{D_2} \cdot e^{\lambda \cdot t} + I_0]}{R_L \cdot C_{load}} \cdot \ln\{[\frac{I_{D_1}^{(j)} + i_{D_1} \cdot e^{\lambda \cdot t}}{I_0} + 1] \cdot [\frac{I_{D_2}^{(j)} + i_{D_2} \cdot e^{\lambda \cdot t}}{I_0} + 1]\}]$$
$$- \frac{R_{DC}}{L_{calc}} \cdot (Z^{(j)} + z \cdot e^{\lambda \cdot t})$$
$$- [-I_{D_1}^{(j)} - i_{D_1} \cdot e^{\lambda \cdot t} + I_{D_2}^{(j)} + i_{D_2} \cdot e^{\lambda \cdot t}] \cdot \frac{R_{DC}}{L_{calc} \cdot C_1}$$
$$- \frac{V_t}{L_{calc} \cdot C_1} \cdot \ln[\frac{I_{D_1}^{(j)} + i_{D_1} \cdot e^{\lambda \cdot t}}{I_0} + 0]$$

We consider $i_{D_1}^2 \to \varepsilon \approx 0; i_{D_2}^2 \to \varepsilon \approx 0; i_{D_2} \cdot i_{D_1} \to \varepsilon \approx 0$

$$z \cdot \lambda \cdot e^{\lambda \cdot t} = \frac{[L_1 - L_{calc}]}{L_{calc} \cdot C_1} \cdot [-(Y^{(j)} + y \cdot e^{\lambda \cdot t}) - \frac{I_{D_2}^{(j)} \cdot (I_{D_2}^{(j)} + I_0) + I_{D_2}^{(j)} \cdot i_{D_2} \cdot e^{\lambda \cdot t} + (I_{D_2}^{(j)} + I_0) \cdot i_{D_2} \cdot e^{\lambda \cdot t}}{C_{load} \cdot V_t}$$
$$- \{\frac{(I_{D_2}^{(j)} + I_0) \cdot (I_{D_1}^{(j)} + I_0) - (I_{D_2}^{(j)} + I_0) \cdot i_{D_1} \cdot e^{\lambda \cdot t} + (I_{D_1}^{(j)} + I_0) \cdot i_{D_2} \cdot e^{\lambda \cdot t}}{(I_{D_1}^{(j)} + I_0)^2}\} \cdot (Y^{(j)} + y \cdot e^{\lambda \cdot t}) \quad (£)$$
$$- \frac{[I_{D_2}^{(j)} + i_{D_2} \cdot e^{\lambda \cdot t} + I_0]}{R_L \cdot C_{load}} \cdot \ln\{[\frac{I_{D_1}^{(j)} + i_{D_1} \cdot e^{\lambda \cdot t}}{I_0} + 1] \cdot [\frac{I_{D_2}^{(j)} + i_{D_2} \cdot e^{\lambda \cdot t}}{I_0} + 1]\}] - \frac{R_{DC}}{L_{calc}} \cdot (Z^{(j)} + z \cdot e^{\lambda \cdot t})$$
$$- [-I_{D_1}^{(j)} - i_{D_1} \cdot e^{\lambda \cdot t} + I_{D_2}^{(j)} + i_{D_2} \cdot e^{\lambda \cdot t}] \cdot \frac{R_{DC}}{L_{calc} \cdot C_1} - \frac{V_t}{L_{calc} \cdot C_1} \cdot \ln[\frac{I_{D_1}^{(j)} + i_{D_1} \cdot e^{\lambda \cdot t}}{I_0} + 1]$$

Calculation No. 1:

$$\ln\{[\frac{I_{D_1}^{(j)} + i_{D_1} \cdot e^{\lambda \cdot t}}{I_0} + 1] \cdot [\frac{I_{D_2}^{(j)} + i_{D_2} \cdot e^{\lambda \cdot t}}{I_0} + 1]\}$$
$$= \ln[([\frac{I_{D_1}^{(j)}}{I_0} + 1] + \frac{1}{I_0} \cdot i_{D_1} \cdot e^{\lambda \cdot t}) \cdot ([\frac{I_{D_2}^{(j)}}{I_0} + 1] + \frac{1}{I_0} \cdot i_{D_2} \cdot e^{\lambda \cdot t})]$$
$$= \ln\{[\frac{I_{D_1}^{(j)}}{I_0} + 1] \cdot [\frac{I_{D_2}^{(j)}}{I_0} + 1] + [\frac{I_{D_1}^{(j)}}{I_0} + 1] \cdot \frac{1}{I_0} \cdot i_{D_2} \cdot e^{\lambda \cdot t}$$
$$+ [\frac{I_{D_2}^{(j)}}{I_0} + 1] \cdot \frac{1}{I_0} \cdot i_{D_1} \cdot e^{\lambda \cdot t} + \frac{1}{I_0^2} \cdot i_{D_1} \cdot i_{D_2} \cdot e^{2 \cdot \lambda \cdot t}\}$$

$i_{D_1} \cdot i_{D_2} \approx 0$

$$\ln\{[\frac{I_{D_1}^{(j)} + i_{D_1} \cdot e^{\lambda \cdot t}}{I_0} + 1] \cdot [\frac{I_{D_2}^{(j)} + i_{D_2} \cdot e^{\lambda \cdot t}}{I_0} + 1]\}$$
$$= \ln\{[\frac{I_{D_1}^{(j)}}{I_0} + 1] \cdot [\frac{I_{D_2}^{(j)}}{I_0} + 1] + [\frac{I_{D_1}^{(j)}}{I_0} + 1] \cdot \frac{1}{I_0} \cdot i_{D_2} \cdot e^{\lambda \cdot t} + [\frac{I_{D_2}^{(j)}}{I_0} + 1] \cdot \frac{1}{I_0} \cdot i_{D_1} \cdot e^{\lambda \cdot t}\}$$

We define:

$$\Omega_1 = \Omega_1(I_{D_1}^{(j)}, I_{D_2}^{(j)}, i_{D_2}, i_{D_1}, \lambda) = [\frac{I_{D_1}^{(j)}}{I_0} + 1] \cdot \frac{1}{I_0} \cdot i_{D_2} \cdot e^{\lambda \cdot t} + [\frac{I_{D_2}^{(j)}}{I_0} + 1] \cdot \frac{1}{I_0} \cdot i_{D_1} \cdot e^{\lambda \cdot t}$$

$$\ln\{[\frac{I_{D_1}^{(j)} + i_{D_1} \cdot e^{\lambda \cdot t}}{I_0} + 1] \cdot [\frac{I_{D_2}^{(j)} + i_{D_2} \cdot e^{\lambda \cdot t}}{I_0} + 1]\} = \ln\{[\frac{I_{D_1}^{(j)}}{I_0} + 1] \cdot [\frac{I_{D_2}^{(j)}}{I_0} + 1] + \Omega_1\}$$

$$\ln\{[\frac{I_{D_1}^{(j)} + i_{D_1} \cdot e^{\lambda \cdot t}}{I_0} + 1] \cdot [\frac{I_{D_2}^{(j)} + i_{D_2} \cdot e^{\lambda \cdot t}}{I_0} + 1]\}$$
$$= \ln\{[\frac{I_{D_1}^{(j)}}{I_0} + 1] \cdot [\frac{I_{D_2}^{(j)}}{I_0} + 1] \cdot (1 + \frac{\Omega_1}{[\frac{I_{D_1}^{(j)}}{I_0} + 1] \cdot [\frac{I_{D_2}^{(j)}}{I_0} + 1]})\}$$

The above is assuming

$$[\frac{I_{D_1}^{(j)}}{I_0} + 1] \cdot [\frac{I_{D_2}^{(j)}}{I_0} + 1] > 0$$

$$\ln\{[\frac{I_{D_1}^{(j)} + i_{D_1} \cdot e^{\lambda \cdot t}}{I_0} + 1] \cdot [\frac{I_{D_2}^{(j)} + i_{D_2} \cdot e^{\lambda \cdot t}}{I_0} + 1]\}$$
$$= \ln\{[\frac{I_{D_1}^{(j)}}{I_0} + 1] \cdot [\frac{I_{D_2}^{(j)}}{I_0} + 1]\} + \ln\{1 + \frac{\Omega_1}{[\frac{I_{D_1}^{(j)}}{I_0} + 1] \cdot [\frac{I_{D_2}^{(j)}}{I_0} + 1]}\}$$

$$\frac{\Omega_1}{[\frac{I_{D_1}^{(j)}}{I_0} + 1] \cdot [\frac{I_{D_2}^{(j)}}{I_0} + 1]} = \frac{[\frac{I_{D_1}^{(j)}}{I_0} + 1] \cdot \frac{1}{I_0} \cdot i_{D_2} \cdot e^{\lambda \cdot t} + [\frac{I_{D_2}^{(j)}}{I_0} + 1] \cdot \frac{1}{I_0} \cdot i_{D_1} \cdot e^{\lambda \cdot t}}{[\frac{I_{D_1}^{(j)}}{I_0} + 1] \cdot [\frac{I_{D_2}^{(j)}}{I_0} + 1]}$$
$$= \{\frac{i_{D_2}}{[\frac{I_{D_2}^{(j)}}{I_0} + 1]} + \frac{i_{D_1}}{[\frac{I_{D_1}^{(j)}}{I_0} + 1]}\} \cdot \frac{1}{I_0} \cdot e^{\lambda \cdot t}$$

$$\ln(1 + x) = x - \frac{x^2}{2} + \frac{x^3}{3} - \frac{x^4}{4} + \ldots = \sum_{n=1}^{\infty} (-1)^{n+1} \ldots \frac{x^n}{n} \Rightarrow \ln(1 + x) \approx x$$

$$\ln\{1+\frac{\Omega_1}{[\frac{I_{D_1}^{(j)}}{I_0}+1]\cdot[\frac{I_{D_2}^{(j)}}{I_0}+1]}\}\approx\frac{\Omega_1}{[\frac{I_{D_1}^{(j)}}{I_0}+1]\cdot[\frac{I_{D_2}^{(j)}}{I_0}+1]}$$
$$=\{\frac{i_{D_2}}{[\frac{I_{D_2}^{(j)}}{I_0}+1]}+\frac{i_{D_1}}{[\frac{I_{D_1}^{(j)}}{I_0}+1]}\}\cdot\frac{1}{I_0}\cdot e^{\lambda\cdot t}$$

$$\ln\{[\frac{I_{D_1}^{(j)}+i_{D_1}\cdot e^{\lambda\cdot t}}{I_0}+1]\cdot[\frac{I_{D_2}^{(j)}+i_{D_2}\cdot e^{\lambda\cdot t}}{I_0}+1]\}$$
$$=\ln\{[\frac{I_{D_1}^{(j)}}{I_0}+1]\cdot[\frac{I_{D_2}^{(j)}}{I_0}+1]\}+\{\frac{i_{D_2}}{[\frac{I_{D_2}^{(j)}}{I_0}+1]}+\frac{i_{D_1}}{[\frac{I_{D_1}^{(j)}}{I_0}+1]}\}\cdot\frac{1}{I_0}\cdot e^{\lambda\cdot t}$$

Calculation No. 2:

$$\frac{[I_{D_2}^{(j)}+i_{D_2}\cdot e^{\lambda\cdot t}+I_0]}{R_L\cdot C_{load}}\cdot\ln\{[\frac{I_{D_1}^{(j)}+i_{D_1}\cdot e^{\lambda\cdot t}}{I_0}+1]\cdot[\frac{I_{D_2}^{(j)}+i_{D_2}\cdot e^{\lambda\cdot t}}{I_0}+1]\}$$
$$=\frac{[I_{D_2}^{(j)}+i_{D_2}\cdot e^{\lambda\cdot t}+I_0]}{R_L\cdot C_{load}}\cdot(\ln\{[\frac{I_{D_1}^{(j)}}{I_0}+1]\cdot[\frac{I_{D_2}^{(j)}}{I_0}+1]\}+\{\frac{i_{D_2}}{[\frac{I_{D_2}^{(j)}}{I_0}+1]}+\frac{i_{D_1}}{[\frac{I_{D_1}^{(j)}}{I_0}+1]}\}\cdot\frac{1}{I_0}\cdot e^{\lambda\cdot t})$$

$$\frac{[I_{D_2}^{(j)}+i_{D_2}\cdot e^{\lambda\cdot t}+I_0]}{R_L\cdot C_{load}}\cdot\ln\{[\frac{I_{D_1}^{(j)}+i_{D_1}\cdot e^{\lambda\cdot t}}{I_0}+1]\cdot[\frac{I_{D_2}^{(j)}+i_{D_2}\cdot e^{\lambda\cdot t}}{I_0}+1]\}$$
$$=\{\frac{I_{D_2}^{(j)}+I_0}{R_L\cdot C_{load}}+\frac{i_{D_2}\cdot e^{\lambda\cdot t}}{R_L\cdot C_{load}}\}\cdot(\ln\{[\frac{I_{D_1}^{(j)}}{I_0}+1]\cdot[\frac{I_{D_2}^{(j)}}{I_0}+1]\}+\{\frac{i_{D_2}}{[\frac{I_{D_2}^{(j)}}{I_0}+1]}+\frac{i_{D_1}}{[\frac{I_{D_1}^{(j)}}{I_0}+1]}\}\cdot\frac{1}{I_0}\cdot e^{\lambda\cdot t})$$
$$=\frac{(I_{D_2}^{(j)}+I_0)}{R_L\cdot C_{load}}\cdot\ln\{[\frac{I_{D_1}^{(j)}}{I_0}+1]\cdot[\frac{I_{D_2}^{(j)}}{I_0}+1]\}+\frac{(I_{D_2}^{(j)}+I_0)}{R_L\cdot C_{load}}\cdot\{\frac{i_{D_2}}{[\frac{I_{D_2}^{(j)}}{I_0}+1]}+\frac{i_{D_1}}{[\frac{I_{D_1}^{(j)}}{I_0}+1]}\}\cdot\frac{1}{I_0}\cdot e^{\lambda\cdot t}$$
$$+\frac{i_{D_2}\cdot e^{\lambda\cdot t}}{R_L\cdot C_{load}}\cdot\ln\{[\frac{I_{D_1}^{(j)}}{I_0}+1]\cdot[\frac{I_{D_2}^{(j)}}{I_0}+1]\}+\frac{e^{2\cdot\lambda\cdot t}}{R_L\cdot C_{load}\cdot I_0}\cdot\{\frac{i_{D_2}\cdot i_{D_2}}{[\frac{I_{D_2}^{(j)}}{I_0}+1]}+\frac{i_{D_2}\cdot i_{D_1}}{[\frac{I_{D_1}^{(j)}}{I_0}+1]}\}$$

$$i_{D_2}\cdot i_{D_2}\approx 0; i_{D_2}\cdot i_{D_1}\approx 0\Rightarrow\frac{e^{2\cdot\lambda\cdot t}}{R_L\cdot C_{load}\cdot I_0}\cdot\{\frac{i_{D_2}\cdot i_{D_2}}{[\frac{I_{D_2}^{(j)}}{I_0}+1]}+\frac{i_{D_2}\cdot i_{D_1}}{[\frac{I_{D_1}^{(j)}}{I_0}+1]}\}\to\varepsilon$$

$$\frac{[I_{D_2}^{(j)} + i_{D_2} \cdot e^{\lambda \cdot t} + I_0]}{R_L \cdot C_{load}} \cdot \ln\{[\frac{I_{D_1}^{(j)} + i_{D_1} \cdot e^{\lambda \cdot t}}{I_0} + 1] \cdot [\frac{I_{D_2}^{(j)} + i_{D_2} \cdot e^{\lambda \cdot t}}{I_0} + 1]\}$$
$$= \frac{(I_{D_2}^{(j)} + I_0)}{R_L \cdot C_{load}} \cdot \ln\{[\frac{I_{D_1}^{(j)}}{I_0} + 1] \cdot [\frac{I_{D_2}^{(j)}}{I_0} + 1]\} + \frac{(I_{D_2}^{(j)} + I_0)}{R_L \cdot C_{load}} \cdot \{\frac{i_{D_2}}{[\frac{I_{D_2}^{(j)}}{I_0} + 1]} + \frac{i_{D_1}}{[\frac{I_{D_1}^{(j)}}{I_0} + 1]}\} \cdot \frac{1}{I_0} \cdot e^{\lambda \cdot t}$$
$$+ \frac{i_{D_2} \cdot e^{\lambda \cdot t}}{R_L \cdot C_{load}} \cdot \ln\{[\frac{I_{D_1}^{(j)}}{I_0} + 1] \cdot [\frac{I_{D_2}^{(j)}}{I_0} + 1]\}$$

<u>Calculation No. 3</u>:

$$\ln[\frac{I_{D_1}^{(j)} + i_{D_1} \cdot e^{\lambda \cdot t}}{I_0} + 1] = \ln[(\frac{I_{D_1}^{(j)}}{I_0} + 1) + \frac{i_{D_1} \cdot e^{\lambda \cdot t}}{I_0}] = \ln[(\frac{I_{D_1}^{(j)}}{I_0} + 1) \cdot \{1 + \frac{i_{D_1} \cdot e^{\lambda \cdot t}}{I_0 \cdot (\frac{I_{D_1}^{(j)}}{I_0} + 1)}\}]$$

$$\ln[\frac{I_{D_1}^{(j)} + i_{D_1} \cdot e^{\lambda \cdot t}}{I_0} + 1] = \ln(\frac{I_{D_1}^{(j)}}{I_0} + 1) + \ln\{1 + \frac{i_{D_1} \cdot e^{\lambda \cdot t}}{I_0 \cdot (\frac{I_{D_1}^{(j)}}{I_0} + 1)}\}$$

$$\ln(1 + x) = x - \frac{x^2}{2} + \frac{x^3}{3} - \frac{x^4}{4} + \cdots = \sum_{n=1}^{\infty} (-1)^{n+1} \cdot \frac{x^n}{n} \Rightarrow \ln(1 + x) \approx x$$

$$\ln[\frac{I_{D_1}^{(j)} + i_{D_1} \cdot e^{\lambda \cdot t}}{I_0} + 1] = \ln(\frac{I_{D_1}^{(j)}}{I_0} + 1) + \ln\{1 + \frac{i_{D_1} \cdot e^{\lambda \cdot t}}{I_0 \cdot (\frac{I_{D_1}^{(j)}}{I_0} + 1)}\}$$
$$= \ln(\frac{I_{D_1}^{(j)}}{I_0} + 1) + \frac{i_{D_1} \cdot e^{\lambda \cdot t}}{I_0 \cdot (\frac{I_{D_1}^{(j)}}{I_0} + 1)}$$

Integrating last results in the next expression:

$$z \cdot \lambda \cdot e^{\lambda \cdot t} = \frac{[L_1 - L_{calc}]}{L_{calc} \cdot C_1} \cdot [-(Y^{(j)} + y \cdot e^{\lambda \cdot t}) - \frac{I_{D_2}^{(j)} \cdot (I_{D_2}^{(j)} + I_0) + I_{D_2}^{(j)} \cdot i_{D_2} \cdot e^{\lambda \cdot t} + (I_{D_2}^{(j)} + I_0) \cdot i_{D_2} \cdot e^{\lambda \cdot t}}{C_{load} \cdot V_t}$$
$$- \{\frac{(I_{D_2}^{(j)} + I_0) \cdot (I_{D_1}^{(j)} + I_0) - (I_{D_2}^{(j)} + I_0) \cdot i_{D_1} \cdot e^{\lambda \cdot t} + (I_{D_1}^{(j)} + I_0) \cdot i_{D_2} \cdot e^{\lambda \cdot t}}{(I_{D_1}^{(j)} + I_0)^2}\} \cdot (Y^{(j)} + y \cdot e^{\lambda \cdot t}) \quad (£)$$
$$- \frac{[I_{D_2}^{(j)} + i_{D_2} \cdot e^{\lambda \cdot t} + I_0]}{R_L \cdot C_{load}} \cdot \ln\{[\frac{I_{D_1}^{(j)} + i_{D_1} \cdot e^{\lambda \cdot t}}{I_0} + 1] \cdot [\frac{I_{D_2}^{(j)} + i_{D_2} \cdot e^{\lambda \cdot t}}{I_0} + 1]\}] - \frac{R_{DC}}{L_{calc}} \cdot (Z^{(j)} + z \cdot e^{\lambda \cdot t})$$
$$-[-I_{D_1}^{(j)} - i_{D_1} \cdot e^{\lambda \cdot t} + I_{D_2}^{(j)} + i_{D_2} \cdot e^{\lambda \cdot t}] \cdot \frac{R_{DC}}{L_{calc} \cdot C_1} - \frac{V_t}{L_{calc} \cdot C_1} \cdot \ln[\frac{I_{D_1}^{(j)} + i_{D_1} \cdot e^{\lambda \cdot t}}{I_0} + 1]$$

$$
\begin{aligned}
z \cdot \lambda \cdot e^{\lambda \cdot t} = & \frac{[L_1 - L_{calc}]}{L_{calc} \cdot C_1} \cdot [-(Y^{(j)} + y \cdot e^{\lambda \cdot t}) \\
& - \frac{I_{D_2}^{(j)} \cdot (I_{D_2}^{(j)} + I_0) + I_{D_2}^{(j)} \cdot i_{D_2} \cdot e^{\lambda \cdot t} + (I_{D_2}^{(j)} + I_0) \cdot i_{D_2} \cdot e^{\lambda \cdot t}}{C_{load} \cdot V_t} \\
& - \{\frac{(I_{D_2}^{(j)} + I_0) \cdot (I_{D_1}^{(j)} + I_0) - (I_{D_2}^{(j)} + I_0) \cdot i_{D_1} \cdot e^{\lambda \cdot t} + (I_{D_1}^{(j)} + I_0) \cdot i_{D_2} \cdot e^{\lambda \cdot t}}{(I_{D_1}^{(j)} + I_0)^2}\} \cdot (Y^{(j)} + y \cdot e^{\lambda \cdot t}) \\
& - \{\frac{(I_{D_2}^{(j)} + I_0)}{R_L \cdot C_{load}} \cdot \ln\{[\frac{I_{D_1}^{(j)}}{I_0} + 1] \cdot [\frac{I_{D_2}^{(j)}}{I_0} + 1]\} \\
& + \frac{(I_{D_2}^{(j)} + I_0)}{R_L \cdot C_{load}} \cdot \{\frac{i_{D_2}}{[\frac{I_{D_2}^{(j)}}{I_0} + 1]} + \frac{i_{D_1}}{[\frac{I_{D_1}^{(j)}}{I_0} + 1]}\} \cdot \frac{1}{I_0} \cdot e^{\lambda \cdot t} \\
& + \frac{i_{D_2} \cdot e^{\lambda \cdot t}}{R_L \cdot C_{load}} \cdot \ln([\frac{I_{D_1}^{(j)}}{I_0} + 1] \cdot [\frac{I_{D_2}^{(j)}}{I_0} + 1])\}] - \frac{R_{DC}}{L_{calc}} \cdot (Z^{(j)} + z \cdot e^{\lambda \cdot t}) \\
& - [-I_{D_1}^{(j)} - i_{D_1} \cdot e^{\lambda \cdot t} + I_{D_2}^{(j)} + i_{D_2} \cdot e^{\lambda \cdot t}] \cdot \frac{R_{DC}}{L_{calc} \cdot C_1} \\
& - \frac{V_t}{L_{calc} \cdot C_1} \cdot \{\ln(\frac{I_{D_1}^{(j)}}{I_0} + 1) + \frac{i_{D_1} \cdot e^{\lambda \cdot t}}{I_0 \cdot (\frac{I_{D_1}^{(j)}}{I_0} + 1)}\}
\end{aligned}
$$

The condition of our system fixed points:

$$
\begin{aligned}
& \frac{dZ}{dt}|_{@I_{D_1}^{(j)}, I_{D_2}^{(j)}, Y^{(j)}, Z^{(j)}} = 0 \\
& \frac{[L_1 - L_{calc}]}{L_{calc} \cdot C_1} \cdot [-Y^{(j)} - \frac{I_{D_2}^{(j)} \cdot (I_{D_2}^{(j)} + I_0)}{C_{load} \cdot V_t} - \{\frac{(I_{D_2}^{(j)} + I_0)}{(I_{D_1}^{(j)} + I_0)}\} \cdot Y^{(j)} \\
& - \frac{(I_{D_2}^{(j)} + I_0)}{R_L \cdot C_{load}} \cdot \ln\{[\frac{I_{D_1}^{(j)}}{I_0} + 1] \cdot [\frac{I_{D_2}^{(j)}}{I_0} + 1]\}] \\
& - \frac{R_{DC}}{L_{calc}} \cdot Z^{(j)} - [-I_{D_1}^{(j)} + I_{D_2}^{(j)}] \cdot \frac{R_{DC}}{L_{calc} \cdot C_1} - \frac{V_t}{L_{calc} \cdot C_1} \cdot \ln(\frac{I_{D_1}^{(j)}}{I_0} + 1) = 0
\end{aligned}
$$

$$
\begin{aligned}
z \cdot \lambda \cdot e^{\lambda \cdot t} = {} & \frac{[L_1 - L_{calc}]}{L_{calc} \cdot C_1} \cdot [-y \cdot e^{\lambda \cdot t} - \frac{I_{D_2}^{(j)} \cdot i_{D_2} \cdot e^{\lambda \cdot t} + (I_{D_2}^{(j)} + I_0) \cdot i_{D_2} \cdot e^{\lambda \cdot t}}{C_{load} \cdot V_t} \\
& - \frac{(I_{D_2}^{(j)} + I_0)}{(I_{D_1}^{(j)} + I_0)} \cdot y \cdot e^{\lambda \cdot t} \\
& + \frac{(I_{D_2}^{(j)} + I_0)}{(I_{D_1}^{(j)} + I_0)^2} \cdot i_{D_1} \cdot e^{\lambda \cdot t} \cdot Y^{(j)} + \frac{(I_{D_2}^{(j)} + I_0)}{(I_{D_1}^{(j)} + I_0)^2} \cdot i_{D_1} \cdot e^{\lambda \cdot t} \cdot y \cdot e^{\lambda \cdot t} \\
& - \frac{i_{D_2} \cdot e^{\lambda \cdot t} \cdot Y^{(j)}}{(I_{D_1}^{(j)} + I_0)} - \frac{i_{D_2} \cdot e^{\lambda \cdot t} \cdot y \cdot e^{\lambda \cdot t}}{(I_{D_1}^{(j)} + I_0)} \\
& - \frac{(I_{D_2}^{(j)} + I_0)}{R_L \cdot C_{load}} \cdot \{\frac{i_{D_2}}{[\frac{I_{D_2}^{(j)}}{I_0} + 1]} + \frac{i_{D_1}}{[\frac{I_{D_1}^{(j)}}{I_0} + 1]}\} \cdot \frac{1}{I_0} \cdot e^{\lambda \cdot t} \\
& - \frac{i_{D_2} \cdot e^{\lambda \cdot t}}{R_L \cdot C_{load}} \cdot \ln([\frac{I_{D_1}^{(j)}}{I_0} + 1] \cdot [\frac{I_{D_2}^{(j)}}{I_0} + 1])] \\
& - \frac{R_{DC}}{L_{calc}} \cdot z \cdot e^{\lambda \cdot t} - [-i_{D_1} \cdot e^{\lambda \cdot t} + i_{D_2} \cdot e^{\lambda \cdot t}] \cdot \frac{R_{DC}}{L_{calc} \cdot C_1} \\
& - \frac{V_t}{L_{calc} \cdot C_1} \cdot \{\frac{i_{D_1} \cdot e^{\lambda \cdot t}}{I_0 \cdot (\frac{I_{D_1}^{(j)}}{I_0} + 1)}\}
\end{aligned}
$$

$$
i_{D_1} \cdot y \approx 0 \Rightarrow \frac{(I_{D_2}^{(j)} + I_0)}{(I_{D_1}^{(j)} + I_0)^2} \cdot i_{D_1} \cdot e^{\lambda \cdot t} \cdot y \cdot e^{\lambda \cdot t} \to \varepsilon;
$$

$$
i_{D_2} \cdot y \approx 0 \Rightarrow \frac{i_{D_2} \cdot e^{\lambda \cdot t} \cdot y \cdot e^{\lambda \cdot t}}{(I_{D_1}^{(j)} + I_0)} \to \varepsilon
$$

$$
\begin{aligned}
z \cdot \lambda \cdot e^{\lambda \cdot t} = {} & \frac{[L_1 - L_{calc}]}{L_{calc} \cdot C_1} \cdot [-y \cdot e^{\lambda \cdot t} - \frac{I_{D_2}^{(j)} \cdot i_{D_2} \cdot e^{\lambda \cdot t} + (I_{D_2}^{(j)} + I_0) \cdot i_{D_2} \cdot e^{\lambda \cdot t}}{C_{load} \cdot V_t} \\
& - \frac{(I_{D_2}^{(j)} + I_0)}{(I_{D_1}^{(j)} + I_0)} \cdot y \cdot e^{\lambda \cdot t} \\
& + \frac{(I_{D_2}^{(j)} + I_0)}{(I_{D_1}^{(j)} + I_0)^2} \cdot i_{D_1} \cdot e^{\lambda \cdot t} \cdot Y^{(j)} - \frac{i_{D_2} \cdot e^{\lambda \cdot t} \cdot Y^{(j)}}{(I_{D_1}^{(j)} + I_0)} \\
& - \frac{(I_{D_2}^{(j)} + I_0)}{R_L \cdot C_{load}} \cdot \{\frac{i_{D_2}}{[\frac{I_{D_2}^{(j)}}{I_0} + 1]} + \frac{i_{D_1}}{[\frac{I_{D_1}^{(j)}}{I_0} + 1]}\} \cdot \frac{1}{I_0} \cdot e^{\lambda \cdot t} \\
& - \frac{i_{D_2} \cdot e^{\lambda \cdot t}}{R_L \cdot C_{load}} \cdot \ln([\frac{I_{D_1}^{(j)}}{I_0} + 1] \cdot [\frac{I_{D_2}^{(j)}}{I_0} + 1])] - \frac{R_{DC}}{L_{calc}} \cdot z \cdot e^{\lambda \cdot t} \\
& - [-i_{D_1} \cdot e^{\lambda \cdot t} + i_{D_2} \cdot e^{\lambda \cdot t}] \cdot \frac{R_{DC}}{L_{calc} \cdot C_1} - \frac{V_t}{L_{calc} \cdot C_1} \cdot \{\frac{i_{D_1} \cdot e^{\lambda \cdot t}}{I_0 \cdot (\frac{I_{D_1}^{(j)}}{I_0} + 1)}\}
\end{aligned}
$$

$$\frac{dI_{D_2}}{dt} = -\frac{I_{D_2} \cdot [I_{D_2} + I_0]}{C_{load} \cdot V_t} - \frac{[I_{D_2} + I_0]}{[I_{D_1} + I_0]} \cdot Y - \frac{[I_{D_2} + I_0]}{R_L \cdot C_{load}} \cdot \ln\{[\frac{I_{D_1}}{I_0} + 1] \cdot [\frac{I_{D_2}}{I_0} + 1]\}$$

$$\begin{aligned}\lambda \cdot i_{D_2} \cdot e^{\lambda \cdot t} = & -\frac{[I_{D_2}^{(j)} + i_{D_2} \cdot e^{\lambda \cdot t}] \cdot [I_{D_2}^{(j)} + i_{D_2} \cdot e^{\lambda \cdot t} + I_0]}{C_{load} \cdot V_t} \\ & -\frac{[I_{D_2}^{(j)} + i_{D_2} \cdot e^{\lambda \cdot t} + I_0]}{[I_{D_1}^{(j)} + i_{D_1} \cdot e^{\lambda \cdot t} + I_0]} \cdot [Y^{(j)} + y \cdot e^{\lambda \cdot t}] \\ & -\frac{[I_{D_2}^{(j)} + i_{D_2} \cdot e^{\lambda \cdot t} + I_0]}{R_L \cdot C_{load}} \cdot \ln\{[\frac{(I_{D_1}^{(j)} + i_{D_1} \cdot e^{\lambda \cdot t})}{I_0} + 1] \cdot [\frac{(I_{D_2}^{(j)} + i_{D_2} \cdot e^{\lambda \cdot t})}{I_0} + 1]\}\end{aligned}$$

$$\begin{aligned}\lambda \cdot i_{D_2} \cdot e^{\lambda \cdot t} = & -\frac{[I_{D_2}^{(j)} + i_{D_2} \cdot e^{\lambda \cdot t}] \cdot [(I_{D_2}^{(j)} + I_0) + i_{D_2} \cdot e^{\lambda \cdot t}]}{C_{load} \cdot V_t} \\ & -\frac{[(I_{D_2}^{(j)} + I_0) + i_{D_2} \cdot e^{\lambda \cdot t}]}{[(I_{D_1}^{(j)} + I_0) + i_{D_1} \cdot e^{\lambda \cdot t}]} \cdot [Y^{(j)} + y \cdot e^{\lambda \cdot t}] \\ & -\frac{[(I_{D_2}^{(j)} + I_0) + i_{D_2} \cdot e^{\lambda \cdot t}]}{R_L \cdot C_{load}} \cdot \ln\{[\frac{(I_{D_1}^{(j)} + i_{D_1} \cdot e^{\lambda \cdot t})}{I_0} + 1] \cdot [\frac{(I_{D_2}^{(j)} + i_{D_2} \cdot e^{\lambda \cdot t})}{I_0} + 1]\}\end{aligned}$$

$$\begin{aligned}\lambda \cdot i_{D_2} \cdot e^{\lambda \cdot t} = & -\frac{I_{D_2}^{(j)} \cdot (I_{D_2}^{(j)} + I_0) + I_{D_2}^{(j)} \cdot i_{D_2} \cdot e^{\lambda \cdot t} + (I_{D_2}^{(j)} + I_0) \cdot i_{D_2} \cdot e^{\lambda \cdot t} + i_{D_2} \cdot i_{D_2} \cdot e^{2 \cdot \lambda \cdot t}}{C_{load} \cdot V_t} \\ & -\frac{[(I_{D_2}^{(j)} + I_0) + i_{D_2} \cdot e^{\lambda \cdot t}]}{[(I_{D_1}^{(j)} + I_0) + i_{D_1} \cdot e^{\lambda \cdot t}]} \cdot \frac{[(I_{D_1}^{(j)} + I_0) - i_{D_1} \cdot e^{\lambda \cdot t}]}{[(I_{D_1}^{(j)} + I_0) - i_{D_1} \cdot e^{\lambda \cdot t}]} \cdot [Y^{(j)} + y \cdot e^{\lambda \cdot t}] \\ & -\frac{[(I_{D_2}^{(j)} + I_0) + i_{D_2} \cdot e^{\lambda \cdot t}]}{R_L \cdot C_{load}} \cdot \ln\{[\frac{(I_{D_1}^{(j)} + i_{D_1} \cdot e^{\lambda \cdot t})}{I_0} + 1] \cdot [\frac{(I_{D_2}^{(j)} + i_{D_2} \cdot e^{\lambda \cdot t})}{I_0} + 1]\}\end{aligned}$$

$$\begin{aligned}\lambda \cdot i_{D_2} \cdot e^{\lambda \cdot t} = & -\frac{I_{D_2}^{(j)} \cdot (I_{D_2}^{(j)} + I_0) + I_{D_2}^{(j)} \cdot i_{D_2} \cdot e^{\lambda \cdot t} + (I_{D_2}^{(j)} + I_0) \cdot i_{D_2} \cdot e^{\lambda \cdot t} + i_{D_2} \cdot i_{D_2} \cdot e^{2 \cdot \lambda \cdot t}}{C_{load} \cdot V_t} \\ & -\{\frac{(I_{D_2}^{(j)} + I_0) \cdot (I_{D_1}^{(j)} + I_0) - (I_{D_2}^{(j)} + I_0) \cdot i_{D_1} \cdot e^{\lambda \cdot t} + (I_{D_1}^{(j)} + I_0) \cdot i_{D_2} \cdot e^{\lambda \cdot t} - i_{D_2} \cdot i_{D_1} \cdot e^{2 \cdot \lambda \cdot t}}{(I_{D_1}^{(j)} + I_0)^2 - i_{D_1}^2 \cdot e^{2 \cdot \lambda \cdot t}}\} \\ & \cdot [Y^{(j)} + y \cdot e^{\lambda \cdot t}] \\ & -\frac{[(I_{D_2}^{(j)} + I_0) + i_{D_2} \cdot e^{\lambda \cdot t}]}{R_L \cdot C_{load}} \cdot \ln\{[\frac{(I_{D_1}^{(j)} + i_{D_1} \cdot e^{\lambda \cdot t})}{I_0} + 1] \cdot [\frac{(I_{D_2}^{(j)} + i_{D_2} \cdot e^{\lambda \cdot t})}{I_0} + 1]\}\end{aligned}$$

$$i_{D_2} \cdot i_{D_2} \approx 0; i_{D_2} \cdot i_{D_1} \approx 0; i_{D_1}^2 \approx 0$$

$$\lambda \cdot i_{D_2} \cdot e^{\lambda \cdot t} = -\frac{I_{D_2}^{(j)} \cdot (I_{D_2}^{(j)} + I_0) + I_{D_2}^{(j)} \cdot i_{D_2} \cdot e^{\lambda \cdot t} + (I_{D_2}^{(j)} + I_0) \cdot i_{D_2} \cdot e^{\lambda \cdot t}}{C_{load} \cdot V_t}$$
$$- \{\frac{(I_{D_2}^{(j)} + I_0) \cdot (I_{D_1}^{(j)} + I_0) - (I_{D_2}^{(j)} + I_0) \cdot i_{D_1} \cdot e^{\lambda \cdot t} + (I_{D_1}^{(j)} + I_0) \cdot i_{D_2} \cdot e^{\lambda \cdot t}}{(I_{D_1}^{(j)} + I_0)^2}\}$$
$$\cdot [Y^{(j)} + y \cdot e^{\lambda \cdot t}]$$
$$- \frac{[(I_{D_2}^{(j)} + I_0) + i_{D_2} \cdot e^{\lambda \cdot t}]}{R_L \cdot C_{load}} \cdot \ln\{[\frac{(I_{D_1}^{(j)} + i_{D_1} \cdot e^{\lambda \cdot t})}{I_0} + 1] \cdot [\frac{(I_{D_2}^{(j)} + i_{D_2} \cdot e^{\lambda \cdot t})}{I_0} + 1]\}$$

$$\lambda \cdot i_{D_2} \cdot e^{\lambda \cdot t} = -\frac{I_{D_2}^{(j)} \cdot (I_{D_2}^{(j)} + I_0) + I_{D_2}^{(j)} \cdot i_{D_2} \cdot e^{\lambda \cdot t} + (I_{D_2}^{(j)} + I_0) \cdot i_{D_2} \cdot e^{\lambda \cdot t}}{C_{load} \cdot V_t}$$
$$- \{\frac{Y^{(j)} \cdot (I_{D_2}^{(j)} + I_0) \cdot (I_{D_1}^{(j)} + I_0) - Y^{(j)} \cdot (I_{D_2}^{(j)} + I_0) \cdot i_{D_1} \cdot e^{\lambda \cdot t} + Y^{(j)} \cdot (I_{D_1}^{(j)} + I_0) \cdot i_{D_2} \cdot e^{\lambda \cdot t}}{(I_{D_1}^{(j)} + I_0)^2}\}$$
$$- \{\frac{(I_{D_2}^{(j)} + I_0) \cdot (I_{D_1}^{(j)} + I_0) \cdot y \cdot e^{\lambda \cdot t} - (I_{D_2}^{(j)} + I_0) \cdot i_{D_1} \cdot y \cdot e^{2 \cdot \lambda \cdot t} + (I_{D_1}^{(j)} + I_0) \cdot i_{D_2} \cdot y \cdot e^{2 \cdot \lambda \cdot t}}{(I_{D_1}^{(j)} + I_0)^2}\}$$
$$- \frac{[(I_{D_2}^{(j)} + I_0) + i_{D_2} \cdot e^{\lambda \cdot t}]}{R_L \cdot C_{load}} \cdot \ln\{[\frac{(I_{D_1}^{(j)} + i_{D_1} \cdot e^{\lambda \cdot t})}{I_0} + 1] \cdot [\frac{(I_{D_2}^{(j)} + i_{D_2} \cdot e^{\lambda \cdot t})}{I_0} + 1]\}$$

$$i_{D_1} \cdot y \approx 0; i_{D_2} \cdot y \approx 0$$

$$\lambda \cdot i_{D_2} \cdot e^{\lambda \cdot t} = -\frac{I_{D_2}^{(j)} \cdot (I_{D_2}^{(j)} + I_0) + I_{D_2}^{(j)} \cdot i_{D_2} \cdot e^{\lambda \cdot t} + (I_{D_2}^{(j)} + I_0) \cdot i_{D_2} \cdot e^{\lambda \cdot t}}{C_{load} \cdot V_t}$$
$$- \{\frac{Y^{(j)} \cdot (I_{D_2}^{(j)} + I_0) \cdot (I_{D_1}^{(j)} + I_0) - Y^{(j)} \cdot (I_{D_2}^{(j)} + I_0) \cdot i_{D_1} \cdot e^{\lambda \cdot t} + Y^{(j)} \cdot (I_{D_1}^{(j)} + I_0) \cdot i_{D_2} \cdot e^{\lambda \cdot t}}{(I_{D_1}^{(j)} + I_0)^2}\}$$
$$- \{\frac{(I_{D_2}^{(j)} + I_0) \cdot (I_{D_1}^{(j)} + I_0) \cdot y \cdot e^{\lambda \cdot t}}{(I_{D_1}^{(j)} + I_0)^2}\}$$
$$- \frac{[(I_{D_2}^{(j)} + I_0) + i_{D_2} \cdot e^{\lambda \cdot t}]}{R_L \cdot C_{load}} \cdot \ln\{[\frac{(I_{D_1}^{(j)} + i_{D_1} \cdot e^{\lambda \cdot t})}{I_0} + 1] \cdot [\frac{(I_{D_2}^{(j)} + i_{D_2} \cdot e^{\lambda \cdot t})}{I_0} + 1]\}$$

We have already approved in calculation No. 1

$$\ln\{[\frac{I_{D_1}^{(j)} + i_{D_1} \cdot e^{\lambda \cdot t}}{I_0} + 1] \cdot [\frac{I_{D_2}^{(j)} + i_{D_2} \cdot e^{\lambda \cdot t}}{I_0} + 1]\} = \ln\{[\frac{I_{D_1}^{(j)}}{I_0} + 1] \cdot [\frac{I_{D_2}^{(j)}}{I_0} + 1]\}$$
$$+ \{\frac{i_{D_2}}{[\frac{I_{D_2}^{(j)}}{I_0} + 1]} + \frac{i_{D_1}}{[\frac{I_{D_1}^{(j)}}{I_0} + 1]}\} \cdot \frac{1}{I_0} \cdot e^{\lambda \cdot t}$$

$$\lambda \cdot i_{D_2} \cdot e^{\lambda \cdot t} = -\frac{I_{D_2}^{(j)} \cdot (I_{D_2}^{(j)} + I_0) + I_{D_2}^{(j)} \cdot i_{D_2} \cdot e^{\lambda \cdot t} + (I_{D_2}^{(j)} + I_0) \cdot i_{D_2} \cdot e^{\lambda \cdot t}}{C_{load} \cdot V_t}$$
$$- \{\frac{Y^{(j)} \cdot (I_{D_2}^{(j)} + I_0) \cdot (I_{D_1}^{(j)} + I_0) - Y^{(j)} \cdot (I_{D_2}^{(j)} + I_0) \cdot i_{D_1} \cdot e^{\lambda \cdot t} + Y^{(j)} \cdot (I_{D_1}^{(j)} + I_0) \cdot i_{D_2} \cdot e^{\lambda \cdot t}}{(I_{D_1}^{(j)} + I_0)^2}\}$$
$$- \{\frac{(I_{D_2}^{(j)} + I_0) \cdot (I_{D_1}^{(j)} + I_0) \cdot y \cdot e^{\lambda \cdot t}}{(I_{D_1}^{(j)} + I_0)^2}\}$$
$$- \{\frac{(I_{D_2}^{(j)} + I_0)}{R_L \cdot C_{load}} + \frac{i_{D_2} \cdot e^{\lambda \cdot t}}{R_L \cdot C_{load}}\} \cdot \{\ln\{[\frac{I_{D_1}^{(j)}}{I_0} + 1] \cdot [\frac{I_{D_2}^{(j)}}{I_0} + 1]\}$$
$$+ \{\frac{i_{D_2}}{[\frac{I_{D_2}^{(j)}}{I_0} + 1]} + \frac{i_{D_1}}{[\frac{I_{D_1}^{(j)}}{I_0} + 1]}\} \cdot \frac{1}{I_0} \cdot e^{\lambda \cdot t}\}$$

$$\lambda \cdot i_{D_2} \cdot e^{\lambda \cdot t} = -\frac{I_{D_2}^{(j)} \cdot (I_{D_2}^{(j)} + I_0)}{C_{load} \cdot V_t} - \frac{I_{D_2}^{(j)} \cdot i_{D_2} \cdot e^{\lambda \cdot t} + (I_{D_2}^{(j)} + I_0) \cdot i_{D_2} \cdot e^{\lambda \cdot t}}{C_{load} \cdot V_t}$$
$$- \{\frac{Y^{(j)} \cdot (I_{D_2}^{(j)} + I_0)}{(I_{D_1}^{(j)} + I_0)} - \frac{Y^{(j)} \cdot (I_{D_2}^{(j)} + I_0) \cdot i_{D_1} \cdot e^{\lambda \cdot t}}{(I_{D_1}^{(j)} + I_0)^2} + \frac{Y^{(j)} \cdot i_{D_2} \cdot e^{\lambda \cdot t}}{(I_{D_1}^{(j)} + I_0)}\}$$
$$- \{\frac{(I_{D_2}^{(j)} + I_0) \cdot (I_{D_1}^{(j)} + I_0) \cdot y \cdot e^{\lambda \cdot t}}{(I_{D_1}^{(j)} + I_0)^2}\}$$
$$- \{\frac{(I_{D_2}^{(j)} + I_0)}{R_L \cdot C_{load}} + \frac{i_{D_2} \cdot e^{\lambda \cdot t}}{R_L \cdot C_{load}}\} \cdot \{\ln\{[\frac{I_{D_1}^{(j)}}{I_0} + 1] \cdot [\frac{I_{D_2}^{(j)}}{I_0} + 1]\}$$
$$+ \{\frac{i_{D_2}}{[\frac{I_{D_2}^{(j)}}{I_0} + 1]} + \frac{i_{D_1}}{[\frac{I_{D_1}^{(j)}}{I_0} + 1]}\} \cdot \frac{1}{I_0} \cdot e^{\lambda \cdot t}\}$$

$$\lambda \cdot i_{D_2} \cdot e^{\lambda \cdot t} = -\frac{I_{D_2}^{(j)} \cdot (I_{D_2}^{(j)} + I_0)}{C_{load} \cdot V_t} - \frac{I_{D_2}^{(j)} \cdot i_{D_2} \cdot e^{\lambda \cdot t} + (I_{D_2}^{(j)} + I_0) \cdot i_{D_2} \cdot e^{\lambda \cdot t}}{C_{load} \cdot V_t}$$
$$- \{\frac{Y^{(j)} \cdot (I_{D_2}^{(j)} + I_0)}{(I_{D_1}^{(j)} + I_0)} - \frac{Y^{(j)} \cdot (I_{D_2}^{(j)} + I_0) \cdot i_{D_1} \cdot e^{\lambda \cdot t}}{(I_{D_1}^{(j)} + I_0)^2} + \frac{Y^{(j)} \cdot i_{D_2} \cdot e^{\lambda \cdot t}}{(I_{D_1}^{(j)} + I_0)}\}$$
$$- \{\frac{(I_{D_2}^{(j)} + I_0) \cdot (I_{D_1}^{(j)} + I_0) \cdot y \cdot e^{\lambda \cdot t}}{(I_{D_1}^{(j)} + I_0)^2}\}$$
$$- \frac{(I_{D_2}^{(j)} + I_0)}{R_L \cdot C_{load}} \cdot \ln\{[\frac{I_{D_1}^{(j)}}{I_0} + 1] \cdot [\frac{I_{D_2}^{(j)}}{I_0} + 1]\}$$
$$- \frac{(I_{D_2}^{(j)} + I_0)}{R_L \cdot C_{load}} \cdot \{\frac{i_{D_2}}{[\frac{I_{D_2}^{(j)}}{I_0} + 1]} + \frac{i_{D_1}}{[\frac{I_{D_1}^{(j)}}{I_0} + 1]}\} \cdot \frac{1}{I_0} \cdot e^{\lambda \cdot t}$$
$$- \frac{i_{D_2} \cdot e^{\lambda \cdot t}}{R_L \cdot C_{load}} \cdot \ln\{[\frac{I_{D_1}^{(j)}}{I_0} + 1] \cdot [\frac{I_{D_2}^{(j)}}{I_0} + 1]\}$$
$$- \frac{1}{R_L \cdot C_{load}} \cdot \{\frac{i_{D_2} \cdot i_{D_2}}{[\frac{I_{D_2}^{(j)}}{I_0} + 1]} + \frac{i_{D_1} \cdot i_{D_2}}{[\frac{I_{D_1}^{(j)}}{I_0} + 1]}\} \cdot \frac{1}{I_0} \cdot e^{2 \cdot \lambda \cdot t}$$

$$i_{D_2} \cdot i_{D_2} \approx 0; i_{D_1} \cdot i_{D_2} \approx 0$$

$$\begin{aligned}
\lambda \cdot i_{D_2} \cdot e^{\lambda \cdot t} = & -\frac{I_{D_2}^{(j)} \cdot (I_{D_2}^{(j)} + I_0)}{C_{load} \cdot V_t} - \frac{I_{D_2}^{(j)} \cdot i_{D_2} \cdot e^{\lambda \cdot t} + (I_{D_2}^{(j)} + I_0) \cdot i_{D_2} \cdot e^{\lambda \cdot t}}{C_{load} \cdot V_t} \\
& - \{\frac{Y^{(j)} \cdot (I_{D_2}^{(j)} + I_0)}{(I_{D_1}^{(j)} + I_0)} - \frac{Y^{(j)} \cdot (I_{D_2}^{(j)} + I_0) \cdot i_{D_1} \cdot e^{\lambda \cdot t}}{(I_{D_1}^{(j)} + I_0)^2} + \frac{Y^{(j)} \cdot i_{D_2} \cdot e^{\lambda \cdot t}}{(I_{D_1}^{(j)} + I_0)}\} \\
& - \{\frac{(I_{D_2}^{(j)} + I_0) \cdot (I_{D_1}^{(j)} + I_0) \cdot y \cdot e^{\lambda \cdot t}}{(I_{D_1}^{(j)} + I_0)^2}\} \\
& - \frac{(I_{D_2}^{(j)} + I_0)}{R_L \cdot C_{load}} \cdot \ln\{[\frac{I_{D_1}^{(j)}}{I_0} + 1] \cdot [\frac{I_{D_2}^{(j)}}{I_0} + 1]\} \\
& - \frac{(I_{D_2}^{(j)} + I_0)}{R_L \cdot C_{load}} \cdot \{\frac{i_{D_2}}{[\frac{I_{D_2}^{(j)}}{I_0} + 1]} + \frac{i_{D_1}}{[\frac{I_{D_1}^{(j)}}{I_0} + 1]}\} \cdot \frac{1}{I_0} \cdot e^{\lambda \cdot t} \\
& - \frac{i_{D_2} \cdot e^{\lambda \cdot t}}{R_L \cdot C_{load}} \cdot \ln\{[\frac{I_{D_1}^{(j)}}{I_0} + 1] \cdot [\frac{I_{D_2}^{(j)}}{I_0} + 1]\}
\end{aligned}$$

At fixed point $\frac{dI_{D_2}}{dt} = 0$

$$\begin{aligned}
\frac{dI_{D_2}}{dt} = 0 \Rightarrow & -\frac{I_{D_2}^{(j)} \cdot (I_{D_2}^{(j)} + I_0)}{C_{load} \cdot V_t} - \frac{Y^{(j)} \cdot (I_{D_2}^{(j)} + I_0)}{(I_{D_1}^{(j)} + I_0)} \\
& - \frac{(I_{D_2}^{(j)} + I_0)}{R_L \cdot C_{load}} \cdot \ln\{[\frac{I_{D_1}^{(j)}}{I_0} + 1] \cdot [\frac{I_{D_2}^{(j)}}{I_0} + 1]\}
\end{aligned}$$

$$\begin{aligned}
\lambda \cdot i_{D_2} \cdot e^{\lambda \cdot t} = & -\frac{I_{D_2}^{(j)} \cdot i_{D_2} \cdot e^{\lambda \cdot t} + (I_{D_2}^{(j)} + I_0) \cdot i_{D_2} \cdot e^{\lambda \cdot t}}{C_{load} \cdot V_t} \\
& - \{-\frac{Y^{(j)} \cdot (I_{D_2}^{(j)} + I_0) \cdot i_{D_1} \cdot e^{\lambda \cdot t}}{(I_{D_1}^{(j)} + I_0)^2} + \frac{Y^{(j)} \cdot i_{D_2} \cdot e^{\lambda \cdot t}}{(I_{D_1}^{(j)} + I_0)}\} \\
& - \{\frac{(I_{D_2}^{(j)} + I_0) \cdot (I_{D_1}^{(j)} + I_0) \cdot y \cdot e^{\lambda \cdot t}}{(I_{D_1}^{(j)} + I_0)^2}\} \\
& - \frac{(I_{D_2}^{(j)} + I_0)}{R_L \cdot C_{load}} \cdot \{\frac{i_{D_2}}{[\frac{I_{D_2}^{(j)}}{I_0} + 1]} + \frac{i_{D_1}}{[\frac{I_{D_1}^{(j)}}{I_0} + 1]}\} \cdot \frac{1}{I_0} \cdot e^{\lambda \cdot t} \\
& - \frac{i_{D_2} \cdot e^{\lambda \cdot t}}{R_L \cdot C_{load}} \cdot \ln\{[\frac{I_{D_1}^{(j)}}{I_0} + 1] \cdot [\frac{I_{D_2}^{(j)}}{I_0} + 1]\}
\end{aligned}$$

Remark: it is reader exercise to build the system Jacobian matrix and analyze the dynamic and stability of the system based on eigenvalues investigation.

We define $I_{D_1}(t-\tau_1) = I_{D_1}^{(j)} + i_{D_1} \cdot e^{\lambda\cdot(t-\tau_1)}$; $I_{D_2}(t-\tau_2) = I_{D_2}^{(j)} + i_{D_2} \cdot e^{\lambda\cdot(t-\tau_2)}$ and $Z(t-\Delta) = Z^{(j)} + z \cdot e^{\lambda\cdot(t-\Delta)}$. Then we get three delayed differential equations with respect to coordinates $[Z, I_{D_1}, I_{D_2}, Y]$ arbitrarily small increments of exponential $[z, i_{D_1}, i_{D_2}, y] \cdot e^{\lambda\cdot t}$. We consider no delay effects on $\frac{dZ(t)}{dt}; \frac{dI_{D_1}(t)}{dt}; \frac{dI_{D_2}(t)}{dt}; Y^{(j)} = 0$.

(i)
$$z\cdot\lambda\cdot e^{\lambda\cdot t} = \frac{[L_1 - L_{calc}]}{L_{calc}\cdot C_1}\cdot[-y\cdot e^{\lambda\cdot t} - \frac{I_{D_2}^{(j)}\cdot i_{D_2}\cdot e^{\lambda\cdot t} + (I_{D_2}^{(j)} + I_0)\cdot i_{D_2}\cdot e^{\lambda\cdot t}}{C_{load}\cdot V_t} - \frac{(I_{D_2}^{(j)} + I_0)}{(I_{D_1}^{(j)} + I_0)}\cdot y\cdot e^{\lambda\cdot t}$$
$$+ \frac{(I_{D_2}^{(j)} + I_0)}{(I_{D_1}^{(j)} + I_0)^2}\cdot i_{D_1}\cdot e^{\lambda\cdot t}\cdot Y^{(j)} - \frac{i_{D_2}\cdot e^{\lambda\cdot t}\cdot Y^{(j)}}{(I_{D_1}^{(j)} + I_0)} + \frac{(I_{D_2}^{(j)} + I_0)}{R_L\cdot C_{load}}\cdot\{\frac{i_{D_2}}{[\frac{I_{D_2}^{(j)}}{I_0} + 1]} + \frac{i_{D_1}}{[\frac{I_{D_1}^{(j)}}{I_0} + 1]}\}\cdot\frac{1}{I_0}\cdot e^{\lambda\cdot t}$$
$$-\frac{i_{D_2}\cdot e^{\lambda\cdot t}}{R_L\cdot C_{load}}\cdot\ln([\frac{I_{D_1}^{(j)}}{I_0} + 1]\cdot[\frac{I_{D_2}^{(j)}}{I_0} + 1])] - \frac{R_{DC}}{L_{calc}}\cdot z\cdot e^{\lambda\cdot t} - [-i_{D_1}\cdot e^{\lambda\cdot t} + i_{D_2}\cdot e^{\lambda\cdot t}]\cdot\frac{R_{DC}}{L_{calc}\cdot C_1}$$
$$-\frac{V_t}{L_{calc}\cdot C_1}\cdot\{\frac{i_{D_1}\cdot e^{\lambda\cdot t}}{I_0\cdot(\frac{I_{D_1}^{(j)}}{I_0} + 1)}\}$$

(ii)
$$\lambda\cdot i_{D_1}\cdot e^{\lambda\cdot t} = Y^{(j)} + y\cdot e^{\lambda\cdot t}$$

(iii)
$$\lambda\cdot i_{D_2}\cdot e^{\lambda\cdot t} = -\frac{I_{D_2}^{(j)}\cdot i_{D_2}\cdot e^{\lambda\cdot t} + (I_{D_2}^{(j)} + I_0)\cdot i_{D_2}\cdot e^{\lambda\cdot t}}{C_{load}\cdot V_t}$$
$$-\{-\frac{Y^{(j)}\cdot(I_{D_2}^{(j)} + I_0)\cdot i_{D_1}\cdot e^{\lambda\cdot t}}{(I_{D_1}^{(j)} + I_0)^2} + \frac{Y^{(j)}\cdot i_{D_2}\cdot e^{\lambda\cdot t}}{(I_{D_1}^{(j)} + I_0)}\}$$
$$-\{\frac{(I_{D_2}^{(j)} + I_0)\cdot(I_{D_1}^{(j)} + I_0)\cdot y\cdot e^{\lambda\cdot t}}{(I_{D_1}^{(j)} + I_0)^2}\}$$
$$-\frac{(I_{D_2}^{(j)} + I_0)}{R_L\cdot C_{load}}\cdot\{\frac{i_{D_2}}{[\frac{I_{D_2}^{(j)}}{I_0} + 1]} + \frac{i_{D_1}}{[\frac{I_{D_1}^{(j)}}{I_0} + 1]}\}\cdot\frac{1}{I_0}\cdot e^{\lambda\cdot t}$$
$$-\frac{i_{D_2}\cdot e^{\lambda\cdot t}}{R_L\cdot C_{load}}\cdot\ln\{[\frac{I_{D_1}^{(j)}}{I_0} + 1]\cdot[\frac{I_{D_2}^{(j)}}{I_0} + 1]\}$$

$$i_{D_1}\cdot e^{\lambda\cdot t} \to i_{D_1}\cdot e^{\lambda\cdot t}\cdot e^{-\lambda\cdot\tau_1}; i_{D_2}\cdot e^{\lambda\cdot t} \to i_{D_2}\cdot e^{\lambda\cdot t}\cdot e^{-\lambda\cdot\tau_2};$$
$$z\cdot e^{\lambda\cdot t} \to z\cdot e^{\lambda\cdot t}\cdot e^{-\lambda\cdot\Delta}; Y^{(j)} = 0$$

Remark: left side of below equation doesn't affect by delay parameter.

(i)

$$z\cdot\lambda\cdot e^{\lambda\cdot t} = \frac{[L_1 - L_{calc}]}{L_{calc}\cdot C_1}\cdot[-y\cdot e^{\lambda\cdot t} - \frac{I_{D_2}^{(j)}\cdot i_{D_2}\cdot e^{\lambda\cdot t}\cdot e^{-\lambda\cdot\tau_2} + (I_{D_2}^{(j)} + I_0)\cdot i_{D_2}\cdot e^{\lambda\cdot t}\cdot e^{-\lambda\cdot\tau_2}}{C_{load}\cdot V_t}$$
$$-\frac{(I_{D_2}^{(j)} + I_0)}{(I_{D_1}^{(j)} + I_0)}\cdot y\cdot e^{\lambda\cdot t}$$
$$+\frac{(I_{D_2}^{(j)} + I_0)}{R_L\cdot C_{load}}\cdot\{\frac{i_{D_2}\cdot e^{-\lambda\cdot\tau_2}}{[\frac{I_{D_2}^{(j)}}{I_0} + 1]} + \frac{i_{D_1}\cdot e^{-\lambda\cdot\tau_1}}{[\frac{I_{D_1}^{(j)}}{I_0} + 1]}\}\cdot\frac{1}{I_0}\cdot e^{\lambda\cdot t}$$
$$-\frac{i_{D_2}\cdot e^{\lambda\cdot t}\cdot e^{-\lambda\cdot\tau_2}}{R_L\cdot C_{load}}\cdot\ln([\frac{I_{D_1}^{(j)}}{I_0} + 1]\cdot[\frac{I_{D_2}^{(j)}}{I_0} + 1])]$$
$$-\frac{R_{DC}}{L_{calc}}\cdot z\cdot e^{\lambda\cdot t}\cdot e^{-\lambda\cdot\Delta} - [-i_{D_1}\cdot e^{\lambda\cdot t}\cdot e^{-\lambda\cdot\tau_1} + i_{D_2}\cdot e^{\lambda\cdot t}\cdot e^{-\lambda\cdot\tau_2}]$$
$$\cdot\frac{R_{DC}}{L_{calc}\cdot C_1} - \frac{V_t}{L_{calc}\cdot C_1}\cdot\{\frac{i_{D_1}\cdot e^{-\lambda\cdot\tau_1}\cdot e^{\lambda\cdot t}}{I_0\cdot(\frac{I_{D_1}^{(j)}}{I_0} + 1)}\}$$

Divide above equations two sides by $e^{\lambda\cdot t}$:

$$z\cdot\lambda = \frac{[L_1 - L_{calc}]}{L_{calc}\cdot C_1}\cdot[-y - \frac{\{2\cdot I_{D_2}^{(j)} + I_0\}\cdot i_{D_2}\cdot e^{-\lambda\cdot\tau_2}}{C_{load}\cdot V_t} - \frac{(I_{D_2}^{(j)} + I_0)}{(I_{D_1}^{(j)} + I_0)}\cdot y$$
$$+\frac{(I_{D_2}^{(j)} + I_0)}{R_L\cdot C_{load}}\cdot\{\frac{i_{D_2}\cdot e^{-\lambda\cdot\tau_2}}{[\frac{I_{D_2}^{(j)}}{I_0} + 1]} + \frac{i_{D_1}\cdot e^{-\lambda\cdot\tau_1}}{[\frac{I_{D_1}^{(j)}}{I_0} + 1]}\}\cdot\frac{1}{I_0}$$
$$-\frac{i_{D_2}\cdot e^{-\lambda\cdot\tau_2}}{R_L\cdot C_{load}}\cdot\ln([\frac{I_{D_1}^{(j)}}{I_0} + 1]\cdot[\frac{I_{D_2}^{(j)}}{I_0} + 1])] - \frac{R_{DC}}{L_{calc}}\cdot z\cdot e^{-\lambda\cdot\Delta}$$
$$-[-i_{D_1}\cdot e^{-\lambda\cdot\tau_1} + i_{D_2}\cdot e^{-\lambda\cdot\tau_2}]\cdot\frac{R_{DC}}{L_{calc}\cdot C_1} - \frac{V_t}{L_{calc}\cdot C_1}\cdot\{\frac{i_{D_1}\cdot e^{-\lambda\cdot\tau_1}}{I_0\cdot(\frac{I_{D_1}^{(j)}}{I_0} + 1)}\}$$

$$-z\cdot\lambda - \frac{[L_1 - L_{calc}]}{L_{calc}\cdot C_1}\cdot y - \frac{[L_1 - L_{calc}]}{L_{calc}\cdot C_1}\cdot\frac{\{2\cdot I_{D_2}^{(j)} + I_0\}\cdot i_{D_2}\cdot e^{-\lambda\cdot\tau_2}}{C_{load}\cdot V_t}$$
$$-\frac{[L_1 - L_{calc}]}{L_{calc}\cdot C_1}\cdot\frac{(I_{D_2}^{(j)} + I_0)}{(I_{D_1}^{(j)} + I_0)}\cdot y$$
$$+\frac{[L_1 - L_{calc}]}{L_{calc}\cdot C_1}\cdot\frac{(I_{D_2}^{(j)} + I_0)}{R_L\cdot C_{load}}\cdot\{\frac{i_{D_2}\cdot e^{-\lambda\cdot\tau_2}}{[\frac{I_{D_2}^{(j)}}{I_0} + 1]} + \frac{i_{D_1}\cdot e^{-\lambda\cdot\tau_1}}{[\frac{I_{D_1}^{(j)}}{I_0} + 1]}\}\cdot\frac{1}{I_0}$$
$$-\frac{[L_1 - L_{calc}]}{L_{calc}\cdot C_1}\cdot\frac{i_{D_2}\cdot e^{-\lambda\cdot\tau_2}}{R_L\cdot C_{load}}\cdot\ln([\frac{I_{D_1}^{(j)}}{I_0} + 1]\cdot[\frac{I_{D_2}^{(j)}}{I_0} + 1])$$
$$-\frac{R_{DC}}{L_{calc}}\cdot z\cdot e^{-\lambda\cdot\Delta} - [-i_{D_1}\cdot e^{-\lambda\cdot\tau_1} + i_{D_2}\cdot e^{-\lambda\cdot\tau_2}]\cdot\frac{R_{DC}}{L_{calc}\cdot C_1}$$
$$-\frac{V_t}{L_{calc}\cdot C_1}\cdot\{\frac{i_{D_1}\cdot e^{-\lambda\cdot\tau_1}}{I_0\cdot(\frac{I_{D_1}^{(j)}}{I_0} + 1)}\} = 0$$

$$- z \cdot \lambda - \frac{[L_1 - L_{calc}]}{L_{calc} \cdot C_1} \cdot y - \frac{[L_1 - L_{calc}]}{L_{calc} \cdot C_1} \cdot \frac{\{2 \cdot I_{D_2}^{(j)} + I_0\} \cdot i_{D_2} \cdot e^{-\lambda \cdot \tau_2}}{C_{load} \cdot V_t}$$
$$- \frac{[L_1 - L_{calc}]}{L_{calc} \cdot C_1} \cdot \frac{(I_{D_2}^{(j)} + I_0)}{(I_{D_1}^{(j)} + I_0)} \cdot y$$
$$+ \frac{[L_1 - L_{calc}]}{L_{calc} \cdot C_1} \cdot \frac{(I_{D_2}^{(j)} + I_0)}{R_L \cdot C_{load} \cdot I_0} \cdot \frac{i_{D_2} \cdot e^{-\lambda \cdot \tau_2}}{[\frac{I_{D_2}^{(j)}}{I_0} + 1]}$$
$$+ \frac{[L_1 - L_{calc}]}{L_{calc} \cdot C_1} \cdot \frac{(I_{D_2}^{(j)} + I_0)}{R_L \cdot C_{load} \cdot I_0} \cdot \frac{i_{D_1} \cdot e^{-\lambda \cdot \tau_1}}{[\frac{I_{D_1}^{(j)}}{I_0} + 1]}$$
$$- \frac{[L_1 - L_{calc}]}{L_{calc} \cdot C_1} \cdot \frac{i_{D_2} \cdot e^{-\lambda \cdot \tau_2}}{R_L \cdot C_{load}} \cdot \ln([\frac{I_{D_1}^{(j)}}{I_0} + 1] \cdot [\frac{I_{D_2}^{(j)}}{I_0} + 1])$$
$$- \frac{R_{DC}}{L_{calc}} \cdot z \cdot e^{-\lambda \cdot \Delta} - [-i_{D_1} \cdot e^{-\lambda \cdot \tau_1} + i_{D_2} \cdot e^{-\lambda \cdot \tau_2}] \cdot \frac{R_{DC}}{L_{calc} \cdot C_1}$$
$$- \frac{V_t}{L_{calc} \cdot C_1} \cdot \{\frac{i_{D_1} \cdot e^{-\lambda \cdot \tau_1}}{I_0 \cdot (\frac{I_{D_1}^{(j)}}{I_0} + 1)}\} = 0$$

$$\{-\frac{R_{DC}}{L_{calc}} \cdot e^{-\lambda \cdot \Delta} - \lambda\} \cdot z + \{\frac{[L_1 - L_{calc}]}{L_{calc} \cdot C_1} \cdot \frac{(I_{D_2}^{(j)} + I_0)}{R_L \cdot C_{load} \cdot I_0} \cdot \frac{e^{-\lambda \cdot \tau_1}}{[\frac{I_{D_1}^{(j)}}{I_0} + 1]}$$
$$- \frac{V_t}{L_{calc} \cdot C_1} \cdot \{\frac{e^{-\lambda \cdot \tau_1}}{I_0 \cdot (\frac{I_{D_1}^{(j)}}{I_0} + 1)}\} + \frac{R_{DC}}{L_{calc} \cdot C_1} \cdot e^{-\lambda \cdot \tau_1}\} \cdot i_{D_1}$$
$$+ \{\frac{[L_1 - L_{calc}]}{L_{calc} \cdot C_1} \cdot \frac{(I_{D_2}^{(j)} + I_0)}{R_L \cdot C_{load} \cdot I_0} \cdot \frac{e^{-\lambda \cdot \tau_2}}{[\frac{I_{D_2}^{(j)}}{I_0} + 1]}$$
$$- \frac{[L_1 - L_{calc}]}{L_{calc} \cdot C_1} \cdot \frac{\{2 \cdot I_{D_2}^{(j)} + I_0\} \cdot e^{-\lambda \cdot \tau_2}}{C_{load} \cdot V_t}$$
$$- \frac{[L_1 - L_{calc}]}{L_{calc} \cdot C_1} \cdot \frac{e^{-\lambda \cdot \tau_2}}{R_L \cdot C_{load}} \cdot \ln([\frac{I_{D_1}^{(j)}}{I_0} + 1] \cdot [\frac{I_{D_2}^{(j)}}{I_0} + 1])$$
$$- \frac{R_{DC}}{L_{calc} \cdot C_1} \cdot e^{-\lambda \cdot \tau_2}\} \cdot i_{D_2} - \frac{[L_1 - L_{calc}]}{L_{calc} \cdot C_1} \cdot \{1 + \frac{(I_{D_2}^{(j)} + I_0)}{(I_{D_1}^{(j)} + I_0)}\} \cdot y = 0$$

$$\{-\frac{R_{DC}}{L_{calc}} \cdot e^{-\lambda \cdot \Delta} - \lambda\} \cdot z + \{\frac{[L_1 - L_{calc}]}{L_{calc} \cdot C_1} \cdot \frac{(I_{D_2}^{(j)} + I_0)}{R_L \cdot C_{load} \cdot I_0} \cdot \frac{1}{[\frac{I_{D_1}^{(j)}}{I_0} + 1]}$$

$$- \frac{V_t}{L_{calc} \cdot C_1} \cdot \{\frac{1}{I_0 \cdot (\frac{I_{D_1}^{(j)}}{I_0} + 1)}\} + \frac{R_{DC}}{L_{calc} \cdot C_1}\} \cdot e^{-\lambda \cdot \tau_1} \cdot i_{D_1}$$

$$+ \{\frac{[L_1 - L_{calc}]}{L_{calc} \cdot C_1} \cdot \frac{(I_{D_2}^{(j)} + I_0)}{R_L \cdot C_{load} \cdot I_0} \cdot \frac{1}{[\frac{I_{D_2}^{(j)}}{I_0} + 1]} - \frac{[L_1 - L_{calc}]}{L_{calc} \cdot C_1} \cdot \frac{\{2 \cdot I_{D_2}^{(j)} + I_0\}}{C_{load} \cdot V_t}$$

$$- \frac{[L_1 - L_{calc}]}{L_{calc} \cdot C_1} \cdot \frac{1}{R_L \cdot C_{load}} \cdot \ln([\frac{I_{D_1}^{(j)}}{I_0} + 1] \cdot [\frac{I_{D_2}^{(j)}}{I_0} + 1]) - \frac{R_{DC}}{L_{calc} \cdot C_1}\} \cdot e^{-\lambda \cdot \tau_2} \cdot i_{D_2}$$

$$- \frac{[L_1 - L_{calc}]}{L_{calc} \cdot C_1} \cdot \{1 + \frac{(I_{D_2}^{(j)} + I_0)}{(I_{D_1}^{(j)} + I_0)}\} \cdot y = 0$$

We define for simplicity the following global parameters:

$$\Upsilon_1 = \frac{[L_1 - L_{calc}]}{L_{calc} \cdot C_1} \cdot \frac{(I_{D_2}^{(j)} + I_0)}{R_L \cdot C_{load} \cdot I_0} \cdot \frac{1}{[\frac{I_{D_1}^{(j)}}{I_0} + 1]} - \frac{V_t}{L_{calc} \cdot C_1} \cdot \{\frac{1}{I_0 \cdot (\frac{I_{D_1}^{(j)}}{I_0} + 1)}\} + \frac{R_{DC}}{L_{calc} \cdot C_1}$$

$$\Upsilon_2 = \frac{[L_1 - L_{calc}]}{L_{calc} \cdot C_1} \cdot \frac{(I_{D_2}^{(j)} + I_0)}{R_L \cdot C_{load} \cdot I_0} \cdot \frac{1}{[\frac{I_{D_2}^{(j)}}{I_0} + 1]} - \frac{[L_1 - L_{calc}]}{L_{calc} \cdot C_1} \cdot \frac{\{2 \cdot I_{D_2}^{(j)} + I_0\}}{C_{load} \cdot V_t}$$

$$- \frac{[L_1 - L_{calc}]}{L_{calc} \cdot C_1} \cdot \frac{1}{R_L \cdot C_{load}} \cdot \ln([\frac{I_{D_1}^{(j)}}{I_0} + 1] \cdot [\frac{I_{D_2}^{(j)}}{I_0} + 1]) - \frac{R_{DC}}{L_{calc} \cdot C_1}$$

$$\Upsilon_3 = \frac{[L_1 - L_{calc}]}{L_{calc} \cdot C_1} \cdot \{1 + \frac{(I_{D_2}^{(j)} + I_0)}{(I_{D_1}^{(j)} + I_0)}\}$$

$$\{-\frac{R_{DC}}{L_{calc}} \cdot e^{-\lambda \cdot \Delta} - \lambda\} \cdot z + \Upsilon_1 \cdot e^{-\lambda \cdot \tau_1} \cdot i_{D_1} + \Upsilon_2 \cdot e^{-\lambda \cdot \tau_2} \cdot i_{D_2} - \Upsilon_3 \cdot y = 0$$

$$i_{D_1} \cdot e^{\lambda \cdot t} \to i_{D_1} \cdot e^{\lambda \cdot t} \cdot e^{-\lambda \cdot \tau_1}; i_{D_2} \cdot e^{\lambda \cdot t} \to i_{D_2} \cdot e^{\lambda \cdot t} \cdot e^{-\lambda \cdot \tau_2};$$

$$z \cdot e^{\lambda \cdot t} \to z \cdot e^{\lambda \cdot t} \cdot e^{-\lambda \cdot \Delta}; Y^{(j)} = 0$$

Remark left side of below equation doesn't affect by delay parameter.

(ii)

$$\lambda \cdot i_{D_1} \cdot e^{\lambda \cdot t} = Y^{(j)} + y \cdot e^{\lambda \cdot t}|_{Y^{(j)}=0} \Rightarrow \lambda \cdot i_{D_1} \cdot e^{\lambda \cdot t} = y \cdot e^{\lambda \cdot t} \Rightarrow -\lambda \cdot i_{D_1} + y = 0$$

$$\lambda \cdot i_{D_2} \cdot e^{\lambda \cdot t} = -\frac{I_{D_2}^{(j)} \cdot i_{D_2} \cdot e^{\lambda \cdot t} + (I_{D_2}^{(j)} + I_0) \cdot i_{D_2} \cdot e^{\lambda \cdot t}}{C_{load} \cdot V_t}$$

(iii)

$$- \{-\frac{Y^{(j)} \cdot (I_{D_2}^{(j)} + I_0) \cdot i_{D_1} \cdot e^{\lambda \cdot t}}{(I_{D_1}^{(j)} + I_0)^2} + \frac{Y^{(j)} \cdot i_{D_2} \cdot e^{\lambda \cdot t}}{(I_{D_1}^{(j)} + I_0)}\}$$
$$- \{\frac{(I_{D_2}^{(j)} + I_0) \cdot (I_{D_1}^{(j)} + I_0) \cdot y \cdot e^{\lambda \cdot t}}{(I_{D_1}^{(j)} + I_0)^2}\}$$
$$- \frac{(I_{D_2}^{(j)} + I_0)}{R_L \cdot C_{load}} \cdot \{\frac{i_{D_2}}{[\frac{I_{D_2}^{(j)}}{I_0} + 1]} + \frac{i_{D_1}}{[\frac{I_{D_1}^{(j)}}{I_0} + 1]}\} \cdot \frac{1}{I_0} \cdot e^{\lambda \cdot t}$$
$$- \frac{i_{D_2} \cdot e^{\lambda \cdot t}}{R_L \cdot C_{load}} \cdot \ln\{[\frac{I_{D_1}^{(j)}}{I_0} + 1] \cdot [\frac{I_{D_2}^{(j)}}{I_0} + 1]\}$$

$$i_{D_1} \cdot e^{\lambda \cdot t} \to i_{D_1} \cdot e^{\lambda \cdot t} \cdot e^{-\lambda \cdot \tau_1}; i_{D_2} \cdot e^{\lambda \cdot t} \to i_{D_2} \cdot e^{\lambda \cdot t} \cdot e^{-\lambda \cdot \tau_2};$$
$$z \cdot e^{\lambda \cdot t} \to z \cdot e^{\lambda \cdot t} \cdot e^{-\lambda \cdot \Delta}; Y^{(j)} = 0$$

Remark left side of below equation doesn't affect by delay parameter.

$$\lambda \cdot i_{D_2} \cdot e^{\lambda \cdot t} = -\frac{I_{D_2}^{(j)} \cdot i_{D_2} \cdot e^{\lambda \cdot t} \cdot e^{-\lambda \cdot \tau_2} + (I_{D_2}^{(j)} + I_0) \cdot i_{D_2} \cdot e^{\lambda \cdot t} \cdot e^{-\lambda \cdot \tau_2}}{C_{load} \cdot V_t}$$
$$- \{\frac{(I_{D_2}^{(j)} + I_0) \cdot (I_{D_1}^{(j)} + I_0) \cdot y \cdot e^{\lambda \cdot t}}{(I_{D_1}^{(j)} + I_0)^2}\}$$
$$- \frac{(I_{D_2}^{(j)} + I_0)}{R_L \cdot C_{load}} \cdot \{\frac{i_{D_2} \cdot e^{-\lambda \cdot \tau_2}}{[\frac{I_{D_2}^{(j)}}{I_0} + 1]} + \frac{i_{D_1} \cdot e^{-\lambda \cdot \tau_1}}{[\frac{I_{D_1}^{(j)}}{I_0} + 1]}\} \cdot \frac{1}{I_0} \cdot e^{\lambda \cdot t}$$
$$- \frac{i_{D_2} \cdot e^{\lambda \cdot t} \cdot e^{-\lambda \cdot \tau_2}}{R_L \cdot C_{load}} \cdot \ln\{[\frac{I_{D_1}^{(j)}}{I_0} + 1] \cdot [\frac{I_{D_2}^{(j)}}{I_0} + 1]\}$$

We divide above two sides by $e^{\lambda \cdot t}$ term.

$$\lambda \cdot i_{D_2} = -\frac{I_{D_2}^{(j)} \cdot i_{D_2} \cdot e^{-\lambda \cdot \tau_2} + (I_{D_2}^{(j)} + I_0) \cdot i_{D_2} \cdot e^{-\lambda \cdot \tau_2}}{C_{load} \cdot V_t} - \frac{(I_{D_2}^{(j)} + I_0)}{(I_{D_1}^{(j)} + I_0)} \cdot y$$

$$- \frac{(I_{D_2}^{(j)} + I_0)}{R_L \cdot C_{load}} \cdot \{\frac{i_{D_2} \cdot e^{-\lambda \cdot \tau_2}}{[\frac{I_{D_2}^{(j)}}{I_0} + 1]} + \frac{i_{D_1} \cdot e^{-\lambda \cdot \tau_1}}{[\frac{I_{D_1}^{(j)}}{I_0} + 1]}\} \cdot \frac{1}{I_0}$$

$$- \frac{i_{D_2} \cdot e^{-\lambda \cdot \tau_2}}{R_L \cdot C_{load}} \cdot \ln\{[\frac{I_{D_1}^{(j)}}{I_0} + 1] \cdot [\frac{I_{D_2}^{(j)}}{I_0} + 1]\}$$

$$\lambda \cdot i_{D_2} = -\frac{[2 \cdot I_{D_2}^{(j)} + I_0] \cdot i_{D_2} \cdot e^{-\lambda \cdot \tau_2}}{C_{load} \cdot V_t} - \frac{(I_{D_2}^{(j)} + I_0)}{(I_{D_1}^{(j)} + I_0)} \cdot y - \frac{(I_{D_2}^{(j)} + I_0)}{R_L \cdot C_{load} \cdot I_0} \cdot \frac{i_{D_2} \cdot e^{-\lambda \cdot \tau_2}}{[\frac{I_{D_2}^{(j)}}{I_0} + 1]}$$

$$- \frac{(I_{D_2}^{(j)} + I_0)}{R_L \cdot C_{load} \cdot I_0} \cdot \frac{i_{D_1} \cdot e^{-\lambda \cdot \tau_1}}{[\frac{I_{D_1}^{(j)}}{I_0} + 1]} - \frac{i_{D_2} \cdot e^{-\lambda \cdot \tau_2}}{R_L \cdot C_{load}} \cdot \ln\{[\frac{I_{D_1}^{(j)}}{I_0} + 1] \cdot [\frac{I_{D_2}^{(j)}}{I_0} + 1]\}$$

$$- \frac{(I_{D_2}^{(j)} + I_0)}{R_L \cdot C_{load} \cdot I_0} \cdot \frac{1}{[\frac{I_{D_1}^{(j)}}{I_0} + 1]} \cdot e^{-\lambda \cdot \tau_1} \cdot i_{D_1} - \lambda \cdot i_{D_2}$$

$$- (\frac{[2 \cdot I_{D_2}^{(j)} + I_0]}{C_{load} \cdot V_t} + \frac{(I_{D_2}^{(j)} + I_0)}{R_L \cdot C_{load} \cdot I_0} \cdot \frac{1}{[\frac{I_{D_2}^{(j)}}{I_0} + 1]}$$

$$+ \frac{1}{R_L \cdot C_{load}} \cdot \ln\{[\frac{I_{D_1}^{(j)}}{I_0} + 1] \cdot [\frac{I_{D_2}^{(j)}}{I_0} + 1]\}) \cdot i_{D_2} \cdot e^{-\lambda \cdot \tau_2} - \frac{(I_{D_2}^{(j)} + I_0)}{(I_{D_1}^{(j)} + I_0)} \cdot y = 0$$

We define the following global parameters for simplicity.

$$\Upsilon_4 = -\frac{(I_{D_2}^{(j)} + I_0)}{R_L \cdot C_{load} \cdot I_0} \cdot \frac{1}{[\frac{I_{D_1}^{(j)}}{I_0} + 1]};$$

$$\Upsilon_5 = \frac{[2 \cdot I_{D_2}^{(j)} + I_0]}{C_{load} \cdot V_t} + \frac{(I_{D_2}^{(j)} + I_0)}{R_L \cdot C_{load} \cdot I_0} \cdot \frac{1}{[\frac{I_{D_2}^{(j)}}{I_0} + 1]} + \frac{1}{R_L \cdot C_{load}} \cdot \ln\{[\frac{I_{D_1}^{(j)}}{I_0} + 1] \cdot [\frac{I_{D_2}^{(j)}}{I_0} + 1]\}$$

$$\Upsilon_6 = \frac{(I_{D_2}^{(j)} + I_0)}{(I_{D_1}^{(j)} + I_0)}$$

$$\Upsilon_4 \cdot e^{-\lambda \cdot \tau_1} \cdot i_{D_1} - \lambda \cdot i_{D_2} - \Upsilon_5 \cdot i_{D_2} \cdot e^{-\lambda \cdot \tau_2} - \Upsilon_6 \cdot y = 0$$

$$\Upsilon_k = \Upsilon_k(Z^{(j)}, I_{D_1}^{(j)}, I_{D_2}^{(j)}, Y^{(j)}, L_1, L_{calc}, R_L, C_{load}, R_{DC}, I_0, V) \,\forall\, k = 1, 2, 3, 4, 5, 6.$$

In the equilibrium fixed points: $Z^{(j)}, I_{D_1}^{(j)}, I_{D_2}^{(j)}, Y^{(j)} = 0$

The small increments Jacobian of our RFID burst switch system is as follows:

$$\{-\frac{R_{DC}}{L_{calc}} \cdot e^{-\lambda \cdot \Delta} - \lambda\} \cdot z + \Upsilon_1 \cdot e^{-\lambda \cdot \tau_1} \cdot i_{D_1}$$
$$+ \Upsilon_2 \cdot e^{-\lambda \cdot \tau_2} \cdot i_{D_2} - \Upsilon_3 \cdot y = 0$$
$$- \lambda \cdot i_{D_1} + y = 0$$
$$\Upsilon_4 \cdot e^{-\lambda \cdot \tau_1} \cdot i_{D_1} - [\lambda + \Upsilon_5 \cdot e^{-\lambda \cdot \tau_2}] \cdot i_{D_2} - \Upsilon_6 \cdot y = 0$$
$$\frac{dI_{D_1}}{dt} = Y \Rightarrow \frac{dY(t)}{dt} = \frac{d^2 I_{D_1}}{dt^2}$$

We consider $\frac{d^2 I_{D_1}}{dt^2} \to \varepsilon$ then $\frac{dY(t)}{dt} = 0$

$$Y(t) = Y^{(j)} + y \cdot e^{\lambda \cdot t} \Rightarrow \frac{dY(t)}{dt} = y \cdot \lambda \cdot e^{\lambda \cdot t};$$
$$\frac{dY(t)}{dt} = 0 \Rightarrow y \cdot \lambda \cdot e^{\lambda \cdot t} = 0 \Rightarrow |_{e^{\lambda \cdot t} \neq 0} y \cdot \lambda = 0 \Rightarrow -y \cdot \lambda = 0$$
$$\begin{pmatrix} \Xi_{11} & \cdots & \Xi_{14} \\ \vdots & \ddots & \vdots \\ \Xi_{41} & \cdots & \Xi_{44} \end{pmatrix} \cdot \begin{pmatrix} z \\ i_{D_1} \\ i_{D_2} \\ y \end{pmatrix} = 0;$$
$$\Xi_{11} = -\frac{R_{DC}}{L_{calc}} \cdot e^{-\lambda \cdot \Delta} - \lambda; \Xi_{12} = \Upsilon_1 \cdot e^{-\lambda \cdot \tau_1}; \Xi_{13} = \Upsilon_2 \cdot e^{-\lambda \cdot \tau_2}$$
$$\Xi_{14} = -\Upsilon_3; \Xi_{21} = 0; \Xi_{22} = -\lambda; \Xi_{23} = 0; \Xi_{24} = 1;$$
$$\Xi_{31} = 0; \Xi_{32} = \Upsilon_4 \cdot e^{-\lambda \cdot \tau_1}; \Xi_{33} = -\lambda - \Upsilon_5 \cdot e^{-\lambda \cdot \tau_2}$$
$$\Xi_{34} = -\Upsilon_6; \Xi_{41} = 0; \Xi_{42} = 0; \Xi_{43} = 0; \Xi_{44} = -\lambda$$
$$A - \lambda \cdot I = \begin{pmatrix} \Xi_{11} & \cdots & \Xi_{14} \\ \vdots & \ddots & \vdots \\ \Xi_{31} & \cdots & \Xi_{34} \end{pmatrix}; \det |A - \lambda \cdot I| = 0$$
$$D(\lambda, \tau_1, \tau_2, \Delta) = \lambda^4 + \lambda^3 \cdot [\frac{R_{DC}}{L_{calc}} \cdot e^{-\lambda \cdot \Delta} + \Upsilon_5 \cdot e^{-\lambda \cdot \tau_2}] + \lambda^2 \cdot \Upsilon_5 \cdot \frac{R_{DC}}{L_{calc}} \cdot e^{-\lambda \cdot (\Delta + \tau_2)}$$

We have three sub cases: (I) $\tau_2 = \tau; \Delta = 0$(II) $\tau_2 = 0; \Delta > 0$(III) $\tau_2 = \Delta = \tau_\Delta$

$$\text{(I)} \qquad D(\lambda, \tau_2 = \tau, \Delta = 0) = \lambda^4 + \lambda^3 \cdot \frac{R_{DC}}{L_{calc}} + [\lambda^3 \cdot \Upsilon_5 + \lambda^2 \cdot \Upsilon_5 \cdot \frac{R_{DC}}{L_{calc}}] \cdot e^{-\lambda \cdot \tau}$$

(II) $$D(\lambda,\tau_2=0,\Delta>0)=\lambda^4+\lambda^3\cdot\Upsilon_5+[\lambda^3\cdot\frac{R_{DC}}{L_{calc}}+\lambda^2\cdot\Upsilon_5\cdot\frac{R_{DC}}{L_{calc}}]\cdot e^{-\lambda\cdot\Delta}$$

(III) $$D(\lambda,\tau_2=\tau_\Delta,\Delta=\tau_\Delta)=\lambda^4+\lambda^3\cdot[\frac{R_{DC}}{L_{calc}}+\Upsilon_5]\cdot e^{-\lambda\cdot\tau_\Delta}+\lambda^2\cdot\Upsilon_5\cdot\frac{R_{DC}}{L_{calc}}\cdot e^{-\lambda\cdot\tau_\Delta}\cdot e^{-\lambda\cdot\tau_\Delta}$$

Under Taylor series approximation: $e^{-\lambda\cdot\tau_\Delta}\approx 1-\lambda\cdot\tau_\Delta$. The Maclaurin series is a Taylor series expansion of a $e^{-\lambda\cdot\tau_\Delta}$ function about zero (0). We get the following general characteristic equation $D(\lambda, \tau_\Delta)$ under Taylor series approximation: $e^{-\lambda\cdot\tau_\Delta}\approx 1-\lambda\cdot\tau_\Delta$[5, 6].

$$D(\lambda,\tau_2=\tau_\Delta,\Delta=\tau_\Delta)=\lambda^4+\{\lambda^3\cdot[\frac{R_{DC}}{L_{calc}}+\Upsilon_5]+\lambda^2\cdot\Upsilon_5\cdot\frac{R_{DC}}{L_{calc}}\cdot(1-\lambda\cdot\tau_\Delta)\}\cdot e^{-\lambda\cdot\tau_\Delta}$$

Possible characteristic equations: (I) $D(\lambda,\tau)=P_n(\lambda,\tau)+Q_m(\lambda,\tau)\cdot e^{-\lambda\cdot\tau}\,\forall n>m$ (II) $D(\lambda,\Delta)=P_n(\lambda,\tau)+Q_m(\lambda,\tau)\cdot e^{-\lambda\cdot\Delta}$(III) $D(\lambda,\tau_\Delta)=P_n(\lambda,\tau)+Q_m(\lambda,\tau)\cdot e^{-\lambda\cdot\tau_\Delta}$.
We summary, our results in the following table:

	$\tau_2=\tau,\Delta=0$(n > m)	$\tau_2=0,\Delta>0$(n > m)	$\tau_2=\tau_\Delta,\Delta=\tau_\Delta$(n > m)
n	4	4	4
m	3	3	3
P_n	$\lambda^4+\lambda^3\cdot\frac{R_{DC}}{L_{calc}}$	$\lambda^4+\lambda^3\cdot\Upsilon_5$	λ^4
Q_m	$\lambda^3\cdot\Upsilon_5+\lambda^2\cdot\Upsilon_5\cdot\frac{R_{DC}}{L_{calc}}$	$\lambda^3\cdot\frac{R_{DC}}{L_{calc}}+\lambda^2\cdot\Upsilon_5\cdot\frac{R_{DC}}{L_{calc}}$	$\lambda^3\cdot[\frac{R_{DC}}{L_{calc}}+\Upsilon_5]+\lambda^2\cdot\Upsilon_5\cdot\frac{R_{DC}}{L_{calc}}\cdot(1-\lambda\cdot\tau_\Delta)$

Our RFID bursts switch homogeneous system for z, i_{D_1}, i_{D_2}, y leads to a characteristic equation for the eigenvalue λ having the form $P(\lambda)+Q(\lambda)\cdot e^{-\lambda\cdot\tau}=0$. First case $\tau_2=\tau$, $\Delta=0$. The general characteristic equation $D(\lambda, \tau)$ is ad follow:

$$D(\lambda,\tau_2=\tau,\Delta=0)=\lambda^4+\lambda^3\cdot\frac{R_{DC}}{L_{calc}}+[\lambda^3\cdot\Upsilon_5+\lambda^2\cdot\Upsilon_5\cdot\frac{R_{DC}}{L_{calc}}]\cdot e^{-\lambda\cdot\tau}$$

The expression for $P_n(\lambda,\tau)$ is

$$P_n(\lambda, \tau) = \sum_{k=0}^{n} P_k(\tau) \cdot \lambda^k = P_0(\tau) + P_1(\tau) \cdot \lambda + P_2(\tau) \cdot \lambda^2 + P_3(\tau) \cdot \lambda^3 + P_4(\tau) \cdot \lambda^4$$

$$P_0(\tau) = 0; P_1(\tau) = 0; P_2(\tau) = 0;$$

$$P_3(\tau) = \frac{R_{DC}}{L_{calc}}; P_4(\tau) = 1$$

The expression for $Q_n(\lambda, \tau)$ is

$$Q_n(\lambda, \tau) = \sum_{k=0}^{M} q_k(\tau) \cdot \lambda^k = q_0(\tau) + q_1(\tau) \cdot \lambda + q_2(\tau) \cdot \lambda^2 + q_3(\tau) \cdot \lambda^3 \; q_0(\tau) = 0; q_1(\tau) = 0; q_2(\tau) = \Upsilon_5 \cdot \frac{R_{DC}}{L_{calc}}; q_3(\tau) = \Upsilon_5$$

The homogeneous system for z, i_{D_1}, i'_{D_2}, y leads to a characteristic equation for the eigenvalue λ having the form

$$P(\lambda, \tau) + Q(\lambda, \tau) \cdot e^{-\lambda \cdot \tau} = 0; P(\lambda) = \sum_{j=0}^{4} a_j \cdot \lambda^j; Q(\lambda) = \sum_{j=0}^{3} c_j \cdot \lambda^j$$

And the coefficients $\{a_j(q_i, q_k, \tau), c_j(q_i, q_k, \tau)\} \in \mathbb{R}$ depend on q_i, q_k and delay q_i, q_k is any RFID burst switching parameters, other parameters keep as a constant.

$$a_0 = 0; a_1 = 0; a_2 = 0; a_3 = \frac{R_{DC}}{L_{calc}};$$

$$a_4 = 1; c_0 = 0; c_1 = 0; c_2 = \Upsilon_5 \cdot \frac{R_{DC}}{L_{calc}}; c_3 = \Upsilon_5$$

Unless strictly necessary, the designation of the varied arguments (q_i, q_k) will subsequently be omitted from P, Q, a_j, c_j. The coefficients a_j, c_j are continuous, and differentiable functions of their arguments, and direct substitution shows that $a_0 + c_0 \neq 0$ (not in sub case I) for $\forall q_i, q_k \in \mathbb{R}_+$, i.e. λ = 0 is not a of $P(\lambda) + Q(\lambda) \cdot e^{-\lambda \cdot \tau} = 0$. Furthermore, P(λ), Q(λ) are analytic functions of λ, for which the following requirements of the analysis [BK] can also be verified in the present case:

(a) If $\lambda = i \cdot \omega$, $\omega \in \mathbb{R}$, then $P(i \cdot \omega) + Q(i \cdot \omega) \neq 0$.
(b) $|Q(\lambda)/P(\lambda)|$ is bounded for $|\lambda| \to \infty$, $\mathrm{Re}\lambda \geq 0$. No roots bifurcation from ∞.
(c) $F(\omega) = |P(i \cdot \omega)|^2 - |Q(i \cdot \omega)|^2$ has a finite number of zeros. Indeed, this is a polynomial in ω.
(d) Each positive root $\omega(q_i, q_k)$ of F (ω) = 0 is continuous and differentiable respect to q_i, q_k.

We assume that $P_n(\lambda,\tau)$ and $Q_m(\lambda,\tau)$ can't have common imaginary roots. That is, for any real number ω,

$$p_n(\lambda = i\cdot\omega,\tau) + Q_m(\lambda = i\cdot\omega,\tau) \neq 0.$$

$$p_n(\lambda = i\cdot\omega,\tau) = \omega^4 - i\cdot\omega^3\cdot\frac{R_{DC}}{L_{calc}};$$

$$Q_m(\lambda = i\cdot\omega,\tau) = -i\cdot\omega^3\cdot\Upsilon_5 - \omega^2\cdot\Upsilon_5\cdot\frac{R_{DC}}{L_{calc}}$$

$$p_n(\lambda = i\cdot\omega,\tau) + Q_m(\lambda = i\cdot\omega,\tau)$$
$$= \omega^4 - i\cdot\omega^3\cdot\frac{R_{DC}}{L_{calc}} - i\cdot\omega^3\cdot\Upsilon_5 - \omega^2\cdot\Upsilon_5\cdot\frac{R_{DC}}{L_{calc}} \neq 0$$

$$|P(i\cdot\omega,\tau)|^2 = P_R^2 + P_I^2 = \omega^8 + \omega^6\cdot[\frac{R_{DC}}{L_{calc}}]^2;$$

$$|Q(i\cdot\omega,\tau)|^2 = Q_R^2 + Q_I^2 = \omega^6\cdot\Upsilon_5^2 + \omega^4\cdot\Upsilon_5^2\cdot[\frac{R_{DC}}{L_{calc}}]^2$$

$$F(\omega,\tau) = |P(i\cdot\omega,\tau)|^2 - |Q(i\cdot\omega,\tau)|^2$$
$$= \omega^8 + \omega^6\cdot[\frac{R_{DC}}{L_{calc}}]^2 - \omega^6\cdot\Upsilon_5^2 - \omega^4\cdot\Upsilon_5^2\cdot[\frac{R_{DC}}{L_{calc}}]^2$$

$$F(\omega,\tau) = |P(i\cdot\omega,\tau)|^2 - |Q(i\cdot\omega,\tau)|^2$$
$$= \omega^8 + \omega^6\cdot\{[\frac{R_{DC}}{L_{calc}}]^2 - \Upsilon_5^2\} - \omega^4\cdot\Upsilon_5^2\cdot[\frac{R_{DC}}{L_{calc}}]^2$$

We define the following parameters for simplicity:

$\Phi_0, \Phi_2, \Phi_4, \Phi_6, \Phi_8$

$$\Phi_0 = 0, \Phi_2 = 0, \Phi_4 = -\Upsilon_5^2\cdot[\frac{R_{DC}}{L_{calc}}]^2, \Phi_6 = [\frac{R_{DC}}{L_{calc}}]^2 - \Upsilon_5^2, \Phi_8 = 1$$

Hence $F(\omega,\tau) = 0$ implies $\sum_{k=0}^{4}\Phi_{2\cdot k}\cdot\omega^{2\cdot k} = 0$. And its roots are given by solving the above polynomial. Furthermore

$$P_R(i\cdot\omega,\tau) = \omega^4; P_I(i\cdot\omega,\tau) = -\omega^3\cdot\frac{R_{DC}}{L_{calc}}; Q_R(i\cdot\omega,\tau) = -\omega^2\cdot\Upsilon_5\cdot\frac{R_{DC}}{L_{calc}}$$

$$Q_I(i\cdot\omega,\tau) = -\omega^3\cdot\Upsilon_5; \sin\theta(\tau) = \frac{-P_R(i\cdot\omega,\tau)\cdot Q_I(i\cdot\omega,\tau) + P_I(i\cdot\omega,\tau)\cdot Q_R(i\cdot\omega,\tau)}{|Q(i\cdot\omega,\tau)|^2}$$

$$\cos\theta(\tau) = -\frac{P_R(i\cdot\omega,\tau)\cdot Q_R(i\cdot\omega,\tau) + P_I(i\cdot\omega,\tau)\cdot Q_I(i\cdot\omega,\tau)}{|Q(i\cdot\omega,\tau)|^2}$$

$$\sin\theta(\tau) = \frac{\omega^7 \cdot \Upsilon_5 + \omega^5 \cdot [\frac{R_{DC}}{L_{calc}}]^2 \cdot \Upsilon_5}{\omega^6 \cdot \Upsilon_5^2 + \omega^4 \cdot \Upsilon_5^2 \cdot [\frac{R_{DC}}{L_{calc}}]^2}; \cos\theta(\tau) = -\frac{[-\frac{R_{DC}}{L_{calc}} + \frac{R_{DC}}{L_{calc}}] \cdot \omega^6 \cdot \Upsilon_5}{\omega^6 \cdot \Upsilon_5^2 + \omega^4 \cdot \Upsilon_5^2 \cdot [\frac{R_{DC}}{L_{calc}}]^2} = 0$$

$$\sin\theta(\tau) = \frac{\omega^7 \cdot \Upsilon_5 + \omega^5 \cdot [\frac{R_{DC}}{L_{calc}}]^2 \cdot \Upsilon_5}{\omega^6 \cdot \Upsilon_5^2 + \omega^4 \cdot \Upsilon_5^2 \cdot [\frac{R_{DC}}{L_{calc}}]^2} = \frac{\omega^5 \cdot \Upsilon_5 \cdot \{\omega^2 + [\frac{R_{DC}}{L_{calc}}]^2\}}{\omega^4 \cdot \Upsilon_5^2 \cdot \{\omega^2 + [\frac{R_{DC}}{L_{calc}}]^2\}} = \frac{\omega}{\Upsilon_5}; \cos\theta(\tau) = 0$$

We use different parameters terminology from our last characteristics parameters definition: $k \to j; p_k(\tau) \to a_j; q_k(\tau) \to c_j; n = 4; m = 3; n > m.$ Additionally $P_n(\lambda,\tau) \to P(\lambda); Q_m(\lambda,\tau) \to Q(\lambda)$ then $P(\lambda) = \sum_{j=0}^{4} a_j \cdot \lambda^j; Q(\lambda) = \sum_{j=0}^{3} c_j \cdot \lambda^j$

$P(\lambda) = \lambda^4 + \lambda^3 \cdot \frac{R_{DC}}{L_{calc}}; Q(\lambda) = \lambda^3 \cdot \Upsilon_5 + \lambda^2 \cdot \Upsilon_5 \cdot \frac{R_{DC}}{L_{calc}}$ $n, m \in \mathbb{N}_0,\ n > m$ and $a_j, c_j : \mathrm{R}_{+0} \to R$ are continuous and differentiable function of τ such that $a_0 + c_0 \neq 0$ (not in sub case I). In the following "—" denotes complex and conjugate. $P(\lambda), Q(\lambda)$ are analytic functions in λ and differentiable in τ. The coefficients $\{a_j(L_{calc}, R_{DC}, C_1, L_1, C_{load}, R_L, \tau, \ldots)$ and $c_j(L_{calc}, R_{DC}, C_1, L_1, C_{load}, R_L, \tau, \ldots)\} \in \mathbb{R}$ depend on RFID burst switch system's $L_{calc}, R_{DC}, C_1, L_1, C_{load}, R_L, \tau, \ldots$ values.

$$a_0 = 0; a_1 = 0; a_2 = 0; a_3 = \frac{R_{DC}}{L_{calc}}; a_4 = 1; c_0 = 0; c_1 = 0; c_2 = \Upsilon_5 \cdot \frac{R_{DC}}{L_{calc}}; c_3 = \Upsilon_5$$

Unless strictly necessary, the designation of the varied arguments $(L_{calc}, R_{DC}, C_1, L_1, C_{load}, R_L, \tau, \ldots)$ will subsequently be omitted from P, Q, a_j, c_j. The coefficients a_j, c_j are continuous, and differentiable functions of their arguments. $\forall L_{calc}, R_{DC}, C_1, L_1, C_{load}, R_L, \tau, \ldots \in \mathbb{R}_+$ I.e. $\lambda = 0$ is not a root of the characteristic equation. Furthermore $P(\lambda)$, $Q(\lambda)$ are analytic functions of λ for which the following requirements of the analysis (see Kuang 1993, Sect. 3.4) can also be verified in the present case [6, 7].

(a) If $\lambda = i \cdot \omega$, $\omega \in \mathbb{R}$ then $P(i \cdot \omega) + Q(i \cdot \omega) \neq 0$, i.e. P and Q have no common imaginary roots. This condition was verified numerically in the entire $(L_{calc}, R_{DC}, C_1, L_1, C_{load}, R_L, \tau, \ldots)$ domain of interest.

(b) $|Q(\lambda)/P(\lambda)|$ is bounded for $|\lambda| \to \infty$, $\mathrm{Re}\lambda \geq 0$. No roots bifurcation from ∞. Indeed, in the limit $|\frac{Q(\lambda)}{P(\lambda)}| = |\frac{\lambda^3 \cdot \Upsilon_5 + \lambda^2 \cdot \Upsilon_5 \frac{R_{DC}}{L_{calc}}}{\lambda^4 + \lambda^3 \cdot \frac{R_{DC}}{L_{calc}}}|$

(c)

$$\begin{aligned} F(\omega,\tau) &= |P(i \cdot \omega,\tau)|^2 - |Q(i \cdot \omega,\tau)|^2 \\ &= \omega^8 + \omega^6 \cdot \{[\frac{R_{DC}}{L_{calc}}]^2 - \Upsilon_5^2\} - \omega^4 \cdot \Upsilon_5^2 \cdot [\frac{R_{DC}}{L_{calc}}]^2 \end{aligned}$$

Has at most a finite number of zeroes. Indeed, this is a polynomial in ω (Degree in ω^8).

(d) Each positive root $\omega(L_{calc}, R_{DC}, C_1, L_1, C_{load}, R_L, \tau, \ldots)$ of $F(\omega) = 0$ is continuous and differentiable with respect to $L_{calc}, R_{DC}, C_1, L_1, C_{load}, R_L, \tau, \ldots$. This condition can only be assessed numerically.

In addition, since the coefficients in P and Q are real, we have, and $\overline{Q(-i \cdot \omega)} = Q(i \cdot \omega)$ thus, $\omega > 0$ may be an eigenvalue of the characteristic equation. The analysis consists in identifying the roots of characteristic equation situated on the imaginary axis of the complex λ-plane, where by increasing the parameters $L_{calc}, R_{DC}, C_1, L_1, C_{load}, R_L, \tau, \ldots$, Reλ may, at the crossing,Change its sign from (−) to (+), i.e. from a stable focus $E^{(j)}(Z^{(j)}, I_{D_1}^{(j)}, I_{D_2}^{(j)}, Y^{(j)} = 0)$ to an unstable one, or vice versa. This feature may be further assessed by examining the sign of the partial derivatives with respect to $L_{calc}, R_{DC}, C_1, L_1, C_{load}, R_L, \tau, \ldots$ and gate antenna parameters.

$$\wedge^{-1}(L_{calc}) = \left(\frac{\partial \mathrm{Re}\lambda}{\partial L_{calc}}\right)_{\lambda = i\cdot\omega}, R_{DC}, C_1, L_1, C_{load}, R_L, \tau, \ldots = const$$

$$\wedge^{-1}(R_{DC}) = \left(\frac{\partial \mathrm{Re}\lambda}{\partial R_{DC}}\right)_{\lambda = i\cdot\omega}, L_{calc}, C_1, L_1, C_{load}, R_L, \tau, \ldots = const$$

$$\wedge^{-1}(C_1) = \left(\frac{\partial \mathrm{Re}\lambda}{\partial C_1}\right)_{\lambda = i\cdot\omega}, L_{calc}, R_{DC}, L_1, C_{load}, R_L, \tau, \ldots = const$$

$$\wedge^{-1}(L_1) = \left(\frac{\partial \mathrm{Re}\lambda}{\partial L_1}\right)_{\lambda = i\cdot\omega}, L_{calc}, R_{DC}, C_1, C_{load}, R_L, \tau, \ldots = const$$

$$\wedge^{-1}(C_{load}) = \left(\frac{\partial \mathrm{Re}\lambda}{\partial C_{load}}\right)_{\lambda = i\cdot\omega}, L_{calc}, R_{DC}, C_1, L_1, R_L, \tau, \ldots = const$$

$$\wedge^{-1}(\tau) = \left(\frac{\partial \mathrm{Re}\lambda}{\partial \tau}\right)_{\lambda = i\cdot\omega}, L_{calc}, R_{DC}, C_1, L_1, C_{load}, R_L, \ldots = const$$

$$\omega \in \mathbb{R}_+ .$$

$$F(\omega, \tau) = |P(i \cdot \omega, \tau)|^2 - |Q(i \cdot \omega, \tau)|^2$$

$$= \Phi_0 + \Phi_2 \cdot \omega^2 + \Phi_4 \cdot \omega^4 + \Phi_6 \cdot \omega^6 + \Phi_8 \cdot \omega^8 = \sum_{k=0}^{4} \Phi_{2\cdot k} \cdot \omega^{2\cdot k}$$

Hence $F(\omega, \tau) = 0$ implies $\sum_{k=0}^{4} \Phi_{2\cdot k} \cdot \omega^{2\cdot k} = 0$ When writing $P(\lambda) = P_R(\lambda) + i \cdot P_I(\lambda)$ and $Q(\lambda) = Q_R(\lambda) + i \cdot Q_I(\lambda)$, and inserting $\lambda = i \cdot \omega$ into RFID burst switch system's characteristic equation, ω must satisfy the following :

$$\sin \omega \cdot \tau = g(\omega) = \frac{-P_R(i \cdot \omega) \cdot Q_I(i \cdot \omega) + P_I(i \cdot \omega) \cdot Q_R(i \cdot \omega)}{|Q(i \cdot \omega)|^2}$$

$$\cos \omega \cdot \tau = h(\omega) = -\frac{P_R(i \cdot \omega) \cdot Q_R(i \cdot \omega) + P_I(i \cdot \omega) \cdot Q_I(i \cdot \omega)}{|Q(i \cdot \omega)|^2}$$

Where $|Q(i \cdot \omega)|^2 \neq 0$ in view of requirement (a) above, and $(g, h) \in R$. Furthermore, it follows above $\sin \omega \cdot \tau$ and $\cos \omega \cdot \tau$ equations that, by squaring and adding the sides, ω must be a positive root of $F(\omega) = |P(i \cdot \omega)|^2 - |Q(i \cdot \omega)|^2 = 0$. Note that $F(\omega)$ is dependent of τ. Now it is important to notice that if $\tau \notin I$ (assume that $I \subseteq R_{+0}$ is the set where $\omega(\tau)$ is a positive root of $F(\omega)$ and for, $\tau \notin I$, $\omega(\tau)$ is not defined. Then for all τ in I $\omega(\tau)$ is satisfied that $F(\omega, \tau) = 0$). Then there are no positive $\omega(\tau)$ solutions for $F(\omega, \tau) = 0$, and we cannot have stability switches. For any $\tau \in I$, where $\omega(\tau)$ is a positive solution of $F(\omega, \tau) = 0$, we can define the angle $\theta(\tau) \in [0, 2 \cdot \pi]$ as the solution of

$$\sin \theta(\tau) = \frac{-P_R(i \cdot \omega) \cdot Q_I(i \cdot \omega) + P_I(i \cdot \omega) \cdot Q_R(i \cdot \omega)}{|Q(i \cdot \omega)|^2};$$

$$\cos \theta(\tau) = -\frac{P_R(i \cdot \omega) \cdot Q_R(i \cdot \omega) + P_I(i \cdot \omega) \cdot Q_I(i \cdot \omega)}{|Q(i \cdot \omega)|^2}$$

And the relation between the argument $\theta(\tau)$ and $\omega(\tau) \cdot \tau$ for $\tau \in I$ must be $\omega(\tau) \cdot \tau = \theta(\tau) + n \cdot 2 \cdot \pi \; \forall \, n \in \mathbb{N}_0$. Hence we can define the maps $\tau_n : I \to R_{+0}$ given by $\tau_n(\tau) = \frac{\theta(\tau) + n \cdot 2 \cdot \pi}{\omega(\tau)}$ $n \in \mathbb{N}_0, \tau \in I$. Let us introduce the functions $I \to R; S_n(\tau) = \tau - \tau_n(\tau) S_n(\tau) = \tau - \tau_n(\tau)$, $\tau \in I$, $n \in \mathbb{N}_0$ that are continuous and differentiable in τ. In the following, the subscripts λ, $\omega, L_{calc}, R_{DC}, C_1, L_1, C_{load}, R_L, \ldots$ indicate the corresponding partial derivatives. Let us first concentrate on $\wedge(x)$, remember in $\lambda(L_{calc}, R_{DC}, C_1, L_1, C_{load}, R_L, \ldots)$ and $\omega(L_{calc}, R_{DC}, C_1, L_1, C_{load}, R_L, \ldots)$, and keeping all parameters except one (x) and τ. The derivation closely follows that in reference [BK]. Differentiating RFID burst switch characteristic equation $P(\lambda) + Q(\lambda) \cdot e^{-\lambda \cdot \tau} = 0$ with respect to specific parameter (x), and inverting the derivative, for convenience, one calculates:

Remark

$$x = L_{calc}, R_{DC}, C_1, L_1, C_{load}, R_L, \ldots, etc.,$$

$$\left(\frac{\partial \lambda}{\partial x}\right)^{-1} = \frac{-P_\lambda(\lambda, x) \cdot Q(\lambda, x) + Q_\lambda(\lambda, x) \cdot P(\lambda, x) - \tau \cdot P(\lambda, x) \cdot Q(\lambda, x)}{P_x(\lambda, x) \cdot Q(\lambda, x) - Q_x(\lambda, x) \cdot P(\lambda, x)}$$

Where $P_\lambda = \frac{\partial P}{\partial \lambda}, \ldots$ etc., Substituting $\lambda = i \cdot \omega$, and bearing i $\overline{P(-i \cdot \omega)} = P(i \cdot \omega)$, $\overline{Q(-i \cdot \omega)} = Q(i \cdot \omega)$ then $i \cdot P_\lambda(i \cdot \omega) = P_\omega(i \cdot \omega); i \cdot Q_\lambda(i \cdot \omega) = Q_\omega(i \cdot \omega)$ and that on the surface $|P(i \cdot \omega)|^2 = |Q(i \cdot \omega)|^2$, one obtains

$$\left(\frac{\partial \lambda}{\partial x}\right)^{-1}\Big|_{\lambda = i \cdot \omega} = \left(\frac{i \cdot P_\omega(i \cdot \omega, x) \cdot \overline{P(i \cdot \omega, x)} + i \cdot Q_\lambda(i \cdot \omega, x) \cdot \overline{Q(\lambda, x)} - \tau \cdot |P(i \cdot \omega, x)|^2}{P_x(i \cdot \omega, x) \cdot \overline{P(i \cdot \omega, x)} - Q_x(i \cdot \omega, x) \cdot \overline{Q(i \cdot \omega, x)}}\right)$$

Upon separating into real and imaginary parts, with $P = P_R + i \cdot P_I$; $Q = Q_R + i \cdot Q_I$; $P_\omega = P_{R\omega} + i \cdot P_{I\omega}$; $Q_\omega = Q_{R\omega} + i \cdot Q_{I\omega}$; $P_x = P_{Rx} + i \cdot P_{Ix}$; $Q_x = Q_{Rx} + i \cdot Q_{Ix}$; $P^2 = P_R^2 + P_I^2$. When (x) can be any RFID burst switch parameters and time delay τ etc. Where for convenience, we have dropped the arguments $(i \cdot \omega, x)$, and where

$$F_{\omega} = 2 \cdot [(P_{R\omega} \cdot P_R + P_{I\omega} \cdot P_I) - (Q_{R\omega} \cdot Q_R + Q_{I\omega} \cdot Q_I)];$$
$$F_x = 2 \cdot [(P_{Rx} \cdot P_R + P_{Ix} \cdot P_I) - (Q_{Rx} \cdot Q_R + Q_{Ix} \cdot Q_I)]$$

$\omega_x = -F_x/F_{\omega}$. We define U and V:

$$U = (P_R \cdot P_{I\omega} - P_I \cdot P_{R\omega}) - (Q_R \cdot Q_{I\omega} - Q_I \cdot Q_{R\omega})$$
$$V = (P_R \cdot P_{Ix} - P_I \cdot P_{Rx}) - (Q_R \cdot Q_{Ix} - Q_I \cdot Q_{Rx}).$$

We choose our specific parameter as time delay x = τ .

$$P_R = \omega^4; P_I = -\omega^3 \cdot \frac{R_{DC}}{L_{calc}}; Q_R = -\omega^2 \cdot \Upsilon_5 \cdot \frac{R_{DC}}{L_{calc}};$$
$$Q_I = -\omega^3 \cdot \Upsilon_5; P_{R\tau} = 0; P_{I\tau} = 0$$
$$Q_{R\tau} = 0; Q_{I\tau} = 0; P_{R\omega} = 4 \cdot \omega^3; P_{I\omega} = -3 \cdot \omega^2 \cdot \frac{R_{DC}}{L_{calc}};$$
$$Q_{R\omega} = -2 \cdot \omega \cdot \Upsilon_5 \cdot \frac{R_{DC}}{L_{calc}}; Q_{I\omega} = -3 \cdot \omega^2 \cdot \Upsilon_5$$
$$Q_I \cdot Q_{R\omega} = 2 \cdot \omega^4 \cdot \Upsilon_5^2 \cdot \frac{R_{DC}}{L_{calc}}; P_R \cdot P_{I\omega} = -3 \cdot \omega^6 \cdot \frac{R_{DC}}{L_{calc}};$$
$$P_I \cdot P_{R\omega} = -4 \cdot \omega^6 \cdot \frac{R_{DC}}{L_{calc}}; Q_R \cdot Q_{I\omega} = 3 \cdot \omega^4 \cdot \Upsilon_5^2 \cdot \frac{R_{DC}}{L_{calc}}$$
$$P_{R\omega} \cdot P_R = 4 \cdot \omega^7; Q_{R\omega} \cdot Q_R = 2 \cdot \omega^3 \cdot \Upsilon_5^2 \cdot [\frac{R_{DC}}{L_{calc}}]^2;$$
$$V = (P_R \cdot P_{I\tau} - P_I \cdot P_{R\tau}) - (Q_R \cdot Q_{I\tau} - Q_I \cdot Q_{R\tau}) = 0$$

$$U = (P_R \cdot P_{I\omega} - P_1 \cdot P_{R\omega}) - (Q_R \cdot Q_{I\omega} - Q_I \cdot Q_{R\omega}) = -3 \cdot \omega^6 \cdot \frac{R_{DC}}{L_{calc}} + 4 \cdot \omega^6 \cdot \frac{R_{DC}}{L_{calc}} - (3 \cdot \omega^4 \cdot \Upsilon_5^2 \cdot \frac{R_{DC}}{L_{calc}} - 2 \cdot \omega^4 \cdot \Upsilon_5^2 \cdot \frac{R_{DC}}{L_{calc}})$$

$$U = -3 \cdot \omega^6 \cdot \frac{R_{DC}}{L_{calc}} + 4 \cdot \omega^6 \cdot \frac{R_{DC}}{L_{calc}} - 3 \cdot \omega^4 \cdot \Upsilon_5^2 \cdot \frac{R_{DC}}{L_{calc}} + 2 \cdot \omega^4 \cdot \Upsilon_5^2 \cdot \frac{R_{DC}}{L_{calc}} = \omega^6 \cdot \frac{R_{DC}}{L_{calc}} - \omega^4 \cdot \Upsilon_5^2 \cdot \frac{R_{DC}}{L_{calc}}$$

$$Q_{I\omega} \cdot Q_I = 3.\omega^5 \cdot \Upsilon_5^2; U = \omega^6 \cdot \frac{R_{DC}}{L_{calc}} - \omega^4 \cdot \Upsilon_5^2 \cdot \frac{R_{DC}}{L_{calc}} = \omega^4 \cdot \frac{R_{DC}}{L_{calc}} \cdot [\omega^2 - \Upsilon_5^2]$$
$$F_{\tau} = 2 \cdot [(P_{R\tau} \cdot P_R + P_{I\tau} \cdot P_I) - (Q_{R\tau} \cdot Q_R + Q_{I\tau} \cdot Q_I)] = 0;$$
$$P_{I\omega} \cdot P_I = 3 \cdot \omega^5 \cdot [\frac{R_{DC}}{L_{calc}}]^2$$

$$F_{\omega} = 2 \cdot [(P_{R\omega} \cdot P_R + P_{I\omega} \cdot P_I) - (Q_{R\omega} \cdot Q_R + Q_{I\omega} \cdot Q_I)]$$
$$= 2 \cdot \{4 \cdot \omega^7 + 3 \cdot \omega^5 \cdot [\frac{R_{DC}}{L_{calc}}]^2 - (2 \cdot \omega^3 \cdot \Upsilon_5^2 \cdot [\frac{R_{DC}}{L_{calc}}]^2 + 3 \cdot \omega^5 \cdot \Upsilon_5^2)\}$$
$$F_{\omega} = 2 \cdot \{4 \cdot \omega^7 - 3 \cdot \omega^5 \cdot \Upsilon_5^2 + 3 \cdot \omega^5 \cdot [\frac{R_{DC}}{L_{calc}}]^2 - 2 \cdot \omega^3 \cdot \Upsilon_5^2 \cdot [\frac{R_{DC}}{L_{calc}}]^2\}$$
$$F_{\omega} = 2 \cdot \{\omega^5 \cdot (4 \cdot \omega^2 - 3 \cdot \Upsilon_5^2) + [\frac{R_{DC}}{L_{calc}}]^2 \cdot \omega^3 \cdot (3 \cdot \omega^2 - 2 \cdot \Upsilon_5^2)\}$$

$$F_{\omega} \cdot \frac{\partial \omega}{\partial \tau} + F_{\tau} = 0; \tau \in I \Rightarrow \frac{\partial \omega}{\partial \tau} = -\frac{F_{\tau}}{F_{\omega}}; \wedge^{-1}(\tau) = \left(\frac{\partial \mathrm{Re}\lambda}{\partial \tau}\right)_{\lambda = i - \omega}; \frac{\partial \omega}{\partial \tau} = \omega_{\tau} = -\frac{F_{\tau}}{F_{\omega}}$$
$$\frac{\partial \omega}{\partial \tau} = \omega_{\tau} = -\frac{F_{\tau}}{F_{\omega}}|_{F_{\tau}=0} = 0; \wedge^{-1}(\tau) = \mathrm{Re}\left\{\frac{-2 \cdot \left[U + \tau \cdot |P|^2 + i \cdot F_{\omega}\right]}{F_{\tau} + i \cdot 2 \cdot \left[V + \omega \cdot |P|^2\right]}\right\}$$

$$sign\{\wedge^{-1}(\tau)\} = sign\{(\frac{\partial \mathrm{Re}\lambda}{\partial \tau})_{\lambda = i - \omega}\};$$
$$sign\{\wedge^{-1}(\tau)\} = sign\{F_{\omega}\} \cdot sign\{\tau \cdot \frac{\partial \omega}{\partial \tau} + \omega + \frac{U \cdot \frac{\partial \omega}{\partial \tau} + V}{|P|^2}\}$$
$$sign\{\wedge^{-1}(\tau)\} = sign\{F_{\omega}\} \cdot sign\{\tau \cdot \frac{\partial \omega}{\partial \tau} + \omega + \frac{U \cdot \frac{\partial \omega}{\partial \tau} + V}{|P|^2}\}|_{\frac{\partial \omega}{\partial \tau}=0}$$
$$= sign\{F_{\omega}\} \cdot sign\{\omega\}$$
$$sign\{\wedge^{-1}(\tau)\} = sign\{4 \cdot \omega^7 + 3 \cdot \omega^5 \cdot [\frac{R_{DC}}{L_{calc}}]^2 - (2 \cdot \omega^3 \cdot \Upsilon_5^2 \cdot [\frac{R_{DC}}{L_{calc}}]^2 + 3 \cdot \omega^5 \cdot \Upsilon_5^2)\} \cdot sign\{\omega\}$$

We shall presently examine the possibility of stability transitions (bifurcations) RFID burst switch system, about the equilibrium point $E^{(j)}(Z^{(j)}, I_{D_1}^{(j)}, I_{D_2}^{(j)}, Y^{(j)} = 0)$ as a result of a variation of delay parameter τ. The analysis consists in identifying the roots of our system characteristic equation situated on the imaginary axis of the complex λ-plane. Where by increasing the delay parameter τ, Re λ may at the crossing, changes its sign from − to +, i.e. from a stable focus $E^{(*)}$ to an unstable one, or vice versa. This feature may be further assessed by examining the sign of the partial derivatives with respect to τ, $\wedge^{-1}(\tau) = (\frac{\partial \mathrm{Re}\lambda}{\partial \tau})_{\lambda = i - \omega}$

$$\wedge^{-1}(\tau) = \left(\frac{\partial \mathrm{Re}\lambda}{\partial \tau}\right)_{\lambda = i - \omega};$$
$$\wedge^{-1}(\tau) = \left(\frac{\partial \mathrm{Re}\lambda}{\partial \tau}\right)_{\lambda = i - \omega}, L_{calc}, R_{DC}, C_1, L_1, C_{load}, R_L, \ldots, etc. = const; \omega \in \mathrm{R}_+$$

We check the sign of $\wedge^{-1}(\tau)$ according the following rule:

$sign[F_\omega]$	$sign\left[\frac{V+\omega_\tau \cdot U}{P^2} + \omega + \omega_\tau \cdot \tau\right]$	$sign[\wedge^{-1}(\tau)]$
$\pm$	$\pm$	$+$
$\pm$	$\mp$	$-$

RFID burst switch system stability switching analysis is done according the below flow chart and based on [BK] geometric stability switch criteria in delay differential systems with delay dependent parameters article [30, 31].

Find our system general characteristic equation D(λ,τ)=0. Find F(ω, τ) for each τ has at most a finite number of real zeros. Find ω, τ values which fulfill F(ω, τ) =0 ; ω(τ), only for those values can be stability switching (first condition).

Those ω, τ values must fulfill the expressions:

$$\sin\theta(\tau) = \sin(\omega\cdot\tau) = g(\omega) = \frac{-P_R(i\cdot\omega)\cdot Q_I(i\cdot\omega) + P_I(i\cdot\omega)\cdot Q_R(i\cdot\omega)}{|Q(i\cdot\omega)|^2}$$

$$\cos\theta(\tau) = \cos(\omega\cdot\tau) = h(\omega) = -\frac{P_R(i\cdot\omega)\cdot Q_R(i\cdot\omega) + P_I(i\cdot\omega)\cdot Q_I(i\cdot\omega)}{|Q(i\cdot\omega)|^2}$$

it is the second condition for stability switching.

If Λ(x)>0 then the crossing proceeds from "-" to "+" respectively (stable to unstable). If Λ(x)<0 then the crossing proceeds from "+" to "-" respectively (unstable to stable).

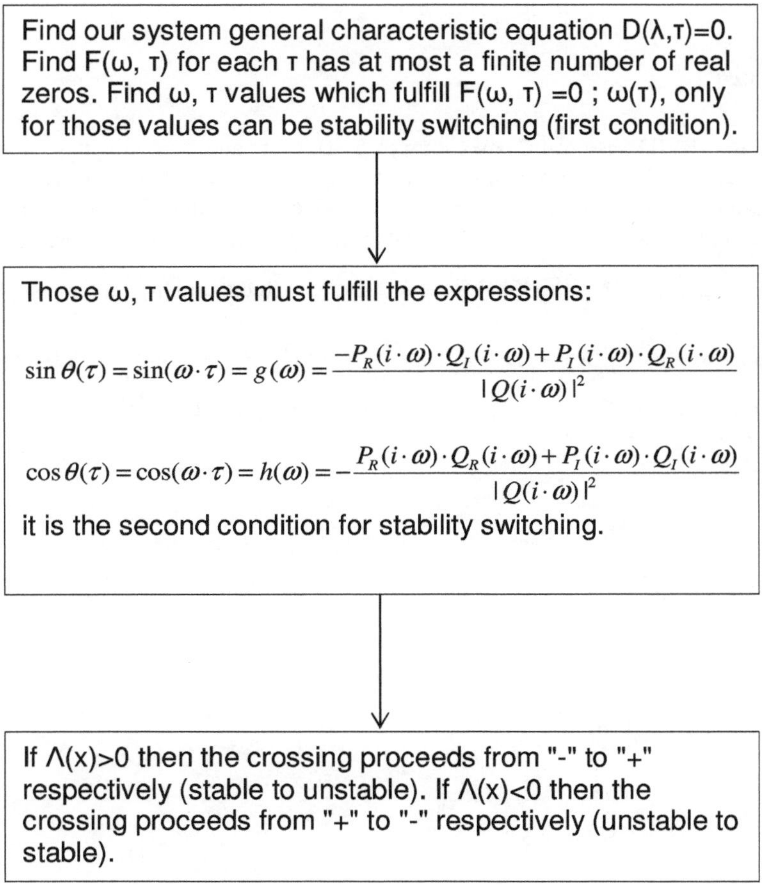

$$x = L_{calc}, R_{DC}, C_1, L_1, C_{load}, R_L, \tau, \ldots, etc.,$$

Remark: The analysis consists in identifying the roots of circuit characteristic equation $P(\lambda) = Q(\lambda) \cdot e^{-\lambda \cdot \tau} = 0$ situated on the imaginary axis of the complex λ-Plane, where, by increasing the RFID burst switch system parameters. Reλ may, at the crossing, change its sign from "−" to "+", i.e. from a stable focus E^* to an unstable one, or vice versa. This feature may be further assessed by examining the sign of the partial derivatives with respect to system parameters. Other sub cases sestability behavior ($\tau_2 = 0; \Delta > 0$ & $\tau_2 = \Delta = \tau_\Delta$) is not discussed and can be good reader exercises [12].

Exercises

1. Active RFID system has two sources $S_1(t)$, $S_2(t)$ and two antennas L_1 and L_2 (rectangular antennas) as appear in the equivalent circuit. L_1 and L_2 configuration structure can be represented as L_2 inductor antenna which is connected in the middle of L_1 antenna. The overall parameters of two antennas are the same. $L_1 = L_{calc1} = [\frac{\mu_0}{\pi} \cdot ([\sum_{i=1}^{2} X_i] - X_3 + X_4) \cdot N_c^p]$. Rectangular antennas. $L_2 = L_{calc2} = [\frac{\mu_0}{\pi} \cdot ([\sum_{i=1}^{2} X_i] - X_3 + X_4) \cdot N_c^{(1+\sqrt{p})}]$. X_1, X_2, X_3, and X_4 global antenna parameters are the same for inductor antenna L_1 and L_2.

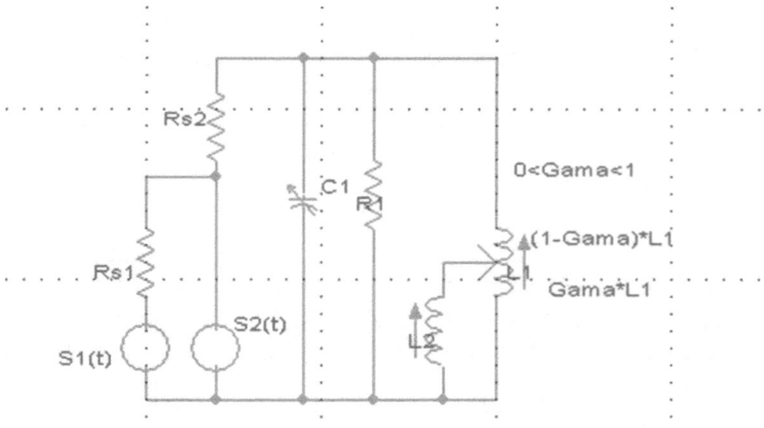

Active RFID's equivalent circuit with two sources and two antennas.

You can neglect the mutual inductance between inductor antennas L_1 and L_2. Γ (Gama) is the two antennas trim parameter ($0 < \Gamma < 1$).

1.1 Write RFID system, differential equations as a matrix representation.
1.2 Write the RFID system analog Van der pol's equations. Find all transformations between Van der pol system parameters and variables to antenna system's variables and parameters.
1.3 Discuss stability, How Γ trim parameter influences our system stability switching?
1.4 How RFID system dynamically changes for (A) $S_1(t)$ – ON, $S_2(t)$ – OFF (B) $S_1(t)$ – OFF, $S_2(t)$ – ON (C) $S_1(t)$ – ON, $S_2(t)$ – ON.
1.5 Find ξ_i (i = 1, 2, 3,...) functions of our RFID system by using regular perturbation or averaging methods.
1.6 RFID TAG IC capacitance C_1 multiple his value $C_1 \to 2 \cdot C_1$, How our RFID system stability change?

2. Active RFID system includes forcing sources $S_i(t)$; i = 1,2,... and antennas inductors (L_1, L_2,...). All antennas are rectangular. The following differential equation describes our RFID system (forced Van der pol equation). R_1 and C_1 are RFID IC parameters.

$$\ddot{V} + (\frac{1}{R_1} + \sum_{i=1}^{k} \frac{1}{R_{S_i}}) \cdot \frac{1}{C_1} \cdot \dot{V} + \frac{1}{[\sqrt{L_1} + \sum_{i=2}^{m} L_i] \cdot C_1} \cdot V = \frac{1}{C_1} \cdot [\sum_{i=1}^{k} \frac{1}{R_{S_i}} \cdot \frac{dV_{S_i}}{dt}]$$

2.1 Express our RFID system as a matrix differential equation system.
2.2 Find fixed point and discuss stability of our system.
2.3 How our Active RFID system behavior is dependent on k and m parameters?
2.4 Write the equivalent Van der pol system parameters Φ(x), α, β when only one forcing source is active. $V_{S_i} - OFF \; \forall \, i \in [1..k] \; \& \; i \neq n$ Except $V_{S_n} - ON; n \notin [1..k]$.
2.5 How the dynamic of our system change for the transformation $\sum_{i=2}^{m} L_i \to \sum_{i=2}^{m} [L_1 + L_i]$. Find fixed points and discuss the stability issue.

3. Our passive RFID TAG contains one RFID IC and two rectangular antennas in the series. The two rectangular antenna parameters are not the same and the definition is related to global parameters: X_{i1}, X_{i2}, X_{i3}, X_{i4}. i = 1for the first antenna and i = 2 for the second antenna. The matrix formulation for RFID differential equation:

$$\begin{bmatrix} \frac{dV_1}{dt} \\ \frac{dV_2}{dt} \\ \frac{dV_3}{dt} \end{bmatrix} = \begin{bmatrix} 0 & 1 & 0 \\ \{-\frac{1}{C_1 \cdot \{\left[\frac{\mu_0}{\pi} \cdot (-X_{13} + \sum_{k=1,k\neq 3}^{4} X_{1k}) \cdot Nc^p\right] + \left[\frac{\mu_0}{\pi} \cdot (-X_{23} + \sum_{k=1,k\neq 3}^{4} X_{2k}) \cdot Nc^{p^2}\right]\}}\} & \{-\frac{1}{C_1 \cdot R_1}\} & 0 \\ 0 & 0 & 0 \end{bmatrix} \cdot \begin{bmatrix} V_1 \\ V_2 \\ V_3 \end{bmatrix} + \begin{bmatrix} 0 \\ 0 \\ 1 \end{bmatrix}$$

R_1 and C_1 are parameters for RFID TAG IC. V_1, V_2, V_3 are system variables. All other antenna parameters are the same as discuss in the chapter.

3.1 Find RFID TAG system fixed points and discuss the stability.
3.2 Discuss the system Eigen direction, Eigen solutions, Eigen vectors, and Eigenvalues behavior for $t \to \infty$.
3.3 How our system stability is affected by different values of "p" parameter? Draw Stable/Unstable diagram.
3.4 Analyze RFID TAG system dynamical behavior for $X_{2k} = X_{1k}^2 - \Gamma$; k = 1,…, 4 RFID TAG antennas global parameter index. Γ is a shifting parameter between the square of first antenna global parameters (X_{1k}^2) and second antenna global parameters (X_{2k}).
3.5 How our RFID TAG system behavior changes for multiple values of rectangular antenna's number of turns (N_C) ; $N_C \to 2 \cdot N_C$. N_C is the same for the first and second RFID TAG antenna.

4. We have delayed in time passive RFID TAG system. Due to electromagnetic interferences, we have RFID TAG's voltage and voltage derivative with delays $\tau + 1$ and $\tau^2 - 1$ respectively in time. $V_1(t) \to V_1(t - [\tau + 1])$; $V_2(t) \to V_2(t - [\tau^2 - 1])$. We consider no delay effect on $\frac{dV_1(t)}{dt}$ and $\frac{dV_2(t)}{dt}$. The RFID TAG antenna is rectangular. X_i; i = 1, 2, 3, 4 are RFID TAG antenna global parameters as discuss in the chapter. R_1 and C_1 are RFID TAG IC parameters.

$$\frac{dV_1}{dt} = V_2(t - [\tau^2 - 1])$$

$$\frac{dV_2}{dt} = \{-\frac{1}{C_1 \cdot \left[\frac{\mu_0}{\pi} \cdot [X_1 + X_2 - X_3 + X_4] \cdot Nc^p\right]}\} \cdot V_1(t - [\tau + 1]) - \frac{1}{C_1 \cdot R_1} \cdot V_2(t - [\tau^2 - 1])$$

4.1 Find system fixed points and discuss stability for $\tau = 0$.

4.2 Find the system characteristic equation ($D(\lambda, \tau)$), τ is our delay parameter. $D(\lambda, \tau) = P_n(\lambda, \tau) + Q_m(\lambda, \tau) \cdot e^{-\lambda\tau}$.

4.3 Find polynomial in ω representation $F(\omega, \tau)$ and sketch 3D function. Find $\sin\theta(\tau)$ and $\cos\theta(\tau)$ expressions.

4.4 Find U, V, ω_τ expressions and define maps $S_n(\tau) = \tau - \tau_n(\tau); \tau \in I, n \in \mathbb{N}_0$.

4.5 Find $\Lambda^{-1}(C_1), \Lambda^{-1}(R_1), \Lambda^{-1}(\tau), sign(\Lambda^{-1}(\tau))$ expressions and discuss stability switching for different values of τ parameters.

5. We have a RFID system with two rectangular antennas (L_1, L_2) in parallel and one RFID IC (R_1 and C_1 parameters). There are parasitic resistances of our RFID system, $r_{p_1}, r_{p_2}; r_{p_1} \neq r_{p_2}$. The following figure is equivalent circuit of our RFID system.

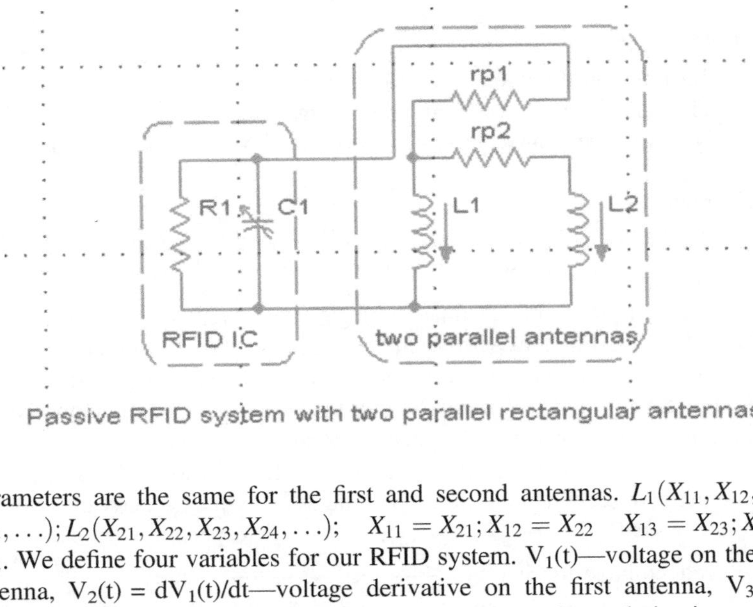

Passive RFID system with two parallel rectangular antennas

Parameters are the same for the first and second antennas. $L_1(X_{11}, X_{12}, X_{13}, X_{14}, \ldots); L_2(X_{21}, X_{22}, X_{23}, X_{24}, \ldots);\quad X_{11} = X_{21}; X_{12} = X_{22}\quad X_{13} = X_{23}; X_{14} = X_{24}$. We define four variables for our RFID system. $V_1(t)$—voltage on the first antenna, $V_2(t) = dV_1(t)/dt$—voltage derivative on the first antenna, $V_3(t)$—voltage on the second antenna, $V_4(t) = dV_3(t)/dt$—voltage derivative on the second antenna.

Remark Voltages on RFID TAG antennas are only on equivalent circuit inductors L_1, L_2 without parasitic resistances. Due to electromagnetic interferences, we get RFID TAG's antenna voltages and voltages derivatives with delays in time:

$$\Delta_1 \neq \Delta_2; V_1(t) \to V_1(t-\tau)$$
$$V_2(t) \to V_2(t-\tau\cdot\Delta_1); V_3(t) \to V_3(t-\sqrt{\tau+1});$$
$$V_4(t) \to V_4(t-|\Delta_1-\Delta_2|\cdot\sqrt{\tau}).$$
$$\Delta_1 > 0; \Delta_2 > 0.$$

5.1 Find RFID system, differential equations, fixed points and discuss stability for $\tau = 0; \Delta_i > 0$ i = 1, 2; $\Delta_1 \neq \Delta_2$.

5.2 Find the system characteristic equation $D(\lambda,\tau,\Delta_1,\Delta_2)$, τ is our delay parameter and Δ_1, Δ_2 are parameters.

$$\xi(\tau,\Delta_1,\Delta_2)$$
$$D(\lambda,\tau,\Delta_1,\Delta_2) = P_n(\lambda,\tau,\Delta_1,\Delta_2) + Q_m(\lambda,\tau,\Delta_1,\Delta_2)\cdot e^{-\lambda\cdot\xi(\tau,\Delta_1,\Delta_2)}.$$

5.3 Find polynomial in ω representation $F(\omega,\tau)|_{\Delta_1,\Delta_2 \text{ parameters}} = 0$ and sketch 3D function. Find $\sin\theta(\tau)$ and $\cos\theta(\tau)$ expressions.

5.4 Find U, V, ω_τ expressions and define maps $S_n(\tau) = \tau - \tau_n(\tau); \tau \in$ I, n $\in \mathbb{N}_0$ for the cases: (A) $\Delta_1 = \Delta$, $\Delta_2 = 0$; (B) $\Delta_1 = 0$, $\Delta_2 = \Delta$; (C) $\Delta_1 = \Delta_2 = \Delta$

5.5 Find $\Lambda^{-1}(\tau), \Lambda^{-1}(\Delta_1), \Lambda^{-1}(\Delta_2), sign(\Lambda^{-1}(\tau)), sign(\Lambda^{-1}(\Delta_1)), sign(\Lambda^{-1}(\Delta_2))$ expressions and discuss stability switching for different values of τ, Δ_1, Δ_2.

6. We have triple loop antennas arranged as a shifted gate in X direction. The RFID TAG is semi passive and contains a battery that enables long reading distance and also enables the tag to operate independently of the reader. The double antenna gate is employed due to the fact that this antenna consists of three parallel loops (primary, secondary, and third loop). Due to electromagnetic interferences there are differences in time delays with respect to gate antenna first, second and third loop voltages and voltages derivatives. The delay voltages are $V_{i1}(t-\tau_1), V_{i2}(t-\tau_2), V_{i3}(t-\tau_3)$ respectively $(\tau_1 \neq \tau_2 \neq \tau_3)$ and delayed voltage derivative $\frac{dV_{i1}(t-\Delta)}{dt}; \frac{dV_{i2}(t-[\Delta+\sqrt{\Delta}])}{dt}$ and $\frac{dV_{i3}(t-[\Delta^2+1])}{dt}; \tau_1 \geq 0;$ $\tau_2 \geq 0; \tau_3 \geq 0; \Delta \geq 0$. Each triple loop gate antenna is defined as a three inductors in series L_{i1}, L_{i2}, L_{i3} with series parasitic resistors r_{p_1}, r_{p_2}; i—index of the first and second gate. First gate: L_{11}, L_{12}, L_{13} is mostly formed by traces on the planar PCB. $2\cdot L_{m,1-2}; 2\cdot L_{m,1-3}; 2\cdot L_{m,2-3}$, elements represent the mutual inductances between each two antenna inductors in the gate. The second loop is within the first loop and third loop is within the second loop. We consider that the triple loop antennas parameter values are the same in the first and second gate $(L_{a1}, L_{a2}, L_{a3}, L_{b1}, L_{b2}, L_{b3}, a_1, a_2, a_3)$.

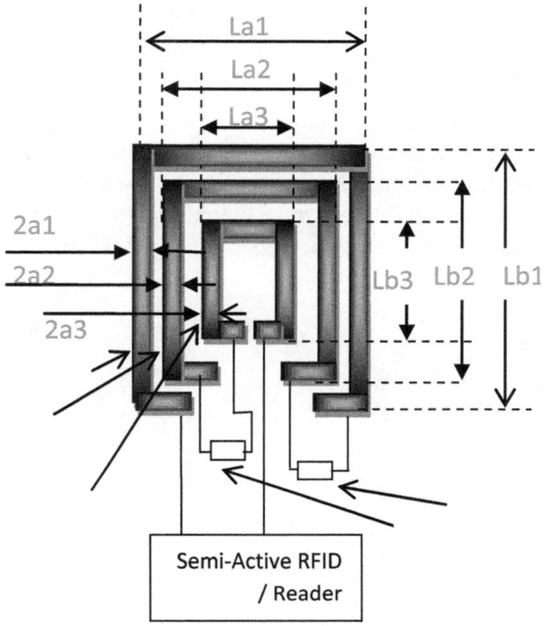

$$L_{1j} = 4 \cdot \{L_{bj} \cdot \ln[\frac{2 \cdot A_j}{a_j \cdot (L_{b1} + l_{cj})}] + L_{aj} \cdot \ln[\frac{2 \cdot A_j}{a_j \cdot (L_{bj} + l_{cj})}] + 2 \cdot [a_j + l_{cj} - (L_{aj} + L_{bj})]\}$$

$$l_{cj} = \sqrt{L_{aj}^2 + L_{bj}^2}; A_j = L_{aj} \cdot L_{bj}$$

j = 1 first loop, j = 2 s loop, j = 3 third loop. Since three inductors (L_{11}, L_{12}, L_{13}) are in series and there are mutual inductances, the total antenna inductance for the first gate:

$$L_T|_{\text{first gate}} = \sum_{k=1}^{3} L_{1k} + 2 \cdot [\sum_{i=1}^{2} L_{m,i-(i+1)} + L_{m,1-3}]; L_{m,1-2} = K_1 \cdot \sqrt{L_{11} \cdot L_{12}}$$

$$L_{m,1-3} = K_2 \cdot \sqrt{L_{11} \cdot L_{13}}; L_{m,2-3} = K_3 \cdot \sqrt{L_{12} \cdot L_{13}}.$$

$L_{m,i\text{-}j}$ is the mutual inductance between inductors i and j. K_1, K_2, and K_3 are the coupling coefficients of two inductors. $0 \le K_1 \le 1; 0 \le K_2 \le 1; 0 \le K_3 \le 1$. We consider the case our RFID shifted gate system is passive (power source is disconnected). Remark: no delay effects on RFID system variables derivatives.

6.1 Find RFID double gate differential equations and fixed points (only one gate).
6.2 Find Jacobian of our RFID shifted gate system and characteristic equation: $D(\lambda, \tau_1, \tau_2, \tau_3, \Delta)$.

6.3 Find F(ω, τ) = 0 and its roots sinθ(τ) and cosθ(τ) expressions.
6.4 Find $\Lambda^{-1}(\tau_1), \Lambda^{-1}(\tau_2), \Lambda^{-1}(\tau_3), \Lambda^{-1}(\Delta)$ expressions.
6.5 Discuss system stability switching for various values of $\tau_1, \tau_2, \tau_3, \Delta$.

7. We have RFID detector system which is represented by the following set of differential equations matrix representation. Ξ_{kl} (k = 1,…5; l = 1,…,5) global parameter expressions are the same as describe in Sect. (1.5). Additional elements are Ω_1, Ω_2 and V(t) second order derivative column matrix element ψ. V (t) represents the RFID tag antenna voltage in time, incoming RF small signal from RFID reader.

$$\begin{pmatrix} \frac{dX}{dt} \\ \frac{dY}{dt} \\ \frac{dI_{L_1}}{dt} \\ \frac{dI_{R_j}}{dt} \\ \frac{dI_{R_S}}{dt} \end{pmatrix} = \begin{pmatrix} \Xi_{11} & \cdots & \Xi_{1n} + \Omega_2 \\ \vdots & \ddots & \vdots \\ \Xi_{m1} + \Omega_1 & \cdots & \Xi_{mn} \end{pmatrix}_{n=m=5} \cdot \begin{pmatrix} X \\ Y \\ I_{L_1} \\ I_{R_j} \\ I_{R_S} \end{pmatrix}$$

$$+ \begin{pmatrix} -\frac{R_{in}}{L_1 \cdot L_P} \\ 0 \\ \frac{1}{L_1} \\ 0 \\ 0 \end{pmatrix} \cdot V(t) + \begin{pmatrix} \frac{1}{L_P} \\ 0 \\ 0 \\ 0 \\ 0 \end{pmatrix} \cdot \frac{dV(t)}{dt} + \begin{pmatrix} 0 \\ \psi \\ 0 \\ 0 \\ 0 \end{pmatrix} \cdot \frac{d^2V(t)}{dt^2}$$

7.1 Draw RFID TAG detector circuit which characterizes by our above differential equations, matrix representation. What are the additional circuit components and their location which represents by matrix's parameters Ω_1, Ω_2 and ψ? <u>Remark</u>: probably they are additional Schottky diode's parasitic elements.
7.2 Find system fixed points and discuss stability in the case of no parasitic delay effects $\tau_i = 0$; i = 1,2,….
7.3 Consider that the Schottky detector diode has a package parasitic inductance L_p delay element in time τ_1 and package parasitic capacitance C_p delay element in time τ_2 . Find fixed points coordinate expressions, consider RF in signal $V(t) = A_0 + B_0 \cdot f^2(t)$.

$$V(t) = A_0 + B_0 \cdot f^2(t) \; ; \; |f(t)| < 1 \; \& A_0 \gg |f(t)| \; ; \; 0 < B_0 < 1.$$

Find Jacobian of our RFID Schotky detector system.
7.4 Find our system characteristic equation $D(\lambda, \tau_1, \tau_2)$ for three cases: (A) $\tau_1 = \tau; \tau_2 = 0$ (B)$\tau_1 = 0; \tau_2 = \tau$ (C) $\tau_1 = \tau_2 = \tau$.

7.5 Find $\wedge^{-1}(\tau) = (\frac{\partial \mathrm{Re}\lambda}{\partial \tau})_{\lambda = i \cdot \omega}, L_P, L_1, C_f, R_{in}, R_s, C_P, R_j, \ldots = const; \omega \in \mathbb{R}_+$ and discuss stability switching for different values of delay parameter τ.

8. We have RFID detector system which is represented by the following set of differential equations matrix representation. Ξ_{kl} (k = 1,…5; l = 1,…,5) global parameter expressions are the same as described in Sect. (1.5). Additional elements are Ω_1, Ω_2 and V(t) second order derivative column matrix element ψ_1 and ψ_2. V(t) represents the RFID tag antenna voltage in time, the incoming RF small signal from the RFID reader.

$$\begin{pmatrix} \frac{dX}{dt} \\ \frac{dY}{dt} \\ \frac{dI_{L_1}}{dt} \\ \frac{dI_{R_j}}{dt} \\ \frac{dI_{R_S}}{dt} \end{pmatrix} = \begin{pmatrix} \Xi_{11} + \Omega_2 & \ldots & \Xi_{1n} \\ \vdots & \ddots & \vdots \\ \Xi_{m1} + \Omega_1 & \cdots & \Xi_{mn} \end{pmatrix}_{n=m=5} \cdot \begin{pmatrix} X \\ Y \\ I_{L_1} \\ I_{R_j} \\ I_{R_S} \end{pmatrix}$$

$$+ \begin{pmatrix} -\frac{R_{in}}{L_1 \cdot L_P} \\ 0 \\ \frac{1}{L_1} \\ 0 \\ 0 \end{pmatrix} \cdot V(t) + \begin{pmatrix} \frac{1}{L_P} \\ 0 \\ 0 \\ 0 \\ 0 \end{pmatrix} \cdot \frac{dV(t)}{dt} + \begin{pmatrix} 0 \\ \psi_1 \\ 0 \\ 0 \\ \psi_2 \end{pmatrix} \cdot \frac{d^2V(t)}{dt^2}$$

8.1 Draw RFID TAG detector circuit which characterizes by our above differential equations, matrix representation. What are the additional circuit components and their location which represents by matrix's parameters Ω_1, Ω_2 and ψ_1, ψ_2? Remark: probably they are additional Schottky diode's parasitic elements.

8.2 Find system fixed points and discuss stability in the case of no parasitic delay effects $\tau_i = 0$; i = 1,2,….

8.3 Consider that the Schottky detector diode has a package parasitic inductance L_p delay element in time τ and package parasitic capacitance C_p delay element in time $\tau^2 + 1$. Find fixed points coordinate expressions, consider RFin signal $V(t) = A_0 + B_0 \cdot f^3(t)$ $V(t) = A_0 + B_0 \cdot f^3(t); |f(t)| < 1 \;\&\; A_0 \gg |f(t)|; 0 < B_0 < 1$. Find Jacobian of our RFID Schottky detector system.

8.4 Find our system characteristic equation $D(\lambda, \tau)$.

8.5 Find $\wedge^{-1}(\tau) = (\frac{\partial \mathrm{Re}\lambda}{\partial \tau})_{\lambda = i \cdot \omega}, L_P, L_1, C_f, R_{in}, R_s, C_P, R_j, \ldots = const; \omega \in \mathbb{R}_+$ and discuss stability switching for different values of delay parameter τ.

9. Active RFID system includes forcing sources $S_i(t)$; i = 1, 2,… and antennas inductors (L_1, L_2,…). All antennas are rectangular. The following differential

equation describes our RFID system (forced Van der pol equation). R_1 and C_1 are RFID IC parameters. Additional parameters are Ω_1 and Ω_2.

$$\ddot{V} + (\frac{1}{R_1} + \Omega_1 \cdot \sum_{i=1}^{k} \frac{1}{R_{S_i}}) \cdot \frac{1}{C_1} \cdot \dot{V} + \frac{1}{[\sqrt{L_1} + \Omega_2 \cdot \sum_{i=2}^{m} L_i] \cdot C_1} \cdot V$$

$$= \frac{1}{C_1} \cdot [\sum_{i=1}^{k} \frac{1}{R_{S_i}} \cdot \frac{dV_{S_i}}{dt}]$$

9.1 Express our RFID system as a matrix differential equation system.

9.2 Find fixed point and discuss stability of our system for different values of Ω_1 and Ω_2 parameters.

9.3 How our Active RFID system behavior is dependent on k, m, Ω_1 and Ω_2 parameters?

9.4 Write the equivalent Van der pol system parameters $\Phi(x)$, α, β when only one forcing source is active. $V_{S_i} - OFF \forall i \in [1 \ldots k]$ & $i \neq n$ Except $V_{S_n} - ON; n \notin [1..k]$.

9.5 How the dynamic of our system change for the transformation $\sum_{i=2}^{m} L_i \to \sum_{i=2}^{m} [L_1 + L_i^2]$. Find fixed points and discuss stability, issue for $\Omega_1 = \Omega; \Omega_2 = 1 + \sqrt{\Omega}$.

10. We have delayed in time passive RFID TAG system. Due to electromagnetic interferences, we have RFID TAG's voltage and voltage derivative with delays $\sqrt{\tau} + 1$ and $\tau^3 - 1$ respectively in time. $V_1(t) \to V_1(t - [\sqrt{\tau} + 1]); V_2(t) \to V_2(t - [\tau^3 - 1])$. We consider no delay effect on $\frac{dV_1(t)}{dt}$ and $\frac{dV_2(t)}{dt}$. The RFID TAG antenna is rectangular. X_i; i = 1, 2, 3, 4 are RFID TAG antenna global parameters as discuss in the chapter. R_1 and C_1 are RFID TAG IC parameters.

$$\frac{dV_1}{dt} = V_2(t - [\tau^3 - 1]);$$

$$\frac{dV_2}{dt} = \{-\frac{1}{C_1 \cdot \left[\frac{\mu_0}{\pi} \cdot [X_1 + X_2 - X_3 + X_4] \cdot Nc^p\right]}\} \cdot V_1(t - [\sqrt{\tau} + 1]) - \frac{1}{C_1 \cdot R_1} \cdot V_2(t - [\tau^3 - 1])$$

10.1 Find system fixed points and discuss stability for $\tau = 0$.

10.2 Find the system characteristic equation ($D(\lambda, \tau)$), τ is our delay parameter. $D(\lambda, \tau) = P_n(\lambda, \tau) + Q_m(\lambda, \tau) \cdot e^{-\lambda\tau}$.

10.3 Find polynomial in ω representation $F(\omega,\tau)$ and sketch 3D function. Find $\sin\theta(\tau)$ and $\cos\theta(\tau)$ expressions.

10.4 Find U, V, ω_τ expressions and define maps $S_n(\tau) = \tau - \tau_n(\tau)$

$$S_n(\tau) = \tau - \tau_n(\tau); \tau \in \mathrm{I}, \mathrm{n} \in \mathbb{N}_0.$$

10.5 Find $\Lambda^{-1}(C_1), \Lambda^{-1}(R_1), \Lambda^{-1}(\tau), sign(\Lambda^{-1}(\tau))$ expressions and discuss stability switching for different value of τ parameters.

Chapter 2
Microwave Elements Description and Stability Analysis

There are three types of microwave circuits which include microwave elements. The first is a discrete circuit; packaged diodes/transistors mounted in coax and waveguide assemblies. Second Hybrid MIC (Microwave Integrated Circuit); diodes/transistors and microstrip fabricated separately and then assembled. The third is MMIC (Monolithic Microwave Integrated Circuit); diodes, transistors and microstrip fabricated simultaneously. The monolithic microwave integrated circuit (MMIC) consists of diodes, transistor, microstrip transmission lines, microstrip circuits, and other circuit elements, such as lumped capacitors, resistors, etc., which have parasitic effects influence on overall system stability behavior. The discrete microwave circuit can be PIN diodes mounted in a coaxial transmission line which characterize by parasitic effects and delay variables in time. Hybrid microwave integrated circuit's wire bonds cause reliability problems and parasitic effects; stability issue can affect every hybrid microwave integrated circuits. Many receivers are often at risk of having their front end burned out by high power RF. Receivers are traditionally protected by a power limiter circuit. The limiter diode is a special type of the PIN diode. Due to the parasitic effects of microstrip transmission lines there is a delay in time for input RF signal result in the end. Power limiters use with transmission line face stability behavior for different delay time values. Reflection Type Phase Shifter (RTPS), employing a circulator. Micro strip transmission lines with three port active circulator, stability analysis under time delayed. Many RF systems are use Active circulator as a passive non-reciprocal three- or four-port device, in which microwave or radio frequency power entering any port is transmitted to the next port in rotation (only). Micro strip transmission lines fid those active circulator ports and face a delay parasitic effect of transferring signals in time. These circulator's micro strip transmission lines, delays cause to system instability. Resonant RF network antennas are important to plasma sources with many applications. The cylindrical resonant RF network antennas run as large volume plasma

O. Aluf, *Microwave RF Antennas and Circuits*,
DOI 10.1007/978-3-319-45427-6_2

sources and have stability switching due to system's copper legs parasitic effects. The cylindrical RF network antennas structure is 16-leg cylindrical (birdcage) RF antenna which has electrical circuit and opposite points of RF feeding and grounding. Due to cylindrical antenna parasitic delayed in time, there is a stability issue by analyzing its operation. Tunnel diode is the p-n junction device that exhibits negative resistance. That means when the voltage increases the current through it decreases. Typical Tunnel Diode (TD) I-V characteristic has two distinct features: (1) it is STRONGLY non-linear (compare to the resistor I-V). Current-Voltage relationships for TDs cannot be described using the Ohm's law (2) it has a negative differential resistance (NDR) region. Tunnel diode can be a microwave oscillator. Transient is in the resonant cavity after turning the bias voltage ON. The resonant circuit with NDR can oscillate. The TD microwave oscillator has parasitic effects in time and delay variables. Stability is a very crucial issue when designing microwave oscillator by using Tunnel Diode (TD) [14, 15].

2.1 Microstrip Transmission Lines Delayed in Time Power Limiters Stability Analysis

Microwave and RF receivers, as well as many instruments, are susceptible to damage from input signals having amplitudes which exceed some danger level. The front end of some receivers can be destroyed by power levels. Avoiding such damage is by using a power limiter which designed around a special type of PIN diode called a limiter diode. A thin epitaxial I-layer is formed on a heavily N^+ doped substrate, after which P^+ top contacts are added by diffusion. Typical limiter diodes have I-layer thickness between 2 and 7 μm, with corresponding values of breakdown voltage. The diode is mounted in shunt across the microstrip transmission line which leads to the receiver front end, and is provided with a DC bias return. For incoming signals which are below the threshold level in amplitude, the diode acts as an ordinary unbiased PIN diode, which is to say that it appears to be a capacitor of relatively small value. When the incident signal exceeds the threshold power level, the diode's I-layer is flooded with carriers during the positive half cycle of the incoming RF signal. Most of these carriers persist through the negative half cycle, DC current begins to flow in the loop formed by the diode and bias return choke, and the diode biases itself to a low value of resistance in a matter of nanoseconds. Under the influence of this self generated bias current, the diode's junction resistance falls to a very low value, shorting out the transmission line. The limiter circuit, then acts as a reflective switch, reflecting the large signal back to its source and protecting the circuitry which is "downstream" from the limiter. Our Microstrip transmission lines with power limiters system delay differential and delay different model can be analytically by using delay differential equations in dynamically. The need of the incorporation of a time delay is often of the existence of any stage structure. It is often difficult to analytically study models with delay dependent parameters, even if only a single discrete delay is present. There is a practical guidelines that combine graphical information with analytical work to effectively study the local stability of models involving delay dependent parameters. The stability of a given steady state is simply determined by the graphs of some function of microstrip delays τ_1, τ_2 which can be expressed, explicitly and thus can be easily depicted by Matlab and other popular software. We need only look at one such function and locate the zeros. This function often has only two zeros, providing thresholds for stability switches. As time delay increases, stability changes from stable to unstable to stable. We emphasize the local stability aspects of some models with delay dependent parameters. Additionally, there is a general geometric criterion that, theoretically speaking, can be applied to models with many delays, or even distributed delays. The simplest case of a first order characteristic equation, providing more user friendly geometric and analytic criteria for stability switches. The analytical criteria provided for the first and second order cases can be used to obtain some insightful analytical statements and can be helpful for conducting simulations. The most obvious way in which to amount limiter diode in shunt across a microstrip line. Two leads of limiter diode are mounted in parallel to the transmission line and the third lead is soldered to the ground pad as shown. D1 is a limiter diode [24–26, 33–35] (Fig. 2.1).

The shunt mounted limiter diode equivalent circuit with microstrip lines delayed in time. The time delay for the first line segment is τ_1 and the second line segment τ_2. See Fig. 2.2.

It is possible to locate several limiter PIN diode on microstrip line, but in the current chapter we focus on one limiter diode with the specific connection structure. We consider for simplicity that the microstrip segments resistances are neglected and either related voltages $V_{\tau 1} \to \varepsilon$, $V_{\tau 2} \to \varepsilon$. Then we can define $V_a(t) = V_i(t - \tau_1)$; $V_o(t) = V_a(t - \tau_2) = V_i(t - (\tau_1 + \tau_2))$. We do our stability analysis of three different cases: $\tau_1 = \tau,\ \tau_2 \to \varepsilon;\ \tau_1 \to \varepsilon,\ \tau_2 = \tau;\ \tau_1 = \tau_2 = \tau$. We defined I_{τ_1}, I_{τ_2} as the current through first and second delay lines respectively. $V_{\tau_1}, V_{\tau_2} \to \varepsilon \Rightarrow V_0(t) = V_i(t - (\tau_1 + \tau_2))$ (V_{τ_1}, V_{τ_2} are voltages of the first and second delay lines). $I_{\tau_1} = \frac{V_i - V_a}{R_{\tau_1}}$; $I_{\tau_2} = \frac{V_a - V_0}{R_{\tau_2}}$; $I_{\tau_1} = I_a + I_{\tau_2}$; $I_a = I_{L_i}$, $i = 0, 1, 2$

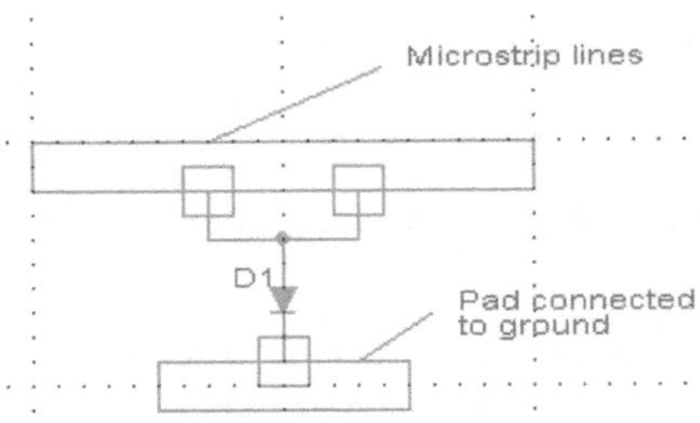

Fig. 2.1 Shunt mounted limiter diode

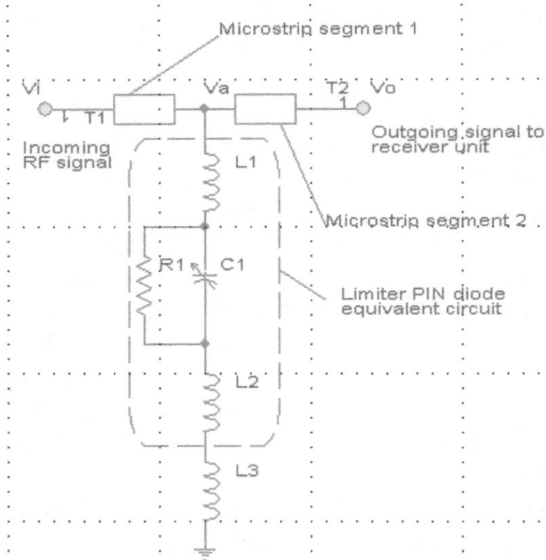

Fig. 2.2 Shunt mounted limiter diode equivalent circuit with microstrip lines delayed in time

$$V = V_{R1} = V_{C1};\ V_a = \frac{dI_a}{dt} \cdot (L_0 + L_1 + L_2) + V;\ I_a = \frac{V}{R1} + C_1 \cdot \frac{dV}{dt}$$

$$\frac{dI_a}{dt} = \frac{1}{R_1} \cdot \frac{dV}{dt} + C_1 \cdot \frac{d^2V}{dt^2};\ V_a = \frac{dI_a}{dt} \cdot \sum_{i=0}^{2} L_i + V;\ V = V(t)$$

$$V_a = [\frac{1}{R_1} \cdot \frac{dV}{dt} + C_1 \cdot \frac{d^2V}{dt^2}] \cdot \sum_{i=0}^{2} L_i + V$$

We consider $V_{\tau_1}, V_{\tau_2} \to \varepsilon$ then $V_i(t - \tau_1) = V_a = [\frac{1}{R_1} \cdot \frac{dV}{dt} + C_1 \cdot \frac{d^2V}{dt^2}] \cdot \sum_{i=0}^{2} L_i + V$. Shifting equation $V_i(t - \tau_1) = V_a = \ldots$ in time by τ_1 gives equation:

$$V_i(t) = [\frac{1}{R_1} \cdot \frac{dV(t + \tau_1)}{dt} + C_1 \cdot \frac{d^2V(t + \tau_1)}{dt^2}] \cdot \sum\nolimits_{i=0}^{2} L_i + V(t + \tau_1).$$

Finally, we get two Power limiter equations (with delays $\boldsymbol{\tau_1}, \boldsymbol{\tau_2}$)

$$V_0(t) = [\frac{1}{R_1} \cdot \frac{dV(t - \tau_2)}{dt} + C_1 \cdot \frac{d^2V(t - \tau_2)}{dt^2}] \cdot \sum_{i=0}^{2} L_i + V(t - \tau_2);\ V_i(t - \tau_1)$$

$$= [\frac{1}{R_1} \cdot \frac{dV}{dt} + C_1 \cdot \frac{d^2V}{dt^2}] \cdot \sum_{i=0}^{2} L_i + V$$

We get two Power limiter equations, one of them is a differential equation which involves input coming signal.

$$V_i(t-\tau_1) = [\frac{1}{R_1}\cdot\frac{dV}{dt} + C_1\cdot\frac{d^2V}{dt^2}]\cdot\sum_{i=0}^{2} L_i + V;\ x(t) = V_i(t-\tau_1);$$

$$\xi_1 = C_1\cdot\sum_{i=0}^{2} L_i;\ \xi_2 = \frac{1}{R_1}\cdot\sum_{i=0}^{2} L_i;\ \xi_3 = 1;\ V = f(\xi_1,\xi_2,\xi_3,x(t))$$

$\ddot{V}\cdot\xi_1 + \dot{V}\cdot\xi_2 + V\cdot\xi_3 = x(t) \Rightarrow \ddot{V} + \dot{V}\cdot\frac{\xi_2}{\xi_1} + V\cdot\frac{\xi_3}{\xi_1} = \frac{1}{\xi_1}x(t)$. This differential equation can recognize as forced Van der Pol equation. The basic Van der Pol equation can be written in the form: $\ddot{V} + \alpha\cdot\phi(V)\cdot\dot{V} + V = \beta\cdot p(t)$

$$\frac{\xi_2}{\xi_1} = \frac{1}{R_1\cdot C_1};\ \frac{\xi_3}{\xi_1} = \frac{1}{C_1\cdot\sum_{i=0}^{2} L_i};\ \frac{\xi_3}{\xi_1} \to (1\pm\varepsilon);\ \beta\cdot p(t) = \frac{1}{\xi_1}x(t) = \frac{1}{C_1\cdot\sum_{i=0}^{2} L_i}\cdot x(t)$$
$$= \frac{1}{C_1\cdot\sum_{i=0}^{2} L_i}\cdot V_i(t-\tau_1)$$

$$\alpha\cdot\phi(V) = \frac{1}{R_1\cdot C_1};\ \alpha = \frac{1}{R_1\cdot C_1};\ \phi(V) = 1;\ \beta = \frac{1}{C_1\cdot\sum_{i=0}^{2} L_i} \to (1\pm\varepsilon);\ p(t)$$
$$= V_i(t-\tau_1)$$

We can consider the incoming signal after delay τ_1 is T periodic and α, β are non-negative parameters. It is convenient to write our power limiter Van der Pol equation as autonomous system when $(V, W, \theta) \in R^2 x\, S^2$.

$\dot{V} = W - \frac{1}{R_1\cdot C_1};\ \dot{W} = -V + \frac{1}{C_1\cdot\sum_{i=0}^{2} L_i}\cdot V_i(\theta-\tau_1);\ \dot{\theta} = 1.$ We suppose $\alpha, \beta \ll 1$, since we are interested in the periodic coming signal response we use $\frac{2\cdot\pi}{\omega}$ periodic transformation.

$$\begin{pmatrix} u_1 \\ u_2 \end{pmatrix} = \begin{pmatrix} \cos\omega\cdot t & -\frac{1}{\omega}\sin\omega\cdot t \\ -\sin\omega\cdot t & -\frac{1}{\omega}\cos\omega\cdot t \end{pmatrix}\cdot\begin{pmatrix} V \\ W \end{pmatrix}$$

$$\frac{du_1}{dt} = -\frac{1}{R_1\cdot C_1}\cdot\phi(V)\cdot\cos\omega\cdot t - (\frac{\omega^2-1}{\omega})\cdot V\cdot\sin\omega\cdot t$$
$$-\frac{1}{\omega\cdot C_1\cdot\sum_{i=0}^{2} L_i}\cdot\sin\omega\cdot t\cdot V_i(t-\tau_1)$$
$$\frac{du_2}{dt} = \frac{1}{R_1\cdot C_1}\cdot\phi(V)\cdot\sin\omega\cdot t - (\frac{\omega^2-1}{\omega})\cdot V\cdot\cos\omega\cdot t$$
$$-\frac{1}{\omega\cdot C_1\cdot\sum_{i=0}^{2} L_i}\cdot\cos\omega\cdot t\cdot V_i(t-\tau_1)$$

First equation solution: $V = u_1 \cdot \cos\omega \cdot t - u_2 \cdot \sin\omega \cdot t$ assuming that we are near resonance, so that $|\omega^2 - 1|, \alpha, \beta$ is all small $(\sigma = (1-\omega^2) \cdot R_1 \cdot C_1/\omega)$, we get

$$\frac{du_1}{dt} = \frac{1}{2 \cdot R_1 \cdot C_1} \cdot [u_1 - \sigma \cdot u_2 - \frac{u_1}{4} \cdot (u_1^2 + u_2^2)];$$
$$\frac{du_2}{dt} = \frac{1}{2 \cdot R_1 \cdot C_1} \cdot [u_2 + \sigma \cdot u_1 - \frac{u_2}{4} \cdot (u_1^2 + u_2^2)] - \frac{1}{2 \cdot \omega \cdot C_1 \cdot \sum_{i=0}^{2} L_i}$$

There is no significant difference if we set one of delays τ_1, τ_2 to zero for that Van der Pol equation. The second Power limiter differential equation involves output voltage.

$$V_0(t) = [\frac{1}{R_1} \cdot \frac{dV(t-\tau_2)}{dt} + C_1 \cdot \frac{d^2V(t-\tau_2)}{dt^2}] \cdot \sum_{i=0}^{2} L_i + V(t-\tau_2)$$

We consider coming signal cause at t = 0 voltage V(t = 0), first power limiter equation (Van der Pol). Additionally $V_0(t = 0) = 0$. Then we get our second delay differential equation: $[\frac{1}{R_1} \cdot \frac{dV(t-\tau_2)}{dt} + C_1 \cdot \frac{d^2V(t-\tau_2)}{dt^2}] \cdot \sum_{i=0}^{2} L_i + V(t-\tau_2) = 0$.

V(t = 0) is calculated from the first power limiter Van der Pol equation, $V(t = 0) = U_1$). We can represent our second power limiter delay differential equation as a general linear real scalar neutral differential equation with single delay τ_2 ($\tau_2 > 0$).

$$\sum_{k=0}^{n} a_k \cdot \frac{d_k}{dt_k} V(t) + \sum_{k=0}^{n} b_k \cdot \frac{d_k}{dt_k} V(t-\tau) = 0; \frac{d^0}{dt^0} V(t) \triangleq V(t)$$

$$\frac{d^0}{dt^0} V(t-\tau_2) \triangleq V(t-\tau_2); n = 2; a_0 = a_1 = a_2; \sum_{k=0}^{2} b_k \cdot \frac{d^k}{dt^k} V(t-\tau_2) = 0$$

$$b_0 = 1; b_1 = \frac{1}{R_1} \cdot \sum_{i=0}^{2} L_i; b_2$$
$$= C_1 \cdot \sum_{i=0}^{2} L_i; b_0 \cdot \frac{d^0}{dt^0} V(t-\tau_2) + b_1 \cdot \frac{d}{dt} V(t-\tau_2) + b_2 \cdot \frac{d^2}{dt^2} V(t-\tau_2) = 0$$

Since $\frac{d^0}{dt^0} V(t-\tau_2) \triangleq V(t-\tau_2)$ then $b_0 \cdot V(t-\tau_2) + b_1 \cdot \frac{d}{dt} V(t-\tau_2) + b_2 \cdot \frac{d^2}{dt^2} V(t-\tau_2) = 0$.

It is well known that if the characteristic equation associated with a linear neutral equation has roots only with negative real parts, and if all the roots are uniformly bounded away from the imaginary axis, then the trivial solution of the linear neutral equation is uniformly asymptotically stable. Thus the stability analysis of power limiter second neural differential equation with single delay τ_2 is very much equivalent to the problem of determining the conditions under which all roots of its characteristic equation lie in the half of the complex plane and are uniformly bounded away from the imaginary axis. In our case $a_k = 0$ then $(\sum_{k=0}^{2} b_k \cdot \lambda^k) \cdot e^{-\lambda \cdot \tau_2} = 0$; $P(\lambda) = 0$; $Q(\lambda) = \sum_{k=0}^{2} b_k \cdot \lambda^k$.

Theorem 1.0 *if* $|b_2| > 1$, *then for all* τ_2, *there is an infinite number of roots of* $Q(\lambda) \cdot e^{-\lambda \cdot \tau_2} = 0$ *whose real parts are positive.* $b_2 = C_1 \cdot \sum_{i=0}^{2} L_i \Rightarrow |C_1 \cdot \sum_{i=0}^{2} L_i| > 1$.

Theorem 2.0 *if* $|b_2| > 1$, *then the trivial solution of power limiter DDE (Delay Differential Equation) is unstable for all* $\tau_2 > 0$.

Theorem 3.0 *Let* $f(\lambda, \tau_2) = \lambda^2 + \alpha \cdot \lambda^2 \cdot e^{-\lambda \cdot \tau_2} + g(\lambda, \tau_2)$ *where* $g(\lambda, \tau_2)$ *is an analytic function. Assume* $|\alpha| > 1$ *and* $\lim_{\substack{\mathrm{Re}\lambda > 0 \\ |\lambda| \to \infty}} \frac{1}{\lambda^2} \cdot g(\lambda, \tau_2) = 0$ *then, for, all* $\tau_2 > 0$, *there is an infinite number of roots of* $f(\lambda, \tau_2) = 0$ *whose real parts are positive. In fact, there is a sequence* $\{\lambda_i\}$ *of the roots of* $f(\lambda, \tau_2) = 0$ *such* $|\lambda_i| \to \infty$, *and* $\lim_{i \to \infty} \mathrm{Re}\, \lambda_i = \frac{1}{\tau_2} \cdot \ln|\alpha| > 0$ *when* $\tau_2 > 0$.

Theorem 4.0 *Let* $f(\lambda, \tau_2) = \lambda^2 + g(\lambda, \tau_2)$ *where* $g(\lambda, \tau_2)$ *is an analytic function. Assume* $\alpha = \lim_{\substack{\mathrm{Re}\lambda > 0 \\ |\lambda| \to \infty}} |\lambda^{-2} \cdot g(\lambda, \tau_2)| < 1$ *when, as* τ_2 *varied, the sum of multiplicities of the roots of* $f(\lambda, \tau_2) = 0$ *in the open right half plane can change only if a root appears on or crosses the imaginary axis. Back to our second power limiter DDE which can be considered as the following second order real scalar linear neutral delay equation:* $\alpha \cdot \frac{d^2V(t-\tau_2)}{dt^2} + \beta \cdot \frac{dV(t-\tau_2)}{dt} + \gamma \cdot V(t - \tau_2) = 0$ *where* $\tau_2, \alpha, \beta, \gamma$ *are real constant. To find the equilibrium points (fixed points) of delayed power limiter circuit is by* $\lim_{t \to \infty} V(t - \tau_2) = \lim_{t \to \infty} V(t)\ \forall\, t \gg \tau_2\ \exists\, (t - \tau_2) \approx t, t \to \infty_{t \to \infty}$. $\frac{dV(t)}{dt} = 0; \frac{dV(t-\tau_2)}{dt} = 0; \frac{d^2V(t-\tau_2)}{dt^2} = 0$ *and we get one fixed point* $\gamma \cdot V^{(i=0)} = 0$; $\gamma \cdot V^{(i=0)} = 0 \Rightarrow \gamma \neq 0 \Rightarrow V^{(i=0)} = 0$.

Stability analysis: The standard local stability analysis about any one of the equilibrium points of delayed circuit consists in adding to coordinate V arbitrarily small increments of exponential form $v \cdot e^{\lambda \cdot t}$. This leads to a polynomial characteristic equation in the eigenvalues λ. The polynomial characteristic equation accepts by set the below voltage and voltage derivative respect to time in delayed power limiter differential equation. The delayed circuit fixed values with an arbitrarily small increment of exponential form $v \cdot e^{\lambda \cdot t}$ is i = 0 (first fixed point), i = 1 (second fixed point), etc., $V(t) = V^{(i)} + v \cdot e^{\lambda \cdot t}$

$$V(t - \tau_2) = V^{(i)} + v \cdot e^{\lambda \cdot (t - \tau_2)};\ \frac{dV(t)}{dt} = v \cdot \lambda \cdot e^{\lambda \cdot t};\ \frac{dV(t - \tau_2)}{dt} = v \cdot \lambda \cdot e^{\lambda \cdot t} \cdot e^{-\lambda \cdot \tau_2};\ \frac{d^2V(t - \tau_2)}{dt^2} = v \cdot \lambda^2 \cdot e^{\lambda \cdot t} \cdot e^{-\lambda \cdot \tau_2}$$

We choose the above expression for our V(t) as small displacement v from our circuit fixed points at time t = 0. $V(t=0) = V^{(i)} + v$. We get the following characteristic equation: $\alpha \cdot \lambda^2 \cdot e^{-\lambda \cdot \tau_2} + \beta \cdot \lambda \cdot e^{-\lambda \cdot \tau_2} + \gamma \cdot e^{-\lambda \cdot \tau_2} = 0$. $(\alpha \cdot \lambda^2 + \beta \cdot \lambda + \gamma) \cdot e^{-\lambda \cdot \tau_2} = 0$. Suppose $\lambda = i \cdot \omega$; $\omega > 0$ is the root of $(\alpha \cdot \lambda^2 + \beta \cdot \lambda + \gamma) \cdot e^{-\lambda \cdot \tau_2} = 0$ for some τ_2. Euler's formulas: $e^{-i \cdot \omega \cdot \tau_2} = \cos(\omega \cdot \tau_2) - i \cdot \sin(\omega \cdot \tau_2)$. We get $[(\gamma - \alpha \cdot \omega^2) + i \cdot \beta \cdot \omega] \cdot e^{-i \cdot \omega \cdot \tau_2} = 0$ and $(\gamma - \alpha \cdot \omega^2) \cdot \cos(\omega \cdot \tau_2) + \beta \cdot \omega \cdot \sin(\omega \cdot \tau_2) + i \cdot (\beta \cdot \omega \cdot \cos(\omega \cdot \tau_2) - (\gamma - \alpha \cdot \omega^2) \cdot \sin(\omega \cdot \tau_2)) = 0$
$\alpha = C_1 \cdot \sum_{i=0}^{2} L_i$; $\beta = \frac{1}{R_1} \cdot \sum_{i=0}^{2} L_i$; $\gamma = 1$; $\omega \neq 0$; $\omega > 0$. We get two equations:

$$(\gamma - \alpha \cdot \omega^2) \cdot \cos(\omega \cdot \tau_2) + \beta \cdot \omega \cdot \sin(\omega \cdot \tau_2)$$
$$= 0; \ (\beta \cdot \omega \cdot \cos(\omega \cdot \tau_2) - (\gamma - \alpha \cdot \omega^2) \cdot \sin(\omega \cdot \tau_2)) = 0$$

Thus $(\gamma - \alpha \cdot \omega^2)^2 + (\beta \cdot \omega)^2 = 0$; Hence $\alpha^2 \cdot \omega^4 + \omega^2 \cdot (\beta^2 - 2 \cdot \gamma \cdot \alpha) + \gamma^2 = 0$. Its roots are $\omega_{\pm}^2 = \frac{1}{2 \cdot \alpha^2} \cdot [2 \cdot \gamma \cdot \alpha - \beta^2 \pm ((\beta^2 - 2 \cdot \gamma \cdot \alpha)^2 - 4 \cdot \gamma^2 \cdot \alpha^2)^{1/2}]$. We have two cases for the above equation ($\omega_{\pm}^2 = \ldots$).

Case (I): $(\beta^2 - 2 \cdot \gamma \cdot \alpha)^2 = 4 \cdot \gamma^2 \cdot \alpha^2 \Rightarrow \beta^2 \cdot (\beta^2 - 4 \cdot \gamma \cdot \alpha) = 0$

$$\beta \to \varepsilon; \frac{1}{R_1} \cdot \sum_{i=0}^{2} L_i \to \varepsilon; \ R_1 \to \infty \ \& \ \sum_{i=0}^{2} L_i \to \varepsilon; \ \beta^2 - 4 \cdot \gamma \cdot \alpha = 0 \Rightarrow \beta^2$$
$$= 4 \cdot \gamma \cdot \alpha \Rightarrow \sum_{i=0}^{2} L_i = 4 \cdot C_1 \cdot R_1^2$$

Then $\omega_{\pm}^2 = \frac{1}{2 \cdot \alpha^2} \cdot [2 \cdot \gamma \cdot \alpha - \beta^2] = \frac{\gamma}{\alpha} - \frac{1}{2} \cdot (\frac{\beta}{\alpha})^2$

Case (II):

$$(\beta^2 - 2 \cdot \gamma \cdot \alpha)^2 > 4 \cdot \gamma^2 \cdot \alpha^2 \Rightarrow \beta^2 \cdot (\beta^2 - 4 \cdot \gamma \cdot \alpha) > 0; \ \beta^2 > 0;$$
$$\beta^2 - 4 \cdot \gamma \cdot \alpha > 0 \Rightarrow \beta^2 > 4 \cdot \gamma \cdot \alpha$$
$$\frac{1}{R_1^2} \cdot (\sum_{i=0}^{2} L_i)^2 > 4 \cdot C_1 \cdot \sum_{i=0}^{2} L_i; \ (\sum_{i=0}^{2} L_i) \cdot [\frac{1}{R_1^2} \cdot \sum_{i=0}^{2} L_i - 4 \cdot C_1] > 0;$$
$$(\sum_{i=0}^{2} L_i) > 0; \ \frac{1}{R_1^2} \cdot \sum_{i=0}^{2} L_i - 4 \cdot C_1 > 0 \Rightarrow \frac{1}{R_1^2 \cdot C_1} \cdot \sum_{i=0}^{2} L_i > 4$$

Thus we have two imaginary solutions $\lambda_{\pm} = i \cdot \omega_{\pm}$ with $\omega_{+} > \omega_{-}$. We need to determine the sign of the derivative of $\mathrm{Re}\,\lambda(\tau_2)$ at the points where $\lambda(\tau_2)$ is purely imaginary. $\frac{d}{d\tau_2} e^{-\lambda(\tau_2) \cdot \tau_2} = -(\frac{d\lambda(\tau_2)}{d\tau_2} \cdot \tau_2 + \lambda) \cdot e^{-\lambda(\tau_2) \cdot \tau_2}$

$\frac{d}{d\tau_2}[\alpha \cdot \lambda^2 + \beta \cdot \lambda + \gamma] = (2 \cdot \alpha \cdot \lambda + \beta) \cdot \frac{d\lambda}{d\tau_2}$. To get the expression for $\frac{d\lambda}{d\tau_2}$ we need to calculate $\frac{d}{d\tau_2}\{[\alpha \cdot \lambda^2(\tau_2) + \beta \cdot \lambda(\tau_2) + \gamma] \cdot e^{-\lambda(\tau_2) \cdot \tau_2}\} = 0$

$$\frac{d\lambda}{d\tau_2} = \frac{\lambda \cdot (\alpha \cdot \lambda^2 + \beta \cdot \lambda + \gamma)}{-\alpha \cdot \tau_2 \cdot \lambda^2 + \lambda \cdot (2 \cdot \alpha - \beta \cdot \tau_2) + (\beta - \gamma \cdot \tau_2)};$$

$$(\frac{d\lambda}{d\tau_2})^{-1} = \frac{-\alpha \cdot \tau_2 \cdot \lambda^2 + \lambda \cdot (2 \cdot \alpha - \beta \cdot \tau_2) + (\beta - \gamma \cdot \tau_2)}{\lambda \cdot (\alpha \cdot \lambda^2 + \beta \cdot \lambda + \gamma)}$$

If $\lambda(\bar{\tau}_2) = i \cdot \omega$ is not simple, then $\frac{d\lambda(\tau_2)}{d\tau_2} = 0$ at $\tau_2 = \bar{\tau}_2$. Since $\omega \neq 0$; $e^{-i \cdot \omega \cdot \tau_2} \neq 0$ hence $\alpha \cdot (i \cdot \omega)^2 + i \cdot \omega \cdot \beta + \gamma = 0$ Which implies $(\gamma - \alpha \cdot \omega^2) + i \cdot \omega \cdot \beta = 0$ and then

$$\beta = 0 \;\&\; \gamma - \alpha \cdot \omega^2 = 0 \Rightarrow \omega^2 \cdot C_1 \cdot \sum_{i=0}^{2} L_i = 1; \frac{1}{R_1} \cdot \sum_{i=0}^{2} L_i \to \varepsilon$$

$$sign\{\mathrm{Re}(\frac{d\lambda}{d\tau_2})^{-1}\}|_{\lambda=i\omega} = sign\{\mathrm{Re}[\frac{-\alpha \cdot \tau_2 \cdot \lambda^2 + \lambda \cdot (2 \cdot \alpha - \beta \cdot \tau_2) + (\beta - \gamma \cdot \tau_2)}{\lambda \cdot (\alpha \cdot \lambda^2 + \beta \cdot \lambda + \gamma)}]\}|_{\lambda=i\omega}$$

$$sign\mathrm{Re}(\frac{d\lambda}{d\tau_2})^{-1}|_{\lambda=i\omega} = sign\{\mathrm{Re}[\frac{-\alpha \cdot \tau_2 \cdot \lambda + (2 \cdot \alpha - \beta \cdot \tau_2)}{(\alpha \cdot \lambda^2 + \beta \cdot \lambda + \gamma)} + \frac{\beta - \gamma \cdot \tau_2}{\lambda \cdot (\alpha \cdot \lambda^2 + \beta \cdot \lambda + \gamma)}]\}|_{\lambda=i\omega}$$

Finally, we get the expression for $sign\{\mathrm{Re}(\frac{d\lambda}{d\tau_2})^{-1}\}|_{\lambda=i\omega}$.

$$sign\{\mathrm{Re}(\frac{d\lambda}{d\tau_2})^{-1}\}|_{\lambda=i\omega} = sign\{\frac{-\alpha \cdot \tau_2 \cdot \omega^2 \cdot \beta + (2 \cdot \alpha - \beta \cdot \tau_2) \cdot (\gamma - \alpha \cdot \omega^2)}{(\gamma - \alpha \cdot \omega^2)^2 + (\omega \cdot \beta)^2} - \frac{\omega^2 \cdot \beta \cdot (\beta - \gamma \cdot \tau_2)}{(\omega^2 \cdot \beta)^2 + \omega^2 \cdot (\gamma - \alpha \cdot \omega^2)^2}\}$$

Since $(\gamma - \alpha \cdot \omega^2)^2 + (\omega \cdot \beta)^2 > 0$ for any $\alpha, \beta, \gamma, \omega$ values we get the expression:

$$sign\{\mathrm{Re}(\frac{d\lambda}{d\tau_2})^{-1}\}|_{\lambda=i\omega} = sign\{-\alpha \cdot \tau_2 \cdot \omega^2 \cdot \beta + (2 \cdot \alpha - \beta \cdot \tau_2) \cdot (\gamma - \alpha \cdot \omega^2) - \omega^2 \cdot \beta \cdot (\beta - \gamma \cdot \tau_2)\}$$

$$sign\{\mathrm{Re}(\frac{d\lambda}{d\tau_2})^{-1}\}|_{\lambda=i\omega} = \{\omega^2 \cdot [-2 \cdot \alpha^2 - \beta^2 + \beta \cdot \gamma \cdot \tau_2] + 2 \cdot \alpha \cdot \gamma - \beta \cdot \tau_2 \cdot \gamma\}$$

By inserting the expression for $\omega_{\pm}^2$, we can check the sign of $\{\mathrm{Re}(\frac{d\lambda}{d\tau_2})^{-1}\}|_{\lambda=i\omega}$. There are two sets of values τ_2 for which there are imaginary roots: $0 \leq \theta_1 < 2\pi$; $0 \leq \theta_2 < 2\pi$. $\tau_{n,1} = \frac{\theta_1}{\omega_+} + \frac{2 \cdot \pi \cdot n}{\omega_+}$; $\tau_{n,2} = \frac{\theta_2}{\omega_-} + \frac{2 \cdot \pi \cdot n}{\omega_-}$.

We choose power limiter critical parameters: L_i, R_1, C_1 and τ_2 delay parameter and examine the possibility of stability transitions (bifurcation) due to parameter values change or τ_2 delay change. The analysis consists in identifying the root of power limiter second DDEs characteristic equation situated on the imaginary axis of the complex λ—plane, where by changing circuits parameters or τ_2 delay,

Reλ may, at crossing, changes its sign from "−" to "+" i.e. from stable focus E^* to an unstable one, or vice versa. This feature may be further assessed by examining the sign of the partial derivatives with respect to circuit parameters or delay τ_2.

$$sign\{\frac{d(\text{Re}\lambda)}{d\tau_2}\}|_{\lambda=i\omega};\ sign\{\frac{d(\text{Re}\lambda)}{dx_i}\}|_{\lambda=i\omega}x_i = R_1, \text{C}_1, \text{L}_i$$

We shall presently examine the possibility of stability transitions (bifurcations) in a power limiter system, about the endemic equilibrium point $V^{(i=0)}$ as a result of the variation of parameters R_1, C_1, L_1 or L_2 in shunt mounted limiter diode equivalent circuit. Our analysis closely follows the procedure described in details in reference [BK] for the time delay variation parameter τ_2. We keep τ_2 fixed and inspect stability switching for variation of parameters R_1, C_1, L_1 or L_2 respectively. We already got the expression for the second power limiter differential equation which lead the characteristic equation for the eigenvalue λ having the form $(\sum_{k=0}^{2} b_k \cdot \lambda^k) \cdot e^{-\lambda \cdot \tau_2} = 0$; $P(\lambda) = 0$; $Q(\lambda) = \sum_{k=0}^{2} b_k \cdot \lambda^k$ where $b_0 = 1$; $b_1 = \frac{1}{R_1} \cdot \sum_{i=0}^{2} L_i$; $b_2 = C_1 \cdot \sum_{i=0}^{2} L_i, a_k = 0$. We do a little parameters terminology change: $k \to j$; $b_k \to c_j$ and we get the following characteristic equation for the eigenvalue $\lambda : (\sum_{j=0}^{2} c_j \cdot \lambda^j) \cdot e^{-\lambda \cdot \tau_2} = 0$; $P(\lambda) = 0$; $Q(\lambda) = \sum_{j=0}^{2} c_j \cdot \lambda^j$ where $c_0 = 1$; $c_1 = \frac{1}{R_1} \cdot \sum_{i=0}^{2} L_i$; $c_2 = C_1 \cdot \sum_{i=0}^{2} L_i, a_j = 0$.

Remark: Do not confuse between c_i parameters and C_1 capacitor element in a power limiter equivalent circuit and the coefficients $\{a_j(R_1, C_1, L_i), c_j(R_1, C_1, L_i)\} \in \mathbb{R}$.

Depend on R_1, C_1, and L_i, but not on τ_2. Unless strictly necessary, the designation of the variable arguments (R_1, C_1, L_i) will subsequently be omitted from P, Q, a_j, c_j.

The coefficients a_j, c_j are continuous, and differentiable functions of their arguments, and direct substitution shows that $a_0 + c_0 \neq 0\ \forall\ R_1, C_1, L_i \in \Re_+$; in our case: $a_0 = 0$; $c_0 = 1 \Rightarrow a_0 + c_0 = 1 \neq 0$. i.e., $\lambda = 0$ is not a root of a power limiter characteristic equation. Furthermore, $P(\lambda), Q(\lambda)$ are analytic functions of λ, for which the following requirements of the analysis [BK] can also be verified in the present case:

(a) If $\lambda = i \cdot \omega$, $\omega \in \mathbb{R}$, then $P(i \cdot \omega) + Q(i \cdot \omega) \neq 0$ i.e. P and Q have no common imaginary roots. This condition was verified numerically in the entire (R_1, C_1, L_i) domain of interest.
(b) $|Q(\lambda)/P(\lambda)|$ is bounded for $|\lambda| \to \infty$, Re $\lambda \geq 0$. No roots bifurcation from ∞. Indeed, in the limit $|Q(\lambda)/P(\lambda)| = O(|c_2/a_3\lambda|) \to 0$.
(c) $F(\omega) = |P(i \cdot \omega)|^2 - |Q(i \cdot \omega)|^2$ has at most a finite number of zeros. Indeed, this is a bi-cubic polynomial in ω.
(d) Each positive root $\omega(R_1, C_1, L_i)$ of $F(\omega) = 0$ being continuous and differentiable with respect to R_1, C_1, L_i. This condition can only be assessed numerically.

In addition, since the coefficients in P and Q are real, we have $\overline{P(-i\cdot\omega)} = P(i\cdot\omega)$, and $\overline{Q(-i\cdot\omega)} = Q(i\cdot\omega)$; thus $\lambda = i\cdot\omega$ may be an eigenvalue of the characteristic equation. The analysis consists in identifying the roots of the characteristic equation situated on the imaginary axis of the complex λ plane, where, by increasing the parameter R_1, and/or C_1 and/or L_i, Re λ may, at the crossing, changes its sign from "−" to "+", i.e. from a stable focus $V^{(*)}$ to an unstable one, or vice versa [5, 6]. This feature may be further assessed by examining the sign of the partial derivatives with respect to R_1, C_1, and L_i.

Reminder: We keep τ_2 fixed.

$$\Lambda^{-1}(R_1) = (\frac{\partial \mathrm{Re}\,\lambda}{\partial R_1})_{\lambda=i\cdot\omega},\ C_1, L_i = const.,;\ \Lambda^{-1}(C_1) = (\frac{\partial \mathrm{Re}\,\lambda}{\partial C_1})_{\lambda=i\cdot\omega},\ R_1, L_i = const.,$$

$\Lambda^{-1}(L_i) = (\frac{\partial \mathrm{Re}\,\lambda}{\partial L_i})_{\lambda=i\cdot\omega},\ R_1, C_1 = const.,$ The subscripts $\lambda, \omega, R_1, C_1, L_i$ indicate the corresponding partial derivatives. Let us first concentrate on $\Lambda(R_1)$, remembering that $\lambda(R_1, C_1, L_i)$, $\omega((R_1, C_1, L_i)$, and keeping C_1, L_i and τ_2 fixed. The derivation closely follows that in reference [BK]. Differentiating characteristic equation with respect to R_1, and inverting the derivative, for convenience, one calculates:

$$(\frac{\partial\lambda}{\partial R_1})^{-1} = (\frac{-P_\lambda(\lambda, R_1)\cdot Q(\lambda, R_1) + Q_\lambda(\lambda, R_1)\cdot P(\lambda, R_1) - \tau_2\cdot P(\lambda, R_1)\cdot Q(\lambda, R_1)}{P_{R_1}(\lambda, R_1)\cdot Q(\lambda, R_1) - Q_{R_1}(\lambda, R_1)\cdot P(\lambda, R_1)})$$

where $P_\lambda = \partial P/\partial\lambda, \ldots,$ etc. Substituting $\lambda = i\cdot\omega$, and bear in mind that $\overline{P(-i\cdot\omega)} = P(i\cdot\omega)$, and $\overline{Q(-i\cdot\omega)} = Q(i\cdot\omega)$, $i\cdot P_\lambda(i\cdot\omega) = P_\omega(i\cdot\omega)$ and $i\cdot Q_\lambda(i\cdot\omega) = Q_\omega(i\cdot\omega)$, and that on the surface exist $|P(i\cdot\omega)|^2 = |Q(i\cdot\omega)|^2$, one obtains:

$$(\frac{\partial\lambda}{\partial R_1})^{-1}|_{\lambda=i\cdot\omega} = (\frac{i\cdot P_\omega(i\cdot\omega, R_1)\cdot\overline{P(i\cdot\omega, R_1)} - i\cdot Q_\lambda(i\cdot\omega, R_1)\cdot\overline{Q(\lambda, R_1)} - \tau_2\cdot|P(i\cdot\omega, R_1)|^2}{P_{R_1}(i\cdot\omega, R_1)\cdot\overline{P(i\cdot\omega, R_1)} - Q_{R_1}(i\cdot\omega, R_1)\cdot\overline{Q(i\cdot\omega, R_1)}})$$

Upon separating into real and imaginary parts, with $P = P_R + i\cdot P_I$; $Q = Q_R + i\cdot Q_I$; $P_\omega = P_{R\omega} + i\cdot P_{I\omega}$; $Q_\omega = Q_{R\omega} + i\cdot Q_{I\omega}$; $P_{R_1} = P_{RR_1} + i\cdot P_{IR_1}$; $Q_{R_1} = Q_{RR_1} + i\cdot Q_{IR_1}$; $P^2 = P_I^2 + P_R^2$, retaining the real part, and noting that the operators ∂ and Re commute, one come up, after some straightforward algebraic manipulations, with the following result:

$$\Lambda(R_1) = \frac{2 \cdot P^2}{F_{R_1}^2 + 4 \cdot V^2} \cdot F_\omega \cdot (\tau_2 \cdot \omega_{R_1} + \frac{U \cdot \omega_{R_1} + V}{P^2});$$

$$P(\lambda) = 0 \Rightarrow P(i \cdot \omega) = 0 \Rightarrow P_R = P_I = 0$$

$$P_{R\omega} = P_{I\omega} = P_{RR_1} = P_{IR_1} = 0;$$

$$Q(\lambda) = \sum_{j=0}^{2} c_j \cdot \lambda^j = 1 + \frac{1}{R_1} \cdot (\sum_{i=0}^{2} L_i) \cdot \lambda + C_1 \cdot (\sum_{i=0}^{2} L_i) \cdot \lambda^2$$

$$Q(\lambda = i \cdot \omega) = \sum_{j=0}^{2} c_j \cdot \lambda^j = 1 + i \frac{1}{R_1} \cdot (\sum_{i=0}^{2} L_i) \cdot \omega - C_1 \cdot (\sum_{i=0}^{2} L_i) \cdot \omega^2;$$

$$Q(i \cdot \omega) = \sum_{j=0}^{2} c_j \cdot \lambda^j = \{1 - C_1 \cdot (\sum_{i=0}^{2} L_i) \cdot \omega^2\} + i \frac{1}{R_1} \cdot (\sum_{i=0}^{2} L_i) \cdot \omega$$

$$Q_R = 1 - C_1 \cdot (\sum_{i=0}^{2} L_i) \cdot \omega^2; \ Q_I = \frac{1}{R_1} \cdot (\sum_{i=0}^{2} L_i) \cdot \omega; \ Q_{R\omega} = \frac{\partial Q_R}{\partial \omega}$$

$$= -2 \cdot \omega \cdot C_1 \cdot (\sum_{i=0}^{2} L_i); \ Q_{I\omega} = \frac{\partial Q_I}{\partial \omega} = \frac{1}{R_1} \cdot (\sum_{i=0}^{2} L_i)$$

$$Q_{RR_1} = \frac{\partial Q_R}{\partial R_1} = 0; \ Q_{IR_1} = -\frac{1}{R_1^2} \cdot (\sum_{i=0}^{2} L_i) \cdot \omega; \ P^2 \to 0; \ \frac{U \cdot \omega_{R_1} + V}{P^2} \gg \tau_2 \cdot \omega_{R_1}$$

$$\Lambda(R_1) \to \frac{2}{F_{R_1}^2 + 4 \cdot V^2} \cdot F_\omega \cdot (U \cdot \omega_{R_1} + V)$$

Where for convenience, we have dropped the arguments ($i \cdot \omega$, R_1), and where
$F_\omega = 2 \cdot [(P_{R\omega} \cdot P_R + P_{I\omega} \cdot P_I) - (Q_{R\omega} \cdot Q_R + Q_{I\omega} \cdot Q_I)]$
$F_{R_1} = 2 \cdot [(P_{RR_1} \cdot P_R + P_{IR_1} \cdot P_I) - (Q_{RR_1} \cdot Q_R + Q_{IR_1} \cdot Q_I)]; \ \omega_{R_1} = -F_{R_1}/F_\omega$

and we get the expressions based on power limiter equivalent parameters:

$$F_\omega = 2 \cdot \omega \cdot (\sum_{i=0}^{2} L_i) \cdot [2 \cdot C_1 - (\sum_{i=0}^{2} L_i) \cdot \{2 \cdot \omega^2 \cdot C_1^2 + \frac{1}{R_1^2}\}];$$

$$F_{R_1} = \frac{2 \cdot \omega^2}{R_1^3} \cdot (\sum_{i=0}^{2} L_i)^2$$

$$\omega_{R_1} = -\frac{F_{R_1}}{F_\omega} = -\frac{\omega \cdot (\sum_{i=0}^{2} L_i)}{R_1^3 \cdot [2 \cdot C_1 - (\sum_{i=0}^{2} L_i) \cdot \{2 \cdot \omega^2 \cdot C_1^2 + \frac{1}{R_1^2}\}]}$$

$$U = (P_R \cdot P_{I\omega} - P_I \cdot P_{R\omega}) - (Q_R \cdot Q_{I\omega} - Q_I \cdot Q_{R\omega});$$

$$V = (P_R \cdot P_{IR_1} - P_I \cdot P_{RR_1}) - (Q_R \cdot Q_{IR_1} - Q_I \cdot Q_{RR_1})$$

$$U = -\frac{1}{R_1} \cdot (\sum_{i=0}^{2} L_i) \cdot \{1 + C_1 \cdot \omega^2 \cdot (\sum_{i=0}^{2} L_i)\};$$

$$V = \frac{1}{R_1^2} \cdot (\sum_{i=0}^{2} L_i) \cdot \omega \cdot [1 - C_1 \cdot (\sum_{i=0}^{2} L_i) \cdot \omega^2]$$

where $\omega \in R_+$. If $\Lambda(R_1) > 0$, $\Lambda(C_1) > 0$, $\Lambda(L_i) > 0$ (or <0), then the crossing proceeds from "−" to "+" (or from "+" to "−"), respectively. Without loss of generality, one may consider only $\lambda = i\omega$, $\omega > 0$ and ignore its complex conjugate. Writing $P(\lambda) = P_R(\lambda) + i \cdot P_I(\lambda)$; $Q(\lambda) = Q_R(\lambda) + i \cdot Q_I(\lambda)$ and inserting $\lambda = i \cdot \omega$ into characteristic equation, ω must satisfy the following:

$$\sin(\omega \cdot \tau_2) = g(\omega) = \frac{-P_R(i \cdot \omega) \cdot Q_I(i \cdot \omega) + P_I(i \cdot \omega) \cdot Q_R(i \cdot \omega)}{|Q(i \cdot \omega)|^2}$$

$$\cos(\omega \cdot \tau_2) = h(\omega) = -\frac{P_R(i \cdot \omega) \cdot Q_R(i \cdot \omega) + P_I(i \cdot \omega) \cdot Q_I(i \cdot \omega)}{|Q(i \cdot \omega)|^2}$$

where $|Q(i \cdot \omega)|^2 \neq 0$ in view of the above requirement, and $(g, h) \in \mathbb{R}$. Furthermore, it follows from the above equations that, by squaring and adding sides, ω must be a positive root of $F(\omega) = |P(i \cdot \omega)|^2 - |Q(i \cdot \omega)|^2 = 0$.

And the above sin/cos equations, of course, are identical to those in reference [BK], except that the variable arguments are R_1, C_1, and L_i instead of τ_2. Note that F (ω) is independent of τ_2; Thus equation (sin/cos) implies F(ω), but not the other way around. The real and imaginary parts of P and Q are discussed, while F and some of its elementary properties are presented. One first solves the polynomial F(ω), retaining only the real positive roots ω, and discarding the others. The result is a 2D manifold (surface) ω = ω(R_1, C_1) in a three dimensional (3D) space (R_1, C_1, ω), where ω is a continuous and differentiable with respect to its arguments, with the possible exception of infinite derivatives on 1D continuous lines. Next, one checks which ω's on the surface also satisfy both (sin/cos) equations for some fixed value of τ_2. This operation results in one, or several continuous lines on the surface. The projection of these lines on the (R_1, C_1) plane gives the loci of possible stability transitions of the dynamical system.

Remark: We can exchange R_1 or C_1 by L_i in the 2D manifold (surface) or in a three dimensional (3D) space.

We can give the sign of $\Lambda(R_1)$, without the leading positive factor by:

$$sign(\frac{\partial \mathrm{Re}\,\lambda}{\partial R_1})^{-1}|_{\lambda = i\omega} = sign\Lambda(R_1);\ sign\Lambda(R_1) = sign(F_\omega) \cdot sign(\tau_2 \cdot \omega_{R1} + \frac{U \cdot \omega_{R1} + V}{P^2})$$

We need now to do the same procedure for C_1 parameter and get the expression:

$$\Lambda^{-1}(C_1) = (\frac{\partial \mathrm{Re}\lambda}{\partial C_1})_{\lambda=i\cdot\omega},\ R_1, L_i = const.,;$$
$$P(\lambda) = 0 \Rightarrow P(i\cdot\omega) = 0 \Rightarrow P_R = P_I = 0$$
$$P_{R\omega} = P_{I\omega} = P_{RC_1} = P_{IC_1} = 0;\ P_{C_1} = P_{RC_1} + i\cdot P_{IC_1};$$
$$Q_R = 1 - C_1\cdot(\sum_{i=0}^{2} L_i)\cdot\omega^2;\ Q_I = \frac{1}{R_1}\cdot(\sum_{i=0}^{2} L_i)\cdot\omega$$
$$Q_{RC_1} = \frac{\partial Q_R}{\partial C_1} = -(\sum_{i=0}^{2} L_i)\cdot\omega^2;\ Q_{IC_1} = \frac{\partial Q_I}{\partial C_1} = 0;$$
$$P = 0 \Rightarrow P^2 = 0;\ Q_{C_1} = \frac{\partial Q}{\partial C_1} = Q_{RC_1} + i\cdot Q_{IC_1}$$

F_ω expression is the same. $F_{C_1} = 2\cdot[(P_{RC_1}\cdot P_R + P_{IC_1}\cdot P_I) - (Q_{RC_1}\cdot Q_R + Q_{IC_1}\cdot Q_I)]$
$\omega_{C_1} = -F_{C_1}/F_\omega$; The expression for F_{C1} is $F_{C_1} = 2\cdot(\sum_{i=0}^{2} L_i)\cdot\omega^2\cdot[1 - C_1\cdot(\sum_{i=0}^{2} L_i)\cdot\omega^2]$

$$\omega_{C_1} = -\frac{F_{C_1}}{F_\omega} = -\frac{\omega\cdot[1 - C_1\cdot(\sum_{i=0}^{2} L_i)\cdot\omega^2]}{2\cdot C_1 - (\sum_{i=0}^{2} L_i)\cdot\{2\cdot\omega^2\cdot C_1^2 + \frac{1}{R_1^2}\}}$$

U expression is the same like our previous calculation.

$$V = (P_R\cdot P_{IC_1} - P_I\cdot P_{RC_1}) - (Q_R\cdot Q_{IC_1} - Q_I\cdot Q_{RC_1});\ V = -(\sum_{i=0}^{2} L_i)\cdot\omega^2\cdot\frac{1}{R_1}\cdot(\sum_{i=0}^{2} L_i)\cdot\omega = -\frac{\omega^3}{R_1}\cdot(\sum_{i=0}^{2} L_i)^2$$

$$sign(\frac{\partial \mathrm{Re}\lambda}{\partial C_1})^{-1}|_{\lambda=i\omega} = sign\Lambda(C_1);\ sign\,\Lambda(C_1) = sign(F_\omega)\cdot sign(\tau_2\cdot\omega_{C1} + \frac{U\cdot\omega_{C1} + V}{P^2})$$

U is always less than zero ($U < 0$). V for parameter R_1: if $\omega > 0$ then $V > 0$ for $C_1\cdot(\sum_{i=0}^{2} L_i)\cdot\omega^2 < 1$ otherwise $V < 0$. V for parameter C_1: If $\omega > 0$ then $V < 0$. Now we choose our parameter L_1.

$$P_I = P_R = 0 \Rightarrow P^2 = 0;\ P_{IL_1} = P_{RL_1} = 0 \sum_{i=1}^{3} L_i = L_1 + L_2 + L_3;$$

$$\frac{\partial \sum_{i=1}^{3} L_i}{\partial L_1} = \frac{\partial \sum_{i=1}^{3} L_i}{\partial L_2} = \frac{\partial \sum_{i=1}^{3} L_i}{\partial L_3} = 1;\ Q_{L_1} = \frac{\partial Q}{\partial L_1} = Q_{RL_1} + i \cdot Q_{IL_1}$$

$$Q_{RL_1} = \frac{\partial Q_R}{\partial L_1} = -C_1 \cdot \omega^2 \cdot \frac{\partial \sum_{i=1}^{3} L_i}{\partial L_1} = -C_1 \cdot \omega^2;\ Q_{IL_1} = \frac{\partial Q_I}{\partial L_1} = \frac{\omega}{R_1} \cdot \frac{\partial \sum_{i=1}^{3} L_i}{\partial L_1} = \frac{\omega}{R_1}$$

$$F_{L1} = 2 \cdot [(P_{RL_1} \cdot P_R + P_{IL_1} \cdot P_I) - (Q_{RL_1} \cdot Q_R + Q_{IL_1} \cdot Q_I)];$$

$$F_{L1} = 2 \cdot \omega^2 \cdot [C_1 - (\sum_{i=1}^{3} L_i) \cdot \{C_1^2 \cdot \omega^2 + \frac{1}{R_1^2}\}]$$

$$\omega_{L_1} = -\frac{F_{L1}}{F_\omega} = \frac{\omega \cdot [C_1 - (\sum_{i=1}^{3} L_i) \cdot \{C_1^2 \cdot \omega^2 + \frac{1}{R_1^2}\}]}{(\sum_{i=1}^{3} L_i) \cdot [2 \cdot C_1 - (\sum_{i=1}^{3} L_i) \cdot \{2 \cdot C_1^2 \cdot \omega^2 + \frac{1}{R_1^2}\}]}$$

U expression is the same like our previous calculation.

$V = (P_R \cdot P_{IL_1} - P_I \cdot P_{RL_1}) - (Q_R \cdot Q_{IL_1} - Q_I \cdot Q_{RL_1}); V = -\frac{\omega}{R_1}$; If $\omega > 0$ then $V < 0$ always. Retaining the real part, and noting that the operators ∂ and Re commute, one come up, after some straightforward algebraic manipulations, with the following result:

$$\Lambda(L_1) = \frac{2 \cdot P^2}{F_{L_1}^2 + 4 \cdot V^2} \cdot F_\omega \cdot (\tau_2 \cdot \omega_{L_1} + \frac{U \cdot \omega_{L_1} + V}{P^2});\ sign(\frac{\partial \mathrm{Re}\,\lambda}{\partial L_1})^{-1}|_{\lambda = i\omega} = sign\,\Lambda(L_1)$$

$$sign\,\Lambda(L_1) = sign(F_\omega) \cdot sign(\tau_2 \cdot \omega_{L1} + \frac{U \cdot \omega_{L1} + V}{P^2})$$

Our switching analysis results are the same if we move from L_1 parameter to L_2 or to L_3 since the partial derivatives are the same.

Summary: We take the assumption that $V_i(t)$ is an incoming signal width $\Delta t < \tau_1, \tau_2$; $\Delta t \to \varepsilon$. There are three time intervals which we analyze our power limiter microstrip line system. The first time interval is $\tau_1 > t > 0$, the coming signal not yet pass the first delay line (τ_1) and V_A, V, V_{out} respectively equal to zero. The second time interval is $\tau_1 + \tau_2 > t \geq \tau_1$, the signal has not yet passed the second delay line (τ_2) and mutual interaction between the signal and power limiter equivalent circuit gives V(t) which is the voltage on resistor R_1 and capacitor C_1. The dynamical analysis is done by using forced Van der Pol equation. The forcing signal X(t) is the coming RF signal. The third time interval is $t \geq \tau_1 + \tau_2$, the incoming signal passes both the first and second delay lines and the dynamical behavior analysis is done by using Delay Differential Equation (DDE) and stability switching analysis. For simplicity, we consider $V_{out} \to \varepsilon$ after $\tau_1 + \tau_2$ second.

In our analysis we consider the incoming RF signal to power limiter input circuit, has time interval Δt. We choose Δt, $\Delta t < \tau_1, \tau_2$. The incoming RF signal time interval is less than delay lines times τ_1, τ_2. First the incoming RF signal cause to the voltage V (on shunt mounted limiter equivalent circuit's R_1, C_1 elements). The analysis is based on the Van der Pol's equation. Second, we analyze output voltage according to Delay Differential Equation (DDE), V is the main equation variable in time [39–41] (Fig. 2.3).

Power limiter with microstrip transmission line system stability switching analysis is done according geometric stability switch criteria [BK] in delay differential system with delay dependent variables. The first analysis is a power limiter microstrip system with incoming RF signal. By using Van der Pol topology we find V(t) voltage after τ_1. The second analysis is power limiter microstrip system output

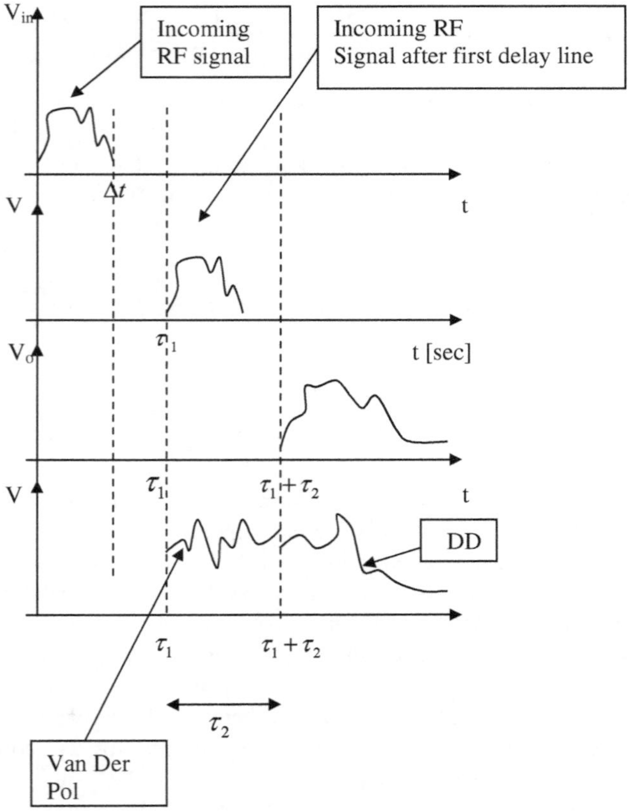

Fig. 2.3 Shunt mounted limiter diode equivalent circuit with microstrip lines delayed time diagram

differential equation with delay variable in time $V(t - \tau_2)$. V is the power limiter equivalent circuit's voltage on R_1 and C_1 . For the second analysis, we find out system general characteristic equation $D(\lambda, \tau_2)$. Find $F(\omega, \tau_2)$ for each τ_2 has at most a finite number of real zeros. Find ω, τ_2 values which fulfill $F(\omega, \tau_2) = 0$; $\omega(\tau_2)$, only for those values can be stability switching (first condition). Next is to find those ω, τ_2 values which fulfill the expressions, it is the second condition for stability switching $\sin(\omega \cdot \tau_2) = \ldots$ $\cos(\omega \cdot \tau_2) = \ldots$. If $\Lambda(R_1) > 0$, $\Lambda(C_1) > 0$, $\Lambda(L_i) > 0$ then the crossing proceeds from "−" to "+" respectively (stable to unstable). If $\Lambda(R_1) < 0$, $\Lambda(C_1) < 0$, $\Lambda(L_i) < 0$ then the crossing proceeds from "+" to "−" respectively (unstable to stable). The analysis consists in identifying the roots of microstrip power limiter circuit characteristic equation $P(\lambda) + Q(\lambda) \cdot \exp(-\lambda \cdot \tau_2) = 0$ situated on the imaginary axis of the complex λ—plane, where, by increasing the circuit parameters R_1, C_1, L_i. Reλ may, at the crossing, change its sign from "−" to "+", i.e. from a stable focus V^* to an unstable one, or vice versa. This feature may be further assessed by examining the sign of the partial derivatives with respect to microstrip power limiter circuit parameters R_1, C_1, L_i.

2.2 Three Ports Active Circulator's Reflection Type Phase Shifter (RTPS) Circuit Transmission Lines Delayed in Time System Stability Analysis

Active circulator consisting of three ports, namely P_1, P_2 and P_3. Active circulator is a three terminal device in which input from one port is transmitted to the next port in rotation. The active circulator acts as an isolator between the input and the output signal so that phase shift is well observed. The RF input signal is given at P_1 of the circulator from the left side. This signal from P_1 is transmitted to P_2. We can connect LC (L_1, C_1) components in series to P_2 port which results in phase shift and helps to reflect the signal to P_3 at the right. At P_3 we get an output RF signal. Each active circulator terminal faces a delay parasitic effect of signal transferring in time [25, 26, 35] (Fig. 2.4).

Our circuit is a Reflection Type Phase Shifter (RTPS), employing a circulator. In the past was little interest in actively circulators since its narrow bandwidth and problems associated with a hybrid realization. We use active circulators since their bandwidths have increased considerably as a result of the advances in transistor technology. Active circulators are ideally suited for realization using MMIC technology. The circuit employs decade bandwidth active circulator which shows very low phase error characteristic. Additionally the phase shifter exhibits an excellent input return loss performance across this decade bandwidth. The circuit configuration of the active circulator used three MESFETs which are the GEC-Marconi standard library cell F20-FET-4x75. As with all the standard library cells, a very accurate, ultra-wideband small signal models of the device. MESFET stands for metal semiconductor field effect transistor. It is similar to a JFET in construction

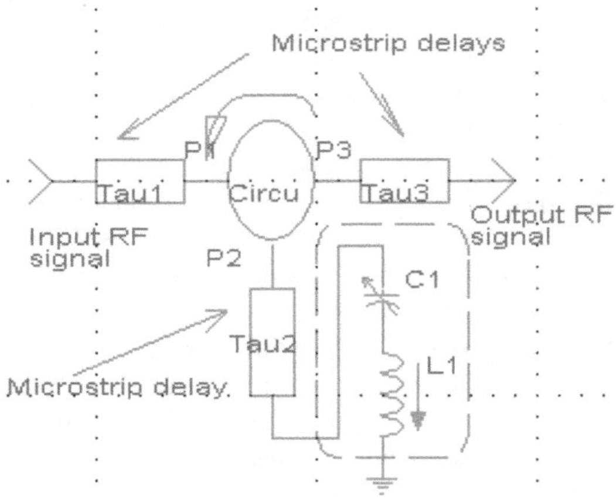

Fig. 2.4 Three ports decade bandwidth active circulator with micro strip delay lines and LC phase shifter in port P_2

and terminology. The difference is that instead of using a p-n junction for the gate, a Schottky (metal semiconductor) junction is used. A typical three ports decade bandwidth active circulator has three MESFETs transistors interconnected with each other. R_F, C_F, L_F, C_C, R_{sb} plays a major role in the working of the circuit. The three feedback branches (R_F, C_F, L_F) are used to link all the three transistors in an end to end fashion. The source resistor (R_{sb}) is shared among all the three MESFETs transistors and one transistor is source coupled with the other two transistors using this source resistor. The circuit works in a symmetric fashion. We consider MESFET high frequency model taking node capacitors into account. Figure 2.5 describes the circuit configuration of the active circulator [35, 36].

In Fig. 2.5 we use N-type MESFET but usually the recommended is a symmetrical bilateral MESFET. All C_c and C_f capacitors are un-polarized. Once we inject RF signal to port P_1, it passes to port P_2 through a feedback branch (R_F, C_F, L_F). The same is between ports P_2 and P_3, ports P_3 and P_1. In case we inject RF signal to port P_2, it reaches the Q1 gate and shorten the Q1's drain and source. Then Port 2's RF signal is shortened to ground through resistor R_{sb} and didn't reach port P_1.

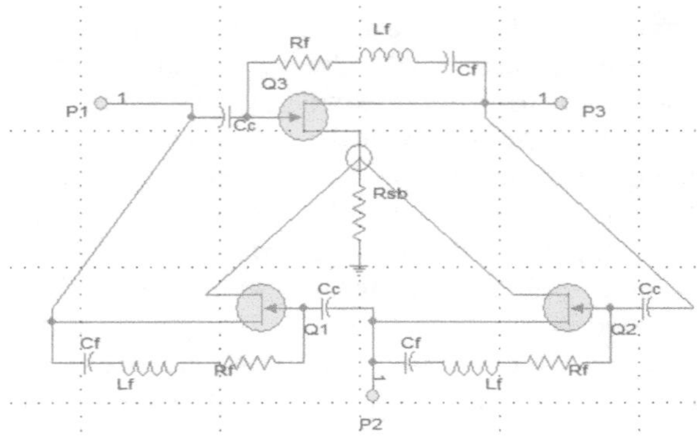

Fig. 2.5 Circuit configuration of the active circulator

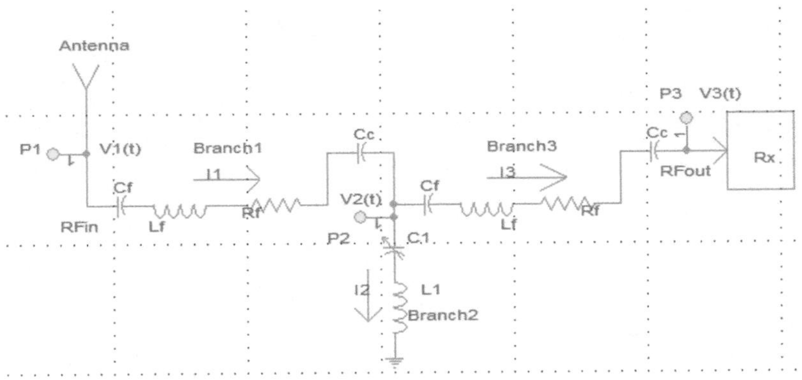

Fig. 2.6 Active circulator system path equivalent circuit

The same is between P_1 to P_3 and P_3 to P_2. We consider a varactor which is realized by connecting together the drain and source terminations of a standard MESFET—resulting in a Schottky junction. The bias potential is then applied across the drain/source and gate terminations. Our three ports decade bandwidth active circulator with micro strip delay lines and LC phase shifter in port P_2 gets his input RF signal from the antenna (port P_1) and feeds receiver unit by active circulator output RF signal (Port P_3) [1, 2]. The active circulator system can be described by the system path from RFin port (P_1) to RFout port (P_3). For simplicity, we ignore MESFET high frequency equivalent model and took it as a cutoff element in our system. Figure 2.6 describes our system path from Antenna RF coming signal to the receiver unit (R_x). Active circulator system path equivalent circuit fulfills current equation: $I_1 = I_2 + I_3$.

We have three main variables in our active circulator system $V_1(t)$, $V_2(t)$, $V_3(t)$. I_1, I_2, I_3 are the currents through related branches. We describe system, differential equations. First branch: $I_1 = C_f \cdot \frac{dV_{c_f}}{dt}$; $V_{L_f} = L_f \cdot \frac{dI_1}{dt}$; $I_1 = C_c \cdot \frac{dV_{c_c}}{dt}$

$$\frac{d}{dt}[V_{c_f} + V_{c_c}] = I_1 \cdot [\frac{1}{C_f} + \frac{1}{C_c}]; \quad V_1 - V_2 = V_{c_f} + V_{L_f} + V_{R_f} + V_{c_c};$$

$$V_{c_f} + V_{c_c} = V_1 - V_2 - L_f \cdot \frac{dI_1}{dt} - I_1 \cdot R_f$$

$$\frac{d}{dt}[V_{c_f} + V_{c_c}] = \frac{dV_1}{dt} - \frac{dV_2}{dt} - L_f \cdot \frac{d^2I_1}{dt^2} - \frac{dI_1}{dt} \cdot R_f;$$

$$I_1 \cdot [\frac{1}{C_f} + \frac{1}{C_c}] = \frac{dV_1}{dt} - \frac{dV_2}{dt} - L_f \cdot \frac{d^2I_1}{dt^2} - \frac{dI_1}{dt} \cdot R_f$$

Second branch: $I_2 = C_1 \cdot \frac{dV_{c_1}}{dt}$; $V_{L_1} = L_1 \cdot \frac{dI_2}{dt}$; $V_{c_1} = V_2 - L_1 \cdot \frac{dI_2}{dt}$; $I_2 = C_1 \cdot \frac{dV_{c_1}}{dt}$ $= C_1 \cdot [\frac{dV_2}{dt} - L_1 \cdot \frac{d^2I_2}{dt^2}]$.

Third branch: $I_3 = C_f \cdot \frac{dV_{c_f}}{dt}$; $V_{L_f} = L_f \cdot \frac{dI_3}{dt}$; $I_3 = C_c \cdot \frac{dV_{c_c}}{dt}$

$$\frac{d}{dt}[V_{c_f} + V_{c_c}] = I_3 \cdot [\frac{1}{C_f} + \frac{1}{C_c}]; \; V_2 - V_3 = V_{c_f} + V_{L_f} + V_{R_f} + V_{Cc};$$

$$V_{c_f} + V_{c_c} = V_2 - V_3 - L_f \cdot \frac{dI_3}{dt} - I_3 \cdot R_f$$

$$\frac{d}{dt}[V_{c_f} + V_{c_c}] = \frac{dV_2}{dt} - \frac{dV_3}{dt} - L_f \cdot \frac{d^2 I_3}{dt^2} - \frac{dI_3}{dt} \cdot R_f;$$

$$I_3 \cdot [\frac{1}{C_f} + \frac{1}{C_c}] = \frac{dV_2}{dt} - \frac{dV_3}{dt} - L_f \cdot \frac{d^2 I_3}{dt^2} - \frac{dI_3}{dt} \cdot R_f$$

We can summarize our system, differential equations:

$$I_1 \cdot [\frac{1}{C_f} + \frac{1}{C_c}] = \frac{dV_1}{dt} - \frac{dV_2}{dt} - L_f \cdot \frac{d^2 I_1}{dt^2} - \frac{dI_1}{dt} \cdot R_f; \; I_2 = C_1 \cdot [\frac{dV_2}{dt} - L_1 \cdot \frac{d^2 I_2}{dt^2}]$$

$$I_3 \cdot [\frac{1}{C_f} + \frac{1}{C_c}] = \frac{dV_2}{dt} - \frac{dV_3}{dt} - L_f \cdot \frac{d^2 I_3}{dt^2} - \frac{dI_3}{dt} \cdot R_f$$

We implement Rx (receiver) unit with an equivalent circuit of the input section of the receiver. The receiver's amplifier is modeled as a noiseless amplifier preceded by noise voltage and noise current generators representing amplifier noise referred to the input. The active circulator's RFout port is connected to the amplifier by a transformer with turns ratio m. We shall assume that this is an ideal transformer. Figure 2.7 describes the receiver input equivalent circuit.

We can consider the above equivalent circuit as resistor R_a, L_a, and L_t in the series. $V_3 = I_3 \cdot R_a + (L_a + L_t) \cdot \frac{dI_3}{dt}$. After we integrated R_x unit differential equation into our system, differential equations we get the following new system differential equations:

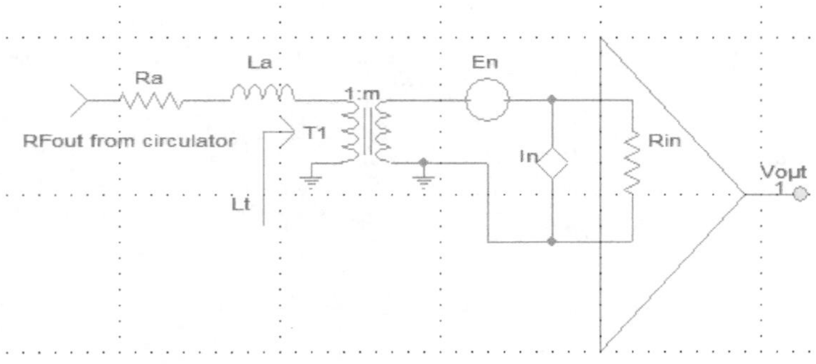

Fig. 2.7 Receiver input equivalent circuit

$$I_1 \cdot \left[\frac{1}{C_f} + \frac{1}{C_c}\right] = \frac{dV_1}{dt} - \frac{dV_2}{dt} - L_f \cdot \frac{d^2 I_1}{dt^2} - \frac{dI_1}{dt} \cdot R_f;\ I_2 = C_1 \cdot \left[\frac{dV_2}{dt} - L_1 \cdot \frac{d^2 I_2}{dt^2}\right]$$

$$I_3 \cdot \left[\frac{1}{C_f} + \frac{1}{C_c}\right] = \frac{dV_2}{dt} - (L_f + L_a + L_t) \cdot \frac{d^2 I_3}{dt^2} - \frac{dI_3}{dt} \cdot (R_f + R_a)$$

$$I_2 = C_1 \cdot \left[\frac{dV_2}{dt} - L_1 \cdot \frac{d^2 I_2}{dt^2}\right] \Rightarrow \frac{dV_2}{dt} = \frac{1}{C_1} \cdot I_2 + L_1 \cdot \frac{d^2 I_2}{dt^2}$$

We get two main systems, differential equations:

$$I_1 \cdot [\frac{1}{C_f} + \frac{1}{C_c}] = \frac{dV_1}{dt} - \frac{1}{C_1} \cdot I_2 - L_1 \cdot \frac{d^2 I_2}{dt^2} - L_f \cdot \frac{d^2 I_1}{dt^2} - \frac{dI_1}{dt} \cdot R_f$$

$$I_3 \cdot [\frac{1}{C_f} + \frac{1}{C_c}] = \frac{1}{C_1} \cdot I_2 + L_1 \cdot \frac{d^2 I_2}{dt^2} - (L_f + L_a + L_t) \cdot \frac{d^2 I_3}{dt^2} - \frac{dI_3}{dt} \cdot (R_f + R_a)$$

Since $I_1 = I_2 + I_3 \Rightarrow I_2 = I_1 - I_3$ we get the following system, differential equations:

$$I_1 \cdot [\frac{1}{C_f} + \frac{1}{C_c}] = \frac{dV_1}{dt} - \frac{1}{C_1} \cdot I_1 + \frac{1}{C_1} \cdot I_3 - L_1 \cdot \frac{d^2 I_1}{dt^2} + L_1 \cdot \frac{d^2 I_3}{dt^2} - L_f \cdot \frac{d^2 I_1}{dt^2} - \frac{dI_1}{dt} \cdot R_f$$

$$I_3 \cdot [\frac{1}{C_f} + \frac{1}{C_c}] = \frac{1}{C_1} \cdot I_1 - \frac{1}{C_1} \cdot I_3 + L_1 \cdot \frac{d^2 I_1}{dt^2} - (L_f + L_a + L_t + L_1) \cdot \frac{d^2 I_3}{dt^2} - \frac{dI_3}{dt} \cdot (R_f + R_a)$$

$$\frac{dI_1}{dt} = I'_1;\ \frac{dI_3}{dt} = I'_3;\ \frac{d^2 I_1}{dt^2} = \frac{dI'_1}{dt};\ \frac{d^2 I_3}{dt^2} = \frac{dI'_3}{dt}$$

We get a new set of system, differential equations:

$$I_1 \cdot [\frac{1}{C_f} + \frac{1}{C_c} + \frac{1}{C_1}] = \frac{dV_1}{dt} + \frac{1}{C_1} \cdot I_3 + L_1 \cdot \frac{dI'_3}{dt} - (L_f + L_1) \cdot \frac{dI'_1}{dt} - I'_1 \cdot R_f$$

$$I_3 \cdot [\frac{1}{C_f} + \frac{1}{C_c} + \frac{1}{C_1}] = \frac{1}{C_1} \cdot I_1 + L_1 \cdot \frac{dI'_1}{dt} - (L_f + L_a + L_t + L_1) \cdot \frac{dI'_3}{dt} - I'_3 \cdot (R_f + R_a);$$

$$\frac{dI_1}{dt} = I'_1;\ \frac{dI_3}{dt} = I'_3$$

For simplicity we define: $C_\Sigma = \frac{1}{C_f} + \frac{1}{C_c} + \frac{1}{C_1}$ and $L_\Sigma = L_f + L_a + L_t + L_1$ and get the following system, differential equations:

$$I_1 \cdot C_\Sigma = \frac{dV_1}{dt} + \frac{1}{C_1} \cdot I_3 + L_1 \cdot \frac{dI'_3}{dt} - (L_f + L_1) \cdot \frac{dI'_1}{dt} - I'_1 \cdot R_f;\ I_3 \cdot C_\Sigma$$
$$= \frac{1}{C_1} \cdot I_1 + L_1 \cdot \frac{dI'_1}{dt} - L_\Sigma \cdot \frac{dI'_3}{dt} - I'_3 \cdot (R_f + R_a);\ \frac{dI_1}{dt} = I'_1;\ \frac{dI_3}{dt} = I'_3$$

From the above differential equation, we get the expressions for $\frac{dI'_1}{dt}$ and $\frac{dI'_3}{dt}$:

$$\frac{dI'_1}{dt} = \frac{dV_1}{dt} \cdot \frac{1}{\{L_f + L_1 \cdot (1 - \frac{L_1}{L_\Sigma})\}} + \frac{(\frac{L_1}{L_\Sigma \cdot C_1} - C_\Sigma)}{\{L_f + L_1 \cdot (1 - \frac{L_1}{L_\Sigma})\}} \cdot I_1 + \frac{(\frac{1}{C_1} - \frac{L_1 \cdot C_\Sigma}{L_\Sigma})}{\{L_f + L_1 \cdot (1 - \frac{L_1}{L_\Sigma})\}}$$
$$\cdot I_3 - \frac{R_f}{\{L_f + L_1 \cdot (1 - \frac{L_1}{L_\Sigma})\}} \cdot I'_1 - \frac{L_1 \cdot (R_f + R_a)}{\{L_f + L_1 \cdot (1 - \frac{L_1}{L_\Sigma})\}} \cdot I'_3$$

For simplicity we define the following parameters:

$$\Gamma_0 = \frac{1}{\{L_f + L_1 \cdot (1 - \frac{L_1}{L_\Sigma})\}};\ \Gamma_1 = \frac{(\frac{L_1}{L_\Sigma \cdot C_1} - C_\Sigma)}{\{L_f + L_1 \cdot (1 - \frac{L_1}{L_\Sigma})\}};\ \Gamma_2 = \frac{(\frac{1}{C_1} - \frac{L_1 \cdot C_\Sigma}{L_\Sigma})}{\{L_f + L_1 \cdot (1 - \frac{L_1}{L_\Sigma})\}};$$
$$\Gamma_3 = -\frac{R_f}{\{L_f + L_1 \cdot (1 - \frac{L_1}{L_\Sigma})\}};\ \Gamma_4 = -\frac{L_1 \cdot (R_f + R_a)}{\{L_f + L_1 \cdot (1 - \frac{L_1}{L_\Sigma})\}}$$
$$\frac{dI'_1}{dt} = \frac{dV_1}{dt} \cdot \Gamma_0 + \Gamma_1 \cdot I_1 + \Gamma_2 \cdot I_3 + \Gamma_3 \cdot I'_1 + \Gamma_4 \cdot I'_3$$
$$\frac{dI'_3}{dt} = \frac{dV_1}{dt} \cdot \frac{L_1 \cdot \Gamma_0}{L_\Sigma} + [\frac{1}{L_\Sigma \cdot C_1} + \frac{L_1 \cdot \Gamma_1}{L_\Sigma}] \cdot I_1 + [\frac{L_1 \cdot \Gamma_2}{L_\Sigma} - \frac{C_\Sigma}{L_\Sigma}] \cdot I_3$$
$$+ \frac{L_1 \cdot \Gamma_3}{L_\Sigma} \cdot I'_1 + [\frac{L_1 \cdot \Gamma_4}{L_\Sigma} - \frac{(R_f + R_a)}{L_\Sigma}] \cdot I'_3$$

$$\frac{dI'_3}{dt} = \frac{dV_1}{dt} \cdot \frac{L_1 \cdot \Gamma_0}{L_\Sigma} + \frac{1}{L_\Sigma} \cdot [\frac{1}{C_1} + L_1 \cdot \Gamma_1] \cdot I_1 + \frac{1}{L_\Sigma} \cdot [L_1 \cdot \Gamma_2 - C_\Sigma] \cdot I_3 + \frac{L_1 \cdot \Gamma_3}{L_\Sigma}$$
$$\cdot I'_1 + \frac{1}{L_\Sigma} \cdot [L_1 \cdot \Gamma_4 - (R_f + R_a)] \cdot I'_3$$

For simplicity we define the following parameters:

$$\Gamma_5 = \frac{L_1 \cdot \Gamma_0}{L_\Sigma};\ \Gamma_6 = \frac{1}{L_\Sigma} \cdot [\frac{1}{C_1} + L_1 \cdot \Gamma_1];\ \Gamma_7 = \frac{1}{L_\Sigma} \cdot [L_1 \cdot \Gamma_2 - C_\Sigma];$$
$$\Gamma_8 = \frac{L_1 \cdot \Gamma_3}{L_\Sigma};\ \Gamma_9 = \frac{1}{L_\Sigma} \cdot [L_1 \cdot \Gamma_4 - (R_f + R_a)]$$
$$\frac{dI'_3}{dt} = \frac{dV_1}{dt} \cdot \Gamma_5 + \Gamma_6 \cdot I_1 + \Gamma_7 \cdot I_3 + \Gamma_8 \cdot I'_1 + \Gamma_9 \cdot I'_3$$

We can summarize our active circulator system, differential equations for coming antenna signal and R_x unit. $\frac{dI_1}{dt} = I'_1; \frac{dI_3}{dt} = I'_3$

$$\frac{dI_1'}{dt} = \frac{dV_1}{dt} \cdot \Gamma_0 + \Gamma_1 \cdot I_1 + \Gamma_2 \cdot I_3 + \Gamma_3 \cdot I_1' + \Gamma_4 \cdot I_3'; \frac{dI_3'}{dt} = \frac{dV_1}{dt} \cdot \Gamma_5 + \Gamma_6 \cdot I_1 + \Gamma_7 \cdot I_3 + \Gamma_8 \cdot I_1' + \Gamma_9 \cdot I_3'$$

We have four system variables: I_1, I_3, I_1', I_3' and coming RF_{in} signal $V_1(t)$. We can write the above system, differential equations in a form of matrix representation: $\Gamma_5 = \frac{L_1}{L_\Sigma} \cdot \Gamma_0$

$$\begin{pmatrix} \frac{dI_1}{dt} \\ \frac{dI_3}{dt} \\ \frac{dI_1'}{dt} \\ \frac{dI_3'}{dt} \end{pmatrix} = \begin{pmatrix} \Upsilon_{11} & \cdots & \Upsilon_{14} \\ \vdots & \ddots & \vdots \\ \Upsilon_{41} & \cdots & \Upsilon_{44} \end{pmatrix} \cdot \begin{pmatrix} I_1 \\ I_3 \\ I_1' \\ I_3' \end{pmatrix} + \begin{pmatrix} 0 \\ 0 \\ 1 \\ \frac{L_1}{L_\Sigma} \end{pmatrix} \cdot \Gamma_0 \cdot \frac{dV_1(t)}{dt}; \ \Upsilon_{11} = 0;$$

$$\Upsilon_{12} = 0; \ \Upsilon_{13} = 1; \ \Upsilon_{14} = 0; \ \Upsilon_{21} = 0; \ \Upsilon_{22} = 0$$

$$\Upsilon_{23} = 0; \Upsilon_{24} = 1; \Upsilon_{31} = \Gamma_1; \Upsilon_{32} = \Gamma_2; \Upsilon_{33} = \Gamma_3; \Upsilon_{34} = \Gamma_4; \Upsilon_{41} = \Gamma_6; \Upsilon_{42} = \Gamma_7; \Upsilon_{43} = \Gamma_8; \Upsilon_{44} = \Gamma_9$$

We consider RF_{in} signal $V_1(t) = A_0 + f(t)$; $|f(t)| <= 1$ and $A_0 \gg |f(t)|$ then $V_1(t) = A_0 + f(t) \approx A_0 \Rightarrow \frac{dV_1(t)}{dt} \to \varepsilon$. We can write our matrix representation: $\varepsilon \to 0$

$$\begin{pmatrix} \frac{dI_1}{dt} \\ \frac{dI_3}{dt} \\ \frac{dI_1'}{dt} \\ \frac{dI_3'}{dt} \end{pmatrix} = \begin{pmatrix} \Upsilon_{11} & \cdots & \Upsilon_{14} \\ \vdots & \ddots & \vdots \\ \Upsilon_{41} & \cdots & \Upsilon_{44} \end{pmatrix} \cdot \begin{pmatrix} I_1 \\ I_3 \\ I_1' \\ I_3' \end{pmatrix} + \varepsilon$$

Due to active circulator's micro strip transmission line delays τ_1 in the first port's current and τ_3 for the third port current. Accordingly, active circulator's micro strip transmission lines, delays Δ_1 for the first port current derivative and Δ_3 for the third port current derivative. $I_1(t) \to I_1(t - \tau_1)$; $I_3(t) \to I_3(t - \tau_3)$ $I_1'(t) \to I_1'(t - \Delta_1)$; $I_3'(t) \to I_3'(t - \Delta_3)$. We consider no delay effect on $\frac{dI_1}{dt}; \frac{dI_3}{dt}; \frac{dI_1'}{dt}; \frac{dI_3'}{dt}$. Active circulator's micro strip transmission lines parasitic effects influence only on P_1 and P_3 current and current derivatives I_1, I_3, I_1', I_3' (I_2, I_2' which are a hidden variable in our analysis).

$$\begin{pmatrix} \frac{dI_1}{dt} \\ \frac{dI_3}{dt} \\ \frac{dI'_1}{dt} \\ \frac{dI'_3}{dt} \end{pmatrix} = \begin{pmatrix} \Upsilon_{11} & \cdots & \Upsilon_{14} \\ \vdots & \ddots & \vdots \\ \Upsilon_{41} & \cdots & \Upsilon_{44} \end{pmatrix} \cdot \begin{pmatrix} I_1(t-\tau_1) \\ I_3(t-\tau_3) \\ I'_1(t-\Delta_1) \\ I'_3(t-\Delta_3) \end{pmatrix} + \varepsilon$$

To find equilibrium points (fixed points) of the active circulator system is by

$$\lim_{t\to\infty} I_1(t-\tau_1) = I_1(t);\ \lim_{t\to\infty} I_3(t-\tau_3) = I_3(t);\ \lim_{t\to\infty} I'_1(t-\tau_1) = I'_1(t);$$
$$\lim_{t\to\infty} I'_3(t-\tau_3) = I'_3(t);\ \frac{dI_1}{dt} = \frac{dI_3}{dt} = 0$$
$$\frac{dI'_1}{dt} = \frac{dI'_3}{dt} = 0\ \forall\, \mathrm{t} \gg \tau_1;\ \mathrm{t} \gg \tau_3; \mathrm{t} \gg \Delta_1;\ \mathrm{t} \gg \Delta_3\ \exists\, (t-\tau_1) \approx t;\ (t-\tau_3) \approx t;$$
$$(t-\Delta_1) \approx t;\ (t-\Delta_3) \approx t,\ \mathrm{t} \to \infty$$

We get four equations and the only one fixed point is $\Gamma_7 - \frac{\Gamma_6 \cdot \Gamma_2}{\Gamma_1} \neq 0$

$$\Gamma_7 - \frac{\Gamma_6 \cdot \Gamma_2}{\Gamma_1} \neq 0 \Rightarrow E^{(0)}(I_1^{(0)}, I_3^{(0)}, I_1'^{(0)}, I_3'^{(0)}) = (0, 0, 0, 0)$$

Stability analysis: The standard local stability analysis about any one of the equilibrium points of the active circulator system consists in adding to coordinate $[I_1, I_3, I'_1, I'_3]$ arbitrarily small increments of exponential form $[i_1, i_3, i'_1, i'_3] \cdot e^{\lambda \cdot t}$, and retaining the first order terms in $[I_1, I_3, I'_1, I'_3]$. The system of four homogeneous equations leads to a polynomial characteristic equation in the eigenvalues λ. The polynomial characteristics equations accept by set the below currents and currents derivative respect to time into active circulator system equations. Active circulator system fixed values with arbitrarily small increments of exponential form $[i_1, i_3, i'_1, i'_3] \cdot e^{\lambda \cdot t}$ are: j = 0 (first fixed point), j = 1 (second fixed point), j = 2 (third fixed point), etc.,

$$I_1(t) = I_1^{(j)} + i_1 \cdot e^{\lambda \cdot t};\ I_3(t) = I_3^{(j)} + i_3 \cdot e^{\lambda \cdot t};\ I'_1(t) = I_1'^{(j)} + i'_1 \cdot e^{\lambda \cdot t};\ I'_3(t)$$
$$= I_3'^{(j)} + i'_3 \cdot e^{\lambda \cdot t}$$

We choose the above expressions for our $I_1(t)$, $I_3(t)$ and $I'_1(t)$, $I'_3(t)$ as a small displacement $[i_1, i_3, i'_1, i'_3]$ from the active circulator system fixed points at time t = 0. $I_1(t=0) = I_1^{(j)} + i_1; I_3(t=0) = I_3^{(j)} + i_3;\ \ I'_1(t=0) = I_1'^{(j)} + i'_1; I'_3(t=0) = I_3'^{(j)} + i'_3$.

For $\lambda > 0$, t > 0 the selected fixed point is stable otherwise $\lambda > 0$, t > 0 is Unstable. Our system tends to the selected fixed point exponentially for $\lambda > 0$, t > 0 otherwise go away from the selected fixed point exponentially. λ is the eigenvalue parameter which establishes if the fixed point is stable or Unstable, additionally his absolute value ($|\lambda|$) establishes the speed of flow toward or away from the selected fixed point [3, 4] (Table 2.1).

The speeds of flow toward or away from the selected fixed point for active circulator system currents and currents derivatives respect to time are

$$\frac{dI_1(t)}{dt} = \lim_{\Delta t \to 0} \frac{I_1(t+\Delta t) - I_1(t)}{\Delta t} = \lim_{\Delta t \to 0} \frac{I_1^{(i)} + i_1 \cdot e^{\lambda \cdot (t+\Delta t)} - [I_1^{(i)} + i_1 \cdot e^{\lambda \cdot t}]}{\Delta t}$$

$$= \lim_{\Delta t \to 0} \frac{i_1 \cdot e^{\lambda \cdot t} \cdot [e^{\lambda \cdot \Delta t} - 1]}{\Delta t} \overset{e^{\lambda \cdot \Delta t} \approx 1 + \lambda \cdot \Delta t}{\longrightarrow} \lambda \cdot i_1 \cdot e^{\lambda \cdot t}$$

$$\frac{dI_3(t)}{dt} = \lambda \cdot i_3 \cdot e^{\lambda \cdot t}; \frac{dI_1'(t)}{dt} = \lambda \cdot i_1' \cdot e^{\lambda \cdot t}; \frac{dI_3'(t)}{dt} = \lambda \cdot i_3' \cdot e^{\lambda \cdot t}$$

$$\frac{dI_1(t-\tau_1)}{dt} = \lambda \cdot i_1 \cdot e^{\lambda \cdot t} \cdot e^{-\lambda \cdot \tau_1}; \frac{dI_3(t-\tau_3)}{dt} = \lambda \cdot i_3 \cdot e^{\lambda \cdot t} \cdot e^{-\lambda \cdot \tau_3}$$

$$\frac{dI_1'(t-\Delta_1)}{dt} = \lambda \cdot i_1' \cdot e^{\lambda \cdot t} \cdot e^{-\lambda \cdot \Delta_1}; \frac{dI_3'(t-\Delta_1)}{dt} = \lambda \cdot i_3' \cdot e^{\lambda \cdot t} \cdot e^{-\lambda \cdot \Delta_3}$$

First, we take the active circulator's currents I_1, I_3 differential equations: $\frac{dI_1}{dt} = I_1'; \frac{dI_3}{dt} = I_3'$ and adding coordinates $[I_1, I_3, I_1', I_3']$ arbitrarily small increments of exponential terms $[i_1, i_3, i_1', i_3'] \cdot e^{\lambda \cdot t}$ and retaining the first order terms in i_1, i_3, i_1', i_3'.

Table 2.1 Active circulator system eigenvalues options

	$\lambda < 0$	$\lambda > 0$
t = 0	$I_1(t=0) = I_1^{(j)} + i_1$ $I_3(t=0) = I_3^{(j)} + i_3$ $I_1'(t=0) = I_1'^{(j)} + i_1'$ $I_3'(t=0) = I_3'^{(j)} + i_3'$	$I_1(t=0) = I_1^{(j)} + i_1$ $I_3(t=0) = I_3^{(j)} + i_3$ $I_1'(t=0) = I_1'^{(j)} + i_1'$ $I_3'(t=0) = I_3'^{(j)} + i_3'$
t > 0	$I_1(t) = I_1^{(j)} + i_1 \cdot e^{-\lvert\lambda\rvert \cdot t}$ $I_3(t) = I_3^{(j)} + i_3 \cdot e^{-\lvert\lambda\rvert \cdot t}$ $I_3'(t) = I_3'^{(j)} + i_3' \cdot e^{-\lvert\lambda\rvert \cdot t}$ $I_3'(t) = I_3'^{(j)} + i_3' \cdot e^{-\lvert\lambda\rvert \cdot t}$	$I_1(t) = I_1^{(j)} + i_1 \cdot e^{\lvert\lambda\rvert \cdot t}$ $I_3(t) = I_3^{(j)} + i_3 \cdot e^{\lvert\lambda\rvert \cdot t}$ $I_3'(t) = I_3'^{(j)} + i_3' \cdot e^{\lvert\lambda\rvert \cdot t}$ $I_3'(t) = I_3'^{(j)} + i_3' \cdot e^{\lvert\lambda\rvert \cdot t}$
t > 0 t → ∞	$I_1(t \to \infty) = I_1^{(i)}$ $I_3(t \to \infty) = I_3^{(i)}$ $I_1'(t \to \infty) = I_1'^{(i)}$ $I_3'(t \to \infty) = I_3'^{(i)}$	$I_1(t \to \infty, \lambda > 0) \approx i_1 \cdot e^{\lvert\lambda\rvert \cdot t}$ $I_3(t \to \infty, \lambda > 0) \approx i_3 \cdot e^{\lvert\lambda\rvert \cdot t}$ $I_1'(t \to \infty, \lambda > 0) \approx i_1' \cdot e^{\lvert\lambda\rvert \cdot t}$ $I_3'(t \to \infty, \lambda > 0) \approx i_3 \cdot e^{\lvert\lambda\rvert \cdot t}$

$$\lambda \cdot i_1 \cdot e^{\lambda \cdot t} = I_1'^{(i)} + i_1' \cdot e^{\lambda \cdot t};\ I_1'^{(i=0)} = 0 \Rightarrow -\lambda \cdot i_1 + i_1' = 0$$
$$\lambda \cdot i_3 \cdot e^{\lambda \cdot t} = I_3'^{(i)} + i_3' \cdot e^{\lambda \cdot t};\ I_3'^{(i=0)} = 0 \Rightarrow -\lambda \cdot i_3 + i_3' = 0$$

Second, we take the active circulator's currents derivative I_1', I_3' differential equations:

$$\frac{dI_1'}{dt} = \frac{dV_1}{dt} \cdot \Gamma_0 + \Gamma_1 \cdot I_1 + \Gamma_2 \cdot I_3 + \Gamma_3 \cdot I_1' + \Gamma_4 \cdot I_3';\ \frac{dI_3'}{dt} = \frac{dV_1}{dt} \cdot \Gamma_5 + \Gamma_6 \cdot I_1 + \Gamma_7 \cdot I_3 + \Gamma_8 \cdot I_1' + \Gamma_9 \cdot I_3'$$

$\frac{dV_1}{dt} \cdot \Gamma_0 \to \varepsilon; \frac{dV_1}{dt} \cdot \Gamma_5 \to \varepsilon$ and adding coordinates $[I_1, I_3, I_1', I_3']$ arbitrarily small increments of exponential terms $[i_1, i_3, i_1', i_3'] \cdot e^{\lambda \cdot t}$ and retaining the first order terms in i_1, i_3, i_1', i_3'.

$$\lambda \cdot i_1' \cdot e^{\lambda \cdot t} = \Gamma_1 \cdot (I_1^{(j)} + i_1 \cdot e^{\lambda \cdot t}) + \Gamma_2 \cdot (I_3^{(j)} + i_3 \cdot e^{\lambda \cdot t}) + \Gamma_3 \cdot (I_1'^{(j)} + i_1' \cdot e^{\lambda \cdot t}) + \Gamma_4 \cdot (I_3'^{(j)} + i_3' \cdot e^{\lambda \cdot t})$$
$$\lambda \cdot i_3' \cdot e^{\lambda \cdot t} = \Gamma_6 \cdot (I_1^{(j)} + i_1 \cdot e^{\lambda \cdot t}) + \Gamma_7 \cdot (I_3^{(j)} + i_3 \cdot e^{\lambda \cdot t}) + \Gamma_8 \cdot (I_1'^{(j)} + i_1' \cdot e^{\lambda \cdot t}) + \Gamma_9 \cdot (I_3'^{(j)} + i_3' \cdot e^{\lambda \cdot t})$$
$$I_1^{(j=0)} = 0\ ;\ I_3^{(j=0)} = 0\ ;\ I_1'^{(j=0)} = 0\ ;\ I_3'^{(j=0)} = 0$$
$$(-\lambda + \Gamma_3) \cdot i_1' + \Gamma_1 \cdot i_1 + \Gamma_2 \cdot i_3 + \Gamma_4 \cdot i_3' = 0\ ;\ (-\lambda + \Gamma_9) \cdot i_3' + \Gamma_6 \cdot i_1 + \Gamma_7 \cdot i_3 + \Gamma_8 \cdot i_1' = 0$$

Remark: It is reader exercise to find system Jacobian matrix and to investigate stability based on system eigenvalues. The system has four eigenvalues: $\lambda_1, \lambda_2, \lambda_3, \lambda_4$.

We define $I_1(t - \tau_1) = I_1^{(i)} + i_1 \cdot e^{\lambda \cdot (t - \tau_1)}$; $I_3(t - \tau_3) = I_3^{(i)} + i_3 \cdot e^{\lambda \cdot (t - \tau_3)}$ $I_1'(t - \Delta_1) = I_1'(i) + i_1' \cdot e^{\lambda \cdot (t - \Delta_1)}$; $I_3'(t - \Delta_3) = I_3'(i) + i_3' \cdot e^{\lambda \cdot (t - \Delta_3)}$ then we get four delayed differential equations with respect to coordinates $[I_1, I_3, I_1', I_3']$ arbitrarily small increments of exponential $[i_1, i_3, i_1', i_3'] \cdot e^{\lambda \cdot t}$. We consider no delay effect on $\frac{dI_1}{dt}; \frac{dI_3}{dt}; \frac{dI_1'}{dt}; \frac{dI_3'}{dt}$ and get the following equations:

$$\lambda \cdot i_1 \cdot e^{\lambda \cdot t} = i_1' \cdot e^{\lambda \cdot (t - \Delta_1)};\ \lambda \cdot i_3 \cdot e^{\lambda \cdot t} = i_3' \cdot e^{\lambda \cdot (t - \Delta_3)}$$
$$\lambda \cdot i_1' \cdot e^{\lambda \cdot t} = \Gamma_1 \cdot i_1 \cdot e^{\lambda \cdot (t - \tau_1)} + \Gamma_2 \cdot i_3 \cdot e^{\lambda \cdot (t - \tau_3)} + \Gamma_3 \cdot i_1' \cdot e^{\lambda \cdot (t - \Delta_1)} + \Gamma_4 \cdot i_3' \cdot e^{\lambda \cdot (t - \Delta_3)}$$
$$\lambda \cdot i_3' \cdot e^{\lambda \cdot t} = \Gamma_6 \cdot i_1 \cdot e^{\lambda \cdot (t - \tau_1)} + \Gamma_7 \cdot i_3 \cdot e^{\lambda \cdot (t - \tau_3)} + \Gamma_8 \cdot i_1' \cdot e^{\lambda \cdot (t - \Delta_1)} + \Gamma_9 \cdot i_3' \cdot e^{\lambda \cdot (t - \Delta_3)}$$
$$-\lambda \cdot i_1 + i_1' \cdot e^{-\lambda \cdot \Delta_1} = 0; -\lambda \cdot i_3 + i_3' \cdot e^{-\lambda \cdot \Delta_3} = 0$$
$$\Gamma_1 \cdot i_1 \cdot e^{-\lambda \cdot \tau_1} + \Gamma_2 \cdot i_3 \cdot e^{-\lambda \cdot \tau_3} + (\Gamma_3 \cdot e^{-\lambda \cdot \Delta_1} - \lambda) \cdot i_1' + \Gamma_4 \cdot i_3' \cdot e^{-\lambda \cdot \Delta_3} = 0$$
$$\Gamma_6 \cdot i_1 \cdot e^{-\lambda \cdot \tau_1} + \Gamma_7 \cdot i_3 \cdot e^{-\lambda \cdot \tau_3} + \Gamma_8 \cdot i_1' \cdot e^{-\lambda \cdot \Delta_1} + (\Gamma_9 \cdot e^{-\lambda \cdot \Delta_3} - \lambda) \cdot i_3' = 0$$

In the equilibrium fixed point $I_1^{(j=0)} = 0; I_3^{(j=0)} = 0; I_1'^{(j=0)} = 0; I_3'^{(j=0)} = 0$.

The small increments Jacobian of our active circulator system is as bellow:

$$\begin{pmatrix} \Xi_{11} & \cdots & \Xi_{14} \\ \vdots & \ddots & \vdots \\ \Xi_{41} & \cdots & \Xi_{44} \end{pmatrix} \cdot \begin{pmatrix} i_1 \\ i_3 \\ i_1' \\ i_3' \end{pmatrix} = 0;\ \Xi_{11} = -\lambda;\ \Xi_{12} = 0;\ \Xi_{13} = e^{-\lambda \cdot \Delta_1};\ \Xi_{14}$$

$$= 0;\ \Xi_{21} = 0;\ \Xi_{22} = -\lambda;\ \Xi_{23} = 0;\ \Xi_{24} = e^{-\lambda \cdot \Delta_3}$$

$$\Xi_{31} = \Gamma_1 \cdot e^{-\lambda \cdot \tau_1};\ \Xi_{32} = \Gamma_2 \cdot e^{-\lambda \cdot \tau_3};\ \Xi_{33} = \Gamma_3 \cdot e^{-\lambda \cdot \Delta_1} - \lambda;\ \Xi_{34} = \Gamma_4 \cdot e^{-\lambda \cdot \Delta_3};$$
$$\Xi_{41} = \Gamma_6 \cdot e^{-\lambda \cdot \tau_1};\ \Xi_{42} = \Gamma_7 \cdot e^{-\lambda \cdot \tau_3}$$
$$\Xi_{43} = \Gamma_8 \cdot e^{-\lambda \cdot \Delta_1};\ \Xi_{44} = \Gamma_9 \cdot e^{-\lambda \cdot \Delta_3} - \lambda$$

$$A - \lambda \cdot I = \begin{pmatrix} \Xi_{11} & \cdots & \Xi_{14} \\ \vdots & \ddots & \vdots \\ \Xi_{41} & \cdots & \Xi_{44} \end{pmatrix};\ \det|A - \lambda \cdot I| = 0$$

$$\begin{aligned} D(\tau_1, \tau_3, \Delta_1, \Delta_3) = {} & \lambda^4 - \lambda^3 \cdot (\Gamma_3 \cdot e^{-\lambda \cdot \Delta_1} + \Gamma_9 \cdot e^{-\lambda \cdot \Delta_3}) + \lambda^2 \cdot \{(\Gamma_3 \cdot \Gamma_9 - \Gamma_8 \cdot \Gamma_4) \\ & \cdot e^{-\lambda \cdot (\Delta_1 + \Delta_3)} - \Gamma_7 \cdot e^{-\lambda \cdot (\Delta_3 + \tau_3)} - \Gamma_1 \cdot e^{-\lambda \cdot (\tau_1 + \Delta_1)}\} \\ & + \lambda \cdot \{(\Gamma_1 \cdot \Gamma_9 - \Gamma_6 \cdot \Gamma_4) \cdot e^{-\lambda \cdot (\tau_1 + \Delta_1 + \Delta_3)} - (\Gamma_2 \cdot \Gamma_8 - \Gamma_7 \cdot \Gamma_3) \cdot e^{-\lambda \cdot (\tau_3 + \Delta_1 + \Delta_3)}\} \\ & + (\Gamma_1 \cdot \Gamma_7 - \Gamma_6 \cdot \Gamma_2) \cdot e^{-\lambda \cdot (\tau_1 + \tau_3 + \Delta_1 + \Delta_3)} \end{aligned}$$

We have three stability cases: $\tau_1 = \tau_3 = \tau \,\&\, \Delta_1 = \Delta_3 = 0$ or $\tau_1 = \tau_3 = 0 \,\&\, \Delta_1 = \Delta_3 = \Delta$ or $\tau_1 = \tau_3 = \Delta_1 = \Delta_3 = \tau_\Delta$ otherwise $\tau_1 \neq \tau_3 \,\&\, \Delta_1 \neq \Delta_3$ and they are positive parameters. There are other possible simple stability cases: $\tau_1 = \tau;\ \tau_3 = 0;\ \Delta_1 = \Delta_3 = 0$ or $\tau_1 = 0;\ \tau_3 = \tau;\ \Delta_1 = \Delta_3 = 0;\ \tau_1 = \tau_3 = 0;\ \Delta_1 = \Delta;\ \Delta_3 = 0$ or $\tau_1 = \tau_3 = 0;\ \Delta_1 = 0;\ \Delta_3 = \Delta$. We need to get characteristics equations for all above stability analysis cases. We study the occurrence of any possible stability switching, resulting from the increase of the value of the time delays $\tau, \Delta, \tau_\Delta$ for the general characteristic equation $D(\lambda, \tau/\Delta/\tau_\Delta)$. If we choose τ parameter, then $D(\lambda, \tau) = P_n(\lambda, \tau) + Q_m(\lambda, \tau) \cdot e^{-\lambda \tau}$. The expression for $P_n(\lambda, \tau)$ is $P_n(\lambda, \tau) = \sum_{k=0}^{n} P_k(\tau) \cdot \lambda^k = P_0(\tau) + P_1(\tau) \cdot \lambda + P_2(\tau) \cdot \lambda^2 + P_3(\tau) \cdot \lambda^3 + \ldots$

The expression for $Q_m(\lambda, \tau)$ is $Q_m(\lambda, \tau) = \sum_{k=0}^{m} q_k(\tau) \cdot \lambda^k = q_0(\tau) + q_1(\tau) \cdot \lambda + q_2(\tau) \cdot \lambda^2 + \ldots$.

The case we analyze is when there is a delay in I_1 and I_3 currents only. The delay is the same for I_1 and I_3 and equal to τ ($\tau_1 = \tau;\ \tau_3 = \tau$) which describe most of active circulator parasitic effects. The general characteristic equation $D(\lambda, \tau)$ is as follow: $D(\tau_1 = \tau_3 = \tau, \Delta_1 = \Delta_3 = 0, \lambda)$

$$\begin{aligned}
D(\tau_1 = \tau_3 = \tau, \Delta_1 = \Delta_3 = 0, \lambda) &= D(\tau, \lambda) \\
&= \lambda^4 - \lambda^3 \cdot (\Gamma_3 + \Gamma_9) + \lambda^2 \cdot (\Gamma_3 \cdot \Gamma_9 - \Gamma_8 \cdot \Gamma_4) + \{-\lambda^2 \cdot (\Gamma_7 + \Gamma_1) \\
&\quad + \lambda \cdot (\Gamma_1 \cdot \Gamma_9 - \Gamma_6 \cdot \Gamma_4 - \Gamma_2 \cdot \Gamma_8 + \Gamma_7 \cdot \Gamma_3) \\
&\quad + (\Gamma_1 \cdot \Gamma_7 - \Gamma_6 \cdot \Gamma_2) \cdot e^{-\lambda \cdot \tau}\} \cdot e^{-\lambda \cdot \tau}
\end{aligned}$$

Under Taylor series approximation: $e^{-\lambda \cdot \tau} \approx 1 - \lambda \cdot \tau$ the Maclaurin series is a Taylor series expansion of a $e^{-\lambda \cdot \tau}$ function about zero (0). We get the following general characteristic equation $D(\lambda, \tau)$ under Taylor series approximation: $e^{-\lambda \cdot \tau} \approx 1 - \lambda \cdot \tau$.

$$\begin{aligned}
& D(\tau_1 = \tau_3 = \tau,\ \Delta_1 = \Delta_3 = 0,\ \lambda) = D(\tau, \lambda) = \lambda^4 - \lambda^3 \cdot (\Gamma_3 + \Gamma_9) \\
&\quad + \lambda^2 \cdot (\Gamma_3 \cdot \Gamma_9 - \Gamma_8 \cdot \Gamma_4) + \{-\lambda^2 \cdot (\Gamma_7 + \Gamma_1) \\
&\quad + \lambda \cdot [\Gamma_1 \cdot \Gamma_9 - \Gamma_6 \cdot \Gamma_4 - \Gamma_2 \cdot \Gamma_8 + \Gamma_7 \cdot \Gamma_3 - \tau \cdot (\Gamma_1 \cdot \Gamma_7 - \Gamma_6 \cdot \Gamma_2)] \\
&\quad + \Gamma_1 \cdot \Gamma_7 - \Gamma_6 \cdot \Gamma_2\} \cdot e^{-\lambda \cdot \tau} \\
& D(\lambda, \tau) = P_n(\lambda, \tau) + Q_m(\lambda, \tau) \cdot e^{-\lambda\tau};\ n = 4;\ m = 2;\ n > m.
\end{aligned}$$

The expression for $P_n(\lambda, \tau)$ is $P_n(\lambda, \tau) = \sum_{k=0}^{n} P_k(\tau) \cdot \lambda^k$

$$\begin{aligned}
P_n(\lambda, \tau) &= \sum_{k=0}^{n} P_k(\tau) \cdot \lambda^k = P_0(\tau) + P_1(\tau) \cdot \lambda + P_2(\tau) \cdot \lambda^2 + P_3(\tau) \cdot \lambda^3 + P_4(\tau) \cdot \lambda^4 \\
&= \lambda^2 \cdot (\Gamma_3 \cdot \Gamma_9 - \Gamma_8 \cdot \Gamma_4) - \lambda^3 \cdot (\Gamma_3 + \Gamma_9) + \lambda^4 \\
& P_0(\tau) = 0;\ P_1(\tau) = 0;\ P_2(\tau) = \Gamma_3 \cdot \Gamma_9 - \Gamma_8 \cdot \Gamma_4;\ P_3(\tau) = -(\Gamma_3 + \Gamma_9); \\
& P_4(\tau) = 1
\end{aligned}$$

The expression for $Q_m(\lambda, \tau)$ is $Q_m(\lambda, \tau) = \sum_{k=0}^{m} q_k(\tau) \cdot \lambda^k = q_0(\tau) + q_1(\tau) \cdot \lambda + q_2(\tau) \cdot \lambda^2$.

$$\begin{aligned}
Q_m(\lambda, \tau) &= \sum_{k=0}^{m} q_k(\tau) \cdot \lambda^k = -\lambda^2 \cdot (\Gamma_7 + \Gamma_1) \\
&\quad + \lambda \cdot [\Gamma_1 \cdot \Gamma_9 - \Gamma_6 \cdot \Gamma_4 - \Gamma_2 \cdot \Gamma_8 + \Gamma_7 \cdot \Gamma_3 - \tau \cdot (\Gamma_1 \cdot \Gamma_7 - \Gamma_6 \cdot \Gamma_2)] \\
&\quad + \Gamma_1 \cdot \Gamma_7 - \Gamma_6 \cdot \Gamma_2 \\
& q_0(\tau) = \Gamma_1 \cdot \Gamma_7 - \Gamma_6 \cdot \Gamma_2; \\
& q_1(\tau) = \Gamma_1 \cdot \Gamma_9 - \Gamma_6 \cdot \Gamma_4 - \Gamma_2 \cdot \Gamma_8 + \Gamma_7 \cdot \Gamma_3 - \tau \cdot (\Gamma_1 \cdot \Gamma_7 - \Gamma_6 \cdot \Gamma_2); \\
& q_2(\tau) = -(\Gamma_7 + \Gamma_1)
\end{aligned}$$

The homogeneous system for I_1, I_3, I_1', I_3' leads to a characteristic equation for the eigenvalue λ having the form $P(\lambda,\tau) + Q(\lambda,\tau)\cdot e^{-\lambda\cdot\tau} = 0$; $P(\lambda) = \sum_{j=0}^{4} a_j \cdot \lambda^j$; $Q(\lambda) = \sum_{j=0}^{2} c_j \cdot \lambda^j$ and the coefficients $\{a_j(q_i, q_k, \tau), c_j(q_i, q_k, \tau)\} \in \mathbb{R}$ depend on q_i, q_k and delay τ. q_i, q_k are any active circulator's parameters, other parameters keep as a constant. $a_0 = 0$; $a_1 = 0$; $a_2 = \Gamma_3 \cdot \Gamma_9 - \Gamma_8 \cdot \Gamma_4$; $a_3 = -(\Gamma_3 + \Gamma_9)$; $a_4 = 1$; $c_0 = \Gamma_1 \cdot \Gamma_7 - \Gamma_6 \cdot \Gamma_2$; $c_2 = -(\Gamma_7 + \Gamma_1)$ $c_1 = \Gamma_1 \cdot \Gamma_9 - \Gamma_6 \cdot \Gamma_4 - \Gamma_2 \cdot \Gamma_8 + \Gamma_7 \cdot \Gamma_3 - \tau \cdot (\Gamma_1 \cdot \Gamma_7 - \Gamma_6 \cdot \Gamma_2)$.

Unless strictly necessary, the designation of the varied arguments (q_i, q_k) will subsequently be omitted from P, Q, a_j, c_j. The coefficients a_j, c_j are continuous, and differentiable functions of their arguments, and direct substitution shows that $a_0 + c_0 \neq 0$ for $\forall\, q_i, q_k \in \mathbb{R}_+$, i.e. $\lambda = 0$ is not a of $P(\lambda) + Q(\lambda)\cdot e^{-\lambda\cdot\tau} = 0$. Furthermore, P(λ), Q(λ) are analytic functions of λ, for which the following requirements of the analysis [BK] can also be verified in the present case.

If $\lambda = i\cdot\omega$, $\omega \in \mathbb{R}$, then $P(i\cdot\omega) + Q(i\cdot\omega) \neq 0$. $|Q(\lambda)/P(\lambda)|$ is bounded for $|\lambda| \to \infty$, $\operatorname{Re}\lambda \geq 0$. No roots bifurcation from ∞. $F(\omega) = |P(i\cdot\omega)|^2 - |Q(i\cdot\omega)|^2$ has a finite number of zeros. Indeed, this is a polynomial in ω. Each positive root $\omega(q_i, q_k)$ of F(ω) = 0 is continuous and differentiable respect to q_i, q_k. We assume that $P_n(\lambda,\tau)$ and $Q_m(\lambda,\tau)$ can't have common imaginary roots. That is for any real number ω.

$$p_n(\lambda = i\cdot\omega,\tau) + Q_m(\lambda = i\cdot\omega,\tau) \neq 0;$$

$$p_n(\lambda = i\cdot\omega,\tau) = \omega^4 - \omega^2\cdot(\Gamma_3\cdot\Gamma_9 - \Gamma_8\cdot\Gamma_4) - i\cdot\omega^3\cdot(\Gamma_3 + \Gamma_9)$$

$$Q_m(\lambda = i\cdot\omega,\tau) = \omega^2\cdot(\Gamma_7 + \Gamma_1) + i\cdot\omega\cdot[\Gamma_1\cdot\Gamma_9 - \Gamma_6\cdot\Gamma_4 - \Gamma_2\cdot\Gamma_8 + \Gamma_7\cdot\Gamma_3 - \tau\cdot(\Gamma_1\cdot\Gamma_7 - \Gamma_6\cdot\Gamma_2)] + \Gamma_1\cdot\Gamma_7 - \Gamma_6\cdot\Gamma_2$$

$$p_n(\lambda = i\cdot\omega,\tau) + Q_m(\lambda = i\cdot\omega,\tau) = \omega^4 + \omega^2\cdot\{\Gamma_7 + \Gamma_1 - \Gamma_3\cdot\Gamma_9 + \Gamma_8\cdot\Gamma_4\} + \Gamma_1\cdot\Gamma_7 - \Gamma_6\cdot\Gamma_2 + i\cdot\omega\cdot\{\Gamma_1\cdot\Gamma_9 - \Gamma_6\cdot\Gamma_4 - \Gamma_2\cdot\Gamma_8 + \Gamma_7\cdot\Gamma_3 - \tau\cdot(\Gamma_1\cdot\Gamma_7 - \Gamma_6\cdot\Gamma_2) - \omega^2\cdot(\Gamma_3 + \Gamma_9)\} \neq 0$$

$$|P(i\cdot\omega,\tau)|^2 = \omega^8 + \omega^6\cdot\{(\Gamma_3 + \Gamma_9)^2 - 2\cdot(\Gamma_3\cdot\Gamma_9 - \Gamma_8\cdot\Gamma_4)\} + \omega^4\cdot(\Gamma_3\cdot\Gamma_9 - \Gamma_8\cdot\Gamma_4)^4$$

For simplicity we define a function: $\Omega(\Gamma_j, \tau;\, 1 \leq j \leq 9)$

$$\Omega = \Omega(\Gamma_j, \tau;\, 1 \le j \le 9) = \Gamma_1 \cdot \Gamma_9 - \Gamma_6 \cdot \Gamma_4 - \Gamma_2 \cdot \Gamma_8 + \Gamma_7 \cdot \Gamma_3 - \tau \cdot (\Gamma_1 \cdot \Gamma_7 - \Gamma_6 \cdot \Gamma_2)$$

$$Q_m(\lambda = i \cdot \omega, \tau) = \omega^2 \cdot (\Gamma_7 + \Gamma_1) + (\Gamma_1 \cdot \Gamma_7 - \Gamma_6 \cdot \Gamma_2) + i \cdot \omega \cdot \Omega(\Gamma_j, \tau;\, 1 \le j \le 9)$$

$$Q_m(\lambda = i \cdot \omega, \tau) = \omega^2 \cdot (\Gamma_7 + \Gamma_1) + \Gamma_1 \cdot \Gamma_7 - \Gamma_6 \cdot \Gamma_2 + i \cdot \omega \cdot \Omega$$

$$|Q(i \cdot \omega, \tau)|^2 = \omega^4 \cdot (\Gamma_7 + \Gamma_1)^2 + \omega^2 \cdot \{\Omega^2 + 2 \cdot (\Gamma_7 + \Gamma_1) \cdot (\Gamma_1 \cdot \Gamma_7 - \Gamma_6 \cdot \Gamma_2)\} + (\Gamma_1 \cdot \Gamma_7 - \Gamma_6 \cdot \Gamma_2)^2$$

$$F(\omega, \tau) = |P(i \cdot \omega, \tau)|^2 - |Q(i \cdot \omega, \tau)|^2 = \omega^8 + \omega^6 \cdot \{(\Gamma_3 + \Gamma_9)^2 - 2 \cdot (\Gamma_3 \cdot \Gamma_9 - \Gamma_8 \cdot \Gamma_4)\} + \omega^4 \cdot \{(\Gamma_3 \cdot \Gamma_9 - \Gamma_8 \cdot \Gamma_4)^4 - (\Gamma_7 + \Gamma_1)^2\} - \omega^2 \cdot \{\Omega^2 + 2 \cdot (\Gamma_7 + \Gamma_1) \cdot (\Gamma_1 \cdot \Gamma_7 - \Gamma_6 \cdot \Gamma_2)\} - (\Gamma_1 \cdot \Gamma_7 - \Gamma_6 \cdot \Gamma_2)^2$$

We define the following parameters for simplicity:

$$\Phi_0 = -(\Gamma_1 \cdot \Gamma_7 - \Gamma_6 \cdot \Gamma_2)^2;\ \Phi_2 = -\{\Omega^2 + 2 \cdot (\Gamma_7 + \Gamma_1) \cdot (\Gamma_1 \cdot \Gamma_7 - \Gamma_6 \cdot \Gamma_2)\}$$
$$\Phi_4 = (\Gamma_3 \cdot \Gamma_9 - \Gamma_8 \cdot \Gamma_4)^4 - (\Gamma_7 + \Gamma_1)^2;$$
$$\Phi_6 = (\Gamma_3 + \Gamma_9)^2 - 2 \cdot (\Gamma_3 \cdot \Gamma_9 - \Gamma_8 \cdot \Gamma_4);\ \Phi_8 = 1$$

Hence $F(\omega, \tau) = 0$ implies $\sum_{k=0}^{4} \Phi_{2 \cdot k} \cdot \omega^{2 \cdot k} = 0$ and its roots are given by solving the above polynomial. Furthermore $P_R(i \cdot \omega, \tau) = \omega^4 - \omega^2 \cdot (\Gamma_3 \cdot \Gamma_9 - \Gamma_8 \cdot \Gamma_4)$

$$P_R(i \cdot \omega, \tau) = \omega^2 \cdot (\omega^2 - \Gamma_3 \cdot \Gamma_9 + \Gamma_8 \cdot \Gamma_4);\ P_I(i \cdot \omega, \tau) = -\omega^3 \cdot (\Gamma_3 + \Gamma_9)$$
$$Q_R(i \cdot \omega, \tau) = \omega^2 \cdot (\Gamma_7 + \Gamma_1) + (\Gamma_1 \cdot \Gamma_7 - \Gamma_6 \cdot \Gamma_2);$$
$$Q_I(i \cdot \omega, \tau) = \omega \cdot \Omega\,(\Gamma_j, \tau;\, 1 \le j \le 9) = \omega \cdot \Omega$$

Hence

$$\sin\theta(\tau) = \frac{-P_R(i \cdot \omega, \tau) \cdot Q_I(i \cdot \omega, \tau) + P_I(i \cdot \omega, \tau) \cdot Q_R(i \cdot \omega, \tau)}{|Q(i \cdot \omega, \tau)|^2}$$

$$\cos\theta(\tau) = -\frac{P_R(i \cdot \omega, \tau) \cdot Q_R(i \cdot \omega, \tau) + P_I(i \cdot \omega, \tau) \cdot Q_I(i \cdot \omega, \tau)}{|Q(i \cdot \omega, \tau)|^2}$$

$$\sin\theta(\tau) = \frac{-\omega^3 \cdot [(\omega^2 - \Gamma_3 \cdot \Gamma_9 + \Gamma_8 \cdot \Gamma_4) \cdot \Omega + (\Gamma_3 + \Gamma_9) \cdot \{\omega^2 \cdot (\Gamma_7 + \Gamma_1) + (\Gamma_1 \cdot \Gamma_7 - \Gamma_6 \cdot \Gamma_2)\}]}{\omega^4 \cdot (\Gamma_7 + \Gamma_1)^2 + \omega^2 \cdot \{\Omega^2 + 2 \cdot (\Gamma_7 + \Gamma_1) \cdot (\Gamma_1 \cdot \Gamma_7 - \Gamma_6 \cdot \Gamma_2)\} + (\Gamma_1 \cdot \Gamma_7 - \Gamma_6 \cdot \Gamma_2)^2}$$

$$\cos\theta(\tau) = -\frac{\omega^2 \cdot (\omega^2 - \Gamma_3 \cdot \Gamma_9 + \Gamma_8 \cdot \Gamma_4) \cdot \{\omega^2 \cdot (\Gamma_7 + \Gamma_1) + (\Gamma_1 \cdot \Gamma_7 - \Gamma_6 \cdot \Gamma_2)\} - \omega^4 \cdot (\Gamma_3 + \Gamma_9) \cdot \Omega}{\omega^4 \cdot (\Gamma_7 + \Gamma_1)^2 + \omega^2 \cdot \{\Omega^2 + 2 \cdot (\Gamma_7 + \Gamma_1) \cdot (\Gamma_1 \cdot \Gamma_7 - \Gamma_6 \cdot \Gamma_2)\} + (\Gamma_1 \cdot \Gamma_7 - \Gamma_6 \cdot \Gamma_2)^2}$$

Which jointly with $F(\omega, \tau) = 0 \Rightarrow \sum_{k=0}^{4} \Phi_{2\cdot k} \cdot \omega^{2\cdot k} = 0$ that is a continuous and differentiable in τ based on Lemma 1.1. Hence we use Theorem 1.2. This proves the Theorem 1.3.

Our active circulator homogeneous system for i_1, i_3, i'_1, i'_3 leads to a characteristic equation for the eigenvalue λ having the form $P(\lambda) + Q(\lambda) \cdot e^{-\lambda\cdot\tau} = 0$; First case $\tau_1 = \tau$; $\tau_3 = \tau$; $\Delta_1 = \Delta_3 = 0$. The general characteristic equation $D(\lambda, \tau)$ is as follow: $D(\tau_1 = \tau_3 = \tau, \Delta_1 = \Delta_3 = 0, \lambda) = D(\tau, \lambda)$

$$\begin{aligned} D(\tau_1 = \tau_3 = \tau, \Delta_1 = \Delta_3 = 0, \lambda) &= D(\tau, \lambda) = \lambda^4 - \lambda^3 \cdot (\Gamma_3 + \Gamma_9) \\ &+ \lambda^2 \cdot (\Gamma_3 \cdot \Gamma_9 - \Gamma_8 \cdot \Gamma_4) + \{-\lambda^2 \cdot (\Gamma_7 + \Gamma_1) \\ &+ \lambda \cdot (\Gamma_1 \cdot \Gamma_9 - \Gamma_6 \cdot \Gamma_4 - \Gamma_2 \cdot \Gamma_8 + \Gamma_7 \cdot \Gamma_3) + (\Gamma_1 \cdot \Gamma_7 - \Gamma_6 \cdot \Gamma_2) \cdot e^{-\lambda\cdot\tau}\} \cdot e^{-\lambda\cdot\tau} \end{aligned}$$

Under Taylor series approximation: $e^{-\lambda\cdot\tau} \approx 1 - \lambda \cdot \tau$. The Maclaurin series is a Taylor series expansion of a $e^{-\lambda\cdot\tau}$ function about zero (0). We get the following general characteristic equation $D(\lambda, \tau)$ under Taylor series approximation: $e^{-\lambda\cdot\tau} \approx 1 - \lambda \cdot \tau$.

$$\begin{aligned} D(\tau_1 = \tau_3 = \tau, \Delta_1 = \Delta_3 = 0, \lambda) &= D(\tau, \lambda) = \lambda^4 - \lambda^3 \cdot (\Gamma_3 + \Gamma_9) \\ &+ \lambda^2 \cdot (\Gamma_3 \cdot \Gamma_9 - \Gamma_8 \cdot \Gamma_4) + \{-\lambda^2 \cdot (\Gamma_7 + \Gamma_1) + \lambda \cdot [\Gamma_1 \cdot \Gamma_9 - \Gamma_6 \cdot \Gamma_4 \\ &- \Gamma_2 \cdot \Gamma_8 + \Gamma_7 \cdot \Gamma_3 - \tau \cdot (\Gamma_1 \cdot \Gamma_7 - \Gamma_6 \cdot \Gamma_2)] + \Gamma_1 \cdot \Gamma_7 - \Gamma_6 \cdot \Gamma_2\} \cdot e^{-\lambda\cdot\tau} \end{aligned}$$

We use different parameters terminology from our last characteristics parameters definition: $k \to j$; $p_k(\tau) \to a_j$; $q_k(\tau) \to c_j$; $n = 4$; $m = 2$; $n > m$.

Additionally $P_n(\lambda, \tau) \to P(\lambda)$; $Q_m(\lambda, \tau) \to Q(\lambda)$ then $P(\lambda) = \sum_{j=0}^{4} a_j \cdot \lambda^j$; $Q(\lambda) = \sum_{j=0}^{2} c_j \cdot \lambda^j$

$$\begin{aligned} P_\lambda &= \lambda^2 \cdot (\Gamma_3 \cdot \Gamma_9 - \Gamma_8 \cdot \Gamma_4) - \lambda^3 \cdot (\Gamma_3 + \Gamma_9) + \lambda^4 \\ Q_\lambda &= -\lambda^2 \cdot (\Gamma_7 + \Gamma_1) + \lambda \cdot [\Gamma_1 \cdot \Gamma_9 - \Gamma_6 \cdot \Gamma_4 - \Gamma_2 \cdot \Gamma_8 + \Gamma_7 \cdot \Gamma_3 \\ &\quad - \tau \cdot (\Gamma_1 \cdot \Gamma_7 - \Gamma_6 \cdot \Gamma_2)] + \Gamma_1 \cdot \Gamma_7 - \Gamma_6 \cdot \Gamma_2 \end{aligned}$$

$n, m \in \mathbb{N}_0$, $n > m$ and $a_j, c_j : \mathrm{R}_{+0} \to R$ are continuous and differentiable function of τ such that $a_0 + c_0 \neq 0$. In the following "–" denotes complex and conjugate. $P(\lambda), Q(\lambda)$ are analytic functions in λ and differentiable in τ. The coefficients $\{a_j(R_f, L_f, C_f, C_c, R_{sb}, \tau, \ldots)$ and $c_j(R_f, L_f, C_f, C_c, R_{sb}, \tau, \ldots)\} \in \mathbb{R}$ depend on active circulator system's $R_f, L_f, C_f, C_c, R_{sb}, \tau, \ldots$ values.

$$a_0 = 0;\ a_1 = 0;\ a_2 = \Gamma_3 \cdot \Gamma_9 - \Gamma_8 \cdot \Gamma_4;\ a_3 = -(\Gamma_3 + \Gamma_9); a_4 = 1$$
$$c_0 = \Gamma_1 \cdot \Gamma_7 - \Gamma_6 \cdot \Gamma_2;\ c_1 = [\Gamma_1 \cdot \Gamma_9 - \Gamma_6 \cdot \Gamma_4 - \Gamma_2 \cdot \Gamma_8 + \Gamma_7 \cdot \Gamma_3 - \tau \cdot (\Gamma_1 \cdot \Gamma_7 - \Gamma_6 \cdot \Gamma_2)];\ c_2 = -(\Gamma_7 + \Gamma_1)$$

Unless strictly necessary, the designation of the varied arguments $(R_f, L_f, C_f, C_c, R_{sb}, \tau, \ldots)$ will subsequently be omitted from P, Q, a_j, c_j. The coefficients a_j, c_j are continuous, and differentiable functions of their arguments, and direct substitution shows that $a_0 + c_0 \neq 0;\ \Gamma_1 \cdot \Gamma_7 - \Gamma_6 \cdot \Gamma_2 \neq 0$

$$\frac{(\frac{L_1}{L_\Sigma \cdot C_1} - C_\Sigma) \cdot \frac{1}{L_\Sigma} \cdot [L_1 \cdot \Gamma_2 - C_\Sigma]}{\{L_f + L_1 \cdot (1 - \frac{L_1}{L_\Sigma})\}} - \frac{\frac{1}{L_\Sigma} \cdot [\frac{1}{C_1} + L_1 \cdot \Gamma_1] \cdot (\frac{1}{C_1} - \frac{L_1 \cdot C_\Sigma}{L_\Sigma})}{\{L_f + L_1 \cdot (1 - \frac{L_1}{L_\Sigma})\}} \neq 0$$

$$\frac{(\frac{L_1}{L_\Sigma \cdot C_1} - C_\Sigma) \cdot \frac{1}{L_\Sigma} \cdot [L_1 \cdot \Gamma_2 - C_\Sigma] - \frac{1}{L_\Sigma} \cdot [\frac{1}{C_1} + L_1 \cdot \Gamma_1] \cdot (\frac{1}{C_1} - \frac{L_1 \cdot C_\Sigma}{L_\Sigma})}{\{L_f + L_1 \cdot (1 - \frac{L_1}{L_\Sigma})\}} \neq 0$$

$\forall R_f, L_f, C_f, C_c, R_{sb}, \tau, \ldots \in \mathbb{R}_+$ i.e. $\lambda = 0$ is not a root of the characteristic equation. Furthermore $P(\lambda)$, $Q(\lambda)$ are analytic function of λ for which the following requirements of the analysis (see Kuang [5], Sect. 3.4) can also be verified in the present case [6, 7].

(a) If $\lambda = i \cdot \omega$, $\omega \in \mathbb{R}$ then $P(i \cdot \omega) + Q(i \cdot \omega) \neq 0$, i.e. P and Q have no common imaginary roots. This condition was verified numerically in the entire $(R_f, L_f, C_f, C_c, R_{sb}, \tau, \ldots)$ domain of interest.

(b) $|Q(\lambda)/P(\lambda)|$ is bounded for $|\lambda| \to \infty$, $\mathrm{Re}\,\lambda \geq 0$. No roots bifurcation from ∞. Indeed, in the limit

$$\left|\frac{Q(\lambda)}{P(\lambda)}\right| = \left|\frac{\{-\lambda^2 \cdot (\Gamma_7 + \Gamma_1) + \lambda \cdot [\Gamma_1 \cdot \Gamma_9 - \Gamma_6 \cdot \Gamma_4 - \Gamma_2 \cdot \Gamma_8 + \Gamma_7 \cdot \Gamma_3 - \tau \cdot (\Gamma_1 \cdot \Gamma_7 - \Gamma_6 \cdot \Gamma_2)] + \Gamma_1 \cdot \Gamma_7 - \Gamma_6 \cdot \Gamma_2\}}{\lambda^2 \cdot (\Gamma_3 \cdot \Gamma_9 - \Gamma_8 \cdot \Gamma_4) - \lambda^3 \cdot (\Gamma_3 + \Gamma_9) + \lambda^4}\right|$$

(c) $F(\omega) = |P(i \cdot \omega)|^2 - |Q(i \cdot \omega)|^2$

$$F(\omega, \tau) = |P(i \cdot \omega, \tau)|^2 - |Q(i \cdot \omega, \tau)|^2 = \omega^8 + \omega^6 \cdot \{(\Gamma_3 + \Gamma_9)^2 - 2 \cdot (\Gamma_3 \cdot \Gamma_9 - \Gamma_8 \cdot \Gamma_4)\} + \omega^4 \cdot \{(\Gamma_3 \cdot \Gamma_9 - \Gamma_8 \cdot \Gamma_4)^4 - (\Gamma_7 + \Gamma_1)^2\} - \omega^2 \cdot \{\Omega^2 + 2 \cdot (\Gamma_7 + \Gamma_1) \cdot (\Gamma_1 \cdot \Gamma_7 - \Gamma_6 \cdot \Gamma_2)\} - (\Gamma_1 \cdot \Gamma_7 - \Gamma_6 \cdot \Gamma_2)^2$$

Has at most a finite number of zeros. Indeed, this is a polynomial in ω (degree in ω^8).

(d) Each positive root $\omega(R_f, L_f, C_f, C_c, R_{sb}, \tau, \ldots)$ of $F(\omega) = 0$ is continuous and differentiable with Respect to $R_f, L_f, C_f, C_c, R_{sb}, \tau, \ldots$. This condition can only be assessed numerically.

In addition, since the coefficients in P and Q are real, we have $\overline{P(-i \cdot \omega)} = P(i \cdot \omega)$, and $\overline{Q(-i \cdot \omega)} = Q(i \cdot \omega)$ thus, $\lambda = i \cdot \omega$, $\omega > 0$ maybe on eigenvalue of characteristic equation. The analysis consists in identifying the roots of the characteristic equation situated on the imaginary axis of the complex λ—plane, whereby increasing the parameters $R_f, L_f, C_f, C_c, R_{sb}, \tau, \ldots$, $\text{Re}\lambda$ may, at the crossing, change its sign from (−) to (+), i.e. from a stable focus $E^{(0)}(I_1^{(0)}, I_3^{(0)}, I_1'^{(0)}, I_3'^{(0)}) = (0,0,0,0)$ to an unstable one, or vice versa. This feature may be further assessed by examining the sign of the partial derivatives with respect to $R_f, L_f, C_f, C_c, R_{sb}, \tau, \ldots$ and gate antenna parameters. $\wedge^{-1}(R_f) = (\frac{\partial \text{Re}\,\lambda}{\partial R_f})_{\lambda = i \cdot \omega}$, $L_f, C_f, C_c, R_{sb}, \tau, \ldots = const$

$$\wedge^{-1}(L_f) = (\frac{\partial \text{Re}\,\lambda}{\partial L_f})_{\lambda = i \cdot \omega},\ R_f, C_f, C_c, R_{sb}, \tau, \ldots = const$$

$$\wedge^{-1}(C_f) = (\frac{\partial \text{Re}\,\lambda}{\partial C_f})_{\lambda = i \cdot \omega},\ R_f, L_f, C_c, R_{sb}, \tau, \ldots = const$$

$$\wedge^{-1}(C_c) = (\frac{\partial \text{Re}\,\lambda}{\partial C_c})_{\lambda = i \cdot \omega},\ R_f, L_f, C_f, R_{sb}, \tau, \ldots = const$$

$$\wedge^{-1}(R_{sb}) = (\frac{\partial \text{Re}\,\lambda}{\partial R_{sb}})_{\lambda = i \cdot \omega},\ R_f, L_f, C_f, C_c, \tau, \ldots = const$$

$$\wedge^{-1}(\tau) = (\frac{\partial \text{Re}\,\lambda}{\partial \tau})_{\lambda = i \cdot \omega},\ R_f, L_f, C_f, C_c, R_{sb}, \ldots = const;\ \omega \in \mathbb{R}_+.$$

In the case $\tau_1 = \tau_3 = \tau \,\&\, \Delta_1 = \Delta_3 = 0$ we get the following results: for simplicity we define a function: $\Omega(\Gamma_j, \tau; 1 \le j \le 9)$

$$\Omega = \Omega(\Gamma_j, \tau; 1 \le j \le 9) = \Gamma_1 \cdot \Gamma_9 - \Gamma_6 \cdot \Gamma_4 - \Gamma_2 \cdot \Gamma_8 + \Gamma_7 \cdot \Gamma_3 - \tau \cdot (\Gamma_1 \cdot \Gamma_7 - \Gamma_6 \cdot \Gamma_2)$$

$$P_R(i \cdot \omega, \tau) = \omega^2 \cdot (\omega^2 - \Gamma_3 \cdot \Gamma_9 + \Gamma_8 \cdot \Gamma_4);\ P_I(i \cdot \omega, \tau) = -\omega^3 \cdot (\Gamma_3 + \Gamma_9)$$

$$Q_R(i \cdot \omega, \tau) = \omega^2 \cdot (\Gamma_7 + \Gamma_1) + (\Gamma_1 \cdot \Gamma_7 - \Gamma_6 \cdot \Gamma_2);\ Q_I(i \cdot \omega, \tau) = \omega \cdot \Omega(\Gamma_j, \tau; 1 \le j \le 9) = \omega \cdot \Omega$$

$$\Phi_0 = -(\Gamma_1 \cdot \Gamma_7 - \Gamma_6 \cdot \Gamma_2)^2;\ \Phi_2 = -\{\Omega^2 + 2 \cdot (\Gamma_7 + \Gamma_1) \cdot (\Gamma_1 \cdot \Gamma_7 - \Gamma_6 \cdot \Gamma_2)\}$$

$$\Phi_4 = (\Gamma_3 \cdot \Gamma_9 - \Gamma_8 \cdot \Gamma_4)^4 - (\Gamma_7 + \Gamma_1)^2;\ \Phi_6 = (\Gamma_3 + \Gamma_9)^2 - 2 \cdot (\Gamma_3 \cdot \Gamma_9 - \Gamma_8 \cdot \Gamma_4);\ \Phi_8 = 1$$

$$F(\omega, \tau) = |P(i \cdot \omega, \tau)|^2 - |Q(i \cdot \omega, \tau)|^2 = \Phi_0 + \Phi_2 \cdot \omega^2 + \Phi_4 \cdot \omega^4 + \Phi_6 \cdot \omega^6 + \Phi_8 \cdot \omega^8 = \sum_{k=0}^{4} \Phi_{2 \cdot k} \cdot \omega^{2 \cdot k}$$

Hence $F(\omega,\tau)=0$ implies $\sum_{k=0}^{4}\Phi_{2\cdot k}\cdot\omega^{2\cdot k}=0$. When writing $P(\lambda)=P_R(\lambda)+i\cdot P_I(\lambda)$ and $Q(\lambda)=Q_R(\lambda)+i\cdot Q_I(\lambda)$, and inserting $\lambda=i\cdot\omega$ into active circulator system's characteristic equation, ω must satisfy the following:

$$\sin\omega\cdot\tau=g(\omega)=\frac{-P_R(i\cdot\omega)\cdot Q_I(i\cdot\omega)+P_I(i\cdot\omega)\cdot Q_R(i\cdot\omega)}{|Q(i\cdot\omega)|^2}$$

$$\cos\omega\cdot\tau=h(\omega)=-\frac{P_R(i\cdot\omega)\cdot Q_R(i\cdot\omega)+P_I(i\cdot\omega)\cdot Q_I(i\cdot\omega)}{|Q(i\cdot\omega)|^2}$$

where $|Q(i\cdot\omega)|^2\neq 0$ in view of requirement (a) above, and $(g,h)\in R$. Furthermore, it follows above $\sin\omega\cdot\tau$ and $\cos\omega\cdot\tau$ equations that, by squaring and adding the sides, ω must be a positive root of $F(\omega)=|P(i\cdot\omega)|^2-|Q(i\cdot\omega)|^2=0$. Note that $F(\omega)$ is dependent of τ. Now it is important to notice that if $\tau\notin I$ (assume that $I\subseteq R_{+0}$ is the set where $\omega(\tau)$ is a positive root of $F(\omega)$ and for $\tau\notin I$, $\omega(\tau)$ is not defined. Then for all τ in I $\omega(\tau)$ is satisfied that $F(\omega,\tau)=0$). Then there are no positive $\omega(\tau)$ solutions for $F(\omega,\tau)=0$, and we cannot have stability switches. For any $\tau\in I$ where $\omega(\tau)$ is a positive solution of $F(\omega,\tau)=0$, we can define the angle $\theta(\tau)\in[0,2\cdot\pi]$ as the solution of $\sin\theta(\tau)=\ldots$; $\cos\theta(\tau)=\ldots$

$$\sin\,\theta(\tau)=\frac{-P_R(i\cdot\omega)\cdot Q_R(i\cdot\omega)+P_I(i\cdot\omega)\cdot Q_I(i\cdot\omega)}{|Q(i\cdot\omega)|^2}$$

$$\cos\,\theta(\tau)=-\frac{P_R(i\cdot\omega)\cdot Q_R(i\cdot\omega)+P_I(i\cdot\omega)\cdot Q_I(i\cdot\omega)}{|Q(i\cdot\omega)|^2}$$

And the relation between the argument $\theta(\tau)$ and $\omega(\tau)\cdot\tau$ for $\omega(\tau)\cdot\tau$ must be $\omega(\tau)\cdot\tau=\theta(\tau)+n\cdot 2\cdot\pi\,\forall n\in\mathbb{N}_0$. Hence we can define the maps $\tau_n: I\to R_{+0}$ given by $\tau_n(\tau)=\frac{\theta(\tau)+n\cdot 2\cdot\pi}{\omega(\tau)}$; $n\in\mathbb{N}_0,\tau\in I$. Let us introduce the functions $I\to R$; $S_n(\tau)=\tau-\tau_n(\tau)$, $\tau\in I$, $n\in\mathbb{N}_0$ that is a continuous and differentiable in τ. In the following, the subscripts $\lambda,\omega,R_f,L_f,C_f,C_c,R_{sb},\ldots$ indicate the corresponding partial derivatives. Let us first concentrate on $\wedge(x)$, remember in $\lambda(R_f,L_f,C_f,C_c,R_{sb},\ldots)$ and $\omega(R_f,L_f,C_f,C_c,R_{sb},\ldots)$, and keeping all parameters except one (x) and τ. The derivation closely follows that in reference [BK]. Differentiating active circulator characteristic equation $P(\lambda)+Q(\lambda)\cdot e^{-\lambda\cdot\tau}=0$ with respect to specific parameter (x), and inverting the derivative, for convenience, one calculates.

Remark: $x=R_f,L_f,C_f,C_c,R_{sb},\ldots, etc.,$

$$\left(\frac{\partial\lambda}{\partial x}\right)^{-1}=\frac{-P_\lambda(\lambda,x)\cdot Q(\lambda,x)+Q_\lambda(\lambda,x)\cdot P(\lambda,x)-\tau\cdot P(\lambda,x)\cdot Q(\lambda,x)}{P_x(\lambda,x)\cdot Q(\lambda,x)-Q_x(\lambda,x)\cdot P(\lambda,x)}$$

where $P_\lambda=\frac{\partial P}{\partial\lambda},\ldots$ etc., Substituting $\lambda=i\cdot\omega$, and bearing i $\overline{P(-i\cdot\omega)}=P(i\cdot\omega)$, $\overline{Q(-i\cdot\omega)}=Q(i\cdot\omega)$ then $i\cdot P_\lambda(i\cdot\omega)=P_\omega(i\cdot\omega)$; $i\cdot Q_\lambda(i\cdot\omega)=Q_\omega(i\cdot\omega)$ and that on the surface $|P(i\cdot\omega)|^2=|Q(i\cdot\omega)|^2$, one obtains

$$\left(\frac{\partial\lambda}{\partial x}\right)^{-1}\Big|_{\lambda=i\cdot\omega} = \left(\frac{i\cdot P_{\omega}(i\cdot\omega,x)\cdot\overline{P(i\cdot\omega,x)} + i\cdot Q_{\lambda}(i\cdot\omega,x)\cdot\overline{Q(\lambda,x)} - \tau\cdot|P(i\cdot\omega,x)|^2}{P_x(i\cdot\omega,x)\cdot\overline{P(i\cdot\omega,x)} - Q_x(i\cdot\omega,x)\cdot\overline{Q(i\cdot\omega,x)}}\right)$$

Upon separating into real and imaginary parts, with $P = P_R + i\cdot P_I$; $Q = Q_R + i\cdot Q_I$; $P_{\omega} = P_{R\omega} + i\cdot P_{I\omega}$; $Q_{\omega} = Q_{R\omega} + i\cdot Q_{I\omega}$; $P_x = P_{Rx} + i\cdot P_{Ix}$; $Q_x = Q_{Rx} + i\cdot Q_{Ix}$

$P^2 = P_R^2 + P_I^2$. When (x) can be any active circulator parameters R_1, C_1, And time delay τ etc. Where for convenience, we have dropped the arguments (i,ω,x), and where $F_{\omega} = 2\cdot[(P_{R\omega}\cdot P_R + P_{I\omega}\cdot P_I) - (Q_{R\omega}\cdot Q_R + Q_{I\omega}\cdot Q_I)]$.

$F_x = 2\cdot[(P_{Rx}\cdot P_R + P_{Ix}\cdot P_I) - (Q_{Rx}\cdot Q_R + Q_{Ix}\cdot Q_I)]$; $\omega_x = -F_x/F_{\omega}$.
We define U and V:

$$U = (P_R\cdot P_{I\omega} - P_I\cdot P_{R\omega}) - (Q_R\cdot Q_{I\omega} - Q_I\cdot Q_{R\omega})$$
$$V = (P_R\cdot P_{Ix} - P_I\cdot P_{Rx}) - (Q_R\cdot Q_{Ix} - Q_I\cdot Q_{Rx})$$

We choose our specific parameter as time delay x = τ

$P_R = \omega^2\cdot(\omega^2 - \Gamma_3\cdot\Gamma_9 + \Gamma_8\cdot\Gamma_4)$; $P_I = -\omega^3\cdot(\Gamma_3+\Gamma_9)$;

$Q_R = \omega^2\cdot(\Gamma_7+\Gamma_1) + (\Gamma_1\cdot\Gamma_7 - \Gamma_6\cdot\Gamma_2)$

$Q_I = \omega\cdot\Omega(\Gamma_j,\tau; 1\le j\le 9) = \omega\cdot\Omega(\tau)$; $P_{R\omega} = 2\cdot\omega\cdot[2\cdot\omega^2 - \Gamma_3\cdot\Gamma_9 + \Gamma_8\cdot\Gamma_4]$

$P_{I\omega} = -3\cdot\omega^2\cdot(\Gamma_3+\Gamma_9)$; $Q_{R\omega} = 2\cdot\omega\cdot(\Gamma_7+\Gamma_1)$; $Q_{I\omega} = \Omega$; $P_{R\tau} = 0$;

$P_{I\tau} = 0$; $Q_{R\tau} = 0$; $\omega_{\tau} = -F_{\tau}/F_{\omega}$

$$Q_{I\tau} = \omega\cdot\frac{\partial\Omega}{\partial\tau} = \omega\cdot(\Gamma_6\cdot\Gamma_2 - \Gamma_1\cdot\Gamma_7)$$

$P_{R\omega}\cdot P_R = 2\cdot\omega^3\cdot(2\cdot\omega^2 - \Gamma_3\cdot\Gamma_9 + \Gamma_8\cdot\Gamma_4)\cdot(\omega^2 - \Gamma_3\cdot\Gamma_9 + \Gamma_8\cdot\Gamma_4)$;

$P_{I\omega}\cdot P_I = 3\cdot\omega^5\cdot(\Gamma_3+\Gamma_9)^2$

$Q_{R\omega}\cdot Q_R = 2\cdot\omega\cdot(\Gamma_7+\Gamma_1)\cdot\{\omega^2\cdot(\Gamma_7+\Gamma_1) + (\Gamma_1\cdot\Gamma_7 - \Gamma_6\cdot\Gamma_2)\}$;

$Q_{I\omega}\cdot Q_I = \omega\cdot\Omega^2(\tau)$

$F_{\tau} = 2\cdot[(P_{R\tau}\cdot P_R + P_{I\tau}\cdot P_I) - (Q_{R\tau}\cdot Q_R + Q_{I\tau}\cdot Q_I)]$;

$F_{\tau} = -2\cdot Q_{I\tau}\cdot Q_I = -2\cdot\omega^2\cdot(\Gamma_6\cdot\Gamma_2 - \Gamma_1\cdot\Gamma_7)\cdot\Omega(\tau)$

$P_R\cdot P_{I\omega} = -3\cdot\omega^4\cdot(\omega^2 - \Gamma_3\cdot\Gamma_9 + \Gamma_8\cdot\Gamma_4)\cdot(\Gamma_3+\Gamma_9)$;

$P_I\cdot P_{R\omega} = -2\cdot\omega^4\cdot(\Gamma_3+\Gamma_9)\cdot[2\cdot\omega^2 - \Gamma_3\cdot\Gamma_9 + \Gamma_8\cdot\Gamma_4]$

$Q_R\cdot Q_{I\omega} = \{\omega^2\cdot(\Gamma_7+\Gamma_1) + (\Gamma_1\cdot\Gamma_7 - \Gamma_6\cdot\Gamma_2)\}\cdot\Omega(\tau)$;

$Q_I\cdot Q_{R\omega} = 2\cdot\omega^2\cdot\Omega(\tau)\cdot(\Gamma_7+\Gamma_1)$

$V = (P_R\cdot P_{I\tau} - P_I\cdot P_{R\tau}) - (Q_R\cdot Q_{I\tau} - Q_I\cdot Q_{R\tau})$

$V = -Q_R\cdot Q_{I\tau} = -\{\omega^2\cdot(\Gamma_7+\Gamma_1) + (\Gamma_1\cdot\Gamma_7 - \Gamma_6\cdot\Gamma_2)\}$
$\quad\cdot\omega\cdot(\Gamma_6\cdot\Gamma_2 - \Gamma_1\cdot\Gamma_7)$; $F(\omega,\tau) = 0$.

Differentiating with respect to τ and we get $F_\omega \cdot \frac{\partial\omega}{\partial\tau} + F_\tau = 0$; $\tau \in I \Rightarrow \frac{\partial\omega}{\partial\tau} = -\frac{F_\tau}{F_\omega}$

$$\wedge^{-1}(\tau) = (\frac{\partial \mathrm{Re}\,\lambda}{\partial\tau})_{\lambda=i\cdot\omega};\ \frac{\partial\omega}{\partial\tau} = \omega_\tau = -\frac{F_\tau}{F_\omega};$$

$$\wedge^{-1}(\tau) = \mathrm{Re}\{\frac{-2\cdot[U+\tau\cdot|P|^2]+i\cdot F_\omega}{F_\tau+i\cdot 2\cdot[V+\omega\cdot|P|^2]}\}$$

$$sign\{\wedge^{-1}(\tau)\} = sign\{(\frac{\partial \mathrm{Re}\lambda}{\partial\tau})_{\lambda=i\cdot\omega}\};$$

$$sign\{\wedge^{-1}(\tau)\} = sign\{F_\omega\}\cdot sign\{\tau\cdot\frac{\partial\omega}{\partial\tau}+\omega+\frac{U\cdot\frac{\partial\omega}{\partial\tau}+V}{|P|^2}\}$$

We shall presently examine the possibility of stability transitions (bifurcations) active circulator system, about the equilibrium point $E^{(0)}(I_0^{(0)}, I_3^{(0)}, I_0'^{(0)}, I_3'^{(0)}) = (0,0,0,0)$ as a result of a variation of delay parameter τ. The analysis consists in identifying the roots of our system characteristic equation situated on the imaginary axis of the complex λ-plane where by increasing the delay parameter τ, Re λ may at the crossing, changes its sign from − to +, i.e. from a stable focus $E^{(*)}$ to an unstable one, or vice versa. This feature may be further assessed by examining the sign of the partial derivatives with respect to τ,

$$\wedge^{-1}(\tau) = (\frac{\partial \mathrm{Re}\,\lambda}{\partial\tau})_{\lambda=i\cdot\omega},\ R_f, L_f, C_f, C_c, R_{sb}, \ldots = const;\ \omega \in \mathbb{R}_+.$$

For our stability switching analysis, we choose typical active circulator parameter values: R_f = 110 Ohm, L_f = 1.4 nH, C_f = 5 pF, C_c = 10 pF, R_{sb} = 115 Ohm, MESFET (F20-FET-4x75), L_1 = 5 nH, C_1 = 5 pF, L_a = 1.6 nH, R_a = 500 Ohm, L_t = 7 nH, L_Σ = 15 nH. $L_\Sigma = L_f + L_a + L_t + L_1 = 1.4\,\mathrm{nH} + 1.6\,\mathrm{nH} + 7\,\mathrm{nH} + 5\,\mathrm{nH} = 15\,\mathrm{nH}$.

$$C_\Sigma = \frac{1}{C_f}+\frac{1}{C_c}+\frac{1}{C_1} = \frac{1}{5\times10^{-12}}+\frac{1}{10\times10^{-12}}+\frac{1}{5\times10^{-12}} = 5\times10^{11};$$

$$\Gamma_0 = 2.11\times10^8;\ \Gamma_1 = \Gamma_0\cdot(\frac{L_1}{L_\Sigma\cdot C_1}-C_\Sigma) = -9.14\times10^{19}$$

$$\Gamma_2 = \Gamma_0\cdot(\frac{1}{C_1}-\frac{L_1\cdot C_\Sigma}{L_\Sigma}) = 0.703\times10^{19};\ \Gamma_3 = -\Gamma_0\cdot R_f = -232.1\times10^8;$$

$$\Gamma_4 = -\Gamma_0\cdot L_1\cdot(R_f+R_a) = -643.55;\ \Gamma_5 = \frac{L_1\cdot\Gamma_0}{L_\Sigma} = 0.7\times10^8$$

$$\Gamma_6 = \frac{1}{L_\Sigma}\cdot[\frac{1}{C_1}+L_1\cdot\Gamma_1] = -0.171\times10^{20};\ \Gamma_7 = \frac{1}{L_\Sigma}\cdot[L_1\cdot\Gamma_2-C_\Sigma] = -3\times10^{19};$$

$$\Gamma_8 = \frac{L_1\times\Gamma_3}{L_\Sigma} = -77.36\times10^8;\ \Gamma_9 = \frac{1}{L_\Sigma}\cdot[L_1\cdot\Gamma_4-(R_f+R_a)] \simeq -40.6\times10^9$$

Then we get the expression for $F(\omega,\tau)$ for an active circulator parameter's value. We find those ω, τ values which fulfill $F(\omega,\tau)=0$. We ignore negative, complex, and imaginary values of ω for specific τ values. $\tau \in [0.001..10]$. And we can be express by 3D function $F(\omega,\tau)=0$. We plot the stability switch diagram based on different delay values of our active circulator system.

$$\wedge^{-1}(\tau) = (\frac{\partial \operatorname{Re}\lambda}{\partial \tau})_{\lambda=i\cdot\omega} = \operatorname{Re}\{\frac{-2\cdot[U+\tau\cdot|P|^2]+i\cdot F_\omega}{F_\tau+i\cdot 2\cdot[V+\omega\cdot|P|^2]}\}$$

$$\wedge^{-1}(\tau) = (\frac{\partial \operatorname{Re}\lambda}{\partial \tau})_{\lambda=i\cdot\omega} = \frac{2\cdot\{F_\omega\cdot(V+\omega\cdot P^2) - F_\tau\cdot(U+\tau\cdot P^2)\}}{F_\tau^2+4\cdot(V+\omega\cdot P^2)^2}$$

The stability switch occurs only on those delay values (τ) which fit the equation: $\tau=\frac{\theta_+(\tau)}{\omega_+(\tau)}$ and $\theta_+(\tau)$ is the solution of $\sin\theta(\tau)=\ldots$; $\cos\theta(\tau)=\ldots$ when $\omega=\omega_+(\tau)$ if only ω_+ is feasible. Additionally When all active circulator's parameters are known and the stability switch due to various time delay values τ is described in the following expression:

$$\begin{aligned} sign\{\wedge^{-1}(\tau)\} &= sign\{F_\omega(\omega(\tau),\tau)\} \\ &\quad \cdot sign\{\tau\cdot\omega_\tau(\omega(\tau))+\omega(\tau)+\frac{U(\omega(\tau))\cdot\omega_\tau(\omega(\tau))+V(\omega(\tau))}{|P(\omega(\tau))|^2}\} \end{aligned}$$

Remark: We know $F(\omega,\tau)=0$ implies its roots $\omega_i(\tau)$ and finding those delays values τ which ω_i is feasible. There are τ values, which ω_i are complex or imaginary numbered, then unable to analyze stability [6, 7].

We find those ω, τ values which fulfill $F(\omega,\tau)=0$. We ignore negative, complex, and imaginary values of ω for specific τ values. $\tau \in [0.001..10]$ and we can be express by 3D function $F(\omega,\tau)=0$. We define new MATLAB script parameters: τ→Tau, Γ_i→G_i (i=0..9), Ω→Omega, Φ_j→Phi_j. Running MATLAB script for τ values ($\tau \in [0.001..10]$) gives the following results.

MATLAB script: Tau=0.1;G0=2.11e8;G1=-9.14e19; G2=0.703e19; G3=-232.1e8;G4=-643.55;G5=0.7e8;G6=-0.171e20;G7=-3e19;G8=-77.36e8;G9=-40.6e9; Omega=G1*G9-G6*G4-G2*G8+G7*G3-Tau*(G1*G7-G6*G2);Phi0=-(G1*G7-G6*G2)^2; Phi2=-(Omega^2+2*(G7+G1)*(G1*G7 G6*G2)); Phi4=(G3*G9-G8*G4)^4-(G7+G1)^2; Phi6=(G3+G9)^2-2*(G3*G9-G8*G4);Phi8=1; p=[Phi8 0 Phi6 0 Phi4 0 Phi2 0 Phi0];r=roots(p).

Results: (Table 2.2).

We plot 3D function $F(\omega,\tau)=0$. τ:0→10; ω:0→1e20. We define additional MATLAB script parameters ω→w, τ→t (Fig. 2.8).

We get two possible real values for ω which fulfil $F(\omega,\tau)=0$; $F(\omega=0\, or\, \omega=1.0e+020,\tau)=0$ $\tau\in[0.001..10]$. Next is to find those ω, τ values which fulfil $\sin\theta(\tau)=\ldots$

Table 2.2 Active circulator roots $\omega_i(\tau)$

τ	τ = 1; τ = 10	τ = [0…0.1]
ω_1	1.0e+020	1.0e+020
ω_2	−6.6468 + 6.6468i	−6.6468 + 6.6468i
ω_3	−6.6468 − 6.6468i	−6.6468 − 6.6468i
ω_4	6.6468 + 6.6468i	6.6468 + 6.6468i
ω_5	6.6468 − 6.6468i	6.6468 − 6.6468i
ω_6	−0.0000	0
ω_7	0.0000	0
ω_8	0.0000 + 0.0000i	0
ω_9	0.0000 − 0.0000i	0

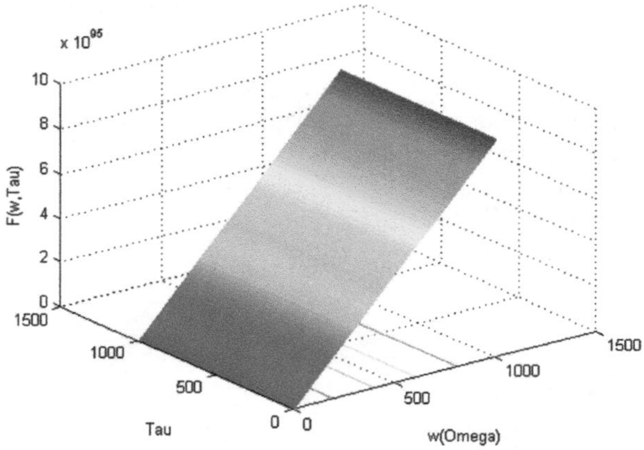

Fig. 2.8 Active circulator F (ω, τ) function

$$\sin(\omega\cdot\tau) = \frac{-P_R\cdot Q_I + P_I\cdot Q_R}{|Q|^2} \text{ and } \cos\theta(\tau) = \ldots$$

$$\cos(\omega\cdot\tau) = -\frac{(P_R\cdot Q_R + P_I\cdot Q_I)}{|Q|^2}; \ |Q|^2 = Q_R^2 + Q_I^2$$

Case I: $\omega = 0$ then $P_R = 0$; $P_I = 0$; $Q_R = \Gamma_1\cdot\Gamma_7 - \Gamma_6\cdot\Gamma_2$; $Q_I = 0$; $\sin(\omega\cdot\tau) = \ldots$ fulfil and $\cos(\omega\cdot\tau) = \ldots$ Can't fulfil since $\cos(\omega\cdot\tau)|_{\omega=0} \neq 0$.

Case II: ω = 1.0e+020 which can fulfil expressions $\sin(\omega\cdot\tau) = \ldots$ and $\cos(\omega\cdot\tau) = \ldots$. Finally, we plot the stability switch diagram based on different delay values of our Active circulator system (ω = 1.0e20). $P_R = \omega^2\cdot(\omega^2 - 9.4233e20)$, $P_I = \omega^3\cdot 6.381e10$, $Q_R = -\omega^2\cdot 1.214e20 + 2.8622e39$, $Q_I = \omega\cdot$ Omega. $Q_{I\tau} = -\omega\cdot 2.86e39$; $V = -Q_R\cdot Q_{I\tau}$; $V = -Q_R\cdot Q_{I\tau} = -\omega^3\cdot 3.473e59$

$$P_{I\omega} = \omega^2 \cdot 19.143e10\,;\; Q_{I\omega} = Omega\,;\; P_{R\omega} = 2 \cdot \omega \cdot [2 \cdot \omega^2 - 9.4233e20]\,;$$
$$Q_{R\omega} = -\omega \cdot 2.428e20$$
$$P_{R\omega} \cdot P_R = 2 \cdot \omega^3 \cdot (2 \cdot \omega^2 - 9.423e20) \cdot (\omega^2 - 9.423e20)\,;$$
$$U = -\omega^4 \cdot (\omega^2 \cdot 6.377e10 + 6.0119e31) - \omega^2 \cdot Omega \cdot 1.214e20$$
$$P_{I\omega} \cdot P_I = \omega^5 \cdot 122.15e20\,;\; Q_{R\omega} \cdot Q_R = \omega^3 \cdot 2.947e40\,;$$
$$Q_{I\omega} \cdot Q_I = \omega \cdot [Omega]^2;\; F_\tau = -2 \cdot Q_{I\tau} \cdot Q_I = -2 \cdot Q_{I\tau} \cdot Q_I$$

$F_\tau = -2 \cdot Q_{I\tau} \cdot Q_I = 5.72e39 \cdot \omega^2 \cdot Omega.$ We plot the function: $g(\tau) = \wedge^{-1}(\tau) = (\frac{\partial \operatorname{Re}\lambda}{\partial \tau})_{\lambda = i\cdot\omega}$

$$g(\tau) = \wedge^{-1}(\tau) = (\frac{\partial \operatorname{Re}\lambda}{\partial \tau})_{\lambda = i\cdot\omega} = \frac{2 \cdot \{F_\omega \cdot (V + \omega \cdot P^2) - F_\tau \cdot (U + \tau \cdot P^2)\}}{F_\tau^2 + 4 \cdot (V + \omega \cdot P^2)^2}$$

$$\begin{aligned} sign\,[g(\tau)] &= sign[\wedge^{-1}(\tau)] = sign[(\frac{\partial \operatorname{Re}\lambda}{\partial \tau})_{\lambda = i\cdot\omega}] \\ &= sign[\frac{2 \cdot \{F_\omega \cdot (V + \omega \cdot P^2) - F_\tau \cdot (U + \tau \cdot P^2)\}}{F_\tau^2 + 4 \cdot (V + \omega \cdot P^2)^2}] \end{aligned}$$

Since $F_\tau^2 + 4 \cdot (V + \omega \cdot P^2)^2 > 0$ then $sign[\wedge^{-1}(\tau)] = sign\{F_\omega \cdot (V + \omega \cdot P^2) - F_\tau \cdot (U + \tau \cdot P^2)\}$

$$sign[\wedge^{-1}(\tau)] = sign\{[F_\omega] \cdot [(V + \omega \cdot P^2) - \frac{F_\tau}{F_\omega} \cdot (U + \tau \cdot P^2)]\}\,;$$
$$\omega_\tau = -\frac{F_\tau}{F_\omega}\,;\; \omega_\tau = (\frac{\partial \omega}{\partial \tau})^{-1} = -\frac{\partial F / \partial \omega}{\partial F / \partial \tau}$$
$$sign\,[\wedge^{-1}(\tau)] = sign\{[F_\omega] \cdot [V + \omega_\tau \cdot U + \omega \cdot P^2 + \omega_\tau \cdot \tau \cdot P^2]\}\,;$$
$$sign\,[\wedge^{-1}(\tau)] = sign\{[F_\omega] \cdot [\frac{1}{P^2}] \cdot [\frac{V + \omega_\tau \cdot U}{P^2} + \omega + \omega_\tau \cdot \tau]\}$$
$$sign\,[\frac{1}{P^2}] > 0 \Rightarrow sign\,[\wedge^{-1}(\tau)] = sign\{[F_\omega] \cdot [\frac{V + \omega_\tau \cdot U}{P^2} + \omega + \omega_\tau \cdot \tau]\}$$
$$sign\,[\wedge^{-1}(\tau)] = sign[F_\omega] \cdot sign\,[\frac{V + \omega_\tau \cdot U}{P^2} + \omega + \omega_\tau \cdot \tau];$$
$$F_\omega = 2 \cdot [(P_{R\omega} \cdot P_R + P_{I\omega} \cdot P_I) - (Q_{R\omega} \cdot Q_R + Q_{I\omega} \cdot Q_I)]$$

We check the sign of $\wedge^{-1}(\tau)$ according the following rule (Table 2.3).

Table 2.3 Active circulator stability switching criteria

$sign\,[F_\omega]$	$sign[\frac{V+\omega_\tau \cdot U}{P^2} + \omega + \omega_\tau \cdot \tau]$	$sign\,[\wedge^{-1}(\tau)]$
$\pm$	$\pm$	+
$\pm$	$\mp$	−

If $\text{sign}[\Lambda^{-1}(\tau)] > 0$ then the crossing proceeds from (−) to (+) respectively (stable to unstable). If $\text{sign}[\Lambda^{-1}(\tau)] < 0$ then the crossing proceeds from (+) to (−) respectively (unstable to stable). Anyway the stability switching can occur only for $\omega = 1.0e + 020$ and $\tau \in [0.001..10]$. Since it is a very complex function, we recommend to solve it numerically rather than analytic. We plot the stability switch diagram based on different delay values of our active circulator system.

We consider Active circulator which connects in a configuration of Reflection Type Phase Shifter (RTPS) circuit. Due to the parasitic effect, there is a delay in time for current which flow in and out Active circulator ports. This delay causes to stability switching for our Active circulator system. We draw our Active circulator (RTPS) equivalent circuit and get system differential equations. Our variables are first and third ports currents and currents derivative. Our system dynamic behavior is dependent on circuit overall parameters and parasitic delay in time. We keep all circuit parameters fix and change, parasitic delay over various values $\tau \in [0.001..10]$. Our analysis results extend that of in the way that it deals with stability switching for different delay values. This implies that our system behavior of the circuit cannot inspect by short analysis and we must study the full system. Several very important issues and possibilities were left out of the present analysis. One possibility is the stability switching by circuit parameters. Every circuit's parameter variation can change our system dynamic and stability behavior. This case can be solved by the same methods combined with alternative and more technical hypotheses. Moreover, numerical simulations for the active circulator model studied in references suggest that this result can be extended to enhance models with more general functions. Still another extension of our results would be to treat the case of delayed Active circulator's port currents derivative in time $\frac{dI_1(t-\Delta_1)}{dt}; \frac{dI_3(t-\Delta_3)}{dt}; \Delta_1 > 0; \Delta_3 > 0$ [5, 6]. It would be extremely desirable to confirm these cases by mathematical proofs. Active circulator transmission lines are characterized by parasitic effects which can influence active circulator system stability in time. There are four main active circulator variables which are affected by transmission lines parasitic effects, first and third branch currents and currents derivatives respectively. Each active circulator currents variable under transmission line parasitic effects is characterized by time delay respectively. The two time delays are not the same, but can be categorized to some sub cases due to interferences behavior. The first case we analyze is when there is delay in active circulator first and third branches current and no delay in active circulator first and third branches current derivative. The second case we analyze is when there is delay in active circulator first and third branches current derivative and no delay in active circulator first and third branches current [6, 7]. The third case we analyze is when there is delay in active circulator first and third branches current and also delay in active circulator first and third branches current derivative.

($\tau_1 = \tau_3 = \Delta_1 = \Delta_3 = \tau_\Delta$) [6, 7]. For simplicity of our analysis we consider in the third case all delays are the same (there is a difference but it is neglected in our analysis). In each case we derive the related characteristic equation. The characteristic equation is dependent on active circulator overall parameters and interferences time delay. Upon mathematics manipulation and [BK] theorems and definitions we derive the expression which gives us a clear picture on active circulator stability map. The stability map gives all possible options for stability segments, each segment belongs to different time delay value segment. Active circulator's stability analysis can be influenced either by system overall parameter values. We left this analysis and do not discuss it in the current chapter [12].

Lemma 1.1 Assume that $\omega(\tau)$ is a positive and real root of $F(\omega, \tau) = 0$.

Defined for $\tau \in I$, which is continuous and differentiable. Assume further that if $\lambda = i \cdot \omega$, $\omega \in R$, then $P_n(i \cdot \omega, \tau) + Q_n(i \cdot \omega, \tau) \neq 0$, $\tau \in \mathrm{R}$ hold true. Then the functions $S_n(\tau)$, $n \in N_0$, are continuous and differentiable on I.

Theorem 1.2 *Assume that $\omega(\tau)$ is a positive real root of $F(\omega, \tau) = 0$ defined for $\tau \in I$, $I \subseteq R_{+0}$, and at some $\tau^* \in I$, $S_n(\tau^*) = 0$. For some $n \in N_0$ then a pair of simple conjugate pure imaginary roots $\lambda_+(\tau^*) = i \cdot \omega(\tau^*)$, $\lambda_-(\tau^*) = -i \cdot \omega(\tau^*)$ of $D(\lambda, \tau) = 0$ exist at $\tau = \tau^*$ which crosses the imaginary* axis *from left to right if $\delta(\tau^*) > 0$ and cross the imaginary axis from right to left if $\delta(\tau^*) < 0$ where*

$$\delta(\tau^*) = sign\{\frac{d\,\mathrm{Re}\,\lambda}{d\tau}|_{\lambda=i\omega(\tau^*)}\} = sign\{F_\omega(\omega(\tau^*), \tau^*)\} \cdot sign\{\frac{dS_n(\tau)}{d\tau}|_{\tau=\tau^*}\}$$

Theorem 1.3 *The characteristic equation has a pair of simple and conjugate pure imaginary roots $\lambda = \pm\omega(\tau^*), \omega(\tau^*)$ real at $\tau^* \in I$ if $S_n(\tau^*) = \tau^* - \tau_n(\tau^*) = 0$ for some $n \in N_0$. If $\omega(\tau^*) = \omega_+(\tau^*)$, this pair of simple conjugate pure imaginary roots crosses the imaginary axis from left to right if $\delta_+(\tau^*) > 0$ and crosses the imaginary axis from right to left if $\delta_+(\tau^*) < 0$ where $\delta_+(\tau^*) = sign\{\frac{d\mathrm{Re}\lambda}{d\tau}|_{\lambda=i\omega_+(\tau^*)}\} = sign\{\frac{dS_n(\tau)}{d\tau}|_{\tau=\tau^*}\}$. If $\omega(\tau^*) = \omega_-(\tau^*)$, this pair of simple conjugate pure imaginary roots crosses the imaginary axis from left to right if $\delta_-(\tau^*) > 0$ and crosses the imaginary axis from right to left If $\delta_-(\tau^*) < 0$ where $\delta_-(\tau^*) = sign\{\frac{d\mathrm{Re}\lambda}{d\tau}|_{\lambda=i\omega_-(\tau^*)}\} = -sign\{\frac{dS_n(\tau)}{d\tau}|_{\tau=\tau^*}\}$.*

If $\omega_+(\tau^*) = \omega_-(\tau^*)$ then $\Delta(\tau^*) = 0$ and $sign\{\frac{d\mathrm{Re}\lambda}{d\tau}|_{\lambda=i\omega(\tau^*)}\} = 0$, the same is true when $S'_n(\tau^*) = 0$. The following result can be useful in identifying values of τ Where stability switches happened.

2.3 Cylindrical RF Network Antennas for Coupled Plasma Sources Copper Legs Delayed in Time System Stability Analysis

In this subchapter, Very Critical and useful subject is discussed: cylindrical (closed) RF network antennas for coupled plasma sources copper legs delayed in time. The resonant RF networks can be arranged to form large-area or large-volume plasma sources with properties similar to Inductive Coupled Plasma (ICP) devices. There are medical applications of Birdcage coils and closed and open configurations of the antenna for plasma production are possible and can be analyzed by using mathematical formulation. There are systems of an open network antenna as a large-area planar plasma source and of a closed network antenna as a cylindrical plasma source. Both are composed of similar electrical meshes. Operation at different normal modes shows the capability of this antenna type of large-volume plasma applications [86].

An important issue of proper antenna operation is the location of the RF feeding and grounding connections on the antenna. There are a large number of different RF antenna arrangements possible in view of the geometry and RF operation and of plasma obtained. In our analysis, we investigated only cylindrical RF antenna which built following a high-pass Birdcage coil. The antenna is mounted outside a glass tube. The RF antenna consists of 16 copper legs (Fig. 2.9), equally spaced interconnected with capacitors, each copper leg current has parasitic time delayed ($\tau_{1-1} \ldots \tau_{1-16}$). We consider for simplicity that all copper legs parasitic time delayed are equal ($\tau_{1-1} = \tau_{1-2} = \cdots = \tau_{1-16}$) and the voltages on delay units (V_ε) are neglected $V_\varepsilon \to \varepsilon$. There is a delay in each Copper leg current $I_1(t - \tau_{1-1}), \ldots, I_{16}(t - \tau_{1-16})$. We consider all interconnected capacitor values are the same (C) and all antenna elements inductance values are the same (L). $C_{A1} = C_{A2} = \cdots = C_{A16} = C$; $C_{B1} = C_{B2} = \cdots = C_{B16} = C L_1 = L_2 = \cdots = L_{16} = L$;

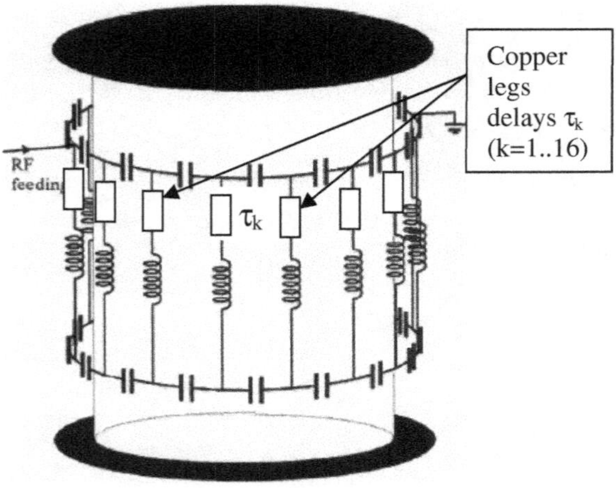

Fig. 2.9 Schematic of the 16-leg cylindrical (Birdcage) RF network antenna (closed)

Fig. 2.10 *Upper* view of 16-leg cylindrical RF antenna

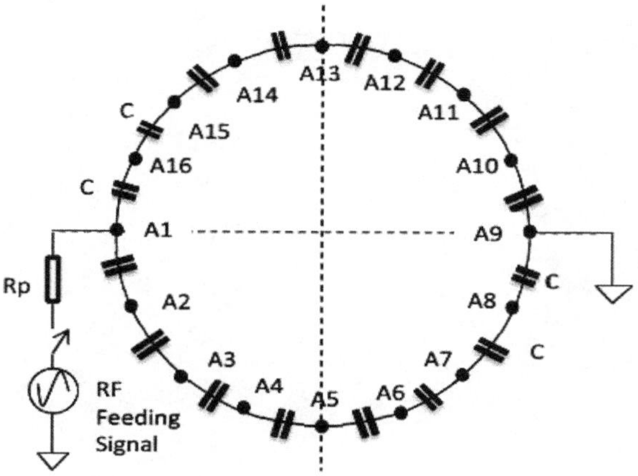

$I_{L1} = I_1$, $I_{L2} = I_2, \ldots, I_{L16} = I_{16}$. We choose first case: antenna network is fed by the transmitter unit (S1 = OFF, no direct RF feeding). The upper view of 16-leg cylindrical RF antenna network described in Fig. 2.10.

The lower view of 16-leg cylindrical RF antenna network described in Fig. 2.11.

Cylindrical RF network antenna system can represent as round strip of capacitors and inductors (Figs. 2.12 and 2.13). The schematic contains RF feeding signal, S1 switch (S1 = ON for direct RF signal feeding, S1 = OFF for RF signal transmitter feeding). The upper network connecting nodes are A1, A2,…,A16 and the lower network connecting nodes are B1, B2,…,B16. Antenna copper leg current parasitic delays are represented by delay units Tau1–1…Tau1–16 ($\tau_{1-1}, \ldots, \tau_{1-16}$). Rp is the parasitic resistance of RF feeding point (A1). The upper system spaced capacitors are CA1,…,CA16 and the lower system spaced capacitors are CB1,…,CB16.

Fig. 2.11 *Lower* view of 16-leg cylindrical RF antenna

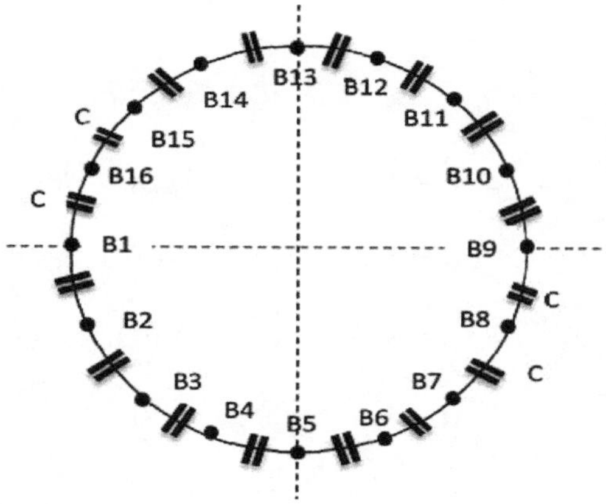

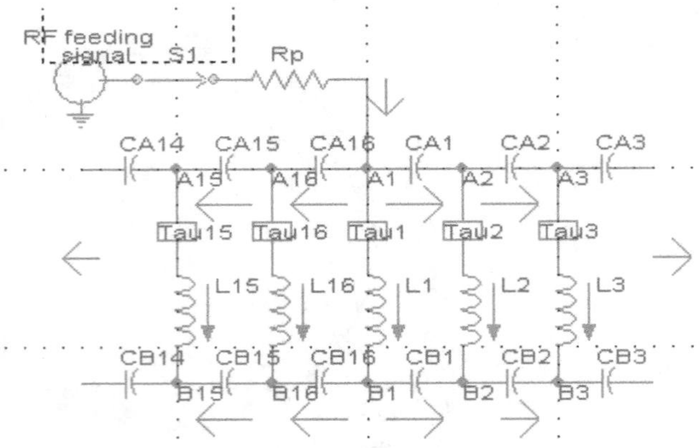

Fig. 2.12 16-leg cylindrical RF antenna strip (feeding side)

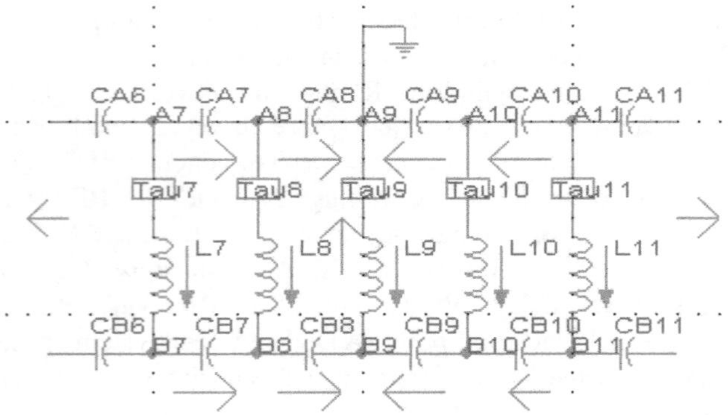

Fig. 2.13 16-leg cylindrical RF antenna strip (ground side)

$$I_{CA1} = C_{A1} \cdot \frac{d}{dt}(V_{A1} - V_{A2}),\ I_{CA2} = C_{A2} \cdot \frac{d}{dt}(V_{A2} - V_{A3})\,;$$

$$I_{CA3} = C_{A3} \cdot \frac{d}{dt}(V_{A3} - V_{A4}), \ldots, I_{CA7} = C_{A7} \cdot \frac{d}{dt}(V_{A7} - V_{A8})$$

$$I_{CA8} = C_{A8} \cdot \frac{dV_{A8}}{dt};\ I_{CAk} = C_{Ak} \cdot \frac{d}{dt}(V_{Ak} - V_{A(k+1)})\,;\ k = 1, \ldots, 7\,;$$

$$I_{CA16} = C_{A16} \cdot \frac{d}{dt}(V_{A1} - V_{A16}),\ I_{CA15} = C_{A15} \cdot \frac{d}{dt}(V_{A16} - V_{A15})$$

$$I_{CA14} = C_{A14} \cdot \frac{d}{dt}(V_{A15} - V_{A14}), \ldots, I_{CA10} = C_{A10} \cdot \frac{d}{dt}(V_{A11} - V_{A10})\,;$$

$$I_{CA9} = C_{A9} \cdot \frac{dV_{A10}}{dt};\ I_{CAl} = C_{Al} \cdot \frac{d}{dt}(V_{A(l+1)} - V_{Al})\,;\ l = 10, \ldots, 15$$

$$I_{CB1} = C_{B1} \cdot \frac{d}{dt}(V_{B1} - V_{B2}), I_{CB2} = C_{B2} \cdot \frac{d}{dt}(V_{B2} - V_{B3})\,;\; k = 1, \ldots, 8\,;\; , \ldots,$$
$$I_{CB8} = C_{B8} \cdot \frac{d}{dt}(V_{B2} - V_{B3})\,;\; I_{CBk} = C_{Bk} \cdot \frac{d}{dt}(V_{Bk} - V_{B(k+1)})$$
$$I_{CB16} = C_{B16} \cdot \frac{d}{dt}(V_{B1} - V_{B16}), I_{CB15} = C_{B15} \cdot \frac{d}{dt}(V_{B16} - V_{B15})\,;\; , \ldots,$$
$$I_{CB9} = C_{B9} \cdot \frac{d}{dt}(V_{B10} - V_{B9})$$
$$I_{CBl} = C_{Bl} \cdot \frac{d}{dt}(V_{B(l+1)} - V_{Bl})\,;\; l = 15, \ldots, 9$$
$$V_{A1} - V_{B1} = L_1 \cdot \frac{dI_{L1}}{dt};\; V_{A2} - V_{B2} = L_2 \cdot \frac{dI_{L2}}{dt};\; V_{A3} - V_{B3} = L_3 \cdot \frac{dI_{L3}}{dt};\; , \ldots,$$
$$V_{A8} - V_{B8} = L_8 \cdot \frac{dI_{L8}}{dt};\, V_{B9} = L_9 \cdot \frac{dI_{L9}}{dt}$$
$$V_{A10} - V_{B10} = L_{10} \cdot \frac{dI_{L10}}{dt}, \ldots, V_{A16} - V_{B16} = L_{16} \cdot \frac{dI_{L16}}{dt};$$
$$V_{Am} - V_{Bm} = L_m \cdot \frac{dI_{Lm}}{dt};\; m = 1, \ldots, 16;\; m \neq 9$$
$$V_{B9} = L_9 \cdot \frac{dI_{L9}}{dt};\; V_{A9} = 0;\quad A9 - ground$$

$$I_{R_P} = I_{CA16} + I_{CA1} + I_{L1};\; I_{CA1} = I_{CA2} + I_{L2};$$
$$I_{CA2} = I_{CA3} + I_{L3}, \ldots, I_{CA7} = I_{CA8} + I_{L8}$$
$$I_{CAl} = I_{CA(l+1)} + I_{L(l+1)};\quad l = 1, \ldots, 7$$
$$I_{CA16} = I_{CA15} + I_{L16};\; I_{CA15} = I_{CA14} + I_{L15};$$
$$I_{CA14} = I_{CA13} + I_{L14}, \ldots, I_{CA10} = I_{CA9} + I_{L10}$$
$$I_{CAk} = I_{CA(k-1)} + I_{Lk};\quad k = 16, \ldots, 10$$
$$I_{L1} = I_{CB1} + I_{CB16};\; I_{CB2} = I_{CB1} + I_{L2};\; I_{CB3} = I_{CB2} + I_{L3};\; I_{CB4} = I_{CB3} + I_{L4}, \ldots,$$
$$I_{CB8} = I_{CB7} + I_{L8};\; I_{L9} = I_{CB8} + I_{CB9};\; I_{CBm} = I_{CB(m-1)} + I_{Lm};\quad m = 2, \ldots, 8$$
$$I_{CB15} = I_{CB16} + I_{L16};\; I_{CB14} = I_{CB15} + I_{L15};\; I_{CB13} = I_{CB14} + I_{L14};$$
$$I_{CB12} = I_{CB13} + I_{L13}, \ldots, I_{CB9} = I_{CB10} + I_{L10}$$
$$I_{CBn} = I_{CB(n+1)} + I_{L(n+1)};\quad n = 15, \ldots, 9.$$

Upon mathematic manipulation we get the following expressions:

$$\frac{I_{C_{A1}}}{C_{A1}} - \frac{I_{C_{B1}}}{C_{B1}} = L_1 \cdot \frac{d^2 I_{L1}}{dt^2} - L_2 \cdot \frac{d^2 I_{L2}}{dt^2};\ C_{A1} = C_{B1} = C;\ L_1 = L_2 = L;$$
$$\frac{1}{LC} \cdot (I_{C_{A1}} - I_{C_{B1}}) = \frac{d^2 I_{L1}}{dt^2} - \frac{d^2 I_{L2}}{dt^2}$$
$$\frac{I_{C_{A2}}}{C_{A2}} - \frac{I_{C_{B2}}}{C_{B2}} = L_2 \cdot \frac{d^2 I_{L2}}{dt^2} - L_3 \cdot \frac{d^2 I_{L3}}{dt^2};\ C_{A2} = C_{B2} = C;\ L_2 = L_3 = L;$$
$$\frac{1}{LC} \cdot (I_{C_{A2}} - I_{C_{B2}}) = \frac{d^2 I_{L2}}{dt^2} - \frac{d^2 I_{L3}}{dt^2} \ldots,$$
$$\frac{I_{C_{A7}}}{C_{A7}} - \frac{I_{C_{B7}}}{C_{B7}} = L_7 \cdot \frac{d^2 I_{L7}}{dt^2} - L_8 \cdot \frac{d^2 I_{L8}}{dt^2};\ L_7 = L_8 = L;$$
$$\frac{1}{LC} \cdot (I_{C_{A7}} - I_{C_{B7}}) = \frac{d^2 I_{L7}}{dt^2} - \frac{d^2 I_{L8}}{dt^2}$$
$$C_{A7} = C_{B7} = C;\ L_1 = L_2 = \cdots = L_{16} = L;\ k = 1,\ldots,7;$$
$$\frac{1}{LC} \cdot (I_{C_{Ak}} - I_{C_{Bk}}) = \frac{d^2 I_{Lk}}{dt^2} - \frac{d^2 I_{L(k+1)}}{dt^2};\ k = 1,\ldots,7$$

$$\frac{1}{LC} \cdot (I_{C_{A10}} - I_{C_{B10}}) = \frac{d^2 I_{L11}}{dt^2} - \frac{d^2 I_{L10}}{dt^2};\ C_{A10} = C_{B10} = C; \frac{1}{LC} \cdot (I_{C_{A15}} - I_{C_{B15}})$$
$$= \frac{d^2 I_{L16}}{dt^2} - \frac{d^2 I_{L15}}{dt^2};\ C_{A15} = C_{B15} = C$$

$$\frac{1}{LC} \cdot (I_{C_{Am}} - I_{C_{Bm}}) = \frac{d^2 I_{L(m+1)}}{dt^2} - \frac{d^2 I_{Lm}}{dt^2};\ m = 10,\ldots,15;\ C_{A8} = C_{B8} = C;$$
$$V_{A9} = 0;\ \frac{1}{LC} \cdot (I_{C_{A8}} - I_{C_{B8}}) = \frac{d^2 I_{L8}}{dt^2} + \frac{d^2 I_{L9}}{dt^2}$$

$$C_{A9} = C_{B9} = C;\ V_{A9} = 0;\ \frac{1}{LC} \cdot (I_{C_{A9}} - I_{C_{B9}}) = \frac{d^2 I_{L9}}{dt^2} + \frac{d^2 I_{L10}}{dt^2};$$
$$C_{A16} = C_{B16} = C;\ V_{A9} = 0;\ \frac{1}{LC} \cdot (I_{C_{A16}} - I_{C_{B16}}) = \frac{d^2 I_{L1}}{dt^2} - \frac{d^2 I_{L16}}{dt^2}$$
$$I_{R_P} = I_{C_{A16}} + I_{C_{A1}} + I_{L1};\ I_{C_{A1}} = I_{C_{A2}} + I_{L2};\ I_{C_{A2}} = I_{C_{A3}} + I_{L3};$$
$$I_{C_{A3}} = I_{C_{A4}} + I_{L4};\ I_{C_{A4}} = I_{C_{A5}} + I_{L5};\ I_{C_{A5}} = I_{C_{A6}} + I_{L6}$$

$$I_{C_{A6}} = I_{C_{A7}} + I_{L7};\ I_{C_{A7}} = I_{C_{A8}} + I_{L8};\ I_{C_{A10}} = I_{C_{A9}} + I_{L10};\ I_{C_{A11}} = I_{C_{A10}} + I_{L11};$$
$$I_{C_{A12}} = I_{C_{A11}} + I_{L12};\ I_{C_{A13}} = I_{C_{A12}} + I_{L13}$$

$$I_{C_{A14}} = I_{C_{A13}} + I_{L14};\ I_{C_{A15}} = I_{C_{A14}} + I_{L15};\ I_{C_{A16}} = I_{C_{A15}} + I_{L16};\ I_{L1} = I_{C_{B1}} + I_{C_{B16}};$$
$$I_{L9} = I_{C_{B8}} + I_{C_{B9}};\ I_{C_{B2}} = I_{C_{B1}} + I_{L2}$$

$$I_{C_{B3}} = I_{C_{B2}} + I_{L3};\ I_{C_{B4}} = I_{C_{B3}} + I_{L4};\ I_{C_{B5}} = I_{C_{B4}} + I_{L5};\ I_{C_{B6}} = I_{C_{B5}} + I_{L6};$$
$$I_{C_{B7}} = I_{C_{B6}} + I_{L7};\ I_{C_{B8}} = I_{C_{B7}} + I_{L8}$$

$I_{C_{B9}} = I_{C_{B10}} + I_{L10}$; $I_{C_{B10}} = I_{C_{B11}} + I_{L11}$; $I_{C_{B11}} = I_{C_{B12}} + I_{L12}$; $I_{C_{B12}} = I_{C_{B13}} + I_{L13}$;
$I_{C_{B13}} = I_{C_{B14}} + I_{L14}$; $I_{C_{B14}} = I_{C_{B15}} + I_{L15}$
$I_{C_{B15}} = I_{C_{B16}} + I_{L16}$. S1 is OFF for RF signal transmitter feeding.
$I_{R_P} = 0 \Rightarrow I_{C_{A16}} + I_{C_{A1}} + I_{L1} = 0$

$$I_{C_{A1}} = I_{C_{A8}} + \sum_{k=2}^{8} I_{Lk};\ I_{C_{A2}} = I_{C_{A8}} + \sum_{k=3}^{8} I_{Lk};\ I_{C_{A3}} = I_{C_{A8}} + \sum_{k=4}^{8} I_{Lk};$$

$$I_{C_{A4}} = I_{C_{A8}} + \sum_{k=5}^{8} I_{Lk};\ I_{C_{A5}} = I_{C_{A8}} + \sum_{k=6}^{8} I_{Lk};\ I_{C_{A6}} = I_{C_{A8}} + \sum_{k=7}^{8} I_{Lk}$$

$$I_{C_{A7}} = I_{C_{A8}} + I_{L8};\ I_{C_{A16}} = I_{C_{A9}} + \sum_{k=10}^{16} I_{Lk};\ I_{C_{A15}} = I_{C_{A9}} + \sum_{k=10}^{15} I_{Lk};$$

$$I_{C_{A14}} = I_{C_{A9}} + \sum_{k=10}^{14} I_{Lk};\ I_{C_{A13}} = I_{C_{A9}} + \sum_{k=10}^{13} I_{Lk};\ I_{C_{A12}} = I_{C_{A9}} + \sum_{k=10}^{12} I_{Lk}$$

$$I_{C_{A11}} = I_{C_{A9}} + \sum_{k=10}^{11} I_{Lk};\ I_{C_{A10}} = I_{C_{A9}} + I_{L_{10}};$$

$$I_{C_{B1}} = I_{L9} - I_{C_{B16}} - \sum_{k=2,k\neq 9}^{16} I_{Lk}; I_{C_{B2}} = I_{L9} - I_{C_{B16}} - \sum_{k=3,k\neq 9}^{16} I_{Lk}$$

$$I_{C_{B3}} = I_{L9} - I_{C_{B16}} - \sum_{k=4,k\neq 9}^{16} I_{Lk};\ I_{C_{B4}} = I_{L9} - I_{C_{B16}} - \sum_{k=5,k\neq 9}^{16} I_{Lk};$$

$$I_{C_{B5}} = I_{L9} - I_{C_{B16}} - \sum_{k=6,k\neq 9}^{16} I_{Lk};\ I_{C_{B6}} = I_{L9} - I_{C_{B16}} - \sum_{k=7,k\neq 9}^{16} I_{Lk}$$

$$I_{C_{B7}} = I_{L9} - I_{C_{B16}} - \sum_{k=8,k\neq 9}^{16} I_{Lk};\ I_{C_{B8}} = I_{L9} - I_{C_{B16}} - \sum_{k=10}^{16} I_{Lk};$$

$$I_{C_{B9}} = I_{C_{B16}} + \sum_{k=10}^{16} I_{Lk};\ I_{C_{B10}} = I_{C_{B16}} + \sum_{k=11}^{16} I_{Lk}$$

$$I_{C_{B11}} = I_{C_{B16}} + \sum_{k=12}^{16} I_{Lk};\ I_{C_{B12}} = I_{C_{B16}} + \sum_{k=13}^{16} I_{Lk};$$

$$I_{C_{B13}} = I_{C_{B16}} + \sum_{k=14}^{16} I_{Lk};\ I_{C_{B14}} = I_{C_{B16}} + \sum_{k=15}^{16} I_{Lk}$$

$I_{C_{B15}} = I_{C_{B16}} + I_{L16}$; $I_{L1} = I_{C_{B1}} + I_{C_{B16}}$. We get the following additional expressions:

$$\frac{2}{L \cdot C} \cdot (I_{L2} - I_{L4}) = \frac{d^2 I_{L1}}{dt^2} - \frac{d^2 I_{L5}}{dt^2} - 2 \cdot [\frac{d^2 I_{L2}}{dt^2} - \frac{d^2 I_{L4}}{dt^2}]; \frac{2}{L \cdot C} \cdot (I_{L6} - I_{L8})$$
$$= \frac{d^2 I_{L5}}{dt^2} + \frac{d^2 I_{L9}}{dt^2} - 2 \cdot [\frac{d^2 I_{L6}}{dt^2} - \frac{d^2 I_{L8}}{dt^2}]$$

$$\frac{-2}{L \cdot C} \cdot (I_{L10} - I_{L12}) = \frac{d^2 I_{L9}}{dt^2} + \frac{d^2 I_{L13}}{dt^2} + 2 \cdot [\frac{d^2 I_{L10}}{dt^2} - \frac{d^2 I_{L12}}{dt^2}]; \frac{-2}{L \cdot C} \cdot (I_{L14} - I_{L16})$$
$$= \frac{d^2 I_{L1}}{dt^2} - \frac{d^2 I_{L13}}{dt^2} + 2 \cdot [\frac{d^2 I_{L14}}{dt^2} - \frac{d^2 I_{L16}}{dt^2}]$$

We add the first and second above equations:

$$[\,*\,]\ \frac{2}{L \cdot C} \cdot \{(I_{L2} - I_{L4}) + (I_{L6} - I_{L8})\} = \frac{d^2 I_{L1}}{dt^2} + \frac{d^2 I_{L9}}{dt^2} - 2 \cdot [\frac{d^2 I_{L2}}{dt^2} - \frac{d^2 I_{L4}}{dt^2} + \frac{d^2 I_{L6}}{dt^2} - \frac{d^2 I_{L8}}{dt^2}]$$

We add the third and forth above equations:

$$[\,*\,*\,]\ \frac{-2}{L \cdot C} \cdot \{(I_{L10} - I_{L12}) + (I_{L14} - I_{L16})\}$$
$$= \frac{d^2 I_{L1}}{dt^2} + \frac{d^2 I_{L9}}{dt^2} + 2 \cdot [\frac{d^2 I_{L10}}{dt^2} - \frac{d^2 I_{L12}}{dt^2} + \frac{d^2 I_{L14}}{dt^2} - \frac{d^2 I_{L16}}{dt^2}]$$

Integrating the last two results ([**]–[*]) gives the following:

$$\frac{-2}{L \cdot C} \cdot \{I_{L10} - I_{L12} + I_{L14} - I_{L16} + I_{L2} - I_{L4} + I_{L6} - I_{L8}\}$$
$$= 2 \cdot [\frac{d^2 I_{L10}}{dt^2} - \frac{d^2 I_{L12}}{dt^2} + \frac{d^2 I_{L14}}{dt^2} - \frac{d^2 I_{L16}}{dt^2} + \frac{d^2 I_{L2}}{dt^2} - \frac{d^2 I_{L4}}{dt^2} + \frac{d^2 I_{L6}}{dt^2} - \frac{d^2 I_{L8}}{dt^2}]$$

We define new global variables for our Cylindrical RF network antennas system.

$$Y = I_{L10} - I_{L12} + I_{L14} - I_{L16} + I_{L2} - I_{L4} + I_{L6} - I_{L8}$$
$$X = \frac{dI_{L10}}{dt} - \frac{dI_{L12}}{dt} + \frac{dI_{L14}}{dt} - \frac{dI_{L16}}{dt} + \frac{dI_{L2}}{dt} - \frac{dI_{L4}}{dt} + \frac{dI_{L6}}{dt} - \frac{dI_{L8}}{dt};$$
$$\frac{dY}{dt} = X; \frac{dX}{dt} = \frac{-1}{L \cdot C} \cdot Y$$

Due to RF antenna copper leg parasitic effect, we get copper leg's current and current derivative with delay τ_{1-k} and τ_{2-k} (k is leg number index, k = 1,…,16). We consider for simplicity $\tau_{1-1} = \tau_{1-2} = \cdots = \tau_{1-16}$; $\tau_{2-1} = \tau_{2-2} = \cdots = \tau_{2-16}$. $I_{Lk}(t) \to I_{Lk}(t - \tau_{1-k})$; $I'_{Lk}(t) = \frac{dI_{Lk}(t)}{dt}$; $I'_{Lk}(t) \to I'_{Lk}(t - \tau_{2-k})$. We consider no delay effect on $\frac{dI'_{Lk}(t)}{dt}$. $Y(t) \to Y(t-\tau_1)$; $X(t) \to X(t-\tau_2)$. $\tau_1 = \tau_{1-1} = \tau_{1-2} = \cdots = \tau_{1-16}$ $\tau_2 = \tau_{2-1} = \tau_{2-2} = \cdots = \tau_{2-16}$. $\frac{dY}{dt} = X(t-\tau_2)$; $\frac{dX}{dt} = \frac{-1}{L \cdot C} \cdot Y(t-\tau_1)$.

To find the Equilibrium points (fixed points) of the Cylindrical RF network antennas system is by $\lim_{t\to\infty} Y(t-\tau_1) = Y(t)$ and $\lim_{t\to\infty} X(t-\tau_2) = X(t)$. $\frac{dY}{dt} = 0$; $\frac{dX}{dt} = 0$; $\forall$ t $\gg \tau_1$; t $\gg \tau_2$ $\exists (t-\tau_1) \approx t$; $(t-\tau_2) \approx t$, t $\to \infty$.

We get two equations and the only fixed point is $E^{(0)}(Y^{(0)}, X^{(0)}) = (0,0)$.

$$Y^{(0)} = I_{L10}^{(0)} - I_{L12}^{(0)} + I_{L14}^{(0)} - I_{L16}^{(0)} + I_{L2}^{(0)} - I_{L4}^{(0)} + I_{L6}^{(0)} - I_{L8}^{(0)} = 0$$
$$X^{(0)} = I_{L10}^{\prime(0)} - I_{L12}^{\prime(0)} + I_{L14}^{\prime(0)} - I_{L16}^{\prime(0)} + I_{L2}^{\prime(0)} - I_{L4}^{\prime(0)} + I_{L6}^{\prime(0)} - I_{L8}^{\prime(0)} = 0$$

Stability analysis: The standard local stability analysis about any one of the equilibrium points of Cylindrical RF network antennas system consists in adding to coordinates [Y X] arbitrarily small increments of exponential form $[y\,x] \cdot e^{\lambda \cdot t}$, and retaining the first order terms in y, x. The system of two homogeneous equations leads to a polynomial characteristics equation in the eigenvalues λ. The polynomial characteristics equations accept by set the below current and current derivative respect to time into two Cylindrical RF network antennas system equations. Cylindrical RF network antennas system fixed values with arbitrarily small increments of exponential form $[y\,x] \cdot e^{\lambda \cdot t}$ are: i = 0 (first fixed point), i = 1 (second fixed point), i = 2 (third fixed point).

$$Y(t) = Y^{(i)} + y \cdot e^{\lambda \cdot t};\ X = X^{(i)} + x \cdot e^{\lambda \cdot t};\ Y(t-\tau_1) = Y^{(i)} + y \cdot e^{\lambda \cdot (t-\tau_1)};\ X(t-\tau_2) = X^{(i)} + x \cdot e^{\lambda \cdot (t-\tau_2)}\ \forall i = 0,1,2$$

We choose the above expressions for our $Y(t)$, $X(t)$ as small displacement [y x] from the system fixed points at time t = 0. $Y(t=0) = Y^{(i)} + y$; $X(t=0) = X^{(i)} + x$ for $\lambda < 0$, t > 0 the selected fixed point is stable otherwise $\lambda > 0$, t > 0 is Unstable. Our Cylindrical RF network antennas system tend to the selected fixed point exponentially for $\lambda < 0$, t > 0 otherwise go away from the selected fixed point exponentially. λ is the eigenvalue parameter which establish if the fixed point is stable or unstable, additionally his absolute value ($|\lambda|$) establish the speed of flow toward or away from the selected fixed point [2–6] (Table 2.4).

Table 2.4 Cylindrical RF network antennas system eigenvalues options

	$\lambda < 0$	$\lambda > 0$
t = 0	$Y(t=0) = Y^{(i)} + y$ $X(t=0) = X^{(i)} + x$	$Y(t=0) = Y^{(i)} + y$ $X(t=0) = X^{(i)} + x$
t > 0	$Y(t) = Y^{(i)} + y \cdot e^{-\lvert\lambda\rvert \cdot t}$ $X(t) = X^{(i)} + x \cdot e^{-\lvert\lambda\rvert \cdot t}$	$Y(t) = Y^{(i)} + y \cdot e^{\lvert\lambda\rvert \cdot t}$ $X(t) = X^{(i)} + x \cdot e^{\lvert\lambda\rvert \cdot t}$
t $\to \infty$	$Y(t \to \infty) = Y^{(i)}$ $X(t \to \infty) = X^{(i)}$	$Y(t \to \infty,\ \lambda > 0) \sim y e^{\lvert\lambda\rvert \cdot t}$ $X(t \to \infty,\ \lambda > 0) \sim x e^{\lvert\lambda\rvert \cdot t}$

The speeds of flow toward or away from the selected fixed point for Cylindrical RF network antennas system currents and currents derivative respect to time are

$$\frac{dY(t)}{dt} = \lim_{\Delta t \to 0} \frac{Y(t+\Delta t) - Y(t)}{\Delta t} = \lim_{\Delta t \to 0} \frac{Y^{(i)} + y \cdot e^{\lambda \cdot (t+\Delta t)} - [Y^{(i)} + y \cdot e^{\lambda \cdot t}]}{\Delta t}$$
$$= \lim_{\Delta t \to 0} \frac{y \cdot e^{\lambda \cdot t} \cdot [e^{\lambda \cdot \Delta t} - 1]}{\Delta t} \overset{e^{\lambda \cdot \Delta t} \approx 1 + \lambda \cdot \Delta t}{\longrightarrow} \lim_{\Delta t \to 0} \frac{y \cdot e^{\lambda \cdot t} \cdot [1 + \lambda \cdot \Delta t - 1]}{\Delta t} = \lambda \cdot y \cdot e^{\lambda \cdot t}$$
$$\frac{dX(t)}{dt} = \lim_{\Delta t \to 0} \frac{X(t+\Delta t) - X(t)}{\Delta t} = \lim_{\Delta t \to 0} \frac{X^{(i)} + x \cdot e^{\lambda \cdot (t+\Delta t)} - [X^{(i)} + x \cdot e^{\lambda \cdot t}]}{\Delta t}$$
$$= \lim_{\Delta t \to 0} \frac{x \cdot e^{\lambda \cdot t} \cdot [e^{\lambda \cdot \Delta t} - 1]}{\Delta t} \overset{e^{\lambda \cdot \Delta t} \approx 1 + \lambda \cdot \Delta t}{\longrightarrow} \lim_{\Delta t \to 0} \frac{x \cdot e^{\lambda \cdot t} \cdot [1 + \lambda \cdot \Delta t - 1]}{\Delta t} = \lambda \cdot x \cdot e^{\lambda \cdot t}$$

and the time derivative of the above equations:

$$\frac{dY(t)}{dt} = y \cdot \lambda \cdot e^{\lambda \cdot t}; \ \frac{dX(t)}{dt} = x \cdot \lambda \cdot e^{\lambda \cdot t};$$
$$\frac{dY(t-\tau_1)}{dt} = y \cdot \lambda \cdot e^{\lambda \cdot (t-\tau_1)} = y \cdot \lambda \cdot e^{\lambda \cdot t} \cdot e^{-\tau_1 \cdot \lambda}$$
$$\frac{dX(t-\tau_2)}{dt} = x \cdot \lambda \cdot e^{\lambda \cdot (t-\tau_2)} = x \cdot \lambda \cdot e^{\lambda \cdot t} \cdot e^{-\tau_2 \cdot \lambda}$$

First we take the Cylindrical RF network antennas (Y) differential equation: $\frac{dY}{dt} = X$ and adding to it coordinates [Y X] arbitrarily small increments of exponential form $[y\,x] \cdot e^{\lambda \cdot t}$ and retaining the first order terms in y, x.

$\lambda \cdot y \cdot e^{\lambda \cdot t} = X^{(i)} + x \cdot e^{\lambda \cdot t}; \ X^{(i=0)} = 0; \ -\lambda \cdot y + x = 0$. Second we take the Cylindrical RF network antennas (X) differential equation: $\frac{dX}{dt} = \frac{-1}{L \cdot C} \cdot Y$ and adding to it coordinates [Y X] arbitrarily small increments of exponential form $[y\,x] \cdot e^{\lambda \cdot t}$ and retaining the first order terms in y, x. $\lambda \cdot x \cdot e^{\lambda \cdot t} = \frac{-1}{L \cdot C} \cdot [Y^{(i)} + y \cdot e^{\lambda \cdot t}]; \ Y^{(i=0)} = 0; \ -\lambda \cdot x = -\frac{1}{L \cdot C} \cdot y = 0$. We define $Y(t-\tau_1) = Y^{(i)} + y \cdot e^{\lambda \cdot (t-\tau_1)}; \ X(t-\tau_2) = X^{(i)} + x \cdot e^{\lambda \cdot (t-\tau_2)}$ then we get two delayed differential equations respect to adding to it coordinates [Y X] arbitrarily small increments of exponential form $[y\,x] \cdot e^{\lambda \cdot t}$. In the equilibrium points: $Y^{(0)} = 0; \ X^{(0)} = 0$

$$\lambda \cdot y \cdot e^{\lambda \cdot t} = X^{(0)} + x \cdot e^{\lambda \cdot (t-\tau_2)}; \ X^{(0)} = 0 \Rightarrow \lambda \cdot y$$
$$= x \cdot e^{-\lambda \cdot \tau_2}; \ \lambda \cdot x \cdot e^{\lambda \cdot t} = \frac{-1}{L \cdot C} \cdot [Y^{(0)} + y \cdot e^{\lambda \cdot (t-\tau_1)}]$$

$Y^{(0)} = 0 \Rightarrow \lambda \cdot x = \frac{-1}{L \cdot C} \cdot y \cdot e^{-\lambda \cdot \tau_1}$. We get the following set of eigenvalues equations: $-\lambda \cdot y + x \cdot e^{-\lambda \cdot \tau_2} = 0; \frac{-1}{L \cdot C} \cdot y \cdot e^{-\lambda \cdot \tau_1} - \lambda \cdot x = 0$

The small increments Jacobian of our Cylindrical RF network antennas.

$$\begin{pmatrix} -\lambda & e^{-\lambda\cdot\tau_2} \\ \frac{-1}{L\cdot C}\cdot e^{-\lambda\cdot\tau_1} & -\lambda \end{pmatrix}\cdot\begin{pmatrix} y \\ x \end{pmatrix} = \begin{pmatrix} 0 \\ 0 \end{pmatrix}$$

$$A-\lambda\cdot I = \begin{pmatrix} -\lambda & e^{-\lambda\cdot\tau_2} \\ \frac{-1}{L\cdot C}\cdot e^{-\lambda\cdot\tau_1} & -\lambda \end{pmatrix};\ \det|A-\lambda\cdot I| = 0;\ D(\lambda,\tau_1,\tau_2)$$
$$= \lambda^2 + \frac{1}{L\cdot C}\cdot e^{-\lambda\cdot\tau_1}\cdot e^{-\lambda\cdot\tau_2}$$

We have three stability analysis cases: $\tau_1 = \tau;\ \tau_2 = 0$ or $\tau_2 = \tau;\ \tau_1 = 0$ or $\tau_1 = \tau_2 = \tau$ otherwise $\tau_1 \neq \tau_2$. We need to get characteristics equations as all above stability analysis cases. We study the occurrence of any possible stability switching resulting from the increase of value of the time delay τ for the general characteristic equation $D(\lambda,\tau)$. $D(\lambda,\tau) = P_n(\lambda,\tau) + Q_m(\lambda,\tau)\cdot e^{-\lambda\tau}$. The expression for $P_n(\lambda,\tau)$ is $P_n(\lambda,\tau) = \sum_{k=0}^{n} P_k(\tau)\cdot\lambda^k = P_0(\tau) + P_1(\tau)\cdot\lambda + P_2(\tau)\cdot\lambda^2 + P_3(\tau)\cdot\lambda^3 + \ldots$

The expression for $Q_m(\lambda,\tau)$ is $Q_m(\lambda,\tau) = \sum_{k=0}^{m} q_k(\tau)\cdot\lambda^k = q_0(\tau) + q_1(\tau)\cdot\lambda + q_2(\tau)\cdot\lambda^2 + \ldots$

The first case we analyze is when there is delay in Cylindrical RF network antennas leg's current and no delay in antennas leg's current derivative or opposite $\tau_1 = \tau;\ \tau_2 = 0\ \&\ \tau_1 = 0;\ \tau_2 = \tau$ [4, 5].

$$D(\lambda,\tau_1=0,\tau_2) = \lambda^2 + \frac{1}{L\cdot C}\cdot e^{-\lambda\cdot\tau_2}_{\tau_2=\tau} = \lambda^2 + \frac{1}{L\cdot C}\cdot e^{-\lambda\cdot\tau};\ D(\lambda,\tau_1,\tau_2=0)$$
$$= \lambda^2 + \frac{1}{L\cdot C}\cdot e^{-\lambda\cdot\tau_1}\Big|_{\tau_1=\tau} = \lambda^2 + \frac{1}{L\cdot C}\cdot e^{-\lambda\cdot\tau}$$

$D(\lambda,\tau) = P_n(\lambda,\tau) + Q_m(\lambda,\tau)\cdot e^{-\lambda\tau}$. The expression for $P_n(\lambda,\tau)$ is

$$P_n(\lambda,\tau) = \sum_{k=0}^{n} P_k(\tau)\cdot\lambda^k = P_0(\tau) + P_1(\tau)\cdot\lambda + P_2(\tau)\cdot\lambda^2 = \lambda^2;\ P_2(\tau) = 1;$$
$$P_1(\tau) = 0;\ P_0(\tau) = 0$$

The expression for $Q_m(\lambda,\tau)$ is $Q_m(\lambda,\tau) = \sum_{k=0}^{m} q_k(\tau)\cdot\lambda^k = q_0(\tau) = \frac{1}{L\cdot C}$.

Our Cylindrical RF network antennas system second order characteristic equation: $D(\lambda,\tau) = \lambda^2 + a(\tau)\cdot\lambda + b(\tau)\cdot\lambda\cdot e^{-\lambda\cdot\tau} + c(\tau) + d(\tau)\cdot e^{-\lambda\cdot\tau}$.

Then $a(\tau) = 0$; $\mathrm{b}(\tau) = 0$; $\mathrm{c}(\tau) = 0$; $\mathrm{d}(\tau) = \frac{1}{L\cdot C}$; $\tau \in R_{+0}$ and $a(\tau), \mathrm{b}(\tau), \mathrm{c}(\tau), \mathrm{d}(\tau) : R_{+0} \to R$ are differentiable functions of class $C^1(R_{+0})$ such that $\mathrm{c}(\tau) + \mathrm{d}(\tau) = \frac{1}{L\cdot C} \neq 0$ for all $\tau \in R_{+0}$ and for any $\tau, \mathrm{b}(\tau), \mathrm{d}(\tau)$ are not simultaneously zero. We have

$$P(\lambda,\tau) = P_n(\lambda,\tau) = \lambda^2 + a(\tau)\cdot\lambda + c(\lambda) = \lambda^2;\ Q(\lambda,\tau) = Q_m(\lambda,\tau) = b(\tau)\cdot\lambda + d(\tau) = \frac{1}{L\cdot C}$$

We assume that $P_n(\lambda,\tau) = P_n(\lambda)$ and $Q_m(\lambda,\tau) = Q_m(\lambda)$ can't have common imaginary roots. That is for any real number ω; $p_n(\lambda = i\cdot\omega,\tau) + Q_m(\lambda = i\cdot\omega,\tau) \neq 0$ $-\omega^2 + \frac{1}{L\cdot C} \neq 0$; $F(\omega,\tau) = |P(i\cdot\omega,\tau)|^2 - |Q(i\cdot\omega,\tau)|^2 = (c-\omega^2)^2 + \omega^2\cdot a^2 - (\omega^2\cdot b^2 + d^2)$

$F(\omega,\tau) = \omega^4 - \frac{1}{(L\cdot C)^2}$; Hence $F(\omega,\tau) = 0$ implies $\omega^4 - \frac{1}{(L\cdot C)^2} = 0$ and its roots are given by $\omega_+^2 = \frac{1}{2}\cdot\{(b^2 + 2\cdot c - a^2) + \sqrt{\Delta}\} = \frac{\sqrt{\Delta}}{2}$; $\omega_-^2 = \frac{1}{2}\cdot\{(b^2 + 2\cdot c - a^2) - \sqrt{\Delta}\} = -\frac{\sqrt{\Delta}}{2}$

$\Delta = (b^2 + 2\cdot c - a^2) - 4\cdot(c^2 - d^2) = \frac{4}{L^2\cdot C^2}$. Therefore the following holds: $2\cdot\omega_\pm^2 - (b^2 + 2\cdot c - a^2) = \pm\sqrt{\Delta}$; $2\cdot\omega_\pm^2 = \pm\sqrt{\Delta}$. Furthermore

$$P_R(i\cdot\omega,\tau) = c(\tau) - \omega^2(\tau) = -\omega^2(\tau);\ P_I(i\cdot\omega,\tau) = \omega(\tau)\cdot a(\tau) = 0;\ Q_R(i\cdot\omega,\tau) = d(\tau) = \frac{1}{L\cdot C}$$

$Q_I(i\cdot\omega,\tau) = \omega(\tau)\cdot b(\tau) = 0$ hence $\sin\theta(\tau) = \frac{-P_R(i\cdot\omega,\tau)\cdot Q_I(i\cdot\omega,\tau) + P_I(i\cdot\omega,\tau)\cdot Q_R(i\cdot\omega,\tau)}{|Q(i\cdot\omega,\tau)|^2}$

$$\cos\theta(\tau) = -\frac{P_R(i\cdot\omega,\tau)\cdot Q_R(i\cdot\omega,\tau) + P_I(i\cdot\omega,\tau)\cdot Q_I(i\cdot\omega,\tau)}{|Q(i\cdot\omega,\tau)|^2};$$

$$\sin\theta(\tau) = \frac{-(c-\omega^2)\cdot\omega\cdot b + \omega\cdot a\cdot d}{\omega^2\cdot b^2 + d^2} = 0$$

$\cos\theta(\tau) = -\frac{(c-\omega^2)\cdot d + \omega^2\cdot a\cdot b}{\omega^2\cdot b^2 + d^2} = \omega^2\cdot L\cdot C$ which jointly with $\omega^4 - \frac{1}{(L\cdot C)^2} = 0$.

Defines the maps $S_n(\tau) = \tau - \tau_n(\tau)$; $\tau \in \mathrm{I}$, $n \in \mathbb{N}_0$ that are continuous and differentiable in τ based on Lemma 1.1. Hence we use Theorem 1.2. This prove the Theorem 1.3 and Theorem 1.4.

Remark: a, b, c, d parameters are independent of delay parameter τ even we use $a(\tau), b(\tau), c(\tau), d(\tau)$.

The second case we analyze is when there is delay both in Cylindrical RF network antennas leg's current and current time derivative $\tau_1 = \tau$; $\tau_2 = \tau$ [4, 5].

$$D(\lambda,\tau_1 = \tau,\tau_2 = \tau) = \lambda^2 + \frac{1}{L\cdot C}\cdot e^{-\lambda\cdot\tau}\cdot e^{-\lambda\cdot\tau};\ D(\lambda,\tau) = P_n(\lambda,\tau) + Q_m(\lambda,\tau)\cdot e^{-\lambda\tau}$$

The expression for $P_n(\lambda,\tau)$ is $P_n(\lambda,\tau) = \sum_{k=0}^{n} P_k(\tau)\cdot\lambda^k = P_0(\tau) + P_1(\tau)\cdot\lambda + P_2(\tau)\cdot\lambda^2 = \lambda^2$

$P_2(\tau) = 1$; $P_1(\tau) = 0$; $P_0(\tau) = 0$. The expression for $Q_m(\lambda, \tau)$; $Q_m(\lambda, \tau) = \sum_{k=0}^{m} q_k(\tau) \cdot \lambda^k = \frac{1}{L \cdot C} \cdot e^{-\lambda \cdot \tau}$.

Taylor expansion: $e^{-\lambda\tau} \approx 1 - \lambda \cdot \tau + \frac{\lambda^2 \cdot \tau^2}{2}$ since we need n > m [BK] analysis we choose $e^{-\lambda\tau} \approx 1 - \lambda \cdot \tau$ then we get $Q_m(\lambda, \tau) = \sum_{k=0}^{m} q_k(\tau) \cdot \lambda^k = \frac{1}{L \cdot C} \cdot (1 - \lambda \cdot \tau) = \frac{1}{L \cdot C} - \frac{1}{L \cdot C} \cdot \lambda \cdot \tau$.

$q_0(\tau, \lambda) = \frac{1}{L \cdot C}$; $q_1(\tau) = -\frac{1}{L \cdot C} \cdot \tau$; $q_2(\tau) = 0$. Our Cylindrical RF network antennas system second order characteristic equation: $D(\lambda, \tau) = \lambda^2 + a(\tau) \cdot \lambda + b(\tau) \cdot \lambda \cdot e^{-\lambda \cdot \tau} + c(\tau) + d(\tau) \cdot e^{-\lambda \cdot \tau}$ then $a(\tau) = 0$; $b(\tau) = \frac{-1}{L \cdot C} \cdot \tau$ $c(\tau) = 0$; $d(\tau) = \frac{1}{L \cdot C}$ and in the same manner like our previous case analysis: $P(\lambda, \tau) = P_n(\lambda, \tau) = \lambda^2$; $Q(\lambda, \tau) = Q_m(\lambda, \tau) = \frac{1}{L \cdot C} - \frac{1}{L \cdot C} \cdot \lambda \cdot \tau$. We assume that $P_n(\lambda, \tau) = P_n(\lambda)$ and $Q_m(\lambda, \tau)$ can't have common imaginary roots. That is for any real number ω; $p_n(\lambda = i \cdot \omega, \tau) + Q_m(\lambda = i \cdot \omega, \tau) \neq 0$; $-\omega^2 - i \cdot \omega \cdot \frac{1}{L \cdot C} \cdot \tau + \frac{1}{L \cdot C} \neq 0$

$$F(\omega, \tau) = |P(i \cdot \omega, \tau)|^2 - |Q(i \cdot \omega, \tau)|^2;\ P(i \cdot \omega, \tau) = -\omega^2;$$
$$P_R(i \cdot \omega, \tau) = -\omega^2; P_I(i \cdot \omega, \tau) = 0$$
$$Q(\lambda = i \cdot \omega, \tau) = -i \cdot \omega \cdot \frac{1}{L \cdot C} \cdot \tau + \frac{1}{L \cdot C};\ Q_I(\lambda = i \cdot \omega, \tau) = -\omega \cdot \frac{1}{L \cdot C} \cdot \tau;$$
$$Q_R(\lambda = i \cdot \omega, \tau) = \frac{1}{L \cdot C}$$
$$|P(i \cdot \omega, \tau)|^2 = P_I^2 + P_R^2;\ |Q(i \cdot \omega, \tau)|^2 = Q_I^2 + Q_R^2;\ |P(i \cdot \omega, \tau)|^2 = P_I^2 + P_R^2 = \omega^4$$
$$|Q(i \cdot \omega, \tau)|^2 = \omega^2 \cdot \frac{\tau^2}{(L \cdot C)^2} + \frac{1}{(L \cdot C)^2};\ F(\omega, \tau) = \omega^4 - \omega^2 \cdot \frac{\tau^2}{(L \cdot C)^2} - \frac{1}{(L \cdot C)^2}$$

Hence $F(\omega, \tau) = 0$ implies $\omega^4 - \omega^2 \cdot \frac{\tau^2}{(L \cdot C)^2} - \frac{1}{(L \cdot C)^2} = 0$; $F_\omega = 4 \cdot \omega^3 - 2 \cdot \omega \cdot \frac{\tau^2}{(L \cdot C)^2} = 2 \cdot \omega \cdot [2 \cdot \omega^2 - \frac{\tau^2}{(L \cdot C)^2}]$

$$F_\tau = \frac{-\omega^2 \cdot 2 \cdot \tau}{(L \cdot C)^2};\ P_{I\omega} = 0;\ P_{R\omega} = -2 \cdot \omega;\ Q_{I\omega} = -\frac{\tau}{L \cdot C};\ Q_{R\omega} = 0;$$
$$P_{I\tau} = 0;\ P_{R\tau} = 0;\ Q_{R\tau} = 0;\ Q_{I\tau} = -\frac{\omega}{L \cdot C}$$

The expressions for U, V can be derive easily [BK]: $x = \tau$

$$U = (P_R \cdot P_{I\omega} - P_I \cdot P_{R\omega}) - (Q_R \cdot Q_{I\omega} - Q_I \cdot Q_{R\omega});$$
$$V = (P_R \cdot P_{Ix} - P_I \cdot P_{Rx}) - (Q_R \cdot Q_{Ix} - Q_I \cdot Q_{Rx})$$

$V = \frac{\omega}{L^2 \cdot C^2}$; $U = \frac{\tau}{L^2 \cdot C^2}$; $\omega_\tau = -\frac{F_\tau}{F_\omega}$ and we get the expression:

$\omega_\tau = -\frac{\frac{-\omega^2 \cdot 2 \cdot \tau}{(L \cdot C)^2}}{2 \cdot \omega \cdot [2 \cdot \omega^2 - \frac{\tau^2}{(L \cdot C)^2}]} = \frac{\frac{-\omega \cdot \tau}{(L \cdot C)^2}}{[2 \cdot \omega^2 - \frac{\tau^2}{(L \cdot C)^2}]}$. Defines the maps $S_n(\tau) = \tau - \tau_n(\tau)$; $\tau \in$ I , n $\in \mathbb{N}_0$.

Defines the maps $S_n(\tau) = \tau - \tau_n(\tau)$; $\tau \in$ I , n $\in \mathbb{N}_0$ that are continuous and differentiable in τ based on Lemma 1.1. Hence we use Theorem 1.2. This prove the Theorem 1.3 and Theorem 1.4.

Remark: Taylor approximation for $e^{-\lambda\tau} \approx 1 - \lambda \cdot \tau$ gives us good stability analysis approximation only for restricted delay time interval.

Our Cylindrical RF network antennas homogeneous system for y, x leads to a characteristic equation for the eigenvalue λ having the for $P(\lambda) + Q(\lambda) \cdot e^{-\lambda \cdot \tau} = 0$; second case $\tau_1 = \tau$; $\tau_2 = \tau$; $D(\lambda, \tau_1 = \tau, \tau_2 = \tau) = \lambda^2 + \frac{1}{L \cdot C} \cdot e^{-\lambda \cdot \tau} \cdot e^{-\lambda \cdot \tau}$. We estimate $e^{-\lambda\tau} \approx 1 - \lambda \cdot \tau$. $D(\lambda, \tau_1 = \tau, \tau_2 = \tau) = \lambda^2 + \frac{1}{L \cdot C} \cdot (1 - \lambda \cdot \tau) \cdot e^{-\lambda \cdot \tau}$ $D(\lambda, \tau_1 = \tau, \tau_2 = \tau) = \lambda^2 + (-\lambda \cdot \frac{1}{L \cdot C} \cdot \tau + \frac{1}{L \cdot C}) \cdot e^{-\lambda \cdot \tau}$. We use different parameters terminology from our last characteristics parameters definition: $k \to j$; $p_k(\tau) \to a_j$; $q_k(\tau) \to c_j$; $n = 2$; $m = 1$. Additionally $P_n(\lambda, \tau) \to P(\lambda)$; $Q_m(\lambda, \tau) \to Q(\lambda)$ then $P(\lambda) = \sum_{j=0}^{2} a_j \cdot \lambda^j$ and $Q(\lambda) = \sum_{j=0}^{1} c_j \cdot \lambda^j$; $P(\lambda) = \lambda^2$; $Q(\lambda, \tau) = -\lambda \cdot \frac{1}{L \cdot C} \cdot \tau + \frac{1}{L \cdot C}$.

$n, m \in \mathbb{N}_0, n > m$ and $a_j, c_j \neq R_{+0} \to R$ are continuous and differentiable function of τ such that $a_0 + c_0 \neq 0$. In the following "–" denotes complex and conjugate. $P(\lambda), Q(\lambda)$ are analytic functions in λ and differentiable in τ. And the coefficients : $\{a_j(C, L), c_j(C, L, \tau)\} \in \mathbb{R}$ depend on Cylindrical RF network antennas C, L, τ values. $a_0 = 0, a_1 = 0, a_2 = 1$; $c_0 = \frac{1}{L \cdot C}, c_1 = -\frac{1}{L \cdot C} \cdot \tau$ unless strictly necessary, the designation of the variation arguments (C, L, τ) will subsequently be omitted from P, Q, a_j, c_j. The coefficients a_j, c_j are continuous, and differentiable functions of their arguments, and direct substitution shows that $a_0 + c_0 = \frac{1}{L \cdot C} \neq 0; \frac{1}{L \cdot C} \neq 0 \, \forall \, C, L, \tau \in \mathbb{R}_+$ i.e. $\lambda = 0$ is not a root of characteristic equation. Furthermore $P(\lambda), Q(\lambda)$ are analytic function of λ for which the following requirements of the analysis (see kuang [5], Sect. 3.4) can also be verified in the present case [4, 5].

(a) If $\lambda = i \cdot \omega, \omega \in \mathbb{R}$ then $P(i \cdot \omega) + Q(i \cdot \omega) \neq 0$, i.e. P and Q have no common imaginary roots. This condition was verified numerically in the entire (C, L, τ) domain of interest.

(b) $|Q(\lambda)/P(\lambda)|$ is bounded for $|\lambda| \to \infty$, Re$\lambda \geq 0$. No roots bifurcation from ∞. Indeed, in the limit $|Q(\lambda)/P(\lambda)| = |\frac{-\lambda \cdot \frac{1}{L \cdot C} \cdot \tau + \frac{1}{L \cdot C}}{\lambda^2}|$

(c) $F(\omega) = |P(i \cdot \omega)|^2 - |Q(i \cdot \omega)|^2$; $F(\omega, \tau) = \omega^4 - \omega^2 \cdot \frac{\tau^2}{(L \cdot C)^2} - \frac{1}{(L \cdot C)^2}$ has at most a finite number of zeros. Indeed, this is a bi-cubic polynomial in ω (second degree in ω^2).

(d) Each positive root $\omega(C, L, \tau)$ of F(ω)= 0 is continuous and differentiable with respect to C, L, τ. This condition can only be assessed numerically.

In addition, since the coefficients in P and Q are real, we have $\overline{P(-i \cdot \omega)} = P(i \cdot \omega)$ and $\overline{Q(-i \cdot \omega)} = Q(i \cdot \omega)$ thus $\lambda = i \cdot \omega$, $\omega > 0$ may be on eigenvalue of characteristic equation. The analysis consists in identifying the roots of characteristic equation situated on the imaginary axis of the complex λ—plane, where by increasing the parameters C, L and delay τ, Reλ may, at the crossing, change its sign from (−) to (+), i.e. from stable focus $E^{(0)}(Y^{(0)}, X^{(0)}) = (0,0)$ to an unstable one, or vice versa. This feature may be further assessed by examining the sign of the partial derivatives with respect to C, L and antenna parameters. $\wedge^{-1}(C) = (\frac{\partial \mathrm{Re}\,\lambda}{\partial C})_{\lambda = i \cdot \omega}$, $L, \tau = const$; $\wedge^{-1}(L) = (\frac{\partial \mathrm{Re}\,\lambda}{\partial L})_{\lambda = i \cdot \omega}$, $C, \tau = const$

$\wedge^{-1}(\tau) = (\frac{\partial \mathrm{Re}\lambda}{\partial \tau})_{\lambda = i \cdot \omega}$, C, L, where $\omega \in \mathbb{R}_+$. For the first case $\tau_1 = \tau$; $\tau_2 = \tau$ we get the following results $P_R(i \cdot \omega) = -\omega^2$; $P_I(i \cdot \omega) = 0$; $Q_R(i \cdot \omega) = \frac{1}{L \cdot C}$; $Q_I(i \cdot \omega) = \frac{-\omega \cdot \tau}{L \cdot C}$.

$F(\omega) = 0$ yield to $\omega^4 - \omega^2 \cdot \frac{\tau^2}{(L \cdot C)^2} - \frac{1}{(L \cdot C)^2} = 0$; $\chi^2 = \omega^4$; $\chi = \omega^2$; $\chi^2 - \chi \cdot \frac{\tau^2}{(L \cdot C)^2} - \frac{1}{(L \cdot C)^2} = 0$

$$\chi = \frac{\tau^2}{2 \cdot (L \cdot C)^2} \pm \frac{1}{2} \cdot \sqrt{\frac{\tau^4}{(L \cdot C)^4} + 4 \cdot \frac{1}{(L \cdot C)^2}}; \; \chi = \omega^2 \Rightarrow \omega = \pm \sqrt{\frac{\tau^2}{2 \cdot (L \cdot C)^2} \pm \frac{1}{2} \cdot \sqrt{\frac{\tau^4}{(L \cdot C)^4} + 4 \cdot \frac{1}{(L \cdot C)^2}}}$$

$\frac{\tau^4}{(L \cdot C)^4} + 4 \cdot \frac{1}{(L \cdot C)^2} > 0$ always and additional for $\omega \in R$; $\omega^2 = \frac{\tau^2}{2 \cdot (L \cdot C)} \pm \frac{1}{2} \sqrt{\frac{\tau^4}{(L \cdot C)} + 4 \cdot \frac{1}{(L \cdot C)}}$ and there are two options: first always exist $\frac{\tau^2}{2 \cdot (L \cdot C)^2} + \frac{1}{2} \cdot \sqrt{\frac{\tau^4}{(L \cdot C)^4} + 4 \cdot \frac{1}{(L \cdot C)^2}} > 0$.

Second $\frac{\tau^2}{2 \cdot (L \cdot C)^2} - \frac{1}{2} \cdot \sqrt{\frac{\tau^4}{(L \cdot C)^4} + 4 \cdot \frac{1}{(L \cdot C)^2}} < 0$; $\omega^2 = \frac{1}{2} \cdot \frac{1}{L \cdot C} \cdot \{\frac{\tau^2}{L \cdot C} \pm \sqrt{\frac{\tau^4}{(L \cdot C)^2} + 4}\}$.

$\sqrt{\frac{\tau^4}{(L \cdot C)^2} + 4} > \frac{\tau^2}{L \cdot C}$, Not exist and always negative for any Cylindrical RF network antennas overall parameters values. We choose only the (+) option (first). Writing $P(\lambda) = P_R(\lambda) + i \cdot P_I(\lambda)$ and $Q(\lambda) = Q_R(\lambda) + i \cdot Q_I(\lambda)$, and inserting $\lambda = i \cdot \omega$ into Cylindrical RF network antennas characteristic equation, ω must satisfy the following: $\sin \omega \cdot \tau = g(\omega) = \frac{-P_R(i \cdot \omega) \cdot Q_I(i \cdot \omega) + P_I(i \cdot \omega) \cdot Q_R(i \cdot \omega)}{|Q(i \cdot \omega)|^2}$.

$\cos \omega \cdot \tau = h(\omega) = -\frac{P_R(i \cdot \omega) \cdot Q_R(i \cdot \omega) + P_I(i \cdot \omega) \cdot Q_I(i \cdot \omega)}{|Q(i \cdot \omega)|^2}$, where $|Q(i \cdot \omega)|^2 \neq 0$ in view of requirement (a) above, and $(g, h) \in R$. Furthermore, it follows above $\sin \omega \cdot \tau$ and $\cos \omega \cdot \tau$ equations that, by squaring and adding the sides, ω must be a positive root of $F(\omega) = |P(i \cdot \omega)|^2 - |Q(i \cdot \omega)|^2 = 0$.

Note that $F(\omega)$ is dependent of τ. Now it is important to notice that if $\tau \notin I$ (assume that $I \subseteq R_{+0}$ is the set where $\omega(\tau)$ is a positive root of $F(\omega)$ and for $\tau \notin I$, $\omega(\tau)$ is not define. Then for all τ in I $\omega(\tau)$ is satisfies that $F(\omega, \tau) = 0$.

Then there are positive $\omega(\tau)$ solutions of $F(\omega,\tau) = 0$, and we analyze stability switches. For any $\tau \in I$ where $\omega(\tau)$ is a positive solution of $F(\omega,\tau) = 0$, we can define the angle $\theta(\tau) \in [0, 2\cdot\pi]$ as the solution of $\sin\theta(\tau) = \ldots$; $\cos\theta(\tau) = \ldots$

$$\sin\theta(\tau) = \frac{-P_R(i\cdot\omega)\cdot Q_I(i\cdot\omega) + P_I(i\cdot\omega)\cdot Q_R(i\cdot\omega)}{|Q(i\cdot\omega)|^2}$$

$$\cos\theta(\tau) = -\frac{P_R(i\cdot\omega)\cdot Q_R(i\cdot\omega) + P_I(i\cdot\omega)\cdot Q_I(i\cdot\omega)}{|Q(i\cdot\omega)|^2}$$

and the relation between the argument $\theta(\tau)$ and $\omega(\tau)\cdot\tau$ for $\tau \in I$ must be $\omega(\tau)\cdot\tau = \theta(\tau) + n\cdot 2\cdot\pi\,\forall n \in \mathbb{N}_0$. Hence we can define the maps $\tau_n : I \to R_{+0}$ given by $\tau_n(\tau) = \frac{\theta(\tau)+n\cdot 2\cdot\pi}{\omega(\tau)}$; $n \in \mathbb{N}_0, \tau \in I$. Let us introduce the functions $I \to R$; $S_n(\tau) = \tau - \tau_n(\tau)$, $\tau \in I$, $n \in \mathbb{N}_0$ that are continuous and differentiable in τ. In the following, the subscripts λ, ω, C, L and Cylindrical RF network antennas parameters $(L, C, \tau\, etc.,)$ indicate the corresponding partial derivatives. Let us first concentrate on $\wedge(x)$, remember in $\lambda(L, C, \tau,\, etc.,)$ and $\omega(L, C, \tau,\, etc.,)$, and keeping all parameters except one (x) and τ. The derivation closely follows that in reference [BK]. Differentiating Cylindrical RF network antennas characteristic equation $P(\lambda) + Q(\lambda)\cdot e^{-\lambda\cdot\tau} = 0$ with respect to specific parameter (x), and inverting the derivative, for convenience, one calculates.

<u>Remark</u>: $x = L, C, \tau,\, etc.,$

$$\left(\frac{\partial\lambda}{\partial x}\right)^{-1} = \frac{-P_\lambda(\lambda,x)\cdot Q(\lambda,x) + Q_\lambda(\lambda,x)\cdot P(\lambda,x) - \tau\cdot P(\lambda,x)\cdot Q(\lambda,x)}{P_x(\lambda,x)\cdot Q(\lambda,x) - Q_x(\lambda,x)\cdot P(\lambda,x)}$$

where $P_\lambda = \frac{\partial P}{\partial\lambda}, \ldots$ etc., Substituting $\lambda = i\cdot\omega$, and bearing i $\overline{P(-i\cdot\omega)} = P(i\cdot\omega)$; $\overline{Q(-i\cdot\omega)} = Q(i\cdot\omega)$,

then $i\cdot P_\lambda(i\cdot\omega) = P_\omega(i\cdot\omega)$ and $i\cdot Q_\lambda(i\cdot\omega) = Q_\omega(i\cdot\omega)$ and that on the surface $|P(i\cdot\omega)|^2 = |Q(i\cdot\omega)|^2$, one obtains

$$\left(\frac{\partial\lambda}{\partial x}\right)^{-1}\Big|_{\lambda=i\cdot\omega} = \left(\frac{i\cdot P_\omega(i\cdot\omega,x)\cdot\overline{P(i\cdot\omega,x)} + i\cdot Q_\lambda(i\cdot\omega,x)\cdot\overline{Q(\lambda,x)} - \tau\cdot|P(i\cdot\omega,x)|^2}{P_x(i\cdot\omega,x)\cdot\overline{P(i\cdot\omega,x)} - Q_x(i\cdot\omega,x)\cdot\overline{Q(i\cdot\omega,x)}}\right).$$

Upon separating into real and imaginary parts, with $P = P_R + i\cdot P_I$; $Q = Q_R + i\cdot Q_I$ $P_\omega = P_{R\omega} + i\cdot P_{I\omega}$; $Q_\omega = Q_{R\omega} + i\cdot Q_{I\omega}$; $P_x = P_{Rx} + i\cdot P_{Ix}$; $Q_x = Q_{Rx} + i\cdot Q_{Ix}$; $P^2 = P_R^2 + P_I^2$.

When (x) can be any Cylindrical RF network antennas parameters L, C, And time delay τ etc. Where for convenience, we have dropped the arguments $(i\cdot\omega, x)$, and where $F_\omega = 2\cdot[(P_{R\omega}\cdot P_R + P_{I\omega}\cdot P_I) - (Q_{R\omega}\cdot Q_R + Q_{I\omega}\cdot Q_I)]$; $\omega_x = -F_x/F_\omega$.

$F_x = 2 \cdot [(P_{Rx} \cdot P_R + P_{Ix} \cdot P_I) - (Q_{Rx} \cdot Q_R + Q_{Ix} \cdot Q_I)]$. We define U and V:

$$U = (P_R \cdot P_{I\omega} - P_I \cdot P_{R\omega}) - (Q_R \cdot Q_{I\omega} - Q_I \cdot Q_{R\omega});$$
$$V = (P_R \cdot P_{Ix} - P_I \cdot P_{Rx}) - (Q_R \cdot Q_{Ix} - Q_I \cdot Q_{Rx})$$

We choose our specific parameter as time delay x = τ. $V = \frac{\omega}{L^2 \cdot C^2}; U = \frac{\tau}{L^2 \cdot C^2}$

$$P^2 = \omega^4 ; F_\tau = \frac{-\omega^2 \cdot 2 \cdot \tau}{(L \cdot C)^2}; \; P_R(\omega,\tau) = -\omega^2; \; P_I(\omega,\tau) = 0;$$
$$Q_I(\omega,\tau) = -\frac{\omega \cdot \tau}{L \cdot C}; \; Q_R(\omega,\tau) = \frac{1}{L \cdot C}$$
$$P_{I\tau} = 0; \; P_{R\tau} = 0; \; Q_{R\tau} = 0; \; Q_{I\tau} = -\frac{\omega}{L \cdot C} \Rightarrow V \neq 0;$$
$$\frac{\partial F}{\partial \omega} = F_\omega = 4 \cdot \omega^3 - 2 \cdot \omega \cdot \frac{\tau^2}{(L \cdot C)^2}$$

$\frac{\partial F}{\partial \omega} = 2 \cdot \omega \cdot [2 \cdot \omega^2 - \frac{\tau^2}{(L \cdot C)^2}]$; $F(\omega,\tau) = 0$ and differentiating with respect to τ and we get $F_\omega \cdot \frac{\partial \omega}{\partial \tau} + F_\tau = 0$; $\tau \in I \Rightarrow \omega_\tau = \frac{\partial \omega}{\partial \tau} = -\frac{F_\tau}{F_\omega}; \frac{\partial \omega}{\partial \tau} = \frac{\frac{\omega \cdot \tau}{(L \cdot C)^2}}{[2 \cdot \omega^2 - \frac{\tau^2}{(L \cdot C)^2}]}$

$$\wedge^{-1}(\tau) = (\frac{\partial \mathrm{Re}\lambda}{\partial \tau})_{\lambda = i \cdot \omega}; \; \omega_\tau = \frac{\partial \omega}{\partial \tau} = \frac{\omega \cdot \tau}{[2 \cdot \omega^2 \cdot (L \cdot C)^2 - \tau^2]}$$
$$\wedge^{-1}(\tau) = \mathrm{Re}\{\frac{-2 \cdot [U + \tau \cdot |P|^2] + i \cdot F_\omega}{F_\tau + i \cdot 2 \cdot [V + \omega \cdot |P|^2]}\}$$
$$= \mathrm{Re}\{\frac{-\tau \cdot [\frac{1}{L^2 \cdot C^2} + \omega^4] + i \cdot \omega \cdot [2 \cdot \omega^2 - \frac{\tau^2}{(L \cdot C)^2}]}{\frac{-\omega^2 \cdot \tau}{(L \cdot C)^2} + i \cdot \omega \cdot [\frac{1}{L^2 \cdot C^2} + \omega^4]}\}$$
$$sign\{\wedge^{-1}(\tau)\} = sign\{(\frac{\partial \mathrm{Re}\lambda}{\partial \tau})_{\lambda = i \cdot \omega}\};$$
$$sign\{\wedge^{-1}(\tau)\} = sign\{F_\omega\} \cdot sign\{\tau \cdot \frac{\partial \omega}{\partial \tau} + \omega + \frac{U \cdot \frac{\partial \omega}{\partial \tau} + V}{|P|^2}\}$$

$$sign\{\wedge^{-1}(\tau)\} = sign\{2 \cdot \omega \cdot [2 \cdot \omega^2 - \frac{\tau^2}{(L \cdot C)^2}]\}$$
$$\cdot sign\{\tau \cdot [\frac{\frac{\omega \cdot \tau}{(L \cdot C)^2}}{[2 \cdot \omega^2 - \frac{\tau^2}{(L \cdot C)^2}]}] + \omega + \frac{\frac{\tau}{L^2 \cdot C^2} \cdot [\frac{\frac{\omega \cdot \tau}{(L \cdot C)^2}}{[2 \cdot \omega^2 - \frac{\tau^2}{(L \cdot C)^2}]}] + \frac{\omega}{L^2 \cdot C^2}}{\omega^4}\}$$

We define new variables: ψ_1, ψ_2, ψ_3: $\psi_1(\omega, \tau, L, C) = 2 \cdot \omega \cdot [2 \cdot \omega^2 - \frac{\tau^2}{(L \cdot C)^2}]$

$$\psi_2(\omega, \tau, L, C) = \tau \cdot [\frac{\frac{\omega \cdot \tau}{(L \cdot C)^2}}{[2 \cdot \omega^2 - \frac{\tau^2}{(L \cdot C)^2}]}]; \; \psi_3(\omega, \tau, L, C) = \frac{\frac{\tau}{L^2 \cdot C^2} \cdot [\frac{\frac{\omega \cdot \tau}{(L \cdot C)^2}}{[2 \cdot \omega^2 - \frac{\tau^2}{(L \cdot C)^2}]}] + \frac{\omega}{L^2 \cdot C^2}}{\omega^4}$$

$sign\{\wedge^{-1}(\tau)\} = sign[\psi_1] \cdot sign[\psi_2 + \omega + \psi_3]$. We check the sign of $\wedge^{-1}(\tau)$ according the following rule.

If sign$[\Lambda^{-1}(\tau)] > 0$ then the crossing proceeds from (−) to (+) respectively (stable to unstable). If sign$[\Lambda^{-1}(\tau)] < 0$ then the crossing proceeds from (+) to (−) respectively (unstable to stable). Anyway the stability switching can occur only for specific ω, τ. Since it is a very complex function, we recommend to solve it numerically rather than analytic. We plot the stability switch diagram based on different delay values of our Cylindrical RF network antennas system. $D(\lambda, \tau_1 = \tau_2 = \tau) = \lambda^2 + \frac{1}{L \cdot C} \cdot e^{-\lambda \cdot \tau} - \lambda \cdot \frac{\tau}{L \cdot C} \cdot e^{-\lambda \cdot \tau}$. Taylor expansion: $e^{-\lambda \tau} \approx 1 - \lambda \cdot \tau + \frac{\lambda^2 \cdot \tau^2}{2}$ since we need n > m [BK] analysis we choose $e^{-\lambda \tau} \approx 1 - \lambda \cdot \tau$ then we get our Cylindrical RF network antennas system second order characteristic equation: $D(\lambda, \tau) = \lambda^2 + a(\tau) \cdot \lambda + b(\tau) \cdot \lambda \cdot e^{-\lambda \cdot \tau} + c(\tau) + d(\tau) \cdot e^{-\lambda \cdot \tau}$ (Table 2.5).

$$a(\tau) = 0; \; b(\tau) = -\frac{\tau}{L \cdot C}; \; c(\tau) = 0; \; d(\tau) = \frac{1}{L \cdot C}; \; F(\omega, \tau)$$
$$= |P(i \cdot \omega, \tau)|^2 - |Q(i \cdot \omega, \tau)|^2 = (c - \omega^2)^2 + \omega^2 \cdot a^2 - (\omega^2 \cdot b^2 + d^2)$$

$F(\omega, \tau) = \omega^4 - \omega^2 \cdot \frac{\tau^2}{(L \cdot C)^2} - \frac{1}{(L \cdot C)^2}$ hence $F(\omega, \tau) = 0$ implies $\omega^4 - \omega^2 \cdot \frac{\tau^2}{(L \cdot C)^2} - \frac{1}{(L \cdot C)^2} = 0$ and its roots are given by $\omega_+^2 = \frac{1}{2} \cdot \{(b^2 + 2 \cdot c - a^2) + \sqrt{\Delta}\} = \frac{1}{2} \cdot \{\sqrt{\Delta} + \frac{\tau^2}{(L \cdot C)^2}\}$

$$\omega_-^2 = \frac{1}{2} \cdot \{(b^2 + 2 \cdot c - a^2) - \sqrt{\Delta}\} = \frac{1}{2} \cdot \{-\sqrt{\Delta} + \frac{\tau^2}{(L \cdot C)^2}\};$$
$$\Delta = (b^2 + 2 \cdot c - a^2) - 4 \cdot (c^2 - d^2) = \frac{\tau^2 + 4}{(L \cdot C)^2}$$

$\Delta = (b^2 + 2 \cdot c - a^2) - 4 \cdot (c^2 - d^2) = \frac{\tau^2 + 4}{(L \cdot C)^2}$ therefore the following holds:

$$2 \cdot \omega_\pm^2 - (b^2 + 2 \cdot c - a^2) = \pm\sqrt{\Delta}; \; \sin\theta(\tau) = \frac{-P_R(i \cdot \omega, \tau) \cdot Q_I(i \cdot \omega, \tau) + P_I(i \cdot \omega, \tau) \cdot Q_R(i \cdot \omega, \tau)}{|Q(i \cdot \omega, \tau)|^2}$$

Table 2.5 Cylindrical RF network antennas system stability switching criteria

$sign[F_\omega]$	$sign[\frac{V + \omega_\tau \cdot U}{P^2} + \omega + \omega_\tau \cdot \tau]$	$sign[\wedge^{-1}(\tau)]$
±	±	+
±	∓	−

$$\cos\theta(\tau) = -\frac{P_R(i\cdot\omega,\tau)\cdot Q_R(i\cdot\omega,\tau) + P_I(i\cdot\omega,\tau)\cdot Q_I(i\cdot\omega,\tau)}{|Q(i\cdot\omega,\tau)|^2};$$

$$\sin\theta(\tau) = \frac{-(c-\omega^2)\cdot\omega\cdot b + \omega\cdot a\cdot d}{\omega^2\cdot b^2 + d^2} = \frac{-\omega^3\cdot\tau\cdot L\cdot C}{(\omega^2\cdot\tau^2+1)}$$

$\cos\theta(\tau) = -\frac{(c-\omega^2)\cdot d + \omega^2\cdot a\cdot b}{\omega^2\cdot b^2 + d^2} = \frac{\omega^2\cdot L\cdot C}{(\omega^2\cdot\tau^2+1)}$. We consider Cylindrical RF antenna which mounted outside a Pyrex glass tube of diameter 32 cm and length 50 cm. The RF antenna consists of 16 copper (Cu) legs equally spaced by 6.7 cm interconnected with capacitors of 2.47nF. Copper leg diameter is equal to 1 mm and length 30 cm = 300 mm (<Pyrex glass tube length, 50 cm). We consider for Copper (Cu), relative permeability is one. f = 10 MHz is the typical testing frequency for cylindrical (birdcage) antenna. L—inductance (nH), l—length of copper leg (mm), d—diameter of copper leg, f—testing frequency. $l > 100\cdot d$ ($300 > 100\cdot 1$ mm), $d^2\cdot f > 1\ \text{mm}^2\cdot\text{MHz}$ ($1\ \text{mm}^2\cdot 10\ \text{MHz} > 1\ \text{mm}^2\cdot\text{MHz}$). L = 365.4 nH. $L = \frac{1}{5}\cdot l\cdot[\ln(\frac{4\cdot l}{d}-1] = 365.4\,\text{nH}$. For our stability switching analysis we choose typical Cylindrical RF network antennas parameters values (as calculated): $C = 2.47\,\text{nF};\ L = 365.4\,\text{nH};\ R_p = 100\text{Ohm}$ then $\frac{1}{L\cdot C} = 0.00110798\times10^{18}$. We find those ω, τ values which fulfill $F(\omega,\tau)=0$. We ignore negative, complex, and imaginary values of ω for specific τ values. The below table gives the list.

Remark: We know $F(\omega,\tau)=0$ implies it roots $\omega_i(\tau)$ and finding those delays values τ which ω_i is feasible. There are τ values, which ω_i are complex or imaginary numbered, then unable to analyze stability [6, 7]. We find those ω, τ values which fulfill $F(\omega,\tau)=0$. We ignore negative, complex, and imaginary values of ω for specific τ values. $\tau\in[0.001..10]$ and we can be express by 3D function $F(\omega,\tau)=0$. $F(\omega,\tau) = \omega^4 - \omega^2\cdot\frac{\tau^2}{(L\cdot C)^2} - \frac{1}{(L\cdot C)^2}$

$$F(\omega,\tau) = |P(i\cdot\omega,\tau)|^2 - |Q(i\cdot\omega,\tau)|^2 = \Phi_0 + \Phi_2\cdot\omega^2 + \Phi_4\cdot\omega^4 = \sum_{k=0}^{2}\Phi_{2\cdot k}\cdot\omega^{2\cdot k}$$

$\Phi_0 = -\frac{1}{(L\cdot C)^2};\ \Phi_2 = -\frac{\tau^2}{(L\cdot C)^2};\ \Phi_4 = 1$ hence $F(\omega,\tau)=0$ implies $\sum_{k=0}^{4}\Phi_{2\cdot k}$ $\tau\omega^{2\cdot k} = 0$.

$\Phi_j \to \text{Phi}_j$. Running MATLAB script for τ values ($\tau\in[0.001..10]$) gives the following results.

MATLAB script: Tau=0.001;C=2.47*1e-9;L=365.4*1e-9;Phi0=-1/(C*L*C*L); Phi2=-(Tau*Tau)/(C*L*C*L); Phi4=1;p=[Phi4 0 Phi2 0 Phi0];r=roots(p) (Tables 2.6, 2.7, 2.8, 2.9, 2.10, 2.11 and 2.12).

We can summery our $\omega_i(\tau)$ results for $\omega_i(\tau) > 0$ and real number (ignore complex, negative, and imaginary values). We exclude from our table the high and real $\omega_i(\tau)$ values (1.0e+007*, 1.0e+012*,…,1.0e+016*) and add results for τ = 15 and τ = 20 s (Figs. 2.14, 2.15 and Table 2.13).

Table 2.6 Cylindrical RF network antennas system roots $\omega_i(\tau)$

τ	τ = 0.01 s	τ = 0.001 s
ω_1	1.0e+013*	1.0e+012*
ω_2	−1.1080	−1.1080
ω_3	1.1080	1.1080
ω_4	0.0000 + 0.0000i	−0.0000 + 0.0000i
ω_5	0.0000 − 0.0000i	−0.0000 − 0.0000i

Table 2.7 Cylindrical RF network antennas system roots $\omega_i(\tau)$

τ	τ = 1 s	τ = 0.1 s
ω_1	1.0e+015*	1.0e+014*
ω_2	−1.1080	−1.1080
ω_3	1.1080	1.1080
ω_4	−0.0000 + 0.0000i	0.0000 + 0.0000i
ω_5	−0.0000 − 0.0000i	0.0000 − 0.0000i

Table 2.8 Cylindrical RF network antennas system roots $\omega_i(\tau)$

τ	τ = 3 s	τ = 2 s
ω_1	1.0e+015*	1.0e+015*
ω_2	3.3240	−2.2160
ω_3	−3.3240	2.2160
ω_4	0 + 0.0000i	−0.0000 + 0.0000i
ω_5	0 − 0.0000i	−0.0000 − 0.0000i

Table 2.9 Cylindrical RF network antennas system roots $\omega_i(\tau)$

τ	τ = 5 s	τ = 4 s
ω_1	1.0e+015*	1.0e+015*
ω_2	−5.5399	4.4319
ω_3	5.5399	−4.4319
ω_4	0.0000 + 0.0000i	0 + 0.0000i
ω_5	0.0000 − 0.0000i	0 − 0.0000i

Table 2.10 Cylindrical RF network antennas system roots $\omega_i(\tau)$

τ	τ = 7 s	τ = 6 s
ω_1	1.0e+015*	1.0e+015*
ω_2	−7.7559	6.6479
ω_3	7.7559	−6.6479
ω_4	0.0000 + 0.0000i	0 + 0.0000i
ω_5	0.0000 − 0.0000i	0 − 0.0000i

Matlab: plot([0 0.001 0.01 0.1 1 2 3 4 5 6 7 8 9 10 15 20], [3.3286 1.1080 1.1080 1.1080 1.1080 1.1080 2.2160 3.3240 4.4319 5.5399 6.6479 7.7559 8.8639 9.9719 1.1080 1.6620 2.2160],'-or'). We plot 3D function $F(\omega, \tau) = 0$. τ:0→10; ω:0→100. We define additional MATLAB script parameters ω→w, τ→t.

Table 2.11 Cylindrical RF network antennas system roots $\omega_i(\tau)$

τ	τ = 9 s	τ = 8 s
ω_1	1.0e+015*	1.0e+015*
ω_2	9.9719	8.8639
ω_3	−9.9719	−8.8639
ω_4	0 + 0.0000i	0 + 0.0000i
ω_5	0 − 0.0000i	0 − 0.0000i

Table 2.12 Cylindrical RF network antennas system roots $\omega_i(\tau)$

τ	τ = 0 s	τ = 10 s
ω_1	1.0e+007*	1.0e+016*
ω_2	−3.3286	−1.1080
ω_3	−0.0000 + 3.3286i	1.1080
ω_4	−0.0000 − 3.3286i	−0.0000 + 0.0000i
ω_5	3.3286	−0.0000 − 0.0000i

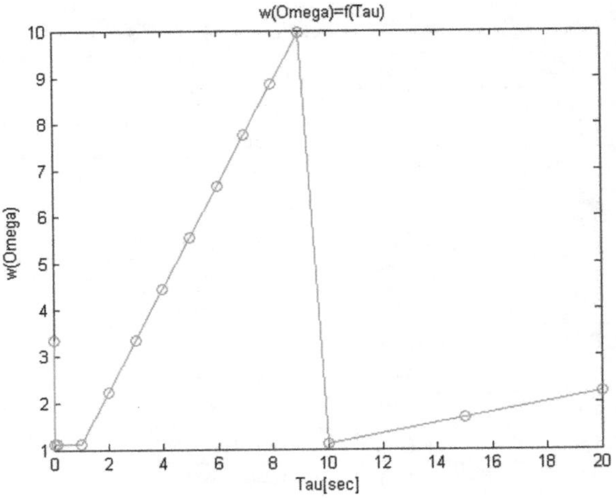

Fig. 2.14 Cylindrical RF network F(ω,τ) function for $\tau_1 = \tau_2 = \tau$

Matlab: [w,t]=meshgrid(1:1:100,0:0.01:10);C=2.47*1e-9; L=365.4*1e-9; f=w.*w.*w.*w-w.*w.*(t.*t)/(C*L*C*L)-1/(C*L*C*L);meshc(f); % $\omega \to w, \tau \to t$.

We get two possible real values for ω which fulfil $F(\omega,\tau) = 0$ $F(\omega = 3.3286$ or $\omega = 1.1080\ldots$ or $\omega = 2.2160, \tau) = 0$; $\tau \in [0.001..10]$. Next is to find those ω, τ values which fulfil $\sin\theta(\tau) = \ldots$; $\sin(\omega \cdot \tau) = \frac{-P_R \cdot Q_I + P_I \cdot Q_R}{|Q|^2}$ and $\cos\theta(\tau) = \ldots$

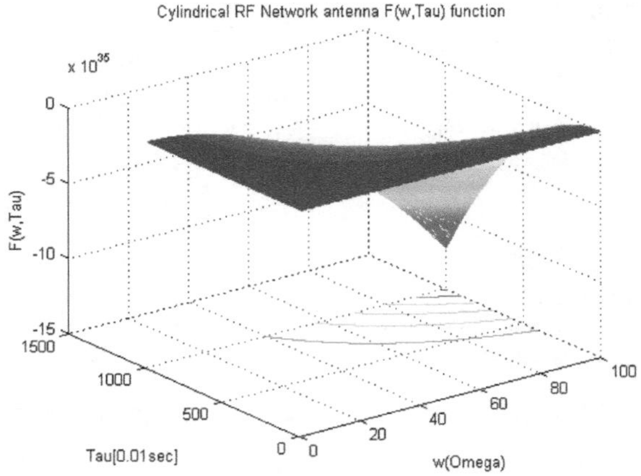

Fig. 2.15 Cylindrical RF network F(ω,τ) function for $\tau_1 = \tau_2 = \tau$

Table 2.13 Cylindrical RF network antennas system positive and real root $\omega_i(\tau)$ values and $\sin(\omega \cdot \tau), \cos(\omega \cdot \tau)$ values

τ[s]	ω	$\sin(\omega \cdot \tau) = \frac{-\omega^3 \cdot \tau \cdot L \cdot C}{(\omega^2 \cdot \tau^2 + 1)}$	$\cos(\omega \cdot \tau) = \frac{\omega^2 \cdot L \cdot C}{(\omega^2 \cdot \tau^2 + 1)}$
0	3.3286	0 = 0	1 ≠ 9.9e−15
0.001..1	1.1080	−1.22e−18 … −5.51e−16	1.108e−15 … 4.973e−16
2	2.2160	−9.5e−16	2.1e−16
3	3.3240	−9.9e−16	9.9e−17
4	4.4319	−9.9e−16	5.62e−17
5	5.5399	−9.9e−16	3.6e−17
6	6.6479	−9.99−16	2.5055e−17
7	7.7559	−9.9966e−16	1.8413e−17
8	8.8639	−9.9980e−16	1.4099e−17
9	9.9719	−9.9988e−16	1.1141e−17
10	1.1080	−9.9193e−17	8.9525e−18
15	1.6620	−9.9841e−17	4.0048e−18
20	2.2160	−9.9950e−17	2.2552e−18

$$\cos(\omega \cdot \tau) = -\frac{(P_R \cdot Q_R + P_I \cdot Q_I)}{|Q|^2}; \ |Q|^2 = Q_R^2 + Q_I^2;$$

$$\sin(\omega \cdot \tau) = \frac{-\omega^3 \cdot \tau \cdot L \cdot C}{(\omega^2 \cdot \tau^2 + 1)}; \ \cos(\omega \cdot \tau) = \frac{\omega^2 \cdot L \cdot C}{(\omega^2 \cdot \tau^2 + 1)}$$

$\frac{-\omega^3 \cdot \tau \cdot L \cdot C}{(\omega^2 \cdot \tau^2 + 1)} < 0$ & $\frac{\omega^2 \cdot L \cdot C}{(\omega^2 \cdot \tau^2 + 1)} > 0$ then $\sin(\omega \cdot \tau) < 0$ and $\cos(\omega \cdot \tau) > 0$; $2 \cdot \pi > \omega \cdot \tau > \frac{\pi}{2} \cdot 3$.

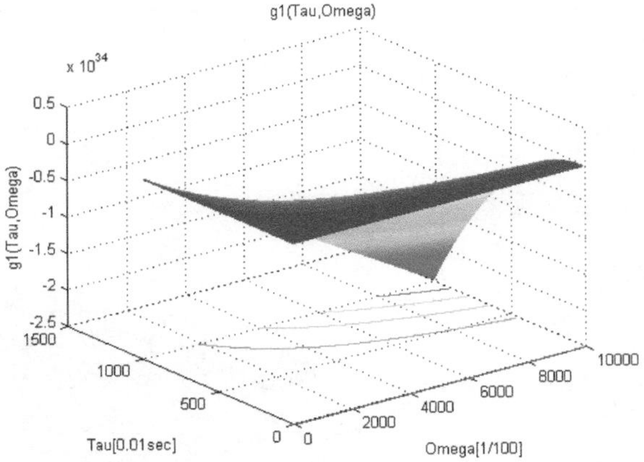

Fig. 2.16 Cylindrical RF network $g_1(\tau,\omega)$ F(ω,τ) function for $\tau_1 = \tau_2 = \tau$

We plot the stability switch diagram based on different delay values of our Cylindrical RF network antennas system. $\wedge^{-1}(\tau) = (\frac{\partial \mathrm{Re}\lambda}{\partial \tau})_{\lambda=i\cdot\omega} = \mathrm{Re}\{\frac{-2\cdot[U+\tau\cdot|P|^2]+i\cdot F_\omega}{F_\tau+i\cdot 2\cdot[V+\omega\cdot|P|^2]}\}$

$$\wedge^{-1}(\tau) = (\frac{\partial \mathrm{Re}\lambda}{\partial \tau})_{\lambda=i\cdot\omega} = \frac{2\cdot\{F_\omega\cdot(V+\omega\cdot P^2) - F_\tau\cdot(U+\tau\cdot P^2)\}}{F_\tau^2+4\cdot(V+\omega\cdot P^2)^2}$$

$$\wedge^{-1}(\tau) = (\frac{\partial \mathrm{Re}\lambda}{\partial \tau})_{\lambda=i\cdot\omega} = \frac{2\cdot\{F_\omega\cdot(V+\omega\cdot P^2) - F_\tau\cdot(U+\tau\cdot P^2)\}}{F_\tau^2+4\cdot(V+\omega\cdot P^2)^2}$$

$sign\{\wedge^{-1}(\tau)\} = sign[\psi_1]\cdot sign[\psi_2+\omega+\psi_3]$. We define the following new functions (Figs. 2.16 and 2.17):

$$g_1 = \psi_1;\; g_2 = \psi_2+\omega+\psi_3;\; sign\{\wedge^{-1}(\tau)\} = sign[g_1]\cdot sign[g_2]$$

Matlab: [w,t]=meshgrid(1:.01:100,0:0.01:10);C=2.47*1e-9; L=365.4*1e-9; f=2*w.*(2*w.*w-(t.*t./(C*L*C*L)));meshc(f) % $\omega \to w, \tau \to t$.

Matlab: [w,t]=meshgrid(1:.1:10,0:0.1:10);C=2.47*1e-9; L=365.4*1e-9;m=w. *t./(2*w.*w.*(L*C*L*C)-t.*t); f=t.*m+w+(m.*t./(L*C*L*C)+w./(L*C*L*C)) ./ (w.*w.*w.*w);meshc(f) % $\omega \to w, \tau \to t$.

$g(Tau) = g_1(Tau)\cdot g_2(Tau) = \wedge^{-1}(\tau) = (\frac{\partial \mathrm{Re}\lambda}{\partial \tau})_{\lambda=i\cdot\omega}$. The stability switch occur only on those delay values (τ) which fit the equation: $\tau = \frac{\theta_+(\tau)}{\omega_+(\tau)}$ and $\theta_+(\tau)$ is the solution of $\sin\theta(\tau) = \frac{-\omega^3\cdot\tau\cdot L\cdot C}{(\omega^2\cdot\tau^2+1)}$; $\cos\theta(\tau) = \frac{\omega^2\cdot L\cdot C}{(\omega^2\cdot\tau^2+1)}$ when $\omega = \omega_+(\tau)$ if only ω_+ is feasible. Additionally When all Cylindrical RF network antennas parameters are known and the stability switch due to various time delay values τ is describe in the below expression (Theorem 1.5):

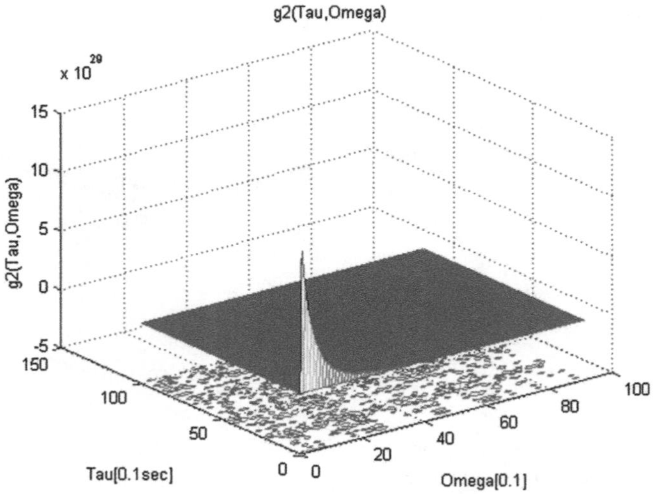

Fig. 2.17 Cylindrical RF network $g_2(\tau, \omega)$ function for $\tau_1 = \tau_2 = \tau$

$$sign\{\wedge^{-1}(\tau)\} = sign\{F_{\omega}(\omega(\tau), \tau)\} \cdot sign\{\tau \cdot \omega_{\tau}(\omega(\tau)) + \omega(\tau) + \frac{U(\omega(\tau)) \cdot \omega_{\tau}(\omega(\tau)) + V(\omega(\tau))}{|P(\omega(\tau))|^2}\}$$

Remark: We know $F(\omega, \tau) = 0$ implies it roots $\omega_i(\tau)$ and finding those delays values τ which ω_i is feasible. There are τ values which ω_i is complex or imaginary number, then unable to analyse stability [4, 5].

Discussion: We consider Cylindrical RF network antennas system. Due to RF antenna copper leg parasitic effect we get copper leg's current and current derivative with delay $\tau_{1\text{-}k}$ and $\tau_{2\text{-}k}$ (k is leg number index, k = 1,…,16). Those delays causes to stability switching for our Cylindrical RF network antennas. We draw our Cylindrical RF network antennas equivalent circuit and get system differential equations. Our variables are Y, X which are function of RF antenna copper leg's current and current derivative. Our system dynamic behavior is dependent on circuit overall parameters and parasitic delay in time. We keep all circuit parameters fix and change, parasitic delay over various values $\tau \in [0.001..10]$. Our analysis results extend that of in the way that it deals with stability switching for different delay values. This implies that our system behavior of the circuit cannot inspect by short analysis and we must study the full system. Several very important issues and possibilities were left out of the present discussion. One possibility is the stability switching by circuit parameters. Every circuit's parameter variation can change our system dynamic and stability behavior. This case can be solved by the same methods combined with alternative and more technical hypotheses. Moreover, numerical simulations for the Cylindrical RF network antennas model studied in references suggest that this result can be extended to enhance models with more general functions. Still another extension of our results would be to treat the case of delayed Cylindrical RF network antennas leg's higher derivative degree of current. It would be extremely desirable to confirm these cases by mathematical proofs.

Conclusion: Cylindrical RF network antennas system is characterized by parasitic effects which can influence Cylindrical RF network antennas system stability in time. There are two main Cylindrical RF network antennas variables which are affected by antenna legs parasitic effects, Y and X functions of antenna leg's currents and currents derivatives respectively. Each Cylindrical RF network antennas system variable under parasitic effects is characterized by time delay respectively. The two time delays are not the same, but can be categorized to some sub cases due to antenna leg parasitic behavior. The first case we analyze is when there is delay in Cylindrical RF network antennas leg's current and no delay in antennas leg's current derivative or opposite. The second case we analyze is when there is delay both in Cylindrical RF network antennas leg's current and current time derivative [4, 5]. For simplicity of our analysis we consider in the second case all delays are the same (there is a difference but it is neglected in our analysis). In each case we derive the related characteristic equation. The characteristic equation is dependent on Cylindrical RF network antennas system overall parameters and parasitic time delay. Upon mathematics manipulation and [BK] theorems and definitions we derive the expression which gives us a clear picture on Cylindrical RF network antennas map. The stability map gives all possible options for stability segments, each segment belongs to different time delay value segment. Cylindrical RF network antennas system's stability analysis can be influenced either by system overall parameter values. We left this analysis and do not discuss it in the current subchapter.

Lemma 1.1 Assume that $\omega(\tau)$ is a positive and real root of $F(\omega, \tau) = 0$.

Defined for $\tau \in I$, which is continuous and differentiable. Assume further that if $\lambda = i \cdot \omega$, $\omega \in R$, then $P_n(i \cdot \omega, \tau) + Q_n(i \cdot \omega, \tau) \neq 0$, $\tau \in \mathrm{R}$ hold true. Then the functions $S_n(\tau)$, $n \in N_0$, are continuous and differentiable on I.

Theorem 1.2 *Assume that $\omega(\tau)$ is a positive real root of $F(\omega, \tau) = 0$ defined for $\tau \in I$, $I \subseteq R_{+0}$, and at some $\tau^* \in I$, $S_n(\tau^*) = 0$ for some $n \in N_0$ then a pair of simple conjugate pure imaginary roots $\lambda_+(\tau^*) = i \cdot \omega(\tau^*)$, $\lambda_-(\tau^*) = -i \cdot \omega(\tau^*)$ of $D(\lambda, \tau) = 0$ exist at $\tau = \tau^*$ which crosses the imaginary axis from left to right if $\delta(\tau^*) > 0$ and cross the imaginary axis from right to left if $\delta(\tau^*) < 0$ where*

$$\delta(\tau^*) = sign\{\frac{d\mathrm{Re}\lambda}{d\tau}|_{\lambda=i\omega(\tau^*)}\} = sign\{F_\omega(\omega(\tau^*), \tau^*)\} \cdot sign\{\frac{dS_n(\tau)}{d\tau}|_{\tau=\tau^*}\}$$

The theorem becomes $sign\{\frac{d\mathrm{Re}\lambda}{d\tau}|_{\lambda=i\omega\pm}\} = sign\{\pm\Delta^{1/2}\} \cdot sign\{\frac{dS_n(\tau)}{d\tau}|_{\tau=\tau^*}\}$.

Theorem 1.3 *The characteristic equation*: $\tau_1 = \tau, \tau_2 = 0$; $\tau_1 = 0, \tau_2 = \tau$

$$D(\lambda,\tau) = \lambda^2 + a(\tau)\cdot\lambda + b(\tau)\cdot\lambda\cdot e^{-\lambda\cdot\tau} + c(\tau) + d(\tau)\cdot e^{-\lambda\cdot\tau};\ D(\lambda,\tau_1,\tau_2) = \lambda^2 + \lambda\cdot\frac{1}{C1\cdot R1}\cdot e^{-\lambda\cdot\tau_2} + \frac{1}{C1\cdot f_{\#}}\cdot e^{-\lambda\cdot(\tau_1+\tau_2)}$$

Has a pair of simple and conjugate pure imaginary roots $\lambda = \pm\omega(\tau^*)$, $\omega(\tau^*)$ *real at* $\tau^* \in I$ *if* $S_n(\tau^*) = \tau^* - \tau_n(\tau^*) = 0$ *for some* $n \in N_0$. *If* $\omega(\tau^*) = \omega_+(\tau^*)$, *this pair of simple conjugate pure imaginary roots crosses the imaginary axis from left to right if* $\delta_+(\tau^*) > 0$ *and crosses the imaginary axis from right to left if* $\delta_+(\tau^*) < 0$ *where* $\delta_+(\tau^*) = sign\{\frac{d\mathrm{Re}\lambda}{d\tau}|_{\lambda=i\omega_+(\tau^*)}\} = sign\{\frac{dS_n(\tau)}{d\tau}|_{\tau=\tau^*}\}$. *If* $\omega(\tau^*) = \omega_-(\tau^*)$, *this pair of simple conjugate pure imaginary roots cross the imaginary axis from left to right if* $\delta_-(\tau^*) > 0$ *and crosses the imaginary axis from right to left If* $\delta_-(\tau^*) < 0$ *where* $\delta_-(\tau^*) = sign\{\frac{d\mathrm{Re}\lambda}{d\tau}|_{\lambda=i\omega_-(\tau^*)}\} = -sign\{\frac{dS_n(\tau)}{d\tau}|_{\tau=\tau^*}\}$. *If* $\omega_+(\tau^*) = \omega_-(\tau^*) = \omega(\tau^*)$ *then* $\Delta(\tau^*) = 0$ *and* $sign\{\frac{d\mathrm{Re}\lambda}{d\tau}|_{\lambda=i\omega(\tau^*)}\} = 0$, *the same is true when* $S_n'(\tau^*) = 0$. *The following result can be useful in identifying values of* τ *where stability switches happened.*

Theorem 1.4 *Assume that for all* $\tau \in I$, $\omega(\tau)$ *is defined as a solution of* $F(\omega,\tau) = 0$ *then* $\delta_\pm(\tau) = sign\{\pm\Delta^{1/2}(\tau)\}\cdot signD_\pm(\tau)$.

$$D_\pm(\tau) = \omega_\pm^2\cdot[(\omega_\pm^2\cdot b^2 + d^2) + a'\cdot(c - \omega_\pm^2) + b\cdot d' - b'\cdot d - a\cdot c'] + \omega_\pm\cdot\omega_\pm'\cdot[\tau\cdot(\omega_\pm^2\cdot b^2 + d^2) - b\cdot d + a\cdot(c - \omega_\pm^2) + 2\cdot\omega_\pm^2\cdot a];\ a' = \frac{da(\tau)}{d\tau};\ b' = \frac{db(\tau)}{d\tau};$$

$$c' = \frac{dc(\tau)}{d\tau};\ d' = \frac{dd(\tau)}{d\tau}$$

Theorem 1.5 *We need to approve the following expression*:

$$sign\{\wedge^{-1}(\tau)\} = sign\{F_\omega(\omega(\tau),\tau)\}\cdot sign\{\tau\cdot\omega_\tau(\omega(\tau)) + \omega(\tau) + \frac{U(\omega(\tau))\cdot\omega_\tau(\omega(\tau)) + V(\omega(\tau))}{|P(\omega(\tau))|^2}\}$$

The basic assumption: $\wedge^{-1}(\tau) = (\frac{\partial\mathrm{Re}\lambda}{\partial\tau})_{\lambda=i\cdot\omega}$

$$\wedge^{-1}(\tau) = (\frac{\partial\mathrm{Re}\lambda}{\partial\tau})_{\lambda=i\cdot\omega};\ \wedge^{-1}(\tau) = (\frac{\partial\mathrm{Re}\lambda}{\partial\tau})_{\lambda=i\cdot\omega} = \frac{2\cdot\{F_\omega\cdot(V+\omega\cdot P^2) - F_\tau\cdot(U+\tau\cdot P^2)\}}{F_\tau^2 + 4\cdot(V+\omega\cdot P^2)^2}$$

$$\wedge^{-1}(\tau) = (\frac{\partial\mathrm{Re}\lambda}{\partial\tau})_{\lambda=i\cdot\omega} = \frac{2\cdot\{F_\omega\cdot(V+\omega\cdot P^2) - F_\tau\cdot(U+\tau\cdot P^2)\}}{F_\tau^2 + 4\cdot(V+\omega\cdot P^2)^2}$$

$$sign\{F_\tau^2 + 4 \cdot (V + \omega \cdot P^2)^2\} > 0 \text{ and } \omega_\tau = -\frac{F_\tau}{F_\omega} \text{ then}$$

$$sign\{\wedge^{-1}(\tau)\} = sign\{(\frac{\partial \mathrm{Re}\lambda}{\partial \tau})_{\lambda = i\cdot\omega}\} = sign\{F_\omega \cdot (V + \omega \cdot P^2) - F_\tau \cdot (U + \tau \cdot P^2)\}$$

$$sign\{\wedge^{-1}(\tau)\} = sign\{F_\omega \cdot \{(V + \omega \cdot P^2) - \frac{F_\tau}{F_\omega} \cdot (U + \tau \cdot P^2)\}\};$$

$$sign\{\wedge^{-1}(\tau)\} = sign\{F_\omega \cdot \{(V + \omega \cdot P^2) + \omega_\tau \cdot (U + \tau \cdot P^2)\}\}$$

$$sign\{\wedge^{-1}(\tau)\} = sign\{F_\omega \cdot \{V + \omega_\tau \cdot U + \omega \cdot P^2 + \omega_\tau \cdot \tau \cdot P^2\}\};$$

$$sign\{\wedge^{-1}(\tau)\} = sign\{P^2 \cdot F_\omega \cdot \{\frac{V + \omega_\tau \cdot U}{P^2} + \omega + \omega_\tau \cdot \tau\}\}$$

$$sign\{\wedge^{-1}(\tau)\} = sign\{P^2\} \cdot sign\{F_\omega\} \cdot sign\{\frac{V + \omega_\tau \cdot U}{P^2} + \omega + \omega_\tau \cdot \tau\};\ sign\{P^2\} > 0$$

$$sign\{\wedge^{-1}(\tau)\} = sign\{F_\omega\} \cdot sign\{\frac{V + \omega_\tau \cdot U}{P^2} + \omega + \omega_\tau \cdot \tau\}.$$

2.4 Tunnel Diode (TD) as a Microwave Oscillator System Cavity Parasitic Elements Stability Analysis

Tunnel diode is used in many engineering applications and specialy as a microwave oscillator. Tunnel diode is the p-n junction device that exhibits negative resistance. That means when the voltage is increased the current through it decreases. We can consider the tunnel diode as an oscillator and high-frequency threshold (trigger) device since it operated at frequencies far greater than the tetrode could, well into the microwave bands. Applications for tunnel diodes included local oscillators for UHF television tuners, trigger circuits in oscilloscopes, high-speed counter circuits, and very fast-rise time pulse generator circuits. The tunnel diode can also be used as low-noise microwave amplifier. The total current that flows through Tunnel Diode (TD) is a summation of three elements, I_{tun}, I_{diode}, I_{excess}. I_{diode} is the P-N junction current $I_{diode} \approx I_s \cdot e^{[(\frac{V_D}{\eta \cdot V_{th}}) - 1]}$; I_s—saturation current, η—ideality factor, $V_{th} = k \cdot T/q$. V_D—tunnel diode voltage. I_{tun} is the tunnel current $I_{tun} = \frac{V_D}{R_0} \cdot e^{[-(\frac{V_D}{V_0})^m]}$; Typically m = 1…3, V_o = 0.1…0.5v, R_o is the tunnel diode resistance in the ohmic region, V_D is the tunnel diode voltage. I_{excess} is an additional tunneling current related to parasitic tunneling via impurities $I_{excess} = \frac{V_D}{R_V} \cdot e^{[\frac{V_D - V_V}{V_{ex}}]}$. This current usually determines the minimum (valley) current I_v, R_V and V_{ex} are the empirical parameters; in high quality diodes, $R_V \gg R_O$, V_{ex} = 1…5v. I_D is the tunnel diode current, $I_D = I_{diode} + I_{tun} + I_{excess} = I_s \cdot e^{[(\frac{V_D}{\eta \cdot V_{th}}) - 1]} + \frac{V_D}{R_0} \cdot e^{[-(\frac{V_D}{V_0})^m]} + \frac{V_D}{R_V} \cdot e^{[\frac{V_D - V_V}{V_{ex}}]}$. If we build a circuit with the Tunnel Diode (TD) and resistor.

Case I: the circuit has three possible operating points. The middle point (intersection between TD's characteristic NDR segment and work line) is typically unstable, depending on parasitic L and C components. The circuit will operate at one of the other two points (intersection between TD's characteristic positive differential resistance segment and work line).

Case II: The circuit has only one operating point (intersection between TD's characteristic NDR segment and work line). The total differential resistance is negative because $R < |R_d|$. Depending on the L and C components, the circuit can be stable (amplifier) or unstable (oscillator). R_d is a TD's differential resistance $R_d = \frac{\partial V_D}{\partial I_D} \approx \frac{\Delta V_D}{\Delta I_D}$. R is a TD's static resistance $R = \frac{V_D}{I_D}$. Typically for linear (Ohmic) components $R = R_d$ and for semiconductor devices $R \neq R_d$ [47–50]. The tunnel diode has a region in its voltage current characteristic where the current decreases with increased forward voltage, known as its negative resistance region. This characteristic makes the tunnel diode useful in oscillators and as a microwave amplifier (case II). The basic circuit structure of Tunnel Diode (TD) as a microwave oscillator includes TD biased voltage and connection to microwave cavity. TD's biasing circuit is constructed from DC voltage V_b, inductor L_b which block oscillation from V_b source and capacitor C_b which shorts to ground V_b bias voltage's oscillations. The DC voltage V_b biases the tunnel diode TD into its negative resistance region and also supplies the power used in amplifying the input signal V_i. We have two topological circuit structures: first, switch S1 is ON and switch and S2 is OFF (growing or decreasing oscillations) and second, switch S1 is ON and S2 in ON (bypass our signal source V_i) and our circuit functions as a oscillator. A microwave cavity or radio frequency (RF) cavity is a special type of resonator, consisting of a closed (or largely closed) metal structure that confines electromagnetic fields in the microwave region of the spectrum. The structure is either hollow or filled with dielectric material. A microwave cavity acts similarly to a resonant circuit with extremely low loss at its frequency of operation. Microwave resonant cavities can be represented and thought of as simple LC circuits. For a microwave cavity, the stored electric energy is equal to the stored magnetic energy at resonance as is the case for a resonant LC circuit. We can represent our microwave cavity as resonant LC circuit. Due to cavity parasitic effects there are delays in the currents which flow through equivalent resonant L and C elements and TD's voltage derivative in time. τ_1 is the time delay for the TD's voltage derivative in time. τ_2 is the time delay for the current flows through C element. R_L is the load resistance. Load resistance (R_L) is chosen so that $R_L < |R_d|$ in the TD's characteristic NDR region. At the TD operating point, the total circuit differential resistance is negative. We have transient in resonant cavity after turning the bias voltage to ON state (switch S1 moves to ON state, S2 is in OFF state) (Fig. 2.18).

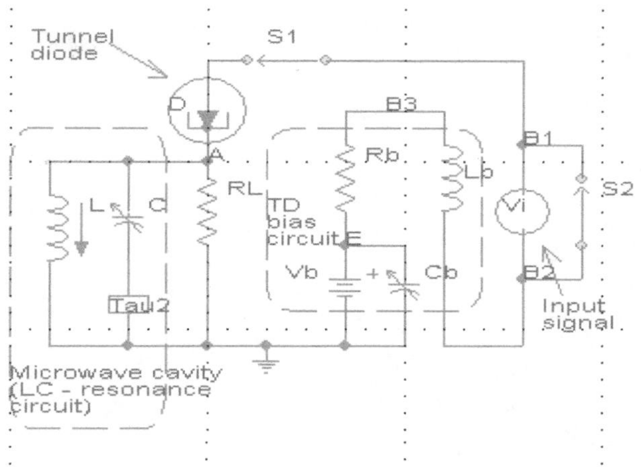

Fig. 2.18 Tunnel diode as a microwave oscillator

If $R_d > 0$ or $R_d < 0$ and $R_L > |R_d|$ then oscillation goes down. If $R_d < 0$ and $R_L < |R_d|$ then the oscillations goes up. The resonator circuit with NDR can oscillate. Maximum frequency of TD-oscillator is limited by the characteristic tunneling time $f_{MAX} \leq (\frac{1}{2 \cdot \pi}) \cdot (\frac{1}{\tau_{tun}})$, tunneling time in TDs is extremely small: $\ll$1ps, $f_{MAX} > 100$ GHz. We represent our Microwave resonant cavities thought of as simple LC circuits. For a microwave cavity, the stored electric energy is equal to the stored magnetic energy at resonance as is the case for a resonant LC circuit. In terms of inductance and capacitance, the resonant frequency for a given *mnl* mode can be written as $L_{mnl} = \mu \cdot k_{mnl}^2 \cdot V_{cavity}$; $C_{mnl} = \frac{\varepsilon}{k_{mnl}^4 \cdot V_{cavity}}$.

V_{cavity} is the cavity volume, k_{mnl} is the mode wavenumber and ε, μ are permittivity and permeability respectively. The resonant frequency (f) for a given mnl mode can be written as $f_{mnl} = \frac{1}{2 \cdot \pi \cdot \sqrt{L_{mnl} \cdot C_{mnl}}} = \frac{1}{2 \cdot \pi \cdot \sqrt{\frac{\mu \cdot \varepsilon}{k_{mnl}^2}}}$.

We consider for our stability analysis the case of switch S1 is ON and S2 in ON (bypass our signal source V_i) and our TD is functions as a microwave oscillator. First we need to locate our TD's voltages segment into negative resistance region $R_d = \frac{\partial V_D}{\partial I_D} < 0$. The expression for tunnel diode current is

$$I_D = \xi(V_D) = \sum_{i=1}^{3} g_i(V_D) = I_s \cdot e^{[(\frac{V_D}{\eta \cdot V_{th}})-1]} + \frac{V_D}{R_0} \cdot e^{[-(\frac{V_D}{V_0})^m]} + \frac{V_D}{R_V} \cdot e^{[\frac{V_D - V_V}{V_{ex}}]};$$

$$g_1(V_D) = I_s \cdot e^{[(\frac{V_D}{\eta \cdot V_{th}})-1]}$$

$$g_2(V_D) = \frac{V_D}{R_0} \cdot e^{[-(\frac{V_D}{V_0})^m]}; \; g_3(V_D) = \frac{V_D}{R_V} \cdot e^{[\frac{V_D - V_V}{V_{ex}}]};$$

$$\frac{\partial g_1(V_D)}{\partial I_D} = I_s \cdot \frac{1}{\eta \cdot V_{th}} \cdot \frac{\partial V_D}{\partial I_D} \cdot e^{[(\frac{V_D}{\eta \cdot V_{th}})-1]}$$

$$\frac{\partial g_2(V_D)}{\partial I_D} = \frac{1}{R_0} \cdot \frac{\partial V_D}{\partial I_D} \cdot e^{[-(\frac{V_D}{V_0})^m]} + \frac{V_D}{R_0} \cdot \{-m \cdot (\frac{V_D}{V_0})^{m-1} \cdot \frac{1}{V_0} \cdot \frac{\partial V_D}{\partial I_D}\} \cdot e^{[-(\frac{V_D}{V_0})^m]}$$

$$\frac{\partial g_2(V_D)}{\partial I_D} = \{1 - m \cdot (\frac{V_D}{V_0})^m\} \cdot \frac{1}{R_0} \cdot \frac{\partial V_D}{\partial I_D} \cdot e^{[-(\frac{V_D}{V_0})^m]};$$

$$\frac{\partial g_3(V_D)}{\partial I_D} = (1 + \frac{V_D}{V_{ex}}) \cdot \frac{1}{R_V} \cdot \frac{\partial V_D}{\partial I_D} \cdot e^{[\frac{V_D - V_V}{V_{ex}}]}$$

$$I_D = \sum_{i=1}^{3} g_i(V_D) \Rightarrow \frac{\partial I_D}{\partial I_D} = \sum_{i=1}^{3} \frac{\partial g_i}{\partial I_D} \Rightarrow \sum_{i=1}^{3} \frac{\partial g_i}{\partial I_D} = \frac{\partial g_1(V_D)}{\partial I_D} + \frac{\partial g_2(V_D)}{\partial I_D} + \frac{\partial g_3(V_D)}{\partial I_D} = 1$$

$$I_s \cdot \frac{1}{\eta \cdot V_{th}} \cdot \frac{\partial V_D}{\partial I_D} \cdot e^{[(\frac{V_D}{\eta \cdot V_{th}})-1]} + \{1 - m \cdot (\frac{V_D}{V_0})^m\} \cdot \frac{1}{R_0} \cdot \frac{\partial V_D}{\partial I_D} \cdot e^{[-(\frac{V_D}{V_0})^m]}$$
$$+ (1 + \frac{V_D}{V_{ex}}) \cdot \frac{1}{R_V} \cdot \frac{\partial V_D}{\partial I_D} \cdot e^{[\frac{V_D - V_V}{V_{ex}}]} = 1$$

$$\frac{\partial V_D}{\partial I_D} \cdot [I_s \cdot \frac{1}{\eta \cdot V_{th}} \cdot e^{[(\frac{V_D}{\eta \cdot V_{th}})-1]} + \{1 - m \cdot (\frac{V_D}{V_0})^m\} \cdot \frac{1}{R_0} \cdot e^{[-(\frac{V_D}{V_0})^m]} + (1 + \frac{V_D}{V_{ex}}) \cdot \frac{1}{R_V} \cdot e^{[\frac{V_D - V_V}{V_{ex}}]}] = 1$$

$$\frac{\partial V_D}{\partial I_D} = \frac{1}{I_s \cdot \frac{1}{\eta \cdot V_{th}} \cdot e^{[(\frac{V_D}{\eta \cdot V_{th}})-1]} + \{1 - m \cdot (\frac{V_D}{V_0})^m\} \cdot \frac{1}{R_0} \cdot e^{[-(\frac{V_D}{V_0})^m]} + (1 + \frac{V_D}{V_{ex}}) \cdot \frac{1}{R_V} \cdot e^{[\frac{V_D - V_V}{V_{ex}}]}}$$

$$\frac{\partial V_D}{\partial I_D} < 0 \Rightarrow \{I_s \cdot \frac{1}{\eta \cdot V_{th}} \cdot e^{[(\frac{V_D}{\eta \cdot V_{th}})-1]} + \{1 - m \cdot (\frac{V_D}{V_0})^m\} \cdot \frac{1}{R_0} \cdot e^{[-(\frac{V_D}{V_0})^m]} + (1 + \frac{V_D}{V_{ex}})$$
$$\cdot \frac{1}{R_V} \cdot e^{[\frac{V_D - V_V}{V_{ex}}]}\} < 0$$

V_V is TD's characteristic valley voltage, V_P is TD's characteristic peak voltage. It is numerical analysis to find our TD's voltages segment into negative resistance region $R_d = \frac{\partial V_D}{\partial I_D} < 0$ by specific TD's parameters ($V_P < V_D < V_V$).

$V_B = V_{B1} = V_{B2}$ (S2 ON). $V_D = V_B - V_A;\ V_A = V_{R_L} = V_C = V_L;\ I_{R_L} = \frac{V_A}{R_L}$

$$I_C = C \cdot \frac{dV_A}{dt} = C \cdot \frac{dV_C}{dt};\ V_A = V_L = L \cdot \frac{dI_L}{dt};\ I_D = I_{L_b} = I_{R_b};\ I_{R_b} = \frac{V_E - V_{B3}}{R_b};\ I_{C_b} = C_b \cdot \frac{dV_E}{dt}\Big|_{\frac{dV_E}{dt}=0} = 0;\ V_E = V_b$$

$V_{L_b} = V_{B3} - V_B = L_b \cdot \frac{dI_{L_b}}{dt} = L_b \cdot \frac{dI_D}{dt};\ I_D = \xi(V_D);\ I_D = I_{R_L} + I_C + I_L;$
V_b—Constant supple voltage, V_D—Tunnel diode voltage, V_C—Capacitor voltage, V_L—Inductor voltage. V_i—Input voltage (is bypass in our case, S2 ON state).

$$\xi(V_D) = \frac{V_A}{R_L} + C \cdot \frac{dV_A}{dt} + \frac{1}{L} \cdot \int V_A \cdot dt;$$
$$V_A = L \cdot \frac{dI_L}{dt} \Rightarrow \frac{V_A}{L} \cdot dt = dI_L \Rightarrow I_L = \frac{1}{L} \cdot \int V_A \cdot dt$$
$$\frac{d\xi(V_D)}{dt} = \frac{1}{R_L} \cdot \frac{dV_A}{dt} + C \cdot \frac{d^2V_A}{dt^2} + \frac{1}{L} \cdot V_A;$$
$$V_b = V_{R_b} + V_{L_b} + V_D + V_A = I_D \cdot R_b + L_b \cdot \frac{dI_D}{dt} + V_D + V_A$$
$$V_A = V_b - I_D \cdot R_b - L_b \cdot \frac{dI_D}{dt} - V_D;\ \frac{dV_A}{dt}\Big|_{\frac{dV_b}{dt}=0} = -R_b \cdot \frac{dI_D}{dt} - L_b \cdot \frac{d^2I_D}{dt^2} - \frac{dV_D}{dt}$$
$$\frac{dV_A}{dt}\Big|_{\frac{dV_b}{dt}=0} = -R_b \cdot \frac{dI_D}{dt} - L_b \cdot \frac{d^2I_D}{dt^2} - \frac{dV_D}{dt} \Rightarrow \frac{d^2V_A}{dt^2} = -R_b \cdot \frac{d^2I_D}{dt^2} - L_b \cdot \frac{d^3I_D}{dt^3} - \frac{d^2V_D}{dt^2}$$

We get the following expression:

$$\frac{d\xi(V_D)}{dt} = \frac{1}{R_L} \cdot \left[-R_b \cdot \frac{dI_D}{dt} - L_b \cdot \frac{d^2I_D}{dt^2} - \frac{dV_D}{dt}\right] + C \cdot \left[-R_b \cdot \frac{d^2I_D}{dt^2} - L_b \cdot \frac{d^3I_D}{dt^3} - \frac{d^2V_D}{dt^2}\right] + \frac{1}{L} \cdot \left[V_b - I_D \cdot R_b - L_b \cdot \frac{dI_D}{dt} - V_D\right]$$

$$\frac{d\xi(V_D)}{dt} = -R_b \cdot \frac{1}{R_L} \cdot \frac{dI_D}{dt} - L_b \cdot \frac{1}{R_L} \cdot \frac{d^2I_D}{dt^2} - \frac{1}{R_L} \cdot \frac{dV_D}{dt} - C \cdot R_b \cdot \frac{d^2I_D}{dt^2} - C \cdot L_b \cdot \frac{d^3I_D}{dt^3} - C \cdot \frac{d^2V_D}{dt^2} + \frac{1}{L} \cdot V_b - \frac{1}{L} \cdot I_D \cdot R_b - \frac{1}{L} \cdot L_b \cdot \frac{dI_D}{dt} - \frac{1}{L} \cdot V_D$$

$$\frac{d\xi(V_D)}{dt} = \frac{1}{L} \cdot V_b - \frac{1}{L} \cdot I_D \cdot R_b - \frac{dI_D}{dt} \cdot [\frac{R_b}{R_L} + \frac{L_b}{L}] - \frac{d^2 I_D}{dt^2} \cdot [L_b \cdot \frac{1}{R_L} + C \cdot R_b]$$
$$- C \cdot L_b \cdot \frac{d^3 I_D}{dt^3} - \frac{1}{L} \cdot V_D - \frac{1}{R_L} \cdot \frac{dV_D}{dt} - C \cdot \frac{d^2 V_D}{dt^2}$$

$I_D = \xi(V_D) = I_s \cdot e^{[(\frac{V_D}{\eta \cdot V_{th}})-1]} + \frac{V_D}{R_0} \cdot e^{[-(\frac{V_D}{V_0})^m]} + \frac{V_D}{R_V} \cdot e^{[\frac{V_D - V_V}{V_{ex}}]}$ and we need to find $\frac{d\xi(V_D)}{dt}$.

$$\frac{d\xi(V_D)}{dt} = I_s \cdot \frac{1}{\eta \cdot V_{th}} \cdot e^{[(\frac{V_D}{\eta \cdot V_{th}})-1]} \cdot \frac{dV_D}{dt} + \frac{1}{R_0} \cdot \frac{dV_D}{dt} \cdot e^{[-(\frac{V_D}{V_0})^m]}$$
$$+ \frac{V_D}{R_0} \cdot (-m \cdot [\frac{V_D}{V_0}]^{m-1} \cdot \frac{1}{V_0} \cdot \frac{dV_D}{dt}) \cdot e^{[-(\frac{V_D}{V_0})^m]}$$
$$+ \frac{1}{R_V} \cdot \frac{dV_D}{dt} \cdot e^{[\frac{V_D - V_V}{V_{ex}}]} + \frac{V_D}{R_V} \cdot \frac{1}{V_{ex}} \cdot \frac{dV_D}{dt} \cdot e^{[\frac{V_D - V_V}{V_{ex}}]}$$

$$\frac{d\xi(V_D)}{dt} = I_s \cdot \frac{1}{\eta \cdot V_{th}} \cdot e^{[(\frac{V_D}{\eta \cdot V_{th}})-1]} \cdot \frac{dV_D}{dt} + \frac{1}{R_0} \cdot e^{[-(\frac{V_D}{V_0})^m]} \cdot \frac{dV_D}{dt}$$
$$- \frac{V_D}{R_0} \cdot [\frac{V_D}{V_0}]^{m-1} \cdot \frac{m}{V_0} \cdot e^{[-(\frac{V_D}{V_0})^m]} \cdot \frac{dV_D}{dt}$$
$$+ \frac{1}{R_V} \cdot e^{[\frac{V_D - V_V}{V_{ex}}]} \cdot \frac{dV_D}{dt} + \frac{V_D}{R_V \cdot V_{ex}} \cdot e^{[\frac{V_D - V_V}{V_{ex}}]} \cdot \frac{dV_D}{dt}$$

$$\frac{d\xi(V_D)}{dt} = I_s \cdot \frac{1}{\eta \cdot V_{th}} \cdot e^{[(\frac{V_D}{\eta \cdot V_{th}})-1]} \cdot \frac{dV_D}{dt} + (1 - m \cdot [\frac{V_D}{V_0}]^m) \cdot \frac{1}{R_0} \cdot e^{[-(\frac{V_D}{V_0})^m]}$$
$$\cdot \frac{dV_D}{dt} + (1 + \frac{V_D}{V_{ex}}) \cdot \frac{1}{R_V} \cdot e^{[\frac{V_D - V_V}{V_{ex}}]} \cdot \frac{dV_D}{dt}$$

$$\frac{d\xi(V_D)}{dt} = \{I_s \cdot \frac{1}{\eta \cdot V_{th}} \cdot e^{[(\frac{V_D}{\eta \cdot V_{th}})-1]} + (1 - m \cdot [\frac{V_D}{V_0}]^m) \cdot \frac{1}{R_0} \cdot e^{[-(\frac{V_D}{V_0})^m]} + (1 + \frac{V_D}{V_{ex}}) \cdot \frac{1}{R_V}$$
$$\cdot e^{[\frac{V_D - V_V}{V_{ex}}]}\} \cdot \frac{dV_D}{dt}$$

$$\psi(V_D) = I_s \cdot \frac{1}{\eta \cdot V_{th}} \cdot e^{[(\frac{V_D}{\eta \cdot V_{th}})-1]} + (1 - m \cdot [\frac{V_D}{V_0}]^m) \cdot \frac{1}{R_0} \cdot e^{[-(\frac{V_D}{V_0})^m]} + (1 + \frac{V_D}{V_{ex}}) \cdot \frac{1}{R_V}$$
$$\cdot e^{[\frac{V_D - V_V}{V_{ex}}]}$$

$\frac{dI_D}{dt} = \frac{d\xi(V_D)}{dt} = \psi(V_D) \cdot \frac{dV_D}{dt}$. We need to find the expression: $\frac{d\psi(V_D)}{dt}$.

$$\frac{d\psi(V_D)}{dt} = I_s \cdot \frac{1}{(\eta \cdot V_{th})^2} \cdot \frac{dV_D}{dt} \cdot e^{[(\frac{V_D}{\eta \cdot V_{th}})-1]} - m^2 \cdot [\frac{V_D}{V_0}]^{m-1} \cdot \frac{1}{V_0} \cdot \frac{dV_D}{dt} \cdot \frac{1}{R_0} \cdot e^{[-(\frac{V_D}{V_0})^m]}$$
$$- (1 - m \cdot [\frac{V_D}{V_0}]^m) \cdot \frac{1}{R_0} \cdot m \cdot (\frac{V_D}{V_0})^{m-1} \cdot \frac{1}{V_0} \cdot \frac{dV_D}{dt} \cdot e^{[-(\frac{V_D}{V_0})^m]}$$
$$+ \frac{1}{V_{ex}} \cdot \frac{dV_D}{dt} \cdot \frac{1}{R_V} \cdot e^{[\frac{V_D - V_V}{V_{ex}}]}$$
$$+ (1 + \frac{V_D}{V_{ex}}) \cdot \frac{1}{R_V} \cdot \frac{1}{V_{ex}} \cdot \frac{dV_D}{dt} \cdot e^{[\frac{V_D - V_V}{V_{ex}}]}$$

$$\frac{d\psi(V_D)}{dt} = I_s \cdot \frac{1}{(\eta \cdot V_{th})^2} \cdot e^{[(\frac{V_D}{\eta \cdot V_{th}})-1]} \cdot \frac{dV_D}{dt} - \{m + (1 - m \cdot [\frac{V_D}{V_0}]^m)\} \cdot m \cdot (\frac{V_D}{V_0})^{m-1}$$
$$\cdot \frac{1}{R_0 \cdot V_0} \cdot e^{[-(\frac{V_D}{V_0})^m]} \cdot \frac{dV_D}{dt} + \{1 + (1 + \frac{V_D}{V_{ex}})\} \cdot \frac{1}{V_{ex} \cdot R_V} \cdot \frac{dV_D}{dt} \cdot e^{[\frac{V_D - V_V}{V_{ex}}]}$$

$$\frac{d\psi(V_D)}{dt} = \{I_s \cdot \frac{1}{(\eta \cdot V_{th})^2} \cdot e^{[(\frac{V_D}{\eta \cdot V_{th}})-1]} - \{1 + m \cdot (1 - [\frac{V_D}{V_0}]^m)\}$$
$$\cdot (\frac{V_D}{V_0})^{m-1} \cdot \frac{m}{R_0 \cdot V_0} \cdot e^{[-(\frac{V_D}{V_0})^m]}$$
$$+ \{2 + \frac{V_D}{V_{ex}}\} \cdot \frac{1}{V_{ex} \cdot R_V} \cdot e^{[\frac{V_D - V_V}{V_{ex}}]}\} \cdot \frac{dV_D}{dt}$$

$$\psi_1(V_D) = I_s \cdot \frac{1}{(\eta \cdot V_{th})^2} \cdot e^{[(\frac{V_D}{\eta \cdot V_{th}})-1]} - \{1 + m \cdot (1 - [\frac{V_D}{V_0}]^m)\}$$
$$\cdot (\frac{V_D}{V_0})^{m-1} \cdot \frac{m}{R_0 \cdot V_0} \cdot e^{[-(\frac{V_D}{V_0})^m]}$$
$$+ \{2 + \frac{V_D}{V_{ex}}\} \cdot \frac{1}{V_{ex} \cdot R_V} \cdot e^{[\frac{V_D - V_V}{V_{ex}}]}; \quad \frac{d\psi(V_D)}{dt} = \psi_1(V_D) \cdot \frac{dV_D}{dt}$$

$$\frac{dI_D}{dt} = \frac{d\xi(V_D)}{dt} = \psi(V_D) \cdot \frac{dV_D}{dt}; \quad \frac{d^2 I_D}{dt^2} = \frac{d^2\xi(V_D)}{dt^2}$$
$$= \frac{d\psi(V_D)}{dt} \cdot [\frac{dV_D}{dt}] + \psi(V_D) \cdot [\frac{d^2 V_D}{dt^2}]$$

$$\frac{d^2 I_D}{dt^2} = \frac{d^2\xi(V_D)}{dt^2} = \psi_1(V_D) \cdot \frac{dV_D}{dt} \cdot [\frac{dV_D}{dt}] + \psi(V_D) \cdot [\frac{d^2 V_D}{dt^2}]$$
$$= \psi_1(V_D) \cdot [\frac{dV_D}{dt}]^2 + \psi(V_D) \cdot [\frac{d^2 V_D}{dt^2}]$$

We define the following new variables: X(t), Y(t). $X = \frac{dV_D}{dt}$; $Y = \frac{dV_A}{dt}$

$$\frac{d\xi(V_D)}{dt}=\frac{1}{R_L}\cdot\frac{dV_A}{dt}+C\cdot\frac{d^2V_A}{dt^2}+\frac{1}{L}\cdot V_A\Rightarrow\psi(V_D)\cdot\frac{dV_D}{dt}=\frac{1}{R_L}\cdot Y+C\cdot\frac{dY}{dt}+\frac{1}{L}\cdot V_A$$

$$\frac{dV_A}{dt}|_{\frac{dV_b}{dt}=0}=-R_b\cdot\frac{dI_D}{dt}-L_b\cdot\frac{d^2I_D}{dt^2}-\frac{dV_D}{dt}\Rightarrow$$

$$Y=-R_b\cdot\psi(V_D)\cdot\frac{dV_D}{dt}-L_b\cdot\{\psi_1(V_D)\cdot[\frac{dV_D}{dt}]^2+\psi(V_D)\cdot[\frac{d^2V_D}{dt^2}]\}-X$$

$$Y=-R_b\cdot\psi(V_D)\cdot\frac{dV_D}{dt}-L_b\cdot\psi_1(V_D)\cdot[\frac{dV_D}{dt}]^2-L_b\cdot\psi(V_D)\cdot[\frac{d^2V_D}{dt^2}]-X$$

$$Y=-R_b\cdot\psi(V_D)\cdot X-L_b\cdot\psi_1(V_D)\cdot X^2-L_b\cdot\psi(V_D)\cdot\frac{dX}{dt}-X$$

$$\psi(V_D)\cdot\frac{dV_D}{dt}=\frac{1}{R_L}\cdot Y+C\cdot\frac{dY}{dt}+\frac{1}{L}\cdot V_A\Rightarrow\psi(V_D)\cdot X=\frac{1}{R_L}\cdot Y+C\cdot\frac{dY}{dt}+\frac{1}{L}\cdot V_A$$

$$\psi(V_D)\cdot X=\frac{1}{R_L}\cdot Y+C\cdot\frac{dY}{dt}+\frac{1}{L}\cdot V_A\Rightarrow\frac{dY}{dt}=\frac{\psi(V_D)}{C}\cdot X-\frac{1}{R_L\cdot C}\cdot Y-\frac{1}{L\cdot C}\cdot V_A$$

$$Y=-R_b\cdot\psi(V_D)\cdot X-L_b\cdot\psi_1(V_D)\cdot X^2-L_b\cdot\psi(V_D)\cdot\frac{dX}{dt}-X\Rightarrow$$

$$\frac{dX}{dt}=-\frac{R_b}{L_b}\cdot X-\frac{L_b\cdot\psi_1(V_D)}{L_b\cdot\psi(V_D)}\cdot X^2-X\cdot\frac{1}{L_b\cdot\psi(V_D)}-Y\cdot\frac{1}{L_b\cdot\psi(V_D)}$$

$$\frac{dX}{dt}=-(R_b+\frac{1}{\psi(V_D)})\cdot\frac{1}{L_b}\cdot X-\frac{\psi_1(V_D)}{\psi(V_D)}\cdot X^2-Y\cdot\frac{1}{L_b\cdot\psi(V_D)}$$

We can summery our Tunnel diode as a microwave oscillator system's differential equations:

$$\frac{dX}{dt}=-(R_b+\frac{1}{\psi(V_D)})\cdot\frac{1}{L_b}\cdot X-\frac{\psi_1(V_D)}{\psi(V_D)}\cdot X^2-Y\cdot\frac{1}{L_b\cdot\psi(V_D)}$$

$$\frac{dY}{dt}=\frac{\psi(V_D)}{C}\cdot X-\frac{1}{R_L\cdot C}\cdot Y-\frac{1}{L\cdot C}\cdot V_A;\ \frac{dV_D}{dt}=X;\ \frac{dV_A}{dt}=Y$$

To find our system equilibrium points (fixed points), we set $\frac{dX}{dt}=0$; $\frac{dY}{dt}=0$ $\frac{dV_D}{dt}=0$; $\frac{dV_A}{dt}=0$. Then our fixed points are $E^*(X^*,Y^*,V_D^*,V_A^*)=(0,0,V_D^*,0)$.

At equilibrium points: $\lim_{t\to\infty}X(t-\tau_1)=\lim_{t\to\infty}X(t)\,\forall\,t\gg\tau_1$; $\lim_{t\to\infty}Y(t-\tau_2)$ $=\lim_{t\to\infty}Y(t)\,\forall\,t\gg\tau_2$.

Under delay parameters we get the following Tunnel diode system's Delay Differential Equation (DDEs).

Remark: Delay parameters don't effect variables derivative in time $\frac{dX}{dt}, \frac{dY}{dt}, \ldots$

$$\frac{dX}{dt} = -(R_b + \frac{1}{\psi(V_D)}) \cdot \frac{1}{L_b} \cdot X(t-\tau_1) - \frac{\psi_1(V_D)}{\psi(V_D)} \cdot X^2(t-\tau_1) - Y(t-\tau_2) \cdot \frac{1}{L_b \cdot \psi(V_D)}$$
$$\frac{dY}{dt} = \frac{\psi(V_D)}{C} \cdot X(t-\tau_1) - \frac{1}{R_L \cdot C} \cdot Y(t-\tau_2) - \frac{1}{L \cdot C} \cdot V_A;$$
$$\frac{dV_D}{dt} = X(t-\tau_1)\,; \frac{dV_A}{dt} = Y(t-\tau_2)$$

Stability analysis: The standard local stability analysis about any one of the equilibrium points of tunnel diode system consists in adding to its coordinates $[X\, Y\, V_D\, V_A]$ arbitrarily small increments of exponential terms $x\, y\, v_D\, v_A$, and retaining the first order terms in $[x\, y\, v_D\, v_A] \cdot e^{\lambda \cdot t}$. The system of four homogeneous equations leads to a polynomial characteristics equation in the eigenvalue λ. The polynomial characteristics equations accept by set the below TD's variables equations (delayed and undelayed) into four tunnel diode equations. Tunnel diode's fixed values with arbitrarily small increments of exponential form $[x\, y\, v_D\, v_A] \cdot e^{\lambda \cdot t}$ are: i = 0 (first fixed point), i = 1 (second fixed point), i = 2 (third fixed point), etc. Under TD's variables X(t) and Y(t), delays in time: $X(t) \to X(t-\tau_1)$; $Y(t) \to Y(t-\tau_2)$ $|_{I_C(t)=C \cdot \frac{dV_A}{dt}=C \cdot Y(t)} \Rightarrow I_C(t) \to I_C(t-\tau_2)$.

$$X(t) = X^{(i)} + x \cdot e^{\lambda \cdot t};\ Y(t) = Y^{(i)} + y \cdot e^{\lambda \cdot t};\ V_D(t) = V_D^{(i)} + v_D \cdot e^{\lambda \cdot t};$$
$$V_A(t) = V_A^{(i)} + v_A \cdot e^{\lambda \cdot t}$$
$$X(t-\tau_1) = X^{(i)} + x \cdot e^{\lambda \cdot (t-\tau_1)};\ Y(t-\tau_2) = Y^{(i)} + y \cdot e^{\lambda \cdot (t-\tau_2)} \text{ for i} = 0, 1, 2, \ldots$$
$$X(t=0) = X^{(i)} + x;\ Y(t=0) = Y^{(i)} + y;\ V_D(t=0) = V_D^{(i)} + v_D;$$
$$V_A(t=0) = V_A^{(i)} + v_A$$
$$X(t-\tau_1)|_{t=0} = X^{(i)} + x \cdot e^{-\lambda \cdot \tau_1};\ Y(t-\tau_2)|_{t=0} = Y^{(i)} + y \cdot e^{-\lambda \cdot \tau_2} \text{ for i} = 0, 1, 2, \ldots$$

for λ < 0, t > 0 the selected fixed point is stable otherwise λ > 0, t > 0 is Unstable. Our system tends to the selected fixed point exponentially for λ < 0, t > 0 otherwise go away from the selected fixed point exponentially. λ is the eigenvalue parameter which establish if the fixed point is stable or Unstable, additionally his absolute value (|λ|) establish the speed of flow toward or away from the selected fixed point [2–5] (Table 2.14).

The speeds of flow toward or away from the selected fixed point for TD system's variables are (k = 1, 2):

Table 2.14 Tunnel Diode (TD) as a microwave oscillator system eigenvalues options

	λ < 0	λ > 0
t = 0	$X(t=0) = X^{(i)} + x$ $Y(t=0) = Y^{(i)} + y$ $V_D(t=0) = V_D^{(i)} + v_D$ $V_A(t=0) = V_A^{(i)} + v_A$	$X(t=0) = X^{(i)} + x$ $Y(t=0) = Y^{(i)} + y$ $V_D(t=0) = V_D^{(i)} + v_D$ $V_A(t=0) = V_A^{(i)} + v_A$
t > 0	$X(t) = X^{(i)} + x \cdot e^{-\|\lambda\| \cdot t}$ $Y(t) = Y^{(i)} + y \cdot e^{-\|\lambda\| \cdot t}$ $V_D(t) = V_D^{(i)} + v_D \cdot e^{-\|\lambda\| \cdot t}$ $V_A(t) = V_A^{(i)} + v_A \cdot e^{-\|\lambda\| \cdot t}$ for i = 0, 1, 2, ...	$X(t) = X^{(i)} + x \cdot e^{\|\lambda\| \cdot t}$ $Y(t) = Y^{(i)} + y \cdot e^{\|\lambda\| \cdot t}$ $V_D(t) = V_D^{(i)} + v_D \cdot e^{\|\lambda\| \cdot t}$ $V_A(t) = V_A^{(i)} + v_A \cdot e^{\|\lambda\| \cdot t}$ for i = 0, 1, 2, ...
t → ∞	$X(t \to \infty) = X^{(i)}$ $Y(t \to \infty) = Y^{(i)}$ $V_D(t \to \infty) = V_D^{(i)}$ $V_A(t \to \infty) = V_A^{(i)}$	$X(t \to \infty) \approx x \cdot e^{\|\lambda\| \cdot t}$ $Y(t \to \infty) \approx y \cdot e^{\|\lambda\| \cdot t}$ $V_D(t \to \infty) \approx v_D \cdot e^{\|\lambda\| \cdot t}$ $V_A(t \to \infty) \approx v_A \cdot e^{\|\lambda\| \cdot t}$

$$\frac{dX(t)}{dt} = \lim_{\Delta t \to \infty} \frac{X(t+\Delta t) - X(t)}{\Delta t} = \lim_{\Delta t \to \infty} \frac{X^{(i)} + x \cdot e^{\lambda \cdot (t+\Delta t)} - [X(t) + x \cdot e^{\lambda \cdot t}]}{\Delta t} \overset{e^{\lambda \cdot \Delta t} \approx 1 + \Delta t \cdot \lambda}{\longrightarrow}$$

$$\lim_{\Delta t \to \infty} \frac{x \cdot e^{\lambda \cdot t} \cdot (e^{\lambda \cdot \Delta t} - 1)}{\Delta t} = \frac{x \cdot e^{\lambda \cdot t} \cdot (1 + \lambda \cdot \Delta t - 1)}{\Delta t} = x \cdot \lambda \cdot e^{\lambda \cdot t}; \frac{dY(t)}{dt} = y \cdot \lambda \cdot e^{\lambda \cdot t}$$

$$\frac{dV_D(t)}{dt} = v_D \cdot \lambda \cdot e^{\lambda \cdot t}; \frac{dV_A(t)}{dt} = v_A \cdot \lambda \cdot e^{\lambda \cdot t}; \frac{dX(t-\tau_1)}{dt} = x \cdot \lambda \cdot e^{\lambda \cdot t} \cdot e^{-\lambda \cdot \tau_1};$$

$$\frac{dY(t-\tau_2)}{dt} = y \cdot \lambda \cdot e^{\lambda \cdot t} \cdot e^{-\lambda \cdot \tau_2}$$

First we take the Tunnel diode's voltage derivative differential equations ($\frac{dX(t)}{dt} = \ldots$). First TD's system differential equation: $\frac{dX}{dt} = -(R_b + \frac{1}{\psi(V_D)}) \cdot \frac{1}{L_b} \cdot X - \frac{\psi_1(V_D)}{\psi(V_D)} \cdot X^2 - Y \cdot \frac{1}{L_b \cdot \psi(V_D)}$ and adding to it's coordinates $[X\, Y\, V_D\, V_A]$ arbitrarily small increments of exponential form $[x\, y\, v_D\, v_A] \cdot e^{\lambda \cdot t}$ and retaining the first order terms in $x\, y\, v_D\, v_A$.

$$\frac{dX}{dt} = -(R_b + \frac{1}{\psi(V_D)}) \cdot \frac{1}{L_b} \cdot X(t-\tau_1) - \frac{\psi_1(V_D)}{\psi(V_D)} \cdot X^2(t-\tau_1) - Y(t-\tau_2) \cdot \frac{1}{L_b \cdot \psi(V_D)}$$

$$x \cdot \lambda \cdot e^{\lambda \cdot t} = -(R_b + \frac{1}{\psi(V_D^{(i)} + v_D \cdot e^{\lambda \cdot t})}) \cdot \frac{1}{L_b} \cdot [X^{(i)} + x \cdot e^{\lambda \cdot (t-\tau_1)}]$$

$$- \frac{\psi_1(V_D^{(i)} + v_D \cdot e^{\lambda \cdot t})}{\psi(V_D^{(i)} + v_D \cdot e^{\lambda \cdot t})} \cdot [X^{(i)} + x \cdot e^{\lambda \cdot (t-\tau_1)}]^2 - [Y^{(i)} + y \cdot e^{\lambda \cdot (t-\tau_2)}] \cdot \frac{1}{L_b \cdot \psi(V_D^{(i)} + v_D \cdot e^{\lambda \cdot t})}$$

$$\psi(V_D) = I_s \cdot \frac{1}{\eta \cdot V_{th}} \cdot e^{[(\frac{V_D}{\eta \cdot V_{th}})-1]} + (1 - m \cdot [\frac{V_D}{V_0}]^m) \cdot \frac{1}{R_0} \cdot e^{[-(\frac{V_D}{V_0})^m]} + (1 + \frac{V_D}{V_{ex}}) \cdot \frac{1}{R_V} \cdot e^{[\frac{V_D - V_V}{V_{ex}}]}$$

$$V_D \to V_D^{(i)} + v_D \cdot e^{\lambda \cdot t} \Rightarrow \psi(V_D) \to \psi(V_D^{(i)} + v_D \cdot e^{\lambda \cdot t})$$

$$\frac{1}{\psi(V_D^{(i)} + v_D \cdot e^{\lambda \cdot t})} = \frac{1}{\{I_s \cdot \frac{1}{\eta \cdot V_{th}} \cdot e^{[(\frac{[V_D^{(i)} + v_D \cdot e^{\lambda \cdot t}]}{\eta \cdot V_{th}})-1]} + (1 - m \cdot [\frac{[V_D^{(i)} + v_D \cdot e^{\lambda \cdot t}]}{V_0}]^m) \cdot \frac{1}{R_0} \cdot e^{[-(\frac{[V_D^{(i)} + v_D \cdot e^{\lambda \cdot t}]}{V_0})^m]} (1 + \frac{[V_D^{(i)} + v_D \cdot e^{\lambda \cdot t}]}{V_{ex}}) \cdot \frac{1}{R_V} \cdot e^{[\frac{[V_D^{(i)} + v_D \cdot e^{\lambda \cdot t}] - V_V}{V_{ex}}]}\}}$$

$\frac{1}{\psi(V_D^{(i)} + v_D \cdot e^{\lambda \cdot t})} = \frac{1}{\psi(V_D^{(i)})} + \Gamma_1(V_D^{(i)}, v_D, \ldots)$. We need to find $\Gamma_1(V_D^{(i)}, v_D, \ldots)$ function.

$$\frac{1}{\psi(V_D^{(i)} + v_D \cdot e^{\lambda \cdot t})} = \frac{1}{\psi(V_D^{(i)})} + \Gamma_1(V_D^{(i)}, v_D, \ldots) \Rightarrow \Gamma_1(V_D^{(i)}, v_D, \ldots) = \frac{1}{\psi(V_D^{(i)} + v_D \cdot e^{\lambda \cdot t})} - \frac{1}{\psi(V_D^{(i)})}$$

$$\frac{1}{\psi(V_D^{(i)})} = \frac{1}{I_s \cdot \frac{1}{\eta \cdot V_{th}} \cdot e^{[(\frac{V_D^{(i)}}{\eta \cdot V_{th}})-1]} + (1 - m \cdot [\frac{V_D^{(i)}}{V_0}]^m) \cdot \frac{1}{R_0} \cdot e^{[-(\frac{V_D^{(i)}}{V_0})^m]} + (1 + \frac{V_D^{(i)}}{V_{ex}}) \cdot \frac{1}{R_V} \cdot e^{[\frac{V_D^{(i)} - V_V}{V_{ex}}]}}$$

<u>Mathematical</u> <u>assumptions</u>: $[\frac{V_D^{(i)} + v_D \cdot e^{\lambda \cdot t}}{V_0}]^m = [\frac{V_D^{(i)}}{V_0} + \frac{v_D \cdot e^{\lambda \cdot t}}{V_0}]^m$; $A = \frac{V_D^{(i)}}{V_0}$; $B = \frac{v_D \cdot e^{\lambda \cdot t}}{V_0}$

$$e^{x_1} \cdot e^{x_2} \cdots e^{x_n} = e^{\sum_{i=1}^{n} x_i}; \ (A+B)^m = A^m + \binom{m}{1} \cdot A^{(m-1)} \cdot B + \binom{m}{2} \cdot A^{(m-2)} \cdot B^2 + \cdots + \binom{m}{m-1} \cdot A \cdot B^{(m-1)} + B^m$$

$$[\frac{V_D^{(i)} + v_D \cdot e^{\lambda \cdot t}}{V_0}]^m = [\frac{V_D^{(i)}}{V_0} + \frac{v_D \cdot e^{\lambda \cdot t}}{V_0}]^m = [\frac{V_D^{(i)}}{V_0}]^m + \begin{pmatrix} m \\ 1 \end{pmatrix} \cdot [\frac{V_D^{(i)}}{V_0}]^{(m-1)} \cdot [\frac{v_D \cdot e^{\lambda \cdot t}}{V_0}]$$
$$+ \begin{pmatrix} m \\ 2 \end{pmatrix} \cdot [\frac{V_D^{(i)}}{V_0}]^{(m-2)} \cdot [\frac{v_D \cdot e^{\lambda \cdot t}}{V_0}]^2$$
$$+ \cdots + \begin{pmatrix} m \\ m-1 \end{pmatrix} \cdot [\frac{V_D^{(i)}}{V_0}] \cdot [\frac{v_D \cdot e^{\lambda \cdot t}}{V_0}]^{(m-1)} + [\frac{v_D \cdot e^{\lambda \cdot t}}{V_0}]^m$$

$$[\frac{V_D^{(i)} + v_D \cdot e^{\lambda \cdot t}}{V_0}]^m = [\frac{V_D^{(i)}}{V_0} + \frac{v_D \cdot e^{\lambda \cdot t}}{V_0}]^m = [\frac{V_D^{(i)}}{V_0} + \frac{v_D \cdot e^{\lambda \cdot t}}{V_0}]^m = [\frac{V_D^{(i)}}{V_0}]^m$$
$$+ \begin{pmatrix} m \\ 1 \end{pmatrix} \cdot [\frac{V_D^{(i)}}{V_0}]^{(m-1)} \cdot [\frac{v_D \cdot e^{\lambda \cdot t}}{V_0}] + \begin{pmatrix} m \\ 2 \end{pmatrix} \cdot [\frac{V_D^{(i)}}{V_0}]^{(m-2)} \cdot \frac{v_D^2 \cdot e^{2 \cdot \lambda \cdot t}}{V_0^2}$$
$$+ \cdots + \begin{pmatrix} m \\ m-1 \end{pmatrix} \cdot [\frac{V_D^{(i)}}{V_0}] \cdot \frac{v_D^{(m-1)} \cdot e^{(m-1) \cdot \lambda \cdot t}}{V_0^{(m-1)}} + \frac{v_D^m \cdot e^{m \cdot \lambda \cdot t}}{V_0^m}$$

We consider $v_D^m \approx 0 \, \forall \, m \geq 2$ then

$$[\frac{V_D^{(i)} + v_D \cdot e^{\lambda \cdot t}}{V_0}]^m = [\frac{V_D^{(i)}}{V_0} + \frac{v_D \cdot e^{\lambda \cdot t}}{V_0}]^m = [\frac{V_D^{(i)}}{V_0}]^m$$
$$+ \begin{pmatrix} m \\ 1 \end{pmatrix} \cdot [\frac{V_D^{(i)}}{V_0}]^{(m-1)} \cdot [\frac{v_D \cdot e^{\lambda \cdot t}}{V_0}] + (\varepsilon \to 0)$$

$$e^{[-(\frac{[V_D^{(i)} + v_D \cdot e^{\lambda \cdot t}]}{V_0})^m]} = e^{[-\{[\frac{V_D^{(i)}}{V_0}]^m + \begin{pmatrix} m \\ 1 \end{pmatrix} \cdot [\frac{V_D^{(i)}}{V_0}]^{(m-1)} \cdot [\frac{v_D \cdot e^{\lambda \cdot t}}{V_0}] + (\varepsilon \to 0)\}]}$$
$$= e^{-[\frac{V_D^{(i)}}{V_0}]^m} \cdot e^{-\{\begin{pmatrix} m \\ 1 \end{pmatrix} \cdot [\frac{V_D^{(i)}}{V_0}]^{(m-1)} \cdot [\frac{v_D \cdot e^{\lambda \cdot t}}{V_0}]\}}$$

$$e^{[(\frac{[V_D^{(i)} + v_D \cdot e^{\lambda \cdot t}]}{\eta \cdot V_{th}}) - 1]} = e^{[\frac{[V_D^{(i)}]}{\eta \cdot V_{th}} - 1]} \cdot e^{[\frac{v_D \cdot e^{\lambda \cdot t}}{\eta \cdot V_{th}}]}; \; e^{[\frac{[V_D^{(i)} + v_D \cdot e^{\lambda \cdot t}] - V_V}{V_{ex}}]} = e^{[\frac{V_D^{(i)} - V_V}{V_{ex}}]} \cdot e^{[\frac{v_D \cdot e^{\lambda \cdot t}}{V_{ex}}]}$$

$$\frac{1}{\psi(V_D^{(i)} + v_D \cdot e^{\lambda \cdot t})} = \frac{1}{\begin{aligned} &\{I_s \cdot \frac{1}{\eta \cdot V_{th}} \cdot e^{[(\frac{[V_D^{(i)} + v_D \cdot e^{\lambda \cdot t}]}{\eta \cdot V_{th}}) - 1]} + (1 - m \cdot [\frac{[V_D^{(i)} + v_D \cdot e^{\lambda \cdot t}]}{V_0}]^m) \\ &\cdot \frac{1}{R_0} \cdot e^{[-(\frac{[V_D^{(i)} + v_D \cdot e^{\lambda \cdot t}]}{V_0})^m]} + (1 + \frac{[V_D^{(i)} + v_D \cdot e^{\lambda \cdot t}]}{V_{ex}}) \\ &\cdot \frac{1}{R_V} \cdot e^{[\frac{[V_D^{(i)} + v_D \cdot e^{\lambda \cdot t}] - V_V}{V_{ex}}]}\} \end{aligned}}$$

$$\frac{1}{\psi(V_D^{(i)} + v_D \cdot e^{\lambda \cdot t})} = \frac{1}{\begin{array}{l} \{I_s \cdot \frac{1}{\eta \cdot V_{th}} \cdot e^{[\frac{[V_D^{(i)}]}{\eta \cdot V_{th}} - 1]} \cdot e^{[\frac{v_D \cdot e^{\lambda \cdot t}}{\eta \cdot V_{th}}]} + (1 - m \cdot \{[\frac{V_D^{(i)}}{V_0}]^m \\ + \binom{m}{1} \cdot [\frac{V_D^{(i)}}{V_0}]^{(m-1)} \cdot [\frac{v_D \cdot e^{\lambda \cdot t}}{V_0}]\}) \\ \cdot \frac{1}{R_0} \cdot e^{-[\frac{V_D^{(i)}}{V_0}]^m} \cdot e^{-\{\binom{m}{1} \cdot [\frac{V_D^{(i)}}{V_0}]^{(m-1)} \cdot [\frac{v_D \cdot e^{\lambda \cdot t}}{V_0}]\}} \\ + \{(1 + \frac{V_D^{(i)}}{V_{ex}}) + \frac{v_D \cdot e^{\lambda \cdot t}}{V_{ex}}\} \cdot \frac{1}{R_V} \cdot e^{[\frac{V_D^{(i)} - V_V}{V_{ex}}]} \cdot e^{[\frac{v_D \cdot e^{\lambda \cdot t}}{V_{ex}}]}\} \end{array}}$$

$$\frac{1}{\psi(V_D^{(i)} + v_D \cdot e^{\lambda \cdot t})} = \frac{1}{\begin{array}{l} \{I_s \cdot \frac{1}{\eta \cdot V_{th}} \cdot e^{[\frac{[V_D^{(i)}]}{\eta \cdot V_{th}} - 1]} \cdot e^{[\frac{v_D \cdot e^{\lambda \cdot t}}{\eta \cdot V_{th}}]} + (1 - m \cdot [\frac{V_D^{(i)}}{V_0}]^m) \\ \cdot \frac{1}{R_0} \cdot e^{-[\frac{V_D^{(i)}}{V_0}]^m} \cdot e^{-\{\binom{m}{1} \cdot [\frac{V_D^{(i)}}{V_0}]^{(m-1)} \cdot [\frac{v_D \cdot e^{\lambda \cdot t}}{V_0}]\}} \\ - m \cdot \binom{m}{1} \cdot [\frac{V_D^{(i)}}{V_0}]^{(m-1)} \cdot [\frac{v_D \cdot e^{\lambda \cdot t}}{V_0}] \\ \cdot \frac{1}{R_0} \cdot e^{-[\frac{V_D^{(i)}}{V_0}]^m} \cdot e^{-\{\binom{m}{1} \cdot [\frac{V_D^{(i)}}{V_0}]^{(m-1)} \cdot [\frac{v_D \cdot e^{\lambda \cdot t}}{V_0}]\}} \\ + (1 + \frac{V_D^{(i)}}{V_{ex}}) \cdot \frac{1}{R_V} \cdot e^{[\frac{V_D^{(i)} - V_V}{V_{ex}}]} \cdot e^{[\frac{v_D \cdot e^{\lambda \cdot t}}{V_{ex}}]} + \frac{v_D \cdot e^{\lambda \cdot t}}{V_{ex}} \\ \cdot \frac{1}{R_V} \cdot e^{[\frac{V_D^{(i)} - V_V}{V_{ex}}]} \cdot e^{[\frac{v_D \cdot e^{\lambda \cdot t}}{V_{ex}}]}\} \end{array}}$$

We define the following global parameters: $B_i = B_i(V_D^{(i)}) \, \forall \, i = 1, 2, 3;$ $B_4 = B_4(V_D^{(i)}, v_D)$

$$B_1 = I_s \cdot \frac{1}{\eta \cdot V_{th}} \cdot e^{[\frac{[V_D^{(i)}]}{\eta \cdot V_{th}} - 1]}; \; B_2 = (1 - m \cdot [\frac{V_D^{(i)}}{V_0}]^m) \cdot \frac{1}{R_0} \cdot e^{-[\frac{V_D^{(i)}}{V_0}]^m}; \; B_3 = (1 + \frac{V_D^{(i)}}{V_{ex}}) \cdot \frac{1}{R_V} \cdot e^{[\frac{V_D^{(i)} - V_V}{V_{ex}}]}$$

$$\frac{1}{\psi(V_D^{(i)}+v_D\cdot e^{\lambda\cdot t})}=\frac{1}{\begin{array}{l}\{B_1\cdot e^{[\frac{v_D\cdot e^{\lambda\cdot t}}{\eta\cdot V_{th}}]}+B_2\cdot e^{-\{\binom{m}{1}\cdot[\frac{V_D^{(i)}}{V_0}]^{(m-1)}\cdot[\frac{v_D\cdot e^{\lambda\cdot t}}{V_0}]\}}+B_3\cdot e^{[\frac{v_D\cdot e^{\lambda\cdot t}}{V_{ex}}]}\\ +\frac{v_D\cdot e^{\lambda\cdot t}}{V_{ex}}\cdot\frac{1}{R_V}\cdot e^{[\frac{V_D^{(i)}-V_V}{V_{ex}}]}\cdot e^{[\frac{v_D\cdot e^{\lambda\cdot t}}{V_{ex}}]}\\ -m\cdot\binom{m}{1}\cdot[\frac{V_D^{(i)}}{V_0}]^{(m-1)}\cdot[\frac{v_D\cdot e^{\lambda\cdot t}}{V_0}]\\ \cdot\frac{1}{R_0}\cdot e^{-[\frac{V_D^{(i)}}{V_0}]^m}\cdot e^{-\{\binom{m}{1}\cdot[\frac{V_D^{(i)}}{V_0}]^{(m-1)}\cdot[\frac{v_D\cdot e^{\lambda\cdot t}}{V_0}]\}}\}\end{array}}$$

We define the following parameter:

$$B_4=\frac{v_D\cdot e^{\lambda\cdot t}}{V_{ex}}\cdot\frac{1}{R_V}\cdot e^{[\frac{V_D^{(i)}-V_V}{V_{ex}}]}\cdot e^{[\frac{v_D\cdot e^{\lambda\cdot t}}{V_{ex}}]}-m\cdot\binom{m}{1}\cdot[\frac{V_D^{(i)}}{V_0}]^{(m-1)}\cdot[\frac{v_D\cdot e^{\lambda\cdot t}}{V_0}]\cdot\frac{1}{R_0}\cdot e^{-[\frac{V_D^{(i)}}{V_0}]^m}\cdot e^{-\{\binom{m}{1}\cdot[\frac{V_D^{(i)}}{V_0}]^{(m-1)}\cdot[\frac{v_D\cdot e^{\lambda\cdot t}}{V_0}]\}}$$

$$\frac{1}{\psi(V_D^{(i)}+v_D\cdot e^{\lambda\cdot t})}=\frac{1}{\{B_1\cdot e^{[\frac{v_D\cdot e^{\lambda\cdot t}}{\eta\cdot V_{th}}]}+B_2\cdot e^{-\{\binom{m}{1}\cdot[\frac{V_D^{(i)}}{V_0}]^{(m-1)}\cdot[\frac{v_D\cdot e^{\lambda\cdot t}}{V_0}]\}}+B_3\cdot e^{[\frac{v_D\cdot e^{\lambda\cdot t}}{V_{ex}}]}+B_4\}};$$

$$\frac{1}{\psi(V_D^{(i)})}=\frac{1}{\sum_{k=1}^{3}B_k}$$

$$\frac{1}{\psi(V_D^{(i)}+v_D\cdot e^{\lambda\cdot t})}=\frac{1}{\psi(V_D^{(i)})}+\Gamma_1(V_D^{(i)},v_D,\ldots)\Rightarrow\Gamma_1(V_D^{(i)},v_D,\ldots)=\frac{1}{\psi(V_D^{(i)}+v_D\cdot e^{\lambda\cdot t})}-\frac{1}{\psi(V_D^{(i)})}$$

$$\Gamma_1(V_D^{(i)},v_D,\ldots)=\frac{1}{\{B_1\cdot e^{[\frac{v_D\cdot e^{\lambda\cdot t}}{\eta\cdot V_{th}}]}+B_2\cdot e^{-\{\binom{m}{1}\cdot[\frac{V_D^{(i)}}{V_0}]^{(m-1)}\cdot[\frac{v_D\cdot e^{\lambda\cdot t}}{V_0}]\}}+B_3\cdot e^{[\frac{v_D\cdot e^{\lambda\cdot t}}{V_{ex}}]}+B_4\}}-\frac{1}{\sum_{k=1}^{3}B_k}$$

$$\Gamma_1(V_D^{(i)}, v_D, \ldots) = \frac{[\sum_{k=1}^{3} B_k] - B_1 \cdot e^{[\frac{v_D \cdot e^{\lambda \cdot t}}{\eta \cdot V_{th}}]} - B_2 \cdot e^{-\{\binom{m}{1} \cdot [\frac{V_D^{(i)}}{V_0}]^{(m-1)} \cdot [\frac{v_D \cdot e^{\lambda \cdot t}}{V_0}]\}} - B_3 \cdot e^{[\frac{v_D \cdot e^{\lambda \cdot t}}{V_{ex}}]} - B_4}{(B_1 \cdot e^{[\frac{v_D \cdot e^{\lambda \cdot t}}{\eta \cdot V_{th}}]} + B_2 \cdot e^{-\{\binom{m}{1} \cdot [\frac{V_D^{(i)}}{V_0}]^{(m-1)} \cdot [\frac{v_D \cdot e^{\lambda \cdot t}}{V_0}]\}} + B_3 \cdot e^{[\frac{v_D \cdot e^{\lambda \cdot t}}{V_{ex}}]} + B_4) \cdot [\sum_{k=1}^{3} B_k]}$$

Binomial coefficients:

$$\binom{n}{k} = \prod_{i=1}^{k} \frac{n-i+1}{k!} = \frac{n \cdot (n-1) \cdots (n-k+1)}{k!} |_{\substack{n=m \\ k=1}}$$
$$= \prod_{i=1}^{1} (n-i+1)|_{\substack{n=m \\ k=1}} = m$$

$$\binom{n}{k} = \frac{n!}{k!(n-k)!} \forall\, 0 \le k \le n;\; n = m \,\&\, k = 1 \Rightarrow \binom{m}{1} = m;$$
$$\binom{n}{k} = \prod_{i=1}^{k} \frac{n-i+1}{k!}$$

$$\Gamma_1(V_D^{(i)}, v_D, \ldots) = \frac{[\sum_{k=1}^{3} B_k] - B_1 \cdot e^{[\frac{v_D \cdot e^{\lambda \cdot t}}{\eta \cdot V_{th}}]} - B_2 \cdot e^{-\{\binom{m}{1} \cdot [\frac{V_D^{(i)}}{V_0}]^{(m-1)} \cdot [\frac{v_D \cdot e^{\lambda \cdot t}}{V_0}]\}} - B_3 \cdot e^{[\frac{v_D \cdot e^{\lambda \cdot t}}{V_{ex}}]} - B_4}{(B_1 \cdot e^{[\frac{v_D \cdot e^{\lambda \cdot t}}{\eta \cdot V_{th}}]} + B_2 \cdot e^{-\{\binom{m}{1} \cdot [\frac{V_D^{(i)}}{V_0}]^{(m-1)} \cdot [\frac{v_D \cdot e^{\lambda \cdot t}}{V_0}]\}} + B_3 \cdot e^{[\frac{v_D \cdot e^{\lambda \cdot t}}{V_{ex}}]} + B_4) \cdot [\sum_{k=1}^{3} B_k]}$$

$$e^{[\frac{v_D \cdot e^{\lambda \cdot t}}{\eta \cdot V_{th}}]} = 1 + \frac{v_D \cdot e^{\lambda \cdot t}}{\eta \cdot V_{th}} + \frac{1}{2} \cdot \frac{v_D^2 \cdot e^{2 \cdot \lambda \cdot t}}{[\eta \cdot V_{th}]^2} + \cdots |_{v_D^k \approx 0 \forall k \ge 2} = 1 + \frac{v_D \cdot e^{\lambda \cdot t}}{\eta \cdot V_{th}} + (\varepsilon \to 0)$$
$$\approx 1 + \frac{v_D \cdot e^{\lambda \cdot t}}{\eta \cdot V_{th}}$$

$$e^{-\{\binom{m}{1} \cdot [\frac{V_D^{(i)}}{V_0}]^{(m-1)} \cdot [\frac{v_D \cdot e^{\lambda \cdot t}}{V_0}]\}} = 1 - \binom{m}{1} \cdot [\frac{V_D^{(i)}}{V_0}]^{(m-1)} \cdot [\frac{v_D \cdot e^{\lambda \cdot t}}{V_0}] + \frac{1}{2}$$
$$\cdot \binom{m}{1}^2 \cdot [\frac{V_D^{(i)}}{V_0}]^{2 \cdot (m-1)} \cdot [\frac{v_D^2 \cdot e^{2 \cdot \lambda \cdot t}}{V_0^2}] + \cdots$$

$$e^{-\{\binom{m}{1} \cdot [\frac{V_D^{(i)}}{V_0}]^{(m-1)} \cdot [\frac{v_D \cdot e^{\lambda \cdot t}}{V_0}]\}} = 1 - \binom{m}{1} \cdot [\frac{V_D^{(i)}}{V_0}]^{(m-1)} \cdot [\frac{v_D \cdot e^{\lambda \cdot t}}{V_0}] + (\varepsilon \to 0)$$
$$\approx 1 - \binom{m}{1} \cdot [\frac{V_D^{(i)}}{V_0}]^{(m-1)} \cdot [\frac{v_D \cdot e^{\lambda \cdot t}}{V_0}]$$

$$e^{[\frac{v_D \cdot e^{\lambda \cdot t}}{V_{ex}}]} = 1 + \frac{v_D \cdot e^{\lambda \cdot t}}{V_{ex}} + \frac{1}{2} \cdot \frac{v_D^2 \cdot e^{2 \cdot \lambda \cdot t}}{V_{ex}^2} + \cdots = 1 + \frac{v_D \cdot e^{\lambda \cdot t}}{V_{ex}} + (\varepsilon \to 0) \approx 1 + \frac{v_D \cdot e^{\lambda \cdot t}}{V_{ex}}$$

$$\Gamma_1(V_D^{(i)}, v_D, \ldots.) = \frac{[\sum_{k=1}^{3} B_k] - B_1 \cdot [1 + \frac{v_D \cdot e^{\lambda \cdot t}}{\eta \cdot V_{th}}] - B_2 \cdot \{1 - \binom{m}{1} \cdot [\frac{V_D^{(i)}}{V_0}]^{(m-1)} \cdot [\frac{v_D \cdot e^{\lambda \cdot t}}{V_0}]\} - B_3 \cdot [1 + \frac{v_D \cdot e^{\lambda \cdot t}}{V_{ex}}] - B_4}{(B_1 \cdot e^{[\frac{v_D \cdot e^{\lambda \cdot t}}{\eta \cdot V_{th}}]} + B_2 \cdot e^{-\{\binom{m}{1} \cdot [\frac{V_D^{(i)}}{V_0}]^{(m-1)} \cdot [\frac{v_D \cdot e^{\lambda \cdot t}}{V_0}]\}} + B_3 \cdot e^{[\frac{v_D \cdot e^{\lambda \cdot t}}{V_{ex}}]} + B_4) \cdot [\sum_{k=1}^{3} B_k]}$$

$$\Gamma_1(V_D^{(i)}, v_D, \ldots) = \frac{-B_1 \frac{v_D \cdot e^{\lambda \cdot t}}{\eta \cdot V_{th}} + B_2 \cdot \binom{m}{1} \cdot [\frac{V_D^{(i)}}{V_0}]^{(m-1)} \cdot [\frac{v_D \cdot e^{\lambda \cdot t}}{V_0}] - B_3 \cdot \frac{v_D \cdot e^{\lambda \cdot t}}{V_{ex}} - B_4}{(B_1 \cdot e^{[\frac{v_D \cdot e^{\lambda \cdot t}}{\eta \cdot V_{th}}]} + B_2 \cdot e^{-\{\binom{m}{1} \cdot [\frac{V_D^{(i)}}{V_0}]^{(m-1)} \cdot [\frac{v_D \cdot e^{\lambda \cdot t}}{V_0}]\}} + B_3 \cdot e^{[\frac{v_D \cdot e^{\lambda \cdot t}}{V_{ex}}]} + B_4) \cdot [\sum_{k=1}^{3} B_k]}$$

$$\Gamma_1(V_D^{(i)}, v_D, \ldots.) = \frac{-B_1 \frac{v_D \cdot e^{\lambda \cdot t}}{\eta \cdot V_{th}} + B_2 \cdot \binom{m}{1} \cdot [\frac{V_D^{(i)}}{V_0}]^{(m-1)} \cdot [\frac{v_D \cdot e^{\lambda \cdot t}}{V_0}] - B_3 \cdot \frac{v_D \cdot e^{\lambda \cdot t}}{V_{ex}} - B_4}{(B_1 \cdot e^{[\frac{v_D \cdot e^{\lambda \cdot t}}{\eta \cdot V_{th}}]} + B_2 \cdot e^{-\{\binom{m}{1} \cdot [\frac{V_D^{(i)}}{V_0}]^{(m-1)} \cdot [\frac{v_D \cdot e^{\lambda \cdot t}}{V_0}]\}} + B_3 \cdot e^{[\frac{v_D \cdot e^{\lambda \cdot t}}{V_{ex}}]}) \cdot [\sum_{k=1}^{3} B_k] + B_4 \cdot [\sum_{k=1}^{3} B_k]}$$

We define: $\Omega_4(v_D) = B_1 \cdot e^{[\frac{v_D \cdot e^{\lambda \cdot t}}{\eta \cdot V_{th}}]} + B_2 \cdot e^{-\{\binom{m}{1} \cdot [\frac{V_D^{(i)}}{V_0}]^{(m-1)} \cdot [\frac{v_D \cdot e^{\lambda \cdot t}}{V_0}]\}} + B_3 \cdot e^{[\frac{v_D \cdot e^{\lambda \cdot t}}{V_{ex}}]}$

$$\Gamma_1(V_D^{(i)}, v_D, \ldots) = \frac{-B_1 \frac{v_D \cdot e^{\lambda \cdot t}}{\eta \cdot V_{th}} + B_2 \cdot \binom{m}{1} \cdot [\frac{V_D^{(i)}}{V_0}]^{(m-1)} \cdot [\frac{v_D \cdot e^{\lambda \cdot t}}{V_0}] - B_3 \cdot \frac{v_D \cdot e^{\lambda \cdot t}}{V_{ex}} - B_4}{\Omega_4(v_D) \cdot [\sum_{k=1}^{3} B_k] + B_4 \cdot [\sum_{k=1}^{3} B_k]}$$

$$B_4 = \frac{v_D \cdot e^{\lambda \cdot t}}{V_{ex}} \cdot \frac{1}{R_V} \cdot e^{[\frac{V_D^{(i)} - V_V}{V_{ex}}]} \cdot e^{[\frac{v_D \cdot e^{\lambda \cdot t}}{V_{ex}}]} - m \cdot \binom{m}{1} \cdot [\frac{V_D^{(i)}}{V_0}]^{(m-1)} \cdot [\frac{v_D \cdot e^{\lambda \cdot t}}{V_0}] \cdot \frac{1}{R_0} \cdot e^{-[\frac{V_D^{(i)}}{V_0}]^m} \cdot e^{-\{\binom{m}{1} \cdot [\frac{V_D^{(i)}}{V_0}]^{(m-1)} \cdot [\frac{v_D \cdot e^{\lambda \cdot t}}{V_0}]\}}$$

We define the following variables: $\Omega_1(v_D)|_{v_D^2 \approx 0} = \Omega_2(v_D)|_{v_D^2 \approx 0} = v_D$

$$\Omega_1(v_D) = v_D \cdot e^{[\frac{v_D \cdot e^{\lambda \cdot t}}{V_{ex}}]} \simeq; v_D \cdot (1 + \frac{v_D \cdot e^{\lambda \cdot t}}{V_{ex}}) = v_D + \frac{v_D^2 \cdot e^{2 \cdot \lambda \cdot t}}{V_{ex}^2}|_{v_D^2 \approx 0} = v_D$$

$$\Omega_2(v_D) = v_D \cdot e^{-\{\binom{m}{1} \cdot [\frac{V_D^{(i)}}{V_0}]^{(m-1)} \cdot [\frac{v_D \cdot e^{\lambda \cdot t}}{V_0}]\}} = v_D \cdot (1 - \binom{m}{1} \cdot [\frac{V_D^{(i)}}{V_0}]^{(m-1)} \cdot [\frac{v_D \cdot e^{\lambda \cdot t}}{V_0}])$$

$$= v_D - \binom{m}{1} \cdot [\frac{V_D^{(i)}}{V_0}]^{(m-1)} \cdot [\frac{v_D^2 \cdot e^{\lambda \cdot t}}{V_0}]$$

$$\Omega_2(v_D) = v_D \cdot e^{-\{\binom{m}{1} \cdot [\frac{V_D^{(i)}}{V_0}]^{(m-1)} \cdot [\frac{v_D \cdot e^{\lambda \cdot t}}{V_0}]\}}|_{v_D^2 \approx 0} = v_D$$

$$B_4 = \frac{e^{\lambda \cdot t}}{V_{ex}} \cdot \frac{1}{R_V} \cdot e^{[\frac{V_D^{(i)} - V_V}{V_{ex}}]} \cdot \Omega_1(v_D) - m \cdot \binom{m}{1} \cdot [\frac{V_D^{(i)}}{V_0}]^{(m-1)} \cdot [\frac{e^{\lambda \cdot t}}{V_0}] \cdot \frac{1}{R_0} \cdot e^{-[\frac{V_D^{(i)}}{V_0}]^m} \cdot \Omega_2(v_D)$$

$$B_4|_{\Omega_1(v_D)|_{v_D^2 \approx 0} = \Omega_2(v_D)|_{v_D^2 \approx 0} = v_D} = \{\frac{1}{V_{ex} \cdot R_V} \cdot e^{[\frac{V_D^{(i)} - V_V}{V_{ex}}]} - m \cdot \binom{m}{1} \cdot [\frac{V_D^{(i)}}{V_0}]^{(m-1)} \cdot \frac{1}{V_0 \cdot R_0} \cdot e^{-[\frac{V_D^{(i)}}{V_0}]^m}\} \cdot v_D \cdot e^{\lambda \cdot t}$$

$$\Omega_3(V_D^{(i)}) = \frac{1}{V_{ex} \cdot R_V} \cdot e^{[\frac{V_D^{(i)} - V_V}{V_{ex}}]} - m \cdot \binom{m}{1} \cdot [\frac{V_D^{(i)}}{V_0}]^{(m-1)} \cdot \frac{1}{V_0 \cdot R_0} \cdot e^{-[\frac{V_D^{(i)}}{V_0}]^m}; B_4|_{\Omega_1(v_D)|_{v_D^2 \approx 0} = \Omega_2(v_D)|_{v_D^2 \approx 0} = v_D} = \Omega_3(V_D^{(i)}) \cdot v_D \cdot e^{\lambda \cdot t}$$

$$\Omega_4(v_D) = B_1 \cdot e^{[\frac{v_D \cdot e^{\lambda \cdot t}}{\eta \cdot V_{th}}]} + B_2 \cdot e^{-\{\binom{m}{1} \cdot [\frac{V_D^{(i)}}{V_0}]^{(m-1)} \cdot [\frac{v_D \cdot e^{\lambda \cdot t}}{V_0}]\}} + B_3 \cdot e^{[\frac{v_D \cdot e^{\lambda \cdot t}}{V_{ex}}]}$$

$$\Omega_4(v_D) = B_1 \cdot [1 + \frac{v_D \cdot e^{\lambda \cdot t}}{\eta \cdot V_{th}}] + B_2 \cdot \{1 - \binom{m}{1} \cdot [\frac{V_D^{(i)}}{V_0}]^{(m-1)} \cdot [\frac{v_D \cdot e^{\lambda \cdot t}}{V_0}]\} + B_3 \cdot [1 + \frac{v_D \cdot e^{\lambda \cdot t}}{V_{ex}}]$$

$$\Omega_4(v_D) = [\sum_{i=1}^{3} B_i] + B_1 \cdot \frac{v_D \cdot e^{\lambda \cdot t}}{\eta \cdot V_{th}} - B_2 \cdot \binom{m}{1} \cdot [\frac{V_D^{(i)}}{V_0}]^{(m-1)} \cdot [\frac{v_D \cdot e^{\lambda \cdot t}}{V_0}] + B_3 \cdot \frac{v_D \cdot e^{\lambda \cdot t}}{V_{ex}}$$

$$\Omega_4(v_D) = [\sum_{i=1}^{3} B_i] + \{\frac{B_1}{\eta \cdot V_{th}} - B_2 \cdot \binom{m}{1} \cdot [\frac{V_D^{(i)}}{V_0}]^{(m-1)} \cdot [\frac{1}{V_0}] + \frac{B_3}{V_{ex}}\} \cdot v_D \cdot e^{\lambda \cdot t}$$

We define: $\Omega_5(V_D^{(i)}) = \frac{B_1}{\eta \cdot V_{th}} - B_2 \cdot \binom{m}{1} \cdot [\frac{V_D^{(i)}}{V_0}]^{(m-1)} \cdot [\frac{1}{V_0}] + \frac{B_3}{V_{ex}}$

$$\Omega_4(v_D) = [\sum_{i=1}^{3} B_i] + \Omega_5(V_D^{(i)}) \cdot v_D \cdot e^{\lambda \cdot t}$$

$$\Gamma_1(V_D^{(i)}, v_D, \ldots) = \frac{-B_1 \frac{v_D \cdot e^{\lambda \cdot t}}{\eta \cdot V_{th}} + B_2 \cdot \binom{m}{1} \cdot [\frac{V_D^{(i)}}{V_0}]^{(m-1)} \cdot [\frac{v_D \cdot e^{\lambda \cdot t}}{V_0}] - B_3 \cdot \frac{v_D \cdot e^{\lambda \cdot t}}{V_{ex}} - \Omega_3(V_D^{(i)}) \cdot v_D \cdot e^{\lambda \cdot t}}{\{[\sum_{i=1}^{3} B_i] + \Omega_5(V_D^{(i)}) \cdot v_D \cdot e^{\lambda \cdot t}\} \cdot [\sum_{k=1}^{3} B_k] + \Omega_3(V_D^{(i)}) \cdot v_D \cdot e^{\lambda \cdot t} \cdot [\sum_{k=1}^{3} B_k]}$$

$$\Gamma_1(V_D^{(i)}, v_D, \ldots) = \frac{\{-B_1 \frac{1}{\eta \cdot V_{th}} + B_2 \cdot \binom{m}{1} \cdot [\frac{V_D^{(i)}}{V_0}]^{(m-1)} \cdot \frac{1}{V_0} - B_3 \cdot \frac{1}{V_{ex}} - \Omega_3(V_D^{(i)})\} \cdot v_D \cdot e^{\lambda \cdot t}}{[\sum_{i=1}^{3} B_i]^2 + [\sum_{k=1}^{3} B_k] \cdot \{\Omega_5(V_D^{(i)}) + \Omega_3(V_D^{(i)})\} \cdot v_D \cdot e^{\lambda \cdot t}}$$

For simplicity we define the following global parameters:

$$\Omega_6(V_D^{(i)}) = -B_1 \frac{1}{\eta \cdot V_{th}} + B_2 \cdot \binom{m}{1} \cdot [\frac{V_D^{(i)}}{V_0}]^{(m-1)} \cdot \frac{1}{V_0} - B_3 \cdot \frac{1}{V_{ex}} - \Omega_3(V_D^{(i)})$$

$$\Omega_7(V_D^{(i)}) = \Omega_5(V_D^{(i)}) + \Omega_3(V_D^{(i)}); \; \Gamma_1(V_D^{(i)}, v_D, \ldots) = \frac{\Omega_6(V_D^{(i)}) \cdot v_D \cdot e^{\lambda \cdot t}}{[\sum_{i=1}^{3} B_i]^2 + [\sum_{k=1}^{3} B_k] \cdot \Omega_7(V_D^{(i)}) \cdot v_D \cdot e^{\lambda \cdot t}}$$

$$\Gamma_1(V_D^{(i)}, v_D, \ldots) = \frac{\Omega_6(V_D^{(i)}) \cdot v_D \cdot e^{\lambda \cdot t}}{[\sum_{k=1}^{3} B_k] \cdot \{[\sum_{k=1}^{3} B_k] + \Omega_7(V_D^{(i)}) \cdot v_D \cdot e^{\lambda \cdot t}\}} \cdot \frac{\{[\sum_{k=1}^{3} B_k] - \Omega_7(V_D^{(i)}) \cdot v_D \cdot e^{\lambda \cdot t}\}}{\{[\sum_{k=1}^{3} B_k] - \Omega_7(V_D^{(i)}) \cdot v_D \cdot e^{\lambda \cdot t}\}}$$

$$\Gamma_1(V_D^{(i)}, v_D, \ldots) = \frac{[\sum_{k=1}^{3} B_k] \cdot \Omega_6(V_D^{(i)}) \cdot v_D \cdot e^{\lambda \cdot t} - [\prod_{i=6}^{7} \Omega_i(V_D^{(i)})] \cdot v_D^2 \cdot e^{2 \cdot \lambda \cdot t}}{[\sum_{k=1}^{3} B_k] \cdot \{[\sum_{k=1}^{3} B_k]^2 - \Omega_7^2 \cdot (V_D^{(i)}) \cdot v_D^2 \cdot e^{2 \cdot \lambda \cdot t}\}}$$

We consider $v_D^2 \to \varepsilon \approx 0$

$$\Gamma_1(V_D^{(i)}, v_D, \ldots)|_{v_D^2 \to \varepsilon \approx 0} = \frac{\Omega_6(V_D^{(i)})}{[\sum_{k=1}^{3} B_k]^2} \cdot v_D \cdot e^{\lambda \cdot t}; \frac{1}{\psi(V_D^{(i)} + v_D \cdot e^{\lambda \cdot t})}$$

$$= \frac{1}{\psi(V_D^{(i)})} + \frac{\Omega_6(V_D^{(i)})}{[\sum_{k=1}^{3} B_k]^2} \cdot v_D \cdot e^{\lambda \cdot t}$$

We can summery our system global parameters in the below Table 2.15:

Table 2.15 Tunnel Diode (TD) as a microwave oscillator system global parameters

Global parameter	Expression
$B_1(V_D^{(i)})$	$I_s \cdot \frac{1}{\eta \cdot V_{th}} \cdot e^{[\frac{[V_D^{(i)}]}{\eta \cdot V_{th}} - 1]}$
$B_2(V_D^{(i)})$	$(1 - m \cdot [\frac{V_D^{(i)}}{V_0}]^m) \cdot \frac{1}{R_0} \cdot e^{-[\frac{V_D^{(i)}}{V_0}]^m}$
$B_3(V_D^{(i)})$	$(1 + \frac{V_D^{(i)}}{V_{ex}}) \cdot \frac{1}{R_V} \cdot e^{[\frac{V_D^{(i)} - V_V}{V_{ex}}]}$
$B_4\|_{\Omega_1(v_D)\|_{v_D^2 \approx 0} = \Omega_2(v_D)\|_{v_D^2 \approx 0} = v_D}$	$\{\frac{1}{V_{ex} \cdot R_V} \cdot e^{[\frac{V_D^{(i)} - V_V}{V_{ex}}]} - m \cdot \binom{m}{1} \cdot [\frac{V_D^{(i)}}{V_0}]^{(m-1)} \cdot \frac{1}{V_0 \cdot R_0} \cdot e^{-[\frac{V_D^{(i)}}{V_0}]^m}\} \cdot v_D \cdot e^{\lambda \cdot t}$ $B_4\|_{\Omega_1(v_D)\|_{v_D^2 \approx 0} = \Omega_2(v_D)\|_{v_D^2 \approx 0} = v_D} = \Omega_3(V_D^{(i)}) \cdot v_D \cdot e^{\lambda \cdot t}$
$\Omega_1(v_D)\|_{v_D^2 \approx 0} = \Omega_2(v_D)\|_{v_D^2 \approx 0}$	v_D
$\Omega_3(V_D^{(i)})$	$\frac{1}{V_{ex} \cdot R_V} \cdot e^{[\frac{V_D^{(i)} - V_V}{V_{ex}}]} - m \cdot \binom{m}{1} \cdot [\frac{V_D^{(i)}}{V_0}]^{(m-1)} \cdot \frac{1}{V_0 \cdot R_0} \cdot e^{-[\frac{V_D^{(i)}}{V_0}]^m}$
$\Omega_4(v_D)$	$[\sum_{i=1}^{3} B_i] + \{\frac{B_1}{\eta \cdot V_{th}} - B_2 \cdot \binom{m}{1} \cdot [\frac{V_D^{(i)}}{V_0}]^{(m-1)} \cdot [\frac{1}{V_0}] + \frac{B_3}{V_{ex}}\} \cdot v_D \cdot e^{\lambda \cdot t}$
$\Omega_5(V_D^{(i)})$	$\frac{B_1}{\eta \cdot V_{th}} - B_2 \cdot \binom{m}{1} \cdot [\frac{V_D^{(i)}}{V_0}]^{(m-1)} \cdot [\frac{1}{V_0}] + \frac{B_3}{V_{ex}}$
$\Omega_6(V_D^{(i)})$	$-B_1 \frac{1}{\eta \cdot V_{th}} + B_2 \cdot \binom{m}{1} \cdot [\frac{V_D^{(i)}}{V_0}]^{(m-1)} \cdot \frac{1}{V_0} - B_3 \cdot \frac{1}{V_{ex}} - \Omega_3(V_D^{(i)})$
$\Omega_7(V_D^{(i)})$	$\Omega_5(V_D^{(i)}) + \Omega_3(V_D^{(i)})$

$\frac{\psi_1(V_D^{(i)}+v_D \cdot e^{\lambda \cdot t})}{\psi(V_D^{(i)}+v_D \cdot e^{\lambda \cdot t})} = \frac{\psi_1(V_D^{(i)})}{\psi(V_D^{(i)})} + \Gamma_2(V_D^{(i)}, v_D, \ldots)$. We need to find $\Gamma_2(V_D^{(i)}, v_D, \ldots)$ function

$$\frac{\psi_1(V_D^{(i)}+v_D \cdot e^{\lambda \cdot t})}{\psi(V_D^{(i)}+v_D \cdot e^{\lambda \cdot t})} = \frac{\begin{array}{l} I_s \cdot \frac{1}{(\eta \cdot V_{th})^2} \cdot e^{[(\frac{V_D^{(i)}+v_D \cdot e^{\lambda \cdot t}}{\eta \cdot V_{th}})-1]} - \{1+m \cdot (1-[\frac{V_D^{(i)}+v_D \cdot e^{\lambda \cdot t}}{V_0}]^m)\} \\ \cdot (\frac{V_D^{(i)}+v_D \cdot e^{\lambda \cdot t}}{V_0})^{m-1} \cdot \frac{m}{R_0 \cdot V_0} \cdot e^{[-(\frac{V_D^{(i)}+v_D \cdot e^{\lambda \cdot t}}{V_0})^m]} \\ + \{2+\frac{V_D^{(i)}+v_D \cdot e^{\lambda \cdot t}}{V_{ex}}\} \cdot \frac{1}{V_{ex} \cdot R_V} \cdot e^{[\frac{V_D^{(i)}+v_D \cdot e^{\lambda \cdot t}-V_V}{V_{ex}}]} \end{array}}{\{B_1 \cdot e^{[\frac{v_D \cdot e^{\lambda \cdot t}}{\eta \cdot V_{th}}]} + B_2 \cdot e^{-\{\binom{m}{1} \cdot [\frac{V_D^{(i)}}{V_0}]^{(m-1)} \cdot [\frac{v_D \cdot e^{\lambda \cdot t}}{V_0}]\}} + B_3 \cdot e^{[\frac{v_D \cdot e^{\lambda \cdot t}}{V_{ex}}]} + B_4\}}$$

Under the Taylor series high order elements tend to zero $(\ldots \cdot v_D^k \to \varepsilon \, \forall k \geq 2)$ assumption for exponent functions: $B_4|_{\Omega_1(v_D)|_{v_D^2 \approx 0} = \Omega_2(v_D)|_{v_D^2 \approx 0} = v_D} = \Omega_3(V_D^{(i)}) \cdot v_D \cdot e^{\lambda \cdot t}$

$$\frac{\psi_1(V_D^{(i)}+v_D \cdot e^{\lambda \cdot t})}{\psi(V_D^{(i)}+v_D \cdot e^{\lambda \cdot t})} = \frac{\begin{array}{l} I_s \cdot \frac{1}{(\eta \cdot V_{th})^2} \cdot e^{[\frac{V_D^{(i)}}{\eta \cdot V_{th}}-1]} \cdot e^{[\frac{v_D \cdot e^{\lambda \cdot t}}{\eta \cdot V_{th}}]} \\ - \{1+m \cdot (1-[\frac{V_D^{(i)}+v_D \cdot e^{\lambda \cdot t}}{V_0}]^m)\} \\ \cdot (\frac{V_D^{(i)}+v_D \cdot e^{\lambda \cdot t}}{V_0})^{m-1} \cdot \frac{m}{R_0 \cdot V_0} \cdot e^{[-(\frac{V_D^{(i)}+v_D \cdot e^{\lambda \cdot t}}{V_0})^m]} \\ + \{(2+\frac{V_D^{(i)}}{V_{ex}}) + \frac{v_D \cdot e^{\lambda \cdot t}}{V_{ex}}\} \cdot \frac{1}{V_{ex} \cdot R_V} \cdot e^{[\frac{V_D^{(i)}-V_V}{V_{ex}}]} \cdot e^{[\frac{v_D \cdot e^{\lambda \cdot t}}{V_{ex}}]} \end{array}}{\begin{array}{l} \{B_1 \cdot e^{[\frac{v_D \cdot e^{\lambda \cdot t}}{\eta \cdot V_{th}}]} + B_2 \cdot e^{-\{\binom{m}{1} \cdot [\frac{V_D^{(i)}}{V_0}]^{(m-1)} \cdot [\frac{v_D \cdot e^{\lambda \cdot t}}{V_0}]\}} \\ + B_3 \cdot e^{[\frac{v_D \cdot e^{\lambda \cdot t}}{V_{ex}}]} + \Omega_3(V_D^{(i)}) \cdot v_D \cdot e^{\lambda \cdot t}\} \end{array}}$$

$$[\frac{V_D^{(i)}+v_D \cdot e^{\lambda \cdot t}}{V_0}]^m \approx [\frac{V_D^{(i)}}{V_0}]^m + \binom{m}{1} \cdot [\frac{V_D^{(i)}}{V_0}]^{(m-1)} \cdot [\frac{v_D \cdot e^{\lambda \cdot t}}{V_0}]$$

$$(\frac{V_D^{(i)}+v_D \cdot e^{\lambda \cdot t}}{V_0})^{m-1} \approx [\frac{V_D^{(i)}}{V_0}]^{m-1} + \binom{m-1}{1} \cdot [\frac{V_D^{(i)}}{V_0}]^{(m-2)} \cdot [\frac{v_D \cdot e^{\lambda \cdot t}}{V_0}]$$

$$e^{[-(\frac{V_D^{(i)}+v_D \cdot e^{\lambda \cdot t}}{V_0})^m]} \approx e^{[-[\frac{V_D^{(i)}}{V_0}]^m - \binom{m}{1} \cdot [\frac{V_D^{(i)}}{V_0}]^{(m-1)} \cdot [\frac{v_D \cdot e^{\lambda \cdot t}}{V_0}]]} = e^{-[\frac{V_D^{(i)}}{V_0}]^m} \cdot e^{-\binom{m}{1} \cdot [\frac{V_D^{(i)}}{V_0}]^{(m-1)} \cdot [\frac{v_D \cdot e^{\lambda \cdot t}}{V_0}]}$$

$$\frac{\psi_1(V_D^{(i)}+v_D\cdot e^{\lambda\cdot t})}{\psi(V_D^{(i)}+v_D\cdot e^{\lambda\cdot t})}=\frac{\begin{array}{l} I_s\cdot\frac{1}{(\eta\cdot V_{th})^2}\cdot e^{[\frac{V_D^{(i)}}{\eta\cdot V_{th}}-1]}\cdot e^{[\frac{v_D\cdot e^{\lambda\cdot t}}{\eta\cdot V_{th}}]}-\{1+m\cdot(1-[\frac{V_D^{(i)}}{V_0}]^m \\ -\binom{m}{1}\cdot[\frac{V_D^{(i)}}{V_0}]^{(m-1)}\cdot[\frac{v_D\cdot e^{\lambda\cdot t}}{V_0}])\} \\ \cdot\{[\frac{V_D^{(i)}}{V_0}]^{m-1}+\binom{m-1}{1}\cdot[\frac{V_D^{(i)}}{V_0}]^{(m-2)} \\ \cdot[\frac{v_D\cdot e^{\lambda\cdot t}}{V_0}]\}\cdot\frac{m}{R_0\cdot V_0}\cdot e^{-[\frac{V_D^{(i)}}{V_0}]^m}\cdot e^{-\binom{m}{1}\cdot[\frac{V_D^{(i)}}{V_0}]^{(m-1)}\cdot[\frac{v_D\cdot e^{\lambda\cdot t}}{V_0}]} \\ +(2+\frac{V_D^{(i)}}{V_{ex}})\cdot\frac{1}{V_{ex}\cdot R_V}\cdot e^{[\frac{V_D^{(i)}-V_V}{V_{ex}}]}\cdot e^{[\frac{v_D\cdot e^{\lambda\cdot t}}{V_{ex}}]} \\ +\frac{v_D\cdot e^{\lambda\cdot t}}{V_{ex}}\cdot\frac{1}{V_{ex}\cdot R_V}\cdot e^{[\frac{V_D^{(i)}-V_V}{V_{ex}}]}\cdot e^{[\frac{v_D\cdot e^{\lambda\cdot t}}{V_{ex}}]}\end{array}}{\begin{array}{l}\{B_1\cdot e^{[\frac{v_D\cdot e^{\lambda\cdot t}}{\eta\cdot V_{th}}]}+B_2\cdot e^{-\{\binom{m}{1}\cdot[\frac{V_D^{(i)}}{V_0}]^{(m-1)}\cdot[\frac{v_D\cdot e^{\lambda\cdot t}}{V_0}]\}} \\ +B_3\cdot e^{[\frac{v_D\cdot e^{\lambda\cdot t}}{V_{ex}}]}+\Omega_3(V_D^{(i)})\cdot v_D\cdot e^{\lambda\cdot t}\}\end{array}}$$

We already define: $\Omega_4(v_D)=B_1\cdot e^{[\frac{v_D\cdot e^{\lambda\cdot t}}{\eta\cdot V_{th}}]}+B_2\cdot e^{-\{\binom{m}{1}\cdot[\frac{V_D^{(i)}}{V_0}]^{(m-1)}\cdot[\frac{v_D\cdot e^{\lambda\cdot t}}{V_0}]\}}+B_3\cdot e^{[\frac{v_D\cdot e^{\lambda\cdot t}}{V_{ex}}]}$

And $\Omega_4(v_D)=[\sum_{i=1}^{3}B_i]+\Omega_5(V_D^{(i)})\cdot v_D\cdot e^{\lambda\cdot t}$

$$\begin{aligned}&B_1\cdot e^{[\frac{v_D\cdot e^{\lambda\cdot t}}{\eta\cdot V_{th}}]}+B_2\cdot e^{-\{\binom{m}{1}\cdot[\frac{V_D^{(i)}}{V_0}]^{(m-1)}\cdot[\frac{v_D\cdot e^{\lambda\cdot t}}{V_0}]\}}+B_3\cdot e^{[\frac{v_D\cdot e^{\lambda\cdot t}}{V_{ex}}]}+\Omega_3(V_D^{(i)})\cdot v_D\cdot e^{\lambda\cdot t}\\ &\quad=\Omega_4(v_D)+\Omega_3(V_D^{(i)})\cdot v_D\cdot e^{\lambda\cdot t}\end{aligned}$$

$$\begin{aligned}\Omega_4(v_D)+\Omega_3(V_D^{(i)})\cdot v_D\cdot e^{\lambda\cdot t}&=[\sum_{i=1}^{3}B_i]+\Omega_5(V_D^{(i)})\cdot v_D\cdot e^{\lambda\cdot t}+\Omega_3(V_D^{(i)})\cdot v_D\cdot e^{\lambda\cdot t}\\ &=[\sum_{i=1}^{3}B_i]+\{\Omega_5(V_D^{(i)})+\Omega_3(V_D^{(i)})\}\cdot v_D\cdot e^{\lambda\cdot t}\end{aligned}$$

$$\frac{\psi_1(V_D^{(i)}+v_D\cdot e^{\lambda\cdot t})}{\psi(V_D^{(i)}+v_D\cdot e^{\lambda\cdot t})}=\frac{\begin{array}{l}I_s\cdot\frac{1}{(\eta\cdot V_{th})^2}\cdot e^{[\frac{V_D^{(i)}}{\eta\cdot V_{th}}-1]}\cdot e^{[\frac{v_D\cdot e^{\lambda\cdot t}}{\eta\cdot V_{th}}]}-\{1+m\cdot(1-[\frac{V_D^{(i)}}{V_0}]^m\\ -\binom{m}{1}\cdot[\frac{V_D^{(i)}}{V_0}]^{(m-1)}\cdot[\frac{v_D\cdot e^{\lambda\cdot t}}{V_0}])\}\\ \cdot\{[\frac{V_D^{(i)}}{V_0}]^{m-1}+\binom{m-1}{1}\cdot[\frac{V_D^{(i)}}{V_0}]^{(m-2)}\\ \cdot[\frac{v_D\cdot e^{\lambda\cdot t}}{V_0}]\}\cdot\frac{m}{R_0\cdot V_0}\cdot e^{-[\frac{V_D^{(i)}}{V_0}]^m}\cdot e^{-\binom{m}{1}\cdot[\frac{V_D^{(i)}}{V_0}]^{(m-1)}\cdot[\frac{v_D\cdot e^{\lambda\cdot t}}{V_0}]}\\ +(2+\frac{V_D^{(i)}}{V_{ex}})\cdot\frac{1}{V_{ex}\cdot R_V}\cdot e^{[\frac{V_D^{(i)}-V_V}{V_{ex}}]}\cdot e^{[\frac{v_D\cdot e^{\lambda\cdot t}}{V_{ex}}]}\\ +\frac{v_D\cdot e^{\lambda\cdot t}}{V_{ex}}\cdot\frac{1}{V_{ex}\cdot R_V}\cdot e^{[\frac{V_D^{(i)}-V_V}{V_{ex}}]}\cdot e^{[\frac{v_D\cdot e^{\lambda\cdot t}}{V_{ex}}]}\end{array}}{[\sum_{i=1}^{3}B_i]+\{\Omega_5(V_D^{(i)})+\Omega_3(V_D^{(i)})\}\cdot v_D\cdot e^{\lambda\cdot t}}$$

We need to get an expression of the above equation's numerator.

$$\begin{aligned}\psi_1(V_D^{(i)}+v_D\cdot e^{\lambda\cdot t})&=I_s\cdot\frac{1}{(\eta\cdot V_{th})^2}\cdot e^{[\frac{V_D^{(i)}}{\eta\cdot V_{th}}-1]}\cdot e^{[\frac{v_D\cdot e^{\lambda\cdot t}}{\eta\cdot V_{th}}]}-\{1+m\cdot(1-[\frac{V_D^{(i)}}{V_0}]^m\\ &-\binom{m}{1}\cdot[\frac{V_D^{(i)}}{V_0}]^{(m-1)}\cdot[\frac{v_D\cdot e^{\lambda\cdot t}}{V_0}])\}\\ &\cdot\{[\frac{V_D^{(i)}}{V_0}]^{m-1}+\binom{m-1}{1}\cdot[\frac{V_D^{(i)}}{V_0}]^{(m-2)}\cdot[\frac{v_D\cdot e^{\lambda\cdot t}}{V_0}]\}\\ &\cdot\frac{m}{R_0\cdot V_0}\cdot e^{-[\frac{V_D^{(i)}}{V_0}]^m}\cdot e^{-\binom{m}{1}\cdot[\frac{V_D^{(i)}}{V_0}]^{(m-1)}\cdot[\frac{v_D\cdot e^{\lambda\cdot t}}{V_0}]}+(2+\frac{V_D^{(i)}}{V_{ex}})\\ &\cdot\frac{1}{V_{ex}\cdot R_V}\cdot e^{[\frac{V_D^{(i)}-V_V}{V_{ex}}]}\cdot e^{[\frac{v_D\cdot e^{\lambda\cdot t}}{V_{ex}}]}+\frac{v_D\cdot e^{\lambda\cdot t}}{V_{ex}}\cdot\frac{1}{V_{ex}\cdot R_V}\cdot e^{[\frac{V_D^{(i)}-V_V}{V_{ex}}]}\cdot e^{[\frac{v_D\cdot e^{\lambda\cdot t}}{V_{ex}}]}\end{aligned}$$

$$\psi_1(V_D^{(i)} + v_D \cdot e^{\lambda \cdot t}) = I_s \cdot \frac{1}{(\eta \cdot V_{th})^2} \cdot e^{[\frac{V_D^{(i)}}{\eta \cdot V_{th}} - 1]} \cdot e^{[\frac{v_D \cdot e^{\lambda \cdot t}}{\eta \cdot V_{th}}]} - \{1 + m \cdot (1 - [\frac{V_D^{(i)}}{V_0}]^m)$$
$$- m \cdot \binom{m}{1} \cdot [\frac{V_D^{(i)}}{V_0}]^{(m-1)} \cdot [\frac{v_D \cdot e^{\lambda \cdot t}}{V_0}]\} \cdot \{[\frac{V_D^{(i)}}{V_0}]^{m-1} + \binom{m-1}{1}$$
$$\cdot [\frac{V_D^{(i)}}{V_0}]^{(m-2)} \cdot [\frac{v_D \cdot e^{\lambda \cdot t}}{V_0}]\}$$
$$\cdot \frac{m}{R_0 \cdot V_0} \cdot e^{-[\frac{V_D^{(i)}}{V_0}]^m} \cdot e^{-\binom{m}{1} \cdot [\frac{V_D^{(i)}}{V_0}]^{(m-1)} \cdot [\frac{v_D \cdot e^{\lambda \cdot t}}{V_0}]} + (2 + \frac{V_D^{(i)}}{V_{ex}})$$
$$\cdot \frac{1}{V_{ex} \cdot R_V} \cdot e^{[\frac{V_D^{(i)} - V_V}{V_{ex}}]} \cdot e^{[\frac{v_D \cdot e^{\lambda \cdot t}}{V_{ex}}]} + \frac{v_D \cdot e^{\lambda \cdot t}}{V_{ex}} \cdot \frac{1}{V_{ex} \cdot R_V} \cdot e^{[\frac{V_D^{(i)} - V_V}{V_{ex}}]} \cdot e^{[\frac{v_D \cdot e^{\lambda \cdot t}}{V_{ex}}]}$$

First we take the internal expression's multiplication:

$$\{[1 + m \cdot (1 - [\frac{V_D^{(i)}}{V_0}]^m)] - m \cdot \binom{m}{1} \cdot [\frac{V_D^{(i)}}{V_0}]^{(m-1)} \cdot [\frac{v_D \cdot e^{\lambda \cdot t}}{V_0}]\} \cdot \{[\frac{V_D^{(i)}}{V_0}]^{m-1}$$
$$+ \binom{m-1}{1} \cdot [\frac{V_D^{(i)}}{V_0}]^{(m-2)} \cdot [\frac{v_D \cdot e^{\lambda \cdot t}}{V_0}]\} = [1 + m \cdot (1 - [\frac{V_D^{(i)}}{V_0}]^m)] \cdot [\frac{V_D^{(i)}}{V_0}]^{m-1}$$
$$+ [1 + m \cdot (1 - [\frac{V_D^{(i)}}{V_0}]^m)] \cdot \binom{m-1}{1} \cdot [\frac{V_D^{(i)}}{V_0}]^{(m-2)} \cdot [\frac{v_D \cdot e^{\lambda \cdot t}}{V_0}]$$
$$- [\frac{V_D^{(i)}}{V_0}]^{m-1} \cdot m \cdot \binom{m}{1} \cdot [\frac{V_D^{(i)}}{V_0}]^{(m-1)} \cdot [\frac{v_D \cdot e^{\lambda \cdot t}}{V_0}] - m \cdot \binom{m}{1} \cdot \binom{m-1}{1}$$
$$\cdot [\frac{V_D^{(i)}}{V_0}]^{(m-1)} \cdot [\frac{V_D^{(i)}}{V_0}]^{(m-2)} \cdot [\frac{v_D^2 \cdot e^{2 \cdot \lambda \cdot t}}{V_0^2}]$$

We consider $\ldots v_D^2 \to \varepsilon = 0$ then

$$m \cdot \binom{m}{1} \cdot \binom{m-1}{1} \cdot [\frac{V_D^{(i)}}{V_0}]^{(m-1)} \cdot [\frac{V_D^{(i)}}{V_0}]^{(m-2)} \cdot [\frac{v_D^2 \cdot e^{2 \cdot \lambda \cdot t}}{V_0^2}] \to \varepsilon$$

$$\{[1 + m \cdot (1 - [\frac{V_D^{(i)}}{V_0}]^m)] - m \cdot \binom{m}{1} \cdot [\frac{V_D^{(i)}}{V_0}]^{(m-1)} \cdot [\frac{v_D \cdot e^{\lambda \cdot t}}{V_0}]\} \cdot \{[\frac{V_D^{(i)}}{V_0}]^{m-1}$$
$$+ \binom{m-1}{1} \cdot [\frac{V_D^{(i)}}{V_0}]^{(m-2)} \cdot [\frac{v_D \cdot e^{\lambda \cdot t}}{V_0}]\} = [1 + m \cdot (1 - [\frac{V_D^{(i)}}{V_0}]^m)] \cdot [\frac{V_D^{(i)}}{V_0}]^{m-1}$$
$$+ [1 + m \cdot (1 - [\frac{V_D^{(i)}}{V_0}]^m)] \cdot \binom{m-1}{1} \cdot [\frac{V_D^{(i)}}{V_0}]^{(m-2)} \cdot [\frac{v_D \cdot e^{\lambda \cdot t}}{V_0}]$$
$$- [\frac{V_D^{(i)}}{V_0}]^{m-1} \cdot m \cdot \binom{m}{1} \cdot [\frac{V_D^{(i)}}{V_0}]^{(m-1)} \cdot [\frac{v_D \cdot e^{\lambda \cdot t}}{V_0}] - (\varepsilon \to 0)$$

$$[1+m\cdot(1-[\frac{V_D^{(i)}}{V_0}]^m)]\cdot[\frac{V_D^{(i)}}{V_0}]^{m-1}+\{[1+m\cdot(1-[\frac{V_D^{(i)}}{V_0}]^m)]\cdot\binom{m-1}{1}\cdot[\frac{V_D^{(i)}}{V_0}]^{(m-2)}$$
$$-[\frac{V_D^{(i)}}{V_0}]^{m-1}\cdot m\cdot\binom{m}{1}\cdot[\frac{V_D^{(i)}}{V_0}]^{(m-1)}\}\cdot[\frac{v_D\cdot e^{\lambda\cdot t}}{V_0}]$$

$$\Xi_1(V_D^{(i)})=[1+m\cdot(1-[\frac{V_D^{(i)}}{V_0}]^m)]\cdot[\frac{V_D^{(i)}}{V_0}]^{m-1};$$
$$\Xi_2(V_D^{(i)})=[1+m\cdot(1-[\frac{V_D^{(i)}}{V_0}]^m)]\cdot\binom{m-1}{1}\cdot[\frac{V_D^{(i)}}{V_0}]^{(m-2)}$$
$$-[\frac{V_D^{(i)}}{V_0}]^{m-1}\cdot m\cdot\binom{m}{1}\cdot[\frac{V_D^{(i)}}{V_0}]^{(m-1)}\Rightarrow\Xi_1(V_D^{(i)})+\Xi_2(V_D^{(i)})\cdot[\frac{v_D\cdot e^{\lambda\cdot t}}{V_0}];$$
$$\Xi_1=\Xi_1(V_D^{(i)});\ \Xi_2=\Xi_2(V_D^{(i)})$$

$$\psi_1(V_D^{(i)}+v_D\cdot e^{\lambda\cdot t})=I_s\cdot\frac{1}{(\eta\cdot V_{th})^2}\cdot e^{[\frac{V_D^{(i)}}{\eta\cdot V_{th}}-1]}\cdot e^{[\frac{v_D\cdot e^{\lambda\cdot t}}{\eta\cdot V_{th}}]}-\{\Xi_1+\Xi_2\cdot[\frac{v_D\cdot e^{\lambda\cdot t}}{V_0}]\}$$
$$\cdot\frac{m}{R_0\cdot V_0}\cdot e^{-[\frac{V_D^{(i)}}{V_0}]^m}\cdot e^{-\binom{m}{1}\cdot[\frac{V_D^{(i)}}{V_0}]^{(m-1)}\cdot[\frac{v_D\cdot e^{\lambda\cdot t}}{V_0}]}+(2+\frac{V_D^{(i)}}{V_{ex}})$$
$$\cdot\frac{1}{V_{ex}\cdot R_V}\cdot e^{[\frac{V_D^{(i)}-V_V}{V_{ex}}]}\cdot e^{[\frac{v_D\cdot e^{\lambda\cdot t}}{V_{ex}}]}+\frac{1}{V_{ex}^2\cdot R_V}\cdot e^{[\frac{V_D^{(i)}-V_V}{V_{ex}}]}\cdot e^{[\frac{v_D\cdot e^{\lambda\cdot t}}{V_{ex}}]}\cdot v_D\cdot e^{\lambda\cdot t}$$

We define for simplicity the following parameters: $\Xi_3=I_s\cdot\frac{1}{(\eta\cdot V_{th})^2}\cdot e^{[\frac{V_D^{(i)}}{\eta\cdot V_{th}}-1]}$

$$\Xi_4=\frac{m}{R_0\cdot V_0}\cdot e^{-[\frac{V_D^{(i)}}{V_0}]^m};\ \Xi_5=(2+\frac{V_D^{(i)}}{V_{ex}})\cdot\frac{1}{V_{ex}\cdot R_V}\cdot e^{[\frac{V_D^{(i)}-V_V}{V_{ex}}]};\ \Xi_6=\frac{1}{V_{ex}^2\cdot R_V}\cdot e^{[\frac{V_D^{(i)}-V_V}{V_{ex}}]}$$

$$\psi_1(V_D^{(i)}+v_D\cdot e^{\lambda\cdot t})=\Xi_3\cdot e^{[\frac{v_D\cdot e^{\lambda\cdot t}}{\eta\cdot V_{th}}]}-\{\Xi_1+\Xi_2\cdot[\frac{v_D\cdot e^{\lambda\cdot t}}{V_0}]\}\cdot\Xi_4\cdot e^{-\binom{m}{1}\cdot[\frac{V_D^{(i)}}{V_0}]^{(m-1)}\cdot[\frac{v_D\cdot e^{\lambda\cdot t}}{V_0}]}$$
$$+\Xi_5\cdot e^{[\frac{v_D\cdot e^{\lambda\cdot t}}{V_{ex}}]}+\Xi_6\cdot e^{[\frac{v_D\cdot e^{\lambda\cdot t}}{V_{ex}}]}\cdot v_D\cdot e^{\lambda\cdot t};\ \Xi_3=\Xi_3(V_D^{(i)});\ \Xi_4=\Xi_4(V_D^{(i)});$$
$$\Xi_5=\Xi_5(V_D^{(i)});\ \Xi_6=\Xi_6(V_D^{(i)})$$

Under the Taylor series high order elements tend to zero $(\ldots \cdot v_D^k \to \varepsilon \forall k \geq 2)$ assumption for exponent functions: $e^{[\frac{v_D \cdot e^{\lambda \cdot t}}{\eta \cdot V_{th}}]} \approx 1 + \frac{v_D \cdot e^{\lambda \cdot t}}{\eta \cdot V_{th}}$; $e^{[\frac{v_D \cdot e^{\lambda \cdot t}}{V_{ex}}]} = 1 + \frac{v_D \cdot e^{\lambda \cdot t}}{V_{ex}}$

$$e^{-\binom{m}{1} \cdot [\frac{V_D^{(i)}}{V_0}]^{(m-1)} \cdot [\frac{v_D \cdot e^{\lambda \cdot t}}{V_0}]} \approx 1 - \binom{m}{1} \cdot [\frac{V_D^{(i)}}{V_0}]^{(m-1)} \cdot [\frac{v_D \cdot e^{\lambda \cdot t}}{V_0}]$$

$$\psi_1(V_D^{(i)} + v_D \cdot e^{\lambda \cdot t}) = \Xi_3 \cdot [1 + \frac{v_D \cdot e^{\lambda \cdot t}}{\eta \cdot V_{th}}] - \{\Xi_1 + \Xi_2 \cdot [\frac{v_D \cdot e^{\lambda \cdot t}}{V_0}]\} \cdot \Xi_4 \cdot \{1 - \binom{m}{1} \cdot [\frac{V_D^{(i)}}{V_0}]^{(m-1)} \cdot [\frac{v_D \cdot e^{\lambda \cdot t}}{V_0}]\} + \Xi_5 \cdot [1 + \frac{v_D \cdot e^{\lambda \cdot t}}{V_{ex}}] + \Xi_6 \cdot [1 + \frac{v_D \cdot e^{\lambda \cdot t}}{V_{ex}}] \cdot v_D \cdot e^{\lambda \cdot t}$$

$$\psi_1(V_D^{(i)} + v_D \cdot e^{\lambda \cdot t}) = \Xi_3 + \Xi_3 \cdot \frac{v_D \cdot e^{\lambda \cdot t}}{\eta \cdot V_{th}} - \{\Xi_4 \cdot \Xi_1 + \Xi_4 \cdot (\Xi_2 - \Xi_1 \cdot \binom{m}{1} \cdot [\frac{V_D^{(i)}}{V_0}]^{(m-1)}) \cdot \frac{v_D \cdot e^{\lambda \cdot t}}{V_0} - \Xi_4 \cdot \Xi_2 \cdot \binom{m}{1} \cdot [\frac{V_D^{(i)}}{V_0}]^{(m-1)} \cdot [\frac{v_D^2 \cdot e^{2 \cdot \lambda \cdot t}}{V_0^2}]\} + \Xi_5 + \Xi_5 \cdot \frac{v_D \cdot e^{\lambda \cdot t}}{V_{ex}} + \Xi_6 \cdot v_D \cdot e^{\lambda \cdot t} + \Xi_6 \cdot \frac{v_D^2 \cdot e^{2 \cdot \lambda \cdot t}}{V_{ex}}$$

All expressions which includes v_D^2 tend to zero $(v_D^2 \to \varepsilon)$.

$$\psi_1(V_D^{(i)} + v_D \cdot e^{\lambda \cdot t}) = \Xi_3 + \Xi_3 \cdot \frac{v_D \cdot e^{\lambda \cdot t}}{\eta \cdot V_{th}} - \Xi_4 \cdot \Xi_1 - \Xi_4 \cdot (\Xi_2 - \Xi_1 \cdot \binom{m}{1} \cdot [\frac{V_D^{(i)}}{V_0}]^{(m-1)}) \cdot \frac{v_D \cdot e^{\lambda \cdot t}}{V_0} + \Xi_5 + \Xi_5 \cdot \frac{v_D \cdot e^{\lambda \cdot t}}{V_{ex}} + \Xi_6 \cdot v_D \cdot e^{\lambda \cdot t}$$

$$\psi_1(V_D^{(i)} + v_D \cdot e^{\lambda \cdot t}) = \{\Xi_5 + \Xi_3 - \Xi_4 \cdot \Xi_1\} + \{\Xi_3 \cdot \frac{1}{\eta \cdot V_{th}} - (\Xi_2 - \Xi_1 \cdot \binom{m}{1} \cdot [\frac{V_D^{(i)}}{V_0}]^{(m-1)}) \cdot \frac{\Xi_4}{V_0} + \frac{\Xi_5}{V_{ex}} + \Xi_6\} \cdot v_D \cdot e^{\lambda \cdot t};$$

$$\psi_1(V_D^{(i)} + v_D \cdot e^{\lambda \cdot t}) = \Phi_1 + \Phi_2 \cdot v_D \cdot e^{\lambda \cdot t}$$

We define: $\Phi_1 = \Xi_5 + \Xi_3 - \Xi_4 \cdot \Xi_1$; $\Phi_1 = \Phi_1(V_D^{(i)})$; $\Phi_2 = \Phi_2(V_D^{(i)})$

$$\Phi_2 = \Xi_3 \cdot \frac{1}{\eta \cdot V_{th}} - (\Xi_2 - \Xi_1 \cdot \binom{m}{1} \cdot [\frac{V_D^{(i)}}{V_0}]^{(m-1)}) \cdot \frac{\Xi_4}{V_0} + \frac{\Xi_5}{V_{ex}} + \Xi_6$$

$$\frac{\psi_1(V_D^{(i)} + v_D \cdot e^{\lambda \cdot t})}{\psi(V_D^{(i)} + v_D \cdot e^{\lambda \cdot t})} = \frac{\Phi_1 + \Phi_2 \cdot v_D \cdot e^{\lambda \cdot t}}{[\sum_{i=1}^{3} B_i] + \{\Omega_5(V_D^{(i)}) + \Omega_3(V_D^{(i)})\} \cdot v_D \cdot e^{\lambda \cdot t}}$$

$$\frac{\psi_1(V_D^{(i)} + v_D \cdot e^{\lambda \cdot t})}{\psi(V_D^{(i)} + v_D \cdot e^{\lambda \cdot t})} = \frac{\psi_1(V_D^{(i)})}{\psi(V_D^{(i)})} + \Gamma_2(V_D^{(i)}, v_D, \ldots) \Rightarrow \Gamma_2(V_D^{(i)}, v_D, \ldots) = \frac{\psi_1(V_D^{(i)} + v_D \cdot e^{\lambda \cdot t})}{\psi(V_D^{(i)} + v_D \cdot e^{\lambda \cdot t})} - \frac{\psi_1(V_D^{(i)})}{\psi(V_D^{(i)})}$$

$$\Gamma_2(V_D^{(i)}, v_D, \ldots.) = \frac{\Phi_1 + \Phi_2 \cdot v_D \cdot e^{\lambda \cdot t}}{[\sum_{i=1}^{3} B_i] + \{\Omega_5(V_D^{(i)}) + \Omega_3(V_D^{(i)})\} \cdot v_D \cdot e^{\lambda \cdot t}} - \frac{\psi_1(V_D^{(i)})}{\psi(V_D^{(i)})}$$

$$= \frac{\Phi_1 \cdot \psi(V_D^{(i)}) + \Phi_2 \cdot \psi(V_D^{(i)}) \cdot v_D \cdot e^{\lambda \cdot t} - \psi_1(V_D^{(i)}) \cdot ([\sum_{i=1}^{3} B_i] + \{\Omega_5(V_D^{(i)}) + \Omega_3(V_D^{(i)})\}) \cdot v_D \cdot e^{\lambda \cdot t}}{[\sum_{i=1}^{3} B_i] \cdot \psi(V_D^{(i)}) + \{\Omega_5(V_D^{(i)}) + \Omega_3(V_D^{(i)})\} \cdot \psi(V_D^{(i)}) \cdot v_D \cdot e^{\lambda \cdot t}}$$

$$\Gamma_2(V_D^{(i)}, v_D, \ldots.) = \frac{\Phi_1 \cdot \psi(V_D^{(i)}) + \{\Phi_2 \cdot \psi(V_D^{(i)}) - \psi_1(V_D^{(i)}) \cdot ([\sum_{i=1}^{3} B_i] + \{\Omega_5(V_D^{(i)}) + \Omega_3(V_D^{(i)})\})\} \cdot v_D \cdot e^{\lambda \cdot t}}{[\sum_{i=1}^{3} B_i] \cdot \psi(V_D^{(i)}) + \{\Omega_5(V_D^{(i)}) + \Omega_3(V_D^{(i)})\} \cdot \psi(V_D^{(i)}) \cdot v_D \cdot e^{\lambda \cdot t}}$$

We define for simplicity the following global parameters: $\Phi_3 = \Phi_1 \cdot \psi(V_D^{(i)})$

$$\Phi_4 = \Phi_2 \cdot \psi(V_D^{(i)}) - \psi_1(V_D^{(i)}) \cdot ([\sum_{i=1}^{3} B_i] + \{\Omega_5(V_D^{(i)}) + \Omega_3(V_D^{(i)})\});$$

$$\Phi_5 = [\sum_{i=1}^{3} B_i] \cdot \psi(V_D^{(i)})$$

$$\Phi_6 = \{\Omega_5(V_D^{(i)}) + \Omega_3(V_D^{(i)})\} \cdot \psi(V_D^{(i)}); \; \Gamma_2(V_D^{(i)}, v_D, \ldots) = \frac{\Phi_3 + \Phi_4 \cdot v_D \cdot e^{\lambda \cdot t}}{\Phi_5 + \Phi_6 \cdot v_D \cdot e^{\lambda \cdot t}}$$

$$\Gamma_2(V_D^{(i)}, v_D, \ldots) = \frac{\Phi_3 + \Phi_4 \cdot v_D \cdot e^{\lambda \cdot t}}{\Phi_5 + \Phi_6 \cdot v_D \cdot e^{\lambda \cdot t}} \cdot \frac{\Phi_5 - \Phi_6 \cdot v_D \cdot e^{\lambda \cdot t}}{\Phi_5 - \Phi_6 \cdot v_D \cdot e^{\lambda \cdot t}}$$
$$= \frac{\Phi_3 \cdot \Phi_5 - \Phi_3 \cdot \Phi_6 \cdot v_D \cdot e^{\lambda \cdot t} + \Phi_5 \cdot \Phi_4 \cdot v_D \cdot e^{\lambda \cdot t} - \Phi_4 \cdot \Phi_6 \cdot v_D^2 \cdot e^{2 \cdot \lambda \cdot t}}{\Phi_5^2 - \Phi_6^2 \cdot v_D^2 \cdot e^{2 \cdot \lambda \cdot t}}$$

All expressions which includes v_D^2 tend to zero ($v_D^2 \to \varepsilon$).

$$\Gamma_2(V_D^{(i)}, v_D, \ldots) = \frac{\Phi_3 \cdot \Phi_5 + (\Phi_5 \cdot \Phi_4 - \Phi_3 \cdot \Phi_6) \cdot v_D \cdot e^{\lambda \cdot t}}{\Phi_5^2}$$
$$= \frac{\Phi_3}{\Phi_5} + \frac{(\Phi_5 \cdot \Phi_4 - \Phi_3 \cdot \Phi_6)}{\Phi_5^2} \cdot v_D \cdot e^{\lambda \cdot t}$$

$$\Phi_3 = \Phi_3(V_D^{(i)});\ \Phi_4 = \Phi_4(V_D^{(i)});\ \Phi_5 = \Phi_5(V_D^{(i)});\ \Phi_6 = \Phi_6(V_D^{(i)})$$

$$\frac{\psi_1(V_D^{(i)} + v_D \cdot e^{\lambda \cdot t})}{\psi(V_D^{(i)} + v_D \cdot e^{\lambda \cdot t})} = \frac{\psi_1(V_D^{(i)})}{\psi(V_D^{(i)})} + \frac{\Phi_3}{\Phi_5} + \frac{(\Phi_5 \cdot \Phi_4 - \Phi_3 \cdot \Phi_6)}{\Phi_5^2} \cdot v_D \cdot e^{\lambda \cdot t}$$

We can summery our last analysis in the next Table 2.16):

Table 2.16 Tunnel Diode (TD) as a microwave oscillator system expressions and equivalent expressions

Expression	Equivalent expression
$\frac{1}{\psi(V_D^{(i)} + v_D \cdot e^{\lambda \cdot t})}$	$\frac{1}{\psi(V_D^{(i)})} + \frac{\Omega_6(V_D^{(i)})}{[\sum_{k=1}^{3} B_k]^2} \cdot v_D \cdot e^{\lambda \cdot t};\ \Gamma_1(V_D^{(i)}, v_D, \ldots)\vert_{v_D^2 \to \varepsilon \approx 0} = \frac{\Omega_6(V_D^{(i)})}{[\sum_{k=1}^{3} B_k]^2} \cdot v_D \cdot e^{\lambda \cdot t}$
$\frac{\psi_1(V_D^{(i)} + v_D \cdot e^{\lambda \cdot t})}{\psi(V_D^{(i)} + v_D \cdot e^{\lambda \cdot t})}$	$\frac{\psi_1(V_D^{(i)})}{\psi(V_D^{(i)})} + \frac{\Phi_3(V_D^{(i)})}{\Phi_5(V_D^{(i)})} + \frac{(\Phi_5(V_D^{(i)}) \cdot \Phi_4(V_D^{(i)}) - \Phi_3(V_D^{(i)}) \cdot \Phi_6(V_D^{(i)}))}{\Phi_5^2(V_D^{(i)})} \cdot v_D \cdot e^{\lambda \cdot t}$ $\Gamma_2(V_D^{(i)}, v_D, \ldots)\vert_{v_D^2 \to \varepsilon \approx 0} = \frac{\Phi_3(V_D^{(i)})}{\Phi_5(V_D^{(i)})} + \frac{(\Phi_5(V_D^{(i)}) \cdot \Phi_4(V_D^{(i)}) - \Phi_3(V_D^{(i)}) \cdot \Phi_6(V_D^{(i)}))}{\Phi_5^2(V_D^{(i)})} \cdot v_D \cdot e^{\lambda \cdot t}$

$$x \cdot \lambda \cdot e^{\lambda \cdot t} = -(R_b + \frac{1}{\psi(V_D^{(i)} + v_D \cdot e^{\lambda \cdot t})}) \cdot \frac{1}{L_b} \cdot [X^{(i)} + x \cdot e^{\lambda \cdot (t-\tau_1)}]$$
$$- \frac{\psi_1(V_D^{(i)} + v_D \cdot e^{\lambda \cdot t})}{\psi(V_D^{(i)} + v_D \cdot e^{\lambda \cdot t})} \cdot [X^{(i)} + x \cdot e^{\lambda \cdot (t-\tau_1)}]^2 - [Y^{(i)} + y \cdot e^{\lambda \cdot (t-\tau_2)}]$$
$$\cdot \frac{1}{L_b \cdot \psi(V_D^{(i)} + v_D \cdot e^{\lambda \cdot t})}$$

$$x \cdot \lambda \cdot e^{\lambda \cdot t} = -(R_b + \{\frac{1}{\psi(V_D^{(i)})} + \frac{\Omega_6(V_D^{(i)})}{[\sum_{k=1}^{3} B_k]^2} \cdot v_D \cdot e^{\lambda \cdot t}\}) \cdot \frac{1}{L_b} \cdot [X^{(i)} + x \cdot e^{\lambda \cdot (t-\tau_1)}]$$
$$- \{\frac{\psi_1(V_D^{(i)})}{\psi(V_D^{(i)})} + \frac{\Phi_3(V_D^{(i)})}{\Phi_5(V_D^{(i)})}$$
$$+ \frac{(\Phi_5(V_D^{(i)}) \cdot \Phi_4(V_D^{(i)}) - \Phi_3(V_D^{(i)}) \cdot \Phi_6(V_D^{(i)}))}{\Phi_5^2(V_D^{(i)})} \cdot v_D \cdot e^{\lambda \cdot t}\}$$
$$\cdot [X^{(i)} + x \cdot e^{\lambda \cdot (t-\tau_1)}]^2 - [Y^{(i)} + y \cdot e^{\lambda \cdot (t-\tau_2)}]$$
$$\cdot \frac{1}{L_b} \cdot \{\frac{1}{\psi(V_D^{(i)})} + \frac{\Omega_6(V_D^{(i)})}{[\sum_{k=1}^{3} B_k]^2} \cdot v_D \cdot e^{\lambda \cdot t}\}$$

$$[X^{(i)} + x \cdot e^{\lambda \cdot (t-\tau_1)}]^2 = [X^{(i)}]^2 + 2 \cdot X^{(i)} \cdot x \cdot e^{\lambda \cdot (t-\tau_1)} + x^2 \cdot e^{2 \cdot \lambda \cdot (t-\tau_1)}|_{x^2 \approx 0}$$
$$= [X^{(i)}]^2 + 2 \cdot X^{(i)} \cdot x \cdot e^{\lambda \cdot (t-\tau_1)}$$

$$x \cdot \lambda \cdot e^{\lambda \cdot t} = -(R_b + \{\frac{1}{\psi(V_D^{(i)})} + \frac{\Omega_6(V_D^{(i)})}{[\sum_{k=1}^{3} B_k]^2} \cdot v_D \cdot e^{\lambda \cdot t}\}) \cdot \frac{1}{L_b} \cdot [X^{(i)} + x \cdot e^{\lambda \cdot (t-\tau_1)}]$$
$$- \{\frac{\psi_1(V_D^{(i)})}{\psi(V_D^{(i)})} + \frac{\Phi_3(V_D^{(i)})}{\Phi_5(V_D^{(i)})}$$
$$+ \frac{(\Phi_5(V_D^{(i)}) \cdot \Phi_4(V_D^{(i)}) - \Phi_3(V_D^{(i)}) \cdot \Phi_6(V_D^{(i)}))}{\Phi_5^2(V_D^{(i)})} \cdot v_D \cdot e^{\lambda \cdot t}\}$$
$$\cdot \{[X^{(i)}]^2 + 2 \cdot X^{(i)} \cdot x \cdot e^{\lambda \cdot (t-\tau_1)}\} - [Y^{(i)} + y \cdot e^{\lambda \cdot (t-\tau_2)}] \cdot \frac{1}{L_b}$$
$$\cdot \{\frac{1}{\psi(V_D^{(i)})} + \frac{\Omega_6(V_D^{(i)})}{[\sum_{k=1}^{3} B_k]^2} \cdot v_D \cdot e^{\lambda \cdot t}\}$$

$$x \cdot \lambda \cdot e^{\lambda \cdot t} = -[R_b + \frac{1}{\psi(V_D^{(i)})}] \cdot \frac{1}{L_b} \cdot X^{(i)} - [R_b + \frac{1}{\psi(V_D^{(i)})}] \cdot \frac{1}{L_b} \cdot x \cdot e^{\lambda \cdot (t-\tau_1)}$$
$$- \frac{X^{(i)}}{L_b} \cdot \frac{\Omega_6(V_D^{(i)})}{[\sum_{k=1}^{3} B_k]^2} \cdot v_D \cdot e^{\lambda \cdot t} - \frac{\Omega_6(V_D^{(i)})}{[\sum_{k=1}^{3} B_k]^2 \cdot L_b} \cdot v_D \cdot x \cdot e^{\lambda \cdot (t-\tau_1)} \cdot e^{\lambda \cdot t}$$
$$- [\frac{\psi_1(V_D^{(i)})}{\psi(V_D^{(i)})} + \frac{\Phi_3(V_D^{(i)})}{\Phi_5(V_D^{(i)})}] \cdot [X^{(i)}]^2 - [\frac{\psi_1(V_D^{(i)})}{\psi(V_D^{(i)})} + \frac{\Phi_3(V_D^{(i)})}{\Phi_5(V_D^{(i)})}] \cdot 2 \cdot X^{(i)} \cdot x \cdot e^{\lambda \cdot (t-\tau_1)}$$
$$- \frac{(\Phi_5(V_D^{(i)}) \cdot \Phi_4(V_D^{(i)}) - \Phi_3(V_D^{(i)}) \cdot \Phi_6(V_D^{(i)})) \cdot [X^{(i)}]^2}{\Phi_5^2(V_D^{(i)})} \cdot v_D \cdot e^{\lambda \cdot t}$$
$$- \frac{(\Phi_5(V_D^{(i)}) \cdot \Phi_4(V_D^{(i)}) - \Phi_3(V_D^{(i)}) \cdot \Phi_6(V_D^{(i)})) \cdot 2 \cdot X^{(i)}}{\Phi_5^2(V_D^{(i)})}$$
$$\cdot v_D \cdot x \cdot e^{\lambda \cdot (t-\tau_1)} \cdot e^{\lambda \cdot t} - \frac{Y^{(i)}}{L_b} \cdot \frac{1}{\psi(V_D^{(i)})} - \frac{Y^{(i)}}{L_b} \cdot \frac{\Omega_6(V_D^{(i)})}{[\sum_{k=1}^{3} B_k]^2} \cdot v_D \cdot e^{\lambda \cdot t}$$
$$- \frac{1}{L_b \cdot \psi(V_D^{(i)})} \cdot y \cdot e^{\lambda \cdot (t-\tau_2)} - \frac{1}{L_b} \cdot \frac{\Omega_6(V_D^{(i)})}{[\sum_{k=1}^{3} B_k]^2} \cdot v_D \cdot y \cdot e^{\lambda \cdot (t-\tau_2)} \cdot e^{\lambda \cdot t}$$

We consider $v_D \cdot x \approx 0;\ v_D \cdot y \approx 0$

$$x \cdot \lambda \cdot e^{\lambda \cdot t} = -[R_b + \frac{1}{\psi(V_D^{(i)})}] \cdot \frac{1}{L_b} \cdot X^{(i)} - [R_b + \frac{1}{\psi(V_D^{(i)})}] \cdot \frac{1}{L_b} \cdot x \cdot e^{\lambda \cdot (t-\tau_1)}$$
$$- \frac{X^{(i)}}{L_b} \cdot \frac{\Omega_6(V_D^{(i)})}{[\sum_{k=1}^{3} B_k]^2} \cdot v_D \cdot e^{\lambda \cdot t} - [\frac{\psi_1(V_D^{(i)})}{\psi(V_D^{(i)})} + \frac{\Phi_3(V_D^{(i)})}{\Phi_5(V_D^{(i)})}] \cdot [X^{(i)}]^2$$
$$- [\frac{\psi_1(V_D^{(i)})}{\psi(V_D^{(i)})} + \frac{\Phi_3(V_D^{(i)})}{\Phi_5(V_D^{(i)})}] \cdot 2 \cdot X^{(i)} \cdot x \cdot e^{\lambda \cdot (t-\tau_1)}$$
$$- \frac{(\Phi_5(V_D^{(i)}) \cdot \Phi_4(V_D^{(i)}) - \Phi_3(V_D^{(i)}) \cdot \Phi_6(V_D^{(i)})) \cdot [X^{(i)}]^2}{\Phi_5^2(V_D^{(i)})}$$
$$\cdot v_D \cdot e^{\lambda \cdot t} - \frac{Y^{(i)}}{L_b} \cdot \frac{1}{\psi(V_D^{(i)})} - \frac{Y^{(i)}}{L_b} \cdot \frac{\Omega_6(V_D^{(i)})}{[\sum_{k=1}^{3} B_k]^2} \cdot v_D \cdot e^{\lambda \cdot t} - \frac{1}{L_b \cdot \psi(V_D^{(i)})} \cdot y \cdot e^{\lambda \cdot (t-\tau_2)}$$

$$x \cdot \lambda \cdot e^{\lambda \cdot t} = -[R_b + \frac{1}{\psi(V_D^{(i)})}] \cdot \frac{1}{L_b} \cdot X^{(i)} - \frac{\psi_1(V_D^{(i)})}{\psi(V_D^{(i)})} \cdot [X^{(i)}]^2 - \frac{Y^{(i)}}{L_b} \cdot \frac{1}{\psi(V_D^{(i)})}$$
$$- [R_b + \frac{1}{\psi(V_D^{(i)})}] \cdot \frac{1}{L_b} \cdot x \cdot e^{\lambda \cdot (t - \tau_1)} - \frac{X^{(i)}}{L_b} \cdot \frac{\Omega_6(V_D^{(i)})}{[\sum_{k=1}^{3} B_k]^2} \cdot v_D \cdot e^{\lambda \cdot t} - \frac{\Phi_3(V_D^{(i)})}{\Phi_5(V_D^{(i)})} \cdot [X^{(i)}]^2$$
$$- [\frac{\psi_1(V_D^{(i)})}{\psi(V_D^{(i)})} + \frac{\Phi_3(V_D^{(i)})}{\Phi_5(V_D^{(i)})}] \cdot 2 \cdot X^{(i)} \cdot x \cdot e^{\lambda \cdot (t - \tau_1)}$$
$$- \frac{(\Phi_5(V_D^{(i)}) \cdot \Phi_4(V_D^{(i)}) - \Phi_3(V_D^{(i)}) \cdot \Phi_6(V_D^{(i)})) \cdot [X^{(i)}]^2}{\Phi_5^2(V_D^{(i)})}$$
$$\cdot v_D \cdot e^{\lambda \cdot t} - \frac{Y^{(i)}}{L_b} \cdot \frac{\Omega_6(V_D^{(i)})}{[\sum_{k=1}^{3} B_k]^2} \cdot v_D \cdot e^{\lambda \cdot t} - \frac{1}{L_b \cdot \psi(V_D^{(i)})} \cdot y \cdot e^{\lambda \cdot (t - \tau_2)}$$

At fixed point: $-(R_b + \frac{1}{\psi(V_D^{(i)})}) \cdot \frac{1}{L_b} \cdot X^{(i)} - \frac{\psi_1(V_D^{(i)})}{\psi(V_D^{(i)})} \cdot [X^{(i)}]^2 - Y^{(i)} \cdot \frac{1}{L_b \cdot \psi(V_D^{(i)})} = 0$

$$x \cdot \lambda \cdot e^{\lambda \cdot t} = -[R_b + \frac{1}{\psi(V_D^{(i)})}] \cdot \frac{1}{L_b} \cdot x \cdot e^{\lambda \cdot (t - \tau_1)} - \frac{X^{(i)}}{L_b} \cdot \frac{\Omega_6(V_D^{(i)})}{[\sum_{k=1}^{3} B_k]^2} \cdot v_D \cdot e^{\lambda \cdot t}$$
$$- \frac{\Phi_3(V_D^{(i)})}{\Phi_5(V_D^{(i)})} \cdot [X^{(i)}]^2 - \frac{1}{L_b \cdot \psi(V_D^{(i)})} \cdot y \cdot e^{\lambda \cdot (t - \tau_2)}$$
$$- [\frac{\psi_1(V_D^{(i)})}{\psi(V_D^{(i)})} + \frac{\Phi_3(V_D^{(i)})}{\Phi_5(V_D^{(i)})}] \cdot 2 \cdot X^{(i)} \cdot x \cdot e^{\lambda \cdot (t - \tau_1)}$$
$$- \frac{(\Phi_5(V_D^{(i)}) \cdot \Phi_4(V_D^{(i)}) - \Phi_3(V_D^{(i)}) \cdot \Phi_6(V_D^{(i)})) \cdot [X^{(i)}]^2}{\Phi_5^2(V_D^{(i)})}$$
$$\cdot v_D \cdot e^{\lambda \cdot t} - \frac{Y^{(i)}}{L_b} \cdot \frac{\Omega_6(V_D^{(i)})}{[\sum_{k=1}^{3} B_k]^2} \cdot v_D \cdot e^{\lambda \cdot t}$$

We need to choose the right parameters which give $\frac{\Phi_3(V_D^{(i)})}{\Phi_5(V_D^{(i)})} \cdot [X^{(i)}]^2 = 0$ since there is no $e^{\lambda \cdot t}$multiplication term. We already approve our fixed points are $E^*(X^*, Y^*, V_D^*, V_A^*) = (0, 0, V_D^*, 0)$. $X^{(i)} = 0 \Rightarrow \frac{\Phi_3(V_D^{(i)})}{\Phi_5(V_D^{(i)})} \cdot [X^{(i)}]^2 = 0$

$$x \cdot \lambda \cdot e^{\lambda \cdot t} = -[R_b + \frac{1}{\psi(V_D^{(i)})}] \cdot \frac{1}{L_b} \cdot x \cdot e^{\lambda \cdot t} \cdot e^{-\lambda \cdot \tau_1} - \frac{X^{(i)}}{L_b} \cdot \frac{\Omega_6(V_D^{(i)})}{[\sum_{k=1}^{3} B_k]^2} \cdot v_D \cdot e^{\lambda \cdot t}$$

$$- \frac{1}{L_b \cdot \psi(V_D^{(i)})} \cdot y \cdot e^{\lambda \cdot t} \cdot e^{-\lambda \cdot \tau_2} - [\frac{\psi_1(V_D^{(i)})}{\psi(V_D^{(i)})} + \frac{\Phi_3(V_D^{(i)})}{\Phi_5(V_D^{(i)})}] \cdot 2 \cdot X^{(i)} \cdot x \cdot e^{\lambda \cdot t} \cdot e^{-\lambda \cdot \tau_1}$$

$$- \frac{(\Phi_5(V_D^{(i)}) \cdot \Phi_4(V_D^{(i)}) - \Phi_3(V_D^{(i)}) \cdot \Phi_6(V_D^{(i)})) \cdot [X^{(i)}]^2}{\Phi_5^2(V_D^{(i)})} \cdot v_D \cdot e^{\lambda \cdot t}$$

$$- \frac{Y^{(i)}}{L_b} \cdot \frac{\Omega_6(V_D^{(i)})}{[\sum_{k=1}^{3} B_k]^2} \cdot v_D \cdot e^{\lambda \cdot t}$$

Dividing two side of above equation by $e^{\lambda \cdot t}$ gives:

$$x \cdot \lambda = -[R_b + \frac{1}{\psi(V_D^{(i)})}] \cdot \frac{1}{L_b} \cdot x \cdot e^{-\lambda \cdot \tau_1} - \frac{X^{(i)}}{L_b} \cdot \frac{\Omega_6(V_D^{(i)})}{[\sum_{k=1}^{3} B_k]^2} \cdot v_D - \frac{1}{L_b \cdot \psi(V_D^{(i)})} \cdot y \cdot e^{-\lambda \cdot \tau_2}$$

$$- [\frac{\psi_1(V_D^{(i)})}{\psi(V_D^{(i)})} + \frac{\Phi_3(V_D^{(i)})}{\Phi_5(V_D^{(i)})}] \cdot 2 \cdot X^{(i)} \cdot x \cdot e^{-\lambda \cdot \tau_1}$$

$$- \frac{(\Phi_5(V_D^{(i)}) \cdot \Phi_4(V_D^{(i)}) - \Phi_3(V_D^{(i)}) \cdot \Phi_6(V_D^{(i)})) \cdot [X^{(i)}]^2}{\Phi_5^2(V_D^{(i)})}$$

$$\cdot v_D - \frac{Y^{(i)}}{L_b} \cdot \frac{\Omega_6(V_D^{(i)})}{[\sum_{k=1}^{3} B_k]^2} \cdot v_D$$

$$\{-\lambda - ([R_b + \frac{1}{\psi(V_D^{(i)})}] \cdot \frac{1}{L_b} + [\frac{\psi_1(V_D^{(i)})}{\psi(V_D^{(i)})} + \frac{\Phi_3(V_D^{(i)})}{\Phi_5(V_D^{(i)})}] \cdot 2 \cdot X^{(i)}) \cdot e^{-\lambda \cdot \tau_1}\} \cdot x$$

$$- \frac{1}{L_b \cdot \psi(V_D^{(i)})} \cdot e^{-\lambda \cdot \tau_2} \cdot y - \{\frac{X^{(i)}}{L_b} \cdot \frac{\Omega_6(V_D^{(i)})}{[\sum_{k=1}^{3} B_k]^2}$$

$$+ \frac{(\Phi_5(V_D^{(i)}) \cdot \Phi_4(V_D^{(i)}) - \Phi_3(V_D^{(i)}) \cdot \Phi_6(V_D^{(i)})) \cdot [X^{(i)}]^2}{\Phi_5^2(V_D^{(i)})}$$

$$+ \frac{Y^{(i)}}{L_b} \cdot \frac{\Omega_6(V_D^{(i)})}{[\sum_{k=1}^{3} B_k]^2}\} \cdot v_D = 0$$

We define for simplicity the following global parameters: $\Pi_1 = \Pi_1(V_D^{(i)}, X^{(i)}, \ldots)$

$$\Pi_1 = [R_b + \frac{1}{\psi(V_D^{(i)})}] \cdot \frac{1}{L_b} + [\frac{\psi_1(V_D^{(i)})}{\psi(V_D^{(i)})} + \frac{\Phi_3(V_D^{(i)})}{\Phi_5(V_D^{(i)})}] \cdot 2 \cdot X^{(i)}; \; \Pi_2 = \Pi_2(V_D^{(i)}, X^{(i)}, Y^{(i)}, \ldots)$$

$$\Pi_2 = \frac{X^{(i)}}{L_b} \cdot \frac{\Omega_6(V_D^{(i)})}{[\sum_{k=1}^{3} B_k]^2} + \frac{(\Phi_5(V_D^{(i)}) \cdot \Phi_4(V_D^{(i)}) - \Phi_3(V_D^{(i)}) \cdot \Phi_6(V_D^{(i)})) \cdot [X^{(i)}]^2}{\Phi_5^2(V_D^{(i)})} + \frac{Y^{(i)}}{L_b} \cdot \frac{\Omega_6(V_D^{(i)})}{[\sum_{k=1}^{3} B_k]^2}$$

$$\{-\lambda - \Pi_1 \cdot e^{-\lambda \cdot \tau_1}\} \cdot x - \frac{1}{L_b \cdot \psi(V_D^{(i)})} \cdot e^{-\lambda \cdot \tau_2} \cdot y - \Pi_2 \cdot v_D = 0$$

Second TD's system differential equation: $\frac{dY}{dt} = \frac{\psi(V_D)}{C} \cdot X - \frac{1}{R_L \cdot C} \cdot Y - \frac{1}{L \cdot C} \cdot V_A$ and adding to it's coordinates $[X \, Y \, V_D \, V_A]$ arbitrarily small increments of exponential form $[x \, y \, v_D \, v_A] \cdot e^{\lambda \cdot t}$ and retaining the first order terms in $x \, y \, v_D \, v_A$.

$$y \cdot \lambda \cdot e^{\lambda \cdot t} = \frac{\psi(V_D^{(i)} + v_D \cdot e^{\lambda \cdot t})}{C} \cdot [X^{(i)} + x \cdot e^{\lambda \cdot (t - \tau_1)}] - \frac{1}{R_L \cdot C} \cdot [Y^{(i)} + y \cdot e^{\lambda \cdot (t - \tau_2)}] - \frac{1}{L \cdot C} \cdot [V_A^{(i)} + v_A \cdot e^{\lambda \cdot t}]$$

$$\psi(V_D^{(i)} + v_D \cdot e^{\lambda \cdot t}) = \psi(V_D^{(i)}) + \Gamma_3(V_D^{(i)}, v_D, \ldots)$$
$$\Rightarrow \Gamma_3(V_D^{(i)}, v_D, \ldots) = \psi(V_D^{(i)} + v_D \cdot e^{\lambda \cdot t}) - \psi(V_D^{(i)})$$

We already approve $B_4|_{\Omega_1(v_D)|_{v_D^2 \approx 0} = \Omega_2(v_D)|_{v_D^2 \approx 0} = v_D} = \Omega_3(V_D^{(i)}) \cdot v_D \cdot e^{\lambda \cdot t}$

$$\psi(V_D^{(i)} + v_D \cdot e^{\lambda \cdot t}) = B_1 \cdot e^{[\frac{v_D \cdot e^{\lambda \cdot t}}{\eta \cdot V_{th}}]} + B_2 \cdot e^{-\{\binom{m}{1} \cdot [\frac{V_D^{(i)}}{V_0}]^{(m-1)} \cdot [\frac{v_D \cdot e^{\lambda \cdot t}}{V_0}]\}} + B_3 \cdot e^{[\frac{v_D \cdot e^{\lambda \cdot t}}{V_{ex}}]} + B_4$$

$$\psi(V_D^{(i)} + v_D \cdot e^{\lambda \cdot t}) = B_1 \cdot e^{[\frac{v_D \cdot e^{\lambda \cdot t}}{\eta \cdot V_{th}}]} + B_2 \cdot e^{-\{\binom{m}{1} \cdot [\frac{V_D^{(i)}}{V_0}]^{(m-1)} \cdot [\frac{v_D \cdot e^{\lambda \cdot t}}{V_0}]\}} + B_3 \cdot e^{[\frac{v_D \cdot e^{\lambda \cdot t}}{V_{ex}}]}$$
$$+ \Omega_3(V_D^{(i)}) \cdot v_D \cdot e^{\lambda \cdot t}$$

$$\psi(V_D^{(i)} + v_D \cdot e^{\lambda \cdot t}) = [\sum_{k=1}^{3} B_k] + \Omega_7(V_D^{(i)}) \cdot v_D \cdot e^{\lambda \cdot t}$$

$$\Gamma_3(V_D^{(i)}, v_D, \ldots) = [\sum_{k=1}^{3} B_k] + \Omega_7(V_D^{(i)}) \cdot v_D \cdot e^{\lambda \cdot t} - \psi(V_D^{(i)})$$
$$= \{[\sum_{k=1}^{3} B_k] - \psi(V_D^{(i)})\} + \Omega_7(V_D^{(i)}) \cdot v_D \cdot e^{\lambda \cdot t}$$

$$[\sum_{k=1}^{3} B_k] - \psi(V_D^{(i)}) = I_s \cdot \frac{1}{\eta \cdot V_{th}} \cdot e^{[\frac{[V_D^{(i)}]}{\eta \cdot V_{th}} - 1]} + (1 - m \cdot [\frac{V_D^{(i)}}{V_0}]^m) \cdot \frac{1}{R_0} \cdot e^{-[\frac{V_D^{(i)}}{V_0}]^m}$$
$$+ (1 + \frac{V_D^{(i)}}{V_{ex}}) \cdot \frac{1}{R_V} \cdot e^{[\frac{V_D^{(i)} - V_V}{V_{ex}}]} - \{I_s \cdot \frac{1}{\eta \cdot V_{th}} \cdot e^{[(\frac{V_D^{(i)}}{\eta \cdot V_{th}}) - 1]}$$
$$+ (1 - m \cdot [\frac{V_D^{(i)}}{V_0}]^m) \cdot \frac{1}{R_0} \cdot e^{[-(\frac{V_D^{(i)}}{V_0})^m]} + (1 + \frac{V_D^{(i)}}{V_{ex}}) \cdot \frac{1}{R_V} \cdot e^{[\frac{V_D^{(i)} - V_V}{V_{ex}}]}\} = 0$$

$$\Gamma_3(V_D^{(i)}, v_D, \ldots) = \Omega_7(V_D^{(i)}) \cdot v_D \cdot e^{\lambda \cdot t} \Rightarrow \psi(V_D^{(i)} + v_D \cdot e^{\lambda \cdot t})$$
$$= \psi(V_D^{(i)}) + \Omega_7(V_D^{(i)}) \cdot v_D \cdot e^{\lambda \cdot t}$$

$$y \cdot \lambda \cdot e^{\lambda \cdot t} = \frac{[\psi(V_D^{(i)}) + \Omega_7(V_D^{(i)}) \cdot v_D \cdot e^{\lambda \cdot t}]}{C} \cdot [X^{(i)} + x \cdot e^{\lambda \cdot (t - \tau_1)}] - \frac{1}{R_L \cdot C} \cdot [Y^{(i)} + y$$
$$\cdot e^{\lambda \cdot (t - \tau_2)}] - \frac{1}{L \cdot C} \cdot [V_A^{(i)} + v_A \cdot e^{\lambda \cdot t}]$$

$$y \cdot \lambda \cdot e^{\lambda \cdot t} = [\frac{\psi(V_D^{(i)})}{C} + \frac{\Omega_7(V_D^{(i)}) \cdot v_D \cdot e^{\lambda \cdot t}}{C}] \cdot [X^{(i)} + x \cdot e^{\lambda \cdot (t - \tau_1)}] - \frac{1}{R_L \cdot C} \cdot Y^{(i)}$$
$$- \frac{1}{R_L \cdot C} \cdot y \cdot e^{\lambda \cdot (t - \tau_2)} - \frac{1}{L \cdot C} \cdot V_A^{(i)} - \frac{1}{L \cdot C} \cdot v_A \cdot e^{\lambda \cdot t}$$

$$y \cdot \lambda \cdot e^{\lambda \cdot t} = \frac{\psi(V_D^{(i)})}{C} \cdot X^{(i)} + \frac{\psi(V_D^{(i)})}{C} \cdot x \cdot e^{\lambda \cdot (t-\tau_1)} + \frac{X^{(i)} \cdot \Omega_7(V_D^{(i)}) \cdot v_D \cdot e^{\lambda \cdot t}}{C}$$
$$+ \frac{\Omega_7(V_D^{(i)}) \cdot v_D \cdot x \cdot e^{\lambda \cdot (t-\tau_1)} \cdot e^{\lambda \cdot t}}{C} - \frac{1}{R_L \cdot C} \cdot Y^{(i)}$$
$$- \frac{1}{R_L \cdot C} \cdot y \cdot e^{\lambda \cdot (t-\tau_2)} - \frac{1}{L \cdot C} \cdot V_A^{(i)} - \frac{1}{L \cdot C} \cdot v_A \cdot e^{\lambda \cdot t}$$

We consider $v_D \cdot x \approx 0$

$$y \cdot \lambda \cdot e^{\lambda \cdot t} = \frac{\psi(V_D^{(i)})}{C} \cdot X^{(i)} + \frac{\psi(V_D^{(i)})}{C} \cdot x \cdot e^{\lambda \cdot (t-\tau_1)} + \frac{X^{(i)} \cdot \Omega_7(V_D^{(i)}) \cdot v_D \cdot e^{\lambda \cdot t}}{C}$$
$$- \frac{1}{R_L \cdot C} \cdot Y^{(i)} - \frac{1}{R_L \cdot C} \cdot y \cdot e^{\lambda \cdot (t-\tau_2)} - \frac{1}{L \cdot C} \cdot V_A^{(i)} - \frac{1}{L \cdot C} \cdot v_A \cdot e^{\lambda \cdot t}$$

$$y \cdot \lambda \cdot e^{\lambda \cdot t} = \frac{\psi(V_D^{(i)})}{C} \cdot X^{(i)} - \frac{1}{R_L \cdot C} \cdot Y^{(i)} - \frac{1}{L \cdot C} \cdot V_A^{(i)} + \frac{\psi(V_D^{(i)})}{C} \cdot x \cdot e^{\lambda \cdot (t-\tau_1)}$$
$$+ \frac{X^{(i)} \cdot \Omega_7(V_D^{(i)}) \cdot v_D \cdot e^{\lambda \cdot t}}{C} - \frac{1}{R_L \cdot C} \cdot y \cdot e^{\lambda \cdot (t-\tau_2)} - \frac{1}{L \cdot C} \cdot v_A \cdot e^{\lambda \cdot t}$$

At fixed point: $\frac{\psi(V_D^{(i)})}{C} \cdot X^{(i)} - \frac{1}{R_L \cdot C} \cdot Y^{(i)} - \frac{1}{L \cdot C} \cdot V_A^{(i)} = 0$

$$y \cdot \lambda \cdot e^{\lambda \cdot t} = \frac{\psi(V_D^{(i)})}{C} \cdot x \cdot e^{\lambda \cdot (t-\tau_1)} + \frac{X^{(i)} \cdot \Omega_7(V_D^{(i)}) \cdot v_D \cdot e^{\lambda \cdot t}}{C} - \frac{1}{R_L \cdot C} \cdot y \cdot e^{\lambda \cdot (t-\tau_2)}$$
$$- \frac{1}{L \cdot C} \cdot v_A \cdot e^{\lambda \cdot t}$$
$$y \cdot \lambda = \frac{\psi(V_D^{(i)})}{C} \cdot x \cdot e^{-\lambda \cdot \tau_1} + \frac{X^{(i)} \cdot \Omega_7(V_D^{(i)})}{C} \cdot v_D - \frac{1}{R_L \cdot C} \cdot y \cdot e^{-\lambda \cdot \tau_2} - \frac{1}{L \cdot C} \cdot v_A$$

$$\frac{\psi(V_D^{(i)})}{C} \cdot x \cdot e^{-\lambda \cdot \tau_1} - y \cdot \lambda + \frac{X^{(i)} \cdot \Omega_7(V_D^{(i)})}{C} \cdot v_D - \frac{1}{R_L \cdot C} \cdot y \cdot e^{-\lambda \cdot \tau_2} - \frac{1}{L \cdot C} \cdot v_A = 0$$

$$\frac{\psi(V_D^{(i)})}{C} \cdot x \cdot e^{-\lambda \cdot \tau_1} - y \cdot \lambda - \frac{1}{R_L \cdot C} \cdot y \cdot e^{-\lambda \cdot \tau_2} + \frac{X^{(i)} \cdot \Omega_7(V_D^{(i)})}{C} \cdot v_D - \frac{1}{L \cdot C} \cdot v_A = 0$$

Third TD's system differential equation: $\frac{dV_D}{dt} = X(t - \tau_1)$ and adding to it's coordinates $[X \; Y \; V_D \; V_A]$ arbitrarily small increments of exponential form $[x \, y \, v_D \, v_A] \cdot e^{\lambda \cdot t}$ and retaining the first order terms in $x \, y \, v_D \, v_A$ [9, 10].

$v_D \cdot \lambda \cdot e^{\lambda \cdot t} = X^{(i)} + x \cdot e^{\lambda \cdot (t-\tau_1)}$. At fixed point: $X(t-\tau_1)|_{t \gg \tau_1} = X(t) \Rightarrow X^{(i)} = 0$.

$v_D \cdot \lambda \cdot e^{\lambda \cdot t} = x \cdot e^{\lambda \cdot t} \cdot e^{-\lambda \cdot \tau_1} \Rightarrow x \cdot e^{-\lambda \cdot \tau_1} - v_D \cdot \lambda = 0$. Fourth TD's system differential equation: $\frac{dV_A}{dt} = Y(t-\tau_2)$ and adding to it's coordinates $[X\, Y\, V_D\, V_A]$ arbitrarily small increments of exponential form $[x\, y\, v_D\, v_A] \cdot e^{\lambda \cdot t}$ and retaining the first order terms in $x\, y\, v_D\, v_A$. $v_A \cdot \lambda \cdot e^{\lambda \cdot t} = Y^{(i)} + y \cdot e^{\lambda \cdot (t-\tau_2)}$. At fixed point: $Y(t-\tau_2)|_{t>>\tau_2} = Y(t) \Rightarrow Y^{(i)} = 0$; $v_A \cdot \lambda \cdot e^{\lambda \cdot t} = Y^{(i)} + y \cdot e^{\lambda \cdot (t-\tau_2)} \Rightarrow v_A \cdot \lambda \cdot e^{\lambda \cdot t} = y \cdot e^{\lambda \cdot (t-\tau_2)}$ $v_A \cdot \lambda \cdot e^{\lambda \cdot t} = y \cdot e^{\lambda \cdot (t-\tau_2)} \Rightarrow y \cdot e^{-\lambda \cdot \tau_2} - v_A \cdot \lambda = 0$.

We summery our TD system's four characteristic equations in the eigenvalue λ with delays:

$$\{-\lambda - \Pi_1(V_D^{(i)}) \cdot e^{-\lambda \cdot \tau_1}\} \cdot x - \frac{1}{L_b \cdot \psi(V_D^{(i)})} \cdot e^{-\lambda \cdot \tau_2} \cdot y - \Pi_2(V_D^{(i)}) \cdot v_D = 0$$

$$\frac{\psi(V_D^{(i)})}{C} \cdot e^{-\lambda \cdot \tau_1} \cdot x - \lambda \cdot y - \frac{1}{R_L \cdot C} \cdot e^{-\lambda \cdot \tau_2} \cdot y + \frac{X^{(i)} \cdot \Omega_7(V_D^{(i)})}{C} \cdot v_D - \frac{1}{L \cdot C} \cdot v_A = 0$$

$$x \cdot e^{-\lambda \cdot \tau_1} - \lambda \cdot v_D = 0; \; y \cdot e^{-\lambda \cdot \tau_2} - \lambda \cdot v_A = 0$$

The small increments Jacobian of our Gradostat system is as bellow:

$$\begin{pmatrix} \Upsilon_{11} & \dots & \Upsilon_{14} \\ \vdots & \ddots & \vdots \\ \Upsilon_{41} & \dots & \Upsilon_{44} \end{pmatrix} \cdot \begin{pmatrix} x \\ y \\ v_D \\ v_A \end{pmatrix} = 0; \; \Upsilon_{11} = -\lambda - \Pi_1(V_D^{(i)}) \cdot e^{-\lambda \cdot \tau_1};$$

$$\Upsilon_{12} = -\frac{1}{L_b \cdot \psi(V_D^{(i)})} \cdot e^{-\lambda \cdot \tau_2}$$

$$\Upsilon_{13} = -\Pi_2(V_D^{(i)}); \; \Upsilon_{14} = 0; \; \Upsilon_{21} = \frac{\psi(V_D^{(i)})}{C} \cdot e^{-\lambda \cdot \tau_1}; \; \Upsilon_{22} = -\lambda - \frac{1}{R_L \cdot C} \cdot e^{-\lambda \cdot \tau_2}$$

$$\Upsilon_{23} = \frac{X^{(i)} \cdot \Omega_7(V_D^{(i)})}{C}; \; \Upsilon_{24} = -\frac{1}{L \cdot C}; \; \Upsilon_{31} = e^{-\lambda \cdot \tau_1}; \; \Upsilon_{32} = 0; \; \Upsilon_{33} = -\lambda; \; \Upsilon_{34} = 0$$

$$\Upsilon_{41} = 0; \; \Upsilon_{42} = e^{-\lambda \cdot \tau_2}; \; \Upsilon_{43} = 0; \; \Upsilon_{44} = -\lambda$$

$$A - \lambda \cdot I = \begin{pmatrix} \Upsilon_{11} & \dots & \Upsilon_{14} \\ \vdots & \ddots & \vdots \\ \Upsilon_{41} & \dots & \Upsilon_{44} \end{pmatrix}; \; \det|A - \lambda \cdot I| = 0$$

$$\det|A-\lambda\cdot I| = -[\lambda+\Pi_1(V_D^{(i)})\cdot e^{-\lambda\cdot\tau_1}]\cdot\det\begin{pmatrix} -\lambda-\frac{1}{R_L\cdot C}\cdot e^{-\lambda\cdot\tau_2} & \frac{X^{(i)}\cdot\Omega_7(V_D^{(i)})}{C} & -\frac{1}{L\cdot C} \\ 0 & -\lambda & 0 \\ e^{-\lambda\cdot\tau_2} & 0 & -\lambda \end{pmatrix}$$

$$+\frac{1}{L_b\cdot\psi(V_D^{(i)})}\cdot e^{-\lambda\cdot\tau_2}\cdot\det\begin{pmatrix} \frac{\psi(V_D^{(i)})}{C}\cdot e^{-\lambda\cdot\tau_1} & \frac{X^{(i)}\cdot\Omega_7(V_D^{(i)})}{C} & -\frac{1}{L\cdot C} \\ e^{-\lambda\cdot\tau_1} & -\lambda & 0 \\ 0 & 0 & -\lambda \end{pmatrix}$$

$$-\Pi_2(V_D^{(i)})\cdot\det\begin{pmatrix} \frac{\psi(V_D^{(i)})}{C}\cdot e^{-\lambda\cdot\tau_1} & -\lambda-\frac{1}{R_L\cdot C}\cdot e^{-\lambda\cdot\tau_2} & -\frac{1}{L\cdot C} \\ e^{-\lambda\cdot\tau_1} & 0 & 0 \\ 0 & e^{-\lambda\cdot\tau_2} & -\lambda \end{pmatrix}$$

$$\det|A-\lambda\cdot I| = -[\lambda+\Pi_1(V_D^{(i)})\cdot e^{-\lambda\cdot\tau_1}]\cdot\{-(\lambda+\frac{1}{R_L\cdot C}\cdot e^{-\lambda\cdot\tau_2})\cdot\lambda^2-\frac{1}{L\cdot C}\cdot e^{-\lambda\cdot\tau_2}\cdot\lambda\}$$
$$+\frac{1}{L_b\cdot\psi(V_D^{(i)})}\cdot e^{-\lambda\cdot\tau_2}\cdot\{\frac{\psi(V_D^{(i)})}{C}\cdot e^{-\lambda\cdot\tau_1}\cdot\lambda^2+\frac{X^{(i)}\cdot\Omega_7(V_D^{(i)})}{C}\cdot e^{-\lambda\cdot\tau_1}\cdot\lambda\}$$
$$-\Pi_2(V_D^{(i)})\cdot\{-(\lambda+\frac{1}{R_L\cdot C}\cdot e^{-\lambda\cdot\tau_2})\cdot e^{-\lambda\cdot\tau_1}\cdot\lambda-\frac{1}{L\cdot C}\cdot e^{-\lambda\cdot(\tau_1+\tau_2)}\}$$

$$D(\lambda,\tau_1,\tau_2) = \lambda^4+\{\lambda^3\cdot\Pi_1(V_D^{(i)})+\lambda^2\cdot\Pi_2(V_D^{(i)})\}\cdot e^{-\lambda\cdot\tau_1}$$
$$+(\lambda^3\cdot\frac{1}{R_L\cdot C}+\lambda^2\cdot\frac{1}{L\cdot C})\cdot e^{-\lambda\cdot\tau_2}+\{\lambda^2\cdot\frac{1}{C}\cdot[\frac{\Pi_1(V_D^{(i)})}{R_L}+\frac{1}{L_b}]$$
$$+\lambda\cdot\frac{1}{C}\cdot[\frac{\Pi_1(V_D^{(i)})}{L}+\frac{X^{(i)}\cdot\Omega_7(V_D^{(i)})}{L_b\cdot\psi(V_D^{(i)})}+\frac{\Pi_2(V_D^{(i)})}{R_L}]+\frac{\Pi_2(V_D^{(i)})}{L\cdot C}\}\cdot e^{-\lambda\cdot\sum_{i=1}^{2}\tau_i}$$

We have three stability cases: (1) $\tau_1=\tau;\ \tau_2=0$ (2) $\tau_1=0;\ \tau_2=\tau$ (3) $\tau_1=\tau_2=\tau$.

We need to get characteristics equations for all above stability analysis cases. We study the occurrence of any possible stability switching resulting from the increase of value of the time delay τ for the general characteristic equation $D(\lambda,\tau)$. If we choose parameter then $D(\lambda,\tau)=P_n(\lambda,\tau)+Q_m(\lambda,\tau)\cdot e^{-\lambda\cdot\tau};\ n>m$.

$$P_n(\lambda,\tau)=\sum_{k=0}^{n}p_k(\tau)\cdot\lambda^k=p_0(\tau)+p_1(\tau)\cdot\lambda+p_2(\tau)\cdot\lambda^2+p_3(\tau)\cdot\lambda^3+\ldots$$
$$Q_m(\lambda,\tau)=\sum_{k=0}^{m}q_k(\tau)\cdot\lambda^k=q_0(\tau)+q_1(\tau)\cdot\lambda+q_2(\tau)\cdot\lambda^2+q_3(\tau)\cdot\lambda^3+\ldots$$

$$D(\lambda, \tau_1 = \tau, \tau_2 = 0) = \lambda^4 + \lambda^3 \cdot \frac{1}{R_L \cdot C} + \lambda^2 \cdot \frac{1}{L \cdot C} + \{\lambda^3 \cdot \Pi_1(V_D^{(i)}) + \lambda^2 \cdot \Pi_2(V_D^{(i)})$$
$$+ \lambda^2 \cdot \frac{1}{C} \cdot [\frac{\Pi_1(V_D^{(i)})}{R_L} + \frac{1}{L_b}] + \lambda \cdot \frac{1}{C} \cdot [\frac{\Pi_1(V_D^{(i)})}{L}$$
$$+ \frac{X^{(i)} \cdot \Omega_7(V_D^{(i)})}{L_b \cdot \psi(V_D^{(i)})} + \frac{\Pi_2(V_D^{(i)})}{R_L}] + \frac{\Pi_2(V_D^{(i)})}{L \cdot C}\} \cdot e^{-\lambda \cdot \tau}$$

$$D(\lambda, \tau_1 = 0, \tau_2 = \tau) = \lambda^4 + \lambda^3 \cdot \Pi_1(V_D^{(i)}) + \lambda^2 \cdot \Pi_2(V_D^{(i)}) + \{\lambda^3 \cdot \frac{1}{R_L \cdot C} + \lambda^2 \cdot \frac{1}{L \cdot C}$$
$$+ \lambda^2 \cdot \frac{1}{C} \cdot [\frac{\Pi_1(V_D^{(i)})}{R_L} + \frac{1}{L_b}] + \lambda \cdot \frac{1}{C} \cdot [\frac{\Pi_1(V_D^{(i)})}{L}$$
$$+ \frac{X^{(i)} \cdot \Omega_7(V_D^{(i)})}{L_b \cdot \psi(V_D^{(i)})} + \frac{\Pi_2(V_D^{(i)})}{R_L}] + \frac{\Pi_2(V_D^{(i)})}{L \cdot C}\} \cdot e^{-\lambda \cdot \tau}$$

$$D(\lambda, \tau_1 = \tau_2 = \tau) = \lambda^4 + \{\lambda^3 \cdot [\Pi_1(V_D^{(i)}) + \frac{1}{R_L \cdot C}] + \lambda^2 \cdot [\Pi_2(V_D^{(i)}) + \frac{1}{L \cdot C}]\} \cdot e^{-\lambda \cdot \tau}$$
$$+ \{\lambda^2 \cdot \frac{1}{C} \cdot [\frac{\Pi_1(V_D^{(i)})}{R_L} + \frac{1}{L_b}] + \lambda \cdot \frac{1}{C} \cdot [\frac{\Pi_1(V_D^{(i)})}{L}$$
$$+ \frac{X^{(i)} \cdot \Omega_7(V_D^{(i)})}{L_b \cdot \psi(V_D^{(i)})} + \frac{\Pi_2(V_D^{(i)})}{R_L}] + \frac{\Pi_2(V_D^{(i)})}{L \cdot C}\} \cdot e^{-2 \cdot \lambda \cdot \tau}$$

Under Taylor series approximation: $e^{-\lambda \cdot \tau} \approx 1 - \lambda \cdot \tau$. The Maclaurin series is a Taylor series expansion of a $e^{-\lambda \cdot \tau}$ function about zero (0). We get the following general characteristic equation D(λ, τ) under Taylor series approximation: $e^{-\lambda \cdot \tau} \approx 1 - \lambda \cdot \tau$. $e^{-2 \cdot \lambda \cdot \tau} \approx e^{-\lambda \cdot \tau} \cdot (1 - \lambda \cdot \tau)$

$$D(\lambda, \tau_1 = \tau_2 = \tau) = \lambda^4 + \{\lambda^3 \cdot [\Pi_1(V_D^{(i)}) + \frac{1}{R_L \cdot C}] + \lambda^2 \cdot [\Pi_2(V_D^{(i)}) + \frac{1}{L \cdot C}]\} \cdot e^{-\lambda \cdot \tau}$$
$$+ \{\lambda^2 \cdot \frac{1}{C} \cdot [\frac{\Pi_1(V_D^{(i)})}{R_L} + \frac{1}{L_b}] + \lambda \cdot \frac{1}{C} \cdot [\frac{\Pi_1(V_D^{(i)})}{L}$$
$$+ \frac{X^{(i)} \cdot \Omega_7(V_D^{(i)})}{L_b \cdot \psi(V_D^{(i)})} + \frac{\Pi_2(V_D^{(i)})}{R_L}] + \frac{\Pi_2(V_D^{(i)})}{L \cdot C}\} \cdot (1 - \lambda \cdot \tau) \cdot e^{-\lambda \cdot \tau}$$

We define for simplicity the following global parameters:

$$B_1(V_D^{(i)}) = \Pi_1(V_D^{(i)}) + \frac{1}{R_L \cdot C}; \; B_2(V_D^{(i)}) = \Pi_2(V_D^{(i)}) + \frac{1}{L \cdot C};$$

$$B_3(V_D^{(i)}) = \frac{1}{C} \cdot [\frac{\Pi_1(V_D^{(i)})}{R_L} + \frac{1}{L_b}]$$

$$B_4(V_D^{(i)}) = \frac{1}{C} \cdot [\frac{\Pi_1(V_D^{(i)})}{L} + \frac{X^{(i)} \cdot \Omega_7(V_D^{(i)})}{L_b \cdot \psi(V_D^{(i)})} + \frac{\Pi_2(V_D^{(i)})}{R_L}]; \; B_5(V_D^{(i)}) = \frac{\Pi_2(V_D^{(i)})}{L \cdot C}$$

$$D(\lambda, \tau_1 = \tau_2 = \tau) = \lambda^4 + \{\lambda^3 \cdot B_1(V_D^{(i)}) + \lambda^2 \cdot B_2(V_D^{(i)})\} \cdot e^{-\lambda \cdot \tau} + \{\lambda^2 \cdot B_3(V_D^{(i)}) + \lambda \cdot B_4(V_D^{(i)}) + B_5(V_D^{(i)})\} \cdot (1 - \lambda \cdot \tau) \cdot e^{-\lambda \cdot \tau}$$

$$D(\lambda, \tau_1 = \tau_2 = \tau) = \lambda^4 + \{\lambda^3 \cdot B_1(V_D^{(i)}) + \lambda^2 \cdot B_2(V_D^{(i)})\} \cdot e^{-\lambda \cdot \tau} + \{\lambda^2 \cdot B_3(V_D^{(i)}) + \lambda \cdot B_4(V_D^{(i)}) + B_5(V_D^{(i)})\} \cdot e^{-\lambda \cdot \tau} - \lambda^3 \cdot B_3(V_D^{(i)}) \cdot \tau \cdot e^{-\lambda \cdot \tau} - \lambda^2 \cdot B_4(V_D^{(i)}) \cdot \tau \cdot e^{-\lambda \cdot \tau} - B_5(V_D^{(i)}) \cdot \lambda \cdot \tau \cdot e^{-\lambda \cdot \tau}$$

$$D(\lambda, \tau_1 = \tau_2 = \tau) = \lambda^4 + \{\lambda^3 \cdot [B_1(V_D^{(i)}) - B_3(V_D^{(i)}) \cdot \tau] + \lambda^2 \cdot [\sum_{k=2}^{3} B_k(V_D^{(i)}) - B_4(V_D^{(i)}) \cdot \tau] + \lambda \cdot [B_4(V_D^{(i)}) - B_5(V_D^{(i)}) \cdot \tau] + B_5(V_D^{(i)})\} \cdot e^{-\lambda \cdot \tau}$$

(Table 2.17)

Table 2.17 Tunnel Diode (TD) as a microwave oscillator system $P_n(\lambda, \tau)$ and $Q_m(\lambda, \tau)$ functions vs τ_1 and τ_2 options

	$\tau_1 = \tau; \tau_2 = 0$	$\tau_1 = 0; \tau_2 = \tau$
$P_n(\lambda, \tau)$	$\lambda^4 + \lambda^3 \cdot \frac{1}{R_L \cdot C} + \lambda^2 \cdot \frac{1}{L \cdot C}$	$\lambda^4 + \lambda^3 \cdot \Pi_1(V_D^{(i)}) + \lambda^2 \cdot \Pi_2(V_D^{(i)})$
$Q_m(\lambda, \tau)$	$\lambda^3 \cdot \Pi_1(V_D^{(i)}) + \lambda^2 \cdot \{\Pi_2(V_D^{(i)}) + \frac{1}{C} \cdot [\frac{\Pi_1(V_D^{(i)})}{R_L} + \frac{1}{L_b}]\} + \lambda \cdot \frac{1}{C} \cdot [\frac{\Pi_1(V_D^{(i)})}{L} + \frac{X^{(i)} \cdot \Omega_7(V_D^{(i)})}{L_b \cdot \psi(V_D^{(i)})} + \frac{\Pi_2(V_D^{(i)})}{R_L}] + \frac{\Pi_2(V_D^{(i)})}{L \cdot C}$	$\lambda^3 \cdot \frac{1}{R_L \cdot C} + \lambda^2 \cdot \{\frac{1}{L \cdot C} + \frac{1}{C} \cdot [\frac{\Pi_1(V_D^{(i)})}{R_L} + \frac{1}{L_b}]\} + \lambda \cdot \frac{1}{C} \cdot [\frac{\Pi_1(V_D^{(i)})}{L} + \frac{X^{(i)} \cdot \Omega_7(V_D^{(i)})}{L_b \cdot \psi(V_D^{(i)})} + \frac{\Pi_2(V_D^{(i)})}{R_L}] + \frac{\Pi_2(V_D^{(i)})}{L \cdot C}$
n	4	4
m	3	3
Status	n > m	n > m

We analyze the TD's system stability for the third case $\tau_1 = \tau_2 = \tau$.

$$D(\lambda, \tau_1 = \tau_2 = \tau) = \lambda^4 + \{\lambda^3 \cdot [B_1(V_D^{(i)}) - B_3(V_D^{(i)}) \cdot \tau] + \lambda^2 \cdot [\sum_{k=2}^{3} B_k(V_D^{(i)}) - B_4(V_D^{(i)}) \cdot \tau] + \lambda \cdot [B_4(V_D^{(i)}) - B_5(V_D^{(i)}) \cdot \tau] + B_5(V_D^{(i)})\} \cdot e^{-\lambda \cdot \tau}; \ n = 4; \ m = 3; \ n > m$$

$$P_n(\lambda, \tau) = \lambda^4; \ Q_m(\lambda, \tau) = \lambda^3 \cdot [B_1(V_D^{(i)}) - B_3(V_D^{(i)}) \cdot \tau] + \lambda^2 \cdot [\sum_{k=2}^{3} B_k(V_D^{(i)}) - B_4(V_D^{(i)}) \cdot \tau] + \lambda \cdot [B_4(V_D^{(i)}) - B_5(V_D^{(i)}) \cdot \tau] + B_5(V_D^{(i)})$$

$$p_0(\tau) = p_1(\tau) = p_2(\tau) = p_3(\tau) = 0; \ p_4(\tau) = 1; \ q_0(\tau) = B_5(V_D^{(i)});$$
$$q_1(\tau) = B_4(V_D^{(i)}) - B_5(V_D^{(i)}) \cdot \tau$$
$$q_2(\tau) = \sum_{k=2}^{3} B_k(V_D^{(i)}) - B_4(V_D^{(i)}) \cdot \tau; \ q_3(\tau) = B_1(V_D^{(i)}) - B_3(V_D^{(i)}) \cdot \tau$$

The homogeneous system for $X \, Y \, V_D \, V_A$ leads to a characteristic equation for the eigenvalue λ having the form $P(\lambda, \tau) + Q(\lambda, \tau) \cdot e^{-\lambda \cdot \tau} = 0$; $P(\lambda, \tau) = \sum_{j=0}^{4} a_j \cdot \lambda^j$ $Q(\lambda, \tau) = \sum_{j=0}^{3} c_j \cdot \lambda^j$ And the coefficients $\{a_j(q_i, q_k, \tau), c_j(q_i, q_k, \tau)\} \in \mathbb{R}$ depend on q_i, q_k and delay τ. q_i, q_k are any TD's parameters, other parameters kept as a constant $a_0(\tau) = a_1(\tau) = a_2(\tau) = a_3(\tau) = 0$; $a_4(\tau) = 1$; $c_0(\tau) = B_5(V_D^{(i)})$

$$c_1(\tau) = B_4(V_D^{(i)}) - B_5(V_D^{(i)}) \cdot \tau; \ c_2(\tau) = \sum_{k=2}^{3} B_k(V_D^{(i)}) - B_4(V_D^{(i)}) \cdot \tau; \ c_3(\tau) = B_1(V_D^{(i)}) - B_3(V_D^{(i)}) \cdot \tau$$

Unless strictly necessary, the designation of the varied arguments (q_i, q_k) will subsequently be omitted from P, Q, a_j, c_j. The coefficients a_j, c_j are continuous, and differentiable functions of their arguments, and direct substitution shows that $a_0 + c_0 \neq 0$ for $\forall q_i, q_k \in \mathbb{R}_+$, i.e. $\lambda = 0$ is not a of $P(\lambda) + Q(\lambda) \cdot e^{-\lambda \cdot \tau} = 0$. Furthermore, $P(\lambda)$, $Q(\lambda)$ are analytic functions of λ, for which the following requirements of the analysis [BK] can also be verified in the present case:

(a) If $\lambda = i \cdot \omega$, $\omega \in \mathbb{R}$, then $P(i \cdot \omega) + Q(i \cdot \omega) \neq 0$.
(b) $|Q(\lambda)/P(\lambda)|$ is bounded for $|\lambda| \to \infty$, $\mathrm{Re}\lambda \geq 0$. No roots bifurcation from ∞.

(c) $F(\omega) = |P(i \cdot \omega)|^2 - |Q(i \cdot \omega)|^2$ has a finite number of zeros. Indeed, this is a polynomial in ω.
(d) Each positive root $\omega(q_i, q_k)$ of F(ω) = 0 is continuous and differentiable respect to q_i, q_k.

We assume that $P_n(\lambda, \tau)$ and $Q_m(\lambda, \tau)$ can't have common imaginary roots. That is for any real number ω; $p_n(\lambda = i \cdot \omega, \tau) + Q_m(\lambda = i \cdot \omega, \tau) \neq 0$. $p_n(\lambda = i \cdot \omega, \tau) = \omega^4$

$$Q_m(\lambda = i \cdot \omega, \tau) = -\omega^2 \cdot [\sum_{k=2}^{3} B_k(V_D^{(i)}) - B_4(V_D^{(i)}) \cdot \tau] + B_5(V_D^{(i)}) + i \cdot \{\omega \cdot [B_4(V_D^{(i)}) - B_5(V_D^{(i)}) \cdot \tau] - \omega^3 \cdot [B_1(V_D^{(i)}) - B_3(V_D^{(i)}) \cdot \tau]\}$$

$$p_n(\lambda = i \cdot \omega, \tau) + Q_m(\lambda = i \cdot \omega, \tau) = \omega^4 - \omega^2 \cdot [\sum_{k=2}^{3} B_k(V_D^{(i)}) - B_4(V_D^{(i)}) \cdot \tau] + B_5(V_D^{(i)}) + i \cdot \{\omega \cdot [B_4(V_D^{(i)}) - B_5(V_D^{(i)}) \cdot \tau] - \omega^3 \cdot [B_1(V_D^{(i)}) - B_3(V_D^{(i)}) \cdot \tau]\} \neq 0$$

$$|P(i \cdot \omega, \tau)|^2 = \omega^8; \; |Q(i \cdot \omega, \tau)|^2 = \{-\omega^2 \cdot [\sum_{k=2}^{3} B_k(V_D^{(i)}) - B_4(V_D^{(i)}) \cdot \tau] + B_5(V_D^{(i)})\}^2 + \{\omega \cdot [B_4(V_D^{(i)}) - B_5(V_D^{(i)}) \cdot \tau] - \omega^3 \cdot [B_1(V_D^{(i)}) - B_3(V_D^{(i)}) \cdot \tau]\}^2$$

$$\{-\omega^2 \cdot [\sum_{k=2}^{3} B_k(V_D^{(i)}) - B_4(V_D^{(i)}) \cdot \tau] + B_5(V_D^{(i)})\}^2 = \omega^4 \cdot [\sum_{k=2}^{3} B_k(V_D^{(i)}) - B_4(V_D^{(i)}) \cdot \tau]^2 + [B_5(V_D^{(i)})]^2 - 2 \cdot \omega^2 \cdot [\sum_{k=2}^{3} B_k(V_D^{(i)}) - B_4(V_D^{(i)}) \cdot \tau] \cdot B_5(V_D^{(i)})$$

$$\{\omega \cdot [B_4(V_D^{(i)}) - B_5(V_D^{(i)}) \cdot \tau] - \omega^3 \cdot [B_1(V_D^{(i)}) - B_3(V_D^{(i)}) \cdot \tau]\}^2 = \omega^2 \cdot [B_4(V_D^{(i)}) - B_5(V_D^{(i)}) \cdot \tau]^2 + \omega^6 \cdot [B_1(V_D^{(i)}) - B_3(V_D^{(i)}) \cdot \tau]^2 - 2 \cdot \omega^4 \cdot [B_4(V_D^{(i)}) - B_5(V_D^{(i)}) \cdot \tau] \cdot [B_1(V_D^{(i)}) - B_3(V_D^{(i)}) \cdot \tau]$$

$$|Q(i \cdot \omega, \tau)|^2 = \omega^6 \cdot [B_1(V_D^{(i)}) - B_3(V_D^{(i)}) \cdot \tau]^2 + \omega^4 \cdot \{[\sum_{k=2}^{3} B_k(V_D^{(i)}) - B_4(V_D^{(i)}) \cdot \tau]^2 - 2 \cdot [B_4(V_D^{(i)}) - B_5(V_D^{(i)}) \cdot \tau] \cdot [B_1(V_D^{(i)}) - B_3(V_D^{(i)}) \cdot \tau]\} + \omega^2 \cdot \{[B_4(V_D^{(i)}) - B_5(V_D^{(i)}) \cdot \tau]^2 - 2 \cdot [\sum_{k=2}^{3} B_k(V_D^{(i)}) - B_4(V_D^{(i)}) \cdot \tau] \cdot B_5(V_D^{(i)})\} + [B_5(V_D^{(i)})]^2$$

$$F(\omega,\tau) = |P(i\cdot\omega,\tau)|^2 - |Q(i\cdot\omega,\tau)|^2 = \omega^8 - \omega^6\cdot[B_1(V_D^{(i)}) - B_3(V_D^{(i)})\cdot\tau]^2 - \omega^4\cdot\{[\sum_{k=2}^{3} B_k(V_D^{(i)}) - B_4(V_D^{(i)})\cdot\tau]^2 - 2\cdot[B_4(V_D^{(i)}) - B_5(V_D^{(i)})\cdot\tau] \cdot[B_1(V_D^{(i)}) - B_3(V_D^{(i)})\cdot\tau]\} - \omega^2\cdot\{[B_4(V_D^{(i)}) - B_5(V_D^{(i)})\cdot\tau]^2 - 2\cdot[\sum_{k=2}^{3} B_k(V_D^{(i)}) - B_4(V_D^{(i)})\cdot\tau]\cdot B_5(V_D^{(i)})\} - [B_5(V_D^{(i)})]^2$$

We define the following parameters for simplicity:

$$\mathrm{H}_0 = -[B_5(V_D^{(i)})]^2\ ;$$

$$\mathrm{H}_2 = -\{[B_4(V_D^{(i)}) - B_5(V_D^{(i)})\cdot\tau]^2 - 2\cdot[\sum_{k=2}^{3} B_k(V_D^{(i)}) - B_4(V_D^{(i)})\cdot\tau]\cdot B_5(V_D^{(i)})\}$$

$$\mathrm{H}_4 = -\{[\sum_{k=2}^{3} B_k(V_D^{(i)}) - B_4(V_D^{(i)})\cdot\tau]^2 - 2\cdot[B_4(V_D^{(i)}) - B_5(V_D^{(i)})\cdot\tau] \cdot[B_1(V_D^{(i)}) - B_3(V_D^{(i)})\cdot\tau]\}$$

$\mathrm{H}_6 = -[B_1(V_D^{(i)}) - B_3(V_D^{(i)})\cdot\tau]^2$; $\mathrm{H}_8 = 1$. Hence $F(\omega,\tau) = 0$ implies $\sum_{k=0}^{4} \mathrm{H}_{2\cdot k}\cdot \omega^{2\cdot k} = 0$.

And its roots are given by solving the above polynomial. Furthermore

$$P_R(i\cdot\omega,\tau) = \omega^4;\ P_I(i\cdot\omega,\tau) = 0;$$

$$Q_R(i\cdot\omega,\tau) = -\omega^2\cdot[\sum_{k=2}^{3} B_k(V_D^{(i)}) - B_4(V_D^{(i)})\cdot\tau] + B_5(V_D^{(i)})$$

$$Q_I(i\cdot\omega,\tau) = \omega\cdot[B_4(V_D^{(i)}) - B_5(V_D^{(i)})\cdot\tau] - \omega^3\cdot[B_1(V_D^{(i)}) - B_3(V_D^{(i)})\cdot\tau]$$

Hence $\sin\theta(\tau) = \frac{-P_R(i\cdot\omega,\tau)\cdot Q_I(i\cdot\omega,\tau) + P_I(i\cdot\omega,\tau)\cdot Q_R(i\cdot\omega,\tau)}{|Q(i\cdot\omega,\tau)|^2}$ and $\cos\theta(\tau) = -\frac{P_R(i\cdot\omega,\tau)\cdot Q_R(i\cdot\omega,\tau) + P_I(i\cdot\omega,\tau)\cdot Q_I(i\cdot\omega,\tau)}{|Q(i\cdot\omega,\tau)|^2}$. We already approve

$$D(\lambda,\tau_1=\tau_2=\tau) = \lambda^4 + \{\lambda^3\cdot[B_1(V_D^{(i)}) - B_3(V_D^{(i)})\cdot\tau] + \lambda^2\cdot[\sum_{k=2}^{3} B_k(V_D^{(i)}) - B_4(V_D^{(i)})\cdot\tau] + \lambda\cdot[B_4(V_D^{(i)}) - B_5(V_D^{(i)})\cdot\tau] + B_5(V_D^{(i)})\}\cdot e^{-\lambda\cdot\tau};\ n=4;\ m=3;\ n>m$$

We use different parameters terminology from our last characteristics parameters definition: $k \to j$; $p_k(\tau) \to a_j$; $q_k(\tau) \to c_j$; $n = 4$; $m = 3$; $n > m$.

Additionally $P_n(\lambda, \tau) \to P(\lambda)$; $Q_m(\lambda, \tau) \to Q(\lambda)$ then $P(\lambda) = \sum_{j=0}^{4} a_j \cdot \lambda^j$; $Q(\lambda) = \sum_{j=0}^{2} c_j \cdot \lambda^j$

$$P_\lambda = \lambda^4;\ Q_\lambda = \lambda^3 \cdot [B_1(V_D^{(i)}) - B_3(V_D^{(i)}) \cdot \tau] + \lambda^2 \cdot [\sum_{k=2}^{3} B_k(V_D^{(i)}) - B_4(V_D^{(i)}) \cdot \tau]$$
$$+ \lambda \cdot [B_4(V_D^{(i)}) - B_5(V_D^{(i)}) \cdot \tau] + B_5(V_D^{(i)})$$

$n, m \in \mathbb{N}_0, n > m$ and $a_j, c_j : R_{+0} \to R$ are continuous and differentiable function of τ such that $a_0 + c_0 \neq 0$. In the following "$-$" denotes complex and conjugate. $P(\lambda), Q(\lambda)$ are analytic functions in λ and differentiable in τ. The coefficients $\{a_j(L, C, R_b, L_b, R_L, V_V, V_{ex}, m, \tau, \ldots)$ and $c_j(L, C, R_b, L_b, R_L, V_V, V_{ex}, m, \tau, \ldots)\} \in \mathbb{R}$ depend on TD system's $L, C, R_b, L_b, R_L, V_V, V_{ex}, m, \tau, \ldots$ values. Unless strictly necessary, the designation of the varied arguments $(L, C, R_b, L_b, R_L, V_V, V_{ex}, m, \tau, \ldots)$ will subsequently be omitted from P, Q, a_j, c_j. The coefficients a_j, c_j are continuous, and differentiable functions of their arguments, and direct substitution shows that $a_0 + c_0 \neq 0$; $B_5(V_D^{(i)}) \neq 0$. $\forall$ $L, C, R_b, L_b, R_L, V_V, V_{ex}, m, \tau, \ldots \in \mathbb{R}_+$ i.e. $\lambda = 0$ is not a root of the characteristic equation. Furthermore $P(\lambda)$, $Q(\lambda)$ are analytic function of λ for which the following requirements of the analysis (see Kuang [5], Sect. 3.4) can also be verified in the present case [4–6].

(a) If $\lambda = i \cdot \omega$, $\omega \in \mathbb{R}$ then $P(i \cdot \omega) + Q(i \cdot \omega) \neq 0$, i.e. P and Q have no common imaginary roots. This condition was verified numerically in the entire $(L, C, R_b, L_b, R_L, V_V, V_{ex}, m, \tau, \ldots)$ domain of interest.

(b) $|Q(\lambda)/P(\lambda)|$ is bounded for $|\lambda| \to \infty$, $\mathrm{Re}\,\lambda \geq 0$. No roots bifurcation from ∞. Indeed, in the limit

$$\left|\frac{Q(\lambda)}{P(\lambda)}\right| = \left|\frac{\lambda^3 \cdot [B_1(V_D^{(i)}) - B_3(V_D^{(i)}) \cdot \tau] + \lambda^2 \cdot [\sum_{k=2}^{3} B_k(V_D^{(i)}) - B_4(V_D^{(i)}) \cdot \tau] + \lambda \cdot [B_4(V_D^{(i)}) - B_5(V_D^{(i)}) \cdot \tau] + B_5(V_D^{(i)})}{\lambda^4}\right|$$

(c) $F(\omega) = |P(i \cdot \omega)|^2 - |Q(i \cdot \omega)|^2$

$$F(\omega,\tau) = |P(i\cdot\omega,\tau)|^2 - |Q(i\cdot\omega,\tau)|^2 = \omega^8 + \omega^6\cdot[B_1(V_D^{(i)}) - B_3(V_D^{(i)})\cdot\tau]^2$$
$$+ \omega^4\cdot\{[\sum_{k=2}^{3} B_k(V_D^{(i)}) - B_4(V_D^{(i)})\cdot\tau]^2 - 2\cdot[B_4(V_D^{(i)}) - B_5(V_D^{(i)})\cdot\tau]$$
$$\cdot[B_1(V_D^{(i)}) - B_3(V_D^{(i)})\cdot\tau]\} + \omega^2\cdot\{[B_4(V_D^{(i)}) - B_5(V_D^{(i)})\cdot\tau]^2$$
$$-2\cdot[\sum_{k=2}^{3} B_k(V_D^{(i)}) - B_4(V_D^{(i)})\cdot\tau]\cdot B_5(V_D^{(i)})\} + [B_5(V_D^{(i)})]^2$$

Has at most a finite number of zeros. Indeed, this is a polynomial in ω .

(d) Each positive root $\omega(L, C, R_b, L_b, R_L, V_V, V_{ex}, m, \tau, \ldots)$ of F(ω)= 0 is continuous and differentiable with respect to $L, C, R_b, L_b, R_L, V_V, V_{ex}, m, \tau, \ldots$. This condition can only be assessed numerically.

In addition, since the coefficients in P and Q are real, we have $\overline{P(-i\cdot\omega)} = P(i\cdot\omega)$ and $\overline{Q(-i\cdot\omega)} = Q(i\cdot\omega)$ thus, $\lambda = i\cdot\omega$, $\omega > 0$ maybe on eigenvalue of characteristic equation. The analysis consists in identifying the roots of the characteristic equation situated on the imaginary axis of the complex λ—plane, whereby increasing the parameters $L, C, R_b, L_b, R_L, V_V, V_{ex}, m, \tau, \ldots$, Reλ may, at the crossing, change its sign from (−) to (+), i.e. from a stable focus $E^*(X^*, Y^*, V_D^*, V_A^*) = (0, 0, V_D^*, 0)$ to an unstable one, or vice versa. This feature may be further assessed by examining the sign of the partial derivatives with respect to $L, C, R_b, L_b, R_L, V_V, V_{ex}, m, \tau, \ldots$ and TD's system parameters. $\omega \in \mathbb{R}_+$.

$$\wedge^{-1}(R_f) = \left(\frac{\partial \mathrm{Re}\lambda}{\partial L}\right)_{\lambda=i\cdot\omega},\ C, R_b, L_b, R_L, V_V, V_{ex}, m, \tau, \ldots = const$$
$$\wedge^{-1}(L_f) = \left(\frac{\partial \mathrm{Re}\lambda}{\partial C}\right)_{\lambda=i\cdot\omega}, L, R_b, L_b, R_L, V_V, V_{ex}, m, \tau, \ldots = const$$
$$\wedge^{-1}(C_f) = \left(\frac{\partial \mathrm{Re}\lambda}{\partial R_b}\right)_{\lambda=i\cdot\omega},\ L, C, L_b, R_L, V_V, V_{ex}, m, \tau, \ldots = const$$
$$\wedge^{-1}(C_c) = \left(\frac{\partial \mathrm{Re}\lambda}{\partial L_b}\right)_{\lambda=i\cdot\omega},\ L, C, R_b, R_L, V_V, V_{ex}, m, \tau, \ldots = const$$
$$\wedge^{-1}(R_{sb}) = \left(\frac{\partial \mathrm{Re}\lambda}{\partial R_L}\right)_{\lambda=i\cdot\omega}, L, C, R_b, L_b, V_V, V_{ex}, m, \tau, \ldots = const$$
$$\wedge^{-1}(\tau) = \left(\frac{\partial \mathrm{Re}\lambda}{\partial \tau}\right)_{\lambda=i\cdot\omega},\ L, C, R_b, L_b, R_L, V_V, V_{ex}, m, \ldots = const$$

When writing $P(\lambda) = P_R(\lambda) + i\cdot P_I(\lambda)$ and $Q(\lambda) = Q_R(\lambda) + i\cdot Q_I(\lambda)$, and inserting $\lambda = i\cdot\omega$.

Into TD system's characteristic equation, ω must satisfy the following:

$$\sin \omega \cdot \tau = g(\omega) = \frac{-P_R(i \cdot \omega) \cdot Q_I(i \cdot \omega) + P_I(i \cdot \omega) \cdot Q_R(i \cdot \omega)}{|Q(i \cdot \omega)|^2}; \cos \omega \cdot \tau = h(\omega)$$
$$= -\frac{P_R(i \cdot \omega) \cdot Q_R(i \cdot \omega) + P_I(i \cdot \omega) \cdot Q_I(i \cdot \omega)}{|Q(i \cdot \omega)|^2}$$

where $|Q(i \cdot \omega)|^2 \neq 0$ in view of requirement (a) above, and $(g, h) \in R$. Furthermore, it follows above $\sin \omega \cdot \tau$ and $\cos \omega \cdot \tau$ equations that, by squaring and adding the sides, ω must be a positive root of $F(\omega) = |P(i \cdot \omega)|^2 - |Q(i \cdot \omega)|^2 = 0$. Note that $F(\omega)$ is dependent of τ. Now it is important to notice that if $\tau \notin I$ (assume that $I \subseteq R_{+0}$ is the set where $\omega(\tau)$ is a positive root of $F(\omega)$ and for $\tau \notin I$, $\omega(\tau)$ is not defined. Then for all τ in I $\omega(\tau)$ is satisfied that $F(\omega, \tau) = 0$). Then there are no positive $\omega(\tau)$ solutions for $F(\omega, \tau) = 0$, and we cannot have stability switches. For any $\tau \in I$ where $\omega(\tau)$ is a positive solution of $F(\omega, \tau) = 0$, we can define the angle $\theta(\tau) \in [0, 2 \cdot \pi]$ as the solution of $\sin \theta(\tau) = \frac{-P_R(i \cdot \omega) \cdot Q_I(i \cdot \omega) + P_I(i \cdot \omega) \cdot Q_R(i \cdot \omega)}{|Q(i \cdot \omega)|^2}$ and $\cos \theta(\tau) = -\frac{P_R(i \cdot \omega) \cdot Q_R(i \cdot \omega) + P_I(i \cdot \omega) \cdot Q_I(i \cdot \omega)}{|Q(i \cdot \omega)|^2}$.

And the relation between the argument $\theta(\tau)$ and $\omega(\tau) \cdot \tau$ for $\tau \in I$ must be $\omega(\tau) \cdot \tau = \theta(\tau) + n \cdot 2 \cdot \pi \quad \forall \quad n \in \mathbb{N}_0$. Hence we can define the maps $\tau_n : I \to R_{+0}$ given by $\tau_n(\tau) = \frac{\theta(\tau) + n \cdot 2 \cdot \pi}{\omega(\tau)}$; $n \in \mathbb{N}_0, \tau \in I$. Let us introduce the functions $I \to R; S_n(\tau) = \tau - \tau_n(\tau), \tau \in I, n \in \mathbb{N}_0$ that is a continuous and differentiable in τ. In the following, the subscripts $\lambda, \omega, L, C, R_b, L_b, R_L, V_V, V_{ex}, m, \tau, \ldots$ indicate the corresponding partial derivatives. Let us first concentrate on $\wedge(x)$, remember in $\lambda(L, C, R_b, L_b, R_L, V_V, V_{ex}, m, \tau, \ldots)$ and $\omega(L, C, R_b, L_b, R_L, V_V, V_{ex}, m, \tau, \ldots)$, and keeping all parameters except one (x) and τ. The derivation closely follows that in reference [BK]. Differentiating TD's system characteristic equation $P(\lambda) + Q(\lambda) \cdot e^{-\lambda \cdot \tau} = 0$ with respect to specific parameter (x), and inverting the derivative, for convenience, one calculates:

Remark: $x = L, C, R_b, L_b, R_L, V_V, V_{ex}, m, \ldots$

$$\left(\frac{\partial \lambda}{\partial x}\right)^{-1} = \frac{-P_\lambda(\lambda, x) \cdot Q(\lambda, x) + Q_\lambda(\lambda, x) \cdot P(\lambda, x) - \tau \cdot P(\lambda, x) \cdot Q(\lambda, x)}{P_x(\lambda, x) \cdot Q(\lambda, x) - Q_x(\lambda, x) \cdot P(\lambda, x)}$$

where $P_\lambda = \frac{\partial P}{\partial \lambda}, \ldots$ etc., Substituting $\lambda = i \cdot \omega$, and bearing i $\overline{P(-i \cdot \omega)} = P(i \cdot \omega)$, $\overline{Q(-i \cdot \omega)} = Q(i \cdot \omega)$ then $i \cdot P_\lambda(i \cdot \omega) = P_\omega(i \cdot \omega)$; $i \cdot Q_\lambda(i \cdot \omega) = Q_\omega(i \cdot \omega)$ and that on the surface $|P(i \cdot \omega)|^2 = |Q(i \cdot \omega)|^2$, one obtains:

$$(\frac{\partial\lambda}{\partial x})^{-1}|_{\lambda=i\cdot\omega} = (\frac{i\cdot P_\omega(i\cdot\omega,x)\cdot\overline{P(i\cdot\omega,x)} + i\cdot Q_\lambda(i\cdot\omega,x)\cdot\overline{Q(\lambda,x)} - \tau\cdot|P(i\cdot\omega,x)|^2}{P_x(i\cdot\omega,x)\cdot\overline{P(i\cdot\omega,x)} - Q_x(i\cdot\omega,x)\cdot\overline{Q(i\cdot\omega,x)}})$$

Upon separating into real and imaginary parts, with $P = P_R + i\cdot P_I$; $Q = Q_R + i\cdot Q_I$; $P_\omega = P_{R\omega} + i\cdot P_{I\omega}$; $Q_\omega = Q_{R\omega} + i\cdot Q_{I\omega}$; $P_x = P_{Rx} + i\cdot P_{Ix}$; $Q_x = Q_{Rx} + i\cdot Q_{Ix}$; $P^2 = P_R^2 + P_I^2$. When (x) can be any TD's system parameters $L, C, R_b, L_b, R_L, V_V, V_{ex}, m, \ldots$, and time delay τ etc. Where for convenience, we have dropped the arguments $(i\cdot\omega, x)$, and where $F_\omega = 2\cdot[(P_{R\omega}\cdot P_R + P_{I\omega}\cdot P_I) - (Q_{R\omega}\cdot Q_R + Q_{I\omega}\cdot Q_I)]$; $\omega_x = -F_x/F_\omega$

$F_x = 2\cdot[(P_{Rx}\cdot P_R + P_{Ix}\cdot P_I) - (Q_{Rx}\cdot Q_R + Q_{Ix}\cdot Q_I)]$ We define U and V:

$$U = (P_R\cdot P_{I\omega} - P_I\cdot P_{R\omega}) - (Q_R\cdot Q_{I\omega} - Q_I\cdot Q_{R\omega});$$
$$V = (P_R\cdot P_{Ix} - P_I\cdot P_{Rx}) - (Q_R\cdot Q_{Ix} - Q_I\cdot Q_{Rx})$$

We choose our specific parameter as time delay x = τ. We already find

$$P_R(i\cdot\omega,\tau) = \omega^4;\ P_I(i\cdot\omega,\tau) = 0;$$

$$Q_R(i\cdot\omega,\tau) = -\omega^2\cdot[\sum_{k=2}^{3} B_k(V_D^{(i)}) - B_4(V_D^{(i)})\cdot\tau] + B_5(V_D^{(i)})$$

$$Q_I(i\cdot\omega,\tau) = \omega\cdot[B_4(V_D^{(i)}) - B_5(V_D^{(i)})\cdot\tau] - \omega^3\cdot[B_1(V_D^{(i)}) - B_3(V_D^{(i)})\cdot\tau]$$

$$P_{R\omega} = 4\cdot\omega^3;\ P_{I\omega} = 0;\ Q_{R\omega} = -2\cdot\omega\cdot[\sum_{k=2}^{3} B_k(V_D^{(i)}) - B_4(V_D^{(i)})\cdot\tau]$$

$$Q_{I\omega} = [B_4(V_D^{(i)}) - B_5(V_D^{(i)})\cdot\tau] - 3\cdot\omega^2\cdot[B_1(V_D^{(i)}) - B_3(V_D^{(i)})\cdot\tau];\ P_{R\tau} = P_{I\tau} = 0$$

$$Q_{R\tau} = \omega^2\cdot B_4(V_D^{(i)});\ Q_{I\tau} = -\omega\cdot B_5(V_D^{(i)}) + \omega^3\cdot B_3(V_D^{(i)})$$

$$P_{R\omega}\cdot P_R = 4\cdot\omega^7;\ P_{I\omega}\cdot P_I = 0;$$

$$Q_{R\omega}\cdot Q_R = 2\cdot\omega^3\cdot[\sum_{k=2}^{3} B_k(V_D^{(i)}) - B_4(V_D^{(i)})\cdot\tau]^2$$
$$- 2\cdot\omega\cdot[\sum_{k=2}^{3} B_k(V_D^{(i)}) - B_4(V_D^{(i)})\cdot\tau]\cdot B_5(V_D^{(i)});$$

$$Q_{I\omega}\cdot Q_I = \{[B_4(V_D^{(i)}) - B_5(V_D^{(i)})\cdot\tau] - 3\cdot\omega^2\cdot[B_1(V_D^{(i)}) - B_3(V_D^{(i)})\cdot\tau]\}$$
$$\cdot\{\omega\cdot[B_4(V_D^{(i)}) - B_5(V_D^{(i)})\cdot\tau] - \omega^3\cdot[B_1(V_D^{(i)}) - B_3(V_D^{(i)})\cdot\tau]\}$$
$$= \omega\cdot[B_4(V_D^{(i)}) - B_5(V_D^{(i)})\cdot\tau]^2 - 4\cdot\omega^3\cdot[B_1(V_D^{(i)}) - B_3(V_D^{(i)})\cdot\tau]$$
$$\cdot[B_4(V_D^{(i)}) - B_5(V_D^{(i)})\cdot\tau] + 3\cdot\omega^5\cdot[B_1(V_D^{(i)}) - B_3(V_D^{(i)})\cdot\tau]^2$$

$$P_{R\tau} \cdot P_R = 0;\ P_{I\tau} \cdot P_I = 0;$$

$$Q_{R\tau} \cdot Q_R = \omega^2 \cdot B_4(V_D^{(i)}) \cdot \{-\omega^2 \cdot [\sum_{k=2}^{3} B_k(V_D^{(i)}) - B_4(V_D^{(i)}) \cdot \tau] + B_5(V_D^{(i)})\}$$

$$\begin{aligned} Q_{I\tau} \cdot Q_I = {} & \{-\omega \cdot B_5(V_D^{(i)}) + \omega^3 \cdot B_3(V_D^{(i)})\} \cdot \{\omega \cdot [B_4(V_D^{(i)}) - B_5(V_D^{(i)}) \cdot \tau] \\ & - \omega^3 \cdot [B_1(V_D^{(i)}) - B_3(V_D^{(i)}) \cdot \tau]\} \end{aligned}$$

$$P_R \cdot P_{I\omega} = 0;\ P_I \cdot P_{R\omega} = 0;\ Q_R \cdot Q_{I\omega} = \{-\omega^2 \cdot [\sum_{k=2}^{3} B_k(V_D^{(i)}) - B_4(V_D^{(i)}) \cdot \tau] + B_5(V_D^{(i)})\}$$
$$\cdot \{[B_4(V_D^{(i)}) - B_5(V_D^{(i)}) \cdot \tau] - 3 \cdot \omega^2 \cdot [B_1(V_D^{(i)}) - B_3(V_D^{(i)}) \cdot \tau]\}$$

$$\begin{aligned} Q_I \cdot Q_{R\omega} = {} & \{\omega \cdot [B_4(V_D^{(i)}) - B_5(V_D^{(i)}) \cdot \tau] - \omega^3 \cdot [B_1(V_D^{(i)}) - B_3(V_D^{(i)}) \cdot \tau]\} \\ & \cdot \{-2 \cdot \omega \cdot [\sum_{k=2}^{3} B_k(V_D^{(i)}) - B_4(V_D^{(i)}) \cdot \tau]\} \end{aligned}$$

$$P_R \cdot P_{I\tau} = 0;\ P_I \cdot P_{R\tau} = 0;$$

$$\begin{aligned} Q_R \cdot Q_{I\tau} = {} & \{-\omega^2 \cdot [\sum_{k=2}^{3} B_k(V_D^{(i)}) - B_4(V_D^{(i)}) \cdot \tau] + B_5(V_D^{(i)})\} \\ & \cdot \{-\omega \cdot B_5(V_D^{(i)}) + \omega^3 \cdot B_3(V_D^{(i)})\}; \\ Q_I \cdot Q_{R\tau} = {} & \{\omega \cdot [B_4(V_D^{(i)}) - B_5(V_D^{(i)}) \cdot \tau] - \omega^3 \cdot [B_1(V_D^{(i)}) - B_3(V_D^{(i)}) \cdot \tau]\} \\ & \cdot \omega^2 \cdot B_4(V_D^{(i)}) \end{aligned}$$

$$U = (P_R \cdot P_{I\omega} - P_I \cdot P_{R\omega}) - (Q_R \cdot Q_{I\omega} - Q_I \cdot Q_{R\omega})|_{\substack{P_R \cdot P_{I\omega} = 0 \\ P_I \cdot P_{R\omega} = 0}} = -Q_R \cdot Q_{I\omega} + Q_I \cdot Q_{R\omega}$$

$$V|_{x=\tau} = (P_R \cdot P_{I\tau} - P_I \cdot P_{R\tau}) - (Q_R \cdot Q_{I\tau} - Q_I \cdot Q_{R\tau})|_{\substack{P_R \cdot P_{I\tau} = 0 \\ P_I \cdot P_{R\tau} = 0}} = -Q_R \cdot Q_{I\tau} + Q_I \cdot Q_{R\tau}$$

$$F_{x=\tau} = 2 \cdot [(P_{R\tau} \cdot P_R + P_{I\tau} \cdot P_I) - (Q_{R\tau} \cdot Q_R + Q_{I\tau} \cdot Q_I)]|_{\substack{P_{R\tau} \cdot P_R = 0 \\ P_{I\tau} \cdot P = 0}} = -2 \cdot (Q_{R\tau} \cdot Q_R + Q_{I\tau} \cdot Q_I)$$

$$F_\omega = 2 \cdot [(P_{R\omega} \cdot P_R + P_{I\omega} \cdot P_I) - (Q_{R\omega} \cdot Q_R + Q_{I\omega} \cdot Q_I)]|_{\substack{P_{R\omega} \cdot P_R = 4 \cdot \omega^7 \\ P_{I\omega} \cdot P_I = 0}} = 2 \cdot [4 \cdot \omega^7 - (Q_{R\omega} \cdot Q_R + Q_{I\omega} \cdot Q_I)]$$

$F(\omega, \tau) = 0$. Differentiating with respect to τ and we get $F_\omega \cdot \frac{\partial \omega}{\partial \tau} + F_\tau = 0$; $\tau \in I \Rightarrow \frac{\partial \omega}{\partial \tau} = -\frac{F_\tau}{F_\omega}$

$$\wedge^{-1}(\tau) = (\frac{\partial \mathrm{Re}\lambda}{\partial \tau})_{\lambda = i \cdot \omega}; \frac{\partial \omega}{\partial \tau} = \omega_\tau = -\frac{F_\tau}{F_\omega};\ \wedge^{-1}(\tau) = \mathrm{Re}\{\frac{-2 \cdot [U + \tau \cdot |P|^2] + i \cdot F_\omega}{F_\tau + i \cdot 2 \cdot [V + \omega \cdot |P|^2]}\}$$

$$sign\{\wedge^{-1}(\tau)\} = sign\{(\frac{\partial \mathrm{Re}\lambda}{\partial \tau})_{\lambda = i \cdot \omega}\};$$

$$sign\{\wedge^{-1}(\tau)\} = sign\{F_\omega\} \cdot sign\{\tau \cdot \frac{\partial \omega}{\partial \tau} + \omega + \frac{U \cdot \frac{\partial \omega}{\partial \tau} + V}{|P|^2}\}$$

We shall presently examine the possibility of stability transitions (bifurcations) TD's system, about the equilibrium point $E^*(X^*, Y^*, V_D^*, V_A^*) = (0, 0, V_D^*, 0)$ as a result of a variation of delay parameter τ. The analysis consists in identifying the roots of our system characteristic equation situated on the imaginary axis of the complex λ-plane. Where by increasing the delay parameter τ, Re λ may at the crossing, changes its sign from − to +, i.e. from a stable focus $E^{(*)}$ to an unstable one, or vice versa. This feature may be further assessed by examining the sign of the partial derivatives with respect to τ, $\wedge^{-1}(\tau) = (\frac{\partial \mathrm{Re}\lambda}{\partial \tau})_{\lambda = i \cdot \omega}$ $\wedge^{-1}(\tau) = (\frac{\partial \mathrm{Re}\lambda}{\partial \tau})_{\lambda = i \cdot \omega}$, $L, C, R_b, L_b, R_L, V_V, V_{ex}, m, \ldots = const$; $\omega \in \mathbb{R}_+$. [12].

Exercises

1. A two-stage limiter circuit is shown in below figure. The limiter PIN diode at the output (D_2), commonly referred to as the "clean-up stage," is the diode with thinner I layer, selected so that the threshold level of the circuit is low enough to protect the remainder of the receiver components. The limiter diode at the input (D_1), often called the "coarse limiter," has a thicker I layer for several reasons. The P layer diameter can be larger for a diode with a thicker I layer while maintaining a capacitance value that produces low insertion loss under small input signal conditions. The circuit components are connected by microstrip segments. We consider for simplicity that the microstrip segments resistances are neglected and either related voltages $V_{\tau k} \to \varepsilon$, k = 1,…,5. Two limiter diode's equivalent circuit parameters are not the same. We consider coming signal cause at t = 0 voltage V(t = 0).

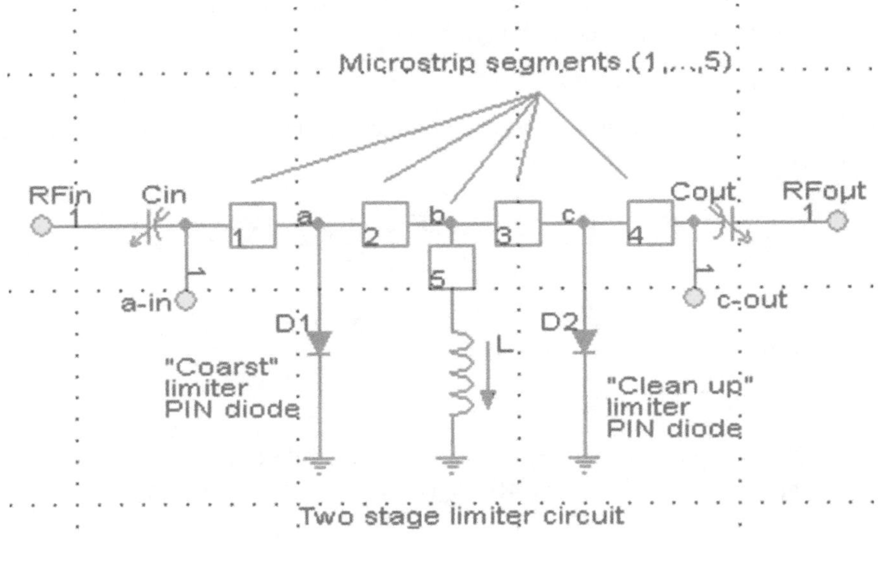

$$V_a(t) = V_{a-in}(t - \tau_1);$$

$$V_b(t) = V_a(t - \tau_2) = V_{a-in}(t - \sum_{i=1}^{2} \tau_i); \; V_c(t) = V_b(t - \tau_3) = V_{a-in}(t - \sum_{i=1}^{3} \tau_i)$$

$$V_{c-out}(t) = V_b(t - \tau_4) = V_{a-in}(t - \sum_{i=1}^{4} \tau_i); \; I_L(t) \to I_L(t - \tau_5)$$

1.1 Write two stage limiter system differential equations.
1.2 Try to recognize our system differential equations as forced Van der Pol equations. Write our two-stage limiter circuit as autonomous system.
1.3 Discuss system stability for the following cases:

(a) $\tau_1 = \tau; \; \sum_{i=1}^{2} \tau_i = \tau^2; \; \sum_{i=1}^{3} \tau_i = \tau^3; \; \sum_{i=1}^{4} \tau_i = \tau^3 + \sqrt{\tau}; \; \tau_5 = \tau$

(b) $\tau_1 = \sqrt{\tau}; \; \sum_{i=1}^{2} \tau_i = \tau \cdot \sqrt{\tau}; \; \sum_{i=1}^{3} \tau_i = \tau \cdot \sqrt{\tau} + \sqrt[3]{\tau}; \; \sum_{i=1}^{4} \tau_i = \tau \cdot \sqrt{\tau} + (\sqrt[3]{\tau})^2; \; \tau_5 = \sqrt{\tau}$

How τ value variations influence our system stability?

1.4 How our system dynamic and stability behavior changes when "Clean up" limiter PIN diode is disconnected?
1.5 How our system dynamic and stability behavior changes when "Coarst" limiter PIN diode is disconnected?
1.6 Microstrip segment No. 5 delay (τ_5) is dependent on other segment's delay summation $\tau_5 = f(\sum_{i=1}^{4} \tau_i); \; \tau_1 = \tau. \; \tau_{i+1} = \sqrt{\tau_i} \, \forall \, i = 1, .., 3$. Try to find f() function which our system is stable for any value of τ parameter. Which possible f() functions there is a stability switching under variation of τ parameter. Investigate stability behavior.

2. We have limiter circuit system which characterize by two differential equations (with delays τ_1, τ_2, τ_3).

$$V_0(t) = [\frac{1}{R_1} \cdot \frac{dV(t - \tau_2)}{dt} + C_1 \cdot \frac{d^2V(t - \sqrt{\tau_2})}{dt^2}] \cdot \sum_{i=0}^{3} L_i + V(t - \sum_{k=2}^{3} \tau_k)$$

$$V_i(t - \tau_1) = [\frac{1}{\sum_{i=1}^{3} R_i} \cdot \frac{dV}{dt} + [\sum_{i=1}^{2} C_i] \cdot \frac{d^2V}{dt^2}] \cdot \sum_{i=0}^{3} L_i + V$$

2.1 Draw possible limiter circuits which can fulfil above system differential equations. V_i—incoming RF signal voltage, V_o—out-going RF signal,

V—circuit internal voltage variable. $\tau_i (i = 1, 2, 3)$ represent microstrip segment delay parameters.

2.2 Find equivalent Van der Pol equation. Represent our system as an autonomous system.

2.3 Discuss stability and stability switching for different values of τ_1, τ_2, τ_3 parameters.

2.4 $\sum_{i=1}^{3} R_i \to \infty$, Draw our limiter circuit. Find equivalent Van der Pol equation. Discuss stability and stability switching for different values of C_i, i = 1,2.

2.5 Discuss system stability behavior for different values of R_1 and L_2.

3. We have active circulator of four ports, namely P_1, P_2, P_3 and P_4. Active circulator is a four terminal device in which input from one port is transmitted to the next port in rotation. The RF input signal is given at P_1 of the circulator from the left side. This signal from P1 is transmitted to P2. We can connect LC (L_1, C_1) components in series to P2 port which results in phase shift and helps to reflect the signal to P3 at the right. We can connect LC (L2, C2) components in series to P3 port which results in phase shift and helps to reflect the signal to P4 at the right. At P4 we get an output RF signal. Each active circulator terminal faces a delay parasitic effect of signal transferring in time. Our circuit is a Reflection Type Phase Shifter (RTPS), employing a circulator. The circuit configuration of the active circulator used four MESFETs which are the GEC-Marconi standard library cell F20-FET-4x75. A typical four ports decade bandwidth active circulator has four MESFETs transistors interconnected with each other. R_F, C_F, L_F, C_C, R_{sb} plays a major role in the working of the circuit. The four feedback branches (R_F, C_F, L_F) are used to link all the four transistors in an end to end fashion. The source resistor (R_{sb}) is shared among all the three MESFETs transistors and one transistor is source coupled with the other two transistors using this source resistor.

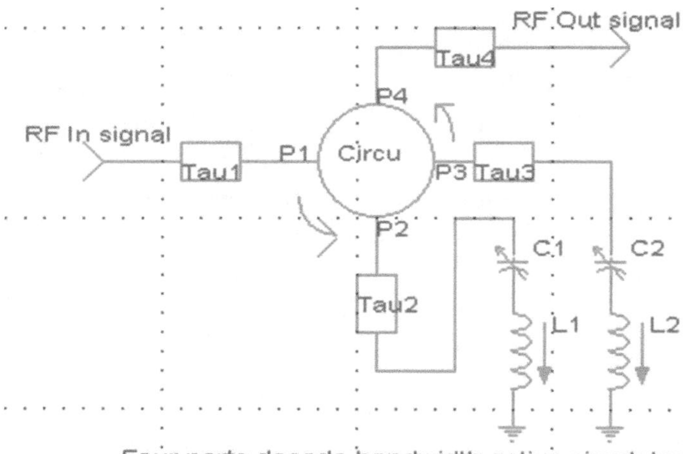

Four ports decade bandwidth active circulator

3.1 Draw four ports active circulator configuration and equivalent circuits. Write system differential equations.
3.2 Find system equilibrium points (fixed points), consider RFin signal

$$V_1(t) = A_0 + f(t) \cdot g(t)\,;\, |f(t)| \Leftarrow 1; |g(t)| \Leftarrow 1\,;\, A_0 \gg |f(t)|;\, A_0 \gg |g(t)|$$

Find small increment Jacobian and characteristic equations.
3.3 Discuss stability behavior and stability switching for different values of τ parameter for the following cases:

(a) $\tau_1 = \tau;\ \sum_{i=1}^{2} \tau_i = \tau^2;\ \sum_{i=1}^{3} \tau_i = \sqrt{\tau^3};\ \sum_{i=1}^{4} \tau_i = \tau^3 + \tau \cdot \sqrt{\tau}.$
(b) $\tau_1 = \sqrt{\tau};\ \sum_{i=1}^{2} \tau_i = \tau^3 + 1;\ \sum_{i=1}^{3} \tau_i = \sqrt{\tau^5};\ \sum_{i=1}^{4} \tau_i = \tau^2 + \tau \cdot \sqrt{\tau}.$

3.4 How our system's dynamical behavior and stability are influenced if port 3 is disconnected? Write system differential equations and discuss stability switching for different values of τ parameter ($\tau_1 = \tau_2 = \tau_3 = \tau$).
3.5 We move port 3 termination components (C_2, L_2) to port 4 and take our RF out signal from port 3. How our system's behavior changes? Write system differential equations and discuss stability behavior.

4. Consider RF system which characterize by Van der Pol equation as autonomous system, when $(V, W, \theta) \in R^2 x S^2$. We suppose $\alpha, \beta \ll 1$, since we are interested in the periodic coming signal response we use $\frac{2\cdot\pi}{\omega}$ periodic transformation. The following equations describe our system:

$$\frac{du_1}{dt} = -\frac{1}{[\sum_{k=1}^{3} R_k] \cdot C_1} \cdot \phi(V) \cdot \cos\omega \cdot t - (\frac{\omega^2 - 1}{\omega}) \cdot V \cdot \sin\omega \cdot t$$
$$- \frac{1}{\omega \cdot [\sum_{k=1}^{3} C_k] \cdot [\sum_{i=0}^{2} L_i]} \cdot \sin\omega \cdot t \cdot V_i(t - \sum_{k=0}^{2} \tau_k)$$

$$\frac{du_2}{dt} = \frac{1}{[\sum_{k=1}^{3} R_k] \cdot C_1} \cdot \phi(V) \cdot \sin\omega \cdot t - (\frac{\omega^2 - 1}{\omega}) \cdot V \cdot \cos\omega \cdot t$$
$$- \frac{1}{\omega \cdot [\sum_{k=1}^{3} C_k] \cdot \sum_{i=0}^{2} L_i} \cdot \cos\omega \cdot t \cdot V_i(t - \sum_{k=0}^{2} \tau_k)$$

4.1 Find the analog basic Van der Pol equation and implement the system by power limiter diodes and discrete components.
4.2 Discuss stability behavior and stability switching for variation of τ_k delay parameter values (k = 0,1,2).
4.3 How the system dynamic and stability are changed for $[\sum_{k=1}^{3} R_k] \to \infty$?
4.4 How the system dynamic and stability are changed for $\sum_{i=0}^{2} L_i \to \varepsilon$?
4.5 Discuss system stability switching for $\sum_{k=0}^{2} \tau_k = \tau^{\Xi}$. How Ξ parameter values influence our system stability switching.

5. We have cylindrical (closed) RF ladder network structure antennas for coupled plasma sources copper legs which delayed in time by parasitic effects. The antenna is mounted outside a glass tube. The RF ladder network antenna consists of 16 copper legs (inductors) equally spaced interconnected with capacitors, each copper leg has parasitic time delay ($\tau_{i+1} = \tau_i + \tau_{i-1}$; $\tau_1 = \tau_2 = \tau \, \forall \, i = 2, .., 15$). We consider for simplicity that all copper legs voltages on delay units (V_ε) are neglected $V_\varepsilon \to \varepsilon$. There is a delay in each copper leg current $I_1(t - \tau_1), \ldots, I_{16}(t - \tau_{16})$. We consider all ladder capacitors are the same (C) and all ending capacitors are the same (C_{end}). We consider all antenna elements inductance values are the same (L). $C_1 = C_2 = \cdots = C_{15} = C_{16} = C$ and $C_{A1} = C_{A2} = \cdots = C_{A16} = C_{end}$; $L_1 = L_2 = \cdots = L_{16} = L$; $I_{L1} = I_1, \cdots, I_{L16} = I_{16}$. The antenna ladder network structure is fed by the transmitter unit (S1 = OFF no direct RF feeding).

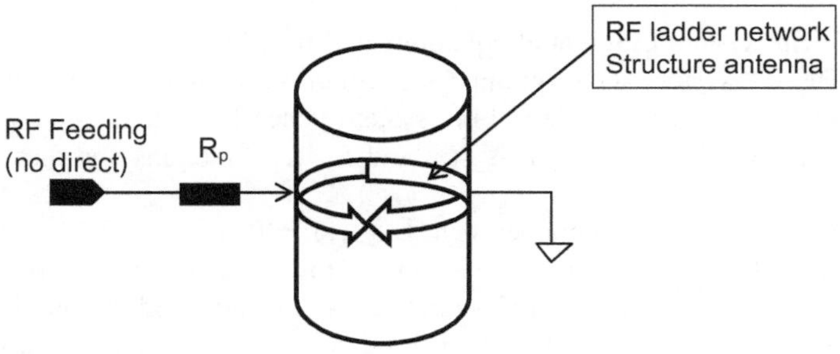

Cylindrical RF ladder network structure can represent as round ladder strip of capacitors and inductors. The schematic contains RF feeding signal, S1 switch (S1 = ON for RF signal feeding, S1 = OFF for RF signal transmitter feeding). R_p is the parasitic resistance of RF feeding direct source.
Remark: Only one ground point exists in our Cylindrical RF ladder network structure, it is connected to point B8.

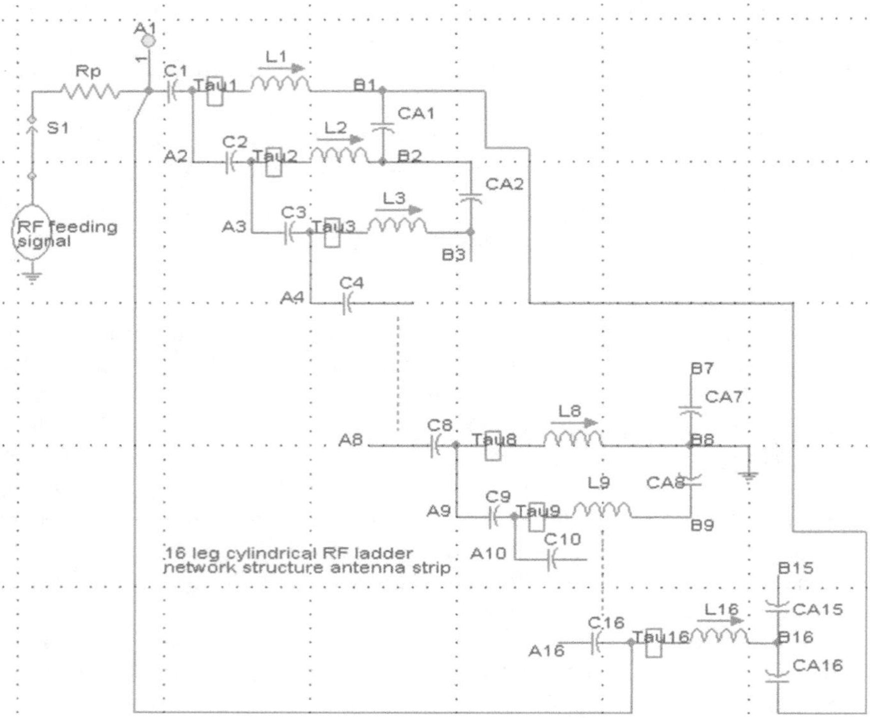

5.1 Write system differential equations and find fixed points.
5.2 Discuss system stability; find Jacobian of our cylindrical RF ladder network structure antenna system. Find system general characteristic equations.
5.3 Find $\Lambda^{-1}(C)$, $\Lambda^{-1}(C_{end})$, $\Lambda^{-1}(\tau)$ and $\Lambda^{-1}(L)$ functions and discuss stability switching.
5.4 We disconnect odd copper legs ($L_{2\cdot k+1}$; k = 0, 1, 2…6, 7) in our cylindrical RF ladder network structure antenna. How our system stability switching is effected? Find system differential equations and fixed points, find Jacobian and discuss stability switching. Find $\Lambda^{-1}(\tau)$ and $\Lambda^{-1}(L)$ functions.
5.5 We disconnect even CA capacitors, disconnect $CA_{2\cdot k}$; k = 1, 2…7, 8) in our cylindrical RF ladder network structure antenna. How our system stability switching is effected? Find system differential equations and fixed points, find Jacobian and discuss stability switching. Find $\Lambda^{-1}()$ and $\Lambda^{-1}(L)$ functions.
5.6 We move our system ground to point B12, How our system behavior changes? Discuss stability switching, $\Lambda^{-1}()$ and $\Lambda^{-1}(L)$ functions.

6. We have the following TD's microwave oscillator which two Tunnel Diodes (TDs) are connected in series. TD's biasing circuit is constructed from DC voltage V_b, inductor L_b which block oscillation from V_b source and capacitor C_b

which shorts to ground V_b bias voltage's oscillations. The DC voltage V_b biases the tunnel diode TD into its negative resistance region and also supplies the power used in amplifying the input signal V_i. Switch S1 is ON and S2 in ON (bypass our signal source V_i) and our circuit functions as a oscillator. S3 can be in OFF state or ON state. Microwave resonant cavities is represented and thought of as simple LC circuits. We represent our microwave cavity as resonant LC circuit. Due to cavity parasitic effects there are delays in the currents which flow through equivalent resonant L and C elements and TD's voltage derivative in time. τ_1, τ_2 are the time delays for the TD's voltage derivative in time respectively (D_1 & D_2). τ_3 is the time delay for the current flows through C element. R_L is the load resistance. Load resistance (R_L) is chosen so that $R_L < |\min(R_{d1}, R_{d2})|$ in the TD's characteristic NDR region. At the Tunnel diodes operating points, the total circuit differential resistance is negative. We consider for simplicity, two Tunnel diodes parameters are the same.

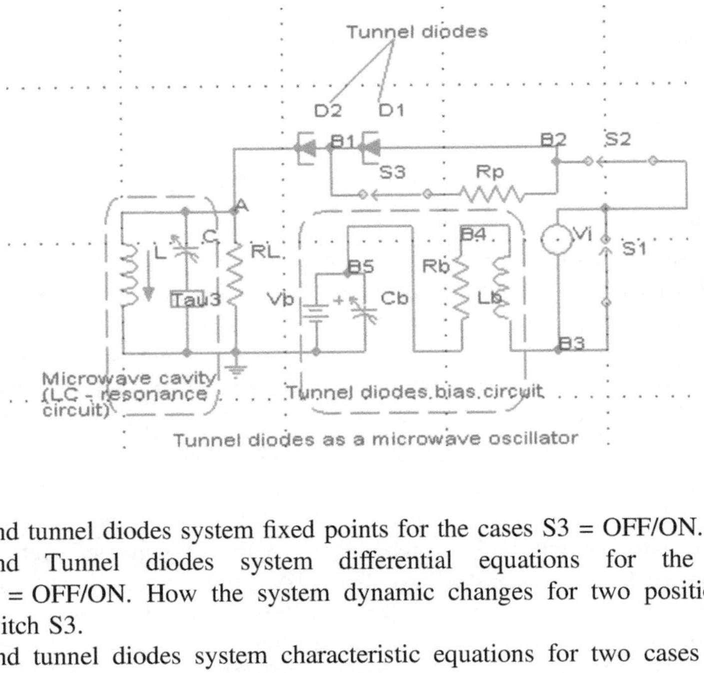

6.1 Find tunnel diodes system fixed points for the cases S3 = OFF/ON.

6.2 Find Tunnel diodes system differential equations for the cases S3 = OFF/ON. How the system dynamic changes for two positions of switch S3.

6.3 Find tunnel diodes system characteristic equations for two cases of S3 switch.

6.4 Discuss stability switching for variation of $\tau_i (i = 1, 2, 3)$ delay parameter values.

6.5 How the system dynamical behavior changes if resistor R_L is disconnected?

6.6 What happened if we bypass D2 tunnel diode (short) and S3 = ON constantly? Discuss stability and stability switching for different values of τ_3 delay parameter.

7. We have planar RTD (Resonant Tunneling Diode) oscillator which eliminate parasitic bias oscillations in an oscillator circuitry by employing a shunt resistor to the NDR device. A non-linear (diode) resistor, Schottky diode. S_d and R_e are a Schottky diode and a resistance, respectively, and form the stabilizing resistance. C_e is a decoupling capacitor (is an RF short circuit), while "TML" is said to be quarter wave transmission line with signal time delay τ. At millimeter wave and low terahertz frequencies (<1 THz), planar RTD oscillator have been integrated at the center of the slot antenna, a location at which the antenna input impedance is infinity.

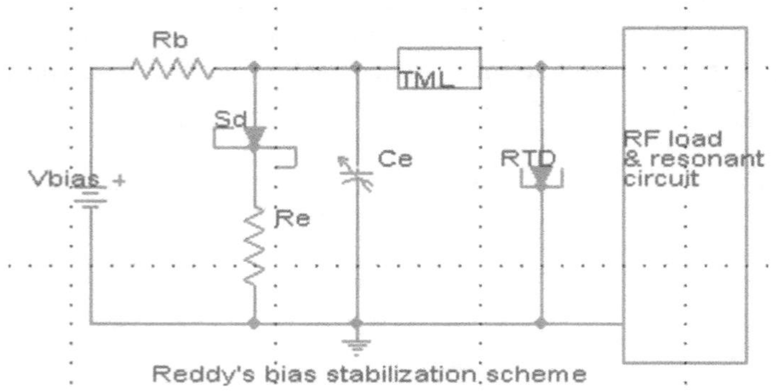

The above schematic is Reddy's bias stabilization scheme for sum mm- wave RTD oscillators. R_b is the resistance of the bias line. S_d, R_e and C_e are the Schottky diode, external resistor and external capacitor, respectively, which form stabilizing circuit. TML is the quarter wave length transmission line at the oscillation frequency. RF load is pure resistive R_{load} and resonant circuit is parallel LC circuit. We consider that Schottky diode current function is $I_D = I_S \cdot [e^{\frac{q \cdot (V - I_D \cdot R_S)}{n \cdot K \cdot T}} - 1] + G_x \cdot V$. n—Ideality factor, I_S—Reverse saturation current, k—Boltzman constant, T—Absolute temperature, V—Bias voltage, $G_X \cdot V$—Edge leakage current, $I_D \cdot R_S$—Voltage drop due to ohmic contact and bulk resistance of semiconductor channel (Ohmic contact is a metal semiconductor contact that has a linear I-V and non-rectifying characteristics). Remark: You should consider RTD equation as in subchapter TD's equation. The input impedance of a transmission line of length L with characteristic impedance Z_0 and connected to load with impedance Z_A (in our case Z_A is RTD element in parallel to RF load and resonant circuit). $Z_{in}(-L) = Z_0 \cdot [\frac{Z_A + j \cdot Z_0 \cdot tg(\beta \cdot L)}{Z_0 + j \cdot Z_A \cdot tg(\beta \cdot L)}]$; $Z_A = Z_{A-R} + j \cdot Z_{A-I}$

$$Z_{in}(L=\frac{\lambda}{4}) = Z_0 \cdot [\frac{Z_A + j \cdot Z_0 \cdot tg(\frac{2\pi}{\lambda} \cdot \frac{\lambda}{4})}{Z_0 + j \cdot Z_A \cdot tg(\frac{2\pi}{\lambda} \cdot \frac{\lambda}{4})}] = \frac{Z_0^2}{Z_A} \cdot Z_{in}(L=\frac{\lambda}{4}) = 50\,\text{Ohm}$$

We know the values for Z_{A-R} and Z_{A-I}. Need to find Z_0 expression.

7.1 Find Reddy's bias stabilization Scheme fixed point and differential equations.
7.2 Discuss stability switching for different values of τ parameter (quarter wave transmission line delay time).
7.3 What happened if C_e is disconnected? How the system dynamic change?
7.4 Find the following functions: $\Lambda(Z_0)$, $\Lambda(Z_A)$, $\Lambda(G_x)$, $\Lambda(V_V)$, $\Lambda(m)$.
Remark: V_V and m parameters are RTD's Internal parameters.
7.5 Discuss stability and dynamical behavior if Schottky diode Sd is shorted.

8. We have tunnel diodes system as a microwave oscillator. D1 and D2 are tunnel diodes which are connected back to back configuration. There are two biasing circuits (V_{b1}, R_{b1} and V_{b2}, R_{b2}). Additionally there is a smart function element which his voltage ($V_A - V_B$) is a complicated function of tunnel diodes currents respectively. TD's parasitic effects cause to current time delays τ_1 and τ_2 respectively $I_{D1}(t) \to I_{D1}(t-\tau_1)$ $I_{D2}(t) \to I_{D2}(t-\tau_2)$; $\tau_1 = \tau$; $\tau_2 = \tau + \sqrt{\tau}$. At time t = 0 switch S1, S2 and S3 move to ON state. We consider for simplicity, two tunnel diodes are identical (internal parameters are the same) and TD's current function is as discuss in the chapter.

$$V_{A-B} = g(I_{D1}, I_{D2}) = \{\max(I_{D1}, I_{D2}) + \sqrt{I_{D1} \cdot I_{D2}}\} \cdot \Gamma; \ \max(I_{D1}, I_{D2}) = \frac{1}{2} \cdot (I_{D1} + I_{D2}) + \frac{1}{2} \cdot |I_{D1} - I_{D2}|$$

Γ—Resistive parameter. L, C—Parallel resonant circuit.

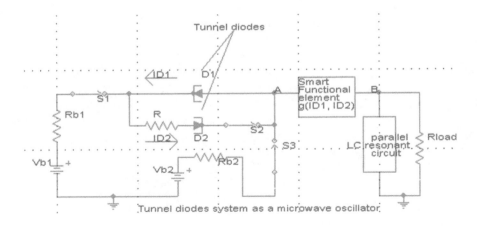

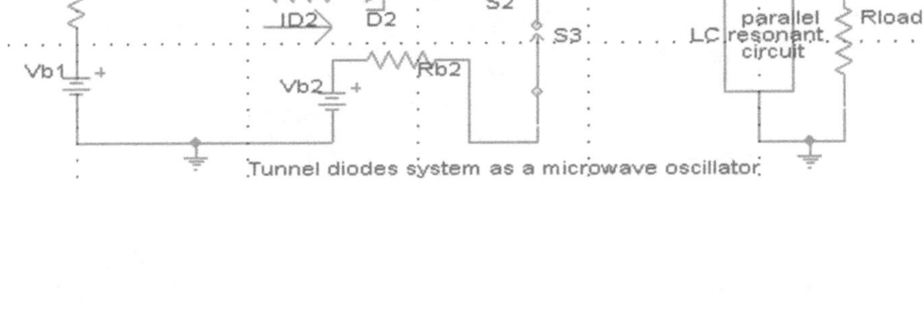

Tunnel diodes system as a microwave oscillator

8.1 Find system fixed points and differential equations.
8.2 Discuss system stability switching for different values of τ time delay parameter.
8.3 How the system dynamic changed if at time t = 0, switch S1 and S2 move to ON state and S3 is constantly in OFF state? Find fixed points and discuss stability switching for different values of τ parameter.
8.4 How the system dynamic changed if at time t = 0, switch S1 and S3 move to ON state and S2 is constantly in OFF state? Find fixed points and discuss stability switching for different values of τ parameter.
8.5 Find the following functions: $\Lambda(\Gamma)$, $\Gamma(R)$, $\Lambda(L)$, and $\Lambda(C)$.
8.6 How the system dynamic changed if at time t = 0, switch S1 move to ON state and S2, S3 is constantly in OFF state? Find fixed points and discuss stability switching for different values of τ parameter.

9. We have tunnel diode's bridge structure system as a microwave oscillator. D1, D2, D3 and D4 are tunnel diodes in our bridge structure. For simplicity all tunnel diodes are identical (they have the same internal parameters values). There are two biasing circuits (V_{b1}, R_{b1} and V_{b2}, R_{b2}). Switches S1 and S2 connect biasing circuits to tunnel diode bridge structure. TD's bridge structure contacts parasitic effects cause to current time delays τ_1 and τ_2 respectively. We have three possible cases: (1) S1 = ON, S2 = OFF, (2) S1 = OFF, S2 = ON, (3) S1 = ON, S2 = ON. TD's current function is as discuss in the chapter. Resonant circuit No. 1 includes Inductor and capacitor in parallel (L_1, C_1). Resonant circuit No. 2 includes inductor and capacitor in parallel (L_2, C_2). Initially all switched are in OFF state, at time t = 0 we move to one of the above cases.

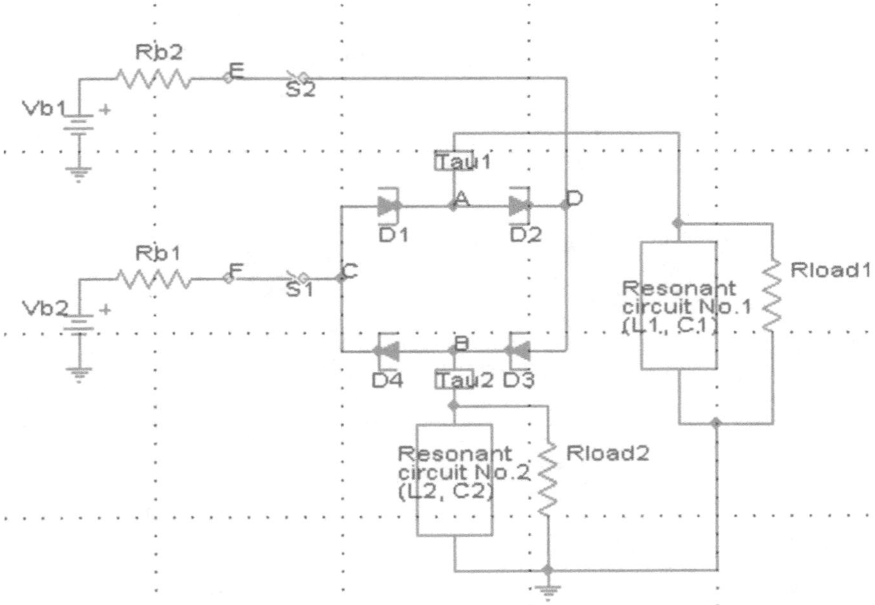

Tunnel diode's bridge structure as a microwave oscillator

9.1 Find system fixed points and differential equations for the three cases. Find the related characteristic equations.
9.2 Discuss stability switching for different values of τ_1 and τ_2 time delay parameters. How system dynamic changes for $\tau_1 = \sqrt{\tau_2 + 1}$?
9.3 Find the following functions: $\Lambda(\tau_1)$, $\Gamma(\tau_2)$, $\Lambda(L_1)$, and $\Lambda(C_2)$.
9.4 D2 tunnel diode is disconnected, how system dynamical behavior changes? Discuss stability and stability switching.
9.5 D1 tunnel diode is disconnected, how system dynamical behavior changes? Discuss stability and stability switching.

10. We have the following custom TD's current expression as a function of voltages V_{Di}. V_{Di} represents specific TD's voltage, the current which flow through all TDs is the same (TDs are in series).

$$I_D = \xi(V_D) = \sum_{k=1}^{3} g_k(V_{Di})$$

$$= I_s \cdot \{\prod_{i=1}^{3} e^{[(\frac{V_{Di}}{\eta \cdot V_{th}})-1]}\} + \frac{\sum_{i=1}^{2} V_{Di}}{R_0} \cdot \{\prod_{i=1}^{2} e^{[-(\frac{V_{Di}+1}{V_0})^m]}\}$$

$$+ \frac{\sum_{i=1}^{2} V_{Di}}{R_V} \cdot \{\prod_{i=2}^{4} e^{[\frac{V_{Di}-V_V}{V_{ex}}]}\}$$

$$g_1(V_{Di}) = I_s \cdot \{\prod_{i=1}^{3} e^{[(\frac{V_{Di}}{\eta \cdot V_{th}})-1]}\};$$

$$g_2(V_{Di}) = \frac{\sum_{i=1}^{2} V_{Di}}{R_0} \cdot \{\prod_{i=1}^{2} e^{[-(\frac{V_{Di}+1}{V_0})^m]}\};\ g_3(V_{Di}) = \frac{\sum_{i=1}^{2} V_{Di}}{R_V} \cdot \{\prod_{i=2}^{4} e^{[\frac{V_{Di}-V_V}{V_{ex}}]}\}$$

We connect our TDs in series to parallel LC resonant circuit and load R_{load}. Additionally, we add biasing circuit (V_b, R_b).

10.1 Find system fixed points and differential equations.
10.2 Discuss system stability switching for different values of L, C resonant circuit values.
10.3 One of the TD in series is shorted, How it influence our system dynamic? Discuss fixed points and stability switching for different values of m parameter.
(Hint: Discuss all possible cases for short TDs).
10.4 We define TD's contact delays as $\tau_l (l = 1, 2, 3)$ respectively, Find $\Lambda(\tau_l)$, $\Lambda(V_V)$, $\Lambda(V_{ex})$ functions.

Chapter 3
Microwave Semiconductor Amplifiers Analysis

Microwave semiconductor amplifiers are widely used in many engineering applications. Among microwave semiconductors there are bipolar transistors which operate at microwave frequencies, microwave Field Effect Transistors (FET) minimize the adverse effects of transit time and internal capacitance and resistance, IMPATT (IMPact ionization Avalanche Transit-Time) amplifiers which widely use at the high end of the microwave band because microwave transistor do not work well above 30 GHz due to transit time limitations. Stability of these microwave amplifiers is affected by internal parameters variation and circuit microstrip parasitic effects. In any microwave system there are parasitic elements with delays in time which affect the performance and stability of the system. There are also IMPATT diodes which are a form of high power diode used in high frequencies electronic and microwave devices. The main advantage of IMPATT diode is the high power capability and frequency band of 3 to 100 GHz and more. FET combined biasing and matching circuit has many stability issues which must be taken for every RF design [14, 15].

3.1 Bipolar Transistor at Microwave Frequencies Description and Stability Analysis

Bipolar transistors are widely in use for many engineering applications. The bipolar transistor consists of an N-doped semiconductors the emitter, a P-doped semiconductor called the base, and an N – doped semiconductor called the collector. The transistor consists of two PN junctions, one between the emitter and the base, the other between the base and the collector. The base collector junction is reverse biased by a rather large voltage. The emitter base junction is forward biased. When the junction is reversed biased no current flows. Current begins to flow when the junction is forward biased about specific voltage (~ 0.6 V). No current flows across

O. Aluf, *Microwave RF Antennas and Circuits*,
DOI 10.1007/978-3-319-45427-6_3

the reverse biased base collector junction. The amplification of AC signal by a bipolar transistor is widely used in every circuit. Initial state is when the emitter base junction is DC biased and there is no AC signal applied. When the input AC signal to be amplified is added to the DC bias, the voltage between the emitter and the base increase (AC signal is positive). The current flows from the emitter into the base. When the input AC signal is negative, it subtracts from the DC bias. There is also an AC current gain. There is a difference in the operation between a microwave bipolar transistor and a low frequency bipolar transistor. The operation is the same except for the degrading effects of transit time, internal capacitance and resistance, and external lead inductance that occur at microwave frequencies. Because the electrons take a finite time to travel across the emitter base junction, through the base, and across the base collector junction, the AC current is reduced at microwave frequencies, compared with low frequencies. The velocity of microwaves, which is the same as the velocity of light $3 \times 10^8 \frac{m}{\sec}$. Electrons in semiconductor travel approximately 1/3000 slower, or 10^8 mm per second. The distance that a microwave or an electron travels in one microwave cycle is the velocity divided by the frequency ($L_{cycle} = \frac{v}{f}$). Next is the bipolar transistor at microwave frequencies.

As the microwave signal enters the transistor, it must flow through the capacitance of the forward biased emitter base junction (C_i). This capacitance exists in low frequency transistor, but its reactance at low frequencies is negligible. Feedback capacitance between the collector and base is important at microwave frequencies, namely the capacitance of the reverse biased base collector junction (C_f). An output

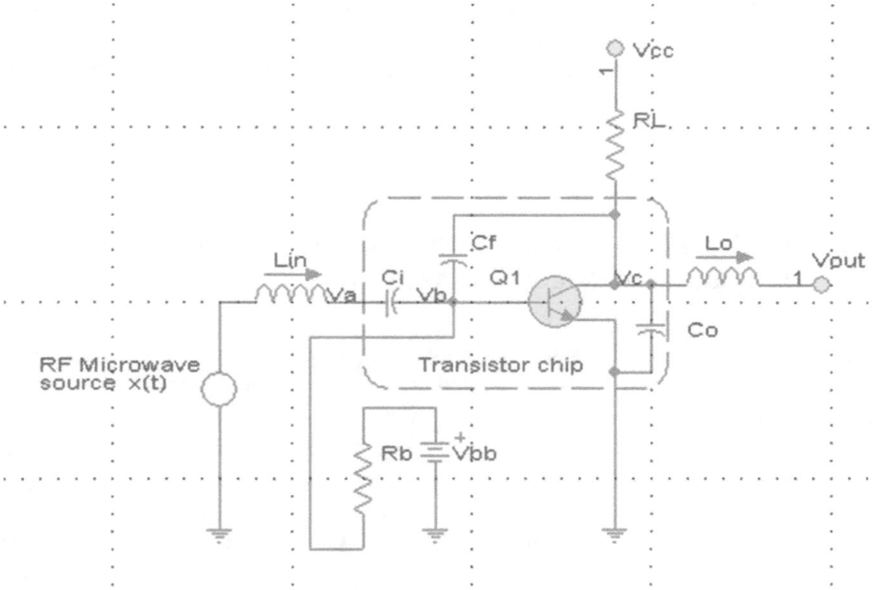

Fig. 3.1 Bipolar transistor at microwave frequencies

collector capacitance between the collector and emitter exists (C_o). In addition, the bonding wire connecting the transistor chip to the input and output microwave transmission lines have inductance (L_{in}, L_o). The lower microwave frequency, the less severe are the transit time effects, because the microwave cycle is longer and the electrons have time to move through the transistor. At low microwave frequencies, the base collector junction thickness can be made large and a higher voltage can be applied and more power can be obtained at low microwave frequencies than at high microwave frequencies. By using Ebers Moll transistor model (see S.M. Sze, *Physics of Semiconductor Devices*), we get the following bipolar transistor at microwave frequency equivalent circuit. X(t) is the microwave source input signal [24, 25, 33, 34]. $I_C = I_c; I_E = I_e$.

$$I_{R_C} = I_{C_f} + I_C + I_{C_0} + I_{L_{out}}; I_{L_{out}} = I_{R_L}; I_{L_{in}} + I_{C_f} + I_{R_b} = I_b; Alfa - f = \alpha_f$$

$$Alfa - R = \alpha_r; I_C + I_{DC} - \alpha_f \cdot I_{DE} = 0; I_b = I_{DC} + I_{DE} - \alpha_r \cdot I_{DC} - \alpha_f \cdot I_{DE}$$

$$I_b = I_{DC} \cdot (1 - \alpha_r) + I_{DE} \cdot (1 - \alpha_f); I_{DE} - \alpha_r \cdot I_{DC} = I_E; I_C = -I_{DC} + \alpha_f \cdot I_{DE}$$

$$I_{L_{in}} = I_{C_{in}}; V_{A2} = V_{BE}; V_{A2} - V_{A1} = V_{BC}; \frac{V_{CC} - V_{A1}}{R_C} = I_{R_C};$$

$$V_{A1} = V_{L_{out}} + V_{R_L} = L_{out} \cdot \frac{dI_{L_{out}}}{dt} + I_{L_{out}} \cdot R_L$$

$$I_{C_0} = C_0 \cdot \frac{dV_{A1}}{dt}; V_{A4} = V_{bb}; V_{A1} - V_{A2} = V_{C_f}; I_{C_f} = C_f \cdot \frac{dV_{C_f}}{dt}; \frac{V_{bb} - V_{A2}}{R_b} = I_{R_b}$$

$$X(t) - V_{A2} = V_{L_{in}} + V_{C_{in}}; V_{L_{in}} = L_{in} \cdot \frac{dI_{Lin}}{dt}; I_{C_{in}} = C_{in} \cdot \frac{dV_{Cin}}{dt}; X(t) - V_{A3} = V_{L_{in}}$$

$$V_{A2} - V_{A3} = V_{Cin}; V_{A4} = V_{bb}; I_{DE} = I_{SE} \cdot [e^{(\frac{V_{be}}{V_T})} - 1]; I_{DC} = I_{SC} \cdot [e^{(\frac{V_{bc}}{V_T})} - 1]$$

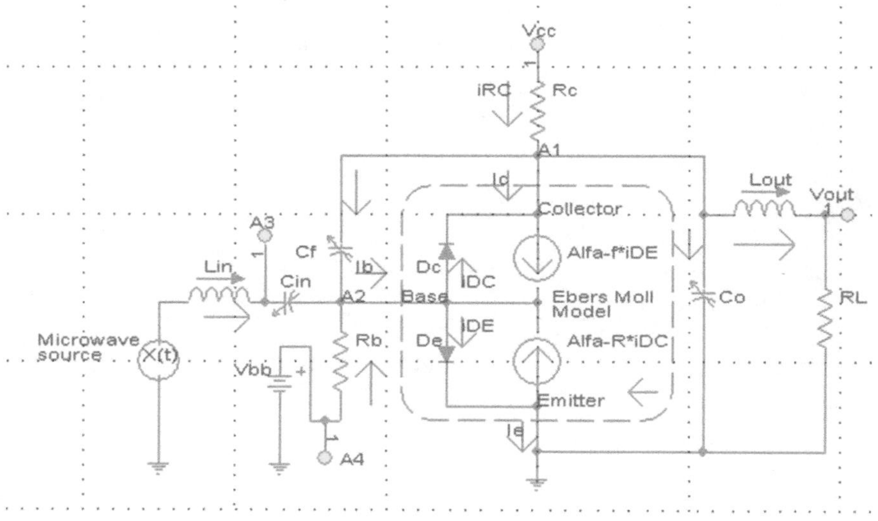

Fig. 3.2 Bipolar transistor at microwave frequencies equivalent circuit

$\alpha_f \cdot I_{SE} = \alpha_r \cdot I_{SC} = I_S$ (See S.M. Sze, Physics of Semiconductor Devices).

$$\beta_f = \frac{\alpha_f}{1-\alpha_f}; \beta_r = \frac{\alpha_r}{1-\alpha_r}; I_E = I_B + I_C \Rightarrow I_B = I_E - I_C;$$

$$\frac{1}{\alpha_f} - 1 = \frac{1-\alpha_f}{\alpha_f} = \frac{1}{\beta_f}; \frac{1}{\alpha_r} - 1 = \frac{1-\alpha_r}{\alpha_r} = \frac{1}{\beta_r}$$

$$I_E = I_{DE} - \alpha_r \cdot I_{DC} = I_{SE} \cdot [e^{(\frac{V_{be}}{V_T})} - 1] - \alpha_r \cdot I_{SC} \cdot [e^{(\frac{V_{bc}}{V_T})} - 1]$$
$$= \frac{I_S}{\alpha_f} \cdot [e^{(\frac{V_{be}}{V_T})} - 1] - I_S \cdot [e^{(\frac{V_{bc}}{V_T})} - 1]$$

$$I_C = -I_{DC} + \alpha_f \cdot I_{DE} = -I_{SC} \cdot [e^{(\frac{V_{bc}}{V_T})} - 1] + \alpha_f \cdot I_{SE} \cdot [e^{(\frac{V_{be}}{V_T})} - 1]$$
$$= -\frac{I_S}{\alpha_r} \cdot [e^{(\frac{V_{bc}}{V_T})} - 1] + I_S \cdot [e^{(\frac{V_{be}}{V_T})} - 1]$$

$$I_B = I_E - I_C = \frac{I_S}{\alpha_f} \cdot [e^{(\frac{V_{be}}{V_T})} - 1] - I_S \cdot [e^{(\frac{V_{bc}}{V_T})} - 1] - \{-\frac{I_S}{\alpha_r} \cdot [e^{(\frac{V_{bc}}{V_T})} - 1] + I_S \cdot [e^{(\frac{V_{be}}{V_T})} - 1]\}$$

$$I_B = I_S \cdot [e^{(\frac{V_{be}}{V_T})} - 1] \cdot (\frac{1}{\alpha_f} - 1) + I_S \cdot [e^{(\frac{V_{bc}}{V_T})} - 1] \cdot (\frac{1}{\alpha_r} - 1) = \frac{I_S}{\beta_f} \cdot [e^{(\frac{V_{be}}{V_T}} - 1] + \frac{I_S}{\beta_r} \cdot [e^{(\frac{V_{bc}}{V_T})} - 1]$$

We can define the following terminology variables and parameters:

I_C—Transistor collector current, I_B—Transistor base current, I_E—Transistor emitter current. β_f—Transistor forward common emitter current gain (20 → 500), β_r- Transistor reverse common emitter current gain (0 → 20), α_f—Transistor common base forward short circuit current gain (0.98 → 0.998). V_T—Transistor thermal voltage $V_T = k \cdot T/q$ (approximately 26 mV at 300 kelvin ≈ room temperature).

$$I_{DC} = \frac{I_C - \alpha_f \cdot I_E}{\alpha_f \cdot \alpha_r - 1}; I_{DE} = \frac{-I_E + \alpha_r \cdot I_C}{\alpha_f \cdot \alpha_r - 1} \Rightarrow I_{SE} \cdot [e^{(\frac{V_{be}}{V_T})} - 1] = \frac{-I_E + \alpha_r \cdot I_C}{\alpha_f \cdot \alpha_r - 1}$$

$$I_{SE} \cdot [e^{(\frac{V_{be}}{V_T})} - 1] = \frac{-I_E + \alpha_r \cdot I_C}{\alpha_f \cdot \alpha_r - 1} \Rightarrow e^{(\frac{V_{be}}{V_T})} = \{\frac{-I_E + \alpha_r \cdot I_C}{I_{SE} \cdot (\alpha_f \cdot \alpha_r - 1)} + 1\}$$

$$\Rightarrow V_{be} = V_T \cdot \ln\{\frac{-I_E + \alpha_r \cdot I_C}{I_{SE} \cdot (\alpha_f \cdot \alpha_r - 1)} + 1\}$$

$$I_{DC} = \frac{I_C - \alpha_f \cdot I_E}{\alpha_f \cdot \alpha_r - 1} \Rightarrow I_{SC} \cdot [e^{(\frac{V_{bc}}{V_T})} - 1] = \frac{I_C - \alpha_f \cdot I_E}{\alpha_f \cdot \alpha_r - 1}$$

$$\Rightarrow V_{bc} = V_T \cdot \ln\{\frac{I_C - \alpha_f \cdot I_E}{I_{SC} \cdot (\alpha_f \cdot \alpha_r - 1)} + 1\}$$

$$V_{ce} = V_{cb} + V_{be}; V_{cb} = -V_{bc} \Rightarrow V_{ce} = -V_{bc} + V_{be}$$

$$V_{ce} = V_T \cdot \ln\{\frac{-I_E + \alpha_r \cdot I_C}{I_{SE} \cdot (\alpha_f \cdot \alpha_r - 1)} + 1\} - V_T \cdot \ln\{\frac{I_C - \alpha_f \cdot I_E}{I_{SC} \cdot (\alpha_f \cdot \alpha_r - 1)} + 1\}$$

$$I_{R_C} = I_{C_f} + I_C + I_{C_0} + I_{L_{out}} \Rightarrow \frac{V_{CC} - V_{A1}}{R_C} = C_f \cdot \frac{dV_{C_f}}{dt} + I_C$$
$$+ C_0 \cdot \frac{dV_{A1}}{dt} + I_{L_{out}}$$

$$\frac{V_{CC} - [L_{out} \cdot \frac{dI_{L_{out}}}{dt} + I_{L_{out}} \cdot R_L]}{R_C} = C_f \cdot \frac{dV_{C_f}}{dt} + I_C + C_0 \cdot \frac{d}{dt}[L_{out} \cdot \frac{dI_{L_{out}}}{dt} + I_{L_{out}} \cdot R_L] + I_{L_{out}}$$

$$\frac{V_{CC} - L_{out} \cdot \frac{dI_{L_{out}}}{dt} - I_{L_{out}} \cdot R_L}{R_C} = C_f \cdot \frac{dV_{C_f}}{dt} + I_C + C_0 \cdot [L_{out} \cdot \frac{d^2 I_{L_{out}}}{dt^2} + \frac{dI_{L_{out}}}{dt} \cdot R_L] + I_{L_{out}}$$

$$I_{L_{in}} + I_{C_f} + I_{R_b} = I_b; I_{L_{in}} = I_{C_{in}} \Rightarrow I_{C_{in}} + I_{C_f} + I_{R_b} = I_b$$

$$\Rightarrow C_{in} \cdot \frac{dV_{Cin}}{dt} + C_f \cdot \frac{dV_{C_f}}{dt} + \frac{V_{bb} - V_{A2}}{R_b} = I_b$$

$$V_{A1} - V_{A2} = V_{C_f} \Rightarrow V_{A2} = V_{A1} - V_{C_f} = L_{out} \cdot \frac{dI_{L_{out}}}{dt} + I_{L_{out}} \cdot R_L - V_{C_f},$$

then we get

$$(*)\, C_{in} \cdot \frac{dV_{Cin}}{dt} + C_f \cdot \frac{dV_{C_f}}{dt} + \frac{V_{bb} - \{L_{out} \cdot \frac{dI_{L_{out}}}{dt} + I_{L_{out}} \cdot R_L - V_{C_f}\}}{R_b} = I_b$$

$$X(t) - V_{A2} = V_{L_{in}} + V_{C_{in}} \Rightarrow V_{C_{in}} = X(t) - V_{A2} - V_{L_{in}}$$

$$= X(t) - \{L_{out} \cdot \frac{dI_{L_{out}}}{dt} + I_{L_{out}} \cdot R_L - V_{C_f}\} - L_{in} \cdot \frac{dI_{Lin}}{dt}$$

$$V_{C_{in}} = X(t) - L_{out} \cdot \frac{dI_{L_{out}}}{dt} - I_{L_{out}} \cdot R_L + V_{C_f} - L_{in} \cdot \frac{dI_{Lin}}{dt}$$

$$(**)\, \frac{dV_{C_{in}}}{dt} = \frac{dX(t)}{dt} - L_{out} \cdot \frac{d^2 I_{L_{out}}}{dt^2} - \frac{dI_{L_{out}}}{dt} \cdot R_L + \frac{dV_{C_f}}{dt} - L_{in} \cdot \frac{d^2 I_{Lin}}{dt^2}$$

$$(**) \to (*) \quad C_{in} \cdot \{\frac{dX(t)}{dt} - L_{out} \cdot \frac{d^2 I_{L_{out}}}{dt^2} - \frac{dI_{L_{out}}}{dt} \cdot R_L + \frac{dV_{C_f}}{dt} - L_{in} \cdot \frac{d^2 I_{Lin}}{dt^2}\} + C_f \cdot \frac{dV_{C_f}}{dt} + \frac{V_{bb} - \{L_{out} \cdot \frac{dI_{L_{out}}}{dt} + I_{L_{out}} \cdot R_L - V_{C_f}\}}{R_b} = I_b$$

We have two global system differential equations:

$$\frac{V_{CC} - L_{out} \cdot \frac{dI_{L_{out}}}{dt} - I_{L_{out}} \cdot R_L}{R_C} = C_f \cdot \frac{dV_{C_f}}{dt} + I_C + C_0 \cdot [L_{out} \cdot \frac{d^2 I_{L_{out}}}{dt^2} + \frac{dI_{L_{out}}}{dt} \cdot R_L] + I_{L_{out}}$$

$$C_{in} \cdot \frac{dX(t)}{dt} - C_{in} \cdot L_{out} \cdot \frac{d^2 I_{L_{out}}}{dt^2} - \frac{dI_{L_{out}}}{dt} \cdot C_{in} \cdot R_L + C_{in} \cdot \frac{dV_{C_f}}{dt} - C_{in} \cdot L_{in} \cdot \frac{d^2 I_{Lin}}{dt^2} + C_f \cdot \frac{dV_{C_f}}{dt} + \frac{V_{bb} - \{L_{out} \cdot \frac{dI_{L_{out}}}{dt} + I_{L_{out}} \cdot R_L - V_{C_f}\}}{R_b} = I_b$$

$$V_{be} = V_{A2} = L_{out} \cdot \frac{dI_{L_{out}}}{dt} + I_{L_{out}} \cdot R_L - V_{C_f}; V_{bc} = V_{A2} - V_{A1} = -V_{C_f}$$

$$I_C = -I_{DC} + \alpha_f \cdot I_{DE} = -\frac{I_S}{\alpha_r} \cdot [e^{(\frac{V_{bc}}{V_T})} - 1] + I_S \cdot [e^{(\frac{V_{be}}{V_T})} - 1]$$

$$= -\frac{I_S}{\alpha_r} \cdot [e^{(\frac{-V_{C_f}}{V_T})} - 1] + I_S \cdot [e^{(\frac{L_{out} \cdot \frac{dI_{L_{out}}}{dt} + I_{L_{out}} \cdot R_L - V_{C_f}}{V_T})} - 1]$$

$$I_B = \frac{I_S}{\beta_f} \cdot [e^{(\frac{V_{be}}{V_T})} - 1] + \frac{I_S}{\beta_r} \cdot [e^{(\frac{V_{bc}}{V_T})} - 1] = \frac{I_S}{\beta_f} \cdot [e^{(\frac{L_{out} \cdot \frac{dI_{L_{out}}}{dt} + I_{L_{out}} \cdot R_L - V_{C_f}}{V_T})} - 1] + \frac{I_S}{\beta_r} \cdot [e^{(\frac{-V_{C_f}}{V_T})} - 1]$$

We define new variables: $Y_1 = \frac{dI_{L_{out}}}{dt}; Y_2 = \frac{dV_{C_f}}{dt}$

$$\frac{V_{CC} - L_{out} \cdot Y_1 - I_{L_{out}} \cdot R_L}{R_C} = C_f \cdot Y_2 + I_C + C_0 \cdot [L_{out} \cdot \frac{dY_1}{dt} + Y_1 \cdot R_L] + I_{L_{out}}$$

$$C_0 \cdot L_{out} \cdot \frac{dY_1}{dt} = \frac{V_{CC} - L_{out} \cdot Y_1 - I_{L_{out}} \cdot R_L}{R_C} - C_f \cdot Y_2 - Y_1 \cdot C_0 \cdot R_L - I_{L_{out}} - I_C$$

$$\frac{dY_1}{dt} = \frac{1}{R_C} \cdot \frac{1}{C_0} \cdot [V_{CC} \cdot \frac{1}{L_{out}} - Y_1 - R_L] - \frac{C_f}{C_0 \cdot L_{out}} \cdot Y_2 - Y_1 \cdot \frac{R_L}{L_{out}} - \frac{1}{C_0 \cdot L_{out}} \cdot I_{L_{out}} - \frac{1}{C_0 \cdot L_{out}} \cdot I_C$$

I_C function: $I_C = \xi_1 = \xi_1(V_{C_f}, Y_1, I_{L_{out}}, I_S, \alpha_r, \ldots)$

$$\frac{dY_1}{dt} = \frac{1}{R_C} \cdot \frac{1}{C_0} \cdot [V_{CC} \cdot \frac{1}{L_{out}} - Y_1 - R_L] - \frac{C_f}{C_0 \cdot L_{out}} \cdot Y_2 - Y_1 \cdot \frac{R_L}{L_{out}} - \frac{1}{C_0 \cdot L_{out}} \cdot I_{L_{out}} - \frac{1}{C_0 \cdot L_{out}} \cdot \xi_1(V_{C_f}, Y_1, I_{L_{out}}, I_S, \alpha_r, \ldots)$$

$$C_{in} \cdot \frac{dX(t)}{dt} - C_{in} \cdot L_{out} \cdot \frac{d^2 I_{L_{out}}}{dt^2} - \frac{dI_{L_{out}}}{dt} \cdot C_{in} \cdot R_L + C_{in} \cdot \frac{dV_{C_f}}{dt} - C_{in} \cdot L_{in} \cdot \frac{d^2 I_{Lin}}{dt^2} + C_f \cdot \frac{dV_{C_f}}{dt} + \frac{V_{bb} - \{L_{out} \cdot \frac{dI_{L_{out}}}{dt} + I_{L_{out}} \cdot R_L - V_{C_f}\}}{R_b} = I_b$$

I_b function: $I_b = \xi_2 = \xi_2(V_{C_f}, Y_1, I_{L_{out}}, I_S, \beta_f, \ldots)$

$$C_{in} \cdot \frac{dX(t)}{dt} - C_{in} \cdot [L_{out} \cdot \frac{d^2 I_{L_{out}}}{dt^2} + L_{in} \cdot \frac{d^2 I_{Lin}}{dt^2}] - \frac{dI_{L_{out}}}{dt} \cdot C_{in} \cdot R_L + [C_{in} + C_f] \cdot \frac{dV_{C_f}}{dt} + \frac{V_{bb} - L_{out} \cdot \frac{dI_{L_{out}}}{dt} - I_{L_{out}} \cdot R_L + V_{C_f}}{R_b} = I_b$$

We define new variable:

$$\frac{dY_3}{dt} = L_{out} \cdot \frac{d^2 I_{L_{out}}}{dt^2} + L_{in} \cdot \frac{d^2 I_{Lin}}{dt^2}; Y_3 = L_{out} \cdot \frac{dI_{L_{out}}}{dt} + L_{in} \cdot \frac{dI_{Lin}}{dt}$$

$$Y_3 = L_{out} \cdot \frac{dI_{L_{out}}}{dt} + L_{in} \cdot \frac{dI_{Lin}}{dt} \Rightarrow Y_3 = L_{out} \cdot Y_1 + L_{in} \cdot \frac{dI_{Lin}}{dt} \Rightarrow \frac{dI_{Lin}}{dt} = Y_3 \cdot \frac{1}{L_{in}} - \frac{L_{out}}{L_{in}} \cdot Y_1$$

$$C_{in} \cdot \frac{dY_3}{dt} = C_{in} \cdot \frac{dX(t)}{dt} - Y_1 \cdot C_{in} \cdot R_L + [C_{in} + C_f] \cdot Y_2 + \frac{V_{bb} - L_{out} \cdot Y_1 - I_{L_{out}} \cdot R_L + V_{C_f}}{R_b} - I_b$$

$$\frac{dY_3}{dt} = \frac{dX(t)}{dt} - Y_1 \cdot R_L + [1 + \frac{C_f}{C_{in}}] \cdot Y_2 + \frac{V_{bb} - L_{out} \cdot Y_1 - I_{L_{out}} \cdot R_L + V_{C_f}}{R_b \cdot C_{in}} - \frac{1}{C_{in}} \cdot I_b$$

$$\frac{dY_3}{dt} = -Y_1 \cdot R_L + [1 + \frac{C_f}{C_{in}}] \cdot Y_2 + \frac{V_{bb} - L_{out} \cdot Y_1 - I_{L_{out}} \cdot R_L + V_{C_f}}{R_b \cdot C_{in}} - \frac{1}{C_{in}} \cdot \xi_2(V_{C_f}, Y_1, I_{L_{out}}, I_S, \beta_f, \ldots) + \frac{dX(t)}{dt}$$

$$\frac{dY_3}{dt} = -[R_L + \frac{L_{out}}{R_b \cdot C_{in}}] \cdot Y_1 + [1 + \frac{C_f}{C_{in}}] \cdot Y_2 - I_{L_{out}} \cdot \frac{R_L}{R_b \cdot C_{in}} + V_{C_f} \cdot \frac{1}{R_b \cdot C_{in}} - \frac{1}{C_{in}} \cdot \xi_2(V_{C_f}, Y_1, I_{L_{out}}, I_S, \beta_f, \ldots) + \frac{V_{bb}}{R_b \cdot C_{in}} + \frac{dX(t)}{dt}$$

We can summery our circuit differential equations:

$$\frac{dY_1}{dt} = \frac{1}{R_C} \cdot \frac{1}{C_0} \cdot [V_{CC} \cdot \frac{1}{L_{out}} - Y_1 - R_L] - \frac{C_f}{C_0 \cdot L_{out}} \cdot Y_2 - Y_1 \cdot \frac{R_L}{L_{out}} - \frac{1}{C_0 \cdot L_{out}} \cdot I_{L_{out}} - \frac{1}{C_0 \cdot L_{out}} \cdot \xi_1(V_{C_f}, Y_1, I_{L_{out}}, I_S, \alpha_r, \ldots)$$

$$\frac{dY_3}{dt} = -[R_L + \frac{L_{out}}{R_b \cdot C_{in}}] \cdot Y_1 + [1 + \frac{C_f}{C_{in}}] \cdot Y_2 - I_{L_{out}} \cdot \frac{R_L}{R_b \cdot C_{in}} + V_{C_f} \cdot \frac{1}{R_b \cdot C_{in}} - \frac{1}{C_{in}} \cdot \xi_2(V_{C_f}, Y_1, I_{L_{out}}, I_S, \beta_f, \ldots) + \frac{V_{bb}}{R_b \cdot C_{in}} + \frac{dX(t)}{dt}$$

$$\frac{dI_{Lin}}{dt} = Y_3 \cdot \frac{1}{L_{in}} - \frac{L_{out}}{L_{in}} \cdot Y_1; \frac{dI_{L_{out}}}{dt} = Y_1; \frac{dV_{C_f}}{dt} = Y_2$$

We find our system equilibrium points by setting $\frac{dY_1}{dt} = 0; \frac{dY_3}{dt} = 0; \frac{dI_{Lin}}{dt} = 0$

$$\frac{dI_{L_{out}}}{dt} = 0; \frac{dV_{C_f}}{dt} = 0. \frac{dI_{L_{out}}}{dt} = 0 \Rightarrow Y_1^* = 0; \frac{dV_{C_f}}{dt} = 0 \Rightarrow Y_2^* = 0; \frac{dI_{Lin}}{dt} = 0 \Rightarrow Y_3^* = 0$$

$$\frac{dY_1}{dt} = 0 \Rightarrow \frac{1}{R_C} \cdot \frac{1}{C_0} \cdot [V_{CC} \cdot \frac{1}{L_{out}} - R_L] - \frac{1}{C_0 \cdot L_{out}} \cdot I^*_{L_{out}} - \frac{1}{C_0 \cdot L_{out}} \cdot \xi_1(V^*_{C_f}, Y^*_1 = 0, I^*_{L_{out}}, I_S, \alpha_r, \ldots) = 0$$

$$\frac{dY_3}{dt} = 0 \Rightarrow -I^*_{L_{out}} \cdot \frac{R_L}{R_b \cdot C_{in}} + V^*_{C_f} \cdot \frac{1}{R_b \cdot C_{in}} - \frac{1}{C_{in}} \cdot \xi_2(V^*_{C_f}, Y^*_1 = 0, I^*_{L_{out}}, I_S, \beta_f, \ldots) + \frac{V_{bb}}{R_b \cdot C_{in}} + \frac{dX(t)}{dt} = 0$$

I_C function: $I_C = \xi_1 = \xi_1(V_{C_f}, Y_1, I_{L_{out}}, I_S, \alpha_r, \ldots)$ and I_b function: $I_b = \xi_2 = \xi_2(V_{C_f}, Y_1, I_{L_{out}}, I_S, \beta_f, \ldots)$

$$\xi_1(V^*_{C_f}, Y^*_1 = 0, I^*_{L_{out}}, I_S, \alpha_r, \ldots) = -\frac{I_S}{\alpha_r} \cdot [e^{(\frac{-V^*_{C_f}}{V_T})} - 1] + I_S \cdot [e^{(\frac{I^*_{L_{out}} \cdot R_L - V^*_{C_f}}{V_T})} - 1]$$

$$\xi_2(V^*_{C_f}, Y^*_1 = 0, I^*_{L_{out}}, I_S, \beta_f, \ldots) = \frac{I_S}{\beta_f} \cdot [e^{(\frac{I^*_{L_{out}} \cdot R_L - V^*_{C_f}}{V_T})} - 1] + \frac{I_S}{\beta_r} \cdot [e^{(\frac{-V^*_{C_f}}{V_T})} - 1]$$

We consider Microwave RF_{in} signal $X(t) = A_0 + f_X(t)$; $|f_X(t)| \leq 1$ and $A_0 \gg |f_X(t)|$ then $X(t) = A_0 + f_X(t) \approx A_0 \Rightarrow \frac{dX(t)}{dt} = \frac{df_X(t)}{dt} \to \varepsilon$. We get the following two equations for our $I^*_{L_{out}}, V^*_{C_f}$ fixed points: $\frac{dX(t)}{dt} \to \varepsilon$

$$\frac{1}{R_C} \cdot \frac{1}{C_0} \cdot [V_{CC} \cdot \frac{1}{L_{out}} - R_L] - \frac{1}{C_0 \cdot L_{out}} \cdot I^*_{L_{out}} - \frac{1}{C_0 \cdot L_{out}} \cdot \{-\frac{I_S}{\alpha_r} \cdot [e^{(\frac{-V^*_{C_f}}{V_T})} - 1] + I_S \cdot [e^{(\frac{I^*_{L_{out}} \cdot R_L - V^*_{C_f}}{V_T})} - 1]\} = 0$$

$$- I^*_{L_{out}} \cdot \frac{R_L}{R_b \cdot C_{in}} + V^*_{C_f} \cdot \frac{1}{R_b \cdot C_{in}} - \frac{1}{C_{in}} \cdot \{\frac{I_S}{\beta_f} \cdot [e^{(\frac{I^*_{L_{out}} \cdot R_L - V^*_{C_f}}{V_T})} - 1] + \frac{I_S}{\beta_r} \cdot [e^{(\frac{-V^*_{C_f}}{V_T})} - 1]\} + \frac{V_{bb}}{R_b \cdot C_{in}} = 0$$

Our system fixed points are $E^{(*)}(Y^*_1, Y^*_2, Y^*_3, I^*_{L_{out}} V^*_{C_f}) = (0, 0, 0, I^*_{L_{out}}, V^*_{C_f})$. We need to find the expressions for $I^*_{L_{out}}, V^*_{C_f}$. First exponent is $e^{(\frac{-V^*_{C_f}}{V_T})}$. We define new exponent argument $\frac{V^*_{C_f}}{V_T} = \Gamma^*_1 \Rightarrow e^{-\Gamma^*_1} \approx 1 - \Gamma^*_1$ by using Taylor series approximation. We define our Taylor series approximation in the interval $0 < \Gamma^*_1 < 1 \Rightarrow 0 < \frac{V^*_{C_f}}{V_T} < 1 \Rightarrow 0 < V^*_{C_f} < V_T \Rightarrow e^{(\frac{-V^*_{C_f}}{V_T})} \approx 1 - \frac{V^*_{C_f}}{V_T}$. Second exponent is

$e^{(\frac{I^*_{L_{out}} \cdot R_L - V^*_{C_f}}{V_T})}$. We define new exponent argument $\frac{I^*_{L_{out}} \cdot R_L - V^*_{C_f}}{V_T} = \Gamma^*_2$ and there are two cases: $\frac{I^*_{L_{out}} \cdot R_L - V^*_{C_f}}{V_T} > 0$ and $\frac{I^*_{L_{out}} \cdot R_L - V^*_{C_f}}{V_T} < 0$. First case $\frac{I^*_{L_{out}} \cdot R_L - V^*_{C_f}}{V_T} > 0 \Rightarrow V_T = 26\,mV \Rightarrow I^*_{L_{out}} \cdot R_L - V^*_{C_f} > 0 \Rightarrow R_L > \frac{V^*_{C_f}}{I^*_{L_{out}}}$ and by using Taylor series approximation: $e^{\Gamma^*_2} = e^{(\frac{I^*_{L_{out}} \cdot R_L - V^*_{C_f}}{V_T})} \approx 1 + \frac{I^*_{L_{out}} \cdot R_L - V^*_{C_f}}{V_T}$. Second case $\frac{I^*_{L_{out}} \cdot R_L - V^*_{C_f}}{V_T} < 0 \Rightarrow V_T = 26\,mV \Rightarrow I^*_{L_{out}} \cdot R_L - V^*_{C_f} < 0 \Rightarrow 0 < R_L < \frac{V^*_{C_f}}{I^*_{L_{out}}}$ and by using Taylor series approximation:

$$e^{\Gamma^*_2}|_{\Gamma^*_2<0} = e^{-|\Gamma^*_2|} = 1 - |\Gamma^*_2| = e^{(\frac{I^*_{L_{out}} \cdot R_L - V^*_{C_f}}{V_T})}$$

$$e^{\Gamma^*_2}|_{\Gamma^*_2<0} = e^{-|\Gamma^*_2|} = 1 - |\Gamma^*_2| = e^{(\frac{I^*_{L_{out}} \cdot R_L - V^*_{C_f}}{V_T})} \approx 1 - |\frac{I^*_{L_{out}} \cdot R_L - V^*_{C_f}}{V_T}|_{\Gamma^*_2<0}$$
$$= 1 + \frac{I^*_{L_{out}} \cdot R_L - V^*_{C_f}}{V_T}$$

We implement our Taylor series approximation expressions into below equations:

$$(\#)\frac{1}{R_C} \cdot \frac{1}{C_0} \cdot [V_{CC} \cdot \frac{1}{L_{out}} - R_L] - \frac{1}{C_0 \cdot L_{out}} \cdot I^*_{L_{out}} - \frac{1}{C_0 \cdot L_{out}} \cdot \{\frac{I_S}{\alpha_r} \cdot \frac{V^*_{C_f}}{V_T} + I_S \cdot [\frac{I^*_{L_{out}} \cdot R_L - V^*_{C_f}}{V_T}]\} = 0$$

$$(\#\#) - I^*_{L_{out}} \cdot \frac{R_L}{R_b \cdot C_{in}} + V^*_{C_f} \cdot \frac{1}{R_b \cdot C_{in}} - \frac{1}{C_{in}} \cdot \{\frac{I_S}{\beta_f} \cdot [\frac{I^*_{L_{out}} \cdot R_L - V^*_{C_f}}{V_T}] - \frac{I_S}{\beta_r} \cdot \frac{V^*_{C_f}}{V_T}\} + \frac{V_{bb}}{R_b \cdot C_{in}} = 0$$

$$(\#) \to \frac{1}{R_C} \cdot L_{out} \cdot [V_{CC} \cdot \frac{1}{L_{out}} - R_L] - I^*_{L_{out}} - \{\frac{I_S}{\alpha_r \cdot V_T} \cdot V^*_{C_f} + \frac{I^*_{L_{out}} \cdot I_S \cdot R_L}{V_T} - \frac{I_S \cdot V^*_{C_f}}{V_T}\} = 0$$

$$\frac{1}{R_C} \cdot [V_{CC} - L_{out} \cdot R_L] + \frac{V^*_{C_f}}{V_T} \cdot [I_S - \frac{I_S}{\alpha_r}] - I^*_{L_{out}} \cdot [\frac{I_S \cdot R_L}{V_T} + 1] = 0 \Rightarrow I^*_{L_{out}} = \frac{\frac{1}{R_C} \cdot [V_{CC} - L_{out} \cdot R_L] + \frac{V^*_{C_f}}{V_T} \cdot [I_S - \frac{I_S}{\alpha_r}]}{[\frac{I_S \cdot R_L}{V_T} + 1]}$$

$$(\#\#) \rightarrow -I^*_{L_{out}} \cdot R_L + V^*_{C_f} - R_b \cdot \{\frac{I_S}{\beta_f} \cdot [\frac{I^*_{L_{out}} \cdot R_L - V^*_{C_f}}{V_T}] - \frac{I_S}{\beta_r} \cdot \frac{V^*_{C_f}}{V_T}\} + V_{bb} = 0$$

$$-I^*_{L_{out}} \cdot R_L + V^*_{C_f} - R_b \cdot \frac{I_S}{\beta_f} \cdot [\frac{I^*_{L_{out}} \cdot R_L - V^*_{C_f}}{V_T}] + R_b \cdot \frac{I_S}{\beta_r \cdot V_T} \cdot V^*_{C_f} + V_{bb} = 0$$

$$V^*_{C_f} \cdot [1 + R_b \cdot \frac{I_S}{\beta_f} \cdot \frac{1}{V_T} + R_b \cdot \frac{I_S}{\beta_r \cdot V_T}] - I^*_{L_{out}} \cdot R_L \cdot [R_b \cdot \frac{I_S}{\beta_f \cdot V_T} + 1] + V_{bb} = 0$$

$I^*_{L_{out}} = \frac{V^*_{C_f} \cdot [1 + R_b \cdot \frac{I_S}{\beta_f \cdot V_T} + R_b \cdot \frac{I_S}{\beta_r \cdot V_T}] + V_{bb}}{R_L \cdot [R_b \cdot \frac{I_S}{\beta_f \cdot V_T} + 1]}$. We have two expressions for $I^*_{L_{out}}$ then $\frac{\frac{1}{R_C} \cdot [V_{CC} - L_{out} \cdot R_L] + \frac{V^*_{C_f}}{V_T} \cdot [I_S - \frac{I_S}{\alpha_r}]}{[\frac{I_S \cdot R_L}{V_T} + 1]} = \frac{V^*_{C_f} \cdot [1 + R_b \cdot \frac{I_S}{\beta_f \cdot V_T} + R_b \cdot \frac{I_S}{\beta_r \cdot V_T}] + V_{bb}}{R_L \cdot [R_b \cdot \frac{I_S}{\beta_f \cdot V_T} + 1]}$

$$\frac{1}{R_C} \cdot [V_{CC} - L_{out} \cdot R_L] \cdot R_L \cdot [R_b \cdot \frac{I_S}{\beta_f \cdot V_T} + 1] + \frac{V^*_{C_f}}{V_T} \cdot I_S \cdot [1 - \frac{1}{\alpha_r}] \cdot R_L \cdot [R_b \cdot \frac{I_S}{\beta_f \cdot V_T} + 1]$$
$$= V^*_{C_f} \cdot [1 + R_b \cdot \frac{I_S}{\beta_f \cdot V_T} + R_b \cdot \frac{I_S}{\beta_r \cdot V_T}] \cdot [\frac{I_S \cdot R_L}{V_T} + 1] + [\frac{I_S \cdot R_L}{V_T} + 1] \cdot V_{bb}$$

$$V^*_{C_f} \cdot \{\frac{1}{V_T} \cdot I_S \cdot [1 - \frac{1}{\alpha_r}] \cdot R_L \cdot [R_b \cdot \frac{I_S}{\beta_f \cdot V_T} + 1] - [1 + \frac{R_b \cdot I_S}{V_T} \cdot (\frac{1}{\beta_f} + \frac{1}{\beta_r})] \cdot [\frac{I_S \cdot R_L}{V_T} + 1]\}$$
$$= [\frac{I_S \cdot R_L}{V_T} + 1] \cdot V_{bb} - \frac{1}{R_C} \cdot [V_{CC} - L_{out} \cdot R_L] \cdot R_L \cdot [R_b \cdot \frac{I_S}{\beta_f \cdot V_T} + 1]$$

$$V^*_{C_f} \simeq \frac{[\frac{I_S \cdot R_L}{V_T} + 1] \cdot V_{bb} - \frac{1}{R_C} \cdot [V_{CC} - L_{out} \cdot R_L] \cdot R_L \cdot [R_b \cdot \frac{I_S}{\beta_f \cdot V_T} + 1]}{\frac{1}{V_T} \cdot I_S \cdot [1 - \frac{1}{\alpha_r}] \cdot R_L \cdot [R_b \cdot \frac{I_S}{\beta_f \cdot V_T} + 1] - [1 + \frac{R_b \cdot I_S}{V_T} \cdot (\frac{1}{\beta_f} + \frac{1}{\beta_r})] \cdot [\frac{I_S \cdot R_L}{V_T} + 1]}$$

$$I^*_{L_{out}} \simeq \frac{V^*_{C_f} \cdot [1 + R_b \cdot \frac{I_S}{\beta_f \cdot V_T} + R_b \cdot \frac{I_S}{\beta_r \cdot V_T}] + V_{bb}}{R_L \cdot [R_b \cdot \frac{I_S}{\beta_f \cdot V_T} + 1]} = \frac{V^*_{C_f} \cdot [1 + R_b \cdot \frac{I_S}{\beta_f \cdot V_T} + R_b \cdot \frac{I_S}{\beta_r \cdot V_T}]}{R_L \cdot [R_b \cdot \frac{I_S}{\beta_f \cdot V_T} + 1]}$$
$$+ \frac{V_{bb}}{R_L \cdot [R_b \cdot \frac{I_S}{\beta_f \cdot V_T} + 1]}$$

$$I^*_{L_{out}} \simeq \{\frac{[\frac{I_S \cdot R_L}{V_T} + 1] \cdot V_{bb} - \frac{1}{R_C} \cdot [V_{CC} - L_{out} \cdot R_L] \cdot R_L \cdot [R_b \cdot \frac{I_S}{\beta_f \cdot V_T} + 1]}{\frac{1}{V_T} \cdot I_S \cdot [1 - \frac{1}{\alpha_r}] \cdot R_L \cdot [R_b \cdot \frac{I_S}{\beta_f \cdot V_T} + 1] - [1 + \frac{R_b \cdot I_S}{V_T} \cdot (\frac{1}{\beta_f} + \frac{1}{\beta_r})] \cdot [\frac{I_S \cdot R_L}{V_T} + 1]}\}$$
$$\cdot \{\frac{[1 + R_b \cdot \frac{I_S}{\beta_f \cdot V_T} + R_b \cdot \frac{I_S}{\beta_r \cdot V_T}]}{R_L \cdot [R_b \cdot \frac{I_S}{\beta_f \cdot V_T} + 1]}\} + \frac{V_{bb}}{R_L \cdot [R_b \cdot \frac{I_S}{\beta_f \cdot V_T} + 1]}$$

There is other way to find our system fixed points. First we write our system differential equations:

$$V_{A1} = L_{out} \cdot \frac{dI_{L_{out}}}{dt} + I_{L_{out}} \cdot R_L \Rightarrow \frac{dI_{L_{out}}}{dt} = \frac{1}{L_{out}} \cdot V_{A1} - I_{L_{out}} \cdot \frac{R_L}{L_{out}}$$

$$I_{C_0} = C_0 \cdot \frac{dV_{A1}}{dt} \Rightarrow \frac{dV_{A1}}{dt} = \frac{1}{C_0} \cdot I_{C_0} = \frac{1}{C_0} \cdot (I_{R_C} - I_{C_f} - I_C - I_{L_{out}}); I_{C_f} = I_b - I_{R_b} - I_{L_{in}}$$

$$\frac{1}{C_0} \cdot I_{C_0} = \frac{1}{C_0} \cdot (I_{R_C} - [I_b - I_{R_b} - I_{L_{in}}] - I_C - I_{L_{out}})$$

$$= \frac{1}{C_0} \cdot (I_{R_C} - [I_b + I_C] + I_{R_b} + I_{L_{in}} - I_{L_{out}})$$

$$\frac{V_{CC} - V_{A1}}{R_C} = I_{R_C} \Rightarrow \frac{1}{C_0} \cdot I_{C_0} = \frac{1}{C_0} \cdot (\frac{V_{CC} - V_{A1}}{R_C} - [I_b + I_C] + I_{R_b} + I_{L_{in}} - I_{L_{out}})$$

$$\frac{V_{bb} - V_{A2}}{R_b} = I_{R_b} \Rightarrow \frac{1}{C_0} \cdot I_{C_0} = \frac{1}{C_0} \cdot (\frac{V_{CC} - V_{A1}}{R_C} - [I_b + I_C] + [\frac{V_{bb} - V_{A2}}{R_b}] + I_{L_{in}} - I_{L_{out}})$$

$$I_C = \xi_1 = \xi_1(V_{C_f}, \frac{dI_{L_{out}}}{dt}, I_{L_{out}}, I_S, \alpha_r, \ldots); I_b = \xi_2 = \xi_2(V_{C_f}, \frac{dI_{L_{out}}}{dt}, I_{L_{out}}, I_S, \beta_f, \ldots)$$

$$\frac{dV_{A1}}{dt} = \frac{1}{C_0} \cdot (\frac{V_{CC} - V_{A1}}{R_C} - [\xi_2(V_{C_f}, \frac{dI_{L_{out}}}{dt}, I_{L_{out}}, I_S, \beta_f, \ldots)$$
$$+ \xi_1(V_{C_f}, \frac{dI_{L_{out}}}{dt}, I_{L_{out}}, I_S, \alpha_r, \ldots)] + [\frac{V_{bb} - V_{A2}}{R_b}] + I_{L_{in}} - I_{L_{out}});$$

$$V_{A1} - V_{A2} = V_{C_f} \Rightarrow V_{A2} = V_{A1} - V_{C_f}$$

$$\frac{dV_{A1}}{dt} = \frac{1}{C_0} \cdot (\frac{V_{CC} - V_{A1}}{R_C} - [\xi_2(V_{C_f}, \frac{dI_{L_{out}}}{dt}, I_{L_{out}}, I_S, \beta_f, \ldots)$$
$$+ \xi_1(V_{C_f}, \frac{dI_{L_{out}}}{dt}, I_{L_{out}}, I_S, \alpha_r, \ldots)] + [\frac{V_{bb} - (V_{A1} - V_{C_f})}{R_b}] + I_{L_{in}} - I_{L_{out}})$$

$$I_{C_f} = C_f \cdot \frac{dV_{C_f}}{dt} \Rightarrow \frac{dV_{C_f}}{dt} = \frac{1}{C_f} \cdot I_{C_f} = \frac{1}{C_f} \cdot (\xi_2(V_{C_f}, \frac{dI_{L_{out}}}{dt}, I_{L_{out}}, I_S, \beta_f, \ldots) - I_{R_b} - I_{L_{in}})$$

$$\frac{dV_{C_f}}{dt} = \frac{1}{C_f} \cdot (\xi_2(V_{C_f}, \frac{dI_{L_{out}}}{dt}, I_{L_{out}}, I_S, \beta_f, \ldots) - [\frac{V_{bb} - V_{A2}}{R_b}] - I_{L_{in}})$$

$$\frac{dV_{C_f}}{dt} = \frac{1}{C_f} \cdot (\xi_2(V_{C_f}, \frac{dI_{L_{out}}}{dt}, I_{L_{out}}, I_S, \beta_f, \ldots) - [\frac{V_{bb} - (V_{A1} - V_{C_f})}{R_b}] - I_{L_{in}})$$

$$V_{L_{in}} = L_{in} \cdot \frac{dI_{Lin}}{dt} \Rightarrow \frac{dI_{Lin}}{dt} = \frac{1}{L_{in}} \cdot (X(t) - [V_{A1} - V_{C_f}] - V_{C_{in}});$$

$$I_{C_{in}} = C_{in} \cdot \frac{dV_{Cin}}{dt} \Rightarrow \frac{dV_{Cin}}{dt} = \frac{1}{C_{in}} \cdot I_{C_{in}} = \frac{1}{C_{in}} \cdot I_{Lin}$$

We can summery our system differential equations:

$$\frac{dI_{L_{out}}}{dt} = \frac{1}{L_{out}} \cdot V_{A1} - I_{L_{out}} \cdot \frac{R_L}{L_{out}}$$
$$\frac{dV_{A1}}{dt} = \frac{1}{C_0} \cdot (\frac{V_{CC} - V_{A1}}{R_C} - [\xi_2(V_{C_f}, \frac{dI_{L_{out}}}{dt}, I_{L_{out}}, I_S, \beta_f, \ldots)$$
$$+ \xi_1(V_{C_f}, \frac{dI_{L_{out}}}{dt}, I_{L_{out}}, I_S, \alpha_r, \ldots)] + [\frac{V_{bb} - (V_{A1} - V_{C_f})}{R_b}] + I_{L_{in}} - I_{L_{out}})$$
$$\frac{dV_{C_f}}{dt} = \frac{1}{C_f} \cdot (\xi_2(V_{C_f}, \frac{dI_{L_{out}}}{dt}, I_{L_{out}}, I_S, \beta_f, \ldots) - [\frac{V_{bb} - (V_{A1} - V_{C_f})}{R_b}] - I_{L_{in}})$$
$$\frac{dI_{Lin}}{dt} = \frac{1}{L_{in}} \cdot (X(t) - [V_{A1} - V_{C_f}] - V_{C_{in}}); \frac{dV_{Cin}}{dt} = \frac{1}{C_{in}} \cdot I_{Lin}$$

At equilibrium points (fixed points):

$$\frac{dI_{L_{out}}}{dt} = 0; \frac{dV_{A1}}{dt} = 0; \frac{dV_{C_f}}{dt} = 0; \frac{dI_{Lin}}{dt} = 0$$
$$\frac{dV_{Cin}}{dt} = 0. \quad I^*_{Lin} = 0; \frac{1}{L_{in}} \cdot (X(t) - [V^*_{A1} - V^*_{C_f}] - V^*_{C_{in}}) = 0; \frac{1}{L_{out}} \cdot V^*_{A1} - I^*_{Lout} \cdot \frac{R_L}{L_{out}} = 0$$
$$\xi_2(V^*_{C_f}, \frac{dI_{L_{out}}}{dt} = 0, I^*_{L_{out}}, I_S, \beta_f, \ldots) - [\frac{V_{bb} - (V^*_{A1} - V^*_{C_f})}{R_b}] = 0$$
$$\frac{V_{CC} - V^*_{A1}}{R_C} - [\xi_2(V^*_{C_f}, \frac{dI_{L_{out}}}{dt} = 0, I^*_{L_{out}}, I_S, \beta_f, \ldots) + \xi_1(V^*_{C_f}, \frac{dI_{L_{out}}}{dt} = 0, I^*_{L_{out}}, I_S, \alpha_r, \ldots)]$$
$$+ [\frac{V_{bb} - (V^*_{A1} - V^*_{C_f})}{R_b}] - I^*_{L_{out}} = 0$$

We consider Microwave RF_{in} signal $X(t) = A_0 + f_X(t)$; $|f_X(t)| <= 1$ and $A_0 >> |f_X(t)|$ then $X(t) = A_0 + f_X(t) \approx A_0$.

$$A_0 - [V^*_{A1} - V^*_{C_f}] - V^*_{C_{in}} = 0; V^*_{A1} - I^*_{Lout} \cdot R_L = 0;$$
$$\xi_2(V^*_{C_f}, \frac{dI_{L_{out}}}{dt} = 0, I^*_{L_{out}}, I_S, \beta_f, \ldots) - [\frac{V_{bb} - (V^*_{A1} - V^*_{C_f})}{R_b}] = 0$$
$$\frac{V_{CC} - V^*_{A1}}{R_C} - [\xi_2(V^*_{C_f}, \frac{dI_{L_{out}}}{dt} = 0, I^*_{L_{out}}, I_S, \beta_f, \ldots) + \xi_1(V^*_{C_f}, \frac{dI_{L_{out}}}{dt} = 0, I^*_{L_{out}}, I_S, \alpha_r, \ldots)]$$
$$+ [\frac{V_{bb} - (V^*_{A1} - V^*_{C_f})}{R_b}] - I^*_{L_{out}} = 0; V^*_{A1} - I^*_{Lout} \cdot R_L = 0 \Rightarrow V^*_{A1} = I^*_{Lout} \cdot R_L$$

Then we get the following equations:

$$A_0 - [I^*_{Lout} \cdot R_L - V^*_{C_f}] - V^*_{C_{in}} = 0; \xi_2(V^*_{C_f}, \frac{dI_{L_{out}}}{dt} = 0, I^*_{L_{out}}, I_S, \beta_f, \ldots)$$
$$- [\frac{V_{bb} - (I^*_{Lout} \cdot R_L - V^*_{C_f})}{R_b}] = 0$$
$$\frac{V_{CC} - I^*_{Lout} \cdot R_L}{R_C} - [\xi_2(V^*_{C_f}, \frac{dI_{L_{out}}}{dt} = 0, I^*_{L_{out}}, I_S, \beta_f, \ldots)$$
$$+ \xi_1(V^*_{C_f}, \frac{dI_{L_{out}}}{dt} = 0, I^*_{L_{out}}, I_S, \alpha_r, \ldots)] + [\frac{V_{bb} - (I^*_{Lout} \cdot R_L - V^*_{C_f})}{R_b}] - I^*_{L_{out}} = 0$$

And we can find analytically or numerically our fixed points (equilibrium points) value: $V^*_{C_f}, I^*_{L_{out}}$. Additionally $I^*_{Lin} = 0; V^*_{A1} = I^*_{Lout} \cdot R_L$.

Stability analysis: We define the following functions:

$$f_1(V_{A1}, I_{L_{out}}, V_{C_f}, I_{Lin}, V_{C_{in}}) = \frac{1}{L_{out}} \cdot V_{A1} - I_{L_{out}} \cdot \frac{R_L}{L_{out}}$$
$$f_2(V_{A1}, I_{L_{out}}, V_{C_f}, I_{Lin}, V_{C_{in}}) = \frac{1}{C_0} \cdot (\frac{V_{CC} - V_{A1}}{R_C} - [\xi_2(V_{C_f}, \frac{dI_{L_{out}}}{dt}, I_{L_{out}}, I_S, \beta_f, \ldots)$$
$$+ \xi_1(V_{C_f}, \frac{dI_{L_{out}}}{dt}, I_{L_{out}}, I_S, \alpha_r, \ldots)] + [\frac{V_{bb} - (V_{A1} - V_{C_f})}{R_b}] + I_{L_{in}} - I_{L_{out}})$$
$$f_3(V_{A1}, I_{L_{out}}, V_{C_f}, I_{Lin}, V_{C_{in}}) = \frac{1}{C_f} \cdot (\xi_2(V_{C_f}, \frac{dI_{L_{out}}}{dt}, I_{L_{out}}, I_S, \beta_f, \ldots)$$
$$- [\frac{V_{bb} - (V_{A1} - V_{C_f})}{R_b}] - I_{L_{in}})$$
$$f_4(V_{A1}, I_{L_{out}}, V_{C_f}, I_{Lin}, V_{C_{in}}) = \frac{1}{L_{in}} \cdot (X(t) - [V_{A1} - V_{C_f}] - V_{C_{in}});$$
$$f_5(V_{A1}, I_{L_{out}}, V_{C_f}, I_{Lin}, V_{C_{in}}) = \frac{1}{C_{in}} \cdot I_{Lin}$$
$$V_{A1} = V_{A1}(t),\ I_{L_{out}} = I_{L_{out}}(t),\ V_{C_f} = V_{C_f}(t),\ I_{Lin} = I_{Lin}(t),\ V_{C_{in}} = V_{C_{in}}(t)$$
$$\xi_1 = \xi_1(V_{C_f}, \frac{dI_{L_{out}}}{dt}, I_{L_{out}}, I_S, \alpha_r, \ldots); \xi_2 = \xi_2(V_{C_f}, \frac{dI_{L_{out}}}{dt}, I_{L_{out}}, I_S, \beta_f, \ldots)$$

To classify our bipolar transistor microwave system fixed points, we need to compute the Jacobian (linearized system) [2–4].

$$A = \begin{pmatrix} \frac{\partial f_1}{\partial V_{A1}} & \cdots & \frac{\partial f_1}{\partial V_{C_{in}}} \\ \vdots & \ddots & \vdots \\ \frac{\partial f_5}{\partial V_{A1}} & \cdots & \frac{\partial f_5}{\partial V_{C_{in}}} \end{pmatrix}; \Xi_{11} = \frac{\partial f_1}{\partial V_{A1}}; \Xi_{12} = \frac{\partial f_1}{\partial I_{L_{out}}}; \Xi_{13} = \frac{\partial f_1}{\partial V_{C_f}}; \Xi_{14} = \frac{\partial f_1}{\partial I_{Lin}}; \Xi_{15}$$
$$= \frac{\partial f_1}{\partial V_{C_{in}}}$$

$$\Xi_{21} = \frac{\partial f_2}{\partial V_{A1}}; \Xi_{22} = \frac{\partial f_2}{\partial I_{L_{out}}}; \Xi_{23} = \frac{\partial f_2}{\partial V_{C_f}}; \Xi_{24} = \frac{\partial f_2}{\partial I_{Lin}};$$

$$\Xi_{25} = \frac{\partial f_2}{\partial V_{C_{in}}}; \Xi_{31} = \frac{\partial f_3}{\partial V_{A1}}; \Xi_{32} = \frac{\partial f_3}{\partial I_{L_{out}}}$$

$$\Xi_{33} = \frac{\partial f_3}{\partial V_{C_f}}; \Xi_{34} = \frac{\partial f_3}{\partial I_{Lin}}; \Xi_{35} = \frac{\partial f_3}{\partial V_{C_{in}}}; \Xi_{41} = \frac{\partial f_4}{\partial V_{A1}};$$

$$\Xi_{42} = \frac{\partial f_4}{\partial I_{L_{out}}}; \Xi_{43} = \frac{\partial f_4}{\partial V_{C_f}}; \Xi_{44} = \frac{\partial f_4}{\partial I_{Lin}}$$

$$\Xi_{45} = \frac{\partial f_4}{\partial V_{C_{in}}}; \Xi_{51} = \frac{\partial f_5}{\partial V_{A1}}; \Xi_{52} = \frac{\partial f_5}{\partial I_{L_{out}}}; \Xi_{53} = \frac{\partial f_5}{\partial V_{C_f}}; \Xi_{54} = \frac{\partial f_5}{\partial I_{Lin}}; \Xi_{55} = \frac{\partial f_5}{\partial V_{C_{in}}}$$

$$\Xi_{k1} = \frac{\partial f_k}{\partial V_{A1}}; \Xi_{k2} = \frac{\partial f_k}{\partial I_{L_{out}}}; \Xi_{k3} = \frac{\partial f_k}{\partial V_{C_f}}; \Xi_{k4} = \frac{\partial f_k}{\partial I_{Lin}}; \Xi_{k5} = \frac{\partial f_k}{\partial V_{C_{in}}}; k = 1, \ldots, 5$$

$$A = \begin{pmatrix} \Xi_{11} & \cdots & \Xi_{15} \\ \vdots & \ddots & \vdots \\ \Xi_{51} & \cdots & \Xi_{55} \end{pmatrix}; \Xi_{11} = \frac{\partial f_1}{\partial V_{A1}} = \frac{1}{L_{out}};$$

$$\Xi_{12} = \frac{\partial f_1}{\partial I_{L_{out}}} = -\frac{R_L}{L_{out}}; \Xi_{13} = \frac{\partial f_1}{\partial V_{C_f}} = 0$$

$$\Xi_{14} = \frac{\partial f_1}{\partial I_{Lin}} = 0; \Xi_{15} = \frac{\partial f_1}{\partial V_{C_{in}}} = 0; \Xi_{21} = \frac{\partial f_2}{\partial V_{A1}} = -\frac{1}{C_0} \cdot \left(\frac{1}{R_C} + \frac{\partial}{\partial V_{A1}}\left[\sum_{k=1}^{2} \xi_k\right] + \frac{1}{R_b}\right)$$

$$\Xi_{21} = \frac{\partial f_2}{\partial V_{A1}} = -\frac{1}{C_0} \cdot \left(\frac{1}{R_C} + \frac{1}{R_b} + \frac{I_S}{V_T} \cdot \left[1 + \frac{1}{\beta_f}\right] \cdot e^{\left(\frac{V_{A1} - V_{C_f}}{V_T}\right)}\right)$$

$$f_2 = \frac{1}{C_0} \cdot \left(\frac{V_{CC} - V_{A1}}{R_C} - [\xi_2 + \xi_1] + \left[\frac{V_{bb} - (V_{A1} - V_{C_f})}{R_b}\right] + I_{L_{in}} - I_{L_{out}}\right);$$

$$\frac{\partial}{\partial V_{A1}}\left[\sum_{k=1}^{2} \xi_k\right] = \frac{I_S}{V_T} \cdot \left[1 + \frac{1}{\beta_f}\right] \cdot e^{\left(\frac{V_{A1} - V_{C_f}}{V_T}\right)}$$

$$\xi_1 = I_C = -\frac{I_S}{\alpha_r} \cdot \left[e^{\left(\frac{-V_{C_f}}{V_T}\right)} - 1\right] + I_S \cdot \left[e^{\left(\frac{L_{out} \cdot \frac{dI_{L_{out}}}{dt} + I_{L_{out}} \cdot R_L - V_{C_f}}{V_T}\right)} - 1\right]$$

$$= -\frac{I_S}{\alpha_r} \cdot \left[e^{\left(\frac{-V_{C_f}}{V_T}\right)} - 1\right] + I_S \cdot \left[e^{\left(\frac{V_{A1} - V_{C_f}}{V_T}\right)} - 1\right]$$

$$\xi_2 = I_B = \frac{I_S}{\beta_f} \cdot \left[e^{\left(\frac{L_{out} \cdot \frac{dI_{L_{out}}}{dt} + I_{L_{out}} \cdot R_L - V_{C_f}}{V_T}\right)} - 1\right] + \frac{I_S}{\beta_r} \cdot \left[e^{\left(\frac{-V_{C_f}}{V_T}\right)} - 1\right]$$

$$= \frac{I_S}{\beta_f} \cdot \left[e^{\left(\frac{V_{A1} - V_{C_f}}{V_T}\right)} - 1\right] + \frac{I_S}{\beta_r} \cdot \left[e^{\left(\frac{-V_{C_f}}{V_T}\right)} - 1\right]$$

Exponent function rules:

$$\frac{d}{dx}[e^{f(x)}] = \frac{df(x)}{dx} \cdot e^{f(x)}; e^{A+B} = e^{A} \cdot e^{B}$$

$$\frac{dI_{L_{out}}}{dt} = \frac{1}{L_{out}} \cdot [V_{A1} - I_{L_{out}} \cdot R_L]; \frac{\partial \xi_1}{\partial V_{C_f}} = \frac{I_S}{V_T} \cdot e^{(\frac{-V_{C_f}}{V_T})} \cdot [\frac{1}{\alpha_r} - e^{(\frac{V_{A1}}{V_T})}]; \frac{\partial \xi_1}{\partial V_{A1}} = \frac{I_S}{V_T} \cdot e^{(\frac{V_{A1} - V_{C_f}}{V_T})}$$

$$\frac{\partial \xi_2}{\partial V_{C_f}} = -\frac{I_S}{V_T} \cdot e^{(\frac{-V_{C_f}}{V_T})} \cdot [\frac{1}{\beta_f} \cdot e^{\frac{V_{A1}}{V_T}} + \frac{1}{\beta_r}]; \frac{\partial \xi_2}{\partial V_{A1}} = \frac{I_S}{\beta_f \cdot V_T} \cdot e^{(\frac{V_{A1} - V_{C_f}}{V_T})}$$

$$\Xi_{22} = \frac{\partial f_2}{\partial I_{L_{out}}} = -\frac{1}{C_0}; \Xi_{23} = \frac{\partial f_2}{\partial V_{C_f}} = \frac{1}{C_0}$$

$$\cdot (-[\frac{\partial \xi_2}{\partial V_{C_f}} + \frac{\partial \xi_1}{\partial V_{C_f}}] + \frac{1}{R_b}) = \frac{1}{C_0} \cdot [\frac{1}{R_b} - \frac{\partial}{\partial V_{C_f}}(\sum_{k=1}^{2} \xi_k)]$$

$$\frac{\partial}{\partial V_{C_f}}(\sum_{k=1}^{2} \xi_k) = \frac{I_S}{V_T} \cdot e^{(\frac{-V_{C_f}}{V_T})} \cdot \{[\frac{1}{\alpha_r} - e^{\frac{V_{A1}}{V_T}}] - [\frac{1}{\beta_f} \cdot e^{\frac{V_{A1}}{V_T}} + \frac{1}{\beta_r}]\}$$

$$= \frac{I_S}{V_T} \cdot e^{(\frac{-V_{C_f}}{V_T})} \cdot [\frac{1}{\alpha_r} - \frac{1}{\beta_r} - e^{\frac{V_{A1}}{V_T}} \cdot (1 + \frac{1}{\beta_f})]$$

$$\Xi_{23} = \frac{\partial f_2}{\partial V_{C_f}} = \frac{1}{C_0} \cdot [\frac{1}{R_b} - \frac{\partial}{\partial V_{C_f}}(\sum_{k=1}^{2} \xi_k)]$$

$$= \frac{1}{C_0} \cdot \{\frac{1}{R_b} - \frac{I_S}{V_T} \cdot e^{(\frac{-V_{C_f}}{V_T})} \cdot [\frac{1}{\alpha_r} - \frac{1}{\beta_r} - e^{\frac{V_{A1}}{V_T}} \cdot (1 + \frac{1}{\beta_f})]\}$$

$$\Xi_{24} = \frac{\partial f_2}{\partial I_{Lin}} = \frac{1}{C_0}; \Xi_{25} = \frac{\partial f_2}{\partial V_{C_{in}}} = 0; \Xi_{31} = \frac{\partial f_3}{\partial V_{A1}}$$

$$= \frac{1}{C_f} \cdot (\frac{\partial \xi_2}{\partial V_{A1}} + \frac{1}{R_b}) = \frac{1}{C_f} \cdot (\frac{I_S}{\beta_f \cdot V_T} \cdot e^{(\frac{V_{A1} - V_{C_f}}{V_T})} + \frac{1}{R_b})$$

$$\Xi_{32} = \frac{\partial f_3}{\partial I_{L_{out}}} = 0; \Xi_{33} = \frac{\partial f_3}{\partial V_{C_f}} = \frac{1}{C_f} \cdot (\frac{\partial \xi_2}{\partial V_{C_f}} - \frac{1}{R_b})$$

$$= -\frac{1}{C_f} \cdot (\frac{I_S}{V_T} \cdot e^{(\frac{-V_{C_f}}{V_T})} \cdot [\frac{1}{\beta_f} \cdot e^{\frac{V_{A1}}{V_T}} + \frac{1}{\beta_r}] + \frac{1}{R_b})$$

$$\Xi_{34} = \frac{\partial f_3}{\partial I_{Lin}} = -\frac{1}{C_f}; \Xi_{35} = \frac{\partial f_3}{\partial V_{C_{in}}} = 0; f_3 = \frac{1}{C_f} \cdot (\xi_2 - [\frac{V_{bb} - (V_{A1} - V_{C_f})}{R_b}] - I_{L_{in}})$$

We consider Microwave RF_{in} signal X(t) = A_0 + $f_X(t)$; $|f_X(t)| \leq 1$ and $A_0 \gg |f_X(t)|$ then $X(t) = A_0 + f_X(t) \approx A_0$.

$$\Xi_{41} = \frac{\partial f_4}{\partial V_{A1}} = -\frac{1}{L_{in}}; \Xi_{42} = \frac{\partial f_4}{\partial I_{L_{out}}} = 0; \Xi_{43} = \frac{\partial f_4}{\partial V_{C_f}} = \frac{1}{L_{in}};$$

$$\Xi_{44} = \frac{\partial f_4}{\partial I_{Lin}} = 0; \Xi_{45} = \frac{\partial f_4}{\partial V_{C_{in}}} = -\frac{1}{L_{in}}$$

$$f_4 \approx \frac{1}{L_{in}} \cdot (A_0 - [V_{A1} - V_{C_f}] - V_{C_{in}}); \Xi_{51} = \frac{\partial f_5}{\partial V_{A1}} = 0;$$

$$\Xi_{52} = \frac{\partial f_5}{\partial I_{L_{out}}} = 0; \Xi_{53} = \frac{\partial f_5}{\partial V_{C_f}} = 0$$

$$\Xi_{54} = \frac{\partial f_5}{\partial I_{Lin}} = \frac{1}{C_{in}}; \Xi_{55} = \frac{\partial f_5}{\partial V_{C_{in}}} = 0; f_5 = \frac{1}{C_{in}} \cdot I_{Lin}$$

We already found our system fixed points: $I^*_{Lin} = 0; V^*_{A1} = I^*_{Lout} \cdot R_L$

$$E^*(V^*_{A1}, I^*_{Lout}, V^*_{Cf}, I^*_{Lin}, V^*_{Cin}) = (I^*_{Lout} \cdot R_L, I^*_{Lout}, V^*_{Cf}, 0, V^*_{Cin})$$

Our system Jacobian elements for our fixed points coordinates are:

$$\Xi_{11} = \frac{1}{L_{out}}; \Xi_{12} = -\frac{R_L}{L_{out}}; \Xi_{13} = 0; \Xi_{14} = 0; \Xi_{15} = 0; \Xi_{22} = -\frac{1}{C_0}; \Xi_{24} = \frac{1}{C_0}$$

$$\Xi_{21} = -\frac{1}{C_0} \cdot (\frac{1}{R_C} + \frac{1}{R_b} + \frac{I_S}{V_T} \cdot [1 + \frac{1}{\beta_f}] \cdot e^{(\frac{I^*_{Lout} \cdot R_L - V^*_{Cf}}{V_T})}); \Xi_{25} = 0; \Xi_{32} = 0$$

$$\Xi_{23} = \frac{1}{C_0} \cdot \{\frac{1}{R_b} - \frac{I_S}{V_T} \cdot e^{(\frac{-V^*_{Cf}}{V_T})} \cdot [\frac{1}{\alpha_r} - \frac{1}{\beta_r} - e^{\frac{I^*_{Lout} \cdot R_L}{V_T}}$$

$$\cdot (1 + \frac{1}{\beta_f})]\}; \Xi_{31} = \frac{1}{C_f} \cdot (\frac{I_S}{\beta_f \cdot V_T} \cdot e^{(\frac{I^*_{Lout} \cdot R_L - V^*_{Cf}}{V_T})} + \frac{1}{R_b})$$

$$\Xi_{33} = -\frac{1}{C_f} \cdot (\frac{I_S}{V_T} \cdot e^{(\frac{-V^*_{Cf}}{V_T})} \cdot [\frac{1}{\beta_f} \cdot e^{\frac{I^*_{Lout} \cdot R_L}{V_T}} + \frac{1}{\beta_r}] + \frac{1}{R_b});$$

$$\Xi_{34} = -\frac{1}{C_f}; \Xi_{35} = 0; \Xi_{54} = \frac{1}{C_{in}}; \Xi_{55} = 0$$

$$\Xi_{41} = -\frac{1}{L_{in}}; \Xi_{42} = 0; \Xi_{43} = \frac{1}{L_{in}}; \Xi_{44} = 0; \Xi_{45} = -\frac{1}{L_{in}}; \Xi_{51} = 0; \Xi_{52} = 0; \Xi_{53} = 0$$

$$A^* = \begin{pmatrix} \Xi_{11} & \cdots & \Xi_{15} \\ \vdots & \ddots & \vdots \\ \Xi_{51} & \cdots & \Xi_{55} \end{pmatrix} \Bigg|_{\substack{@E^*(V^*_{A1}, I^*_{Lout}, V^*_{Cf}, I^*_{Lin}, V^*_{Cin}) \\ = (I^*_{Lout} \cdot R_L, I^*_{Lout}, V^*_{Cf}, 0, V^*_{Cin})}} = \begin{pmatrix} \frac{\partial f_1}{\partial V_{A1}} & \cdots & \frac{\partial f_1}{\partial V_{C_{in}}} \\ \vdots & \ddots & \vdots \\ \frac{\partial f_5}{\partial V_{A1}} & \cdots & \frac{\partial f_5}{\partial V_{C_{in}}} \end{pmatrix} \Bigg|_{\substack{@E^*(V^*_{A1}, I^*_{Lout}, V^*_{Cf}, I^*_{Lin}, V^*_{Cin}) \\ = (I^*_{Lout} \cdot R_L, I^*_{Lout}, V^*_{Cf}, 0, V^*_{Cin})}}$$

$(A^* - \lambda \cdot I) = (A - \lambda \cdot I)^* \Rightarrow \det|A^* - \lambda \cdot I| = 0$ To classify our system fixed points.

We define our system Jacobian elements at fixed points:

$$\Xi_{kl}@(I^*_{Lout}\cdot R_L,\ I^*_{Lout},\ V^*_{Cf},\ 0,\ V^*_{Cin}) \to \Xi^*_{kl}\ \forall k=1,\ldots,5; l=1,\ldots,5$$

$$\det|A^*-\lambda\cdot I| = \left(\frac{1}{L_{out}}-\lambda\right)\cdot(\Xi^*_{22}-\lambda)\cdot\det\begin{pmatrix}(\Xi^*_{33}-\lambda) & -\frac{1}{C_f} & 0\\ \frac{1}{L_{in}} & -\lambda & -\frac{1}{L_{in}}\\ 0 & \frac{1}{C_{in}} & -\lambda\end{pmatrix}$$

$$+\frac{R_L}{L_{out}}\cdot\left\{\Xi^*_{21}\cdot\det\begin{pmatrix}(\Xi^*_{33}-\lambda) & -\frac{1}{C_f} & 0\\ \frac{1}{L_{in}} & -\lambda & -\frac{1}{L_{in}}\\ 0 & \frac{1}{C_{in}} & -\lambda\end{pmatrix}\right.$$

$$\left.-\Xi^*_{23}\cdot\det\begin{pmatrix}\Xi^*_{31} & -\frac{1}{C_f} & 0\\ -\frac{1}{L_{in}} & -\lambda & -\frac{1}{L_{in}}\\ 0 & \frac{1}{C_{in}} & -\lambda\end{pmatrix}+\frac{1}{C_0}\cdot\det\begin{pmatrix}\Xi^*_{31} & (\Xi^*_{33}-\lambda) & 0\\ -\frac{1}{L_{in}} & \frac{1}{L_{in}} & -\frac{1}{L_{in}}\\ 0 & 0 & -\lambda\end{pmatrix}\right\}$$

$$\det\begin{pmatrix}(\Xi^*_{33}-\lambda) & -\frac{1}{C_f} & 0\\ \frac{1}{L_{in}} & -\lambda & -\frac{1}{L_{in}}\\ 0 & \frac{1}{C_{in}} & -\lambda\end{pmatrix} = (\Xi^*_{33}-\lambda)\cdot\det\begin{pmatrix}-\lambda & -\frac{1}{L_{in}}\\ \frac{1}{C_{in}} & -\lambda\end{pmatrix}+\frac{1}{C_f}\cdot\det\begin{pmatrix}\frac{1}{L_{in}} & -\frac{1}{L_{in}}\\ 0 & -\lambda\end{pmatrix}$$

$$=(\Xi^*_{33}-\lambda)\cdot\left(\lambda^2+\frac{1}{C_{in}\cdot L_{in}}\right)-\frac{1}{C_f}\cdot\frac{1}{L_{in}}\cdot\lambda$$

$$=-\lambda^3+\Xi^*_{33}\cdot\lambda^2-\lambda\cdot\frac{1}{L_{in}}\cdot\left(\frac{1}{C_{in}}+\frac{1}{C_f}\right)+\frac{\Xi^*_{33}}{C_{in}\cdot L_{in}}$$

$$\det\begin{pmatrix}\Xi^*_{31} & -\frac{1}{C_f} & 0\\ -\frac{1}{L_{in}} & -\lambda & -\frac{1}{L_{in}}\\ 0 & \frac{1}{C_{in}} & -\lambda\end{pmatrix} = \Xi^*_{31}\cdot\det\begin{pmatrix}-\lambda & -\frac{1}{L_{in}}\\ \frac{1}{C_{in}} & -\lambda\end{pmatrix}+\frac{1}{C_f}\cdot\begin{pmatrix}-\frac{1}{L_{in}} & -\frac{1}{L_{in}}\\ 0 & -\lambda\end{pmatrix}$$

$$=\Xi^*_{31}\cdot\left(\lambda^2+\frac{1}{C_{in}\cdot L_{in}}\right)+\frac{1}{C_f\cdot L_{in}}\cdot\lambda=\Xi^*_{31}\cdot\lambda^2+\frac{1}{C_f\cdot L_{in}}\cdot\lambda+\frac{\Xi^*_{31}}{C_{in}\cdot L_{in}}$$

$$\det\begin{pmatrix}\Xi^*_{31} & (\Xi^*_{33}-\lambda) & 0\\ -\frac{1}{L_{in}} & \frac{1}{L_{in}} & -\frac{1}{L_{in}}\\ 0 & 0 & -\lambda\end{pmatrix} = \Xi^*_{31}\cdot\begin{pmatrix}\frac{1}{L_{in}} & -\frac{1}{L_{in}}\\ 0 & -\lambda\end{pmatrix}-(\Xi^*_{33}-\lambda)\cdot\begin{pmatrix}-\frac{1}{L_{in}} & -\frac{1}{L_{in}}\\ 0 & -\lambda\end{pmatrix}$$

$$=-\frac{\Xi^*_{31}}{L_{in}}\cdot\lambda-(\Xi^*_{33}-\lambda)\cdot\frac{1}{L_{in}}\cdot\lambda=\frac{1}{L_{in}}\cdot\lambda^2-\lambda\cdot\frac{1}{L_{in}}\cdot(\Xi^*_{31}+\Xi^*_{33})$$

We can summery our last results:

$$\det|A^* - \lambda \cdot I| = (\frac{1}{L_{out}} - \lambda) \cdot (\Xi_{22}^* - \lambda) \cdot \{-\lambda^3 + \Xi_{33}^* \cdot \lambda^2 - \lambda \cdot \frac{1}{L_{in}} \cdot (\frac{1}{C_{in}} + \frac{1}{C_f}) + \frac{\Xi_{33}^*}{C_{in} \cdot L_{in}}\}$$
$$+ \frac{R_L}{L_{out}} \cdot \{\Xi_{21}^* \cdot [-\lambda^3 + \Xi_{33}^* \cdot \lambda^2 - \lambda \cdot \frac{1}{L_{in}} \cdot (\frac{1}{C_{in}} + \frac{1}{C_f}) + \frac{\Xi_{33}^*}{C_{in} \cdot L_{in}}]$$
$$- \Xi_{23}^* \cdot (\Xi_{31}^* \cdot \lambda^2 + \frac{1}{C_f \cdot L_{in}} \cdot \lambda$$
$$+ \frac{\Xi_{31}^*}{C_{in} \cdot L_{in}}) + \frac{1}{C_0} \cdot [\frac{1}{L_{in}} \cdot \lambda^2 - \lambda \cdot \frac{1}{L_{in}} \cdot (\Xi_{31}^* + \Xi_{33}^*)]\}$$

$$\det|A^* - \lambda \cdot I| = [\lambda^2 - \lambda \cdot (\frac{1}{L_{out}} + \Xi_{22}^*) + \frac{\Xi_{22}^*}{L_{out}}]$$
$$\cdot [-\lambda^3 + \Xi_{33}^* \cdot \lambda^2 - \lambda \cdot \frac{1}{L_{in}} \cdot (\frac{1}{C_{in}} + \frac{1}{C_f}) + \frac{\Xi_{33}^*}{C_{in} \cdot L_{in}}]$$
$$+ \frac{R_L}{L_{out}} \cdot \{-\lambda^3 \cdot \Xi_{21}^* + \Xi_{21}^* \cdot \Xi_{33}^* \cdot \lambda^2 - \lambda \cdot \Xi_{21}^* \cdot \frac{1}{L_{in}} \cdot (\frac{1}{C_{in}} + \frac{1}{C_f})$$
$$+ \frac{\Xi_{21}^* \cdot \Xi_{33}^*}{C_{in} \cdot L_{in}} - \Xi_{23}^* \cdot \Xi_{31}^* \cdot \lambda^2 - \frac{\Xi_{23}^*}{C_f \cdot L_{in}} \cdot \lambda$$
$$- \frac{\Xi_{31}^* \cdot \Xi_{23}^*}{C_{in} \cdot L_{in}} + \frac{1}{L_{in} \cdot C_0} \cdot \lambda^2 - \lambda \cdot \frac{1}{L_{in} \cdot C_0} \cdot (\Xi_{31}^* + \Xi_{33}^*)\}$$

$$\det|A^* - \lambda \cdot I| = [\lambda^2 - \lambda \cdot (\frac{1}{L_{out}} + \Xi_{22}^*) + \frac{\Xi_{22}^*}{L_{out}}]$$
$$\cdot [-\lambda^3 + \Xi_{33}^* \cdot \lambda^2 - \lambda \cdot \frac{1}{L_{in}} \cdot (\frac{1}{C_{in}} + \frac{1}{C_f}) + \frac{\Xi_{33}^*}{C_{in} \cdot L_{in}}]$$
$$+ \frac{R_L}{L_{out}} \cdot \{-\lambda^3 \cdot \Xi_{21}^* + \lambda^2 \cdot [\Xi_{21}^* \cdot \Xi_{33}^* - \Xi_{23}^* \cdot \Xi_{31}^* + \frac{1}{L_{in} \cdot C_0}]$$
$$- \lambda \cdot \frac{1}{L_{in}} \cdot [\Xi_{21}^* \cdot (\frac{1}{C_{in}} + \frac{1}{C_f}) + \frac{\Xi_{23}^*}{C_f} + \frac{1}{C_0} \cdot (\Xi_{31}^* + \Xi_{33}^*)]$$
$$+ \frac{1}{C_{in} \cdot L_{in}} \cdot [\Xi_{21}^* \cdot \Xi_{33}^* - \Xi_{31}^* \cdot \Xi_{23}^*]\}$$

$$\det|A^{*}-\lambda\cdot I| = -\lambda^{5}+\Xi_{33}^{*}\cdot\lambda^{4}-\lambda^{3}\cdot\frac{1}{L_{in}}\cdot(\frac{1}{C_{in}}+\frac{1}{C_{f}})+\lambda^{2}\cdot\frac{\Xi_{33}^{*}}{C_{in}\cdot L_{in}}$$
$$+\lambda^{4}\cdot(\frac{1}{L_{out}}+\Xi_{22}^{*})-\lambda^{3}\cdot(\frac{1}{L_{out}}+\Xi_{22}^{*})\cdot\Xi_{33}^{*}+\lambda^{2}\cdot\frac{1}{L_{in}}\cdot(\frac{1}{C_{in}}+\frac{1}{C_{f}})\cdot(\frac{1}{L_{out}}+\Xi_{22}^{*})$$
$$-\lambda\cdot(\frac{1}{L_{out}}+\Xi_{22}^{*})\cdot\frac{\Xi_{33}^{*}}{C_{in}\cdot L_{in}}-\lambda^{3}\cdot\frac{\Xi_{22}^{*}}{L_{out}}+\frac{\Xi_{22}^{*}}{L_{out}}\cdot\Xi_{33}^{*}\cdot\lambda^{2}-\lambda$$
$$\cdot\frac{1}{L_{in}}\cdot\frac{\Xi_{22}^{*}}{L_{out}}\cdot(\frac{1}{C_{in}}+\frac{1}{C_{f}})+\frac{\Xi_{33}^{*}}{C_{in}\cdot L_{in}}\cdot\frac{\Xi_{22}^{*}}{L_{out}}$$
$$+-\lambda^{3}\cdot\frac{R_{L}}{L_{out}}\cdot\Xi_{21}^{*}+\lambda^{2}\cdot\frac{R_{L}}{L_{out}}\cdot[\Xi_{21}^{*}\cdot\Xi_{33}^{*}-\Xi_{23}^{*}\cdot\Xi_{31}^{*}+\frac{1}{L_{in}\cdot C_{0}}]$$
$$-\lambda\cdot\frac{R_{L}}{L_{out}}\cdot\frac{1}{L_{in}}\cdot[\Xi_{21}^{*}\cdot(\frac{1}{C_{in}}+\frac{1}{C_{f}})+\frac{\Xi_{23}^{*}}{C_{f}}+\frac{1}{C_{0}}\cdot(\Xi_{31}^{*}+\Xi_{33}^{*})]$$
$$+\frac{1}{C_{in}\cdot L_{in}}\cdot\frac{R_{L}}{L_{out}}\cdot[\Xi_{21}^{*}\cdot\Xi_{33}^{*}-\Xi_{31}^{*}\cdot\Xi_{23}^{*}]$$

$$\det|A^{*}-\lambda\cdot I| = -\lambda^{5}+\lambda^{4}\cdot[\Xi_{33}^{*}+\frac{1}{L_{out}}+\Xi_{22}^{*}]-\lambda^{3}$$
$$\cdot[\frac{1}{L_{in}}\cdot(\frac{1}{C_{in}}+\frac{1}{C_{f}})+(\frac{1}{L_{out}}+\Xi_{22}^{*})\cdot\Xi_{33}^{*}+\frac{\Xi_{22}^{*}}{L_{out}}+\frac{R_{L}}{L_{out}}\cdot\Xi_{21}^{*}]$$
$$+\lambda^{2}\cdot\{\frac{\Xi_{33}^{*}}{C_{in}\cdot L_{in}}+\frac{1}{L_{in}}\cdot(\frac{1}{C_{in}}+\frac{1}{C_{f}})\cdot(\frac{1}{L_{out}}+\Xi_{22}^{*})+\frac{\Xi_{22}^{*}}{L_{out}}$$
$$\cdot\Xi_{33}^{*}+\frac{R_{L}}{L_{out}}\cdot[\Xi_{21}^{*}\cdot\Xi_{33}^{*}-\Xi_{23}^{*}\cdot\Xi_{31}^{*}+\frac{1}{L_{in}\cdot C_{0}}]\}$$
$$-\lambda\cdot\{(\frac{1}{L_{out}}+\Xi_{22}^{*})\cdot\frac{\Xi_{33}^{*}}{C_{in}\cdot L_{in}}+\frac{1}{L_{in}}\cdot\frac{\Xi_{22}^{*}}{L_{out}}\cdot(\frac{1}{C_{in}}+\frac{1}{C_{f}})$$
$$+\frac{R_{L}}{L_{out}}\cdot\frac{1}{L_{in}}\cdot[\Xi_{21}^{*}\cdot(\frac{1}{C_{in}}+\frac{1}{C_{f}})+\frac{\Xi_{23}^{*}}{C_{f}}+\frac{1}{C_{0}}\cdot(\Xi_{31}^{*}+\Xi_{33}^{*})]\}$$
$$+\frac{\Xi_{33}^{*}}{C_{in}\cdot L_{in}}\cdot\frac{\Xi_{22}^{*}}{L_{out}}+\frac{1}{C_{in}\cdot L_{in}}\cdot\frac{R_{L}}{L_{out}}\cdot[\Xi_{21}^{*}\cdot\Xi_{33}^{*}-\Xi_{31}^{*}\cdot\Xi_{23}^{*}]$$

We get fifth degree polynomial in λ (eigenvalue) and define the following equilibrium parameters:

$$\Omega_5^* = -1; \Omega_4^* = \Xi_{33}^* + \frac{1}{L_{out}} + \Xi_{22}^*; \Omega_3^* = -[\frac{1}{L_{in}} \cdot (\frac{1}{C_{in}} + \frac{1}{C_f})$$
$$+ (\frac{1}{L_{out}} + \Xi_{22}^*) \cdot \Xi_{33}^* + \frac{\Xi_{22}^*}{L_{out}} + \frac{R_L}{L_{out}} \cdot \Xi_{21}^*]$$
$$\Omega_2^* = \frac{\Xi_{33}^*}{C_{in} \cdot L_{in}} + \frac{1}{L_{in}} \cdot (\frac{1}{C_{in}} + \frac{1}{C_f}) \cdot (\frac{1}{L_{out}} + \Xi_{22}^*)$$
$$+ \frac{\Xi_{22}^*}{L_{out}} \cdot \Xi_{33}^* + \frac{R_L}{L_{out}} \cdot [\Xi_{21}^* \cdot \Xi_{33}^* - \Xi_{23}^* \cdot \Xi_{31}^* + \frac{1}{L_{in} \cdot C_0}]$$
$$\Omega_1^* = -\{(\frac{1}{L_{out}} + \Xi_{22}^*) \cdot \frac{\Xi_{33}^*}{C_{in} \cdot L_{in}} + \frac{1}{L_{in}} \cdot \frac{\Xi_{22}^*}{L_{out}} \cdot (\frac{1}{C_{in}} + \frac{1}{C_f})$$
$$+ \frac{R_L}{L_{out}} \cdot \frac{1}{L_{in}} \cdot [\Xi_{21}^* \cdot (\frac{1}{C_{in}} + \frac{1}{C_f}) + \frac{\Xi_{23}^*}{C_f} + \frac{1}{C_0} \cdot (\Xi_{31}^* + \Xi_{33}^*)]\};$$
$$\Omega_0^* = \frac{\Xi_{33}^*}{C_{in} \cdot L_{in}} \cdot \frac{\Xi_{22}^*}{L_{out}} + \frac{1}{C_{in} \cdot L_{in}} \cdot \frac{R_L}{L_{out}} \cdot [\Xi_{21}^* \cdot \Xi_{33}^* - \Xi_{31}^* \cdot \Xi_{23}^*]$$

$$\det |A^* - \lambda \cdot I| = \sum_{i=0}^{5} \lambda^i \cdot \Omega_i^*; \det |A^* - \lambda \cdot I| = 0 \Rightarrow \sum_{i=0}^{5} \lambda^i \cdot \Omega_i^* = 0 \Rightarrow \lambda_1, \lambda_2, \ldots$$

We need to classify our system stability fixed points according to eigenvalues:

Table 3.1 Bipolar transistor at microwave frequencies system stability fixed points and eigenvalues

	System Eigen values $(\sum_{i=0}^{5} \lambda^i \cdot \Omega_i^* = 0 \Rightarrow \lambda_1, \lambda_2, \ldots, \lambda_n)$ Number of eigenvalues is n	System fixed point classification
1	$\lambda_k > 0$ and real $\forall k \in [1, \ldots, n]$ n, k are integers	Unstable node
2	$\lambda_k < 0$ and real $\forall k \in [1, \ldots, n]$ n, k are integers	Stable node
3	At least one Eigen value is negative real number ($\lambda_l < 0$) and all other Eigenvalues are positive real number $\lambda_k > 0$ $\forall k \in [1, \ldots, n]$; n, k, l are integers, $\lambda_l < 0$; $0 \le l \le n; n, l \in [0, \ldots, n]$	Saddle point
4	$\lambda_k < 0$ and real $\forall k \in [1, \ldots, n]$ n, k are integers except λ_l, λ_m $\lambda_l = \gamma_1 + j \cdot \gamma_2; \lambda_m = \gamma_1 - j \cdot \gamma_2$ $0 \le l \le n; 0 \le m \le n; l, m \in [1, \ldots, n]$ n, m are integer numbers $\gamma_1 < 0, \gamma_2 > 0$ and real number $l \ne m \& \gamma_1 = \mathrm{Re}(\lambda_{m,n}) < 0$	Stable spiral (decay oscillation spiral). If at list one of our Eigenvalues is positive then we have Saddle point spiral
5	$\lambda_k < 0$ and real $\forall k \in [1, \ldots, n]$ n, k are integers except λ_l, λ_m $\lambda_l = \gamma_1 + j \cdot \gamma_2; \lambda_m = \gamma_1 - j \cdot \gamma_2$ $0 \le l \le n;\ 0 \le m \le n;\ l, m \in [1, \ldots, n]$	Unstable spiral (growing oscillation spiral)

(continued)

Table 3.1 (continued)

	System Eigen values $(\sum_{i=0}^{5} \lambda^i \cdot \Omega_i^* = 0 \Rightarrow \lambda_1, \lambda_2, \ldots, \lambda_n)$ Number of eigenvalues is n	System fixed point classification
	n, m are integer numbers $\gamma_1 > 0, \gamma_2 > 0$ and real number $l \neq m \,\&\, \gamma_1 = \mathrm{Re}(\lambda_{m,n}) > 0$	
6	$\lambda_l = \gamma_1 + j \cdot \gamma_2$; $\lambda_m = \gamma_1 - j \cdot \gamma_2$ $0 \leq l \leq n$; $0 \leq m \leq n$; $l, m \in [1, \ldots, n]$ l, n, m are integer numbers $\gamma_1 = 0, \gamma_2 > 0$ and real number $l \neq m \,\&\, \gamma_1 = \mathrm{Re}(\lambda_{m,n}) = 0$, $\lambda_{m,n}$ are pure imaginary	Solutions are periodic with period $T = \frac{2 \cdot \pi}{\gamma_2}$

3.2 Field Effect Transistor (FETs) at Microwave Frequencies Description

There are three major types of FETs. First type is Junction FET (JFET). The second type is the metal oxide semiconductor FET (MOSFET). MOSFETs transistor is widely use as discrete devices in UHF band communications. The third type is Schottky barrier type FETs made of gallium arsenide. It is known as the Gallium Arsenide Metal semiconductor FET (GaAs MESFET) which showed performances better than bipolar transistor. GaAs MESFET provide lower noise and higher gain for solid state applications, frequency characteristics previously unavailable from bipolar transistor. It is made by using gallium arsenide. The electron mobility of gallium arsenide is five to seven times that of silicon. The GaAs FET is different from the MOSFET by the use of schottky barrier at the gate instead of an oxide layer. GaAs FET are called "Normally ON" type device, the maximum gate voltage must be zero. The design of microwave circuits includes active components GaAs MESFET's and GaAs or InP based MOD-FET's. Small Signal Model (SSM's) are used as a building block for large signal modeling nonlinear circuits like power amplifiers, mixers, oscillators, etc., There are equivalent circuit elements which required if small signal broadband behavior of microwave and millimeter wave FET's has to be modeled. Optimization is done by using nonlinear dynamic. The FET is fabricated on a semi insulating substrate, which serves as the transistor support. An epitaxial layer of N-doped semiconductor material is deposited on top of the substrate, and practically the FET is built into epitaxial layer. The FET's ports are source, gate and drain. The source is at one end of the transistor, and the drain is at other end. We connect positive voltage to the drain, and electrons are drawn from the source to the drain. The gate is between the source and drain on the surface of the epitaxial layer. Microwave FET is constructed from a metal to semiconductor junction (Schottky junction) at the gate. Another name to microwave FET is MESFET (gate is a metal to semiconductor junction). FET semiconductor material is GaAs and silicon not recommended since electrons travel twice as fast in GaAs as in silicon. Better high frequency performance is obtained in GaAs FET. Microwave FETs are made with GaAs and called GaAs FETs. When the source to gate voltage is zero, the electrons move through the entire thickness of the epitaxial layer and the FET draws the maximum current, saturated drain to source current I_{DSS}. A microwave FET operated with its gate voltage negative with respect to its source voltage. A reversed biased Schottky junction is formed around the gate. Increasing the negative voltage on the gate causes the size of the insulating barrier region increases, and reducing the current flow from the source to the drain. If the gate voltage is negative enough, the insulating region around the gate extend across the entire epitaxial layer and cut off the current flow. FET amplification is performing because a small voltage applied to the gate controls a large amount of current which flowing through the transistor. This current used to generate a large voltage in the output circuit. We interested on the properties of a GaAs MESFET. A linear amplifier circuit biases the GaAs MESFET. Considering gate bias alone, the range must be from I_{DSS}, $V_G = 0$ to $I_{DS} = 0$ at pinch off, $V_G = V_P$. FET square law characteristic is done according to the formula $I_D = I_{DSS} \cdot (1 - \frac{V_{GS}}{V_P})^2$. In this range, the voltage V_{DS} between the drain and the source has little effect on the current I_{DS} flowing through the channel. By changing the gate voltage, V_G, the drain to source current can be controlled. There is a transfer characteristic of a GaAs FET with n channels. This FET transfer characteristic is an important basic parameter in circuit design because it sets the bias conditions and operating point. The operating point line is directly related to the mutual conductance g_m. Mutual conductance is defined as the ratio of the change in direct current to the minor change in voltage between gate sources. We differentiate I_D expression with respect to V_{GS}: $g_m = \frac{\partial I_D}{\partial V_{GS}} = 2 \cdot I_{DSS} \cdot (1 - \frac{V_{GS}}{V_P}) \cdot (-\frac{1}{V_P}) = \frac{-2 \cdot I_{DSS}}{V_P} \cdot (1 - \frac{V_{GS}}{V_P})$. The most important characteristic when designing a bias circuit for small signal GaAs FETs is transfer characteristic. There are two methods to bias a GaAs FET: dual power source method and self-bias method (Auto-bias) [37, 62].

Dual power source: $V_P < V_{GS} < 0$ must always apply to a GaAs FET, and we get the expression for V_{GS}, $V_{GS} = V_P \cdot (1 - \sqrt{\frac{I_D}{I_{DSS}}})$.

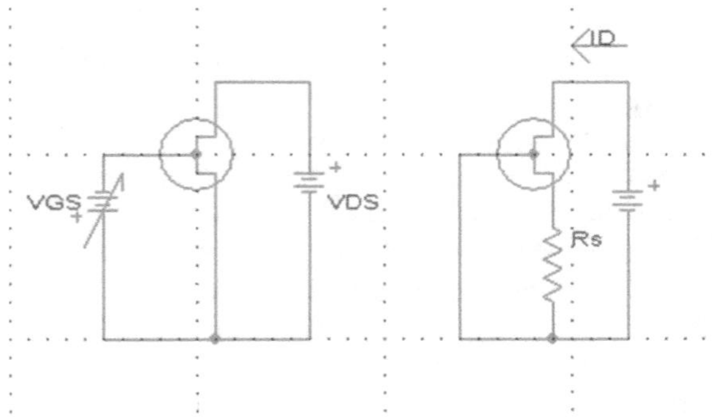

Fig. 3.3 FET dual source bias and FET self (auto) bias method

Self-bias method (Auto-Bias): This is the most universal method for reducing electrical potential between a gate and the source when there is only one power source. If the source resistance is R_S, and the operating current is I_D, then the drop in electric potential caused by R_S is $I_D \cdot R_S$ and the actual electrical potential between the gate and the source is $V_{GS} = -I_D \cdot R_S$. V_{GS} is negative and the FET can be turned on. We can get the value for R_S:

$R_S = \frac{-V_{GS}}{I_D} = -\frac{1}{I_D} \cdot V_P \cdot (1 - \sqrt{\frac{I_D}{I_{DSS}}})$. We implement a system of RF FET typical Band 2 amplifier. We have two bias voltages: $V_G < 0$, $V_D > 0$, input and output capacitors C_{in}, C_{out}, Microwave RF source X(t) and series resistance R_{in}. The dual power source bias method ($V_G < 0$, $V_D > 0$) is appropriate for use in higher frequencies. When we connect the source to the ground terminal, source inductance can be made relatively small. By using this method, higher gain can be obtained and a lower noise factor anticipated in the higher frequencies. A large DC voltage is applied between the source and the drain, and the drain is positive with respect to the source. The gate is biased at a DC negative voltage, which shows I_{DS} as a function of the gate voltage, current flows through the FET and through R_D drain resistor. In the FET none of the transistor current flows into the gate circuit, because the gate junction is reverse biased. In a bipolar transistor, a small of the emitter current flows into the base. The current flowing from the drain to the source, which is opposite to electron current flow. The gate length determines the transit time of the FET. The increased power is obtained by using multiple sources, gates, and drains. We consider Microwave RF_{in} signal $X(t) = A_0 + f_X(t)$; $|f_X(t)| \leq 1$ and $A_0 \gg |f_X(t)|$ then $X(t) = A_0 + f_X(t) \approx A_0$.

Remark: The Microwave applications implementation for GaAs FET is recommended since it has a greatest advantage in the higher frequency band. GaAs FETs are far better in term of noise, gain and output power saturation characteristics compare to silicon bipolar transistor and tunnel diodes. Most small signal FETs are in low noise amplifiers. They are used over the horizon microwave

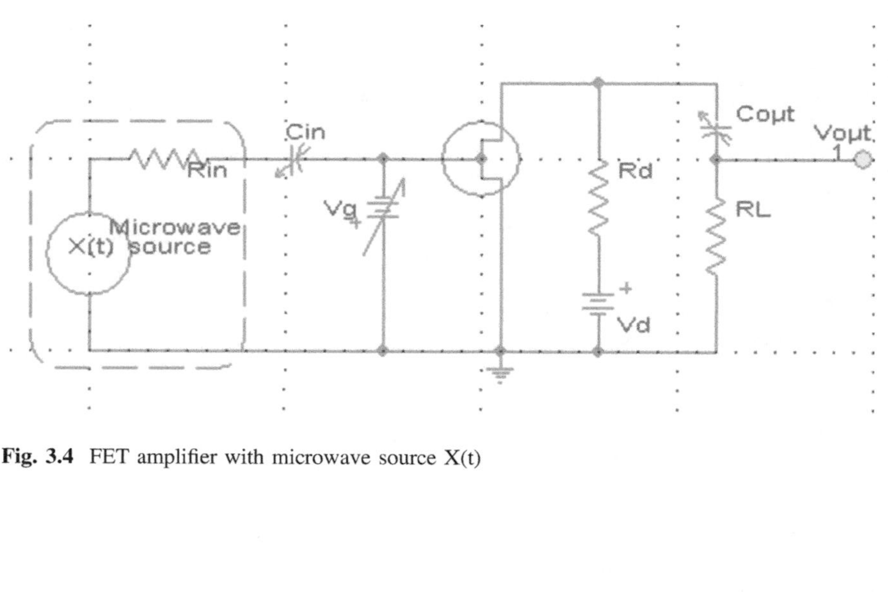

Fig. 3.4 FET amplifier with microwave source X(t)

communications, and in earth stations communicating with satellites. A low noise amplifier is designed by minimizing the noise measure (M) which can show by the expression $M = \frac{NF-1}{1-\frac{1}{G}}$. NF is the amplifier noise factor and G is the amplifier gain. The input output impedances of a GaAs FET, there is a difference in impedance between maximum gain and minimum NF (Noise Figure). This difference is particularly apparent at lower frequencies. As the frequencies go higher, the difference seems to decrease. The NF (Noise Factor) will be low when the gain is maximized at high frequencies. The Noise Factor (NF) is a function of device gain. There are two possible equivalent circuit topologies for FET small signal modeling analysis. The first equivalent circuit topology is present in the below figure. For FET pad capacitance we define C_{pgs}, C_{pds}. R_g, R_s, and R_d are FET ports resistances respectively. L_g, L_s, and L_d are FET ports inductances respectively. C_{gd} is the capacitance between the gate and the drain. C_{ds} is the capacitance between the drain and the source. C_{gs} is the capacitance between the gate and the source. R_i is the FET input resistance. The second equivalent circuit topology 19 parameters small signal equivalent circuit topology that is implemented in our analysis. The broad band modeling of millimeter wave FET's requires that the parasitic elements such as pad capacitances C_{pgs}, C_{pds}, and C_{pgd} are taken into account. Additionally high performance devices such as InP-based MODFET's often have rather leaky gates, the reverse currents of which must be modeled by the resistances R_{gs}, and R_{gd}. All other FET parameters are the same like in the first equivalent circuit.

We define circuit node in below schematic as A_1, A_2 …

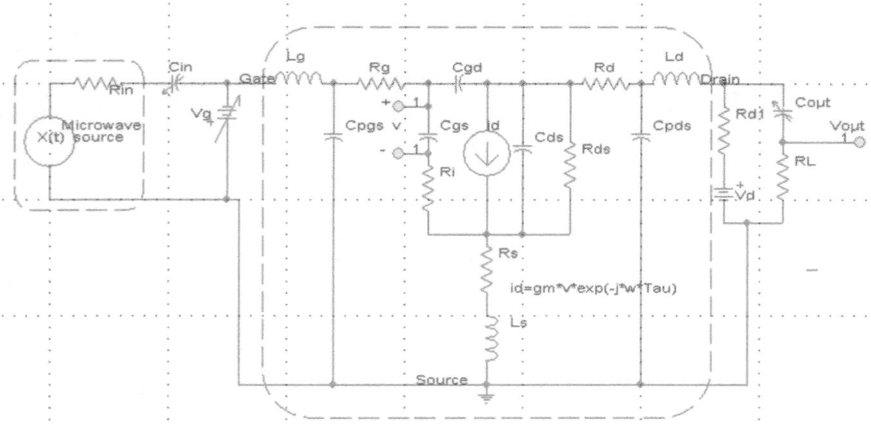

Fig. 3.5 Small signal equivalent circuit of FET

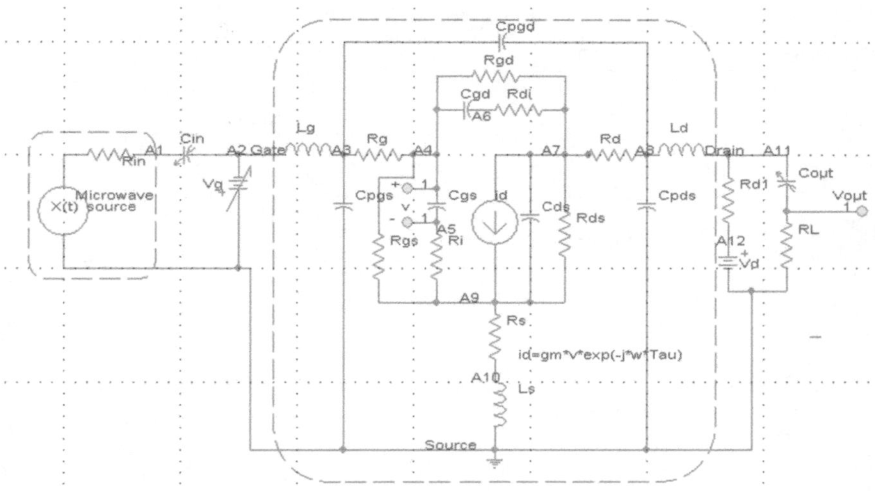

Fig. 3.6 Small signal enhance equivalent circuit for FET (version 1)

The above FET equivalent circuits are for high frequency model and operation, taking the node capacitors and other elements into account. If we switch to low frequency small signal FET model, all capacitors in the above model disconnected and all inductors are short. Capacitor impedance is $Z_c = \frac{1}{\omega \cdot C}$; $\lim_{\omega \to \varepsilon} Z_c \Rightarrow \lim_{\omega \to \varepsilon} \frac{1}{\omega \cdot C} \to \infty$ and for inductance impedance is $Z_l = \omega \cdot L$ $Z_l = \omega \cdot L \Rightarrow \lim_{\omega \to \varepsilon} \omega \cdot L \to \varepsilon$. We get the low frequency small signal FET model.

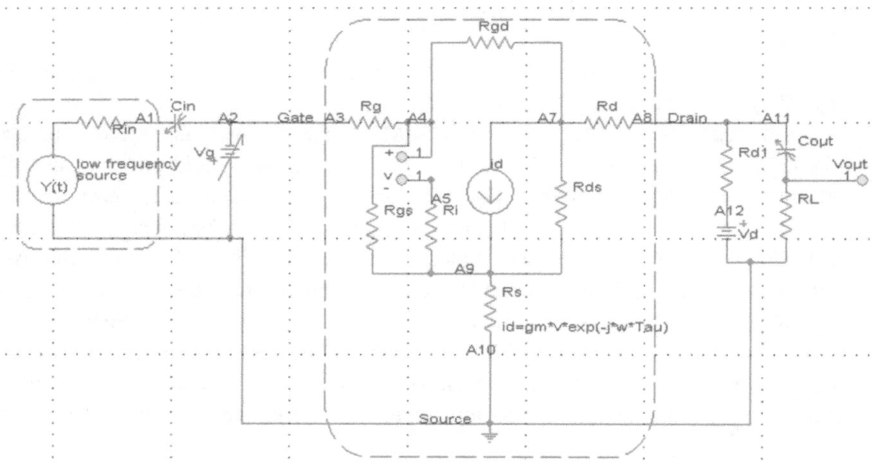

Fig. 3.7 Small signal enhance equivalent circuit for FET (version 2)

Table 3.2 Small signal FET model parameters (JFET, MOSFET)

Parameter	JFET	MOSFET
g_m	0.1–10 mA/V	0.1–20 mA/V or more
R_{ds}	0.1–1 MΩ	1–50 KΩ
C_{ds}	0.1–1 pF	0.1–1 pF
C_{gs}, C_{gd}	1–10 pF	1–10 pF
R_{gs}	$>10^8$ Ω	$>10^{10}$ Ω
R_{gd}	$>10^8$ Ω	$>10^{10}$ Ω

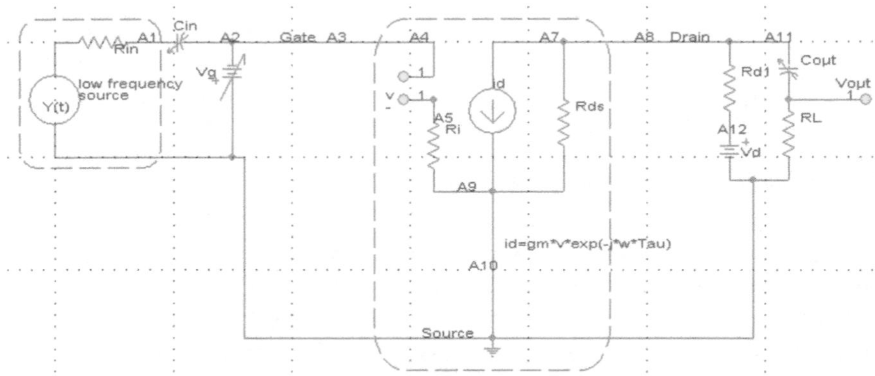

Fig. 3.8 Small signal equivalent circuit for FET (R_{gs} and R_{gd} disconnected)

More restricted low frequency small signal FET model is taking Rs, Rg, and Rd small and we neglect them. The range of parameter values for an FET is present in the below table:

R_{gs} and R_{gd} are high resistance elements which can be taken as disconnected in our low frequency small signal model.

A High-electron-mobility transistor (HEMT), also known as hetero-structure FET (HFET) or modulation-doped FET (MODFET), is a field-effect transistor incorporating a junction between two materials with different band gaps (i.e. a hetero-junction) as the channel instead of a doped region (as is generally the case for MOSFET). A commonly used material combination is GaAs with AlGaAs, though there is wide variation, dependent on the application of the device. Devices incorporating more indium generally show better high-frequency performance, while in recent years, gallium nitride HEMTs have attracted attention due to their high-power performance. HEMT transistors are able to operate at higher frequencies than ordinary transistors, up to millimeter wave frequencies, and are used in high-frequency products. The minimum, maximum and optimal values for each parameter of the model used for PHEMT SSM (Small Signal Model) Optimization ([62], Fundamental theory and applications, Vol. 43, No. 10, October 1996) is describe in the below table. Femtofarads (1 fF = 0.001 pF = 10^{-15} F). Ω—Ohm. The unit for the conductance (g) is defined by $S = \Omega^{-1} = A/V$ (A—ampere, V—voltage). Mho is an alternative name of the same unit, the reciprocal of one ohm. Mho is derived from spelling ohm backwards and is written with an upside-down capital Greek letter Omega [62].

We do our circuit analysis only for second equivalent circuit topology (full version) which is high frequency model and good for Input microwave source X(t). $i_d = g_m \cdot v \cdot e^{-j \cdot \omega \cdot Tau}$; $\lim_{Tau \to \varepsilon} i_d = g_m \cdot v$; $\lim_{Tau \to \varepsilon} e^{-j \cdot \omega \cdot Tau} = 1$. We describe our circuit nodes Kirchhoff's current law and elements equations in the below tables.

$\sum_{k=1}^{n} I_k = 0$; n is the total number of branches with currents flowing towards or away from the node.

Table 3.3 PHEMT SSM (Small Signal Model) parameters, minimum, maximum and optimal values

No.	Parameter	Minimum	Maximum	Optimal
1	L_g[nH]	0.005	0.5	0.0479
2	R_g[Ω]	0.5	50	7.97
3	L_s[nH]	0.005	0.5	0.011
4	R_s[Ω]	0.5	50	2.68
5	L_d[nH]	0.005	0.5	0.0257
6	R_d[Ω]	0.5	50	4.14
7	C_{pgs}[fF]	1	100	24.5
8	C_{pds}[fF]	1	100	18.2
9	C_{pgd}[fF]	1	100	3.01
10	R_{gs}[MΩ]	5	500	394
11	C_{gs}[fF]	10	1000	86.4
12	R_i[Ω]	0.5	50	1.55
13	R_{gd}[MΩ]	5	500	485
14	C_{gd}[fF]	1	100	18.8
15	R_{di}[Ω]	0.5	50	10.4
16	g_m[mS]	10	1000	65.8
17	τ [ps]	0.001	0.1	0.0977
18	R_{ds}[Ω]	10	1000	227
19	C_{ds}[fF]	1	100	30.8

Table 3.4 PHEMT SSM (Small Signal Model) circuit nodes and Kirchhoff's Current Law (KCL)

Circuit node	Kirchhoff's current law (KCL) - $\sum_{k=1}^{n} I_k = 0$
A_1	$I_{Rin} = I_{Cin}$
A_2	$V_{A2} = -V_g$
A_3	$I_{Lg} = I_{Cpgd} + I_{Rg} + I_{Cpgs}$
A_4	$I_{Rg} = I_{Rgs} + I_{Cgs} + I_{Rgd} + I_{Cgd}$
A_5	$I_{Cgs} = I_{Ri}$
A_6	$I_{Cgd} = I_{Rdi}$
A_7	$I_{Rgd} + I_{Rdi} = I_{Rd} + I_{Rds} + I_{Cds} + g_m \cdot v$
A_8	$I_{Rd} + I_{Cpgd} = I_{Ld} + I_{Cpds}$
A_9	$I_{Rgs} + I_{Cgs} + g_m \cdot v + I_{Cds} + I_{Rds} = I_{Rs}$
A_{10}	$I_{Rs} = I_{Ls}$
A_{11}	$I_{Ld} + I_{Rd1} = I_{Cout} = I_{R_L}$
A_{12}	$V_{A12} = V_d$

Below table describes circuit element and related equation.

Table 3.5 PHEMT SSM (Small Signal Model) circuit elements and elements related equations

Circuit element	Element related equation	Circuit element	Element related equation
R_{in}	$\frac{X(t)-V_{A1}}{R_{in}} = I_{Rin}$	R_i	$I_{Ri} = \frac{V_{A5}-V_{A9}}{R_i}$
C_{in}	$I_{Cin} = C_{in} \cdot \frac{d}{dt}(V_{A1} + V_g)$	C_{ds}	$I_{Cds} = C_{ds} \cdot \frac{d}{dt}(V_{A7} - V_{A9})$
L_g	$-V_g - V_{A3} = L_g \cdot \frac{dI_{Lg}}{dt}$	R_{ds}	$I_{Rds} = \frac{V_{A7}-V_{A9}}{R_{ds}}$
C_{pgs}	$I_{Cpgs} = C_{pgs} \cdot \frac{dV_{Cpgs}}{dt} = C_{pgs} \cdot \frac{dV_{A3}}{dt}$	R_d	$I_{Rd} = \frac{V_{A7}-V_{A8}}{R_d}$
C_{pgd}	$I_{Cpgd} = C_{pgd} \cdot \frac{d}{dt}(V_{A3} - V_{A8})$	C_{pds}	$I_{Cpds} = C_{pds} \cdot \frac{dV_{A8}}{dt}$
R_{gd}	$I_{Rgd} = \frac{V_{A4}-V_{A7}}{R_{gd}}$	R_S	$I_{Rs} = \frac{V_{A9}-V_{A10}}{R_S}$
C_{gd}	$I_{Cgd} = C_{gd} \cdot \frac{d}{dt}(V_{A4} - V_{A6})$	L_S	$V_{A10} = L_S \cdot \frac{dI_{Ls}}{dt}$
R_{di}	$I_{Rdi} = \frac{V_{A6}-V_{A7}}{R_{di}}$	L_d	$V_{A8} - V_{A11} = L_d \cdot \frac{dI_{Ld}}{dt}$
R_g	$I_{Rg} = \frac{V_{A3}-V_{A4}}{R_g}$	R_{d1}	$I_{Rd1} = \frac{V_d-V_{A11}}{R_{d1}}$
R_{gs}	$I_{Rgs} = \frac{V_{A4}-V_{A9}}{R_{gs}}$	C_{out}	$I_{Cout} = C_{out} \cdot \frac{d}{dt}(V_{A11} - V_{out})$
C_{gs}	$I_{Cgs} = C_{gs} \cdot \frac{dv}{dt} = C_{gs} \cdot \frac{d}{dt}(V_{A4} - V_{A5})$	R_L	$I_{R_L} = \frac{V_{out}}{R_L}$

We have two methods to investigate our system fixed points.

Method A:

V_g and V_d are constant circuit biasing voltages $V_g > 0$; $V_d > 0$; $\frac{dV_g}{dt} = 0$; $\frac{dV_d}{dt} = 0$

$$\frac{X(t) - V_{A1}}{R_{in}} = I_{Rin} \Rightarrow X(t) - V_{A1} = R_{in} \cdot I_{Rin} \Rightarrow V_{A1} = X(t) - R_{in} \cdot I_{Rin}$$

$$I_{Cin} = C_{in} \cdot \frac{d}{dt}(V_{A1} + V_g) \Rightarrow I_{Cin} = C_{in} \cdot \frac{dV_{A1}}{dt} \Rightarrow \frac{dV_{A1}}{dt} = \frac{1}{C_{in}} \cdot I_{Cin}$$

$$-V_g - V_{A3} = L_g \cdot \frac{dI_{Lg}}{dt} \Rightarrow \frac{dI_{Lg}}{dt} = -\frac{1}{L_g} \cdot (V_g + V_{A3});$$

$$I_{Cpgs} = C_{pgs} \cdot \frac{dV_{A3}}{dt} \Rightarrow \frac{dV_{A3}}{dt} = \frac{1}{C_{pgs}} \cdot I_{Cpgs}$$

$$I_{Cpgd} = C_{pgd} \cdot \frac{d}{dt}(V_{A3} - V_{A8}) \Rightarrow \frac{d}{dt}(V_{A3} - V_{A8}) = \frac{1}{C_{pgd}} \cdot I_{Cpgd};$$

$$I_{Rgd} = \frac{V_{A4} - V_{A7}}{R_{gd}} \Rightarrow V_{A4} - V_{A7} = I_{Rgd} \cdot R_{gd}$$

$$I_{Cgd} = C_{gd} \cdot \frac{d}{dt}(V_{A4} - V_{A6}) \Rightarrow \frac{d}{dt}(V_{A4} - V_{A6}) = \frac{1}{C_{gd}} \cdot I_{Cgd};$$

$$I_{Rdi} = \frac{V_{A6} - V_{A7}}{R_{di}} \Rightarrow V_{A6} - V_{A7} = I_{Rdi} \cdot R_{di}$$

$$I_{Rg} = \frac{V_{A3} - V_{A4}}{R_g} \Rightarrow V_{A3} - V_{A4} = I_{Rg} \cdot R_g; I_{Rgs} = \frac{V_{A4} - V_{A9}}{R_{gs}} \Rightarrow V_{A4} - V_{A9} = I_{Rgs} \cdot R_{gs};$$

$$I_{Cgs} = C_{gs} \cdot \frac{dv}{dt} \Rightarrow \frac{dv}{dt} = \frac{1}{C_{gs}} \cdot I_{Cgs}$$

$$I_{Ri} = \frac{V_{A5} - V_{A9}}{R_i} \Rightarrow V_{A5} - V_{A9} = I_{Ri} \cdot R_i;$$

$$I_{Cds} = C_{ds} \cdot \frac{d}{dt}(V_{A7} - V_{A9}) \Rightarrow \frac{d}{dt}(V_{A7} - V_{A9}) = \frac{1}{C_{ds}} \cdot I_{Cds}$$

$$I_{Rds} = \frac{V_{A7} - V_{A9}}{R_{ds}} \Rightarrow V_{A7} - V_{A9} = I_{Rds} \cdot R_{ds}; I_{Rd} = \frac{V_{A7} - V_{A8}}{R_d} \Rightarrow V_{A7} - V_{A8} = I_{Rd} \cdot R_d$$

$$I_{Cpds} = C_{pds} \cdot \frac{dV_{A8}}{dt} \Rightarrow \frac{dV_{A8}}{dt} = \frac{1}{C_{pds}} \cdot I_{Cpds}; I_{Rs} = \frac{V_{A9} - V_{A10}}{R_S} \Rightarrow V_{A9} - V_{A10} = I_{Rs} \cdot R_S$$

$$V_{A10} = L_S \cdot \frac{dI_{Ls}}{dt} \Rightarrow \frac{dI_{Ls}}{dt} = \frac{1}{L_S} \cdot V_{A10}; V_{A8} - V_{A11} = L_d \cdot \frac{dI_{Ld}}{dt} \Rightarrow \frac{dI_{Ld}}{dt} = \frac{1}{L_d} \cdot (V_{A8} - V_{A11})$$

$$I_{Rd1} = \frac{V_d - V_{A11}}{R_{d1}} \Rightarrow V_d - V_{A11} = I_{Rd1} \cdot R_{d1} \Rightarrow V_{A11} = V_d - I_{Rd1} \cdot R_{d1}$$

$$I_{Cout} = C_{out} \cdot \frac{d}{dt}(V_{A11} - V_{out}) \Rightarrow \frac{d}{dt}(V_{A11} - V_{out}) = \frac{1}{C_{out}} \cdot I_{Cout};$$

$$I_{R_L} = \frac{V_{out}}{R_L} \Rightarrow V_{out} = I_{R_L} \cdot R_L$$

We do some mathematical manipulations:

$$\frac{dV_{A1}}{dt}=\frac{1}{C_{in}}\cdot I_{Cin}\Rightarrow\frac{d[X(t)-R_{in}\cdot I_{Rin}]}{dt}=\frac{1}{C_{in}}\cdot I_{Cin}\Rightarrow\frac{dX(t)}{dt}-R_{in}\cdot\frac{dI_{Rin}}{dt}=\frac{1}{C_{in}}\cdot I_{Cin}$$

$$X(t)=A_0+f_X(t)\approx A_0\Rightarrow\frac{dX(t)}{dt}\to\varepsilon;\frac{dA_0}{dt}=0;\frac{dI_{Rin}}{dt}=-\frac{1}{C_{in}\cdot R_{in}}\cdot I_{Cin};I_{Rin}=I_{Cin}$$

$$\frac{dI_{Rin}}{dt}=-\frac{1}{C_{in}\cdot R_{in}}\cdot I_{Rin};\frac{dI_{Lg}}{dt}=-\frac{1}{L_g}\cdot(V_g+V_{A3});\frac{dV_{A3}}{dt}=\frac{1}{C_{pgs}}\cdot I_{Cpgs}$$

$$\frac{d}{dt}(V_{A3}-V_{A8})=\frac{1}{C_{pgd}}\cdot I_{Cpgd}\Rightarrow\frac{dV_{A3}}{dt}-\frac{dV_{A8}}{dt}=\frac{1}{C_{pgd}}\cdot I_{Cpgd}$$

$$\Rightarrow\frac{dV_{A8}}{dt}=\frac{1}{C_{pgs}}\cdot I_{Cpgs}-\frac{1}{C_{pgd}}\cdot I_{Cpgd}$$

$$V_{A4}-V_{A7}=I_{Rgd}\cdot R_{gd}\ \&\ V_{A6}-V_{A7}=I_{Rdi}\cdot R_{di}\Rightarrow V_{A4}-V_{A6}=I_{Rgd}\cdot R_{gd}-I_{Rdi}\cdot R_{di}$$

$$\frac{d}{dt}(V_{A4}-V_{A6})=\frac{1}{C_{gd}}\cdot I_{Cgd}\Rightarrow\frac{d}{dt}(I_{Rgd}\cdot R_{gd}-I_{Rdi}\cdot R_{di})=\frac{1}{C_{gd}}\cdot I_{Cgd}$$

$$\frac{d}{dt}(I_{Rgd}\cdot R_{gd}-I_{Rdi}\cdot R_{di})=\frac{1}{C_{gd}}\cdot I_{Cgd}\Rightarrow R_{gd}\cdot\frac{dI_{Rgd}}{dt}-R_{di}\cdot\frac{dI_{Rdi}}{dt}=\frac{1}{C_{gd}}\cdot I_{Cgd}$$

$$V_{A3}-V_{A4}=I_{Rg}\cdot R_g\ \&\ V_{A4}-V_{A9}=I_{Rgs}\cdot R_{gs}\Rightarrow V_{A3}-V_{A9}=I_{Rg}\cdot R_g+I_{Rgs}\cdot R_{gs}$$

$$\frac{dv}{dt}=\frac{1}{C_{gs}}\cdot I_{Cgs};V_{A5}-V_{A9}=I_{Ri}\cdot R_i;V_{A7}-V_{A9}=I_{Rds}\cdot R_{ds}\ \&\ \frac{d}{dt}(V_{A7}-V_{A9})$$

$$=\frac{1}{C_{ds}}\cdot I_{Cds}\Rightarrow\frac{dI_{Rds}}{dt}=\frac{1}{C_{ds}\cdot R_{ds}}\cdot I_{Cds}$$

$$V_{A7}-V_{A8}=I_{Rd}\cdot R_d;V_{A9}-V_{A10}=I_{Rs}\cdot R_S;\frac{dV_{A8}}{dt}=\frac{1}{C_{pds}}\cdot I_{Cpds};\frac{dI_{Ls}}{dt}=\frac{1}{L_S}\cdot V_{A10}$$

$$\frac{d}{dt}(V_{A3}-V_{A8})=\frac{1}{C_{pgd}}\cdot I_{Cpgd}\Rightarrow\frac{dV_{A3}}{dt}-\frac{dV_{A8}}{dt}=\frac{1}{C_{pgd}}$$

$$\cdot I_{Cpgd}\Rightarrow\frac{dV_{A8}}{dt}=\frac{1}{C_{pgs}}\cdot I_{Cpgs}-\frac{1}{C_{pgd}}\cdot I_{Cpgd}$$

$$V_{A4}-V_{A7}=I_{Rgd}\cdot R_{gd}\ \&\ V_{A6}-V_{A7}=I_{Rdi}\cdot R_{di}\Rightarrow V_{A4}-V_{A6}=I_{Rgd}\cdot R_{gd}-I_{Rdi}\cdot R_{di}$$

$$\frac{dI_{Ld}}{dt}=\frac{1}{L_d}\cdot(V_{A8}-V_{A11})\ \&\ V_{A11}=V_d-I_{Rd1}$$

$$\cdot R_{d1}\Rightarrow\frac{dI_{Ld}}{dt}=\frac{1}{L_d}\cdot(V_{A8}-V_d+I_{Rd1}\cdot R_{d1})$$

$$\frac{d}{dt}(V_{A11}-V_{out})=\frac{1}{C_{out}}\cdot I_{Cout}\ \&\ V_{out}=I_{R_L}\cdot R_L\Rightarrow\frac{d}{dt}(V_{A11}-I_{R_L}\cdot R_L)=\frac{1}{C_{out}}\cdot I_{Cout}$$

$$\frac{d}{dt}(V_{A11}-I_{R_L}\cdot R_L)=\frac{1}{C_{out}}\cdot I_{Cout}\Rightarrow\frac{d}{dt}(V_d-I_{Rd1}\cdot R_{d1}-I_{R_L}\cdot R_L)=\frac{1}{C_{out}}\cdot I_{Cout}$$

$$\frac{d}{dt}(V_d - I_{Rd1} \cdot R_{d1} - I_{R_L} \cdot R_L) = \frac{1}{C_{out}} \cdot I_{Cout};$$

$$I_{Cout} = I_{R_L} \Rightarrow \frac{dV_d}{dt} - \frac{dI_{Rd1}}{dt} \cdot R_{d1} - \frac{dI_{R_L}}{dt} \cdot R_L = \frac{1}{C_{out}} \cdot I_{R_L}$$

$$\frac{dV_d}{dt} = 0 \Rightarrow \frac{dI_{Rd1}}{dt} \cdot R_{d1} + \frac{dI_{R_L}}{dt} \cdot R_L = -\frac{1}{C_{out}}$$

$$\cdot I_{R_L}; I_{Ld} + I_{Rd1} = I_{R_L} \Rightarrow I_{Rd1} = I_{R_L} - I_{Ld}$$

$$\frac{d(I_{R_L} - I_{Ld})}{dt} \cdot R_{d1} + \frac{dI_{R_L}}{dt} \cdot R_L = -\frac{1}{C_{out}} \cdot I_{R_L}$$

$$\Rightarrow \frac{dI_{R_L}}{dt} \cdot R_{d1} - \frac{dI_{Ld}}{dt} \cdot R_{d1} + \frac{dI_{R_L}}{dt} \cdot R_L = -\frac{1}{C_{out}} \cdot I_{R_L}$$

$$\frac{dI_{R_L}}{dt} \cdot R_{d1} - \frac{dI_{Ld}}{dt} \cdot R_{d1} + \frac{dI_{R_L}}{dt} \cdot R_L = -\frac{1}{C_{out}} \cdot I_{R_L}$$

$$\Rightarrow \frac{dI_{R_L}}{dt} \cdot (R_{d1} + R_L) - \frac{dI_{Ld}}{dt} \cdot R_{d1} = -\frac{1}{C_{out}} \cdot I_{R_L}$$

$$V_{A7} - V_{A9} = I_{Rds} \cdot R_{ds} \Rightarrow V_{A9} = V_{A7} - I_{Rds} \cdot R_{ds};$$

$$V_{A9} - V_{A10} = I_{Rs} \cdot R_S \Rightarrow V_{A10} = V_{A9} - I_{Rs} \cdot R_S$$

$$V_{A10} = V_{A7} - I_{Rds} \cdot R_{ds} - I_{Rs} \cdot R_S; V_{A7} = I_{Rd} \cdot R_d + V_{A8};$$

$$V_{A10} = I_{Rd} \cdot R_d + V_{A8} - I_{Rds} \cdot R_{ds} - I_{Rs} \cdot R_S$$

$$\frac{dI_{Ls}}{dt} = \frac{1}{L_S} \cdot V_{A10} = \frac{1}{L_S} \cdot (I_{Rd} \cdot R_d + V_{A8} - I_{Rds} \cdot R_{ds} - I_{Rs} \cdot R_S)$$

$$\frac{dI_{Ls}}{dt} = \frac{1}{L_S} \cdot (I_{Rd} \cdot R_d + V_{A8} - I_{Rds} \cdot R_{ds} - I_{Ls} \cdot R_S)$$

$$\frac{dI_{R_L}}{dt} \cdot (R_{d1} + R_L) - \frac{dI_{Ld}}{dt} \cdot R_{d1} = -\frac{1}{C_{out}} \cdot I_{R_L}$$

$$\Rightarrow \frac{dI_{R_L}}{dt} = \frac{1}{(R_{d1} + R_L)} \cdot [\frac{dI_{Ld}}{dt} \cdot R_{d1} - \frac{1}{C_{out}} \cdot I_{R_L}]$$

$$\frac{dI_{R_L}}{dt} = \frac{1}{(R_{d1} + R_L)} \cdot [\frac{1}{L_d} \cdot (V_{A8} - V_d + I_{Rd1} \cdot R_{d1}) \cdot R_{d1} - \frac{1}{C_{out}} \cdot I_{R_L}]$$

The condition to find our system fixed points is no variation with time for our variables. $\frac{d''Variable''}{dt} = 0$.

Table 3.6 Small signal equivalent circuit of FET system differential equations, fixed points and outcome

System differential equation	@ Fixed points	Outcome
$\frac{dI_{Rin}}{dt} = -\frac{1}{C_{in} \cdot R_{in}} \cdot I_{Rin}$	$\frac{dI_{Rin}}{dt} = 0$	$I^*_{Rin} = 0$
$\frac{dI_{Lg}}{dt} = -\frac{1}{L_g} \cdot (V_g + V_{A3})$	$\frac{dI_{Lg}}{dt} = 0$	$V^*_{A3} = -V_g$
$\frac{dV_{A3}}{dt} = \frac{1}{C_{pgs}} \cdot I_{Cpgs}$	$\frac{dV_{A3}}{dt} = 0$	$I^*_{Cpgs} = 0$
$\frac{dV_{A8}}{dt} = \frac{1}{C_{pgs}} \cdot I_{Cpgs} - \frac{1}{C_{pgd}} \cdot I_{Cpgd}$	$\frac{dV_{A8}}{dt} = 0$	$\frac{I^*_{Cpgd}}{I^*_{Cpgs}} = \frac{C_{pgd}}{C_{pgs}}$

(continued)

Table 3.6 (continued)

System differential equation	@ Fixed points	Outcome
$R_{gd}\cdot\frac{dI_{Rgd}}{dt}-R_{di}\cdot\frac{dI_{Rdi}}{dt}=\frac{1}{C_{gd}}\cdot I_{Cgd}$	$\frac{dI_{Rgd}}{dt}=0;\frac{dI_{Rdi}}{dt}=0$	$I^*_{Cgd}=0$
$\frac{dv}{dt}=\frac{1}{C_{gs}}\cdot I_{Cgs}$	$\frac{dv}{dt}=0$	$I^*_{Cgs}=0$
$\frac{dI_{Rds}}{dt}=\frac{1}{C_{ds}\cdot R_{ds}}\cdot I_{Cds}$	$\frac{dI_{Rds}}{dt}=0$	$I^*_{Cds}=0$
$\frac{dV_{A8}}{dt}=\frac{1}{C_{pds}}\cdot I_{Cpds}$	$\frac{dV_{A8}}{dt}=0$	$I^*_{Cpds}=0$
$\frac{dI_{Ls}}{dt}=\frac{1}{L_S}\cdot V_{A10}$	$\frac{dI_{Ls}}{dt}=0$	$V^*_{A10}=0$
$\frac{dI_{Ld}}{dt}=\frac{1}{L_d}\cdot(V_{A8}-V_d+I_{Rd1}\cdot R_{d1})$	$\frac{dI_{Ld}}{dt}=0$	$V^*_{A8}+I^*_{Rd1}\cdot R_{d1}=V_d$
$\frac{dI_{Rd1}}{dt}\cdot R_{d1}+\frac{dI_{R_L}}{dt}\cdot R_L=-\frac{1}{C_{out}}\cdot I_{R_L}$	$\frac{dI_{Rd1}}{dt}=0;\frac{dI_{R_L}}{dt}=0$	$I^*_{R_L}=0$
$\frac{dI_{Ls}}{dt}=\frac{1}{L_S}\cdot(I_{Rd}\cdot R_d+V_{A8}-I_{Rds}\cdot R_{ds}$ $-I_{Ls}\cdot R_S)$	$\frac{dI_{Ls}}{dt}=0$	$I^*_{Rd}\cdot R_d+V^*_{A8}-I^*_{Rds}\cdot R_{ds}$ $-I^*_{Ls}\cdot R_S=0$
$\frac{dI_{R_L}}{dt}=\frac{1}{(R_{d1}+R_L)}\cdot[\frac{1}{L_d}\cdot(V_{A8}-V_d$ $+I_{Rd1}\cdot R_{d1})\cdot R_{d1}-\frac{1}{C_{out}}\cdot I_{R_L}]$	$\frac{dI_{R_L}}{dt}=0$	$\frac{R_{d1}}{L_d}\cdot(V^*_{A8}-V_d+I^*_{Rd1}\cdot R_{d1})=0$

Method B:

V_g and V_d are constant circuit biasing voltages $V_g>0; V_d>0; \frac{dV_g}{dt}=0; \frac{dV_d}{dt}=0$

(1) KCL @ circuit node A_1: $\frac{X(t)-V_{A1}}{R_{in}}=C_{in}\cdot\frac{d}{dt}(V_{A1}+V_g)$.

(2) KCL @ circuit node A_3: $I_{Lg}=C_{pgd}\cdot\frac{d}{dt}(V_{A3}-V_{A8})+\frac{V_{A3}-V_{A4}}{R_g}+C_{pgs}\cdot\frac{dV_{A3}}{dt}$.

(3) KCL @ circuit node A_4: $\frac{V_{A3}-V_{A4}}{R_g}=\frac{V_{A4}-V_{A9}}{R_{gs}}+C_{gs}\cdot\frac{dv}{dt}+\frac{V_{A4}-V_{A7}}{R_{gd}}+C_{gd}\cdot\frac{d}{dt}(V_{A4}-V_{A6})$.

(4) KCL @ circuit node A_5: $C_{gs}\cdot\frac{dv}{dt}=\frac{V_{A5}-V_{A9}}{R_i}$.

(5) KCL @ circuit node A_6: $C_{gd}\cdot\frac{d}{dt}(V_{A4}-V_{A6})=\frac{V_{A6}-V_{A7}}{R_{di}}$

(6) KCL @ circuit node A_7:

$$\frac{V_{A4}-V_{A7}}{R_{gd}}+\frac{V_{A6}-V_{A7}}{R_{di}}=\frac{V_{A7}-V_{A8}}{R_d}+\frac{V_{A7}-V_{A9}}{R_{ds}}+C_{ds}\cdot\frac{d}{dt}(V_{A7}-V_{A9})+g_m\cdot v$$

(7) KCL @ circuit node A_8: $\frac{V_{A7}-V_{A8}}{R_d}+C_{pgd}\cdot\frac{d}{dt}(V_{A3}-V_{A8})=I_{Ld}+C_{pds}\cdot\frac{dV_{A8}}{dt}$

(8) KCL @ circuit node A_9:

$$\frac{V_{A4}-V_{A9}}{R_{gs}}+C_{gs}\cdot\frac{dv}{dt}+g_m\cdot v+C_{ds}\cdot\frac{d}{dt}(V_{A7}-V_{A9})+\frac{V_{A7}-V_{A9}}{R_{ds}}=\frac{V_{A9}-V_{A10}}{R_s}$$

(9) KCL @ circuit node A_{10}: $\frac{V_{A9}-V_{A10}}{R_s} = I_{Ls}$.

(10) KCL @ circuit node A_{11}:

$$I_{Ld} + \frac{V_d - V_{A11}}{R_{d1}} = C_{out} \cdot \frac{d}{dt}(V_{A11} - V_{out}) = \frac{V_{out}}{R_L}$$

$$-V_g - V_{A3} = L_g \cdot \frac{dI_{Lg}}{dt}\ ;\ V_{A10} = L_S \cdot \frac{dI_{Ls}}{dt}\ ;\ V_{A8} - V_{A11} = L_d \cdot \frac{dI_{Ld}}{dt}$$

$$(2)\ \begin{matrix} I_{Lg} = C_{pgd} \cdot \frac{d}{dt}(V_{A3} - V_{A8}) + \frac{V_{A3}-V_{A4}}{R_g} + C_{pgs} \cdot \frac{dV_{A3}}{dt} \Rightarrow \\ I_{Lg} = [C_{pgs} + C_{pgd}] \cdot \frac{dV_{A3}}{dt} - C_{pgd} \cdot \frac{dV_{A8}}{dt} + \frac{V_{A3}}{R_g} - \frac{V_{A4}}{R_g} \end{matrix}$$

$$C_{pgd} \cdot \frac{dV_{A8}}{dt} = [C_{pgs} + C_{pgd}] \cdot \frac{dV_{A3}}{dt} - I_{Lg} + \frac{V_{A3}}{R_g} - \frac{V_{A4}}{R_g} \Rightarrow$$

$$\frac{dV_{A8}}{dt} = [\frac{C_{pgs}}{C_{pgd}} + 1] \cdot \frac{dV_{A3}}{dt} - \frac{1}{C_{pgd}} \cdot I_{Lg} + \frac{V_{A3}}{C_{pgd} \cdot R_g} - \frac{V_{A4}}{C_{pgd} \cdot R_g}$$

$$(7)\ \frac{V_{A7} - V_{A8}}{R_d} + C_{pgd} \cdot \frac{d}{dt}(V_{A3} - V_{A8}) = I_{Ld} + C_{pds} \cdot \frac{dV_{A8}}{dt}$$

$$\frac{V_{A7} - V_{A8}}{R_d} + C_{pgd} \cdot \frac{dV_{A3}}{dt} - I_{Ld} = C_{pds} \cdot \frac{dV_{A8}}{dt} + C_{pgd} \cdot \frac{dV_{A8}}{dt}$$

$$\frac{dV_{A8}}{dt} = \frac{V_{A7} - V_{A8}}{R_d \cdot [C_{pds} + C_{pgd}]} + \frac{C_{pgd}}{[C_{pds} + C_{pgd}]} \cdot \frac{dV_{A3}}{dt} - \frac{1}{[C_{pds} + C_{pgd}]} \cdot I_{Ld}$$

$$(2) = (7) \rightarrow \begin{matrix} [\frac{C_{pgs}}{C_{pgd}} + 1] \cdot \frac{dV_{A3}}{dt} - \frac{1}{C_{pgd}} \cdot I_{Lg} + \frac{V_{A3}}{C_{pgd} \cdot R_g} - \frac{V_{A4}}{C_{pgd} \cdot R_g} \\ = \frac{V_{A7}-V_{A8}}{R_d \cdot [C_{pds} + C_{pgd}]} + \frac{C_{pgd}}{[C_{pds} + C_{pgd}]} \cdot \frac{dV_{A3}}{dt} - \frac{1}{[C_{pds} + C_{pgd}]} \cdot I_{Ld} \end{matrix}$$

$$\{[\frac{C_{pgs}}{C_{pgd}} + 1] - \frac{C_{pgd}}{[C_{pds} + C_{pgd}]}\} \cdot \frac{dV_{A3}}{dt}$$
$$= \frac{V_{A7} - V_{A8}}{R_d \cdot [C_{pds} + C_{pgd}]} - \frac{1}{[C_{pds} + C_{pgd}]} \cdot I_{Ld} + \frac{1}{C_{pgd}} \cdot I_{Lg} - \frac{V_{A3}}{C_{pgd} \cdot R_g} + \frac{V_{A4}}{C_{pgd} \cdot R_g}$$

We define for simplicity: $C_{T1} = \frac{C_{pgs}}{C_{pgd}} - \frac{C_{pgd}}{[C_{pds} + C_{pgd}]} + 1$

$$C_{T1} = \frac{C_{pgs}}{C_{pgd}} - \frac{C_{pgd}}{[C_{pds} + C_{pgd}]} + 1 = \frac{C_{pgs} \cdot [C_{pds} + C_{pgd}] + [C_{pds} + C_{pgd}] \cdot C_{pgd} - [C_{pgd}]^2}{[C_{pds} + C_{pgd}] \cdot C_{pgd}}$$

$$C_{T1} = \frac{C_{pgs} \cdot C_{pds} + C_{pgs} \cdot C_{pgd} + C_{pds} \cdot C_{pgd}}{[C_{pds} + C_{pgd}] \cdot C_{pgd}}$$

$$\{\frac{C_{pgs} \cdot C_{pds} + C_{pgs} \cdot C_{pgd} + C_{pds} \cdot C_{pgd}}{[C_{pds} + C_{pgd}] \cdot C_{pgd}}\} \cdot \frac{dV_{A3}}{dt}$$
$$= \frac{V_{A7} - V_{A8}}{R_d \cdot [C_{pds} + C_{pgd}]} - \frac{1}{[C_{pds} + C_{pgd}]} \cdot I_{Ld} + \frac{1}{C_{pgd}} \cdot I_{Lg} - \frac{V_{A3}}{C_{pgd} \cdot R_g} + \frac{V_{A4}}{C_{pgd} \cdot R_g}$$

$$\frac{dV_{A3}}{dt} = \frac{V_{A7} - V_{A8}}{R_d \cdot [C_{pds} + C_{pgd}] \cdot C_{T1}} - \frac{1}{[C_{pds} + C_{pgd}] \cdot C_{T1}} \cdot I_{Ld} + \frac{1}{C_{pgd} \cdot C_{T1}} \cdot I_{Lg}$$
$$- \frac{V_{A3}}{C_{pgd} \cdot R_g \cdot C_{T1}} + \frac{V_{A4}}{C_{pgd} \cdot R_g \cdot C_{T1}}$$

(1)
$$\frac{X(t) - V_{A1}}{R_{in}} = C_{in} \cdot \frac{d}{dt}(V_{A1} + V_g) \Rightarrow \frac{X(t) - V_{A1}}{R_{in}}$$
$$= C_{in} \cdot \frac{dV_{A1}}{dt} - C_{in} \cdot \frac{dV_g}{dt}; \frac{dV_g}{dt} = 0$$
$$\frac{dV_{A1}}{dt} = \frac{X(t) - V_{A1}}{R_{in} \cdot C_{in}}.$$

(3)
$$\frac{V_{A3} - V_{A4}}{R_g} = \frac{V_{A4} - V_{A9}}{R_{gs}} + C_{gs} \cdot \frac{dv}{dt} + \frac{V_{A4} - V_{A7}}{R_{gd}}$$
$$+ C_{gd} \cdot \frac{d}{dt}(V_{A4} - V_{A6}); C_{gs} \cdot \frac{dv}{dt} = \frac{V_{A5} - V_{A9}}{R_i}$$
$$\frac{V_{A3} - V_{A4}}{R_g} = \frac{V_{A4} - V_{A9}}{R_{gs}} + \frac{V_{A5} - V_{A9}}{R_i} + \frac{V_{A4} - V_{A7}}{R_{gd}} + C_{gd} \cdot \frac{d}{dt}(V_{A4} - V_{A6})$$

$$C_{gd} \cdot \frac{d}{dt}(V_{A4} - V_{A6}) = \frac{V_{A3} - V_{A4}}{R_g} - \frac{(V_{A4} - V_{A9})}{R_{gs}} - \frac{(V_{A5} - V_{A9})}{R_i} - \frac{(V_{A4} - V_{A7})}{R_{gd}}$$
$$\frac{dV_{A4}}{dt} = \frac{dV_{A6}}{dt} + \frac{V_{A3} - V_{A4}}{C_{gd} \cdot R_g} - \frac{(V_{A4} - V_{A9})}{C_{gd} \cdot R_{gs}} - \frac{(V_{A5} - V_{A9})}{C_{gd} \cdot R_i} - \frac{(V_{A4} - V_{A7})}{C_{gd} \cdot R_{gd}}$$

(5)
$$C_{gd} \cdot \frac{d}{dt}(V_{A4} - V_{A6}) = \frac{V_{A6} - V_{A7}}{R_{di}} \Rightarrow \frac{dV_{A4}}{dt} = \frac{dV_{A6}}{dt} + \frac{V_{A6} - V_{A7}}{C_{gd} \cdot R_{di}}$$

(3) = (5) →

$$\frac{V_{A3} - V_{A4}}{C_{gd} \cdot R_g} - \frac{(V_{A4} - V_{A9})}{C_{gd} \cdot R_{gs}} - \frac{(V_{A5} - V_{A9})}{C_{gd} \cdot R_i} - \frac{(V_{A4} - V_{A7})}{C_{gd} \cdot R_{gd}} = \frac{V_{A6} - V_{A7}}{C_{gd} \cdot R_{di}}$$

(4)
$$C_{gs} \cdot \frac{dv}{dt} = \frac{V_{A5} - V_{A9}}{R_i} \Rightarrow \frac{dv}{dt} = \frac{V_{A5} - V_{A9}}{C_{gs} \cdot R_i}$$

$$(6)\quad \frac{V_{A4}-V_{A7}}{R_{gd}}+\frac{V_{A6}-V_{A7}}{R_{di}}=\frac{V_{A7}-V_{A8}}{R_d}+\frac{V_{A7}-V_{A9}}{R_{ds}}+C_{ds}\cdot\frac{d}{dt}(V_{A7}-V_{A9})+g_m\cdot v$$

$$C_{ds}\cdot\frac{d}{dt}(V_{A7}-V_{A9})=\frac{V_{A4}-V_{A7}}{R_{gd}}+\frac{V_{A6}-V_{A7}}{R_{di}}-\frac{(V_{A7}-V_{A8})}{R_d}-\frac{(V_{A7}-V_{A9})}{R_{ds}}-g_m\cdot v$$

$$(8)\quad \frac{V_{A4}-V_{A9}}{R_{gs}}+C_{gs}\cdot\frac{dv}{dt}+g_m\cdot v+C_{ds}\cdot\frac{d}{dt}(V_{A7}-V_{A9})+\frac{V_{A7}-V_{A9}}{R_{ds}}=\frac{V_{A9}-V_{A10}}{R_s}$$

$$C_{gs}\cdot\frac{dv}{dt}=\frac{V_{A5}-V_{A9}}{R_i};\frac{V_{A4}-V_{A9}}{R_{gs}}+\frac{V_{A5}-V_{A9}}{R_i}+g_m\cdot v+C_{ds}\cdot\frac{d}{dt}(V_{A7}-V_{A9})+\frac{V_{A7}-V_{A9}}{R_{ds}}=\frac{V_{A9}-V_{A10}}{R_s}$$

$$C_{ds}\cdot\frac{d}{dt}(V_{A7}-V_{A9})=\frac{V_{A9}-V_{A10}}{R_s}-\frac{(V_{A4}-V_{A9})}{R_{gs}}-\frac{(V_{A5}-V_{A9})}{R_i}-\frac{(V_{A7}-V_{A9})}{R_{ds}}-g_m\cdot v$$

(6) = (8) →

$$\frac{V_{A4}-V_{A7}}{R_{gd}}+\frac{V_{A6}-V_{A7}}{R_{di}}-\frac{(V_{A7}-V_{A8})}{R_d}=\frac{V_{A9}-V_{A10}}{R_s}-\frac{(V_{A4}-V_{A9})}{R_{gs}}-\frac{(V_{A5}-V_{A9})}{R_i}$$

We can summery our last results (circuit node voltages equations) in the below table:

Table 3.7 Small signal equivalent circuit of FET nodes KCL and circuit nodes voltages equations

Nodes KCL	Circuit node voltages equation
(3) = (5) (*)	$\frac{V_{A3}-V_{A4}}{C_{gd}\cdot R_g}-\frac{(V_{A4}-V_{A9})}{C_{gd}\cdot R_{gs}}-\frac{(V_{A5}-V_{A9})}{C_{gd}\cdot R_i}-\frac{(V_{A4}-V_{A7})}{C_{gd}\cdot R_{gd}}=\frac{V_{A6}-V_{A7}}{C_{gd}\cdot R_{di}}$
(6) = (8) (**)	$\frac{V_{A4}-V_{A7}}{R_{gd}}+\frac{V_{A6}-V_{A7}}{R_{di}}-\frac{(V_{A7}-V_{A8})}{R_d}=\frac{V_{A9}-V_{A10}}{R_s}-\frac{(V_{A4}-V_{A9})}{R_{gs}}-\frac{(V_{A5}-V_{A9})}{R_i}$

$$V_{A10} = L_S \cdot \frac{dI_{Ls}}{dt} \; ; \; I_{Rs} = I_{Ls} \; ; \; I_{Rs} = \frac{V_{A9} - V_{A10}}{R_S} \; ; \; V_{A10} = \frac{L_S}{R_S} \cdot \frac{d}{dt}(V_{A9} - V_{A10})$$

$$-V_g - V_{A3} = L_g \cdot \frac{dI_{Lg}}{dt} \Rightarrow \frac{dI_{Lg}}{dt} = -\frac{1}{L_g} \cdot (V_g + V_{A3}) \; ;$$

$$V_{A8} - V_{A11} = L_d \cdot \frac{dI_{Ld}}{dt} \Rightarrow \frac{dI_{Ld}}{dt} = \frac{1}{L_d} \cdot (V_{A8} - V_{A11})$$

$$I_{Ld} + \frac{V_d - V_{A11}}{R_{d1}} = \frac{V_{out}}{R_L} \Rightarrow V_{out} = I_{Ld} \cdot R_L + \frac{(V_d - V_{A11}) \cdot R_L}{R_{d1}} \; ;$$

$$I_{Ld} + \frac{V_d - V_{A11}}{R_{d1}} = C_{out} \cdot \frac{d}{dt}(V_{A11} - V_{out})$$

$$V_{out} = I_{Ld} \cdot R_L + \frac{(V_d - V_{A11}) \cdot R_L}{R_{d1}} \Rightarrow V_{out} - V_{A11} = I_{Ld} \cdot R_L + \frac{(V_d - V_{A11}) \cdot R_L}{R_{d1}} - V_{A11}$$

$$V_{A11} - V_{out} = -I_{Ld} \cdot R_L + \frac{(-V_d + V_{A11}) \cdot R_L}{R_{d1}} + V_{A11} \; ;$$

$$\frac{d}{dt}(V_{A11} - V_d) = \frac{dV_{A11}}{dt} \; ; \; \frac{dV_d}{dt} = 0$$

$$\frac{dV_{A11}}{dt} - \frac{dV_{out}}{dt} = -\frac{dI_{Ld}}{dt} \cdot R_L + \frac{R_L}{R_{d1}} \cdot \frac{d}{dt}(V_{A11} - V_d) + \frac{dV_{A11}}{dt}$$

$$\frac{d}{dt}(V_{A11} - V_{out}) = -\frac{dI_{Ld}}{dt} \cdot R_L + [\frac{R_L}{R_{d1}} + 1] \cdot \frac{dV_{A11}}{dt} ; \frac{dI_{Ld}}{dt} = \frac{1}{L_d} \cdot [V_{A8} - V_{A11}]$$

$$\frac{d}{dt}(V_{A11} - V_{out}) = -\frac{R_L}{L_d} \cdot [V_{A8} - V_{A11}] + [\frac{R_L}{R_{d1}} + 1] \cdot \frac{dV_{A11}}{dt}$$

$$I_{Ld} + \frac{V_d - V_{A11}}{R_{d1}} = C_{out} \cdot \frac{d}{dt}(V_{A11} - V_{out}) \Rightarrow I_{Ld} + \frac{V_d - V_{A11}}{R_{d1}} = C_{out} \cdot \{[\frac{R_L}{R_{d1}} + 1] \cdot \frac{dV_{A11}}{dt} - \frac{R_L}{L_d} \cdot [V_{A8} - V_{A11}]\}$$

$$C_{out} \cdot [\frac{R_L}{R_{d1}} + 1] \cdot \frac{dV_{A11}}{dt} = I_{Ld} + \frac{V_d - V_{A11}}{R_{d1}} + C_{out} \cdot \frac{R_L}{L_d} \cdot [V_{A8} - V_{A11}]$$

$$\frac{dV_{A11}}{dt} = \frac{1}{C_{out} \cdot [\frac{R_L}{R_{d1}} + 1]} \cdot I_{Ld} + \frac{V_d}{C_{out} \cdot [\frac{R_L}{R_{d1}} + 1] \cdot R_{d1}} - \frac{V_{A11}}{C_{out} \cdot [\frac{R_L}{R_{d1}} + 1] \cdot R_{d1}} + \frac{R_L}{[\frac{R_L}{R_{d1}} + 1] \cdot L_d} \cdot [V_{A8} - V_{A11}]$$

$$\frac{dV_{A11}}{dt} = \frac{1}{C_{out} \cdot [\frac{R_L}{R_{d1}} + 1]} \cdot I_{Ld} + \frac{V_d}{C_{out} \cdot [\frac{R_L}{R_{d1}} + 1] \cdot R_{d1}} + \frac{R_L}{[\frac{R_L}{R_{d1}} + 1] \cdot L_d} \cdot V_{A8} - [\frac{R_L}{L_d} + \frac{1}{C_{out} \cdot R_{d1}}] \cdot \frac{1}{[\frac{R_L}{R_{d1}} + 1]} \cdot V_{A11}$$

We can summery our system new differential equations representation:

$$\frac{dV_{A3}}{dt} = \frac{V_{A7} - V_{A8}}{R_d \cdot [C_{pds} + C_{pgd}] \cdot C_{T1}} - \frac{1}{[C_{pds} + C_{pgd}] \cdot C_{T1}} \cdot I_{Ld} + \frac{1}{C_{pgd} \cdot C_{T1}} \cdot I_{Lg} - \frac{V_{A3}}{C_{pgd} \cdot R_g \cdot C_{T1}} + \frac{V_{A4}}{C_{pgd} \cdot R_g \cdot C_{T1}}$$

$$\frac{d}{dt}(V_{A4} - V_{A6}) = \frac{V_{A6} - V_{A7}}{C_{gd} \cdot R_{di}}; \frac{dv}{dt} = \frac{1}{C_{gs}} \cdot \frac{(V_{A5} - V_{A9})}{R_i}$$

$$\frac{d}{dt}(V_{A7} - V_{A9}) = \frac{V_{A4} - V_{A7}}{C_{ds} \cdot R_{gd}} + \frac{V_{A6} - V_{A7}}{C_{ds} \cdot R_{di}} - \frac{(V_{A7} - V_{A8})}{C_{ds} \cdot R_d} - \frac{(V_{A7} - V_{A9})}{C_{ds} \cdot R_{ds}} - \frac{g_m}{C_{ds}} \cdot v$$

$$V_{A10} = L_S \cdot \frac{dI_{Ls}}{dt}; \frac{V_{A9} - V_{A10}}{R_s} = I_{Ls} \Rightarrow \frac{d}{dt}(V_{A9} - V_{A10}) = \frac{R_s}{L_S} \cdot V_{A10}; \frac{dI_{Lg}}{dt} = -\frac{1}{L_g} \cdot (V_g + V_{A3})$$

$$\frac{dI_{Ld}}{dt} = \frac{1}{L_d} \cdot (V_{A8} - V_{A11})$$

$$\frac{dV_{A11}}{dt} = \frac{1}{C_{out} \cdot [\frac{R_L}{R_{d1}} + 1]} \cdot I_{Ld} + \frac{V_d}{C_{out} \cdot [\frac{R_L}{R_{d1}} + 1] \cdot R_{d1}} + \frac{R_L}{[\frac{R_L}{R_{d1}} + 1] \cdot L_d} \cdot V_{A8} - [\frac{R_L}{L_d} + \frac{1}{C_{out} \cdot R_{d1}}] \cdot \frac{1}{[\frac{R_L}{R_{d1}} + 1]} \cdot V_{A11}$$

We need to find our system fixed points. First we analyze our circuit node voltages equations.

$$(*) \frac{V_{A3} - V_{A4}}{C_{gd} \cdot R_g} - \frac{(V_{A4} - V_{A9})}{C_{gd} \cdot R_{gs}} - \frac{(V_{A5} - V_{A9})}{C_{gd} \cdot R_i} - \frac{(V_{A4} - V_{A7})}{C_{gd} \cdot R_{gd}} = \frac{V_{A6} - V_{A7}}{C_{gd} \cdot R_{di}}$$

$$\frac{V_{A3}}{C_{gd} \cdot R_g} - \frac{V_{A4}}{C_{gd} \cdot R_g} - \frac{V_{A4}}{C_{gd} \cdot R_{gs}} + \frac{V_{A9}}{C_{gd} \cdot R_{gs}} - \frac{V_{A5}}{C_{gd} \cdot R_i} + \frac{V_{A9}}{C_{gd} \cdot R_i} - \frac{V_{A4}}{C_{gd} \cdot R_{gd}} + \frac{V_{A7}}{C_{gd} \cdot R_{gd}} = \frac{V_{A6}}{C_{gd} \cdot R_{di}} - \frac{V_{A7}}{C_{gd} \cdot R_{di}}$$

$$\frac{V_{A3}}{C_{gd}\cdot R_g}-V_{A4}\cdot\{\frac{1}{C_{gd}\cdot R_g}+\frac{1}{C_{gd}\cdot R_{gs}}+\frac{1}{C_{gd}\cdot R_{gd}}\}-\frac{V_{A5}}{C_{gd}\cdot R_i}-\frac{V_{A6}}{C_{gd}\cdot R_{di}}$$
$$+V_{A7}\cdot\{\frac{1}{C_{gd}\cdot R_{di}}+\frac{1}{C_{gd}\cdot R_{gd}}\}+V_{A9}\cdot\{\frac{1}{C_{gd}\cdot R_{gs}}+\frac{1}{C_{gd}\cdot R_i}\}=0$$

$$\frac{V_{A3}}{C_{gd}\cdot R_g}-V_{A4}\cdot\{\frac{1}{C_{gd}\cdot R_g}+\frac{1}{C_{gd}\cdot R_{gs}}+\frac{1}{C_{gd}\cdot R_{gd}}\}-\frac{V_{A5}}{C_{gd}\cdot R_i}-\frac{V_{A6}}{C_{gd}\cdot R_{di}}$$
$$+V_{A7}\cdot\{\frac{1}{C_{gd}\cdot R_{di}}+\frac{1}{C_{gd}\cdot R_{gd}}\}+V_{A8}\cdot 0+V_{A9}$$
$$\cdot\{\frac{1}{C_{gd}\cdot R_{gs}}+\frac{1}{C_{gd}\cdot R_i}\}+V_{A10}\cdot 0=0$$

$$\Gamma_{11}=\frac{1}{C_{gd}\cdot R_g};\Gamma_{12}=-\{\frac{1}{C_{gd}\cdot R_g}+\frac{1}{C_{gd}\cdot R_{gs}}+\frac{1}{C_{gd}\cdot R_{gd}}\};\Gamma_{13}$$
$$=-\frac{1}{C_{gd}\cdot R_i};\Gamma_{14}=-\frac{1}{C_{gd}\cdot R_{di}}$$

$$\Gamma_{15}=\frac{1}{C_{gd}\cdot R_{di}}+\frac{1}{C_{gd}\cdot R_{gd}};\Gamma_{16}=0;\Gamma_{17}=\frac{1}{C_{gd}\cdot R_{gs}}+\frac{1}{C_{gd}\cdot R_i};\Gamma_{18}$$
$$=0;\sum_{k=1}^{8}\Gamma_{1k}\cdot V_{A_{k+2}}=0$$

$$(**)\frac{V_{A4}-V_{A7}}{R_{gd}}+\frac{V_{A6}-V_{A7}}{R_{di}}-\frac{(V_{A7}-V_{A8})}{R_d}$$
$$=\frac{V_{A9}-V_{A10}}{R_s}-\frac{(V_{A4}-V_{A9})}{R_{gs}}-\frac{(V_{A5}-V_{A9})}{R_i}$$

$$\frac{V_{A4}}{R_{gd}}-\frac{V_{A7}}{R_{gd}}+\frac{V_{A6}}{R_{di}}-\frac{V_{A7}}{R_{di}}-\frac{V_{A7}}{R_d}+\frac{V_{A8}}{R_d}=\frac{V_{A9}}{R_s}-\frac{V_{A10}}{R_s}-\frac{V_{A4}}{R_{gs}}+\frac{V_{A9}}{R_{gs}}-\frac{V_{A5}}{R_i}+\frac{V_{A9}}{R_i}$$

$$V_{A4}\cdot(\frac{1}{R_{gd}}+\frac{1}{R_{gs}})+\frac{V_{A5}}{R_i}+\frac{V_{A6}}{R_{di}}-V_{A7}\cdot\{\frac{1}{R_{gd}}+\frac{1}{R_{di}}+\frac{1}{R_d}\}+\frac{V_{A8}}{R_d}-V_{A9}$$
$$\cdot\{\frac{1}{R_s}+\frac{1}{R_{gs}}+\frac{1}{R_i}\}+\frac{V_{A10}}{R_s}$$
$$=0$$

$$V_{A3}\cdot 0+V_{A4}\cdot(\frac{1}{R_{gd}}+\frac{1}{R_{gs}})+\frac{V_{A5}}{R_i}+\frac{V_{A6}}{R_{di}}-V_{A7}\cdot\{\frac{1}{R_{gd}}+\frac{1}{R_{di}}+\frac{1}{R_d}\}$$
$$+\frac{V_{A8}}{R_d}-V_{A9}\cdot\{\frac{1}{R_s}+\frac{1}{R_{gs}}+\frac{1}{R_i}\}+\frac{V_{A10}}{R_s}=0$$

$$\Gamma_{21} = 0; \Gamma_{22} = \frac{1}{R_{gd}} + \frac{1}{R_{gs}}; \Gamma_{23} = \frac{1}{R_i}; \Gamma_{24} = \frac{1}{R_{di}};$$
$$\Gamma_{25} = -\{\frac{1}{R_{gd}} + \frac{1}{R_{di}} + \frac{1}{R_d}\}; \Gamma_{26} = \frac{1}{R_d}$$

$$\Gamma_{27} = -\{\frac{1}{R_s} + \frac{1}{R_{gs}} + \frac{1}{R_i}\}; \Gamma_{28} = \frac{1}{R_s}; \sum_{k=1}^{8} \Gamma_{2k} \cdot V_{A_{k+2}} = 0$$

$(*)-(**)$

$$\sum_{k=1}^{8} \Gamma_{1k} \cdot V_{A_{k+2}} - \sum_{k=1}^{8} \Gamma_{2k} \cdot V_{A_{k+2}} = 0; \sum_{k=1}^{8} \Gamma_{1k} \cdot V_{A_{k+2}}$$
$$- \sum_{k=1}^{8} \Gamma_{2k} \cdot V_{A_{k+2}} = \sum_{k=1}^{8} (\Gamma_{1k} - \Gamma_{2k}) \cdot V_{A_{k+2}}$$

$$V_{A3} \cdot \Gamma_{11} + V_{A4} \cdot (\Gamma_{12} - \Gamma_{22}) + V_{A5} \cdot (\Gamma_{13} - \Gamma_{23}) + V_{A6} \cdot (\Gamma_{14} - \Gamma_{24}) + V_{A7} \cdot (\Gamma_{15} - \Gamma_{25})$$
$$- V_{A8} \cdot \Gamma_{26} + V_{A9} \cdot (\Gamma_{17} - \Gamma_{27}) - V_{A10} \cdot \Gamma_{28} = 0$$

At fixed points:

$$\frac{dV_{A3}}{dt} = 0; \frac{d}{dt}(V_{A4} - V_{A6}) = 0; \frac{dv}{dt} = 0; \frac{d}{dt}(V_{A7} - V_{A9}) = 0$$
$$\frac{d}{dt}(V_{A9} - V_{A10}) = 0; \frac{dI_{Lg}}{dt} = 0; \frac{dI_{Ld}}{dt} = 0; \frac{dV_{A11}}{dt} = 0$$

Fixed points: $E^* = (V_{A3}^*, \ldots, V_{A11}^*, I_{Ld}^*, I_{Lg}^*, v^*)$

$$\frac{V_{A7}^* - V_{A8}^*}{R_d \cdot [C_{pds} + C_{pgd}] \cdot C_{T1}} - \frac{1}{[C_{pds} + C_{pgd}] \cdot C_{T1}} \cdot I_{Ld}^* + \frac{1}{C_{pgd} \cdot C_{T1}} \cdot I_{Lg}^*$$
$$- \frac{V_{A3}^*}{C_{pgd} \cdot R_g \cdot C_{T1}} + \frac{V_{A4}^*}{C_{pgd} \cdot R_g \cdot C_{T1}}$$
$$= 0$$

$$V_{A6}^* - V_{A7}^* = 0; V_{A5}^* - V_{A9}^* = 0; V_{A10}^* = 0; V_{A3}^* = -V_g; V_{A8}^* - V_{A11}^* = 0$$
$$\frac{V_{A4}^*}{C_{ds} \cdot R_{gd}} + \frac{V_{A6}^*}{C_{ds} \cdot R_{di}} - V_{A7}^* \cdot [\frac{1}{C_{ds} \cdot R_{gd}} - \frac{1}{C_{ds} \cdot R_{di}} - \frac{1}{C_{ds} \cdot R_d} - \frac{1}{C_{ds} \cdot R_{ds}}]$$
$$+ \frac{V_{A8}^*}{C_{ds} \cdot R_d} + \frac{V_{A9}^*}{C_{ds} \cdot R_{ds}} - \frac{g_m}{C_{ds}} \cdot v^* = 0$$

$$\frac{1}{C_{out} \cdot [\frac{R_L}{R_{d1}} + 1]} \cdot I_{Ld}^* + \frac{V_d}{C_{out} \cdot [\frac{R_L}{R_{d1}} + 1] \cdot R_{d1}} + \frac{R_L}{[\frac{R_L}{R_{d1}} + 1] \cdot L_d} \cdot V_{A8}^* - [\frac{R_L}{L_d} + \frac{1}{C_{out} \cdot R_{d1}}] \cdot \frac{1}{[\frac{R_L}{R_{d1}} + 1]} \cdot V_{A11}^* = 0$$

We can minimize the above fixed points equations:

$$\frac{V_{A7}^* - V_{A8}^*}{R_d \cdot [C_{pds} + C_{pgd}] \cdot C_{T1}} - \frac{1}{[C_{pds} + C_{pgd}] \cdot C_{T1}} \cdot I_{Ld}^* + \frac{1}{C_{pgd} \cdot C_{T1}} \cdot I_{Lg}^* + \frac{V_g}{C_{pgd} \cdot R_g \cdot C_{T1}} + \frac{V_{A4}^*}{C_{pgd} \cdot R_g \cdot C_{T1}} = 0$$

$$\frac{V_{A4}^*}{C_{ds} \cdot R_{gd}} - V_{A7}^* \cdot [\frac{1}{C_{ds} \cdot R_{gd}} + \frac{1}{C_{ds} \cdot R_{di}} + \frac{1}{C_{ds} \cdot R_d} + \frac{1}{C_{ds} \cdot R_{ds}} - \frac{1}{C_{ds} \cdot R_{di}}] + \frac{V_{A8}^*}{C_{ds} \cdot R_d} + \frac{V_{A9}^*}{C_{ds} \cdot R_{ds}} - \frac{g_m}{C_{ds}} \cdot v^* = 0$$

$$\frac{1}{C_{out} \cdot [\frac{R_L}{R_{d1}} + 1]} \cdot I_{Ld}^* + \frac{V_d}{C_{out} \cdot [\frac{R_L}{R_{d1}} + 1] \cdot R_{d1}} + V_{A8}^* \cdot \{\frac{R_L}{[\frac{R_L}{R_{d1}} + 1] \cdot L_d} - [\frac{R_L}{L_d} + \frac{1}{C_{out} \cdot R_{d1}}] \cdot \frac{1}{[\frac{R_L}{R_{d1}} + 1]}\} = 0$$

$$V_{A6}^* = V_{A7}^*; V_{A5}^* = V_{A9}^*; V_{A10}^* = 0; V_{A3}^* = -V_g; V_{A8}^* = V_{A11}^*; \sum_{k=1}^{8} \Gamma_{1k} \cdot V_{A_{k+2}}^* = 0; \sum_{k=1}^{8} \Gamma_{2k} \cdot V_{A_{k+2}}^* = 0$$

3.3 Field Effect Transistor (FETs) at Microwave Frequencies Stability Analysis

In Sect. 3.2, we present FET system's differential equations representation:

$$\frac{dV_{A3}}{dt} = \frac{V_{A7} - V_{A8}}{R_d \cdot [C_{pds} + C_{pgd}] \cdot C_{T1}} - \frac{1}{[C_{pds} + C_{pgd}] \cdot C_{T1}} \cdot I_{Ld} + \frac{1}{C_{pgd} \cdot C_{T1}} \cdot I_{Lg} - \frac{V_{A3}}{C_{pgd} \cdot R_g \cdot C_{T1}} + \frac{V_{A4}}{C_{pgd} \cdot R_g \cdot C_{T1}}$$

$$\frac{d}{dt}(V_{A4} - V_{A6}) = \frac{V_{A6} - V_{A7}}{C_{gd} \cdot R_{di}}; \frac{dv}{dt} = \frac{1}{C_{gs}} \cdot \frac{(V_{A5} - V_{A9})}{R_i}$$

$$\frac{d}{dt}(V_{A7} - V_{A9}) = \frac{V_{A4} - V_{A7}}{C_{ds} \cdot R_{gd}} + \frac{V_{A6} - V_{A7}}{C_{ds} \cdot R_{di}} - \frac{(V_{A7} - V_{A8})}{C_{ds} \cdot R_d} - \frac{(V_{A7} - V_{A9})}{C_{ds} \cdot R_{ds}} - \frac{g_m}{C_{ds}} \cdot v$$

$$V_{A10} = L_S \cdot \frac{dI_{Ls}}{dt}; \frac{V_{A9} - V_{A10}}{R_s} = I_{Ls} \Rightarrow \frac{d}{dt}(V_{A9} - V_{A10})$$
$$= \frac{R_s}{L_S} \cdot V_{A10}; \frac{dI_{Lg}}{dt} = -\frac{1}{L_g} \cdot (V_g + V_{A3})$$
$$\frac{dI_{Ld}}{dt} = \frac{1}{L_d} \cdot (V_{A8} - V_{A11})$$

$$\frac{dV_{A11}}{dt} = \frac{1}{C_{out} \cdot [\frac{R_L}{R_{d1}} + 1]} \cdot I_{Ld} + \frac{V_d}{C_{out} \cdot [\frac{R_L}{R_{d1}} + 1] \cdot R_{d1}} + \frac{R_L}{[\frac{R_L}{R_{d1}} + 1] \cdot L_d} \cdot V_{A8}$$
$$- [\frac{R_L}{L_d} + \frac{1}{C_{out} \cdot R_{d1}}] \cdot \frac{1}{[\frac{R_L}{R_{d1}} + 1]} \cdot V_{A11}$$

We need to discuss its stability analysis under parameter variation.

We derivate the first equation $\frac{dV_{A3}}{dt} = \cdots; \frac{dI_{Ld}}{dt} = \frac{1}{L_d} \cdot (V_{A8} - V_{A11}); \frac{dI_{Lg}}{dt} = -\frac{1}{L_g} \cdot (V_g + V_{A3})$

$$\frac{d^2 V_{A3}}{dt^2} = \frac{d}{dt}\{\frac{V_{A7} - V_{A8}}{R_d \cdot [C_{pds} + C_{pgd}] \cdot C_{T1}} - \frac{1}{[C_{pds} + C_{pgd}] \cdot C_{T1}} \cdot I_{Ld} + \frac{1}{C_{pgd} \cdot C_{T1}} \cdot I_{Lg}$$
$$- \frac{V_{A3}}{C_{pgd} \cdot R_g \cdot C_{T1}} + \frac{V_{A4}}{C_{pgd} \cdot R_g \cdot C_{T1}}\}$$

$$\frac{d^2 V_{A3}}{dt^2} = \frac{1}{R_d \cdot [C_{pds} + C_{pgd}] \cdot C_{T1}} \cdot \frac{d(V_{A7} - V_{A8})}{dt}$$
$$- \frac{1}{[C_{pds} + C_{pgd}] \cdot C_{T1}} \cdot \frac{dI_{Ld}}{dt} + \frac{1}{C_{pgd} \cdot C_{T1}} \cdot \frac{dI_{Lg}}{dt}$$
$$- \frac{1}{C_{pgd} \cdot R_g \cdot C_{T1}} \cdot \frac{dV_{A3}}{dt} + \frac{1}{C_{pgd} \cdot R_g \cdot C_{T1}} \cdot \frac{dV_{A4}}{dt}$$

Inserting expressions: $\frac{dI_{Ld}}{dt} = \cdots; \frac{dI_{Lg}}{dt} = \cdots$

$$\frac{d^2V_{A3}}{dt^2} = \frac{1}{R_d \cdot [C_{pds} + C_{pgd}] \cdot C_{T1}} \cdot \frac{d(V_{A7} - V_{A8})}{dt} - \frac{1}{[C_{pds} + C_{pgd}] \cdot C_{T1}} \cdot \frac{1}{L_d} \cdot (V_{A8} - V_{A11}) - \frac{1}{C_{pgd} \cdot C_{T1}} \cdot \frac{1}{L_g} \cdot (V_g + V_{A3}) - \frac{1}{C_{pgd} \cdot R_g \cdot C_{T1}} \cdot \frac{dV_{A3}}{dt} + \frac{1}{C_{pgd} \cdot R_g \cdot C_{T1}} \cdot \frac{dV_{A4}}{dt}$$

$$\frac{d^2V_{A3}}{dt^2} - \frac{1}{R_d \cdot [C_{pds} + C_{pgd}] \cdot C_{T1}} \cdot \frac{d(V_{A7} - V_{A8})}{dt} + \frac{1}{[C_{pds} + C_{pgd}] \cdot C_{T1}} \cdot \frac{1}{L_d} \cdot (V_{A8} - V_{A11}) + \frac{1}{C_{pgd} \cdot C_{T1}} \cdot \frac{1}{L_g} \cdot (V_g + V_{A3}) + \frac{1}{C_{pgd} \cdot R_g \cdot C_{T1}} \cdot \frac{dV_{A3}}{dt} - \frac{1}{C_{pgd} \cdot R_g \cdot C_{T1}} \cdot \frac{dV_{A4}}{dt} = 0$$

$$\frac{d^2V_{A3}}{dt^2} - \frac{1}{R_d \cdot [C_{pds} + C_{pgd}] \cdot C_{T1}} \cdot \frac{dV_{A7}}{dt} + \frac{1}{R_d \cdot [C_{pds} + C_{pgd}] \cdot C_{T1}} \cdot \frac{dV_{A8}}{dt} + \frac{1}{[C_{pds} + C_{pgd}] \cdot C_{T1}} \cdot \frac{1}{L_d} \cdot V_{A8} - \frac{1}{[C_{pds} + C_{pgd}] \cdot C_{T1}} \cdot \frac{1}{L_d} \cdot V_{A11} + \frac{1}{C_{pgd} \cdot C_{T1}} \cdot \frac{1}{L_g} \cdot V_g + \frac{1}{C_{pgd} \cdot C_{T1}} \cdot \frac{1}{L_g} \cdot V_{A3} + \frac{1}{C_{pgd} \cdot R_g \cdot C_{T1}} \cdot \frac{dV_{A3}}{dt} - \frac{1}{C_{pgd} \cdot R_g \cdot C_{T1}} \cdot \frac{dV_{A4}}{dt} = 0$$

We define the following new variables: $\frac{dV_{A3}}{dt} = Y_1$; $\frac{d^2V_{A3}}{dt^2} = \frac{dY_1}{dt}$; $\frac{dV_{A4}}{dt} = Y_2$

$$\frac{dV_{A7}}{dt} = Y_3 ; \frac{dV_{A8}}{dt} = Y_4$$

$$\frac{dY_1}{dt} - \frac{1}{R_d \cdot [C_{pds} + C_{pgd}] \cdot C_{T1}} \cdot Y_3 + \frac{1}{R_d \cdot [C_{pds} + C_{pgd}] \cdot C_{T1}} \cdot Y_4 + \frac{1}{[C_{pds} + C_{pgd}] \cdot C_{T1}} \cdot \frac{1}{L_d} \cdot V_{A8} - \frac{1}{[C_{pds} + C_{pgd}] \cdot C_{T1}} \cdot \frac{1}{L_d} \cdot V_{A11} + \frac{1}{C_{pgd} \cdot C_{T1}} \cdot \frac{1}{L_g} \cdot V_g + \frac{1}{C_{pgd} \cdot C_{T1}} \cdot \frac{1}{L_g} \cdot V_{A3} + \frac{1}{C_{pgd} \cdot R_g \cdot C_{T1}} \cdot Y_1 - \frac{1}{C_{pgd} \cdot R_g \cdot C_{T1}} \cdot Y_2 = 0$$

$$\frac{dY_1}{dt} = \frac{1}{R_d \cdot [C_{pds} + C_{pgd}] \cdot C_{T1}} \cdot Y_3 - \frac{1}{R_d \cdot [C_{pds} + C_{pgd}] \cdot C_{T1}} \cdot Y_4 - \frac{1}{[C_{pds} + C_{pgd}] \cdot C_{T1}} \cdot \frac{1}{L_d} \cdot V_{A8} + \frac{1}{[C_{pds} + C_{pgd}] \cdot C_{T1}} \cdot \frac{1}{L_d} \cdot V_{A11} - \frac{1}{C_{pgd} \cdot C_{T1}} \cdot \frac{1}{L_g} \cdot V_g - \frac{1}{C_{pgd} \cdot C_{T1}} \cdot \frac{1}{L_g} \cdot V_{A3} - \frac{1}{C_{pgd} \cdot R_g \cdot C_{T1}} \cdot Y_1 + \frac{1}{C_{pgd} \cdot R_g \cdot C_{T1}} \cdot Y_2$$

We need to derivative equation: $\frac{dV_{A11}}{dt} = \cdots ; \frac{dI_{Ld}}{dt} = \frac{1}{L_d} \cdot (V_{A8} - V_{A11}) ; \frac{dV_d}{dt} = 0$

$$\frac{dV_{A11}}{dt} = \frac{1}{C_{out} \cdot [\frac{R_L}{R_{d1}} + 1]} \cdot I_{Ld} + \frac{V_d}{C_{out} \cdot [\frac{R_L}{R_{d1}} + 1] \cdot R_{d1}} + \frac{R_L}{[\frac{R_L}{R_{d1}} + 1] \cdot L_d} \cdot V_{A8} - [\frac{R_L}{L_d} + \frac{1}{C_{out} \cdot R_{d1}}] \cdot \frac{1}{[\frac{R_L}{R_{d1}} + 1]} \cdot V_{A11}$$

$$\frac{d^2V_{A11}}{dt^2} = \frac{1}{C_{out} \cdot [\frac{R_L}{R_{d1}} + 1]} \cdot \frac{dI_{Ld}}{dt} + \frac{R_L}{[\frac{R_L}{R_{d1}} + 1] \cdot L_d} \cdot \frac{dV_{A8}}{dt} - [\frac{R_L}{L_d} + \frac{1}{C_{out} \cdot R_{d1}}] \cdot \frac{1}{[\frac{R_L}{R_{d1}} + 1]} \cdot \frac{dV_{A11}}{dt}$$

$$\frac{d^2V_{A11}}{dt^2} = \frac{1}{C_{out} \cdot [\frac{R_L}{R_{d1}} + 1]} \cdot \frac{1}{L_d} \cdot (V_{A8} - V_{A11}) + \frac{R_L}{[\frac{R_L}{R_{d1}} + 1] \cdot L_d} \cdot \frac{dV_{A8}}{dt} - [\frac{R_L}{L_d} + \frac{1}{C_{out} \cdot R_{d1}}] \cdot \frac{1}{[\frac{R_L}{R_{d1}} + 1]} \cdot \frac{dV_{A11}}{dt}$$

We define the following new variables: $\frac{dV_{A8}}{dt} = Y_4 ; \frac{dV_{A11}}{dt} = Y_5 ; \frac{d^2V_{A11}}{dt^2} = \frac{dY_5}{dt}$

$$\frac{dY_5}{dt} = \frac{1}{C_{out} \cdot [\frac{R_L}{R_{d1}} + 1]} \cdot \frac{1}{L_d} \cdot (V_{A8} - V_{A11}) + \frac{R_L}{[\frac{R_L}{R_{d1}} + 1] \cdot L_d} \cdot Y_4 - [\frac{R_L}{L_d} + \frac{1}{C_{out} \cdot R_{d1}}] \cdot \frac{1}{[\frac{R_L}{R_{d1}} + 1]} \cdot Y_5$$

$$\frac{dY_5}{dt} = \frac{1}{C_{out} \cdot [\frac{R_L}{R_{d1}} + 1]} \cdot \frac{1}{L_d} \cdot V_{A8} - \frac{1}{C_{out} \cdot [\frac{R_L}{R_{d1}} + 1]} \cdot \frac{1}{L_d} \cdot V_{A11} + \frac{R_L}{[\frac{R_L}{R_{d1}} + 1] \cdot L_d} \cdot Y_4 - [\frac{R_L}{L_d} + \frac{1}{C_{out} \cdot R_{d1}}] \cdot \frac{1}{[\frac{R_L}{R_{d1}} + 1]} \cdot Y_5$$

We need to derivative equation: $\frac{d}{dt}(V_{A7} - V_{A9}) = \cdots$; $\frac{dv}{dt} = \frac{1}{C_{gs}} \cdot \frac{(V_{A5} - V_{A9})}{R_i}$

$$\frac{d}{dt}(V_{A7} - V_{A9}) = \frac{V_{A4} - V_{A7}}{C_{ds} \cdot R_{gd}} + \frac{V_{A6} - V_{A7}}{C_{ds} \cdot R_{di}} - \frac{(V_{A7} - V_{A8})}{C_{ds} \cdot R_d} - \frac{(V_{A7} - V_{A9})}{C_{ds} \cdot R_{ds}} - \frac{g_m}{C_{ds}} \cdot v$$

$$\frac{d^2}{dt^2}(V_{A7} - V_{A9}) = \frac{d}{dt}\{\frac{V_{A4} - V_{A7}}{C_{ds} \cdot R_{gd}} + \frac{V_{A6} - V_{A7}}{C_{ds} \cdot R_{di}} - \frac{(V_{A7} - V_{A8})}{C_{ds} \cdot R_d} - \frac{(V_{A7} - V_{A9})}{C_{ds} \cdot R_{ds}} - \frac{g_m}{C_{ds}} \cdot v\}$$

$$\begin{aligned}\frac{d^2}{dt^2}(V_{A7} - V_{A9}) = & \frac{1}{C_{ds} \cdot R_{gd}} \cdot \frac{dV_{A4}}{dt} - \frac{1}{C_{ds} \cdot R_{gd}} \cdot \frac{1}{dt} \cdot \frac{dV_{A7}}{dt} \\ & + \frac{1}{C_{ds} \cdot R_{di}} \cdot \frac{dV_{A6}}{dt} - \frac{1}{C_{ds} \cdot R_{di}} \cdot \frac{dV_{A7}}{dt} \\ & - \frac{1}{C_{ds} \cdot R_d} \cdot \frac{dV_{A7}}{dt} + \frac{1}{C_{ds} \cdot R_d} \cdot \frac{dV_{A8}}{dt} - \frac{1}{C_{ds} \cdot R_{ds}} \cdot \frac{dV_{A7}}{dt} \\ & + \frac{1}{C_{ds} \cdot R_{ds}} \cdot \frac{dV_{A9}}{dt} - \frac{g_m}{C_{ds}} \cdot \frac{dv}{dt}\end{aligned}$$

$$\frac{d^2}{dt^2}(V_{A7} - V_{A9}) = \frac{1}{C_{ds} \cdot R_{gd}} \cdot \frac{dV_{A4}}{dt} - \frac{1}{C_{ds} \cdot R_{gd}} \cdot \frac{1}{dt} \cdot \frac{dV_{A7}}{dt} + \frac{1}{C_{ds} \cdot R_{di}} \cdot \frac{dV_{A6}}{dt} - \frac{1}{C_{ds} \cdot R_{di}} \cdot \frac{dV_{A7}}{dt} - \frac{1}{C_{ds} \cdot R_d} \cdot \frac{dV_{A7}}{dt} + \frac{1}{C_{ds} \cdot R_d} \cdot \frac{dV_{A8}}{dt} - \frac{1}{C_{ds} \cdot R_{ds}} \cdot \frac{dV_{A7}}{dt} + \frac{1}{C_{ds} \cdot R_{ds}} \cdot \frac{dV_{A9}}{dt} - \frac{g_m}{C_{ds}} \cdot \frac{1}{C_{gs}} \cdot \frac{(V_{A5} - V_{A9})}{R_i}$$

$$\frac{d^2}{dt^2}(V_{A7} - V_{A9}) = \frac{1}{C_{ds} \cdot R_{ds}} \cdot [\frac{dV_{A9}}{dt} - \frac{dV_{A7}}{dt}] + \frac{1}{C_{ds} \cdot R_{gd}} \cdot \frac{dV_{A4}}{dt} + \frac{1}{C_{ds} \cdot R_{di}} \cdot \frac{dV_{A6}}{dt} - \frac{dV_{A7}}{dt} \cdot [\frac{1}{C_{ds} \cdot R_{gd}} + \frac{1}{C_{ds} \cdot R_{di}} + \frac{1}{C_{ds} \cdot R_d}] + \frac{1}{C_{ds} \cdot R_d} \cdot \frac{dV_{A8}}{dt} - \frac{g_m}{C_{ds}} \cdot \frac{1}{C_{gs}} \cdot \frac{V_{A5}}{R_i} + \frac{g_m}{C_{ds}} \cdot \frac{1}{C_{gs}} \cdot \frac{V_{A9}}{R_i}$$

We define the following new variables: $\frac{dV_{A6}}{dt} = Y_6$; $Y_7 = \frac{dV_{A9}}{dt} - \frac{dV_{A7}}{dt}$; $\frac{dV_{A7}}{dt} = Y_3$

$$Y_7 = \frac{dV_{A9}}{dt} - Y_3 \Rightarrow \frac{dV_{A9}}{dt} = Y_7 + Y_3$$

$$\begin{aligned}\frac{dY_7}{dt} = & \frac{1}{C_{ds} \cdot R_{ds}} \cdot Y_7 + \frac{1}{C_{ds} \cdot R_{gd}} \cdot Y_2 + \frac{1}{C_{ds} \cdot R_{di}} \cdot Y_6 - Y_3 \\ & \cdot [\frac{1}{C_{ds} \cdot R_{gd}} + \frac{1}{C_{ds} \cdot R_{di}} + \frac{1}{C_{ds} \cdot R_d}] \\ & + \frac{1}{C_{ds} \cdot R_d} \cdot Y_4 - \frac{g_m}{C_{ds}} \cdot \frac{1}{C_{gs}} \cdot \frac{V_{A5}}{R_i} + \frac{g_m}{C_{ds}} \cdot \frac{1}{C_{gs}} \cdot \frac{V_{A9}}{R_i}\end{aligned}$$

We can summery our system new differential equations representation:

$$\frac{dV_{A3}}{dt} = Y_1 = f_1(Y_1); \frac{dV_{A4}}{dt} = Y_2 = f_3(Y_2); \frac{dV_{A7}}{dt} = f_4(Y_3) = Y_3; \frac{dV_{A8}}{dt} = f_5(Y_4) = Y_4$$

$$\begin{aligned}
\frac{dY_1}{dt} &= f_2(Y_1, Y_2, Y_3, Y_4, V_{A3}, V_{A8}, V_{A11}) = \frac{1}{R_d \cdot [C_{pds} + C_{pgd}] \cdot C_{T1}} \\
&\quad \cdot Y_3 - \frac{1}{R_d \cdot [C_{pds} + C_{pgd}] \cdot C_{T1}} \cdot Y_4 \\
&\quad - \frac{1}{[C_{pds} + C_{pgd}] \cdot C_{T1}} \cdot \frac{1}{L_d} \cdot V_{A8} + \frac{1}{[C_{pds} + C_{pgd}] \cdot C_{T1}} \cdot \frac{1}{L_d} \cdot V_{A11} \\
&\quad - \frac{1}{C_{pgd} \cdot C_{T1}} \cdot \frac{1}{L_g} \cdot V_g - \frac{1}{C_{pgd} \cdot C_{T1}} \cdot \frac{1}{L_g} \cdot V_{A3} \\
&\quad - \frac{1}{C_{pgd} \cdot R_g \cdot C_{T1}} \cdot Y_1 + \frac{1}{C_{pgd} \cdot R_g \cdot C_{T1}} \cdot Y_2
\end{aligned}$$

$$\begin{aligned}
\frac{dV_{A11}}{dt} &= f_6(Y_5) = Y_5; \frac{dY_5}{dt} = f_7(V_{A8}, V_{A11}, Y_4, Y_5); \frac{dV_{A6}}{dt} = f_8(Y_6) = Y_6; \frac{dV_{A9}}{dt} \\
&= f_9(Y_7, Y_3) = Y_7 + Y_3
\end{aligned}$$

$$\begin{aligned}
\frac{dY_5}{dt} &= \frac{1}{C_{out} \cdot [\frac{R_L}{R_{d1}} + 1]} \cdot \frac{1}{L_d} \cdot V_{A8} - \frac{1}{C_{out} \cdot [\frac{R_L}{R_{d1}} + 1]} \cdot \frac{1}{L_d} \cdot V_{A11} + \frac{R_L}{[\frac{R_L}{R_{d1}} + 1] \cdot L_d} \cdot Y_4 \\
&\quad - [\frac{R_L}{L_d} + \frac{1}{C_{out} \cdot R_{d1}}] \cdot \frac{1}{[\frac{R_L}{R_{d1}} + 1]} \cdot Y_5
\end{aligned}$$

$$\begin{aligned}
\frac{dY_7}{dt} &= f_{10}(V_{A5}, V_{A9}, Y_2, Y_6, Y_7, Y_3, Y_4) = \frac{1}{C_{ds} \cdot R_{ds}} \cdot Y_7 + \frac{1}{C_{ds} \cdot R_{gd}} \cdot Y_2 + \frac{1}{C_{ds} \cdot R_{di}} \cdot Y_6 \\
&\quad - Y_3 \cdot [\frac{1}{C_{ds} \cdot R_{gd}} + \frac{1}{C_{ds} \cdot R_{di}} + \frac{1}{C_{ds} \cdot R_d}] + \frac{1}{C_{ds} \cdot R_d} \\
&\quad \cdot Y_4 - \frac{g_m}{C_{ds}} \cdot \frac{1}{C_{gs}} \cdot \frac{V_{A5}}{R_i} + \frac{g_m}{C_{ds}} \cdot \frac{1}{C_{gs}} \cdot \frac{V_{A9}}{R_i}
\end{aligned}$$

We have ten differential equations which represent our system.

$$\begin{aligned}
&\frac{dV_{A3}}{dt} = f_1(Y_1)\,; \frac{dY_1}{dt} = f_2(Y_1, Y_2, Y_3, Y_4, V_{A3}, V_{A8}, V_{A11})\,; \frac{dV_{A4}}{dt} = f_3(Y_2); \frac{dV_{A7}}{dt} = f_4(Y_3) \\
&\frac{dV_{A8}}{dt} = f_5(Y_4)\,; \frac{dV_{A11}}{dt} = f_6(Y_5); \frac{dY_5}{dt} = f_7(V_{A8}, V_{A11}, Y_4, Y_5)\,; \frac{dV_{A6}}{dt} = f_8(Y_6) \\
&\frac{dV_{A9}}{dt} = f_9(Y_7, Y_3)\,; \frac{dY_7}{dt} = f_{10}(V_{A5}, V_{A9}, Y_2, Y_6, Y_7, Y_3, Y_4)
\end{aligned}$$

@ Fixed points:

$$\frac{dV_{A3}}{dt} = 0;\ \frac{dY_1}{dt} = 0;\ \frac{dV_{A4}}{dt} = 0;\ \frac{dV_{A7}}{dt} = 0;\ \frac{dV_{A8}}{dt} = 0$$

$$\frac{dV_{A11}}{dt} = 0;\ \frac{dY_5}{dt} = 0;\ \frac{dV_{A6}}{dt} = 0;\ \frac{dV_{A9}}{dt} = 0;\ \frac{dY_7}{dt} = 0$$

$$Y_1^* = 0;\ Y_2^* = 0;\ Y_3^* = 0;\ Y_4^* = 0;\ Y_5^* = 0;\ Y_6^* = 0;\ Y_7^* = 0$$

$$-\frac{1}{[C_{pds} + C_{pgd}] \cdot C_{T1}} \cdot \frac{1}{L_d} \cdot V_{A8}^* + \frac{1}{[C_{pds} + C_{pgd}] \cdot C_{T1}} \cdot \frac{1}{L_d} \cdot V_{A11}^* - \frac{1}{C_{pgd} \cdot C_{T1}} \cdot \frac{1}{L_g} \cdot V_g - \frac{1}{C_{pgd} \cdot C_{T1}} \cdot \frac{1}{L_g} \cdot V_{A3}^* = 0$$

$$\frac{1}{C_{out} \cdot [\frac{R_L}{R_{d1}} + 1]} \cdot \frac{1}{L_d} \cdot V_{A8}^* - \frac{1}{C_{out} \cdot [\frac{R_L}{R_{d1}} + 1]} \cdot \frac{1}{L_d} \cdot V_{A11}^* = 0$$

$$-\frac{g_m}{C_{ds}} \cdot \frac{1}{C_{gs}} \cdot \frac{V_{A5}^*}{R_i} + \frac{g_m}{C_{ds}} \cdot \frac{1}{C_{gs}} \cdot \frac{V_{A9}^*}{R_i} = 0$$

<u>Stability analysis</u>: The standard local stability analysis about any one of the equilibrium points of our system consists in adding to coordinates $[V_{A3}, V_{A4}, V_{A7}, V_{A8}, Y_1, V_{A11}, Y_5, V_{A6}, V_{A9}, \ldots]$ arbitrarily small increments of exponential form $[v_{A3}, v_{A4}, v_{A7}, v_{A8}, y_1, v_{A11}, y_5, v_{A6}, v_{A9}, \ldots] \cdot e^{\lambda \cdot t}$, and retaining the first order terms in $V_{A3}, V_{A4}, V_{A7}, V_{A8}, Y_1, V_{A11}, Y_5, V_{A6}, V_{A9}, \ldots$. The system of homogeneous equations leads to a polynomial characteristics equation in the eigenvalues λ. Our system fixed values with arbitrarily small increments of exponential form $[v_{A3}, v_{A4}, v_{A7}, v_{A8}, y_1, v_{A11}, y_5, v_{A6}, v_{A9}, \ldots] \cdot e^{\lambda \cdot t}$ are: j = 0 (first fixed point), j = 1 (second fixed point), j = 2 (third fixed point), etc. $V_{A3}(t) = V_{A3}^{(j)} + v_{A3} \cdot e^{\lambda \cdot t}$; $V_{A4}(t) = V_{A4}^{(j)} + v_{A4} \cdot e^{\lambda \cdot t}$

$$V_{A7}(t) = V_{A7}^{(j)} + v_{A7} \cdot e^{\lambda \cdot t};\ V_{A8}(t) = V_{A8}^{(j)} + v_{A8} \cdot e^{\lambda \cdot t};$$

$$Y_1(t) = Y_1^{(j)} + y_1 \cdot e^{\lambda \cdot t};\ V_{A11}(t) = V_{A11}^{(j)} + v_{A11} \cdot e^{\lambda \cdot t}$$

$$Y_5(t) = Y_5^{(j)} + y_5 \cdot e^{\lambda \cdot t};\ V_{A6}(t) = V_{A6}^{(j)} + v_{A6} \cdot e^{\lambda \cdot t};\ V_{A9}(t) = V_{A9}^{(j)} + v_{A9} \cdot e^{\lambda \cdot t} \ldots$$

We choose the above expressions for our $V_{A3}(t), V_{A4}(t), V_{A7}(t), V_{A8}(t), Y_1(t), V_{A11}(t), Y_5(t), V_{A6}(t), V_{A9}(t), \ldots$ as small displacement $[v_{A3}, v_{A4}, v_{A7}, v_{A8}, y_1, v_{A11}, y_5, v_{A6}, v_{A9}, \ldots]$

from our system fixed points at time t = 0. $V_{A3}(t = 0) = V_{A3}^{(j)} + v_{A3}$; $V_{A4}(t = 0) = V_{A4}^{(j)} + v_{A4}$

$$V_{A7}(t = 0) = V_{A7}^{(j)} + v_{A7};\ V_{A8}(t = 0) = V_{A8}^{(j)} + v_{A8};$$

$$Y_1(t = 0) = Y_1^{(j)} + y_1;\ V_{A11}(t = 0) = V_{A11}^{(j)} + v_{A11}$$

$$Y_5(t = 0) = Y_5^{(j)} + y_5;\ V_{A6}(t = 0) = V_{A6}^{(j)} + v_{A6};\ V_{A9}(t = 0) = V_{A9}^{(j)} + v_{A9} \ldots\ldots$$

For λ < 0, t > 0 the selected fixed point is stable otherwise λ > 0, t > 0 is unstable. Our system tends to the selected fixed point exponentially for λ < 0, t > 0 otherwise go away from the selected fixed point exponentially. λ is the eigenvalue parameter which is established if the fixed point is stable or unstable; additionally, his absolute value |λ| establishes the speed of flow toward or away from the selected fixed point (Yuri 1995; Jack and Huseyin 1991) [2–4]. The speeds of flow toward or away from the selected fixed point for system variables derivatives with respect to time are:

$$\frac{dV_{A3}(t)}{dt} = v_{A3} \cdot \lambda \cdot e^{\lambda \cdot t}; \frac{dV_{A4}(t)}{dt} = v_{A4} \cdot \lambda \cdot e^{\lambda \cdot t}$$
$$\frac{dV_{A7}(t)}{dt} = v_{A7} \cdot \lambda \cdot e^{\lambda \cdot t}; \frac{dV_{A8}(t)}{dt} = v_{A8} \cdot \lambda \cdot e^{\lambda \cdot t};$$
$$\frac{dY_1(t)}{dt} = y_1 \cdot \lambda \cdot e^{\lambda \cdot t}; \frac{dV_{A11}(t)}{dt} = v_{A11} \cdot \lambda \cdot e^{\lambda \cdot t}$$
$$\frac{dY_5(t)}{dt} = y_5 \cdot \lambda \cdot e^{\lambda \cdot t}; \frac{dV_{A6}(t)}{dt} = v_{A6} \cdot \lambda \cdot e^{\lambda \cdot t}; \frac{dV_{A9}(t)}{dt} = v_{A9} \cdot \lambda \cdot e^{\lambda \cdot t}\ldots$$
$$\frac{dV_{A3}}{dt} = Y_1 \Rightarrow v_{A3} \cdot \lambda \cdot e^{\lambda \cdot t} = Y_1^{(j)} + y_1 \cdot e^{\lambda \cdot t}; \text{ @ fixed point } Y_1^{(j)} = 0$$
$$Y_1^{(j)} = 0;\ y_1 - v_{A3} \cdot \lambda = 0; \frac{dV_{A4}}{dt} = Y_2 \Rightarrow v_{A4} \cdot \lambda \cdot e^{\lambda \cdot t} = Y_2^{(j)} + y_2 \cdot e^{\lambda \cdot t}$$
$$\text{@ fixed point } Y_2^{(j)} = 0; y_2 - v_{A4} \cdot \lambda = 0$$
$$\frac{dV_{A7}}{dt} = Y_3 \Rightarrow v_{A7} \cdot \lambda \cdot e^{\lambda \cdot t} = Y_3^{(j)} + y_3 \cdot e^{\lambda \cdot t} \text{@ fixed point } Y_3^{(j)} = 0$$
$$y_3 - v_{A7} \cdot \lambda = 0;\ \frac{dV_{A8}}{dt} = Y_4 \Rightarrow v_{A8} \cdot \lambda \cdot e^{\lambda \cdot t} = Y_4^{(j)} + y_4 \cdot e^{\lambda \cdot t}$$
$$\text{@ fixed point } Y_4^{(j)} = 0 \Rightarrow y_4 - v_{A8} \cdot \lambda = 0$$

$$\begin{aligned}
\frac{dY_1}{dt} = & \frac{1}{R_d \cdot [C_{pds} + C_{pgd}] \cdot C_{T1}} \cdot Y_3 - \frac{1}{R_d \cdot [C_{pds} + C_{pgd}] \cdot C_{T1}} \cdot Y_4 \\
& - \frac{1}{[C_{pds} + C_{pgd}] \cdot C_{T1}} \cdot \frac{1}{L_d} \cdot V_{A8} + \frac{1}{[C_{pds} + C_{pgd}] \cdot C_{T1}} \cdot \frac{1}{L_d} \\
& \cdot V_{A11} - \frac{1}{C_{pgd} \cdot C_{T1}} \cdot \frac{1}{L_g} \cdot V_g - \frac{1}{C_{pgd} \cdot C_{T1}} \cdot \frac{1}{L_g} \cdot V_{A3} \\
& - \frac{1}{C_{pgd} \cdot R_g \cdot C_{T1}} \cdot Y_1 + \frac{1}{C_{pgd} \cdot R_g \cdot C_{T1}} \cdot Y_2
\end{aligned}$$

$$y_1 \cdot \lambda \cdot e^{\lambda \cdot t} = \frac{1}{R_d \cdot [C_{pds} + C_{pgd}] \cdot C_{T1}} \cdot [Y_3^{(j)} + y_3 \cdot e^{\lambda \cdot t}] - \frac{1}{R_d \cdot [C_{pds} + C_{pgd}] \cdot C_{T1}} \cdot [Y_4^{(j)} + y_4 \cdot e^{\lambda \cdot t}] - \frac{1}{[C_{pds} + C_{pgd}] \cdot C_{T1}} \cdot \frac{1}{L_d} \cdot [V_{A8}^{(j)} + v_{A8} \cdot e^{\lambda \cdot t}] + \frac{1}{[C_{pds} + C_{pgd}] \cdot C_{T1}} \cdot \frac{1}{L_d} \cdot [V_{A11}^{(j)} + v_{A11} \cdot e^{\lambda \cdot t}] - \frac{1}{C_{pgd} \cdot C_{T1}} \cdot \frac{1}{L_g} \cdot V_g - \frac{1}{C_{pgd} \cdot C_{T1}} \cdot \frac{1}{L_g} \cdot [V_{A3}^{(j)} + v_{A3} \cdot e^{\lambda \cdot t}] - \frac{1}{C_{pgd} \cdot R_g \cdot C_{T1}} \cdot [Y_1^{(j)} + y_1 \cdot e^{\lambda \cdot t}] + \frac{1}{C_{pgd} \cdot R_g \cdot C_{T1}} \cdot [Y_2^{(j)} + y_2 \cdot e^{\lambda \cdot t}]$$

$$y_1 \cdot \lambda \cdot e^{\lambda \cdot t} = \{\frac{1}{R_d \cdot [C_{pds} + C_{pgd}] \cdot C_{T1}} \cdot Y_3^{(j)} - \frac{1}{R_d \cdot [C_{pds} + C_{pgd}] \cdot C_{T1}} \cdot Y_4^{(j)} - \frac{1}{[C_{pds} + C_{pgd}] \cdot C_{T1}} \cdot \frac{1}{L_d} \cdot V_{A8}^{(j)} + \frac{1}{[C_{pds} + C_{pgd}] \cdot C_{T1}} \cdot \frac{1}{L_d} \cdot V_{A11}^{(j)} - \frac{1}{C_{pgd} \cdot C_{T1}} \cdot \frac{1}{L_g} \cdot V_g - \frac{1}{C_{pgd} \cdot C_{T1}} \cdot \frac{1}{L_g} \cdot V_{A3}^{(j)} - \frac{1}{C_{pgd} \cdot R_g \cdot C_{T1}} \cdot Y_1^{(j)} + \frac{1}{C_{pgd} \cdot R_g \cdot C_{T1}} \cdot Y_2^{(j)}\} \frac{1}{R_d \cdot [C_{pds} + C_{pgd}] \cdot C_{T1}} \cdot y_3 \cdot e^{\lambda \cdot t} - \frac{1}{R_d \cdot [C_{pds} + C_{pgd}] \cdot C_{T1}} \cdot y_4 \cdot e^{\lambda \cdot t} - \frac{1}{[C_{pds} + C_{pgd}] \cdot C_{T1}} \cdot \frac{1}{L_d} \cdot v_{A8} \cdot e^{\lambda \cdot t} \frac{1}{[C_{pds} + C_{pgd}] \cdot C_{T1}} \cdot \frac{1}{L_d} \cdot v_{A11} \cdot e^{\lambda \cdot t} - \frac{1}{C_{pgd} \cdot C_{T1}} \cdot \frac{1}{L_g} \cdot v_{A3} \cdot e^{\lambda \cdot t} - \frac{1}{C_{pgd} \cdot R_g \cdot C_{T1}} \cdot y_1 \cdot e^{\lambda \cdot t} + \frac{1}{C_{pgd} \cdot R_g \cdot C_{T1}} \cdot y_2 \cdot e^{\lambda \cdot t}$$

@ fixed point

$$\{\frac{1}{R_d \cdot [C_{pds} + C_{pgd}] \cdot C_{T1}} \cdot Y_3^{(j)} - \frac{1}{R_d \cdot [C_{pds} + C_{pgd}] \cdot C_{T1}} \cdot Y_4^{(j)} - \frac{1}{[C_{pds} + C_{pgd}] \cdot C_{T1}} \cdot \frac{1}{L_d} \cdot V_{A8}^{(j)} + \frac{1}{[C_{pds} + C_{pgd}] \cdot C_{T1}} \cdot \frac{1}{L_d} \cdot V_{A11}^{(j)} - \frac{1}{C_{pgd} \cdot C_{T1}} \cdot \frac{1}{L_g} \cdot V_g - \frac{1}{C_{pgd} \cdot C_{T1}} \cdot \frac{1}{L_g} \cdot V_{A3}^{(j)} - \frac{1}{C_{pgd} \cdot R_g \cdot C_{T1}} \cdot Y_1^{(j)} + \frac{1}{C_{pgd} \cdot R_g \cdot C_{T1}} \cdot Y_2^{(j)}\} = 0$$

$$y_1 \cdot \lambda \cdot e^{\lambda \cdot t} = \frac{1}{R_d \cdot [C_{pds} + C_{pgd}] \cdot C_{T1}} \cdot y_3 \cdot e^{\lambda \cdot t} - \frac{1}{R_d \cdot [C_{pds} + C_{pgd}] \cdot C_{T1}} \cdot y_4 \cdot e^{\lambda \cdot t} - \frac{1}{[C_{pds} + C_{pgd}] \cdot C_{T1}} \cdot \frac{1}{L_d} \cdot v_{A8} \cdot e^{\lambda \cdot t} + \frac{1}{[C_{pds} + C_{pgd}] \cdot C_{T1}} \cdot \frac{1}{L_d} \cdot v_{A11} \cdot e^{\lambda \cdot t} - \frac{1}{C_{pgd} \cdot C_{T1}} \cdot \frac{1}{L_g} \cdot v_{A3} \cdot e^{\lambda \cdot t} - \frac{1}{C_{pgd} \cdot R_g \cdot C_{T1}} \cdot y_1 \cdot e^{\lambda \cdot t} + \frac{1}{C_{pgd} \cdot R_g \cdot C_{T1}} \cdot y_2 \cdot e^{\lambda \cdot t}$$

$$y_1 \cdot \lambda = \frac{1}{R_d \cdot [C_{pds} + C_{pgd}] \cdot C_{T1}} \cdot y_3 - \frac{1}{R_d \cdot [C_{pds} + C_{pgd}] \cdot C_{T1}} \cdot y_4 - \frac{1}{[C_{pds} + C_{pgd}] \cdot C_{T1}} \cdot \frac{1}{L_d} \cdot v_{A8} + \frac{1}{[C_{pds} + C_{pgd}] \cdot C_{T1}} \cdot \frac{1}{L_d} \cdot v_{A11} - \frac{1}{C_{pgd} \cdot C_{T1}} \cdot \frac{1}{L_g} \cdot v_{A3} - \frac{1}{C_{pgd} \cdot R_g \cdot C_{T1}} \cdot y_1 + \frac{1}{C_{pgd} \cdot R_g \cdot C_{T1}} \cdot y_2$$

$$\frac{1}{R_d \cdot [C_{pds} + C_{pgd}] \cdot C_{T1}} \cdot y_3 - \frac{1}{R_d \cdot [C_{pds} + C_{pgd}] \cdot C_{T1}} \cdot y_4 - \frac{1}{[C_{pds} + C_{pgd}] \cdot C_{T1}} \cdot \frac{1}{L_d} \cdot v_{A8} + \frac{1}{[C_{pds} + C_{pgd}] \cdot C_{T1}} \cdot \frac{1}{L_d} \cdot v_{A11} - \frac{1}{C_{pgd} \cdot C_{T1}} \cdot \frac{1}{L_g} \cdot v_{A3} - \frac{1}{C_{pgd} \cdot R_g \cdot C_{T1}} \cdot y_1 + \frac{1}{C_{pgd} \cdot R_g \cdot C_{T1}} \cdot y_2 - y_1 \cdot \lambda = 0$$

$\frac{dV_{A11}}{dt} = Y_5 \Rightarrow v_{A11} \cdot \lambda \cdot e^{\lambda \cdot t} = Y_5^{(j)} + y_5 \cdot e^{\lambda \cdot t}$; @ fixed point $Y_5^{(j)} = 0 \Rightarrow v_{A11} \cdot \lambda \cdot e^{\lambda \cdot t} = y_5 \cdot e^{\lambda \cdot t}$

$$v_{A11} \cdot \lambda \cdot e^{\lambda \cdot t} = y_5 \cdot e^{\lambda \cdot t} \Rightarrow y_5 - v_{A11} \cdot \lambda = 0; \frac{dV_{A6}}{dt} = Y_6 \Rightarrow v_{A6} \cdot \lambda \cdot e^{\lambda \cdot t} = Y_6^{(j)} + y_6 \cdot e^{\lambda \cdot t}$$

@ fixed point $Y_6^{(j)} = 0 \Rightarrow v_{A6} \cdot \lambda \cdot e^{\lambda \cdot t} = y_6 \cdot e^{\lambda \cdot t} \Rightarrow y_6 - v_{A6} \cdot \lambda = 0$

$\frac{dV_{A9}}{dt} = Y_7 + Y_3 \Rightarrow v_{A9} \cdot \lambda \cdot e^{\lambda \cdot t} = Y_7^{(j)} + y_7 \cdot e^{\lambda \cdot t} + Y_3^{(j)} + y_3 \cdot e^{\lambda \cdot t}$ @ fixed point $Y_7^{(j)} + Y_3^{(j)} = 0$

$$v_{A9} \cdot \lambda \cdot e^{\lambda \cdot t} = y_7 \cdot e^{\lambda \cdot t} + y_3 \cdot e^{\lambda \cdot t} \Rightarrow v_{A9} \cdot \lambda = y_7 + y_3 \Rightarrow y_7 + y_3 - v_{A9} \cdot \lambda = 0$$

$$\frac{dY_5}{dt} = \frac{1}{C_{out} \cdot [\frac{R_L}{R_{d1}} + 1]} \cdot \frac{1}{L_d} \cdot V_{A8} - \frac{1}{C_{out} \cdot [\frac{R_L}{R_{d1}} + 1]} \cdot \frac{1}{L_d} \cdot V_{A11} + \frac{R_L}{[\frac{R_L}{R_{d1}} + 1] \cdot L_d} \cdot Y_4 - [\frac{R_L}{L_d} + \frac{1}{C_{out} \cdot R_{d1}}] \cdot \frac{1}{[\frac{R_L}{R_{d1}} + 1]} \cdot Y_5$$

$$y_5 \cdot \lambda \cdot e^{\lambda \cdot t} + \frac{R_L}{[\frac{R_L}{R_{d1}} + 1] \cdot L_d} \cdot [Y_4^{(j)} + y_4 \cdot e^{\lambda \cdot t}] - [\frac{R_L}{L_d} + \frac{1}{C_{out} \cdot R_{d1}}] \cdot \frac{1}{[\frac{R_L}{R_{d1}} + 1]} \cdot [Y_5^{(j)} + y_5 \cdot e^{\lambda \cdot t}]$$

$$y_5 \cdot \lambda \cdot e^{\lambda \cdot t} = \{\frac{1}{C_{out} \cdot [\frac{R_L}{R_{d1}} + 1]} \cdot \frac{1}{L_d} \cdot V_{A8}^{(j)} - \frac{1}{C_{out} \cdot [\frac{R_L}{R_{d1}} + 1]} \cdot \frac{1}{L_d} \cdot V_{A11}^{(j)} + \frac{R_L}{[\frac{R_L}{R_{d1}} + 1] \cdot L_d} \cdot Y_4^{(j)} - [\frac{R_L}{L_d} + \frac{1}{C_{out} \cdot R_{d1}}] \cdot \frac{1}{[\frac{R_L}{R_{d1}} + 1]} \cdot Y_5^{(j)}\} + \frac{1}{C_{out} \cdot [\frac{R_L}{R_{d1}} + 1]} \cdot \frac{1}{L_d} \cdot v_{A8} \cdot e^{\lambda \cdot t} - \frac{1}{C_{out} \cdot [\frac{R_L}{R_{d1}} + 1]} \cdot \frac{1}{L_d} \cdot v_{A11} \cdot e^{\lambda \cdot t} + \frac{R_L}{[\frac{R_L}{R_{d1}} + 1] \cdot L_d} \cdot y_4 \cdot e^{\lambda \cdot t} - [\frac{R_L}{L_d} + \frac{1}{C_{out} \cdot R_{d1}}] \cdot \frac{1}{[\frac{R_L}{R_{d1}} + 1]} \cdot y_5 \cdot e^{\lambda \cdot t}$$

@ fixed point

$$\{\frac{1}{C_{out} \cdot [\frac{R_L}{R_{d1}} + 1]} \cdot \frac{1}{L_d} \cdot V_{A8}^{(j)} - \frac{1}{C_{out} \cdot [\frac{R_L}{R_{d1}} + 1]} \cdot \frac{1}{L_d} \cdot V_{A11}^{(j)} + \frac{R_L}{[\frac{R_L}{R_{d1}} + 1] \cdot L_d} \cdot Y_4^{(j)} - [\frac{R_L}{L_d} + \frac{1}{C_{out} \cdot R_{d1}}] \cdot \frac{1}{[\frac{R_L}{R_{d1}} + 1]} \cdot Y_5^{(j)}\} = 0$$

$$y_5 \cdot \lambda = \frac{1}{C_{out} \cdot [\frac{R_L}{R_{d1}} + 1]} \cdot \frac{1}{L_d} \cdot v_{A8} - \frac{1}{C_{out} \cdot [\frac{R_L}{R_{d1}} + 1]} \cdot \frac{1}{L_d} \cdot v_{A11} + \frac{R_L}{[\frac{R_L}{R_{d1}} + 1] \cdot L_d} \cdot y_4 - [\frac{R_L}{L_d} + \frac{1}{C_{out} \cdot R_{d1}}] \cdot \frac{1}{[\frac{R_L}{R_{d1}} + 1]} \cdot y_5$$

$$\frac{1}{C_{out} \cdot [\frac{R_L}{R_{d1}} + 1]} \cdot \frac{1}{L_d} \cdot v_{A8} - \frac{1}{C_{out} \cdot [\frac{R_L}{R_{d1}} + 1]} \cdot \frac{1}{L_d} \cdot v_{A11} + \frac{R_L}{[\frac{R_L}{R_{d1}} + 1] \cdot L_d} \cdot y_4 - [\frac{R_L}{L_d} + \frac{1}{C_{out} \cdot R_{d1}}] \cdot \frac{1}{[\frac{R_L}{R_{d1}} + 1]} \cdot y_5 - y_5 \cdot \lambda = 0$$

$$\frac{dY_7}{dt} = \frac{1}{C_{ds} \cdot R_{ds}} \cdot Y_7 + \frac{1}{C_{ds} \cdot R_{gd}} \cdot Y_2 + \frac{1}{C_{ds} \cdot R_{di}} \cdot Y_6 - Y_3 \cdot [\frac{1}{C_{ds} \cdot R_{gd}} + \frac{1}{C_{ds} \cdot R_{di}} + \frac{1}{C_{ds} \cdot R_d}] + \frac{1}{C_{ds} \cdot R_d} \cdot Y_4 - \frac{g_m}{C_{ds}} \cdot \frac{1}{C_{gs}} \cdot \frac{V_{A5}}{R_i} + \frac{g_m}{C_{ds}} \cdot \frac{1}{C_{gs}} \cdot \frac{V_{A9}}{R_i}$$

$$y_7 \cdot \lambda \cdot e^{\lambda \cdot t} = \frac{1}{C_{ds} \cdot R_{ds}} \cdot [Y_7^{(j)} + y_7 \cdot e^{\lambda \cdot t}] + \frac{1}{C_{ds} \cdot R_{gd}} \cdot [Y_2^{(j)} + y_2 \cdot e^{\lambda \cdot t}] + \frac{1}{C_{ds} \cdot R_{di}} \cdot [Y_6^{(j)} + y_6 \cdot e^{\lambda \cdot t}] - [Y_3^{(j)} + y_3 \cdot e^{\lambda \cdot t}] \cdot [\frac{1}{C_{ds} \cdot R_{gd}} + \frac{1}{C_{ds} \cdot R_{di}} + \frac{1}{C_{ds} \cdot R_d}] + \frac{1}{C_{ds} \cdot R_d} \cdot [Y_4^{(j)} + y_4 \cdot e^{\lambda \cdot t}] - \frac{g_m}{C_{ds}} \cdot \frac{1}{C_{gs}} \cdot \frac{[V_{A5}^{(j)} + v_{A5} \cdot e^{\lambda \cdot t}]}{R_i} + \frac{g_m}{C_{ds}} \cdot \frac{1}{C_{gs}} \cdot \frac{[V_{A9}^{(j)} + v_{A9} \cdot e^{\lambda \cdot t}]}{R_i}$$

$$y_7 \cdot \lambda \cdot e^{\lambda \cdot t} = \{\frac{1}{C_{ds} \cdot R_{ds}} \cdot Y_7^{(j)} + \frac{1}{C_{ds} \cdot R_{gd}} \cdot Y_2^{(j)} + \frac{1}{C_{ds} \cdot R_{di}} \cdot Y_6^{(j)} - Y_3^{(j)} \cdot [\frac{1}{C_{ds} \cdot R_{gd}} + \frac{1}{C_{ds} \cdot R_{di}} + \frac{1}{C_{ds} \cdot R_d}] + \frac{1}{C_{ds} \cdot R_d} \cdot Y_4^{(j)} - \frac{g_m}{C_{ds}} \cdot \frac{1}{C_{gs}} \cdot \frac{V_{A5}^{(j)}}{R_i} + \frac{g_m}{C_{ds}} \cdot \frac{1}{C_{gs}} \cdot \frac{V_{A9}^{(j)}}{R_i}\} + \frac{1}{C_{ds} \cdot R_{ds}} \cdot y_7 \cdot e^{\lambda \cdot t} + \frac{1}{C_{ds} \cdot R_{gd}} \cdot y_2 \cdot e^{\lambda \cdot t} + \frac{1}{C_{ds} \cdot R_{di}} \cdot y_6 \cdot e^{\lambda \cdot t} - [\frac{1}{C_{ds} \cdot R_{gd}} + \frac{1}{C_{ds} \cdot R_{di}} + \frac{1}{C_{ds} \cdot R_d}] \cdot y_3 \cdot e^{\lambda \cdot t} + \frac{1}{C_{ds} \cdot R_d} \cdot y_4 \cdot e^{\lambda \cdot t} - \frac{g_m}{C_{ds}} \cdot \frac{1}{C_{gs} \cdot R_i} \cdot v_{A5} \cdot e^{\lambda \cdot t} + \frac{g_m}{C_{ds}} \cdot \frac{1}{C_{gs} \cdot R_i} \cdot v_{A9} \cdot e^{\lambda \cdot t}$$

@ fixed point

$$\{\frac{1}{C_{ds}\cdot R_{ds}}\cdot Y_7^{(j)}+\frac{1}{C_{ds}\cdot R_{gd}}\cdot Y_2^{(j)}+\frac{1}{C_{ds}\cdot R_{di}}\cdot Y_6^{(j)}-Y_3^{(j)}$$

$$\cdot[\frac{1}{C_{ds}\cdot R_{gd}}+\frac{1}{C_{ds}\cdot R_{di}}+\frac{1}{C_{ds}\cdot R_d}]$$

$$+\frac{1}{C_{ds}\cdot R_d}\cdot Y_4^{(j)}-\frac{g_m}{C_{ds}}\cdot\frac{1}{C_{gs}}\cdot\frac{V_{A5}^{(j)}}{R_i}+\frac{g_m}{C_{ds}}\cdot\frac{1}{C_{gs}}\cdot\frac{V_{A9}^{(j)}}{R_i}\}=0$$

$$y_7\cdot\lambda=\frac{1}{C_{ds}\cdot R_{ds}}\cdot y_7+\frac{1}{C_{ds}\cdot R_{gd}}\cdot y_2+\frac{1}{C_{ds}\cdot R_{di}}\cdot y_6$$

$$-[\frac{1}{C_{ds}\cdot R_{gd}}+\frac{1}{C_{ds}\cdot R_{di}}+\frac{1}{C_{ds}\cdot R_d}]\cdot y_3$$

$$+\frac{1}{C_{ds}\cdot R_d}\cdot y_4-\frac{g_m}{C_{ds}}\cdot\frac{1}{C_{gs}\cdot R_i}\cdot v_{A5}+\frac{g_m}{C_{ds}}\cdot\frac{1}{C_{gs}\cdot R_i}\cdot v_{A9}$$

$$\frac{1}{C_{ds}\cdot R_{ds}}\cdot y_7+\frac{1}{C_{ds}\cdot R_{gd}}\cdot y_2+\frac{1}{C_{ds}\cdot R_{di}}\cdot y_6-[\frac{1}{C_{ds}\cdot R_{gd}}+\frac{1}{C_{ds}\cdot R_{di}}+\frac{1}{C_{ds}\cdot R_d}]\cdot y_3$$

$$+\frac{1}{C_{ds}\cdot R_d}\cdot y_4-\frac{g_m}{C_{ds}}\cdot\frac{1}{C_{gs}\cdot R_i}\cdot v_{A5}+\frac{g_m}{C_{ds}}\cdot\frac{1}{C_{gs}\cdot R_i}\cdot v_{A9}-y_7\cdot\lambda=0$$

Summary of our results, we get arbitrarily small increments $(v_{A3}, v_{A4}, v_{A7}, v_{A8}, y_1, v_{A11}, y_5, v_{A6}, v_{A9}, \ldots)$ ten equations:

$$y_1-v_{A3}\cdot\lambda=0\,;\;y_2-v_{A4}\cdot\lambda=0\,;\;y_3-v_{A7}\cdot\lambda=0\,;\;y_4-v_{A8}\cdot\lambda=0$$

$$\frac{1}{R_d\cdot[C_{pds}+C_{pgd}]\cdot C_{T1}}\cdot y_3-\frac{1}{R_d\cdot[C_{pds}+C_{pgd}]\cdot C_{T1}}\cdot y_4-\frac{1}{[C_{pds}+C_{pgd}]\cdot C_{T1}}\cdot\frac{1}{L_d}\cdot v_{A8}$$

$$+\frac{1}{[C_{pds}+C_{pgd}]\cdot C_{T1}}\cdot\frac{1}{L_d}\cdot v_{A11}-\frac{1}{C_{pgd}\cdot C_{T1}}\cdot\frac{1}{L_g}\cdot v_{A3}$$

$$-\frac{1}{C_{pgd}\cdot R_g\cdot C_{T1}}\cdot y_1+\frac{1}{C_{pgd}\cdot R_g\cdot C_{T1}}\cdot y_2-y_1\cdot\lambda=0$$

$$y_5-v_{A11}\cdot\lambda=0\,;\;y_6-v_{A6}\cdot\lambda=0\,;\;y_7+y_3-v_{A9}\cdot\lambda=0$$

$$\frac{1}{C_{out}\cdot[\frac{R_L}{R_{d1}}+1]}\cdot\frac{1}{L_d}\cdot v_{A8}-\frac{1}{C_{out}\cdot[\frac{R_L}{R_{d1}}+1]}\cdot\frac{1}{L_d}\cdot v_{A11}+\frac{R_L}{[\frac{R_L}{R_{d1}}+1]\cdot L_d}\cdot y_4$$

$$-[\frac{R_L}{L_d}+\frac{1}{C_{out}\cdot R_{d1}}]\cdot\frac{1}{[\frac{R_L}{R_{d1}}+1]}\cdot y_5-y_5\cdot\lambda=0$$

$$\frac{1}{C_{ds}\cdot R_{ds}}\cdot y_7+\frac{1}{C_{ds}\cdot R_{gd}}\cdot y_2+\frac{1}{C_{ds}\cdot R_{di}}\cdot y_6-[\frac{1}{C_{ds}\cdot R_{gd}}+\frac{1}{C_{ds}\cdot R_{di}}+\frac{1}{C_{ds}\cdot R_d}]\cdot y_3$$
$$+\frac{1}{C_{ds}\cdot R_d}\cdot y_4-\frac{g_m}{C_{ds}}\cdot\frac{1}{C_{gs}\cdot R_i}\cdot v_{A5}+\frac{g_m}{C_{ds}}\cdot\frac{1}{C_{gs}\cdot R_i}\cdot v_{A9}-y_7\cdot\lambda=0$$

$$\begin{pmatrix}\iota_{1\_1} & \cdots & \iota_{1\_10}\\ \vdots & \ddots & \vdots\\ \iota_{10\_1} & \cdots & \iota_{10\_10}\end{pmatrix}\cdot\begin{pmatrix}v_{A_3}\\ v_{A_4}\\ v_{A_7}\\ v_{A_8}\\ y_1\\ v_{A_{11}}\\ v_{A_6}\\ v_{A_9}\\ y_5\\ y_7\end{pmatrix}+\begin{pmatrix}v_{11} & \cdots & v_{15}\\ \vdots & \ddots & \vdots\\ v_{10\_1} & \cdots & v_{10\_5}\end{pmatrix}\cdot\begin{pmatrix}y_2\\ y_3\\ y_4\\ y_6\\ v_{A_5}\end{pmatrix}=0;\ \iota_{1\_1}$$
$$=-\lambda;\ \iota_{1\_2}=\iota_{1\_3}=\iota_{1\_4}=0$$

$$\iota_{1\_6}=\cdots=\iota_{1\_10}=0;\ \iota_{2\_1}=0;\ \iota_{2\_2}=-\lambda;$$
$$\iota_{2\_3}=\ldots=\iota_{2\_10};\ \iota_{3\_1}=\iota_{3\_2}=0;\ \iota_{3\_3}=-\lambda$$
$$\iota_{3\_4}=\ldots=\iota_{3\_10}=0;\ \iota_{4\_1}=\iota_{4\_2}=\iota_{4\_3}=0;\ \iota_{4\_4}=-\lambda;$$
$$\iota_{4\_5}=\ldots=\iota_{4\_10}=0;\ \iota_{5\_1}=-\frac{1}{C_{pgd}\cdot C_{T1}}\cdot\frac{1}{L_g}$$
$$\iota_{5\_2}=0;\ \iota_{5\_3}=0;\ \iota_{5\_4}=-\frac{1}{[C_{pds}+C_{pgd}]\cdot C_{T1}}\cdot\frac{1}{L_d};$$
$$\iota_{5\_5}=-\lambda-\frac{1}{C_{pgd}\cdot R_g\cdot C_{T1}}$$
$$\iota_{5\_6}=\frac{1}{[C_{pds}+C_{pgd}]\cdot C_{T1}}\cdot\frac{1}{L_d};\ \iota_{5\_7}=\iota_{5\_8}=\iota_{5\_9}=\iota_{5\_10}=0$$
$$\iota_{6\_1}=\ldots=\iota_{6\_5}=0;\ \iota_{6\_6}=-\lambda$$
$$\iota_{6\_7}=0;\ \iota_{6\_8}=0;\ \iota_{6\_9}=1;\ \iota_{6\_10}=0;\ \iota_{7\_1}=\ldots=\iota_{7\_6}=0;$$
$$\iota_{7\_7}=-\lambda;\ \iota_{7\_8}=\iota_{7\_9}=\iota_{7\_10}=0$$
$$\iota_{8\_1}=\ldots=\iota_{8\_7}=0;\ \iota_{8\_8}=-\lambda;\ \iota_{8\_9}=0;\ \iota_{8\_10}=1;$$
$$\iota_{9\_1}=\iota_{9\_2}=\iota_{9\_3}=0\ \iota_{9\_4}=\frac{1}{C_{out}\cdot[\frac{R_L}{R_{d1}}+1]}\cdot\frac{1}{L_d}$$
$$\iota_{9\_5}=0;\ \iota_{9\_6}=-\frac{1}{C_{out}\cdot[\frac{R_L}{R_{d1}}+1]}\cdot\frac{1}{L_d};\ \iota_{9\_7}=\iota_{9\_8}=0$$
$$\iota_{9\_9}=-\lambda-[\frac{R_L}{L_d}+\frac{1}{C_{out}\cdot R_{d1}}]\cdot\frac{1}{[\frac{R_L}{R_{d1}}+1]}$$

$$\iota_{9\_10} = 0\,;\ \iota_{10\_1} = \iota_{10\_2} = \iota_{10\_3} = \iota_{10\_4} = \iota_{10\_5} = \iota_{10\_6}$$
$$= \iota_{10\_7} = 0; \iota_{10\_8} = \frac{g_m}{C_{ds}} \cdot \frac{1}{C_{gs} \cdot R_i}$$
$$\iota_{10\_9} = 0\,;\ \iota_{10\_10} = -\lambda + \frac{1}{C_{ds} \cdot R_{ds}}\,;\ v_{11} = \ldots = v_{15} = 0\,;$$
$$v_{21} = 1\,;\ v_{22} = \ldots = v_{25} = 0$$
$$v_{31} = 0\,;\ v_{32} = 1\,;\ v_{33} = v_{34} = v_{35} = 0\,;$$
$$v_{41} = v_{42} = 0\,;\ v_{43} = 1\,;\ v_{44} = 0\,;\ v_{45} = 0$$
$$v_{51} = \frac{1}{C_{pgd} \cdot R_g \cdot C_{T1}}\,;\ v_{52} = \frac{1}{R_d \cdot [C_{pds} + C_{pgd}] \cdot C_{T1}}\,;$$
$$v_{53} = -\frac{1}{R_d \cdot [C_{pds} + C_{pgd}] \cdot C_{T1}}$$
$$v_{54} = v_{55} = 0\,;\ v_{61} = \ldots = v_{65} = 0\,;\ v_{71} = v_{72} = v_{73} = 0;$$
$$v_{74} = 1\,;\ v_{75} = 0\,;\ v_{81} = 0\,;\ v_{82} = 1$$
$$v_{83} = v_{84} = v_{85} = 0\,;\ v_{91} = v_{92} = 0;$$
$$v_{93} = \frac{R_L}{[\frac{R_L}{R_{d1}} + 1] \cdot L_d}\,;\ v_{94} = v_{95} = 0$$
$$v_{10\_1} = \frac{1}{C_{ds} \cdot R_{gd}}\,;\ v_{10\_2} = -[\frac{1}{C_{ds} \cdot R_{gd}} + \frac{1}{C_{ds} \cdot R_{di}} + \frac{1}{C_{ds} \cdot R_d}]$$
$$v_{10\_3} = \frac{1}{C_{ds} \cdot R_d} v_{10\_4} = \frac{1}{C_{ds} \cdot R_{di}}\,;\ v_{10\_5} = -\frac{g_m}{C_{ds}} \cdot \frac{1}{C_{gs} \cdot R_i}$$

Assumption:

$$\begin{pmatrix} v_{11} & \cdots & v_{15} \\ \vdots & \ddots & \vdots \\ v_{10\_1} & \cdots & v_{10\_5} \end{pmatrix} \rightarrow \varepsilon; \det(A - \lambda \cdot I) = 0\,;\ A - \lambda \cdot I = \begin{pmatrix} \iota_{1\_1} & \cdots & \iota_{1\_10} \\ \vdots & \ddots & \vdots \\ \iota_{10\_1} & \cdots & \iota_{10\_10} \end{pmatrix}$$

$$\begin{pmatrix} \iota_{1\_1} & \cdots & \iota_{1\_10} \\ \vdots & \ddots & \vdots \\ \iota_{10\_1} & \cdots & \iota_{10\_10} \end{pmatrix} \cdot \begin{pmatrix} v_{A_3} \\ v_{A_4} \\ v_{A_7} \\ v_{A_8} \\ y_1 \\ v_{A_{11}} \\ v_{A_6} \\ v_{A_9} \\ y_5 \\ y_7 \end{pmatrix} \approx 0\,;\ \det(A - \lambda \cdot I) = 0 \Rightarrow \det \begin{pmatrix} \iota_{1\_1} & \cdots & \iota_{1\_10} \\ \vdots & \ddots & \vdots \\ \iota_{10\_1} & \cdots & \iota_{10\_10} \end{pmatrix} \approx 0$$

To effectively apply the stability criterion of Lipunov to our system, we require a criterion for when the equation $\det(A - \lambda \cdot I) = 0$ has a zero in the left half plane, without calculating the eigenvalues explicitly. We use criterion of Routh-Hurwitz [2–4].

3.4 IMPATT Amplifier Stability Analysis

A wide variety of solid state diodes and transistor have been developed for microwave use. IMPact ionization Avalanche Transit-Time (IMPATT) diode functions as microwave oscillator. It used to produce carrier signal for microwave transmission system. IMPATT can operate from a few GHz to a few hundred GHz. The diode is operated in reverse bias near breakdown, and both the N and N-regions are completely depleted. Because of the difference in doping between the "drift region" and "avalanche region", the electric field is highly peaked in the avalanche region and nearly flat in drift region. In operation, avalanche breakdown occurs at the point of highest electric field, and this generates a large number of hole-electron pairs by impact ionization. The holes are swept into the cathode, but the electrons travel across the drift region toward anode. As they drift, they induce image charges on the anode, giving rise to a displacement current in the external circuit that is 180° out of phase with the nearly sinusoidal voltage waveform. It's buildup of microwave oscillations in the diode current and voltage when the diode is embedded in a resonant cavity and biased at breakdown. The IMPATT diode has a negative resistance from DC through microwave frequencies. Consequently, it is prone to oscillate at low frequencies, with the lead inductance from bias circuit connections. The voltage due to bias circuit oscillations may be large enough to burn the device out if adequate precautions are not observed. It is prudent practice to suppress the bias circuit oscillation. Adequate heat sink must be provided for the diode to operate properly. These IMPATT diodes have been designed to operate in the pre-collection mode. As the diode is tuned up from a low operating current from a constant current source, it will be noticed that at the onset of pre-collection mode, the diode voltage falls down. The power output will increase by several dBs with a slight shift in the operating frequency. When the circuit is detuned in such a fashion that the diode falls out of the pre-collection mode, the diode voltage will increase. The power dissipation will increase as the power output falls down. If the diode is not adequately heat sink, the diode may burn out. A main advantage is their high power capability. These diodes are used in a variety of applications from low power radar systems to alarms. IMPATT oscillator is for higher-power output, higher efficiency, and higher frequency range of operation. The effect of negative resistance of IMPATT diode in the amplification of microwave signals. Nonlinear effects are dominant considerations in power amplifier design because of efficiency and economy consideration of the device. IMPATT amplifier at lower frequency band is a very interesting area. Ka-band reflection type IMPATT amplifier has been developed using a Ka-band IMPATT diode oscillator as an input signal source. These amplifiers have small size, simple arrangement, and sufficient power addition for various applications in the field of high frequency communication and radar. IMPATT amplifiers are used mostly at the high end of the microwave band because microwave transistors do not work well above 30 GHz due to transit time limitations. IMPATT uses transit time effects to generate microwaves. IMPATT amplifier consist circulator (three ports circulator) which is connected to an IMPATT diode, mounted in a resonant circuit or cavity. A circulator must be used to separate the input and output powers. Since IMPATT is a single port device, circulator must be used to separate the input and the output power RF signal. The microwave power to be amplified enters one port of the circulator and is routed into the IMPATT diode in cavity. The incoming microwave RF signal is amplified and leaves the cavity from the same port that it entered. It is then routed by the circulator into the output transmission line [67–74].

IMPATT diode construction: IMPATT diode consists of a PN junction between the P^+ and the N regions, a drift region of intrinsic (I) material, and an N^+ connection. IMPATT diode has negative resistance characteristics. The microwave negative resistance of an IMPATT diode arises out of a phase difference between the RF voltage and RF current. This phase difference is produced by the lagging RF current generated in the space charge layer with respect to the applied RF voltage.

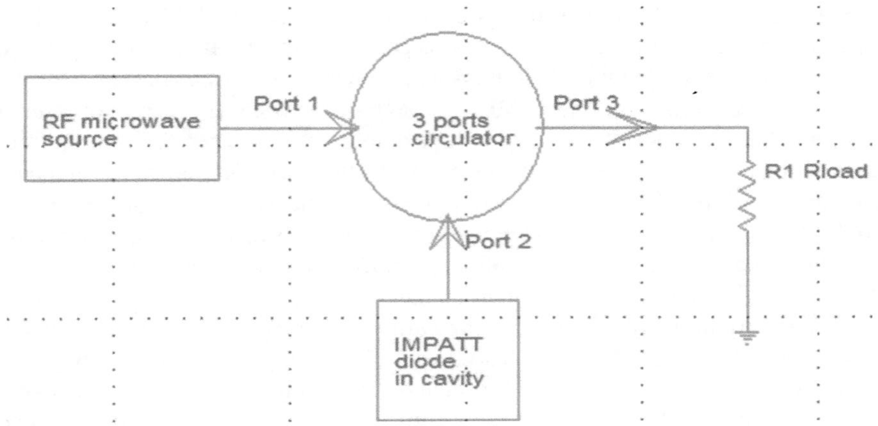

Fig. 3.9 IMPATT amplifier circuit with R_1 load

An IMPATT diode can function as an amplifier if the load resistance presented to it is larger in magnitude than the diode's negative resistance.

The intrinsic region thickness is controlled and the electron transit time through it is a half a microwave cycle at the selected operational frequency. Basically IMPATT diode is a form of high power diode used in high frequency electronics and microwave devices. They operate at frequencies between 3 GHz to 100GHz or more. The main advantage of IMPATT diode is their high power capability. These diodes are used in a variety of applications from low power radar systems to alarms. These diodes make excellent microwave generators for many applications. An IMPATT diode is mounted in a microwave package. The diode is mounted with its high field region close to a copper heat sink so that the heat generated at the diode junction can be readily dissipated.

An IMPATT diode can function as an amplifier if the load resistance presented to it is larger in magnitude than the diode's negative resistance. A Bias T is required for operation of the amplifier cavity. The IMPATT diode requires a DC bias current. It is necessary to block the DC bias current from the rest of the circuitry. This has been accomplished using a microstrip inter-digital DC blocking capacitor circuit. The DC bias current is applied to the diode through a very high RF impedance (quarter-wavelength long, very narrow piece of copper) terminated in a low RF impedance (wide piece of copper). The next figure shows IMPATT reflection amplifier. A circulator is used to separate input and output signals. The maximum possible DC bias current above which the diode breaks into oscillations can be readily observed also. An IMPATT diode exhibited the same characteristics of tending toward saturation with increasing input power levels and a corresponding increase in bandwidth.

IMPATT diode negative resistance typically varies as a function of the diode RF current amplitude. R_D is the terminal (negative) resistance of the packaged diode and R_L is the diode's load resistance. IR_DI decreases with signal level. R_D also varies with DC bias current and thus the upper limit of bias current is established at the value that causes IR_DI to equal R_L. Exceeding this maximum value of bias current will cause the diode to act as an oscillator instead of an amplifier because the diodes load resistance, R_L, is no longer greater than the magnitude of the diode's negative resistance, IR_DI. An IMPATT amplifier requires that R_L be larger than IR_DI for all values of the RF current through the diode. Since R_D varies not only with different types of diodes, but also with DC bias current and signal level, the selection of R_L for optimum power gain is of prime importance in the design of the amplifier.

If $R_L > |R_D|$ then IMPATT diode acts as an amplifier. If $R_L < |R_D|$ then IMPATT diode acts as an oscillator ($R_L = R_{load}$).

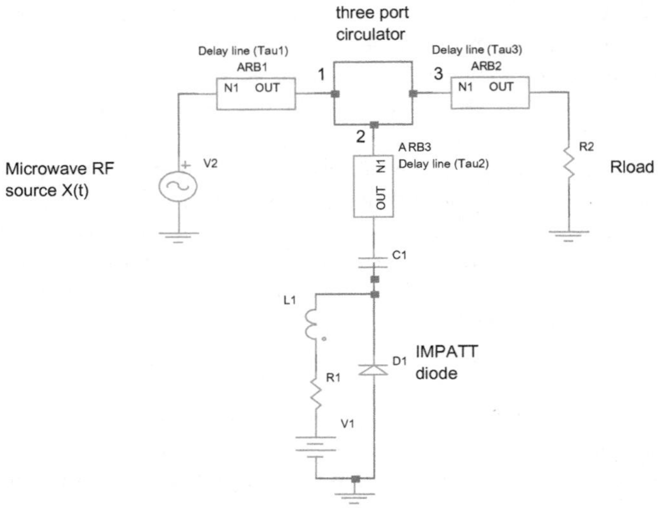

V_1 — DC voltage source (bias voltage to IMPATT diode).
R_1 — parasitic resistance of DC voltage source.
V_2 — Microwave RF source X(t).
L_1, C_1 — inductor and capacitor of BiasT circuit.
ARB1, ARB2, ARB3 — circuit micro strip delay lines.
R_2 — circuit load resistance.
D_1 — IMPATT diode.

Fig. 3.10 IMPATT amplifier diode negative resistance circuit

IMPATT diode is current controlled negative resistance (CCNR, open circuit stable, or "S" type). In this type, the voltage is a single valued function of current, but the current is a multivalued function of voltage. The graph is a curve shaped like the letter "S". Negative differential resistance devices such as IMPATT diode are used to make amplifiers, particularly at microwave frequencies, but not as commonly as oscillators. IMPATT diode (negative resistance device) has only one port (two terminals), unlike two ports devices such as transistors, the outgoing amplified signal has to leave the IMPATT diode by the same terminals as the incoming signal enter it. If we do not use the circulator the IMPATT diode negative resistance amplifier is bilateral. It amplifiers in both directions then there is high sensitivity to load impedance and feedback problems. To separate the input and output signals, IMPATT diode negative resistance amplifier use nonreciprocal device such as isolator and directional couplers. In our case we use reflection amplifier in which the separation is accomplished by an active circulator. The IMPATT diode chip RF equivalent circuit includes the active part of the diode (the chip, excluding the

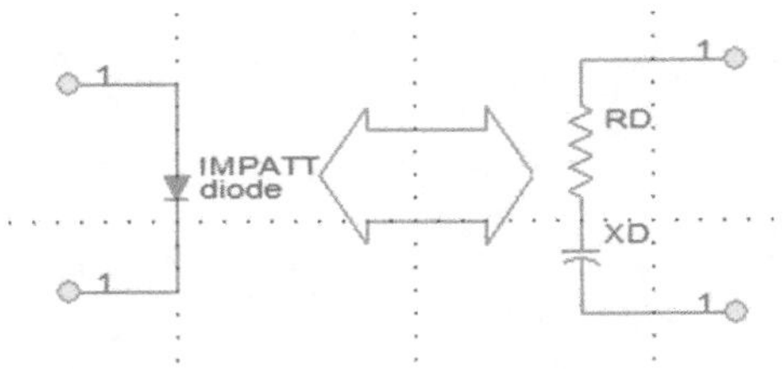

Fig. 3.11 IMPATT diode chip RF equivalent circuit

package) as a negative resistance R_D, and a reactance X_D. Included in R_D is the unavoidable parasitic series resistance, R_S, contributed by contacts and the un depleted portion of the N region.

$R_D = -R_\gamma$; $R_\gamma > 0$; $R_D < 0$; $|R_D| = R_\gamma$; $X_D = \frac{1}{j \cdot \omega \cdot C_D}$. If we connect IMPATT diode directly to bias voltage V_b which locate our IMPATT diode working point DC characteristic in the negative resistance region. V_b is IMPATT diode bias voltage source. R_b is voltage source series resistance. S_1 is a bypass RF microwave source switch. X(t) is RF microwave signal source.

$$I = I_{C_d} + I_{R_d} = I_{R_b} \text{ ; } I_{C_d} = C_d \cdot \frac{dV_{C_d}}{dt} \text{ ; } V_{C_d} = \frac{1}{C_d} \cdot \int I_{C_d} \cdot dt \text{ ; } V_b = I \cdot R_b + V_{C_d} + V_{R_d}$$

$$V_b = I \cdot R_b + \frac{1}{C_d} \cdot \int I_{C_d} \cdot dt + V_{R_d} \text{ ; } \frac{d}{dt}\{V_b = I \cdot R_b + \frac{1}{C_d} \cdot \int I_{C_d} \cdot dt + V_{R_d}\}$$

$$0 = \frac{dI}{dt} \cdot R_b + \frac{1}{C_d} \cdot I_{C_d} + \frac{dV_{R_d}}{dt} \Rightarrow \frac{dI}{dt} \cdot R_b + \frac{1}{C_d} \cdot I_{C_d} + \frac{dI}{dt} \cdot \frac{dV_{R_d}}{dI} = 0 \text{ ; } \frac{dV_{R_d}}{dI} = R_d = -R_\gamma$$

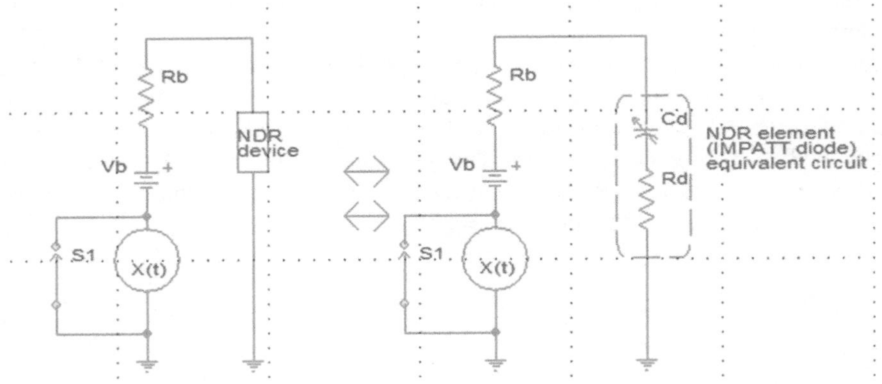

Fig. 3.12 IMPATT diode and NDR element equivalent circuit

$$\frac{dI}{dt}\cdot R_b + \frac{1}{C_d}\cdot I_{C_d} - \frac{dI}{dt}\cdot R_\gamma = 0 \Rightarrow \frac{dI}{dt}\cdot R_\gamma - \frac{dI}{dt}\cdot R_b = \frac{1}{C_d}\cdot I_{C_d};\; R_\gamma - R_b > 0$$

$$\frac{dI}{dt}\cdot (R_\gamma - R_b) = \frac{1}{C_d}\cdot I \Rightarrow \frac{dI/dt}{I} = \frac{1}{C_d\cdot (R_\gamma - R_b)} \Rightarrow \frac{d}{dt}\{\ln I(t)\} = \frac{1}{C_d\cdot (R_\gamma - R_b)}$$

$$\int \{\frac{d}{dt}\{\ln I(t)\} = \frac{1}{C_d\cdot (R_\gamma - R_b)}\} \Rightarrow \ln I(t) = \frac{1}{C_d\cdot (R_\gamma - R_b)}\cdot t \Rightarrow I(t) = e^{\frac{1}{C_d\cdot (R_\gamma - R_b)}\cdot t}$$

$\frac{1}{C_d\cdot (R_\gamma - R_b)}$ is the exponential coefficient.

We use active circulator in our circuit. Active circulators are ideally suited for realization using monolithic microwave integrated circuit (MMIC) technology. The circuit employs decade bandwidth active circulator which shows very low phase error characteristic. The circuit configuration of the active circulator used three metal–semiconductor field effect transistors (MESFETs) which are the GEC-Marconi standard library cell F20-FET-4 × 75. With all the standard library cells, it is a very accurate ultra-wideband small signal model for the device. It is similar to a junction gate field-effect transistor (JFET) in construction and terminology. The difference is that instead of using a p-n junction for gate, a schottky (metal semiconductor) junction is used. A typical three ports decade bandwidth active circulator has three MESFETs transistors interconnected with each other. R_F, C_F, L_F, C_C, Rsb play a major role in the working of the circuit. The three feedback branches (RF, CF, LF) are used to link all the three transistors in an end to end fashion. The source resistor (Rsb) is shared among all the three MESFETs transistors and one transistor is source coupled with the other two transistors using this source resistor. The circuit works in a symmetric fashion. We consider MESFET high frequency model taking node capacitors into account. Next figure describes the circuit configuration of the active circulator [36, 37]. We use N-type MESFET but usually the recommended is a symmetrical bilateral MESFET. All Cc and Cf capacitors are un-polarized. Once we inject RF signal to port P_1, it passes to port P_2 through feedback branch (RF, CF, LF). The same is between ports P_2 and P_3, ports P_3 and P_1. In case we inject RF signal to port P_2, it reaches Q1 gate and shorten Q1's drain and source. Then Port 2's RF signal is shortened to ground through resistor Rsb and did not reach port P_1. The same is between P_1 to P_3 and P_3 to P_2. We consider a varactor which is realized by connecting together the drain and source terminations of a standard MESFET, resulting in a Schottky junction. The bias potential is then applied across the drain/source and gate terminations. Our three ports decade bandwidth active circulator with micro strip delay lines and IMPATT diode circuit in port P_2 gets his input RF signal from microwave RF source (port P_1) and feeds antenna unit by active circulator output RF signal (Port P_3) [36, 37].

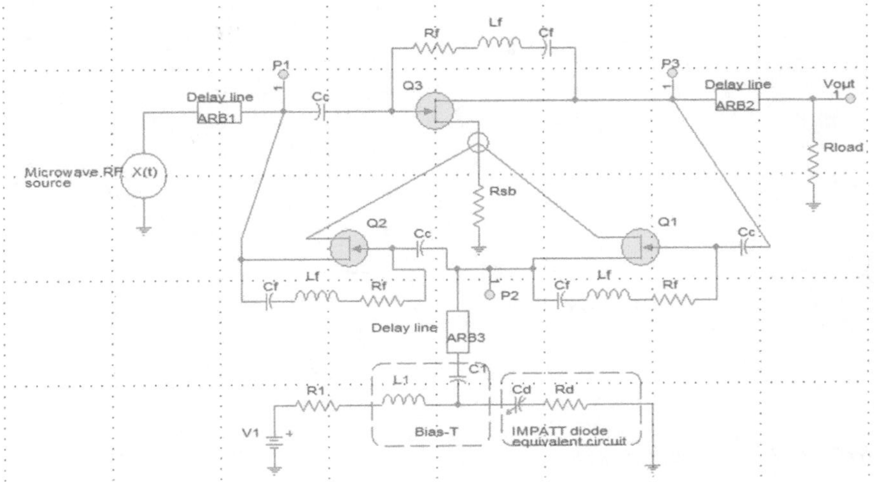

Fig. 3.13 Active circulator circuit system (version 1)

Active circulator system can be described by system path from RFin port (P_1) to RFout port (P_3). For simplicity we ignore MESFET high frequency equivalent model and took it as cutoff element in our system. Next figure describes our IMPATT amplifier system path from microwave RF source X(t) to antenna unit (R_{load}). We ignore the signal path from P_3 to P_1 since our amplifier RF signal is feed directly to load antenna. R_{load} is a pure resistive but can be taken with additional reactance part. We consider fully matching between antenna load resistance and three power active circulator IMPATT diode circuit, no signal reflections. ARB1, ARB2, and ARB3 are circuit micro strip delay line, $V_{ARB1}(t) \to \varepsilon$; $V_{ARB2}(t) \to \varepsilon$; $V_{ARB3}(t) \to \varepsilon$. Due to active circulator's micro strip transmission lines delays, τ_1 for the first port current, τ_2 for the second port current, and τ_3 for the third port current. V_1 is IMPATT diode bias voltage.

$$I_1(t) \to I_1(t-\tau_1)\,;\; I_2(t) \to I_2(t-\tau_2)\,;\; I_3(t) \to I_3(t-\tau_3)\,;\; \frac{dV_{R_D}}{dI_{R_D}} < 0\,;\; \frac{dV_{R_D}}{dI_{R_D}} = R_D\,;\; \left|\frac{dV_{R_D}}{dI_{R_D}}\right| = R_\gamma$$

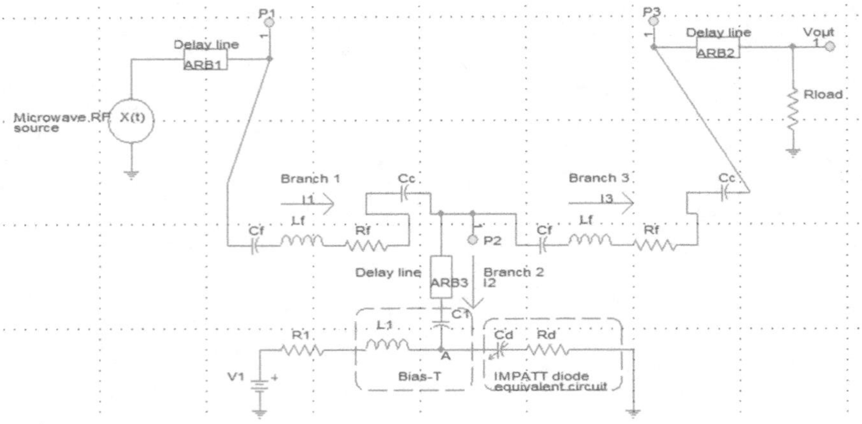

Fig. 3.14 Active circulator circuit system (version 2)

<u>First branch</u>:

$$I_1 = C_f \cdot \frac{dV_{C_f}}{dt}\ ;\ V_{L_f} = L_f \cdot \frac{dI_1}{dt}\ ; I_1 = C_C \cdot \frac{dV_{C_C}}{dt}\ ;$$

$$\mathrm{C}_{eq} \cdot \frac{d}{dt}[V_{c_f} + V_{C_C}] = I_1\ ; \mathrm{C}_{eq} = C_f || C_C$$

$$\frac{1}{\mathrm{C}_{eq}} = \frac{1}{C_f} + \frac{1}{C_C}\ ; \frac{d}{dt}[V_{c_f} + V_{C_C}] = I_1 \cdot (\frac{1}{C_f} + \frac{1}{C_C});$$

$$V_{P_1} - V_{P_2} = V_{c_f} + V_{L_f} + V_{R_f} + V_{C_C}$$

$$V_{C_f} + V_{C_C} = V_{P_1} - V_{P_2} - V_{R_f} - V_{L_f}\ ;\ V_{C_f} + V_{C_C} = V_{P_1} - V_{P_2} - I_1 \cdot R_f - L_f \cdot \frac{dI_1}{dt}$$

$$\frac{d}{dt}(V_{C_f} + V_{C_C}) = \frac{dV_{P_1}}{dt} - \frac{dV_{P_2}}{dt} - \frac{dI_1}{dt} \cdot R_f - L_f \cdot \frac{d^2 I_1}{dt^2}\ ;\ V_{R_f} = I_1 \cdot R_f\ ;\ R_d = R_D$$

$$I_1 \cdot (\frac{1}{C_f} + \frac{1}{C_C}) = \frac{dV_{P_1}}{dt} - \frac{dV_{P_2}}{dt} - \frac{dI_1}{dt} \cdot R_f - L_f \cdot \frac{d^2 I_1}{dt^2}$$

<u>Second branch</u>:

$$I_2 = C_1 \cdot \frac{dV_{C_1}}{dt}\ ;\ V_{L_1} = L_1 \cdot \frac{dI_{L_1}}{dt}\ ;\ I_{C_d} = C_d \cdot \frac{dV_{C_d}}{dt};$$

$$\frac{dV_{R_d}}{dI_{R_d}} = R_d\ ;\ I_2 + I_{L_1} = I_{C_d}\ ;\ I_{R_1} = I_{L_1}$$

$$I_{Cd} = I_{RD}\ ;\ V_1 - V_A = V_{R_1} + V_{L_1}\ ;\ V_A = V_{C_d} + V_{R_d};$$

$$V_{P_2} - V_A = V_{C_1}\ ;\ C_1 \cdot \frac{dV_{C_1}}{dt} + I_{L_1} = C_d \cdot \frac{dV_{C_d}}{dt}$$

$$C_1 \cdot \frac{dV_{C_1}}{dt} + \frac{1}{L_1} \cdot \int V_{L_1} \cdot dt = C_d \cdot \frac{dV_{C_d}}{dt} \Rightarrow C_1 \cdot \frac{d^2 V_{C_1}}{dt^2}$$

$$+ \frac{1}{L_1} \cdot V_{L_1} = C_d \cdot \frac{d^2 V_{C_d}}{dt^2}\ ;\ V_{R_1} = I_{L_1} \cdot R_1$$

$$V_1 - V_{C_d} - V_{R_d} = V_{R_1} + V_{L_1}\ ;\ \frac{dV_{R_d}}{dI_{R_d}} = R_d\ ;\ \frac{dV_{C_d}}{dt} = \frac{I_{C_d}}{C_d} = \frac{I_{Rd}}{C_d}$$

$$V_{C_1} = V_{P_2} - V_A = V_{P_2} - V_{C_d} - V_{R_d} \Rightarrow \frac{dV_{C_1}}{dt} = \frac{dV_{P_2}}{dt} - \frac{dV_{C_d}}{dt}$$

$$- \frac{dV_{R_d}}{dt}\ ;\ \frac{dV_{C_1}}{dt} = \frac{dV_{P_2}}{dt} - \frac{I_{Rd}}{C_d} - \frac{dV_{R_d}}{dt}$$

$$I_2 = C_1 \cdot \frac{dV_{C_1}}{dt} = C_1 \cdot \left(\frac{dV_{P_2}}{dt} - \frac{I_{Rd}}{C_d} - \frac{dV_{R_d}}{dt}\right)$$

$$= C_1 \cdot \left(\frac{dV_{P_2}}{dt} - \frac{I_{Rd}}{C_d} - \frac{dI_{R_d}}{dt} \cdot \frac{dV_{R_d}}{dI_{R_d}}\right)\ ;\ \frac{dV_{R_d}}{dI_{R_d}} = -R_\gamma$$

$$I_2 = C_1 \cdot \left(\frac{dV_{P_2}}{dt} - \frac{I_{Rd}}{C_d} + \frac{dI_{R_d}}{dt} \cdot R_\gamma\right)\ ;\ I_2 + I_{L_1} = I_{C_d} = I_{Rd} \Rightarrow I_2$$

$$= C_1 \cdot \left(\frac{dV_{P_2}}{dt} - \frac{I_2 + I_{L_1}}{C_d} + \frac{d(I_2 + I_{L_1})}{dt} \cdot R_\gamma\right)$$

<u>Third branch</u>:

$$I_3 = C_f \cdot \frac{dV_{C_f}}{dt}\ ;\ V_{L_f} = L_f \cdot \frac{dI_3}{dt}\ ;\ V_{R_f} = I_3 \cdot R_f;$$

$$I_3 = C_C \cdot \frac{dV_{C_C}}{dt}\ ;\ I_{R_{load}} = I_3(t - \tau_3)\ ;\ \frac{dV_{C_f}}{dt} = \frac{I_3}{C_f}$$

$$\frac{dV_{C_C}}{dt} = \frac{I_3}{C_C}\ ;\ \frac{dV_{C_C}}{dt} + \frac{dV_{C_f}}{dt} = \frac{I_3}{C_f} + \frac{I_3}{C_C} \Rightarrow \frac{d}{dt}(V_{C_C} + V_{C_f}) = I_3 \cdot \left(\frac{1}{C_f} + \frac{1}{C_C}\right)$$

$$V_{P_2} - V_{P_3} = V_{C_f} + V_{L_f} + V_{R_f} + V_{C_C}\ ;\ V_{P_3} = V_{R_{load}} = R_{load} \cdot I_{R_{load}} = R_{load} \cdot I_3(t - \tau_3)$$

$$V_{P_2} - V_{P_3} - V_{L_f} - V_{R_f} = V_{C_f} + V_{C_C}\ ;\ V_{C_f} + V_{C_C} = V_{P_2} - V_{P_3} - L_f \cdot \frac{dI_3}{dt} - I_3 \cdot R_f$$

$$\frac{d}{dt}(V_{C_f} + V_{C_C}) = \frac{dV_{P_2}}{dt} - \frac{dV_{P_3}}{dt} - L_f \cdot \frac{d^2 I_3}{dt^2} - \frac{dI_3}{dt} \cdot R_f;$$

$$I_3 \cdot \left(\frac{1}{C_f} + \frac{1}{C_C}\right) = \frac{dV_{P_2}}{dt} - \frac{dV_{P_3}}{dt} - L_f \cdot \frac{d^2 I_3}{dt^2} - \frac{dI_3}{dt} \cdot R_f$$

We can summarize our system differential equation:

$$I_1 \cdot (\frac{1}{C_f} + \frac{1}{C_C}) = \frac{dV_{P_1}}{dt} - \frac{dV_{P_2}}{dt} - \frac{dI_1}{dt} \cdot R_f - L_f \cdot \frac{d^2 I_1}{dt^2};$$

$$I_2 = C_1 \cdot (\frac{dV_{P_2}}{dt} - \frac{(I_2 + I_{L_1})}{C_d} + \frac{d(I_2 + I_{L_1})}{dt} \cdot R_\gamma)$$

$$I_3 \cdot (\frac{1}{C_f} + \frac{1}{C_C}) = \frac{dV_{P_2}}{dt} - \frac{dV_{P_3}}{dt} - L_f \cdot \frac{d^2 I_3}{dt^2} - \frac{dI_3}{dt} \cdot R_f$$

$$I_2 = C_1 \cdot (\frac{dV_{P_2}}{dt} - \frac{I_2 + I_{L_1}}{C_d} + \frac{d(I_2 + I_{L_1})}{dt} \cdot R_\gamma)$$

$$\Rightarrow \frac{dV_{P_2}}{dt} = \frac{I_2}{C_1} + \frac{(I_2 + I_{L_1})}{C_d} - \frac{d(I_2 + I_{L_1})}{dt} \cdot R_\gamma$$

We get two main system differential equations:

$$V_{P_3} = R_{load} \cdot I_3(t - \tau_3) \Rightarrow \frac{dV_{P_3}}{dt} = R_{load} \cdot \frac{dI_3(t - \tau_3)}{dt}$$

$$I_1 \cdot (\frac{1}{C_f} + \frac{1}{C_C}) = \frac{dV_{P_1}}{dt} - [\frac{I_2}{C_1} + \frac{(I_2 + I_{L_1})}{C_d} - \frac{d(I_2 + I_{L_1})}{dt} \cdot R_\gamma]$$

$$- \frac{dI_1}{dt} \cdot R_f - L_f \cdot \frac{d^2 I_1}{dt^2}$$

$$I_3 \cdot (\frac{1}{C_f} + \frac{1}{C_C}) = \frac{I_2}{C_1} + \frac{(I_2 + I_{L_1})}{C_d} - \frac{d(I_2 + I_{L_1})}{dt} \cdot R_\gamma - R_{load}$$

$$\cdot \frac{dI_3(t - \tau_3)}{dt} - L_f \cdot \frac{d^2 I_3}{dt^2} - \frac{dI_3}{dt} \cdot R_f$$

Since $I_1 = I_2 + I_3 \Rightarrow I_2 = I_1 - I_3$ we get the following system differential equations:

$$I_1 \cdot (\frac{1}{C_f} + \frac{1}{C_C}) = \frac{dV_{P_1}}{dt} - [\frac{I_1 - I_3}{C_1} + \frac{(I_1 - I_3 + I_{L_1})}{C_d}$$

$$- \frac{d(I_1 - I_3 + I_{L_1})}{dt} \cdot R_\gamma] - \frac{dI_1}{dt} \cdot R_f - L_f \cdot \frac{d^2 I_1}{dt^2}$$

$$I_3 \cdot (\frac{1}{C_f} + \frac{1}{C_C}) = \frac{I_1 - I_3}{C_1} + \frac{(I_1 - I_3 + I_{L_1})}{C_d} - \frac{d(I_1 - I_3 + I_{L_1})}{dt} \cdot R_\gamma$$

$$- R_{load} \cdot \frac{dI_3(t - \tau_3)}{dt} - L_f \cdot \frac{d^2 I_3}{dt^2} - \frac{dI_3}{dt} \cdot R_f$$

We define the following new variables: $I'_1 = \frac{dI_1}{dt}; I'_3 = \frac{dI_3}{dt}; I'_{L_1} = \frac{dI_{L_1}}{dt}$

$$\frac{d(I_1 - I_3 + I_{L_1})}{dt} = I_1' - I_3' + I_{L_1}' \; ; \; \frac{d^2 I_1}{dt^2} = \frac{dI_1'}{dt} \; ; \; \frac{d^2 I_3}{dt^2} = \frac{dI_3'}{dt}$$

$$I_1 \cdot \left(\frac{1}{C_f} + \frac{1}{C_C}\right) = \frac{dV_{P_1}}{dt} - \left[\frac{I_1 - I_3}{C_1} + \frac{(I_1 - I_3 + I_{L_1})}{C_d} - (I_1' - I_3' + I_{L_1}') \cdot R_\gamma\right] - I_1' \cdot R_f - L_f \cdot \frac{dI_1'}{dt}$$

$$L_f \cdot \frac{dI_1'}{dt} = \frac{dV_{P_1}}{dt} - \left[\frac{I_1 - I_3}{C_1} + \frac{(I_1 - I_3 + I_{L_1})}{C_d} - (I_1' - I_3' + I_{L_1}') \cdot R_\gamma\right] - I_1' \cdot R_f - I_1 \cdot \left(\frac{1}{C_f} + \frac{1}{C_C}\right)$$

$$L_f \cdot \frac{dI_1'}{dt} = \frac{dV_{P_1}}{dt} - \left[\frac{I_1}{C_1} - \frac{I_3}{C_1} + \frac{I_1}{C_d} - \frac{I_3}{C_d} + \frac{I_{L_1}}{C_d} - I_1' \cdot R_\gamma + I_3' \cdot R_\gamma - I_{L_1}' \cdot R_\gamma\right] - I_1' \cdot R_f - I_1 \cdot \left(\frac{1}{C_f} + \frac{1}{C_C}\right)$$

$$\frac{dI_1'}{dt} = \frac{1}{L_f} \cdot \frac{dV_{P_1}}{dt} - I_1 \cdot \left[\frac{1}{C_1} + \frac{1}{C_d} + \frac{1}{C_f} + \frac{1}{C_C}\right] \cdot \frac{1}{L_f} + I_3 \cdot \left[\frac{1}{C_1} + \frac{1}{C_d}\right] \cdot \frac{1}{L_f} - \frac{I_{L_1}}{C_d \cdot L_f} + I_1' \cdot [R_\gamma - R_f] \cdot \frac{1}{L_f} - I_3' \cdot \frac{R_\gamma}{L_f} + I_{L_1}' \cdot \frac{R_\gamma}{L_f}$$

$$I_3 \cdot \left(\frac{1}{C_f} + \frac{1}{C_C}\right) = \frac{I_1 - I_3}{C_1} + \frac{(I_1 - I_3 + I_{L_1})}{C_d} - [I_1' - I_3' + I_{L_1}'] \cdot R_\gamma - R_{load} \cdot \frac{dI_3(t - \tau_3)}{dt} - L_f \cdot \frac{dI_3'}{dt} - I_3' \cdot R_f$$

$$I_3 \cdot \left(\frac{1}{C_f} + \frac{1}{C_C}\right) = \frac{I_1}{C_1} - \frac{I_3}{C_1} + \frac{I_1}{C_d} - \frac{I_3}{C_d} + \frac{I_{L_1}}{C_d} - I_1' \cdot R_\gamma + I_3' \cdot R_\gamma - I_{L_1}' \cdot R_\gamma - R_{load} \cdot \frac{dI_3(t - \tau_3)}{dt} - L_f \cdot \frac{dI_3'}{dt} - I_3' \cdot R_f$$

$$L_f \cdot \frac{dI_3'}{dt} = I_1 \cdot \left[\frac{1}{C_1} + \frac{1}{C_d}\right] - I_3 \cdot \left[\frac{1}{C_1} + \frac{1}{C_d} + \left(\frac{1}{C_f} + \frac{1}{C_C}\right)\right] + \frac{I_{L_1}}{C_d} - I_1' \cdot R_\gamma + I_3' \cdot [R_\gamma - R_f] - I_{L_1}' \cdot R_\gamma - R_{load} \cdot \frac{dI_3(t - \tau_3)}{dt}$$

$$\frac{dI_3'}{dt} = I_1 \cdot \left[\frac{1}{C_1} + \frac{1}{C_d}\right] \cdot \frac{1}{L_f} - I_3 \cdot \left[\frac{1}{C_1} + \frac{1}{C_d} + \frac{1}{C_f} + \frac{1}{C_C}\right] \cdot \frac{1}{L_f} + \frac{I_{L_1}}{C_d \cdot L_f} - I_1' \cdot \frac{R_\gamma}{L_f} + I_3' \cdot \frac{[R_\gamma - R_f]}{L_f} - I_{L_1}' \cdot \frac{R_\gamma}{L_f} - \frac{R_{load}}{L_f} \cdot \frac{dI_3(t - \tau_3)}{dt}$$

Finally we get system set of differential equations: $I_1' = \frac{dI_1}{dt}; I_3' = \frac{dI_3}{dt}; I_{L_1}' = \frac{dI_{L_1}}{dt}$

$$\frac{dI_1'}{dt} = \frac{1}{L_f}\cdot\frac{dV_{P_1}}{dt} - I_1\cdot\left[\frac{1}{C_1}+\frac{1}{C_d}+\frac{1}{C_f}+\frac{1}{C_C}\right]\cdot\frac{1}{L_f} + I_3\cdot\left[\frac{1}{C_1}+\frac{1}{C_d}\right]\cdot\frac{1}{L_f} - \frac{I_{L_1}}{C_d\cdot L_f}$$
$$+ I_1'\cdot[R_\gamma - R_f]\cdot\frac{1}{L_f} - I_3'\cdot\frac{R_\gamma}{L_f} + I_{L_1}'\cdot\frac{R_\gamma}{L_f}$$

$$\frac{dI_3'}{dt} = I_1\cdot\left[\frac{1}{C_1}+\frac{1}{C_d}\right]\cdot\frac{1}{L_f} - I_3\cdot\left[\frac{1}{C_1}+\frac{1}{C_d}+\frac{1}{C_f}+\frac{1}{C_C}\right]$$
$$\cdot\frac{1}{L_f} + \frac{I_{L_1}}{C_d\cdot L_f} - I_1'\cdot\frac{R_\gamma}{L_f} + I_3'\cdot\frac{[R_\gamma - R_f]}{L_f}$$
$$- I_{L_1}'\cdot\frac{R_\gamma}{L_f} - \frac{R_{load}}{L_f}\cdot\frac{dI_3(t-\tau_3)}{dt}$$

We add the above two differential equations: $\{\frac{dI_1'}{dt} = \ldots\} + \{\frac{dI_3'}{dt} = \ldots\}$

$$\frac{dI_1'}{dt} + \frac{dI_3'}{dt} = \frac{1}{L_f}\cdot\frac{dV_{P_1}}{dt} - I_1\cdot\left[\frac{1}{C_f}+\frac{1}{C_C}\right]\cdot\frac{1}{L_f} - I_3\cdot\left[\frac{1}{C_f}+\frac{1}{C_C}\right]\cdot\frac{1}{L_f} - I_1'\cdot\frac{R_f}{L_f}$$
$$- I_3'\cdot\frac{R_f}{L_f} - \frac{R_{load}}{L_f}\cdot\frac{dI_3(t-\tau_3)}{dt}$$
$$\frac{dI_1'}{dt} + \frac{dI_3'}{dt} = \frac{1}{L_f}\cdot\frac{dV_{P_1}}{dt} - \left[\frac{1}{C_f}+\frac{1}{C_C}\right]\cdot\frac{1}{L_f}\cdot(I_1+I_3) - \frac{R_f}{L_f}\cdot(I_1'+I_3') - \frac{R_{load}}{L_f}\cdot\frac{dI_3(t-\tau_3)}{dt}$$

We define for simplicity new variables: $X = I_1 + I_3; Y = I_1' + I_3'; \frac{dX}{dt} = Y$.

$$\frac{dY}{dt} = \frac{1}{L_f}\cdot\frac{dV_{P_1}}{dt} - \left[\frac{1}{C_f}+\frac{1}{C_C}\right]\cdot\frac{1}{L_f}\cdot X - \frac{R_f}{L_f}\cdot Y - \frac{R_{load}}{L_f}\cdot\frac{dI_3(t-\tau_3)}{dt}; \quad \frac{dX}{dt} = Y$$

I_1 flows through delay line (ARB1) before it enters to port 1 then $I_1(t) \to I_1(t-\tau_1)$.

I_2 flows through delay line (ARB3) before entering to Bias-T and IMPATT diode circuit. Then $I_2(t) \to I_2(t-\tau_2)$; accordingly, active circulator's microstrip transmission lines delays Δ_1 for the first current derivative and Δ_3 for the third port current derivative $I_3'(t) \to I_3'(t-\Delta_3); I_1'(t) \to I_1'(t-\Delta_1)$. I_3 is the current which flows through active circulator port 3 and it flows through delay line (ABR2) then $I_{R_{load}} = I_3(t-\tau_3)$. I_2, I_2' are hidden variables in our analysis [6, 12, 13]. There is no time delay in I_{L_1}', I_{L_1}. To find our system fixed point, $\frac{dY}{dt} = 0$ $\frac{dX}{dt} = 0$. There is no effect of variables delay in time since $t \gg \tau_i \Rightarrow t - \tau_i \approx t; i = 1,2,3$ and $t \gg \Delta_i \Rightarrow t - \Delta_i \approx t; i = 1,2,3$ then $\frac{dI_3(t-\tau_3)}{dt} \to \frac{dI_3(t)}{dt} = 0$ .We get two main fixed points: X^*, Y^*. $Y^* = 0$ and $\frac{1}{L_f}\cdot\frac{dV_{P_1}}{dt} - [\frac{1}{C_f}+\frac{1}{C_C}]\cdot\frac{1}{L_f}\cdot X^* = 0 \Rightarrow X^* = \frac{\frac{dV_{P_1}}{dt}}{[\frac{1}{C_f}+\frac{1}{C_C}]\frac{1}{L_f}}$.

We consider the assumption which the IMPATT/Circulator amplifier input voltage $V_{P_1} = \Gamma + \xi(t)$; Γ is constant voltage and $\xi(t)$ is RF signal in time.

$$\frac{dV_{P_1}}{dt} = \frac{d\Gamma}{dt} + \frac{d\xi(t)}{dt}\,;\ \frac{d\Gamma}{dt} \to 0\,;\ \frac{d\xi(t)}{dt} \to \varepsilon \text{ or } \frac{d\xi(t)}{dt} \to \Omega_0 \cdot sgn[\frac{d\xi(t)}{dt}]$$

$$\frac{d\xi(t)}{dt} = \text{sgn}[\frac{d\xi(t)}{dt}] \cdot |\frac{d\xi(t)}{dt}|\,;\ \forall \text{real } \frac{d\xi(t)}{dt} \ \exists\, |\frac{d\xi(t)}{dt}| = \text{sgn}[\frac{d\xi(t)}{dt}] \cdot \frac{d\xi(t)}{dt}$$

$$\frac{d|\frac{d\xi(t)}{dt}|}{d[\frac{d\xi(t)}{dt}]} = \text{sgn}[\frac{d\xi(t)}{dt}]\ \forall\ \frac{d\xi(t)}{dt} \neq 0\,;\ \frac{dV_{P_1}}{dt} = 0 \text{ for } \frac{d\xi(t)}{dt} = 0$$

$\frac{dV_{P_1}}{dt} = \Omega_0$ for $\frac{d\xi(t)}{dt} > 0$; $\frac{dV_{P_1}}{dt} = -\Omega_0$ for $\frac{d\xi(t)}{dt} < 0$. Then we get some possibilities for X^*. $X^* = 0\ \forall\ \frac{d\xi(t)}{dt} = 0$. $X^* = \frac{\Omega_0}{[\frac{1}{C_f}+\frac{1}{C_C}]\cdot\frac{1}{L_f}}\ \forall\ \frac{d\xi(t)}{dt} > 0$ $X^* = \frac{-\Omega_0}{[\frac{1}{C_f}+\frac{1}{C_C}]\cdot\frac{1}{L_f}}\ \forall\ \frac{d\xi(t)}{dt} < 0$.

We get the following options for system fixed points:

$$E^{(0)}(X^{(0)}, Y^{(0)}) = (0,0)\ \forall\ \frac{d\xi(t)}{dt} = 0\,;\ E^{(1)}(X^{(1)}, Y^{(1)}) = (\frac{\Omega_0}{[\frac{1}{C_f} + \frac{1}{C_C}] \cdot \frac{1}{L_f}}, 0)\ \forall\ \frac{d\xi(t)}{dt} > 0$$

$$E^{(2)}(X^{(2)}, Y^{(2)}) = (\frac{-\Omega_0}{[\frac{1}{C_f} + \frac{1}{C_C}] \cdot \frac{1}{L_f}}, 0)\ \forall\ \frac{d\xi(t)}{dt} < 0.$$

Stability analysis: We got system five differential equations:

$$I_1' = \frac{dI_1}{dt}\,;\ I_3' = \frac{dI_3}{dt}\,;\ I_{L_1}' = \frac{dI_{L_1}}{dt}$$

$$\frac{dI_1'}{dt} = \frac{1}{L_f} \cdot \frac{dV_{P_1}}{dt} - I_1 \cdot [\frac{1}{C_1} + \frac{1}{C_d} + \frac{1}{C_f} + \frac{1}{C_C}] \cdot \frac{1}{L_f} + I_3 \cdot [\frac{1}{C_1} + \frac{1}{C_d}] \cdot \frac{1}{L_f} - \frac{I_{L_1}}{C_d \cdot L_f}$$
$$+ I_1' \cdot [R_\gamma - R_f] \cdot \frac{1}{L_f} - I_3' \cdot \frac{R_\gamma}{L_f} + I_{L_1}' \cdot \frac{R_\gamma}{L_f}$$

$$\frac{dI_3'}{dt} = I_1 \cdot [\frac{1}{C_1} + \frac{1}{C_d}] \cdot \frac{1}{L_f} - I_3 \cdot [\frac{1}{C_1} + \frac{1}{C_d} + \frac{1}{C_f} + \frac{1}{C_C}]$$
$$\cdot \frac{1}{L_f} + \frac{I_{L_1}}{C_d \cdot L_f} - I_1' \cdot \frac{R_\gamma}{L_f} + I_3' \cdot \frac{[R_\gamma - R_f]}{L_f}$$
$$- I_{L_1}' \cdot \frac{R_\gamma}{L_f} - \frac{R_{load}}{L_f} \cdot \frac{dI_3(t - \tau_3)}{dt}$$

We define for simplicity two global parameters:

$$C_{\sum 1} = \frac{1}{C_1} + \frac{1}{C_d} + \frac{1}{C_f} + \frac{1}{C_C}$$

$$C_{\sum 2} = \frac{1}{C_1} + \frac{1}{C_d} \cdot \frac{dI_1}{dt} = I'_1 ; \frac{dI_3}{dt} = I'_3 ; \frac{dI_{L_1}}{dt} = I'_{L_1}$$

$$\frac{dI'_1}{dt} = \frac{1}{L_f} \cdot \frac{dV_{P_1}}{dt} - I_1 \cdot C_{\sum 1} \cdot \frac{1}{L_f} + I_3 \cdot C_{\sum 2} \cdot \frac{1}{L_f} - \frac{I_{L_1}}{C_d \cdot L_f} + I'_1 \cdot [R_\gamma - R_f] \cdot \frac{1}{L_f} - I'_3 \cdot \frac{R_\gamma}{L_f} + I'_{L_1} \cdot \frac{R_\gamma}{L_f}$$

$$\frac{dI'_3}{dt} = I_1 \cdot C_{\sum 2} \cdot \frac{1}{L_f} - I_3 \cdot C_{\sum 1} \cdot \frac{1}{L_f} + \frac{I_{L_1}}{C_d \cdot L_f} - I'_1 \cdot \frac{R_\gamma}{L_f} + I'_3 \cdot \frac{[R_\gamma - R_f]}{L_f} - I'_{L_1} \cdot \frac{R_\gamma}{L_f} - \frac{R_{load}}{L_f} \cdot \frac{dI_3(t - \tau_3)}{dt}$$

$$\Xi_{11} = [R_\gamma - R_f] \cdot \frac{1}{L_f} ; \Xi_{12} = -\frac{R_\gamma}{L_f} ; \Xi_{13} = -C_{\sum 1} \cdot \frac{1}{L_f} ;$$

$$\Xi_{14} = C_{\sum 2} \cdot \frac{1}{L_f} ; \Xi_{15} = -\frac{1}{C_d \cdot L_f}$$

$$\Xi_{21} = -\frac{R_\gamma}{L_f} ; \Xi_{22} = \frac{[R_\gamma - R_f]}{L_f} ; \Xi_{23} = C_{\sum 2} \cdot \frac{1}{L_f} ;$$

$$\Xi_{24} = -C_{\sum 1} \cdot \frac{1}{L_f} ; \Xi_{25} = \frac{1}{C_d \cdot L_f}$$

$$\Xi_{31} = 1 ; \Xi_{32} = 0 ; \Xi_{33} = 0 ; \Xi_{34} = 0 ; \Xi_{35} = 0 ; \Xi_{41} = 0;$$

$$\Xi_{42} = 1 ; \Xi_{43} = 0 ; \Xi_{44} = 0$$

$$\Xi_{45} = 0 ; \Xi_{51} = \cdots \Xi_{55} = 0.$$

We can write our system differential equations matrix representation:

$$\begin{pmatrix} \frac{dI'_1}{dt} \\ \frac{dI'_3}{dt} \\ \frac{dI_1}{dt} \\ \frac{dI_3}{dt} \\ \frac{dI_{L_1}}{dt} \end{pmatrix} = \begin{pmatrix} \Xi_{11} & \cdots & \Xi_{15} \\ \vdots & \ddots & \vdots \\ \Xi_{51} & \cdots & \Xi_{55} \end{pmatrix} \cdot \begin{pmatrix} I'_1 \\ I'_3 \\ I_1 \\ I_3 \\ I_{L_1} \end{pmatrix} + \begin{pmatrix} 1 \\ 0 \\ 0 \\ 0 \\ 0 \end{pmatrix} \cdot \frac{1}{L_f} \cdot \frac{dV_{P_1}}{dt} + \begin{pmatrix} \frac{R_\gamma}{L_f} \\ -\frac{R_\gamma}{L_f} \\ 0 \\ 0 \\ 0 \end{pmatrix} \cdot I'_{L_1} + \begin{pmatrix} 0 \\ -\frac{R_{load}}{L_f} \\ 0 \\ 0 \\ 0 \end{pmatrix} \cdot \frac{dI_3(t - \tau_3)}{dt}$$

We consider no delay effect on $\frac{dI'_1}{dt}, \frac{dI'_3}{dt}, \frac{dI_1}{dt}, \frac{dI_3}{dt}, \frac{dI_{L_1}}{dt}$. $I'_{L_1} \to \varepsilon; \tau_2, \Delta_2 \to \varepsilon$.

$$I_1(t) \to I_1(t-\tau_1)\,;\; I_1'(t) \to I_1'(t-\Delta_1).$$

$$\begin{pmatrix} \frac{dI_1'}{dt} \\ \frac{dI_3'}{dt} \\ \frac{dI_1}{dt} \\ \frac{dI_3}{dt} \\ \frac{dI_{L_1}}{dt} \end{pmatrix} = \begin{pmatrix} \Xi_{11} & \cdots & \Xi_{15} \\ \vdots & \ddots & \vdots \\ \Xi_{51} & \cdots & \Xi_{55} \end{pmatrix} \cdot \begin{pmatrix} I_1'(t-\Delta_1) \\ I_3'(t) \\ I_1(t-\tau_1) \\ I_3(t) \\ I_{L_1}(t) \end{pmatrix} + \begin{pmatrix} 1 \\ 0 \\ 0 \\ 0 \\ 0 \end{pmatrix}$$

$$\cdot \frac{1}{L_f} \cdot \frac{dV_{P_1}}{dt} + \begin{pmatrix} 0 \\ -\frac{R_{load}}{L_f} \\ 0 \\ 0 \\ 0 \end{pmatrix} \cdot \frac{dI_3(t-\tau_3)}{dt} + \varepsilon$$

To find equilibrium points (fixed points) of active circulator IMPATT amplifier system by the following assumptions:

$$\lim_{t\to\infty} I_1'(t-\Delta_1) = I_1'(t)\,;\; \lim_{t\to\infty} I_1(t-\tau_1) = I_1(t)\,;$$

$$\lim_{t\to\infty} I_3(t-\tau_3) = I_3(t)\,;\; \frac{dI_1'}{dt} = 0\,;\; \frac{dI_3'}{dt} = 0$$

$$\frac{dI_1}{dt} = 0\,;\; \frac{dI_3}{dt} = 0\,;\; \frac{dI_{L_1}}{dt} = 0\,;\; I_3'^{(*)} = 0\,;\; I_1'^{(*)} = 0$$

$$\frac{1}{L_f} \cdot \frac{dV_{P_1}}{dt} - I_1^* \cdot C_{\sum 1} \cdot \frac{1}{L_f} + I_3^* \cdot C_{\sum 2} \cdot \frac{1}{L_f} - \frac{I_{L_1}^*}{C_d \cdot L_f} = 0$$

$$I_1^* \cdot C_{\sum 2} \cdot \frac{1}{L_f} - I_3^* \cdot C_{\sum 1} \cdot \frac{1}{L_f} + \frac{I_{L_1}^*}{C_d \cdot L_f} = 0\,;\; \frac{dV_{P_1}}{dt} + (I_1^* + I_3^*) \cdot [C_{\sum 2} - C_{\sum 1}] = 0$$

$$\frac{dV_{P_1}}{dt} + (I_1^* + I_3^*) \cdot [C_{\sum 2} - C_{\sum 1}] = 0 \Rightarrow I_1^* + I_3^* = \frac{\frac{dV_{P_1}}{dt}}{C_{\sum 1} - C_{\sum 2}}.$$

The standard local stability analysis about anyone of the equilibrium points of active circulator IMPATT amplifier system consist in adding to coordinates $[I_1', I_3', I_1, I_3, I_{L_1}]$ arbitrarily small increments of exponential form $[i_1', i_3', i_1, i_3, i_{L_1}] \cdot e^{\lambda \cdot t}$ and retaining the first order terms in $I_1', I_3', I_1, I_3, I_{L_1}$. The system of five homogeneous equations leads to a polynomial characteristics equation in the eigenvalues-λ. The polynomial characteristics equations accept by set the below currents and currents derivative with respect to time into active circulator IMPATT diode system equations. Active circulator IMPATT diode system fixed values with arbitrarily small increments of exponential form $[i_1', i_3', i_1, i_3, i_{L_1}] \cdot e^{\lambda \cdot t}$ are: j = 0 (first fixed point), j = 1 (second fixed point), j = 2 (third fixed point), etc.

$I_1'(t) = I_1'^{(j)} + i_1' \cdot e^{\lambda \cdot t}$; $I_3'(t) = I_3'^{(j)} + i_3' \cdot e^{\lambda \cdot t}$; $I_1(t) = I_1^{(j)} + i_1 \cdot e^{\lambda \cdot t}$
$I_3(t) = I_3^{(j)} + i_3 \cdot e^{\lambda \cdot t}$; $I_{L_1}(t) = I_{L_1}^{(j)} + i_{L_1} \cdot e^{\lambda \cdot t}$.

We choose the expressions for our $I_1'(t), I_3'(t), I_1(t), I_3(t), I_{L_1}(t)$ as small displacement $[i_1', i_3', i_1, i_3, i_{L_1}]$ from the active circulator IMPATT diode system fixed points at time t = 0. $I_1'(t=0) = I_1'^{(j)} + i_1'$

$$I_3'(t=0) = I_3'^{(j)} + i_3' \,;\, I_1(t=0) = I_1^{(j)} + i_1 \,;\, I_3(t=0) = I_3^{(j)} + i_3 \,;\, I_{L_1}(t=0) = I_{L_1}^{(j)} + i_{L_1}.$$

For λ < 0, t > 0, the selected fixed point is stable otherwise λ > 0, t > 0 is unstable (Table 1). Our system tends to the selected fixed point exponentially for λ < 0, t > 0 otherwise go away from the selected fixed point exponentially. λ is the eigenvalue parameter which is established if the fixed point is stable or unstable; additionally, his absolute value (|λ|) establishes the speed of flow toward or away from the selected fixed point [2] (Jack and Huseyin 1991). The speeds of flow toward or away from the selected fixed point for active circulator IMPATT diode amplifier system currents and currents derivatives with respect to time are:

$$\frac{dI_1'(t)}{dt} = \lim_{\Delta t \to 0} \frac{I_1'(t+\Delta t) - I_1'(t)}{\Delta t} = \lim_{\Delta t \to 0} \frac{I_1'^{(j)} + i_1' \cdot e^{\lambda \cdot (t+\Delta t)} - (I_1'^{(j)} + i_1' \cdot e^{\lambda \cdot t})}{\Delta t}$$
$$= \lim_{\Delta t \to 0} \frac{i_1' \cdot e^{\lambda \cdot t}[e^{\lambda \cdot \Delta t} - 1]}{\Delta t} \overset{e^{\lambda \cdot \Delta t} \approx 1 + \lambda \cdot \Delta t}{\longrightarrow} \lambda \cdot i_1' \cdot e^{\lambda \cdot t} \,;\, \frac{dI_3'(t)}{dt} = \lambda \cdot i_3' \cdot e^{\lambda \cdot t}$$
$$\frac{dI_1(t)}{dt} = \lambda \cdot i_1 \cdot e^{\lambda \cdot t} \,;\, \frac{dI_3(t)}{dt} = \lambda \cdot i_3 \cdot e^{\lambda \cdot t} \,;$$
$$\frac{dI_{L_1}(t)}{dt} = \lambda \cdot i_{L_1} \cdot e^{\lambda \cdot t} \,;\, \frac{dI_1'(t-\Delta_1)}{dt} = \lambda \cdot i_1' \cdot e^{\lambda \cdot t} \cdot e^{-\lambda \cdot \Delta_1}$$
$$\frac{dI_3'(t-\Delta_3)}{dt} = \lambda \cdot i_3' \cdot e^{\lambda \cdot t} \cdot e^{-\lambda \cdot \Delta_1} \,;\, \frac{dI_1'(t-\Delta_1)}{dt} = \lambda \cdot i_1' \cdot e^{\lambda \cdot t} \cdot e^{-\lambda \cdot \Delta_1} \,;$$
$$\frac{dI_1(t-\tau_1)}{dt} = \lambda \cdot i_1 \cdot e^{\lambda \cdot t} \cdot e^{-\lambda \cdot \tau_1}$$
$$\frac{dI_3(t-\tau_3)}{dt} = \lambda \cdot i_3 \cdot e^{\lambda \cdot t} \cdot e^{-\lambda \cdot \tau_3}.$$

First we take The IMPATT amplifier system's current differential equations: $\frac{dI_1}{dt} = I_1'$; $\frac{dI_3}{dt} = I_3'$ and adding coordinates $[I_1', I_3', I_1, I_3, I_{L_1}]$ arbitrarily small increments of exponential terms $[i_1', i_3', i_1, i_3, i_{L_1}] \cdot e^{\lambda \cdot t}$ and retaining the first order terms in $i_1', i_3', i_1, i_3, i_{L_1}$: $\lambda \cdot i_1 \cdot e^{\lambda \cdot t} = I_1'^{(j)} + i_1' \cdot e^{\lambda \cdot t}$; $I_1'^{(j)} = 0 \Rightarrow i_1' - \lambda \cdot i_1 = 0$.

$\lambda \cdot i_3 \cdot e^{\lambda \cdot t} = I_3'^{(j)} + i_3' \cdot e^{\lambda \cdot t}$; $I_3'^{(j)} = 0 \Rightarrow i_3' - \lambda \cdot i_3 = 0$. Second we take the active circulator IMPATT diode's current derivatives I_1', I_3' differential equations:

$$\frac{dI'_1}{dt} = \frac{1}{L_f} \cdot \frac{dV_{P_1}}{dt} - I_1 \cdot C_{\sum 1} \cdot \frac{1}{L_f} + I_3 \cdot C_{\sum 2} \cdot \frac{1}{L_f} - \frac{I_{L_1}}{C_d \cdot L_f} + I'_1 \cdot [R_\gamma - R_f] \cdot \frac{1}{L_f} - I'_3 \cdot \frac{R_\gamma}{L_f} + I'_{L_1} \cdot \frac{R_\gamma}{L_f}$$

$$\frac{dI'_3}{dt} = I_1 \cdot C_{\sum 2} \cdot \frac{1}{L_f} - I_3 \cdot C_{\sum 1} \cdot \frac{1}{L_f} + \frac{I_{L_1}}{C_d \cdot L_f} - I'_1 \cdot \frac{R_\gamma}{L_f} + I'_3 \cdot \frac{[R_\gamma - R_f]}{L_f} - I'_{L_1} \cdot \frac{R_\gamma}{L_f} - \frac{R_{load}}{L_f} \cdot \frac{dI_3(t-\tau_3)}{dt}$$

We already get $\frac{dV_{P_1}}{dt} = \Omega_0 \cdot sgn[\frac{d\xi(t)}{dt}]$. We add coordinates $[I'_1, I'_3, I_1, I_3, I_{L_1}]$ arbitrarily small increments of exponential terms $[i'_1, i'_3, i_1, i_3, i_{L_1}] \cdot e^{\lambda \cdot t}$ and retaining the first order terms in $i'_1, i'_3, i_1, i_3, i_{L_1}$.

$$\lambda \cdot i'_1 \cdot e^{\lambda \cdot t} = \frac{1}{L_f} \cdot \Omega_0 \cdot sgn[\frac{d\xi(t)}{dt}] - (I_1^{(j)} + i_1 \cdot e^{\lambda \cdot t}) \cdot C_{\sum 1} \cdot \frac{1}{L_f} + (I_3^{(j)} + i_3 \cdot e^{\lambda \cdot t}) \cdot C_{\sum 2} \cdot \frac{1}{L_f} - \frac{(I_{L_1}^{(j)} + i_{L_1} \cdot e^{\lambda \cdot t})}{C_d \cdot L_f} + (I_1'^{(j)} + i'_1 \cdot e^{\lambda \cdot t}) \cdot [R_\gamma - R_f] \cdot \frac{1}{L_f} - (I_3'^{(j)} + i'_3 \cdot e^{\lambda \cdot t}) \cdot \frac{R_\gamma}{L_f} + (I'_{L_1} \to \varepsilon) \cdot \frac{R_\gamma}{L_f}$$

$$I_3'^{(j)} = 0\,;\; I_1'^{(j)} = 0\,;\; I_1^{(j)} + I_3^{(j)} = \frac{\frac{dV_{P_1}}{dt}}{C_{\sum 1} - C_{\sum 2}}\,;\; (I'_{L_1} \to \varepsilon) \cdot \frac{R_\gamma}{L_f} \approx 0$$

$$(\#)\; \lambda \cdot i'_1 \cdot e^{\lambda \cdot t} = \frac{1}{L_f} \cdot \Omega_0 \cdot sgn[\frac{d\xi(t)}{dt}] - I_1^{(j)} \cdot C_{\sum 1} \cdot \frac{1}{L_f} + I_3^{(j)} \cdot C_{\sum 2} \cdot \frac{1}{L_f} - i_1 \cdot C_{\sum 1} \cdot \frac{1}{L_f} \cdot e^{\lambda \cdot t} + i_3 \cdot C_{\sum 2} \cdot \frac{1}{L_f} \cdot e^{\lambda \cdot t} - \frac{(I_{L_1}^{(j)} + i_{L_1} \cdot e^{\lambda \cdot t})}{C_d \cdot L_f} + i'_1 \cdot e^{\lambda \cdot t} \cdot [R_\gamma - R_f] \cdot \frac{1}{L_f} - i'_3 \cdot e^{\lambda \cdot t} \cdot \frac{R_\gamma}{L_f}$$

$$\lambda \cdot i_3' \cdot e^{\lambda \cdot t} = (I_1^{(j)} + i_1 \cdot e^{\lambda \cdot t}) \cdot C_{\sum 2} \cdot \frac{1}{L_f} - (I_3^{(j)} + i_3 \cdot e^{\lambda \cdot t}) \cdot C_{\sum 1}$$
$$\cdot \frac{1}{L_f} + \frac{(I_{L_1}^{(j)} + i_{L_1} \cdot e^{\lambda \cdot t})}{C_d \cdot L_f} - (I_1'^{(j)} + i_1' \cdot e^{\lambda \cdot t}) \cdot \frac{R_\gamma}{L_f}$$
$$+ (I_3'^{(j)} + i_3' \cdot e^{\lambda \cdot t}) \cdot \frac{[R_\gamma - R_f]}{L_f} - (I_{L_1}' \to \varepsilon) \cdot \frac{R_\gamma}{L_f} - \frac{R_{load}}{L_f}$$
$$\cdot \lambda \cdot i_3 \cdot e^{\lambda \cdot t} \cdot e^{-\lambda \cdot \tau_3}; e^{-\lambda \cdot \tau_3} \to 1$$

$$(\#\#) \quad \begin{aligned} \lambda \cdot i_3' \cdot e^{\lambda \cdot t} = {} & I_1^{(j)} \cdot C_{\sum 2} \cdot \tfrac{1}{L_f} + i_1 \cdot C_{\sum 2} \cdot \tfrac{1}{L_f} \\ & \cdot e^{\lambda \cdot t} - I_3^{(j)} \cdot C_{\sum 1} \cdot \tfrac{1}{L_f} - i_3 \cdot C_{\sum 1} \cdot \tfrac{1}{L_f} \cdot e^{\lambda \cdot t} \\ & + \tfrac{(I_{L_1}^{(j)} + i_{L_1} \cdot e^{\lambda \cdot t})}{C_d \cdot L_f} - i_1' \cdot e^{\lambda \cdot t} \cdot \tfrac{R_\gamma}{L_f} + i_3' \\ & \cdot \tfrac{[R_\gamma - R_f]}{L_f} - \tfrac{R_{load}}{L_f} \cdot \lambda \cdot i_3 \cdot e^{\lambda \cdot t} \end{aligned}$$

Adding two expressions: (#) + (##)

$$\lambda \cdot i_1' \cdot e^{\lambda \cdot t} + \lambda \cdot i_3' \cdot e^{\lambda \cdot t} = \frac{1}{L_f} \cdot \Omega_0 \cdot sgn[\frac{d\xi(t)}{dt}] - I_1^{(j)} \cdot C_{\sum 1} \cdot \frac{1}{L_f}$$
$$+ I_3^{(j)} \cdot C_{\sum 2} \cdot \frac{1}{L_f} - i_1 \cdot C_{\sum 1} \cdot \frac{1}{L_f} \cdot e^{\lambda \cdot t} + i_3 \cdot C_{\sum 2}$$
$$\cdot \frac{1}{L_f} \cdot e^{\lambda \cdot t} - \frac{(I_{L_1}^{(j)} + i_{L_1} \cdot e^{\lambda \cdot t})}{C_d \cdot L_f} + i_1' \cdot e^{\lambda \cdot t} \cdot [R_\gamma - R_f]$$
$$\cdot \frac{1}{L_f} - i_3' \cdot e^{\lambda \cdot t} \cdot \frac{R_\gamma}{L_f} + \{I_1^{(j)} \cdot C_{\sum 2} \cdot \frac{1}{L_f} + i_1 \cdot C_{\sum 2} \cdot \frac{1}{L_f} \cdot e^{\lambda \cdot t} - I_3^{(j)}$$
$$\cdot C_{\sum 1} \cdot \frac{1}{L_f} - i_3 \cdot C_{\sum 1} \cdot \frac{1}{L_f} \cdot e^{\lambda \cdot t} + \frac{(I_{L_1}^{(j)} + i_{L_1} \cdot e^{\lambda \cdot t})}{C_d \cdot L_f}$$
$$- i_1' \cdot e^{\lambda \cdot t} \cdot \frac{R_\gamma}{L_f} + i_3' \cdot e^{\lambda \cdot t} \cdot \frac{[R_\gamma - R_f]}{L_f} - \frac{R_{load}}{L_f} \cdot \lambda \cdot i_3 \cdot e^{\lambda \cdot t}\}$$

$$\lambda \cdot [i_1' + i_3'] \cdot e^{\lambda \cdot t} = \frac{1}{L_f} \cdot \Omega_0 \cdot sgn[\frac{d\xi(t)}{dt}] + I_1^{(j)} \cdot [C_{\sum 2} - C_{\sum 1}]$$
$$\cdot \frac{1}{L_f} + I_3^{(j)} \cdot [C_{\sum 2} - C_{\sum 1}] \cdot \frac{1}{L_f}$$
$$+ i_1 \cdot [C_{\sum 2} - C_{\sum 1}] \cdot \frac{1}{L_f} \cdot e^{\lambda \cdot t} + i_3 \cdot [C_{\sum 2} - C_{\sum 1} - R_{load} \cdot \lambda] \cdot \frac{1}{L_f} \cdot e^{\lambda \cdot t}$$
$$+ i_1' \cdot e^{\lambda \cdot t} \cdot [R_\gamma - R_f] \cdot \frac{1}{L_f} - i_1' \cdot e^{\lambda \cdot t} \cdot \frac{R_\gamma}{L_f} + i_3' \cdot e^{\lambda \cdot t} \cdot \frac{[R_\gamma - R_f]}{L_f} - i_3' \cdot e^{\lambda \cdot t} \cdot \frac{R_\gamma}{L_f}$$

$$\lambda \cdot [i'_1 + i'_3] \cdot e^{\lambda \cdot t} = \frac{1}{L_f} \cdot \Omega_0 \cdot sgn[\frac{d\xi(t)}{dt}] + [C_{\sum 2} - C_{\sum 1}] \cdot \frac{1}{L_f} \cdot (I_1^{(j)} + I_3^{(j)})$$
$$+ i_1 \cdot [C_{\sum 2} - C_{\sum 1}] \cdot \frac{1}{L_f} \cdot e^{\lambda \cdot t} + i_3 \cdot [C_{\sum 2} - C_{\sum 1} - R_{load} \cdot \lambda]$$
$$\cdot \frac{1}{L_f} \cdot e^{\lambda \cdot t} - i'_1 \cdot e^{\lambda \cdot t} \cdot \frac{R_f}{L_f} - i'_3 \cdot e^{\lambda \cdot t} \cdot \frac{R_f}{L_f}$$

We already found the system fixed points condition: $I_1^{(j)} + I_3^{(j)} = \frac{\frac{dV_{P_1}}{dt}}{C_{\sum 1} - C_{\sum 2}}$

$$\lambda \cdot [i'_1 + i'_3] \cdot e^{\lambda \cdot t} = \frac{1}{L_f} \cdot \Omega_0 \cdot sgn[\frac{d\xi(t)}{dt}] + [C_{\sum 2} - C_{\sum 1}] \cdot \frac{1}{L_f} \cdot \frac{\frac{dV_{P_1}}{dt}}{C_{\sum 1} - C_{\sum 2}}$$
$$+ i_1 \cdot [C_{\sum 2} - C_{\sum 1}] \cdot \frac{1}{L_f} \cdot e^{\lambda \cdot t} + i_3 \cdot [C_{\sum 2} - C_{\sum 1} - R_{load} \cdot \lambda]$$
$$\cdot \frac{1}{L_f} \cdot e^{\lambda \cdot t} - i'_1 \cdot e^{\lambda \cdot t} \cdot \frac{R_f}{L_f} - i'_3 \cdot e^{\lambda \cdot t} \cdot \frac{R_f}{L_f}$$

$$\lambda \cdot [i'_1 + i'_3] \cdot e^{\lambda \cdot t} = \frac{1}{L_f} \cdot [\Omega_0 \cdot sgn[\frac{d\xi(t)}{dt}] - \frac{dV_{P_1}}{dt}] + i_1 \cdot [C_{\sum 2} - C_{\sum 1}] \cdot \frac{1}{L_f} \cdot e^{\lambda \cdot t}$$
$$+ i_3 \cdot [C_{\sum 2} - C_{\sum 1} - R_{load} \cdot \lambda] \cdot \frac{1}{L_f} \cdot e^{\lambda \cdot t} - i'_1 \cdot e^{\lambda \cdot t}$$
$$\cdot \frac{R_f}{L_f} - i'_3 \cdot e^{\lambda \cdot t} \cdot \frac{R_f}{L_f}$$

$$\frac{dV_{P_1}}{dt} = \Omega_0 \cdot sgn[\frac{d\xi(t)}{dt}] \Rightarrow \lambda \cdot [i'_1 + i'_3] \cdot e^{\lambda \cdot t} = i_1 \cdot [C_{\sum 2} - C_{\sum 1}] \cdot \frac{1}{L_f} \cdot e^{\lambda \cdot t}$$
$$+ i_3 \cdot [C_{\sum 2} - C_{\sum 1} - R_{load} \cdot \lambda] \cdot \frac{1}{L_f} \cdot e^{\lambda \cdot t} - i'_1 \cdot e^{\lambda \cdot t} \cdot \frac{R_f}{L_f} - i'_3 \cdot e^{\lambda \cdot t} \cdot \frac{R_f}{L_f}$$

We divide two sides of the above equation by $e^{\lambda \cdot t}$

$$\lambda \cdot [i'_1 + i'_3] = i_1 \cdot [C_{\sum 2} - C_{\sum 1}] \cdot \frac{1}{L_f} + i_3 \cdot [C_{\sum 2} - C_{\sum 1} - R_{load} \cdot \lambda] \cdot \frac{1}{L_f} - i'_1 \cdot \frac{R_f}{L_f}$$
$$- i'_3 \cdot \frac{R_f}{L_f}$$

$$i_1 \cdot [C_{\sum 2} - C_{\sum 1}] \cdot \frac{1}{L_f} + i_3 \cdot [C_{\sum 2} - C_{\sum 1} - R_{load} \cdot \lambda] \cdot \frac{1}{L_f} - i_1' \cdot \frac{R_f}{L_f} - i_3' \cdot \frac{R_f}{L_f} - \lambda \cdot [i_1' + i_3'] = 0$$

$$i_1 \cdot [C_{\sum 2} - C_{\sum 1}] \cdot \frac{1}{L_f} + i_3 \cdot (C_{\sum 2} - C_{\sum 1}) \cdot \frac{1}{L_f} - i_1' \cdot \frac{R_f}{L_f} - i_3' \cdot \frac{R_f}{L_f} - (i_1' + i_3' + i_3 \cdot R_{load} \cdot \frac{1}{L_f}) \cdot \lambda = 0$$

Three cases: $C_{\sum 2} - C_{\sum 1} > R_f$; $C_{\sum 2} - C_{\sum 1} < R_f$. $C_{\sum 2} - C_{\sum 1} = R_f$.
We can summery our equations of arbitrarily small increments:

$$i_1' - \lambda \cdot i_1 = 0; i_3' - \lambda \cdot i_3 = 0$$

$$i_1 \cdot [C_{\sum 2} - C_{\sum 1}] \cdot \frac{1}{L_f} + i_3 \cdot (C_{\sum 2} - C_{\sum 1}) \cdot \frac{1}{L_f} - i_1' \cdot \frac{R_f}{L_f} - i_3' \cdot \frac{R_f}{L_f} - (i_1' + i_3' + i_3 \cdot R_{load} \cdot \frac{1}{L_f}) \cdot \lambda = 0$$

The active circulator IMPATT diode amplifier system eigenvalues options are describe in the below table.

Table 3.8 Active circulator IMPATT diode amplifier system eigenvalues options

	$\lambda < 0$	$\lambda > 0$
t = 0	$I_1'(t=0) = I_1'^{(j)} + i_1'$ $I_3'(t=0) = I_3'^{(j)} + i_3'$ $I_1(t=0) = I_1^{(j)} + i_1$ $I_3(t=0) = I_3^{(j)} + i_3$ $I_{L_1}(t=0) = I_{L_1}^{(j)} + i_{L_1}$	$I_1'(t=0) = I_1'^{(j)} + i_1'$ $I_3'(t=0) = I_3'^{(j)} + i_3'$ $I_1(t=0) = I_1^{(j)} + i_1$ $I_3(t=0) = I_3^{(j)} + i_3$ $I_{L_1}(t=0) = I_{L_1}^{(j)} + i_{L_1}$
t > 0	$I_1'(t) = I_1'^{(j)} + i_1' \cdot e^{-\vert\lambda\vert \cdot t}$ $I_3'(t) = I_3'^{(j)} + i_3' \cdot e^{-\vert\lambda\vert \cdot t}$ $I_1(t) = I_1^{(j)} + i_1 \cdot e^{-\vert\lambda\vert \cdot t}$ $I_3(t) = I_3^{(j)} + i_3 \cdot e^{-\vert\lambda\vert \cdot t}$ $I_{L_1}(t) = I_{L_1}^{(j)} + i_{L_1} \cdot e^{-\vert\lambda\vert \cdot t}$	$I_1'(t) = I_1'^{(j)} + i_1' \cdot e^{\vert\lambda\vert \cdot t}$ $I_3'(t) = I_3'^{(j)} + i_3' \cdot e^{\vert\lambda\vert \cdot t}$ $I_1(t) = I_1^{(j)} + i_1 \cdot e^{\vert\lambda\vert \cdot t}$ $I_3(t) = I_3^{(j)} + i_3 \cdot e^{\vert\lambda\vert \cdot t}$ $I_{L_1} = I_{L_1}^{(j)} + i_{L_1} \cdot e^{\vert\lambda\vert \cdot t}$
t > 0; t → ∞	$I_1'(t \to \infty) = I_1'^{(j)}$ $I_3'(t \to \infty) = I_3'^{(j)}$ $I_1(t \to \infty) = I_1^{(j)}$ $I_3(t \to \infty) = I_3^{(j)}$ $I_{L_1}(t \to \infty) = I_{L_1}^{(j)}$	$I_1'(t \to \infty) \approx i_1' \cdot e^{\vert\lambda\vert \cdot t}$ $I_3'(t \to \infty) \approx i_3' \cdot e^{\vert\lambda\vert \cdot t}$ $I_1(t \to \infty) \approx i_1 \cdot e^{\vert\lambda\vert \cdot t}$ $I_3(t \to \infty) \approx i_3 \cdot e^{\vert\lambda\vert \cdot t}$ $I_{L_1}(t \to \infty) \approx i_{L_1} \cdot e^{\vert\lambda\vert \cdot t}$

We already define

$$I_1(t-\tau_1) = I_1^{(j)} + i_1 \cdot e^{\lambda\cdot(t-\tau_1)}\,;\; I_3(t-\tau_3) = I_3^{(j)} + i_3 \cdot e^{\lambda\cdot(t-\tau_3)};$$
$$I'_1(t-\Delta_1) = I_1'^{(j)} + i'_1 \cdot e^{\lambda\cdot(t-\Delta_1)}$$
$$I'_3(t-\Delta_3) = I_3'^{(j)} + i'_3 \cdot e^{\lambda\cdot(t-\Delta_3)}\,;\; I_{L_1} = I_{L_1}^{(j)} + i_{L_1} \cdot e^{\lambda\cdot t}$$

Then we get four delayed differential equations with respect to coordinates $[I'_1, I'_3, I_1, I_3, I_{L_1}]$ arbitrarily small increments of exponential $[i'_1, i'_3, i_1, i_3, i_{L_1}] \cdot e^{\lambda\cdot t}$. We consider no delay effect on $\frac{dI'_1}{dt}, \frac{dI'_3}{dt}, \frac{dI_1}{dt}, \frac{dI_3}{dt}, \frac{dI_{L_1}}{dt}$. We neglect the time delay τ_2 which related to active circulator branch 2.

$$I'_1(t) = I_1'^{(j)} + i'_1 \cdot e^{\lambda\cdot t}\,;\; I'_3(t) = I_3'^{(j)} + i'_3 \cdot e^{\lambda\cdot t}\,;\; I_1(t) = I_1^{(j)} + i_1 \cdot e^{\lambda\cdot t}$$
$$I_3(t) = I_3^{(j)} + i_3 \cdot e^{\lambda\cdot t}\,;\; I_{L_1}(t) = I_{L_1}^{(j)} + i_{L_1} \cdot e^{\lambda\cdot t}$$
$$\Xi_{11} = [R_\gamma - R_f] \cdot \frac{1}{L_f}\,;\; \Xi_{12} = -\frac{R_\gamma}{L_f}\,;\; \Xi_{13} = -C_{\sum 1} \cdot \frac{1}{L_f};$$
$$\Xi_{14} = C_{\sum 2} \cdot \frac{1}{L_f}\,;\; \Xi_{15} = -\frac{1}{C_d \cdot L_f}$$
$$\Xi_{21} = -\frac{R_\gamma}{L_f}\,;\; \Xi_{22} = \frac{[R_\gamma - R_f]}{L_f}\,;\; \Xi_{23} = C_{\sum 2} \cdot \frac{1}{L_f}\,;$$
$$\Xi_{24} = -C_{\sum 1} \cdot \frac{1}{L_f}\,;\; \Xi_{25} = \frac{1}{C_d \cdot L_f}$$
$$\Xi_{31} = 1\,;\; \Xi_{32} = 0\,;\; \Xi_{33} = 0\,;\; \Xi_{34} = 0\,;\; \Xi_{35} = 0;$$
$$\Xi_{41} = 0\,;\; \Xi_{42} = 1\,;\; \Xi_{43} = 0\,;\; \Xi_{44} = 0$$
$$\Xi_{45} = 0\,;\; \Xi_{51} = \cdots \Xi_{55} = 0.$$

$$\begin{pmatrix} \frac{dI'_1}{dt} \\ \frac{dI'_3}{dt} \\ \frac{dI_1}{dt} \\ \frac{dI_3}{dt} \\ \frac{dI_{L_1}}{dt} \end{pmatrix} = \begin{pmatrix} \Xi_{11} & \cdots & \Xi_{15} \\ \vdots & \ddots & \vdots \\ \Xi_{51} & \cdots & \Xi_{55} \end{pmatrix} \cdot \begin{pmatrix} I'_1(t-\Delta_1) \\ I'_3(t) \\ I_1(t-\tau_1) \\ I_3(t) \\ I_{L_1}(t) \end{pmatrix} + \begin{pmatrix} 1 \\ 0 \\ 0 \\ 0 \\ 0 \end{pmatrix} \cdot \frac{1}{L_f} \cdot \frac{dV_{P_1}}{dt} + \begin{pmatrix} 0 \\ -\frac{R_{load}}{L_f} \\ 0 \\ 0 \\ 0 \end{pmatrix} \cdot \frac{dI_3(t-\tau_3)}{dt} + \varepsilon$$

$$i'_1 \cdot \lambda \cdot e^{\lambda\cdot t} = \Xi_{11} \cdot (I_1'^{(j)} + i'_1 \cdot e^{\lambda\cdot(t-\Delta_1)}) + \Xi_{12} \cdot (I_3'^{(j)} + i'_3 \cdot e^{\lambda\cdot t}) + \Xi_{13} \cdot (I_1^{(j)} + i_1 \cdot e^{\lambda\cdot(t-\tau_1)})$$
$$+ \Xi_{14} \cdot (I_3^{(j)} + i_3 \cdot e^{\lambda\cdot t}) + \Xi_{15} \cdot (I_{L_1}^{(j)} + i_{L_1} \cdot e^{\lambda\cdot t}) + \frac{1}{L_f} \cdot \frac{dV_{P_1}}{dt}$$

$$i_1' \cdot \lambda \cdot e^{\lambda \cdot t} = \Xi_{11} \cdot I_1'^{(j)} + \Xi_{12} \cdot I_3'^{(j)} + \Xi_{13} \cdot I_1^{(j)} + \Xi_{14} \cdot I_3^{(j)} + \Xi_{15} \cdot I_{L_1}^{(j)} + \frac{1}{L_f} \cdot \frac{dV_{P_1}}{dt} + \Xi_{11} \cdot i_1' \cdot e^{\lambda \cdot (t - \Delta_1)}$$
$$+ \Xi_{12} \cdot i_3' \cdot e^{\lambda \cdot t} + \Xi_{13} \cdot i_1 \cdot e^{\lambda \cdot (t - \tau_1)} + \Xi_{14} \cdot i_3 \cdot e^{\lambda \cdot t} + \Xi_{15} \cdot i_{L_1} \cdot e^{\lambda \cdot t}$$

We already know that at fixed point: $\frac{dI_1'}{dt} = 0$

$$\frac{dI_1'}{dt} = 0 \Rightarrow \Xi_{11} \cdot I_1'^{(j)} + \Xi_{12} \cdot I_3'^{(j)} + \Xi_{13} \cdot I_1^{(j)} + \Xi_{14} \cdot I_3^{(j)} + \Xi_{15} \cdot I_{L_1}^{(j)} + \frac{1}{L_f} \cdot \frac{dV_{P_1}}{dt} = 0$$

$$i_1' \cdot \lambda \cdot e^{\lambda \cdot t} = \Xi_{11} \cdot i_1' \cdot e^{\lambda \cdot (t - \Delta_1)} + \Xi_{12} \cdot i_3' \cdot e^{\lambda \cdot t}$$
$$+ \Xi_{13} \cdot i_1 \cdot e^{\lambda \cdot (t - \tau_1)} + \Xi_{14} \cdot i_3 \cdot e^{\lambda \cdot t} + \Xi_{15} \cdot i_{L_1} \cdot e^{\lambda \cdot t}$$
$$i_1' \cdot [\Xi_{11} \cdot e^{-\lambda \cdot \Delta_1} - \lambda] \cdot e^{\lambda \cdot t} + \Xi_{12} \cdot i_3' \cdot e^{\lambda \cdot t} + \Xi_{13} \cdot i_1 \cdot e^{\lambda \cdot (t - \tau_1)}$$
$$+ \Xi_{14} \cdot i_3 \cdot e^{\lambda \cdot t} + \Xi_{15} \cdot i_{L_1} \cdot e^{\lambda \cdot t} = 0$$
$$i_1' \cdot [\Xi_{11} \cdot e^{-\lambda \cdot \Delta_1} - \lambda] + \Xi_{12} \cdot i_3' + \Xi_{13} \cdot i_1 \cdot e^{-\lambda \cdot \tau_1} + \Xi_{14} \cdot i_3 + \Xi_{15} \cdot i_{L_1} = 0$$

$$\lambda \cdot i_3' \cdot e^t = \Xi_{21} \cdot (I_1'^{(j)} + i_1' \cdot e^{\lambda \cdot (t - \Delta_1)}) + \Xi_{22} \cdot (I_3'^{(j)} + i_3' \cdot e^{\lambda \cdot t})$$
$$+ \Xi_{23} \cdot (I_1^{(j)} + i_1 \cdot e^{\lambda \cdot (t - \tau_1)}) + \Xi_{24} \cdot (I_3^{(j)} + i_3 \cdot e^{\lambda \cdot t}) + \Xi_{25} \cdots (I_{L_1}^{(j)} + i_{L_1} \cdot e^{\lambda \cdot t})$$
$$- \frac{R_{load}}{L_f} \cdot \frac{dI_3(t - \tau_3)}{dt}$$

<u>Assumption</u>: since $\frac{dI_3(t)}{dt} = I_3'(t)$ then $I_3'(t - \Delta_3) = \frac{dI_3(t - \tau_3)}{dt}$.
$I_3'(t) = I_3'^{(j)} + i_3' \cdot e^{\lambda \cdot t} \Rightarrow I_3'(t - \Delta_3) = I_3'^{(j)} + i_3' \cdot e^{\lambda \cdot (t - \Delta_3)}$. At fixed point $I_3'^{(j)} = 0$ since $\frac{dI_3'}{dt} = 0$ then $\frac{dI_3(t - \tau_3)}{dt}|_{@\,fixed\,point} = I_3'(t - \Delta_3)|_{@fixed\,point} = i_3' \cdot e^{\lambda \cdot (t - \Delta_3)}$.

$$\lambda \cdot i_3' \cdot e^t = \Xi_{21} \cdot (I_1'^{(j)} + i_1' \cdot e^{\lambda \cdot (t - \Delta_1)}) + \Xi_{22} \cdot (I_3'^{(j)} + i_3' \cdot e^{\lambda \cdot t}) + \Xi_{23} \cdot (I_1^{(j)} + i_1 \cdot e^{\lambda \cdot (t - \tau_1)})$$
$$+ \Xi_{24} \cdot (I_3^{(j)} + i_3 \cdot e^{\lambda \cdot t}) + \Xi_{25} \cdots (I_{L_1}^{(j)} + i_{L_1} \cdot e^{\lambda \cdot t}) - \frac{R_{load}}{L_f} \cdot i_3' \cdot e^{\lambda \cdot (t - \Delta_3)}$$

$$\lambda \cdot i_3' \cdot e^t = \Xi_{21} \cdot I_1'^{(j)} + \Xi_{22} \cdot I_3'^{(j)} + \Xi_{23} \cdot I_1^{(j)} + \Xi_{24} \cdot I_3^{(j)}$$
$$+ \Xi_{25} \cdot I_{L_1}^{(j)} + i_1' \cdot \Xi_{21} \cdot e^{\lambda \cdot (t - \Delta_1)}$$
$$+ i_3' \cdot \Xi_{22} \cdot e^{\lambda \cdot t} + i_1 \cdot \Xi_{23} \cdot e^{\lambda \cdot (t - \tau_1)} + i_3 \cdot \Xi_{24} \cdot e^{\lambda \cdot t} + i_{L_1}$$
$$\cdot \Xi_{25} \cdot e^{\lambda \cdot t} - \frac{R_{load}}{L_f} \cdot i_3' \cdot e^{\lambda \cdot (t - \Delta_3)}$$

We already know that at fixed point: $\frac{dI_3'}{dt} = 0$

$$\frac{dI'_3}{dt} = 0 \Rightarrow \Xi_{21} \cdot I_1'^{(j)} + \Xi_{22} \cdot I_3'^{(j)} + \Xi_{23} \cdot I_1^{(j)} + \Xi_{24} \cdot I_3^{(j)} + \Xi_{25} \cdot I_{L_1}^{(j)} = 0$$

$$\lambda \cdot i'_3 \cdot e^t = i'_1 \cdot \Xi_{21} \cdot e^{\lambda \cdot (t-\Delta_1)} + i'_3 \cdot \Xi_{22} \cdot e^{\lambda \cdot t} + i_1 \cdot \Xi_{23} \cdot e^{\lambda \cdot (t-\tau_1)} + i_3 \cdot \Xi_{24} \cdot e^{\lambda \cdot t} + i_{L_1} \cdot \Xi_{25} \cdot e^{\lambda \cdot t} - \frac{R_{load}}{L_f} \cdot \lambda \cdots i'_3 \cdot e^{\lambda \cdot (t-\Delta_3)}$$

$$i'_1 \cdot \Xi_{21} \cdot e^{\lambda \cdot t} \cdot e^{-\lambda \cdot \Delta_1} + i'_3 \cdot \Xi_{22} \cdot e^{\lambda \cdot t} - \lambda \cdot i'_3 \cdot e^t + i_1 \cdot \Xi_{23} \cdot e^{\lambda \cdot t} \cdot e^{-\lambda \cdot \tau_1} + i_3 \cdot \Xi_{24} \cdot e^{\lambda \cdot t} + i_{L_1} \cdot \Xi_{25} \cdot e^{\lambda \cdot t} - \frac{R_{load}}{L_f} \cdot i'_3 \cdot e^{\lambda \cdot (t-\Delta_3)} = 0$$

$$i'_1 \cdot \Xi_{21} \cdot e^{-\lambda \cdot \Delta_1} + i'_3 \cdot [\Xi_{22} - \frac{R_{load}}{L_f} \cdot e^{-\lambda \cdot \Delta_3} - \lambda] + i_1 \cdot \Xi_{23} \cdot e^{-\lambda \cdot \tau_1} + i_3 \cdot \Xi_{24} + i_{L_1} \cdot \Xi_{25} = 0$$

$$\frac{dI_1}{dt} = \Xi_{31} \cdot I'_1(t-\Delta_1) \Rightarrow \lambda \cdot i_1 \cdot e^{\lambda \cdot t} = \Xi_{31} \cdot (I_1'^{(j)} + i'_1 \cdot e^{\lambda \cdot (t-\Delta_1)});$$

$$I_1'^{(j)} = 0;\ \Xi_{31} \cdot i'_1 \cdot e^{-\lambda \cdot \Delta_1} - \lambda \cdot i_1 = 0$$

$$\frac{dI_3}{dt} = \Xi_{42} \cdot I'_3(t) \Rightarrow \lambda \cdot i_3 \cdot e^{\lambda \cdot t} = \Xi_{42} \cdot (I_3'^{(j)} + i'_3 \cdot e^{\lambda \cdot t});$$

$$I_3'^{(j)} = 0;\ \Xi_{42} \cdot i'_3 - \lambda \cdot i_3 = 0$$

L_1 is an element of IMPATT diode Bias-T circuit. It forwards DC current to bias the IMPATT diode in the negative resistance characteristic region. It blocks any RF signal which comes from V_1 (DC voltage source). The IMPATT diode work point is stable and the DC current which flows through inductor L_1 is fixing. $\frac{dI_{L_1}}{dt} = 0 \Rightarrow \lambda \cdot i_{L_1} \cdot e^{\lambda \cdot t} = 0 \Rightarrow \lambda \cdot i_{L_1} = 0$. We can summery our small increments equations of our active circulator IMPATT diode amplifier system.

$$i'_1 \cdot [\Xi_{11} \cdot e^{-\lambda \cdot \Delta_1} - \lambda] + \Xi_{12} \cdot i'_3 + \Xi_{13} \cdot i_1 \cdot e^{-\lambda \cdot \tau_1} + \Xi_{14} \cdot i_3 + \Xi_{15} \cdot i_{L_1} = 0$$

$$i'_1 \cdot \Xi_{21} \cdot e^{-\lambda \cdot \Delta_1} + i'_3 \cdot [\Xi_{22} - \frac{R_{load}}{L_f} \cdot e^{-\lambda \cdot \Delta_3} - \lambda] + i_1 \cdot \Xi_{23} \cdot e^{-\lambda \cdot \tau_1} + i_3 \cdot \Xi_{24} + i_{L_1} \cdot \Xi_{25} = 0$$

$$\Xi_{31} \cdot i'_1 \cdot e^{-\lambda \cdot \Delta_1} - \lambda \cdot i_1 = 0;\ \Xi_{42} \cdot i'_3 - \lambda \cdot i_3 = 0;\ \lambda \cdot i_{L_1} = 0 \Rightarrow -\lambda \cdot i_{L_1} = 0$$

The small increments Jacobian of our active circulator IMPATT diode amplifier system is as follow:

$$\begin{pmatrix} \Upsilon_{11} & \dots & \Upsilon_{15} \\ \vdots & \ddots & \vdots \\ \Upsilon_{51} & \cdots & \Upsilon_{55} \end{pmatrix} \cdot \begin{pmatrix} i'_1 \\ i'_3 \\ i_1 \\ i_3 \\ i_{L_1} \end{pmatrix} = 0\,;\; \Upsilon_{11} = \Xi_{11} \cdot e^{-\lambda \cdot \Delta_1} - \lambda;$$

$$\Upsilon_{12} = \Xi_{12}\,;\; \Upsilon_{13} = \Xi_{13} \cdot e^{-\lambda \cdot \tau_1}\,;\; \Upsilon_{14} = \Xi_{14}$$

$$\Upsilon_{15} = \Xi_{15}\,;\; \Upsilon_{21} = \Xi_{21} \cdot e^{-\lambda \cdot \Delta_1}\,;\; \Upsilon_{22} = \Xi_{22} - \frac{R_{load}}{L_f}$$

$$\cdot e^{-\lambda \cdot \Delta_3} - \lambda\,;\; \Upsilon_{23} = \Xi_{23} \cdot e^{-\lambda \cdot \tau_1}\,;\; \Upsilon_{24} = \Xi_{24}$$

$$\Upsilon_{25} = \Xi_{25}\,;\; \Upsilon_{31} = \Xi_{31} \cdot e^{-\lambda \cdot \Delta_1}\,;\; \Upsilon_{32} = 0\,;\; \Upsilon_{33} = -\lambda\,;$$

$$\Upsilon_{34} = \Upsilon_{35} = 0\,;\; \Upsilon_{41} = 0\,;\; \Upsilon_{42} = \Xi_{42}$$

$$\Upsilon_{43} = 0\,;\; \Upsilon_{44} = -\lambda\,;\; \Upsilon_{45} = 0\,;\; \Upsilon_{51} = \Upsilon_{52} = \Upsilon_{53} = \Upsilon_{54} = 0\,;\; \Upsilon_{55} = -\lambda$$

$$A - \lambda \cdot I = \begin{pmatrix} \Upsilon_{11} & \dots & \Upsilon_{15} \\ \vdots & \ddots & \vdots \\ \Upsilon_{51} & \cdots & \Upsilon_{55} \end{pmatrix}\,;\; \det|A - \lambda \cdot I| = 0\,;\; D(\tau_1, \Delta_1, \Delta_3) = \det|A - \lambda \cdot I|$$

We inspect the occurrence of any possible stability switching resulting from the increase of value of time delays $\tau_1, \Delta_1, \Delta_3$ for the active circulator IMPATT diode amplifier system general characteristic equation $D(\tau_1, \Delta_1, \Delta_3) = 0$[6].

$$D(\tau_1, \Delta_1, \Delta_3) = \det|A - \lambda \cdot I| = (\Xi_{11} \cdot e^{-\lambda \cdot \Delta_1} - \lambda)$$

$$\cdot \det \begin{pmatrix} (\Xi_{22} - \frac{R_{load}}{L_f} \cdot e^{-\lambda \cdot \Delta_3} - \lambda) & \dots & \Xi_{25} \\ \vdots & \ddots & \vdots \\ 0 & \cdots & -\lambda \end{pmatrix}$$

$$- \Xi_{12} \cdot \det \begin{pmatrix} \Xi_{21} \cdot e^{-\lambda \cdot \Delta_1} & \dots & \Xi_{25} \\ \vdots & \ddots & \vdots \\ 0 & \cdots & -\lambda \end{pmatrix} + \Xi_{13} \cdot e^{-\lambda \cdot \tau_1}$$

$$\cdot \det \begin{pmatrix} \Xi_{21} \cdot e^{-\lambda \cdot \Delta_1} & \dots & \Xi_{25} \\ \vdots & \ddots & \vdots \\ 0 & \cdots & -\lambda \end{pmatrix} - \Xi_{14} \cdot \det \begin{pmatrix} \Xi_{21} \cdot e^{-\lambda \cdot \Delta_1} & \dots & \Xi_{25} \\ \vdots & \ddots & \vdots \\ 0 & \cdots & -\lambda \end{pmatrix}$$

$$+ \Xi_{15} \cdot \det \begin{pmatrix} \Xi_{21} \cdot e^{-\lambda \cdot \Delta_1} & \dots & \Xi_{24} \\ \vdots & \ddots & \vdots \\ 0 & \cdots & 0 \end{pmatrix}$$

We define $\eta_i; i = 1, 2, \ldots, 5$ functions. $D(\tau_1, \Delta_1, \Delta_3) = \det|A - \lambda \cdot I| = \sum_{i=1}^{5} \eta_i$

$$\eta_i = \eta_i(\Xi_{lk}, R_{load}, L_f, \lambda, \Delta_1, \Delta_3, \tau_1); l = 1, \ldots, 5\,;\ k = 1, \ldots, 5$$
$$\eta_1 = \eta_1(\Xi_{lk}, R_{load}, L_f, \lambda, \Delta_1, \Delta_3, \tau_1); \eta_2 = \eta_2(\Xi_{lk}, R_{load}, L_f, \lambda, \Delta_1, \Delta_3, \tau_1)$$
$$\eta_3 = \eta_3(\Xi_{lk}, R_{load}, L_f, \lambda, \Delta_1, \Delta_3, \tau_1); \eta_4 = \eta_4(\Xi_{lk}, R_{load}, L_f, \lambda, \Delta_1, \Delta_3, \tau_1)\,;\ \eta_5 = 0$$

$$\eta_1 = \eta_1(\Xi_{lk}, R_{load}, L_f, \lambda, \Delta_1, \Delta_3, \tau_1) = (\Xi_{11} \cdot e^{-\lambda \cdot \Delta_1} - \lambda) \cdot \det\begin{pmatrix} (\Xi_{22} - \frac{R_{load}}{L_f} \cdot e^{-\lambda \cdot \Delta_3} - \lambda) & \ldots & \Xi_{25} \\ \vdots & \ddots & \vdots \\ 0 & \ldots & -\lambda \end{pmatrix}$$

$$\eta_2 = \eta_2(\Xi_{lk}, R_{load}, L_f, \lambda, \Delta_1, \Delta_3, \tau_1) = -\Xi_{12} \cdot \det\begin{pmatrix} \Xi_{21} \cdot e^{-\lambda \cdot \Delta_1} & \ldots & \Xi_{25} \\ \vdots & \ddots & \vdots \\ 0 & \ldots & -\lambda \end{pmatrix}$$

$$\eta_3 = \eta_3(\Xi_{lk}, R_{load}, L_f, \lambda, \Delta_1, \Delta_3, \tau_1) = \Xi_{13} \cdot e^{-\lambda \cdot \tau_1} \cdot \det\begin{pmatrix} \Xi_{21} \cdot e^{-\lambda \cdot \Delta_1} & \ldots & \Xi_{25} \\ \vdots & \ddots & \vdots \\ 0 & \ldots & -\lambda \end{pmatrix}$$

$$\eta_4 = \eta_4(\Xi_{lk}, R_{load}, L_f, \lambda, \Delta_1, \Delta_3, \tau_1) = -\Xi_{14} \cdot \det\begin{pmatrix} \Xi_{21} \cdot e^{-\lambda \cdot \Delta_1} & \ldots & \Xi_{25} \\ \vdots & \ddots & \vdots \\ 0 & \ldots & -\lambda \end{pmatrix}$$

$$\eta_5 = \Xi_{15} \cdot \det\begin{pmatrix} \Xi_{21} \cdot e^{-\lambda \cdot \Delta_1} & \ldots & \Xi_{24} \\ \vdots & \ddots & \vdots \\ 0 & \ldots & 0 \end{pmatrix} = 0$$

$$\begin{aligned}\eta_1 &= (\Xi_{11} \cdot e^{-\lambda \cdot \Delta_1} - \lambda) \cdot (-\Xi_{22} \cdot \lambda^3 + \lambda^3 \cdot \frac{R_{load}}{L_f} \cdot e^{-\lambda \cdot \Delta_3} + \lambda^4 - \lambda^2 \cdot \Xi_{24} \cdot \Xi_{42}) \\ &= -\Xi_{11} \cdot \Xi_{24} \cdot \Xi_{42} \cdot e^{-\lambda \cdot \Delta_1} \cdot \lambda^2 + \{\Xi_{11} \cdot \frac{R_{load}}{L_f} \cdot e^{-\lambda \cdot (\Delta_3 + \Delta_1)} \\ &\quad - \Xi_{11} \cdot \Xi_{22} \cdot e^{-\lambda \cdot \Delta_1}\} \cdot \lambda^3 + (\Xi_{11} \cdot e^{-\lambda \cdot \Delta_1} - \frac{R_{load}}{L_f} \cdot e^{-\lambda \cdot \Delta_3}\} \\ &\quad \cdot \lambda^4 + \Xi_{24} \cdot \Xi_{42} \cdot \lambda^3 + \Xi_{22} \cdot \lambda^4 - \lambda^5\end{aligned}$$

$$\eta_2 = -\Xi_{12} \cdot \{-\lambda^3 \cdot \Xi_{21} \cdot e^{-\lambda \cdot \Delta_1} - \Xi_{23} \cdot \Xi_{31} \cdot \lambda^2 \cdot e^{-\lambda \cdot (\tau_1 + \Delta_1)}\} = \lambda^3 \cdot \Xi_{12} \cdot \Xi_{21} \cdot e^{-\lambda \cdot \Delta_1} + \Xi_{23} \cdot \Xi_{31} \cdot \Xi_{12} \cdot \lambda^2 \cdot e^{-\lambda \cdot (\tau_1 + \Delta_1)}$$

$$\begin{aligned}\eta_3 &= \Xi_{13} \cdot e^{-\lambda \cdot \tau_1} \cdot \{-\Xi_{22} \cdot \Xi_{31} \cdot e^{-\lambda \cdot \Delta_1} \\ &\quad \cdot \lambda^2 + \frac{R_{load}}{L_f} \cdot e^{-\lambda \cdot (\Delta_3 + \Delta_1)} \cdot \Xi_{31} \cdot \lambda^2 + \lambda^3 \cdot \Xi_{31} \cdot e^{-\lambda \cdot \Delta_1} \\ &\quad - \Xi_{24} \cdot \Xi_{31} \cdot \Xi_{42} \cdot \lambda \cdot e^{-\lambda \cdot \Delta_1}\} = \lambda^3 \cdot \Xi_{31} \cdot \Xi_{13} \cdot e^{-\lambda \cdot (\Delta_1 + \tau_1)} \\ &\quad + \lambda^2 \cdot \{\Xi_{31} \cdot \Xi_{13} \cdot \frac{R_{load}}{L_f} \cdot e^{-\lambda \cdot (\Delta_3 + \Delta_1 + \tau_1)} \\ &\quad - \Xi_{31} \cdot \Xi_{13} \cdot \Xi_{22} \cdot e^{-\lambda \cdot (\Delta_1 + \tau_1)}\} - \Xi_{24} \cdot \Xi_{31} \cdot \Xi_{42} \cdot \Xi_{13} \cdot e^{-\lambda \cdot (\Delta_1 + \tau_1)} \cdot \lambda\end{aligned}$$

$$\begin{aligned}\eta_4 &= -\Xi_{14} \cdot (-\Xi_{21} \cdot \Xi_{42} \cdot e^{-\lambda \cdot \Delta_1} \cdot \lambda^2 - \Xi_{23} \cdot \Xi_{31} \cdot \Xi_{42} \cdot e^{-\lambda \cdot (\tau_1 + \Delta_1)} \cdot \lambda) \\ &= \Xi_{14} \cdot \Xi_{21} \cdot \Xi_{42} \cdot e^{-\lambda \cdot \Delta_1} \cdot \lambda^2 \\ &\quad + \Xi_{14} \cdot \Xi_{23} \cdot \Xi_{31} \cdot \Xi_{42} \cdot e^{-\lambda \cdot (\tau_1 + \Delta_1)} \cdot \lambda\end{aligned}$$

We analyze our system stability switching for the following cases:
(I) $\tau_1 = \tau\,;\ \Delta_1 = \Delta_2 = 0$ (II) $\tau_1 = 0\,;\ \Delta_1 = \Delta_2 = \Delta$
(III) $\tau_1 = \tau\,;\ \Delta_1 = k \cdot \tau\,;\ \Delta_2 = (1 - k) \cdot \tau\,;\ 0<k<1$
We summery our results in the following table: $D(\tau_1, \Delta_1, \Delta_3) = \det|A - \lambda \cdot I| = \sum_{i=1}^{5} \eta_i$

We need to get characteristics equations for all above stability analysis cases. We study the occurrence of any possible stability switching resulting from the increase of value of the time delays Δ, τ and balance parameter k (0 < k < 1) for the general characteristic equation $D(\Delta, \tau, k)$. If we choose τ parameter then $D(\lambda, \tau, k) = \sum_{i=1}^{5} \eta_i = P_n(\lambda, \tau, k) + Q_m(\lambda, \tau, k) \cdot e^{-\lambda \cdot \tau}$. The expression for $P_n(\lambda, \tau)$:

Table 3.9 IMPATT amplifier system stability switching cases

	Case (I) $\tau_1 = \tau\,;\ \Delta_1 = \Delta_2 = 0$	Case (II) $\tau_1 = 0; \Delta_1 = \Delta_2 = \Delta$	Case (III) $\tau_1 = \tau; \Delta_1 = k \cdot \tau$; $\Delta_2 = (1-k) \cdot \tau; 0<k<1$
η_1	$\eta_1 = -\Xi_{11} \cdot \Xi_{24} \cdot \Xi_{42} \cdot \lambda^2 + \{\Xi_{11} \cdot \frac{R_{load}}{L_f} - \Xi_{11} \cdot \Xi_{22} + \Xi_{24} \cdot \Xi_{42}\} \cdot \lambda^3 + \{\Xi_{11} + \Xi_{22} - \frac{R_{load}}{L_f}\} \cdot \lambda^4 - \lambda^5$	$\eta_1 = [-\Xi_{11} \cdot \Xi_{24} \cdot \Xi_{42} \cdot \lambda^2 + (\frac{R_{load}}{L_f} \cdot e^{-\lambda \cdot \Delta} - \Xi_{22}) \cdot \Xi_{11} \cdot \lambda^3 + (\Xi_{11} - \frac{R_{load}}{L_f}) \cdot \lambda^4] \cdot e^{-\lambda \cdot \Delta} + \Xi_{24} \cdot \Xi_{42} \cdot \lambda^3 + \Xi_{22} \cdot \lambda^4 - \lambda^5$	$\eta_1 = -\Xi_{11} \cdot \Xi_{24} \cdot \Xi_{42} \cdot e^{-\lambda \cdot k \cdot \tau} \cdot \lambda^2 + (\Xi_{24} \cdot \Xi_{42} - \Xi_{11} \cdot \Xi_{22} \cdot e^{-\lambda \cdot k \cdot \tau}) \cdot \lambda^3 + (\Xi_{11} \cdot e^{-\lambda \cdot k \cdot \tau} + \Xi_{22}) \cdot \lambda^4 - \lambda^5 + (\Xi_{11} \cdot \frac{R_{load}}{L_f} \cdot \lambda^3 - \frac{R_{load}}{L_f} \cdot e^{\lambda \cdot k \cdot \tau} \cdot \lambda^4) \cdot e^{-\lambda \cdot \tau}$
η_2	$\eta_2 = \lambda^3 \cdot \Xi_{12} \cdot \Xi_{21} + \lambda^2 \cdot \Xi_{12} \cdot \Xi_{23} \cdot \Xi_{31} \cdot e^{-\lambda \cdot \tau}$	$\eta_2 = (\Xi_{12} \cdot \Xi_{23} \cdot \Xi_{31} \cdot \lambda^2 + \Xi_{12} \cdot \Xi_{21} \cdot \lambda^3) \cdot e^{-\lambda \cdot \Delta}$	$\eta_2 = \Xi_{12} \cdot \Xi_{21} \cdot e^{-\lambda \cdot k \cdot \tau} \cdot \lambda^3 + \Xi_{12} \cdot \Xi_{23} \cdot \Xi_{31} \cdot e^{-\lambda \cdot k \cdot \tau} \cdot e^{-\lambda \cdot \tau} \cdot \lambda^2$
η_3	$\eta_3 = \{-\Xi_{24} \cdot \Xi_{31} \cdot \Xi_{42} \cdot \Xi_{13} \cdot \lambda + (\Xi_{13} \cdot \Xi_{31} \cdot \frac{R_{load}}{L_f} - \Xi_{13} \cdot \Xi_{22} \cdot \Xi_{31}) \cdot \lambda^2 + \Xi_{13} \cdot \Xi_{31} \cdot \lambda^3\} \cdot e^{-\lambda \cdot \tau}$	$\eta_3 = \{-\Xi_{24} \cdot \Xi_{31} \cdot \Xi_{42} \cdot \Xi_{13} \cdot \lambda + (\Xi_{13} \cdot \Xi_{31} \cdot \frac{R_{load}}{L_f} \cdot e^{-\lambda \cdot \Delta} - \Xi_{13} \cdot \Xi_{22} \cdot \Xi_{31}) \cdot \lambda^2 + \Xi_{13} \cdot \Xi_{31} \cdot \lambda^3\} \cdot e^{-\lambda \cdot \Delta}$	$\eta_3 = \Xi_{13} \cdot \Xi_{31} \cdot e^{-\lambda \cdot k \cdot \tau} \cdot e^{-\lambda \cdot \tau} \cdot \lambda^3 + \{\Xi_{13} \cdot \Xi_{31} \cdot \frac{R_{load}}{L_f} \cdot e^{-\lambda \cdot \tau} - \Xi_{13} \cdot \Xi_{22} \cdot \Xi_{31} \cdot e^{-\lambda \cdot k \cdot \tau}\} \cdot \lambda^2 \cdot e^{-\lambda \cdot \tau} - \Xi_{24} \cdot \Xi_{31} \cdot \Xi_{42} \cdot \Xi_{13} \cdot e^{-\lambda \cdot k \cdot \tau} \cdot e^{-\lambda \cdot \tau} \cdot \lambda$
η_4	$\eta_4 = \Xi_{14} \cdot \Xi_{21} \cdot \Xi_{42} \cdot \lambda^2 + \Xi_{14} \cdot \Xi_{23} \cdot \Xi_{31} \cdot \Xi_{42} \cdot e^{-\lambda \cdot \tau} \cdot \lambda$	$\eta_4 = \Xi_{14} \cdot \Xi_{21} \cdot \Xi_{42} \cdot \lambda^2 + \Xi_{14} \cdot \Xi_{23} \cdot \Xi_{31} \cdot \Xi_{42} \cdot e^{-\lambda \cdot \tau} \cdot \lambda$	$\eta_4 = \Xi_{14} \cdot \Xi_{21} \cdot \Xi_{42} \cdot e^{-\lambda \cdot k \cdot \tau} \cdot \lambda^2 + \Xi_{14} \cdot \Xi_{23} \cdot \Xi_{31} \cdot \Xi_{42} \cdot e^{-\lambda \cdot \tau} \cdot e^{-\lambda \cdot k \cdot \tau} \cdot \lambda$

$$P_n(\lambda,\tau) = \sum_{j=0}^{n} P_j(\tau)\cdot\lambda^j = P_0(\tau) + P_1(\tau)\cdot\lambda + P_2(\tau)\cdot\lambda^2 + P_3(\tau)\cdot\lambda^3 + \cdots$$

The expression for $Q_m(\lambda,\tau)$:

$$Q_m(\lambda,\tau) = \sum_{j=0}^{m} q_j(\tau)\cdot\lambda^j = q_0(\tau) + q_1(\tau)\cdot\lambda + q_2(\tau)\cdot\lambda^2 + q_3(\tau)\cdot\lambda^3 + \cdots$$

If we choose Δ parameter then $D(\lambda,\Delta) = \sum_{i=1}^{5}\eta_i = P_n(\lambda,\Delta) + Q_m(\lambda,\Delta)\cdot e^{-\lambda\cdot\Delta}$.

The expression for $P_n(\lambda,\Delta)$:

$$P_n(\lambda,\Delta) = \sum_{j=0}^{n} P_j(\Delta)\cdot\lambda^j = P_0(\Delta) + P_1(\Delta)\cdot\lambda + P_2(\Delta)\cdot\lambda^2 + P_3(\Delta)\cdot\lambda^3 + \cdots$$

The expression for $Q_m(\lambda,\Delta)$ [38, 39]:

$$Q_m(\lambda,\Delta) = \sum_{j=0}^{m} q_j(\Delta)\cdot\lambda^j = q_0(\Delta) + q_1(\Delta)\cdot\lambda + q_2(\Delta)\cdot\lambda^2 + q_3(\Delta)\cdot\lambda^3 + \cdots$$

Remark: Balance parameter k (0 < k < 1) appears only in the third case. For all cases which τ and/or k or Δ include in P_n and Q_m expressions, we take the following assumptions for power Taylor approximation series only for these expressions.

$$e^{-\lambda\cdot k\cdot\tau} = \sum_{n=0}^{\infty}\frac{(-\lambda\cdot k\cdot\tau)^n}{n!} \approx 1 - \lambda\cdot k\cdot\tau + \frac{\lambda^2\cdot k^2\cdot\tau^2}{2};$$

$$e^{-\lambda\cdot\tau} = \sum_{n=0}^{\infty}\frac{(-\lambda\cdot\tau)^n}{n!} \approx 1 - \lambda\cdot\tau + \frac{\lambda^2\cdot\tau^2}{2}$$

$$e^{\lambda\cdot k\cdot\tau} = \sum_{n=0}^{\infty}\frac{(\lambda\cdot k\cdot\tau)^n}{n!} \approx 1 + \lambda\cdot k\cdot\tau + \frac{\lambda^2\cdot k^2\cdot\tau^2}{2};$$

$$e^{\lambda\cdot\tau} = \sum_{n=0}^{\infty}\frac{(\lambda\cdot\tau)^n}{n!} \approx 1 + \lambda\cdot\tau + \frac{\lambda^2\cdot\tau^2}{2}$$

$$e^{-\lambda\cdot\Delta} = \sum_{n=0}^{\infty}\frac{(-\lambda\cdot\Delta)^n}{n!} \approx 1 - \lambda\cdot\Delta + \frac{\lambda^2\cdot\Delta^2}{2};$$

$$e^{\lambda\cdot\Delta} = \sum_{n=0}^{\infty}\frac{(\lambda\cdot\Delta)^n}{n!} \approx 1 + \lambda\cdot\Delta + \frac{\lambda^2\cdot\Delta^2}{2}$$

Balance parameter, k only appears in the third case (Case III). We use the general geometric criterion [BK] and investigate the occurrence of any possible stability switching resulting from the increase of value of time delay parameters τ, Δ for the general characteristic equation $D(\Delta, \tau, k) = 0$ [6].

$$D(\lambda, \tau) = P_n(\lambda, \tau) + Q_m(\lambda, \tau) \cdot e^{-\lambda \cdot \tau}\,;\; D(\lambda, \Delta) = P_n(\lambda, \Delta) + Q_m(\lambda, \Delta) \cdot e^{-\lambda \cdot \Delta}$$

In the case our time delay parameter is τ: $P_n(\lambda, \tau) = \sum_{j=0}^{n} P_j(\tau) \cdot \lambda^j Q_m(\lambda, \tau) = \sum_{j=0}^{m} q_j(\tau) \cdot \lambda^j$. In the case our time delay parameter is Δ:

$P_n(\lambda, \Delta) = \sum_{j=0}^{n} P_j(\Delta) \cdot \lambda^j; Q_m(\lambda, \Delta) = \sum_{j=0}^{m} q_j(\Delta) \cdot \lambda^j$. $n, m \in \mathbb{N}_0; n > m; p_j(\cdot), q_j(\cdot): \mathbb{R}_{+0} \to \mathbb{R}$ The expressions are continuous and differentiable functions of τ or Δ.

We summery our $P_n(\lambda, \tau, k)$ and $Q_m(\lambda, \tau, k)$ expressions in the below table:

There are three cases to analyze stability switching under delay parameter variation. We choose to analyze the second case (II). It is reader exercise to do the same analysis for cases I and III. According stability switch criteria [BK], $n, m \in \mathbb{N}_0$, $n > m$.

$$P_n(\lambda, \Delta) = \Xi_{24} \cdot \Xi_{42} \cdot \lambda^3 + \Xi_{22} \cdot \lambda^4 - \lambda^5; e^{-\lambda \cdot \Delta} \approx \sum_{n=0}^{\infty} \frac{(-\lambda \cdot \Delta)^n}{n!} \approx 1 - \lambda \cdot \Delta$$

Table 3.10 IMPATT amplifier system $P_n(\lambda, \tau, k)$ and $Q_m(\lambda, \tau, k)$ expressions

	Case (I) $\tau_1 = \tau; \Delta_1 = \Delta_2 = 0$ n = 5; m = 3; n > m	Case (II) $\tau_1 = 0; \Delta_1 = \Delta_2 = \Delta$ n = 5; m = 4; n > m	Case (III) $\tau_1 = \tau; \Delta_1 = k \cdot \tau$ $\Delta_2 = (1 - k) \cdot \tau; 0 < k < 1$
$P_n(\lambda, \tau)$ Or $P_n(\lambda, \tau, k)$ Or $P_n(\lambda, \Delta)$	$P_n(\lambda, \tau) = \lambda^2 \cdot (-\Xi_{11} \cdot \Xi_{24} \cdot \Xi_{42}$ $+ \Xi_{14} \cdot \Xi_{21} \cdot \Xi_{42})$ $+ \lambda^3 \cdot (\Xi_{11} \cdot \frac{R_{load}}{L_f} - \Xi_{11} \cdot \Xi_{22}$ $+ \Xi_{24} \cdot \Xi_{42} + \Xi_{12} \cdot \Xi_{21})$ $+ \lambda^4 \cdot (\Xi_{11} + \Xi_{22} - \frac{R_{load}}{L_f}) - \lambda^5$	$P_n(\lambda, \Delta) = \Xi_{24} \cdot \Xi_{42} \cdot \lambda^3$ $+ \Xi_{22} \cdot \lambda^4 - \lambda^5$	$P_n(\lambda, \tau, k) = [\Xi_{14} \cdot \Xi_{21}$ $- \Xi_{11} \cdot \Xi_{24}] \cdot \Xi_{42} \cdot e^{-\lambda \cdot k \cdot \tau} \cdot \lambda^2$ $+ [\Xi_{24} \cdot \Xi_{42} - \Xi_{11} \cdot \Xi_{22} \cdot e^{-\lambda \cdot k \cdot \tau}$ $+ \Xi_{12} \cdot \Xi_{21} \cdot e^{-\lambda \cdot k \cdot \tau}] \cdot \lambda^3$ $+ (\Xi_{11} \cdot e^{-\lambda \cdot k \cdot \tau} + \Xi_{22}) \cdot \lambda^4 - \lambda^5$
$Q_m(\lambda, \tau)$ Or $Q_m(\lambda, \tau, k)$ Or $Q_m(\lambda, \Delta)$	$Q_m(\lambda, \tau) = (-\Xi_{24} \cdot \Xi_{31} \cdot \Xi_{42} \cdot \Xi_{13}$ $+ \Xi_{14} \cdot \Xi_{23} \cdot \Xi_{31} \cdot \Xi_{42}) \cdot \lambda$ $+ (\Xi_{12} \cdot \Xi_{23} \cdot \Xi_{31}$ $+ \Xi_{13} \cdot \Xi_{31} \cdot \frac{R_{load}}{L_f}$ $- \Xi_{13} \cdot \Xi_{22} \cdot \Xi_{31}) \cdot \lambda^2$ $+ \Xi_{13} \cdot \Xi_{31} \cdot \lambda^3$	$Q_m(\lambda, \Delta) = (-\Xi_{24} \cdot \Xi_{31} \cdot \Xi_{42} \cdot \Xi_{13}$ $+ \Xi_{14} \cdot \Xi_{23} \cdot \Xi_{31} \cdot \Xi_{42}) \cdot \lambda$ $+ (-\Xi_{11} \cdot \Xi_{24} \cdot \Xi_{42} + \Xi_{12} \cdot \Xi_{23} \cdot \Xi_{31}$ $+ \Xi_{13} \cdot \Xi_{31} \cdot \frac{R_{load}}{L_f} \cdot e^{-\lambda \cdot \Delta}$ $- \Xi_{13} \cdot \Xi_{22} \cdot \Xi_{31} + \Xi_{14} \cdot \Xi_{21} \cdot \Xi_{42}) \cdot \lambda^2$ $+ \{(\frac{R_{load}}{L_f} \cdot e^{-\lambda \cdot \Delta} - \Xi_{22}) \cdot \Xi_{11}$ $+ \Xi_{12} \cdot \Xi_{21} + \Xi_{13} \cdot \Xi_{31}\} \cdot \lambda^3$ $+ (\Xi_{11} - \frac{R_{load}}{L_f}) \cdot \lambda^4$	$Q_m(\lambda, \tau, k) = -\frac{R_{load}}{L_f} \cdot e^{\lambda \cdot k \cdot \tau} \cdot \lambda^4$ $+ [\Xi_{11} \cdot \frac{R_{load}}{L_f} + \Xi_{13} \cdot \Xi_{31} \cdot e^{-\lambda \cdot k \cdot \tau}] \cdot \lambda^3$ $+ [\Xi_{12} \cdot \Xi_{23} \cdot \Xi_{31} \cdot e^{-\lambda \cdot k \cdot \tau}$ $+ \Xi_{13} \cdot \Xi_{31} \cdot \frac{R_{load}}{L_f} \cdot e^{-\lambda \cdot \tau}$ $- \Xi_{13} \cdot \Xi_{22} \cdot \Xi_{31} \cdot e^{-\lambda \cdot k \cdot \tau}] \cdot \lambda^2$ $+ [\Xi_{14} \cdot \Xi_{23} \cdot \Xi_{42}$ $- \Xi_{24} \cdot \Xi_{42} \cdot \Xi_{13}] \cdot \Xi_{31} \cdot e^{-\lambda \cdot k \cdot \tau} \cdot \lambda$

$$Q_m(\lambda,\Delta) = (-\Xi_{24}\cdot\Xi_{31}\cdot\Xi_{42}\cdot\Xi_{13} + \Xi_{14}\cdot\Xi_{23}\cdot\Xi_{31}\cdot\Xi_{42})\cdot\lambda$$
$$+ (-\Xi_{11}\cdot\Xi_{24}\cdot\Xi_{42} + \Xi_{12}\cdot\Xi_{23}\cdot\Xi_{31} + \Xi_{13}\cdot\Xi_{31}\cdot\frac{R_{load}}{L_f}$$
$$\cdot e^{-\lambda\cdot\Delta} - \Xi_{13}\cdot\Xi_{22}\cdot\Xi_{31} + \Xi_{14}\cdot\Xi_{21}\cdot\Xi_{42})\cdot\lambda^2$$
$$+ \{(\frac{R_{load}}{L_f}\cdot e^{-\lambda\cdot\Delta} - \Xi_{22})\cdot\Xi_{11} + \Xi_{12}\cdot\Xi_{21}$$
$$+ \Xi_{13}\cdot\Xi_{31}\}\cdot\lambda^3 + (\Xi_{11} - \frac{R_{load}}{L_f})\cdot\lambda^4$$

$$Q_m(\lambda,\Delta) = (-\Xi_{24}\cdot\Xi_{31}\cdot\Xi_{42}\cdot\Xi_{13} + \Xi_{14}\cdot\Xi_{23}\cdot\Xi_{31}\cdot\Xi_{42})\cdot\lambda$$
$$+ (-\Xi_{11}\cdot\Xi_{24}\cdot\Xi_{42} + \Xi_{12}\cdot\Xi_{23}\cdot\Xi_{31} + \Xi_{13}\cdot\Xi_{31}$$
$$\cdot\frac{R_{load}}{L_f}\cdot[1 - \lambda\cdot\Delta] - \Xi_{13}\cdot\Xi_{22}\cdot\Xi_{31} + \Xi_{14}\cdot\Xi_{21}\cdot\Xi_{42})\cdot\lambda^2$$
$$+ \{(\frac{R_{load}}{L_f}\cdot[1 - \lambda\cdot\Delta] - \Xi_{22})\cdot\Xi_{11} + \Xi_{12}$$
$$\cdot\Xi_{21} + \Xi_{13}\cdot\Xi_{31}\}\cdot\lambda^3 + (\Xi_{11} - \frac{R_{load}}{L_f})\cdot\lambda^4$$

$$Q_m(\lambda,\Delta) = (-\Xi_{24}\cdot\Xi_{31}\cdot\Xi_{42}\cdot\Xi_{13} + \Xi_{14}\cdot\Xi_{23}\cdot\Xi_{31}\cdot\Xi_{42})\cdot\lambda$$
$$+ (-\Xi_{11}\cdot\Xi_{24}\cdot\Xi_{42} + \Xi_{12}\cdot\Xi_{23}\cdot\Xi_{31} + \Xi_{13}\cdot\Xi_{31}$$
$$\cdot\frac{R_{load}}{L_f} - \Xi_{13}\cdot\Xi_{22}\cdot\Xi_{31} + \Xi_{14}\cdot\Xi_{21}\cdot\Xi_{42})\cdot\lambda^2$$
$$+ \{\frac{R_{load}}{L_f}\cdot\Xi_{11} - \Xi_{22}\cdot\Xi_{11} + \Xi_{12}\cdot\Xi_{21} + \Xi_{13}\cdot\Xi_{31} - \Xi_{13}\cdot\Xi_{31}\cdot\frac{R_{load}}{L_f}\cdot\Delta\}$$
$$\cdot\lambda^3 + \{\Xi_{11} - \frac{R_{load}}{L_f} - \frac{R_{load}}{L_f}\cdot\Delta\cdot\Xi_{11}\}\cdot\lambda^4$$

Result: n = 5, m = 4; n > m. $n, m \in \mathbb{N}_0$, $n > m$
The expression $P_n(\lambda,\tau,k)$ is as follow:

$$P_n(\Delta) = \sum_{j=0}^{n} P_j(\Delta)\cdot\lambda^j = P_0(\Delta) + P_1(\Delta)\cdot\lambda + P_2(\Delta)$$
$$\cdot\lambda^2 + P_3(\Delta)\cdot\lambda^3 + P_4(\Delta)\cdot\lambda^4 + P_5(\Delta)\cdot\lambda^5$$

$$P_0(\Delta) = 0; P_1(\Delta) = 0; P_2(\Delta) = 0; P_5(\Delta) = -1; P_3(\Delta) = \Xi_{24}\cdot\Xi_{42}; P_4(\Delta) = \Xi_{22}$$

The expression for $Q_m(\Delta)$ is as follow:

$$Q_m(\Delta) = \sum_{j=0}^{m} q_j(\Delta) \cdot \lambda^j = q_0(\Delta) + q_1(\Delta) \cdot \lambda + q_2(\Delta) \cdot \lambda^2 + q_3(\Delta) \cdot \lambda^3 + q_4(\Delta) \cdot \lambda^4$$

$$q_0(\Delta) = 0; q_1(\Delta) = -\Xi_{24} \cdot \Xi_{31} \cdot \Xi_{42} \cdot \Xi_{13} + \Xi_{14} \cdot \Xi_{23} \cdot \Xi_{31} \cdot \Xi_{42}$$

$$q_2(\Delta) = -\Xi_{11} \cdot \Xi_{24} \cdot \Xi_{42} + \Xi_{12} \cdot \Xi_{23} \cdot \Xi_{31} + \Xi_{13} \cdot \Xi_{31} \cdot \frac{R_{load}}{L_f} - \Xi_{13} \cdot \Xi_{22} \cdot \Xi_{31} + \Xi_{14} \cdot \Xi_{21} \cdot \Xi_{42}$$

$$q_3(\Delta) = \frac{R_{load}}{L_f} \cdot \Xi_{11} - \Xi_{22} \cdot \Xi_{11} + \Xi_{12} \cdot \Xi_{21} + \Xi_{13} \cdot \Xi_{31} - \Xi_{13} \cdot \Xi_{31} \cdot \frac{R_{load}}{L_f} \cdot \Delta$$

$$q_4(\Delta) = \Xi_{11} - \frac{R_{load}}{L_f} - \frac{R_{load}}{L_f} \cdot \Delta \cdot \Xi_{11}$$

The homogeneous system for $I_1', I_3', I_1, I_3, I_{L_1}$ leads to a characteristic equation for the eigenvalue λ having the form $P(\Delta) + Q(\Delta) \cdot e^{-\lambda \cdot \Delta} = 0$; $P(\Delta) = \sum_{j=0}^{5} a_j \cdot \lambda^j$; $Q(\Delta) = \sum_{j=0}^{4} c_j \cdot \lambda^j$ and the coefficients $\{a_j(q_i, q_l, \tau), c_j(q_i, q_l, \tau)\} \in \mathbb{R}$ depend on q_i, q_l and delay τ. q_i, q_l are any system's parameters, other parameters kept as a constant $a_0 = 0; a_1 = 0; a_5 = -1$

$$a_2 = 0; a_3 = \Xi_{24} \cdot \Xi_{42}; a_4 = \Xi_{22}; c_0 = 0;$$

$$c_1 = -\Xi_{24} \cdot \Xi_{31} \cdot \Xi_{42} \cdot \Xi_{13} + \Xi_{14} \cdot \Xi_{23} \cdot \Xi_{31} \cdot \Xi_{42}$$

$$c_2 = -\Xi_{11} \cdot \Xi_{24} \cdot \Xi_{42} + \Xi_{12} \cdot \Xi_{23} \cdot \Xi_{31} + \Xi_{13} \cdot \Xi_{31} \cdot \frac{R_{load}}{L_f} - \Xi_{13} \cdot \Xi_{22} \cdot \Xi_{31} + \Xi_{14} \cdot \Xi_{21} \cdot \Xi_{42}$$

$$c_3 = \frac{R_{load}}{L_f} \cdot \Xi_{11} - \Xi_{22} \cdot \Xi_{11} + \Xi_{12} \cdot \Xi_{21} + \Xi_{13} \cdot \Xi_{31} - \Xi_{13} \cdot \Xi_{31} \cdot \frac{R_{load}}{L_f} \cdot \Delta; c_4 = \Xi_{11} - \frac{R_{load}}{L_f} - \frac{R_{load}}{L_f} \cdot \Delta \cdot \Xi_{11}$$

Unless strictly necessary, the designation of the variation arguments (q_i, q_l) will subsequently be omitted from P, Q, a_j, c_j. The coefficients a_j, c_j are continuous, and differentiable functions of their arguments, and direct substitution shows that $a_0 + c_0 = 0 \, \forall \, q_i, q_l \in \mathbb{R}_+$; that is $\lambda = 0$ is of $P(\lambda, \Delta) + Q(\lambda, \Delta) \cdot e^{-\lambda \cdot \tau} = 0$. Furthermore, $P(\lambda, \Delta)$, $Q(\lambda, \Delta)$ are analytic functions of λ, for which the following requirements of the analysis [5, 41] can also be verified in the present case:

(a) If $\lambda = i \cdot \omega$, $\omega \in \mathbb{R}$, then $P(i \cdot \omega) + Q(i \cdot \omega) \neq 0$.
(b) $|Q(\lambda)/P(\lambda)|$ is bounded for $|\lambda| \to \infty$, $\mathrm{Re}\lambda \geq 0$. No roots bifurcation from ∞.
(c) $F(\omega) = |P(i \cdot \omega)|^2 - |Q(i \cdot \omega)|^2$ has a finite number of zeros. Indeed, this is a polynomial in ω.
(d) Each positive root $\omega(q_i, q_l)$ of $F(\omega) = 0$ is continuous and differentiable with respect to q_i, q_l.

We assume that $P_n(\Delta)$ and $Q_m(\Delta)$ can't have common imaginary roots. That is for any real number ω;

$$P_n(\lambda = i\cdot\omega,\Delta) + Q_m(\lambda = i\cdot\omega,\Delta) \neq 0;\ \lambda^2 = -\omega^2;\ \lambda^3 = -i\cdot\omega^3$$
$$\lambda^4 = \omega^4;\ \lambda^5 = i\cdot\omega^5 . P_n(\lambda = i\cdot\omega,\Delta) = \Xi_{22}\cdot\omega^4 - i\cdot[\Xi_{24}\cdot\Xi_{42}\cdot\omega^3 + \omega^5]$$

$$\begin{aligned} Q_m(\lambda = i\cdot\omega,\Delta) &= (-\Xi_{24}\cdot\Xi_{31}\cdot\Xi_{42}\cdot\Xi_{13} + \Xi_{14}\cdot\Xi_{23}\cdot\Xi_{31}\cdot\Xi_{42})\cdot i\cdot\omega \\ &- (-\Xi_{11}\cdot\Xi_{24}\cdot\Xi_{42} + \Xi_{12}\cdot\Xi_{23}\cdot\Xi_{31} + \Xi_{13}\cdot\Xi_{31}\cdot\frac{R_{load}}{L_f} - \Xi_{13}\cdot\Xi_{22}\cdot\Xi_{31} + \Xi_{14}\cdot\Xi_{21}\cdot\Xi_{42})\cdot\omega^2 \\ &- \{\frac{R_{load}}{L_f}\cdot\Xi_{11} - \Xi_{22}\cdot\Xi_{11} + \Xi_{12}\cdot\Xi_{21} + \Xi_{13}\cdot\Xi_{31} - \Xi_{13}\cdot\Xi_{31}\cdot\frac{R_{load}}{L_f}\cdot\Delta\}\cdot i\cdot\omega^3 \\ &+ \{\Xi_{11} - \frac{R_{load}}{L_f} - \frac{R_{load}}{L_f}\cdot\Delta\cdot\Xi_{11}\}\cdot\omega^4 \end{aligned}$$

$$\begin{aligned} Q_m(\lambda = i\cdot\omega,\Delta) = &-(-\Xi_{11}\cdot\Xi_{24}\cdot\Xi_{42} + \Xi_{12}\cdot\Xi_{23}\cdot\Xi_{31} + \Xi_{13}\cdot\Xi_{31} \\ &\cdot\frac{R_{load}}{L_f} - \Xi_{13}\cdot\Xi_{22}\cdot\Xi_{31} + \Xi_{14}\cdot\Xi_{21}\cdot\Xi_{42})\cdot\omega^2 \\ &+ \{\Xi_{11} - \frac{R_{load}}{L_f} - \frac{R_{load}}{L_f}\cdot\Delta\cdot\Xi_{11}\}\cdot\omega^4 \\ &+ \{(-\Xi_{24}\cdot\Xi_{31}\cdot\Xi_{42}\cdot\Xi_{13} + \Xi_{14}\cdot\Xi_{23}\cdot\Xi_{31}\cdot\Xi_{42})\cdot\omega \\ &- (\frac{R_{load}}{L_f}\cdot\Xi_{11} - \Xi_{22}\cdot\Xi_{11} + \Xi_{12}\cdot\Xi_{21} + \Xi_{13} \\ &\cdot\Xi_{31} - \Xi_{13}\cdot\Xi_{31}\cdot\frac{R_{load}}{L_f}\cdot\Delta)\cdot\omega^3\}\cdot i \end{aligned}$$

$$\begin{aligned} P_n(\lambda = i\cdot\omega,\Delta) + Q_m(\lambda = i\cdot\omega,\Delta) = &-(-\Xi_{11}\cdot\Xi_{24}\cdot\Xi_{42} + \Xi_{12} \\ &\cdot\Xi_{23}\cdot\Xi_{31} + \Xi_{13}\cdot\Xi_{31}\cdot\frac{R_{load}}{L_f} \\ &- \Xi_{13}\cdot\Xi_{22}\cdot\Xi_{31} + \Xi_{14}\cdot\Xi_{21}\cdot\Xi_{42})\cdot\omega^2 + \{\Xi_{11} - \frac{R_{load}}{L_f} \\ &- \frac{R_{load}}{L_f}\cdot\Delta\cdot\Xi_{11} + \Xi_{22}\}\cdot\omega^4 \\ &+ i\cdot\{(-\Xi_{24}\cdot\Xi_{31}\cdot\Xi_{42}\cdot\Xi_{13} + \Xi_{14}\cdot\Xi_{23}\cdot\Xi_{31}\cdot\Xi_{42})\cdot\omega \\ &- (\frac{R_{load}}{L_f}\cdot\Xi_{11} - \Xi_{22}\cdot\Xi_{11} + \Xi_{12}\cdot\Xi_{21} + \Xi_{13}\cdot\Xi_{31} - \Xi_{13}\cdot\Xi_{31} \\ &\cdot\frac{R_{load}}{L_f}\cdot\Delta + \Xi_{24}\cdot\Xi_{42})\cdot\omega^3 - \omega^5\} \neq 0 \end{aligned}$$

$|P(i\cdot\omega,\Delta)|^2 = \Xi_{24}^2\cdot\Xi_{42}^2\cdot\omega^6 + [\Xi_{22}^2 + 2\cdot\Xi_{24}\cdot\Xi_{42}]\cdot\omega^8 + \omega^{10}$. We define for simplicity the following global parameters:

$$\Omega_1 = -\Xi_{11} \cdot \Xi_{24} \cdot \Xi_{42} + \Xi_{12} \cdot \Xi_{23} \cdot \Xi_{31} + \Xi_{13} \cdot \Xi_{31} \cdot \frac{R_{load}}{L_f} - \Xi_{13} \cdot \Xi_{22} \cdot \Xi_{31} + \Xi_{14} \cdot \Xi_{21} \cdot \Xi_{42}$$

$$\Omega_2(\Delta) = \Xi_{11} - \frac{R_{load}}{L_f} - \frac{R_{load}}{L_f} \cdot \Delta \cdot \Xi_{11};$$

$$\Omega_3 = -\Xi_{24} \cdot \Xi_{31} \cdot \Xi_{42} \cdot \Xi_{13} + \Xi_{14} \cdot \Xi_{23} \cdot \Xi_{31} \cdot \Xi_{42}$$

$$\Omega_4(\Delta) = \frac{R_{load}}{L_f} \cdot \Xi_{11} - \Xi_{22} \cdot \Xi_{11} + \Xi_{12} \cdot \Xi_{21} + \Xi_{13} \cdot \Xi_{31} - \Xi_{13} \cdot \Xi_{31} \cdot \frac{R_{load}}{L_f} \cdot \Delta$$

$$Q_m(\lambda = i \cdot \omega, \Delta) = -\Omega_1 \cdot \omega^2 + \Omega_2(\Delta) \cdot \omega^4 + \{\Omega_3 \cdot \omega - \Omega_4(\Delta) \cdot \omega^3\} \cdot i$$

$$|Q(i \cdot \omega, \Delta)|^2 = \{-\Omega_1 \cdot \omega^2 + \Omega_2(\Delta) \cdot \omega^4\}^2 + \{\Omega_3 \cdot \omega - \Omega_4(\Delta) \cdot \omega^3\}^2$$

$$|Q(i \cdot \omega, \Delta)|^2 = \Omega_1^2 \cdot \omega^4 + \Omega_2^2(\Delta) \cdot \omega^8 - 2 \cdot \Omega_1 \cdot \Omega_2(\Delta) \cdot \omega^6 + \Omega_3^2 \cdot \omega^2 + \Omega_4^2(\Delta) \cdot \omega^6 - 2 \cdot \Omega_3 \cdot \Omega_4(\Delta) \cdot \omega^4$$

$$|Q(i \cdot \omega, \Delta)|^2 = \Omega_3^2 \cdot \omega^2 + [\Omega_1^2 - 2 \cdot \Omega_3 \cdot \Omega_4(\Delta)] \cdot \omega^4 + [\Omega_4^2(\Delta) - 2 \cdot \Omega_1 \cdot \Omega_2(\Delta)] \cdot \omega^6 + \Omega_2^2(\Delta) \cdot \omega^8$$

$$\begin{aligned} F(\omega, \Delta) = |P(i \cdot \omega, \Delta)|^2 - |Q(i \cdot \omega, \Delta)|^2 &= \Xi_{24}^2 \cdot \Xi_{42}^2 \cdot \omega^6 \\ &+ [\Xi_{22}^2 + 2 \cdot \Xi_{24} \cdot \Xi_{42}] \cdot \omega^8 + \omega^{10} \\ &- \Omega_3^2 \cdot \omega^2 - [\Omega_1^2 - 2 \cdot \Omega_3 \cdot \Omega_4(\Delta)] \cdot \omega^4 \\ &- [\Omega_4^2(\Delta) - 2 \cdot \Omega_1 \cdot \Omega_2(\Delta)] \cdot \omega^6 - \Omega_2^2(\Delta) \cdot \omega^8 \end{aligned}$$

$$\begin{aligned} F(\omega, \Delta) = |P(i \cdot \omega, \Delta)|^2 - |Q(i \cdot \omega, \Delta)|^2 &= -\Omega_3^2 \cdot \omega^2 - [\Omega_1^2 - 2 \cdot \Omega_3 \cdot \Omega_4(\Delta)] \cdot \omega^4 \\ &+ \Xi_{24}^2 \cdot \Xi_{42}^2 \cdot \omega^6 - [\Omega_4^2(\Delta) - 2 \cdot \Omega_1 \cdot \Omega_2(\Delta)] \cdot \omega^6 \\ &+ [\Xi_{22}^2 + 2 \cdot \Xi_{24} \cdot \Xi_{42}] \cdot \omega^8 - \Omega_2^2(\Delta) \cdot \omega^8 + \omega^{10} \end{aligned}$$

$$\begin{aligned} F(\omega, \Delta) = |P(i \cdot \omega, \Delta)|^2 - |Q(i \cdot \omega, \Delta)|^2 &= -\Omega_3^2 \cdot \omega^2 - [\Omega_1^2 - 2 \cdot \Omega_3 \cdot \Omega_4(\Delta)] \cdot \omega^4 \\ &+ \{\Xi_{24}^2 \cdot \Xi_{42}^2 - \Omega_4^2(\Delta) + 2 \cdot \Omega_1 \cdot \Omega_2(\Delta)\} \cdot \omega^6 \\ &+ \{\Xi_{22}^2 + 2 \cdot \Xi_{24} \cdot \Xi_{42} - \Omega_2^2(\Delta)\} \cdot \omega^8 + \omega^{10} \end{aligned}$$

We define the following parameters for simplicity: $\Phi_0 = 0; \Phi_2 = -\Omega_3^2$

$$\Phi_4 = -[\Omega_1^2 - 2 \cdot \Omega_3 \cdot \Omega_4(\Delta)]; \Phi_6 = \Xi_{24}^2 \cdot \Xi_{42}^2 - \Omega_4^2(\Delta) + 2 \cdot \Omega_1 \cdot \Omega_2(\Delta)$$

$\Phi_8 = \Xi_{22}^2 + 2 \cdot \Xi_{24} \cdot \Xi_{42} - \Omega_2^2(\Delta); \Phi_{10} = 1.$ Hence $F(\omega, \Delta) = 0$ implies $\sum_{k=0}^{5} \Phi_{2 \cdot k} \cdot \omega^{2 \cdot k} = 0$ and its roots are given by solving the polynomial.

Furthermore

$$P_R(i\cdot\omega,\Delta)=\Xi_{22}\cdot\omega^4;P_I(i\cdot\omega,\Delta)=-[\Xi_{24}\cdot\Xi_{42}\cdot\omega^3+\omega^5]$$
$$Q_R(i\cdot\omega,\Delta)=-\Omega_1\cdot\omega^2+\Omega_2(\Delta)\cdot\omega^4;Q_I(i\cdot\omega,\Delta)=\Omega_3\cdot\omega-\Omega_4(\Delta)\cdot\omega^3$$

Hence

$$\sin\theta(\Delta)=\frac{-P_R(i\cdot\omega,\Delta)\cdot Q_I(i\cdot\omega,\Delta)+P_I(i\cdot\omega,\Delta)\cdot Q_R(i\cdot\omega,\Delta)}{|Q(i\cdot\omega,\Delta)|^2}$$
$$\cos\theta(\Delta)=-\frac{P_R(i\cdot\omega,\Delta)\cdot Q_R(i\cdot\omega,\Delta)+P_I(i\cdot\omega,\Delta)\cdot Q_I(i\cdot\omega,\Delta)}{|Q(i\cdot\omega,\Delta)|^2}$$
$$\sin\theta(\Delta)=\frac{-\Xi_{22}\cdot\omega^4\cdot[\Omega_3\cdot\omega-\Omega_4(\Delta)\cdot\omega^3]-[\Xi_{24}\cdot\Xi_{42}\cdot\omega^3+\omega^5]\cdot[-\Omega_1\cdot\omega^2+\Omega_2(\Delta)\cdot\omega^4]}{\Omega_3^2\cdot\omega^2+[\Omega_1^2-2\cdot\Omega_3\cdot\Omega_4(\Delta)]\cdot\omega^4+[\Omega_4^2(\Delta)-2\cdot\Omega_1\cdot\Omega_2(\Delta)]\cdot\omega^6+\Omega_2^2(\Delta)\cdot\omega^8}$$
$$\cos\theta(\Delta)=-\frac{\Xi_{22}\cdot\omega^4\cdot[-\Omega_1\cdot\omega^2+\Omega_2(\Delta)\cdot\omega^4]-[\Xi_{24}\cdot\Xi_{42}\cdot\omega^3+\omega^5]\cdot[\Omega_3\cdot\omega-\Omega_4(\Delta)\cdot\omega^3]}{\Omega_3^2\cdot\omega^2+[\Omega_1^2-2\cdot\Omega_3\cdot\Omega_4(\Delta)]\cdot\omega^4+[\Omega_4^2(\Delta)-2\cdot\Omega_1\cdot\Omega_2(\Delta)]\cdot\omega^6+\Omega_2^2(\Delta)\cdot\omega^8}$$

Which jointly with $F(\omega,\Delta)=0\Rightarrow\sum_{k=0}^{5}\Phi_{2\cdot k}\cdot\omega^{2\cdot k}=0$ that are continuous and differentiable in Δ, based on Lema 1and Hence we use theorem 1. This proves the theorem 2.

Lemma 1 *Assume that* $\omega(\Delta)$ *is a positive and real root of* $F(\omega,\Delta)=0$ *defined for* $\Delta\in I$*, which is continuous and differentiable. Assume further that if* $\lambda=i\cdot\omega$ $\omega\in\mathbb{R}$ *then* $P_n(\lambda=i\cdot\omega,\Delta)+Q_m(\lambda=i\cdot\omega,\Delta)\neq0$, $\Delta\in\mathbb{R}$ *hold true. Then the functions* $S_n(\Delta)$, $n\in\mathbb{N}_0$ *are continuous and differentiable on* I.

Theorem 1 *Assume that* $\omega(\Delta)$ *is a positive real root of* $F(\omega,\Delta)=0$ *defined for* $\Delta\in I$, $I\subseteq\mathbb{R}_{+0}$*, and at some* $\Delta^*\in I$, $S_n(\Delta^*)=0$*. For some* $n\in\mathbb{N}_0$ *then a pair of simple conjugate pure imaginary roots* $\lambda_+(\Delta^*)=i\cdot\omega(\Delta^*)$, $\lambda_-(\Delta^*)=-i\cdot\omega(\Delta^*)$. $D(\lambda,\Delta)=0$ *exist at* $\Delta=\Delta^*$ *which crosses the imaginary axis from left to right if* $\delta(\Delta^*)>0$ *and cross the imaginary axis from right to left if* $\delta(\Delta^*)<0$ *where*

$$\delta(\Delta^*)=sign\{\frac{d\mathrm{Re}\lambda}{d\Delta}|_{\lambda=i\omega(\Delta^*)}\}=sign\{F_\omega(\omega(\Delta^*),\Delta^*)\}\cdot sign\{\frac{dS_n(\Delta)}{d\Delta}|_{\Delta=\Delta^*}\}.$$

Theorem 2 *The characteristic equation has a pair of simple and conjugate pure imaginary roots* $\lambda=\pm\omega(\Delta^*)$, $\omega(\Delta^*)$ *real at* $\Delta^*\in I$ if $S_n(\Delta^*)=\Delta^*-\Delta_n(\Delta^*)=0$ *for some* $n\in\mathbb{N}_0$*. If* $\omega(\Delta^*)=\omega_+(\Delta^*)$ *this pair of simple conjugate pure imaginary roots crosses the imaginary axis from left to right if* $\delta_+(\Delta^*)>0$ *and crosses the imaginary axis from right to left if* $\delta_+(\Delta^*)<0$ *where* $\delta_+(\Delta^*)=sign\{\frac{d\mathrm{Re}\lambda}{d\Delta}|_{\lambda=i\omega_+(\Delta^*)}\}$

$\delta_+(\Delta^*)=sign\{\frac{d\mathrm{Re}\lambda}{d\Delta}|_{\lambda=i\omega_+(\Delta^*)}\}=sign\{\frac{dS_n(\Delta)}{d\Delta}|_{\Delta=\Delta^*}\}$*. If* $\omega(\Delta^*)=\omega_-(\Delta^*)$ *these pair of simple conjugates pure imaginary roots cross the imaginary axis from left to right if* $\delta_-(\Delta^*)>0$ *and crosses the imaginary axis from right to left if* $\delta_-(\Delta^*)<0$ *where* $\delta_-(\Delta^*)=sign\{\frac{d\mathrm{Re}\lambda}{d\Delta}|_{\lambda=i\omega_-(\Delta^*)}\}=-sign\{\frac{dS_n(\Delta)}{d\Delta}|_{\Delta=\Delta^*}\}$*. If* $\omega_+(\Delta^*)=\omega_-(\Delta^*)=\omega(\Delta^*)$ *then* $\Delta(\Delta^*)=0$ *and* $sign\{\frac{d\mathrm{Re}\lambda}{d\Delta}|_{\lambda=i\omega(\Delta^*)}\}=0$*, the same is true* *when* $S'_n(\Delta^*) = 0$. *The following result can be useful in identifying values of* Δ *where stability switches happened.*

Our IMPATT *amplifier homogenous system for* $i'_1, i'_3, i_1, i_3, i_{L_1}$ *leads to a characteristic equation for the eigenvalue* λ *having the form (second case)* $P(\lambda, \Delta) + Q(\lambda, \Delta) \cdot e^{-\lambda \cdot \Delta} = 0$. $D(\tau_1 = 0; \Delta_1 = \Delta_2 = \Delta, \lambda) = D(\Delta, \lambda)$ *and we use Taylor series approximation:* $e^{-\lambda \cdot \Delta} \approx \sum_{n=0}^{\infty} \frac{(-\lambda \cdot \Delta)^n}{n!} \approx 1 - \lambda \cdot \Delta$ *the Maclaurin series is a Taylor series expansion of a* $e^{-\lambda \cdot \Delta}$ *function about zero* (0). *We get the following general characteristic equation* $D(\Delta, \lambda)$ *under Taylor series approximation*:

$e^{-\lambda \cdot \Delta} \approx 1 - \lambda \cdot \Delta.$

$$\begin{aligned} D(\tau_1 = 0; \Delta_1 = \Delta_2 = \Delta, \lambda) = D(\Delta, \lambda) &= \Xi_{24} \cdot \Xi_{42} \cdot \lambda^3 + \Xi_{22} \cdot \lambda^4 - \lambda^5 \\ &+ \{(-\Xi_{24} \cdot \Xi_{31} \cdot \Xi_{42} \cdot \Xi_{13} + \Xi_{14} \cdot \Xi_{23} \cdot \Xi_{31} \cdot \Xi_{42}) \\ &\cdot \lambda + (-\Xi_{11} \cdot \Xi_{24} \cdot \Xi_{42} + \Xi_{12} \cdot \Xi_{23} \cdot \Xi_{31} \\ &+ \Xi_{13} \cdot \Xi_{31} \cdot \frac{R_{load}}{L_f} - \Xi_{13} \cdot \Xi_{22} \cdot \Xi_{31} + \Xi_{14} \cdot \Xi_{21} \cdot \Xi_{42}) \\ &\cdot \lambda^2 + \{\frac{R_{load}}{L_f} \cdot \Xi_{11} - \Xi_{22} \cdot \Xi_{11} + \Xi_{12} \cdot \Xi_{21} + \Xi_{13} \cdot \Xi_{31} - \Xi_{13} \cdot \Xi_{31} \\ &\cdot \frac{R_{load}}{L_f} \cdot \Delta\} \cdot \lambda^3 + \{\Xi_{11} - \frac{R_{load}}{L_f} - \frac{R_{load}}{L_f} \cdot \Delta \cdot \Xi_{11}\} \cdot \lambda^4\} \cdot e^{-\lambda \cdot \Delta} \end{aligned}$$

We use different parameters terminology from our last characteristics parameters definition: $p_j(\Delta) \to a_j; q_j(\Delta) \to c_j; n = 5; m = 4; n > m$.

$$P_n(\lambda, \Delta) \to P(\lambda); Q_m(\lambda, \Delta) \to Q(\lambda); P(\lambda) = \sum_{j=0}^{5} a_j \cdot \lambda^j; Q(\lambda) = \sum_{j=0}^{4} c_j \cdot \lambda^j$$

$$\begin{aligned} P(\lambda) &= \Xi_{24} \cdot \Xi_{42} \cdot \lambda^3 + \Xi_{22} \cdot \lambda^4 - \lambda^5 \\ Q(\lambda) &= (-\Xi_{24} \cdot \Xi_{31} \cdot \Xi_{42} \cdot \Xi_{13} + \Xi_{14} \cdot \Xi_{23} \cdot \Xi_{31} \cdot \Xi_{42}) \cdot \lambda \\ &+ (-\Xi_{11} \cdot \Xi_{24} \cdot \Xi_{42} + \Xi_{12} \cdot \Xi_{23} \cdot \Xi_{31} + \Xi_{13} \cdot \Xi_{31} \\ &\cdot \frac{R_{load}}{L_f} - \Xi_{13} \cdot \Xi_{22} \cdot \Xi_{31} + \Xi_{14} \cdot \Xi_{21} \cdot \Xi_{42}) \cdot \lambda^2 \\ &+ \{\frac{R_{load}}{L_f} \cdot \Xi_{11} - \Xi_{22} \cdot \Xi_{11} + \Xi_{12} \cdot \Xi_{21} + \Xi_{13} \\ &\cdot \Xi_{31} - \Xi_{13} \cdot \Xi_{31} \cdot \frac{R_{load}}{L_f} \cdot \Delta\} \cdot \lambda^3 + \{\Xi_{11} - \frac{R_{load}}{L_f} - \frac{R_{load}}{L_f} \cdot \Delta \cdot \Xi_{11}\} \cdot \lambda^4 \end{aligned}$$

$n, m \in \mathbb{N}_0; n > m$ *and* $a_j, c_j : \mathbb{R}_{+0} \to \mathbb{R}$ *are continuous and differentiable function of* Δ *such that* $a_0 + c_0 = 0$. *In the following "—" denotes complex conjugate.* P(λ), Q(λ) *are analytic functions in* λ *and differentiable in* Δ. *The coefficients*:

$\{a_j(C_f, L_f, R_f, C_d, R_d, R_1, L_1, \Delta, \ldots)$ *and* $c_j(C_f, L_f, R_f, C_d, R_d, R_1, L_1, \Delta, \ldots)\} \in \mathbb{R}$ *are dependent on IMPATT amplifier system's* $C_f, L_f, R_f, C_d, R_d, R_1, L_1, \Delta, \ldots$ *values.*

We already got the following expressions: $a_0 = 0; a_1 = 0; a_5 = -1$

$$a_2 = 0; a_3 = \Xi_{24} \cdot \Xi_{42}; a_4 = \Xi_{22}; c_0 = 0; c_1 = -\Xi_{24} \cdot \Xi_{31} \cdot \Xi_{42} \cdot \Xi_{13} + \Xi_{14} \cdot \Xi_{23} \cdot \Xi_{31} \cdot \Xi_{42}$$

$$c_2 = -\Xi_{11} \cdot \Xi_{24} \cdot \Xi_{42} + \Xi_{12} \cdot \Xi_{23} \cdot \Xi_{31} + \Xi_{13} \cdot \Xi_{31} \cdot \frac{R_{load}}{L_f} - \Xi_{13} \cdot \Xi_{22} \cdot \Xi_{31} + \Xi_{14} \cdot \Xi_{21} \cdot \Xi_{42}$$

$$c_3 = \frac{R_{load}}{L_f} \cdot \Xi_{11} - \Xi_{22} \cdot \Xi_{11} + \Xi_{12} \cdot \Xi_{21} + \Xi_{13} \cdot \Xi_{31} - \Xi_{13} \cdot \Xi_{31} \cdot \frac{R_{load}}{L_f} \cdot \Delta; c_4 = \Xi_{11} - \frac{R_{load}}{L_f} - \frac{R_{load}}{L_f} \cdot \Delta \cdot \Xi_{11}$$

Unless strictly necessary, the designation of the variation arguments $(C_f, L_f, R_f, C_d, R_d, R_1, L_1, \Delta, \ldots)$ *will subsequently be omitted from* P, Q, a_j, c_j *. The coefficients* a_j, c_j *are continuous, and differentiable functions of their* arguments, *and direct substitution shows that* $a_0 + c_0 = 0$. *In our case* $\lambda = 0$ *is a root of characteristic equation. Furthermore,* $P(\lambda)$, $Q(\lambda)$ *are analytic function of* λ *of the analysis* [5] *can also be verified in the present case* [6].

Remark: In our case $P_n(\lambda = 0, \Delta) + Q_m(\lambda = 0, \Delta) = p_0(\Delta) + q_0(\Delta) = a_0 + c_0 = 0 \forall \Delta \in \mathbb{R}_{+0}$ and $\lambda = 0$ is a characteristic root of $D(\lambda, \Delta) = 0$. It is against general geometric criterion [BK]. But we inspect our analysis for $\lambda \neq 0$.

(a) If $\lambda = i \cdot \omega$, $\omega \in \mathbb{R}$ then $P(i \cdot \omega) + Q(i \cdot \omega) \neq 0$, that is P and Q have no common imaginary roots. This condition was verified numerically in the entire $(C_f, L_f, R_f, C_d, R_d, R_1, L_1, \Delta, \ldots)$ domain interest.

(b) $|Q(\lambda)/P(\lambda)|$ is bounded for $|\lambda| \to \infty$, $\mathrm{Re}\lambda \geq 0$, No roots bifurcation from ∞. Indeed, in the limit:

$$\left|\frac{Q(\lambda)}{P(\lambda)}\right| = \left|\frac{\begin{array}{l}(-\Xi_{24} \cdot \Xi_{31} \cdot \Xi_{42} \cdot \Xi_{13} + \Xi_{14} \cdot \Xi_{23} \cdot \Xi_{31} \cdot \Xi_{42}) \cdot \lambda \\ + (-\Xi_{11} \cdot \Xi_{24} \cdot \Xi_{42} + \Xi_{12} \cdot \Xi_{23} \cdot \Xi_{31} + \Xi_{13} \cdot \Xi_{31} \cdot \frac{R_{load}}{L_f} - \Xi_{13} \cdot \Xi_{22} \cdot \Xi_{31} + \Xi_{14} \cdot \Xi_{21} \cdot \Xi_{42}) \cdot \lambda^2 \\ + \{\frac{R_{load}}{L_f} \cdot \Xi_{11} - \Xi_{22} \cdot \Xi_{11} + \Xi_{12} \cdot \Xi_{21} + \Xi_{13} \cdot \Xi_{31} - \Xi_{13} \cdot \Xi_{31} \cdot \frac{R_{load}}{L_f} \cdot \Delta\} \cdot \lambda^3 \\ + \{\Xi_{11} - \frac{R_{load}}{L_f} - \frac{R_{load}}{L_f} \cdot \Delta \cdot \Xi_{11}\} \cdot \lambda^4\end{array}}{\Xi_{24} \cdot \Xi_{42} \cdot \lambda^3 + \Xi_{22} \cdot \lambda^4 - \lambda^5}\right|$$

$$|\frac{Q(\lambda)}{P(\lambda)}| = |\frac{\begin{array}{l}(-\Xi_{24}\cdot\Xi_{31}\cdot\Xi_{42}\cdot\Xi_{13}+\Xi_{14}\cdot\Xi_{23}\cdot\Xi_{31}\cdot\Xi_{42}) \\ +(-\Xi_{11}\cdot\Xi_{24}\cdot\Xi_{42}+\Xi_{12}\cdot\Xi_{23}\cdot\Xi_{31}+\Xi_{13}\cdot\Xi_{31} \\ \cdot\frac{R_{load}}{L_f}-\Xi_{13}\cdot\Xi_{22}\cdot\Xi_{31}+\Xi_{14}\cdot\Xi_{21}\cdot\Xi_{42})\cdot\lambda \\ +\{\frac{R_{load}}{L_f}\cdot\Xi_{11}-\Xi_{22}\cdot\Xi_{11}+\Xi_{12}\cdot\Xi_{21}+\Xi_{13} \\ \cdot\Xi_{31}-\Xi_{13}\cdot\Xi_{31}\cdot\frac{R_{load}}{L_f}\cdot\Delta\}\cdot\lambda^2 \\ +\{\Xi_{11}-\frac{R_{load}}{L_f}-\frac{R_{load}}{L_f}\cdot\Delta\cdot\Xi_{11}\}\cdot\lambda^3\end{array}}{\Xi_{24}\cdot\Xi_{42}\cdot\lambda^2+\Xi_{22}\cdot\lambda^3-\lambda^4}|$$

(c)

$$\begin{aligned} F(\omega) &= |P(i\cdot\omega)|^2 - |Q(i\cdot\omega)|^2. \\ F(\omega,\Delta) &= |P(i\cdot\omega,\Delta)|^2 - |Q(i\cdot\omega,\Delta)|^2 \\ &= -\Omega_3^2\cdot\omega^2 - [\Omega_1^2 - 2\cdot\Omega_3\cdot\Omega_4(\Delta)]\cdot\omega^4 \\ &\quad + \{\Xi_{24}^2\cdot\Xi_{42}^2 - \Omega_4^2(\Delta) + 2\cdot\Omega_1\cdot\Omega_2(\Delta)\}\cdot\omega^6 \\ &\quad + \{\Xi_{22}^2 + 2\cdot\Xi_{24}\cdot\Xi_{42} - \Omega_2^2(\Delta)\}\cdot\omega^8 + \omega^{10} \end{aligned}$$

It has at most a finite number of zeros. Indeed, this is a polynomial in ω (degree in ω^{10}).

(d) Each positive root $\omega(C_f, L_f, R_f, C_d, R_d, R_1, L_1, \Delta, \ldots)$ of $F(\omega) = 0$ is continuous and differentiable with respect to $C_f, L_f, R_f, C_d, R_d, R_1, L_1, \Delta, \ldots$ This condition can only be assessed numerically.

In addition, since the coefficients in P and Q are real, we have $\overline{P(-i\cdot\omega)} = P(i\cdot\omega)$.

And $\overline{Q(-i\cdot\omega)} = Q(i\cdot\omega)$ thus $\lambda = i\cdot\omega$, $\omega > 0$ may be on eigenvalue of characteristic equation. The analysis consists in identifying the roots of characteristic equation situated on the imaginary axis of the complex λ-plane, whereby increasing the parameters $C_f, L_f, R_f, C_d, R_d, R_1, L_1, \Delta, \ldots$ Reλ may, at the crossing, change its sign from (-) to (+), that is, from a stable focus $E^{(*)}(I_1'^{(*)}, I_3'^{(*)}, I_1^{(*)}, I_3^{(*)}, I_{L1}^{(*)}) = (0, 0, I_1^{(*)}, \frac{\frac{dV_{P_1}}{dt}}{C_{\sum_1} - C_{\sum_2}} - I_1^*, I_{L1}^{(*)})$ to an unstable one, or vice versa. This feature may be further assessed by examining the sign of the partial derivatives with respect to $C_f, L_f, R_f, C_d, R_d, R_1, L_1, \Delta, \ldots$ and system parameters.

$$\Lambda^{-1}(R_f) = (\frac{\partial \mathrm{Re}\lambda}{\partial R_f})_{\lambda=i\cdot\omega}, C_f, L_f, C_d, R_d, R_1, L_1, \Delta, \ldots = const$$

$$\Lambda^{-1}(L_f) = (\frac{\partial \mathrm{Re}\lambda}{\partial L_f})_{\lambda=i\cdot\omega}, C_f, R_f, C_d, R_d, R_1, L_1, \Delta, \ldots = const$$

$$\Lambda^{-1}(C_f) = (\frac{\partial \mathrm{Re}\lambda}{\partial C_f})_{\lambda=i\cdot\omega}, L_f, R_f, C_d, R_d, R_1, L_1, \Delta, \ldots = const$$

$$\Lambda^{-1}(C_d) = (\frac{\partial \mathrm{Re}\lambda}{\partial C_d})_{\lambda=i\cdot\omega}, L_f, R_f, C_f, R_d, R_1, L_1, \Delta, \ldots = const$$

$$\Lambda^{-1}(R_d) = (\frac{\partial \mathrm{Re}\lambda}{\partial R_d})_{\lambda=i\cdot\omega}, L_f, R_f, C_f, C_d, R_1, L_1, \Delta, \ldots = const$$

$$\Lambda^{-1}(\Delta) = (\frac{\partial \mathrm{Re}\lambda}{\partial \Delta})_{\lambda=i\cdot\omega}, L_f, R_f, C_f, C_d, R_1, L_1, R_d, \ldots = const$$

For the second case $\tau_1 = 0; \Delta_1 = \Delta_2 = \Delta$ we got the following results:

$$P_R(i\cdot\omega,\Delta) = \Xi_{22}\cdot\omega^4; P_I(i\cdot\omega,\Delta) = -[\Xi_{24}\cdot\Xi_{42}\cdot\omega^3 + \omega^5]$$

$$Q_R(i\cdot\omega,\Delta) = -\Omega_1\cdot\omega^2 + \Omega_2(\Delta)\cdot\omega^4; Q_I(i\cdot\omega,\Delta) = \Omega_3\cdot\omega - \Omega_4(\Delta)\cdot\omega^3$$

$$\Phi_0 = 0; \Phi_2 = -\Omega_3^2; \Phi_4 = -[\Omega_1^2 - 2\cdot\Omega_3\cdot\Omega_4(\Delta)];$$

$$\Phi_6 = \Xi_{24}^2\cdot\Xi_{42}^2 - \Omega_4^2(\Delta) + 2\cdot\Omega_1\cdot\Omega_2(\Delta)$$

$$\Phi_8 = \Xi_{22}^2 + 2\cdot\Xi_{24}\cdot\Xi_{42} - \Omega_2^2(\Delta); \Phi_{10} = 1;$$

$$F(\omega,\Delta) = |P(i\cdot\omega,\Delta)|^2 - |Q(i\cdot\omega,\Delta)|^2 = \sum_{k=0}^{5}\Phi_{2\cdot k}\cdot\omega^{2\cdot k}$$

Hence $F(\omega,\Delta) = 0$ implies $\sum_{k=0}^{5}\Phi_{2\cdot k}\cdot\omega^{2\cdot k} = 0$. When writing $P(\lambda) = P_R(\lambda) + i\cdot P_I(\lambda)$ and $Q(\lambda) = Q_R(\lambda) + i\cdot Q_I(\lambda)$, inserting $\lambda = i\cdot\omega$ into IMPATT amplifier system's characteristic equation, ω must satisfy the following:

$$\sin\omega\cdot\Delta = g(\omega) = \frac{-P_R(i\cdot\omega,\Delta)\cdot Q_I(i\cdot\omega,\Delta) + P_I(i\cdot\omega,\Delta)\cdot Q_R(i\cdot\omega,\Delta)}{|Q(i\cdot\omega,\Delta)|^2}$$

$$\cos\omega\cdot\Delta = h(\omega) = -\frac{P_R(i\cdot\omega,\Delta)\cdot Q_R(i\cdot\omega,\Delta) + P_I(i\cdot\omega,\Delta)\cdot Q_I(i\cdot\omega,\Delta)}{|Q(i\cdot\omega,\Delta)|^2}$$

where $|Q(i\cdot\omega,\Delta)|^2 \neq 0$ in view of requirement (a) above, and $(g,h)\in\mathbb{R}$. Furthermore, it follows above $\sin\omega\cdot\Delta$ and $\cos\omega\cdot\Delta$ equations that, by squaring and adding the sides, ω must be a positive root of $F(\omega) = |P(i\cdot\omega)|^2 - |Q(i\cdot\omega)|^2 = 0$.

Note that $F(\omega)$ is dependent of Δ. Now it is important to notice that if $\Delta \notin I$ (assume that $I \subseteq \mathbb{R}_{+0}$ is the set where $\omega(\Delta)$ is a positive root of $F(\omega)$ and for, $\Delta \notin I, \omega(\Delta)$ is not defined, then for all Δ in I $\omega(\Delta)$ satisfies that $F(\omega) = 0$) . Then

There are no positive $\omega(\Delta)$ solutions for $F(\omega,\Delta)=0$, and we cannot have stability switches. For any $\Delta \in I$, where $\omega(\Delta)$ is a positive solution of $F(\omega,\Delta)=0$, we can define the angle $\theta(\Delta) \in [0, 2\cdot\pi]$ as the solution of $\sin\theta(\Delta) = \cdots; \cos\theta(\Delta) = \cdots$ And the relation between the argument $\theta(\Delta)$ and $\omega(\Delta)\cdot\Delta$ for $\Delta \in I$ must be $\omega(\Delta)\cdot\Delta = \theta(\Delta) + n\cdot 2\cdot\pi \, \forall n \in \mathbb{N}_0$. Hence we can define the maps $\Delta_n : I \to \mathbb{R}_{+0}$ given by $\Delta_n(\Delta) = \frac{\theta(\Delta)+n\cdot 2\cdot\pi}{\omega(\Delta)}; n \in \mathbb{N}_0; \Delta \in I$. Let as introduce the functions $I \to \mathbb{R}; S_n(\Delta) = \Delta - \Delta_n(\Delta); \Delta \in I; n \in \mathbb{N}_0$ that are continuous and differentiable in Δ. In the following, the subscripts $\lambda, \omega, L_f, R_f, C_f, R_d, R_1, L_1, \ldots$ indicate the corresponding partial derivatives. Let us first concentrate on $\Lambda(x)$, remember in $\lambda(L_f, R_f, C_f, R_d, R_1, L_1, \ldots)$ and $\omega(L_f, R_f, C_f, R_d, R_1, L_1, \ldots)$, and keeping all parameters except (x) and Δ. The derivation closely follows that in reference [BK]. Differentiating IMPATT amplifier system characteristic equation $P(\lambda) + Q(\lambda)\cdot e^{-\lambda\cdot\Delta} = 0$ with respect to specific parameter (x), and inverting the derivative, for convenience, one calculates:

Remark:

$$x = L_f, R_f, C_f, R_d, R_1, L_1, \ldots, etc.,$$

$$(\frac{\partial\lambda}{\partial x})^{-1} = \frac{-P_\lambda(\lambda,x)\cdot Q(\lambda,x) + Q_\lambda(\lambda,x)\cdot P(\lambda,x) - \Delta\cdot P(\lambda,x)\cdot Q(\lambda,x)}{P_x(\lambda,x)\cdot Q(\lambda,x) - Q_x(\lambda,x)\cdot P(\lambda,x)}$$

where $P_\lambda = \frac{\partial P}{\partial\lambda}$ ...etc., Substituting $\lambda = i\cdot\omega$ and bearing $i\cdot\overline{P(-i\cdot\omega)} = P(i\cdot\omega)$ $\overline{Q(-i\cdot\omega)} = Q(i\cdot\omega)$; $i\cdot P_\lambda(i\cdot\omega) = P_\omega(i\cdot\omega)$; $i\cdot Q_\lambda(i\cdot\omega) = Q_\omega(i\cdot\omega)$ and that on the surface $|P(i\cdot\omega)|^2 = |Q(i\cdot\omega)|^2$, one obtain the following expression:

$$(\frac{\partial\lambda}{\partial x})^{-1}|_{\lambda=i\cdot\omega} = \frac{i\cdot P_\omega(i\cdot\omega,x)\cdot\overline{P(i\cdot\omega,x)} + i\cdot Q_\lambda(i\cdot\omega,x)\cdot\overline{Q(\lambda,x)} - \Delta\cdot|P(i\cdot\omega,x)|^2}{P_x(i\cdot\omega,x)\cdot\overline{P(i\cdot\omega,x)} - Q_x(i\cdot\omega,x)\cdot\overline{Q(i\cdot\omega,x)}}$$

Upon separating into real and imaginary parts, with $P = P_R + i\cdot P_I$; $Q = Q_R + i\cdot Q_I$

$$P_\omega = P_{R\omega} + i\cdot P_{I\omega}; Q_\omega = Q_{R\omega} + i\cdot Q_{I\omega}; P_x = P_{Rx} + i\cdot P_{Ix}; Q_x = Q_{Rx} + i\cdot Q_{Ix}; P^2 = P_R^2 + P_I^2$$

When (x) can be any IMPATT diode active circulator system's parameter L_f, R_f, C_f,... and time delay Δ etc.,. Where for convenience, we dropped the arguments (i•ω, x), and where $F_\omega = 2\cdot[(P_{R\omega}\cdot P_R + P_{I\omega}\cdot P_I) - (Q_{R\omega}\cdot Q_R + Q_{I\omega}\cdot Q_I)]$

$F_x = 2\cdot[(P_{Rx}\cdot P_R + P_{Ix}\cdot P_I) - (Q_{Rx}\cdot Q_R + Q_{Ix}\cdot Q_I)]$; $\omega_x = -F_x/F_\omega$ We define U and V:

$$U = (P_R \cdot P_{I\omega} - P_I \cdot P_{R\omega}) - (Q_R \cdot Q_{I\omega} - Q_I \cdot Q_{R\omega})$$
$$V = (P_R \cdot P_{Ix} - P_I \cdot P_{Rx}) - (Q_R \cdot Q_{Ix} - Q_I \cdot Q_{Rx})$$

We choose our specific parameter as time delay x = Δ.

$$P_R = P_R(i \cdot \omega, \Delta) = \Xi_{22} \cdot \omega^4; P_I = P_I(i \cdot \omega, \Delta) = -[\Xi_{24} \cdot \Xi_{42} \cdot \omega^3 + \omega^5]$$
$$Q_R = Q_R(i \cdot \omega, \Delta) = -\Omega_1 \cdot \omega^2 + \Omega_2(\Delta) \cdot \omega^4; Q_I = Q_I(i \cdot \omega, \Delta) = \Omega_3 \cdot \omega - \Omega_4(\Delta) \cdot \omega^3$$

$$P_{R\omega} = 4 \cdot \Xi_{22} \cdot \omega^3; P_{I\omega} = -[\Xi_{24} \cdot \Xi_{42} \cdot 3 \cdot \omega^2 + 5 \cdot \omega^4];$$
$$Q_{R\omega} = -\Omega_1 \cdot 2 \cdot \omega + \Omega_2(\Delta) \cdot 4 \cdot \omega^3$$
$$Q_{I\omega} = \Omega_3 - \Omega_4(\Delta) \cdot 3 \cdot \omega^2; P_{R\Delta} = 0; P_{I\Delta} = 0;$$
$$Q_{R\Delta} = \frac{\partial \Omega_2(\Delta)}{\partial \Delta} \cdot \omega^4; Q_{I\Delta} = -\frac{\partial \Omega_4(\Delta)}{\partial \Delta} \cdot \omega^3$$
$$\omega_\Delta = -F_\Delta / F_\omega; \frac{\partial \Omega_2(\Delta)}{\partial \Delta} = -\frac{R_{load}}{L_f} \cdot \Xi_{11}; \frac{\partial \Omega_4(\Delta)}{\partial \Delta} = -\Xi_{13} \cdot \Xi_{31} \cdot \frac{R_{load}}{L_f}$$

$$Q_{R\Delta} = -\frac{R_{load}}{L_f} \cdot \Xi_{11} \cdot \omega^4; Q_{I\Delta} = \Xi_{13} \cdot \Xi_{31} \cdot \frac{R_{load}}{L_f} \cdot \omega^3; F_\Delta = -2 \cdot (Q_{R\Delta} \cdot Q_R + Q_{I\Delta} \cdot Q_I)$$
$$P_{R\omega} \cdot P_R = 4 \cdot \Xi_{22}^2 \cdot \omega^7; P_{I\omega} \cdot P_I = [\Xi_{24} \cdot \Xi_{42} \cdot 3 + 5 \cdot \omega^2] \cdot [\Xi_{24} \cdot \Xi_{42} + \omega^2] \cdot \omega^5$$
$$Q_{R\omega} \cdot Q_R = 2 \cdot [-\Omega_1 + \Omega_2(\Delta) \cdot 2 \cdot \omega^2] \cdot [-\Omega_1 + \Omega_2(\Delta) \cdot \omega^2] \cdot \omega^3$$
$$Q_{I\omega} \cdot Q_I = [\Omega_3 - \Omega_4(\Delta) \cdot 3 \cdot \omega^2] \cdot [\Omega_3 - \Omega_4(\Delta) \cdot \omega^2] \cdot \omega; P_{R\Delta} \cdot P_R + P_{I\Delta} \cdot P_I = 0$$
$$P_{I\Delta} \cdot P_I = 0; P_{R\Delta} \cdot P_R = 0; F_\Delta = 2 \cdot [(P_{R\Delta} \cdot P_R + P_{I\Delta} \cdot P_I) - (Q_{R\Delta} \cdot Q_R + Q_{I\Delta} \cdot Q_I)]$$

$$Q_{R\Delta} \cdot Q_R = -\frac{R_{load}}{L_f} \cdot \Xi_{11} \cdot (-\Omega_1 + \Omega_2(\Delta) \cdot \omega^2) \cdot \omega^6;$$
$$Q_{I\Delta} \cdot Q_I = \Xi_{13} \cdot \Xi_{31} \cdot \frac{R_{load}}{L_f} \cdot [\Omega_3 - \Omega_4(\Delta) \cdot \omega^2] \cdot \omega^4$$
$$F_\Delta = -2 \cdot \{-\frac{R_{load}}{L_f} \cdot \Xi_{11} \cdot (-\Omega_1 + \Omega_2(\Delta) \cdot \omega^2) \cdot \omega^2$$
$$+ \Xi_{13} \cdot \Xi_{31} \cdot \frac{R_{load}}{L_f} \cdot [\Omega_3 - \Omega_4(\Delta) \cdot \omega^2]\} \cdot \omega^4$$
$$P_R \cdot P_{I\omega} = -\Xi_{22} \cdot [\Xi_{24} \cdot \Xi_{42} \cdot 3 + 5 \cdot \omega^2] \cdot \omega^6;$$
$$P_I \cdot P_{R\omega} = -4 \cdot [\Xi_{24} \cdot \Xi_{42} + \omega^2] \cdot \Xi_{22} \cdot \omega^6$$
$$Q_R \cdot Q_{I\omega} = [-\Omega_1 + \Omega_2(\Delta) \cdot \omega^2] \cdot [\Omega_3 - \Omega_4(\Delta) \cdot 3 \cdot \omega^2] \cdot \omega^2$$
$$Q_I \cdot Q_{R\omega} = [\Omega_3 - \Omega_4(\Delta) \cdot \omega^2] \cdot [-\Omega_1 + \Omega_2(\Delta) \cdot 2 \cdot \omega^2] \cdot 2 \cdot \omega^2$$

$$V = (P_R \cdot P_{I\Delta} - P_I \cdot P_{R\Delta}) - (Q_R \cdot Q_{I\Delta} - Q_I \cdot Q_{R\Delta}); P_R \cdot P_{I\Delta} - P_I \cdot P_{R\Delta} = 0$$

$$V = -(Q_R \cdot Q_{I\Delta} - Q_I \cdot Q_{R\Delta}) = -([-\Omega_1 \cdot \omega^2 + \Omega_2(\Delta) \cdot \omega^4] \cdot \Xi_{13} \cdot \Xi_{31} \cdot \frac{R_{load}}{L_f} \cdot \omega^3$$
$$- [\Omega_3 \cdot \omega - \Omega_4(\Delta) \cdot \omega^3] \cdot [-\frac{R_{load}}{L_f} \cdot \Xi_{11} \cdot \omega^4])$$

$$V = -([-\Omega_1 + \Omega_2(\Delta) \cdot \omega^2] \cdot \Xi_{13} \cdot \Xi_{31} \cdot \frac{R_{load}}{L_f} \cdot \omega^5 + [\Omega_3 - \Omega_4(\Delta) \cdot \omega^2] \cdot \frac{R_{load}}{L_f} \cdot \Xi_{11} \cdot \omega^5)$$

F(ω, Δ) = 0. Differentiating with respect to Δ and we get

$$F_\omega \cdot \frac{\partial \omega}{\partial \Delta} + F_\Delta = 0; \Delta \in I \Rightarrow \frac{\partial \omega}{\partial \Delta} = -\frac{F_\Delta}{F_\omega};$$
$$\Lambda^{-1}(\Delta) = (\frac{\partial \mathrm{Re}\lambda}{\partial \Delta})_{\lambda = i \cdot \omega}; \frac{\partial \omega}{\partial \Delta} = \omega_\Delta = -\frac{F_\Delta}{F_\omega}$$
$$\Lambda^{-1}(\Delta) = \mathrm{Re}\{\frac{-2 \cdot [U + \Delta \cdot |P|^2] + i \cdot F_\omega}{F_\Delta + i \cdot 2 \cdot [V + \omega \cdot |P|^2]}; sign\ \Lambda^{-1}(\Delta)\} = sign\{(\frac{\partial \mathrm{Re}\lambda}{\partial \Delta})_{\lambda = i \cdot \omega}\}$$

$sign\ \Lambda^{-1}(\Delta)\} = sign\{F_\omega\} \cdot\ sign\{\Delta \cdot \frac{\partial \omega}{\partial \Delta} + \omega + \frac{U \cdot \frac{\partial \omega}{\partial \Delta} + V}{|P|^2}\}$. We shall presently examine the possibility of stability transition (bifurcations) of our system, about the equilibrium point $E^{(*)}(I_1'^{(*)}, I_3'^{(*)}, I_1^{(*)}, I_3^{(*)}, I_{L1}^{(*)}) = (0, 0, I_1^{(*)}, \frac{\frac{dV_{P_1}}{dt}}{C_{\sum_1} - C_{\sum_2}} - I_1^*, I_{L1}^{(*)})$ as a result of a variation of delay parameter Δ. The analysis consists in identifying the roots of our system characteristic equation situated on the imaginary axis of the complex λ-plane whereby increasing the delay parameter Δ, Reλ may at the crossing, change its sign from – to +, that is, from a stable focus $E^{(*)}$ to an unstable one, or vice versa. This feature may be further assessed by examining the sign of the partial derivatives with respect to Δ.

$$\Lambda^{-1}(\Delta) = (\frac{\partial \mathrm{Re}\lambda}{\partial \Delta})_{\lambda = i \cdot \omega}; \Lambda^{-1}(\Delta) = (\frac{\partial \mathrm{Re}\lambda}{\partial \Delta})_{\lambda = i \cdot \omega}, L_f, R_f, C_f, R_d, R_1, L_1, \ldots$$
$$= const, \omega \in \mathbb{R}_+$$

Numerical analysis: We get the expression for $F(\omega, \Delta)$ system parameters values. We find those ω, Δ values which fulfil $F(\omega, \Delta) = 0$. We ignore negative, complex, and imaginary values of ω for specific Δ values. $\Delta \in [0.001 \ldots 10]$ and we can express by 3D function $F(\omega, \Delta) = 0$. We plot the stability switch diagram based on different delay values of our system.

$$\Lambda^{-1}(\Delta) = (\frac{\partial \mathrm{Re}\lambda}{\partial \Delta})_{\lambda = i\cdot\omega} = \mathrm{Re}\{\frac{-2\cdot[U+\Delta\cdot|P|^2]+i\cdot F_\omega}{F_\Delta + i\cdot 2\cdot[V+\omega\cdot|P|^2]}\}$$

$$\Lambda^{-1}(\Delta) = (\frac{\partial \mathrm{Re}\lambda}{\partial \Delta})_{\lambda = i\cdot\omega} = \frac{2\cdot\{F_\omega\cdot(V+\omega\cdot P^2) - F_\Delta\cdot(U+\Delta\cdot P^2)\}}{F_\Delta^2 + 4\cdot(V+\omega\cdot P^2)^2}$$

The stability switch occur only on those delay values (Δ) which fit the equation: $\Delta = \frac{\theta_+(\Delta)}{\omega_+(\Delta)}$ and $\theta_+(\Delta)$ is the solution of $\sin\theta(\Delta) = \cdots; \cos\theta(\Delta) = \ldots$ when $\omega = \omega_+(\Delta)$ if only ω_+ is feasible. Additionally, when system's parameters are known and the stability switch due to various time delay values Δ is described in the following expression: (Steven 1994).

$$sign\{\Lambda^{-1}(\Delta)\} = sign\{F_\omega(\omega(\Delta),\Delta)\}$$

$$sign\{\Delta\cdot\omega_\Delta(\omega(\Delta)) + \omega(\Delta) + \frac{U(\omega(\Delta))\cdot\omega_\Delta(\omega(\Delta)) + V(\omega(\Delta))}{|P(\omega(\Delta))|^2}\}$$

Remark: We know $F(\omega,\Delta) = 0$ implies it roots $\omega_i(\Delta)$ and finding those delays values Δ which ω_i is feasible. There are Δ values which ω_i is complex or imaginary number, then unable to analyze stability [6, 19, 32].

3.5 Multistage IMPATT Amplifier System Microstrip Delayed in Time Stability Switching Analysis

In many applications there is a use of multistage IMPATT amplifier. Multistage IMPATT amplifier is constructed from many single circulator/IMPATT diode amplifier which are connected as a chain structure. Each circulator/IMPATT diode amplifier gets the RF signal from the previous amplifier unit and after amplification feeds the RF signal to the next amplifier unit. In that multistage IMPATT amplifier structure we use two modes of operation as an amplifier. The first is the negative resistance mode, where the input signal entering the IMPATT through the circulator is amplified due to the negative resistance phenomena in the IMPATT. The amplified signal passes out of the diode through the same port at which the input signal entered, and because of the circulator, passes into the output line. The second is the injection locked mode. The IMPATT is biases so that it is oscillating all the time, but the frequency is locked to the input frequency and the power leaving the IMPATT is at the same frequency as the input. If we compare the two modes bandwidth, the negative resistance mode provides the optimal bandwidth and the bandwidth of an injection locked amplifier is only few percent. The efficiency of the injection locked mode is greater than the efficiency of negative resistance mode. In multistage IMPATT amplifier, all the amplifier unit's mode of operation is negative resistance except the last amplifier unit (output stage) which the mode of operation is injection locked mode.

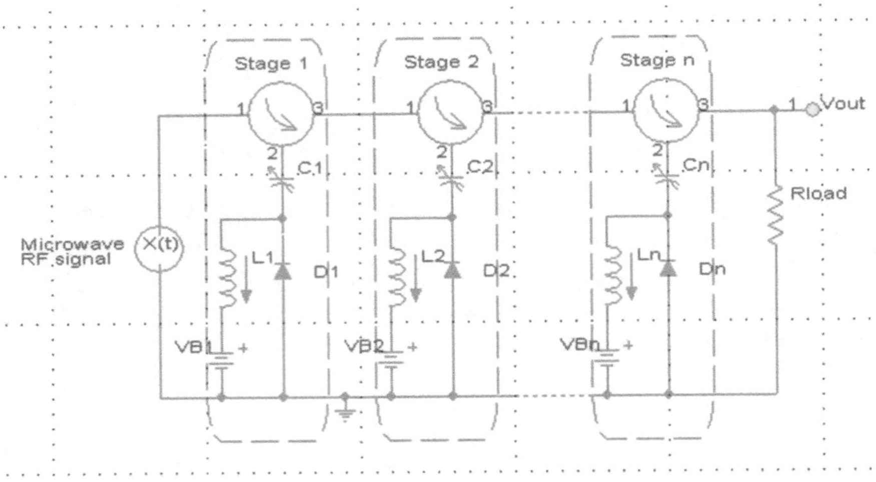

Fig. 3.15 Multistage IMPATT amplifier circuit

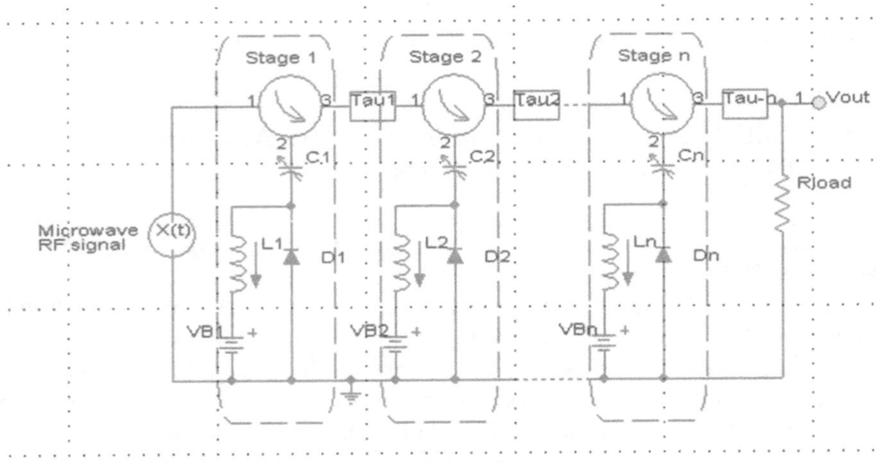

Fig. 3.16 Multistage IMPATT amplifier circuit with microstrip delay lines

Remark: we neglect the microstrip time delay which is connected between each stage circulator port 2 to IMPATT diode [70–72].

D_1, D_2, …, D_n are IMPATT diodes. Each single circulator/IMPATT diode amplifier is connected to the next amplifier unit by microstrip line. We characterize each microstrip segment as a delay line. We define our multistage IMPATT amplifier with additional n delay lines, $\tau_1, \ldots, \tau_n$ which represent (n) microstrip segments. We neglect the voltages on delay lines $V_{\tau_i} \rightarrow \varepsilon \, \forall \, \tau_1, \ldots, \tau_n \, ; \, 1 \le i \le n$. There is a delay in the current which flow through each microstrip delay line $I(t) \rightarrow I(t - \tau_i) \, \forall \, 1 \le i \le n$.

The input current to circuit stage (i) is define as $I_{in(i)}(t)$ and the output current to circuit stage (i) is define as $I_{out(i)}(t)$. The transfer function from circuit stage (i) input current $I_{in(i)}(t)$ to output current $I_{out(i)}(t)$ is $I_{out(i)}(t) = f_i(I_{in(i)}(t), D_i, L_i, C_i, V_{B_i})$

$$\forall 1 \le i \le n; I_{in(i)}(t) = I_{out(i-1)}(t - \tau_{i-1}); I_{out(i)}(t) = f_i(I_{out(i-1)}(t - \tau_{i-1}), D_i, L_i, C_i, V_{B_i}) \forall$$
$$2 \le i \le n. I_{out(1)}(t) = f_1(I_{in(1)}(t), D_1, L_1, C_1, V_{B_1}); I_{in(1)}(t) = g(X(t)).$$
$$I_{in(2)}(t) = I_{out(1)}(t - \tau_1); I_{in(3)}(t) = I_{out(2)}(t - \tau_2); I_{in(n-1)}(t) = I_{out(n-2)}(t - \tau_{n-2})$$
$$I_{in(n)}(t) = I_{out(n-1)}(t - \tau_{n-1}); I_{R_L}(t) = I_{out(n)}(t - \tau_n);$$
$$I_{out(1)}(t) = f_1(g(X(t)), D_1, L_1, C_1, V_{B_1}).$$
$$I_{out(1)}(t) = f_1(I_{in(1)}(t), D_1, C_1, L_1, V_{B_1}); I_{in(2)}(t) = I_{out(1)}(t - \tau_1);$$
$$I_{out(2)}(t) = f_2(I_{in(2)}(t), D_2, C_2, L_2, V_{B_2})$$
$$I_{out(2)}(t) = f_2(I_{out(1)}(t - \tau_1), D_2, C_2, L_2, V_{B_2}); I_{in(3)}(t) = I_{out(2)}(t - \tau_2);$$
$$I_{out(3)}(t) = f_3(I_{in(3)}(t), D_3, C_3, L_3, V_{B_3})$$
$$I_{in(n-1)}(t) = I_{out(n-2)}(t - \tau_{n-2}); I_{out(n-1)}(t) = f_{n-1}(I_{in(n-1)}(t), D_{n-1}, L_{n-1}, C_{n-1}, V_{B_{n-1}})$$
$$I_{out(n-1)}(t) = f_{n-1}(I_{out(n-2)}(t - \tau_{n-2}), D_{n-1}, L_{n-1}, C_{n-1}, V_{B_{n-1}});$$
$$I_{in(n)}(t) = I_{out(n-1)}(t - \tau_{n-1})$$

$$I_{out(n)}(t) = f_n(I_{out(n-1)}(t - \tau_{n-1}), D_n, L_n, C_n, V_{B_n});$$
$$I_{out(n)}(t) = f_n(I_{in(n)}(t), D_n, L_n, C_n, V_{B_n})$$
$$I_{R_L}(t) = I_{out(n)}(t - \tau_n); I_{R_L}(t) = f_n(I_{out(n-1)}(t - \tau_{n-1} - \tau_n), D_n, L_n, C_n, V_{B_n})$$
$$I_{R_L}(t) = f_n(f_{n-1}(I_{out(n-2)}(t - \tau_{n-2} - \tau_{n-1} - \tau_n), D_{n-1}, L_{n-1}, C_{n-1}, V_{B_{n-1}}), D_n, L_n, C_n, V_{B_n})$$
$$I_{out(n-1)}(t - \tau_{n-1} - \tau_n) = f_{n-1}(I_{in(n-1)}(t - \tau_{n-1} - \tau_n), D_{n-1}, L_{n-1}, C_{n-1}, V_{B_{n-1}})$$
$$I_{out(n-1)}(t - \tau_{n-1} - \tau_n) = f_{n-1}(I_{out(n-2)}(t - \tau_{n-2} - \tau_{n-1} - \tau_n), D_{n-1}, L_{n-1}, C_{n-1}, V_{B_{n-1}})$$

$$I_{out(n-2)}(t) = f_{n-2}(I_{in(n-2)}(t), D_{n-2}, L_{n-2}, C_{n-2}, V_{B_{n-2}});$$
$$I_{in(n-2)}(t) = I_{out(n-3)}(t - \tau_{n-3})$$
$$I_{out(n-2)}(t) = f_{n-2}(I_{out(n-3)}(t - \tau_{n-3}), D_{n-2}, L_{n-2}, C_{n-2}, V_{B_{n-2}})$$
$$I_{out(n-2)}(t - \tau_{n-2} - \tau_{n-1} - \tau_n) = f_{n-2}(I_{out(n-3)}(t - \tau_{n-3} - \tau_{n-2} - \tau_{n-1} - \tau_n), D_{n-2}, L_{n-2}, C_{n-2}, V_{B_{n-2}})$$
$$I_{out(n-2)}(t - \tau_{n-2} - \tau_{n-1} - \tau_n) = f_{n-2}(I_{out(n-3)}(t - \sum_{k=n-3}^{n} \tau_k), D_{n-2}, L_{n-2}, C_{n-2}, V_{B_{n-2}})$$

$$I_{R_L}(t) = f_n(f_{n-1}(f_{n-2}(I_{out(n-3)}(t - \tau_{n-3} - \tau_{n-2} - \tau_{n-1} - \tau_n), D_{n-2}, L_{n-2}, C_{n-2}, V_{B_{n-2}}), D_{n-1}, L_{n-1}, C_{n-1}, V_{B_{n-1}}), D_n, L_n, C_n, V_{B_n})$$

$$I_{R_L}(t) = f_n(f_{n-1}(f_{n-2}(I_{out(n-3)}(t - \sum_{k=n-3}^{n} \tau_k), D_{n-2}, L_{n-2}, C_{n-2}, V_{B_{n-2}}), D_{n-1}, L_{n-1}, C_{n-1}, V_{B_{n-1}}), D_n, L_n, C_n, V_{B_n})$$

$$I_{R_L}(t) = f_n(f_{n-1}(f_{n-2}(\ldots(f_1(I_{in(1)}(t - \tau_1 - \tau_2 \ldots - \tau_n), D_1, L_1, C_1, V_{B_1})\ldots) D_{n-1}, L_{n-1}, C_{n-1}, V_{B_{n-1}}), D_n, L_n, C_n, V_{B_n})$$

$$I_{R_L}(t) = f_n(f_{n-1}(f_{n-2}(\ldots(f_1(I_{in(1)}(t - \sum_{k=1}^{n} \tau_k), D_1, L_1, C_1, V_{B_1})\ldots) D_{n-1}, L_{n-1}, C_{n-1}, V_{B_{n-1}}), D_n, L_n, C_n, V_{B_n})$$

$$I_{in(n-k)}(t) = I_{out(n-(k+1))}(t - \tau_{n-(k+1)})\ \forall k = 0, 1, 2, \ldots, n-2$$
$$I_{in(n)}(t) = I_{out(n-1)}(t - \tau_{n-1}), \ldots, I_{in(2)}(t) = I_{out(1)}(t - \tau_1)$$

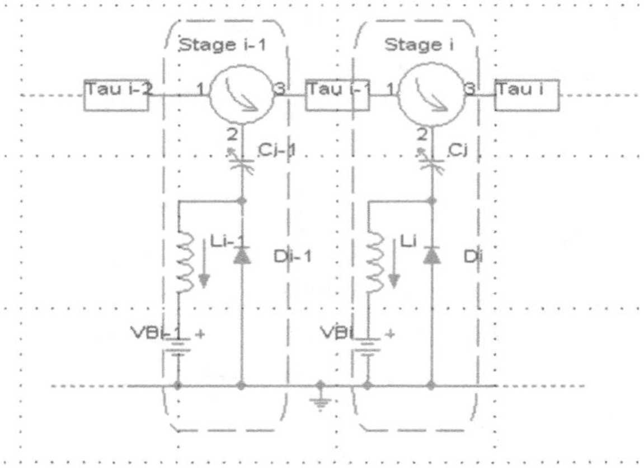

Fig. 3.17 Multistage IMPATT amplifier circuit follow stages

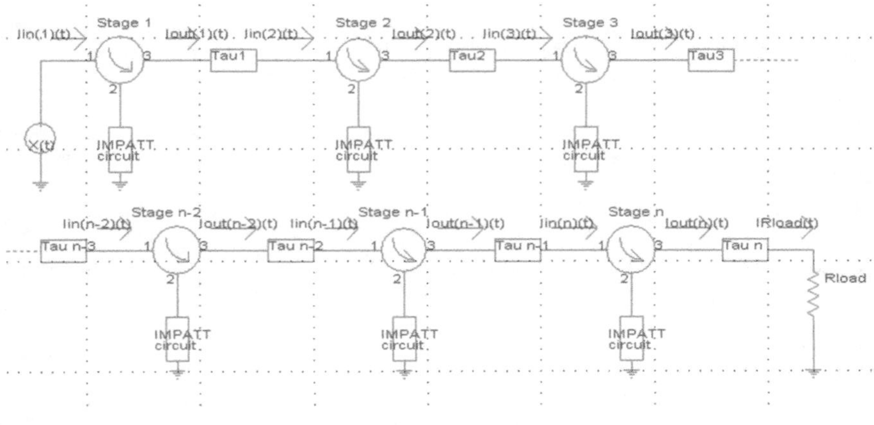

Fig. 3.18 Multistage IMPATT amplifier circuit – n stages

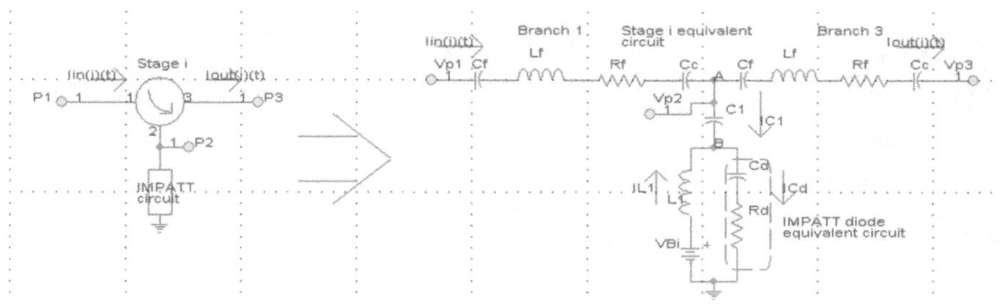

Fig. 3.19 IMPATT amplifier circuit and equivalent circuit

Multistage IMPATT amplifier one stage current transfer function: The transfer function from circuit stage (i) input current $I_{in(i)}(t)$ to output current $I_{out(i)}(t)$ is $I_{out(i)}(t) = f_i(I_{in(i)}(t), D_i, L_i, C_i, V_{B_i})$. Circuit stage active circulator can be described by system path from RFin port (P_1) to RFout port (P_3). For simplicity we ignore MESFET high frequency equivalent model and took it as cutoff element in our system. The equivalent circuit for one stage current transfer function is present in the below figure. We ignore the circulator signal path from P_3 to P_1 since our amplifier RF signal is feed directly to the next IMPATT amplifier stage. We consider fully matching between two follow amplifier stages (stages i − 1 and i), no signal reflections.

Branch 1: $I_{C_C} = I_{L_f} = I_{R_f} = I_{C_f} = I_{in(i)}(t)$; Branch 3: $I_{C_C} = I_{L_f} = I_{R_f} = I_{C_f} = I_{out(i)}(t)$

$$R_D = -R_\gamma; R_\gamma > 0; R_D < 0; |R_D| = R_\gamma; \frac{dV_{R_D}}{dI_{R_D}} < 0; \frac{dV_{R_D}}{dI_{R_D}} = R_D = R_d; \left|\frac{dV_{R_D}}{dI_{R_D}}\right| = R_\gamma$$

$$I_{in(i)}(t) = C_f \cdot \frac{dV_{C_f}}{dt}; I_{in(i)}(t) = C_C \cdot \frac{dV_{C_C}}{dt}; V_{R_f} = I_{in(i)}(t) \cdot R_f; V_{L_f} = L_f \cdot \frac{dI_{in(i)}(t)}{dt}$$

$$I_{in(i)}(t) = I_{C_1} + I_{out(i)}(t); I_{C_d} = I_{R_d} = I_{IMPATT}; V_{L_1} = L_1 \cdot \frac{dI_{L_1}}{dt} \Rightarrow I_{L_1} = \frac{1}{L_1}\int V_{L_1} \cdot dt$$

Branch 1: $V_{P_1} = V_{C_f} + V_{L_f} + V_{R_f} + V_{C_C} + V_A$; Branch 3: $V_A = V_{C_f} + V_{L_f} + V_{R_f} + V_{C_C} + V_{P_3}$

$$V_A - V_B = V_{C_1}; I_{C_1} = C_1 \cdot \frac{dV_{C_1}}{dt}; I_{C_d} = C_d \cdot \frac{dV_{C_d}}{dt}; I_{L_1} + I_{C_1} = I_{R_d}; I_{C_1} = I_{R_d} - I_{L_1}$$

$$\frac{dV_{R_d}}{dI_{R_d}} = \frac{dV_{R_d}}{dI_{C_d}} = R_d; I_{out(i)}(t) = I_{in(i)}(t) - I_{C_1};$$

$$I_{out(i)}(t) = I_{in(i)}(t) - I_{C_1} = I_{in(i)}(t) - I_{R_d} + I_{L_1}$$

$$I_{C_d} = C_d \cdot \frac{dV_{C_d}}{dt} \Rightarrow V_{C_d} = \frac{1}{C_d} \cdot \int I_{C_d} \cdot dt;$$

$$I_{out(i)}(t) = I_{in(i)}(t) - I_{R_d} + \frac{1}{L_1} \int (V_{B_i} - V_B) \cdot dt$$

$$I_{out(i)}(t) = I_{in(i)}(t) - I_{R_d} + \frac{1}{L_1} \int V_{L_1} \cdot dt; V_{L_1} = V_{B_i} - V_B;$$

$$V_B = V_{C_d} + V_{R_d} = \frac{1}{C_d} \cdot \int I_{C_d} \cdot dt + V_{R_d}$$

$$V_A - V_B = V_{C_1}; I_{C_1} = C_1 \cdot \frac{dV_{C_1}}{dt}; I_{C_1} = C_1 \cdot \frac{d(V_A - V_B)}{dt} \Rightarrow V_A - V_B = \frac{1}{C_1} \cdot \int I_{C_1} \cdot dt$$

$$V_B = V_A - \frac{1}{C_1} \cdot \int I_{C_1} \cdot dt; I_{out(i)}(t) = I_{in(i)}(t) - I_{R_d} + \frac{1}{L_1} \cdot \int (V_{B_i} - V_B) \cdot dt; V_{B_i} - const$$

$$\begin{aligned} I_{out(i)}(t) &= I_{in(i)}(t) - I_{R_d} + \frac{1}{L_1} \cdot \int V_{B_i} \cdot dt - \frac{1}{L_1} \cdot \int V_B \cdot dt \\ &= I_{in(i)}(t) - I_{R_d} + \frac{1}{L_1} \cdot V_{B_i} \cdot t - \frac{1}{L_1} \cdot \int V_B \cdot dt \end{aligned}$$

$$I_{out(i)}(t) = I_{in(i)}(t) - I_{R_d} + \frac{1}{L_1} \cdot V_{B_i} \cdot t - \frac{1}{L_1} \cdot \int [V_A - \frac{1}{C_1} \cdot \int I_{C_1} \cdot dt] \cdot dt; I_{C_1}(t) = I_{in(i)}(t) - I_{out(i)}(t)$$

$$V_A = V_{P_1} - (V_{C_f} + V_{L_f} + V_{R_f} + V_{Cc})_{@Branch1};$$

$$I_{in(i)}(t) = C_f \cdot \frac{dV_{C_f}}{dt} \Rightarrow V_{C_f} = \frac{1}{C_f} \cdot \int I_{in(i)}(t) \cdot dt$$

$$V_A = V_{P_1} - (\frac{1}{C_f} \cdot \int I_{in(i)}(t) \cdot dt + L_f \cdot \frac{dI_{in(i)}(t)}{dt} + I_{in(i)}(t) \cdot R_f + \frac{1}{C_C} \cdot \int I_{in(i)}(t) \cdot dt)$$

$$\frac{dV_A}{dt} = \frac{dV_{P_1}}{dt} - [\frac{1}{C_f} \cdot I_{in(i)}(t) + L_f \cdot \frac{d^2 I_{in(i)}(t)}{dt^2} + \frac{dI_{in(i)}(t)}{dt} \cdot R_f + \frac{1}{C_C} \cdot I_{in(i)}(t)]$$

$$\frac{dV_A}{dt} = \frac{dV_{P_1}}{dt} - [(\frac{1}{C_C} + \frac{1}{C_f}) \cdot I_{in(i)}(t) + L_f \cdot \frac{d^2 I_{in(i)}(t)}{dt^2} + \frac{dI_{in(i)}(t)}{dt} \cdot R_f]$$

We are interested to find out our $I_{out(i)}(t)$ expression.

$$I_{out(i)}(t) = I_{in(i)}(t) - I_{R_d} + \frac{1}{L_1} \cdot V_{B_i} \cdot t - \frac{1}{L_1} \cdot \int V_A \cdot dt + \frac{1}{L_1 \cdot C_1} \cdot \iint I_{C_1} \cdot dt \cdot dt$$

$$\frac{dI_{out(i)}(t)}{dt} = \frac{dI_{in(i)}(t)}{dt} - \frac{dI_{R_d}}{dt} + \frac{1}{L_1} \cdot V_{B_i} - \frac{1}{L_1} \cdot V_A + \frac{1}{L_1 \cdot C_1} \cdot \int I_{C_1} \cdot dt$$

$$\frac{d}{dt}\{\frac{dI_{out(i)}(t)}{dt} = \frac{dI_{in(i)}(t)}{dt} - \frac{dI_{R_d}}{dt} + \frac{1}{L_1} \cdot V_{B_i} - \frac{1}{L_1} \cdot V_A + \frac{1}{L_1 \cdot C_1} \cdot \int I_{C_1} \cdot dt\}$$

$$\frac{d^2 I_{out(i)}(t)}{dt^2} = \frac{d^2 I_{in(i)}(t)}{dt^2} - \frac{d^2 I_{R_d}}{dt^2} + \frac{1}{L_1} \cdot \frac{dV_{B_i}}{dt} - \frac{1}{L_1} \cdot \frac{dV_A}{dt} + \frac{1}{L_1 \cdot C_1} \cdot I_{C_1}; \frac{dV_{B_i}}{dt} = 0$$

$$(*)\frac{d^2 I_{out(i)}(t)}{dt^2} = \frac{d^2 I_{in(i)}(t)}{dt^2} - \frac{d^2 I_{R_d}}{dt^2} - \frac{1}{L_1} \cdot \frac{dV_A}{dt} + \frac{1}{L_1 \cdot C_1} \cdot [I_{in(i)}(t) - I_{out(i)}(t)]$$

$$(**)\frac{dV_A}{dt} = \frac{dV_{P_1}}{dt} - [(\frac{1}{C_C} + \frac{1}{C_f}) \cdot I_{in(i)}(t) + L_f \cdot \frac{d^2 I_{in(i)}(t)}{dt^2} + \frac{dI_{in(i)}(t)}{dt} \cdot R_f]$$

We consider the assumption which the IMPATT/Circulator amplifier input voltage $V_{P_1} = \Gamma + \xi(t)$; Γ is constant voltage and $\xi(t)$ is RF signal in time.

$$\frac{dV_{P_1}}{dt} = \frac{d\Gamma}{dt} + \frac{d\xi(t)}{dt}; \frac{d\Gamma}{dt} \to 0; \frac{d\xi(t)}{dt} \to \varepsilon \text{ Or } \frac{d\xi(t)}{dt} \to \Omega_0 \cdot sgn[\frac{d\xi(t)}{dt}]$$

$$\frac{dV_A}{dt} = \Omega_0 \cdot sgn[\frac{d\xi(t)}{dt}] - [(\frac{1}{C_C} + \frac{1}{C_f}) \cdot I_{in(i)}(t) + L_f \cdot \frac{d^2 I_{in(i)}(t)}{dt^2} + \frac{dI_{in(i)}(t)}{dt} \cdot R_f]$$

$$\frac{d\xi(t)}{dt} = \text{sgn}[\frac{d\xi(t)}{dt}] \cdot |\frac{d\xi(t)}{dt}|; \forall \text{ real } \frac{d\xi(t)}{dt} \exists |\frac{d\xi(t)}{dt}| = \text{sgn}[\frac{d\xi(t)}{dt}] \cdot \frac{d\xi(t)}{dt}$$

$$\frac{d|\frac{d\xi(t)}{dt}|}{d[\frac{d\xi(t)}{dt}]} = \text{sgn}[\frac{d\xi(t)}{dt}] \forall \frac{d\xi(t)}{dt} \neq 0; \frac{dV_{P_1}}{dt} = 0 \text{ for } \frac{d\xi(t)}{dt} = 0$$

$$\frac{dV_{P_1}}{dt} = \Omega_0 \text{ for } \frac{d\xi(t)}{dt} > 0; \frac{dV_{P_1}}{dt} = -\Omega_0 \text{ for } \frac{d\xi(t)}{dt} < 0$$

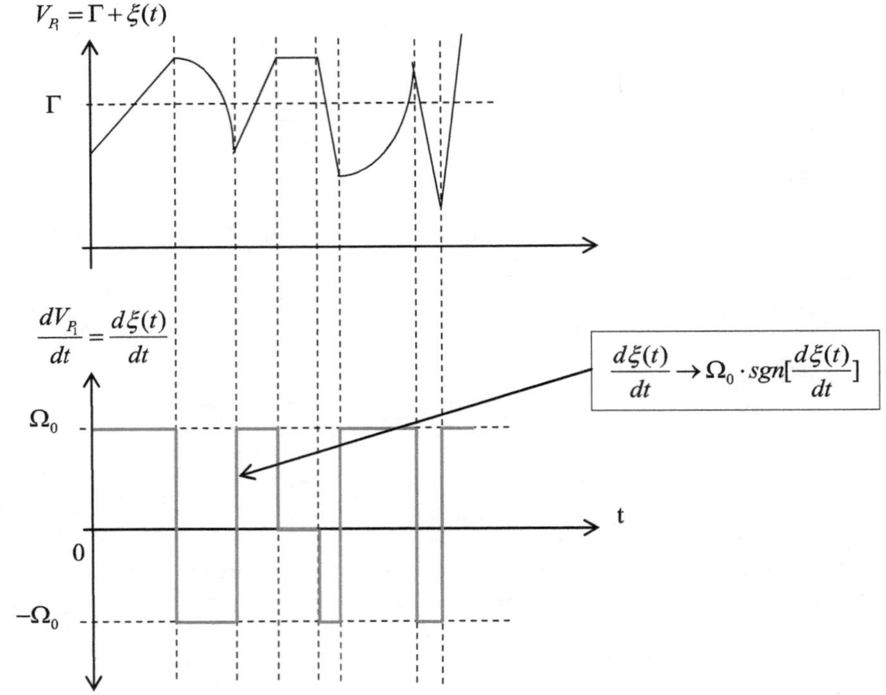

Fig. 3.20 IMPATT amplifier circuit V_{P_1} and $\frac{dV_{P_1}}{dt}$ functions of time

We neglect the IMPATT diode reciprocal negative resistance ($\frac{d}{dt}[\frac{1}{R_d}] \to \varepsilon$) characteristic slope changes over time.

$$\frac{d^2 I_{R_d}}{dt^2} = \frac{d}{dt}[\frac{dI_{R_d}}{dt}] = \frac{d}{dt}[\frac{dI_{R_d}}{dV_{R_d}} \cdot \frac{dV_{R_d}}{dt}]; \frac{dI_{R_d}}{dV_{R_d}} = \frac{1}{R_d}; \frac{d^2 I_{R_d}}{dt^2} = \frac{d}{dt}[\frac{1}{R_d} \cdot \frac{dV_{R_d}}{dt}]$$

$$\frac{d^2 I_{R_d}}{dt^2} = \frac{d}{dt}[\frac{1}{R_d} \cdot \frac{dV_{R_d}}{dt}] = \frac{d}{dt}[\frac{1}{R_d}] \cdot \frac{dV_{R_d}}{dt} + \frac{1}{R_d} \cdot \frac{d^2 V_{R_d}}{dt^2}; \frac{d}{dt}[\frac{1}{R_d}] \to \varepsilon$$

$$\frac{d^2 I_{R_d}}{dt^2} = \frac{1}{R_d} \cdot \frac{d^2 V_{R_d}}{dt^2} = \frac{1}{R_d} \cdot \frac{d}{dt}[\frac{dV_{R_d}}{dt}]; \frac{dV_{R_d}}{dt} = \Upsilon_{R_d}; \frac{d\Upsilon_{R_d}}{dt} \to \varepsilon; \frac{d^2 I_{R_d}}{dt^2} \to \varepsilon$$

$$(*) \frac{d^2 I_{out(i)}(t)}{dt^2} = \frac{d^2 I_{in(i)}(t)}{dt^2} - \frac{1}{L_1} \cdot \frac{dV_A}{dt} + \frac{1}{L_1 \cdot C_1} \cdot [I_{in(i)}(t) - I_{out(i)}(t)]$$

$(**) \rightarrow (*)$

$$\frac{d^2 I_{out(i)}(t)}{dt^2} = \frac{d^2 I_{in(i)}(t)}{dt^2} - \frac{1}{L_1} \cdot \{\Omega_0 \cdot sgn[\frac{d\xi(t)}{dt}] - [(\frac{1}{C_C} + \frac{1}{C_f}) \cdot I_{in(i)}(t) + L_f \cdot \frac{d^2 I_{in(i)}(t)}{dt^2}$$
$$+ \frac{dI_{in(i)}(t)}{dt} \cdot R_f]\} + \frac{1}{L_1 \cdot C_1} \cdot [I_{in(i)}(t) - I_{out(i)}(t)]$$

$$\frac{d^2 I_{out(i)}(t)}{dt^2} = \frac{d^2 I_{in(i)}(t)}{dt^2} - \frac{\Omega_0}{L_1} \cdot sgn[\frac{d\xi(t)}{dt}] + \frac{1}{L_1} \cdot (\frac{1}{C_C} + \frac{1}{C_f}) \cdot I_{in(i)}(t) + \frac{L_f}{L_1} \cdot \frac{d^2 I_{in(i)}(t)}{dt^2}$$
$$+ \frac{dI_{in(i)}(t)}{dt} \cdot \frac{R_f}{L_1} + \frac{1}{L_1 \cdot C_1} \cdot I_{in(i)}(t) - \frac{1}{L_1 \cdot C_1} \cdot I_{out(i)}(t)$$

$$\frac{d^2 I_{out(i)}(t)}{dt^2} + \frac{1}{L_1 \cdot C_1} \cdot I_{out(i)}(t) = (1 + \frac{L_f}{L_1}) \cdot \frac{d^2 I_{in(i)}(t)}{dt^2} + \frac{R_f}{L_1} \cdot \frac{dI_{in(i)}(t)}{dt}$$
$$+ \frac{1}{L_1} \cdot [(\frac{1}{C_C} + \frac{1}{C_f}) + \frac{1}{C_1}] \cdot I_{in(i)}(t) - \frac{\Omega_0}{L_1} \cdot sgn[\frac{d\xi(t)}{dt}]$$

<u>Case I</u>: $\frac{dV_{P_1}}{dt} = 0$ for $\frac{d\xi(t)}{dt} = 0$

$$\frac{d^2 I_{out(i)}(t)}{dt^2} + \frac{1}{L_1 \cdot C_1} \cdot I_{out(i)}(t) = (1 + \frac{L_f}{L_1}) \cdot \frac{d^2 I_{in(i)}(t)}{dt^2} + \frac{R_f}{L_1} \cdot \frac{dI_{in(i)}(t)}{dt} + \frac{1}{L_1} \cdot [(\frac{1}{C_C} + \frac{1}{C_f}) + \frac{1}{C_1}] \cdot I_{in(i)}(t)$$

<u>Case II</u>: $\frac{dV_{P_1}}{dt} = \Omega_0$ for $\frac{d\xi(t)}{dt} > 0$

$$\frac{d^2 I_{out(i)}(t)}{dt^2} + \frac{1}{L_1 \cdot C_1} \cdot I_{out(i)}(t) = (1 + \frac{L_f}{L_1}) \cdot \frac{d^2 I_{in(i)}(t)}{dt^2} + \frac{R_f}{L_1} \cdot \frac{dI_{in(i)}(t)}{dt}$$
$$+ \frac{1}{L_1} \cdot [(\frac{1}{C_C} + \frac{1}{C_f}) + \frac{1}{C_1}] \cdot I_{in(i)}(t) - \frac{\Omega_0}{L_1}$$

<u>Case III</u>: $\frac{dV_{P_1}}{dt} = -\Omega_0$ for $\frac{d\xi(t)}{dt} < 0$

$$\frac{d^2 I_{out(i)}(t)}{dt^2} + \frac{1}{L_1 \cdot C_1} \cdot I_{out(i)}(t) = (1 + \frac{L_f}{L_1}) \cdot \frac{d^2 I_{in(i)}(t)}{dt^2} + \frac{R_f}{L_1} \cdot \frac{dI_{in(i)}(t)}{dt}$$
$$+ \frac{1}{L_1} \cdot [(\frac{1}{C_C} + \frac{1}{C_f}) + \frac{1}{C_1}] \cdot I_{in(i)}(t) + \frac{\Omega_0}{L_1}$$

We define for simplicity a new function: $\psi = \psi(\frac{d^2 I_{in(i)}(t)}{dt^2}, \frac{dI_{in(i)}(t)}{dt}, I_{in(i)}(t), \ldots)$

$$\psi = (1 + \frac{L_f}{L_1}) \cdot \frac{d^2 I_{in(i)}(t)}{dt^2} + \frac{R_f}{L_1} \cdot \frac{dI_{in(i)}(t)}{dt} + \frac{1}{L_1} \cdot [(\frac{1}{C_C} + \frac{1}{C_f}) + \frac{1}{C_1}] \cdot I_{in(i)}(t) - \frac{\Omega_0}{L_1}$$
$$\cdot sgn[\frac{d\xi(t)}{dt}]$$

$$\frac{d^2 I_{out(i)}(t)}{dt^2} + \frac{1}{L_1 \cdot C_1} \cdot I_{out(i)}(t) = \psi(\frac{d^2 I_{in(i)}(t)}{dt^2}, \frac{dI_{in(i)}(t)}{dt}, I_{in(i)}(t), \ldots)$$

Next it is reader exercise to find $I_{out(i)}(t) = f_i(I_{in(i)}(t), D_i, L_i, C_i, V_{B_i})$ function.

3.6 FET Combined Biasing and Matching Circuit Stability Analysis

FET RF transistor is biased by using two power supplies, one for V_{DS} and the other for V_{GS}. Another way to bias microwave FET is to use source resistor. The source resistor has the advantage of providing feedback to stabilize the FET performance and requires only one power supply. Once the transistor has been properly biased, it must be matched to microstrip transmission line. By implementing biasing and matching elements to RF microwave FET, each electrode must be simultaneously connected to an RF circuit and a DC circuit, and the two circuits must not interfere. The required isolation between the biasing and the matching circuit is done with RF chocks (which pass the DC and block the RF) and coupling capacitors which pass RF and block DC [33, 34, 62].

C_{in}—Input coupling capacitor. It allows the input microwave signal X(t) to enter the transistor gate but prevents the input microstrip line from shorting out the gate bias voltage.

L_{in}—We use RF chock because the gate must be connected to DC ground, but the RF must not leak through this ground.

The RF FET source port is connected to RF ground through coupling capacitor C_s which allows the FET source to be at RF ground. It allows the biasing source resistor R_s to be used between the source and DC ground.

The FET drain port is connected to the drain resistor R_d through RF choke L_d, which is connected to the positive supply voltage V_{dd}. The RF choke L_d presents

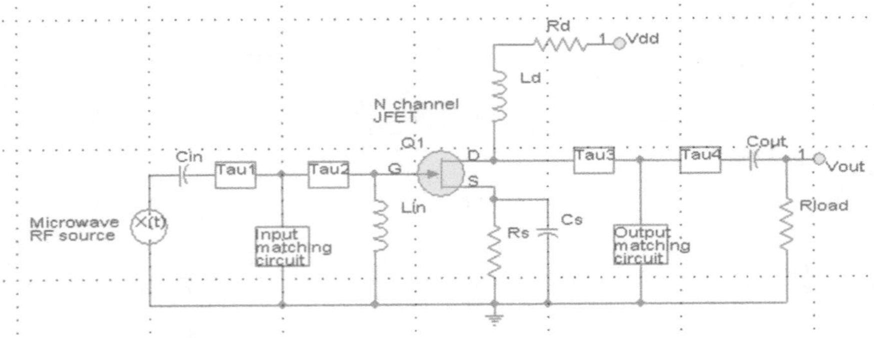

Fig. 3.21 FET combined biasing and matching circuit

the microwave signal from being shorted out by the drain resistor R_d and the power supply V_{dd}. Coupling capacitor C_{out} allows the RF microwave signal to pass into the output microstrip line and the load R_{load}, but prevents the output microstrip line, which is a DC ground from shorting out the drain voltage. We represent our circuit microstrip elements as a delay lines Tau1, Tau2, Tau3, Tau4 $(\tau_1, \tau_2, \tau_3, \tau_4)$. We neglect the voltages on delay lines $V_{\tau_i} \to \varepsilon \forall \tau_1, \ldots, \tau_4; 1 \leq i \leq 4$. There is a delay in the current which flow through each microstrip delay line $I(t) \to I(t - \tau_i) \forall 1 \leq i \leq 4$. Next is to investigate how these delay line elements influence our circuit performance and stability. Input and output matching circuits can be π or T models. We use for our analysis the FET small signal equivalent circuit (reduced version) and input and output T matching network. We give different name to RF chocke L_d, L_{dd} (L_d is the FET small signal equivalent circuit drain inductance) and drain resistor R_d, R_{dd} (R_d is the FET small signal equivalent circuit drain resistance). $L_d \to L_{dd}; R_d \to R_{dd}$. Since we neglect the voltages on the delay lines $V_{\tau_i} \to \varepsilon \forall \tau_1, \ldots, \tau_4; 1 \leq i \leq 4$ the voltages on each delay line's ports is the same (consider it like a short contact). We write the FET small signal equivalent circuit for our analysis:

$$I_{Cin1} = C_{in1} \cdot \frac{d(V_{A_2} - V_{A_3})}{dt}; \; I_{R_d} = \frac{V_{A_7} - V_{A_8}}{R_d}; \; I_{R_i} = I_{C_{gs}}$$

$$I_{Cin} = C_{in} \cdot \frac{d(V_{A_1} - V_{A_2})}{dt}; I_{C_{in1}}(t) = I_{C_{in}}(t - \tau_1); V_{A_1} = X(t);$$

$$V_{A_3} = L_{in1} \cdot \frac{dI_{Lin1}}{dt}; I_{C_{in2}} = C_{in2} \cdot \frac{d(V_{A_3} - V_{A_4})}{dt}; V_{A_4} = L_{in} \cdot \frac{dI_{Lin}}{dt}$$

$$V_{A_4} - V_{A_5} = V_{L_g} = L_g \cdot \frac{dI_{L_g}}{dt}; I_{C_{in}}(t - \tau_1) = I_{C_{in1}}(t)$$

$$= I_{Lin1}(t) + I_{C_{in2}}(t); I_{C_{in2}}(t - \tau_2) = I_{Lin}(t) + I_{L_g}(t)$$

$$I_{C_{pgs}} = C_{pgs} \cdot \frac{dV_{A_5}}{dt}; I_{L_g} = I_{R_g} + I_{C_{pgs}}; I_{C_{gs}} = I_{R_i};$$

$$I_{R_g} = \frac{V_{A_5} - V_{A_6}}{R_g}; I_{C_{gd}} = C_{gd} \cdot \frac{d(V_{A_6} - V_{A_7})}{dt}$$

$$I_{R_g} = I_{C_{gd}} + I_{C_{gs}}; V = V_{A_6} - V_{A_9}; I_{C_{gs}} = C_{gs} \cdot \frac{dV}{dt};$$

$$I_{R_i} = \frac{V_{A_9} - V_{A_{15}}}{R_i}; I_{C_{gd}} = I_d + I_{C_{ds}} + I_{R_{ds}} + I_{R_d}$$

$$I_{C_{ds}} = C_{ds} \cdot \frac{d(V_{A_7} - V_{A_{15}})}{dt}; I_{R_{ds}} = \frac{V_{A_7} - V_{A_{15}}}{R_{ds}};$$

$$I_{R_d} = I_{L_d} + I_{C_{pds}}; I_{R_s} = \frac{V_{A_{15}} - V_{A_{16}}}{R_S}; V_{A_{16}} = V_{L_S} = L_S \cdot \frac{dI_{L_S}}{dt}$$

$$I_{R_S} = I_{L_S}; I_{C_{pds}} = C_{pds} \cdot \frac{dV_{A_8}}{dt}; V_{L_d} = V_{A_8} - V_{A_{10}} = L_d \cdot \frac{dI_{L_d}}{dt};$$

$$I_{R_{dd}} = I_{L_{dd}}; \frac{V_{dd} - V_{A_{11}}}{R_{dd}} = I_{R_{dd}}$$

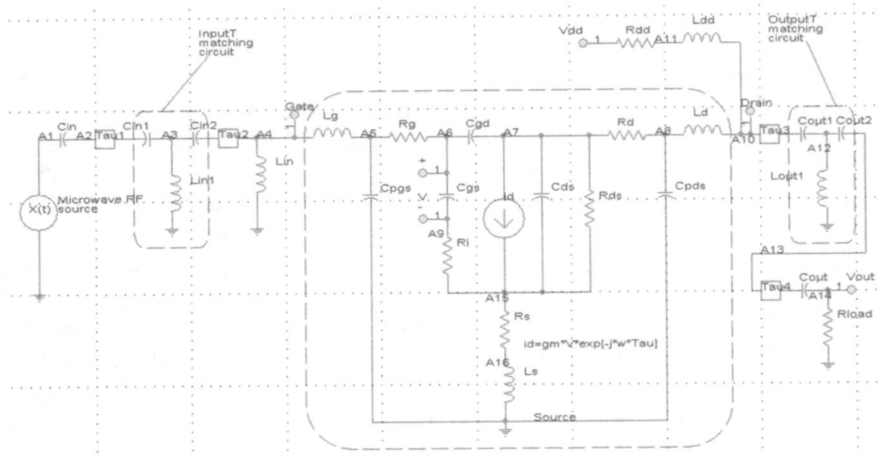

Fig. 3.22 FET combined biasing and matching equivalent circuit

$$V_{A_{11}} - V_{A_{10}} = L_{dd} \cdot \frac{dI_{L_{dd}}}{dt} \; ; \; I_{L_d}(t - \tau_3) + I_{L_{dd}}(t - \tau_3)$$
$$= I_{C_{out1}}(t) \; ; \; I_{C_{out1}} = C_{out1} \cdot \frac{d(V_{A_{10}} - V_{A_{12}})}{dt}$$
$$V_{A_{12}} = L_{out1} \cdot \frac{dI_{L_{out1}}}{dt};$$
$$I_{C_{out1}} = I_{L_{out1}} + I_{C_{out2}}; I_{C_{out2}} = C_{out2} \cdot \frac{d(V_{A_{12}} - V_{A_{13}})}{dt}; I_{C_{out}}(t) = I_{C_{out2}}(t - \tau_4)$$
$$I_{C_{out}} = I_{R_{load}}; I_{C_{out}} = C_{out} \cdot \frac{d(V_{A_{13}} - V_{A_{14}})}{dt}; I_{R_{load}} = \frac{V_{A_{14}}}{R_{load}}; V_O = V_{A_{14}}$$

We write FET small signal equivalent circuit's Kirchhoff's current law (KCL) for circuit's node A_2, A_3,…,A_{16}.

Table 3.11 FET combined biasing and matching equivalent circuit's Kirchhoff's Current Law (KCL) and expressions

	KCL @ A_x (x = 2,…,16)	Expression of Kirchhoff's current law
1	A_2	$I_{C_{in1}}(t) = I_{C_{in}}(t - \tau_1)$
2	A_3	$I_{C_{in}}(t - \tau_1) = I_{C_{in1}}(t) = I_{Lin1}(t) + I_{C_{in2}}(t)$
3	A_4	$I_{C_{in2}}(t - \tau_2) = I_{Lin}(t) + I_{L_g}(t)$
4	A_5	$I_{L_g} = I_{R_g} + I_{C_{pgs}}$
5	A_6	$I_{R_g} = I_{C_{gd}} + I_{C_{gs}}$
6	A_7	$I_{C_{gd}} = I_d + I_{C_{ds}} + I_{R_{ds}} + I_{R_d}$
7	A_8	$I_{R_d} = I_{L_d} + I_{C_{pds}}$
8	A_9	$I_{C_{gs}} = I_{R_i}$
9	A_{10}	$I_{L_d}(t - \tau_3) + I_{L_{dd}}(t - \tau_3) = I_{C_{out1}}(t)$
10	A_{11}	$I_{R_{dd}} = I_{L_{dd}}$

(continued)

Table 3.11 (continued)

	KCL @ A_x (x = 2,…,16)	Expression of Kirchhoff's current law
11	A_{12}	$I_{C_{out1}} = I_{L_{out1}} + I_{C_{out2}}$
12	A_{13}	$I_{C_{out}}(t) = I_{C_{out2}}(t - \tau_4)$
13	A_{14}	$I_{C_{out}} = I_{R_{load}}$
14	A_{15}	$I_{R_S} = I_{R_i} + I_d + I_{C_{ds}} + I_{R_{ds}}$
15	A_{16}	$I_{R_S} = I_{L_S}$

$$I_{C_{in}} = C_{in} \cdot \frac{d}{dt}(V_{A_1} - V_{A_2}) = C_{in} \cdot \frac{d}{dt}(X(t) - V_{A_2});$$

$$I_{C_{in1}} = C_{in1} \cdot \frac{d}{dt}(V_{A_2} - V_{A_3}) = C_{in1} \cdot \frac{d}{dt}(V_{A_2} - L_{in1} \cdot \frac{dI_{L_{in1}}}{dt})$$

$$I_{C_{in}} = C_{in} \cdot (\frac{dX(t)}{dt} - \frac{dV_{A_2}}{dt}); I_{C_{in1}} = C_{in1} \cdot (\frac{dV_{A_2}}{dt} - L_{in1} \cdot \frac{d^2 I_{L_{in1}}}{dt^2});$$

$$\frac{1}{C_{in}} \cdot I_{C_{in}} = \frac{dX(t)}{dt} - \frac{dV_{A_2}}{dt}$$

$$\frac{dV_{A_2}}{dt} = \frac{dX(t)}{dt} - \frac{1}{C_{in}} \cdot I_{C_{in}}; \frac{1}{C_{in1}} \cdot I_{C_{in1}} = \frac{dV_{A_2}}{dt} - L_{in1} \cdot \frac{d^2 I_{L_{in1}}}{dt^2}$$

$$\Rightarrow \frac{dV_{A_2}}{dt} = \frac{1}{C_{in1}} \cdot I_{C_{in1}} + L_{in1} \cdot \frac{d^2 I_{L_{in1}}}{dt^2}$$

$$\frac{dX(t)}{dt} - \frac{1}{C_{in}} \cdot I_{C_{in}} = \frac{1}{C_{in1}} \cdot I_{C_{in1}} + L_{in1} \cdot \frac{d^2 I_{L_{in1}}}{dt^2};$$

$$I_{C_{in2}} = C_{in2} \cdot \frac{d}{dt}(V_{A_3} - V_{A_4}) = C_{in2} \cdot \frac{d}{dt}(L_{in1} \cdot \frac{dI_{L_{in1}}}{dt} - L_{in} \cdot \frac{dI_{L_{in}}}{dt})$$

$$I_{C_{in2}} = C_{in2} \cdot (L_{in1} \cdot \frac{d^2 I_{L_{in1}}}{dt^2} - L_{in} \cdot \frac{d^2 I_{L_{in}}}{dt^2});$$

$$V_{A_4} - V_{A_5} = L_g \cdot \frac{dI_{L_g}}{dt} \Rightarrow L_{in} \cdot \frac{dI_{L_{in}}}{dt} - \frac{1}{C_{pgs}} \cdot \int I_{C_{pgs}} \cdot dt = L_g \cdot \frac{dI_{L_g}}{dt}$$

$$L_{in} \cdot \frac{d^2 I_{L_{in}}}{dt^2} - \frac{1}{C_{pgs}} \cdot I_{C_{pgs}} = L_g \cdot \frac{d^2 I_{L_g}}{dt^2}; I_{C_{ds}} = C_{ds} \cdot \frac{d}{dt}(V_{A_7} - V_{A_{15}});$$

$$I_{R_{ds}} = \frac{V_{A_7} - V_{A_{15}}}{R_{ds}} \Rightarrow I_{R_{ds}} \cdot R_{ds} = V_{A_7} - V_{A_{15}}$$

$$I_{C_{ds}} = C_{ds} \cdot R_{ds} \cdot \frac{dI_{R_{ds}}}{dt}; V_{A_5} = \frac{1}{C_{pgs}} \cdot \int I_{C_{pgs}} \cdot dt; V_{A_8} = \frac{1}{C_{pds}} \cdot \int I_{C_{pds}} \cdot dt; V_{A_6} = \frac{1}{C_{pgs}} \cdot \int I_{C_{pgs}} \cdot dt - I_{R_g} \cdot R_g$$

$$V_{A_7} = I_{R_d} \cdot R_d + \frac{1}{C_{pds}} \cdot \int I_{C_{pds}} \cdot dt; I_{C_{gd}} = C_{gd} \cdot \frac{d}{dt}[\frac{1}{C_{pgs}} \cdot \int I_{C_{pgs}} \cdot dt - I_{R_g} \cdot R_g - I_{R_d} \cdot R_d - \frac{1}{C_{pds}} \cdot \int I_{C_{pds}} \cdot dt]$$

$$I_{C_{gd}} = C_{gd} \cdot [\frac{1}{C_{pgs}} \cdot I_{C_{pgs}} - \frac{1}{C_{pds}} \cdot I_{C_{pds}} - R_g \cdot \frac{dI_{R_g}}{dt} - R_d \cdot \frac{dI_{R_d}}{dt}];$$

$$V_{A_{15}} = I_{R_s} \cdot R_s + L_S \cdot \frac{dI_{R_S}}{dt}$$

$$V = \frac{1}{C_{gs}} \cdot \int I_{C_{gs}} \cdot dt; R_i \cdot \frac{dI_{R_i}}{dt} = \frac{1}{C_{pgs}} \cdot I_{C_{pgs}} - R_g \cdot \frac{dI_{R_g}}{dt} - R_S \cdot \frac{dI_{R_S}}{dt} - L_S \cdot \frac{d^2 I_{R_S}}{dt^2} - \frac{1}{C_{gs}} \cdot I_{C_{gs}}$$

$$I_{R_i} = I_{C_{gs}}; R_i \cdot \frac{dI_{C_{gs}}}{dt} = \frac{1}{C_{pgs}} \cdot I_{C_{pgs}} - R_g \cdot \frac{dI_{R_g}}{dt} - R_S \cdot \frac{dI_{R_S}}{dt} - L_S \cdot \frac{d^2 I_{R_S}}{dt^2} - \frac{1}{C_{gs}} \cdot I_{C_{gs}}$$

$$V_{A_8} = \frac{1}{C_{pds}} \cdot \int I_{C_{pds}} \cdot dt; V_{A_{10}} = \frac{1}{C_{pds}} \cdot \int I_{C_{pds}} \cdot dt - L_d \cdot \frac{dI_{L_d}}{dt}; V_{A_{11}} = V_{dd} - R_{dd} \cdot I_{R_{dd}}$$

$$V_{dd} - R_{dd} \cdot I_{R_{dd}} - \frac{1}{C_{pds}} \cdot \int I_{C_{pds}} \cdot dt + L_d \cdot \frac{dI_{L_d}}{dt} = L_{dd} \cdot \frac{dI_{L_{dd}}}{dt}$$

$$\frac{d}{dt}\{V_{dd} - R_{dd} \cdot I_{R_{dd}} - \frac{1}{C_{pds}} \cdot \int I_{C_{pds}} \cdot dt + L_d \cdot \frac{dI_{L_d}}{dt}\} = \frac{d}{dt}\{L_{dd} \cdot \frac{dI_{L_{dd}}}{dt}\}$$

$$\frac{dV_{dd}}{dt} - R_{dd} \cdot \frac{dI_{R_{dd}}}{dt} - \frac{1}{C_{pds}} \cdot I_{C_{pds}} + L_d \cdot \frac{d^2 I_{L_d}}{dt^2} = L_{dd} \cdot \frac{d^2 I_{L_{dd}}}{dt^2}; \frac{dV_{dd}}{dt} = 0$$

$$- R_{dd} \cdot \frac{dI_{R_{dd}}}{dt} - \frac{1}{C_{pds}} \cdot I_{C_{pds}} + L_d \cdot \frac{d^2 I_{L_d}}{dt^2} = L_{dd} \cdot \frac{d^2 I_{L_{dd}}}{dt^2}$$

$$V_{A_8} = \frac{1}{C_{pds}} \cdot \int I_{C_{pds}} \cdot dt; V_{A_{10}} = \frac{1}{C_{pds}} \cdot \int I_{C_{pds}} \cdot dt - L_d \cdot \frac{dI_{L_d}}{dt}$$

$$I_{C_{out1}} = C_{out1} \cdot \frac{d}{dt}[\frac{1}{C_{pds}} \cdot \int I_{C_{pds}} \cdot dt - L_d \cdot \frac{dI_{L_d}}{dt} - L_{out1} \cdot \frac{dI_{L_{out1}}}{dt}]$$
$$= C_{out1} \cdot [\frac{1}{C_{pds}} \cdot I_{C_{pds}} - L_d \cdot \frac{d^2 I_{L_d}}{dt^2} - L_{out1} \cdot \frac{d^2 I_{L_{out1}}}{dt^2}]$$

$$V_{A_{12}} = L_{out1} \cdot \frac{dI_{L_{out1}}}{dt}; V_{A_{14}} = I_{R_{load}} \cdot R_{load}; V_{A_{13}} = \frac{1}{C_{out}} \cdot \int I_{C_{out}} \cdot dt + I_{R_{load}} \cdot R_{load}$$

$$I_{C_{out2}} = C_{out2} \cdot \frac{d}{dt}[L_{out1} \cdot \frac{dI_{L_{out1}}}{dt} - \frac{1}{C_{out}} \cdot \int I_{C_{out}} \cdot dt - I_{R_{load}} \cdot R_{load}]$$
$$= C_{out2} \cdot [L_{out1} \cdot \frac{d^2 I_{L_{out1}}}{dt^2} - \frac{1}{C_{out}} \cdot I_{C_{out}} - R_{load} \cdot \frac{dI_{R_{load}}}{dt}]$$

$$I_{C_{out2}} = I_{C_{out}} = I_{R_{load}}; I_{C_{out2}} = C_{out2} \cdot [L_{out1} \cdot \frac{d^2 I_{L_{out1}}}{dt^2} - \frac{1}{C_{out}} \cdot I_{C_{out}} - R_{load} \cdot \frac{dI_{R_{load}}}{dt}]$$

$$I_{R_{load}} = C_{out2} \cdot [L_{out1} \cdot \frac{d^2 I_{L_{out1}}}{dt^2} - \frac{1}{C_{out}} \cdot I_{R_{load}} - R_{load} \cdot \frac{dI_{R_{load}}}{dt}]$$

Our system equations with delays:

$$I_{C_{in1}}(t) = I_{C_{in}}(t-\tau_1);\ I_{L_d}(t-\tau_3) + I_{L_{dd}}(t-\tau_3) = I_{C_{out1}}(t)$$
$$I_{C_{in}}(t-\tau_1) = I_{C_{in1}}(t) = I_{Lin1}(t) + I_{C_{in2}}(t);$$
$$I_{C_{in2}}(t-\tau_2) = I_{Lin}(t) + I_{L_g}(t);\ I_{C_{out}}(t) = I_{C_{out2}}(t-\tau_4)$$

We can summery our system differential equations:

$$\frac{dX(t)}{dt} - \frac{1}{C_{in}} \cdot I_{C_{in}} = \frac{1}{C_{in1}} \cdot I_{C_{in1}} + L_{in1} \cdot \frac{d^2 I_{L_{in1}}}{dt^2};\ I_{C_{in2}} = C_{in2} \cdot (L_{in1} \cdot \frac{d^2 I_{L_{in1}}}{dt^2} - L_{in} \cdot \frac{d^2 I_{L_{in}}}{dt^2})$$

$$L_{in} \cdot \frac{d^2 I_{L_{in}}}{dt^2} - \frac{1}{C_{pgs}} \cdot I_{C_{pgs}} = L_g \cdot \frac{d^2 I_{L_g}}{dt^2};\ I_{C_{ds}} = C_{ds} \cdot R_{ds} \cdot \frac{dI_{R_{ds}}}{dt}$$

$$I_{C_{gd}} = C_{gd} \cdot [\frac{1}{C_{pgs}} \cdot I_{C_{pgs}} - \frac{1}{C_{pds}} \cdot I_{C_{pds}} - R_g \cdot \frac{dI_{R_g}}{dt} - R_d \cdot \frac{dI_{R_d}}{dt}]$$

$$R_i \cdot \frac{dI_{C_{gs}}}{dt} = \frac{1}{C_{pgs}} \cdot I_{C_{pgs}} - R_g \cdot \frac{dI_{R_g}}{dt} - R_S \cdot \frac{dI_{R_S}}{dt} - L_S \cdot \frac{d^2 I_{R_S}}{dt^2} - \frac{1}{C_{gs}} \cdot I_{C_{gs}}$$
$$- R_{dd} \cdot \frac{dI_{R_{dd}}}{dt} - \frac{1}{C_{pds}} \cdot I_{C_{pds}} + L_d \cdot \frac{d^2 I_{L_d}}{dt^2} = L_{dd} \cdot \frac{d^2 I_{L_{dd}}}{dt^2}$$

$$I_{C_{out1}} = C_{out1} \cdot [\frac{1}{C_{pds}} \cdot I_{C_{pds}} - L_d \cdot \frac{d^2 I_{L_d}}{dt^2} - L_{out1} \cdot \frac{d^2 I_{L_{out1}}}{dt^2}];$$

$$I_{R_{load}} = C_{out2} \cdot [L_{out1} \cdot \frac{d^2 I_{L_{out1}}}{dt^2} - \frac{1}{C_{out}} \cdot I_{R_{load}} - R_{load} \cdot \frac{dI_{R_{load}}}{dt}]$$

We implement delay variables in the above system differential equations:

$$\frac{dX(t)}{dt} - \frac{1}{C_{in}} \cdot I_{C_{in}} = \frac{1}{C_{in1}} \cdot I_{C_{in}}(t-\tau_1) + L_{in1} \cdot \frac{d^2 I_{L_{in1}}}{dt^2};\ I_{C_{in2}}$$
$$= C_{in2} \cdot (L_{in1} \cdot \frac{d^2[I_{C_{in}}(t-\tau_1) - I_{C_{in2}}(t)]}{dt^2} - L_{in} \cdot \frac{d^2 I_{L_{in}}}{dt^2})$$

$$L_{in} \cdot \frac{d^2 I_{L_{in}}}{dt^2} - \frac{1}{C_{pgs}} \cdot I_{C_{pgs}} = L_g \cdot \frac{d^2[I_{C_{in2}}(t-\tau_2) - I_{Lin}(t)]}{dt^2};\ I_{C_{ds}} = C_{ds} \cdot R_{ds} \cdot \frac{dI_{R_{ds}}}{dt}$$

$$I_{C_{gd}} = C_{gd} \cdot [\frac{1}{C_{pgs}} \cdot I_{C_{pgs}} - \frac{1}{C_{pds}} \cdot I_{C_{pds}} - R_g \cdot \frac{dI_{R_g}}{dt} - R_d \cdot \frac{dI_{R_d}}{dt}]$$

$$R_i \cdot \frac{dI_{C_{gs}}}{dt} = \frac{1}{C_{pgs}} \cdot I_{C_{pgs}} - R_g \cdot \frac{dI_{R_g}}{dt} - R_S \cdot \frac{dI_{R_S}}{dt} - L_S \cdot \frac{d^2 I_{R_S}}{dt^2} - \frac{1}{C_{gs}} \cdot I_{C_{gs}}$$

$$-R_{dd} \cdot \frac{dI_{R_{dd}}}{dt} - \frac{1}{C_{pds}} \cdot I_{C_{pds}} + L_d \cdot \frac{d^2 I_{L_d}}{dt^2} = L_{dd} \cdot \frac{d^2 I_{L_{dd}}}{dt^2}; I_{C_{out}} = I_{R_{load}}; I_{R_{load}}(t) = I_{C_{out}}(t) = I_{C_{out2}}(t - \tau_4)$$

$$I_{L_d}(t - \tau_3) + I_{L_{dd}}(t - \tau_3) = C_{out1} \cdot [\frac{1}{C_{pds}} \cdot I_{C_{pds}} - L_d \cdot \frac{d^2 I_{L_d}}{dt^2} - L_{out1} \cdot \frac{d^2 I_{L_{out1}}}{dt^2}]$$

$$I_{C_{out2}}(t - \tau_4) = C_{out2} \cdot [L_{out1} \cdot \frac{d^2 I_{L_{out1}}}{dt^2} - \frac{1}{C_{out}} \cdot I_{C_{out2}}(t - \tau_4) - R_{load} \cdot \frac{dI_{C_{out2}}(t - \tau_4)}{dt}]$$

To find equilibrium points (fixed points) of our system we define

$$\lim_{t \to \infty} I_{C_{in}}(t - \tau_1) = I_{C_{in}}(t); \lim_{t \to \infty} I_{L_d}(t - \tau_3) = I_{L_d}(t); \lim_{t \to \infty} I_{L_{dd}}(t - \tau_3) = I_{L_{dd}}(t)$$
$$\lim_{t \to \infty} I_{C_{in2}}(t - \tau_2) = I_{C_{in2}}(t); \lim_{t \to \infty} I_{C_{out2}}(t - \tau_4) = I_{C_{out2}}(t)$$

$$\frac{d^2 I_{L_{in1}}}{dt^2} = 0; \frac{d^2 [I_{C_{in}}(t - \tau_1) - I_{C_{in2}}(t)]}{dt^2} = 0;$$
$$\frac{d^2 I_{L_{in}}}{dt^2} = 0; \frac{d^2 [I_{C_{in2}}(t - \tau_2) - I_{Lin}(t)]}{dt^2} = 0; \frac{dI_{R_{ds}}}{dt} = 0$$
$$\frac{dI_{R_d}}{dt} = 0; R_i \cdot \frac{dI_{C_{gs}}}{dt} = 0; \frac{dI_{R_g}}{dt} = 0; \frac{dI_{R_S}}{dt} = 0; \frac{d^2 I_{R_S}}{dt^2} = 0;$$
$$\frac{dI_{R_{dd}}}{dt} = 0; \frac{d^2 I_{L_d}}{dt^2} = 0; \frac{d^2 I_{L_{dd}}}{dt^2} = 0$$
$$\frac{d^2 I_{L_d}}{dt^2} = 0; \frac{d^2 I_{L_{out1}}}{dt^2} = 0; \frac{dI_{C_{out2}}(t - \tau_4)}{dt} = 0$$

We get our system fixed points: $\frac{dX(t)}{dt} = [\frac{1}{C_{in1}} + \frac{1}{C_{in}}] \cdot I^*_{C_{in}}; I^*_{C_{in2}} = 0; I^*_{C_{pgs}} = 0; I^*_{C_{ds}} = 0$

$$I^*_{C_{gd}} = 0; I^*_{C_{pds}} = 0; I^*_{C_{gd}} = 0; I^*_{C_{gs}} = 0; I^*_{L_d} + I^*_{L_{dd}} = 0; I^*_{C_{out2}} = 0$$

We consider microwave RF source $X(t) = A_0 + f(t)$; $|f(t)| < 1 \,\&\, A_0 \gg |f(t)|$ then $X(t)|_{A_0 \gg |f(t)|}$ $X(t)|_{A_0 \gg |f(t)|} = A_0 + f(t) \approx A_0$; $\frac{dX(t)}{dt}|_{A_0 \gg |f(t)|} = \frac{df(t)}{dt} \to \varepsilon$.

Stability analysis:

The standard local stability analysis about any one of the equilibrium points of the small signal equivalent circuit for FET consists in adding to coordinate

$[I_{C_{in}}, I_{L_{in1}}, I_{C_{in2}}, I_{L_{in}}, I_{C_{pgs}}, I_{C_{ds}}, I_{R_{ds}}, I_{C_{gd}}, I_{C_{pds}}, I_{R_g}, I_{R_d}, I_{C_{gs}}, I_{R_s}, I_{R_{dd}}, I_{L_d}, I_{L_{dd}}, I_{L_{out1}}, I_{C_{out2}}]$ arbitrarily small increments of exponential form $[x, y, i_{L_1}, i_{R_j}, i_{R_s}] \cdot e^{\lambda \cdot t}$ and retaining the first order terms in $I_{C_{in}}, I_{L_{in1}}, I_{C_{in2}}, I_{L_{in}}, I_{C_{pgs}}, I_{C_{ds}}, I_{R_{ds}}, I_{C_{gd}}, \ldots$ The system of homogeneous equations leads to a polynomial characteristic equation in the eigenvalues. The polynomial characteristic equations accept by set the below circuit variables and circuit variables derivative with respect to time into equivalent circuit for FET equations. FET circuit fixed values with arbitrarily small increments of exponential form $[i_{C_{in}}, i_{L_{in1}}, i_{C_{in2}}, i_{L_{in}}, i_{C_{pgs}}, i_{C_{ds}}, i_{R_{ds}}, i_{C_{gd}}, \ldots] \cdot e^{\lambda \cdot t}$ are: j = 0 (first fixed point), j = 1 (second fixed point), j = 2 (third fixed point), etc.

$$I_{C_{in}}(t) = I_{C_{in}}^{(j)} + i_{C_{in}} \cdot e^{\lambda \cdot t}; I_{C_{in}}(t - \tau_1) = I_{C_{in}}^{(j)} + i_{C_{in}} \cdot e^{\lambda \cdot (t - \tau_1)};$$

$$I_{L_{in1}}(t) = I_{L_{in1}}^{(j)} + i_{L_{in1}} \cdot e^{\lambda \cdot t}; \frac{dI_{L_{in1}}(t)}{dt} = i_{L_{in1}} \cdot \lambda \cdot e^{\lambda \cdot t}$$

$$\frac{d^2 I_{L_{in1}}(t)}{dt^2} = i_{L_{in1}} \cdot \lambda^2 \cdot e^{\lambda \cdot t}; \frac{dI_{C_{in}}(t - \tau_1)}{dt} = i_{C_{in}} \cdot \lambda \cdot e^{\lambda \cdot (t - \tau_1)};$$

$$\frac{d^2 I_{C_{in}}(t - \tau_1)}{dt^2} = i_{C_{in}} \cdot \lambda^2 \cdot e^{\lambda \cdot (t - \tau_1)}$$

$$I_{L_{in}}(t) = I_{L_{in}}^{(j)} + i_{L_{in}} \cdot e^{\lambda \cdot t}; \frac{dI_{L_{in}}(t)}{dt} = i_{L_{in}} \cdot \lambda \cdot e^{\lambda \cdot t}; \frac{d^2 I_{L_{in}}(t)}{dt^2}$$
$$= i_{L_{in}} \cdot \lambda^2 \cdot e^{\lambda \cdot t}; I_{C_{in2}}(t) = I_{C_{in2}}^{(j)} + i_{C_{in2}} \cdot e^{\lambda \cdot t}$$

$$\frac{d^2 I_{C_{in2}}(t)}{dt^2} = i_{C_{in2}} \cdot \lambda^2 \cdot e^{\lambda \cdot t}; I_{C_{pgs}}(t) = I_{C_{pgs}}^{(j)} + i_{C_{pgs}} \cdot e^{\lambda \cdot t};$$

$$I_{C_{in2}}(t - \tau_2) = I_{C_{in2}}^{(j)} + i_{C_{in2}} \cdot e^{\lambda \cdot (t - \tau_2)}$$

$$\frac{d^2 I_{C_{in2}}(t - \tau_2)}{dt^2} = i_{C_{in2}} \cdot \lambda^2 \cdot e^{\lambda \cdot (t - \tau_2)}; I_{C_{ds}} = I_{C_{ds}}^{(j)} + i_{C_{ds}} \cdot e^{\lambda \cdot t};$$

$$I_{R_{ds}} = I_{R_{ds}}^{(j)} + i_{R_{ds}} \cdot e^{\lambda \cdot t}; \frac{dI_{R_{ds}}}{dt} = i_{R_{ds}} \cdot \lambda \cdot e^{\lambda \cdot t}$$

$$I_{C_{gd}}(t) = I_{C_{gd}}^{(j)} + i_{C_{gd}} \cdot e^{\lambda \cdot t}; I_{C_{pds}}(t) = I_{C_{pds}}^{(j)} + i_{C_{pds}} \cdot e^{\lambda \cdot t};$$

$$I_{R_g}(t) = I_{R_g}^{(j)} + i_{R_g} \cdot e^{\lambda \cdot t}; \frac{dI_{R_g}(t)}{dt} = i_{R_g} \cdot \lambda \cdot e^{\lambda \cdot t}$$

$$I_{R_d}(t) = I_{R_d}^{(j)} + i_{R_d} \cdot e^{\lambda \cdot t}; \frac{dI_{R_d}(t)}{dt} = i_{R_d} \cdot \lambda \cdot e^{\lambda \cdot t};$$

$$I_{C_{gs}}(t) = I_{C_{gs}}^{(j)} + i_{C_{gs}} \cdot e^{\lambda \cdot t}; \frac{dI_{C_{gs}}(t)}{dt} = i_{C_{gs}} \cdot \lambda \cdot e^{\lambda \cdot t}$$

$$I_{R_s}(t) = I_{R_s}^{(j)} + i_{R_s} \cdot e^{\lambda \cdot t}; \frac{dI_{R_s}(t)}{dt} = i_{R_s} \cdot \lambda \cdot e^{\lambda \cdot t};$$

$$\frac{d^2 I_{R_s}(t)}{dt^2} = i_{R_s} \cdot \lambda^2 \cdot e^{\lambda \cdot t}; I_{R_{dd}}(t) = I_{R_{dd}}^{(j)} + i_{R_{dd}} \cdot e^{\lambda \cdot t}$$

$$\frac{dI_{R_{dd}}(t)}{dt} = i_{R_{dd}} \cdot \lambda \cdot e^{\lambda \cdot t}; I_{L_d}(t) = I_{L_d}^{(j)} + i_{L_d} \cdot e^{\lambda \cdot t}; \frac{d^2 I_{L_d}(t)}{dt^2} = i_{L_d}$$

$$\cdot \lambda^2 \cdot e^{\lambda \cdot t}; I_{L_{dd}}(t) = I_{L_{dd}}^{(j)} + i_{L_{dd}} \cdot e^{\lambda \cdot t}$$

$$\frac{d^2 I_{L_{dd}}(t)}{dt^2} = i_{L_{dd}} \cdot \lambda^2 \cdot e^{\lambda \cdot t}; I_{L_d}(t - \tau_3) = I_{L_d}^{(j)} + i_{L_d} \cdot e^{\lambda \cdot (t - \tau_3)};$$

$$I_{L_{dd}}(t - \tau_3) = I_{L_{dd}}^{(j)} + i_{L_{dd}} \cdot e^{\lambda \cdot (t - \tau_3)}$$

$$I_{L_{out1}}(t) = I_{L_{out1}}^{(j)} + i_{L_{out1}} \cdot e^{\lambda \cdot t}; \frac{d^2 I_{L_{out1}}(t)}{dt^2} = i_{L_{out1}} \cdot \lambda^2 \cdot e^{\lambda \cdot t};$$

$$I_{C_{out2}}(t - \tau_4) = I_{C_{out2}}^{(j)} + i_{C_{out2}} \cdot e^{\lambda \cdot (t - \tau_4)}$$

$$\frac{dI_{C_{out2}}(t - \tau_4)}{dt} = i_{C_{out2}} \cdot \lambda^2 \cdot e^{\lambda \cdot (t - \tau_4)}$$

By implementing the above delay equations, we get the following system eigenvalues equations:

$$[\frac{dX(t)}{dt} \to \varepsilon] - \frac{1}{C_{in}} \cdot (I_{C_{in}}^{(j)} + i_{C_{in}} \cdot e^{\lambda \cdot t}) = \frac{1}{C_{in1}} \cdot (I_{C_{in}}^{(j)} + i_{C_{in}} \cdot e^{\lambda \cdot (t - \tau_1)}) + L_{in1} \cdot i_{L_{in1}} \cdot \lambda^2 \cdot e^{\lambda \cdot t};$$

$$I_{C_{in2}}^{(j)} + i_{C_{in2}} \cdot e^{\lambda \cdot t} = C_{in2} \cdot (L_{in1} \cdot [i_{C_{in}} \cdot e^{-\lambda \cdot \tau_1} - i_{C_{in2}}] \cdot \lambda^2 \cdot e^{\lambda \cdot t} - L_{in} \cdot i_{L_{in}} \cdot \lambda^2 \cdot e^{\lambda \cdot t})$$

$$L_{in} \cdot i_{L_{in}} \cdot \lambda^2 \cdot e^{\lambda \cdot t} - \frac{1}{C_{pgs}} \cdot [I_{C_{pgs}}^{(j)} + i_{C_{pgs}} \cdot e^{\lambda \cdot t}] = L_g \cdot [i_{C_{in2}} \cdot e^{-\lambda \cdot \tau_2} - i_{L_{in}}] \cdot \lambda^2 \cdot e^{\lambda \cdot t}$$

$$I_{C_{ds}}^{(j)} + i_{C_{ds}} \cdot e^{\lambda \cdot t} = C_{ds} \cdot R_{ds} \cdot i_{R_{ds}} \cdot \lambda \cdot e^{\lambda \cdot t}$$

$$I_{C_{gd}}^{(j)} + i_{C_{gd}} \cdot e^{\lambda \cdot t} = C_{gd} \cdot \{\frac{1}{C_{pgs}} \cdot [I_{C_{pgs}}^{(j)} + i_{C_{pgs}} \cdot e^{\lambda \cdot t}]$$

$$- \frac{1}{C_{pds}} \cdot [I_{C_{pds}}^{(j)} + i_{C_{pds}} \cdot e^{\lambda \cdot t}] - R_g \cdot i_{R_g} \cdot \lambda \cdot e^{\lambda \cdot t} - R_d \cdot i_{R_d} \cdot \lambda \cdot e^{\lambda \cdot t}\}$$

$$R_i \cdot i_{C_{gs}} \cdot \lambda \cdot e^{\lambda \cdot t} = \frac{1}{C_{pgs}} \cdot [I_{C_{pgs}}^{(j)} + i_{C_{pgs}} \cdot e^{\lambda \cdot t}] - R_g \cdot i_{R_g} \cdot \lambda \cdot e^{\lambda \cdot t} - R_S \cdot i_{R_s} \cdot \lambda \cdot e^{\lambda \cdot t}$$

$$- L_S \cdot i_{R_s} \cdot \lambda^2 \cdot e^{\lambda \cdot t} - \frac{1}{C_{gs}} \cdot [I_{C_{gs}}^{(j)} + i_{C_{gs}} \cdot e^{\lambda \cdot t}]$$

$$- R_{dd} \cdot i_{R_{dd}} \cdot \lambda \cdot e^{\lambda \cdot t} - \frac{1}{C_{pds}} \cdot [I_{C_{pds}}^{(j)} + i_{C_{pds}} \cdot e^{\lambda \cdot t}] + L_d \cdot i_{L_d} \cdot \lambda^2 \cdot e^{\lambda \cdot t} = L_{dd} \cdot i_{L_{dd}} \cdot \lambda^2 \cdot e^{\lambda \cdot t}$$

$$I_{L_d}^{(j)} + i_{L_d} \cdot e^{\lambda \cdot (t-\tau_3)} + I_{L_{dd}}^{(j)} + i_{L_{dd}} \cdot e^{\lambda \cdot (t-\tau_3)} = C_{out1} \cdot [\frac{1}{C_{pds}} \cdot (I_{C_{pds}}^{(j)} + i_{C_{pds}} \cdot e^{\lambda \cdot t})$$
$$- L_d \cdot i_{L_d} \cdot \lambda^2 \cdot e^{\lambda \cdot t} - L_{out1} \cdot i_{L_{out1}} \cdot \lambda^2 \cdot e^{\lambda \cdot t}]$$
$$I_{C_{out2}}^{(j)} + i_{C_{out2}} \cdot e^{\lambda \cdot (t-\tau_4)} = C_{out2} \cdot [L_{out1} \cdot i_{L_{out1}} \cdot \lambda^2 \cdot e^{\lambda \cdot t}$$
$$- \frac{1}{C_{out}} \cdot (I_{C_{out2}}^{(j)} + i_{C_{out2}} \cdot e^{\lambda \cdot (t-\tau_4)}) - R_{load} \cdot i_{C_{out2}} \cdot \lambda^2 \cdot e^{\lambda \cdot (t-\tau_4)}]$$

We implement our system fixed points values:

$$\frac{dX(t)}{dt} \to \varepsilon \Rightarrow I_{C_{in}}^{(j)} = I_{C_{in}}^{*} = 0; I_{C_{in2}}^{(j)} = I_{C_{in2}}^{*} = 0; I_{C_{pgs}}^{(j)} = I_{C_{pgs}}^{*} = 0; I_{C_{ds}}^{(j)} = I_{C_{ds}}^{*} = 0$$
$$I_{C_{pgs}}^{(j)} = I_{C_{pgs}}^{*}; I_{C_{gd}}^{(j)} = I_{C_{gd}}^{*}; I_{C_{gd}}^{*} = 0; I_{C_{gs}}^{(j)} = I_{C_{gs}}^{*} = 0; I_{C_{pds}}^{(j)} = I_{C_{pds}}^{*} = 0$$
$$I_{L_{dd}}^{(j)} = I_{L_{dd}}^{*}; I_{L_d}^{(j)} = I_{L_d}^{*}; I_{L_d}^{*} + I_{L_{dd}}^{*} = 0; I_{C_{out2}}^{(j)} = I_{C_{out2}}^{*} = 0$$

System set of eigenvalues equations (eliminating $e^{\lambda \cdot t}$ term):

$$-\frac{1}{C_{in}} \cdot i_{C_{in}} = \frac{1}{C_{in1}} \cdot i_{C_{in}} \cdot e^{-\lambda \cdot \tau_1} + L_{in1} \cdot i_{L_{in1}} \cdot \lambda^2; i_{C_{in2}}$$
$$= C_{in2} \cdot (L_{in1} \cdot [i_{C_{in}} \cdot e^{-\lambda \cdot \tau_1} - i_{C_{in2}}] - L_{in} \cdot i_{L_{in}}) \cdot \lambda^2$$

$$L_{in} \cdot i_{L_{in}} \cdot \lambda^2 - \frac{1}{C_{pgs}} \cdot i_{C_{pgs}} = L_g \cdot [i_{C_{in2}} \cdot e^{-\lambda \cdot \tau_2} - i_{L_{in}}] \cdot \lambda^2; i_{C_{ds}} = C_{ds} \cdot R_{ds} \cdot i_{R_{ds}} \cdot \lambda$$

$$i_{C_{gd}} = \frac{1}{C_{pgs}} \cdot C_{gd} \cdot i_{C_{pgs}} - C_{gd} \cdot \frac{1}{C_{pds}} \cdot i_{C_{pds}} - C_{gd} \cdot [R_g \cdot i_{R_g} + R_d \cdot i_{R_d}] \cdot \lambda$$

$$R_i \cdot i_{C_{gs}} \cdot \lambda = \frac{1}{C_{pgs}} \cdot i_{C_{pgs}} - R_g \cdot i_{R_g} \cdot \lambda - R_S \cdot i_{R_s} \cdot \lambda - L_S \cdot i_{R_s} \cdot \lambda^2 - \frac{1}{C_{gs}} \cdot i_{C_{gs}}$$

$$-R_{dd} \cdot i_{R_{dd}} \cdot \lambda - \frac{1}{C_{pds}} \cdot i_{C_{pds}} + L_d \cdot i_{L_d} \cdot \lambda^2 = L_{dd} \cdot i_{L_{dd}} \cdot \lambda^2$$

$$i_{L_d} \cdot e^{-\lambda \cdot \tau_3} + i_{L_{dd}} \cdot e^{-\lambda \cdot \tau_3} = C_{out1} \cdot [\frac{1}{C_{pds}} \cdot i_{C_{pds}} - L_d \cdot i_{L_d} \cdot \lambda^2 - L_{out1} \cdot i_{L_{out1}} \cdot \lambda^2]$$

$$i_{C_{out2}} \cdot e^{-\lambda \cdot \tau_4} = C_{out2} \cdot [L_{out1} \cdot i_{L_{out1}} \cdot \lambda^2 - \frac{1}{C_{out}} \cdot i_{C_{out2}} \cdot e^{-\lambda \cdot \tau_4} - R_{load} \cdot i_{C_{out2}} \cdot \lambda^2 \cdot e^{-\lambda \cdot \tau_4}]$$

Remark: it a reader task to analyze system stability under delay parameters variation based on eigenvalues equations.

Exercises

1. We have system of a bipolar transistor that activated at microwave frequency and include feedback loop (C_{f_x}, R_{f_x}, R_A). R_A is a variable resistor (3 ports) which established the strength of our RF circuit feedback. We can divide our variable resistor R_A to two resistive sections: $R_A \cdot \delta; R_A \cdot (1-\delta)$. Delta parameter ($\delta$) is the feedback balance resistor parameter $0<\delta<1$. Consider the base diagram of bipolar transistor at microwave frequency circuit (Sect. 3.1) with additional feedback loop which describe in the below figure.

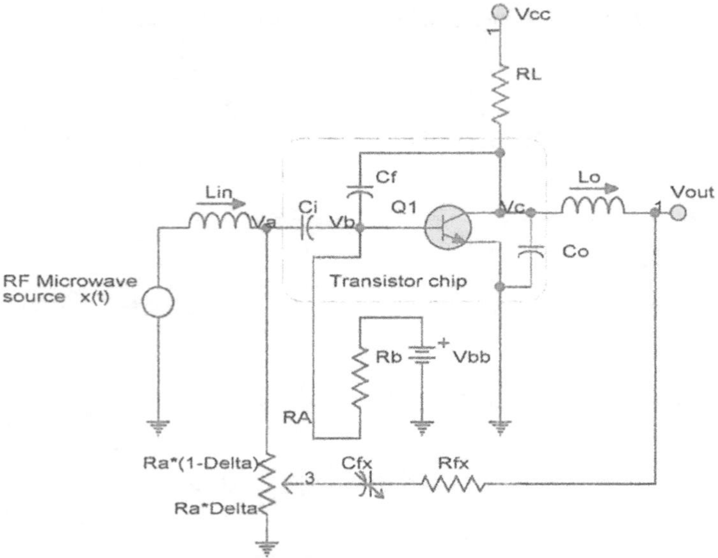

Bipolar transistor at microwave frequencies with feedback loop

1.1 Draw the circuit of bipolar transistor with feedback loop at microwave frequencies equivalent circuit. Input Microwave RF source X(t). We consider Microwave RF_{in} signal $X(t) = A_0 + f_X(t)$; $|f_X(t)| \leq 1$ and $A_0 \gg |f_X(t)|$ then $X(t) = A_0 + f_X(t) \approx A_0 \Rightarrow \frac{dX(t)}{dt} = \frac{df_X(t)}{dt} \to \varepsilon$. Find system differential equations and fixed points.

1.2 Discuss stability analysis, linearization, and find system Jacobian elements at fixed points. How Delta (δ; $0<\delta<1$) feedback balance resistor parameter influences our system stability?

1.3 Classify system stability fixed points according to eigenvalues variation. How Delta (δ ; $0<\delta<1$) parameter influences our eigenvalues variation?
1.4 What happened if our feedback loop Delta (δ ; $0<\delta<1$) parameter is constantly equal to one $\delta = 1$? How our system behaviors change?
1.5 What happened if our feedback loop Delta ($\delta; 0<\delta<1$) parameter is constantly equal to zero $\delta = 0$? How our system behaviors change?

2. We have system of a bipolar transistor that activated at microwave frequency and includes two feedback loops, first loop C_{f_A}, R_{f_A}, R_A and second loop C_{f_B}, R_{f_B}, R_B. R_A and R_B are variable resistors (3 ports) which establish the strength of our RF circuit feedback. We can divide our system circuit feedback variable resistors to two sections respectively. $\delta_A \cdot R_A, (1 - \delta_A) \cdot R_A$ and $\delta_B \cdot R_B, (1 - \delta_B) \cdot R_B$. They δ_A, δ_B are two feedback balance resistors parameters $0<\delta_A<1, 0<\delta_B<1; \delta_A \neq \delta_B$.
Consider the base diagram of bipolar transistor at microwave frequency circuit (Sect. 3.1) with two additional feedback loops which describe in the below figure.

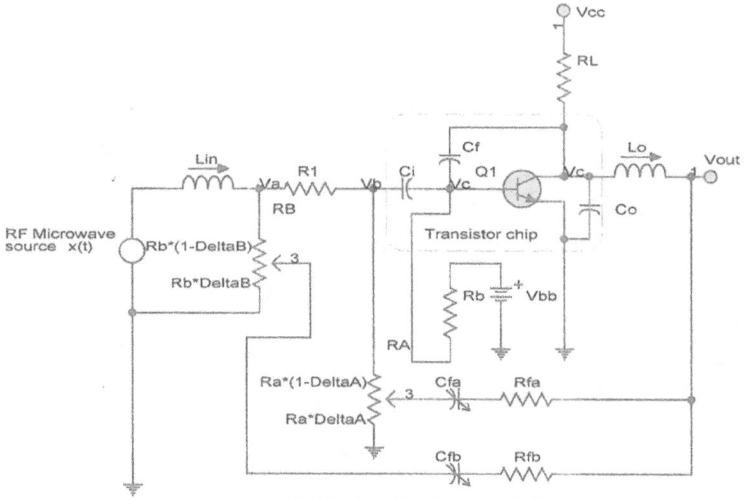

Bipolar transistor at microwave frequencies with two feedback loops

2.1 Draw system circuit bipolar transistor with two feedback loops equivalent circuit at microwave frequencies. Input Microwave RF source X(t). We consider Microwave RF_{in} signal $X(t) = A_0 + f_X(t)$; $|f_X(t)| \leq 1$ and $A_0 \gg |f_X(t)|$ then $X(t) = A_0 + f_X(t) \approx A_0 \Rightarrow \frac{dX(t)}{dt} = \frac{df_X(t)}{dt} \to \varepsilon$
Find system differential equations and fixed points.
2.2 Discuss stability analysis, linearization, and find Jacobian elements at fixed points. How DeltaA ($\delta_A; 0<\delta_A<1$) and DeltaB (δ_B ; $0<\delta_B<1$) feedback balance resistor parameters influence our system stability? $\delta_A \neq \delta_B$

2.3 Classify system stability fixed points according to eigenvalues. How δ_A, δ_B parameters influence our system eigenvalues variation?

2.4 What happened if our feedback loop δ_A parameter is constantly equal to zero ($\delta_A = 0$) or one ($\delta_A = 1$); $\delta_B; 0 < \delta_B < 1$? How our system behavior changes?

2.5 What happened if our feedback loop δ_B parameter is constantly equal to zero ($\delta_B = 0$) or one ($\delta_B = 1$);$\delta_A; 0 < \delta_A < 1$? How our system behavior change?

3. Consider RF FETs amplifier system which includes two RF FETs with peripheral components. We consider Microwave RF_{in} signal $X(t) = A_0 + f_X(t)$; $|f_X(t)| <= 1$ and $A_0 >> |f_X(t)|$ then $X(t) = A_0 + f_X(t) \approx A_0$.
The FET equivalent circuits are for high frequency model and operation, taking the node capacitors and other elements into account. If we switch to low frequency small signal FET model, all capacitors in the above model disconnected and all inductors are short.

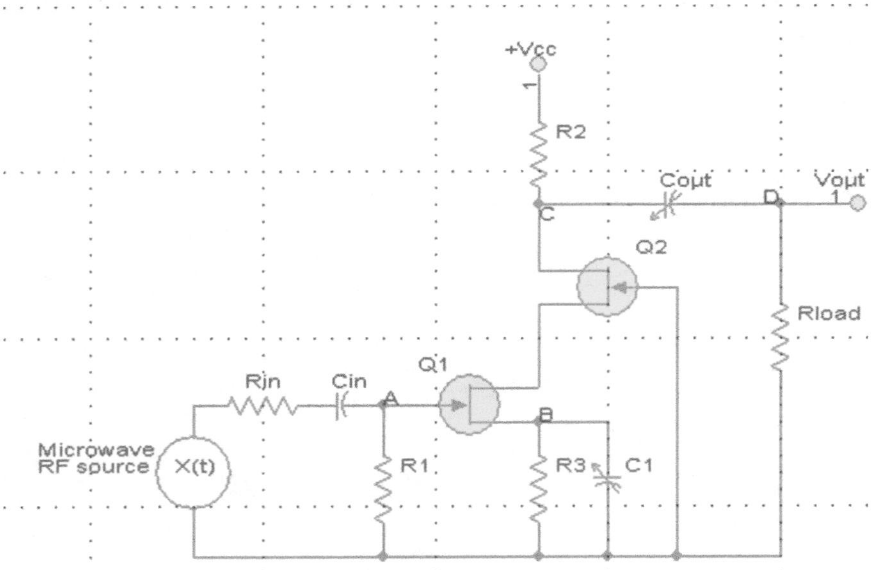

3.1. Draw full small signal equivalent circuit for FETs system and write FETs system differential Eqs.

3.2. Find FETs system fixed points, How our system fixed points change if $R_3 = 0$? How our system fixed points change if capacitor C_1 is disconnected?

3.3. Discuss RF FETs system standard local stability analysis about anyone of the equilibrium points. How the stability changes if $R_3 = 0$?

3.4. Classify RF FETs system fixed points and discuss bifurcation for different values of R_1, R_2, R_3 resistors.

3.5. Resistor R_3 is disconnected, How our RF FETs system dynamical behavior changes?
3.6. Capacitor C_1 is disconnected, How our system bifurcation behavior changes?

4. We have a schematic of an RF amplifier using a JFET as the active element. The configuration of the JFET amplifier is common gate. The circuit includes two iron-core inductors L_1, L_2 (RF chockes). The iron core inductors block the high frequency AC signals from getting to the DC power supply. +V is positive DC voltage source and –V is negative DC voltage source. +V and –V are biasing voltages to our FET circuit.

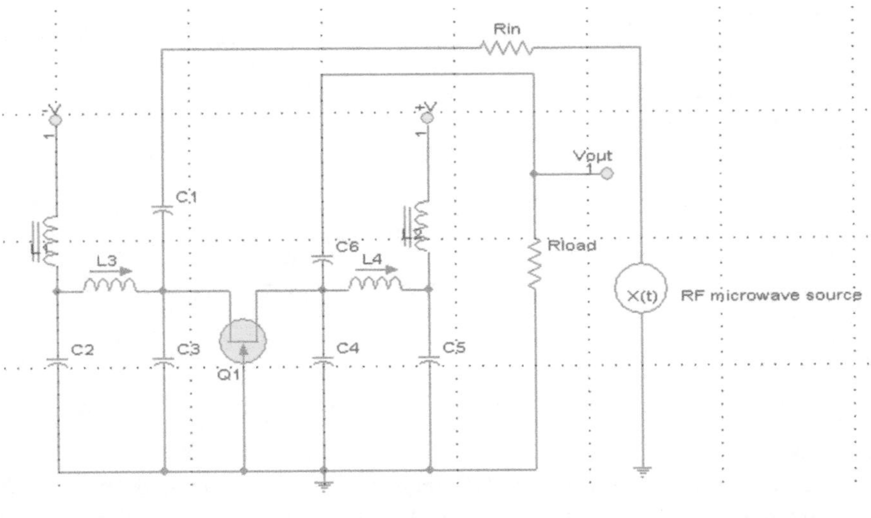

We consider RFin signal $X(t) = A_0 + f_x(t); |f_x(t)| <= 1$ and $A_0 \gg |f_x(t)|$ then $X(t) = A_0 + f_x(t)|_{A_0 \gg |f_x(t)|} \approx A_0; |f_x(t)| \leq 1$.

4.1 Draw full small signal equivalent circuit for FET RF amplifier and write system differential equations.
4.2 Find JFET amplifier circuit fixed points, How our system fixed points change if C_4 is disconnected?
4.3 Discuss JFET amplifier standard local stability analysis about anyone of the equilibrium points. How the stability changes if C_3 is disconnected?
4.4 Classify JFET amplifier circuit fixed points and discuss bifurcation for different values of L_3, L_4.
4.5 L_3 is disconnected, How our JFET amplifier circuit behavior changes?
4.6 Capacitor C_5 is disconnected, How our system bifurcation behavior changes?

5. We have IMPATT amplifier system which is constructed from three ports active circulator and two IMPATT diodes. The circulator is used to separate input and output signals. We consider IMPATT diodes chip RF equivalent circuit which can be represent as a series resistor and capacitor R_{D1}, C_{D1} and R_{D2}, C_{D2} respectively for the first IMPATT diode D_1 and second IMPATT diode D_2. $R_{D1} < 0$, $R_{D2} < 0$. R_{D1} and R_{D2} are the terminal negative resistances of the packaged diodes (D_1 and D_2). IMPATT diodes act as an amplifier in the negative differential resistance characteristics.

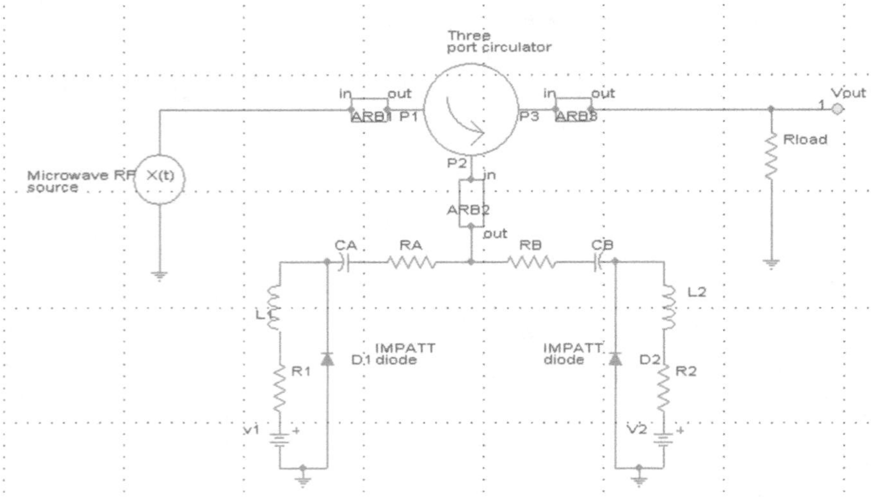

τ_1 is the time delay for ARB1 microstrip. τ_2 is the time delay for ARB2 microstrip. τ_3 is the time delay for ARB3 microstrip. R_{load} is the circuit load resistance. V_1, V_2 are DC voltage sources (bias voltages to IMPATT diodes D_1 and D_2 respectively). $V_1 \neq V_2$ is different biasing voltages. R_1, R_2 are parasitic resistances of DC voltage sources. L_1, C_A—inductor and capacitor of Bias-T circuit for D_1. L_2, C_B—inductor and capacitor of Bias-T circuit for D_2. ARB1, ARB2, and ARB3 are circuit microstrip delay lines (τ_1, τ_2, τ_3). D_1, D_2—first and second IMPATT diodes.

5.1 Draw system amplifier full equivalent circuit and find differential equations. Find fixed points and consider amplifier system input voltage $V_{P_1} = \Gamma + \xi(t)$, Γ is constant voltage and ξ(t) is RF signal in time.

5.2 Write system differential equations in matrix representation. Discuss system eigenvalues and related fixed points classification.

5.3 If we short inductor L_1 in our amplifier system, How it influences system stability?

5.4 If we short R_A in our amplifier system, How it influences system stability?

5.5 If we short diode D_2, How it influences system differential equations, fixed points and stability?

5.6 If we disconnect diode D_1, How it influences system differential equations, fixed points and stability?

Hint: The delay parameters τ_1, τ_2 and τ_3 are related to currents which flows through micro strips ARB1, ARB2, and ARB3. We can consider additional delay parameters $\Delta_1, \Delta_2, \Delta_3$ for the current derivatives which flows through micro strips respectively.

6. We have two sets of multistage IMPATT amplifiers. The output of our system is a summation of two set's outputs. The first set of multistage IMPATT amplifier has additional k_1 delay lines $\tau_1, \tau_3, \tau_5, \ldots, \tau_{2 \cdot n-1}$ for n $n = 1, 2, \ldots, k_1$ which represents $(2 \cdot k_1 - 1)$ IMPATT diode/active circulator circuit stages. The delay lines for the first set are exist only on odd stages outputs. The second set of multistage IMPATT amplifiers has additional k_2 delay lines $\tau_2, \tau_4, \tau_6, \ldots, \tau_{2 \cdot n}$; $n = 1, 2, \ldots, k_2$ which represents $(2 \cdot k_2)$ IMPATT diode/active circulator circuit stages. The delay lines for the second set are exist only on even stages outputs ($k_1 \neq k_2$). The first set of multistage IMPATT amplifiers is feed by microwave RF signal $X_1(t)$ and the second set of multistage IMPATT amplifiers is feed by microwave RF signal $X_2(t)$; $X_1(t) \neq X_2(t)$. $X_1(t) = \Gamma_1 + \xi_1(t)$ $X_2(t) = \Gamma_2 + \xi_2(t)$. Γ_1, Γ_2 are constant voltages ($\Gamma_1 \neq \Gamma_2$) and $\xi_1(t)$, $\xi_2(t)$ are RF signals in time $\xi_1(t) \neq \xi_2(t)$.

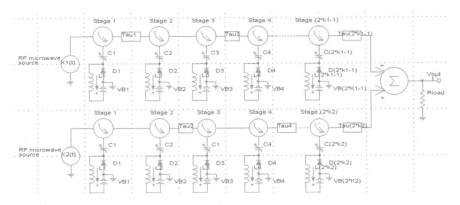

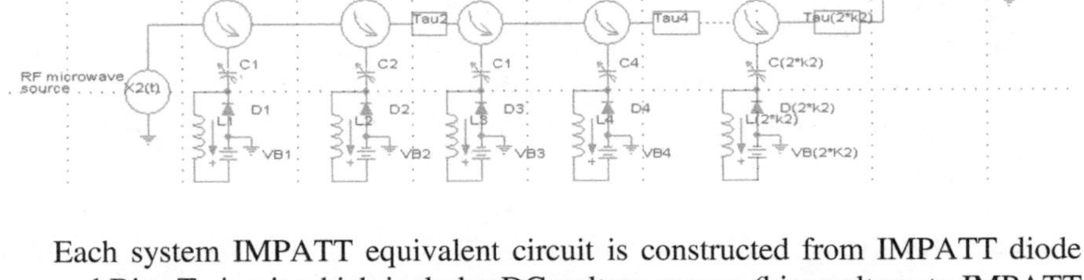

Each system IMPATT equivalent circuit is constructed from IMPATT diode and Bias-T circuit which includes DC voltage source (bias voltage to IMPATT diode), parasitic resistance of DC voltage source, inductor and capacitor. We neglect the microstrip time delay which is connected between each stage circulator's port 2 to IMPATT diode. We neglect the voltages on system delay lines $V_{\tau_{2 \cdot n-1}} \to \varepsilon \forall n = 1, 2, \ldots, k_1$ and $V_{\tau_{2 \cdot n}} \to \varepsilon \forall n = 1, 2, \ldots, k_2$. There is a delay in the current that flow through each microstrip delay line in the two sets of multistage IMPATT amplifiers. $1 \leq (2 \cdot n - 1) \leq (2 \cdot k_1 - 1)$ for the first set and $1 \leq 2 \cdot n \leq (2 \cdot k_2)$ for the second set. The input current for the circuit stage

(i) is defined as $I_{in(i)}(t)$ in the two sets respectively and the output current for circuit stage (i) is defined as $I_{out(i)}(t)$ in the two sets respectively. The transfer function from circuit stage (i) input current $I_{in(i)}(t)$ to output current $I_{out(i)}(t)$ is define as $I_{out(i)}(t) = \ldots; I_{out(i)}(t) = g_{1i}(I_{in(i)}(t), D_i, L_i, C_i, V_{Bi})$ for the first set and $I_{out(i)}(t) = \cdots; I_{out(i)}(t) = g_{2i}(I_{in(i)}(t), D_i, L_i, C_i, V_{Bi})$ for the second set respectively. $g_{1i}(I_{in(i)}(t), D_i, L_i, C_i, V_{Bi}) \neq g_{2i}(I_{in(i)}(t), D_i, L_i, C_i, V_{Bi})$

6.1 Find the expression for $I_{Rload}(t)$ as a function of the two set's outputs functions.
6.2 Find the two functions for the first and second sets: $g_{1i}(I_{in(i)}(t), D_i, L_i, C_i, V_{Bi})$ and $g_{2i}(I_{in(i)}(t), D_i, L_i, C_i, V_{Bi})$
6.3 Find the expressions for $I_{Rload}(t)$ if $X_1(t) = 0$ or $X_2(t) = 0$.
6.4 How our system behavior changes if each IMPATT/active circulator amplifier stage in the two sets suffers from short Bias-T's inductor or shorted Bias-T's capacitor?
6.5 How our system behavior changes if two set's input RF signals are as follow: $X_1(t) = \Gamma + \xi(t); X_2(t) = a_1 \cdot X_1(t) + a_2 \cdot X_1^2(t)$. Γ is a constant voltage. $\xi(t)$ is a RF signal. a_1, a_2 are constants.

7. We have system of three sets of multistage IMPATT amplifiers. The RF microwave signal input to the third set is a summation of two signals: first is the output signal from the first set and second is the output signal from the second set. The first set is constructed from multistage IMPATT amplifier which has a delay line in each output stage. The number of IMPATT/active circulator stages in the first set is k_1 and the number of delay lines is k_1 $(\tau_1, \tau_2, \ldots, \tau_{k_1})$. k_1 can be odd or even number. The second set of multistage IMPATT amplifier has additional k_2 delay lines $\tau_1, \tau_3, \tau_5, \ldots, \tau_{2 \cdot n-1}$; $n = 1, 2, \ldots, k_2$ which represent $(2 \cdot k_2 - 1)$ IMPATT/ active circulator circuit stages. The delay lines for the second set are exist only on odd stages outputs. The third set of multistage IMPATT amplifier has additional k_3 delay lines $\tau_2, \tau_4, \tau_6, \ldots, \tau_{2 \cdot n}$; $n = 1, 2, \ldots, k_3$ which represent $2 \cdot k_3$ IMPATT/active circulator stages. The delay lines for the third set are exist only on even stages outputs $(k_1 \neq k_2 \neq k_3)$. The first set of multistage IMPATT amplifiers is feed by microwave RF signal $X_1(t)$; and the second set of multistage IMPATT amplifier is feed by microwave RF signal $X_2(t)$; $X_1(t) \neq X_2(t)$. $X_1(t) = \Gamma + \xi(t); X_2(t) = \sqrt{1 + X_1(t)}$. Γ is constant voltage and $\xi(t)$ is RF signal in time. Each system IMPATT equivalent circuit is constructed from IMPATT diode and Bias-T circuit which includes DC voltage source (bias voltage to IMPATT diode), parasitic resistance of the DC voltage source, inductor and capacitor. We consider IMPATT diode acts as an amplifier. We neglect the microstrip time delay which is connected between each stage circulator port 2 to IMPATT diode. We neglect the voltages on system delay lines $V_{\tau_n} \to \varepsilon \, \forall n = 1, 2, \ldots, k_1$; $V_{\tau_{2 \cdot n-1}} \to \varepsilon \, \forall n = 1, 2, \ldots, k_2$ and $V_{\tau_{2 \cdot n}} \to \varepsilon \, \forall n = 1, 2, \ldots, k_3$.

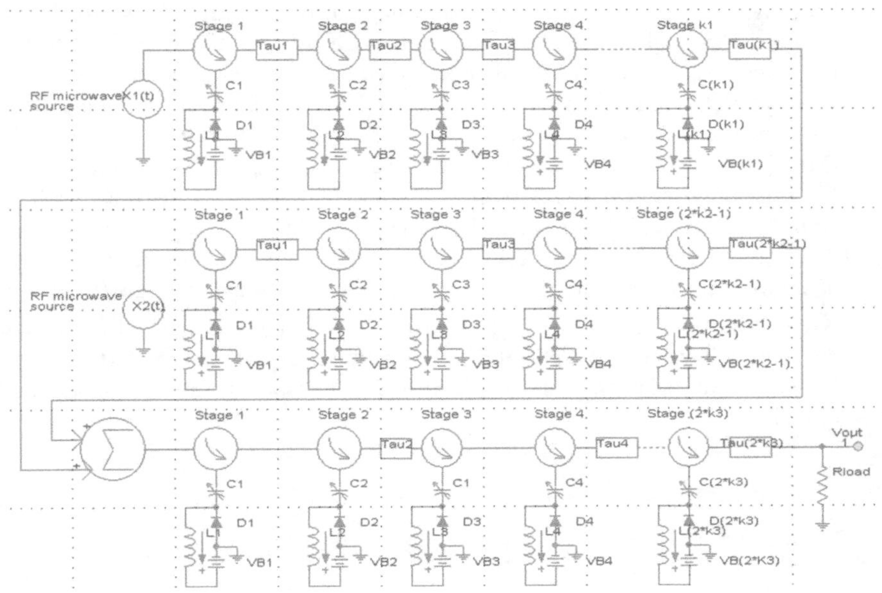

There is a delay in the current that flow through each microstrip delay line in the three sets of multistage IMPATT amplifier in our system. $\forall\, 1 \leq n \leq k_1$ first set ; $\forall\, 1 \leq n \leq k_2$ second set ; $\forall\, 1 \leq n \leq k_3$ third set

The input current for circuit stage (i) is defined as $I_{in(i)}(t)$ in all sets and the output current for circuit stage (i) is defined as $I_{out}(t)$ in all sets. The transfer function from circuit stage (i) input current $I_{in(i)}(t)$ to output current $I_{out(i)}(t)$ in set m (m = 1,2,3 stand for set number). $I_{out(i)}(t) = g_{mi}(I_{in(i)}(t), D_i, L_i, C_i, V_{Bi})\ \forall\, m = 1, 2, 3$ (set number).

$$g_{l_1 i}(I_{in(i)}(t), \ldots) \neq g_{l_2 i}(I_{in(i)}(t), \ldots)\ ;\ l_1 \neq l_2\ ;\ l_1 = 1, 2, 3\ ;\ l_2 = 1, 2, 3$$

7.1 Find the expression for $I_{Rload}(t)$ as a function of the three set's output functions.
7.2 Find the three functions (m = 1,2,3), $g_{mi}(I_{in(i)}(t), D_i, L_i, C_i, V_{Bi})$.
7.3 Find the expression for $I_{Rload}(t)$ if $X_1(t) = 0$.
7.4 Find the expression for $I_{Rload}(t)$ if $X_2(t) = 0$.
7.5 How our system behavior changes if each IMPATT diode/active circulator amplifier stage in the second set is suffered from short Bias-T's inductor or short Bias-T's capacitor?
7.6 How our system behavior changes if each IMPATT diode/active circulator amplifier stage in the third set is suffered from short Bias-T's inductor or short Bias-T's capacitor?

8. We have system of two sets of multistage IMPATT amplifiers which are connected through isolators to loads network (R_A, R_B, and R − Q). The RF microwave signal input to the first set is $X_1(t)$ and for the second set is $X_2(t)$. The first set of multistage IMPATT amplifier has additional k_1 delay lines $\tau_1, \tau_3, \tau_5, \ldots, \tau_{2\cdot n-1}$ for n $n = 1, 2, \ldots, k_1$ which represents $(2 \cdot k_1 - 1)$ IMPATT diode/active circulator circuit stages. The delay lines for the first set are exist only on odd stages outputs. The second set of multistage IMPATT amplifiers has additional k2 delay lines $\tau_2, \tau_4, \tau_6, \ldots, \tau_{2\cdot n}$; $n = 1, 2, \ldots, k_2$ which represents $(2 \cdot k_2)$ IMPATT diode/active circulator circuit stages. The delay lines for the second set are exist only on even stages outputs ($k_1 \neq k_2$). $X_1(t) \neq X_2(t)$. $X_1(t) = \Gamma_1 + \xi_1^2(t)$; $X_2(t) = \xi_2(t) \cdot \Gamma_2 + \sqrt{X_1(t)}$. Γ_1, Γ_2 are constant voltages ($\Gamma_1 \neq \Gamma_2$) and $\xi_1(t)$, $\xi_2(t)$ are RF signals in time $\xi_1(t) \neq \xi_2(t)$. Consider that isolators A and B are ideal and transparent to RF signal in one direction.

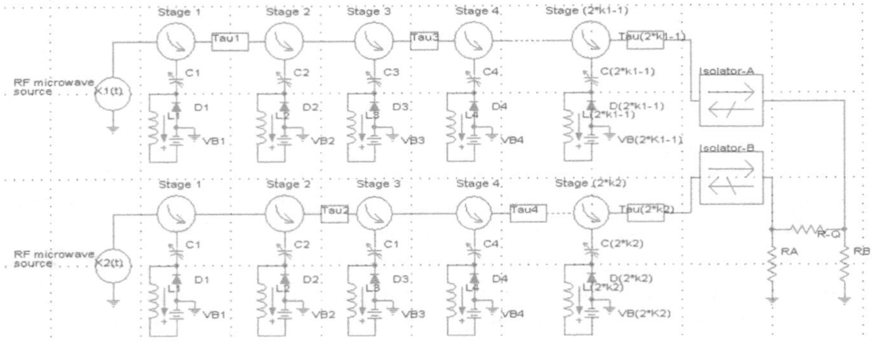

Each system IMPATT equivalent circuit is constructed from IMPATT diode and Bias-T circuit which includes DC voltage source (bias voltage to IMPATT diode), parasitic resistance of DC voltage source, inductor and capacitor. We neglect the microstrip time delay which is connected between each stage circulator's port 2 to IMPATT diode. We neglect the voltages on system delay lines $V_{\tau_{2\cdot n-1}} \to \varepsilon \forall n = 1, 2, \ldots, k_1$ and $V_{\tau_{2\cdot n}} \to \varepsilon \forall n = 1, 2, \ldots, k_2$. There is a delay in the current that flow through each microstrip delay line in the two sets of multistage IMPATT amplifiers. $1 \leq (2 \cdot n - 1) \leq (2 \cdot k_1 - 1)$ for the first set and $1 \leq 2 \cdot n \leq (2 \cdot k_2)$ for the second set. The input current for the circuit stage (i) is defined as $I_{in(i)}(t)$ in the two sets respectively and the output current for circuit stage (i) is defined as $I_{out(i)}(t)$ in the two sets respectively. The transfer function from circuit stage (i) input current $I_{in(i)}(t)$ to output current $I_{out(i)}(t)$ is define as $I_{out(i)}(t) = \ldots$; $I_{out(i)}(t) = g_{1i}(I_{in(i)}(t), D_i, L_i, C_i, V_{Bi})$ for the first set and $I_{out(i)}(t) = \ldots$; $I_{out(i)}(t) = g_{2i}(I_{in(i)}(t), D_i, L_i, C_i, V_{Bi})$ for the second set respectively. $g_{1i}(I_{in(i)}(t), D_i, L_i, C_i, V_{Bi}) \neq g_{2i}(I_{in(i)}(t), D_i, L_i, C_i, V_{Bi})$

8.1 Find the expression for $I_{RA}(t)$, $I_{RB}(t)$, $I_{R\text{-}Q}(t)$, as a function of the two set's outputs functions.
8.2 Find the two functions for the first and second sets: $g_{1i}(I_{in(i)}(t), D_i, L_i, C_i, V_{Bi})$ and $g_{2i}(I_{in(i)}(t), D_i, L_i, C_i, V_{Bi})$
8.3 Find the expressions for $I_{R\text{-}Q}(t)$if $X_1(t) = 0$ or $X_2(t) = 0$.
8.4 How our system behavior changes if each IMPATT/active circulator amplifier stage in the two sets suffers from short Bias-T's inductor or shorted Bias-T's capacitor and additionally R_A is disconnected?
8.5 How our system behavior changes if two set's input RF signals are as follow: $X_1(t) = \Gamma + \xi(t); X_2(t) = a_1 \cdot X_1(t) + a_2 \cdot X_1^2(t)$. Γ is a constant voltage. ξ(t) is a RF signal. a_1, a_2 are constants and R_B is disconnected?

9. We have a system of two sets of multistage IMPATT amplifier which feed load resistance R_L. Each circulator/IMPATT diode amplifier stage gets RF signal from the previous amplifier unit and after amplification feeds the RF signal to the next amplifier unit. Input RF microwave signal to the first set is defined as $X_1(t)$ and to the second set $X_2(t)$. The function expression which characterize our load resistance current in time is as follow:

$$I_{R_L} = f_n(f_{n-1}(f_{n-2}(\ldots(f_1(I_{in(1)}(t - \sum_{k=3}^{n} \tau_k), D_1, L_1, C_1, V_{B1})\ldots), D_{n-1}, L_{n-1}, C_{n-1}, V_{B_{n-1}}),$$
$$D_n, L_n, C_n, V_{B_n}) + g_2(g_1(I_{in(1)}(t - \tau_1 - \tau_2), D_1, L_1, C_1, V_{B1}), D_2, L_2, C_2, V_{B2})$$

The first set is characterized by the f functions and the second set is characterized by the g function.

First set: $\tau_k = |\tau_{k-1} \cdot (-1)^k + \tau_{k-2} \cdot (-1)^{k+1}| \forall n \geq k \geq 3$ is recursive function of micro strip delay line parameter in the first set.

$$\tau_3 = |\tau_1 - \tau_2|\,;\ \tau_4 = |\tau_3 - \tau_2|\,;\ \tau_5 = |\tau_3 - \tau_4|\,;\ \tau_6 = |\tau_5 - \tau_4|\ldots$$

Second set:

$$\tau_1 > 0; \tau_2 > 0; I'_{out(2)}(t) = I_{out(2)}(t - \tau_2); I_{in(1)}(t) = \xi_2(X_2(t))$$
$$I_{out(2)}(t) = g_2(I_{out(1)}(t - \tau_1), D_2, L_2, C_2, V_{B2}); I_{in(2)}(t) = I_{out(1)}(t - \tau_1)$$
$$I_{out(1)}(t) = g_1(I_{in(1)}(t), D_1, L_1, C_1, V_{B1}); I'_{out(2)}(t) = g_2(I_{out(1)}(t - \tau_1 - \tau_2), D_2, L_2, C_2, V_{B2})$$
$$I'_{out(2)}(t) = g_2(g_1(I_{in(1)}(t - \tau_1 - \tau_2), D_1, L_1, C_1, V_{B1}), D_2, L_2, C_2, V_{B2}).$$

The system's summation operator inputs are $I'_{out(2)}(t)$ and $f_n(f_{n-1}(f_{n-2}(\ldots)$

9.1 Draw our two sets amplifier system.
9.2 How our system dynamic change for $\tau_{k-1} > \tau_{k-2}$ and for $\tau_{k-1} < \tau_{k-2}$? (First set micro strip delay lines parameters, n ≥ k ≥ 3).
9.3 The system's first set of IMPATT multistage amplifiers has micro strip delay lines only on odd places of IMPATT diode/active circulator stages.

Write the related functions for $I_{RL}(t)$ and draw our system. How the dynamic of the system changes?

9.4 System's first set stages is suffered from Bias-T's short capacitor. How our system functionality changes? Find the related f functions for that case.

9.5 We change the direction polarity of each IMPATT diode in the first set. How it influences our system functionality. Find the expression for $I_{RL}(t)$ and transformation function in each multistage amplifier in the first set.

9.6 The recursive function of micro strip delay lines in the first set is as follow:

$$\tau_k = |\tau_{k-1} \cdot (-1)^k + |\tau_{k-2} \cdot (-1)^{k+1} - \tau_{k-3} \cdot (-1)^k|| \forall n \geq k \geq 4; \tau_3 \geq 0$$
$$\eta_1(\tau, k) = \tau_{k-2} \cdot (-1)^{k+1} - \tau_{k-3} \cdot (-1)^k;$$
$$\eta_2(\tau, k) = \tau_{k-1} \cdot (-1)^k; \tau_k = |\eta_2(\tau, k) + |\eta_1(\tau, k)||$$

How our system dynamic changes? Write all possible options for $\tau_3, \tau_4, \tau_5, \tau_6, \ldots, \tau_n$. Analyze the $I_{RL}(t)$ expression for that case.

10. We have a system of RF FET transistor combined biasing and matching circuit. The RF FET transistor is N channel JFET and the input and output matching circuits are Pi type. We represent our circuit microstrip elements as a delay lines τ_1, …,τ_6 ($\tau_{k+1} = \tau_k + \Gamma \cdot \tau_{k-1}$) for $\Gamma \in [1, \ldots, 10]$; $k = 2, \ldots, 5$ and $\tau_1 = \tau$, $\tau_2 = \tau^2$. We neglect the voltages on the delay lines $V_{\tau_i} \to \varepsilon$; $\tau_1, \ldots, \tau_6$; $1 \leq i \leq 6$. There is a delay in the current which flows through each microstrip delay line $I(t) \to I(t - \tau_i) \forall 1 \leq i \leq 6$. We use for our analysis the FET small signal equivalent circuit (reduced version). The RF choke L_{dd} presents the microwave signal from being shorted out by the drain resistor R_{dd} through RF choke L_{dd}, which is connected to the positive supply voltage V_{dd}. Input and output capacitors are C_{in} and C_{out}. X(t) is a input microwave RF source.

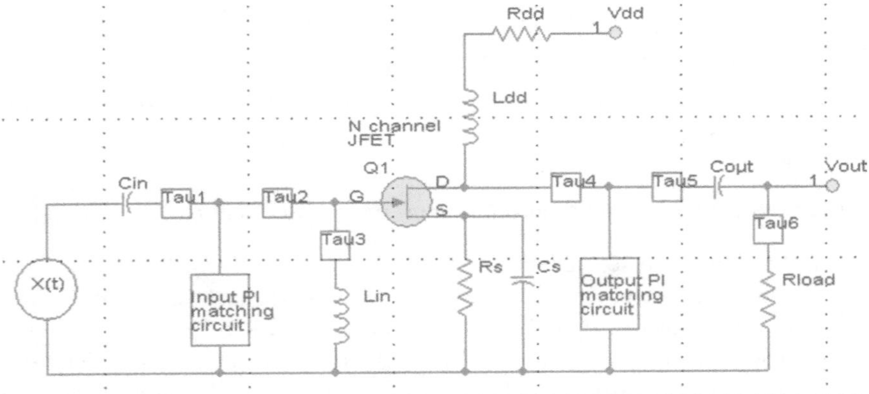

N channel JFET combined biasing and matching circuit

10.1 Draw our system complete circuit which includes small signal equivalent circuit for FET.
10.2 Write system represented differential equations with delay variables in time ($\tau_1, \ldots, \tau_6$ delay parameters).
10.3 Find our system small increments Jacobian and characteristic Eq.
10.4 Discuss stability switching under delay parameter (τ) variation and Γ parameter variation.
10.5 How our system dynamic changes if $\tau_{k+1} = \tau_k \cdot \Gamma + \Gamma^{\tau_k} \cdot \tau_{k-1}$ $\tau_{k+1} = \tau_k \cdot \Gamma + \Gamma^{\tau_k} \cdot \quad \tau_{k-1} \forall \Gamma \in [1, \ldots, 10]$; $k = 2, \ldots, 5$. Discuss stability Switching under delay parameter (τ) variation and Γ parameter variation.
10.6 Resistor R_s is disconnected. How our system dynamic changes? Discuss stability switching under delay parameter (τ) variation and Γ parameter variation.

Chapter 4
Small Signal (SS) Amplifiers and Matching Network Stability Analysis

An amplifier is an active device that has the ability to amplify voltage, amplify current or amplify both voltage and current. There are some types of amplifiers. Amplifiers types: zero frequency amplifiers (DC amplifiers), low frequency amplifiers (audio amplifiers), and high frequency amplifiers (RF amplifiers). Power is $P = V \cdot I$, when current (I) or voltage (V) is raising then will create power amplification. Amplifiers come in three basic flavors: Common Base (CB) amplifiers, Common Collector (CC) amplifiers, and Common Emitters (CE) amplifiers. It depends whether the base, collector, or emitter is common to both the input and output of the amplifier. Common Base (CB) amplifier, input signal inserted at emitter (E) and output signal taken from the collector (C). The CB amplifier can operate as a voltage amplifier for low input impedance circuits. The most popular amplifier circuit is Common Emitter (CE). The CE amplifier has a greater current gain and voltage gain combination than any other type. CE amplifier makes excellent power amplifiers. Common Emitter (CE) amplifier, input signal inserted at the base (B) and output signal taken from the collector (C). The CE amplifier's output voltage is shifted by $\sim 180°$ in phase compared to CE amplifier's input signal. At RF frequencies there is an effect of "positive feedback", which creates amplifier instability and oscillation. "Positive feed-back" is when there is internal feedback capacitance between the transistor's collector and its base. It cause to undesired CE oscillations. At a specific frequency this capacitance will send an in phase signal back into the base input from the collector's output. This back in phase signal creates an "oscillator". The Common Collector (CC) amplifier is emitter follower amplifier. The CC amplifier has the input signal inserted into the base and output signal from the emitter. The CC amplifier has current and power gain, voltage gain less than one. CC amplifier's common use buffer or active impedance matching circuit. The CC

O. Aluf, *Microwave RF Antennas and Circuits*,
DOI 10.1007/978-3-319-45427-6_4

amplifier has high input impedance and low output impedance. There is no phase inversion between CC amplifier's input and output. When an amplifier's output impedance matches the load impedance maximum power is transferred to the load and all reflections are eliminated. When an amplifier's output impedance unmatched the load impedance there are reflections and less than maximum power is transferred to the load. There are instabilities behaviors in these three types of amplifiers causes by circuit micro-strip delays in time parasitic effects. We use RF matching network which able and facilitate impedance matching and filtering of signal, coupling between RF stages. There are typical amplifiers matching networks: L matching network, T matching network, and PI matching network. In design of microwave matching network, device parasitic effects of length on RF circuit matching and stability. Many RF circuits contain Bias-T three ports network. The function of the Bias-T is to simultaneously allow a DC bias voltage and RF test signal to be applied to the port of a transistor during measurement and operation. Bias-T three ports network suffers from instability under delayed micro-strip in time. The passive filter with Bias-T suffers from instability under parameters variation. Many RF circuits include a PIN diode. A PIN diode is a diode with a wide, lightly doped 'near' intrinsic semiconductor region between a p-type semiconductor and a n-type semiconductor region. A PIN diode suitable for many applications: Attenuators, fast switches, photo detectors, and high voltage power electronics applications. A PIN diode operates under high level injection. The PIN diode suffers from instability under parameters variations.

4.1 Small Signal (SS) Amplifiers and Matching Network

Amplifier can be categorized in two manner, first according to signal level (small signal amplifier, power/large signal amplifier) and second according to DC biasing scheme of the active component (Classes A, B, AB, C). There are also other classes, such as class D (D stands for digital), class E and class F. These all uses the transistor/FET as a switch [24–26]. Input and output voltage relation of the amplifier can be modeled simply as: $V_{out}(t) = \sum_{k=1}^{3} a_k \cdot v_{in}^k + H.O.T$; $V_{out}(t) = a_1 \cdot v_i(t) + a_2 \cdot v_i^2(t) + a_3 \cdot v_i^3(t) + H.O.T$. The general block diagram is described below (Fig. 4.1):

If we compare small signal versus Large-signal operation amplifier output voltage expression: Large signal $V_{out}(t) = a_1 \cdot v_i(t) + a_2 \cdot v_i^2(t) + a_3 \cdot v_i^3(t) + H.O.T; V_{out}(t) = \sum_{k=1}^{3} a_k \cdot v_{in}^k + H.O.T$. We get usually non-sinusoidal waveform

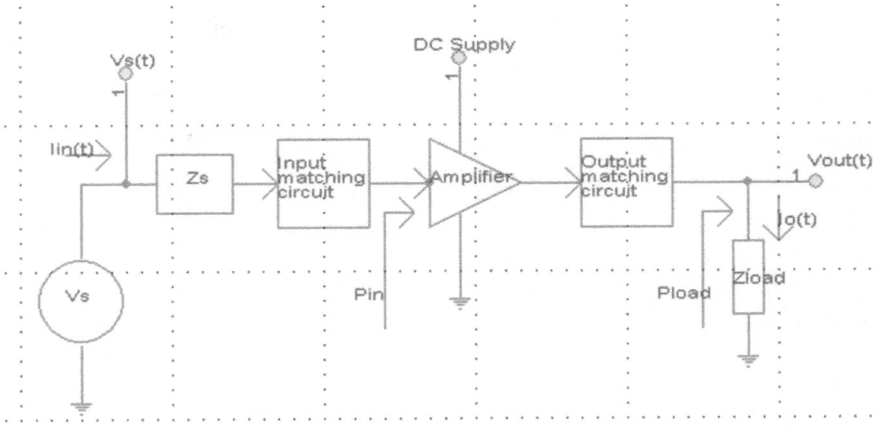

Fig. 4.1 Small Signal (SS) amplifier and matching network

amplifier output signal. Small signal, output voltage expression is linear $V_{out}(t) \simeq a_1 \cdot v_i(t)$ and the output signal is sinusoidal waveform. All amplifiers are inherently nonlinear. However when the input signal is small, the input and output relationship of the amplifier is approximately linear.

$V_{out}(t) = a_1 \cdot v_i(t) + a_2 \cdot v_i^2(t) + a_3 \cdot v_i^3(t) + H.O.T \approx a_1 \cdot v_i(t)$. When $v_i(t) \to \varepsilon$ ($<$2.6 mV) Then $V_{out}(t) \approx a_1 \cdot v_i(t)$ (Fig. 4.2).

The active component can be BJT transistor (Fig. 4.3):

The linear relationship applies also to current and power. An amplifier that fulfills these conditions: Small signal operation and linearity is called Small Signal Amplifier (SSA). If a SSA amplifier contains BJT and FET, these components can

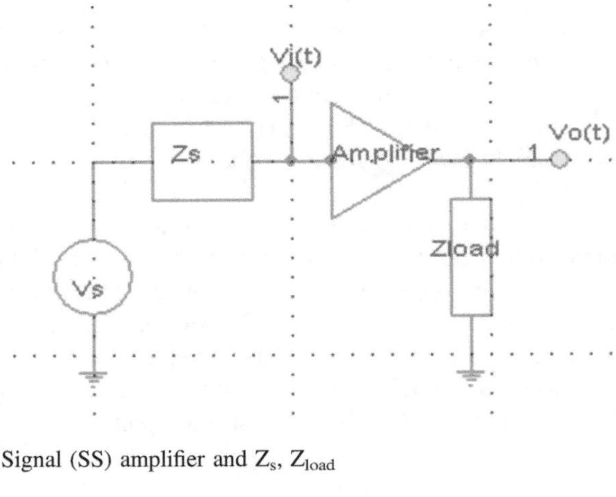

Fig. 4.2 Small Signal (SS) amplifier and Z_s, Z_{load}

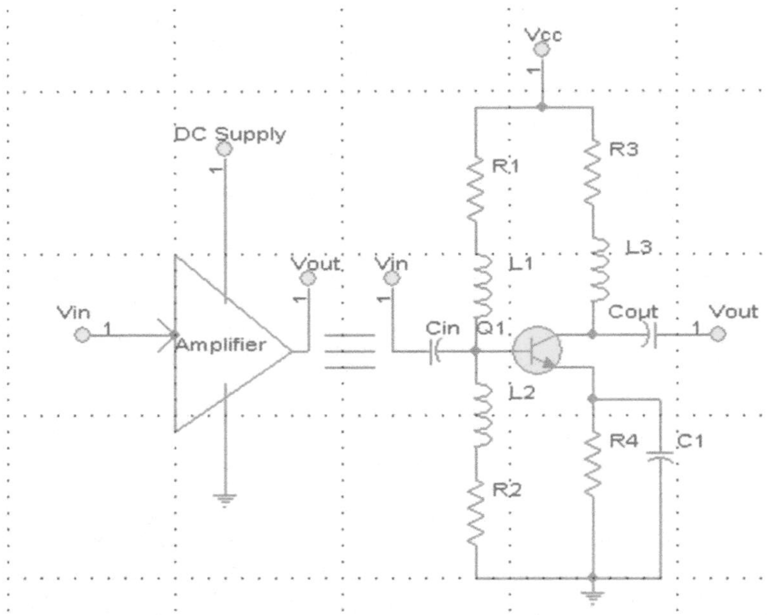

Fig. 4.3 Small Signal (SS) amplifier BJT transistor active component

be replaced by their respective small signal model, for instance the Hybrid Pi model of BJT. To determine the performance of an amplifier, the following characteristics are typically observed: power gain, bandwidth (operation frequency range), noise figure, phase response, gain compression, dynamic range, harmonic distortion, intermodulation distortion and Third Order Intercept point (TOI). The characteristics which are important to small signal amplifier are bandwidth, noise figure, and phase response. The important parameters of Large-signal amplifier (Related to linearity) are gain compression, dynamic range, harmonic distortion, intermodulation distortion, and Third Order Intercept point (TOI).

Typical RF amplifier schematics (Fig. 4.4):

Under AC and Small Signal (SS) conditions, the BJT can be replaced with linear Hybrid Pi model (Fig. 4.5).

At low frequencies it is assumed that the transistor responds instantly to charges of input voltage or current but actually because the mechanism of the transport of charge carriers from emitter to collector is one of diffusion. The transistor behavior at high frequencies is inspected by examine this diffusion mechanism in more details. The Hybrid Pi model gives a reasonable compromise between accuracy and simplicity. Using this model, a detailed analysis of a single stage CE transistor amplifier is made. The Common Emitter (CE) is the most important practical configuration.

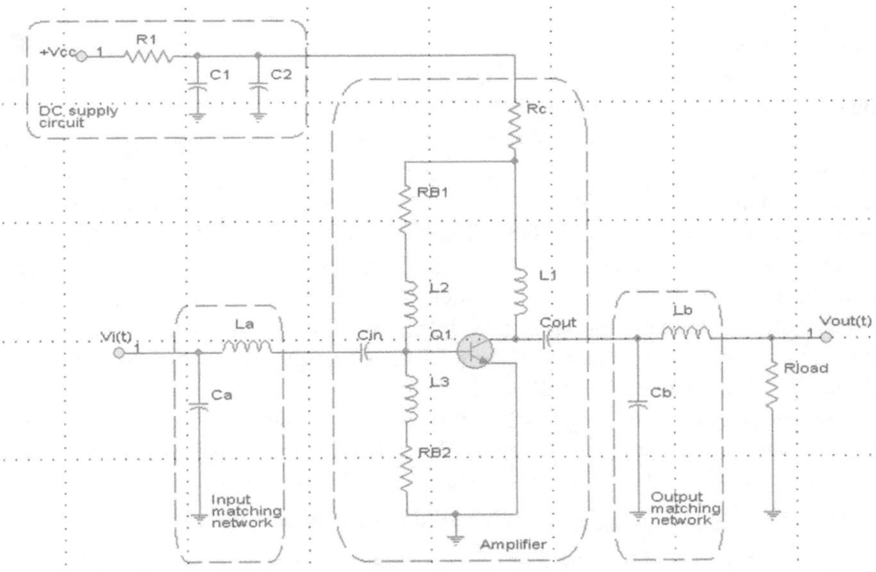

Fig. 4.4 Typical RF amplifier schematics

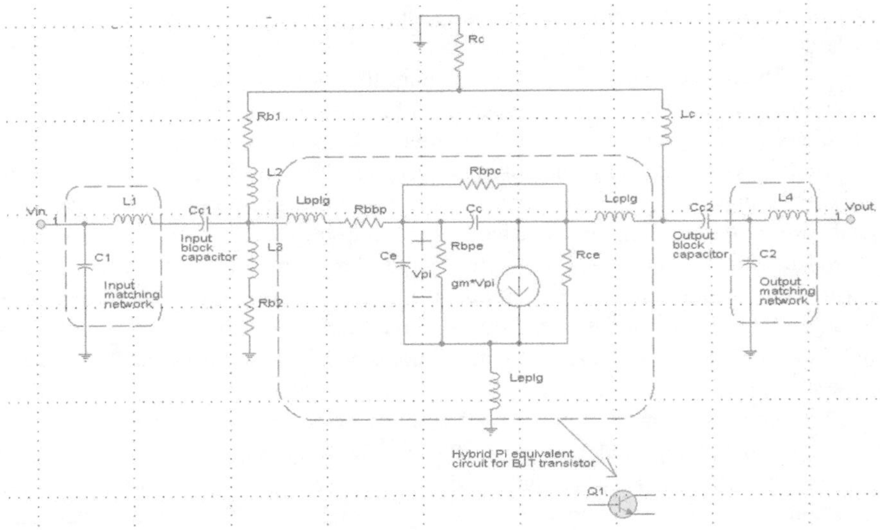

Fig. 4.5 AC Small Signal (SS) amplifier Hybrid Pi model

The CE model is valid at high frequencies. The circuit called the Hybrid Pi, Giacoletto or model. The analyses of circuits using this model are not too difficult and give results which are excellent with experiment at all frequencies for which the transistor gives reasonable amplification. All parameters (resistances and capacitances, inductances) in the model are assumed to be independent of frequency. They may vary with the quiescent operation point, but under given bias conditions are reasonably constant for small signal swings [92,93].

Remark: like a PN diode, we can break the BJT up into a large signal analysis and small signal analysis and "linearize" the non-linear behavior of the Ebers-Moll model. Small signal models are only useful for forward active mode and thus, are derived under this condition. Saturation and cutoff are used for switches which involve very large voltage/current swings from the on to off states. Small signal models are used to determine amplifier characteristics. When a diode voltage exceeds a certain value, the non-linear behavior of the diode leads to distortion of the current/voltage curves, and if the BJT transistor inputs/outputs exceed certain limits, the full Ebers-Moll model must be used.

For amplifiers functioning at RF and microwave frequencies, usually of interest is the input and output power relation. The ratio of output power over input power is called power gain (G), usually expressed in dB. Power gain:

$G = 10 \cdot \log_{10}(\frac{\text{output power}}{\text{input power}})\ dB$. There are a number of definitions for power gain. Furthermore G is a function of frequency and the input signal level. Power gain is preferred for high frequency amplifiers as the impedance encountered is usually low (due to presence of parasitic capacitance).

Power = Voltage × Current. If the amplifier is required to drive 50 Ohm load the voltage across the load may be small, although the corresponding current may be large (there is a current gain). For amplifiers functioning at lower frequency (such as IF frequency), it is the voltage gain that is of interest, since impedance encounter is usually higher (less parasitic). If the output of an IF amplifier drives the modulator circuits, which are usually digital systems, the impedance looking into the digital system is high and large voltage can be developed across it. Thus working with voltage gain is more convenient. When the input driving signal is small, the amplifier is linear and harmonic components are almost non-existent. When the input driving signal is too large the amplifier becomes nonlinear. Harmonics are introduced at the output. Harmonic generation reduces the gain of the amplifier, as some of the output power at the fundamental frequency is shifted to higher harmonics. This result is in gain compression. The amplifier also introduces noise into the output in addition to the noise from the environment.

Phase consideration is important for amplifier working with wideband signals. For signal to be amplified with no distortion, two requirements are needed from linear systems theory. First the magnitude of the power gain transfer function must be a constant with respect to frequency (f). Second the phase of the power gain transfer function must be a linear function of (f). A linear phase produces a constant time delay for all signal frequencies, and a linear phase shift produces different time delay for different frequencies. Property means that all frequency components will be amplified by similar amount and implies that all frequency components will be delayed by similar amount. Essence of Small Signal Amplifier (SSA) design: In essence, designing a small signal amplifier with transistor or Monolithic Microwave Integrated Circuit (MMIC) implies finding the suitable load and source impedance to be connected to the output and input port, and getting the required transducer power gain G_T, bandwidth and other characteristics. An amplifier is a circuit designed to enlarged electrical signals. When there is no input, there should be no output; this condition is known as stable. On the contrary, if the amplifier produces an output when there is no input, it is unstable. In fact the amplifier becomes an oscillator. Thus a stability analysis is required to determine whether an amplifier circuit is stable or not. Stability analysis is also carried out by assuming a small-signal amplifier, since the initial signal that causes oscillation is always very small. Stability of an amplifier is affected by the load and source impedance connected to its two ports. An unstable or marginally stable amplifier can be made more stable. When amplifier is unstable, or stable region is too small there are some steps which need to be done: use negative feedback to reduce amplifier gain, redesign DC biasing, finding new operating point (or Q point) that will result in more stable amplifier, add some resistive loss to the circuit to improve stability, and use a new component with better stability.

When an amplifier's output impedance matches the load impedance, maximum power is transferred to the load. When amplifier's output impedance matches the load impedance all reflections are eliminated. When an amplifier's output impedance unmatched the load impedance there are reflections and less than maximum power is transferred to the load. In order to develop maximum power the Z_{out} of the amplifier must be complex conjugate of the Z_{in} of the load. Amplifier matching allows to amplifier maximum power transfer and attenuation of harmonics to be achieved between stages. RF matching network able to facilitates impedance matching, filtering of signal and coupling between RF stages. The amplifier matching networks types are L matching network, T matching network, and PI matching network.

L matchingnetwork: LC matching topology is especially for narrowband impedance matching. L network name is due to its L shape. L network can furnish low pass filtering to decrease harmonic output. Two stages (amplifier, load) are constructed with simple low pass L network between stages. Low pass L network can matches a higher output impedance source (Z_{high}) to a lower input impedance load (Z_{low}). Low pass L network can matches a lower output impedance source (Z_{low}) to a higher input impedance load (Z_{high}). We need to analyze the stability of typical RF amplifier under RF and small signal conditions, the BJT can be replaced with linear Hybrid Pi model and additional input and output matching circuits. We define $X_s(t)$ as the RF source and Rs RF source parasitic resistance (Fig. 4.6).

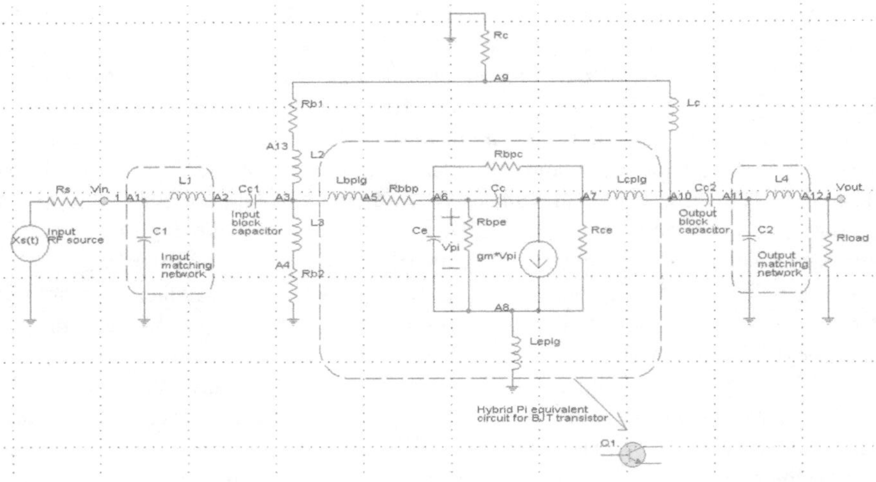

Fig. 4.6 Amplifier Hybrid Pi model and additional input and output matching circuits

$$I_{R_s} = \frac{X_s(t) - V_{A_1}}{R_s}; I_{C_1} = C_1 \cdot \frac{dV_{A_1}}{dt}; V_{A_1} - V_{A_2} = L_1 \cdot \frac{dI_{L_1}}{dt}; I_{C_{C_1}} = C_{C_1} \cdot \frac{d}{dt}(V_{A_2} - V_{A_3})$$

$$I_{L_1} = I_{C_{C_1}}; V_{A_3} - V_{A_4} = L_3 \cdot \frac{dI_{L_3}}{dt}; I_{R_{b2}} = \frac{V_{A_4}}{R_{b_2}}; I_{L_3} = I_{R_{b2}}; V_{A_3} - V_{A_{13}} = L_2 \cdot \frac{dI_{L_2}}{dt};$$

$$I_{R_{b1}} = \frac{V_{A_{13}} - V_{A_9}}{R_{b1}} V_{A_3} - V_{A_5} = L_{bplg} \cdot \frac{dI_{L_{bplg}}}{dt}; I_{R_{bbp}} = \frac{V_{A_5} - V_{A_6}}{R_{bbp}}; I_{L_{bplg}} = I_{R_{bbp}};$$

$$I_{R_c} = \frac{V_{A_9}}{R_c}; V_{pi} = V_{A_6} - V_{A_8} = V_{C_e}$$

$$I_{C_e} = C_e \cdot \frac{dV_{pi}}{dt}; I_{R_{bpe}} = \frac{V_{A_6} - V_{A_8}}{R_{bpe}}; I_{C_c} = C_c \cdot \frac{d(V_{A_6} - V_{A_7})}{dt}; I_{R_{bpc}} = \frac{V_{A_6} - V_{A_7}}{R_{bpc}};$$

$$I_{R_{ce}} = \frac{V_{A_7} - V_{A_8}}{R_{ce}}$$

$$V_{A_8} = L_{eplg} \cdot \frac{dI_{L_{eplg}}}{dt}; V_{A_7} - V_{A_{10}} = L_{cplg} \cdot \frac{dI_{L_{cplg}}}{dt}; V_{A_9} - V_{A_{10}} = L_c \cdot \frac{dL_{L_c}}{dt};$$

$$I_{C_{c2}} = C_{c2} \cdot \frac{d(V_{A_{10}} - V_{A_{11}})}{dt}$$

$$I_{C_2} = C_2 \cdot \frac{dV_{A_{11}}}{dt}; V_{A_{11}} - V_{A_{12}} = L_4 \cdot \frac{dI_{L_4}}{dt}; V_{A_{12}} = V_{out}; I_{L_4} = I_{R_{load}};$$

$$I_{R_{load}} = \frac{V_{out}}{R_{load}} = \frac{V_{A_{12}}}{R_{load}}$$

Table 4.1 Amplifier Hybrid Pi model and matching circuit Kirchhoff Current Law (KCL) for all nodes

Node number	KCL @ A_i	Node number	KCL @ A_i
A_1	$I_{R_s} = I_{L_1} + I_{C_1}$	A_8	$I_{C_e} + I_{R_{bpe}} + g_m \cdot V_{pi} + I_{R_{ce}} = I_{L_{eplg}}$
A_2	$I_{L_1} = I_{C_{c1}}$	A_9	$I_{R_{b1}} = I_{R_c} + I_{L_c}$
A_3	$I_{C_{c1}} = I_{L_2} + I_{L_{bplg}} + I_{L_3}$	A_{10}	$I_{L_c} + I_{L_{cplg}} = I_{C_{c2}}$
A_4	$I_{L_3} = I_{R_{b2}}$	A_{11}	$I_{C_{c2}} = I_{L_4} + I_{C_2}$
A_5	$I_{L_{bplg}} = I_{R_{bbp}}$	A_{12}	$I_{L_4} = I_{R_{load}}$
A_6	$I_{R_{bbp}} = I_{R_{bpc}} + I_{C_c} + I_{R_{bpe}} + I_{C_e}$	A_{13}	$I_{L_2} = I_{R_{b1}}$
A_7	$I_{C_c} + I_{R_{bpc}} = g_m \cdot V_{pi} + I_{R_{ce}} + I_{L_{cplg}}$		

We can write our circuit Kirchhoff Current Law (KCL) for all nodes (Table 4.1):

$$I_{R_s} = \frac{X_s(t) - V_{A_1}}{R_s} \Rightarrow I_{R_s} \cdot R_s = X_s(t) - V_{A_1} \Rightarrow V_{A_1} = X_s(t) - I_{R_s} \cdot R_s$$

$$I_{C_1} = C_1 \cdot \frac{dV_{A_1}}{dt} = C_1 \cdot \left(\frac{dX_s(t)}{dt} - \frac{dI_{R_s}}{dt} \cdot R_s\right); V_{A_1} - V_{A_2} = L_1 \cdot \frac{dI_{L_1}}{dt} \Rightarrow V_{A_2} = V_{A_1} - L_1 \cdot \frac{dI_{L_1}}{dt}$$

$$V_{A_2} = X_s(t) - I_{R_s} \cdot R_s - L_1 \cdot \frac{dI_{L_1}}{dt}; I_{C_{C_1}} = C_{C_1} \cdot \left(\frac{dV_{A_2}}{dt} - \frac{dV_{A_3}}{dt}\right) = C_{C_1} \cdot \left(\frac{dX_s(t)}{dt} - \frac{dI_{R_s}}{dt} \cdot R_s - L_1 \cdot \frac{d^2 I_{L_1}}{dt^2} - \frac{dV_{A_3}}{dt}\right)$$

$$I_{R_{b2}} = \frac{V_{A_4}}{R_{b_2}} \Rightarrow V_{A_4} = I_{R_{b2}} \cdot R_{b_2}; V_{A_3} - V_{A_4} = L_3 \cdot \frac{dI_{L_3}}{dt} \Rightarrow V_{A_3} = I_{R_{b2}} \cdot R_{b_2} + L_3 \cdot \frac{dI_{L_3}}{dt}; \frac{dV_{A_3}}{dt} = \frac{dI_{R_{b2}}}{dt} \cdot R_{b_2} + L_3 \cdot \frac{d^2 I_{L_3}}{dt^2}$$

System equation No. 1: $I_{C_{C_1}} = C_{C_1} \cdot \left(\frac{dX_s(t)}{dt} - \frac{dI_{R_s}}{dt} \cdot R_s - L_1 \cdot \frac{d^2 I_{L_1}}{dt^2} - \frac{dI_{R_{b2}}}{dt} \cdot R_{b_2} - L_3 \cdot \frac{d^2 I_{L_3}}{dt^2}\right)$

$$V_{A_3} - V_{A_{13}} = L_2 \cdot \frac{dI_{L_2}}{dt} \Rightarrow V_{A_{13}} = V_{A_3} - L_2 \cdot \frac{dI_{L_2}}{dt} = I_{R_{b2}} \cdot R_{b_2} + L_3 \cdot \frac{dI_{L_3}}{dt} - L_2 \cdot \frac{dI_{L_2}}{dt}$$

$$I_{R_{b1}} = \frac{V_{A_{13}} - V_{A_9}}{R_{b1}} \Rightarrow I_{R_{b1}} \cdot R_{b1} = V_{A_{13}} - V_{A_9} \Rightarrow V_{A_9} = V_{A_{13}} - I_{R_{b1}} \cdot R_{b1}$$

$$V_{A_9} = V_{A_{13}} - I_{R_{b1}} \cdot R_{b1} = I_{R_{b2}} \cdot R_{b_2} + L_3 \cdot \frac{dI_{L_3}}{dt} - L_2 \cdot \frac{dI_{L_2}}{dt} - I_{R_{b1}} \cdot R_{b1}$$

$$V_{A_3} - V_{A_5} = L_{bplg} \cdot \frac{dI_{L_{bplg}}}{dt} \Rightarrow V_{A_5} = I_{R_{b2}} \cdot R_{b_2} + L_3 \cdot \frac{dI_{L_3}}{dt} - L_{bplg} \cdot \frac{dI_{L_{bplg}}}{dt}$$

$$I_{R_{bbp}} = \frac{V_{A_5} - V_{A_6}}{R_{bbp}} \Rightarrow I_{R_{bbp}} \cdot R_{bbp} = V_{A_5} - V_{A_6} \Rightarrow V_{A_6} = I_{R_{b2}} \cdot R_{b_2} + L_3 \cdot \frac{dI_{L_3}}{dt} - L_{bplg} \cdot \frac{dI_{L_{bplg}}}{dt} - I_{R_{bbp}} \cdot R_{bbp}$$

$$I_{R_c} = \frac{V_{A_9}}{R_c} \Rightarrow V_{A_9} = I_{R_c} \cdot R_c; I_{R_{bpe}} = \frac{V_{A_6} - V_{A_8}}{R_{bpe}} \Rightarrow I_{R_{bpe}} \cdot R_{bpe} = V_{A_6} - V_{A_8} \Rightarrow V_{A_8} = V_{A_6} - I_{R_{bpe}} \cdot R_{bpe}$$

$$V_{A_8} = V_{A_6} - I_{R_{bpe}} \cdot R_{bpe} = I_{R_{b2}} \cdot R_{b_2} + L_3 \cdot \frac{dI_{L_3}}{dt} - L_{bplg} \cdot \frac{dI_{L_{bplg}}}{dt} - I_{R_{bbp}} \cdot R_{bbp} - I_{R_{bpe}} \cdot R_{bpe}$$

$$I_{C_c} = C_c \cdot \frac{d(V_{A_6} - V_{A_7})}{dt}; I_{R_{bpc}} = \frac{V_{A_6} - V_{A_7}}{R_{bpc}} \Rightarrow I_{C_c} = C_c \cdot R_{bpc} \cdot \frac{dI_{R_{bpc}}}{dt}; V_{A_7} - V_{A_8} = I_{R_{ce}} \cdot R_{ce}$$

$$V_{A_7} = V_{A_8} + I_{R_{ce}} \cdot R_{ce} = I_{R_{b2}} \cdot R_{b_2} + L_3 \cdot \frac{dI_{L_3}}{dt} - L_{bplg} \cdot \frac{dI_{L_{bplg}}}{dt} - I_{R_{bbp}} \cdot R_{bbp} - I_{R_{bpe}} \cdot R_{bpe} + I_{R_{ce}} \cdot R_{ce}$$

System equation No. 2:

$$V_{A_8} = L_{eplg} \cdot \frac{dI_{L_{eplg}}}{dt} \Rightarrow I_{R_{b2}} \cdot R_{b_2} + L_3 \cdot \frac{dI_{L_3}}{dt} - L_{bplg} \cdot \frac{dI_{L_{bplg}}}{dt} - I_{R_{bbp}} \cdot R_{bbp} - I_{R_{bpe}} \cdot R_{bpe} = L_{eplg} \cdot \frac{dI_{L_{eplg}}}{dt}$$

$$V_{A_7} - V_{A_{10}} = L_{cplg} \cdot \frac{dI_{L_{cplg}}}{dt} \Rightarrow V_{A_{10}} = V_{A_7} - L_{cplg} \cdot \frac{dI_{L_{cplg}}}{dt}$$

$$V_{A_{10}} = I_{R_{b2}} \cdot R_{b_2} + L_3 \cdot \frac{dI_{L_3}}{dt} - L_{bplg} \cdot \frac{dI_{L_{bplg}}}{dt} - I_{R_{bbp}} \cdot R_{bbp} - I_{R_{bpe}} \cdot R_{bpe} + I_{R_{ce}} \cdot R_{ce} - L_{cplg} \cdot \frac{dI_{L_{cplg}}}{dt}$$

$$V_{A_9} - V_{A_{10}} = L_c \cdot \frac{dL_{L_c}}{dt} \Rightarrow V_{A_9} = V_{A_{10}} + L_c \cdot \frac{dL_{L_c}}{dt}; \frac{dV_{A_{10}}}{dt} = \frac{I_{C_{c2}}}{C_{c2}} + \frac{I_{C_2}}{C_2}$$

$$V_{A_9} = I_{R_{b2}} \cdot R_{b_2} + L_3 \cdot \frac{dI_{L_3}}{dt} - L_{bplg} \cdot \frac{dI_{L_{bplg}}}{dt} - I_{R_{bbp}} \cdot R_{bbp} - I_{R_{bpe}} \cdot R_{bpe} + I_{R_{ce}} \cdot R_{ce} - L_{cplg} \cdot \frac{dI_{L_{cplg}}}{dt} + L_c \cdot \frac{dL_{L_c}}{dt}$$

$$I_{C_{c2}} = C_{c2} \cdot \frac{d(V_{A_{10}} - V_{A_{11}})}{dt} \Rightarrow I_{C_{c2}} = C_{c2} \cdot \left(\frac{dV_{A_{10}}}{dt} - \frac{dV_{A_{11}}}{dt} \right) \Rightarrow I_{C_{c2}} = C_{c2} \cdot \left(\frac{dV_{A_{10}}}{dt} - \frac{I_{C_2}}{C_2} \right)$$

System equation No. 3:

$$\frac{I_{C_{c2}}}{C_{c2}} + \frac{I_{C_2}}{C_2} = \frac{d}{dt} \left\{ I_{R_{b2}} \cdot R_{b_2} + L_3 \cdot \frac{dI_{L_3}}{dt} - L_{bplg} \cdot \frac{dI_{L_{bplg}}}{dt} - I_{R_{bbp}} \cdot R_{bbp} - I_{R_{bpe}} \cdot R_{bpe} + I_{R_{ce}} \cdot R_{ce} - L_{cplg} \cdot \frac{dI_{L_{cplg}}}{dt} \right\}$$

$$\frac{I_{C_{c2}}}{C_{c2}} + \frac{I_{C_2}}{C_2} = \frac{dI_{R_{b2}}}{dt} \cdot R_{b_2} + L_3 \cdot \frac{d^2 I_{L_3}}{dt^2} - L_{bplg} \cdot \frac{d^2 I_{L_{bplg}}}{dt^2} - \frac{dI_{R_{bbp}}}{dt} \cdot R_{bbp} - \frac{dI_{R_{bpe}}}{dt} \cdot R_{bpe} + \frac{dI_{R_{ce}}}{dt} \cdot R_{ce} - L_{cplg} \cdot \frac{d^2 I_{L_{cplg}}}{dt^2}$$

$$V_{A_{11}} - V_{A_{12}} = L_4 \cdot \frac{dI_{L_4}}{dt} \Rightarrow V_{A_{11}} = V_{A_{12}} + L_4 \cdot \frac{dI_{L_4}}{dt}; I_{R_{load}} = \frac{V_{A_{12}}}{R_{load}} \Rightarrow V_{A_{12}} = I_{R_{load}} \cdot R_{load}$$

System equation No. 4:

$$V_{A_{11}} = I_{R_{load}} \cdot R_{load} + L_4 \cdot \frac{dI_{L_4}}{dt}; I_{C_2} = C_2 \cdot \left(\frac{dI_{R_{load}}}{dt} \cdot R_{load} + L_4 \cdot \frac{d^2 I_{L_4}}{dt^2} \right)$$

We can summery our system equations:

$$I_{C_{C_1}} = C_{C_1} \cdot \left(\frac{dX_s(t)}{dt} - \frac{dI_{R_s}}{dt} \cdot R_s - L_1 \cdot \frac{d^2 I_{L_1}}{dt^2} - \frac{dI_{R_{b2}}}{dt} \cdot R_{b_2} - L_3 \cdot \frac{d^2 I_{L_3}}{dt^2} \right)$$

$$I_{R_{b2}} \cdot R_{b_2} + L_3 \cdot \frac{dI_{L_3}}{dt} - L_{bplg} \cdot \frac{dI_{L_{bplg}}}{dt} - I_{R_{bbp}} \cdot R_{bbp} - I_{R_{bpe}} \cdot R_{bpe} = L_{eplg} \cdot \frac{dI_{L_{eplg}}}{dt}$$

$$\frac{I_{C_{c2}}}{C_{c2}} + \frac{I_{C_2}}{C_2} = \frac{dI_{R_{b2}}}{dt} \cdot R_{b_2} + L_3 \cdot \frac{d^2 I_{L_3}}{dt^2} - L_{bplg} \cdot \frac{d^2 I_{L_{bplg}}}{dt^2} - \frac{dI_{R_{bbp}}}{dt} \cdot R_{bbp} - \frac{dI_{R_{bpe}}}{dt} \cdot R_{bpe} + \frac{dI_{R_{ce}}}{dt} \cdot R_{ce} - L_{cplg} \cdot \frac{d^2 I_{L_{cplg}}}{dt^2} I_{C_2} = C_2 \cdot \left(\frac{dI_{R_{load}}}{dt} \cdot R_{load} + L_4 \cdot \frac{d^2 I_{L_4}}{dt^2} \right)$$

Since $I_{C_{c2}} = I_{L_4} + I_{C_2} = I_{R_{load}} + I_{C_2} \Rightarrow I_{C_2} = I_{C_{c2}} - I_{R_{load}}$ then

$$I_{C_{C_1}} = C_{C_1} \cdot \left(\frac{dX_s(t)}{dt} - \frac{dI_{R_s}}{dt} \cdot R_s - L_1 \cdot \frac{d^2 I_{L_1}}{dt^2} - \frac{dI_{R_{b2}}}{dt} \cdot R_{b_2} - L_3 \cdot \frac{d^2 I_{L_3}}{dt^2} \right)$$

$$I_{R_{b2}} \cdot R_{b_2} + L_3 \cdot \frac{dI_{L_3}}{dt} - L_{bplg} \cdot \frac{dI_{L_{bplg}}}{dt} - I_{R_{bbp}} \cdot R_{bbp} - I_{R_{bpe}} \cdot R_{bpe} = L_{eplg} \cdot \frac{dI_{L_{eplg}}}{dt}$$

$$\frac{I_{C_{c2}}}{C_{c2}} + \frac{I_{C_2}}{C_2} = \frac{dI_{R_{b2}}}{dt} \cdot R_{b_2} + L_3 \cdot \frac{d^2 I_{L_3}}{dt^2} - L_{bplg} \cdot \frac{d^2 I_{L_{bplg}}}{dt^2} - \frac{dI_{R_{bbp}}}{dt} \cdot R_{bbp} - \frac{dI_{R_{bpe}}}{dt} \cdot R_{bpe} + \frac{dI_{R_{ce}}}{dt} \cdot R_{ce} - L_{cplg} \cdot \frac{d^2 I_{L_{cplg}}}{dt^2}$$

$I_{C_{c2}} - I_{R_{load}} = C_2 \cdot \left(\frac{dI_{R_{load}}}{dt} \cdot R_{load} + L_4 \cdot \frac{d^2 I_{L_4}}{dt^2} \right)$. We can restrict our two differential equations to one differential equation:

$$\frac{d}{dt} \left\{ I_{R_{b2}} \cdot R_{b_2} + L_3 \cdot \frac{dI_{L_3}}{dt} - L_{bplg} \cdot \frac{dI_{L_{bplg}}}{dt} - I_{R_{bbp}} \cdot R_{bbp} - I_{R_{bpe}} \cdot R_{bpe} \right\} = L_{eplg} \cdot \frac{d^2 I_{L_{eplg}}}{dt^2}$$

$$\frac{dI_{R_{b2}}}{dt} \cdot R_{b_2} + L_3 \cdot \frac{d^2 I_{L_3}}{dt^2} - L_{bplg} \cdot \frac{d^2 I_{L_{bplg}}}{dt^2} - \frac{dI_{R_{bbp}}}{dt} \cdot R_{bbp} - \frac{dI_{R_{bpe}}}{dt} \cdot R_{bpe} = L_{eplg} \cdot \frac{d^2 I_{L_{eplg}}}{dt^2}$$

$$\frac{I_{C_{c2}}}{C_{c2}} + \frac{I_{C_2}}{C_2} = \frac{dI_{R_{b2}}}{dt} \cdot R_{b_2} + L_3 \cdot \frac{d^2 I_{L_3}}{dt^2} - L_{bplg} \cdot \frac{d^2 I_{L_{bplg}}}{dt^2} - \frac{dI_{R_{bbp}}}{dt} \cdot R_{bbp} - \frac{dI_{R_{bpe}}}{dt} \cdot R_{bpe} + \frac{dI_{R_{ce}}}{dt} \cdot R_{ce} - L_{cplg} \cdot \frac{d^2 I_{L_{cplg}}}{dt^2}$$

One differential equation: $\frac{I_{C_{c2}}}{C_{c2}} + \frac{I_{C_2}}{C_2} - \frac{dI_{R_{ce}}}{dt} \cdot R_{ce} + L_{cplg} \cdot \frac{d^2 I_{L_{cplg}}}{dt^2} = L_{eplg} \cdot \frac{d^2 I_{L_{eplg}}}{dt^2}$
We get system three differential equations:

$$I_{L_3} = I_{R_{b2}}; I_{L_1} = I_{C_{c1}}; I_{C_{c2}} = I_{L_4} + I_{C_2} \Rightarrow I_{C_2} = I_{C_{c2}} - I_{R_{load}}$$

$$I_{L_1} = C_{C_1} \cdot \left(\frac{dX_s(t)}{dt} - \frac{dI_{R_s}}{dt} \cdot R_s - L_1 \cdot \frac{d^2 I_{L_1}}{dt^2} - \frac{dI_{R_{b2}}}{dt} \cdot R_{b_2} - L_3 \cdot \frac{d^2 I_{L_3}}{dt^2} \right)$$

$$\frac{I_{C_{c2}}}{C_{c2}} + \frac{I_{C_2}}{C_2} - \frac{dI_{R_{ce}}}{dt} \cdot R_{ce} + L_{cplg} \cdot \frac{d^2 I_{L_{cplg}}}{dt^2} = L_{eplg} \cdot \frac{d^2 I_{L_{eplg}}}{dt^2}$$

$$I_{C_{c2}} - I_{R_{load}} = C_2 \cdot \left(\frac{dI_{R_{load}}}{dt} \cdot R_{load} + L_4 \cdot \frac{d^2 I_{R_{load}}}{dt^2} \right)$$

We need to find our system equilibrium points (fixed points). It is done by setting $\frac{dI_{R_s}}{dt} = 0; \frac{dI_{R_{b2}}}{dt} = 0; \frac{dI_{L_1}}{dt} = 0 \Rightarrow \frac{d^2 I_{L_1}}{dt^2} = 0; \frac{dI_{L_3}}{dt} = 0 \Rightarrow \frac{d^2 I_{L_3}}{dt^2} = 0; \frac{dI_{R_{ce}}}{dt} = 0$

$$\frac{dI_{L_{cplg}}}{dt} = 0 \Rightarrow \frac{d^2 I_{L_{cplg}}}{dt^2} = 0; \frac{dI_{L_{eplg}}}{dt} = 0 \Rightarrow \frac{d^2 I_{L_{eplg}}}{dt^2} = 0; \frac{dI_{R_{load}}}{dt} = 0 \Rightarrow \frac{d^2 I_{R_{load}}}{dt^2} = 0$$

We get the following system fixed points:

$$I^*_{L_1} = C_{C_1} \cdot \frac{dX_s(t)}{dt}; I^*_{C_{c2}} \cdot (\frac{1}{C_{c2}} + \frac{1}{C_2}) - \frac{I^*_{R_{load}}}{C_2} = 0; I^*_{C_{c2}} - I^*_{R_{load}} = 0$$

$$I^*_{C_{c2}} - I^*_{R_{load}} = 0 \Rightarrow I^*_{C_2} = 0; I^*_{L_1} = C_{C_1} \cdot \frac{dX_s(t)}{dt} \Rightarrow I^*_{C_{c1}} = C_{C_1} \cdot \frac{dX_s(t)}{dt}$$

$$I^*_{L_1} = C_{C_1} \cdot \frac{dX_s(t)}{dt} \Rightarrow C_{C_1} \cdot \frac{dX_s(t)}{dt} = I^*_{R_s} - I^*_{C_1}; C_{C_1} \cdot \frac{dX_s(t)}{dt} = I^*_{L_2} + I^*_{L_{bplg}} + I^*_{L_3}$$

We consider input RF source $X_s(t) = A_0 + \xi(t)$; $|\xi(t)| < 1 \& A_0 \gg |\xi(t)|$Then $X_s(t)|_{A_0 \gg |\xi(t)|} = A_0 + \xi(t) \approx A_0; \frac{dX_s(t)|_{A_0 \gg |\xi(t)|}}{dt} = \frac{d\xi(t)}{dt} \to \varepsilon$.

Based on the above assumption, we get the following system fixed points:

$$I^*_{L_1} = 0; I^*_{C_{c2}} \cdot \left(\frac{1}{C_{c2}} + \frac{1}{C_2}\right) - \frac{I^*_{R_{load}}}{C_2} = 0; I^*_{C_{c2}} - I^*_{R_{load}} = 0; I^*_{R_s} - I^*_{C_1} \Rightarrow I^*_{R_s} = I^*_{C_1}$$

$$I^*_{C_{c2}} - I^*_{R_{load}} = 0 \Rightarrow I^*_{C_2} = 0; I^*_{L_1} = 0 \Rightarrow I^*_{C_{c1}} = 0; I^*_{L_1} = 0 \Rightarrow I^*_{C_1} = I^*_{R_s}; I^*_{L_2} + I^*_{L_{bplg}} + I^*_{L_3} = 0$$

$$E^*(I^*_{L_1}, I^*_{C_{c2}}, I^*_{R_{load}}, I^*_{R_{ce}}, I^*_{C_2}, I^*_{C_1}, I^*_{R_s}, I^*_{L_2}, I^*_{L_{bplg}}, I^*_{L_3}) = (0, I^*_{R_{load}}, I^*_{R_{load}}, I^*_{R_{ce}}, 0, I^*_{R_s}, I^*_{R_s}, -(I^*_{L_{bplg}} + I^*_{L_3}), I^*_{L_{bplg}}, I^*_{L_3})$$

Stability analysis: The standard local stability analysis about any one of the equilibrium points of the Small Signal (SS) amplifier with matching network consists in adding to coordinate $[I_{L_1}, I_{C_{c2}}, I_{R_{load}}, I_{R_{ce}}, I_{C_2}, I_{C_1}, I_{R_s}, I_{L_2}, I_{L_{bplg}}, I_{L_3}]$ arbitrarily small increments of exponentially form $[i_{L_1}, i_{C_{c2}}, i_{R_{load}}, i_{R_{ce}}, i_{C_2}, i_{C_1}, i_{R_s}, i_{L_2}, i_{L_{bplg}}, i_{L_3}] \cdot e^{\lambda \cdot t}$ and retaining the first order terms in $I_{L_1}, I_{C_{c2}}, I_{R_{load}}, I_{R_{ce}}, I_{C_2}, I_{C_1}, I_{R_s}, I_{L_2}, I_{L_{bplg}}, I_{L_3}$. The system of homogenous equations leads to a polynomial characteristic equation in the eigenvalues. The polynomial characteristic equations accept by set of the below circuit variables, circuit variables derivative and circuit variables second order derivative with respect to time into Small Signal (SS) amplifier with matching network equivalent circuit [2–4]. Our Small Signal (SS) amplifier with matching network equivalent circuit fixed values with arbitrarily small increments of exponential form $[i_{L_1}, i_{C_{c2}}, i_{R_{load}}, i_{R_{ce}}, i_{C_2}, i_{C_1}, i_{R_s}, i_{L_2}, i_{L_{bplg}}, i_{L_3}] \cdot e^{\lambda \cdot t}$ are: j = 0 (first fixed point), j = 1 (second fixed point), j = 2 (third fixed point), etc.,

We define new variables: $Y_1 = \frac{dI_{L_1}}{dt} \Rightarrow \frac{dY_1}{dt} = \frac{d^2 I_{L_1}}{dt^2}; Y_2 = \frac{dI_{L_3}}{dt} \Rightarrow \frac{dY_2}{dt} = \frac{d^2 I_{L_3}}{dt^2}$

$$Y_3 = \frac{dI_{R_{load}}}{dt} \Rightarrow \frac{dY_3}{dt} = \frac{d^2 I_{R_{load}}}{dt^2}; I_{C_{c2}} - I_{R_{load}} = C_2 \cdot \left(Y_3 \cdot R_{load} + L_4 \cdot \frac{dY_3}{dt} \right)$$

$$I_{L_1} = C_{C_1} \cdot \left(\left[\frac{dX_s(t)}{dt} \to \varepsilon \right] - \frac{dI_{R_s}}{dt} \cdot R_s - L_1 \cdot \frac{dY_1}{dt} - \frac{dI_{R_{b2}}}{dt} \cdot R_{b_2} - L_3 \cdot \frac{dY_2}{dt} \right)$$

$$I_{C_{c2}} \cdot \left(\frac{1}{C_{c2}} + \frac{1}{C_2}\right) - \frac{I_{R_{load}}}{C_2} - \frac{dI_{R_{ce}}}{dt} \cdot R_{ce} + L_{cplg} \cdot \frac{d^2 I_{L_{cplg}}}{dt^2} = L_{eplg} \cdot \frac{d^2 I_{L_{eplg}}}{dt^2}$$

$$Y_4 = \frac{dI_{L_{cplg}}}{dt} \Rightarrow \frac{dY_4}{dt} = \frac{d^2 I_{L_{cplg}}}{dt^2}; Y_5 = \frac{dI_{L_{eplg}}}{dt} \Rightarrow \frac{dY_5}{dt} = \frac{d^2 I_{L_{eplg}}}{dt^2}$$

$$I_{C_{c2}} \cdot \left(\frac{1}{C_{c2}} + \frac{1}{C_2}\right) - \frac{I_{R_{load}}}{C_2} - \frac{dI_{R_{ce}}}{dt} \cdot R_{ce} + L_{cplg} \cdot \frac{dY_4}{dt} = L_{eplg} \cdot \frac{dY_5}{dt}$$

$$Y_1(t) = Y_1^{(j)} + y_1 \cdot e^{\lambda \cdot t}; Y_2(t) = Y_2^{(j)} + y_2 \cdot e^{\lambda \cdot t}; Y_3(t) = Y_3^{(j)} + y_3 \cdot e^{\lambda \cdot t}; I_{L_1}(t) = I_{L_1}^{(j)} + i_{L_1} \cdot e^{\lambda \cdot t}$$

$$Y_4(t) = Y_4^{(j)} + y_4 \cdot e^{\lambda \cdot t}; Y_5(t) = Y_5^{(j)} + y_5 \cdot e^{\lambda \cdot t}; \frac{dY_4(t)}{dt} = y_4 \cdot \lambda \cdot e^{\lambda \cdot t}; \frac{dY_5(t)}{dt} = y_5 \cdot \lambda \cdot e^{\lambda \cdot t}$$

$$I_{L_3}(t) = I_{L_3}^{(j)} + i_{L_3} \cdot e^{\lambda \cdot t}; I_{R_{load}}(t) = I_{R_{load}}^{(j)} + i_{R_{load}} \cdot e^{\lambda \cdot t}; \frac{dI_{L_1}(t)}{dt} = i_{L_1} \cdot \lambda \cdot e^{\lambda \cdot t}; \frac{dI_{L_3}(t)}{dt} = i_{L_3} \cdot \lambda \cdot e^{\lambda \cdot t}$$

$$\frac{dI_{R_{load}}(t)}{dt} = i_{R_{load}} \cdot \lambda \cdot e^{\lambda \cdot t}; Y_1 = \frac{dI_{L_1}}{dt} \Rightarrow Y_1^{(j)} + y_1 \cdot e^{\lambda \cdot t} = i_{L_1} \cdot \lambda \cdot e^{\lambda \cdot t}; I_{C_{c2}}(t) = I_{C_{c2}}^{(j)} + i_{C_{c2}} \cdot e^{\lambda \cdot t}$$

At fixed point $\frac{dI_{L_1}}{dt} = 0 \Rightarrow Y_1^{(j)} = 0; \frac{dI_{L_3}}{dt} = 0 \Rightarrow Y_2^{(j)} = 0; \frac{dI_{R_{load}}}{dt} = 0 \Rightarrow Y_3^{(j)} = 0$

$$\frac{dI_{L_{cplg}}}{dt} = 0 \Rightarrow Y_4^{(j)} = 0; \frac{dI_{L_{eplg}}}{dt} = 0 \Rightarrow Y_5^{(j)} = 0; \frac{dI_{L_{cplg}}}{dt}\Big|_{Y_4^{(j)}=0} \Rightarrow -i_{L_{cplg}} \cdot \lambda + y_4 = 0$$

$$\frac{dI_{L_{eplg}}}{dt}\Big|_{Y_5^{(j)}=0} \Rightarrow i_{L_{eplg}} \cdot \lambda \cdot e^{\lambda \cdot t} = y_5 \cdot e^{\lambda \cdot t}; -i_{L_{eplg}} \cdot \lambda + y_5 = 0$$

$$I_{R_s}(t) = I_{R_s}^{(j)} + i_{R_s} \cdot e^{\lambda \cdot t}; I_{R_{b2}}(t) = I_{R_{b2}}^{(j)} + i_{R_{b2}} \cdot e^{\lambda \cdot t}; I_{R_{Ce}}(t) = I_{R_{Ce}}^{(j)} + i_{R_{Ce}} \cdot e^{\lambda \cdot t}$$

$$I_{L_{cplg}}(t) = I_{L_{cplg}}^{(j)} + i_{L_{cplg}} \cdot e^{\lambda \cdot t}; I_{L_{eplg}}(t) = I_{L_{eplg}}^{(j)} + i_{L_{eplg}} \cdot e^{\lambda \cdot t}; Y_1 = \frac{dI_{L_1}}{dt}\Big|_{Y_1^{(j)}=0} \Rightarrow y_1 \cdot e^{\lambda \cdot t} = i_{L_1} \cdot \lambda \cdot e^{\lambda \cdot t}$$

$$-i_{L_1} \cdot \lambda + y_1 = 0; Y_2 = \frac{dI_{L_3}}{dt}\Big|_{Y_2^{(j)}=0} \Rightarrow y_2 \cdot e^{\lambda \cdot t} = i_{L_3} \cdot \lambda \cdot e^{\lambda \cdot t}; -i_{L_3} \cdot \lambda + y_2 = 0$$

$$Y_3 = \frac{dI_{R_{load}}}{dt}|_{Y_3^{(j)}=0} \Rightarrow y_3 \cdot e^{\lambda \cdot t} = i_{R_{load}} \cdot \lambda \cdot e^{\lambda \cdot t}; -i_{R_{load}} \cdot \lambda + y_3 = 0$$

$$I_{C_{c2}} - I_{R_{load}} = C_2 \cdot \left(Y_3 \cdot R_{load} + L_4 \cdot \frac{dY_3}{dt} \right)$$

$$I_{C_{c2}}^{(j)} + i_{C_{c2}} \cdot e^{\lambda \cdot t} - I_{R_{load}}^{(j)} - i_{R_{load}} \cdot e^{\lambda \cdot t} = C_2 \cdot (Y_3^{(j)} + y_3 \cdot e^{\lambda \cdot t}) \cdot R_{load} + L_4 \cdot y_3 \cdot \lambda \cdot e^{\lambda \cdot t})$$

@ fixed point $\frac{dY_3}{dt} = 0 \Rightarrow I_{C_{c2}}^{(j)} - I_{R_{load}}^{(j)} - C_2 \cdot R_{load} \cdot Y_3^{(j)} = 0$

$$I_{C_{c2}}^{(j)} - I_{R_{load}}^{(j)} - C_2 \cdot Y_3^{(j)} \cdot R_{load} + i_{C_{c2}} \cdot e^{\lambda \cdot t} - i_{R_{load}} \cdot e^{\lambda \cdot t} = y_3 \cdot C_2 \cdot R_{load} \cdot e^{\lambda \cdot t} + C_2 \cdot L_4 \cdot y_3 \cdot \lambda \cdot e^{\lambda \cdot t}$$

$$i_{C_{c2}} \cdot e^{\lambda \cdot t} - i_{R_{load}} \cdot e^{\lambda \cdot t} = y_3 \cdot C_2 \cdot R_{load} \cdot e^{\lambda \cdot t} + C_2 \cdot L_4 \cdot y_3 \cdot \lambda \cdot e^{\lambda \cdot t}$$

$$y_3 \cdot C_2 \cdot R_{load} + C_2 \cdot L_4 \cdot y_3 \cdot \lambda = i_{C_{c2}} - i_{R_{load}} \Rightarrow -y_3 \cdot C_2 \cdot L_4 \cdot \lambda - y_3 \cdot C_2 \cdot R_{load} + i_{C_{c2}} - i_{R_{load}} = 0$$

$$\frac{dX_s(t)}{dt} \to \varepsilon; I_{L_1} = C_{C_1} \cdot - \left(\frac{dI_{R_s}}{dt} \cdot R_s - L_1 \cdot \frac{dY_1}{dt} - \frac{dI_{R_{b2}}}{dt} \cdot R_{b_2} - L_3 \cdot \frac{dY_2}{dt} \right)$$

$$I_{L_1}^{(j)} + i_{L_1} \cdot e^{\lambda \cdot t} = C_{C_1} \cdot (-i_{R_s} \cdot \lambda \cdot e^{\lambda \cdot t} \cdot R_s - L_1 \cdot y_1 \cdot \lambda \cdot e^{\lambda \cdot t} - i_{R_{b2}} \cdot \lambda \cdot e^{\lambda \cdot t} \cdot R_{b_2} - L_3 \cdot y_2 \cdot \lambda \cdot e^{\lambda \cdot t})$$

@ fixed point $I_{L_1}^{(j)} = 0; i_{L_1} = C_{C_1} \cdot (-i_{R_s} \cdot R_s - L_1 \cdot y_1 - i_{R_{b2}} \cdot R_{b_2} - L_3 \cdot y_2) \cdot \lambda$

$$I_{C_{c2}} \cdot \left(\frac{1}{C_{c2}} + \frac{1}{C_2} \right) - \frac{I_{R_{load}}}{C_2} - \frac{dI_{R_{ce}}}{dt} \cdot R_{ce} + L_{cplg} \cdot \frac{d^2 I_{L_{cplg}}}{dt^2} = L_{eplg} \cdot \frac{d^2 I_{L_{eplg}}}{dt^2}$$

$$I_{C_{c2}} \cdot \left(\frac{1}{C_{c2}} + \frac{1}{C_2} \right) - \frac{I_{R_{load}}}{C_2} - \frac{dI_{R_{ce}}}{dt} \cdot R_{ce} + L_{cplg} \cdot \frac{dY_4}{dt} = L_{eplg} \cdot \frac{dY_5}{dt}$$

$$\left(I_{C_{c2}}^{(j)} + i_{C_{c2}} \cdot e^{\lambda \cdot t} \right) \cdot \left(\frac{1}{C_{c2}} + \frac{1}{C_2} \right) - \frac{\left(I_{R_{load}}^{(j)} + i_{R_{load}} \cdot e^{\lambda \cdot t} \right)}{C_2} - i_{R_{Ce}} \cdot \lambda \cdot e^{\lambda \cdot t} \cdot R_{ce} + L_{cplg} \cdot y_4 \cdot \lambda \cdot e^{\lambda \cdot t} = L_{eplg} \cdot y_5 \cdot \lambda \cdot e^{\lambda \cdot t}$$

@ fixed point $I_{C_{c2}}^{(j)} \cdot (\frac{1}{C_{c2}} + \frac{1}{C_2}) - \frac{I_{R_{load}}^{(j)}}{C_2} = 0$

$$I_{C_{c2}}^{(j)} \cdot (\frac{1}{C_{c2}} + \frac{1}{C_2}) + i_{C_{c2}} \cdot (\frac{1}{C_{c2}} + \frac{1}{C_2}) \cdot e^{\lambda \cdot t} - \frac{I_{R_{load}}^{(j)}}{C_2} - \frac{i_{R_{load}}}{C_2} \cdot e^{\lambda \cdot t} - i_{R_{Ce}} \cdot \lambda \cdot e^{\lambda \cdot t} \cdot R_{ce}$$
$$+ L_{cplg} \cdot y_4 \cdot \lambda \cdot e^{\lambda \cdot t} = L_{eplg} \cdot y_5 \cdot \lambda \cdot e^{\lambda \cdot t}$$

$$\{I_{C_{c2}}^{(j)} \cdot (\frac{1}{C_{c2}} + \frac{1}{C_2}) - \frac{I_{R_{load}}^{(j)}}{C_2}\} + i_{C_{c2}} \cdot (\frac{1}{C_{c2}} + \frac{1}{C_2}) \cdot e^{\lambda \cdot t} - \frac{i_{R_{load}}}{C_2} \cdot e^{\lambda \cdot t} - i_{R_{Ce}} \cdot \lambda \cdot e^{\lambda \cdot t} \cdot R_{ce}$$
$$+ L_{cplg} \cdot y_4 \cdot \lambda \cdot e^{\lambda \cdot t} = L_{eplg} \cdot y_5 \cdot \lambda \cdot e^{\lambda \cdot t}$$
$$i_{C_{c2}} \cdot \left(\frac{1}{C_{c2}} + \frac{1}{C_2}\right) \cdot e^{\lambda \cdot t} - \frac{i_{R_{load}}}{C_2} \cdot e^{\lambda \cdot t} - i_{R_{Ce}} \cdot \lambda \cdot e^{\lambda \cdot t} \cdot R_{ce} + L_{cplg} \cdot y_4 \cdot \lambda \cdot e^{\lambda \cdot t}$$
$$= L_{eplg} \cdot y_5 \cdot \lambda \cdot e^{\lambda \cdot t}$$

$$i_{C_{c2}} \cdot (\frac{1}{C_{c2}} + \frac{1}{C_2}) - \frac{i_{R_{load}}}{C_2} - i_{R_{Ce}} \cdot \lambda \cdot R_{ce} + L_{cplg} \cdot y_4 \cdot \lambda = L_{eplg} \cdot y_5 \cdot \lambda$$

Small Signal (SS) amplifier with matching network system matrix $\left(i_{L_{cplg}}, i_{L_{eplg}}, i_{L_1}, i_{L_3}, i_{R_{load}}, i_{C_{c2}}, i_{R_s}, i_{R_{b2}}, i_{R_{Ce}}, y_1, y_2, y_3, y_4, y_5\right)$ can be constructed from the below list of equations:

$$- i_{L_{cplg}} \cdot \lambda + y_4 = 0; -i_{L_{eplg}} \cdot \lambda + y_5 = 0; -i_{L_1} \cdot \lambda + y_1 = 0; -i_{L_3} \cdot \lambda + y_2 = 0;$$
$$- i_{R_{load}} \cdot \lambda + y_3 = 0$$

$$-y_3 \cdot C_2 \cdot L_4 \cdot \lambda - y_3 \cdot C_2 \cdot R_{load} + i_{C_{c2}} - i_{R_{load}} = 0$$

$$i_{L_1} = C_{C_1} \cdot (-i_{R_s} \cdot R_s - L_1 \cdot y_1 - i_{R_{b2}} \cdot R_{b_2} - L_3 \cdot y_2) \cdot \lambda$$

$$i_{C_{c2}} \cdot \left(\frac{1}{C_{c2}} + \frac{1}{C_2}\right) - \frac{i_{R_{load}}}{C_2} - i_{R_{Ce}} \cdot \lambda \cdot R_{ce} + L_{cplg} \cdot y_4 \cdot \lambda = L_{eplg} \cdot y_5 \cdot \lambda$$

Assumption: We consider for simplicity that arbitrarily small increments elements $i_{R_s}, y_1, i_{R_{b2}}, y_2$ are proximally the same $i_{R_s} \approx y_1 \approx i_{R_{b2}} \approx y_2 \rightarrow z_1$ and we represent them as small increment element z_1. Additionally, arbitrarily small increments elements $i_{R_{Ce}}, y_4, y_5$ are proximally the same $i_{R_{Ce}} \approx y_4 \approx y_5 \rightarrow z_2$ and we represent them as small increment element z_2.

$$- i_{L_{cplg}} \cdot \lambda + z_2 = 0; -i_{L_{eplg}} \cdot \lambda + z_2 = 0; -i_{L_1} \cdot \lambda + z_1 = 0; -i_{L_3} \cdot \lambda + z_1 = 0;$$
$$- i_{R_{load}} \cdot \lambda + y_3 = 0$$

$$-y_3 \cdot \lambda - y_3 \cdot \frac{R_{load}}{L_4} + i_{C_{c2}} \cdot \frac{1}{C_2 \cdot L_4} - i_{R_{load}} \cdot \frac{1}{C_2 \cdot L_4} = 0; -z_1 \cdot \lambda - i_{L_1} \cdot \frac{1}{C_{C_1} \cdot (R_s + L_1 + R_{b_2} + L_3)} = 0$$

$$-z_2 \cdot \lambda + i_{C_{c2}} \cdot \frac{\left(\frac{1}{C_{c2}} + \frac{1}{C_2}\right)}{(R_{ce} - L_{cplg} + L_{eplg})} - \frac{i_{R_{load}}}{C_2 \cdot (R_{ce} - L_{cplg} + L_{eplg})} = 0$$

Small Signal (SS) amplifier with matching network system matrixes:

$$\begin{pmatrix} \iota_{11} & \cdots & \iota_{18} \\ \vdots & \ddots & \vdots \\ \iota_{81} & \cdots & \iota_{88} \end{pmatrix} \cdot \begin{pmatrix} i_{L_{cplg}} \\ i_{L_{eplg}} \\ i_{L_1} \\ i_{L_3} \\ i_{R_{load}} \\ y_3 \\ z_1 \\ z_2 \end{pmatrix} + \begin{pmatrix} 0 \\ 0 \\ 0 \\ 0 \\ 0 \\ \frac{1}{C_2 \cdot L_4} \\ 0 \\ \frac{\left(\frac{1}{C_{c2}} + \frac{1}{C_2}\right)}{(R_{ce} - L_{cplg} + L_{eplg})} \end{pmatrix} \cdot i_{C_{c2}} = 0$$

$$\iota_{11} = -\lambda; \iota_{12} = \ldots = \iota_{17} = 0; \iota_{18} = 1; \iota_{21} = 0; \iota_{22} = -\lambda; \iota_{23} = \ldots = \iota_{27} = 0; \iota_{28} = 1$$

$$\iota_{31} = \iota_{32} = 0; \iota_{33} = -\lambda; \iota_{34} = \iota_{35} = \iota_{36} = 0; \iota_{37} = 1; \iota_{38} = 0; \iota_{41} = \iota_{42} = \iota_{43} = 0; \iota_{44} = -\lambda$$

$$\iota_{45} = \iota_{46} = 0; \iota_{47} = 1; \iota_{48} = 0; \iota_{51} = \ldots = \iota_{54} = 0; \iota_{55} = -\lambda; \iota_{56} = 1; \iota_{57} = \iota_{58} = 0$$

$$\iota_{61} = \ldots = \iota_{64} = 0; \iota_{65} = -\frac{1}{C_2 \cdot L_4}; \iota_{66} = -\lambda - \frac{R_{load}}{L_4}; \iota_{67} = \iota_{68} = 0$$

$$\iota_{71} = \iota_{72} = 0; \iota_{73} = -\frac{1}{C_{C_1} \cdot (R_s + L_1 + R_{b_2} + L_3)}; \iota_{74} = \iota_{75} = \iota_{76} = 0; \iota_{77} = -\lambda; \iota_{78} = 0$$

$$\iota_{81} = \ldots = \iota_{84} = 0; \iota_{85} = -\frac{1}{C_2 \cdot (R_{ce} - L_{cplg} + L_{eplg})}; \iota_{86} = \iota_{87} = 0; \iota_{88} = -\lambda$$

$$\text{We consider}\begin{pmatrix} 0 \\ 0 \\ 0 \\ 0 \\ 0 \\ \frac{1}{C_2 \cdot L_4} \\ 0 \\ \frac{\left(\frac{1}{C_{c2}}+\frac{1}{C_2}\right)}{(R_{ce}-L_{cplg}+L_{eplg})} \end{pmatrix} \to \varepsilon \begin{pmatrix} \imath_{11} & \cdots & \imath_{18} \\ \vdots & \ddots & \vdots \\ \imath_{81} & \cdots & \imath_{88} \end{pmatrix} \cdot \begin{pmatrix} i_{L_{cplg}} \\ i_{L_{eplg}} \\ i_{L_1} \\ i_{L_3} \\ i_{R_{load}} \\ y_3 \\ z_1 \\ z_2 \end{pmatrix} \approx 0$$

$$A - \lambda \cdot I = \begin{pmatrix} \imath_{11} & \cdots & \imath_{18} \\ \vdots & \ddots & \vdots \\ \imath_{81} & \cdots & \imath_{88} \end{pmatrix}; \det(A - \lambda \cdot I) = 0 \Rightarrow \det \begin{pmatrix} \imath_{11} & \cdots & \imath_{18} \\ \vdots & \ddots & \vdots \\ \imath_{81} & \cdots & \imath_{88} \end{pmatrix} = 0$$

To effectively apply the stability criterion of Lipunov to our system, we require a criterion for when the equation $\det(A - \lambda \cdot I) = 0$ has a zero in the left half plane, without calculating the eigenvalues explicitly. We use criterion of Routh-Hurwitz [2–4].

4.2 Small Signal (SS) Amplifiers PI & T's Matching Network and Transformation

In (4.1) we discuss the stability and fixed points (equilibrium points) analysis of Small Signal (SS) amplifiers with input and output L-matching network. Another way is to use Small Signal (SS) amplifier with input and output PI & T's matching networks. The PI or T matching networks constructed from three elements impedance matching and are used in many narrow band applications. The narrow band is due to the higher loaded Q over what the L network possesses. PI and T networks also permit any Q to be selected and always PI and T Q's (loaded quality factor) is bigger or equal to the L network Q's. The Q is desired for a particular applications and is calculated with the formula $\left(Q = \frac{f_c}{f_2 - f_1}\right)$ when utilizing with high Q inductor. Q is the loaded quality factor of the matching circuit. f_c is the center frequency of the circuit. f_2 is the upper frequency and need to pass with little loss. f_1 is the lower frequency and need to pass with little loss. We consider PI network as a two L networks which attached back to back. Any PI network can be transformed to an equivalent T network. This is also known as the Wye-Delta transformation, which is the terminology used in power distribution and electrical engineering. The PI matching network is equivalent to the Delta and the T matching network is equivalent to the Wye (or star) form [25, 26, 33]. The typical RF amplifier schematic with input and output matching circuit is as follow (Fig. 4.7):

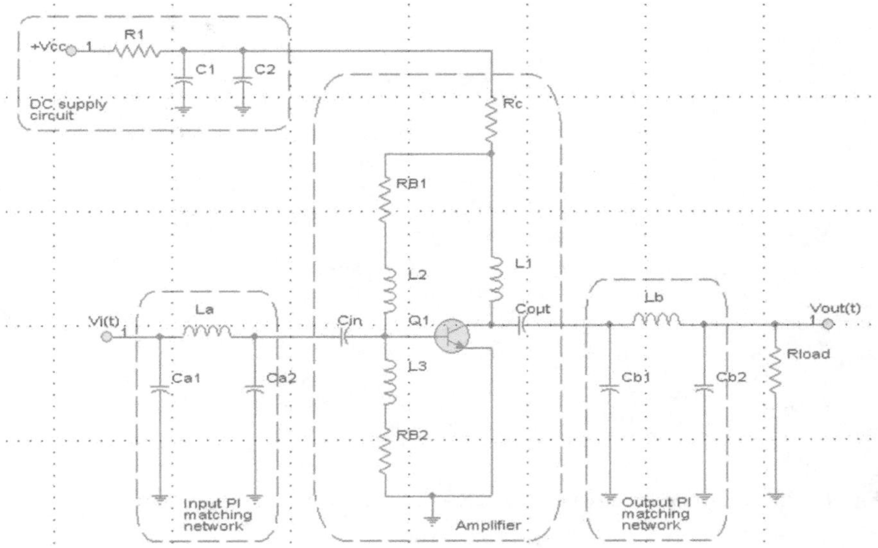

Fig. 4.7 Small Signal (SS) amplifier with PI matching networks

Under AC and Small Signal (SS) conditions, the BJT can be replaced with linear Hybrid PI model (Fig. 4.8):

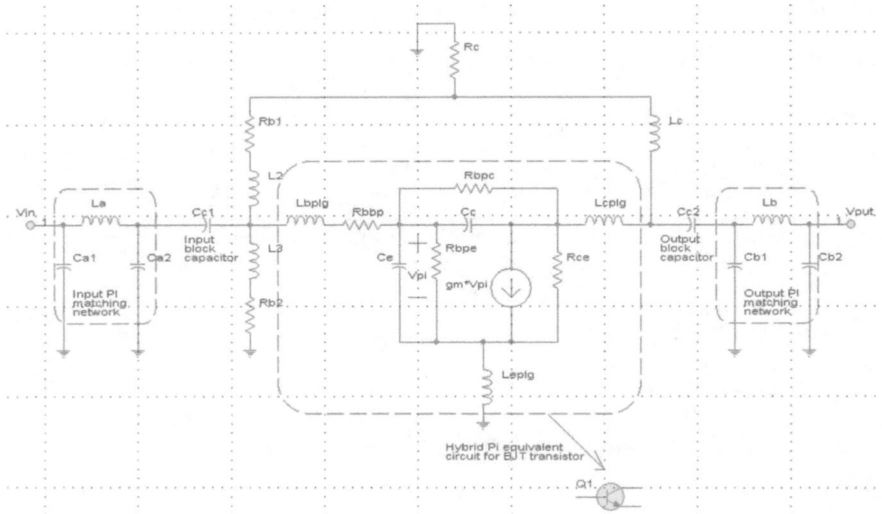

Fig. 4.8 Small Signal (SS) amplifier Hybrid PI model with PI matching networks

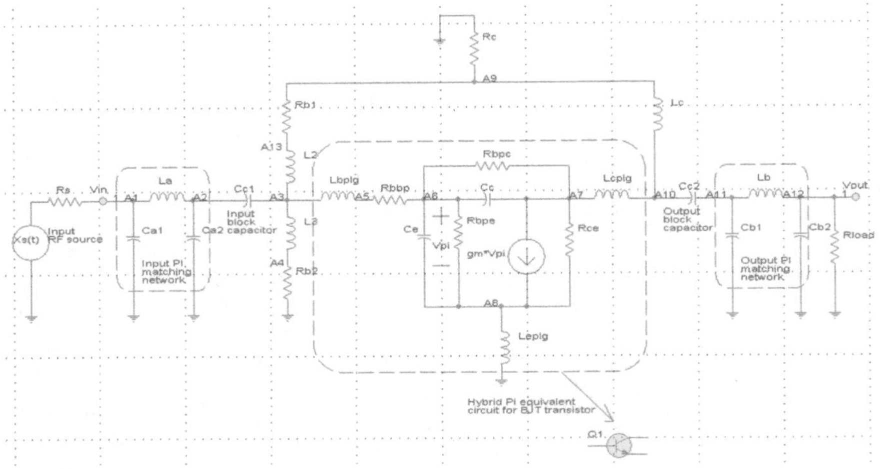

Fig. 4.9 Linear Hybrid PI model and additional input and output matching circuits

We need to analyze the stability of typical RF amplifier under RF and small signal conditions, the BJT can be replaced with linear Hybrid Pi model and additional input and output matching circuits. We define $X_s(t)$ as the RF source and Rs RF source parasitic resistance (Fig. 4.9).

$$I_{R_s} = \frac{X_s(t) - V_{A_1}}{R_s}; I_{C_{a1}} = C_{a1} \cdot \frac{dV_{A_1}}{dt}; I_{C_{a2}} = C_{a2} \cdot \frac{dV_{A_2}}{dt}; V_{L_a} = V_{A_1} - V_{A_2} = L_a \cdot \frac{dI_{L_a}}{dt}$$

$$I_{C_{C_1}} = C_{C_1} \cdot \frac{d}{dt}(V_{A_2} - V_{A_3}); V_{A_1} = V_{C_{a1}}; V_{A_2} = V_{C_{a2}}; I_{L_a} = I_{C_{a2}} + I_{C_{C_1}}; I_{R_s} = I_{L_a} + I_{C_{a1}}$$

$$V_{A_3} - V_{A_4} = L_3 \cdot \frac{dI_{L_3}}{dt}; I_{R_{b2}} = \frac{V_{A_4}}{R_{b_2}}; I_{L_3} = I_{R_{b2}}; V_{A_3} - V_{A_{13}} = L_2 \cdot \frac{dI_{L_2}}{dt}; I_{R_{b1}} = \frac{V_{A_{13}} - V_{A_9}}{R_{b1}}$$

$$V_{A_3} - V_{A_5} = L_{bplg} \cdot \frac{dI_{L_{bplg}}}{dt}; I_{R_{bbp}} = \frac{V_{A_5} - V_{A_6}}{R_{bbp}}; I_{L_{bplg}} = I_{R_{bbp}}; I_{R_c} = \frac{V_{A_9}}{R_c}; V_{pi} = V_{A_6} - V_{A_8} = V_{C_e}$$

$$I_{C_e} = C_e \cdot \frac{dV_{pi}}{dt}; I_{R_{bpe}} = \frac{V_{A_6} - V_{A_8}}{R_{bpe}}; I_{C_c} = C_c \cdot \frac{d(V_{A_6} - V_{A_7})}{dt}; I_{R_{bpc}} = \frac{V_{A_6} - V_{A_7}}{R_{bpc}}; I_{R_{ce}} = \frac{V_{A_7} - V_{A_8}}{R_{ce}}$$

$$V_{A_8} = L_{eplg} \cdot \frac{dI_{L_{eplg}}}{dt}; V_{A_7} - V_{A_{10}} = L_{cplg} \cdot \frac{dI_{L_{cplg}}}{dt}; V_{A_9} - V_{A_{10}} = L_c \cdot \frac{dL_{L_c}}{dt}; I_{C_{c2}} = C_{c2} \cdot \frac{d(V_{A_{10}} - V_{A_{11}})}{dt}$$

$$I_{C_{b1}} = C_{b1} \cdot \frac{dV_{A_{11}}}{dt}; V_{L_b} = V_{A_{11}} - V_{A_{12}} = L_b \cdot \frac{dI_{L_b}}{dt}; V_{A_{12}} = V_{out}; I_{C_{b2}} = C_{b2} \cdot \frac{dV_{A_{12}}}{dt};$$

$$I_{R_{load}} = \frac{V_{out}}{R_{load}} = \frac{V_{A_{12}}}{R_{load}}$$

$$V_{A_{11}} = V_{C_{b1}}; V_{A_{12}} = V_{C_{b2}}; I_{C_{c2}} = I_{L_b} + I_{C_{b1}}; I_{L_b} = I_{C_{b2}} + I_{R_{load}}$$

Table 4.2 Linear Hybrid PI model and additional matching circuits Kirchhoff's Current Law (KCL) for all nodes

Node number	KCL @ A_i	Node number	KCL @ A_i
A_1	$I_{R_s} = I_{L_a} + I_{C_{a1}}$	A_8	$I_{C_e} + I_{R_{bpe}} + g_m \cdot V_{pi} + I_{R_{ce}} = I_{L_{eplg}}$
A_2	$I_{L_a} = I_{C_{a2}} + I_{C_{C_1}}$	A_9	$I_{R_{b1}} = I_{R_c} + I_{L_c}$
A_3	$I_{C_{c1}} = I_{L_2} + I_{L_{bplg}} + I_{L_3}$	A_{10}	$I_{L_c} + I_{L_{cplg}} = I_{C_{c2}}$
A_4	$I_{L_3} = I_{R_{b2}}$	A_{11}	$I_{C_{c2}} = I_{L_b} + I_{C_{b1}}$
A_5	$I_{L_{bplg}} = I_{R_{bbp}}$	A_{12}	$I_{L_b} = I_{C_{b2}} + I_{R_{load}}$
A_6	$I_{R_{bbp}} = I_{R_{bpc}} + I_{C_c} + I_{R_{bpe}} + I_{C_e}$	A_{13}	$I_{L_2} = I_{R_{b1}}$
A_7	$I_{C_c} + I_{R_{bpc}} = g_m \cdot V_{pi} + I_{R_{ce}} + I_{L_{cplg}}$		

We can write our circuit Kirchhoff's Current Law (KCL) for all nodes:

$$I_{R_s} = \frac{X_s(t) - V_{A_1}}{R_s} \Rightarrow I_{R_s} \cdot R_s = X_s(t) - V_{A_1} \Rightarrow V_{A_1} = X_s(t) - I_{R_s} \cdot R_s$$

$$I_{C_{a1}} = C_{a1} \cdot \frac{dV_{A_1}}{dt} = C_{a1} \cdot \frac{d}{dt}(X_s(t) - I_{R_s} \cdot R_s) = C_{a1} \cdot \frac{dX_s(t)}{dt} - C_{a1} \cdot R_s \cdot \frac{dI_{R_s}}{dt}$$

$$V_{L_a} = V_{A_1} - V_{A_2} = L_a \cdot \frac{dI_{L_a}}{dt} \Rightarrow V_{A_2} = V_{A_1} - L_a \cdot \frac{dI_{L_a}}{dt} = X_s(t) - I_{R_s} \cdot R_s - L_a \cdot \frac{dI_{L_a}}{dt}$$

$$I_{C_{a2}} = C_{a2} \cdot \frac{dV_{A_2}}{dt} = C_{a2} \cdot \frac{d}{dt}(X_s(t) - I_{R_s} \cdot R_s - L_a \cdot \frac{dI_{L_a}}{dt})$$

$$= C_{a2} \cdot \frac{dX_s(t)}{dt} - C_{a2} \cdot R_s \cdot \frac{dI_{R_s}}{dt} - C_{a2} \cdot L_a \cdot \frac{d^2 I_{L_a}}{dt^2}$$

$$I_{C_{C_1}} = C_{C_1} \cdot \frac{d}{dt}(V_{A_2} - V_{A_3}) = C_{C_1} \cdot \frac{dV_{A_2}}{dt} - C_{C_1} \cdot \frac{dV_{A_3}}{dt}$$

$$= C_{C_1} \cdot \frac{d}{dt}(X_s(t) - I_{R_s} \cdot R_s - L_a \cdot \frac{dI_{L_a}}{dt}) - C_{C_1} \cdot \frac{dV_{A_3}}{dt}$$

$$I_{C_{C_1}} = C_{C_1} \cdot \frac{d}{dt}(V_{A_2} - V_{A_3}) = C_{C_1} \cdot \frac{dX_s(t)}{dt} - \frac{dI_{R_s}}{dt} \cdot C_{C_1} \cdot R_s - C_{C_1} \cdot L_a \cdot \frac{d^2 I_{L_a}}{dt^2}$$
$$- C_{C_1} \cdot \frac{dV_{A_3}}{dt}$$

$$I_{R_{b2}} = \frac{V_{A_4}}{R_{b_2}} \Rightarrow V_{A_4} = I_{R_{b2}} \cdot R_{b_2}; V_{A_3} - V_{A_4} = L_3 \cdot \frac{dI_{L_3}}{dt} \Rightarrow V_{A_3} = V_{A_4} + L_3 \cdot \frac{dI_{L_3}}{dt}$$
$$= I_{R_{b2}} \cdot R_{b_2} + L_3 \cdot \frac{dI_{L_3}}{dt}$$

$$\frac{dV_{A_3}}{dt} = \frac{dV_{A_4}}{dt} + L_3 \cdot \frac{d^2 I_{L_3}}{dt^2} = \frac{dI_{R_{b2}}}{dt} \cdot R_{b_2} + L_3 \cdot \frac{d^2 I_{L_3}}{dt^2}$$

System equation No. 1:

$$I_{C_{C_1}} = C_{C_1} \cdot \frac{d}{dt}(V_{A_2} - V_{A_3})$$
$$= C_{C_1} \cdot \frac{dX_s(t)}{dt} - \frac{dI_{R_s}}{dt} \cdot C_{C_1} \cdot R_s - C_{C_1} \cdot L_a \cdot \frac{d^2 I_{L_a}}{dt^2} - C_{C_1} \cdot (\frac{dI_{R_{b2}}}{dt} \cdot R_{b_2} + L_3 \cdot \frac{d^2 I_{L_3}}{dt^2})$$

$$I_{C_{C_1}} = C_{C_1} \cdot \frac{dX_s(t)}{dt} - \frac{dI_{R_s}}{dt} \cdot C_{C_1} \cdot R_s - C_{C_1} \cdot L_a \cdot \frac{d^2 I_{L_a}}{dt^2} - \frac{dI_{R_{b2}}}{dt} \cdot C_{C_1} \cdot R_{b_2} - C_{C_1} \cdot L_3 \cdot \frac{d^2 I_{L_3}}{dt^2}$$

$$V_{A_3} - V_{A_{13}} = L_2 \cdot \frac{dI_{L_2}}{dt} \Rightarrow V_{A_{13}} = V_{A_3} - L_2 \cdot \frac{dI_{L_2}}{dt} = I_{R_{b2}} \cdot R_{b_2} + L_3 \cdot \frac{dI_{L_3}}{dt} - L_2 \cdot \frac{dI_{L_2}}{dt}$$

$$I_{R_{b1}} = \frac{V_{A_{13}} - V_{A_9}}{R_{b1}} \Rightarrow I_{R_{b1}} \cdot R_{b1} = V_{A_{13}} - V_{A_9} \Rightarrow V_{A_9} = V_{A_{13}} - I_{R_{b1}} \cdot R_{b1}$$

$$V_{A_9} = V_{A_{13}} - I_{R_{b1}} \cdot R_{b1} = I_{R_{b2}} \cdot R_{b_2} + L_3 \cdot \frac{dI_{L_3}}{dt} - L_2 \cdot \frac{dI_{L_2}}{dt} - I_{R_{b1}} \cdot R_{b1}$$

$$V_{A_3} - V_{A_5} = L_{bplg} \cdot \frac{dI_{L_{bplg}}}{dt} \Rightarrow V_{A_5} = I_{R_{b2}} \cdot R_{b_2} + L_3 \cdot \frac{dI_{L_3}}{dt} - L_{bplg} \cdot \frac{dI_{L_{bplg}}}{dt}$$

$$I_{R_{bbp}} = \frac{V_{A_5} - V_{A_6}}{R_{bbp}} \Rightarrow I_{R_{bbp}} \cdot R_{bbp} = V_{A_5} - V_{A_6} \Rightarrow V_{A_6}$$
$$= I_{R_{b2}} \cdot R_{b_2} + L_3 \cdot \frac{dI_{L_3}}{dt} - L_{bplg} \cdot \frac{dI_{L_{bplg}}}{dt} - I_{R_{bbp}} \cdot R_{bbp}$$

$$I_{R_c} = \frac{V_{A_9}}{R_c} \Rightarrow V_{A_9} = I_{R_c} \cdot R_c; I_{R_{bpe}} = \frac{V_{A_6} - V_{A_8}}{R_{bpe}} \Rightarrow I_{R_{bpe}} \cdot R_{bpe}$$
$$= V_{A_6} - V_{A_8} \Rightarrow V_{A_8} = V_{A_6} - I_{R_{bpe}} \cdot R_{bpe}$$

$$V_{A_8} = V_{A_6} - I_{R_{bpe}} \cdot R_{bpe} = I_{R_{b2}} \cdot R_{b_2} + L_3 \cdot \frac{dI_{L_3}}{dt} - L_{bplg} \cdot \frac{dI_{L_{bplg}}}{dt} - I_{R_{bbp}} \cdot R_{bbp} - I_{R_{bpe}} \cdot R_{bpe}$$

$$I_{C_c} = C_c \cdot \frac{d(V_{A_6} - V_{A_7})}{dt}; I_{R_{bpc}} = \frac{V_{A_6} - V_{A_7}}{R_{bpc}} \Rightarrow I_{C_c}$$
$$= C_c \cdot R_{bpc} \cdot \frac{dI_{R_{bpc}}}{dt}; V_{A_7} - V_{A_8} = I_{R_{ce}} \cdot R_{ce}$$

$$V_{A_7} = V_{A_8} + I_{R_{ce}} \cdot R_{ce} = I_{R_{b2}} \cdot R_{b_2} + L_3 \cdot \frac{dI_{L_3}}{dt} - L_{bplg} \cdot \frac{dI_{L_{bplg}}}{dt} - I_{R_{bbp}} \cdot R_{bbp} - I_{R_{bpe}} \cdot R_{bpe} + I_{R_{ce}} \cdot R_{ce}$$

System equation No. 2:

$$V_{A_8} = L_{eplg} \cdot \frac{dI_{L_{eplg}}}{dt} \Rightarrow I_{R_{b2}} \cdot R_{b_2} + L_3 \cdot \frac{dI_{L_3}}{dt} - L_{bplg} \cdot \frac{dI_{L_{bplg}}}{dt} - I_{R_{bbp}} \cdot R_{bbp} - I_{R_{bpe}} \cdot R_{bpe} = L_{eplg} \cdot \frac{dI_{L_{eplg}}}{dt}$$

$$V_{A_7} - V_{A_{10}} = L_{cplg} \cdot \frac{dI_{L_{cplg}}}{dt} \Rightarrow V_{A_{10}} = V_{A_7} - L_{cplg} \cdot \frac{dI_{L_{cplg}}}{dt}$$

$$V_{A_{10}} = I_{R_{b2}} \cdot R_{b_2} + L_3 \cdot \frac{dI_{L_3}}{dt} - L_{bplg} \cdot \frac{dI_{L_{bplg}}}{dt} - I_{R_{bbp}} \cdot R_{bbp} - I_{R_{bpe}} \cdot R_{bpe} + I_{R_{ce}} \cdot R_{ce} - L_{cplg} \cdot \frac{dI_{L_{cplg}}}{dt}$$

$$V_{A_9} - V_{A_{10}} = L_c \cdot \frac{dL_{L_c}}{dt} \Rightarrow V_{A_9} = V_{A_{10}} + L_c \cdot \frac{dL_{L_c}}{dt}; \frac{dV_{A_{10}}}{dt} = \frac{I_{C_{c2}}}{C_{c2}} + \frac{I_{C_{b1}}}{C_{b1}}$$

$$V_{A_9} = I_{R_{b2}} \cdot R_{b_2} + L_3 \cdot \frac{dI_{L_3}}{dt} - L_{bplg} \cdot \frac{dI_{L_{bplg}}}{dt} - I_{R_{bbp}} \cdot R_{bbp} - I_{R_{bpe}} \cdot R_{bpe} + I_{R_{ce}} \cdot R_{ce} - L_{cplg} \cdot \frac{dI_{L_{cplg}}}{dt} + L_c \cdot \frac{dL_{L_c}}{dt}$$

$$I_{C_{c2}} = C_{c2} \cdot \frac{d(V_{A_{10}} - V_{A_{11}})}{dt} \Rightarrow I_{C_{c2}} = C_{c2} \cdot \left(\frac{dV_{A_{10}}}{dt} - \frac{dV_{A_{11}}}{dt}\right) \Rightarrow I_{C_{c2}} = C_{c2} \cdot \left(\frac{dV_{A_{10}}}{dt} - \frac{I_{C_{b1}}}{C_{b1}}\right)$$

System equation No. 3:

$$\frac{I_{C_{c2}}}{C_{c2}} + \frac{I_{C_{b1}}}{C_{b1}} = \frac{d}{dt}\left\{I_{R_{b2}} \cdot R_{b_2} + L_3 \cdot \frac{dI_{L_3}}{dt} - L_{bplg} \cdot \frac{dI_{L_{bplg}}}{dt} - I_{R_{bbp}} \cdot R_{bbp} - I_{R_{bpe}} \cdot R_{bpe} + I_{R_{ce}} \cdot R_{ce} - L_{cplg} \cdot \frac{dI_{L_{cplg}}}{dt}\right\}$$

$$\frac{I_{C_{c2}}}{C_{c2}} + \frac{I_{C_{b1}}}{C_{b1}} = \frac{dI_{R_{b2}}}{dt} \cdot R_{b_2} + L_3 \cdot \frac{d^2 I_{L_3}}{dt^2} - L_{bplg} \cdot \frac{d^2 I_{L_{bplg}}}{dt^2} - \frac{dI_{R_{bbp}}}{dt} \cdot R_{bbp} - \frac{dI_{R_{bpe}}}{dt} \cdot R_{bpe} + \frac{dI_{R_{ce}}}{dt} \cdot R_{ce} - L_{cplg} \cdot \frac{d^2 I_{L_{cplg}}}{dt^2}$$

$$V_{A_{11}} - V_{A_{12}} = L_b \cdot \frac{dI_{L_b}}{dt} \Rightarrow V_{A_{11}} = V_{A_{12}} + L_b \cdot \frac{dI_{L_b}}{dt}; I_{R_{load}} = \frac{V_{A12}}{R_{load}} \Rightarrow V_{A_{12}} = I_{R_{load}} \cdot R_{load}$$

System equation No. 4:

$$V_{A_{11}} = I_{R_{load}} \cdot R_{load} + L_b \cdot \frac{dI_{L_b}}{dt}; V_{A_{12}} = V_{C_{b2}}; I_{C_{b2}} = C_{b2} \cdot \frac{dV_{A_{12}}}{dt} = C_{b2} \cdot R_{load} \cdot \frac{dI_{R_{load}}}{dt}$$

$$I_{L_b} = I_{C_{b2}} + I_{R_{load}} \Rightarrow I_{L_b} = C_{b2} \cdot R_{load} \cdot \frac{dI_{R_{load}}}{dt} + I_{R_{load}}$$

We can summery our system equations:

$$I_{C_{C_1}} = C_{C_1} \cdot \frac{dX_s(t)}{dt} - \frac{dI_{R_s}}{dt} \cdot C_{C_1} \cdot R_s - C_{C_1} \cdot L_a \cdot \frac{d^2 I_{L_a}}{dt^2} - \frac{dI_{R_{b2}}}{dt} \cdot C_{C_1} \cdot R_{b_2} - C_{C_1} \cdot L_3 \cdot \frac{d^2 I_{L_3}}{dt^2}$$

$$I_{R_{b2}} \cdot R_{b_2} + L_3 \cdot \frac{dI_{L_3}}{dt} - L_{bplg} \cdot \frac{dI_{L_{bplg}}}{dt} - I_{R_{bbp}} \cdot R_{bbp} - I_{R_{bpe}} \cdot R_{bpe} = L_{eplg} \cdot \frac{dI_{L_{eplg}}}{dt}$$

$$\frac{I_{C_{c2}}}{C_{c2}} + \frac{I_{C_{b1}}}{C_{b1}} = \frac{dI_{R_{b2}}}{dt} \cdot R_{b_2} + L_3 \cdot \frac{d^2 I_{L_3}}{dt^2} - L_{bplg} \cdot \frac{d^2 I_{L_{bplg}}}{dt^2} - \frac{dI_{R_{bbp}}}{dt} \cdot R_{bbp} - \frac{dI_{R_{bpe}}}{dt} \cdot R_{bpe} + \frac{dI_{R_{ce}}}{dt} \cdot R_{ce} - L_{cplg} \cdot \frac{d^2 I_{L_{cplg}}}{dt^2}$$

$$I_{L_b} = C_{b2} \cdot R_{load} \cdot \frac{dI_{R_{load}}}{dt} + I_{R_{load}}$$

Since
$I_{C_{c2}} = I_{L_b} + I_{C_{b1}} \Rightarrow I_{C_{b1}} = I_{C_{c2}} - I_{L_b}$; $I_{L_b} = I_{C_{b2}} + I_{R_{load}}$; $I_{C_{b1}} = I_{C_{c2}} - I_{C_{b2}} - I_{R_{load}}$

$$I_{C_{C_1}} = C_{C_1} \cdot \frac{dX_s(t)}{dt} - \frac{dI_{R_s}}{dt} \cdot C_{C_1} \cdot R_s - C_{C_1} \cdot L_a \cdot \frac{d^2 I_{L_a}}{dt^2} - \frac{dI_{R_{b2}}}{dt} \cdot C_{C_1} \cdot R_{b_2} - C_{C_1} \cdot L_3 \cdot \frac{d^2 I_{L_3}}{dt^2}$$

$$I_{R_{b2}} \cdot R_{b_2} + L_3 \cdot \frac{dI_{L_3}}{dt} - L_{bplg} \cdot \frac{dI_{L_{bplg}}}{dt} - I_{R_{bbp}} \cdot R_{bbp} - I_{R_{bpe}} \cdot R_{bpe} = L_{eplg} \cdot \frac{dI_{L_{eplg}}}{dt}$$

$$\frac{I_{C_{c2}}}{C_{c2}} + \frac{I_{C_{b1}}}{C_{b1}} = \frac{dI_{R_{b2}}}{dt} \cdot R_{b_2} + L_3 \cdot \frac{d^2 I_{L_3}}{dt^2} - L_{bplg} \cdot \frac{d^2 I_{L_{bplg}}}{dt^2} - \frac{dI_{R_{bbp}}}{dt} \cdot R_{bbp} - \frac{dI_{R_{bpe}}}{dt} \cdot R_{bpe} + \frac{dI_{R_{ce}}}{dt} \cdot R_{ce} - L_{cplg} \cdot \frac{d^2 I_{L_{cplg}}}{dt^2}$$

$$\frac{d}{dt}\left\{ I_{R_{b2}} \cdot R_{b_2} + L_3 \cdot \frac{dI_{L_3}}{dt} - L_{bplg} \cdot \frac{dI_{L_{bplg}}}{dt} - I_{R_{bbp}} \cdot R_{bbp} - I_{R_{bpe}} \cdot R_{bpe} \right\} = L_{eplg} \cdot \frac{d^2 I_{L_{eplg}}}{dt^2}$$

We get one differential equation: $\frac{I_{C_{c2}}}{C_{c2}} + \frac{I_{C_{b1}}}{C_{b1}} = L_{eplg} \cdot \frac{d^2 I_{L_{eplg}}}{dt^2} + \frac{dI_{R_{ce}}}{dt} \cdot R_{ce} - L_{cplg} \cdot \frac{d^2 I_{L_{cplg}}}{dt^2}$

$$I_{C_{b1}} = I_{C_{c2}} - I_{L_b} \Rightarrow I_{C_{c2}} \cdot \left(\frac{1}{C_{c2}} + \frac{1}{C_{b1}}\right) - \frac{I_{L_b}}{C_{b1}} = L_{eplg} \cdot \frac{d^2 I_{L_{eplg}}}{dt^2} + \frac{dI_{R_{ce}}}{dt} \cdot R_{ce} - L_{cplg} \cdot \frac{d^2 I_{L_{cplg}}}{dt^2}$$

We get for our system three differential equations:

$$I_{C_{C_1}} = C_{C_1} \cdot \frac{dX_s(t)}{dt} - \frac{dI_{R_s}}{dt} \cdot C_{C_1} \cdot R_s - C_{C_1} \cdot L_a \cdot \frac{d^2 I_{L_a}}{dt^2} - \frac{dI_{R_{b2}}}{dt} \cdot C_{C_1} \cdot R_{b_2} - C_{C_1} \cdot L_3 \cdot \frac{d^2 I_{L_3}}{dt^2}$$

$$I_{C_{c2}} \cdot \left(\frac{1}{C_{c2}} + \frac{1}{C_{b1}}\right) - \frac{I_{L_b}}{C_{b1}} = L_{eplg} \cdot \frac{d^2 I_{L_{eplg}}}{dt^2} + \frac{dI_{R_{ce}}}{dt} \cdot R_{ce} - L_{cplg} \cdot \frac{d^2 I_{L_{cplg}}}{dt^2};$$

$$I_{L_b} = C_{b2} \cdot R_{load} \cdot \frac{dI_{R_{load}}}{dt} + I_{R_{load}}$$

We need to find our system equilibrium points (fixed points). It is done by setting $\frac{dI_{R_s}}{dt} = 0; \frac{dI_{L_a}}{dt} = 0 \Rightarrow \frac{d^2 I_{L_a}}{dt^2} = 0; \frac{dI_{R_{b2}}}{dt} = 0; \frac{dI_{L_3}}{dt} \Rightarrow \frac{d^2 I_{L_3}}{dt^2} = 0$

$$\frac{dI_{L_{eplg}}}{dt} = 0 \Rightarrow \frac{d^2 I_{L_{eplg}}}{dt^2} = 0; \frac{dI_{R_{ce}}}{dt} = 0; \frac{dI_{L_{cplg}}}{dt} \Rightarrow \frac{d^2 I_{L_{cplg}}}{dt^2} = 0; \frac{dI_{R_{load}}}{dt} = 0$$

We get the system fixed points (equilibrium points):

$$I^*_{C_{C_1}} = C_{C_1} \cdot \frac{dX_s(t)}{dt}; I^*_{C_{c2}} \cdot \left(\frac{1}{C_{c2}} + \frac{1}{C_{b1}}\right) - \frac{I^*_{L_b}}{C_{b1}} = 0 \Rightarrow I^*_{L_b} = I^*_{C_{c2}} \cdot \left(\frac{1}{C_{c2}} + \frac{1}{C_{b1}}\right) \cdot C_{b1}; I^*_{L_b} = I^*_{R_{load}}$$

We consider input RF source $X_s(t) = A_0 + \xi(t); |\xi(t)| < 1 \& A_0 \gg |\xi(t)|$

Then. $X_s(t)|_{A_0 \gg |\xi(t)|} = A_0 + \xi(t) \approx A_0; \frac{dX_s(t)|_{A_0 \gg |\xi(t)|}}{dt} = \frac{d\xi(t)}{dt} \to \varepsilon.$

Based on the above assumption, we get the following system fixed points:

$$I^*_{C_{C_1}} = 0; I^*_{C_{c2}} \cdot \left(\frac{1}{C_{c2}} + \frac{1}{C_{b1}}\right) - \frac{I^*_{L_b}}{C_{b1}} = 0 \Rightarrow I^*_{L_b} = I^*_{C_{c2}} \cdot \left(\frac{1}{C_{c2}} + \frac{1}{C_{b1}}\right) \cdot C_{b1}; I^*_{L_b} = I^*_{R_{load}}$$

$$E^*\left(I^*_{C_{C_1}}, I^*_{R_s}, I^*_{L_a}, I^*_{R_{b2}}, I^*_{L_3}, I^*_{C_{c2}}, I^*_{L_b}, I^*_{L_{eplg}}, I^*_{R_{ce}}, I^*_{L_{cplg}}, I^*_{R_{load}}\right) = \left(0, I^*_{R_s}, I^*_{L_a}, I^*_{R_{b2}}, I^*_{L_3}, I^*_{C_{c2}}, I^*_{C_{c2}} \cdot \left(\frac{1}{C_{c2}} + \frac{1}{C_{b1}}\right) \cdot C_{b1}, I^*_{L_{eplg}}, I^*_{R_{ce}}, I^*_{L_{cplg}}, I^*_{C_{c2}} \cdot \left(\frac{1}{C_{c2}} + \frac{1}{C_{b1}}\right) \cdot C_{b1}\right)$$

Stability analysis: The standard local stability analysis about any one of the equilibrium points of the Small Signal (SS) amplifier with matching network consists in adding to coordinate $\left[I_{C_{C_1}}, I_{R_s}, I_{L_a}, I_{R_{b2}}, I_{L_3}, I_{C_{c2}}, I_{L_b}, I_{L_{eplg}}, I_{R_{ce}}, I_{L_{cplg}}, I_{R_{load}}\right]$ arbitrarily small increments of exponentially form $[i_{C_{C_1}}, i_{R_s}, i_{L_a}, i_{R_{b2}}, i_{L_3}, i_{C_{c2}}, i_{L_b}, i_{L_{eplg}}, i_{R_{ce}}, i_{L_{cplg}}, i_{R_{load}}] \cdot e^{\lambda \cdot t}$ and retaining the first order terms in $I_{C_{C_1}}, I_{R_s}, I_{L_a}, I_{R_{b2}}, I_{L_3}, I_{C_{c2}}, I_{L_b}, I_{L_{eplg}}, I_{R_{ce}}, I_{L_{cplg}}, I_{R_{load}}$ [3, 4]. The system of homogenous equations leads to a polynomial characteristic equation in the eigenvalues. The polynomial characteristic equations accept by set of the below circuit variables, circuit variables derivative and circuit variables second order derivative with respect to time into Small Signal (SS) amplifier with matching network equivalent circuit. Our Small Signal (SS) amplifier with matching network equivalent circuit fixed values with arbitrarily small increments of exponential form $[i_{C_{C_1}}, i_{R_s}, i_{L_a}, i_{R_{b2}}, i_{L_3}, i_{C_{c2}}, i_{L_b}, i_{L_{eplg}}, i_{R_{ce}}, i_{L_{cplg}}, i_{R_{load}}] \cdot e^{\lambda \cdot t}$ are: j = 0 (first fixed point), j = 1 (second fixed point), j = 2 (third fixed point), etc.,

We define new variables: $Y_1 = \frac{dI_{L_a}}{dt} \Rightarrow \frac{dY_1}{dt} = \frac{d^2 I_{L_a}}{dt^2}; Y_2 = \frac{dI_{L_3}}{dt} \Rightarrow \frac{dY_2}{dt} = \frac{d^2 I_{L_3}}{dt^2}$

$$I_{C_{C_1}} = C_{C_1} \cdot \left(\frac{dX_s(t)}{dt} \to \varepsilon\right) - \frac{dI_{R_s}}{dt} \cdot C_{C_1} \cdot R_s - C_{C_1} \cdot L_a \cdot \frac{dY_1}{dt} - \frac{dI_{R_{b2}}}{dt} \cdot C_{C_1} \cdot R_{b_2} - C_{C_1} \cdot L_3 \cdot \frac{dY_2}{dt}$$

$$Y_3 = \frac{dI_{L_{eplg}}}{dt} \Rightarrow \frac{dY_3}{dt} = \frac{d^2 I_{L_{eplg}}}{dt^2}; Y_4 = \frac{dI_{L_{cplg}}}{dt} \Rightarrow \frac{dY_4}{dt} = \frac{d^2 I_{L_{cplg}}}{dt^2}$$

$$I_{C_{c2}} \cdot \left(\frac{1}{C_{c2}} + \frac{1}{C_{b1}}\right) - \frac{I_{L_b}}{C_{b1}} = L_{eplg} \cdot \frac{dY_3}{dt} + \frac{dI_{R_{ce}}}{dt} \cdot R_{ce} - L_{cplg} \cdot \frac{dY_4}{dt}$$

$$Y_1(t) = Y_1^{(j)} + y_1 \cdot e^{\lambda \cdot t}; Y_2(t) = Y_2^{(j)} + y_2 \cdot e^{\lambda \cdot t}; Y_3(t) = Y_3^{(j)} + y_3 \cdot e^{\lambda \cdot t}; Y_4(t) = Y_4^{(j)} + y_4 \cdot e^{\lambda \cdot t}$$

$$I_{C_{c1}}(t) = I_{C_{c1}}^{(j)} + i_{C_{c1}} \cdot e^{\lambda \cdot t}; I_{R_s}(t) = I_{R_s}^{(j)} + i_{R_s} \cdot e^{\lambda \cdot t}; I_{R_{b2}}(t) = I_{R_{b2}}^{(j)} + i_{R_{b2}} \cdot e^{\lambda \cdot t}; I_{C_{c2}}(t) = I_{C_{c2}}^{(j)} + i_{C_{c2}} \cdot e^{\lambda \cdot t}$$

$$I_{L_b}(t) = I_{L_b}^{(j)} + i_{L_b} \cdot e^{\lambda \cdot t}; I_{R_{ce}}(t) = I_{R_{ce}}^{(j)} + i_{R_{ce}} \cdot e^{\lambda \cdot t}; \frac{dI_{R_s}(t)}{dt} = i_{R_s} \cdot \lambda \cdot e^{\lambda \cdot t}; \frac{dY_1(t)}{dt} = y_1 \cdot \lambda \cdot e^{\lambda \cdot t}$$

$$\frac{dI_{R_{b2}}(t)}{dt} = i_{R_{b2}} \cdot \lambda \cdot e^{\lambda \cdot t}; \frac{dY_2(t)}{dt} = y_2 \cdot \lambda \cdot e^{\lambda \cdot t}; I_{R_{load}}(t) = I_{R_{load}}^{(j)} + i_{R_{load}} \cdot e^{\lambda \cdot t}; \frac{dI_{R_{load}}(t)}{dt} = i_{R_{load}} \cdot \lambda \cdot e^{\lambda \cdot t}$$

At fixed point $\frac{dI_{La}}{dt} = 0 \Rightarrow Y_1^{(j)} = 0; \frac{dI_{L_3}}{dt} = 0 \Rightarrow Y_2^{(j)} = 0; \frac{dI_{L_{eplg}}}{dt} = 0 \Rightarrow Y_3^{(j)} = 0$

$$\frac{dI_{L_{cplg}}}{dt} = 0 \Rightarrow Y_4^{(j)} = 0; Y_1 = \frac{dI_{L_a}}{dt}|_{Y_1^{(j)}=0} \Rightarrow y_1 \cdot e^{\lambda \cdot t} = i_{L_a} \cdot \lambda \cdot e^{\lambda \cdot t}; y_1 - i_{L_a} \cdot \lambda = 0$$

$$Y_2 = \frac{dI_{L_3}}{dt}|_{Y_2^{(j)}=0} \Rightarrow y_2 \cdot e^{\lambda \cdot t} = i_{L_3} \cdot \lambda \cdot e^{\lambda \cdot t}; y_2 - i_{L_3} \cdot \lambda = 0; Y_3 = \frac{dI_{L_{eplg}}}{dt}|_{Y_3^{(j)}=0}$$
$$\Rightarrow y_3 \cdot e^{\lambda \cdot t} = i_{L_{eplg}} \cdot \lambda \cdot e^{\lambda \cdot t}$$

$$y_3 - i_{L_{eplg}} \cdot \lambda = 0; Y_4 = \frac{dI_{L_{cplg}}}{dt}|_{Y_4^{(j)}=0} \Rightarrow y_4 \cdot e^{\lambda \cdot t} = i_{L_{cplg}} \cdot \lambda \cdot e^{\lambda \cdot t};$$
$$y_4 - i_{L_{cplg}} \cdot \lambda = 0; \frac{dX_s(t)}{dt} \rightarrow \varepsilon = 0$$

$$I_{C_{C_1}} = -\frac{dI_{R_s}}{dt} \cdot C_{C_1} \cdot R_s - C_{C_1} \cdot L_a \cdot \frac{dY_1}{dt} - \frac{dI_{R_{b2}}}{dt} \cdot C_{C_1} \cdot R_{b_2} - C_{C_1} \cdot L_3 \cdot \frac{dY_2}{dt}$$

$$I_{C_{c1}}^{(j)} + i_{C_{c1}} \cdot e^{\lambda \cdot t} = -i_{R_s} \cdot \lambda \cdot e^{\lambda \cdot t} \cdot C_{C_1} \cdot R_s - C_{C_1} \cdot L_a \cdot y_1 \cdot \lambda \cdot e^{\lambda \cdot t} - i_{R_{b2}} \cdot \lambda \cdot e^{\lambda \cdot t} \cdot C_{C_1}$$
$$\cdot R_{b_2} - C_{C_1} \cdot L_3 \cdot y_2 \cdot \lambda \cdot e^{\lambda \cdot t}$$

@ fixed point $I_{C_{c1}}^{(j)} = 0$; $i_{C_{c1}} = -(i_{R_s} \cdot R_s + L_a \cdot y_1 + i_{R_{b2}} \cdot R_{b_2} + L_3 \cdot y_2) \cdot C_{C_1} \cdot \lambda$
We divide the two side of the above equation by $i_{C_{c1}}$ term

$$i_{C_{c1}} = -i_{R_s} \cdot \lambda \cdot C_{C_1} \cdot R_s - C_{C_1} \cdot L_a \cdot y_1 \cdot \lambda - i_{R_{b2}} \cdot \lambda \cdot C_{C_1} \cdot R_{b_2} - C_{C_1} \cdot L_3 \cdot y_2 \cdot \lambda$$

$$I_{C_{c2}} \cdot \left(\frac{1}{C_{c2}} + \frac{1}{C_{b1}}\right) - \frac{I_{L_b}}{C_{b1}} = L_{eplg} \cdot \frac{dY_3}{dt} + \frac{dI_{R_{ce}}}{dt} \cdot R_{ce} - L_{cplg} \cdot \frac{dY_4}{dt}$$

$$\left(I_{C_{c2}}^{(j)} + i_{C_{c2}} \cdot e^{\lambda \cdot t}\right) \cdot \left(\frac{1}{C_{c2}} + \frac{1}{C_{b1}}\right) - \frac{\left(I_{L_b}^{(j)} + i_{L_b} \cdot e^{\lambda \cdot t}\right)}{C_{b1}}$$
$$= L_{eplg} \cdot y_3 \cdot \lambda \cdot e^{\lambda \cdot t} + i_{R_{ce}} \cdot \lambda \cdot e^{\lambda \cdot t} \cdot R_{ce} - L_{cplg} \cdot y_4 \cdot \lambda \cdot e^{\lambda \cdot t}$$

@ fixed point $I_{C_{c2}}^{*} \cdot \left(\frac{1}{C_{c2}} + \frac{1}{C_{b1}}\right) - \frac{I_{L_b}^{*}}{C_{b1}} = 0 \Rightarrow I_{L_b}^{*} = I_{C_{c2}}^{*} \cdot \left(\frac{1}{C_{c2}} + \frac{1}{C_{b1}}\right) \cdot C_{b1}$

$$\left\{I_{C_{c2}}^{(j)} \cdot \left(\frac{1}{C_{c2}} + \frac{1}{C_{b1}}\right) - \frac{I_{L_b}^{(j)}}{C_{b1}}\right\} + i_{C_{c2}} \cdot \left(\frac{1}{C_{c2}} + \frac{1}{C_{b1}}\right) \cdot e^{\lambda \cdot t} - \frac{i_{L_b} \cdot e^{\lambda \cdot t}}{C_{b1}}$$
$$= L_{eplg} \cdot y_3 \cdot \lambda \cdot e^{\lambda \cdot t} + i_{R_{ce}} \cdot \lambda \cdot e^{\lambda \cdot t} \cdot R_{ce} - L_{cplg} \cdot y_4 \cdot \lambda \cdot e^{\lambda \cdot t}$$

$$i_{C_{c2}} \cdot \left(\frac{1}{C_{c2}} + \frac{1}{C_{b1}}\right) \cdot e^{\lambda \cdot t} - \frac{i_{L_b} \cdot e^{\lambda \cdot t}}{C_{b1}} = L_{eplg} \cdot y_3 \cdot \lambda \cdot e^{\lambda \cdot t} + i_{R_{ce}} \cdot \lambda \cdot e^{\lambda \cdot t} \cdot R_{ce} - L_{cplg} \cdot y_4 \cdot \lambda \cdot e^{\lambda \cdot t}$$

We divide the two side of the above equation by $e^{\lambda \cdot t}$ term.

$$i_{C_{c2}} \cdot \left(\frac{1}{C_{c2}} + \frac{1}{C_{b1}}\right) - \frac{i_{L_b}}{C_{b1}} = L_{eplg} \cdot y_3 \cdot \lambda + i_{R_{ce}} \cdot \lambda \cdot R_{ce} - L_{cplg} \cdot y_4 \cdot \lambda$$

$$I_{L_b} = C_{b2} \cdot R_{load} \cdot \frac{dI_{R_{load}}}{dt} + I_{R_{load}} \Rightarrow I_{L_b}^{(j)} + i_{L_b} \cdot e^{\lambda \cdot t}$$
$$= C_{b2} \cdot R_{load} \cdot i_{R_{load}} \cdot \lambda \cdot e^{\lambda \cdot t} + I_{R_{load}}^{(j)} + i_{R_{load}} \cdot e^{\lambda \cdot t}$$

@ fixed point $I_{L_b}^{*} = I_{R_{load}}^{*}; I_{L_b}^{*} = I_{C_{c2}}^{*} \cdot (\frac{1}{C_{c2}} + \frac{1}{C_{b1}}) \cdot C_{b1}$

$$\left\{I_{L_b}^{(j)} - I_{R_{load}}^{(j)}\right\} + i_{L_b} \cdot e^{\lambda \cdot t} = C_{b2} \cdot R_{load} \cdot i_{R_{load}} \cdot \lambda \cdot e^{\lambda \cdot t} + i_{R_{load}} \cdot e^{\lambda \cdot t}; \left\{I_{L_b}^{(j)} - I_{R_{load}}^{(j)}\right\}|_{I_{L_b}^{*} = I_{R_{load}}^{*}} = 0$$

$$C_{b2} \cdot R_{load} \cdot i_{R_{load}} \cdot \lambda \cdot e^{\lambda \cdot t} = i_{L_b} \cdot e^{\lambda \cdot t} - i_{R_{load}} \cdot e^{\lambda \cdot t} \Rightarrow i_{L_b} - i_{R_{load}} - C_{b2} \cdot R_{load} \cdot i_{R_{load}} \cdot \lambda = 0$$

We can summery our Small Signal (SS) amplifier with matching network (PI input and output matching networks) small increments elements:

$$y_1 - i_{L_a} \cdot \lambda = 0; y_2 - i_{L_3} \cdot \lambda = 0; y_3 - i_{L_{eplg}} \cdot \lambda = 0; y_4 - i_{L_{cplg}} \cdot \lambda = 0$$

$$i_{C_{c1}} = -i_{R_s} \cdot \lambda \cdot C_{C_1} \cdot R_s - C_{C_1} \cdot L_a \cdot y_1 \cdot \lambda - i_{R_{b2}} \cdot \lambda \cdot C_{C_1} \cdot R_{b_2} - C_{C_1} \cdot L_3 \cdot y_2 \cdot \lambda$$

$$i_{C_{c2}} \cdot \left(\frac{1}{C_{c2}} + \frac{1}{C_{b1}}\right) - \frac{i_{L_b}}{C_{b1}} = L_{eplg} \cdot y_3 \cdot \lambda + i_{R_{ce}} \cdot \lambda \cdot R_{ce} - L_{cplg} \cdot y_4 \cdot \lambda$$

$$i_{L_b} - i_{R_{load}} - C_{b2} \cdot R_{load} \cdot i_{R_{load}} \cdot \lambda = 0$$

Assumption: We consider for simplicity that arbitrarily small increments elements $i_{R_s}, y_1, i_{R_{b2}}, y_2$ are proximally the same $i_{R_s} \approx y_1 \approx i_{R_{b2}} \approx y_2 \to z_1$ and we represent them as small increment element z_1. Additionally, arbitrarily small increments elements $y_3, i_{R_{Ce}}, y_4$ are proximally the same $y_3 \approx i_{R_{Ce}} \approx y_4 \to z_2$ and we represent them as small increment element z_2.

$$z_1 - i_{L_a} \cdot \lambda = 0; z_2 - i_{L_3} \cdot \lambda = 0; z_3 - i_{L_{eplg}} \cdot \lambda = 0; z_4 - i_{L_{cplg}} \cdot \lambda = 0$$

$$-i_{C_{c1}} \cdot \frac{1}{(C_{C_1} \cdot R_s + C_{C_1} \cdot L_a + C_{C_1} \cdot R_{b_2} + C_{C_1} \cdot L_3)} - z_1 \cdot \lambda = 0$$

$$i_{C_{c2}} \cdot \frac{\left(\frac{1}{C_{c2}} + \frac{1}{C_{b1}}\right)}{(L_{eplg} + R_{ce} - L_{cplg})} - \frac{i_{L_b}}{C_{b1} \cdot (L_{eplg} + R_{ce} - L_{cplg})} - z_2 \cdot \lambda = 0$$

$$i_{L_b} - i_{R_{load}} - C_{b2} \cdot R_{load} \cdot i_{R_{load}} \cdot \lambda = 0$$
$$\Rightarrow i_{L_b} \cdot \frac{1}{C_{b2} \cdot R_{load}} - i_{R_{load}} \cdot \frac{1}{C_{b2} \cdot R_{load}} - i_{R_{load}} \cdot \lambda = 0$$

Small Signal (SS) amplifier with matching network system matrixes:

$$\begin{pmatrix} \iota_{11} & \cdots & \iota_{17} \\ \vdots & \ddots & \vdots \\ \iota_{71} & \cdots & \iota_{77} \end{pmatrix} \cdot \begin{pmatrix} i_{L_a} \\ i_{L_3} \\ i_{L_{eplg}} \\ i_{L_{cplg}} \\ z_1 \\ z_2 \\ i_{R_{load}} \end{pmatrix} + \begin{pmatrix} \varphi_{11} & \cdots & \varphi_{13} \\ \vdots & \ddots & \vdots \\ \varphi_{71} & \cdots & \varphi_{73} \end{pmatrix} \cdot \begin{pmatrix} i_{C_{c1}} \\ i_{C_{c2}} \\ i_{L_b} \end{pmatrix} = 0$$

$$\imath_{11} = -\lambda; \imath_{12} = \imath_{13} = \imath_{14} = 0; \imath_{15} = 1; \imath_{16} = \imath_{17} = 0; \imath_{21} = 0; \imath_{22} = -\lambda;$$
$$\imath_{23} = \imath_{24} = 0; \imath_{25} = 1; \imath_{26} = \imath_{27} = 0$$

$$\imath_{31} = \imath_{32} = 0; \imath_{33} = -\lambda; \imath_{34} = \imath_{35} = 0; \imath_{36} = 1; \imath_{37} = 0; \imath_{41} = \imath_{42} = \imath_{43} = 0$$

$$\imath_{44} = -\lambda; \imath_{45} = 0; \imath_{46} = 1; \imath_{47} = 0; \imath_{51} = \ldots = \imath_{54} = 0; \imath_{55} = -\lambda; \imath_{56} = \imath_{57} = 0$$

$$\imath_{61} = \imath_{65} = 0; \imath_{66} = -\lambda; \imath_{67} = 0; \imath_{71} = \ldots = \imath_{76} = 0; \imath_{77} = -\lambda - \frac{1}{C_{b2} \cdot R_{load}}$$

$$\varphi_{11} = \varphi_{12} = \varphi_{13} = 0; \varphi_{21} = \varphi_{22} = \varphi_{23} = 0; \varphi_{31} = \varphi_{32} = \varphi_{33} = 0; \varphi_{41} = \varphi_{42} = \varphi_{43} = 0$$

$$\varphi_{51} = -\frac{1}{(C_{C_1} \cdot R_s + C_{C_1} \cdot L_a + C_{C_1} \cdot R_{b_2} + C_{C_1} \cdot L_3)}; \varphi_{52} = \varphi_{53} = 0$$

$$\varphi_{61} = 0; \varphi_{62} = \frac{(\frac{1}{C_{c2}} + \frac{1}{C_{b1}})}{(L_{eplg} + R_{ce} - L_{cplg})}; \varphi_{63} = -\frac{1}{C_{b1} \cdot (L_{eplg} + R_{ce} - L_{cplg})}$$

$$\varphi_{71} = \varphi_{72} = 0; \varphi_{73} = \frac{1}{C_{b2} \cdot R_{load}}$$

$$\text{We consider } \begin{pmatrix} \varphi_{11} & \cdots & \varphi_{13} \\ \vdots & \ddots & \vdots \\ \varphi_{71} & \cdots & \varphi_{73} \end{pmatrix} \rightarrow \varepsilon; \begin{pmatrix} \imath_{11} & \cdots & \imath_{17} \\ \vdots & \ddots & \vdots \\ \imath_{71} & \cdots & \imath_{77} \end{pmatrix} \cdot \begin{pmatrix} i_{L_a} \\ i_{L_3} \\ i_{L_{eplg}} \\ i_{L_{cplg}} \\ z_1 \\ z_2 \\ i_{R_{load}} \end{pmatrix} \approx 0$$

$$A - \lambda \cdot I = \begin{pmatrix} \imath_{11} & \cdots & \imath_{17} \\ \vdots & \ddots & \vdots \\ \imath_{71} & \cdots & \imath_{77} \end{pmatrix}; \det(A - \lambda \cdot I) = 0 \Rightarrow \det \begin{pmatrix} \imath_{11} & \cdots & \imath_{17} \\ \vdots & \ddots & \vdots \\ \imath_{71} & \cdots & \imath_{77} \end{pmatrix} = 0$$

To effectively apply the stability criterion of Lipunov to our system, we require a criterion for when the equation $\det(A - \lambda \cdot I) = 0$ has a zero in the left half plane, without calculating the eigenvalues explicitly. We use criterion of Routh-Hurwitz [2–4].

4.3 Small Signal (SS) Amplifiers Matching Network Stability Analysis Under Microstrip Parasitic Parameters Variation

In our stability analysis of Small Signal (SS) amplifiers which include input and output matching networks, we need to consider the microstrip lines that connect our input RF source to input matching network and input matching network to Small Signal (SS) amplifier. Additionally there is a microstrip lines between Small Signal (SS) amplifier to the output matching network and between output matching network to the load (R_{load}). We represent in our stability analysis the microstrip lines as a parasitic delay lines in time. We define our microstrip line's delay parameters as τ_1, τ_2, τ_3, τ_4 respectively. Under AC and Small Signal (SS) conditions, the BJT can be replaced with linear Hybrid PI model, input and output matching circuits, RF input source, load resistance and microstrip delay lines. The amplifier matching networks types are L matching network in our analysis [25, 26, 33] (Fig. 4.10).

Remark: microstrip lines have many parasitic effects. We neglect all those effects and concentrate on representation of microstrip line as a delay line. We consider that the voltage on microstrip delay line is very small and the assumption that the current flow through each microstrip is delay in time.

If we define the voltage on (i) delay line as V_{τ_i} (i = 1, 2, 3…). $V_{\tau_i} \to \varepsilon$. If we define the current that flow through microstrip as a $I(t)$ then the effect of parasitic delay in time is $I(t-\tau_i).I(t) \to I(t-\tau_i)$. We consider the following:

$\mathrm{Tau1} = \tau_1$, $\mathrm{Tau2} = \tau_2$, $\mathrm{Tau3} = \tau_3$, $\mathrm{Tau4} = \tau_4$, $V_{\tau_1} = (V_{A_{1-1}} - V_{A_{1-2}}) \to \varepsilon$; $V_{\tau_2} = (V_{A_{2-1}} - V_{A_{2-2}}) \to \varepsilon$

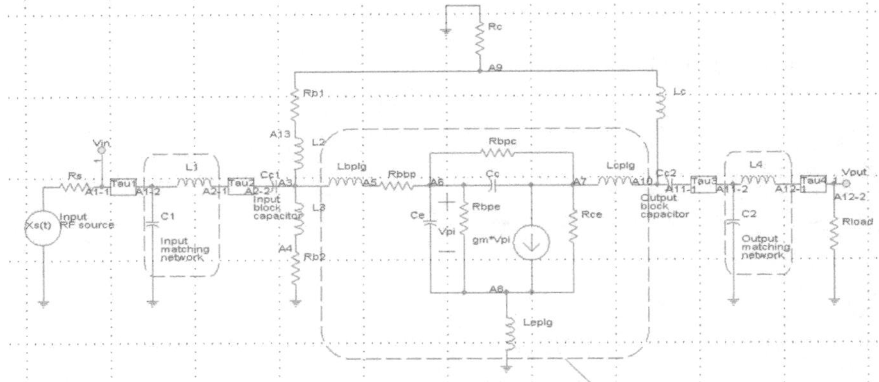

Fig. 4.10 Small Signal (SS) amplifier with matching networks (Hybrid PI equivalent circuit for BJT transistor)

$$V_{\tau 3} = (V_{A_{11-1}} - V_{A_{11-2}}) \rightarrow \varepsilon;\ V_{\tau 4} = (V_{A_{12-1}} - V_{A_{12-2}}) \rightarrow \varepsilon;$$
$$(V_{A_{1-1}} - V_{A_{1-2}}) \rightarrow \varepsilon \Rightarrow V_{A_1} = V_{A_{1-1}} = V_{A_{1-2}}$$
$$(V_{A_{2-1}} - V_{A_{2-2}}) \rightarrow \varepsilon \Rightarrow V_{A_2} = V_{A_{2-1}} = V_{A_{2-2}};$$
$$(V_{A_{11-1}} - V_{A_{11-2}}) \rightarrow \varepsilon \Rightarrow V_{A_{11}} = V_{A_{11-1}} = V_{A_{11-2}}$$
$$(V_{A_{12-1}} - V_{A_{12-2}}) \rightarrow \varepsilon \Rightarrow V_{A_{12}} = V_{A_{12-1}} = V_{A_{12-2}}$$

$$I_{R_s} = \frac{X_s(t) - V_{A_1}}{R_s}; I_{C_1} = C_1 \cdot \frac{dV_{A_1}}{dt}; V_{A_1} - V_{A_2} = L_1 \cdot \frac{dI_{L_1}}{dt}; I_{Cc_1} = C_{C_1} \cdot \frac{d}{dt}(V_{A_2} - V_{A_3})$$

$$I_{L_1}(t - \tau_2) = I_{C_{c1}}(t); V_{A_3} - V_{A_4} = L_3 \cdot \frac{dI_{L_3}}{dt}; I_{R_{b2}} = \frac{V_{A_4}}{R_{b_2}}; I_{L_3} = I_{R_{b2}};$$

$$V_{A_3} - V_{A_{13}} = L_2 \cdot \frac{dI_{L_2}}{dt}; I_{R_{b1}} = \frac{V_{A_{13}} - V_{A_9}}{R_{b1}}$$

$$V_{A_3} - V_{A_5} = L_{bplg} \cdot \frac{dI_{L_{bplg}}}{dt}; I_{R_{bbp}} = \frac{V_{A_5} - V_{A_6}}{R_{bbp}}; I_{L_{bplg}} = I_{R_{bbp}}; I_{R_c} = \frac{V_{A_9}}{R_c};$$

$$V_{pi} = V_{A_6} - V_{A_8} = V_{C_e}$$

$$I_{C_e} = C_e \cdot \frac{dV_{pi}}{dt}; I_{R_{bpe}} = \frac{V_{A_6} - V_{A_8}}{R_{bpe}}; I_{C_c} = C_c \cdot \frac{d(V_{A_6} - V_{A_7})}{dt}; I_{R_{bpc}} = \frac{V_{A_6} - V_{A_7}}{R_{bpc}};$$

$$I_{R_{ce}} = \frac{V_{A_7} - V_{A_8}}{R_{ce}}$$

$$V_{A_8} = L_{eplg} \cdot \frac{dI_{L_{eplg}}}{dt}; V_{A_7} - V_{A_{10}} = L_{cplg} \cdot \frac{dI_{L_{cplg}}}{dt}; V_{A_9} - V_{A_{10}} = L_c \cdot \frac{dL_{L_c}}{dt};$$

$$I_{C_{c2}} = C_{c2} \cdot \frac{d(V_{A_{10}} - V_{A_{11}})}{dt}$$

$$I_{C_2} = C_2 \cdot \frac{dV_{A_{11}}}{dt}; V_{A_{11}} - V_{A_{12}} = L_4 \cdot \frac{dI_{L_4}}{dt}; V_{A_{12}} = V_{out}; I_{L_4} = I_{R_{load}}; I_{R_{load}} = \frac{V_{out}}{R_{load}} = \frac{V_{A_{12}}}{R_{load}}$$

We can write our circuit Kirchhoff's Current Law (KCL) for all nodes:

Table 4.3 Small Signal (SS) amplifier with matching networks circuit Kirchhoff's Current Law (KCL) for all nodes

Node number	KCL @ A_i	Node number	KCL @ A_i
A_1 ($A_1 = A_{1\text{-}1} = A_{1\text{-}2}$)	$I_{R_s}(t-\tau_1) = I_{L_1}(t) + I_{C_1}(t)$	A_8	$I_{C_e} + I_{R_{bpe}} + g_m \cdot V_{pi} + I_{R_{ce}} = I_{L_{eplg}}$
A_2 ($A_2 = A_{2\text{-}1} = A_{2\text{-}2}$)	$I_{L_1}(t-\tau_2) = I_{C_{c1}}(t)$	A_9	$I_{R_{b1}} = I_{R_c} + I_{L_c}$
A_3	$I_{C_{c1}} = I_{L_2} + I_{L_{bplg}} + I_{L_3}$	A_{10}	$I_{L_c} + I_{L_{cplg}} = I_{C_{c2}}$
A_4	$I_{L_3} = I_{R_{b2}}$	A_{11} ($A_{11} = A_{11\text{-}1} = A_{11\text{-}2}$)	$I_{C_{c2}}(t-\tau_3) = I_{L_4}(t) + I_{C_2}(t)$
A_5	$I_{L_{bplg}} = I_{R_{bbp}}$	A_{12} ($A_{12} = A_{12\text{-}1} = A_{12\text{-}2}$)	$I_{L_4}(t-\tau_4) = I_{R_{load}}(t)$
A_6	$I_{R_{bbp}} = I_{R_{bpc}} + I_{C_c} + I_{R_{bpe}} + I_{C_e}$	A_{13}	$I_{L_2} = I_{R_{b1}}$
A_7	$I_{C_c} + I_{R_{bpc}} = g_m \cdot V_{pi} + I_{R_{ce}} + I_{L_{cplg}}$		

$$I_{R_s} = \frac{X_s(t) - V_{A_1}}{R_s} \Rightarrow I_{R_s} \cdot R_s = X_s(t) - V_{A_1} \Rightarrow V_{A_1} = X_s(t) - I_{R_s} \cdot R_s$$

$$I_{C_1} = C_1 \cdot \frac{dV_{A_1}}{dt} = C_1 \cdot (\frac{dX_s(t)}{dt} - \frac{dI_{R_s}}{dt} \cdot R_s); V_{A_1} - V_{A_2}$$
$$= L_1 \cdot \frac{dI_{L_1}}{dt} \Rightarrow V_{A_2} = V_{A_1} - L_1 \cdot \frac{dI_{L_1}}{dt}$$

$$V_{A_2} = X_s(t) - I_{R_s} \cdot R_s - L_1 \cdot \frac{dI_{L_1}}{dt}; I_{C_{C_1}} = C_{C_1} \cdot \left(\frac{dV_{A_2}}{dt} - \frac{dV_{A_3}}{dt}\right)$$
$$= C_{C_1} \cdot \left(\frac{dX_s(t)}{dt} - \frac{dI_{R_s}}{dt} \cdot R_s - L_1 \cdot \frac{d^2 I_{L_1}}{dt^2} - \frac{dV_{A_3}}{dt}\right)$$

$$I_{R_{b2}} = \frac{V_{A_4}}{R_{b_2}} \Rightarrow V_{A_4} = I_{R_{b2}} \cdot R_{b_2}; V_{A_3} - V_{A_4} = L_3 \cdot \frac{dI_{L_3}}{dt} \Rightarrow V_{A_3}$$
$$= I_{R_{b2}} \cdot R_{b_2} + L_3 \cdot \frac{dI_{L_3}}{dt}; \frac{dV_{A_3}}{dt} = \frac{dI_{R_{b2}}}{dt} \cdot R_{b_2} + L_3 \cdot \frac{d^2 I_{L_3}}{dt^2}$$
$$I_{C_{C_1}} = C_{C_1} \cdot (\frac{dX_s(t)}{dt} - \frac{dI_{R_s}}{dt} \cdot R_s - L_1 \cdot \frac{d^2 I_{L_1}}{dt^2} - \frac{dI_{R_{b2}}}{dt} \cdot R_{b_2} - L_3 \cdot \frac{d^2 I_{L_3}}{dt^2})$$
$$I_{L_1}(t - \tau_2) = I_{C_{c1}}(t); I_{L_1}(t) = I_{R_s}(t - \tau_1) - I_{C_1}(t); I_{C_{c1}}(t) = I_{L_1}(t - \tau_2)$$
$$= I_{R_s}(t - \tau_1 - \tau_2) - I_{C_1}(t - \tau_2)$$

$I_{C_{c1}}(t) = I_{L_1}(t - \tau_2) = I_{R_s}(t - \sum_{i=1}^{2} \tau_i) - I_{C_1}(t - \tau_2)$ System equation No. 1:

$$I_{R_s}\left(t - \sum_{i=1}^{2} \tau_i\right) - I_{C_1}(t - \tau_2)$$
$$= C_{C_1} \cdot \left(\frac{dX_s(t)}{dt} - \frac{dI_{R_s}}{dt} \cdot R_s - L_1 \cdot \frac{d^2 I_{L_1}}{dt^2} - \frac{dI_{R_{b2}}}{dt} \cdot R_{b_2} - L_3 \cdot \frac{d^2 I_{L_3}}{dt^2}\right)$$

$$V_{A_3} - V_{A_{13}} = L_2 \cdot \frac{dI_{L_2}}{dt} \Rightarrow V_{A_{13}} = V_{A_3} - L_2 \cdot \frac{dI_{L_2}}{dt} = I_{R_{b2}} \cdot R_{b_2} + L_3 \cdot \frac{dI_{L_3}}{dt} - L_2 \cdot \frac{dI_{L_2}}{dt}$$

$$I_{R_{b1}} = \frac{V_{A_{13}} - V_{A_9}}{R_{b1}} \Rightarrow I_{R_{b1}} \cdot R_{b1} = V_{A_{13}} - V_{A_9} \Rightarrow V_{A_9} = V_{A_{13}} - I_{R_{b1}} \cdot R_{b1}$$

$$V_{A_9} = V_{A_{13}} - I_{R_{b1}} \cdot R_{b1} = I_{R_{b2}} \cdot R_{b_2} + L_3 \cdot \frac{dI_{L_3}}{dt} - L_2 \cdot \frac{dI_{L_2}}{dt} - I_{R_{b1}} \cdot R_{b1}$$

$$V_{A_3} - V_{A_5} = L_{bplg} \cdot \frac{dI_{L_{bplg}}}{dt} \Rightarrow V_{A_5} = I_{R_{b2}} \cdot R_{b_2} + L_3 \cdot \frac{dI_{L_3}}{dt} - L_{bplg} \cdot \frac{dI_{L_{bplg}}}{dt}$$

$$I_{R_{bbp}} = \frac{V_{A_5} - V_{A_6}}{R_{bbp}} \Rightarrow I_{R_{bbp}} \cdot R_{bbp} = V_{A_5} - V_{A_6} \Rightarrow V_{A_6} = I_{R_{b2}} \cdot R_{b_2} + L_3 \cdot \frac{dI_{L_3}}{dt} - L_{bplg} \cdot \frac{dI_{L_{bplg}}}{dt} - I_{R_{bbp}} \cdot R_{bbp}$$

$$I_{R_c} = \frac{V_{A_9}}{R_c} \Rightarrow V_{A_9} = I_{R_c} \cdot R_c; I_{R_{bpe}} = \frac{V_{A_6} - V_{A_8}}{R_{bpe}} \Rightarrow I_{R_{bpe}} \cdot R_{bpe} = V_{A_6} - V_{A_8} \Rightarrow V_{A_8} = V_{A_6} - I_{R_{bpe}} \cdot R_{bpe}$$

$$V_{A_8} = V_{A_6} - I_{R_{bpe}} \cdot R_{bpe} = I_{R_{b2}} \cdot R_{b_2} + L_3 \cdot \frac{dI_{L_3}}{dt} - L_{bplg} \cdot \frac{dI_{L_{bplg}}}{dt} - I_{R_{bbp}} \cdot R_{bbp} - I_{R_{bpe}} \cdot R_{bpe}$$

$$I_{C_c} = C_c \cdot \frac{d(V_{A_6} - V_{A_7})}{dt}; I_{R_{bpc}} = \frac{V_{A_6} - V_{A_7}}{R_{bpc}} \Rightarrow I_{C_c} = C_c \cdot R_{bpc} \cdot \frac{dI_{R_{bpc}}}{dt}; V_{A_7} - V_{A_8} = I_{R_{ce}} \cdot R_{ce}$$

$$V_{A_7} = V_{A_8} + I_{R_{ce}} \cdot R_{ce} = I_{R_{b2}} \cdot R_{b_2} + L_3 \cdot \frac{dI_{L_3}}{dt} - L_{bplg} \cdot \frac{dI_{L_{bplg}}}{dt} - I_{R_{bbp}} \cdot R_{bbp} - I_{R_{bpe}} \cdot R_{bpe} + I_{R_{ce}} \cdot R_{ce}$$

System equation No. 2:

$$V_{A_8} = L_{eplg} \cdot \frac{dI_{L_{eplg}}}{dt} \Rightarrow I_{R_{b2}} \cdot R_{b_2} + L_3 \cdot \frac{dI_{L_3}}{dt} - L_{bplg} \cdot \frac{dI_{L_{bplg}}}{dt} - I_{R_{bbp}} \cdot R_{bbp} - I_{R_{bpe}} \cdot R_{bpe} = L_{eplg} \cdot \frac{dI_{L_{eplg}}}{dt}$$

$$V_{A_7} - V_{A_{10}} = L_{cplg} \cdot \frac{dI_{L_{cplg}}}{dt} \Rightarrow V_{A_{10}} = V_{A_7} - L_{cplg} \cdot \frac{dI_{L_{cplg}}}{dt}$$

$$V_{A_{10}} = I_{R_{b2}} \cdot R_{b_2} + L_3 \cdot \frac{dI_{L_3}}{dt} - L_{bplg} \cdot \frac{dI_{L_{bplg}}}{dt} - I_{R_{bbp}} \cdot R_{bbp} - I_{R_{bpe}} \cdot R_{bpe} + I_{R_{ce}} \cdot R_{ce} - L_{cplg} \cdot \frac{dI_{L_{cplg}}}{dt}$$

$$V_{A_9} - V_{A_{10}} = L_c \cdot \frac{dL_{L_c}}{dt} \Rightarrow V_{A_9} = V_{A_{10}} + L_c \cdot \frac{dL_{L_c}}{dt}; \frac{dV_{A_{10}}}{dt} = \frac{I_{C_{c2}}}{C_{c2}} + \frac{I_{C_2}}{C_2}$$

$$V_{A_9} = I_{R_{b2}} \cdot R_{b_2} + L_3 \cdot \frac{dI_{L_3}}{dt} - L_{bplg} \cdot \frac{dI_{L_{bplg}}}{dt} - I_{R_{bbp}} \cdot R_{bbp} - I_{R_{bpe}} \cdot R_{bpe} + I_{R_{ce}} \cdot R_{ce} - L_{cplg} \cdot \frac{dI_{L_{cplg}}}{dt} + L_c \cdot \frac{dL_{L_c}}{dt}$$

$$I_{C_{c2}} = C_{c2} \cdot \frac{d(V_{A_{10}} - V_{A_{11}})}{dt} \Rightarrow I_{C_{c2}} = C_{c2} \cdot (\frac{dV_{A_{10}}}{dt} - \frac{dV_{A_{11}}}{dt}) \Rightarrow I_{C_{c2}} = C_{c2} \cdot (\frac{dV_{A_{10}}}{dt} - \frac{I_{C_2}}{C_2})$$

$$\frac{I_{C_{c2}}}{C_{c2}} + \frac{I_{C_2}}{C_2} = \frac{d}{dt}\left\{I_{R_{b2}} \cdot R_{b_2} + L_3 \cdot \frac{dI_{L_3}}{dt} - L_{bplg} \cdot \frac{dI_{L_{bplg}}}{dt} - I_{R_{bbp}} \cdot R_{bbp} - I_{R_{bpe}} \cdot R_{bpe} + I_{R_{ce}} \cdot R_{ce} - L_{cplg} \cdot \frac{dI_{L_{cplg}}}{dt}\right\}$$

$$\frac{I_{C_{c2}}}{C_{c2}} + \frac{I_{C_2}}{C_2} = \frac{dI_{R_{b2}}}{dt} \cdot R_{b_2} + L_3 \cdot \frac{d^2 I_{L_3}}{dt^2} - L_{bplg} \cdot \frac{d^2 I_{L_{bplg}}}{dt^2} - \frac{dI_{R_{bbp}}}{dt} \cdot R_{bbp} - \frac{dI_{R_{bpe}}}{dt} \cdot R_{bpe} + \frac{dI_{R_{ce}}}{dt} \cdot R_{ce} - L_{cplg} \cdot \frac{d^2 I_{L_{cplg}}}{dt^2}$$

$$I_{C_{c2}}(t - \tau_3) = I_{L_4}(t) + I_{C_2}(t) \Rightarrow I_{C_2}(t) = I_{C_{c2}}(t - \tau_3) - I_{L_4}(t)$$

System equation No. 3:

$$\frac{I_{C_{c2}}}{C_{c2}} + \frac{I_{C_{c2}}(t - \tau_3)}{C_2} - \frac{I_{L_4}(t)}{C_2} = \frac{dI_{R_{b2}}}{dt} \cdot R_{b_2} + L_3 \cdot \frac{d^2 I_{L_3}}{dt^2} - L_{bplg} \cdot \frac{d^2 I_{L_{bplg}}}{dt^2} - \frac{dI_{R_{bbp}}}{dt} \cdot R_{bbp} - \frac{dI_{R_{bpe}}}{dt} \cdot R_{bpe} + \frac{dI_{R_{ce}}}{dt} \cdot R_{ce} - L_{cplg} \cdot \frac{d^2 I_{L_{cplg}}}{dt^2}$$

$$V_{A_{11}} - V_{A_{12}} = L_4 \cdot \frac{dI_{L_4}}{dt} \Rightarrow V_{A_{11}} = V_{A_{12}} + L_4 \cdot \frac{dI_{L_4}}{dt}; I_{R_{load}} = \frac{V_{A_{12}}}{R_{load}} \Rightarrow V_{A_{12}} = I_{R_{load}} \cdot R_{load}$$

System equation No. 4: $I_{R_{load}}(t) = I_{L_4}(t - \tau_4); I_{C_2}(t) = I_{C_{c2}}(t - \tau_3) - I_{L_4}(t)$

$$V_{A_{11}} = I_{R_{load}} \cdot R_{load} + L_4 \cdot \frac{dI_{L_4}}{dt}; I_{C_2} = C_2 \cdot \left(\frac{dI_{R_{load}}}{dt} \cdot R_{load} + L_4 \cdot \frac{d^2 I_{L_4}}{dt^2}\right)$$

$$I_{C_{c2}}(t - \tau_3) - I_{L_4}(t) = C_2 \cdot \left(\frac{dI_{L_4}(t - \tau_4)}{dt} \cdot R_{load} + L_4 \cdot \frac{d^2 I_{L_4}}{dt^2}\right)$$

We can summery our system equations:

$$I_{R_s}\left(t - \sum_{i=1}^{2} \tau_i\right) - I_{C_1}(t - \tau_2) = C_{C_1} \cdot \left(\frac{dX_s(t)}{dt} - \frac{dI_{R_s}}{dt} \cdot R_s - L_1 \cdot \frac{d^2 I_{L_1}}{dt^2} - \frac{dI_{R_{b2}}}{dt} \cdot R_{b_2} - L_3 \cdot \frac{d^2 I_{L_3}}{dt^2}\right)$$

$$I_{R_{b2}} \cdot R_{b_2} + L_3 \cdot \frac{dI_{L_3}}{dt} - L_{bplg} \cdot \frac{dI_{L_{bplg}}}{dt} - I_{R_{bbp}} \cdot R_{bbp} - I_{R_{bpe}} \cdot R_{bpe} = L_{eplg} \cdot \frac{dI_{L_{eplg}}}{dt}$$

$$\frac{I_{C_{c2}}}{C_{c2}} + \frac{I_{C_{c2}}(t-\tau_3)}{C_2} - \frac{I_{L_4}(t)}{C_2} = \frac{dI_{R_{b2}}}{dt} \cdot R_{b_2} + L_3 \cdot \frac{d^2 I_{L_3}}{dt^2} - L_{bplg} \cdot \frac{d^2 I_{L_{bplg}}}{dt^2}$$
$$- \frac{dI_{R_{bbp}}}{dt} \cdot R_{bbp} - \frac{dI_{R_{bpe}}}{dt} \cdot R_{bpe} + \frac{dI_{R_{ce}}}{dt} \cdot R_{ce} - L_{cplg} \cdot \frac{d^2 I_{L_{cplg}}}{dt^2}$$
$$I_{C_{c2}}(t-\tau_3) - I_{L_4}(t) = C_2 \cdot \left(\frac{dI_{L_4}(t-\tau_4)}{dt} \cdot R_{load} + L_4 \cdot \frac{d^2 I_{L_4}}{dt^2} \right)$$

$$\frac{d}{dt}\left\{ I_{R_{b2}} \cdot R_{b_2} + L_3 \cdot \frac{dI_{L_3}}{dt} - L_{bplg} \cdot \frac{dI_{L_{bplg}}}{dt} - I_{R_{bbp}} \cdot R_{bbp} - I_{R_{bpe}} \cdot R_{bpe} \right\} = L_{eplg} \cdot \frac{d^2 I_{L_{eplg}}}{dt^2}$$

$$\frac{dI_{R_{b2}}}{dt} \cdot R_{b_2} + L_3 \cdot \frac{d^2 I_{L_3}}{dt^2} - L_{bplg} \cdot \frac{d^2 I_{L_{bplg}}}{dt^2} - \frac{dI_{R_{bbp}}}{dt} \cdot R_{bbp} - \frac{dI_{R_{bpe}}}{dt} \cdot R_{bpe} = L_{eplg} \cdot \frac{d^2 I_{L_{eplg}}}{dt^2}$$

We can merge our system two differential equations to one differential equation:

$$\frac{I_{C_{c2}}}{C_{c2}} + \frac{I_{C_{c2}}(t-\tau_3)}{C_2} - \frac{I_{L_4}(t)}{C_2} = L_{eplg} \cdot \frac{d^2 I_{L_{eplg}}}{dt^2} + \frac{dI_{R_{ce}}}{dt} \cdot R_{ce} - L_{cplg} \cdot \frac{d^2 I_{L_{cplg}}}{dt^2}$$

We can summery our system three differential equations:

$$I_{R_s}(t - \sum_{i=1}^{2} \tau_i) - I_{C_1}(t-\tau_2)$$
$$= C_{C_1} \cdot \left(\frac{dX_s(t)}{dt} - \frac{dI_{R_s}}{dt} \cdot R_s - L_1 \cdot \frac{d^2 I_{L_1}}{dt^2} - \frac{dI_{R_{b2}}}{dt} \cdot R_{b_2} - L_3 \cdot \frac{d^2 I_{L_3}}{dt^2} \right)$$
$$\frac{I_{C_{c2}}}{C_{c2}} + \frac{I_{C_{c2}}(t-\tau_3)}{C_2} - \frac{I_{L_4}(t)}{C_2} = L_{eplg} \cdot \frac{d^2 I_{L_{eplg}}}{dt^2} + \frac{dI_{R_{ce}}}{dt} \cdot R_{ce} - L_{cplg} \cdot \frac{d^2 I_{L_{cplg}}}{dt^2}$$
$$I_{C_{c2}}(t-\tau_3) - I_{L_4}(t) = C_2 \cdot \left(\frac{dI_{L_4}(t-\tau_4)}{dt} \cdot R_{load} + L_4 \cdot \frac{d^2 I_{L_4}}{dt^2} \right)$$

To find our equilibrium points (fixed points) of the Small Signal (SS) amplifier with L—matching networks is by $\lim_{t \to \infty} I_{R_s}\left(t - \sum_{i=1}^{2} \tau_i\right) = I_{R_s}(t)$; $\lim_{t \to \infty} I_{C_{c2}}(t-\tau_3) = I_{C_{c2}}(t)$

$$\lim_{t\to\infty} I_{C_1}(t-\tau_2) = I_{C_1}(t); \lim_{t\to\infty} I_{L_4}(t-\tau_4) = I_{L_4}(t); t \gg \tau_2; t \gg \tau_4; t \gg \sum_{i=1}^{2}\tau_i; t \gg \tau_3$$

$$\frac{dI_{R_s}}{dt} = 0; \frac{dI_{R_{b2}}}{dt} = 0; \frac{dI_{R_{ce}}}{dt} = 0; \frac{dI_{L_4}(t-\tau_4)}{dt} = 0; \frac{dI_{L_1}}{dt} = 0 \Rightarrow \frac{d^2I_{L_1}}{dt^2} = 0$$

$$\frac{dI_{L_3}}{dt} = 0 \Rightarrow \frac{d^2I_{L_3}}{dt^2} = 0; \frac{dI_{L_{eplg}}}{dt} = 0 \Rightarrow \frac{d^2I_{L_{eplg}}}{dt^2} = 0; \frac{dI_{L_{cplg}}}{dt} \Rightarrow \frac{d^2I_{L_{cplg}}}{dt^2} = 0$$

$$\frac{dI_{L_4}}{dt} = 0 \Rightarrow \frac{d^2I_{L_4}}{dt^2} = 0; I^*_{R_s} - I^*_{C_1} = C_{C_1}\cdot\frac{dX_s(t)}{dt}; I^*_{C_{c2}}\cdot\left(\frac{1}{C_{c2}} + \frac{1}{C_2}\right) - \frac{I^*_{L_4}}{C_2} = 0$$

$$I^*_{C_{c2}} - I^*_{L_4} = 0 \Rightarrow I^*_{C_{c2}} = I^*_{L_4}; \frac{dX_s(t)}{dt} \to \varepsilon$$

$$I_{C_{c1}} = I_{L_2} + I_{L_{bplg}} + I_{L_3} = I_{L_2} + I_{R_{bbp}} + I_{L_3} = I_{L_2} + (I_{R_{bpc}} + I_{C_c}) + I_{R_{bpe}} + I_{C_e} + I_{L_3}$$

$$I_{C_c} + I_{R_{bpc}} = g_m\cdot V_{pi} + I_{R_{ce}} + I_{L_{cplg}}; I_{C_{c1}} = I_{L_2} + (I_{R_{bpc}} + I_{C_c}) + I_{R_{bpe}} + I_{C_e} + I_{L_3}$$

$$I_{C_{c1}} = I_{L_2} + g_m\cdot V_{pi} + I_{R_{ce}} + I_{L_{cplg}} + I_{R_{bpe}} + I_{C_e} + I_{L_3}; I_{C_e} + I_{R_{bpe}} + g_m\cdot V_{pi} + I_{R_{ce}} = I_{L_{eplg}}$$

$$I_{C_e} + I_{R_{bpe}} + g_m\cdot V_{pi} = I_{L_{eplg}} - I_{R_{ce}}; I_{C_{c1}} = I_{L_2} + I_{L_{eplg}} - I_{R_{ce}} + I_{R_{ce}} + I_{L_{cplg}} + I_{L_3}$$

$$I_{C_{c1}} = I_{L_2} + I_{L_{eplg}} + I_{L_{cplg}} + I_{L_3}; I_{C_{c2}} = I_{L_c} + I_{L_{cplg}} = I_{R_{b1}} - I_{R_c} + I_{L_{cplg}};$$

$$I_{R_{b1}} = I_{R_c} + I_{L_c} \Rightarrow I_{L_c} = I_{R_{b1}} - I_{R_c}$$

$$I_{C_{c2}} = I_{L_c} + I_{L_{cplg}} = I_{R_{b1}} - I_{R_c} + I_{L_{cplg}} = I_{L_2} - I_{R_c} + I_{L_{cplg}}; I_{L_2} = I_{C_{c1}} - I_{L_{bplg}} - I_{L_3}$$

$$I_{C_{c2}} = I_{L_2} - I_{R_c} + I_{L_{cplg}} = I_{C_{c1}} - I_{L_{bplg}} - I_{L_3} - I_{R_c} + I_{L_{cplg}}; I_{C_{c1}}(t) = I_{L_1}(t-\tau_2)$$

$$I_{C_{c2}} = I_{L_1}(t-\tau_2) - I_{L_{bplg}} - I_{L_3} - I_{R_c} + I_{L_{cplg}}; I_{L_{bplg}} = I_{R_{bbp}} = I_{R_{bpc}} + I_{C_c} + I_{R_{bpe}} + I_{C_e}$$

$$I_{L_{bplg}} = (I_{R_{bpc}} + I_{C_c}) + I_{R_{bpe}} + I_{C_e}; I_{C_c} + I_{R_{bpc}} = g_m\cdot V_{pi} + I_{R_{ce}} + I_{L_{cplg}}$$

$$I_{L_{bplg}} = (g_m\cdot V_{pi} + I_{R_{ce}} + I_{L_{cplg}}) + I_{R_{bpe}} + I_{C_e}; I_{C_e} + I_{R_{bpe}} + g_m\cdot V_{pi} + I_{R_{ce}} = I_{L_{eplg}}$$

$$I_{L_{bplg}} = (g_m\cdot V_{pi} + I_{R_{ce}} + I_{R_{bpe}} + I_{C_e}) + I_{L_{cplg}} = I_{L_{eplg}} + I_{L_{cplg}}; I_{L_{bplg}} = I_{L_{eplg}} + I_{L_{cplg}}$$

$$I_{C_{c2}} = I_{L_1}(t-\tau_2) - I_{L_{bplg}} - I_{L_3} - I_{R_c} + I_{L_{cplg}} = I_{L_1}(t-\tau_2) - I_{L_{eplg}} - I_{L_{cplg}} - I_{L_3} - I_{R_c} + I_{L_{cplg}}$$

$$I_{C_{c2}} = I_{L_1}(t-\tau_2) - I_{L_{eplg}} - I_{L_3} - I_{R_c}; I_{C_{c2}}(t-\tau_3) = I_{L_1}(t-\tau_2-\tau_3) - I_{L_{eplg}}(t-\tau_3) - I_{L_3}(t-\tau_3) - I_{R_c}(t-\tau_3)$$

$$I_{C_{c2}}(t-\tau_3) = I_{L_1}\left(t - \sum_{i=2}^{3}\tau_i\right) - I_{L_{eplg}}(t-\tau_3) - I_{L_3}(t-\tau_3) - I_{R_c}(t-\tau_3); I_{R_{b2}} = I_{L_3}$$

We can write our system set of differential equations:

$$I_{R_s}\left(t-\sum_{i=1}^{2}\tau_i\right)-I_{C_1}(t-\tau_2)=C_{C_1}\cdot\left(\frac{dX_s(t)}{dt}-\frac{dI_{R_s}}{dt}\cdot R_s-L_1\cdot\frac{d^2I_{L_1}}{dt^2}-\frac{dI_{L_3}}{dt}\cdot R_{b_2}-L_3\cdot\frac{d^2I_{L_3}}{dt^2}\right)$$

$$\frac{1}{C_{c2}}\cdot[I_{L_1}(t-\tau_2)-I_{L_{eplg}}-I_{L_3}-I_{R_c}]+\frac{1}{C_2}\cdot[I_{L_1}(t-\sum_{i=2}^{3}\tau_i)-I_{L_{eplg}}(t-\tau_3)-I_{L_3}(t-\tau_3)$$
$$-I_{R_c}(t-\tau_3)]-\frac{I_{L_4}(t)}{C_2}=L_{eplg}\cdot\frac{d^2I_{L_{eplg}}}{dt^2}+\frac{dI_{R_{ce}}}{dt}\cdot R_{ce}-L_{cplg}\cdot\frac{d^2I_{L_{cplg}}}{dt^2}$$

$$I_{L_1}(t-\sum_{i=2}^{3}\tau_i)-I_{L_{eplg}}(t-\tau_3)-I_{L_3}(t-\tau_3)-I_{R_c}(t-\tau_3)-I_{L_4}(t)$$
$$=C_2\cdot\left(\frac{dI_{L_4}(t-\tau_4)}{dt}\cdot R_{load}+L_4\cdot\frac{d^2I_{L_4}}{dt^2}\right)$$

Remark: Some of our system variables include time argument X(t) and other without time argument. Always consider these system variables as a function of time.

We consider $L_{13}=L_1\approx L_3; L_{e-cplg}=L_{eplg}\approx L_{cplg}$. We define for simplicity of our analysis new variables: $Y_1=\frac{dI_{R_s}}{dt}; Y_2=\frac{dI_{L_1}}{dt}+\frac{dI_{L_3}}{dt}; \frac{d^2Y_2}{dt^2}=\frac{d^2I_{L_1}}{dt^2}+\frac{d^2I_{L_3}}{dt^2}; Y_3=\frac{dI_{L_3}}{dt}$.

$Y_4=\frac{dI_{R_{ce}}}{dt}; Y_5=\frac{dI_{L_{eplg}}}{dt}-\frac{dI_{L_{cplg}}}{dt}; \frac{dY_5}{dt}=\frac{d^2I_{L_{eplg}}}{dt^2}-\frac{d^2I_{L_{cplg}}}{dt^2}$. The system new set of delay differential equations:

$$I_{R_s}(t-\sum_{i=1}^{2}\tau_i)-I_{C_1}(t-\tau_2)=C_{C_1}\cdot(\frac{dX_s(t)}{dt}-Y_1\cdot R_s-L_{13}\cdot\frac{dY_2}{dt}-Y_3\cdot R_{b_2})$$

$$\frac{1}{C_{c2}}\cdot[I_{L_1}(t-\tau_2)-I_{L_{eplg}}-I_{L_3}-I_{R_c}]+\frac{1}{C_2}\cdot[I_{L_1}(t-\sum_{i=2}^{3}\tau_i)-I_{L_{eplg}}(t-\tau_3)-I_{L_3}(t-\tau_3)$$
$$-I_{R_c}(t-\tau_3)]-\frac{I_{L_4}(t)}{C_2}=L_{e-cplg}\cdot\frac{dY_5}{dt}+Y_4\cdot R_{ce}$$

$$I_{L_1}(t-\sum_{i=2}^{3}\tau_i)-I_{L_{eplg}}(t-\tau_3)-I_{L_3}(t-\tau_3)-I_{R_c}(t-\tau_3)-I_{L_4}(t)$$
$$=C_2\cdot(\frac{dI_{L_4}(t-\tau_4)}{dt}\cdot R_{load}+L_4\cdot\frac{d^2I_{L_4}}{dt^2})$$

We can rewrite our new set of system delay differential equations:

$$Y_6 = \frac{dI_{L_4}}{dt}; \frac{dY_6}{dt} = \frac{d^2 I_{L_4}}{dt^2}; Y_6(t-\tau_4) = \frac{dI_{L_4}(t-\tau_4)}{dt}$$

$$\frac{dY_2}{dt} = -\frac{1}{C_{C_1} \cdot L_{13}} \cdot I_{R_s}\left(t - \sum_{i=1}^{2} \tau_i\right) + \frac{1}{C_{C_1} \cdot L_{13}} \cdot I_{C_1}(t-\tau_2) - \frac{1}{L_{13}} \cdot Y_1 \cdot R_s - \frac{1}{L_{13}} \cdot Y_3 \cdot R_{b_2} + \frac{1}{L_{13}} \cdot \frac{dX_s(t)}{dt}$$

$$\frac{dY_5}{dt} = \frac{1}{C_{c2} \cdot L_{e-cplg}} \cdot [I_{L_1}(t-\tau_2) - I_{L_{eplg}} - I_{L_3} - I_{R_c}] + \frac{1}{C_2 \cdot L_{e-cplg}} \cdot [I_{L_1}(t - \sum_{i=2}^{3} \tau_i) - I_{L_{eplg}}(t-\tau_3) - I_{L_3}(t-\tau_3) - I_{R_c}(t-\tau_3)] - \frac{I_{L_4}(t)}{C_2 \cdot L_{e-cplg}} - Y_4 \cdot \frac{R_{ce}}{L_{e-cplg}}$$

$$\frac{dI_{R_s}}{dt} = Y_1; \frac{dI_{L_1}}{dt} + \frac{dI_{L_3}}{dt} = Y_2; \frac{dI_{L_3}}{dt} = Y_3; \frac{dI_{R_{ce}}}{dt} = Y_4; \frac{dI_{L_{eplg}}}{dt} - \frac{dI_{L_{cplg}}}{dt} = Y_5; \frac{dI_{L_4}}{dt} = Y_6$$

We need to find our system equilibrium points (fixed points) based on the new set of delay differential equations. It is done by setting $\frac{dY_2}{dt} = 0; \frac{dY_5}{dt} = 0; \frac{dY_6}{dt} = 0$

$$\frac{dI_{R_s}}{dt} = 0; \frac{dI_{L_1}}{dt} + \frac{dI_{L_3}}{dt} = 0; \frac{dI_{L_3}}{dt} = 0; \frac{dI_{R_{ce}}}{dt} = 0; \frac{dI_{L_{eplg}}}{dt} - \frac{dI_{L_{cplg}}}{dt} = 0; \frac{dI_{L_4}}{dt} = 0$$

@ fixed points

$$\lim_{t\to\infty} I_{R_s}(t - \sum_{i=1}^{2} \tau_i) \approx I_{R_s}(t); \lim_{t\to\infty} I_{L_1}(t-\tau_2) = I_{L_1}(t); t \gg \sum_{i=1}^{2} \tau_i; t \gg \tau_2$$

$$\lim_{t\to\infty} I_{C_1}(t-\tau_2) = I_{C_1}(t); \lim_{t\to\infty} I_{L_1}(t - \sum_{i=2}^{3} \tau_i) = I_{L_1}(t); t \gg \sum_{i=2}^{3} \tau_i; \lim_{t\to\infty} I_{L_{eplg}}(t-\tau_3) = I_{L_{eplg}}(t)$$

$$\lim_{t\to\infty} I_{L_3}(t-\tau_3) = I_{L_3}(t); t \gg \tau_3; \lim_{t\to\infty} I_{R_c}(t-\tau_3) = I_{R_c}(t); \lim_{t\to\infty} I_{L_1}(t - \sum_{i=2}^{3} \tau_i) = I_{L_1}(t); t \gg \sum_{i=2}^{3} \tau_i$$

$$\lim_{t\to\infty} Y_6(t-\tau_4) = Y_6(t); t \gg \tau_4$$

$$-\frac{1}{C_{C_1}\cdot L_{13}}I^*_{R_s}+\frac{1}{C_{C_1}\cdot L_{13}}\cdot I^*_{C_1}+\frac{1}{L_{13}}\cdot\frac{dX_s(t)}{dt}=0;$$
$$(\frac{1}{C_{c2}}+\frac{1}{C_2})\cdot\frac{[I^*_{L_1}-I^*_{L_{eplg}}-I^*_{L_3}-I^*_{R_c}]}{L_{e-cplg}}-\frac{I^*_{L_4}}{C_2\cdot L_{e-cplg}}=0$$
$$\frac{1}{C_2\cdot L_4}\cdot(I^*_{L_1}-I^*_{L_{eplg}}-I^*_{L_3}-I^*_{R_c}-I^*_{L_4})=0;Y^*_1=0;$$
$$Y^*_2=0;Y^*_3=0;Y^*_4=0;Y^*_5=0;Y^*_6=0$$

Stability analysis: The standard local stability analysis about any one of the equilibrium points of the Small Signal (SS) amplifier with L matching networks consists in adding to coordinate $[I_{R_s}, I_{C_1}, I_{L_1}, I_{L_3}, I_{L_{eplg}}, I_{R_c}, I_{L_4}, Y_1, Y_2, Y_3, Y_4, Y_5, Y_6]$ arbitrarily small increments of exponentially form $[i_{R_s}, i_{C_1}, i_{L_1}, i_{L_3}, i_{L_{eplg}}, i_{R_c}, i_{L_4}, y_1, y_2, y_3, y_4, y_5, y_6]\cdot e^{\lambda\cdot t}$ and retaining the first order terms in $I_{R_s}, I_{C_1}, I_{L_1}, I_{L_3}, I_{L_{eplg}}, I_{R_c}, I_{L_4}, Y_1, Y_2, Y_3, Y_4, Y_5, Y_6$ [2, 3]. The system of homogenous equations leads to a polynomial characteristic equation in the eigenvalues. The polynomial characteristic equations accept by set of the below circuit variables, circuit variables derivative and circuit variables second order derivative with respect to time into Small Signal (SS) amplifier with L matching networks equivalent circuit. Our Small Signal (SS) amplifier with matching L networks equivalent circuit fixed values with arbitrarily small increments of exponential form $[i_{R_s}, i_{C_1}, i_{L_1}, i_{L_3}, i_{L_{eplg}}, i_{R_c}, i_{L_4}, y_1, y_2, y_3, y_4, y_5, y_6]\cdot e^{\lambda\cdot t}$ are: j = 0(first fixed point), j = 1(second fixed point), j = 2(third fixed point), etc.,

$$Y_1(t)=Y_1^{(j)}+y_1\cdot e^{\lambda\cdot t};Y_2(t)=Y_2^{(j)}+y_2\cdot e^{\lambda\cdot t};Y_3(t)=Y_3^{(j)}+y_3\cdot e^{\lambda\cdot t};Y_4(t)=Y_4^{(j)}+y_4\cdot e^{\lambda\cdot t}$$
$$Y_5(t)=Y_5^{(j)}+y_5\cdot e^{\lambda\cdot t};Y_6(t)=Y_6^{(j)}+y_6\cdot e^{\lambda\cdot t};I_{R_s}(t-\sum_{i=1}^{2}\tau_i)=I_{R_s}^{(j)}+i_{R_s}\cdot e^{\lambda\cdot(t-\sum_{i=1}^{2}\tau_i)}$$

$$I_{C_1}(t-\tau_2)=I_{C_1}^{(j)}+i_{C_1}\cdot e^{\lambda\cdot(t-\tau_2)};I_{L_1}(t-\tau_2)=I_{L_1}^{(j)}+i_{L_1}\cdot e^{\lambda\cdot(t-\tau_2)};$$
$$I_{L_{eplg}}(t)=I_{L_{eplg}}^{(j)}+i_{L_{eplg}}\cdot e^{\lambda\cdot t}$$

$$I_{L_3}(t)=I_{L_3}^{(j)}+i_{L_3}\cdot e^{\lambda\cdot t};\ I_{R_c}(t)=I_{R_c}^{(j)}+i_{R_c}\cdot e^{\lambda\cdot t};I_{L_1}(t-\sum_{i=2}^{3}\tau_i)=I_{L_1}^{(j)}+i_{L_1}\cdot e^{\lambda\cdot(t-\sum_{i=2}^{3}\tau_i)}$$
$$I_{L_{eplg}}(t-\tau_3)=I_{L_{eplg}}^{(j)}+i_{L_{eplg}}\cdot e^{\lambda\cdot(t-\tau_3)};\ I_{L_3}(t-\tau_3)=I_{L_3}^{(j)}+i_{L_3}\cdot e^{\lambda\cdot(t-\tau_3)};$$
$$I_{R_c}(t-\tau_3)=I_{R_c}^{(j)}+i_{R_c}\cdot e^{\lambda\cdot(t-\tau_3)}$$
$$I_{L_4}(t)$$
$$\frac{dI_{R_s}}{dt}=i_{R_s}\cdot\lambda\cdot e^{\lambda\cdot t};\frac{dI_{L_1}}{dt}+\frac{dI_{L_3}}{dt}=[i_{L_1}+i_{L_1}]\cdot\lambda\cdot e^{\lambda\cdot t}$$

$$\frac{dI_{L_3}}{dt} = i_{L_3} \cdot \lambda \cdot e^{\lambda \cdot t}; \frac{dI_{R_{ce}}}{dt} = i_{R_{ce}} \cdot \lambda \cdot e^{\lambda \cdot t}; \frac{dI_{L_{eplg}}}{dt} - \frac{dI_{L_{cplg}}}{dt} = [i_{L_{eplg}} - i_{L_{cplg}}] \cdot \lambda \cdot e^{\lambda \cdot t};$$
$$\frac{dI_{L_4}}{dt} = i_{L_4} \cdot \lambda \cdot e^{\lambda \cdot t} I_{L_{eplg}} = I_{L_{eplg}}^{(j)} + i_{L_{eplg}} \cdot e^{\lambda \cdot t}; I_{L_{eplg}}(t - \tau_3) = I_{L_{eplg}}^{(j)} + i_{L_{eplg}} \cdot e^{\lambda \cdot (t - \tau_3)}$$

$$\frac{dY_2}{dt} = -\frac{1}{C_{C_1} \cdot L_{13}} \cdot I_{R_s}(t - \sum_{i=1}^{2} \tau_i) + \frac{1}{C_{C_1} \cdot L_{13}} \cdot I_{C_1}(t - \tau_2) - \frac{1}{L_{13}} \cdot Y_1 \cdot R_s - \frac{1}{L_{13}} \cdot Y_3 \cdot R_{b_2} + \frac{1}{L_{13}} \cdot \frac{dX_s(t)}{dt}$$

$$y_2 \cdot \lambda \cdot e^{\lambda \cdot t} = -\frac{1}{C_{C_1} \cdot L_{13}} \cdot (I_{R_s}^{(j)} + i_{R_s} \cdot e^{\lambda \cdot (t - \sum_{i=1}^{2} \tau_i)}) + \frac{1}{C_{C_1} \cdot L_{13}} \cdot (I_{C_1}^{(j)} + i_{C_1} \cdot e^{\lambda \cdot (t - \tau_2)}) - \frac{1}{L_{13}} \cdot (Y_1^{(j)} + y_1 \cdot e^{\lambda \cdot t}) \cdot R_s - \frac{1}{L_{13}} \cdot (Y_3^{(j)} + y_3 \cdot e^{\lambda \cdot t}) \cdot R_{b_2} + \frac{1}{L_{13}} \cdot \frac{dX_s(t)}{dt}$$

We consider input RF source $X_s(t) = A_0 + \xi(t); |\xi(t)| < 1 \,\&\, A_0 \gg |\xi(t)|$
Then $X_s(t)|_{A_0 \gg |\xi(t)|} = A_0 + \xi(t) \approx A_0; \frac{dX_s(t)|_{A_0 \gg |\xi(t)|}}{dt} = \frac{d\xi(t)}{dt} \to \varepsilon$.

$$y_2 \cdot \lambda \cdot e^{\lambda \cdot t} = -\frac{1}{C_{C_1} \cdot L_{13}} \cdot \left(I_{R_s}^{(j)} + i_{R_s} \cdot e^{\lambda \cdot \left(t - \sum_{i=1}^{2} \tau_i \right)} \right) + \frac{1}{C_{C_1} \cdot L_{13}} \cdot \left(I_{C_1}^{(j)} + i_{C_1} \cdot e^{\lambda \cdot (t - \tau_2)} \right) - \frac{1}{L_{13}} \cdot (Y_1^{(j)} + y_1 \cdot e^{\lambda \cdot t}) \cdot R_s - \frac{1}{L_{13}} \cdot \left(Y_3^{(j)} + y_3 \cdot e^{\lambda \cdot t} \right) \cdot R_{b_2}$$

$$y_2 \cdot \lambda \cdot e^{\lambda \cdot t} = -\frac{1}{C_{C_1} \cdot L_{13}} \cdot I_{R_s}^{(j)} - \frac{1}{C_{C_1} \cdot L_{13}} \cdot i_{R_s} \cdot e^{\lambda \cdot \left(t - \sum_{i=1}^{2} \tau_i \right)} + \frac{1}{C_{C_1} \cdot L_{13}} \cdot I_{C_1}^{(j)} + \frac{1}{C_{C_1} \cdot L_{13}} \cdot i_{C_1} \cdot e^{\lambda \cdot (t - \tau_2)} - \frac{1}{L_{13}} \cdot Y_1^{(j)} \cdot R_s - \frac{R_s}{L_{13}} \cdot y_1 \cdot e^{\lambda \cdot t} - \frac{R_{b_2}}{L_{13}} \cdot Y_3^{(j)} - \frac{R_{b_2}}{L_{13}} \cdot y_3 \cdot e^{\lambda \cdot t}$$

$$y_2 \cdot \lambda \cdot e^{\lambda \cdot t} = -\frac{1}{C_{C_1} \cdot L_{13}} \cdot I_{R_s}^{(j)} + \frac{1}{C_{C_1} \cdot L_{13}} \cdot I_{C_1}^{(j)} - \frac{1}{L_{13}} \cdot Y_1^{(j)} \cdot R_s - \frac{R_{b_2}}{L_{13}} \cdot Y_3^{(j)} - \frac{1}{C_{C_1} \cdot L_{13}} \cdot i_{R_s} \cdot e^{\lambda \cdot \left(t - \sum_{i=1}^{2} \tau_i \right)} + \frac{1}{C_{C_1} \cdot L_{13}} \cdot i_{C_1} \cdot e^{\lambda \cdot (t - \tau_2)} - \frac{R_s}{L_{13}} \cdot y_1 \cdot e^{\lambda \cdot t} - \frac{R_{b_2}}{L_{13}} \cdot y_3 \cdot e^{\lambda \cdot t}$$

@ fixed point $-\frac{1}{C_{C_1} \cdot L_{13}} \cdot I_{R_s}^{(j)} + \frac{1}{C_{C_1} \cdot L_{13}} \cdot I_{C_1}^{(j)} - \frac{1}{L_{13}} \cdot Y_1^{(j)} \cdot R_s - \frac{R_{b_2}}{L_{13}} \cdot Y_3^{(j)} = 0$

$$y_2 \cdot \lambda \cdot e^{\lambda \cdot t} = -\frac{1}{C_{C_1} \cdot L_{13}} \cdot i_{R_s} \cdot e^{\lambda \cdot (t - \sum_{i=1}^{2} \tau_i)} + \frac{1}{C_{C_1} \cdot L_{13}} \cdot i_{C_1} \cdot e^{\lambda \cdot (t - \tau_2)} - \frac{R_s}{L_{13}} \cdot y_1 \cdot e^{\lambda \cdot t} - \frac{R_{b_2}}{L_{13}} \cdot y_3 \cdot e^{\lambda \cdot t}$$

Dividing two sides of the above by $e^{\lambda \cdot t}$ term

$$-y_2 \cdot \lambda - \frac{1}{C_{C_1} \cdot L_{13}} \cdot i_{R_s} \cdot e^{-\lambda \cdot \sum_{i=1}^{2} \tau_i} + \frac{1}{C_{C_1} \cdot L_{13}} \cdot i_{C_1} \cdot e^{-\lambda \cdot \tau_2} - \frac{R_s}{L_{13}} \cdot y_1 - \frac{R_{b_2}}{L_{13}} \cdot y_3 = 0$$

$$\frac{dY_5}{dt} = \frac{1}{C_{c2} \cdot L_{e-cplg}} \cdot [I_{L_1}(t - \tau_2) - I_{L_{eplg}} - I_{L_3} - I_{R_c}] + \frac{1}{C_2 \cdot L_{e-cplg}} \cdot [I_{L_1}(t - \sum_{i=2}^{3} \tau_i) - I_{L_{eplg}}(t - \tau_3) - I_{L_3}(t - \tau_3) - I_{R_c}(t - \tau_3)] - \frac{I_{L_4}(t)}{C_2 \cdot L_{e-cplg}} - Y_4 \cdot \frac{R_{ce}}{L_{e-cplg}}$$

$$\begin{aligned} y_5 \cdot \lambda \cdot e^{\lambda \cdot t} = {} & \frac{1}{C_{c2} \cdot L_{e-cplg}} \cdot \left[\left(I_{L_1}^{(j)} + i_{L_1} \cdot e^{\lambda \cdot (t - \tau_2)} \right) - \left(I_{L_{ep\,\lg}}^{(j)} + i_{L_{ep\,\lg}} \cdot e^{\lambda \cdot t} \right) \right. \\ & \left. - \left(I_{L_3}^{(j)} + i_{L_3} \cdot e^{\lambda \cdot t} \right) - \left(I_{R_c}^{(j)} + i_{R_c} \cdot e^{\lambda \cdot t} \right) \right] \\ & + \frac{1}{C_2 \cdot L_{e-cplg}} \cdot \left[\left(I_{L_1}^{(j)} + i_{L_1} \cdot e^{\lambda \cdot (t - \sum_{i=2}^{3} \tau_i)} \right) - \left(I_{L_{ep\,\lg}}^{(j)} + i_{L_{ep\,\lg}} \cdot e^{\lambda \cdot (t - \tau_3)} \right) \right. \\ & - \left(I_{L_3}^{(j)} + i_{L_3} \cdot e^{\lambda \cdot (t - \tau_3)} \right) \\ & \left. - \left(I_{R_c}^{(j)} + i_{R_c} \cdot e^{\lambda \cdot (t - \tau_3)} \right) \right] - \frac{\left(I_{L_4}^{(j)} + i_{L_4} \cdot e^{\lambda \cdot t} \right)}{C_2 \cdot L_{e-cplg}} - \left(Y_4^{(j)} + y_4 \cdot e^{\lambda \cdot t} \right) \cdot \frac{R_{ce}}{L_{e-cplg}} \end{aligned}$$

$$y_5 \cdot \lambda \cdot e^{\lambda \cdot t} = \left(\frac{1}{C_{c2} \cdot L_{e-cplg}} \cdot I_{L_1}^{(j)} + \frac{1}{C_{c2} \cdot L_{e-cplg}} \cdot i_{L_1} \cdot e^{\lambda \cdot (t - \tau_2)} \right)$$
$$- \left(\frac{1}{C_{c2} \cdot L_{e-cplg}} \cdot I_{L_{eplg}}^{(j)} + \frac{1}{C_{c2} \cdot L_{e-cplg}} \cdot i_{L_{eplg}} \cdot e^{\lambda \cdot t} \right)$$
$$- \left(\frac{1}{C_{c2} \cdot L_{e-cplg}} \cdot I_{L_3}^{(j)} + \frac{1}{C_{c2} \cdot L_{e-cplg}} \cdot i_{L_3} \cdot e^{\lambda \cdot t} \right)$$
$$- \left(\frac{1}{C_{c2} \cdot L_{e-cplg}} \cdot I_{R_c}^{(j)} + \frac{1}{C_{c2} \cdot L_{e-cplg}} \cdot i_{R_c} \cdot e^{\lambda \cdot t} \right)$$
$$+ \left(\frac{1}{C_2 \cdot L_{e-cplg}} \cdot I_{L_1}^{(j)} + \frac{1}{C_2 \cdot L_{e-cplg}} \cdot i_{L_1} \cdot e^{\lambda \cdot (t - \sum_{i=2}^{3} \tau_i)} \right)$$
$$- \left(\frac{1}{C_2 \cdot L_{e-cplg}} \cdot I_{L_{eplg}}^{(j)} + \frac{1}{C_2 \cdot L_{e-cplg}} \cdot i_{L_{eplg}} \cdot e^{\lambda \cdot (t - \tau_3)} \right)$$
$$- \left(\frac{1}{C_2 \cdot L_{e-cplg}} \cdot I_{L_3}^{(j)} + \frac{1}{C_2 \cdot L_{e-cplg}} \cdot i_{L_3} \cdot e^{\lambda \cdot (t - \tau_3)} \right)$$
$$- \left(\frac{1}{C_2 \cdot L_{e-cplg}} \cdot I_{R_c}^{(j)} + \frac{1}{C_2 \cdot L_{e-cplg}} \cdot i_{R_c} \cdot e^{\lambda \cdot (t - \tau_3)} \right)$$
$$- \frac{I_{L_4}^{(j)}}{C_2 \cdot L_{e-cplg}} - \frac{i_{L_4} \cdot e^{\lambda \cdot t}}{C_2 \cdot L_{e-cplg}} - \left(Y_4^{(j)} \cdot \frac{R_{ce}}{L_{e-cplg}} + y_4 \cdot \frac{R_{ce}}{L_{e-cplg}} \cdot e^{\lambda \cdot t} \right)$$

$$y_5 \cdot \lambda \cdot e^{\lambda \cdot t} = \{ \frac{1}{C_{c2} \cdot L_{e-cplg}} \cdot I_{L_1}^{(j)} - \frac{1}{C_{c2} \cdot L_{e-cplg}} \cdot I_{L_{eplg}}^{(j)} - \frac{1}{C_{c2} \cdot L_{e-cplg}} \cdot I_{L_3}^{(j)}$$
$$- \frac{1}{C_{c2} \cdot L_{e-cplg}} \cdot I_{R_c}^{(j)} + \frac{1}{C_2 \cdot L_{e-cplg}} \cdot I_{L_1}^{(j)}$$
$$- \frac{1}{C_2 \cdot L_{e-cplg}} \cdot I_{L_{eplg}}^{(j)} - \frac{1}{C_2 \cdot L_{e-cplg}} \cdot I_{L_3}^{(j)} - \frac{1}{C_2 \cdot L_{e-cplg}} \cdot I_{R_c}^{(j)}$$
$$- \frac{I_{L_4}^{(j)}}{C_2 \cdot L_{e-cplg}} - Y_4^{(j)} \cdot \frac{R_{ce}}{L_{e-cplg}} \} + \frac{1}{C_{c2} \cdot L_{e-cplg}} \cdot i_{L_1} \cdot e^{\lambda \cdot (t - \tau_2)}$$
$$- \frac{1}{C_{c2} \cdot L_{e-cplg}} \cdot i_{L_{eplg}} \cdot e^{\lambda \cdot t} - \frac{1}{C_{c2} \cdot L_{e-cplg}} \cdot i_{L_3} \cdot e^{\lambda \cdot t}$$
$$- \frac{1}{C_{c2} \cdot L_{e-cplg}} \cdot i_{R_c} \cdot e^{\lambda \cdot t} + \frac{1}{C_2 \cdot L_{e-cplg}} \cdot i_{L_1} \cdot e^{\lambda \cdot (t - \sum_{i=2}^{3} \tau_i)}$$
$$- \frac{1}{C_2 \cdot L_{e-cplg}} \cdot i_{L_{eplg}} \cdot e^{\lambda \cdot (t - \tau_3)} - \frac{1}{C_2 \cdot L_{e-cplg}} \cdot i_{L_3} \cdot e^{\lambda \cdot (t - \tau_3)}$$
$$- \frac{1}{C_2 \cdot L_{e-cplg}} \cdot i_{R_c} \cdot e^{\lambda \cdot (t - \tau_3)}$$
$$- \frac{i_{L_4} \cdot e^{\lambda \cdot t}}{C_2 \cdot L_{e-cplg}} - y_4 \cdot \frac{R_{ce}}{L_{e-cplg}} \cdot e^{\lambda \cdot t}$$

@ fixed point

$$\{\frac{1}{C_{c2} \cdot L_{e-cplg}} \cdot I_{L_1}^{(j)} - \frac{1}{C_{c2} \cdot L_{e-cplg}} \cdot I_{L_{eplg}}^{(j)} - \frac{1}{C_{c2} \cdot L_{e-cplg}} \cdot I_{L_3}^{(j)} - \frac{1}{C_{c2} \cdot L_{e-cplg}} \cdot I_{R_c}^{(j)} + \frac{1}{C_2 \cdot L_{e-cplg}} \cdot I_{L_1}^{(j)} - \frac{1}{C_2 \cdot L_{e-cplg}} \cdot I_{L_{eplg}}^{(j)} - \frac{1}{C_2 \cdot L_{e-cplg}} \cdot I_{L_3}^{(j)} - \frac{1}{C_2 \cdot L_{e-cplg}} \cdot I_{R_c}^{(j)} - \frac{I_{L_4}^{(j)}}{C_2 \cdot L_{e-cplg}} - Y_4^{(j)} \cdot \frac{R_{ce}}{L_{e-cplg}}\} = 0$$

$$\begin{aligned} y_5 \cdot \lambda \cdot e^{\lambda \cdot t} = {} & \frac{1}{C_{c2} \cdot L_{e-cplg}} \cdot i_{L_1} \cdot e^{\lambda \cdot (t-\tau_2)} - \frac{1}{C_{c2} \cdot L_{e-cplg}} \cdot i_{L_{eplg}} \cdot e^{\lambda \cdot t} \\ & - \frac{1}{C_{c2} \cdot L_{e-cplg}} \cdot i_{L_3} \cdot e^{\lambda \cdot t} - \frac{1}{C_{c2} \cdot L_{e-cplg}} \cdot i_{R_c} \cdot e^{\lambda \cdot t} \\ & + \frac{1}{C_2 \cdot L_{e-cplg}} \cdot i_{L_1} \cdot e^{\lambda \cdot (t - \sum_{i=2}^{3} \tau_i)} - \frac{1}{C_2 \cdot L_{e-cplg}} \cdot i_{L_{eplg}} \cdot e^{\lambda \cdot (t-\tau_3)} \\ & - \frac{1}{C_2 \cdot L_{e-cplg}} \cdot i_{L_3} \cdot e^{\lambda \cdot (t-\tau_3)} - \frac{1}{C_2 \cdot L_{e-cplg}} \cdot i_{R_c} \cdot e^{\lambda \cdot (t-\tau_3)} \\ & - \frac{i_{L_4} \cdot e^{\lambda \cdot t}}{C_2 \cdot L_{e-cplg}} - y_4 \cdot \frac{R_{ce}}{L_{e-cplg}} \cdot e^{\lambda \cdot t} \end{aligned}$$

Dividing two sides of the above by $e^{\lambda \cdot t}$ term

$$\begin{aligned} y_5 \cdot \lambda = {} & \frac{1}{C_{c2} \cdot L_{e-cplg}} \cdot i_{L_1} \cdot e^{-\lambda \cdot \tau_2} - \frac{1}{C_{c2} \cdot L_{e-cplg}} \cdot i_{L_{eplg}} - \frac{1}{C_{c2} \cdot L_{e-cplg}} \cdot i_{L_3} \\ & - \frac{1}{C_{c2} \cdot L_{e-cplg}} \cdot i_{R_c} + \frac{1}{C_2 \cdot L_{e-cplg}} \cdot i_{L_1} \cdot e^{-\lambda \cdot \sum_{i=2}^{3} \tau_i} \\ & - \frac{1}{C_2 \cdot L_{e-cplg}} \cdot i_{L_{eplg}} \cdot e^{-\lambda \cdot \tau_3} - \frac{1}{C_2 \cdot L_{e-cplg}} \cdot i_{L_3} \cdot e^{-\lambda \cdot \tau_3} \\ & - \frac{1}{C_2 \cdot L_{e-cplg}} \cdot i_{R_c} \cdot e^{-\lambda \cdot \tau_3} - \frac{i_{L_4}}{C_2 \cdot L_{e-cplg}} - y_4 \cdot \frac{R_{ce}}{L_{e-cplg}} \end{aligned}$$

$$\begin{aligned} & - y_5 \cdot \lambda + \frac{1}{C_{c2} \cdot L_{e-cplg}} \cdot i_{L_1} \cdot e^{-\lambda \cdot \tau_2} - \frac{1}{C_{c2} \cdot L_{e-cplg}} \cdot i_{L_{eplg}} - \frac{1}{C_{c2} \cdot L_{e-cplg}} \cdot i_{L_3} \\ & - \frac{1}{C_{c2} \cdot L_{e-cplg}} \cdot i_{R_c} + \frac{1}{C_2 \cdot L_{e-cplg}} \cdot i_{L_1} \cdot e^{-\lambda \cdot \sum_{i=2}^{3} \tau_i} - \frac{1}{C_2 \cdot L_{e-cplg}} \cdot i_{L_{eplg}} \cdot e^{-\lambda \cdot \tau_3} \\ & - \frac{1}{C_2 \cdot L_{e-cplg}} \cdot i_{L_3} \cdot e^{-\lambda \cdot \tau_3} - \frac{1}{C_2 \cdot L_{e-cplg}} \cdot i_{R_c} \cdot e^{-\lambda \cdot \tau_3} - \frac{i_{L_4}}{C_2 \cdot L_{e-cplg}} - y_4 \cdot \frac{R_{ce}}{L_{e-cplg}} = 0 \end{aligned}$$

$$- y_5 \cdot \lambda + \frac{1}{C_{c2} \cdot L_{e-cplg}} \cdot (e^{-\lambda \cdot \tau_2} + e^{-\lambda \cdot \sum_{i=2}^{3} \tau_i}) \cdot i_{L_1} - \frac{1}{C_{c2} \cdot L_{e-cplg}} \cdot (1 + e^{-\lambda \cdot \tau_3}) \cdot i_{L_{eplg}}$$
$$- \frac{1}{C_2 \cdot L_{e-cplg}} \cdot (1 + e^{-\lambda \cdot \tau_3}) \cdot i_{L_3} - \frac{1}{C_2 \cdot L_{e-cplg}} \cdot (1 + e^{-\lambda \cdot \tau_3}) \cdot i_{R_c} - \frac{i_{L_4}}{C_2 \cdot L_{e-cplg}}$$
$$- y_4 \cdot \frac{R_{ce}}{L_{e-cplg}} = 0$$

$$\frac{dY_6}{dt} = \frac{1}{C_2 \cdot L_4} \cdot I_{L_1}(t - \sum_{i=2}^{3} \tau_i) - \frac{1}{C_2 \cdot L_4} \cdot I_{L_{eplg}}(t - \tau_3) - \frac{1}{C_2 \cdot L_4} \cdot I_{L_3}(t - \tau_3)$$
$$- \frac{1}{C_2 \cdot L_4} \cdot I_{R_c}(t - \tau_3) - \frac{1}{C_2 \cdot L_4} \cdot I_{L_4}(t) - \frac{1}{L_4} \cdot Y_6(t - \tau_4) \cdot R_{load}$$

$$y_6 \cdot \lambda \cdot e^{\lambda \cdot t} = \frac{1}{C_2 \cdot L_4} \cdot (I_{L_1}^{(j)} + i_{L_1} \cdot e^{\lambda \cdot (t - \sum_{i=2}^{3} \tau_i)}) - \frac{1}{C_2 \cdot L_4} \cdot (I_{L_{eplg}}^{(j)} + i_{L_{eplg}} \cdot e^{\lambda \cdot (t - \tau_3)})$$
$$- \frac{1}{C_2 \cdot L_4} \cdot (I_{L_3}^{(j)} + i_{L_3} \cdot e^{\lambda \cdot (t - \tau_3)}) - \frac{1}{C_2 \cdot L_4} \cdot (I_{R_c}^{(j)} + i_{R_c} \cdot e^{\lambda \cdot (t - \tau_3)})$$
$$- \frac{1}{C_2 \cdot L_4} \cdot (I_{L_4}^{(j)} + i_{L_4} \cdot e^{\lambda \cdot t}) - \frac{1}{L_4} \cdot (Y_6^{(j)} + y_6 \cdot e^{\lambda \cdot (t - \tau_4)}) \cdot R_{load}$$

$$y_6 \cdot \lambda \cdot e^{\lambda \cdot t} = \frac{1}{C_2 \cdot L_4} \cdot I_{L_1}^{(j)} + \frac{1}{C_2 \cdot L_4} \cdot i_{L_1} \cdot e^{\lambda \cdot (t - \sum_{i=2}^{3} \tau_i)} - \frac{1}{C_2 \cdot L_4} \cdot I_{L_{eplg}}^{(j)}$$
$$- \frac{1}{C_2 \cdot L_4} \cdot i_{L_{eplg}} \cdot e^{\lambda \cdot (t - \tau_3)} - \frac{1}{C_2 \cdot L_4} \cdot I_{L_3}^{(j)} - \frac{1}{C_2 \cdot L_4} \cdot i_{L_3} \cdot e^{\lambda \cdot (t - \tau_3)}$$
$$- \frac{1}{C_2 \cdot L_4} \cdot I_{R_c}^{(j)} - \frac{1}{C_2 \cdot L_4} \cdot i_{R_c} \cdot e^{\lambda \cdot (t - \tau_3)} - \frac{1}{C_2 \cdot L_4} \cdot I_{L_4}^{(j)} - \frac{1}{C_2 \cdot L_4} \cdot i_{L_4} \cdot e^{\lambda \cdot t}$$
$$- \frac{R_{load}}{L_4} \cdot Y_6^{(j)} - \frac{R_{load}}{L_4} \cdot y_6 \cdot e^{\lambda \cdot (t - \tau_4)}$$

$$y_6 \cdot \lambda \cdot e^{\lambda \cdot t} = \left\{ \frac{1}{C_2 \cdot L_4} \cdot I_{L_1}^{(j)} - \frac{1}{C_2 \cdot L_4} \cdot I_{L_{eplg}}^{(j)} - \frac{1}{C_2 \cdot L_4} \cdot I_{L_3}^{(j)} - \frac{1}{C_2 \cdot L_4} \cdot I_{R_c}^{(j)} \right.$$
$$\left. - \frac{1}{C_2 \cdot L_4} \cdot I_{L_4}^{(j)} - \frac{R_{load}}{L_4} \cdot Y_6^{(j)} \right\}$$
$$+ \frac{1}{C_2 \cdot L_4} \cdot i_{L_1} \cdot e^{\lambda \cdot (t - \sum_{i=2}^{3} \tau_i)} - \frac{1}{C_2 \cdot L_4} \cdot i_{L_{eplg}} \cdot e^{\lambda \cdot (t - \tau_3)}$$
$$- \frac{1}{C_2 \cdot L_4} \cdot i_{L_3} \cdot e^{\lambda \cdot (t - \tau_3)} - \frac{1}{C_2 \cdot L_4} \cdot i_{R_c} \cdot e^{\lambda \cdot (t - \tau_3)}$$
$$- \frac{1}{C_2 \cdot L_4} \cdot i_{L_4} \cdot e^{\lambda \cdot t} - \frac{R_{load}}{L_4} \cdot y_6 \cdot e^{\lambda \cdot (t - \tau_4)}$$

@ fixed point

$$\frac{1}{C_2 \cdot L_4} \cdot I_{L_1}^{(j)} - \frac{1}{C_2 \cdot L_4} \cdot I_{L_{eplg}}^{(j)} - \frac{1}{C_2 \cdot L_4} \cdot I_{L_3}^{(j)} - \frac{1}{C_2 \cdot L_4} \cdot I_{R_c}^{(j)} - \frac{1}{C_2 \cdot L_4} \cdot I_{L_4}^{(j)} - \frac{R_{load}}{L_4} \cdot Y_6^{(j)} = 0$$

$$y_6 \cdot \lambda \cdot e^{\lambda \cdot t} = \frac{1}{C_2 \cdot L_4} \cdot i_{L_1} \cdot e^{\lambda \cdot (t - \sum_{i=2}^{3} \tau_i)} - \frac{1}{C_2 \cdot L_4} \cdot i_{L_{eplg}} \cdot e^{\lambda \cdot (t - \tau_3)} - \frac{1}{C_2 \cdot L_4} \cdot i_{L_3} \cdot e^{\lambda \cdot (t - \tau_3)} - \frac{1}{C_2 \cdot L_4} \cdot i_{R_c} \cdot e^{\lambda \cdot (t - \tau_3)} - \frac{1}{C_2 \cdot L_4} \cdot i_{L_4} \cdot e^{\lambda \cdot t} - \frac{R_{load}}{L_4} \cdot y_6 \cdot e^{\lambda \cdot (t - \tau_4)}$$

$$- y_6 \cdot \lambda \cdot e^{\lambda \cdot t} + \frac{1}{C_2 \cdot L_4} \cdot i_{L_1} \cdot e^{\lambda \cdot (t - \sum_{i=2}^{3} \tau_i)} - \frac{1}{C_2 \cdot L_4} \cdot i_{L_{eplg}} \cdot e^{\lambda \cdot (t - \tau_3)} - \frac{1}{C_2 \cdot L_4} \cdot i_{L_3} \cdot e^{\lambda \cdot (t - \tau_3)} - \frac{1}{C_2 \cdot L_4} \cdot i_{R_c} \cdot e^{\lambda \cdot (t - \tau_3)} - \frac{1}{C_2 \cdot L_4} \cdot i_{L_4} \cdot e^{\lambda \cdot t} - \frac{R_{load}}{L_4} \cdot y_6 \cdot e^{\lambda \cdot (t - \tau_4)} = 0$$

Dividing two sides of the above by $e^{\lambda \cdot t}$ term

$$- y_6 \cdot \lambda + \frac{1}{C_2 \cdot L_4} \cdot i_{L_1} \cdot e^{-\lambda \cdot \sum_{i=2}^{3} \tau_i} - \frac{1}{C_2 \cdot L_4} \cdot i_{L_{eplg}} \cdot e^{-\lambda \cdot \tau_3} - \frac{1}{C_2 \cdot L_4} \cdot i_{L_3} \cdot e^{-\lambda \cdot \tau_3} - \frac{1}{C_2 \cdot L_4} \cdot i_{R_c} \cdot e^{-\lambda \cdot \tau_3} - \frac{1}{C_2 \cdot L_4} \cdot i_{L_4} - \frac{R_{load}}{L_4} \cdot y_6 \cdot e^{-\lambda \cdot \tau_4} = 0$$

$$\frac{dI_{R_s}}{dt} = Y_1; \frac{dI_{L_1}}{dt} + \frac{dI_{L_3}}{dt} = Y_2; \frac{dI_{L_3}}{dt} = Y_3; \frac{dI_{R_{ce}}}{dt} = Y_4; \frac{dI_{L_{eplg}}}{dt} - \frac{dI_{L_{cplg}}}{dt} = Y_5; \frac{dI_{L_4}}{dt} = Y_6$$

$$\frac{dI_{R_s}}{dt} = Y_1 \Rightarrow i_{R_s} \cdot \lambda \cdot e^{\lambda \cdot t} = Y_1^{(j)} + y_1 \cdot e^{\lambda \cdot t} \Rightarrow_{Y_1^{(j)}=0} -i_{R_s} \cdot \lambda + y_1 = 0$$

$$\frac{dI_{L_1}}{dt} + \frac{dI_{L_3}}{dt} = Y_2 \Rightarrow i_{L_1} \cdot \lambda \cdot e^{\lambda \cdot t} + i_{L_3} \cdot \lambda \cdot e^{\lambda \cdot t} = Y_2^{(j)} + y_2 \cdot e^{\lambda \cdot t} \Rightarrow_{Y_2^{(j)}=0}$$
$$-(i_{L_1} + i_{L_3}) \cdot \lambda + y_2 = 0$$

$$\frac{dI_{L_3}}{dt} = Y_3 \Rightarrow i_{L_3} \cdot \lambda \cdot e^{\lambda t} = Y_3^{(j)} + y_3 \cdot e^{\lambda t} \Rightarrow_{Y_3^{(j)}=0} -i_{L_3} \cdot \lambda + y_3 = 0$$

$$\frac{dI_{R_{ce}}}{dt} = Y_4 \Rightarrow i_{R_{ce}} \cdot \lambda \cdot e^{\lambda \cdot t} = Y_4^{(j)} + y_4 \cdot e^{\lambda \cdot t} \Rightarrow_{Y_4^{(j)}=0} -i_{R_{ce}} \cdot \lambda + y_4 = 0$$

$$\frac{dI_{L_{eplg}}}{dt} - \frac{dI_{L_{cplg}}}{dt} = Y_5 \Rightarrow i_{L_{eplg}} \cdot \lambda \cdot e^{\lambda \cdot t} - i_{L_{cplg}} \cdot \lambda \cdot e^{\lambda \cdot t} = Y_5^{(j)} + y_5 \cdot e^{\lambda \cdot t} \Rightarrow_{Y_5^{(j)}=0}$$
$$-(i_{L_{eplg}} - i_{L_{cplg}}) \cdot \lambda + y_5 = 0$$

$$\frac{dI_{L_4}}{dt} = Y_6 \Rightarrow i_{L_4} \cdot \lambda \cdot e^{\lambda \cdot t} = Y_6^{(j)} + y_6 \cdot e^{\lambda \cdot t} \Rightarrow_{Y_6^{(j)}=0} -i_{L_4} \cdot \lambda + y_6 = 0$$

We can summery our results:

$$-y_2 \cdot \lambda - \frac{1}{C_{C_1} \cdot L_{13}} \cdot i_{R_s} \cdot e^{-\lambda \cdot \sum_{i=1}^{2} \tau_i} + \frac{1}{C_{C_1} \cdot L_{13}} \cdot i_{C_1} \cdot e^{-\lambda \cdot \tau_2} - \frac{R_s}{L_{13}} \cdot y_1 - \frac{R_{b_2}}{L_{13}} \cdot y_3 = 0$$

$$-y_5 \cdot \lambda + \frac{1}{C_{c2} \cdot L_{e-cplg}} \cdot (e^{-\lambda \cdot \tau_2} + e^{-\lambda \cdot \sum_{i=2}^{3} \tau_i}) \cdot i_{L_1} - \frac{1}{C_{c2} \cdot L_{e-cplg}} \cdot (1 + e^{-\lambda \cdot \tau_3}) \cdot i_{L_{eplg}}$$
$$-\frac{1}{C_2 \cdot L_{e-cplg}} \cdot (1 + e^{-\lambda \cdot \tau_3}) \cdot i_{L_3}$$
$$-\frac{1}{C_2 \cdot L_{e-cplg}} \cdot (1 + e^{-\lambda \cdot \tau_3}) \cdot i_{R_c} - \frac{i_{L_4}}{C_2 \cdot L_{e-cplg}} - y_4 \cdot \frac{R_{ce}}{L_{e-cplg}} = 0$$

$$-(\lambda + \frac{R_{load}}{L_4} \cdot e^{-\lambda \cdot \tau_4}) \cdot y_6 + \frac{1}{C_2 \cdot L_4} \cdot i_{L_1} \cdot e^{-\lambda \cdot \sum_{i=2}^{3} \tau_i} - \frac{1}{C_2 \cdot L_4} \cdot i_{L_{eplg}} \cdot e^{-\lambda \cdot \tau_3}$$
$$-\frac{1}{C_2 \cdot L_4} \cdot i_{L_3} \cdot e^{-\lambda \cdot \tau_3}$$
$$-\frac{1}{C_2 \cdot L_4} \cdot i_{R_c} \cdot e^{-\lambda \cdot \tau_3} - \frac{1}{C_2 \cdot L_4} \cdot i_{L_4} = 0$$

$$-i_{R_s}\cdot\lambda+y_1=0;-(i_{L_1}+i_{L_3})\cdot\lambda+y_2=0;-i_{L_3}\cdot\lambda+y_3=0;-i_{R_{ce}}\cdot\lambda+y_4=0$$
$$-(i_{L_{eplg}}-i_{L_{cplg}})\cdot\lambda+y_5=0;-i_{L_4}\cdot\lambda+y_6=0;-i_{L_3}\cdot\lambda+y_3=0\Rightarrow y_3=i_{L_3}\cdot\lambda$$
$$-(i_{L_1}+i_{L_3})\cdot\lambda+y_2=0\Rightarrow -i_{L_1}\cdot\lambda-i_{L_3}\cdot\lambda+y_2=0\Rightarrow -i_{L_1}\cdot\lambda-y_3+y_2=0$$

The small increments of our Small Signal (SS) amplifier with L matching networks can be divided to two matrixes. The first matrix is (8x8) and the second matrix is (6x8).

$$\begin{pmatrix}\Upsilon_{11} & \dots & \Upsilon_{18}\\ \vdots & \ddots & \vdots\\ \Upsilon_{81} & \dots & \Upsilon_{88}\end{pmatrix}\cdot\begin{pmatrix}y_2\\ i_{R_s}\\ y_5\\ i_{L_1}\\ i_{L_3}\\ i_{L_4}\\ y_6\\ i_{R_{ce}}\end{pmatrix}+\begin{pmatrix}\Xi_{11} & \dots & \Xi_{16}\\ \vdots & \ddots & \vdots\\ \Xi_{81} & \dots & \Xi_{86}\end{pmatrix}\cdot\begin{pmatrix}i_{C_1}\\ y_1\\ y_3\\ i_{Leplg}\\ i_{R_c}\\ y_4\end{pmatrix}=0;\Upsilon_{11}$$
$$=-\lambda;\Upsilon_{12}=-\frac{1}{C_{C_1}\cdot L_{13}}\cdot e^{-\lambda\cdot\sum_{i=1}^{2}\tau_i}$$

$$\Upsilon_{13}=\Upsilon_{14}=\Upsilon_{15}=\Upsilon_{16}=\Upsilon_{17}=\Upsilon_{18}=0;\Upsilon_{21}=0;\Upsilon_{22}=-\lambda;\Upsilon_{23}=\Upsilon_{24}=\Upsilon_{25}$$
$$=\Upsilon_{26}=\Upsilon_{27}=\Upsilon_{28}=0$$

$$\Upsilon_{31}=\Upsilon_{32}=0;\Upsilon_{33}=-\lambda;\Upsilon_{34}=\frac{1}{C_{c2}\cdot L_{e-cplg}}\cdot(e^{-\lambda\cdot\tau_2}+e^{-\lambda\cdot\sum_{i=2}^{3}\tau_i});$$
$$\Upsilon_{35}=-\frac{1}{C_2\cdot L_{e-cplg}}\cdot(1+e^{-\lambda\cdot\tau_3})$$
$$\Upsilon_{36}=-\frac{1}{C_2\cdot L_{e-cplg}};\Upsilon_{37}=\Upsilon_{38}=0;\Upsilon_{41}=1;\Upsilon_{42}=\Upsilon_{43}=0;$$
$$\Upsilon_{44}=-\lambda;\Upsilon_{45}=\Upsilon_{46}=\Upsilon_{47}=\Upsilon_{48}=0$$

$$\Upsilon_{51} = \Upsilon_{52} = \Upsilon_{53} = \Upsilon_{54} = 0; \Upsilon_{55} = -\lambda; \Upsilon_{56} = \Upsilon_{57} = \Upsilon_{58} = 0;$$
$$\Upsilon_{61} = \Upsilon_{62} = \Upsilon_{63} = \Upsilon_{64} = \Upsilon_{65} = 0$$

$$\Upsilon_{66} = -\lambda; \Upsilon_{67} = 1; \Upsilon_{68} = 0; \Upsilon_{71} = \Upsilon_{72} = \Upsilon_{73} = 0; \Upsilon_{74} = \frac{1}{C_2 \cdot L_4} \cdot e^{-\lambda \cdot \sum_{i=2}^{3} \tau_i};$$

$$\Upsilon_{75} = -\frac{1}{C_2 \cdot L_4} \cdot e^{-\lambda \cdot \tau_3}$$

$$\Upsilon_{76} = -\frac{1}{C_2 \cdot L_4}; \Upsilon_{77} = -(\lambda + \frac{R_{load}}{L_4} \cdot e^{-\lambda \cdot \tau_4}); \Upsilon_{78} = 0; \Upsilon_{81} = \ldots = \Upsilon_{87} = 0;$$

$$\Upsilon_{88} = -\lambda$$

$$\Xi_{11} = \frac{1}{C_{C_1} \cdot L_{13}} \cdot e^{-\lambda \cdot \tau_2}; \Xi_{12} = -\frac{R_s}{L_{13}}; \Xi_{13} = -\frac{R_{b_2}}{L_{13}}; \Xi_{14} = \Xi_{15} = \Xi_{16} = 0$$

$$\Xi_{21} = 0; \Xi_{22} = 1; \Xi_{23} = \Xi_{24} = \Xi_{25} = \Xi_{26} = 0; \Xi_{31} = \Xi_{32} = \Xi_{33} = 0;$$

$$\Xi_{34} = -\frac{1}{C_{c2} \cdot L_{e-cplg}} \cdot (1 + e^{-\lambda \cdot \tau_3})$$

$$\Xi_{35} = -\frac{1}{C_2 \cdot L_{e-cplg}} \cdot (1 + e^{-\lambda \cdot \tau_3}); \Xi_{36} = -\frac{R_{ce}}{L_{e-cplg}}; \Xi_{41} = \Xi_{42} = 0;$$

$$\Xi_{43} = -1; \Xi_{44} = \Xi_{45} = \Xi_{46} = 0$$

$$\Xi_{51} = \Xi_{52} = 0; \Xi_{53} = 1; \Xi_{54} = \Xi_{55} = \Xi_{56} = 0; \Xi_{61} = \ldots = \Xi_{66} = 0;$$

$$\Xi_{71} = \Xi_{72} = \Xi_{73} = 0$$

$$\Xi_{74} = -\frac{1}{C_2 \cdot L_4} \cdot e^{-\lambda \cdot \tau_3}; \Xi_{75} = -\frac{1}{C_2 \cdot L_4} \cdot e^{-\lambda \cdot \tau_3}; \Xi_{76} = 0; \Xi_{81} = \ldots = \Xi_{85} = 0;$$

$$\Xi_{86} = 1$$

We consider in our analysis the following: $\begin{pmatrix} \Xi_{11} & \cdots & \Xi_{16} \\ \vdots & \ddots & \vdots \\ \Xi_{61} & \cdots & \Xi_{66} \end{pmatrix} \cdot \begin{pmatrix} i_{C_1} \\ y_1 \\ y_3 \\ i_{Leplg} \\ i_{R_c} \\ y_4 \end{pmatrix} \to \varepsilon$

Since $\Xi_{61} = \ldots = \Xi_{66} = 0$ then $\det \begin{pmatrix} \Xi_{11} & \cdots & \Xi_{16} \\ \vdots & \ddots & \vdots \\ \Xi_{61} & \cdots & \Xi_{66} \end{pmatrix} = 0$

We consider in our stability analysis small increments Jacobian of our Small Signal (SS) amplifier with L matching networks, first matrix is (8 × 8) [5, 6].

$$\begin{pmatrix} \Upsilon_{11} & \cdots & \Upsilon_{18} \\ \vdots & \ddots & \vdots \\ \Upsilon_{81} & \cdots & \Upsilon_{88} \end{pmatrix} \cdot \begin{pmatrix} y_2 \\ i_{R_s} \\ y_5 \\ i_{L_1} \\ i_{L_3} \\ i_{L_4} \\ y_6 \\ i_{R_{ce}} \end{pmatrix} = 0; A - \lambda \cdot I$$

$$= \begin{pmatrix} \Upsilon_{11} & \cdots & \Upsilon_{18} \\ \vdots & \ddots & \vdots \\ \Upsilon_{81} & \cdots & \Upsilon_{88} \end{pmatrix}; \ \det|A - \lambda \cdot I| = 0$$

$$D(\lambda, \tau_1, \tau_2, \tau_3, \tau_4) = \lambda^8 + \lambda^6 \cdot \frac{1}{C_2 \cdot L_4} + \lambda^7 \cdot \frac{R_{load}}{L_4} \cdot e^{-\lambda \cdot \tau_4}; \tau_4 = \tau$$

$$D(\lambda, \tau) = \lambda^8 + \lambda^6 \cdot \frac{1}{C_2 \cdot L_4} + \lambda^7 \cdot \frac{R_{load}}{L_4} \cdot e^{-\lambda \cdot \tau}$$

We need to get the characteristic equation for stability analysis. We study the occurrence of any possible stability switching resulting from the increase of value of the time delay τ parameter then $D(\lambda, \tau) = P_n(\lambda, \tau) + Q_m(\lambda, \tau) \cdot e^{-\lambda \cdot \tau}$.

The expression for $P_n(\lambda, \tau)$ is $P_n(\lambda, \tau) = \sum_{k=0}^{n} p_k(\tau) \cdot \lambda^k$. The expression for $Q_m(\lambda, \tau)$ is $Q_m(\lambda, \tau) = \sum_{k=0}^{m} q_k(\tau) \cdot \lambda^k$. $P_{n=8}(\lambda) = \lambda^8 + \lambda^6 \cdot \frac{1}{C_2 \cdot L_4}; Q_{m=7}(\lambda) = \lambda^7 \cdot \frac{R_{load}}{L_4}$

$$n = 8; m = 7; n > m; p_0(\tau) = \ldots = p_5(\tau) = 0; p_6(\tau) = \frac{1}{C_2 \cdot L_4}; p_7(\tau) = 0; p_8(\tau) = 1$$

$q_0(\tau) = \ldots = q_6(\tau); q_7(\tau) = \frac{R_{load}}{L_4}$. The homogeneous system for $I_{R_s}, I_{C_1}, I_{L_1}, I_{L_3}, I_{L_{eplg}}, I_{R_c}, I_{L_4}, Y_1, Y_2, Y_3, Y_4, Y_5, Y_6$ leads to a characteristic equation for the eigenvalue λ having the form $P(\lambda, \tau) + Q(\lambda, \tau) \cdot e^{-\lambda \cdot \tau} = 0; P(\lambda, \tau) = \sum_{j=0}^{8} a_j \cdot \lambda^j Q(\lambda, \tau) = \sum_{j=0}^{7} c_j \cdot \lambda^j$ and the coefficients $\{a_j(q_i, q_k, \tau), c_j(q_i, q_k, \tau)\} \in R$ depend on q_i, q_k and delay τ. q_i, q_k are any Small Signal (SS) amplifier with L matching network parameters, other parameters kept as a constant. Unless strictly necessary, the designation of variation arguments (q_i, q_k) will subsequently be omitted from P, Q, a_j and c_j. The coefficients a_j, c_j are continuous, and differentiable functions of their arguments. Furthermore P(λ), Q(λ) are analytic functions of λ, for which the following requirements of the analysis (Kuang and Cong 2005; Kuang 1993) can also be verified in the present case.

(a) If $\lambda = i \cdot \omega$, $\omega \in R$ then $P(i \cdot \omega) + Q(i \cdot \omega) \neq 0$.
(b) $|Q(\lambda)/P(\lambda)|$ is bounded for $|\lambda| \to \infty, \mathrm{Re}\lambda \geq 0$. No roots bifurcation from ∞.
(c) $F(\omega) = |P(i \cdot \omega)|^2 - |Q(i \cdot \omega)|^2$. It has a finite number of zeros; indeed, this is a polynomial in ω.
(d) Each positive root $\omega(q_i, q_k)$ of $F(\omega) = 0$ is continuous and differentiable with respect to q_i, q_k.

We assume that $P_n(\lambda, \tau)$ and $Q_m(\lambda, \tau)$ cannot have common imaginary roots. That is for any real number ω; $P_n(\lambda = i \cdot \omega, \tau) + Q_m(\lambda = i \cdot \omega, \tau) \neq 0$.

$$P_n(\lambda = i \cdot \omega, \tau) = \omega^8 - \omega^6 \cdot \frac{1}{C_2 \cdot L_4}; Q_m(\lambda = i \cdot \omega, \tau) = -i \cdot \omega^7 \cdot \frac{R_{load}}{L_4}$$

$$P_n(\lambda = i \cdot \omega, \tau) + Q_m(\lambda = i \cdot \omega, \tau) = \omega^8 - \omega^6 \cdot \frac{1}{C_2 \cdot L_4} - i \cdot \omega^7 \cdot \frac{R_{load}}{L_4} \neq 0$$

$$|P(i \cdot \omega)|^2 = \omega^{16} + \omega^{12} \cdot \frac{1}{C_2^2 \cdot L_4^2} - 2 \cdot \omega^{14} \cdot \frac{1}{C_2 \cdot L_4}; |Q(i \cdot \omega)|^2 = \omega^{14} \cdot \frac{R_{load}^2}{L_4^2}$$

$$F(\omega) = |P(i \cdot \omega)|^2 - |Q(i \cdot \omega)|^2 = \omega^{16} + \omega^{12} \cdot \frac{1}{C_2^2 \cdot L_4^2} - 2 \cdot \omega^{14} \cdot \frac{1}{C_2 \cdot L_4} - \omega^{14} \cdot \frac{R_{load}^2}{L_4^2}$$

$$F(\omega) = |P(i \cdot \omega)|^2 - |Q(i \cdot \omega)|^2 = \omega^{16} - [2 \cdot \frac{1}{C_2 \cdot L_4} + \frac{R_{load}^2}{L_4^2}] \cdot \omega^{14} + \omega^{12} \cdot \frac{1}{C_2^2 \cdot L_4^2}$$

We define the following parameters for simplicity:

$$\Phi_{16} = 1; \Phi_{14} = -[2 \cdot \frac{1}{C_2 \cdot L_4} + \frac{R_{load}^2}{L_4^2}]; \Phi_{12} = \frac{1}{C_2^2 \cdot L_4^2}; \Phi_{2 \cdot k} = 0 \, \forall \, k = 0, \ldots, 5$$

Hence $F(\omega) = 0$ implies $\sum_{k=0}^{8} \Phi_{2 \cdot k} \cdot \omega^{2 \cdot k} = 0$ and its roots are given by solving the polynomial. Furthermore

$$P_R(i \cdot \omega, \tau) = \omega^8 - \omega^6 \cdot \frac{1}{C_2 \cdot L_4}; P_I(i \cdot \omega, \tau) = 0; Q_R(i \cdot \omega, \tau) = 0;$$

$$Q_I(i \cdot \omega, \tau) = -\omega^7 \cdot \frac{R_{load}}{L_4}$$

$$\sin \theta(\tau) = \frac{-P_R(i \cdot \omega, \tau) \cdot Q_I(i \cdot \omega, \tau) + P_I(i \cdot \omega, \tau) \cdot Q_R(i \cdot \omega, \tau)}{|Q(i \cdot \omega)|^2}$$

$$\cos \theta(\tau) = -\frac{P_R(i \cdot \omega, \tau) \cdot Q_R(i \cdot \omega, \tau) + P_I(i \cdot \omega, \tau) \cdot Q_I(i \cdot \omega, \tau)}{|Q(i \cdot \omega)|^2}$$

$$\sin\theta(\tau) = \frac{[\omega^8 - \omega^6 \cdot \frac{1}{C_2 \cdot L_4}] \cdot \omega^7 \cdot \frac{R_{load}}{L_4}}{\omega^{14} \cdot \frac{R_{load}^2}{L_4^2}} = \left[\omega - \frac{1}{\omega} \cdot \frac{1}{C_2 \cdot L_4}\right] \cdot \frac{L_4}{R_{load}}; \cos\theta(\tau) = 0$$

We can use different parameters terminology from our last characteristics parameters definition: $k \to j; p_k(\tau) \to a_j; q_k(\tau) \to c_j; n = 8; m = 7; n > m$.

Additionally

$$P_n(\lambda,\tau) \to P(\lambda); Q_m(\lambda,\tau) \to Q(\lambda); P(\lambda) = \sum_{j=0}^{8} a_j \cdot \lambda^j; Q(\lambda) = \sum_{j=0}^{7} c_j \cdot \lambda^j$$

$$P(\lambda) + Q(\lambda) \cdot e^{-\lambda \cdot \tau} = 0; P(\lambda) = \lambda^8 + \lambda^6 \cdot \frac{1}{C_2 \cdot L_4}; Q(\lambda) = \lambda^7 \cdot \frac{R_{load}}{L_4}$$

$n, m \in \mathbb{N}_0; n > m; a_j, c_j : \mathbb{R}_{+0} \to \mathbb{R}$ are continuous and differentiable function of τ. In the following "—" denotes complex and conjugate. $P(\lambda), Q(\lambda)$ are analytic functions in λ and differentiable in τ. The coefficients a_j, c_j as follow $\{a_j(C_2, L_4, R_{load,\ldots}) \& c_j(C_2, L_4, R_{load,\ldots})\} \in \mathbb{R}$ depend on Small Signal (SS) amplifier with L matching network system's C_2, L_4, $R_{load}\cdots$ values. $a_0 = \ldots = a_5 = 0; a_6 = \frac{1}{C_2 \cdot L_4}$

$a_7 = 0; a_8 = 1; c_0 = c_1 = \ldots = c_6 = 0; c_7 = \frac{R_{load}}{L_4}$. Unless strictly necessary, the designation of the variation arguments $(C_2, L_4, R_{load,\ldots})$ will subsequently be omitted from P, Q, a_j, c_j. The coefficients a_j, c_j are continuous and differential functions of their arguments. Furthermore P(λ), Q(λ) are analytic function of τ for which the following requirements of the analysis (Kuang 1993) can also be verified in the present case (Beretta and Kuang 2002) [5, 6].

(a) If $\lambda = i \cdot \omega$, $\omega \in \mathbb{R}$ then $P(i \cdot \omega) + Q(i \cdot \omega) \neq 0$ that is P and Q have no common imaginary roots. This condition can verified numerically in the entire $(C_2, L_4, R_{load,\ldots})$ domain of interest.

(b) $|\frac{Q(\lambda)}{P(\lambda)}|$ is bounded for $|\lambda| \to \infty, \mathrm{Re}\lambda \geq 0$. No roots bifurcation from ∞. Indeed in the limit $|\frac{Q(\lambda)}{P(\lambda)}| = |\frac{\lambda^7 \cdot \frac{R_{load}}{L_4}}{\lambda^8 + \lambda^6 \cdot \frac{1}{C_2 \cdot L_4}}|$.

(c) $F(\omega) = |P(i \cdot \omega)|^2 - |Q(i \cdot \omega)|^2 = \omega^{16} - [2 \cdot \frac{1}{C_2 \cdot L_4} + \frac{R_{load}^2}{L_4^2}] \cdot \omega^{14} + \omega^{12} \cdot \frac{1}{C_2^2 \cdot L_4^2}$
It has at most a finite number of zeros. Indeed, this is a polynomial in ω (degree in ω^{16}).

(d) Each positive root $\omega(C_2, L_4, R_{load,\ldots})$ of $F(\omega) = 0$ is continuous and differentiable with respect to $C_2, L_4, R_{load,\ldots}$ this condition can only be assessed numerically.

In addition, since the coefficients in P and Q are real we have $\overline{P(-i \cdot \omega)} = P(i \cdot \omega); \overline{Q(-i \cdot \omega)} = Q(i \cdot \omega)$. $\lambda = i \cdot \omega$, $\omega > 0$ may be on eigenvalue of characteristic equation. The analysis consists in identifying the roots of characteristic equation situated on the imaginary axis of the complex λ plane, whereby increasing the parameters $C_2, L_4, R_{load,...}$ Re λ may at the crossing, change its sign from (−) to (+), that is, from a stable focus $E^*\left(I^*_{R_s}, I^*_{C_1}, I^*_{L_1}, I^*_{L_3}, \ldots\right)$ to an unstable one, or vice versa. This feature may be further assessed by examining the sign of the partial derivatives with respect to $C_2, L_4, R_{load,...}$ and Small Signal (SS) amplifier with L matching networks parameters.

$$\Lambda^{-1}(C_2) = \left(\frac{\partial \mathrm{Re}\lambda}{\partial C_2}\right)_{\lambda=i\cdot\omega}, L_4,\ R_{load}, \ldots = const;\ \Lambda^{-1}(L_4) = \left(\frac{\partial \mathrm{Re}\lambda}{\partial L_4}\right)_{\lambda=i\cdot\omega},$$
$$C_2,\ R_{load}, \ldots = const$$
$$\Lambda^{-1}(R_{load}) = \left(\frac{\partial \mathrm{Re}\lambda}{\partial R_{load}}\right)_{\lambda=i\cdot\omega}, C_2,\ L_4, \ldots = const.\ F(\omega,\tau) = 0 \Rightarrow \sum_{k=0}^{8} \Phi_{2\cdot k}\cdot\omega^{2\cdot k} = 0.$$

When writing $P(\lambda) = P_R(\lambda) + i\cdot P_I(\lambda); Q(\lambda) = Q_R(\lambda) + i\cdot Q_I(\lambda)$ and inserting $\lambda = i\cdot\omega,\ \omega\in\mathbb{R}$ into Small Signal (SS) amplifier with L matching networks system's characteristic equation, ω must satisfy the following:

$$\sin\omega\cdot\tau = g(\omega) = \frac{-P_R(i\cdot\omega)\cdot Q_I(i\cdot\omega) + P_I(i\cdot\omega)\cdot Q_R(i\cdot\omega)}{|Q(i\cdot\omega)|^2}$$

$$\cos\omega\cdot\tau = h(\omega) = -\frac{P_R(i\cdot\omega)\cdot Q_R(i\cdot\omega) + P_I(i\cdot\omega)\cdot Q_I(i\cdot\omega)}{|Q(i\cdot\omega)|^2}$$

Where $|Q(i\cdot\omega)|^2 \neq 0$ in view of requirement (a) above, and $(g,h)\in\mathbb{R}$. Furthermore, it follows above $\sin\omega\cdot\tau$ and $\cos\omega\cdot\tau$ equations that, by squaring and adding the sides, ω must be a positive root of $F(\omega) = |P(i\cdot\omega)|^2 - |Q(i\cdot\omega)|^2 = 0$. Note that $F(\omega)$ can be dependent of τ. If $\tau \notin I$ (assume $I \subseteq \mathbb{R}_{+0}$ is the set where $\omega(\tau)$ is a positive root of $F(\omega)$ and for $\tau\notin I$, $\omega(\tau)$ is not defined then for all τ in I, $\omega(\tau)$ satisfies that $F(\omega,\tau)$) then there are no positive $\omega(\tau)$ solutions for $F(\omega,\tau) = 0$ and we cannot have stability switches. For any $\tau\in I$, where $\omega(\tau)$ is a positive solution of $F(\omega,\tau) = 0$, we can define the angle $\theta(\tau)\in[0, 2\cdot\pi]$ as the solution of

$$\sin\theta(\tau) = \frac{-P_R(i\cdot\omega,\tau)\cdot Q_I(i\cdot\omega,\tau) + P_I(i\cdot\omega,\tau)\cdot Q_R(i\cdot\omega,\tau)}{|Q(i\cdot\omega)|^2}$$

$$\cos\theta(\tau) = -\frac{P_R(i\cdot\omega,\tau)\cdot Q_R(i\cdot\omega,\tau) + P_I(i\cdot\omega,\tau)\cdot Q_I(i\cdot\omega,\tau)}{|Q(i\cdot\omega)|^2}$$

And the relation between the argument $\theta(\tau)$ and $\omega(\tau)\cdot\tau$ for $\tau\in I$ must be $\omega(\tau)\cdot\tau = \theta(\tau) + n\cdot 2\cdot\pi\,\forall n\in\mathbb{N}_0$. Hence we can define the maps $\tau_n : I\to\mathbb{R}_{+0}$ given by $\tau_n(\tau) = \frac{\theta(\tau)+n\cdot 2\cdot\pi}{\omega(\tau)}; n\in\mathbb{N}_0,\ \tau\in I$. Let us introduce the functions

$I\to\mathbb{R};\ S_n(\tau) = \tau - \tau_n(\tau),\ \tau\in I,\ n\in\mathbb{N}_0$ that are continuous and differentiable in τ. In the following, the subscripts $\lambda, \omega, C_2, L_4, R_{load,...}$ indicate the corresponding partial derivatives. Let us first concentrate on $\Lambda(x)$, remember in $\lambda(C_2, L_4, R_{load,\ldots})$ and $\omega(C_2, L_4, R_{load,\ldots})$, and keeping all parameters except one (x) and τ. The derivation closely follows that in reference [BK]. Differentiation Small Signal (SS) amplifier with L matching networks characteristic equation: $P(\lambda) + Q(\lambda) \cdot e^{-\lambda \cdot \tau} = 0$ with respect to specific parameter (x), and inverting the derivative, for convenience one calculates:

Remark: $x = L_4, R_{load}, C_2, \ldots etc$

$$(\frac{\partial \lambda}{\partial x})^{-1} = \frac{-P_{\lambda}(\lambda, x) \cdot Q(\lambda, x) + Q_{\lambda}(\lambda, x) \cdot P(\lambda, x) - \tau \cdot P(\lambda, x) \cdot Q(\lambda, x)}{P_x(\lambda, x) \cdot Q(\lambda, x) - Q_x(\lambda, x) \cdot P(\lambda, x)}$$

where $P_{\lambda} = \frac{\partial P}{\partial \lambda}$ …etc., substituting $\lambda = i \cdot \omega$ and bearing $\overline{P(-i \cdot \omega)} = P(i \cdot \omega)$ and $\overline{Q(-i \cdot \omega)} = Q(i \cdot \omega)$. Then $i \cdot P_{\lambda}(i \cdot \omega) = P_{\omega}(i \cdot \omega); i \cdot Q_{\lambda}(i \cdot \omega) = Q_{\omega}(i \cdot \omega)$ and that on the surface $|P(i \cdot \omega)|^2 = |Q(i \cdot \omega)|^2$, one obtain:

$$(\frac{\partial \lambda}{\partial x})^{-1}|_{\lambda = i \cdot \omega} = \frac{i \cdot P_{\omega}(i \cdot \omega, x) \cdot \overline{P(i \cdot \omega, x)} + i \cdot Q_{\lambda}(i \cdot \omega, x) \cdot \overline{Q(\lambda, x)} - \tau \cdot |P(i \cdot \omega, x)|^2}{P_x(i \cdot \omega, x) \cdot \overline{P(i \cdot \omega, x)} - Q_x(i \cdot \omega, x) \cdot \overline{Q(i \cdot \omega, x)}}$$

Upon separating into real and imaginary parts, with $P = P_R + i \cdot P_I$; $Q = Q_R + i \cdot Q_I$

$P_{\omega} = P_{R\omega} + i \cdot P_{I\omega}; Q_{\omega} = Q_{R\omega} + i \cdot Q_{I\omega}; P_x = P_{Rx} + i \cdot P_{Ix}; Q_x = Q_{Rx} + i \cdot Q_{Ix};$
$P^2 = P_R^2 + P_I^2$

When (x) can be any Small Signal (SS) amplifier with L matching network parameters $C_2, L_4, R_{load,\ldots}$ and any time delay τ etc. Where for convenience, we dropped the arguments $(i \cdot \omega, x)$, and where

$$P_R = \omega^8 - \omega^6 \cdot \frac{1}{C_2 \cdot L_4}; P_I = 0; Q_R = 0; Q_I = -\omega^7 \cdot \frac{R_{load}}{L_4};$$

$$P_{R\omega} = 8 \cdot \omega^7 - 6 \cdot \omega^5 \cdot \frac{1}{C_2 \cdot L_4}$$

$$P_{I\omega} = 0; Q_{R\omega} = 0; Q_{I\omega} = -7 \cdot \omega^6 \cdot \frac{R_{load}}{L_4}; P_{R\tau} = 0; P_{I\tau} = 0; Q_{R\tau} = 0; Q_{I\tau} = 0$$

$$F_{\omega} = 2 \cdot [(P_{R\omega} \cdot P_R + P_{I\omega} \cdot P_I) - (Q_{R\omega} \cdot Q_R + Q_{I\omega} \cdot Q_I)];$$

$$P^2 = P_R^2 + P_I^2 = \omega^{12} \cdot [\omega^2 - \frac{1}{C_2 \cdot L_4}]^2$$

$$F_x = 2 \cdot [(P_{Rx} \cdot P_R + P_{Ix} \cdot P_I) - (Q_{Rx} \cdot Q_R + Q_{Ix} \cdot Q_I)]; \omega_x = -\frac{F_x}{F_{\omega}}$$

We define U and V: $U = (P_R \cdot P_{I\omega} - P_I \cdot P_{R\omega}) - (Q_R \cdot Q_{I\omega} - Q_I \cdot Q_{R\omega})$

$V = (P_R \cdot P_{Ix} - P_I \cdot P_{Rx}) - (Q_R \cdot Q_{Ix} - Q_I \cdot Q_{Rx})$. First we choose our specific parameter as time delay τ. $P_{R\omega} \cdot P_R = [8 \cdot \omega^7 - 6 \cdot \omega^5 \cdot \frac{1}{C_2 \cdot L_4}] \cdot [\omega^8 - \omega^6 \cdot \frac{1}{C_2 \cdot L_4}]$

$$P_{R\omega} \cdot P_R = 2 \cdot \omega^{11} \cdot [4 \cdot \omega^4 - 7 \cdot \omega^2 \cdot \frac{1}{C_2 \cdot L_4} + 3 \cdot \frac{1}{(C_2 \cdot L_4)^2}]; Q_{R\omega} \cdot Q_R = 0; P_{I\omega} \cdot P_I = 0$$

$$Q_{I\omega} \cdot Q_I = 7 \cdot \omega^{13} \cdot \left[\frac{R_{load}}{L_4}\right]^2; P_R \cdot P_{I\omega} = 0; P_I \cdot P_{R\omega} = 0; Q_R \cdot Q_{I\omega} = 0; Q_I \cdot Q_{R\omega} = 0$$

$$F_\tau = 0; F_\omega = 2 \cdot \left\{2 \cdot \omega^{11} \cdot \left[4 \cdot \omega^4 - 7 \cdot \omega^2 \cdot \frac{1}{C_2 \cdot L_4} + 3 \cdot \frac{1}{(C_2 \cdot L_4)^2}\right] - 7 \cdot \omega^{13} \cdot \left[\frac{R_{load}}{L_4}\right]^2\right\}$$

$$F_\omega = 2 \cdot \omega^{11} \cdot \left\{8 \cdot \omega^4 - 7 \cdot \omega^2 \cdot \left(\frac{2}{C_2 \cdot L_4} + \left[\frac{R_{load}}{L_4}\right]^2\right) + \frac{6}{(C_2 \cdot L_4)^2}\right\}; \omega_\tau = 0; V = 0; U = 0$$

$$F_\omega \cdot \frac{\partial \omega}{\partial \tau} + F_\tau = 0; \tau \in I \Rightarrow \omega_\tau = \frac{\partial \omega}{\partial \tau} = -\frac{F_\tau}{F_\omega}; \Lambda^{-1}(\tau) = (\frac{\partial \mathrm{Re}\lambda}{\partial \tau})_{\lambda = i \cdot \omega}$$

$$\Lambda^{-1}(\tau) = \mathrm{Re}\left\{\frac{-2 \cdot [U + \tau \cdot |P|^2] + i \cdot F_\omega}{F_\tau + i \cdot 2 \cdot [V + \omega \cdot |P|^2]}\right\}; sign\{\Lambda^{-1}(\tau)\} = sign\left\{\left(\frac{\partial \mathrm{Re}\lambda}{\partial \tau}\right)_{\lambda = i \cdot \omega}\right\}$$

$$sign\{\Lambda^{-1}(\tau)\} = sign\{F_\omega\} \cdot sign\left\{\tau \cdot \frac{\partial \omega}{\partial \tau} + \omega + \frac{U \cdot \frac{\partial \omega}{\partial \tau} + V}{|P|^2}\right\}$$

$$sign\{\Lambda^{-1}(\tau)\} = sign\left[2 \cdot \omega^{11} \cdot \left\{8 \cdot \omega^4 - 7 \cdot \omega^2 \cdot (\frac{2}{C_2 \cdot L_4} + [\frac{R_{load}}{L_4}]^2) + \frac{6}{(C_2 \cdot L_4)^2}\right\}\right] \cdot sign\{\omega\}$$

Remark: Since P and Q are independent on τ parameter, $sign\{\Lambda^{-1}(\tau)\}$ function Is independent on τ parameter and it is only a function of ω.

We shall presently examine the possibility of stability transitions (bifurcation) of Small Signal (SS) amplifier with L matching networks system, about the equilibrium point $E^*(I_{R_s}^*, I_{C_1}^*, I_{L_1}^*, I_{L_3}^*, \ldots)$ as a result of a variation of delay parameter τ or any other system parameter $(C_2, L_4, R_{load,\ldots})$. The analysis consists in identifying the roots of our system characteristic equation situated on the imaginary axis of the complex λ-plane whereby increasing the delay parameter τ or other system's parameter, Re λ may at the crossing, change its sign from—to +, that is, from a stable focus $E^{(*)}$ to an unstable one, or vice versa. This feature may be further assessed by examining the sign of the partial derivatives with respect to τ or other system parameter $(C_2, L_4, R_{load,\ldots})$.

It is a reader exercise to find the expressions for $sign\{\Lambda^{-1}(C_2)\}$, $sign\{\Lambda^{-1}(L_4)\}, \ldots$

And discuss stability switching for different value of system parameter.

Remark: Our system Jacobian is related to part of small increments, so our stability analysis is not fully implemented to the actual behavior of our system. Our stability switching analysis is under these assumptions.

4.4 Bias—T Three Port Network Stability Switching Under Delayed Micro Strip in Time

The function of the bias T is to simultaneously allow a DC bias voltage and an RF test signal to be applied to the port of a transistor during measurement. In S—parameter measurement system, the DC bias is applied at the port labeled "DC", and the RF test signal from the Vector Network Analyzer (VNA)) is applied to the port labeled "RF". At the RF + DC port, both RF and DC voltages are applied to the device. Basic Bias T schematics: capacitor transfers only RF signal and block DC. Inductor transfers DC signal and block RF [91–93] (Fig. 4.11)..

A bias T is a three ports network designed to provide power to remote devices, such as amplifiers, over the same coaxial cable that RF signals are conveyed. Commercially available bias Ts are available in both "connectorized" and surface mount versions. These units are typically expensive and, although designed to be wideband, often suffer in performance at frequencies below 50 MHz. It is consisting of one inductor and one capacitor the bias T circuit is simple, but particular consideration must be given to component selection. The basic topology and means of operation, of a bias T network is described in the below figure (Fig. 4.12).

The shunt capacitor (C2) on the DC port should not be considered optional. It increases isolation between the RF ports and the DC supply connection by routing any remaining RF leakage on the supply side of the inductor to ground. The circuit is evaluated both as a single unit and in the intended configuration with two bias Ts connected together and transferring power. Bias T design considerations: Finding a

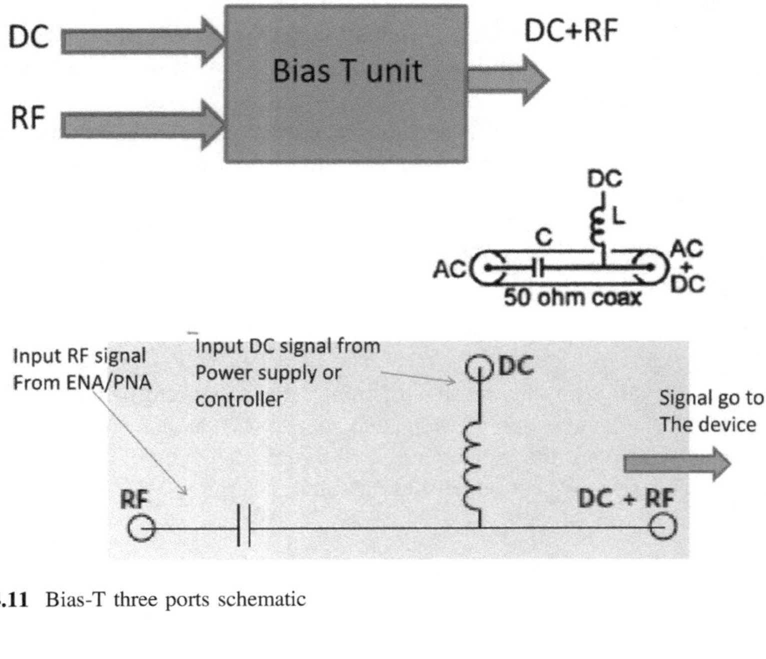

Fig. 4.11 Bias-T three ports schematic

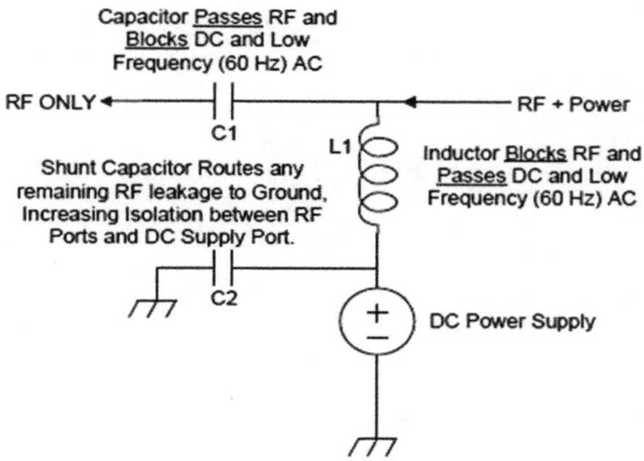

Fig. 4.12 Basic topology of Bias-T network

DC/RF isolation inductor is the main challenge. The inductor must provide high reactance across the bands of interest ($X_L \to \infty$; $Z_L = X_L$) and it must carry the required DC current. Inductor's Q must be high to minimize inductor power dissipation due to the RF signal

$(Q \to \infty \Rightarrow P_L(average) \to \varepsilon \Rightarrow R \to \varepsilon)$. The real component of an inductor has loss due to the resistive component. The stored energy in the inductor is marked as E_L. The average power dissipated in an inductor is marked as P_{avg}.

We analyze the stability of Bias T circuit which feed coax cable. For Bias T's microstrip lines circuit connections, we represent microstrip as a delay lines in time. We neglect the voltage on delay lines $V_{\tau_i} \to \varepsilon$ then $V_{\tau_i} \to \varepsilon \,\forall\, i = 1, 2, \ldots, 5$ (Fig. 4.13).

The delay is in the current that flows through microstrip represented delay line $I(t) \to I(t - \tau_i)$ (Fig. 4.14).

The purpose of the inductor L_1 is to prevent the RF signal from entering DC path, and the purpose of the capacitor C_1 is to keep the DC signal from entering the RF path. The inductor and capacitor should be designed such that the upper cut-off frequency of the low pass DC path is lower than the lower cut-off frequency of the high pass RF path. We define R_{load} as the total resistance seen at the RF + DC port (purely resistive). The equivalent circuit for the proposed system is as follow (Fig. 4.15):

Terminology: $Tau1 \leftrightarrow \tau_1; Tau2 \leftrightarrow \tau_2; Tau3 \leftrightarrow \tau_3; Tau4 \leftrightarrow \tau_4; Tau5 \leftrightarrow \tau_5$

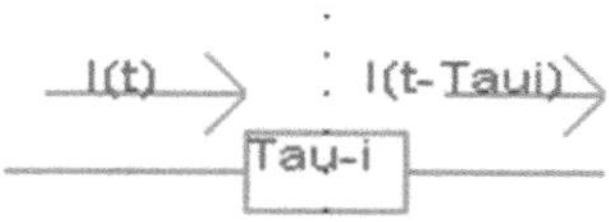

Fig. 4.13 Bias-T microstrip delay line in time

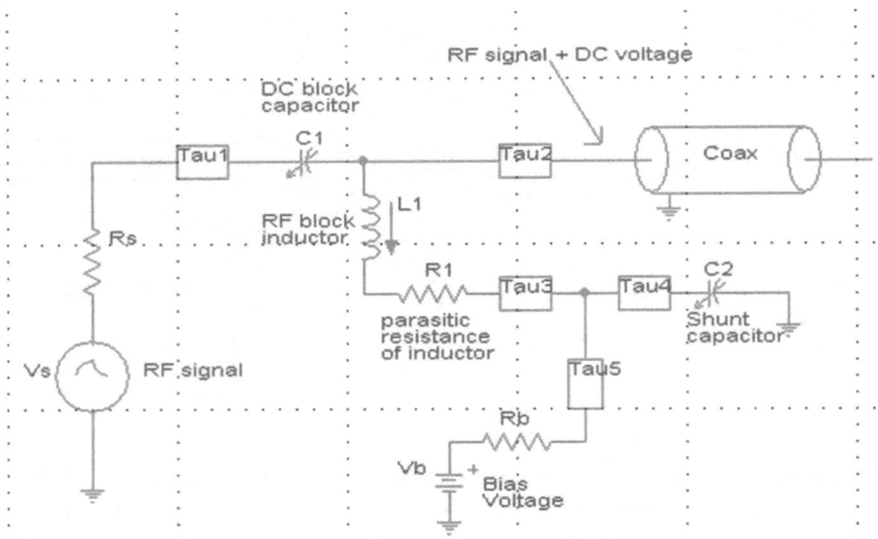

Fig. 4.14 Full Bias-T circuit implementation with delay lines

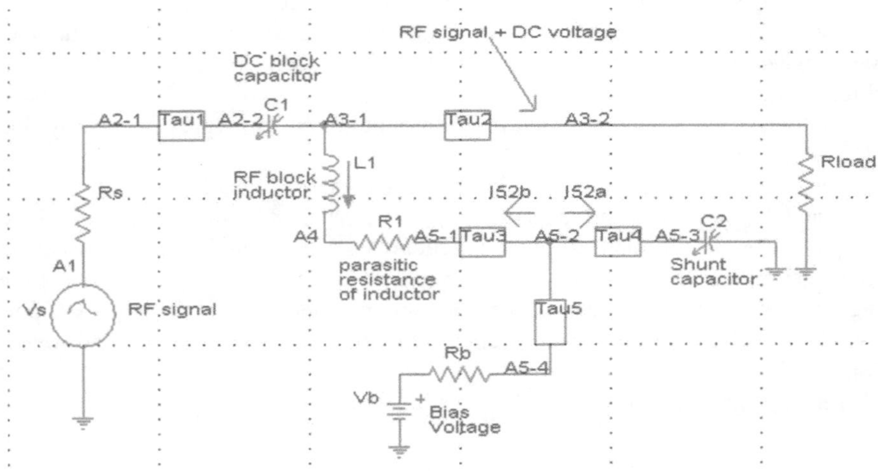

Fig. 4.15 Full Bias-T circuit implementation with delay lines and nodes index

$$V_{\tau_1} \to \varepsilon \Rightarrow A_2 = A_{2-1} = A_{2-2}; V_{\tau_2} \to \varepsilon \Rightarrow A_3 = A_{3-1} = A_{3-2}$$
$$\{V_{\tau_3} \to \varepsilon\}\&\{V_{\tau_4} \to \varepsilon\}\&\{V_{\tau_5} \to \varepsilon\} \Rightarrow A_5 = A_{5-1} = A_{5-2} = A_{5-3} = A_{5-4}$$
$$V_{A_1}=V_s(t); I_{R_s} = \frac{V_s(t) - V_{A_2}}{R_s}; I_{C_1}(t) = I_{R_s}(t-\tau_1);$$
$$\text{KCL @ } A_{3-1} \Rightarrow I_{C_1}(t-\tau_2) + I_{L_1}(t-\tau_2) = I_{R_{load}}(t)$$

$$\text{KCL @ } A_{5-2} \Rightarrow I_{R_b}(t-\tau_5) = I_{52a}(t) + I_{52b}(t); I_{C_2} = I_{52a}(t-\tau_4); I_{R_1}(t) = I_{52b}(t-\tau_3); I_{L_1}(t) = I_{R_1}(t)$$

$$I_{C_1} = C_1 \cdot \frac{d}{dt}(V_{A_{2-2}} - V_{A_{3-1}}); V_{A_2} = V_{A_{2-2}}; V_{A_3} = V_{A_{3-1}}; I_{C_1} = C_1 \cdot \frac{d}{dt}(V_{A_2} - V_{A_3})$$

$I_{C_2} = C_2 \cdot \frac{dV_{A_{5-2}}}{dt}$, since $A_5 = A_{5-1} = A_{5-2}$, $I_{C_2} = C_2 \cdot \frac{dV_{A_5}}{dt}; I_{R_b} = \frac{V_b - V_{A_{5-4}}}{R_b}$ and since $A_5 = A_{5-4} \Rightarrow I_{R_b} = \frac{V_b - V_{A_5}}{R_b}; I_{R_1} = \frac{V_{A_{5-1}} - V_{A_4}}{R_1}$ Since $A_5 = A_{5-1} \Rightarrow I_{R_1} = \frac{V_{A_5} - V_{A_4}}{R_1}$ $V_{L_1} = L_1 \cdot \frac{dI_{L_1}}{dt}; V_{L_1} = V_{A_4} - V_{A_{3-1}}$, since $V_{A_3} = V_{A_{3-1}}; V_{L_1} = V_{A_4} - V_{A_3};$ $V_{A_4} - V_{A_3} = L_1 \cdot \frac{dI_{L_1}}{dt}$

$$V_{A_3} = V_{A_{3-1}} = V_{A_{3-2}} \Rightarrow I_{R_{load}} = \frac{V_{A_{3-2}}}{R_{load}} = \frac{V_{A_3}}{R_{load}}$$

Remark: we consider in our analysis $V_{A_{5-2}} > V_{A_{3-1}}$ then the current flow is from node A_{5-2} to node A_{3-1} otherwise $(V_{A_{5-2}} < V_{A_{3-1}})$ the current flow is from node A_{3-1} to node A_{5-2}.

$$I_{R_s} = \frac{V_s(t) - V_{A_2}}{R_s} \Rightarrow V_s(t) - V_{A_2} = I_{R_s} \cdot R_s \Rightarrow V_{A_2} = V_s(t) - I_{R_s} \cdot R_s$$
$$I_{C_1}(t) = I_{R_s}(t-\tau_1); I_{C_1}(t-\tau_2) + I_{L_1}(t-\tau_2) = I_{R_{load}}(t); I_{R_s}(t-\tau_1-\tau_2) + I_{L_1}(t-\tau_2) = I_{R_{load}}(t)$$
$$I_{C_2} = I_{52a}(t-\tau_4); I_{C_2} = C_2 \cdot \frac{dV_{A_5}}{dt}; I_{52a}(t-\tau_4) = C_2 \cdot \frac{dV_{A_5}}{dt} \Rightarrow V_{A_5} = \frac{1}{C_2} \cdot \int I_{52a}(t-\tau_4) \cdot dt$$
$$I_{R_b} = \frac{V_b - V_{A_5}}{R_b} \Rightarrow V_b - V_{A_5} = I_{R_b} \cdot R_b \Rightarrow V_{A_5} = V_b - I_{R_b} \cdot R_b; I_{R_1} = \frac{V_{A_5} - V_{A_4}}{R_1} \Rightarrow V_{A_5} - V_{A_4} = I_{R_1} \cdot R_1$$

$$I_{R_{load}} = \frac{V_{A_3}}{R_{load}} \Rightarrow V_{A_3} = I_{R_{load}} \cdot R_{load}; V_{A_3} = R_{load} \cdot [I_{R_s}(t - \sum_{i=1}^{2} \tau_i) + I_{L_1}(t - \tau_2)]$$

$$I_{C_1} = C_1 \cdot \frac{d}{dt}(V_{A_2} - V_{A_3}) \Rightarrow V_{A_2} - V_{A_3} = \frac{1}{C_1} \cdot \int I_{C_1} \cdot dt; V_{A_2} - V_{A_3}$$
$$= \frac{1}{C_1} \cdot \int I_{R_s}(t - \tau_1) \cdot dt I_{R_b}(t - \tau_5) = I_{52a}(t) + I_{52b}(t) \Rightarrow I_{52b}(t)$$
$$= I_{R_b}(t - \tau_5) - I_{52a}(t); I_{52b}(t - \tau_3) = I_{R_b}(t - \tau_5 - \tau_3) - I_{52a}(t - \tau_3) I_{R_1}$$
$$= \frac{V_{A_5} - V_{A_4}}{R_1} \Rightarrow V_{A_5} - V_{A_4} = I_{R_1} \cdot R_1; V_{A_5} - V_{A_4} = I_{R_1} \cdot R_1;$$

$$I_{R_1}(t) = I_{52b}(t - \tau_3); I_{R_1}(t) = I_{52b}(t - \tau_3)$$
$$= I_{R_b}(t - \tau_5 - \tau_3) - I_{52a}(t - \tau_3);$$
$$V_{A_5} - V_{A_4} = [I_{R_b}(t - \tau_5 - \tau_3) - I_{52a}(t - \tau_3)] \cdot R_1$$

We can summery our intermediate equations:

$$V_{A_2} = V_s(t) - I_{R_s} \cdot R_s; V_{A_5} = \frac{1}{C_2} \cdot \int I_{52a}(t - \tau_4) \cdot dt; V_{A_5} = V_b - I_{R_b} \cdot R_b$$

$$V_{A_5} - V_{A_4} = I_{R_1} \cdot R_1 = I_{52b}(t - \tau_3) \cdot R_1; V_{A_3} = R_{load} \cdot \left[I_{R_s}(t - \sum_{i=1}^{2} \tau_i) + I_{L_1}(t - \tau_2)\right]$$

$$V_{A_2} - V_{A_3} = \frac{1}{C_1} \cdot \int I_{R_s}(t - \tau_1) \cdot dt; V_{A_5} - V_{A_4} = [I_{R_b}(t - \tau_5 - \tau_3) - I_{52a}(t - \tau_3)] \cdot R_1$$

$$V_{A_5} - V_{A_4} = \left[I_{R_b}\left(t - \sum_{\substack{i=3 \\ i \neq 4}}^{5} \tau_i\right) - I_{52a}(t - \tau_3)\right] \cdot R_1$$

$$V_{A_5} = \frac{1}{C_2} \cdot \int I_{52a}(t - \tau_4) \cdot dt; V_{A_5} = V_b - I_{R_b} \cdot R_b \Rightarrow \frac{1}{C_2} \cdot \int I_{52a}(t - \tau_4) \cdot dt = V_b - I_{R_b} \cdot R_b$$

$$\frac{d}{dt}\left\{\frac{1}{C_2} \cdot \int I_{52a}(t - \tau_4) \cdot dt\right\} = \frac{d}{dt}\{V_b - I_{R_b} \cdot R_b\} \Rightarrow \frac{1}{C_2} \cdot I_{52a}(t - \tau_4) = \frac{dV_b}{dt} - \frac{dI_{R_b}}{dt} \cdot R_b; \frac{dV_b}{dt} = 0$$

$$\frac{1}{C_2} \cdot I_{52a}(t - \tau_4) = -\frac{dI_{R_b}}{dt} \cdot R_b \Rightarrow \frac{dI_{R_b}}{dt} = -\frac{1}{C_2 \cdot R_b} \cdot I_{52a}(t - \tau_4)$$

$$V_s(t) - I_{R_s} \cdot R_s - V_{A_3} = \frac{1}{C_1} \cdot \int I_{R_s}(t - \tau_1) \cdot dt \Rightarrow V_{A_2}$$
$$- R_{load} \cdot [I_{R_s}(t - \sum_{i=1}^{2} \tau_i) + I_{L_1}(t - \tau_2)] = \frac{1}{C_1} \cdot \int I_{R_s}(t - \tau_1) \cdot dt$$

$$V_s(t) - I_{R_s} \cdot R_s - R_{load} \cdot [I_{R_s}(t - \sum_{i=1}^{2} \tau_i) + I_{L_1}(t - \tau_2)] = \frac{1}{C_1} \cdot \int I_{R_s}(t - \tau_1) \cdot dt$$

$$\frac{d}{dt}\{V_s(t) - I_{R_s} \cdot R_s - R_{load} \cdot [I_{R_s}(t - \sum_{i=1}^{2} \tau_i) + I_{L_1}(t - \tau_2)]\} = \frac{d}{dt}\{\frac{1}{C_1} \cdot \int I_{R_s}(t - \tau_1) \cdot dt\}$$

$$\frac{dV_s(t)}{dt} - \frac{dI_{R_s}}{dt} \cdot R_s - R_{load} \cdot [\frac{dI_{R_s}(t - \sum_{i=1}^{2} \tau_i)}{dt} + \frac{dI_{L_1}(t - \tau_2)}{dt}] = \frac{I_{R_s}(t - \tau_1)}{C_1}$$

$$V_{A_4} = V_{A_5} - \left[I_{R_b}\left(t - \sum_{\substack{i=3\\i\neq 4}}^{5} \tau_i\right) - I_{52a}(t - \tau_3) \right] \cdot R_1; V_{A_4} = V_{A_3} + L_1 \cdot \frac{dI_{L_1}}{dt}$$

$$V_{A_3} + L_1 \cdot \frac{dI_{L_1}}{dt} = V_{A_5} - \left[I_{R_b}\left(t - \sum_{\substack{i=3\\i\neq 4}}^{5} \tau_i\right) - I_{52a}(t - \tau_3 \Bigg)\right] \cdot R_1$$

$$R_{load} \cdot \left[I_{R_s}\left(t - \sum_{i=1}^{2} \tau_i\right) + I_{L_1}(t - \tau_2) \right] + L_1 \cdot \frac{dI_{L_1}}{dt} = \frac{1}{C_2} \cdot \int I_{52a}(t - \tau_4) \cdot dt$$
$$- \left[I_{R_b}\left(t - \sum_{\substack{i=3\\i\neq 4}}^{5} \tau_i) - I_{52a}(t - \tau_3 \right)\right] \cdot R_1$$

$$\frac{1}{C_2} \cdot \int I_{52a}(t - \tau_4) \cdot dt = R_{load} \cdot \left[I_{R_s}\left(t - \sum_{i=1}^{2} \tau_i\right) + I_{L_1}(t - \tau_2) \right]$$
$$+ L_1 \cdot \frac{dI_{L_1}}{dt} + \left[I_{R_b}\left(t - \sum_{\substack{i=3\\i\neq 4}}^{5} \tau_i) - I_{52a}(t - \tau_3 \right)\right] \cdot R_1$$

$$\frac{d}{dt}\left\{ \frac{1}{C_2} \cdot \int I_{52a}(t - \tau_4) \cdot dt \right\} = \frac{d}{dt}\{R_{load} \cdot \left[I_{R_s}\left(t - \sum_{i=1}^{2} \tau_i\right) + I_{L_1}(t - \tau_2) \right]$$
$$+ L_1 \cdot \frac{dI_{L_1}}{dt} + [I_{R_b}(t - \sum_{\substack{i=3\\i\neq 4}}^{5} \tau_i) - I_{52a}(t - \tau_3)] \cdot R_1\}$$

$$\frac{I_{52a}(t - \tau_4)}{C_2} = R_{load} \cdot \frac{dI_{R_s}(t - \sum_{i=1}^{2} \tau_i)}{dt} + R_{load} \cdot \frac{dI_{L_1}(t - \tau_2)}{dt} + L_1 \cdot \frac{d^2 I_{L_1}}{dt^2}$$
$$+ \frac{dI_{R_b}(t - \sum_{\substack{i=3\\i\neq 4}}^{5} \tau_i)}{dt} \cdot R_1 - \frac{dI_{52a}(t - \tau_3)}{dt} \cdot R_1$$

$$L_1 \cdot \frac{d^2 I_{L_1}}{dt^2} = \frac{I_{52a}(t - \tau_4)}{C_2} - R_{load} \cdot \frac{dI_{R_s}(t - \sum_{i=1}^{2} \tau_i)}{dt} - R_{load} \cdot \frac{dI_{L_1}(t - \tau_2)}{dt}$$
$$- \frac{dI_{R_b}(t - \sum_{\substack{i=3\\i\neq 4}}^{5} \tau_i)}{dt} \cdot R_1 + \frac{dI_{52a}(t - \tau_3)}{dt} \cdot R_1$$

We can summery our equations as follow: $\frac{dI_{R_b}}{dt} = -\frac{1}{C_2 \cdot R_b} \cdot I_{52a}(t - \tau_4)$

$$\frac{dV_s(t)}{dt} - \frac{dI_{R_s}}{dt} \cdot R_s - R_{load} \cdot [\frac{dI_{R_s}(t - \sum_{i=1}^{2} \tau_i)}{dt} + \frac{dI_{L_1}(t - \tau_2)}{dt}] = \frac{I_{R_s}(t - \tau_1)}{C_1}$$

$$L_1 \cdot \frac{d^2 I_{L_1}}{dt^2} = \frac{I_{52a}(t - \tau_4)}{C_2} - R_{load} \cdot \frac{dI_{R_s}(t - \sum_{i=1}^{2} \tau_i)}{dt} - R_{load} \cdot \frac{dI_{L_1}(t - \tau_2)}{dt}$$
$$- \frac{dI_{R_b}(t - \sum_{\substack{i=3 \\ i \neq 4}}^{5} \tau_i)}{dt} \cdot R_1 + \frac{dI_{52a}(t - \tau_3)}{dt} \cdot R_1$$

We can merge two differential equations:

$$R_{load} \cdot \left[\frac{dI_{R_s}\left(t - \sum_{i=1}^{2} \tau_i\right)}{dt} + \frac{dI_{L_1}(t - \tau_2)}{dt} \right] = \frac{dV_s(t)}{dt} - \frac{dI_{R_s}}{dt} \cdot R_s - \frac{I_{R_s}(t - \tau_1)}{C_1}$$

$$L_1 \cdot \frac{d^2 I_{L_1}}{dt^2} = \frac{I_{52a}(t - \tau_4)}{C_2} - R_{load} \cdot \left[\frac{dI_{R_s}\left(t - \sum_{i=1}^{2} \tau_i\right)}{dt} + \frac{dI_{L_1}(t - \tau_2)}{dt} \right]$$
$$- \frac{dI_{R_b}\left(t - \sum_{\substack{i=3 \\ i \neq 4}}^{5} \tau_i\right)}{dt} \cdot R_1 + \frac{dI_{52a}(t - \tau_3)}{dt} \cdot R_1$$

$$L_1 \cdot \frac{d^2 I_{L_1}}{dt^2} = \frac{I_{52a}(t - \tau_4)}{C_2} - \left[\frac{dV_s(t)}{dt} - \frac{dI_{R_s}}{dt} \cdot R_s - \frac{I_{R_s}(t - \tau_1)}{C_1} \right]$$
$$- \frac{dI_{R_b}\left(t - \sum_{\substack{i=3 \\ i \neq 4}}^{5} \tau_i\right)}{dt} \cdot R_1 + \frac{dI_{52a}(t - \tau_3)}{dt} \cdot R_1$$

We can summery our system two differential equations:

$$\frac{dI_{R_b}}{dt} = -\frac{1}{C_2 \cdot R_b} \cdot I_{52a}(t - \tau_4)$$

$$L_1 \cdot \frac{d^2 I_{L_1}}{dt^2} = \frac{I_{52a}(t - \tau_4)}{C_2} - \left[\frac{dV_s(t)}{dt} - \frac{dI_{R_s}}{dt} \cdot R_s - \frac{I_{R_s}(t - \tau_1)}{C_1}\right] - \frac{dI_{R_b}\left(t - \sum_{\substack{i=3 \\ i \neq 4}}^{5} \tau_i\right)}{dt} \cdot R_1 + \frac{dI_{52a}(t - \tau_3)}{dt} \cdot R_1$$

Some definitions for new variables: $\frac{dI_{R_b}}{dt} = -\frac{1}{C_2 \cdot R_b} \cdot I_{52a}(t - \tau_4); \frac{dY_1}{dt} = \frac{d^2 I_{L_1}}{dt^2};$ $\frac{dI_{L_1}}{dt} = Y_1$

$$L_1 \cdot \frac{dY_1}{dt} = \frac{I_{52a}(t - \tau_4)}{C_2} - \left[\frac{dV_s(t)}{dt} - \frac{dI_{R_s}}{dt} \cdot R_s - \frac{I_{R_s}(t - \tau_1)}{C_1}\right] - \frac{dI_{R_b}\left(t - \sum_{\substack{i=3 \\ i \neq 4}}^{5} \tau_i\right)}{dt} \cdot R_1 + \frac{dI_{52a}(t - \tau_3)}{dt} \cdot R_1$$

At fixed point: $\frac{dI_{L_1}}{dt} = 0; \frac{dY_1}{dt} = 0; \frac{dI_{R_s}}{dt} = 0; \frac{dI_{R_b}(t - \sum_{\substack{i=3 \\ i \neq 4}}^{5} \tau_i)}{dt} = 0; \frac{dI_{52a}(t - \tau_3)}{dt} = 0$

$$\lim_{t\to\infty} I_{52a}(t - \tau_4) = I_{52a}(t); \lim_{t\to\infty} I_{R_s}(t - \tau_1) = I_{R_s}(t); \lim_{t\to\infty} I_{R_b}(t - \sum_{\substack{i=3 \\ i \neq 4}}^{5} \tau_i) = I_{R_b}(t)$$

$$\lim_{t\to\infty} I_{52a}(t - \tau_3) = I_{52a}(t); t > > \tau_4; t > > \tau_1; t > > \sum_{\substack{i=3 \\ i \neq 4}}^{5} \tau_i; t > > \tau_3$$

j is the Index of system fixed points, first fixed point j = 0, second fixed point j = 1, third fixed point j = 2, etc.,

$$\frac{dI_{R_b}}{dt} = 0 \Rightarrow -\frac{1}{C_2 \cdot R_b} \cdot I_{52a}(t-\tau_4) = 0 \Rightarrow I_{52a}^{(j)} = 0; \frac{dI_{L_1}}{dt} = 0 \Rightarrow Y_1^{(j)} = 0$$

$$\frac{I_{52a}^{(j)}}{C_2} - \left[\frac{dV_s(t)}{dt} - \frac{I_{R_s}^{(j)}}{C_1}\right] = 0 \Rightarrow I_{52a}^{(j)} + I_{R_s}^{(j)} = \frac{dV_s(t)}{dt} \cdot C_1; Y_2 = \frac{dI_{R_s}}{dt}; Y_3 = \frac{dI_{52a}}{dt}$$

$$\frac{dI_{R_s}}{dt} = 0 \Rightarrow Y_2^{(j)} = 0; \frac{dI_{52a}}{dt} = 0 \Rightarrow Y_3^{(j)} = 0$$

Stability analysis: The standard local stability analysis about any one of the equilibrium points of the Bias-T circuit which feed coax cable consists in adding to coordinate $[Y_1, Y_2, Y_{3,} I_{52a}, I_{R_s}]$ arbitrarily small increments of exponentially form $[y_1, y_2, y_3, i_{52a}, i_{R_s}] \cdot e^{\lambda \cdot t}$ and retaining the first order terms in $Y_1, Y_2, Y_{3,} I_{52a}, I_{R_s}$. The system of homogenous equations leads to a polynomial characteristic equation in the eigenvalues. The polynomial characteristic equations accept by set of the below circuit variables, circuit variables derivative and circuit variables second order derivative with respect to time into Bias-T circuit which feed coax cable equivalent circuit. Our Bias-T circuit which feed coax cable equivalent circuit fixed values with arbitrarily small increments of exponential form $[y_1, y_2, y_3, i_{52a}, i_{R_s}] \cdot e^{\lambda \cdot t}$ are: j = 0 (first fixed point), j = 1 (second fixed point), j = 2 (third fixed point), etc.,

$$I_{L_1} = I_{L_1}^{(j)} + i_{L_1} \cdot e^{\lambda \cdot t}, Y_1 = Y_1^{(j)} + y_1 \cdot e^{\lambda \cdot t}, I_{52a} = I_{52a}^{(j)} + i_{52a} \cdot e^{\lambda \cdot t}, I_{R_s} = I_{R_s}^{(j)} + i_{R_s} \cdot e^{\lambda \cdot t}, I_{R_b} = I_{R_b}^{(j)} + i_{R_b} \cdot e^{\lambda \cdot t}$$

$$I_{52a}(t-\tau_4) = I_{52a}^{(j)} + i_{52a} \cdot e^{\lambda \cdot (t-\tau_4)}; I_{R_s}(t-\tau_1) = I_{R_s}^{(j)} + i_{R_s} \cdot e^{\lambda \cdot (t-\tau_1)};$$

$$I_{R_b}(t - \sum_{\substack{i=3 \\ i \neq 4}}^{5} \tau_i) = I_{R_b}^{(j)} + i_{R_b} \cdot \exp[\lambda \cdot (t - \sum_{\substack{i=3 \\ i \neq 4}}^{5} \tau_i)]$$

$$\frac{dI_{L_1}}{dt} = i_{L_1} \cdot \lambda \cdot e^{\lambda \cdot t}, \frac{dY_1}{dt} = y_1 \cdot \lambda \cdot e^{\lambda \cdot t}, \frac{dI_{52a}}{dt} = i_{52a} \cdot \lambda \cdot e^{\lambda \cdot t}, \frac{dI_{R_s}}{dt} = i_{R_s} \cdot \lambda \cdot e^{\lambda \cdot t}, \frac{dI_{R_b}}{dt} = i_{R_b} \cdot \lambda \cdot e^{\lambda \cdot t}$$

$$\frac{dI_{52a}(t-\tau_4)}{dt} = i_{52a} \cdot \lambda \cdot e^{\lambda \cdot t} \cdot e^{-\lambda \cdot \tau_4}; \frac{dI_{R_s}(t-\tau_1)}{dt} = i_{R_s} \cdot \lambda \cdot e^{\lambda \cdot t} \cdot e^{-\lambda \cdot \tau_1}; \frac{dI_{R_b}\left(t - \sum_{\substack{i=3 \\ i \neq 4}}^{5} \tau_i\right)}{dt} = i_{R_b} \cdot \lambda \cdot e^{\lambda \cdot t} \cdot \exp[\lambda \cdot (t - \sum_{\substack{i=3 \\ i \neq 4}}^{5} \tau_i)]$$

$$I_{52a}(t-\tau_3) = I_{52a}^{(j)} + i_{52a} \cdot e^{\lambda \cdot (t-\tau_3)}; \frac{dI_{52a}(t-\tau_3)}{dt}$$
$$= i_{52a} \cdot \lambda \cdot e^{\lambda \cdot t} \cdot e^{-\lambda \cdot \tau_3}; Y_2 = Y_2^{(j)} + y_2 \cdot e^{\lambda \cdot t}$$

$$\frac{dI_{R_b}\left(t - \sum_{\substack{i=3 \\ i \neq 4}}^{5} \tau_i\right)}{dt} = \frac{dI_{R_b}}{dt} \cdot \exp[\lambda \cdot (t - \sum_{\substack{i=3 \\ i \neq 4}}^{5} \tau_i)]$$

$$= -\frac{1}{C_2 \cdot R_b} \cdot I_{52a}(t-\tau_4) \cdot \exp[\lambda \cdot (t - \sum_{\substack{i=3 \\ i \neq 4}}^{5} \tau_i)]; \frac{dI_{52a}(t-\tau_3)}{dt} = \frac{dI_{52a}}{dt} \cdot e^{-\lambda \cdot \tau_3}$$

For $\lambda < 0$, $t > 0$, the selected fixed point is stable otherwise $\lambda > 0$, $t > 0$ is unstable. Our system tends to the selected fixed point exponentially for $\lambda < 0$, $t > 0$ otherwise go away from the selected fixed point exponentially. λ is the eigenvalue parameter which is established if the fixed point is stable or unstable; additionally, his absolute value $|\lambda|$ established the speed of flow toward or away from the selected fixed point [5, 6].

We can rewrite our system differential equations: $\frac{dI_{R_b}}{dt} = -\frac{1}{C_2 \cdot R_b} \cdot I_{52a}(t-\tau_4)$; $\frac{dI_{L_1}}{dt} = Y_1$

$$L_1 \cdot \frac{dY_1}{dt} = \frac{I_{52a}(t-\tau_4)}{C_2} - \frac{dV_s(t)}{dt} + \frac{dI_{R_s}}{dt} \cdot R_s + \frac{I_{R_s}(t-\tau_1)}{C_1} + \frac{1}{C_2 \cdot R_b} \cdot I_{52a}(t-\tau_4)$$
$$\cdot \exp[\lambda \cdot (t - \sum_{\substack{i=3 \\ i \neq 4}}^{5} \tau_i)] \cdot R_1 + \frac{dI_{52a}}{dt} \cdot e^{-\lambda \cdot \tau_3} \cdot R_1$$

$$L_1 \cdot \frac{dY_1}{dt} = I_{52a}(t-\tau_4) \cdot \frac{1}{C_2} \cdot \left[1 + \frac{R_1}{R_b} \cdot \exp[\lambda \cdot (t - \sum_{\substack{i=3 \\ i \neq 4}}^{5} \tau_i)]\right]$$
$$- \frac{dV_s(t)}{dt} + \frac{dI_{R_s}}{dt} \cdot R_s + \frac{I_{R_s}(t-\tau_1)}{C_1} + \frac{dI_{52a}}{dt} \cdot e^{-\lambda \cdot \tau_3} \cdot R_1$$

$$\frac{dY_1}{dt} = I_{52a}(t-\tau_4) \cdot \frac{1}{L_1 \cdot C_2} \cdot [1 + \frac{R_1}{R_b} \cdot \exp[\lambda \cdot (t - \sum_{\substack{i=3 \\ i \neq 4}}^{5} \tau_i)]]$$
$$- \frac{1}{L_1} \cdot \frac{dV_s(t)}{dt} + \frac{dI_{R_s}}{dt} \cdot \frac{R_s}{L_1} + \frac{I_{R_s}(t-\tau_1)}{C_1 \cdot L_1} + \frac{dI_{52a}}{dt} \cdot e^{-\lambda \cdot \tau_3} \cdot \frac{R_1}{L_1}$$

We define two new variables: $Y_2 = \frac{dI_{R_s}}{dt}$; $Y_3 = \frac{dI_{52a}}{dt}$

$$\frac{dY_1}{dt} = I_{52a}(t-\tau_4)\cdot\frac{1}{L_1\cdot C_2}\cdot[1+\frac{R_1}{R_b}\cdot\exp[\lambda\cdot(t-\sum_{\substack{i=3\\i\neq4}}^{5}\tau_i)]]+Y_2$$
$$\cdot\frac{R_s}{L_1}+\frac{I_{R_s}(t-\tau_1)}{C_1\cdot L_1}+Y_3\cdot e^{-\lambda\cdot\tau_3}\cdot\frac{R_1}{L_1}-\frac{1}{L_1}\cdot\frac{dV_s(t)}{dt}$$

Expression for differential equation: $\frac{dY_2}{dt} = \ldots\ldots$

$$V_{A_3} = R_{load}\cdot\left[I_{R_s}\left(t-\sum_{i=1}^{2}\tau_i\right)+I_{L_1}(t-\tau_2)\right];\ V_{A_2}-V_{A_3}$$
$$=\frac{1}{C_1}\cdot\int I_{C_1}\cdot dt;\ I_{C_1}(t)=I_{R_s}(t-\tau_1)$$
$$V_{A_2}=\frac{1}{C_1}\cdot\int I_{R_s}(t-\tau_1)\cdot dt+R_{load}\cdot[I_{R_s}(t-\sum_{i=1}^{2}\tau_i)+I_{L_1}(t-\tau_2)];$$
$$I_{R_s}=\frac{V_s(t)-V_{A_2}}{R_s}=\frac{1}{R_s}\cdot V_s(t)-\frac{1}{R_s}\cdot V_{A_2}$$
$$I_{R_s}=\frac{1}{R_s}\cdot V_s(t)-\frac{1}{R_s}\cdot\left\{\frac{1}{C_1}\cdot\int I_{R_s}(t-\tau_1)\cdot dt+R_{load}\cdot\left[I_{R_s}\left(t-\sum_{i=1}^{2}\tau_i\right)+I_{L_1}(t-\tau_2)\right]\right\}$$

$$\frac{dI_{R_s}}{dt}=\frac{1}{R_s}\cdot\frac{dV_s(t)}{dt}-\frac{1}{R_s}\cdot\left\{\frac{1}{C_1}\cdot I_{R_s}(t-\tau_1)+R_{load}\cdot\left[\frac{dI_{R_s}(t-\sum_{i=1}^{2}\tau_i)}{dt}+\frac{dI_{L_1}(t-\tau_2)}{dt}\right]\right\}$$

$$\frac{dI_{R_s}(t-\sum_{i=1}^{2}\tau_i)}{dt}=\frac{dI_{R_s}(t)}{dt}\cdot\exp[\lambda\cdot(t-\sum_{\substack{i=3\\i\neq4}}^{5}\tau_i)];\frac{dI_{R_s}(t-\tau_1)}{dt}=\frac{dI_{R_s}(t)}{dt}\cdot e^{-\lambda\cdot\tau_1}$$

$$\frac{dI_{R_s}}{dt}=\frac{1}{R_s}\cdot\frac{dV_s(t)}{dt}-\frac{1}{C_1R_s}\cdot I_{R_s}(t-\tau_1)-\frac{R_{load}}{R_s}\cdot\frac{dI_{R_s}(t-\sum_{i=1}^{2}\tau_i)}{dt}-\frac{R_{load}}{R_s}\cdot\frac{dI_{L_1}(t-\tau_2)}{dt}$$

$$\frac{dI_{R_s}}{dt}=\frac{1}{R_s}\cdot\frac{dV_s(t)}{dt}-\frac{1}{C_1R_s}\cdot I_{R_s}(t-\tau_1)-\frac{R_{load}}{R_s}\cdot\frac{dI_{R_s}(t)}{dt}\cdot\exp[\lambda\cdot(t-\sum_{\substack{i=3\\i\neq4}}^{5}\tau_i)]-\frac{R_{load}}{R_s}\cdot\frac{dI_{L_1}(t-\tau_2)}{dt}$$

$$\frac{dI_{R_s}}{dt}\cdot[1+\frac{R_{load}}{R_s}\cdot\exp[\lambda\cdot(t-\sum_{\substack{i=3\\i\neq4}}^{5}\tau_i)]]=\frac{1}{R_s}\cdot\frac{dV_s(t)}{dt}-\frac{1}{C_1\cdot R_s}\cdot I_{R_s}(t-\tau_1)-\frac{R_{load}}{R_s}\cdot\frac{dI_{L_1}(t-\tau_2)}{dt}$$

We derive the two side of the above equation and get.

$$\frac{d^2 I_{R_s}}{dt^2} \cdot [1 + \frac{R_{load}}{R_s} \cdot \exp[\lambda \cdot (t - \sum_{\substack{i=3 \\ i \neq 4}}^{5} \tau_i)]] = \frac{1}{R_s} \cdot \frac{d^2 V_s(t)}{dt^2} - \frac{1}{C_1 \cdot R_s} \cdot \frac{dI_{R_s}(t)}{dt} \cdot e^{-\lambda \cdot \tau_1} - \frac{R_{load}}{R_s} \cdot \frac{d^2 I_{L_1}(t - \tau_2)}{dt^2}$$

$$\frac{dY_2}{dt} \cdot [1 + \frac{R_{load}}{R_s} \cdot \exp[\lambda \cdot (t - \sum_{\substack{i=3 \\ i \neq 4}}^{5} \tau_i)]] = \frac{1}{R_s} \cdot \frac{d^2 V_s(t)}{dt^2} - \frac{1}{C_1 \cdot R_s} \cdot Y_2 \cdot e^{-\lambda \cdot \tau_1} - \frac{R_{load}}{R_s} \cdot \frac{d^2 I_{L_1}(t - \tau_2)}{dt^2}$$

We consider $\frac{d^2 I_{L_1}(t-\tau_2)}{dt^2} \to \varepsilon = 0$ then $\frac{dY_2}{dt} \cdot [1 + \frac{R_{load}}{R_s} \cdot \exp[\lambda \cdot (t - \sum_{\substack{i=3 \\ i \neq 4}}^{5} \tau_i)]] = \frac{1}{R_s} \cdot \frac{d^2 V_s(t)}{dt^2} - \frac{1}{C_1 \cdot R_s} \cdot Y_2 \cdot e^{-\lambda \cdot \tau_1}$

$$\frac{dY_2}{dt} = \frac{1}{R_s \cdot [1 + \frac{R_{load}}{R_s} \cdot \exp[\lambda \cdot (t - \sum_{\substack{i=3 \\ i \neq 4}}^{5} \tau_i)]]} \cdot \frac{d^2 V_s(t)}{dt^2} - \frac{1}{C_1 \cdot R_s \cdot [1 + \frac{R_{load}}{R_s} \cdot \exp[\lambda \cdot (t - \sum_{\substack{i=3 \\ i \neq 4}}^{5} \tau_i)]]} \cdot Y_2 \cdot e^{-\lambda \cdot \tau_1}$$

Case I: $\tau_1 \to \varepsilon(=0); \tau_2 \to \varepsilon(=0); \sum_{i=1}^{2} \tau_i \to \varepsilon$

Assumption: Our Bias-T circuit's first and second microstrips parasitic effect (delay parameters in time) is minor, $\tau_1 \to \varepsilon(=0); \tau_2 \to \varepsilon(=0); \sum_{i=1}^{2} \tau_i \to \varepsilon$

$$\lim_{\substack{\tau_i \to \varepsilon \\ i=1,2}} \exp[\lambda \cdot (t - \sum_{\substack{i=3 \\ i \neq 4}}^{5} \tau_i)] = 1; 1 + \frac{R_{load}}{R_s} \cdot \exp[\lambda \cdot (t - \sum_{\substack{i=3 \\ i \neq 4}}^{5} \tau_i)] = 1 + \frac{R_{load}}{R_s};$$

$$\frac{dY_2}{dt} = \frac{1}{R_s \cdot [1 + \frac{R_{load}}{R_s}]} \cdot \frac{d^2 V_s(t)}{dt^2} - \frac{1}{C_1 \cdot R_s \cdot [1 + \frac{R_{load}}{R_s}]} \cdot Y_2 \cdot e^{-\lambda \cdot \tau_1}$$

$$\frac{dY_1}{dt} = I_{52a}(t - \tau_4) \cdot \frac{1}{L_1 \cdot C_2} \cdot [1 + \frac{R_1}{R_b} \cdot \exp[\lambda \cdot (t - \sum_{\substack{i=3 \\ i \neq 4}}^{5} \tau_i)]]$$

$$+ Y_2 \cdot \frac{R_s}{L_1} + \frac{I_{R_s}(t)}{C_1 \cdot L_1} + Y_3 \cdot e^{-\lambda \cdot \tau_3} \cdot \frac{R_1}{L_1} - \frac{1}{L_1} \cdot \frac{dV_s(t)}{dt}$$

Expression for differential equation: $\frac{dY_3}{dt} = \ldots\ldots$

$$I_{C_2} = I_{52a}(t-\tau_4) \Rightarrow \frac{dI_{C_2}}{dt} = \frac{dI_{52a}(t-\tau_4)}{dt}; \frac{dI_{52a}(t-\tau_4)}{dt} = \frac{dI_{52a}(t)}{dt} \cdot e^{-\lambda \cdot \tau_4}$$

$$I_{C_2} = C_2 \cdot \frac{dV_{A_5}}{dt}; I_{R_b} = \frac{V_b - V_{A_5}}{R_b} \Rightarrow V_{A_5} = V_b - I_{R_b} \cdot R_b; \frac{dV_{A_5}}{dt} = \frac{dV_b}{dt} - \frac{dI_{R_b}}{dt} \cdot R_b; \frac{dV_b}{dt} = 0$$

$$I_{C_2} = -C_2 \cdot R_b \cdot \frac{dI_{R_b}}{dt} \Rightarrow \frac{dI_{C_2}}{dt} = -C_2 \cdot R_b \cdot \frac{d^2 I_{R_b}}{dt^2}; -C_2 \cdot R_b \cdot \frac{d^2 I_{R_b}}{dt^2} = \frac{dI_{52a}(t)}{dt} \cdot e^{-\lambda \cdot \tau_4}$$

$$-C_2 \cdot R_b \cdot \frac{d^2 I_{R_b}}{dt^2} = \frac{dI_{52a}(t)}{dt} \cdot e^{-\lambda \cdot \tau_4} \Rightarrow -C_2 \cdot R_b \cdot \frac{d^3 I_{R_b}}{dt^3} = \frac{d^2 I_{52a}(t)}{dt^2} \cdot e^{-\lambda \cdot \tau_4};$$

$$\frac{d^2 I_{52a}(t)}{dt^2} = \frac{dY_3}{dt}$$

$$-C_2 \cdot R_b \cdot \frac{d^3 I_{R_b}}{dt^3} = \frac{dY_3}{dt} \cdot e^{-\lambda \cdot \tau_4}; \frac{d^3 I_{R_b}}{dt^3} \to (\varepsilon = 0) \Rightarrow \frac{dY_3}{dt} = 0$$

We can summery our system five Delay Differential Equations (DDE):

$$\frac{dY_1}{dt} = I_{52a}(t-\tau_4) \cdot \frac{1}{L_1 \cdot C_2} \cdot [1 + \frac{R_1}{R_b} \cdot \exp[\lambda \cdot (t - \sum_{\substack{i=3 \\ i \neq 4}}^{5} \tau_i)]] + Y_2 \cdot \frac{R_s}{L_1} + \frac{I_{R_s}(t)}{C_1 \cdot L_1} + Y_3 \cdot e^{-\lambda \cdot \tau_3} \cdot \frac{R_1}{L_1} - \frac{1}{L_1} \cdot \frac{dV_s(t)}{dt}$$

$$\frac{dY_2}{dt} = \frac{1}{R_s \cdot [1 + \frac{R_{load}}{R_s}]} \cdot \frac{d^2 V_s(t)}{dt^2} - \frac{1}{C_1 \cdot R_s \cdot [1 + \frac{R_{load}}{R_s}]} \cdot Y_2 \cdot e^{-\lambda \cdot \tau_1}; \frac{dY_3}{dt} = 0; \frac{dI_{R_s}}{dt} = Y_2; \frac{dI_{52a}}{dt} = Y_3$$

Remark: Some of our system variables include time argument X(t) and other without time argument. Always consider these system variables as a function of time.

We consider RF signal source $V_s(t) = A_0 + \xi(t); |\xi(t)| < 1 \,\&\, A_0 \gg |\xi(t)|]] >$

$$V_s(t)|_{A_0 \gg |\xi(t)|} = A_0 + \xi(t) \approx A_0; \frac{dV_s(t)|_{A_0 \gg |\xi(t)|}}{dt} = \frac{d\xi(t)}{dt} \to \varepsilon; \frac{d^2 V_s(t)|_{A_0 \gg |\xi(t)|}}{dt^2} = \frac{d^2 \xi(t)}{dt^2} \to \varepsilon.$$

Under the above consideration we can rewrite our system five Delay Differential Equations (DDE): $\tau_1 \to \varepsilon(=0); \tau_2 \to \varepsilon(=0); \sum_{i=1}^{2} \tau_i \to \varepsilon; e^{-\lambda \cdot \tau_1} \to 1$

$$\frac{dY_1}{dt} = I_{52a}(t - \tau_4) \cdot \frac{1}{L_1 \cdot C_2} \cdot \left[1 + \frac{R_1}{R_b} \cdot \exp[\lambda \cdot (t - \sum_{\substack{i=3 \\ i \neq 4}}^{5} \tau_i)] \right] + Y_2$$
$$\cdot \frac{R_s}{L_1} + \frac{I_{R_s}(t)}{C_1 \cdot L_1} + Y_3 \cdot e^{-\lambda \cdot \tau_3} \cdot \frac{R_1}{L_1}$$

$$\frac{dY_2}{dt} = -\frac{1}{C_1 \cdot R_s \cdot \left[1 + \frac{R_{load}}{R_s}\right]} \cdot Y_2; \frac{dY_3}{dt} = 0; \frac{dI_{R_s}}{dt} = Y_2; \frac{dI_{52a}}{dt} = Y_3$$

&&&

$$\frac{dY_1}{dt} = I_{52a}(t - \tau_4) \cdot \frac{1}{L_1 \cdot C_2} \cdot \left[1 + \frac{R_1}{R_b} \cdot \exp[\lambda \cdot (t - \sum_{\substack{i=3 \\ i \neq 4}}^{5} \tau_i)] \right] + Y_2$$
$$\cdot \frac{R_s}{L_1} + \frac{I_{R_s}(t)}{C_1 \cdot L_1} + Y_3 \cdot e^{-\lambda \cdot \tau_3} \cdot \frac{R_1}{L_1}$$

$$y_1 \cdot \lambda \cdot e^{\lambda \cdot t} = [I_{52a}^{(j)} + i_{52a} \cdot e^{\lambda \cdot (t - \tau_4)}] \cdot \frac{1}{L_1 \cdot C_2} \cdot \left[1 + \frac{R_1}{R_b} \cdot \exp[\lambda \cdot (t - \sum_{\substack{i=3 \\ i \neq 4}}^{5} \tau_i)] \right]$$
$$+ [Y_2^{(j)} + y_2 \cdot e^{\lambda \cdot t}] \cdot \frac{R_s}{L_1} + \frac{1}{C_1 \cdot L_1} \cdot [I_{R_s}^{(j)} + i_{R_s} \cdot e^{\lambda \cdot t}]$$
$$+ [Y_3^{(j)} + y_3 \cdot e^{\lambda \cdot t}] \cdot e^{-\lambda \cdot \tau_3} \cdot \frac{R_1}{L_1}$$

$$y_1 \cdot \lambda \cdot e^{\lambda \cdot t} = I_{52a}^{(j)} \cdot \frac{1}{L_1 \cdot C_2} \cdot \left[1 + \frac{R_1}{R_b} \cdot \exp[\lambda \cdot (t - \sum_{\substack{i=3 \\ i \neq 4}}^{5} \tau_i)] \right]$$
$$+ Y_2^{(j)} \cdot \frac{R_s}{L_1} + \frac{1}{C_1 \cdot L_1} \cdot I_{R_s}^{(j)} + Y_3^{(j)} \cdot e^{-\lambda \cdot \tau_3} \cdot \frac{R_1}{L_1}$$
$$+ \frac{1}{L_1 \cdot C_2} \cdot [1 + \frac{R_1}{R_b} \cdot \exp[\lambda \cdot (t - \sum_{\substack{i=3 \\ i \neq 4}}^{5} \tau_i)]] \cdot i_{52a} \cdot e^{\lambda \cdot (t - \tau_4)}$$
$$+ \frac{R_s}{L_1} \cdot y_2 \cdot e^{\lambda \cdot t} + \frac{1}{C_1 \cdot L_1} \cdot i_{R_s} \cdot e^{\lambda \cdot t} + y_3 \cdot e^{\lambda \cdot t} \cdot e^{-\lambda \cdot \tau_3} \cdot \frac{R_1}{L_1}$$

At fixed point $I_{52a}^{(j)} \cdot \frac{1}{L_1 \cdot C_2} \cdot [1 + \frac{R_1}{R_b} \cdot \exp[\lambda \cdot (t - \sum_{\substack{i=3 \\ i \neq 4}}^{5} \tau_i)]] + Y_2^{(j)} \cdot \frac{R_s}{L_1} + \frac{1}{C_1 \cdot L_1} \cdot I_{R_s}^{(j)} + Y_3^{(j)} \cdot e^{-\lambda \cdot \tau_3} \cdot \frac{R_1}{L_1} = 0$

$$y_1 \cdot \lambda \cdot e^{\lambda \cdot t} = \frac{1}{L_1 \cdot C_2} \cdot [1 + \frac{R_1}{R_b} \cdot \exp[\lambda \cdot (t - \sum_{\substack{i=3 \\ i \neq 4}}^{5} \tau_i)]] \cdot i_{52a} \cdot e^{\lambda \cdot t} \cdot e^{-\lambda \cdot \tau_4} + \frac{R_s}{L_1} \cdot y_2 \cdot e^{\lambda \cdot t} + \frac{1}{C_1 \cdot L_1} \cdot i_{R_s} \cdot e^{\lambda \cdot t} + y_3 \cdot e^{\lambda \cdot t} \cdot e^{-\lambda \cdot \tau_3} \cdot \frac{R_1}{L_1}$$

Dividing the two side of the above equation by $e^{\lambda \cdot t}$ term gives the equation:

$$-y_1 \cdot \lambda + \frac{1}{L_1 \cdot C_2} \cdot \left[1 + \frac{R_1}{R_b} \cdot \exp[\lambda \cdot (t - \sum_{\substack{i=3 \\ i \neq 4}}^{5} \tau_i)] \right] \cdot i_{52a} \cdot e^{-\lambda \cdot \tau_4} + \frac{R_s}{L_1} \cdot y_2 + \frac{1}{C_1 \cdot L_1} \cdot i_{R_s} + y_3 \cdot e^{-\lambda \cdot \tau_3} \cdot \frac{R_1}{L_1} = 0$$

$$\frac{dY_2}{dt} = -\frac{1}{C_1 \cdot R_s \cdot \left[1 + \frac{R_{load}}{R_s}\right]} \cdot Y_2; y_2 \cdot \lambda \cdot e^{\lambda \cdot t} = -\frac{1}{C_1 \cdot R_s \cdot \left[1 + \frac{R_{load}}{R_s}\right]} \cdot [Y_2^{(j)} + y_2 \cdot e^{\lambda \cdot t}]$$

At fixed point $-\frac{1}{C_1 \cdot R_s \cdot [1 + \frac{R_{load}}{R_s}]} \cdot Y_2^{(j)} = 0; \left\{ -\lambda - \frac{1}{C_1 \cdot R_s \cdot [1 + \frac{R_{load}}{R_s}]} \right\} \cdot y_2 = 0$

$$\frac{dY_3}{dt} = 0 \Rightarrow -y_3 \cdot \lambda \cdot e^{\lambda \cdot t} = 0 \Rightarrow -y_3 \cdot \lambda = 0; \frac{dI_{R_s}}{dt} = Y_2 \Rightarrow i_{R_s} \cdot \lambda \cdot e^{\lambda \cdot t} = Y_2^{(j)} + y_2 \cdot e^{\lambda \cdot t}$$

At fixed point $Y_2^{(j)} = 0 \Rightarrow i_{R_s} \cdot \lambda \cdot e^{\lambda \cdot t} = y_2 \cdot e^{\lambda \cdot t} \Rightarrow -i_{R_s} \cdot \lambda + y_2 = 0$

$\frac{dI_{52a}}{dt} = Y_3 \Rightarrow i_{52a} \cdot \lambda \cdot e^{\lambda \cdot t} = Y_3^{(j)} + y_3 \cdot e^{\lambda \cdot t}$. At fixed point $Y_3^{(j)} = 0 \Rightarrow -i_{52a} \cdot \lambda + y_3 = 0$

The small increments Jacobian of our Bias-T circuit is as follow:

$$\begin{pmatrix} \Upsilon_{11} & \cdots & \Upsilon_{15} \\ \vdots & \ddots & \vdots \\ \Upsilon_{51} & \cdots & \Upsilon_{55} \end{pmatrix} \cdot \begin{pmatrix} y_1 \\ y_2 \\ y_3 \\ i_{R_s} \\ i_{52a} \end{pmatrix} = 0; \Upsilon_{11} = -\lambda; \Upsilon_{12} = \frac{R_s}{L_1};$$

$$\Upsilon_{13} = \frac{R_1}{L_1} \cdot e^{-\lambda \cdot \tau_3}; \Upsilon_{14} = \frac{1}{C_1 \cdot L_1}$$

$$\Upsilon_{15} = \frac{1}{L_1 \cdot C_2} \cdot \left[1 + \frac{R_1}{R_b} \cdot \exp[\lambda \cdot (t - \sum_{\substack{i=3 \\ i \neq 4}}^{5} \tau_i)] \right] \cdot e^{-\lambda \cdot \tau_4}; \Upsilon_{21} = 0;$$

$$\Upsilon_{22} = -\lambda - \frac{1}{C_1 \cdot R_s \cdot \left[1 + \frac{R_{load}}{R_s} \right]}$$

$$\Upsilon_{23} = 0; \Upsilon_{24} = 0; \Upsilon_{25} = 0; \Upsilon_{31} = 0; \Upsilon_{32} = 0; \Upsilon_{33} = -\lambda; \Upsilon_{34} = 0; \Upsilon_{35} = 0$$
$$\Upsilon_{41} = 0; \Upsilon_{42} = 1; \Upsilon_{43} = 0; \Upsilon_{44} = -\lambda; \Upsilon_{45} = 0; \Upsilon_{51} = 0; \Upsilon_{52} = 0; \Upsilon_{53} = 1;$$
$$\Upsilon_{54} = 0; \Upsilon_{55} = -\lambda$$

$$|A - \lambda \cdot I| = \begin{pmatrix} \Upsilon_{11} & \cdots & \Upsilon_{15} \\ \vdots & \ddots & \vdots \\ \Upsilon_{51} & \cdots & \Upsilon_{55} \end{pmatrix}$$; $\det|A - \lambda \cdot I| = 0$. We define for simplicity the following parameters: $\sigma_1 = \frac{R_s}{L_1}; \sigma_2 = \frac{R_1}{L_1}; \sigma_3 = \frac{1}{C_1 \cdot L_1}; \sigma_4 = \frac{1}{L_1 \cdot C_2}; \sigma_5 = \frac{R_1}{R_b}$

$$\sigma_6 = -\frac{1}{C_1 \cdot R_s \cdot \left[1 + \frac{R_{load}}{R_s} \right]}; \Upsilon_{12} = \sigma_1; \Upsilon_{13} = \sigma_2 \cdot e^{-\lambda \cdot \tau_3};$$

$$\Upsilon_{14} = \sigma_3; \Upsilon_{15} = \sigma_4 \cdot \left[1 + \sigma_5 \cdot \exp[\lambda \cdot (t - \sum_{\substack{i=3 \\ i \neq 4}}^{5} \tau_i)] \right] \cdot e^{-\lambda \cdot \tau_4}$$

$$\Upsilon_{22} = -\lambda + \sigma_6; \det|A - \lambda \cdot I| = \lambda^4 \cdot (-\lambda + \sigma_6) = 0;$$

$$\lambda_1 = 0; \lambda_2 = \sigma_6 = -\frac{1}{C_1 \cdot R_s \cdot [1 + \frac{R_{load}}{R_s}]}$$

We get a stability solution which is independent on Bias-T microstrip delay lines parameters, $\lambda_1 = 0$, $\lambda_2 < 0$ then our stability map is attracting line [26].

Case II: $1 \gg \tau_1 > 0; 1 \gg \tau_2 > 0; 1 \gg \sum_{i=1}^{2} \tau_i > 0$ then our system delay differential equations are as below. $\frac{d^2 V_s(t)}{dt^2} \to \varepsilon; \frac{d^2 I_{L_1}(t-\tau_2)}{dt^2} \to \varepsilon$

$$\frac{dY_1}{dt} = I_{52a}(t-\tau_4) \cdot \frac{1}{L_1 \cdot C_2} \cdot \left[1 + \frac{R_1}{R_b} \cdot \exp[\lambda \cdot (t - \sum_{\substack{i=3 \\ i \neq 4}}^{5} \tau_i)] \right] + Y_2$$

$$\cdot \frac{R_s}{L_1} + \frac{I_{R_s}(t-\tau_1)}{C_1 \cdot L_1} + Y_3 \cdot e^{-\lambda \cdot \tau_3} \cdot \frac{R_1}{L_1} - \frac{1}{L_1} \cdot \frac{dV_s(t)}{dt}$$

$$\frac{dY_2}{dt} = -\frac{1}{C_1 \cdot R_s \cdot \left[1 + \frac{R_{load}}{R_s} \cdot \exp[\lambda \cdot (t - \sum_{\substack{i=3 \\ i \neq 4}}^{5} \tau_i)] \right]} \cdot Y_2 \cdot e^{-\lambda \cdot \tau_1};$$

$$\frac{dY_3}{dt} = 0; \frac{dI_{R_s}}{dt} = Y_2; \frac{dI_{52a}}{dt} = Y_3$$

Small increment equations: $-i_{52a} \cdot \lambda + y_3 = 0$

$$-y_1 \cdot \lambda + \frac{1}{L_1 \cdot C_2} \cdot \left[1 + \frac{R_1}{R_b} \cdot \exp[\lambda \cdot (t - \sum_{\substack{i=3 \\ i \neq 4}}^{5} \tau_i)] \right] \cdot i_{52a} \cdot e^{-\lambda \cdot \tau_4} + \frac{R_s}{L_1} \cdot y_2$$

$$+ \frac{1}{C_1 \cdot L_1} \cdot i_{R_s} \cdot e^{-\lambda \cdot \tau_1} + y_3 \cdot e^{-\lambda \cdot \tau_3} \cdot \frac{R_1}{L_1} = 0$$

$$-y_2 \cdot \lambda - \frac{1}{C_1 \cdot R_s \cdot \left[1 + \frac{R_{load}}{R_s} \cdot \exp[\lambda \cdot (t - \sum_{\substack{i=3 \\ i \neq 4}}^{5} \tau_i)] \right]} \cdot y_2 \cdot e^{-\lambda \cdot \tau_1} = 0;$$

$$-y_3 \cdot \lambda = 0; -i_{R_s} \cdot \lambda + y_2 = 0$$

The small increments Jacobian of our Bias-T circuit is as follow:

$$\begin{pmatrix} \Upsilon_{11} & \dots & \Upsilon_{15} \\ \vdots & \ddots & \vdots \\ \Upsilon_{51} & \dots & \Upsilon_{55} \end{pmatrix} \cdot \begin{pmatrix} y_1 \\ y_2 \\ y_3 \\ i_{R_s} \\ i_{52a} \end{pmatrix} = 0;$$

$$\Upsilon_{11} = -\lambda; \Upsilon_{12} = \frac{R_s}{L_1}; \Upsilon_{13} = \frac{R_1}{L_1} \cdot e^{-\lambda \cdot \tau_3}; \Upsilon_{14} = \frac{1}{C_1 \cdot L_1} \cdot e^{-\lambda \cdot \tau_1}$$

$$\Upsilon_{15} = \frac{1}{L_1 \cdot C_2} \cdot \left[1 + \frac{R_1}{R_b} \cdot e^{-\lambda \cdot \sum_{i=3, i \neq 4}^{5} \tau_i} \right] \cdot e^{-\lambda \cdot \tau_4}; \Upsilon_{21} = 0;$$

$$\Upsilon_{22} = -\lambda - \frac{1}{C_1 \cdot R_s \cdot \left[1 + \frac{R_{load}}{R_s} \cdot e^{-\lambda \cdot \sum_{i=1}^{2} \tau_i} \right]} \cdot e^{-\lambda \cdot \tau_1}$$

$$\Upsilon_{23} = 0; \Upsilon_{24} = 0; \Upsilon_{25} = 0; \Upsilon_{31} = 0;$$
$$\Upsilon_{32} = 0; \Upsilon_{33} = -\lambda; \Upsilon_{34} = 0; \Upsilon_{35} = 0$$
$$\Upsilon_{41} = 0; \Upsilon_{42} = 1; \Upsilon_{43} = 0; \Upsilon_{44} = -\lambda;$$
$$\Upsilon_{45} = 0; \Upsilon_{51} = 0; \Upsilon_{52} = 0; \Upsilon_{53} = 1; \Upsilon_{54} = 0; \Upsilon_{55} = -\lambda$$

$|A - \lambda \cdot I| = \begin{pmatrix} \Upsilon_{11} & \dots & \Upsilon_{15} \\ \vdots & \ddots & \vdots \\ \Upsilon_{51} & \dots & \Upsilon_{55} \end{pmatrix}$; $\det|A - \lambda \cdot I| = 0$. We define for simplicity the following parameters: $\sigma_1 = \frac{R_s}{L_1}; \sigma_2 = \frac{R_1}{L_1}; \sigma_3 = \frac{1}{C_1 \cdot L_1}; \sigma_4 = \frac{1}{L_1 \cdot C_2}; \sigma_5 = \frac{R_1}{R_b}$

$$\sigma_6(\tau_1, \tau_2) = -\frac{1}{C_1 \cdot R_s \cdot \left[1 + \frac{R_{load}}{R_s} \cdot e^{-\lambda \cdot \sum_{i=1}^{2} \tau_i} \right]} \cdot e^{-\lambda \cdot \tau_1}; \Upsilon_{12} = \sigma_1; \Upsilon_{13} = \sigma_2 \cdot e^{-\lambda \cdot \tau_3};$$

$$\Upsilon_{14} = \sigma_3 \cdot e^{-\lambda \cdot \tau_1}$$

$$\Upsilon_{15} = \sigma_4 \cdot \left[1 + \sigma_5 \cdot e^{-\lambda \cdot \sum_{i=3, i \neq 4}^{5} \tau_i} \right] \cdot e^{-\lambda \cdot \tau_4}; \Upsilon_{22} = -\lambda + \sigma_6; \det|A - \lambda \cdot I| = \lambda^4 \cdot (-\lambda + \sigma_6) = 0$$

$$\det|A - \lambda \cdot I| = \lambda^4 \cdot (-\lambda + \sigma_6) = -\lambda^5 + \lambda^4 \cdot \sigma_6(\tau_1, \tau_2); D(\tau_1, \tau_2) = -\lambda^5 + \lambda^4 \cdot \sigma_6(\tau_1, \tau_2)$$

We need to find $D(\tau_1, \tau_2)$ for the following cases: (A) $\tau_1 = \tau; \tau_2 = 0$ (B) $\tau_1 = 0; \tau_2 = \tau$ (C) $\tau_1 = \tau; \tau_2 = \tau$. We need to get characteristics equations for all above stability analysis cases. We study the occurrence of any possible stability switching, resulting from the increase of the value of the time delays τ_1, τ_2 for the general characteristic equation $D(\tau_1, \tau_2)$. If we choose τ as a parameter, then the expression: $D(\lambda, \tau) = P_n(\lambda, \tau) + Q_m(\lambda, \tau) \cdot e^{-\lambda \cdot \tau}; n, m \in R_0; n > m$.

We analyze the stability switching for the third case (C) $\tau_1 = \tau; \tau_2 = \tau$.

$$D(\tau_1, \tau_2) = -\lambda^5 - \lambda^4 \cdot \frac{1}{C_1 \cdot R_s \cdot [1 + \frac{R_{load}}{R_s} \cdot e^{-2 \cdot \lambda \cdot \tau}]} \cdot e^{-\lambda \cdot \tau};$$

$$P_n(\lambda, \tau) = -\lambda^5; Q_m(\lambda, \tau) = -\lambda^4 \cdot f(\lambda, \tau)$$

$$Q_m(\lambda, \tau) = -\lambda^4 \cdot \frac{1}{C_1 \cdot R_s \cdot \left[1 + \frac{R_{load}}{R_s} \cdot e^{-2 \cdot \lambda \cdot \tau}\right]}; f(\lambda, \tau) = \frac{1}{C_1 \cdot R_s \cdot \left[1 + \frac{R_{load}}{R_s} \cdot e^{-2 \cdot \lambda \cdot \tau}\right]}$$

The exponential function $e^{-\varsigma(\lambda, \tau)}; \varsigma(\lambda, \tau) = 2 \cdot \lambda \cdot \tau$ can be characterized in a variety of equivalent ways. In particular it may be defined by the following power series: $\varsigma = \varsigma(\lambda, \tau); e^{-\varsigma(\lambda, \tau)} = 1 - \varsigma + \frac{\varsigma^2}{2!} - \frac{\varsigma^3}{3!} + \ldots = \sum_{n=0}^{\infty} \frac{\varsigma^n}{n!} \cdot (-1)^n$.

We take it as approximation expression: $e^{-\varsigma(\lambda, \tau)} \approx 1 - \varsigma + \frac{\varsigma^2}{2!} - \frac{\varsigma^3}{3!}$

$$e^{-2 \cdot \lambda \cdot \tau} \approx 1 - \lambda \cdot 2 \cdot \tau + \lambda^2 \cdot 2 \cdot \tau^2 - \lambda^3 \cdot \frac{4 \cdot \tau^3}{3}; C_1 \cdot R_s \cdot \left[1 + \frac{R_{load}}{R_s} \cdot e^{-2 \cdot \lambda \cdot \tau}\right]$$
$$= C_1 \cdot R_s + C_1 \cdot R_{load} \cdot e^{-2 \cdot \lambda \cdot \tau}$$
$$C_1 \cdot R_s + C_1 \cdot R_{load} \cdot e^{-2 \cdot \lambda \cdot \tau} = C_1 \cdot [R_s + R_{load}] - \lambda \cdot 2 \cdot C_1 \cdot R_{load} \cdot \tau$$
$$+ \lambda^2 \cdot 2 \cdot C_1 \cdot R_{load} \cdot \tau^2 - \lambda^3 \cdot C_1 \cdot R_{load} \cdot \frac{4 \cdot \tau^3}{3}$$

$$f(\lambda, \tau) \approx \frac{1}{C_1 \cdot [R_s + R_{load}] - \lambda \cdot 2 \cdot C_1 \cdot R_{load} \cdot \tau + \lambda^2 \cdot 2 \cdot C_1 \cdot R_{load} \cdot \tau^2 - \lambda^3 \cdot C_1 \cdot R_{load} \cdot \frac{4 \cdot \tau^3}{3}}$$

$$Q_m(\lambda, \tau) = -\lambda^4 \cdot f(\lambda, \tau)$$
$$= \frac{-\lambda^4}{C_1 \cdot [R_s + R_{load}] - \lambda \cdot 2 \cdot C_1 \cdot R_{load} \cdot \tau + \lambda^2 \cdot 2 \cdot C_1 \cdot R_{load} \cdot \tau^2 - \lambda^3 \cdot C_1 \cdot R_{load} \cdot \frac{4 \cdot \tau^3}{3}}$$

$$Q_m(\lambda, \tau) \cdot C_1 \cdot [R_s + R_{load}] - \lambda \cdot 2 \cdot C_1 \cdot R_{load} \cdot Q_m(\lambda, \tau) \cdot \tau$$
$$+ \lambda^2 \cdot 2 \cdot C_1 \cdot R_{load} \cdot Q_m(\lambda, \tau) \cdot \tau^2$$
$$- \lambda^3 \cdot C_1 \cdot R_{load} \cdot Q_m(\lambda, \tau) \cdot \frac{4 \cdot \tau^3}{3} + \lambda^4 = 0; \sum_{i=0}^{4} \psi_i(\lambda, \tau) \cdot \lambda^i = 0$$

$$\psi_0(\lambda,\tau) = Q_m(\lambda,\tau)\cdot C_1\cdot[R_s + R_{load}];\psi_1(\lambda,\tau) = -2\cdot C_1\cdot R_{load}\cdot Q_m(\lambda,\tau)\cdot\tau$$
$$\psi_2(\lambda,\tau) = 2\cdot C_1\cdot R_{load}\cdot Q_m(\lambda,\tau)\cdot\tau^2;\psi_3(\lambda,\tau)$$
$$= -C_1\cdot R_{load}\cdot Q_m(\lambda,\tau)\cdot\frac{4\cdot\tau^3}{3};\psi_4(\lambda,\tau) = 1$$

We can solve a quartic function by factoring it into a product of two quadratic equations: $\sum_{i=0}^{4}\psi_i(\lambda,\tau)\cdot\lambda^i = (\lambda^2 + \lambda\cdot\Gamma_1 + \Gamma_2)\cdot(\lambda^2 + \lambda\cdot\Gamma_3 + \Gamma_4)$.

$$\sum_{i=0}^{4}\psi_i(\lambda,\tau)\cdot\lambda^i = \lambda^4 + \lambda^3\cdot(\Gamma_3 + \Gamma_1) + \lambda^2\cdot(\Gamma_4 + \Gamma_1\cdot\Gamma_3 + \Gamma_2)$$
$$+\lambda\cdot(\Gamma_1\cdot\Gamma_4 + \Gamma_3\cdot\Gamma_2) + \Gamma_2\cdot\Gamma_4$$

$$\psi_0(\lambda,\tau) = \Gamma_2\cdot\Gamma_4;\psi_1(\lambda,\tau) = \Gamma_1\cdot\Gamma_4 + \Gamma_3\cdot\Gamma_2;\psi_2(\lambda,\tau)$$
$$= \Gamma_4 + \Gamma_1\cdot\Gamma_3 + \Gamma_2$$
$$\psi_3(\lambda,\tau) = \Gamma_3 + \Gamma_1;\psi_4(\lambda,\tau) = 1$$

Remark: it is easier to solve the above equations numerically rather than analytically. The target is to find the two quadratic equations parameters as a function of $Q_m(\lambda,\tau).\Gamma_k(Q_m(\lambda,\tau),\tau,\ldots);k = 1,2,3,4$.

We have two possible solutions: $\sum_{i=0}^{4}\psi_i(\lambda,\tau)\cdot\lambda^i = (\lambda^2 + \lambda\cdot\Gamma_1 + \Gamma_2)\cdot(\lambda^2 + \lambda\cdot\Gamma_3 + \Gamma_4) = 0$

$\lambda^2 + \lambda\cdot\Gamma_1(Q_m(\lambda,\tau),\tau,\ldots) + \Gamma_2(Q_m(\lambda,\tau),\tau,\ldots) = 0$ or $\lambda^2 + \lambda\cdot\Gamma_3(Q_m(\lambda,\tau),\tau,\ldots) + \Gamma_4(Q_m(\lambda,\tau),\tau,\ldots) = 0$.

$$Q_m(\lambda,\tau) = \sum_{k=0}^{m} q_k(\tau)\cdot\lambda^k; m<n = 5; P_n(\lambda,\tau) = -\lambda^5; P_{n=5}(\lambda,\tau) = \sum_{k=0}^{n=5} p_k(\tau)\cdot\lambda^k$$

$$p_0(\tau) = 0; p_1(\tau) = 0; p_2(\tau) = 0; p_3(\tau) = 0; p_4(\tau) = 0; p_5(\tau) = -1$$

The homogenous system for $Y_1, Y_2, Y_3, I_{R_s}, I_{52a}$ leads to a characteristic equation for the eigenvalue λ having the form $P(\lambda,\tau) + Q(\lambda,\tau)\cdot e^{-\lambda\cdot\tau} = 0; P(\lambda) = \sum_{j=0}^{5} a_j\cdot\lambda^j$

$Q(\lambda) = \sum_{j=0}^{m<5} c_j\cdot\lambda^j$. The coefficients $\{a_j(q_i,q_k,\tau), c_j(q_i,q_k,\tau)\}\in R$ depend on q_i, q_k and delay parameter τ. q_i, q_k are any Bias-T circuit's global parameter, other parameters kept as a constant. Unless strictly necessary, the designation of the varied arguments (q_i, q_k) will subsequently be omitted from P, Q, a_j, and c_j. The coefficients a_j, c_j are continuous, and differentiable functions of their arguments, and direct substitution shows that $a_0 + c_0 \neq 0$ for $q_i, q_k \in R_+$; that is λ = 0 is not of

$P(\lambda,\tau)+Q(\lambda,\tau)\cdot e^{-\lambda\cdot\tau}=0$. Furthermore $P(\lambda,\tau),Q(\lambda,\tau)$ are analytic functions of λ, for which the following requirements of the analysis (Kuang and Cong 2005; Kuang 1993) can also be verified in the present case.

(a) If $\lambda=i\cdot\omega,\omega\in R$ then $P(i\cdot\omega)+Q(i\cdot\omega)\neq 0$.
(b) $|\frac{Q(\lambda)}{P(\lambda)}|$ is bounded for $|\lambda|\to\infty$. $\mathrm{Re}\lambda\geq 0$ no roots bifurcation from ∞.
(c) $F(\omega)=|P(i\cdot\omega)|^2-|Q(i\cdot\omega)|^2$ has a finite number of zeros. Indeed, this is a polynomial in ω.
(d) Each positive root $\omega(q_i,q_k)$ of $F(\omega)=0$ is continuous and differentiable with respect to q_i,q_k.

We assume that $P_n(\lambda,\tau)$ and $Q_m(\lambda,\tau)$ cannot have common imaginary roots. That is for any real number ω. $P_n(\lambda=i\cdot\omega,\tau)+Q_m(\lambda=i\cdot\omega,\tau)\neq 0$.

$$P_n(\lambda=i\cdot\omega,\tau)=-i\cdot\omega^5;Q_m(\lambda=i\cdot\omega,\tau)=\sum_{k=0}^{m<5}q_k(\tau)\cdot i^k\cdot\omega^k;0\leq k\leq m<n=5$$

$$i^k;i^0=1;i^1=i;i^2=-1;i^3=-i;i^4=1;P_n(\lambda=i\cdot\omega,\tau)+Q_m(\lambda=i\cdot\omega,\tau)\neq 0$$

$$-i\cdot\omega^5+\sum_{k=0}^{m<5}q_k(\tau)\cdot i^k\cdot\omega^k\neq 0;|P(i\omega,\tau)|^2=\omega^{10}$$

$$|Q(i\omega,\tau)|^2_{m=4}=\left[\sum_{k=0}^{2}q_{2\cdot k}(\tau)\cdot(-1)^k\cdot\omega^{2\cdot k}\right]^2+\left[\sum_{k=0}^{1}q_{2\cdot k+1}(\tau)\cdot(-1)^k\cdot\omega^{2\cdot k+1}\right]^2$$

$$F(\omega)=|P(i\cdot\omega)|^2-|Q(i\cdot\omega)|^2=\omega^{10}-\left[\sum_{k=0}^{2}q_{2\cdot k}(\tau)\cdot(-1)^k\cdot\omega^{2\cdot k}\right]^2$$
$$-\left[\sum_{k=0}^{1}q_{2\cdot k+1}(\tau)\cdot(-1)^k\cdot\omega^{2\cdot k+1}\right]^2$$

$$\sum_{k=0}^{2}q_{2\cdot k}(\tau)\cdot(-1)^k\cdot\omega^{2\cdot k}=q_0(\tau)-q_2(\tau)\cdot\omega^2+q_4(\tau)\cdot\omega^4;\sum_{k=0}^{1}q_{2\cdot k+1}(\tau)\cdot(-1)^k$$
$$\cdot\,\omega^{2\cdot k+1}=q_1(\tau)\cdot\omega-q_3(\tau)\cdot\omega^3$$

$$\left[\sum_{k=0}^{2}q_{2\cdot k}(\tau)\cdot(-1)^k\cdot\omega^{2\cdot k}\right]^2=q_0^2(\tau)-2\cdot q_0(\tau)\cdot q_2(\tau)\cdot\omega^2+[q_2^2(\tau)+2\cdot q_0(\tau)\cdot q_4(\tau)]\cdot\omega^4$$
$$-2\cdot q_2(\tau)\cdot q_4(\tau)\cdot\omega^6+q_4^2(\tau)\cdot\omega^8$$

$$\left[\sum_{k=0}^{1} q_{2\cdot k+1}(\tau)\cdot(-1)^k\cdot\omega^{2\cdot k+1}\right]^2 = q_1^2(\tau)\cdot\omega^2 - 2\cdot q_1(\tau)\cdot q_3(\tau)\cdot\omega^4 + q_3^2(\tau)\cdot\omega^6$$

$$\begin{aligned}&\left[\sum_{k=0}^{2} q_{2\cdot k}(\tau)\cdot(-1)^k\cdot\omega^{2\cdot k}\right]^2 + \left[\sum_{k=0}^{1} q_{2\cdot k+1}(\tau)\cdot(-1)^k\cdot\omega^{2\cdot k+1}\right]^2\\ &= q_0^2(\tau) + [q_1^2(\tau) - 2\cdot q_0(\tau)\cdot q_2(\tau)]\cdot\omega^2\\ &\quad + [q_2^2(\tau) + 2\cdot q_0(\tau)\cdot q_4(\tau) - 2\cdot q_1(\tau)\cdot q_3(\tau)]\cdot\omega^4 + [q_3^2(\tau)\\ &- 2\cdot q_2(\tau)\cdot q_4(\tau)]\cdot\omega^6 + q_4^2(\tau)\cdot\omega^8\end{aligned}$$

$$\begin{aligned}F(\omega) = |P(i\cdot\omega)|^2 - |Q(i\cdot\omega)|^2 &= -q_0^2(\tau) - [q_1^2(\tau) - 2\cdot q_0(\tau)\cdot q_2(\tau)]\cdot\omega^2\\ &- [q_2^2(\tau) + 2\cdot q_0(\tau)\cdot q_4(\tau) - 2\cdot q_1(\tau)\cdot q_3(\tau)]\cdot\omega^4\\ &- [q_3^2(\tau) - 2\cdot q_2(\tau)\cdot q_4(\tau)]\cdot\omega^6 - q_4^2(\tau)\cdot\omega^8 + \omega^{10}\end{aligned}$$

We define the following parameters for simplicity $\Pi_0, \Pi_2, \Pi_4, \Pi_6, \Pi_8, \Pi_{10}$

$$\begin{aligned}\Pi_0(\tau) &= -q_0^2(\tau); \Pi_2(\tau) = -[q_1^2(\tau) - 2\cdot q_0(\tau)\cdot q_2(\tau)]; \Pi_4(\tau)\\ &= -[q_2^2(\tau) + 2\cdot q_0(\tau)\cdot q_4(\tau) - 2\cdot q_1(\tau)\cdot q_3(\tau)]\\ \Pi_6(\tau) &= -[q_3^2(\tau) - 2\cdot q_2(\tau)\cdot q_4(\tau)]; \Pi_8(\tau) = -q_4^2(\tau); \Pi_{10}(\tau) = 1.\end{aligned}$$

Hence $F(\omega) = 0$ implies $\sum_{k=0}^{5} \Pi_{2\cdot k}\cdot\omega^{2\cdot k} = 0$ and its roots are given by solving the above polynomial. $Q_I(i\cdot\omega,\tau) = q_1(\tau)\cdot\omega - q_3(\tau)\cdot\omega^3$

$$P_R(i\cdot\omega,\tau) = 0; P_I(i\cdot\omega,\tau) = -\omega^5; Q_R(i\cdot\omega,\tau) = q_0(\tau) - q_2(\tau)\cdot\omega^2 + q_4(\tau)\cdot\omega^4$$

$$\sin\theta(\tau) = \frac{-P_R(i\cdot\omega,\tau)\cdot Q_I(i\cdot\omega,\tau) + P_I(i\cdot\omega,\tau)\cdot Q_R(i\cdot\omega,\tau)}{|Q(i\cdot\omega,\tau)|^2}$$

$$\cos\theta(\tau) = -\frac{P_R(i\cdot\omega,\tau)\cdot Q_R(i\cdot\omega,\tau) + P_I(i\cdot\omega,\tau)\cdot Q_I(i\cdot\omega,\tau)}{|Q(i\cdot\omega,\tau)|^2}$$

We use different parameters terminology from our last characteristics parameters definition: $k \to j; p_k(\tau) \to a_j; q_k(\tau) \to c_j; n = 5; m < 5; n > m$

Additionally $P_n(\lambda,\tau) \to P(\lambda); Q_m(\lambda,\tau) \to Q(\lambda,\tau)$ then $P(\lambda) = \sum_{j=0}^{5} a_j\cdot\lambda^j$

$$Q(\lambda,\tau) = \sum_{j=0}^{m<5} c_j \cdot \lambda^j; P(\lambda) = -\lambda^5 \; ; \mathrm{a}_0 = a_1 = a_2 = a_3 = a_4 = 0; a_5 = -1$$

$$Q(\lambda,\tau) = \sum_{j=0}^{m<5} c_j \cdot \lambda^j = c_0 + c_1 \cdot \lambda + c_2 \cdot \lambda^2 + c_3 \cdot \lambda^3 + c_4 \cdot \lambda^4$$

$n, m \in R_0$, $n > m$ and a_j, c_j:$R_{+0} \to R$ are continuous and differentiable function of τ such that $a_0 + c_0 \neq 0$. In the following "—" denotes complex and conjugate. $P(\lambda), Q(\lambda,\tau)$ are analytic functions in λ and differentiable in τ. The coefficients $\{a_j(R_s, C_1, L_1, R_1, C_2, R_{load}, R_b, \tau)$ and $c_j(R_s, C_1, L_1, R_1, C_2, R_{load}, R_b, \tau)\} \in R$ depend on Bias-T's $R_s, C_1, L_1, R_1, C_2, R_{load}, R_b, \tau$ values. Unless strictly necessary, the designation of the varied arguments: $(R_s, C_1, L_1, R_1, C_2, R_{load}, R_b, \tau)$ will subsequently be omitted from P, Q, a_j, c_j. The coefficients a_j, c_j are continuous and differentiable functions of their arguments, and direct substitution shows that $a_0 + c_0 \neq 0. a_0 = 0$; $c_0 = q_0(\tau); a_0 + c_0 \neq 0 \Rightarrow q_0(\tau) \neq 0$.
$\forall R_s, C_1, L_1, R_1, C_2, R_{load}, R_b, \tau \in R_+$ i.e. λ = 0 is not a root of the characteristic equation. Furthermore $P(\lambda), Q(\lambda,\tau)$ are analytic functions of λ for which the following requirements of the analysis (see Kuang 1993, Sect. 3.4) can also be verified in the present case.

(a) If $\lambda = i \cdot \omega$, $\omega \in R$ then $P(i \cdot \omega) + Q(i \cdot \omega) \neq 0$ i.e. P and Q have no common imaginary roots. This condition was verified numerically in the entire $(R_s, C_1, L_1, R_1, C_2, R_{load}, R_b, \tau)$ domain of interest.
(b) $|\frac{Q(\lambda,\tau)}{P(\lambda)}|$ is bounded for $|\lambda| \to \infty$, $\mathrm{Re}\lambda \geq 0$. No roots bifurcation from ∞. Indeed, in the limit: $|\frac{Q(\lambda,\tau)}{P(\lambda)}| = |\frac{\sum_{j=0}^{m<5} c_j \cdot \lambda^j}{-\lambda^5}| = |-\sum_{j=0}^{m<5} c_j \cdot \lambda^{j-5}|$
(c) $F(\omega) = |P(i \cdot \omega)|^2 - |Q(i \cdot \omega)|^2; F(\omega) = \sum_{k=0}^{5} \Pi_{2 \cdot k} \cdot \omega^{2 \cdot k}$ has at most a finite number of zeros. Indeed, this is a polynomial in ω (degree in ω^{10}).
(d) Each positive root $\omega(R_s, C_1, L_1, R_1, C_2, R_{load}, R_b, \tau)$ of $F(\omega) = 0$ is continuous and differentiable with respect to $R_s, C_1, L_1, R_1, C_2, R_{load}, R_b, \tau$. This condition can be assessed numerically.

In addition, since the coefficients in P and Q are real, we have $\overline{P(-i \cdot \omega)} = P(i \cdot \omega)$ and $\overline{Q(-i \cdot \omega)} = Q(i \cdot \omega)$ thus, ω > 0 maybe on eigenvalue of characteristic equations. The analysis consists in identifying the roots of the characteristic equation situated on the imaginary axis of the complex λ—plane, whereby increasing the parameters: $R_s, C_1, L_1, R_1, C_2, R_{load}, R_b, \tau$ Reλ may, at the crossing, change its sign from (−) to (+), i.e. from stable focus $E^{(j)}(Y_1^{(j)}, Y_2^{(j)}, Y_3^{(j)}, I_{52a}^{(j)}, I_{R_s}^{(j)})$ $j = 0, 1, 2, \ldots$ to an unstable one, or vice versa. This feature may be further assessed by examining the sign of the partial derivatives with respect to $R_s, C_1, L_1, R_1, C_2, R_{load}, R_b, \tau$.

$$\Lambda^{-1}(\tau) = (\frac{\partial \mathrm{Re}\lambda}{\partial \tau})_{\lambda=i\cdot\omega}, R_s, C_1, L_1, R_1, C_2, R_{load}, R_b = Const; \omega \in \mathbb{R}_+$$

When writing $P(\lambda) = P_R(\lambda) + i \cdot P_I(\lambda); Q(\lambda, \tau) = Q_R(\lambda, \tau) + i \cdot Q_I(\lambda, \tau)$ and inserting $\lambda = i \cdot \omega$ into Bias-T circuit's characteristic equation, ω must satisfy the following:

$$\sin \omega \cdot \tau = g(\omega) = \frac{-P_R(i \cdot \omega, \tau) \cdot Q_I(i \cdot \omega, \tau) + P_I(i \cdot \omega, \tau) \cdot Q_R(i \cdot \omega, \tau)}{|Q(i \cdot \omega, \tau)|^2}$$

$$\cos \omega \cdot \tau = h(\omega) = -\frac{P_R(i \cdot \omega, \tau) \cdot Q_R(i \cdot \omega, \tau) + P_I(i \cdot \omega, \tau) \cdot Q_I(i \cdot \omega, \tau)}{|Q(i \cdot \omega, \tau)|^2}$$

where $|Q(i \cdot \omega, \tau)|^2 \neq 0$ in view of requirement (a) above, and $(g, h) \in R$. Furthermore, it follows above $\sin \omega \cdot \tau$ and $\cos \omega \cdot \tau$ equations that, by squaring and adding the sides, ω must be a positive root of $F(\omega) = |P(i \cdot \omega)|^2 - |Q(i \cdot \omega)|^2 = 0$.

Note: F(ω) is dependent on τ. Now it is important to notice that if $\tau \notin I$(assume that $I \subseteq R_{+0}$ is the set where $\omega(\tau)$ is a positive root of F(ω) and for $\tau \notin I, \omega(\tau)$ is not defined. Then for all τ in I, $\omega(\tau)$ is satisfied that $F(\omega) = 0$. Then there are no positive $\omega(\tau)$ solutions for $F(\omega) = 0$, and we cannot have stability switches. For $\tau \in I$ where $\omega(\tau)$ is a positive solution of $F(\omega) = 0$, we can define the angle $\theta(\tau) \in [0, 2 \cdot \pi]$ as the solution of $\sin \theta(\tau) = \ldots; \cos \theta(\tau) = \ldots$

And the relation between the argument $\theta(\tau)$ and $\omega(\tau) \cdot \tau$ for $\tau \in I$ must be $\omega(\tau) \cdot \tau = \theta(\tau) + 2 \cdot n \cdot \pi \, \forall n \in R_0$. Hence we can define the maps $\tau_n : I \to R_{+0}$ given by $\tau_n(\tau) = \frac{\theta(\tau) + 2 \cdot n \cdot \pi}{\omega(\tau)}; n \in R_0, \tau \in I$. Let us introduce the functions: $I \to R; S_n(\tau) = \tau - \tau_n(\tau), \tau \in I, n \in R_0$ that is continuous and differentiable in τ. In the following the subscripts $\lambda, \omega, R_s, C_1, L_1, R_1, C_2, R_{load}, R_b, ..$ indicate the corresponding partial derivatives. Let us first concentrate on Λ(x), remember in $\lambda(R_s, C_1, L_1, R_1, C_2, R_{load}, R_b, ..)$ and $\omega(R_s, C_1, L_1, R_1, C_2, R_{load}, R_b, ..)$, and keeping all parameters except one (x) and τ. The derivation closely follows that in reference [BK]. Differentiating Bias-T circuit characteristic equation $P(\lambda) + Q(\lambda, \tau) \cdot e^{-\lambda \cdot \tau} = 0$ with respect to specific parameter (x), and inverting the derivative, for convenience, one calculates:

Remark: $x = R_s, C_1, L_1, R_1, C_2, R_{load}, R_b, ..$

$$(\frac{\partial \lambda}{\partial x})^{-1} = \frac{-P_\lambda(\lambda, x) \cdot Q(\lambda, x) + Q_\lambda(\lambda, x) \cdot P(\lambda, x) - \tau \cdot P(\lambda, x) \cdot Q(\lambda, x)}{P_x(\lambda, x) \cdot Q(\lambda, x) - Q_x(\lambda, x) \cdot P(\lambda, x)}$$

where $P_\lambda = \frac{\partial P}{\partial \lambda}; Q_\lambda = \frac{\partial Q}{\partial \lambda}; P_x = \frac{\partial P}{\partial x}; Q_x = \frac{\partial Q}{\partial x}$, substituting $\lambda = i \cdot \omega$ and bearing $\overline{P(-i \cdot \omega)} = P(i \cdot \omega); \overline{Q(-i \cdot \omega)} = Q(i \cdot \omega)$. Then $\frac{\partial P(\lambda, x)}{\partial \lambda} = \frac{\partial P(\lambda, x)}{\partial [i \cdot \omega]} = \frac{\partial P(\lambda, x)}{i \cdot \partial \omega} = -i \cdot \frac{\partial P(\lambda, x)}{\partial \omega}$

$i \cdot \frac{\partial P(\lambda,x)}{\partial \lambda} = \frac{\partial P(\lambda,x)}{\partial \omega}; i \cdot P_\lambda(i \cdot \omega) = P_\omega(i \cdot \omega); i \cdot Q_\lambda(i \cdot \omega) = Q_\omega(i \cdot \omega)$ and that on the surface $|P(i \cdot \omega)|^2 = |Q(i \cdot \omega)|^2$, one obtains:

$$(\frac{\partial \lambda}{\partial x})^{-1}|_{\lambda=i \cdot \omega} = (\frac{i \cdot P_\omega(i \cdot \omega, x) \cdot \overline{P(i \cdot \omega, x)} + i \cdot Q_\lambda(i \cdot \omega, x) \cdot \overline{Q(\lambda, x)} - \tau \cdot |P(i \cdot \omega)|^2}{P_x(i \cdot \omega, x) \cdot \overline{P(i \cdot \omega, x)} - Q_x(i \cdot \omega, x) \cdot \overline{Q(i \cdot \omega, x)}})$$

Upon separating into real and imaginary parts, with $P = P_R + i \cdot P_I$; $Q = Q_R + i \cdot Q_I$

$$P_\omega = P_{R\omega} + i \cdot P_{I\omega}; Q_\omega = Q_{R\omega} + i \cdot Q_{I\omega}; P_x = P_{Rx} + i \cdot P_{Ix}; Q_x = Q_{Rx} + i \cdot Q_{Ix}; P^2 = P_R^2 + P_I^2 = \omega^{10}$$

When (x) can be any Bias-T circuit parameter's $R_s, C_1, L_1, R_1, C_2, R_{load}, R_b$ and τ etc. Where for convenience, we have dropped the arguments $(i \cdot \omega, x)$, and where $F_\omega = 2 \cdot [(P_{R\omega} \cdot P_R + P_{I\omega} \cdot P_I) - (Q_{R\omega} \cdot Q_R + Q_{I\omega} \cdot Q_I)]; \omega_x = \frac{-F_x}{F_\omega}$
$F_x = 2 \cdot [(P_{Rx} \cdot P_R + P_{Ix} \cdot P_I) - (Q_{Rx} \cdot Q_R + Q_{Ix} \cdot Q_I)]$. We define U and V:

$$U = (P_R \cdot P_{I\omega} - P_I \cdot P_{R\omega}) - (Q_R \cdot Q_{I\omega} - Q_I \cdot Q_{R\omega})$$
$$V = (P_R \cdot P_{Ix} - P_I \cdot P_{Rx}) - (Q_R \cdot Q_{Ix} - Q_I \cdot Q_{Rx})$$

We choose our specific parameter as time delay x = τ.

$$Q_I = q_1(\tau) \cdot \omega - q_3(\tau) \cdot \omega^3; P_R = 0; P_I = -\omega^5; Q_R = q_0(\tau) - q_2(\tau) \cdot \omega^2 + q_4(\tau) \cdot \omega^4$$
$$P_{R\omega} = 0; P_{I\omega} = -5 \cdot \omega^4; Q_{I\omega} = q_1(\tau) - q_3(\tau) \cdot 3 \cdot \omega^2;$$
$$Q_{R\omega} = -2 \cdot q_2(\tau) \cdot \omega + 4 \cdot q_4(\tau) \cdot \omega^3$$
$$P_{R\tau} = 0; P_{I\tau} = 0; Q_{I\tau} = \frac{\partial q_1(\tau)}{\partial \tau} \cdot \omega - \frac{\partial q_3(\tau)}{\partial \tau} \cdot \omega^3;$$
$$Q_{R\tau} = \frac{\partial q_0(\tau)}{\partial \tau} - \frac{\partial q_2(\tau)}{\partial \tau} \cdot \omega^2 + \frac{\partial q_4(\tau)}{\partial \tau} \cdot \omega^4$$
$$\omega_\tau = \frac{-F_\tau}{F_\omega}; P_{R\omega} \cdot P_R = 0; Q_{R\omega} \cdot Q_R = [-2 \cdot q_2(\tau) \cdot \omega + 4 \cdot q_4(\tau) \cdot \omega^3] \cdot [q_0(\tau) - q_2(\tau) \cdot \omega^2 + q_4(\tau) \cdot \omega^4]$$
$$Q_{R\omega} \cdot Q_R = -2 \cdot q_0(\tau) \cdot q_2(\tau) \cdot \omega + 2 \cdot [q_2^2(\tau) + 2 \cdot q_0(\tau) \cdot q_4(\tau)] \cdot \omega^3 - 6 \cdot q_2(\tau) \cdot q_4(\tau) \cdot \omega^5 + 4 \cdot q_4^2(\tau) \cdot \omega^7$$
$$\Upsilon_1(\tau) = -2 \cdot q_0(\tau) \cdot q_2(\tau); \Upsilon_3(\tau) = 2 \cdot [q_2^2(\tau) + 2 \cdot q_0(\tau) \cdot q_4(\tau)];$$
$$\Upsilon_5(\tau) = -6 \cdot q_2(\tau) \cdot q_4(\tau)$$
$$\Upsilon_7(\tau) = 4 \cdot q_4^2(\tau); Q_{R\omega} \cdot Q_R = \sum_{k=1}^{4} \Upsilon_{2 \cdot k-1} \cdot \omega^{2 \cdot k-1}; P_R \cdot P_{I\omega} = 0; P_I \cdot P_{R\omega} = 0;$$
$$P_{I\omega} \cdot P_I = 5 \cdot \omega^9$$
$$Q_{I\omega} \cdot Q_I = [q_1(\tau) - q_3(\tau) \cdot 3 \cdot \omega^2] \cdot [q_1(\tau) \cdot \omega - q_3(\tau) \cdot \omega^3]$$
$$= q_1^2(\tau) \cdot \omega - 4 \cdot q_1(\tau) \cdot q_3(\tau) \cdot \omega^3 + 3 \cdot q_3^2(\tau) \cdot \omega^5$$

$$\psi_1(\tau) = q_1^2(\tau); \psi_3(\tau) = -4 \cdot q_1(\tau) \cdot q_3(\tau); \psi_5(\tau) = 3 \cdot q_3^2(\tau); Q_{I\omega} \cdot Q_I = \sum_{k=1}^{3} \psi_{2\cdot k-1}(\tau) \cdot \omega^{2\cdot k-1}$$

$$Q_R \cdot Q_{I\omega} = [q_0(\tau) - q_2(\tau) \cdot \omega^2 + q_4(\tau) \cdot \omega^4] \cdot [q_1(\tau) - q_3(\tau) \cdot 3 \cdot \omega^2] = q_0(\tau) \cdot q_1(\tau) - [3 \cdot q_0(\tau) \cdot q_3(\tau) + q_2(\tau) \cdot q_1(\tau)] \cdot \omega^2 + [3 \cdot q_3(\tau) \cdot q_2(\tau) + q_4(\tau) \cdot q_1(\tau)] \cdot \omega^4 - 3 \cdot q_4(\tau) \cdot q_3(\tau) \cdot \omega^6$$

$$\mathrm{A}_0(\tau) = q_0(\tau) \cdot q_1(\tau); \mathrm{A}_2(\tau) = -[3 \cdot q_0(\tau) \cdot q_3(\tau) + q_2(\tau) \cdot q_1(\tau)]$$

$$\mathrm{A}_4(\tau) = 3 \cdot q_3(\tau) \cdot q_2(\tau) + q_4(\tau) \cdot q_1(\tau); \mathrm{A}_6(\tau) = -3 \cdot q_4(\tau) \cdot q_3(\tau);$$

$$Q_R \cdot Q_{I\omega} = \sum_{k=0}^{3} \mathrm{A}_{2\cdot k}(\tau) \cdot \omega^{2\cdot k}$$

$$Q_I \cdot Q_{R\omega} = [q_1(\tau) \cdot \omega - q_3(\tau) \cdot \omega^3] \cdot [-2 \cdot q_2(\tau) \cdot \omega + 4 \cdot q_4(\tau) \cdot \omega^3] = -2 \cdot q_1(\tau) \cdot q_2(\tau) \cdot \omega^2 + 2 \cdot [2 \cdot q_4(\tau) \cdot q_1(\tau) + q_2(\tau) \cdot q_3(\tau)] \cdot \omega^4 - 4 \cdot q_3(\tau) \cdot q_4(\tau) \cdot \omega^6$$

$$\xi_2(\tau) = -2 \cdot q_1(\tau) \cdot q_2(\tau); \xi_4(\tau) = 2 \cdot [2 \cdot q_4(\tau) \cdot q_1(\tau) + q_2(\tau) \cdot q_3(\tau)];$$

$$\xi_6(\tau) = -4 \cdot q_3(\tau) \cdot q_4(\tau)$$

$$Q_I \cdot Q_{R\omega} = \sum_{k=1}^{3} \xi_{2\cdot k}(\tau) \cdot \omega^{2\cdot k}; P_R \cdot P_{I\tau} = 0; P_I \cdot P_{R\tau} = 0$$

$$Q_R \cdot Q_{I\tau} = [q_0(\tau) - q_2(\tau) \cdot \omega^2 + q_4(\tau) \cdot \omega^4] \cdot [\frac{\partial q_1(\tau)}{\partial \tau} \cdot \omega - \frac{\partial q_3(\tau)}{\partial \tau} \cdot \omega^3] = q_0(\tau) \cdot \frac{\partial q_1(\tau)}{\partial \tau} \cdot \omega - [q_0(\tau) \cdot \frac{\partial q_3(\tau)}{\partial \tau} + q_2(\tau) \cdot \frac{\partial q_1(\tau)}{\partial \tau}] \cdot \omega^3 + [q_2(\tau) \cdot \frac{\partial q_3(\tau)}{\partial \tau} + q_4(\tau) \cdot \frac{\partial q_1(\tau)}{\partial \tau}] \cdot \omega^5 - q_4(\tau) \cdot \frac{\partial q_3(\tau)}{\partial \tau} \cdot \omega^7$$

$$\zeta_1(\tau) = q_0(\tau) \cdot \frac{\partial q_1(\tau)}{\partial \tau}; \zeta_3(\tau) = -\left[q_0(\tau) \cdot \frac{\partial q_3(\tau)}{\partial \tau} + q_2(\tau) \cdot \frac{\partial q_1(\tau)}{\partial \tau}\right]$$

$$\zeta_5(\tau) = q_2(\tau) \cdot \frac{\partial q_3(\tau)}{\partial \tau} + q_4(\tau) \cdot \frac{\partial q_1(\tau)}{\partial \tau}; \zeta_7(\tau) = -q_4(\tau) \cdot \frac{\partial q_3(\tau)}{\partial \tau};$$

$$Q_R \cdot Q_{I\tau} = \sum_{K=0}^{3} \zeta_{2\cdot k+1} \cdot \omega^{2\cdot k+1}$$

$$Q_I \cdot Q_{R\tau} = [q_1(\tau) \cdot \omega - q_3(\tau) \cdot \omega^3] \cdot [\frac{\partial q_0(\tau)}{\partial \tau} - \frac{\partial q_2(\tau)}{\partial \tau} \cdot \omega^2 + \frac{\partial q_4(\tau)}{\partial \tau} \cdot \omega^4]$$
$$= q_1(\tau) \cdot \frac{\partial q_0(\tau)}{\partial \tau} \cdot \omega - [q_1(\tau) \cdot \frac{\partial q_2(\tau)}{\partial \tau} + q_3(\tau) \cdot \frac{\partial q_0(\tau)}{\partial \tau}] \cdot \omega^3$$
$$+ [q_1(\tau) \cdot \frac{\partial q_4(\tau)}{\partial \tau} + q_3(\tau) \cdot \frac{\partial q_2(\tau)}{\partial \tau}] \cdot \omega^5 - q_3(\tau) \cdot \frac{\partial q_4(\tau)}{\partial \tau} \cdot \omega^7$$
$$\eta_1(\tau) = q_1(\tau) \cdot \frac{\partial q_0(\tau)}{\partial \tau}; \eta_3(\tau) = -[q_1(\tau) \cdot \frac{\partial q_2(\tau)}{\partial \tau} + q_3(\tau) \cdot \frac{\partial q_0(\tau)}{\partial \tau}]$$
$$\eta_5(\tau) = q_1(\tau) \cdot \frac{\partial q_4(\tau)}{\partial \tau} + q_3(\tau) \cdot \frac{\partial q_2(\tau)}{\partial \tau}; \eta_7(\tau) = -q_3(\tau) \cdot \frac{\partial q_4(\tau)}{\partial \tau};$$
$$Q_I \cdot Q_{R\tau} = \sum_{k=0}^{3} \eta_{2\cdot k+1}(\tau) \cdot \omega^{2\cdot k+1}$$

$$F_\omega = 10 \cdot \omega^9 - 2 \cdot [\sum_{k=1}^{3} [\psi_{2\cdot k-1}(\tau) + \Upsilon_{2\cdot k-1}] \cdot \omega^{2\cdot k-1} + \Upsilon_7 \cdot \omega^7]$$
$$F_\tau = -2 \cdot (Q_{R\tau} \cdot Q_R + Q_{I\tau} \cdot Q_I) = -2 \cdot ([\frac{\partial q_0(\tau)}{\partial \tau} - \frac{\partial q_2(\tau)}{\partial \tau} \cdot \omega^2 + \frac{\partial q_4(\tau)}{\partial \tau} \cdot \omega^4] \cdot$$
$$[q_0(\tau) - q_2(\tau) \cdot \omega^2 + q_4(\tau) \cdot \omega^4] + [\frac{\partial q_1(\tau)}{\partial \tau} \cdot \omega - \frac{\partial q_3(\tau)}{\partial \tau} \cdot \omega^3] \cdot$$
$$[q_1(\tau) \cdot \omega - q_3(\tau) \cdot \omega^3])$$

$$F_\tau = -2 \cdot (Q_{R\tau} \cdot Q_R + Q_{I\tau} \cdot Q_I) = -2 \cdot \{\frac{\partial q_0(\tau)}{\partial \tau} \cdot q_0(\tau) - [\frac{\partial q_0(\tau)}{\partial \tau} \cdot q_2(\tau)$$
$$+ \frac{\partial q_2(\tau)}{\partial \tau} \cdot q_0(\tau) - \frac{\partial q_1(\tau)}{\partial \tau} \cdot q_1(\tau)] \cdot \omega^2$$
$$+ [\frac{\partial q_0(\tau)}{\partial \tau} \cdot q_4(\tau) + \frac{\partial q_2(\tau)}{\partial \tau} \cdot q_2(\tau) + \frac{\partial q_4(\tau)}{\partial \tau} \cdot q_0(\tau) - \frac{\partial q_1(\tau)}{\partial \tau} \cdot q_3(\tau)$$
$$- \frac{\partial q_3(\tau)}{\partial \tau} \cdot q_1(\tau)] \cdot \omega^4 - [\frac{\partial q_2(\tau)}{\partial \tau} \cdot q_4(\tau) + \frac{\partial q_4(\tau)}{\partial \tau} \cdot q_2(\tau) - \frac{\partial q_3(\tau)}{\partial \tau} \cdot q_3(\tau)]$$
$$\cdot \omega^6 + \frac{\partial q_4(\tau)}{\partial \tau} \cdot q_4(\tau) \cdot \omega^8\}$$

$$F_\tau = -2 \cdot (Q_{R\tau} \cdot Q_R + Q_{I\tau} \cdot Q_I) = -2 \cdot \frac{\partial q_0(\tau)}{\partial \tau} \cdot q_0(\tau) + 2 \cdot [\frac{\partial q_0(\tau)}{\partial \tau} \cdot q_2(\tau)$$
$$+ \frac{\partial q_2(\tau)}{\partial \tau} \cdot q_0(\tau) - \frac{\partial q_1(\tau)}{\partial \tau} \cdot q_1(\tau)] \cdot \omega^2 - 2 \cdot [\frac{\partial q_0(\tau)}{\partial \tau} \cdot q_4(\tau) + \frac{\partial q_2(\tau)}{\partial \tau} \cdot q_2(\tau)$$
$$+ \frac{\partial q_4(\tau)}{\partial \tau} \cdot q_0(\tau) - \frac{\partial q_1(\tau)}{\partial \tau} \cdot q_3(\tau) - \frac{\partial q_3(\tau)}{\partial \tau} \cdot q_1(\tau)] \cdot \omega^4$$
$$+ 2 \cdot [\frac{\partial q_2(\tau)}{\partial \tau} \cdot q_4(\tau) + \frac{\partial q_4(\tau)}{\partial \tau} \cdot q_2(\tau) - \frac{\partial q_3(\tau)}{\partial \tau} \cdot q_3(\tau)] \cdot \omega^6 - 2 \cdot \frac{\partial q_4(\tau)}{\partial \tau} \cdot q_4(\tau) \cdot \omega^8$$

We define for simplicity the following functions: $B_0(\tau) = -2 \cdot \frac{\partial q_0(\tau)}{\partial \tau} \cdot q_0(\tau)$

$$B_2(\tau) = 2 \cdot \left[\frac{\partial q_0(\tau)}{\partial \tau} \cdot q_2(\tau) + \frac{\partial q_2(\tau)}{\partial \tau} \cdot q_0(\tau) - \frac{\partial q_1(\tau)}{\partial \tau} \cdot q_1(\tau)\right]$$

$$B_4(\tau) = -2 \cdot \left[\frac{\partial q_0(\tau)}{\partial \tau} \cdot q_4(\tau) + \frac{\partial q_2(\tau)}{\partial \tau} \cdot q_2(\tau) + \frac{\partial q_4(\tau)}{\partial \tau} \cdot q_0(\tau) - \frac{\partial q_1(\tau)}{\partial \tau} \cdot q_3(\tau) - \frac{\partial q_3(\tau)}{\partial \tau} \cdot q_1(\tau)\right]$$

$$B_6(\tau) = 2 \cdot \left[\frac{\partial q_2(\tau)}{\partial \tau} \cdot q_4(\tau) + \frac{\partial q_4(\tau)}{\partial \tau} \cdot q_2(\tau) - \frac{\partial q_3(\tau)}{\partial \tau} \cdot q_3(\tau)\right];$$

$$B_8(\tau) = -2 \cdot \frac{\partial q_4(\tau)}{\partial \tau} \cdot q_4(\tau)$$

$$F_\tau = -2 \cdot (Q_{R\tau} \cdot Q_R + Q_{I\tau} \cdot Q_I) = \sum_{k=0}^{4} B_{2\cdot k}(\tau) \cdot \omega^{2\cdot k}$$

$$U = (P_R \cdot P_{I\omega} - P_I \cdot P_{R\omega}) - (Q_R \cdot Q_{I\omega} - Q_I \cdot Q_{R\omega}) = -\left(\sum_{k=0}^{3} A_{2\cdot k}(\tau) \cdot \omega^{2\cdot k} - \sum_{k=1}^{3} \xi_{2\cdot k}(\tau) \cdot \omega^{2\cdot k}\right)$$

$$U = \sum_{k=1}^{3} \xi_{2\cdot k}(\tau) \cdot \omega^{2\cdot k} - \sum_{k=0}^{3} A_{2\cdot k}(\tau) \cdot \omega^{2\cdot k} = \sum_{k=1}^{3} [\xi_{2\cdot k}(\tau) - A_{2\cdot k}(\tau)] \cdot \omega^{2\cdot k} - A_0(\tau)$$

$$V|_{x=\tau} = -(Q_R \cdot Q_{I\tau} - Q_I \cdot Q_{R\tau}) = -\left(\sum_{k=0}^{3} \zeta_{2\cdot k+1} \cdot \omega^{2\cdot k+1} - \sum_{k=0}^{3} \eta_{2\cdot k+1}(\tau) \cdot \omega^{2\cdot k+1}\right) = \sum_{k=0}^{3} [\eta_{2\cdot k+1}(\tau) - \zeta_{2\cdot k+1}] \cdot \omega^{2\cdot k+1}$$

$F(\omega, \tau) = 0$. Differentiating with respect to τ and we get $F_\omega \cdot \frac{\partial \omega}{\partial \tau} + F_\tau = 0$

$$\tau \in I \Rightarrow \frac{\partial \omega}{\partial \tau} = \omega_\tau = -\frac{F_\tau}{F_\omega}; \Lambda^{-1}(\tau) = \left(\frac{\partial \operatorname{Re}\lambda}{\partial \tau}\right)_{\lambda = i\cdot\omega};$$

$$\Lambda^{-1}(\tau) = \operatorname{Re}\left\{\frac{-2 \cdot [U + \tau \cdot |P|^2] + i \cdot F_\omega}{F_\tau + i \cdot 2 \cdot [V + \omega \cdot |P|^2]}\right\}$$

$$sign\{\Lambda^{-1}(\tau)\} = sign\left\{\left(\frac{\partial \operatorname{Re}\lambda}{\partial \tau}\right)_{\lambda = i\cdot\omega}\right\};$$

$$sign\{\Lambda^{-1}(\tau)\} = sign\{F_\omega\} \cdot sign\left\{\tau \cdot \frac{\partial \omega}{\partial \tau} + \omega + \frac{U \cdot \frac{\partial \omega}{\partial \tau} + V}{|P|^2}\right\}$$

We shall presently examine the possibility of stability transitions (bifurcations) Bias-T circuit, about the equilibrium points $E^{(j)}(Y_1^{(j)}, Y_2^{(j)}, Y_3^{(j)}, I_{52a}^{(j)}, I_{R_s}^{(j)})$, j = 0, 1,…

As a result of a variation of delay parameter τ. The analysis consists in identifying the roots of our system characteristic equation situated on the imaginary axis of the complex λ-plane where by increasing the delay parameter τ, Reλ may at the crossing, changes its sign from − to +, i.e. from stable focus $E^{(j)}$ to an unstable one, or vice versa. This feature may be further assessed by examining the sign of the partial derivatives with respect to τ,

$$\Lambda^{-1}(\tau) = \left(\frac{\partial \mathrm{Re}\lambda}{\partial \tau}\right)_{\lambda = i\cdot\omega}, R_s, C_1, L_1, R_1, C_2, R_{load}, R_b, \ldots = const; \omega \in \mathbb{R}_+$$

$$sign\{\Lambda^{-1}(\tau)\} = sign\left\{10\cdot\omega^9 - 2\cdot\left[\sum_{k=1}^{3}[\psi_{2\cdot k-1}(\tau) + \Upsilon_{2\cdot k-1}]\cdot\omega^{2\cdot k-1} + \Upsilon_7\cdot\omega^7\right]\right\}$$
$$\cdot\, sign\{-\tau\cdot\left(\frac{\sum_{k=0}^{4}\mathrm{B}_{2\cdot k}(\tau)\cdot\omega^{2\cdot k}}{10\cdot\omega^9 - 2\cdot\left[\sum_{k=1}^{3}[\psi_{2\cdot k-1}(\tau) + \Upsilon_{2\cdot k-1}]\cdot\omega^{2\cdot k-1} + \Upsilon_7\cdot\omega^7\right]}\right) + \omega$$
$$+\frac{\{-\left[\sum_{k=1}^{3}[\xi_{2\cdot k}(\tau) - \mathrm{A}_{2\cdot k}(\tau)]\cdot\omega^{2\cdot k} - \mathrm{A}_0(\tau)\right]\cdot\left(\frac{\sum_{k=0}^{4}\mathrm{B}_{2\cdot k}(\tau)\cdot\omega^{2\cdot k}}{10\cdot\omega^9 - 2\cdot\left[\sum_{k=1}^{3}[\psi_{2\cdot k-1}(\tau) + \Upsilon_{2\cdot k-1}\right]\cdot\omega^{2\cdot k-1} + \Upsilon_7\cdot\omega^7]}\right) + \sum_{k=0}^{3}[\eta_{2\cdot k+1}(\tau) - \zeta_{2\cdot k+1}]\cdot\omega^{2\cdot k+1}\}}{\omega^{10}}\}$$

The stability switch occurs only on those delay values (τ) which fit the equation: $\tau = \frac{\theta_+(\tau)}{\omega_+(\tau)}$ and $\theta_+(\tau)$ is the solution of $\sin\theta(\tau) = \ldots; \cos\theta(\tau) = \ldots$ when $\omega = \omega_+(\tau)$ if only ω_+ is feasible. Additionally, when all Bias-T circuit parameters are known and the stability switch due to various time delay values τ is described in the following expression:

$$sign\{\Lambda^{-1}(\tau)\} = sign\{F_\omega(\omega(\tau), \tau)\}$$
$$\cdot\, sign\left\{\tau\cdot\omega_\tau(\omega(\tau)) + \omega(\tau) + \frac{U(\omega(\tau))\cdot\omega_\tau(\omega(\tau)) + V(\omega(\tau))}{|P(\omega(\tau))|^2}\right\}$$

Remark: We know $F_\omega(\omega(\tau), \tau) = 0$ implies its roots $\omega_i(\tau)$ and finding those delays values τ which ω_i is feasible. There are τ values which give complex ω_i or imaginary number, then unable to analyze the stability.

4.5 PIN Diode Stability Analysis Under Parameters Variation

A PIN diode is a diode with a wide, lightly doped 'near' intrinsic semiconductor region between a p-type semiconductors an n-type semiconductor region. The p-type and n-type regions are typically heavily doped because they are used for ohmic contacts. The wide intrinsic region is in contrast to an ordinary PN diode. The wide intrinsic region makes the PIN diode an inferior rectifier (one typical function of a diode). The wide intrinsic region makes the PIN diode suitable for many applications [33,91,92]. The PIN diode suitable for many applications: attenuators, fast switches, photo detectors and high voltage power electronics applications. A PIN diode operates under what is known as high level injection. PIN's intrinsic "i" region is flooded with charge carriers from the "p" and "n" regions. Its function can be likened to filling up a water bucket with a hole on the side. Once the water reaches the hole's level it will begin to pour out. A PIN diode obeys the standard diode equation for low frequency signals. At higher frequencies, the diode looks like an almost perfect (very linear, even for large signals) resistor. At low frequencies, the charge can be removed and the diode turns off. At higher frequencies, there is not enough time to remove the charge, so the diode never turns off. The high frequency resistance is inversely proportional to the DC bias current through the diode. A PIN diode, suitably biased, therefore acts as a variable resistor. The high frequency resistance may vary over a wide range (from 0.1 to 10 k-ohm in some cases; the useful range is smaller, though). The wide intrinsic region also means the diode will have a low capacitance when reverse biased.

PIN diode fundamentals: A PIN diode is a semiconductor device that operates as a variable resistor at RF and microwave frequencies. The resistance value of the PIN diode is determined only by the forward biased DC current. In switch and attenuator applications, the PIN diode should ideally control the RF signal level without introduction distortion which might change the shape of the RF signal. An important additional feature of the PIN diode is its ability to control large RF signals while using much smaller level of DC excitation. A model of a PIN diode chip is presented and the chip is prepared by starting with a wafer of almost intrinsically pure silicon, having high resistivity and long lifetime. A P-region is then diffused into one diode surface and an N-region is diffused into the other surface. The resulting intrinsic or I-region thickness (W) is a function of the thickness of the original silicon wafer.

The area of the chip (A) depends upon how many small sections are defined from the original wafer. The performance of the PIN diode primarily depends on chip geometry and nature of the semiconductor material in the finished diode, particularly in the I-region. The characteristics of PIN diodes are controlled thickness I-regions having long carrier lifetimes and very high resistivity. These characteristics enhance the ability to control RF signals with a minimum of distortion while requiring low DC supply. When a PIN diode is forward biased, holes and electrons are injected from the P and N regions into the I-region. These charges do not recombine immediately. Instead, a finite quantity of charge always remains stored and results in a lowering of the resistivity of the I-region. The quantity of stored charge, Q depends on the recombination time (τ)—the carrier life time, and the forward bias current (I_F), $Q = I_F \cdot \tau$. The resistance of the I-region under forward bias, R_s is inversely proportional to Q. $R_s = \frac{W^2}{(\mu_N + \mu_p) \cdot Q}$ (ohm), W—I-region width, μ_N—electron mobility, μ_p- hole mobility. We get the expression for R_s as an inverse function of current $R_s = \frac{W^2}{(\mu_N + \mu_p) \cdot I_F \cdot \tau}$ (ohm). The equation is independent of area (A). R_s is slightly dependent upon area because the effective lifetime varies with area and thickness $\tau(A, W)$ due to edge recombination effects. Typically, PIN diodes display a resistance characteristic consistent with this model. Resistance of the order of 0.1 Ohm at 1A forward bias increasing to about 10000 Ohm (10Kohm) at 1 μA. The forward bias represents a realistic range for a PIN diode. The maximum forward resistance, R_s(max), of a PIN diode is generally specified at 100 mA forward bias current. Some PIN diodes suppliers specifies not only the R_s(max) but also the R_s(min) at a lower forward bias current ($\sim$ 10 mA). It ensures a wide range of diode resistance which is particularly important in attenuator applications. At the lower frequencies R_s is not constant but increases as the frequency is lowered. The normal PIN diodes which are designed to operate in RF/Microwave frequencies exhibit this increase in R_s in the 1–10 MHz range.

A properly designed PIN will maintain constant Rs well into the 10 kHz region. The results obtained are valid over an extremely broad frequency range. The practical low resistance limitations result from package parasitic inductances and junction contact resistances. Both of which are minimized in the construction of PIN diodes. The high resistance range of PIN diodes is usually limited by the effect of the diode capacitance (C_t). The maximum dynamic range of the PIN diode at high frequencies, this diode reactance may have to be tuned out. The “skin effect” is much less pronounced in relatively poor conductors such as silicon, than with good metallic conductors. The “skin depth” is proportional to the square root of the resistivity of the conducting material. RF signals penetrate deeply into the semiconductor and “skin effect” is not a significant factor in PIN diodes below X-band frequencies. At DC and very low frequencies, the PIN diode is similar to a PN diode. The diode resistance is described by the dynamic resistance of the I–V characteristics at any quiescent bias point. The DC dynamic resistance point is not, however, valid in PIN diodes at frequencies above which the period is shorter than the transit time of the I-region. The frequency at which this occurs, f_T is called transit time frequency and may be considered the lower frequency limit. The lower frequency limit is primarily a function of W, the I-region thickness and can be expressed at $f_T = 1300/W$, where W is the I-region thickness in microns.

The conductance of the diode is proportional to the stored charge and the charge is in turn related to the diode current by $I_d = \frac{dQ_d}{dt} + \frac{Q_d}{\tau}$ where I_d is the diode current, Q_d is the charge stored in the diode, τ is diode recombination lifetime. If the diode is biased with only a constant current, the stored charge is constant and is equal to $Q_d = I_d \cdot \tau$. The PIN diode store charge equation:

$I_d(t) = \frac{dQ_d(t)}{dt} + \frac{Q_d(t)}{\tau}$, $I_d(t)$ is a function of time. First we consider the simpler equation $I_d(t) = 0$. The equation is $\frac{dQ_d(t)}{dt} + \frac{Q_d(t)}{\tau} = 0$ or $\frac{\frac{dQ_d(t)}{dt}}{Q_d(t)} = -\frac{1}{\tau}$

$$\frac{dQ_d}{Q_d} = -\frac{1}{\tau} \cdot dt; \ln Q_d = -\int \frac{1}{\tau} \cdot dt + const; Q_d = e^{-\int \frac{1}{\tau} \cdot dt + const} = A \cdot e^{-\int \frac{1}{\tau} \cdot dt};$$

$$A = e^{const}$$

$I_\tau = \int \frac{1}{\tau} \cdot dt \Rightarrow \frac{dI_\tau}{dt} = \frac{1}{\tau}; Q_d = A \cdot e^{-I_\tau}; Q_d \cdot e^{I_\tau} = A$. We can see how to solve our equation. If we differentiate equation $Q_d \cdot e^{I_\tau} = A$ respect to time t and use $\frac{dI_\tau}{dt} = \frac{1}{\tau}$, we get $\frac{d}{dt}(Q_d \cdot e^{I_\tau}) = \frac{dQ_d}{dt} \cdot e^{I_\tau} + Q_d \cdot e^{I_\tau} \cdot \frac{dI_\tau}{dt}; \frac{d}{dt}(Q_d \cdot e^{I_\tau}) = e^{I_\tau} \cdot [\frac{dQ_d}{dt} + Q_d \cdot \frac{dI_\tau}{dt}]$ e^{I_τ} is the integrating factor. Since $\frac{dQ_d}{dt} + Q_d \cdot \frac{dI_\tau}{dt}|_{\frac{dI_\tau}{dt} = \frac{1}{\tau}} = I_d$ then

$\frac{d}{dt}(Q_d \cdot e^{I_\tau}) = e^{I_\tau} \cdot [\frac{dQ_d}{dt} + Q_d \cdot \frac{dI_\tau}{dt}] = e^{I_\tau} \cdot I_d$. I_d and e^{I_τ} are functions of t only, we can now integrate both sides of $\frac{d}{dt}(Q_d \cdot e^{I_\tau}) = e^{I_\tau} \cdot I_d$ with respect to time t to get $Q_d \cdot e^{I_\tau} = \int e^{I_\tau} \cdot I_d \cdot dt + const; Q_d = e^{-I_\tau} \cdot \int e^{I_\tau} \cdot I_d \cdot dt + const; I_\tau = \int \frac{1}{\tau} \cdot dt$. If we consider that recombination lifetime parameter τ is independent on time then

$I_\tau = \int \frac{1}{\tau} \cdot dt = \frac{1}{\tau} \cdot t + const$. If the bias consists of both a constant current and a low frequency RF or time varying signal, then the DC component of stored charge will be modulated by the presence of an AC component. The degree of modulation depends on the relative level the two charge components and the frequency of the RF signal. At signal frequency below $f_c = \frac{1}{2 \cdot \pi \cdot \tau}$ the RF signal has about the same effect as the DC bias. Above f_c, the modulation effect decreases. The lifetime of PIN diodes is determined by design and is based on the desired switching speed. Typically, diode recombination lifetime τ can be in the range of 0.005 μsec to over 3 μsec. At frequencies bellow f_c, the PIN diode behaves as an ordinary PN junction diode. The RF signal incident on the diode will be rectified and considerable distortion of the signal will occur. In the vicinity of f_c, the diode begins to behave as a linear resistor with a small nonlinear component. At frequencies well above f_c, the

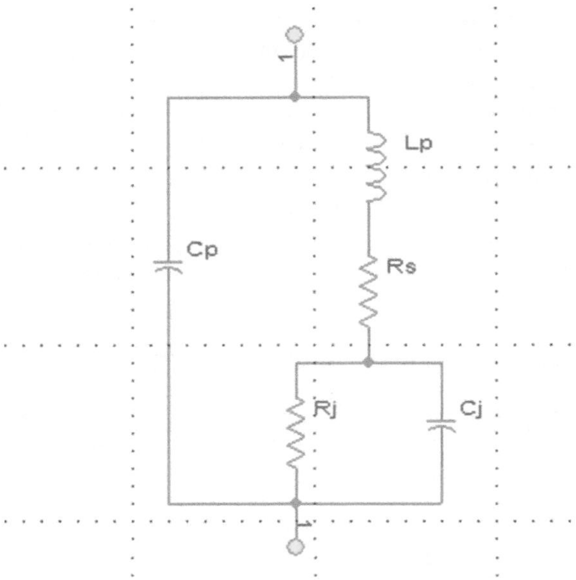

Fig. 4.16 PIN diode low frequency equivalent circuit

diode appears essentially as a pure linear resistance whose value can be controlled by the DC or a low frequency control signal. The equivalent circuit of the PIN diode also depends on the frequency. At frequencies much less than f_c the equivalent circuit is as shown in the below figure (normal PN junction) [24] (Fig. 4.16).

In this circuit L_P is the package inductance, C_P is the package capacitance, R_s is the series resistance, and R_j is the junction resistance $\left(R_j = \frac{n \cdot k \cdot T}{q \cdot I_{dc}}\right)$. Typical value for n is 1.8 then at room temperature and $R_j = \frac{48}{I_{dc}(mA)}$. I_{dc} is the forward DC bias current, $C_{j(V)}$ is the junction capacitance which is a function of the applied voltage. At frequencies much higher than f_c, we can draw equivalent circuit is as shown in the below figure. L_P, C_P, and R_s are the same as in the low frequency equivalent circuit. The element C_I represents the I-layer capacitance which is constant and dependent only on the geometry of the I-layer (typical values of C_I are between 0.02 and 2 pF and are dependent on diode design). The element R_I represents the effective RF resistance of the I-layer. This resistance is constant with respect to RF signal, providing the signal frequency much higher than f_c. It is variable by the DC or very low frequency control current (Fig. 4.17).

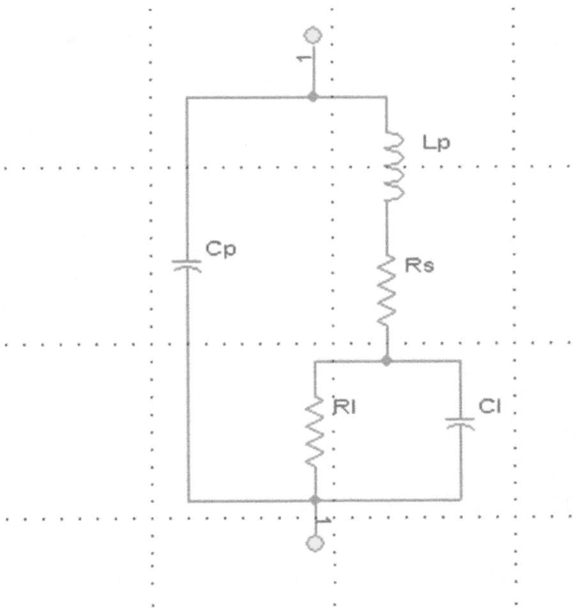

Fig. 4.17 PIN diode high frequency equivalent circuit

We analyze the stability of Single Pole Single Throw (SPST)) PIN switches. The RF switch circuit requires a few components and a very simple to implement. Nevertheless it is able to act as an RF switch for radio frequency or RF applications and is adequate for many applications. The RF switch circuit comprises a single PIN diode (D_1), an RF inductor or choke (L_c), a current limiting resistor or RF choke (L_d) and a DC block capacitor (C_c). In operation, when a positive potential is applied to the control point current, this forward biases the diode and as a result the radio frequency signal is able to pass through the circuit. When a negative bias is applied to the circuit, the diode become reverse biased and is effectively switched off. Under these conditions the depletion layer in the diode becomes wide and does not allow signal to pass. PIN diodes have a number of advantages as switches. In the first place they are more linear than ordinary PN junction diodes. This means that in their action as a radio frequency switch they do not create as many spurious products. Secondly when reverse biased and switched off, the depletion layer is wider than with an ordinary diode and this provides for greater isolation when switching. By varying the amount of bias on the PIN diode it is possible to vary the level of attenuation provided. In this way the circuit can be used as a very simple RF attenuator. Although the circuit can be used as an RF attenuator, more effective RF attenuator circuits are available for more demanding applications [92, 93] (Fig. 4.18).

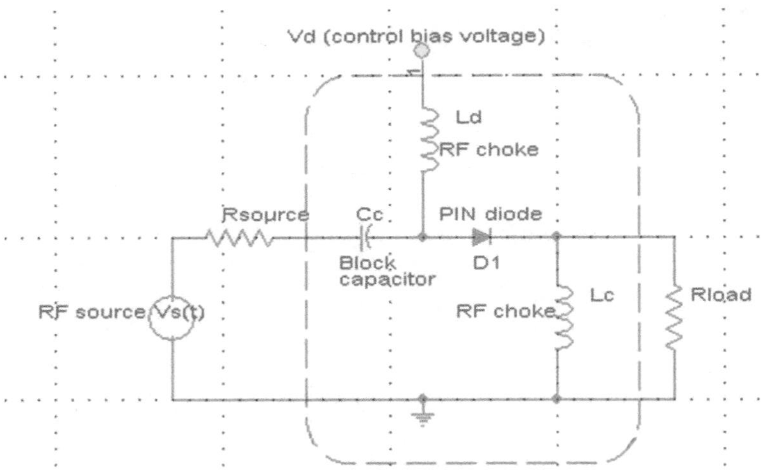

Fig. 4.18 PIN diode attenuator and switch

Applications of PIN diode:

- A variable resistor in a variable attenuator, a function that few other components can achieve as effectively. The fact that when it is forward biased, the diode is linear, behaving like a resistor, can be put to good use in a variety of applications.
- The PIN diode can also be used as an RF switch. In the forward direction it can be biased sufficiently to ensure it has a low resistance to the RF that needs to be passed, and when a reverse bias is applied it acts as an open circuit, with only a relatively small level of capacitance.
- PIN diode is for use in RF protection circuits. When used with RF, the diode normally behaves like a resistor when a small bias is applied. Hover this is only true for RF levels below a certain level. Above this the resistance drops considerably. Thus it can be used to protect a sensitive receiver from the effects of a large transmitter if it is placed across the receiver input.

We consider the RF source as a high frequency signal and use PIN diode high frequency equivalent circuit.

$$V_{A_1} = V_s(t); I_{R_{source}} = \frac{V_{A_1} - V_{A_2}}{R_{source}}; I_{R_{source}} = I_{C_c}; I_{C_c} = C_c \cdot \frac{d}{dt}(V_{A_2} - V_{A_3}); V_{L_d} = V_d - V_{A_3} = L_d \cdot \frac{dI_{L_d}}{dt}$$

(Fig. 4.19).

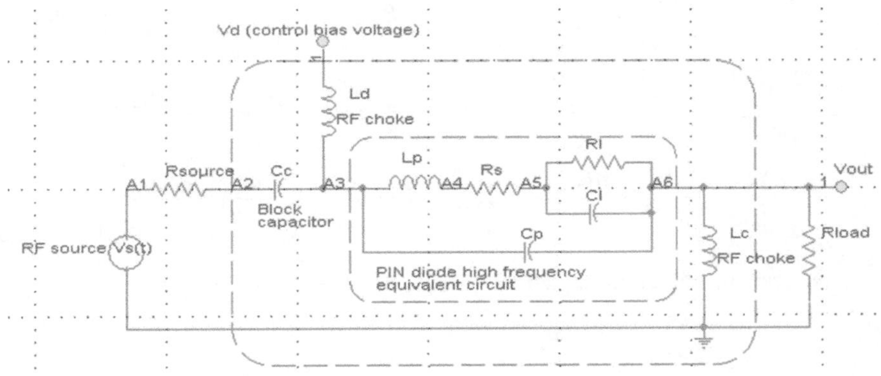

Fig. 4.19 PIN diode attenuator and switch equivalent circuit

$$V_{L_p} = V_{A_3} - V_{A_4} = L_p \cdot \frac{dI_{L_p}}{dt}; I_{L_p} = I_{R_s}; I_{R_s} = \frac{V_{A_4} - V_{A_5}}{R_s}; I_{R_I} = \frac{V_{A_5} - V_{A_6}}{R_I}; I_{C_I} = C_I \cdot \frac{d}{dt}(V_{A_5} - V_{A_6})$$

$$I_{C_p} = C_p \cdot \frac{d}{dt}(V_{A_3} - V_{A_6}); V_{A_6} = V_{L_c} = L_c \cdot \frac{dI_{L_c}}{dt}; I_{R_{load}} = \frac{V_{A_6}}{R_{load}}; V_{L_c} = V_{R_{load}} = V_{out}$$

We write our circuit Kirchhoff's Current Law (KCL) for all nodes:

$$I_{R_{source}} = \frac{V_s(t) - V_{A_2}}{R_{source}} \Rightarrow V_s(t) - V_{A_2} = I_{R_{source}} \cdot R_{source}; V_{A_2} = V_s(t) - I_{R_{source}} \cdot R_{source}$$

$$I_{C_c} = C_c \cdot \frac{d}{dt}(V_{A_2} - V_{A_3}) \Rightarrow I_{C_c} \cdot \frac{1}{C_c} = \frac{d}{dt}(V_{A_2} - V_{A_3}); V_{A_2} - V_{A_3} = \frac{1}{C_c} \cdot \int I_{C_c} \cdot dt$$

$$V_{A_3} = V_{A_2} - \frac{1}{C_c} \cdot \int I_{C_c} \cdot dt; V_{A_3} = V_s(t) - I_{R_{source}} \cdot R_{source} - \frac{1}{C_c} \cdot \int I_{C_c} \cdot dt$$

Table 4.4 PIN diode attenuator and switch equivalent circuit Kirchhoff's Current Law (KCL) for all nodes

Node number	KCL @ A_i
A_2	$I_{R_{source}} = I_{C_c}$
A_3	$I_{C_c} + I_{L_d} = I_{L_p} + I_{C_p}$
A_4	$I_{L_p} = I_{R_s}$
A_5	$I_{R_s} = I_{R_I} + I_{C_I}$
A_6	$I_{R_I} + I_{C_I} + I_{C_p} = I_{L_c} + I_{R_{load}}$

$$V_d - V_{A_3} = L_d \cdot \frac{dI_{L_d}}{dt}; V_d - [V_s(t) - I_{R_{source}} \cdot R_{source} - \frac{1}{C_c} \cdot \int I_{C_c} \cdot dt] = L_d \cdot \frac{dI_{L_d}}{dt}$$

$$\frac{d}{dt}\{V_d - [V_s(t) - I_{R_{source}} \cdot R_{source} - \frac{1}{C_c} \cdot \int I_{C_c} \cdot dt]\} = L_d \cdot \frac{d^2 I_{L_d}}{dt^2}$$

$$\frac{dV_d}{dt} - \frac{dV_s(t)}{dt} + \frac{dI_{R_{source}}}{dt} \cdot R_{source} + \frac{1}{C_c} \cdot I_{C_c} = L_d \cdot \frac{d^2 I_{L_d}}{dt^2}; \frac{dV_d}{dt} \to \varepsilon$$

First system differential equation: $-\frac{dV_s(t)}{dt} + \frac{dI_{R_{source}}}{dt} \cdot R_{source} + \frac{1}{C_c} \cdot I_{C_c} = L_d \cdot \frac{d^2 I_{L_d}}{dt^2}$

$$V_{A_3} - V_{A_4} = L_p \cdot \frac{dI_{L_p}}{dt} \Rightarrow V_{A_4} = V_{A_3} - \frac{dI_{L_p}}{dt}; V_{A_4} = V_s(t) - I_{R_{source}} \cdot R_{source} - \frac{1}{C_c} \cdot \int I_{C_c} \cdot dt - \frac{dI_{L_p}}{dt}$$

$$V_{A_3} - V_{A_4} = L_p \cdot \frac{dI_{L_p}}{dt} \Rightarrow V_{A_4} = V_{A_3} - L_p \cdot \frac{dI_{L_p}}{dt}; V_{A_4} = V_s(t) - I_{R_{source}} \cdot R_{source} - \frac{1}{C_c} \cdot \int I_{C_c} \cdot dt - L_p \cdot \frac{dI_{L_p}}{dt}$$

$$I_{R_s} = \frac{V_{A_4} - V_{A_5}}{R_s} \Rightarrow V_{A_4} - V_{A_5} = I_{R_s} \cdot R_s; V_{A_5} = V_{A_4} - I_{R_s} \cdot R_s$$

$$V_{A_5} = V_s(t) - I_{R_{source}} \cdot R_{source} - \frac{1}{C_c} \cdot \int I_{C_c} \cdot dt - L_p \cdot \frac{dI_{L_p}}{dt} - I_{R_s} \cdot R_s$$

$$I_{R_I} = \frac{V_{A_5} - V_{A_6}}{R_I} \Rightarrow V_{A_5} - V_{A_6} = I_{R_I} \cdot R_I; V_{A_6} = V_{A_5} - I_{R_I} \cdot R_I$$

$$V_{A_5} = V_s(t) - I_{R_{source}} \cdot R_{source} - \frac{1}{C_c} \cdot \int I_{C_c} \cdot dt - L_p \cdot \frac{dI_{L_p}}{dt} - I_{R_s} \cdot R_s$$

$$I_{R_I} = \frac{V_{A_5} - V_{A_6}}{R_I} \Rightarrow V_{A_5} - V_{A_6} = I_{R_I} \cdot R_I; V_{A_6} = V_{A_5} - I_{R_I} \cdot R_I$$

$$V_{A_6} = V_s(t) - I_{R_{source}} \cdot R_{source} - \frac{1}{C_c} \cdot \int I_{C_c} \cdot dt - L_p \cdot \frac{dI_{L_p}}{dt} - I_{R_s} \cdot R_s - I_{R_I} \cdot R_I$$

$$I_{C_I} = C_I \cdot \frac{d}{dt}(V_{A_5} - V_{A_6}); I_{R_I} = \frac{V_{A_5} - V_{A_6}}{R_I} \Rightarrow V_{A_5} - V_{A_6} = I_{R_I} \cdot R_I; I_{C_I} = C_I \cdot R_I \cdot \frac{dI_{R_I}}{dt}$$

$$I_{C_p} = C_p \cdot \frac{d}{dt}(V_{A_3} - V_{A_6}) \Rightarrow \frac{I_{C_p}}{C_p} = \frac{d}{dt}(V_{A_3} - V_{A_6}); V_{A_3} - V_{A_6} = \frac{1}{C_p} \cdot \int I_{C_p} \cdot dt$$

$$V_s(t) - I_{R_{source}} \cdot R_{source} - \frac{1}{C_c} \cdot \int I_{C_c} \cdot dt - \{V_s(t) - I_{R_{source}} \cdot R_{source}$$
$$- \frac{1}{C_c} \cdot \int I_{C_c} \cdot dt - L_p \cdot \frac{dI_{L_p}}{dt}$$
$$- I_{R_s} \cdot R_s - I_{R_I} \cdot R_I\} = \frac{1}{C_p} \cdot \int I_{C_p} \cdot dt$$

$$L_p \cdot \frac{dI_{L_p}}{dt} + I_{R_s} \cdot R_s + I_{R_I} \cdot R_I = \frac{1}{C_p} \cdot \int I_{C_p} \cdot dt; \frac{d}{dt}\left\{L_p \cdot \frac{dI_{L_p}}{dt} + I_{R_s} \cdot R_s + I_{R_I} \cdot R_I\right\} = \frac{1}{C_p} \cdot I_{C_p}$$

$$L_p \cdot \frac{d^2 I_{L_p}}{dt^2} + \frac{dI_{R_s}}{dt} \cdot R_s + \frac{dI_{R_I}}{dt} \cdot R_I = \frac{1}{C_p} \cdot I_{C_p}; \frac{1}{C_I} \cdot I_{C_I} = R_I \cdot \frac{dI_{R_I}}{dt}$$

Second system differential equation: $L_p \cdot \frac{d^2 I_{L_p}}{dt^2} + \frac{dI_{R_s}}{dt} \cdot R_s + \frac{1}{C_I} \cdot I_{C_I} = \frac{1}{C_p} \cdot I_{C_p}$

$$V_{A_6} = L_c \cdot \frac{dI_{L_c}}{dt}; I_{R_{load}} = \frac{V_{A6}}{R_{load}}; L_c \cdot \frac{dI_{L_c}}{dt} = I_{R_{load}} \cdot R_{load}$$

Third system differential equation: $L_c \cdot \frac{dI_{L_c}}{dt} = I_{R_{load}} \cdot R_{load}$

We can summery our system differential equations:

$$\frac{d^2 I_{L_d}}{dt^2} = -\frac{dV_s(t)}{dt} \cdot \frac{1}{L_d} + \frac{dI_{R_{source}}}{dt} \cdot \frac{R_{source}}{L_d} + \frac{1}{C_c \cdot L_d} \cdot I_{C_c}$$

$$\frac{d^2 I_{L_p}}{dt^2} = \frac{1}{C_p \cdot L_p} \cdot I_{C_p} - \frac{1}{C_I \cdot L_p} \cdot I_{C_I} - \frac{dI_{R_s}}{dt} \cdot \frac{R_s}{L_p}; \frac{dI_{L_c}}{dt} = \frac{R_{load}}{L_c} \cdot I_{R_{load}}$$

Since $I_{R_s} = I_{L_p}; \frac{dI_{R_s}}{dt} = \frac{dI_{L_p}}{dt}; I_{R_{source}} = I_{C_c}; \frac{dI_{R_{source}}}{dt} = \frac{dI_{C_c}}{dt}$

$$\frac{d^2 I_{L_d}}{dt^2} = -\frac{dV_s(t)}{dt} \cdot \frac{1}{L_d} + \frac{dI_{C_c}}{dt} \cdot \frac{R_{source}}{L_d} + \frac{1}{C_c \cdot L_d} \cdot I_{C_c}$$

$$\frac{d^2 I_{L_p}}{dt^2} = \frac{1}{C_p \cdot L_p} \cdot I_{C_p} - \frac{1}{C_I \cdot L_p} \cdot I_{C_I} - \frac{dI_{L_p}}{dt} \cdot \frac{R_s}{L_p}; \frac{dI_{L_c}}{dt} = \frac{R_{load}}{L_c} \cdot I_{R_{load}}$$

We define new variables:

$$Y_1 = \frac{dI_{C_c}}{dt}; Y_2 = \frac{dI_{L_d}}{dt} \Leftrightarrow \frac{dY_2}{dt} = \frac{d^2 I_{L_d}}{dt^2}; Y_3 = \frac{dI_{L_p}}{dt} \Rightarrow \frac{dY_3}{dt} = \frac{d^2 I_{L_p}}{dt^2}$$

$$\frac{dY_2}{dt} = -\frac{dV_s(t)}{dt} \cdot \frac{1}{L_d} + Y_1 \cdot \frac{R_{source}}{L_d} + \frac{1}{C_c \cdot L_d} \cdot I_{C_c}$$

$$\frac{dY_3}{dt} = \frac{1}{C_p \cdot L_p} \cdot I_{C_p} - \frac{1}{C_I \cdot L_p} \cdot I_{C_I} - Y_3 \cdot \frac{R_s}{L_p}; \frac{dI_{L_c}}{dt} = \frac{R_{load}}{L_c} \cdot I_{R_{load}}$$

We can summery our new system differential equations:

$$\frac{dI_{C_c}}{dt} = Y_1; \frac{dI_{L_d}}{dt} = Y_2; \frac{dI_{L_p}}{dt} = Y_3; \frac{dY_2}{dt} = -\frac{dV_s(t)}{dt} \cdot \frac{1}{L_d} + Y_1 \cdot \frac{R_{source}}{L_d} + \frac{1}{C_c \cdot L_d} \cdot I_{C_c}$$

$$\frac{dY_3}{dt} = \frac{1}{C_p \cdot L_p} \cdot I_{C_p} - \frac{1}{C_I \cdot L_p} \cdot I_{C_I} - Y_3 \cdot \frac{R_s}{L_p}; \frac{dI_{L_c}}{dt} = \frac{R_{load}}{L_c} \cdot I_{R_{load}}$$

At fixed points: $\frac{dI_{C_c}}{dt} = 0; \frac{dI_{L_d}}{dt} = 0; \frac{dI_{L_p}}{dt} = 0; \frac{dY_2}{dt} = 0; \frac{dY_3}{dt} = 0; \frac{dI_{L_c}}{dt} = 0$

$$Y_1^* = 0; Y_2^* = 0; Y_3^* = 0; -\frac{dV_s(t)}{dt} + \frac{1}{C_c} \cdot I_{C_c}^* = 0; \frac{1}{C_p} \cdot I_{C_P}^* - \frac{1}{C_I} \cdot I_{C_I}^* = 0; I_{R_{load}}^* = 0$$

Stability analysis: The standard local stability analysis about any one of the equilibrium points of PIN diode attenuator and switch circuit in adding to coordinate $[Y_1, Y_2, Y_3, I_{C_c}, I_{L_d}, I_{L_p}, I_{C_p}, I_{C_I}, I_{L_c}, I_{Rload}]$ arbitrarily small increments of exponentially form $[y_1, y_2, y_3, i_{C_c}, i_{L_d}, i_{L_p}, i_{C_p}, i_{C_I}, i_{L_c}, i_{Rload}] \cdot e^{\lambda \cdot t}$ and retaining the first order terms in $Y_1, Y_2, Y_3, I_{C_c}, I_{L_d}, I_{L_p}, I_{C_p}, I_{C_I}, I_{L_c}, I_{Rload}$. The system of homogenous equations leads to a polynomial characteristic equation in the eigenvalues. The polynomial characteristic equations accept by set of the below circuit variables, circuit variables derivative and circuit variables second order derivative with respect to time into PIN diode attenuator and switch circuit equivalent circuit. Our PIN diode attenuator and switch circuit equivalent circuit fixed values with arbitrarily small increments of exponential form $[y_1, y_2, y_3, i_{C_c}, i_{L_d}, i_{L_p}, i_{C_p}, i_{C_I}, i_{L_c}, i_{Rload}] \cdot e^{\lambda \cdot t}$ are: j = 0 (first fixed point), j = 1 (second fixed point), j = 2 (third fixed point), etc.,

$$Y_1 = Y_1^{(j)} + y_1 \cdot e^{\lambda \cdot t}; Y_2 = Y_2^{(j)} + y_2 \cdot e^{\lambda \cdot t}; Y_3 = Y_3^{(j)} + y_3 \cdot e^{\lambda \cdot t}; I_{C_c} = I_{C_c}^{(j)} + i_{C_c} \cdot e^{\lambda \cdot t}$$

$$I_{L_d} = I_{L_d}^{(j)} + i_{L_d} \cdot e^{\lambda \cdot t}; I_{L_p} = I_{L_p}^{(j)} + i_{L_p} \cdot e^{\lambda \cdot t}; I_{C_p} = I_{C_p}^{(j)} + i_{C_p} \cdot e^{\lambda \cdot t}; I_{C_I} = I_{C_I}^{(j)} + i_{C_I} \cdot e^{\lambda \cdot t}$$

$$I_{L_c} = I_{L_c}^{(j)} + i_{L_c} \cdot e^{\lambda \cdot t}; I_{Rload} = I_{Rload}^{(j)} + i_{Rload} \cdot e^{\lambda \cdot t}; \frac{dI_{C_c}}{dt} = i_{C_c} \cdot \lambda \cdot e^{\lambda \cdot t}; \frac{dI_{L_d}}{dt} = i_{L_d} \cdot \lambda \cdot e^{\lambda \cdot t}$$

$$\frac{dI_{L_p}}{dt} = i_{L_p} \cdot \lambda \cdot e^{\lambda \cdot t}; \frac{dY_2}{dt} = y_2 \cdot \lambda \cdot e^{\lambda \cdot t}; \frac{dY_3}{dt} = y_3 \cdot \lambda \cdot e^{\lambda \cdot t}; \frac{dI_{L_c}}{dt} = i_{L_c} \cdot \lambda \cdot e^{\lambda \cdot t}$$

For $\lambda < 0$, $t > 0$, the selected fixed point is stable otherwise $\lambda > 0$, $t > 0$ is unstable. Our system tends to the selected fixed point exponentially for $\lambda < 0$, $t > 0$

otherwise go away from the selected fixed point exponentially. λ is the eigenvalue parameter which is established if the fixed point is stable or unstable; additionally, his absolute value |λ| established the speed of flow toward or away from the selected fixed point [2–4].

We can rewrite our system differential equations:

$i_{C_c} \cdot \lambda \cdot e^{\lambda \cdot t} = Y_1^{(j)} + y_1 \cdot e^{\lambda \cdot t}$ At fixed point $Y_1^{(j)} = 0$ then $y_1 - i_{C_c} \cdot \lambda = 0$

$i_{L_d} \cdot \lambda \cdot e^{\lambda \cdot t} = Y_2^{(j)} + y_2 \cdot e^{\lambda \cdot t}$ At fixed point $Y_2^{(j)} = 0$ then $y_2 - i_{L_d} \cdot \lambda = 0$

$i_{L_p} \cdot \lambda \cdot e^{\lambda \cdot t} = Y_3^{(j)} + y_3 \cdot e^{\lambda \cdot t}$ At fixed point $Y_3^{(j)} = 0$ then $y_3 - i_{L_p} \cdot \lambda = 0$

$$\frac{dY_2}{dt} = -\frac{dV_s(t)}{dt} \cdot \frac{1}{L_d} + Y_1 \cdot \frac{R_{source}}{L_d} + \frac{1}{C_c \cdot L_d} \cdot I_{C_c}; \frac{dV_s(t)}{dt} \to \varepsilon$$

$$y_2 \cdot \lambda \cdot e^{\lambda \cdot t} = [Y_1^{(j)} + y_1 \cdot e^{\lambda \cdot t}] \cdot \frac{R_{source}}{L_d} + \frac{1}{C_c \cdot L_d} \cdot [I_{C_c}^{(j)} + i_{C_c} \cdot e^{\lambda \cdot t}]$$

$$y_2 \cdot \lambda \cdot e^{\lambda \cdot t} = Y_1^{(j)} \cdot \frac{R_{source}}{L_d} + \frac{1}{C_c \cdot L_d} \cdot I_{C_c}^{(j)} + y_1 \cdot \frac{R_{source}}{L_d} \cdot e^{\lambda \cdot t} + i_{C_c} \cdot \frac{1}{C_c \cdot L_d} \cdot e^{\lambda \cdot t}$$

At fixed point $Y_1^{(j)} \cdot \frac{R_{source}}{L_d} + \frac{1}{C_c \cdot L_d} \cdot I_{C_c}^{(j)} = 0$ then $y_1 \cdot \frac{R_{source}}{L_d} + i_{C_c} \cdot \frac{1}{C_c \cdot L_d} - y_2 \cdot \lambda = 0$

$$y_3 \cdot \lambda \cdot e^{\lambda \cdot t} = \frac{1}{C_p \cdot L_p} \cdot [I_{C_p}^{(j)} + i_{C_p} \cdot e^{\lambda \cdot t}] - \frac{1}{C_I \cdot L_p} \cdot [I_{C_I}^{(j)} + i_{C_I} \cdot e^{\lambda \cdot t}] - [Y_3^{(j)} + y_3 \cdot e^{\lambda \cdot t}] \cdot \frac{R_s}{L_p}$$

$$y_3 \cdot \lambda \cdot e^{\lambda \cdot t} = [\frac{1}{C_p \cdot L_p} \cdot I_{C_p}^{(j)} - \frac{1}{C_I \cdot L_p} \cdot I_{C_I}^{(j)} - Y_3^{(j)} \cdot \frac{R_s}{L_p}] + i_{C_p} \cdot \frac{1}{C_p \cdot L_p} \cdot e^{\lambda \cdot t} - i_{C_I} \cdot \frac{1}{C_I \cdot L_p} \cdot e^{\lambda \cdot t} - y_3 \cdot \frac{R_s}{L_p} \cdot e^{\lambda \cdot t}$$

At fixed point $\frac{1}{C_p \cdot L_p} \cdot I_{C_p}^{(j)} - \frac{1}{C_I \cdot L_p} \cdot I_{C_I}^{(j)} - Y_3^{(j)} \cdot \frac{R_s}{L_p} = 0$

$$i_{C_p} \cdot \frac{1}{C_p \cdot L_p} - i_{C_I} \cdot \frac{1}{C_I \cdot L_p} - y_3 \cdot \frac{R_s}{L_p} - y_3 \cdot \lambda = 0$$

$$i_{L_c} \cdot \lambda \cdot e^{\lambda \cdot t} = \frac{R_{load}}{L_c} \cdot [I_{Rload}^{(j)} + i_{Rload} \cdot e^{\lambda \cdot t}]; i_{L_c} \cdot \lambda \cdot e^{\lambda \cdot t} = \frac{R_{load}}{L_c} \cdot I_{Rload}^{(j)} + i_{Rload} \cdot \frac{R_{load}}{L_c} \cdot e^{\lambda \cdot t}$$

At fixed point $\frac{R_{load}}{L_c} \cdot I_{Rload}^{(j)} = 0$ then $i_{Rload} \cdot \frac{R_{load}}{L_c} - i_{L_c} \cdot \lambda = 0$

PIN diode attenuator and switch circuit system matrix $(y_1, y_2, y_3, i_{C_c}, i_{L_d}, i_{L_p}, i_{C_p}, i_{C_I}, i_{L_c}, i_{Rload})$ can be constructed from the below list of equations:

$$y_1 - i_{C_c} \cdot \lambda = 0; y_2 - i_{L_d} \cdot \lambda = 0; y_3 - i_{L_p} \cdot \lambda = 0; y_1 \cdot \frac{R_{source}}{L_d} + i_{C_c} \cdot \frac{1}{C_c \cdot L_d} - y_2 \cdot \lambda = 0$$

$$i_{C_p} \cdot \frac{1}{C_p \cdot L_p} - i_{C_I} \cdot \frac{1}{C_I \cdot L_p} - y_3 \cdot \frac{R_s}{L_p} - y_3 \cdot \lambda = 0; i_{Rload} \cdot \frac{R_{load}}{L_c} - i_{L_c} \cdot \lambda = 0$$

PIN diode attenuator and switch circuit system matrixes:

$$\begin{pmatrix} \imath_{11} & \cdots & \imath_{16} \\ \vdots & \ddots & \vdots \\ \imath_{61} & \cdots & \imath_{66} \end{pmatrix} \cdot \begin{pmatrix} i_{C_c} \\ i_{L_d} \\ i_{L_p} \\ y_2 \\ y_3 \\ i_{L_c} \end{pmatrix} + \begin{pmatrix} \varphi_{11} & \cdots & \varphi_{14} \\ \vdots & \ddots & \vdots \\ \varphi_{61} & \cdots & \varphi_{64} \end{pmatrix} \cdot \begin{pmatrix} y_1 \\ i_{C_p} \\ i_{C_I} \\ i_{R_{load}} \end{pmatrix} = 0$$

$$\imath_{11} = -\lambda; \imath_{12} = \ldots = \imath_{16} = 0; \imath_{21} = 0; \imath_{22} = -\lambda; \imath_{23} = 0; \imath_{24} = 1; \imath_{25} = \imath_{26} = 0$$

$$\imath_{31} = \imath_{32} = 0; \imath_{33} = -\lambda; \imath_{34} = 0; \imath_{35} = 1; \imath_{36} = 0; \imath_{41} = \frac{1}{C_c \cdot L_d}; \imath_{42} = \imath_{43} = 0; \imath_{44} = -\lambda; \imath_{45} = \imath_{46} = 0$$

$$\imath_{51} = \ldots = \imath_{54} = 0; \imath_{55} = -\lambda - \frac{R_s}{L_p}; \imath_{56} = 0; \imath_{61} = \ldots = \imath_{65} = 0; \imath_{66} = -\lambda; \varphi_{11} = 1; \varphi_{12} = \varphi_{13} = \varphi_{14} = 0$$

$$\varphi_{21} = \ldots = \varphi_{24} = 0; \varphi_{31} = \ldots = \varphi_{34} = 0; \varphi_{41} = \frac{R_{source}}{L_d}; \varphi_{42} = \varphi_{43} = \varphi_{44} = 0$$

$$\varphi_{51} = 0; \varphi_{52} = \frac{1}{C_p \cdot L_p}; \varphi_{53} = -\frac{1}{C_I \cdot L_p}; \varphi_{54} = 0; \varphi_{61} = \varphi_{62} = \varphi_{63} = 0; \varphi_{64} = \frac{R_{load}}{L_c}$$

We consider

$$\begin{pmatrix} \varphi_{11} & \cdots & \varphi_{14} \\ \vdots & \ddots & \vdots \\ \varphi_{61} & \cdots & \varphi_{64} \end{pmatrix} \to \varepsilon; \begin{pmatrix} \imath_{11} & \cdots & \imath_{16} \\ \vdots & \ddots & \vdots \\ \imath_{61} & \cdots & \imath_{66} \end{pmatrix} \cdot \begin{pmatrix} i_{C_c} \\ i_{L_d} \\ i_{L_p} \\ y_2 \\ y_3 \\ i_{L_c} \end{pmatrix} \approx 0$$

$$A - \lambda \cdot I = \begin{pmatrix} \imath_{11} & \cdots & \imath_{16} \\ \vdots & \ddots & \vdots \\ \imath_{61} & \cdots & \imath_{66} \end{pmatrix}; \det(A - \lambda \cdot I) = 0 \Rightarrow \det \begin{pmatrix} \imath_{11} & \cdots & \imath_{16} \\ \vdots & \ddots & \vdots \\ \imath_{61} & \cdots & \imath_{66} \end{pmatrix} = 0$$

To effectively apply the stability criterion of Lipunov to our system, we require a criterion for when the equation $\det(A - \lambda \cdot I) = 0$ has a zero in the left half plane, without calculating the eigenvalues explicitly. We use criterion of Routh-Hurwitz [2–4].

Exercises

1. We have amplifier system which contains BJT transistor amplifier. The amplifier is operated as a Small Signal Amplifier. We represent our BJT transistor by Hybrid Pi model of BJT. Our amplifier input and output matching networks are T-type. Amplifier's load is represented by parallel resistor (R_{load}) and capacitor (C_{load}). We feed the amplifier by input RF source.

   1.1 Draw our amplifier system by using BJT transistor linear Hybrid Pi model, input and output block capacitors, input and output T-type matching networks, amplifier load circuit and input RF source.

   1.2 Write our amplifier system differential equations. Find fixed points and discuss stability. *Remark*: We consider input RF source $X_s(t) = A_0 + \xi(t); |\xi(t)| < 1 \& A_0 >> |\xi(t)|$ then $X_s(t)|_{A_0 >> |\xi(t)|} = A_0 + \xi(t) \approx A_0$; $\frac{dX_s(t)|_{A_0 >> |\xi(t)|}}{dt} = \frac{d\xi(t)}{dt} \to \varepsilon$.

   1.3 How our amplifier system stability and dynamical behavior change if the input matching network is T-type and output matching network is Pi-type?

   1.4 We change our amplifier load circuit to parallel resistor (R_{load}) and inductance (L_{load}). How the dynamical behavior and stability of the circuit change? Find fixed points.

2. We have amplifier system which contains two BJT transistors amplifiers in cascade. The amplifiers are operated as a Small Signal Amplifiers. We represent our BJT transistors by two Hybrid Pi model of BJTs. Our first amplifier input

and output matching networks are Pi-type and second amplifier input and output matching networks are T-type. Last amplifier's load is represented by parallel inductance (L_{load}) and capacitor (C_{load}). We feed the two amplifiers by input RF source.

2.1 Draw our amplifiers system by using BJT transistors linear Hybrid Pi models, input and output block capacitors, first and second amplifiers matching networks, last amplifier load circuit and input RF source.
2.2 Write our amplifiers system differential equations. Find fixed points and discuss stability. *Remark*: We consider first amplifier circuit input RF source $X_s(t) = A_0 + \xi(t); ; |\xi(t)| < 1 \,\&\, A_0 > > |\xi(t)|$ then $X_s(t)|_{A_0 > > |\xi(t)|} = A_0 + \xi(t) \approx A_0; \frac{dX_s(t)|_{A_0 > > |\xi(t)|}}{dt} = \frac{d\xi(t)}{dt} \to \varepsilon$
2.3 How our amplifier system stability and dynamical behavior change if the load is pure resistive (R_{load})?
2.4 How our amplifier system stability and dynamical behavior change if all input and output matching networks are L-type?

3. We have Small Signal (SS) amplifier with PI and T's matching networks equivalent circuit. Under AC and Small Signal (SS) conditions, we replace the BJT transistor with linear Hybrid PI model. Input matching network constructed from two Pi-type matching networks in cascade. Output matching network can be L-type (switch S_1 position A) or PI-type (switch S_1 position B).

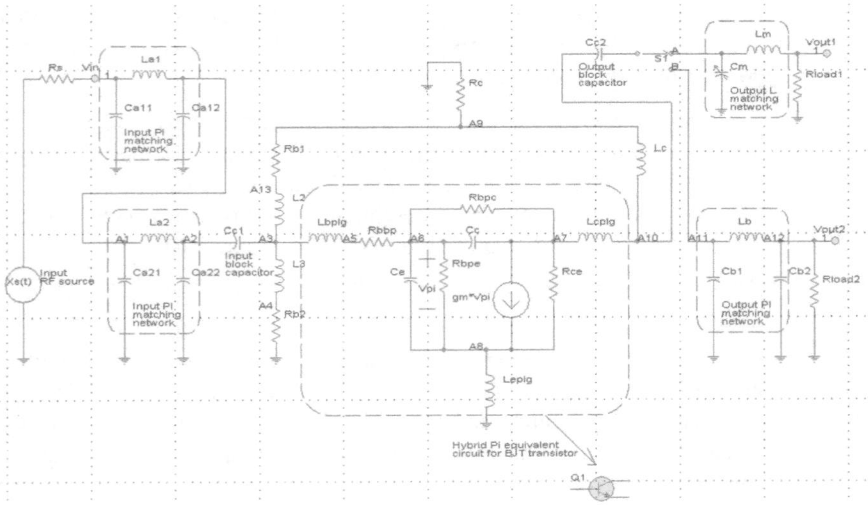

3.1 Switch S_1 is in (A) position, Find circuit differential equations and fixed points. Discuss stability of the circuit.
3.2 Switch S_2 is in (B) position, Find circuit differential equations and fixed point s. Discuss stability of the circuit.
3.3 Capacitors C_{a11} and C_{a12} are disconnected. How the circuit dynamical behavior and stability is changed?
3.4 Capacitor C_m is a function of C_{a11} and C_{a12}. $C_m = C_{a11} \cdot \Gamma_1 + C_{a12} \cdot \Gamma_2$ $\Gamma_1 = \alpha \cdot \Gamma_2; \Gamma_1, \Gamma_2, \alpha \in R_+$. How the dynamical behavior and stability of the circuit change for different values of α parameter?

Remark: We consider amplifier circuit input RF source $X_s(t) = A_0 + \xi(t); |\xi(t)| < 1 \,\&\, A_0 \gg |\xi(t)|$ then $X_s(t)|_{A_0 >> |\xi(t)|} = A_0 + \xi(t) \approx A_0; \frac{dX_s(t)|_{A_0 >> |\xi(t)|}}{dt} = \frac{d\xi(t)}{dt} \to \varepsilon$

4. We have balanced amplifier which contains two BJT NPN transistors (Q_1 and Q_2), input and output matching networks (R_{M1}, C_{c1}, R_{M2}, C_{c2}, R_{M3}, C_{c3}, R_{M4}, C_{c4}), feedback resistor (R_f), two input RF sources ($V_{s1}(t)$ and $V_{s2}(t)$), biasing resistors (R_{b1}, R_{c1}, R_{b2}, R_{c2}), and output load (R_{load})

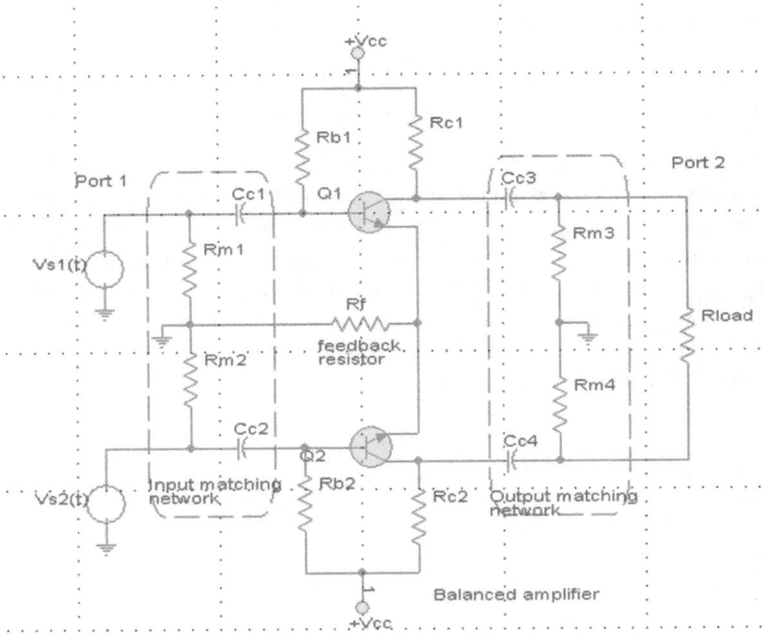

Use BJT NPN transistor's Small Signal (SS) Hybrid Pi equivalent circuit in your analysis.

4.1 Find circuit differential equations and fixed points. Discuss stability of the circuit.
4.2 Resistor R_{m4} is disconnected, how it influences on the circuit dynamical behavior and stability.
4.3 We short capacitors C_{c2} and C_{c3}. How the circuit dynamical behavior and stability is changed?
4.4. Capacitors C_{c3} and C_{c4} are functions of capacitors C_{c1} and C_{c2}.

$$C_{c3} = C_{c1} \cdot \Gamma_1 + C_{c2} \cdot (1 - \Gamma_1); C_{c4} = C_{c1} \cdot \Gamma_2 + C_{c2} \cdot (1 - \Gamma_2)$$
$$\Gamma_1 \neq \Gamma_2; 0 < \Gamma_1, \Gamma_2 < 1; \Gamma_1, \Gamma_2 \in R_+$$

How the dynamical behavior and stability of the circuit change for different values of Γ_1 and Γ_2 parameters?
4.5 Circuit feedback resistor (R_f) is disconnected, How the dynamical behavior and stability of the circuit is changed?

Remarks: We consider balance amplifier circuit input RF sources $V_{s1}(t) = A_{01} + \xi_1(t); |\xi_1(t)| < 1; A_{01} \gg |\xi_1(t)|$ then $V_{s1}(t)|_{A_{01} \gg |\xi_1(t)|} = A_{01} + \xi_1(t) \approx A_{01}$; $\frac{dV_{s1}(t)|_{A_{01} \gg |\xi_1(t)|}}{dt} = \frac{d\xi_1(t)}{dt} \to \varepsilon$; $V_{s2}(t) = A_{02} + \xi_2(t); |\xi_2(t)| < 1; A_{02} \gg |\xi_2(t)|$ then $V_{s2}(t)|_{A_{02} \gg |\xi_2(t)|} = A_{02} + \xi_2(t) \approx A_{02}; \frac{dV_{s2}(t)|_{A_{02} \gg |\xi_2(t)|}}{dt} = \frac{d\xi_2(t)}{dt} \to \varepsilon$

5. We have Common Emitter (CE) and Common Base (CB) BJT transistors amplifier circuit. Q_1 is connected as CE and Q_2 is connected as CB. Common Emitter (CE) amplifier, input signal inserted at base (B) and output signal is taken from the collector (C). The CE amplifier's output voltage is shifted by $\sim 180°$ in phase compared to CE amplifier's input signal. Common Base (CB) amplifier, input signal inserted at emitter (E) and output signal taken from the collector (C). The CB amplifier can operate as a voltage amplifier for low input impedance.

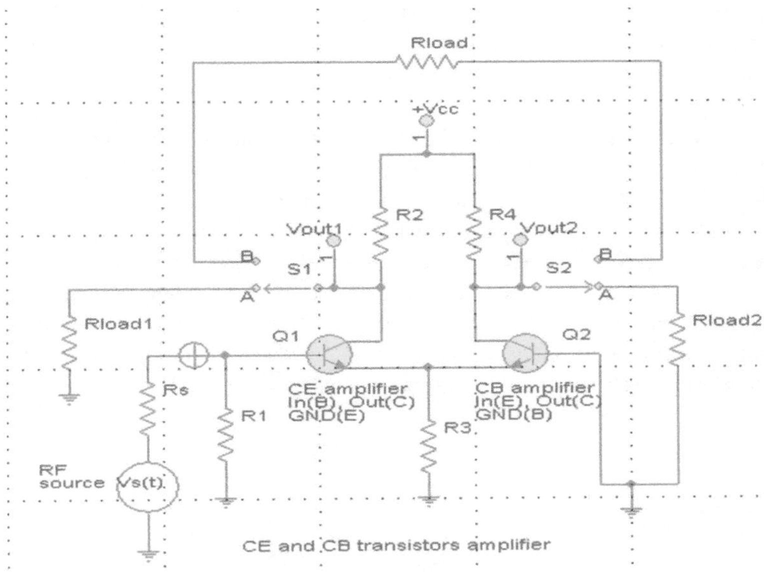

We define two circuit outputs. First circuit output (V_{out1}) is a voltage phase shift by 180° and second circuit output (V_{out2}) has no voltage phase shift. The input RF source is $V_s(t)$ with serial resistor R_s and is injected to Q_1 base. There are three possible loads connections to out circuit, R_{load1}, R_{load2}, and R_{load}. R_1, R_2, R_3, and R_4 are circuit biasing voltage resistors. Under AC and Small Signal (SS) conditions, we replace the BJT transistors (Q_1 and Q_2) with linear Hybrid PI model. Switches S_1 and S_2 connect the loads to our circuit. In the current circuit there are no matching networks between the circuit amplifier and loads (R_{load1}, R_{load2}, and R_{load}).

5.1 Find circuit differential equations and fixed points. Discuss stability of the circuit for the cases: (1) S_1 is in position A and S_2 is in position A, (2) S_1 is in position A and S_2 is in position B, (3) S_1 and S_2 are in position B, (4) S_1 is in position B and S_2 in position A.

5.2 Resistor R_1 is disconnected, how it influences on the circuit dynamical behavior and stability?

5.3 We add L—matching network between output one and R_{load1}. Switches S_1 and S_2 are in position A. Find circuit differential equations and fixed points. Discuss stability of the circuit.

5.4 We add PI matching network between output two and R_{load2}. Switches S_1 and S_2 are in position A. Find circuit differential equations and fixed points. Discuss stability of the circuit.

Remarks: We consider CE and CB transistor amplifier circuit input RF sources $V_s(t) = A_0 + \xi(t); |\xi(t)| < 1; A_0 \gg |\xi(t)|$ then $V_s(t)|_{A_0 \gg |\xi(t)|} = A_0 + \xi(t) \approx A_0$; $\frac{dV_s(t)|_{A_0 \gg |\xi(t)|}}{dt} = \frac{d\xi(t)}{dt} \to \varepsilon$

6. We have Bias-T circuit with 4th order filter. Bias-T's RF choke is L_2 and capacitor is C_1. We consider our 4th order filter with Bias-T feed coax cable with RF + DC signal. The coax and the remote device which is connected to the coax cable are taken as a pure resistive load (R_{load}). Due to parasitic effects of the circuit microstrip lines there are some delay elements in time in our circuit (τ_1, τ_2, τ_3). *Assumptions*: The voltages on the microstrip represented delay lines are neglected $V_{\tau_i} \to \varepsilon(i = 1, 2, 3)$ and the delay is in the current that flows through each delay line $I(t) \to I(t - \tau)$. V_1—DC voltage source, V_2—RF voltage source. $Tau1 \leftrightarrow \tau_1; Tau2 \leftrightarrow \tau_2; Tau3 \leftrightarrow \tau_3$

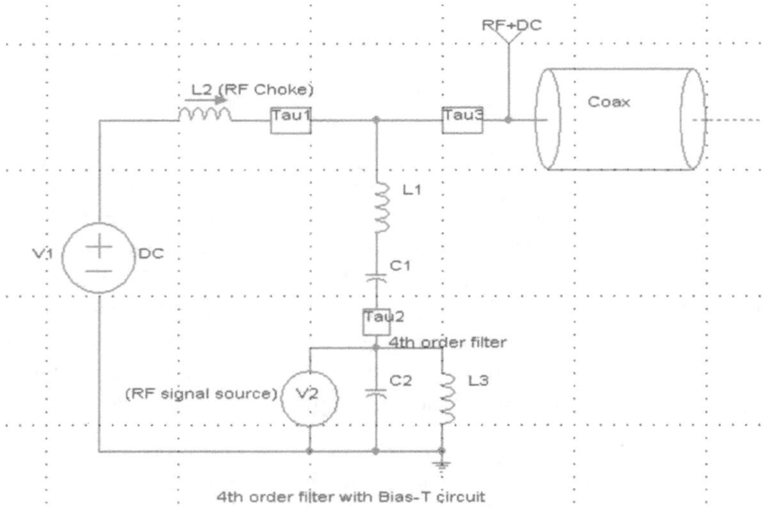

4th order filter with Bias-T circuit

6.1 Write 4th order filter with Bias-T circuit differential Eqs.
6.2 Discuss stability and dynamical behavior of the circuit for $\tau_i \to \varepsilon(=0)\ \forall\ i = 1, 2, 3$. Find circuit fixed points and characteristic Eq.
6.3 Discuss stability and stability switching for $\tau_i > 0\ \forall\ i = 1, 2, 3$ due to different values of τ parameter $\left(\tau_1 = \tau; \tau_2 = \tau^2; \tau_3 = \sqrt{\tau + 1}\right)$.
6.4 How the circuit dynamical behavior changes if L_3 is disconnected?
6.5 How the circuit dynamical behavior changes if we short inductor L_1?
6.6. Discuss stability and stability switching for $\tau_i > 0\,\forall\, i = 1, 2, 3$ due to different values of τ parameter $\left(\tau_1 = \tau; \tau_2 = \tau^2; \tau_3 = \left\{\begin{matrix} 0 \text{ for } \tau < \tau_\Gamma \\ \tau^3 \text{ for } \tau \geq \tau_\Gamma \end{matrix}\right\}\right)$

τ_Γ is a critical delay parameter value. $\tau_3(\tau) = \tau^3 \cdot U(\tau - \tau_\Gamma)$. How the circuit stability and stability switching is dependent on the critical delay parameter (τ_Γ)?

7. We have Bias-T circuit which is driving VCSEL (Vertical Cavity Surface Emitting Laser) diode (or other laser diode). It can be done with no IC. The VCSEL or laser diode is biased with a DC current until it just begins to lase. An RF sine wave then applied to this laser through a Bias-T network. The VCSEL or laser diode is biased with a DC current until it just begins to lase. An RF sine wave is then applied to this laser through a Bias-T network. The internal capacitance and structure of these small lasers then does something. Instead of slowly turning ON and OFF with the application of the RF, the diode is driven for threshold during one part of the sine wave and then begin to store energy on the opposite swing of the wave. When it has reached a certain level of stored energy (gain), it "snaps" on and releases all of this energy as laser light. The laser then turns off because all of the gain was extracted. The phenomenon is called "gain-switching". Tuning the laser diode: if DC is too high, the laser diode may produce light all the time. If the RF is too high, the laser diode may produce an "after pulse". If either is too low, the laser diode won't produce the desired amplitude. Due to parasitic effects of the circuit microstrip lines there are some delay elements in time in our circuit (τ_1, τ_2). *Assumptions*: The voltages on the microstrip represented delay lines are neglected $V_{\tau_i} \to \varepsilon(i = 1, 2)$ and the delay is in the current that flows through each delay line $I(t) \to I(t - \tau)$.

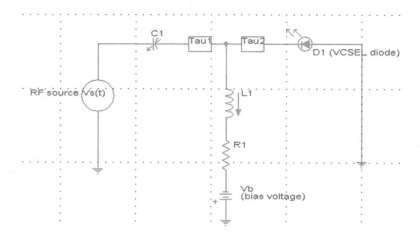

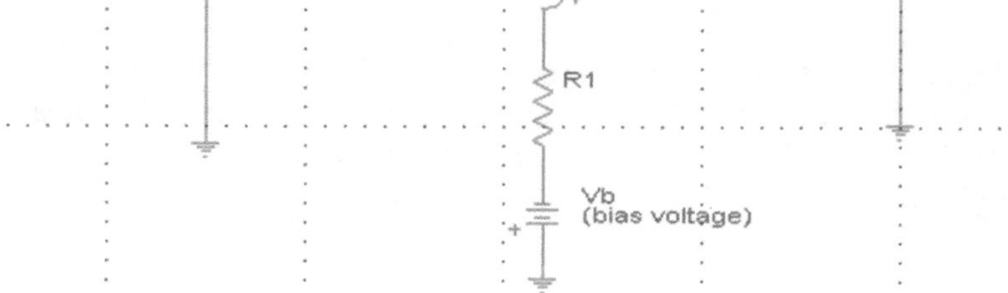

7.1 Write VCSEL diode driving Bias-T circuit differential equations. Take into consideration in your differential equations the full equivalent circuit of VCSEL diode.

7.2 Discuss stability and dynamical behavior for the circuit $\tau_i = 0; i = 1,2$.

7.3 Discuss stability and stability switching for $\tau_i \neq 0; \tau_i > 0 \, \forall \, i = 1,2$ due to different values of τ parameter $\left(\tau_1 = \tau^2; \tau_2 = \sqrt{\tau + \tau^2 + \Gamma \cdot \tau}; \Gamma \in R\right)$. How the stability changes for different values of Γ parameter (τ is constant)?

7.4 Discuss stability and stability switching for $\tau_i > 0 \, \forall \, i = 1,2$ due to different values of τ parameter $\left(\tau_1 = \tau^4; \tau_2 = \begin{Bmatrix} \sqrt{\tau + 1} \text{ for } 0 < \tau < \tau_\Gamma \\ \tau^3 \text{ for } \tau \geq \tau_\Gamma \end{Bmatrix}\right)$. τ_Γ is a critical delay parameter value. How the circuit stability and stability switching is dependent on the critical delay parameter (τ_Γ)? $\tau_2(\tau) = (\sqrt{\tau + 1}) \cdot [U(\tau) - U(\tau - \tau_\Gamma)] + \tau^3 \cdot U(\tau - \tau_\Gamma)$

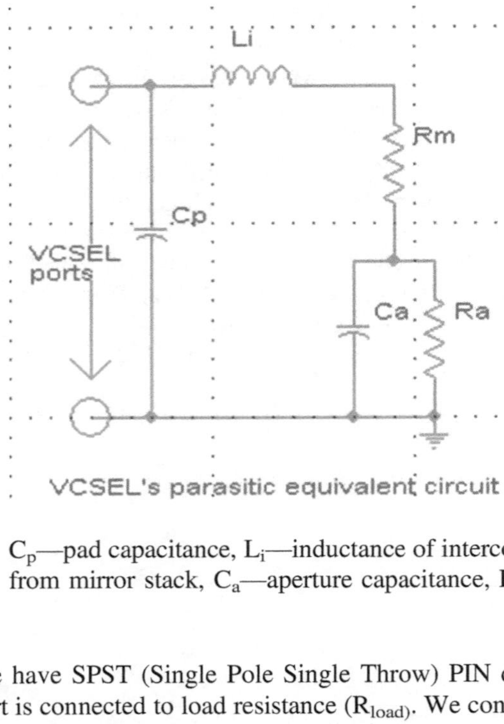

C_p—pad capacitance, L_i—inductance of interconnect metal, R_m—resistance from mirror stack, C_a—aperture capacitance, R_a—aperture resistance.

8. We have SPST (Single Pole Single Throw) PIN diode circuit. The output RF port is connected to load resistance (R_{load}). We consider the RF source as a high frequency signal and use PIN diode high frequency equivalent circuit. Consider $\frac{dV_s(t)}{dt} \to \varepsilon$

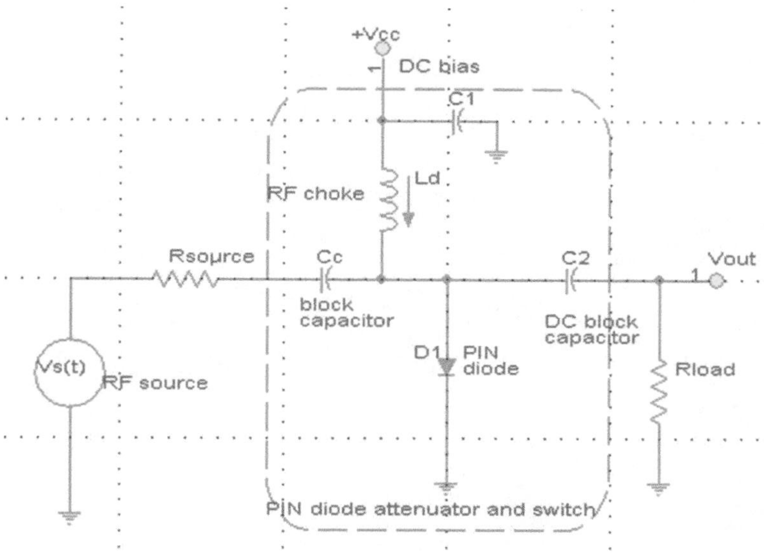

8.1 Draw PIN diode SPST circuit equivalent circuit.
8.2 Write circuit differential equations and find fixed points.
8.3 Find circuit eigenvalues expressions and discuss stability switching for parameter values variation.
8.4 We replace inductor L_d by resistor R_d. How the circuit dynamic changes? Find circuit differential equations and fixed points. Discuss stability.
8.5 We add PIN diode D_2 in parallel to D_1. How the circuit dynamical behavior changes? Find circuit differential equations and fixed point. Discuss stability.

9. We have SPST with series and shunt PIN didoes (D_1 and D_2). The output RF port is connected to load resistance (R_{load}). We consider the RF source as a high frequency signal and use PIN diode high frequency equivalent circuit. Consider $\frac{dV_s(t)}{dt} \to \varepsilon$.

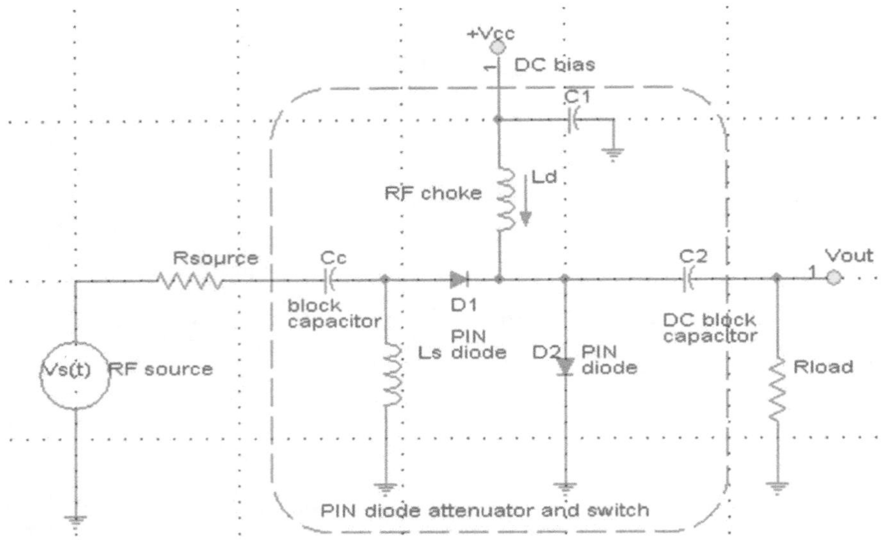

9.1 Draw PIN diodes (D_1 and D_2) SPST circuit equivalent circuit.
9.2 Write circuit differential equations and find fixed points.
9.3 Find circuit eigenvalues expressions and discuss stability switching for parameters variation.
9.4 Inductor L_s is disconnected, How the circuit dynamic changes? Find circuit differential equations and fixed points. Discuss stability.
9.5 We disconnect diode D_2, How the circuit dynamic changes? Find circuit differential equations and fixed points. Discuss stability.

10. We have high isolation generic PIN SPST circuit. PIN diode switches can handle very large power signals. The important diode parameters for switches that must handle power levels higher than 1 w include the diode's voltage rating and thermal resistance. Other diode parameters, such as series resistance, capacitance and I layer thickness, are also contributing factors to the determination of maximum power handling. We inject the RF signal ($V_s(t)$) through circuit RF common port and can switch it to RF Out1 or RF Out1. We have two bias voltage sources to our circuit (V_{ss1} and V_{ss2}). Circuit RF chokes are L_{d1}, L_{d2}, and L_{d3}. Three C_{block} capacitors (C_{b1}, C_{b2}, and C_{b3}). Two C_{filter} capacitors (C_{f1} and C_{f2}). Two PIN diodes are low capacitance and the other two PIN diodes are low resistance. R_{load1} and R_{load2} are our circuit's load resistances. We consider for simplicity $\frac{dV_s(t)}{dt} \to \varepsilon$. Circuit PIN diodes parameters are not the

same and there is slightly differences. Use PIN diode high frequency equivalent circuit in your analysis.

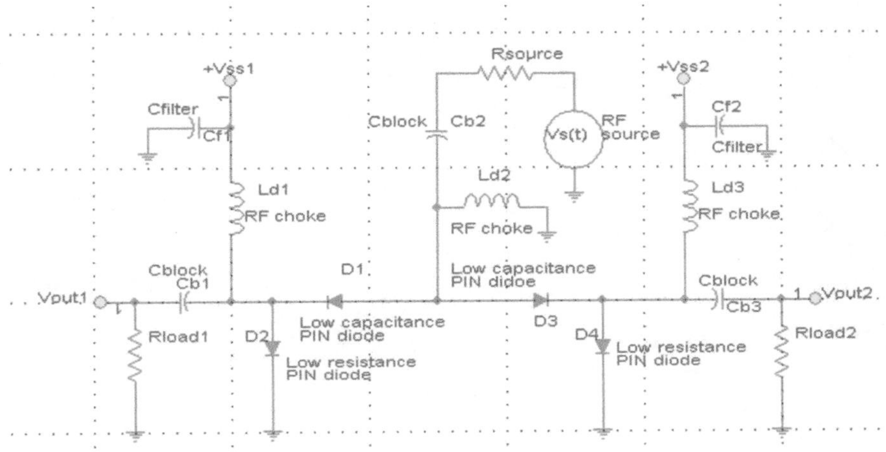

$$L_d = L_{d1}; L_{d2} = \Gamma_1 \cdot L_{d1}; L_{d3} = (1 - \Gamma_1) \cdot L_{d1}; 0 < \Gamma_1 < 1; C_f = C_{f1}; C_{f2} = (1 - \Gamma_2) \cdot C_{f1}; 0 < \Gamma_2 < 1$$

10.1 Draw PIN diodes ($D_1,\ldots,D_4$) SPST circuit equivalent circuit.

10.2 Write circuit differential equations and find fixed points.

10.3 Find circuit eigenvalues expressions and discuss stability switching for parameters Γ_1, Γ_2 variation.

10.4 Diode D_2 is disconnected, how the dynamical of the circuit is changes? Find circuit differential equations and discuss stability.

10.5 We short diode D_3, how the dynamical of the circuit is changed? Find circuit differential equations and discuss stability.

10.6 We short R_{load2}, How the dynamical of the circuit is changed? Discuss stability switching for parameters Γ_1, Γ_2 variations.

10.7 We disconnect C_{filter} capacitors, How the dynamical of the circuit is changed? Discuss stability switching for circuit parameters variations.

Chapter 5
Power Amplifier (PA) System Stability Analysis

Generally, large signal or Power Amplifier (PA) are used in the output stages of audio amplifier systems to derive a loudspeaker load. Power amplifier must be able to supply the high peak currents required to drive the low impedance speaker. One method used to distinguish the electrical characteristics of different types of amplifiers is by "class", and as such amplifiers are classified according to their circuit configuration and method of operation. Then Amplifier Classes is the term used to differentiate between the different amplifier types. Amplifier Classes represent the amount of the output signal which varies within the amplifier circuit over one cycle of operation when excited by a sinusoidal input signal. The classification of amplifiers range from entirely linear operation (for use in high-fidelity signal amplification) with very low efficiency, to entirely non-linear (where a faithful signal reproduction is not so important) operation but with a much higher efficiency, while others are a compromise between the two. Amplifier classes are mainly lumped into two basic groups. The first are the classically controlled conduction angle amplifiers forming the more common amplifier classes of A, B, AB and C, which are defined by the length of their conduction state over some portion of the output waveform, such that the output stage transistor operation lies somewhere between being "fully-ON" and "fully-OFF". The second set of amplifiers are the newer so-called "switching" amplifier classes of D, E, F, G, S, T etc., which use digital circuits and pulse width modulation (PWM) to constantly switch the signal between "fully-ON" and "fully-OFF" driving the output hard into the transistors saturation and cut-off regions. The most commonly constructed amplifier classes are those that are used as audio amplifiers, mainly class A, B, AB and C. Class A Amplifiers are the most common type of amplifier class due mainly to their simple design. Class A, literally means "the best class" of amplifier due mainly to their low signal distortion levels and are probably the best sounding of all the amplifier classes mentioned here. The class A amplifier has the highest linearity over the other amplifier classes and as such operates in the linear portion of the characteristics curve. Class B amplifiers were invented as a solution to the efficiency and heating problems associated with the previous class A amplifier. The basic class B

O. Aluf, *Microwave RF Antennas and Circuits*,
DOI 10.1007/978-3-319-45427-6_5

amplifier uses two complimentary transistors either bipolar of FET for each half of the waveform with its output stage configured in a "push-pull" type arrangement, so that each transistor device amplifies only half of the output waveform. Class AB Amplifier is a combination of the "Class A" and the "Class B" type amplifiers. The AB classification of amplifier is currently one of the most common used types of audio power amplifier design. The class AB amplifier is a variation of a class B amplifier, except that both devices are allowed to conduct at the same time around the waveforms crossover point eliminating the crossover distortion problems of the previous class B amplifier. The Class C Amplifier design has the greatest efficiency but the poorest linearity of the classes of amplifiers mentioned here. The previous classes, A, B and AB are considered linear amplifiers, as the output signals amplitude and phase are linearly related to the input signals amplitude and phase. Class C amplifier is heavily biased so that the output current is zero for more than one half of an input sinusoidal signal cycle with the transistor idling at its cut-off point. The conduction angle for the transistor is significantly less than 180°, and is generally around the 90° area. We analyse the stability of these amplifiers by inspecting the equivalent circuit differential equations, fixed points, bifurcation and stability switching for circuit parameters variation. BJT transistor is replaced by large signal model in our analysis.

We use in our analysis the Bipolar transistor model for large signal circuit simulation: The BJT model used in circuit simulation can accurately represent the DC and dynamic currents of the transistor in response to $V_{BE}(t)$ and $V_{CE}(t)$. A typical circuit simulation model or compact model is made of the Ebers-Moll model when V_{BE} and V_{BC} are two driving forces for I_C and I_B, plus additional enhancements for high level injection, voltage dependent capacitances that accurately represent the charge storage in the transistor, and parasitic resistances. This BJT model is known as the Gummel-Poon model. The Ebers-Moll BJT model is a good large signal. if the inputs/outputs exceed certain limits, the full Ebers-Moll model must be used. When certain parameters are omitted, the Gummel–Poon model reduces to the simpler Ebers–Moll model. Gummel-Poon nonlinear model is the "large signal model". Large signal models is closer to reality but is computationally complex or even intractable. Additionally we discuss the stability of wideband LNA with negative feedback.

5.1 Class AB Push-Pull Power Amplifiers Stability Analysis Under Parameters Variation

Class AB Amplifier is a combination of the "Class A" and the "Class B" type amplifiers. The AB classification of amplifier is currently one of the most common used types of audio power amplifier design. Push-pull mechanism is essential for realizing practical class AB power amplifiers. Circuit diagram of a typical class AB push-pull amplifier is shown in the next figure. The technical designation is "complementary symmetry class AB power amplifier". The active elements used in this circuit (Q_1 and Q_2) are complementary symmetric transistors and it means the transistors are similar in all aspects except one is NPN and the other is PNP. The use of this complementary pair eliminates the bulky transformer for phase splitting the input signal for driving the individual transistor. The NPN transistor alone will conduct the positive half cycle and PNP transistor alone will conduct the negative half cycle. Pre-biasing is given to the transistors using the network comprising of resistors R_1, R_2 and biasing diodes D_1 and D_2. The NPN transistor will start conducting when its base voltage is above the base emitter voltage ($V_{BE} \approx 0.7$ v) and a PNP transistor will start conducting when its base voltage is below the base emitter voltage ($V_{BE} \approx -0.7$ v). A forward biased diode will drop approximately 0.7 v across it and the biasing diodes used here will keep the transistor slightly forward biased even if there is no input signal. Compensating diodes (D_1, D_2) characteristics must match as close as possible to the transistors. Resistors R_1 and R_2 are used for forward biasing the diodes and they drop 0.7 v across it for biasing the individual transistors (Q_1, Q_2). C_1 and C_2 are input DC decoupling capacitors. The advantages of class AB power amplifier: no cross over distortion, no need for the bulky coupling transformers, and no hum in the output. The disadvantages of class AB power amplifier: efficiency is slightly less when compared to class B configuration, there will be some DC components in the output as the load is directly coupled, and capacitive coupling can eliminate DC components but it is not practical in case of heavy loads. For simplicity in our circuit analysis we reduces BJT transistors Gummel–Poon model to the simpler Ebers–Moll model [24–26] (Fig. 5.1).

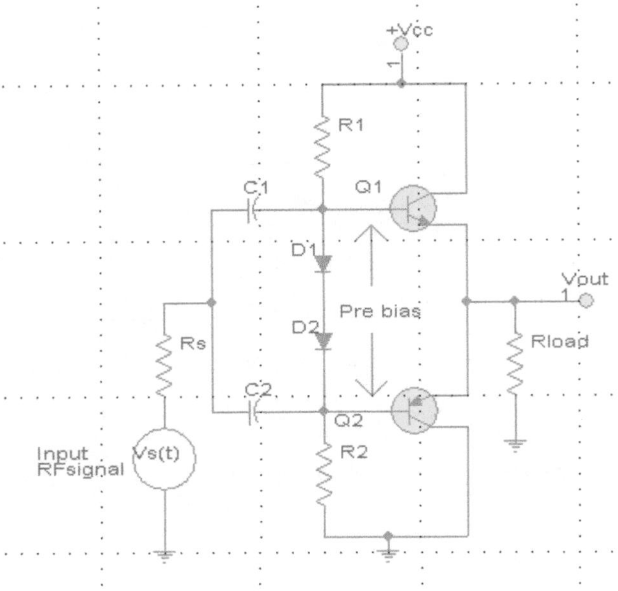

Fig. 5.1 Class AB push-pull amplifier

$$I_{EQ_1} = \frac{I_s}{\alpha_f} \cdot (e^{\frac{V_{BEQ_1}}{V_T}} - 1) - I_s \cdot (e^{\frac{V_{BCQ_1}}{V_T}} - 1); \quad I_{CQ_1} = I_s \cdot (e^{\frac{V_{BEQ_1}}{V_T}} - 1) - \frac{I_s}{\alpha_r} \cdot (e^{\frac{V_{BCQ_1}}{V_T}} - 1)$$

$$I_{BQ_1} = \frac{I_s}{\beta_f} \cdot (e^{\frac{V_{BEQ_1}}{V_T}} - 1) + \frac{I_s}{\beta_r} \cdot (e^{\frac{V_{BCQ_1}}{V_T}} - 1); \quad I_{EQ_2} = \frac{I_s}{\alpha_f} \cdot (e^{\frac{V_{EBQ_2}}{V_T}} - 1) - I_s \cdot (e^{\frac{V_{CBQ_2}}{V_T}} - 1)$$

$$I_{CQ_2} = I_s \cdot (e^{\frac{V_{EBQ_2}}{V_T}} - 1) - \frac{I_s}{\alpha_r} \cdot (e^{\frac{V_{CBQ_2}}{V_T}} - 1); \quad I_{BQ_2} = \frac{I_s}{\beta_f} \cdot (e^{\frac{V_{EBQ_2}}{V_T}} - 1) + \frac{I_s}{\beta_r} \cdot (e^{\frac{V_{CBQ_2}}{V_T}} - 1)$$

$$V_{D_1} = V_t \cdot \ln(\frac{I_{D_1}}{I_0} + 1); \quad V_{D_2} = V_t \cdot \ln(\frac{I_{D_2}}{I_0} + 1); \quad I_D = I_{D_1} = I_{D_2}; \quad I_{EQ_1} = I_{R_{load}} + I_{EQ_2}$$

It can be shown that $\alpha_F \cdot I_{SE} = \alpha_R \cdot I_{SC} = I_S$ (see S.M. Sze, *Physics of Semiconductor Devices*) (Fig. 5.2).

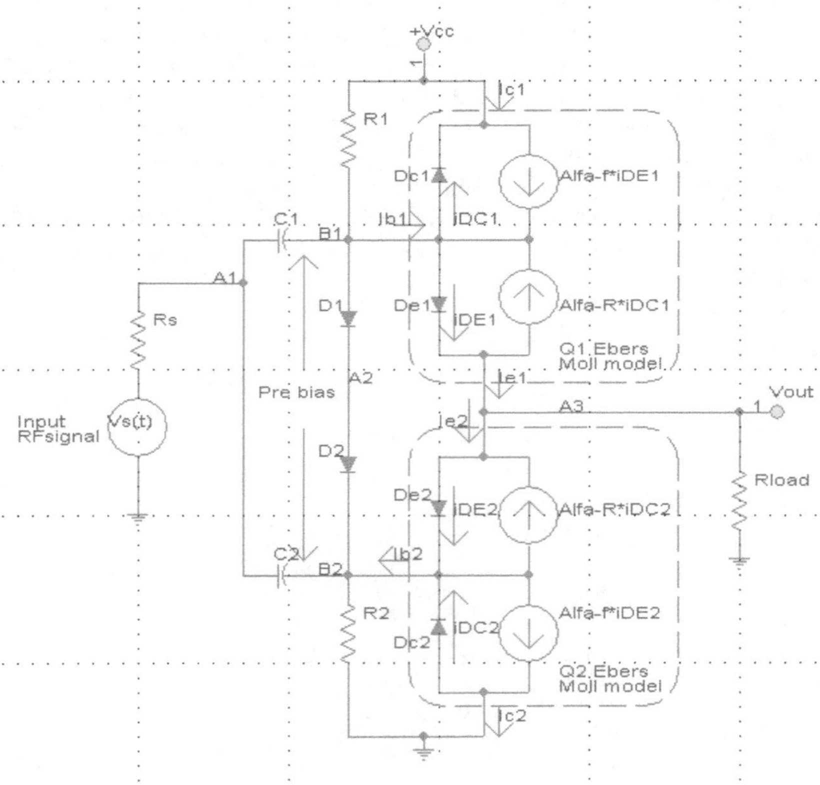

Fig. 5.2 Class AB push-pull amplifier equivalent circuit

$$I_{CQ_1} + I_{BQ_1} = I_{EQ_1}; I_{EQ_2} = I_{CQ_2} + I_{BQ_2}; I_{R_1} = \frac{V_{cc} - V_{B_1}}{R_1}; I_{R_2} = \frac{V_{B_2}}{R_2}; I_{C_1} = C_1 \cdot \frac{d}{dt}(V_{A_1} - V_{B_1})$$
$$I_{C_2} = C_2 \cdot \frac{d}{dt}(V_{A_1} - V_{B_2}); I_{R_s} = I_{C_1} + I_{C_2}; I_{R_1} + I_{C_1} = I_{BQ_1} + I_{D_1} = I_{BQ_1} + I_D; \ln(\frac{I_{sc}}{I_{se}}) \approx 0$$
$$I_{D_2} + I_{C_2} + I_{BQ_2} = I_{R_2}; I_{R_{load}} = \frac{V_{A_3}}{R_{load}}; V_{ECQ_2} = V_{A_3}; V_{CEQ_1} + V_{ECQ_2} = V_{cc}; I_{R_s} = \frac{V_s(t) - V_{A_1}}{R_s}$$

Collector emitter voltage expression for BJT NPN and PNP transistor

$$V_{CE-NPN} \approx V_T \cdot \ln\{\frac{[\alpha_r \cdot I_C - I_E + (\alpha_r \cdot \alpha_f - 1) \cdot I_{se}]}{[I_C - \alpha_f \cdot I_E + (\alpha_f \cdot \alpha_r - 1) \cdot I_{sc}]}\};$$
$$V_{CE-PNP} \approx V_T \cdot \ln\{\frac{I_C - \alpha_f \cdot I_E + (\alpha_f \cdot \alpha_r - 1) \cdot I_{sc}}{\alpha_r \cdot I_C - I_E + (\alpha_r \cdot \alpha_f - 1) \cdot I_{se}}\}$$

we can write the following circuit collector emitter equations for Q_1 and Q_2:

$$V_{CEQ_1} \approx V_T \cdot \ln\{\frac{[\alpha_r \cdot I_{CQ_1} - I_{EQ_1} + (\alpha_r \cdot \alpha_f - 1) \cdot I_{se}]}{[I_{CQ_1} - \alpha_f \cdot I_{EQ_1} + (\alpha_f \cdot \alpha_r - 1) \cdot I_{sc}]}\};$$
$$V_{CEQ_2} \approx V_T \cdot \ln\{\frac{I_{CQ_2} - \alpha_f \cdot I_{EQ_2} + (\alpha_f \cdot \alpha_r - 1) \cdot I_{sc}}{\alpha_r \cdot I_{CQ_2} - I_{EQ_2} + (\alpha_r \cdot \alpha_f - 1) \cdot I_{se}}\}$$
$$V_{ECQ_2} = -V_{CEQ_2}; V_{ECQ_2} \approx V_T \cdot \ln\{\frac{\alpha_r \cdot I_{CQ_2} - I_{EQ_2} + (\alpha_r \cdot \alpha_f - 1) \cdot I_{se}}{I_{CQ_2} - \alpha_f \cdot I_{EQ_2} + (\alpha_f \cdot \alpha_r - 1) \cdot I_{sc}}\};$$
$$V_{CEQ_1} + V_{ECQ_2} = V_{cc}$$

$$V_T \cdot \ln\{\frac{[\alpha_r \cdot I_{CQ_1} - I_{EQ_1} + (\alpha_r \cdot \alpha_f - 1) \cdot I_{se}]}{[I_{CQ_1} - \alpha_f \cdot I_{EQ_1} + (\alpha_f \cdot \alpha_r - 1) \cdot I_{sc}]}\} + V_T \cdot \ln\{\frac{\alpha_r \cdot I_{CQ_2} - I_{EQ_2} + (\alpha_r \cdot \alpha_f - 1) \cdot I_{se}}{I_{CQ_2} - \alpha_f \cdot I_{EQ_2} + (\alpha_f \cdot \alpha_r - 1) \cdot I_{sc}}\} = V_{cc}$$

We can rewrite transistors collector and emitter currents:

$$I_{CQ_1} = \beta_f \cdot I_{BQ_1} + (\beta_f + 1) \cdot I_{CBQ_10}; \quad I_{EQ_1} = (\beta_f + 1) \cdot I_{BQ_1} + (\beta_f + 1) \cdot I_{CBQ_10}$$
$$I_{CQ_2} = \beta_f \cdot I_{BQ_2} + (\beta_f + 1) \cdot I_{CBQ_20}; \quad I_{EQ_2} = (\beta_f + 1) \cdot I_{BQ_2} + (\beta_f + 1) \cdot I_{CBQ_20}$$

Remark Q_1 and Q_2 are complementary symmetric transistors and it means the transistors are similar in all aspects (parameters are the same, β_f, α_f,...) except one is NPN and the other is PNP.

$$\beta_f : 20 \rightarrow 500;\ \beta_f = \frac{\alpha_f}{1-\alpha_f};\ \alpha_f = \frac{\beta_f}{1+\beta_f};\ \alpha_f : 0.95 \rightarrow 0.99; \beta_f > \beta_r$$

$$\beta_r : 0 \rightarrow 20;\ \beta_r = \frac{\alpha_r}{1-\alpha_r};\ \alpha_r = \frac{\beta_r}{1+\beta_r};\ \alpha_r : 0 \rightarrow 0.95; \alpha_f > \alpha_r$$

$$I_{EQ_1} = (\beta_f+1)\cdot I_{BQ_1} + (\beta_f+1)\cdot I_{CBQ_10};\ \ I_{BQ_1} = \frac{I_{EQ_1}}{(\beta_f+1)} - I_{CBQ_10}$$

$$I_{EQ_2} = (\beta_f+1)\cdot I_{BQ_2} + (\beta_f+1)\cdot I_{CBQ_20};\ \ I_{BQ_2} = \frac{I_{EQ_2}}{(\beta_f+1)} - I_{CBQ_20}$$

$$I_{CQ_1} = \frac{\beta_f\cdot I_{EQ_1}}{(\beta_f+1)} + I_{CBQ_10};\ \ I_{CQ_2} = \frac{\beta_f\cdot I_{EQ_2}}{(\beta_f+1)} + I_{CBQ_20}$$

$$I_{R_1} = \frac{V_{cc}-V_{B_1}}{R_1} \Rightarrow I_{R_1}\cdot R_1 = V_{cc} - V_{B_1}; V_{B_1} = V_{cc} - I_{R_1}\cdot R_1; I_{R_2} = \frac{V_{B_2}}{R_2} \Rightarrow V_{B_2} = I_{R_2}\cdot R_2$$

$$I_{C_1} = C_1\cdot\frac{d}{dt}(V_{A_1} - V_{B_1}) \Rightarrow V_{A_1} - V_{B_1} = \frac{1}{C_1}\cdot\int I_{C_1}\cdot dt;$$
$$V_{A_1} = V_{B_1} + \frac{1}{C_1}\cdot\int I_{C_1}\cdot dt$$

$$I_{C_1} = C_1\cdot\frac{d}{dt}(V_{A_1} - V_{B_1}) \Rightarrow V_{A_1} - V_{B_1} = \frac{1}{C_1}\cdot\int I_{C_1}\cdot dt; V_{A_1} = V_{B_1} + \frac{1}{C_1}\cdot\int I_{C_1}\cdot dt$$
$$I_{C_2} = C_2\cdot\frac{d}{dt}(V_{A_1} - V_{B_2}) \Rightarrow V_{A_1} - V_{B_2} = \frac{1}{C_2}\cdot\int I_{C_2}\cdot dt; V_{A_1} = V_{B_2} + \frac{1}{C_2}\cdot\int I_{C_2}\cdot dt$$
$$V_{B_1} + \frac{1}{C_1}\cdot\int I_{C_1}\cdot dt = V_{B_2} + \frac{1}{C_2}\cdot\int I_{C_2}\cdot dt;$$
$$V_{cc} - I_{R_1}\cdot R_1 + \frac{1}{C_1}\cdot\int I_{C_1}\cdot dt = I_{R_2}\cdot R_2 + \frac{1}{C_2}\cdot\int I_{C_2}\cdot dt$$

$$\frac{d}{dt}\{V_{cc} - I_{R_1}\cdot R_1 + \frac{1}{C_1}\cdot\int I_{C_1}\cdot dt\} = \frac{d}{dt}\{I_{R_2}\cdot R_2 + \frac{1}{C_2}\cdot\int I_{C_2}\cdot dt\}$$

$$\frac{dV_{cc}}{dt} - \frac{dI_{R_1}}{dt}\cdot R_1 + \frac{1}{C_1}\cdot I_{C_1} = \frac{dI_{R_2}}{dt}\cdot R_2 + \frac{1}{C_2}\cdot I_{C_2}; \frac{dV_{cc}}{dt} \rightarrow \varepsilon$$

We get the equation: $-\frac{dI_{R_1}}{dt}\cdot R_1 + \frac{1}{C_1}\cdot I_{C_1} = \frac{dI_{R_2}}{dt}\cdot R_2 + \frac{1}{C_2}\cdot I_{C_2}$

$$I_{R_{load}} = \frac{V_{A_3}}{R_{load}} \Rightarrow V_{A_3} = I_{R_{load}}\cdot R_{load}; V_{A_3} = V_{ECQ_2}$$
$$= V_T\cdot\ln\{\frac{\alpha_r\cdot I_{CQ_2} - I_{EQ_2} + (\alpha_r\cdot\alpha_f - 1)\cdot I_{se}}{I_{CQ_2} - \alpha_f\cdot I_{EQ_2} + (\alpha_f\cdot\alpha_r - 1)\cdot I_{sc}}\}$$

$$I_{R_{load}} = \frac{V_T}{R_{load}} \cdot \ln\{\frac{\alpha_r \cdot I_{CQ_2} - I_{EQ_2} + (\alpha_r \cdot \alpha_f - 1) \cdot I_{se}}{I_{CQ_2} - \alpha_f \cdot I_{EQ_2} + (\alpha_f \cdot \alpha_r - 1) \cdot I_{sc}}\};$$

$$V_{cc} = I_{R_1} \cdot R_1 + 2 \cdot \ln(\frac{I_D}{I_0} + 1) + I_{R_2} \cdot R_2$$

$$I_{EQ_1} = I_{R_{load}} + I_{EQ_2} \Rightarrow \frac{I_s}{\alpha_f} \cdot (e^{\frac{V_{BEQ_1}}{V_T}} - 1) - I_s \cdot (e^{\frac{V_{BCQ_1}}{V_T}} - 1)$$
$$= I_{R_{load}} + \frac{I_s}{\alpha_f} \cdot (e^{\frac{V_{EBQ_2}}{V_T}} - 1) - I_s \cdot (e^{\frac{V_{CBQ_2}}{V_T}} - 1)$$

$$I_{R_1} + I_{C_1} = I_{BQ_1} + I_D \Rightarrow I_{R_1} = I_{BQ_1} + I_D - I_{C_1}; I_{R_2} = I_D + I_{C_2} + I_{BQ_2}$$

We add one equation to the other and get the following expression:

$$(*) \quad I_{R_1} = I_{BQ_1} + I_D - I_{C_1}; (**) \quad I_{R_2} = I_D + I_{C_2} + I_{BQ_2}; (*) + (**) \Rightarrow I_{R_1} + I_{R_2}$$
$$= I_{BQ_1} + I_{BQ_2} + I_{C_2} - I_{C_1} + 2 \cdot I_D$$

$$-\frac{dI_{R_1}}{dt} \cdot R_1 + \frac{1}{C_1} \cdot I_{C_1} = \frac{dI_{R_2}}{dt} \cdot R_2 + \frac{1}{C_2} \cdot I_{C_2} \Rightarrow -(\frac{dI_{R_1}}{dt} \cdot R_1 + \frac{dI_{R_2}}{dt} \cdot R_2)$$
$$= \frac{1}{C_2} \cdot I_{C_2} - \frac{1}{C_1} \cdot I_{C_1}$$

If C = C_1 = C_2 then $-(\frac{dI_{R_1}}{dt} \cdot R_1 + \frac{dI_{R_2}}{dt} \cdot R_2) = \frac{1}{C} \cdot (I_{C_2} - I_{C_1})$

$$I_{R_1} + I_{R_2} = I_{BQ_1} + I_{BQ_2} + I_{C_2} - I_{C_1} + 2 \cdot I_D \Rightarrow I_{C_2} - I_{C_1}$$
$$= I_{R_1} + I_{R_2} - (I_{BQ_1} + I_{BQ_2}) - 2 \cdot I_D$$

$$I_{C_2} - I_{C_1} = \sum_{k=1}^{2} I_{R_k} - \sum_{k=1}^{2} I_{BQ_k} - 2 \cdot I_D; -(\frac{dI_{R_1}}{dt} \cdot R_1 + \frac{dI_{R_2}}{dt} \cdot R_2)$$
$$= \frac{1}{C} \cdot (\sum_{k=1}^{2} I_{R_k} - \sum_{k=1}^{2} I_{BQ_k} - 2 \cdot I_D)$$

$$V_{cc} = I_{R_1} \cdot R_1 + 2 \cdot \ln(\frac{I_D}{I_0} + 1) + I_{R_2} \cdot R_2;$$

$$I_D = I_0 \cdot \{\exp[\frac{1}{2}(V_{cc} - I_{R_1} \cdot R_1 - I_{R_2} \cdot R_2)] - 1\}$$

$$-(\frac{dI_{R_1}}{dt} \cdot R_1 + \frac{dI_{R_2}}{dt} \cdot R_2) = \frac{1}{C} \cdot (\sum_{k=1}^{2} I_{R_k} - \sum_{k=1}^{2} I_{BQ_k} - 2 \cdot I_0$$
$$\cdot \{\exp[\frac{1}{2}(V_{cc} - I_{R_1} \cdot R_1 - I_{R_2} \cdot R_2)] - 1\})$$

$$I_{R_{load}} = I_{EQ_1} - I_{EQ_2}; I_{R_{load}} = \frac{V_T}{R_{load}} \cdot \ln\{\frac{\alpha_r \cdot I_{CQ_2} - I_{EQ_2} + (\alpha_r \cdot \alpha_f - 1) \cdot I_{se}}{I_{CQ_2} - \alpha_f \cdot I_{EQ_2} + (\alpha_f \cdot \alpha_r - 1) \cdot I_{sc}}\}$$

$$I_{EQ_1} - I_{EQ_2} = \frac{V_T}{R_{load}} \cdot \ln\{\frac{\alpha_r \cdot I_{CQ_2} - I_{EQ_2} + (\alpha_r \cdot \alpha_f - 1) \cdot I_{se}}{I_{CQ_2} - \alpha_f \cdot I_{EQ_2} + (\alpha_f \cdot \alpha_r - 1) \cdot I_{sc}}\}$$

$$I_{EQ_1} - I_{EQ_2} = \frac{V_T}{R_{load}} \cdot \ln\{\frac{\alpha_r \cdot [\frac{\beta_f \cdot I_{EQ_2}}{(\beta_f + 1)} + I_{CBQ_2 0}] - I_{EQ_2} + (\alpha_r \cdot \alpha_f - 1) \cdot I_{se}}{\frac{\beta_f \cdot I_{EQ_2}}{(\beta_f + 1)} + I_{CBQ_2 0} - \alpha_f \cdot I_{EQ_2} + (\alpha_f \cdot \alpha_r - 1) \cdot I_{sc}}\};$$

$$V_{CEQ_1} + V_{ECQ_2} = V_{cc}$$

$$\xi_1(I_{EQ_1}, I_{EQ_2}) = \frac{V_T}{R_{load}} \cdot \ln\{\frac{\alpha_r \cdot [\frac{\beta_f \cdot I_{EQ_2}}{(\beta_f + 1)} + I_{CBQ_2 0}] - I_{EQ_2} + (\alpha_r \cdot \alpha_f - 1) \cdot I_{se}}{\frac{\beta_f \cdot I_{EQ_2}}{(\beta_f + 1)} + I_{CBQ_2 0} - \alpha_f \cdot I_{EQ_2} + (\alpha_f \cdot \alpha_r - 1) \cdot I_{sc}}\} - I_{EQ_1} + I_{EQ_2} = 0$$

$$V_T \cdot \ln\{\frac{[\alpha_r \cdot I_{CQ_1} - I_{EQ_1} + (\alpha_r \cdot \alpha_f - 1) \cdot I_{se}]}{[I_{CQ_1} - \alpha_f \cdot I_{EQ_1} + (\alpha_f \cdot \alpha_r - 1) \cdot I_{sc}]}\} + V_T \cdot \ln\{\frac{\alpha_r \cdot I_{CQ_2} - I_{EQ_2} + (\alpha_r \cdot \alpha_f - 1) \cdot I_{se}}{I_{CQ_2} - \alpha_f \cdot I_{EQ_2} + (\alpha_f \cdot \alpha_r - 1) \cdot I_{sc}}\} = V_{cc}$$

$$\ln\{\frac{\alpha_r \cdot [\frac{\beta_f \cdot I_{EQ_1}}{(\beta_f + 1)} + I_{CBQ_1 0}] - I_{EQ_1} + (\alpha_r \cdot \alpha_f - 1) \cdot I_{se}}{[\frac{\beta_f \cdot I_{EQ_1}}{(\beta_f + 1)} + I_{CBQ_1 0}] - \alpha_f \cdot I_{EQ_1} + (\alpha_f \cdot \alpha_r - 1) \cdot I_{sc}}\} + \ln\{\frac{\alpha_r \cdot [\frac{\beta_f \cdot I_{EQ_2}}{(\beta_f + 1)} + I_{CBQ_2 0}] - I_{EQ_2} + (\alpha_r \cdot \alpha_f - 1) \cdot I_{se}}{[\frac{\beta_f \cdot I_{EQ_2}}{(\beta_f + 1)} + I_{CBQ_2 0}] - \alpha_f \cdot I_{EQ_2} + (\alpha_f \cdot \alpha_r - 1) \cdot I_{sc}}\} = \frac{V_{cc}}{V_T}$$

$$\ln[\{\frac{\alpha_r \cdot [\frac{\beta_f \cdot I_{EQ_1}}{(\beta_f + 1)} + I_{CBQ_1 0}] - I_{EQ_1} + (\alpha_r \cdot \alpha_f - 1) \cdot I_{se}}{[\frac{\beta_f \cdot I_{EQ_1}}{(\beta_f + 1)} + I_{CBQ_1 0}] - \alpha_f \cdot I_{EQ_1} + (\alpha_f \cdot \alpha_r - 1) \cdot I_{sc}}\} \cdot \{\frac{\alpha_r \cdot [\frac{\beta_f \cdot I_{EQ_2}}{(\beta_f + 1)} + I_{CBQ_2 0}] - I_{EQ_2} + (\alpha_r \cdot \alpha_f - 1) \cdot I_{se}}{[\frac{\beta_f \cdot I_{EQ_2}}{(\beta_f + 1)} + I_{CBQ_2 0}] - \alpha_f \cdot I_{EQ_2} + (\alpha_f \cdot \alpha_r - 1) \cdot I_{sc}}\}] = \frac{V_{cc}}{V_T}$$

$$e^{[\frac{V_{cc}}{V_T}]} = \{\frac{\alpha_r \cdot [\frac{\beta_f \cdot I_{EQ_1}}{(\beta_f + 1)} + I_{CBQ_1 0}] - I_{EQ_1} + (\alpha_r \cdot \alpha_f - 1) \cdot I_{se}}{[\frac{\beta_f \cdot I_{EQ_1}}{(\beta_f + 1)} + I_{CBQ_1 0}] - \alpha_f \cdot I_{EQ_1} + (\alpha_f \cdot \alpha_r - 1) \cdot I_{sc}}\} \cdot \{\frac{\alpha_r \cdot [\frac{\beta_f \cdot I_{EQ_2}}{(\beta_f + 1)} + I_{CBQ_2 0}] - I_{EQ_2} + (\alpha_r \cdot \alpha_f - 1) \cdot I_{se}}{[\frac{\beta_f \cdot I_{EQ_2}}{(\beta_f + 1)} + I_{CBQ_2 0}] - \alpha_f \cdot I_{EQ_2} + (\alpha_f \cdot \alpha_r - 1) \cdot I_{sc}}\}$$

$$\xi_2(I_{EQ_1}, I_{EQ_2}) = \{\frac{\alpha_r \cdot [\frac{\beta_f \cdot I_{EQ_1}}{(\beta_f+1)} + I_{CBQ_1 0}] - I_{EQ_1} + (\alpha_r \cdot \alpha_f - 1) \cdot I_{se}}{[\frac{\beta_f \cdot I_{EQ_1}}{(\beta_f+1)} + I_{CBQ_1 0}] - \alpha_f \cdot I_{EQ_1} + (\alpha_f \cdot \alpha_r - 1) \cdot I_{sc}}\}$$
$$\cdot \{\frac{\alpha_r \cdot [\frac{\beta_f \cdot I_{EQ_2}}{(\beta_f+1)} + I_{CBQ_2 0}] - I_{EQ_2} + (\alpha_r \cdot \alpha_f - 1) \cdot I_{se}}{[\frac{\beta_f \cdot I_{EQ_2}}{(\beta_f+1)} + I_{CBQ_2 0}] - \alpha_f \cdot I_{EQ_2} + (\alpha_f \cdot \alpha_r - 1) \cdot I_{sc}}\};$$

$$\xi_2(I_{EQ_1}, I_{EQ_2}) = e^{[\frac{V_{cc}}{V_T}]}$$

We need to solve the following two equations:

$$\xi_1(I_{EQ_1}, I_{EQ_2}) = 0; \xi_2(I_{EQ_1}, I_{EQ_2}) = e^{[\frac{V_{cc}}{V_T}]}$$

Assumptions: $\alpha_r : 0 \to 0.95; \alpha_r = 0.95 \approx 1; \beta_f : 20 \to 500; \beta_f \gg 1 \Rightarrow \beta_f + 1 \approx \beta_f$

$$\frac{\beta_f}{(\beta_f+1)} \approx 1; \alpha_f : 0.95 \to 0.99; \alpha_f \approx 1; (\frac{\beta_f}{(\beta_f+1)} - \alpha_f) \to \varepsilon \to 0; \Omega(\alpha_r, \beta_f) = \frac{\alpha_r \cdot \beta_f}{(\beta_f+1)} - 1$$

$$\xi_1(I_{EQ_1}, I_{EQ_2}) = \frac{V_T}{R_{load}} \cdot \ln\{\frac{(\frac{\alpha_r \cdot \beta_f}{(\beta_f+1)} - 1) \cdot I_{EQ_2} + \alpha_r \cdot I_{CBQ_2 0} + (\alpha_r \cdot \alpha_f - 1) \cdot I_{se}}{(\frac{\beta_f}{(\beta_f+1)} - \alpha_f) \cdot I_{EQ_2} + I_{CBQ_2 0} + (\alpha_f \cdot \alpha_r - 1) \cdot I_{sc}}\} - I_{EQ_1} + I_{EQ_2} = 0$$

$$\xi_1(I_{EQ_1}, I_{EQ_2}) = \frac{V_T}{R_{load}} \cdot \ln\{\frac{(\frac{\alpha_r \cdot \beta_f}{(\beta_f+1)} - 1) \cdot I_{EQ_2} + \alpha_r \cdot I_{CBQ_2 0} + (\alpha_r \cdot \alpha_f - 1) \cdot I_{se}}{I_{CBQ_2 0} + (\alpha_f \cdot \alpha_r - 1) \cdot I_{sc}}\} - I_{EQ_1} + I_{EQ_2} = 0$$

$$\Gamma_1(\alpha_r, \alpha_f, I_{se}, I_{CBQ_2 0}) = \alpha_r \cdot I_{CBQ_2 0} + (\alpha_r \cdot \alpha_f - 1) \cdot I_{se}; \Gamma_2(\alpha_r, \alpha_f, I_{sc}, I_{CBQ_2 0}) = I_{CBQ_2 0} + (\alpha_f \cdot \alpha_r - 1) \cdot I_{sc}$$

$$\xi_1(I_{EQ_1}, I_{EQ_2}) = \frac{V_T}{R_{load}} \cdot \ln\{\frac{\Omega(\alpha_r, \beta_f) \cdot I_{EQ_2} + \Gamma_1(\alpha_r, \alpha_f, I_{se}, I_{CBQ_2 0})}{\Gamma_2(\alpha_r, \alpha_f, I_{sc}, I_{CBQ_2 0})}\} - I_{EQ_1} + I_{EQ_2} = 0$$

$$\ln\{\frac{\Omega(\alpha_r, \beta_f) \cdot I_{EQ_2} + \Gamma_1(\alpha_r, \alpha_f, I_{se}, I_{CBQ_2 0})}{\Gamma_2(\alpha_r, \alpha_f, I_{sc}, I_{CBQ_2 0})}\} = (I_{EQ_1} - I_{EQ_2}) \cdot \frac{R_{load}}{V_T}$$

$$\exp\{(I_{EQ_1} - I_{EQ_2}) \cdot \frac{R_{load}}{V_T}\} = \frac{\Omega(\alpha_r, \beta_f) \cdot I_{EQ_2} + \Gamma_1(\alpha_r, \alpha_f, I_{se}, I_{CBQ_2 0})}{\Gamma_2(\alpha_r, \alpha_f, I_{sc}, I_{CBQ_2 0})}$$

$$\exp\{(I_{EQ_1} - I_{EQ_2}) \cdot \frac{R_{load}}{V_T}\} = \sum_{n=0}^{\infty} \frac{(I_{EQ_1} - I_{EQ_2})^n \cdot \frac{R_{load}^n}{V_T^n}}{n!}$$

$$= 1 + (I_{EQ_1} - I_{EQ_2}) \cdot \frac{R_{load}}{V_T} + \frac{(I_{EQ_1} - I_{EQ_2})^2 \cdot \frac{R_{load}^2}{V_T^2}}{2} + \frac{(I_{EQ_1} - I_{EQ_2})^3 \cdot \frac{R_{load}^3}{V_T^3}}{6} + \cdots$$

$$\exp\{(I_{EQ_1} - I_{EQ_2}) \cdot \frac{R_{load}}{V_T}\} \approx 1 + (I_{EQ_1} - I_{EQ_2}) \cdot \frac{R_{load}}{V_T} + \frac{(I_{EQ_1} - I_{EQ_2})^2 \cdot \frac{R_{load}^2}{V_T^2}}{2}$$

$$\Gamma_1 = \Gamma_1(\alpha_r, \alpha_f, I_{se}, I_{CBQ_2 0}); \quad \Gamma_2 = \Gamma_2(\alpha_r, \alpha_f, I_{sc}, I_{CBQ_2 0}); \quad \Omega = \Omega(\alpha_r, \beta_f)$$

$$1 + (I_{EQ_1} - I_{EQ_2}) \cdot \frac{R_{load}}{V_T} + \frac{(I_{EQ_1} - I_{EQ_2})^2 \cdot \frac{R_{load}^2}{V_T^2}}{2} \approx \frac{\Omega(\alpha_r, \beta_f) \cdot I_{EQ_2} + \Gamma_1(\alpha_r, \alpha_f, I_{se}, I_{CBQ_2 0})}{\Gamma_2(\alpha_r, \alpha_f, I_{sc}, I_{CBQ_2 0})}$$

$$(*) \quad 1 + (I_{EQ_1} - I_{EQ_2}) \cdot \frac{R_{load}}{V_T} + \frac{(I_{EQ_1} - I_{EQ_2})^2 \cdot \frac{R_{load}^2}{V_T^2}}{2} \approx \frac{\Omega}{\Gamma_2} \cdot I_{EQ_2} + \frac{\Gamma_1}{\Gamma_2}$$

$$\xi_2(I_{EQ_1}, I_{EQ_2}) = \{\frac{\alpha_r \cdot [\frac{\beta_f \cdot I_{EQ_1}}{(\beta_f + 1)} + I_{CBQ_1 0}] - I_{EQ_1} + (\alpha_r \cdot \alpha_f - 1) \cdot I_{se}}{[\frac{\beta_f \cdot I_{EQ_1}}{(\beta_f + 1)} + I_{CBQ_1 0}] - \alpha_f \cdot I_{EQ_1} + (\alpha_f \cdot \alpha_r - 1) \cdot I_{sc}}\} \cdot \{\frac{\alpha_r \cdot [\frac{\beta_f \cdot I_{EQ_2}}{(\beta_f + 1)} + I_{CBQ_2 0}] - I_{EQ_2} + (\alpha_r \cdot \alpha_f - 1) \cdot I_{se}}{[\frac{\beta_f \cdot I_{EQ_2}}{(\beta_f + 1)} + I_{CBQ_2 0}] - \alpha_f \cdot I_{EQ_2} + (\alpha_f \cdot \alpha_r - 1) \cdot I_{sc}}\};$$

$$\xi_2(I_{EQ_1}, I_{EQ_2}) = e^{[\frac{V_{cc}}{V_T}]}$$

$$\{\frac{(\alpha_r \cdot \frac{\beta_f}{(\beta_f + 1)} - 1) \cdot I_{EQ_1} + \alpha_r \cdot I_{CBQ_1 0} + (\alpha_r \cdot \alpha_f - 1) \cdot I_{se}}{(\frac{\beta_f}{(\beta_f + 1)} - \alpha_f) \cdot I_{EQ_1} + I_{CBQ_1 0} + (\alpha_f \cdot \alpha_r - 1) \cdot I_{sc}}\} \cdot \{\frac{(\alpha_r \cdot \frac{\beta_f}{(\beta_f + 1)} - 1) \cdot I_{EQ_2} + \alpha_r \cdot I_{CBQ_2 0} + (\alpha_r \cdot \alpha_f - 1) \cdot I_{se}}{(\frac{\beta_f}{(\beta_f + 1)} - \alpha_f) \cdot I_{EQ_2} + I_{CBQ_2 0} + (\alpha_f \cdot \alpha_r - 1) \cdot I_{sc}}\} = e^{[\frac{V_{cc}}{V_T}]}$$

Assumptions:

$(\frac{\beta_f}{(\beta_f + 1)} - \alpha_f) \to \varepsilon; \Omega = \alpha_r \cdot \frac{\beta_f}{(\beta_f + 1)} - 1; \Gamma_3 = \alpha_r \cdot I_{CBQ_1 0} + (\alpha_r \cdot \alpha_f - 1) \cdot I_{se}$

$$\Gamma_4 = I_{CBQ_1 0} + (\alpha_f \cdot \alpha_r - 1) \cdot I_{sc}; \quad \Gamma_1 = \alpha_r \cdot I_{CBQ_2 0} + (\alpha_r \cdot \alpha_f - 1) \cdot I_{se};$$

$$\Gamma_2 = I_{CBQ_2 0} + (\alpha_f \cdot \alpha_r - 1) \cdot I_{sc}$$

$(**)\quad (\frac{\Omega\cdot I_{EQ_1}+\Gamma_3}{\Gamma_4})\cdot(\frac{\Omega\cdot I_{EQ_2}+\Gamma_1}{\Gamma_2}) = e^{[\frac{V_{cc}}{V_T}]}$

We can summary our intermediate results:

$$(*)\quad 1+(I_{EQ_1}-I_{EQ_2})\cdot\frac{R_{load}}{V_T}+\frac{(I_{EQ_1}-I_{EQ_2})^2\cdot\frac{R_{load}^2}{V_T^2}}{2}\approx\frac{\Omega}{\Gamma_2}\cdot I_{EQ_2}+\frac{\Gamma_1}{\Gamma_2}$$

$$(**)\quad (\frac{\Omega\cdot I_{EQ_1}+\Gamma_3}{\Gamma_4})\cdot(\frac{\Omega\cdot I_{EQ_2}+\Gamma_1}{\Gamma_2}) = e^{[\frac{V_{cc}}{V_T}]}$$

$$I_{EQ_1}=\frac{1}{\Omega}\cdot(\frac{\Gamma_4\cdot\Gamma_2\cdot e^{[\frac{V_{cc}}{V_T}]}}{\Omega\cdot I_{EQ_2}+\Gamma_1}-\Gamma_3)$$

$$1+(\frac{1}{\Omega}\cdot(\frac{\Gamma_4\cdot\Gamma_2\cdot e^{[\frac{V_{cc}}{V_T}]}}{\Omega\cdot I_{EQ_2}+\Gamma_1}-\Gamma_3)-I_{EQ_2})\cdot\frac{R_{load}}{V_T}$$
$$+\frac{(\frac{1}{\Omega}\cdot(\frac{\Gamma_4\cdot\Gamma_2\cdot e^{[\frac{V_{cc}}{V_T}]}}{\Omega\cdot I_{EQ_2}+\Gamma_1}-\Gamma_3)-I_{EQ_2})^2\cdot\frac{R_{load}^2}{V_T^2}}{2}\approx\frac{\Omega}{\Gamma_2}\cdot I_{EQ_2}+\frac{\Gamma_1}{\Gamma_2}$$

$$\frac{\Gamma_4\cdot\Gamma_2\cdot R_{load}\cdot e^{[\frac{V_{cc}}{V_T}]}}{\Omega^2\cdot V_T\cdot I_{EQ_2}+\Omega\cdot\Gamma_1\cdot V_T}-I_{EQ_2}\cdot\frac{R_{load}}{V_T}+1-\frac{\Gamma_3\cdot R_{load}}{\Omega\cdot V_T}+([\frac{\Gamma_4\cdot\Gamma_2\cdot e^{[\frac{V_{cc}}{V_T}]}}{\Omega^2\cdot I_{EQ_2}+\Omega\cdot\Gamma_1}-I_{EQ_2}]^2$$
$$+[\frac{\Gamma_3}{\Omega}]^2-2\cdot\frac{\Gamma_3}{\Omega}\cdot[\frac{\Gamma_4\cdot\Gamma_2\cdot e^{[\frac{V_{cc}}{V_T}]}}{\Omega^2\cdot I_{EQ_2}+\Omega\cdot\Gamma_1}-I_{EQ_2}])\cdot\frac{R_{load}^2}{2\cdot V_T^2}\approx\frac{\Omega}{\Gamma_2}\cdot I_{EQ_2}+\frac{\Gamma_1}{\Gamma_2}$$

$$\frac{\Gamma_4\cdot\Gamma_2\cdot R_{load}\cdot e^{[\frac{V_{cc}}{V_T}]}}{\Omega^2\cdot V_T\cdot I_{EQ_2}+\Omega\cdot\Gamma_1\cdot V_T}-I_{EQ_2}\cdot\frac{R_{load}}{V_T}+1-\frac{\Gamma_3\cdot R_{load}}{\Omega\cdot V_T}$$
$$+(\frac{\Gamma_4^2\cdot\Gamma_2^2\cdot e^{2\cdot[\frac{V_{cc}}{V_T}]}}{[\Omega^2\cdot I_{EQ_2}+\Omega\cdot\Gamma_1]^2}+I_{EQ_2}^2-2\cdot\frac{\Gamma_4\cdot\Gamma_2\cdot e^{[\frac{V_{cc}}{V_T}]}}{[\Omega^2\cdot I_{EQ_2}+\Omega\cdot\Gamma_1]}\cdot I_{EQ_2}$$
$$+[\frac{\Gamma_3}{\Omega}]^2-2\cdot\frac{\Gamma_3}{\Omega}\cdot[\frac{\Gamma_4\cdot\Gamma_2\cdot e^{[\frac{V_{cc}}{V_T}]}}{\Omega^2\cdot I_{EQ_2}+\Omega\cdot\Gamma_1}-I_{EQ_2}])\cdot\frac{R_{load}^2}{2\cdot V_T^2}\approx\frac{\Omega}{\Gamma_2}\cdot I_{EQ_2}+\frac{\Gamma_1}{\Gamma_2}$$

$$\frac{\Gamma_4 \cdot \Gamma_2 \cdot R_{load} \cdot e^{[\frac{V_{cc}}{V_T}]}}{\Omega^2 \cdot V_T \cdot I_{EQ_2} + \Omega \cdot \Gamma_1 \cdot V_T} - I_{EQ_2} \cdot \frac{R_{load}}{V_T} + 1 - \frac{\Gamma_3 \cdot R_{load}}{\Omega \cdot V_T}$$
$$+ \frac{\Gamma_4^2 \cdot \Gamma_2^2 \cdot e^{2 \cdot [\frac{V_{cc}}{V_T}]} \cdot \frac{R_{load}^2}{2 \cdot V_T^2}}{[\Omega^2 \cdot I_{EQ_2} + \Omega \cdot \Gamma_1]^2} + I_{EQ_2}^2 \cdot \frac{R_{load}^2}{2 \cdot V_T^2} - 2 \cdot \frac{\Gamma_4 \cdot \Gamma_2 \cdot e^{[\frac{V_{cc}}{V_T}]} \cdot \frac{R_{load}^2}{2 \cdot V_T^2}}{[\Omega^2 \cdot I_{EQ_2} + \Omega \cdot \Gamma_1]} \cdot I_{EQ_2}$$
$$+ [\frac{\Gamma_3}{\Omega}]^2 \cdot \frac{R_{load}^2}{2 \cdot V_T^2} - 2 \cdot \frac{\Gamma_3}{\Omega} \cdot [\frac{\Gamma_4 \cdot \Gamma_2 \cdot e^{[\frac{V_{cc}}{V_T}]} \cdot \frac{R_{load}^2}{2 \cdot V_T^2}}{\Omega^2 \cdot I_{EQ_2} + \Omega \cdot \Gamma_1} - I_{EQ_2} \cdot \frac{R_{load}^2}{2 \cdot V_T^2}] \approx \frac{\Omega}{\Gamma_2} \cdot I_{EQ_2} + \frac{\Gamma_1}{\Gamma_2}$$

$$\frac{\Gamma_4 \cdot \Gamma_2 \cdot R_{load} \cdot e^{[\frac{V_{cc}}{V_T}]}}{\Omega^2 \cdot V_T \cdot I_{EQ_2} + \Omega \cdot \Gamma_1 \cdot V_T} + I_{EQ_2} \cdot [2 \cdot \frac{\Gamma_3}{\Omega} \cdot \frac{R_{load}^2}{2 \cdot V_T^2} - \frac{R_{load}}{V_T} - \frac{\Omega}{\Gamma_2}] + I_{EQ_2}^2 \cdot \frac{R_{load}^2}{2 \cdot V_T^2}$$
$$+ \frac{\Gamma_4^2 \cdot \Gamma_2^2 \cdot e^{2 \cdot [\frac{V_{cc}}{V_T}]} \cdot \frac{R_{load}^2}{2 \cdot V_T^2}}{[\Omega^2 \cdot I_{EQ_2} + \Omega \cdot \Gamma_1]^2} - 2 \cdot \frac{\Gamma_3}{\Omega} \cdot (\frac{\Gamma_4 \cdot \Gamma_2 \cdot e^{[\frac{V_{cc}}{V_T}]} \cdot \frac{R_{load}^2}{2 \cdot V_T^2}}{\Omega^2 \cdot I_{EQ_2} + \Omega \cdot \Gamma_1})$$
$$- 2 \cdot \frac{\Gamma_4 \cdot \Gamma_2 \cdot e^{[\frac{V_{cc}}{V_T}]} \cdot \frac{R_{load}^2}{2 \cdot V_T^2}}{[\Omega^2 \cdot I_{EQ_2} + \Omega \cdot \Gamma_1]} \cdot I_{EQ_2} \approx \frac{\Gamma_1}{\Gamma_2} - [\frac{\Gamma_3}{\Omega}]^2 \cdot \frac{R_{load}^2}{2 \cdot V_T^2} + \frac{\Gamma_3 \cdot R_{load}}{\Omega \cdot V_T} - 1$$

The above equation can be solve numerically and we get some options for I_{EQ_2} values. We ignore complex and negative values [33].

$$\psi_1(I_{EQ_2}) \approx \frac{\Gamma_1}{\Gamma_2} - [\frac{\Gamma_3}{\Omega}]^2 \cdot \frac{R_{load}^2}{2 \cdot V_T^2} + \frac{\Gamma_3 \cdot R_{load}}{\Omega \cdot V_T} - 1;$$
$$I_{EQ_2} = f_n(\Gamma_1, \ldots, \Gamma_4, \Omega, R_{load}, V_T, \ldots); \quad n = 1, 2, \ldots$$

$$I_{EQ_1} = \frac{1}{\Omega} \cdot (\frac{\Gamma_4 \cdot \Gamma_2 \cdot e^{[\frac{V_{cc}}{V_T}]}}{\Omega \cdot f_n(\Gamma_1, \ldots, \Gamma_4, \Omega, R_{load}, V_T, \ldots) + \Gamma_1} - \Gamma_3);$$
$$I_{EQ_1} = g_n(\Gamma_1, \ldots, \Gamma_4, \Omega, R_{load}, V_T, \ldots)$$

Summary: We get some options for I_{EQ_1}, I_{EQ_2} values, and ignore negative and complex results.

$$I_{EQ_1} = g_n(\Gamma_1, \ldots, \Gamma_4, \Omega, R_{load}, V_T, \ldots);$$
$$I_{EQ_2} = f_n(\Gamma_1, \ldots, \Gamma_4, \Omega, R_{load}, V_T, \ldots); \quad n = 1, 2, \ldots$$

Back to our previous differential equation:

$$- (\frac{dI_{R_1}}{dt} \cdot R_1 + \frac{dI_{R_2}}{dt} \cdot R_2) = \frac{1}{C} \cdot (\sum_{k=1}^{2} I_{R_k} - \sum_{k=1}^{2} I_{BQ_k} - 2 \cdot I_0$$
$$\cdot \{\exp[\frac{1}{2}(V_{cc} - I_{R_1} \cdot R_1 - I_{R_2} \cdot R_2)] - 1\})$$

We consider $R_1 \approx R_2$ and $R = R_2; R = R_1$ then we can write the above circuit differential equation:

$$-(\frac{dI_{R_1}}{dt} + \frac{dI_{R_2}}{dt}) \cdot R = \frac{1}{C} \cdot (\sum_{k=1}^{2} I_{R_k} - \sum_{k=1}^{2} I_{BQ_k} - 2 \cdot I_0 \cdot \{\exp[\frac{1}{2}(V_{cc} - R \cdot \sum_{k=1}^{2} I_{R_k})] - 1\})$$

We define new variable

$$X = \sum_{k=1}^{2} I_{R_k}; \quad \frac{dX}{dt} = \sum_{k=1}^{2} \frac{dI_{R_k}}{dt}$$

$$I_{BQ_1} = \frac{I_{EQ_1}}{(\beta_f + 1)} - I_{CBQ_1 0} = \frac{g_n(\Gamma_1, \ldots, \Gamma_4, \Omega, R_{load}, V_T, \ldots)}{(\beta_f + 1)} - I_{CBQ_1 0}$$

$$I_{BQ_2} = \frac{I_{EQ_2}}{(\beta_f + 1)} - I_{CBQ_2 0} = \frac{f_n(\Gamma_1, \ldots, \Gamma_4, \Omega, R_{load}, V_T, \ldots)}{(\beta_f + 1)} - I_{CBQ_2 0}$$

$$\sum_{k=1}^{2} I_{BQ_k} = \frac{g_n(\Gamma_1, \ldots, \Gamma_4, \Omega, R_{load}, V_T, \ldots)}{(\beta_f + 1)} + \frac{f_n(\Gamma_1, \ldots, \Gamma_4, \Omega, R_{load}, V_T, \ldots)}{(\beta_f + 1)} - \sum_{k=1}^{2} I_{CBQ_k 0}$$

$$g_n = g_n(\Gamma_1, \ldots, \Gamma_4, \Omega, R_{load}, V_T, \ldots); \; f_n = f_n(\Gamma_1, \ldots, \Gamma_4, \Omega, R_{load}, V_T, \ldots)$$

$$\sum_{k=1}^{2} I_{BQ_k} = \frac{g_n + f_n}{(\beta_f + 1)} - \sum_{k=1}^{2} I_{CBQ_k 0}; \quad n = 1, 2, \ldots$$

$$-R \cdot \sum_{k=1}^{2} \frac{dI_{R_k}}{dt} = \frac{1}{C} \cdot (\sum_{k=1}^{2} I_{R_k} - \frac{g_n + f_n}{(\beta_f + 1)} + \sum_{k=1}^{2} I_{CBQ_k 0} - 2 \cdot I_0 \cdot \{\exp[\frac{1}{2}(V_{cc} - R \cdot \sum_{k=1}^{2} I_{R_k})] - 1\})$$

$$\frac{dX}{dt} = -\frac{1}{C \cdot R} \cdot (X - \frac{g_n + f_n}{(\beta_f + 1)} + \sum_{k=1}^{2} I_{CBQ_k 0} - 2 \cdot I_0 \cdot \{\exp[\frac{1}{2}(V_{cc} - R \cdot X)] - 1\})$$

The first stage is to find our circuit fixed point: $\frac{dX}{dt} = 0$

$$X^{(j)} - \frac{g_n + f_n}{(\beta_f + 1)} + \sum_{k=1}^{2} I_{CBQ_k0} - 2 \cdot I_0 \cdot \{\exp[\frac{1}{2}(V_{cc} - R \cdot X^{(j)})] - 1\} = 0;$$
$$j = 0, 1, 2, \ldots$$

$$X^{(j)} - 2 \cdot I_0 \cdot \{\exp[\frac{1}{2}(V_{cc} - R \cdot X^{(j)})] - 1\} = \frac{g_n + f_n}{(\beta_f + 1)} - \sum_{k=1}^{2} I_{CBQ_k0}$$

$$X^{(j)} - 2 \cdot I_0 \cdot \{\exp[\frac{1}{2}(V_{cc} - R \cdot X^{(j)})] - 1\} = \frac{g_n + f_n}{(\beta_f + 1)} - \sum_{k=1}^{2} I_{CBQ_k0}$$

$$\exp[\frac{1}{2} \cdot (V_{cc} - R \cdot X^{(j)})] = \sum_{n=0}^{\infty} \frac{\frac{1}{2^n} \cdot (V_{cc} - R \cdot X^{(j)})^n}{n!}$$
$$= 1 + \frac{1}{2} \cdot (V_{cc} - R \cdot X^{(j)}) + \frac{1}{8} \cdot (V_{cc} - R \cdot X^{(j)})^2 + \cdots$$

$$\exp[\frac{1}{2} \cdot (V_{cc} - R \cdot X^{(j)})] \approx 1 + \frac{1}{2} \cdot (V_{cc} - R \cdot X^{(j)}) + \frac{1}{8} \cdot (V_{cc} - R \cdot X^{(j)})^2$$

$$X^{(j)} - I_0 \cdot \{(V_{cc} - R \cdot X^{(j)}) + \frac{1}{4} \cdot (V_{cc} - R \cdot X^{(j)})^2\} = \frac{g_n + f_n}{(\beta_f + 1)} - \sum_{k=1}^{2} I_{CBQ_k0}$$

$$(1 + I_0 \cdot R) \cdot X^{(j)} - I_0 \cdot V_{cc} - \frac{1}{4} \cdot I_0 \cdot [V_{cc}^2 - 2 \cdot V_{cc} \cdot R \cdot X^{(j)} + R^2 \cdot [X^{(j)}]^2]$$
$$= \frac{g_n + f_n}{(\beta_f + 1)} - \sum_{k=1}^{2} I_{CBQ_k0}$$

$$\frac{1}{4} \cdot I_0 \cdot R^2 \cdot [X^{(j)}]^2 - (1 + I_0 \cdot R + \frac{1}{2} \cdot I_0 \cdot V_{cc} \cdot R) \cdot X^{(j)}$$
$$+ [I_0 \cdot V_{cc} + \frac{1}{4} \cdot I_0 \cdot V_{cc}^2 + \frac{g_n + f_n}{(\beta_f + 1)} - \sum_{k=1}^{2} I_{CBQ_k0}] = 0$$

$$X^{(j)} = \frac{(1 + I_0 \cdot R + \frac{1}{2} \cdot I_0 \cdot V_{cc} \cdot R) \pm \sqrt{\begin{array}{l}(1 + I_0 \cdot R + \frac{1}{2} \cdot I_0 \cdot V_{cc} \cdot R)^2 \\ - I_0 \cdot R^2 \cdot [I_0 \cdot V_{cc} + \frac{1}{4} \cdot I_0 \cdot V_{cc}^2 + \frac{g_n + f_n}{(\beta_f + 1)} - \sum_{k=1}^{2} I_{CBQ_k0}]\end{array}}}{\frac{1}{2} \cdot I_0 \cdot R^2}$$

We get two groups of fixed points for our circuit: $g_n, f_n \in \mathbb{R}_+; n = 1, 2, \ldots$

$$X^{(j)}_{group_1} = \frac{(1+I_0\cdot R+\frac{1}{2}\cdot I_0\cdot V_{cc}\cdot R)+\sqrt{\begin{array}{l}(1+I_0\cdot R+\frac{1}{2}\cdot I_0\cdot V_{cc}\cdot R)^2\\ -I_0\cdot R^2\cdot[I_0\cdot V_{cc}+\frac{1}{4}\cdot I_0\cdot V_{cc}^2+\frac{g_n+f_n}{(\beta_f+1)}-\sum_{k=1}^{2}I_{CBQ_k0}]\end{array}}}{\frac{1}{2}\cdot I_0\cdot R^2}$$

$$X^{(j)}_{group_2} = \frac{(1+I_0\cdot R+\frac{1}{2}\cdot I_0\cdot V_{cc}\cdot R)-\sqrt{\begin{array}{l}(1+I_0\cdot R+\frac{1}{2}\cdot I_0\cdot V_{cc}\cdot R)^2\\ -I_0\cdot R^2\cdot[I_0\cdot V_{cc}+\frac{1}{4}\cdot I_0\cdot V_{cc}^2+\frac{g_n+f_n}{(\beta_f+1)}-\sum_{k=1}^{2}I_{CBQ_k0}]\end{array}}}{\frac{1}{2}\cdot I_0\cdot R^2}$$

We ignore in our analysis negative and complex fixed points values. We get a set of one dimension systems $\frac{dX}{dt} = \xi_n(X); n = 1,2,\ldots$

$$\xi_n(X) = -\frac{1}{C\cdot R}\cdot(X - \frac{g_n+f_n}{(\beta_f+1)} + \sum_{k=1}^{2} I_{CBQ_k0} - 2\cdot I_0\cdot\{\exp[\frac{1}{2}(V_{cc}-R\cdot X)]-1\})$$

$$\frac{dX}{dt} = -\frac{1}{C\cdot R}\cdot(X - \frac{g_n+f_n}{(\beta_f+1)} + \sum_{k=1}^{2} I_{CBQ_k0} - 2\cdot I_0\cdot\{\exp[\frac{1}{2}(V_{cc}-R\cdot X)]-1\})$$

We can draw the graphs $\xi_n(X); n = 1,2,\ldots$ and then use it to sketch the vector fields on the real line. A fluid is flowing along the real line with a local velocities $\xi_n(X)$. This imaginary fluid is called the phase fluid of our class AB push-pull amplifier system, and the real line is the phase space. The flow is to the right where $\xi_n(X) > 0$ and to the left where $\xi_n(X) < 0$. To find the solutions to $\frac{dX}{dt} = \xi_n(X); n = 1,2,\ldots$ starting from an arbitrary initial condition X_0 As time goes, the phase point moves along the X-axis according to some functions $\xi_n(X); n = 1,2,\ldots$ This function is called the trajectory based at X_0, and it represents the solutions of the differential equation starting from the initial conditions X_0. A picture which shows all the qualitatively different trajectories of our class AB push-pull amplifier system, is called a phase portrait. The appearance of the phase portrait is controlled by the fixed points $X^{(j)}_{group_1}$ or $X^{(j)}_{group_2}; j = 0,1,\ldots,$ defined by $\xi_n(X^{(j)}_{group_1}) = 0$, $\xi_n(X^{(j)}_{group_2}) = 0; n = 1,2,\ldots$; they correspond to stagnation points of the flow. Our system fixed points represent equilibrium solutions (steady state, constant, rest solutions), since if $X = X^{(j)}_{group_1}$ or $X = X^{(j)}_{group_2}$; $j = 0,1,\ldots$ initially, then $X(t) = X^{(j)}_{group_1}$ or $X(t) = X^{(j)}_{group_2}; j = 0,1,\ldots$ for all time. An class AB push-pull amplifier system equilibrium is defined to be stable if all sufficiently small disturbances away from it damp out in time. Stable system equilibria are represented geometrically by stable fixed points. Conversely, unstable equilibria, in which disturbances grow in time, are represented by unstable fixed points [2–4].

5.2 Class C Power Amplifier (PA) with Parallel Resonance Circuit Stability Analysis Under Parameters Variation

The Class C Amplifier design has the greatest efficiency but the poorest linearity of the classes of amplifiers. The class C amplifier is heavily biased so that the output current is zero for more than one half of an input sinusoidal signal cycle with the transistor idling at its cut-off point. Due to its heavy audio distortion, class C amplifiers are commonly used in high frequency sine wave oscillators and certain types of radio frequency amplifiers. The class C amplifier conduction angle is slightly less than 180°. The most common application of the Class C amplifier is the RF (radio frequency) circuits like RF oscillator, RF amplifier etc. where there are additional tuned circuits for retrieving the original input signal from the pulsed output of the Class C amplifier and so the distortion caused by the amplifier has little effect on the final output. Biasing resistor Rb pulls the base of Q1 further downwards and the Q-point will be set some way below the cut-off point in the DC load line. As a result the transistor will start conducting only after the input signal amplitude has risen above the base emitter voltage (Vbe ~ 0.7 V) plus the downward bias voltage caused by Rb. That is the reason why the major portion of the input signal is absent in the output signal. Inductor L_1 and capacitor C_1 forms a tank circuit which aids in the extraction of the required signal from the pulsed output of the transistor. Class C operation means that the collector current flows for less than 180° of the ac cycle. This implies that the collector current of a class C amplifier is highly non-sinusoidal because current flows in pulses. To avoid distortion, class C amplifier makes use of a resonant tank circuit. This results in a sinusoidal output voltage. Actual job of the active element (transistor Q1) here is to produce a series of current pulses according to the input and make it flow through the resonant circuit. Values of L_1 and C_1 are so selected that the resonant circuit oscillates in the frequency of the input signal. Since the resonant circuit oscillates in one frequency (generally the carrier frequency) all other frequencies are attenuated and the required frequency can be squeezed out using a suitably tuned load. Harmonics or noise present in the output signal can be eliminated using additional filters. A coupling transformer can be used for transferring the power to the load. The Class C amplifier has high efficiency and it is excellent in RF applications [24–26] (Fig. 5.3).

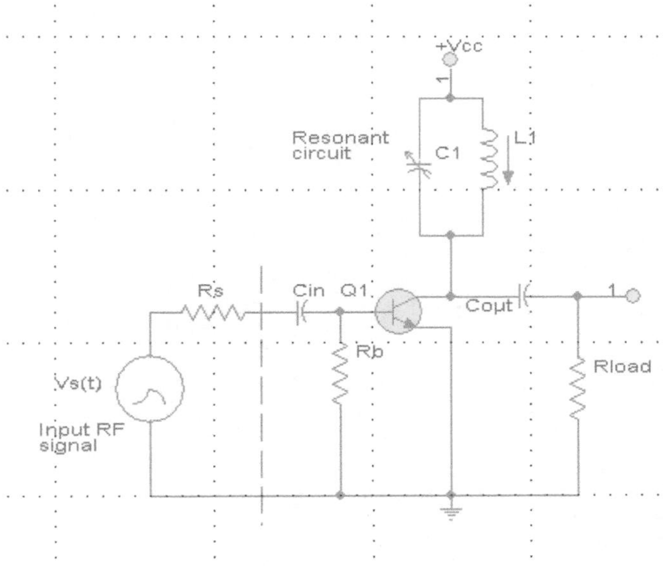

Fig. 5.3 Class C power amplifier

Inductor L_1 and capacitor C_1 forms a tank circuit which aids in the extraction of the required signal from the pulsed output of the transistor. Class C operation means that the collector current flows for less than 180° of the ac cycle. This implies that the collector current of a class C amplifier is highly non-sinusoidal because current flows in pulses. To avoid distortion, class C amplifier makes use of a resonant tank circuit. This results in a sinusoidal output voltage. Actual job of the active element (transistor Q1) here is to produce a series of current pulses according to the input and make it flow through the resonant circuit. Values of L_1 and C_1 are so selected that the resonant circuit oscillates in the frequency of the input signal. Since the resonant circuit oscillates in one frequency (generally the carrier frequency) all other frequencies are attenuated and the required frequency can be squeezed out using a suitably tuned load. Harmonics or noise present in the output signal can be eliminated using additional filters. A coupling transformer can be used for transferring the power to the load. The Class C amplifier has high efficiency and it is excellent in RF applications.

Since the input RF signal is a large signal we use in our analysis the Ebers-Moll BJT model for Q_1. The Ebers-Moll BJT model is a good large signal, steady state model of the transistor and allows the state of conduction of the device to be easily determined for different modes of operation of the device. The different modes of operation are determined by the manner in which the junctions are biased [91–93] (Fig. 5.4).

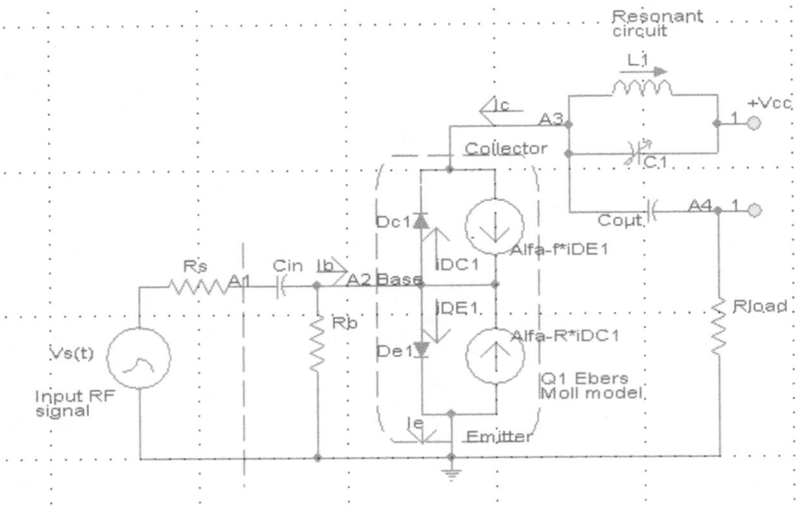

Fig. 5.4 Class C power amplifier equivalent circuit

$$I_{CQ_1} + I_{BQ_1} = I_{EQ_1}; I_{R_s} = \frac{V_s(t) - V_{A_1}}{R_s}; I_{C_{in}} = C_{in} \cdot \frac{d}{dt}(V_{A_1} - V_{A_2}); I_{R_s} = I_{C_{in}}; I_{C_{in}} = I_{BQ1} + I_{R_b}$$

Collector emitter voltage expression for BJT NPN:

$$V_{CE-NPN} \approx V_T \cdot \ln\{\frac{[\alpha_r \cdot I_C - I_E + (\alpha_r \cdot \alpha_f - 1) \cdot I_{se}]}{[I_C - \alpha_f \cdot I_E + (\alpha_f \cdot \alpha_r - 1) \cdot I_{sc}]}\};$$

$$V_{CEQ_1} \approx V_T \cdot \ln\{\frac{[\alpha_r \cdot I_{CQ_1} - I_{EQ_1} + (\alpha_r \cdot \alpha_f - 1) \cdot I_{se}]}{[I_{CQ_1} - \alpha_f \cdot I_{EQ_1} + (\alpha_f \cdot \alpha_r - 1) \cdot I_{sc}]}\}$$

$$I_{R_b} = \frac{V_{A_2}}{R_b}; I_{L_1} + I_{C_1} = I_{CQ_1} + I_{C_{out}}; I_{C_{out}} = I_{R_{load}}; V_{cc} - V_{A_3} = L_1 \cdot \frac{dI_{L_1}}{dt};$$

$$I_{C_1} = C_1 \cdot \frac{d}{dt}(V_{cc} - V_{A_3})$$

$$V_{A_2} = V_{BEQ_1}; V_{CEQ_1} = V_{C_{out}} + V_{R_{load}}; I_{C_{out}} = C_{out} \cdot \frac{d}{dt}(V_{A_3} - V_{A_4}); I_{R_{load}} = \frac{V_{A_4}}{R_{load}}$$

$$I_{R_s} = \frac{V_s(t) - V_{A_1}}{R_s} \Rightarrow V_{A_1} = V_s(t) - I_{R_s} \cdot R_s; I_{C_{in}} = C_{in} \cdot \frac{d}{dt}(V_{A_1} - V_{A_2})$$

$$\Rightarrow V_{A_1} - V_{A_2} = \frac{1}{C_{in}} \cdot \int I_{C_{in}} \cdot dt$$

$$V_{A_1} = V_{A_2} + \frac{1}{C_{in}} \cdot \int I_{C_{in}} \cdot dt; V_s(t) - I_{R_s} \cdot R_s = V_{A_2} + \frac{1}{C_{in}} \cdot \int I_{C_{in}} \cdot dt;$$

$$V_{A_2} = V_s(t) - I_{R_s} \cdot R_s - \frac{1}{C_{in}} \cdot \int I_{C_{in}} \cdot dt$$

$$I_{R_b} = \frac{V_{A_2}}{R_b} \Rightarrow V_{A_2} = I_{R_b} \cdot R_b; I_{R_b} \cdot R_b$$
$$= V_s(t) - I_{R_s} \cdot R_s - \frac{1}{C_{in}} \cdot \int I_{C_{in}} \cdot dt; V_{CEQ_1} = V_{A_3}$$

$$\frac{d}{dt}\{I_{R_b} \cdot R_b = V_s(t) - I_{R_s} \cdot R_s - \frac{1}{C_{in}} \cdot \int I_{C_{in}} \cdot dt\};$$

$$\frac{dI_{R_b}}{dt} \cdot R_b = \frac{dV_s(t)}{dt} - \frac{dI_{R_s}}{dt} \cdot R_s - \frac{1}{C_{in}} \cdot I_{C_{in}}$$

$$V_{cc} - V_{A_3} = L_1 \cdot \frac{dI_{L_1}}{dt}; I_{C_1} = C_1 \cdot \frac{d}{dt}(V_{cc} - V_{A_3}); I_{C_1} = C_1 \cdot L_1 \cdot \frac{d^2 I_{L_1}}{dt^2}$$

$$I_{C_{out}} = C_{out} \cdot \frac{d}{dt}(V_{A_3} - V_{A_4}) \Rightarrow V_{A_3} - V_{A_4} = \frac{1}{C_{out}} \cdot \int I_{C_{out}} \cdot dt; V_{A_3} - I_{R_{load}} \cdot R_{load}$$
$$= \frac{1}{C_{out}} \cdot \int I_{C_{out}} \cdot dt$$

$$V_{CEQ_1} = V_{A_3}; V_T \cdot \ln\{\frac{[\alpha_r \cdot I_{CQ_1} - I_{EQ_1} + (\alpha_r \cdot \alpha_f - 1) \cdot I_{se}]}{[I_{CQ_1} - \alpha_f \cdot I_{EQ_1} + (\alpha_f \cdot \alpha_r - 1) \cdot I_{sc}]}\}$$
$$= I_{R_{load}} \cdot R_{load} + \frac{1}{C_{out}} \cdot \int I_{C_{out}} \cdot dt$$

$$\frac{d}{dt}(\ln\{\frac{[\alpha_r \cdot I_{CQ_1} - I_{EQ_1} + (\alpha_r \cdot \alpha_f - 1) \cdot I_{se}]}{[I_{CQ_1} - \alpha_f \cdot I_{EQ_1} + (\alpha_f \cdot \alpha_r - 1) \cdot I_{sc}]}\})$$
$$= \frac{(\alpha_r \cdot \frac{dI_{CQ_1}}{dt} - \frac{dI_{EQ_1}}{dt}) \cdot [I_{CQ_1} - \alpha_f \cdot I_{EQ_1} + (\alpha_f \cdot \alpha_r - 1) \cdot I_{sc}] - (\frac{dI_{CQ_1}}{dt} - \alpha_f \cdot \frac{dI_{EQ_1}}{dt}) \cdot [\alpha_r \cdot I_{CQ_1} - I_{EQ_1} + (\alpha_r \cdot \alpha_f - 1) \cdot I_{se}]}{[I_{CQ_1} - \alpha_f \cdot I_{EQ_1} + (\alpha_f \cdot \alpha_r - 1) \cdot I_{sc}] \cdot [\alpha_r \cdot I_{CQ_1} - I_{EQ_1} + (\alpha_r \cdot \alpha_f - 1) \cdot I_{se}]}$$

$$V_T \cdot \{\frac{(\alpha_r \cdot \frac{dI_{CQ_1}}{dt} - \frac{dI_{EQ_1}}{dt}) \cdot [I_{CQ_1} - \alpha_f \cdot I_{EQ_1} + (\alpha_f \cdot \alpha_r - 1) \cdot I_{sc}] - (\frac{dI_{CQ_1}}{dt} - \alpha_f \cdot \frac{dI_{EQ_1}}{dt}) \cdot [\alpha_r \cdot I_{CQ_1} - I_{EQ_1} + (\alpha_r \cdot \alpha_f - 1) \cdot I_{se}]}{[I_{CQ_1} - \alpha_f \cdot I_{EQ_1} + (\alpha_f \cdot \alpha_r - 1) \cdot I_{sc}] \cdot [\alpha_r \cdot I_{CQ_1} - I_{EQ_1} + (\alpha_r \cdot \alpha_f - 1) \cdot I_{se}]}\}$$
$$= \frac{dI_{R_{load}}}{dt} \cdot R_{load} + \frac{1}{C_{out}} \cdot I_{C_{out}}$$

$$I_{C_{out}} = I_{R_{load}}; V_T \cdot \{\frac{(\alpha_r \cdot \frac{dI_{CQ_1}}{dt} - \frac{dI_{EQ_1}}{dt}) \cdot [I_{CQ_1} - \alpha_f \cdot I_{EQ_1} + (\alpha_f \cdot \alpha_r - 1) \cdot I_{sc}] - (\frac{dI_{CQ_1}}{dt} - \alpha_f \cdot \frac{dI_{EQ_1}}{dt}) \cdot [\alpha_r \cdot I_{CQ_1} - I_{EQ_1} + (\alpha_r \cdot \alpha_f - 1) \cdot I_{se}]}{[I_{CQ_1} - \alpha_f \cdot I_{EQ_1} + (\alpha_f \cdot \alpha_r - 1) \cdot I_{sc}] \cdot [\alpha_r \cdot I_{CQ_1} - I_{EQ_1} + (\alpha_r \cdot \alpha_f - 1) \cdot I_{se}]}\}$$
$$= \frac{dI_{C_{out}}}{dt} \cdot R_{load} + \frac{1}{C_{out}} \cdot I_{C_{out}}$$

We can summary our equations: $I_{R_s} = I_{C_{in}}; I_{C_1} = C_1 \cdot L_1 \cdot \frac{d^2 I_{L_1}}{dt^2}$

$$\frac{dI_{R_b}}{dt} \cdot R_b = \frac{dV_s(t)}{dt} - \frac{dI_{C_{in}}}{dt} \cdot R_s - \frac{1}{C_{in}} \cdot I_{C_{in}};$$
$$Y_1 = \frac{dI_{C_{in}}}{dt}; \frac{dI_{R_b}}{dt} = \frac{1}{R_b} \cdot \frac{dV_s(t)}{dt} - Y_1 \cdot \frac{R_s}{R_b} - \frac{1}{C_{in} \cdot R_b} \cdot I_{C_{in}}$$
$$I_{CQ_1} = I_{L_1} + I_{C_1} - I_{C_{out}} = I_{L_1} + C_1 \cdot L_1 \cdot \frac{d^2 I_{L_1}}{dt^2} - I_{C_{out}}; I_{BQ1} = I_{C_{in}} - I_{R_b}$$
$$I_{EQ1} = I_{BQ1} + I_{CQ1} = I_{C_{in}} - I_{R_b} + I_{L_1} + C_1 \cdot L_1 \cdot \frac{d^2 I_{L_1}}{dt^2} - I_{C_{out}}$$

$$\frac{dI_{CQ_1}}{dt} = \frac{dI_{L_1}}{dt} + C_1 \cdot L_1 \cdot \frac{d^3 I_{L_1}}{dt^3} - \frac{dI_{C_{out}}}{dt};$$
$$\frac{dI_{EQ_1}}{dt} = \frac{dI_{C_{in}}}{dt} - \frac{dI_{R_b}}{dt} + \frac{dI_{L_1}}{dt} + C_1 \cdot L_1 \cdot \frac{d^3 I_{L_1}}{dt^3} - \frac{dI_{C_{out}}}{dt}$$

$$I_{CQ_1} - \alpha_f \cdot I_{EQ_1} = (1 - \alpha_f) \cdot I_{L_1} + (1 - \alpha_f) \cdot C_1 \cdot L_1 \cdot \frac{d^2 I_{L_1}}{dt^2} + (\alpha_f - 1) \cdot I_{C_{out}} - \alpha_f \cdot I_{C_{in}} + \alpha_f \cdot I_{R_b}$$

$$\alpha_r \cdot I_{CQ_1} - I_{EQ_1} = (\alpha_r - 1) \cdot I_{L_1} + (\alpha_r - 1) \cdot C_1 \cdot L_1 \cdot \frac{d^2 I_{L_1}}{dt^2} + (1 - \alpha_r) \cdot I_{C_{out}} - I_{C_{in}} + I_{R_b}$$

$$\alpha_r \cdot \frac{dI_{CQ_1}}{dt} - \frac{dtI_{EQ_1}}{dt} = (\alpha_r - 1) \cdot \frac{dI_{L_1}}{dt} + (\alpha_r - 1) \cdot C_1 \cdot L_1 \cdot \frac{d^3 I_{L_1}}{dt^3} + (1 - \alpha_r) \cdot \frac{dI_{C_{out}}}{dt} - \frac{dI_{C_{in}}}{dt} + \frac{dI_{R_b}}{dt}$$

$$\frac{dI_{CQ_1}}{dt} - \alpha_f \cdot \frac{dI_{EQ_1}}{dt} = (1 - \alpha_f) \cdot \frac{dI_{L_1}}{dt} + (1 - \alpha_f) \cdot C_1 \cdot L_1 \cdot \frac{d^3 I_{L_1}}{dt^3} + (\alpha_f - 1) \cdot \frac{dI_{C_{out}}}{dt} - \alpha_f \cdot \frac{dI_{C_{in}}}{dt} + \alpha_f \cdot \frac{dI_{R_b}}{dt}$$

We define new variables in our system:

$$Y_1 = \frac{dI_{C_{in}}}{dt}; Y_2 = \frac{dI_{C_{out}}}{dt}; Y_3 = \frac{dI_{L_1}}{dt}; Y_4 = \frac{dY_3}{dt} = \frac{d^2 I_{L_1}}{dt^2}$$

$$I_{CQ_1} - \alpha_f \cdot I_{EQ_1} = (1 - \alpha_f) \cdot I_{L_1} + (1 - \alpha_f) \cdot C_1 \cdot L_1 \cdot Y_4 + (\alpha_f - 1) \cdot I_{C_{out}} - \alpha_f \cdot I_{C_{in}} + \alpha_f \cdot I_{R_b}$$

$$\alpha_r \cdot I_{CQ_1} - I_{EQ_1} = (\alpha_r - 1) \cdot I_{L_1} + (\alpha_r - 1) \cdot C_1 \cdot L_1 \cdot Y_4 + (1 - \alpha_r) \cdot I_{C_{out}} - I_{C_{in}} + I_{R_b}$$

$$\alpha_r \cdot \frac{dI_{CQ_1}}{dt} - \frac{dtI_{EQ_1}}{dt} = (\alpha_r - 1) \cdot Y_3 + (\alpha_r - 1) \cdot C_1 \cdot L_1 \cdot \frac{dY_4}{dt} + (1 - \alpha_r) \cdot Y_2 - Y_1 + \frac{1}{R_b} \cdot \frac{dV_s(t)}{dt} - Y_1 \cdot \frac{R_s}{R_b} - \frac{1}{C_{in} \cdot R_b} \cdot I_{C_{in}}$$

$$\frac{dI_{CQ_1}}{dt} - \alpha_f \cdot \frac{dI_{EQ_1}}{dt} = (1 - \alpha_f) \cdot Y_3 + (1 - \alpha_f) \cdot C_1 \cdot L_1 \cdot \frac{dY_4}{dt} + (\alpha_f - 1) \cdot Y_2 - \alpha_f \cdot Y_1 + \alpha_f \cdot (\frac{1}{R_b} \cdot \frac{dV_s(t)}{dt} - Y_1 \cdot \frac{R_s}{R_b} - \frac{1}{C_{in} \cdot R_b} \cdot I_{C_{in}})$$

We define for simplicity of our analysis four functions:

$$I_{CQ_1} - \alpha_f \cdot I_{EQ_1} = (1 - \alpha_f) \cdot I_{L_1} + (1 - \alpha_f) \cdot C_1 \cdot L_1 \cdot Y_4 + (\alpha_f - 1) \cdot I_{C_{out}} - \alpha_f \cdot I_{C_{in}} + \alpha_f \cdot I_{R_b}$$

$$g_1(I_{L_1}, Y_4, I_{C_{out}}, I_{C_{in}}, I_{R_b}) = (1 - \alpha_f) \cdot I_{L_1} + (1 - \alpha_f) \cdot C_1 \cdot L_1 \cdot Y_4 + (\alpha_f - 1) \cdot I_{C_{out}} - \alpha_f \cdot I_{C_{in}} + \alpha_f \cdot I_{R_b}$$

$$I_{CQ_1} - \alpha_f \cdot I_{EQ_1} = g_1(I_{L_1}, Y_4, I_{C_{out}}, I_{C_{in}}, I_{R_b}); g_1 = g_1(I_{L_1}, Y_4, I_{C_{out}}, I_{C_{in}}, I_{R_b})$$

$$\alpha_r \cdot I_{CQ_1} - I_{EQ_1} = (\alpha_r - 1) \cdot I_{L_1} + (\alpha_r - 1) \cdot C_1 \cdot L_1 \cdot Y_4 + (1 - \alpha_r) \cdot I_{C_{out}} - I_{C_{in}} + I_{R_b}$$

$$g_2(I_{L_1}, Y_4, I_{C_{out}}, I_{C_{in}}, I_{R_b}) = (\alpha_r - 1) \cdot I_{L_1} + (\alpha_r - 1) \cdot C_1 \cdot L_1 \cdot Y_4 + (1 - \alpha_r) \cdot I_{C_{out}} - I_{C_{in}} + I_{R_b}$$

$$\alpha_r \cdot I_{CQ_1} - I_{EQ_1} = g_2(I_{L_1}, Y_4, I_{C_{out}}, I_{C_{in}}, I_{R_b}); g_2 = g_2(I_{L_1}, Y_4, I_{C_{out}}, I_{C_{in}}, I_{R_b})$$

$$\alpha_r \cdot \frac{dI_{CQ_1}}{dt} - \frac{dtI_{EQ_1}}{dt} = (\alpha_r - 1) \cdot C_1 \cdot L_1 \cdot \frac{dY_4}{dt} + g_3(Y_3, Y_2, Y_1, I_{C_{in}}, \frac{dV_s(t)}{dt});$$

$$g_3 = g_3(Y_3, Y_2, Y_1, I_{C_{in}}, \frac{dV_s(t)}{dt})$$

$$g_3(Y_3, Y_2, Y_1, I_{C_{in}}, \frac{dV_s(t)}{dt}) = (\alpha_r - 1) \cdot Y_3 + (1 - \alpha_r) \cdot Y_2 - Y_1 \cdot (1 + \frac{R_s}{R_b}) + \frac{1}{R_b} \cdot \frac{dV_s(t)}{dt} - \frac{1}{C_{in} \cdot R_b} \cdot I_{C_{in}}$$

$$\frac{dI_{CQ_1}}{dt} - \alpha_f \cdot \frac{dI_{EQ_1}}{dt} = (1 - \alpha_f) \cdot C_1 \cdot L_1 \cdot \frac{dY_4}{dt} + g_4(Y_3, Y_2, Y_1, I_{C_{in}}, \frac{dV_s(t)}{dt});$$

$$g_4 = g_4(Y_3, Y_2, Y_1, I_{C_{in}}, \frac{dV_s(t)}{dt})$$

$$g_4(Y_3, Y_2, Y_1, I_{C_{in}}, \frac{dV_s(t)}{dt}) = (1 - \alpha_f) \cdot Y_3 + (\alpha_f - 1) \cdot Y_2 - Y_1 \cdot \alpha_f \cdot (1 + \frac{R_s}{R_b}) + \frac{\alpha_f}{R_b} \cdot \frac{dV_s(t)}{dt} - \frac{\alpha_f}{C_{in} \cdot R_b} \cdot I_{C_{in}}$$

The main system differential equation can be present as follow:

$$V_T \cdot \{\frac{[(\alpha_r - 1) \cdot C_1 \cdot L_1 \cdot \frac{dY_4}{dt} + g_3] \cdot [g_1 + (\alpha_f \cdot \alpha_r - 1) \cdot I_{sc}] - [(1 - \alpha_f) \cdot C_1 \cdot L_1 \cdot \frac{dY_4}{dt} + g_4] \cdot [g_2 + (\alpha_r \cdot \alpha_f - 1) \cdot I_{se}]}{[g_1 + (\alpha_f \cdot \alpha_r - 1) \cdot I_{sc}] \cdot [g_2 + (\alpha_r \cdot \alpha_f - 1) \cdot I_{se}]}\} = Y_2 \cdot R_{load} + \frac{1}{C_{out}} \cdot I_{C_{out}}$$

$$[(\alpha_r - 1) \cdot C_1 \cdot L_1 \cdot \frac{dY_4}{dt} + g_3] \cdot [g_1 + (\alpha_f \cdot \alpha_r - 1) \cdot I_{sc}] - [(1 - \alpha_f) \cdot C_1 \cdot L_1 \cdot \frac{dY_4}{dt} + g_4] \cdot [g_2 + (\alpha_r \cdot \alpha_f - 1) \cdot I_{se}] = \frac{1}{V_T} \cdot [Y_2 \cdot R_{load} + \frac{1}{C_{out}} \cdot I_{C_{out}}] \cdot [g_1 + (\alpha_f \cdot \alpha_r - 1) \cdot I_{sc}] \cdot [g_2 + (\alpha_r \cdot \alpha_f - 1) \cdot I_{se}]$$

$$\{(\alpha_r - 1) \cdot [g_1 + (\alpha_f \cdot \alpha_r - 1) \cdot I_{sc}] - (1 - \alpha_f) \cdot [g_2 + (\alpha_r \cdot \alpha_f - 1) \cdot I_{se}]\} \cdot C_1 \cdot L_1 \cdot \frac{dY_4}{dt} = \frac{1}{V_T} \cdot [Y_2 \cdot R_{load} + \frac{1}{C_{out}} \cdot I_{C_{out}}] \cdot [g_1 + (\alpha_f \cdot \alpha_r - 1) \cdot I_{sc}] \cdot [g_2 + (\alpha_r \cdot \alpha_f - 1) \cdot I_{se}] - g_3 \cdot [g_1 + (\alpha_f \cdot \alpha_r - 1) \cdot I_{sc}] - g_4 \cdot [g_2 + (\alpha_r \cdot \alpha_f - 1) \cdot I_{se}]$$

$$\frac{dY_4}{dt} = \frac{\frac{1}{V_T} \cdot [Y_2 \cdot R_{load} + \frac{1}{C_{out}} \cdot I_{C_{out}}] \cdot [g_1 + (\alpha_f \cdot \alpha_r - 1) \cdot I_{sc}] \cdot [g_2 + (\alpha_r \cdot \alpha_f - 1) \cdot I_{se}] - g_3 \cdot [g_1 + (\alpha_f \cdot \alpha_r - 1) \cdot I_{sc}] - g_4 \cdot [g_2 + (\alpha_r \cdot \alpha_f - 1) \cdot I_{se}]}{\{(\alpha_r - 1) \cdot [g_1 + (\alpha_f \cdot \alpha_r - 1) \cdot I_{sc}] - (1 - \alpha_f) \cdot [g_2 + (\alpha_r \cdot \alpha_f - 1) \cdot I_{se}]\} \cdot C_1 \cdot L_1}$$

We can summary our system differential equations:

$$\frac{dI_{C_{in}}}{dt} = Y_1; \ \frac{dI_{C_{out}}}{dt} = Y_2; \ \frac{dI_{L_1}}{dt} = Y_3; \ \frac{dY_3}{dt} = Y_4$$

$$\frac{dY_4}{dt} = \frac{\frac{1}{V_T} \cdot [Y_2 \cdot R_{load} + \frac{1}{C_{out}} \cdot I_{C_{out}}] \cdot [g_1 + (\alpha_f \cdot \alpha_r - 1) \cdot I_{sc}] \cdot [g_2 + (\alpha_r \cdot \alpha_f - 1) \cdot I_{se}] - g_3 \cdot [g_1 + (\alpha_f \cdot \alpha_r - 1) \cdot I_{sc}] - g_4 \cdot [g_2 + (\alpha_r \cdot \alpha_f - 1) \cdot I_{se}]}{\{(\alpha_r - 1) \cdot [g_1 + (\alpha_f \cdot \alpha_r - 1) \cdot I_{sc}] - (1 - \alpha_f) \cdot [g_2 + (\alpha_r \cdot \alpha_f - 1) \cdot I_{se}]\} \cdot C_1 \cdot L_1}$$

The first stage is to find our system fixed points: It is done by letting

$$\frac{dI_{C_{in}}}{dt} = 0 \Rightarrow Y_1^* = 0; \ \frac{dI_{C_{out}}}{dt} = 0 \Rightarrow Y_2^* = 0; \ \frac{dI_{L_1}}{dt} = 0 \Rightarrow Y_3^* = 0; \ \frac{dY_3}{dt} = 0 \Rightarrow Y_4^* = 0; \ \frac{dY_4}{dt} = 0$$

$$g_1(I_{L_1}^*, Y_4^*, I_{C_{out}}^*, I_{C_{in}}^*, I_{R_b}^*) = (1 - \alpha_f) \cdot I_{L_1}^* + (1 - \alpha_f) \cdot C_1 \cdot L_1 \cdot Y_4^* + (\alpha_f - 1) \cdot I_{C_{out}}^* - \alpha_f \cdot I_{C_{in}}^* + \alpha_f \cdot I_{R_b}^*$$

$$g_2(I_{L_1}^*, Y_4^*, I_{C_{out}}^*, I_{C_{in}}^*, I_{R_b}^*) = (\alpha_r - 1) \cdot I_{L_1}^* + (\alpha_r - 1) \cdot C_1 \cdot L_1 \cdot Y_4^* + (1 - \alpha_r) \cdot I_{C_{out}}^* - I_{C_{in}}^* + I_{R_b}^*$$

$$g_3^* = g_3(Y_3^*, Y_2^*, Y_1^*, I_{C_{in}}^*, \frac{dV_s(t)}{dt}) = \frac{1}{R_b} \cdot \frac{dV_s(t)}{dt} - \frac{1}{C_{in} \cdot R_b} \cdot I_{C_{in}}^*$$

$$g_4^* = g_4(Y_3^*, Y_2^*, Y_1^*, I_{C_{in}}^*, \frac{dV_s(t)}{dt}) = \frac{\alpha_f}{R_b} \cdot \frac{dV_s(t)}{dt} - \frac{\alpha_f}{C_{in} \cdot R_b} \cdot I_{C_{in}}$$

$$\frac{dY_4}{dt} = 0 \Rightarrow \{(\alpha_r - 1) \cdot [g_1^* + (\alpha_f \cdot \alpha_r - 1) \cdot I_{sc}] - (1 - \alpha_f) \cdot [g_2^* + (\alpha_r \cdot \alpha_f - 1) \cdot I_{se}]\} \cdot C_1 \cdot L_1 \neq 0$$

$$C_1 \cdot L_1 \neq 0 \Rightarrow (\alpha_r - 1) \cdot [g_1^* + (\alpha_f \cdot \alpha_r - 1) \cdot I_{sc}] - (1 - \alpha_f) \cdot [g_2^* + (\alpha_r \cdot \alpha_f - 1) \cdot I_{se}] \neq 0$$

$$\frac{1}{V_T} \cdot \frac{1}{C_{out}} \cdot I_{C_{out}}^* \cdot [g_1^* + (\alpha_f \cdot \alpha_r - 1) \cdot I_{sc}] \cdot [g_2^* + (\alpha_r \cdot \alpha_f - 1) \cdot I_{se}] - g_3^* \cdot [g_1^* + (\alpha_f \cdot \alpha_r - 1) \cdot I_{sc}] - g_4^* \cdot [g_2^* + (\alpha_r \cdot \alpha_f - 1) \cdot I_{se}] = 0$$

Assumption $\frac{dV_s(t)}{dt} \to \varepsilon; g_3^* = -\frac{1}{C_{in} \cdot R_b} \cdot I_{C_{in}}^*; g_4^* = -\frac{\alpha_f}{C_{in} \cdot R_b} \cdot I_{C_{in}}$

Stability analysis: The standard local stability analysis about any one of the equilibrium points of the class C power amplifier equivalent circuit consists in adding to coordinate $[I_{L_1}, I_{R_b}, I_{R_{load}}, I_{C_{out}}, I_{C_{in}}, Y_1, Y_2, Y_3, Y_4]$ arbitrarily small increments of exponentially form $[i_{L_1}, i_{R_b}, i_{R_{load}}, i_{C_{out}}, i_{C_{in}}, y_1, y_2, y_3, y_4] \cdot e^{\lambda \cdot t}$ and retaining the first order terms in $I_{L_1}, I_{R_b}, I_{R_{load}}, I_{C_{out}}, I_{C_{in}}, Y_1, Y_2, Y_3, Y_4$. The system of homogenous equations leads to a polynomial characteristic equation in the eigenvalues [4]. The polynomial characteristic equations accept by set of the below circuit variables, circuit variables derivative and circuit variables second order derivative with respect to time into class C power amplifier [2–4]. Our class C power amplifier equivalent circuit fixed values with arbitrarily small increments of exponential form $[i_{L_1}, i_{R_b}, i_{R_{load}}, i_{C_{out}}, i_{C_{in}}, y_1, y_2, y_3, y_4] \cdot e^{\lambda \cdot t}$ are: j = 0 (first fixed point), j = 1 (second fixed point), j = 2 (third fixed point), etc.,

$$Y_1(t) = Y_1^{(j)} + y_1 \cdot e^{\lambda \cdot t}; Y_2(t) = Y_2^{(j)} + y_2 \cdot e^{\lambda \cdot t}; Y_3(t) = Y_3^{(j)} + y_3 \cdot e^{\lambda \cdot t};$$
$$I_{L_1}(t) = I_{L_1}^{(j)} + i_{L_1} \cdot e^{\lambda \cdot t}$$

$$Y_4(t) = Y_4^{(j)} + y_4 \cdot e^{\lambda \cdot t}; I_{C_{in}}(t) = I_{C_{in}}^{(j)} + i_{C_{in}} \cdot e^{\lambda \cdot t}; I_{C_{out}}(t) = I_{C_{out}}^{(j)} + i_{C_{out}} \cdot e^{\lambda \cdot t};$$
$$I_{R_b}(t) = I_{R_b}^{(j)} + i_{R_b} \cdot e^{\lambda \cdot t}$$

$$\frac{dI_{C_{in}}}{dt} = i_{C_{in}} \cdot \lambda \cdot e^{\lambda \cdot t}; \frac{dI_{C_{out}}}{dt} = i_{C_{out}} \cdot \lambda \cdot e^{\lambda \cdot t}; \frac{dI_{L_1}}{dt} = i_{L_1} \cdot \lambda \cdot e^{\lambda \cdot t}; \frac{dY_3}{dt} = y_3 \cdot \lambda \cdot e^{\lambda \cdot t};$$
$$\frac{dY_4}{dt} = y_4 \cdot \lambda \cdot e^{\lambda \cdot t}$$

$$I_{R_{load}}(t) = I_{R_{load}}^{(j)} + i_{R_{load}} \cdot e^{\lambda \cdot t}; \frac{dI_{R_{load}}(t)}{dt} = i_{R_{load}} \cdot \lambda \cdot e^{\lambda \cdot t}; \frac{dI_{R_b}(t)}{dt} = i_{R_b} \cdot \lambda \cdot e^{\lambda \cdot t};$$
$$\frac{dV_s(t)}{dt} \to \varepsilon$$

$$i_{R_b} \cdot \lambda \cdot e^{\lambda \cdot t} = -[Y_1^{(j)} + y_1 \cdot e^{\lambda \cdot t}] \cdot \frac{R_s}{R_b} - \frac{1}{C_{in} \cdot R_b} \cdot [I_{C_{in}}^{(j)} + i_{C_{in}} \cdot e^{\lambda \cdot t}]$$

$$i_{R_b} \cdot \lambda \cdot e^{\lambda \cdot t} = -Y_1^{(j)} \cdot \frac{R_s}{R_b} - \frac{1}{C_{in} \cdot R_b} \cdot I_{C_{in}}^{(j)} - y_1 \cdot \frac{R_s}{R_b} \cdot e^{\lambda \cdot t} - i_{C_{in}} \cdot \frac{1}{C_{in} \cdot R_b} \cdot e^{\lambda \cdot t}$$

At fixed point $-Y_1^{(j)} \cdot \frac{R_s}{R_b} - \frac{1}{C_{in}} \cdot R_b \cdot I_{C_{in}}^{(j)} = 0$. $-i_{R_b} \cdot \lambda - y_1 \cdot \frac{R_s}{R_b} - i_{C_{in}} \cdot \frac{1}{C_{in}} \cdot R_b = 0$

$$\frac{dI_{C_{in}}}{dt} = Y_1; i_{C_{in}} \cdot \lambda \cdot e^{\lambda \cdot t} = Y_1^{(j)} + y_1 \cdot e^{\lambda \cdot t}; Y_1^{(j)} = 0; -i_{C_{in}} \cdot \lambda + y_1 = 0$$

$$\frac{dI_{C_{out}}}{dt} = Y_2; i_{C_{out}} \cdot \lambda \cdot e^{\lambda \cdot t} = Y_2^{(j)} + y_2 \cdot e^{\lambda \cdot t}; Y_2^{(j)} = 0; -i_{C_{out}} \cdot \lambda + y_2 = 0$$

$$\frac{dI_{L_1}}{dt} = Y_3; i_{L_1} \cdot \lambda \cdot e^{\lambda \cdot t} = Y_3(t) = Y_3^{(j)} + y_3 \cdot e^{\lambda \cdot t}; Y_3^{(j)} = 0; -i_{L_1} \cdot \lambda + y_3 = 0$$

$$\frac{dY_3}{dt} = Y_4; y_3 \cdot \lambda \cdot e^{\lambda \cdot t} = Y_4^{(j)} + y_4 \cdot e^{\lambda \cdot t}; Y_4^{(j)} = 0; -y_3 \cdot \lambda + y_4 = 0$$

$$\begin{aligned} g_1(I_{L_1}(t), Y_4(t), I_{C_{out}}(t), I_{C_{in}}(t), I_{R_b}(t)) &= (1-\alpha_f) \cdot (I_{L_1}^{(j)} + i_{L_1} \cdot e^{\lambda \cdot t}) \\ &+ (1-\alpha_f) \cdot C_1 \cdot L_1 \cdot (Y_4^{(j)} + y_4 \cdot e^{\lambda \cdot t}) + (\alpha_f - 1) \cdot (I_{C_{out}}^{(j)} + i_{C_{out}} \cdot e^{\lambda \cdot t}) \\ &- \alpha_f \cdot (I_{C_{in}}^{(j)} + i_{C_{in}} \cdot e^{\lambda \cdot t}) + \alpha_f \cdot (I_{R_b}^{(j)} + i_{R_b} \cdot e^{\lambda \cdot t}) \end{aligned}$$

$$\begin{aligned} g_1(I_{L_1}(t), Y_4(t), I_{C_{out}}(t), I_{C_{in}}(t), I_{R_b}(t)) &= (1-\alpha_f) \cdot I_{L_1}^{(j)} + (1-\alpha_f) \cdot C_1 \cdot L_1 \cdot Y_4^{(j)} \\ &+ (\alpha_f - 1) \cdot I_{C_{out}}^{(j)} - \alpha_f \cdot I_{C_{in}}^{(j)} + \alpha_f \cdot I_{R_b}^{(j)} + i_{L_1} \cdot (1-\alpha_f) \cdot e^{\lambda \cdot t} + y_4 \cdot (1-\alpha_f) \cdot C_1 \cdot L_1 \cdot e^{\lambda \cdot t} \\ &+ i_{C_{out}} \cdot (\alpha_f - 1) \cdot e^{\lambda \cdot t} - i_{C_{in}} \cdot \alpha_f \cdot e^{\lambda \cdot t} + i_{R_b} \cdot \alpha_f \cdot e^{\lambda \cdot t} \end{aligned}$$

At fixed point:

$$\begin{aligned} g_1^* = g_1(I_{L_1}^*, Y_4^*, I_{C_{out}}^*, I_{C_{in}}^*, I_{R_b}^*) &= (1-\alpha_f) \cdot I_{L_1}^{(j)} + (1-\alpha_f) \cdot C_1 \cdot L_1 \cdot Y_4^{(j)} \\ &+ (\alpha_f - 1) \cdot I_{C_{out}}^{(j)} - \alpha_f \cdot I_{C_{in}}^{(j)} + \alpha_f \cdot I_{R_b}^{(j)} \end{aligned}$$

$$\begin{aligned} g_1(t) = g_1(I_{L_1}(t), Y_4(t), I_{C_{out}}(t), I_{C_{in}}(t), I_{R_b}(t)) &= g_1(I_{L_1}^*, Y_4^*, I_{C_{out}}^*, I_{C_{in}}^*, I_{R_b}^*) + i_{L_1} \cdot (1-\alpha_f) \cdot e^{\lambda \cdot t} \\ &+ y_4 \cdot (1-\alpha_f) \cdot C_1 \cdot L_1 \cdot e^{\lambda \cdot t} + i_{C_{out}} \cdot (\alpha_f - 1) \cdot e^{\lambda \cdot t} - i_{C_{in}} \cdot \alpha_f \cdot e^{\lambda \cdot t} + i_{R_b} \cdot \alpha_f \cdot e^{\lambda \cdot t} \end{aligned}$$

$$\begin{aligned} g_1(t) = g_1^* + [&i_{L_1} \cdot (1-\alpha_f) + y_4 \cdot (1-\alpha_f) \cdot C_1 \cdot L_1 + i_{C_{out}} \cdot (\alpha_f - 1) - i_{C_{in}} \cdot \alpha_f \\ &+ i_{R_b} \cdot \alpha_f] \cdot e^{\lambda \cdot t} \end{aligned}$$

$$\begin{aligned} g_2(I_{L_1}(t), Y_4(t), I_{C_{out}}(t), I_{C_{in}}(t), I_{R_b}(t)) &= (\alpha_r - 1) \cdot I_{L_1}^{(j)} + i_{L_1} \cdot (\alpha_r - 1) \cdot e^{\lambda \cdot t} \\ &+ (\alpha_r - 1) \cdot C_1 \cdot L_1 \cdot Y_4^{(j)} + y_4 \cdot (\alpha_r - 1) \cdot C_1 \cdot L_1 \cdot e^{\lambda \cdot t} + (1-\alpha_r) \cdot I_{C_{out}}^{(j)} \\ &+ i_{C_{out}} \cdot (1-\alpha_r) \cdot e^{\lambda \cdot t} - I_{C_{in}}^{(j)} - i_{C_{in}} \cdot e^{\lambda \cdot t} + I_{R_b}^{(j)} + i_{R_b} \cdot e^{\lambda \cdot t} \end{aligned}$$

$$\begin{aligned} g_2(I_{L_1}(t), Y_4(t), I_{C_{out}}(t), I_{C_{in}}(t), I_{R_b}(t)) &= (\alpha_r - 1) \cdot I_{L_1}^{(j)} + (\alpha_r - 1) \cdot C_1 \cdot L_1 \cdot Y_4^{(j)} \\ &+ (1-\alpha_r) \cdot I_{C_{out}}^{(j)} - I_{C_{in}}^{(j)} + I_{R_b}^{(j)} + i_{L_1} \cdot (\alpha_r - 1) \cdot e^{\lambda \cdot t} + y_4 \cdot (\alpha_r - 1) \cdot C_1 \cdot L_1 \cdot e^{\lambda \cdot t} \\ &+ i_{C_{out}} \cdot (1-\alpha_r) \cdot e^{\lambda \cdot t} - i_{C_{in}} \cdot e^{\lambda \cdot t} + i_{R_b} \cdot e^{\lambda \cdot t} \end{aligned}$$

At fixed point: $(\alpha_r - 1) \cdot I_{L_1}^{(j)} + (\alpha_r - 1) \cdot C_1 \cdot L_1 \cdot Y_4^{(j)} + (1-\alpha_r) \cdot I_{C_{out}}^{(j)} - I_{C_{in}}^{(j)} + I_{R_b}^{(j)} = 0$

$$g_2^* = g_2(I_{L_1}^*, Y_4^*, I_{C_{out}}^*, I_{C_{in}}^*, I_{R_b}^*) = (\alpha_r - 1) \cdot I_{L_1}^{(j)} + (\alpha_r - 1) \cdot C_1 \cdot L_1 \cdot Y_4^{(j)} + (1 - \alpha_r) \cdot I_{C_{out}}^{(j)} - I_{C_{in}}^{(j)} + I_{R_b}^{(j)} = 0$$

$$g_2(t) = g_2^* + [i_{L_1} \cdot (\alpha_r - 1) + y_4 \cdot (\alpha_r - 1) \cdot C_1 \cdot L_1 + i_{C_{out}} \cdot (1 - \alpha_r) - i_{C_{in}} + i_{R_b}] \cdot e^{\lambda \cdot t}$$

$$\begin{aligned} g_3(Y_3(t), Y_2(t), Y_1(t), I_{C_{in}}(t), \frac{dV_s(t)}{dt}) &= (\alpha_r - 1) \cdot [Y_3^{(j)} + y_3 \cdot e^{\lambda \cdot t}] \\ &+ (1 - \alpha_r) \cdot [Y_2^{(j)} + y_2 \cdot e^{\lambda \cdot t}] \\ &- [Y_1^{(j)} + y_1 \cdot e^{\lambda \cdot t}] \cdot (1 + \frac{R_s}{R_b}) + \frac{1}{R_b} \cdot \frac{dV_s(t)}{dt} \\ &- \frac{1}{C_{in} \cdot R_b} \cdot [I_{C_{in}}^{(j)} + i_{C_{in}} \cdot e^{\lambda \cdot t}] \end{aligned}$$

$$g_3(Y_3(t), Y_2(t), Y_1(t), I_{C_{in}}(t), \frac{dV_s(t)}{dt}) = (\alpha_r - 1) \cdot Y_3^{(j)} + (1 - \alpha_r) \cdot Y_2^{(j)} - Y_1^{(j)} \cdot (1 + \frac{R_s}{R_b}) - \frac{1}{C_{in} \cdot R_b} \cdot I_{C_{in}}^{(j)} + y_3 \cdot (\alpha_r - 1) \cdot e^{\lambda \cdot t} + y_2 \cdot (1 - \alpha_r) \cdot e^{\lambda \cdot t} - y_1 \cdot (1 + \frac{R_s}{R_b}) \cdot e^{\lambda \cdot t} - i_{C_{in}} \cdot \frac{1}{C_{in} \cdot R_b} \cdot e^{\lambda \cdot t} + \frac{1}{R_b} \cdot \frac{dV_s(t)}{dt}$$

$$g_3^* = g_3(Y_3^*, Y_2^*, Y_1^*, I_{C_{in}}^*, \frac{dV_s(t)}{dt}) = (\alpha_r - 1) \cdot Y_3^{(j)} + (1 - \alpha_r) \cdot Y_2^{(j)} - Y_1^{(j)} \cdot (1 + \frac{R_s}{R_b}) - \frac{1}{C_{in} \cdot R_b} \cdot I_{C_{in}}^{(j)}$$

$$g_3(Y_3(t), Y_2(t), Y_1(t), I_{C_{in}}(t), \frac{dV_s(t)}{dt}) = g_3^* + y_3 \cdot (\alpha_r - 1) \cdot e^{\lambda \cdot t} + y_2 \cdot (1 - \alpha_r) \cdot e^{\lambda \cdot t} - y_1 \cdot (1 + \frac{R_s}{R_b}) \cdot e^{\lambda \cdot t} - i_{C_{in}} \cdot \frac{1}{C_{in} \cdot R_b} \cdot e^{\lambda \cdot t} + \frac{1}{R_b} \cdot \frac{dV_s(t)}{dt}; \frac{dV_s(t)}{dt} \to \varepsilon$$

$$g_3(\frac{dV_s(t)}{dt} \to \varepsilon) = g_3^* + [y_3 \cdot (\alpha_r - 1) + y_2 \cdot (1 - \alpha_r) - y_1 \cdot (1 + \frac{R_s}{R_b}) - i_{C_{in}} \cdot \frac{1}{C_{in} \cdot R_b}] \cdot e^{\lambda \cdot t}$$

$$g_4(Y_3(t), Y_2(t), Y_1(t), I_{C_{in}}, \frac{dV_s(t)}{dt}) = (1 - \alpha_f) \cdot [Y_3^{(j)} + y_3 \cdot e^{\lambda \cdot t}] + (\alpha_f - 1) \cdot [Y_2^{(j)} + y_2 \cdot e^{\lambda \cdot t}] - [Y_1^{(j)} + y_1 \cdot e^{\lambda \cdot t}] \cdot \alpha_f \cdot (1 + \frac{R_s}{R_b}) + \frac{\alpha_f}{R_b} \cdot \frac{dV_s(t)}{dt} - \frac{\alpha_f}{C_{in} \cdot R_b} \cdot [I_{C_{in}}^{(j)} + i_{C_{in}} \cdot e^{\lambda \cdot t}]$$

$$g_4(Y_3(t), Y_2(t), Y_1(t), I_{C_{in}}, \frac{dV_s(t)}{dt}) = (1-\alpha_f)\cdot Y_3^{(j)} + y_3\cdot(1-\alpha_f)\cdot e^{\lambda\cdot t} + (\alpha_f - 1)\cdot Y_2^{(j)} + y_2\cdot(\alpha_f - 1)\cdot e^{\lambda\cdot t} - Y_1^{(j)}\cdot\alpha_f\cdot(1+\frac{R_s}{R_b}) - y_1\cdot\alpha_f\cdot(1+\frac{R_s}{R_b})\cdot e^{\lambda\cdot t} + \frac{\alpha_f}{R_b}\cdot\frac{dV_s(t)}{dt} - \frac{\alpha_f}{C_{in}\cdot R_b}\cdot I_{C_{in}}^{(j)} - i_{C_{in}}\cdot\frac{\alpha_f}{C_{in}\cdot R_b}\cdot e^{\lambda\cdot t}$$

$$g_4(Y_3(t), Y_2(t), Y_1(t), I_{C_{in}}, \frac{dV_s(t)}{dt}) = (1-\alpha_f)\cdot Y_3^{(j)} + (\alpha_f - 1)\cdot Y_2^{(j)} - Y_1^{(j)}\cdot\alpha_f\cdot(1+\frac{R_s}{R_b}) - \frac{\alpha_f}{C_{in}\cdot R_b}\cdot I_{C_{in}}^{(j)} + y_3\cdot(1-\alpha_f)\cdot e^{\lambda\cdot t} + y_2\cdot(\alpha_f - 1)\cdot e^{\lambda\cdot t} - y_1\cdot\alpha_f\cdot(1+\frac{R_s}{R_b})\cdot e^{\lambda\cdot t} - i_{C_{in}}\cdot\frac{\alpha_f}{C_{in}\cdot R_b}\cdot e^{\lambda\cdot t} + \frac{\alpha_f}{R_b}\cdot\frac{dV_s(t)}{dt}$$

$$g_4^* = g_4(Y_3^*, Y_2^*, Y_1^*, I_{C_{in}}^*, \frac{dV_s(t)}{dt}) = (1-\alpha_f)\cdot Y_3^{(j)} + (\alpha_f - 1)\cdot Y_2^{(j)} - Y_1^{(j)}\cdot\alpha_f\cdot(1+\frac{R_s}{R_b}) - \frac{\alpha_f}{C_{in}\cdot R_b}\cdot I_{C_{in}}^{(j)}$$

$$g_4(Y_3(t), Y_2(t), Y_1(t), I_{C_{in}}(t), \frac{dV_s(t)}{dt}) = g_4^* + y_3\cdot(1-\alpha_f)\cdot e^{\lambda\cdot t} + y_2\cdot(\alpha_f - 1)\cdot e^{\lambda\cdot t} - y_1\cdot\alpha_f\cdot(1+\frac{R_s}{R_b})\cdot e^{\lambda\cdot t} - i_{C_{in}}\cdot\frac{\alpha_f}{C_{in}\cdot R_b}\cdot e^{\lambda\cdot t} + \frac{\alpha_f}{R_b}\cdot\frac{dV_s(t)}{dt};\ \frac{dV_s(t)}{dt} \to \varepsilon$$

$$g_4(\frac{dV_s(t)}{dt} \to \varepsilon) = g_4^* + [y_3\cdot(1-\alpha_f) + y_2\cdot(\alpha_f - 1) - y_1\cdot\alpha_f\cdot(1+\frac{R_s}{R_b}) - i_{C_{in}}\cdot\frac{\alpha_f}{C_{in}\cdot R_b}]\cdot e^{\lambda\cdot t}$$

Back to our last differential equation:

$$\frac{dY_4}{dt} = \frac{\frac{1}{V_T}\cdot[Y_2\cdot R_{load} + \frac{1}{C_{out}}\cdot I_{C_{out}}]\cdot[g_1 + (\alpha_f\cdot\alpha_r - 1)\cdot I_{sc}]\cdot[g_2 + (\alpha_r\cdot\alpha_f - 1)\cdot I_{se}] - g_3\cdot[g_1 + (\alpha_f\cdot\alpha_r - 1)\cdot I_{sc}] - g_4\cdot[g_2 + (\alpha_r\cdot\alpha_f - 1)\cdot I_{se}]}{\{(\alpha_r - 1)\cdot[g_1 + (\alpha_f\cdot\alpha_r - 1)\cdot I_{sc}] - (1-\alpha_f)\cdot[g_2 + (\alpha_r\cdot\alpha_f - 1)\cdot I_{se}]\}\cdot C_1\cdot L_1}$$

At fixed point: $\frac{dY_4}{dt} = 0$

$$\frac{\frac{1}{V_T}\cdot[Y_2^{(j)}\cdot R_{load}+\frac{1}{C_{out}}\cdot I_{C_{out}}^{(j)}]\cdot[g_1^*+(\alpha_f\cdot\alpha_r-1)\cdot I_{sc}]\cdot[g_2^*+(\alpha_r\cdot\alpha_f-1)\cdot I_{se}] -g_3^*\cdot[g_1+(\alpha_f\cdot\alpha_r-1)\cdot I_{sc}]-g_4^*\cdot[g_2^*+(\alpha_r\cdot\alpha_f-1)\cdot I_{se}]}{\{(\alpha_r-1)\cdot[g_1^*+(\alpha_f\cdot\alpha_r-1)\cdot I_{sc}]-(1-\alpha_f)\cdot[g_2^*+(\alpha_r\cdot\alpha_f-1)\cdot I_{se}]\}\cdot C_1\cdot L_1} = 0$$

We define function: $\frac{dY_4}{dt}=\upsilon(Y_2, I_{C_{out}}, \ldots)$

$$\upsilon(Y_2, I_{C_{out}}, \ldots)=\frac{\frac{1}{V_T}\cdot[Y_2\cdot R_{load}+\frac{1}{C_{out}}\cdot I_{C_{out}}]\cdot[g_1+(\alpha_f\cdot\alpha_r-1)\cdot I_{sc}]\cdot[g_2+(\alpha_r\cdot\alpha_f-1)\cdot I_{se}] -g_3\cdot[g_1+(\alpha_f\cdot\alpha_r-1)\cdot I_{sc}]-g_4\cdot[g_2+(\alpha_r\cdot\alpha_f-1)\cdot I_{se}]}{\{(\alpha_r-1)\cdot[g_1+(\alpha_f\cdot\alpha_r-1)\cdot I_{sc}]-(1-\alpha_f)\cdot[g_2+(\alpha_r\cdot\alpha_f-1)\cdot I_{se}]\}\cdot C_1\cdot L_1}$$

First we extract the above function $\upsilon(Y_2, I_{C_{out}}, \ldots)$ *denominator*.

$$\begin{aligned}&\{(\alpha_r-1)\cdot[g_1+(\alpha_f\cdot\alpha_r-1)\cdot I_{sc}]-(1-\alpha_f)\cdot[g_2+(\alpha_r\cdot\alpha_f-1)\cdot I_{se}]\}\cdot C_1\cdot L_1\\&=\{(\alpha_r-1)\cdot g_1-(1-\alpha_f)\cdot g_2+(\alpha_r-1)\cdot(\alpha_f\cdot\alpha_r-1)\cdot I_{sc}\\&\quad-(1-\alpha_f)\cdot(\alpha_r\cdot\alpha_f-1)\cdot I_{se}\}\cdot C_1\cdot L_1\end{aligned}$$

We define global parameter:

$$\Gamma_1=\Gamma_1(\alpha_r,\alpha_f,I_{sc},I_{se})=(\alpha_r-1)\cdot(\alpha_f\cdot\alpha_r-1)\cdot I_{sc}-(1-\alpha_f)\cdot(\alpha_r\cdot\alpha_f-1)\cdot I_{se}$$

$$\begin{aligned}&\{(\alpha_r-1)\cdot[g_1+(\alpha_f\cdot\alpha_r-1)\cdot I_{sc}]-(1-\alpha_f)\cdot[g_2+(\alpha_r\cdot\alpha_f-1)\cdot I_{se}]\}\cdot C_1\cdot L_1\\&=\{(\alpha_r-1)\cdot g_1-(1-\alpha_f)\cdot g_2+\Gamma_1(\alpha_r,\alpha_f,I_{sc},I_{se})\}\cdot C_1\cdot L_1\end{aligned}$$

$$\begin{aligned}&\{(\alpha_r-1)\cdot[g_1+(\alpha_f\cdot\alpha_r-1)\cdot I_{sc}]-(1-\alpha_f)\cdot[g_2+(\alpha_r\cdot\alpha_f-1)\cdot I_{se}]\}\cdot C_1\cdot L_1\\&=\{(\alpha_r-1)\cdot(g_1^*+[i_{L_1}\cdot(1-\alpha_f)+y_4\cdot(1-\alpha_f)\cdot C_1\cdot L_1+i_{C_{out}}\cdot(\alpha_f-1)\\&\quad-i_{C_{in}}\cdot\alpha_f+i_{R_b}\cdot\alpha_f]\cdot e^{\lambda\cdot t})-(1-\alpha_f)\cdot(g_2^*+[i_{L_1}\cdot(\alpha_r-1)\\&\quad+y_4\cdot(\alpha_r-1)\cdot C_1\cdot L_1+i_{C_{out}}\cdot(1-\alpha_r)-i_{C_{in}}+i_{R_b}]\cdot e^{\lambda\cdot t})\\&\quad+\Gamma_1(\alpha_r,\alpha_f,I_{sc},I_{se})\}\cdot C_1\cdot L_1\end{aligned}$$

$$\begin{aligned}&\{(\alpha_r-1)\cdot[g_1+(\alpha_f\cdot\alpha_r-1)\cdot I_{sc}]-(1-\alpha_f)\cdot[g_2+(\alpha_r\cdot\alpha_f-1)\cdot I_{se}]\}\cdot C_1\cdot L_1\\&=\{(\alpha_r-1)\cdot g_1^*+[i_{L_1}\cdot(\alpha_r-1)\cdot(1-\alpha_f)+y_4\cdot(\alpha_r-1)\cdot(1-\alpha_f)\cdot C_1\cdot L_1\\&\quad+i_{C_{out}}\cdot(\alpha_r-1)\cdot(\alpha_f-1)-i_{C_{in}}\cdot(\alpha_r-1)\cdot\alpha_f+i_{R_b}\cdot(\alpha_r-1)\cdot\alpha_f]\cdot e^{\lambda\cdot t}\\&\quad-(1-\alpha_f)\cdot g_2^*-[i_{L_1}\cdot(1-\alpha_f)\cdot(\alpha_r-1)+y_4\cdot(1-\alpha_f)\cdot(\alpha_r-1)\\&\quad\cdot C_1\cdot L_1+i_{C_{out}}\cdot(1-\alpha_f)\cdot(1-\alpha_r)-(1-\alpha_f)\cdot i_{C_{in}}+(1-\alpha_f)\cdot i_{R_b}]\cdot e^{\lambda\cdot t}\\&\quad+\Gamma_1(\alpha_r,\alpha_f,I_{sc},I_{se})\}\cdot C_1\cdot L_1\end{aligned}$$

$$\begin{aligned}&\{(\alpha_r-1)\cdot[g_1+(\alpha_f\cdot\alpha_r-1)\cdot I_{sc}]-(1-\alpha_f)\cdot[g_2+(\alpha_r\cdot\alpha_f-1)\cdot I_{se}]\}\cdot C_1\cdot L_1\\&\quad=\{(\alpha_r-1)\cdot g_1^*-(1-\alpha_f)\cdot g_2^*+\Gamma_1(\alpha_r,\alpha_f,I_{sc},I_{se})+[i_{L_1}\cdot(\alpha_r-1)\cdot(1-\alpha_f)\\&\qquad+y_4\cdot(\alpha_r-1)\cdot(1-\alpha_f)\cdot C_1\cdot L_1+i_{C_{out}}\cdot(\alpha_r-1)\cdot(\alpha_f-1)\\&\qquad-i_{C_{in}}\cdot(\alpha_r-1)\cdot\alpha_f+i_{R_b}\cdot(\alpha_r-1)\cdot\alpha_f]\cdot e^{\lambda\cdot t}\\&\qquad-[i_{L_1}\cdot(1-\alpha_f)\cdot(\alpha_r-1)+y_4\cdot(1-\alpha_f)\cdot(\alpha_r-1)\cdot C_1\cdot L_1\\&\qquad+i_{C_{out}}\cdot(1-\alpha_f)\cdot(1-\alpha_r)\\&\qquad-(1-\alpha_f)\cdot i_{C_{in}}+(1-\alpha_f)\cdot i_{R_b}]\cdot e^{\lambda\cdot t}\}\cdot C_1\cdot L_1\end{aligned}$$

$$\begin{aligned}&\{(\alpha_r-1)\cdot[g_1+(\alpha_f\cdot\alpha_r-1)\cdot I_{sc}]-(1-\alpha_f)\cdot[g_2+(\alpha_r\cdot\alpha_f-1)\cdot I_{se}]\}\cdot C_1\cdot L_1\\&\quad=\{(\alpha_r-1)\cdot g_1^*-(1-\alpha_f)\cdot g_2^*+\Gamma_1(\alpha_r,\alpha_f,I_{sc},I_{se})\\&\qquad+[(1-\alpha_f)-(\alpha_r-1)\cdot\alpha_f]\cdot i_{C_{in}}\cdot e^{\lambda\cdot t}+[(\alpha_r-1)\cdot\alpha_f-(1-\alpha_f)]\cdot i_{R_b}\cdot e^{\lambda\cdot t}\}\cdot C_1\cdot L_1\end{aligned}$$

We define the following new system parameters for simplicity.

$$\Gamma_2(g_1^*,g_2^*,\Gamma_1,\alpha_r,\alpha_f)=(\alpha_r-1)\cdot g_1^*-(1-\alpha_f)\cdot g_2^*+\Gamma_1(\alpha_r,\alpha_f,I_{sc},I_{se})$$
$$\Gamma_3(\alpha_f,\alpha_r)=(1-\alpha_f)-(\alpha_r-1)\cdot\alpha_f;-\Gamma_3(\alpha_f,\alpha_r)=(\alpha_r-1)\cdot\alpha_f-(1-\alpha_f)$$

$$\begin{aligned}&\{(\alpha_r-1)\cdot[g_1+(\alpha_f\cdot\alpha_r-1)\cdot I_{sc}]-(1-\alpha_f)\cdot[g_2+(\alpha_r\cdot\alpha_f-1)\cdot I_{se}]\}\cdot C_1\cdot L_1\\&\quad=\{\Gamma_2(g_1^*,g_2^*,\Gamma_1,\alpha_r,\alpha_f)+\Gamma_3(\alpha_f,\alpha_r)\cdot(i_{C_{in}}-i_{R_b})\cdot e^{\lambda\cdot t}\}\cdot C_1\cdot L_1\end{aligned}$$

Second we extract the above function $\upsilon(Y_2,I_{C_{out}},\ldots)$ *numerator.*

We define for simplicity three functions which there summation gives the function $\upsilon(Y_2,I_{C_{out}},\ldots)$ numerator.

$$\begin{aligned}\Phi_1(Y_2,g_1,g_2,\alpha_f,\alpha_r,\ldots)&=\frac{1}{V_T}\cdot[Y_2\cdot R_{load}+\frac{1}{C_{out}}\cdot I_{C_{out}}]\cdot[g_1+(\alpha_f\cdot\alpha_r-1)\cdot I_{sc}]\\&\quad\cdot[g_2+(\alpha_r\cdot\alpha_f-1)\cdot I_{se}]\\\Phi_2(g_1,g_3,\alpha_f,\alpha_r,I_{sc})&=-g_3\cdot[g_1+(\alpha_f\cdot\alpha_r-1)\cdot I_{sc}];\Phi_3(g_2,g_4,\alpha_f,\alpha_r,I_{se})\\&=-g_4\cdot[g_2+(\alpha_r\cdot\alpha_f-1)\cdot I_{se}]\end{aligned}$$

Function $\upsilon(Y_2,I_{C_{out}},\ldots)$ numerator is define as $\sum_{k=1}^{3}\Phi_k.\Phi_1(Y_2,g_1,g_2,\alpha_f,\alpha_r,\ldots)=\frac{1}{V_T}\cdot\sum_{j=1}^{8}\Phi_{1j}$.

$$\begin{aligned}\Phi_{1j=1}=g_1\cdot g_2\cdot Y_2\cdot R_{load}&=\{g_1^*+[i_{L_1}\cdot(1-\alpha_f)+y_4\cdot(1-\alpha_f)\cdot C_1\cdot L_1\\&\quad+i_{C_{out}}\cdot(\alpha_f-1)-i_{C_{in}}\cdot\alpha_f+i_{R_b}\cdot\alpha_f]\cdot e^{\lambda\cdot t}\}\\&\quad\cdot\{g_2^*+[i_{L_1}\cdot(\alpha_r-1)+y_4\cdot(\alpha_r-1)\cdot C_1\cdot L_1\\&\quad+i_{C_{out}}\cdot(1-\alpha_r)-i_{C_{in}}+i_{R_b}]\cdot e^{\lambda\cdot t}\}\cdot\{Y_2^{(j)}+y_2\cdot e^{\lambda\cdot t}\}\cdot R_{load}\end{aligned}$$

$$\Phi_{1j=1} = \{g_1^* \cdot g_2^* + g_1^* \cdot [i_{L_1} \cdot (\alpha_r - 1) + y_4 \cdot (\alpha_r - 1) \cdot C_1 \cdot L_1 + i_{C_{out}} \cdot (1 - \alpha_r) - i_{C_{in}} + i_{R_b}] \cdot e^{\lambda \cdot t}$$
$$+ g_2^* \cdot [i_{L_1} \cdot (1 - \alpha_f) + y_4 \cdot (1 - \alpha_f) \cdot C_1 \cdot L_1 + i_{C_{out}} \cdot (\alpha_f - 1) - i_{C_{in}} \cdot \alpha_f + i_{R_b} \cdot \alpha_f] \cdot e^{\lambda \cdot t}$$
$$+ [i_{L_1} \cdot (1 - \alpha_f) + y_4 \cdot (1 - \alpha_f) \cdot C_1 \cdot L_1 + i_{C_{out}} \cdot (\alpha_f - 1) - i_{C_{in}} \cdot \alpha_f + i_{R_b} \cdot \alpha_f] \cdot e^{\lambda \cdot t}$$
$$\cdot [i_{L_1} \cdot (\alpha_r - 1) + y_4 \cdot (\alpha_r - 1) \cdot C_1 \cdot L_1 + i_{C_{out}} \cdot (1 - \alpha_r) - i_{C_{in}} + i_{R_b}] \cdot e^{\lambda \cdot t}\} \cdot \{Y_2^{(j)} + y_2 \cdot e^{\lambda \cdot t}\} \cdot R_{load}$$

Since $i_{L_1} \cdot i_{L_1} \to \varepsilon; y_4 \cdot i_{L_1} \to \varepsilon; i_{C_{out}} \cdot i_{L_1} \to \varepsilon \ldots; \frac{dI_{C_{out}}}{dt} = 0 \Rightarrow Y_2^{(j)} = 0$ then

$$\Phi_{1j=1} = \{g_1^* \cdot g_2^* + g_1^* \cdot [i_{L_1} \cdot (\alpha_r - 1) + y_4 \cdot (\alpha_r - 1) \cdot C_1 \cdot L_1$$
$$+ i_{C_{out}} \cdot (1 - \alpha_r) - i_{C_{in}} + i_{R_b}] \cdot e^{\lambda \cdot t} + g_2^* \cdot [i_{L_1} \cdot (1 - \alpha_f)$$
$$+ y_4 \cdot (1 - \alpha_f) \cdot C_1 \cdot L_1 + i_{C_{out}} \cdot (\alpha_f - 1) - i_{C_{in}} \cdot \alpha_f + i_{R_b} \cdot \alpha_f] \cdot e^{\lambda \cdot t}\} \cdot y_2 \cdot e^{\lambda \cdot t} \cdot R_{load}$$

$$\Phi_{1j=1} = \{g_1^* \cdot g_2^* \cdot y_2 \cdot e^{\lambda \cdot t} + g_1^* \cdot [i_{L_1} \cdot (\alpha_r - 1) + y_4 \cdot (\alpha_r - 1) \cdot C_1 \cdot L_1$$
$$+ i_{C_{out}} \cdot (1 - \alpha_r) - i_{C_{in}} + i_{R_b}] \cdot y_2 \cdot e^{\lambda \cdot t} \cdot e^{\lambda \cdot t} + g_2^* \cdot [i_{L_1} \cdot (1 - \alpha_f)$$
$$+ y_4 \cdot (1 - \alpha_f) \cdot C_1 \cdot L_1 + i_{C_{out}} \cdot (\alpha_f - 1) - i_{C_{in}} \cdot \alpha_f + i_{R_b} \cdot \alpha_f] \cdot y_2 \cdot e^{\lambda \cdot t} \cdot e^{\lambda \cdot t}\} \cdot R_{load}$$

Since $i_{L_1} \cdot y_2 \to \varepsilon; y_4 \cdot y_2 \to \varepsilon; \ldots; i_{L_1} \cdot y_2 \to \varepsilon; y_4 \cdot y_2 \to \varepsilon \ldots; \Phi_{1j=1} = g_1^* \cdot g_2^* \cdot R_{load} \cdot y_2 \cdot e^{\lambda \cdot t}$

$$\Phi_{1j=2} = Y_2 \cdot g_1 \cdot (\alpha_r \cdot \alpha_f - 1) \cdot I_{se} \cdot R_{load}$$
$$= (Y_2^{(j)} + y_2 \cdot e^{\lambda \cdot t}) \cdot \{g_1^* + [i_{L_1} \cdot (1 - \alpha_f)$$
$$+ y_4 \cdot (1 - \alpha_f) \cdot C_1 \cdot L_1 + i_{C_{out}} \cdot (\alpha_f - 1)$$
$$- i_{C_{in}} \cdot \alpha_f + i_{R_b} \cdot \alpha_f] \cdot e^{\lambda \cdot t}\} \cdot (\alpha_r \cdot \alpha_f - 1) \cdot I_{se} \cdot R_{load}$$

Since $\frac{dI_{C_{out}}}{dt} = 0 \Rightarrow Y_2^{(j)} = 0; y_2 \cdot i_{L_1} \to \varepsilon; y_2 \cdot y_4 \to \varepsilon \ldots$ then

$$\Phi_{1j=2} = y_2 \cdot e^{\lambda \cdot t} \cdot g_1^* \cdot (\alpha_r \cdot \alpha_f - 1) \cdot I_{se} \cdot R_{load}$$
$$= g_1^* \cdot (\alpha_r \cdot \alpha_f - 1) \cdot I_{se} \cdot R_{load} \cdot y_2 \cdot e^{\lambda \cdot t}$$

$$\Phi_{1j=3} = Y_2 \cdot g_2 \cdot (\alpha_r \cdot \alpha_f - 1) \cdot I_{sc} \cdot R_{load}$$
$$= (Y_2^{(j)} + y_2 \cdot e^{\lambda \cdot t}) \cdot \{g_2^* + [i_{L_1} \cdot (\alpha_r - 1) + y_4 \cdot (\alpha_r - 1)$$
$$\cdot C_1 \cdot L_1 + i_{C_{out}} \cdot (1 - \alpha_r) - i_{C_{in}} + i_{R_b}] \cdot e^{\lambda \cdot t}\}$$
$$\cdot (\alpha_r \cdot \alpha_f - 1) \cdot I_{sc} \cdot R_{load}$$

Since $\frac{dI_{C_{out}}}{dt} = 0 \Rightarrow Y_2^{(j)} = 0; y_2 \cdot i_{L_1} \to \varepsilon; y_2 \cdot y_4 \to \varepsilon\ldots$ then

$$\Phi_{1j=3} = Y_2 \cdot g_2 \cdot (\alpha_r \cdot \alpha_f - 1) \cdot I_{sc} \cdot R_{load} = g_2^* \cdot (\alpha_r \cdot \alpha_f - 1) \cdot I_{sc} \cdot R_{load} \cdot y_2 \cdot e^{\lambda \cdot t}$$

$$\Phi_{1j=4} = Y_2 \cdot R_{load} \cdot (\alpha_r \cdot \alpha_f - 1)^2 \cdot I_{sc} \cdot I_{se}; \frac{dI_{C_{out}}}{dt} = 0 \Rightarrow Y_2^{(j)} = 0;$$
$$\Phi_{1j=4} = R_{load} \cdot (\alpha_r \cdot \alpha_f - 1)^2 \cdot I_{sc} \cdot I_{se} \cdot y_2 \cdot e^{\lambda \cdot t}$$

$$\begin{aligned}\Phi_{1j=5} = I_{C_{out}} \cdot g_1 \cdot g_2 \cdot \frac{1}{C_{out}} &= (I_{C_{out}}^{(j)} + i_{C_{out}} \cdot e^{\lambda \cdot t}) \cdot \{g_1^* + [i_{L_1} \cdot (1 - \alpha_f) + y_4 \cdot (1 - \alpha_f) \cdot C_1 \cdot L_1 \\ &+ i_{C_{out}} \cdot (\alpha_f - 1) - i_{C_{in}} \cdot \alpha_f + i_{R_b} \cdot \alpha_f] \cdot e^{\lambda \cdot t}\} \cdot \{g_2^* + [i_{L_1} \cdot (\alpha_r - 1) + y_4 \cdot (\alpha_r - 1) \cdot C_1 \cdot L_1 \\ &+ i_{C_{out}} \cdot (1 - \alpha_r) - i_{C_{in}} + i_{R_b}] \cdot e^{\lambda \cdot t}\} \cdot \frac{1}{C_{out}}\end{aligned}$$

$$\begin{aligned}\Phi_{1j=5} = (I_{C_{out}}^{(j)} + i_{C_{out}} \cdot e^{\lambda \cdot t}) \cdot \{&g_1^* \cdot g_2^* + g_1^* \cdot [i_{L_1} \cdot (\alpha_r - 1) + y_4 \cdot (\alpha_r - 1) \cdot C_1 \cdot L_1 \\ &+ i_{C_{out}} \cdot (1 - \alpha_r) - i_{C_{in}} + i_{R_b}] \cdot e^{\lambda \cdot t} + g_2^* \cdot [i_{L_1} \cdot (1 - \alpha_f) + y_4 \cdot (1 - \alpha_f) \cdot C_1 \cdot L_1 \\ &+ i_{C_{out}} \cdot (\alpha_f - 1) - i_{C_{in}} \cdot \alpha_f + i_{R_b} \cdot \alpha_f] \cdot e^{\lambda \cdot t} + [i_{L_1} \cdot (1 - \alpha_f) + y_4 \cdot (1 - \alpha_f) \cdot C_1 \cdot L_1 \\ &+ i_{C_{out}} \cdot (\alpha_f - 1) - i_{C_{in}} \cdot \alpha_f + i_{R_b} \cdot \alpha_f] \cdot e^{\lambda \cdot t} \cdot [i_{L_1} \cdot (\alpha_r - 1) + y_4 \cdot (\alpha_r - 1) \cdot C_1 \cdot L_1 \\ &+ i_{C_{out}} \cdot (1 - \alpha_r) - i_{C_{in}} + i_{R_b}] \cdot e^{\lambda \cdot t}\} \cdot \frac{1}{C_{out}}\end{aligned}$$

Since $i_{L_1} \cdot i_{L_1} \to \varepsilon; y_4 \cdot i_{L_1} \to \varepsilon; i_{C_{out}} \cdot i_{L_1} \to \varepsilon\ldots; \frac{dI_{C_{out}}}{dt} = 0 \Rightarrow Y_2^{(j)} = 0$ then

$$\begin{aligned}\Phi_{1j=5} = (I_{C_{out}}^{(j)} + i_{C_{out}} \cdot e^{\lambda \cdot t}) \cdot \{&g_1^* \cdot g_2^* + g_1^* \cdot [i_{L_1} \cdot (\alpha_r - 1) + y_4 \cdot (\alpha_r - 1) \cdot C_1 \cdot L_1 \\ &+ i_{C_{out}} \cdot (1 - \alpha_r) - i_{C_{in}} + i_{R_b}] \cdot e^{\lambda \cdot t} + g_2^* \cdot [i_{L_1} \cdot (1 - \alpha_f) + y_4 \cdot (1 - \alpha_f) \cdot C_1 \cdot L_1 \\ &+ i_{C_{out}} \cdot (\alpha_f - 1) - i_{C_{in}} \cdot \alpha_f + i_{R_b} \cdot \alpha_f] \cdot e^{\lambda \cdot t}\} \cdot \frac{1}{C_{out}}\end{aligned}$$

$$\begin{aligned}\Phi_{1j=5} = (I_{C_{out}}^{(j)} + i_{C_{out}} \cdot e^{\lambda \cdot t}) \cdot (&g_1^* \cdot g_2^* + \{i_{L_1} \cdot [g_1^* \cdot (\alpha_r - 1) + g_2^* \cdot (1 - \alpha_f)] \\ &+ [g_1^* \cdot (\alpha_r - 1) + g_2^* \cdot (1 - \alpha_f)] \cdot C_1 \cdot L_1 \cdot y_4 + i_{C_{out}} \cdot [g_1^* \cdot (1 - \alpha_r) + g_2^* \cdot (\alpha_f - 1)] \\ &- i_{C_{in}} \cdot [g_1^* + g_2^* \cdot \alpha_f] + i_{R_b} \cdot [g_1^* + g_2^* \cdot \alpha_f]\} \cdot e^{\lambda \cdot t}) \cdot \frac{1}{C_{out}}\end{aligned}$$

$$\Phi_{1j=5} = (I_{C_{out}}^{(j)} \cdot g_1^* \cdot g_2^* + I_{C_{out}}^{(j)} \cdot \{i_{L_1} \cdot [g_1^* \cdot (\alpha_r - 1) + g_2^* \cdot (1 - \alpha_f)]$$
$$+ [g_1^* \cdot (\alpha_r - 1) + g_2^* \cdot (1 - \alpha_f)] \cdot C_1 \cdot L_1 \cdot y_4$$
$$+ i_{C_{out}} \cdot [g_1^* \cdot (1 - \alpha_r) + g_2^* \cdot (\alpha_f - 1)] - i_{C_{in}} \cdot [g_1^* + g_2^* \cdot \alpha_f]$$
$$+ i_{R_b} \cdot [g_1^* + g_2^* \cdot \alpha_f]\} \cdot e^{\lambda \cdot t} + g_1^* \cdot g_2^* \cdot i_{C_{out}} \cdot e^{\lambda \cdot t}$$
$$+ \{i_{L_1} \cdot [g_1^* \cdot (\alpha_r - 1) + g_2^* \cdot (1 - \alpha_f)]$$
$$+ [g_1^* \cdot (\alpha_r - 1) + g_2^* \cdot (1 - \alpha_f)] \cdot C_1 \cdot L_1 \cdot y_4$$
$$+ i_{C_{out}} \cdot [g_1^* \cdot (1 - \alpha_r) + g_2^* \cdot (\alpha_f - 1)] - i_{C_{in}} \cdot [g_1^* + g_2^* \cdot \alpha_f]$$
$$+ i_{R_b} \cdot [g_1^* + g_2^* \cdot \alpha_f]\} \cdot i_{C_{out}} \cdot e^{\lambda \cdot t} \cdot e^{\lambda \cdot t}) \cdot \frac{1}{C_{out}}$$

Since $i_{L_1} \cdot i_{C_{out}} \to \varepsilon; y_4 \cdot i_{C_{out}} \to \varepsilon \ldots$

$$\Phi_{1j=5} = (I_{C_{out}}^{(j)} \cdot g_1^* \cdot g_2^* + I_{C_{out}}^{(j)} \cdot \{i_{L_1} \cdot [g_1^* \cdot (\alpha_r - 1) + g_2^* \cdot (1 - \alpha_f)]$$
$$+ [g_1^* \cdot (\alpha_r - 1) + g_2^* \cdot (1 - \alpha_f)] \cdot C_1 \cdot L_1 \cdot y_4 + i_{C_{out}} \cdot [g_1^* \cdot (1 - \alpha_r) + g_2^* \cdot (\alpha_f - 1)]$$
$$- i_{C_{in}} \cdot [g_1^* + g_2^* \cdot \alpha_f] + i_{R_b} \cdot [g_1^* + g_2^* \cdot \alpha_f]\} \cdot e^{\lambda \cdot t} + g_1^* \cdot g_2^* \cdot i_{C_{out}} \cdot e^{\lambda \cdot t}) \cdot \frac{1}{C_{out}}$$

$$f_1(i_{L_1}, y_4, \ldots) = i_{L_1} \cdot [g_1^* \cdot (\alpha_r - 1) + g_2^* \cdot (1 - \alpha_f)]$$
$$+ [g_1^* \cdot (\alpha_r - 1) + g_2^* \cdot (1 - \alpha_f)] \cdot C_1 \cdot L_1 \cdot y_4$$
$$+ i_{C_{out}} \cdot [g_1^* \cdot (1 - \alpha_r) + g_2^* \cdot (\alpha_f - 1)]$$
$$- i_{C_{in}} \cdot [g_1^* + g_2^* \cdot \alpha_f] + i_{R_b} \cdot [g_1^* + g_2^* \cdot \alpha_f]$$

$$\Phi_{1j=5} = (I_{C_{out}}^{(j)} \cdot g_1^* \cdot g_2^* + I_{C_{out}}^{(j)} \cdot f_1(i_{L_1}, y_4, \ldots) \cdot e^{\lambda \cdot t} + g_1^* \cdot g_2^* \cdot i_{C_{out}} \cdot e^{\lambda \cdot t}) \cdot \frac{1}{C_{out}}$$

$$\Phi_{1j=6} = I_{C_{out}} \cdot g_1 \cdot (\alpha_r \cdot \alpha_f - 1) \cdot \frac{1}{C_{out}} \cdot I_{se}$$
$$= (I_{C_{out}}^{(j)} + i_{C_{out}} \cdot e^{\lambda \cdot t}) \cdot \{g_1^* + [i_{L_1} \cdot (1 - \alpha_f) + y_4 \cdot (1 - \alpha_f) \cdot C_1 \cdot L_1$$
$$+ i_{C_{out}} \cdot (\alpha_f - 1) - i_{C_{in}} \cdot \alpha_f + i_{R_b} \cdot \alpha_f] \cdot e^{\lambda \cdot t}\} \cdot (\alpha_r \cdot \alpha_f - 1) \cdot \frac{1}{C_{out}} \cdot I_{se}$$

$$\Phi_{1j=6} = I_{C_{out}} \cdot g_1 \cdot (\alpha_r \cdot \alpha_f - 1) \cdot \frac{1}{C_{out}} \cdot I_{se}$$
$$= \{I_{C_{out}}^{(j)} \cdot g_1^* + I_{C_{out}}^{(j)} \cdot [i_{L_1} \cdot (1 - \alpha_f) + y_4 \cdot (1 - \alpha_f) \cdot C_1 \cdot L_1$$
$$+ i_{C_{out}} \cdot (\alpha_f - 1) - i_{C_{in}} \cdot \alpha_f + i_{R_b} \cdot \alpha_f] \cdot e^{\lambda \cdot t} + i_{C_{out}} \cdot e^{\lambda \cdot t} \cdot g_1^*$$
$$+ i_{C_{out}} \cdot e^{\lambda \cdot t} \cdot [i_{L_1} \cdot (1 - \alpha_f)$$
$$+ y_4 \cdot (1 - \alpha_f) \cdot C_1 \cdot L_1 + i_{C_{out}} \cdot (\alpha_f - 1)$$
$$- i_{C_{in}} \cdot \alpha_f + i_{R_b} \cdot \alpha_f] \cdot e^{\lambda \cdot t}\} \cdot (\alpha_r \cdot \alpha_f - 1) \cdot \frac{1}{C_{out}} \cdot I_{se}$$

Since $i_{L_1} \cdot i_{C_{out}} \to \varepsilon; y_4 \cdot i_{C_{out}} \to \varepsilon \ldots$

$$\begin{aligned}\Phi_{1j=6} &= I_{C_{out}} \cdot g_1 \cdot (\alpha_r \cdot \alpha_f - 1) \cdot \frac{1}{C_{out}} \cdot I_{se} \\ &= \{I_{C_{out}}^{(j)} \cdot g_1^* + I_{C_{out}}^{(j)} \cdot [i_{L_1} \cdot (1 - \alpha_f) + y_4 \cdot (1 - \alpha_f) \cdot C_1 \cdot L_1 \\ &\quad + i_{C_{out}} \cdot (\alpha_f - 1) - i_{C_{in}} \cdot \alpha_f + i_{R_b} \cdot \alpha_f] \cdot e^{\lambda \cdot t} + i_{C_{out}} \cdot e^{\lambda \cdot t} \cdot g_1^*\} \cdot (\alpha_r \cdot \alpha_f - 1) \cdot \frac{1}{C_{out}} \cdot I_{se}\end{aligned}$$

$$\begin{aligned}f_2(i_{L_1}, y_4, \ldots) = i_{L_1} \cdot (1 - \alpha_f) + y_4 \cdot (1 - \alpha_f) \cdot C_1 \cdot L_1 + i_{C_{out}} \cdot (\alpha_f - 1) - i_{C_{in}} \\ \cdot \alpha_f + i_{R_b} \cdot \alpha_f\end{aligned}$$

$$\begin{aligned}\Phi_{1j=6} = I_{C_{out}} \cdot g_1 \cdot (\alpha_r \cdot \alpha_f - 1) \cdot \frac{1}{C_{out}} \cdot I_{se} = \{I_{C_{out}}^{(j)} \cdot g_1^* + I_{C_{out}}^{(j)} \cdot f_2(i_{L_1}, y_4, \ldots) \cdot e^{\lambda \cdot t} \\ + g_1^* \cdot i_{C_{out}} \cdot e^{\lambda \cdot t}\} \cdot (\alpha_r \cdot \alpha_f - 1) \cdot \frac{1}{C_{out}} \cdot I_{se}\end{aligned}$$

$$\begin{aligned}\Phi_{1j=7} &= I_{C_{out}} \cdot g_2 \cdot (\alpha_r \cdot \alpha_f - 1) \cdot \frac{1}{C_{out}} \cdot I_{sc} \\ &= (I_{C_{out}}^{(j)} + i_{C_{out}} \cdot e^{\lambda \cdot t}) \cdot \{g_2^* + [i_{L_1} \cdot (\alpha_r - 1) + y_4 \cdot (\alpha_r - 1) \cdot C_1 \cdot L_1 \\ &\quad + i_{C_{out}} \cdot (1 - \alpha_r) - i_{C_{in}} + i_{R_b}] \cdot e^{\lambda \cdot t}\} \cdot (\alpha_r \cdot \alpha_f - 1) \cdot \frac{1}{C_{out}} \cdot I_{sc}\end{aligned}$$

$$\begin{aligned}\Phi_{1j=7} = I_{C_{out}} \cdot g_2 \cdot (\alpha_r \cdot \alpha_f - 1) \cdot \frac{1}{C_{out}} \cdot I_{sc} = \{I_{C_{out}}^{(j)} \cdot g_2^* + I_{C_{out}}^{(j)} \cdot [i_{L_1} \cdot (\alpha_r - 1) \\ + y_4 \cdot (\alpha_r - 1) \cdot C_1 \cdot L_1 + i_{C_{out}} \cdot (1 - \alpha_r) - i_{C_{in}} + i_{R_b}] \cdot e^{\lambda \cdot t} \\ + g_2^* \cdot i_{C_{out}} \cdot e^{\lambda \cdot t} + [i_{L_1} \cdot (\alpha_r - 1) \\ + y_4 \cdot (\alpha_r - 1) \cdot C_1 \cdot L_1 + i_{C_{out}} \cdot (1 - \alpha_r) - i_{C_{in}} + i_{R_b}] \\ \cdot i_{C_{out}} \cdot e^{\lambda \cdot t} \cdot e^{\lambda \cdot t}\} \cdot (\alpha_r \cdot \alpha_f - 1) \cdot \frac{1}{C_{out}} \cdot I_{sc}\end{aligned}$$

Since $i_{L_1} \cdot i_{C_{out}} \to \varepsilon; y_4 \cdot i_{C_{out}} \to \varepsilon \ldots$

$$\begin{aligned}\Phi_{1j=7} &= I_{C_{out}} \cdot g_2 \cdot (\alpha_r \cdot \alpha_f - 1) \cdot \frac{1}{C_{out}} \cdot I_{sc} \\ &= \{I_{C_{out}}^{(j)} \cdot g_2^* + I_{C_{out}}^{(j)} \cdot [i_{L_1} \cdot (\alpha_r - 1) \\ &\quad + y_4 \cdot (\alpha_r - 1) \cdot C_1 \cdot L_1 + i_{C_{out}} \cdot (1 - \alpha_r) - i_{C_{in}} + i_{R_b}] \cdot e^{\lambda \cdot t} \\ &\quad + g_2^* \cdot i_{C_{out}} \cdot e^{\lambda \cdot t}\} \cdot (\alpha_r \cdot \alpha_f - 1) \cdot \frac{1}{C_{out}} \cdot I_{sc}\end{aligned}$$

$$f_3(i_{L_1}, y_4, \ldots) = i_{L_1} \cdot (\alpha_r - 1) + y_4 \cdot (\alpha_r - 1) \cdot C_1 \cdot L_1 + i_{C_{out}} \cdot (1 - \alpha_r) - i_{C_{in}} + i_{R_b}$$

$$\Phi_{1j=7} = I_{C_{out}} \cdot g_2 \cdot (\alpha_r \cdot \alpha_f - 1) \cdot \frac{1}{C_{out}} \cdot I_{sc} = \{I_{C_{out}}^{(j)} \cdot g_2^* + I_{C_{out}}^{(j)} \cdot f_3(i_{L_1}, y_4, \ldots) \cdot e^{\lambda \cdot t} + g_2^* \cdot i_{C_{out}} \cdot e^{\lambda \cdot t}\} \cdot (\alpha_r \cdot \alpha_f - 1) \cdot \frac{1}{C_{out}} \cdot I_{sc}$$

$$\Phi_{1j=8} = I_{C_{out}} \cdot \frac{1}{C_{out}} \cdot (\alpha_r \cdot \alpha_f - 1)^2 \cdot I_{sc} \cdot I_{se} = (I_{C_{out}}^{(j)} + i_{C_{out}} \cdot e^{\lambda \cdot t}) \cdot \frac{1}{C_{out}} \cdot (\alpha_r \cdot \alpha_f - 1)^2 \cdot I_{sc} \cdot I_{se}$$

$$\Phi_{1j=8} = I_{C_{out}} \cdot \frac{1}{C_{out}} \cdot (\alpha_r \cdot \alpha_f - 1)^2 \cdot I_{sc} \cdot I_{se} = I_{C_{out}}^{(j)} \cdot \frac{1}{C_{out}} \cdot (\alpha_r \cdot \alpha_f - 1)^2 \cdot I_{sc} \cdot I_{se} + \frac{1}{C_{out}} \cdot (\alpha_r \cdot \alpha_f - 1)^2 \cdot I_{sc} \cdot I_{se} \cdot i_{C_{out}} \cdot e^{\lambda \cdot t}$$

We can summary our last results in the following Table 5.1.

Table 5.1 Class C power amplifier (PA) with parallel resonance, Φ_{1j} expressions

Φ_{1j}	Expression
$j = 1\ (\Phi_{1j=1})$	$\Phi_{1j=1} = g_1^* \cdot g_2^* \cdot R_{load} \cdot y_2 \cdot e^{\lambda \cdot t}$
$j = 2\ (\Phi_{1j=2})$	$\Phi_{1j=2} = g_1^* \cdot (\alpha_r \cdot \alpha_f - 1) \cdot I_{se} \cdot R_{load} \cdot y_2 \cdot e^{\lambda \cdot t}$
$j = 3\ (\Phi_{1j=3})$	$\Phi_{1j=3} = g_2^* \cdot (\alpha_r \cdot \alpha_f - 1) \cdot I_{sc} \cdot R_{load} \cdot y_2 \cdot e^{\lambda \cdot t}$
$j = 4\ (\Phi_{1j=4})$	$\Phi_{1j=4} = R_{load} \cdot (\alpha_r \cdot \alpha_f - 1)^2 \cdot I_{sc} \cdot I_{se} \cdot y_2 \cdot e^{\lambda \cdot t}$
$j = 5\ (\Phi_{1j=5})$	$\Phi_{1j=5} = (I_{C_{out}}^{(j)} \cdot g_1^* \cdot g_2^* + I_{C_{out}}^{(j)} \cdot f_1(i_{L_1}, y_4, \ldots) \cdot e^{\lambda \cdot t} + g_1^* \cdot g_2^* \cdot i_{C_{out}} \cdot e^{\lambda \cdot t}) \cdot \frac{1}{C_{out}}$
$j = 6\ (\Phi_{1j=6})$	$\Phi_{1j=6} = \{I_{C_{out}}^{(j)} \cdot g_1^* + I_{C_{out}}^{(j)} \cdot f_2(i_{L_1}, y_4, \ldots) \cdot e^{\lambda \cdot t} + g_1^* \cdot i_{C_{out}} \cdot e^{\lambda \cdot t}\} \cdot (\alpha_r \cdot \alpha_f - 1) \cdot \frac{1}{C_{out}} \cdot I_{se}$
$j = 7\ (\Phi_{1j=7})$	$\Phi_{1j=7} = \{I_{C_{out}}^{(j)} \cdot g_2^* + I_{C_{out}}^{(j)} \cdot f_3(i_{L_1}, y_4, \ldots) \cdot e^{\lambda \cdot t} + g_2^* \cdot i_{C_{out}} \cdot e^{\lambda \cdot t}\} \cdot (\alpha_r \cdot \alpha_f - 1) \cdot \frac{1}{C_{out}} \cdot I_{sc}$
$j = 8\ (\Phi_{1j=8})$	$\Phi_{1j=8} = I_{C_{out}}^{(j)} \cdot \frac{1}{C_{out}} \cdot (\alpha_r \cdot \alpha_f - 1)^2 \cdot I_{sc} \cdot I_{se} + \frac{1}{C_{out}} \cdot (\alpha_r \cdot \alpha_f - 1)^2 \cdot I_{sc} \cdot I_{se} \cdot i_{C_{out}} \cdot e^{\lambda \cdot t}$

$$\Phi_2(g_1, g_3, \alpha_f, \alpha_r, I_{sc}) = -g_3 \cdot [g_1 + (\alpha_f \cdot \alpha_r - 1) \cdot I_{sc}] = -g_3(\frac{dV_s(t)}{dt} \to \varepsilon) \cdot [g_1 + (\alpha_f \cdot \alpha_r - 1) \cdot I_{sc}]$$

$$g_1(t) = g_1^* + [i_{L_1} \cdot (1 - \alpha_f) + y_4 \cdot (1 - \alpha_f) \cdot C_1 \cdot L_1 + i_{C_{out}} \cdot (\alpha_f - 1) - i_{C_{in}} \cdot \alpha_f + i_{R_b} \cdot \alpha_f] \cdot e^{\lambda \cdot t}$$

$$g_2(t) = g_2^* + [i_{L_1} \cdot (\alpha_r - 1) + y_4 \cdot (\alpha_r - 1) \cdot C_1 \cdot L_1 + i_{C_{out}} \cdot (1 - \alpha_r) - i_{C_{in}} + i_{R_b}] \cdot e^{\lambda \cdot t}$$

$$g_3(\frac{dV_s(t)}{dt} \to \varepsilon) = g_3^* + [y_3 \cdot (\alpha_r - 1) + y_2 \cdot (1 - \alpha_r) - y_1 \cdot (1 + \frac{R_s}{R_b}) - i_{C_{in}} \cdot \frac{1}{C_{in} \cdot R_b}] \cdot e^{\lambda \cdot t}$$

$$g_4(\frac{dV_s(t)}{dt} \to \varepsilon) = g_4^* + [y_3 \cdot (1 - \alpha_f) + y_2 \cdot (\alpha_f - 1) - y_1 \cdot \alpha_f \cdot (1 + \frac{R_s}{R_b}) - i_{C_{in}} \cdot \frac{\alpha_f}{C_{in} \cdot R_b}] \cdot e^{\lambda \cdot t}$$

$$\Phi_2(g_1, g_3(\frac{dV_s(t)}{dt} \to \varepsilon), \alpha_f, \alpha_r, I_{sc}) = -\{g_3^* + [y_3 \cdot (\alpha_r - 1) + y_2 \cdot (1 - \alpha_r) - y_1 \cdot (1 + \frac{R_s}{R_b}) - i_{C_{in}} \cdot \frac{1}{C_{in} \cdot R_b}] \cdot e^{\lambda \cdot t}\} \cdot \{[g_1^* + (\alpha_f \cdot \alpha_r - 1) \cdot I_{sc}] + [i_{L_1} \cdot (1 - \alpha_f) + y_4 \cdot (1 - \alpha_f) \cdot C_1 \cdot L_1 + i_{C_{out}} \cdot (\alpha_f - 1) - i_{C_{in}} \cdot \alpha_f + i_{R_b} \cdot \alpha_f] \cdot e^{\lambda \cdot t}\}$$

$$\Phi_2(g_1, g_3(\frac{dV_s(t)}{dt} \to \varepsilon), \alpha_f, \alpha_r, I_{sc}) = -\{g_3^* \cdot [g_1^* + (\alpha_f \cdot \alpha_r - 1) \cdot I_{sc}] + g_3^* \cdot [i_{L_1} \cdot (1 - \alpha_f) + y_4 \cdot (1 - \alpha_f) \cdot C_1 \cdot L_1 + i_{C_{out}} \cdot (\alpha_f - 1) - i_{C_{in}} \cdot \alpha_f + i_{R_b} \cdot \alpha_f] \cdot e^{\lambda \cdot t} + [g_1^* + (\alpha_f \cdot \alpha_r - 1) \cdot I_{sc}] \cdot [y_3 \cdot (\alpha_r - 1) + y_2 \cdot (1 - \alpha_r) - y_1 \cdot (1 + \frac{R_s}{R_b}) - i_{C_{in}} \cdot \frac{1}{C_{in} \cdot R_b}] \cdot e^{\lambda \cdot t} + [y_3 \cdot (\alpha_r - 1) + y_2 \cdot (1 - \alpha_r) - y_1 \cdot (1 + \frac{R_s}{R_b}) - i_{C_{in}} \cdot \frac{1}{C_{in} \cdot R_b}] \cdot e^{\lambda \cdot t} \cdot [i_{L_1} \cdot (1 - \alpha_f) + y_4 \cdot (1 - \alpha_f) \cdot C_1 \cdot L_1 + i_{C_{out}} \cdot (\alpha_f - 1) - i_{C_{in}} \cdot \alpha_f + i_{R_b} \cdot \alpha_f] \cdot e^{\lambda \cdot t}\}$$

Since $y_3 \cdot i_{L_1} \to \varepsilon; y_3 \cdot y_4 \to \varepsilon \ldots$

$$\Phi_2(g_1, g_3(\frac{dV_s(t)}{dt} \to \varepsilon), \alpha_f, \alpha_r, I_{sc}) = -\{g_3^* \cdot [g_1^* + (\alpha_f \cdot \alpha_r - 1) \cdot I_{sc}] + g_3^* \cdot [i_{L_1} \cdot (1 - \alpha_f) + y_4 \cdot (1 - \alpha_f) \cdot C_1 \cdot L_1 + i_{C_{out}} \cdot (\alpha_f - 1) - i_{C_{in}} \cdot \alpha_f + i_{R_b} \cdot \alpha_f] \cdot e^{\lambda \cdot t} + [g_1^* + (\alpha_f \cdot \alpha_r - 1) \cdot I_{sc}] \cdot [y_3 \cdot (\alpha_r - 1) + y_2 \cdot (1 - \alpha_r) - y_1 \cdot (1 + \frac{R_s}{R_b}) - i_{C_{in}} \cdot \frac{1}{C_{in} \cdot R_b}] \cdot e^{\lambda \cdot t}\}$$

We define two functions:

$$f_4(i_{L_1}, y_4, \ldots) = i_{L_1} \cdot (1 - \alpha_f) + y_4 \cdot (1 - \alpha_f) \cdot C_1 \cdot L_1 + i_{C_{out}} \cdot (\alpha_f - 1) - i_{C_{in}} \cdot \alpha_f + i_{R_b} \cdot \alpha_f$$

$$f_5(y_3, y_2, \ldots) = y_3 \cdot (\alpha_r - 1) + y_2 \cdot (1 - \alpha_r) - y_1 \cdot (1 + \frac{R_s}{R_b}) - i_{C_{in}} \cdot \frac{1}{C_{in} \cdot R_b}$$

$$\Phi_2(g_1, g_3(\frac{dV_s(t)}{dt} \to \varepsilon), \alpha_f, \alpha_r, I_{sc}) = -\{g_3^* \cdot [g_1^* + (\alpha_f \cdot \alpha_r - 1) \cdot I_{sc}] + g_3^* \cdot f_4(i_{L_1}, y_4, \ldots) \cdot e^{\lambda \cdot t} + [g_1^* + (\alpha_f \cdot \alpha_r - 1) \cdot I_{sc}] \cdot f_5(y_3, y_2, \ldots) \cdot e^{\lambda \cdot t}\}$$

$$\Phi_3(g_2, g_4(\frac{dV_s(t)}{dt} \to \varepsilon), \alpha_f, \alpha_r, I_{se}) = -g_4(\frac{dV_s(t)}{dt} \to \varepsilon) \cdot [g_2 + (\alpha_r \cdot \alpha_f - 1) \cdot I_{se}]$$

$$\Phi_3(g_2, g_4(\frac{dV_s(t)}{dt} \to \varepsilon), \alpha_f, \alpha_r, I_{se}) = -\{g_4^* + [y_3 \cdot (1 - \alpha_f) + y_2 \cdot (\alpha_f - 1) - y_1 \cdot \alpha_f \cdot (1 + \frac{R_s}{R_b}) - i_{C_{in}} \cdot \frac{\alpha_f}{C_{in} \cdot R_b}] \cdot e^{\lambda \cdot t}\} \cdot \{g_2^* + [i_{L_1} \cdot (\alpha_r - 1) + y_4 \cdot (\alpha_r - 1) \cdot C_1 \cdot L_1 + i_{C_{out}} \cdot (1 - \alpha_r) - i_{C_{in}} + i_{R_b}] \cdot e^{\lambda \cdot t} + (\alpha_r \cdot \alpha_f - 1) \cdot I_{se}\}$$

$$\Phi_3(g_2, g_4(\frac{dV_s(t)}{dt} \to \varepsilon), \alpha_f, \alpha_r, I_{se}) = -\{g_4^* \cdot [g_2^* + (\alpha_r \cdot \alpha_f - 1) \cdot I_{se}]$$
$$+ g_4^* \cdot [i_{L_1} \cdot (\alpha_r - 1) + y_4 \cdot (\alpha_r - 1) \cdot C_1 \cdot L_1 + i_{C_{out}} \cdot (1 - \alpha_r) - i_{C_{in}} + i_{R_b}] \cdot e^{\lambda \cdot t}$$
$$+ [g_2^* + (\alpha_r \cdot \alpha_f - 1) \cdot I_{se}] \cdot [y_3 \cdot (1 - \alpha_f) + y_2 \cdot (\alpha_f - 1) - y_1 \cdot \alpha_f \cdot (1 + \frac{R_s}{R_b})$$
$$- i_{C_{in}} \cdot \frac{\alpha_f}{C_{in} \cdot R_b}] \cdot e^{\lambda \cdot t} + [y_3 \cdot (1 - \alpha_f) + y_2 \cdot (\alpha_f - 1)$$
$$- y_1 \cdot \alpha_f \cdot (1 + \frac{R_s}{R_b}) - i_{C_{in}} \cdot \frac{\alpha_f}{C_{in} \cdot R_b}] \cdot e^{\lambda \cdot t} \cdot [i_{L_1} \cdot (\alpha_r - 1) + y_4 \cdot (\alpha_r - 1) \cdot C_1 \cdot L_1$$
$$+ i_{C_{out}} \cdot (1 - \alpha_r) - i_{C_{in}} + i_{R_b}] \cdot e^{\lambda \cdot t}\}$$

Since $y_3 \cdot i_{L_1} \to \varepsilon; y_3 \cdot y_4 \to \varepsilon \ldots$

$$\Phi_3(g_2, g_4(\frac{dV_s(t)}{dt} \to \varepsilon), \alpha_f, \alpha_r, I_{se}) = -\{g_4^* \cdot [g_2^* + (\alpha_r \cdot \alpha_f - 1) \cdot I_{se}]$$
$$+ g_4^* \cdot [i_{L_1} \cdot (\alpha_r - 1) + y_4 \cdot (\alpha_r - 1) \cdot C_1 \cdot L_1$$
$$+ i_{C_{out}} \cdot (1 - \alpha_r) - i_{C_{in}} + i_{R_b}] \cdot e^{\lambda \cdot t} + [g_2^* + (\alpha_r \cdot \alpha_f - 1) \cdot I_{se}]$$
$$\cdot [y_3 \cdot (1 - \alpha_f) + y_2 \cdot (\alpha_f - 1) - y_1 \cdot \alpha_f \cdot (1 + \frac{R_s}{R_b})$$
$$- i_{C_{in}} \cdot \frac{\alpha_f}{C_{in} \cdot R_b}] \cdot e^{\lambda \cdot t}\}$$

We define two functions:

$$f_6(i_{L_1}, y_4, \ldots) = i_{L_1} \cdot (\alpha_r - 1) + y_4 \cdot (\alpha_r - 1) \cdot C_1 \cdot L_1 + i_{C_{out}} \cdot (1 - \alpha_r) - i_{C_{in}} + i_{R_b}$$
$$f_7(y_3, y_2, \ldots) = y_3 \cdot (1 - \alpha_f) + y_2 \cdot (\alpha_f - 1) - y_1 \cdot \alpha_f \cdot (1 + \frac{R_s}{R_b}) - i_{C_{in}} \cdot \frac{\alpha_f}{C_{in} \cdot R_b}$$

$$\Phi_3(g_2, g_4(\frac{dV_s(t)}{dt} \to \varepsilon), \alpha_f, \alpha_r, I_{se}) = -\{g_4^* \cdot [g_2^* + (\alpha_r \cdot \alpha_f - 1) \cdot I_{se}]$$
$$+ g_4^* \cdot f_6(i_{L_1}, y_4, \ldots) \cdot e^{\lambda \cdot t} + [g_2^* + (\alpha_r \cdot \alpha_f - 1) \cdot I_{se}] \cdot f_7(y_3, y_2, \ldots) \cdot e^{\lambda \cdot t}\}$$

Finally we get the enhance expression for $\upsilon(Y_2, I_{C_{out}}, \ldots)$

$$\upsilon(Y_2, I_{C_{out}}, \ldots) = \frac{\sum_{k=1}^{3} \Phi_k}{\{\Gamma_2(g_1^*, g_2^*, \Gamma_1, \alpha_r, \alpha_f) + \Gamma_3(\alpha_f, \alpha_r) \cdot (i_{C_{in}} - i_{R_b}) \cdot e^{\lambda \cdot t}\} \cdot C_1 \cdot L_1}$$

$$\upsilon(Y_2, I_{C_{out}}, \ldots) = \frac{\frac{1}{V_T} \cdot \sum_{j=1}^{8} \Phi_{1j} + \sum_{k=2}^{3} \Phi_k}{\{\Gamma_2(g_1^*, g_2^*, \Gamma_1, \alpha_r, \alpha_f) + \Gamma_3(\alpha_f, \alpha_r) \cdot (i_{C_{in}} - i_{R_b}) \cdot e^{\lambda \cdot t}\} \cdot C_1 \cdot L_1}$$

$$\upsilon(Y_2, I_{C_{out}}, \ldots = \frac{1}{C_1 \cdot L_1} \cdot \{\frac{1}{V_T} \cdot \frac{\sum_{j=1}^{8} \Phi_{1j}}{[\Gamma_2(g_1^*, g_2^*, \Gamma_1, \alpha_r, \alpha_f) + \Gamma_3(\alpha_f, \alpha_r) \cdot (i_{C_{in}} - i_{R_b}) \cdot e^{\lambda \cdot t}]} + \frac{\sum_{k=2}^{3} \Phi_k}{[\Gamma_2(g_1^*, g_2^*, \Gamma_1, \alpha_r, \alpha_f) + \Gamma_3(\alpha_f, \alpha_r) \cdot (i_{C_{in}} - i_{R_b}) \cdot e^{\lambda \cdot t}]}\}$$

$$\upsilon(Y_2, I_{C_{out}}, \ldots) = \frac{1}{C_1 \cdot L_1} \cdot \{\frac{1}{V_T} \cdot \upsilon_1(Y_2, I_{C_{out}}, \ldots) + \upsilon_2(Y_2, I_{C_{out}}, \ldots)\}$$

$$\upsilon(Y_2, I_{C_{out}}, \ldots) = \frac{1}{C_1 \cdot L_1} \cdot \{\frac{1}{V_T} \cdot \sum_{k=1}^{2} \upsilon_{1k}(Y_2, I_{C_{out}}, \ldots) + \sum_{k=1}^{2} \upsilon_{2k}(Y_2, I_{C_{out}}, \ldots)\}$$

$$\upsilon_1(Y_2, I_{C_{out}}, \ldots) = \frac{\sum_{j=1}^{8} \Phi_{1j}}{[\Gamma_2(g_1^*, g_2^*, \Gamma_1, \alpha_r, \alpha_f) + \Gamma_3(\alpha_f, \alpha_r) \cdot (i_{C_{in}} - i_{R_b}) \cdot e^{\lambda \cdot t}]}$$

$$\upsilon_2(Y_2, I_{C_{out}}, \ldots) = \frac{\sum_{k=2}^{3} \Phi_k}{[\Gamma_2(g_1^*, g_2^*, \Gamma_1, \alpha_r, \alpha_f) + \Gamma_3(\alpha_f, \alpha_r) \cdot (i_{C_{in}} - i_{R_b}) \cdot e^{\lambda \cdot t}]}$$

$$\upsilon(Y_2, I_{C_{out}}, \ldots) = \frac{1}{C_1 \cdot L_1} \cdot \{\frac{1}{V_T} \cdot \upsilon_1(Y_2, I_{C_{out}}, \ldots) + \upsilon_2(Y_2, I_{C_{out}}, \ldots)\}$$

Stage 1:

$$\upsilon_1(Y_2, I_{C_{out}}, \ldots) = \frac{\sum_{j=1}^{8} \Phi_{1j}}{\begin{matrix}[\Gamma_2(g_1^*, g_2^*, \Gamma_1, \alpha_r, \alpha_f) \\ + \Gamma_3(\alpha_f, \alpha_r) \cdot (i_{C_{in}} - i_{R_b}) \cdot e^{\lambda \cdot t}]\end{matrix}} \cdot \frac{\begin{matrix}[\Gamma_2(g_1^*, g_2^*, \Gamma_1, \alpha_r, \alpha_f) \\ - \Gamma_3(\alpha_f, \alpha_r) \cdot (i_{C_{in}} - i_{R_b}) \cdot e^{\lambda \cdot t}]\end{matrix}}{\begin{matrix}[\Gamma_2(g_1^*, g_2^*, \Gamma_1, \alpha_r, \alpha_f) \\ - \Gamma_3(\alpha_f, \alpha_r) \cdot (i_{C_{in}} - i_{R_b}) \cdot e^{\lambda \cdot t}]\end{matrix}}$$

$$\upsilon_1(Y_2, I_{C_{out}}, \ldots) = \frac{[\Gamma_2(g_1^*, g_2^*, \Gamma_1, \alpha_r, \alpha_f) - \Gamma_3(\alpha_f, \alpha_r) \cdot (i_{C_{in}} - i_{R_b}) \cdot e^{\lambda \cdot t}] \cdot \sum_{j=1}^{8} \Phi_{1j}}{[\Gamma_2^2(g_1^*, g_2^*, \Gamma_1, \alpha_r, \alpha_f) - \Gamma_3^2(\alpha_f, \alpha_r) \cdot (i_{C_{in}} - i_{R_b})^2 \cdot e^{2 \cdot \lambda \cdot t}]}$$

$$(i_{C_{in}} - i_{R_b})^2 = i_{C_{in}}^2 - 2 \cdot i_{C_{in}} \cdot i_{R_b} + i_{R_b}^2; i_{C_{in}}^2 \to \varepsilon; i_{C_{in}} \cdot i_{R_b} \to \varepsilon; i_{R_b}^2 \to \varepsilon; (i_{C_{in}} - i_{R_b})^2 \to \varepsilon$$

$$\upsilon_1(Y_2, I_{C_{out}}, \ldots) = \frac{[\Gamma_2(g_1^*, g_2^*, \Gamma_1, \alpha_r, \alpha_f) - \Gamma_3(\alpha_f, \alpha_r) \cdot (i_{C_{in}} - i_{R_b}) \cdot e^{\lambda \cdot t}] \cdot \sum_{j=1}^{8} \Phi_{1j}}{[\Gamma_2^2(g_1^*, g_2^*, \Gamma_1, \alpha_r, \alpha_f)]}$$

$$\upsilon_1(Y_2, I_{C_{out}}, \ldots) = \frac{\sum_{j=1}^{8} \Phi_{1j}}{\Gamma_2(g_1^*, g_2^*, \Gamma_1, \alpha_r, \alpha_f)} - \frac{\Gamma_3(\alpha_f, \alpha_r) \cdot (i_{C_{in}} - i_{R_b}) \cdot e^{\lambda \cdot t} \cdot \sum_{j=1}^{8} \Phi_{1j}}{\Gamma_2^2(g_1^*, g_2^*, \Gamma_1, \alpha_r, \alpha_f)}$$

$$\upsilon_{11}(Y_2, I_{C_{out}}, \ldots) = \frac{\sum_{j=1}^{8} \Phi_{1j}}{\Gamma_2(g_1^*, g_2^*, \Gamma_1, \alpha_r, \alpha_f)}; \upsilon_{12}(Y_2, I_{C_{out}}, \ldots) = \frac{\Gamma_3(\alpha_f, \alpha_r) \cdot (i_{C_{in}} - i_{R_b}) \cdot e^{\lambda \cdot t} \cdot \sum_{j=1}^{8} \Phi_{1j}}{\Gamma_2^2(g_1^*, g_2^*, \Gamma_1, \alpha_r, \alpha_f)}$$

$$\upsilon_1(Y_2, I_{C_{out}}, \ldots) = \sum_{k=1}^{2} \upsilon_{1k}(Y_2, I_{C_{out}}, \ldots) = \upsilon_{11}(Y_2, I_{C_{out}}, \ldots) + \upsilon_{12}(Y_2, I_{C_{out}}, \ldots)$$

$$\upsilon_{11}(Y_2, I_{C_{out}}, \ldots) = \frac{\begin{array}{l}[g_1^* \cdot g_2^* \cdot R_{load} \cdot y_2 \cdot e^{\lambda \cdot t} + g_1^* \cdot (\alpha_r \cdot \alpha_f - 1) \cdot I_{se} \cdot R_{load} \cdot y_2 \cdot e^{\lambda \cdot t} \\ + g_2^* \cdot (\alpha_r \cdot \alpha_f - 1) \cdot I_{sc} \cdot R_{load} \cdot y_2 \cdot e^{\lambda \cdot t} + R_{load} \cdot (\alpha_r \cdot \alpha_f - 1)^2 \cdot I_{sc} \cdot I_{se} \cdot y_2 \cdot e^{\lambda \cdot t} \\ + I_{C_{out}}^{(j)} \cdot g_1^* \cdot g_2^* \cdot \frac{1}{C_{out}} + I_{C_{out}}^{(j)} \cdot \frac{1}{C_{out}} \cdot f_1(i_{L_1}, y_4, \ldots) \cdot e^{\lambda \cdot t} + g_1^* \cdot g_2^* \cdot \frac{1}{C_{out}} \cdot i_{C_{out}} \cdot e^{\lambda \cdot t} \\ + \{I_{C_{out}}^{(j)} \cdot g_1^* + I_{C_{out}}^{(j)} \cdot f_2(i_{L_1}, y_4, \ldots) \cdot e^{\lambda \cdot t} + g_1^* \cdot i_{C_{out}} \cdot e^{\lambda \cdot t}\} \cdot (\alpha_r \cdot \alpha_f - 1) \cdot \frac{1}{C_{out}} \cdot I_{se} \\ + \{I_{C_{out}}^{(j)} \cdot g_2^* + I_{C_{out}}^{(j)} \cdot f_3(i_{L_1}, y_4, \ldots) \cdot e^{\lambda \cdot t} + g_2^* \cdot i_{C_{out}} \cdot e^{\lambda \cdot t}\} \cdot (\alpha_r \cdot \alpha_f - 1) \cdot \frac{1}{C_{out}} \cdot I_{sc} \\ + I_{C_{out}}^{(j)} \cdot \frac{1}{C_{out}} \cdot (\alpha_r \cdot \alpha_f - 1)^2 \cdot I_{sc} \cdot I_{se} + \frac{1}{C_{out}} \cdot (\alpha_r \cdot \alpha_f - 1)^2 \cdot I_{sc} \cdot I_{se} \cdot i_{C_{out}} \cdot e^{\lambda \cdot t}]\end{array}}{\Gamma_2(g_1^*, g_2^*, \Gamma_1, \alpha_r, \alpha_f)}$$

$$\upsilon_{11}(Y_2, I_{C_{out}}, \ldots) = \frac{\begin{array}{l}[I_{C_{out}}^{(j)} \cdot \frac{1}{C_{out}} \cdot [(\alpha_r \cdot \alpha_f - 1)^2 \cdot I_{sc} \cdot I_{se} + g_1^* \cdot g_2^* + g_1^* \cdot (\alpha_r \cdot \alpha_f - 1) \cdot I_{se} + g_2^* \cdot (\alpha_r \cdot \alpha_f - 1) \cdot I_{sc}] \\ + [g_1^* \cdot g_2^* \cdot R_{load} \cdot y_2 + g_1^* \cdot (\alpha_r \cdot \alpha_f - 1) \cdot I_{se} \cdot R_{load} \cdot y_2 \\ + g_2^* \cdot (\alpha_r \cdot \alpha_f - 1) \cdot I_{sc} \cdot R_{load} \cdot y_2 + R_{load} \cdot (\alpha_r \cdot \alpha_f - 1)^2 \cdot I_{sc} \cdot I_{se} \cdot y_2 \\ + I_{C_{out}}^{(j)} \cdot \frac{1}{C_{out}} \cdot f_1(i_{L_1}, y_4, \ldots) + g_1^* \cdot g_2^* \cdot \frac{1}{C_{out}} \cdot i_{C_{out}} + I_{C_{out}}^{(j)} \cdot (\alpha_r \cdot \alpha_f - 1) \cdot \frac{1}{C_{out}} \cdot I_{se} \cdot f_2(i_{L_1}, y_4, \ldots) \\ + g_1^* \cdot (\alpha_r \cdot \alpha_f - 1) \cdot \frac{1}{C_{out}} \cdot I_{se} \cdot i_{C_{out}} + I_{C_{out}}^{(j)} \cdot (\alpha_r \cdot \alpha_f - 1) \cdot \frac{1}{C_{out}} \cdot I_{sc} \cdot f_3(i_{L_1}, y_4, \ldots) \\ + g_2^* \cdot (\alpha_r \cdot \alpha_f - 1) \cdot \frac{1}{C_{out}} \cdot I_{sc} \cdot i_{C_{out}} + \frac{1}{C_{out}} \cdot (\alpha_r \cdot \alpha_f - 1)^2 \cdot I_{sc} \cdot I_{se} \cdot i_{C_{out}}] \cdot e^{\lambda \cdot t}\end{array}}{\Gamma_2(g_1^*, g_2^*, \Gamma_1, \alpha_r, \alpha_f)}$$

$$\Omega_1(I_{C_{out}}^{(j)}, g_1^*, g_2^*, \ldots) = I_{C_{out}}^{(j)} \cdot \frac{1}{C_{out}} \cdot [(\alpha_r \cdot \alpha_f - 1)^2 \cdot I_{sc} \cdot I_{se} + g_1^* \cdot g_2^* + g_1^* \cdot (\alpha_r \cdot \alpha_f - 1) \cdot I_{se} + g_2^* \cdot (\alpha_r \cdot \alpha_f - 1) \cdot I_{sc}]$$

$$\upsilon_{11}(Y_2, I_{C_{out}}, \ldots) = \frac{\Omega_1(I_{C_{out}}^{(j)}, g_1^*, g_2^*, \ldots)}{\Gamma_2(g_1^*, g_2^*, \Gamma_1, \alpha_r, \alpha_f)} + \frac{\begin{array}{l}[g_1^* \cdot g_2^* \cdot R_{load} \cdot y_2 + g_1^* \cdot (\alpha_r \cdot \alpha_f - 1) \cdot I_{se} \cdot R_{load} \cdot y_2 \\ + g_2^* \cdot (\alpha_r \cdot \alpha_f - 1) \cdot I_{sc} \cdot R_{load} \cdot y_2 + R_{load} \cdot (\alpha_r \cdot \alpha_f - 1)^2 \cdot I_{sc} \cdot I_{se} \cdot y_2 \\ + I_{C_{out}}^{(j)} \cdot \frac{1}{C_{out}} \cdot f_1(i_{L_1}, y_4, \ldots) + g_1^* \cdot g_2^* \cdot \frac{1}{C_{out}} \cdot i_{C_{out}} + I_{C_{out}}^{(j)} \cdot (\alpha_r \cdot \alpha_f - 1) \cdot \frac{1}{C_{out}} \cdot I_{se} \cdot f_2(i_{L_1}, y_4, \ldots) \\ + g_1^* \cdot (\alpha_r \cdot \alpha_f - 1) \cdot \frac{1}{C_{out}} \cdot I_{se} \cdot i_{C_{out}} + I_{C_{out}}^{(j)} \cdot (\alpha_r \cdot \alpha_f - 1) \cdot \frac{1}{C_{out}} \cdot I_{sc} \cdot f_3(i_{L_1}, y_4, \ldots) \\ + g_2^* \cdot (\alpha_r \cdot \alpha_f - 1) \cdot \frac{1}{C_{out}} \cdot I_{sc} \cdot i_{C_{out}} + \frac{1}{C_{out}} \cdot (\alpha_r \cdot \alpha_f - 1)^2 \cdot I_{sc} \cdot I_{se} \cdot i_{C_{out}}] \cdot e^{\lambda \cdot t}\end{array}}{\Gamma_2(g_1^*, g_2^*, \Gamma_1, \alpha_r, \alpha_f)}$$

We define $\upsilon_{11}(Y_2, I_{C_{out}}, \ldots) = \frac{\Omega_1(I_{C_{out}}^{(j)}, g_1^*, g_2^*, \ldots)}{\Gamma_2(g_1^*, g_2^*, \Gamma_1, \alpha_r, \alpha_f)} + \frac{\Upsilon_1(y_2, f_1(i_{L_1}, y_4, \ldots), i_{C_{out}}, \ldots) \cdot e^{\lambda \cdot t}}{\Gamma_2(g_1^*, g_2^*, \Gamma_1, \alpha_r, \alpha_f)}$

$$\Upsilon_1(y_2, f_1(i_{L_1}, y_4, \ldots), i_{C_{out}}, \ldots) = [g_1^* \cdot g_2^* \cdot R_{load} \cdot y_2 + g_1^* \cdot (\alpha_r \cdot \alpha_f - 1) \cdot I_{se} \cdot R_{load} \cdot y_2$$
$$+ g_2^* \cdot (\alpha_r \cdot \alpha_f - 1) \cdot I_{sc} \cdot R_{load} \cdot y_2 + R_{load} \cdot (\alpha_r \cdot \alpha_f - 1)^2 \cdot I_{sc} \cdot I_{se} \cdot y_2$$
$$+ I_{C_{out}}^{(j)} \cdot \frac{1}{C_{out}} \cdot f_1(i_{L_1}, y_4, \ldots) + g_1^* \cdot g_2^* \cdot \frac{1}{C_{out}} \cdot i_{C_{out}} + I_{C_{out}}^{(j)} \cdot (\alpha_r \cdot \alpha_f - 1) \cdot \frac{1}{C_{out}} \cdot I_{se} \cdot f_2(i_{L_1}, y_4, \ldots)$$
$$+ g_1^* \cdot (\alpha_r \cdot \alpha_f - 1) \cdot \frac{1}{C_{out}} \cdot I_{se} \cdot i_{C_{out}} + I_{C_{out}}^{(j)} \cdot (\alpha_r \cdot \alpha_f - 1) \cdot \frac{1}{C_{out}} \cdot I_{sc} \cdot f_3(i_{L_1}, y_4, \ldots)$$
$$+ g_2^* \cdot (\alpha_r \cdot \alpha_f - 1) \cdot \frac{1}{C_{out}} \cdot I_{sc} \cdot i_{C_{out}} + \frac{1}{C_{out}} \cdot (\alpha_r \cdot \alpha_f - 1)^2 \cdot I_{sc} \cdot I_{se} \cdot i_{C_{out}}]$$

$$\upsilon_{12}(Y_2, I_{C_{out}}, \ldots) = \frac{\Gamma_3(\alpha_f, \alpha_r) \cdot (i_{C_{in}} - i_{R_b}) \cdot e^{\lambda \cdot t} \cdot \sum_{j=1}^{8} \Phi_{1j}}{\Gamma_2^2(g_1^*, g_2^*, \Gamma_1, \alpha_r, \alpha_f)}$$

$$\upsilon_{12}(Y_2, I_{C_{out}}, \ldots) = \frac{\begin{array}{r} \Gamma_3(\alpha_f, \alpha_r) \cdot (i_{C_{in}} - i_{R_b}) \cdot e^{\lambda \cdot t} \cdot [g_1^* \cdot g_2^* \cdot R_{load} \cdot y_2 \cdot e^{\lambda \cdot t} + g_1^* \cdot (\alpha_r \cdot \alpha_f - 1) \cdot I_{se} \cdot R_{load} \cdot y_2 \cdot e^{\lambda \cdot t} \\ + g_2^* \cdot (\alpha_r \cdot \alpha_f - 1) \cdot I_{sc} \cdot R_{load} \cdot y_2 \cdot e^{\lambda \cdot t} + R_{load} \cdot (\alpha_r \cdot \alpha_f - 1)^2 \cdot I_{sc} \cdot I_{se} \cdot y_2 \cdot e^{\lambda \cdot t} \\ + I_{C_{out}}^{(j)} \cdot g_1^* \cdot g_2^* \cdot \frac{1}{C_{out}} + I_{C_{out}}^{(j)} \cdot \frac{1}{C_{out}} \cdot f_1(i_{L_1}, y_4, \ldots) \cdot e^{\lambda \cdot t} + g_1^* \cdot g_2^* \cdot \frac{1}{C_{out}} \cdot i_{C_{out}} \cdot e^{\lambda \cdot t} \\ + \{I_{C_{out}}^{(j)} \cdot g_1^* + I_{C_{out}}^{(j)} \cdot f_2(i_{L_1}, y_4, \ldots) \cdot e^{\lambda \cdot t} + g_1^* \cdot i_{C_{out}} \cdot e^{\lambda \cdot t}\} \cdot (\alpha_r \cdot \alpha_f - 1) \cdot \frac{1}{C_{out}} \cdot I_{se} \\ + \{I_{C_{out}}^{(j)} \cdot g_2^* + I_{C_{out}}^{(j)} \cdot f_3(i_{L_1}, y_4, \ldots) \cdot e^{\lambda \cdot t} + g_2^* \cdot i_{C_{out}} \cdot e^{\lambda \cdot t}\} \cdot (\alpha_r \cdot \alpha_f - 1) \cdot \frac{1}{C_{out}} \cdot I_{sc} \\ + I_{C_{out}}^{(j)} \cdot \frac{1}{C_{out}} \cdot (\alpha_r \cdot \alpha_f - 1)^2 \cdot I_{sc} \cdot I_{se} + \frac{1}{C_{out}} \cdot (\alpha_r \cdot \alpha_f - 1)^2 \cdot I_{sc} \cdot I_{se} \cdot i_{C_{out}} \cdot e^{\lambda \cdot t}] \end{array}}{\Gamma_2^2(g_1^*, g_2^*, \Gamma_1, \alpha_r, \alpha_f)}$$

$$\upsilon_{12}(Y_2, I_{C_{out}}, \ldots) = \frac{\begin{array}{r} \Gamma_3(\alpha_f, \alpha_r) \cdot (i_{C_{in}} - i_{R_b}) \cdot e^{\lambda \cdot t} \cdot [I_{C_{out}}^{(j)} \cdot \frac{1}{C_{out}} \cdot [(\alpha_r \cdot \alpha_f - 1)^2 \cdot I_{sc} \cdot I_{se} + g_1^* \cdot g_2^* \\ + g_1^* \cdot (\alpha_r \cdot \alpha_f - 1) \cdot I_{se} + g_2^* \cdot (\alpha_r \cdot \alpha_f - 1) \cdot I_{sc}] \\ + \Gamma_3(\alpha_f, \alpha_r) \cdot (i_{C_{in}} - i_{R_b}) \cdot e^{\lambda \cdot t} \cdot [g_1^* \cdot g_2^* \cdot R_{load} \cdot y_2 + g_1^* \cdot (\alpha_r \cdot \alpha_f - 1) \cdot I_{se} \cdot R_{load} \cdot y_2 \\ + g_2^* \cdot (\alpha_r \cdot \alpha_f - 1) \cdot I_{sc} \cdot R_{load} \cdot y_2 + R_{load} \cdot (\alpha_r \cdot \alpha_f - 1)^2 \cdot I_{sc} \cdot I_{se} \cdot y_2 \\ + I_{C_{out}}^{(j)} \cdot \frac{1}{C_{out}} \cdot f_1(i_{L_1}, y_4, \ldots) + g_1^* \cdot g_2^* \cdot \frac{1}{C_{out}} \cdot i_{C_{out}} + I_{C_{out}}^{(j)} \cdot (\alpha_r \cdot \alpha_f - 1) \cdot \frac{1}{C_{out}} \cdot I_{se} \cdot f_2(i_{L_1}, y_4, \ldots) \\ + g_1^* \cdot (\alpha_r \cdot \alpha_f - 1) \cdot \frac{1}{C_{out}} \cdot I_{se} \cdot i_{C_{out}} + I_{C_{out}}^{(j)} \cdot (\alpha_r \cdot \alpha_f - 1) \cdot \frac{1}{C_{out}} \cdot I_{sc} \cdot f_3(i_{L_1}, y_4, \ldots) \\ + g_2^* \cdot (\alpha_r \cdot \alpha_f - 1) \cdot \frac{1}{C_{out}} \cdot I_{sc} \cdot i_{C_{out}} + \frac{1}{C_{out}} \cdot (\alpha_r \cdot \alpha_f - 1)^2 \cdot I_{sc} \cdot I_{se} \cdot i_{C_{out}}] \cdot e^{\lambda \cdot t} \end{array}}{\Gamma_2^2(g_1^*, g_2^*, \Gamma_1, \alpha_r, \alpha_f)}$$

Since $(i_{C_{in}} - i_{R_b}) \cdot y_2 \to \varepsilon; (i_{C_{in}} - i_{R_b}) \cdot f_1(i_{L_1}, y_4, \ldots) \to \varepsilon \ldots$

$$\upsilon_{12}(Y_2, I_{C_{out}}, \ldots) = \frac{\begin{array}{r} \Gamma_3(\alpha_f, \alpha_r) \cdot (i_{C_{in}} - i_{R_b}) \cdot e^{\lambda \cdot t} \cdot [I_{C_{out}}^{(j)} \cdot \frac{1}{C_{out}} \cdot [(\alpha_r \cdot \alpha_f - 1)^2 \cdot I_{sc} \cdot I_{se} + g_1^* \cdot g_2^* \\ + g_1^* \cdot (\alpha_r \cdot \alpha_f - 1) \cdot I_{se} + g_2^* \cdot (\alpha_r \cdot \alpha_f - 1) \cdot I_{sc}] \end{array}}{\Gamma_2^2(g_1^*, g_2^*, \Gamma_1, \alpha_r, \alpha_f)}$$

We define $\upsilon_{12}(Y_2, I_{C_{out}}, \ldots) = \frac{\Upsilon_2(g_1^*, g_2^*, \alpha_r, \alpha_f, \ldots) \cdot \Gamma_3(\alpha_f, \alpha_r) \cdot (i_{C_{in}} - i_{R_b}) \cdot e^{\lambda \cdot t}}{\Gamma_2^2(g_1^*, g_2^*, \Gamma_1, \alpha_r, \alpha_f)}$

$$\Upsilon_2(g_1^*, g_2^*, \alpha_r, \alpha_f, \ldots) = [I_{C_{out}}^{(j)} \cdot \frac{1}{C_{out}} \cdot [(\alpha_r \cdot \alpha_f - 1)^2 \cdot I_{sc} \cdot I_{se} + g_1^* \cdot g_2^* + g_1^* \cdot (\alpha_r \cdot \alpha_f - 1) \cdot I_{se} + g_2^* \cdot (\alpha_r \cdot \alpha_f - 1) \cdot I_{sc}]$$

Stage 2:

$$\upsilon_2(Y_2, I_{C_{out}}, \ldots) = \frac{\sum_{k=2}^{3} \Phi_k}{[\Gamma_2(g_1^*, g_2^*, \Gamma_1, \alpha_r, \alpha_f) + \Gamma_3(\alpha_f, \alpha_r) \cdot (i_{C_{in}} - i_{R_b}) \cdot e^{\lambda \cdot t}]}$$

$$\Phi_2(g_1, g_3(\frac{dV_s(t)}{dt} \to \varepsilon), \alpha_f, \alpha_r, I_{sc}) = -\{g_3^* \cdot [g_1^* + (\alpha_f \cdot \alpha_r - 1) \cdot I_{sc}] + g_3^* \cdot f_4(i_{L_1}, y_4, \ldots) \cdot e^{\lambda \cdot t} + [g_1^* + (\alpha_f \cdot \alpha_r - 1) \cdot I_{sc}] \cdot f_5(y_3, y_2, \ldots) \cdot e^{\lambda \cdot t}\}$$

$$\Phi_3(g_2, g_4(\frac{dV_s(t)}{dt} \to \varepsilon), \alpha_f, \alpha_r, I_{se}) = -\{g_4^* \cdot [g_2^* + (\alpha_r \cdot \alpha_f - 1) \cdot I_{se}] + g_4^* \cdot f_6(i_{L_1}, y_4, \ldots) \cdot e^{\lambda \cdot t} + [g_2^* + (\alpha_r \cdot \alpha_f - 1) \cdot I_{se}] \cdot f_7(y_3, y_2, \ldots) \cdot e^{\lambda \cdot t}\}$$

$$\upsilon_2(Y_2, I_{C_{out}}, \ldots) = \frac{\Phi_2(g_1, g_3(\frac{dV_s(t)}{dt} \to \varepsilon), \alpha_f, \alpha_r, I_{sc}) + \Phi_3(g_2, g_4(\frac{dV_s(t)}{dt} \to \varepsilon), \alpha_f, \alpha_r, I_{se})}{[\Gamma_2(g_1^*, g_2^*, \Gamma_1, \alpha_r, \alpha_f) + \Gamma_3(\alpha_f, \alpha_r) \cdot (i_{C_{in}} - i_{R_b}) \cdot e^{\lambda \cdot t}]}$$

$$\upsilon_2(Y_2, I_{C_{out}}, \ldots) = \frac{\Phi_2(g_1, g_3(\frac{dV_s(t)}{dt} \to \varepsilon), \alpha_f, \alpha_r, I_{sc})}{[\Gamma_2(g_1^*, g_2^*, \Gamma_1, \alpha_r, \alpha_f) + \Gamma_3(\alpha_f, \alpha_r) \cdot (i_{C_{in}} - i_{R_b}) \cdot e^{\lambda \cdot t}]} + \frac{\Phi_3(g_2, g_4(\frac{dV_s(t)}{dt} \to \varepsilon), \alpha_f, \alpha_r, I_{se})}{[\Gamma_2(g_1^*, g_2^*, \Gamma_1, \alpha_r, \alpha_f) + \Gamma_3(\alpha_f, \alpha_r) \cdot (i_{C_{in}} - i_{R_b}) \cdot e^{\lambda \cdot t}]}$$

$$\upsilon_{21}(Y_2, I_{C_{out}}, \ldots) = \frac{\Phi_2(g_1, g_3(\frac{dV_s(t)}{dt} \to \varepsilon), \alpha_f, \alpha_r, I_{sc})}{[\Gamma_2(g_1^*, g_2^*, \Gamma_1, \alpha_r, \alpha_f) + \Gamma_3(\alpha_f, \alpha_r) \cdot (i_{C_{in}} - i_{R_b}) \cdot e^{\lambda \cdot t}]}$$

$$\upsilon_{22}(Y_2, I_{C_{out}}, \ldots) = \frac{\Phi_3(g_2, g_4(\frac{dV_s(t)}{dt} \to \varepsilon), \alpha_f, \alpha_r, I_{se})}{[\Gamma_2(g_1^*, g_2^*, \Gamma_1, \alpha_r, \alpha_f) + \Gamma_3(\alpha_f, \alpha_r) \cdot (i_{C_{in}} - i_{R_b}) \cdot e^{\lambda \cdot t}]}$$

$$\upsilon_2(Y_2, I_{C_{out}}, \ldots) = \sum_{k=1}^{2} \upsilon_{2k}(Y_2, I_{C_{out}}, \ldots) = \upsilon_{21}(Y_2, I_{C_{out}}, \ldots) + \upsilon_{22}(Y_2, I_{C_{out}}, \ldots)$$

$$\upsilon_{21}(Y_2, I_{C_{out}}, \ldots) = \{\frac{\Phi_2(g_1, g_3(\frac{dV_s(t)}{dt} \to \varepsilon), \alpha_f, \alpha_r, I_{sc})}{[\Gamma_2(g_1^*, g_2^*, \Gamma_1, \alpha_r, \alpha_f) + \Gamma_3(\alpha_f, \alpha_r) \cdot (i_{C_{in}} - i_{R_b}) \cdot e^{\lambda \cdot t}]}\} \cdot \{\frac{[\Gamma_2(g_1^*, g_2^*, \Gamma_1, \alpha_r, \alpha_f) - \Gamma_3(\alpha_f, \alpha_r) \cdot (i_{C_{in}} - i_{R_b}) \cdot e^{\lambda \cdot t}]}{[\Gamma_2(g_1^*, g_2^*, \Gamma_1, \alpha_r, \alpha_f) - \Gamma_3(\alpha_f, \alpha_r) \cdot (i_{C_{in}} - i_{R_b}) \cdot e^{\lambda \cdot t}]}\}$$

$$\upsilon_{21}(Y_2, I_{C_{out}}, \ldots) = \frac{\Phi_2(g_1, g_3(\frac{dV_s(t)}{dt} \to \varepsilon), \alpha_f, \alpha_r, I_{sc}) \cdot [\Gamma_2(g_1^*, g_2^*, \Gamma_1, \alpha_r, \alpha_f) - \Gamma_3(\alpha_f, \alpha_r) \cdot (i_{C_{in}} - i_{R_b}) \cdot e^{\lambda \cdot t}]}{\Gamma_2^2(g_1^*, g_2^*, \Gamma_1, \alpha_r, \alpha_f) - \Gamma_3^2(\alpha_f, \alpha_r) \cdot (i_{C_{in}} - i_{R_b})^2 \cdot e^{2 \cdot \lambda \cdot t}}$$

Assume $(i_{C_{in}} - i_{R_b})^2 \to \varepsilon$ then

$$\upsilon_{21}(Y_2, I_{C_{out}}, \ldots) = \frac{\Phi_2(g_1, g_3(\frac{dV_s(t)}{dt} \to \varepsilon), \alpha_f, \alpha_r, I_{sc}) \cdot [\Gamma_2(g_1^*, g_2^*, \Gamma_1, \alpha_r, \alpha_f) - \Gamma_3(\alpha_f, \alpha_r) \cdot (i_{C_{in}} - i_{R_b}) \cdot e^{\lambda \cdot t}]}{\Gamma_2^2(g_1^*, g_2^*, \Gamma_1, \alpha_r, \alpha_f)}$$

$$\upsilon_{21}(Y_2, I_{C_{out}}, \ldots) = \frac{\Phi_2(g_1, g_3(\frac{dV_s(t)}{dt} \to \varepsilon), \alpha_f, \alpha_r, I_{sc})}{\Gamma_2(g_1^*, g_2^*, \Gamma_1, \alpha_r, \alpha_f)} - \frac{\Phi_2(g_1, g_3(\frac{dV_s(t)}{dt} \to \varepsilon), \alpha_f, \alpha_r, I_{sc}) \cdot \Gamma_3(\alpha_f, \alpha_r) \cdot (i_{C_{in}} - i_{R_b}) \cdot e^{\lambda \cdot t}}{\Gamma_2^2(g_1^*, g_2^*, \Gamma_1, \alpha_r, \alpha_f)}$$

$$\upsilon_{21}(Y_2, I_{C_{out}}, \ldots) = \frac{\Phi_2(g_1, g_3(\frac{dV_s(t)}{dt} \to \varepsilon), \alpha_f, \alpha_r, I_{sc})}{\Gamma_2(g_1^*, g_2^*, \Gamma_1, \alpha_r, \alpha_f)} + \frac{\{g_3^* \cdot [g_1^* + (\alpha_f \cdot \alpha_r - 1) \cdot I_{sc}] + g_3^* \cdot f_4(i_{L_1}, y_4, \ldots) \cdot e^{\lambda \cdot t} + [g_1^* + (\alpha_f \cdot \alpha_r - 1) \cdot I_{sc}] \cdot f_5(y_3, y_2, \ldots) \cdot e^{\lambda \cdot t}\} \cdot \Gamma_3(\alpha_f, \alpha_r) \cdot (i_{C_{in}} - i_{R_b}) \cdot e^{\lambda \cdot t}}{\Gamma_2^2(g_1^*, g_2^*, \Gamma_1, \alpha_r, \alpha_f)}$$

Since $f_4(i_{L_1}, y_4, \ldots) \cdot (i_{C_{in}} - i_{R_b}) \to \varepsilon; f_5(y_3, y_2, \ldots) \cdot (i_{C_{in}} - i_{R_b}) \to \varepsilon$

$$\upsilon_{21}(Y_2, I_{C_{out}}, \ldots) = \frac{\Phi_2(g_1, g_3(\frac{dV_s(t)}{dt} \to \varepsilon), \alpha_f, \alpha_r, I_{sc})}{\Gamma_2(g_1^*, g_2^*, \Gamma_1, \alpha_r, \alpha_f)} + \frac{g_3^* \cdot [g_1^* + (\alpha_f \cdot \alpha_r - 1) \cdot I_{sc}] \cdot \Gamma_3(\alpha_f, \alpha_r) \cdot (i_{C_{in}} - i_{R_b}) \cdot e^{\lambda \cdot t}}{\Gamma_2^2(g_1^*, g_2^*, \Gamma_1, \alpha_r, \alpha_f)}$$

$$\upsilon_{21}(Y_2, I_{C_{out}}, \ldots) = -\frac{g_3^* \cdot [g_1^* + (\alpha_f \cdot \alpha_r - 1) \cdot I_{sc}]}{\Gamma_2(g_1^*, g_2^*, \Gamma_1, \alpha_r, \alpha_f)} - \frac{\{g_3^* \cdot f_4(i_{L_1}, y_4, \ldots) + [g_1^* + (\alpha_f \cdot \alpha_r - 1) \cdot I_{sc}] \cdot f_5(y_3, y_2, \ldots)\} \cdot e^{\lambda \cdot t}}{\Gamma_2(g_1^*, g_2^*, \Gamma_1, \alpha_r, \alpha_f)} + \frac{g_3^* \cdot [g_1^* + (\alpha_f \cdot \alpha_r - 1) \cdot I_{sc}] \cdot \Gamma_3(\alpha_f, \alpha_r) \cdot (i_{C_{in}} - i_{R_b}) \cdot e^{\lambda \cdot t}}{\Gamma_2^2(g_1^*, g_2^*, \Gamma_1, \alpha_r, \alpha_f)}$$

We define $\upsilon_{21}(Y_2, I_{C_{out}}, \ldots) = -\frac{g_3^* \cdot [g_1^* + (\alpha_f \cdot \alpha_r - 1) \cdot I_{sc}]}{\Gamma_2(g_1^*, g_2^*, \Gamma_1, \alpha_r, \alpha_f)} + \Upsilon_4(i_{C_{in}}, i_{R_b}, f_4(i_{L_1}, y_4, \ldots), \ldots) \cdot e^{\lambda \cdot t}$

$$\Upsilon_4(i_{C_{in}}, i_{R_b}, f_4(i_{L_1}, y_4, \ldots), \ldots) = \frac{g_3^* \cdot [g_1^* + (\alpha_f \cdot \alpha_r - 1) \cdot I_{sc}] \cdot \Gamma_3(\alpha_f, \alpha_r) \cdot (i_{C_{in}} - i_{R_b})}{\Gamma_2^2(g_1^*, g_2^*, \Gamma_1, \alpha_r, \alpha_f)} - \frac{\{g_3^* \cdot f_4(i_{L_1}, y_4, \ldots) + [g_1^* + (\alpha_f \cdot \alpha_r - 1) \cdot I_{sc}] \cdot f_5(y_3, y_2, \ldots)\}}{\Gamma_2(g_1^*, g_2^*, \Gamma_1, \alpha_r, \alpha_f)}$$

$$\upsilon_{22}(Y_2, I_{C_{out}}, \ldots) = \frac{\Phi_3(g_2, g_4(\frac{dV_s(t)}{dt} \to \varepsilon), \alpha_f, \alpha_r, I_{se})}{[\Gamma_2(g_1^*, g_2^*, \Gamma_1, \alpha_r, \alpha_f) + \Gamma_3(\alpha_f, \alpha_r) \cdot (i_{C_{in}} - i_{R_b}) \cdot e^{\lambda \cdot t}]} \cdot \{\frac{[\Gamma_2(g_1^*, g_2^*, \Gamma_1, \alpha_r, \alpha_f) - \Gamma_3(\alpha_f, \alpha_r) \cdot (i_{C_{in}} - i_{R_b}) \cdot e^{\lambda \cdot t}]}{[\Gamma_2(g_1^*, g_2^*, \Gamma_1, \alpha_r, \alpha_f) - \Gamma_3(\alpha_f, \alpha_r) \cdot (i_{C_{in}} - i_{R_b}) \cdot e^{\lambda \cdot t}]}\}$$

$$\upsilon_{22}(Y_2, I_{C_{out}}, \ldots) = \frac{\Phi_3(g_2, g_4(\frac{dV_s(t)}{dt} \to \varepsilon), \alpha_f, \alpha_r, I_{se}) \cdot [\Gamma_2(g_1^*, g_2^*, \Gamma_1, \alpha_r, \alpha_f) - \Gamma_3(\alpha_f, \alpha_r) \cdot (i_{C_{in}} - i_{R_b}) \cdot e^{\lambda \cdot t}]}{\Gamma_2^2(g_1^*, g_2^*, \Gamma_1, \alpha_r, \alpha_f) - \Gamma_3^2(\alpha_f, \alpha_r) \cdot (i_{C_{in}} - i_{R_b})^2 \cdot e^{2 \cdot \lambda \cdot t}}$$

Assume $(i_{C_{in}} - i_{R_b})^2 \to \varepsilon$ then

$$\upsilon_{22}(Y_2, I_{C_{out}}, \ldots) = \frac{\Phi_3(g_2, g_4(\frac{dV_s(t)}{dt} \to \varepsilon), \alpha_f, \alpha_r, I_{se}) \cdot [\Gamma_2(g_1^*, g_2^*, \Gamma_1, \alpha_r, \alpha_f) - \Gamma_3(\alpha_f, \alpha_r) \cdot (i_{C_{in}} - i_{R_b}) \cdot e^{\lambda \cdot t}]}{\Gamma_2^2(g_1^*, g_2^*, \Gamma_1, \alpha_r, \alpha_f)}$$

$$\upsilon_{22}(Y_2, I_{C_{out}}, \ldots) = \frac{\Phi_3(g_2, g_4(\frac{dV_s(t)}{dt} \to \varepsilon), \alpha_f, \alpha_r, I_{se})}{\Gamma_2(g_1^*, g_2^*, \Gamma_1, \alpha_r, \alpha_f)} - \frac{\Phi_3(g_2, g_4(\frac{dV_s(t)}{dt} \to \varepsilon), \alpha_f, \alpha_r, I_{se}) \cdot \Gamma_3(\alpha_f, \alpha_r) \cdot (i_{C_{in}} - i_{R_b}) \cdot e^{\lambda \cdot t}}{\Gamma_2^2(g_1^*, g_2^*, \Gamma_1, \alpha_r, \alpha_f)}$$

$$\upsilon_{22}(Y_2, I_{C_{out}}, \ldots) = \frac{\Phi_3(g_2, g_4(\frac{dV_s(t)}{dt} \to \varepsilon), \alpha_f, \alpha_r, I_{se})}{\Gamma_2(g_1^*, g_2^*, \Gamma_1, \alpha_r, \alpha_f)}$$
$$+ \frac{\{g_4^* \cdot [g_2^* + (\alpha_r \cdot \alpha_f - 1) \cdot I_{se}] + g_4^* \cdot f_6(i_{L_1}, y_4, \ldots) \cdot e^{\lambda \cdot t} + [g_2^* + (\alpha_r \cdot \alpha_f - 1) \cdot I_{se}] \cdot f_7(y_3, y_2, \ldots) \cdot e^{\lambda \cdot t}\} \cdot \Gamma_3(\alpha_f, \alpha_r) \cdot (i_{C_{in}} - i_{R_b}) \cdot e^{\lambda \cdot t}}{\Gamma_2^2(g_1^*, g_2^*, \Gamma_1, \alpha_r, \alpha_f)}$$

Since $f_6(i_{L_1}, y_4, \ldots) \cdot (i_{C_{in}} - i_{R_b}) \to \varepsilon; f_7(y_3, y_2, \ldots) \cdot (i_{C_{in}} - i_{R_b}) \to \varepsilon$

$$\upsilon_{22}(Y_2, I_{C_{out}}, \ldots) = \frac{\Phi_3(g_2, g_4(\frac{dV_s(t)}{dt} \to \varepsilon), \alpha_f, \alpha_r, I_{se})}{\Gamma_2(g_1^*, g_2^*, \Gamma_1, \alpha_r, \alpha_f)} + \frac{g_4^* \cdot [g_2^* + (\alpha_r \cdot \alpha_f - 1) \cdot I_{se}] \cdot \Gamma_3(\alpha_f, \alpha_r) \cdot (i_{C_{in}} - i_{R_b}) \cdot e^{\lambda \cdot t}}{\Gamma_2^2(g_1^*, g_2^*, \Gamma_1, \alpha_r, \alpha_f)}$$

$$\upsilon_{22}(Y_2, I_{C_{out}}, \ldots) = \frac{-g_4^* \cdot [g_2^* + (\alpha_r \cdot \alpha_f - 1) \cdot I_{se}]}{\Gamma_2(g_1^*, g_2^*, \Gamma_1, \alpha_r, \alpha_f)} - \frac{\{g_4^* \cdot f_6(i_{L_1}, y_4, \ldots) + [g_2^* + (\alpha_r \cdot \alpha_f - 1) \cdot I_{se}] \cdot f_7(y_3, y_2, \ldots)\} \cdot e^{\lambda \cdot t}}{\Gamma_2(g_1^*, g_2^*, \Gamma_1, \alpha_r, \alpha_f)}$$
$$+ \frac{g_4^* \cdot [g_2^* + (\alpha_r \cdot \alpha_f - 1) \cdot I_{se}] \cdot \Gamma_3(\alpha_f, \alpha_r) \cdot (i_{C_{in}} - i_{R_b}) \cdot e^{\lambda \cdot t}}{\Gamma_2^2(g_1^*, g_2^*, \Gamma_1, \alpha_r, \alpha_f)}$$

We define $\upsilon_{22}(Y_2, I_{C_{out}}, \ldots) = \frac{-g_4^* \cdot [g_2^* + (\alpha_r \cdot \alpha_f - 1) \cdot I_{se}]}{\Gamma_2(g_1^*, g_2^*, \Gamma_1, \alpha_r, \alpha_f)} + \Upsilon_3(i_{C_{in}}, i_{R_b}, \ldots) \cdot e^{\lambda \cdot t}$

$$\Upsilon_3(i_{C_{in}}, i_{R_b}, f_6(i_{L_1}, y_4, \ldots), f_7(y_3, y_2, \ldots)) = \frac{g_4^* \cdot [g_2^* + (\alpha_r \cdot \alpha_f - 1) \cdot I_{se}] \cdot \Gamma_3(\alpha_f, \alpha_r) \cdot (i_{C_{in}} - i_{R_b})}{\Gamma_2^2(g_1^*, g_2^*, \Gamma_1, \alpha_r, \alpha_f)} - \frac{\{g_4^* \cdot f_6(i_{L_1}, y_4, \ldots) + [g_2^* + (\alpha_r \cdot \alpha_f - 1) \cdot I_{se}] \cdot f_7(y_3, y_2, \ldots)\}}{\Gamma_2(g_1^*, g_2^*, \Gamma_1, \alpha_r, \alpha_f)}$$

Summary: We define function: $\frac{dY_4}{dt} = \upsilon(Y_2, I_{C_{out}}, \ldots)$

$$\frac{dY_4}{dt} = \frac{1}{C_1 \cdot L_1} \cdot \{\frac{1}{V_T} \cdot \sum_{k=1}^{2} \upsilon_{1k}(Y_2, I_{C_{out}}, \ldots) + \sum_{k=1}^{2} \upsilon_{2k}(Y_2, I_{C_{out}}, \ldots)\}$$
$$\frac{dY_4}{dt} = \frac{1}{C_1 \cdot L_1 \cdot V_T} \cdot \sum_{k=1}^{2} \upsilon_{1k}(Y_2, I_{C_{out}}, \ldots) + \frac{1}{C_1 \cdot L_1} \cdot \sum_{k=1}^{2} \upsilon_{2k}(Y_2, I_{C_{out}}, \ldots)$$

$$\upsilon_{11}(Y_2, I_{C_{out}}, \ldots) = \frac{\Omega_1(I_{C_{out}}^{(j)}, g_1^*, g_2^*, \ldots)}{\Gamma_2(g_1^*, g_2^*, \Gamma_1, \alpha_r, \alpha_f)} + \frac{\Upsilon_1(y_2, f_1(i_{L_1}, y_4, \ldots), i_{C_{out}}, \ldots) \cdot e^{\lambda \cdot t}}{\Gamma_2(g_1^*, g_2^*, \Gamma_1, \alpha_r, \alpha_f)}$$

$$\upsilon_{12}(Y_2, I_{C_{out}}, \ldots) = \frac{\Upsilon_2(g_1^*, g_2^*, \alpha_r, \alpha_f, \ldots) \cdot \Gamma_3(\alpha_f, \alpha_r) \cdot (i_{C_{in}} - i_{R_b}) \cdot e^{\lambda \cdot t}}{\Gamma_2^2(g_1^*, g_2^*, \Gamma_1, \alpha_r, \alpha_f)}$$

$$\upsilon_{21}(Y_2, I_{C_{out}}, \ldots) = -\frac{g_3^* \cdot [g_1^* + (\alpha_f \cdot \alpha_r - 1) \cdot I_{sc}]}{\Gamma_2(g_1^*, g_2^*, \Gamma_1, \alpha_r, \alpha_f)} + \Upsilon_4(i_{C_{in}}, i_{R_b}, f_4(i_{L_1}, y_4, \ldots), \ldots \cdot e^{\lambda \cdot t}$$

$$\upsilon_{22}(Y_2, I_{C_{out}}, \ldots) = \frac{-g_4^* \cdot [g_2^* + (\alpha_r \cdot \alpha_f - 1) \cdot I_{se}]}{\Gamma_2(g_1^*, g_2^*, \Gamma_1, \alpha_r, \alpha_f)} + \Upsilon_3(i_{C_{in}}, i_{R_b}, \ldots) \cdot e^{\lambda \cdot t}$$

$$\begin{aligned}
\upsilon(Y_2, I_{C_{out}}, \ldots) = {} & \frac{1}{C_1 \cdot L_1 \cdot V_T} \cdot \{\frac{\Omega_1(I_{C_{out}}^{(j)}, g_1^*, g_2^*, \ldots)}{\Gamma_2(g_1^*, g_2^*, \Gamma_1, \alpha_r, \alpha_f)} + \frac{\Upsilon_1(y_2, f_1(i_{L_1}, y_4, \ldots), i_{C_{out}}, \ldots) \cdot e^{\lambda \cdot t}}{\Gamma_2(g_1^*, g_2^*, \Gamma_1, \alpha_r, \alpha_f)} \\
& + \frac{\Upsilon_2(g_1^*, g_2^*, \alpha_r, \alpha_f, \ldots) \cdot \Gamma_3(\alpha_f, \alpha_r) \cdot (i_{C_{in}} - i_{R_b}) \cdot e^{\lambda \cdot t}}{\Gamma_2^2(g_1^*, g_2^*, \Gamma_1, \alpha_r, \alpha_f)}\} \\
& + \frac{1}{C_1 \cdot L_1} \cdot \{-\frac{g_3^* \cdot [g_1^* + (\alpha_f \cdot \alpha_r - 1) \cdot I_{sc}]}{\Gamma_2(g_1^*, g_2^*, \Gamma_1, \alpha_r, \alpha_f)} \\
& + \Upsilon_4(i_{C_{in}}, i_{R_b}, f_4(i_{L_1}, y_4, \ldots), \ldots) \cdot e^{\lambda \cdot t} + \{\frac{-g_4^* \cdot [g_2^* + (\alpha_r \cdot \alpha_f - 1) \cdot I_{se}]}{\Gamma_2(g_1^*, g_2^*, \Gamma_1, \alpha_r, \alpha_f)}\} \\
& + \Upsilon_3(i_{C_{in}}, i_{R_b}, \ldots) \cdot e^{\lambda \cdot t}\}
\end{aligned}$$

$$\begin{aligned}
\upsilon(Y_2, I_{C_{out}}, \ldots) = {} & \frac{1}{C_1 \cdot L_1 \cdot V_T} \cdot \frac{\Omega_1(I_{C_{out}}^{(j)}, g_1^*, g_2^*, \ldots)}{\Gamma_2(g_1^*, g_2^*, \Gamma_1, \alpha_r, \alpha_f)} \\
& + \frac{1}{C_1 \cdot L_1 \cdot V_T} \cdot \frac{\Upsilon_1(y_2, f_1(i_{L_1}, y_4, \ldots), i_{C_{out}}, \ldots) \cdot e^{\lambda \cdot t}}{\Gamma_2(g_1^*, g_2^*, \Gamma_1, \alpha_r, \alpha_f)} \\
& + \frac{1}{C_1 \cdot L_1 \cdot V_T} \cdot \frac{\Upsilon_2(g_1^*, g_2^*, \alpha_r, \alpha_f, \ldots) \cdot \Gamma_3(\alpha_f, \alpha_r) \cdot (i_{C_{in}} - i_{R_b}) \cdot e^{\lambda \cdot t}}{\Gamma_2^2(g_1^*, g_2^*, \Gamma_1, \alpha_r, \alpha_f)} \\
& - \frac{1}{C_1 \cdot L_1} \cdot \frac{g_3^* \cdot [g_1^* + (\alpha_f \cdot \alpha_r - 1) \cdot I_{sc}]}{\Gamma_2(g_1^*, g_2^*, \Gamma_1, \alpha_r, \alpha_f)} + \frac{1}{C_1 \cdot L_1} \cdot \Upsilon_4(i_{C_{in}}, i_{R_b}, f_4(i_{L_1}, y_4, \ldots), \ldots) \cdot e^{\lambda \cdot t} \\
& - \frac{1}{C_1 \cdot L_1} \cdot \frac{g_4^* \cdot [g_2^* + (\alpha_r \cdot \alpha_f - 1) \cdot I_{se}]}{\Gamma_2(g_1^*, g_2^*, \Gamma_1, \alpha_r, \alpha_f)} + \frac{1}{C_1 \cdot L_1} \cdot \Upsilon_3(i_{C_{in}}, i_{R_b}, \ldots) \cdot e^{\lambda \cdot t}
\end{aligned}$$

$$\begin{aligned}
\upsilon(Y_2, I_{C_{out}}, \ldots) = {} & \frac{\{\frac{1}{V_T} \cdot \Omega_1(I_{C_{out}}^{(j)}, g_1^*, g_2^*, \ldots) - g_3^* \cdot [g_1^* + (\alpha_f \cdot \alpha_r - 1) \cdot I_{sc}] - g_4^* \cdot [g_2^* + (\alpha_r \cdot \alpha_f - 1) \cdot I_{se}]\}}{C_1 \cdot L_1 \cdot \Gamma_2(g_1^*, g_2^*, \Gamma_1, \alpha_r, \alpha_f)} \\
& + \{\frac{1}{C_1 \cdot L_1 \cdot V_T} \cdot \frac{\Upsilon_1(y_2, f_1(i_{L_1}, y_4, \ldots), i_{C_{out}}, \ldots)}{\Gamma_2(g_1^*, g_2^*, \Gamma_1, \alpha_r, \alpha_f)} + \frac{1}{C_1 \cdot L_1 \cdot V_T} \\
& \cdot \frac{\Upsilon_2(g_1^*, g_2^*, \alpha_r, \alpha_f, \ldots) \cdot \Gamma_3(\alpha_f, \alpha_r) \cdot (i_{C_{in}} - i_{R_b})}{\Gamma_2^2(g_1^*, g_2^*, \Gamma_1, \alpha_r, \alpha_f)} \\
& + \frac{1}{C_1 \cdot L_1} \cdot \Upsilon_4(i_{C_{in}}, i_{R_b}, f_4(i_{L_1}, y_4, \ldots), \ldots) + \frac{1}{C_1 \cdot L_1} \cdot \Upsilon_3(i_{C_{in}}, i_{R_b}, \ldots)\} \cdot e^{\lambda \cdot t}
\end{aligned}$$

At fixed point:

$$\frac{\{\frac{1}{V_T}\cdot\Omega_1(I_{C_{out}}^{(j)}, g_1^*, g_2^*,\ldots) - g_3^*\cdot[g_1^* + (\alpha_f\cdot\alpha_r - 1)\cdot I_{sc}] - g_4^*\cdot[g_2^* + (\alpha_r\cdot\alpha_f - 1)\cdot I_{se}]\}}{C_1\cdot L_1\cdot\Gamma_2(g_1^*, g_2^*, \Gamma_1, \alpha_r, \alpha_f)} = 0$$

$$\begin{aligned}\upsilon(Y_2, I_{C_{out}},\ldots) = \{&\frac{1}{C_1\cdot L_1\cdot V_T}\cdot\frac{\Upsilon_1(y_2, f_1(i_{L_1}, y_4,\ldots), i_{C_{out}},\ldots)}{\Gamma_2(g_1^*, g_2^*, \Gamma_1, \alpha_r, \alpha_f)}\\ &+\frac{1}{C_1\cdot L_1\cdot V_T}\cdot\frac{\Upsilon_2(g_1^*, g_2^*, \alpha_r, \alpha_f,\ldots)\cdot\Gamma_3(\alpha_f, \alpha_r)\cdot(i_{C_{in}} - i_{R_b})}{\Gamma_2^2(g_1^*, g_2^*, \Gamma_1, \alpha_r, \alpha_f)}\\ &+\frac{1}{C_1\cdot L_1}\cdot\Upsilon_4(i_{C_{in}}, i_{R_b}, f_4(i_{L_1}, y_4,\ldots),\ldots) + \frac{1}{C_1\cdot L_1}\cdot\Upsilon_3(i_{C_{in}}, i_{R_b},\ldots)\}\cdot e^{\lambda\cdot t}\end{aligned}$$

$$\frac{dY_4}{dt} = \upsilon(Y_2, I_{C_{out}},\ldots); \frac{dY_4}{dt} = y_4\cdot\lambda\cdot e^{\lambda\cdot t}$$

$$\begin{aligned}&-y_4\cdot\lambda + \{\frac{1}{C_1\cdot L_1\cdot V_T}\cdot\frac{\Upsilon_1(y_2, f_1(i_{L_1}, y_4,\ldots), i_{C_{out}},\ldots)}{\Gamma_2(g_1^*, g_2^*, \Gamma_1, \alpha_r, \alpha_f)}\\ &+\frac{1}{C_1\cdot L_1\cdot V_T}\cdot\frac{\Upsilon_2(g_1^*, g_2^*, \alpha_r, \alpha_f,\ldots)\cdot\Gamma_3(\alpha_f, \alpha_r)\cdot(i_{C_{in}} - i_{R_b})}{\Gamma_2^2(g_1^*, g_2^*, \Gamma_1, \alpha_r, \alpha_f)}\\ &+\frac{1}{C_1\cdot L_1}\cdot\Upsilon_4(i_{C_{in}}, i_{R_b}, f_4(i_{L_1}, y_4,\ldots),\ldots) + \frac{1}{C_1\cdot L_1}\cdot\Upsilon_3(i_{C_{in}}, i_{R_b},\ldots)\} = 0\end{aligned}$$

We build our system matrices $(i_{R_b}, i_{C_{in}}, i_{C_{out}}, i_{L_1}, y_3, y_4)$ and (y_1, y_2)

$$-i_{R_b}\cdot\lambda - y_1\cdot\frac{R_s}{R_b} - i_{C_{in}}\cdot\frac{1}{C_{in}\cdot R_b} = 0; -i_{C_{in}}\cdot\lambda + y_1 = 0; -i_{C_{out}}\cdot\lambda + y_2 = 0$$

$$-i_{L_1}\cdot\lambda + y_3 = 0; -y_3\cdot\lambda + y_4 = 0$$

$$\begin{aligned}&-y_4\cdot\lambda + \{\frac{1}{C_1\cdot L_1\cdot V_T}\cdot\frac{\Upsilon_1(y_2, f_1(i_{L_1}, y_4,\ldots), i_{C_{out}},\ldots)}{\Gamma_2(g_1^*, g_2^*, \Gamma_1, \alpha_r, \alpha_f)}\\ &+\frac{1}{C_1\cdot L_1\cdot V_T}\cdot\frac{\Upsilon_2(g_1^*, g_2^*, \alpha_r, \alpha_f,\ldots)\cdot\Gamma_3(\alpha_f, \alpha_r)\cdot(i_{C_{in}} - i_{R_b})}{\Gamma_2^2(g_1^*, g_2^*, \Gamma_1, \alpha_r, \alpha_f)}\\ &+\frac{1}{C_1\cdot L_1}\cdot\Upsilon_4(i_{C_{in}}, i_{R_b}, f_4(i_{L_1}, y_4,\ldots),\ldots) + \frac{1}{C_1\cdot L_1}\cdot\Upsilon_3(i_{C_{in}}, i_{R_b},\ldots)\} = 0\end{aligned}$$

$$\begin{pmatrix} \imath_{11} & \cdots & \imath_{16} \\ \vdots & \ddots & \vdots \\ \imath_{61} & \cdots & \imath_{66} \end{pmatrix} \cdot \begin{pmatrix} i_{R_b} \\ i_{C_{in}} \\ i_{C_{out}} \\ i_{L_1} \\ y_3 \\ y_4 \end{pmatrix} + \begin{pmatrix} v_{11} & v_{12} \\ v_{21} & v_{22} \\ v_{31} & v_{32} \\ v_{41} & v_{42} \\ v_{51} & v_{52} \\ v_{61} & v_{62} \end{pmatrix} \cdot \begin{pmatrix} y_1 \\ y_2 \end{pmatrix} = 0; \imath_{11} = -\lambda;$$

$$\imath_{12} = -\frac{1}{C_{in} \cdot R_b}$$

$$\imath_{13} = 0; \imath_{14} = \imath_{15} = \imath_{16} = 0; \imath_{21} = 0; \imath_{22} = -\lambda; \imath_{23} = \imath_{24} = \imath_{25} = \imath_{26} = 0$$

$$\imath_{31} = \imath_{32} = 0; \imath_{33} = -\lambda; \imath_{34} = \imath_{35} = \imath_{36} = 0; \imath_{41} = \imath_{42} = \imath_{43} = 0; \imath_{44} = -\lambda; \imath_{45} = 1; \imath_{46} = 0$$

$\imath_{51} = \imath_{52} = \imath_{53} = \imath_{54} = 0; \imath_{55} = -\lambda; \imath_{56} = 1; \imath_{66} = -\lambda$. To find $\imath_{16}, \ldots, \imath_{56}, v_{61}, v_{62}$, we need to do some analytic work with the expression:

$$\frac{1}{C_1 \cdot L_1 \cdot V_T} \cdot \frac{\Upsilon_1(y_2, f_1(i_{L_1}, y_4, \ldots), i_{C_{out}}, \ldots)}{\Gamma_2(g_1^*, g_2^*, \Gamma_1, \alpha_r, \alpha_f)} + \frac{1}{C_1 \cdot L_1 \cdot V_T} \cdot \frac{\Upsilon_2(g_1^*, g_2^*, \alpha_r, \alpha_f, \ldots) \cdot \Gamma_3(\alpha_f, \alpha_r) \cdot (i_{C_{in}} - i_{R_b})}{\Gamma_2^2(g_1^*, g_2^*, \Gamma_1, \alpha_r, \alpha_f)} + \frac{1}{C_1 \cdot L_1} \cdot \Upsilon_4(i_{C_{in}}, i_{R_b}, f_4(i_{L_1}, y_4, \ldots), \ldots) + \frac{1}{C_1 \cdot L_1} \cdot \Upsilon_3(i_{C_{in}}, i_{R_b}, \ldots)$$

$$v_{11} = -\frac{R_s}{R_b}; v_{12} = 0; v_{21} = 1; v_{22} = 0; v_{31} = 0; v_{32} = 1; v_{41} = v_{42} = 0; v_{51} = v_{52} = 0$$

Assumption $v_{lk}; l = 1, \ldots, 6; k = 1, 2$ elements are neglected compare to $\imath_{lk}$ elements $l = 1, \ldots, 6; k = 1, \ldots, 6$.

$$(A - \lambda \cdot I) = \begin{pmatrix} \imath_{11} & \cdots & \imath_{16} \\ \vdots & \ddots & \vdots \\ \imath_{61} & \cdots & \imath_{66} \end{pmatrix}; \det(A - \lambda \cdot I) = 0 \Rightarrow \det \begin{pmatrix} \imath_{11} & \cdots & \imath_{16} \\ \vdots & \ddots & \vdots \\ \imath_{61} & \cdots & \imath_{66} \end{pmatrix} = 0$$

To effectively apply the stability criterion of Lipunov to our system, we require a criterion for when the equation $\det(A - \lambda \cdot I) = 0$ has a zero in the left half plane, without calculating the eigenvalues explicit. We use criterion of Routh-Hurwitz [2–4].

5.3 Single Ended Class B Amplifier Gummel-Poon Model Analysis Under Parameters Variation

Class B amplifier is a type of power amplifier where the active device (transistor) conducts only for one half cycle of the input signal. That means the conduction angle is 180° for a Class B amplifier. Since the active device is switched off for half the input cycle, the active device dissipates less power and hence the efficiency is improved. Theoretical maximum efficiency of Class B power amplifier is 78.5 %. it improves the power efficiency, it creates a lot of distortion. You can find class B amplifier in the RF power amplifiers where the distortion is not a matter of major concern. We use in our stability analysis The BJT NPN transistor Gummel-Poon model since the input signal is large. The small signal S-parameters are not useful for large-signal or high power circuit design such as power amplifier, mixers, frequency converters because the active devices (transistor/FET/diode) in these circuits usually operate in the nonlinear regions. In large signal circuits the voltage and current variation will be large, for BJT this means the variation of the transistor terminals voltages will be greater than V_T. We must use large signal model of the transistor, such as Ebers-Molls model, the Gummel-Poon model, the VBIC model. The most popular large-signal model for BJT is the Spice Gummel Poon (SGP) model. A more recent alternative to the SGP model is the Vertical Bipolar Intercompany Model (VBIC) model which offers more accuracies as compared to SGP model. The Spice Gummel Poon model is based on the device physics of bipolar junction transistor. The Gummel Poon model is a compact model for bipolar junction transistor which also takes into account effects of low currents and at high level injection signal [91–93] (Fig. 5.5).

The base of the transistor Q1 is not biased and the negative half cycle of the input waveform is missing in the output. Even though it improves the power

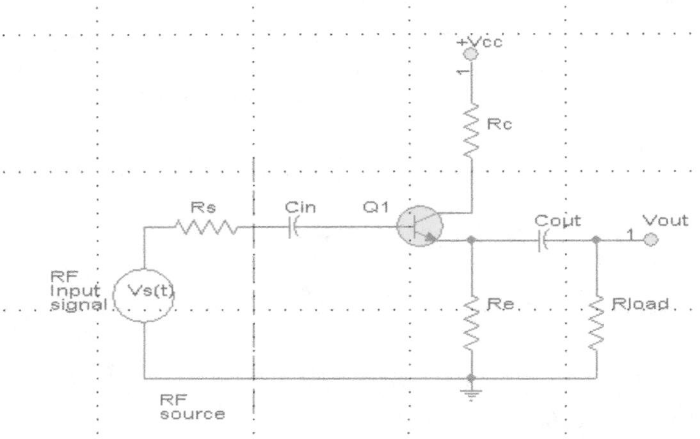

Fig. 5.5 Single ended class B amplifier

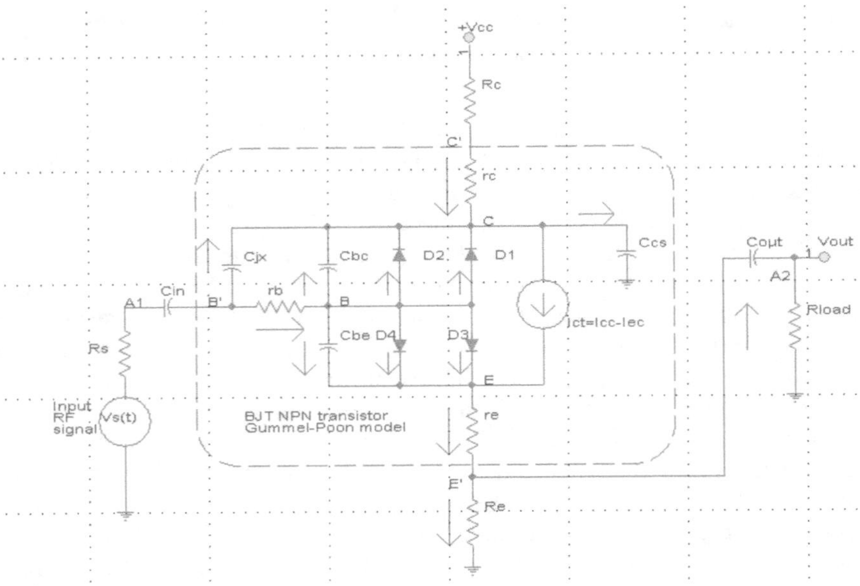

Fig. 5.6 Single ended class B amplifier equivalent circuit with BJT transistor Gummel-Poon model

efficiency, it creates a lot of distortion. Only half the information present in the input will be available in the output and that is a bad thing. Single ended Class B amplifiers are not used in present day practical audio amplifier application and they can be found only in some earlier gadgets. Another place where you can find them is the RF power amplifiers where the distortion is not a matter of major concern. The single ended class B amplifier equivalent circuit with BJT transistor Gummel-Poon model is present in the below schematic. (Fig. 5.6).

The Gummel-Poon schematic equivalent circuit shows the large signal schematic of the Gummel-Poon model. It represents the physical transistor, a current-controlled output current sink, and two diode structures including their capacitors.

$$I_{ec} = \frac{I_{ss}}{q_b} \cdot (e^{\frac{q \cdot V_{BC}}{n_r \cdot k_B \cdot T}} - 1); I_{cc} = \frac{I_{ss}}{q_b} \cdot (e^{\frac{q \cdot V_{BE}}{n_f \cdot k_B \cdot T}} - 1); I_{ct} = I_{cc} - I_{ec} = \frac{I_{ss}}{q_b} [e^{\frac{q \cdot V_{BE}}{n_f \cdot k_B \cdot T}} - e^{\frac{q \cdot V_{BC}}{n_r \cdot k_B \cdot T}}]$$

$$I_{D_1} = \frac{I_{ec}}{\beta_r} = \frac{I_{ss}}{q_b \cdot \beta_r} \cdot (e^{\frac{q \cdot V_{BC}}{n_r \cdot k_B \cdot T}} - 1); I_{D_3} = \frac{I_{cc}}{\beta_f} = \frac{I_{ss}}{q_b \cdot \beta_f} \cdot (e^{\frac{q \cdot V_{BE}}{n_f \cdot k_B \cdot T}} - 1)$$

$$I_{D_2} = C_4 \cdot I_{ss} \cdot (e^{\frac{q \cdot V_{BC}}{n_{cl} \cdot k_B \cdot T}} - 1); I_{D_4} = C_2 \cdot I_{ss} \cdot (e^{\frac{q \cdot V_{BE}}{n_{el} \cdot k_B \cdot T}} - 1); I_{R_c} = I_{r_c}; I_{R_s} = I_{C_{in}}; I_{C_{out}} = I_{R_{load}}$$

$$KCL\ @\ (C){:}\ C = C'; I_{r_c} \to \varepsilon; V_{C'} - V_C \to \varepsilon; I_{R_c} + I_{C_{jx}} + I_{C_{bc}} + \sum_{k=1}^{2} I_{D_k} = I_{C_t} + I_{C_{cs}}$$

$$KCL\ @\ (B){:}\ B = B'; I_{r_b} \to \varepsilon; V_{B'} - V_B \to \varepsilon; I_{C_{in}} = I_{C_{jx}} + I_{C_{bc}} + I_{C_{be}} + \sum_{k=1}^{4} I_{D_k}$$

KCL @ (E): $E = E'; I_{r_e} \to \varepsilon; V_E - V_{E'} \to \varepsilon; I_{C_{be}} + \sum_{k=3}^{4} I_{D_k} + I_{C_t} = I_{R_e} + I_{C_{out}}$

$$I_{C_{jx}} = C_{jx} \cdot \frac{d}{dt}(V_{B'} - V_C)|_{V_{B'}=V_B} = C_{jx} \cdot \frac{d}{dt}(V_B - V_C);$$
$$I_{C_{in}} = C_{in} \cdot \frac{d}{dt}(V_{A_1} - V_{B'})|_{V_{B'}=V_B} = C_{in} \cdot \frac{d}{dt}(V_{A_1} - V_B)$$

$$I_{C_{bc}} = C_{bc} \cdot \frac{d}{dt}(V_B - V_C);\ I_{C_{be}} = C_{be} \cdot \frac{d}{dt}(V_B - V_E);$$
$$I_{C_{out}} = C_{out} \cdot \frac{d}{dt}(V_{E'} - V_{A_2})|_{V_{E'}=V_E} = C_{out} \cdot \frac{d}{dt}(V_E - V_{A_2})$$

$$I_{C_{cs}} = C_{cs} \cdot \frac{dV_c}{dt}; I_{R_s} = \frac{V_s(t) - V_{A_1}}{R_s}; V_{A_2} = V_{out}; I_{R_{load}} = \frac{V_{A_2}}{R_{load}} = \frac{V_{out}}{R_{load}}; I_{C_{out}} = I_{R_{load}}$$

$$I_{R_c} = \frac{V_{cc} - V_{c'}}{R_c}|_{V_c=V_{c'}} = \frac{V_{cc} - V_c}{R_c}; I_{R_e} = \frac{V_{E'}}{R_e}|_{V_E=V_{E'}} = \frac{V_E}{R_e}$$

Gummel-Poon model parameters:

n_f	forward current emission coefficient.
n_r	reverse current emission coefficient.
q_b	variable model parameter (early and Kirk effects).
n_{el}, n_{cl}, C_2, C_4	model parameters responsible for low current effects.
$\frac{k_B \cdot T}{q}$	thermal voltage. β_f—forward current gain. β_r— reverse current gain.

KCL @ (C): $C = C'; I_{r_c} \to \varepsilon; V_{C'} - V_C \to \varepsilon; I_{R_c} + I_{C_{jx}} + I_{C_{bc}} + \sum_{k=1}^{2} I_{D_k} = I_{C_t} + I_{C_{cs}}$

$$\frac{V_{cc} - V_c}{R_c} + [C_{jx} + C_{bc}] \cdot \frac{d}{dt}(V_B - V_C) + I_{ss} \cdot [\frac{1}{q_b \cdot \beta_r} \cdot (e^{\frac{q \cdot V_{BC}}{n_r \cdot k_B \cdot T}} - 1) + C_4 \cdot (e^{\frac{q \cdot V_{BC}}{n_{cl} \cdot k_B \cdot T}} - 1)]$$
$$= \frac{I_{ss}}{q_b}[e^{\frac{q \cdot V_{BE}}{n_f \cdot k_B \cdot T}} - e^{\frac{q \cdot V_{BC}}{n_r \cdot k_B \cdot T}})] + C_{cs} \cdot \frac{dV_c}{dt}; V_{BC} = V_B - V_C$$

We define for simplicity new functions:

$$\xi_1(V_{BC}) = I_{ss} \cdot [\frac{1}{q_b \cdot \beta_r} \cdot (e^{\frac{q \cdot V_{BC}}{n_r \cdot k_B \cdot T}} - 1) + C_4 \cdot (e^{\frac{q \cdot V_{BC}}{n_{cl} \cdot k_B \cdot T}} - 1)]; \xi_2(V_{BE}, V_{BC}) = I_{C_t}$$
$$= \frac{I_{ss}}{q_b} \cdot [e^{\frac{q \cdot V_{BE}}{n_f \cdot k_B \cdot T}} - e^{\frac{q \cdot V_{BC}}{n_r \cdot k_B \cdot T}}]$$

$$\frac{V_{cc} - V_c}{R_c} + [C_{jx} + C_{bc}] \cdot \frac{dV_{BC}}{dt} + \xi_1(V_{BC}) = \xi_2(V_{BE}, V_{BC}) + C_{cs} \cdot \frac{dV_c}{dt}$$

KCL @ (B): $B = B'; I_{r_b} \to \varepsilon; V_{B'} - V_B \to \varepsilon; I_{C_{in}} = I_{C_{jx}} + I_{C_{bc}} + I_{C_{be}} + \sum_{k=1}^{4} I_{D_k}$

$$\begin{aligned} C_{in} \cdot \frac{d}{dt}(V_{A_1} - V_B) &= C_{jx} \cdot \frac{d}{dt}(V_B - V_C) + C_{bc} \cdot \frac{d}{dt}(V_B - V_C) \\ &\quad + C_{be} \cdot \frac{d}{dt}(V_B - V_E) + \frac{I_{ss}}{q_b \cdot \beta_r} \cdot (e^{\frac{q \cdot V_{BC}}{n_r \cdot k_B \cdot T}} - 1) \\ &\quad + C_4 \cdot I_{ss} \cdot (e^{\frac{q \cdot V_{BC}}{n_{cl} \cdot k_B \cdot T}} - 1) + \frac{I_{ss}}{q_b \cdot \beta_f} \cdot (e^{\frac{q \cdot V_{BE}}{n_f \cdot k_B \cdot T}} - 1) \\ &\quad + C_2 \cdot I_{ss} \cdot (e^{\frac{q \cdot V_{BE}}{n_{el} \cdot k_B \cdot T}} - 1) \end{aligned}$$

$$C_{in} \cdot \frac{d}{dt}(V_{A_1} - V_B) = [C_{jx} + C_{bc}] \cdot \frac{dV_{BC}}{dt} + C_{be} \cdot \frac{dV_{BE}}{dt} + \xi_3(V_{BE}, V_{BC})$$

$$\begin{aligned} \xi_3(V_{BE}, V_{BC}) &= I_{ss} \cdot [\frac{1}{q_b \cdot \beta_r} \cdot (e^{\frac{q \cdot V_{BC}}{n_r \cdot k_B \cdot T}} - 1) + C_4 \cdot (e^{\frac{q \cdot V_{BC}}{n_{cl} \cdot k_B \cdot T}} - 1)] \\ &\quad + I_{ss} \cdot [\frac{1}{q_b \cdot \beta_f} \cdot (e^{\frac{q \cdot V_{BE}}{n_f \cdot k_B \cdot T}} - 1) + C_2 \cdot (e^{\frac{q \cdot V_{BE}}{n_{el} \cdot k_B \cdot T}} - 1)] \end{aligned}$$

$$\xi_4(V_{BE}) = I_{ss} \cdot [\frac{1}{q_b \cdot \beta_f} \cdot (e^{\frac{q \cdot V_{BE}}{n_f \cdot k_B \cdot T}} - 1) + C_2 \cdot (e^{\frac{q \cdot V_{BE}}{n_{el} \cdot k_B \cdot T}} - 1)];$$

$$\xi_3(V_{BE}, V_{BC}) = \xi_1(V_{BC}) + \xi_4(V_{BE})$$

KCL @ (E): $E = E'; I_{r_e} \to \varepsilon; V_E - V_{E'} \to \varepsilon; I_{C_{be}} + \sum_{k=3}^{4} I_{D_k} + I_{C_t} = I_{R_e} + I_{C_{out}}$

$$\begin{aligned} C_{be} \cdot \frac{d}{dt}(V_B - V_E) + I_{ss} \cdot [\frac{1}{q_b \cdot \beta_f} \cdot (e^{\frac{q \cdot V_{BE}}{n_f \cdot k_B \cdot T}} - 1) + C_2 \cdot (e^{\frac{q \cdot V_{BE}}{n_{el} \cdot k_B \cdot T}} - 1)] \\ + \frac{I_{ss}}{q_b}[e^{\frac{q \cdot V_{BE}}{n_f \cdot k_B \cdot T}} - e^{\frac{q \cdot V_{BC}}{n_r \cdot k_B \cdot T}})] = \frac{V_E}{R_e} + \frac{V_{A_2}}{R_{load}} \end{aligned}$$

$$C_{be} \cdot \frac{dV_{BE}}{dt} + \xi_4(V_{BE}) + \xi_2(V_{BE}, V_{BC}) = \frac{V_E}{R_e} + \frac{V_{A_2}}{R_{load}};$$
$$C_{be} \cdot \frac{dV_{BE}}{dt} + \xi_4(V_{BE}) + \xi_2(V_{BE}, V_{BC}) = \frac{V_E}{R_e} + C_{out} \cdot \frac{d}{dt}(V_E - V_{A_2})$$

$$I_{C_{out}} = I_{R_{load}} \Rightarrow \frac{V_{A_2}}{R_{load}} = C_{out} \cdot \frac{d}{dt}(V_E - V_{A_2}) \Rightarrow V_{A_2} = R_{load} \cdot C_{out} \cdot \frac{d}{dt}(V_E - V_{A_2})$$

$$\begin{aligned} I_{R_s} = I_{C_{in}} \Rightarrow \frac{V_s(t) - V_{A_1}}{R_s} &= C_{in} \cdot \frac{d}{dt}(V_{A_1} - V_B) \Rightarrow V_s(t) - V_{A_1} \\ &= R_s \cdot C_{in} \cdot \frac{d}{dt}(V_{A_1} - V_B) \end{aligned}$$

Summary of our system differential equations:

$$\frac{V_{cc}-V_c}{R_c}+[C_{jx}+C_{bc}]\cdot\frac{dV_{BC}}{dt}+\xi_1(V_{BC})=\xi_2(V_{BE},V_{BC})+C_{cs}\cdot\frac{dV_c}{dt}$$

$$C_{in}\cdot\frac{d}{dt}(V_{A_1}-V_B)=[C_{jx}+C_{bc}]\cdot\frac{dV_{BC}}{dt}+C_{be}\cdot\frac{dV_{BE}}{dt}+\xi_3(V_{BE},V_{BC})$$

$$C_{be}\cdot\frac{dV_{BE}}{dt}+\xi_4(V_{BE})+\xi_2(V_{BE},V_{BC})=\frac{V_E}{R_e}+\frac{V_{A_2}}{R_{load}};$$
$$C_{be}\cdot\frac{dV_{BE}}{dt}+\xi_4(V_{BE})+\xi_2(V_{BE},V_{BC})=\frac{V_E}{R_e}+C_{out}\cdot\frac{d}{dt}(V_E-V_{A_2})$$

$$V_{A_2}=R_{load}\cdot C_{out}\cdot\frac{d}{dt}(V_E-V_{A_2});\ V_s(t)-V_{A_1}=R_s\cdot C_{in}\cdot\frac{d}{dt}(V_{A_1}-V_B)$$

&&

$$V_B=V_{BE}+V_E;\ V_C=V_{CB}+V_{BE}+V_E;\ V_{BC}=-V_{CB};\ V_{CB}=-V_{BC};\ V_C=V_{BE}-V_{BC}+V_E$$

$$\frac{V_{cc}-[V_{BE}-V_{BC}+V_E]}{R_c}+[C_{jx}+C_{bc}]\cdot\frac{dV_{BC}}{dt}+\xi_1(V_{BC})=\xi_2(V_{BE},V_{BC})+C_{cs}\cdot\frac{d}{dt}(V_{BE}-V_{BC}+V_E)$$

$$\frac{V_{cc}-[V_{BE}-V_{BC}+V_E]}{R_c}+[C_{jx}+C_{bc}+C_{cs}]\cdot\frac{dV_{BC}}{dt}+\xi_1(V_{BC})=\xi_2(V_{BE},V_{BC})+C_{cs}\cdot\frac{dV_{BE}}{dt}+C_{cs}\cdot\frac{dV_E}{dt}$$

$$C_{in}\cdot\frac{d}{dt}(V_{A_1}-V_{BE}-V_E)=[C_{jx}+C_{bc}]\cdot\frac{dV_{BC}}{dt}+C_{be}\cdot\frac{dV_{BE}}{dt}+\xi_3(V_{BE},V_{BC})$$

$$C_{in}\cdot\frac{dV_{A_1}}{dt}-C_{in}\cdot\frac{dV_E}{dt}=[C_{jx}+C_{bc}]\cdot\frac{dV_{BC}}{dt}+[C_{be}+C_{in}]\cdot\frac{dV_{BE}}{dt}+\xi_3(V_{BE},V_{BC})$$

$$C_{be}\cdot\frac{dV_{BE}}{dt}+\xi_4(V_{BE})+\xi_2(V_{BE},V_{BC})=\frac{V_E}{R_e}+\frac{V_{A_2}}{R_{load}};$$
$$V_{A_2}=R_{load}\cdot C_{out}\cdot\frac{dV_E}{dt}-R_{load}\cdot C_{out}\cdot\frac{dV_{A_2}}{dt}$$

$$V_s(t)-V_{A_1}=R_s\cdot C_{in}\cdot\frac{d}{dt}(V_{A_1}-V_{BE}-V_E)\Rightarrow V_s(t)-V_{A_1}=R_s\cdot C_{in}\cdot\frac{dV_{A_1}}{dt}-R_s\cdot C_{in}\cdot\frac{dV_{BE}}{dt}-R_s\cdot C_{in}\cdot\frac{dV_E}{dt}$$

$$V_s(t) = A_0 + \xi(t)|_{A_0 \gg \xi(t)} \approx A_0; \frac{dV_s(t)}{dt} = \frac{dA_0}{dt} + \frac{d\xi(t)}{dt}|_{A_0 \gg \xi(t)} \approx \varepsilon; \frac{d\xi(t)}{dt} \rightarrow \varepsilon; \frac{dV_s(t)}{dt} \rightarrow \varepsilon$$

Assumption $I_{R_s} \cdot R_s \rightarrow \varepsilon \Rightarrow V_{A_1} \approx V_s(t); \frac{dV_s(t)}{dt} \rightarrow \varepsilon \Rightarrow \frac{dV_{A_1}}{dt} \rightarrow \varepsilon; V_s(t) - V_{A_1} \rightarrow \varepsilon$ and we get the following system differential equations:

$$\frac{V_{cc} - [V_{BE} - V_{BC} + V_E]}{R_c} + [C_{jx} + C_{bc} + C_{cs}] \cdot \frac{dV_{BC}}{dt} + \xi_1(V_{BC}) = \xi_2(V_{BE}, V_{BC}) + C_{cs} \cdot \frac{dV_{BE}}{dt} + C_{cs} \cdot \frac{dV_E}{dt}$$

$$-C_{in} \cdot \frac{dV_E}{dt} = [C_{jx} + C_{bc}] \cdot \frac{dV_{BC}}{dt} + [C_{be} + C_{in}] \cdot \frac{dV_{BE}}{dt} + \xi_3(V_{BE}, V_{BC}); -R_s \cdot C_{in} \cdot \frac{dV_{BE}}{dt} - R_s \cdot C_{in} \cdot \frac{dV_E}{dt} = 0$$

$$C_{be} \cdot \frac{dV_{BE}}{dt} + \xi_4(V_{BE}) + \xi_2(V_{BE}, V_{BC}) = \frac{V_E}{R_e} + \frac{V_{A_2}}{R_{load}}; V_{A_2} = R_{load} \cdot C_{out} \cdot \frac{dV_E}{dt} - R_{load} \cdot C_{out} \cdot \frac{dV_{A_2}}{dt}$$

$$R_s \cdot C_{in} \cdot (\frac{dV_{A_1}}{dt} \rightarrow \varepsilon) - R_s \cdot C_{in} \cdot \frac{dV_{BE}}{dt} - R_s \cdot C_{in} \cdot \frac{dV_E}{dt} = 0; \frac{dV_{BE}}{dt} = -\frac{dV_E}{dt}$$

We get the following set of system differential equations:

$$\frac{V_{cc} - [V_{BE} - V_{BC} + V_E]}{R_c} + [C_{jx} + C_{bc} + C_{cs}] \cdot \frac{dV_{BC}}{dt} + \xi_1(V_{BC}) = \xi_2(V_{BE}, V_{BC})$$

$$C_{in} \cdot \frac{dV_{BE}}{dt} = [C_{jx} + C_{bc}] \cdot \frac{dV_{BC}}{dt} + [C_{be} + C_{in}] \cdot \frac{dV_{BE}}{dt} + \xi_3(V_{BE}, V_{BC})$$

$$C_{be} \cdot \frac{dV_{BE}}{dt} + \xi_4(V_{BE}) + \xi_2(V_{BE}, V_{BC}) = \frac{V_E}{R_e} + \frac{V_{A_2}}{R_{load}};$$

$$V_{A_2} = -R_{load} \cdot C_{out} \cdot \frac{dV_{BE}}{dt} - R_{load} \cdot C_{out} \cdot \frac{dV_{A_2}}{dt}$$

First differential equation:

$$\frac{V_{cc} - [V_{BE} - V_{BC} + V_E]}{R_c} + [C_{jx} + C_{bc} + C_{cs}] \cdot \frac{dV_{BC}}{dt} + \xi_1(V_{BC}) = \xi_2(V_{BE}, V_{BC})$$

$$V_E = V_{cc} - V_{BE} + V_{BC} - R_c \cdot \xi_2(V_{BE}, V_{BC}) + R_c \cdot \xi_1(V_{BC}) + R_c \cdot [C_{jx} + C_{bc} + C_{cs}] \cdot \frac{dV_{BC}}{dt}$$

Third differential equation:

$$C_{be} \cdot \frac{dV_{BE}}{dt} + \xi_4(V_{BE}) + \xi_2(V_{BE}, V_{BC}) = \frac{V_E}{R_e} + \frac{V_{A_2}}{R_{load}}$$

$$V_E = C_{be} \cdot R_e \cdot \frac{dV_{BE}}{dt} + \xi_4(V_{BE}) \cdot R_e + \xi_2(V_{BE}, V_{BC}) \cdot R_e - \frac{V_{A_2}}{R_{load}} \cdot R_e$$

Combine of the first and third differential equations:

$$V_{cc} - V_{BE} + V_{BC} - R_c \cdot \xi_2(V_{BE}, V_{BC}) + R_c \cdot \xi_1(V_{BC}) + R_c \cdot [C_{jx} + C_{bc} + C_{cs}] \cdot \frac{dV_{BC}}{dt}$$
$$= C_{be} \cdot R_e \cdot \frac{dV_{BE}}{dt} + \xi_4(V_{BE}) \cdot R_e + \xi_2(V_{BE}, V_{BC}) \cdot R_e - \frac{V_{A_2}}{R_{load}} \cdot R_e$$

$$V_{cc} - V_{BE} + V_{BC} + R_c \cdot \xi_1(V_{BC}) + R_c \cdot [C_{jx} + C_{bc} + C_{cs}] \cdot \frac{dV_{BC}}{dt}$$
$$= C_{be} \cdot R_e \cdot \frac{dV_{BE}}{dt} + \xi_4(V_{BE}) \cdot R_e + [R_e + R_c] \cdot \xi_2(V_{BE}, V_{BC}) - \frac{V_{A_2}}{R_{load}} \cdot R_e$$

We can summary our system reduced three differential equations:

(1) $$V_{cc} - V_{BE} + V_{BC} + R_c \cdot \xi_1(V_{BC}) + R_c \cdot [C_{jx} + C_{bc} + C_{cs}] \cdot \frac{dV_{BC}}{dt} = C_{be} \cdot R_e \cdot \frac{dV_{BE}}{dt} + \xi_4(V_{BE}) \cdot R_e + [R_e + R_c] \cdot \xi_2(V_{BE}, V_{BC}) - \frac{V_{A_2}}{R_{load}} \cdot R_e$$

(2) $C_{in} \cdot \frac{dV_{BE}}{dt} = [C_{jx} + C_{bc}] \cdot \frac{dV_{BC}}{dt} + [C_{be} + C_{in}] \cdot \frac{dV_{BE}}{dt} + \xi_3(V_{BE}, V_{BC})$

(3) $V_{A_2} = -R_{load} \cdot C_{out} \cdot \frac{dV_{BE}}{dt} - R_{load} \cdot C_{out} \cdot \frac{dV_{A_2}}{dt}$

(3) $V_{A_2} = -R_{load} \cdot C_{out} \cdot \frac{dV_{BE}}{dt} - R_{load} \cdot C_{out} \cdot \frac{dV_{A_2}}{dt} \Rightarrow \frac{dV_{A_2}}{dt} = -\frac{dV_{BE}}{dt} - \frac{1}{R_{load} \cdot C_{out}} \cdot V_{A_2}$

(2) $C_{in} \cdot \frac{dV_{BE}}{dt} = [C_{jx} + C_{bc}] \cdot \frac{dV_{BC}}{dt} + [C_{be} + C_{in}] \cdot \frac{dV_{BE}}{dt} + \xi_3(V_{BE}, V_{BC})$

$$[C_{jx} + C_{bc}] \cdot \frac{dV_{BC}}{dt} + C_{be} \cdot \frac{dV_{BE}}{dt} + \xi_3(V_{BE}, V_{BC}) = 0$$
$$\Rightarrow C_{be} \cdot \frac{dV_{BE}}{dt} = -[C_{jx} + C_{bc}] \cdot \frac{dV_{BC}}{dt} - \xi_3(V_{BE}, V_{BC})$$

$$\frac{dV_{BE}}{dt} = -[\frac{C_{jx} + C_{bc}}{C_{be}}] \cdot \frac{dV_{BC}}{dt} - \frac{1}{C_{be}} \cdot \xi_3(V_{BE}, V_{BC})$$

$$(1)\quad \begin{aligned} &V_{cc} - V_{BE} + V_{BC} + R_c \cdot \xi_1(V_{BC}) + R_c \cdot [C_{jx} + C_{bc} + C_{cs}] \cdot \frac{dV_{BC}}{dt} \\ &= C_{be} \cdot R_e \cdot \frac{dV_{BE}}{dt} + \xi_4(V_{BE}) \cdot R_e + [R_e + R_c] \cdot \xi_2(V_{BE}, V_{BC}) - \frac{V_{A_2}}{R_{load}} \cdot R_e \end{aligned}$$

$$\begin{aligned} &V_{cc} - V_{BE} + V_{BC} + R_c \cdot \xi_1(V_{BC}) + R_c \cdot [C_{jx} + C_{bc} + C_{cs}] \cdot \frac{dV_{BC}}{dt} \\ &= -C_{be} \cdot R_e \cdot \{[\frac{C_{jx} + C_{bc}}{C_{be}}] \cdot \frac{dV_{BC}}{dt} \\ &+ \frac{1}{C_{be}} \cdot \xi_3(V_{BE}, V_{BC})\} + \xi_4(V_{BE}) \cdot R_e + [R_e + R_c] \cdot \xi_2(V_{BE}, V_{BC}) - \frac{V_{A_2}}{R_{load}} \cdot R_e \end{aligned}$$

$$\begin{aligned} &(R_c \cdot [C_{jx} + C_{bc} + C_{cs}] + R_e \cdot [C_{jx} + C_{bc}]) \cdot \frac{dV_{BC}}{dt} = -R_e \cdot \xi_3(V_{BE}, V_{BC}) + \xi_4(V_{BE}) \cdot R_e \\ &+ [R_e + R_c] \cdot \xi_2(V_{BE}, V_{BC}) - R_c \cdot \xi_1(V_{BC}) - \frac{V_{A_2}}{R_{load}} \cdot R_e - V_{cc} + V_{BE} - V_{BC} \end{aligned}$$

We define for simplicity global parameter $\Gamma_1 = \Gamma_1(R_c, R_e, \ldots)$

$$\Gamma_1(R_c, R_e, \ldots) = (R_c \cdot [C_{jx} + C_{bc} + C_{cs}] + R_e \cdot [C_{jx} + C_{bc}])$$

$$\begin{aligned} \frac{dV_{BC}}{dt} = &-\frac{R_e}{\Gamma_1(R_c, R_e, \ldots)} \cdot \xi_3(V_{BE}, V_{BC}) + \xi_4(V_{BE}) \cdot \frac{R_e}{\Gamma_1(R_c, R_e, \ldots)} \\ &+ [\frac{R_e + R_c}{\Gamma_1(R_c, R_e, \ldots)}] \cdot \xi_2(V_{BE}, V_{BC}) - \frac{R_c}{\Gamma_1(R_c, R_e, \ldots)} \cdot \xi_1(V_{BC}) \\ &- \frac{V_{A_2}}{R_{load}} \cdot \frac{R_e}{\Gamma_1(R_c, R_e, \ldots)} + \frac{V_{BE} - V_{BC} - V_{cc}}{\Gamma_1(R_c, R_e, \ldots)} \end{aligned}$$

$$\begin{aligned} \Omega_1(R_c, R_e, \ldots) &= \frac{R_e}{\Gamma_1(R_c, R_e, \ldots)}; \Omega_2(R_c, R_e, \ldots) = \frac{R_e + R_c}{\Gamma_1(R_c, R_e, \ldots)}; \Omega_3(R_c, R_e, \ldots) \\ &= \frac{R_c}{\Gamma_1(R_c, R_e, \ldots)} \end{aligned}$$

$$\begin{aligned} \frac{dV_{BC}}{dt} = &-\Omega_1(R_c, R_e, \ldots) \cdot \xi_3(V_{BE}, V_{BC}) + \xi_4(V_{BE}) \cdot \Omega_1(R_c, R_e, \ldots) \\ &+ \Omega_2(R_c, R_e, \ldots) \cdot \xi_2(V_{BE}, V_{BC}) - \Omega_3(R_c, R_e, \ldots) \cdot \xi_1(V_{BC}) \\ &- \frac{V_{A_2}}{R_{load}} \cdot \Omega_1(R_c, R_e, \ldots) + \frac{V_{BE} - V_{BC} - V_{cc}}{\Gamma_1(R_c, R_e, \ldots)} \end{aligned}$$

$$\frac{dV_{BE}}{dt} = -[\frac{C_{jx}+C_{bc}}{C_{be}}] \cdot \{-\Omega_1(R_c,R_e,\ldots)\cdot\xi_3(V_{BE},V_{BC}) + \xi_4(V_{BE})\cdot\Omega_1(R_c,R_e,\ldots) + \Omega_2(R_c,R_e,\ldots)\cdot\xi_2(V_{BE},V_{BC}) - \Omega_3(R_c,R_e,\ldots)\cdot\xi_1(V_{BC}) - \frac{V_{A_2}}{R_{load}}\cdot\Omega_1(R_c,R_e,\ldots) + \frac{V_{BE}-V_{BC}-V_{cc}}{\Gamma_1(R_c,R_e,\ldots)}\} - \frac{1}{C_{be}}\cdot\xi_3(V_{BE},V_{BC})$$

$$\frac{dV_{A_2}}{dt} = [\frac{C_{jx}+C_{bc}}{C_{be}}] \cdot \{-\Omega_1(R_c,R_e,\ldots)\cdot\xi_3(V_{BE},V_{BC}) + \xi_4(V_{BE})\cdot\Omega_1(R_c,R_e,\ldots) + \Omega_2(R_c,R_e,\ldots)\cdot\xi_2(V_{BE},V_{BC}) - \Omega_3(R_c,R_e,\ldots)\cdot\xi_1(V_{BC}) - \frac{V_{A_2}}{R_{load}}\cdot\Omega_1(R_c,R_e,\ldots) + \frac{V_{BE}-V_{BC}-V_{cc}}{\Gamma_1(R_c,R_e,\ldots)}\} + \frac{1}{C_{be}}\cdot\xi_3(V_{BE},V_{BC}) - \frac{1}{R_{load}\cdot C_{out}}\cdot V_{A_2}$$

We define three functions:

$$\psi_1(V_{BE},V_{BC},V_{A_2},\ldots) = -\Omega_1(R_c,R_e,\ldots)\cdot\xi_3(V_{BE},V_{BC}) + \xi_4(V_{BE})\cdot\Omega_1(R_c,R_e,\ldots) + \Omega_2(R_c,R_e,\ldots)\cdot\xi_2(V_{BE},V_{BC}) - \Omega_3(R_c,R_e,\ldots)\cdot\xi_1(V_{BC}) - \frac{V_{A_2}}{R_{load}}\cdot\Omega_1(R_c,R_e,\ldots) + \frac{V_{BE}-V_{BC}-V_{cc}}{\Gamma_1(R_c,R_e,\ldots)}$$

$$\psi_2(V_{BE},V_{BC},V_{A_2},\ldots) = -[\frac{C_{jx}+C_{bc}}{C_{be}}] \cdot \{-\Omega_1(R_c,R_e,\ldots)\cdot\xi_3(V_{BE},V_{BC}) + \xi_4(V_{BE})\cdot\Omega_1(R_c,R_e,\ldots) + \Omega_2(R_c,R_e,\ldots)\cdot\xi_2(V_{BE},V_{BC}) - \Omega_3(R_c,R_e,\ldots)\cdot\xi_1(V_{BC}) - \frac{V_{A_2}}{R_{load}}\cdot\Omega_1(R_c,R_e,\ldots) + \frac{V_{BE}-V_{BC}-V_{cc}}{\Gamma_1(R_c,R_e,\ldots)}\} - \frac{1}{C_{be}}\cdot\xi_3(V_{BE},V_{BC})$$

$$\psi_3(V_{BE},V_{BC},V_{A_2},\ldots) = [\frac{C_{jx}+C_{bc}}{C_{be}}] \cdot \{-\Omega_1(R_c,R_e,\ldots)\cdot\xi_3(V_{BE},V_{BC}) + \xi_4(V_{BE})\cdot\Omega_1(R_c,R_e,\ldots) + \Omega_2(R_c,R_e,\ldots)\cdot\xi_2(V_{BE},V_{BC}) - \Omega_3(R_c,R_e,\ldots)\cdot\xi_1(V_{BC}) - \frac{V_{A_2}}{R_{load}}\cdot\Omega_1(R_c,R_e,\ldots) + \frac{V_{BE}-V_{BC}-V_{cc}}{\Gamma_1(R_c,R_e,\ldots)}\} + \frac{1}{C_{be}}\cdot\xi_3(V_{BE},V_{BC}) - \frac{1}{R_{load}\cdot C_{out}}\cdot V_{A_2}$$

Our system differential equations: $\frac{dV_{BC}}{dt} = \psi_1(V_{BE},V_{BC},V_{A_2},\ldots)$

$$\frac{dV_{BE}}{dt} = \psi_2(V_{BE}, V_{BC}, V_{A_2}, \ldots); \frac{dV_{A_2}}{dt} = \psi_3(V_{BE}, V_{BC}, V_{A_2}, \ldots)$$

To find system fixed points: $\frac{dV_{BC}}{dt} = 0$; $\frac{dV_{BE}}{dt} = 0$; $\frac{dV_{A_2}}{dt} = 0$

$$\psi_1(V_{BE}^*, V_{BC}^*, V_{A_2}^*, \ldots) = 0; \psi_2(V_{BE}^*, V_{BC}^*, V_{A_2}^*, \ldots) = 0; \psi_3(V_{BE}^*, V_{BC}^*, V_{A_2}^*, \ldots) = 0$$

Fixed point and linearization: We approximate our system phase portrait near a fixed point $V_{BE}^*, V_{BC}^*, V_{A_2}^*$ by corresponding it as a linear system.

We consider the system $\frac{dV_{BC}}{dt} = \psi_1(V_{BE}, V_{BC}, V_{A_2})$; $\frac{dV_{BE}}{dt} = \psi_2(V_{BE}, V_{BC}, V_{A_2})$ $\frac{dV_{A_2}}{dt} = \psi_3(V_{BE}, V_{BC}, V_{A_2})$ and suppose that $(V_{BE}^*, V_{BC}^*, V_{A_2}^*)$ is a fixed point, i.e., $\psi_1(V_{BE}^*, V_{BC}^*, V_{A_2}^*) = 0; \psi_2(V_{BE}^*, V_{BC}^*, V_{A_2}^*) = 0; \psi_3(V_{BE}^*, V_{BC}^*, V_{A_2}^*) = 0$. Let $u = V_{BC} - V_{BC}^*$ $v = V_{BE} - V_{BE}^*$; $w = V_{A_2} - V_{A_2}^*$ denote the components of a small disturbance from the fixed point. To see whether the disturbance grows or decays, we need to derive differential equations for u, v, and w. $\frac{du}{dt} = \frac{dV_{BC}}{dt}; \frac{dv}{dt} = \frac{dV_{BE}}{dt}; \frac{dw}{dt} = \frac{dV_{A_2}}{dt}$

Since $V_{BE}^*, V_{BC}^*, V_{A_2}^*$ are constants. By substitution:

$$\frac{du}{dt} = \frac{dV_{BC}}{dt} = \psi_1(v + V_{BE}^*, u + V_{BC}^*, w + V_{A_2}^*) = \psi_1(V_{BE}^*, V_{BC}^*, V_{A_2}^*) + u \cdot \frac{\partial \psi_1}{\partial V_{BC}} + v \cdot \frac{\partial \psi_1}{\partial V_{BE}} + w \cdot \frac{\partial \psi_1}{\partial V_{A_2}} + O(u^2, v^2, w^2, uvw, \ldots)$$

Since $\psi_1(V_{BE}^*, V_{BC}^*, V_{A_2}^*) = 0$ then $\frac{du}{dt} = \frac{dV_{BC}}{dt} = u \cdot \frac{\partial \psi_1}{\partial V_{BC}} + v \cdot \frac{\partial \psi_1}{\partial V_{BE}} + w \cdot \frac{\partial \psi_1}{\partial V_{A_2}} + O(u^2, v^2, w^2, \ldots)$.

The partial derivatives are to be evaluated at the fixed point $(V_{BE}^*, V_{BC}^*, V_{A_2}^*)$ and they are numbers and not functions. Also the shorthand notation $O(u^2, v^2, w^2, \ldots)$.

Denotes quadratic terms in u, v, and w and it extremely small. Similarly we find $\frac{dv}{dt} = \frac{dV_{BE}}{dt} = u \cdot \frac{\partial \psi_2}{\partial V_{BC}} + v \cdot \frac{\partial \psi_2}{\partial V_{BE}} + w \cdot \frac{\partial \psi_3}{\partial V_{A_2}} + O(u^2, v^2, w^2, \ldots)$ and the expression $\frac{dw}{dt} = \frac{dV_{A_2}}{dt} = u \cdot \frac{\partial \psi_3}{\partial V_{BC}} + v \cdot \frac{\partial \psi_3}{\partial V_{BE}} + w \cdot \frac{\partial \psi_3}{\partial V_{A_2}} + O(u^2, v^2, w^2, \ldots)$. We denote (u, v, w) a disturbance [2–4].

$$\begin{pmatrix} \frac{du}{dt} \\ \frac{dv}{dt} \\ \frac{dw}{dt} \end{pmatrix} = \begin{pmatrix} \frac{\partial \psi_1}{\partial V_{BC}} & \frac{\partial \psi_1}{\partial V_{BE}} & \frac{\partial \psi_1}{\partial V_{A_2}} \\ \frac{\partial \psi_2}{\partial V_{BC}} & \frac{\partial \psi_2}{\partial V_{BE}} & \frac{\partial \psi_2}{\partial V_{A_2}} \\ \frac{\partial \psi_3}{\partial V_{BC}} & \frac{\partial \psi_3}{\partial V_{BE}} & \frac{\partial \psi_3}{\partial V_{A_2}} \end{pmatrix} \cdot \begin{pmatrix} u \\ v \\ w \end{pmatrix} + \text{quadratic term}$$

The matrix $A = \begin{pmatrix} \frac{\partial\psi_1}{\partial V_{BC}} & \frac{\partial\psi_1}{\partial V_{BE}} & \frac{\partial\psi_1}{\partial V_{A_2}} \\ \frac{\partial\psi_2}{\partial V_{BC}} & \frac{\partial\psi_2}{\partial V_{BE}} & \frac{\partial\psi_2}{\partial V_{A_2}} \\ \frac{\partial\psi_3}{\partial V_{BC}} & \frac{\partial\psi_3}{\partial V_{BE}} & \frac{\partial\psi_3}{\partial V_{A_2}} \end{pmatrix}_{(V_{BE}^*, V_{BC}^*, V_{A_2}^*)}$ is called the Jacobian matrix at the fixed point $(V_{BE}^*, V_{BC}^*, V_{A_2}^*)$. The quadratic terms are tiny and we neglect them altogether. We obtain the linearized system.

$$\begin{pmatrix} \frac{du}{dt} \\ \frac{dv}{dt} \\ \frac{dw}{dt} \end{pmatrix} = \begin{pmatrix} \frac{\partial\psi_1}{\partial V_{BC}} & \frac{\partial\psi_1}{\partial V_{BE}} & \frac{\partial\psi_1}{\partial V_{A_2}} \\ \frac{\partial\psi_2}{\partial V_{BC}} & \frac{\partial\psi_2}{\partial V_{BE}} & \frac{\partial\psi_2}{\partial V_{A_2}} \\ \frac{\partial\psi_3}{\partial V_{BC}} & \frac{\partial\psi_3}{\partial V_{BE}} & \frac{\partial\psi_3}{\partial V_{A_2}} \end{pmatrix}$$

$$\frac{\partial\psi_1}{\partial V_{BC}} = -\Omega_1(R_c, R_e, \ldots)\cdot\frac{\partial\xi_3(V_{BE}, V_{BC})}{\partial V_{BC}} + \Omega_2(R_c, R_e, \ldots)\cdot\frac{\partial\xi_2(V_{BE}, V_{BC})}{\partial V_{BC}} - \Omega_3(R_c, R_e, \ldots)\cdot\frac{\partial\xi_1(V_{BC})}{\partial V_{BC}} - \frac{1}{\Gamma_1(R_c, R_e, \ldots)}$$

$$\frac{\partial\xi_1(V_{BC})}{\partial V_{BC}} = I_{ss}\cdot\frac{q}{k_B\cdot T}\cdot\left[\frac{1}{q_b\cdot\beta_r\cdot n_r}\cdot e^{\frac{q\cdot V_{BC}}{n_r\cdot k_B\cdot T}} + \frac{C_4}{n_{cl}}\cdot e^{\frac{q\cdot V_{BC}}{n_{cl}\cdot k_B\cdot T}}\right]; \frac{\partial\xi_1(V_{BC})}{\partial V_{BC}} = \frac{\partial\xi_3(V_{BE}, V_{BC})}{\partial V_{BC}}$$

$$\frac{\partial\xi_2(V_{BE}, V_{BC})}{\partial V_{BC}} = -\frac{I_{ss}}{q_b}\cdot\frac{q}{n_r\cdot k_B\cdot T}\cdot e^{\frac{q\cdot V_{BC}}{n_r\cdot k_B\cdot T}};$$

$$\frac{\partial\xi_3(V_{BE}, V_{BC})}{\partial V_{BC}} = I_{ss}\cdot\frac{q}{k_B\cdot T}\cdot\left[\frac{1}{q_b\cdot\beta_r\cdot n_r}\cdot e^{\frac{q\cdot V_{BC}}{n_r\cdot k_B\cdot T}} + \frac{C_4}{n_{cl}}\cdot e^{\frac{q\cdot V_{BC}}{n_{cl}\cdot k_B\cdot T}}\right]$$

$$\frac{\partial\psi_1}{\partial V_{BE}} = -\Omega_1(R_c, R_e, \ldots)\cdot\frac{\partial\xi_3(V_{BE}, V_{BC})}{\partial V_{BE}} + \frac{\partial\xi_4(V_{BE})}{\partial V_{BE}}\cdot\Omega_1(R_c, R_e, \ldots) + \Omega_2(R_c, R_e, \ldots\cdot\frac{\partial\xi_2(V_{BE}, V_{BC})}{\partial V_{BE}} + \frac{1}{\Gamma_1(R_c, R_e, \ldots)}$$

$$\frac{\partial\xi_3(V_{BE}, V_{BC})}{\partial V_{BE}} = I_{ss}\cdot\frac{q}{k_B\cdot T}\cdot\left[\frac{1}{q_b\cdot\beta_f\cdot n_f}\cdot e^{\frac{q\cdot V_{BE}}{n_f\cdot k_B\cdot T}} + \frac{C_2}{n_{el}}\cdot e^{\frac{q\cdot V_{BE}}{n_{el}\cdot k_B\cdot T}}\right]$$

$$\frac{\partial\xi_4(V_{BE})}{\partial V_{BE}} = I_{ss}\cdot\frac{q}{k_B\cdot T}\cdot\left[\frac{1}{q_b\cdot\beta_f\cdot n_f}\cdot e^{\frac{q\cdot V_{BE}}{n_f\cdot k_B\cdot T}} + \frac{C_2}{n_{el}}\cdot e^{\frac{q\cdot V_{BE}}{n_{el}\cdot k_B\cdot T}}\right]; \frac{\partial\xi_3(V_{BE}, V_{BC})}{\partial V_{BE}} = \frac{\partial\xi_4(V_{BE})}{\partial V_{BE}}$$

$$\frac{\partial \xi_2(V_{BE}, V_{BC})}{\partial V_{BE}} = \frac{I_{ss}}{q_b} \cdot \frac{q}{n_f \cdot k_B \cdot T} \cdot e^{\frac{q \cdot V_{BE}}{n_f \cdot k_B \cdot T}}; \frac{\partial \psi_1}{\partial V_{A_2}} = -\frac{1}{R_{load}} \cdot \Omega_1(R_c, R_e, \ldots)$$

$$\frac{\partial \psi_2}{\partial V_{BC}} = -[\frac{C_{jx} + C_{bc}}{C_{be}}] \cdot \{-\Omega_1(R_c, R_e, \ldots) \cdot \frac{\partial \xi_3(V_{BE}, V_{BC})}{\partial V_{BC}} + \Omega_2(R_c, R_e, \ldots) \cdot \frac{\partial \xi_2(V_{BE}, V_{BC})}{\partial V_{BC}}$$
$$- \Omega_3(R_c, R_e, \ldots) \cdot \frac{\partial \xi_1(V_{BC})}{\partial V_{BC}} - \frac{1}{\Gamma_1(R_c, R_e, \ldots)}\} - \frac{1}{C_{be}} \cdot \frac{\partial \xi_3(V_{BE}, V_{BC})}{\partial V_{BC}}$$

$$\frac{\partial \psi_2}{\partial V_{BE}} = -[\frac{C_{jx} + C_{bc}}{C_{be}}] \cdot \{-\Omega_1(R_c, R_e, \ldots) \cdot \frac{\partial \xi_3(V_{BE}, V_{BC})}{\partial V_{BE}} + \frac{\partial \xi_4(V_{BE})}{\partial V_{BE}} \cdot \Omega_1(R_c, R_e, \ldots)$$
$$+ \Omega_2(R_c, R_e, \ldots) \cdot \frac{\partial \xi_2(V_{BE}, V_{BC})}{\partial V_{BE}} + \frac{1}{\Gamma_1(R_c, R_e, \ldots)}\} - \frac{1}{C_{be}} \cdot \partial V_{BE} \frac{\partial \xi_3(V_{BE}, V_{BC})}{\partial V_{BE}}$$

$$\frac{\partial \psi_2}{\partial V_{A_2}} = [\frac{C_{jx} + C_{bc}}{C_{be}}] \cdot \frac{1}{R_{load}} \cdot \Omega_1(R_c, R_e, \ldots)$$

$$\frac{\partial \psi_3}{\partial V_{BC}} = [\frac{C_{jx} + C_{bc}}{C_{be}}] \cdot \{-\Omega_1(R_c, R_e, \ldots) \cdot \frac{\partial \xi_3(V_{BE}, V_{BC})}{\partial V_{BC}} + \Omega_2(R_c, R_e, \ldots) \cdot \frac{\partial \xi_2(V_{BE}, V_{BC})}{\partial V_{BC}}$$
$$- \Omega_3(R_c, R_e, \ldots) \cdot \frac{\partial \xi_1(V_{BC})}{\partial V_{BC}} - \frac{1}{\Gamma_1(R_c, R_e, \ldots)}\} + \frac{1}{C_{be}} \cdot \frac{\partial \xi_3(V_{BE}, V_{BC})}{\partial V_{BC}}$$

$$\frac{\partial \psi_3}{\partial V_{BE}} = [\frac{C_{jx} + C_{bc}}{C_{be}}] \cdot \{-\Omega_1(R_c, R_e, \ldots) \cdot \frac{\partial \xi_3(V_{BE}, V_{BC})}{\partial V_{BE}} + \frac{\partial \xi_4(V_{BE})}{\partial V_{BE}} \cdot \Omega_1(R_c, R_e, \ldots)$$
$$+ \Omega_2(R_c, R_e, \ldots) \cdot \frac{\partial \xi_2(V_{BE}, V_{BC})}{\partial V_{BE}} + \frac{1}{\Gamma_1(R_c, R_e, \ldots)}\} + \frac{1}{C_{be}} \cdot \frac{\partial \xi_3(V_{BE}, V_{BC})}{\partial V_{BE}}$$

$$\frac{\partial \psi_3}{\partial V_{A_2}} = -[\frac{C_{jx} + C_{bc}}{C_{be}}] \cdot \frac{1}{R_{load}} \cdot \Omega_1(R_c, R_e, \ldots) - \frac{1}{R_{load} \cdot C_{out}}$$

We already found matrix A. The eigenvalues of a matrix A are given by the characteristic equation $\det(A - \lambda \cdot I) = 0$, where I is the identity matrix 3 × 3.

$$A - \lambda \cdot I = \begin{pmatrix} \frac{\partial \psi_1}{\partial V_{BC}} & \frac{\partial \psi_1}{\partial V_{BE}} & \frac{\partial \psi_1}{\partial V_{A_2}} \\ \frac{\partial \psi_2}{\partial V_{BC}} & \frac{\partial \psi_2}{\partial V_{BE}} & \frac{\partial \psi_2}{\partial V_{A_2}} \\ \frac{\partial \psi_3}{\partial V_{BC}} & \frac{\partial \psi_3}{\partial V_{BE}} & \frac{\partial \psi_3}{\partial V_{A_2}} \end{pmatrix}_{(V_{BE}^*, V_{BC}^*, V_{A_2}^*)} - \begin{pmatrix} \lambda & 0 & 0 \\ 0 & \lambda & 0 \\ 0 & 0 & \lambda \end{pmatrix}$$

$$A - \lambda \cdot I = \begin{pmatrix} \frac{\partial \psi_1}{\partial V_{BC}} - \lambda & \frac{\partial \psi_1}{\partial V_{BE}} & \frac{\partial \psi_1}{\partial V_{A_2}} \\ \frac{\partial \psi_2}{\partial V_{BC}} & \frac{\partial \psi_2}{\partial V_{BE}} - \lambda & \frac{\partial \psi_2}{\partial V_{A_2}} \\ \frac{\partial \psi_3}{\partial V_{BC}} & \frac{\partial \psi_3}{\partial V_{BE}} & \frac{\partial \psi_3}{\partial V_{A_2}} - \lambda \end{pmatrix}_{(V_{BE}^*, V_{BC}^*, V_{A_2}^*)}$$

$$\det(A - \lambda \cdot I) = (\frac{\partial \psi_1}{\partial V_{BC}} - \lambda)_{(V_{BE}^*, V_{BC}^*, V_{A_2}^*)} \cdot \det\begin{pmatrix} \frac{\partial \psi_2}{\partial V_{BE}} - \lambda & \frac{\partial \psi_2}{\partial V_{A_2}} \\ \frac{\partial \psi_3}{\partial V_{BE}} & \frac{\partial \psi_3}{\partial V_{A_2}} - \lambda \end{pmatrix}_{(V_{BE}^*, V_{BC}^*, V_{A_2}^*)}$$
$$- (\frac{\partial \psi_1}{\partial V_{BE}})_{(V_{BE}^*, V_{BC}^*, V_{A_2}^*)} \cdot \det\begin{pmatrix} \frac{\partial \psi_2}{\partial V_{BC}} & \frac{\partial \psi_2}{\partial V_{A_2}} \\ \frac{\partial \psi_3}{\partial V_{BC}} & \frac{\partial \psi_3}{\partial V_{A_2}} - \lambda \end{pmatrix}_{(V_{BE}^*, V_{BC}^*, V_{A_2}^*)}$$
$$+ (\frac{\partial \psi_1}{\partial V_{A_2}})_{(V_{BE}^*, V_{BC}^*, V_{A_2}^*)} \cdot \det\begin{pmatrix} \frac{\partial \psi_2}{\partial V_{BC}} & \frac{\partial \psi_2}{\partial V_{BE}} - \lambda \\ \frac{\partial \psi_3}{\partial V_{BC}} & \frac{\partial \psi_3}{\partial V_{BE}} \end{pmatrix}_{(V_{BE}^*, V_{BC}^*, V_{A_2}^*)}$$

$$\det(A - \lambda \cdot I) = (\frac{\partial \psi_1}{\partial V_{BC}} - \lambda)_{(V_{BE}^*, V_{BC}^*, V_{A_2}^*)} \cdot [(\frac{\partial \psi_2}{\partial V_{BE}} - \lambda) \cdot (\frac{\partial \psi_3}{\partial V_{A_2}} - \lambda) - \frac{\partial \psi_3}{\partial V_{BE}} \cdot \frac{\partial \psi_2}{\partial V_{A_2}}]_{(V_{BE}^*, V_{BC}^*, V_{A_2}^*)}$$
$$- (\frac{\partial \psi_1}{\partial V_{BE}})_{(V_{BE}^*, V_{BC}^*, V_{A_2}^*)} \cdot [\frac{\partial \psi_2}{\partial V_{BC}} \cdot (\frac{\partial \psi_3}{\partial V_{A_2}} - \lambda) - \frac{\partial \psi_3}{\partial V_{BC}} \cdot \frac{\partial \psi_2}{\partial V_{A_2}}]_{(V_{BE}^*, V_{BC}^*, V_{A_2}^*)}$$
$$+ (\frac{\partial \psi_1}{\partial V_{A_2}})_{(V_{BE}^*, V_{BC}^*, V_{A_2}^*)} \cdot [\frac{\partial \psi_2}{\partial V_{BC}} \cdot \frac{\partial \psi_3}{\partial V_{BE}} - \frac{\partial \psi_3}{\partial V_{BC}} \cdot (\frac{\partial \psi_2}{\partial V_{BE}} - \lambda)]_{(V_{BE}^*, V_{BC}^*, V_{A_2}^*)}$$

$$\det(A - \lambda \cdot I) = (\frac{\partial \psi_1}{\partial V_{BC}} - \lambda)_{(V_{BE}^*, V_{BC}^*, V_{A_2}^*)} \cdot [(\frac{\partial \psi_2}{\partial V_{BE}} \cdot \frac{\partial \psi_3}{\partial V_{A_2}} - \frac{\partial \psi_3}{\partial V_{BE}} \cdot \frac{\partial \psi_2}{\partial V_{A_2}})$$
$$- (\frac{\partial \psi_2}{\partial V_{BE}} + \frac{\partial \psi_3}{\partial V_{A_2}}) \cdot \lambda + \lambda^2]_{(V_{BE}^*, V_{BC}^*, V_{A_2}^*)} - (\frac{\partial \psi_1}{\partial V_{BE}})_{(V_{BE}^*, V_{BC}^*, V_{A_2}^*)}$$
$$\cdot [(\frac{\partial \psi_2}{\partial V_{BC}} \cdot \frac{\partial \psi_3}{\partial V_{A_2}} - \frac{\partial \psi_3}{\partial V_{BC}} \cdot \frac{\partial \psi_2}{\partial V_{A_2}}) - \frac{\partial \psi_2}{\partial V_{BC}} \cdot \lambda]_{(V_{BE}^*, V_{BC}^*, V_{A_2}^*)}$$
$$+ (\frac{\partial \psi_1}{\partial V_{A_2}})_{(V_{BE}^*, V_{BC}^*, V_{A_2}^*)} \cdot [(\frac{\partial \psi_2}{\partial V_{BC}} \cdot \frac{\partial \psi_3}{\partial V_{BE}} - \frac{\partial \psi_3}{\partial V_{BC}} \cdot \frac{\partial \psi_2}{\partial V_{BE}}) + \frac{\partial \psi_3}{\partial V_{BC}} \cdot \lambda]_{(V_{BE}^*, V_{BC}^*, V_{A_2}^*)}$$

$$\det(A - \lambda \cdot I) = [\frac{\partial \psi_1}{\partial V_{BC}} \cdot (\frac{\partial \psi_2}{\partial V_{BE}} \cdot \frac{\partial \psi_3}{\partial V_{A_2}} - \frac{\partial \psi_3}{\partial V_{BE}} \cdot \frac{\partial \psi_2}{\partial V_{A_2}}) - \frac{\partial \psi_1}{\partial V_{BC}} \cdot (\frac{\partial \psi_2}{\partial V_{BE}} + \frac{\partial \psi_3}{\partial V_{A_2}}) \cdot \lambda$$
$$+ \frac{\partial \psi_1}{\partial V_{BC}} \cdot \lambda^2 - (\frac{\partial \psi_2}{\partial V_{BE}} \cdot \frac{\partial \psi_3}{\partial V_{A_2}} - \frac{\partial \psi_3}{\partial V_{BE}} \cdot \frac{\partial \psi_2}{\partial V_{A_2}}) \cdot \lambda$$
$$+ (\frac{\partial \psi_2}{\partial V_{BE}} + \frac{\partial \psi_3}{\partial V_{A_2}}) \cdot \lambda^2 - \lambda^3]_{(V_{BE}^*, V_{BC}^*, V_{A_2}^*)} - [(\frac{\partial \psi_1}{\partial V_{BE}}) \cdot (\frac{\partial \psi_2}{\partial V_{BC}} \cdot \frac{\partial \psi_3}{\partial V_{A_2}}$$
$$- \frac{\partial \psi_3}{\partial V_{BC}} \cdot \frac{\partial \psi_2}{\partial V_{A_2}})]_{(V_{BE}^*, V_{BC}^*, V_{A_2}^*)} + [(\frac{\partial \psi_1}{\partial V_{BE}}) \cdot \frac{\partial \psi_2}{\partial V_{BC}}]_{(V_{BE}^*, V_{BC}^*, V_{A_2}^*)} \cdot \lambda$$
$$+ [(\frac{\partial \psi_1}{\partial V_{A_2}}) \cdot (\frac{\partial \psi_2}{\partial V_{BC}} \cdot \frac{\partial \psi_3}{\partial V_{BE}} - \frac{\partial \psi_3}{\partial V_{BC}} \cdot \frac{\partial \psi_2}{\partial V_{BE}}) + (\frac{\partial \psi_1}{\partial V_{A_2}}) \cdot \frac{\partial \psi_3}{\partial V_{BC}} \cdot \lambda]_{(V_{BE}^*, V_{BC}^*, V_{A_2}^*)}$$

$$\det(A - \lambda \cdot I) = -\lambda^3 + [\frac{\partial \psi_1}{\partial V_{BC}} + (\frac{\partial \psi_2}{\partial V_{BE}} + \frac{\partial \psi_3}{\partial V_{A_2}})]_{(V_{BE}^*, V_{BC}^*, V_{A_2}^*)} \cdot \lambda^2$$
$$+ [(\frac{\partial \psi_1}{\partial V_{A_2}}) \cdot \frac{\partial \psi_3}{\partial V_{BC}} - \frac{\partial \psi_1}{\partial V_{BC}} \cdot (\frac{\partial \psi_2}{\partial V_{BE}} + \frac{\partial \psi_3}{\partial V_{A_2}})$$
$$- (\frac{\partial \psi_2}{\partial V_{BE}} \cdot \frac{\partial \psi_3}{\partial V_{A_2}} - \frac{\partial \psi_3}{\partial V_{BE}} \cdot \frac{\partial \psi_2}{\partial V_{A_2}}) + (\frac{\partial \psi_1}{\partial V_{BE}}) \cdot \frac{\partial \psi_2}{\partial V_{BC}}]_{(V_{BE}^*, V_{BC}^*, V_{A_2}^*)} \cdot \lambda$$
$$+ [\frac{\partial \psi_1}{\partial V_{BC}} \cdot (\frac{\partial \psi_2}{\partial V_{BE}} \cdot \frac{\partial \psi_3}{\partial V_{A_2}} - \frac{\partial \psi_3}{\partial V_{BE}} \cdot \frac{\partial \psi_2}{\partial V_{A_2}})$$
$$- (\frac{\partial \psi_1}{\partial V_{BE}}) \cdot (\frac{\partial \psi_2}{\partial V_{BC}} \cdot \frac{\partial \psi_3}{\partial V_{A_2}} - \frac{\partial \psi_3}{\partial V_{BC}} \cdot \frac{\partial \psi_2}{\partial V_{A_2}})$$
$$+ (\frac{\partial \psi_1}{\partial V_{A_2}}) \cdot (\frac{\partial \psi_2}{\partial V_{BC}} \cdot \frac{\partial \psi_3}{\partial V_{BE}} - \frac{\partial \psi_3}{\partial V_{BC}} \cdot \frac{\partial \psi_2}{\partial V_{BE}})]_{(V_{BE}^*, V_{BC}^*, V_{A_2}^*)}$$

The eigenvalues of a matrix A are given by the characteristic equation $\det(A - \lambda \cdot I) = 0$; $\det(A - \lambda \cdot I) = \sum_{k=0}^{3} \Xi_k \cdot \lambda^k = 0$.

$$\Xi_3 = -1; \Xi_2 = [\frac{\partial \psi_1}{\partial V_{BC}} + (\frac{\partial \psi_2}{\partial V_{BE}} + \frac{\partial \psi_3}{\partial V_{A_2}})]_{(V_{BE}^*, V_{BC}^*, V_{A_2}^*)}$$
$$\Xi_1 = [(\frac{\partial \psi_1}{\partial V_{A_2}}) \cdot \frac{\partial \psi_3}{\partial V_{BC}} - \frac{\partial \psi_1}{\partial V_{BC}} \cdot (\frac{\partial \psi_2}{\partial V_{BE}} + \frac{\partial \psi_3}{\partial V_{A_2}})$$
$$- (\frac{\partial \psi_2}{\partial V_{BE}} \cdot \frac{\partial \psi_3}{\partial V_{A_2}} - \frac{\partial \psi_3}{\partial V_{BE}} \cdot \frac{\partial \psi_2}{\partial V_{A_2}}) + (\frac{\partial \psi_1}{\partial V_{BE}}) \cdot \frac{\partial \psi_2}{\partial V_{BC}}]_{(V_{BE}^*, V_{BC}^*, V_{A_2}^*)}$$

$$\Xi_0 = [\frac{\partial \psi_1}{\partial V_{BC}} \cdot (\frac{\partial \psi_2}{\partial V_{BE}} \cdot \frac{\partial \psi_3}{\partial V_{A_2}} - \frac{\partial \psi_3}{\partial V_{BE}} \cdot \frac{\partial \psi_2}{\partial V_{A_2}}) - (\frac{\partial \psi_1}{\partial V_{BE}}) \cdot (\frac{\partial \psi_2}{\partial V_{BC}} \cdot \frac{\partial \psi_3}{\partial V_{A_2}} - \frac{\partial \psi_3}{\partial V_{BC}} \cdot \frac{\partial \psi_2}{\partial V_{A_2}})$$
$$+ (\frac{\partial \psi_1}{\partial V_{A_2}}) \cdot (\frac{\partial \psi_2}{\partial V_{BC}} \cdot \frac{\partial \psi_3}{\partial V_{BE}} - \frac{\partial \psi_3}{\partial V_{BC}} \cdot \frac{\partial \psi_2}{\partial V_{BE}})]_{(V_{BE}^*, V_{BC}^*, V_{A_2}^*)}$$

We get three eigenvalues (λ_1, λ_2, λ_3) for our system and need to classify them. If $\lambda_1, \lambda_2, \lambda_3 \in \mathbb{R}$ then our fixed point $V_{BE}^*, V_{BC}^*, V_{A_2}^*$ is classify in the below Table 5.2

Table 5.2 Single ended class B amplifier system eigenvalues and stability classification

System eigenvalues	Stability classification $V_{BE}^*, V_{BC}^*, V_{A_2}^*$
$\lambda_1 > 0, \lambda_2 > 0, \lambda_3 > 0$	Unstable node
$\lambda_1 > 0, \lambda_2 > 0, \lambda_3 < 0$	Saddle point
$\lambda_1 > 0, \lambda_2 < 0, \lambda_3 < 0$	Saddle point
$\lambda_1 < 0, \lambda_2 < 0, \lambda_3 < 0$	Stable node

If $\lambda_1, \lambda_2 \in \mathbb{C}; \lambda_1 = \eta_1 + i \cdot \eta_2; \lambda_2 = \eta_1 - i \cdot \eta_2; \eta_1, \eta_2 \in \mathbb{R}; \lambda_3 \in \mathbb{R}$ then our fixed point $V_{BE}^*, V_{BC}^*, V_{A_2}^*$ is classify in the below table

System eigenvalues	Stability classification $V_{BE}^{*}, V_{BC}^{*}, V_{A_2}^{*}$
$\eta_1 > 0; \lambda_3 > 0$	Unstable spiral node
$\eta_1 > 0; \lambda_3 < 0$	Unstable spiral saddle
$\eta_1 < 0; \lambda_3 > 0$	Unstable spiral saddle
$\eta_1 < 0; \lambda_3 < 0$	Stable spiral node

If $\lambda_1, \lambda_2 < 0; \lambda_3 = 0$ or $\lambda_1, \lambda_3 < 0; \lambda_2 = 0$ or $\lambda_2, \lambda_3 < 0; \lambda_1 = 0$ then we get attracting line. If $\lambda_1, \lambda_2 > 0; \lambda_3 = 0$ or $\lambda_1, \lambda_3 > 0; \lambda_2 = 0$ or $\lambda_2, \lambda_3 > 0; \lambda_1 = 0$ then we get repelling line.

5.4 Wideband Low Noise Amplifier (LNA) with Negative Feedback Circuit Stability Analysis Under Circuit's Parameters Variation

When we want to amplify a very low power signal, we use Low Noise Amplifier (LNA). It is done without degrading its signal to noise ratio (SNR). LNA device is a crucial element in every RF receiver system and it amplifies the signal that comes from the antenna. Regular amplifier will increase the power of both the signal and the noise which come from the antenna and present at the amplifier's input. Additionally amplifiers are not ideal and they add noise to the input signal. Low Noise Amplifiers (LNAs) are designed to minimize the additional noise. The target is to minimize the additional noise by considering tradeoffs that include impedance matching. choosing the amplifier technology, and selecting low-noise biasing conditions. Low-noise amplifiers are found in many radio communications systems, medical instruments, and electronic equipments. The conventional LNA operates on a single band, while wideband LNA operate typically from 100 MHz to 1GHz and it hard to design. It is a challenge to design broadband amplifier with the best performances. One architecture is the combination of several narrows band LNA circuit into a single wideband LNA circuit. LNA feedback technique is proposed to simultaneously achieve improvement in bandwidth and on its gain, noise figure and return loss. The negative feedback technique can be used in wideband amplifier to provide a flat gain response and to reduce the input and output VSWR. It controls the amplifier performance due to technical specifications variation from transistor to transistor and in band stability is also improved by employing negative feedback. The LNA is most important block in any receiving system because the receiving system sensitivity is generally determined by its gain and noise figure. LNAs figures of merit are reduced Noise Figure (NF), moderated gain, good input/output impedance matching, low power consumption, isolation between input and output, acceptable linearity (low distortion), and stability. There are many ways to design LNAs. It can be single ended or differential, single stage, multistage, depending on type of engineering application and applications. Typical single end LNA system is a two stage single ended LNA. The first stage cascode amplifier is chosen for its simple input matching, its higher gain compared to an inductively degenerated common emitter amplifier, and its high reverse isolation and higher stability compared to a common base counterpart. The second stage consists of a common emitter cascode amplifier without emitter degeneration for higher gain. Differential LNA is composed of two stages. The first stage is a differential cascode amplifier using LC impedance peaking network as load. This load impedance can be made very large across the desired frequency band in order to force the output current to flow into the following stage. In practice the series resistance in the transmission line inductor will limit the impedance peaking effect. The second stage is a conventional emitter coupled differential amplifier with high common mode rejection. It is used to amplify the desired signal and compress the common mode signal. The two stages are connected by a coupling capacitor C. The single ended architecture has one disadvantage that it is very sensitive to parasitic ground inductance. A differential LNA can beneficial while the noise figure is higher than single ended design. We get higher gain by using multi stage LNA but the problem is that is difficult to maintain stability than single stage LNA. The selection of design option depends on type of application and specific design targets. The wideband LNA is required to be in single stage, low power consumption, and minimum components. The most important design considerations in a LNA design are stability, noise, power gain, bandwidth, and DC requirements. The DC biasing circuit is used to bias the selected transistor and the input and output matching network is important for maximum power transfer in the circuit. LNAs operate in class A mode, characterized by a bias point at the center of maximum voltage and current of the bias supply for the transistor. The biasing point for the LNA should have high gain, low noise figure, linear, good input and output matching and stable at the lowest current drain from the supply. In designing LNAs, stability of the circuit is important parameter. This stability characteristic means that the device does not oscillate over a range of frequencies with any combination of source and load impedance. The next figure describes the block diagram of LNA [121, 122] (Fig. 5.7).

BJT technology is selected to design the wideband low noise amplifier due to the higher gain at low power consumption, with reasonable low noise figure.

The RC feedback is the one of the most popular techniques to be used in amplifiers circuit for its wideband input match and good linearity. The schematic of the LNA is shown in the below figure. The transistor is self biased with the biasing resistor of R_1 and R_2 and designed to low power product application. (Fig. 5.8).

The LNA design has implemented RLC feedback (L_1, R_3, C_1) in order to lower the gain at the lower frequencies and hence improve the stability of the circuit. We need to tune the RLC feedback banch in order to meet the design specifications. The LNA design also employs output resistive loading in stabilizing the circuit. The initial output resistor value R_4 is set to less than 50 Ω because high output resistor value may result in huge decrease of gain and P1 dB point. L_2 and L_3 in the circuit acts as RF choke which separate RF and DC path in the circuit. The LNA is matched using lumped element as it is simple and compact (L matching network). Typical LC matching network include the use of capacitors and inductors in either series or shunt configuration. Circuit stability analysis is done by considering BJT Small Signal (SS) equivalent circuit model. We consider "AC ground" in the circuit. Since the voltage at this terminal is held constant at V_{cc}, there is no time

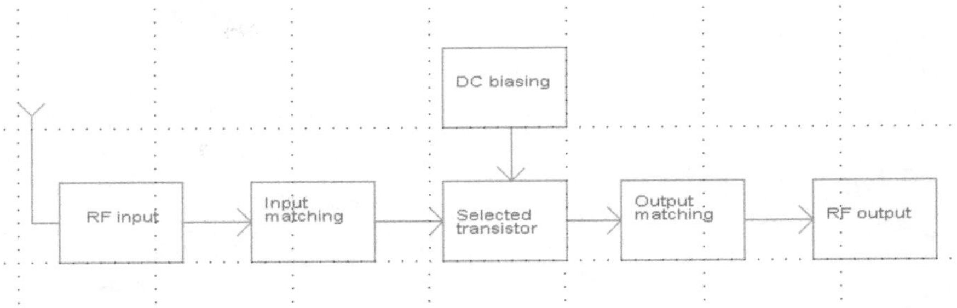

Fig. 5.7 Block diagram of LNA

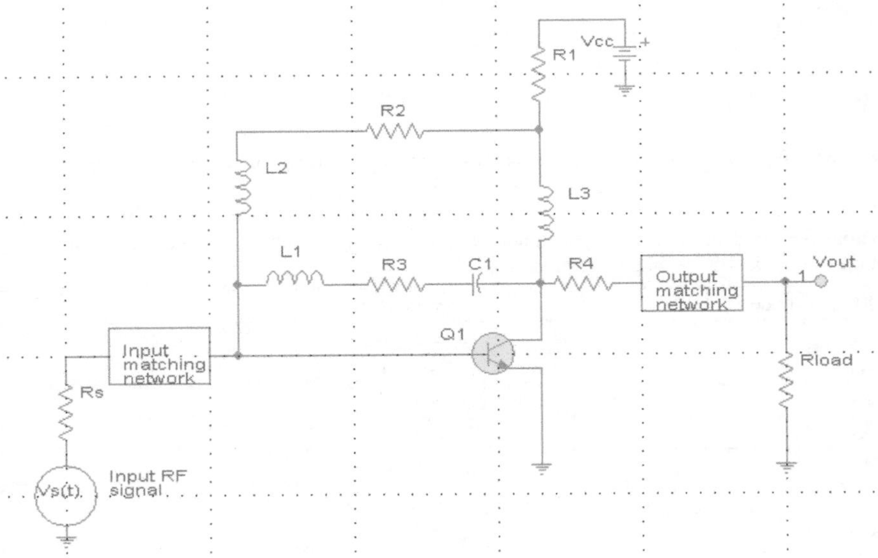

Fig. 5.8 Wideband Low Noise Amplifier (LNA) with input and output matching network schematics

variation of the voltage. Consequently. We can set this terminal to be an "AC ground" in the small signal circuit. For AC ground, we "kill" the DC sources at that terminal (short circuit voltage sources and open circuit current sources). Input and output matching networks are L-type. Under AC and Small Signal (SS) conditions, the BJT can be replaced with linear Hybrid Pi model. Let's verify that this circuit incorporates all the necessary Small Signal (SS) characteristics of BJT: $i_b = v_{be}/r_{\pi}; i_c = g_m \cdot v_{be}; i_b + i_c = i_e$ [24–26] (Fig. 5.9, Table 5.3).

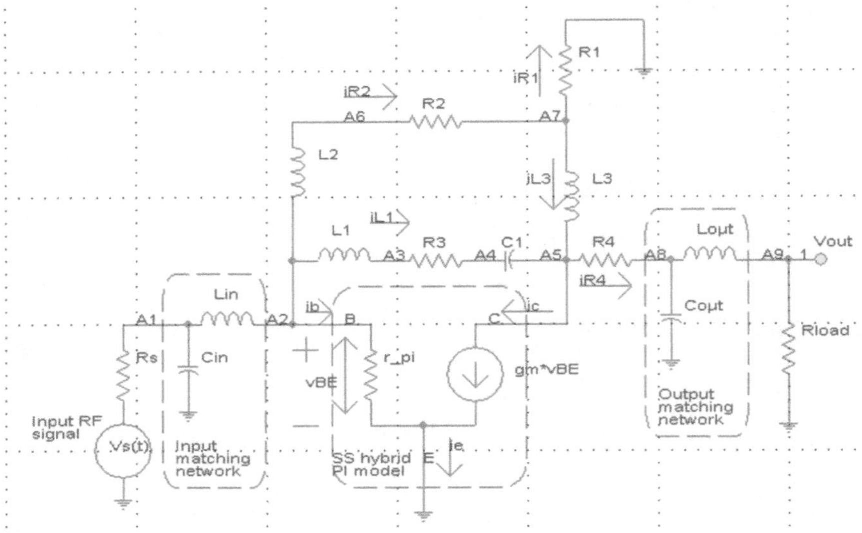

Fig. 5.9 Wideband Low Noise Amplifier (LNA) with SS hybrid PI model and matching networks

Table 5.3 Wideband Low Noise Amplifier (LNA) with SS hybrid PI model and matching networks KCL@ node and expressions

KCL @ node	Expression
A_1	$I_{R_s} = I_{C_{in}} + I_{L_{in}}$
A_2	$I_{L_{in}} = I_{r_\pi} + I_{L_1} + I_{L_2}$
A_3	$I_{L_1} = I_{R_3}$
A_4	$I_{R_3} = I_{C_1}$
A_5	$I_{C_1} + I_{L_3} = I_{R_4} + g_m \cdot v_{BE}$
A_6	$I_{L_2} = I_{R_2}$
A_7	$I_{R_2} = I_{R_1} + I_{L_3}$
A_8	$I_{R_4} = I_{C_{out}} + I_{L_{out}}$
A_9	$I_{L_{out}} = I_{R_{load}}$

$$I_{R_s} = \frac{V_s(t) - V_{A_1}}{R_s}; I_{C_{in}} = C_{in} \cdot \frac{dV_{A_1}}{dt}; V_{L_{in}} = V_{A_1} - V_{A_2} = L_{in} \cdot \frac{dI_{L_{in}}}{dt}; I_{r_\pi} = \frac{V_{A_2}}{r_\pi}$$

$$V_{L_1} = V_{A_2} - V_{A_3} = L_1 \cdot \frac{dI_{L_1}}{dt}; I_{R_3} = \frac{V_{A_3} - V_{A_4}}{R_3}; I_{C_1} = C_1 \cdot \frac{d}{dt}(V_{A_4} - V_{A_5}) = C_1 \cdot \frac{dV_{C_1}}{dt}$$

$$V_{L_2} = V_{A_2} - V_{A_6} = L_2 \cdot \frac{dI_{L_2}}{dt}; I_{R_2} = \frac{V_{A_6} - V_{A_7}}{R_2}; I_{R_1} = \frac{V_{A_7}}{R_1}; V_{L_3} = V_{A_7} - V_{A_5} = L_3 \cdot \frac{dI_{L_3}}{dt}$$

$$I_{L_1} = I_{R_3} = I_{C_1}; I_{L_2} = I_{R_2}; I_{R_4} = \frac{V_{A_5} - V_{A_8}}{R_4}; I_{C_{out}} = C_{out} \cdot \frac{dV_{A_8}}{dt}; V_{L_{out}} = V_{A_8} - V_{A_9} = L_{out} \cdot \frac{dI_{L_{out}}}{dt}$$

$$I_{R_{load}} = \frac{V_{out}}{R_{load}} = \frac{V_{A_9}}{R_{load}}; I_{L_{out}} = I_{R_{load}}$$

$$I_{R_s} = \frac{V_s(t) - V_{A_1}}{R_s} \Rightarrow I_{R_s} \cdot R_s = V_s(t) - V_{A_1} \Rightarrow V_{A_1} = V_s(t) - I_{R_s} \cdot R_s$$

$$I_{C_{in}} = C_{in} \cdot \frac{d}{dt}(V_s(t) - I_{R_s} \cdot R_s) = C_{in} \cdot (\frac{dV_s(t)}{dt} - \frac{dI_{R_s}}{dt} \cdot R_s); V_{A_1} - V_{A_2} = L_{in} \cdot \frac{dI_{L_{in}}}{dt}$$

$$V_{A_2} = V_{A_1} - L_{in} \cdot \frac{dI_{L_{in}}}{dt} = V_s(t) - I_{R_s} \cdot R_s - L_{in} \cdot \frac{dI_{L_{in}}}{dt}; I_{r_\pi} = \frac{V_{A_2}}{r_\pi} \Rightarrow V_{A_2} = I_{r_\pi} \cdot r_\pi$$

$$I_{r_\pi} \cdot r_\pi = V_s(t) - I_{R_s} \cdot R_s - L_{in} \cdot \frac{dI_{L_{in}}}{dt}; V_{A_2} - V_{A_3} = L_1 \cdot \frac{dI_{L_1}}{dt} \Rightarrow V_{A_3} = V_{A_2} - L_1 \cdot \frac{dI_{L_1}}{dt}$$

$$V_{A_3} = I_{r_\pi} \cdot r_\pi - L_1 \cdot \frac{dI_{L_1}}{dt}; I_{R_3} = \frac{V_{A_3} - V_{A_4}}{R_3}$$

$$\Rightarrow I_{R_3} \cdot R_3 = V_{A_3} - V_{A_4} \Rightarrow V_{A_4} = V_{A_3} - I_{R_3} \cdot R_3$$

$$V_{A_4} = I_{r_\pi} \cdot r_\pi - L_1 \cdot \frac{dI_{L_1}}{dt} - I_{R_3} \cdot R_3; I_{C_1} = C_1 \cdot \frac{d}{dt}(V_{A_4} - V_{A_5})$$

$$= C_1 \cdot \frac{d}{dt}(I_{r_\pi} \cdot r_\pi - L_1 \cdot \frac{dI_{L_1}}{dt} - I_{R_3} \cdot R_3 - V_{A_5})$$

$$I_{C_1} = C_1 \cdot (\frac{dI_{r_\pi}}{dt} \cdot r_\pi - L_1 \cdot \frac{d^2 I_{L_1}}{dt^2} - \frac{dI_{R_3}}{dt} \cdot R_3 - \frac{dV_{A_5}}{dt}); V_{A_2} - V_{A_6} = L_2 \cdot \frac{dI_{L_2}}{dt} \Rightarrow V_{A_6}$$
$$= V_{A_2} - L_2 \cdot \frac{dI_{L_2}}{dt}$$

$$V_{A_6} = I_{r_\pi} \cdot r_\pi - L_2 \cdot \frac{dI_{L_2}}{dt}; I_{R_2} = \frac{V_{A_6} - V_{A_7}}{R_2}$$

$$\Rightarrow I_{R_2} \cdot R_2 = V_{A_6} - V_{A_7} = I_{r_\pi} \cdot r_\pi - L_2 \cdot \frac{dI_{L_2}}{dt} - I_{R_1} \cdot R_1$$

$$I_{R_1} = \frac{V_{A_7}}{R_1} \Rightarrow V_{A_7} = I_{R_1} \cdot R_1; I_{R_2} \cdot R_2$$

$$= I_{r_\pi} \cdot r_\pi - L_2 \cdot \frac{dI_{L_2}}{dt} - I_{R_1} \cdot R_1; V_{A_7} - V_{A_5} = L_3 \cdot \frac{dI_{L_3}}{dt}$$

$$V_{A_5} = V_{A_7} - L_3 \cdot \frac{dI_{L_3}}{dt}; I_{C_1} = C_1 \cdot (\frac{dI_{r_\pi}}{dt} \cdot r_\pi - L_1 \cdot \frac{d^2 I_{L_1}}{dt^2} - \frac{dI_{R_3}}{dt} \cdot R_3 - \frac{dV_{A_7}}{dt} + L_3 \cdot \frac{d^2 I_{L_3}}{dt^2})$$

$$V_{A_7} = I_{R_1} \cdot R_1 \Rightarrow I_{C_1} = C_1 \cdot (\frac{dI_{r_\pi}}{dt} \cdot r_\pi - L_1 \cdot \frac{d^2 I_{L_1}}{dt^2} - \frac{dI_{R_3}}{dt} \cdot R_3 - \frac{dI_{R_1}}{dt} \cdot R_1 + L_3 \cdot \frac{d^2 I_{L_3}}{dt^2})$$

$$I_{R_4} = \frac{V_{A_5} - V_{A_8}}{R_4} \Rightarrow I_{R_4} \cdot R_4 = V_{A_5} - V_{A_8} \Rightarrow V_{A_8} = V_{A_5} - I_{R_4} \cdot R_4; I_{C_{out}} = C_{out} \cdot \frac{dV_{A_8}}{dt}$$
$$= C_{out} \cdot (\frac{dV_{A_5}}{dt} - \frac{dI_{R_4}}{dt} \cdot R_4)$$

$$V_{A_5} = I_{R_1} \cdot R_1 - L_3 \cdot \frac{dI_{L_3}}{dt} \Rightarrow I_{C_{out}} = C_{out} \cdot (\frac{dI_{R_1}}{dt} \cdot R_1 - L_3 \cdot \frac{d^2 I_{L_3}}{dt^2} - \frac{dI_{R_4}}{dt} \cdot R_4)$$

$$V_{A_8} - V_{A_9} = L_{out} \cdot \frac{dI_{L_{out}}}{dt} \Rightarrow V_{A_8} = V_{A_9} + L_{out} \cdot \frac{dI_{L_{out}}}{dt}; I_{R_{load}} = \frac{V_{A_9}}{R_{load}} \Rightarrow V_{A_9} = I_{R_{load}} \cdot R_{load}$$

$$V_{A_8} = I_{R_{load}} \cdot R_{load} + L_{out} \cdot \frac{dI_{L_{out}}}{dt}; I_{C_{out}} = C_{out} \cdot \frac{dV_{A_8}}{dt} = C_{out} \cdot (\frac{dI_{R_{load}}}{dt} \cdot R_{load} + L_{out} \cdot \frac{d^2 I_{L_{out}}}{dt^2})$$

$$I_{R_{load}} = I_{L_{out}} \Rightarrow I_{C_{out}} = C_{out} \cdot \frac{dV_{A_8}}{dt} = C_{out} \cdot (\frac{dI_{L_{out}}}{dt} \cdot R_{load} + L_{out} \cdot \frac{d^2 I_{L_{out}}}{dt^2})$$

Additional equations: $I_{L_1} = I_{R_3} = I_{C_1}; I_{L_2} = I_{R_2}$.

We can summary our wideband LNA with negative feedback circuit differential equation:

[I] $I_{C_{in}} = C_{in} \cdot (\frac{dV_s(t)}{dt} - \frac{dI_{R_s}}{dt} \cdot R_s)$, [II] $I_{r_\pi} \cdot r_\pi = V_s(t) - I_{R_s} \cdot R_s - L_{in} \cdot \frac{dI_{L_{in}}}{dt}$.

[III] $I_{R_2} \cdot R_2 = I_{r_\pi} \cdot r_\pi - L_2 \cdot \frac{dI_{R_2}}{dt} - I_{R_1} \cdot R_1$

[IV] $I_{C_1} = C_1 \cdot (\frac{dI_{r_\pi}}{dt} \cdot r_\pi - L_1 \cdot \frac{d^2 I_{C_1}}{dt^2} - \frac{dI_{C_1}}{dt} \cdot R_3 - \frac{dI_{R_1}}{dt} \cdot R_1 + L_3 \cdot \frac{d^2 I_{L_3}}{dt^2})$

[V] $I_{C_{out}} = C_{out} \cdot (\frac{dI_{R_1}}{dt} \cdot R_1 - L_3 \cdot \frac{d^2 I_{L_3}}{dt^2} - \frac{dI_{R_4}}{dt} \cdot R_4)$, [VI] $I_{C_{out}} = C_{out} \cdot (\frac{dI_{L_{out}}}{dt} \cdot R_{load} + L_{out} \cdot \frac{d^2 I_{L_{out}}}{dt^2})$

[IV] + [V] $L_3 \cdot \frac{d^2 I_{L_3}}{dt^2} = \frac{dI_{R_1}}{dt} \cdot R_1 - \frac{dI_{R_4}}{dt} \cdot R_4 - I_{C_{out}} \cdot \frac{1}{C_{out}}$

$$I_{C_1} = C_1 \cdot (\frac{dI_{r_\pi}}{dt} \cdot r_\pi - L_1 \cdot \frac{d^2 I_{C_1}}{dt^2} - \frac{dI_{C_1}}{dt} \cdot R_3 - \frac{dI_{R_4}}{dt} \cdot R_4 - I_{C_{out}} \cdot \frac{1}{C_{out}})$$

$$I_{R_s} = I_{C_{in}} + I_{L_{in}} \Rightarrow I_{C_{in}} = I_{R_s} - I_{L_{in}}; I_{C_1} + I_{R_2} - I_{R_1} = I_{R_4} + g_m \cdot I_{r_\pi} \cdot r_\pi$$

$$I_{R_1} = I_{C_1} + I_{R_2} - g_m \cdot I_{r_\pi} \cdot r_\pi - I_{R_4}; I_{R_4} = I_{C_{out}} + I_{L_{out}} \Rightarrow I_{C_{out}} = I_{R_4} - I_{L_{out}}$$

We can summary our system differential equations:

[I] $I_{R_s} - I_{L_{in}} = C_{in} \cdot (\frac{dV_s(t)}{dt} - \frac{dI_{R_s}}{dt} \cdot R_s)$, [II] $I_{r_\pi} \cdot r_\pi = V_s(t) - I_{R_s} \cdot R_s - L_{in} \cdot \frac{dI_{L_{in}}}{dt}$.

[III] $I_{R_2} \cdot R_2 = I_{r_\pi} \cdot r_\pi - L_2 \cdot \frac{dI_{R_2}}{dt} - [I_{C_1} + I_{R_2} - g_m \cdot I_{r_\pi} \cdot r_\pi - I_{R_4}] \cdot R_1$

[IV] + [V] $I_{C_1} = C_1 \cdot (\frac{dI_{r_\pi}}{dt} \cdot r_\pi - L_1 \cdot \frac{d^2 I_{C_1}}{dt^2} - \frac{dI_{C_1}}{dt} \cdot R_3 - \frac{dI_{R_4}}{dt} \cdot R_4 - [I_{R_4} - I_{L_{out}}] \cdot \frac{1}{C_{out}})$

[VI] $I_{R_4} - I_{L_{out}} = C_{out} \cdot (\frac{dI_{L_{out}}}{dt} \cdot R_{load} + L_{out} \cdot \frac{d^2 I_{L_{out}}}{dt^2})$

We define new variables: $Y_1 = \frac{dI_{r_\pi}}{dt}; Y_2 = \frac{dI_{C_1}}{dt}; \frac{dY_2}{dt} = \frac{d^2 I_{C_1}}{dt^2}; Y_3 = \frac{dI_{R_4}}{dt}; Y_4 = \frac{dI_{L_{out}}}{dt}$

[I] $I_{R_s} - I_{L_{in}} = C_{in} \cdot (\frac{dV_s(t)}{dt} - \frac{dI_{R_s}}{dt} \cdot R_s)$, [II] $I_{r_\pi} \cdot r_\pi = V_s(t) - I_{R_s} \cdot R_s - L_{in} \cdot \frac{dI_{L_{in}}}{dt}$.

[III] $I_{R_2} \cdot R_2 = I_{r_\pi} \cdot r_\pi - L_2 \cdot \frac{dI_{R_2}}{dt} - [I_{C_1} + I_{R_2} - g_m \cdot I_{r_\pi} \cdot r_\pi - I_{R_4}] \cdot R_1$

[IV] + [V] $I_{C_1} = C_1 \cdot (Y_1 \cdot r_\pi - L_1 \cdot \frac{dY_2}{dt} - Y_2 \cdot R_3 - Y_3 \cdot R_4 - I_{R_4} \cdot \frac{1}{C_{out}} + I_{L_{out}} \cdot \frac{1}{C_{out}})$

[VI] $I_{R_4} - I_{L_{out}} = C_{out} \cdot (Y_4 \cdot R_{load} + L_{out} \cdot \frac{dY_4}{dt})$

We can summary our system differential equations:

$$\frac{dI_{r_\pi}}{dt} = Y_1; \frac{dI_{C_1}}{dt} = Y_2; \frac{dI_{R_4}}{dt} = Y_3; \frac{dI_{L_{out}}}{dt} = Y_4; \frac{dI_{R_s}}{dt} = \frac{1}{R_s} \cdot \frac{dV_s(t)}{dt} - (I_{R_s} - I_{L_{in}}) \cdot \frac{1}{C_{in} \cdot R_s}$$

$$\frac{dI_{L_{in}}}{dt} = \frac{1}{L_{in}} \cdot V_s(t) - I_{R_s} \cdot \frac{R_s}{L_{in}} - I_{r_\pi} \cdot \frac{r_\pi}{L_{in}}; \frac{dY_4}{dt} = (I_{R_4} - I_{L_{out}}) \cdot \frac{1}{C_{out} \cdot L_{out}} - Y_4 \cdot \frac{R_{load}}{L_{out}}$$

$$\frac{dI_{R_2}}{dt} = -I_{R_2} \cdot \frac{R_2}{L_2} + I_{r_\pi} \cdot \frac{r_\pi}{L_2} - [I_{C_1} + I_{R_2} - g_m \cdot I_{r_\pi} \cdot r_\pi - I_{R_4}] \cdot \frac{R_1}{L_2}$$

$$\frac{dY_2}{dt} = -I_{C_1} \cdot \frac{1}{C_1 \cdot L_1} + Y_1 \cdot \frac{r_\pi}{L_1} - Y_2 \cdot \frac{R_3}{L_1} - Y_3 \cdot \frac{R_4}{L_1} - I_{R_4} \cdot \frac{1}{C_{out} \cdot L_1} + I_{L_{out}} \cdot \frac{1}{C_{out} \cdot L_1}$$

At fixed points (equilibrium points): $\frac{dI_{r_\pi}}{dt} = 0$; $\frac{dI_{C_1}}{dt} = 0$; $\frac{dI_{R_4}}{dt} = 0$; $\frac{dI_{L_{out}}}{dt} = 0$; $\frac{dI_{R_s}}{dt} = 0$

$$\frac{dI_{L_{in}}}{dt} = 0; \frac{dI_{R_2}}{dt} = 0; \frac{dY_2}{dt} = 0; Y_1^* = 0; Y_2^* = 0; Y_3^* = 0; Y_4^* = 0; I_{R_s}^* - I_{L_{in}}^* = C_{in} \cdot \frac{dV_s(t)}{dt}$$

$$I_{R_s}^* \cdot R_s + I_{r_\pi}^* \cdot r_\pi = V_s(t); I_{R_4}^* - I_{L_{out}}^* = Y_4^* \cdot R_{load} \cdot C_{out}; -I_{C_1}^* \cdot \frac{1}{C_1} - I_{R_4}^* \cdot \frac{1}{C_{out}} + I_{L_{out}}^* \cdot \frac{1}{C_{out}} = 0$$

$$-I_{R_2}^* \cdot R_2 + I_{r_\pi}^* \cdot r_\pi \cdot (1 + g_m \cdot R_1) - I_{C_1}^* \cdot R_1 - I_{R_2}^* \cdot R_1 + I_{R_4}^* \cdot R_1 = 0$$

If $\frac{dV_s(t)}{dt} \to \varepsilon$ then $I_{R_s}^* = I_{L_{in}}^*$.

Stability analysis: The standard local stability analysis about any one of the equilibrium points of the wideband Low Noise Amplifier (LNA) with negative feedback circuit equivalent circuit consists in adding to coordinate $[I_{R_s}, I_{L_{in}}, I_{r_\pi}, I_{R_2}, I_{C_1}, I_{R_4}, I_{L_{out}}, Y_1, Y_2, Y_3, Y_4]$ arbitrarily small increments of exponentially form $[i_{R_s}, i_{L_{in}}, i_{r_\pi}, i_{R_2}, i_{C_1}, i_{R_4}, i_{L_{out}}, y_1, y_2, y_3, y_4] \cdot e^{\lambda \cdot t}$ and retaining the first order terms in $I_{R_s}, I_{L_{in}}, I_{r_\pi}, I_{R_2}, I_{C_1}, I_{R_4}, I_{L_{out}}, Y_1, Y_2, Y_3, Y_4$. The system of homogenous equations leads to a polynomial characteristic equation in the eigenvalues. The polynomial characteristic equations accept by set of the below circuit variables, circuit variables derivative and circuit variables second order derivative with respect to time into wideband Low Noise Amplifier (LNA) with negative feedback [2–4]. Our wideband Low Noise Amplifier (LNA) with negative feedback equivalent circuit fixed values with arbitrarily small increments of exponential form $[i_{L_1}, i_{R_b}, i_{R_{load}}, i_{C_{out}}, i_{C_{in}}, y_1, y_2, y_3, y_4] \cdot e^{\lambda \cdot t}$ are: j = 0 (first fixed point), j = 1 (second fixed point), j = 2 (third fixed point), etc.,

$$Y_1(t) = Y_1^{(j)} + y_1 \cdot e^{\lambda \cdot t}; Y_2(t) = Y_2^{(j)} + y_2 \cdot e^{\lambda \cdot t}; Y_3(t) = Y_3^{(j)} + y_3 \cdot e^{\lambda \cdot t};$$
$$Y_4(t) = Y_4^{(j)} + y_4 \cdot e^{\lambda \cdot t}$$

$$I_{R_s}(t) = I_{R_s}^{(j)} + i_{R_s} \cdot e^{\lambda \cdot t}; I_{L_{in}}(t) = I_{L_{in}}^{(j)} + i_{L_{in}} \cdot e^{\lambda \cdot t}; I_{r_\pi}(t) = I_{r_\pi}^{(j)} + i_{R_\pi} \cdot e^{\lambda \cdot t}; I_{R_2}(t) = I_{R_2}^{(j)} + i_{R_2} \cdot e^{\lambda \cdot t}$$
$$I_{C_1}(t) = I_{C_1}^{(j)} + i_{C_1} \cdot e^{\lambda \cdot t}; I_{R_4}(t) = I_{R_4}^{(j)} + i_{R_4} \cdot e^{\lambda \cdot t}; I_{L_{out}}(t) = I_{L_{out}}^{(j)} + i_{L_{out}} \cdot e^{\lambda \cdot t}; \frac{dI_{r_\pi}(t)}{dt} = i_{R_\pi} \cdot \lambda \cdot e^{\lambda \cdot t}$$
$$\frac{dI_{L_{in}}(t)}{dt} = i_{L_{in}} \cdot \lambda \cdot e^{\lambda \cdot t}; \frac{dI_{R_2}(t)}{dt} = i_{R_2} \cdot \lambda \cdot e^{\lambda \cdot t}; \frac{dY_2(t)}{dt} = y_2 \cdot \lambda \cdot e^{\lambda \cdot t}; \frac{dY_4(t)}{dt} = y_4 \cdot \lambda \cdot e^{\lambda \cdot t}$$
$$\frac{dI_{C_1}(t)}{dt} = i_{C_1} \cdot \lambda \cdot e^{\lambda \cdot t}; \frac{dI_{R_4}(t)}{dt} = i_{R_4} \cdot \lambda \cdot e^{\lambda \cdot t}; \frac{dI_{L_{out}}(t)}{dt} = i_{L_{out}} \cdot \lambda \cdot e^{\lambda \cdot t}; \frac{dI_{r_s}(t)}{dt} = i_{R_s} \cdot \lambda \cdot e^{\lambda \cdot t}$$

$$i_{R_\pi} \cdot \lambda \cdot e^{\lambda \cdot t} = Y_1^{(j)} + y_1 \cdot e^{\lambda \cdot t}; Y_1^{(j)} = 0 \Rightarrow -i_{R_\pi} \cdot \lambda + y_1 = 0; i_{C_1} \cdot \lambda \cdot e^{\lambda \cdot t} = Y_2^{(j)} + y_2 \cdot e^{\lambda \cdot t};$$
$$Y_2^{(j)} = 0 \Rightarrow -i_{C_1} \cdot \lambda + y_2 = 0$$

$$i_{R_4} \cdot \lambda \cdot e^{\lambda \cdot t} = Y_3^{(j)} + y_3 \cdot e^{\lambda \cdot t}; Y_3^{(j)} = 0 \Rightarrow -i_{R_4} \cdot \lambda + y_3 = 0; i_{L_{out}} \cdot \lambda \cdot e^{\lambda \cdot t} = Y_4^{(j)} + y_4 \cdot e^{\lambda \cdot t};$$
$$Y_4^{(j)} = 0 \Rightarrow -i_{L_{out}} \cdot \lambda + y_4 = 0$$

$$i_{R_s} \cdot \lambda \cdot e^{\lambda \cdot t} = \frac{1}{R_s} \cdot \frac{dV_s(t)}{dt} - [I_{R_s}^{(j)} + i_{R_s} \cdot e^{\lambda \cdot t}] \cdot \frac{1}{C_{in} \cdot R_s} + [I_{L_{in}}^{(j)} + i_{L_{in}} \cdot e^{\lambda \cdot t}] \cdot \frac{1}{C_{in} \cdot R_s}$$

$$i_{R_s} \cdot \lambda \cdot e^{\lambda \cdot t} = \frac{1}{R_s} \cdot \frac{dV_s(t)}{dt} - I_{R_s}^{(j)} \cdot \frac{1}{C_{in} \cdot R_s} + I_{L_{in}}^{(j)} \cdot \frac{1}{C_{in} \cdot R_s} - i_{R_s} \cdot \frac{1}{C_{in} \cdot R_s} \cdot e^{\lambda \cdot t} + i_{L_{in}} \cdot \frac{1}{C_{in} \cdot R_s} \cdot e^{\lambda \cdot t}$$

At fixed point: $\frac{1}{R_s} \cdot \frac{dV_s(t)}{dt} - I_{R_s}^{(j)} \cdot \frac{1}{C_{in} \cdot R_s} + I_{L_{in}}^{(j)} \cdot \frac{1}{C_{in} \cdot R_s} = 0$

$$-i_{R_s} \cdot \lambda - i_{R_s} \cdot \frac{1}{C_{in} \cdot R_s} + i_{L_{in}} \cdot \frac{1}{C_{in} \cdot R_s} = 0$$

$$i_{L_{in}} \cdot \lambda \cdot e^{\lambda \cdot t} = \frac{1}{L_{in}} \cdot V_s(t) - [I_{R_s}^{(j)} + i_{R_s} \cdot e^{\lambda \cdot t}] \cdot \frac{R_s}{L_{in}} - [I_{r_\pi}^{(j)} + i_{R_\pi} \cdot e^{\lambda \cdot t}] \cdot \frac{r_\pi}{L_{in}}$$

$$i_{L_{in}} \cdot \lambda \cdot e^{\lambda \cdot t} = \frac{1}{L_{in}} \cdot V_s(t) - I_{R_s}^{(j)} \cdot \frac{R_s}{L_{in}} - I_{r_\pi}^{(j)} \cdot \frac{r_\pi}{L_{in}} - i_{R_s} \cdot \frac{R_s}{L_{in}} \cdot e^{\lambda \cdot t} - i_{R_\pi} \cdot \frac{r_\pi}{L_{in}} \cdot e^{\lambda \cdot t}$$

At fixed point: $\frac{1}{L_{in}} \cdot V_s(t) - I_{R_s}^{(j)} \cdot \frac{R_s}{L_{in}} - I_{r_\pi}^{(j)} \cdot \frac{r_\pi}{L_{in}} = 0; -i_{L_{in}} \cdot \lambda - i_{R_s} \cdot \frac{R_s}{L_{in}} - i_{R_\pi} \cdot \frac{r_\pi}{L_{in}} = 0$

$$y_4 \cdot \lambda \cdot e^{\lambda \cdot t} = (I_{R_4}^{(j)} + i_{R_4} \cdot e^{\lambda \cdot t} - I_{L_{out}}^{(j)} - i_{L_{out}} \cdot e^{\lambda \cdot t}) \cdot \frac{1}{C_{out} \cdot L_{out}} - [Y_4^{(j)} + y_4 \cdot e^{\lambda \cdot t}] \cdot \frac{R_{load}}{L_{out}}$$

$$y_4 \cdot \lambda \cdot e^{\lambda \cdot t} = (I_{R_4}^{(j)} - I_{L_{out}}^{(j)}) \cdot \frac{1}{C_{out} \cdot L_{out}} - Y_4^{(j)} \cdot \frac{R_{load}}{L_{out}} + (i_{R_4} \cdot e^{\lambda \cdot t} - i_{L_{out}} \cdot e^{\lambda \cdot t}) \cdot \frac{1}{C_{out} \cdot L_{out}} - y_4 \cdot \frac{R_{load}}{L_{out}} \cdot e^{\lambda \cdot t}$$

At fixed point: $(I_{R_4}^{(j)} - I_{L_{out}}^{(j)}) \cdot \frac{1}{C_{out} \cdot L_{out}} - Y_4^{(j)} \cdot \frac{R_{load}}{L_{out}} = 0$

$$-y_4 \cdot \lambda + i_{R_4} \cdot \frac{1}{C_{out} \cdot L_{out}} - i_{L_{out}} \cdot \frac{1}{C_{out} \cdot L_{out}} - y_4 \cdot \frac{R_{load}}{L_{out}} = 0$$

$$i_{R_2} \cdot \lambda \cdot e^{\lambda \cdot t} = -[I_{R_2}^{(j)} + i_{R_2} \cdot e^{\lambda \cdot t}] \cdot \frac{R_2}{L_2} + [I_{r_\pi}^{(j)} + i_{R_\pi} \cdot e^{\lambda \cdot t}] \cdot \frac{r_\pi}{L_2} - [I_{C_1}^{(j)} + i_{C_1} \cdot e^{\lambda \cdot t} + I_{R_2}^{(j)} + i_{R_2} \cdot e^{\lambda \cdot t} - g_m \cdot (I_{r_\pi}^{(j)} + i_{R_\pi} \cdot e^{\lambda \cdot t}) \cdot r_\pi - I_{R_4}^{(j)} - i_{R_4} \cdot e^{\lambda \cdot t}] \cdot \frac{R_1}{L_2}$$

$$i_{R_2} \cdot \lambda \cdot e^{\lambda \cdot t} = -I_{R_2}^{(j)} \cdot \frac{R_2}{L_2} + I_{r_\pi}^{(j)} \cdot \frac{r_\pi}{L_2} - [I_{C_1}^{(j)} + I_{R_2}^{(j)} - g_m \cdot I_{r_\pi}^{(j)} \cdot r_\pi - I_{R_4}^{(j)}] \cdot \frac{R_1}{L_2} - i_{R_2} \cdot \frac{R_2}{L_2} \cdot e^{\lambda \cdot t} + i_{R_\pi} \cdot \frac{r_\pi}{L_2} \cdot e^{\lambda \cdot t} - [i_{C_1} \cdot e^{\lambda \cdot t} + i_{R_2} \cdot e^{\lambda \cdot t} - g_m \cdot i_{R_\pi} \cdot e^{\lambda \cdot t} \cdot r_\pi - i_{R_4} \cdot e^{\lambda \cdot t}] \cdot \frac{R_1}{L_2}$$

At fixed point: $-I_{R_2}^{(j)} \cdot \frac{R_2}{L_2} + I_{r_\pi}^{(j)} \cdot \frac{r_\pi}{L_2} - [I_{C_1}^{(j)} + I_{R_2}^{(j)} - g_m \cdot I_{r_\pi}^{(j)} \cdot r_\pi - I_{R_4}^{(j)}] \cdot \frac{R_1}{L_2} = 0$

$$-i_{R_2} \cdot \lambda - i_{R_2} \cdot \frac{R_2}{L_2} + i_{R_\pi} \cdot \frac{r_\pi}{L_2} - [i_{C_1} + i_{R_2} - g_m \cdot i_{R_\pi} \cdot r_\pi - i_{R_4}] \cdot \frac{R_1}{L_2} = 0$$

$$-i_{R_2} \cdot \lambda - i_{R_2} \cdot \frac{R_2}{L_2} - i_{R_2} \cdot \frac{R_1}{L_2} + i_{R_\pi} \cdot \frac{r_\pi}{L_2} + i_{R_\pi} \cdot g_m \cdot r_\pi \cdot \frac{R_1}{L_2} - i_{C_1} \cdot \frac{R_1}{L_2} + \frac{R_1}{L_2} \cdot i_{R_4} = 0$$

$$-i_{R_2} \cdot \lambda - i_{R_2} \cdot \frac{1}{L_2} \cdot \sum_{k=1}^{2} R_k + i_{R_\pi} \cdot \frac{1}{L_2} \cdot r_\pi \cdot (1 + g_m \cdot R_1) - i_{C_1} \cdot \frac{R_1}{L_2} + \frac{R_1}{L_2} \cdot i_{R_4} = 0$$

$$y_2 \cdot \lambda \cdot e^{\lambda \cdot t} = -[I_{C_1}^{(j)} + i_{C_1} \cdot e^{\lambda \cdot t}] \cdot \frac{1}{C_1 \cdot L_1} + [Y_1^{(j)} + y_1 \cdot e^{\lambda \cdot t}] \cdot \frac{r_\pi}{L_1} - [Y_2^{(j)} + y_2 \cdot e^{\lambda \cdot t}] \cdot \frac{R_3}{L_1} - [Y_3^{(j)} + y_3 \cdot e^{\lambda \cdot t}] \cdot \frac{R_4}{L_1} - [I_{R_4}^{(j)} + i_{R_4} \cdot e^{\lambda \cdot t}] \cdot \frac{1}{C_{out} \cdot L_1} + [I_{L_{out}}^{(j)} + i_{L_{out}} \cdot e^{\lambda \cdot t}] \cdot \frac{1}{C_{out} \cdot L_1}$$

$$y_2 \cdot \lambda \cdot e^{\lambda \cdot t} = -I_{C_1}^{(j)} \cdot \frac{1}{C_1 \cdot L_1} + Y_1^{(j)} \cdot \frac{r_\pi}{L_1} - Y_2^{(j)} \cdot \frac{R_3}{L_1} - Y_3^{(j)} \cdot \frac{R_4}{L_1} - I_{R_4}^{(j)} \cdot \frac{1}{C_{out} \cdot L_1} + I_{L_{out}}^{(j)} \cdot \frac{1}{C_{out} \cdot L_1} - i_{C_1} \cdot \frac{1}{C_1 \cdot L_1} \cdot e^{\lambda \cdot t} + y_1 \cdot \frac{r_\pi}{L_1} \cdot e^{\lambda \cdot t} - y_2 \cdot \frac{R_3}{L_1} \cdot e^{\lambda \cdot t} - y_3 \cdot \frac{R_4}{L_1} \cdot e^{\lambda \cdot t} - i_{R_4} \cdot \frac{1}{C_{out} \cdot L_1} \cdot e^{\lambda \cdot t} + i_{L_{out}} \cdot \frac{1}{C_{out} \cdot L_1} \cdot e^{\lambda \cdot t}$$

At fixed point:

$$-I_{C_1}^{(j)} \cdot \frac{1}{C_1 \cdot L_1} + Y_1^{(j)} \cdot \frac{r_\pi}{L_1} - Y_2^{(j)} \cdot \frac{R_3}{L_1} - Y_3^{(j)} \cdot \frac{R_4}{L_1} - I_{R_4}^{(j)} \cdot \frac{1}{C_{out} \cdot L_1} + I_{L_{out}}^{(j)} \cdot \frac{1}{C_{out} \cdot L_1} = 0$$

$$y_2 \cdot \lambda \cdot e^{\lambda \cdot t} = -i_{C_1} \cdot \frac{1}{C_1 \cdot L_1} \cdot e^{\lambda \cdot t} + y_1 \cdot \frac{r_\pi}{L_1} \cdot e^{\lambda \cdot t} - y_2 \cdot \frac{R_3}{L_1} \cdot e^{\lambda \cdot t} - y_3 \cdot \frac{R_4}{L_1} \cdot e^{\lambda \cdot t} - i_{R_4} \cdot \frac{1}{C_{out} \cdot L_1} \cdot e^{\lambda \cdot t} + i_{L_{out}} \cdot \frac{1}{C_{out} \cdot L_1} \cdot e^{\lambda \cdot t}$$

$$-y_2 \cdot \lambda - i_{C_1} \cdot \frac{1}{C_1 \cdot L_1} + y_1 \cdot \frac{r_\pi}{L_1} - y_2 \cdot \frac{R_3}{L_1} - y_3 \cdot \frac{R_4}{L_1} - i_{R_4} \cdot \frac{1}{C_{out} \cdot L_1} + i_{L_{out}} \cdot \frac{1}{C_{out} \cdot L_1} = 0$$

We build our system matrices $(i_{R_\pi}, i_{C_1}, i_{R_4}, i_{L_{out}}, i_{R_s}, i_{L_{in}}, y_4, i_{R_2}, y_2)$ and (y_1, y_2, y_3)

$$-i_{R_\pi} \cdot \lambda + y_1 = 0; -i_{C_1} \cdot \lambda + y_2 = 0; -i_{R_4} \cdot \lambda + y_3 = 0; -i_{L_{out}} \cdot \lambda + y_4 = 0$$

$$-i_{R_s} \cdot \lambda - i_{R_s} \cdot \frac{1}{C_{in} \cdot R_s} + i_{L_{in}} \cdot \frac{1}{C_{in} \cdot R_s} = 0; -i_{L_{in}} \cdot \lambda - i_{R_s} \cdot \frac{R_s}{L_{in}} - i_{R_\pi} \cdot \frac{r_\pi}{L_{in}} = 0$$

$$-y_4 \cdot \lambda - y_4 \cdot \frac{R_{load}}{L_{out}} + i_{R_4} \cdot \frac{1}{C_{out} \cdot L_{out}} - i_{L_{out}} \cdot \frac{1}{C_{out} \cdot L_{out}} = 0$$

$$-i_{R_2} \cdot \lambda - i_{R_2} \cdot \frac{1}{L_2} \cdot \sum_{k=1}^{2} R_k + i_{R_\pi} \cdot \frac{1}{L_2} \cdot r_\pi \cdot (1 + g_m \cdot R_1) - i_{C_1} \cdot \frac{R_1}{L_2} + \frac{R_1}{L_2} \cdot i_{R_4} = 0$$

$$-y_2 \cdot \lambda - y_2 \cdot \frac{R_3}{L_1} - i_{C_1} \cdot \frac{1}{C_1 \cdot L_1} + y_1 \cdot \frac{r_\pi}{L_1} - y_3 \cdot \frac{R_4}{L_1} - i_{R_4} \cdot \frac{1}{C_{out} \cdot L_1} + i_{L_{out}} \cdot \frac{1}{C_{out} \cdot L_1} = 0$$

$$\begin{pmatrix} \iota_{11} & \cdots & \iota_{19} \\ \vdots & \ddots & \vdots \\ \iota_{91} & \cdots & \iota_{99} \end{pmatrix} \cdot \begin{pmatrix} i_{R_\pi} \\ i_{C_1} \\ i_{R_4} \\ i_{L_{out}} \\ i_{R_s} \\ i_{L_{in}} \\ y_4 \\ i_{R_2} \\ y_2 \end{pmatrix} + \begin{pmatrix} v_{11} & v_{12} \\ v_{21} & v_{22} \\ v_{31} & v_{32} \\ v_{41} & v_{42} \\ v_{51} & v_{52} \\ v_{61} & v_{62} \\ v_{71} & v_{72} \\ v_{81} & v_{82} \\ v_{91} & v_{92} \end{pmatrix} \cdot \begin{pmatrix} y_1 \\ y_3 \end{pmatrix} = 0; \iota_{11} = -\lambda; \iota_{12} = \cdots$$
$$= \iota_{19} = 0; \iota_{21} = 0; \iota_{22} = -\lambda$$

$$\iota_{23} = \cdots = \iota_{28} = 0; \iota_{29} = 1; \iota_{31} = \iota_{32} = 0; \iota_{33} = -\lambda; \iota_{34} = \cdots = \iota_{39} = 0; \iota_{41} = \iota_{42} = \iota_{43} = 0; \iota_{44} = -\lambda$$

$$\iota_{45} = \iota_{46} = 0; \iota_{47} = 1; \iota_{48} = \iota_{49} = 0; \iota_{51} = \iota_{54} = 0; \iota_{55} = -\lambda - \frac{1}{C_{in} \cdot R_s}; \iota_{56} = -\frac{1}{C_{in} \cdot R_s}; \iota_{57} = \iota_{58} = \iota_{59} = 0$$

$$\iota_{61} = -\frac{r_\pi}{L_{in}}; \iota_{62} = \cdots = \iota_{64} = 0; \iota_{65} = -\frac{R_s}{L_{in}}; \iota_{66} = -\lambda; \iota_{67} = \cdots = \iota_{69} = 0; \iota_{71} = \iota_{72} = 0$$

$$\iota_{73} = \frac{1}{C_{out} \cdot L_{out}}; \iota_{74} = -\frac{1}{C_{out} \cdot L_{out}}; \iota_{75} = \iota_{76} = 0; \iota_{77} = -\lambda - \frac{R_{load}}{L_{out}}; \iota_{78} = \iota_{79} = 0$$

$$\iota_{81} = \frac{1}{L_2} \cdot r_\pi \cdot (1 + g_m \cdot R_1); \iota_{82} = -\frac{R_1}{L_2}; \iota_{83} = \frac{R_1}{L_2}; \iota_{84} = \iota_{85} = \iota_{86} = \iota_{87} = 0; \iota_{88} = -\lambda - \frac{1}{L_2} \cdot \sum_{k=1}^{2} R_k$$

$$\iota_{89} = 0; \iota_{91} = 0; \iota_{92} = -\frac{1}{C_1 \cdot L_1}; \iota_{93} = -\frac{1}{C_{out} \cdot L_1}; \iota_{94} = \frac{1}{C_{out} \cdot L_1}; \iota_{95} = \iota_{96} = \iota_{97} = 0$$

$$\iota_{98} = 0; \iota_{99} = -\lambda - \frac{R_3}{L_1}; v_{11} = 1; v_{12} = 0; v_{21} = 0; v_{22} = 0; v_{31} = 0; v_{32} = 1$$

$$v_{41} = 0; v_{42} = 0; v_{51} = 0; v_{52} = 0; v_{61} = 0; v_{62} = 0; v_{71} = 0; v_{72} = 0$$

$$v_{81} = 0; v_{82} = 0; v_{91} = \frac{r_\pi}{L_1}; v_{92} = -\frac{R_4}{L_1}$$

Assumption v_{lk}; $l = 1,\ldots,9$; $k = 1,2$ elements are neglected compar to $\imath_{lk}$ elements $l = 1,\ldots,9$; $k = 1,\ldots,9$.

$$(A - \lambda \cdot I) = \begin{pmatrix} \imath_{11} & \ldots & \imath_{19} \\ \vdots & \ddots & \vdots \\ \imath_{91} & \ldots & \imath_{99} \end{pmatrix}; \det(A - \lambda \cdot I) = 0 \Rightarrow \det \begin{pmatrix} \imath_{11} & \ldots & \imath_{19} \\ \vdots & \ddots & \vdots \\ \imath_{91} & \ldots & \imath_{99} \end{pmatrix} = 0$$

To effectively apply the stability criterion of Lipunov to our system, we require a criterion for when the equation $\det(A - \lambda \cdot I) = 0$ has a zero in the left half plane, without calculating the eigenvalues explicit. We use criterion of Routh-Hurwitz [2–4].

BJT Small Signal (SS) equivalent circuit models:

In order to develop these BJT small signal models, there are two small signal resistances that determine. These are r_π the small signal, active mode input resistance between the base and emitter, as "seen looking into the base" and r_e the small signal, active mode output resistance between the base and emitter, "as looking into the emitter". These resistances are not the same because the transistor is not a reciprocal device. The behavior of the BJT in the circuit changes if we interchange the terminals. Determine r_π: assuming the transistor in this circuit is operating in the active mode, then we get BJT base current $i_B = \frac{i_C}{\beta_f} = \frac{1}{\beta_f} \cdot (I_C + \frac{I_C}{V_T} \cdot v_{be}); i_b = \frac{I_C}{\beta_f \cdot V_T} \cdot v_{be} = \frac{g_m}{\beta_f} \cdot v_{be}$. The i_C DC section is I_C and the AC section is $\frac{I_C}{V_T} \cdot v_{be}$ [33, 34] (Fig. 5.10).

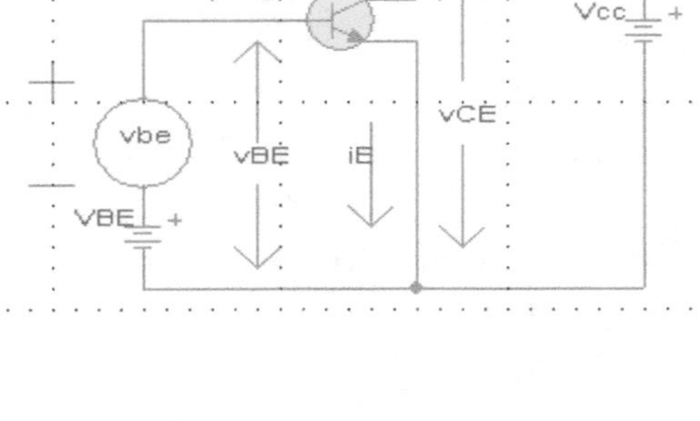

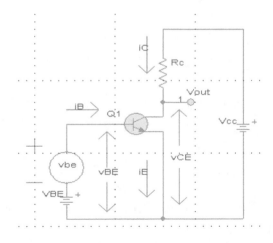

Fig. 5.10 BJT transistor circuit with biasing voltages

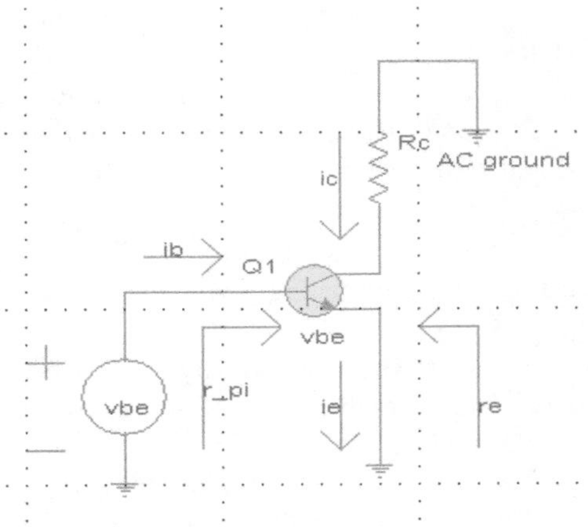

Fig. 5.11 BJT transistor AC small signal equivalent circuit (version 1)

The AC small signal equivalent circuit is as follow: (Fig. 5.11).

Since the voltage at circuit terminal is held constant at V_{cc}, there is no time variation of the voltage. Consequently, this terminal ca be an "AC ground" in the small signal circuit. For AC grounds, we "kill" the DC sources at that terminal: short circuit voltage sources an open current sources.

We get $r_\pi = \frac{v_{be}}{i_b} = \frac{\beta_f}{g_m}[\Omega]$, this r_π is the BJT active mode small signal input resistance of the BJT between the base and the emitter as seen looking into the base terminals. Determine r_e: We determine r_e following a similar procedure as for r_π but start with $i_E = \frac{i_C}{\alpha_f} = \frac{I_C}{\alpha_f} + \frac{i_c}{\alpha_f}; i_e = \frac{i_c}{\alpha_f} = \frac{I_C}{\alpha_f \cdot V_T} \cdot v_{be}$; $I_E = \frac{I_C}{\alpha_f}; i_e = \frac{I_E}{V_T} \cdot v_{be}; i_E = I_E + i_e$. r_e is the BJT small signal resistance between the emitter and base seen looking into the emitter. Mathematically, this is stated as $r_e \equiv \frac{v_e}{-i_e} \Rightarrow v_e = -v_{be}; r_e \equiv \frac{v_{be}}{i_e}; r_e = \frac{V_T}{I_E}; g_m = \frac{I_C}{V_T} = \frac{\alpha_f \cdot I_E}{V_T} \Rightarrow \frac{V_T}{I_E} = \frac{\alpha_f}{g_m}; r_e = \frac{\alpha_f}{g_m} \approx \frac{1}{g_m}[\Omega]$.

It can be shown that $r_\pi = (\beta_f + 1) \cdot r_e[\Omega]; r_\pi \neq r_e$. The active mode BJT is a non reciprocal device (Fig. 5.12).

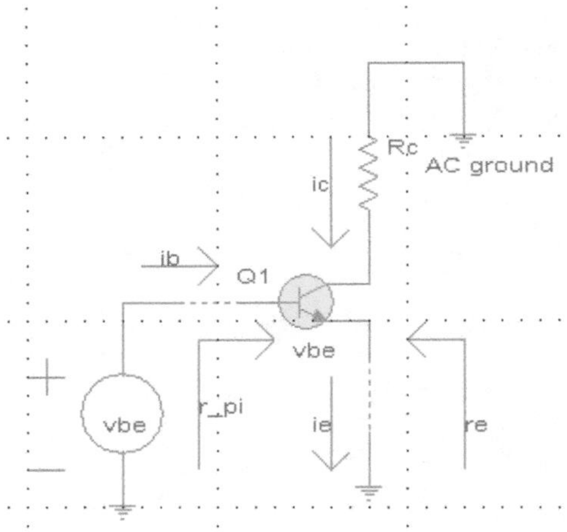

Fig. 5.12 BJT transistor AC small signal (version 2)

There are two families of equivalent active mode BJT small signal circuit: Hybrid Pi model and T model. Both are equally valid models but choosing one over the other sometimes leads to simpler analysis of certain circuits.

Hybrid Pi model Version A: The circuit incorporates all of the necessary small signal characteristics of the BJT $i_b = \frac{v_{be}}{r_\pi}; i_c = g_m \cdot v_{be}; i_b + i_c = i_e; i_e = \frac{v_{be}}{r_e}$ (Fig. 5.13).

Hybrid Pi model Version B: The second equivalent circuit is constructed by using the following notation: $g_m \cdot v_{be} = g_m \cdot (i_b \cdot r_\pi) = g_m \cdot r_\pi \cdot i_b = \beta_f \cdot i_b$ (Fig. 5.14).

T model: The hybrid Pi model is the most popular small signal model for the BJT. The alternative is the T model, which is useful in certain situations.

T model Version A: (Fig. 5.15).

T model Version B: (Fig. 5.16).

The small signal model for PNP BJTs are identically the same as for NPN transistors. There is no change in any polarities (voltage or current) for the PNP models relative to the NPN models. These small signal models are identically the same.

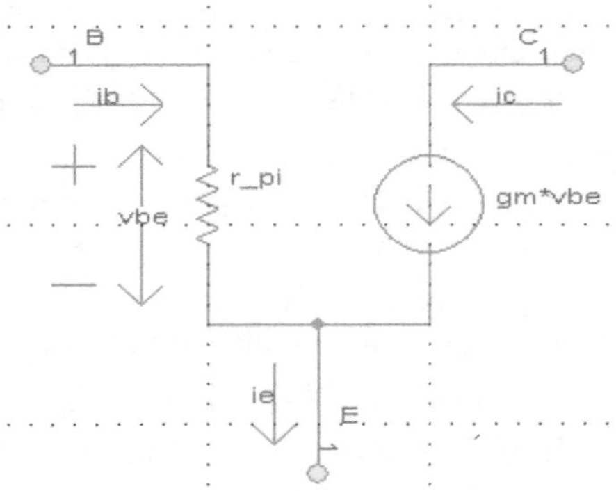

Fig. 5.13 Equivalent active mode BJT small signal circuit Hybrid Pi model (version A)

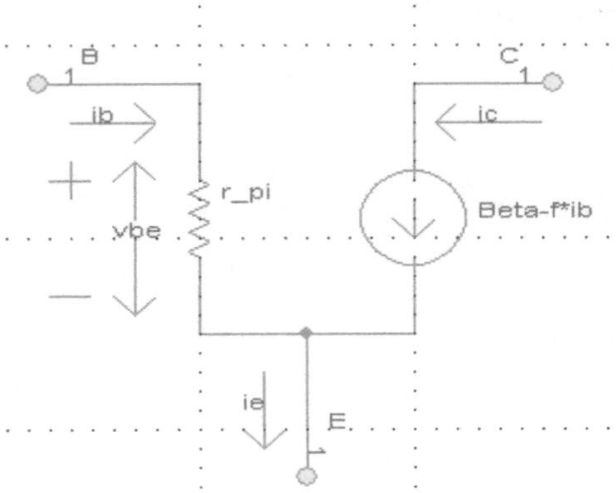

Fig. 5.14 Equivalent active mode BJT small signal circuit Hybrid Pi model (version B)

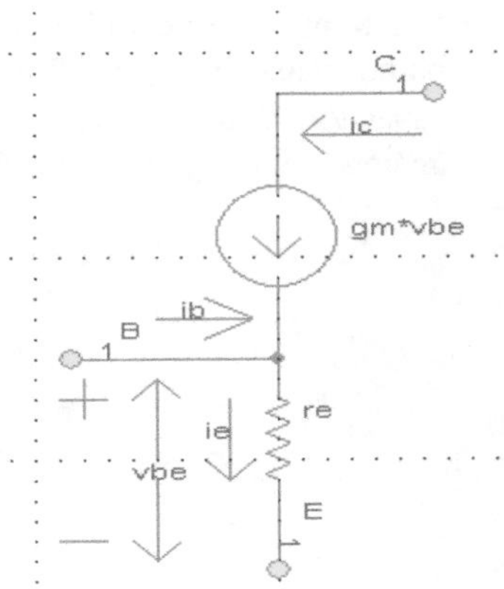

Fig. 5.15 Equivalent active model BJT small signal circuit, T model (version A)

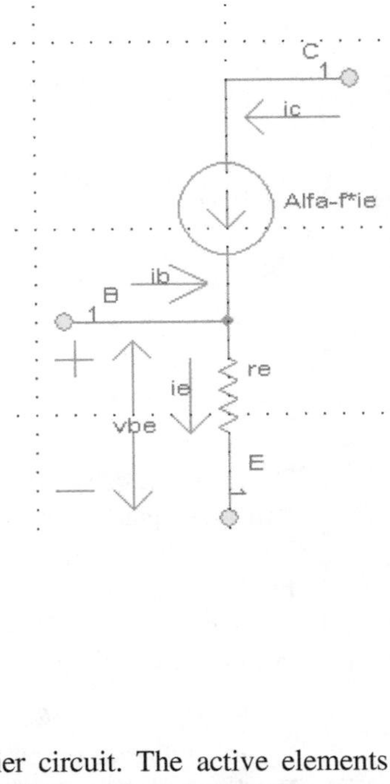

Fig. 5.16 Equivalent active model BJT small signal circuit, T model (version B)

Exercises

1. We have class AB push-pull variation amplifier circuit. The active elements used in this circuit (Q_1, D_3 (Opto-coupler), and Q_2) are almost complementary symmetric transistors. Q_1 is NPN phototransistor and Q_2 is PNP transistor. Pre-biasing is given to the transistor Q_2 and LED D_3 using the network comprising of resistors R_1, R_2 and the biasing diodes D_1 and D_2. Resistor r_p is a parasitic resistance between diodes D_1 and D_2. We use in our circuit analysis

BJT transistor Ebers-Moll model. The coupling coefficient between LED D_3 and photo transistor Q_1 is "k" $(I_{BQ_1} = k \cdot I_{D_3})$. C_1 and C_2 are input DC decoupling capacitors and input RF source $V_s(t)$ with series resistance R_s. *Hint*: we use analog optocoupler in our circuit.

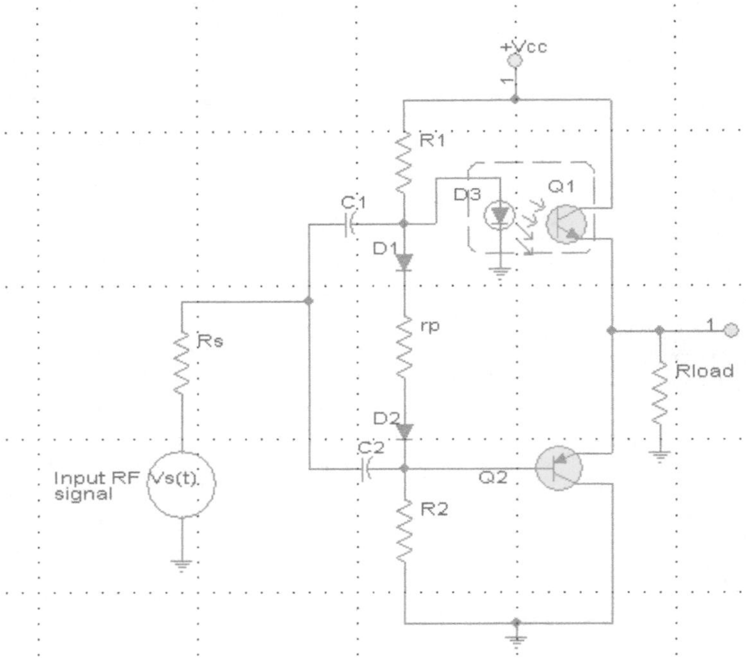

1.1 Draw circuit class AB push-pull variation amplifier equivalent circuit and write the related circuit differential equations (Use transistor Q_1 and Q_2 Ebers-Moll model).
1.2 Find circuit fixed points and discuss stability by parameters variation.
1.3 We short diode D_1, How it influences circuit behavior and stability?
1.4 We define coupling function between LED D_3 current and photo transistor Q_1 base current $\left(I_{BQ_1} = \sum_{k=1}^{M} I_{D_3}^k \cdot a_k; M \in \mathbb{N}\right)$. How parameters $a_1, a_2, \ldots, a_M$ influence on circuit behavior stability?
1.5 Diode D_3 is disconnected, Find circuit differential equations and fixed points. Discuss stability.

2. We have class AB push-pull variation amplifier circuit. The active elements used in this circuit are Q_1, Q_2, and D_3, Q_3 (Opto coupler). Q_1 is NPN transistor and Q_2 is NPN photo transistor, Q_3 is PNP transistor. Pre-biasing is given to the transistor Q_1, LED D_3, and transistor Q_3 using the network comprising of resistors R_1, R_2, R_3 and the biasing diodes D_1 and D_2. Resistor r_p is the parasitic resistance between diodes D_1 and D_2. We use in our analysis BJT transistor Ebers-Moll model. The coupling coefficient between LED D_3 and photo transistor Q_2 is "k" parameter $(I_{BQ_2} = k \cdot I_{D_3})$.$C_1$ and C_2 are input DC decoupling capacitors and input RF source $V_s(t)$ with series resistance R_s. *Hint*: we use analog optocoupler in our circuit.

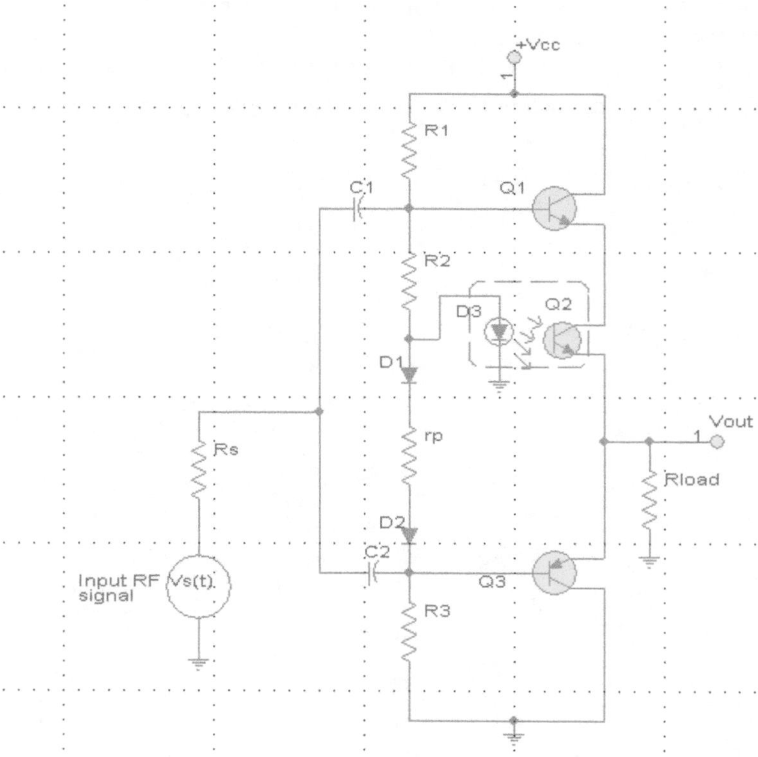

2.1 Draw circuit class AB push-pull variation amplifier equivalent circuit and write the related circuit differential equations. (Use transistors Q_1, Q_2, and Q_3 Ebers-Moll model.)
2.2 Find circuit fixed points and discuss stability by circuit parameters variation.
2.3 Resistor $r_p \rightarrow \varepsilon$, How it influences on circuit behavior and stability?

2.4 We define coupling function between diode D_3 and photo transistor Q_2 base current $\left(I_{BQ_2} = \sum_{l=1}^{P} [b_l \cdot \prod_{k=1}^{l} I_{D_3}^k \cdot a_k];\ P \in \mathbb{N}\right)$. How parameters a_k, b_l influence on circuit behavior and stability?

2.5 We short diode D_1, How it influence circuit behavior and stability? Find circuit differential equations and fixed points.

3. We have power amplifier with two internal resonant circuits. The first resonant circuit is connected to transistor Q_1's emitter (C_1 and L_1) and the second resonant circuit is connected to transistor Q_1's collector (C_2 and L_2). Resistor R_b is transistor Q_1's base resistor and C_{in}, C_{out} are input and output blocking capacitors. Input RF source is $V_s(t)$ with series resistor R_s. Since the input RF signal is a large signal, we use in our analysis the Ebers-Moll BJT model for transistor Q_1. The values of L_1 and C_1 or L_2 and C_2 are so selected that the resonant circuit oscillates in the frequency of the input signal. Capacitors C_{in}, C_{out} are input and output blocking capacitors. Assumption: $\frac{dV_s(t)}{dt} \to \varepsilon$.

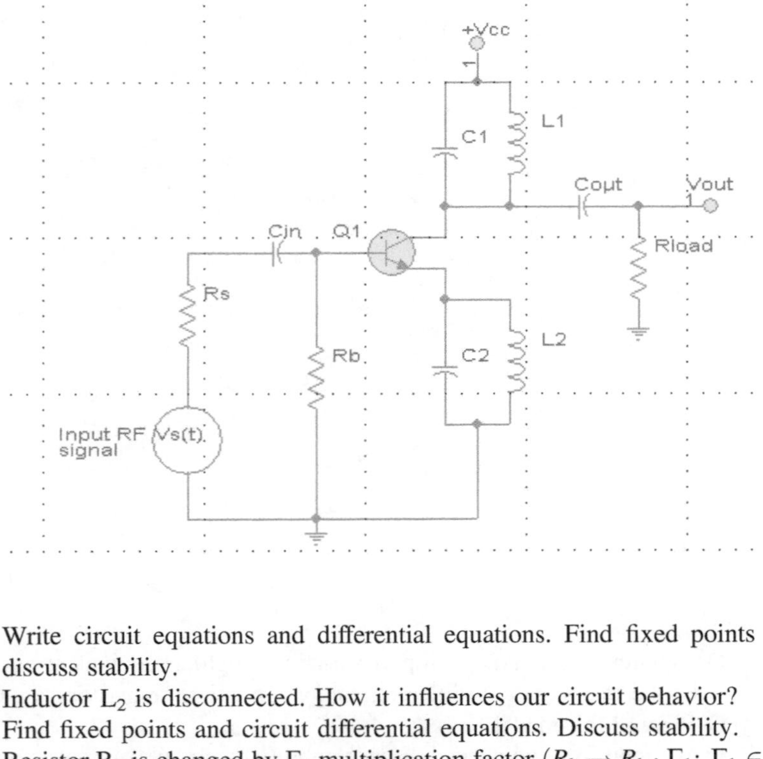

3.1 Write circuit equations and differential equations. Find fixed points and discuss stability.

3.2 Inductor L_2 is disconnected. How it influences our circuit behavior? Find fixed points and circuit differential equations. Discuss stability.

3.3 Resistor R_b is changed by Γ_1 multiplication factor ($R_b \to R_b \cdot \Gamma_1;\ \Gamma_1 \in \mathbb{R}$). We have two cases: (a) $0 < \Gamma_1 < 1$(b) $\Gamma_1 > 1$. How the dynamical behavior of the circuit is changed for cases (a) and (b)? Discuss stability for both cases.

3.4 Capacitor C_1 is disconnected. How it influences our circuit behavior? Find fixed points and circuit differential equations. Discuss stability.

3.5 Capacitor C_1 and inductor L_1 are changed according the following transformation: $C_1 \to C_1 \cdot \Gamma_2; L_1 \to L_1 \cdot \Gamma_2^2; \Gamma_2 > 1; \Gamma_2 > 0; \Gamma_2 \in \mathbb{R}$. How the dynamical behavior of the circuit is changed for different values of parameter Γ_2? Discuss stability.

4. We have power amplifier with two internal resonant circuits. The first resonant circuit is connected between the V_{cc} and Q_1's emitter (C_1 and L_1) and the second resonant circuit is connected between Q_1's emitter and ground (C_2 and L_2). R_b is transistor Q_1's base resistor and C_{in}, C_{out} are input and output blocking capacitors. Input RF source is $V_s(t)$ with series resistance R_s. Since the input RF signal is a large signal, we use in our analysis the Ebers-Moll BJT model of Q_1 transistor. Assumption: $\frac{dV_s(t)}{dt} \to \varepsilon$.

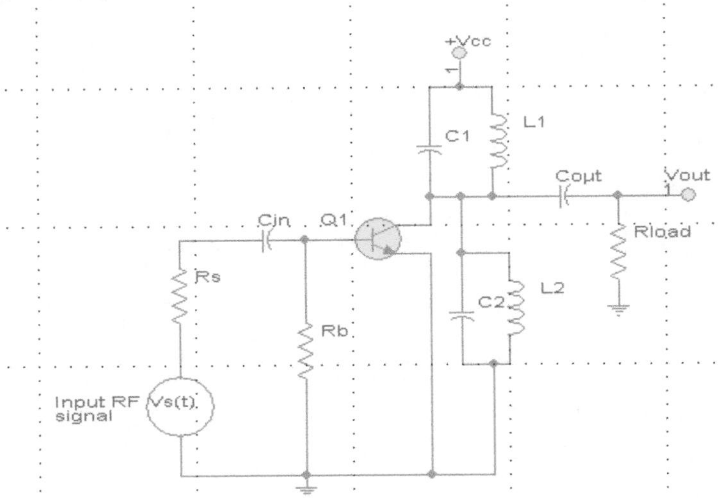

4.1 Write circuit equations and differential equations. Find fixed points and discuss stability.

4.2 Capacitor C_2 is disconnected. Find fixed points and circuit differential equations. Discuss stability.

4.3 Transistor Q_1's α_f, α_r parameters are changed by multiplication factors Γ_1 and Γ_2 respectively $\alpha_f \to \alpha_f \cdot \Gamma_1$; $\alpha_r \to \alpha_r \cdot \Gamma_2$; $\Gamma_1, \Gamma_2 \in \mathbb{R}$. α_f: $0.95 \to 0.99$; $0.95 \leq \alpha_f \cdot \Gamma_1 \leq 0.99$; $\alpha_r : 0 \to 0.95$; $0 < \alpha_r \cdot \Gamma_2 < 0.95$; $\alpha_f > \alpha_r$. How the dynamical behavior of the circuit is changed for different values of

Γ_1 and Γ_2 factors? Discuss stability for different values of Γ_1 and Γ_2 factors $\Gamma_1, \Gamma_2 > 0$.

4.4 Inductor L_1 is disconnected. How it influences our circuit behavior? Find fixed points and circuit differential equations. Discuss stability.

4.5 Capacitors C_2 and inductor L_2 are changed according to the following transformations: $C_2 \rightarrow C_2 \cdot \Gamma_3$; $L_1 \rightarrow L_1 \cdot (\sqrt{\Gamma_3} + \Gamma_3^3)$; $\Gamma_3 > 0$; $\Gamma_3 \in \mathbb{R}$. How the dynamical behavior of the circuit is changed for different values of parameter Γ_3? Discuss stability.

5. We have power amplifier with resonant circuit (C_1 and L_1) between transistor Q_1 emitter-collector, Q_1's emitter branch can be connected to resistor R_{e1} or resistor R_{e2} and inductor L_e (you choose it by S_1 switch). R_b is the transistor Q_1 base resistor and C_{in}, C_{out} are input and output blocking capacitors. Input RF source is $V_s(t)$ with series resistance R_s. Since the input RF signal is a large signal we use in our analysis the Ebers-Moll BJT model for Q_1. The values of capacitors C_1 and inductor L_1 are selected that the resonant circuit is oscillated in the frequency of the input signal. We have two cases in our circuit, case (a): switch S_1 at position (1), case (b): switch S_2 at position (2). Assumption: $\frac{dV_s(t)}{dt} \rightarrow \varepsilon$

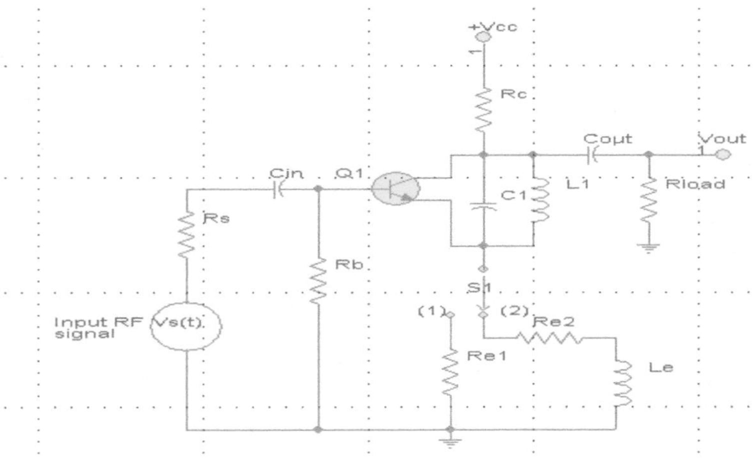

5.1 Write circuit equations and differential equations. Find fixed points and discuss stability [two cases, S_1 in position (1) and (2)].

5.2 Resistor R_c is changed by factor Ω_1 ($R_c \rightarrow R_c \cdot \Omega_1$; $\Omega_1 > 0$; $\Omega_1 \in \mathbb{R}$). How the dynamical behavior of the circuit is changed for different values of Ω_1 parameter? [Two cases: S_1 in position (1) and (2)].

5.3 Transistor Q_1's α_f, α_r parameters are changed by multiplication Γ_1 and Γ_2 factors respectively $\alpha_f \to \alpha_f \cdot \Gamma_1$; $\alpha_r \to \alpha_r \cdot \sqrt{\Gamma_2}$; $\Gamma_1, \Gamma_2 \in \mathbb{R}$. α_f: $0.95 \to 0.99$; $0.95 \leq \alpha_f \cdot \Gamma_1 \leq 0.99$; α_r: $0 \to 0.95$; $0 < \alpha_r \cdot \Gamma_2 < 0.95$; $\alpha_f > \alpha_r$. How the dynamical behavior of the circuit is changed for different values of Γ_1 and Γ_2 factors? Discuss stability for different values of Γ_1 and Γ_2 factors $\Gamma_1, \Gamma_2 > 0$.

5.4 Capacitors C_1 and inductor L_1 are changed according to the following transformation: $C_1 \to C_1 \cdot \sqrt{\Gamma_3}$; $L_1 \to L_1 \cdot (\sqrt[3]{\Gamma_3} + \Gamma_3^2)$ $\Gamma_3 > 0$; $\Gamma_3 \in \mathbb{R}$. How the dynamical behavior of the circuit is changed for different values of parameter Γ_3? Discuss stability.

5.5 Inductor L_1 is disconnected. How the dynamical behavior of the circuit is changed? [Cases: (1) and (2)]. Discuss stability.

6. We have power amplifier system which constructed from two BJT NPN transistors Q_1 and Q_2, peripheral components. Capacitors C_{in} and C_{out} are input and output blocking capacitors. Transistor Q_2 is RF transistor, L_2 is radio frequency choke which isolated the RF form the DC source V_{cc}. DC voltage source V_{cc} is between 2v to 48v. The bias input at junction A to transistor Q_2 having the desired DC and RF impedance characteristics which will allow linear amplification of RF frequencies over a wide power range. The biasing circuit has two portions: one functioning essentially only at low power level and another portion with the one portion functioning at increasing to high power level with a smooth and continuous transition between them. Diode D_2 is a constant current diode, resistor R_3 (low power level portion). Additionally resistor R_1, inductor L_1, and Ferrite bead F_1. When we increase to high power level portion of the biasing circuit, its done additionally to low power level portion. *Q_1 biasing series circuit*: resistor R_2, diode D_1, and resistor R_{b1}. The series circuit provides bias to the Q1's base. The emitter-collector of transistor Q_1 is a bypass around resistor R_1 and at these two components provides variable impedance for giving the varying bias needed, as between low power inputs and high power inputs. The purpose of diode D_2 is to provide a high impedance constant current through resistor R_1. The circuit of constant current is through diode D_2 connected to V_{cc} through resistor R_3, ferrite bead F_1, RF inductor L_1, and resistor R_1. Diode D_2 provides a voltage of about 0.5v at point A under DC static condition. The voltage at point A is the base voltage V_b of the transistor Q_1. The transistor Q_1 having turn ON base voltage of about 0.6v. The DC bias voltage value 0.5v initially maintains the transistor Q_1 in a non-conducting state and the amplifier is, in effect, operating in a class B mode. Resistor R_3 is a power dissipation limiting resistor and has a typical value of 1.8 kΩ in order to provide large impedance.

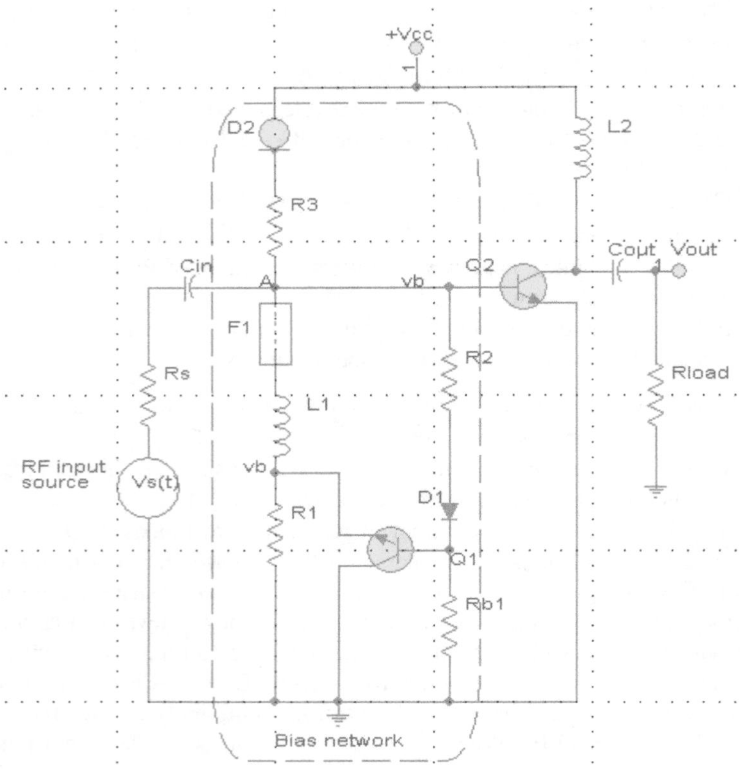

Biasing circuit of Q_2 at high power level: Diode D_1, resistor R_2, and resistor R_{b1} rectify the RF supplied from the input and develops DC bias from transistor Q_1 which is proportional to the RF input drive level. At low RF input level, Q_1 is biased OFF and allowing resistor R_1 to develop the bias for RF amplifier Q_2 which operating in a class B. When RF input is increased sufficiently, transistor Q_1 begin to saturate, shunting resistor R_1 and changing the operating mode of amplifier Q_2 from class B to class C. At high drive level conditions, the DC dynamic impedance between the base-emitter of Q_2 is very low and promoting maximum transistor gain and efficiency. As the drive level is increased or decreased from an intermediate level ($V \pm \Delta V$), the dynamic impedance changes in a nonlinear manner. At low power input RF transistor Q_2 operates at class B and at higher power levels transistor Q_2 operates at class C.

Constant current diode: Constant current diode is an electric device that limits current to a maximum specified value for device (CLD—Current Limiting Diode, CRD—Current Regulating Diode). This diode consist of a n-channel JFET transistor with the gate shorted to the source, which functions like a two terminal current limiter or current source (analog to voltage limiting Zener diode). It allows a current through to rise to a certain value, and then level OFF at specific value. This diode keeps the current constant. This device keeps the current flowing through it unchanged when the voltage changes.

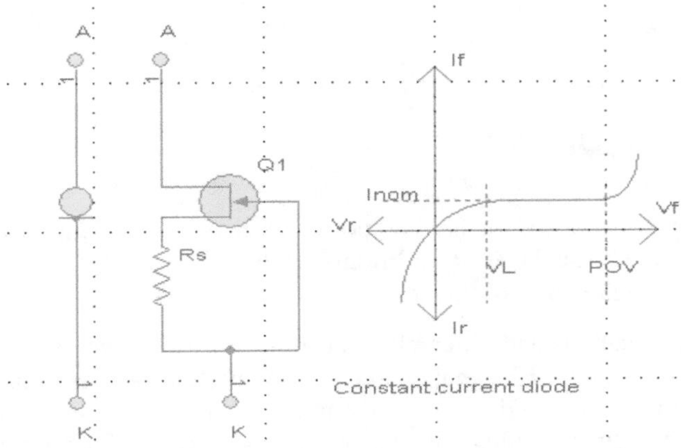

V_L—Limiting voltage: measured at I_L, V_L together with knee AC impedance (Z_k), indicates the knee characteristics of the device.

POV = Peak Operating Voltage: maximum voltage to be applied to device.

In operation the CLD regulates the amount of current that can flow over a voltage range of about 1–100 V. The equivalent circuit of the CLD is a current generator in series with a parallel combination of the dynamic impedance and the junction capacitance. The shunt capacitance of Central's CLD is about 4–10 pF over the useful operating voltage range.

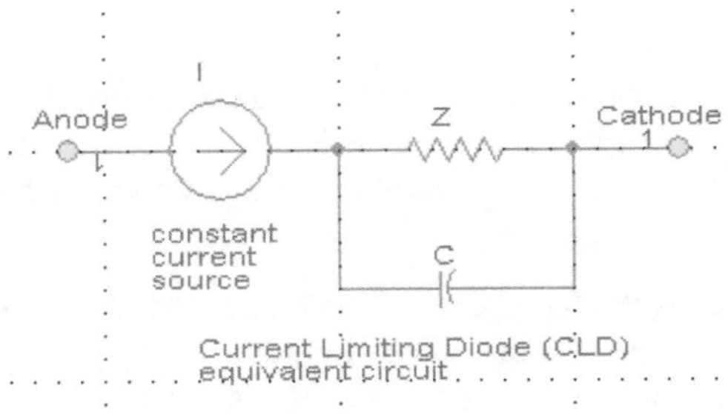

Ferrite bead circuit model: A ferrite bead is a passive device that suppresses high frequency noise in electric circuit. It is a specific type of electronic choke. Ferrite beads prevent interference in two directions, from a device or to a device. The equivalent ferrite bead is as follow:

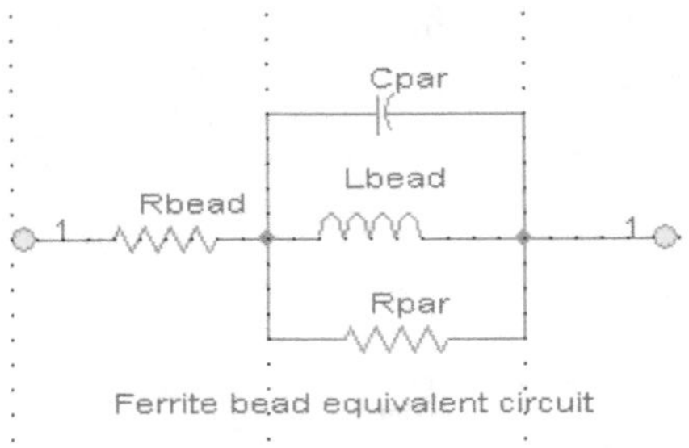

In circuit analysis we need to use for low power level input RF signal—BJT transistor Small Signal (SS) model (Hybrid Pi) and for high power level input RF signal—BJT transistor Gummel-Poon model.

6.1 Write circuit equations and differential equations for low power level and high power level input RF signals. Find fixed points and discuss stability.
6.2 Resistor R_1 is disconnected. How the dynamical behavior of the circuit is changed? Find circuit differential equations and fixed points. Discuss stability.
6.3 We replace constant current diode (D_2) by regular diode. How the dynamical behavior of the circuit is changed? Find circuit differential equations and fixed points. Discuss stability.
6.4 We replace Ferrite bead F_1 by regular RF choke (L_F). How the dynamical behavior of the circuit is changed? Find circuit differential equations and fixed points. Discuss stability.
6.5 We replace diode D_1 by D_2 and diode D_2 by D_1. How the dynamical behavior of the circuit is changed? Find circuit differential equations and fixed points. Discuss stability.

7. We have wideband Low Noise Amplifier (LNA) with negative feedback network. The feedback network consists of capacitors, resistors, and inductors as describe in the below circuit. Input and output matching networks are Pi-type. Input RF signal is $V_s(t)$, $V_s(t) = A_0 + \xi(t)$; $A_0 \gg \xi(t)$.

Inductors L_2, L_3 in the circuit act as the RF choke which blocks the DC current from entering the RF path. The LNA design also employs output resistive loading in stabilizing the circuit. The initial output resistor value, R_4 is set to less than 100 Ω ($R_4 < 100$ Ω) because high output resistor value may result in huge decrease of gain and P1 dB point. The BJT transistor (Q_1) is self-biased with the biasing resistors of R_2 and R_3.

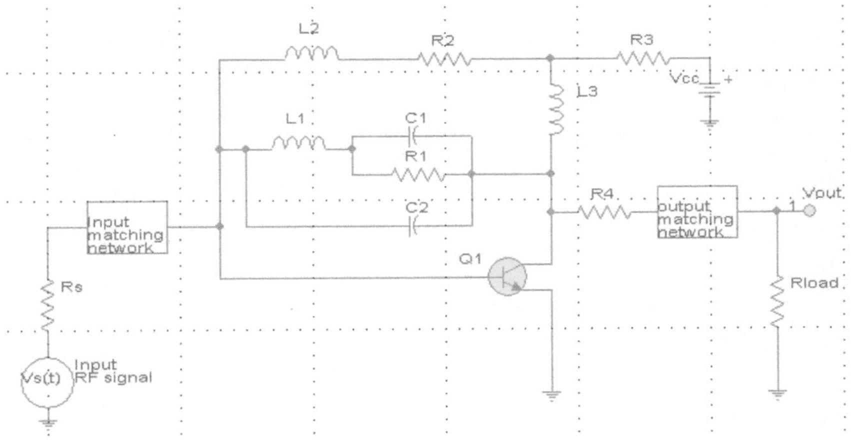

7.1 Write circuit equations and differential equations. Find fixed points and discuss stability.

7.2 Capacitor C_2 is disconnected. How the dynamical behavior of the circuit is changed? Discuss stability and find circuit differential equations and fixed points.

7.3 We change the input and output matching networks from Pi-type to T-type. How the dynamical behavior of the circuit is changed? Discuss stability.

7.4 Resistor R_4 is changed by factor Γ ($R_4 \to R_4 \cdot \Gamma, \Gamma > 0, \Gamma \in \mathbb{R}_+$). How the dynamical behavior of the circuit is changed for different values of Γ parameter? Discuss stability for the cases: (a) $0 < \Gamma \leq 1$, (b) $\Gamma > 0$.

7.5 Transistor Q_1's α_f and α_r parameters are changed by factor Ω_1 and $\sqrt[3]{\Omega_1^2}$ respectively ($\alpha_f \to \alpha_f \cdot \Omega_1$; $\alpha_r \to \alpha_r \cdot \sqrt[3]{\Omega_1^2}$; $\Omega_1 > 0$; $\Omega_1 \in \mathbb{R}_+$). How the dynamical behavior of the circuit is changed for different values of Ω_1 parameter? Discuss stability for different value of Ω_1 parameter.

$$\alpha_f\colon 0.95 \to 0.99;\ 0.95 \leq \alpha_f \cdot \Omega_1 \leq 0.99;\ \alpha_r\colon 0$$
$$\to 0.95;\ 0 < \alpha_r \cdot \sqrt[3]{\Omega_1^2} < 0.95;\ \alpha_f > \alpha_r$$

8. We have Gummel-Poon model equivalent circuit for BJT. Model's diode D_1 is disconnected, how it influences our circuit dynamic? Consider single ended class B amplifier. Write circuit differential equations and find fixed points, discuss stability. Model's capacitors C_{jx} and C_{cs} are disconnected. Find circuit differential equations, fixed points and discuss stability. Consider single ended class B amplifier.

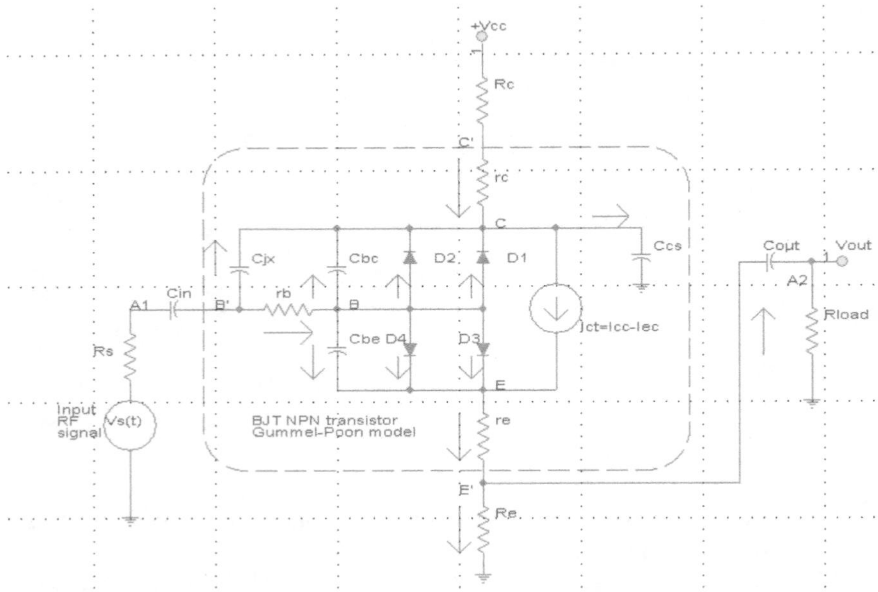

9. We have two wideband LNAs with negative feedback networks which are connected in cascade. The first wideband LNA circuit feeds the second wideband LNA circuit. The feedback networks include mixture of capacitors, inductors, and resistors. There are three circuit's matching networks, input matching network Pi-type, cascade matching network T-type, and output matching network Pi-type. Input RF signal is $V_s(t)$, $V_s(t) = A_0 + \xi(t)$; $A_0 \gg \xi(t)$. Inductors L_3, L_4 in the first LNA circuit and inductors L_7, L_8 in the second LNA circuit act as the RF choke which blocks the DC current from entering the RF part in each LNA circuit. The LNAs design also employs output resistive loadings (R_4.R_5 and R_9) in stabilizing the circuit. The initial output resistors values, R_4.R_5 and R_9 are set to less than 100 Ω ($R_4, R_5, R_9 < 100\,\Omega$) because high output resistor values may result in huge decrease of gain and P_1dB point. The BJT transistors (Q_1 and Q_2) are self-biased with the biasing resistors R_1, R_2 and R_8, R_7 respectively. Switch S_1 connects and disconnects the second LNA negative feedback network.

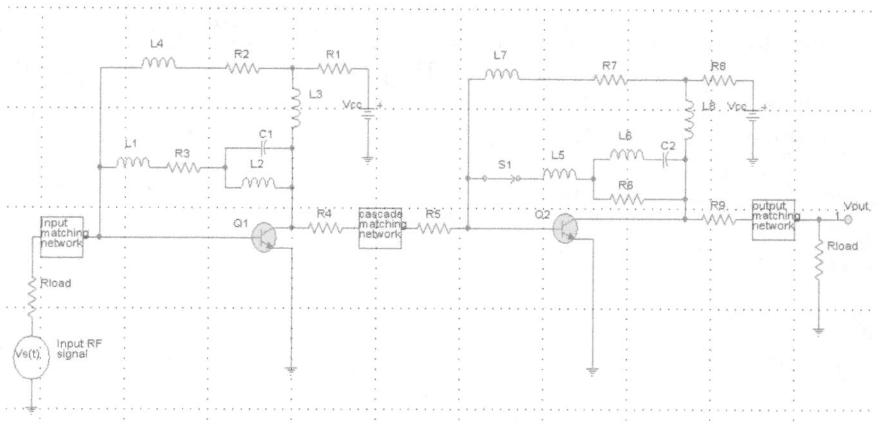

9.1 Write circuit equations and differential equations. Find fixed points and discuss stability (switch S_1 is close).

9.2 How the circuit dynamic is changed if S_1 switch is open? Find circuit differential equations and fixed points. Discuss stability.

9.3 We change the input and output matching networks to T-type. The cascade matching network is L-type. Switch S_1 is in close state. Find circuit differential equations and fixed points. Discuss stability.

9.4 Transistor Q_2's α_f and α_r parameters are changed by factor Ω_1 and $\sqrt[5]{\Omega_1^3+1}$ respectively $(\alpha_f \to \alpha_f \cdot \Omega_1;\ \alpha_r \to \alpha_r \cdot \sqrt[5]{\Omega_1^3+1};\ \Omega_1 > 0;\ \Omega_1 \in \mathbb{R}_+)$.

Switch S_1 is open. How the dynamical behavior of the circuit is changed for different values of Ω_1 parameter. Discuss stability for different values of Ω_1 parameter.

$$\alpha_f\!: 0.95 \to 0.99;\ 0.95 \le \alpha_f \cdot \Omega_1 \le 0.99;\ \alpha_r\!: 0 \to 0.95;\ 0 < \alpha_r \cdot \sqrt[5]{\Omega_1^3+1} < 0.95;\ \alpha_f > \alpha_r$$

9.5 We disconnect capacitor C_1. How the dynamical behavior of the circuit is changed? Discuss stability and find circuit differential equations, fixed points (switch S_1 is close).

10. We have wideband Low Noise Amplifier (LNA) with possible four options of negative feedback networks. The feedback network consists of capacitors, resistors, inductors, and two SPD2 switches. Input and output matching networks are T-type. Input RF signal is $V_s(t)$, $V_s(t) = A_0 + \xi(t);\ A_0 \gg \xi(t)$. Inductors L_1, L_2 in the circuit act as the RF choke which blocks the DC current from entering the RF path. The LNA design also employs output resistive loading in stabilizing the circuit. The initial output resistor value, R_3 is set to

less than 100 Ω ($R_3 < 100\ \Omega$) because high output resistor value may result in huge decrease of gain and P1 dB point. The BJT transistor (Q_1) is self-biased with the biasing resistors R_1 and R_2. The possible negative feedback networks options are as follow: (1) $S_1(a_1)$ & $S_2(a_2)$, (2) $S_1(a_1)$ & $S_2(b_2)$, (3) $S_1(b_1)$ & $S_2(a_2)$, (4) $S_1(b_1)$ & $S_2(b_2)$.

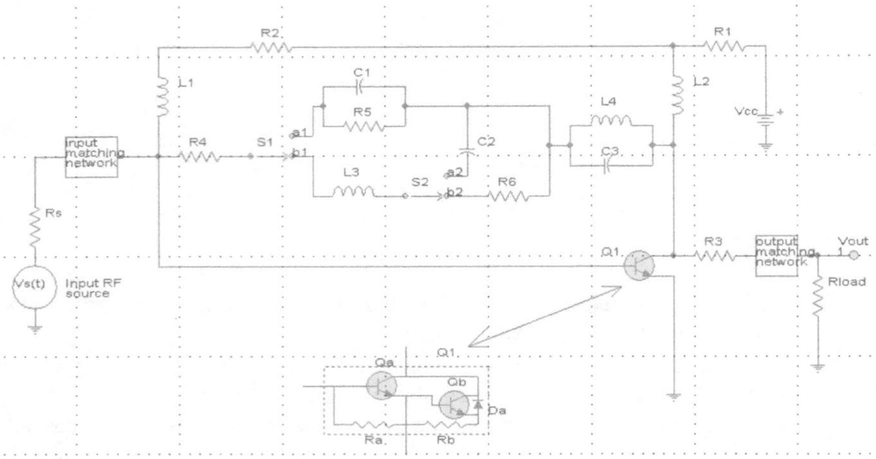

10.1 Write circuit equations and circuit differential equations for each of the negative feedback network options. Find fixed points and discuss stability for each network option.

10.2 We short inductor L_1, How it influences the circuit dynamics? Find circuit differential equations and fixed points. Discuss stability (all network options cases).

10.3 We short resistor R_3, How it influences circuit dynamics? Find circuit differential equations, fixed points and discuss stability (all network options cases).

10.4 Return (10.1), (10.2), and (10.3) when the input and output matching networks are L-type.

10.5 Return (10.1), (10.2), and (10.3) when we replace Q_1 by n-Darlington transistor (two BJT transistors (Q_a, Q_b), two resistors R_a, R_b and diode D_a). Consider that Q_a and Q_b transistors internal parameters are the same $\left(\alpha_{f_a} = \alpha_{f_b}; \alpha_{r_a} = \alpha_{r_b}, \text{etc.}\ldots\right)$.

Chapter 6
Microwave/RF Oscillator Systems Stability Analysis

An electronic oscillator is an electronic circuit that produces a periodic, oscillating electronic signal, often a sine wave or a square wave. Oscillators are class of circuits with one terminal or port, which produce a periodic electrical output upon power up. Oscillators can be classified into two types: (a) Relaxation and (b) Harmonic oscillators. Relaxation oscillators (also called unstable multi-vibrator) is a class of circuits with two unstable states. The circuit switches back and forth between these states. The output is generally square waves. Harmonic oscillators are capable of producing near sinusoidal output, and are based on positive feedback approach. In microwave the purpose of a microwave oscillator is to generate a microwave signal. An oscillator consists of two parts: An active device to generate the power and a resonator to control the frequency of the microwave signal. The oscillations are made using feedback or negative resistance. Important issues in oscillators are frequency stability, frequency tuning, and phase noise. Oscillator is a non-linear circuit, initially upon power up the condition of oscillation to start up will prevail. As the magnitudes of voltages and currents in the circuit increase, the amplifier in the oscillator begins to saturate, reducing the gain, until the loop gain becomes one. A steady state condition is reached when loop gain is equal to one. We can view an oscillator as an amplifier that produces an output when there is no input. Thus is an unstable amplifier that becomes an oscillator. An amplifier can be made unstable by providing some kind of local positive feedback. Two favorite transistor amplifier configurations used for oscillator design are the Common-Base (CB) configuration with base feedback and Common-Emitter (CE) configuration with emitter degeneration. Oscillator performance requirements are frequency, frequency stability, tuning, phase noise, power and efficiency.

O. Aluf, *Microwave RF Antennas and Circuits*,
DOI 10.1007/978-3-319-45427-6_6

6.1 A Resonator Circuit 180° Phase Shift from Its Input to Output Stability Analysis Under Delayed Variables in Time

A phase-shift oscillator is a linear electronic oscillator circuit that produces a sine wave output. It consists of an inverting amplifier element such as a transistor or op amp with its output fed back to its input through a phase-shift network consisting of resistors and capacitors in a ladder network. The feedback network 'shifts' the phase of the amplifier output by 180° at the oscillation frequency to give positive feedback. Phase-shift oscillators are often used at audio frequency as audio oscillators. A phase shift oscillator consists of a single stage of amplifier that amplifies the input signal and produces a phase shift of 180° the input and its output signal. If a part of this output is taken and feedback to input, it results in negative feedback causing the output voltage to decrease. We require positive feedback which means that the voltage signal feedback should be in phase with the input signal. The output of the amplifier should take through a phase shift network to provide it an additional phase shift of 180°. Amplifier provides a phase shift of 180° and the phase shift network also gives a 180° and therefore, a total phase-shift of 360° (which is equivalent to 0°) results [25, 26] (Fig. 6.1).

The RC network provides the required phase shift by using three RC. Each having some value of R and C. These values are selected so as to produce 60° phase shift per section, resulting in total of 180° phase shift as desired. Practically each RC section does not provide the same phase shift because each section leads the previous one but the overall phase shift is 180° which is the requirement. The frequency at which phase shift

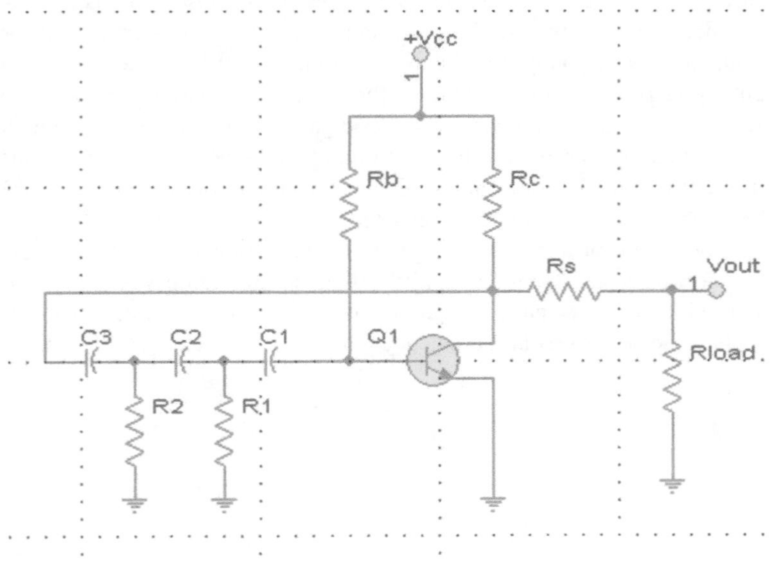

Fig. 6.1 Resonator circuit 180° (degree) phase shift

is 180° is $f = \frac{1}{2 \cdot \pi \cdot R \cdot C \cdot \sqrt{6}}$. The circuit generates a sinusoidal wave at its output. The RC phase shift network is used to form a positive feedback loop. R_b and R_c resistors polarize the transistor Q_1 for DC current, R_s is a separating resistor. $R_1 = R_2 = R$; $C_1 = C_2 = C_3 = C$. Phase shift resonant circuit stability analysis is done by considering BJT Small Signal (SS) equivalent circuit model. We consider "AC ground" in the circuit. Since the voltage at this terminal is held constant at V_{cc}, there is no time variation of the voltage consequently. We can set this terminal to be an "AC ground" in the small signal circuit. For AC ground, we "kill" the DC sources at that terminal (short circuit voltage sources and open circuit current sources) [33, 34] (Fig. 6.2).

$$I_{R_c} = \frac{V_{cc} - V_{A_4}}{R_c}; \; I_{R_b} = \frac{V_{cc} - V_{A_1}}{R_b}; \; I_{R_s} = \frac{V_{A_4} - V_{A_5}}{R_s}; \; I_{R_{load}} = \frac{V_{A_5}}{R_{load}}; \; I_{r\_pi} = \frac{V_{A_1}}{r\_pi}; \; I_{C_3} = C_3 \cdot \frac{d}{dt}(V_{A_4} - V_{A_3})$$

$$I_{R_2} = \frac{V_{A_3}}{R_3}; \; I_{C_3} = I_{C_2} + I_{R_2}; \; I_{C_2} = C_2 \cdot \frac{d}{dt}(V_{A_3} - V_{A_2}); \; I_{R_1} = \frac{V_{A_2}}{R_1}; \; I_{C_2} = I_{C_1} + I_{R_1}$$

$$I_{C_1} = C_1 \cdot \frac{d}{dt}(V_{A_2} - V_{A_1}); \; I_{R_b} + I_{C_1} = I_{r\_pi}; \; I_{R_c} = I_{C_3} + I_{R_s} + g_m \cdot v_{BE}; \; v_{BE} = V_{A_1}$$

$$I_{R_c} = I_{C_3} + I_{R_s} + g_m \cdot V_{A_1} = I_{C_3} + I_{R_s} + g_m \cdot [V_{cc} - I_{R_b} \cdot R_b]$$

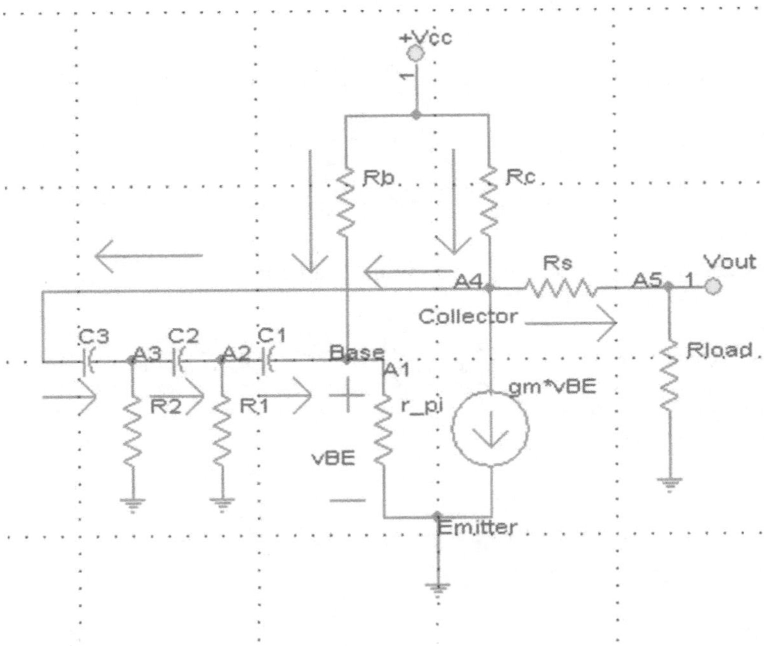

Fig. 6.2 Phase shift resonator equivalent circuit (BJT small signal)

$$I_{R_c} = \frac{V_{cc} - V_{A_4}}{R_c} \Rightarrow I_{R_c} \cdot R_c = V_{cc} - V_{A_4} \Rightarrow V_{A_4} = V_{cc} - I_{R_c} \cdot R_c$$
$$I_{R_b} = \frac{V_{cc} - V_{A_1}}{R_b} \Rightarrow I_{R_b} \cdot R_b = V_{cc} - V_{A_1} \Rightarrow V_{A_1} = V_{cc} - I_{R_b} \cdot R_b$$
$$I_{R_{load}} = \frac{V_{A_5}}{R_{load}};\ I_{R_{load}} = I_{R_s};\ V_{A_5} = I_{R_{load}} \cdot R_{load} = I_{R_s} \cdot R_{load}$$
$$I_{R_s} = \frac{V_{A_4} - V_{A_5}}{R_s} \Rightarrow I_{R_s} \cdot R_s = V_{A_4} - V_{A_5} = V_{cc} - I_{R_c} \cdot R_c - I_{R_s} \cdot R_{load}$$

$$I_{R_s} \cdot (R_s + R_{load}) = V_{cc} - I_{R_c} \cdot R_c; I_{r\_pi} = \frac{V_{A_1}}{r\_pi} \Rightarrow V_{A_1} = I_{r\_pi} \cdot r\_pi; I_{r\_pi} \cdot r\_pi = V_{cc} - I_{R_b} \cdot R_b$$
$$I_{R_2} = \frac{V_{A_3}}{R_3} \Rightarrow V_{A_3} = I_{R_2} \cdot R_3; I_{R_1} = \frac{V_{A_2}}{R_1} \Rightarrow V_{A_2} = I_{R_1} \cdot R_1; I_{C_3} = C_3 \cdot \frac{d}{dt}(V_{cc} - I_{R_c} \cdot R_c - I_{R_2} \cdot R_3)$$

$$I_{C_3} = C_3 \cdot (\frac{dV_{cc}}{dt} - \frac{dI_{R_c}}{dt} \cdot R_c - \frac{dI_{R_2}}{dt} \cdot R_3); \frac{dV_{cc}}{dt} \to \varepsilon;$$
$$I_{C_3} = -C_3 \cdot (\frac{dI_{R_c}}{dt} \cdot R_c + \frac{dI_{R_2}}{dt} \cdot R_3)$$

$$I_{C_2} = C_2 \cdot \frac{d}{dt}(V_{A_3} - V_{A_2}) = C_2 \cdot \frac{d}{dt}(I_{R_2} \cdot R_3 - I_{R_1} \cdot R_1);$$
$$I_{C_2} = C_2 \cdot (\frac{dI_{R_2}}{dt} \cdot R_3 - \frac{dI_{R_1}}{dt} \cdot R_1)$$

$$I_{C_1} = C_1 \cdot \frac{d}{dt}(V_{A_2} - V_{A_1}) = C_1 \cdot \frac{d}{dt}(I_{R_1} \cdot R_1 - V_{cc} + I_{R_b} \cdot R_b);$$
$$I_{C_1} = C_1 \cdot (\frac{dI_{R_1}}{dt} \cdot R_1 - \frac{dV_{cc}}{dt} + \frac{dI_{R_b}}{dt} \cdot R_b)$$
$$\frac{dV_{cc}}{dt} \to \varepsilon; I_{C_1} = C_1 \cdot (\frac{dI_{R_1}}{dt} \cdot R_1 + \frac{dI_{R_b}}{dt} \cdot R_b)$$

We can write KCL for the above circuit (Table 6.1):

Table 6.1 Phase shift resonator equivalent circuit (BJT small signal) KCL@ nodes and expressions

KCL @ node	Expression
A_1	$I_{R_b} + I_{C_1} = I_{r\_pi}$
A_2	$I_{C_2} = I_{C_1} + I_{R_1}$
A_3	$I_{C_3} = I_{C_2} + I_{R_2}$
A_4	$I_{R_c} = I_{C_3} + I_{R_s} + g_m \cdot v_{BE}$
A_5	$I_{R_{load}} = I_{R_s}$

$$I_{R_b} + I_{C_1} = I_{r\_pi} \Rightarrow I_{C_1} = I_{r\_pi} - I_{R_b}; I_{C_2} = I_{C_1} + I_{R_1} \Rightarrow I_{C_2} = I_{r\_pi} - I_{R_b} + I_{R_1}$$

$$I_{C_3} = I_{C_2} + I_{R_2} = I_{r\_pi} - I_{R_b} + I_{R_1} + I_{R_2} = I_{r\_pi} - I_{R_b} + \sum_{k=1}^{2} I_{R_k};$$

$$I_{C_3} = I_{r\_pi} - I_{R_b} + \sum_{k=1}^{2} I_{R_k}$$

We can summery our system three differential equations:

$$I_{r\_pi} - I_{R_b} = C_1 \cdot (\frac{dI_{R_1}}{dt} \cdot R_1 + \frac{dI_{R_b}}{dt} \cdot R_b); I_{r\_pi} - I_{R_b} + I_{R_1} = C_2 \cdot (\frac{dI_{R_2}}{dt} \cdot R_3 - \frac{dI_{R_1}}{dt} \cdot R_1)$$

$$I_{r\_pi} - I_{R_b} + \sum_{k=1}^{2} I_{R_k} = -C_3 \cdot (\frac{dI_{R_c}}{dt} \cdot R_c + \frac{dI_{R_2}}{dt} \cdot R_3)$$

$$I_{R_s} \cdot (R_s + R_{load}) = V_{cc} - I_{R_c} \cdot R_c \Rightarrow I_{R_s} = \frac{V_{cc}}{(R_s + R_{load})} - I_{R_c} \cdot \frac{R_c}{(R_s + R_{load})}$$

$$I_{R_c} = I_{r\_pi} - I_{R_b} + \sum_{k=1}^{2} I_{R_k} + \frac{V_{cc}}{(R_s + R_{load})} - I_{R_c} \cdot \frac{R_c}{(R_s + R_{load})} + g_m \cdot [V_{cc} - I_{R_b} \cdot R_b]$$

$$I_{R_c} \cdot (1 + \frac{R_c}{(R_s + R_{load})}) = I_{r\_pi} - I_{R_b} + \sum_{k=1}^{2} I_{R_k} + \frac{V_{cc}}{(R_s + R_{load})} + g_m \cdot [V_{cc} - I_{R_b} \cdot R_b]$$

$$I_{r\_pi} \cdot r\_pi = V_{cc} - I_{R_b} \cdot R_b \Rightarrow I_{r\_pi} = \frac{V_{cc}}{r\_pi} - I_{R_b} \cdot \frac{R_b}{r\_pi}$$

$$I_{R_c} \cdot (1 + \frac{R_c}{(R_s + R_{load})}) = \frac{V_{cc}}{r\_pi} - I_{R_b} \cdot \frac{R_b}{r\_pi} - I_{R_b} + \sum_{k=1}^{2} I_{R_k} + \frac{V_{cc}}{(R_s + R_{load})} + g_m \cdot [V_{cc} - I_{R_b} \cdot R_b]$$

$$I_{R_c} \cdot (1 + \frac{R_c}{(R_s + R_{load})}) = V_{cc} \cdot [\frac{1}{r\_pi} + \frac{1}{(R_s + R_{load})} + g_m] - I_{R_b} \cdot [\frac{R_b}{r\_pi} + 1 + g_m \cdot R_b] + \sum_{k=1}^{2} I_{R_k}$$

$$I_{R_c} = V_{cc} \cdot \frac{[\frac{1}{r\_pi} + \frac{1}{(R_s + R_{load})} + g_m]}{(1 + \frac{R_c}{(R_s + R_{load})})} - I_{R_b} \cdot \frac{[\frac{R_b}{r\_pi} + 1 + g_m \cdot R_b]}{(1 + \frac{R_c}{(R_s + R_{load})})} + \frac{1}{(1 + \frac{R_c}{(R_s + R_{load})})} \cdot \sum_{k=1}^{2} I_{R_k}$$

We define for simplicity the following global parameters:

$$\Gamma_1(r\_pi, R_s, R_{load}, \ldots) = \frac{[\frac{1}{r\_pi} + \frac{1}{(R_s + R_{load})} + g_m]}{(1 + \frac{R_c}{(R_s + R_{load})})}; \; \Gamma_1 = \Gamma_1(r\_pi, R_s, R_{load}, \ldots)$$

$$\Gamma_2(R_b, r\_pi, R_b, \ldots) = \frac{[\frac{R_b}{r\_pi} + 1 + g_m \cdot R_b]}{(1 + \frac{R_c}{(R_s + R_{load})})}; \; \Gamma_2 = \Gamma_2(R_b, r\_pi, R_b, \ldots)$$

$$\Gamma_3(R_c, R_s, \ldots) = \frac{1}{(1 + \frac{R_c}{(R_s + R_{load})})}; \; \Gamma_3 = \Gamma_3(R_c, R_s, \ldots)$$

$$I_{R_c} = V_{cc} \cdot \Gamma_1(r\_pi, R_s, R_{load}, \ldots) - I_{R_b} \cdot \Gamma_2(R_b, r\_pi, R_b, \ldots) + \Gamma_3(R_c, R_s, \ldots) \cdot \sum_{k=1}^{2} I_{R_k}$$

$$\frac{dI_{R_c}}{dt} = \frac{dV_{cc}}{dt} \cdot \Gamma_1(r\_pi, R_s, R_{load}, \ldots) - \frac{dI_{R_b}}{dt} \cdot \Gamma_2(R_b, r\_pi, R_b, \ldots) + \Gamma_3(R_c, R_s, \ldots) \cdot \sum_{k=1}^{2} \frac{dI_{R_k}}{dt}$$

$$\frac{dV_{cc}}{dt} \to \varepsilon \Rightarrow \frac{dV_{cc}}{dt} \cdot \Gamma_1(r\_pi, R_s, R_{load}, \ldots) \to \varepsilon; \; \frac{dI_{R_c}}{dt} = -\frac{dI_{R_b}}{dt} \cdot \Gamma_2 + \Gamma_3 \cdot \sum_{k=1}^{2} \frac{dI_{R_k}}{dt}$$

We can summery our system three differential equations with variables: R_1, R_2, R_b.

$$(*) \frac{V_{cc}}{r\_pi} - I_{R_b} \cdot \frac{R_b}{r\_pi} - I_{R_b} = C_1 \cdot (\frac{dI_{R_1}}{dt} \cdot R_1 + \frac{dI_{R_b}}{dt} \cdot R_b)$$

$$(**) \frac{V_{cc}}{r\_pi} - I_{R_b} \cdot [\frac{R_b}{r\_pi} + 1] + I_{R_1} = C_2 \cdot (\frac{dI_{R_2}}{dt} \cdot R_3 - \frac{dI_{R_1}}{dt} \cdot R_1)$$

$$(***)\frac{V_{cc}}{r\_pi} - I_{R_b} \cdot \frac{R_b}{r\_pi} - I_{R_b} + \sum_{k=1}^{2} I_{R_k} = -C_3 \cdot (-\frac{dI_{R_b}}{dt} \cdot \Gamma_2 \cdot R_c + \Gamma_3 \cdot R_c \cdot \sum_{k=1}^{2} \frac{dI_{R_k}}{dt} + \frac{dI_{R_2}}{dt} \cdot R_3)$$

Another representation of system three differential equations:

$$(*)\frac{V_{cc}}{r\_pi} - I_{R_b} \cdot \frac{R_b}{r\_pi} - I_{R_b} = C_1 \cdot (\frac{dI_{R_1}}{dt} \cdot R_1 + \frac{dI_{R_b}}{dt} \cdot R_b)$$

$$\frac{V_{cc}}{r\_pi} - I_{R_b} \cdot [\frac{R_b}{r\_pi} + 1] - \frac{dI_{R_b}}{dt} \cdot C_1 \cdot R_b = \frac{dI_{R_1}}{dt} \cdot C_1 \cdot R_1$$

$$(**)\frac{V_{cc}}{r\_pi} - I_{R_b} \cdot [\frac{R_b}{r\_pi} + 1] + I_{R_1} = C_2 \cdot (\frac{dI_{R_2}}{dt} \cdot R_3 - \frac{dI_{R_1}}{dt} \cdot R_1)$$

$$\frac{V_{cc}}{r\_pi} - I_{R_b} \cdot [\frac{R_b}{r\_pi} + 1] + I_{R_1} = C_2 \cdot (\frac{dI_{R_2}}{dt} \cdot R_3 - \frac{dI_{R_1}}{dt} \cdot R_1)$$

$$(***)\frac{V_{cc}}{r\_pi} - I_{R_b} \cdot \frac{R_b}{r\_pi} - I_{R_b} + \sum_{k=1}^{2} I_{R_k} = -C_3 \cdot (-\frac{dI_{R_b}}{dt} \cdot \Gamma_2 \cdot R_c + \Gamma_3 \cdot R_c \cdot \sum_{k=1}^{2} \frac{dI_{R_k}}{dt} + \frac{dI_{R_2}}{dt} \cdot R_3)$$

$$\frac{V_{cc}}{r\_pi} - I_{R_b} \cdot [\frac{R_b}{r\_pi} + 1] + \sum_{k=1}^{2} I_{R_k} = \frac{dI_{R_b}}{dt} \cdot \Gamma_2 \cdot C_3 \cdot R_c - \Gamma_3 \cdot R_c \cdot C_3 \cdot \frac{dI_{R_1}}{dt} - \frac{dI_{R_2}}{dt} \cdot C_3 \cdot [\Gamma_3 \cdot R_c + R_3]$$

And we get the following differential equations:

$$(*)\frac{dI_{R_1}}{dt} = \frac{V_{cc}}{r\_pi \cdot C_1 \cdot R_1} - I_{R_b} \cdot \frac{[\frac{R_b}{r\_pi} + 1]}{C_1 \cdot R_1} - \frac{dI_{R_b}}{dt} \cdot \frac{R_b}{R_1}$$

$(*) \to (**)$

$$\frac{V_{cc}}{r\_pi} - I_{R_b} \cdot [\frac{R_b}{r\_pi} + 1] + I_{R_1}$$

$$= C_2 \cdot (\frac{dI_{R_2}}{dt} \cdot R_3 - [\frac{V_{cc}}{r\_pi \cdot C_1 \cdot R_1} - I_{R_b} \cdot \frac{[\frac{R_b}{r\_pi} + 1]}{C_1 \cdot R_1} - \frac{dI_{R_b}}{dt} \cdot \frac{R_b}{R_1}] \cdot R_1)$$

$$\frac{V_{cc}}{r\_pi} - I_{R_b} \cdot [\frac{R_b}{r\_pi} + 1] + I_{R_1} = \frac{dI_{R_2}}{dt} \cdot C_2 \cdot R_3 - \frac{V_{cc} \cdot C_2}{r\_pi \cdot C_1} + I_{R_b} \cdot \frac{[\frac{R_b}{r\_pi} + 1] \cdot C_2}{C_1} + \frac{dI_{R_b}}{dt} \cdot R_b \cdot C_2$$

$$\frac{V_{cc}}{r\_pi} + I_{R_1} = \frac{dI_{R_2}}{dt} \cdot C_2 \cdot R_3 - \frac{V_{cc} \cdot C_2}{r\_pi \cdot C_1} + I_{R_b} \cdot [\frac{R_b}{r\_pi} + 1] \cdot [\frac{C_2}{C_1} + 1] + \frac{dI_{R_b}}{dt} \cdot R_b \cdot C_2$$

$(*) \rightarrow (***)$

$$\frac{V_{cc}}{r\_pi} - I_{R_b} \cdot [\frac{R_b}{r\_pi} + 1] + \sum_{k=1}^{2} I_{R_k} = \frac{dI_{R_b}}{dt} \cdot \Gamma_2 \cdot C_3 \cdot R_c - \Gamma_3 \cdot R_c \cdot C_3 \cdot [\frac{V_{cc}}{r\_pi \cdot C_1 \cdot R_1} - I_{R_b} \cdot \frac{[\frac{R_b}{r\_pi} + 1]}{C_1 \cdot R_1} - \frac{dI_{R_b}}{dt} \cdot \frac{R_b}{R_1}] - \frac{dI_{R_2}}{dt} \cdot C_3 \cdot [\Gamma_3 \cdot R_c + R_3]$$

$$\frac{V_{cc}}{r\_pi} - I_{R_b} \cdot [\frac{R_b}{r\_pi} + 1] + \sum_{k=1}^{2} I_{R_k} = \frac{dI_{R_b}}{dt} \cdot R_c \cdot C_3 \cdot [\Gamma_2 + \frac{\Gamma_3 \cdot R_b}{R_1}] - \frac{\Gamma_3 \cdot R_c \cdot C_3 \cdot V_{cc}}{r\_pi \cdot C_1 \cdot R_1} + I_{R_b} \cdot \frac{\Gamma_3 \cdot R_c \cdot C_3 \cdot [\frac{R_b}{r\_pi} + 1]}{C_1 \cdot R_1} - \frac{dI_{R_2}}{dt} \cdot C_3 \cdot [\Gamma_3 \cdot R_c + R_3]$$

We define new global parameters for simplicity:

$$\Omega_1 = \frac{C_2}{r\_pi \cdot C_1}; \ \Omega_2 = [\frac{R_b}{r\_pi} + 1] \cdot [\frac{C_2}{C_1} + 1]; \ \Omega_3 = [\frac{R_b}{r\_pi} + 1];$$
$$\Omega_4 = R_c \cdot C_3 \cdot [\Gamma_2 + \frac{\Gamma_3 \cdot R_b}{R_1}]$$

$$\Omega_5 = \frac{\Gamma_3 \cdot R_c \cdot C_3}{r\_pi \cdot C_1 \cdot R_1}; \Omega_6 = \frac{\Gamma_3 \cdot R_c \cdot C_3 \cdot [\frac{R_b}{r\_pi} + 1]}{C_1 \cdot R_1}; \ \Omega_7 = C_3 \cdot [\Gamma_3 \cdot R_c + R_3]$$

$$\frac{V_{cc}}{r\_pi} + I_{R_1} = \frac{dI_{R_2}}{dt} \cdot C_2 \cdot R_3 - V_{cc} \cdot \Omega_1 + I_{R_b} \cdot \Omega_2 + \frac{dI_{R_b}}{dt} \cdot R_b \cdot C_2$$

$$\frac{V_{cc}}{r\_pi} - I_{R_b} \cdot \Omega_3 + \sum_{k=1}^{2} I_{R_k} = \frac{dI_{R_b}}{dt} \cdot \Omega_4 - V_{cc} \cdot \Omega_5 + I_{R_b} \cdot \Omega_6 - \frac{dI_{R_2}}{dt} \cdot \Omega_7$$

$$\frac{V_{cc}}{r\_pi} + I_{R_1} = \frac{dI_{R_2}}{dt} \cdot C_2 \cdot R_3 - V_{cc} \cdot \Omega_1 + I_{R_b} \cdot \Omega_2 + \frac{dI_{R_b}}{dt} \cdot R_b \cdot C_2$$

$$\frac{dI_{R_b}}{dt} = V_{cc} \cdot \frac{[\frac{1}{r\_pi} + \Omega_1]}{R_b \cdot C_2} + \frac{1}{R_b \cdot C_2} \cdot I_{R_1} - \frac{R_3}{R_b} \cdot \frac{dI_{R_2}}{dt} - I_{R_b} \cdot \frac{\Omega_2}{R_b \cdot C_2}$$

$$\frac{V_{cc}}{r\_pi} - I_{R_b} \cdot \Omega_3 + \sum_{k=1}^{2} I_{R_k} = \frac{dI_{R_b}}{dt} \cdot \Omega_4 - V_{cc} \cdot \Omega_5 + I_{R_b} \cdot \Omega_6 - \frac{dI_{R_2}}{dt} \cdot \Omega_7$$

$$\frac{V_{cc}}{r\_pi} - I_{R_b} \cdot \Omega_3 + \sum_{k=1}^{2} I_{R_k} = [V_{cc} \cdot \frac{[\frac{1}{r\_pi} + \Omega_1]}{R_b \cdot C_2} + \frac{1}{R_b \cdot C_2} \cdot I_{R_1} - \frac{R_3}{R_b} \cdot \frac{dI_{R_2}}{dt} - I_{R_b} \cdot \frac{\Omega_2}{R_b \cdot C_2}] \cdot \Omega_4 - V_{cc} \cdot \Omega_5 + I_{R_b} \cdot \Omega_6 - \frac{dI_{R_2}}{dt} \cdot \Omega_7$$

$$-[\frac{R_3 \cdot \Omega_4}{R_b} + \Omega_7] \cdot \frac{dI_{R_2}}{dt} = V_{cc} \cdot (\frac{1}{r\_pi} - \frac{[\frac{1}{r\_pi} + \Omega_1] \cdot \Omega_4}{R_b \cdot C_2} + \Omega_5) + I_{R_b} \cdot [\frac{\Omega_2 \cdot \Omega_4}{R_b \cdot C_2} - \Omega_6 - \Omega_3] + [1 - \frac{\Omega_4}{R_b \cdot C_2}] \cdot I_{R_1} + I_{R_2}$$

$$\frac{dI_{R_2}}{dt} = V_{cc} \cdot \frac{(\frac{[\frac{1}{r\_pi} + \Omega_1] \cdot \Omega_4}{R_b \cdot C_2} - \frac{1}{r\_pi} - \Omega_5)}{[\frac{R_3 \cdot \Omega_4}{R_b} + \Omega_7]} + I_{R_b} \cdot \frac{[\Omega_6 + \Omega_3 - \frac{\Omega_2 \cdot \Omega_4}{R_b \cdot C_2}]}{[\frac{R_3 \cdot \Omega_4}{R_b} + \Omega_7]} + \frac{[\frac{\Omega_4}{R_b \cdot C_2} - 1]}{[\frac{R_3 \cdot \Omega_4}{R_b} + \Omega_7]} \cdot I_{R_1} - \frac{1}{[\frac{R_3 \cdot \Omega_4}{R_b} + \Omega_7]} \cdot I_{R_2}$$

We define for simplicity the following global parameters:

$$\Xi_1 = \Xi_1(r\_pi, R_3, R_b, \Omega_1, \Omega_4, \ldots) = \frac{(\frac{[\frac{1}{r\_pi} + \Omega_1] \cdot \Omega_4}{R_b \cdot C_2} - \frac{1}{r\_pi} - \Omega_5)}{[\frac{R_3 \cdot \Omega_4}{R_b} + \Omega_7]}$$

$$\Xi_2 = \Xi_2(R_b, R_3, \Omega_2, \Omega_3, \Omega_4, \ldots) = \frac{[\Omega_6 + \Omega_3 - \frac{\Omega_2 \cdot \Omega_4}{R_b \cdot C_2}]}{[\frac{R_3 \cdot \Omega_4}{R_b} + \Omega_7]}; \; \Xi_3 = \Xi_3(R_b, R_3, \Omega_4, \ldots) = \frac{[\frac{\Omega_4}{R_b \cdot C_2} - 1]}{[\frac{R_3 \cdot \Omega_4}{R_b} + \Omega_7]}$$

$$\Xi_4 = \Xi_4(R_3, R_b, \Omega_4, \ldots) = -\frac{1}{[\frac{R_3 \cdot \Omega_4}{R_b} + \Omega_7]}$$

$$\frac{dI_{R_2}}{dt} = V_{cc} \cdot \Xi_1 + I_{R_b} \cdot \Xi_2 + \Xi_3 \cdot I_{R_1} + \Xi_4 \cdot I_{R_2}$$

$$\frac{dI_{R_b}}{dt} = V_{cc} \cdot \frac{[\frac{1}{r\_pi} + \Omega_1]}{R_b \cdot C_2} + \frac{1}{R_b \cdot C_2} \cdot I_{R_1} - \frac{R_3}{R_b} \cdot [V_{cc} \cdot \Xi_1 + I_{R_b} \cdot \Xi_2 + \Xi_3 \cdot I_{R_1} + \Xi_4 \cdot I_{R_2}] - I_{R_b} \cdot \frac{\Omega_2}{R_b \cdot C_2}$$

$$\frac{dI_{R_b}}{dt} = V_{cc} \cdot \frac{1}{R_b} \cdot [\frac{[\frac{1}{r\_pi} + \Omega_1]}{C_2} - R_3 \cdot \Xi_1] + [\frac{1}{C_2} - \Xi_3 \cdot R_3] \cdot \frac{1}{R_b} \cdot I_{R_1} - \Xi_4 \cdot \frac{R_3}{R_b} \cdot I_{R_2} - I_{R_b} \cdot \frac{1}{R_b} \cdot [R_3 \cdot \Xi_2 + \frac{\Omega_2}{C_2}]$$

We define for simplicity the following global parameters:

$$\Xi_5 = \Xi_5(r\_pi, R_b, R_3, \ldots) = \frac{1}{R_b} \cdot [\frac{[\frac{1}{r\_pi} + \Omega_1]}{C_2} - R_3 \cdot \Xi_1]$$

$$\Xi_6 = \Xi_6(\Xi_3, R_3, R_b, \ldots) = [\frac{1}{C_2} - \Xi_3 \cdot R_3] \cdot \frac{1}{R_b}; \; \Xi_7 = \Xi_7(\Xi_4, R_3, \ldots) = -\Xi_4 \cdot \frac{R_3}{R_b}$$

$$\Xi_8 = \Xi_8(R_3, \Xi_2, \ldots) = -\frac{1}{R_b} \cdot [R_3 \cdot \Xi_2 + \frac{\Omega_2}{C_2}];$$

$$\frac{dI_{R_b}}{dt} = V_{cc} \cdot \Xi_5 + \Xi_6 \cdot I_{R_1} + \Xi_7 \cdot I_{R_2} + I_{R_b} \cdot \Xi_8$$

$$\frac{dI_{R_1}}{dt} = \frac{V_{cc}}{r\_pi \cdot C_1 \cdot R_1} - I_{R_b} \cdot \frac{[\frac{R_b}{r\_pi} + 1]}{C_1 \cdot R_1} - \frac{dI_{R_b}}{dt} \cdot \frac{R_b}{R_1}$$

$$\frac{dI_{R_1}}{dt} = \frac{V_{cc}}{r\_pi \cdot C_1 \cdot R_1} - I_{R_b} \cdot \frac{[\frac{R_b}{r\_pi} + 1]}{C_1 \cdot R_1} - [V_{cc} \cdot \Xi_5 + \Xi_6 \cdot I_{R_1} + \Xi_7 \cdot I_{R_2} + I_{R_b} \cdot \Xi_8] \cdot \frac{R_b}{R_1}$$

$$\frac{dI_{R_1}}{dt} = V_{cc} \cdot [\frac{1}{r\_pi \cdot C_1 \cdot R_1} - \Xi_5 \cdot \frac{R_b}{R_1}] - \Xi_6 \cdot \frac{R_b}{R_1} \cdot I_{R_1} - \Xi_7 \cdot \frac{R_b}{R_1} \cdot I_{R_2} - I_{R_b} \cdot \frac{1}{R_1} \cdot [\Xi_8 \cdot R_b + \frac{[\frac{R_b}{r\_pi} + 1]}{C_1}]$$

$$\Xi_9 = \Xi_9(R_1, R_b, r\_pi, \ldots) = \frac{1}{r\_pi \cdot C_1 \cdot R_1} - \Xi_5 \cdot \frac{R_b}{R_1}; \; \Xi_{10} = \Xi_{10}(R_b, R_1, \ldots) = -\Xi_6 \cdot \frac{R_b}{R_1}$$

$$\Xi_{11} = \Xi_{11}(\Xi_7, R_b, \ldots) = -\Xi_7 \cdot \frac{R_b}{R_1}; \; \Xi_{12} = \Xi_{12}(\Xi_8, R_1, R_b, \ldots) = -\frac{1}{R_1} \cdot [\Xi_8 \cdot R_b + \frac{[\frac{R_b}{r\_pi} + 1]}{C_1}]$$

$$\frac{dI_{R_1}}{dt} = V_{cc} \cdot \Xi_9 + \Xi_{10} \cdot I_{R_1} + \Xi_{11} \cdot I_{R_2} + I_{R_b} \cdot \Xi_{12}$$

We can summery our system set of differential equations:

$$\frac{dI_{R_1}}{dt} = V_{cc} \cdot \Xi_9 + \Xi_{10} \cdot I_{R_1} + \Xi_{11} \cdot I_{R_2} + I_{R_b} \cdot \Xi_{12}; \frac{dI_{R_2}}{dt} = V_{cc} \cdot \Xi_1 + I_{R_b} \cdot \Xi_2 + \Xi_3 \cdot I_{R_1} + \Xi_4 \cdot I_{R_2}$$

$$\frac{dI_{R_b}}{dt} = V_{cc} \cdot \Xi_5 + \Xi_6 \cdot I_{R_1} + \Xi_7 \cdot I_{R_2} + I_{R_b} \cdot \Xi_8$$

Phase shift resonator circuit stability analysis is done by considering BJT Small Signal (SS) equivalent circuit model. We consider "AC ground" in the circuit. Since the voltage at this terminal is held constant at V_{cc}, there is no time variation of the voltage. Consequently, we can set this terminal to be an "AC ground" in the small signal circuit ($V_{cc} = 0$). Under AC and Small Signal (SS) conditions, the BJT is replaced with linear Hybrid Pi model [26, 27].

$$\frac{dI_{R_1}}{dt} = \Xi_{10} \cdot I_{R_1} + \Xi_{11} \cdot I_{R_2} + I_{R_b} \cdot \Xi_{12}; \frac{dI_{R_2}}{dt} = I_{R_b} \cdot \Xi_2 + \Xi_3 \cdot I_{R_1} + \Xi_4 \cdot I_{R_2}$$

$$\frac{dI_{R_b}}{dt} = \Xi_6 \cdot I_{R_1} + \Xi_7 \cdot I_{R_2} + I_{R_b} \cdot \Xi_8$$

At fixed points: $\frac{dI_{R_1}}{dt} = 0; \frac{dI_{R_2}}{dt} = 0; \frac{dI_{R_b}}{dt} = 0$

$$\Xi_{10} \cdot I^*_{R_1} + \Xi_{11} \cdot I^*_{R_2} + I^*_{R_b} \cdot \Xi_{12} = 0; \ \Xi_3 \cdot I^*_{R_1} + \Xi_4 \cdot I^*_{R_2} + I^*_{R_b} \cdot \Xi_2 = 0;$$
$$\Xi_6 \cdot I^*_{R_1} + \Xi_7 \cdot I^*_{R_2} + I^*_{R_b} \cdot \Xi_8 = 0$$

$E(I^*_{R_1}, I^*_{R_2}, I^*_{R_b}) = (0, 0, 0)$. If $V_{cc} \neq 0$ then we get the following three equations for system fixed points (equilibrium points):

$$\Xi_{10} \cdot I^*_{R_1} + \Xi_{11} \cdot I^*_{R_2} + I^*_{R_b} \cdot \Xi_{12} = -V_{cc} \cdot \Xi_9; \Xi_3 \cdot I^*_{R_1} + \Xi_4 \cdot I^*_{R_2} + I^*_{R_b} \cdot \Xi_2 = -V_{cc} \cdot \Xi_1$$
$$\Xi_6 \cdot I^*_{R_1} + \Xi_7 \cdot I^*_{R_2} + I^*_{R_b} \cdot \Xi_8 = -V_{cc} \cdot \Xi_5$$

To use determinants to solve our system fixed points equations ($V_{cc} \neq 0$) with three fixed points (Cramer's Rule), $I^*_{R_1}, I^*_{R_2}, I^*_{R_b}$, four determinants must be formed following the procedure:

(I) Write all fixed points equations in standard form.

(II) Create the denominator determinant, D, by using the coefficients of $I^*_{R_1}, I^*_{R_2}, I^*_{R_b}$ from the equations and evaluate it.

(III) Create the $I^*_{R_1}$—numerator determinant, $D_{I^*_{R_1}}$, the $I^*_{R_2}$—numerator determinant, $D_{I^*_{R_2}}$, and the $I^*_{R_b}$—numerator determinant, $D_{I^*_{R_b}}$, by replacing the respective $I^*_{R_1}, I^*_{R_2}$, and $I^*_{R_b}$ coefficients with the constants from the equations in standard form and evaluate each determinant.

The answers for $I_{R_1}^*, I_{R_2}^*$, and $I_{R_b}^*$ are as follow: $I_{R_1}^* = \frac{D_{I_{R_1}^*}}{D}; I_{R_2}^* = \frac{D_{I_{R_2}^*}}{D}; I_{R_b}^* = \frac{D_{I_{R_b}^*}}{D}$. We solve this system of fixed point's equations, using Cramer's Rule.

$$D = \det\begin{pmatrix} \Xi_{10} & \Xi_{11} & \Xi_{12} \\ \Xi_3 & \Xi_4 & \Xi_2 \\ \Xi_6 & \Xi_7 & \Xi_8 \end{pmatrix} = \Xi_{10} \cdot \det\begin{pmatrix} \Xi_4 & \Xi_2 \\ \Xi_7 & \Xi_8 \end{pmatrix} - \Xi_{11} \cdot \det\begin{pmatrix} \Xi_3 & \Xi_2 \\ \Xi_6 & \Xi_8 \end{pmatrix} + \Xi_{12} \cdot \det\begin{pmatrix} \Xi_3 & \Xi_4 \\ \Xi_6 & \Xi_7 \end{pmatrix}$$

$$D = \Xi_{10} \cdot (\Xi_4 \cdot \Xi_8 - \Xi_7 \cdot \Xi_2) - \Xi_{11} \cdot (\Xi_3 \cdot \Xi_8 - \Xi_6 \cdot \Xi_2) + \Xi_{12} \cdot (\Xi_3 \cdot \Xi_7 - \Xi_6 \cdot \Xi_4)$$

$$D = \Xi_{10} \cdot \Xi_4 \cdot \Xi_8 - \Xi_{10} \cdot \Xi_7 \cdot \Xi_2 - \Xi_{11} \cdot \Xi_3 \cdot \Xi_8 + \Xi_{11} \cdot \Xi_6 \cdot \Xi_2 + \Xi_{12} \cdot \Xi_3 \cdot \Xi_7 - \Xi_{12} \cdot \Xi_6 \cdot \Xi_4$$

We use the constants to replace the $I_{R_1}^*$—coefficients.

$$D_{I_{R_1}^*} = \det\begin{pmatrix} -V_{cc} \cdot \Xi_9 & \Xi_{11} & \Xi_{12} \\ -V_{cc} \cdot \Xi_1 & \Xi_4 & \Xi_2 \\ -V_{cc} \cdot \Xi_5 & \Xi_7 & \Xi_8 \end{pmatrix} = -V_{cc} \cdot \Xi_9 \cdot \det\begin{pmatrix} \Xi_4 & \Xi_2 \\ \Xi_7 & \Xi_8 \end{pmatrix}$$
$$+ V_{cc} \cdot \Xi_1 \cdot \det\begin{pmatrix} \Xi_{11} & \Xi_{12} \\ \Xi_7 & \Xi_8 \end{pmatrix} - V_{cc} \cdot \Xi_5 \cdot \det\begin{pmatrix} \Xi_{11} & \Xi_{12} \\ \Xi_4 & \Xi_2 \end{pmatrix}$$

$$D_{I_{R_1}^*} = -V_{cc} \cdot \Xi_9 \cdot (\Xi_4 \cdot \Xi_8 - \Xi_7 \cdot \Xi_2) + V_{cc} \cdot \Xi_1 \cdot (\Xi_{11} \cdot \Xi_8 - \Xi_7 \cdot \Xi_{12}) - V_{cc} \cdot \Xi_5 \cdot (\Xi_{11} \cdot \Xi_2 - \Xi_4 \cdot \Xi_{12})$$

$$D_{I_{R_1}^*} = -V_{cc} \cdot \Xi_9 \cdot \Xi_4 \cdot \Xi_8 + V_{cc} \cdot \Xi_9 \cdot \Xi_7 \cdot \Xi_2 + V_{cc} \cdot \Xi_1 \cdot \Xi_{11} \cdot \Xi_8 - V_{cc} \cdot \Xi_1 \cdot \Xi_7 \cdot \Xi_{12}$$
$$- V_{cc} \cdot \Xi_5 \cdot \Xi_{11} \cdot \Xi_2 + V_{cc} \cdot \Xi_5 \cdot \Xi_4 \cdot \Xi_{12}$$

We use the constants to replace the $I_{R_2}^*$—coefficients.

$$D_{I_{R_2}^*} = \det\begin{pmatrix} \Xi_{10} & -V_{cc} \cdot \Xi_9 & \Xi_{12} \\ \Xi_3 & -V_{cc} \cdot \Xi_1 & \Xi_2 \\ \Xi_6 & -V_{cc} \cdot \Xi_5 & \Xi_8 \end{pmatrix} = \Xi_{10} \cdot \det\begin{pmatrix} -V_{cc} \cdot \Xi_1 & \Xi_2 \\ -V_{cc} \cdot \Xi_5 & \Xi_8 \end{pmatrix}$$
$$+ V_{cc} \cdot \Xi_9 \cdot \begin{pmatrix} \Xi_3 & \Xi_2 \\ \Xi_6 & \Xi_8 \end{pmatrix} - \Xi_{12} \cdot \begin{pmatrix} \Xi_3 & -V_{cc} \cdot \Xi_1 \\ \Xi_6 & -V_{cc} \cdot \Xi_5 \end{pmatrix}$$

$$D_{I_{R_2}^*} = \Xi_{10} \cdot (-V_{cc} \cdot \Xi_1 \cdot \Xi_8 + V_{cc} \cdot \Xi_5 \cdot \Xi_2) + V_{cc} \cdot \Xi_9 \cdot (\Xi_3 \cdot \Xi_8 - \Xi_6 \cdot \Xi_2)$$
$$- \Xi_{12} \cdot (-V_{cc} \cdot \Xi_5 \cdot \Xi_3 + V_{cc} \cdot \Xi_1 \cdot \Xi_6)$$

$$D_{I_{R_2}^*} = -V_{cc} \cdot \Xi_1 \cdot \Xi_{10} \cdot \Xi_8 + V_{cc} \cdot \Xi_5 \cdot \Xi_{10} \cdot \Xi_2 + V_{cc} \cdot \Xi_9 \cdot \Xi_3 \cdot \Xi_8 - V_{cc} \cdot \Xi_9 \cdot \Xi_6 \cdot \Xi_2$$
$$+ V_{cc} \cdot \Xi_5 \cdot \Xi_{12} \cdot \Xi_3 - V_{cc} \cdot \Xi_1 \cdot \Xi_{12} \cdot \Xi_6$$

We use the constants to replace the $I_{R_b}^*$—coefficients.

$$D_{I^*_{R_b}} = \det\begin{pmatrix} \Xi_{10} & \Xi_{11} & -V_{cc}\cdot\Xi_9 \\ \Xi_3 & \Xi_4 & -V_{cc}\cdot\Xi_1 \\ \Xi_6 & \Xi_7 & -V_{cc}\cdot\Xi_5 \end{pmatrix} = \Xi_{10}\cdot\det\begin{pmatrix} \Xi_4 & -V_{cc}\cdot\Xi_1 \\ \Xi_7 & -V_{cc}\cdot\Xi_5 \end{pmatrix}$$
$$-\,\Xi_{11}\cdot\begin{pmatrix} \Xi_3 & -V_{cc}\cdot\Xi_1 \\ \Xi_6 & -V_{cc}\cdot\Xi_5 \end{pmatrix} - V_{cc}\cdot\Xi_9\cdot\begin{pmatrix} \Xi_3 & \Xi_4 \\ \Xi_6 & \Xi_7 \end{pmatrix}$$

$$D_{I^*_{R_b}} = \Xi_{10}\cdot(-V_{cc}\cdot\Xi_5\cdot\Xi_4 + V_{cc}\cdot\Xi_1\cdot\Xi_7) - \Xi_{11}\cdot(-V_{cc}\cdot\Xi_5\cdot\Xi_3 + V_{cc}\cdot\Xi_1\cdot\Xi_6)$$
$$-\,V_{cc}\cdot\Xi_9\cdot(\Xi_3\cdot\Xi_7 - \Xi_6\cdot\Xi_4)$$

$$D_{I^*_{R_b}} = -V_{cc}\cdot\Xi_5\cdot\Xi_{10}\cdot\Xi_4 + V_{cc}\cdot\Xi_1\cdot\Xi_{10}\cdot\Xi_7 + V_{cc}\cdot\Xi_5\cdot\Xi_{11}\cdot\Xi_3 - V_{cc}\cdot\Xi_1\cdot\Xi_{11}\cdot\Xi_6$$
$$-\,V_{cc}\cdot\Xi_9\cdot\Xi_3\cdot\Xi_7 + V_{cc}\cdot\Xi_9\cdot\Xi_6\cdot\Xi_4$$

Therefore, $I^*_{R_1} = \frac{D_{I^*_{R_1}}}{D}; I^*_{R_2} = \frac{D_{I^*_{R2}}}{D}; I^*_{R_b} = \frac{D_{I^*_{R_b}}}{D}$. If the denominator determinant, D, has a value of zero, then system is either inconsistent or dependent. The system is dependent if all the determinants have a value of zero. The system is inconsistent if at least one of the determinants, $D_{I^*_{R_1}}, D_{I^*_{R_2}}$ or $D_{I^*_{R_b}}$ has a value not equal to zero and the denominator determinant has a value of zero.

Stability discussion:

$$\begin{pmatrix} \frac{dI_{R_1}}{dt} \\ \frac{dI_{R_2}}{dt} \\ \frac{dI_{R_b}}{dt} \end{pmatrix} = \begin{pmatrix} \Xi_{10} & \Xi_{11} & \Xi_{12} \\ \Xi_3 & \Xi_4 & \Xi_2 \\ \Xi_6 & \Xi_7 & \Xi_8 \end{pmatrix}\cdot\begin{pmatrix} I_{R_1} \\ I_{R_2} \\ I_{R_b} \end{pmatrix} + \begin{pmatrix} b_1(I_{R_1},t) \\ b_2(I_{R_2},t) \\ b_3(I_{R_b},t) \end{pmatrix}; \begin{pmatrix} I_{R_1}(t=0) \\ I_{R_2}(t=0) \\ I_{R_b}(t=0) \end{pmatrix} = a_0$$

We define the following notation: $X = \begin{pmatrix} I_{R_1} \\ I_{R_2} \\ I_{R_b} \end{pmatrix}; X = (I_{R_1}, I_{R_2}, I_{R_b})^{\perp}$

$$A = \begin{pmatrix} \Xi_{10} & \Xi_{11} & \Xi_{12} \\ \Xi_3 & \Xi_4 & \Xi_2 \\ \Xi_6 & \Xi_7 & \Xi_8 \end{pmatrix}; b = (array{*}20cb_1b_2b_3); b = (b_1, b_2, b_3)^{\perp}$$

A is a real, time independent (3 × 3)-matrix and the components of $b = (b_1, b_2, b_3)^{\perp}$

Are real C^1—functions for all $X = \begin{pmatrix} I_{R_1} \\ I_{R_2} \\ I_{R_b} \end{pmatrix}$ in a neighborhood of the origin for all times t ≥ 0. Moreover, let $\begin{pmatrix} b_1(I_{R_1}=0,t) \\ b_2(I_{R_2}=0,t) \\ b_3(I_{R_b}=0,t) \end{pmatrix} = 0$. Hence $X = \begin{pmatrix} I_{R_1} \\ I_{R_2} \\ I_{R_b} \end{pmatrix} \equiv 0$ is a solution of our system linear differential equation which corresponds to a point of equilibrium of the system. The equilibrium point is at X = 0.

Stability: The equilibrium point $E(I^*_{R_1}, I^*_{R_2}, I^*_{R_b}) = (0,0,0)$ is stable if and only if for each ε > 0 there is a number δ > 0 such that from $|t| < \delta$, the existence of a unique solution $X = X(t) \Rightarrow (I_{R_1} I_{R_2} I_{R_b}) = (I_{R_1}(t) I_{R_2}(t) I_{R_b}(t))$ of our system linear differential equation follows, with $|X(t)| < \varepsilon$ for all times t ≥ 0. This means that sufficiently small perturbations of the equilibrium configuration at $X = (I_{R_1} I_{R_2} I_{R_b}) = 0$ remain small for all t ≥ 0.

Asymptotic stability: The equilibrium point $E(I^*_{R_1}, I^*_{R_2}, I^*_{R_b}) = (0,0,0)$ is asymptotically stable if and only if it is stable and in addition there is a number $\delta_* > 0$ such that, for each solution with $|X(0)| < \delta_*$, we have the limit relation $\lim_{t \to \infty} X(t) = 0$. Sufficient small perturbations of the equilibrium configuration at time t = 0 return to their starting configuration after a sufficiently long term.

Instability: The equilibrium point $E(I^*_{R_1}, I^*_{R_2}, I^*_{R_b}) = (0,0,0)$ is instable if and only if it is not stable.

We can implement in our system the theorem on stability by Liapunov:

Suppose a perturbation b of the linear system $(\frac{dI_{R_1}}{dt} \frac{dI_{R_2}}{dt} \frac{dI_{R_b}}{dt}) = \begin{pmatrix} \Xi_{10} & \Xi_{11} & \Xi_{12} \\ \Xi_3 & \Xi_4 & \Xi_2 \\ \Xi_6 & \Xi_7 & \Xi_8 \end{pmatrix} \cdot (I_{R_1} I_{R_2} I_{R_b})$

With constant coefficients is sufficiently small, i.e., one has $\lim_{|X| \to \infty} (\sup_{t \geq 0} \frac{|b(X,t)|}{|X|}) = 0$

The equilibrium point $E(I^*_{R_1}, I^*_{R_2}, I^*_{R_b}) = (0,0,0)$ is asymptotically stable, if all eigenvalues $\lambda_1, \lambda_2, \ldots$ of the matrix A are in the left half plane, have negative real part for all j. The equilibrium point $E(I^*_{R_1}, I^*_{R_2}, I^*_{R_b}) = (0,0,0)$ is unstable, if an eigenvalue of A is in the right half plane, i.e., one has $\text{Re}\,\lambda_j > 0$ for some j. If an eigenvalue of A is on the imaginary axis, then the method of the center manifold must be applied. To apply the stability criterion of Liapunov to our system, the equation $\det(A - \lambda \cdot I) = 0$ has a zero in the left half plane, without calculating the eigenvalues explicitly.

Eigenvalues stability discussion: Our phase shift oscillation system involving N variables (N > 2, N = 3), the characteristic equation is of degree N = 3 and must often be solved numerically. Expect in some particular cases, such an equation has (N = 3) distinct roots that can be real or complex. These values are the eigenvalues of the 3 × 3 Jacobian matrix (A). The general rule is that the Steady State (SS) is stable if there is no eigenvalue with positive real part. It is sufficient that one eigenvalue is positive for the steady state to be unstable. Our 3-variables $(I_{R_1}, I_{R_2}, I_{R_b})$ system has three eigenvalues. The type of behavior can be characterized as a function of the position of these eigenvalues in the Re/Im plane. Five non-degenerated cases can be distinguished: (1) the three eigenvalues are real and negative (stable steady state), (2) the three eigenvalues are real, two of them are negative (unstable steady state), (3) and (4) two eigenvalues are complex conjugates with a negative real part and the third one real is negative (stable steady state), two cases can be distinguished depending on the relative value of the real part of the complex eigenvalues and of the real one, (5) two eigenvalues are complex conjugates with a negative real part and the third one real is positive (unstable steady state) [2, 3, 4].

$$A = \begin{pmatrix} \Xi_{10} & \Xi_{11} & \Xi_{12} \\ \Xi_3 & \Xi_4 & \Xi_2 \\ \Xi_6 & \Xi_7 & \Xi_8 \end{pmatrix}; \det(A - \lambda \cdot I) = 0$$

$$\Rightarrow \det \begin{pmatrix} \Xi_{10} - \lambda & \Xi_{11} & \Xi_{12} \\ \Xi_3 & \Xi_4 - \lambda & \Xi_2 \\ \Xi_6 & \Xi_7 & \Xi_8 - \lambda \end{pmatrix} = 0$$

$$\det \begin{pmatrix} \Xi_{10} - \lambda & \Xi_{11} & \Xi_{12} \\ \Xi_3 & \Xi_4 - \lambda & \Xi_2 \\ \Xi_6 & \Xi_7 & \Xi_8 - \lambda \end{pmatrix} = (\Xi_{10} - \lambda) \cdot \det \begin{pmatrix} \Xi_4 - \lambda & \Xi_2 \\ \Xi_7 & \Xi_8 - \lambda \end{pmatrix} - \Xi_{11} \cdot \det \begin{pmatrix} \Xi_3 & \Xi_2 \\ \Xi_6 & \Xi_8 - \lambda \end{pmatrix} + \Xi_{12} \cdot \det \begin{pmatrix} \Xi_3 & \Xi_4 - \lambda \\ \Xi_6 & \Xi_7 \end{pmatrix}$$

$$\det(A - \lambda \cdot I) = (\Xi_{10} - \lambda) \cdot [(\Xi_4 - \lambda) \cdot (\Xi_8 - \lambda) - \Xi_7 \cdot \Xi_2] - \Xi_{11} \cdot [\Xi_3 \cdot (\Xi_8 - \lambda) - \Xi_6 \cdot \Xi_2] + \Xi_{12} \cdot [\Xi_3 \cdot \Xi_7 - \Xi_6 \cdot (\Xi_4 - \lambda)]$$

$$\det(A - \lambda \cdot I) = (\Xi_{10} - \lambda) \cdot [(\Xi_4 \cdot \Xi_8 - \Xi_7 \cdot \Xi_2) - (\Xi_4 + \Xi_8) \cdot \lambda + \lambda^2] - \Xi_{11} \cdot [(\Xi_3 \cdot \Xi_8 - \Xi_6 \cdot \Xi_2) - \Xi_3 \cdot \lambda] + \Xi_{12} \cdot [(\Xi_3 \cdot \Xi_7 - \Xi_6 \cdot \Xi_4) + \Xi_6 \cdot \lambda]$$

$$\det(A - \lambda \cdot I) = \Xi_{10} \cdot (\Xi_4 \cdot \Xi_8 - \Xi_7 \cdot \Xi_2) - \Xi_{10} \cdot (\Xi_4 + \Xi_8) \cdot \lambda + \Xi_{10} \cdot \lambda^2 - \lambda \cdot (\Xi_4 \cdot \Xi_8 - \Xi_7 \cdot \Xi_2) + (\Xi_4 + \Xi_8) \cdot \lambda^2 - \lambda^3 - \Xi_{11} \cdot (\Xi_3 \cdot \Xi_8 - \Xi_6 \cdot \Xi_2) + \Xi_{11} \cdot \Xi_3 \cdot \lambda + \Xi_{12} \cdot (\Xi_3 \cdot \Xi_7 - \Xi_6 \cdot \Xi_4) + \Xi_6 \cdot \Xi_{12} \cdot \lambda$$

$$\det(A - \lambda \cdot I) = -\lambda^3 + [\Xi_{10} + \Xi_4 + \Xi_8] \cdot \lambda^2 + [\Xi_{11} \cdot \Xi_3 + \Xi_6 \cdot \Xi_{12} - (\Xi_4 \cdot \Xi_8 - \Xi_7 \cdot \Xi_2) - \Xi_{10} \cdot (\Xi_4 + \Xi_8)] \cdot \lambda + \Xi_{10} \cdot (\Xi_4 \cdot \Xi_8 - \Xi_7 \cdot \Xi_2) - \Xi_{11} \cdot (\Xi_3 \cdot \Xi_8 - \Xi_6 \cdot \Xi_2) + \Xi_{12} \cdot (\Xi_3 \cdot \Xi_7 - \Xi_6 \cdot \Xi_4)$$

We define new parameters: $\Upsilon_3 = -1; \Upsilon_2 = \Xi_{10} + \Xi_4 + \Xi_8$

$$\Upsilon_1 = \Xi_{11} \cdot \Xi_3 + \Xi_6 \cdot \Xi_{12} - (\Xi_4 \cdot \Xi_8 - \Xi_7 \cdot \Xi_2) - \Xi_{10} \cdot (\Xi_4 + \Xi_8)$$
$$\Upsilon_0 = \Xi_{10} \cdot (\Xi_4 \cdot \Xi_8 - \Xi_7 \cdot \Xi_2) - \Xi_{11} \cdot (\Xi_3 \cdot \Xi_8 - \Xi_6 \cdot \Xi_2) + \Xi_{12} \cdot (\Xi_3 \cdot \Xi_7 - \Xi_6 \cdot \Xi_4)$$

$$\det(A - \lambda \cdot I) = \sum_{k=0}^{3} \Upsilon_k \cdot \lambda^k; \det(A - \lambda \cdot I) = 0 \Rightarrow \sum_{k=0}^{3} \Upsilon_k \cdot \lambda^k = 0$$

We get zeros of the polynomial $\sum_{k=0}^{3} \Upsilon_k \cdot \lambda^k = 0$ with real coefficients Υ_k. The characteristic equation is a polynomial of degree N = 3: $\sum_{k=0}^{3} \Upsilon_k \cdot \lambda^k = 0$. We suppose $\lambda_1, \lambda_2, \lambda_3$ all (known) eigenvalues of the linearized system: $\frac{dX(t)}{dt} = A \cdot X(t); X(t) = (I_{R_1}(t) I_{R_2}(t) I_{R_b}(t))$. If one eigenvalue λ_k ; $k = 1, 2, 3$ has a real part greater than zero $\lambda_k > 0$; $k = 1, 2, 3$, the perturbation will ultimately increase and the steady state is thus unstable. To determine if all the eigenvalue have a negative real part can be done by checking some conditions, known as the Routh-Hurwitz criteria. In our case we define N = 3 matrices as follow:

Coefficients Υ_k are real and fulfilling $\Upsilon_{k=3}$ lie in the left half plane, if and only if, all determinants: H_1, H_2, and H_3 (with $\Upsilon_m = 0$ for $m > 3$, then $\Upsilon_4 = 0, \Upsilon_5 = 0$) are positive. It is the condition for stability.

$$H_1 = (\Upsilon_1); H_2 = \begin{pmatrix} \Upsilon_1 & \Upsilon_0 \\ \Upsilon_3 & \Upsilon_2 \end{pmatrix}; H_3 = \begin{pmatrix} \Upsilon_1 & \Upsilon_0 & 0 \\ \Upsilon_3 & \Upsilon_2 & \Upsilon_1 \\ \Upsilon_5 = 0 & \Upsilon_4 = 0 & \Upsilon_3 \end{pmatrix}$$

Stability criteria:

$$\det(H_1) = \det(\Upsilon_1) > 0 \Rightarrow \Upsilon_1 > 0 \text{ ; } \det(H_2) = \det\begin{pmatrix} \Upsilon_1 & \Upsilon_0 \\ \Upsilon_3 & \Upsilon_2 \end{pmatrix} > 0$$
$$\Rightarrow \Upsilon_1 \cdot \Upsilon_2 - \Upsilon_3 \cdot \Upsilon_0 > 0 \Rightarrow \Upsilon_1 \cdot \Upsilon_2 > \Upsilon_3 \cdot \Upsilon_0$$

$$\det(H_3) = \det\begin{pmatrix} \Upsilon_1 & \Upsilon_0 & 0 \\ \Upsilon_3 & \Upsilon_2 & \Upsilon_1 \\ \Upsilon_5 = 0 & \Upsilon_4 = 0 & \Upsilon_3 \end{pmatrix} > 0 \Rightarrow \Upsilon_1 \cdot \Upsilon_2 \cdot \Upsilon_3 - \Upsilon_0 \cdot \Upsilon_3^2$$
$$= \prod_{k=1}^{3} \Upsilon_k - \Upsilon_0 \cdot \Upsilon_3^2 > 0$$
$$\prod_{k=1}^{3} \Upsilon_k - \Upsilon_0 \cdot \Upsilon_3^2 = \Upsilon_3 \cdot (\Upsilon_1 \cdot \Upsilon_2 - \Upsilon_0 \cdot \Upsilon_3) > 0$$

We have two cases for $\det(H_3) > 0$: Case (I) $\Upsilon_3 > 0 \,\&\, \Upsilon_1 \cdot \Upsilon_2 > \Upsilon_0 \cdot \Upsilon_3$ Case (II) $\Upsilon_3 < 0$ and $\Upsilon_1 \cdot \Upsilon_2 < \Upsilon_0 \cdot \Upsilon_3$.

We are interesting to get oscillations from the phase shift resonator circuit. To get oscillations, we need eigenvalue of A is on imaginary axis, and then the method of the center manifold is applied. The center manifold of an equilibrium point (fixed point) of our dynamical system consists of orbits whose behavior around the equilibrium point is not controlled by either the attraction of the stable manifold or the repulsion of the unstable manifold. Phase shift resonator eigenvalues corresponding to eigenvalues with negative real part form the stable Eigen space, which gives rise to the stable manifold. Eigenvalue with positive real part yield the unstable manifold. If there are eigenvalues whose real part is zero, then these give rise to the center manifold. If the eigenvalues are precisely zero, then these more specifically give rise to a slow manifold [3, 4].

The system matrix A defines three main subspaces:

(I) Stable subspace, which is spanned by our system generalized eigenvectors corresponding to the eigenvalues λ with Re $\lambda < 0$.
(II) Unstable subspace, which is spanned by our system generalized eigenvectors corresponding to the eigenvalues λ with Re $\lambda > 0$.
(III) Center subspace, which is spanned by our generalized eigenvectors corresponding to the eigenvalues λ with Re $\lambda = 0$.

The options for sub spaces of interest include center-stable, center-unstable, sub-center, slow, and fast subspaces. These subspaces are all invariant subspaces of the system's linear equations.

Center manifold theorem: The neighborhood may be chosen so that all solutions of the system staying in the neighborhood tend exponentially to some solution on the center manifold. A wide variety of initial conditions yields to solutions of the full system which decay exponentially quickly to a solution on the relatively low dimensional center manifold.

6.2 Closed Loop Functioning Oscillator Stability Analysis Under Parameters Variations

A good oscillator is stable in that its frequency and amplitude of oscillation do not vary appreciably with temperature, process, power supply and external disturbances. The amplitude of oscillation is particularly stable, always returning to the same value. LC tank oscillator is not a good oscillator. Due to loss, no matter how small, the amplitude of the oscillator decays. Many oscillators can be viewed as feedback systems. The oscillation is sustained by feeding back a fraction of the output signal, using an amplifier to gain the signal, and then injecting the energy back into the tank. The transistor "pushes" the LC tank with just about enough energy to compensate for the loss. Typical oscillator feedback system is describes in the below figure (Fig. 6.3).

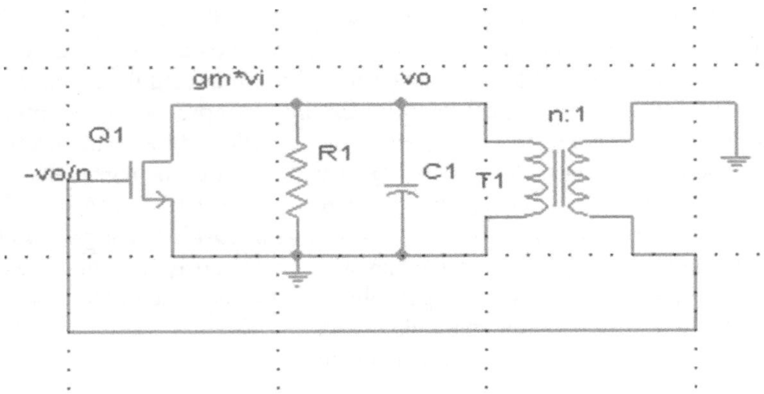

Fig. 6.3 Typical closed loop oscillator feedback system

Fig. 6.4 LC tank circuit and active device as a negative resistance generator

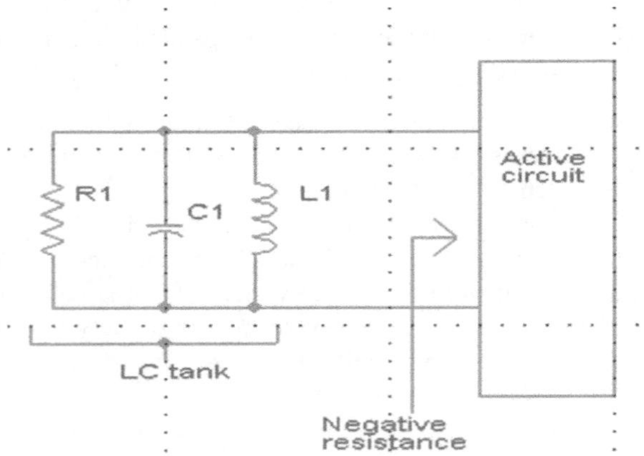

Another option is to view the active device as a negative resistance generator. In steady state, the losses in the tank due to conductance G are balanced by the power drawn from the active device through the negative conductance—G (Fig. 6.4).

In a real oscillator, the amplitude of oscillation initially grows exponentially as linear system theory predicts. The oscillator amplitude is initially very small. But as the oscillations become more vigorous, the non-linearity of the system comes into play. We will analyze by using nonlinear dynamic the steady-state behavior, where the system is non-linear but periodically time-varying. Typical circuit is BJT NPN transistor LC oscillator. The base of the NPN transistor (Q_1) is conveniently biased through the transformer windings. The transistor Q_1's emitter resistor is bypassed by a large capacitor at AC frequencies. The LC oscillator uses a transformer for feedback. Since the amplifier has a phase shift of 180°, the feedback transformer needs to provide an additional phase shift of 180° to provide positive feedback [33, 34]. (Fig. 6.5).

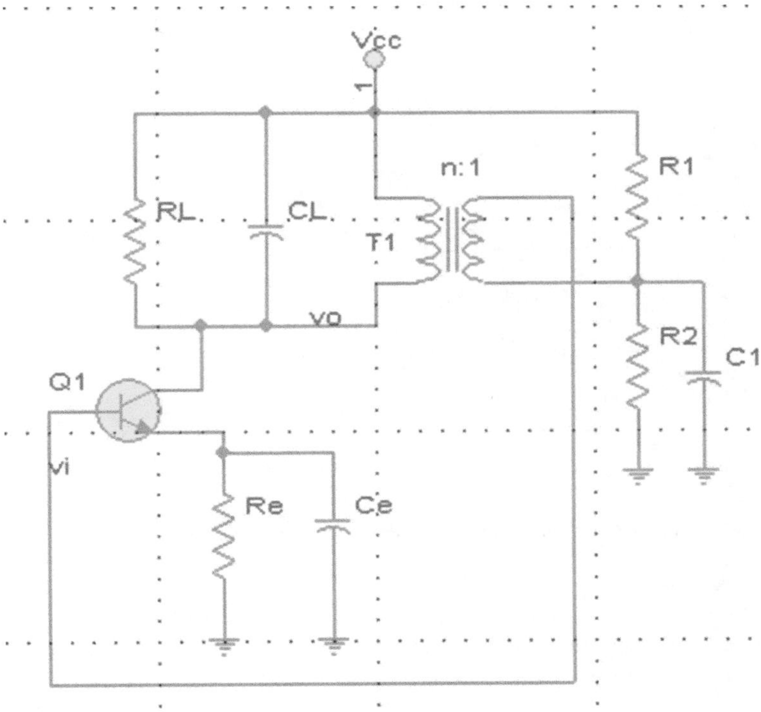

Fig. 6.5 Closed loop functioning oscillator with transformer T_1

We use for our stability analysis the BJT NPN transistor Small Signal (SS) model equivalent circuit (Hybrid Pi model Version A). We consider "AC ground" in the circuit. Since the voltage at this terminal is held constant at V_{cc}, there is no time variation of the voltage. Consequently, we can set this terminal to be an "AC ground" in the small signal circuit. For AC ground, we "kill" the DC sources at that terminal (short circuit voltage sources and open circuit current sources).

Transformer T_1: v_p—transformer primary voltage, v_s—transformer secondary voltage. ϕ is the flux through a one turn coil located anywhere on the transformer core. n_p and n_s are the number of turns of the coil 1 (primary) and 2 (secondary), respectively, then the total flux ϕ_p and ϕ_s through coils 1 and 2 respectively are $\phi_p = n_p \cdot \phi; \phi_s = n_s \cdot \phi; v_p = \frac{d\phi_p}{dt}; v_s = \frac{d\phi_s}{dt}; v_p = n_p \cdot \frac{d\phi}{dt}; \; v_s = n_s \cdot \frac{d\phi}{dt}; v_p(t) = v_p; v_s(t) = v_s; \frac{v_p(t)}{v_s(t)} = \frac{n_p}{n_s}; a = \frac{n_p}{n_s} \cdot \frac{v_p(t)}{v_s(t)} = \frac{n_p}{n_s}$ for all times t and for all voltages v_p and v_s. We define the ratio between n_p and n_s as "a" $(a = \frac{n_p}{n_s})$. Step down transformer a > 1 and step up transformer a < 1. By law of conservation of energy, apparent real and reactive powers are each conserved in the input and output $(S = I_p \cdot v_p = I_s \cdot v_s)$. Ideal transformer identity: $\frac{v_p}{v_s} = \frac{I_s}{I_p} = \frac{n_p}{n_s} = \sqrt{\frac{L_p}{L_s}} = a$. $n_p = n$; $n_s = 1; \frac{v_p}{v_s} = n; n > 1 \Rightarrow v_s = \frac{v_p}{n} = \frac{-v_0}{n} = \frac{-v_{A_2}}{n}$.

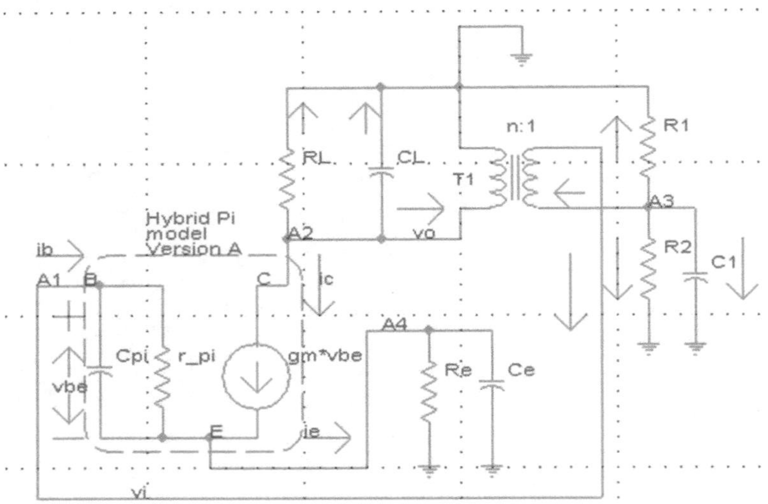

Fig. 6.6 Closed loop functioning oscillator with transformer T_1 and Hybrid Pi model (version A)

$v_p = -v_0 = -v_{A_2}; v_s = v_i - v_{A_3}; v_i - v_{A_3} = \frac{-v_0}{n} = \frac{-v_{A_2}}{n}$ (Reference directions are in the top of each transformer's coil). $n_p \cdot i_p + n_s \cdot i_s = 0; i_p = i_p(t); i_s = i_s(t);$ $\frac{i_p(t)}{i_s(t)} = -\frac{n_s}{n_p}$.

For all t and all currents i_p and i_s. The voltage v_p across coil 1 does not depend on i_p or on i_s; it depends only on v_s. Similarly the current i_p depends only on i_s and is independent of v_p, v_s (Fig. 6.6).

Remark: When we have two coils of wire in close physical proximity to one another, no importance whether or not the coils are wrapped around a common core of magnetic material. We assume that the coils do not move with respect to one another or with respect to a core they might be wrapped around. If we have some ferromagnetic material in the magnetic circuit of the two coils, then when the current sufficiently large, the relation between the fluxes ϕ_p, ϕ_s and the currents i_p, i_s are no longer linear. In this case the equations have the following form: $\phi_p = f_p(i_p, i_s)$; $\phi_s = f_s(i_p, i_s)$, where $f_p(i_p, i_s), f_s(i_p, i_s)$ are nonlinear functions of the currents i_p, i_s, By Faraday's law we get the following:

$$v_p = \frac{d\phi_p}{dt} = \frac{\partial f_p}{\partial i_p} \cdot \frac{di_p}{dt} + \frac{\partial f_p}{\partial i_s} \cdot \frac{di_s}{dt}; f_p = f_p(i_p, i_s); f_s = f_s(i_p, i_s)$$

$$v_s = \frac{d\phi_s}{dt} = \frac{\partial f_s}{\partial i_p} \cdot \frac{di_p}{dt} + \frac{\partial f_s}{\partial i_s} \cdot \frac{di_s}{dt}; f_p = f_p(i_p, i_s); f_s = f_s(i_p, i_s)$$

The four partial derivatives $(\frac{\partial f_p}{\partial i_p}, \frac{\partial f_p}{\partial i_s}, \frac{\partial f_s}{\partial i_p}, \frac{\partial f_s}{\partial i_s})$ are function of i_p, i_s (Fig. 6.7).

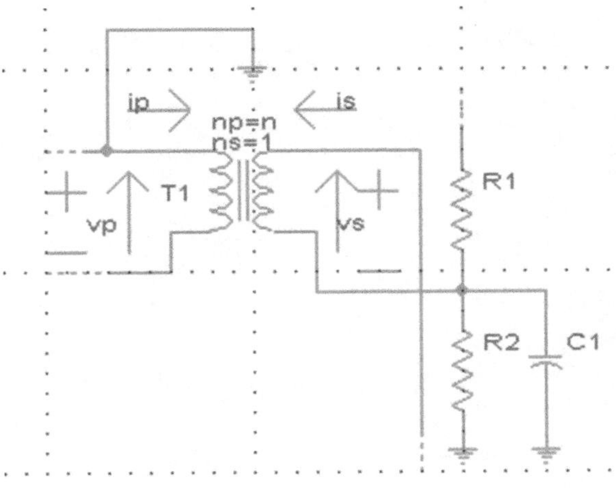

Fig. 6.7 Closed loop functioning oscillator's T_1 transformer circuit

$$I_{C_{pi}} = C_{pi} \cdot \frac{d(V_{A_1} - V_{A_4})}{dt}; I_{r\_pi} = \frac{V_{A_1} - V_{A_4}}{r\_pi}; I_{R_L} = \frac{V_{A_2}}{R_L}; I_{C_L} = C_L \cdot \frac{dV_{A_2}}{dt};$$

$$v_{be} = V_{A_1} - V_{A_4}$$

$$I_{R_e} = \frac{V_{A_4}}{R_e}; I_{C_e} = C_e \cdot \frac{dV_{A_4}}{dt}; I_{R_1} = \frac{V_{A_3}}{R_1}; I_{R_2} = \frac{V_{A_3}}{R_2}; I_{C_1} = C_1 \cdot \frac{dV_{A_3}}{dt}$$

KCL at nodes A_1, A_2, A_3, and A_4 gives the below results (Table 6.2):

$$n_p \cdot i_p + n_s \cdot i_s = 0; n_p = n; n_s = 1; n \cdot i_p + i_s = 0 \Rightarrow i_s = -n \cdot i_p; n > 1$$

$$I_{C_{pi}} = C_{pi} \cdot \frac{d(V_{A_1} - V_{A_4})}{dt}; I_{r\_pi} = \frac{V_{A_1} - V_{A_4}}{r\_pi} \Rightarrow V_{A_1} - V_{A_4} = I_{r\_pi} \cdot r\_pi;$$

$$I_{C_{pi}} = C_{pi} \cdot \frac{d(I_{r\_pi} \cdot r\_pi)}{dt}$$

$$I_{C_{pi}} = C_{pi} \cdot r\_pi \cdot \frac{dI_{r\_pi}}{dt}; I_{R_L} = \frac{V_{A_2}}{R_L} \Rightarrow V_{A_2} = I_{R_L} \cdot R_L; I_{C_L} = C_L \cdot \frac{d(I_{R_L} \cdot R_L)}{dt}$$

Table 6.2 Closed loop Hybrid Pi model microwave RF oscillator KCL@ nodes and expressions

KCL @ node	Expression
A_1	$i_s + I_{C_{pi}} + I_{r\_pi} = 0$
A_2	$g_m \cdot v_{be} + I_{R_L} + I_{C_L} - i_p = 0$
A_3	$I_{R_2} + I_{C_1} + I_{R_1} - i_s = 0$
A_4	$g_m \cdot v_{be} + I_{r\_pi} + I_{C_{pi}} = I_{R_e} + I_{C_e}$

$$I_{C_L} = C_L \cdot R_L \cdot \frac{dI_{R_L}}{dt}; I_{R_e} = \frac{V_{A_4}}{R_e} \Rightarrow V_{A_4} = I_{R_e} \cdot R_e; I_{C_e} = C_e \cdot \frac{dV_{A_4}}{dt} = \frac{d(I_{R_e} \cdot R_e)}{dt}$$

$$I_{C_e} = C_e \cdot R_e \cdot \frac{dI_{R_e}}{dt}; I_{R_1} = \frac{V_{A_3}}{R_1} \Rightarrow V_{A_3} = I_{R_1} \cdot R_1; I_{R_2} = \frac{V_{A_3}}{R_2} \Rightarrow V_{A_3} = I_{R_2} \cdot R_2$$

$$I_{C_1} = C_1 \cdot \frac{dV_{A_3}}{dt} = C_1 \cdot \frac{d(I_{R_1} \cdot R_1)}{dt} = C_1 \cdot \frac{d(I_{R_2} \cdot R_2)}{dt}; I_{C_1} = C_1 \cdot R_1 \cdot \frac{dI_{R_1}}{dt};$$
$$I_{C_1} = C_1 \cdot R_2 \cdot \frac{dI_{R_2}}{dt}$$

$$\frac{v_p}{v_s} = \frac{n_p}{n_s} = n \Rightarrow v_p = v_s \cdot n; v_s = v_i - V_{A_3} = V_{A_1} - V_{A_3}; v_p = -V_{A_2}; -V_{A_2} = (V_{A_1} - V_{A_3}) \cdot n$$

$$-I_{R_L} \cdot R_L = (V_{A_1} - I_{R_1} \cdot R_1) \cdot n; -I_{R_L} \cdot R_L = (V_{A_1} - I_{R_2} \cdot R_2) \cdot n; I_{r\_pi} = \frac{V_{A_1} - I_{R_e} \cdot R_e}{r\_pi}$$

$$I_{r\_pi} = \frac{V_{A_1} - I_{R_e} \cdot R_e}{r\_pi} \Rightarrow V_{A_1} = I_{r\_pi} \cdot r\_pi + I_{R_e} \cdot R_e; -I_{R_L} \cdot R_L = (I_{r\_pi} \cdot r\_pi + I_{R_e} \cdot R_e - I_{R_2} \cdot R_2) \cdot n$$

$$I_{R_2} = I_{r\_pi} \cdot \frac{r\_pi}{R_2} + I_{R_e} \cdot \frac{R_e}{R_2} + I_{R_L} \cdot \frac{R_L}{n \cdot R_2}; i_s = -n \cdot i_p; n > 1$$

KCL @ node 1: $i_s + I_{C_{pi}} + I_{r\_pi} = 0$

$$-n \cdot i_p + C_{pi} \cdot r\_pi \cdot \frac{dI_{r\_pi}}{dt} + I_{r\_pi} = 0 \Rightarrow \frac{dI_{r\_pi}}{dt} = \frac{n}{C_{pi} \cdot r\_pi} \cdot i_p - \frac{1}{C_{pi} \cdot r\_pi} \cdot I_{r\_pi}$$

KCL @ node 2: $g_m \cdot v_{be} + I_{R_L} + I_{C_L} - i_p = 0; v_{be} = V_{A_1} - V_{A_4}; v_{be} = I_{r\_pi} \cdot r\_pi$

$$g_m \cdot I_{r\_pi} \cdot r\_pi + I_{R_L} + C_L \cdot R_L \cdot \frac{dI_{R_L}}{dt} - i_p = 0 \Rightarrow \frac{dI_{R_L}}{dt} = -\frac{g_m \cdot r\_pi}{C_L \cdot R_L} \cdot I_{r\_pi} - \frac{1}{C_L \cdot R_L} \cdot I_{R_L} + \frac{1}{C_L \cdot R_L} \cdot i_p$$

KCL @ node 3: $I_{R_2} + I_{C_1} + I_{R_1} - i_s = 0$

$$I_{R_1} + I_{r\_pi} \cdot \frac{r\_pi}{R_2} + I_{R_e} \cdot \frac{R_e}{R_2} + I_{R_L} \cdot \frac{R_L}{n \cdot R_2} + C_1 \cdot R_1 \cdot \frac{dI_{R_1}}{dt} + n \cdot i_p = 0$$

$$C_1 \cdot R_1 \cdot \frac{dI_{R_1}}{dt} = -I_{R_1} - I_{r\_pi} \cdot \frac{r\_pi}{R_2} - I_{R_e} \cdot \frac{R_e}{R_2} - I_{R_L} \cdot \frac{R_L}{n \cdot R_2} - n \cdot i_p$$

$$\frac{dI_{R_1}}{dt} = -\frac{1}{C_1 \cdot R_1} \cdot I_{R_1} - I_{r\_pi} \cdot \frac{r\_pi}{C_1 \cdot \prod\limits_{k=1}^{2} R_k} - I_{R_e} \cdot \frac{R_e}{C_1 \cdot \prod\limits_{k=1}^{2} R_k} - I_{R_L} \cdot \frac{R_L}{n \cdot C_1 \cdot \prod\limits_{k=1}^{2} R_k}$$
$$- \frac{n}{C_1 \cdot R_1} \cdot i_p$$

KCL @ node 4: $g_m \cdot v_{be} + I_{r\_pi} + I_{C_{pi}} = I_{R_e} + I_{C_e}$

$$g_m \cdot I_{r\_pi} \cdot r\_pi + I_{r\_pi} + C_{pi} \cdot r\_pi \cdot \frac{dI_{r\_pi}}{dt} = I_{R_e} + C_e \cdot R_e \cdot \frac{dI_{R_e}}{dt}$$

$$g_m \cdot I_{r\_pi} \cdot r\_pi + I_{r\_pi} + C_{pi} \cdot r\_pi \cdot \left[\frac{n}{C_{pi} \cdot r\_pi} \cdot i_p - \frac{1}{C_{pi} \cdot r\_pi} \cdot I_{r\_pi} \right]$$
$$= I_{R_e} + C_e \cdot R_e \cdot \frac{dI_{R_e}}{dt}$$

$$\frac{dI_{R_e}}{dt} = \frac{g_m \cdot r\_pi}{C_e \cdot R_e} \cdot I_{r\_pi} + \frac{n}{C_e \cdot R_e} \cdot i_p - \frac{1}{C_e \cdot R_e} \cdot I_{R_e}$$

We can summery our system differential equations:

$$\frac{dI_{r\_pi}}{dt} = \frac{n}{C_{pi} \cdot r\_pi} \cdot i_p - \frac{1}{C_{pi} \cdot r\_pi} \cdot I_{r\_pi}$$

$$\frac{dI_{R_L}}{dt} = -\frac{g_m \cdot r\_pi}{C_L \cdot R_L} \cdot I_{r\_pi} - \frac{1}{C_L \cdot R_L} \cdot I_{R_L} + \frac{1}{C_L \cdot R_L} \cdot i_p$$

$$\frac{dI_{R_1}}{dt} = -\frac{1}{C_1 \cdot R_1} \cdot I_{R_1} - I_{r\_pi} \cdot \frac{r\_pi}{C_1 \cdot \prod\limits_{k=1}^{2} R_k} - I_{R_e} \cdot \frac{R_e}{C_1 \cdot \prod\limits_{k=1}^{2} R_k} - I_{R_L} \cdot \frac{R_L}{n \cdot C_1 \cdot \prod\limits_{k=1}^{2} R_k}$$
$$- \frac{n}{C_1 \cdot R_1} \cdot i_p$$

$$\frac{dI_{R_e}}{dt} = \frac{g_m \cdot r\_pi}{C_e \cdot R_e} \cdot I_{r\_pi} + \frac{n}{C_e \cdot R_e} \cdot i_p - \frac{1}{C_e \cdot R_e} \cdot I_{R_e}$$

At fixed points (equilibrium points): $\frac{dI_{r\_pi}}{dt} = 0; \frac{dI_{R_L}}{dt} = 0; \frac{dI_{R_1}}{dt} = 0; \frac{dI_{R_e}}{dt} = 0$

$$I_{r\_pi}^* = n \cdot i_p^*; -\frac{g_m \cdot r\_pi}{C_L \cdot R_L} \cdot I_{r\_pi}^* - \frac{1}{C_L \cdot R_L} \cdot I_{R_L}^* + \frac{1}{C_L \cdot R_L} \cdot i_p^* = 0$$

$$I_{R_L}^* = i_p^* \cdot (1 - g_m \cdot r\_pi \cdot n); -I_{R_1}^* - I_{r\_pi}^* \cdot \frac{r\_pi}{R_2} - I_{R_e}^* \cdot \frac{R_e}{R_2} - I_{R_L}^* \cdot \frac{R_L}{n \cdot R_2} - n \cdot i_p^* = 0$$

$$I_{R_1}^* + I_{R_e}^* \cdot \frac{R_e}{R_2} = -i_p^* \cdot [n \cdot \frac{r\_pi}{R_2} + (1 - g_m \cdot r\_pi \cdot n) \cdot \frac{R_L}{n \cdot R_2} + n]$$

$$\frac{g_m \cdot r\_pi}{C_e \cdot R_e} \cdot I_{r\_pi}^* + \frac{n}{C_e \cdot R_e} \cdot i_p^* - \frac{1}{C_e \cdot R_e} \cdot I_{R_e}^* = 0$$

$$\frac{g_m \cdot r\_pi}{C_e \cdot R_e} \cdot n \cdot i_p^* + \frac{n}{C_e \cdot R_e} \cdot i_p^* - \frac{1}{C_e \cdot R_e} \cdot I_{R_e}^* = 0 \Rightarrow I_{R_e}^* = i_p^* \cdot n \cdot [g_m \cdot r\_pi + 1]$$

$$I_{R_1}^* + I_{R_e}^* \cdot \frac{R_e}{R_2} = -i_p^* \cdot [n \cdot \frac{r\_pi}{R_2} + (1 - g_m \cdot r\_pi \cdot n) \cdot \frac{R_L}{n \cdot R_2} + n]$$

$$I_{R_1}^* = -i_p^* \cdot \{n \cdot \frac{r\_pi}{R_2} + (1 - g_m \cdot r\_pi \cdot n) \cdot \frac{R_L}{n \cdot R_2} + n + n \cdot [g_m \cdot r\_pi + 1] \cdot \frac{R_e}{R_2}\}$$

We can summery our system fixed points: $E(i_p^*, I_{r\_pi}^*, I_{R_L}^*, I_{R_1}^*, I_{R_e}^*)$

$$\begin{aligned} E(i_p^*, I_{r\_pi}^*, I_{R_L}^*, I_{R_1}^*, I_{R_e}^*) = \{ & i_p^*, n \cdot i_p^*, i_p^* \cdot (1 - g_m \cdot r\_pi \cdot n), \\ & - i_p^* \cdot \{n \cdot \frac{r\_pi}{R_2} + (1 - g_m \cdot r\_pi \cdot n) \cdot \frac{R_L}{n \cdot R_2} + n + n \cdot [g_m \cdot r\_pi + 1] \cdot \frac{R_e}{R_2}\}, \\ & i_p^* \cdot n \cdot [g_m \cdot r\_pi + 1]\} \end{aligned}$$

$$\begin{aligned} E(i_p^*, I_{r\_pi}^*, I_{R_L}^*, I_{R_1}^*, I_{R_e}^*) = i_p^* \cdot \{ & 1, n, (1 - g_m \cdot r\_pi \cdot n), \\ & - \{n \cdot \frac{r\_pi}{R_2} + (1 - g_m \cdot r\_pi \cdot n) \cdot \frac{R_L}{n \cdot R_2} + n + n \cdot [g_m \cdot r\_pi + 1] \cdot \frac{R_e}{R_2}\}, \\ & n \cdot [g_m \cdot r\_pi + 1]\} \end{aligned}$$

Stability analysis: The standard local stability analysis about any one of the equilibrium points of the closed loop functioning oscillator circuit (BJT transistor small signal model) consists in adding to coordinate $[i_p, I_{r\_pi}, I_{R_L}, I_{R_1}, I_{R_e}]$ arbitrarily small increments of exponentially form $[i'_p, i_{r\_pi}, i_{R_L}, i_{R_1}, i_{R_e}] \cdot e^{\lambda \cdot t}$ and retaining the first order terms in $i_p, I_{r\_pi}, I_{R_L}, I_{R_1}, I_{R_e}$. The system of homogenous equations leads to a polynomial characteristic equation in the eigenvalues [4]. The polynomial characteristic equations accept by set of the below circuit variables, circuit variables derivative and circuit variables second order derivative with respect to time into closed loop functioning oscillator circuit [2, 3, 4]. Our closed loop functioning oscillator circuit fixed values with arbitrarily small increments of exponential form $[i'_p, i_{r\_pi}, i_{R_L}, i_{R_1}, i_{R_e}] \cdot e^{\lambda \cdot t}$ are: j = 0 (first fixed point), j = 1 (second fixed point), j = 2(third fixed point), etc.,

$$i_p(t) = i_p^{(j)} + i'_p \cdot e^{\lambda \cdot t}; I_{r\_pi}(t) = I_{r\_pi}^{(j)} + i_{r\_pi} \cdot e^{\lambda \cdot t}; I_{R_L}(t) = I_{R_L}^{(j)} + i_{R_L} \cdot e^{\lambda \cdot t}; I_{R_1}(t) = I_{R_1}^{(j)} + i_{R_1} \cdot e^{\lambda \cdot t}$$

$$I_{R_e}(t) = I_{R_e}^{(j)} + i_{R_e} \cdot e^{\lambda \cdot t}; \frac{dI_{r\_pi}(t)}{dt} = i_{r\_pi} \cdot \lambda \cdot e^{\lambda \cdot t}; \frac{dI_{R_L}(t)}{dt} = i_{R_L} \cdot \lambda \cdot e^{\lambda \cdot t}; \frac{dI_{R_1}(t)}{dt} = i_{R_1} \cdot \lambda \cdot e^{\lambda \cdot t}$$

$$\frac{dI_{R_e}(t)}{dt} = i_{R_e} \cdot \lambda \cdot e^{\lambda \cdot t}$$

$$\frac{dI_{r\_pi}}{dt} = \frac{n}{C_{pi} \cdot r\_pi} \cdot i_p - \frac{1}{C_{pi} \cdot r\_pi} \cdot I_{r\_pi}$$

$$i_{r\_pi} \cdot \lambda \cdot e^{\lambda \cdot t} = \frac{n}{C_{pi} \cdot r\_pi} \cdot [i_p^{(j)} + i'_p \cdot e^{\lambda \cdot t}] - \frac{1}{C_{pi} \cdot r\_pi} \cdot [I_{r\_pi}^{(j)} + i_{r\_pi} \cdot e^{\lambda \cdot t}]$$

$$i_{r\_pi} \cdot \lambda \cdot e^{\lambda \cdot t} = \frac{n}{C_{pi} \cdot r\_pi} \cdot i_p^{(j)} - \frac{1}{C_{pi} \cdot r\_pi} \cdot I_{r\_pi}^{(j)} + \frac{n}{C_{pi} \cdot r\_pi} \cdot i'_p \cdot e^{\lambda \cdot t} - \frac{1}{C_{pi} \cdot r\_pi} \cdot i_{r\_pi} \cdot e^{\lambda \cdot t}$$

At fixed point $\frac{n}{C_{pi} \cdot r\_pi} \cdot i_p^{(j)} - \frac{1}{C_{pi} \cdot r\_pi} \cdot I_{r\_pi}^{(j)} = 0$

$$\frac{n}{C_{pi} \cdot r\_pi} \cdot i'_p - \frac{1}{C_{pi} \cdot r\_pi} \cdot i_{r\_pi} - i_{r\_pi} \cdot \lambda = 0$$

$$\frac{dI_{R_L}}{dt} = -\frac{g_m \cdot r\_pi}{C_L \cdot R_L} \cdot I_{r\_pi} - \frac{1}{C_L \cdot R_L} \cdot I_{R_L} + \frac{1}{C_L \cdot R_L} \cdot i_p$$

$$i_{R_L} \cdot \lambda \cdot e^{\lambda \cdot t} = -\frac{g_m \cdot r\_pi}{C_L \cdot R_L} \cdot [I_{r\_pi}^{(j)} + i_{r\_pi} \cdot e^{\lambda \cdot t}] - \frac{1}{C_L \cdot R_L} \cdot [I_{R_L}^{(j)} + i_{R_L} \cdot e^{\lambda \cdot t}] + \frac{1}{C_L \cdot R_L} \cdot [i_p^{(j)} + i'_p \cdot e^{\lambda \cdot t}]$$

$$i_{R_L} \cdot \lambda \cdot e^{\lambda \cdot t} = -\frac{g_m \cdot r\_pi}{C_L \cdot R_L} \cdot I_{r\_pi}^{(j)} - \frac{1}{C_L \cdot R_L} \cdot I_{R_L}^{(j)} + \frac{1}{C_L \cdot R_L} \cdot i_p^{(j)} - \frac{g_m \cdot r\_pi}{C_L \cdot R_L} \cdot i_{r\_pi} \cdot e^{\lambda \cdot t} - \frac{1}{C_L \cdot R_L} \cdot i_{R_L} \cdot e^{\lambda \cdot t} + \frac{1}{C_L \cdot R_L} \cdot i'_p \cdot e^{\lambda \cdot t}$$

At fixed point $-\frac{g_m \cdot r\_pi}{C_L \cdot R_L} \cdot I_{r\_pi}^{(j)} - \frac{1}{C_L \cdot R_L} \cdot I_{R_L}^{(j)} + \frac{1}{C_L \cdot R_L} \cdot i_p^{(j)} = 0$

$$-\frac{g_m \cdot r\_pi}{C_L \cdot R_L} \cdot i_{r\_pi} + \frac{1}{C_L \cdot R_L} \cdot i'_p - \frac{1}{C_L \cdot R_L} \cdot i_{R_L} - i_{R_L} \cdot \lambda = 0$$

$$\frac{dI_{R_1}}{dt} = -\frac{1}{C_1 \cdot R_1} \cdot I_{R_1} - I_{r\_pi} \cdot \frac{r\_pi}{C_1 \cdot \prod_{k=1}^{2} R_k} - I_{R_e} \cdot \frac{R_e}{C_1 \cdot \prod_{k=1}^{2} R_k} - I_{R_L} \cdot \frac{R_L}{n \cdot C_1 \cdot \prod_{k=1}^{2} R_k} - \frac{n}{C_1 \cdot R_1} \cdot i_p$$

$$i_{R_1} \cdot \lambda \cdot e^{\lambda \cdot t} = -\frac{1}{C_1 \cdot R_1} \cdot [I_{R_1}^{(j)} + i_{R_1} \cdot e^{\lambda \cdot t}] - [I_{r\_pi}^{(j)} + i_{r\_pi} \cdot e^{\lambda \cdot t}] \cdot \frac{r\_pi}{C_1 \cdot \prod_{k=1}^{2} R_k} - [I_{R_e}^{(j)} + i_{R_e} \cdot e^{\lambda \cdot t}] \cdot \frac{R_e}{C_1 \cdot \prod_{k=1}^{2} R_k} - [I_{R_L}^{(j)} + i_{R_L} \cdot e^{\lambda \cdot t}] \cdot \frac{R_L}{n \cdot C_1 \cdot \prod_{k=1}^{2} R_k} - \frac{n}{C_1 \cdot R_1} \cdot [i_p^{(j)} + i'_p \cdot e^{\lambda \cdot t}]$$

$$i_{R_1} \cdot \lambda \cdot e^{\lambda \cdot t} = -\frac{1}{C_1 \cdot R_1} \cdot I_{R_1}^{(j)} - \frac{1}{C_1 \cdot R_1} \cdot i_{R_1} \cdot e^{\lambda \cdot t} - I_{r\_pi}^{(j)} \cdot \frac{r\_pi}{C_1 \cdot \prod_{k=1}^{2} R_k} - i_{r\_pi} \cdot \frac{r\_pi}{C_1 \cdot \prod_{k=1}^{2} R_k} \cdot e^{\lambda \cdot t} - I_{R_e}^{(j)} \cdot \frac{R_e}{C_1 \cdot \prod_{k=1}^{2} R_k} - i_{R_e} \cdot \frac{R_e}{C_1 \cdot \prod_{k=1}^{2} R_k} \cdot e^{\lambda \cdot t} - I_{R_L}^{(j)} \cdot \frac{R_L}{n \cdot C_1 \cdot \prod_{k=1}^{2} R_k} - i_{R_L} \cdot \frac{R_L}{n \cdot C_1 \cdot \prod_{k=1}^{2} R_k} \cdot e^{\lambda \cdot t} - \frac{n}{C_1 \cdot R_1} \cdot i_p^{(j)} - \frac{n}{C_1 \cdot R_1} \cdot i'_p \cdot e^{\lambda \cdot t}$$

$$i_{R_1} \cdot \lambda \cdot e^{\lambda \cdot t} = -\frac{1}{C_1 \cdot R_1} \cdot I_{R_1}^{(j)} - I_{r\_pi}^{(j)} \cdot \frac{r\_pi}{C_1 \cdot \prod_{k=1}^{2} R_k} - I_{R_e}^{(j)} \cdot \frac{R_e}{C_1 \cdot \prod_{k=1}^{2} R_k} - I_{R_L}^{(j)} \cdot \frac{R_L}{n \cdot C_1 \cdot \prod_{k=1}^{2} R_k} - \frac{n}{C_1 \cdot R_1} \cdot i_p^{(j)} - \frac{1}{C_1 \cdot R_1} \cdot i_{R_1} \cdot e^{\lambda \cdot t} - i_{r\_pi} \cdot \frac{r\_pi}{C_1 \cdot \prod_{k=1}^{2} R_k} \cdot e^{\lambda \cdot t} - i_{R_e} \cdot \frac{R_e}{C_1 \cdot \prod_{k=1}^{2} R_k} \cdot e^{\lambda \cdot t} - i_{R_L} \cdot \frac{R_L}{n \cdot C_1 \cdot \prod_{k=1}^{2} R_k} \cdot e^{\lambda \cdot t} - \frac{n}{C_1 \cdot R_1} \cdot i'_p \cdot e^{\lambda \cdot t}$$

At fixed point

$$-\frac{1}{C_1 \cdot R_1} \cdot I_{R_1}^{(j)} - I_{r\_pi}^{(j)} \cdot \frac{r\_pi}{C_1 \cdot \prod_{k=1}^{2} R_k} - I_{R_e}^{(j)} \cdot \frac{R_e}{C_1 \cdot \prod_{k=1}^{2} R_k} - I_{R_L}^{(j)} \cdot \frac{R_L}{n \cdot C_1 \cdot \prod_{k=1}^{2} R_k} - \frac{n}{C_1 \cdot R_1} \cdot i_p^{(j)} = 0$$

$$-\frac{1}{C_1 \cdot R_1} \cdot i_{R_1} - i_{R_1} \cdot \lambda - i_{r\_pi} \cdot \frac{r\_pi}{C_1 \cdot \prod_{k=1}^{2} R_k} - i_{R_e} \cdot \frac{R_e}{C_1 \cdot \prod_{k=1}^{2} R_k} - i_{R_L} \cdot \frac{R_L}{n \cdot C_1 \cdot \prod_{k=1}^{2} R_k} - \frac{n}{C_1 \cdot R_1} \cdot i'_p = 0$$

$$\frac{dI_{R_e}}{dt} = \frac{g_m \cdot r\_pi}{C_e \cdot R_e} \cdot I_{r\_pi} + \frac{n}{C_e \cdot R_e} \cdot i_p - \frac{1}{C_e \cdot R_e} \cdot I_{R_e}$$

$$i_{R_e} \cdot \lambda \cdot e^{\lambda \cdot t} = \frac{g_m \cdot r\_pi}{C_e \cdot R_e} \cdot [I_{r\_pi}^{(j)} + i_{r\_pi} \cdot e^{\lambda \cdot t}] + \frac{n}{C_e \cdot R_e} \cdot [i_p^{(j)} + i'_p \cdot e^{\lambda \cdot t}] - \frac{1}{C_e \cdot R_e} \cdot [I_{R_e}^{(j)} + i_{R_e} \cdot e^{\lambda \cdot t}]$$

$$i_{R_e} \cdot \lambda \cdot e^{\lambda \cdot t} = \frac{g_m \cdot r\_pi}{C_e \cdot R_e} \cdot I_{r\_pi}^{(j)} + \frac{g_m \cdot r\_pi}{C_e \cdot R_e} \cdot i_{r\_pi} \cdot e^{\lambda \cdot t} + \frac{n}{C_e \cdot R_e} \cdot i_p^{(j)} + \frac{n}{C_e \cdot R_e} \cdot i'_p \cdot e^{\lambda \cdot t} - \frac{1}{C_e \cdot R_e} \cdot I_{R_e}^{(j)} - \frac{1}{C_e \cdot R_e} \cdot i_{R_e} \cdot e^{\lambda \cdot t}$$

$$i_{R_e} \cdot \lambda \cdot e^{\lambda \cdot t} = \frac{g_m \cdot r\_pi}{C_e \cdot R_e} \cdot I_{r\_pi}^{(j)} + \frac{n}{C_e \cdot R_e} \cdot i_p^{(j)} - \frac{1}{C_e \cdot R_e} \cdot I_{R_e}^{(j)} + \frac{g_m \cdot r\_pi}{C_e \cdot R_e} \cdot i_{r\_pi} \cdot e^{\lambda \cdot t} + \frac{n}{C_e \cdot R_e} \cdot i'_p \cdot e^{\lambda \cdot t} - \frac{1}{C_e \cdot R_e} \cdot i_{R_e} \cdot e^{\lambda \cdot t}$$

At fixed point $\frac{g_m \cdot r\_pi}{C_e \cdot R_e} \cdot I_{r\_pi}^{(j)} + \frac{n}{C_e \cdot R_e} \cdot i_p^{(j)} - \frac{1}{C_e \cdot R_e} \cdot I_{R_e}^{(j)} = 0$

$$\frac{g_m \cdot r\_pi}{C_e \cdot R_e} \cdot i_{r\_pi} + \frac{n}{C_e \cdot R_e} \cdot i'_p - \frac{1}{C_e \cdot R_e} \cdot i_{R_e} - i_{R_e} \cdot \lambda = 0$$

We can summery our arbitrarily small increments equations:

$$\frac{n}{C_{pi} \cdot r\_pi} \cdot i'_p - \frac{1}{C_{pi} \cdot r\_pi} \cdot i_{r\_pi} - i_{r\_pi} \cdot \lambda = 0$$

$$-\frac{g_m \cdot r\_pi}{C_L \cdot R_L} \cdot i_{r\_pi} + \frac{1}{C_L \cdot R_L} \cdot i'_p - \frac{1}{C_L \cdot R_L} \cdot i_{R_L} - i_{R_L} \cdot \lambda = 0$$

$$-\frac{1}{C_1 \cdot R_1} \cdot i_{R_1} - i_{R_1} \cdot \lambda - i_{r\_pi} \cdot \frac{r\_pi}{C_1 \cdot \prod_{k=1}^{2} R_k} - i_{R_e} \cdot \frac{R_e}{C_1 \cdot \prod_{k=1}^{2} R_k}$$
$$- i_{R_L} \cdot \frac{R_L}{n \cdot C_1 \cdot \prod_{k=1}^{2} R_k} - \frac{n}{C_1 \cdot R_1} \cdot i'_p = 0$$

$$\frac{g_m \cdot r\_pi}{C_e \cdot R_e} \cdot i_{r\_pi} + \frac{n}{C_e \cdot R_e} \cdot i'_p - \frac{1}{C_e \cdot R_e} \cdot i_{R_e} - i_{R_e} \cdot \lambda = 0$$

$$\begin{pmatrix} \Xi_{11} & \cdots & \Xi_{14} \\ \vdots & \ddots & \vdots \\ \Xi_{41} & \cdots & \Xi_{44} \end{pmatrix} \cdot \left(i_{r\_pi} i_{R_L} i_{R_1} i_{R_e} \right) + \begin{pmatrix} \frac{n}{C_{pi} \cdot r\_pi} \\ \frac{1}{C_L \cdot R_L} \\ -\frac{n}{C_1 \cdot R_1} \\ \frac{n}{C_e \cdot R_e} \end{pmatrix} \cdot i'_p = 0;$$

$$\begin{pmatrix} \Xi_{11} & \cdots & \Xi_{14} \\ \vdots & \ddots & \vdots \\ \Xi_{41} & \cdots & \Xi_{44} \end{pmatrix} \cdot \begin{pmatrix} i_{r\_pi} \\ i_{R_L} \\ i_{R_1} \\ i_{R_e} \end{pmatrix} \approx 0$$

$$\Xi_{11} = -\frac{1}{C_{pi} \cdot r\_pi} - \lambda; \Xi_{12} = \Xi_{13} = \Xi_{14} = 0; \Xi_{21} = -\frac{g_m \cdot r\_pi}{C_L \cdot R_L};$$
$$\Xi_{22} = -\frac{1}{C_L \cdot R_L} - \lambda$$

$$\Xi_{23} = \Xi_{24} = 0; \Xi_{31} = -\frac{r\_pi}{C_1 \cdot \prod_{k=1}^{2} R_k}; \Xi_{32} = -\frac{R_L}{n \cdot C_1 \cdot \prod_{k=1}^{2} R_k}; \Xi_{33} = -\frac{1}{C_1 \cdot R_1} - \lambda$$

$$\Xi_{34} = -\frac{R_e}{C_1 \cdot \prod_{k=1}^{2} R_k}; \Xi_{41} = \frac{g_m \cdot r\_pi}{C_e \cdot R_e}; \Xi_{42} = \Xi_{43} = 0; \Xi_{44} = -\frac{1}{C_e \cdot R_e} - \lambda$$

Assumption: Arbitrarily small increment i'_p is very small compare to other system arbitrarily small increments $(i_{r\_pi}, i_{R_L}, i_{R_1}, i_{R_e})$.

$$\begin{pmatrix} \frac{n}{C_{pi}\cdot r\_pi} \\ \frac{1}{C_L\cdot R_L} \\ -\frac{n}{C_1\cdot R_1} \\ \frac{n}{C_e\cdot R_e} \end{pmatrix} \cdot i'_p \to \varepsilon; (A - \lambda \cdot I) = \begin{pmatrix} \Xi_{11} & \cdots & \Xi_{14} \\ \vdots & \ddots & \vdots \\ \Xi_{41} & \cdots & \Xi_{44} \end{pmatrix};$$

$$\det(A - \lambda \cdot I) = \det\begin{pmatrix} \Xi_{11} & \cdots & \Xi_{14} \\ \vdots & \ddots & \vdots \\ \Xi_{41} & \cdots & \Xi_{44} \end{pmatrix}$$

$$\begin{aligned} \det(A - \lambda \cdot I) = {} & \lambda^4 + \lambda^3 \cdot [\frac{1}{C_1 \cdot R_1} + \frac{1}{C_e \cdot R_e} + \frac{1}{C_{pi} \cdot r\_pi} + \frac{1}{C_L \cdot R_L}] \\ & + \lambda^2 \cdot [\frac{1}{C_1 \cdot R_1 \cdot C_e \cdot R_e} + (\frac{1}{C_{pi} \cdot r\_pi} + \frac{1}{C_L \cdot R_L}) \cdot (\frac{1}{C_1 \cdot R_1} + \frac{1}{C_e \cdot R_e}) \\ & + \frac{1}{C_{pi} \cdot r\_pi \cdot C_L \cdot R_L}] + \lambda \cdot \{[\frac{1}{C_{pi} \cdot r\_pi} + \frac{1}{C_L \cdot R_L}] \cdot \frac{1}{C_1 \cdot R_1 \cdot C_e \cdot R_e} \\ & + [\frac{1}{C_1 \cdot R_1} + \frac{1}{C_e \cdot R_e}] \cdot \frac{1}{C_{pi} \cdot r\_pi \cdot C_L \cdot R_L}\} \\ & + \frac{1}{C_{pi} \cdot r\_pi \cdot C_L \cdot R_L \cdot C_1 \cdot R_1 \cdot C_e \cdot R_e} \end{aligned}$$

$$\Upsilon_4 = 1; \Upsilon_3 = \frac{1}{C_1 \cdot R_1} + \frac{1}{C_e \cdot R_e} + \frac{1}{C_{pi} \cdot r\_pi} + \frac{1}{C_L \cdot R_L}$$

$$\begin{aligned} \Upsilon_2 = {} & \frac{1}{C_1 \cdot R_1 \cdot C_e \cdot R_e} + (\frac{1}{C_{pi} \cdot r\_pi} + \frac{1}{C_L \cdot R_L}) \\ & \cdot (\frac{1}{C_1 \cdot R_1} + \frac{1}{C_e \cdot R_e}) + \frac{1}{C_{pi} \cdot r\_pi \cdot C_L \cdot R_L} \end{aligned}$$

$$\begin{aligned} \Upsilon_1 = {} & [\frac{1}{C_{pi} \cdot r\_pi} + \frac{1}{C_L \cdot R_L}] \cdot \frac{1}{C_1 \cdot R_1 \cdot C_e \cdot R_e} + [\frac{1}{C_1 \cdot R_1} + \frac{1}{C_e \cdot R_e}] \\ & \cdot \frac{1}{C_{pi} \cdot r\_pi \cdot C_L \cdot R_L} \end{aligned}$$

$$\Upsilon_0 = \frac{1}{C_{pi} \cdot r\_pi \cdot C_L \cdot R_L \cdot C_1 \cdot R_1 \cdot C_e \cdot R_e}; \det(A - \lambda \cdot I) = \sum_{k=0}^{4} \Upsilon_k \cdot \lambda^k$$

Eigenvalues stability discussion: Our closed loop functioning oscillator circuit (BJT transistor small signal model) involving N variables (N > 2, N = 5, arbitrarily small increments), the characteristic equation is of degree N = 4 (we exclude small increment i'_p) and must often be solved numerically. Expect in some particular cases, such an equation has (N = 4) distinct roots that can be real or complex. These values are the eigenvalues of the 4 × 4 Jacobian matrix (A).

The general rule is that the closed loop functioning oscillator circuit is stable if there is no eigenvalue with positive real part. It is sufficient that one eigenvalue is positive for the steady state to be unstable. Our 5-variables $(i_p, I_{r\_pi}, I_{R_L}, I_{R_1}, I_{R_e})$ system has four eigenvalues (reduce to four system arbitrarily small increments). The type of behavior can be characterized as a function of the position of these eigenvalues in the Re/Im plane. Five non-degenerated cases can be distinguished: (1) the four eigenvalues are real and negative (stable steady state), (2) the four eigenvalues are real, two of them are negative (unstable steady state), (3) and (4) two eigenvalues are complex conjugates with a negative real part and other two real are negative (stable steady state), two cases can be distinguished depending on the relative value of the real part of the complex eigenvalues and of the real one, (5) two eigenvalues are complex conjugates with a negative real part and the other two real are positive (unstable steady state) [2, 3].

$$\det(A - \lambda \cdot I) = \sum_{k=0}^{4} \Upsilon_k \cdot \lambda^k; \det(A - \lambda \cdot I) = 0 \Rightarrow \sum_{k=0}^{4} \Upsilon_k \cdot \lambda^k = 0$$

We suppose $\lambda_1, \lambda_2, \lambda_3, \lambda_4$ all (known) eigenvalues of the linearized system.

We are interesting to get oscillations from the closed loop functioning oscillator circuit. To get oscillations, we need eigenvalue of A is on imaginary axis, and then the method of the center manifold is applied. The center manifold of an equilibrium point (fixed point) of our dynamical system consists of orbits whose behavior around the equilibrium point is not controlled by either the attraction of the stable manifold or the repulsion of the unstable manifold. Closed loop functioning oscillator circuit eigenvalues which corresponding to eigenvalues with negative real part form the stable Eigen space, which gives rise to the stable manifold. Eigenvalue with positive real part yield the unstable manifold. If there are eigenvalues whose real part is zero, then these give rise to the center manifold. If the eigenvalues are precisely zero, then these more specifically give rise to a slow manifold.

6.3 Hartley Oscillator Stability Analysis

There are types of transistor oscillators which use feedback and lumped inductance and capacitance resonators. These oscillators are like their low frequency counterparts and very small values of inductance and capacitance must be used to make them resonate at microwave frequencies. A common base configuration is used for oscillators, although common emitter configurations can be used. The transistor input and output are matched and feedback is supplied from the output to the input. The feedback circuit contains the resonator, which controls the oscillation frequency. There are three types of transistor LC oscillators, Colpitts, Hartley and Clapp. These oscillators differ only in the way that the feedback is applied. In the

Hartley oscillator, the feedback is supplied by the inductive divider formed by two inductors. The advantages of the Hartley oscillator are oscillator's frequency may be adjusted using a single variable capacitor (one side of which can be earthed), oscillator's output amplitude remains constant over the frequency range, either a tapped coil or two fixed inductors are needed, and easy to create an accurate fixed frequency crystal oscillator variation (it is done by replacing the capacitor with a quartz crystal). The disadvantage of Hartley oscillator is the fact that harmonic rich output if taken from the amplifier and not directly from the LC circuit. Basic LC oscillator circuits have no means of controlling the amplitude of the oscillations and it is difficult to tune the oscillator to the required frequency. Hartley Oscillator configuration has a tuned tank circuit with its resonant coil tapped to feed a fraction of the output signal back to the emitter of the transistor. The output of the transistors emitter is always "in-phase" with the output at the collector, this feedback signal is positive. The oscillating frequency which is a sine-wave voltage is determined by the resonance frequency of the tank circuit [9, 10] (Fig. 6.8).

Resistor R_1 provide the usual stabilizing DC bias for the transistor. L_1 is the Radio Frequency Coil (RFC). It is an RF Choke which has a high reactance at the frequency of oscillations so most of the RF current is applied to the LC tanking circuit

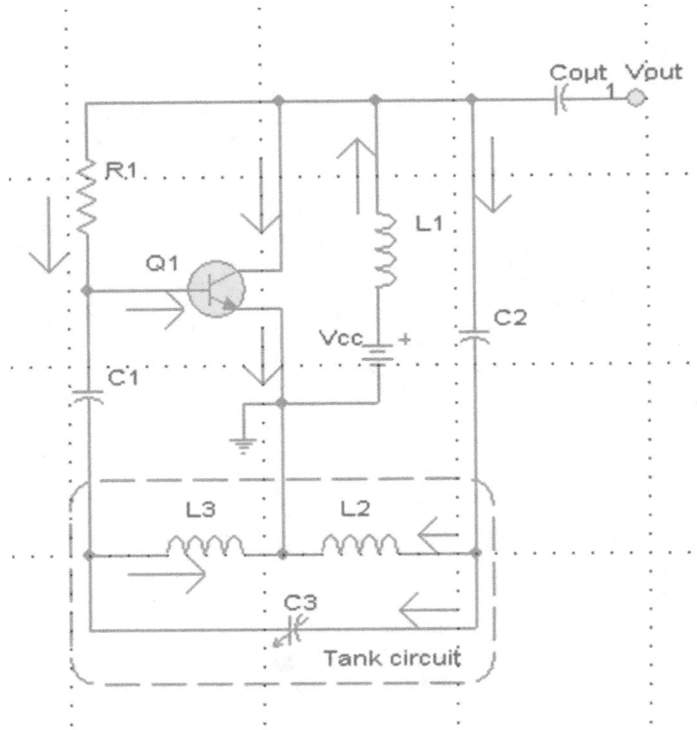

Fig. 6.8 Hartley oscillator circuit

via capacitor C_2 and the DC component passes through L_1 to the power supply. We use for our stability analysis the BJT NPN transistor Small Signal (SS) model equivalent circuit (Hybrid Pi model Version A). We consider "AC ground" in the circuit. Since the voltage at this terminal is held constant at V_{cc}, there is no time variation of the voltage. Consequently, we can set this terminal to be an "AC ground" in the small signal circuit. For AC ground, we "kill" the DC sources at that terminal (short circuit voltage sources and open circuit current sources). (Fig. 6.9).

$$V_{L_1} = V_{A_1}; V_{L_3} = V_{A_3}; V_{L_2} = V_{A_4}; V_{C_3} = V_{A_4} - V_{A_3}; v_{be} = V_{A_2};$$

$$V_{A_1} = V_{L_1} = L_1 \cdot \frac{dI_{L_1}}{dt}; V_{C_2} = V_{A_1} - V_{A_4}$$

$$I_{C_2} = C_2 \cdot \frac{d(V_{A_1} - V_{A_4})}{dt}; V_{L_2} = V_{A_4} = L_2 \cdot \frac{dI_{L_2}}{dt};$$

$$V_{L_3} = V_{A_3} = L_3 \cdot \frac{dI_{L_3}}{dt}; V_{C_1} = V_{A_2} - V_{A_3}$$

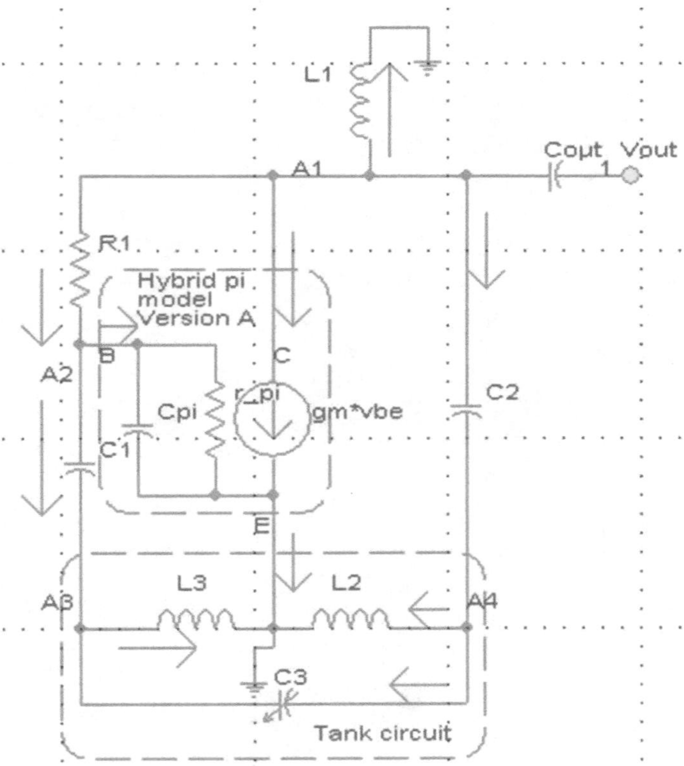

Fig. 6.9 Hartley oscillator Hybrid Pi model (Version A) schematic

$$I_{C_1} = C_1 \cdot \frac{d(V_{A_2} - V_{A_3})}{dt}; V_{Cpi} = V_{A_2}; I_{Cpi} = C_{pi} \cdot \frac{dV_{A_2}}{dt};$$
$$I_{r\_pi} = \frac{V_{A_2}}{r\_pi}; V_{R_1} = V_{A_1} - V_{A_2}; I_{R_1} = \frac{V_{A_1} - V_{A_2}}{R_1}$$

KCL @ node A_1: $I_{R_1} + g_m \cdot v_{be} + I_{C_2} + I_{L_1} = 0$, KCL @ node A_2: $I_{R_1} = I_{C_1} + I_{Cpi} + I_{r\_pi}$

KCL @ node A_3: $I_{C_1} + I_{C_3} = I_{L_3}$, KCL @ node A_4: $I_{C_2} = I_{L_2} + I_{C_3}$

$$V_{A_1} = V_{L_1} = L_1 \cdot \frac{dI_{L_1}}{dt}; I_{C_2} = C_2 \cdot \frac{d(V_{A_1} - V_{A_4})}{dt} \Rightarrow V_{A_1} - V_{A_4} = \frac{1}{C_2} \cdot \int I_{C_2} \cdot dt;$$
$$V_{A_4} = V_{A_1} - \frac{1}{C_2} \cdot \int I_{C_2} \cdot dt$$

$$\frac{d}{dt}\{V_{A_4} = V_{A_1} - \frac{1}{C_2} \cdot \int I_{C_2} \cdot dt\} \Rightarrow \frac{dV_{A_4}}{dt} = \frac{dV_{A_1}}{dt} - \frac{1}{C_2} \cdot I_{C_2};$$
$$V_{A_4} = L_1 \cdot \frac{dI_{L_1}}{dt} - \frac{1}{C_2} \cdot \int I_{C_2} \cdot dt$$

$$V_{A_4} = L_2 \cdot \frac{dI_{L_2}}{dt} \Rightarrow L_2 \cdot \frac{dI_{L_2}}{dt} = L_1 \cdot \frac{dI_{L_1}}{dt} - \frac{1}{C_2} \cdot \int I_{C_2} \cdot dt; L_2 \cdot \frac{d^2 I_{L_2}}{dt^2}$$
$$= L_1 \cdot \frac{d^2 I_{L_1}}{dt^2} - \frac{1}{C_2} \cdot I_{C_2}$$

$$V_{A_3} = L_3 \cdot \frac{dI_{L_3}}{dt}; I_{C_1} = C_1 \cdot \frac{d(V_{A_2} - V_{A_3})}{dt} \Rightarrow V_{A_2} - V_{A_3} = \frac{1}{C_1} \cdot \int I_{C_1} \cdot dt;$$
$$V_{A_2} = V_{A_3} + \frac{1}{C_1} \cdot \int I_{C_1} \cdot dt$$

$$V_{A_2} = L_3 \cdot \frac{dI_{L_3}}{dt} + \frac{1}{C_1} \cdot \int I_{C_1} \cdot dt; I_{Cpi} = C_{pi} \cdot \frac{dV_{A_2}}{dt} \Rightarrow V_{A_2} = \frac{1}{C_{pi}} \cdot \int I_{Cpi} \cdot dt$$

$$\frac{1}{C_{pi}} \cdot \int I_{Cpi} \cdot dt = L_3 \cdot \frac{dI_{L_3}}{dt} + \frac{1}{C_1} \cdot \int I_{C_1} \cdot dt; \frac{d}{dt}\{\frac{1}{C_{pi}} \cdot \int I_{Cpi} \cdot dt$$
$$= L_3 \cdot \frac{dI_{L_3}}{dt} + \frac{1}{C_1} \cdot \int I_{C_1} \cdot dt\}$$

$$\frac{1}{C_{pi}} \cdot I_{Cpi} = L_3 \cdot \frac{d^2 I_{L_3}}{dt^2} + \frac{1}{C_1} \cdot I_{C_1}; I_{r\_pi} = \frac{V_{A_2}}{r\_pi} \Rightarrow V_{A_2} = I_{r\_pi} \cdot r\_pi;$$
$$I_{R_1} = \frac{V_{A_1} - V_{A_2}}{R_1}$$

$V_{A_1} - V_{A_2} = I_{R_1} \cdot R_1 \Rightarrow V_{A_1} = V_{A_2} + I_{R_1} \cdot R_1; V_{A_1} = I_{r\_pi} \cdot r\_pi + I_{R_1} \cdot R_1;$
$V_{A_1} = I_{r\_pi} \cdot r\_pi + I_{R_1} \cdot R_1$

$$V_{A_1} = L_1 \cdot \frac{dI_{L_1}}{dt}; L_1 \cdot \frac{dI_{L_1}}{dt} = I_{r\_pi} \cdot r\_pi + I_{R_1} \cdot R_1; \frac{dI_{L_1}}{dt} = I_{r\_pi} \cdot \frac{r\_pi}{L_1} + I_{R_1} \cdot \frac{R_1}{L_1}$$

$$\frac{d^2 I_{L_1}}{dt^2} = \frac{dI_{r\_pi}}{dt} \cdot \frac{r\_pi}{L_1} + \frac{dI_{R_1}}{dt} \cdot \frac{R_1}{L_1}; L_2 \cdot \frac{d^2 I_{L_2}}{dt^2} = L_1 \cdot [\frac{dI_{r\_pi}}{dt} \cdot \frac{r\_pi}{L_1} + \frac{dI_{R_1}}{dt} \cdot \frac{R_1}{L_1}] - \frac{1}{C_2} \cdot I_{C_2}$$

$$L_2 \cdot \frac{d^2 I_{L_2}}{dt^2} = \frac{dI_{r\_pi}}{dt} \cdot r\_pi + \frac{dI_{R_1}}{dt} \cdot R_1 - \frac{1}{C_2} \cdot I_{C_2}; \frac{d^2 I_{L_2}}{dt^2} = \frac{dI_{r\_pi}}{dt} \cdot \frac{r\_pi}{L_2} + \frac{dI_{R_1}}{dt} \cdot \frac{R_1}{L_2} - \frac{1}{C_2 \cdot L_2} \cdot I_{C_2}$$

We can summery our Hartley oscillator differential equation:

$$L_2 \cdot \frac{d^2 I_{L_2}}{dt^2} = \frac{dI_{r\_pi}}{dt} \cdot r\_pi + \frac{dI_{R_1}}{dt} \cdot R_1 - \frac{1}{C_2} \cdot I_{C_2}$$

$$\frac{dI_{L_1}}{dt} = I_{r\_pi} \cdot \frac{r\_pi}{L_1} + I_{R_1} \cdot \frac{R_1}{L_1}; \frac{1}{C_{pi}} \cdot I_{Cpi} = L_3 \cdot \frac{d^2 I_{L_3}}{dt^2} + \frac{1}{C_1} \cdot I_{C_1}$$

We define for simplicity the following new variables: $\frac{dI_{L_2}}{dt} = Y_1; \frac{dI_{r\_pi}}{dt} = Y_2$

$$\frac{dI_{R_1}}{dt} = Y_3; \frac{dI_{L_3}}{dt} = Y_4; \frac{d^2 I_{L_2}}{dt^2} = \frac{dY_1}{dt}; \frac{d^2 I_{L_3}}{dt^2} = \frac{dY_4}{dt}$$

$$\frac{dY_1}{dt} = Y_2 \cdot \frac{r\_pi}{L_2} + Y_3 \cdot \frac{R_1}{L_2} - \frac{1}{C_2 \cdot L_2} \cdot I_{C_2}; \frac{dY_4}{dt} = \frac{1}{C_{pi} \cdot L_3} \cdot I_{Cpi} - \frac{1}{C_1 \cdot L_3} \cdot I_{C_1}$$

Our circuit KCLs @ nodes $A_1 - A_4$: $I_{R_1} + g_m \cdot I_{r\_pi} \cdot r\_pi + I_{C_2} + I_{L_1} = 0$

$$I_{R_1} = I_{C_1} + I_{Cpi} + I_{r\_pi}; I_{C_1} + I_{C_3} = I_{L_3}; I_{C_2} = I_{L_2} + I_{C_3}$$

$I_{C_2} = -I_{L_1} - g_m \cdot r\_pi \cdot I_{r\_pi} - I_{R_1}; I_{Cpi} = I_{R_1} - I_{C_1} - I_{r\_pi}; I_{C_1} = I_{L_3} - I_{C_3}; I_{C_3} = I_{C_2} - I_{L_2}$
$I_{C_1} = I_{L_3} - [I_{C_2} - I_{L_2}] = I_{L_3} - I_{C_2} + I_{L_2}; I_{C_1} = I_{L_3} + I_{L_2} - [-I_{L_1} - g_m \cdot r\_pi \cdot I_{r\_pi} - I_{R_1}]$

$$I_{C_1} = I_{L_3} + I_{L_2} + I_{L_1} + g_m \cdot r\_pi \cdot I_{r\_pi} + I_{R_1}; I_{C_1} = \sum_{k=1}^{3} I_{L_k} + g_m \cdot r\_pi \cdot I_{r\_pi} + I_{R_1}$$

$$I_{Cpi} = I_{R_1} - I_{r\_pi} - [\sum_{k=1}^{3} I_{L_k} + g_m \cdot r\_pi \cdot I_{r\_pi} + I_{R_1}];$$

$$I_{Cpi} = -I_{r\_pi} \cdot [1 + g_m \cdot r\_pi] - \sum_{k=1}^{3} I_{L_k}$$

We can summery our intermediate results:

$$I_{Cpi} = -I_{r\_pi} \cdot [1 + g_m \cdot r\_pi] - \sum_{k=1}^{3} I_{L_k}; I_{C_1} = \sum_{k=1}^{3} I_{L_k} + g_m \cdot r\_pi \cdot I_{r\_pi} + I_{R_1}$$

$$I_{C_2} = -I_{L_1} - g_m \cdot r\_pi \cdot I_{r\_pi} - I_{R_1}$$

&&&

$$\frac{dY_1}{dt} = Y_2 \cdot \frac{r\_pi}{L_2} + Y_3 \cdot \frac{R_1}{L_2} - \frac{1}{C_2 \cdot L_2} \cdot I_{C_2}$$
$$= Y_2 \cdot \frac{r\_pi}{L_2} + Y_3 \cdot \frac{R_1}{L_2} - \frac{1}{C_2 \cdot L_2} \cdot [-I_{L_1} - g_m \cdot r\_pi \cdot I_{r\_pi} - I_{R_1}]$$

$$\frac{dY_1}{dt} = Y_2 \cdot \frac{r\_pi}{L_2} + Y_3 \cdot \frac{R_1}{L_2} + \frac{1}{C_2 \cdot L_2} \cdot I_{L_1} + \frac{1}{C_2 \cdot L_2} \cdot g_m \cdot r\_pi \cdot I_{r\_pi} + \frac{1}{C_2 \cdot L_2} \cdot I_{R_1}$$

$$\frac{dY_4}{dt} = \frac{1}{C_{pi} \cdot L_3} \cdot [-I_{r\_pi} \cdot [1 + g_m \cdot r\_pi] - \sum_{k=1}^{3} I_{L_k}]$$
$$- \frac{1}{C_1 \cdot L_3} \cdot [\sum_{k=1}^{3} I_{L_k} + g_m \cdot r\_pi \cdot I_{r\_pi} + I_{R_1}]$$

$$\frac{dY_4}{dt} = -\frac{I_{r\_pi}}{L_3} \cdot (\frac{[1 + g_m \cdot r\_pi]}{C_{pi}} + \frac{g_m \cdot r\_pi}{C_1}) - [\frac{1}{C_{pi}} + \frac{1}{C_1}] \cdot \frac{1}{L_3} \cdot \sum_{k=1}^{3} I_{L_k}$$
$$- \frac{1}{C_1 \cdot L_3} \cdot I_{R_1}$$

We can summery our system differential equations:

$$\frac{dY_1}{dt} = Y_2 \cdot \frac{r\_pi}{L_2} + Y_3 \cdot \frac{R_1}{L_2} + \frac{1}{C_2 \cdot L_2} \cdot I_{L_1} + \frac{1}{C_2 \cdot L_2} \cdot g_m \cdot r\_pi \cdot I_{r\_pi} + \frac{1}{C_2 \cdot L_2} \cdot I_{R_1}$$

$$\frac{dY_4}{dt} = -\frac{I_{r\_pi}}{L_3} \cdot (\frac{[1 + g_m \cdot r\_pi]}{C_{pi}} + \frac{g_m \cdot r\_pi}{C_1}) - [\frac{1}{C_{pi}} + \frac{1}{C_1}] \cdot \frac{1}{L_3} \cdot \sum_{k=1}^{3} I_{L_k}$$
$$- \frac{1}{C_1 \cdot L_3} \cdot I_{R_1}$$

$$\frac{dI_{L_2}}{dt} = Y_1; \frac{dI_{r\_pi}}{dt} = Y_2; \frac{dI_{R_1}}{dt} = Y_3; \frac{dI_{L_3}}{dt} = Y_4; \frac{dI_{L_1}}{dt} = I_{r\_pi} \cdot \frac{r\_pi}{L_1} + I_{R_1} \cdot \frac{R_1}{L_1}$$

At fixed points:

$$\frac{dY_1}{dt} = 0; \frac{dY_4}{dt} = 0; \frac{dI_{L_2}}{dt} = 0; \frac{dI_{r\_pi}}{dt} = 0; \frac{dI_{R_1}}{dt} = 0; \frac{dI_{L_3}}{dt} = 0; \frac{dI_{L_1}}{dt} = 0$$

$$Y_1^* = 0;\ Y_2^* = 0;\ Y_3^* = 0;\ Y_4^* = 0;\ r\_pi \cdot I_{r\_pi}^* + I_{R_1}^* \cdot R_1 = 0 \Rightarrow I_{r\_pi}^* = -I_{R_1}^* \cdot \frac{R_1}{r\_pi}$$

$$\frac{1}{C_2 \cdot L_2} \cdot I_{L_1}^* + \frac{1}{C_2 \cdot L_2} \cdot g_m \cdot r\_pi \cdot I_{r\_pi}^* + \frac{1}{C_2 \cdot L_2} \cdot I_{R_1}^* = 0$$

$$-\frac{I_{r\_pi}^*}{L_3} \cdot \left(\frac{[1 + g_m \cdot r\_pi]}{C_{pi}} + \frac{g_m \cdot r\_pi}{C_1}\right) - \left[\frac{1}{C_{pi}} + \frac{1}{C_1}\right] \cdot \frac{1}{L_3} \cdot \sum_{k=1}^{3} I_{L_k}^* - \frac{1}{C_1 \cdot L_3} \cdot I_{R_1}^* = 0$$

We get two fixed points equations:

$$I_{L_1}^* + I_{R_1}^* \cdot (1 - g_m \cdot R_1) = 0$$

$$I_{R_1}^* \cdot \left\{\frac{R_1}{r\_pi \cdot L_3} \cdot \left(\frac{[1 + g_m \cdot r\_pi]}{C_{pi}} + \frac{g_m \cdot r\_pi}{C_1}\right) - \frac{1}{C_1 \cdot L_3}\right\} - \left[\frac{1}{C_{pi}} + \frac{1}{C_1}\right] \cdot \frac{1}{L_3} \cdot \sum_{k=1}^{3} I_{L_k}^* = 0$$

$$I_{R_1}^* \cdot \left\{\frac{R_1}{r\_pi \cdot L_3} \cdot \left(\frac{[1 + g_m \cdot r\_pi]}{C_{pi}} + \frac{g_m \cdot r\_pi}{C_1}\right) - \frac{1}{C_1 \cdot L_3}\right\} - \left[\frac{1}{C_{pi}} + \frac{1}{C_1}\right] \cdot \frac{1}{L_3} \cdot \left[I_{L_1}^* + \sum_{k=2}^{3} I_{L_k}^*\right] = 0$$

$$\sum_{k=1}^{3} I_{L_k}^* = I_{L_1}^* + \sum_{k=2}^{3} I_{L_k}^*; I_{L_1}^* + I_{R_1}^* \cdot (1 - g_m \cdot R_1) = 0 \Rightarrow I_{L_1}^* = -I_{R_1}^* \cdot (1 - g_m \cdot R_1)$$

$$I_{R_1}^* \cdot \left\{\frac{R_1}{r\_pi \cdot L_3} \cdot \left(\frac{[1 + g_m \cdot r\_pi]}{C_{pi}} + \frac{g_m \cdot r\_pi}{C_1}\right) - \frac{1}{C_1 \cdot L_3}\right\} - \left[\frac{1}{C_{pi}} + \frac{1}{C_1}\right] \cdot \frac{1}{L_3} \cdot I_{L_1}^* - \left[\frac{1}{C_{pi}} + \frac{1}{C_1}\right] \cdot \frac{1}{L_3} \cdot \sum_{k=2}^{3} I_{L_k}^* = 0$$

$$I_{R_1}^* \cdot \{\frac{R_1}{r\_pi \cdot L_3} \cdot (\frac{[1 + g_m \cdot r\_pi]}{C_{pi}} + \frac{g_m \cdot r\_pi}{C_1}) - \frac{1}{C_1 \cdot L_3}\} + [\frac{1}{C_{pi}} + \frac{1}{C_1}] \cdot \frac{(1 - g_m \cdot R_1)}{L_3} \cdot I_{R_1}^*$$
$$- [\frac{1}{C_{pi}} + \frac{1}{C_1}] \cdot \frac{1}{L_3} \cdot \sum_{k=2}^{3} I_{L_k}^* = 0$$

$$I_{R_1}^* \cdot \{\frac{R_1}{r\_pi \cdot L_3} \cdot (\frac{[1 + g_m \cdot r\_pi]}{C_{pi}} + \frac{g_m \cdot r\_pi}{C_1}) - \frac{1}{C_1 \cdot L_3} + [\frac{1}{C_{pi}} + \frac{1}{C_1}] \cdot \frac{(1 - g_m \cdot R_1)}{L_3}\}$$
$$- [\frac{1}{C_{pi}} + \frac{1}{C_1}] \cdot \frac{1}{L_3} \cdot \sum_{k=2}^{3} I_{L_k}^* = 0$$

$$I_{R_1}^* = \frac{[\frac{1}{C_{pi}} + \frac{1}{C_1}] \cdot \frac{1}{L_3} \cdot \sum_{k=2}^{3} I_{L_k}^*}{\frac{R_1}{r\_pi \cdot L_3} \cdot (\frac{[1 + g_m \cdot r\_pi]}{C_{pi}} + \frac{g_m \cdot r\_pi}{C_1}) - \frac{1}{C_1 \cdot L_3} + [\frac{1}{C_{pi}} + \frac{1}{C_1}] \cdot \frac{(1 - g_m \cdot R_1)}{L_3}}$$

$$I_{L_1}^* = \frac{-(1 - g_m \cdot R_1) \cdot [\frac{1}{C_{pi}} + \frac{1}{C_1}] \cdot \frac{1}{L_3} \cdot \sum_{k=2}^{3} I_{L_k}^*}{\frac{R_1}{r\_pi \cdot L_3} \cdot (\frac{[1 + g_m \cdot r\_pi]}{C_{pi}} + \frac{g_m \cdot r\_pi}{C_1}) - \frac{1}{C_1 \cdot L_3} + [\frac{1}{C_{pi}} + \frac{1}{C_1}] \cdot \frac{(1 - g_m \cdot R_1)}{L_3}}$$

$$I_{r\_pi}^* = \frac{-\frac{R_1}{r\_pi} \cdot [\frac{1}{C_{pi}} + \frac{1}{C_1}] \cdot \frac{1}{L_3} \cdot \sum_{k=2}^{3} I_{L_k}^*}{\frac{R_1}{r\_pi \cdot L_3} \cdot (\frac{[1 + g_m \cdot r\_pi]}{C_{pi}} + \frac{g_m \cdot r\_pi}{C_1}) - \frac{1}{C_1 \cdot L_3} + [\frac{1}{C_{pi}} + \frac{1}{C_1}] \cdot \frac{(1 - g_m \cdot R_1)}{L_3}}$$

We can define our system fixed points as $E^*(Y_1^*, Y_2^*, Y_3^*, Y_4^*, I_{r\_pi}^*, I_{R_1}^*, I_{L_1}^*, I_{L_2}^*, I_{L_3}^*)$
We define three global parameters in our system: $\Gamma_1, \Gamma_2, \Gamma_3$

$$\Gamma_1 = \Gamma_1(R_1, r\_pi, C_1, C_{pi}, \ldots); \Gamma_2 = \Gamma_1(R_1, r\_pi, C_1, C_{pi}, \ldots);$$
$$\Gamma_3 = \Gamma_3(R_1, r\_pi, C_1, C_{pi}, \ldots)$$

$$\Gamma_1 = \frac{[\frac{1}{C_{pi}} + \frac{1}{C_1}] \cdot \frac{1}{L_3}}{\frac{R_1}{r\_pi \cdot L_3} \cdot (\frac{[1 + g_m \cdot r\_pi]}{C_{pi}} + \frac{g_m \cdot r\_pi}{C_1}) - \frac{1}{C_1 \cdot L_3} + [\frac{1}{C_{pi}} + \frac{1}{C_1}] \cdot \frac{(1 - g_m \cdot R_1)}{L_3}}$$

$$\Gamma_2 = \frac{-(1 - g_m \cdot R_1) \cdot [\frac{1}{C_{pi}} + \frac{1}{C_1}] \cdot \frac{1}{L_3}}{\frac{R_1}{r\_pi \cdot L_3} \cdot (\frac{[1 + g_m \cdot r\_pi]}{C_{pi}} + \frac{g_m \cdot r\_pi}{C_1}) - \frac{1}{C_1 \cdot L_3} + [\frac{1}{C_{pi}} + \frac{1}{C_1}] \cdot \frac{(1 - g_m \cdot R_1)}{L_3}}$$

$$\Gamma_3 = \frac{-\frac{R_1}{r\_pi} \cdot [\frac{1}{C_{pi}} + \frac{1}{C_1}] \cdot \frac{1}{L_3}}{\frac{R_1}{r\_pi \cdot L_3} \cdot (\frac{[1 + g_m \cdot r\_pi]}{C_{pi}} + \frac{g_m \cdot r\_pi}{C_1}) - \frac{1}{C_1 \cdot L_3} + [\frac{1}{C_{pi}} + \frac{1}{C_1}] \cdot \frac{(1 - g_m \cdot R_1)}{L_3}}$$

$$I_{R_1}^* = \Gamma_1(R_1, r\_pi, C_1, C_{pi}, \ldots) \cdot \sum_{k=2}^{3} I_{L_k}^*; I_{L_1}^* = \Gamma_2(R_1, r\_pi, C_1, C_{pi}, \ldots) \cdot \sum_{k=2}^{3} I_{L_k}^*$$

$$I_{r\_pi}^* = \Gamma_3(R_1, r\_pi, C_1, C_{pi}, \ldots) \cdot \sum_{k=2}^{3} I_{L_k}^*$$

$$E^*(Y_1^*, Y_2^*, Y_3^*, Y_4^*, I_{r\_pi}^*, I_{R_1}^*, I_{L_1}^*, I_{L_2}^*, I_{L_3}^*) = (0, 0, 0, 0, \Gamma_3(R_1, r\_pi, C_1, C_{pi}, \ldots) \cdot \sum_{k=2}^{3} I_{L_k}^*,$$

$$\Gamma_1(R_1, r\_pi, C_1, C_{pi}, \ldots) \cdot \sum_{k=2}^{3} I_{L_k}^*, \Gamma_2(R_1, r\_pi, C_1, C_{pi}, \ldots) \cdot \sum_{k=2}^{3} I_{L_k}^*, I_{L_2}^*, I_{L_3}^*)$$

Stability analysis: The standard local stability analysis about any one of the equilibrium points of the Hartley oscillator circuit (BJT transistor small signal model) consists in adding to coordinate $[Y_1, Y_2, Y_3, Y_4, I_{r\_pi}, I_{R_1}, I_{L_1}, I_{L_2}, I_{L_3}]$ arbitrarily small increments of exponentially form $[y_1, y_2, y_3, y_4, i_{r\_pi}, i_{R_1}, i_{L_1}, i_{L_2}, i_{L_3}] \cdot e^{\lambda \cdot t}$ and retaining the first order terms in $Y_1, Y_2, Y_3, Y_4, I_{r\_pi}, I_{R_1}, I_{L_1}, I_{L_2}, I_{L_3}$. The system of homogenous equations leads to a polynomial characteristic equation in the eigenvalues [4]. The polynomial characteristic equations accept by set of the below circuit variables, circuit variables derivative and circuit variables second order derivative with respect to time into Hartley oscillator circuit [2, 3, 4]. Our Hartley oscillator fixed values with arbitrarily small increments of exponential form $[y_1, y_2, y_3, y_4, i_{r\_pi}, i_{R_1}, i_{L_1}, i_{L_2}, i_{L_3}] \cdot e^{\lambda \cdot t}$ are: j = 0(first fixed point), j = 1(second fixed point), j = 2(third fixed point), etc.,

$$I_{r\_pi}(t) = I_{r\_pi}^{(j)} + i_{r\_pi} \cdot e^{\lambda \cdot t}; I_{R_1}(t) = I_{R_1}^{(j)} + i_{R_1} \cdot e^{\lambda \cdot t};$$
$$I_{L_1}(t) = I_{L_1}^{(j)} + i_{L_1} \cdot e^{\lambda \cdot t}; I_{L_2}(t) = I_{L_2}^{(j)} + i_{L_2} \cdot e^{\lambda \cdot t}$$

$$I_{L_3}(t) = I_{L_3}^{(j)} + i_{L_3} \cdot e^{\lambda \cdot t}; Y_1(t) = Y_1^{(j)} + y_1 \cdot e^{\lambda \cdot t}; Y_2(t) = Y_2^{(j)} + y_2 \cdot e^{\lambda \cdot t};$$
$$Y_3(t) = Y_3^{(j)} + y_3 \cdot e^{\lambda \cdot t}$$

$$Y_4(t) = Y_4^{(j)} + y_4 \cdot e^{\lambda \cdot t}; \frac{dY_1(t)}{dt} = y_1 \cdot \lambda \cdot e^{\lambda \cdot t};$$
$$\frac{dY_4(t)}{dt} = y_4 \cdot \lambda \cdot e^{\lambda \cdot t}; \frac{dI_{r\_pi}(t)}{dt} = i_{r\_pi} \cdot \lambda \cdot e^{\lambda \cdot t}$$

$$\frac{dI_{R_1}(t)}{dt} = i_{R_1} \cdot \lambda \cdot e^{\lambda \cdot t}; \frac{dI_{L_1}(t)}{dt} = i_{L_1} \cdot \lambda \cdot e^{\lambda \cdot t};$$
$$\frac{dI_{L_2}(t)}{dt} = i_{L_2} \cdot \lambda \cdot e^{\lambda \cdot t}; \frac{dI_{L_3}(t)}{dt} = i_{L_3} \cdot \lambda \cdot e^{\lambda \cdot t}$$

&&&

$$\frac{dY_1}{dt} = Y_2 \cdot \frac{r\_pi}{L_2} + Y_3 \cdot \frac{R_1}{L_2} + \frac{1}{C_2 \cdot L_2} \cdot I_{L_1} + \frac{1}{C_2 \cdot L_2} \cdot g_m \cdot r\_pi \cdot I_{r\_pi} + \frac{1}{C_2 \cdot L_2} \cdot I_{R_1}$$

$$y_1 \cdot \lambda \cdot e^{\lambda \cdot t} = [Y_2^{(j)} + y_2 \cdot e^{\lambda \cdot t}] \cdot \frac{r\_pi}{L_2} + [Y_3^{(j)} + y_3 \cdot e^{\lambda \cdot t}] \cdot \frac{R_1}{L_2} + \frac{1}{C_2 \cdot L_2} \cdot [I_{L_1}^{(j)} + i_{L_1} \cdot e^{\lambda \cdot t}]$$
$$+ \frac{1}{C_2 \cdot L_2} \cdot g_m \cdot r\_pi \cdot [I_{r\_pi}^{(j)} + i_{r\_pi} \cdot e^{\lambda \cdot t}] + \frac{1}{C_2 \cdot L_2} \cdot [I_{R_1}^{(j)} + i_{R_1} \cdot e^{\lambda \cdot t}]$$

$$y_1 \cdot \lambda \cdot e^{\lambda \cdot t} = Y_2^{(j)} \cdot \frac{r\_pi}{L_2} + Y_3^{(j)} \cdot \frac{R_1}{L_2} + \frac{1}{C_2 \cdot L_2} \cdot I_{L_1}^{(j)} + \frac{1}{C_2 \cdot L_2} \cdot g_m \cdot r\_pi \cdot I_{r\_pi}^{(j)}$$
$$+ \frac{1}{C_2 \cdot L_2} \cdot I_{R_1}^{(j)} + y_2 \cdot \frac{r\_pi}{L_2} \cdot e^{\lambda \cdot t} + y_3 \cdot \frac{R_1}{L_2} \cdot e^{\lambda \cdot t} + \frac{1}{C_2 \cdot L_2} \cdot i_{L_1} \cdot e^{\lambda \cdot t}$$
$$+ \frac{1}{C_2 \cdot L_2} \cdot g_m \cdot r\_pi \cdot i_{r\_pi} \cdot e^{\lambda \cdot t} + \frac{1}{C_2 \cdot L_2} \cdot i_{R_1} \cdot e^{\lambda \cdot t}$$

At fixed point:

$$Y_2^{(j)} \cdot \frac{r\_pi}{L_2} + Y_3^{(j)} \cdot \frac{R_1}{L_2} + \frac{1}{C_2 \cdot L_2} \cdot I_{L_1}^{(j)} + \frac{1}{C_2 \cdot L_2} \cdot g_m \cdot r\_pi \cdot I_{r\_pi}^{(j)} + \frac{1}{C_2 \cdot L_2} \cdot I_{R_1}^{(j)} = 0$$

$$- y_1 \cdot \lambda \cdot e^{\lambda \cdot t} + y_2 \cdot \frac{r\_pi}{L_2} \cdot e^{\lambda \cdot t} + y_3 \cdot \frac{R_1}{L_2} \cdot e^{\lambda \cdot t} + \frac{1}{C_2 \cdot L_2} \cdot i_{L_1} \cdot e^{\lambda \cdot t}$$
$$+ \frac{1}{C_2 \cdot L_2} \cdot g_m \cdot r\_pi \cdot i_{r\_pi} \cdot e^{\lambda \cdot t} + \frac{1}{C_2 \cdot L_2} \cdot i_{R_1} \cdot e^{\lambda \cdot t} = 0$$

Dividing the two sides of the above equation by $e^{\lambda \cdot t}$ term:

$$- y_1 \cdot \lambda + y_2 \cdot \frac{r\_pi}{L_2} + y_3 \cdot \frac{R_1}{L_2} + \frac{1}{C_2 \cdot L_2} \cdot i_{L_1} + \frac{1}{C_2 \cdot L_2} \cdot g_m \cdot r\_pi \cdot i_{r\_pi}$$
$$+ \frac{1}{C_2 \cdot L_2} \cdot i_{R_1} = 0$$

$$\frac{dY_4}{dt} = - \frac{I_{r\_pi}}{L_3} \cdot (\frac{[1 + g_m \cdot r\_pi]}{C_{pi}} + \frac{g_m \cdot r\_pi}{C_1}) - [\frac{1}{C_{pi}} + \frac{1}{C_1}] \cdot \frac{1}{L_3} \cdot \sum_{k=1}^{3} I_{L_k}$$
$$- \frac{1}{C_1 \cdot L_3} \cdot I_{R_1}$$

$$y_4 \cdot \lambda \cdot e^{\lambda \cdot t} = -\frac{1}{L_3} \cdot [I_{r\_pi}^{(j)} + i_{r\_pi} \cdot e^{\lambda \cdot t}] \cdot (\frac{[1 + g_m \cdot r\_pi]}{C_{pi}} + \frac{g_m \cdot r\_pi}{C_1})$$
$$- [\frac{1}{C_{pi}} + \frac{1}{C_1}] \cdot \frac{1}{L_3} \cdot \sum_{k=1}^{3} (I_{L_k}^{(j)} + i_{L_k} \cdot e^{\lambda \cdot t}) - \frac{1}{C_1 \cdot L_3} \cdot [I_{R_1}^{(j)} + i_{R_1} \cdot e^{\lambda \cdot t}]$$

$$y_4 \cdot \lambda \cdot e^{\lambda \cdot t} = -\frac{1}{L_3} \cdot I_{r\_pi}^{(j)} \cdot (\frac{[1 + g_m \cdot r\_pi]}{C_{pi}} + \frac{g_m \cdot r\_pi}{C_1})$$
$$- \frac{1}{L_3} \cdot i_{r\_pi} \cdot (\frac{[1 + g_m \cdot r\_pi]}{C_{pi}} + \frac{g_m \cdot r\_pi}{C_1}) \cdot e^{\lambda \cdot t}$$
$$- [\frac{1}{C_{pi}} + \frac{1}{C_1}] \cdot \frac{1}{L_3} \cdot \sum_{k=1}^{3} I_{L_k}^{(j)} - [\frac{1}{C_{pi}} + \frac{1}{C_1}] \cdot \frac{1}{L_3} \cdot \sum_{k=1}^{3} i_{L_k} \cdot e^{\lambda \cdot t}$$
$$- \frac{1}{C_1 \cdot L_3} \cdot I_{R_1}^{(j)} - \frac{1}{C_1 \cdot L_3} \cdot i_{R_1} \cdot e^{\lambda \cdot t}$$

$$y_4 \cdot \lambda \cdot e^{\lambda \cdot t} = -\frac{1}{L_3} \cdot I_{r\_pi}^{(j)} \cdot (\frac{[1 + g_m \cdot r\_pi]}{C_{pi}} + \frac{g_m \cdot r\_pi}{C_1})$$
$$- [\frac{1}{C_{pi}} + \frac{1}{C_1}] \cdot \frac{1}{L_3} \cdot \sum_{k=1}^{3} I_{L_k}^{(j)} - \frac{1}{C_1 \cdot L_3} \cdot I_{R_1}^{(j)}$$
$$- \frac{1}{L_3} \cdot i_{r\_pi} \cdot (\frac{[1 + g_m \cdot r\_pi]}{C_{pi}} + \frac{g_m \cdot r\_pi}{C_1}) \cdot e^{\lambda \cdot t}$$
$$- [\frac{1}{C_{pi}} + \frac{1}{C_1}] \cdot \frac{1}{L_3} \cdot \sum_{k=1}^{3} i_{L_k} \cdot e^{\lambda \cdot t} - \frac{1}{C_1 \cdot L_3} \cdot i_{R_1} \cdot e^{\lambda \cdot t}$$

At fixed point:

$$-\frac{1}{L_3} \cdot I_{r\_pi}^{(j)} \cdot (\frac{[1 + g_m \cdot r\_pi]}{C_{pi}} + \frac{g_m \cdot r\_pi}{C_1}) - [\frac{1}{C_{pi}} + \frac{1}{C_1}] \cdot \frac{1}{L_3} \cdot \sum_{k=1}^{3} I_{L_k}^{(j)} - \frac{1}{C_1 \cdot L_3} \cdot I_{R_1}^{(j)}$$
$$= 0$$

$$y_4 \cdot \lambda \cdot e^{\lambda \cdot t} = -\frac{1}{L_3} \cdot i_{r\_pi} \cdot (\frac{[1 + g_m \cdot r\_pi]}{C_{pi}} + \frac{g_m \cdot r\_pi}{C_1}) \cdot e^{\lambda \cdot t} - [\frac{1}{C_{pi}} + \frac{1}{C_1}] \cdot \frac{1}{L_3}$$
$$\cdot \sum_{k=1}^{3} i_{L_k} \cdot e^{\lambda \cdot t} - \frac{1}{C_1 \cdot L_3} \cdot i_{R_1} \cdot e^{\lambda \cdot t}$$

Dividing the two sides of the above equation by $e^{\lambda \cdot t}$ term:

$$-y_4 \cdot \lambda - \frac{1}{L_3} \cdot i_{r\_pi} \cdot (\frac{[1 + g_m \cdot r\_pi]}{C_{pi}} + \frac{g_m \cdot r\_pi}{C_1}) - [\frac{1}{C_{pi}} + \frac{1}{C_1}] \cdot \frac{1}{L_3} \cdot \sum_{k=1}^{3} i_{L_k}$$
$$-\frac{1}{C_1 \cdot L_3} \cdot i_{R_1} = 0$$

$$\frac{dI_{L_2}}{dt} = Y_1; \frac{dI_{r\_pi}}{dt} = Y_2; \frac{dI_{R_1}}{dt} = Y_3; \frac{dI_{L_3}}{dt} = Y_4$$

$$i_{L_2} \cdot \lambda \cdot e^{\lambda \cdot t} = Y_1^{(j)} + y_1 \cdot e^{\lambda \cdot t}; i_{r\_pi} \cdot \lambda \cdot e^{\lambda \cdot t} = Y_2^{(j)} + y_2 \cdot e^{\lambda \cdot t}; i_{R_1} \cdot \lambda \cdot e^{\lambda \cdot t} = Y_3^{(j)} + y_3 \cdot e^{\lambda \cdot t}$$
$$i_{L_3} \cdot \lambda \cdot e^{\lambda \cdot t} = Y_4^{(j)} + y_4 \cdot e^{\lambda \cdot t}$$

At fixed point:

$$Y_1^* = 0; Y_2^* = 0; Y_3^* = 0; Y_4^* = 0$$

$$-i_{L_2} \cdot \lambda + y_1 = 0; -i_{r\_pi} \cdot \lambda + y_2 = 0; -i_{R_1} \cdot \lambda + y_3 = 0; -i_{L_3} \cdot \lambda + y_4 = 0$$

$$\frac{dI_{L_1}}{dt} = I_{r\_pi} \cdot \frac{r\_pi}{L_1} + I_{R_1} \cdot \frac{R_1}{L_1}; i_{L_1} \cdot \lambda \cdot e^{\lambda \cdot t}$$
$$= [I_{r\_pi}^{(j)} + i_{r\_pi} \cdot e^{\lambda \cdot t}] \cdot \frac{r\_pi}{L_1} + [I_{R_1}^{(j)} + i_{R_1} \cdot e^{\lambda \cdot t}] \cdot \frac{R_1}{L_1}$$

$$i_{L_1} \cdot \lambda \cdot e^{\lambda \cdot t} = I_{r\_pi}^{(j)} \cdot \frac{r\_pi}{L_1} + I_{R_1}^{(j)} \cdot \frac{R_1}{L_1} + i_{r\_pi} \cdot \frac{r\_pi}{L_1} \cdot e^{\lambda \cdot t} + i_{R_1} \cdot \frac{R_1}{L_1} \cdot e^{\lambda \cdot t}$$

At fixed point :

$$I_{r\_pi}^{(j)} \cdot \frac{r\_pi}{L_1} + I_{R_1}^{(j)} \cdot \frac{R_1}{L_1} = 0.$$

We get $-i_{L_1} \cdot \lambda + i_{r\_pi} \cdot \frac{r\_pi}{L_1} + i_{R_1} \cdot \frac{R_1}{L_1} = 0.$

We can summery our Hartley oscillator circuit (BJT transistor small signal model) arbitrarily small increments equations:

$$-y_1 \cdot \lambda + y_2 \cdot \frac{r\_pi}{L_2} + y_3 \cdot \frac{R_1}{L_2} + \frac{1}{C_2 \cdot L_2} \cdot i_{L_1} + \frac{1}{C_2 \cdot L_2} \cdot g_m \cdot r\_pi \cdot i_{r\_pi}$$
$$+ \frac{1}{C_2 \cdot L_2} \cdot i_{R_1} = 0$$

$$- y_4 \cdot \lambda - \frac{1}{L_3} \cdot i_{r\_pi} \cdot (\frac{[1 + g_m \cdot r\_pi]}{C_{pi}} + \frac{g_m \cdot r\_pi}{C_1}) - [\frac{1}{C_{pi}} + \frac{1}{C_1}] \cdot \frac{1}{L_3} \cdot \sum_{k=1}^{3} i_{L_k}$$

$$- \frac{1}{C_1 \cdot L_3} \cdot i_{R_1} = 0$$

$$-i_{L_2} \cdot \lambda + y_1 = 0; - i_{r\_pi} \cdot \lambda + y_2 = 0; - i_{R_1} \cdot \lambda + y_3 = 0; -i_{L_3} \cdot \lambda + y_4 = 0$$

$$-i_{L_1} \cdot \lambda + i_{r\_pi} \cdot \frac{r\_pi}{L_1} + i_{R_1} \cdot \frac{R_1}{L_1} = 0$$

$$\begin{pmatrix} \Xi_{11} & \cdots & \Xi_{17} \\ \vdots & \ddots & \vdots \\ \Xi_{71} & \cdots & \Xi_{77} \end{pmatrix} \cdot \begin{pmatrix} y_1 \\ y_4 \\ i_{r\_pi} \\ i_{R_1} \\ i_{L_1} \\ i_{L_2} \\ i_{L_3} \end{pmatrix} + \begin{pmatrix} \frac{r\_pi}{L_2} & \frac{R_1}{L_2} \\ 0 & 0 \\ 1 & 0 \\ 0 & 1 \\ 0 & 0 \\ 0 & 0 \\ 0 & 0 \end{pmatrix} \cdot \begin{pmatrix} y_2 \\ y_3 \end{pmatrix} = 0;$$

$$\begin{pmatrix} \Xi_{11} & \cdots & \Xi_{17} \\ \vdots & \ddots & \vdots \\ \Xi_{71} & \cdots & \Xi_{77} \end{pmatrix} \cdot \begin{pmatrix} y_1 \\ y_4 \\ i_{r\_pi} \\ i_{R_1} \\ i_{L_1} \\ i_{L_2} \\ i_{L_3} \end{pmatrix} \approx 0$$

Assumption: arbitrarily small increments: $y_2 \to \varepsilon; y_3 \to \varepsilon$

$$\begin{pmatrix} \frac{r\_pi}{L_2} & \frac{R_1}{L_2} \\ 0 & 0 \\ 1 & 0 \\ 0 & 1 \\ 0 & 0 \\ 0 & 0 \\ 0 & 0 \end{pmatrix} \cdot \begin{pmatrix} y_2 \\ y_3 \end{pmatrix} \to \varepsilon; \Xi_{11} = -\lambda; \Xi_{12} = 0;$$

$$\Xi_{13} = \frac{1}{C_2 \cdot L_2} \cdot g_m \cdot r\_pi; \Xi_{14} = \frac{1}{C_2 \cdot L_2}$$

$$\Xi_{15} = \frac{1}{C_2 \cdot L_2}; \Xi_{16} = \Xi_{17} = 0; \Xi_{21} = 0; \Xi_{22} = -\lambda;$$
$$\Xi_{23} = -\frac{1}{L_3} \cdot (\frac{[1 + g_m \cdot r\_pi]}{C_{pi}} + \frac{g_m \cdot r\_pi}{C_1})$$

$$\Xi_{24} = -\frac{1}{C_1 \cdot L_3}; \Xi_{25} = \Xi_{26} = \Xi_{27} = -[\frac{1}{C_{pi}} + \frac{1}{C_1}] \cdot \frac{1}{L_3}; \Xi_{31} = \Xi_{32} = 0; \Xi_{33} = -\lambda$$

$$\Xi_{34} = \Xi_{35} = \Xi_{36} = \Xi_{37} = 0; \Xi_{41} = \Xi_{42} = \Xi_{43} = 0; \Xi_{44} = -\lambda; \Xi_{45} = \Xi_{46} = \Xi_{47} = 0$$

$$\Xi_{51} = \Xi_{52} = 0; \Xi_{53} = \frac{r\_pi}{L_1}; \Xi_{54} = \frac{R_1}{L_1}; \Xi_{55} = -\lambda; \Xi_{56} = \Xi_{57} = 0; \Xi_{61} = 1; \Xi_{62} = 0$$

$$\Xi_{63} = \Xi_{64} = \Xi_{65} = 0; \Xi_{66} = -\lambda; \Xi_{67} = 0; \Xi_{71} = 0; \Xi_{72} = 1; \Xi_{73} = \Xi_{74} = \Xi_{75} = \Xi_{76} = 0$$

$$\Xi_{77} = -\lambda; (A - \lambda \cdot I) = \begin{pmatrix} \Xi_{11} & \cdots & \Xi_{17} \\ \vdots & \ddots & \vdots \\ \Xi_{71} & \cdots & \Xi_{77} \end{pmatrix}; \det(A - \lambda \cdot I) = \det \begin{pmatrix} \Xi_{11} & \cdots & \Xi_{17} \\ \vdots & \ddots & \vdots \\ \Xi_{71} & \cdots & \Xi_{77} \end{pmatrix}$$

$$\det(A - \lambda \cdot I) = -\lambda^7 - \lambda^5 \cdot [\frac{1}{C_{pi}} + \frac{1}{C_1}] \cdot \frac{1}{L_3}; \det(A - \lambda \cdot I) = -\lambda^5 \cdot (\lambda^2 + [\frac{1}{C_{pi}} + \frac{1}{C_1}] \cdot \frac{1}{L_3})$$

If an eigenvalue of A is on the imaginary axis, then the method of the center manifold must be applied. To apply the stability criterion of Liapunov to our system, the equation $\det(A - \lambda \cdot I) = 0$ has a zero in the left half plane, without calculating the eigenvalues explicitly [2, 3].

$$\lambda_1 = \lambda_2 = \ldots = \lambda_5 = 0; \lambda^2 + [\frac{1}{C_{pi}} + \frac{1}{C_1}] \cdot \frac{1}{L_3} = 0 \Rightarrow \lambda_{6,7} = \pm j \cdot \sqrt{[\frac{1}{C_{pi}} + \frac{1}{C_1}] \cdot \frac{1}{L_3}}$$

6.4 Colpitts Oscillator Stability Analysis

The Colpitt's oscillator is designed for generation of high frequency sinusoidal oscillations (radio frequencies ranging from 10 kHz to 100 MHz). They are widely used in commercial signal generators up to 100 MHz. Colpitt's oscillator is same as Hartley oscillator except for one difference. Instead of using a tapped inductance, Colpitt's oscillator uses a tapped capacitance. The circuit diagram of Colpitt's oscillator using BJT. It consists of an R-C coupled amplifier using an NPN transistor in CE configuration. R_1 is resistor which forms a voltage bias to the transistor. We can connect resistor R_E which stabilizes the circuit against temperature variation (not in our circuit). If we connect R_E resistor then a capacitor C_E is connected parallel with R_E, acts as a bypass capacitor provides a low reactive path to the amplified AC signal. The coupling capacitor C_2 blocks DC and provides an AC path from collector to the tank circuit [25, 26] (Fig. 6.10).

The feedback network (tank circuit) consists of two capacitors C_3, C_4 (in series) which placed across a common inductor L_2. The Centre of the two capacitors is tapped (grounded). The feedback network (C_3, C_4, and L_2) determines the frequency of oscillation of the oscillator. The two capacitors C_3, C_4 form the potential divider led for providing the feedback voltage. The voltage developed across the capacitor

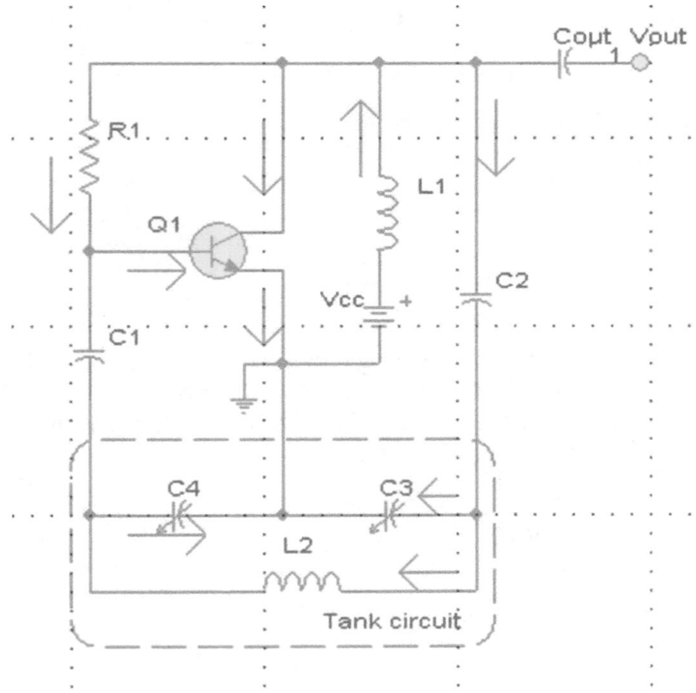

Fig. 6.10 Colpitts oscillator circuit

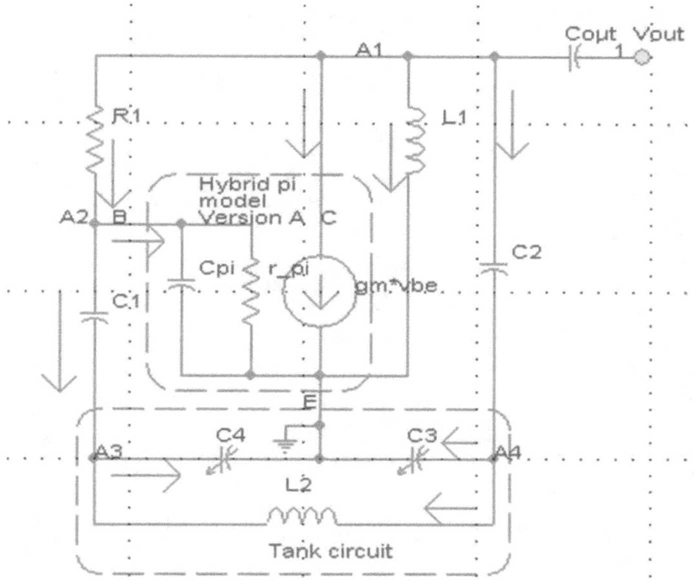

Fig. 6.11 Colpitts oscillator Hybrid Pi model (Version A) circuit

C_4 provides regenerative feedback which is essential for sustained oscillators. We use for our stability analysis the BJT NPN transistor Small Signal (SS) model equivalent circuit (Hybrid Pi model Version A). We consider "AC ground" in the circuit. Since the voltage at this terminal is held constant at V_{cc}, there is no time variation of the voltage. Consequently, we can set this terminal to be an "AC ground" in the small signal circuit. For AC ground, we "kill" the DC sources at that terminal (short circuit voltage sources and open circuit current sources) (Fig. 6.11)..

$$I_{R_1} = \frac{V_{A_1} - V_{A_2}}{R_1}; V_{L_1} = V_{A_1} = L_1 \cdot \frac{dI_{L_1}}{dt}; v_{be} = V_{A_2}; I_{r\_pi} = \frac{V_{A_2}}{r\_pi}; I_{C_{pi}} = C_{pi} \cdot \frac{dV_{C_{pi}}}{dt} = C_{pi} \cdot \frac{dV_{A_2}}{dt}$$

$$I_{C_2} = C_2 \cdot \frac{d(V_{A_1} - V_{A_4})}{dt}; I_{C_4} = C_4 \cdot \frac{dV_{A_3}}{dt}; I_{C_3} = C_3 \cdot \frac{dV_{A_4}}{dt}; V_{C_3} = V_{A_4}; V_{C_4} = V_{A_3}; V_{L_2} = L_2 \cdot \frac{dI_{L_2}}{dt}$$

$$V_{L_2} = V_{A_4} - V_{A_3}; V_{C_1} = V_{A_2} - V_{A_3}; V_{A_2} = V_{C_{pi}}; V_{R_1} = V_{A_1} - V_{A_2}; V_{L_1} = V_{A_1}; I_{C_1} = C_1 \cdot \frac{d(V_{A_2} - V_{A_3})}{dt}$$

KCL @ node A_1: $I_{R_1} + g_m \cdot v_{be} + I_{L_1} + I_{C_2} = 0$; KCL @ node A_2: $I_{R_1} = I_{C_1} + I_{C_{pi}} + I_{r\_pi}$

KCL @ node A_3: $I_{C_1} + I_{L_2} = I_{C_4}$; KCL @ node A_4: $I_{C_2} = I_{C_3} + I_{L_2}$

$$I_{r\_pi} = \frac{V_{A_2}}{r\_pi} \Rightarrow V_{A_2} = I_{r\_pi} \cdot r\_pi; I_{R_1} = \frac{V_{A_1} - V_{A_2}}{R_1} \Rightarrow V_{A_1} - V_{A_2} = I_{R_1} \cdot R_1; V_{A_1} = L_1 \cdot \frac{dI_{L_1}}{dt}$$

$$L_1 \cdot \frac{dI_{L_1}}{dt} - I_{r\_pi} \cdot r\_pi = I_{R_1} \cdot R_1; I_{C_{pi}} = C_{pi} \cdot \frac{dV_{A_2}}{dt} \Rightarrow V_{A_2} = \frac{1}{C_{pi}} \cdot \int I_{C_{pi}} \cdot dt$$

$$I_{C_2} = C_2 \cdot \frac{d(V_{A_1} - V_{A_4})}{dt} \Rightarrow V_{A_1} - V_{A_4} = \frac{1}{C_2} \cdot \int I_{C_2} \cdot dt; I_{C_4} = C_4 \cdot \frac{dV_{A_3}}{dt} \Rightarrow V_{A_3} = \frac{1}{C_4} \cdot \int I_{C_4} \cdot dt$$

$$I_{C_3} = C_3 \cdot \frac{dV_{A_4}}{dt} \Rightarrow V_{A_4} = \frac{1}{C_3} \cdot \int I_{C_3} \cdot dt; V_{A_4} - V_{A_3} = L_2 \cdot \frac{dI_{L_2}}{dt} \Rightarrow \frac{1}{C_3} \cdot \int I_{C_3} \cdot dt - \frac{1}{C_4} \cdot \int I_{C_4} \cdot dt = L_2 \cdot \frac{dI_{L_2}}{dt}$$

$$\frac{d}{dt}\{\frac{1}{C_3} \cdot \int I_{C_3} \cdot dt - \frac{1}{C_4} \cdot \int I_{C_4} \cdot dt = L_2 \cdot \frac{dI_{L_2}}{dt}\} \Rightarrow \frac{1}{C_3} \cdot I_{C_3} - \frac{1}{C_4} \cdot I_{C_4} = L_2 \cdot \frac{d^2 I_{L_2}}{dt^2}$$

$$I_{C_2} = C_2 \cdot \frac{d(V_{A_1} - V_{A_4})}{dt} \Rightarrow V_{A_1} - V_{A_4} = \frac{1}{C_2} \cdot \int I_{C_2} \cdot dt; V_{A_1} - V_{A_4} = \frac{1}{C_2} \cdot \int I_{C_2} \cdot dt$$

$$L_1 \cdot \frac{dI_{L_1}}{dt} - \frac{1}{C_3} \cdot \int I_{C_3} \cdot dt = \frac{1}{C_2} \cdot \int I_{C_2} \cdot dt; \frac{d}{dt}\{L_1 \cdot \frac{dI_{L_1}}{dt} - \frac{1}{C_3} \cdot \int I_{C_3} \cdot dt = \frac{1}{C_2} \cdot \int I_{C_2} \cdot dt\}$$

$$L_1 \cdot \frac{d^2 I_{L_1}}{dt^2} - \frac{1}{C_3} \cdot I_{C_3} = \frac{1}{C_2} \cdot I_{C_2}; I_{C_1} = C_1 \cdot \frac{d(V_{A_2} - V_{A_3})}{dt} \Rightarrow V_{A_2} - V_{A_3} = \frac{1}{C_1} \cdot \int I_{C_1} \cdot dt$$

$$I_{r\_pi} \cdot r\_pi - \frac{1}{C_4} \cdot \int I_{C_4} \cdot dt = \frac{1}{C_1} \cdot \int I_{C_1} \cdot dt; \frac{d}{dt}\{I_{r\_pi} \cdot r\_pi - \frac{1}{C_4} \cdot \int I_{C_4} \cdot dt = \frac{1}{C_1} \cdot \int I_{C_1} \cdot dt\}$$

$$\frac{dI_{r\_pi}}{dt} \cdot r\_pi - \frac{1}{C_4} \cdot I_{C_4} = \frac{1}{C_1} \cdot I_{C_1}$$

We can summery our Colpitt's oscillator differential equations:

$$L_1 \cdot \frac{dI_{L_1}}{dt} - I_{r\_pi} \cdot r\_pi = I_{R_1} \cdot R_1; \frac{1}{C_3} \cdot I_{C_3} - \frac{1}{C_4} \cdot I_{C_4} = L_2 \cdot \frac{d^2 I_{L_2}}{dt^2}$$

$$L_1 \cdot \frac{d^2 I_{L_1}}{dt^2} - \frac{1}{C_3} \cdot I_{C_3} = \frac{1}{C_2} \cdot I_{C_2}; \frac{dI_{r\_pi}}{dt} \cdot r\_pi - \frac{1}{C_4} \cdot I_{C_4} = \frac{1}{C_1} \cdot I_{C_1}$$

KCL @ nodes A_1 – A_4: $I_{R_1} + g_m \cdot I_{r\_pi} \cdot r\_pi + I_{L_1} + I_{C_2} = 0; I_{R_1} = I_{C_1} + I_{C_{pi}} + I_{r\_pi}$

$$I_{C_1} + I_{L_2} = I_{C_4}; I_{C_2} = I_{C_3} + I_{L_2}.$$

$$\frac{dI_{L_1}}{dt} = I_{R_1} \cdot \frac{R_1}{L_1} + I_{r\_pi} \cdot \frac{r\_pi}{L_1} \Rightarrow \frac{d^2 I_{L_1}}{dt^2} = \frac{dI_{R_1}}{dt} \cdot \frac{R_1}{L_1} + \frac{dI_{r\_pi}}{dt} \cdot \frac{r\_pi}{L_1}$$

$$\frac{d^2 I_{L_1}}{dt^2} = \frac{1}{C_2 \cdot L_1} \cdot I_{C_2} + \frac{1}{C_3 \cdot L_1} \cdot I_{C_3}; \frac{dI_{R_1}}{dt} \cdot \frac{R_1}{L_1} + \frac{dI_{r\_pi}}{dt} \cdot \frac{r\_pi}{L_1}$$
$$= \frac{1}{C_2 \cdot L_1} \cdot I_{C_2} + \frac{1}{C_3 \cdot L_1} \cdot I_{C_3}$$

$$\frac{dI_{R_1}}{dt} \cdot R_1 + \frac{dI_{r\_pi}}{dt} \cdot r\_pi = \frac{1}{C_2} \cdot I_{C_2} + \frac{1}{C_3} \cdot I_{C_3}; \frac{dI_{r\_pi}}{dt}$$
$$= \frac{1}{C_1 \cdot r\_pi} \cdot I_{C_1} + \frac{1}{C_4 \cdot r\_pi} \cdot I_{C_4}$$

$$\frac{dI_{R_1}}{dt} \cdot R_1 + [\frac{1}{C_1 \cdot r\_pi} \cdot I_{C_1} + \frac{1}{C_4 \cdot r\_pi} \cdot I_{C_4}] \cdot r\_pi = \frac{1}{C_2} \cdot I_{C_2} + \frac{1}{C_3} \cdot I_{C_3}$$

$$\frac{dI_{R_1}}{dt} = \frac{1}{C_2 \cdot R_1} \cdot I_{C_2} + \frac{1}{C_3 \cdot R_1} \cdot I_{C_3} - \frac{1}{C_1 \cdot R_1} \cdot I_{C_1} - \frac{1}{C_4 \cdot R_1} \cdot I_{C_4}$$

$$L_2 \cdot \frac{d^2 I_{L_2}}{dt^2} = \frac{1}{C_3} \cdot I_{C_3} - \frac{1}{C_4} \cdot I_{C_4}; \frac{dI_{L_2}}{dt} = Y_1; \frac{dY_1}{dt} = \frac{1}{C_3 \cdot L_2} \cdot I_{C_3} - \frac{1}{C_4 \cdot L_2} \cdot I_{C_4}$$

We can summery our system differential equations (version 1):

$$\frac{dI_{R_1}}{dt} = \frac{1}{C_2 \cdot R_1} \cdot I_{C_2} + \frac{1}{C_3 \cdot R_1} \cdot I_{C_3} - \frac{1}{C_1 \cdot R_1} \cdot I_{C_1} - \frac{1}{C_4 \cdot R_1} \cdot I_{C_4}; \frac{dI_{L_1}}{dt}$$
$$= I_{R_1} \cdot \frac{R_1}{L_1} + I_{r\_pi} \cdot \frac{r\_pi}{L_1}$$

$$\frac{dI_{r\_pi}}{dt} = \frac{1}{C_1 \cdot r\_pi} \cdot I_{C_1} + \frac{1}{C_4 \cdot r\_pi} \cdot I_{C_4}; \frac{dI_{L_2}}{dt} = Y_1; \frac{dY_1}{dt} = \frac{1}{C_3 \cdot L_2} \cdot I_{C_3} - \frac{1}{C_4 \cdot L_2} \cdot I_{C_4}$$

KCL @ nodes A₁ – A₄: $I_{C_2} = -I_{R_1} - g_m \cdot I_{r\_pi} \cdot r\_pi - I_{L_1}; I_{C_1} = I_{R_1} - I_{C_{pi}} - I_{r\_pi}$

$$I_{C_4} = I_{R_1} - I_{C_{pi}} - I_{r\_pi} + I_{L_2}; I_{C_3} = -I_{R_1} - g_m \cdot I_{r\_pi} \cdot r\_pi - \sum_{k=1}^{2} I_{L_k}.$$

We can summery our system differential equations (version 2):

$$\frac{dI_{R_1}}{dt} = \frac{1}{C_2 \cdot R_1} \cdot [-I_{R_1} - g_m \cdot I_{r\_pi} \cdot r\_pi - I_{L_1}] + \frac{1}{C_3 \cdot R_1} \cdot [-I_{R_1} - g_m \cdot I_{r\_pi} \cdot r\_pi - \sum_{k=1}^{2} I_{L_k}] - \frac{1}{C_1 \cdot R_1} \cdot [I_{R_1} - I_{C_{pi}} - I_{r\_pi}] - \frac{1}{C_4 \cdot R_1} \cdot [I_{R_1} - I_{C_{pi}} - I_{r\_pi} + I_{L_2}]$$

$$\frac{dI_{R_1}}{dt} = -\frac{1}{C_2 \cdot R_1} \cdot I_{R_1} - \frac{1}{C_2 \cdot R_1} \cdot g_m \cdot I_{r\_pi} \cdot r\_pi - \frac{1}{C_2 \cdot R_1} \cdot I_{L_1} - \frac{1}{C_3 \cdot R_1} \cdot I_{R_1} - \frac{1}{C_3 \cdot R_1} \cdot g_m \cdot I_{r\_pi} \cdot r\_pi - \frac{1}{C_3 \cdot R_1} \cdot \sum_{k=1}^{2} I_{L_k} - \frac{1}{C_1 \cdot R_1} \cdot I_{R_1} + \frac{1}{C_1 \cdot R_1} \cdot I_{C_{pi}} + \frac{1}{C_1 \cdot R_1} \cdot I_{r\_pi} - \frac{1}{C_4 \cdot R_1} \cdot I_{R_1} + \frac{1}{C_4 \cdot R_1} \cdot I_{C_{pi}} + \frac{1}{C_4 \cdot R_1} \cdot I_{r\_pi} - \frac{1}{C_4 \cdot R_1} \cdot I_{L_2}$$

$$\frac{dI_{R_1}}{dt} = -\frac{1}{C_2 \cdot R_1} \cdot I_{R_1} - \frac{1}{C_1 \cdot R_1} \cdot I_{R_1} - \frac{1}{C_3 \cdot R_1} \cdot I_{R_1} - \frac{1}{C_4 \cdot R_1} \cdot I_{R_1} - \frac{1}{C_2 \cdot R_1} \cdot g_m \cdot I_{r\_pi} \cdot r\_pi - \frac{1}{C_3 \cdot R_1} \cdot g_m \cdot I_{r\_pi} \cdot r\_pi + \frac{1}{C_1 \cdot R_1} \cdot I_{r\_pi} + \frac{1}{C_4 \cdot R_1} \cdot I_{r\_pi} + \frac{1}{C_1 \cdot R_1} \cdot I_{C_{pi}} + \frac{1}{C_4 \cdot R_1} \cdot I_{C_{pi}} - \frac{1}{C_2 \cdot R_1} \cdot I_{L_1} - \frac{1}{C_4 \cdot R_1} \cdot I_{L_2} - \frac{1}{C_3 \cdot R_1} \cdot \sum_{k=1}^{2} I_{L_k}$$

$$\frac{dI_{R_1}}{dt} = -[\sum_{k=1}^{4} \frac{1}{C_k}] \cdot \frac{1}{R_1} \cdot I_{R_1} + [\sum_{\substack{k=1 \\ k\neq 2 \\ k\neq 3}}^{4} \frac{1}{C_k} - (\sum_{k=2}^{3} \frac{1}{C_k}) \cdot g_m \cdot r\_pi] \cdot \frac{1}{R_1} \cdot I_{r\_pi} + (\sum_{\substack{k=1 \\ k\neq 2 \\ k\neq 3}}^{4} \frac{1}{C_k}) \cdot \frac{1}{R_1} \cdot I_{C_{pi}} - [\sum_{k=2}^{3} \frac{1}{C_k}] \cdot \frac{1}{R_1} \cdot I_{L_1} - [\sum_{k=3}^{4} \frac{1}{C_k}] \cdot \frac{1}{R_1} \cdot I_{L_2}; \frac{dI_{L_1}}{dt} = I_{R_1} \cdot \frac{R_1}{L_1} + I_{r\_pi} \cdot \frac{r\_pi}{L_1}$$

$$\frac{dI_{r\_pi}}{dt} = \frac{1}{C_1 \cdot r\_pi} \cdot [I_{R_1} - I_{C_{pi}} - I_{r\_pi}] + \frac{1}{C_4 \cdot r\_pi} \cdot [I_{R_1} - I_{C_{pi}} - I_{r\_pi} + I_{L_2}]$$

$$\frac{dI_{r\_pi}}{dt} = [\sum_{\substack{k=1\\k\neq 2\\k\neq 3}}^{4} \frac{1}{C_k}] \cdot \frac{1}{r\_pi} \cdot I_{R_1} - [\sum_{\substack{k=1\\k\neq 2\\k\neq 3}}^{4} \frac{1}{C_k}] \cdot \frac{1}{r\_pi} \cdot I_{C_{pi}} - [\sum_{\substack{k=1\\k\neq 2\\k\neq 3}}^{4} \frac{1}{C_k}] \cdot \frac{1}{r\_pi} \cdot I_{r\_pi} + \frac{1}{C_4 \cdot r\_pi} \cdot I_{L_2}$$

$$\frac{dI_{L_2}}{dt} = Y_1; \frac{dY_1}{dt} = \frac{1}{C_3 \cdot L_2} \cdot [-I_{R_1} - g_m \cdot I_{r\_pi} \cdot r\_pi - \sum_{k=1}^{2} I_{L_k}] - \frac{1}{C_4 \cdot L_2} \cdot [I_{R_1} - I_{C_{pi}} - I_{r\_pi} + I_{L_2}]$$

$$\frac{dY_1}{dt} = -\frac{1}{C_3 \cdot L_2} \cdot I_{R_1} - \frac{1}{C_4 \cdot L_2} \cdot I_{R_1} - \frac{1}{C_3 \cdot L_2} \cdot g_m \cdot I_{r\_pi} \cdot r\_pi + \frac{1}{C_4 \cdot L_2} \cdot I_{r\_pi} - \frac{1}{C_3 \cdot L_2} \cdot I_{L_1} - \frac{1}{C_3 \cdot L_2} \cdot I_{L_2} - \frac{1}{C_4 \cdot L_2} \cdot I_{L_2} + \frac{1}{C_4 \cdot L_2} \cdot I_{C_{pi}}$$

$$\frac{dY_1}{dt} = -[\sum_{k=3}^{4} \frac{1}{C_k}] \cdot \frac{1}{L_2} \cdot I_{R_1} + [\frac{1}{C_4} - \frac{1}{C_3} \cdot g_m \cdot r\_pi] \cdot \frac{1}{L_2} \cdot I_{r\_pi} - \frac{1}{C_3 \cdot L_2} \cdot I_{L_1} - [\sum_{k=3}^{4} \frac{1}{C_k}] \cdot \frac{1}{L_2} \cdot I_{L_2} + \frac{1}{C_4 \cdot L_2} \cdot I_{C_{pi}}$$

We can summery our system differential equations (version 2):

$$\frac{dI_{R_1}}{dt} = -[\sum_{k=1}^{4} \frac{1}{C_k}] \cdot \frac{1}{R_1} \cdot I_{R_1} + [\sum_{\substack{k=1\\k\neq 2\\k\neq 3}}^{4} \frac{1}{C_k} - (\sum_{k=2}^{3} \frac{1}{C_k}) \cdot g_m \cdot r\_pi] \cdot \frac{1}{R_1} \cdot I_{r\_pi} + (\sum_{\substack{k=1\\k\neq 2\\k\neq 3}}^{4} \frac{1}{C_k}) \cdot \frac{1}{R_1} \cdot I_{C_{pi}} - [\sum_{k=2}^{3} \frac{1}{C_k}] \cdot \frac{1}{R_1} \cdot I_{L_1} - [\sum_{k=3}^{4} \frac{1}{C_k}] \cdot \frac{1}{R_1} \cdot I_{L_2}$$

$$\frac{dI_{r\_pi}}{dt} = [\sum_{\substack{k=1\\k\neq 2\\k\neq 3}}^{4} \frac{1}{C_k}] \cdot \frac{1}{r\_pi} \cdot I_{R_1} - [\sum_{\substack{k=1\\k\neq 2\\k\neq 3}}^{4} \frac{1}{C_k}] \cdot \frac{1}{r\_pi} \cdot I_{C_{pi}} - [\sum_{\substack{k=1\\k\neq 2\\k\neq 3}}^{4} \frac{1}{C_k}] \cdot \frac{1}{r\_pi} \cdot I_{r\_pi} + \frac{1}{C_4 \cdot r\_pi} \cdot I_{L_2}$$

$$\frac{dY_1}{dt} = -[\sum_{k=3}^{4} \frac{1}{C_k}] \cdot \frac{1}{L_2} \cdot I_{R_1} + [\frac{1}{C_4} - \frac{1}{C_3} \cdot g_m \cdot r\_pi] \cdot \frac{1}{L_2} \cdot I_{r\_pi} - \frac{1}{C_3 \cdot L_2} \cdot I_{L_1} - [\sum_{k=3}^{4} \frac{1}{C_k}] \cdot \frac{1}{L_2} \cdot I_{L_2} + \frac{1}{C_4 \cdot L_2} \cdot I_{C_{pi}}$$

$$\frac{dI_{L_1}}{dt} = I_{R_1} \cdot \frac{R_1}{L_1} + I_{r\_pi} \cdot \frac{r\_pi}{L_1}; \frac{dI_{L_2}}{dt} = Y_1$$

At fixed points (equilibrium points):

$$\frac{dI_{R_1}}{dt} = 0; \frac{dI_{r\_pi}}{dt} = 0; \frac{dY_1}{dt} = 0; \frac{dI_{L_1}}{dt} = 0; \frac{dI_{L_2}}{dt} = 0$$

$$\frac{dI_{L_2}}{dt} = 0 \Rightarrow Y_1^* = 0; \frac{dI_{L_1}}{dt} = 0 \Rightarrow I_{R_1}^* \cdot \frac{R_1}{L_1} + I_{r\_pi}^* \cdot \frac{r\_pi}{L_1} = 0$$

$$\frac{dY_1}{dt} = 0 \Rightarrow -[\sum_{k=3}^{4} \frac{1}{C_k}] \cdot \frac{1}{L_2} \cdot I_{R_1}^* + [\frac{1}{C_4} - \frac{1}{C_3} \cdot g_m \cdot r\_pi] \cdot \frac{1}{L_2} \cdot I_{r\_pi}^* - \frac{1}{C_3 \cdot L_2} \cdot I_{L_1}^*$$
$$- [\sum_{k=3}^{4} \frac{1}{C_k}] \cdot \frac{1}{L_2} \cdot I_{L_2}^* + \frac{1}{C_4 \cdot L_2} \cdot I_{C_{pi}}^* = 0$$

$$\frac{dI_{r\_pi}}{dt} = 0$$
$$\Rightarrow [\sum_{\substack{k=1 \\ k\neq 2 \\ k\neq 3}}^{4} \frac{1}{C_k}] \cdot \frac{1}{r\_pi} \cdot I_{R_1}^* - [\sum_{\substack{k=1 \\ k\neq 2 \\ k\neq 3}}^{4} \frac{1}{C_k}] \cdot \frac{1}{r\_pi} \cdot I_{C_{pi}}^* - [\sum_{\substack{k=1 \\ k\neq 2 \\ k\neq 3}}^{4} \frac{1}{C_k}] \cdot \frac{1}{r\_pi}$$
$$\cdot I_{r\_pi}^* + \frac{1}{C_4 \cdot r\_pi} \cdot I_{L_2}^*$$
$$= 0$$

$$\frac{dI_{R_1}}{dt} = 0 \Rightarrow -[\sum_{k=1}^{4} \frac{1}{C_k}] \cdot \frac{1}{R_1} \cdot I_{R_1}^* + [\sum_{\substack{k=1 \\ k\neq 2 \\ k\neq 3}}^{4} \frac{1}{C_k} - (\sum_{k=2}^{3} \frac{1}{C_k}) \cdot g_m \cdot r\_pi] \cdot \frac{1}{R_1} \cdot I_{r\_pi}^*$$
$$+ (\sum_{\substack{k=1 \\ k\neq 2 \\ k\neq 3}}^{4} \frac{1}{C_k}) \cdot \frac{1}{R_1} \cdot I_{C_{pi}}^* - [\sum_{k=2}^{3} \frac{1}{C_k}] \cdot \frac{1}{R_1} \cdot I_{L_1}^* - [\sum_{k=3}^{4} \frac{1}{C_k}] \cdot \frac{1}{R_1} \cdot I_{L_2}^* = 0$$

We can summery our system fixed points equations (version 1):

$$Y_1^* = 0; I_{R_1}^* \cdot R_1 + I_{r\_pi}^* \cdot r\_pi = 0 \Rightarrow I_{R_1}^* = -I_{r\_pi}^* \cdot \frac{r\_pi}{R_1}$$

$$-[\sum_{k=3}^{4}\frac{1}{C_k}]\cdot\frac{1}{L_2}\cdot I^*_{R_1}+[\frac{1}{C_4}-\frac{1}{C_3}\cdot g_m\cdot r\_pi]\cdot\frac{1}{L_2}\cdot I^*_{r\_pi}-\frac{1}{C_3\cdot L_2}\cdot I^*_{L_1}-[\sum_{k=3}^{4}\frac{1}{C_k}]\cdot\frac{1}{L_2}\cdot I^*_{L_2}$$
$$+\frac{1}{C_4\cdot L_2}\cdot I^*_{C_{pi}}=0$$

$$[\sum_{\substack{k=1\\k\neq2\\k\neq3}}^{4}\frac{1}{C_k}]\cdot\frac{1}{r\_pi}\cdot I^*_{R_1}-[\sum_{\substack{k=1\\k\neq2\\k\neq3}}^{4}\frac{1}{C_k}]\cdot\frac{1}{r\_pi}\cdot I^*_{C_{pi}}-[\sum_{\substack{k=1\\k\neq2\\k\neq3}}^{4}\frac{1}{C_k}]\cdot\frac{1}{r\_pi}\cdot I^*_{r\_pi}+\frac{1}{C_4\cdot r\_pi}\cdot I^*_{L_2}=0$$

$$-[\sum_{k=1}^{4}\frac{1}{C_k}]\cdot\frac{1}{R_1}\cdot I^*_{R_1}+[\sum_{\substack{k=1\\k\neq2\\k\neq3}}^{4}\frac{1}{C_k}-(\sum_{k=2}^{3}\frac{1}{C_k})\cdot g_m\cdot r\_pi]\cdot\frac{1}{R_1}\cdot I^*_{r\_pi}+(\sum_{\substack{k=1\\k\neq2\\k\neq3}}^{4}\frac{1}{C_k})\cdot\frac{1}{R_1}\cdot I^*_{C_{pi}}$$
$$-[\sum_{k=2}^{3}\frac{1}{C_k}]\cdot\frac{1}{R_1}\cdot I^*_{L_1}-[\sum_{k=3}^{4}\frac{1}{C_k}]\cdot\frac{1}{R_1}\cdot I^*_{L_2}=0$$

We can summery our system fixed points equations (version 2):

$$(*)\{[\sum_{k=3}^{4}\frac{1}{C_k}]\cdot\frac{r\_pi}{R_1\cdot L_2}+[\frac{1}{C_4}-\frac{1}{C_3}\cdot g_m\cdot r\_pi]\cdot\frac{1}{L_2}\}\cdot I^*_{r\_pi}-\frac{1}{C_3\cdot L_2}\cdot I^*_{L_1}$$
$$-[\sum_{k=3}^{4}\frac{1}{C_k}]\cdot\frac{1}{L_2}\cdot I^*_{L_2}+\frac{1}{C_4\cdot L_2}\cdot I^*_{C_{pi}}=0$$

$$(**)-[\sum_{\substack{k=1\\k\neq2\\k\neq3}}^{4}\frac{1}{C_k}]\cdot(\frac{1}{R_1}+\frac{1}{r\_pi})\cdot I^*_{r\_pi}+\frac{1}{C_4\cdot r\_pi}\cdot I^*_{L_2}-[\sum_{\substack{k=1\\k\neq2\\k\neq3}}^{4}\frac{1}{C_k}]\cdot\frac{1}{r\_pi}\cdot I^*_{C_{pi}}=0$$

$$(***)\{[\sum_{k=1}^{4}\frac{1}{C_k}]\cdot\frac{r\_pi}{R_1}+[\sum_{\substack{k=1\\k\neq2\\k\neq3}}^{4}\frac{1}{C_k}-(\sum_{k=2}^{3}\frac{1}{C_k})\cdot g_m\cdot r\_pi]\}\cdot\frac{1}{R_1}\cdot I^*_{r\_pi}+(\sum_{\substack{k=1\\k\neq2\\k\neq3}}^{4}\frac{1}{C_k})\cdot\frac{1}{R_1}\cdot I^*_{C_{pi}}$$
$$-[\sum_{k=2}^{3}\frac{1}{C_k}]\cdot\frac{1}{R_1}\cdot I^*_{L_1}-[\sum_{k=3}^{4}\frac{1}{C_k}]\cdot\frac{1}{R_1}\cdot I^*_{L_2}=0$$

We can summery our system fixed points equations (version 3):

$$(**)I^*_{L_2}=[\sum_{\substack{k=1\\k\neq2\\k\neq3}}^{4}\frac{1}{C_k}]\cdot C_4\cdot I^*_{C_{pi}}+[\sum_{\substack{k=1\\k\neq2\\k\neq3}}^{4}\frac{1}{C_k}]\cdot(\frac{C_4\cdot r\_pi}{R_1}+C_4)\cdot I^*_{r\_pi}$$

$(**) \rightarrow (*)$

$$\{[\sum_{k=3}^{4}\frac{1}{C_k}]\cdot\frac{r\_pi}{R_1\cdot L_2}+[\frac{1}{C_4}-\frac{1}{C_3}\cdot g_m\cdot r\_pi]\cdot\frac{1}{L_2}\}\cdot I^*_{r\_pi}-\frac{1}{C_3\cdot L_2}\cdot I^*_{L_1}$$
$$-[\sum_{k=3}^{4}\frac{1}{C_k}]\cdot\frac{1}{L_2}\cdot\{[\sum_{\substack{k=1\\k\neq 2\\k\neq 3}}^{4}\frac{1}{C_k}]\cdot C_4\cdot I^*_{C_{pi}}+[\sum_{\substack{k=1\\k\neq 2\\k\neq 3}}^{4}\frac{1}{C_k}]\cdot(\frac{C_4\cdot r\_pi}{R_1}+C_4)\cdot I^*_{r\_pi}\}$$
$$+\frac{1}{C_4\cdot L_2}\cdot I^*_{C_{pi}}=0$$

$$\{[\sum_{k=3}^{4}\frac{1}{C_k}]\cdot\frac{r\_pi}{R_1\cdot L_2}+[\frac{1}{C_4}-\frac{1}{C_3}\cdot g_m\cdot r\_pi]\cdot\frac{1}{L_2}-[\sum_{k=3}^{4}\frac{1}{C_k}]\cdot\frac{1}{L_2}\cdot[\sum_{\substack{k=1\\k\neq 2\\k\neq 3}}^{4}\frac{1}{C_k}]\cdot(\frac{C_4\cdot r\_pi}{R_1}+C_4)\cdot I^*_{r\_pi}\}$$
$$-\frac{1}{C_3\cdot L_2}\cdot I^*_{L_1}+\{\frac{1}{C_4}-[\sum_{k=3}^{4}\frac{1}{C_k}]\cdot[\sum_{\substack{k=1\\k\neq 2\\k\neq 3}}^{4}\frac{1}{C_k}]\cdot C_4\}\cdot\frac{1}{L_2}\cdot I^*_{C_{pi}}=0$$

$(**) \rightarrow (***)$

$$\{[\sum_{k=1}^{4}\frac{1}{C_k}]\cdot\frac{r\_pi}{R_1}+[\sum_{\substack{k=1\\k\neq 2\\k\neq 3}}^{4}\frac{1}{C_k}-(\sum_{k=2}^{3}\frac{1}{C_k})\cdot g_m\cdot r\_pi]\}\cdot\frac{1}{R_1}\cdot I^*_{r\_pi}+(\sum_{\substack{k=1\\k\neq 2\\k\neq 3}}^{4}\frac{1}{C_k})\cdot\frac{1}{R_1}\cdot I^*_{C_{pi}}$$
$$-[\sum_{k=2}^{3}\frac{1}{C_k}]\cdot\frac{1}{R_1}\cdot I^*_{L_1}-[\sum_{k=3}^{4}\frac{1}{C_k}]\cdot\frac{1}{R_1}\cdot\{[\sum_{\substack{k=1\\k\neq 2\\k\neq 3}}^{4}\frac{1}{C_k}]\cdot C_4\cdot I^*_{C_{pi}}+[\sum_{\substack{k=1\\k\neq 2\\k\neq 3}}^{4}\frac{1}{C_k}]\cdot(\frac{C_4\cdot r\_pi}{R_1}+C_4)\cdot I^*_{r\_pi}\}=0$$

$$\{\{[\sum_{k=1}^{4}\frac{1}{C_k}]\cdot\frac{r\_pi}{R_1}+[\sum_{\substack{k=1\\k\neq 2\\k\neq 3}}^{4}\frac{1}{C_k}-(\sum_{k=2}^{3}\frac{1}{C_k})\cdot g_m\cdot r\_pi]\}\cdot\frac{1}{R_1}$$
$$-[\sum_{k=3}^{4}\frac{1}{C_k}]\cdot\frac{1}{R_1}\cdot[\sum_{\substack{k=1\\k\neq 2\\k\neq 3}}^{4}\frac{1}{C_k}]\cdot(\frac{C_4\cdot r\_pi}{R_1}+C_4)\}\cdot I^*_{r\_pi}$$
$$+\{(\sum_{\substack{k=1\\k\neq 2\\k\neq 3}}^{4}\frac{1}{C_k})\cdot\frac{1}{R_1}-[\sum_{k=3}^{4}\frac{1}{C_k}]\cdot\frac{1}{R_1}\cdot[\sum_{\substack{k=1\\k\neq 2\\k\neq 3}}^{4}\frac{1}{C_k}]\cdot C_4\}\cdot I^*_{C_{pi}}-[\sum_{k=2}^{3}\frac{1}{C_k}]\cdot\frac{1}{R_1}\cdot I^*_{L_1}=0$$

Remark: it is reader exercise to get the exact fixed points options in our case.

Stability analysis: The standard local stability analysis about any one of the equilibrium points of the Colpitt's oscillator circuit (BJT transistor small signal model) consists in adding to coordinate $[Y_1, I_{r\_pi}, I_{R_1}, I_{L_1}, I_{L_2}]$ arbitrarily small increments of exponentially form $[y_1, i_{r\_pi}, i_{R_1}, i_{L_1}, i_{L_2}]\cdot e^{\lambda\cdot t}$ and retaining the first order terms in $Y_1, I_{r\_pi}, I_{R_1}, I_{L_1}, I_{L_2}$. The system of homogenous equations leads to a polynomial characteristic equation in the eigenvalues [4]. The polynomial characteristic equations accept by set of the below circuit variables, circuit variables derivative and circuit variables second order derivative with respect to time into Colpitt's oscillator circuit [2, 3, 4]. Our Colpitt's oscillator fixed values with arbitrarily small increments of exponential form $[y_1, i_{r\_pi}, i_{R_1}, i_{L_1}, i_{L_2}] \cdot e^{\lambda \cdot t}$ are: j = 0 (first fixed point), j = 1(second fixed point), j = 2(third fixed point), etc.,

$$I_{r\_pi}(t) = I_{r\_pi}^{(j)} + i_{r\_pi} \cdot e^{\lambda \cdot t}; I_{R_1}(t) = I_{R_1}^{(j)} + i_{R_1} \cdot e^{\lambda \cdot t}; I_{L_1}(t) = I_{L_1}^{(j)} + i_{L_1} \cdot e^{\lambda \cdot t}; I_{L_2}(t) = I_{L_2}^{(j)} + i_{L_2} \cdot e^{\lambda \cdot t}$$

$$Y_1(t) = Y_1^{(j)} + y_1 \cdot e^{\lambda \cdot t}; \frac{dY_1(t)}{dt} = y_1 \cdot \lambda \cdot e^{\lambda \cdot t}; \frac{dI_{r\_pi}(t)}{dt} = i_{r\_pi} \cdot \lambda \cdot e^{\lambda \cdot t}; I_{C_{pi}}(t) = I_{C_{pi}}^{(j)} + i_{C_{pi}} \cdot e^{\lambda \cdot t}$$

$$\frac{dI_{R_1}(t)}{dt} = i_{R_1} \cdot \lambda \cdot e^{\lambda \cdot t}; \frac{dI_{L_1}(t)}{dt} = i_{L_1} \cdot \lambda \cdot e^{\lambda \cdot t}; \frac{dI_{L_2}(t)}{dt} = i_{L_2} \cdot \lambda \cdot e^{\lambda \cdot t}; \frac{dI_{C_{pi}}(t)}{dt} = i_{C_{pi}} \cdot \lambda \cdot e^{\lambda \cdot t}$$

&&&

$$\frac{dI_{R_1}}{dt} = -\left[\sum_{k=1}^{4} \frac{1}{C_k}\right] \cdot \frac{1}{R_1} \cdot I_{R_1} + \left[\sum_{\substack{k=1 \\ k \neq 2 \\ k \neq 3}}^{4} \frac{1}{C_k} - \left(\sum_{k=2}^{3} \frac{1}{C_k}\right) \cdot g_m \cdot r\_pi\right] \cdot \frac{1}{R_1} \cdot I_{r\_pi} + \left(\sum_{\substack{k=1 \\ k \neq 2 \\ k \neq 3}}^{4} \frac{1}{C_k}\right) \cdot \frac{1}{R_1} \cdot I_{C_{pi}} - \left[\sum_{k=2}^{3} \frac{1}{C_k}\right] \cdot \frac{1}{R_1} \cdot I_{L_1} - \left[\sum_{k=3}^{4} \frac{1}{C_k}\right] \cdot \frac{1}{R_1} \cdot I_{L_2}$$

$$\begin{aligned} i_{R_1} \cdot \lambda \cdot e^{\lambda \cdot t} = & -\left[\sum_{k=1}^{4} \frac{1}{C_k}\right] \cdot \frac{1}{R_1} \cdot [I_{R_1}^{(j)} + i_{R_1} \cdot e^{\lambda \cdot t}] \\ & + \left[\sum_{\substack{k=1 \\ k \neq 2 \\ k \neq 3}}^{4} \frac{1}{C_k} - \left(\sum_{k=2}^{3} \frac{1}{C_k}\right) \cdot g_m \cdot r\_pi\right] \cdot \frac{1}{R_1} \cdot [I_{r\_pi}^{(j)} + i_{r\_pi} \cdot e^{\lambda \cdot t}] \\ & + \left(\sum_{\substack{k=1 \\ k \neq 2 \\ k \neq 3}}^{4} \frac{1}{C_k}\right) \cdot \frac{1}{R_1} \cdot [I_{C_{pi}}^{(j)} + i_{C_{pi}} \cdot e^{\lambda \cdot t}] - \left[\sum_{k=2}^{3} \frac{1}{C_k}\right] \cdot \frac{1}{R_1} \cdot [I_{L_1}^{(j)} + i_{L_1} \cdot e^{\lambda \cdot t}] \\ & - \left[\sum_{k=3}^{4} \frac{1}{C_k}\right] \cdot \frac{1}{R_1} \cdot [I_{L_2}^{(j)} + i_{L_2} \cdot e^{\lambda \cdot t}] \end{aligned}$$

$$i_{R_1} \cdot \lambda \cdot e^{\lambda \cdot t} = \{-[\sum_{k=1}^{4} \frac{1}{C_k}] \cdot \frac{1}{R_1} \cdot I_{R_1}^{(j)} + [\sum_{\substack{k=1 \\ k \neq 2 \\ k \neq 3}}^{4} \frac{1}{C_k} - (\sum_{k=2}^{3} \frac{1}{C_k}) \cdot g_m \cdot r\_pi] \cdot \frac{1}{R_1} \cdot I_{r\_pi}^{(j)}$$

$$+ (\sum_{\substack{k=1 \\ k \neq 2 \\ k \neq 3}}^{4} \frac{1}{C_k}) \cdot \frac{1}{R_1} \cdot I_{C_{pi}}^{(j)} - [\sum_{k=2}^{3} \frac{1}{C_k}] \cdot \frac{1}{R_1} \cdot I_{L_1}^{(j)} - [\sum_{k=3}^{4} \frac{1}{C_k}] \cdot \frac{1}{R_1} \cdot I_{L_2}^{(j)}\}$$

$$- [\sum_{k=1}^{4} \frac{1}{C_k}] \cdot \frac{1}{R_1} \cdot i_{R_1} \cdot e^{\lambda \cdot t} + [\sum_{\substack{k=1 \\ k \neq 2 \\ k \neq 3}}^{4} \frac{1}{C_k} - (\sum_{k=2}^{3} \frac{1}{C_k}) \cdot g_m \cdot r\_pi] \cdot \frac{1}{R_1} \cdot i_{r\_pi} \cdot e^{\lambda \cdot t}$$

$$+ (\sum_{\substack{k=1 \\ k \neq 2 \\ k \neq 3}}^{4} \frac{1}{C_k}) \cdot \frac{1}{R_1} \cdot i_{C_{pi}} \cdot e^{\lambda \cdot t} - [\sum_{k=2}^{3} \frac{1}{C_k}] \cdot \frac{1}{R_1} \cdot i_{L_1} \cdot e^{\lambda \cdot t} - [\sum_{k=3}^{4} \frac{1}{C_k}] \cdot \frac{1}{R_1} \cdot i_{L_2} \cdot e^{\lambda \cdot t}$$

At fixed points:

$$- [\sum_{k=1}^{4} \frac{1}{C_k}] \cdot \frac{1}{R_1} \cdot I_{R_1}^{(j)} + [\sum_{\substack{k=1 \\ k \neq 2 \\ k \neq 3}}^{4} \frac{1}{C_k} - (\sum_{k=2}^{3} \frac{1}{C_k}) \cdot g_m \cdot r\_pi] \cdot \frac{1}{R_1} \cdot I_{r\_pi}^{(j)} + (\sum_{\substack{k=1 \\ k \neq 2 \\ k \neq 3}}^{4} \frac{1}{C_k}) \cdot \frac{1}{R_1} \cdot I_{C_{pi}}^{(j)}$$

$$- [\sum_{k=2}^{3} \frac{1}{C_k}] \cdot \frac{1}{R_1} \cdot I_{L_1}^{(j)} - [\sum_{k=3}^{4} \frac{1}{C_k}] \cdot \frac{1}{R_1} \cdot I_{L_2}^{(j)} = 0$$

$$- i_{R_1} \cdot \lambda - [\sum_{k=1}^{4} \frac{1}{C_k}] \cdot \frac{1}{R_1} \cdot i_{R_1} + [\sum_{\substack{k=1 \\ k \neq 2 \\ k \neq 3}}^{4} \frac{1}{C_k} - (\sum_{k=2}^{3} \frac{1}{C_k}) \cdot g_m \cdot r\_pi] \cdot \frac{1}{R_1} \cdot i_{r\_pi} + (\sum_{\substack{k=1 \\ k \neq 2 \\ k \neq 3}}^{4} \frac{1}{C_k}) \cdot \frac{1}{R_1} \cdot i_{C_{pi}}$$

$$- [\sum_{k=2}^{3} \frac{1}{C_k}] \cdot \frac{1}{R_1} \cdot i_{L_1} - [\sum_{k=3}^{4} \frac{1}{C_k}] \cdot \frac{1}{R_1} \cdot i_{L_2} = 0$$

$$\frac{dI_{r\_pi}}{dt} = [\sum_{\substack{k=1 \\ k \neq 2 \\ k \neq 3}}^{4} \frac{1}{C_k}] \cdot \frac{1}{r\_pi} \cdot I_{R_1} - [\sum_{\substack{k=1 \\ k \neq 2 \\ k \neq 3}}^{4} \frac{1}{C_k}] \cdot \frac{1}{r\_pi} \cdot I_{C_{pi}} - [\sum_{\substack{k=1 \\ k \neq 2 \\ k \neq 3}}^{4} \frac{1}{C_k}] \cdot \frac{1}{r\_pi}$$

$$\cdot I_{r\_pi} + \frac{1}{C_4 \cdot r\_pi} \cdot I_{L_2}$$

$$i_{r\_pi} \cdot \lambda \cdot e^{\lambda \cdot t} = [\sum_{\substack{k=1 \\ k \neq 2 \\ k \neq 3}}^{4} \frac{1}{C_k}] \cdot \frac{1}{r\_pi} \cdot [I_{R_1}^{(j)} + i_{R_1} \cdot e^{\lambda \cdot t}] - [\sum_{\substack{k=1 \\ k \neq 2 \\ k \neq 3}}^{4} \frac{1}{C_k}] \cdot \frac{1}{r\_pi} \cdot [I_{C_{pi}}^{(j)} + i_{C_{pi}} \cdot e^{\lambda \cdot t}]$$

$$- [\sum_{\substack{k=1 \\ k \neq 2 \\ k \neq 3}}^{4} \frac{1}{C_k}] \cdot \frac{1}{r\_pi} \cdot [I_{r\_pi}^{(j)} + i_{r\_pi} \cdot e^{\lambda \cdot t}] + \frac{1}{C_4 \cdot r\_pi} \cdot [I_{L_2}^{(j)} + i_{L_2} \cdot e^{\lambda \cdot t}]$$

$$i_{r\_pi} \cdot \lambda \cdot e^{\lambda \cdot t} = [\sum_{\substack{k=1 \\ k \neq 2 \\ k \neq 3}}^{4} \frac{1}{C_k}] \cdot \frac{1}{r\_pi} \cdot I_{R_1}^{(j)} - [\sum_{\substack{k=1 \\ k \neq 2 \\ k \neq 3}}^{4} \frac{1}{C_k}] \cdot \frac{1}{r\_pi} \cdot I_{C_{pi}}^{(j)} - [\sum_{\substack{k=1 \\ k \neq 2 \\ k \neq 3}}^{4} \frac{1}{C_k}] \cdot \frac{1}{r\_pi} \cdot I_{r\_pi}^{(j)}$$

$$+ \frac{1}{C_4 \cdot r\_pi} \cdot I_{L_2}^{(j)} + [\sum_{\substack{k=1 \\ k \neq 2 \\ k \neq 3}}^{4} \frac{1}{C_k}] \cdot \frac{1}{r\_pi} \cdot i_{R_1} \cdot e^{\lambda \cdot t}$$

$$- [\sum_{\substack{k=1 \\ k \neq 2 \\ k \neq 3}}^{4} \frac{1}{C_k}] \cdot \frac{1}{r\_pi} \cdot i_{C_{pi}} \cdot e^{\lambda \cdot t} - [\sum_{\substack{k=1 \\ k \neq 2 \\ k \neq 3}}^{4} \frac{1}{C_k}] \cdot \frac{1}{r\_pi} \cdot i_{r\_pi} \cdot e^{\lambda \cdot t} + \frac{1}{C_4 \cdot r\_pi} \cdot i_{L_2} \cdot e^{\lambda \cdot t}$$

At fixed points:

$$[\sum_{\substack{k=1 \\ k \neq 2 \\ k \neq 3}}^{4} \frac{1}{C_k}] \cdot \frac{1}{r\_pi} \cdot I_{R_1}^{(j)} - [\sum_{\substack{k=1 \\ k \neq 2 \\ k \neq 3}}^{4} \frac{1}{C_k}] \cdot \frac{1}{r\_pi} \cdot I_{C_{pi}}^{(j)} - [\sum_{\substack{k=1 \\ k \neq 2 \\ k \neq 3}}^{4} \frac{1}{C_k}] \cdot \frac{1}{r\_pi} \cdot I_{r\_pi}^{(j)} + \frac{1}{C_4 \cdot r\_pi} \cdot I_{L_2}^{(j)} = 0$$

$$- i_{r\_pi} \cdot \lambda + [\sum_{\substack{k=1 \\ k \neq 2 \\ k \neq 3}}^{4} \frac{1}{C_k}] \cdot \frac{1}{r\_pi} \cdot i_{R_1} - [\sum_{\substack{k=1 \\ k \neq 2 \\ k \neq 3}}^{4} \frac{1}{C_k}] \cdot \frac{1}{r\_pi} \cdot i_{C_{pi}} - [\sum_{\substack{k=1 \\ k \neq 2 \\ k \neq 3}}^{4} \frac{1}{C_k}] \cdot \frac{1}{r\_pi} \cdot i_{r\_pi}$$

$$+ \frac{1}{C_4 \cdot r\_pi} \cdot i_{L_2} = 0$$

$$\frac{dY_1}{dt} = -[\sum_{k=3}^{4} \frac{1}{C_k}] \cdot \frac{1}{L_2} \cdot I_{R_1} + [\frac{1}{C_4} - \frac{1}{C_3} \cdot g_m \cdot r\_pi] \cdot \frac{1}{L_2} \cdot I_{r\_pi} - \frac{1}{C_3 \cdot L_2} \cdot I_{L_1}$$

$$- [\sum_{k=3}^{4} \frac{1}{C_k}] \cdot \frac{1}{L_2} \cdot I_{L_2} + \frac{1}{C_4 \cdot L_2} \cdot I_{C_{pi}}$$

$$y_1 \cdot \lambda \cdot e^{\lambda \cdot t} = -[\sum_{k=3}^{4} \frac{1}{C_k}] \cdot \frac{1}{L_2} \cdot [I_{R_1}^{(j)} + i_{R_1} \cdot e^{\lambda \cdot t}] + [\frac{1}{C_4} - \frac{1}{C_3} \cdot g_m \cdot r\_pi] \cdot \frac{1}{L_2} \cdot [I_{r\_pi}^{(j)} + i_{r\_pi} \cdot e^{\lambda \cdot t}]$$
$$- \frac{1}{C_3 \cdot L_2} \cdot [I_{L_1}^{(j)} + i_{L_1} \cdot e^{\lambda \cdot t}] - [\sum_{k=3}^{4} \frac{1}{C_k}] \cdot \frac{1}{L_2} \cdot [I_{L_2}^{(j)} + i_{L_2} \cdot e^{\lambda \cdot t}]$$
$$+ \frac{1}{C_4 \cdot L_2} \cdot [I_{C_{pi}}^{(j)} + i_{C_{pi}} \cdot e^{\lambda \cdot t}]$$

$$y_1 \cdot \lambda \cdot e^{\lambda \cdot t} = -[\sum_{k=3}^{4} \frac{1}{C_k}] \cdot \frac{1}{L_2} \cdot I_{R_1}^{(j)} - [\sum_{k=3}^{4} \frac{1}{C_k}] \cdot \frac{1}{L_2} \cdot i_{R_1} \cdot e^{\lambda \cdot t} + [\frac{1}{C_4} - \frac{1}{C_3} \cdot g_m \cdot r\_pi] \cdot \frac{1}{L_2} \cdot I_{r\_pi}^{(j)}$$
$$+ [\frac{1}{C_4} - \frac{1}{C_3} \cdot g_m \cdot r\_pi] \cdot \frac{1}{L_2} \cdot i_{r\_pi} \cdot e^{\lambda \cdot t} - \frac{1}{C_3 \cdot L_2} \cdot I_{L_1}^{(j)} - \frac{1}{C_3 \cdot L_2} \cdot i_{L_1} \cdot e^{\lambda \cdot t} - [\sum_{k=3}^{4} \frac{1}{C_k}] \cdot \frac{1}{L_2} \cdot I_{L_2}^{(j)}$$
$$- [\sum_{k=3}^{4} \frac{1}{C_k}] \cdot \frac{1}{L_2} \cdot i_{L_2} \cdot e^{\lambda \cdot t} + \frac{1}{C_4 \cdot L_2} \cdot I_{C_{pi}}^{(j)} + \frac{1}{C_4 \cdot L_2} \cdot i_{C_{pi}} \cdot e^{\lambda \cdot t}$$

$$y_1 \cdot \lambda \cdot e^{\lambda \cdot t} = \{-[\sum_{k=3}^{4} \frac{1}{C_k}] \cdot \frac{1}{L_2} \cdot I_{R_1}^{(j)} + [\frac{1}{C_4} - \frac{1}{C_3} \cdot g_m \cdot r\_pi] \cdot \frac{1}{L_2} \cdot I_{r\_pi}^{(j)} - \frac{1}{C_3 \cdot L_2} \cdot I_{L_1}^{(j)}$$
$$- [\sum_{k=3}^{4} \frac{1}{C_k}] \cdot \frac{1}{L_2} \cdot I_{L_2}^{(j)} + \frac{1}{C_4 \cdot L_2} \cdot I_{C_{pi}}^{(j)}\} - [\sum_{k=3}^{4} \frac{1}{C_k}] \cdot \frac{1}{L_2} \cdot i_{R_1} \cdot e^{\lambda \cdot t}$$
$$+ [\frac{1}{C_4} - \frac{1}{C_3} \cdot g_m \cdot r\_pi] \cdot \frac{1}{L_2} \cdot i_{r\_pi} \cdot e^{\lambda \cdot t}$$
$$- \frac{1}{C_3 \cdot L_2} \cdot i_{L_1} \cdot e^{\lambda \cdot t} - [\sum_{k=3}^{4} \frac{1}{C_k}] \cdot \frac{1}{L_2} \cdot i_{L_2} \cdot e^{\lambda \cdot t} + \frac{1}{C_4 \cdot L_2} \cdot i_{C_{pi}} \cdot e^{\lambda \cdot t}$$

At fixed points:

$$- [\sum_{k=3}^{4} \frac{1}{C_k}] \cdot \frac{1}{L_2} \cdot I_{R_1}^{(j)} + [\frac{1}{C_4} - \frac{1}{C_3} \cdot g_m \cdot r\_pi] \cdot \frac{1}{L_2} \cdot I_{r\_pi}^{(j)} - \frac{1}{C_3 \cdot L_2} \cdot I_{L_1}^{(j)}$$
$$- [\sum_{k=3}^{4} \frac{1}{C_k}] \cdot \frac{1}{L_2} \cdot I_{L_2}^{(j)} + \frac{1}{C_4 \cdot L_2} \cdot I_{C_{pi}}^{(j)} = 0$$

$$- y_1 \cdot \lambda - [\sum_{k=3}^{4} \frac{1}{C_k}] \cdot \frac{1}{L_2} \cdot i_{R_1} + [\frac{1}{C_4} - \frac{1}{C_3} \cdot g_m \cdot r\_pi] \cdot \frac{1}{L_2} \cdot i_{r\_pi} - \frac{1}{C_3 \cdot L_2} \cdot i_{L_1}$$
$$- [\sum_{k=3}^{4} \frac{1}{C_k}] \cdot \frac{1}{L_2} \cdot i_{L_2} + \frac{1}{C_4 \cdot L_2} \cdot i_{C_{pi}} = 0$$

$$\frac{dI_{L_1}}{dt} = I_{R_1} \cdot \frac{R_1}{L_1} + I_{r\_pi} \cdot \frac{r\_pi}{L_1}; i_{L_1} \cdot \lambda \cdot e^{\lambda \cdot t} = [I_{R_1}^{(j)} + i_{R_1} \cdot e^{\lambda \cdot t}] \cdot \frac{R_1}{L_1}$$
$$+ [I_{r\_pi}^{(j)} + i_{r\_pi} \cdot e^{\lambda \cdot t}] \cdot \frac{r\_pi}{L_1} i_{L_1} \cdot \lambda \cdot e^{\lambda \cdot t} = I_{R_1}^{(j)} \cdot \frac{R_1}{L_1} + I_{r\_pi}^{(j)} \cdot \frac{r\_pi}{L_1} + i_{R_1} \cdot \frac{R_1}{L_1} \cdot e^{\lambda \cdot t}$$
$$+ i_{r\_pi} \cdot \frac{r\_pi}{L_1} \cdot e^{\lambda \cdot t}$$

At fixed points:

$$I_{R_1}^{(j)} \cdot \frac{R_1}{L_1} + I_{r\_pi}^{(j)} \cdot \frac{r\_pi}{L_1} = 0; -i_{L_1} \cdot \lambda + i_{R_1} \cdot \frac{R_1}{L_1} + i_{r\_pi} \cdot \frac{r\_pi}{L_1} = 0$$

$\frac{dI_{L_2}}{dt} = Y_1; i_{L_2} \cdot \lambda \cdot e^{\lambda \cdot t} = Y_1^{(j)} + y_1 \cdot e^{\lambda \cdot t}$. At fixed points: $Y_1^{(j)} = 0; -i_{L_2} \cdot \lambda + y_1 = 0$

We can summery our Colpitt's oscillator circuit arbitrarily small increments equations:

$$-i_{R_1} \cdot \lambda - [\sum_{k=1}^{4} \frac{1}{C_k}] \cdot \frac{1}{R_1} \cdot i_{R_1} + [\sum_{\substack{k=1 \\ k \neq 2 \\ k \neq 3}}^{4} \frac{1}{C_k} - (\sum_{k=2}^{3} \frac{1}{C_k}) \cdot g_m \cdot r\_pi] \cdot \frac{1}{R_1} \cdot i_{r\_pi} + (\sum_{\substack{k=1 \\ k \neq 2 \\ k \neq 3}}^{4} \frac{1}{C_k}) \cdot \frac{1}{R_1} \cdot i_{C_{pi}}$$
$$-[\sum_{k=2}^{3} \frac{1}{C_k}] \cdot \frac{1}{R_1} \cdot i_{L_1} - [\sum_{k=3}^{4} \frac{1}{C_k}] \cdot \frac{1}{R_1} \cdot i_{L_2} = 0$$

$$-i_{r\_pi} \cdot \lambda + [\sum_{\substack{k=1 \\ k \neq 2 \\ k \neq 3}}^{4} \frac{1}{C_k}] \cdot \frac{1}{r\_pi} \cdot i_{R_1} - [\sum_{\substack{k=1 \\ k \neq 2 \\ k \neq 3}}^{4} \frac{1}{C_k}] \cdot \frac{1}{r\_pi} \cdot i_{C_{pi}} - [\sum_{\substack{k=1 \\ k \neq 2 \\ k \neq 3}}^{4} \frac{1}{C_k}] \cdot \frac{1}{r\_pi} \cdot i_{r\_pi}$$
$$+ \frac{1}{C_4 \cdot r\_pi} \cdot i_{L_2} = 0$$

$$-y_1 \cdot \lambda - [\sum_{k=3}^{4} \frac{1}{C_k}] \cdot \frac{1}{L_2} \cdot i_{R_1} + [\frac{1}{C_4} - \frac{1}{C_3} \cdot g_m \cdot r\_pi] \cdot \frac{1}{L_2} \cdot i_{r\_pi} - \frac{1}{C_3 \cdot L_2} \cdot i_{L_1}$$
$$-[\sum_{k=3}^{4} \frac{1}{C_k}] \cdot \frac{1}{L_2} \cdot i_{L_2} + \frac{1}{C_4 \cdot L_2} \cdot i_{C_{pi}} = 0$$

$$-i_{L_1} \cdot \lambda + i_{R_1} \cdot \frac{R_1}{L_1} + i_{r\_pi} \cdot \frac{r\_pi}{L_1} = 0; -i_{L_2} \cdot \lambda + y_1 = 0$$

$$\begin{pmatrix} \Xi_{11} & \cdots & \Xi_{15} \\ \vdots & \ddots & \vdots \\ \Xi_{51} & \cdots & \Xi_{55} \end{pmatrix} \cdot \begin{pmatrix} i_{R_1} \\ i_{r\_pi} \\ y_1 \\ i_{L_1} \\ i_{L_2} \end{pmatrix} + \begin{pmatrix} [(\sum_{\substack{k=1 \\ k\neq 2 \\ k\neq 3}}^{4} \frac{1}{C_k}) \cdot \frac{1}{R_1}] \\ \{-[\sum_{\substack{k=1 \\ k\neq 2 \\ k\neq 3}}^{4} \frac{1}{C_k}] \cdot \frac{1}{r\_pi}\} \\ \frac{1}{C_4 \cdot L_2} \\ 0 \\ 0 \end{pmatrix} \cdot (i_{C_{pi}}) = 0;$$

$$\begin{pmatrix} \Xi_{11} & \cdots & \Xi_{15} \\ \vdots & \ddots & \vdots \\ \Xi_{51} & \cdots & \Xi_{55} \end{pmatrix} \cdot \begin{pmatrix} i_{R_1} \\ i_{r\_pi} \\ y_1 \\ i_{L_1} \\ i_{L_2} \end{pmatrix} \approx 0$$

Assumption: $$\begin{pmatrix} [(\sum_{\substack{k=1 \\ k\neq 2 \\ k\neq 3}}^{4} \frac{1}{C_k}) \cdot \frac{1}{R_1}] \\ \{-[\sum_{\substack{k=1 \\ k\neq 2 \\ k\neq 3}}^{4} \frac{1}{C_k}] \cdot \frac{1}{r\_pi}\} \\ \frac{1}{C_4 \cdot L_2} \\ 0 \\ 0 \end{pmatrix} \cdot (i_{C_{pi}}) \to \varepsilon; \Xi_{11} = -\lambda - [\sum_{k=1}^{4} \frac{1}{C_k}] \cdot \frac{1}{R_1}$$

$$\Xi_{12} = [\sum_{\substack{k=1 \\ k\neq 2 \\ k\neq 3}}^{4} \frac{1}{C_k} - (\sum_{k=2}^{3} \frac{1}{C_k}) \cdot g_m \cdot r\_pi] \cdot \frac{1}{R_1}; \ \Xi_{13} = 0;$$

$$\Xi_{14} = -[\sum_{k=2}^{3} \frac{1}{C_k}] \cdot \frac{1}{R_1}; \ \Xi_{15} = -[\sum_{k=3}^{4} \frac{1}{C_k}] \cdot \frac{1}{R_1}$$

$$\Xi_{21} = [\sum_{\substack{k=1 \\ k\neq 2 \\ k\neq 3}}^{4} \frac{1}{C_k}] \cdot \frac{1}{r\_pi}; \ \Xi_{22} = -\lambda - [\sum_{\substack{k=1 \\ k\neq 2 \\ k\neq 3}}^{4} \frac{1}{C_k}] \cdot \frac{1}{r\_pi};$$

$$\Xi_{23} = 0; \ \Xi_{24} = 0; \ \Xi_{25} = \frac{1}{C_4 \cdot r\_pi}$$

$$\Xi_{31} = -[\sum_{k=3}^{4} \frac{1}{C_k}] \cdot \frac{1}{L_2}; \Xi_{32} = [\frac{1}{C_4} - \frac{1}{C_3} \cdot g_m \cdot r\_pi] \cdot \frac{1}{L_2}; \Xi_{33} = -\lambda; \Xi_{34} = -\frac{1}{C_3 \cdot L_2}$$

$$\Xi_{35} = -[\sum_{k=3}^{4} \frac{1}{C_k}] \cdot \frac{1}{L_2}; \Xi_{41} = \frac{R_1}{L_1}; \Xi_{42} = \frac{r\_pi}{L_1}; \Xi_{43} = 0; \Xi_{44} = -\lambda; \Xi_{45} = 0$$

$$\Xi_{51} = \Xi_{52} = 0; \Xi_{53} = 1; \Xi_{54} = 0; \Xi_{55} = -\lambda$$

$$(A - \lambda \cdot I) = \begin{pmatrix} \Xi_{11} & \cdots & \Xi_{15} \\ \vdots & \ddots & \vdots \\ \Xi_{51} & \cdots & \Xi_{55} \end{pmatrix}; \det(A - \lambda \cdot I) = \det \begin{pmatrix} \Xi_{11} & \cdots & \Xi_{15} \\ \vdots & \ddots & \vdots \\ \Xi_{51} & \cdots & \Xi_{55} \end{pmatrix}$$

If an eigenvalue of A is on the imaginary axis, then the method of the center manifold must be applied. To apply the stability criterion of Liapunov to our system, the equation $\det(A - \lambda \cdot I) = 0$ has a zero in the left half plane, without calculating the eigenvalues explicitly. We define new system parameters functions:

$$\xi_1 = \xi_1(\underset{[k=1,2,3,4]}{C_k}, R_1) = -[\sum_{k=1}^{4} \frac{1}{C_k}] \cdot \frac{1}{R_1}; \xi_1 = \xi_2(\underset{[k=1,2,3,4]}{C_k},$$

$$R_1, r\_pi, g_m) = [\sum_{\substack{k=1 \\ k\neq 2 \\ k\neq 3}}^{4} \frac{1}{C_k} - (\sum_{k=2}^{3} \frac{1}{C_k}) \cdot g_m \cdot r\_pi] \cdot \frac{1}{R_1}$$

$$\xi_3 = \xi_3(C_2, C_3, R_1) = -[\sum_{k=2}^{3} \frac{1}{C_k}] \cdot \frac{1}{R_1}; \xi_4 = \xi_4(C_3, C_4, R_1) = -[\sum_{k=3}^{4} \frac{1}{C_k}] \cdot \frac{1}{R_1}$$

$$\xi_5 = \xi_5(C_1, C_4, r\_pi) = [\sum_{\substack{k=1 \\ k\neq 2 \\ k\neq 3}}^{4} \frac{1}{C_k}] \cdot \frac{1}{r\_pi}; \xi_6 = \xi_6(C_1, C_4, r\_pi) = -[\sum_{\substack{k=1 \\ k\neq 2 \\ k\neq 3}}^{4} \frac{1}{C_k}] \cdot \frac{1}{r\_pi}$$

$$\xi_7 = \xi_7(C_4, r\_pi) = \frac{1}{C_4 \cdot r\_pi}; \xi_8 = \xi_8(C_3, C_4, L_2) = -[\sum_{k=3}^{4} \frac{1}{C_k}] \cdot \frac{1}{L_2}$$

$$\xi_9 = \xi_9(C_3, C_4, r\_pi, L_2) = [\frac{1}{C_4} - \frac{1}{C_3} \cdot g_m \cdot r\_pi] \cdot \frac{1}{L_2}; \xi_{10} = \xi_{10}(C_3, L_2) = -\frac{1}{C_3 \cdot L_2}$$

$$\xi_{11} = \xi_{11}(C_3, C_4, L_2) = -[\sum_{k=3}^{4} \frac{1}{C_k}] \cdot \frac{1}{L_2}; \xi_{12} = \xi_{12}(R_1, L_1) = \frac{R_1}{L_1};$$

$$\xi_{13} = \xi_{13}(r\_pi, L_1) = \frac{r\_pi}{L_1}$$

$$\det(A - \lambda \cdot I) = -\lambda \cdot \xi_1 \cdot \{\lambda^4 - \lambda^3 \cdot \xi_6 - \lambda^2 \cdot \xi_{11} + \lambda \cdot [\xi_{11} \cdot \xi_6 - \xi_7 \cdot \xi_9] - \xi_7 \cdot \xi_{10} \cdot \xi_{13}\} - \xi_2 \cdot \{-\lambda^3 \cdot \xi_5 - \lambda \cdot [\xi_{11} \cdot \xi_5 + \xi_7 \cdot \xi_8] - \xi_7 \cdot \xi_{10} \cdot \xi_{12}\} - \xi_3 \cdot \{-\lambda^3 \cdot \xi_{12} + \lambda^2 \cdot [\xi_6 \cdot \xi_{12} - \xi_5 \cdot \xi_{13}] + \lambda \cdot \xi_{11} \cdot \xi_{12} + [\xi_{11} \cdot \xi_{13} \cdot \xi_5 - \xi_{11} \cdot \xi_6 \cdot \xi_{12} - \xi_7 \cdot \xi_8 \cdot \xi_{13} + \xi_7 \cdot \xi_9 \cdot \xi_{12}]\} + \xi_4 \cdot \{\lambda^2 \cdot \xi_8 + \lambda \cdot [\xi_5 \cdot \xi_9 + \xi_{10} \cdot \xi_{12} - \xi_6 \cdot \xi_8] + [\xi_{10} \cdot \xi_{13} \cdot \xi_5 - \xi_6 \cdot \xi_{10} \cdot \xi_{12}]\}$$

$$\det(A - \lambda \cdot I) = -\xi_1 \cdot \lambda^5 + \lambda^4 \cdot \xi_6 \cdot \xi_1 + \lambda^3 \cdot \xi_1 \cdot \xi_{11} - \lambda^2 \cdot \xi_1 \cdot [\xi_{11} \cdot \xi_6 - \xi_7 \cdot \xi_9] + \lambda \cdot \xi_1 \cdot \xi_7 \cdot \xi_{10} \cdot \xi_{13} + \lambda^3 \cdot \xi_2 \cdot \xi_5 + \lambda \cdot \xi_2 \cdot [\xi_{11} \cdot \xi_5 + \xi_7 \cdot \xi_8] + \xi_2 \cdot \xi_7 \cdot \xi_{10} \cdot \xi_{12} + \lambda^3 \cdot \xi_3 \cdot \xi_{12} - \lambda^2 \cdot \xi_3 \cdot [\xi_6 \cdot \xi_{12} - \xi_5 \cdot \xi_{13}] - \lambda \cdot \xi_{11} \cdot \xi_3 \cdot \xi_{12} - \xi_3 \cdot [\xi_{11} \cdot \xi_{13} \cdot \xi_5 - \xi_{11} \cdot \xi_6 \cdot \xi_{12} - \xi_7 \cdot \xi_8 \cdot \xi_{13} + \xi_7 \cdot \xi_9 \cdot \xi_{12}] + \lambda^2 \cdot \xi_4 \cdot \xi_8 + \lambda \cdot \xi_4 \cdot [\xi_5 \cdot \xi_9 + \xi_{10} \cdot \xi_{12} - \xi_6 \cdot \xi_8] + \xi_4 \cdot [\xi_{10} \cdot \xi_{13} \cdot \xi_5 - \xi_6 \cdot \xi_{10} \cdot \xi_{12}]$$

$$\det(A - \lambda \cdot I) = -\xi_1 \cdot \lambda^5 + \lambda^4 \cdot \xi_6 \cdot \xi_1 + \lambda^3 \cdot [\xi_1 \cdot \xi_{11} + \xi_2 \cdot \xi_5 + \xi_3 \cdot \xi_{12}] + \lambda^2\{\xi_4 \cdot \xi_8 - \xi_1 \cdot [\xi_{11} \cdot \xi_6 - \xi_7 \cdot \xi_9] - \xi_3 \cdot [\xi_6 \cdot \xi_{12} - \xi_5 \cdot \xi_{13}]\} + \lambda \cdot \{\xi_1 \cdot \xi_7 \cdot \xi_{10} \cdot \xi_{13} + \lambda \cdot \xi_2 \cdot [\xi_{11} \cdot \xi_5 + \xi_7 \cdot \xi_8] - \xi_{11} \cdot \xi_3 \cdot \xi_{12} + \xi_4 \cdot [\xi_5 \cdot \xi_9 + \xi_{10} \cdot \xi_{12} - \xi_6 \cdot \xi_8]\} + \xi_2 \cdot \xi_7 \cdot \xi_{10} \cdot \xi_{12} - \xi_3 \cdot [\xi_{11} \cdot \xi_{13} \cdot \xi_5 - \xi_{11} \cdot \xi_6 \cdot \xi_{12} - \xi_7 \cdot \xi_8 \cdot \xi_{13} + \xi_7 \cdot \xi_9 \cdot \xi_{12}] + \xi_4 \cdot [\xi_{10} \cdot \xi_{13} \cdot \xi_5 - \xi_6 \cdot \xi_{10} \cdot \xi_{12}]$$

We define new global parameters: $\Upsilon_5 = -\xi_1; \Upsilon_4 = \xi_6 \cdot \xi_1; \Upsilon_3 = \xi_1 \cdot \xi_{11} + \xi_2 \cdot \xi_5 + \xi_3 \cdot \xi_{12}$

$$\Upsilon_2 = \xi_4 \cdot \xi_8 - \xi_1 \cdot [\xi_{11} \cdot \xi_6 - \xi_7 \cdot \xi_9] - \xi_3 \cdot [\xi_6 \cdot \xi_{12} - \xi_5 \cdot \xi_{13}]$$

$$\Upsilon_1 = \xi_1 \cdot \xi_7 \cdot \xi_{10} \cdot \xi_{13} + \lambda \cdot \xi_2 \cdot [\xi_{11} \cdot \xi_5 + \xi_7 \cdot \xi_8] - \xi_{11} \cdot \xi_3 \cdot \xi_{12} + \xi_4 \cdot [\xi_5 \cdot \xi_9 + \xi_{10} \cdot \xi_{12} - \xi_6 \cdot \xi_8]$$

$$\Upsilon_0 = \xi_2 \cdot \xi_7 \cdot \xi_{10} \cdot \xi_{12} - \xi_3 \cdot [\xi_{11} \cdot \xi_{13} \cdot \xi_5 - \xi_{11} \cdot \xi_6 \cdot \xi_{12} - \xi_7 \cdot \xi_8 \cdot \xi_{13} + \xi_7 \cdot \xi_9 \cdot \xi_{12}] + \xi_4 \cdot [\xi_{10} \cdot \xi_{13} \cdot \xi_5 - \xi_6 \cdot \xi_{10} \cdot \xi_{12}]$$

$\det(A - \lambda \cdot I) = \sum_{k=0}^{5} \Upsilon_k \cdot \lambda^k = 0$. Next is to find zeros of $\sum_{k=0}^{5} \Upsilon_k \cdot \lambda^k = 0$ and establish stability according to criterion of Liapunov [2, 3, 4].

Exercises

1. We have resonator circuit $\theta(0<\theta<\infty)$ degree phase shift. The output of the amplifier is taken through a phase shift network to provide it an additional phase shift of $\theta(0<\theta<\infty)$ degree. Amplifier provides a phase shift of 180° and the phase shift network also gives a θ°. Therefore total phases shift of (180 + θ) degree (it is a 60° phase shift per RC section). The phase shift network is constructed from n RC sections ($n > 3; n \in \mathbb{N}$).

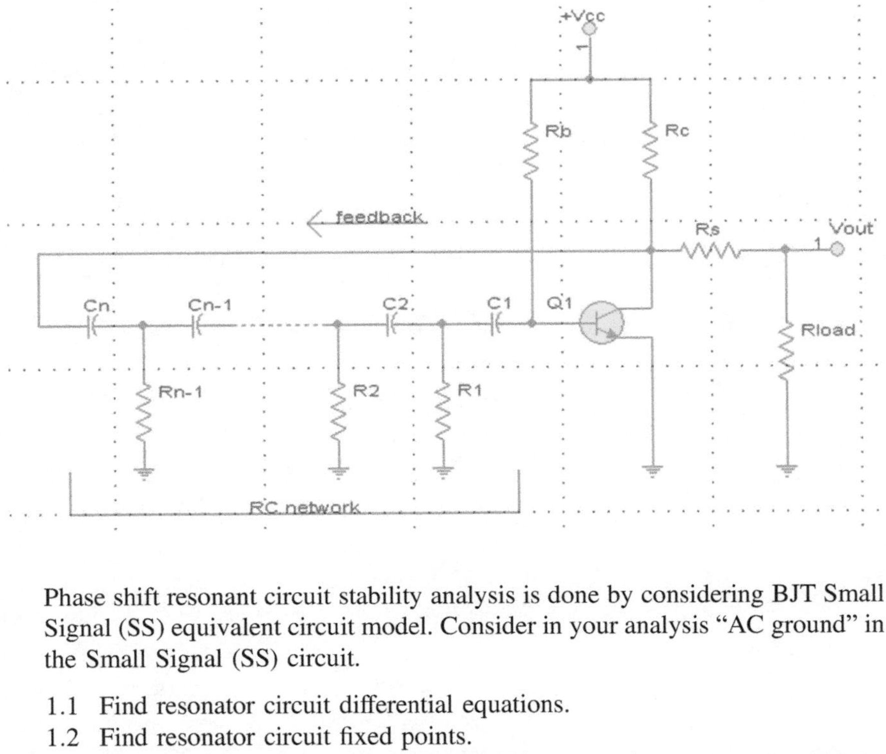

Phase shift resonant circuit stability analysis is done by considering BJT Small Signal (SS) equivalent circuit model. Consider in your analysis "AC ground" in the Small Signal (SS) circuit.

1.1 Find resonator circuit differential equations.
1.2 Find resonator circuit fixed points.
1.3 Discuss stability of resonator circuit $\theta(0<\theta<\infty)$ degree phase shift by using Liapunov theory. How the stability is dependent on the number (n) of RC sections?
1.4 Discuss the circuit behavior which characterized as a function of the position of eigenvalues in the Re/Im plane. How the position of eigenvalues in the Re/Im plane is changed for different numbers (n) of RC sections?
1.5 Discuss center manifold theorem in our circuit for different number (n) of RC sections.

2. We have resonator circuit $\varphi(0 < \varphi < \infty)$ degree phase shift. The output of the amplifier is taken through two phase shift networks in cascade (series). There is intermediate RF choke (L) between the first and the second RC networks. The phase shift of the first network is $\varphi_1(0 < \varphi_1 < \infty)$ degree and the phase shift of the second network is $\varphi_2(0 < \varphi_2 < \infty)$ degree $\varphi = \sum_{k=1}^{2} \varphi_k; 0 < \varphi < \infty$. In the first RC network the number of RC sections is n. In the second RC network the number of RC sections is m $(n \neq m; n, m > 3; n, m \in \mathbb{N})$.

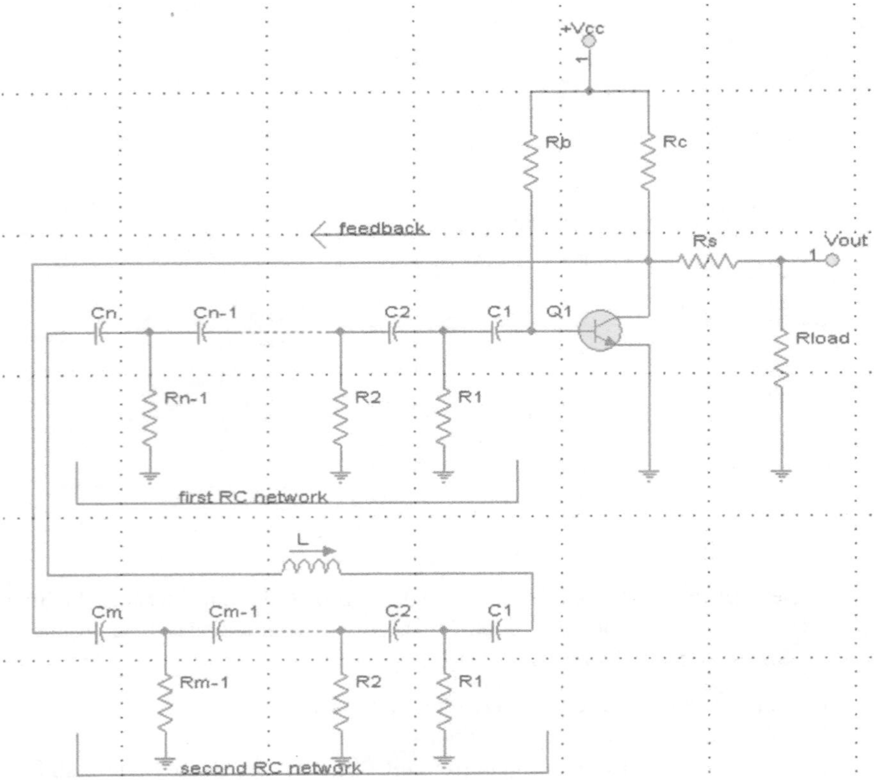

Phase shift resonator circuit stability analysis is done by considering BJT Small Signal (SS) equivalent circuit model. Consider in your analysis "AC signal" in the small signal circuit model.

2.1 Find resonator circuit differential equations.
2.2 Find resonator circuit fixed points.
2.3 Discuss stability of resonator circuit $\varphi(0 < \varphi < \infty)$ phase shift by using Liapunov theory. How the stability is dependent on the number (n) of RC

sections in the first network? and the number (m) of RC sections in the second network? How the stability of resonator circuit is dependent on the value of intermediate RF choke (L)?

2.4 Discuss the circuit behavior which characterized as a function of the position of eigenvalues in the Re/Im plane. How the position of eigenvalues in the Re/Im plane is changed for different number of RC sections in the first network (n) and second network (m)?

2.5 Discuss center manifold theorem in our circuit for different number of RC sections in the first network (n) and second network (m).

3. We have parallel resonator crystal oscillator circuit (BJT crystal oscillator). The BJT's collector is connected to RFC (Radio Frequency Coil) L_1. Additionally we have XTAL which is connected to circuit's output. The circuit contains biasing resistors R_1 and R_2 and additional elements (C_B, R_E, C_1, C_2). Our circuit stability analysis is done by considering BJT Small Signal (SS) equivalent model. Consider in your analysis "AC ground" in the Small Signal (SS) circuit.

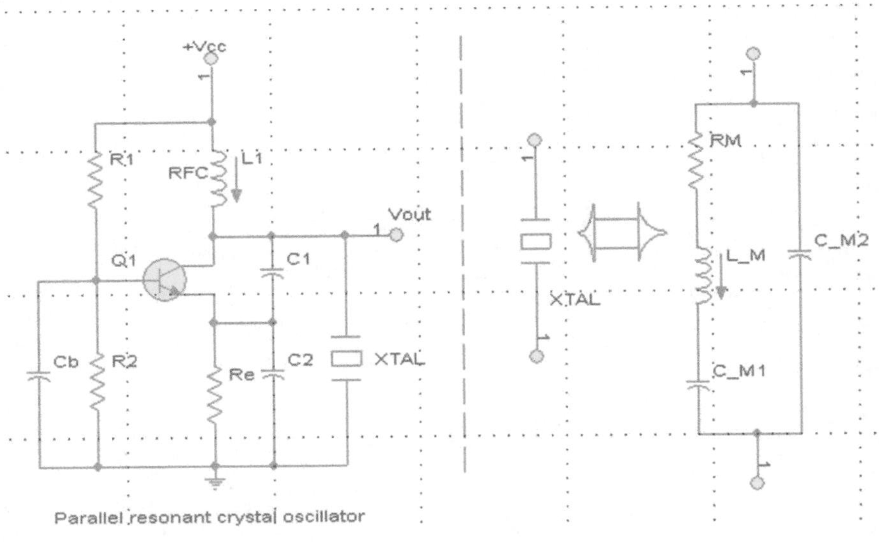

Hint: Replace the circuit's crystal by equivalent circuit (resonator circuit).

3.1 Find parallel resonator crystal oscillator circuit differential equations.

3.2 Find parallel resonator crystal oscillator circuit fixed points.

3.3 Discuss stability of parallel resonator crystal oscillator circuit by using Liapunov theorem. How the stability is changed if capacitor Cb is disconnected? How the stability is changed if we disconnect resistor Re?

3.4 Discuss circuit behavior which characterized as a function of the position of eigenvalues in the Re/Im plane. How the position of eigenvalues in the Re/Im plane changed if we connect two parallel crystals (XTAL1 and XTAL2) to our circuit's output? Consider that two parallel crystals (XTAL1 and XTAL2) are not same.

3.5 Discuss center manifold theorem in our circuit for two cases. First case: resistor R_2 is disconnected and Second case: capacitor C_2 is disconnected.

4. We have BJT based Hartley oscillator circuit. The circuit is constructed from sub Tank circuit and peripheral components. The sub tank circuit is constructed from capacitor C and primary transformer (T_1) inductances (L_1 and L_2), total $L_p = \sum_{i=1}^{2} L_i$. The secondary transformer (T_1) inductance is L_s ($L_p \neq L_s$). Additionally there are L_3 (RFC = Radio Frequency Coil), BJT transistor bias resistors R_1 and R_2, Q_1 transistor emitter resistor (R_e) and capacitor (C_e), feedback capacitor (C_c).

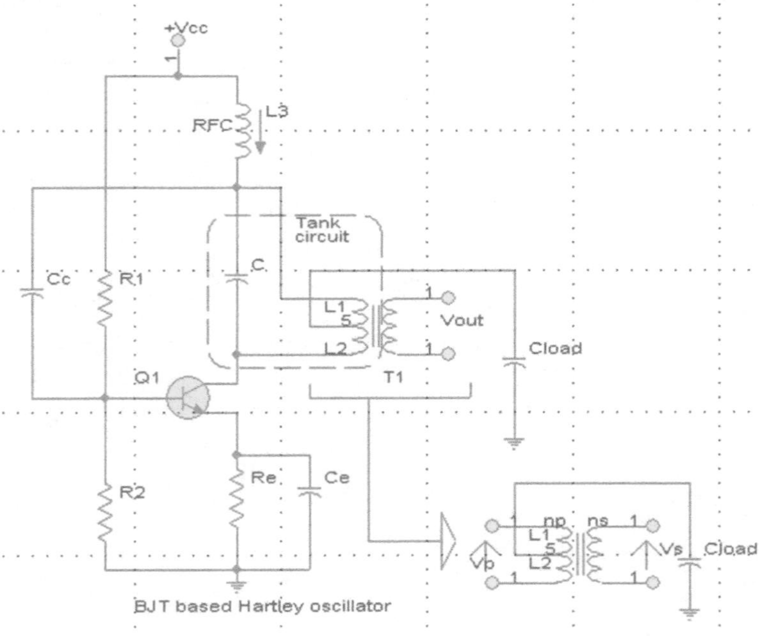

Transformer T_1: v_p—transformer primary voltage, v_s—transformer secondary voltage. ϕ is the flux through a one turn coil located anywhere on the transformer core. n_p and n_s are the number of turns of the coil 1 (primary) and 2 (secondary), respectively, then the total flux ϕ_p and ϕ_s through coils 1 and 2 respectively are $\phi_p = n_p \cdot \phi; \phi_s = n_s \cdot \phi; v_p = \frac{d\phi_p}{dt}; v_s = \frac{d\phi_s}{dt}; v_p = n_p \cdot \frac{d\phi}{dt}; L_p = \sum_{i=1}^{2} L_i$

$v_s = n_s \cdot \frac{d\phi}{dt}; v_p(t) = v_p; v_s(t) = v_s; \frac{v_p(t)}{v_s(t)} = \frac{n_p}{n_s}; a = \frac{n_p}{n_s} \cdot \frac{v_p(t)}{v_s(t)} = \frac{n_p}{n_s}$ for all times t and for all voltages v_p and v_s. We define the ratio between n_p and n_s as "a" $(a = \frac{n_p}{n_s})$. Step down transformer a > 1 and step up transformer a < 1. By law of conservation of energy, apparent real and reactive powers are each conserved in the input and output $(S = I_p \cdot v_p = I_s \cdot v_s; v_p = V_{L_1} + V_{L_2})$.
Remark: Coil 1 (primary) is constructed from two coils (L_1 and L_2 in series). Hartley oscillator circuit stability analysis is done by considering BJT Small Signal (SS) equivalent circuit model. Consider in your analysis "AC signal" in the small signal circuit model.

4.1 Find Hartley oscillator circuit differential equations.
4.2 Find Hartley oscillator circuit fixed points.
4.3 Discuss stability of Hartley oscillator circuit by using Liapunov theorem. How the stability is dependent on the number of turns of the coil 1 (primary) and 2 (secondary) of T_1 transformer?
4.4 Discuss the circuit behavior which characterized as a function of the position of eigenvalues in the Re/Im plane. How the position of the eigenvalues in the Re/Im plane is changed if we short resistor R_e? And if we disconnect capacitor C_e?
4.5 Discuss center manifold theorem in our circuit if we multiple the capacitance C by factor Γ $(C \rightarrow \Gamma \cdot C; \Gamma \in \mathbb{R}_+)$. How the circuit stability is dependent on Γ parameter?

5. We have transistor Colpitts oscillator circuit. The circuit is constructed from sub tank circuit and peripheral components. The sub tank circuit is constructed from two capacitors C_1 and C_2, primary transformer (T_1)—inductance L ($L_p = L$). The secondary transformer (T_1) inductance is L_s ($L_p \neq L_s$). Additionally there are L_1 (RFC = Radio Frequency Coil), BJT transistor bias resistors R_1 and R_2, Q_1 transistor emitter resistor (R_e) and capacitor (C_e), feedback capacitor (C_c). Colpitts oscillator circuit stability analysis is done by considering BJT Small Signal (SS) equivalent circuit model. Consider in your analysis "AC signal" in the small signal circuit model. Transformer T_1: v_p—transformer primary voltage, v_s—transformer secondary voltage. ϕ is the flux through a one turn coil located anywhere on the transformer core. n_p and n_s are the number of turns of the coil 1 (primary) and 2 (secondary), respectively, then the total flux ϕ_p and ϕ_s through coils 1 and 2 respectively are $\phi_p = n_p \cdot \phi; \phi_s = n_s \cdot \phi; v_p = \frac{d\phi_p}{dt}; v_s = \frac{d\phi_s}{dt}; v_p = n_p \cdot \frac{d\phi}{dt}$ $v_s = n_s \cdot \frac{d\phi}{dt}; v_p(t) = v_p; v_s(t) = v_s; \frac{v_p(t)}{v_s(t)} = \frac{n_p}{n_s}; a = \frac{n_p}{n_s} \cdot \frac{v_p(t)}{v_s(t)} = \frac{n_p}{n_s}$ for all times t and for all voltages v_p and v_s. We define the ratio between n_p and n_s as "a" $(a = \frac{n_p}{n_s})$. Step down transformer a > 1 and step up transformer a < 1. By law of conservation of energy, apparent real and reactive powers are each conserved in the input and output $(S = I_p \cdot v_p = I_s \cdot v_s)$.

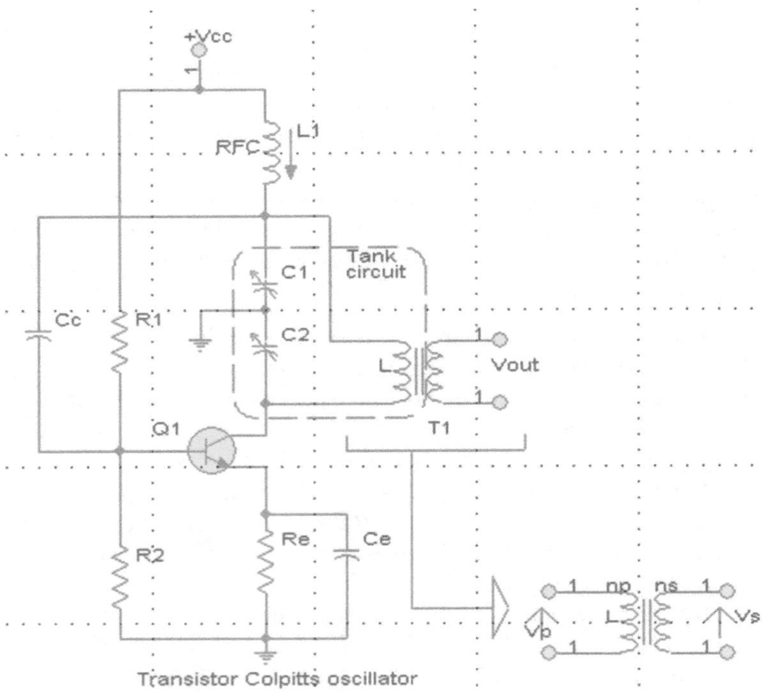

5.1 Find Colpitts oscillator circuit differential equations.
5.2 Find Colpitts oscillator circuit fixed points.
5.3 Discuss stability of Colpitts oscillator circuit by using Liapunov theorem. How the stability is dependent on the number of turns of the coil 1 (primary) and 2 (secondary) of T_1 transformer?
5.4 Discuss the circuit behavior which characterized as a function of the position of eigenvalues in the Re/Im plane. How the position of the eigenvalues in the Re/Im plane is changed if we disconnect resistor R_e? And if we disconnect capacitor C_e?
5.5 Discuss center manifold theorem in our circuit if we multiple the capacitance C_c by factor Γ ($C_c \to \Gamma \cdot C_c; \Gamma \in \mathbb{R}_+$). How the circuit stability is dependent on Γ parameter?

6. We have Colpitts crystal oscillator circuit. The crystal oscillator (XTAL) is designed around a CC (Common Collector), emitter-follower amplifier. The R_1 and R_2 resistor network sets the DC bias level on the base while emitter resistor R_e sets the output voltage level. Resistor R_2 is set as large as possible to prevent loading to the parallel connected crystal. The type of transistor is NPN connected in a common collector configuration and is capable of operating at high switching speeds. Capacitors C_1 and C_2 shunt the output of the transistor which

reduces the feedback signal. Therefore, the gain of the transistor limits the maximum value of C_1 and C_2. The output amplitude should be kept low in order to avoid excessive power dissipation in the crystal.

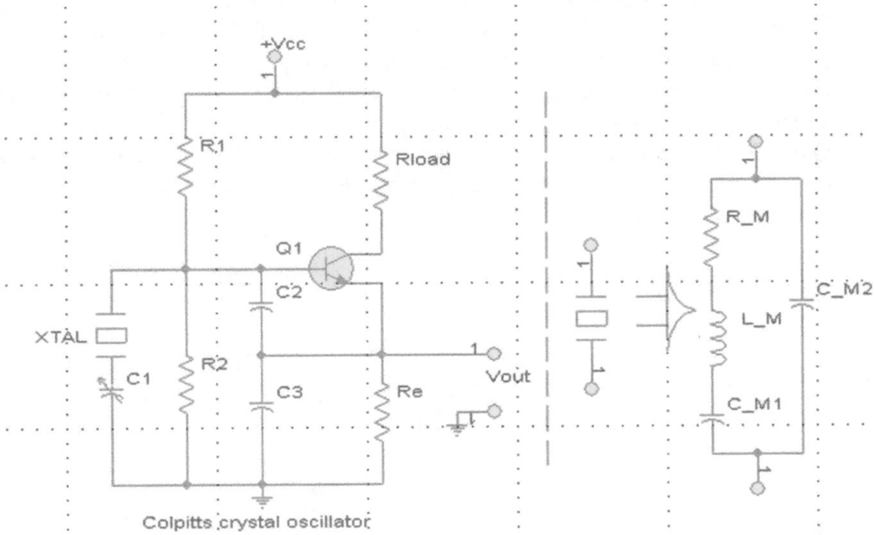

Our circuit stability analysis is done by considering BJT Small Signal (SS) equivalent model. Consider in your analysis "AC ground" in the Small Signal (SS) circuit. Hint: Replace the circuit's crystal by equivalent circuit (resonator circuit).

6.1 Find Colpitts crystal oscillator circuit differential equations.
6.2 Find Colpitts crystal oscillator circuit fixed points.
6.3 Discuss stability of Colpitts crystal oscillator circuit by using Liapunov theorem. How the stability is dependent on values of capacitors C_1, C_2, and C_3?
6.4 Discuss the circuit behavior which characterized as a function of the position of eigenvalues in the Re/Im plane. How the position of the eigenvalues in the Re/Im plane is changed if we short resistor R_e? And if we disconnect capacitor C_3?
6.5 Discuss center manifold theorem in our circuit if we multiple the capacitance C_1 by factor Γ ($C_1 \to \Gamma \cdot C_1; \Gamma \in \mathbb{R}_+$). How the circuit stability is dependent on Γ parameter?

7. The Colpitts oscillator, like Hartley is capable of giving an excellent sine wave shape, and also has the advantage of better stability at very high frequencies. It can be recognized by always having a "tapped capacitor". The circuit is a buffered Colpitts oscillator which is a common solution to feed the oscillator

output into an emitter follower buffer amplifier. The oscillator section of this circuit is a slightly different version of typical Colpitts oscillator. The RF choke (L_1) is the load impedance for Q_1 and the tank circuit is isolated from Q_1 by two DC blocking capacitors, C_1 and C_4. The Colpitts oscillator circuit uses a tuned feedback path rather than a tuned amplifier. The emitter follower stage (R_4, Q_2 and R_5) has very high input impedance, thus having little loading effect on the oscillator, and very low output impedance allowing it to drive loads of only a few tens of ohms impedance. The frequency stability of oscillators can be affected by variations in supply voltage.
Our circuit stability analysis is done by considering BJT Small Signal (SS) equivalent model for Q_1 and Q_2. Consider in your analysis "AC ground" in the Small Signal (SS) circuit.

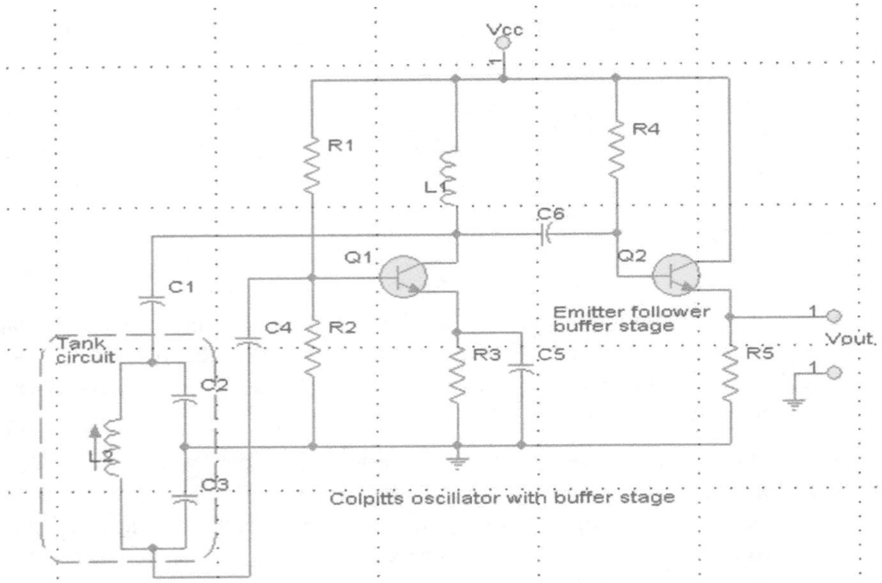

7.1 Find buffered Colpitts oscillator circuit differential equations
7.2 Find buffered Colpitts oscillator circuit fixed points.
7.3 Discuss stability of buffered Colpitts oscillator circuit by using Liapunov theorem. How the stability is dependent on values of capacitors C_6, C_2, and C_3?
7.4 Discuss the circuit behavior which characterized as a function of the position of eigenvalues in the Re/Im plane. How the position of the eigenvalues in the Re/Im plane is changed if we short resistor R_3? And if we disconnect capacitor C_5?

7.5 Discuss center manifold theorem in our circuit if we multiple the capacitance C_2 by factor Γ ($C_2 \rightarrow \Gamma \cdot C_2; \Gamma \in \mathbb{R}_+$). How the circuit stability is dependent on Γ parameter?

8. We have Colpitts double crystals oscillator circuit. The crystals oscillator (XTAL1 and XTAL2) is designed around a CC (Common Collector), emitter-follower amplifier. The R_1 resistor sets the DC bias level on the base while emitter resistor R_e sets the output voltage level. The type of transistor is NPN connected in a common collector configuration and is capable of operating at high switching speeds. Capacitor C_1 can be tune in his value.

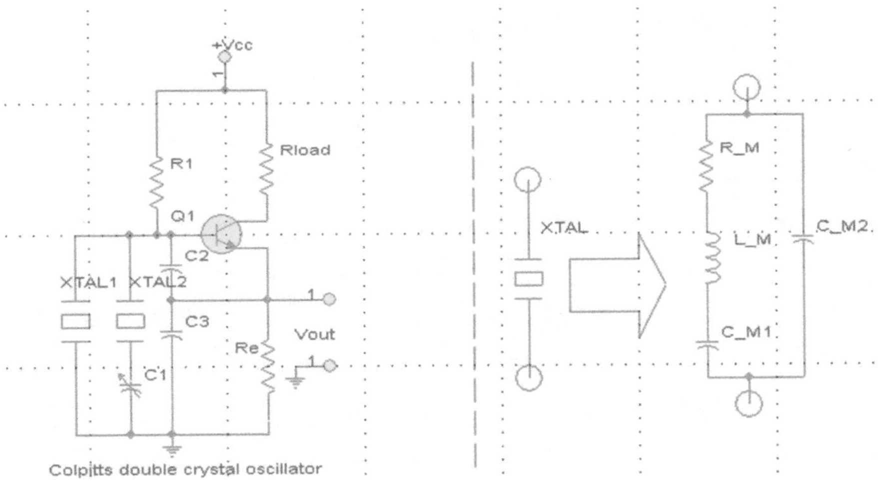

Our circuit stability analysis is done by considering BJT Small Signal (SS) equivalent model. Consider in your analysis "AC ground" in the Small Signal (SS) circuit. Hint: Replace the circuit's crystal by equivalent circuit (resonator circuit).

8.1 Find Colpitts double crystals oscillator circuit differential equations.
8.2 Find Colpitts double crystals circuit fixed points.
8.3 Discuss stability of Colpitts double crystals circuit by using Liapunov theorem. How the stability is dependent on values of capacitors C_1, R_e?
8.4 Discuss the circuit behavior which characterized as a function of the position of eigenvalues in the Re/Im plane. How the position of the eigenvalues in the Re/Im plane is changed if we disconnect resistor R_e? And if we disconnect capacitor C_3?
8.5 Discuss center manifold theorem in our circuit if capacitance C_1 transform according to the function $C_1 \rightarrow C_1 = \xi(C_1, \Gamma)$ ($\xi(C_1, \Gamma) = C_1 \cdot \Gamma + \sqrt{C_1 \cdot \Gamma^3}; \Gamma \in \mathbb{R}_+$). How the circuit stability is dependent on Γ parameter?

9. We have voltage controlled RF oscillator (VCO) circuit. The VCO circuit may be considered as an amplifier and a feedback loop. For the circuit to oscillate the total phase shift around the loop must be 360° and the gain must be unity. VCO circuit uses a common emitter circuit. This is itself produces a phase shift of 180°, leaving the feedback network to provide a further 180°. VCO circuit which uses a common base circuit where there is no phase shift between the emitter and collector signals (using bipolar transistor) and the phase shift network must provide either 0° or 360°. The system includes a resonator circuit to ensure that the oscillation occurs on a given frequency. The resonator circuit can be an LC resonator circuit in either series or parallel resonance depending upon the circuit, or a quartz crystal. Resistors R_{b1} and R_{b2} are Q_1's transistor biasing elements.

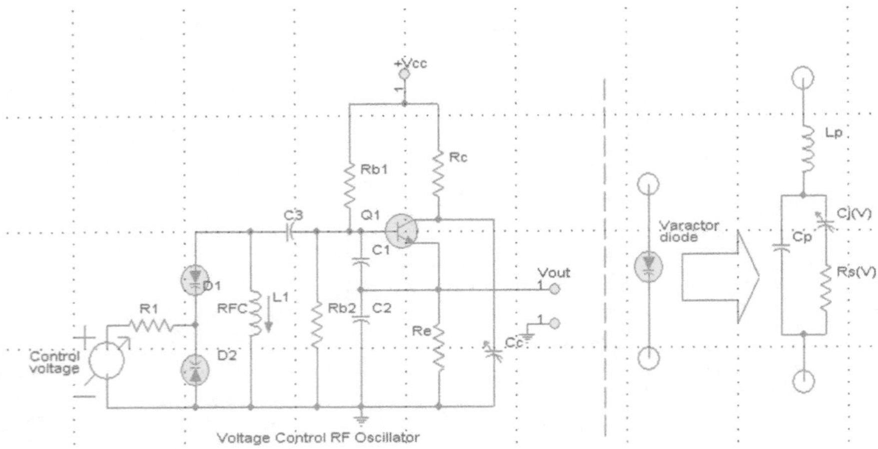

One of the options for VCO active device is bipolar transistor. The bipolar transistor has a low input impedance and is current driven. To make a VCO, the oscillator needs to be tuned by a voltage. This is done by variable capacitor from varactor diodes. The tune voltage for the VCO can then be applied to the varactors. The control line from the phase detector is isolated from the varactor diodes using a resistor R_1. Inductor L_1 is the RF choke (RFC). The series capacitor C_3 is used to block the DC from the inductor otherwise it would provide a direct short to ground and upset the bias arrangements of the circuit. The value of C_3 is normally large in comparison with C_1 and C_2 and can be ignored from the resonance perspective. If we use a varactor diodes within a voltage controlled oscillator, care must be taken in the design of the circuit to ensure that the drive level in the tuned circuit is not too high. In that case the varactor diodes may be driven into forward conduction, reducing the Q and increasing the level of spurious signals.

Varactor diodes: In your stability analysis use the simple model of a packaged varactor diode. A varactor diode is a P-N diode that changes its capacitance and the series resistance as the bias applied to the diode is varied. The property of

capacitance change is utilized to achieve a change in the frequency and/or the phase of an electrical circuit. In the simple model of a packaged varactor diode $C_j(V)$ is the variable junction capacitance of the diode die, $R_s(V)$ is the variable series resistance of the diode die, and C_p is the fixed parasitic capacitance arising from the installation of the die in a package. Package parasitic inductance L_p. The contribution to the series resistance from the packaging is very small and may be ignored. Similarly, the inductance associated with the die itself is very small and may be ignored. Variation of the junction capacitance and the junction series resistance as a function of applied reverse voltage is reported in the individual varactor data sheets. D_1 and D_2 are varactor diodes, consider that they identical for circuit analysis.
Our circuit stability analysis is done by considering BJT Small Signal (SS) equivalent model. Consider in your analysis "AC ground" in the Small Signal (SS) circuit.

9.1 Find voltage controlled RF oscillator (VCO) circuit differential equations.
9.2 Find voltage controlled RF oscillator (VCO) circuit fixed points.
9.3 Discuss stability of voltage controlled RF oscillator (VCO) circuit by using Liapunov theorem. How the stability is dependent on values of capacitors C_c, R_e?
9.4 We disconnect resistor R_{b2}, How the stability and dynamics of our circuit is changed?
9.5 Discuss center manifold theorem in our circuit if capacitance C_1 and C_2 transform according to the functions $C_1 \to C_1 = \xi_1(C_1, \Gamma)$

$$C_2 \to C_2 = \xi_2(C_2, \Omega); (\xi_1(C_1, \Gamma) = C_1 \cdot \sqrt{\Gamma} + \sqrt{C_1 \cdot \Gamma^3}\,;\ \Gamma \in \mathbb{R}_+).$$

$(\xi_2(C_2, \Omega) = C_2 \cdot \sqrt{\Omega^3} + \sqrt{C_2 \cdot \Omega}\,;\ \Omega \in \mathbb{R}_+)$. How the circuit stability is dependent on Γ and Ω parameters?

10. We have Common Base (CB) bipolar colpitts oscillator with varactor diode. Common Base (CB) amplifier, input signal is inserted at emitter (E) and output signal taken from the collector (C). The CB amplifier can operate as a voltage amplifier for low input impedance circuits. The transistor is matched to the load. One capacitor of the divider circuit is the varactor, whose voltage can be changed by applying a tuning voltage through RF choke (L_3) and a bypass conductor. Note that the capacitance variation with voltage is nonlinear, with the capacitance changing more at low values of bias voltage than at high values. The typical capacitance variation of a varactor diode (D_1) is as a function of the reverse bias voltage. This leads to a nonlinear frequency versus voltage tuning curve for the oscillator. If linearity is required, a compensation network must be used to modify the tuning voltage before it is applied to varactor. Resistors R_1... R4 are circuit biasing elements. Output matching network can be Pi or T type. Varactor diodes: In your stability analysis use the simple model of a

packaged varactor diode. A varactor diode is a P-N diode that changes its capacitance and the series resistance as the bias applied to the diode is varied.

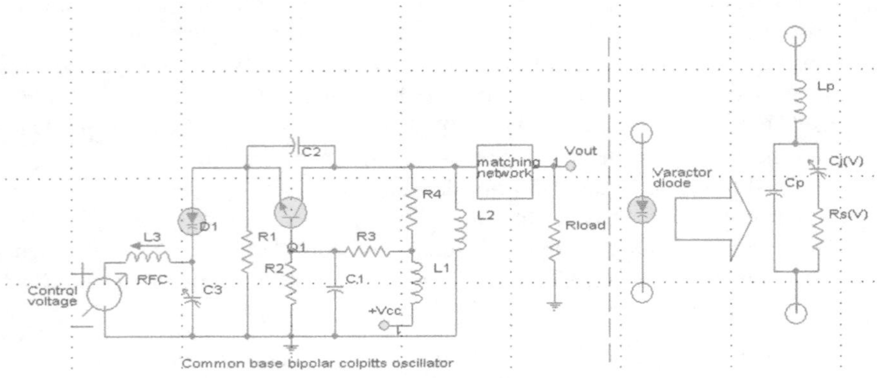

Our circuit stability analysis is done by considering BJT Small Signal (SS) equivalent model. Consider in your analysis "AC ground" in the Small Signal (SS) circuit.

10.1 Find voltage Common Base (CB) bipolar colpitts oscillator circuit differential equations for Pi and T matching networks.

10.2 Find Common Base (CB) bipolar colpitts oscillator circuit fixed points for Pi and T matching networks.

10.3 Discuss stability of Common Base (CB) bipolar colpitts circuit by using Liapunov theorem. How the stability is dependent on values of capacitors C_2, R_4?

10.4 We disconnect capacitor C_1, How the stability and dynamics of our circuit is changed?

10.5 Discuss center manifold theorem in our circuit if capacitance C_1 and C_2 transform according to the functions $C_1 \to C_1 = \xi_1(C_1, \Gamma)$ $C_2 \to C_2 = \xi_2(C_2, \Omega)$; $(\xi_1(C_1, \Gamma) = C_1 \cdot \sqrt{\Gamma} + \sqrt{C_1 \cdot \sqrt{\Gamma^3}}; \Gamma \in \mathbb{R}_+)$. $(\xi_2(C_2, \Omega) = C_2 \cdot \sqrt{\Omega^3} + \sqrt{C_2 \cdot \sqrt{\Omega}}; \Omega \in \mathbb{R}_+)$. How the circuit stability Is dependent on Γ and Ω parameters?

Chapter 7
Filters Systems Stability Analysis

The target of analog and RF filtering is to modify the magnitude and phase of signal frequency components. Many analog or radio frequency (RF) circuits perform filtering on the signals passing through them. The analog and RF filters types are defined on the criteria how they modify the magnitude and/or phase of sinusoidal frequency components. The primary issue is magnitude response. In other cases filters concerned with phase modifications. Filters are typically classified based on how they modify the frequency spectrum. The four basic types of filters are; the low pass filter, High pass filter, bandpass filter and band stop filter. Microwave and RF filters pass a range of frequencies and reject other frequencies. Filters are widely used at the input of a microwave receiver. The RF and microwave system's frequencies are picked up by the receiving antenna system and enter the radio receiver. The first target of the filter is to pass only those frequencies in the assigned operation range. The second target of the filter is to reject all other frequencies. Important graphs when analyzing the functionality of filter is the attenuation of a microwave signal passing through the filter as a function of frequency. Good filters have a large out of band attenuation and a low in band insertion loss. The selectivity feature defines the frequency range over which the filter characteristics change from passing the signal to blocking it. A Diplexer is a three port network that splits the incoming signal on one end and directs it through two outputs to different lines, dependent on frequency. A diplexer is the simplest form of a multiplexer, which can split the signals from one common port into many different paths. Quartz crystals are widely used In many filters applications. Quartz crystals have piezo electric properties; they develop an electric potential upon the application of mechanical stress. One of the most common piezoelectric uses of quartz is as a crystal oscillator. The resonant frequency of a quartz crystal oscillator is changed by mechanically loading it. A quartz crystal has two resonant frequencies, a series resonant frequency and a parallel resonant frequency. A quartz crystal operating on its parallel resonant frequency behaves like a parallel LC circuit. It has high impedance at one frequency and other frequency only. The practical advantage of quartz crystal is that it has an extremely high Q and result of an excellent high

O. Aluf, *Microwave RF Antennas and Circuits*,
DOI 10.1007/978-3-319-45427-6_7

selectivity filter. The crystal lattice filter can be a band pass filter, suitable for use as a sideband filter in an SSB transmitter receiver or transmitter. The two resonant frequencies of a quartz crystal are important when designing many RF systems. A tunable third order bandpass filter using varactors is commonly used in many RF applications. A tunable filter has greater functionality, better channel selectivity, reduced size, and lower weight since the same hardware can be employed at multiple bands. Practically tunable frequency filters are used as tracking filters for multi band telecommunication systems, wideband radar systems and radiometers. Tracking filters are mechanically tuned by adjusting the cavity dimensions of the resonators or magnetically altering the resonant frequency. The tuning element is a reverse-biased varactor diode [26, 33, 34, 42].

7.1 BPF Diplexer Without a Series Input Stability Analysis

The terminology duplexers and diplexers are very important for wireless communication. Duplexer is when two band pass filters are duplexed, meaning one common input, and two outputs (reverse is also correct, two inputs and one output). Duplexer will duplex a receiving and transmitting signal using two band pass filter, one common input (or output), and two outputs (or inputs), and is a three port device. A diplexer will refer to a duplexed high pass and low pass where broad bands transmit and receive is necessary, it is also a three port device with a common input and two outputs. A diplexer is a passive device that implements frequency domain multiplexing. Two ports (e.g., L and H) are multiplexed onto a third port (e.g., S). The signals on ports L and H occupy disjoint frequency bands. Consequently, the signals on L and H can coexist on port S without interfering with each other. The signal on port L will occupy a single low frequency band and the signal on port H will occupy a higher frequency band. In that situation, the diplexer consists of a low pass filter connecting ports L and S and high pass filter connecting ports H and S. Ideally, all the signal power on port L is transferred to the S port and vice versa. None of the low band signal is transferred from S port to the H port. Some power will be lost, and some signal power will leak to the wrong port. The diplexer, being a passive device, is reciprocal; the device itself doesn't have a notion of input or output. The diplexer is a different device than a passive combiner or splitter. The ports of a diplexer are frequency selective; the ports of a combiner are not. This is also a power "loss" difference—a combiner takes all the power delivered to the S port and divides it between the A and B ports. A diplexer multiplexes two ports onto one port, but more than two ports may be multiplexed: a three port to one port multiplexer is known as a triplexer. Our BPF diplexer without a series input circuits splits transmit from receive frequency in an FDD (Frequency Division Duplex) transceiver. A diplexer can be placed at the output of a frequency RF source, where it functions as an absorptive filter [91, 92].

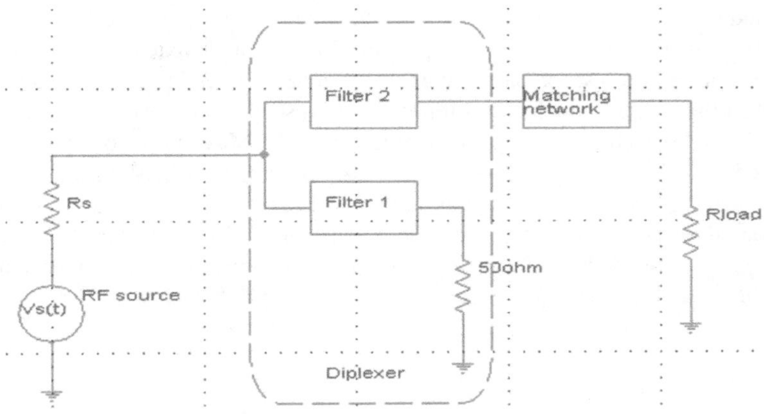

Fig. 7.1 BPF diplexer without a series input block diagram with RF source, output matching network and R_{load}

Filter 1 of the duplexer has a BPF that corresponds with the undesired frequency band, which can pass right through with little attenuation, and is terminated within the 50 Ω load. These undesired frequencies are blocked from entering filter 2 by that filter's stop band. Filter 2 BPF passes all the desired signals onto the load through matching network with little attenuation. The undesired signals through the diplexer are absorbed instead of being reflected as they would be in a typical filter. This absorption will prevent any undesired frequency products that were created by the RF source nonlinearity, from being bounced off of a reflective filters stopbands, which would return to the RF source and cause increased Inter Modulation Distortion (IMD) levels. The design of a diplexer is as two different frequency filters with non-overlapping band pass. The full BPF diplexer without a series input circuit is presented in the next figure. The matching network between diplexer unit and load can be L-type, Pi-type, and T-type.

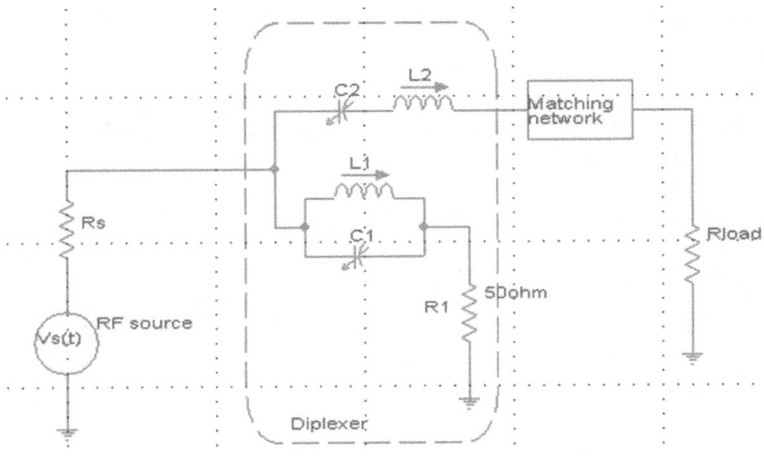

Fig. 7.2 BPF diplexer without a series input circuit

In actual microwave and RF system the diplexer unit is connected through micro strips lines to RF source, 50 Ω resistor, and load through matching network. We represent these micro strips parasitic effects as a delay lines in time. The delays are related to the current which flows through micro strips and are τ_{in}, τ_{out1}, and τ_{out2} respectively. We choose T-type matching network. Maximum power transfer is achieved by using T matching network (passive) connected between diplexer unit and load R_{load}. The circuit matching network no only designed to meet the requirement of minimum power loss but to additional targets. The additional targets of matching network are minimizing noise influence, maximizing power handling capabilities, and linearizing the frequency response [107, 108].

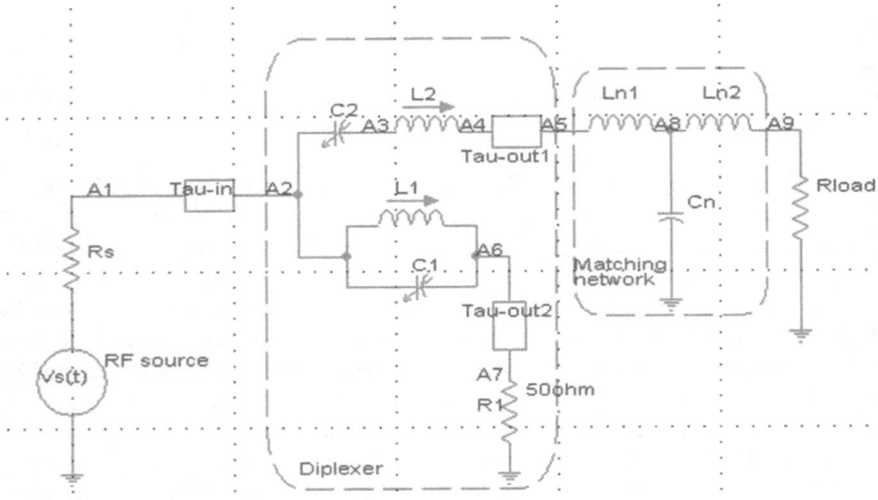

Fig. 7.3 BPF diplexer without a series input full schematic with delay line elements

$$V_{\tau_{in}} = V_{A_1} - V_{A_2};\ V_{\tau_{out1}} = V_{A_4} - V_{A_5};\ V_{\tau_{out2}} = V_{A_6} - V_{A_7};\ V_{\tau_{in}} \to \varepsilon;\ V_{\tau_{out1}} \to \varepsilon;\ V_{\tau_{out2}} \to \varepsilon$$

$$I_{R_1}(t) = I_{L_1}(t - \tau_{out2}) + I_{C_1}(t - \tau_{out2});\ I_{L_{n1}}(t) = I_{L_2}(t - \tau_{out1});\ I_{R_s}(t - \tau_{in}) = I_{C_2}(t) + I_{L_1}(t) + I_{C_1}(t)$$

$$I_{R_s} = \frac{V_s(t) - V_{A_1}}{R_s};\ I_{C_2} = C_2 \cdot \frac{d(V_{A_2} - V_{A_3})}{dt};\ V_{L_2} = V_{A_3} - V_{A_4} = L_2 \cdot \frac{dI_{L_2}}{dt}$$

$$V_{L_1} = V_{A_2} - V_{A_6} = L_1 \cdot \frac{dI_{L_1}}{dt};\ I_{C_1} = C_1 \cdot \frac{d(V_{A_2} - V_{A_6})}{dt};\ I_{R_1} = \frac{V_{A_7}}{R_1};\ V_{L_{n1}} = V_{A_5} - V_{A_8} = L_{n1} \cdot \frac{dI_{L_{n1}}}{dt}$$

$$V_{L_{n2}} = V_{A_8} - V_{A_9} = L_{n2} \cdot \frac{dI_{L_{n2}}}{dt};\ I_{C_n} = C_n \cdot \frac{dV_{A_8}}{dt};\ I_{R_{load}} = \frac{V_{A_9}}{R_{load}};\ I_{L_{n2}} = I_{R_{load}};\ I_{C_2} = I_{L_2}$$

$$L_{n1} = I_{L_{n2}} + I_{C_n};\ I_{R_s} = \frac{V_s(t) - V_{A_1}}{R_s} \Rightarrow V_{A_1} = V_s(t) - I_{R_s}R_s;$$

$$I_{C_2} = C_2 \cdot \frac{d(V_{A_2} - V_{A_3})}{dt} \Rightarrow V_{A_2} - V_{A_3} = \frac{1}{C_2} \cdot \int I_{C_2} \cdot dt$$

$$V_{A_3} - V_{A_4} = L_2 \cdot \frac{dI_{L_2}}{dt};\ V_{A_2} - V_{A_6} = L_1 \cdot \frac{dI_{L_1}}{dt};$$

$$I_{C_1} = C_1 \cdot \frac{d(V_{A_2} - V_{A_6})}{dt} \Rightarrow V_{A_2} - V_{A_6} = \frac{1}{C_1} \cdot \int I_{C_1} \cdot dt$$

$$I_{R_1} = \frac{V_{A_7}}{R_1} \Rightarrow V_{A_7} = I_{R_1} \cdot R_1;\ V_{A_5} - V_{A_8} = L_{n1} \cdot \frac{dI_{L_{n1}}}{dt};\ V_{A_8} - V_{A_9} = L_{n2} \cdot \frac{dI_{L_{n2}}}{dt}$$

$$I_{C_n} = C_n \cdot \frac{dV_{A_8}}{dt} \Rightarrow V_{A_8} = \frac{1}{C_n} \cdot \int I_{C_n} \cdot dt;\ I_{R_{load}} = \frac{V_{A_9}}{R_{load}} \Rightarrow V_{A_9} = I_{R_{load}} \cdot R_{load}$$

$$V_{A_1} \approx V_{A_2};\ V_{A_4} \approx V_{A_5};\ V_{A_6} \approx V_{A_7};\ V_{A_2} = V_s(t) - I_{R_s}R_s;\ V_{A_3} - V_{A_5} = L_2 \cdot \frac{dI_{L_2}}{dt}$$

$$V_{A_2} - V_{A_7} = L_1 \cdot \frac{dI_{L_1}}{dt};\ V_{A_2} - V_{A_7} = \frac{1}{C_1} \cdot \int I_{C_1} \cdot dt;\ V_{A_3} = V_{A_2} - \frac{1}{C_2} \cdot \int I_{C_2} \cdot dt$$

$$V_{A_3} = V_s(t) - I_{R_s}R_s - \frac{1}{C_2} \cdot \int I_{C_2} \cdot dt;\ V_{A_3} - V_{A_5} = L_2 \cdot \frac{dI_{L_2}}{dt} \Rightarrow V_{A_5} = V_{A_3} - L_2 \cdot \frac{dI_{L_2}}{dt}$$

$$V_{A_5} = V_s(t) - I_{R_s}R_s - \frac{1}{C_2} \cdot \int I_{C_2} \cdot dt - L_2 \cdot \frac{dI_{L_2}}{dt};\ V_{A_7} = V_{A_2} - L_1 \cdot \frac{dI_{L_1}}{dt}$$

$$V_{A_7} = V_s(t) - I_{R_s}R_s - L_1 \cdot \frac{dI_{L_1}}{dt};\ V_{A_7} = V_{A_2} - \frac{1}{C_1} \cdot \int I_{C_1} \cdot dt;\ V_{A_7} = V_s(t) - I_{R_s}R_s - \frac{1}{C_1} \cdot \int I_{C_1} \cdot dt$$

&&&

$$I_{R_1} \cdot R_1 = V_s(t) - I_{R_s}R_s - \frac{1}{C_1} \cdot \int I_{C_1} \cdot dt;\ I_{R_1} \cdot R_1 = V_s(t) - I_{R_s}R_s - L_1 \cdot \frac{dI_{L_1}}{dt};\ \frac{1}{C_1} \cdot \int I_{C_1} \cdot dt = L_1 \cdot \frac{dI_{L_1}}{dt}$$

$$V_{A_5} - V_{A_8} = L_{n1} \cdot \frac{dI_{L_{n1}}}{dt} \Rightarrow V_{A_8} = V_{A_5} - L_{n1} \cdot \frac{dI_{L_{n1}}}{dt};\ V_{A_8} = V_s(t) - I_{R_s}R_s - \frac{1}{C_2} \cdot \int I_{C_2} \cdot dt - L_2 \cdot \frac{dI_{L_2}}{dt} - L_{n1} \cdot \frac{dI_{L_{n1}}}{dt}$$

$$\frac{1}{C_n} \cdot \int I_{C_n} \cdot dt = V_s(t) - I_{R_s}R_s - \frac{1}{C_2} \cdot \int I_{C_2} \cdot dt - L_2 \cdot \frac{dI_{L_2}}{dt} - L_{n1} \cdot \frac{dI_{L_{n1}}}{dt};\ \frac{1}{C_n} \cdot \int I_{C_n} \cdot dt - I_{R_{load}} \cdot R_{load} = L_{n2} \cdot \frac{dI_{L_{n2}}}{dt}$$

We can summery our system differential equations (Version 1):

$$I_{R_1} \cdot R_1 = V_s(t) - I_{R_s}R_s - \frac{1}{C_1} \cdot \int I_{C_1} \cdot dt;\ I_{R_1} \cdot R_1 = V_s(t) - I_{R_s}R_s - L_1 \cdot \frac{dI_{L_1}}{dt};\ \frac{1}{C_1} \cdot \int I_{C_1} \cdot dt = L_1 \cdot \frac{dI_{L_1}}{dt}$$

$$\frac{1}{C_n} \cdot \int I_{C_n} \cdot dt = V_s(t) - I_{R_s}R_s - \frac{1}{C_2} \cdot \int I_{C_2} \cdot dt - L_2 \cdot \frac{dI_{L_2}}{dt} - L_{n1} \cdot \frac{dI_{L_{n1}}}{dt}$$

$$\frac{1}{C_n} \cdot \int I_{C_n} \cdot dt - I_{R_{load}} \cdot R_{load} = L_{n2} \cdot \frac{dI_{L_{n2}}}{dt};\ I_{L_{n2}} = I_{R_{load}};\ I_{C_2} = I_{L_2}$$

We can summery our system differential equations (Version 2):

$$\frac{dI_{R_1}}{dt} \cdot R_1 = \frac{dV_s(t)}{dt} - \frac{dI_{R_s}}{dt} \cdot R_s - \frac{1}{C_1} \cdot I_{C_1};\ \frac{dI_{R_1}}{dt} \cdot R_1 = \frac{dV_s(t)}{dt} - \frac{dI_{R_s}}{dt} \cdot R_s - L_1 \cdot \frac{d^2I_{L_1}}{dt^2};\ \frac{1}{C_1} \cdot I_{C_1} = L_1 \cdot \frac{d^2I_{L_1}}{dt^2}$$

$$\frac{1}{C_n} \cdot I_{C_n} = \frac{dV_s(t)}{dt} - \frac{dI_{R_s}}{dt} \cdot R_s - \frac{1}{C_2} \cdot I_{L_2} - L_2 \cdot \frac{d^2I_{L_2}}{dt^2} - L_{n1} \cdot \frac{d^2I_{L_{n1}}}{dt^2};\ \frac{1}{C_n} \cdot I_{C_n} - \frac{dI_{L_{n2}}}{dt} \cdot R_{load} = L_{n2} \cdot \frac{d^2I_{L_{n2}}}{dt^2}$$

We can summery our system differential equations (Version 3):

$$I_{R_1}(t) = I_{L_1}(t - \tau_{out2}) + I_{C_1}(t - \tau_{out2});\ I_{L_{n1}}(t) = I_{L_2}(t - \tau_{out1});\ I_{R_s}(t - \tau_{in}) = I_{C_2}(t) + I_{L_1}(t) + I_{C_1}(t)$$

$$I_{R_1} \cdot R_1 = V_s(t) - I_{R_s} R_s - L_1 \cdot \frac{dI_{L_1}}{dt}; \; [I_{L_1}(t - \tau_{out2}) + I_{C_1}(t - \tau_{out2})] \cdot R_1 = V_s(t) - I_{R_s} R_s - L_1 \cdot \frac{dI_{L_1}}{dt}$$

$$\frac{d}{dt}[I_{L_1}(t - \tau_{out2}) + I_{C_1}(t - \tau_{out2})] \cdot R_1 = \frac{dV_s(t)}{dt} - \frac{dI_{R_s}}{dt} \cdot R_s - \frac{1}{C_1} \cdot I_{C_1}; \; \frac{1}{C_1} \cdot I_{C_1} = L_1 \cdot \frac{d^2 I_{L_1}}{dt^2}$$

$$\frac{1}{C_n} \cdot I_{C_n} = \frac{dV_s(t)}{dt} - \frac{dI_{R_s}}{dt} \cdot R_s - \frac{1}{C_2} \cdot I_{L_2} - L_2 \cdot \frac{d^2 I_{L_2}}{dt^2} - L_{n1} \cdot \frac{d^2 I_{L_2}(t - \tau_{out1})}{dt^2}$$

$$\frac{1}{C_n} \cdot I_{C_n} - \frac{dI_{L_{n2}}}{dt} \cdot R_{load} = L_{n2} \cdot \frac{d^2 I_{L_{n2}}}{dt^2}; \; I_{R_s}(t - \tau_{in}) = I_{L_2}(t) + I_{L_1}(t) + I_{C_1}(t)$$

We define new system variables: $\frac{dI_{L_1}}{dt} = X_1$; $\frac{dI_{L_2}}{dt} = X_2$; $\frac{dI_{L_{n2}}}{dt} = X_3$
We can summery our system differential equations (Version 4):

$$[I_{L_1}(t - \tau_{out2}) + I_{C_1}(t - \tau_{out2})] \cdot R_1 = V_s(t) - I_{R_s} R_s - L_1 \cdot X_1$$

$$\frac{dI_{L_1}(t - \tau_{out2})}{dt} \cdot R_1 + \frac{dI_{C_1}(t - \tau_{out2})}{dt} \cdot R_1 = \frac{dV_s(t)}{dt} - \frac{dI_{R_s}}{dt} \cdot R_s - \frac{1}{C_1} \cdot I_{C_1}; \; \frac{1}{C_1} \cdot I_{C_1} = L_1 \cdot \frac{dX_1}{dt}$$

$$\frac{1}{C_n} \cdot I_{C_n} = \frac{dV_s(t)}{dt} - \frac{dI_{R_s}}{dt} \cdot R_s - \frac{1}{C_2} \cdot I_{L_2} - L_2 \cdot \frac{dX_2}{dt} - L_{n1} \cdot \frac{d^2 I_{L_2}(t - \tau_{out1})}{dt^2}$$

$$\frac{1}{C_n} \cdot I_{C_n} - X_3 \cdot R_{load} = L_{n2} \cdot \frac{dX_3}{dt} \Rightarrow \frac{dX_3}{dt} = \frac{1}{C_n \cdot L_{n2}} \cdot I_{C_n} - X_3 \cdot \frac{R_{load}}{L_{n2}}$$

$$I_{R_s}(t - \tau_{in}) = I_{C_2}(t) + I_{L_1}(t) + I_{C_1}(t); \; I_{L_{n1}} = I_{L_{n2}} + I_{C_n}$$

Remark 1.0 $I_{L_1}(t - \tau_{out2}) = I_{L_1}^* + i_{L_1} \cdot e^{\lambda \cdot t} \cdot e^{-\lambda \cdot \tau_{out2}}$; $I_{L_1}(t) = I_{L_1}^* + i_{L_1} \cdot e^{\lambda \cdot t}$

$$\frac{dI_{L_1}(t)}{dt} = i_{L_1} \cdot \lambda \cdot e^{\lambda \cdot t}; \; \frac{dI_{L_1}(t - \tau_{out2})}{dt} = i_{L_1} \cdot \lambda \cdot e^{\lambda \cdot t} \cdot e^{-\lambda \cdot \tau_{out2}}; \; \frac{dI_{L_1}(t - \tau_{out2})}{dt} = \frac{dI_{L_1}(t)}{dt} \cdot e^{-\lambda \cdot \tau_{out2}}$$

Remark 1.1 $I_{L_2}(t - \tau_{out1}) = I_{L_2}^* + i_{L_2} \cdot e^{\lambda \cdot t} \cdot e^{-\lambda \cdot \tau_{out1}}$; $I_{L_2}(t) = I_{L_2}^* + i_{L_2} \cdot e^{\lambda \cdot t}$

$$\frac{dI_{L_2}(t)}{dt} = i_{L_2} \cdot \lambda \cdot e^{\lambda \cdot t}; \frac{dI_{L_2}(t - \tau_{out1})}{dt} = i_{L_2} \cdot \lambda \cdot e^{\lambda \cdot t} \cdot e^{-\lambda \cdot \tau_{out1}}; \frac{dI_{L_2}(t - \tau_{out1})}{dt} = \frac{dI_{L_2}(t)}{dt} \cdot e^{-\lambda \cdot \tau_{out1}}$$

Remark 1.2 $I_{L_2}(t - \tau_{out1}) = I^*_{L_2} + i_{L_2} \cdot e^{\lambda \cdot t} \cdot e^{-\lambda \cdot \tau_{out1}}; \; I_{L_2}(t) = I^*_{L_2} + i_{L_2} \cdot e^{\lambda \cdot t}$

$$\frac{d^2 I_{L_2}(t)}{dt^2} = i_{L_2} \cdot \lambda^2 \cdot e^{\lambda \cdot t}; \frac{d^2 I_{L_2}(t - \tau_{out1})}{dt^2} = i_{L_2} \cdot \lambda^2 \cdot e^{\lambda \cdot t} \cdot e^{-\lambda \cdot \tau_{out1}}; \frac{d^2 I_{L_2}(t - \tau_{out1})}{dt^2} = \frac{d^2 I_{L_2}(t)}{dt^2} \cdot e^{-\lambda \cdot \tau_{out1}}$$

Remark 1.3 $I_{C_1}(t - \tau_{out2}) = I^*_{C_1} + i_{C_1} \cdot e^{\lambda \cdot t} \cdot e^{-\lambda \cdot \tau_{out2}}; \; I_{C_1}(t) = I^*_{C_1} + i_{C_1} \cdot e^{\lambda \cdot t}$

$$\frac{dI_{C_1}(t)}{dt} = i_{C_1} \cdot \lambda \cdot e^{\lambda \cdot t}; \frac{dI_{C_1}(t - \tau_{out2})}{dt} = i_{C_1} \cdot \lambda \cdot e^{\lambda \cdot t} \cdot e^{-\lambda \cdot \tau_{out2}}; \frac{dI_{C_1}(t - \tau_{out2})}{dt} = \frac{dI_{C_1}(t)}{dt} \cdot e^{-\lambda \cdot \tau_{out2}}$$

Remark 1.4 $I_{R_s}(t - \tau_{in}) = I^*_{R_s} + i_{R_s} \cdot e^{\lambda \cdot t} \cdot e^{-\lambda \cdot \tau_{in}}; \; I_{R_s}(t) = I^*_{R_s} + i_{R_s} \cdot e^{\lambda \cdot t}$

$$\frac{dI_{R_s}(t)}{dt} = i_{R_s} \cdot \lambda \cdot e^{\lambda \cdot t}; \frac{dI_{R_s}(t - \tau_{in})}{dt} = i_{R_s} \cdot \lambda \cdot e^{\lambda \cdot t} \cdot e^{-\lambda \cdot \tau_{in}}; \frac{dI_{R_s}(t - \tau_{in})}{dt} = \frac{dI_{R_s}(t)}{dt} \cdot e^{-\lambda \cdot \tau_{in}}$$

We can summery our system differential equations (Version 5):

$$\frac{dI_{L_1}}{dt} = \frac{dI_{L_1}(t)}{dt}; \frac{dI_{L_2}}{dt} = \frac{dI_{L_2}(t)}{dt}; \frac{dX_2}{dt} = \frac{d^2 I_{L_2}}{dt^2} = \frac{d^2 I_{L_2}(t)}{dt^2}; \frac{dI_{C_1}}{dt} = \frac{dI_{C_1}(t)}{dt}$$

&&& $\frac{dX_3}{dt} = \frac{1}{C_n \cdot L_{n2}} \cdot I_{C_n} - X_3 \cdot \frac{R_{load}}{L_{n2}}$

$$\frac{dI_{R_s}(t - \tau_{in})}{dt} = \frac{dI_{L_2}(t)}{dt} + \frac{dI_{L_1}(t)}{dt} + \frac{dI_{C_1}(t)}{dt}; \frac{dI_{R_s}(t)}{dt} = [\sum_{k=1}^{2} X_k] \cdot e^{\lambda \cdot \tau_{in}} + \frac{dI_{C_1}}{dt} \cdot e^{\lambda \cdot \tau_{in}}$$

$$\frac{dI_{L_1}(t - \tau_{out2})}{dt} \cdot R_1 + \frac{dI_{C_1}(t - \tau_{out2})}{dt} \cdot R_1 = \frac{dV_s(t)}{dt} - \frac{dI_{R_s}}{dt} \cdot R_s - \frac{1}{C_1} \cdot I_{C_1}; \frac{1}{C_1} \cdot I_{C_1} = L_1 \cdot \frac{dX_1}{dt}$$

$$\frac{dI_{L_1}(t)}{dt} \cdot e^{-\lambda \cdot \tau_{out2}} \cdot R_1 + \frac{dI_{C_1}(t)}{dt} \cdot e^{-\lambda \cdot \tau_{out2}} \cdot R_1 = \frac{dV_s(t)}{dt} - \frac{dI_{R_s}}{dt} \cdot R_s - \frac{1}{C_1} \cdot I_{C_1}; \frac{dX_1}{dt} = \frac{1}{L_1 \cdot C_1} \cdot I_{C_1}$$

$$X_1 \cdot e^{-\lambda \cdot \tau_{out2}} \cdot R_1 + \frac{dI_{C_1}}{dt} \cdot e^{-\lambda \cdot \tau_{out2}} \cdot R_1 = \frac{dV_s(t)}{dt} - \{[\sum_{k=1}^{2} X_k] \cdot e^{\lambda \cdot \tau_{in}} + \frac{dI_{C_1}}{dt} \cdot e^{\lambda \cdot \tau_{in}}\} \cdot R_s - \frac{1}{C_1} \cdot I_{C_1}$$

$$[e^{-\lambda \cdot \tau_{out2}} \cdot R_1 + R_s \cdot e^{\lambda \cdot \tau_{in}}] \cdot \frac{dI_{C_1}}{dt} = \frac{dV_s(t)}{dt} - [\sum_{k=1}^{2} X_k] \cdot R_s \cdot e^{\lambda \cdot \tau_{in}} - \frac{1}{C_1} \cdot I_{C_1} - X_1 \cdot e^{-\lambda \cdot \tau_{out2}} \cdot R_1$$

$$\frac{dI_{C_1}}{dt} = \frac{1}{(e^{-\lambda \cdot \tau_{out2}} \cdot R_1 + R_s \cdot e^{\lambda \cdot \tau_{in}})} \cdot \frac{dV_s(t)}{dt} - X_1 - \frac{X_{2a} \cdot R_s}{(e^{-\lambda \cdot \tau_{out2}} \cdot R_1 + R_s \cdot e^{\lambda \cdot \tau_{in}})} \cdot e^{\lambda \cdot \tau_{in}} - \frac{1}{C_1 \cdot (e^{-\lambda \cdot \tau_{out2}} \cdot R_1 + R_s \cdot e^{\lambda \cdot \tau_{in}})} \cdot I_{C_1}; \; \xi_1(R_1, R_s, \tau_{out2}, \tau_{in}, \lambda) = e^{-\lambda \cdot \tau_{out2}} \cdot R_1 + R_s \cdot e^{\lambda \cdot \tau_{in}}$$

$$\xi_1 = \xi_1(R_1, R_s, \tau_{out2}, \tau_{in}, \lambda); \; \frac{dI_{C_1}}{dt} = \frac{1}{\xi_1} \cdot \frac{dV_s(t)}{dt} - X_1 - \frac{X_2 \cdot R_s}{\xi_1} \cdot e^{\lambda \cdot \tau_{in}} - \frac{1}{C_1 \cdot \xi_1} \cdot I_{C_1}$$

$$\frac{1}{C_n} \cdot I_{C_n} = \frac{dV_s(t)}{dt} - \frac{dI_{R_s}}{dt} \cdot R_s - \frac{1}{C_2} \cdot I_{L_2} - L_2 \cdot \frac{dX_2}{dt} - L_{n1} \cdot \frac{d^2 I_{L_2}(t - \tau_{out1})}{dt^2}$$

$$\frac{1}{C_n} \cdot I_{C_n} = \frac{dV_s(t)}{dt} - \frac{dI_{R_s}}{dt} \cdot R_s - \frac{1}{C_2} \cdot I_{L_2} - L_2 \cdot \frac{dX_2}{dt} - L_{n1} \cdot \frac{d^2 I_{L_2}(t)}{dt^2} \cdot e^{-\lambda \cdot \tau_{out1}}$$

$$\frac{1}{C_n} \cdot I_{C_n} = \frac{dV_s(t)}{dt} - \frac{dI_{R_s}}{dt} \cdot R_s - \frac{1}{C_2} \cdot I_{L_2} - L_2 \cdot \frac{dX_2}{dt} - L_{n1} \cdot \frac{dX_2}{dt} \cdot e^{-\lambda \cdot \tau_{out1}}$$

$$[L_2 + L_{n1} \cdot e^{-\lambda \cdot \tau_{out1}}] \cdot \frac{dX_2}{dt} = \frac{dV_s(t)}{dt} - \frac{dI_{R_s}}{dt} \cdot R_s - \frac{1}{C_2} \cdot I_{L_2} - \frac{1}{C_n} \cdot I_{C_n}$$

$$[L_2 + L_{n1} \cdot e^{-\lambda \cdot \tau_{out1}}] \cdot \frac{dX_2}{dt} = \frac{dV_s(t)}{dt} - [\sum_{k=1}^{2} X_k] \cdot e^{\lambda \cdot \tau_{in}} \cdot R_s - \frac{dI_{C_1}}{dt} \cdot e^{\lambda \cdot \tau_{in}} \cdot R_s - \frac{1}{C_2} \cdot I_{L_2} - \frac{1}{C_n} \cdot I_{C_n}$$

We can summery our system differential equations (Version 6):

$$\frac{dI_{L_1}}{dt} = X_1; \; \frac{dI_{L_2}}{dt} = X_2; \; \frac{dI_{L_{n2}}}{dt} = X_3; \; \frac{dX_1}{dt} = \frac{1}{L_1 \cdot C_1} \cdot I_{C_1}$$

$$\frac{dI_{C_1}}{dt} = \frac{1}{\xi_1} \cdot \frac{dV_s(t)}{dt} - X_1 - \frac{X_2 \cdot R_s}{\xi_1} \cdot e^{\lambda \cdot \tau_{in}} - \frac{1}{C_1 \cdot \xi_1} \cdot I_{C_1}; \; \frac{dX_3}{dt} = \frac{1}{C_n \cdot L_{n2}} \cdot I_{C_n} - X_3 \cdot \frac{R_{load}}{L_{n2}}$$

$$\frac{dX_2}{dt} = \frac{1}{[L_2 + L_{n1} \cdot e^{-\lambda \cdot \tau_{out1}}]} \cdot \frac{dV_s(t)}{dt} - \frac{[\sum_{k=1}^{2} X_k] \cdot e^{\lambda \cdot \tau_{in}}}{[L_2 + L_{n1} \cdot e^{-\lambda \cdot \tau_{out1}}]} \cdot R_s - \frac{dI_{C_1}}{dt} \cdot \frac{R_s \cdot e^{\lambda \cdot \tau_{in}}}{[L_2 + L_{n1} \cdot e^{-\lambda \cdot \tau_{out1}}]} - \frac{1}{C_2 \cdot [L_2 + L_{n1} \cdot e^{-\lambda \cdot \tau_{out1}}]} \cdot I_{L_2} - \frac{1}{C_n \cdot [L_2 + L_{n1} \cdot e^{-\lambda \cdot \tau_{out1}}]} \cdot I_{C_n}$$

$$\frac{dI_{R_s}}{dt} = [\sum_{k=1}^{2} X_k] \cdot e^{\lambda \cdot \tau_{in}} + [\frac{1}{\xi_1} \cdot \frac{dV_s(t)}{dt} - X_1 - \frac{X_2 \cdot R_s}{\xi_1} \cdot e^{\lambda \cdot \tau_{in}} - \frac{1}{C_1 \cdot \xi_1} \cdot I_{C_1}] \cdot e^{\lambda \cdot \tau_{in}}$$

At fixed points: $\frac{dI_{L_1}}{dt} = 0$; $\frac{dI_{L_2}}{dt} = 0$; $\frac{dI_{L_{n2}}}{dt} = 0$; $\frac{dX_1}{dt} = 0$; $\frac{dX_3}{dt} = 0$; $\frac{dX_2}{dt} = 0$; $\frac{dI_{R_s}}{dt} = 0$

Assumption $\frac{dV_s(t)}{dt} \to \varepsilon$

$$X_1^* = 0;\ X_2^* = 0;\ X_3^* = 0;\ I_{C_1}^* = 0;\ I_{C_1}^* = C_1 \cdot \frac{dV_s(t)}{dt} \to \varepsilon;\ I_{C_n}^* = 0;\ I_{L_2}^* = C_2 \cdot \frac{dV_s(t)}{dt} \to \varepsilon$$

$$I_{L_{n1}} = I_{L_{n2}} + I_{C_n} \Rightarrow I_{C_n} = I_{L_{n1}} - I_{L_{n2}};\ I_{C_n} = I_{L_2}(t - \tau_{out1}) - I_{L_{n2}}$$

We can summery our system differential equations (Version 7):

$$\frac{dI_{L_1}}{dt} = X_1;\ \frac{dI_{L_2}}{dt} = X_2;\ \frac{dI_{L_{n2}}}{dt} = X_3;\ \frac{dX_1}{dt} = \frac{1}{L_1 \cdot C_1} \cdot I_{C_1}$$

$$\frac{dI_{C_1}}{dt} = \frac{1}{\xi_1} \cdot \frac{dV_s(t)}{dt} - X_1 - \frac{X_2 \cdot R_s}{\xi_1} \cdot e^{\lambda \cdot \tau_{in}} - \frac{1}{C_1 \cdot \xi_1} \cdot I_{C_1};$$

$$\frac{dX_3}{dt} = \frac{1}{C_n \cdot L_{n2}} \cdot [I_{L_2}(t - \tau_{out1}) - I_{L_{n2}}] - X_3 \cdot \frac{R_{load}}{L_{n2}}$$

$$\frac{dX_2}{dt} = \frac{1}{[L_2 + L_{n1} \cdot e^{-\lambda \cdot \tau_{out1}}]} \cdot \frac{dV_s(t)}{dt} - \frac{[\sum_{k=1}^{2} X_k] \cdot e^{\lambda \cdot \tau_{in}}}{[L_2 + L_{n1} \cdot e^{-\lambda \cdot \tau_{out1}}]} \cdot R_s - [\frac{1}{\xi_1} \cdot \frac{dV_s(t)}{dt} - X_1 - \frac{X_2 \cdot R_s}{\xi_1} \cdot e^{\lambda \cdot \tau_{in}} - \frac{1}{C_1 \cdot \xi_1} \cdot I_{C_1}] \cdot \frac{R_s \cdot e^{\lambda \cdot \tau_{in}}}{[L_2 + L_{n1} \cdot e^{-\lambda \cdot \tau_{out1}}]} - \frac{1}{C_2 \cdot [L_2 + L_{n1} \cdot e^{-\lambda \cdot \tau_{out1}}]} \cdot I_{L_2} - \frac{1}{C_n \cdot [L_2 + L_{n1} \cdot e^{-\lambda \cdot \tau_{out1}}]} \cdot [I_{L_2}(t - \tau_{out1}) - I_{L_{n2}}]$$

$$\frac{dI_{R_s}}{dt} = [\sum_{k=1}^{2} X_k] \cdot e^{\lambda \cdot \tau_{in}} + [\frac{1}{\xi_1} \cdot \frac{dV_s(t)}{dt} - X_1 - \frac{X_2 \cdot R_s}{\xi_1} \cdot e^{\lambda \cdot \tau_{in}} - \frac{1}{C_1 \cdot \xi_1} \cdot I_{C_1}] \cdot e^{\lambda \cdot \tau_{in}}$$

Stability analysis: The standard local stability analysis about any one of the equilibrium points of BPF diplexer circuit consists in adding to its coordinated $[X_1\ X_2\ X_3\ I_{L_1}\ I_{L_2}\ I_{L_{n2}}\ I_{C_1}\ I_{R_S}]$ arbitrarily small increments of exponential terms $[x_1\ x_2\ x_3\ i_{L_1}\ i_{L_2}\ i_{L_{n2}}\ i_{C_1}\ i_{R_S}] \cdot e^{\lambda \cdot t}$, and retaining the first order terms in $x_1\ x_2\ x_3\ i_{L_1}\ i_{L_2}\ i_{L_{n2}}\ i_{C_1}\ i_{R_S}$.

The system of eight homogeneous equations leads to a polynomial characteristic equation in the eigenvalue λ. The polynomial characteristic equation accepts by set the BPF diplexer circuit equations. The BPF diplexer circuit fixed values with arbitrarily small increments of exponential form $[x_1\ x_2\ x_3\ i_{L_1}\ i_{L_2}\ i_{L_{n2}}\ i_{C_1}\ i_{R_S}] \cdot e^{\lambda \cdot t}$ are; i = 0 (first fixed point), i = 1 (second fixed point), i = 2 (third fixed point), etc., [2–4].

$$X_1(t) = X_1^{(i)} + x_1 \cdot e^{\lambda \cdot t};\ X_2(t) = X_2^{(i)} + x_2 \cdot e^{\lambda \cdot t};$$
$$X_3(t) = X_3^{(i)} + x_3 \cdot e^{\lambda \cdot t};\ I_{L_1}(t) = I_{L_1}^{(i)} + i_{L_1} \cdot e^{\lambda \cdot t}$$
$$I_{L_2}(t) = I_{L_2}^{(i)} + i_{L_2} \cdot e^{\lambda \cdot t};\ I_{L_{n2}}(t) = I_{L_{n2}}^{(i)} + i_{L_{n2}} \cdot e^{\lambda \cdot t};$$
$$I_{C_1}(t) = I_{C_1}^{(i)} + i_{C_1} \cdot e^{\lambda \cdot t};\ I_{R_s} = I_{R_s}^{(i)} + i_{R_s} \cdot e^{\lambda \cdot t}$$
$$I_{L_2}(t - \tau_{out1}) = I_{L_2}^{(i)} + i_{L_2} \cdot e^{\lambda \cdot (t - \tau_{out1})};\ \frac{dI_{L_1}}{dt} = i_{L_1} \cdot \lambda \cdot e^{\lambda \cdot t};$$
$$\frac{dI_{L_2}}{dt} = i_{L_2} \cdot \lambda \cdot e^{\lambda \cdot t};\ \frac{dI_{L_{n2}}}{dt} = i_{L_{n2}} \cdot \lambda \cdot e^{\lambda \cdot t}$$

$$\frac{dX_1}{dt} = x_1 \cdot \lambda \cdot e^{\lambda \cdot t};\ \frac{dX_2}{dt} = x_2 \cdot \lambda \cdot e^{\lambda \cdot t};\ \frac{dX_3}{dt} = x_3 \cdot \lambda \cdot e^{\lambda \cdot t};\ \frac{dI_{C_1}}{dt} = i_{C_1} \cdot \lambda \cdot e^{\lambda \cdot t};\ \forall i = 0, 1, 2, \ldots$$

For λ < 0, t > 0 the selected fixed point is stable otherwise λ > 0, t > 0 unstable. Our BPF diplexer circuit tends to the selected fixed point exponentially for λ < 0, t > 0 otherwise go away from the selected fixed point exponentially. λ is the eigenvalue parameter which establish if the fixed point is stable or unstable, additionally his absolute value |λ| establish the speed of flow toward or away from the selected fixed point [4].

$$\frac{dI_{L_1}}{dt} = X_1 \Rightarrow i_{L_1} \cdot \lambda \cdot e^{\lambda \cdot t} = X_1^{(i)} + x_1 \cdot e^{\lambda \cdot t};\ X_1^{(i)} = X_1^* = 0;\ -i_{L_1} \cdot \lambda + x_1 = 0$$
$$\frac{dI_{L_2}}{dt} = X_2 \Rightarrow i_{L_2} \cdot \lambda \cdot e^{\lambda \cdot t} = X_2^{(i)} + x_2 \cdot e^{\lambda \cdot t};\ X_2^{(i)} = X_2^* = 0;\ -i_{L_2} \cdot \lambda + x_2 = 0$$

$$\frac{dI_{L_{n2}}}{dt} = X_3 \Rightarrow i_{L_{n2}} \cdot \lambda \cdot e^{\lambda \cdot t} = X_3^{(i)} + x_3 \cdot e^{\lambda \cdot t};\ X_3^{(i)} = X_3^* = 0;\ -i_{L_{n2}} \cdot \lambda + x_3 = 0$$

$$\frac{dX_1}{dt} = \frac{1}{L_1 \cdot C_1} \cdot I_{C_1} \Rightarrow x_1 \cdot \lambda \cdot e^{\lambda \cdot t} = \frac{1}{L_1 \cdot C_1} \cdot [I_{C_1}^{(i)} + i_{C_1} \cdot e^{\lambda \cdot t}];\ I_{C_1}^{(i)} = I_{C_1}^* = 0;\ -x_1 \cdot \lambda + \frac{1}{L_1 \cdot C_1} \cdot i_{C_1} = 0$$

$$\frac{dI_{C_1}}{dt} = \frac{1}{\xi_1} \cdot \frac{dV_s(t)}{dt} - X_1 - \frac{X_2 \cdot R_s}{\xi_1} \cdot e^{\lambda \cdot \tau_{in}} - \frac{1}{C_1 \cdot \xi_1} \cdot I_{C_1}$$

$$i_{C_1} \cdot \lambda \cdot e^{\lambda \cdot t} = \frac{1}{\xi_1} \cdot \frac{dV_s(t)}{dt} - [X_1^{(i)} + x_1 \cdot e^{\lambda \cdot t}] - \frac{[X_2^{(i)} + x_2 \cdot e^{\lambda \cdot t}] \cdot R_s}{\xi_1} \cdot e^{\lambda \cdot \tau_{in}} - \frac{1}{C_1 \cdot \xi_1} \cdot [I_{C_1}^{(i)} + i_{C_1} \cdot e^{\lambda \cdot t}]$$

$$i_{C_1} \cdot \lambda \cdot e^{\lambda \cdot t} = \frac{1}{\xi_1} \cdot \frac{dV_s(t)}{dt} - X_1^{(i)} - \frac{X_2^{(i)} \cdot R_s}{\xi_1} \cdot e^{\lambda \cdot \tau_{in}} - \frac{1}{C_1 \cdot \xi_1} \cdot I_{C_1}^{(i)} - x_1 \cdot e^{\lambda \cdot t} - \frac{x_2 \cdot e^{\lambda \cdot t} \cdot R_s}{\xi_1} \cdot e^{\lambda \cdot \tau_{in}} - \frac{1}{C_1 \cdot \xi_1} \cdot i_{C_1} \cdot e^{\lambda \cdot t}$$

At fixed point: $\frac{1}{\xi_1} \cdot \frac{dV_s(t)}{dt} - X_1^{(i)} - \frac{X_2^{(i)} \cdot R_s}{\xi_1} \cdot e^{\lambda \cdot \tau_{in}} - \frac{1}{C_1 \cdot \xi_1} \cdot I_{C_1}^{(i)} = 0$

$$-i_{C_1} \cdot \lambda - x_1 - x_2 \cdot \frac{R_s}{\xi_1} \cdot e^{\lambda \cdot \tau_{in}} - \frac{1}{C_1 \cdot \xi_1} \cdot i_{C_1} = 0$$

$$\frac{dX_3}{dt} = \frac{1}{C_n \cdot L_{n2}} \cdot [I_{L_2}(t - \tau_{out1}) - I_{L_{n2}}(t)] - X_3 \cdot \frac{R_{load}}{L_{n2}}; \ \lim_{t \to \infty} I_{L_2}(t - \tau_{out1}) = I_{L_{n2}}(t); \ t \gg \tau_{out1}$$

$$x_3 \cdot \lambda \cdot e^{\lambda \cdot t} = \frac{1}{C_n \cdot L_{n2}} \cdot [I_{L_2}^{(i)} + i_{L_2} \cdot e^{\lambda \cdot (t - \tau_{out1})} - I_{L_{n2}}^{(i)} - i_{L_{n2}} \cdot e^{\lambda \cdot t}] - [X_3^{(i)} + x_3 \cdot e^{\lambda \cdot t}] \cdot \frac{R_{load}}{L_{n2}}$$

$$x_3 \cdot \lambda \cdot e^{\lambda \cdot t} = \frac{1}{C_n \cdot L_{n2}} \cdot [I_{L_2}^{(i)} - I_{L_{n2}}^{(i)}] - X_3^{(i)} \cdot \frac{R_{load}}{L_{n2}} + \frac{1}{C_n \cdot L_{n2}} \cdot [i_{L_2} \cdot e^{\lambda \cdot (t - \tau_{out1})} - i_{L_{n2}} \cdot e^{\lambda \cdot t}] - x_3 \cdot \frac{R_{load}}{L_{n2}} \cdot e^{\lambda \cdot t}$$

At fixed point: $\frac{1}{C_n \cdot L_{n2}} \cdot [I_{L_2}^{(i)} - I_{L_{n2}}^{(i)}] - X_3^{(i)} \cdot \frac{R_{load}}{L_{n2}} = 0$

$$-x_3 \cdot \lambda - x_3 \cdot \frac{R_{load}}{L_{n2}} + \frac{1}{C_n \cdot L_{n2}} \cdot i_{L_2} \cdot e^{-\lambda \cdot \tau_{out1}} - \frac{1}{C_n \cdot L_{n2}} \cdot i_{L_{n2}} = 0$$

$$\frac{dX_2}{dt} = \frac{1}{[L_2 + L_{n1} \cdot e^{-\lambda \cdot \tau_{out1}}]} \cdot \frac{dV_s(t)}{dt} - \frac{[\sum_{k=1}^{2} X_k] \cdot e^{\lambda \cdot \tau_{in}}}{[L_2 + L_{n1} \cdot e^{-\lambda \cdot \tau_{out1}}]} \cdot R_s - [\frac{1}{\xi_1} \cdot \frac{dV_s(t)}{dt} - X_1 - \frac{X_2 \cdot R_s}{\xi_1} \cdot e^{\lambda \cdot \tau_{in}} - \frac{1}{C_1 \cdot \xi_1} \cdot I_{C_1}] \cdot \frac{R_s \cdot e^{\lambda \cdot \tau_{in}}}{[L_2 + L_{n1} \cdot e^{-\lambda \cdot \tau_{out1}}]} - \frac{1}{C_2 \cdot [L_2 + L_{n1} \cdot e^{-\lambda \cdot \tau_{out1}}]} \cdot I_{L_2} - \frac{1}{C_n \cdot [L_2 + L_{n1} \cdot e^{-\lambda \cdot \tau_{out1}}]} \cdot [I_{L_2}(t - \tau_{out1}) - I_{L_{n2}}]$$

$$x_2 \cdot \lambda \cdot e^{\lambda \cdot t} = \frac{1}{[L_2 + L_{n1} \cdot e^{-\lambda \cdot \tau_{out1}}]} \cdot \frac{dV_s(t)}{dt} - \frac{[\sum_{k=1}^{2} (X_k^{(i)} + x_k \cdot e^{\lambda \cdot t})] \cdot e^{\lambda \cdot \tau_{in}}}{[L_2 + L_{n1} \cdot e^{-\lambda \cdot \tau_{out1}}]} \cdot R_s$$
$$- \{\frac{1}{\xi_1} \cdot \frac{dV_s(t)}{dt} - [X_1^{(i)} + x_1 \cdot e^{\lambda \cdot t}] - \frac{[X_2^{(i)} + x_2 \cdot e^{\lambda \cdot t}] \cdot R_s}{\xi_1} \cdot e^{\lambda \cdot \tau_{in}}$$
$$- \frac{1}{C_1 \cdot \xi_1} \cdot [I_{C_1}^{(i)} + i_{C_1} \cdot e^{\lambda \cdot t}]\} \cdot \frac{R_s \cdot e^{\lambda \cdot \tau_{in}}}{[L_2 + L_{n1} \cdot e^{-\lambda \cdot \tau_{out1}}]}$$
$$- \frac{1}{C_2 \cdot [L_2 + L_{n1} \cdot e^{-\lambda \cdot \tau_{out1}}]} \cdot [I_{L_2}^{(i)} + i_{L_2} \cdot e^{\lambda \cdot t}]$$
$$- \frac{1}{C_n \cdot [L_2 + L_{n1} \cdot e^{-\lambda \cdot \tau_{out1}}]} \cdot [I_{L_2}^{(i)} + i_{L_2} \cdot e^{\lambda \cdot (t - \tau_{out1})} - I_{L_{n2}}^{(i)} - i_{L_{n2}} \cdot e^{\lambda \cdot t}]$$

$$x_2 \cdot \lambda \cdot e^{\lambda \cdot t} = \frac{1}{[L_2 + L_{n1} \cdot e^{-\lambda \cdot \tau_{out1}}]} \cdot \frac{dV_s(t)}{dt} - \frac{[\sum_{k=1}^{2} X_k^{(i)}] \cdot e^{\lambda \cdot \tau_{in}}}{[L_2 + L_{n1} \cdot e^{-\lambda \cdot \tau_{out1}}]} \cdot R_s$$
$$- \{\frac{1}{\xi_1} \cdot \frac{dV_s(t)}{dt} - X_1^{(i)} - \frac{X_2^{(i)} \cdot R_s}{\xi_1} \cdot e^{\lambda \cdot \tau_{in}} - \frac{1}{C_1 \cdot \xi_1} \cdot I_{C_1}^{(i)}\} \cdot \frac{R_s \cdot e^{\lambda \cdot \tau_{in}}}{[L_2 + L_{n1} \cdot e^{-\lambda \cdot \tau_{out1}}]}$$
$$- \frac{1}{C_2 \cdot [L_2 + L_{n1} \cdot e^{-\lambda \cdot \tau_{out1}}]} \cdot I_{L_2}^{(i)} - \frac{1}{C_n \cdot [L_2 + L_{n1} \cdot e^{-\lambda \cdot \tau_{out1}}]} \cdot [I_{L_2}^{(i)} - I_{L_{n2}}^{(i)}]$$
$$- \frac{[\sum_{k=1}^{2} x_k] \cdot e^{\lambda \cdot \tau_{in}} \cdot e^{\lambda \cdot t}}{[L_2 + L_{n1} \cdot e^{-\lambda \cdot \tau_{out1}}]} \cdot R_s - \{-x_1 \cdot e^{\lambda \cdot t} - \frac{x_2 \cdot e^{\lambda \cdot t} \cdot R_s}{\xi_1} \cdot e^{\lambda \cdot \tau_{in}}$$
$$- \frac{1}{C_1 \cdot \xi_1} \cdot i_{C_1} \cdot e^{\lambda \cdot t}\} \cdot \frac{R_s \cdot e^{\lambda \cdot \tau_{in}}}{[L_2 + L_{n1} \cdot e^{-\lambda \cdot \tau_{out1}}]}$$
$$- \frac{1}{C_2 \cdot [L_2 + L_{n1} \cdot e^{-\lambda \cdot \tau_{out1}}]} \cdot i_{L_2} \cdot e^{\lambda \cdot t}$$
$$- \frac{1}{C_n \cdot [L_2 + L_{n1} \cdot e^{-\lambda \cdot \tau_{out1}}]} \cdot [i_{L_2} \cdot e^{-\lambda \cdot \tau_{out1}} - i_{L_{n2}}] \cdot e^{\lambda \cdot t}$$

At fixed point:

$$\frac{1}{[L_2 + L_{n1} \cdot e^{-\lambda \cdot \tau_{out1}}]} \cdot \frac{dV_s(t)}{dt} - \frac{[\sum_{k=1}^{2} X_k^{(i)}] \cdot e^{\lambda \cdot \tau_{in}}}{[L_2 + L_{n1} \cdot e^{-\lambda \cdot \tau_{out1}}]} \cdot R_s$$
$$- \{\frac{1}{\xi_1} \cdot \frac{dV_s(t)}{dt} - X_1^{(i)} - \frac{X_2^{(i)} \cdot R_s}{\xi_1} \cdot e^{\lambda \cdot \tau_{in}} - \frac{1}{C_1 \cdot \xi_1} \cdot I_{C_1}^{(i)}\} \cdot \frac{R_s \cdot e^{\lambda \cdot \tau_{in}}}{[L_2 + L_{n1} \cdot e^{-\lambda \cdot \tau_{out1}}]}$$
$$- \frac{1}{C_2 \cdot [L_2 + L_{n1} \cdot e^{-\lambda \cdot \tau_{out1}}]} \cdot I_{L_2}^{(i)} - \frac{1}{C_n \cdot [L_2 + L_{n1} \cdot e^{-\lambda \cdot \tau_{out1}}]} \cdot [I_{L_2}^{(i)} - I_{L_{n2}}^{(i)}] = 0$$

$$- x_2 \cdot \lambda - \frac{[\sum_{k=1}^{2} x_k] \cdot e^{\lambda \cdot \tau_{in}}}{[L_2 + L_{n1} \cdot e^{-\lambda \cdot \tau_{out1}}]} \cdot R_s - \{-x_1 - \frac{x_2 \cdot R_s}{\xi_1} \cdot e^{\lambda \cdot \tau_{in}}$$
$$- \frac{1}{C_1 \cdot \xi_1} \cdot i_{C_1}\} \cdot \frac{R_s \cdot e^{\lambda \cdot \tau_{in}}}{[L_2 + L_{n1} \cdot e^{-\lambda \cdot \tau_{out1}}]}$$
$$- \frac{1}{C_2 \cdot [L_2 + L_{n1} \cdot e^{-\lambda \cdot \tau_{out1}}]} \cdot i_{L_2}$$
$$- \frac{1}{C_n \cdot [L_2 + L_{n1} \cdot e^{-\lambda \cdot \tau_{out1}}]} \cdot [i_{L_2} \cdot e^{-\lambda \cdot \tau_{out1}} - i_{L_{n2}}] = 0$$

$$- x_2 \cdot \lambda + x_2 \cdot \frac{(\frac{R_s}{\xi_1} \cdot e^{\lambda \cdot \tau_{in}} - 1) \cdot e^{\lambda \cdot \tau_{in}}}{[L_2 + L_{n1} \cdot e^{-\lambda \cdot \tau_{out1}}]} \cdot R_s + \frac{1}{C_1 \cdot \xi_1} \cdot \frac{R_s \cdot e^{\lambda \cdot \tau_{in}}}{[L_2 + L_{n1} \cdot e^{-\lambda \cdot \tau_{out1}}]} \cdot i_{C_1}$$
$$- (\frac{1}{C_2} + \frac{1}{C_n} \cdot e^{-\lambda \cdot \tau_{out1}}) \cdot \frac{1}{[L_2 + L_{n1} \cdot e^{-\lambda \cdot \tau_{out1}}]} \cdot i_{L_2} + \frac{1}{C_n \cdot [L_2 + L_{n1} \cdot e^{-\lambda \cdot \tau_{out1}}]} \cdot i_{L_{n2}} = 0$$

Remark: The last differential equation $\frac{dI_{R_s}}{dt} = \ldots$ is not essential for our circuit stability analysis.

We can summery our BPF diplexer arbitrarily small increments equations.

$$- i_{L_1} \cdot \lambda + x_1 = 0; -i_{L_2} \cdot \lambda + x_2 = 0; -i_{L_{n2}} \cdot \lambda + x_3 = 0;$$
$$- i_{C_1} \cdot \lambda - \frac{1}{C_1 \cdot \xi_1} \cdot i_{C_1} - x_1 - x_2 \cdot \frac{R_s}{\xi_1} \cdot e^{\lambda \cdot \tau_{in}} = 0$$
$$- x_1 \cdot \lambda + \frac{1}{L_1 \cdot C_1} \cdot i_{C_1} = 0; -x_3 \cdot \lambda - x_3 \cdot \frac{R_{load}}{L_{n2}}$$
$$+ \frac{1}{C_n \cdot L_{n2}} \cdot i_{L_2} \cdot e^{-\lambda \cdot \tau_{out1}} - \frac{1}{C_n \cdot L_{n2}} \cdot i_{L_{n2}} = 0$$

$$- x_2 \cdot \lambda + x_2 \cdot \frac{(\frac{R_s}{\xi_1} \cdot e^{\lambda \cdot \tau_{in}} - 1) \cdot e^{\lambda \cdot \tau_{in}}}{[L_2 + L_{n1} \cdot e^{-\lambda \cdot \tau_{out1}}]} \cdot R_s + \frac{1}{C_1 \cdot \xi_1} \cdot \frac{R_s \cdot e^{\lambda \cdot \tau_{in}}}{[L_2 + L_{n1} \cdot e^{-\lambda \cdot \tau_{out1}}]} \cdot i_{C_1}$$
$$- (\frac{1}{C_2} + \frac{1}{C_n} \cdot e^{-\lambda \cdot \tau_{out1}}) \cdot \frac{1}{[L_2 + L_{n1} \cdot e^{-\lambda \cdot \tau_{out1}}]} \cdot i_{L_2} + \frac{1}{C_n \cdot [L_2 + L_{n1} \cdot e^{-\lambda \cdot \tau_{out1}}]} \cdot i_{L_{n2}} = 0$$

$$\begin{pmatrix} \Xi_{11} & \cdots & \Xi_{17} \\ \vdots & \ddots & \vdots \\ \Xi_{71} & \cdots & \Xi_{77} \end{pmatrix} \cdot \begin{pmatrix} i_{L_1} \\ i_{L_2} \\ i_{L_{n2}} \\ i_{C_1} \\ x_1 \\ x_2 \\ x_3 \end{pmatrix} = 0 \; ; \; \Xi_{11} = -\lambda \; ; \; \Xi_{12} = \Xi_{13} = \Xi_{14} = 0 \; ; \; \Xi_{15}$$
$$= 1 \; ; \; \Xi_{16} = \Xi_{17} = 0$$

$$\Xi_{21} = 0;\ \Xi_{22} = -\lambda;\ \Xi_{23} = \Xi_{24} = \Xi_{25} = 0;\ \Xi_{26} = 1;\ \Xi_{27} = 0;\ \Xi_{31} = \Xi_{32} = 0;\ \Xi_{33} = -\lambda$$

$$\Xi_{34} = \Xi_{35} = \Xi_{36} = 0;\ \Xi_{37} = 1;\ \Xi_{41} = \Xi_{42} = \Xi_{43} = 0;$$

$$\Xi_{44} = -\lambda - \frac{1}{C_1 \cdot \xi_1};\ \Xi_{45} = -1$$

$$\Xi_{46} = -\frac{R_s}{\xi_1} \cdot e^{\lambda \cdot \tau_{in}};\ \Xi_{47} = 0;\ \Xi_{51} = \Xi_{52} = \Xi_{53} = 0;\ \Xi_{54} = \frac{1}{L_1 \cdot C_1};$$

$$\Xi_{55} = -\lambda;\ \Xi_{56} = \Xi_{57} = 0$$

$$\Xi_{61} = 0;\ \Xi_{62} = -(\frac{1}{C_2} + \frac{1}{C_n} \cdot e^{-\lambda \cdot \tau_{out1}}) \cdot \frac{1}{[L_2 + L_{n1} \cdot e^{-\lambda \cdot \tau_{out1}}]};\ \Xi_{63} = \frac{1}{C_n \cdot [L_2 + L_{n1} \cdot e^{-\lambda \cdot \tau_{out1}}]}$$

$$\Xi_{64} = \frac{1}{C_1 \cdot \xi_1} \cdot \frac{R_s \cdot e^{\lambda \cdot \tau_{in}}}{[L_2 + L_{n1} \cdot e^{-\lambda \cdot \tau_{out1}}]};\ \Xi_{65} = 0;\ \Xi_{66} = -\lambda + \frac{(\frac{R_s}{\xi_1} \cdot e^{\lambda \cdot \tau_{in}} - 1) \cdot e^{\lambda \cdot \tau_{in}}}{[L_2 + L_{n1} \cdot e^{-\lambda \cdot \tau_{out1}}]} \cdot R_s;\ \Xi_{67} = 0$$

$$\Xi_{71} = 0;\ \Xi_{72} = \frac{1}{C_n \cdot L_{n2}} \cdot e^{-\lambda \cdot \tau_{out1}};\ \Xi_{73} = -\frac{1}{C_n \cdot L_{n2}};\ \Xi_{74} = \Xi_{75} = \Xi_{76} = 0;\ \Xi_{77} = -\lambda - \frac{R_{load}}{L_{n2}}$$

We define for simplicity global parameters in our system:

$$\xi_1 = \xi_1(R_1, R_s, \tau_{out2}, \tau_{in}, \lambda) = e^{-\lambda \cdot \tau_{out2}} \cdot R_1 + R_s \cdot e^{\lambda \cdot \tau_{in}}$$

$$\psi_1 = \psi_1(C_2, C_n, L_2, L_{n1}, \tau_{out1}, \lambda) = -\frac{(\frac{1}{C_2} + \frac{1}{C_n} \cdot e^{-\lambda \cdot \tau_{out1}})}{[L_2 + L_{n1} \cdot e^{-\lambda \cdot \tau_{out1}}]}$$

$$\psi_2 = \psi_2(C_n, L_2, L_{n1}, \tau_{out1}, \lambda) = \frac{1}{C_n \cdot [L_2 + L_{n1} \cdot e^{-\lambda \cdot \tau_{out1}}]}$$

$$\psi_3 = \psi_3(C_1, R_1, R_s, L_2, L_{n1}, \tau_{in}, \tau_{out1}, \tau_{out2}, \lambda) = \frac{1}{C_1 \cdot \xi_1} \cdot \frac{R_s \cdot e^{\lambda \cdot \tau_{in}}}{[L_2 + L_{n1} \cdot e^{-\lambda \cdot \tau_{out1}}]} = \frac{1}{C_1 \cdot [e^{-\lambda \cdot \tau_{out2}} \cdot R_1 + R_s \cdot e^{\lambda \cdot \tau_{in}}]} \cdot \frac{R_s \cdot e^{\lambda \cdot \tau_{in}}}{[L_2 + L_{n1} \cdot e^{-\lambda \cdot \tau_{out1}}]}$$

$$\psi_4 = \psi_4(R_s, R_1, L_2, L_{n1}, \tau_{out1}, \tau_{out2}, \lambda) = \frac{(\frac{R_s}{\xi_1} \cdot e^{\lambda \cdot \tau_{in}} - 1) \cdot e^{\lambda \cdot \tau_{in}}}{[L_2 + L_{n1} \cdot e^{-\lambda \cdot \tau_{out1}}]} \cdot R_s$$
$$= \frac{(\frac{R_s}{[e^{-\lambda \cdot \tau_{out2}} \cdot R_1 + R_s \cdot e^{\lambda \cdot \tau_{in}}]} \cdot e^{\lambda \cdot \tau_{in}} - 1) \cdot e^{\lambda \cdot \tau_{in}}}{[L_2 + L_{n1} \cdot e^{-\lambda \cdot \tau_{out1}}]} \cdot R_s$$

$$\Xi_{62} = \psi_1 = \psi_1(C_2, C_n, L_2, L_{n1}, \tau_{out1}, \lambda); \; \Xi_{63} = \psi_2 = \psi_2(C_n, L_2, L_{n1}, \tau_{out1}, \lambda)$$

$$\Xi_{64} = \psi_3 = \psi_3(C_1, R_1, R_s, L_2, L_{n1}, \tau_{in}, \tau_{out1}, \tau_{out2}, \lambda); \; \Xi_{66} = \psi_4$$
$$= \psi_4(R_s, R_1, L_2, L_{n1}, \tau_{out1}, \tau_{out2}, \lambda)$$

$$(A - \lambda \cdot I) = \begin{pmatrix} \Xi_{11} & \cdots & \Xi_{17} \\ \vdots & \ddots & \vdots \\ \Xi_{71} & \cdots & \Xi_{77} \end{pmatrix}; \det(A - \lambda \cdot I) = \det \begin{pmatrix} \Xi_{11} & \cdots & \Xi_{17} \\ \vdots & \ddots & \vdots \\ \Xi_{71} & \cdots & \Xi_{77} \end{pmatrix} = 0$$

$$\det(A - \lambda \cdot I) = \det \begin{pmatrix} \Xi_{11} & \cdots & \Xi_{17} \\ \vdots & \ddots & \vdots \\ \Xi_{71} & \cdots & \Xi_{77} \end{pmatrix} = (-\lambda) \cdot \{(-\lambda) \cdot [(-\lambda) \cdot P_1(\lambda) + P_2(\lambda)]$$
$$- [\lambda \cdot P_3(\lambda) + P_4(\lambda)]\}$$

$$P_1(\lambda) = \lambda^4 + \lambda^3 \cdot [\frac{R_{load}}{L_{n2}} - \psi_4 + \frac{1}{C_1 \cdot \xi_1}] + \lambda^2$$
$$\times [\frac{1}{C_1 \cdot \xi_1} \cdot (\frac{R_{load}}{L_{n2}} - \psi_4) - \psi_4 \cdot \frac{R_{load}}{L_{n2}} + \frac{1}{L_1 \cdot C_1} + \frac{R_s}{\xi_1} \cdot e^{\tau_{in} \cdot \lambda} \cdot \psi_3]$$
$$+ \lambda \cdot [(\frac{R_{load}}{L_{n2}} - \psi_4) \cdot \frac{1}{L_1 \cdot C_1} - \frac{\psi_4 \cdot R_{load}}{C_1 \cdot \xi_1 \cdot L_{n2}} + \frac{R_{load} \cdot R_s}{\xi_1 \cdot L_{n2}} \cdot e^{\lambda \cdot \tau_{in}} \cdot \psi_3]$$
$$- \psi_4 \cdot \frac{R_{load}}{L_{n2} \cdot L_1 \cdot C_1}$$

$$P_2(\lambda) = \frac{1}{C_n \cdot L_{n2}} \cdot [-\lambda^3 + \lambda^2 \cdot (\psi_4 - \frac{1}{C_1 \cdot \xi_1}) + \lambda \cdot (\frac{\psi_4}{C_1 \cdot \xi_1} - \frac{1}{L_1 \cdot C_1} - \frac{R_s}{\xi_1} \cdot e^{\lambda \cdot \tau_{in}} \cdot \psi_3)$$
$$+ \psi_4 \cdot \frac{1}{L_1 \cdot C_1}]$$

$$P_2(\lambda) = -\frac{1}{C_n \cdot L_{n2}} \cdot \lambda^3 + \lambda^2 \cdot \frac{1}{C_n \cdot L_{n2}} \cdot (\psi_4 - \frac{1}{C_1 \cdot \xi_1})$$
$$+ \lambda \cdot \frac{1}{C_n \cdot L_{n2}} \cdot (\frac{\psi_4}{C_1 \cdot \xi_1} - \frac{1}{L_1 \cdot C_1} - \frac{R_s}{\xi_1} \cdot e^{\lambda \cdot \tau_{in}} \cdot \psi_3)$$
$$+ \frac{1}{C_n \cdot L_{n2}} \cdot \psi_4 \cdot \frac{1}{L_1 \cdot C_1}$$

Table 7.1 BPF diplexer circuit cases for τ_{in}, τ_{out1}, τ_{out2}

Case no.	τ_{in}	τ_{out1}	τ_{out2}
1	0	0	0
2	τ	0	0
3	0	τ	0
4	0	0	τ
5	τ	τ	0
6	0	τ	τ
7	τ	0	τ
8	τ	τ	τ

$$P_3(\lambda) = \psi_1 \cdot \lambda^3 + \lambda^2 \cdot \psi_1 \cdot \left(\frac{R_{load}}{L_{n2}} + \frac{1}{C_1 \cdot \xi_1}\right) + \lambda \cdot \psi_1 \cdot \left(\frac{R_{load}}{C_1 \cdot \xi_1 \cdot L_{n2}} + \frac{1}{L_1 \cdot C_1}\right) + \frac{\psi_1 \cdot R_{load}}{L_1 \cdot C_1 \cdot L_{n2}}$$

$$P_4(\lambda) = \lambda^2 \cdot \left[\frac{1}{C_n \cdot L_{n2}} \cdot \psi_1 + \psi_2 \cdot \frac{1}{C_n \cdot L_{n2}} \cdot e^{-\lambda \cdot \tau_{out1}}\right] + \lambda \cdot \frac{1}{C_1 \cdot \xi_1} \cdot \frac{1}{C_n \cdot L_{n2}} \cdot \left[\psi_1 + \psi_2 \cdot e^{-\lambda \cdot \tau_{out1}}\right] + \frac{1}{L_1 \cdot C_1} \cdot \frac{1}{C_n \cdot L_{n2}} \cdot \left[\psi_1 + \psi_2 \cdot e^{-\lambda \cdot \tau_{out1}}\right]$$

$$\det(A - \lambda \cdot I) = -\lambda^3 \cdot P_1(\lambda, \tau_{in}, \tau_{out1}, \tau_{out2}) + \lambda^2 \cdot P_2(\lambda, \tau_{in}, \tau_{out1}, \tau_{out2}) + \lambda^2 \cdot P_3(\lambda, \tau_{in}, \tau_{out1}, \tau_{out2}) + \lambda \cdot P_4(\lambda, \tau_{in}, \tau_{out1}, \tau_{out2})$$

We analyze our BPF diplexer circuit for the following cases:
Case No. 1: $\tau_{in} = 0$; $\tau_{out1} = 0$; $\tau_{out2} = 0$.

$$\xi_1(R_1, R_s, \tau_{out2} = 0, \tau_{in} = 0, \lambda) = R_1 + R_s; \; \psi_1 = \psi_1(C_2, C_n, L_2, L_{n1}, \tau_{out1} = 0, \lambda) = -\frac{\left(\frac{1}{C_2} + \frac{1}{C_n}\right)}{[L_2 + L_{n1}]}$$

$$\psi_2 = \psi_2(C_n, L_2, L_{n1}, \tau_{out1} = 0, \lambda) = \frac{1}{C_n \cdot [L_2 + L_{n1}]}$$

$$\psi_3 = \psi_3(C_1, R_1, R_s, L_2, L_{n1}, \tau_{in} = 0, \tau_{out1} = 0, \tau_{out2} = 0, \lambda) = \frac{1}{C_1 \cdot \xi_1(\tau_{out2} = 0, \tau_{in} = 0)} \cdot \frac{R_s}{[L_2 + L_{n1}]} = \frac{1}{C_1 \cdot [R_1 + R_s]} \cdot \frac{R_s}{[L_2 + L_{n1}]}$$

$$\psi_4 = \psi_4(R_s, R_1, L_2, L_{n1}, \tau_{out1} = 0, \tau_{out2} = 0, \lambda) = \frac{(\frac{R_s}{\xi_1} - 1)}{[L_2 + L_{n1}]} \cdot R_s$$
$$= \frac{(\frac{R_s}{[R_1 + R_s]} - 1)}{[L_2 + L_{n1}]} \cdot R_s$$

$$P_2(\lambda, \tau_{in} = 0, \tau_{out1} = 0, \tau_{out2} = 0) = -\frac{1}{C_n \cdot L_{n2}} \cdot \lambda^3 + \lambda^2 \cdot \frac{1}{C_n \cdot L_{n2}}$$
$$\times (\psi_4(, \tau_{in} = 0, \tau_{out1} = 0, \tau_{out2} = 0) - \frac{1}{C_1 \cdot \xi_1(, \tau_{in} = 0, \tau_{out1} = 0, \tau_{out2} = 0)})$$
$$+ \lambda \cdot \frac{1}{C_n \cdot L_{n2}} \cdot (\frac{\psi_4(, \tau_{in} = 0, \tau_{out1} = 0, \tau_{out2} = 0)}{C_1 \cdot \xi_1(, \tau_{in} = 0, \tau_{out1} = 0, \tau_{out2} = 0)} - \frac{1}{L_1 \cdot C_1}$$
$$- \frac{R_s}{\xi_1(, \tau_{in} = 0, \tau_{out1} = 0, \tau_{out2} = 0)} \cdot \psi_3(, \tau_{in} = 0, \tau_{out1} = 0, \tau_{out2} = 0))$$
$$+ \frac{1}{C_n \cdot L_{n2}} \cdot \psi_4(, \tau_{in} = 0, \tau_{out1} = 0, \tau_{out2} = 0) \cdot \frac{1}{L_1 \cdot C_1}$$

$$P_1(\lambda, \tau_{in} = 0, \tau_{out1} = 0, \tau_{out2} = 0) = \lambda^4 + \lambda^3 \cdot [\frac{R_{load}}{L_{n2}} - \psi_4(\tau_{in} = 0, \tau_{out1} = 0, \tau_{out2} = 0)$$
$$+ \frac{1}{C_1 \cdot \xi_1(\tau_{in} = 0, \tau_{out1} = 0, \tau_{out2} = 0)}] + \lambda^2 \cdot [\frac{1}{C_1 \cdot \xi_1(\tau_{in} = 0, \tau_{out1} = 0, \tau_{out2} = 0)}$$
$$\cdot (\frac{R_{load}}{L_{n2}} - \psi_4(\tau_{in} = 0, \tau_{out1} = 0, \tau_{out2} = 0)) - \psi_4(\tau_{in} = 0, \tau_{out1} = 0, \tau_{out2} = 0)$$
$$\cdot \frac{R_{load}}{L_{n2}} + \frac{1}{L_1 \cdot C_1} + \frac{R_s}{\xi_1(\tau_{in} = 0, \tau_{out1} = 0, \tau_{out2} = 0)}$$
$$\cdot e^{\tau_{in} \cdot \lambda} \cdot \psi_3(\tau_{in} = 0, \tau_{out1} = 0, \tau_{out2} = 0)] + \lambda$$
$$\cdot [(\frac{R_{load}}{L_{n2}} - \psi_4(\tau_{in} = 0, \tau_{out1} = 0, \tau_{out2} = 0))$$
$$\cdot \frac{1}{L_1 \cdot C_1} - \frac{\psi_4(\tau_{in} = 0, \tau_{out1} = 0, \tau_{out2} = 0) \cdot R_{load}}{C_1 \cdot \xi_1(\tau_{in} = 0, \tau_{out1} = 0, \tau_{out2} = 0) \cdot L_{n2}}$$
$$+ \frac{R_{load} \cdot R_s}{\xi_1(\tau_{in} = 0, \tau_{out1} = 0, \tau_{out2} = 0) \cdot L_{n2}}$$
$$\cdot e^{\lambda \cdot \tau_{in}} \cdot \psi_3(\tau_{in} = 0, \tau_{out1} = 0, \tau_{out2} = 0)] - \psi_4(\tau_{in} = 0, \tau_{out1} = 0, \tau_{out2} = 0)$$
$$\cdot \frac{R_{load}}{L_{n2} \cdot L_1 \cdot C_1}$$

$$P_3(\lambda, \tau_{in} = 0, \tau_{out1} = 0, \tau_{out2} = 0) = \psi_1(\tau_{in} = 0, \tau_{out1} = 0, \tau_{out2} = 0) \cdot \lambda^3$$
$$+ \lambda^2 \cdot \psi_1(\tau_{in} = 0, \tau_{out1} = 0, \tau_{out2} = 0)$$
$$\cdot \left(\frac{R_{load}}{L_{n2}} + \frac{1}{C_1 \cdot \xi_1(\tau_{in} = 0, \tau_{out1} = 0, \tau_{out2} = 0)}\right)$$
$$+ \lambda \cdot \psi_1(\tau_{in} = 0, \tau_{out1} = 0, \tau_{out2} = 0)$$
$$\cdot \left(\frac{R_{load}}{C_1 \cdot \xi_1(\tau_{in} = 0, \tau_{out1} = 0, \tau_{out2} = 0) \cdot L_{n2}} + \frac{1}{L_1 \cdot C_1}\right)$$
$$+ \frac{\psi_1(\tau_{in} = 0, \tau_{out1} = 0, \tau_{out2} = 0) \cdot R_{load}}{L_1 \cdot C_1 \cdot L_{n2}}$$

$$P_4(\lambda, \tau_{in} = 0, \tau_{out1} = 0, \tau_{out2} = 0) = \lambda^2 \cdot \left[\frac{1}{C_n \cdot L_{n2}} \cdot \psi_1(\tau_{in} = 0, \tau_{out1} = 0, \tau_{out2} = 0)\right.$$
$$\left.+ \psi_2(\tau_{in} = 0, \tau_{out1} = 0, \tau_{out2} = 0) \cdot \frac{1}{C_n \cdot L_{n2}}\right]$$
$$+ \lambda \cdot \frac{1}{C_1 \cdot \xi_1(\tau_{in} = 0, \tau_{out1} = 0, \tau_{out2} = 0)} \cdot \frac{1}{C_n \cdot L_{n2}}$$
$$\cdot [\psi_1(\tau_{in} = 0, \tau_{out1} = 0, \tau_{out2} = 0)$$
$$+ \psi_2(\tau_{in} = 0, \tau_{out1} = 0, \tau_{out2} = 0)] + \frac{1}{L_1 \cdot C_1} \cdot \frac{1}{C_n \cdot L_{n2}}$$
$$\cdot [\psi_1(\tau_{in} = 0, \tau_{out1} = 0, \tau_{out2} = 0)$$
$$+ \psi_2(\tau_{in} = 0, \tau_{out1} = 0, \tau_{out2} = 0)]$$

We can summery our $\psi_1, \psi_2, \psi_3, \psi_4$ expression for $\tau_{in} = 0, \tau_{out1} = 0, \tau_{out2} = 0$:

$$\psi_1(\tau_{out1} = 0) = -\frac{(\frac{1}{C_2} + \frac{1}{C_n})}{[L_2 + L_{n1}]}; \; \psi_2(\tau_{out1} = 0) = \frac{1}{C_n \cdot [L_2 + L_{n1}]}; \xi_1(\tau_{out2} = 0, \tau_{in} = 0) = R_1 + R_s$$

$$\psi_3(\tau_{in} = 0, \tau_{out1} = 0, \tau_{out2} = 0) = \frac{1}{C_1 \cdot [R_1 + R_s]} \cdot \frac{R_s}{[L_2 + L_{n1}]}; \; \psi_4(\tau_{out1} = 0, \tau_{out2} = 0) = \frac{(\frac{R_s}{[R_1 + R_s]} - 1)}{[L_2 + L_{n1}]} \cdot R_s$$

&&&

$$P_1(\lambda, \tau_{in} = 0, \tau_{out1} = 0, \tau_{out2} = 0) = \lambda^4 + \lambda^3 \cdot [\frac{R_{load}}{L_{n2}} - (\frac{(\frac{R_s}{[R_1 + R_s]} - 1)}{[L_2 + L_{n1}]} \cdot R_s)$$
$$+ \frac{1}{C_1 \cdot (R_1 + R_s)}] + \lambda^2 \cdot [\frac{1}{C_1 \cdot (R_1 + R_s)} \cdot (\frac{R_{load}}{L_{n2}} - (\frac{(\frac{R_s}{[R_1 + R_s]} - 1)}{[L_2 + L_{n1}]} \cdot R_s))$$
$$- (\frac{(\frac{R_s}{[R_1 + R_s]} - 1)}{[L_2 + L_{n1}]} \cdot R_s) \cdot \frac{R_{load}}{L_{n2}} + \frac{1}{L_1 \cdot C_1} + \frac{R_s}{(R_1 + R_s)} \cdot (\frac{1}{C_1 \cdot [R_1 + R_s]} \cdot \frac{R_s}{[L_2 + L_{n1}]})]$$
$$+ \lambda \cdot [(\frac{R_{load}}{L_{n2}} - (\frac{(\frac{R_s}{[R_1 + R_s]} - 1)}{[L_2 + L_{n1}]} \cdot R_s)) \cdot \frac{1}{L_1 \cdot C_1} - \frac{(\frac{(\frac{R_s}{[R_1 + R_s]} - 1)}{[L_2 + L_{n1}]} \cdot R_s) \cdot R_{load}}{C_1 \cdot (R_1 + R_s) \cdot L_{n2}}$$
$$+ \frac{R_{load} \cdot R_s}{(R_1 + R_s) \cdot L_{n2}} \cdot (\frac{1}{C_1 \cdot [R_1 + R_s]} \cdot \frac{R_s}{[L_2 + L_{n1}]})]$$
$$- (\frac{(\frac{R_s}{[R_1 + R_s]} - 1)}{[L_2 + L_{n1}]} \cdot R_s) \cdot \frac{R_{load}}{L_{n2} \cdot L_1 \cdot C_1}$$

We define for simplicity new global parameters:

$$A_4 = 1; A_3 = \frac{R_{load}}{L_{n2}} - (\frac{(\frac{R_s}{[R_1 + R_s]} - 1)}{[L_2 + L_{n1}]} \cdot R_s) + \frac{1}{C_1 \cdot (R_1 + R_s)}; A_0$$
$$= -(\frac{(\frac{R_s}{[R_1 + R_s]} - 1)}{[L_2 + L_{n1}]} \cdot R_s) \cdot \frac{R_{load}}{L_{n2} \cdot L_1 \cdot C_1}$$

$$A_2 = \frac{1}{C_1 \cdot (R_1 + R_s)} \cdot (\frac{R_{load}}{L_{n2}} - (\frac{(\frac{R_s}{[R_1 + R_s]} - 1)}{[L_2 + L_{n1}]} \cdot R_s)) - (\frac{(\frac{R_s}{[R_1 + R_s]} - 1)}{[L_2 + L_{n1}]} \cdot R_s) \cdot \frac{R_{load}}{L_{n2}}$$
$$+ \frac{1}{L_1 \cdot C_1} + \frac{R_s}{(R_1 + R_s)} \cdot (\frac{1}{C_1 \cdot [R_1 + R_s]} \cdot \frac{R_s}{[L_2 + L_{n1}]})$$

$$A_1 = (\frac{R_{load}}{L_{n2}} - (\frac{(\frac{R_s}{[R_1 + R_s]} - 1)}{[L_2 + L_{n1}]} \cdot R_s)) \cdot \frac{1}{L_1 \cdot C_1} - \frac{(\frac{(\frac{R_s}{[R_1 + R_s]} - 1)}{[L_2 + L_{n1}]} \cdot R_s) \cdot R_{load}}{C_1 \cdot (R_1 + R_s) \cdot L_{n2}}$$
$$+ \frac{R_{load} \cdot R_s}{(R_1 + R_s) \cdot L_{n2}} \cdot (\frac{1}{C_1 \cdot [R_1 + R_s]} \cdot \frac{R_s}{[L_2 + L_{n1}]})$$

$$P_1(\lambda, \tau_{in} = 0, \tau_{out1} = 0, \tau_{out2} = 0) = \sum_{k=0}^{4} \lambda^k \cdot A_k$$
$$= \lambda^4|_{A_4=1} + \lambda^3 \cdot A_3 + \lambda^2 \cdot A_2 + \lambda \cdot A_1 + A_0$$

$$P_2(\lambda, \tau_{in} = 0, \tau_{out1} = 0, \tau_{out2} = 0) = -\frac{1}{C_n \cdot L_{n2}} \cdot \lambda^3 + \lambda^2 \cdot \frac{1}{C_n \cdot L_{n2}} \cdot \left(\frac{\left(\frac{R_s}{[R_1 + R_s]} - 1\right)}{[L_2 + L_{n1}]} \cdot R_s - \frac{1}{C_1 \cdot (R_1 + R_s)}\right) + \lambda \cdot \frac{1}{C_n \cdot L_{n2}} \cdot \left(\frac{\frac{\left(\frac{R_s}{[R_1 + R_s]} - 1\right)}{[L_2 + L_{n1}]} \cdot R_s}{C_1 \cdot (R_1 + R_s)} - \frac{1}{L_1 \cdot C_1} - \frac{R_s}{(R_1 + R_s)} \cdot \frac{1}{C_1 \cdot [R_1 + R_s]} \cdot \frac{R_s}{[L_2 + L_{n1}]}\right) + \frac{1}{C_n \cdot L_{n2}} \cdot \frac{\left(\frac{R_s}{[R_1 + R_s]} - 1\right)}{[L_2 + L_{n1}]} \cdot R_s \cdot \frac{1}{L_1 \cdot C_1}$$

We define for simplicity new global parameters:

$$B_3 = -\frac{1}{C_n \cdot L_{n2}}; \; B_2 = \frac{1}{C_n \cdot L_{n2}} \cdot \left(\frac{\left(\frac{R_s}{[R_1 + R_s]} - 1\right)}{[L_2 + L_{n1}]} \cdot R_s - \frac{1}{C_1 \cdot (R_1 + R_s)}\right)$$

$$B_1 = \frac{1}{C_n \cdot L_{n2}} \cdot \left(\frac{\frac{\left(\frac{R_s}{[R_1 + R_s]} - 1\right)}{[L_2 + L_{n1}]} \cdot R_s}{C_1 \cdot (R_1 + R_s)} - \frac{1}{L_1 \cdot C_1} - \frac{R_s}{(R_1 + R_s)} \cdot \frac{1}{C_1 \cdot [R_1 + R_s]} \cdot \frac{R_s}{[L_2 + L_{n1}]}\right)$$

$$B_0 = \frac{1}{C_n \cdot L_{n2}} \cdot \frac{\left(\frac{R_s}{[R_1 + R_s]} - 1\right)}{[L_2 + L_{n1}]} \cdot R_s \cdot \frac{1}{L_1 \cdot C_1}$$

$$P_2(\lambda, \tau_{in} = 0, \tau_{out1} = 0, \tau_{out2} = 0) = \sum_{k=0}^{3} \lambda^k \cdot B_k = \lambda^3 \cdot B_3 + \lambda^2 \cdot B_2 + \lambda \cdot B_1 + B_0$$

$$P_3(\lambda, \tau_{in} = 0, \tau_{out1} = 0, \tau_{out2} = 0) = -\frac{\left(\frac{1}{C_2} + \frac{1}{C_n}\right)}{[L_2 + L_{n1}]} \cdot \lambda^3 - \lambda^2 \cdot \frac{\left(\frac{1}{C_2} + \frac{1}{C_n}\right)}{[L_2 + L_{n1}]} \cdot \left(\frac{R_{load}}{L_{n2}} + \frac{1}{C_1 \cdot (R_1 + R_s)}\right) - \lambda \cdot \frac{\left(\frac{1}{C_2} + \frac{1}{C_n}\right)}{[L_2 + L_{n1}]} \cdot \left(\frac{R_{load}}{C_1 \cdot (R_1 + R_s) \cdot L_{n2}} + \frac{1}{L_1 \cdot C_1}\right) - \frac{\frac{\left(\frac{1}{C_2} + \frac{1}{C_n}\right)}{[L_2 + L_{n1}]} \cdot R_{load}}{L_1 \cdot C_1 \cdot L_{n2}}$$

$$C_3 = -\frac{\left(\frac{1}{C_2} + \frac{1}{C_n}\right)}{[L_2 + L_{n1}]}; \; C_2 = -\frac{\left(\frac{1}{C_2} + \frac{1}{C_n}\right)}{[L_2 + L_{n1}]} \cdot \left(\frac{R_{load}}{L_{n2}} + \frac{1}{C_1 \cdot (R_1 + R_s)}\right)$$

$$C_1 = -\frac{\left(\frac{1}{C_2} + \frac{1}{C_n}\right)}{[L_2 + L_{n1}]} \cdot \left(\frac{R_{load}}{C_1 \cdot (R_1 + R_s) \cdot L_{n2}} + \frac{1}{L_1 \cdot C_1}\right); \; C_0 = -\frac{\frac{\left(\frac{1}{C_2} + \frac{1}{C_n}\right)}{[L_2 + L_{n1}]} \cdot R_{load}}{L_1 \cdot C_1 \cdot L_{n2}}$$

$$P_3(\lambda, \tau_{in} = 0, \tau_{out1} = 0, \tau_{out2} = 0) = \sum_{k=0}^{3} \lambda^k \cdot C_k = \lambda^3 \cdot C_3 + \lambda^2 \cdot C_2 + \lambda \cdot C_1 + C_0$$

$$P_4(\lambda, \tau_{in} = 0, \tau_{out1} = 0, \tau_{out2} = 0) = \lambda^2 \cdot [-\frac{1}{C_n \cdot L_{n2}} \cdot (\frac{(\frac{1}{C_2} + \frac{1}{C_n})}{[L_2 + L_{n1}]}) + \frac{1}{C_n \cdot [L_2 + L_{n1}]} \cdot \frac{1}{C_n \cdot L_{n2}}] + \lambda \cdot \frac{1}{C_1 \cdot (R_1 + R_s)} \cdot \frac{1}{C_n \cdot L_{n2}} \cdot [-\frac{(\frac{1}{C_2} + \frac{1}{C_n})}{[L_2 + L_{n1}]} + \frac{1}{C_n \cdot [L_2 + L_{n1}]}] + \frac{1}{L_1 \cdot C_1} \cdot \frac{1}{C_n \cdot L_{n2}} \cdot [-\frac{(\frac{1}{C_2} + \frac{1}{C_n})}{[L_2 + L_{n1}]} + \frac{1}{C_n \cdot [L_2 + L_{n1}]}]$$

$$D_2 = -\frac{1}{C_n \cdot L_{n2}} \cdot (\frac{(\frac{1}{C_2} + \frac{1}{C_n})}{[L_2 + L_{n1}]}) + \frac{1}{C_n \cdot [L_2 + L_{n1}]} \cdot \frac{1}{C_n \cdot L_{n2}}$$

$$D_1 = \frac{1}{C_1 \cdot (R_1 + R_s)} \cdot \frac{1}{C_n \cdot L_{n2}} \cdot [-\frac{(\frac{1}{C_2} + \frac{1}{C_n})}{[L_2 + L_{n1}]} + \frac{1}{C_n \cdot [L_2 + L_{n1}]}]$$

$$D_0 = \frac{1}{L_1 \cdot C_1} \cdot \frac{1}{C_n \cdot L_{n2}} \cdot [-\frac{(\frac{1}{C_2} + \frac{1}{C_n})}{[L_2 + L_{n1}]} + \frac{1}{C_n \cdot [L_2 + L_{n1}]}]$$

$$P_4(\lambda, \tau_{in} = 0, \tau_{out1} = 0, \tau_{out2} = 0) = \sum_{k=0}^{2} \lambda^k \cdot D_k = \lambda^2 \cdot D_2 + \lambda \cdot D_1 + D_0$$

$$\det(A - \lambda \cdot I)|_{\tau_{in}=0, \tau_{out1}=0, \tau_{out2}=0} = -\lambda^3 \cdot \sum_{k=0}^{4} \lambda^k \cdot \mathrm{A}_k + \lambda^2 \cdot \sum_{k=0}^{3} \lambda^k \cdot \mathrm{B}_k + \lambda^2 \cdot \sum_{k=0}^{3} \lambda^k \cdot C_k + \lambda \cdot \sum_{k=0}^{2} \lambda^k \cdot D_k$$

$$\det(A - \lambda \cdot I)|_{\tau_{in}=0, \tau_{out1}=0, \tau_{out2}=0} = -\sum_{k=0}^{4} \lambda^{k+3} \cdot \mathrm{A}_k + \sum_{k=0}^{3} \lambda^{k+2} \cdot \mathrm{B}_k + \sum_{k=0}^{3} \lambda^{k+2} \cdot C_k + \sum_{k=0}^{2} \lambda^{k+1} \cdot D_k$$

$$\sum_{k=0}^{3} \lambda^{k+2} \cdot C_k = \lambda^5 \cdot C_3 + \sum_{k=0}^{2} \lambda^{k+2} \cdot C_k; \; \sum_{k=0}^{3} \lambda^{k+2} \cdot \mathrm{B}_k = \lambda^5 \cdot \mathrm{B}_3 + \sum_{k=0}^{2} \lambda^{k+2} \cdot \mathrm{B}_k$$

$$\sum_{k=0}^{4} \lambda^{k+3} \cdot \mathrm{A}_k = \lambda^7 \cdot \mathrm{A}_4 + \lambda^6 \cdot \mathrm{A}_3 + \sum_{k=0}^{2} \lambda^{k+3} \cdot \mathrm{A}_k$$

$$\det(A - \lambda \cdot I)|_{\tau_{in}=0,\tau_{out1}=0,\tau_{out2}=0} = -\lambda^7 \cdot \mathrm{A}_4 - \lambda^6 \cdot \mathrm{A}_3 - \sum_{k=0}^{2} \lambda^{k+3} \cdot \mathrm{A}_k + \lambda^5 \cdot \mathrm{B}_3 + \sum_{k=0}^{2} \lambda^{k+2} \cdot \mathrm{B}_k + \lambda^5 \cdot C_3 + \sum_{k=0}^{2} \lambda^{k+2} \cdot C_k + \sum_{k=0}^{2} \lambda^{k+1} \cdot D_k$$

$$\det(A - \lambda \cdot I)|_{\tau_{in}=0,\tau_{out1}=0,\tau_{out2}=0} = -\lambda^7 \cdot \mathrm{A}_4 - \lambda^6 \cdot \mathrm{A}_3 + \lambda^5 \cdot (\mathrm{B}_3 + C_3) \sum_{k=0}^{2} [\lambda^{k+2} \cdot \mathrm{B}_k + \lambda^{k+2} \cdot C_k + \lambda^{k+1} \cdot D_k - \lambda^{k+3} \cdot \mathrm{A}_k]$$

$$\det(A - \lambda \cdot I)|_{\tau_{in}=0,\tau_{out1}=0,\tau_{out2}=0} = -\lambda^7 \cdot \mathrm{A}_4 - \lambda^6 \cdot \mathrm{A}_3 - \sum_{k=0}^{2} \lambda^{k+3} \cdot \mathrm{A}_k + \lambda^5 \cdot \mathrm{B}_3 + \sum_{k=0}^{2} \lambda^{k+2} \cdot \mathrm{B}_k + \lambda^5 \cdot C_3 + \sum_{k=0}^{2} \lambda^{k+2} \cdot C_k + \sum_{k=0}^{2} \lambda^{k+1} \cdot D_k$$

$$\sum_{k=0}^{2} \lambda^{k+3} \cdot \mathrm{A}_k = \lambda^5 \cdot \mathrm{A}_2 + \lambda^4 \cdot \mathrm{A}_1 + \lambda^3 \cdot \mathrm{A}_0;$$

$$\sum_{k=0}^{2} \lambda^{k+2} \cdot \mathrm{B}_k = \lambda^4 \cdot \mathrm{B}_2 + \lambda^3 \cdot \mathrm{B}_1 + \lambda^2 \cdot \mathrm{B}_0$$

$$\sum_{k=0}^{2} \lambda^{k+2} \cdot C_k = \lambda^4 \cdot C_2 + \lambda^3 \cdot C_1 + \lambda^2 \cdot C_0;$$

$$\sum_{k=0}^{2} \lambda^{k+1} \cdot D_k = \lambda^3 \cdot D_2 + \lambda^2 \cdot D_1 + \lambda \cdot D_0$$

$$\det(A - \lambda \cdot I)|_{\tau_{in}=0,\tau_{out1}=0,\tau_{out2}=0} = -\lambda^7 \cdot \mathrm{A}_4 - \lambda^6 \cdot \mathrm{A}_3 - [\lambda^5 \cdot \mathrm{A}_2 + \lambda^4 \cdot \mathrm{A}_1 + \lambda^3 \cdot \mathrm{A}_0] + \lambda^5 \cdot \mathrm{B}_3 + \lambda^4 \cdot \mathrm{B}_2 + \lambda^3 \cdot \mathrm{B}_1 + \lambda^2 \cdot \mathrm{B}_0 + \lambda^5 \cdot C_3 + \lambda^4 \cdot C_2 + \lambda^3 \cdot C_1 + \lambda^2 \cdot C_0 + \lambda^3 \cdot D_2 + \lambda^2 \cdot D_1 + \lambda \cdot D_0$$

$$\det(A - \lambda \cdot I)|_{\tau_{in}=0,\tau_{out1}=0,\tau_{out2}=0} = -\lambda^7 \cdot \mathrm{A}_4 - \lambda^6 \cdot \mathrm{A}_3 + \lambda^5 \cdot [\mathrm{B}_3 - \mathrm{A}_2 + C_3] + \lambda^4 \cdot [\mathrm{B}_2 - \mathrm{A}_1 + C_2] + \lambda^3 \cdot [\mathrm{B}_1 - \mathrm{A}_0 + C_1 + D_2] + \lambda^2 \cdot [\mathrm{B}_0 + C_0 + D_1] + \lambda \cdot D_0$$

$$\det(A - \lambda \cdot I) = \sum_{k=0}^{7} \Upsilon_k \cdot \lambda^k; \ \det(A - \lambda \cdot I) = 0 \Rightarrow \sum_{k=0}^{7} \Upsilon_k \cdot \lambda^k = 0$$

$$\Upsilon_7 = -1; \ \Upsilon_6 = -\mathrm{A}_3; \ \Upsilon_5 = \mathrm{B}_3 - \mathrm{A}_2 + C_3; \ \Upsilon_4 = \mathrm{B}_2 - \mathrm{A}_1 + C_2; \ \Upsilon_3 = \mathrm{B}_1 - \mathrm{A}_0 + C_1 + D_2$$

$$\Upsilon_2 = B_0 + C_0 + D_1;\ \Upsilon_1 = D_0;\ \Upsilon_0 = 0$$

Eigenvalues stability discussion: Our BPF diplexer circuit involving N variables (N > 2, N = 7), the characteristic equation is of degree N = 7 and must often be solved numerically. Expect in some particular cases, such an equation has (N = 7) distinct roots that can be real or complex. These values are the eigenvalues of the 7 × 7 Jacobian matrix (A). The general rule is that the Steady State (SS) is stable if there is no eigenvalue with positive real part. It is sufficient that one eigenvalue is positive for the steady state to be unstable. Our 7-variables ($X_1\ X_2\ X_3\ I_{L_1}\ I_{L_2}\ I_{L_{m2}}\ I_{C_1}$) BPF diplexer circuit has seven eigenvalues. The type of behavior can be characterized as a function of the position of these eigenvalues in the Re/Im plane. Five non-degenerated cases can be distinguished: (1) the seven eigenvalues are real and negative (stable steady state), (2) the seven eigenvalues are real, six of them are negative (unstable steady state), (3) and (4) two eigenvalues are complex conjugates with a negative real part and the other eigenvalues are real and negative (stable steady state), two cases can be distinguished depending on the relative value of the real part of the complex eigenvalues and of the real one, (5) two eigenvalues are complex conjugates with a negative real part and at least one eigenvalue is positive (unstable steady state) [12, 13].

Remark It is reader exercise to analyze BPF diplexer circuit stability for cases 2–8 (at least one delay parameter $\tau_{in}, \tau_{out1}, \tau_{out2}$ is positive and real number). The stability analysis is done by using geometric stability switch criteria in delay differential systems (E. Beretta and Y. Kuang). It is a practical guideline that combines graphical information with analytical work to effectively study the local stability of BPF diplexer circuit model involving delay dependent parameters. The stability of BPF diplexer circuit steady state is determined by the graphs of some functions of τ which can be expressed explicitly.

The general geometric criterion: The occurrence of any possible stability switching resulting from the increase of value of the time delay τ for our BPF diplexer circuit characteristic equation.

$$D(\lambda, \tau_{in}, \tau_{out1}, \tau_{out2}) = \det(A - \lambda \cdot I);\ D(\lambda, \tau_{in}, \tau_{out1}, \tau_{out2}) = 0;\ D(\lambda, \tau) = P_n(\lambda, \tau) + Q_m(\lambda, \tau) \cdot e^{-\lambda \cdot \tau}$$

$$P_n(\lambda, \tau) = \sum_{k=0}^{n} p_k(\tau) \cdot \lambda^k;\ Q_m(\lambda, \tau) = \sum_{k=0}^{m} q_k(\tau) \cdot \lambda^k;\ n, m \in \mathbb{N}_0;\ n > m$$

$p_k(\bullet), q_k(\bullet) : \mathbb{R}_{+0} \to \mathbb{R}$ are continuous and differentiable functions of τ.

$$P_n(\lambda = 0, \tau) + Q_m(\lambda = 0, \tau) = p_0(\tau) + q_0(\tau) \neq 0\ \forall \tau \in \mathbb{R}_{+0}$$

$P_n(\lambda, \tau), Q_m(\lambda, \tau)$ are analytic functions in λ and differentiable in τ for which we assume:

(I) If $\lambda = i \cdot \omega$; $\omega \in \mathbb{R}$ then $P_n(i \cdot \omega, \tau) + Q_m(i \cdot \omega, \tau) \neq 0$; $\tau \in \mathbb{R}$.
(II) $\limsup\{|Q_m(\lambda, \tau)/P_n(\lambda, \tau)| \, : \, |\lambda| \to \infty, \ \mathrm{Re}\lambda \geq 0\} < 1$ for any τ.
(III) $F(\omega, \tau) = |P_n(i \cdot \omega, \tau)|^2 - |Q_m(i \cdot \omega, \tau)|^2$ for each τ has at most a finite number of real zeros.
(IV) Each positive root $\omega(\tau)$ of $F(\omega, \tau) = 0$ is continuous and differentiable in τ whenever it exists.

7.2 Dual Band Diplexer Filter Stability Analysis Under Parameters Variation

A diplexer filters to pass two bands to separate ports, and its stability analysis under parameters variation. In our RF and microwave system, two desired frequencies are relative close together; design each separate filter to have its own band edge as far as possible from the other filter. The desired frequency we want to pass is located at the upper band edge of BPF1, while the desired frequency of BPF2 is at its lower band edge.

The circuit of dual band diplexer filter contains capacitors and inductors, RF source $V_s(t)$ and series resistor R_s, two matching networks, and resistive loads R_{load1}, R_{load2}. Three delay lines $\tau_{in}, \tau_{out1}, \tau_{out2}$ represent circuit micro strip lines parasitic effects. In our stability analysis we consider fully match between R_{load1} and BPF1 and between R_{load2} and BPF2 (no matching networks) [25, 26].

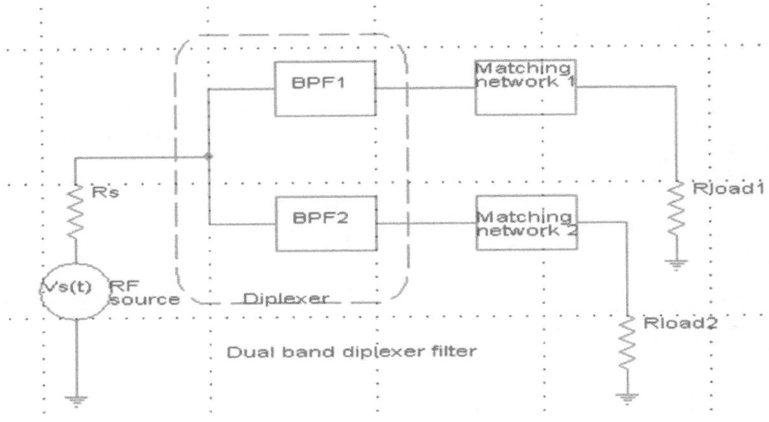

Fig. 7.4 Dual band diplexer filter block diagram

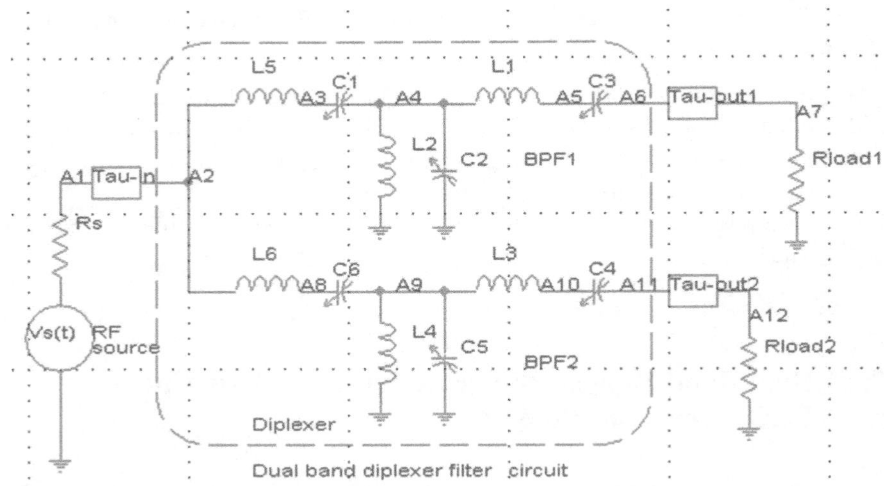

Fig. 7.5 Dual band diplexer filter full schematic with delay lines

$$V_{\tau_{in}} \to \varepsilon;\ V_{\tau_{out1}} \to \varepsilon;\ V_{\tau_{out2}} \to \varepsilon;\ V_{\tau_{in}} = V_{A_1} - V_{A_2};\ V_{\tau_{out1}} = V_{A_6} - V_{A_7};\ V_{\tau_{out2}} = V_{A_{11}} - V_{A_{12}}$$

$$I_{R_s} = \frac{V_s(t) - V_{A_1}}{R_s}; V_{L_5} = V_{A_2} - V_{A_3} = L_5 \cdot \frac{dI_{L_5}}{dt}; I_{C_1} = C_1 \cdot \frac{d(V_{A_3} - V_{A_4})}{dt}; V_{A_4} = V_{L_2} = L_2 \cdot \frac{dI_{L_2}}{dt}$$

$$I_{C_2} = C_2 \cdot \frac{dV_{A_4}}{dt}; V_{L_1} = V_{A_4} - V_{A_5} = L_1 \cdot \frac{dI_{L_1}}{dt}; I_{C_3} = C_3 \cdot \frac{d(V_{A_5} - V_{A_6})}{dt};\ I_{R_{load1}} = \frac{V_{A_7}}{R_{load1}}; V_{A_1} \approx V_{A_2}$$

$$V_{A_6} \approx V_{A_7}; V_{L_6} = V_{A_2} - V_{A_8} = L_6 \cdot \frac{dI_{L_6}}{dt}; I_{C_6} = C_6 \cdot \frac{d(V_{A_8} - V_{A_9})}{dt}; V_{A_9} = V_{L_4} = L_4 \cdot \frac{dI_{L_4}}{dt}$$

$$I_{C_5} = C_5 \cdot \frac{dV_{A_9}}{dt}; V_{L_3} = V_{A_9} - V_{A_{10}} = L_3 \cdot \frac{dI_{L_3}}{dt};\ I_{C_4} = C_4 \cdot \frac{d(V_{A_{10}} - V_{A_{11}})}{dt};\ I_{R_{load2}} = \frac{V_{A_{12}}}{R_{load2}}; V_{A_{11}} \approx V_{A_{12}}$$

$$I_{R_s}(t - \tau_{in}) = I_{L_5} + I_{L_6} = \sum_{k=5}^{6} I_{L_k};\ I_{L_5} = I_{C_1};\ I_{L_6} = I_{C_6};\ I_{L_1} = I_{C_3};\ I_{L_3} = I_{C_4};\ I_{R_{load1}} = I_{C_3}(t - \tau_{out1})$$

$$I_{R_{load1}} = I_{L_1}(t - \tau_{out1});\ I_{R_{load2}} = I_{C_4}(t - \tau_{out2});\ I_{R_{load2}} = I_{L_3}(t - \tau_{out2})$$

KCL @ node A_2: $I_{R_s}(t-\tau_{in}) = I_{L_5} + I_{L_6} = \sum_{k=5}^{6} I_{L_k}$, KCL @ node A_4: $I_{C_1} = I_{L_2} + I_{C_2} + I_{L_1} = I_{C_2} + \sum_{k=1}^{2} I_{L_k}$

KCL @ node A_9: $I_{C_6} = I_{L_4} + I_{C_5} + I_{L_3} = I_{C_5} + \sum_{k=3}^{4} I_{L_k}$

&&&

$$I_{R_s} = \frac{V_s(t) - V_{A_1}}{R_s} \Rightarrow V_{A_1} = V_s(t) - I_{R_s} \cdot R_s; \; V_{A_2} - V_{A_3} = L_5 \cdot \frac{dI_{L_5}}{dt}$$

$$I_{C_2} = C_2 \cdot \frac{dV_{A_4}}{dt} \Rightarrow V_{A_4} = \frac{1}{C_2} \cdot \int I_{C_2} \cdot dt$$

$$I_{C_1} = C_1 \cdot \frac{d(V_{A_3} - V_{A_4})}{dt} \Rightarrow V_{A_3} - V_{A_4} = \frac{1}{C_1} \cdot \int I_{C_1} \cdot dt; \; V_{A_4} - V_{A_5} = L_1 \cdot \frac{dI_{L_1}}{dt}$$
$$\Rightarrow V_{A_5} = V_{A_4} - L_1 \cdot \frac{dI_{L_1}}{dt}$$

$$V_{A_5} = \frac{1}{C_2} \cdot \int I_{C_2} \cdot dt - L_1 \cdot \frac{dI_{L_1}}{dt}; \; I_{C_3} = C_3 \cdot \frac{d(V_{A_5} - V_{A_6})}{dt} \Rightarrow V_{A_5} - V_{A_6}$$
$$= \frac{1}{C_3} \cdot \int I_{C_3} \cdot dt; \; V_{A_7} = I_{R_{load1}} \cdot R_{load}1$$

$$V_{A_2} - V_{A_8} = L_6 \cdot \frac{dI_{L_6}}{dt}; \; I_{C_6} = C_6 \cdot \frac{d(V_{A_8} - V_{A_9})}{dt} \Rightarrow V_{A_8} - V_{A_9}$$
$$= \frac{1}{C_6} \cdot \int I_{C_6} \cdot dt; \; V_{A_9} = L_4 \cdot \frac{dI_{L_4}}{dt}$$

$$I_{C_5} = C_5 \cdot \frac{dV_{A_9}}{dt} \Rightarrow V_{A_9} = \frac{1}{C_5} \cdot \int I_{C_5} \cdot dt; \; V_{A_9} - V_{A_{10}} = L_3 \cdot \frac{dI_{L_3}}{dt}$$
$$I_{C_4} = C_4 \cdot \frac{d(V_{A_{10}} - V_{A_{11}})}{dt} \Rightarrow V_{A_{10}} - V_{A_{11}} = \frac{1}{C_4} \cdot \int I_{C_4} \cdot dt$$

$$I_{R_{load2}} = \frac{V_{A_{12}}}{R_{load2}}; \; V_{A_{11}} \approx V_{A_{12}}; \; I_{R_{load2}} = \frac{V_{A_{11}}}{R_{load2}} \Rightarrow V_{A_{11}} = I_{R_{load2}} \cdot R_{load2}$$

$$V_{\tau_{in}} \to \varepsilon; \; V_{A_1} \to V_{A_2}; \; V_{A_2} = V_s(t) - I_{R_s} \cdot R_s;$$
$$V_{A_3} = V_{A_2} - L_5 \cdot \frac{dI_{L_5}}{dt} = V_s(t) - I_{R_s} \cdot R_s - L_5 \cdot \frac{dI_{L_5}}{dt}$$
$$V_{A_2} - V_{A_3} = L_5 \cdot \frac{dI_{L_5}}{dt} \Rightarrow V_{A_3} = V_{A_2} - L_5 \cdot \frac{dI_{L_5}}{dt} = V_s(t) - I_{R_s} \cdot R_s - L_5 \cdot \frac{dI_{L_5}}{dt}$$

$$V_{A_3} - V_{A_4} = \frac{1}{C_1} \cdot \int I_{C_1} \cdot dt \Rightarrow V_s(t) - I_{R_s} \cdot R_s - L_5 \cdot \frac{dI_{L_5}}{dt} - \frac{1}{C_2} \cdot \int I_{C_2} \cdot dt = \frac{1}{C_1} \cdot \int I_{C_1} \cdot dt$$

$$\frac{d}{dt}\{V_s(t) - I_{R_s} \cdot R_s - L_5 \cdot \frac{dI_{L_5}}{dt} - \frac{1}{C_2} \cdot \int I_{C_2} \cdot dt = \frac{1}{C_1} \cdot \int I_{C_1} \cdot dt\}$$

$$\frac{dV_s(t)}{dt} - \frac{dI_{R_s}}{dt} \cdot R_s - L_5 \cdot \frac{d^2 I_{L_5}}{dt^2} - \frac{1}{C_2} \cdot I_{C_2} = \frac{1}{C_1} \cdot I_{C_1}$$

$$V_{A_5} - V_{A_6} = \frac{1}{C_3} \cdot \int I_{C_3} \cdot dt \Rightarrow V_{A_6} = V_{A_5} - \frac{1}{C_3} \cdot \int I_{C_3} \cdot dt = \frac{1}{C_2} \cdot \int I_{C_2} \cdot dt - L_1 \cdot \frac{dI_{L_1}}{dt} - \frac{1}{C_3} \cdot \int I_{C_3} \cdot dt$$

$$V_{\tau_{out1}} \rightarrow \varepsilon;\ V_{A_6} \rightarrow V_{A_7};\ I_{R_{load1}} = \frac{V_{A_6}}{R_{load1}} \Rightarrow V_{A_6} = I_{R_{load1}} \cdot R_{load1}$$

$$\frac{1}{C_2} \cdot \int I_{C_2} \cdot dt - L_1 \cdot \frac{dI_{L_1}}{dt} - \frac{1}{C_3} \cdot \int I_{C_3} \cdot dt = I_{R_{load1}} \cdot R_{load1};$$

$$\frac{1}{C_2} \cdot I_{C_2} - L_1 \cdot \frac{d^2 I_{L_1}}{dt^2} - \frac{1}{C_3} \cdot I_{C_3} = \frac{dI_{R_{load1}}}{dt} \cdot R_{load1}$$

$$V_{A_2} - V_{A_8} = L_6 \cdot \frac{dI_{L_6}}{dt} \Rightarrow V_{A_8} = V_{A_2} - L_6 \cdot \frac{dI_{L_6}}{dt} = V_s(t) - I_{R_s} \cdot R_s - L_6 \cdot \frac{dI_{L_6}}{dt}$$

$$V_{A_8} - V_{A_9} = \frac{1}{C_6} \cdot \int I_{C_6} \cdot dt \Rightarrow V_{A_9} = V_{A_8} - \frac{1}{C_6} \cdot \int I_{C_6} \cdot dt = V_s(t) - I_{R_s} \cdot R_s - L_6 \cdot \frac{dI_{L_6}}{dt} - \frac{1}{C_6} \cdot \int I_{C_6} \cdot dt$$

$$V_{A_9} = L_4 \cdot \frac{dI_{L_4}}{dt};\ V_{A_9} = \frac{1}{C_5} \cdot \int I_{C_5} \cdot dt;\ L_4 \cdot \frac{dI_{L_4}}{dt} = V_s(t) - I_{R_s} \cdot R_s - L_6 \cdot \frac{dI_{L_6}}{dt} - \frac{1}{C_6} \cdot \int I_{C_6} \cdot dt$$

$$L_4 \cdot \frac{d^2 I_{L_4}}{dt^2} = \frac{dV_s(t)}{dt} - \frac{dI_{R_s}}{dt} \cdot R_s - L_6 \cdot \frac{d^2 I_{L_6}}{dt^2} - \frac{1}{C_6} \cdot I_{C_6}$$

$$\frac{1}{C_5} \cdot \int I_{C_5} \cdot dt = V_s(t) - I_{R_s} \cdot R_s - L_6 \cdot \frac{dI_{L_6}}{dt} - \frac{1}{C_6} \cdot \int I_{C_6} \cdot dt;\ \frac{1}{C_5} \cdot I_{C_5} = \frac{dV_s(t)}{dt} - \frac{dI_{R_s}}{dt} \cdot R_s - L_6 \cdot \frac{d^2 I_{L_6}}{dt^2} - \frac{1}{C_6} \cdot I_{C_6}$$

$$V_{A_9} - V_{A_{10}} = L_3 \cdot \frac{dI_{L_3}}{dt} \Rightarrow V_{A_{10}} = V_{A_9} - L_3 \cdot \frac{dI_{L_3}}{dt} = L_4 \cdot \frac{dI_{L_4}}{dt} - L_3 \cdot \frac{dI_{L_3}}{dt};$$

$$\frac{1}{C_5} \cdot I_{C_5} = L_4 \cdot \frac{d^2 I_{L_4}}{dt^2}$$

$$V_{A_{10}} - V_{A_{11}} = \frac{1}{C_4} \cdot \int I_{C_4} \cdot dt \Rightarrow V_{A_{11}} = V_{A_{10}} - \frac{1}{C_4} \cdot \int I_{C_4} \cdot dt$$
$$= L_4 \cdot \frac{dI_{L_4}}{dt} - L_3 \cdot \frac{dI_{L_3}}{dt} - \frac{1}{C_4} \cdot \int I_{C_4} \cdot dt$$

$$V_{\tau_{out2}} \to \varepsilon;\ V_{A_{11}} \to V_{A_{12}};\ V_{A_{11}} = I_{R_{load2}} \cdot R_{load2};\ I_{R_{load2}} \cdot R_{load2}$$
$$= L_4 \cdot \frac{dI_{L_4}}{dt} - L_3 \cdot \frac{dI_{L_3}}{dt} - \frac{1}{C_4} \cdot \int I_{C_4} \cdot dt$$

$$\frac{d}{dt}\{I_{R_{load2}} \cdot R_{load2} = L_4 \cdot \frac{dI_{L_4}}{dt} - L_3 \cdot \frac{dI_{L_3}}{dt} - \frac{1}{C_4} \cdot \int I_{C_4} \cdot dt\};\ \frac{dI_{R_{load2}}}{dt} \cdot R_{load2}$$
$$= L_4 \cdot \frac{d^2 I_{L_4}}{dt^2} - L_3 \cdot \frac{d^2 I_{L_3}}{dt^2} - \frac{1}{C_4} \cdot I_{C_4}$$

Summary of our circuit differential equations (Version 1):

$$\frac{dV_s(t)}{dt} - \frac{dI_{R_s}}{dt} \cdot R_s - L_5 \cdot \frac{d^2 I_{L_5}}{dt^2} - \frac{1}{C_2} \cdot I_{C_2} = \frac{1}{C_1} \cdot I_{C_1};\ \frac{1}{C_2} \cdot I_{C_2} - L_1 \cdot \frac{d^2 I_{L_1}}{dt^2} - \frac{1}{C_3} \cdot I_{C_3}$$
$$= \frac{dI_{R_{load1}}}{dt} \cdot R_{load1}$$

$$L_4 \cdot \frac{d^2 I_{L_4}}{dt^2} = \frac{dV_s(t)}{dt} - \frac{dI_{R_s}}{dt} \cdot R_s - L_6 \cdot \frac{d^2 I_{L_6}}{dt^2} - \frac{1}{C_6} \cdot I_{C_6};\ \frac{1}{C_5} \cdot I_{C_5}$$
$$= \frac{dV_s(t)}{dt} - \frac{dI_{R_s}}{dt} \cdot R_s - L_6 \cdot \frac{d^2 I_{L_6}}{dt^2} - \frac{1}{C_6} \cdot I_{C_6}$$

$$\frac{dI_{R_{load2}}}{dt} \cdot R_{load2} = L_4 \cdot \frac{d^2 I_{L_4}}{dt^2} - L_3 \cdot \frac{d^2 I_{L_3}}{dt^2} - \frac{1}{C_4} \cdot I_{C_4};\ \frac{1}{C_5} \cdot I_{C_5} = L_4 \cdot \frac{d^2 I_{L_4}}{dt^2}$$

Summary of our circuit differential equations (Version 2):

$$I_{C_1} \to I_{L_5};\ I_{C_6} \to I_{L_6};\ I_{L_1} \to I_{C_3};\ I_{L_3} \to I_{C_4}$$

$$\frac{dV_s(t)}{dt} - \frac{dI_{R_s}}{dt} \cdot R_s - L_5 \cdot \frac{d^2 I_{L_5}}{dt^2} - \frac{1}{C_2} \cdot I_{C_2} = \frac{1}{C_1} \cdot I_{L_5};\ \frac{1}{C_2} \cdot I_{C_2} - L_1 \cdot \frac{d^2 I_{C_3}}{dt^2} - \frac{1}{C_3} \cdot I_{C_3}$$
$$= \frac{dI_{R_{load1}}}{dt} \cdot R_{load1}$$

$$\frac{1}{C_5} \cdot I_{C_5} = \frac{dV_s(t)}{dt} - \frac{dI_{R_s}}{dt} \cdot R_s - L_6 \cdot \frac{d^2 I_{L_6}}{dt^2} - \frac{1}{C_6} \cdot I_{L_6}; \frac{1}{C_5} \cdot I_{C_5} = L_4 \cdot \frac{d^2 I_{L_4}}{dt^2}$$

$$\frac{dI_{R_{load2}}}{dt} \cdot R_{load2} = L_4 \cdot \frac{d^2 I_{L_4}}{dt^2} - L_3 \cdot \frac{d^2 I_{C_4}}{dt^2} - \frac{1}{C_4} \cdot I_{C_4}$$

Remark 1.0 $I_{R_s}(t - \tau_{in}) = I_{L_5} + I_{L_6}; I_{R_{load2}} = I_{C_4}(t - \tau_{out2}); I_{R_{load1}} = I_{C_3}(t - \tau_{out1})$

$$\frac{dI_{R_{load2}}}{dt} = \frac{dI_{C_4}(t - \tau_{out2})}{dt}; I_{C_4}(t - \tau_{out2}) = I^*_{C_4} + i_{C_4} \cdot e^{\lambda \cdot t} \cdot e^{-\lambda \cdot \tau_{out2}}; I_{C_4}(t) = I^*_{C_4} + i_{C_4} \cdot e^{\lambda \cdot t}$$

$$\frac{dI_{C_4}(t - \tau_{out2})}{dt} = i_{C_4} \cdot \lambda \cdot e^{\lambda \cdot t} \cdot e^{-\lambda \cdot \tau_{out2}}; \frac{dI_{C_4}(t)}{dt} = i_{C_4} \cdot \lambda \cdot e^{\lambda \cdot t}; \frac{dI_{C_4}(t - \tau_{out2})}{dt} = \frac{dI_{C_4}(t)}{dt} \cdot e^{-\lambda \cdot \tau_{out2}}$$

Remark 1.1 $\frac{dI_{R_{load1}}}{dt} = \frac{dI_{C_3}(t-\tau_{out1})}{dt}; I_{C_3}(t - \tau_{out1}) = I^*_{C_3} + i_{C_3} \cdot e^{\lambda \cdot t} \cdot e^{-\lambda \cdot \tau_{out1}}; I_{C_3}(t) = I^*_{C_3} + i_{C_3} \cdot e^{\lambda \cdot t}$

$$\frac{dI_{C_3}(t - \tau_{out1})}{dt} = i_{C_3} \cdot \lambda \cdot e^{\lambda \cdot t} \cdot e^{-\lambda \cdot \tau_{out1}}; \frac{dI_{C_3}(t)}{dt} = i_{C_3} \cdot \lambda \cdot e^{\lambda \cdot t}; \frac{dI_{C_3}(t - \tau_{out1})}{dt} = \frac{dI_{C_3}(t)}{dt} \cdot e^{-\lambda \cdot \tau_{out1}}$$

$$\frac{dI_{R_{load1}}}{dt} = \frac{dI_{C_3}(t - \tau_{out1})}{dt} = \frac{dI_{C_3}(t)}{dt} \cdot e^{-\lambda \cdot \tau_{out1}}$$

Remark 1.2 $I_{R_s}(t - \tau_{in}) = I_{L_5} + I_{L_6}; \frac{dI_{R_s}(t-\tau_{in})}{dt} = \frac{dI_{L_5}}{dt} + \frac{dI_{L_6}}{dt}$

$$I_{R_s}(t - \tau_{in}) = I^*_{R_s} + i_{R_s} \cdot e^{\lambda \cdot t} \cdot e^{-\lambda \cdot \tau_{in}}; I_{R_s}(t) = I^*_{R_s} + i_{R_s} \cdot e^{\lambda \cdot t}; \frac{dI_{R_s}(t - \tau_{in})}{dt} = i_{R_s} \cdot \lambda \cdot e^{\lambda \cdot t} \cdot e^{-\lambda \cdot \tau_{in}}$$

$$\frac{dI_{R_s}(t)}{dt} = i_{R_s} \cdot \lambda \cdot e^{\lambda \cdot t}; \frac{dI_{R_s}(t - \tau_{in})}{dt} = \frac{dI_{R_s}(t)}{dt} \cdot e^{-\lambda \cdot \tau_{in}}; \frac{dI_{R_s}(t)}{dt} \cdot e^{-\lambda \cdot \tau_{in}} = \frac{dI_{L_5}}{dt} + \frac{dI_{L_6}}{dt}$$

$$\frac{dI_{R_s}(t)}{dt} = \frac{dI_{L_5}}{dt} \cdot e^{\lambda \cdot \tau_{in}} + \frac{dI_{L_6}}{dt} \cdot e^{\lambda \cdot \tau_{in}}; \frac{dI_{R_{load2}}}{dt} = \frac{dI_{C_4}(t - \tau_{out2})}{dt} = \frac{dI_{C_4}(t)}{dt} \cdot e^{-\lambda \cdot \tau_{out2}}$$

Summary of our circuit differential equations (Version 3):

$$\frac{dX_2}{dt} = \frac{d^2 I_{L_5}}{dt^2}; X_2 = \frac{dI_{L_5}}{dt}; \frac{dX_3}{dt} = \frac{d^2 I_{C_3}}{dt^2}; X_3 = \frac{dI_{C_3}}{dt}; \frac{dX_4}{dt} = \frac{d^2 I_{L_6}}{dt^2}; X_4 = \frac{dI_{L_6}}{dt}$$

$$\frac{dX_5}{dt} = \frac{d^2 I_{L_4}}{dt^2}; \; X_5 = \frac{dI_{L_4}}{dt}; \; \frac{dX_6}{dt} = \frac{d^2 I_{C_4}}{dt^2}; \; X_6 = \frac{dI_{C_4}}{dt}; \; \frac{dI_{R_s}(t)}{dt} = X_2 \cdot e^{\lambda \cdot \tau_{in}} + X_4 \cdot e^{\lambda \cdot \tau_{in}}$$

$$\frac{dV_s(t)}{dt} - [X_2 \cdot e^{\lambda \cdot \tau_{in}} + X_4 \cdot e^{\lambda \cdot \tau_{in}}] \cdot R_s - L_5 \cdot \frac{dX_2}{dt} - \frac{1}{C_2} \cdot I_{C_2}$$
$$= \frac{1}{C_1} \cdot I_{L_5}; \; \frac{1}{C_2} \cdot I_{C_2} - L_1 \cdot \frac{dX_3}{dt} - \frac{1}{C_3} \cdot I_{C_3} = X_3 \cdot e^{-\lambda \cdot \tau_{out1}} \cdot R_{load1}$$

$$\frac{1}{C_5} \cdot I_{C_5} = \frac{dV_s(t)}{dt} - [X_2 \cdot e^{\lambda \cdot \tau_{in}} + X_4 \cdot e^{\lambda \cdot \tau_{in}}] \cdot R_s - L_6 \cdot \frac{dX_4}{dt} - \frac{1}{C_6} \cdot I_{L_6}; \; \frac{1}{C_5} \cdot I_{C_5}$$
$$= L_4 \cdot \frac{dX_5}{dt} \Rightarrow \frac{dX_5}{dt} = \frac{1}{C_5 \cdot L_4} \cdot I_{C_5}$$

$$X_6 \cdot e^{-\lambda \cdot \tau_{out2}} \cdot R_{load2} = L_4 \cdot \frac{dX_5}{dt} - L_3 \cdot \frac{dX_6}{dt} - \frac{1}{C_4} \cdot I_{C_4}$$

$$X_6 \cdot e^{-\lambda \cdot \tau_{out2}} \cdot R_{load2} = L_4 \cdot \frac{1}{C_5 \cdot L_4} \cdot I_{C_5} - L_3 \cdot \frac{dX_6}{dt} - \frac{1}{C_4} \cdot I_{C_4}$$

Summary of our circuit differential equations (Version 4):

$$\frac{dI_{L_5}}{dt} = X_2; \frac{dI_{C_3}}{dt} = X_3; \frac{dI_{L_6}}{dt} = X_4; \frac{dI_{L_4}}{dt} = X_5; \frac{dI_{C_4}}{dt} = X_6$$

$$\frac{dX_2}{dt} = \frac{1}{L_5} \cdot \frac{dV_s(t)}{dt} - [X_2 \cdot e^{\lambda \cdot \tau_{in}} + X_4 \cdot e^{\lambda \cdot \tau_{in}}] \cdot \frac{R_s}{L_5} - \frac{1}{C_2 \cdot L_5} \cdot I_{C_2} - \frac{1}{C_1 \cdot L_5} \cdot I_{L_5}$$
$$\frac{dX_3}{dt} = \frac{1}{C_2 \cdot L_1} \cdot I_{C_2} - \frac{1}{C_3 \cdot L_1} \cdot I_{C_3} - X_3 \cdot e^{-\lambda \cdot \tau_{out1}} \cdot \frac{R_{load1}}{L_1}; \; \frac{dX_5}{dt} = \frac{1}{C_5 \cdot L_4} \cdot I_{C_5}$$

$$\frac{dX_2}{dt} = \frac{1}{L_5} \cdot \frac{dV_s(t)}{dt} - [X_2 \cdot e^{\lambda \cdot \tau_{in}} + X_4 \cdot e^{\lambda \cdot \tau_{in}}] \cdot \frac{R_s}{L_5} - \frac{1}{C_2 \cdot L_5} \cdot I_{C_2} - \frac{1}{C_1 \cdot L_5} \cdot I_{L_5}$$

$$\frac{dX_3}{dt} = \frac{1}{C_2 \cdot L_1} \cdot I_{C_2} - \frac{1}{C_3 \cdot L_1} \cdot I_{C_3} - X_3 \cdot e^{-\lambda \cdot \tau_{out1}} \cdot \frac{R_{load1}}{L_1}; \; \frac{dX_5}{dt} = \frac{1}{C_5 \cdot L_4} \cdot I_{C_5}$$

At fixed points (equilibrium points): $\frac{dV_s(t)}{dt} \to \varepsilon$

$$\frac{dI_{L_5}}{dt} = 0 \Rightarrow X_2^{(i)} = 0; \frac{dI_{C_3}}{dt} = 0 \Rightarrow X_3^{(i)} = 0; \frac{dI_{L_6}}{dt} = 0 \Rightarrow X_4^{(i)} = 0; \frac{dI_{L_4}}{dt} = 0 \Rightarrow X_5^{(i)} = 0$$

$$\frac{dI_{C_4}}{dt} = 0 \Rightarrow X_6^{(i)} = 0; \; \frac{dX_2}{dt} = 0 \Rightarrow \{\frac{dV_s(t)}{dt} \to \varepsilon\} - \frac{1}{C_2} \cdot I_{C_2}^{(i)} - \frac{1}{C_1} \cdot I_{L_5}^{(i)} = 0$$

$$\frac{dX_3}{dt} = 0 \Rightarrow \frac{1}{C_2} \cdot I_{C_2}^{(i)} - \frac{1}{C_3} \cdot I_{C_3}^{(i)} = 0; \frac{dX_5}{dt} = 0 \Rightarrow \frac{1}{C_5 \cdot L_4} \cdot I_{C_5}^{(i)} = 0 \Rightarrow I_{C_5}^{(i)} = 0$$

$$\frac{dX_4}{dt} = 0 \Rightarrow \{\frac{dV_s(t)}{dt} \to \varepsilon\} - \frac{1}{C_6} \cdot I_{L_6}^{(i)} - \frac{1}{C_5} \cdot I_{C_5}^{(i)} = 0; I_{C_5}^{(i)} = 0 \Rightarrow I_{L_6}^{(i)} = 0$$

$$\frac{dX_6}{dt} = 0 \Rightarrow \frac{1}{C_5} \cdot I_{C_5}^{(i)} - \frac{1}{C_4} \cdot I_{C_4}^{(i)} = 0; I_{C_5}^{(i)} = 0; I_{C_4}^{(i)} = 0; I_{L_6}^{(i)} = 0$$

We can summery our system fixed points:

$$\underset{\forall k=2,3,4,5,6}{X_k^{(i)}} = 0; I_{C_5}^{(i)} = 0; I_{C_4}^{(i)} = 0; I_{L_6}^{(i)} = 0$$

$$\frac{1}{C_2} \cdot I_{C_2}^{(i)} + \frac{1}{C_1} \cdot I_{L_5}^{(i)} = 0; \frac{1}{C_2} \cdot I_{C_2}^{(i)} - \frac{1}{C_3} \cdot I_{C_3}^{(i)} = 0$$

Stability analysis: The standard local stability analysis about any one of the equilibrium points of dual band diplexer filter circuit consists in adding to its coordinated $[X_2\ X_3\ X_4\ X_5\ X_6\ I_{L_5}\ I_{C_3}\ I_{L_6}\ I_{L_4}\ I_{C_4}\ I_{C_2}\ I_{C_5}]$ arbitrarily small increments of exponential terms $[x_2\ x_3\ x_4\ x_5\ x_6\ i_{L_5}\ i_{C_3}\ i_{L_6}\ i_{L_4}\ i_{C_4}\ i_{C_2}\ i_{C_5}] \cdot e^{\lambda \cdot t}$, and retaining the first order terms in $x_2\ x_3\ x_4\ x_5\ x_6\ i_{L_5}\ i_{C_3}\ i_{L_6}\ i_{L_4}\ i_{C_4}\ i_{C_2}\ i_{C_5}$. The system of ten homogeneous equations leads to a polynomial characteristic equation in the eigenvalueλ. The polynomial characteristic equation accepts by set the dual band diplexer filter circuit equations. The dual band diplexer filter circuit fixed values with arbitrarily small increments of exponential form $[x_2\ x_3\ x_4\ x_5\ x_6\ i_{L_5}\ i_{C_3}\ i_{L_6}\ i_{L_4}\ i_{C_4}\ i_{C_2}\ i_{C_5}] \cdot e^{\lambda \cdot t}$ are; i = 0 (first fixed point), i = 1 (second fixed point), i = 2 (third fixed point), etc., [2–4].

$$X_2(t) = X_2^{(i)} + x_2 \cdot e^{\lambda \cdot t}; X_3(t) = X_3^{(i)} + x_3 \cdot e^{\lambda \cdot t};$$
$$X_4(t) = X_4^{(i)} + x_4 \cdot e^{\lambda \cdot t}; X_5(t) = X_5^{(i)} + x_5 \cdot e^{\lambda \cdot t}$$
$$X_6(t) = X_6^{(i)} + x_6 \cdot e^{\lambda \cdot t}; I_{L_5}(t) = I_{L_5}^{(i)} + i_{L_5} \cdot e^{\lambda \cdot t};$$
$$I_{C_3}(t) = I_{C_3}^{(i)} + i_{C_3} \cdot e^{\lambda \cdot t}; I_{L_6}(t) = I_{L_6}^{(i)} + i_{L_6} \cdot e^{\lambda \cdot t}$$

$$I_{L_4}(t) = I_{L_4}^{(i)} + i_{L_4} \cdot e^{\lambda \cdot t}; I_{C_4}(t) = I_{C_4}^{(i)} + i_{C_4} \cdot e^{\lambda \cdot t}; I_{C_2}(t) = I_{C_2}^{(i)} + i_{C_2} \cdot e^{\lambda \cdot t}; I_{C_5}(t) = I_{C_5}^{(i)} + i_{C_5} \cdot e^{\lambda \cdot t}$$

$$\frac{dX_2(t)}{dt} = x_2 \cdot \lambda \cdot e^{\lambda \cdot t}; \frac{dX_3(t)}{dt} = x_3 \cdot \lambda \cdot e^{\lambda \cdot t}; \frac{dX_4(t)}{dt} = x_4 \cdot \lambda \cdot e^{\lambda \cdot t}; \frac{dX_5(t)}{dt} = x_5 \cdot \lambda \cdot e^{\lambda \cdot t}$$

$$\frac{dX_6(t)}{dt} = x_6 \cdot \lambda \cdot e^{\lambda \cdot t}; \frac{dI_{L_5}(t)}{dt} = i_{L_5} \cdot \lambda \cdot e^{\lambda \cdot t}; \frac{dI_{C_3}(t)}{dt} = i_{C_3} \cdot \lambda \cdot e^{\lambda \cdot t}; \frac{dI_{L_6}(t)}{dt} = i_{L_6} \cdot \lambda \cdot e^{\lambda \cdot t}$$

$$\frac{dI_{L_4}(t)}{dt} = i_{L_4} \cdot \lambda \cdot e^{\lambda \cdot t}; \frac{dI_{C_4}(t)}{dt} = i_{C_4} \cdot \lambda \cdot e^{\lambda \cdot t}$$

For λ < 0, t > 0 the selected fixed point is stable otherwise λ > 0, t > 0 unstable. Our dual band diplexer filter circuit tends to the selected fixed point exponentially for λ < 0, t > 0 otherwise go away from the selected fixed point exponentially. λ is the eigenvalue parameter which establish if the fixed point is stable or unstable, additionally his absolute value |λ| establish the speed of flow toward or away from the selected fixed point [4].

$$\frac{dI_{L_5}}{dt} = X_2 \Rightarrow i_{L_5} \cdot \lambda \cdot e^{\lambda \cdot t} = X_2^{(i)} + x_2 \cdot e^{\lambda \cdot t};\ X_2^{(i)} = 0; -i_{L_5} \cdot \lambda + x_2 = 0$$

$$\frac{dI_{C_3}}{dt} = X_3 \Rightarrow i_{C_3} \cdot \lambda \cdot e^{\lambda \cdot t} = X_3^{(i)} + x_3 \cdot e^{\lambda \cdot t};\ X_3^{(i)} = 0; -i_{C_3} \cdot \lambda + x_3 = 0$$

$$\frac{dI_{L_6}}{dt} = X_4 \Rightarrow i_{L_6} \cdot \lambda \cdot e^{\lambda \cdot t} = X_4^{(i)} + x_4 \cdot e^{\lambda \cdot t};\ X_4^{(i)} = 0;\ -i_{L_6} \cdot \lambda + x_4 = 0$$

$$\frac{dI_{L_4}}{dt} = X_5 \Rightarrow i_{L_4} \cdot \lambda \cdot e^{\lambda \cdot t} = X_5^{(i)} + x_5 \cdot e^{\lambda \cdot t};\ X_5^{(i)} = 0; -i_{L_4} \cdot \lambda + x_5 = 0$$

$$\frac{dI_{C_4}}{dt} = X_6 \Rightarrow i_{C_4} \cdot \lambda \cdot e^{\lambda \cdot t} = X_6^{(i)} + x_6 \cdot e^{\lambda \cdot t};\ X_6^{(i)} = 0; -i_{C_4} \cdot \lambda + x_6 = 0$$

$$\frac{dX_2}{dt} = -[X_2 + X_4] \cdot e^{\lambda \cdot \tau_{in}} \cdot \frac{R_s}{L_5} - \frac{1}{C_2 \cdot L_5} \cdot I_{C_2} - \frac{1}{C_1 \cdot L_5} \cdot I_{L_5}$$

$$x_2 \cdot \lambda \cdot e^{\lambda \cdot t} = -[X_2^{(i)} + x_2 \cdot e^{\lambda \cdot t} + X_4^{(i)} + x_4 \cdot e^{\lambda \cdot t}] \cdot e^{\lambda \cdot \tau_{in}} \cdot \frac{R_s}{L_5} - \frac{1}{C_2 \cdot L_5} \cdot [I_{C_2}^{(i)} + i_{C_2} \cdot e^{\lambda \cdot t}] - \frac{1}{C_1 \cdot L_5} \cdot [I_{L_5}^{(i)} + i_{L_5} \cdot e^{\lambda \cdot t}]$$

$$x_2 \cdot \lambda \cdot e^{\lambda \cdot t} = -[X_2^{(i)} + X_4^{(i)}] \cdot e^{\lambda \cdot \tau_{in}} \cdot \frac{R_s}{L_5} - \frac{1}{C_2 \cdot L_5} \cdot I_{C_2}^{(i)} - \frac{1}{C_1 \cdot L_5} \cdot I_{L_5}^{(i)} - [x_2 \cdot e^{\lambda \cdot t} + x_4 \cdot e^{\lambda \cdot t}] \cdot e^{\lambda \cdot \tau_{in}} \cdot \frac{R_s}{L_5} - \frac{1}{C_2 \cdot L_5} \cdot i_{C_2} \cdot e^{\lambda \cdot t} - \frac{1}{C_1 \cdot L_5} \cdot i_{L_5} \cdot e^{\lambda \cdot t}$$

At fixed points: $-[X_2^{(i)} + X_4^{(i)}] \cdot e^{\lambda \cdot \tau_{in}} \cdot \frac{R_s}{L_5} - \frac{1}{C_2 \cdot L_5} \cdot I_{C_2}^{(i)} - \frac{1}{C_1 \cdot L_5} \cdot I_{L_5}^{(i)} = 0$

$$x_2 \cdot \lambda \cdot e^{\lambda \cdot t} = -[x_2 \cdot e^{\lambda \cdot t} + x_4 \cdot e^{\lambda \cdot t}] \cdot e^{\lambda \cdot \tau_{in}} \cdot \frac{R_s}{L_5} - \frac{1}{C_2 \cdot L_5} \cdot i_{C_2} \cdot e^{\lambda \cdot t} - \frac{1}{C_1 \cdot L_5} \cdot i_{L_5} \cdot e^{\lambda \cdot t}$$

$$-x_2 \cdot \lambda - x_2 \cdot e^{\lambda \cdot \tau_{in}} \cdot \frac{R_s}{L_5} - x_4 \cdot e^{\lambda \cdot \tau_{in}} \cdot \frac{R_s}{L_5} - \frac{1}{C_2 \cdot L_5} \cdot i_{C_2} - \frac{1}{C_1 \cdot L_5} \cdot i_{L_5} = 0$$

$$\frac{dX_3}{dt} = \frac{1}{C_2 \cdot L_1} \cdot I_{C_2} - \frac{1}{C_3 \cdot L_1} \cdot I_{C_3} - X_3 \cdot e^{-\lambda \cdot \tau_{out1}} \cdot \frac{R_{load1}}{L_1}$$

$$x_3 \cdot \lambda \cdot e^{\lambda \cdot t} = \frac{1}{C_2 \cdot L_1} \cdot [I_{C_2}^{(i)} + i_{C_2} \cdot e^{\lambda \cdot t}] - \frac{1}{C_3 \cdot L_1} \cdot [I_{C_3}^{(i)} + i_{C_3} \cdot e^{\lambda \cdot t}] - [X_3^{(i)} + x_3 \cdot e^{\lambda \cdot t}] \cdot e^{-\lambda \cdot \tau_{out1}} \cdot \frac{R_{load1}}{L_1}$$

$$x_3 \cdot \lambda \cdot e^{\lambda \cdot t} = \frac{1}{C_2 \cdot L_1} \cdot I_{C_2}^{(i)} - \frac{1}{C_3 \cdot L_1} \cdot I_{C_3}^{(i)} - X_3^{(i)} \cdot e^{-\lambda \cdot \tau_{out1}} \cdot \frac{R_{load1}}{L_1} + \frac{1}{C_2 \cdot L_1} \cdot i_{C_2} \cdot e^{\lambda \cdot t} - \frac{1}{C_3 \cdot L_1} \cdot i_{C_3} \cdot e^{\lambda \cdot t} - x_3 \cdot e^{\lambda \cdot t} \cdot e^{-\lambda \cdot \tau_{out1}} \cdot \frac{R_{load1}}{L_1}$$

At fixed points: $\frac{1}{C_2 \cdot L_1} \cdot I_{C_2}^{(i)} - \frac{1}{C_3 \cdot L_1} \cdot I_{C_3}^{(i)} - X_3^{(i)} \cdot e^{-\lambda \cdot \tau_{out1}} \cdot \frac{R_{load1}}{L_1} = 0$

$$x_3 \cdot \lambda \cdot e^{\lambda \cdot t} = \frac{1}{C_2 \cdot L_1} \cdot i_{C_2} \cdot e^{\lambda \cdot t} - \frac{1}{C_3 \cdot L_1} \cdot i_{C_3} \cdot e^{\lambda \cdot t} - x_3 \cdot e^{\lambda \cdot t} \cdot e^{-\lambda \cdot \tau_{out1}} \cdot \frac{R_{load1}}{L_1}$$

$$-x_3 \cdot \lambda + \frac{1}{C_2 \cdot L_1} \cdot i_{C_2} - \frac{1}{C_3 \cdot L_1} \cdot i_{C_3} - x_3 \cdot e^{-\lambda \cdot \tau_{out1}} \cdot \frac{R_{load1}}{L_1} = 0$$

$$\frac{dX_5}{dt} = \frac{1}{C_5 \cdot L_4} \cdot I_{C_5} \Rightarrow x_5 \cdot \lambda \cdot e^{\lambda \cdot t} = \frac{1}{C_5 \cdot L_4} \cdot [I_{C_5}^{(i)} + i_{C_5} \cdot e^{\lambda \cdot t}];\ x_5 \cdot \lambda \cdot e^{\lambda \cdot t} = \frac{1}{C_5 \cdot L_4} \cdot I_{C_5}^{(i)} + \frac{1}{C_5 \cdot L_4} \cdot i_{C_5} \cdot e^{\lambda \cdot t}$$

At fixed points: $\frac{1}{C_5 \cdot L_4} \cdot I_{C_5}^{(i)} = 0;\ -x_5 \cdot \lambda + \frac{1}{C_5 \cdot L_4} \cdot i_{C_5} = 0$

$$\frac{dX_4}{dt} = \frac{1}{L_6} \cdot \frac{dV_s(t)}{dt} - [X_2 \cdot e^{\lambda \cdot \tau_{in}} + X_4 \cdot e^{\lambda \cdot \tau_{in}}] \cdot \frac{R_s}{L_6} - \frac{1}{C_6 \cdot L_6} \cdot I_{L_6} - \frac{1}{C_5 \cdot L_6} \cdot I_{C_5}$$

$$x_4 \cdot \lambda \cdot e^{\lambda \cdot t} = \frac{1}{L_6} \cdot \frac{dV_s(t)}{dt} - [X_2^{(i)} + x_2 \cdot e^{\lambda \cdot t} + X_4^{(i)} + x_4 \cdot e^{\lambda \cdot t}] \cdot \frac{R_s}{L_6} \cdot e^{\lambda \cdot \tau_{in}} - \frac{1}{C_6 \cdot L_6} \cdot [I_{L_6}^{(i)} + i_{L_6} \cdot e^{\lambda \cdot t}] - \frac{1}{C_5 \cdot L_6} \cdot [I_{C_5}^{(i)} + i_{C_5} \cdot e^{\lambda \cdot t}]$$

$$x_4 \cdot \lambda \cdot e^{\lambda \cdot t} = \frac{1}{L_6} \cdot \frac{dV_s(t)}{dt} - [X_2^{(i)} + X_4^{(i)}] \cdot \frac{R_s}{L_6} \cdot e^{\lambda \cdot \tau_{in}} - \frac{1}{C_6 \cdot L_6} \cdot I_{L_6}^{(i)} - \frac{1}{C_5 \cdot L_6} \cdot I_{C_5}^{(i)} - [x_2 \cdot e^{\lambda \cdot t} + x_4 \cdot e^{\lambda \cdot t}] \cdot \frac{R_s}{L_6} \cdot e^{\lambda \cdot \tau_{in}} - \frac{1}{C_6 \cdot L_6} \cdot i_{L_6} \cdot e^{\lambda \cdot t} - \frac{1}{C_5 \cdot L_6} \cdot i_{C_5} \cdot e^{\lambda \cdot t}$$

At fixed points: $\frac{1}{L_6} \cdot \{\frac{dV_s(t)}{dt} \to \varepsilon\} - [X_2^{(i)} + X_4^{(i)}] \cdot \frac{R_s}{L_6} \cdot e^{\lambda \cdot \tau_{in}} - \frac{1}{C_6 \cdot L_6} \cdot I_{L_6}^{(i)} - \frac{1}{C_5 \cdot L_6} \cdot I_{C_5}^{(i)} = 0$

$$x_4 \cdot \lambda \cdot e^{\lambda \cdot t} = -[x_2 \cdot e^{\lambda \cdot t} + x_4 \cdot e^{\lambda \cdot t}] \cdot \frac{R_s}{L_6} \cdot e^{\lambda \cdot \tau_{in}} - \frac{1}{C_6 \cdot L_6} \cdot i_{L_6} \cdot e^{\lambda \cdot t} - \frac{1}{C_5 \cdot L_6} \cdot i_{C_5} \cdot e^{\lambda \cdot t}$$

$$-x_4 \cdot \lambda - x_2 \cdot \frac{R_s}{L_6} \cdot e^{\lambda \cdot \tau_{in}} - x_4 \cdot \frac{R_s}{L_6} \cdot e^{\lambda \cdot \tau_{in}} - \frac{1}{C_6 \cdot L_6} \cdot i_{L_6} - \frac{1}{C_5 \cdot L_6} \cdot i_{C_5} = 0$$

$$\frac{dX_6}{dt} = \frac{1}{C_5 \cdot L_3} \cdot I_{C_5} - \frac{1}{C_4 \cdot L_3} \cdot I_{C_4} - X_6 \cdot e^{-\lambda \cdot \tau_{out2}} \cdot \frac{R_{load2}}{L_3}$$

$$x_6 \cdot \lambda \cdot e^{\lambda \cdot t} = \frac{1}{C_5 \cdot L_3} \cdot [I_{C_5}^{(i)} + i_{C_5} \cdot e^{\lambda \cdot t}] - \frac{1}{C_4 \cdot L_3} \cdot [I_{C_4}^{(i)} + i_{C_4} \cdot e^{\lambda \cdot t}] - [X_6^{(i)} + x_6 \cdot e^{\lambda \cdot t}] \cdot e^{-\lambda \cdot \tau_{out2}} \cdot \frac{R_{load2}}{L_3}$$

$$x_6 \cdot \lambda \cdot e^{\lambda \cdot t} = \frac{1}{C_5 \cdot L_3} \cdot I_{C_5}^{(i)} - \frac{1}{C_4 \cdot L_3} \cdot I_{C_4}^{(i)} - X_6^{(i)} \cdot e^{-\lambda \cdot \tau_{out2}} \cdot \frac{R_{load2}}{L_3} + \frac{1}{C_5 \cdot L_3} \cdot i_{C_5} \cdot e^{\lambda \cdot t} - \frac{1}{C_4 \cdot L_3} \cdot i_{C_4} \cdot e^{\lambda \cdot t} - x_6 \cdot e^{\lambda \cdot t} \cdot e^{-\lambda \cdot \tau_{out2}} \cdot \frac{R_{load2}}{L_3}$$

At fixed points: $\frac{1}{C_5 \cdot L_3} \cdot I_{C_5}^{(i)} - \frac{1}{C_4 \cdot L_3} \cdot I_{C_4}^{(i)} - X_6^{(i)} \cdot e^{-\lambda \cdot \tau_{out2}} \cdot \frac{R_{load2}}{L_3} = 0$

$$x_6 \cdot \lambda \cdot e^{\lambda \cdot t} = \frac{1}{C_5 \cdot L_3} \cdot i_{C_5} \cdot e^{\lambda \cdot t} - \frac{1}{C_4 \cdot L_3} \cdot i_{C_4} \cdot e^{\lambda \cdot t} - x_6 \cdot e^{\lambda \cdot t} \cdot e^{-\lambda \cdot \tau_{out2}} \cdot \frac{R_{load2}}{L_3}$$

$$-x_6 \cdot \lambda + \frac{1}{C_5 \cdot L_3} \cdot i_{C_5} - \frac{1}{C_4 \cdot L_3} \cdot i_{C_4} - x_6 \cdot e^{-\lambda \cdot \tau_{out2}} \cdot \frac{R_{load2}}{L_3} = 0$$

We can summery our dual band diplexer filter circuit arbitrarily small increments equations:

$$-i_{L_5} \cdot \lambda + x_2 = 0; -i_{C_3} \cdot \lambda + x_3 = 0; -i_{L_6} \cdot \lambda + x_4 = 0; -i_{L_4} \cdot \lambda + x_5 = 0; -i_{C_4} \cdot \lambda + x_6 = 0$$

$$-x_2 \cdot \lambda - x_2 \cdot e^{\lambda \cdot \tau_{in}} \cdot \frac{R_s}{L_5} - x_4 \cdot e^{\lambda \cdot \tau_{in}} \cdot \frac{R_s}{L_5} - \frac{1}{C_2 \cdot L_5} \cdot i_{C_2} - \frac{1}{C_1 \cdot L_5} \cdot i_{L_5} = 0$$

$$-x_3 \cdot \lambda + \frac{1}{C_2 \cdot L_1} \cdot i_{C_2} - \frac{1}{C_3 \cdot L_1} \cdot i_{C_3} - x_3 \cdot e^{-\lambda \cdot \tau_{out1}} \cdot \frac{R_{load1}}{L_1} = 0; -x_5 \cdot \lambda + \frac{1}{C_5 \cdot L_4} \cdot i_{C_5} = 0$$

$$-x_4 \cdot \lambda - x_2 \cdot \frac{R_s}{L_6} \cdot e^{\lambda \cdot \tau_{in}} - x_4 \cdot \frac{R_s}{L_6} \cdot e^{\lambda \cdot \tau_{in}} - \frac{1}{C_6 \cdot L_6} \cdot i_{L_6} - \frac{1}{C_5 \cdot L_6} \cdot i_{C_5} = 0$$

$$-x_6 \cdot \lambda + \frac{1}{C_5 \cdot L_3} \cdot i_{C_5} - \frac{1}{C_4 \cdot L_3} \cdot i_{C_4} - x_6 \cdot e^{-\lambda \cdot \tau_{out2}} \cdot \frac{R_{load2}}{L_3} = 0$$

$$\begin{pmatrix} \Xi_{11} & \cdots & \Xi_{1\_10} \\ \vdots & \ddots & \vdots \\ \Xi_{10\_1} & \cdots & \Xi_{10\_10} \end{pmatrix} \cdot \begin{pmatrix} i_{L_5} \\ i_{C_3} \\ i_{L_6} \\ i_{L_4} \\ i_{C_5} \\ x_2 \\ x_3 \\ x_4 \\ x_5 \\ x_6 \end{pmatrix} + \begin{pmatrix} \Upsilon_{11} & \Upsilon_{12} \\ \Upsilon_{21} & \Upsilon_{22} \\ \Upsilon_{31} & \Upsilon_{32} \\ \Upsilon_{41} & \Upsilon_{42} \\ \Upsilon_{51} & \Upsilon_{52} \\ \Upsilon_{61} & \Upsilon_{62} \\ \Upsilon_{71} & \Upsilon_{72} \\ \Upsilon_{81} & \Upsilon_{82} \\ \Upsilon_{91} & \Upsilon_{92} \\ \Upsilon_{10\_1} & \Upsilon_{10\_2} \end{pmatrix} \cdot \begin{pmatrix} i_{C_2} \\ i_{C_5} \end{pmatrix}$$

$$= 0;\ \Upsilon_{kl}\Big|_{\forall k=1,2,3,4,5;l=1,2};\ \Upsilon_{61} = -\frac{1}{C_2 \cdot L_5};\ \Upsilon_{62} = 0$$

$$\Upsilon_{71} = \frac{1}{C_2 \cdot L_1};\ \Upsilon_{72} = 0;\ \Upsilon_{81} = 0;\ \Upsilon_{82} = -\frac{1}{C_5 \cdot L_6};\ \Upsilon_{91} = 0;\ \Upsilon_{92} = \frac{1}{C_5 \cdot L_4};\ \Upsilon_{10\_1} = 0;\ \Upsilon_{10\_2} = \frac{1}{C_5 \cdot L_3}$$

Assumption

$$\begin{pmatrix} \Upsilon_{11} & \Upsilon_{12} \\ \Upsilon_{21} & \Upsilon_{22} \\ \Upsilon_{31} & \Upsilon_{32} \\ \Upsilon_{41} & \Upsilon_{42} \\ \Upsilon_{51} & \Upsilon_{52} \\ \Upsilon_{61} & \Upsilon_{62} \\ \Upsilon_{71} & \Upsilon_{72} \\ \Upsilon_{81} & \Upsilon_{82} \\ \Upsilon_{91} & \Upsilon_{92} \\ \Upsilon_{10\_1} & \Upsilon_{10\_2} \end{pmatrix} \cdot \begin{pmatrix} i_{C_2} \\ i_{C_5} \end{pmatrix} \to \varepsilon;\ \begin{pmatrix} \Xi_{11} & \cdots & \Xi_{1\_10} \\ \vdots & \ddots & \vdots \\ \Xi_{10\_1} & \cdots & \Xi_{10\_10} \end{pmatrix} \cdot \begin{pmatrix} i_{L_5} \\ i_{C_3} \\ i_{L_6} \\ i_{L_4} \\ i_{C_5} \\ x_2 \\ x_3 \\ x_4 \\ x_5 \\ x_6 \end{pmatrix} \approx 0$$

Arbitrarily small increments i_{C_2}, i_{C_5} are very small and $\begin{pmatrix} i_{C_2} \\ i_{C_5} \end{pmatrix} \to \varepsilon$.

$$\Xi_{11} = -\lambda;\ \Xi_{12} = \ldots = \Xi_{15} = 0;\ \Xi_{16} = 1;\ \Xi_{17} = \ldots \Xi_{1\_10} = 0;\ \Xi_{21} = 0;\ \Xi_{22} = -\lambda$$

$$\Xi_{23} = \ldots \Xi_{26} = 0;\ \Xi_{27} = 1;\ \Xi_{28} = \Xi_{29} = \Xi_{2\_10} = 0;\ \Xi_{31} = \Xi_{32} = 0;\ \Xi_{33} = -\lambda$$

$$\Xi_{34} = \ldots \Xi_{37} = 0;\ \Xi_{38} = 1;\ \Xi_{39} = \Xi_{3\_10} = 0;\ \Xi_{41} = \Xi_{42} = \Xi_{43} = 0;\ \Xi_{44} = -\lambda$$

$$\Xi_{45} = \ldots = \Xi_{48} = 0;\ \Xi_{49} = 1;\ \Xi_{4\_10} = 0;\ \Xi_{51} = \ldots = \Xi_{54} = 0;\ \Xi_{55} = -\lambda;\ \Xi_{56} = \ldots \Xi_{59} = 0$$

$$\Xi_{5\_10} = 1;\ \Xi_{61} = -\frac{1}{C_1 \cdot L_5};\ \Xi_{62} = 0;\ \Xi_{63} = \Xi_{64} = \Xi_{65} = 0;\ \Xi_{66} = -\lambda - e^{\lambda \cdot \tau_{in}} \cdot \frac{R_s}{L_5}$$

$$\Xi_{67} = 0;\ \Xi_{68} = -e^{\lambda \cdot \tau_{in}} \cdot \frac{R_s}{L_5};\ \Xi_{69} = 0;\ \Xi_{6\_10} = 0;\ \Xi_{71} = 0;\ \Xi_{72} = -\frac{1}{C_3 \cdot L_1};\ \Xi_{73} = 0$$

$$\Xi_{74} = \Xi_{75} = 0;\ \Xi_{76} = 0;\ \Xi_{77} = -\lambda - e^{-\lambda \cdot \tau_{out1}} \cdot \frac{R_{load1}}{L_1};\ \Xi_{78} = \Xi_{79} = \Xi_{7\_10} = 0$$

$$\Xi_{81} = \Xi_{82} = 0;\ \Xi_{83} = -\frac{1}{C_6 \cdot L_6};\ \Xi_{84} = \Xi_{85} = 0;\ \Xi_{86} = -\frac{R_s}{L_6} \cdot e^{\lambda \cdot \tau_{in}};\ \Xi_{87} = 0$$

$$\Xi_{88} = -\lambda - \frac{R_s}{L_6} \cdot e^{\lambda \cdot \tau_{in}};\ \Xi_{89} = \Xi_{8\_10} = 0;\ \Xi_{91} = \ldots = \Xi_{98} = 0;\ \Xi_{99} = -\lambda;\ \Xi_{9\_10} = 0$$

$$\Xi_{10\_1} = \Xi_{10\_2} = \Xi_{10\_3} = \Xi_{10\_4} = 0;\ \Xi_{10\_5} = -\frac{1}{C_4 \cdot L_3};\ \Xi_{10\_6} = \ldots = \Xi_{10\_9} = 0$$

$$\Xi_{10\_10} = -\lambda - e^{-\lambda \cdot \tau_{out2}} \cdot \frac{R_{load2}}{L_3}$$

We analyze dual band diplexer filter circuit for the following cases:

Table 7.2 Dual band diplexer filter circuit for cases of τ_{in}, τ_{out1}, τ_{out2}

Case no.	τ_{in}	τ_{out1}	τ_{out2}
1	0	0	0
2	τ	0	0
3	0	τ	0
4	0	0	τ
5	τ	τ	0
6	0	τ	τ
7	τ	0	τ
8	τ	τ	τ

Case No. 1: $\tau_{in} = 0; \tau_{out1} = 0; \tau_{out2} = 0; \Xi_{66} = -\lambda - \frac{R_s}{L_5}; \Xi_{68} = -\frac{R_s}{L_5}.$

$$\Xi_{77} = -\lambda - \frac{R_{load1}}{L_1}; \Xi_{86} = -\frac{R_s}{L_6}; \Xi_{88} = -\lambda - \frac{R_s}{L_6}; \Xi_{10\_10} = -\lambda - \frac{R_{load2}}{L_3}$$

$$\det(A - \lambda \cdot I)|_{\substack{\tau_{in}=0 \\ \tau_{out1}=0 \\ \tau_{out2}=0}} = \sum_{k=0}^{10} \Upsilon_k \cdot \lambda^k; \det(A - \lambda \cdot I)|_{\substack{\tau_{in}=0 \\ \tau_{out1}=0 \\ \tau_{out2}=0}} = 0 \Rightarrow \sum_{k=0}^{10} \Upsilon_k \cdot \lambda^k|_{\substack{\tau_{in}=0 \\ \tau_{out1}=0 \\ \tau_{out2}=0}} = 0$$

Remark It is reader exercise to find above polynomial parameters $\Upsilon_1, \ldots, \Upsilon_{10}$.

Eigenvalues stability discussion: Our dual band diplexer filter circuit involving N variables (N > 2, N = 10), the characteristic equation is of degree N = 10 and must often be solved numerically. Expect in some particular cases, such an equation has (N = 10) distinct roots that can be real or complex. These values are the eigenvalues of the 10 × 10 Jacobian matrix (A). The general rule is that the Steady State (SS) is stable if there is no eigenvalue with positive real part. It is sufficient that one eigenvalue is positive for the steady state to be unstable. Our 10-variables (X_2 X_3 X_4 X_5 X_6 I_{L_5} I_{C_3} I_{L_6} I_{L_4} I_{C_4}) dual band diplexer filter circuit has ten eigenvalues. The type of behavior can be characterized as a function of the position of these eigenvalues in the Re/Im plane. Five non-degenerated cases can be distinguished: (1) the ten eigenvalues are real and negative (stable steady state), (2) the ten eigenvalues are real, nine of them are negative (unstable steady state), (3) and (4) two eigenvalues are complex conjugates with a negative real part and the other eigenvalues are real and negative (stable steady state), two cases can be distinguished depending on the relative value of the real part of the complex eigenvalues and of the real one, (5) two eigenvalues are complex conjugates with a negative real part and at least one eigenvalue is positive (unstable steady state).

Remark It is reader exercise to analyze dual band diplexer filter circuit stability for cases 2 to 8 (at least one delay parameter $\tau_{in}, \tau_{out1}, \tau_{out2}$ is positive and real number). The stability analysis is done by using geometric stability switch criteria in delay differential systems (E. Beretta and Y. Kuang). It is a practical guideline that combines graphical information with analytical work to effectively study the local stability of dual band diplexer filter circuit model involving delay dependent parameters. The stability of dual band diplexer filter circuit steady state is determined by the graphs of some functions of τ which can be expressed explicitly [5, 6].

The general geometric criterion: The occurrence of any possible stability switching resulting from the increase of value of the time delay τ for our dual band diplexer filter circuit characteristic equation.

$$D(\lambda, \tau_{in}, \tau_{out1}, \tau_{out2}) = \det(A - \lambda \cdot I);\ D(\lambda, \tau_{in}, \tau_{out1}, \tau_{out2}) = 0;\ D(\lambda, \tau) = P_n(\lambda, \tau) + Q_m(\lambda, \tau) \cdot e^{-\lambda \cdot \tau}$$

$$P_n(\lambda, \tau) = \sum_{k=0}^{n} p_k(\tau) \cdot \lambda^k;\ Q_m(\lambda, \tau) = \sum_{k=0}^{m} q_k(\tau) \cdot \lambda^k;\ n, m \in \mathbb{N}_0;\ n > m$$

$p_k(\bullet), q_k(\bullet) : \mathbb{R}_{+0} \to \mathbb{R}$ are continuous and differentiable functions of τ.

$$P_n(\lambda = 0, \tau) + Q_m(\lambda = 0, \tau) = p_0(\tau) + q_0(\tau) \neq 0 \,\forall \tau \in \mathbb{R}_{+0}$$

$P_n(\lambda, \tau), Q_m(\lambda, \tau)$ are analytic functions in λ and differentiable in τ for which we assume:

(I) If $\lambda = i \cdot \omega; \omega \in \mathbb{R}$ then $P_n(i \cdot \omega, \tau) + Q_m(i \cdot \omega, \tau) \neq 0; \tau \in \mathbb{R}$.
(II) $\limsup\{|Q_m(\lambda, \tau)/P_n(\lambda, \tau)| : |\lambda| \to \infty, \operatorname{Re}\lambda \geq 0\} < 1$ for any τ.
(III) $F(\omega, \tau) = |P_n(i \cdot \omega, \tau)|^2 - |Q_m(i \cdot \omega, \tau)|^2$ for each τ has at most a finite number of real zeros.
(IV) Each positive root $\omega(\tau)$ of $F(\omega, \tau) = 0$ is continuous and differentiable in τ whenever it exists.

7.3 A Crystal-Lattice BPF Circuit Stability Analysis

We use crystal in place of LC filter for low frequency applications. It is emphasis in narrow bandwidth filtering. Crystals have a series and parallel resonant mode. Other crystal mode is overtone or harmonic mode. There are lattice crystal filter, half latticex and cascaded half lattice filters. The bandwidth of these filters is a function of the frequency separation of the crystals. Another form of filter is the ladder crystal filter. It has an asymmetrical response and it is called the "lower-sideband ladder" configuration. Ladder filters have some advantages: no need to pick crystals for proper frequency separation, no need to match crystal pairs, simple filter topology, simple construction methods, no adjustable components are required after alignment is completed, benefits by the absence of coils, compact assembly, high number of poles, and shape the filter response with great accuracy. The equivalent circuit of a quartz crystal is describe in the below figure.

We need to measure the crystal parameters. The important parameters in our design are: ΔF—frequency offset or deviation from the specific center frequency, r—series resistance of the crystal, f_L and f_H—3 dB points required for the Q calculation, L_m—motional inductance which is derived from the Q and r, C_p—parallel capacitance of the crystal's holder, C_m—motional capacitance. In our design the crystals can be matched for Q, L_m and ΔF. Several factors influence the choice of bandwidth of a crystal filter: the desired selectivity (narrow filters for contest work and wider filters for casual rag-chewing), receiver sensitivity, dynamic range, and personal preference. The value of the terminating resistance should be as low as possible to minimize the transformation ratio of the impedance matching transformers. The crystal can be forced to resonate efficiency at odd harmonic intervals of its fundamental frequency. The overtone modes force a crystal filter to have undesired reentrance modes (odd multiples of the series resonant frequency). Crystals can be in larger combinations, within RF filter packages. The crystal lattice filters contain several crystals within a single circuit (adopted for use as a very sharp bandpass filter). The input and outputs employ RF transformers (T_1 and T_2) with shunt capacitors (C_1 and C_2). Each set of crystals XTAL1 plus XTAL2 and XTAL3 plus XTAL4 are cut to different frequencies. The matched set of XTAL1 and XTAL2 having a lower resonant frequency than the other

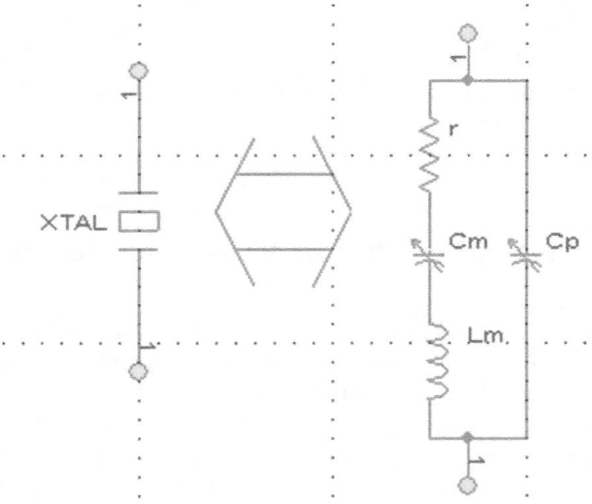

Fig. 7.6 Equivalent circuit of a quartz crystal

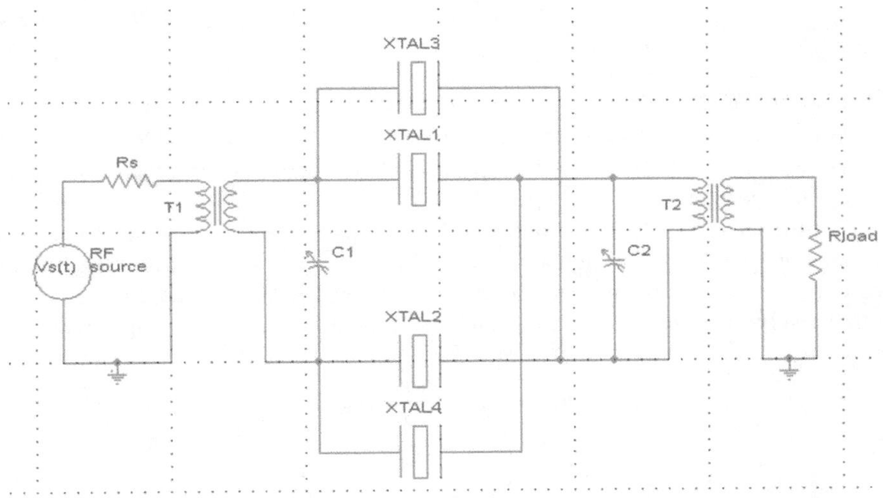

Fig. 7.7 Lattice crystal filter

matched set of XTAL3 and XTAL4. This structure circuit attains the desired bandwidth and selectivity [107, 108].

Terminology: i_{p_1}-input current to transformer T_1's primary coil, i_{s_1}-input current to transformer T_1's secondary coil, i_{p_2}-input current to transformer T_2's primary coil, i_{s_2}-input current to transformer T_2's secondary coil

Transformer T_1: v_{p1}—transformer primary voltage, v_{s1}—transformer secondary voltage. ϕ_1 is the flux through a one turn coil located anywhere on the transformer core. n_{p1} and n_{s1} are the number of turns of the coil 1 (primary) and 2 (secondary), respectively, then the total flux ϕ_{p1} and ϕ_{s1} through coils 1 and 2 respectively are $\phi_{p_1} = n_{p_1} \cdot \phi_1; \phi_{s_1} = n_{s_1} \cdot \phi; v_{p_1} = \frac{d\phi_{p_1}}{dt}; v_{s_1} = \frac{d\phi_{s_1}}{dt}; v_{p_1} = n_{p_1} \cdot \frac{d\phi_1}{dt}$
$v_{s_1} = n_{s_1} \cdot \frac{d\phi_1}{dt}; v_{p_1}(t) = v_{p_1}; v_{s_1}(t) = v_{s_1}; \frac{v_{p_1}(t)}{v_{s_1}(t)} = \frac{n_{p_1}}{n_{s_1}}; a_1 = \frac{n_{p_1}}{n_{s_1}}. \frac{v_{p_1}(t)}{v_{s_1}(t)} = \frac{n_{p_1}}{n_{s_1}}$ for all times t and for all voltages v_{p_1} and v_{s_1}. We define the ratio between n_{p_1} and n_{s_1} as "a_1" ($a_1 = \frac{n_{p_1}}{n_{s_1}}$). Step down transformer $a_1 > 1$ and step up transformer $a_1 < 1$. By law of conservation of energy, apparent real and reactive powers are each conserved in the input and output ($S_1 = I_{p_1} \cdot v_{p_1} = I_{s_1} \cdot v_{s_1}$). Ideal transformer identity: $\frac{v_{p_1}}{v_{s_1}} = \frac{I_{s_1}}{I_{p_1}} = \frac{n_{p_1}}{n_{s_1}} = \sqrt{\frac{L_{p_1}}{L_{s_1}}} = a_1$.

$$n_{p_1} = n_1; n_{s_1} = n_2; \frac{v_{p_1}}{v_{s_1}} = \frac{n_1}{n_2}; n_1, n_2 > 1 \Rightarrow v_{s_1} = v_{p_1} \cdot \frac{n_2}{n_1} = V_{A_1} \cdot \frac{n_2}{n_1}$$
$$= [V_s(t) - R_s \cdot I_{R_s}] \cdot \frac{n_2}{n_1}$$

$v_{s_1} = V_{A_1} \cdot \frac{n_2}{n_1}; v_{p_1} = V_s(t) - R_s \cdot I_{R_s}; v_{s_1} = [V_s(t) - R_s \cdot I_{R_s}] \cdot \frac{n_2}{n_1}$ (Reference directions are in the top of each transformer's coil). $n_{p_1} \cdot i_{p_1} + n_{s_1} \cdot i_{s_1} = 0; i_{p_1} = i_{p_1}(t)$; $i_{s_1} = i_{s_1}(t); \frac{i_{p_1}(t)}{i_{s_1}(t)} = -\frac{n_{s_1}}{n_{p_1}}$

For all t and all currents i_{p_1} and i_{s_1}. The voltage v_{p_1} across coil 1 does not depend on i_{p_1} or on i_{s_1}; it depends only on v_{s_1}. Similarly the current i_{p_1} depends only on i_{s_1} and is independent of v_{p_1}, v_{s_1}. $v_{s_1} = V_{A_2} - V_{A_9}$

Transformer T_2: v_{p2}—transformer primary voltage, v_{s2}—transformer secondary voltage. ϕ_2 is the flux through a one turn coil located anywhere on the transformer core. n_{p2} and n_{s2} are the number of turns of the coil 1 (primary) and 2 (secondary), respectively, then the total flux ϕ_{p2} and ϕ_{s2} through coils 1 and 2 respectively are

$$\phi_{p_2} = n_{p_2} \cdot \phi_2; \phi_{s_2} = n_{s_2} \cdot \phi; v_{p_2} = \frac{d\phi_{p_2}}{dt}; v_{s_2} = \frac{d\phi_{s_2}}{dt}; v_{p_2} = n_{p_2} \cdot \frac{d\phi_2}{dt}$$

$$v_{s_2} = n_{s_2} \cdot \frac{d\phi_2}{dt}; v_{p_2}(t) = v_{p_2}; v_{s_2}(t) = v_{s_2}; \frac{v_{p_2}(t)}{v_{s_2}(t)} = \frac{n_{p_2}}{n_{s_2}}; a_2 = \frac{n_{p_2}}{n_{s_2}}.$$

$\frac{v_{p_2}(t)}{v_{s_2}(t)} = \frac{n_{p_2}}{n_{s_2}}$ for all times t and for all voltages v_{p_2} and v_{s_2}. We define the ratio between n_{p_2} and n_{s_2} as "a_2" ($a_2 = \frac{n_{p_2}}{n_{s_2}}$). Step down transformer $a_2 > 1$ and step up transformer $a_2 < 1$. By law of conservation of energy, apparent real and reactive powers are each conserved in the input and output ($S_2 = I_{p_2} \cdot v_{p_2} = I_{s_2} \cdot v_{s_2}$). Ideal transformer identity: $\frac{v_{p_2}}{v_{s_2}} = \frac{I_{s_2}}{I_{p_2}} = \frac{n_{p_2}}{n_{s_2}} = \sqrt{\frac{L_{p_2}}{L_{s_2}}} = a_2$.

$$v_{s_2} = V_{A_{14}} = I_{R_{load}} \cdot R_{load}$$

$$n_{p_2} = n_3; n_{s_2} = n_4; \frac{v_{p_2}}{v_{s_2}} = \frac{n_3}{n_4}; n_3, n_4 > 1 \Rightarrow v_{s_2} = v_{p_2} \cdot \frac{n_4}{n_3} = [V_{A_6} - V_{A_3}] \cdot \frac{n_4}{n_3}$$

$v_{s_2} = [V_{A_6} - V_{A_3}] \cdot \frac{n_4}{n_3}$; $v_{p_2} = V_{A_6} - V_{A_3}; v_{s_2} = v_{p_2} \cdot \frac{n_4}{n_3}$ (Reference directions are in the top of each transformer's coil). $n_{p_2} \cdot i_{p_2} + n_{s_2} \cdot i_{s_2} = 0; i_{p_2} = i_{p_2}(t); i_{s_2} = i_{s_2}(t)$; $\frac{i_{p_2}(t)}{i_{s_2}(t)} = -\frac{n_{s_2}}{n_{p_2}}$.

For all t and all currents i_{p_2} and i_{s_2}. The voltage v_{p_2} across coil 1 does not depend on i_{p_2} or on i_{s_2}; it depends only on v_{s_2}. Similarly the current i_{p_2} depends only on i_{s_2} and is independent of v_{p_2}, v_{s_2}. We consider for two transformers $a_1 \neq a_2$ [24].

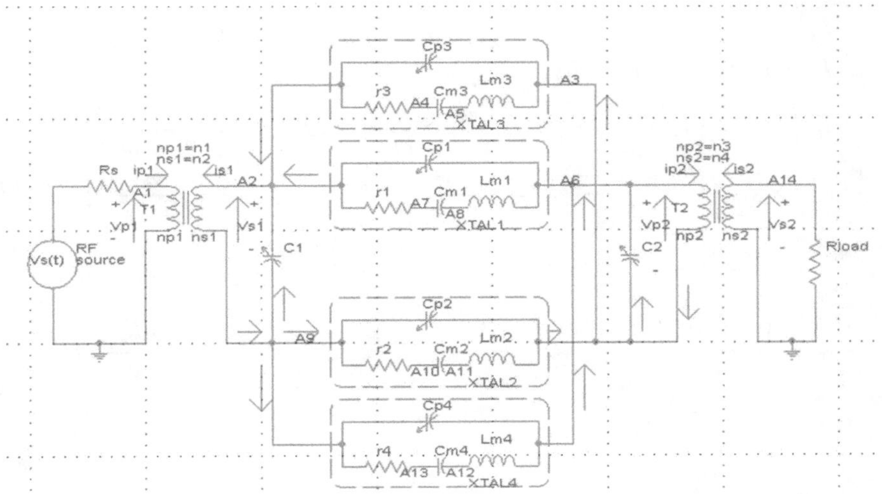

Fig. 7.8 Lattice crystal filter equivalent circuit

Remark When we have two coils of wire in close physical proximity to one another, no importance whether or not the coils are wrapped around a common core of magnetic material. We assume that the coils do not move with respect to one another or with respect to a core they might be wrapped around. If we have some ferromagnetic material in the magnetic circuit of the two coils, then when the current sufficiently large, the relation between the fluxes ϕ_p, ϕ_s and the currents i_p, i_s are no longer linear. In this case the equations have the following form: $\phi_p = f_p(i_p, i_s)$; $\phi_s = f_s(i_p, i_s)$, where $f_p(i_p, i_s)$,$f_s(i_p, i_s)$ are nonlinear functions of the currents i_p, i_s, By Faraday's law we get the following:

$$v_p = \frac{d\phi_p}{dt} = \frac{\partial f_p}{\partial i_p} \cdot \frac{di_p}{dt} + \frac{\partial f_p}{\partial i_s} \cdot \frac{di_s}{dt}; f_p = f_p(i_p, i_s); f_s = f_s(i_p, i_s)$$

$$v_s = \frac{d\phi_s}{dt} = \frac{\partial f_s}{\partial i_p} \cdot \frac{di_p}{dt} + \frac{\partial f_s}{\partial i_s} \cdot \frac{di_s}{dt}; f_p = f_p(i_p, i_s); f_s = f_s(i_p, i_s)$$

The four partial derivatives $(\frac{\partial f_p}{\partial i_p}, \frac{\partial f_p}{\partial i_s}, \frac{\partial f_s}{\partial i_p}, \frac{\partial f_s}{\partial i_s})$ are function of i_p, i_s. It is applicable for the first transformer (T_1) and the second transformer (T_2) in our Lattice crystal filter circuit.

$$I_{r_1} = I_{C_{m_1}} = I_{L_{m_1}};\ I_{r_2} = I_{C_{m_2}} = I_{L_{m_2}};\ I_{r_3} = I_{C_{m_3}} = I_{L_{m_3}};$$
$$I_{r_4} = I_{C_{m_4}} = I_{L_{m_4}};\ I_{r_k} = I_{C_{m_k}} = I_{L_{m_k}} \quad \forall k = 1, 2, 3, 4$$
$$I_{R_{load}} = -i_{s_2};\ I_{R_s} = i_{p_1};\ I_{C_1} = C_1 \cdot \frac{d(V_{A_9} - V_{A_2})}{dt};\ I_{C_2} = C_2 \cdot \frac{d(V_{A_3} - V_{A_6})}{dt};$$
$$I_{C_{p_3}} = C_{p_3} \cdot \frac{d(V_{A_3} - V_{A_2})}{dt} I_{r_3} = \frac{V_{A_4} - V_{A_2}}{r_3};\ I_{C_{m_3}} = C_{m_3} \cdot \frac{d(V_{A_5} - V_{A_4})}{dt};$$
$$V_{A_3} - V_{A_5} = L_{m_3} \cdot \frac{dI_{L_{m_3}}}{dt};\ I_{C_{p_1}} = C_{p_1} \cdot \frac{d(V_{A_6} - V_{A_2})}{dt}$$

$$I_{r_1} = \frac{V_{A_7} - V_{A_2}}{r_1};\ I_{C_{m_1}} = C_{m_1} \cdot \frac{d(V_{A_8} - V_{A_7})}{dt};\ V_{A_6} - V_{A_8} = L_{m_1} \cdot \frac{dI_{L_{m_1}}}{dt};\ I_{C_{p_2}} = C_{p_2} \cdot \frac{d(V_{A_9} - V_{A_3})}{dt}$$

$$I_{r_2} = \frac{V_{A_9} - V_{A_{10}}}{r_2};\ I_{C_{m_2}} = C_{m_2} \cdot \frac{d(V_{A_{10}} - V_{A_{11}})}{dt};\ V_{A_{11}} - V_{A_3} = L_{m_2} \cdot \frac{dI_{L_{m_2}}}{dt};\ I_{C_{p_4}} = C_{p_4} \cdot \frac{d(V_{A_9} - V_{A_6})}{dt}$$

$$I_{r_4} = \frac{V_{A_9} - V_{A_{13}}}{r_4};\ I_{C_{m4}} = C_{m_4} \cdot \frac{d(V_{A_{13}} - V_{A_{12}})}{dt};\ V_{A_{12}} - V_{A_6} = L_{m_4} \cdot \frac{dI_{L_{m4}}}{dt}$$

$$v_{s2} = V_{A_{14}} = I_{R_{load}} \cdot R_{load};\ v_{s_1} = V_{A_2} - V_{A_9};\ v_{p2} = V_{A_6} - V_{A_3};\ v_{p_1} = V_s(t) - I_{R_s} \cdot R_s$$

KCL @ A_2: $I_{C_{p_3}} + I_{r_3} + I_{C_{p_1}} + I_{r_1} + I_{C_1} = i_{s_1}$; KCL @ A_9: $i_{s_1} = I_{C_1} + I_{C_{p_2}} + I_{r_2} + I_{C_{p_4}} + I_{r_4}$

KCL @ A_3: $I_{C_{p_3}} + I_{L_{m_3}} + I_{C_2} = i_{p_2} + I_{C_{p_2}} + I_{L_{m_2}}$ KCL @ A_6: $I_{C_{p_4}} + I_{L_{m_4}} + I_{C_2} = i_{p_2} + I_{C_{p_1}} + I_{L_{m_1}}$

$$I_{C_2} = C_2 \cdot \frac{d(V_{A_3} - V_{A_6})}{dt} \Rightarrow V_{A_3} - V_{A_6} = \frac{1}{C_2} \cdot \int I_{C_2} \cdot dt;\ I_{C_1} = C_1 \cdot \frac{d(V_{A_9} - V_{A_2})}{dt} \Rightarrow V_{A_9} - V_{A_2} = \frac{1}{C_1} \cdot \int I_{C_1} \cdot dt$$

$$I_{C_{p_3}} = C_{p_3} \cdot \frac{d(V_{A_3} - V_{A_2})}{dt} \Rightarrow V_{A_3} - V_{A_2} = \frac{1}{C_{p_3}} \cdot \int I_{C_{p_3}} \cdot dt;\ I_{r_3} = \frac{V_{A_4} - V_{A_2}}{r_3} \Rightarrow V_{A_4} - V_{A_2} = I_{r_3} \cdot r_3$$

$$I_{C_{m_3}} = C_{m_3} \cdot \frac{d(V_{A_5} - V_{A_4})}{dt} \Rightarrow V_{A_5} - V_{A_4} = \frac{1}{C_{m_3}} \cdot \int I_{C_{m_3}} \cdot dt;\ V_{A_3} - V_{A_5} = L_{m_3} \cdot \frac{dI_{L_{m_3}}}{dt}$$

$$I_{C_{p_1}} = C_{p_1} \cdot \frac{d(V_{A_6} - V_{A_2})}{dt} \Rightarrow V_{A_6} - V_{A_2} = \frac{1}{C_{p_1}} \cdot \int I_{C_{p_1}} \cdot dt;\ I_{r_1} = \frac{V_{A_7} - V_{A_2}}{r_1}$$
$$\Rightarrow V_{A_7} - V_{A_2} = I_{r_1} \cdot r_1$$

$$I_{C_{m_1}} = C_{m_1} \cdot \frac{d(V_{A_8} - V_{A_7})}{dt} \Rightarrow V_{A_8} - V_{A_7} = \frac{1}{C_{m_1}} \cdot \int I_{C_{m_1}} \cdot dt;\ V_{A_6} - V_{A_8}$$
$$= L_{m_1} \cdot \frac{dI_{L_{m_1}}}{dt}$$

$$I_{C_{p_2}} = C_{p_2} \cdot \frac{d(V_{A_9} - V_{A_3})}{dt} \Rightarrow V_{A_9} - V_{A_3} = \frac{1}{C_{p_2}} \cdot \int I_{C_{p_2}} \cdot dt;$$
$$I_{r_2} = \frac{V_{A_9} - V_{A_{10}}}{r_2} \Rightarrow V_{A_9} - V_{A_{10}} = I_{r_2} \cdot r_2$$
$$I_{C_{m_2}} = C_{m_2} \cdot \frac{d(V_{A_{10}} - V_{A_{11}})}{dt} \Rightarrow V_{A_{10}} - V_{A_{11}} = \frac{1}{C_{m_2}} \cdot \int I_{C_{m_2}} \cdot dt;$$
$$V_{A_{11}} - V_{A_3} = L_{m_2} \cdot \frac{dI_{L_{m_2}}}{dt}$$

$$I_{C_{p_4}} = C_{p_4} \cdot \frac{d(V_{A_9} - V_{A_6})}{dt} \Rightarrow V_{A_9} - V_{A_6} = \frac{1}{C_{p_4}} \cdot \int I_{C_{p_4}} \cdot dt;\ I_{r_4} = \frac{V_{A_9} - V_{A_{13}}}{r_4} \Rightarrow V_{A_9} - V_{A_{13}} = I_{r_4} \cdot r_4$$
$$I_{C_{m_4}} = C_{m_4} \cdot \frac{d(V_{A_{13}} - V_{A_{12}})}{dt} \Rightarrow V_{A_{13}} - V_{A_{12}} = \frac{1}{C_{m_4}} \cdot \int I_{C_{m_4}} \cdot dt;\ V_{A_{12}} - V_{A_6} = L_{m_4} \cdot \frac{dI_{L_{m_4}}}{dt}$$

Lattice crystal filter differential equations group No. 1:

$$V_{A_9} - V_{A_2} = \frac{1}{C_1} \cdot \int I_{C_1} \cdot dt;\ V_{A_3} - V_{A_6} = \frac{1}{C_2} \cdot \int I_{C_2} \cdot dt;\ V_{A_3} - V_{A_2}$$
$$= \frac{1}{C_{p_3}} \cdot \int I_{C_{p3}} \cdot dt$$

$$V_{A_5} - V_{A_4} = \frac{1}{C_{m_3}} \cdot \int I_{C_{m3}} \cdot dt;\ V_{A_6} - V_{A_2} = \frac{1}{C_{p_1}} \cdot \int I_{C_{p_1}} \cdot dt;\ V_{A_8} - V_{A_7}$$
$$= \frac{1}{C_{m_1}} \cdot \int I_{C_{m1}} \cdot dt$$

$$V_{A_9} - V_{A_3} = \frac{1}{C_{p_2}} \cdot \int I_{C_{p2}} \cdot dt;\ V_{A_{10}} - V_{A_{11}} = \frac{1}{C_{m_2}} \cdot \int I_{C_{m2}} \cdot dt;\ V_{A_9} - V_{A_6}$$
$$= \frac{1}{C_{p_4}} \cdot \int I_{C_{p4}} \cdot dt$$

$$V_{A_{13}} - V_{A_{12}} = \frac{1}{C_{m_4}} \cdot \int I_{C_{m4}} \cdot dt;\ V_{A_4} - V_{A_2} = I_{r_3} \cdot r_3;\ V_{A_3} - V_{A_5}$$
$$= L_{m_3} \cdot \frac{dI_{L_{m3}}}{dt};\ V_{A_7} - V_{A_2} = I_{r_1} \cdot r_1$$

$$V_{A_6} - V_{A_8} = L_{m_1} \cdot \frac{dI_{L_{m1}}}{dt}; V_{A_9} - V_{A_{10}} = I_{r_2} \cdot r2; V_{A_{11}} - V_{A_3} = L_{m_2} \cdot \frac{dI_{L_{m2}}}{dt}; V_{A_9} - V_{A_{13}} = I_{r_4} \cdot r4$$

$$V_{A_{12}} - V_{A_6} = L_{m_4} \cdot \frac{dI_{L_{m4}}}{dt}$$

&&&

$$V_{A_9} - V_{A_2} = \frac{1}{C_1} \cdot \int I_{C_1} \cdot dt \Rightarrow V_{A_2} = V_{A_9} - \frac{1}{C_1} \cdot \int I_{C_1} \cdot dt$$

$$V_{A_3} - V_{A_2} = \frac{1}{C_{p_3}} \cdot \int I_{C_{p3}} \cdot dt \Rightarrow V_{A_3} = V_{A_2} + \frac{1}{C_{p_3}} \cdot \int I_{C_{p3}} \cdot dt$$

$$V_{A_3} = V_{A_9} - \frac{1}{C_1} \cdot \int I_{C_1} \cdot dt + \frac{1}{C_{p3}} \cdot \int I_{C_{p3}} \cdot dt; \; V_{A_6} - V_{A_2} = \frac{1}{C_{p_1}} \cdot \int I_{C_{p1}} \cdot dt$$

$$V_{A_6} = V_{A_2} + \frac{1}{C_{p1}} \cdot \int I_{C_{p1}} \cdot dt = V_{A_9} - \frac{1}{C_1} \cdot \int I_{C_1} \cdot dt + \frac{1}{C_{p1}} \cdot \int I_{C_{p1}} \cdot dt$$

$$V_{A_3} - V_{A_6} = \frac{1}{C_2} \cdot \int I_{C_2} \cdot dt \Rightarrow \frac{1}{C_{p3}} \cdot \int I_{C_{p3}} \cdot dt - \frac{1}{C_{p_1}} \cdot \int I_{C_{p1}} \cdot dt = \frac{1}{C_2} \cdot \int I_{C_2} \cdot dt$$

$$\frac{d}{dt}\{\frac{1}{C_{p_3}} \cdot \int I_{C_{p3}} \cdot dt - \frac{1}{C_{p1}} \cdot \int I_{C_{p1}} \cdot dt = \frac{1}{C_2} \cdot \int I_{C_2} \cdot dt\} \Rightarrow \frac{1}{C_{p_3}} \cdot I_{C_{p3}} - \frac{1}{C_{p1}} \cdot I_{C_{p1}} = \frac{1}{C_2} \cdot I_{C_2}$$

$$V_{A_9} - V_{A_3} = \frac{1}{C_{p_2}} \cdot \int I_{C_{p_2}} \cdot dt \Rightarrow V_{A_9} = V_{A_3} + \frac{1}{C_{p_2}} \cdot \int I_{C_{p_2}} \cdot dt$$

$$V_{A_9} = V_{A_9} - \frac{1}{C_1} \cdot \int I_{C_1} \cdot dt + \frac{1}{C_{p_3}} \cdot \int I_{C_{p_3}} \cdot dt + \frac{1}{C_{p_2}} \cdot \int I_{C_{p_2}} \cdot dt$$

$$-\frac{1}{C_1} \cdot \int I_{C_1} \cdot dt + \frac{1}{C_{p_3}} \cdot \int I_{C_{p_3}} \cdot dt + \frac{1}{C_{p_2}} \cdot \int I_{C_{p_2}} \cdot dt = 0$$

$$\frac{d}{dt}\{-\frac{1}{C_1} \cdot \int I_{C_1} \cdot dt + \frac{1}{C_{p_3}} \cdot \int I_{C_{p_3}} \cdot dt + \frac{1}{C_{p_2}} \cdot \int I_{C_{p_2}} \cdot dt = 0\}; \frac{1}{C_1} \cdot I_{C_1} = \frac{1}{C_{p_3}} \cdot I_{C_{p_3}} + \frac{1}{C_{p_2}} \cdot I_{C_{p_2}}$$

$$V_{A_9} - V_{A_6} = \frac{1}{C_{p4}} \cdot \int I_{C_{p_4}} \cdot dt \Rightarrow \frac{1}{C_1} \cdot \int I_{C_1} \cdot dt - \frac{1}{C_{p_1}} \cdot \int I_{C_{p_1}} \cdot dt = \frac{1}{C_{p4}} \cdot \int I_{C_{p_4}} \cdot dt$$

$$\frac{d}{dt}\{\frac{1}{C_1} \cdot \int I_{C_1} \cdot dt - \frac{1}{C_{p_1}} \cdot \int I_{C_{p_1}} \cdot dt = \frac{1}{C_{p_4}} \cdot \int I_{C_{p_4}} \cdot dt\} \Rightarrow \frac{1}{C_1} \cdot I_{C_1} - \frac{1}{C_{p_1}} \cdot I_{C_{p_1}} = \frac{1}{C_{p_4}} \cdot I_{C_{p_4}}$$

$$V_{A_4} - V_{A_2} = I_{r_3} \cdot r_3 \Rightarrow V_{A_4} = V_{A_2} + I_{r_3} \cdot r_3 = V_{A_9} - \frac{1}{C_1} \cdot \int I_{C_1} \cdot dt + I_{r_3} \cdot r_3$$

$$V_{A_3} - V_{A_5} = L_{m_3} \cdot \frac{dI_{L_{m_3}}}{dt} \Rightarrow V_{A_5} = V_{A_3} - L_{m_3} \cdot \frac{dI_{L_{m_3}}}{dt} = V_{A_9} - \frac{1}{C_1} \cdot \int I_{C_1} \cdot dt + \frac{1}{C_{p_3}} \cdot \int I_{C_{p_3}} \cdot dt - L_{m_3} \cdot \frac{dI_{L_{m_3}}}{dt}$$

$$V_{A_7} - V_{A_2} = I_{r_1} \cdot r_1 \Rightarrow V_{A_7} = V_{A_2} + I_{r_1} \cdot r_1 = V_{A_9} - \frac{1}{C_1} \cdot \int I_{C_1} \cdot dt + I_{r_1} \cdot r_1$$

$$V_{A_6} - V_{A_8} = L_{m_1} \cdot \frac{dI_{L_{m_1}}}{dt} \Rightarrow V_{A_8} = V_{A_6} - L_{m_1} \cdot \frac{dI_{L_{m_1}}}{dt} = V_{A_9} - \frac{1}{C_1} \cdot \int I_{C_1} \cdot dt + \frac{1}{C_{p_1}} \cdot \int I_{C_{p_1}} \cdot dt - L_{m_1} \cdot \frac{dI_{L_{m_1}}}{dt}$$

$$V_{A_9} - V_{A_{10}} = I_{r_2} \cdot r_2 \Rightarrow V_{A_{10}} = V_{A_9} - I_{r_2} \cdot r_2;$$

$$V_{A_{11}} - V_{A_3} = L_{m_2} \cdot \frac{dI_{L_{m_2}}}{dt} \Rightarrow V_{A_{11}} = V_{A_3} + L_{m_2} \cdot \frac{dI_{L_{m_2}}}{dt}$$

$$V_{A_{11}} = V_{A_9} - \frac{1}{C_1} \cdot \int I_{C_1} \cdot dt + \frac{1}{C_{p_3}} \cdot \int I_{C_{p_3}} \cdot dt + L_{m_2} \cdot \frac{dI_{L_{m_2}}}{dt};$$

$$V_{A_9} - V_{A_{13}} = I_{r_4} \cdot r_4 \Rightarrow V_{A_{13}} = V_{A_9} - I_{r_4} \cdot r_4$$

$$V_{A_{12}} - V_{A_6} = L_{m_4} \cdot \frac{dI_{L_{m_4}}}{dt} \Rightarrow V_{A_{12}} = V_{A_6} + L_{m_4} \cdot \frac{dI_{L_{m_4}}}{dt} = V_{A_9} - \frac{1}{C_1} \cdot \int I_{C_1} \cdot dt + \frac{1}{C_{p_1}} \cdot \int I_{C_{p_1}} \cdot dt + L_{m_4} \cdot \frac{dI_{L_{m_4}}}{dt}$$

Remark A

$$V_{A_5} - V_{A_4} = \frac{1}{C_{m_3}} \cdot \int I_{C_{m_3}} \cdot dt; \frac{1}{C_{p_3}} \cdot \int I_{C_{p_3}} \cdot dt - L_{m_3} \cdot \frac{dI_{L_{m_3}}}{dt} - I_{r_3} \cdot r_3 = \frac{1}{C_{m_3}} \cdot \int I_{C_{m_3}} \cdot dt$$

$$V_{A_9} - \frac{1}{C_1} \cdot \int I_{C_1} \cdot dt + \frac{1}{C_{p3}} \cdot \int I_{C_{p3}} \cdot dt - L_{m3} \cdot \frac{dI_{L_{m3}}}{dt}$$
$$- [V_{A_9} - \frac{1}{C_1} \cdot \int I_{C_1} \cdot dt + I_{r_3} \cdot r_3]$$
$$= \frac{1}{C_{m3}} \cdot \int I_{C_{m3}} \cdot dt$$

$$\frac{d}{dt}\{\frac{1}{C_{p3}} \cdot \int I_{C_{p3}} \cdot dt - L_{m3} \cdot \frac{dI_{L_{m3}}}{dt} - I_{r_3} \cdot r_3 = \frac{1}{C_{m3}} \cdot \int I_{C_{m3}} \cdot dt\}$$
$$\Rightarrow \frac{1}{C_{p3}} \cdot I_{C_{p3}} - L_{m3} \cdot \frac{d^2 I_{L_{m3}}}{dt^2} - \frac{dI_{r3}}{dt} \cdot r_3 = \frac{1}{C_{m3}} \cdot I_{C_{m3}}$$

Remark B

$$V_{A_8} - V_{A_7} = \frac{1}{C_{m1}} \cdot \int I_{C_{m1}} \cdot dt; \frac{1}{C_{p1}} \cdot \int I_{C_{p1}} \cdot dt - L_{m1} \cdot \frac{dI_{L_{m1}}}{dt} - I_{r_1} \cdot r_1$$
$$= \frac{1}{C_{m1}} \cdot \int I_{C_{m1}} \cdot dt$$

$$V_{A_9} - \frac{1}{C_1} \cdot \int I_{C_1} \cdot dt + \frac{1}{C_{p1}} \cdot \int I_{C_{p1}} \cdot dt - L_{m1} \cdot \frac{dI_{L_{m1}}}{dt}$$
$$- [V_{A_9} - \frac{1}{C_1} \cdot \int I_{C_1} \cdot dt + I_{r_1} \cdot r_1]$$
$$= \frac{1}{C_{m1}} \cdot \int I_{C_{m1}} \cdot dt$$

$$\frac{d}{dt}\{\frac{1}{C_{p1}} \cdot \int I_{C_{p1}} \cdot dt - L_{m1} \cdot \frac{dI_{L_{m1}}}{dt} - I_{r_1} \cdot r_1 = \frac{1}{C_{m1}} \cdot \int I_{C_{m1}} \cdot dt\}$$
$$\Rightarrow \frac{1}{C_{p1}} \cdot I_{C_{p1}} - L_{m1} \cdot \frac{d^2 I_{L_{m1}}}{dt^2} - \frac{dI_{r_1}}{dt} \cdot r_1 = \frac{1}{C_{m1}} \cdot I_{C_{m1}}$$

Remark C

$$V_{A_{10}} - V_{A_{11}} = \frac{1}{C_{m2}} \cdot \int I_{C_{m2}} \cdot dt; -\frac{dI_{r_2}}{dt} \cdot r_2 + \frac{1}{C_1} \cdot I_{C_1} - \frac{1}{C_{p3}} \cdot I_{C_{p3}} - L_{m2} \cdot \frac{d^2 I_{L_{m2}}}{dt^2}$$
$$= \frac{1}{C_{m2}} \cdot I_{C_{m2}}$$

$$V_{A_9} - I_{r_2} \cdot r_2 - [V_{A_9} - \frac{1}{C_1} \cdot \int I_{C_1} \cdot dt + \frac{1}{C_{p3}} \cdot \int I_{C_{p3}} \cdot dt + L_{m2} \cdot \frac{dI_{L_{m2}}}{dt}]$$
$$= \frac{1}{C_{m2}} \cdot \int I_{C_{m2}} \cdot dt$$

$$-I_{r_2}\cdot r_2+\frac{1}{C_1}\cdot\int I_{C_1}\cdot dt-\frac{1}{C_{p3}}\cdot\int I_{C_{p3}}\cdot dt-L_{m2}\cdot\frac{dI_{L_{m_2}}}{dt}=\frac{1}{C_{m_2}}\cdot\int I_{C_{m_2}}\cdot dt$$

$$\frac{d}{dt}\{-I_{r_2}\cdot r_2+\frac{1}{C_1}\cdot\int I_{C_1}\cdot dt-\frac{1}{C_{p3}}\cdot\int I_{C_{p3}}\cdot dt-L_{m2}\cdot\frac{dI_{L_{m_2}}}{dt}=\frac{1}{C_{m_2}}\cdot\int I_{C_{m_2}}\cdot dt\}$$

$$-\frac{dI_{r_2}}{dt}\cdot r_2+\frac{1}{C_1}\cdot I_{C_1}-\frac{1}{C_{p3}}\cdot I_{C_{p3}}-L_{m2}\cdot\frac{d^2I_{L_{m_2}}}{dt^2}=\frac{1}{C_{m_2}}\cdot I_{C_{m_2}}$$

Remark D

$$V_{A_{13}}-V_{A_{12}}=\frac{1}{C_{m4}}\cdot\int I_{C_{m_4}}\cdot dt;-\frac{dI_{r_4}}{dt}\cdot r_4+\frac{1}{C_1}\cdot I_{C_1}-\frac{1}{C_{p_1}}\cdot I_{C_{p_1}}-L_{m4}\cdot\frac{d^2I_{L_{m_4}}}{dt^2}$$
$$=\frac{1}{C_{m4}}\cdot I_{C_{m_4}}$$

$$V_{A_9}-I_{r_4}\cdot r_4-[V_{A_9}-\frac{1}{C_1}\cdot\int I_{C_1}\cdot dt+\frac{1}{C_{p_1}}\cdot\int I_{C_{p_1}}\cdot dt+L_{m4}\cdot\frac{dI_{L_{m_4}}}{dt}]$$
$$=\frac{1}{C_{m4}}\cdot\int I_{C_{m_4}}\cdot dt$$

$$-I_{r_4}\cdot r_4+\frac{1}{C_1}\cdot\int I_{C_1}\cdot dt-\frac{1}{C_{p_1}}\cdot\int I_{C_{p_1}}\cdot dt-L_{m4}\cdot\frac{dI_{L_{m_4}}}{dt}=\frac{1}{C_{m4}}\cdot\int I_{C_{m_4}}\cdot dt$$

$$\frac{d}{dt}\{-I_{r_4}\cdot r_4+\frac{1}{C_1}\cdot\int I_{C_1}\cdot dt-\frac{1}{C_{p_1}}\cdot\int I_{C_{p_1}}\cdot dt-L_{m4}\cdot\frac{dI_{L_{m_4}}}{dt}=\frac{1}{C_{m4}}\cdot\int I_{C_{m_4}}\cdot dt\}$$

$$-\frac{dI_{r_4}}{dt}\cdot r_4+\frac{1}{C_1}\cdot I_{C_1}-\frac{1}{C_{p_1}}\cdot I_{C_{p_1}}-L_{m4}\cdot\frac{d^2I_{L_{m_4}}}{dt^2}=\frac{1}{C_{m4}}\cdot I_{C_{m_4}}$$

Lattice crystal filter differential equations group No. 2:

$$\frac{1}{C_{p3}}\cdot I_{C_{p3}}-\frac{1}{C_{p1}}\cdot I_{C_{p1}}=\frac{1}{C_2}\cdot I_{C_2};\frac{1}{C_1}\cdot I_{C_1}$$
$$=\frac{1}{C_{p3}}\cdot I_{C_{p3}}+\frac{1}{C_{p2}}\cdot I_{C_{p2}};\frac{1}{C_1}\cdot I_{C_1}-\frac{1}{C_{p1}}\cdot I_{C_{p1}}=\frac{1}{C_{p4}}\cdot I_{C_{p4}}$$

$$\frac{1}{C_{p3}}\cdot I_{C_{p3}}-L_{m3}\cdot\frac{d^2I_{L_{m3}}}{dt^2}-\frac{dI_{r_3}}{dt}\cdot r_3=\frac{1}{C_{m3}}\cdot I_{C_{m3}};\frac{1}{C_{p_1}}\cdot I_{C_{p_1}}-L_{m_1}\cdot\frac{d^2I_{L_{m_1}}}{dt^2}-\frac{dI_{r_1}}{dt}\cdot r_1$$
$$=\frac{1}{C_{m_1}}\cdot I_{C_{m_1}}$$

$$-\frac{dI_{r_2}}{dt}\cdot r_2+\frac{1}{C_1}\cdot I_{C_1}-\frac{1}{C_{p_3}}\cdot I_{C_{p_3}}-L_{m_2}\cdot\frac{d^2I_{L_{m_2}}}{dt^2}=\frac{1}{C_{m_2}}\cdot I_{C_{m_2}}$$

$$-\frac{dI_{r_4}}{dt}\cdot r_4+\frac{1}{C_1}\cdot I_{C_1}-\frac{1}{C_{p_1}}\cdot I_{C_{p_1}}-L_{m_4}\cdot\frac{d^2I_{L_{m_4}}}{dt^2}=\frac{1}{C_{m_4}}\cdot I_{C_{m_4}}$$

Lattice crystal filters Variables transformation:

$$I_{r_1}=I_{C_{m_1}}=I_{L_{m_1}};\ I_{r_2}=I_{C_{m_2}}=I_{L_{m_2}};\ I_{r_3}=I_{C_{m_3}}=I_{L_{m_3}};$$
$$I_{r_4}=I_{C_{m_4}}=I_{L_{m_4}};I_{r_k}=I_{C_{m_k}}=I_{L_{m_k}}\quad\forall k=1,2,3,4$$
$$I_{r_1}\to I_{L_{m_1}};\ I_{C_{m_1}}\to I_{L_{m_1}};\ I_{r_2}\to I_{L_{m_2}};\ I_{C_{m_2}}\to I_{L_{m_2}};\ I_{r_3}\to I_{L_{m_3}};$$
$$I_{C_{m_3}}\to I_{L_{m_3}};\ I_{r_4}\to I_{L_{m_4}};\ I_{C_{m_4}}\to I_{L_{m_4}}$$

Lattice crystal filter differential equations group No. 3:

$$\frac{1}{C_{p_3}}\cdot I_{C_{p_3}}-\frac{1}{C_{p_1}}\cdot I_{C_{p_1}}=\frac{1}{C_2}\cdot I_{C_2};\ \frac{1}{C_1}\cdot I_{C_1}=\frac{1}{C_{p_3}}\cdot I_{C_{p_3}}+\frac{1}{C_{p_2}}\cdot I_{C_{p_2}};\ \frac{1}{C_1}\cdot I_{C_1}-\frac{1}{C_{p_1}}\cdot I_{C_{p_1}}=\frac{1}{C_{p_4}}\cdot I_{C_{p_4}}$$

$$\frac{1}{C_{p_3}}\cdot I_{C_{p_3}}-L_{m_3}\cdot\frac{d^2I_{L_{m_3}}}{dt^2}-\frac{dI_{L_{m_3}}}{dt}\cdot r_3=\frac{1}{C_{m_3}}\cdot I_{L_{m_3}};\ \frac{1}{C_{p_1}}\cdot I_{C_{p_1}}-L_{m_1}\cdot\frac{d^2I_{L_{m_1}}}{dt^2}-\frac{dI_{L_{m_1}}}{dt}\cdot r_1=\frac{1}{C_{m_1}}\cdot I_{L_{m_1}}$$

$$-\frac{dI_{L_{m_2}}}{dt}\cdot r_2+\frac{1}{C_1}\cdot I_{C_1}-\frac{1}{C_{p_3}}\cdot I_{C_{p_3}}-L_{m_2}\cdot\frac{d^2I_{L_{m_2}}}{dt^2}=\frac{1}{C_{m_2}}\cdot I_{L_{m_2}}$$

$$-\frac{dI_{L_{m_4}}}{dt}\cdot r_4+\frac{1}{C_1}\cdot I_{C_1}-\frac{1}{C_{p_1}}\cdot I_{C_{p_1}}-L_{m_4}\cdot\frac{d^2I_{L_{m_4}}}{dt^2}=\frac{1}{C_{m_4}}\cdot I_{L_{m_4}}$$

KCL @ A_2: $I_{C_{p_3}}+I_{L_{m_3}}+I_{C_{p_1}}+I_{L_{m_1}}+I_{C_1}=i_{s_1}$; KCL @ A_9: $i_{s_1}=I_{C_1}+I_{C_{p_2}}+I_{L_{m_2}}+I_{C_{p_4}}+I_{L_{m_4}}$

KCL @ A_3: $I_{C_{p_3}}+I_{L_{m_3}}+I_{C_2}=i_{p_2}+I_{C_{p_2}}+I_{L_{m_2}}$ KCL @ A_6: $I_{C_{p_4}}+I_{L_{m_4}}+I_{C_2}=i_{p_2}+I_{C_{p_1}}+I_{L_{m_1}}$

Lattice crystal filter circuit transformer T_1's secondary coil current differential equation:

$$\frac{i_{p_1}(t)}{i_{s_1}(t)} = -\frac{n_{s_1}}{n_{p_1}} = -\frac{n_2}{n_1} \Rightarrow i_{s_1}(t) = -i_{p_1}(t) \cdot \frac{n_1}{n_2};\ i_{s_1} = i_{s_1}(t);\ i_{p_1} = i_{p_1}(t);\ i_{s_1} = -i_{p_1} \cdot \frac{n_1}{n_2}$$

$$v_{s_1} = -V_{C_1} = -(V_{A_9} - V_{A_2}) \Rightarrow V_{A_9} - V_{A_2} = -v_{s_1};\ I_{C_1} = C_1 \cdot \frac{d}{dt}(V_{A_9} - V_{A_2});\ I_{C_1} = -C_1 \cdot \frac{dv_{s_1}}{dt}$$

$$i_{p_1} = I_{R_s};\ v_{s_1} = [V_s(t) - R_s \cdot i_{p_1}] \cdot \frac{n_2}{n_1};\ i_{p_1} = -i_{s_1} \cdot \frac{n_2}{n_1} \Rightarrow \frac{di_{p_1}}{dt} = -\frac{di_{s_1}}{dt} \cdot \frac{n_2}{n_1}$$

$$I_{C_1} = -C_1 \cdot \frac{d[V_s(t) - R_s \cdot i_{p_1}] \cdot \frac{n_2}{n_1}}{dt} = -C_1 \cdot \frac{n_2}{n_1} \cdot [\frac{dV_s(t)}{dt} - R_s \cdot \frac{di_{p_1}}{dt}];$$

$$I_{C_1} = -C_1 \cdot \frac{n_2}{n_1} \cdot [\frac{dV_s(t)}{dt} - R_s \cdot (-\frac{di_{s_1}}{dt} \cdot \frac{n_2}{n_1})]$$

$$I_{C_1} = -C_1 \cdot \frac{n_2}{n_1} \cdot [\frac{dV_s(t)}{dt} + R_s \cdot \frac{n_2}{n_1} \cdot \frac{di_{s_1}}{dt}];$$

$$R_s \cdot \frac{n_2}{n_1} \cdot \frac{di_{s_1}}{dt} = -I_{C_1} \cdot \frac{1}{C_1} \cdot \frac{n_1}{n_2} - \frac{dV_s(t)}{dt}$$

$$\frac{di_{s_1}}{dt} = -I_{C_1} \cdot \frac{1}{C_1 \cdot R_s} \cdot \frac{n_1^2}{n_2^2} - \frac{n_1}{n_2} \cdot \frac{1}{R_s} \cdot \frac{dV_s(t)}{dt}$$

Lattice crystal filter circuit transformer T_2's primary coil current differential equation:

$$I_{R_{load}} = -i_{s_2};\ v_{p_2} = -V_{C_2};\ \frac{i_{p_2}(t)}{i_{s_2}(t)} = -\frac{n_{s_2}}{n_{p_2}} = -\frac{n_4}{n_3};\ i_{p_2} = i_{p_2}(t);\ i_{s_2} = i_{s_2}(t);\ \frac{i_{p_2}}{i_{s_2}} = -\frac{n_4}{n_3}$$

$$v_{s_2} = v_{p_2} \cdot \frac{n_4}{n_3} = [V_{A_6} - V_{A_3}] \cdot \frac{n_4}{n_3};\ V_{A_6} - V_{A_3} = v_{s_2} \cdot \frac{n_3}{n_4};\ V_{A_3} - V_{A_6} = -v_{s_2} \cdot \frac{n_3}{n_4};\ v_{s_2} = -i_{s_2} \cdot R_{load}$$

$$I_{C_2} = C_2 \cdot \frac{d}{dt}[V_{A_3} - V_{A_6}];\ v_{s_2} = V_{R_{load}};\ I_{C_2} = C_2 \cdot \frac{d}{dt}[-v_{s_2} \cdot \frac{n_3}{n_4}];\ I_{C_2} = -C_2 \cdot \frac{n_3}{n_4} \cdot \frac{dv_{s_2}}{dt}$$

$$\frac{i_{p_2}}{i_{s_2}} = -\frac{n_4}{n_3} \Rightarrow i_{s_2} = -i_{p_2} \cdot \frac{n_3}{n_4};\ \frac{di_{s_2}}{dt} = -\frac{di_{p_2}}{dt} \cdot \frac{n_3}{n_4};\ I_{C_2} = C_2 \cdot \frac{n_3}{n_4} \cdot R_{load} \cdot \frac{di_{s_2}}{dt}$$

$$\frac{di_{s2}}{dt} = \frac{n_4}{n_3 \cdot C_2 \cdot R_{load}} \cdot I_{C_2}; \; -\frac{di_{p2}}{dt} \cdot \frac{n_3}{n_4} = \frac{n_4}{n_3 \cdot C_2 \cdot R_{load}} \cdot I_{C_2}; \; \frac{di_{p2}}{dt} = -\frac{n_4^2}{n_3^2} \cdot \frac{1}{C_2 \cdot R_{load}} \cdot I_{C_2}$$

&&&

$$\frac{1}{C_{p3}} \cdot I_{C_{p3}} - \frac{1}{C_{p1}} \cdot I_{C_{p1}} = \frac{1}{C_2} \cdot I_{C_2}; \; \frac{1}{C_1} \cdot I_{C_1} = \frac{1}{C_{p3}} \cdot I_{C_{p3}} + \frac{1}{C_{p2}} \cdot I_{C_{p2}}; \; \frac{1}{C_1} \cdot I_{C_1} - \frac{1}{C_{p1}} \cdot I_{C_{p1}} = \frac{1}{C_{p4}} \cdot I_{C_{p4}}$$

KCL @ A_2: $I_{C_{p3}} + I_{L_{m3}} + I_{C_{p1}} + I_{L_{m1}} + I_{C_1} = i_{s_1}$; KCL @ A_9: $i_{s_1} = I_{C_1} + I_{C_{p2}} + I_{L_{m2}} + I_{C_{p4}} + I_{L_{m4}}$

KCL @ A_3: $I_{C_{p3}} + I_{L_{m3}} + I_{C_2} = i_{p2} + I_{C_{p2}} + I_{L_{m2}}$ KCL @ A_6: $I_{C_{p4}} + I_{L_{m4}} + I_{C_2} = i_{p2} + I_{C_{p1}} + I_{L_{m1}}$

Find circuit variables $I_{C_{p1}}, I_{C_{p2}}, I_{C_{p3}}, I_{C_{p4}}$ as a function of circuit variables $I_{L_{m1}}, I_{L_{m2}}, I_{L_{m3}}, I_{L_{m4}}$ and circuit parameters Version No. 1:

$$\frac{1}{C_{p1}} \cdot I_{C_{p1}} - \frac{1}{C_{p2}} \cdot I_{C_{p2}} - \frac{1}{C_{p3}} \cdot I_{C_{p3}} + \frac{1}{C_{p4}} \cdot I_{C_{p4}} = 0;$$

$$I_{C_{p3}} + I_{C_{p1}} - I_{C_{p2}} - I_{C_{p4}} = -I_{L_{m1}} + I_{L_{m2}} - I_{L_{m3}} + I_{L_{m4}}$$

$$I_{C_2} = \frac{C_2}{C_{p3}} \cdot I_{C_{p3}} - \frac{C_2}{C_{p1}} \cdot I_{C_{p1}}; \; \frac{C_2}{C_{p1}} \cdot I_{C_{p1}} + I_{C_{p2}} - [\frac{C_2}{C_{p3}} + 1] \cdot I_{C_{p3}} = -i_{p2} - I_{L_{m2}} + I_{L_{m3}}$$

$$I_{C_2} = \frac{C_2}{C_{p3}} \cdot I_{C_{p3}} - \frac{C_2}{C_{p1}} \cdot I_{C_{p1}}; \; [\frac{C_2}{C_{p1}} + 1] \cdot I_{C_{p1}} - \frac{C_2}{C_{p3}} \cdot I_{C_{p3}} - I_{C_{p4}} = -i_{p2} - I_{L_{m1}} + I_{L_{m4}}$$

$$\sum_{k=1}^{4} I_{L_{m_k}} \cdot (-1)^k = -I_{L_{m1}} + I_{L_{m2}} - I_{L_{m3}} + I_{L_{m4}}; \; -i_{p2} - I_{L_{m2}} + I_{L_{m3}} = -i_{p2} + \sum_{k=2}^{3} I_{L_{m_k}} \cdot (-1)^{k+1}$$

$$-i_{p2} - I_{L_{m1}} + I_{L_{m4}} = -i_{p2} + \sum_{\substack{k=1 \\ k \neq 2 \\ k \neq 3}}^{4} I_{L_{m_k}} \cdot (-1)^k$$

Find circuit variables $I_{C_{p1}}, I_{C_{p2}}, I_{C_{p3}}, I_{C_{p4}}$ as a function of circuit variables $I_{L_{m_1}}, I_{L_{m_2}}, I_{L_{m_3}}, I_{L_{m_4}}$ and circuit parameters Version No. 2:

$$\frac{1}{C_{p1}} \cdot I_{C_{p1}} - \frac{1}{C_{p2}} \cdot I_{C_{p2}} - \frac{1}{C_{p3}} \cdot I_{C_{p3}} + \frac{1}{C_{p4}} \cdot I_{C_{p4}} = 0;$$

$$I_{C_{p3}} + I_{C_{p1}} - I_{C_{p2}} - I_{C_{p4}} = \sum_{k=1}^{4} I_{L_{m_k}} \cdot (-1)^k$$

$$\frac{C_2}{C_{p1}} \cdot I_{C_{p1}} + I_{C_{p2}} - [\frac{C_2}{C_{p3}} + 1] \cdot I_{C_{p3}} = -i_{p2} + \sum_{k=2}^{3} I_{L_{m_k}} \cdot (-1)^{k+1}$$

$$[\frac{C_2}{C_{p1}} + 1] \cdot I_{C_{p1}} - \frac{C_2}{C_{p3}} \cdot I_{C_{p3}} - I_{C_{p4}} = -i_{p2} + \sum_{\substack{k=1 \\ k \neq 2 \\ k \neq 3}}^{4} I_{L_{m_k}} \cdot (-1)^k$$

Find circuit variables $I_{C_{p1}}, I_{C_{p2}}, I_{C_{p3}}, I_{C_{p4}}$ as a function of circuit variables $I_{L_{m_1}}, I_{L_{m_2}}, I_{L_{m_3}}, I_{L_{m_4}}$ and circuit parameters Version No. 3:

$$I_{C_{p1}} \cdot \begin{pmatrix} \frac{1}{C_{p1}} \\ 1 \\ \frac{C_2}{C_{p1}} \\ \frac{C_2}{C_{p1}} + 1 \end{pmatrix} + I_{C_{p2}} \cdot \begin{pmatrix} -\frac{1}{C_{p2}} \\ -1 \\ 1 \\ 0 \end{pmatrix} + I_{C_{p3}} \cdot \begin{pmatrix} -\frac{1}{C_{p3}} \\ 1 \\ -[\frac{C_2}{C_{p3}} + 1] \\ -\frac{C_2}{C_{p3}} \end{pmatrix} + I_{C_{p4}} \cdot \begin{pmatrix} \frac{1}{C_{p4}} \\ -1 \\ 0 \\ -1 \end{pmatrix}$$
$$= \begin{pmatrix} 0 \\ \sum_{k=1}^{4} I_{L_{m_k}} \cdot (-1)^k \\ -i_{p2} + \sum_{k=2}^{3} I_{L_{m_k}} \cdot (-1)^{k+1} \\ -i_{p2} + \sum_{\substack{k=1 \\ k \neq 2 \\ k \neq 3}}^{4} I_{L_{m_k}} \cdot (-1)^k \end{pmatrix}$$

$$\frac{1}{C_{p1}} \cdot I_{C_{p1}} - \frac{1}{C_{p2}} \cdot I_{C_{p2}} - \frac{1}{C_{p3}} \cdot I_{C_{p3}} + \frac{1}{C_{p4}} \cdot I_{C_{p4}} = 0 \Rightarrow I_{C_{p1}}$$
$$= \frac{C_{p1}}{C_{p2}} \cdot I_{C_{p2}} + \frac{C_{p1}}{C_{p3}} \cdot I_{C_{p3}} - \frac{C_{p1}}{C_{p4}} \cdot I_{C_{p4}}$$

$$[\frac{C_{p1}}{C_{p2}} - 1] \cdot I_{C_{p2}} + [\frac{C_{p1}}{C_{p3}} + 1] \cdot I_{C_{p3}} - [\frac{C_{p1}}{C_{p4}} + 1] \cdot I_{C_{p4}} = \sum_{k=1}^{4} I_{L_{m_k}} \cdot (-1)^k$$

$$[\frac{C_2}{C_{p2}} + 1] \cdot I_{C_{p2}} - I_{C_{p3}} - \frac{C_2}{C_{p4}} \cdot I_{C_{p4}} = -i_{p2} + \sum_{k=2}^{3} I_{L_{m_k}} \cdot (-1)^{k+1}$$

$$[\frac{C_2}{C_{p_1}} + 1] \cdot \frac{C_{p_1}}{C_{p_2}} \cdot I_{C_{p_2}} + \frac{C_{p_1}}{C_{p_3}} \cdot I_{C_{p_3}} - ([\frac{C_2}{C_{p_1}} + 1] \cdot \frac{C_{p_1}}{C_{p_4}} + 1) \cdot I_{C_{p_4}} = -i_{p_2} + \sum_{\substack{k=1 \\ k \neq 2 \\ k \neq 3}}^{4} I_{L_{m_k}} \cdot (-1)^k$$

We can summery our circuit variables $I_{C_{p_1}}, I_{C_{p_2}}, I_{C_{p_3}}, I_{C_{p_4}}$ as a function of circuit variables $I_{L_{m_1}}, I_{L_{m_2}}, I_{L_{m_3}}, I_{L_{m_4}}$ and circuit parameters Version No. 4:

$$I_{C_{p_1}} = \frac{C_{p_1}}{C_{p_2}} \cdot I_{C_{p_2}} + \frac{C_{p_1}}{C_{p_3}} \cdot I_{C_{p_3}} - \frac{C_{p_1}}{C_{p_4}} \cdot I_{C_{p_4}}$$

$$\begin{pmatrix} [\frac{C_{p1}}{C_{p2}} - 1] \\ [\frac{C_2}{C_{p2}} + 1] \\ [\frac{C_2}{C_{p1}} + 1] \cdot \frac{C_{p1}}{C_{p2}} \end{pmatrix} \cdot I_{C_{p_2}} + \begin{pmatrix} [\frac{C_{p1}}{C_{p3}} + 1] \\ -1 \\ \frac{C_{p1}}{C_{p3}} \end{pmatrix} \cdot I_{C_{p_3}} + \begin{pmatrix} -[\frac{C_{p1}}{C_{p4}} + 1] \\ -\frac{C_2}{C_{p4}} \\ -([\frac{C_2}{C_{p1}} + 1] \cdot \frac{C_{p1}}{C_{p4}} + 1) \end{pmatrix} \cdot I_{C_{p_4}} = \begin{pmatrix} \sum_{k=1}^{4} I_{Lm_k} \cdot (-1)^k \\ -i_{p2} + \sum_{k=2}^{3} I_{Lm_k} \cdot (-1)^{k+1} \\ -i_{p2} + \sum_{\substack{k=1 \\ k \neq 2 \\ k \neq 3}}^{4} I_{Lm_k} \cdot (-1)^k \end{pmatrix}$$

We use Cramer's rule for the solution of above linear equations. Each variable $(C_{p_k};\ k = 2, 3, 4)$ given by a quotient two determinants.

$$\Delta = \det \begin{pmatrix} [\frac{C_{p1}}{C_{p2}} - 1] & [\frac{C_{p1}}{C_{p3}} + 1] & -[\frac{C_{p1}}{C_{p4}} + 1] \\ [\frac{C_2}{C_{p2}} + 1] & -1 & -\frac{C_2}{C_{p4}} \\ [\frac{C_2}{C_{p1}} + 1] \cdot \frac{C_{p1}}{C_{p2}} & \frac{C_{p1}}{C_{p3}} & -([\frac{C_2}{C_{p1}} + 1] \cdot \frac{C_{p1}}{C_{p4}} + 1) \end{pmatrix}$$

$$\Delta = [\frac{C_{p_1}}{C_{p_2}} - 1] \cdot \det \begin{pmatrix} -1 & -\frac{C_2}{C_{p4}} \\ \frac{C_{p1}}{C_{p3}} & -([\frac{C_2}{C_{p1}} + 1] \cdot \frac{C_{p1}}{C_{p4}} + 1) \end{pmatrix} - [\frac{C_{p_1}}{C_{p_3}} + 1] \cdot \det \begin{pmatrix} [\frac{C_2}{C_{p2}} + 1] & -\frac{C_2}{C_{p4}} \\ [\frac{C_2}{C_{p1}} + 1] \cdot \frac{C_{p1}}{C_{p2}} & -([\frac{C_2}{C_{p1}} + 1] \cdot \frac{C_{p1}}{C_{p4}} + 1) \end{pmatrix} - [\frac{C_{p_1}}{C_{p_4}} + 1] \cdot \det \begin{pmatrix} [\frac{C_2}{C_{p2}} + 1] & -1 \\ [\frac{C_2}{C_{p1}} + 1] \cdot \frac{C_{p1}}{C_{p2}} & \frac{C_{p1}}{C_{p3}} \end{pmatrix}; \Delta = \Delta(C_{p_1}, \ldots, C_{p_4}, C_2)$$

$$\Delta_{I_{C_{p_2}}} = \det\begin{pmatrix} \sum_{k=1}^{4} I_{L_{m_k}} \cdot (-1)^k & [\frac{C_{p_1}}{C_{p_3}} + 1] & -[\frac{C_{p_1}}{C_{p_4}} + 1] \\ -i_{p_2} + \sum_{k=2}^{3} I_{L_{m_k}} \cdot (-1)^{k+1} & -1 & -\frac{C_2}{C_{p_4}} \\ -i_{p_2} + \sum_{\substack{k=1 \\ k\neq 2 \\ k\neq 3}}^{4} I_{L_{m_k}} \cdot (-1)^k & \frac{C_{p_1}}{C_{p_3}} & -([\frac{C_2}{C_{p_1}} + 1] \cdot \frac{C_{p_1}}{C_{p_4}} + 1) \end{pmatrix}$$

$$\Delta_{I_{C_{p_2}}} = (\sum_{k=1}^{4} I_{L_{m_k}} \cdot (-1)^k) \cdot \det\begin{pmatrix} -1 & -\frac{C_2}{C_{p_4}} \\ \frac{C_{p_1}}{C_{p_3}} & -([\frac{C_2}{C_{p_1}} + 1] \cdot \frac{C_{p_1}}{C_{p_4}} + 1) \end{pmatrix}$$

$$- [\frac{C_{p_1}}{C_{p_3}} + 1] \cdot \det\begin{pmatrix} -i_{p_2} + \sum_{k=2}^{3} I_{L_{m_k}} \cdot (-1)^{k+1} & -\frac{C_2}{C_{p_4}} \\ -i_{p_2} + \sum_{\substack{k=1 \\ k\neq 2 \\ k\neq 3}}^{4} I_{L_{m_k}} \cdot (-1)^k & -([\frac{C_2}{C_{p_1}} + 1] \cdot \frac{C_{p_1}}{C_{p_4}} + 1) \end{pmatrix}$$

$$-[\frac{C_{p_1}}{C_{p_4}} + 1] \cdot \det\begin{pmatrix} -i_{p_2} + \sum_{k=2}^{3} I_{L_{m_k}} \cdot (-1)^{k+1} & -1 \\ -i_{p_2} + \sum_{\substack{k=1 \\ k\neq 2 \\ k\neq 3}}^{4} I_{L_{m_k}} \cdot (-1)^k & \frac{C_{p_1}}{C_{p_3}} \end{pmatrix}$$

$$\begin{aligned} \Delta_{I_{C_{p_2}}} &= (\sum_{k=1}^{4} I_{L_{m_k}} \cdot (-1)^k) \cdot [([\frac{C_2}{C_{p_1}} + 1] \cdot \frac{C_{p_1}}{C_{p_4}} + 1) + \frac{C_{p_1}}{C_{p_3}} \cdot \frac{C_2}{C_{p_4}}] \\ &- [\frac{C_{p_1}}{C_{p_3}} + 1] \cdot [i_{p_2} \cdot ([\frac{C_2}{C_{p_1}} + 1] \cdot \frac{C_{p_1}}{C_{p_4}} + 1) - ([\frac{C_2}{C_{p_1}} + 1] \cdot \frac{C_{p_1}}{C_{p_4}} + 1) \\ &\cdot \sum_{k=2}^{3} I_{L_{m_k}} \cdot (-1)^{k+1} - i_{p_2} \cdot \frac{C_2}{C_{p_4}} + \frac{C_2}{C_{p_4}} \cdot \sum_{\substack{k=1 \\ k\neq 2 \\ k\neq 3}}^{4} I_{L_{m_k}} \cdot (-1)^k] - [\frac{C_{p_1}}{C_{p_4}} + 1] \\ &\cdot [-i_{p_2} \cdot \frac{C_{p_1}}{C_{p_3}} + \frac{C_{p_1}}{C_{p_3}} \cdot \sum_{k=2}^{3} I_{L_{m_k}} \cdot (-1)^{k+1} - i_{p_2} + \sum_{\substack{k=1 \\ k\neq 2 \\ k\neq 3}}^{4} I_{L_{m_k}} \cdot (-1)^k] \end{aligned}$$

$$\Delta_{I_{C_{p2}}} = (\sum_{k=1}^{4} I_{L_{m_k}} \cdot (-1)^k) \cdot [(([\frac{C_2}{C_{p_1}} + 1] \cdot \frac{C_{p_1}}{C_{p_4}} + 1) + \frac{C_{p_1}}{C_{p_3}} \cdot \frac{C_2}{C_{p_4}}]$$
$$- i_{p_2} \cdot [\frac{C_{p_1}}{C_{p_3}} + 1] \cdot ([\frac{C_2}{C_{p_1}} + 1] \cdot \frac{C_{p_1}}{C_{p_4}} + 1) + [\frac{C_{p_1}}{C_{p_3}} + 1] \cdot ([\frac{C_2}{C_{p_1}} + 1] \cdot \frac{C_{p_1}}{C_{p_4}} + 1)$$
$$\cdot \sum_{k=2}^{3} I_{L_{m_k}} \cdot (-1)^{k+1} + i_{p_2} \cdot [\frac{C_{p_1}}{C_{p_3}} + 1] \cdot \frac{C_2}{C_{p_4}} - \frac{C_2}{C_{p_4}} \cdot [\frac{C_{p_1}}{C_{p_3}} + 1] \cdot \sum_{\substack{k=1 \\ k \neq 2 \\ k \neq 3}}^{4} I_{L_{m_k}} \cdot (-1)^k$$
$$+ i_{p_2} \cdot [\frac{C_{p_1}}{C_{p_4}} + 1] \cdot \frac{C_{p_1}}{C_{p_3}} - \frac{C_{p_1}}{C_{p_3}} \cdot [\frac{C_{p_1}}{C_{p_4}} + 1] \cdot \sum_{k=2}^{3} I_{L_{m_k}} \cdot (-1)^{k+1} + [\frac{C_{p_1}}{C_{p_4}} + 1] \cdot i_{p_2}$$
$$-[\frac{C_{p_1}}{C_{p_4}} + 1] \cdot \sum_{\substack{k=1 \\ k \neq 2 \\ k \neq 3}}^{4} I_{L_{m_k}} \cdot (-1)^k$$

$$\Delta_{I_{C_{p2}}} = (\sum_{k=1}^{4} I_{L_{m_k}} \cdot (-1)^k) \cdot [(([\frac{C_2}{C_{p_1}} + 1] \cdot \frac{C_{p_1}}{C_{p_4}} + 1) + \frac{C_{p_1}}{C_{p_3}} \cdot \frac{C_2}{C_{p_4}}] + [\frac{C_{p_1}}{C_{p_3}} + 1]$$
$$\cdot ([\frac{C_2}{C_{p_1}} + 1] \cdot \frac{C_{p_1}}{C_{p_4}} + 1) \cdot \sum_{k=2}^{3} I_{L_{m_k}} \cdot (-1)^{k+1} - \frac{C_2}{C_{p_4}} \cdot [\frac{C_{p_1}}{C_{p_3}} + 1]$$
$$\cdot \sum_{\substack{k=1 \\ k \neq 2 \\ k \neq 3}}^{4} I_{L_{m_k}} \cdot (-1)^k - \frac{C_{p_1}}{C_{p_3}} \cdot [\frac{C_{p_1}}{C_{p_4}} + 1] \cdot \sum_{k=2}^{3} I_{L_{m_k}} \cdot (-1)^{k+1} - [\frac{C_{p_1}}{C_{p_4}} + 1]$$
$$\cdot \sum_{\substack{k=1 \\ k \neq 2 \\ k \neq 3}}^{4} I_{L_{m_k}} \cdot (-1)^k + i_{p_2} \cdot \{[\frac{C_{p_1}}{C_{p_3}} + 1] \cdot \frac{C_2}{C_{p_4}} - [\frac{C_{p_1}}{C_{p_3}} + 1] \cdot ([\frac{C_2}{C_{p_1}} + 1] \cdot \frac{C_{p_1}}{C_{p_4}} + 1)$$
$$+ [\frac{C_{p_1}}{C_{p_4}} + 1] \cdot \frac{C_{p_1}}{C_{p_3}} + [\frac{C_{p_1}}{C_{p_4}} + 1]\}$$

We can define $\Delta_{I_{C_{p2}}}$ as $\Delta_{I_{C_{p2}}} = i_{p_2} \cdot \Gamma_1 + \psi_1(I_{L_{m_k}}; k = 1, 2, 3, 4; C_{p_1}, C_{p_1}, \ldots)$

$$\Gamma_1 = [\frac{C_{p_1}}{C_{p_3}} + 1] \cdot \frac{C_2}{C_{p_4}} - [\frac{C_{p_1}}{C_{p_3}} + 1] \cdot ((\frac{C_2}{C_{p_1}} + 1) \cdot \frac{C_{p_1}}{C_{p_4}} + 1) + [\frac{C_{p_1}}{C_{p_4}} + 1]$$
$$\cdot \frac{C_{p_1}}{C_{p_3}} + [\frac{C_{p_1}}{C_{p_4}} + 1]$$

$$\psi_1 = \psi_1(I_{L_{m_k}};\ k = 1, 2, 3, 4; C_{p_1}, C_{p_1}, \ldots)$$

$$\psi_1(I_{L_{m_k}}; k = 1,2,3,4; C_{p_1}, C_{p_1}, \ldots) = (\sum_{k=1}^{4} I_{L_{m_k}} \cdot (-1)^k) \cdot [([\frac{C_2}{C_{p_1}} + 1] \cdot \frac{C_{p_1}}{C_{p_4}} + 1) + \frac{C_{p_1}}{C_{p_3}} \cdot \frac{C_2}{C_{p_4}}]$$
$$+ [\frac{C_{p_1}}{C_{p_3}} + 1] \cdot ([\frac{C_2}{C_{p_1}} + 1] \cdot \frac{C_{p_1}}{C_{p_4}} + 1) \cdot \sum_{k=2}^{3} I_{L_{m_k}} \cdot (-1)^{k+1} - \frac{C_2}{C_{p_4}} \cdot [\frac{C_{p_1}}{C_{p_3}} + 1] \cdot \sum_{\substack{k=1 \\ k\neq 2 \\ k\neq 3}}^{4} I_{L_{m_k}} \cdot (-1)^k$$
$$- \frac{C_{p_1}}{C_{p_3}} \cdot [\frac{C_{p_1}}{C_{p_4}} + 1] \cdot \sum_{k=2}^{3} I_{L_{m_k}} \cdot (-1)^{k+1} - [\frac{C_{p_1}}{C_{p_4}} + 1] \cdot \sum_{\substack{k=1 \\ k\neq 2 \\ k\neq 3}}^{4} I_{L_{m_k}} \cdot (-1)^k$$

$$\Delta_{I_{C_{p_3}}} = \det\begin{pmatrix} [\frac{C_{p_1}}{C_{p_2}} - 1] & \sum_{k=1}^{4} I_{L_{m_k}} \cdot (-1)^k & -[\frac{C_{p_1}}{C_{p_4}} + 1] \\ [\frac{C_2}{C_{p_2}} + 1] & -i_{p_2} + \sum_{k=2}^{3} I_{L_{m_k}} \cdot (-1)^{k+1} & -\frac{C_2}{C_{p_4}} \\ [\frac{C_2}{C_{p_1}} + 1] \cdot \frac{C_{p_1}}{C_{p_2}} & -i_{p_2} + \sum_{\substack{k=1 \\ k\neq 2 \\ k\neq 3}}^{4} I_{L_{m_k}} \cdot (-1)^k & -([\frac{C_2}{C_{p_1}} + 1] \cdot \frac{C_{p_1}}{C_{p_4}} + 1) \end{pmatrix}$$

$$\Delta_{I_{C_{p_3}}} = [\frac{C_{p_1}}{C_{p_2}} - 1] \cdot \det\begin{pmatrix} -i_{p_2} + \sum_{k=2}^{3} I_{L_{m_k}} \cdot (-1)^{k+1} & -\frac{C_2}{C_{p_4}} \\ -i_{p_2} + \sum_{\substack{k=1 \\ k\neq 2 \\ k\neq 3}}^{4} I_{L_{m_k}} \cdot (-1)^k & -([\frac{C_2}{C_{p_1}} + 1] \cdot \frac{C_{p_1}}{C_{p_4}} + 1) \end{pmatrix}$$
$$- (\sum_{k=1}^{4} I_{L_{m_k}} \cdot (-1)^k) \cdot \det\begin{pmatrix} [\frac{C_2}{C_{p_2}} + 1] & -\frac{C_2}{C_{p_4}} \\ [\frac{C_2}{C_{p_1}} + 1] \cdot \frac{C_{p_1}}{C_{p_2}} & -([\frac{C_2}{C_{p_1}} + 1] \cdot \frac{C_{p_1}}{C_{p_4}} + 1) \end{pmatrix}$$
$$- [\frac{C_{p_1}}{C_{p_4}} + 1] \cdot \det\begin{pmatrix} [\frac{C_2}{C_{p_2}} + 1] & -i_{p_2} + \sum_{k=2}^{3} I_{L_{m_k}} \cdot (-1)^{k+1} \\ [\frac{C_2}{C_{p_1}} + 1] \cdot \frac{C_{p_1}}{C_{p_2}} & -i_{p_2} + \sum_{\substack{k=1 \\ k\neq 2 \\ k\neq 3}}^{4} I_{L_{m_k}} \cdot (-1)^k \end{pmatrix}$$

$$\Delta_{I_{C_{p_3}}} = [\frac{C_{p_1}}{C_{p_2}} - 1] \cdot \{i_{p_2} \cdot ([\frac{C_2}{C_{p_1}} + 1] \cdot \frac{C_{p_1}}{C_{p_4}} + 1) - ([\frac{C_2}{C_{p_1}} + 1] \cdot \frac{C_{p_1}}{C_{p_4}} + 1) \cdot \sum_{k=2}^{3} I_{L_{m_k}} \cdot (-1)^{k+1}$$
$$-i_{p_2} \cdot \frac{C_2}{C_{p_4}} + \frac{C_2}{C_{p_4}} \cdot \sum_{\substack{k=1 \\ k\neq 2 \\ k\neq 3}}^{4} I_{L_{m_k}} \cdot (-1)^k\} - (\sum_{k=1}^{4} I_{L_{m_k}} \cdot (-1)^k) \cdot \{-([\frac{C_2}{C_{p_1}} + 1] \cdot \frac{C_{p_1}}{C_{p_4}} + 1) \cdot [\frac{C_2}{C_{p_2}} + 1] + \frac{C_2}{C_{p_4}} \cdot [\frac{C_2}{C_{p_1}} + 1] \cdot \frac{C_{p_1}}{C_{p_2}}\}$$
$$- [\frac{C_{p_1}}{C_{p_4}} + 1] \cdot \{-i_{p_2} \cdot [\frac{C_2}{C_{p_2}} + 1] + [\frac{C_2}{C_{p_2}} + 1] \cdot \sum_{\substack{k=1 \\ k\neq 2 \\ k\neq 3}}^{4} I_{L_{m_k}} \cdot (-1)^k + [\frac{C_2}{C_{p_1}} + 1] \cdot \frac{C_{p_1}}{C_{p_2}} \cdot i_{p_2} - [\frac{C_2}{C_{p_1}} + 1] \cdot \frac{C_{p_1}}{C_{p_2}} \cdot \sum_{k=2}^{3} I_{L_{m_k}} \cdot (-1)^{k+1}\}$$

$$\Delta_{I_{C_{p3}}} = i_{p_2} \cdot [\frac{C_{p1}}{C_{p2}} - 1] \cdot ([\frac{C_2}{C_{p1}} + 1] \cdot \frac{C_{p1}}{C_{p4}} + 1) - [\frac{C_{p1}}{C_{p2}} - 1] \cdot ([\frac{C_2}{C_{p1}} + 1] \cdot \frac{C_{p1}}{C_{p4}} + 1)$$
$$\cdot \sum_{k=2}^{3} I_{L_{m_k}} \cdot (-1)^{k+1} - i_{p_2} \cdot [\frac{C_{p1}}{C_{p2}} - 1] \cdot \frac{C_2}{C_{p4}} + [\frac{C_{p1}}{C_{p2}} - 1] \cdot \frac{C_2}{C_{p4}} \cdot \sum_{\substack{k=1\\k\neq 2\\k\neq 3}}^{4} I_{L_{m_k}} \cdot (-1)^k$$
$$- (\sum_{k=1}^{4} I_{L_{m_k}} \cdot (-1)^k) \cdot \{-([\frac{C_2}{C_{p1}} + 1] \cdot \frac{C_{p1}}{C_{p4}} + 1) \cdot [\frac{C_2}{C_{p2}} + 1] + \frac{C_2}{C_{p4}} \cdot [\frac{C_2}{C_{p1}} + 1] \cdot \frac{C_{p1}}{C_{p2}}\}$$
$$+ i_{p_2} \cdot [\frac{C_{p1}}{C_{p4}} + 1] \cdot [\frac{C_2}{C_{p2}} + 1] - [\frac{C_{p1}}{C_{p4}} + 1] \cdot [\frac{C_2}{C_{p2}} + 1] \cdot \sum_{\substack{k=1\\k\neq 2\\k\neq 3}}^{4} I_{L_{m_k}} \cdot (-1)^k$$
$$- [\frac{C_{p1}}{C_{p4}} + 1] \cdot [\frac{C_2}{C_{p1}} + 1] \cdot \frac{C_{p1}}{C_{p2}} \cdot i_{p_2} + [\frac{C_{p1}}{C_{p4}} + 1] \cdot [\frac{C_2}{C_{p1}} + 1] \cdot \frac{C_{p1}}{C_{p2}} \cdot \sum_{k=2}^{3} I_{L_{m_k}} \cdot (-1)^{k+1}$$

$$\Delta_{I_{C_{p3}}} = i_{p_2} \cdot \{[\frac{C_{p1}}{C_{p2}} - 1] \cdot ([\frac{C_2}{C_{p1}} + 1] \cdot \frac{C_{p1}}{C_{p4}} + 1) - [\frac{C_{p1}}{C_{p2}} - 1] \cdot \frac{C_2}{C_{p4}} + [\frac{C_{p1}}{C_{p4}} + 1]$$
$$\cdot [\frac{C_2}{C_{p2}} + 1] - [\frac{C_{p1}}{C_{p4}} + 1] \cdot [\frac{C_2}{C_{p1}} + 1] \cdot \frac{C_{p1}}{C_{p2}}\} - [\frac{C_{p1}}{C_{p2}} - 1] \cdot ([\frac{C_2}{C_{p1}} + 1] \cdot \frac{C_{p1}}{C_{p4}} + 1)$$
$$\cdot \sum_{k=2}^{3} I_{L_{m_k}} \cdot (-1)^{k+1} + [\frac{C_{p1}}{C_{p2}} - 1] \cdot \frac{C_2}{C_{p4}} \cdot \sum_{\substack{k=1\\k\neq 2\\k\neq 3}}^{4} I_{L_{m_k}} \cdot (-1)^k - (\sum_{k=1}^{4} I_{L_{m_k}} \cdot (-1)^k)$$
$$\cdot \{-([\frac{C_2}{C_{p1}} + 1] \cdot \frac{C_{p1}}{C_{p4}} + 1) \cdot [\frac{C_2}{C_{p2}} + 1] + \frac{C_2}{C_{p4}} \cdot [\frac{C_2}{C_{p1}} + 1] \cdot \frac{C_{p1}}{C_{p2}}\} - [\frac{C_{p1}}{C_{p4}} + 1] \cdot [\frac{C_2}{C_{p2}} + 1]$$
$$\cdot \sum_{\substack{k=1\\k\neq 2\\k\neq 3}}^{4} I_{L_{m_k}} \cdot (-1)^k + [\frac{C_{p1}}{C_{p4}} + 1] \cdot [\frac{C_2}{C_{p1}} + 1] \cdot \frac{C_{p1}}{C_{p2}} \cdot \sum_{k=2}^{3} I_{L_{m_k}} \cdot (-1)^{k+1}$$

We can define $\Delta_{I_{C_{p3}}}$ as $\Delta_{I_{C_{p3}}} = i_{p_2} \cdot \Gamma_2 + \psi_2(I_{L_{m_k}}; k = 1, 2, 3, 4; C_{p_1}, C_{p_1}, \ldots)$.

$$\Gamma_2 = [\frac{C_{p1}}{C_{p2}} - 1] \cdot ([\frac{C_2}{C_{p1}} + 1] \cdot \frac{C_{p1}}{C_{p4}} + 1) - [\frac{C_{p1}}{C_{p2}} - 1] \cdot \frac{C_2}{C_{p4}} + [\frac{C_{p1}}{C_{p4}} + 1] \cdot [\frac{C_2}{C_{p2}} + 1]$$
$$- [\frac{C_{p1}}{C_{p4}} + 1] \cdot [\frac{C_2}{C_{p1}} + 1] \cdot \frac{C_{p1}}{C_{p2}}$$

$$\psi_2 = \psi_2(I_{L_{m_k}}; k = 1, 2, 3, 4; C_{p_1}, C_{p_1}, \ldots)$$

$$\psi_2(I_{L_{mk}}; k = 1,2,3,4; C_{p_1}, C_{p_1}, \ldots) = -[\frac{C_{p_1}}{C_{p_2}} - 1] \cdot ([\frac{C_2}{C_{p_1}} + 1] \cdot \frac{C_{p_1}}{C_{p_4}} + 1)$$

$$\cdot \sum_{k=2}^{3} I_{L_{m_k}} \cdot (-1)^{k+1} + [\frac{C_{p_1}}{C_{p_2}} - 1] \cdot \frac{C_2}{C_{p_4}} \cdot \sum_{\substack{k=1 \\ k\neq 2 \\ k\neq 3}}^{4} I_{L_{m_k}} \cdot (-1)^k - (\sum_{k=1}^{4} I_{L_{m_k}} \cdot (-1)^k)$$

$$\cdot \{-([\frac{C_2}{C_{p_1}} + 1] \cdot \frac{C_{p_1}}{C_{p_4}} + 1) \cdot [\frac{C_2}{C_{p_2}} + 1] + \frac{C_2}{C_{p_4}} \cdot [\frac{C_2}{C_{p_1}} + 1] \cdot \frac{C_{p_1}}{C_{p_2}}\} - [\frac{C_{p_1}}{C_{p_4}} + 1]$$

$$\cdot [\frac{C_2}{C_{p_2}} + 1] \cdot \sum_{\substack{k=1 \\ k\neq 2 \\ k\neq 3}}^{4} I_{L_{m_k}} \cdot (-1)^k + [\frac{C_{p_1}}{C_{p_4}} + 1] \cdot [\frac{C_2}{C_{p_1}} + 1] \cdot \frac{C_{p_1}}{C_{p_2}} \cdot \sum_{k=2}^{3} I_{L_{m_k}} \cdot (-1)^{k+1}$$

$$\Delta_{I_{C_{p_4}}} = \det\begin{pmatrix} [\frac{C_{p_1}}{C_{p_2}} - 1] & [\frac{C_{p_1}}{C_{p_3}} + 1] & \sum_{k=1}^{4} I_{L_{m_k}} \cdot (-1)^k \\ [\frac{C_2}{C_{p_2}} + 1] & -1 & -i_{p_2} + \sum_{k=2}^{3} I_{L_{m_k}} \cdot (-1)^{k+1} \\ [\frac{C_2}{C_{p_1}} + 1] \cdot \frac{C_{p_1}}{C_{p_2}} & \frac{C_{p_1}}{C_{p_3}} & -i_{p_2} + \sum_{\substack{k=1 \\ k\neq 2 \\ k\neq 3}}^{4} I_{L_{m_k}} \cdot (-1)^k \end{pmatrix}$$

$$\Delta_{I_{C_{p_4}}} = [\frac{C_{p_1}}{C_{p_2}} - 1] \cdot \det\begin{pmatrix} -1 & -i_{p_2} + \sum_{k=2}^{3} I_{L_{m_k}} \cdot (-1)^{k+1} \\ \frac{C_{p_1}}{C_{p_3}} & -i_{p_2} + \sum_{\substack{k=1 \\ k\neq 2 \\ k\neq 3}}^{4} I_{L_{m_k}} \cdot (-1)^k \end{pmatrix}$$

$$- [\frac{C_{p_1}}{C_{p_3}} + 1] \cdot \det\begin{pmatrix} [\frac{C_2}{C_{p_2}} + 1] & -i_{p_2} + \sum_{k=2}^{3} I_{L_{m_k}} \cdot (-1)^{k+1} \\ [\frac{C_2}{C_{p_1}} + 1] \cdot \frac{C_{p_1}}{C_{p_2}} & -i_{p_2} + \sum_{\substack{k=1 \\ k\neq 2 \\ k\neq 3}}^{4} I_{L_{m_k}} \cdot (-1)^k \end{pmatrix}$$

$$+ \left(\sum_{k=1}^{4} I_{L_{m_k}} \cdot (-1)^k\right) \cdot \det\begin{pmatrix} [\frac{C_2}{C_{p_2}} + 1] & -1 \\ [\frac{C_2}{C_{p_1}} + 1] \cdot \frac{C_{p_1}}{C_{p_2}} & \frac{C_{p_1}}{C_{p_3}} \end{pmatrix}$$

$$\Delta_{I_{C_{p_4}}} = [\frac{C_{p_1}}{C_{p_2}} - 1] \cdot \{i_{p_2} - \sum_{\substack{k=1 \\ k \neq 2 \\ k \neq 3}}^{4} I_{L_{m_k}} \cdot (-1)^k + i_{p_2} \cdot \frac{C_{p_1}}{C_{p_3}} - \frac{C_{p_1}}{C_{p_3}} \cdot \sum_{k=2}^{3} I_{L_{m_k}} \cdot (-1)^{k+1}\}$$
$$- [\frac{C_{p_1}}{C_{p_3}} + 1] \cdot \{-i_{p_2} \cdot [\frac{C_2}{C_{p_2}} + 1] + [\frac{C_2}{C_{p_2}} + 1] \cdot \sum_{\substack{k=1 \\ k \neq 2 \\ k \neq 3}}^{4} I_{L_{m_k}} \cdot (-1)^k + i_{p_2} \cdot [\frac{C_2}{C_{p_1}} + 1] \cdot \frac{C_{p_1}}{C_{p_2}}$$
$$- [\frac{C_2}{C_{p_1}} + 1] \cdot \frac{C_{p_1}}{C_{p_2}} \cdot \sum_{k=2}^{3} I_{L_{m_k}} \cdot (-1)^{k+1}\} + (\sum_{k=1}^{4} I_{L_{m_k}} \cdot (-1)^k) \cdot \{[\frac{C_2}{C_{p_2}} + 1] \cdot \frac{C_{p_1}}{C_{p_3}} + [\frac{C_2}{C_{p_1}} + 1] \cdot \frac{C_{p_1}}{C_{p_2}}\}$$

$$\Delta_{I_{C_{p_4}}} = [\frac{C_{p_1}}{C_{p_2}} - 1] \cdot i_{p_2} - [\frac{C_{p_1}}{C_{p_2}} - 1] \cdot \sum_{\substack{k=1 \\ k \neq 2 \\ k \neq 3}}^{4} I_{L_{m_k}} \cdot (-1)^k + i_{p_2} \cdot [\frac{C_{p_1}}{C_{p_2}} - 1] \cdot \frac{C_{p_1}}{C_{p_3}} - [\frac{C_{p_1}}{C_{p_2}} - 1]$$
$$\cdot \frac{C_{p_1}}{C_{p_3}} \cdot \sum_{k=2}^{3} I_{L_{m_k}} \cdot (-1)^{k+1} + i_{p_2} \cdot [\frac{C_{p_1}}{C_{p_3}} + 1] \cdot [\frac{C_2}{C_{p_2}} + 1] - [\frac{C_{p_1}}{C_{p_3}} + 1] \cdot [\frac{C_2}{C_{p_2}} + 1]$$
$$\cdot \sum_{\substack{k=1 \\ k \neq 2 \\ k \neq 3}}^{4} I_{L_{m_k}} \cdot (-1)^k - i_{p_2} \cdot [\frac{C_{p_1}}{C_{p_3}} + 1] \cdot [\frac{C_2}{C_{p_1}} + 1] \cdot \frac{C_{p_1}}{C_{p_2}} + [\frac{C_{p_1}}{C_{p_3}} + 1] \cdot [\frac{C_2}{C_{p_1}} + 1] \cdot \frac{C_{p_1}}{C_{p_2}}$$
$$\cdot \sum_{k=2}^{3} I_{L_{m_k}} \cdot (-1)^{k+1} + (\sum_{k=1}^{4} I_{L_{m_k}} \cdot (-1)^k) \cdot \{[\frac{C_2}{C_{p_2}} + 1] \cdot \frac{C_{p_1}}{C_{p_3}} + [\frac{C_2}{C_{p_1}} + 1] \cdot \frac{C_{p_1}}{C_{p_2}}\}$$

$$\Delta_{I_{C_{p_4}}} = i_{p_2} \cdot \{[\frac{C_{p_1}}{C_{p_2}} - 1] + [\frac{C_{p_1}}{C_{p_2}} - 1] \cdot \frac{C_{p_1}}{C_{p_3}} + [\frac{C_{p_1}}{C_{p_3}} + 1] \cdot [\frac{C_2}{C_{p_2}} + 1] - [\frac{C_{p_1}}{C_{p_3}} + 1]$$
$$\cdot [\frac{C_2}{C_{p_1}} + 1] \cdot \frac{C_{p_1}}{C_{p_2}}\} - [\frac{C_{p_1}}{C_{p_2}} - 1] \cdot \sum_{\substack{k=1 \\ k \neq 2 \\ k \neq 3}}^{4} I_{L_{m_k}} \cdot (-1)^k - [\frac{C_{p_1}}{C_{p_2}} - 1]$$
$$\cdot \frac{C_{p_1}}{C_{p_3}} \cdot \sum_{k=2}^{3} I_{L_{m_k}} \cdot (-1)^{k+1} - [\frac{C_{p_1}}{C_{p_3}} + 1] \cdot [\frac{C_2}{C_{p_2}} + 1] \cdot \sum_{\substack{k=1 \\ k \neq 2 \\ k \neq 3}}^{4} I_{L_{m_k}} \cdot (-1)^k$$
$$+ [\frac{C_{p_1}}{C_{p_3}} + 1] \cdot [\frac{C_2}{C_{p_1}} + 1] \cdot \frac{C_{p_1}}{C_{p_2}} \cdot \sum_{k=2}^{3} I_{L_{m_k}} \cdot (-1)^{k+1} + (\sum_{k=1}^{4} I_{L_{m_k}} \cdot (-1)^k)$$
$$\cdot \{[\frac{C_2}{C_{p_2}} + 1] \cdot \frac{C_{p_1}}{C_{p_3}} + [\frac{C_2}{C_{p_1}} + 1] \cdot \frac{C_{p_1}}{C_{p_2}}\}$$

We can define $\Delta_{I_{C_{p_4}}}$ as $\Delta_{I_{C_{p_4}}} = i_{p_2} \cdot \Gamma_3 + \psi_3(I_{L_{m_k}}; k = 1, 2, 3, 4; C_{p_1}, C_{p_1}, \ldots)$.

$$\Gamma_3 = [\frac{C_{p_1}}{C_{p_2}} - 1] + [\frac{C_{p_1}}{C_{p_2}} - 1] \cdot \frac{C_{p_1}}{C_{p_3}} + [\frac{C_{p_1}}{C_{p_3}} + 1] \cdot [\frac{C_2}{C_{p_2}} + 1] - [\frac{C_{p_1}}{C_{p_3}} + 1] \cdot [\frac{C_2}{C_{p_1}} + 1]$$
$$\cdot \frac{C_{p_1}}{C_{p_2}}$$

$$\psi_3 = \psi_3(I_{L_{m_k}}; k = 1,2,3,4; C_{p_1}, C_{p_1}, \ldots)$$

$$\psi_3(I_{L_{m_k}}; k = 1,2,3,4; C_{p_1}, C_{p_1}, \ldots) = -[\frac{C_{p_1}}{C_{p_2}} - 1] \cdot \sum_{\substack{k=1 \\ k\neq 2 \\ k\neq 3}}^{4} I_{L_{m_k}} \cdot (-1)^k - [\frac{C_{p_1}}{C_{p_2}} - 1]$$

$$\cdot \frac{C_{p_1}}{C_{p_3}} \cdot \sum_{k=2}^{3} I_{L_{m_k}} \cdot (-1)^{k+1} - [\frac{C_{p_1}}{C_{p_3}} + 1] \cdot [\frac{C_2}{C_{p_2}} + 1] \cdot \sum_{\substack{k=1 \\ k\neq 2 \\ k\neq 3}}^{4} I_{L_{m_k}} \cdot (-1)^k + [\frac{C_{p_1}}{C_{p_3}} + 1] \cdot [\frac{C_2}{C_{p_1}} + 1]$$

$$\cdot \frac{C_{p_1}}{C_{p_2}} \cdot \sum_{k=2}^{3} I_{L_{m_k}} \cdot (-1)^{k+1} + (\sum_{k=1}^{4} I_{L_{m_k}} \cdot (-1)^k) \cdot \{[\frac{C_2}{C_{p_2}} + 1] \cdot \frac{C_{p_1}}{C_{p_3}} + [\frac{C_2}{C_{p_1}} + 1] \cdot \frac{C_{p_1}}{C_{p_2}}\}$$

We can summery our expressions for $I_{L_{m_k}}; k = 1,2,3,4$:

$$I_{C_{p_2}} = \frac{\Delta_{I_{C_{p_2}}}}{\Delta} = i_{p_2} \cdot \frac{\Gamma_1}{\Delta} + \frac{\psi_1}{\Delta}; \; I_{C_{p_3}} = \frac{\Delta_{I_{C_{p_3}}}}{\Delta} = i_{p_2} \cdot \frac{\Gamma_2}{\Delta} + \frac{\psi_2}{\Delta}; \; I_{C_{p_4}} = \frac{\Delta_{I_{C_{p_4}}}}{\Delta}$$
$$= i_{p_2} \cdot \frac{\Gamma_3}{\Delta} + \frac{\psi_3}{\Delta}$$

$$I_{C_{p_1}} = \frac{C_{p_1}}{C_{p_2}} \cdot [i_{p_2} \cdot \frac{\Gamma_1}{\Delta} + \frac{\psi_1}{\Delta}] + \frac{C_{p_1}}{C_{p_3}} \cdot [i_{p_2} \cdot \frac{\Gamma_2}{\Delta} + \frac{\psi_2}{\Delta}] - \frac{C_{p_1}}{C_{p_4}} \cdot [i_{p_2} \cdot \frac{\Gamma_3}{\Delta} + \frac{\psi_3}{\Delta}]$$

$$I_{C_{p_1}} = i_{p_2} \cdot \frac{C_{p_1}}{\Delta} \cdot [\frac{\Gamma_1}{C_{p_2}} + \frac{\Gamma_2}{C_{p_3}} - \frac{\Gamma_3}{C_{p_4}}] + \frac{C_{p_1}}{\Delta} \cdot [\frac{\psi_1}{C_{p_2}} + \frac{\psi_2}{C_{p_3}} - \frac{\psi_3}{C_{p_4}}]$$

We define for simplicity new parameter Γ_4 and function $\psi_4(\psi_1, \psi_2, \psi_{3,\ldots})$

$$\Gamma_4 = \frac{\Gamma_1}{C_{p_2}} + \frac{\Gamma_2}{C_{p_3}} - \frac{\Gamma_3}{C_{p_4}}; \psi_4 = \frac{\psi_1}{C_{p_2}} + \frac{\psi_2}{C_{p_3}} - \frac{\psi_3}{C_{p_4}}; \; I_{C_{p_1}} = i_{p_2} \cdot \frac{C_{p_1}}{\Delta} \cdot \Gamma_4 + \frac{C_{p_1}}{\Delta} \cdot \psi_4$$

Lattice crystal filter differential equations group No. 4:

$$\frac{1}{C_{p_3}} \cdot I_{C_{p_3}} - L_{m_3} \cdot \frac{d^2 I_{L_{m_3}}}{dt^2} - \frac{dI_{L_{m_3}}}{dt} \cdot r_3 = \frac{1}{C_{m_3}} \cdot I_{L_{m_3}}; \; \frac{1}{C_{p_1}} \cdot I_{C_{p_1}} - L_{m_1} \cdot \frac{d^2 I_{L_{m_1}}}{dt^2} - \frac{dI_{L_{m_1}}}{dt}$$
$$\cdot r_1 = \frac{1}{C_{m_1}} \cdot I_{L_{m_1}}$$

$$-\frac{dI_{L_{m_2}}}{dt} \cdot r_2 + \frac{1}{C_1} \cdot I_{C_1} - \frac{1}{C_{p_3}} \cdot I_{C_{p_3}} - L_{m_2} \cdot \frac{d^2 I_{L_{m_2}}}{dt^2} = \frac{1}{C_{m_2}} \cdot I_{L_{m_2}}$$

$$-\frac{dI_{L_{m_4}}}{dt} \cdot r_4 + \frac{1}{C_1} \cdot I_{C_1} - \frac{1}{C_{p_1}} \cdot I_{C_{p_1}} - L_{m_4} \cdot \frac{d^2 I_{L_{m_4}}}{dt^2} = \frac{1}{C_{m_4}} \cdot I_{L_{m_4}}$$

$$\frac{di_{s_1}}{dt} = -I_{C_1} \cdot \frac{1}{C_1 \cdot R_s} \cdot \frac{n_1^2}{n_2^2} - \frac{n_1}{n_2} \cdot \frac{1}{R_s} \cdot \frac{dV_s(t)}{dt}; \frac{di_{p_2}}{dt} = -\frac{n_4^2}{n_3^2} \cdot \frac{1}{C_2 \cdot R_{load}} \cdot I_{C_2}$$

We define for simplicity new variables $Y_1 = \frac{dI_{Lm_1}}{dt}$; $Y_2 = \frac{dI_{Lm_2}}{dt}$; $Y_3 = \frac{dI_{Lm_3}}{dt}$; $Y_4 = \frac{dI_{Lm_4}}{dt}$

$$\frac{dY_1}{dt} = \frac{d^2 I_{L_{m_1}}}{dt^2}; \frac{dY_2}{dt} = \frac{d^2 I_{L_{m_2}}}{dt^2}; \frac{dY_3}{dt} = \frac{d^2 I_{L_{m_3}}}{dt^2}; \frac{dY_4}{dt} = \frac{d^2 I_{L_{m_4}}}{dt^2}$$

Lattice crystal filter differential equations group No. 5:

$$\frac{1}{C_{p_3}} \cdot I_{C_{p_3}} - L_{m_3} \cdot \frac{dY_3}{dt} - Y_3 \cdot r_3 = \frac{1}{C_{m_3}} \cdot I_{L_{m_3}}; \frac{1}{C_{p_1}} \cdot I_{C_{p_1}} - L_{m_1} \cdot \frac{dY_1}{dt} - Y_1 \cdot r_1 = \frac{1}{C_{m_1}} \cdot I_{L_{m_1}}$$

$$-Y_2 \cdot r_2 + \frac{1}{C_1} \cdot I_{C_1} - \frac{1}{C_{p_3}} \cdot I_{C_{p_3}} - L_{m_2} \cdot \frac{dY_2}{dt} = \frac{1}{C_{m_2}} \cdot I_{L_{m_2}}; \frac{dI_{L_{m_1}}}{dt} = Y_1; \frac{dI_{L_{m_2}}}{dt} = Y_2; \frac{dI_{L_{m_3}}}{dt} = Y_3; \frac{dI_{L_{m_4}}}{dt} = Y_4$$

$$-Y_4 \cdot r_4 + \frac{1}{C_1} \cdot I_{C_1} - \frac{1}{C_{p_1}} \cdot I_{C_{p_1}} - L_{m_4} \cdot \frac{dY_4}{dt} = \frac{1}{C_{m_4}} \cdot I_{L_{m_4}}$$

$$\frac{di_{s_1}}{dt} = -I_{C_1} \cdot \frac{1}{C_1 \cdot R_s} \cdot \frac{n_1^2}{n_2^2} - \frac{n_1}{n_2} \cdot \frac{1}{R_s} \cdot \frac{dV_s(t)}{dt}; \frac{di_{p_2}}{dt} = -\frac{n_4^2}{n_3^2} \cdot \frac{1}{C_2 \cdot R_{load}} \cdot I_{C_2}$$

Lattice crystal filter differential equations group No. 6:

$$\frac{dY_3}{dt} = \frac{1}{C_{p_3} \cdot L_{m_3}} \cdot I_{C_{p_3}} - Y_3 \cdot \frac{r_3}{L_{m_3}} - \frac{1}{C_{m_3} \cdot L_{m_3}} \cdot I_{L_{m_3}}; \frac{dY_1}{dt} = \frac{1}{C_{p_1} \cdot L_{m_1}} \cdot I_{C_{p_1}} - Y_1 \cdot \frac{r_1}{L_{m_1}} - \frac{1}{C_{m_1} \cdot L_{m_1}} \cdot I_{L_{m_1}}$$

$$\frac{dY_2}{dt} = -Y_2 \cdot \frac{r_2}{L_{m_2}} + \frac{1}{C_1 \cdot L_{m_2}} \cdot I_{C_1} - \frac{1}{C_{p_3} \cdot L_{m_2}} \cdot I_{C_{p_3}} - \frac{1}{C_{m_2} \cdot L_{m_2}} \cdot I_{L_{m_2}}$$

$$\frac{dY_4}{dt} = -Y_4 \cdot \frac{r_4}{L_{m_4}} + \frac{1}{C_1 \cdot L_{m_4}} \cdot I_{C_1} - \frac{1}{C_{p_1} \cdot L_{m_4}} \cdot I_{C_{p_1}} - \frac{1}{C_{m_4} \cdot L_{m_4}} \cdot I_{L_{m_4}}$$

$$\frac{dI_{L_{m_1}}}{dt} = Y_1; \frac{dI_{L_{m_2}}}{dt} = Y_2; \frac{dI_{L_{m_3}}}{dt} = Y_3; \frac{dI_{L_{m_4}}}{dt} = Y_4$$

$$\frac{di_{s_1}}{dt} = -I_{C_1} \cdot \frac{1}{C_1 \cdot R_s} \cdot \frac{n_1^2}{n_2^2} - \frac{n_1}{n_2} \cdot \frac{1}{R_s} \cdot \frac{dV_s(t)}{dt}; \frac{di_{p_2}}{dt} = -\frac{n_4^2}{n_3^2} \cdot \frac{1}{C_2 \cdot R_{load}} \cdot I_{C_2}$$

Lattice crystal filter differential equations group No. 7:

$$\frac{dY_3}{dt} = \frac{1}{C_{p_3} \cdot L_{m_3}} \cdot [i_{p_2} \cdot \frac{\Gamma_2}{\Delta} + \frac{\psi_2}{\Delta}] - Y_3 \cdot \frac{r_3}{L_{m_3}} - \frac{1}{C_{m_3} \cdot L_{m_3}} \cdot I_{L_{m_3}}$$

$$\frac{dY_1}{dt} = \frac{1}{C_{p_1} \cdot L_{m_1}} \cdot [i_{p_2} \cdot \frac{C_{p_1}}{\Delta} \cdot \Gamma_4 + \frac{C_{p_1}}{\Delta} \cdot \psi_4] - Y_1 \cdot \frac{r_1}{L_{m_1}} - \frac{1}{C_{m_1} \cdot L_{m_1}} \cdot I_{L_{m_1}}$$

$$\frac{dY_2}{dt} = -Y_2 \cdot \frac{r_2}{L_{m_2}} + \frac{1}{C_1 \cdot L_{m_2}} \cdot I_{C_1} - \frac{1}{C_{p_3} \cdot L_{m_2}} \cdot [i_{p_2} \cdot \frac{\Gamma_2}{\Delta} + \frac{\psi_2}{\Delta}] - \frac{1}{C_{m_2} \cdot L_{m_2}} \cdot I_{L_{m_2}}$$

$$\frac{dY_4}{dt} = -Y_4 \cdot \frac{r_4}{L_{m_4}} + \frac{1}{C_1 \cdot L_{m_4}} \cdot I_{C_1} - \frac{1}{C_{p_1} \cdot L_{m_4}} \cdot [i_{p_2} \cdot \frac{C_{p_1}}{\Delta} \cdot \Gamma_4 + \frac{C_{p_1}}{\Delta} \cdot \psi_4] - \frac{1}{C_{m_4} \cdot L_{m_4}} \cdot I_{L_{m_4}}$$

$$\frac{dI_{L_{m_1}}}{dt} = Y_1; \frac{dI_{L_{m_2}}}{dt} = Y_2; \frac{dI_{L_{m_3}}}{dt} = Y_3; \frac{dI_{L_{m_4}}}{dt} = Y_4$$

$$\frac{di_{s_1}}{dt} = -I_{C_1} \cdot \frac{1}{C_1 \cdot R_s} \cdot \frac{n_1^2}{n_2^2} - \frac{n_1}{n_2} \cdot \frac{1}{R_s} \cdot \frac{dV_s(t)}{dt}; \frac{di_{p_2}}{dt} = -\frac{n_4^2}{n_3^2} \cdot \frac{1}{C_2 \cdot R_{load}} \cdot I_{C_2}$$

At fixed points (equilibrium points): $\frac{dY_k}{dt} = 0 \, \forall \, \text{k} = 1,2,3,4; \frac{dI_{L_{m_1}}}{dt} = 0$

$$\frac{dI_{L_{m_2}}}{dt} = 0; \frac{dI_{L_{m_3}}}{dt} = 0; \frac{dI_{L_{m_4}}}{dt} = 0; \frac{di_{s_1}}{dt} = 0; \frac{di_{p_2}}{dt} = 0; Y_k^* = 0 \, \forall \, k = 1,2,3,4$$

Assumption $\frac{dV_s(t)}{dt} \to \varepsilon. -I_{C_1} \cdot \frac{1}{C_1 \cdot R_s} \cdot \frac{n_1^2}{n_2^2} \to \varepsilon \Rightarrow I_{C_1}^* = 0; \; I_{C_2}^* = 0.$

$$\frac{dY_1}{dt} = 0 \Rightarrow \frac{1}{C_{p_1} \cdot L_{m_1}} \cdot [i_{p_2}^* \cdot \frac{C_{p_1}}{\Delta} \cdot \Gamma_4 + \frac{C_{p_1}}{\Delta} \cdot \psi_4(I_{L_{m_k}}^*; k = 1,2,3,4; C_{p_1}, C_{p_1}, \ldots)] - \frac{1}{C_{m_1} \cdot L_{m_1}} \cdot I_{L_{m_1}}^* = 0$$

$$\frac{dY_2}{dt} = 0 \Rightarrow -\frac{1}{C_{p_3} \cdot L_{m_2}} \cdot [i_{p_2}^* \cdot \frac{\Gamma_2}{\Delta} + \frac{\psi_2(I_{L_{m_k}}^*; k = 1,2,3,4; C_{p_1}, C_{p_1}, \ldots)}{\Delta}] - \frac{1}{C_{m_2} \cdot L_{m_2}} \cdot I_{L_{m_2}}^* = 0$$

$$\frac{dY_3}{dt} = 0 \Rightarrow \frac{1}{C_{p_3} \cdot L_{m_3}} \cdot [i_{p_2}^* \cdot \frac{\Gamma_2}{\Delta} + \frac{\psi_2(I_{L_{m_k}}^*; \; k = 1,2,3,4; \; C_{p_1}, C_{p_1}, \ldots)}{\Delta}] - \frac{1}{C_{m_3} \cdot L_{m_3}} \cdot I_{L_{m_3}}^* = 0$$

$$\frac{dY_4}{dt} = 0 \Rightarrow -\frac{1}{C_{p_1} \cdot L_{m_4}} \cdot [i^*_{p_2} \cdot \frac{C_{p_1}}{\Delta} \cdot \Gamma_4 + \frac{C_{p_1}}{\Delta} \cdot \psi_4(I^*_{L_{m_k}}; k = 1,2,3,4; C_{p_1}, C_{p_1}, \ldots)] - \frac{1}{C_{m_4} \cdot L_{m_4}} \cdot I^*_{L_{m_4}} = 0$$

Stability analysis: The standard local stability analysis about any one of the equilibrium points of Lattice crystal filter circuit consists in adding to its coordinated $[Y_1\ Y_2\ Y_3\ Y_4\ I_{L_{m_1}}\ I_{L_{m_2}}\ I_{L_{m_3}}\ I_{L_{m_4}}\ i_{s_1}\ i_{p_2}]$ arbitrarily small increments of exponential terms $[y_1\ y_2\ y_3\ y_4\ i_{L_{m_1}}\ i_{L_{m_2}}\ i_{L_{m_3}}\ i_{L_{m_4}}\ i'_{s_1}\ i'_{p_2}] \cdot e^{\lambda \cdot t}$, and retaining the first order terms in $y_1\ y_2\ y_3\ y_4\ i_{L_{m_1}}\ i_{L_{m_2}}\ i_{L_{m_3}}\ i_{L_{m_4}}\ i'_{s_1}\ i'_{p_2}$. The system of ten homogeneous equations leads to a polynomial characteristic equation in the eigenvalueλ. The polynomial characteristic equation accepts by set the Lattice crystal filter circuit equations. The Lattice crystal filter circuit fixed values with arbitrarily small increments of exponential form $[y_1\ y_2\ y_3\ y_4\ i_{L_{m_1}}\ i_{L_{m_2}}\ i_{L_{m_3}}\ i_{L_{m_4}}\ i'_{s_1}\ i'_{p_2}] \cdot e^{\lambda \cdot t}$ are; i = 0 (first fixed point), i = 1 (second fixed point), i = 2 (third fixed point), etc.,

$$Y_1(t) = Y_1^{(i)} + y_1 \cdot e^{\lambda \cdot t}; Y_2(t) = Y_2^{(i)} + y_2 \cdot e^{\lambda \cdot t};$$
$$Y_3(t) = Y_3^{(i)} + y_3 \cdot e^{\lambda \cdot t}; Y_4(t) = Y_4^{(i)} + y_4 \cdot e^{\lambda \cdot t}$$
$$I_{L_{m_1}}(t) = I^{(i)}_{L_{m_1}} + i_{L_{m_1}} \cdot e^{\lambda \cdot t}; I_{L_{m_2}}(t) = I^{(i)}_{L_{m_2}} + i_{L_{m_2}} \cdot e^{\lambda \cdot t};$$
$$I_{L_{m_3}}(t) = I^{(i)}_{L_{m_3}} + i_{L_{m_3}} \cdot e^{\lambda \cdot t}; I_{L_{m_4}}(t) = I^{(i)}_{L_{m_4}} + i_{L_{m_4}} \cdot e^{\lambda \cdot t}$$
$$i_{s_1}(t) = i^{(i)}_{s_1} + i'_{s_1} \cdot e^{\lambda \cdot t}; i_{p_2}(t) = i^{(i)}_{p_2} + i'_{p_2} \cdot e^{\lambda \cdot t};$$
$$I_{C_2}(t) = I^{(i)}_{C_2} + i_{C_2} \cdot e^{\lambda \cdot t}; I_{C_1}(t) = I^{(i)}_{C_1} + i_{C_1} \cdot e^{\lambda \cdot t}$$

$$\frac{dY_1(t)}{dt} = y_1 \cdot \lambda \cdot e^{\lambda \cdot t}; \frac{dY_2(t)}{dt} = y_2 \cdot \lambda \cdot e^{\lambda \cdot t}; \frac{dY_3(t)}{dt} = y_3 \cdot \lambda \cdot e^{\lambda \cdot t}; \frac{dY_4(t)}{dt} = y_4 \cdot \lambda \cdot e^{\lambda \cdot t}$$

$$\frac{dI_{L_{m_1}}(t)}{dt} = i_{L_{m_1}} \cdot \lambda \cdot e^{\lambda \cdot t}; \frac{dI_{L_{m_2}}(t)}{dt} = i_{L_{m_2}} \cdot \lambda \cdot e^{\lambda \cdot t}; \frac{dI_{L_{m_3}}(t)}{dt} = i_{L_{m_3}} \cdot \lambda \cdot e^{\lambda \cdot t}; \frac{dI_{L_{m_4}}(t)}{dt}$$
$$= i_{L_{m_4}} \cdot \lambda \cdot e^{\lambda \cdot t}$$

$$\frac{di_{s_1}(t)}{dt} = i'_{s_1} \cdot \lambda \cdot e^{\lambda \cdot t}; \frac{di_{p_2}(t)}{dt} = i'_{p_2} \cdot \lambda \cdot e^{\lambda \cdot t}$$

$$I_{C_2} = \frac{C_2}{C_{p_3}} \cdot I_{C_{p_3}} - \frac{C_2}{C_{p_1}} \cdot I_{C_{p_1}} = i_{p_2} \cdot \frac{C_2}{\Delta} \cdot [\frac{\Gamma_2}{C_{p_3}} - \Gamma_4]$$
$$+ \frac{C_2}{\Delta} \cdot [\frac{1}{C_{p_3}} \cdot \psi_2(I_{L_{m_k}}; k = 1,2,3,4; C_{p_1}, C_{p_1}, \ldots)$$
$$- \psi_4(I_{L_{m_k}}; k = 1,2,3,4; C_{p_1}, C_{p_1}, \ldots)]$$

$$I_{C_1} = \frac{C_1}{C_{p3}} \cdot I_{C_{p3}} + \frac{C_1}{C_{p2}} \cdot I_{C_{p2}} = i_{p_2} \cdot \frac{C_1}{\Delta} \cdot [\frac{\Gamma_2}{C_{p_3}} + \frac{\Gamma_1}{C_{p_2}}]$$
$$+ \frac{C_1}{\Delta} \cdot [\frac{1}{C_{p_3}} \cdot \psi_2(I_{L_{m_k}}; k = 1,2,3,4; C_{p_1}, C_{p_1}, \ldots) + \frac{1}{C_{p_2}}$$
$$\cdot \psi_1(I_{L_{m_k}}; k = 1,2,3,4;\ C_{p_1}, C_{p_1}, \ldots)]$$
$$I_{C_1} = \frac{C_1}{C_{p_4}} \cdot I_{C_{p_4}} + \frac{C_1}{C_{p_1}} \cdot I_{C_{p_1}} = i_{p_2} \cdot \frac{C_1}{\Delta} \cdot [\frac{\Gamma_3}{C_{p_4}} + \Gamma_4]$$
$$+ \frac{C_1}{\Delta} \cdot [\frac{1}{C_{p_4}} \cdot \psi_3(I_{L_{m_k}}; k = 1,2,3,4; C_{p_1}, C_{p_1}, \ldots)$$
$$+ \psi_4(I_{L_{m_k}}; k = 1,2,3,4; C_{p_1}, C_{p_1}, \ldots)]$$

Lattice crystal filter differential equations group No. 8: $\frac{dV_s(t)}{dt} \to \varepsilon$

$$\frac{dY_3}{dt} = \frac{1}{C_{p_3} \cdot L_{m_3}} \cdot [i_{p_2} \cdot \frac{\Gamma_2}{\Delta} + \frac{\psi_2}{\Delta}] - Y_3 \cdot \frac{r_3}{L_{m_3}} - \frac{1}{C_{m_3} \cdot L_{m_3}} \cdot I_{L_{m_3}}$$
$$\frac{dY_1}{dt} = \frac{1}{C_{p_1} \cdot L_{m_1}} \cdot [i_{p_2} \cdot \frac{C_{p_1}}{\Delta} \cdot \Gamma_4 + \frac{C_{p_1}}{\Delta} \cdot \psi_4] - Y_1 \cdot \frac{r_1}{L_{m_1}} - \frac{1}{C_{m_1} \cdot L_{m_1}} \cdot I_{L_{m_1}}$$

$$\frac{dY_2}{dt} = -Y_2 \cdot \frac{r_2}{L_{m_2}} + \frac{1}{C_1 \cdot L_{m_2}} \cdot (i_{p_2} \cdot \frac{C_1}{\Delta} \cdot [\frac{\Gamma_2}{C_{p_3}} + \frac{\Gamma_1}{C_{p_2}}] + \frac{C_1}{\Delta} \cdot [\frac{1}{C_{p_3}} \cdot \psi_2 + \frac{1}{C_{p_2}} \cdot \psi_1])$$
$$- \frac{1}{C_{p_3} \cdot L_{m_2}} \cdot [i_{p_2} \cdot \frac{\Gamma_2}{\Delta} + \frac{\psi_2}{\Delta}] - \frac{1}{C_{m_2} \cdot L_{m_2}} \cdot I_{L_{m_2}}$$

$$\frac{dY_4}{dt} = -Y_4 \cdot \frac{r_4}{L_{m_4}} + \frac{1}{C_1 \cdot L_{m_4}} \cdot (i_{p_2} \cdot \frac{C_1}{\Delta} \cdot [\frac{\Gamma_2}{C_{p_3}} + \frac{\Gamma_1}{C_{p_2}}] + \frac{C_1}{\Delta} \cdot [\frac{1}{C_{p_3}} \cdot \psi_2 + \frac{1}{C_{p_2}} \cdot \psi_1])$$
$$- \frac{1}{C_{p_1} \cdot L_{m_4}} \cdot [i_{p_2} \cdot \frac{C_{p_1}}{\Delta} \cdot \Gamma_4 + \frac{C_{p_1}}{\Delta} \cdot \psi_4] - \frac{1}{C_{m_4} \cdot L_{m_4}} \cdot I_{L_{m_4}}$$

$$\frac{dI_{L_{m_1}}}{dt} = Y_1; \frac{dI_{L_{m_2}}}{dt} = Y_2; \frac{dI_{L_{m_3}}}{dt} = Y_3; \frac{dI_{L_{m_4}}}{dt} = Y_4$$

$$\frac{di_{s_1}}{dt} = -\{i_{p_2} \cdot \frac{C_1}{\Delta} \cdot [\frac{\Gamma_2}{C_{p_3}} + \frac{\Gamma_1}{C_{p_2}}] + \frac{C_1}{\Delta} \cdot [\frac{1}{C_{p_3}} \cdot \psi_2 + \frac{1}{C_{p_2}} \cdot \psi_1]\} \cdot \frac{1}{C_1 \cdot R_s}$$
$$\cdot \frac{n_1^2}{n_2^2} - \frac{n_1}{n_2} \cdot \frac{1}{R_s} \cdot \frac{dV_s(t)}{dt}$$
$$\frac{di_{p_2}}{dt} = -\frac{n_4^2}{n_3^2} \cdot \frac{1}{C_2 \cdot R_{load}} \cdot \{i_{p_2} \cdot \frac{C_2}{\Delta} \cdot [\frac{\Gamma_2}{C_{p_3}} - \Gamma_4] + \frac{C_2}{\Delta} \cdot [\frac{1}{C_{p_3}} \cdot \psi_2 - \psi_4]\}$$

Remark

$$\begin{aligned}
&\psi_k(I_{L_{m_k}}(t) = I^*_{L_{m_k}} + i_{L_{m_k}} \cdot e^{\lambda \cdot t};\ k = 1,2,3,4; C_{p_1}, C_{p_1}, \ldots) \\
&\quad = \psi_k(I^*_{L_{m_k}};\ k = 1,2,3,4;\ C_{p_1}, C_{p_1}, \ldots) \\
&\quad + \psi_k(i_{L_{m_k}} \cdot e^{\lambda \cdot t};\ k = 1,2,3,4; C_{p_1}, C_{p_1}, \ldots); \\
&\psi_k(I_{L_{m_k}}) = \psi_k(I^*_{L_{m_k}}) + \psi_k(i_{L_{m_k}} \cdot e^{\lambda \cdot t});\ k = 1,2,3,4 \\
&\psi_k(i_{L_{m_k}} \cdot e^{\lambda \cdot t}) = e^{\lambda \cdot t} \cdot \psi_k(i_{L_{m_k}})\, \forall k = 1,2,3,4
\end{aligned}$$

For $\lambda < 0$, $t > 0$ the selected fixed point is stable otherwise $\lambda > 0$, $t > 0$ unstable. Our Lattice crystal filter circuit tends to the selected fixed point exponentially for $\lambda < 0$, $t > 0$ otherwise go away from the selected fixed point exponentially. λ is the eigenvalue parameter which establish if the fixed point is stable or unstable, additionally his absolute value $|\lambda|$ establish the speed of flow toward or away from the selected fixed point [2–4].

$$\frac{dY_1}{dt} = \frac{1}{C_{p_1} \cdot L_{m_1}} \cdot [i_{p_2} \cdot \frac{C_{p_1}}{\Delta} \cdot \Gamma_4 + \frac{C_{p_1}}{\Delta} \cdot \psi_4] - Y_1 \cdot \frac{r_1}{L_{m_1}} - \frac{1}{C_{m_1} \cdot L_{m_1}} \cdot I_{L_{m_1}}$$

$$\begin{aligned}
y_1 \cdot \lambda \cdot e^{\lambda \cdot t} &= \frac{1}{C_{p_1} \cdot L_{m_1}} \cdot [(i^{(i)}_{p_2} + i'_{p_2} \cdot e^{\lambda \cdot t}) \cdot \frac{C_{p_1}}{\Delta} \cdot \Gamma_4 + \frac{C_{p_1}}{\Delta} \cdot (\psi_4(I^*_{L_{m_k}}) + \psi_4(i_{L_{m_k}} \cdot e^{\lambda \cdot t}))] \\
&\quad - [Y^{(i)}_1 + y_1 \cdot e^{\lambda \cdot t}] \cdot \frac{r_1}{L_{m_1}} - \frac{1}{C_{m_1} \cdot L_{m_1}} \cdot [I^{(i)}_{L_{m_1}} + i_{L_{m_1}} \cdot e^{\lambda \cdot t}]
\end{aligned}$$

$$\begin{aligned}
y_1 \cdot \lambda \cdot e^{\lambda \cdot t} &= \frac{1}{C_{p_1} \cdot L_{m_1}} \cdot [i^{(i)}_{p_2} \cdot \frac{C_{p_1}}{\Delta} \cdot \Gamma_4 + i'_{p_2} \cdot \frac{C_{p_1}}{\Delta} \cdot \Gamma_4 \cdot e^{\lambda \cdot t} + \frac{C_{p_1}}{\Delta} \cdot \psi_4(I^*_{L_{m_k}}) \\
&\quad + \frac{C_{p_1}}{\Delta} \cdot \psi_4(i_{L_{m_k}} \cdot e^{\lambda \cdot t})] - Y^{(i)}_1 \cdot \frac{r_1}{L_{m_1}} - y_1 \cdot \frac{r_1}{L_{m_1}} \cdot e^{\lambda \cdot t} - \frac{1}{C_{m_1} \cdot L_{m_1}} \\
&\quad \cdot I^{(i)}_{L_{m_1}} - \frac{1}{C_{m_1} \cdot L_{m_1}} \cdot i_{L_{m_1}} \cdot e^{\lambda \cdot t}
\end{aligned}$$

$$\begin{aligned}
y_1 \cdot \lambda \cdot e^{\lambda \cdot t} &= \frac{1}{C_{p_1} \cdot L_{m_1}} \cdot [i^{(i)}_{p_2} \cdot \frac{C_{p_1}}{\Delta} \cdot \Gamma_4 + \frac{C_{p_1}}{\Delta} \cdot \psi_4(I^*_{L_{m_k}})] - Y^{(i)}_1 \cdot \frac{r_1}{L_{m_1}} - \frac{1}{C_{m_1} \cdot L_{m_1}} \cdot I^{(i)}_{L_{m_1}} \\
&\quad + \frac{1}{C_{p_1} \cdot L_{m_1}} \cdot [i'_{p_2} \cdot \frac{C_{p_1}}{\Delta} \cdot \Gamma_4 \cdot e^{\lambda \cdot t} + \frac{C_{p_1}}{\Delta} \cdot \psi_4(i_{L_{m_k}} \cdot e^{\lambda \cdot t})] \\
&\quad - y_1 \cdot \frac{r_1}{L_{m_1}} \cdot e^{\lambda \cdot t} - \frac{1}{C_{m_1} \cdot L_{m_1}} \cdot i_{L_{m_1}} \cdot e^{\lambda \cdot t}
\end{aligned}$$

At fixed points: $\frac{1}{C_{p_1} \cdot L_{m_1}} \cdot [i^{(i)}_{p_2} \cdot \frac{C_{p_1}}{\Delta} \cdot \Gamma_4 + \frac{C_{p_1}}{\Delta} \cdot \psi_4(I^*_{L_{m_k}})] - Y^{(i)}_1 \cdot \frac{r_1}{L_{m_1}} - \frac{1}{C_{m_1} \cdot L_{m_1}} \cdot I^{(i)}_{L_{m_1}} = 0$

$$-y_1 \cdot \lambda + \frac{1}{C_{p_1} \cdot L_{m_1}} \cdot [i'_{p_2} \cdot \frac{C_{p_1}}{\Delta} \cdot \Gamma_4 + \frac{C_{p_1}}{\Delta} \cdot \psi_4(i_{L_{m_k}})] - y_1 \cdot \frac{r_1}{L_{m_1}} - \frac{1}{C_{m_1} \cdot L_{m_1}} \cdot i_{L_{m_1}} = 0$$

$$\frac{dY_2}{dt} = -Y_2 \cdot \frac{r_2}{L_{m_2}} + \frac{1}{C_1 \cdot L_{m_2}} \cdot (i_{p_2} \cdot \frac{C_1}{\Delta} \cdot [\frac{\Gamma_2}{C_{p_3}} + \frac{\Gamma_1}{C_{p_2}}] + \frac{C_1}{\Delta} \cdot [\frac{1}{C_{p_3}} \cdot \psi_2 + \frac{1}{C_{p_2}} \cdot \psi_1]) - \frac{1}{C_{p_3} \cdot L_{m_2}} \cdot [i_{p_2} \cdot \frac{\Gamma_2}{\Delta} + \frac{\psi_2}{\Delta}] - \frac{1}{C_{m_2} \cdot L_{m_2}} \cdot I_{L_{m_2}}$$

$$y_2 \cdot \lambda \cdot e^{\lambda \cdot t} = -[Y_2^{(i)} + y_2 \cdot e^{\lambda \cdot t}] \cdot \frac{r_2}{L_{m_2}} + \frac{1}{C_1 \cdot L_{m_2}} \cdot ([i_{p_2}^{(i)} + i'_{p_2} \cdot e^{\lambda \cdot t}] \cdot \frac{C_1}{\Delta} \cdot [\frac{\Gamma_2}{C_{p_3}} + \frac{\Gamma_1}{C_{p_2}}] + \frac{C_1}{\Delta} \cdot [\frac{1}{C_{p_3}} \cdot [\psi_2(I^*_{L_{m_k}}) + \psi_2(i_{L_{m_k}} \cdot e^{\lambda \cdot t})] + \frac{1}{C_{p_2}} \cdot \{\psi_1(I^*_{L_{m_k}}) + \psi_1(i_{L_{m_k}} \cdot e^{\lambda \cdot t})\}]) - \frac{1}{C_{p_3} \cdot L_{m_2}} \cdot [(i_{p_2}^{(i)} + i'_{p_2} \cdot e^{\lambda \cdot t}) \cdot \frac{\Gamma_2}{\Delta} + \frac{[\psi_2(I^*_{L_{m_k}}) + \psi_2(i_{L_{m_k}} \cdot e^{\lambda \cdot t})]}{\Delta}] - \frac{1}{C_{m_2} \cdot L_{m_2}} \cdot [I_{L_{m_2}}^{(i)} + i_{L_{m_2}} \cdot e^{\lambda \cdot t}]$$

$$y_2 \cdot \lambda \cdot e^{\lambda \cdot t} = -Y_2^{(i)} \cdot \frac{r_2}{L_{m_2}} + \frac{1}{C_1 \cdot L_{m_2}} \cdot (i_{p_2}^{(i)} \cdot \frac{C_1}{\Delta} \cdot [\frac{\Gamma_2}{C_{p_3}} + \frac{\Gamma_1}{C_{p_2}}] + \frac{C_1}{\Delta} \cdot [\frac{1}{C_{p_3}} \cdot \psi_2(I^*_{L_{m_k}}) + \frac{1}{C_{p_2}} \cdot \psi_1(I^*_{L_{m_k}})]) - \frac{1}{C_{p_3} \cdot L_{m_2}} \cdot [i_{p_2}^{(i)} \cdot \frac{\Gamma_2}{\Delta} + \frac{\psi_2(I^*_{L_{m_k}})}{\Delta}] - \frac{1}{C_{m_2} \cdot L_{m_2}} \cdot I_{L_{m_2}}^{(i)} - y_2 \cdot e^{\lambda \cdot t} \cdot \frac{r_2}{L_{m_2}} + \frac{1}{C_1 \cdot L_{m_2}} \cdot (i'_{p_2} \cdot e^{\lambda \cdot t} \cdot \frac{C_1}{\Delta} \cdot [\frac{\Gamma_2}{C_{p_3}} + \frac{\Gamma_1}{C_{p_2}}] + \frac{C_1}{\Delta} \cdot [\frac{1}{C_{p_3}} \cdot \psi_2(i_{L_{m_k}} \cdot e^{\lambda \cdot t}) + \frac{1}{C_{p_2}} \cdot \psi_1(i_{L_{m_k}} \cdot e^{\lambda \cdot t})]) - \frac{1}{C_{p_3} \cdot L_{m_2}} \cdot [i'_{p_2} \cdot e^{\lambda \cdot t} \cdot \frac{\Gamma_2}{\Delta} + \frac{\psi_2(i_{L_{m_k}} \cdot e^{\lambda \cdot t})}{\Delta}] - \frac{1}{C_{m_2} \cdot L_{m_2}} \cdot i_{L_{m_2}} \cdot e^{\lambda \cdot t}$$

At fixed points:

$$-Y_2^{(i)} \cdot \frac{r_2}{L_{m_2}} + \frac{1}{C_1 \cdot L_{m_2}} \cdot (i_{p_2}^{(i)} \cdot \frac{C_1}{\Delta} \cdot [\frac{\Gamma_2}{C_{p_3}} + \frac{\Gamma_1}{C_{p_2}}] + \frac{C_1}{\Delta} \cdot [\frac{1}{C_{p_3}} \cdot \psi_2(I^*_{L_{m_k}}) + \frac{1}{C_{p_2}} \cdot \psi_1(I^*_{L_{m_k}})]) - \frac{1}{C_{p_3} \cdot L_{m_2}} \cdot [i_{p_2}^{(i)} \cdot \frac{\Gamma_2}{\Delta} + \frac{\psi_2(I^*_{L_{m_k}})}{\Delta}] - \frac{1}{C_{m_2} \cdot L_{m_2}} \cdot I_{L_{m_2}}^{(i)} = 0$$

$$y_2 \cdot \lambda \cdot e^{\lambda \cdot t} = -y_2 \cdot e^{\lambda \cdot t} \cdot \frac{r_2}{L_{m_2}} + \frac{1}{C_1 \cdot L_{m_2}} \cdot (i'_{p_2} \cdot e^{\lambda \cdot t} \cdot \frac{C_1}{\Delta} \cdot [\frac{\Gamma_2}{C_{p_3}} + \frac{\Gamma_1}{C_{p_2}}] + \frac{C_1}{\Delta} \cdot [\frac{1}{C_{p_3}} \cdot \psi_2(i_{L_{m_k}} \cdot e^{\lambda \cdot t}) + \frac{1}{C_{p_2}} \cdot \psi_1(i_{L_{m_k}} \cdot e^{\lambda \cdot t})]) - \frac{1}{C_{p_3} \cdot L_{m_2}} \cdot [i'_{p_2} \cdot e^{\lambda \cdot t} \cdot \frac{\Gamma_2}{\Delta} + \frac{\psi_2(i_{L_{m_k}} \cdot e^{\lambda \cdot t})}{\Delta}] - \frac{1}{C_{m_2} \cdot L_{m_2}} \cdot i_{L_{m_2}} \cdot e^{\lambda \cdot t}$$

$$\psi_1(i_{L_{m_k}} \cdot e^{\lambda \cdot t}) = e^{\lambda \cdot t} \cdot \psi_1(i_{L_{m_k}});\ \psi_2(i_{L_{m_k}} \cdot e^{\lambda \cdot t}) = e^{\lambda \cdot t} \cdot \psi_2(i_{L_{m_k}})$$

$$- y_2 \cdot \lambda - y_2 \cdot \frac{r_2}{L_{m_2}} + i'_{p_2} \cdot \frac{1}{\Delta \cdot L_{m_2}} \cdot \frac{\Gamma_1}{C_{p_2}} + \frac{1}{L_{m_2} \cdot \Delta} \cdot [\frac{1}{C_{p_3}} \cdot \psi_2(i_{L_{m_k}}) + \frac{1}{C_{p_2}} \cdot \psi_1(i_{L_{m_k}})]$$
$$- \frac{1}{C_{p_3} \cdot L_{m_2}} \cdot \frac{\psi_2(i_{L_{m_k}})}{\Delta} - \frac{1}{C_{m_2} \cdot L_{m_2}} \cdot i_{L_{m_2}} = 0$$

$$\frac{dY_3}{dt} = \frac{1}{C_{p_3} \cdot L_{m_3}} \cdot [i_{p_2} \cdot \frac{\Gamma_2}{\Delta} + \frac{\psi_2}{\Delta}] - Y_3 \cdot \frac{r_3}{L_{m_3}} - \frac{1}{C_{m_3} \cdot L_{m_3}} \cdot I_{L_{m_3}}$$

$$y_3 \cdot \lambda \cdot e^{\lambda \cdot t} = \frac{1}{C_{p_3} \cdot L_{m_3}} \cdot [(i_{p_2}^{(i)} + i'_{p_2} \cdot e^{\lambda \cdot t}) \cdot \frac{\Gamma_2}{\Delta} + \frac{[\psi_2(I^*_{L_{m_k}}) + \psi_2(i_{L_{m_k}} \cdot e^{\lambda \cdot t})]}{\Delta}]$$
$$- (Y_3^{(i)} + y_3 \cdot e^{\lambda \cdot t}) \cdot \frac{r_3}{L_{m_3}} - \frac{1}{C_{m_3} \cdot L_{m_3}} \cdot (I_{L_{m_3}}^{(i)} + i_{L_{m_3}} \cdot e^{\lambda \cdot t})$$

$$y_3 \cdot \lambda \cdot e^{\lambda \cdot t} = \frac{1}{C_{p_3} \cdot L_{m_3}} \cdot [i_{p_2}^{(i)} \cdot \frac{\Gamma_2}{\Delta} + \frac{\psi_2(I^*_{L_{m_k}})}{\Delta}] - Y_3^{(i)} \cdot \frac{r_3}{L_{m_3}} - \frac{1}{C_{m_3} \cdot L_{m_3}} \cdot I_{L_{m_3}}^{(i)}$$
$$+ \frac{1}{C_{p_3} \cdot L_{m_3}} \cdot [i'_{p_2} \cdot e^{\lambda \cdot t} \cdot \frac{\Gamma_2}{\Delta} + \frac{\psi_2(i_{L_{m_k}} \cdot e^{\lambda \cdot t})}{\Delta}] - y_3 \cdot e^{\lambda \cdot t} \cdot \frac{r_3}{L_{m_3}}$$
$$- \frac{1}{C_{m_3} \cdot L_{m_3}} \cdot i_{L_{m_3}} \cdot e^{\lambda \cdot t}$$

At fixed points: $\frac{1}{C_{p_3} \cdot L_{m_3}} \cdot [i_{p_2}^{(i)} \cdot \frac{\Gamma_2}{\Delta} + \frac{\psi_2(I^*_{L_{m_k}})}{\Delta}] - Y_3^{(i)} \cdot \frac{r_3}{L_{m_3}} - \frac{1}{C_{m_3} \cdot L_{m_3}} \cdot I_{L_{m_3}}^{(i)} = 0$

$$-y_3 \cdot \lambda + \frac{1}{C_{p_3} \cdot L_{m_3}} \cdot [i'_{p_2} \cdot \frac{\Gamma_2}{\Delta} + \frac{\psi_2(i_{L_{m_k}})}{\Delta}] - y_3 \cdot \frac{r_3}{L_{m_3}} - \frac{1}{C_{m_3} \cdot L_{m_3}} \cdot i_{L_{m_3}} = 0$$

$$\frac{dY_4}{dt} = -Y_4 \cdot \frac{r_4}{L_{m_4}} + \frac{1}{C_1 \cdot L_{m_4}} \cdot (i_{p_2} \cdot \frac{C_1}{\Delta} \cdot [\frac{\Gamma_2}{C_{p_3}} + \frac{\Gamma_1}{C_{p_2}}] + \frac{C_1}{\Delta} \cdot [\frac{1}{C_{p_3}} \cdot \psi_2 + \frac{1}{C_{p_2}} \cdot \psi_1])$$
$$- \frac{1}{C_{p_1} \cdot L_{m_4}} \cdot [i_{p_2} \cdot \frac{C_{p_1}}{\Delta} \cdot \Gamma_4 + \frac{C_{p_1}}{\Delta} \cdot \psi_4] - \frac{1}{C_{m_4} \cdot L_{m_4}} \cdot I_{L_{m_4}}$$

$$y_4 \cdot \lambda \cdot e^{\lambda \cdot t} = -(Y_4^{(i)} + y_4 \cdot e^{\lambda \cdot t}) \cdot \frac{r_4}{L_{m_4}} + \frac{1}{C_1 \cdot L_{m_4}} \cdot ([i_{p_2}^{(i)} + i'_{p_2} \cdot e^{\lambda \cdot t}] \cdot \frac{C_1}{\Delta} \cdot [\frac{\Gamma_2}{C_{p_3}} + \frac{\Gamma_1}{C_{p_2}}]$$
$$+ \frac{C_1}{\Delta} \cdot [\frac{1}{C_{p_3}} \cdot [\psi_2(I^*_{L_{m_k}}) + \psi_2(i_{L_{m_k}} \cdot e^{\lambda \cdot t})] + \frac{1}{C_{p_2}} \cdot \{\psi_1(I^*_{L_{m_k}}) + \psi_1(i_{L_{m_k}} \cdot e^{\lambda \cdot t})\}])$$
$$- \frac{1}{C_{p_1} \cdot L_{m_4}} \cdot [(i_{p_2}^{(i)} + i'_{p_2} \cdot e^{\lambda \cdot t}) \cdot \frac{C_{p_1}}{\Delta} \cdot \Gamma_4 + \frac{C_{p_1}}{\Delta} \cdot \{\psi_4(I^*_{L_{m_k}}) + \psi_4(i_{L_{m_k}} \cdot e^{\lambda \cdot t})\}]$$
$$- \frac{1}{C_{m_4} \cdot L_{m_4}} \cdot (I_{L_{m_4}}^{(i)} + i_{L_{m_4}} \cdot e^{\lambda \cdot t})$$

$$y_4 \cdot \lambda \cdot e^{\lambda \cdot t} = -Y_4^{(i)} \cdot \frac{r_4}{L_{m_4}} + \frac{1}{C_1 \cdot L_{m_4}} \cdot (i_{p2}^{(i)} \cdot \frac{C_1}{\Delta} \cdot [\frac{\Gamma_2}{C_{p_3}} + \frac{\Gamma_1}{C_{p_2}}] + \frac{C_1}{\Delta} \cdot [\frac{1}{C_{p_3}} \cdot \psi_2(I_{L_{m_k}}^*) + \frac{1}{C_{p_2}} \cdot \psi_1(I_{L_{m_k}}^*)])$$
$$- \frac{1}{C_{p_1} \cdot L_{m_4}} \cdot [i_{p2}^{(i)} \cdot \frac{C_{p_1}}{\Delta} \cdot \Gamma_4 + \frac{C_{p_1}}{\Delta} \cdot \psi_4(I_{L_{m_k}}^*)] - \frac{1}{C_{m_4} \cdot L_{m_4}} \cdot I_{L_{m_4}}^{(i)} - y_4 \cdot e^{\lambda \cdot t} \cdot \frac{r_4}{L_{m_4}}$$
$$+ \frac{1}{C_1 \cdot L_{m_4}} \cdot (i'_{p_2} \cdot e^{\lambda \cdot t} \cdot \frac{C_1}{\Delta} \cdot [\frac{\Gamma_2}{C_{p_3}} + \frac{\Gamma_1}{C_{p_2}}] + \frac{C_1}{\Delta} \cdot [\frac{1}{C_{p_3}} \cdot \psi_2(i_{L_{m_k}} \cdot e^{\lambda \cdot t}) + \frac{1}{C_{p_2}} \cdot \psi_1(i_{L_{m_k}} \cdot e^{\lambda \cdot t})])$$
$$- \frac{1}{C_{p_1} \cdot L_{m_4}} \cdot [i'_{p_2} \cdot e^{\lambda \cdot t} \cdot \frac{C_{p_1}}{\Delta} \cdot \Gamma_4 + \frac{C_{p_1}}{\Delta} \cdot \psi_4(i_{L_{m_k}} \cdot e^{\lambda \cdot t})] - \frac{1}{C_{m_4} \cdot L_{m_4}} \cdot i_{L_{m_4}} \cdot e^{\lambda \cdot t}$$

At fixed points:

$$- Y_4^{(i)} \cdot \frac{r_4}{L_{m_4}} + \frac{1}{C_1 \cdot L_{m_4}} \cdot (i_{p2}^{(i)} \cdot \frac{C_1}{\Delta} \cdot [\frac{\Gamma_2}{C_{p_3}} + \frac{\Gamma_1}{C_{p_2}}] + \frac{C_1}{\Delta} \cdot [\frac{1}{C_{p_3}} \cdot \psi_2(I_{L_{m_k}}^*) + \frac{1}{C_{p_2}} \cdot \psi_1(I_{L_{m_k}}^*)])$$
$$- \frac{1}{C_{p_1} \cdot L_{m_4}} \cdot [i_{p2}^{(i)} \cdot \frac{C_{p_1}}{\Delta} \cdot \Gamma_4 + \frac{C_{p_1}}{\Delta} \cdot \psi_4(I_{L_{m_k}}^*)] - \frac{1}{C_{m_4} \cdot L_{m_4}} \cdot I_{L_{m_4}}^{(i)} = 0$$

$$y_4 \cdot \lambda \cdot e^{\lambda \cdot t} = -y_4 \cdot e^{\lambda \cdot t} \cdot \frac{r_4}{L_{m_4}} + \frac{1}{C_1 \cdot L_{m_4}} \cdot (i'_{p_2} \cdot e^{\lambda \cdot t} \cdot \frac{C_1}{\Delta} \cdot [\frac{\Gamma_2}{C_{p_3}} + \frac{\Gamma_1}{C_{p_2}}] + \frac{C_1}{\Delta}$$
$$\cdot [\frac{1}{C_{p_3}} \cdot \psi_2(i_{L_{m_k}} \cdot e^{\lambda \cdot t}) + \frac{1}{C_{p_2}} \cdot \psi_1(i_{L_{m_k}} \cdot e^{\lambda \cdot t})]) - \frac{1}{C_{p_1} \cdot L_{m_4}}$$
$$\cdot [i'_{p_2} \cdot e^{\lambda \cdot t} \cdot \frac{C_{p_1}}{\Delta} \cdot \Gamma_4 + \frac{C_{p_1}}{\Delta} \cdot \psi_4(i_{L_{m_k}} \cdot e^{\lambda \cdot t})] - \frac{1}{C_{m_4} \cdot L_{m_4}} \cdot i_{L_{m_4}} \cdot e^{\lambda \cdot t}$$

$$- y_4 \cdot \lambda - y_4 \cdot \frac{r_4}{L_{m_4}} + \frac{1}{C_1 \cdot L_{m_4}} \cdot (i'_{p_2} \cdot \frac{C_1}{\Delta} \cdot [\frac{\Gamma_2}{C_{p_3}} + \frac{\Gamma_1}{C_{p_2}}] + \frac{C_1}{\Delta} \cdot [\frac{1}{C_{p_3}} \cdot \psi_2(i_{L_{m_k}})$$
$$+ \frac{1}{C_{p_2}} \cdot \psi_1(i_{L_{m_k}})]) - \frac{1}{C_{p_1} \cdot L_{m_4}} \cdot [i'_{p_2} \cdot \frac{C_{p_1}}{\Delta} \cdot \Gamma_4 + \frac{C_{p_1}}{\Delta} \cdot \psi_4(i_{L_{m_k}})] - \frac{1}{C_{m_4} \cdot L_{m_4}} \cdot i_{L_{m_4}} = 0$$

$$i_{L_{m_1}} \cdot \lambda \cdot e^{\lambda \cdot t} = Y_1^{(i)} + y_1 \cdot e^{\lambda \cdot t};\ Y_1^{(i)} = 0;\ -i_{L_{m_1}} \cdot \lambda + y_1 = 0$$
$$i_{L_{m_2}} \cdot \lambda \cdot e^{\lambda \cdot t} = Y_2^{(i)} + y_2 \cdot e^{\lambda \cdot t};\ Y_2^{(i)} = 0;\ -i_{L_{m_2}} \cdot \lambda + y_2 = 0$$
$$i_{L_{m_3}} \cdot \lambda \cdot e^{\lambda \cdot t} = Y_3^{(i)} + y_3 \cdot e^{\lambda \cdot t};\ Y_3^{(i)} = 0;\ -i_{L_{m_3}} \cdot \lambda + y_3 = 0$$
$$i_{L_{m_4}} \cdot \lambda \cdot e^{\lambda \cdot t} = Y_4^{(i)} + y_4 \cdot e^{\lambda \cdot t};\ Y_4^{(i)} = 0;\ -i_{L_{m_4}} \cdot \lambda + y_4 = 0$$

$$\frac{di_{s_1}}{dt} = -[i_{p_2} \cdot \frac{C_1}{\Delta} \cdot [\frac{\Gamma_2}{C_{p_3}} + \frac{\Gamma_1}{C_{p_2}}] + \frac{C_1}{\Delta} \cdot [\frac{1}{C_{p_3}} \cdot \psi_2 + \frac{1}{C_{p_2}} \cdot \psi_1]] \cdot \frac{1}{C_1 \cdot R_s} \cdot \frac{n_1^2}{n_2^2} - \frac{n_1}{n_2} \cdot \frac{1}{R_s}$$
$$\cdot \frac{dV_s(t)}{dt}$$

$$i'_{s_1}\cdot\lambda\cdot e^{\lambda\cdot t}=-\{(i_{p_2}^{(i)}+i'_{p_2}\cdot e^{\lambda\cdot t})\cdot\frac{C_1}{\Delta}\cdot[\frac{\Gamma_2}{C_{p_3}}+\frac{\Gamma_1}{C_{p_2}}]+\frac{C_1}{\Delta}\cdot[\frac{1}{C_{p_3}}\cdot\{\psi_2(I^*_{L_{m_k}}) +\psi_2(i_{L_{m_k}}\cdot e^{\lambda\cdot t})\}+\frac{1}{C_{p_2}}\cdot\{\psi_1(I^*_{L_{m_k}})+\psi_1(i_{L_{m_k}}\cdot e^{\lambda\cdot t})\}]\}\cdot\frac{1}{C_1\cdot R_s} \cdot\frac{n_1^2}{n_2^2}-\frac{n_1}{n_2}\cdot\frac{1}{R_s}\cdot\frac{dV_s(t)}{dt}$$

$$i'_{s_1}\cdot\lambda\cdot e^{\lambda\cdot t}=-\{(i_{p_2}^{(i)})\cdot\frac{C_1}{\Delta}\cdot[\frac{\Gamma_2}{C_{p_3}}+\frac{\Gamma_1}{C_{p_2}}]+\frac{C_1}{\Delta}\cdot[\frac{1}{C_{p_3}}\cdot\psi_2(I^*_{L_{m_k}})+\frac{1}{C_{p_2}}\cdot\psi_1(I^*_{L_{m_k}})]\} \cdot\frac{1}{C_1\cdot R_s}\cdot\frac{n_1^2}{n_2^2}-\frac{n_1}{n_2}\cdot\frac{1}{R_s}\cdot\{\frac{dV_s(t)}{dt}\to\varepsilon\}-\{i'_{p_2}\cdot e^{\lambda\cdot t}\cdot\frac{C_1}{\Delta}\cdot[\frac{\Gamma_2}{C_{p_3}}+\frac{\Gamma_1}{C_{p_2}}] +\frac{C_1}{\Delta}\cdot[\frac{1}{C_{p_3}}\cdot\psi_2(i_{L_{m_k}}\cdot e^{\lambda\cdot t})+\frac{1}{C_{p_2}}\cdot\psi_1(i_{L_{m_k}}\cdot e^{\lambda\cdot t})]\}\cdot\frac{1}{C_1\cdot R_s}\cdot\frac{n_1^2}{n_2^2}$$

At fixed points: $-\{(i_{p_2}^{(i)})\cdot\frac{C_1}{\Delta}\cdot[\frac{\Gamma_2}{C_{p_3}}+\frac{\Gamma_1}{C_{p_2}}]+\frac{C_1}{\Delta}\cdot[\frac{1}{C_{p_3}}\cdot\psi_2(I^*_{L_{m_k}})+\frac{1}{C_{p_2}}\cdot\psi_1(I^*_{L_{m_k}})]\}\cdot\frac{1}{C_1\cdot R_s}\cdot\frac{n_1^2}{n_2^2}=0$

$$i'_{s_1}\cdot\lambda\cdot e^{\lambda\cdot t}=-\{i'_{p_2}\cdot e^{\lambda\cdot t}\cdot\frac{C_1}{\Delta}\cdot[\frac{\Gamma_2}{C_{p_3}}+\frac{\Gamma_1}{C_{p_2}}]+\frac{C_1}{\Delta}\cdot[\frac{1}{C_{p_3}}\cdot\psi_2(i_{L_{m_k}}\cdot e^{\lambda\cdot t})+\frac{1}{C_{p_2}}\cdot\psi_1(i_{L_{m_k}}\cdot e^{\lambda\cdot t})]\}\cdot\frac{1}{C_1\cdot R_s}\cdot\frac{n_1^2}{n_2^2}$$

$$-i'_{s_1}\cdot\lambda-\{i'_{p_2}\cdot\frac{C_1}{\Delta}\cdot[\frac{\Gamma_2}{C_{p_3}}+\frac{\Gamma_1}{C_{p_2}}]+\frac{C_1}{\Delta}\cdot[\frac{1}{C_{p_3}}\cdot\psi_2(i_{L_{m_k}})+\frac{1}{C_{p_2}}\cdot\psi_1(i_{L_{m_k}})]\}\cdot\frac{1}{C_1\cdot R_s}\cdot\frac{n_1^2}{n_2^2}=0$$

$$\frac{di_{p_2}}{dt}=-\frac{n_4^2}{n_3^2}\cdot\frac{1}{C_2\cdot R_{load}}\cdot\{i_{p_2}\cdot\frac{C_2}{\Delta}\cdot[\frac{\Gamma_2}{C_{p_3}}-\Gamma_4]+\frac{C_2}{\Delta}\cdot[\frac{1}{C_{p_3}}\cdot\psi_2-\psi_4]\}$$

$$i'_{p_2}\cdot\lambda\cdot e^{\lambda\cdot t}=-\frac{n_4^2}{n_3^2}\cdot\frac{1}{C_2\cdot R_{load}}\cdot\{(i_{p_2}^{(i)}+i'_{p_2}\cdot e^{\lambda\cdot t})\cdot\frac{C_2}{\Delta}\cdot[\frac{\Gamma_2}{C_{p_3}}-\Gamma_4]+\frac{C_2}{\Delta}\cdot[\frac{1}{C_{p_3}}\cdot\{\psi_2(I^*_{L_{m_k}}) +\psi_2(i_{L_{m_k}}\cdot e^{\lambda\cdot t})\}-\{\psi_4(I^*_{L_{m_k}})+\psi_4(i_{L_{m_k}}\cdot e^{\lambda\cdot t})\}]\}$$

$$i'_{p_2}\cdot\lambda\cdot e^{\lambda\cdot t}=-\frac{n_4^2}{n_3^2}\cdot\frac{1}{C_2\cdot R_{load}}\cdot\{i_{p_2}^{(i)}\cdot\frac{C_2}{\Delta}\cdot[\frac{\Gamma_2}{C_{p_3}}-\Gamma_4]+\frac{C_2}{\Delta}\cdot[\frac{1}{C_{p_3}}\cdot\psi_2(I^*_{L_{m_k}})-\psi_4(I^*_{L_{m_k}})]\} -\frac{n_4^2}{n_3^2}\cdot\frac{1}{C_2\cdot R_{load}}\cdot\{i'_{p_2}\cdot e^{\lambda\cdot t}\cdot\frac{C_2}{\Delta}\cdot[\frac{\Gamma_2}{C_{p_3}}-\Gamma_4]+\frac{C_2}{\Delta}\cdot[\frac{1}{C_{p_3}}\cdot\psi_2(i_{L_{m_k}}\cdot e^{\lambda\cdot t}) -\psi_4(i_{L_{m_k}}\cdot e^{\lambda\cdot t})]\}$$

At fixed points: $-\frac{n_4^2}{n_3^2}\cdot\frac{1}{C_2\cdot R_{load}}\cdot\{i_{p_2}^{(i)}\cdot\frac{C_2}{\Delta}\cdot[\frac{\Gamma_2}{C_{p_3}}-\Gamma_4]+\frac{C_2}{\Delta}\cdot[\frac{1}{C_{p_3}}\cdot\psi_2(I^*_{L_{m_k}})-\psi_4(I^*_{L_{m_k}})]\}=0$

$$-i'_{p_2} \cdot \lambda - \frac{n_4^2}{n_3^2} \cdot \frac{1}{C_2 \cdot R_{load}} \cdot \{i'_{p_2} \cdot \frac{C_2}{\Delta} \cdot [\frac{\Gamma_2}{C_{p_3}} - \Gamma_4] + \frac{C_2}{\Delta} \cdot [\frac{1}{C_{p_3}} \cdot \psi_2(i_{L_{m_k}}) - \psi_4(i_{L_{m_k}})]\}$$
$$= 0$$

We can summery our Lattice crystal filter arbitrarily small increments equations:

$$-y_1 \cdot \lambda + \frac{1}{C_{p_1} \cdot L_{m_1}} \cdot [i'_{p_2} \cdot \frac{C_{p_1}}{\Delta} \cdot \Gamma_4 + \frac{C_{p_1}}{\Delta} \cdot \psi_4(i_{L_{m_k}})] - y_1 \cdot \frac{r_1}{L_{m_1}} - \frac{1}{C_{m_1} \cdot L_{m_1}} \cdot i_{L_{m_1}} = 0$$

$$-y_2 \cdot \lambda - y_2 \cdot \frac{r_2}{L_{m_2}} + i'_{p_2} \cdot \frac{1}{\Delta \cdot L_{m_2}} \cdot \frac{\Gamma_1}{C_{p_2}} + \frac{1}{L_{m_2} \cdot \Delta} \cdot [\frac{1}{C_{p_3}} \cdot \psi_2(i_{L_{m_k}}) + \frac{1}{C_{p_2}} \cdot \psi_1(i_{L_{m_k}})]$$
$$- \frac{1}{C_{p_3} \cdot L_{m_2}} \cdot \frac{\psi_2(i_{L_{m_k}})}{\Delta} - \frac{1}{C_{m_2} \cdot L_{m_2}} \cdot i_{L_{m_2}} = 0$$

$$-y_3 \cdot \lambda + \frac{1}{C_{p_3} \cdot L_{m_3}} \cdot [i'_{p_2} \cdot \frac{\Gamma_2}{\Delta} + \frac{\psi_2(i_{L_{m_k}})}{\Delta}] - y_3 \cdot \frac{r_3}{L_{m_3}} - \frac{1}{C_{m_3} \cdot L_{m_3}} \cdot i_{L_{m_3}} = 0$$

$$-y_4 \cdot \lambda - y_4 \cdot \frac{r_4}{L_{m_4}} + \frac{1}{C_1 \cdot L_{m_4}} \cdot (i'_{p_2} \cdot \frac{C_1}{\Delta} \cdot [\frac{\Gamma_2}{C_{p_3}} + \frac{\Gamma_1}{C_{p_2}}] + \frac{C_1}{\Delta} \cdot [\frac{1}{C_{p_3}} \cdot \psi_2(i_{L_{m_k}})$$
$$+ \frac{1}{C_{p_2}} \cdot \psi_1(i_{L_{m_k}})]) - \frac{1}{C_{p_1} \cdot L_{m_4}} \cdot [i'_{p_2} \cdot \frac{C_{p_1}}{\Delta} \cdot \Gamma_4 + \frac{C_{p_1}}{\Delta} \cdot \psi_4(i_{L_{m_k}})] - \frac{1}{C_{m_4} \cdot L_{m_4}} \cdot i_{L_{m_4}} = 0$$

$$-i_{L_{m_1}} \cdot \lambda + y_1 = 0; \quad -i_{L_{m_2}} \cdot \lambda + y_2 = 0; \quad -i_{L_{m_3}} \cdot \lambda + y_3 = 0; \quad -i_{L_{m_4}} \cdot \lambda + y_4 = 0$$

$$-i'_{s_1} \cdot \lambda - \{i'_{p_2} \cdot \frac{C_1}{\Delta} \cdot [\frac{\Gamma_2}{C_{p_3}} + \frac{\Gamma_1}{C_{p_2}}] + \frac{C_1}{\Delta} \cdot [\frac{1}{C_{p_3}} \cdot \psi_2(i_{L_{m_k}}) + \frac{1}{C_{p_2}} \cdot \psi_1(i_{L_{m_k}})]\} \cdot \frac{1}{C_1 \cdot R_s} \cdot \frac{n_1^2}{n_2^2} = 0$$

$$-i'_{p_2} \cdot \lambda - \frac{n_4^2}{n_3^2} \cdot \frac{1}{C_2 \cdot R_{load}} \cdot \{i'_{p_2} \cdot \frac{C_2}{\Delta} \cdot [\frac{\Gamma_2}{C_{p_3}} - \Gamma_4] + \frac{C_2}{\Delta} \cdot [\frac{1}{C_{p_3}} \cdot \psi_2(i_{L_{m_k}}) - \psi_4(i_{L_{m_k}})]\} = 0$$

$$\begin{pmatrix} \Xi_{11} & \cdots & \Xi_{1\_10} \\ \vdots & \ddots & \vdots \\ \Xi_{10\_1} & \cdots & \Xi_{10\_10} \end{pmatrix} \cdot \begin{pmatrix} y_1 \\ y_2 \\ y_3 \\ y_4 \\ i_{L_{m_1}} \\ i_{L_{m_2}} \\ i_{L_{m_3}} \\ i_{L_{m_4}} \\ i'_{s_1} \\ i'_{p_2} \end{pmatrix} = 0; \quad \Xi_{11} = -\lambda - \frac{r_1}{L_{m_1}}; \quad \Xi_{12} = \Xi_{13} = \Xi_{14} = 0$$

Remark Reader exercise to find $\Xi_{15}, \Xi_{16}, \Xi_{17}, \Xi_{18}$. $\Xi_{19} = 0$; $\Xi_{1\_10} = \frac{1}{L_{m_1} \cdot \Delta} \cdot \Gamma_4$

$$\Xi_{21} = 0;\ \Xi_{22} = -\lambda - \frac{r_2}{L_{m_2}};\ \Xi_{23} = \Xi_{24} = 0;\ \Xi_{29} = 0;\ \Xi_{2\_10} = \frac{1}{\Delta \cdot L_{m_2}} \cdot \frac{\Gamma_1}{C_{p_2}}$$

Remark Reader exercise to find $\Xi_{25}, \Xi_{26}, \Xi_{27}, \Xi_{28}$.

$$\Xi_{31} = \Xi_{32} = 0;\ \Xi_{33} = -\lambda - \frac{r_3}{L_{m_3}};\ \Xi_{34} = 0;\ \Xi_{39} = 0;\ \Xi_{3\_10} = \frac{1}{C_{p_3} \cdot L_{m_3}} \cdot \frac{\Gamma_2}{\Delta}$$

Remark Reader exercise to find $\Xi_{35}, \Xi_{36}, \Xi_{37}, \Xi_{38}$.

$$\Xi_{41} = \Xi_{42} = \Xi_{43} = 0;\ \Xi_{44} = -\lambda - \frac{r_4}{L_{m_4}};\ \Xi_{49} = 0;\ \Xi_{4\_10}$$
$$= \frac{1}{L_{m_4} \cdot \Delta} \cdot [\frac{\Gamma_2}{C_{p_3}} + \frac{\Gamma_1}{C_{p_2}}] - \frac{1}{L_{m_4} \cdot \Delta} \cdot \Gamma_4$$

Remark Reader exercise to find $\Xi_{45}, \Xi_{46}, \Xi_{47}, \Xi_{48}$.

$$\Xi_{51} = 1;\ \Xi_{52} = \Xi_{53} = \Xi_{54} = 0;\ \Xi_{55} = -\lambda;\ \Xi_{56} = \ldots = \Xi_{5\_10} = 0;\ \Xi_{61} = 0;\ \Xi_{62} = 1$$

$$\Xi_{63} = \Xi_{64} = 0;\ \Xi_{65} = 0;\ \Xi_{66} = -\lambda;\ \Xi_{67} = \ldots = \Xi_{6\_10} = 0;\ \Xi_{71} = \Xi_{72} = 0;\ \Xi_{73} = 1;\ \Xi_{74} = 0$$
$$\Xi_{75} = \Xi_{76} = 0;\ \Xi_{77} = -\lambda;\ \Xi_{78} = \Xi_{79} = \Xi_{7\_10} = 0;\ \Xi_{81} = \Xi_{82} = \Xi_{83} = 0;\ \Xi_{84} = 1$$

$$\Xi_{85} = \Xi_{86} = \Xi_{87} = 0;\ \Xi_{88} = -\lambda;\ \Xi_{89} = \Xi_{8\_10} = 0;\ \Xi_{91} = \ldots = \Xi_{94} = 0;\ \Xi_{99} = -\lambda$$

$$\Xi_{9\_10} = -\frac{C_1}{\Delta} \cdot [\frac{\Gamma_2}{C_{p_3}} + \frac{\Gamma_1}{C_{p_2}}] \cdot \frac{1}{C_1 \cdot R_s} \cdot \frac{n_1^2}{n_2^2}.$$

Remark: Reader exercise to find $\Xi_{95}, \Xi_{96}, \Xi_{97}, \Xi_{98}$.

$$\Xi_{10\_1} = \ldots = \Xi_{10\_4} = 0;\ \Xi_{10\_9} = 0;\ \Xi_{10\_10} = -\lambda - \frac{n_4^2}{n_3^2} \cdot \frac{1}{R_{load} \cdot \Delta} \cdot [\frac{\Gamma_2}{C_{p_3}} - \Gamma_4].$$

Remark Reader exercise to find $\Xi_{10\_5}, \Xi_{10\_6}, \Xi_{10\_7}, \Xi_{10\_8}$.

$$(A - \lambda \cdot I) = \begin{pmatrix} \Xi_{11} & \cdots & \Xi_{1\_10} \\ \vdots & \ddots & \vdots \\ \Xi_{10\_1} & \cdots & \Xi_{10\_10} \end{pmatrix};\ \det(A - \lambda \cdot I)$$
$$= \det\begin{pmatrix} \Xi_{11} & \cdots & \Xi_{1\_10} \\ \vdots & \ddots & \vdots \\ \Xi_{10\_1} & \cdots & \Xi_{10\_10} \end{pmatrix} = 0$$

$$\det(A - \lambda \cdot I) = \sum_{k=0}^{10} \Upsilon_k \cdot \lambda^k; \sum_{k=0}^{10} \Upsilon_k \cdot \lambda^k = 0.$$

Remark It is reader exercise to find the expressions for $\Upsilon_k \, \forall k = 0, 1, \ldots, 10$.

Eigenvalues stability discussion: Our Lattice crystal filter circuit involving N variables (N > 2, N = 10, arbitrarily small increments), the characteristic equation is of degree N = 10 and must often be solved numerically. Expect in some particular cases, such an equation has (N = 10) distinct roots that can be real or complex. These values are the eigenvalues of the (10 × 10) Jacobian matrix (A). The general rule is that the Lattice crystal filter circuit is stable if there is no eigenvalue with positive real part. It is sufficient that one eigenvalue is positive for the steady state to be unstable [3, 4]. Our 10-variables $(y_1, y_2, y_3, y_4, i_{L_{m_1}}, i_{L_{m_2}}, i_{L_{m_3}}, i_{L_{m_4}}, i'_{s_1}, i'_{p_2})$ system has ten eigenvalues (ten system's arbitrarily small increments). The type of behavior can be characterized as a function of the position of these eigenvalues in the Re/Im plane. Five non-degenerated cases can be distinguished: (1) the ten eigenvalues are real and negative (stable steady state), (2) the ten eigenvalues are real, at least one of them is positive (unstable steady state), (3) and (4) two eigenvalues are complex conjugates with a negative real part and other eigenvalues real are negative (stable steady state), two cases can be distinguished depending on the relative value of the real part of the complex eigenvalues and of the real one, (5) two eigenvalues are complex conjugates with a negative real part and at least one of the other eigenvalues real is positive (unstable steady state).

$$\det(A - \lambda \cdot I) = \sum_{k=0}^{10} \Upsilon_k \cdot \lambda^k; \ \det(A - \lambda \cdot I) = 0 \Rightarrow \sum_{k=0}^{10} \Upsilon_k \cdot \lambda^k = 0$$

7.4 A Tunable BPF Employing Varactor Diodes Stability Analysis

We have circuit which represents a tunable BPF employing varactor diodes. It is ideal for many diverse wireless applications. There are two types of tunable BPF employing varactor diodes; top inductively coupled variable BPF and capacitive coupled variable band pass filter. The best for wideband applications is the top inductively coupled variable band pass filter. In that subchapter we discuss the stability analysis of capacitive coupled variable BPF (employing varactor diodes). The design of the basic top capacitive coupled BPF is based on selecting a center frequency for the top capacitive coupled BPF at either the high and, low end, or middle of the tunable range of the desired bandpass frequencies, depending on the initial tuning voltages we supply to the tuning capacitors. The circuit includes varactor diodes (C_v) and C_T capacitors, bias resistors R_1 and R_2, R_1 resistor isolates the two varactors (C_v) from the effects of each other, and resistor R_2 represents a direct RF short to ground through V_{tune}. Capacitor C_T blocks the DC inserted by V_{tune} from being shorted by L_1 or L_2. Varactor diodes (C_v) supply the variable tuning capacitance. Capacitor C_c couples the two tank circuits consisting of L_1 and C_T/C_v, and L_2 and C_T/C_v. The capacitance of the series combination of C_T and C_v in series is $\frac{C_T \cdot C_v}{C_T + C_v}$. The capacitor C_T mainly functioning as a DC blocking capacitor, while C_v, the varactor is supplying all of the tuning capacitance for the filter's tanks. By applying positive tuning voltage V_{tune}, we allow the varactor to either linearly tune the filter to its maximum and minimum values or, by supplying V_{tune} with discrete voltages it filter the incoming RF/Microwave signal in discrete steps. Due to parasitic capacitances and inductances at these frequencies we need to optimize our filter and investigate his stability. Input RF/Microwave signal is $V_s(t)$ and his series resistance R_s. Typical varactors are limited in the value of their maximum capacitance, and in case we want to operate our tunable BPF at low frequencies region, then we need to increase the capacitance of C_T and C_v combination and it is done by adding capacitor C_s in shunt with C_T and C_v which will increase the capacitance in each leg to $C_s + \frac{C_T \cdot C_v}{C_T + C_v}$. The circuit load resistance R_{load} is connected in our tunable BPF circuit (parallel to inductor L_2) [25, 26].

Varactor diode: Varactor diode is an electronic component whose reactance can be varied, usually electronically. Varactor diode is a variable capacitance diode. Other names of varactors are varicap, tuning diode, and voltage variable capacitor. Varactors are integral part in many RF and microwave circuits. The capacitance of a varactor can be controlled electronically and automatic circuit tuning becomes practical. A varactor is a diode in which P and N regions are doped in such a manner that the capacitance that normally forms near the PN junction can be precisely controlled by a reverse bias voltage. There is an inverse relationship between the capacitance of the varactor and the applied voltage; a small reverse bias

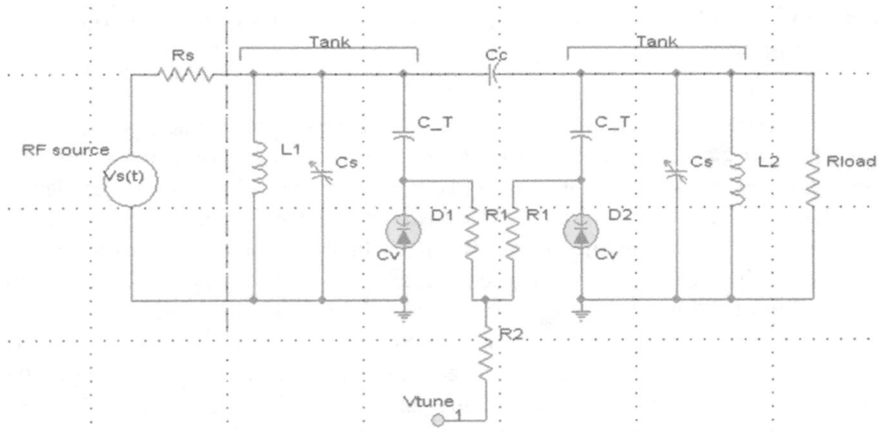

Fig. 7.9 Tunable BPF employing varactor diodes

voltage gives a big capacitance, and a large reverse bias voltage gives a small capacitance. The varactor capacitance is changed as a function of the bias voltage. There are two circuits' models of varactor diode, high frequency circuit model and low frequency circuit model. High frequency circuit model: L_s is the lead inductance and C_c is the package inductance. These components of the model only become significant at very high frequencies. C_j is the junction capacitance which varies with applied voltage according to equation $C_j(V) = \frac{C_0}{(1+\frac{V}{V_0})^n}$. V_0 is the junction potential with no bias voltage applied and is usually in the range of 0.5–0.7 v. It can be determined by measuring the voltage drop across the diode when it is conducting under forward bias. C_0 is the capacitance with zero bias voltage. The exponent n is dependent on the doping profile: n = 1/3 for a graded junction, n = 1/2 for an abrupt junction, and n = 1 to n = 2 for hyperabrupt junction. V is the applied voltage and is positive for reverse bias and negative for forward bias.

The $C_j(V) = \ldots$ equation does apply for forward bias voltages but only up to about $V = \frac{V_0}{2}$. It is possible to derive an expression for C_0 but in practice the value of C_0 usually has to be determined experimentally. The series resistance, R_s, is due to the resistance of the semiconductor material of which the diode is made as well as any lead and contact resistance. The part of the diode which is not part of the depletion region contributes to this resistance. R_s is a function of the bias voltage. As reverse bias is increased, the depletion region gets larger and R_s gets smaller and vice versa. The parallel resistance R_p represents the reverse leakage current. It is in general varying somewhat with the applied voltage and becoming smaller very rapidly near the reverse breakdown voltage. The low frequency model for a varactor is as follow:

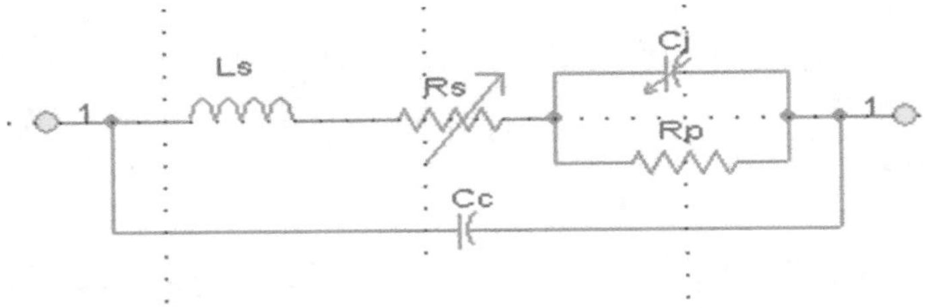

Fig. 7.10 High frequency circuit model for a varactor

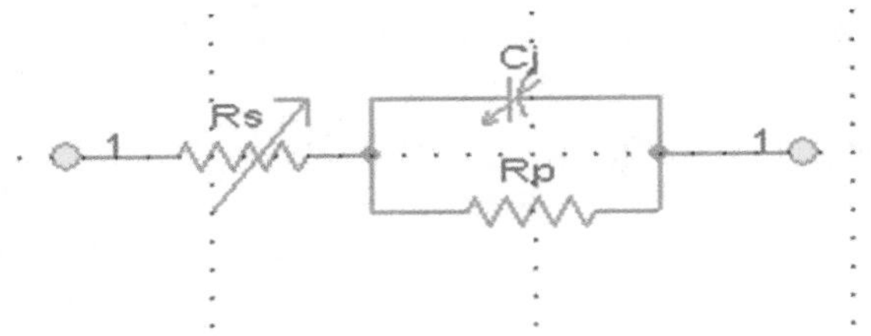

Fig. 7.11 Low frequency model for a varactor

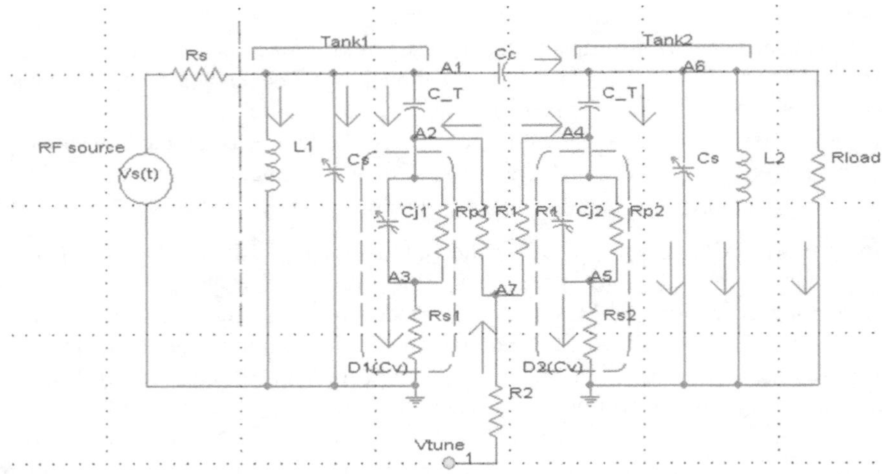

Fig. 7.12 Tunable BPF employing varactor diodes equivalent circuit

The Tunable BPF (varactor diodes) equivalent circuit is present in the below figure. We consider $V_s(t)$ low frequency RF source. We use varactor low frequency model in our analysis.

$$I_{R_s} = \frac{V_s(t) - V_{A_1}}{R_s};\ V_{A_1} = L_1 \cdot \frac{dI_{L_1}}{dt};$$
$$I_{C_s(\mathrm{tank1})} = C_s \cdot \frac{dV_{A_1}}{dt};\ I_{C_T(\mathrm{tank1})} = C_T \cdot \frac{d(V_{A_1} - V_{A_2})}{dt}$$
$$I_{C_{j1}} = C_{j1} \cdot \frac{d(V_{A_2} - V_{A_3})}{dt};\ I_{R_{p_1}} = \frac{V_{A_2} - V_{A_3}}{R_{p_1}};$$
$$I_{R_{s_1}} = \frac{V_{A_3}}{R_{s_1}};\ I_{C_c} = C_c \cdot \frac{d(V_{A_1} - V_{A_6})}{dt}$$

$$I_{C_T(\mathrm{tank2})} = C_T \cdot \frac{d(V_{A_6} - V_{A_4})}{dt};\ I_{C_{j2}} = C_{j2} \cdot \frac{d(V_{A_4} - V_{A_5})}{dt};\ I_{R_{p_2}} = \frac{V_{A_4} - V_{A_5}}{R_{p_2}};\ I_{R_{s_2}} = \frac{V_{A_5}}{R_{s_2}}$$

$$I_{C_s(\mathrm{tank2})} = C_s \cdot \frac{dV_{A_6}}{dt};\ V_{A_6} = L_2 \cdot \frac{dI_{L_2}}{dt};\ I_{R_{load}} = \frac{V_{A_6}}{R_{load}};\ V_{L_1} = V_{C_s(\mathrm{tank1})} = V_{A_1}$$

$$V_{C_s(\mathrm{tank2})} = V_{L_2} = V_{R_{load}} = V_{A_6};\ I_{R_2} = \frac{V_{tune} - V_{A_7}}{R_2};\ I_{R_1(\mathrm{tank1})} = \frac{V_{A_7} - V_{A_2}}{R_1};\ I_{R_1(\mathrm{tank2})} = \frac{V_{A_7} - V_{A_4}}{R_1}$$

V_{A2} is the applied voltage to varactor diode D_1. V_{A4} is the applied voltage to varactor diode D_2.

$$C_{j1}(V_{A_2}) = \frac{C_0}{(1+\frac{V_{A2}}{V_0})^n}; C_{j1}(V_{A_4}) = \frac{C_0}{(1+\frac{V_{A4}}{V_0})^n}; I_{C_{j1}} + I_{R_{p1}} = I_{R_{s1}}; I_{C_{j2}} + I_{R_{p2}} = I_{R_{s2}}$$

$$I_{R_2} = I_{R_1(\text{tank1})} + I_{R_1(\text{tank2})} = \sum_{i=1}^{2} I_{R_1(\text{tanki})}$$

KCL @ node A_1: $I_{R_s} = I_{L_1} + I_{C_s(\text{tank1})} + I_{C_T(\text{tank1})} + I_{C_c}$.
KCL @ node A_6: $I_{C_c} = I_{C_T(\text{tank2})} + I_{C_s(\text{tank2})} + I_{L_2} + I_{R_{load}}$
KCL @ node A_2: $I_{C_T(\text{tank1})} + I_{R_1(\text{tank1})} = I_{C_{j_1}} + I_{R_{p_1}}$
KCL @ node A_4: $I_{C_T(\text{tank2})} + I_{R_1(\text{tank2})} = I_{C_{j2}} + I_{R_{p_2}}$

$$I_{C_s(\text{tank1})} = C_s \cdot \frac{dV_{A_1}}{dt} = C_s \cdot \frac{d}{dt}(L_1 \cdot \frac{dI_{L_1}}{dt}) = C_s \cdot L_1 \cdot \frac{d^2 I_{L_1}}{dt^2}; V_{A_1} = \frac{1}{C_s} \cdot \int I_{C_s(\text{tank1})} \cdot dt$$

$$I_{C_T(\text{tank1})} = C_T \cdot \frac{d(V_{A_1} - V_{A_2})}{dt} \Rightarrow V_{A_1} - V_{A_2} = \frac{1}{C_T} \cdot \int I_{C_T(\text{tank1})} \cdot dt;$$

$$V_{A_2} = V_{A_1} - \frac{1}{C_T} \cdot \int I_{C_T(\text{tank1})} \cdot dt$$

$$V_{A_2} = L_1 \cdot \frac{dI_{L_1}}{dt} - \frac{1}{C_T} \cdot \int I_{C_T(\text{tank1})} \cdot dt; I_{C_{j1}} = C_{j1} \cdot \frac{d(V_{A_2} - V_{A_3})}{dt}$$

$$\Rightarrow V_{A_2} - V_{A_3} = \frac{1}{C_{j1}} \cdot \int I_{C_{j1}} \cdot dt$$

$$V_{A_3} = V_{A_2} - \frac{1}{C_{j1}} \cdot \int I_{C_{j1}} \cdot dt = L_1 \cdot \frac{dI_{L_1}}{dt} - \frac{1}{C_T} \cdot \int I_{C_T(\text{tank1})} \cdot dt - \frac{1}{C_{j1}} \cdot \int I_{C_{j1}} \cdot dt$$

$$I_{R_{p_1}} = \frac{V_{A_2} - V_{A_3}}{R_{p_1}} \Rightarrow V_{A_2} - V_{A_3} = I_{R_{p_1}} \cdot R_{p_1}; I_{R_{p_1}} \cdot R_{p_1} = \frac{1}{C_{j1}} \cdot \int I_{C_{j1}} \cdot dt$$

$$\frac{d}{dt}\{I_{R_{p_1}} \cdot R_{p_1} = \frac{1}{C_{j1}} \cdot \int I_{C_{j1}} \cdot dt\} \Rightarrow \frac{dI_{R_{p_1}}}{dt} \cdot R_{p_1} = \frac{1}{C_{j1}} \cdot I_{C_{j1}} \Rightarrow \frac{dI_{R_{p_1}}}{dt} = \frac{1}{C_{j1} \cdot R_{p_1}} \cdot I_{C_{j1}}$$

$$I_{R_{s_1}} = \frac{V_{A_3}}{R_{s_1}} \Rightarrow V_{A_3} = I_{R_{s_1}} \cdot R_{s_1}; I_{R_{s_1}} \cdot R_{s_1} = L_1 \cdot \frac{dI_{L_1}}{dt} - \frac{1}{C_T} \cdot \int I_{C_T(\text{tank1})} \cdot dt - \frac{1}{C_{j1}} \cdot \int I_{C_{j1}} \cdot dt$$

$$\frac{dI_{R_{s1}}}{dt}\cdot R_{s1} = L_1\cdot\frac{d^2I_{L_1}}{dt^2} - \frac{1}{C_T}\cdot I_{C_T(\text{tank1})} - \frac{1}{C_{j1}}\cdot I_{C_{j1}}; \frac{dI_{R_{s1}}}{dt} = \frac{1}{C_s\cdot R_{s1}}\cdot I_{C_s(\text{tank1})} - \frac{1}{C_T\cdot R_{s1}}\cdot I_{C_T(\text{tank1})} - \frac{1}{C_{j1}\cdot R_{s1}}\cdot I_{C_{j1}}$$

$$I_{C_c} = C_c\cdot\frac{d(V_{A_1}-V_{A_6})}{dt} \Rightarrow V_{A_1}-V_{A_6} = \frac{1}{C_c}\cdot\int I_{C_c}\cdot dt$$

$$V_{A_6} = V_{A_1} - \frac{1}{C_c}\cdot\int I_{C_c}\cdot dt = \frac{1}{C_s}\cdot\int I_{C_s(\text{tank1})}\cdot dt - \frac{1}{C_c}\cdot\int I_{C_c}\cdot dt$$

$$I_{C_T(\text{tank2})} = C_T\cdot\frac{d(V_{A_6}-V_{A_4})}{dt} \Rightarrow V_{A_6}-V_{A_4} = \frac{1}{C_T}\cdot\int I_{C_T(\text{tank2})}\cdot dt;$$

$$V_{A_4} = V_{A_6} - \frac{1}{C_T}\cdot\int I_{C_T(\text{tank2})}\cdot dt$$

$$V_{A_4} = \frac{1}{C_s}\cdot\int I_{C_s(\text{tank1})}\cdot dt - \frac{1}{C_c}\cdot\int I_{C_c}\cdot dt - \frac{1}{C_T}\cdot\int I_{C_T(\text{tank2})}\cdot dt$$

$$I_{C_{j2}} = C_{j2}\cdot\frac{d(V_{A_4}-V_{A_5})}{dt} \Rightarrow V_{A_4}-V_{A_5} = \frac{1}{C_{j2}}\cdot\int I_{C_{j2}}\cdot dt \Rightarrow V_{A_5} = V_{A_4} - \frac{1}{C_{j2}}\cdot\int I_{C_{j2}}\cdot dt$$

$$V_{A_5} = \frac{1}{C_s}\cdot\int I_{C_s(\text{tank1})}\cdot dt - \frac{1}{C_c}\cdot\int I_{C_c}\cdot dt - \frac{1}{C_T}\cdot\int I_{C_T(\text{tank2})}\cdot dt - \frac{1}{C_{j2}}\cdot\int I_{C_{j2}}\cdot dt$$

$$I_{R_{p2}} = \frac{V_{A_4}-V_{A_5}}{R_{p2}} \Rightarrow V_{A_4}-V_{A_5} = I_{R_{p2}}\cdot R_{p2} \Rightarrow I_{R_{p2}}\cdot R_{p2} = \frac{1}{C_{j2}}\cdot\int I_{C_{j2}}\cdot dt; \frac{dI_{R_{p2}}}{dt} = \frac{1}{C_{j2}\cdot R_{p2}}\cdot I_{C_{j2}}$$

$$I_{R_{s2}} = \frac{V_{A_5}}{R_{s2}} \Rightarrow V_{A_5} = I_{R_{s2}}\cdot R_{s2}; I_{R_{s2}}\cdot R_{s2} = \frac{1}{C_s}\cdot\int I_{C_s(\text{tank1})}\cdot dt - \frac{1}{C_c}\cdot\int I_{C_c}\cdot dt - \frac{1}{C_T}\cdot\int I_{C_T(\text{tank2})}\cdot dt - \frac{1}{C_{j2}}\cdot\int I_{C_{j2}}\cdot dt$$

$$\frac{d}{dt}\{I_{R_{s2}}\cdot R_{s2} = \frac{1}{C_s}\cdot\int I_{C_s(\text{tank1})}\cdot dt - \frac{1}{C_c}\cdot\int I_{C_c}\cdot dt - \frac{1}{C_T}\cdot\int I_{C_T(\text{tank2})}\cdot dt - \frac{1}{C_{j2}}\cdot\int I_{C_{j2}}\cdot dt\}$$

$$\frac{dI_{R_{s2}}}{dt}\cdot R_{s2} = \frac{1}{C_s}\cdot I_{C_s(\text{tank1})} - \frac{1}{C_c}\cdot I_{C_c} - \frac{1}{C_T}\cdot I_{C_T(\text{tank2})} - \frac{1}{C_{j2}}\cdot I_{C_{j2}}$$

$$I_{C_s(\text{tank2})} = C_s \cdot \frac{dV_{A_6}}{dt} \Rightarrow V_{A_6} = \frac{1}{C_s} \cdot \int I_{C_s(\text{tank2})} \cdot dt; \frac{1}{C_s} \cdot \int I_{C_s(\text{tank2})} \cdot dt$$
$$= \frac{1}{C_s} \cdot \int I_{C_s(\text{tank1})} \cdot dt - \frac{1}{C_c} \cdot \int I_{C_c} \cdot dt$$
$$\frac{d}{dt}\{\frac{1}{C_s} \cdot \int I_{C_s(\text{tank2})} \cdot dt = \frac{1}{C_s} \cdot \int I_{C_s(\text{tank1})} \cdot dt - \frac{1}{C_c} \cdot \int I_{C_c} \cdot dt\};$$
$$\frac{1}{C_s} \cdot I_{C_s(\text{tank2})} = \frac{1}{C_s} \cdot I_{C_s(\text{tank1})} - \frac{1}{C_c} \cdot I_{C_c}$$
$$I_{C_s(\text{tank2})} = C_s \cdot \frac{dV_{A_6}}{dt} = C_s \cdot \frac{d}{dt}(L_2 \cdot \frac{dI_{L_2}}{dt}) = C_s \cdot L_2 \cdot \frac{d^2 I_{L_2}}{dt^2};$$
$$I_{R_{load}} = \frac{V_{A_6}}{R_{load}} \Rightarrow V_{A_6} = I_{R_{load}} \cdot R_{load}$$

$$V_{A_6} = I_{R_{load}} \cdot R_{load} = L_2 \cdot \frac{dI_{L_2}}{dt} \Rightarrow \frac{dI_{L_2}}{dt} = I_{R_{load}} \cdot \frac{R_{load}}{L_2}; I_{R_2} = \frac{V_{tune} - V_{A_7}}{R_2} \Rightarrow V_{A_7}$$
$$= V_{tune} - I_{R_2} \cdot R_2$$

$$\frac{1}{C_s} \cdot I_{C_s(\text{tank2})} = L_2 \cdot \frac{d^2 I_{L_2}}{dt^2}; \frac{dI_{R_{load}}}{dt} \cdot R_{load} = L_2 \cdot \frac{d^2 I_{L_2}}{dt^2}; \frac{dI_{R_{load}}}{dt} = \frac{1}{C_s \cdot R_{load}} \cdot I_{C_s(\text{tank2})}$$

$$I_{R_1(\text{tank1})} = \frac{V_{A_7} - V_{A_2}}{R_1} \Rightarrow I_{R_1(\text{tank1})} \cdot R_1 = V_{A_7} - V_{A_2};$$
$$I_{R_1(\text{tank1})} \cdot R_1 = V_{tune} - I_{R_2} \cdot R_2 - L_1 \cdot \frac{dI_{L_1}}{dt} + \frac{1}{C_T} \cdot \int I_{C_T(\text{tank1})} \cdot dt$$
$$\frac{d}{dt}\{I_{R_1(\text{tank1})} \cdot R_1 = V_{tune} - I_{R_2} \cdot R_2 - L_1 \cdot \frac{dI_{L_1}}{dt} + \frac{1}{C_T} \cdot \int I_{C_T(\text{tank1})} \cdot dt\}$$

$$\frac{dI_{R_1(\text{tank1})}}{dt} \cdot R_1 = \frac{dV_{tune}}{dt} - \frac{dI_{R_2}}{dt} \cdot R_2 - L_1 \cdot \frac{d^2 I_{L_1}}{dt^2} + \frac{1}{C_T} \cdot I_{C_T(\text{tank1})}; I_{R_1(\text{tank2})}$$
$$= \frac{V_{A_7} - V_{A_4}}{R_1}$$

$$I_{R_1(\text{tank2})} = \frac{V_{A_7} - V_{A_4}}{R_1} \Rightarrow V_{A_7} - V_{A_4} = I_{R_1(\text{tank2})} \cdot R_1$$

$$\frac{dI_{R_1(\text{tank2})}}{dt} \cdot R_1 = \frac{dV_{tune}}{dt} - \frac{dI_{R_2}}{dt} \cdot R_2 - \frac{1}{C_s} \cdot I_{C_s(\text{tank1})} + \frac{1}{C_c} \cdot I_{C_c} + \frac{1}{C_T} \cdot I_{C_T(\text{tank2})}$$

We consider that $I_{R_1(\text{tank1})} \approx I_{R_1(\text{tank2})}; \frac{dI_{R_1(\text{tank1})}}{dt} \approx \frac{dI_{R_1(\text{tank2})}}{dt}$. V_{tune} Is a DC voltage then $\frac{dV_{tune}}{dt} = 0$. $\frac{dI_{R_1(\text{tank1})}}{dt} \cdot R_1 + \frac{dI_{R_2}}{dt} \cdot R_2 = -L_1 \cdot \frac{d^2 I_{L_1}}{dt^2} + \frac{1}{C_T} \cdot I_{C_T(\text{tank1})}$.

$$\frac{dI_{R_1(\text{tank2})}}{dt}\cdot R_1+\frac{dI_{R_2}}{dt}\cdot R_2=-\frac{1}{C_s}\cdot I_{C_s(\text{tank1})}+\frac{1}{C_c}\cdot I_{C_c}+\frac{1}{C_T}\cdot I_{C_T(\text{tank2})}$$

$$\frac{dI_{R_1(\text{tank1})}}{dt}\cdot R_1+\frac{dI_{R_2}}{dt}\cdot R_2\approx\frac{dI_{R_1(\text{tank2})}}{dt}\cdot R_1+\frac{dI_{R_2}}{dt}\cdot R_2$$
$$\Rightarrow -L_1\cdot\frac{d^2I_{L_1}}{dt^2}+\frac{1}{C_T}\cdot I_{C_T(\text{tank1})}=-\frac{1}{C_s}\cdot I_{C_s(\text{tank1})}+\frac{1}{C_c}\cdot I_{C_c}+\frac{1}{C_T}\cdot I_{C_T(\text{tank2})}$$

$$L_1\cdot\frac{d^2I_{L_1}}{dt^2}-\frac{1}{C_T}\cdot I_{C_T(\text{tank1})}=\frac{1}{C_s}\cdot I_{C_s(\text{tank1})}-\frac{1}{C_c}\cdot I_{C_c}-\frac{1}{C_T}\cdot I_{C_T(\text{tank2})}$$

We define new variables: $I_{C_s(\text{tank1})}=C_s\cdot L_1\cdot\frac{d^2I_{L_1}}{dt^2};\, Y_1=\frac{dI_{L_1}}{dt};\, \frac{dY_1}{dt}=\frac{d^2I_{L_1}}{dt^2}$

$$\frac{dY_1}{dt}=\frac{1}{C_s\cdot L_1}\cdot I_{C_s(\text{tank1})}$$

Tunable BPF (varactor diodes) differential equations group No. 1:

$$\frac{dI_{L_1}}{dt}=Y_1;\frac{dY_1}{dt}=\frac{1}{C_s\cdot L_1}\cdot I_{C_s(\text{tank1})};\frac{dI_{R_{p_1}}}{dt}=\frac{1}{C_{j1}\cdot R_{p_1}}\cdot I_{C_{j1}}$$

$$\frac{dI_{R_{s_1}}}{dt}=\frac{1}{C_s\cdot R_{s_1}}\cdot I_{C_s(\text{tank1})}-\frac{1}{C_T\cdot R_{s_1}}\cdot I_{C_T(\text{tank1})}-\frac{1}{C_{j1}\cdot R_{s_1}}\cdot I_{C_{j1}}$$

$$\frac{dI_{R_{p_2}}}{dt}=\frac{1}{C_{j2}\cdot R_{p_2}}\cdot I_{C_{j2}};\frac{dI_{R_{s_2}}}{dt}=\frac{1}{C_s\cdot R_{s_2}}\cdot I_{C_s(\text{tank1})}$$
$$-\frac{1}{C_c\cdot R_{s_2}}\cdot I_{C_c}-\frac{1}{C_T\cdot R_{s_2}}\cdot I_{C_T(\text{tank2})}-\frac{1}{C_{j2}\cdot R_{s_2}}\cdot I_{C_{j2}}$$
$$\frac{dI_{L_2}}{dt}=I_{R_{load}}\cdot\frac{R_{load}}{L_2};\frac{dI_{R_{load}}}{dt}=\frac{1}{C_s\cdot R_{load}}\cdot I_{C_s(\text{tank2})}$$

Tunable BPF (varactor diodes) KCL equations:

$$\frac{1}{C_s}\cdot I_{C_s(\text{tank2})}=\frac{1}{C_s}\cdot I_{C_s(\text{tank1})}-\frac{1}{C_c}\cdot I_{C_c};\; I_{R_s}=I_{L_1}+I_{C_s(\text{tank1})}+I_{C_T(\text{tank1})}+I_{C_c}$$

$$I_{C_c}=I_{C_T(\text{tank2})}+I_{C_s(\text{tank2})}+I_{L_2}+I_{R_{load}};\; I_{C_T(\text{tank1})}+I_{R_1(\text{tank1})}=I_{C_{j_1}}+I_{R_{p_1}}$$

$$I_{C_T(\text{tank2})}+I_{R_1(\text{tank2})}=I_{C_{j2}}+I_{R_{p_2}};\; I_{C_{j_1}}+I_{R_{p_1}}=I_{R_{s_1}};\; I_{C_{j_2}}+I_{R_{p_2}}=I_{R_{s_2}}$$

$$I_{R_2} = I_{R_1(\text{tank1})} + I_{R_1(\text{tank2})} = \sum_{i=1}^{2} I_{R_1(\text{tanki})}$$

$$I_{C_{j1}} + I_{R_{p1}} = I_{R_{s1}} \Rightarrow I_{C_{j1}} = I_{R_{s1}} - I_{R_{p1}};\ I_{C_{j2}} + I_{R_{p2}} = I_{R_{s2}} \Rightarrow I_{C_{j2}} = I_{R_{s2}} - I_{R_{p2}}$$

$$I_{C_T(\text{tank1})} + I_{R_1(\text{tank1})} = I_{C_{j1}} + I_{R_{p1}} = I_{R_{s1}};\ I_{C_T(\text{tank2})} + I_{R_1(\text{tank2})} = I_{C_{j2}} + I_{R_{p2}} = I_{R_{s2}}$$

Tunable BPF (varactor diodes) differential equations group No. 2:

$$\frac{dI_{L_1}}{dt} = Y_1;\ \frac{dY_1}{dt} = \frac{1}{C_s \cdot L_1} \cdot I_{C_s(\text{tank1})};\ \frac{dI_{R_{p1}}}{dt} = \frac{1}{C_{j1} \cdot R_{p1}} \cdot (I_{R_{s1}} - I_{R_{p1}})$$

$$\frac{dI_{R_{s1}}}{dt} = \frac{1}{C_s \cdot R_{s1}} \cdot I_{C_s(\text{tank1})} - \frac{1}{C_T \cdot R_{s1}} \cdot I_{C_T(\text{tank1})} - \frac{1}{C_{j1} \cdot R_{s1}} \cdot (I_{R_{s1}} - I_{R_{p1}});$$

$$\frac{dI_{R_{p2}}}{dt} = \frac{1}{C_{j2} \cdot R_{p2}} \cdot (I_{R_{s2}} - I_{R_{p2}})$$

$$\frac{dI_{R_{s2}}}{dt} = \frac{1}{C_s \cdot R_{s2}} \cdot I_{C_s(\text{tank1})} - \frac{1}{C_c \cdot R_{s2}} \cdot I_{C_c} - \frac{1}{C_T \cdot R_{s2}} \cdot I_{C_T(\text{tank2})} - \frac{1}{C_{j2} \cdot R_{s2}} \cdot (I_{R_{s2}} - I_{R_{p2}})$$

$$\frac{dI_{L_2}}{dt} = I_{R_{load}} \cdot \frac{R_{load}}{L_2};\ \frac{dI_{R_{load}}}{dt} = \frac{1}{C_s \cdot R_{load}} \cdot I_{C_s(\text{tank2})}$$

$$I_{C_T(\text{tank1})} + I_{R_1(\text{tank1})} = I_{R_{s1}} \Rightarrow I_{C_T(\text{tank1})} = I_{R_{s1}} - I_{R_1(\text{tank1})};\ I_{C_T(\text{tank2})} + I_{R_1(\text{tank2})} = I_{R_{s2}} \Rightarrow I_{C_T(\text{tank2})} = I_{R_{s2}} - I_{R_1(\text{tank2})}$$

Tunable BPF (varactor diodes) differential equations group No. 3:

$$\frac{dI_{L_1}}{dt} = Y_1;\ \frac{dY_1}{dt} = \frac{1}{C_s \cdot L_1} \cdot I_{C_s(\text{tank1})};\ \frac{dI_{R_{p1}}}{dt} = \frac{1}{C_{j1} \cdot R_{p1}} \cdot (I_{R_{s1}} - I_{R_{p1}})$$

$$\frac{dI_{R_{s1}}}{dt} = \frac{1}{C_s \cdot R_{s1}} \cdot I_{C_s(\text{tank1})} - \frac{1}{C_T \cdot R_{s1}} \cdot (I_{R_{s1}} - I_{R_1(\text{tank1})}) - \frac{1}{C_{j1} \cdot R_{s1}} \cdot (I_{R_{s1}} - I_{R_{p1}});\ \frac{dI_{R_{p2}}}{dt} = \frac{1}{C_{j2} \cdot R_{p2}} \cdot (I_{R_{s2}} - I_{R_{p2}})$$

$$\frac{dI_{R_{s2}}}{dt} = \frac{1}{C_s \cdot R_{s2}} \cdot I_{C_s(\text{tank1})} - \frac{1}{C_c \cdot R_{s2}} \cdot I_{C_c} - \frac{1}{C_T \cdot R_{s2}} \cdot (I_{R_{s2}} - I_{R_1(\text{tank2})}) - \frac{1}{C_{j2} \cdot R_{s2}} \cdot (I_{R_{s2}} - I_{R_{p2}})$$

$$\frac{dI_{L_2}}{dt} = I_{R_{load}} \cdot \frac{R_{load}}{L_2}; \frac{dI_{R_{load}}}{dt} = \frac{1}{C_s \cdot R_{load}} \cdot I_{C_s(\text{tank2})}$$

&&&

$$I_{R_s} = I_{L_1} + I_{C_s(\text{tank1})} + I_{C_T(\text{tank1})} + I_{C_c}; \; I_{C_c} = I_{C_T(\text{tank2})} + I_{C_s(\text{tank2})} + I_{L_2} + I_{R_{load}};$$

$$\frac{1}{C_s} \cdot I_{C_s(\text{tank2})} = \frac{1}{C_s} \cdot I_{C_s(\text{tank1})} - \frac{1}{C_c} \cdot I_{C_c}$$

$$I_{C_c} = I_{R_s} - I_{L_1} - I_{C_s(\text{tank1})} - I_{C_T(\text{tank1})} = I_{R_s} - I_{L_1} - I_{C_s(\text{tank1})} - I_{R_{s_1}} + I_{R_1(\text{tank1})}$$

$$I_{R_s} - I_{L_1} - I_{C_s(\text{tank1})} - I_{R_{s_1}} + I_{R_1(\text{tank1})} = I_{R_{s_2}} - I_{R_1(\text{tank2})} + I_{C_s(\text{tank2})} + I_{L_2} + I_{R_{load}}$$

$$I_{R_s} - I_{L_1} - I_{L_2} - I_{R_{s_1}} - I_{R_{s_2}} + I_{R_1(\text{tank1})} + I_{R_1(\text{tank2})} - I_{R_{load}} = I_{C_s(\text{tank1})} + I_{C_s(\text{tank2})}$$

$$I_{R_s} - \sum_{k=1}^{2} I_{L_k} - \sum_{i=1}^{2} I_{R_{s_i}} + \sum_{i=1}^{2} I_{R_1(\text{tanki})} - I_{R_{load}} = I_{C_s(\text{tank1})} + I_{C_s(\text{tank2})}$$

$$\frac{1}{C_s} \cdot I_{C_s(\text{tank2})} = \frac{1}{C_s} \cdot I_{C_s(\text{tank1})} - \frac{1}{C_c} \cdot (I_{R_s} - I_{L_1} - I_{C_s(\text{tank1})} - I_{R_{s_1}} + I_{R_1(\text{tank1})})$$

$$-\frac{1}{C_c} \cdot I_{R_s} + \frac{1}{C_c} \cdot I_{L_1} + \frac{1}{C_c} \cdot I_{R_{s_1}} - \frac{1}{C_c} \cdot I_{R_1(\text{tank1})}$$
$$= \frac{1}{C_s} \cdot I_{C_s(\text{tank2})} - (\frac{1}{C_s} + \frac{1}{C_c}) \cdot I_{C_s(\text{tank1})}$$

$$I_{C_s(\text{tank1})} = I_{R_s} - \sum_{k=1}^{2} I_{L_k} - \sum_{i=1}^{2} I_{R_{s_i}} + \sum_{i=1}^{2} I_{R_1(\text{tanki})} - I_{R_{load}} - I_{C_s(\text{tank2})}$$

$$-\frac{1}{C_c} \cdot I_{R_s} + \frac{1}{C_c} \cdot I_{L_1} + \frac{1}{C_c} \cdot I_{R_{s_1}} - \frac{1}{C_c} \cdot I_{R_1(\text{tank1})} = \frac{1}{C_s} \cdot I_{C_s(\text{tank2})}$$
$$- (\frac{1}{C_s} + \frac{1}{C_c}) \cdot [I_{R_s} - \sum_{k=1}^{2} I_{L_k} - \sum_{i=1}^{2} I_{R_{s_i}} + \sum_{i=1}^{2} I_{R_1(\text{tanki})} - I_{R_{load}} - I_{C_s(\text{tank2})}]$$

$$-\frac{1}{C_c} \cdot I_{R_s} + \frac{1}{C_c} \cdot I_{L_1} + \frac{1}{C_c} \cdot I_{R_{s_1}} - \frac{1}{C_c} \cdot I_{R_1(\text{tank1})} = \frac{1}{C_s} \cdot I_{C_s(\text{tank2})}$$
$$- (\frac{1}{C_s} + \frac{1}{C_c}) \cdot I_{R_s} + (\frac{1}{C_s} + \frac{1}{C_c}) \cdot \sum_{k=1}^{2} I_{L_k} + (\frac{1}{C_s} + \frac{1}{C_c}) \cdot \sum_{i=1}^{2} I_{R_{s_i}} - (\frac{1}{C_s} + \frac{1}{C_c})$$
$$\cdot \sum_{i=1}^{2} I_{R_1(\text{tanki})} + (\frac{1}{C_s} + \frac{1}{C_c}) \cdot I_{R_{load}} + (\frac{1}{C_s} + \frac{1}{C_c}) \cdot I_{C_s(\text{tank2})}$$

$$(\frac{2}{C_s}+\frac{1}{C_c})\cdot I_{C_s(\text{tank2})} = -\frac{1}{C_c}\cdot I_{R_s}+\frac{1}{C_c}\cdot I_{L_1}+\frac{1}{C_c}\cdot I_{R_{s_1}}-\frac{1}{C_c}\cdot I_{R_1(\text{tank1})}+(\frac{1}{C_s}+\frac{1}{C_c})$$
$$\cdot\sum_{i=1}^{2} I_{R_1(\text{tanki})}+(\frac{1}{C_s}+\frac{1}{C_c})\cdot I_{R_s}-(\frac{1}{C_s}+\frac{1}{C_c})\cdot\sum_{k=1}^{2} I_{L_k}$$
$$-(\frac{1}{C_s}+\frac{1}{C_c})\cdot\sum_{i=1}^{2} I_{R_{s_i}}-(\frac{1}{C_s}+\frac{1}{C_c})\cdot I_{R_{load}}$$

$$(\frac{2}{C_s}+\frac{1}{C_c})\cdot I_{C_s(\text{tank2})} = \frac{1}{C_s}\cdot I_{R_s}+I_{R_1(\text{tank1})}\cdot\frac{1}{C_s}+I_{R_1(\text{tank2})}\cdot(\frac{1}{C_s}+\frac{1}{C_c})$$
$$-\frac{1}{C_s}\cdot I_{L_1}-(\frac{1}{C_s}+\frac{1}{C_c})\cdot I_{L_2}-\frac{1}{C_s}\cdot I_{R_{s_1}}-(\frac{1}{C_s}+\frac{1}{C_c})$$
$$\cdot I_{R_{s_2}}-(\frac{1}{C_s}+\frac{1}{C_c})\cdot I_{R_{load}}$$

$$I_{C_s(\text{tank2})} = \frac{1}{(\frac{2}{C_s}+\frac{1}{C_c})\cdot C_s}\cdot I_{R_s}+I_{R_1(\text{tank1})}\cdot\frac{1}{(\frac{2}{C_s}+\frac{1}{C_c})\cdot C_s}+I_{R_1(\text{tank2})}\cdot\frac{(\frac{1}{C_s}+\frac{1}{C_c})}{(\frac{2}{C_s}+\frac{1}{C_c})}$$
$$-\frac{1}{C_s\cdot(\frac{2}{C_s}+\frac{1}{C_c})}\cdot I_{L_1}-\frac{(\frac{1}{C_s}+\frac{1}{C_c})}{(\frac{2}{C_s}+\frac{1}{C_c})}\cdot I_{L_2}-\frac{1}{C_s\cdot(\frac{2}{C_s}+\frac{1}{C_c})}\cdot I_{R_{s_1}}-\frac{(\frac{1}{C_s}+\frac{1}{C_c})}{(\frac{2}{C_s}+\frac{1}{C_c})}$$
$$\cdot I_{R_{s_2}}-\frac{(\frac{1}{C_s}+\frac{1}{C_c})}{(\frac{2}{C_s}+\frac{1}{C_c})}\cdot I_{R_{load}}$$

$$I_{C_s(\text{tank2})} = \frac{1}{(2+\frac{C_s}{C_c})}\cdot I_{R_s}+I_{R_1(\text{tank1})}\cdot\frac{1}{(2+\frac{C_s}{C_c})}+I_{R_1(\text{tank2})}\cdot\frac{(C_c+C_s)}{(2\cdot C_c+C_s)}$$
$$-\frac{1}{(2+\frac{C_s}{C_c})}\cdot I_{L_1}-\frac{(C_c+C_s)}{(2\cdot C_c+C_s)}\cdot I_{L_2}-\frac{1}{(2+\frac{C_s}{C_c})}\cdot I_{R_{s_1}}$$
$$-\frac{(C_c+C_s)}{(2\cdot C_c+C_s)}\cdot I_{R_{s_2}}-\frac{(C_c+C_s)}{(2\cdot C_c+C_s)}\cdot I_{R_{load}}$$

$$\Gamma_1 = \Gamma_1(C_c, C_s) = \frac{(\frac{1}{C_s} + \frac{1}{C_c})}{(\frac{2}{C_s} + \frac{1}{C_c})} = \frac{(\frac{1}{C_s} + \frac{1}{C_c})}{(\frac{2}{C_s} + \frac{1}{C_c})} \cdot \frac{C_s \cdot C_c}{C_s \cdot C_c} = \frac{(C_c + C_s)}{(2 \cdot C_c + C_s)}$$

$$I_{C_s(\text{tank2})} = \frac{1}{(2+\frac{C_s}{C_c})} \cdot I_{R_s} + I_{R_1(\text{tank1})} \cdot \frac{1}{(2+\frac{C_s}{C_c})} + I_{R_1(\text{tank2})} \cdot \Gamma_1(C_c, C_s) - \frac{1}{(2+\frac{C_s}{C_c})} \cdot I_{L_1} - \Gamma_1(C_c, C_s) \cdot I_{L_2} - \frac{1}{(2+\frac{C_s}{C_c})} \cdot I_{R_{s_1}} - \Gamma_1(C_c, C_s) \cdot I_{R_{s_2}} - \Gamma_1(C_c, C_s) \cdot I_{R_{load}}$$

$$I_{C_s(\text{tank1})} = I_{R_s} + [\frac{1}{(2+\frac{C_s}{C_c})} - 1] \cdot I_{L_1} + [\Gamma_1(C_c, C_s) - 1] \cdot I_{L_2} + [\frac{1}{(2+\frac{C_s}{C_c})} - 1] \cdot I_{R_{s_1}} + [\Gamma_1(C_c, C_s) - 1] \cdot I_{R_{s_2}} + I_{R_1(\text{tank1})} \cdot [1 - \frac{1}{(2+\frac{C_s}{C_c})}] + I_{R_1(\text{tank2})} \cdot [1 - \Gamma_1(C_c, C_s)] + [\Gamma_1(C_c, C_s) - 1] \cdot I_{R_{load}} - \frac{1}{(2+\frac{C_s}{C_c})} \cdot I_{R_s}$$

$$\Gamma_2 = \Gamma_2(C_s, C_c) = 1 - \frac{1}{(2+\frac{C_s}{C_c})}; -\Gamma_2 = -\Gamma_2(C_s, C_c) = \frac{1}{(2+\frac{C_s}{C_c})} - 1$$

$$\Gamma_3 = \Gamma_3(C_c, C_s) = 1 - \Gamma_1(C_c, C_s); -\Gamma_3 = -\Gamma_3(C_c, C_s) = \Gamma_1(C_c, C_s) - 1$$

$$I_{C_s(\text{tank1})} = I_{R_s} - \Gamma_2 \cdot I_{L_1} - \Gamma_3 \cdot I_{L_2} - \Gamma_2 \cdot I_{R_{s_1}} - \Gamma_3 \cdot I_{R_{s_2}} + I_{R_1(\text{tank1})} \cdot \Gamma_2 + I_{R_1(\text{tank2})} \cdot \Gamma_3 - \Gamma_3 \cdot I_{R_{load}} - \frac{1}{(2+\frac{C_s}{C_c})} \cdot I_{R_s}$$

$$I_{C_s(\text{tank2})} = \frac{1}{(2+\frac{C_s}{C_c})} \cdot I_{R_s} + I_{R_1(\text{tank1})} \cdot \frac{1}{(2+\frac{C_s}{C_c})} + I_{R_1(\text{tank2})} \cdot \Gamma_1 - \frac{1}{(2+\frac{C_s}{C_c})} \cdot I_{L_1} - \Gamma_1 \cdot I_{L_2} - \frac{1}{(2+\frac{C_s}{C_c})} \cdot I_{R_{s_1}} - \Gamma_1 \cdot I_{R_{s_2}} - \Gamma_1 \cdot I_{R_{load}}$$

Tunable BPF (varactor diodes) differential equations group No. 4:

$$\frac{dI_{L_1}}{dt} = Y_1; \frac{dY_1}{dt} = \frac{1}{C_s \cdot L_1} \cdot \{I_{R_s} - \Gamma_2 \cdot I_{L_1} - \Gamma_3 \cdot I_{L_2} - \Gamma_2 \cdot I_{R_{s_1}} - \Gamma_3 \cdot I_{R_{s_2}} + I_{R_1(\text{tank1})} \cdot \Gamma_2 + I_{R_1(\text{tank2})} \cdot \Gamma_3 - \Gamma_3 \cdot I_{R_{load}} - \frac{1}{(2+\frac{C_s}{C_c})} \cdot I_{R_s}\};$$

$$\frac{dI_{R_{p_1}}}{dt} = \frac{1}{C_{j1} \cdot R_{p_1}} \cdot (I_{R_{s_1}} - I_{R_{p_1}})$$

$$\frac{dI_{R_{s_1}}}{dt} = \frac{1}{C_s \cdot R_{s_1}} \cdot (I_{R_s} - \Gamma_2 \cdot I_{L_1} - \Gamma_3 \cdot I_{L_2} - \Gamma_2 \cdot I_{R_{s_1}} - \Gamma_3 \cdot I_{R_{s_2}} + I_{R_1(\text{tank1})} \cdot \Gamma_2$$
$$+ I_{R_1(\text{tank2})} \cdot \Gamma_3 - \Gamma_3 \cdot I_{R_{load}} - \frac{1}{(2 + \frac{C_s}{C_c})} \cdot I_{R_s}) - \frac{1}{C_T \cdot R_{s_1}} \cdot (I_{R_{s_1}} - I_{R_1(\text{tank1})})$$
$$- \frac{1}{C_{j1} \cdot R_{s_1}} \cdot (I_{R_{s_1}} - I_{R_{p_1}}); \frac{dI_{R_{p_2}}}{dt} = \frac{1}{C_{j2} \cdot R_{p_2}} \cdot (I_{R_{s_2}} - I_{R_{p_2}})$$

$$\frac{dI_{R_{s_2}}}{dt} = \frac{1}{C_s \cdot R_{s_2}} \cdot (I_{R_s} - \Gamma_2 \cdot I_{L_1} - \Gamma_3 \cdot I_{L_2} - \Gamma_2 \cdot I_{R_{s_1}} - \Gamma_3 \cdot I_{R_{s_2}} + I_{R_1(\text{tank1})} \cdot \Gamma_2 + I_{R_1(\text{tank2})}$$
$$\cdot \Gamma_3 - \Gamma_3 \cdot I_{R_{load}} - \frac{1}{(2 + \frac{C_s}{C_c})} \cdot I_{R_s}) - \frac{1}{C_c \cdot R_{s_2}} \cdot I_{C_c} - \frac{1}{C_T \cdot R_{s_2}} \cdot (I_{R_{s_2}} - I_{R_1(\text{tank2})})$$
$$- \frac{1}{C_{j2} \cdot R_{s_2}} \cdot (I_{R_{s_2}} - I_{R_{p_2}})$$

$$\frac{dI_{L_2}}{dt} = I_{R_{load}} \cdot \frac{R_{load}}{L_2}; \frac{dI_{R_{load}}}{dt} = \frac{1}{C_s \cdot R_{load}} \cdot \{\frac{1}{(2 + \frac{C_s}{C_c})} \cdot I_{R_s} + I_{R_1(\text{tank1})} \cdot \frac{1}{(2 + \frac{C_s}{C_c})}$$
$$+ I_{R_1(\text{tank2})} \cdot \Gamma_1 - \frac{1}{(2 + \frac{C_s}{C_c})} \cdot I_{L_1} - \Gamma_1 \cdot I_{L_2} - \frac{1}{(2 + \frac{C_s}{C_c})}$$
$$\cdot I_{R_{s_1}} - \Gamma_1 \cdot I_{R_{s_2}} - \Gamma_1 \cdot I_{R_{load}}\}$$

$$I_{C_c} = I_{R_s} - I_{L_1} - I_{C_s(\text{tank1})} - I_{R_{s_1}} + I_{R_1(\text{tank1})}$$
$$= I_{R_s} - I_{L_1} - \{I_{R_s} - \Gamma_2 \cdot I_{L_1} - \Gamma_3 \cdot I_{L_2} - \Gamma_2 \cdot I_{R_{s_1}}$$
$$- \Gamma_3 \cdot I_{R_{s_2}} + I_{R_1(\text{tank1})} \cdot \Gamma_2 + I_{R_1(\text{tank2})} \cdot \Gamma_3 - \Gamma_3 \cdot I_{R_{load}}$$
$$- \frac{1}{(2 + \frac{C_s}{C_c})} \cdot I_{R_s}\} - I_{R_{s_1}} + I_{R_1(\text{tank1})}$$

$$I_{C_c} = I_{R_s} - I_{L_1} - I_{R_s} + \Gamma_2 \cdot I_{L_1} + \Gamma_3 \cdot I_{L_2} + \Gamma_2 \cdot I_{R_{s1}} + \Gamma_3 \cdot I_{R_{s2}} - I_{R_1(\text{tank1})} \cdot \Gamma_2$$
$$- I_{R_1(\text{tank2})} \cdot \Gamma_3 + \Gamma_3 \cdot I_{R_{load}} + \frac{1}{(2 + \frac{C_s}{C_c})} \cdot I_{R_s} - I_{R_{s1}} + I_{R_1(\text{tank1})}$$

$$I_{C_c} = [1 + \frac{1}{(2 + \frac{C_s}{C_c})}] \cdot I_{R_s} + [\Gamma_2 - 1] \cdot I_{L_1} - I_{R_s} + \Gamma_3 \cdot I_{L_2} + [\Gamma_2 - 1] \cdot I_{R_{s1}} + \Gamma_3 \cdot I_{R_{s2}}$$
$$- I_{R_1(\text{tank2})} \cdot \Gamma_3 + \Gamma_3 \cdot I_{R_{load}} + I_{R_1(\text{tank1})} \cdot [1 - \Gamma_2]$$

Tunable BPF (varactor diodes) differential equations group No. 5:

$$\frac{dI_{L_1}}{dt} = Y_1; \frac{dY_1}{dt} = \frac{1}{C_s \cdot L_1} \cdot \{I_{R_s} - \Gamma_2 \cdot I_{L_1} - \Gamma_3 \cdot I_{L_2} - \Gamma_2 \cdot I_{R_{s_1}} - \Gamma_3 \cdot I_{R_{s_2}} + I_{R_1(\text{tank1})} \cdot \Gamma_2$$
$$+ I_{R_1(\text{tank2})} \cdot \Gamma_3 - \Gamma_3 \cdot I_{R_{load}} - \frac{1}{(2 + \frac{C_s}{C_c})} \cdot I_{R_s}\}; \frac{dI_{R_{p_1}}}{dt} = \frac{1}{C_{j1} \cdot R_{p_1}} \cdot (I_{R_{s_1}} - I_{R_{p_1}})$$

$$\frac{dI_{R_{s_1}}}{dt} = \frac{1}{C_s \cdot R_{s_1}} \cdot (I_{R_s} - \Gamma_2 \cdot I_{L_1} - \Gamma_3 \cdot I_{L_2} - \Gamma_2 \cdot I_{R_{s_1}} - \Gamma_3 \cdot I_{R_{s_2}} + I_{R_1(\text{tank1})}$$
$$\cdot \Gamma_2 + I_{R_1(\text{tank2})} \cdot \Gamma_3 - \Gamma_3 \cdot I_{R_{load}} - \frac{1}{(2 + \frac{C_s}{C_c})} \cdot I_{R_s}) - \frac{1}{C_T \cdot R_{s_1}} \cdot (I_{R_{s_1}} - I_{R_1(\text{tank1})}) - \frac{1}{C_{j1} \cdot R_{s_1}}$$
$$\cdot (I_{R_{s_1}} - I_{R_{p_1}}); \frac{dI_{R_{p_2}}}{dt} = \frac{1}{C_{j2} \cdot R_{p_2}} \cdot (I_{R_{s_2}} - I_{R_{p_2}})$$

$$\frac{dI_{R_{s_2}}}{dt} = \frac{1}{C_s \cdot R_{s_2}} \cdot (I_{R_s} - \Gamma_2 \cdot I_{L_1} - \Gamma_3 \cdot I_{L_2} - \Gamma_2 \cdot I_{R_{s_1}} - \Gamma_3 \cdot I_{R_{s_2}} + I_{R_1(\text{tank1})}$$
$$\cdot \Gamma_2 + I_{R_1(\text{tank2})} \cdot \Gamma_3 - \Gamma_3 \cdot I_{R_{load}} - \frac{1}{(2 + \frac{C_s}{C_c})} \cdot I_{R_s}) - \frac{1}{C_c \cdot R_{s_2}} \cdot \{[1 + \frac{1}{(2 + \frac{C_s}{C_c})}] \cdot I_{R_s}$$
$$+ [\Gamma_2 - 1] \cdot I_{L_1} - I_{R_s} + \Gamma_3 \cdot I_{L_2} + [\Gamma_2 - 1] \cdot I_{R_{s_1}} + \Gamma_3 \cdot I_{R_{s_2}}$$
$$- I_{R_1(\text{tank2})} \cdot \Gamma_3 + \Gamma_3 \cdot I_{R_{load}} + I_{R_1(\text{tank1})} \cdot [1 - \Gamma_2]\} - \frac{1}{C_T \cdot R_{s_2}} \cdot (I_{R_{s_2}} - I_{R_1(\text{tank2})})$$
$$- \frac{1}{C_{j2} \cdot R_{s_2}} \cdot (I_{R_{s_2}} - I_{R_{p_2}})$$

$$\frac{dI_{L_2}}{dt} = I_{R_{load}} \cdot \frac{R_{load}}{L_2}; \frac{dI_{R_{load}}}{dt} = \frac{1}{C_s \cdot R_{load}} \cdot \{\frac{1}{(2 + \frac{C_s}{C_c})} \cdot I_{R_s} + I_{R_1(\text{tank1})} \cdot \frac{1}{(2 + \frac{C_s}{C_c})}$$
$$+ I_{R_1(\text{tank2})} \cdot \Gamma_1 - \frac{1}{(2 + \frac{C_s}{C_c})} \cdot I_{L_1} - \Gamma_1 \cdot I_{L_2} - \frac{1}{(2 + \frac{C_s}{C_c})} \cdot I_{R_{s_1}} - \Gamma_1 \cdot I_{R_{s_2}} - \Gamma_1 \cdot I_{R_{load}}\}$$

Tunable BPF (varactor diodes) differential equations group No. 6:

$$\frac{dY_1}{dt} = \frac{1}{C_s \cdot L_1} \cdot I_{R_s} - \frac{1}{C_s \cdot L_1} \cdot \Gamma_2 \cdot I_{L_1} - \frac{1}{C_s \cdot L_1} \cdot \Gamma_3 \cdot I_{L_2} - \frac{1}{C_s \cdot L_1} \cdot \Gamma_2$$
$$\cdot I_{R_{s_1}} - \frac{1}{C_s \cdot L_1} \cdot \Gamma_3 \cdot I_{R_{s_2}} + \frac{1}{C_s \cdot L_1} \cdot I_{R_1(\text{tank1})} \cdot \Gamma_2 + \frac{1}{C_s \cdot L_1} \cdot I_{R_1(\text{tank2})}$$
$$\cdot \Gamma_3 - \frac{1}{C_s \cdot L_1} \cdot \Gamma_3 \cdot I_{R_{load}} - \frac{1}{C_s \cdot L_1} \cdot \frac{1}{(2 + \frac{C_s}{C_c})} \cdot I_{R_s}$$

$$\frac{dI_{R_{p_1}}}{dt} = \frac{1}{C_{j1} \cdot R_{p_1}} \cdot I_{R_{s_1}} - \frac{1}{C_{j1} \cdot R_{p_1}} \cdot I_{R_{p_1}}; \frac{dI_{L_1}}{dt} = Y_1$$

$$\frac{dI_{R_{s_1}}}{dt} = [1 - \frac{1}{(2+\frac{C_s}{C_c})}] \cdot \frac{1}{C_s \cdot R_{s_1}} \cdot I_{R_s} - \frac{1}{C_s \cdot R_{s_1}} \cdot \Gamma_2 \cdot I_{L_1} - \frac{1}{C_s \cdot R_{s_1}} \cdot \Gamma_3 \cdot I_{L_2}$$
$$- [\frac{1}{C_s} \cdot \Gamma_2 + \frac{1}{C_T} + \frac{1}{C_{j1}}] \cdot \frac{1}{R_{s_1}} \cdot I_{R_{s_1}} - \frac{1}{C_s \cdot R_{s_1}} \cdot \Gamma_3 \cdot I_{R_{s_2}} + [\frac{1}{C_s} \cdot \Gamma_2 + \frac{1}{C_T}]$$
$$\cdot \frac{1}{R_{s_1}} \cdot I_{R_1(\text{tank1})} + I_{R_1(\text{tank2})}$$
$$\cdot \frac{1}{C_s \cdot R_{s_1}} \cdot \Gamma_3 - \frac{1}{C_s \cdot R_{s_1}} \cdot \Gamma_3 \cdot I_{R_{load}} + \frac{1}{C_{j1} \cdot R_{s_1}} \cdot I_{R_{p_1}}$$

$$\frac{dI_{R_{p_2}}}{dt} = \frac{1}{C_{j2} \cdot R_{p_2}} \cdot I_{R_{s_2}} - \frac{1}{C_{j2} \cdot R_{p_2}} \cdot I_{R_{p_2}}$$

$$\frac{dI_{R_{s_2}}}{dt} = \{\frac{1}{C_s} - \frac{1}{C_s} \cdot \frac{1}{(2+\frac{C_s}{C_c})} - \frac{1}{C_c} \cdot [1 + \frac{1}{(2+\frac{C_s}{C_c})}] + \frac{1}{C_c}\} \cdot \frac{1}{R_{s_2}} \cdot I_{R_s}$$
$$- (\frac{1}{C_s} \cdot \Gamma_2 + \frac{1}{C_c} \cdot [\Gamma_2 - 1]) \cdot \frac{1}{R_{s_2}} \cdot I_{L_1} - (\frac{1}{C_s} + \frac{1}{C_c}) \cdot \frac{1}{R_{s_2}} \cdot \Gamma_3 \cdot I_{L_2}$$
$$- (\frac{1}{C_s} \cdot \Gamma_2 + \frac{1}{C_c} \cdot [\Gamma_2 - 1]) \cdot \frac{1}{R_{s_2}} \cdot I_{R_{s_1}} - (\frac{1}{C_s} \cdot \Gamma_3 + \frac{1}{C_c} \cdot \Gamma_3 + \frac{1}{C_{j2}} + \frac{1}{C_T})$$
$$\cdot \frac{1}{R_{s_2}} \cdot I_{R_{s_2}} + (\frac{1}{C_s} \cdot \Gamma_2 - \frac{1}{C_c} \cdot [1 - \Gamma_2]) \cdot \frac{1}{R_{s_2}} \cdot I_{R_1(\text{tank1})} + (\frac{1}{C_c} \cdot \Gamma_3 + \frac{1}{C_T} + \frac{1}{C_s} \cdot \Gamma_3)$$
$$\cdot \frac{1}{R_{s_2}} \cdot I_{R_1(\text{tank2})} - (\frac{1}{C_c} + \frac{1}{C_s}) \cdot \Gamma_3 \cdot \frac{1}{R_{s_2}} \cdot I_{R_{load}} + \frac{1}{C_{j2} \cdot R_{s_2}} \cdot I_{R_{p_2}}$$

$$\frac{dI_{L_2}}{dt} = I_{R_{load}} \cdot \frac{R_{load}}{L_2}; \frac{dI_{R_{load}}}{dt} = \frac{1}{C_s \cdot R_{load}} \cdot \frac{1}{(2+\frac{C_s}{C_c})} \cdot I_{R_s} + I_{R_1(\text{tank1})} \cdot \frac{1}{C_s \cdot R_{load}} \cdot \frac{1}{(2+\frac{C_s}{C_c})}$$
$$+ I_{R_1(\text{tank2})} \cdot \frac{1}{C_s \cdot R_{load}} \cdot \Gamma_1 - \frac{1}{C_s \cdot R_{load}} \cdot \frac{1}{(2+\frac{C_s}{C_c})} \cdot I_{L_1} - \frac{1}{C_s \cdot R_{load}} \cdot \Gamma_1 \cdot I_{L_2}$$
$$- \frac{1}{C_s \cdot R_{load}} \cdot \frac{1}{(2+\frac{C_s}{C_c})} \cdot I_{R_{s_1}} - \frac{1}{C_s \cdot R_{load}} \cdot \Gamma_1 \cdot I_{R_{s_2}} - \frac{1}{C_s \cdot R_{load}} \cdot \Gamma_1 \cdot I_{R_{load}}$$

We define for simplicity new global parameters:

$$\Omega_1 = \frac{1}{C_s \cdot L_1}; \ \Omega_2 = \frac{1}{C_{j1} \cdot R_{p_1}}; \ \Omega_3 = [1 - \frac{1}{(2+\frac{C_s}{C_c})}] \cdot \frac{1}{C_s \cdot R_{s_1}}; \ \Omega_4 = \frac{1}{C_s \cdot R_{s_1}} \cdot \Gamma_2;$$
$$\Omega_5 = \frac{1}{C_s \cdot R_{s_1}} \cdot \Gamma_3; \ \Omega_6 = [\frac{1}{C_s} \cdot \Gamma_2 + \frac{1}{C_T} + \frac{1}{C_{j1}}] \cdot \frac{1}{R_{s_1}}; \Omega_7 = [\frac{1}{C_s} \cdot \Gamma_2 + \frac{1}{C_T}] \cdot \frac{1}{R_{s_1}};$$
$$\Omega_8 = \frac{1}{C_{j1} \cdot R_{s_1}}$$

$$\Omega_9 = \frac{1}{C_{j2} \cdot R_{p_2}}; \Omega_{10} = \{\frac{1}{C_s} - \frac{1}{C_s} \cdot \frac{1}{(2 + \frac{C_s}{C_c})} - \frac{1}{C_c} \cdot [1 + \frac{1}{(2 + \frac{C_s}{C_c})}] + \frac{1}{C_c}\} \cdot \frac{1}{R_{s_2}}$$

$$\Omega_{11} = (\frac{1}{C_s} \cdot \Gamma_2 + \frac{1}{C_c} \cdot [\Gamma_2 - 1]) \cdot \frac{1}{R_{s_2}}; \Omega_{12} = (\frac{1}{C_s} + \frac{1}{C_c}) \cdot \frac{1}{R_{s_2}} \cdot \Gamma_3$$

$$\Omega_{13} = (\frac{1}{C_s} \cdot \Gamma_3 + \frac{1}{C_c} \cdot \Gamma_3 + \frac{1}{C_{j2}} + \frac{1}{C_T}) \cdot \frac{1}{R_{s_2}}; \Omega_{14} = (\frac{1}{C_c} \cdot \Gamma_3 + \frac{1}{C_T} + \frac{1}{C_s} \cdot \Gamma_3) \cdot \frac{1}{R_{s_2}}$$

$$\Omega_{15} = \frac{1}{C_{j2} \cdot R_{s_2}}; \Omega_{16} = \frac{1}{C_s \cdot R_{load}} \cdot \frac{1}{(2 + \frac{C_s}{C_c})}; \Omega_{17} = \frac{1}{C_s \cdot R_{load}} \cdot \Gamma_1$$

Tunable BPF (varactor diodes) differential equations group No. 7:

$$\frac{dY_1}{dt} = \Omega_1 \cdot I_{R_s} - \Omega_1 \cdot \Gamma_2 \cdot I_{L_1} - \Omega_1 \cdot \Gamma_3 \cdot I_{L_2} - \Omega_1 \cdot \Gamma_2 \cdot I_{R_{s_1}} - \Omega_1 \cdot \Gamma_3 \cdot I_{R_{s_2}}$$
$$+ \Omega_1 \cdot I_{R_1(\text{tank1})} \cdot \Gamma_2 + \Omega_1 \cdot I_{R_1(\text{tank2})} \cdot \Gamma_3 - \Omega_1 \cdot \Gamma_3 \cdot I_{R_{load}} - \Omega_1 \cdot \frac{1}{(2 + \frac{C_s}{C_c})} \cdot I_{R_s}$$

$$\frac{dI_{R_{p_1}}}{dt} = \Omega_2 \cdot I_{R_{s_1}} - \Omega_2 \cdot I_{R_{p_1}}; \frac{dI_{L_1}}{dt} = Y_1; \frac{dI_{R_{p_2}}}{dt} = \Omega_9 \cdot I_{R_{s_2}} - \Omega_9 \cdot I_{R_{p_2}}; \frac{dI_{L_2}}{dt}$$
$$= I_{R_{load}} \cdot \frac{R_{load}}{L_2}$$

$$\frac{dI_{R_{s_1}}}{dt} = \Omega_3 \cdot I_{R_s} - \Omega_4 \cdot I_{L_1} - \Omega_5 \cdot I_{L_2} - \Omega_6 \cdot I_{R_{s_1}} - \Omega_5 \cdot I_{R_{s_2}} + \Omega_7$$
$$\cdot I_{R_1(\text{tank1})} + I_{R_1(\text{tank2})} \cdot \Omega_5 - \Omega_5 \cdot I_{R_{load}} + \Omega_8 \cdot I_{R_{p_1}}$$

$$\frac{dI_{R_{s_2}}}{dt} = \Omega_{10} \cdot I_{R_s} - \Omega_{11} \cdot I_{L_1} - \Omega_{12} \cdot I_{L_2} - \Omega_{11} \cdot I_{R_{s_1}} - \Omega_{13} \cdot I_{R_{s_2}} + \Omega_{11} \cdot I_{R_1(\text{tank1})}$$
$$+ \Omega_{14} \cdot I_{R_1(\text{tank2})} - \Omega_{12} \cdot I_{R_{load}} + \Omega_{15} \cdot I_{R_{p_2}}$$

$$\frac{dI_{R_{load}}}{dt} = \Omega_{16} \cdot I_{R_s} + I_{R_1(\text{tank1})} \cdot \Omega_{16} + I_{R_1(\text{tank2})} \cdot \Omega_{17} - \Omega_{16} \cdot I_{L_1} - \Omega_{17} \cdot I_{L_2}$$
$$- \Omega_{16} \cdot I_{R_{s_1}} - \Omega_{17} \cdot I_{R_{s_2}} - \Omega_{17} \cdot I_{R_{load}}$$

At fixed points: $\frac{dY_1}{dt} = 0; \frac{dI_{R_{p_1}}}{dt} = 0; \frac{dI_{L_1}}{dt} = 0; \frac{dI_{R_{p_2}}}{dt} = 0; \frac{dI_{L_2}}{dt} = 0; \frac{dI_{R_{s_1}}}{dt} = 0$

$$\frac{dI_{R_{s_2}}}{dt} = 0; \frac{dI_{R_{load}}}{dt} = 0$$

$$I^*_{R_{p_1}} = I^*_{R_{s_1}}; Y^*_1 = 0; I^*_{R_{p_2}} = I^*_{R_{s_2}}; I^*_{R_{load}} = 0$$

$$[1-\frac{1}{(2+\frac{C_s}{C_c})}]\cdot\Omega_1\cdot I^*_{R_s}-\Omega_1\cdot\Gamma_2\cdot I^*_{L_1}-\Omega_1\cdot\Gamma_3\cdot I^*_{L_2}-\Omega_1\cdot\Gamma_2\cdot I^*_{R_{s_1}}-\Omega_1\cdot\Gamma_3\cdot I^*_{R_{s_2}}$$
$$+\Omega_1\cdot\Gamma_2\cdot I^*_{R_1(\text{tank1})}+\Omega_1\cdot\Gamma_3\cdot I^*_{R_1(\text{tank2})}=0$$

$$\Omega_3\cdot I^*_{R_s}-\Omega_4\cdot I^*_{L_1}-\Omega_5\cdot I^*_{L_2}+[\Omega_8-\Omega_6]\cdot I^*_{R_{s_1}}-\Omega_5\cdot I^*_{R_{s_2}}+\Omega_7\cdot I^*_{R_1(\text{tank1})}+I^*_{R_1(\text{tank2})}$$
$$\cdot\Omega_5=0$$

$$\Omega_{10}\cdot I^*_{R_s}-\Omega_{11}\cdot I^*_{L_1}-\Omega_{12}\cdot I^*_{L_2}-\Omega_{11}\cdot I^*_{R_{s_1}}+[\Omega_{15}-\Omega_{13}]\cdot I^*_{R_{s_2}}+\Omega_{11}$$
$$\cdot I^*_{R_1(\text{tank1})}+\Omega_{14}\cdot I^*_{R_1(\text{tank2})}$$
$$=0$$

$$\Omega_{16}\cdot I^*_{R_s}+I^*_{R_1(\text{tank1})}\cdot\Omega_{16}+I^*_{R_1(\text{tank2})}\cdot\Omega_{17}-\Omega_{16}\cdot I^*_{L_1}-\Omega_{17}\cdot I^*_{L_2}-\Omega_{16}\cdot I^*_{R_{s_1}}-\Omega_{17}$$
$$\cdot I^*_{R_{s_2}}$$
$$=0$$

Stability analysis: The standard local stability analysis about any one of the equilibrium points of Tunable BPF (varactor diodes) circuit consists in adding to its coordinated $[Y_1\ I_{R_{p_1}}\ I_{L_1}\ I_{R_{s_1}}\ I_{R_{p_2}}\ I_{R_{s_2}}\ I_{L_2}\ I_{R_{load}}\ I_{R_s}\ I_{R_1(\text{tank1})}\ I_{R_1(\text{tank2})}]$ arbitrarily small increments of exponential terms $[y_1\ i_{R_{p_1}}\ i_{L_1}\ i_{R_{s_1}}\ i_{R_{p_2}}\ i_{R_{s_2}}\ i_{L_2}\ i_{R_{load}}\ i_{R_s}\ i_{R_1(\text{tank1})}\ i_{R_1(\text{tank2})}]\cdot e^{\lambda\cdot t}$, and retaining the first order terms in $y_1\ i_{R_{p_1}}\ i_{L_1}\ i_{R_{s_1}}\ i_{R_{p_2}}\ i_{R_{s_2}}\ i_{L_2}\ i_{R_{load}}\ i_{R_s}\ i_{R_1(\text{tank1})}\ i_{R_1(\text{tank2})}$. The system of eight homogeneous equations leads to a polynomial characteristic equation in the eigenvalue λ. The polynomial characteristic equation accepts by set the tunable BPF (varactor diodes) circuit equations. The tunable BPF (varactor diodes) circuit fixed values with arbitrarily small increments of exponential form $[y_1\ i_{R_{p_1}}\ i_{L_1}\ i_{R_{s_1}}\ i_{R_{p_2}}\ i_{R_{s_2}}\ i_{L_2}\ i_{R_{load}}\ i_{R_s}\ i_{R_1(\text{tank1})}\ i_{R_1(\text{tank2})}]\cdot e^{\lambda\cdot t}$ are; i = 0 (first fixed point), i = 1 (second fixed point), i = 2 (third fixed point), etc.,

$$Y_1(t)=Y_1^{(i)}+y_1\cdot e^{\lambda\cdot t};\ I_{R_{p_1}}(t)=I^{(i)}_{R_{p_1}}+i_{R_{p_1}}\cdot e^{\lambda\cdot t};\ I_{L_1}(t)=I^{(i)}_{L_1}+i_{L_1}\cdot e^{\lambda\cdot t};\ I_{R_{s_1}}(t)=I^{(i)}_{R_{s_1}}+i_{R_{s_1}}\cdot e^{\lambda\cdot t}$$
$$I_{R_{p_2}}(t)=I^{(i)}_{R_{p_2}}+i_{R_{p_2}}\cdot e^{\lambda\cdot t};\ I_{R_{s_2}}(t)=I^{(i)}_{R_{s_2}}+i_{R_{s_2}}\cdot e^{\lambda\cdot t};\ I_{L_2}(t)=I^{(i)}_{L_2}+i_{L_2}\cdot e^{\lambda\cdot t};\ I_{R_{load}}(t)=I^{(i)}_{R_{load}}+i_{R_{load}}\cdot e^{\lambda\cdot t}$$
$$I_{R_s}(t)=I^{(i)}_{R_s}+i_{R_s}\cdot e^{\lambda\cdot t};\ I_{R_1(\text{tank1})}(t)=I^{(i)}_{R_1(\text{tank1})}+i_{R_1(\text{tank1})}\cdot e^{\lambda\cdot t};\ I_{R_1(\text{tank2})}(t)=I^{(i)}_{R_1(\text{tank2})}+i_{R_1(\text{tank2})}\cdot e^{\lambda\cdot t}$$

$$\frac{dY_1(t)}{dt}=y_1\cdot\lambda\cdot e^{\lambda\cdot t};\frac{dI_{R_{p_1}}(t)}{dt}=i_{R_{p_1}}\cdot\lambda\cdot e^{\lambda\cdot t};\frac{dI_{L_1}(t)}{dt}=i_{L_1}\cdot\lambda\cdot e^{\lambda\cdot t};\frac{dI_{R_{s_1}}(t)}{dt}$$
$$=i_{R_{s_1}}\cdot\lambda\cdot e^{\lambda\cdot t}$$

$$\frac{dI_{R_{p_2}}(t)}{dt}=i_{R_{p_2}}\cdot\lambda\cdot e^{\lambda\cdot t};\frac{dI_{R_{s_2}}(t)}{dt}=i_{R_{s_2}}\cdot\lambda\cdot e^{\lambda\cdot t};\frac{dI_{L_2}(t)}{dt}=i_{L_2}\cdot\lambda\cdot e^{\lambda\cdot t};\frac{dI_{R_{load}}(t)}{dt}$$
$$=i_{R_{load}}\cdot\lambda\cdot e^{\lambda\cdot t}$$

For λ < 0, t > 0 the selected fixed point is stable otherwise λ > 0, t > 0 unstable. Our BPF (varactor diodes) circuit tends to the selected fixed point exponentially for λ < 0, t > 0 otherwise go away from the selected fixed point exponentially. λ is the eigenvalue parameter which establish if the fixed point is stable or unstable, additionally his absolute value |λ| establish the speed of flow toward or away from the selected fixed point [2–4].

$$\frac{dY_1}{dt} = \Omega_1 \cdot I_{R_s} - \Omega_1 \cdot \Gamma_2 \cdot I_{L_1} - \Omega_1 \cdot \Gamma_3 \cdot I_{L_2} - \Omega_1 \cdot \Gamma_2 \cdot I_{R_{s_1}} - \Omega_1 \cdot \Gamma_3 \cdot I_{R_{s_2}} + \Omega_1 \cdot I_{R_1(\text{tank1})} \cdot \Gamma_2 + \Omega_1 \cdot I_{R_1(\text{tank2})} \cdot \Gamma_3 - \Omega_1 \cdot \Gamma_3 \cdot I_{R_{load}} - \Omega_1 \cdot \frac{1}{(2 + \frac{C_s}{C_c})} \cdot I_{R_s}$$

$$y_1 \cdot \lambda \cdot e^{\lambda \cdot t} = \Omega_1 \cdot (I_{R_s}^{(i)} + i_{R_s} \cdot e^{\lambda \cdot t}) - \Omega_1 \cdot \Gamma_2 \cdot (I_{L_1}^{(i)} + i_{L_1} \cdot e^{\lambda \cdot t}) - \Omega_1 \cdot \Gamma_3 \cdot (I_{L_2}^{(i)} + i_{L_2} \cdot e^{\lambda \cdot t}) - \Omega_1 \cdot \Gamma_2 \cdot (I_{R_{s_1}}^{(i)} + i_{R_{s_1}} \cdot e^{\lambda \cdot t}) - \Omega_1 \cdot \Gamma_3 \cdot (I_{R_{s_2}}^{(i)} + i_{R_{s_2}} \cdot e^{\lambda \cdot t}) + \Omega_1 \cdot \Gamma_2 \cdot (I_{R_1(\text{tank1})}^{(i)} + i_{R_1(\text{tank1})} \cdot e^{\lambda \cdot t}) + \Omega_1 \cdot \Gamma_3 \cdot (I_{R_1(\text{tank2})}^{(i)} + i_{R_1(\text{tank2})} \cdot e^{\lambda \cdot t}) - \Omega_1 \cdot \Gamma_3 \cdot (I_{R_{load}}^{(i)} + i_{R_{load}} \cdot e^{\lambda \cdot t}) - \Omega_1 \cdot \frac{1}{(2 + \frac{C_s}{C_c})} \cdot (I_{R_s}^{(i)} + i_{R_s} \cdot e^{\lambda \cdot t})$$

$$y_1 \cdot \lambda \cdot e^{\lambda \cdot t} = \Omega_1 \cdot I_{R_s}^{(i)} - \Omega_1 \cdot \Gamma_2 \cdot I_{L_1}^{(i)} - \Omega_1 \cdot \Gamma_3 \cdot I_{L_2}^{(i)} - \Omega_1 \cdot \Gamma_2 \cdot I_{R_{s_1}}^{(i)} - \Omega_1 \cdot \Gamma_3 \cdot I_{R_{s_2}}^{(i)} + \Omega_1 \cdot \Gamma_2 \cdot I_{R_1(\text{tank1})}^{(i)} + \Omega_1 \cdot \Gamma_3 \cdot I_{R_1(\text{tank2})}^{(i)} - \Omega_1 \cdot \Gamma_3 \cdot I_{R_{load}}^{(i)} - \Omega_1 \cdot \frac{1}{(2 + \frac{C_s}{C_c})} \cdot I_{R_s}^{(i)} + \Omega_1 \cdot i_{R_s} \cdot e^{\lambda \cdot t} - \Omega_1 \cdot \Gamma_2 \cdot i_{L_1} \cdot e^{\lambda \cdot t} - \Omega_1 \cdot \Gamma_3 \cdot i_{L_2} \cdot e^{\lambda \cdot t} - \Omega_1 \cdot \Gamma_2 \cdot i_{R_{s_1}} \cdot e^{\lambda \cdot t} - \Omega_1 \cdot \Gamma_3 \cdot i_{R_{s_2}} \cdot e^{\lambda \cdot t} + \Omega_1 \cdot \Gamma_2 \cdot i_{R_1(\text{tank1})} \cdot e^{\lambda \cdot t} + \Omega_1 \cdot \Gamma_3 \cdot i_{R_1(\text{tank2})} \cdot e^{\lambda \cdot t} - \Omega_1 \cdot \Gamma_3 \cdot i_{R_{load}} \cdot e^{\lambda \cdot t} - \Omega_1 \cdot \frac{1}{(2 + \frac{C_s}{C_c})} \cdot i_{R_s} \cdot e^{\lambda \cdot t}$$

At fixed points:

$$\Omega_1 \cdot I_{R_s}^{(i)} - \Omega_1 \cdot \Gamma_2 \cdot I_{L_1}^{(i)} - \Omega_1 \cdot \Gamma_3 \cdot I_{L_2}^{(i)} - \Omega_1 \cdot \Gamma_2 \cdot I_{R_{s_1}}^{(i)} - \Omega_1 \cdot \Gamma_3 \cdot I_{R_{s_2}}^{(i)} + \Omega_1 \cdot \Gamma_2 \cdot I_{R_1(\text{tank1})}^{(i)} + \Omega_1 \cdot \Gamma_3 \cdot I_{R_1(\text{tank2})}^{(i)} - \Omega_1 \cdot \Gamma_3 \cdot I_{R_{load}}^{(i)} - \Omega_1 \cdot \frac{1}{(2 + \frac{C_s}{C_c})} \cdot I_{R_s}^{(i)} = 0$$

$$-y_1 \cdot \lambda - \Omega_1 \cdot \Gamma_2 \cdot i_{L_1} - \Omega_1 \cdot \Gamma_3 \cdot i_{L_2} - \Omega_1 \cdot \Gamma_2 \cdot i_{R_{s_1}} - \Omega_1 \cdot \Gamma_3 \cdot i_{R_{s_2}} + \Omega_1 \cdot \Gamma_2 \cdot i_{R_1(\text{tank1})} + \Omega_1 \cdot \Gamma_3 \cdot i_{R_1(\text{tank2})} - \Omega_1 \cdot \Gamma_3 \cdot i_{R_{load}} + \Omega_1 \cdot i_{R_s} \cdot [1 - \frac{1}{(2 + \frac{C_s}{C_c})}] = 0$$

$$\frac{dI_{R_{p1}}}{dt} = \Omega_2 \cdot I_{R_{s1}} - \Omega_2 \cdot I_{R_{p1}}; i_{R_{p1}} \cdot \lambda \cdot e^{\lambda \cdot t}$$
$$= \Omega_2 \cdot (I^{(i)}_{R_{s1}} + i_{R_{s1}} \cdot e^{\lambda \cdot t}) - \Omega_2 \cdot (I^{(i)}_{R_{p1}} + i_{R_{p1}} \cdot e^{\lambda \cdot t})$$

$$i_{R_{p1}} \cdot \lambda \cdot e^{\lambda \cdot t} = \Omega_2 \cdot I^{(i)}_{R_{s1}} - \Omega_2 \cdot I^{(i)}_{R_{p1}} + \Omega_2 \cdot i_{R_{s1}} \cdot e^{\lambda \cdot t} - \Omega_2 \cdot i_{R_{p1}} \cdot e^{\lambda \cdot t}$$

At fixed points: $\Omega_2 \cdot I^{(i)}_{R_{s1}} - \Omega_2 \cdot I^{(i)}_{R_{p1}} = 0$ then $-i_{R_{p1}} \cdot \lambda + \Omega_2 \cdot i_{R_{s1}} - \Omega_2 \cdot i_{R_{p1}} = 0$

$$\frac{dI_{L_1}}{dt} = Y_1; i_{L_1} \cdot \lambda \cdot e^{\lambda \cdot t} = Y^{(i)}_1 + y_1 \cdot e^{\lambda \cdot t}; Y^{(i)}_1 = 0 \Rightarrow -i_{L_1} \cdot \lambda + y_1 = 0$$

$$\frac{dI_{R_{p2}}}{dt} = \Omega_9 \cdot I_{R_{s2}} - \Omega_9 \cdot I_{R_{p2}}; i_{R_{p2}} \cdot \lambda \cdot e^{\lambda \cdot t}$$
$$= \Omega_9 \cdot (I^{(i)}_{R_{s2}} + i_{R_{s2}} \cdot e^{\lambda \cdot t}) - \Omega_9 \cdot (I^{(i)}_{R_{p2}} + i_{R_{p2}} \cdot e^{\lambda \cdot t})$$

$$i_{R_{p2}} \cdot \lambda \cdot e^{\lambda \cdot t} = \Omega_9 \cdot I^{(i)}_{R_{s2}} - \Omega_9 \cdot I^{(i)}_{R_{p2}} + \Omega_9 \cdot i_{R_{s2}} \cdot e^{\lambda \cdot t} - \Omega_9 \cdot i_{R_{p2}} \cdot e^{\lambda \cdot t}$$

At fixed points: $\Omega_9 \cdot I^{(i)}_{R_{s2}} - \Omega_9 \cdot I^{(i)}_{R_{p2}} = 0$ then $-i_{R_{p2}} \cdot \lambda + \Omega_9 \cdot i_{R_{s2}} - \Omega_9 \cdot i_{R_{p2}} = 0$

$$\frac{dI_{L_2}}{dt} = I_{R_{load}} \cdot \frac{R_{load}}{L_2}; i_{L_2} \cdot \lambda \cdot e^{\lambda \cdot t} = (I^{(i)}_{R_{load}} + i_{R_{load}} \cdot e^{\lambda \cdot t}) \cdot \frac{R_{load}}{L_2}; i_{L_2} \cdot \lambda \cdot e^{\lambda \cdot t}$$
$$= I^{(i)}_{R_{load}} \cdot \frac{R_{load}}{L_2} + i_{R_{load}} \cdot \frac{R_{load}}{L_2} \cdot e^{\lambda \cdot t}$$

At fixed points: $I^{(i)}_{R_{load}} \cdot \frac{R_{load}}{L_2} = 0$ then $-i_{L_2} \cdot \lambda + i_{R_{load}} \cdot \frac{R_{load}}{L_2} = 0$

$$\frac{dI_{R_{s1}}}{dt} = \Omega_3 \cdot I_{R_s} - \Omega_4 \cdot I_{L_1} - \Omega_5 \cdot I_{L_2} - \Omega_6 \cdot I_{R_{s1}} - \Omega_5 \cdot I_{R_{s2}} + \Omega_7$$
$$\cdot I_{R_1(\text{tank1})} + I_{R_1(\text{tank2})} \cdot \Omega_5 - \Omega_5 \cdot I_{R_{load}} + \Omega_8 \cdot I_{R_{p1}}$$

$$i_{R_{s1}} \cdot \lambda \cdot e^{\lambda \cdot t} = \Omega_3 \cdot (I^{(i)}_{R_s} + i_{R_s} \cdot e^{\lambda \cdot t}) - \Omega_4 \cdot (I^{(i)}_{L_1} + i_{L_1} \cdot e^{\lambda \cdot t}) - \Omega_5 \cdot (I^{(i)}_{L_2} + i_{L_2} \cdot e^{\lambda \cdot t})$$
$$- \Omega_6 \cdot (I^{(i)}_{R_{s1}} + i_{R_{s1}} \cdot e^{\lambda \cdot t}) - \Omega_5 \cdot (I^{(i)}_{R_{s2}} + i_{R_{s2}} \cdot e^{\lambda \cdot t}) + \Omega_7$$
$$\cdot (I^{(i)}_{R_1(\text{tank1})} + i_{R_1(\text{tank1})} \cdot e^{\lambda \cdot t}) + (I^{(i)}_{R_1(\text{tank2})} + i_{R_1(\text{tank2})} \cdot e^{\lambda \cdot t}) \cdot \Omega_5$$
$$- \Omega_5 \cdot (I^{(i)}_{R_{load}} + i_{R_{load}} \cdot e^{\lambda \cdot t}) + \Omega_8 \cdot (I^{(i)}_{R_{p1}} + i_{R_{p1}} \cdot e^{\lambda \cdot t})$$

$$i_{R_{s_1}} \cdot \lambda \cdot e^{\lambda \cdot t} = \Omega_3 \cdot I_{R_s}^{(i)} - \Omega_4 \cdot I_{L_1}^{(i)} - \Omega_5 \cdot I_{L_2}^{(i)} - \Omega_6 \cdot I_{R_{s_1}}^{(i)} - \Omega_5 \cdot I_{R_{s_2}}^{(i)} + \Omega_7 \cdot I_{R_1(\text{tank1})}^{(i)} + I_{R_1(\text{tank2})}^{(i)} \cdot \Omega_5 - \Omega_5 \cdot I_{R_{load}}^{(i)} + \Omega_8 \cdot I_{R_{p_1}}^{(i)} + \Omega_3 \cdot i_{R_s} \cdot e^{\lambda \cdot t} - \Omega_4 \cdot i_{L_1} \cdot e^{\lambda \cdot t} - \Omega_5 \cdot i_{L_2} \cdot e^{\lambda \cdot t} - \Omega_6 \cdot i_{R_{s_1}} \cdot e^{\lambda \cdot t} - \Omega_5 \cdot i_{R_{s_2}} \cdot e^{\lambda \cdot t} + \Omega_7 \cdot i_{R_1(\text{tank1})} \cdot e^{\lambda \cdot t} + i_{R_1(\text{tank2})} \cdot e^{\lambda \cdot t} \cdot \Omega_5 - \Omega_5 \cdot i_{R_{load}} \cdot e^{\lambda \cdot t} + \Omega_8 \cdot i_{R_{p_1}} \cdot e^{\lambda \cdot t}$$

At fixed points:

$$\Omega_3 \cdot I_{R_s}^{(i)} - \Omega_4 \cdot I_{L_1}^{(i)} - \Omega_5 \cdot I_{L_2}^{(i)} - \Omega_6 \cdot I_{R_{s_1}}^{(i)} - \Omega_5 \cdot I_{R_{s_2}}^{(i)} + \Omega_7 \cdot I_{R_1(\text{tank1})}^{(i)} + I_{R_1(\text{tank2})}^{(i)} \cdot \Omega_5 - \Omega_5 \cdot I_{R_{load}}^{(i)} + \Omega_8 \cdot I_{R_{p_1}}^{(i)} = 0$$

$$-i_{R_{s_1}} \cdot \lambda + \Omega_3 \cdot i_{R_s} - \Omega_4 \cdot i_{L_1} - \Omega_5 \cdot i_{L_2} - \Omega_6 \cdot i_{R_{s_1}} - \Omega_5 \cdot i_{R_{s_2}} + \Omega_7 \cdot i_{R_1(\text{tank1})} + i_{R_1(\text{tank2})} \cdot \Omega_5 - \Omega_5 \cdot i_{R_{load}} + \Omega_8 \cdot i_{R_{p_1}} = 0$$

$$\frac{dI_{R_{s_2}}}{dt} = \Omega_{10} \cdot I_{R_s} - \Omega_{11} \cdot I_{L_1} - \Omega_{12} \cdot I_{L_2} - \Omega_{11} \cdot I_{R_{s_1}} - \Omega_{13} \cdot I_{R_{s_2}} + \Omega_{11} \cdot I_{R_1(\text{tank1})} + \Omega_{14} \cdot I_{R_1(\text{tank2})} - \Omega_{12} \cdot I_{R_{load}} + \Omega_{15} \cdot I_{R_{p_2}}$$

$$i_{R_{s_2}} \cdot \lambda \cdot e^{\lambda \cdot t} = \Omega_{10} \cdot (I_{R_s}^{(i)} + i_{R_s} \cdot e^{\lambda \cdot t}) - \Omega_{11} \cdot (I_{L_1}^{(i)} + i_{L_1} \cdot e^{\lambda \cdot t}) - \Omega_{12} \cdot (I_{L_2}^{(i)} + i_{L_2} \cdot e^{\lambda \cdot t}) - \Omega_{11} \cdot (I_{R_{s_1}}^{(i)} + i_{R_{s_1}} \cdot e^{\lambda \cdot t}) - \Omega_{13} \cdot (I_{R_{s_2}}^{(i)} + i_{R_{s_2}} \cdot e^{\lambda \cdot t}) + \Omega_{11} \cdot (I_{R_1(\text{tank1})}^{(i)} + i_{R_1(\text{tank1})} \cdot e^{\lambda \cdot t}) + \Omega_{14} \cdot (I_{R_1(\text{tank2})}^{(i)} + i_{R_1(\text{tank2})} \cdot e^{\lambda \cdot t}) - \Omega_{12} \cdot (I_{R_{load}}^{(i)} + i_{R_{load}} \cdot e^{\lambda \cdot t}) + \Omega_{15} \cdot (I_{R_{p_2}}^{(i)} + i_{R_{p_2}} \cdot e^{\lambda \cdot t})$$

$$i_{R_{s_2}} \cdot \lambda \cdot e^{\lambda \cdot t} = \Omega_{10} \cdot I_{R_s}^{(i)} - \Omega_{11} \cdot I_{L_1}^{(i)} - \Omega_{12} \cdot I_{L_2}^{(i)} - \Omega_{11} \cdot I_{R_{s_1}}^{(i)} - \Omega_{13} \cdot I_{R_{s_2}}^{(i)} + \Omega_{11} \cdot I_{R_1(\text{tank1})}^{(i)} + \Omega_{14} \cdot I_{R_1(\text{tank2})}^{(i)} - \Omega_{12} \cdot I_{R_{load}}^{(i)} + \Omega_{15} \cdot I_{R_{p_2}}^{(i)} + \Omega_{10} \cdot i_{R_s} \cdot e^{\lambda \cdot t} - \Omega_{11} \cdot i_{L_1} \cdot e^{\lambda \cdot t} - \Omega_{12} \cdot i_{L_2} \cdot e^{\lambda \cdot t} - \Omega_{11} \cdot i_{R_{s_1}} \cdot e^{\lambda \cdot t} - \Omega_{13} \cdot i_{R_{s_2}} \cdot e^{\lambda \cdot t} + \Omega_{11} \cdot i_{R_1(\text{tank1})} \cdot e^{\lambda \cdot t} + \Omega_{14} \cdot i_{R_1(\text{tank2})} \cdot e^{\lambda \cdot t} - \Omega_{12} \cdot i_{R_{load}} \cdot e^{\lambda \cdot t} + \Omega_{15} \cdot i_{R_{p_2}} \cdot e^{\lambda \cdot t}$$

At fixed points:

$$\Omega_{10} \cdot I_{R_s}^{(i)} - \Omega_{11} \cdot I_{L_1}^{(i)} - \Omega_{12} \cdot I_{L_2}^{(i)} - \Omega_{11} \cdot I_{R_{s_1}}^{(i)} - \Omega_{13} \cdot I_{R_{s_2}}^{(i)} + \Omega_{11} \cdot I_{R_1(\text{tank1})}^{(i)} + \Omega_{14} \cdot I_{R_1(\text{tank2})}^{(i)} - \Omega_{12} \cdot I_{R_{load}}^{(i)} + \Omega_{15} \cdot I_{R_{p_2}}^{(i)} = 0$$

$$-i_{R_{s_2}} \cdot \lambda + \Omega_{10} \cdot i_{R_s} - \Omega_{11} \cdot i_{L_1} - \Omega_{12} \cdot i_{L_2} - \Omega_{11} \cdot i_{R_{s_1}} - \Omega_{13} \cdot i_{R_{s_2}} + \Omega_{11} \cdot i_{R_1(\text{tank1})} + \Omega_{14} \cdot i_{R_1(\text{tank2})} - \Omega_{12} \cdot i_{R_{load}} + \Omega_{15} \cdot i_{R_{p_2}} = 0$$

$$\frac{dI_{R_{load}}}{dt} = \Omega_{16} \cdot I_{R_s} + I_{R_1(\text{tank1})} \cdot \Omega_{16} + I_{R_1(\text{tank2})} \cdot \Omega_{17} - \Omega_{16} \cdot I_{L_1} - \Omega_{17} \cdot I_{L_2} - \Omega_{16} \cdot I_{R_{s_1}} - \Omega_{17} \cdot I_{R_{s_2}} - \Omega_{17} \cdot I_{R_{load}}$$

$$i_{R_{load}} \cdot \lambda \cdot e^{\lambda \cdot t} = \Omega_{16} \cdot (I_{R_s}^{(i)} + i_{R_s} \cdot e^{\lambda \cdot t}) + (I_{R_1(\text{tank1})}^{(i)} + i_{R_1(\text{tank1})} \cdot e^{\lambda \cdot t}) \cdot \Omega_{16} + (I_{R_1(\text{tank2})}^{(i)} + i_{R_1(\text{tank2})} \cdot e^{\lambda \cdot t}) \cdot \Omega_{17} - \Omega_{16} \cdot (I_{L_1}^{(i)} + i_{L_1} \cdot e^{\lambda \cdot t}) - \Omega_{17} \cdot (I_{L_2}^{(i)} + i_{L_2} \cdot e^{\lambda \cdot t}) - \Omega_{16} \cdot (I_{R_{s_1}}^{(i)} + i_{R_{s_1}} \cdot e^{\lambda \cdot t}) - \Omega_{17} \cdot (I_{R_{s_2}}^{(i)} + i_{R_{s_2}} \cdot e^{\lambda \cdot t}) - \Omega_{17} \cdot (I_{R_{load}}^{(i)} + i_{R_{load}} \cdot e^{\lambda \cdot t})$$

$$i_{R_{load}} \cdot \lambda \cdot e^{\lambda \cdot t} = \Omega_{16} \cdot I_{R_s}^{(i)} + I_{R_1(\text{tank1})}^{(i)} \cdot \Omega_{16} + I_{R_1(\text{tank2})}^{(i)} \cdot \Omega_{17} - \Omega_{16} \cdot I_{L_1}^{(i)} - \Omega_{17} \cdot I_{L_2}^{(i)} - \Omega_{16} \cdot I_{R_{s_1}}^{(i)} - \Omega_{17} \cdot I_{R_{s_2}}^{(i)} - \Omega_{17} \cdot I_{R_{load}}^{(i)} + \Omega_{16} \cdot i_{R_s} \cdot e^{\lambda \cdot t} + i_{R_1(\text{tank1})} \cdot e^{\lambda \cdot t} \cdot \Omega_{16} + i_{R_1(\text{tank2})} \cdot e^{\lambda \cdot t} \cdot \Omega_{17} - \Omega_{16} \cdot i_{L_1} \cdot e^{\lambda \cdot t} - \Omega_{17} \cdot i_{L_2} \cdot e^{\lambda \cdot t} - \Omega_{16} \cdot i_{R_{s_1}} \cdot e^{\lambda \cdot t} - \Omega_{17} \cdot i_{R_{s_2}} \cdot e^{\lambda \cdot t} - \Omega_{17} \cdot i_{R_{load}} \cdot e^{\lambda \cdot t}$$

At fixed points:

$$\Omega_{16} \cdot I_{R_s}^{(i)} + I_{R_1(\text{tank1})}^{(i)} \cdot \Omega_{16} + I_{R_1(\text{tank2})}^{(i)} \cdot \Omega_{17} - \Omega_{16} \cdot I_{L_1}^{(i)} - \Omega_{17} \cdot I_{L_2}^{(i)} - \Omega_{16} \cdot I_{R_{s_1}}^{(i)} - \Omega_{17} \cdot I_{R_{s_2}}^{(i)} - \Omega_{17} \cdot I_{R_{load}}^{(i)} = 0$$

$$-i_{R_{load}} \cdot \lambda + \Omega_{16} \cdot i_{R_s} + i_{R_1(\text{tank1})} \cdot \Omega_{16} + i_{R_1(\text{tank2})} \cdot \Omega_{17} - \Omega_{16} \cdot i_{L_1} - \Omega_{17} \cdot i_{L_2} - \Omega_{16} \cdot i_{R_{s_1}} - \Omega_{17} \cdot i_{R_{s_2}} - \Omega_{17} \cdot i_{R_{load}} = 0$$

We can summery our BPF (varactor diodes) circuit arbitrarily small increments equations:

$$-y_1 \cdot \lambda - \Omega_1 \cdot \Gamma_2 \cdot i_{L_1} - \Omega_1 \cdot \Gamma_3 \cdot i_{L_2} - \Omega_1 \cdot \Gamma_2 \cdot i_{R_{s_1}} - \Omega_1 \cdot \Gamma_3 \cdot i_{R_{s_2}} + \Omega_1 \cdot \Gamma_2 \cdot i_{R_1(\text{tank1})} + \Omega_1 \cdot \Gamma_3 \cdot i_{R_1(\text{tank2})} - \Omega_1 \cdot \Gamma_3 \cdot i_{R_{load}} + \Omega_1 \cdot i_{R_s} - \Omega_1 \cdot \frac{1}{(2 + \frac{C_s}{C_c})} \cdot i_{R_s} = 0$$

$$-i_{R_{p_1}} \cdot \lambda - \Omega_2 \cdot i_{R_{p_1}} + \Omega_2 \cdot i_{R_{s_1}} = 0; -i_{L_1} \cdot \lambda + y_1 = 0$$

$$- i_{R_{s_1}} \cdot \lambda - \Omega_6 \cdot i_{R_{s_1}} + \Omega_3 \cdot i_{R_s} - \Omega_4 \cdot i_{L_1} - \Omega_5 \cdot i_{L_2} - \Omega_5 \cdot i_{R_{s_2}} + \Omega_7 \cdot i_{R_1(\text{tank1})} + i_{R_1(\text{tank2})} \cdot \Omega_5 - \Omega_5 \cdot i_{R_{load}} + \Omega_8 \cdot i_{R_{p_1}} = 0$$

$$-i_{R_{p_2}} \cdot \lambda - \Omega_9 \cdot i_{R_{p_2}} + \Omega_9 \cdot i_{R_{s_2}} = 0$$

$$- i_{R_{s_2}} \cdot \lambda - \Omega_{13} \cdot i_{R_{s_2}} + \Omega_{10} \cdot i_{R_s} - \Omega_{11} \cdot i_{L_1} - \Omega_{12} \cdot i_{L_2} - \Omega_{11} \cdot i_{R_{s_1}} + \Omega_{11} \cdot i_{R_1(\text{tank1})} + \Omega_{14} \cdot i_{R_1(\text{tank2})} - \Omega_{12} \cdot i_{R_{load}} + \Omega_{15} \cdot i_{R_{p_2}} = 0$$

$$-i_{L_2} \cdot \lambda + i_{R_{load}} \cdot \frac{R_{load}}{L_2} = 0$$

$$-i_{R_{load}} \cdot \lambda - \Omega_{17} \cdot i_{R_{load}} + \Omega_{16} \cdot i_{R_s} + i_{R_1(\text{tank1})} \cdot \Omega_{16} + i_{R_1(\text{tank2})} \cdot \Omega_{17} - \Omega_{16} \cdot i_{L_1} - \Omega_{17} \cdot i_{L_2} - \Omega_{16} \cdot i_{R_{s_1}} - \Omega_{17} \cdot i_{R_{s_2}} = 0$$

$$\begin{pmatrix} \Xi_{11} & \cdots & \Xi_{18} \\ \vdots & \ddots & \vdots \\ \Xi_{81} & \cdots & \Xi_{88} \end{pmatrix} \cdot \begin{pmatrix} y_1 \\ i_{R_{p_1}} \\ i_{L_1} \\ i_{R_{s_1}} \\ i_{R_{p_2}} \\ i_{R_{s_2}} \\ i_{L_2} \\ i_{R_{load}} \end{pmatrix} + \begin{pmatrix} \imath_{11} & \cdots & \imath_{13} \\ \vdots & \ddots & \vdots \\ \imath_{81} & \cdots & \imath_{83} \end{pmatrix} \cdot \begin{pmatrix} i_{R_s} \\ i_{R_1(\text{tank1})} \\ i_{R_1(\text{tank2})} \end{pmatrix} = 0; \ \Xi_{11} = -\lambda; \ \Xi_{12} = 0; \ \Xi_{13} = -\Omega_1 \cdot \Gamma_2$$

$$\Xi_{14} = -\Omega_1 \cdot \Gamma_2; \Xi_{15} = 0; \Xi_{16} = -\Omega_1 \cdot \Gamma_3; \Xi_{17} = -\Omega_1 \cdot \Gamma_3; \Xi_{18} = -\Omega_1 \cdot \Gamma_3; \Xi_{21} = 0$$

$$\Xi_{22} = -\lambda - \Omega_2; \Xi_{23} = 0; \Xi_{24} = \Omega_2; \Xi_{25} = \Xi_{26} = \Xi_{27} = \Xi_{28} = 0; \Xi_{31} = 1; \Xi_{32} = 0$$

$$\Xi_{33} = -\lambda; \Xi_{34} = \ldots \Xi_{38} = 0; \Xi_{41} = 0; \Xi_{42} = \Omega_8; \Xi_{43} = -\Omega_4; \Xi_{44} = -\lambda - \Omega_6$$

$$\Xi_{45} = 0; \Xi_{46} = -\Omega_5; \Xi_{47} = -\Omega_5; \Xi_{48} = -\Omega_5; \Xi_{51} = \Xi_{52} = \Xi_{53} = \Xi_{54} = 0; \Xi_{55} = -\lambda - \Omega_9$$

$$\Xi_{56} = \Omega_9; \Xi_{57} = \Xi_{58} = 0; \Xi_{61} = 0; \Xi_{62} = 0; \Xi_{63} = -\Omega_{11}; \Xi_{64} = -\Omega_{11}; \Xi_{65} = \Omega_{15}$$

$$\Xi_{66} = -\lambda - \Omega_{13}; \Xi_{67} = -\Omega_{12}; \Xi_{68} = -\Omega_{12}; \Xi_{71} = \ldots = \Xi_{76} = 0; \Xi_{77} = -\lambda; \Xi_{78} = \frac{R_{load}}{L_2}$$

$$\Xi_{81} = \Xi_{82} = 0; \Xi_{83} = -\Omega_{16}; \Xi_{84} = -\Omega_{16}; \Xi_{85} = 0; \Xi_{86} = -\Omega_{17}; \Xi_{87} = -\Omega_{17}; \Xi_{88} = -\lambda - \Omega_{17}$$

$$\imath_{11} = \Omega_1 \cdot [1 - \frac{1}{(2 + \frac{C_s}{C_c})}]; \imath_{12} = \Omega_1 \cdot \Gamma_2; \imath_{13} = \Omega_1 \cdot \Gamma_3; \imath_{21} = \imath_{22} = \imath_{23} = 0; \imath_{31} = \imath_{32} = \imath_{33} = 0$$

$$\imath_{41} = \Omega_3; \imath_{42} = \Omega_7; \imath_{43} = \Omega_5; \imath_{51} = \imath_{52} = \imath_{53} = 0; \imath_{61} = \Omega_{10}; \imath_{62} = \Omega_{11}; \imath_{63} = \Omega_{14}$$

$$\imath_{71} = \imath_{72} = \imath_{73} = 0; \imath_{81} = \Omega_{16}; \imath_{82} = \Omega_{16}; \imath_{83} = \Omega_{17}$$

Assumption $\begin{pmatrix} \imath_{11} & \cdots & \imath_{13} \\ \vdots & \ddots & \vdots \\ \imath_{81} & \cdots & \imath_{83} \end{pmatrix} \cdot \begin{pmatrix} i_{R_s} \\ i_{R_1(\text{tank1})} \\ i_{R_1(\text{tank2})} \end{pmatrix} \to \varepsilon$

$$(A - \lambda \cdot I) = \begin{pmatrix} \Xi_{11} & \cdots & \Xi_{18} \\ \vdots & \ddots & \vdots \\ \Xi_{81} & \cdots & \Xi_{88} \end{pmatrix}; \det(A - \lambda \cdot I) = \det \begin{pmatrix} \Xi_{11} & \cdots & \Xi_{18} \\ \vdots & \ddots & \vdots \\ \Xi_{81} & \cdots & \Xi_{88} \end{pmatrix} = 0$$

$$\det(A - \lambda \cdot I) = \sum_{k=0}^{8} \Upsilon_k \cdot \lambda^k; \sum_{k=0}^{8} \Upsilon_k \cdot \lambda^k = 0.$$

Remark It is reader exercise to find the expressions for $\Upsilon_k \, \forall k = 0, 1, \ldots, 8$.

Eigenvalues stability discussion: Our BPF (varactor diodes) circuit involving N variables (N > 2, N = 11, arbitrarily small increments), the characteristic equation is of degree N = 8 (reduced) and must often be solved numerically. Expect in some particular cases, such an equation has (N = 8) distinct roots that can be real or complex. These values are the eigenvalues of the (10 × 10) Jacobian matrix (A). The general rule is that the BPF (varactor diodes) circuit is stable if there is no eigenvalue with positive real part. It is sufficient that one eigenvalue is positive for the steady state to be unstable. Our 8-variables (y_1 $i_{R_{p1}}$ i_{L_1} $i_{R_{s_1}}$ $i_{R_{p2}}$ $i_{R_{s_2}}$ i_{L_2} $i_{R_{load}}$) system has eight eigenvalues (eight system's arbitrarily small increments). The type of behavior can be characterized as a function of the position of these eigenvalues in the Re/Im plane. Five non-degenerated cases can be distinguished: (1) the eight eigenvalues are real and negative (stable steady state), (2) the eight eigenvalues are real, at least one of them is positive (unstable steady state), (3) and (4) two eigenvalues are complex conjugates with a negative real part and other eigenvalues real are negative (stable steady state), two cases can be distinguished depending on the relative value of the real part of the complex eigenvalues and of the real one, (5) two eigenvalues are complex conjugates with a negative real part and at least one of the other eigenvalues real is positive (unstable steady state) [12, 13].

$$\det(A - \lambda \cdot I) = \sum_{k=0}^{8} \Upsilon_k \cdot \lambda^k; \det(A - \lambda \cdot I) = 0 \Rightarrow \sum_{k=0}^{8} \Upsilon_k \cdot \lambda^k = 0$$

Exercises

1. We have triplexer circuit. It is a four ports device with a common input and three outputs. A triplexer is a passive device that implements frequency domain multiplexing. Three ports (e.g., L, I, H) are multiplexed onto a fourth port (e.g., S). The signals on L, I, and H occupy disjoint frequency bands. Consequently, the signals on L, I, and H can coexist on port S without interfering with each other. The signal on port L (low band) will occupy a single low frequency band, the signal on port I (intermediate band) will occupy a single intermediate frequency band and the signal on port H will occupy a higher frequency band. In that situation, the triplexer consists of a low pass filter connecting ports L and S, and an intermediate pass filter connecting ports I and S, and a high pass filter connecting ports H and S. The triplexer, being a passive device, is reciprocal; the device itself doesn't have a nothing of input or output. The ports of a diplexer are frequency selective. A triplexer multiplexes three ports onto one port. A triplexer can be placed at the output of a frequency RF source, where it functions as an absorptive filter.

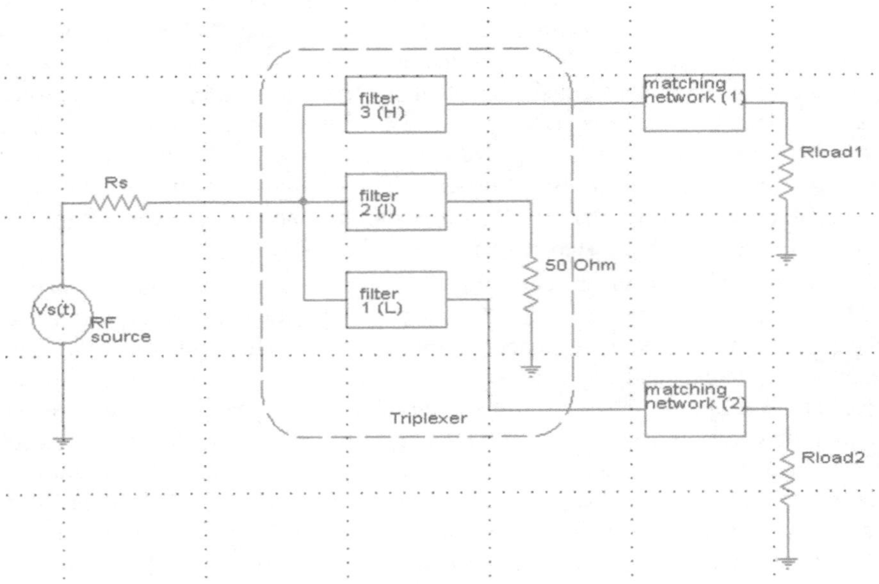

Filter 2 (I) of the triplexer has a BPF that corresponds with the undesired frequency band, which can pass right through with little attenuation, and is terminated within to 50 Ω load. These undesired frequencies are blocked from entering filter 2 (I) by that filter's stop band. Filters 1(L) and 3(H) BPFs passes all the desired signals onto load 1 or load 2 through matching networks (1 and 2) with little attenuation. The undesired signals through the triplexer are absorbed instead of being reflected as they would be in a typical filter. The design of a triplexer is as three different frequency filters with non-overlapping bandpass. The matching network between triplexer circuit unit and loads (R_{load1} and R_{load2}) can be L-type, Pi-type, and T-type.

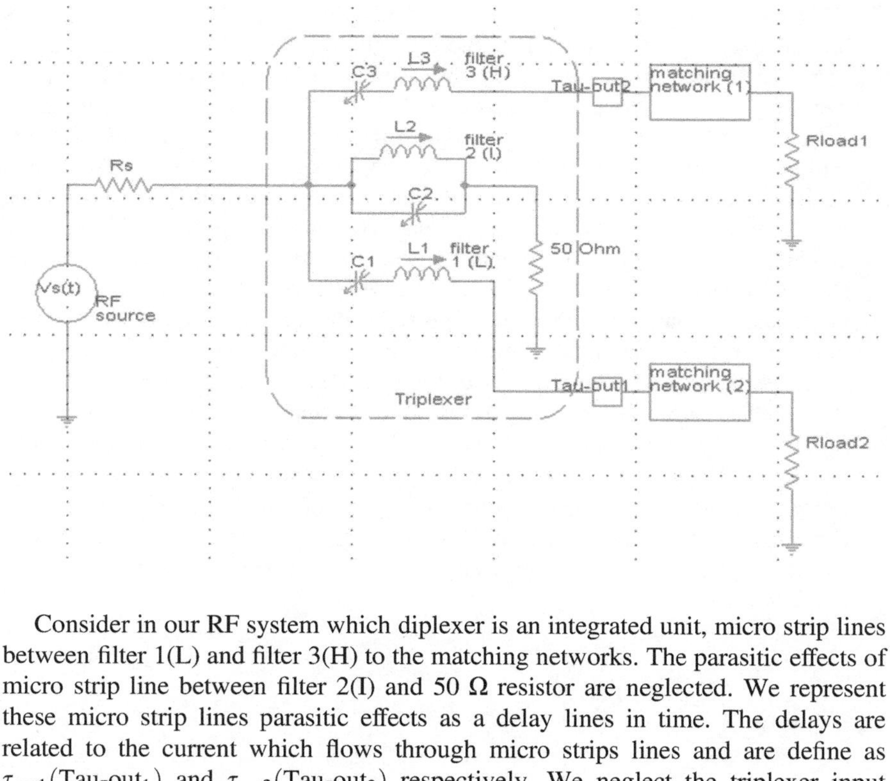

Consider in our RF system which diplexer is an integrated unit, micro strip lines between filter 1(L) and filter 3(H) to the matching networks. The parasitic effects of micro strip line between filter 2(I) and 50 Ω resistor are neglected. We represent these micro strip lines parasitic effects as a delay lines in time. The delays are related to the current which flows through micro strips lines and are define as τ_{out1}(Tau-out$_1$) and τ_{out2}(Tau-out$_2$) respectively. We neglect the triplexer input micro strip line parasitic effects. $\tau > 0$; $\tau \in \mathbb{R}$

1.1 Write circuit differential equations and find fixed points (matching network 1: T type and matching network 2: L type). Assumption: $\frac{dV_s(t)}{dt} \to \varepsilon$.

1.2 Discuss stability of our circuit for the following cases: (a) $\tau_{out_1} = \tau$; $\tau_{out_2} = 0$(b) $\tau_{out_1} = 0$; $\tau_{out_2} = \tau$(c) $\tau_{out_1} = \Gamma \cdot \tau_{out_2}$; $\tau_{out_2} = \tau$; $\Gamma \in \mathbb{R}_+$ How our circuit behavior changes for different values of Γ and τ Parameters?

1.3 How our circuit dynamic is changed if we disconnect inductor L_2? Write circuit differential equations and find fixed points. Discuss stability and stability switching.

1.4 We choose both circuits' matching networks Pi type. Assumption: $\frac{dV_s(t)}{dt} \to \varepsilon$. Write circuit differential equations and find fixed points. Discuss stability and stability switching.

1.5 We add additional filter to our triplexer (inductor L_4 and capacitor C_4 in series) which terminated by 50 Ω resistor. Our circuit is Quad-plexer. Write circuit differential equations and find fixed points (matching networks are L type). Discuss stability and stability switching.

2. We have a system of two diplexers in series. *Diplexer (I)*: Two band pass filters are duplexed, one common input, and two outputs. A diplexer (I) is referring to a duplex high pass and low pass where broad bands transmit and receive is necessary. It is a three ports device with common input and two outputs. It implements frequency domain multiplexing two ports (e.g., L_I and H_I) are multiplexed onto a third port (e.g., S_I). The signal on port L_I and H_I occupy disjoint frequency bands. Consequently, the signals on L_I and H_I can coexist on port S_I without interfering with each other. The signal on port L_I will occupy a single low frequency band and the signal on port H_I will occupy a higher frequency band. *Diplexer (II)*: Two band pass filters are duplexed, one common input, and two outputs. A diplexer (II) is referring to a duplex high pass and low pass where broad bands transmit and receive is necessary. It is a three ports device with common input and two outputs. It implements frequency domain multiplexing two ports (e.g., L_{II} and H_{II}) are multiplexed onto a third port (e.g., S_{II}). The signal on port L_{II} and H_{II} occupy disjoint frequency bands. Consequently, the signals on L_{II} and H_{II} can coexist on port S_{II} without interfering with each other. The signal on port L_{II} will occupy a single low frequency band and the signal on port H_{II} will occupy a higher frequency band. There is an overlap between diplexer (I) port H_I frequency band and diplexer (II) port H_{II} frequency band.

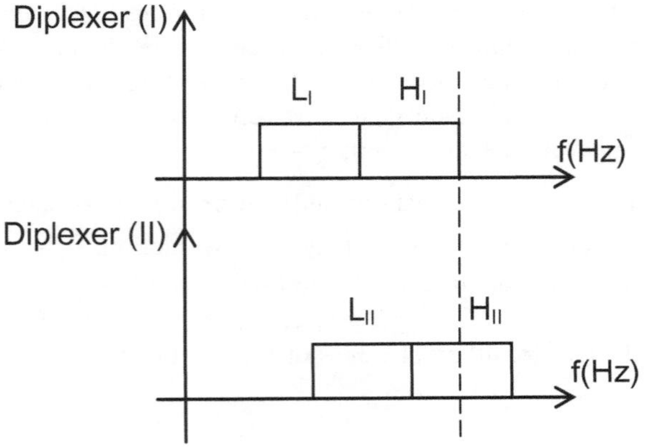

Filter (II) of diplexer (I)—low band and filter (IV) of diplexer (II)—low band have a BPFs that correspond with the undesired frequency bands, and are terminated within the 50 Ω load. These undesired frequencies are blocked from entering filters II and IV by that filter's stop band. Diplexers (I) and (II) are connected in series by micro strip lines and matching networks (I). The output load (R_{load}) is connected to diplexer II's output through matching network (II).

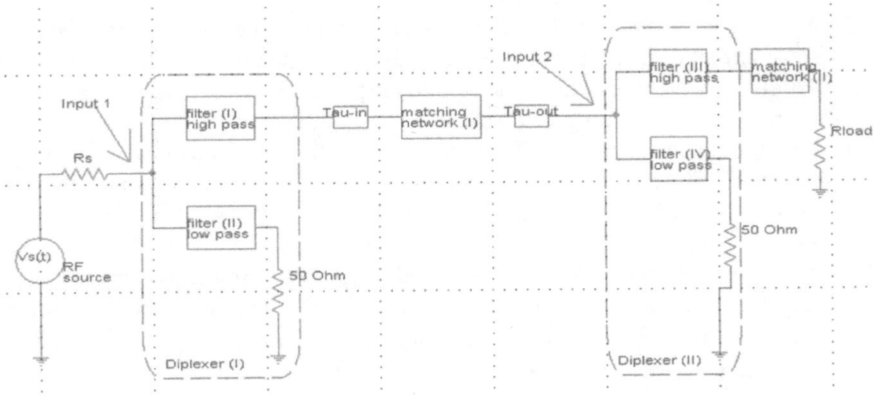

We represent these micro strips parasitic effects as a delay lines in time. The delays are related to the current which flows through micro strips and are τ_{in}(Tau-in) and τ_{out}(Tau-out) respectively. We choose T type matching network (I) and L type matching network (II). We neglect in our analysis other circuit micro strips parasitic effects and consider them as an ideal elements. $\tau > 0$; $\tau \in \mathbb{R}$.

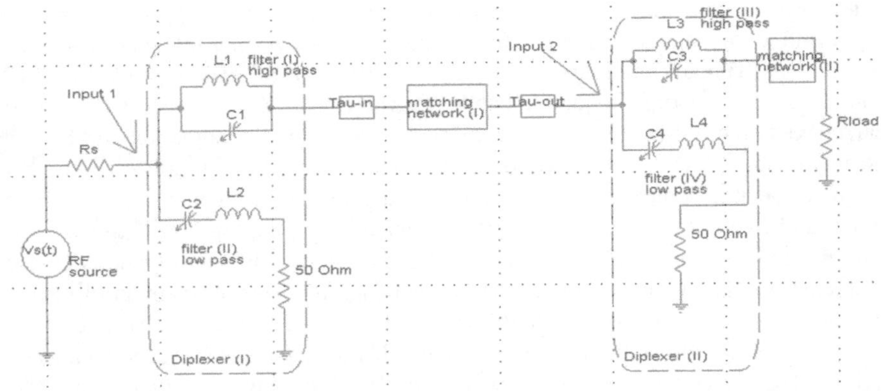

2.1 Write circuit differential equations and find fixed points.

Assumption $\frac{dV_s(t)}{dt} \to \varepsilon$.

2.2 Discuss stability of our circuit for the following cases: (a) $\tau_{in} = \tau; \tau_{out} = 0$, (b) $\tau_{in} = \tau; \tau_{out} = \tau^2$, (c) $\tau_{in} = \tau^2; \tau_{out} = \tau - \sqrt{\tau}$. How our circuit behavior changes for different values of τ parameter? Discuss stability switching for different values of τ parameter.

2.3 We disconnect diplexer (I)'s filter (II). How it influences our circuit behavior? Write circuit differential equations and find fixed points. Discuss stability and stability switching for the following cases: (a) $\tau_{in} = \tau; \tau_{out} = 0$, (b) $\tau_{in} = 0; \tau_{out} = \tau$, (c) $\tau_{in} = \tau_{out} = \tau$.

2.4 We disconnect diplexer (II)'s filter (IV). How it influences our circuit behavior? Write circuit differential equations and find fixed points. Discuss stability and stability switching for the following cases: (a) $\tau_{in} = \sqrt{\tau}; \tau_{out} = 0$, (b) $\tau_{in} = 0; \tau_{out} = \tau$, (c) $\tau_{in} = \tau_{out} = \tau \cdot \sqrt[3]{\tau}$.

2.5 We disconnect inductor L_3. How it influences our circuit dynamics? Write circuit differential equations and find fixed points. Discuss stability and stability switching for the following cases: (a) $\tau_{in} = \tau \cdot \sqrt{\tau}; \tau_{out} = 0$, (b) $\tau_{in} = 0; \tau_{out} = \sqrt{\tau}$, (c) $\tau_{in} = \tau_{out} = \sqrt{\tau \cdot \sqrt[3]{\tau}}$.

3. We have a system of two diplexers (I and II), two RF/Microwave sources, and summation operator. The summation operator is implemented by using ideal op amps (summing amplifier follows by inverting amplifier). $R_1 = R_2 = R_{f_1}; \frac{R_{f_2}}{R_{in}} = 1; V_C = V_A + V_B; R_2 \gg R_{s_2}; R_2 + R_{s_2} \approx R_2$

Diplexer (I): Two band pass filters are duplexed, one common input, and two outputs. A diplexer (I) is referring to a duplex high pass and low pass where broad bands transmit and receive is necessary. It is a three ports device with common input and two outputs. It implements frequency domain multiplexing two ports (e.g., LI and HI) are multiplexed onto a third port (e.g., SI). The signal on port LI and HI occupy disjoint frequency bands. Consequently, the signals on LI and HI can coexist on port SI without interfering with each other. The signal on port LI will occupy a single low frequency band and the signal on port HI will occupy a higher frequency band. Diplexer (II): Two band pass filters are duplexed, one common input, and two outputs. A diplexer (II) is referring to a duplex high pass and low pass where broad bands transmit and receive is necessary. It is a three ports device with common input and two outputs. It implements frequency domain multiplexing two ports (e.g., LII and HII) are multiplexed onto a third port (e.g., SII). The signal on port LII and HII occupy disjoint frequency bands. Consequently, the signals on LII and HII can coexist on port SII without interfering with each other. The signal on port LII will occupy a single low frequency band and the signal on port HII will occupy a higher frequency band. There is an overlap between diplexer (I) port HI frequency band and diplexer (II) port HII frequency band. The output of diplexer

(I)'s filter (I) is connected to summation operator (port A). RF/Microwave source $V_{s2}(t)$ is connected through resistor R_{s2} to summation operator (port B). Load resistor R_{load} is connected to diplexer (II)'s filter (IV) through T type matching network.

Filter (II) of diplexer (I)—low band and filter (III) of diplexer (II)—high band have a BPFs that correspond with the undesired frequency bands, and are terminated within the 50 Ω load. These undesired frequencies are blocked from entering filters II and III by that filter's stop band. Diplexers (I) and (II) are connected to summation operator by micro strip lines (τ_{in}(Tau-in); τ_{out}(Tau-out)) .Matching networks (I) is connected to diplexer (II)'s filter (IV) by micro strip line (τ_m(Tau-m)).

Assumption There is an overlaps between all circuit's filters frequency bands, the output signal from diplexer (II)'s filter (IV) is within specific frequency band and target for our load resistance (R_{load}).

Assumption $\frac{dV_{s_1}(t)}{dt} \to \varepsilon$; $\frac{dV_{s_2}(t)}{dt} \to \varepsilon$. All other micro strip lines in our system are ideal with no parasitic effects in our circuit. $\tau > 0; \tau \in \mathbb{R}$

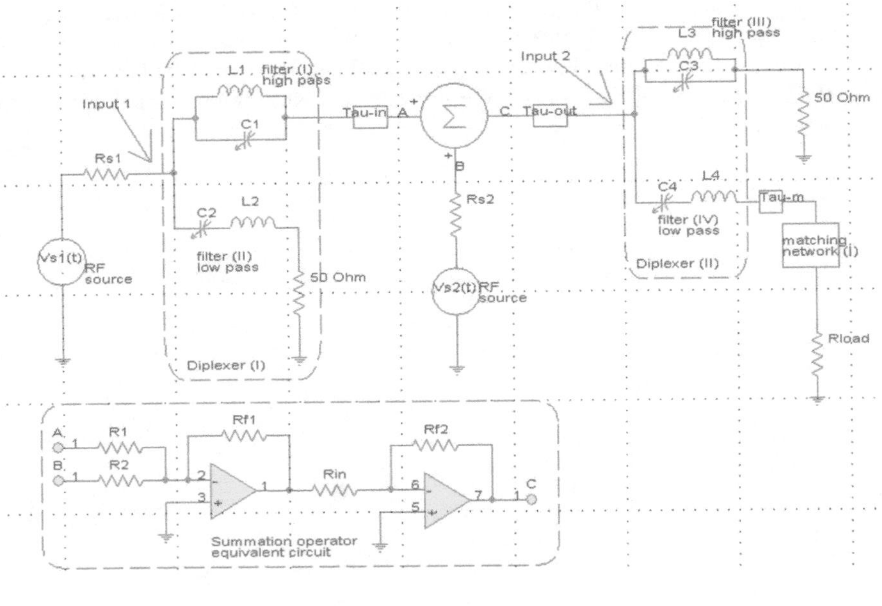

3.1 Write circuit differential equations and find fixed points.

3.2 Discuss stability of our circuit for the following cases: $\tau_m = \tau$ (a) $\tau_{in} = \tau^2; \tau_{out} = 0$, (b) $\tau_{in} = \tau; \tau_{out} = \sqrt[3]{\tau^2}$, (c) $\tau_{in} = \tau^2; \tau_{out} = \sqrt[3]{\tau} - \sqrt{\tau}$.

How our circuit behavior changes for different values of τ parameter? Discuss stability switching for different values of τ parameter.

3.3 We disconnect diplexer (I)'s filter (II). How it influences our circuit behavior? Write circuit differential equations and find fixed points. Discuss stability and stability switching for the following cases: $\tau_m = \tau$ (a) $\tau_{in} = \tau; \tau_{out} = 0$, (b) $\tau_{in} = 0; \tau_{out} = \tau$, (c) $\tau_{in} = \tau_{out} = \tau$.

3.4 We disconnect diplexer (II)'s filter (III). How it influences our circuit behavior? Write circuit differential equations and find fixed points. Discuss stability and stability switching for the following cases: $\tau_m = \sqrt{\tau}$ (a) $\tau_{in} = \tau \cdot \sqrt{\tau}$; $\tau_{out} = 0$, (b) $\tau_{in} = 0$; $\tau_{out} = \sqrt{4}\tau$, (c) $\tau_{in} = \tau_{out} = \tau \cdot \sqrt[3]{\tau}$.

3.5 We disconnect diplexer (I) from summation operator. How it influences our circuit dynamics? Write circuit differential equations and find fixed points. Discuss stability and stability switching for the following cases: $\tau_m = \tau \cdot \sqrt{\tau}$ (a) $\tau_{in} = \tau \cdot \sqrt{\tau}; \tau_{out} = 0$, (b) $\tau_{in} = 0; \tau_{out} = \sqrt{\tau}$, (c) $\tau_{in} = \tau_{out} = \sqrt{\tau \cdot \sqrt[3]{\tau}}$.

4. We have a system of triple band triplexer filter. The circuit of triple band triplexer contains capacitors and inductors, RF source $V_s(t)$ and series resistor R_s, and resistive loads R_{load1}, R_{load2} and R_{load3}. Three delays lines τ_{out_1}, τ_{out_2} and τ_{out_3} represent circuit micro strip lines parasitic effects. In our analysis we consider full matching between R_{load1} and BPF1, R_{load2} and BPF2, R_{load3} and BPF3 (no matching networks). We consider all other micro strip lines in our circuit are an ideal micro strips lines.

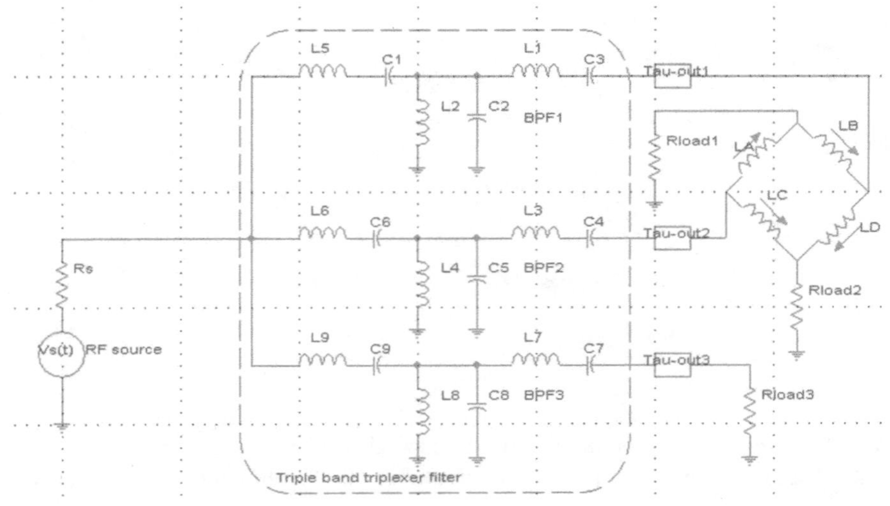

Load resistors R_{load1} and R_{load2} are connected to BPF1 and BPF2 through inductors bridge (L_A, L_B, L_C, L_D; $L_A \neq L_B \neq L_C \neq L_D$).

4.1 Write circuit differential equations and find fixed points,

Assumption $\frac{dV_s(t)}{dt} \to \varepsilon$.

4.2 Discuss stability of our circuit for the following cases:

(a) $\tau_{out_1} = \tau; \tau_{out_2} = \sqrt{\tau}; \tau_{out_3} = 0$
(b) $\tau_{out_1} = 0; \tau_{out_2} = \tau^2; \tau_{out_3} = \tau \cdot \sqrt{\tau}$
(c) $\tau_{out_1} = \tau^2; \tau_{out_2} = \tau; \tau_{out_3} = \sqrt{\tau}$.

How our circuit behavior changes for different values of τ parameter? Discuss stability switching for different values of τ parameter.

4.3 We disconnect capacitor C_2. How it influences our circuit dynamics? Write circuit differential equations and find fixed points. Discuss stability of our circuit for the following cases:

(a) $\tau_{out_1} = \tau \cdot \Gamma_1; \tau_{out_2} = \tau^2_{out_1}; \tau_{out_3} = \sum_{k=1}^{2} \tau_{out_k}$

(b) $\tau_{out_1} = \tau^2 \cdot \Gamma_2; \tau_{out_2} = \sqrt{\tau}; \tau_{out_3} = \sum_{k=1}^{2} \tau^2_{out_k}$

$\Gamma_1, \Gamma_2 \in R_+; \Gamma_1 \neq \Gamma_2; \tau > 0; \tau \in \mathbb{R}$. Discuss stability switching for different values of Γ_1, Γ_2, τ parameters.

4.4 Return (4.3) if we disconnect inductors L_2 and L_4.

4.5 We disconnect Inductors Bridge's L_B, how it influences our circuit dynamics? Write circuit differential equations and find fixed points. Discuss stability of our circuit for the following cases:

(a) $\tau_{out_1} = \sqrt{\tau} \cdot \Gamma_1; \tau_{out_2} = \tau^2_{out_1}; \tau_{out_3} = \sum_{k=1}^{2} \Gamma_k \cdot \tau_{out_k}$

(b) $\tau_{out_1} = \sqrt[3]{\tau^2} \cdot \Gamma_2; \tau_{out_2} = \tau \cdot \sqrt{\tau}; \tau_{out_3} = \sum_{k=1}^{2} \Gamma_k \cdot \tau^2_{out_k}$

$\Gamma_1, \Gamma_2 \in R_+; \Gamma_1 \neq \Gamma_2; \tau > 0; \tau \in \mathbb{R}$. Discuss stability switching for different values of Γ_1, Γ_2, τ parameters.

5. We have a system of triple band triplexer filter. The circuit of triple band triplexer contains capacitors and inductors, RF source $V_s(t)$ and series resistor R_s, and resistive loads R_{load1}, R_{load2}. Three delays lines $\tau_{out_1}, \tau_{out_2}$ and τ_{out_3} represent circuit micro strip lines parasitic effects. In our analysis we consider full matching between R_{load1} and BPF1, R_{load2} and BPF2 (no matching networks). We consider all other micro strip lines in our circuit are an ideal micro strips lines. RF signals from BPF2 and BPF3 are added by using summation operator and feed to capacitors and inductors bridge. Load resistors R_{load1} and R_{load2} are connected to BPF1, BPF2, and BPF3 through inductors and capacitors bridge ($L_A, L_B, C_A, C_B; L_A \neq L_B; C_A \neq C_B$). The summation operator is implemented by using ideal op amps (summing amplifier follows by

inverting amplifier). $R_1 = R_2 = R_{f_1}; \frac{R_{f_2}}{R_{in}} = 1; V_C = V_A + V_B; R_2 \gg R_{s_2}; R_2 + R_{s_2} \approx R_2$ (see question 3).

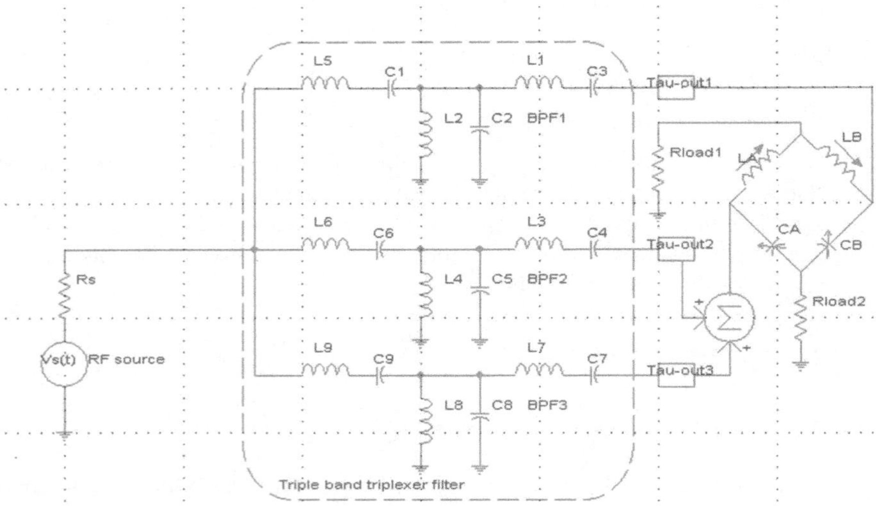

5.1 Write circuit differential equations and find fixed points, Assumption: $\frac{dV_s(t)}{dt} \to \varepsilon$.

5.2 Discuss stability of our circuit for the following cases:

(a) $\tau_{out_1} = \sqrt[3]{\tau}; \tau_{out_2} = \sqrt{\tau}; \tau_{out_3} = 0$

(b) $\tau_{out_1} = 0; \tau_{out_2} = \sqrt[3]{\tau^2}; \tau_{out_3} = \tau^2 \cdot \sqrt{\tau}$

(c) $\tau_{out_1} = \sqrt[3]{\tau^2}; \tau_{out_2} = \tau; \tau_{out_3} = \tau \cdot \sqrt{\tau}$.

How our circuit behavior changes for different values of τ parameter? Discuss stability switching for different values of τ parameter.

5.3 We disconnect capacitor C_5. How it influences our circuit dynamics? Write circuit differential equations and find fixed points. Discuss stability of our circuit for the following cases:

(a) $\tau_{out_1} = \tau \cdot \Gamma_1; \tau_{out_2} = \sqrt[3]{\tau_{out_1}^2}; \tau_{out_3} = \sum_{k=1}^{2} \Gamma_1^2 \cdot \tau_{out_k}$

(b) $\tau_{out_1} = \sqrt[3]{\tau^2 \cdot \Gamma_2}; \tau_{out_2} = \tau \cdot \sqrt{\tau}; \tau_{out_3} = \sum_{k=1}^{2} \tau_{out_k}^2$

$\Gamma_1, \Gamma_2 \in R_+; \Gamma_1 \neq \Gamma_2; \tau > 0; \tau \in \mathbb{R}$. Discuss stability switching for different values of Γ_1, Γ_2, τ parameters.

5.4 Return (5.3) if we disconnect inductors L_2 and L_9.

5.5 We disconnect Inductors Bridge's C_B, how it influences our circuit dynamics? Write circuit differential equations and find fixed points. Discuss stability of our circuit for the following cases:

(a) $\tau_{out_1} = \sqrt{\sqrt{\tau} \cdot \Gamma_1}; \tau_{out_2} = \tau_{out_1}^2; \tau_{out_3} = \sum_{k=1}^{2} \Gamma_k \cdot \tau_{out_k}$

(b) $\tau_{out_1} = \sqrt[3]{\tau^2} \cdot \Gamma_2; \tau_{out_2} = \sqrt{\tau \cdot \sqrt{\tau}}; \tau_{out_3} = \sum_{k=1}^{2} \Gamma_k \cdot \tau_{out_k}^2$

$\Gamma_1, \Gamma_2 \in R_+; \Gamma_1 \neq \Gamma_2; \tau > 0; \tau \in \mathbb{R}$. Discuss stability switching for different values of Γ_1, Γ_2, τ parameters.

6. We have crystal-lattice BPF circuit with two output loads resistances R_{load1} and R_{load2}. Crystals have series and parallel resonant mode. The crystal lattice filter contains several crystals within a single circuit. The input employs RF transformer (T_1) with shunt capacitor (C_1). The output employs five terminals transformer (T_2) with shunt capacitor (C_2). Each set of crystals XTAL1 plus XTAL2 and XTAL3 plus XTAL4 are cut to different frequencies. The matched set of XTAL1 and XTAL2 having a lower resonant frequency than the other matched set of XTAL3 and XTAL4. *Transformer T_1*: Parameters n_{p1} and n_{s1} are the number of turns of coil 1 (primary) and 2 (secondary), respectively. *Transformer T_2*: Parameter n_{p2} is the number of turns of coil 1 (primary). Parameters n_{s2a} and n_{s2b} are the number of turns of coil 2 (secondary). In your analysis use the equivalent circuit of a quartz crystal and use typical crystal parameters.

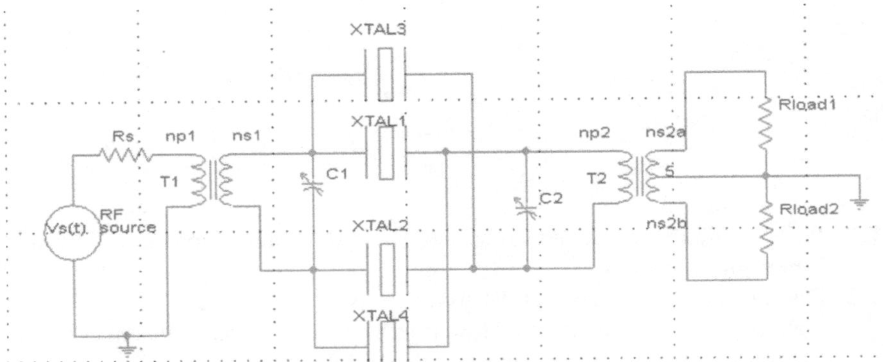

6.1 Write circuit differential equations and find fixed points. Find arbitrarily small increments equations.

6.2 Find circuit eigenvalues and discuss stability.

6.3 Capacitor C_2 is disconnected. How it influences our circuit dynamics? Write circuit differential equations and find fixed points. Discuss stability.

6.4 Load resistance R_{load2} is disconnected. How it influences our circuit dynamic? Write circuit differential equations and find fixed points. Discuss stability.

6.5 We multiple the value of capacitor C_1 by Γ parameter $(C_1 \rightarrow C_1 \cdot \Gamma)$, $\Gamma > 0; \Gamma \in \mathbb{R}_+$. How it influences our circuit dynamics. Discuss stability and stability switching for different values of Γ parameter.

7. We have Half-lattice crystal filter circuit with two crystals and one load R_{load}. Crystals have a series and parallel resonant mode. The Half lattice crystal filter contains two crystals within a single circuit. The input employs RF transformer (T_1) with shunt capacitor (C_1). The set of crystals XTAL1 and XTAL2 are cut to different frequencies. Half-lattice crystal filter offers a flatter in band response. The two crystals have different resonant frequencies. The response has a small peak at either side of the center frequency and a small dip in the middle. Transformer T_1: n_p is the number of turns of coil 1 (primary). N_{sa} and n_{sb} are number of turns of coil 2 (secondary) and coil 3 (secondary). In our analysis, we use the equivalent circuit of a quartz crystal and use typical crystal parameters.

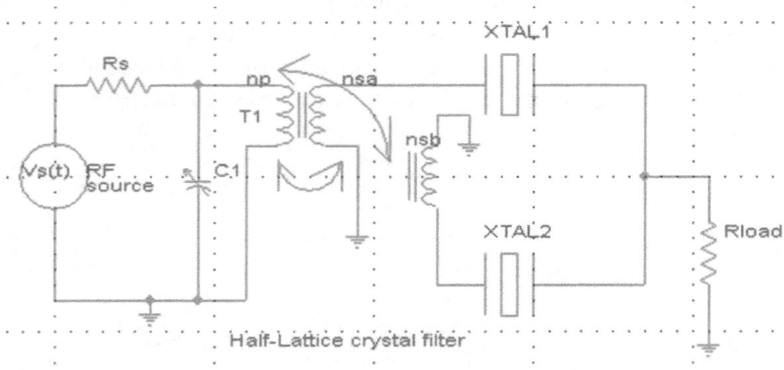

7.1 Write circuit differential equations and find fixed points. Find arbitrarily small increments equations.

7.2 Find circuit eigenvalues and discuss stability.

7.3 Capacitor C_1 is disconnected. How it influences our circuit dynamics? Write circuit differential equations and find fixed points. Discuss stability.

7.4 We increase the number of coil 3 turns by Γ factor $(n_{s_b} \rightarrow n_{s_b} \cdot \Gamma)$, $\Gamma > 0; \Gamma \in \mathbb{R}_+$. How it influences our circuit dynamics? Discuss stability switching for different values of Γ parameter.

7.5 We short crystal XTAL2. How it influences circuit behavior? Write circuit differential equation and find fixed points. Discuss stability.

8. We have Half-lattice crystal filter circuit with two crystals and one load R_{load}. Crystals have a series and parallel resonant mode. The Half lattice crystal filter contains two crystals within a single circuit. The input employs RF transformer (T_1) with shunt capacitor (C_1). The RF transformer (T_1)'s secondary is connected to two capacitors (C_a and C_b). The set of crystals XTAL1 and XTAL2 are cut to different frequencies. Half-lattice crystal filter offers a flatter in band response. The two crystals have different resonant frequencies. The response has a small peak at either side of the center frequency and a small dip in the middle. Transformer T_1: n_p is the number of turns of coil 1 (primary). n_s is number of turns of coil 2 (secondary). In our analysis, we use the equivalent circuit of a quartz crystal and use typical crystal parameters.

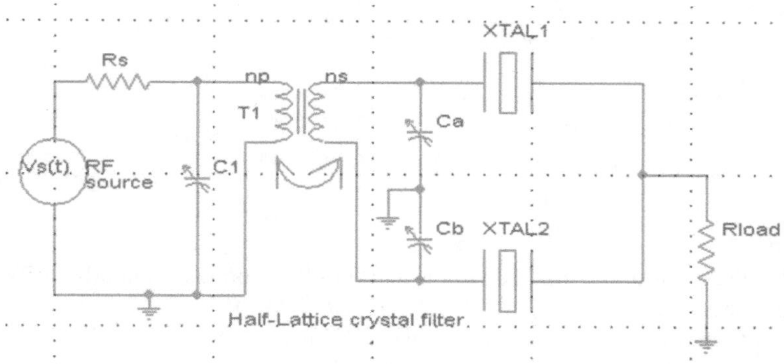

8.1 Write circuit differential equations and find fixed points. Find arbitrarily small increments Eqs.
8.2 Find circuit eigenvalues and discuss stability.
8.3 Capacitor C_b is disconnected. How it influences our circuit dynamics? Write circuit differential equations and find fixed points. Discuss stability.
8.4 We increase the number of coil 1 turns by Γ factor ($n_p \to n_p \cdot \Gamma$), $\Gamma > 0; \Gamma \in \mathbb{R}_+$. How it influences our circuit dynamics? Discuss stability switching for different values of Γ parameter.
8.5 We short crystal XTAL1. How it influences circuit behavior? Write circuit differential equation and find fixed points. Discuss stability.

9. We have circuit of a single crystal filter. It employs the very high Q of the crystal. Its response is asymmetric and it is too narrow for most applications, having a bandwidth of a hundred Hz or less. In the circuit there is a variable capacitor (C_x) that is used to compensate for the parasitic capacitance in the

crystal. This capacitor was normally included as a front panel control. The input employs RF transformer (T_1) with shunt capacitor (C_1). The RF transformer (T_1)'s secondary is connected to two capacitors (C_a and C_b). Transformer T1: np is the number of turns of coil 1 (primary). ns is number of turns of coil 2 (secondary). In our analysis, we use the equivalent circuit of a quartz crystal and use typical crystal parameters. The crystals XTAL1 is cut to different frequency.

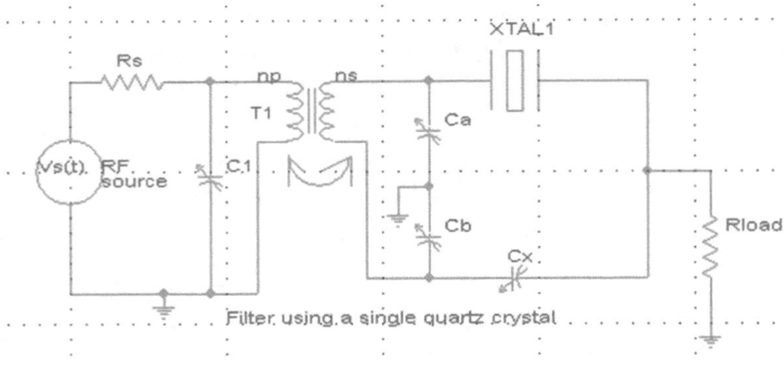

9.1 Write circuit differential equations and find fixed points. Find arbitrarily small increments equations
9.2 Find circuit eigenvalues and discuss stability.
9.3 Capacitor C_a is disconnected. How it influences our circuit dynamics? Write circuit differential equations and find fixed points. Discuss stability.
9.4 We increase the number of coil 1 turns by Γ factor ($n_p \to n_p \cdot \Gamma$), $\Gamma > 0; \Gamma \in \mathbb{R}_+$. How it influences our circuit dynamics? Discuss stability switching for different values of Γ parameter.
9.5 We increase the C_x capacitor value by ψ parameter ($C_x \to C_x \cdot \psi$), $\psi > 0; \psi \in \mathbb{R}_+$. Write circuit differential equations and find fixed points. Discuss stability and stability switching for different values of ψ parameter.

10. We have circuit of top inductively coupled variable BPF. For wideband applications the top inductively coupled variable band pass filter is the best. The circuit includes varactor diodes (C_v) and inductors $L_1, \ldots, L_5$, bias resistors R_1 and R_2 isolated the two varactors (C_v) from the effects of each other, and resistor R_3 represents a direct RF short to ground through V_{tune}. Capacitors C_2 and C_3 block the DC inserted by V_{tune} from being shorted by L_2 and L_4. Varactor diodes (C_v) supply the variable tuning capacitance. The capacitance of the series combination of C_2 and C_v in series is $\frac{C_2 \cdot C_v}{C_2 + C_v}$. The capacitance of the series combination of C_3 and C_v in series is $\frac{C_3 \cdot C_v}{C_3 + C_v}$. The capacitors C_2 and C_3 mainly functioning as a DC blocking capacitor, while C_v, the varactor is

supplying all of the tuning capacitance for the filter tanks. By applying positive V_{tune} we allow the varactor to either linearly tune the filter to its maximum and minimum values or, by supplying V_{tune} with discrete voltages it filter the incoming RF/Microwave signal in discrete steps. The circuit load resistance R_{load} is connected in our voltage tunable inductor coupled bandpass filter. Input RF/microwave signal voltage $V_s(t)$ and his series resistance R_s.

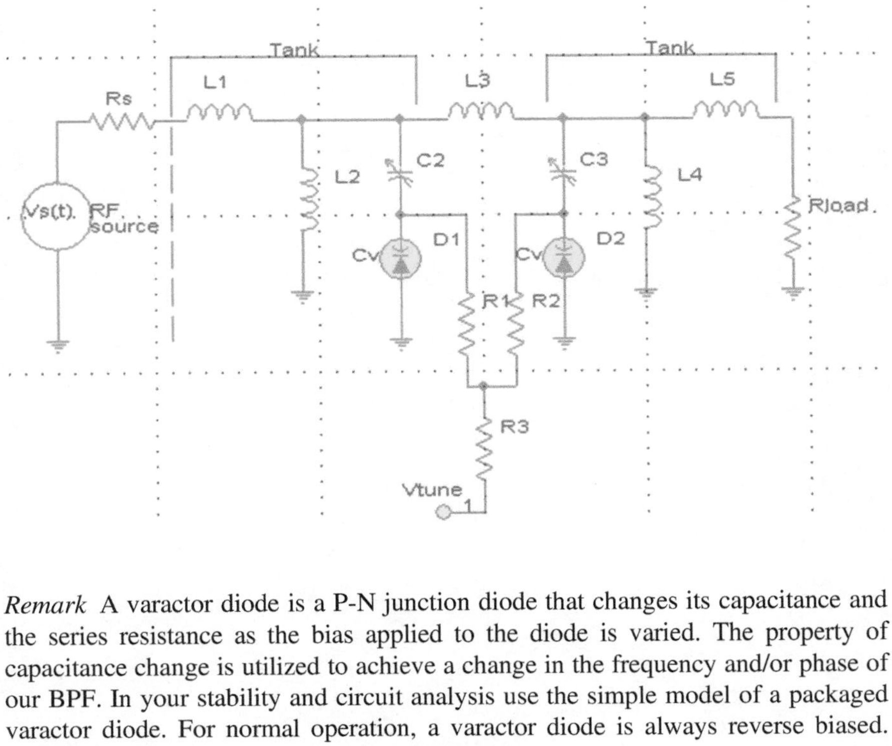

Remark A varactor diode is a P-N junction diode that changes its capacitance and the series resistance as the bias applied to the diode is varied. The property of capacitance change is utilized to achieve a change in the frequency and/or phase of our BPF. In your stability and circuit analysis use the simple model of a packaged varactor diode. For normal operation, a varactor diode is always reverse biased. Varactor diode is called voltage controlled capacitor.

10.1 Write circuit differential equations and find fixed points.

10.2 Find arbitrarily small increments equations and discuss stability.

10.3 Resistor R_1 is disconnected. How it influences our circuit dynamics? Write circuit differential equations and find fixed points. Discuss stability.

10.4 We increase the value of L_3 inductor by Γ parameter ($L_3 \rightarrow L_3 \cdot \Gamma$), $\Gamma > 0; \Gamma \in \mathbb{R}_+$. Write circuit differential equations and find fixed points. Discuss stability and stability switching for different values of Γ parameter.

10.5 Inductor L_4 is disconnected. How it influences our circuit dynamics? Find circuit differential equations and fixed points. Discuss stability.

Chapter 8
Antennas System Stability Analysis

An antenna is a conductor or group of conductors used for radiating electromagnetic energy into space or collecting electromagnetic energy from space. The radio signal is generated in a transmitter and radiates through space to a receiver by antenna. The transmitter signal energy is sent into space by a transmitting antenna and the radio frequency energy is picked up from space by the receiving antenna. As the electromagnetic field arrives at the receiving antenna, a voltage is induced into the antenna and passed into the receiver. There are many types of antennas and we discuss those antennas that operate at microwave frequencies. Microwave refer to radio waves with wavelength ranging from as long as 1 m to as short as 1 mm with frequencies between 300 MHz to 300 GHz. Microwave antenna is used for radiating microwave signal into space and receiving microwave signal from space. Microwave antenna is the transition region between free space and guiding structure. Antenna requirements include gain, receiving area, beam width, polarization, bandwidth, and side lobes. The major function of the antenna used at the receiver end of a RF/Microwave system is to collect as much of this transmitted power as possible. It is important in many applications to make the receiving area or aperture of the antenna as large as possible. Another antennas area is for RFID applications. Complete RFID system includes RFID reader and transponder units. Electrical current flowing through a conductor generates electromagnetic fields. We distinguish two fields regions related to RFID applications. The first region is far-field region. In this region, the generated fields are radiated fields which energy propagates through the space with an energy density proportional to the inverse of the distance. The second region is the near-field region. In this region, radiated fields are not prevalent. Instead, attenuating fields, in which the strength of the field decreases with $(1/r^3)$ are dominant. Furthermore, the power in this region is reactive. The boundaries between far-field and near-field regions depend on the relationship between the physical dimensions of the antenna and the wavelength of the

O. Aluf, *Microwave RF Antennas and Circuits*,
DOI 10.1007/978-3-319-45427-6_8

propagating signal. In particular, the dimensions of the antenna should be compatible to the wavelength of the signal in order to achieve optimal performance. The wavelength of signals operating in the LF (Low Frequency) region is 2.4 km, while the wavelength for signals operating in the HF (High Frequency) region is 24 m. Therefore, at these frequencies it would not be practical to build antennas with dimensions similar to the wavelength of their signals. Any antennas that can be built in a practical manner for RFID transponders operating in the LF or HF ranges will be electrically small and therefore highly inefficient. RFID transponders operating in the LF or HF frequencies cannot use dipole antennas because of the mismatch in dimensions. The solution is to use a small loop antenna instead of a dipole. A small loop antenna is a closed loop with a maximum dimension that is less than about a tenth of the wavelength of the signal. The small loop antenna is the dual equivalent of an ideal dipole and it is suitable for antennas incorporated in transponders operating at LF or HF frequencies. A conductor of infinite length carrying a magnitude of current of (I) amps. The magnetic field (B_ϕ) measured at a distance of (r) meters from the conductor can be found using Ampere's law as $B_\phi = \frac{\mu_0 \cdot I}{2 \cdot \pi \cdot r} (\frac{\text{Wb}}{\text{m}^2})$, where μ_0 is the permeability of the free space $\mu_0 = 4 \cdot \pi \times 10^{-7} \frac{\text{H}}{\text{m}}$. A conductor of infinite length is not realistic. Its practical implementation is based on building a loop antenna by bending the original, finite wire, which carries a current of (I) amps in a circle with a radius of (a) meters. In practice, the wire is bent in such a way that produces a total of N turns as this allows using a longer wire with a relatively small diameter. In this situation, the value of the magnetic field in the (z) coordinate direction (B_z) for a point located at a distance of (r) meters from the plane of the coil and located along the axis of the coil can be found as $B_z = \frac{\mu_0 \cdot I \cdot N \cdot a^2}{2 \cdot (a^2 + r^2)^{3/2}} (\frac{\text{Wb}}{\text{m}^2})$, where (a) is the radius of the loop in meters. The other kind of N turns antenna is rectangular spiral antenna. Both antennas are sensitive to their parameters variation and stability need to be investigated. Additionally, micro strips lines in RFID system cause to system's parasitic delays and influence stability. Special antenna to many RF and microwave applications is N-turn multilayer circular coil and there is an expression which define its inductance as a function of overall parameters [7, 8]. The stability is inspected for parameters variations and optimization under delayed electromagnetic interferences. Some antennas systems are straight thin film inductors antennas structure (single turn square planar straight thin film inductors antenna system) and its stability is inspected for many RF applications. Helix (Helical) antenna is consisting of a conducting wire wound in the form of a helix. Helical antennas are mounted over a ground plane. The feed line is connected between the bottom of the helix and the ground plane. Helical antennas can operate in one of two modes, normal mode or axial mode. In each operation mode we can represent helical antenna as equivalent circuit and inspect stability for parameters variations.

8.1 N-Turn Multilayer Circular Coil Antennas Transceiver System Stability Optimization Under Delayed Electromagnetic Interferences

N-turn multilayer circular coil antennas can be integrated with RFID IC for complete RFID tags. We investigate the system stability optimization under delayed electromagnetic interference and parasitic effects. An N-turn multilayer circular coil antenna is constructed from N-turn of circular coil with multilayer. Our system is constructed from two antenna; each one, N-turn multilayer circular coil antenna. Antennas are connected in series with micro strip line and to the RFID IC. An N-turn multilayer circular coil antennas system is influenced by electromagnetic interference, which effect their stability behavior. Additionally, micro strip line which connected each antenna in series, has a parasitic effect, a delay in time Δ_{μ}. We inspect our system performances under electromagnetic interferences and micro strip parasitic effects. Generally, N-turn multilayer circular coil antennas system is good for many RF and microwave applications. The micro strip line feed technique enhances the bandwidth of the simple micro strip antenna. Every N-turn multilayer circular coil antenna has a parasitic DC resistance which needs to be calculated. Index (i) indicates the first N-turn multilayer circular coil antenna (i = 1) or second N-turn multilayer circular coil antenna (i = 2). We define RFID's N-turn multilayer coil antenna parameters, a_i—Average radius of the coil in cm, N_i—number of turns, b_i—winding thickness in cm, S_i—wire cross section area, m_i—radius of the wire and h_i—winding height in cm. Integrating all those parameters gives the equations for N-turn multilayer circular coil antenna inductance calculation [85].

$$L_{calc-i} = \frac{0.31 \cdot (a_i \cdot N_i)^2}{6 \cdot a_i + 9 \cdot h_i + 10 \cdot b_i} (\mu\text{H}).$$

The N turn multilayer circular coil antenna length is calculated as follows: l_i is the length of one turn $l_i = 2 \cdot \pi \cdot a_i$. l_N is the length of N turn $l_{N_i} = N_i \cdot l_i = 2 \cdot \pi \cdot a_i \cdot N_i$. Assumption: $a_i \gg b_i; a_i + b_i \approx a_i$ (Fig. 8.1).

We consider system's two N-turn multilayer circular coil antennas are not identical $(a_1 \neq a_2; N_1 \neq N_2; h_1 \neq h_2; b_1 \neq b_2; a_i, N_i, h_i, b_i \in \mathbb{R}_+)$. The DC resistance of the N-turn multilayer circular coil antenna:

$$R_{DC-i} = \frac{l_{N_i}}{\sigma_i \cdot S_i} = \frac{2 \cdot \pi \cdot a_i \cdot N_i}{\sigma_i \cdot S_i} = \frac{2 \cdot a_i \cdot N_i}{\sigma_i \cdot m_i^2}.$$

l_{N_i}—total length of the wire, σ_i—conductivity of the wire (S/m), S_i—wire cross section area $(\pi \cdot m_i^2)$, m_i—radius of the wire. Due to electromagnetic interference there are differences in time delays with respect to the first (i = 1) and second (i = 2) N-turn multilayer circular coil antenna voltages and voltages derivatives. The delayed voltages are $V_1(t - \tau_1)$ and $V_2(t - \tau_2)$ respectively $(\tau_1 \neq \tau_2)$ and delayed voltages derivatives are $\frac{dV_1(t-\Delta_1)}{dt}; \frac{dV_2(t-\Delta_2)}{dt}$ respectively. Assumption:

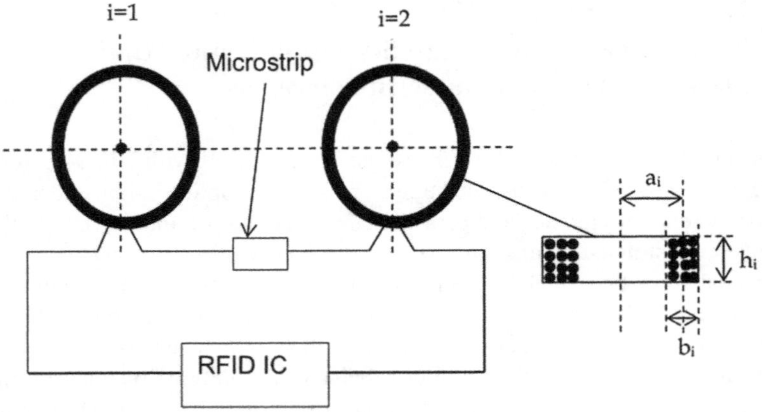

Fig. 8.1 N-turn multilayer circular coil antennas transceiver system

$\Delta_1 \neq \Delta_2; \tau_1 > 0; \tau_2 > 0; \Delta_1, \Delta_2 \geq 0$. Additionally, there is a delay in time for the micro strip parasitic effects Δ_μ. The stability of a given steady state is simply determined by the graphs of some function of τ_1, τ_2 which can be expressed, explicitly and thus can be easily depicted by MATLAB and other popular software. We need only look at one such function and locate the zeros. At time delay increases, stability changes from stable to unstable to stable. The analytical criteria provided for the first and second order cases can be used to obtain some insightful analytical statements and can be helpful for conducting simulations. N-turn multilayer circular coil antennas transceiver (RFID system) system can be represented as two inductors (L_{calc-1} and L_{calc-2}), parasitic resistances (R_{DC-1} and R_{DC-2}) and microstrip delay line. The N-turn multilayer circular coil antennas in series are connected in parallel to RFID IC. The equivalent circuit of N-turn multilayer circular coil antennas transceiver (RFID system) is capacitor (C_1) and resistor (R_1) in parallel with N-turn multilayer circular coil antennas in series. Element $2 \cdot L_m$ represents the mutual inductance between L_{calc-1} and L_{calc-2}. Since two inductors (L_{calc-1} and L_{calc-2}) are in series and there is a mutual inductance between L_{calc-1} and L_{calc-2}, the total antenna inductance L_T: $L_T = L_{calc-1} + L_{calc-2} + 2 \cdot L_m$ and $L_m = K \cdot \sqrt{\prod_{i=1}^{2} L_{calc-i}}$, L_m is the mutual inductance between L_{calc-1} and L_{calc-2}, K is the coupling coefficient of two inductors $0 \leq K \leq 1$. Variable I(t) is the current that flows through a N-turn multilayer circular coil antennas transceiver system for $\Delta_\mu \to \varepsilon$. The V_1 and V_2 are the voltages on L_{calc-1} and L_{calc-2} respectively, V_m is the voltage on N-turn multilayer circular coil antennas mutual inductance element. We neglect the voltage on microstrip delay line $V_\mu \to \varepsilon$. The delay which is related to microstrip element is on current $I_{L_{calc-2}}(t) = I_{L_{calc-1}}(t - \Delta_\mu)$ and $I_{R_{DC-i}} = I_{L_{calc-i}}; i = 1, 2$. N-turn multilayer circular coil antennas RFID system equivalent circuit is present in the next figure (Fig. 8.2).

Fig. 8.2 N-turn multilayer circular coil antennas equivalent circuit

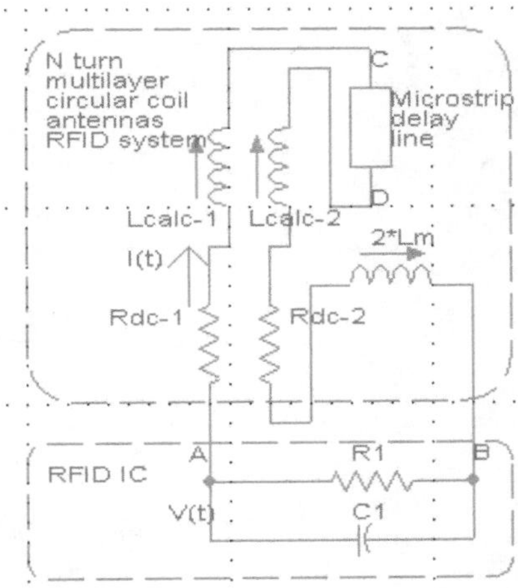

$$I_{L_{calc-2}}(t) = I(t-\Delta_\mu); I_{R_{DC-2}} = I(t-\Delta_\mu); I_{L_m} = I(t-\Delta_\mu)$$

$$I_{R_1} = \frac{V_{AB}}{R_1} = \frac{V_A - V_B}{R_1}; I_{C_1} = C_1 \cdot \frac{dV_{AB}}{dt} = C_1 \cdot \frac{d(V_A - V_B)}{dt}; I(t) + I_{R_1} + I_{C_1} = 0$$

$$V_{AB} = V_{R_1} = V_{C_1}; V_{AB} = V_{R_{DC-1}} + V_{L_{calc-1}} + V_{R_{DC-2}} + V_{L_{calc-2}} + V_{L_m};$$

$$V_1(t) = V_{L_{calc-1}} = V_{L_{calc-1}}(t)$$

$$V_2(t) = V_{L_{calc-2}} = V_{L_{calc-2}}(t); V_{AB} = V_A - V_B = R_1 \cdot I_{R_1}; V_{AB} = \frac{1}{C_1} \cdot \int I_{C_1} \cdot dt;$$

$$I(t) + \frac{V_{AB}}{R_1} + C_1 \cdot \frac{dV_{AB}}{dt} = 0$$

$$V_{AB} = V_{R_{DC-1}} + V_{L_{calc-1}} + V_{R_{DC-2}} + V_{L_{calc-2}} + V_m + (V_\mu \to \varepsilon); V_m = V_{L_m};$$

$$V_{AB} = \sum_{i=1}^{2} V_{R_{DC-i}} + \sum_{i=1}^{2} V_{L_{calc-i}} + V_m + (V_\mu \to \varepsilon)$$

$$V_{R_{DC-1}} = I(t) \cdot R_{DC-1} = I(t) \cdot \frac{2 \cdot a_1 \cdot N_1}{\sigma_1 \cdot m_1^2};$$

$$V_{R_{DC-2}} = I(t-\Delta_\mu) \cdot R_{DC-2} = I(t-\Delta_\mu) \cdot \frac{2 \cdot a_2 \cdot N_2}{\sigma_2 \cdot m_2^2}$$

$$V_{L_{calc-1}} = L_{calc-1} \cdot \frac{dI(t)}{dt} = \frac{0.31 \cdot (a_1 \cdot N_1)^2}{6 \cdot a_1 + 9 \cdot h_1 + 10 \cdot b_1} \cdot \frac{dI(t)}{dt};$$

$$V_{L_{calc-2}} = L_{calc-2} \cdot \frac{dI(t-\Delta_\mu)}{dt} = \frac{0.31 \cdot (a_2 \cdot N_2)^2}{6 \cdot a_2 + 9 \cdot h_2 + 10 \cdot b_2} \cdot \frac{dI(t-\Delta_\mu)}{dt}$$

$$V_m = 2 \cdot L_m \cdot \frac{dI(t-\Delta_\mu)}{dt} = 2 \cdot K \cdot (\sqrt{\prod_{i=1}^{2} L_{calc-i}}) \cdot \frac{dI(t-\Delta_\mu)}{dt}; L_{calc-1} \neq L_{calc-2}$$

$$I(t) + \frac{V_{C_1}}{R_1} + C_1 \cdot \frac{dV_{C_1}}{dt} = 0;$$

$$V_{C_1} = \sum_{i=1}^{2} V_{R_{DC-i}} + \sum_{i=1}^{2} V_{L_{calc-i}} + 2 \cdot K \cdot (\sqrt{\prod_{i=1}^{2} L_{calc-i}}) \cdot \frac{dI(t-\Delta_\mu)}{dt} + (V_\mu \to \varepsilon)$$

$$\frac{dV_{R_1}}{dt} = \frac{dV_{C_1}}{dt} = \sum_{i=1}^{2} \frac{dV_{R_{DC-i}}}{dt} + \sum_{i=1}^{2} \frac{dV_{L_{calc-i}}}{dt} + 2 \cdot K \cdot (\sqrt{\prod_{i=1}^{2} L_{calc-i}}) \cdot \frac{d^2 I(t-\Delta_\mu)}{dt^2} + (\frac{dV_\mu}{dt} \to \varepsilon)$$

$$V_{L_{calc-1}} = L_{calc-1} \cdot \frac{dI(t)}{dt} \Rightarrow I(t) = \frac{1}{L_{calc-1}} \cdot \int V_{L_{calc-1}} \cdot dt;$$

$$I(t) = (\frac{6 \cdot a_1 + 9 \cdot h_1 + 10 \cdot b_1}{0.31 \cdot (a_1 \cdot N_1)^2}) \cdot \int V_{L_{calc-1}} \cdot dt$$

$$V_{L_{calc-2}} = L_{calc-2} \cdot \frac{dI(t-\Delta_\mu)}{dt} \Rightarrow I(t-\Delta_\mu) = \frac{1}{L_{calc-2}} \cdot \int V_{L_{calc-2}} \cdot dt;$$

$$I(t-\Delta_\mu) = \frac{6 \cdot a_2 + 9 \cdot h_2 + 10 \cdot b_2}{0.31 \cdot (a_2 \cdot N_2)^2} \cdot \int V_{L_{calc-2}} \cdot dt$$

$$\frac{dI(t)}{dt} = \frac{V_{L_{calc-1}}}{L_{calc-1}}; \frac{dI(t-\Delta_\mu)}{dt} = \frac{V_{L_{calc-2}}}{L_{calc-2}};$$

$$\left|\frac{dI(t-\Delta_\mu)}{dt} - \frac{dI(t)}{dt}\right| \leq \Omega; \Omega \to \varepsilon; \frac{dI(t-\Delta_\mu)}{dt} \approx \frac{dI(t)}{dt}$$

$$\frac{V_{L_{calc-1}}}{L_{calc-1}} \approx \frac{V_{L_{calc-2}}}{L_{calc-2}}; V_m = 2 \cdot L_m \cdot \frac{dI(t-\Delta_\mu)}{dt} = 2 \cdot L_m \cdot \frac{V_{L_{calc-2}}}{L_{calc-2}};$$

$$V_m = 2 \cdot K \cdot (\sqrt{\prod_{i=1}^{2} L_{calc-i}}) \cdot \frac{V_{L_{calc-2}}}{L_{calc-2}}$$

$$V_m = 2 \cdot K \cdot (\sqrt{\frac{L_{calc-1}}{L_{calc-2}}}) \cdot V_{L_{calc-2}}; \frac{dV_m}{dt} = 2 \cdot K \cdot (\sqrt{\frac{L_{calc-1}}{L_{calc-2}}}) \cdot \frac{dV_{L_{calc-2}}}{dt}$$

We get the following differential equation respect to $V_{L_{calc-1}}$ variable:

$$\frac{1}{L_{calc-1}} \cdot \int V_{L_{calc-1}} \cdot dt + \frac{V_{AB}}{R_1} + C_1 \cdot \frac{dV_{AB}}{dt} = 0;$$

$$V_{R_{DC-1}} = I(t) \cdot R_{dc-1}; V_{R_{DC-2}} = I(t-\Delta_\mu) \cdot R_{dc-2}$$

$$\frac{1}{L_{calc-1}} \cdot \int V_{L_{calc-1}} \cdot dt + \frac{1}{R_1} \cdot (\sum_{i=1}^{2} V_{R_{DC-i}} + \sum_{i=1}^{2} V_{L_{calc-i}} + V_m) + C_1 \cdot \frac{d}{dt}(\sum_{i=1}^{2} V_{R_{DC-i}} + \sum_{i=1}^{2} V_{L_{calc-i}} + V_m) = 0$$

$$\frac{1}{L_{calc-1}} \cdot \int V_{L_{calc-1}} \cdot dt + \frac{1}{R_1} \cdot (\sum_{i=1}^{2} V_{R_{DC-i}} + \sum_{i=1}^{2} V_{L_{calc-i}} + V_m) + C_1 \cdot (\sum_{i=1}^{2} \frac{dV_{R_{DC-i}}}{dt} + \sum_{i=1}^{2} \frac{dV_{L_{calc-i}}}{dt} + \frac{dV_m}{dt}) = 0$$

$$\frac{1}{L_{calc-1}} \cdot \int V_{L_{calc-1}} \cdot dt + \frac{1}{R_1} \cdot \{I(t) \cdot R_{dc-1} + I(t-\Delta_\mu) \cdot R_{dc-2} + \sum_{i=1}^{2} V_{L_{calc-i}} + 2 \cdot K \cdot (\sqrt{\frac{L_{calc-1}}{L_{calc-2}}}) \cdot V_{L_{calc-2}}\} + C_1 \cdot \{\frac{dI(t)}{dt} \cdot R_{dc-1} + \frac{dI(t-\Delta_\mu)}{dt} \cdot R_{dc-2} + \sum_{i=1}^{2} \frac{dV_{L_{calc-i}}}{dt} + 2 \cdot K \cdot (\sqrt{\frac{L_{calc-1}}{L_{calc-2}}}) \cdot \frac{dV_{L_{calc-2}}}{dt}\} = 0$$

$$\frac{dI(t)}{dt} = \frac{V_{L_{calc-1}}}{L_{calc-1}}; \frac{dI(t-\Delta_\mu)}{dt} = \frac{V_{L_{calc-2}}}{L_{calc-2}}$$

$$\frac{1}{L_{calc-1}} \cdot \int V_{L_{calc-1}} \cdot dt + \frac{1}{R_1} \cdot \{\frac{1}{L_{calc-1}} \cdot R_{dc-1} \cdot \int V_{L_{calc-1}} \cdot dt + \frac{1}{L_{calc-2}} \cdot R_{dc-2} \cdot \int V_{L_{calc-2}} \cdot dt + \sum_{i=1}^{2} V_{L_{calc-i}} + 2 \cdot K \cdot (\sqrt{\frac{L_{calc-1}}{L_{calc-2}}}) \cdot V_{L_{calc-2}}\} + C_1 \cdot \{\frac{V_{L_{calc-1}}}{L_{calc-1}} \cdot R_{dc-1} + \frac{V_{L_{calc-2}}}{L_{calc-2}} \cdot R_{dc-2} + \sum_{i=1}^{2} \frac{dV_{L_{calc-i}}}{dt} + 2 \cdot K \cdot (\sqrt{\frac{L_{calc-1}}{L_{calc-2}}}) \cdot \frac{dV_{L_{calc-2}}}{dt}\} = 0$$

We derive in time the two sides of the above equations:

$$\frac{1}{L_{calc-1}} \cdot V_{L_{calc-1}} + \frac{1}{R_1} \cdot \{\frac{1}{L_{calc-1}} \cdot R_{dc-1} \cdot V_{L_{calc-1}} + \frac{1}{L_{calc-2}} \cdot R_{dc-2} \cdot V_{L_{calc-2}} + \sum_{i=1}^{2} \frac{dV_{L_{calc-i}}}{dt} + 2 \cdot K \cdot (\sqrt{\frac{L_{calc-1}}{L_{calc-2}}}) \cdot \frac{dV_{L_{calc-2}}}{dt}\} + C_1 \cdot \{\frac{dV_{L_{calc-1}}}{dt} \cdot \frac{R_{dc-1}}{L_{calc-1}} + \frac{dV_{L_{calc-2}}}{dt} \cdot \frac{R_{dc-2}}{L_{calc-2}} + \sum_{i=1}^{2} \frac{d^2 V_{L_{calc-i}}}{dt^2} + 2 \cdot K \cdot (\sqrt{\frac{L_{calc-1}}{L_{calc-2}}}) \cdot \frac{d^2 V_{L_{calc-2}}}{dt^2}\} = 0$$

$$\frac{V_{L_{calc-1}}}{L_{calc-1}} \simeq \frac{V_{L_{calc-2}}}{L_{calc-2}} \Rightarrow V_{L_{calc-2}} = V_{L_{calc-1}} \cdot \frac{L_{calc-2}}{L_{calc-1}}$$

Then we get the following expression:

$$\frac{1}{L_{calc-1}}\cdot V_{L_{calc-1}}+\frac{1}{R_1}\cdot\{\frac{1}{L_{calc-1}}\cdot R_{dc-1}\cdot V_{L_{calc-1}}+\frac{1}{L_{calc-2}}\cdot R_{dc-2}\cdot V_{L_{calc-1}}\cdot\frac{L_{calc-2}}{L_{calc-1}}+\sum_{i=1}^{2}\frac{dV_{L_{calc-i}}}{dt}$$
$$+2\cdot K\cdot(\sqrt{\frac{L_{calc-1}}{L_{calc-2}}})\cdot\frac{dV_{L_{calc-1}}}{dt}\cdot\frac{L_{calc-2}}{L_{calc-1}}\}+C_1\cdot\{\frac{dV_{L_{calc-1}}}{dt}\cdot\frac{R_{dc-1}}{L_{calc-1}}+\frac{dV_{L_{calc-1}}}{dt}\cdot\frac{R_{dc-2}}{L_{calc-1}}$$
$$+\sum_{i=1}^{2}\frac{d^2V_{L_{calc-i}}}{dt^2}+2\cdot K\cdot(\sqrt{\frac{L_{calc-1}}{L_{calc-2}}})\cdot\frac{L_{calc-2}}{L_{calc-1}}\cdot\frac{d^2V_{L_{calc-1}}}{dt^2}\}=0$$

$$\sum_{i=1}^{2}\frac{d^2V_{L_{calc-i}}}{dt^2}=\frac{d^2V_{L_{calc-1}}}{dt^2}+\frac{d^2V_{L_{calc-2}}}{dt^2}=\frac{d^2V_{L_{calc-1}}}{dt^2}\cdot(1+\frac{L_{calc-2}}{L_{calc-1}});$$
$$\sum_{i=1}^{2}\frac{dV_{L_{calc-i}}}{dt}=\frac{dV_{L_{calc-1}}}{dt}+\frac{dV_{L_{calc-2}}}{dt}=\frac{dV_{L_{calc-1}}}{dt}\cdot(1+\frac{L_{calc-2}}{L_{calc-1}})$$

$$\frac{1}{L_{calc-1}}\cdot V_{L_{calc-1}}+\frac{1}{R_1}\cdot\{\frac{1}{L_{calc-1}}\cdot R_{dc-1}\cdot V_{L_{calc-1}}+R_{dc-2}\cdot V_{L_{calc-1}}\cdot\frac{1}{L_{calc-1}}$$
$$+\frac{dV_{L_{calc-1}}}{dt}\cdot(1+\frac{L_{calc-2}}{L_{calc-1}})+2\cdot K\cdot(\sqrt{\frac{L_{calc-1}}{L_{calc-2}}})\cdot\frac{dV_{L_{calc-1}}}{dt}\cdot\frac{L_{calc-2}}{L_{calc-1}}\}$$
$$+C_1\cdot\{\frac{dV_{L_{calc-1}}}{dt}\cdot\frac{R_{dc-1}}{L_{calc-1}}+\frac{dV_{L_{calc-1}}}{dt}\cdot\frac{R_{dc-2}}{L_{calc-1}}+\frac{d^2V_{L_{calc-1}}}{dt^2}\cdot(1+\frac{L_{calc-2}}{L_{calc-1}})$$
$$+2\cdot K\cdot(\sqrt{\frac{L_{calc-1}}{L_{calc-2}}})\cdot\frac{L_{calc-2}}{L_{calc-1}}\cdot\frac{d^2V_{L_{calc-1}}}{dt^2}\}=0$$

$$[\frac{1}{L_{calc-1}}+\frac{1}{R_1}\cdot\frac{(R_{dc-1}+R_{dc-2})}{L_{calc-1}}]\cdot V_{L_{calc-1}}+[\frac{1}{R_1}\cdot(1+\frac{L_{calc-2}}{L_{calc-1}})+2\cdot K\cdot\frac{1}{R_1}\cdot\sqrt{\frac{L_{calc-2}}{L_{calc-1}}}$$
$$+C_1\cdot(\frac{R_{dc-1}+R_{dc-2}}{L_{calc-1}})]\cdot\frac{dV_{L_{calc-1}}}{dt}+\frac{d^2V_{L_{calc-1}}}{dt^2}\cdot C_1\cdot[1+\frac{L_{calc-2}}{L_{calc-1}}+2\cdot K\cdot\sqrt{\frac{L_{calc-2}}{L_{calc-1}}}]=0$$

We define the following global parameters:

$$\eta_1=\frac{1}{L_{calc-1}}+\frac{1}{R_1}\cdot\frac{(R_{dc-1}+R_{dc-2})}{L_{calc-1}};$$
$$\eta_2=\frac{1}{R_1}\cdot(1+\frac{L_{calc-2}}{L_{calc-1}})+2\cdot K\cdot\frac{1}{R_1}\cdot\sqrt{\frac{L_{calc-2}}{L_{calc-1}}}+C_1\cdot(\frac{R_{dc-1}+R_{dc-2}}{L_{calc-1}})$$

$$\eta_3=C_1\cdot[1+\frac{L_{calc-2}}{L_{calc-1}}+2\cdot K\cdot\sqrt{\frac{L_{calc-2}}{L_{calc-1}}}];$$
$$\eta_1\cdot V_{L_{calc-1}}+\eta_2\cdot\frac{dV_{L_{calc-1}}}{dt}+\eta_3\cdot\frac{d^2V_{L_{calc-1}}}{dt^2}=0$$

We get the following differential equation respect to $V_{L_{calc-1}}$:

$$\eta_1 \cdot V_{L_{calc-1}} + \eta_2 \cdot \frac{dV_{L_{calc-1}}}{dt} + \eta_3 \cdot \frac{d^2 V_{L_{calc-1}}}{dt^2} = 0; \eta_1 = \eta_1(R_{dc-1}, R_{dc-2}, L_{calc-1}, R_1)$$

$$\eta_2 = \eta_2(R_{dc-1}, R_{dc-2}, L_{calc-1}, L_{calc-2}, R_1, K, C_1); \eta_3 = \eta_3(L_{calc-1}, L_{calc-2}, C_1, K)$$

We define new variables: $Y_1 = \frac{dV_{L_{calc-1}}}{dt}; \frac{dY_1}{dt} = \frac{d^2 V_{L_{calc-1}}}{dt^2}$

$$\eta_1 \cdot V_{L_{calc-1}} + \eta_2 \cdot Y_1 + \eta_3 \cdot \frac{dY_1}{dt} = 0; \frac{dY_1}{dt} = -\frac{\eta_1}{\eta_3} \cdot V_{L_{calc-1}} - \frac{\eta_2}{\eta_3} \cdot Y_1;$$

$$\eta_1 = \eta_1(R_{dc-1}, R_{dc-2}, L_{calc-1}, R_1)$$

In the same manner we find $V_{L_{calc-2}}$ differential equation:

$$\frac{1}{L_{calc-1}} \cdot V_{L_{calc-1}} + \frac{1}{R_1} \cdot \{\frac{1}{L_{calc-1}} \cdot R_{dc-1} \cdot V_{L_{calc-1}} + \frac{1}{L_{calc-2}} \cdot R_{dc-2} \cdot V_{L_{calc-2}}$$
$$+ \sum_{i=1}^{2} \frac{dV_{L_{calc-i}}}{dt} + 2 \cdot K \cdot (\sqrt{\frac{L_{calc-1}}{L_{calc-2}}}) \cdot \frac{dV_{L_{calc-2}}}{dt}\} + C_1 \cdot \{\frac{dV_{L_{calc-1}}}{dt} \cdot \frac{R_{dc-1}}{L_{calc-1}}$$
$$+ \frac{dV_{L_{calc-2}}}{dt} \cdot \frac{R_{dc-2}}{L_{calc-2}} + \sum_{i=1}^{2} \frac{d^2 V_{L_{calc-i}}}{dt^2} + 2 \cdot K \cdot (\sqrt{\frac{L_{calc-1}}{L_{calc-2}}}) \cdot \frac{d^2 V_{L_{calc-2}}}{dt^2}\} = 0$$

$$\frac{V_{L_{calc-1}}}{L_{calc-1}} \simeq \frac{V_{L_{calc-2}}}{L_{calc-2}} \Rightarrow V_{L_{calc-1}} = V_{L_{calc-2}} \cdot \frac{L_{calc-1}}{L_{calc-2}}$$

Then we get the following expression:

$$\frac{V_{L_{calc-2}}}{L_{calc-2}} + \frac{1}{R_1} \cdot \{V_{L_{calc-2}} \cdot \frac{R_{dc-1}}{L_{calc-2}} + \frac{R_{dc-2}}{L_{calc-2}} \cdot V_{L_{calc-2}} + \sum_{i=1}^{2} \frac{dV_{L_{calc-i}}}{dt}$$
$$+ 2 \cdot K \cdot (\sqrt{\frac{L_{calc-1}}{L_{calc-2}}}) \cdot \frac{dV_{L_{calc-2}}}{dt}\} + C_1 \cdot \{\frac{dV_{L_{calc-2}}}{dt} \cdot \frac{R_{dc-1}}{L_{calc-2}} + \frac{dV_{L_{calc-2}}}{dt} \cdot \frac{R_{dc-2}}{L_{calc-2}}$$
$$+ \sum_{i=1}^{2} \frac{d^2 V_{L_{calc-i}}}{dt^2} + 2 \cdot K \cdot (\sqrt{\frac{L_{calc-1}}{L_{calc-2}}}) \cdot \frac{d^2 V_{L_{calc-2}}}{dt^2}\} = 0$$

$$\sum_{i=1}^{2} \frac{d^2 V_{L_{calc-i}}}{dt^2} = \frac{d^2 V_{L_{calc-1}}}{dt^2} + \frac{d^2 V_{L_{calc-2}}}{dt^2} = \frac{d^2 V_{L_{calc-2}}}{dt^2} \cdot (\frac{L_{calc-1}}{L_{calc-2}} + 1);$$

$$\sum_{i=1}^{2} \frac{dV_{L_{calc-i}}}{dt} = \frac{dV_{L_{calc-1}}}{dt} + \frac{dV_{L_{calc-2}}}{dt} = \frac{dV_{L_{calc-2}}}{dt} \cdot (\frac{L_{calc-1}}{L_{calc-2}} + 1)$$

$$\frac{V_{L_{calc-2}}}{L_{calc-2}} + \frac{1}{R_1} \cdot \{V_{L_{calc-2}} \cdot \frac{R_{dc-1}}{L_{calc-2}} + \frac{R_{dc-2}}{L_{calc-2}} \cdot V_{L_{calc-2}} + \frac{dV_{L_{calc-2}}}{dt} \cdot (\frac{L_{calc-1}}{L_{calc-2}} + 1)$$
$$+ 2 \cdot K \cdot (\sqrt{\frac{L_{calc-1}}{L_{calc-2}}}) \cdot \frac{dV_{L_{calc-2}}}{dt}\} + C_1 \cdot \{\frac{dV_{L_{calc-2}}}{dt} \cdot \frac{R_{dc-1}}{L_{calc-2}} + \frac{dV_{L_{calc-2}}}{dt} \cdot \frac{R_{dc-2}}{L_{calc-2}}$$
$$+ \frac{d^2V_{L_{calc-2}}}{dt^2} \cdot (\frac{L_{calc-1}}{L_{calc-2}} + 1) + 2 \cdot K \cdot (\sqrt{\frac{L_{calc-1}}{L_{calc-2}}}) \cdot \frac{d^2V_{L_{calc-2}}}{dt^2}\} = 0$$

$$\frac{V_{L_{calc-2}}}{L_{calc-2}} + V_{L_{calc-2}} \cdot \frac{1}{R_1} \cdot \frac{R_{dc-1}}{L_{calc-2}} + \frac{1}{R_1} \cdot \frac{R_{dc-2}}{L_{calc-2}} \cdot V_{L_{calc-2}} + \frac{dV_{L_{calc-2}}}{dt} \cdot \frac{1}{R_1} \cdot (\frac{L_{calc-1}}{L_{calc-2}} + 1)$$
$$+ 2 \cdot K \cdot \frac{1}{R_1} \cdot (\sqrt{\frac{L_{calc-1}}{L_{calc-2}}}) \cdot \frac{dV_{L_{calc-2}}}{dt} + \frac{dV_{L_{calc-2}}}{dt} \cdot \frac{C_1 \cdot R_{dc-1}}{L_{calc-2}} + \frac{dV_{L_{calc-2}}}{dt} \cdot \frac{C_1 \cdot R_{dc-2}}{L_{calc-2}}$$
$$+ \frac{d^2V_{L_{calc-2}}}{dt^2} \cdot C_1 \cdot (\frac{L_{calc-1}}{L_{calc-2}} + 1) + 2 \cdot K \cdot C_1 \cdot (\sqrt{\frac{L_{calc-1}}{L_{calc-2}}}) \cdot \frac{d^2V_{L_{calc-2}}}{dt^2} = 0$$

$$[\frac{1}{L_{calc-2}} + \frac{1}{R_1} \cdot (\frac{R_{dc-1} + R_{dc-2}}{L_{calc-2}})] \cdot V_{L_{calc-2}} + [\frac{1}{R_1} \cdot (\frac{L_{calc-1}}{L_{calc-2}} + 1) + 2 \cdot K \cdot \frac{1}{R_1} \cdot \sqrt{\frac{L_{calc-1}}{L_{calc-2}}}$$
$$+ C_1 \cdot (\frac{R_{dc-1} + R_{dc-2}}{L_{calc-2}})] \cdot \frac{dV_{L_{calc-2}}}{dt} + \frac{d^2V_{L_{calc-2}}}{dt^2} \cdot C_1 \cdot [\frac{L_{calc-1}}{L_{calc-2}} + 1 + 2 \cdot K \cdot \sqrt{\frac{L_{calc-1}}{L_{calc-2}}}] = 0$$

We define the following global parameters: $\xi_1 = \frac{1}{L_{calc-2}} + \frac{1}{R_1} \cdot (\frac{R_{dc-1} + R_{dc-2}}{L_{calc-2}})$

$$\xi_2 = \frac{1}{R_1} \cdot (\frac{L_{calc-1}}{L_{calc-2}} + 1) + 2 \cdot K \cdot \frac{1}{R_1} \cdot \sqrt{\frac{L_{calc-1}}{L_{calc-2}}} + C_1 \cdot (\frac{R_{dc-1} + R_{dc-2}}{L_{calc-2}});$$
$$\xi_3 = C_1 \cdot [\frac{L_{calc-1}}{L_{calc-2}} + 1 + 2 \cdot K \cdot \sqrt{\frac{L_{calc-1}}{L_{calc-2}}}]$$

$\xi_1 = \xi_1(R_{dc-1}, R_{dc-2}, L_{calc-2}, R_1)$; $\xi_2 = \xi_2(L_{calc-1}, L_{calc-2}, R_{dc-1}, R_{dc-2}, R_1, K, C_1)$; $\xi_3 = \xi_3(L_{calc-1}, L_{calc-2}, K, C_1)$

$\xi_1 \cdot V_{L_{calc-2}} + \xi_2 \cdot \frac{dV_{L_{calc-2}}}{dt} + \frac{d^2V_{L_{calc-2}}}{dt^2} \cdot \xi_3 = 0$. We define new variables: $Y_2 = \frac{dV_{L_{calc-2}}}{dt}$; $\frac{dY_2}{dt} = \frac{d^2V_{L_{calc-2}}}{dt^2}$

$$\xi_1 \cdot V_{L_{calc-2}} + \xi_2 \cdot \frac{dV_{L_{calc-2}}}{dt} + \frac{d^2V_{L_{calc-2}}}{dt^2} \cdot \xi_3 = 0;$$
$$\xi_1 \cdot V_{L_{calc-2}} + \xi_2 \cdot Y_2 + \frac{dY_2}{dt} \cdot \xi_3 = 0 \Rightarrow \frac{dY_2}{dt} = -\frac{\xi_1}{\xi_3} \cdot V_{L_{calc-2}} - \frac{\xi_2}{\xi_3} \cdot Y_2$$

Summary: We can get our N-turn multilayer circular coil antennas (RFID system) system.

$$\frac{dV_{L_{calc-1}}}{dt} = Y_1; \frac{dY_1}{dt} = -\frac{\eta_1}{\eta_3} \cdot V_{L_{calc-1}} - \frac{\eta_2}{\eta_3} \cdot Y_1;$$
$$\frac{dV_{L_{calc-2}}}{dt} = Y_2; \frac{dY_2}{dt} = -\frac{\xi_1}{\xi_3} \cdot V_{L_{calc-2}} - \frac{\xi_2}{\xi_3} \cdot Y_2$$

$$\begin{pmatrix} \frac{dY_1}{dt} \\ \frac{dV_{L_{calc-1}}}{dt} \\ \frac{dY_2}{dt} \\ \frac{dV_{L_{calc-2}}}{dt} \end{pmatrix} = \begin{pmatrix} \Gamma_{11} & \dots & \Gamma_{14} \\ \vdots & \ddots & \vdots \\ \Gamma_{41} & \dots & \Gamma_{44} \end{pmatrix} \cdot \begin{pmatrix} Y_1 \\ V_{L_{calc-1}} \\ Y_2 \\ V_{L_{calc-2}} \end{pmatrix}; \Gamma_{11} = -\frac{\eta_2}{\eta_3}; \Gamma_{12} = -\frac{\eta_1}{\eta_3}; \Gamma_{13} = 0; \Gamma_{14} = 0$$

$$\Gamma_{21} = 1; \Gamma_{22} = \Gamma_{23} = \Gamma_{24} = 0; \Gamma_{31} = \Gamma_{32} = 0; \Gamma_{33} = -\frac{\xi_2}{\xi_3}; \Gamma_{34} = -\frac{\xi_1}{\xi_3}; \Gamma_{41} = \Gamma_{42} = 0$$

$$\Gamma_{43} = 1; \Gamma_{44} = 0$$

N-turn multilayer circular coil antennas (RFID system) system is composed from two N-turn thin wire multilayer and circular. Units are all in centimeters (cm), $a_i, h_i, b_i \; \forall \; i = 1, 2; N_i \in \mathbb{R}_+$. Inductors L_{calc-1}, L_{calc-2} units are in μH. Due to electromagnetic interferences there are different in time delays respect to first and second N-turn thin wire multilayer and circular antennas voltages and voltages derivatives. Additionally, there is a delay in time Δ_μ for the current that flows through inductor L_{calc-2} and mutual inductance. The delay voltages are $V_1(t - \tau_1); V_2(t - \tau_2)$ respectively $(\tau_1 \neq \tau_2)$ and delayed voltages derivatives are $\frac{dV_1(t-\tau_1)}{dt}; \frac{dV_2(t-\tau_2)}{dt}$ respectively $(\Delta_1 \neq \Delta_2; \tau_1 \geq 0; \tau_2 \geq 0; \Delta_1, \Delta_2 \in \mathbb{R}_+)$.

$$V_1(t) = V_{L_{calc-1}} = V_{L_{calc-1}}(t); V_2(t) = V_{L_{calc-2}} = V_{L_{calc-2}}(t);$$
$$V_{L_{calc-1}}(t) \to V_{L_{calc-1}}(t - \tau_1); V_{L_{calc-2}}(t) \to V_{L_{calc-2}}(t - \tau_2)$$

We consider no delay effect on $\frac{dY_1}{dt}, \frac{dV_{L_{calc-1}}}{dt}, \frac{dY_2}{dt}, \frac{dV_{L_{calc-2}}}{dt}$. The N-turn multilayer circular coil antennas system differential equations under electromagnetic interferences (delay terms) influence only N-turn multilayer circular coil antennas system first and second N-turn multilayer circular coil voltages $V_{L_{calc-1}}(t), V_{L_{calc-2}}(t)$ and voltages derivatives $Y_1(t)$ and $Y_2(t)$ respect to time, there is no influence on $\frac{dY_1(t)}{dt}, \frac{dV_{L_{calc-1}}(t)}{dt}, \frac{dY_2(t)}{dt}, \frac{dV_{L_{calc-2}}(t)}{dt}$ [85].

$$\begin{pmatrix} \frac{dY_1}{dt} \\ \frac{dV_{L_{calc-1}}}{dt} \\ \frac{dY_2}{dt} \\ \frac{dV_{L_{calc-2}}}{dt} \end{pmatrix} = \begin{pmatrix} \Gamma_{11} & \dots & \Gamma_{14} \\ \vdots & \ddots & \vdots \\ \Gamma_{41} & \dots & \Gamma_{44} \end{pmatrix} \cdot \begin{pmatrix} Y_1(t - \Delta_1) \\ V_{L_{calc-1}}(t - \tau_1) \\ Y_2(t - \Delta_2) \\ V_{L_{calc-2}}(t - \tau_2) \end{pmatrix}$$

To find equilibrium points (fixed points) of N-turn multilayer circular coil antennas RFID system is by

$$\lim_{t\to\infty} Y_1(t-\Delta_1) = Y_1(t); \lim_{t\to\infty} V_{L_{calc-1}}(t-\tau_1) = V_{L_{calc-1}}(t)$$
$$\lim_{t\to\infty} Y_2(t-\Delta_2) = Y_2(t); \lim_{t\to\infty} V_{L_{calc-2}}(t-\tau_2) = V_{L_{calc-2}}(t).$$

$$\frac{dY_1}{dt} = 0; \frac{dV_{L_{calc-1}}}{dt} = 0; \frac{dY_2}{dt} = 0; \frac{dV_{L_{calc-2}}}{dt} = 0; t \gg \tau_1; t \gg \tau_2; t \gg \Delta_1; t \gg \Delta_2$$

We get four equations and the only fixed point is $E^{(0)}(Y_1^{(0)}, V_{L_{calc-1}}^{(0)}, Y_2^{(0)}, V_{L_{calc-2}}^{(0)}) = (0,0,0,0)$.

$$\eta_3 \neq 0 \,\&\, \eta_2 \neq 0 \Rightarrow \frac{1}{R_1} \cdot (1 + \frac{L_{calc-2}}{L_{calc-1}}) + 2 \cdot K \cdot \frac{1}{R_1} \cdot \sqrt{\frac{L_{calc-2}}{L_{calc-1}}} + C_1 \cdot (\frac{R_{dc-1} + R_{dc-2}}{L_{calc-1}}) \neq 0; \Gamma_{11} \neq 0$$

$$\eta_3 \neq 0 \,\&\, \eta_1 \neq 0 \Rightarrow \Gamma_{12} \neq 0; \frac{1}{L_{calc-1}} + \frac{1}{R_1} \cdot \frac{(R_{dc-1} + R_{dc-2})}{L_{calc-1}} \neq 0$$

$$\xi_3 \neq 0; \xi_1 \neq 0 \Rightarrow \frac{1}{L_{calc-2}} + \frac{1}{R_1} \cdot (\frac{R_{dc-1} + R_{dc-2}}{L_{calc-2}}) \neq 0; \Gamma_{34} \neq 0$$

$$\xi_3 \neq 0; \xi_2 \neq 0 \Rightarrow \frac{1}{R_1} \cdot (\frac{L_{calc-1}}{L_{calc-2}} + 1) + 2 \cdot K \cdot \frac{1}{R_1} \cdot \sqrt{\frac{L_{calc-1}}{L_{calc-2}}} + C_1 \cdot (\frac{R_{dc-1} + R_{dc-2}}{L_{calc-2}}) \neq 0; \Gamma_{33} \neq 0$$

Stability analysis: The standard local stability analysis about any one of the equilibrium points (fixed points) of N-turn multilayer circular coil antennas RFID system consists in adding to coordinates $[Y_1 V_{L_{calc-1}} Y_2 V_{L_{calc-2}}]$ arbitrarily small increments of exponential form $[y_1 v_{L_{calc-1}} y_2 v_{L_{calc-2}}] \cdot e^{\lambda \cdot t}$, and retaining the first order terms in $Y_1 V_{L_{calc-1}} Y_2 V_{L_{calc-2}}$. The system of four homogeneous equations leads to a polynomial characteristics equation in the eigenvalues λ. The polynomial characteristics equations accept by set the below voltages and voltages derivative respect to the time into N-turn multilayer circular coil antennas RFID system equations. N-turn multilayer circular coil antennas RFID system fixed values with arbitrarily small increments of exponential form $[Y_1 V_{L_{calc-1}} Y_2 V_{L_{calc-2}}] \cdot e^{\lambda \cdot t}$ are: i = 1 (first fixed point), i = 2 (second fixed point), i = 3 (third fixed point), etc.,

$$Y_1(t) = Y_1^{(i)} + y_1 \cdot e^{\lambda \cdot t}; V_{L_{calc-1}}(t) = V_{L_{calc-1}}^{(i)} + v_{L_{calc-1}} \cdot e^{\lambda \cdot t};$$
$$Y_2(t) = Y_2^{(i)} + y_2 \cdot e^{\lambda \cdot t}; V_{L_{calc-2}}(t) = V_{L_{calc-2}}^{(i)} + v_{L_{calc-2}} \cdot e^{\lambda \cdot t}$$

We choose the above expressions for our Y_1(t),$V_{L_{calc-1}}$(t),Y_2(t),$V_{L_{calc-2}}$(t) as small displacement $[y_1 v_{L_{calc-1}} y_2 v_{L_{calc-2}}]$ from the system fixed points at time t = 0.

$$Y_1(t=0) = Y_1^{(i)} + y_1;\ V_{L_{calc-1}}(t=0) = V_{L_{calc-1}}^{(i)} + v_{L_{calc-1}};$$
$$Y_2(t=0) = Y_2^{(i)} + y_2;\ V_{L_{calc-2}}(t=0) = V_{L_{calc-2}}^{(i)} + v_{L_{calc-2}}$$

For t > 0, λ < 0 the selected fixed point is stable otherwise t > 0, λ > 0 is unstable. Our system tends to the selected fixed point exponentially for t > 0, λ < 0 otherwise go away from the selected fixed point exponentially. Eigenvalue λ is the parameter which establish if the fixed point is stable or unstable, additionally his absolute value (|λ|) establish the speed of flow toward or away from the selected fixed point. Table 8.1 describes N-turn multilayer circular coil antennas RFID system variables for different eigenvalue λ and t values.

The speeds of flow toward or away from the selected fixed point for N-turn multilayer circular coil antennas RFID system voltages and voltages derivatives respect to time are as follow:

$$\frac{dY_1(t)}{dt} = y_1 \cdot \lambda \cdot e^{\lambda \cdot t}; \frac{dV_{L_{calc-1}}(t)}{dt} = v_{L_{calc-1}} \cdot \lambda \cdot e^{\lambda \cdot t};$$
$$\frac{dY_2(t)}{dt} = y_2 \cdot \lambda \cdot e^{\lambda \cdot t}; \frac{dV_{L_{calc-2}}(t)}{dt} = v_{L_{calc-2}} \cdot \lambda \cdot e^{\lambda \cdot t}$$

Table 8.1 N-turn multilayer circular coil antennas RFID system variables for different eigenvalue λ and time

	λ < 0	λ > 0
t = 0	$Y_1(t=0) = Y_1^{(i)} + y_1$ $V_{L_{calc-1}}(t=0) = V_{L_{calc-1}}^{(i)} + v_{L_{calc-1}}$ $Y_2(t=0) = Y_2^{(i)} + y_2$ $V_{L_{calc-2}}(t=0) = V_{L_{calc-2}}^{(i)} + v_{L_{calc-2}}$	$Y_1(t=0) = Y_1^{(i)} + y_1$ $V_{L_{calc-1}}(t=0) = V_{L_{calc-1}}^{(i)} + v_{L_{calc-1}}$ $Y_2(t=0) = Y_2^{(i)} + y_2$ $V_{L_{calc-2}}(t=0) = V_{L_{calc-2}}^{(i)} + v_{L_{calc-2}}$
t > 0	$Y_1(t) = Y_1^{(i)} + y_1 \cdot e^{-\lvert\lambda\rvert \cdot t}$ $V_{L_{calc-1}}(t) = V_{L_{calc-1}}^{(i)} + v_{L_{calc-1}} \cdot e^{-\lvert\lambda\rvert \cdot t}$ $Y_2(t) = Y_2^{(i)} + y_2 \cdot e^{-\lvert\lambda\rvert \cdot t}$ $V_{L_{calc-2}}(t) = V_{L_{calc-2}}^{(i)} + v_{L_{calc-2}} \cdot e^{-\lvert\lambda\rvert \cdot t}$	$Y_1(t) = Y_1^{(i)} + y_1 \cdot e^{\lvert\lambda\rvert \cdot t}$ $V_{L_{calc-1}}(t) = V_{L_{calc-1}}^{(i)} + v_{L_{calc-1}} \cdot e^{\lvert\lambda\rvert \cdot t}$ $Y_2(t) = Y_2^{(i)} + y_2 \cdot e^{\lvert\lambda\rvert \cdot t}$ $V_{L_{calc-2}}(t) = V_{L_{calc-2}}^{(i)} + v_{L_{calc-2}} \cdot e^{\lvert\lambda\rvert \cdot t}$
t → ∞	$Y_1(t \to \infty) = Y_1^{(i)}$ $V_{L_{calc-1}}(t \to \infty) = V_{L_{calc-1}}^{(i)}$ $Y_2(t \to \infty) = Y_2^{(i)}$ $V_{L_{calc-2}}(t \to \infty) = V_{L_{calc-2}}^{(i)}$	$Y_1(t \to \infty) \simeq y_1 \cdot e^{\lvert\lambda\rvert \cdot t}$ $V_{L_{calc-1}}(t \to \infty) \simeq v_{L_{calc-1}} \cdot e^{\lvert\lambda\rvert \cdot t}$ $Y_2(t \to \infty) \simeq y_2 \cdot e^{\lvert\lambda\rvert \cdot t}$ $V_{L_{calc-2}}(t \to \infty) \simeq v_{L_{calc-2}} \cdot e^{\lvert\lambda\rvert \cdot t}$

$$\frac{dV_{L_{calc-1}}}{dt} = Y_1 \Rightarrow v_{L_{calc-1}} \cdot \lambda \cdot e^{\lambda \cdot t} = Y_1^{(i)} + y_1 \cdot e^{\lambda \cdot t}; Y_1^{(i)} = 0; -v_{L_{calc-1}} \cdot \lambda + y_1 = 0$$

$$\frac{dY_1}{dt} = -\frac{\eta_1}{\eta_3} \cdot V_{L_{calc-1}} - \frac{\eta_2}{\eta_3} \cdot Y_1$$

$$\Rightarrow y_1 \cdot \lambda \cdot e^{\lambda \cdot t} = -\frac{\eta_1}{\eta_3} \cdot (V_{L_{calc-1}}^{(i)} + v_{L_{calc-1}} \cdot e^{\lambda \cdot t}) - \frac{\eta_2}{\eta_3} \cdot (Y_1^{(i)} + y_1 \cdot e^{\lambda \cdot t})$$

$$y_1 \cdot \lambda \cdot e^{\lambda \cdot t} = -\frac{\eta_1}{\eta_3} \cdot V_{L_{calc-1}}^{(i)} - \frac{\eta_2}{\eta_3} \cdot Y_1^{(i)} - \frac{\eta_1}{\eta_3} \cdot v_{L_{calc-1}} \cdot e^{\lambda \cdot t} - \frac{\eta_2}{\eta_3} \cdot y_1 \cdot e^{\lambda \cdot t};$$

$$-\frac{\eta_1}{\eta_3} \cdot V_{L_{calc-1}}^{(i)} - \frac{\eta_2}{\eta_3} \cdot Y_1^{(i)} = 0$$

$$-y_1 \cdot \lambda - \frac{\eta_1}{\eta_3} \cdot v_{L_{calc-1}} - \frac{\eta_2}{\eta_3} \cdot y_1 = 0;$$

$$\frac{dV_{L_{calc-2}}}{dt} = Y_2 \Rightarrow v_{L_{calc-2}} \cdot \lambda \cdot e^{\lambda \cdot t} = Y_2^{(i)} + y_2 \cdot e^{\lambda \cdot t}; Y_2^{(i)} = 0$$

$$-v_{L_{calc-2}} \cdot \lambda + y_2 = 0; \frac{dY_2}{dt} = -\frac{\xi_1}{\xi_3} \cdot V_{L_{calc-2}} - \frac{\xi_2}{\xi_3} \cdot Y_2$$

$$\Rightarrow y_2 \cdot \lambda \cdot e^{\lambda \cdot t} = -\frac{\xi_1}{\xi_3} \cdot (V_{L_{calc-2}}^{(i)} + v_{L_{calc-2}} \cdot e^{\lambda \cdot t}) - \frac{\xi_2}{\xi_3} \cdot (Y_2^{(i)} + y_2 \cdot e^{\lambda \cdot t})$$

$$y_2 \cdot \lambda \cdot e^{\lambda \cdot t} = -\frac{\xi_1}{\xi_3} \cdot V_{L_{calc-2}}^{(i)} - \frac{\xi_2}{\xi_3} \cdot Y_2^{(i)} - \frac{\xi_1}{\xi_3} \cdot v_{L_{calc-2}} \cdot e^{\lambda \cdot t} - \frac{\xi_2}{\xi_3} \cdot y_2 \cdot e^{\lambda \cdot t};$$

$$-\frac{\xi_1}{\xi_3} \cdot V_{L_{calc-2}}^{(i)} - \frac{\xi_2}{\xi_3} \cdot Y_2^{(i)} = 0$$

$$-y_2 \cdot \lambda - \frac{\xi_1}{\xi_3} \cdot v_{L_{calc-2}} - \frac{\xi_2}{\xi_3} \cdot y_2 = 0$$

We can summary our N-turn multilayer circular coil antennas RFID system arbitrarily small increments equations:

$$-v_{L_{calc-1}} \cdot \lambda + y_1 = 0; -y_1 \cdot \lambda - \frac{\eta_2}{\eta_3} \cdot y_1 - \frac{\eta_1}{\eta_3} \cdot v_{L_{calc-1}} = 0; -v_{L_{calc-2}} \cdot \lambda + y_2 = 0;$$

$$-y_2 \cdot \lambda - \frac{\xi_2}{\xi_3} \cdot y_2 - \frac{\xi_1}{\xi_3} \cdot v_{L_{calc-2}} = 0$$

$$\begin{pmatrix} \Upsilon_{11} & \dots & \Upsilon_{14} \\ \vdots & \ddots & \vdots \\ \Upsilon_{41} & \cdots & \Upsilon_{44} \end{pmatrix} \cdot \begin{pmatrix} y_1 \\ v_{L_{calc-1}} \\ y_2 \\ v_{L_{calc-2}} \end{pmatrix} = 0; \Upsilon_{11} = -\lambda - \frac{\eta_2}{\eta_3}; \Upsilon_{12} = -\frac{\eta_1}{\eta_3};$$

$$\Upsilon_{13} = 0; \Upsilon_{14} = 0; \Upsilon_{21} = 1$$

$$\Upsilon_{22} = -\lambda; \Upsilon_{23} = 0; \Upsilon_{24} = 0; \Upsilon_{31} = 0; \Upsilon_{32} = 0; \Upsilon_{33} = -\lambda - \frac{\xi_2}{\xi_3};$$

$$\Upsilon_{34} = -\frac{\xi_1}{\xi_3}; \Upsilon_{41} = 0$$

$$\Upsilon_{42} = 0; \Upsilon_{43} = 1; \Upsilon_{44} = -\lambda$$

$$A - \lambda \cdot I = \begin{pmatrix} \Upsilon_{11} & \dots & \Upsilon_{14} \\ \vdots & \ddots & \vdots \\ \Upsilon_{41} & \cdots & \Upsilon_{44} \end{pmatrix}; \det(A - \lambda \cdot I) = 0 \Rightarrow \sum_{k=0}^{4} \lambda^k \cdot \Pi_k = 0;$$

$$\Pi_0 = \frac{\eta_1 \cdot \xi_1}{\eta_3 \cdot \xi_3}; \Pi_1 = \frac{\eta_2 \cdot \xi_1}{\eta_3 \cdot \xi_3} + \frac{\eta_1 \cdot \xi_2}{\eta_3 \cdot \xi_3}$$

$$\Pi_2 = \frac{\xi_1}{\xi_3} + \frac{\eta_2 \cdot \xi_2}{\eta_3 \cdot \xi_3} + \frac{\eta_1}{\eta_3}; \Pi_3 = \frac{\xi_2}{\xi_3} + \frac{\eta_2}{\eta_3}; \Pi_4 = 1$$

Eigenvalues stability discussion: Our N-turn multilayer circular coil antennas RFID system involving N_a variables ($N_a > 2$, $N_a = 4$), the characteristic equation is of degree $N_a = 4$ ($\sum_{k=0}^{4} \lambda^k \cdot \Pi_k = 0$) and must often be solved numerically. Expect in some particular cases, such an equation has ($N_a = 4$) distinct roots that can be real or complex. These values are the eigenvalues of the 4×4 Jacobian matrix (A). The general rule is that the Steady State (SS) is stable if there is no eigenvalue with positive real part. It is sufficient that one eigenvalue is positive for the steady state to be unstable. Our 4-variables ($Y_1 V_{L_{calc-1}} Y_2 V_{L_{calc-2}}$) N-turn multilayer circular coil antennas RFID system has four eigenvalues. The type of behavior can be characterized as a function of the position of these eigenvalues in the Re/Im plane. Five non-degenerated cases can be distinguished: (1) the four eigenvalues are real and negative (stable steady state), (2) the four eigenvalues are real, three of them are negative (unstable steady state), (3) and (4) two eigenvalues are complex conjugates with a negative real part and the other eigenvalues are real and negative (stable steady state), two cases can be distinguished depending on the relative value of the real part of the complex eigenvalues and of the real one, (5) two eigenvalues are complex conjugates with a negative real part and at least one eigenvalue is positive (unstable steady state) [2, 3].

We define the following N-turn multilayer circular coil antennas RFID system variables delayed in time:

$$Y_1(t-\Delta_1) = Y_1^{(i)} + y_1 \cdot e^{\lambda\cdot(t-\Delta_1)}; V_{L_{calc-1}}(t-\tau_1) = V_{L_{calc-1}}^{(i)} + v_{L_{calc-1}} \cdot e^{\lambda\cdot(t-\tau_1)};$$
$$Y_2(t-\Delta_2) = Y_2^{(i)} + y_2 \cdot e^{\lambda\cdot(t-\Delta_2)}; V_{L_{calc-2}}(t-\tau_2) = V_{L_{calc-2}}^{(i)} + v_{L_{calc-2}} \cdot e^{\lambda\cdot(t-\tau_2)}$$

We get four delayed differential equations respect to coordinates $[Y_1 V_{L_{calc-1}} Y_2 V_{L_{calc-2}}]$ arbitrarily small increments of exponential $[y_1 v_{L_{calc-1}} y_2 v_{L_{calc-2}}] \cdot e^{\lambda\cdot t}$.

$$v_{L_{calc-1}} \cdot \lambda \cdot e^{\lambda\cdot t} = Y_1^{(i)} + y_1 \cdot e^{\lambda\cdot(t-\Delta_1)}; Y_1^{(i)} = 0; -v_{L_{calc-1}} \cdot \lambda + y_1 \cdot e^{-\lambda\cdot\Delta_1} = 0$$
$$y_1 \cdot \lambda \cdot e^{\lambda\cdot t} = -\frac{\eta_1}{\eta_3} \cdot (V_{L_{calc-1}}^{(i)} + v_{L_{calc-1}} \cdot e^{\lambda\cdot(t-\tau_1)}) - \frac{\eta_2}{\eta_3} \cdot (Y_1^{(i)} + y_1 \cdot e^{\lambda\cdot(t-\Delta_1)})$$
$$y_1 \cdot \lambda \cdot e^{\lambda\cdot t} = -\frac{\eta_1}{\eta_3} \cdot V_{L_{calc-1}}^{(i)} - \frac{\eta_2}{\eta_3} \cdot Y_1^{(i)} - \frac{\eta_1}{\eta_3} \cdot v_{L_{calc-1}} \cdot e^{\lambda\cdot(t-\tau_1)} - \frac{\eta_2}{\eta_3} \cdot y_1 \cdot e^{\lambda\cdot(t-\Delta_1)}$$
$$-\frac{\eta_1}{\eta_3} \cdot V_{L_{calc-1}}^{(i)} - \frac{\eta_2}{\eta_3} \cdot Y_1^{(i)} = 0; -y_1 \cdot \lambda - \frac{\eta_1}{\eta_3} \cdot v_{L_{calc-1}} \cdot e^{-\lambda\cdot\tau_1} - \frac{\eta_2}{\eta_3} \cdot y_1 \cdot e^{-\lambda\cdot\Delta_1} = 0$$
$$\frac{dV_{L_{calc-2}}}{dt} = Y_2; v_{L_{calc-2}} \cdot \lambda \cdot e^{\lambda\cdot t} = Y_2^{(i)} + y_2 \cdot e^{\lambda\cdot(t-\Delta_2)}; Y_2^{(i)} = 0; -v_{L_{calc-2}} \cdot \lambda + y_2 \cdot e^{-\lambda\cdot\Delta_2} = 0$$
$$y_2 \cdot \lambda \cdot e^{\lambda\cdot t} = -\frac{\xi_1}{\xi_3} \cdot (V_{L_{calc-2}}^{(i)} + v_{L_{calc-2}} \cdot e^{\lambda\cdot(t-\tau_2)}) - \frac{\xi_2}{\xi_3} \cdot Y_2^{(i)} - \frac{\xi_2}{\xi_3} \cdot y_2 \cdot e^{\lambda\cdot(t-\Delta_2)}$$
$$y_2 \cdot \lambda \cdot e^{\lambda\cdot t} = -\frac{\xi_1}{\xi_3} \cdot V_{L_{calc-2}}^{(i)} - \frac{\xi_2}{\xi_3} \cdot Y_2^{(i)} - \frac{\xi_1}{\xi_3} \cdot v_{L_{calc-2}} \cdot e^{\lambda\cdot(t-\tau_2)} - \frac{\xi_2}{\xi_3} \cdot y_2 \cdot e^{\lambda\cdot(t-\Delta_2)}$$
$$-\frac{\xi_1}{\xi_3} \cdot V_{L_{calc-2}}^{(i)} - \frac{\xi_2}{\xi_3} \cdot Y_2^{(i)} = 0; -y_2 \cdot \lambda - \frac{\xi_1}{\xi_3} \cdot v_{L_{calc-2}} \cdot e^{-\lambda\cdot\tau_2} - \frac{\xi_2}{\xi_3} \cdot y_2 \cdot e^{-\lambda\cdot\Delta_2} = 0$$

The small increments Jacobian of our N-turn multilayer circular coil antennas RFID system is as follow:

$$[-\lambda - \frac{\eta_2}{\eta_3} \cdot e^{-\lambda\cdot\Delta_1}] \cdot y_1 - \frac{\eta_1}{\eta_3} \cdot e^{-\lambda\cdot\tau_1} \cdot v_{L_{calc-1}} = 0; -v_{L_{calc-1}} \cdot \lambda + y_1 \cdot e^{-\lambda\cdot\Delta_1} = 0$$
$$[-\lambda - \frac{\xi_2}{\xi_3} \cdot e^{-\lambda\cdot\Delta_2}] \cdot y_2 - \frac{\xi_1}{\xi_3} \cdot e^{-\lambda\cdot\tau_2} \cdot v_{L_{calc-2}} = 0; -v_{L_{calc-2}} \cdot \lambda + y_2 \cdot e^{-\lambda\cdot\Delta_2} = 0$$
$$\begin{pmatrix} \iota_{11} & \cdots & \iota_{14} \\ \vdots & \ddots & \vdots \\ \iota_{41} & \cdots & \iota_{44} \end{pmatrix} \cdot \begin{pmatrix} y_1 \\ v_{L_{calc-1}} \\ y_2 \\ v_{L_{calc-2}} \end{pmatrix} = 0; \iota_{11} = -\lambda - \frac{\eta_2}{\eta_3} \cdot e^{-\lambda\cdot\Delta_1};$$
$$\iota_{12} = -\frac{\eta_1}{\eta_3} \cdot e^{-\lambda\cdot\tau_1}; \iota_{13} = \iota_{14} = 0$$

$$\iota_{21} = e^{-\lambda\cdot\Delta_1}; \iota_{22} = -\lambda; \iota_{23} = \iota_{24} = 0; \iota_{31} = \iota_{32} = 0; \iota_{33} = -\lambda - \frac{\xi_2}{\xi_3} \cdot e^{-\lambda\cdot\Delta_2}; \iota_{34} = -\frac{\xi_1}{\xi_3} \cdot e^{-\lambda\cdot\tau_2}$$

$$\iota_{41} = \iota_{42} = 0; \iota_{43} = e^{-\lambda\cdot\Delta_2}; \iota_{44} = -\lambda$$

$$A - \lambda \cdot I = \begin{pmatrix} \iota_{11} & \cdots & \iota_{14} \\ \vdots & \ddots & \vdots \\ \iota_{41} & \cdots & \iota_{44} \end{pmatrix}; \det|A - \lambda \cdot I| = 0; D = D(\lambda, \tau_1, \tau_2, \Delta_1, \Delta_2)$$

$$\Gamma_{33} = -\frac{\xi_2}{\xi_3}; \Gamma_{34} = -\frac{\xi_1}{\xi_3}; \Gamma_{11} = -\frac{\eta_2}{\eta_3}; \Gamma_{12} = -\frac{\eta_1}{\eta_3}; \iota_{33} = -\lambda + \Gamma_{33} \cdot e^{-\lambda \cdot \Delta_2};$$

$$\iota_{34} = \Gamma_{34} \cdot e^{-\lambda \cdot \tau_2}$$

$$\iota_{11} = -\lambda + \Gamma_{11} \cdot e^{-\lambda \cdot \Delta_1}; \iota_{12} = \Gamma_{12} \cdot e^{-\lambda \cdot \tau_1}$$

$$D(\lambda, \tau_1, \tau_2, \Delta_1, \Delta_2) = \lambda^4 - \lambda^3 \cdot \Gamma_{33} \cdot e^{-\lambda \cdot \Delta_2} - \lambda^2 \cdot \Gamma_{34} \cdot e^{-\lambda \cdot (\Delta_2 + \tau_2)} - \lambda^3 \cdot \Gamma_{11} \cdot e^{-\lambda \cdot \Delta_1}$$
$$+ \lambda^2 \cdot \Gamma_{11} \cdot \Gamma_{33} \cdot e^{-\lambda \cdot \sum_{i=1}^{2} \Delta_i} + \lambda \cdot \Gamma_{11} \cdot \Gamma_{34} \cdot e^{-\lambda \cdot (\tau_2 + \sum_{i=1}^{2} \Delta_i)} - \lambda^2 \cdot \Gamma_{12} \cdot e^{-\lambda \cdot (\tau_1 + \Delta_1)}$$
$$+ \lambda \cdot \Gamma_{12} \cdot \Gamma_{33} \cdot e^{-\lambda \cdot (\tau_1 + \sum_{i=1}^{2} \Delta_i)} + \Gamma_{12} \cdot \Gamma_{34} \cdot e^{-\lambda \cdot (\sum_{i=1}^{2} \tau_i + \sum_{i=1}^{2} \Delta_i)}$$

$$D(\lambda, \tau_1, \tau_2, \Delta_1, \Delta_2) = \lambda^4 + \Gamma_{12} \cdot \Gamma_{34} \cdot e^{-\lambda \cdot (\sum_{i=1}^{2} \tau_i + \sum_{i=1}^{2} \Delta_i)} + \lambda \cdot [\Gamma_{11} \cdot \Gamma_{34} \cdot e^{-\lambda \cdot (\tau_2 + \sum_{i=1}^{2} \Delta_i)}$$
$$+ \Gamma_{12} \cdot \Gamma_{33} \cdot e^{-\lambda \cdot (\tau_1 + \sum_{i=1}^{2} \Delta_i)}] + \lambda^2 \cdot [\Gamma_{11} \cdot \Gamma_{33} \cdot e^{-\lambda \cdot \sum_{i=1}^{2} \Delta_i} - \Gamma_{12} \cdot e^{-\lambda \cdot (\tau_1 + \Delta_1)} - \Gamma_{34} \cdot e^{-\lambda \cdot (\Delta_2 + \tau_2)}]$$
$$- \lambda^3 \cdot [\Gamma_{33} \cdot e^{-\lambda \cdot \Delta_2} + \Gamma_{11} \cdot e^{-\lambda \cdot \Delta_1}]$$

We have three stability cases: (1) $\tau_1 = \tau_2 = \tau \;\&\; \Delta_1 = 0, \Delta_2 = 0$ (2) $\tau_1 = \tau_2 = 0 \;\&\; \Delta_1 = \Delta, \Delta_2 = \Delta$ (3) $\tau_1 = \tau_2 = \tau_\Delta \;\&\; \Delta_1 = \tau_\Delta, \Delta_2 = \tau_\Delta$. Otherwise $\tau_1 \neq \tau_2$ and $\Delta_1 \neq \Delta_2$, they are positive parameters. There are other possible simple stability cases: (4) $\tau_1 = \tau; \tau_2 = 0; \Delta_1 = \Delta_2 = 0$ (5) $\tau_1 = 0; \tau_2 = \tau; \Delta_1 = \Delta_2 = 0$ (6) $\tau_1 = 0; \tau_2 = 0; \Delta_1 = \Delta; \Delta_2 = 0$ (7) $\tau_1 = 0; \tau_2 = 0; \Delta_1 = 0; \Delta_2 = \Delta$.

We need to get characteristics equations for all above stability analysis cases. We study the occurrence of any possible stability switching resulting from the increase the value of time delays $\tau, \Delta, \tau_\Delta$ for the general characteristic equation $D(\lambda, \tau/\Delta/\tau_\Delta)$. If we choose τ parameter then $D(\lambda, \tau) = P_n(\lambda, \tau) + Q_m(\lambda, \tau) \cdot e^{-\lambda \cdot \tau}$. The expression for $P_n(\lambda, \tau)$ is

$$P_n(\lambda, \tau) = \sum_{k=0}^{n} p_k(\tau) \cdot \lambda^k = p_0(\tau) + p_1(\tau) \cdot \lambda + p_2(\tau) \cdot \lambda^2 + \ldots$$

The expression for $Q_m(\lambda, \tau)$ is

$$Q_m(\lambda, \tau) = \sum_{k=0}^{m} q_k(\tau) \cdot \lambda^k = q_0(\tau) + q_1(\tau) \cdot \lambda + q_2(\tau) \cdot \lambda^2 + \ldots$$

First we analyze the case when there is delay in first (i = 1) RFID's N-turn multilayer coil antenna voltage ($\tau_1 = \tau$) and there is no delay in second (i = 2) RFID's N-turn multilayer coil antenna voltage ($\tau_2 = \tau$). Additionally there is no delay in RFID's N-turn multilayer coil antenna voltages derivatives ($\Delta_1 = 0, \Delta_2 = 0$).

The general characteristic equation $D(\lambda, \tau)$ is as follow:

$$D(\lambda, \tau_1 = \tau, \tau_2 = 0, \Delta_{1=0}, \Delta_2 = 0) = \lambda^4 + \Gamma_{12} \cdot \Gamma_{34} \cdot e^{-\lambda \cdot \tau} + \lambda \cdot [\Gamma_{11} \cdot \Gamma_{34} + \Gamma_{12} \cdot \Gamma_{33} \cdot e^{-\lambda \cdot \tau}] + \lambda^2 \cdot [\Gamma_{11} \cdot \Gamma_{33} - \Gamma_{34} - \Gamma_{12} \cdot e^{-\lambda \cdot \tau}] - \lambda^3 \cdot [\Gamma_{33} + \Gamma_{11}]$$

$$D(\lambda, \tau_1 = \tau, \tau_2 = 0, \Delta_{1=0}, \Delta_2 = 0) = \lambda^4 - \lambda^3 \cdot [\Gamma_{33} + \Gamma_{11}] + \lambda^2 \cdot (\Gamma_{11} \cdot \Gamma_{33} - \Gamma_{34}) + \lambda \cdot \Gamma_{11} \cdot \Gamma_{34} + (-\lambda^2 \cdot \Gamma_{12} + \lambda \cdot \Gamma_{12} \cdot \Gamma_{33} + \Gamma_{12} \cdot \Gamma_{34}) \cdot e^{-\lambda \cdot \tau}$$

$$D(\lambda, \tau) = P_n(\lambda, \tau) + Q_m(\lambda, \tau) \cdot e^{-\lambda \cdot \tau}; n = 4; m = 2; n > m$$

The expression for $P_n(\lambda, \tau)$:

$$P_{n=4}(\lambda, \tau) = \sum_{k=0}^{n=4} p_k(\tau) \cdot \lambda^k = p_0(\tau) + p_1(\tau) \cdot \lambda + p_2(\tau) \cdot \lambda^2 + p_3(\tau) \cdot \lambda^3 + p_4(\tau) \cdot \lambda^4$$

$$p_0(\tau) = 0; p_1(\tau) = \Gamma_{11} \cdot \Gamma_{34}; p_2(\tau) = \Gamma_{11} \cdot \Gamma_{33} - \Gamma_{34}; p_3(\tau) = -[\Gamma_{33} + \Gamma_{11}]; p_4(\tau) = 1$$

The expression for $Q_m(\lambda, \tau)$:

$$Q_{m=2}(\lambda, \tau) = \sum_{k=0}^{m=2} q_k(\tau) \cdot \lambda^k = q_0(\tau) + q_1(\tau) \cdot \lambda + q_2(\tau) \cdot \lambda^2$$

$q_0(\tau) = \Gamma_{12} \cdot \Gamma_{34}; q_1(\tau) = \Gamma_{12} \cdot \Gamma_{33}; q_2(\tau) = -\Gamma_{12}$. The homogeneous system for $Y_1 V_{L_{calc-1}} Y_2 V_{L_{calc-2}}$ leads to a characteristic equation for the eigenvalue λ having the form $P(\lambda) + Q(\lambda) \cdot e^{-\lambda \cdot \tau}; P(\lambda) = \sum_{j=0}^{4} a_j \cdot \lambda^j; Q(\lambda) = \sum_{j=0}^{2} c_j \cdot \lambda^j$ and the coefficients $\{a_j(q_i, q_k), c_j(q_i, q_k)\} \in \mathbb{R}$ depend on q_i, q_k but not on τ, q_i, q_k are any RFID's N-turn multilayer coil antennas system's parameters, other parameters keep as a constant.

$$a_0 = 0; a_1 = \Gamma_{11} \cdot \Gamma_{34}; a_2 = \Gamma_{11} \cdot \Gamma_{33} - \Gamma_{34}; a_3 = -(\Gamma_{33} + \Gamma_{11}); a_4 = 1;$$
$$c_0 = \Gamma_{12} \cdot \Gamma_{34}; c_1 = \Gamma_{33} \cdot \Gamma_{12}; c_2 = -\Gamma_{12}.$$

Unless strictly necessary, the designation of the variation arguments (q_i, q_k) will subsequently be omitted from P, Q, a_j, and c_j. The coefficients a_j, and c_j are continuous, and differential functions of their arguments, and direct substitution shows that $a_0 + c_0 \neq 0 \; \forall \; q_i, q_k \in \mathbb{R}_+$ $(\Gamma_{12} \cdot \Gamma_{34} \neq 0)$. $\lambda = 0$ is not a of $P(\lambda) + Q(\lambda) \cdot e^{-\lambda \cdot \tau} = 0$. Furthermore, $P(\lambda), Q(\lambda)$ are analytic functions of λ, for which the following requirements of the analysis [BK] can also be verified in the present case [4–6]:

(a) If $\lambda = i \cdot \omega, \omega \in \mathbb{R}$, then $P(i \cdot \omega) + Q(i \cdot \omega) \neq 0$.
(b) $|\frac{Q(\lambda)}{P(\lambda)}|$ is bounded for $|\lambda| \to \infty$, $\mathrm{Re}\lambda \geq 0$. No roots bifurcation from ∞.
(c) $F(\omega) = |P(i \cdot \omega)|^2 - |Q(i \cdot \omega)|^2$ has a finite number of zeros. Indeed, this is a polynomial in ω.
(d) Each positive root $\omega(q_i, q_k)$ of $F(\omega) = 0$ is continuous and differentiable respect to q_i, q_k.

We assume that $P_n(\lambda, \tau) = P_n(\lambda)$ and $Q_m(\lambda, \tau) = Q_m(\lambda)$ cannot have common imaginary roots. That is for any real number ω: $P_n(\lambda = i \cdot \omega, \tau) + Q_m(\lambda = i \cdot \omega, \tau) \neq 0$.

$$P_n(\lambda = i \cdot \omega, \tau) = (\Gamma_{11} \cdot \Gamma_{34} \cdot \omega + [\Gamma_{33} + \Gamma_{11}] \cdot \omega^3) \cdot i - (\Gamma_{11} \cdot \Gamma_{33} - \Gamma_{34}) \cdot \omega^2 + \omega^4$$
$$Q_m(\lambda = i \cdot \omega, \tau) = \Gamma_{12} \cdot \Gamma_{34} + \Gamma_{12} \cdot \omega^2 + \Gamma_{12} \cdot \Gamma_{33} \cdot i \cdot \omega$$

$$\begin{aligned} P_n(\lambda = i \cdot \omega, \tau) + Q_m(\lambda = i \cdot \omega, \tau) &= \Gamma_{12} \cdot \Gamma_{34} + (\Gamma_{12} - \Gamma_{11} \cdot \Gamma_{33} + \Gamma_{34}) \cdot \omega^2 + \omega^4 \\ &+ (\Gamma_{11} \cdot \Gamma_{34} + \Gamma_{12} \cdot \Gamma_{33}) \cdot i \cdot \omega + i \cdot [\Gamma_{33} + \Gamma_{11}] \cdot \omega^3 \neq 0 \end{aligned}$$

$$\begin{aligned} |P_n(\lambda = i \cdot \omega, \tau)|^2 &= (\Gamma_{11} \cdot \Gamma_{34} \cdot \omega + [\Gamma_{33} + \Gamma_{11}] \cdot \omega^3)^2 + (-(\Gamma_{11} \cdot \Gamma_{33} - \Gamma_{34}) \cdot \omega^2 + \omega^4)^2 \\ &= \Gamma_{11}^2 \cdot \Gamma_{34}^2 \cdot \omega^2 + [\Gamma_{33} + \Gamma_{11}]^2 \cdot \omega^6 + 2 \cdot \Gamma_{11} \cdot \Gamma_{34} \cdot [\Gamma_{33} + \Gamma_{11}] \cdot \omega^4 \\ &+ (\Gamma_{11} \cdot \Gamma_{33} - \Gamma_{34})^2 \cdot \omega^4 + \omega^8 - 2 \cdot (\Gamma_{11} \cdot \Gamma_{33} - \Gamma_{34}) \cdot \omega^6 \end{aligned}$$

$$\begin{aligned} |P_n(\lambda = i \cdot \omega, \tau)|^2 = |P(i \cdot \omega)|^2 &= \omega^8 + \{[\Gamma_{33} + \Gamma_{11}]^2 - 2 \cdot (\Gamma_{11} \cdot \Gamma_{33} - \Gamma_{34})\} \cdot \omega^6 \\ &+ (2 \cdot \Gamma_{11} \cdot \Gamma_{34} \cdot [\Gamma_{33} + \Gamma_{11}] + (\Gamma_{11} \cdot \Gamma_{33} - \Gamma_{34})^2) \cdot \omega^4 + \Gamma_{11}^2 \cdot \Gamma_{34}^2 \cdot \omega^2 \end{aligned}$$

$$\begin{aligned} |Q_m(\lambda = i \cdot \omega, \tau)|^2 &= (\Gamma_{12} \cdot \Gamma_{34} + \Gamma_{12} \cdot \omega^2)^2 + \Gamma_{12}^2 \cdot \Gamma_{33}^2 \cdot \omega^2 = \Gamma_{12}^2 \cdot \Gamma_{34}^2 + \Gamma_{12}^2 \cdot \omega^4 \\ &+ 2 \cdot \Gamma_{12}^2 \cdot \Gamma_{34} \cdot \omega^2 + \Gamma_{12}^2 \cdot \Gamma_{33}^2 \cdot \omega^2 \end{aligned}$$

$$|Q_m(\lambda = i \cdot \omega, \tau)|^2 = |Q(i \cdot \omega)|^2 = \Gamma_{12}^2 \cdot \Gamma_{34}^2 + (2 \cdot \Gamma_{34} + \Gamma_{33}^2) \cdot \Gamma_{12}^2 \cdot \omega^2 + \Gamma_{12}^2 \cdot \omega^4$$

$$\begin{aligned} F(\omega) = |P(i \cdot \omega)|^2 - |Q(i \cdot \omega)|^2 &= \omega^8 + \{[\Gamma_{33} + \Gamma_{11}]^2 - 2 \cdot (\Gamma_{11} \cdot \Gamma_{33} - \Gamma_{34})\} \cdot \omega^6 \\ &+ (2 \cdot \Gamma_{11} \cdot \Gamma_{34} \cdot [\Gamma_{33} + \Gamma_{11}] + (\Gamma_{11} \cdot \Gamma_{33} - \Gamma_{34})^2) \cdot \omega^4 + \Gamma_{11}^2 \cdot \Gamma_{34}^2 \cdot \omega^2 - \Gamma_{12}^2 \cdot \Gamma_{34}^2 \\ &- (2 \cdot \Gamma_{34} + \Gamma_{33}^2) \cdot \Gamma_{12}^2 \cdot \omega^2 - \Gamma_{12}^2 \cdot \omega^4 \end{aligned}$$

$$\begin{aligned} F(\omega) = |P(i \cdot \omega)|^2 - |Q(i \cdot \omega)|^2 &= \omega^8 + \{[\Gamma_{33} + \Gamma_{11}]^2 - 2 \cdot (\Gamma_{11} \cdot \Gamma_{33} - \Gamma_{34})\} \cdot \omega^6 \\ &+ \{(2 \cdot \Gamma_{11} \cdot \Gamma_{34} \cdot [\Gamma_{33} + \Gamma_{11}] + (\Gamma_{11} \cdot \Gamma_{33} - \Gamma_{34})^2) - \Gamma_{12}^2\} \cdot \omega^4 \\ &+ \{\Gamma_{11}^2 \cdot \Gamma_{34}^2 - (2 \cdot \Gamma_{34} + \Gamma_{33}^2) \cdot \Gamma_{12}^2\} \cdot \omega^2 - \Gamma_{12}^2 \cdot \Gamma_{34}^2 \end{aligned}$$

We define the following parameters for simplicity: $\Xi_0 = -\Gamma_{12}^2 \cdot \Gamma_{34}^2$

$$\Xi_2 = \Gamma_{11}^2 \cdot \Gamma_{34}^2 - (2 \cdot \Gamma_{34} + \Gamma_{33}^2) \cdot \Gamma_{12}^2; \Xi_4 = 2 \cdot \Gamma_{11} \cdot \Gamma_{34} \cdot [\Gamma_{33} + \Gamma_{11}] + (\Gamma_{11} \cdot \Gamma_{33} - \Gamma_{34})^2 - \Gamma_{12}^2$$

$$\Xi_6 = [\Gamma_{33} + \Gamma_{11}]^2 - 2 \cdot (\Gamma_{11} \cdot \Gamma_{33} - \Gamma_{34}); \Xi_8 = 1$$

$$F(\omega) = |P(i \cdot \omega)|^2 - |Q(i \cdot \omega)|^2 = \Xi_8 \cdot \omega^8 + \Xi_6 \cdot \omega^6 + \Xi_4 \cdot \omega^4 + \Xi_2 \cdot \omega^2 + \Xi_0 = \sum_{k=0}^{4} \Xi_{2 \cdot k} \cdot \omega^{2 \cdot k}$$

Hence $F(\omega) = 0$ implies $\sum_{k=0}^{4} \Xi_{2 \cdot k} \cdot \omega^{2 \cdot k} = 0$ and its roots are given by solving the above polynomial. Furthermore $P_R(i \cdot \omega, \tau) = -(\Gamma_{11} \cdot \Gamma_{33} - \Gamma_{34}) \cdot \omega^2 + \omega^4$

$$P_I(i \cdot \omega, \tau) = \Gamma_{11} \cdot \Gamma_{34} \cdot \omega + [\Gamma_{33} + \Gamma_{11}] \cdot \omega^3; Q_R(i \cdot \omega, \tau) = \Gamma_{12} \cdot \Gamma_{34} + \Gamma_{12} \cdot \omega^2; Q_I(i \cdot \omega, \tau) = \Gamma_{12} \cdot \Gamma_{33} \cdot \omega$$

Hence

$$\sin \theta(\tau) = \frac{-P_R(i \cdot \omega, \tau) \cdot Q_I(i \cdot \omega, \tau) + P_I(i \cdot \omega, \tau) \cdot Q_R(i \cdot \omega, \tau)}{|Q(i \cdot \omega, \tau)|^2}$$

And

$$\cos \theta(\tau) = -\frac{P_R(i \cdot \omega, \tau) \cdot Q_R(i \cdot \omega, \tau) + P_I(i \cdot \omega, \tau) \cdot Q_I(i \cdot \omega, \tau)}{|Q(i \cdot \omega, \tau)|^2}$$

$$\sin \theta(\tau) = \frac{\begin{array}{c} -\{-(\Gamma_{11} \cdot \Gamma_{33} - \Gamma_{34}) \cdot \omega^2 + \omega^4\} \cdot \Gamma_{12} \cdot \Gamma_{33} \cdot \omega \\ + \{\Gamma_{11} \cdot \Gamma_{34} \cdot \omega + [\Gamma_{33} + \Gamma_{11}] \cdot \omega^3\} \cdot \{\Gamma_{12} \cdot \Gamma_{34} + \Gamma_{12} \cdot \omega^2\} \end{array}}{\Gamma_{12}^2 \cdot \Gamma_{34}^2 + \Gamma_{12}^2 \cdot \omega^4 + 2 \cdot \Gamma_{12}^2 \cdot \Gamma_{34} \cdot \omega^2 + \Gamma_{12}^2 \cdot \Gamma_{33}^2 \cdot \omega^2}$$

$$\cos \theta(\tau) = -\frac{\begin{array}{c} \{-(\Gamma_{11} \cdot \Gamma_{33} - \Gamma_{34}) \cdot \omega^2 + \omega^4\} \cdot \{\Gamma_{12} \cdot \Gamma_{34} + \Gamma_{12} \cdot \omega^2\} \\ + \{\Gamma_{11} \cdot \Gamma_{34} \cdot \omega + [\Gamma_{33} + \Gamma_{11}] \cdot \omega^3\} \cdot \Gamma_{12} \cdot \Gamma_{33} \cdot \omega \end{array}}{\Gamma_{12}^2 \cdot \Gamma_{34}^2 + \Gamma_{12}^2 \cdot \omega^4 + 2 \cdot \Gamma_{12}^2 \cdot \Gamma_{34} \cdot \omega^2 + \Gamma_{12}^2 \cdot \Gamma_{33}^2 \cdot \omega^2}$$

Which jointly with $F(\omega) = 0 \Rightarrow \sum_{k=0}^{4} \Xi_{2 \cdot k} \cdot \omega^{2 \cdot k} = 0$ that are continuous and differentiable in τ based on Lemma 1.1. Hence we use Theorem 1.2 and this prove the Theorem 1.3.

The second case we analyze is when there is a delay in first (i = 1) and second (i = 2) RFID's N-turn multilayer coil antennas voltages ($\tau_1 = \tau; \tau_2 = \tau$) and no delay in RFID's N-turn multilayer coil antennas voltages derivatives $\Delta_1 = 0$ $\Delta_2 = 0$. The general characteristic equation $D(\lambda, \tau)$ is as follow:

$$D(\lambda, \tau_1, \tau_2, \Delta_1, \Delta_2) = \lambda^4 + \Gamma_{12} \cdot \Gamma_{34} \cdot e^{-\lambda \cdot (\sum_{i=1}^{2} \tau_i + \sum_{i=1}^{2} \Delta_i)} + \lambda \cdot [\Gamma_{11} \cdot \Gamma_{34} \cdot e^{-\lambda \cdot (\tau_2 + \sum_{i=1}^{2} \Delta_i)} + \Gamma_{12} \cdot \Gamma_{33} \cdot e^{-\lambda \cdot (\tau_1 + \sum_{i=1}^{2} \Delta_i)}] + \lambda^2 \cdot [\Gamma_{11} \cdot \Gamma_{33} \cdot e^{-\lambda \cdot \sum_{i=1}^{2} \Delta_i} - \Gamma_{12} \cdot e^{-\lambda \cdot (\tau_1 + \Delta_1)} - \Gamma_{34} \cdot e^{-\lambda \cdot (\Delta_2 + \tau_2)}] - \lambda^3 \cdot [\Gamma_{33} \cdot e^{-\lambda \cdot \Delta_2} + \Gamma_{11} \cdot e^{-\lambda \cdot \Delta_1}]$$

$$D(\lambda, \tau_1 = \tau, \tau_2 = \tau, \Delta_1 = 0, \Delta_2 = 0) = \lambda^4 + \Gamma_{12} \cdot \Gamma_{34} \cdot e^{-\lambda \cdot 2 \cdot \tau} + \lambda \cdot [\Gamma_{11} \cdot \Gamma_{34} \cdot e^{-\lambda \cdot \tau} + \Gamma_{12} \cdot \Gamma_{33} \cdot e^{-\lambda \cdot \tau}] + \lambda^2 \cdot [\Gamma_{11} \cdot \Gamma_{33} - \Gamma_{12} \cdot e^{-\lambda \cdot \tau} - \Gamma_{34} \cdot e^{-\lambda \cdot \tau}] - \lambda^3 \cdot [\Gamma_{33} + \Gamma_{11}]$$

$$D(\lambda, \tau_1 = \tau, \tau_2 = \tau, \Delta_1 = 0, \Delta_2 = 0) = \lambda^4 - \lambda^3 \cdot [\Gamma_{33} + \Gamma_{11}] + \lambda^2 \cdot \Gamma_{11} \cdot \Gamma_{33} + (\Gamma_{12} \cdot \Gamma_{34} \cdot e^{-\lambda \cdot \tau} + \lambda \cdot [\Gamma_{11} \cdot \Gamma_{34} + \Gamma_{12} \cdot \Gamma_{33}] - \lambda^2 \cdot [\Gamma_{12} + \Gamma_{34}]) \cdot e^{-\lambda \cdot \tau}$$

Under Taylor series approximation: $e^{-\lambda \cdot \tau} \approx 1 - \lambda \cdot \tau + \frac{1}{2} \cdot \lambda^2 \cdot \tau^2$. The Maclaurin series is a Taylor series expansion of a $\exp(-\lambda \cdot \tau)$ function about zero (0). We get the following general characteristic equation D(λ,τ) under Taylor series approximation: $e^{-\lambda \cdot \tau} \approx 1 - \lambda \cdot \tau + \frac{1}{2} \cdot \lambda^2 \cdot \tau^2$.

$$D(\lambda, \tau_1 = \tau, \tau_2 = \tau, \Delta_1 = 0, \Delta_2 = 0) = \lambda^4 - \lambda^3 \cdot [\Gamma_{33} + \Gamma_{11}] + \lambda^2 \cdot \Gamma_{11} \cdot \Gamma_{33} + (\Gamma_{12} \cdot \Gamma_{34} \cdot [1 - \lambda \cdot \tau + \frac{1}{2} \cdot \lambda^2 \cdot \tau^2] + \lambda \cdot [\Gamma_{11} \cdot \Gamma_{34} + \Gamma_{12} \cdot \Gamma_{33}] - \lambda^2 \cdot [\Gamma_{12} + \Gamma_{34}]) \cdot e^{-\lambda \cdot \tau}$$

$$D(\lambda, \tau_1 = \tau, \tau_2 = \tau, \Delta_1 = 0, \Delta_2 = 0) = \lambda^4 - \lambda^3 \cdot [\Gamma_{33} + \Gamma_{11}] + \lambda^2 \cdot \Gamma_{11} \cdot \Gamma_{33} + \{\lambda^2 \cdot (\frac{1}{2} \cdot \Gamma_{12} \cdot \Gamma_{34} \cdot \tau^2 - \Gamma_{12} - \Gamma_{34}) + \lambda \cdot (\Gamma_{11} \cdot \Gamma_{34} + \Gamma_{12} \cdot \Gamma_{33} - \Gamma_{12} \cdot \Gamma_{34} \cdot \tau) + \Gamma_{12} \cdot \Gamma_{34}\} \cdot e^{-\lambda \cdot \tau}$$

$$D(\lambda, \tau) = P_n(\lambda, \tau) + Q_m(\lambda, \tau) \cdot e^{-\lambda \cdot \tau}; n = 4; m = 2; n > m.$$

The expression for $P_n(\lambda, \tau)$:

$$P_n(\lambda, \tau) = \lambda^4 - \lambda^3 \cdot [\Gamma_{33} + \Gamma_{11}] + \lambda^2 \cdot \Gamma_{11} \cdot \Gamma_{33}; P_{n=4}(\lambda, \tau) = \sum_{k=0}^{n=4} p_k(\tau) \cdot \lambda^k$$

$$\sum_{k=0}^{n=4} p_k(\tau) \cdot \lambda^k = p_0(\tau) + p_1(\tau) \cdot \lambda + p_2(\tau) \cdot \lambda^2 + p_3(\tau) \cdot \lambda^3 + p_4(\tau) \cdot \lambda^4$$

$$p_0(\tau) = 0; p_1(\tau) = 0; p_2(\tau) = \Gamma_{11} \cdot \Gamma_{33}; p_3(\tau) = -[\Gamma_{33} + \Gamma_{11}]; p_4(\tau) = 1$$

$$Q_{m=2}(\lambda, \tau) = \lambda^2 \cdot (\frac{1}{2} \cdot \Gamma_{12} \cdot \Gamma_{34} \cdot \tau^2 - \Gamma_{12} - \Gamma_{34}) + \lambda \cdot (\Gamma_{11} \cdot \Gamma_{34} + \Gamma_{12} \cdot \Gamma_{33} - \Gamma_{12} \cdot \Gamma_{34} \cdot \tau) + \Gamma_{12} \cdot \Gamma_{34}$$

$$Q_{m=2}(\lambda, \tau) = \sum_{k=0}^{k=2} q_k(\tau) \cdot \lambda^k = q_0 + q_1 \cdot \lambda + q_2 \cdot \lambda^2; q_0(\tau) = \Gamma_{12} \cdot \Gamma_{34}$$

$$q_2(\tau) = \frac{1}{2} \cdot \Gamma_{12} \cdot \Gamma_{34} \cdot \tau^2 - \Gamma_{12} - \Gamma_{34}; q_1(\tau) = \Gamma_{11} \cdot \Gamma_{34} + \Gamma_{12} \cdot \Gamma_{33} - \Gamma_{12} \cdot \Gamma_{34} \cdot \tau$$

The homogeneous system for $Y_1 V_{L_{calc-1}} Y_2 V_{L_{calc-2}}$ leads to characteristic equation for the eigenvalue λ having the form $P(\lambda, \tau) + Q(\lambda, \tau) \cdot e^{-\lambda \cdot \tau}$, $P(\lambda, \tau) = \sum_{j=0}^{4} a_j(\tau) \cdot \lambda^j Q(\lambda) = \sum_{j=0}^{2} c_j(\tau) \cdot \lambda^j$ and the coefficients $\{a_j(q_i, q_k, \tau), c_j(q_i, q_k, \tau)\} \in \mathbb{R}$ depend on q_i, q_k and delay τ, q_i, q_k are any RFID's N-turn multilayer coil antennas system's parameters, other parameters keep as a constant.

$$a_0 = 0; a_1 = 0; a_2 = \Gamma_{11} \cdot \Gamma_{33}; a_3 = -[\Gamma_{33} + \Gamma_{11}]; a_4 = 1; c_0 = \Gamma_{12} \cdot \Gamma_{34}$$
$$c_1 = \Gamma_{11} \cdot \Gamma_{34} + \Gamma_{12} \cdot \Gamma_{33} - \Gamma_{12} \cdot \Gamma_{34} \cdot \tau; c_2 = \frac{1}{2} \cdot \Gamma_{12} \cdot \Gamma_{34} \cdot \tau^2 - \Gamma_{34} - \Gamma_{12}$$

Unless strictly necessary, the designation of the variation arguments (q_i, q_k, τ) will subsequently be omitted from P, Q, a_j, and c_j. The coefficients a_j, c_j are continuous and differentiable functions of their arguments, and direct substitution shows that $a_0 + c_0 \neq 0; \Gamma_{12} \cdot \Gamma_{34} \neq 0 \; \forall \, q_i, q_k \in \mathbb{R}_+$, i.e. $\lambda = 0$ is not a $P(\lambda, \tau) + Q(\lambda, \tau) \cdot e^{-\lambda \cdot \tau}$. We assume that $P_n(\lambda, \tau)$ and $Q_m(\lambda, \tau)$ can't have common imaginary roots. That is for any real number ω: $P_n(\lambda = i \cdot \omega, \tau) + Q_m(\lambda = i \cdot \omega, \tau) \neq 0$.

$$P_n(\lambda = i \cdot \omega, \tau) = \omega^4 - \omega^2 \cdot \Gamma_{11} \cdot \Gamma_{33} + i \cdot \omega^3 \cdot [\Gamma_{33} + \Gamma_{11}]$$
$$Q_m(\lambda = i \cdot \omega, \tau) = -\omega^2 \cdot (\frac{1}{2} \cdot \Gamma_{12} \cdot \Gamma_{34} \cdot \tau^2 - \Gamma_{12} - \Gamma_{34}) + \Gamma_{12} \cdot \Gamma_{34} + i \cdot \omega \cdot (\Gamma_{11} \cdot \Gamma_{34} + \Gamma_{12} \cdot \Gamma_{33} - \Gamma_{12} \cdot \Gamma_{34} \cdot \tau)$$

$$P_n(\lambda = i \cdot \omega, \tau) + Q_m(\lambda = i \cdot \omega, \tau) = \omega^4 - \omega^2 \cdot [\frac{1}{2} \cdot \Gamma_{12} \cdot \Gamma_{34} \cdot \tau^2 - \Gamma_{12} - \Gamma_{34} + \Gamma_{11} \cdot \Gamma_{33}] + \Gamma_{12} \cdot \Gamma_{34} + i \cdot \{\omega^3 \cdot [\Gamma_{33} + \Gamma_{11}] + \omega \cdot (v_{11} \cdot \Gamma_{34} + \Gamma_{12} \cdot \Gamma_{33} - \Gamma_{12} \cdot \Gamma_{34} \cdot \tau)\} \neq 0$$

$$|P_n(\lambda = i \cdot \omega, \tau)|^2 = |P(i \cdot \omega)|^2 = (\omega^4 - \omega^2 \cdot \Gamma_{11} \cdot \Gamma_{33})^2 + \omega^6 \cdot [\Gamma_{33} + \Gamma_{11}]^2 = \omega^8 + \omega^6 \cdot ([\Gamma_{33} + \Gamma_{11}]^2 - 2 \cdot \Gamma_{11} \cdot \Gamma_{33}) + \omega^4 \cdot \Gamma_{11}^2 \cdot \Gamma_{33}^2$$

$$|P_n(\lambda = i \cdot \omega, \tau)|^2 = |P(i \cdot \omega)|^2 = \omega^8 + \omega^6 \cdot ([\Gamma_{33} + \Gamma_{11}]^2 - 2 \cdot \Gamma_{11} \cdot \Gamma_{33}) + \omega^4 \cdot \Gamma_{11}^2 \cdot \Gamma_{33}^2$$

$$|Q_m(\lambda = i \cdot \omega, \tau)|^2 = |Q(i \cdot \omega)|^2 = (-\omega^2 \cdot (\frac{1}{2} \cdot \Gamma_{12} \cdot \Gamma_{34} \cdot \tau^2 - \Gamma_{12} - \Gamma_{34}) + \Gamma_{12} \cdot \Gamma_{34})^2 + \omega^2 \cdot (\Gamma_{11} \cdot \Gamma_{34} + \Gamma_{12} \cdot \Gamma_{33} - \Gamma_{12} \cdot \Gamma_{34} \cdot \tau)^2$$

$$|Q_m(\lambda = i \cdot \omega, \tau)|^2 = |Q(i \cdot \omega)|^2 = \omega^4 \cdot (\frac{1}{2} \cdot \Gamma_{12} \cdot \Gamma_{34} \cdot \tau^2 - \Gamma_{12} - \Gamma_{34})^2 + \omega^2 \cdot \{(\Gamma_{11} \cdot \Gamma_{34} + \Gamma_{12} \cdot \Gamma_{33} - \Gamma_{12} \cdot \Gamma_{34} \cdot \tau)^2 - 2 \cdot (\frac{1}{2} \cdot \Gamma_{12} \cdot \Gamma_{34} \cdot \tau^2 - \Gamma_{12} - \Gamma_{34}) \cdot \Gamma_{12} \cdot \Gamma_{34}\} + \Gamma_{12}^2 \cdot \Gamma_{34}^2$$

We need to find the expression for $F(\omega) = |P(i \cdot \omega)|^2 - |Q(i \cdot \omega)|^2$

$$F(\omega) = |P(i \cdot \omega)|^2 - |Q(i \cdot \omega)|^2 = \omega^8 + \omega^6 \cdot ([\Gamma_{33} + \Gamma_{11}]^2 - 2 \cdot \Gamma_{11} \cdot \Gamma_{33}) + \omega^4 \cdot \{\Gamma_{11}^2 \cdot \Gamma_{33}^2 - (\frac{1}{2} \cdot \Gamma_{12} \cdot \Gamma_{34} \cdot \tau^2 - \Gamma_{12} - \Gamma_{34})^2\} - \omega^2 \cdot \{(\Gamma_{11} \cdot \Gamma_{34} + \Gamma_{12} \cdot \Gamma_{33} - \Gamma_{12} \cdot \Gamma_{34} \cdot \tau)^2 - 2 \cdot (\frac{1}{2} \cdot \Gamma_{12} \cdot \Gamma_{34} \cdot \tau^2 - \Gamma_{12} - \Gamma_{34}) \cdot \Gamma_{12} \cdot \Gamma_{34}\} - \Gamma_{12}^2 \cdot \Gamma_{34}^2$$

We define the following parameters for simplicity:

$$\Xi_0 = -\Gamma_{12}^2 \cdot \Gamma_{34}^2; \Xi_2 = -(\Gamma_{11} \cdot \Gamma_{34} + \Gamma_{12} \cdot \Gamma_{33} - \Gamma_{12} \cdot \Gamma_{34} \cdot \tau)^2 + 2 \cdot (\frac{1}{2} \cdot \Gamma_{12} \cdot \Gamma_{34} \cdot \tau^2 - \Gamma_{12} - \Gamma_{34}) \cdot \Gamma_{12} \cdot \Gamma_{34}$$

$$\Xi_4 = \Gamma_{11}^2 \cdot \Gamma_{33}^2 - (\frac{1}{2} \cdot \Gamma_{12} \cdot \Gamma_{34} \cdot \tau^2 - \Gamma_{12} - \Gamma_{34})^2;$$

$$\Xi_6 = [\Gamma_{33} + \Gamma_{11}]^2 - 2 \cdot \Gamma_{11} \cdot \Gamma_{33}; \Xi_8 = 1$$

$$F(\omega, \tau) = |P(i \cdot \omega)|^2 - |Q(i \cdot \omega)|^2 = \sum_{k=0}^{4} \Xi_{2 \cdot k} \cdot \omega^{2 \cdot k} = \Xi_0 + \Xi_2 \cdot \omega^2 + \Xi_4 \cdot \omega^4 + \Xi_6 \cdot \omega^6 + \Xi_8 \cdot \omega^8$$

Hence $F(\omega, \tau) = 0$ implies $\sum_{k=0}^{4} \Xi_{2 \cdot k} \cdot \omega^{2 \cdot k} = 0$ and its roots are given by solving the above polynomial. Furthermore $P_R(i \cdot \omega, \tau) = \omega^4 - \omega^2 \cdot \Gamma_{11} \cdot \Gamma_{33}; P_I(i \cdot \omega, \tau) = \omega^3 \cdot [\Gamma_{33} + \Gamma_{11}]$

$$Q_R(i\cdot\omega,\tau) = -\omega^2\cdot(\frac{1}{2}\cdot\Gamma_{12}\cdot\Gamma_{34}\cdot\tau^2 - \Gamma_{12} - \Gamma_{34}) + \Gamma_{12}\cdot\Gamma_{34};$$

$$Q_I(i\cdot\omega,\tau) = \omega\cdot(\Gamma_{11}\cdot\Gamma_{34} + \Gamma_{12}\cdot\Gamma_{33} - \Gamma_{12}\cdot\Gamma_{34}\cdot\tau)$$

Hence

$$\sin\theta(\tau) = \frac{-P_R(i\cdot\omega,\tau)\cdot Q_I(i\cdot\omega,\tau) + P_I(i\cdot\omega,\tau)\cdot Q_R(i\cdot\omega,\tau)}{|Q(i\cdot\omega,\tau)|^2}$$

And

$$\cos\theta(\tau) = -\frac{P_R(i\cdot\omega,\tau)\cdot Q_R(i\cdot\omega,\tau) + P_I(i\cdot\omega,\tau)\cdot Q_I(i\cdot\omega,\tau)}{|Q(i\cdot\omega,\tau)|^2}$$

$$\sin\theta(\tau) = \frac{\begin{array}{c}-(\omega^4 - \omega^2\cdot\Gamma_{11}\cdot\Gamma_{33})\cdot[\omega\cdot(\Gamma_{11}\cdot\Gamma_{34} + \Gamma_{12}\cdot\Gamma_{33} - \Gamma_{12}\cdot\Gamma_{34}\cdot\tau)] \\ +\omega^3\cdot[\Gamma_{33}+\Gamma_{11}]\cdot[-\omega^2\cdot(\frac{1}{2}\cdot\Gamma_{12}\cdot\Gamma_{34}\cdot\tau^2 - \Gamma_{12} - \Gamma_{34}) + \Gamma_{12}\cdot\Gamma_{34}]\end{array}}{\begin{array}{c}\omega^4\cdot(\frac{1}{2}\cdot\Gamma_{12}\cdot\Gamma_{34}\cdot\tau^2 - \Gamma_{12} - \Gamma_{34})^2 + \omega^2\cdot\{(\Gamma_{11}\cdot\Gamma_{34} + \Gamma_{12}\cdot\Gamma_{33} - \Gamma_{12}\cdot\Gamma_{34}\cdot\tau)^2 \\ -2\cdot(\frac{1}{2}\cdot\Gamma_{12}\cdot\Gamma_{34}\cdot\tau^2 - \Gamma_{12} - \Gamma_{34})\cdot\Gamma_{12}\cdot\Gamma_{34}\} + \Gamma_{12}^2\cdot\Gamma_{34}^2\end{array}}$$

$$\cos\theta(\tau) = -\frac{\begin{array}{c}(\omega^4 - \omega^2\cdot\Gamma_{11}\cdot\Gamma_{33})\cdot[-\omega^2\cdot(\frac{1}{2}\cdot\Gamma_{12}\cdot\Gamma_{34}\cdot\tau^2 - \Gamma_{12} - \Gamma_{34}) + \Gamma_{12}\cdot\Gamma_{34}] \\ +\omega^4\cdot[\Gamma_{33}+\Gamma_{11}]\cdot(\Gamma_{11}\cdot\Gamma_{34} + \Gamma_{12}\cdot\Gamma_{33} - \Gamma_{12}\cdot\Gamma_{34}\cdot\tau)\end{array}}{\begin{array}{c}\omega^4\cdot(\frac{1}{2}\cdot\Gamma_{12}\cdot\Gamma_{34}\cdot\tau^2 - \Gamma_{12} - \Gamma_{34})^2 + \omega^2\cdot\{(\Gamma_{11}\cdot\Gamma_{34} + \Gamma_{12}\cdot\Gamma_{33} - \Gamma_{12}\cdot\Gamma_{34}\cdot\tau)^2 \\ -2\cdot(\frac{1}{2}\cdot\Gamma_{12}\cdot\Gamma_{34}\cdot\tau^2 - \Gamma_{12} - \Gamma_{34})\cdot\Gamma_{12}\cdot\Gamma_{34}\} + \Gamma_{12}^2\cdot\Gamma_{34}^2\end{array}}$$

Those are continuous and differentiable in τ based on Lemma 1.1. Hence we use Theorem 1.2 and this approve the Theorem 1.3.

The third case we analyze is when there is a delay in first (i = 1) and second (i = 2) RFID's N-turn multilayer coil antennas voltages and antennas voltages derivatives ($\tau_1 = \tau_2 = \tau_\Delta; \Delta_1 = \Delta_2 = \tau_\Delta$). The general characteristic equation $D(\lambda,\tau_\Delta)$ is as follow:

$$\begin{aligned} D(\lambda,\tau_1,\tau_2,\Delta_1,\Delta_2) &= \lambda^4 + \Gamma_{12}\cdot\Gamma_{34}\cdot e^{-\lambda\cdot(\sum_{i=1}^{2}\tau_i + \sum_{i=1}^{2}\Delta_i)} + \lambda\cdot[\Gamma_{11}\cdot\Gamma_{34}\cdot e^{-\lambda\cdot(\tau_2 + \sum_{i=1}^{2}\Delta_i)} \\ &+ \Gamma_{12}\cdot\Gamma_{33}\cdot e^{-\lambda\cdot(\tau_1 + \sum_{i=1}^{2}\Delta_i)}] + \lambda^2\cdot[\Gamma_{11}\cdot\Gamma_{33}\cdot e^{-\lambda\cdot\sum_{i=1}^{2}\Delta_i} - \Gamma_{12}\cdot e^{-\lambda\cdot(\tau_1+\Delta_1)} \\ &- \Gamma_{34}\cdot e^{-\lambda\cdot(\Delta_2+\tau_2)}] \\ &- \lambda^3\cdot[\Gamma_{33}\cdot e^{-\lambda\cdot\Delta_2} + \Gamma_{11}\cdot e^{-\lambda\cdot\Delta_1}] \end{aligned}$$

$$D(\lambda, \tau_1 = \tau_2 = \tau_\Delta; \Delta_1 = \Delta_2 = \tau_\Delta) = \lambda^4 + \Gamma_{12} \cdot \Gamma_{34} \cdot e^{-\lambda \cdot 4 \cdot \tau_\Delta} + \lambda \cdot [\Gamma_{11} \cdot \Gamma_{34} \cdot e^{-\lambda \cdot 3 \cdot \tau_\Delta} + \Gamma_{12} \cdot \Gamma_{33} \cdot e^{-\lambda \cdot 3 \cdot \tau_\Delta}] + \lambda^2 \cdot [\Gamma_{11} \cdot \Gamma_{33} \cdot e^{-\lambda \cdot 2 \cdot \tau_\Delta} - \Gamma_{12} \cdot e^{-\lambda \cdot 2 \cdot \tau_\Delta} - \Gamma_{34} \cdot e^{-\lambda \cdot 2 \cdot \tau_\Delta}] - \lambda^3 \cdot [\Gamma_{33} \cdot e^{-\lambda \cdot \tau_\Delta} + \Gamma_{11} \cdot e^{-\lambda \cdot \tau_\Delta}]$$

$$D(\lambda, \tau_1 = \tau_2 = \tau_\Delta; \Delta_1 = \Delta_2 = \tau_\Delta) = \lambda^4 + \Gamma_{12} \cdot \Gamma_{34} \cdot e^{-\lambda \cdot 4 \cdot \tau_\Delta} + \lambda \cdot [\Gamma_{11} \cdot \Gamma_{34} + \Gamma_{12} \cdot \Gamma_{33}] \cdot e^{-\lambda \cdot 3 \cdot \tau_\Delta} + \lambda^2 \cdot [\Gamma_{11} \cdot \Gamma_{33} - \Gamma_{12} - \Gamma_{34}] \cdot e^{-\lambda \cdot 2 \cdot \tau_\Delta} - \lambda^3 \cdot [\Gamma_{33} + \Gamma_{11}] \cdot e^{-\lambda \cdot \tau_\Delta}$$

$$D(\lambda, \tau_1 = \tau_2 = \tau_\Delta; \Delta_1 = \Delta_2 = \tau_\Delta) = \lambda^4 + (\Gamma_{12} \cdot \Gamma_{34} \cdot e^{-\lambda \cdot 3 \cdot \tau_\Delta} + \lambda \cdot [\Gamma_{11} \cdot \Gamma_{34} + \Gamma_{12} \cdot \Gamma_{33}] \cdot e^{-\lambda \cdot 2 \cdot \tau_\Delta} + \lambda^2 \cdot [\Gamma_{11} \cdot \Gamma_{33} - \Gamma_{12} - \Gamma_{34}] \cdot e^{-\lambda \cdot \tau_\Delta} - \lambda^3 \cdot [\Gamma_{33} + \Gamma_{11}]) \cdot e^{-\lambda \cdot \tau_\Delta}$$

The Maclaurin series is a Taylor series expansion of $e^{-\lambda \cdot 3 \cdot \tau_\Delta}, e^{-\lambda \cdot 2 \cdot \tau_\Delta}, e^{-\lambda \cdot \tau_\Delta}$ functions about zero (0). We get the following general characteristic equation $D(\lambda, \tau_\Delta)$ under Taylor series approximation:

$$e^{-\lambda \cdot 3 \cdot \tau_\Delta} \simeq 1 - \lambda \cdot 3 \cdot \tau_\Delta; e^{-\lambda \cdot 2 \cdot \tau_\Delta} \simeq 1 - \lambda \cdot 2 \cdot \tau_\Delta;$$
$$e^{-\lambda \cdot \tau_\Delta} \simeq 1 - \lambda \cdot \tau_\Delta.$$

$$D(\lambda, \tau_1 = \tau_2 = \tau_\Delta; \Delta_1 = \Delta_2 = \tau_\Delta) = \lambda^4 + (\Gamma_{12} \cdot \Gamma_{34} \cdot (1 - \lambda \cdot 3 \cdot \tau_\Delta) + \lambda \cdot [\Gamma_{11} \cdot \Gamma_{34} + \Gamma_{12} \cdot \Gamma_{33}] \cdot (1 - \lambda \cdot 2 \cdot \tau_\Delta) + \lambda^2 \cdot [\Gamma_{11} \cdot \Gamma_{33} - \Gamma_{12} - \Gamma_{34}] \cdot (1 - \lambda \cdot \tau_\Delta) - \lambda^3 \cdot [\Gamma_{33} + \Gamma_{11}]) \cdot e^{-\lambda \cdot \tau_\Delta}$$

$$D(\lambda, \tau_1 = \tau_2 = \tau_\Delta; \Delta_1 = \Delta_2 = \tau_\Delta) = \lambda^4 + (\Gamma_{12} \cdot \Gamma_{34} + \lambda \cdot (\Gamma_{11} \cdot \Gamma_{34} + \Gamma_{12} \cdot \Gamma_{33} - \Gamma_{12} \cdot \Gamma_{34} \cdot 3 \cdot \tau_\Delta) + \lambda^2 \cdot (\Gamma_{11} \cdot \Gamma_{33} - \Gamma_{12} - \Gamma_{34} - [\Gamma_{11} \cdot \Gamma_{34} + \Gamma_{12} \cdot \Gamma_{33}] \cdot 2 \cdot \tau_\Delta) - \lambda^3 \cdot \{[\Gamma_{11} \cdot \Gamma_{33} - \Gamma_{12} - \Gamma_{34}] \cdot \tau_\Delta + \Gamma_{33} + \Gamma_{11}\}) \cdot e^{-\lambda \cdot \tau_\Delta}$$

$$D(\lambda, \tau_1 = \tau_2 = \tau_\Delta; \Delta_1 = \Delta_2 = \tau_\Delta) = \lambda^4 + (\Gamma_{12} \cdot \Gamma_{34} + \lambda \cdot (\Gamma_{11} \cdot \Gamma_{34} + \Gamma_{12} \cdot \Gamma_{33} - \Gamma_{12} \cdot \Gamma_{34} \cdot 3 \cdot \tau_\Delta) + \lambda^2 \cdot (\Gamma_{11} \cdot \Gamma_{33} - \Gamma_{12} - \Gamma_{34} - [\Gamma_{11} \cdot \Gamma_{34} + \Gamma_{12} \cdot \Gamma_{33}] \cdot 2 \cdot \tau_\Delta) + \lambda^3 \cdot \{[-\Gamma_{11} \cdot \Gamma_{33} + \Gamma_{12} + \Gamma_{34}] \cdot \tau_\Delta - \Gamma_{33} - \Gamma_{11}\}) \cdot e^{-\lambda \cdot \tau_\Delta}$$

$$D(\lambda, \tau_\Delta) = P_n(\lambda, \tau_\Delta) + Q_m(\lambda, \tau_\Delta) \cdot e^{-\lambda \cdot \tau_\Delta}; n = 4; m = 3; n > m$$

The expression for $P_n(\lambda, \tau_\Delta)$:

$$p_0(\tau_\Delta) = 0; p_1(\tau_\Delta) = 0; p_2(\tau_\Delta) = 0; p_3(\tau_\Delta) = 0; p_4(\tau_\Delta) = 1$$

$$P_{n=4}(\lambda, \tau_\Delta) = \sum_{k=0}^{n=4} p_k(\tau_\Delta) \cdot \lambda^k = p_0(\tau_\Delta) + p_1(\tau_\Delta) \cdot \lambda + p_2(\tau_\Delta) \cdot \lambda^2 + p_3(\tau_\Delta) \cdot \lambda^3 + p_4(\tau_\Delta) \cdot \lambda^4$$

$$P_n(\lambda, \tau_\Delta) = \lambda^4$$

The expression for $Q_m(\lambda, \tau_\Delta)$:

$$q_0(\tau_\Delta) = \Gamma_{12} \cdot \Gamma_{34}; q_1(\tau_\Delta) = \Gamma_{11} \cdot \Gamma_{34} + \Gamma_{12} \cdot \Gamma_{33} - \Gamma_{12} \cdot \Gamma_{34} \cdot 3 \cdot \tau_\Delta$$

$$q_2(\tau_\Delta) = \Gamma_{11} \cdot \Gamma_{33} - \Gamma_{12} - \Gamma_{34} - [\Gamma_{11} \cdot \Gamma_{34} + \Gamma_{12} \cdot \Gamma_{33}] \cdot 2 \cdot \tau_\Delta;$$
$$q_3(\tau_\Delta) = [-\Gamma_{11} \cdot \Gamma_{33} + \Gamma_{12} + \Gamma_{34}] \cdot \tau_\Delta - \Gamma_{33} - \Gamma_{11}$$

$$Q_{m=3}(\lambda, \tau_\Delta) = \sum_{k=0}^{m=3} q_k(\tau_\Delta) \cdot \lambda^k = q_0(\tau_\Delta) + q_1(\tau_\Delta) \cdot \lambda + q_2(\tau_\Delta) \cdot \lambda^2 + q_3(\tau_\Delta) \cdot \lambda^3$$

$$\begin{aligned} Q_m(\lambda, \tau_\Delta) = {} & \Gamma_{12} \cdot \Gamma_{34} + \lambda \cdot (\Gamma_{11} \cdot \Gamma_{34} + \Gamma_{12} \cdot \Gamma_{33} - \Gamma_{12} \cdot \Gamma_{34} \cdot 3 \cdot \tau_\Delta) \\ & + \lambda^2 \cdot (\Gamma_{11} \cdot \Gamma_{33} - \Gamma_{12} - \Gamma_{34} - [\Gamma_{11} \cdot \Gamma_{34} + \Gamma_{12} \cdot \Gamma_{33}] \cdot 2 \cdot \tau_\Delta) \\ & + \lambda^3 \cdot \{[-\Gamma_{11} \cdot \Gamma_{33} + \Gamma_{12} + \Gamma_{34}] \cdot \tau_\Delta - \Gamma_{33} - \Gamma_{11}\} \end{aligned}$$

The homogenous system for $Y_1 V_{L_{calc-1}} Y_2 V_{L_{calc-2}}$ lead to a characteristic equation for the eigenvalue λ having the form $P(\lambda, \tau_\Delta) + Q(\lambda, \tau_\Delta) \cdot e^{-\lambda \cdot \tau_\Delta} = 0$ $P(\lambda, \tau_\Delta) = \sum_{j=0}^{4} a_j(\tau_\Delta) \cdot \lambda^j; Q(\lambda, \tau_\Delta) = \sum_{j=0}^{3} c_j(\tau_\Delta) \cdot \lambda^j$ and the coefficients $\{a_j(q_i, q_k, \tau),$ $c_j(q_i, q_k, \tau)\} \in \mathbb{R}$ depend on q_i, q_k and delay τ_Δ, q_i, q_k are any RFID's N-turn multilayer coil antennas system's parameters, other parameters keep as a constant.

$$\begin{aligned} & a_0 = 0; a_1 = 0; a_2 = 0; a_3 = 0; a_4 = 1; c_0 = \Gamma_{12} \cdot \Gamma_{34}; \\ & c_1 = \Gamma_{11} \cdot \Gamma_{34} + \Gamma_{12} \cdot \Gamma_{33} - \Gamma_{12} \cdot \Gamma_{34} \cdot 3 \cdot \tau_\Delta; \\ & c_2 = \Gamma_{11} \cdot \Gamma_{33} - \Gamma_{12} - \Gamma_{34} - [\Gamma_{11} \cdot \Gamma_{34} + \Gamma_{12} \cdot \Gamma_{33}] \cdot 2 \cdot \tau_\Delta; \\ & c_3 = [-\Gamma_{11} \cdot \Gamma_{33} + \Gamma_{12} + \Gamma_{34}] \cdot \tau_\Delta - \Gamma_{33} - \Gamma_{11} \end{aligned}$$

Unless strictly necessary, the designation of the variation arguments (q_i, q_k) will subsequently be omitted from P, Q, a_j, and c_j. The coefficients a_j, c_j are continuous and differentiable functions of their arguments, and direct substitution shows that $a_0 + c_0 \neq 0 \Rightarrow \Gamma_{12} \cdot \Gamma_{34} \neq 0 \; \forall \, q_i, q_k \in \mathbb{R}_+$, $\lambda = 0$ is not a $P(\lambda, \tau_\Delta) + Q(\lambda, \tau_\Delta) \cdot e^{-\lambda \cdot \tau_\Delta} = 0$. We assume that $P_n(\lambda, \tau_\Delta), Q_m(\lambda, \tau_\Delta)$ can't have common imaginary roots. That is for any real number ω: $P_n(\lambda = i \cdot \omega, \tau_\Delta)$, $Q_m(\lambda = i \cdot \omega, \tau_\Delta) \neq 0$.

$$\begin{aligned} P_n(\lambda, \tau_\Delta) = \omega^4; Q_m(\lambda, \tau_\Delta) = {} & \Gamma_{12} \cdot \Gamma_{34} - \omega^2 \cdot (\Gamma_{11} \cdot \Gamma_{33} - \Gamma_{12} - \Gamma_{34} \\ & - [\Gamma_{11} \cdot \Gamma_{34} + \Gamma_{12} \cdot \Gamma_{33}] \cdot 2 \cdot \tau_\Delta) + i \cdot \{\omega \cdot (\Gamma_{11} \cdot \Gamma_{34} + \Gamma_{12} \cdot \Gamma_{33} - \Gamma_{12} \cdot \Gamma_{34} \cdot 3 \cdot \tau_\Delta) \\ & - \omega^3 \cdot ([-\Gamma_{11} \cdot \Gamma_{33} + \Gamma_{12} + \Gamma_{34}] \cdot \tau_\Delta - \Gamma_{33} - \Gamma_{11})\} \end{aligned}$$

$$P_n(\lambda,\tau_\Delta)+Q_m(\lambda,\tau_\Delta)=\Gamma_{12}\cdot\Gamma_{34}-\omega^2\cdot(\Gamma_{11}\cdot\Gamma_{33}-\Gamma_{12}-\Gamma_{34} - [\Gamma_{11}\cdot\Gamma_{34}+\Gamma_{12}\cdot\Gamma_{33}]\cdot 2\cdot\tau_\Delta)+\omega^4+i\cdot\{\omega\cdot(\Gamma_{11}\cdot\Gamma_{34}+\Gamma_{12}\cdot\Gamma_{33}-\Gamma_{12}\cdot\Gamma_{34}\cdot 3\cdot\tau_\Delta) - \omega^3\cdot([-\Gamma_{11}\cdot\Gamma_{33}+\Gamma_{12}+\Gamma_{34}]\cdot\tau_\Delta-\Gamma_{33}-\Gamma_{11})\}\neq 0$$

$$|P(i\cdot\omega,\tau_\Delta)|^2=\omega^8$$

$$|Q(i\cdot\omega,\tau_\Delta)|^2=\{\Gamma_{12}\cdot\Gamma_{34}-\omega^2\cdot(\Gamma_{11}\cdot\Gamma_{33}-\Gamma_{12}-\Gamma_{34} - [\Gamma_{11}\cdot\Gamma_{34}+\Gamma_{12}\cdot\Gamma_{33}]\cdot 2\cdot\tau_\Delta)\}^2+\{\omega\cdot(\Gamma_{11}\cdot\Gamma_{34}+\Gamma_{12}\cdot\Gamma_{33} - \Gamma_{12}\cdot\Gamma_{34}\cdot 3\cdot\tau_\Delta)-\omega^3\cdot([-\Gamma_{11}\cdot\Gamma_{33}+\Gamma_{12}+\Gamma_{34}]\cdot\tau_\Delta-\Gamma_{33}-\Gamma_{11})\}^2$$

$$|Q(i\cdot\omega,\tau_\Delta)|^2=\Gamma_{12}^2\cdot\Gamma_{34}^2+\omega^2\cdot\{(\Gamma_{11}\cdot\Gamma_{34}+\Gamma_{12}\cdot\Gamma_{33}-\Gamma_{12}\cdot\Gamma_{34}\cdot 3\cdot\tau_\Delta)^2 - 2\cdot\Gamma_{12}\cdot\Gamma_{34}\cdot(\Gamma_{11}\cdot\Gamma_{33}-\Gamma_{12}-\Gamma_{34}-[\Gamma_{11}\cdot\Gamma_{34}+\Gamma_{12}\cdot\Gamma_{33}]\cdot 2\cdot\tau_\Delta)\} + \omega^4\cdot\{(\Gamma_{11}\cdot\Gamma_{33}-\Gamma_{12}-\Gamma_{34}-[\Gamma_{11}\cdot\Gamma_{34}+\Gamma_{12}\cdot\Gamma_{33}]\cdot 2\cdot\tau_\Delta)^2 - 2\cdot(\Gamma_{11}\cdot\Gamma_{34}+\Gamma_{12}\cdot\Gamma_{33}-\Gamma_{12}\cdot\Gamma_{34}\cdot 3\cdot\tau_\Delta)\cdot([-\Gamma_{11}\cdot\Gamma_{33}+\Gamma_{12} + \Gamma_{34}]\cdot\tau_\Delta-\Gamma_{33}-\Gamma_{11})\}+\omega^6\cdot([-\Gamma_{11}\cdot\Gamma_{33}+\Gamma_{12}+\Gamma_{34}]\cdot\tau_\Delta-\Gamma_{33}-\Gamma_{11})^2$$

We need to find the expression: $F(\omega,\tau_\Delta)=|P(i\cdot\omega)|^2-|Q(i\cdot\omega)|^2$

$$F(\omega,\tau_\Delta)=|P(i\cdot\omega)|^2-|Q(i\cdot\omega)|^2=-\Gamma_{12}^2\cdot\Gamma_{34}^2-\omega^2\cdot\{(\Gamma_{11}\cdot\Gamma_{34}+\Gamma_{12}\cdot\Gamma_{33}-\Gamma_{12}\cdot\Gamma_{34}\cdot 3\cdot\tau_\Delta)^2 - 2\cdot\Gamma_{12}\cdot\Gamma_{34}\cdot(\Gamma_{11}\cdot\Gamma_{33}-\Gamma_{12}-\Gamma_{34}-[\Gamma_{11}\cdot\Gamma_{34}+\Gamma_{12}\cdot\Gamma_{33}]\cdot 2\cdot\tau_\Delta)\} - \omega^4\cdot\{(\Gamma_{11}\cdot\Gamma_{33}-\Gamma_{12}-\Gamma_{34}-[\Gamma_{11}\cdot\Gamma_{34}+\Gamma_{12}\cdot\Gamma_{33}]\cdot 2\cdot\tau_\Delta)^2 - 2\cdot(\Gamma_{11}\cdot\Gamma_{34}+\Gamma_{12}\cdot\Gamma_{33}-\Gamma_{12}\cdot\Gamma_{34}\cdot 3\cdot\tau_\Delta)\cdot([-\Gamma_{11}\cdot\Gamma_{33}+\Gamma_{12}+\Gamma_{34}]\cdot\tau_\Delta-\Gamma_{33}-\Gamma_{11})\} - \omega^6\cdot([-\Gamma_{11}\cdot\Gamma_{33}+\Gamma_{12}+\Gamma_{34}]\cdot\tau_\Delta-\Gamma_{33}-\Gamma_{11})^2+\omega^8$$

We define the following parameters for simplicity: $\Xi_0=-\Gamma_{12}^2\cdot\Gamma_{34}^2$

$$\Xi_2=-\{(\Gamma_{11}\cdot\Gamma_{34}+\Gamma_{12}\cdot\Gamma_{33}-\Gamma_{12}\cdot\Gamma_{34}\cdot 3\cdot\tau_\Delta)^2-2\cdot\Gamma_{12}\cdot\Gamma_{34}\cdot(\Gamma_{11}\cdot\Gamma_{33}-\Gamma_{12}-\Gamma_{34} - [\Gamma_{11}\cdot\Gamma_{34}+\Gamma_{12}\cdot\Gamma_{33}]\cdot 2\cdot\tau_\Delta)\}$$

$$\Xi_4=-\{(\Gamma_{11}\cdot\Gamma_{33}-\Gamma_{12}-\Gamma_{34}-[\Gamma_{11}\cdot\Gamma_{34}+\Gamma_{12}\cdot\Gamma_{33}]\cdot 2\cdot\tau_\Delta)^2 - 2\cdot(\Gamma_{11}\cdot\Gamma_{34}+\Gamma_{12}\cdot\Gamma_{33}-\Gamma_{12}\cdot\Gamma_{34}\cdot 3\cdot\tau_\Delta)\cdot([-\Gamma_{11}\cdot\Gamma_{33}+\Gamma_{12}+\Gamma_{34}]\cdot\tau_\Delta-\Gamma_{33}-\Gamma_{11})\}$$

$$\Xi_6=-([-\Gamma_{11}\cdot\Gamma_{33}+\Gamma_{12}+\Gamma_{34}]\cdot\tau_\Delta-\Gamma_{33}-\Gamma_{11})^2;\ \Xi_8=1$$

$$F(\omega,\tau_\Delta)=|P(i\cdot\omega)|^2-|Q(i\cdot\omega)|^2=\sum_{k=0}^{4}\Xi_{2\cdot k}\cdot\omega^{2\cdot k} =\Xi_0+\Xi_2\cdot\omega^2+\Xi_4\cdot\omega^4+\Xi_6\cdot\omega^6+\Xi_8\cdot\omega^8$$

Hence $F(\omega,\tau_\Delta)=0$ implies $\sum_{k=0}^{4}\Xi_{2\cdot k}\cdot\omega^{2\cdot k}=0$ and its roots are given by solving the above polynomial. Furthermore, $P_R(i\cdot\omega,\tau_\Delta)=\omega^4; P_I(i\cdot\omega,\tau_\Delta)=0$

$$Q_R(i\cdot\omega,\tau_\Delta)=\Gamma_{12}\cdot\Gamma_{34}-\omega^2\cdot(\Gamma_{11}\cdot\Gamma_{33}-\Gamma_{12}-\Gamma_{34}-[\Gamma_{11}\cdot\Gamma_{34}+\Gamma_{12}\cdot\Gamma_{33}]\cdot 2\cdot\tau_\Delta)$$
$$\begin{aligned}Q_I(i\cdot\omega,\tau_\Delta)&=\omega\cdot(\Gamma_{11}\cdot\Gamma_{34}+\Gamma_{12}\cdot\Gamma_{33}-\Gamma_{12}\cdot\Gamma_{34}\cdot 3\cdot\tau_\Delta)\\&\quad-\omega^3\cdot([-\Gamma_{11}\cdot\Gamma_{33}+\Gamma_{12}+\Gamma_{34}]\cdot\tau_\Delta-\Gamma_{33}-\Gamma_{11})\end{aligned}$$

Hence

$$\sin\theta(\tau_\Delta)=\frac{-P_R(i\cdot\omega,\tau_\Delta)\cdot Q_I(i\cdot\omega,\tau_\Delta)+P_I(i\cdot\omega,\tau_\Delta)\cdot Q_R(i\cdot\omega,\tau_\Delta)}{|Q(i\cdot\omega,\tau_\Delta)|^2}$$

And

$$\cos\theta(\tau_\Delta)=-\frac{P_R(i\cdot\omega,\tau_\Delta)\cdot Q_R(i\cdot\omega,\tau_\Delta)+P_I(i\cdot\omega,\tau_\Delta)\cdot Q_I(i\cdot\omega,\tau_\Delta)}{|Q(i\cdot\omega,\tau_\Delta)|^2}$$

$$\sin\theta(\tau_\Delta)=\frac{\begin{array}{c}-\omega^4\cdot\{\omega\cdot(\Gamma_{11}\cdot\Gamma_{34}+\Gamma_{12}\cdot\Gamma_{33}-\Gamma_{12}\cdot\Gamma_{34}\cdot 3\cdot\tau_\Delta)\\-\omega^3\cdot([-\Gamma_{11}\cdot\Gamma_{33}+\Gamma_{12}+\Gamma_{34}]\cdot\tau_\Delta-\Gamma_{33}-\Gamma_{11})\}\end{array}}{\begin{array}{l}\Gamma_{12}^2\cdot\Gamma_{34}^2+\omega^2\cdot\{(\Gamma_{11}\cdot\Gamma_{34}+\Gamma_{12}\cdot\Gamma_{33}-\Gamma_{12}\cdot\Gamma_{34}\cdot 3\cdot\tau_\Delta)^2\\-2\cdot\Gamma_{12}\cdot\Gamma_{34}\cdot(\Gamma_{11}\cdot\Gamma_{33}-\Gamma_{12}-\Gamma_{34}-[\Gamma_{11}\cdot\Gamma_{34}+\Gamma_{12}\cdot\Gamma_{33}]\cdot 2\cdot\tau_\Delta)\}\\+\omega^4\cdot\{(\Gamma_{11}\cdot\Gamma_{33}-\Gamma_{12}-\Gamma_{34}-[\Gamma_{11}\cdot\Gamma_{34}+\Gamma_{12}\cdot\Gamma_{33}]\cdot 2\cdot\tau_\Delta)^2\\-2\cdot(\Gamma_{11}\cdot\Gamma_{34}+\Gamma_{12}\cdot\Gamma_{33}-\Gamma_{12}\cdot\Gamma_{34}\cdot 3\cdot\tau_\Delta)\cdot([-\Gamma_{11}\cdot\Gamma_{33}+\Gamma_{12}+\Gamma_{34}]\cdot\tau_\Delta-\Gamma_{33}-\Gamma_{11})\}\\+\omega^6\cdot([-\Gamma_{11}\cdot\Gamma_{33}+\Gamma_{12}+\Gamma_{34}]\cdot\tau_\Delta-\Gamma_{33}-\Gamma_{11})^2\end{array}}$$

$$\cos\theta(\tau_\Delta)=-\frac{\omega^4\cdot\{\Gamma_{12}\cdot\Gamma_{34}-\omega^2\cdot(\Gamma_{11}\cdot\Gamma_{33}-\Gamma_{12}-\Gamma_{34}-[\Gamma_{11}\cdot\Gamma_{34}+\Gamma_{12}\cdot\Gamma_{33}]\cdot 2\cdot\tau_\Delta)\}}{\begin{array}{l}\Gamma_{12}^2\cdot\Gamma_{34}^2+\omega^2\cdot\{(\Gamma_{11}\cdot\Gamma_{34}+\Gamma_{12}\cdot\Gamma_{33}-\Gamma_{12}\cdot\Gamma_{34}\cdot 3\cdot\tau_\Delta)^2\\-2\cdot\Gamma_{12}\cdot\Gamma_{34}\cdot(\Gamma_{11}\cdot\Gamma_{33}-\Gamma_{12}-\Gamma_{34}-[\Gamma_{11}\cdot\Gamma_{34}+\Gamma_{12}\cdot\Gamma_{33}]\cdot 2\cdot\tau_\Delta)\}\\+\omega^4\cdot\{(\Gamma_{11}\cdot\Gamma_{33}-\Gamma_{12}-\Gamma_{34}-[\Gamma_{11}\cdot\Gamma_{34}+\Gamma_{12}\cdot\Gamma_{33}]\cdot 2\cdot\tau_\Delta)^2\\-2\cdot(\Gamma_{11}\cdot\Gamma_{34}+\Gamma_{12}\cdot\Gamma_{33}-\Gamma_{12}\cdot\Gamma_{34}\cdot 3\cdot\tau_\Delta)\cdot([-\Gamma_{11}\cdot\Gamma_{33}+\Gamma_{12}+\Gamma_{34}]\cdot\tau_\Delta-\Gamma_{33}-\Gamma_{11})\}\\+\omega^6\cdot([-\Gamma_{11}\cdot\Gamma_{33}+\Gamma_{12}+\Gamma_{34}]\cdot\tau_\Delta-\Gamma_{33}-\Gamma_{11})^2\end{array}}$$

Above expressions are continuous and differentiable in τ_Δ based on Lemma 1.1. Hence we use Theorem 1.2 and this prove Theorem 1.3.

We use different parameters terminology from our last characteristics parameters definition $k\to j; p_k(\tau_\Delta)\to a_j; q_k(\tau_\Delta)\to c_j; n=4; m=3; n>m$ additionally $P_n(\lambda,\tau_\Delta)\to P(\lambda,\tau_\Delta); Q_m(\lambda,\tau_\Delta)\to Q(\lambda,\tau_\Delta)$ then $P(\lambda,\tau_\Delta)=\sum_{j=0}^{4}a_j\cdot\lambda^j; Q(\lambda,\tau_\Delta)=\sum_{j=0}^{2}c_j\cdot\lambda^j$.

$$\begin{aligned}P(\lambda)=\lambda^4; Q(\lambda,\tau_\Delta)&=\Gamma_{12}\cdot\Gamma_{34}+\lambda\cdot(\Gamma_{11}\cdot\Gamma_{34}+\Gamma_{12}\cdot\Gamma_{33}-\Gamma_{12}\cdot\Gamma_{34}\cdot 3\cdot\tau_\Delta)\\&\quad+\lambda^2\cdot(\Gamma_{11}\cdot\Gamma_{33}-\Gamma_{12}-\Gamma_{34}-[\Gamma_{11}\cdot\Gamma_{34}+\Gamma_{12}\cdot\Gamma_{33}]\cdot 2\cdot\tau_\Delta)\\&\quad+\lambda^3\cdot\{[-\Gamma_{11}\cdot\Gamma_{33}+\Gamma_{12}+\Gamma_{34}]\cdot\tau_\Delta-\Gamma_{33}-\Gamma_{11}\}\end{aligned}$$

$n, m \in \mathbb{N}_0, n > m$ and $a_j, c_j : \mathrm{R}_{+0} \to R$ are continuous and differentiable function of τ_Δ such that $a_0 + c_0 \neq 0$. In the following "_" denotes complex and conjugate. $P(\lambda), Q(\lambda, \tau_\Delta)$ are analytic functions in λ and differentiable in τ_Δ. The coefficients:

$$\{a_j(C_1, R_1, \text{RFID'sN-turn multilayer coil antennas system/s parameters})\ \&$$
$$c_j(C_1, R_1, \tau_\Delta, \text{RFID'sN-turn multilayer coil antennas system/s parameters})\ \} \in \mathbb{R}$$

depend on RFID's N-turn multilayer coil antennas system's parameters C_1, R_1, τ_Δ values and antennas parameters. Unless strictly necessary, the designation of the variation arguments. Resistor R_1, capacitor C_1, τ_Δ and N-turn multilayer coil antennas system's parameters will subsequently be omitted from P, Q, a_j, c_j. The coefficients a_j, c_j are continuous, and differentiable functions of their arguments, and direct substitution shows that $a_0 + c_0 \neq 0; \Gamma_{12} \cdot \Gamma_{34} \neq 0$.

$$\Gamma_{12} \cdot \Gamma_{34} \neq \frac{\eta_1}{\eta_3} \cdot \frac{\xi_1}{\xi_3}$$

$$= [\frac{\frac{1}{L_{calc-1}} + \frac{1}{R_1} \cdot \frac{(\sum_{i=1}^{2} R_{dc-i})}{L_{calc-1}}}{C_1 \cdot [1 + \frac{L_{calc-2}}{L_{calc-1}} + 2 \cdot K \cdot \sqrt{\frac{L_{calc-2}}{L_{calc-1}}}]}] \cdot [\frac{\frac{1}{L_{calc-2}} + \frac{1}{R_1} \cdot (\frac{\sum_{i=1}^{2} R_{dc-i}}{L_{calc-2}})}{C_1 \cdot [\frac{L_{calc-1}}{L_{calc-2}} + 1 + 2 \cdot K \cdot \sqrt{\frac{L_{calc-1}}{L_{calc-2}}}]}] \neq 0$$

$$\forall C_1, R_1, \text{RFID'sN-turn multilayer coil antennas system's parameters} \in \mathbb{R}_+$$

i.e. $\lambda = 0$ is not a root of characteristic equation. Furthermore $P(\lambda), Q(\lambda, \tau_\Delta)$ are analytic function of λ for which the following requirements of the analysis (see kuang, 1993, Sect. 3.4) can also be verified in the present case [2–4].

(a) If $\lambda = i \cdot \omega$, $\omega \in \mathbb{R}$ then $P(i \cdot \omega) + Q(i \cdot \omega) \neq 0$,
i.e. P and Q have no common imaginary roots. This condition was verified numerically in the entire (R_1, C_1, N-turn multilayer coil antennas system parameters) domain of interest.

(b) $|Q(\lambda)/P(\lambda)|$ is bounded for $|\lambda| \to \infty$, $\mathrm{Re}\lambda \geq 0$. No roots bifurcation from ∞. Indeed, in the limit:

$$|\frac{Q(\lambda)}{P(\lambda)}| = |\frac{\begin{array}{c}\Gamma_{12} \cdot \Gamma_{34} + \lambda \cdot (\Gamma_{11} \cdot \Gamma_{34} + \Gamma_{12} \cdot \Gamma_{33} - \Gamma_{12} \cdot \Gamma_{34} \cdot 3 \cdot \tau_\Delta) + \lambda^2 \cdot (\Gamma_{11} \cdot \Gamma_{33} \\ -\Gamma_{12} - \Gamma_{34} - [\Gamma_{11} \cdot \Gamma_{34} + \Gamma_{12} \cdot \Gamma_{33}] \cdot 2 \cdot \tau_\Delta) + \lambda^3 \cdot \{[-\Gamma_{11} \cdot \Gamma_{33} + \Gamma_{12} + \Gamma_{34}] \cdot \tau_\Delta - \Gamma_{33} - \Gamma_{11}\}\end{array}}{\lambda^4}|$$

(c) $F(\omega) = |P(i \cdot \omega)|^2 - |Q(i \cdot \omega)|^2$

$$\begin{aligned} F(\omega, \tau_\Delta) &= |P(i \cdot \omega)|^2 - |Q(i \cdot \omega)|^2 \\ &= -\Gamma_{12}^2 \cdot \Gamma_{34}^2 - \omega^2 \cdot \{(\Gamma_{11} \cdot \Gamma_{34} + \Gamma_{12} \cdot \Gamma_{33} - \Gamma_{12} \cdot \Gamma_{34} \cdot 3 \cdot \tau_\Delta)^2 \\ &\quad - 2 \cdot \Gamma_{12} \cdot \Gamma_{34} \cdot (\Gamma_{11} \cdot \Gamma_{33} - \Gamma_{12} - \Gamma_{34} - [\Gamma_{11} \cdot \Gamma_{34} + \Gamma_{12} \cdot \Gamma_{33}] \cdot 2 \cdot \tau_\Delta)\} \\ &\quad - \omega^4 \cdot \{(\Gamma_{11} \cdot \Gamma_{33} - \Gamma_{12} - \Gamma_{34} - [\Gamma_{11} \cdot \Gamma_{34} + \Gamma_{12} \cdot \Gamma_{33}] \cdot 2 \cdot \tau_\Delta)^2 \\ &\quad - 2 \cdot (\Gamma_{11} \cdot \Gamma_{34} + \Gamma_{12} \cdot \Gamma_{33} - \Gamma_{12} \cdot \Gamma_{34} \cdot 3 \cdot \tau_\Delta) \\ &\quad \cdot ([-\Gamma_{11} \cdot \Gamma_{33} + \Gamma_{12} + \Gamma_{34}] \cdot \tau_\Delta - \Gamma_{33} - \Gamma_{11})\} \\ &\quad - \omega^6 \cdot ([-\Gamma_{11} \cdot \Gamma_{33} + \Gamma_{12} + \Gamma_{34}] \cdot \tau_\Delta - \Gamma_{33} - \Gamma_{11})^2 + \omega^8 \end{aligned}$$

Has at most a finite number of zeros. Indeed, this is a polynomial in ω (Degree in ω^8).

(d) Each positive root ω (R_1, C_1, τ_Δ, N-turn multilayer coil antennas system parameters) of $F(\omega) = 0$ is continuous and differentiable with respect to R_1, C_1, τ_Δ, N-turn multilayer coil antennas system parameters. This condition can only be assessed numerically.

In addition, since the coefficients in P and Q are real, we have $\overline{P(-i \cdot \omega)} = P(i \cdot \omega)$, and $\overline{Q(-i \cdot \omega)} = Q(i \cdot \omega)$ thus $\lambda = i \cdot \omega$, $\omega > 0$ may be on eigenvalue of characteristic equation. The analysis consists in identifying the roots of characteristic equation situated on the imaginary axis of the complex λ-plane, where by increasing the parameters R_1, C_1, τ_Δ, and N-turn multilayer coil antennas system parameters, Reλ may, at the crossing, Change its sign from (−) to (+), i.e. from a stable focus $E^{(0)}(Y_1^{(0)}, V_{L_{calc-1}}^{(0)}, Y_2^{(0)}, V_{L_{calc-2}}^{(0)}) = (0, 0, 0, 0)$ to an unstable one, or vice versa. This feature may be further assessed by examining the sign of the partial derivatives with respect to C_1, R_1, τ_Δ and N-turn multilayer coil antennas parameters. $\wedge^{-1}(R_1) = (\frac{\partial \mathrm{Re}\lambda}{\partial R_1})_{\lambda = i \cdot \omega}$, C_1, τ_Δ, N-turn multilayer coil antennas system parameters are constant.

$$\wedge^{-1}(L_{calc-1}) = (\frac{\partial \mathrm{Re}\lambda}{\partial L_{calc-1}})_{\lambda = i \cdot \omega}, C_1, R_1, \tau_\Delta = const;$$

$$\wedge^{-1}(L_{calc-2}) = (\frac{\partial \mathrm{Re}\lambda}{\partial L_{calc-2}})_{\lambda = i \cdot \omega}, C_1, R_1, \tau_\Delta = const$$

$$\begin{aligned} \wedge^{-1}(\tau_\Delta) &= (\frac{\partial \mathrm{Re}\lambda}{\partial \tau_\Delta})_{\lambda = i \cdot \omega}, C_1, R_1, \text{N-turn multilayer coil antennas system's parameters} \\ &= const \text{ where } \omega \in \mathbb{R}_+. \end{aligned}$$

When writing $P(\lambda) = P_R(\lambda) + i \cdot P_I(\lambda)$ and $Q(\lambda) = Q_R(\lambda) + i \cdot Q_I(\lambda)$, and inserting $\lambda = i \cdot \omega$ Into RFID N-turn multilayer coil antennas system's characteristic equation, ω must satisfy the following:

$$\sin \omega \cdot \tau_{\Delta} = g(\omega) = \frac{-P_R(i \cdot \omega) \cdot Q_I(i \cdot \omega) + P_I(i \cdot \omega) \cdot Q_R(i \cdot \omega)}{|Q(i \cdot \omega)|^2}$$

and

$$\cos \omega \cdot \tau_{\Delta} = h(\omega) = -\frac{P_R(i \cdot \omega) \cdot Q_R(i \cdot \omega) + P_I(i \cdot \omega) \cdot Q_I(i \cdot \omega)}{|Q(i \cdot \omega)|^2}.$$

Where $|Q(i \cdot \omega)|^2 \neq 0$ in view of requirement (a) above, $(g, h) \in R$. Furthermore, it follows above equations $\sin \omega \cdot \tau_{\Delta}$ and $\cos \omega \cdot \tau_{\Delta}$ that, by squaring and adding the sides, ω must be a positive root of $F(\omega) = |P(i \cdot \omega)|^2 - |Q(i \cdot \omega)|^2 = 0$. Note $F(\omega)$ is dependent of τ_{Δ}. Now it is important to notice that if $\tau_{\Delta} \notin I$ (assume that $I \subseteq R_{+0}$ is the set where $\omega(\tau_{\Delta})$ is a positive root of $F(\omega)$ and for $\tau_{\Delta} \notin I$; $\omega(\tau_{\Delta})$ is not define. Then for all τ_{Δ} in I $\omega(\tau_{\Delta})$ is satisfies that $F(\omega, \tau_{\Delta}) = 0$). Then there are no positive $\omega(\tau_{\Delta})$ solutions for $F(\omega, \tau_{\Delta}) = 0$, and we cannot have stability switches. For any $\tau_{\Delta} \in I$, where $\omega(\tau_{\Delta})$ is a positive solution of $F(\omega, \tau_{\Delta}) = 0$, we can define the angle $\theta(\tau_{\Delta}) \in [0, 2 \cdot \pi]$ as the solution of $\sin \theta(\tau_{\Delta}) = \frac{-P_R(i \cdot \omega) \cdot Q_I(i \cdot \omega) + P_I(i \cdot \omega) \cdot Q_R(i \cdot \omega)}{|Q(i \cdot \omega)|^2}$; $\cos \theta(\tau_{\Delta}) = -\frac{P_R(i \cdot \omega) \cdot Q_R(i \cdot \omega) + P_I(i \cdot \omega) \cdot Q_I(i \cdot \omega)}{|Q(i \cdot \omega)|^2}$ and the relation between the argument $\theta(\tau_{\Delta})$ and $\omega(\tau_{\Delta}) \cdot \tau_{\Delta}$ for $\tau_{\Delta} \in I$ must be $\omega(\tau_{\Delta}) \cdot \tau_{\Delta} = \theta(\tau_{\Delta}) + n \cdot 2 \cdot \pi \; \forall \, n \in \mathbb{N}_0$. Hence we can define the maps $\tau_{\Delta n} : I \to R_{+0}$ given by $\tau_{\Delta n}(\tau_{\Delta}) = \frac{\theta(\tau_{\Delta}) + n \cdot 2 \cdot \pi}{\omega(\tau_{\Delta})}$; $n \in \mathbb{N}_0, \tau_{\Delta} \in I$. Let us introduce the functions $I \to R$; $S_n(\tau) = \tau_{\Delta} - \tau_{\Delta n}(\tau_{\Delta}), \tau_{\Delta} \in I, n \in \mathbb{N}_0$ that are continuous and differentiable in τ. In the following, the subscripts $\lambda, \omega, R_1, C_1$ and N-turn multilayer coil antennas system parameters $(a_1 \neq a_2; N_1 \neq N_2; h_1 \neq h_2; b_1 \neq b_2; a_i, N_i, h_i, b_i \in \mathbb{R}_+)$ indicate the corresponding partial derivatives. Let us first concentrate on $\wedge(x)$, remember in

$$\lambda(a_1 \neq a_2; N_1 \neq N_2; h_1 \neq h_2; b_1 \neq b_2; a_i, N_i, h_i, b_i \in \mathbb{R}_+)$$
$$\omega(a_1 \neq a_2; N_1 \neq N_2; h_1 \neq h_2; b_1 \neq b_2; a_i, N_i, h_i, b_i \in \mathbb{R}_+)$$

and keeping all parameters except one (x) and τ_{Δ}. The derivation closely follows that in reference [BK]. Differentiating N-turn multilayer coil antennas system characteristic equation $P(\lambda) + Q(\lambda) \cdot e^{-\lambda \cdot \tau_{\Delta}} = 0$ with respect to specific parameter (x), and inverting the derivative, for convenience, one calculates:

Remark:
$x = a_i, N_i, h_i, b_i; a_1 \neq a_2; N_1 \neq N_2; h_1 \neq h_2; b_1 \neq b_2; a_i, N_i, h_i, b_i \in \mathbb{R}_+$

$$\left(\frac{\partial \lambda}{\partial x}\right)^{-1} = \frac{-P_{\lambda}(\lambda, x) \cdot Q(\lambda, x) + Q_{\lambda}(\lambda, x) \cdot P(\lambda, x) - \tau_{\Delta} \cdot P(\lambda, x) \cdot Q(\lambda, x)}{P_x(\lambda, x) \cdot Q(\lambda, x) - Q_x(\lambda, x) \cdot P(\lambda, x)}$$

Where $P_\lambda = \frac{\partial P}{\partial \lambda}, \ldots$ etc., Substituting $\lambda = i \cdot \omega$, and bearing $\overline{P(-i \cdot \omega)} = P(i \cdot \omega)$; $\overline{Q(-i \cdot \omega)} = Q(i \cdot \omega)$ then $i \cdot P_\lambda(i \cdot \omega) = P_\omega(i \cdot \omega)$ and $i \cdot Q_\lambda(i \cdot \omega) = Q_\omega(i \cdot \omega)$ that on the surface $|P(i \cdot \omega)|^2 = |Q(i \cdot \omega)|^2$, one obtains

$$(\frac{\partial \lambda}{\partial x})^{-1}|_{\lambda = i \cdot \omega} = (\frac{i \cdot P_\omega(i \cdot \omega, x) \cdot \overline{P(i \cdot \omega, x)} + i \cdot Q_\lambda(i \cdot \omega, x) \cdot \overline{Q(\lambda, x)} - \tau_\Delta \cdot |P(i \cdot \omega, x)|^2}{P_x(i \cdot \omega, x) \cdot \overline{P(i \cdot \omega, x)} - Q_x(i \cdot \omega, x) \cdot \overline{Q(i \cdot \omega, x)}}).$$

Upon separating into real and imaginary parts, with $P = P_R + i \cdot P_I$; $Q = Q_R + i \cdot Q_I$

$$P_\omega = P_{R\omega} + i \cdot P_{I\omega}; Q_\omega = Q_{R\omega} + i \cdot Q_{I\omega}; P_x = P_{Rx} + i \cdot P_{Ix}; Q_x = Q_{Rx} + i \cdot Q_{Ix}; P^2 = P_R^2 + P_I^2.$$

When (x) can be any N-turn multilayer coil antennas system parameters R_1, C_1, and time delay τ_Δ etc. Where for convenience, we have dropped the arguments $(i \cdot \omega, x)$, and where $F_\omega = 2 \cdot [(P_{R\omega} \cdot P_R + P_{I\omega} \cdot P_I) - (Q_{R\omega} \cdot Q_R + Q_{I\omega} \cdot Q_I)]$
$F_x = 2 \cdot [(P_{Rx} \cdot P_R + P_{Ix} \cdot P_I) - (Q_{Rx} \cdot Q_R + Q_{Ix} \cdot Q_I)]; \omega_x = -F_x / F_\omega$.
We define U and V:

$$U = (P_R \cdot P_{I\omega} - P_I \cdot P_{R\omega}) - (Q_R \cdot Q_{I\omega} - Q_I \cdot Q_{R\omega})$$
$$V = (P_R \cdot P_{Ix} - P_I \cdot P_{Rx}) - (Q_R \cdot Q_{Ix} - Q_I \cdot Q_{Rx})$$

We choose our specific parameter as time delay x = τ_Δ.

$$\begin{aligned} P_R(i \cdot \omega, \tau_\Delta) &= \omega^4; P_I(i \cdot \omega, \tau_\Delta) = 0 \\ Q_R(i \cdot \omega, \tau_\Delta) &= \Gamma_{12} \cdot \Gamma_{34} - \omega^2 \cdot (\Gamma_{11} \cdot \Gamma_{33} - \Gamma_{12} - \Gamma_{34} \\ &\quad - [\Gamma_{11} \cdot \Gamma_{34} + \Gamma_{12} \cdot \Gamma_{33}] \cdot 2 \cdot \tau_\Delta) \\ Q_I(i \cdot \omega, \tau_\Delta) &= \omega \cdot (\Gamma_{11} \cdot \Gamma_{34} + \Gamma_{12} \cdot \Gamma_{33} - \Gamma_{12} \cdot \Gamma_{34} \cdot 3 \cdot \tau_\Delta) \\ &\quad - \omega^3 \cdot ([-\Gamma_{11} \cdot \Gamma_{33} + \Gamma_{12} + \Gamma_{34}] \cdot \tau_\Delta - \Gamma_{33} - \Gamma_{11}) \end{aligned}$$

$$P_{R\omega} = 4 \cdot \omega^3; P_{I\omega} = 0; P_{R\tau_\Delta} = 0; P_{I\tau_\Delta} = 0; Q_{R\tau_\Delta} = \omega^2 \cdot [\Gamma_{11} \cdot \Gamma_{34} + \Gamma_{12} \cdot \Gamma_{33}] \cdot 2$$

$Q_{I\tau_\Delta} = -\omega \cdot \Gamma_{12} \cdot \Gamma_{34} \cdot 3 - \omega^3 \cdot [-\Gamma_{11} \cdot \Gamma_{33} + \Gamma_{12} + \Gamma_{34}]; P_{R\omega} \cdot P_R = 4 \cdot \omega^7$; $P_{I\omega} \cdot P_I = 0; \omega_{\tau_\Delta} = -F_{\tau_\Delta} / F_\omega$

$$Q_{R\omega} = -2 \cdot \omega \cdot (\Gamma_{11} \cdot \Gamma_{33} - \Gamma_{12} - \Gamma_{34} - [\Gamma_{11} \cdot \Gamma_{34} + \Gamma_{12} \cdot \Gamma_{33}] \cdot 2 \cdot \tau_\Delta)$$

$$Q_{I\omega} = (\Gamma_{11} \cdot \Gamma_{34} + \Gamma_{12} \cdot \Gamma_{33} - \Gamma_{12} \cdot \Gamma_{34} \cdot 3 \cdot \tau_\Delta) - 3 \cdot \omega^2 \cdot ([-\Gamma_{11} \cdot \Gamma_{33} + \Gamma_{12} + \Gamma_{34}] \cdot \tau_\Delta - \Gamma_{33} - \Gamma_{11})$$

$$Q_{R\omega} \cdot Q_R = -2 \cdot \omega \cdot (\Gamma_{11} \cdot \Gamma_{33} - \Gamma_{12} - \Gamma_{34} - [\Gamma_{11} \cdot \Gamma_{34} + \Gamma_{12} \cdot \Gamma_{33}] \cdot 2 \cdot \tau_\Delta) \cdot \{\Gamma_{12} \cdot \Gamma_{34} - \omega^2 \cdot (\Gamma_{11} \cdot \Gamma_{33} - \Gamma_{12} - \Gamma_{34} - [\Gamma_{11} \cdot \Gamma_{34} + \Gamma_{12} \cdot \Gamma_{33}] \cdot 2 \cdot \tau_\Delta)\}$$

$$Q_{I\omega} \cdot Q_I = \{(\Gamma_{11} \cdot \Gamma_{34} + \Gamma_{12} \cdot \Gamma_{33} - \Gamma_{12} \cdot \Gamma_{34} \cdot 3 \cdot \tau_\Delta) - 3 \cdot \omega^2 \cdot ([-\Gamma_{11} \cdot \Gamma_{33} + \Gamma_{12} + \Gamma_{34}] \cdot \tau_\Delta - \Gamma_{33} - \Gamma_{11})\} \cdot \{\omega \cdot (\Gamma_{11} \cdot \Gamma_{34} + \Gamma_{12} \cdot \Gamma_{33} - \Gamma_{12} \cdot \Gamma_{34} \cdot 3 \cdot \tau_\Delta) - \omega^3 \cdot ([-\Gamma_{11} \cdot \Gamma_{33} + \Gamma_{12} + \Gamma_{34}] \cdot \tau_\Delta - \Gamma_{33} - \Gamma_{11})\}$$

$$F_{\tau_\Delta} = 2 \cdot [(P_{R\tau_\Delta} \cdot P_R + P_{I\tau_\Delta} \cdot P_I) - (Q_{R\tau_\Delta} \cdot Q_R + Q_{I\tau_\Delta} \cdot Q_I)]; P_{R\tau_\Delta} = 0; P_{I\tau_\Delta} = 0$$

$$F_{\tau_\Delta} = -2 \cdot \{\omega^2 \cdot [\Gamma_{11} \cdot \Gamma_{34} + \Gamma_{12} \cdot \Gamma_{33}] \cdot 2 \cdot [\Gamma_{12} \cdot \Gamma_{34} - \omega^2 \cdot (\Gamma_{11} \cdot \Gamma_{33} - \Gamma_{12} - \Gamma_{34} - [\Gamma_{11} \cdot \Gamma_{34} + \Gamma_{12} \cdot \Gamma_{33}] \cdot 2 \cdot \tau_\Delta)] + (-\omega \cdot \Gamma_{12} \cdot \Gamma_{34} \cdot 3 - \omega^3 \cdot [-\Gamma_{11} \cdot \Gamma_{33} + \Gamma_{12} + \Gamma_{34}]) \cdot [\omega \cdot (\Gamma_{11} \cdot \Gamma_{34} + \Gamma_{12} \cdot \Gamma_{33} - \Gamma_{12} \cdot \Gamma_{34} \cdot 3 \cdot \tau_\Delta) - \omega^3 \cdot ([-\Gamma_{11} \cdot \Gamma_{33} + \Gamma_{12} + \Gamma_{34}] \cdot \tau_\Delta - \Gamma_{33} - \Gamma_{11})]\}$$

$$P_R \cdot P_{I\omega} = 0; P_I \cdot P_{R\omega} = 0; Q_R \cdot Q_{I\omega} = \{\Gamma_{12} \cdot \Gamma_{34} - \omega^2 \cdot (\Gamma_{11} \cdot \Gamma_{33} - \Gamma_{12} - \Gamma_{34} - [\Gamma_{11} \cdot \Gamma_{34} + \Gamma_{12} \cdot \Gamma_{33}] \cdot 2 \cdot \tau_\Delta)\} \cdot \{(\Gamma_{11} \cdot \Gamma_{34} + \Gamma_{12} \cdot \Gamma_{33} - \Gamma_{12} \cdot \Gamma_{34} \cdot 3 \cdot \tau_\Delta) - 3 \cdot \omega^2 \cdot ([-\Gamma_{11} \cdot \Gamma_{33} + \Gamma_{12} + \Gamma_{34}] \cdot \tau_\Delta - \Gamma_{33} - \Gamma_{11})\}$$

$$Q_I \cdot Q_{R\omega} = -2 \cdot \omega \cdot \{\omega \cdot (\Gamma_{11} \cdot \Gamma_{34} + \Gamma_{12} \cdot \Gamma_{33} - \Gamma_{12} \cdot \Gamma_{34} \cdot 3 \cdot \tau_\Delta) - \omega^3 \cdot ([-\Gamma_{11} \cdot \Gamma_{33} + \Gamma_{12} + \Gamma_{34}] \cdot \tau_\Delta - \Gamma_{33} - \Gamma_{11})\} \cdot \{(\Gamma_{11} \cdot \Gamma_{33} - \Gamma_{12} - \Gamma_{34} - [\Gamma_{11} \cdot \Gamma_{34} + \Gamma_{12} \cdot \Gamma_{33}] \cdot 2 \cdot \tau_\Delta)\}$$

$$V = (P_R \cdot P_{I\tau_\Delta} - P_I \cdot P_{R\tau_\Delta}) - (Q_R \cdot Q_{I\tau_\Delta} - Q_I \cdot Q_{R\tau_\Delta}); P_R \cdot P_{I\tau_\Delta} = 0; P_I \cdot P_{R\tau_\Delta} = 0$$

$$Q_R \cdot Q_{I\tau_\Delta} = -\{\Gamma_{12} \cdot \Gamma_{34} - \omega^2 \cdot (\Gamma_{11} \cdot \Gamma_{33} - \Gamma_{12} - \Gamma_{34} - [\Gamma_{11} \cdot \Gamma_{34} + \Gamma_{12} \cdot \Gamma_{33}] \cdot 2 \cdot \tau_\Delta)\} \cdot \omega \cdot \{\Gamma_{12} \cdot \Gamma_{34} \cdot 3 + \omega^2 \cdot [-\Gamma_{11} \cdot \Gamma_{33} + \Gamma_{12} + \Gamma_{34}]\}$$

$$Q_I \cdot Q_{R\tau_\Delta} = \{\omega \cdot (\Gamma_{11} \cdot \Gamma_{34} + \Gamma_{12} \cdot \Gamma_{33} - \Gamma_{12} \cdot \Gamma_{34} \cdot 3 \cdot \tau_\Delta) - \omega^3 \cdot ([-\Gamma_{11} \cdot \Gamma_{33} + \Gamma_{12} + \Gamma_{34}] \cdot \tau_\Delta - \Gamma_{33} - \Gamma_{11})\} \cdot \{\omega^2 \cdot [\Gamma_{11} \cdot \Gamma_{34} + \Gamma_{12} \cdot \Gamma_{33}] \cdot 2\}; F(\omega, \tau_\Delta) = 0$$

Differentiating with respect to τ_Δ and we get

$$F_\omega \cdot \frac{\partial \omega}{\partial \tau_\Delta} + F_{\tau_\Delta} = 0; \tau_\Delta \in I \Rightarrow \frac{\partial \omega}{\partial \tau_\Delta} = -\frac{F_{\tau_\Delta}}{F_\omega}$$

$$\wedge^{-1}(\tau_\Delta) = (\frac{\partial \mathrm{Re}\lambda}{\partial \tau_\Delta})_{\lambda=i\cdot\omega}; \wedge^{-1}(\tau_\Delta) = \mathrm{Re}\{\frac{-2\cdot[U+\tau_\Delta\cdot|P|^2]+i\cdot F_\omega}{F_{\tau_\Delta}+i\cdot 2\cdot[V+\omega\cdot|P|^2]}\}; \frac{\partial \omega}{\partial \tau_\Delta} = \omega_{\tau_\Delta} = -\frac{F_{\tau_\Delta}}{F_\omega}$$

$$sign\{\wedge^{-1}(\tau_\Delta)\} = sign\{(\frac{\partial \mathrm{Re}\lambda}{\partial \tau_\Delta})_{\lambda=i\cdot\omega}\};$$

$$sign\{\wedge^{-1}(\tau_\Delta)\} = sign\{F_\omega\}\cdot sign\{\tau_\Delta\cdot\frac{\partial \omega}{\partial \tau_\Delta} + \omega + \frac{U\cdot\frac{\partial \omega}{\partial \tau_\Delta}+V}{|P|^2}\}$$

We shall presently examine the possibility of stability transitions (bifurcations) in a N-turn multilayer coil antennas system, about the equilibrium point $E^{(0)}(Y_1^{(0)}, V_{L_{calc-1}}^{(0)}, Y_2^{(0)}, V_{L_{calc-2}}^{(0)})$ as a result of a variation of delay parameter τ_Δ. The analysis consists in identifying the roots of our system characteristic equation situated on the imaginary axis of the complex λ-plane where by increasing the delay parameter τ_Δ, Re λ may at the crossing, change its sign from − to +, i.e. from a stable focus $E^{(*)}$ to an unstable one, or vice versa. This feature may be further assessed by examining the sign of the partial derivatives with respect to τ_Δ, $\wedge^{-1}(\tau_\Delta) = (\frac{\partial \mathrm{Re}\lambda}{\partial \tau_\Delta})_{\lambda=i\cdot\omega}$, C_1, R_1 and N-turn multilayer coil antennas system parameters are constant where $\omega \in \mathbb{R}_+$. We need to plot the stability switch diagram based on different delay values of our N-turn multilayer coil antennas system. Since it is a very complex function we recommend to solve it numerically rather than analytic.

$$\wedge^{-1}(\tau_\Delta) = (\frac{\partial \mathrm{Re}\lambda}{\partial \tau_\Delta})_{\lambda=i\cdot\omega} = \mathrm{Re}\{\frac{-2\cdot[U+\tau_\Delta\cdot|P|^2]+i\cdot F_\omega}{F_{\tau_\Delta}+i\cdot 2\cdot[V+\omega\cdot|P|^2]}\}$$

$$\wedge^{-1}(\tau_\Delta) = (\frac{\partial \mathrm{Re}\lambda}{\partial \tau_\Delta})_{\lambda=i\cdot\omega} = \frac{2\cdot\{F_\omega\cdot(V+\omega\cdot P^2) - F_{\tau_\Delta}\cdot(U+\tau_\Delta\cdot P^2)\}}{F_{\tau_\Delta}^2 + 4\cdot(V+\omega\cdot P^2)^2}$$

The stability switch occurs only on those delay values (τ_Δ) which fit the equation: $\tau_\Delta = \frac{\theta_+(\tau_\Delta)}{\omega_+(\tau_\Delta)}$ and $\theta_+(\tau_\Delta)$ is the solution of $\sin\theta(\tau) = \ldots; \cos\theta(\tau) = \ldots$ when $\omega = \omega_+(\tau_\Delta)$ if only ω_+ is feasible. Additionally when all N-turn multilayer coil antennas system parameters are known and the stability switch due to various time delay values τ_Δ is describe in the following expression:

$$sign\{\wedge^{-1}(\tau_\Delta)\} = sign\{F_\omega(\omega(\tau_\Delta), \tau_\Delta)\}\cdot sign\{\tau_\Delta\cdot\omega_{\tau_\Delta}(\omega(\tau_\Delta)) + \omega(\tau_\Delta) + \frac{U(\omega(\tau_\Delta))\cdot\omega_{\tau_\Delta}(\omega(\tau_\Delta)) + V(\omega(\tau_\Delta))}{|P(\omega(\tau_\Delta))|^2}\}$$

Remark: we know $F(\omega, \tau_\Delta) = 0$ implies it roots $\omega_i(\tau_\Delta)$ and finding those delays values τ_Δ which ω_i is feasible. There are τ_Δ values which ω_i are complex or imaginary numbers, then unable to analyse stability.

Lemma 1.1 *Assume that $\omega(\tau)$ is a positive and real root of $F(\omega, \tau) = 0$ defined for $\tau \in I$, which is continuous and differentiable. Assume further that if $\lambda = i \cdot \omega$, $\omega \in R$, then $P_n(i \cdot \omega, \tau) + Q_n(i \cdot \omega, \tau) \neq 0, \tau \in \mathrm{R}$ hold true. The functions $S_n(\tau), n \in N_0$, are continuous and differentiable on I.*

Theorem 1.2 *Assume that $\omega(\tau)$ is a positive real root of $F(\omega, \tau) = 0$ defined for $\tau \in I, I \subseteq R_{+0}$, and at some $\tau^* \in I$, $S_n(\tau^*) = 0$ for some $n \in N_0$ then a pair of simple conjugate pure imaginary roots $\lambda_+(\tau^*) = i \cdot \omega(\tau^*), \lambda_-(\tau^*) = -i \cdot \omega(\tau^*)$ of $D(\lambda, \tau) = 0$ exist at $\tau = \tau^*$ which crosses the imaginary axis from left to right if $\delta(\tau^*) > 0$ and cross the imaginary axis from right to left if $\delta(\tau^*) < 0$ where*

$$\delta(\tau^*) = sign\{\frac{d\mathrm{Re}\lambda}{d\tau}|_{\lambda=i\omega(\tau^*)}\} = sign\{F_\omega(\omega(\tau^*), \tau^*)\} \cdot sign\{\frac{dS_n(\tau)}{d\tau}|_{\tau=\tau^*}\}$$

Theorem 1.3 *The characteristic equation has a pair of simple and conjugate pure imaginary roots $\lambda = \pm\omega(\tau^*), \omega(\tau^*)$ real at $\tau^* \in I$ if $S_n(\tau^*) = \tau^* - \tau_n(\tau^*) = 0$ for some $n \in N_0$. If $\omega(\tau^*) = \omega_+(\tau^*)$, this pair of simple conjugate pure imaginary roots crosses the imaginary axis from left to right if $\delta_+(\tau^*) > 0$ and crosses the imaginary axis from right to left if $\delta_+(\tau^*) < 0$ where $\delta_+(\tau^*) = sign\{\frac{d\mathrm{Re}\lambda}{d\tau}|_{\lambda=i\omega_+(\tau^*)}\}$*

$$\delta_+(\tau^*) = sign\{\frac{d\mathrm{Re}\lambda}{d\tau}|_{\lambda=i\omega_+(\tau^*)}\} = sign\{\frac{dS_n(\tau)}{d\tau}|_{\tau=\tau^*}\}$$

If $\omega(\tau^*) = \omega_-(\tau^*)$, this pair of simple conjugate pure imaginary roots cross the imaginary axis from left to right, if $\delta_-(\tau^*) > 0$ and crosses the imaginary axis from right to left. If $\delta_-(\tau^*) < 0$ where $\delta_-(\tau^*) = sign\{\frac{d\mathrm{Re}\lambda}{d\tau}|_{\lambda=i\omega_-(\tau^*)}\} = -sign\{\frac{dS_n(\tau)}{d\tau}|_{\tau=\tau^*}\}$

If $\omega_+(\tau^*) = \omega_-(\tau^*) = \omega(\tau^*)$ then $\Delta(\tau^*) = 0$ and $sign\{\frac{d\mathrm{Re}\lambda}{d\tau}|_{\lambda=i\omega(\tau^*)}\} = 0$, the same is true when $S'_n(\tau^*) = 0$ the following result can be useful in identifying values of τ where stability switches happened.

Remark: Lemma 1.1 and Theorems 1.2, 1.3: In the first and second cases we discuss delay parameter τ and in the third case we discuss delay parameter τ_Δ [18, 19].

8.2 Double Rectangular Spiral Coils Antennas System Stability Optimization Under Delayed Electromagnetic Interferences and Parasitic Effects

A double rectangular spiral antennas are constructed from two antennas, each antenna is a rectangular spiral antenna. Antennas are connected in series with micro strip line and to the RFID IC. A double rectangular spiral antennas system influence by electromagnetic interferences which effect there stability behavior. Additionally, micro strip line which connected antennas in the series has parasitic effect, delay in time Δ_{μ}. We inspect our system performances under electromagnetic interferences and micro strip parasitic effects. Generally double spiral micro strip antenna over rectangular patch improved the bandwidth of Mobile, WiMAX applications. Micro strip line feed technique enhances the bandwidth of the simple micro strip antenna. Every rectangular spiral antenna has a parasitic DC resistance which needs to be calculated. The below figure describes the double rectangular spiral antenna system [1, 7, 8] (Fig. 8.3).

Index (i) indicates first rectangular spiral antenna (i = 1) or second rectangular spiral antenna (i = 2). We define RFID's coil dimensional parameters. A_{0i}, B_{0i}—Overal dimensions of the coil. $Aavg_i$, $Bavg_i$—Average dimensions of the coil. t_i—track thickness. w_i—track width. g_i—gaps between tracks. Nc_i—number of turns. d_i—equivalent diameter of the track. Average coil area $Ac_i = Aavg_i \cdot Bavg_i$. P_i—coil manufacturing technology parameter. Integrating all those parameters gives the equations for rectangular spiral antenna inductance calculation:

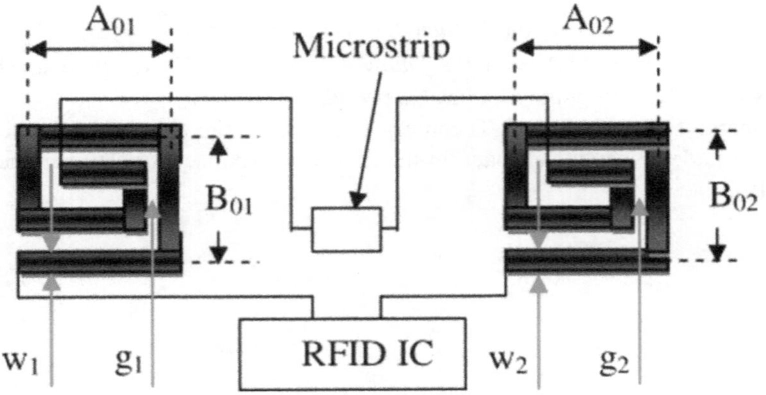

Fig. 8.3 Double rectangular spiral coils antennas system

$$L_{calc-i} = \frac{\mu_0}{\pi} \cdot \left(\sum_{k=1,k\neq 3}^{4} X_{ki} - X_3 \right) \cdot N_{ci}^{P_i}; X_{1i} = A_{avgi} \cdot \ln\left(\frac{2 \cdot A_{avgi} \cdot B_{avgi}}{d_i \cdot (A_{avgi} + \sqrt{A_{avgi}^2 + B_{avgi}^2})} \right)$$

$$X_{2i} = B_{avgi} \cdot \ln\left(\frac{2 \cdot A_{avgi} \cdot B_{avgi}}{d_i \cdot (B_{avgi} + \sqrt{A_{avgi}^2 + B_{avgi}^2})} \right); X_{3i} = 2 \cdot (A_{avgi} + B_{avgi} - \sqrt{A_{avgi}^2 + B_{avgi}^2})$$

$$X_{4i} = \frac{(A_{avgi} + B_{avgi})}{4}; d_i = \frac{2 \cdot (t_i + w_i)}{\pi}; A_{avgi} = A_{0i} - N_{ci} \cdot (g_i + w_i);$$

$$B_{avgi} = B_{0i} - N_{ci} \cdot (g_i + w_i)$$

The rectangular spiral antenna length is calculated as follows: l_{0i} is the length of the first turn $l_{0i} = 2 \cdot (A_{0i} + B_{0i}) - (w_i + g_i)$. l_k is the length of turn k + 1. We define the following:

$$l_{Ti} = l_{0i} + \sum_{k=1}^{N_{ci}-1} \{A_{0i} - [1 + (k-1) \cdot 2] \cdot (w_i + g_i) + B_{0i} - [2 + (k-1) \cdot 2] \cdot (w_i + g_i)$$
$$+ A_{0i} - [2 + (k-1) \cdot 2] \cdot (w_i + g_i) + B_{0i} - [3 + (k-1) \cdot 2] \cdot (w_i + g_i)\}$$

$$L_{Ti} = L_{0i} + 2 \cdot (A_{0i} + B_{0i}) \cdot (N_{Ci} - 1) - 8 \cdot (w_i + g_i) \cdot \sum_{k=1}^{N_C-1} k; \sum_{k=1}^{N_C-1} k = N_{Ci} - 1$$

$$L_{Ti} = 2 \cdot \{(A_{0i} + B_{0i}) \cdot (1 + N_{Ci}) - (w_i + g_i) \cdot [4 \cdot N_{Ci} - 3]\}$$

The DC resistance of rectangular spiral RFID antenna: $R_{DC-i} = \frac{l_{Ti}}{\sigma_i \cdot S_i} = \frac{l_{Ti}}{\sigma_i \cdot \pi \cdot a_i^2}$. l_{Ti}—total length of the wire. σ_i—conductivity of the wire (mΩ/m). S_i—Cross section area $\pi \cdot a_i^2$. a_i—radius of the wire.

Remark: $a_i^2 = w_i^2$.

$$R_{DC-i} = \frac{2 \cdot \{(A_{0i} + B_{0i}) \cdot (1 + N_{Ci}) - (w_i + g_i) \cdot [4 \cdot N_{Ci} - 3]\}}{\sigma_i \cdot \pi \cdot w_i^2}$$

Due to electromagnetic interferences there are different in time delays respect to first and second rectangular spiral antennas voltages and voltages derivatives. The delayed voltages are $V_1(t - \tau_1)$ and $V_2(t - \tau_2)$ respectively ($\tau_1 \neq \tau_2$) and delayed voltages derivatives are $dV_1(t - \Delta_1)/dt$, $dV_2(t - \Delta_2)/dt$ respectively $(\Delta_1 \neq \Delta_2; \tau_1 \geq 0);(\tau_2 \geq 0; \Delta_1, \Delta_2 \geq 0)$. Additionally, there is a delay in time for the micro strip parasitic effects Δ_μ. The double rectangular spiral antenna system equivalent circuit can represent as delayed differential equations which, depending on variable parameters and delays. Our double rectangular spiral antennas system delay differential and delay different model can be analytically by using delay differential equations in dynamically. The need of the incorporation of a time delay is often of the existence of any stage structure. It is often difficult to analytically study models with delay dependent parameters, even if only a single discrete delay is present. There are practical guidelines that combine graphical information with analytical work to effectively study the local stability of models involving delay dependent parameters. The stability of a given steady state is simply determined by the graphs of some function of $\tau_{1,}$ τ_2 which can be expressed, explicitly and thus can be easily depicted by MATLAB and other popular software. We need only look at one such function and locate the zeros. This function often has only two zeros, providing thresholds for stability switches. As time delay increases, stability changes from stable to unstable to stable. We emphasize the local stability aspects of some models with delay dependent parameters, additionally there is a general geometric criterion that, theoretically speaking, can be applied to models with many delays, or even distributed delays. The simplest case of a first order characteristic equation, providing more user friendly geometric and analytic criteria for stability switches. The analytical criteria provided for the first and second order cases can be used to obtain some insightful analytical statements and can be helpful for conducting simulations. Double rectangular spiral antennas can be represented as a two inductors in series (L_{calc-1} and L_{calc-2}), parasitic resistances (R_{DC-1} and R_{DC-2}) and micro strip delay line. The rectangular spiral antennas in series are connected in parallel to RFID TAG IC. The Equivalent Circuit of Passive RFID TAG with double rectangular antennas is Capacitor (C_1) and Resistor (R_1) in parallel with double rectangular antennas in the series [85] (Fig. 8.4).

L_{calc-1} and L_{calc-2} are mostly formed by traces on planar PCB. $2 \cdot L_m$ element represents the mutual inductance between L_{calc-1} and L_{calc-2}. Since two inductors (L_{calc-1}, L_{calc-2}) are in series and there is a mutual inductance between L_{calc-1} and L_{calc-2}, the total antenna inductance L_T: $L_T = L_{calc-1} + L_{calc-2} + 2 \cdot L_m$ and $L_m = K \cdot \sqrt{L_{calc-1} \cdot L_{calc-2}}$. L_m is the mutual inductance between L_{calc-1} and L_{calc-2}. K is the coupling coefficient of two inductors $0 \leq K \leq 1$. I(t) is the current that flow through a double rectangular spiral antenna. V_1 and V_1 are the voltages on L_{calc-1} and L_{calc-2} respectively. V_m is the voltage on double loop antenna mutual inductance element. We neglect the voltage on microstrip delay line $V\mu \to \varepsilon$. $I_{L_{calc-1}} = I(t)$ $I_{L_{calc-2}} = I(t - \Delta_\mu); I_{Rdc-2} = I(t - \Delta_\mu); I_{L_m} = I(t - \Delta_\mu)$.

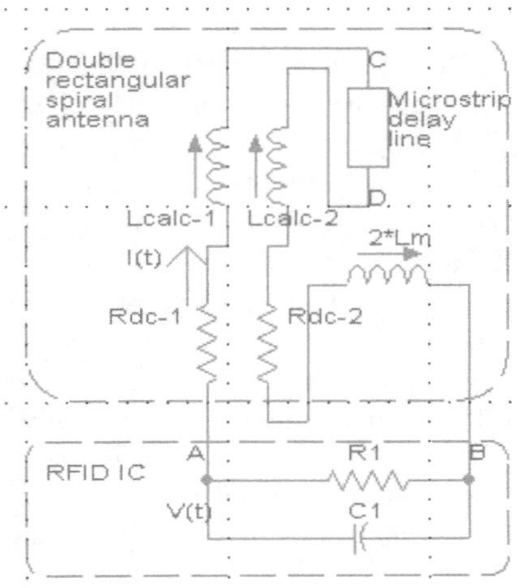

Fig. 8.4 Double rectangular spiral coils antennas system equivalent circuit

$$V_1(t) = V_{L_{calc-1}} = V_{L_{calc-1}}(t); V_2(t) = V_{L_{calc-2}} = V_{L_{calc-2}}(t)$$
$$V_{AB} = V_{R_1} = V_{C_1}; V_{AB} = R_1 \cdot I_{R_1}; V_{AB} = \frac{1}{C_1} \cdot \int I_{C_1} \cdot dt; I_{C_1} = C_1 \cdot \frac{dV_{AB}}{dt}; I(t) + I_{C_1} + I_{R_1} = 0$$
$$I(t) + \frac{V_{AB}}{R_1} + C_1 \cdot \frac{dV_{AB}}{dt} = 0; V_{AB} = V_{L_{calc-1}} + V_{Rdc-1} + V_{L_{calc-2}} + V_{Rdc-2} + V_m + (V_\mu \to \varepsilon)$$
$$V_{Rdc-1} = I(t) \cdot R_{dc-1} = I(t) \cdot \frac{l_{T1}}{\sigma_1 \cdot \pi \cdot a_1^2}; V_{Rdc-2} = I(t - \Delta_\mu) \cdot R_{dc-2} = I(t - \Delta_\mu) \cdot \frac{l_{T2}}{\sigma_2 \cdot \pi \cdot a_2^2}$$
$$V_{L_{calc-1}} = L_{calc-1} \cdot \frac{dI(t)}{dt}; V_{L_{calc-2}} = L_{calc-2} \cdot \frac{dI(t - \Delta_\mu)}{dt};$$
$$V_m = 2 \cdot L_m \cdot \frac{dI(t - \Delta_\mu)}{dt}; L_{calc-1} \neq L_{calc-2}$$

$$C_1 \cdot \frac{dV_{C_1}}{dt} + \frac{V_{C_1}}{R_1} + I(t) = 0; V_{C_1} = V_{L_{calc-1}} + V_{Rdc-1} + V_{L_{calc-2}} + V_{Rdc-2} + V_m$$
$$\frac{dV_{R_1}}{dt} = \frac{dV_{C_1}}{dt} = \frac{dV_{L_{calc-1}}}{dt} + \frac{dV_{Rdc-1}}{dt} + \frac{dV_{L_{calc-2}}}{dt} + \frac{dV_{Rdc-2}}{dt} + \frac{dV_m}{dt}$$
$$I(t) = \frac{1}{L_{calc-1}} \cdot \int V_{L_{calc-1}} \cdot dt; I(t - \Delta_\mu) = \frac{1}{L_{calc-2}} \cdot \int V_{L_{calc-2}} \cdot dt$$
$$\frac{dI(t)}{dt} = \frac{V_{L_{calc-1}}}{L_{calc-1}}; \frac{dI(t - \Delta_\mu)}{dt} = \frac{V_{L_{calc-2}}}{L_{calc-2}}; \left|\frac{dI(t - \Delta_\mu)}{dt} - \frac{dI(t)}{dt}\right| \leq \Omega$$
$$\Omega \to \varepsilon; \frac{dI(t - \Delta_\mu)}{dt} \approx \frac{dI(t)}{dt} \Rightarrow \frac{V_{L_{calc-1}}}{L_{calc-1}} = \frac{V_{L_{calc-2}}}{L_{calc-2}};$$
$$V_m = 2 \cdot L_m \cdot \frac{dI(t - \Delta_\mu)}{dt} = 2 \cdot L_m \cdot \frac{V_{L_{calc-2}}}{L_{calc-2}}$$

$$V_m = 2 \cdot K \cdot \{\sqrt{L_{calc-1} \cdot L_{calc-2}}\} \cdot \frac{V_{L_{calc-2}}}{L_{calc-2}} = 2 \cdot K \cdot \sqrt{\frac{L_{calc-1}}{L_{calc-2}}} \cdot V_{L_{calc-2}}$$
$$\frac{dV_m}{dt} = 2 \cdot K \cdot \{\sqrt{L_{calc-1} \cdot L_{calc-2}}\} \cdot \frac{dVL_{calc-2}/dt}{L_{calc-2}} = 2 \cdot K \cdot \sqrt{\frac{L_{calc-1}}{L_{calc-2}}} \cdot \frac{dV_{L_{calc-2}}}{dt}$$

We get the following differential equation respect to $V_{L_{calc-1}}$ variable:

$$\frac{1}{L_{calc-1}} \cdot \int V_{L_{calc-1}} \cdot dt + \frac{V_{AB}}{R_1} + C_1 \cdot \frac{dV_{AB}}{dt} = 0$$
$$\frac{1}{L_{calc-1}} \cdot \int V_{L_{calc-1}} \cdot dt + \frac{1}{R_1} \cdot [V_{L_{calc-1}} + V_{Rdc-1} + V_{L_{calc-2}} + V_{Rdc-2} + V_m]$$
$$+ C_1 \cdot [\frac{dV_{L_{calc-1}}}{dt} + \frac{dV_{Rdc-1}}{dt} + \frac{dV_{L_{calc-2}}}{dt} + \frac{dV_{Rdc-2}}{dt} + \frac{dV_m}{dt}] = 0$$

$$V_{Rdc-1} = I(t) \cdot R_{dc-1}; V_{Rdc-2} = I(t - \Delta_\mu) \cdot R_{dc-2}$$

$$\frac{1}{L_{calc-1}} \cdot \int V_{L_{calc-1}} \cdot dt + \frac{1}{R_1} \cdot [V_{L_{calc-1}} + I(t) \cdot R_{dc-1} + V_{L_{calc-2}} + I(t-\Delta_\mu) \cdot R_{dc-2} + V_m] + C_1 \cdot [\frac{dV_{L_{calc-1}}}{dt} + \frac{dI(t)}{dt} \cdot R_{dc-1} + \frac{dV_{L_{calc-2}}}{dt} + \frac{dI(t-\Delta_\mu)}{dt} \cdot R_{dc-2} + 2 \cdot K \cdot \sqrt{\frac{L_{calc-1}}{L_{calc-2}}} \cdot \frac{dV_{L_{calc-2}}}{dt}] = 0$$

$$\frac{V_{L_{calc-1}}}{L_{calc-1}} = \frac{dI(t)}{dt}; \frac{V_{L_{calc-2}}}{L_{calc-2}} = \frac{dI(t-\Delta_\mu)}{dt}$$

$$\frac{1}{L_{calc-1}} \cdot \int V_{L_{calc-1}} \cdot dt + \frac{1}{R_1} \cdot [V_{L_{calc-1}} + \frac{R_{dc-1}}{L_{calc-1}} \cdot \int V_{L_{calc-1}} \cdot dt + V_{L_{calc-2}} + \frac{R_{dc-2}}{L_{calc-2}} \cdot \int V_{L_{calc-2}} \cdot dt + 2 \cdot K \cdot \sqrt{\frac{L_{calc-1}}{L_{calc-2}}} \cdot V_{L_{calc-2}}] + C_1 \cdot [\frac{dV_{L_{calc-1}}}{dt} + \frac{V_{L_{calc-1}}}{L_{calc-1}} \cdot R_{dc-1} + \frac{dV_{L_{calc-2}}}{dt} + \frac{V_{L_{calc-2}}}{L_{calc-2}} \cdot R_{dc-2} + 2 \cdot K \cdot \sqrt{\frac{L_{calc-1}}{L_{calc-2}}} \cdot \frac{dV_{L_{calc-2}}}{dt}] = 0$$

We derivative in time the two sides of the above equation.

$$\frac{V_{L_{calc-1}}}{L_{calc-1}} + \frac{1}{R_1} \cdot [\frac{dV_{L_{calc-1}}}{dt} + \frac{R_{dc-1}}{L_{calc-1}} \cdot V_{L_{calc-1}} + \frac{dV_{L_{calc-2}}}{dt} + \frac{R_{dc-2}}{L_{calc-2}} \cdot V_{L_{calc-2}} + 2 \cdot K \cdot \sqrt{\frac{L_{calc-1}}{L_{calc-2}}} \cdot \frac{dV_{L_{calc-2}}}{dt}] + C_1 \cdot [\frac{d^2 V_{L_{calc-1}}}{dt^2} + \frac{R_{dc-1}}{L_{calc-1}} \cdot \frac{dV_{L_{calc-1}}}{dt} + \frac{d^2 V_{L_{calc-2}}}{dt^2} + \frac{R_{dc-2}}{L_{calc-2}} \cdot \frac{dV_{L_{calc-2}}}{dt} + 2 \cdot K \cdot \sqrt{\frac{L_{calc-1}}{L_{calc-2}}} \cdot \frac{d^2 V_{L_{calc-2}}}{dt^2}] = 0$$

$$V_{L_{calc-2}} = \frac{L_{calc-2}}{L_{calc-1}} \cdot V_{L_{calc-1}}$$

Then we get the following expression:

$$\frac{V_{L_{calc-1}}}{L_{calc-1}} + \frac{1}{R_1} \cdot \frac{dV_{L_{calc-1}}}{dt} + \frac{R_{dc-1}}{R_1 \cdot L_{calc-1}} \cdot V_{L_{calc-1}} + \frac{L_{calc-2}}{R_1 \cdot L_{calc-1}} \cdot \frac{dV_{L_{calc-1}}}{dt} + \frac{R_{dc-2}}{R_1 \cdot L_{calc-1}} \cdot V_{L_{calc-1}} + 2 \cdot K \cdot \frac{L_{calc-2}}{R_1 \cdot L_{calc-1}} \cdot \sqrt{\frac{L_{calc-1}}{L_{calc-2}}} \cdot \frac{dV_{L_{calc-1}}}{dt} + C_1 \cdot \frac{d^2 V_{L_{calc-1}}}{dt^2} + \frac{C_1 \cdot R_{dc-1}}{L_{calc-1}} \cdot \frac{dV_{L_{calc-1}}}{dt} + \frac{C_1 \cdot L_{calc-2}}{L_{calc-1}} \cdot \frac{d^2 V_{L_{calc-1}}}{dt^2} + \frac{C_1 \cdot L_{calc-2}}{L_{calc-1}} \cdot \frac{R_{dc-2}}{L_{calc-2}} \cdot \frac{dV_{L_{calc-1}}}{dt} + 2 \cdot K \cdot \frac{C_1 \cdot L_{calc-2}}{L_{calc-1}} \cdot \sqrt{\frac{L_{calc-1}}{L_{calc-2}}} \cdot \frac{d^2 V_{L_{calc-1}}}{dt^2} = 0$$

$$[\frac{1}{L_{calc-1}} + \frac{(R_{dc-1} + R_{dc-2})}{R_1 \cdot L_{calc-1}}] \cdot V_{L_{calc-1}} + [\frac{1}{R_1} + \frac{L_{calc-2}}{R_1 \cdot L_{calc-1}} + 2 \cdot K \cdot \frac{1}{R_1} \cdot \sqrt{\frac{L_{calc-2}}{L_{calc-1}}} + \frac{C_1 \cdot (R_{dc-1} + R_{dc-2})}{L_{calc-1}}] \cdot \frac{dV_{L_{calc-1}}}{dt} + C_1 \cdot [1 + \frac{L_{calc-2}}{L_{calc-1}} + 2 \cdot K \cdot \sqrt{\frac{L_{calc-2}}{L_{calc-1}}}] \cdot \frac{d^2 V_{L_{calc-1}}}{dt^2} = 0$$

We get the following differential equation respect to $V_{L_{calc-1}}$ variable, η_1, η_2, η_3 are global parameters.

$$\eta_1 \cdot V_{L_{calc-1}} + \eta_2 \cdot \frac{dV_{L_{calc-1}}}{dt} + \eta_3 \cdot \frac{d^2 V_{L_{calc-1}}}{dt^2} = 0$$

$$\eta_1 = \frac{1}{L_{calc-1}} + \frac{(R_{dc-1} + R_{dc-2})}{R_1 \cdot L_{calc-1}};$$

$$\eta_2 = \frac{L_{calc-2}}{R_1 \cdot L_{calc-1}} + 2 \cdot K \cdot \frac{1}{R_1} \cdot \sqrt{\frac{L_{calc-2}}{L_{calc-1}}} + \frac{C_1 \cdot (R_{dc-1} + R_{dc-2})}{L_{calc-1}} + \frac{1}{R_1}$$

$$\eta_3 = C_1 \cdot [1 + \frac{L_{calc-2}}{L_{calc-1}} + 2 \cdot K \cdot \sqrt{\frac{L_{calc-2}}{L_{calc-1}}}\}];$$

$$\eta_1 = \eta_1(R_1, L_{calc-1} R_{dc-1}, R_{dc-2})$$

$$\eta_2 = \eta_2(L_{calc-1}, L_{calc-2}, R_1, C_1, K, R_{dc-1}, R_{dc-2});$$

$$\eta_3 = \eta_3(L_{calc-1}, L_{calc-2}, C_1, K)$$

$$X_1 = \frac{dV_{L_{calc-1}}}{dt}; \frac{dX_1}{dt} = \frac{d^2 V_{L_{calc-1}}}{dt^2}; \frac{dX_1}{dt} = -\frac{\eta_1}{\eta_3} \cdot V_{L_{calc-1}} - \frac{\eta_2}{\eta_3} \cdot X_1; \frac{dV_{L_{calc-1}}}{dt} = X_1$$

In the same manner we find our $V_{L_{calc-2}}$ differential equation. We get the following differential equation respect to $V_{L_{calc-1}}$ variable, ξ_1, ξ_2, ξ_3 are global parameters.

$$\frac{V_{L_{calc-1}}}{L_{calc-1}} + \frac{1}{R_1} \cdot [\frac{dV_{L_{calc-1}}}{dt} + \frac{R_{dc-1}}{L_{calc-1}} \cdot V_{L_{calc-1}} + \frac{dV_{L_{calc-2}}}{dt} + \frac{R_{dc-2}}{L_{calc-2}} \cdot V_{L_{calc-2}}$$
$$+ 2 \cdot K \cdot \sqrt{\frac{L_{calc-1}}{L_{calc-2}}} \cdot \frac{dV_{L_{calc-2}}}{dt}] + C_1 \cdot [\frac{d^2 V_{L_{calc-1}}}{dt^2} + \frac{R_{dc-1}}{L_{calc-1}} \cdot \frac{dV_{L_{calc-1}}}{dt}$$
$$+ \frac{d^2 V_{L_{calc-2}}}{dt^2} + \frac{R_{dc-2}}{L_{calc-2}} \cdot \frac{dV_{L_{calc-2}}}{dt} + 2 \cdot K \cdot \sqrt{\frac{L_{calc-1}}{L_{calc-2}}} \cdot \frac{d^2 V_{L_{calc-2}}}{dt^2}] = 0$$

$$V_{L_{calc-1}} = \frac{L_{calc-1}}{L_{calc-2}} \cdot V_{L_{calc-2}}$$

$$\frac{1}{L_{calc-2}} \cdot V_{L_{calc-2}} + \frac{1}{R_1} \cdot [\frac{L_{calc-1}}{L_{calc-2}} \cdot \frac{dV_{L_{calc-2}}}{dt} + \frac{R_{dc-1}}{L_{calc-2}} \cdot V_{L_{calc-2}} + \frac{dV_{L_{calc-2}}}{dt} + \frac{R_{dc-2}}{L_{calc-2}} \cdot V_{L_{calc-2}}$$
$$+ 2 \cdot K \cdot \sqrt{\frac{L_{calc-1}}{L_{calc-2}}} \cdot \frac{dV_{L_{calc-2}}}{dt}] + C_1 \cdot [\frac{L_{calc-1}}{L_{calc-2}} \cdot \frac{d^2 V_{L_{calc-2}}}{dt^2} + \frac{R_{dc-1}}{L_{calc-2}} \cdot \frac{dV_{L_{calc-2}}}{dt} + \frac{d^2 V_{L_{calc-2}}}{dt^2}$$
$$+ \frac{R_{dc-2}}{L_{calc-2}} \cdot \frac{dV_{L_{calc-2}}}{dt} + 2 \cdot K \cdot \sqrt{\frac{L_{calc-1}}{L_{calc-2}}} \cdot \frac{d^2 V_{L_{calc-2}}}{dt^2}] = 0$$

$$\frac{1}{L_{calc-2}} \cdot V_{L_{calc-2}} + \frac{L_{calc-1}}{R_1 \cdot L_{calc-2}} \cdot \frac{dV_{L_{calc-2}}}{dt} + \frac{R_{dc-1}}{R_1 \cdot L_{calc-2}} \cdot V_{L_{calc-2}} + \frac{1}{R_1} \cdot \frac{dV_{L_{calc-2}}}{dt} + \frac{R_{dc-2}}{R_1 \cdot L_{calc-2}} \cdot V_{L_{calc-2}}$$
$$+ \frac{2 \cdot K}{R_1} \cdot \sqrt{\frac{L_{calc-1}}{L_{calc-2}}} \cdot \frac{dV_{L_{calc-2}}}{dt} + \frac{C_1 \cdot L_{calc-1}}{L_{calc-2}} \cdot \frac{d^2 V_{L_{calc-2}}}{dt^2} + \frac{C_1 \cdot R_{dc-1}}{L_{calc-2}} \cdot \frac{dV_{L_{calc-2}}}{dt} + C_1 \cdot \frac{d^2 V_{L_{calc-2}}}{dt^2}$$
$$+ \frac{C_1 \cdot R_{dc-2}}{L_{calc-2}} \cdot \frac{dV_{L_{calc-2}}}{dt} + 2 \cdot K \cdot C_1 \cdot \sqrt{\frac{L_{calc-1}}{L_{calc-2}}} \cdot \frac{d^2 V_{L_{calc-2}}}{dt^2} = 0$$

$$[\frac{1}{L_{calc-2}} + \frac{(R_{dc-1} + R_{dc-2})}{R_1 \cdot L_{calc-2}}] \cdot V_{L_{calc-2}} + [\frac{L_{calc-1}}{R_1 \cdot L_{calc-2}} + \frac{1}{R_1} + \frac{2 \cdot K}{R_1} \cdot \sqrt{\frac{L_{calc-1}}{L_{calc-2}}}$$
$$+ \frac{C_1 \cdot (R_{dc-1} + R_{dc-2})}{L_{calc-2}}] \cdot \frac{dV_{L_{calc-2}}}{dt} + C_1 \cdot [1 + \frac{L_{calc-1}}{L_{calc-2}} + 2 \cdot K \cdot \sqrt{\frac{L_{calc-1}}{L_{calc-2}}}] \cdot \frac{d^2 V_{L_{calc-2}}}{dt^2} = 0$$

$$\xi_1 = \frac{1}{L_{calc-2}} + \frac{(R_{dc-1} + R_{dc-2})}{R_1 \cdot L_{calc-2}}; \xi_2 = \frac{L_{calc-1}}{R_1 \cdot L_{calc-2}} + \frac{1}{R_1} + \frac{2 \cdot K}{R_1} \cdot \sqrt{\frac{L_{calc-1}}{L_{calc-2}}}$$
$$+ \frac{C_1 \cdot (R_{dc-1} + R_{dc-2})}{L_{calc-2}}$$
$$\xi_3 = C_1 \cdot [1 + \frac{L_{calc-1}}{L_{calc-2}} + 2 \cdot K \cdot \sqrt{\frac{L_{calc-1}}{L_{calc-2}}}]; \xi_1 = (L_{calc-1}, L_{calc-2}, R_1, R_{dc-1}, R_{dc-2})$$
$$\xi_2 = (L_{calc-1}, L_{calc-2}, R_1, R_{dc-1}, R_{dc-2}, C_1, K); \xi_3 = (L_{calc-1}, L_{calc-2}, C_1, K)$$

We get the following differential equation respect to $V_{L_{calc-2}}$ variable, ξ_1, ξ_2, ξ_3 are global parameters.

$$\xi_1 \cdot V_{L_{calc-2}} + \xi_2 \cdot \frac{dV_{L_{calc-2}}}{dt} + \xi_3 \cdot \frac{d^2 V_{L_{calc-2}}}{dt^2} = 0; X_2 = \frac{dV_{L_{calc-2}}}{dt}; \frac{dX_2}{dt} = \frac{d^2 V_{L_{calc-2}}}{dt^2}$$
$$\frac{dX_2}{dt} = -\frac{\xi_1}{\xi_3} \cdot V_{L_{calc-2}} - \frac{\xi_2}{\xi_3} \cdot X_2; \frac{dV_{L_{calc-2}}}{dt} = X_2$$

Summary: We get our RFID TAGs with double rectangular spiral antenna system's four differential equations.

$$\frac{dX_1}{dt} = -\frac{\eta_1}{\eta_3} \cdot V_{L_{calc-1}} - \frac{\eta_2}{\eta_3} \cdot X_1; \frac{dV_{L_{calc-1}}}{dt} = X_1; \frac{dX_2}{dt}$$
$$= -\frac{\xi_1}{\xi_3} \cdot V_{L_{calc-2}} - \frac{\xi_2}{\xi_3} \cdot X_2; \frac{dV_{L_{calc-2}}}{dt} = X_2$$

$$\begin{pmatrix} \frac{dX_1}{dt} \\ \frac{dV_{L_{calc-1}}}{dt} \\ \frac{dX_2}{dt} \\ \frac{dV_{L_{calc-2}}}{dt} \end{pmatrix} = \begin{pmatrix} \Gamma_{11} & \cdots & \Gamma_{14} \\ \vdots & \ddots & \vdots \\ \Gamma_{41} & \cdots & \Gamma_{44} \end{pmatrix} \cdot \begin{pmatrix} X_1 \\ V_{L_{calc-1}} \\ X_2 \\ V_{L_{calc-2}} \end{pmatrix}; \Gamma_{11} = -\frac{\eta_2}{\eta_3}; \Gamma_{12} = -\frac{\eta_1}{\eta_3}; \Gamma_{33} = -\frac{\xi_2}{\xi_3};$$

$$\Gamma_34 = -\frac{\xi_1}{\xi_3}; \Gamma_{21} = \Gamma_{43} = 1$$

$$\Gamma_{13} = \Gamma_{14} = \Gamma_{22} = \Gamma_{23} = \Gamma_{24} = \Gamma_{31} = \Gamma_{32} = \Gamma_{41} = \Gamma_{42} = \Gamma_{44} = 0$$

The RFID TAGs with double rectangular spiral antenna system's first and second rectangular spiral antenna are composed of a thin wire or a thin plate element. Units are all in cm, and $a_{1,}$ a_2 are radiuses of the first and second wires in cm. There inductances can be calculated by the following formulas:

$$L_{calc-1} = \frac{\mu_0}{\pi} \cdot \left(\sum_{k=1,k\neq 3}^{4} X_{k1} - X_3 \right) \cdot N_{c1}^{P_1}; L_{calc-2} = \frac{\mu_0}{\pi} \cdot \left(\sum_{k=1,k\neq 3}^{4} X_{k2} - X_3 \right) \cdot N_{c2}^{P_2}$$

Due to electromagnetic interferences there are different in time delays respect to first and second rectangular spiral antennas voltages and voltages derivatives. The delayed voltages are $V_1(t - \tau_1)$ and $V_2(t - \tau_2)$ respectively ($\tau_1 \neq \tau_2$) and delayed voltages derivatives are $dV_1(t - \Delta_1)/dt$, $dV_2(t - \Delta_2)/dt$ respectively.

$$\begin{aligned} &(\Delta_1 \neq \Delta_2; \tau_1 \geq 0; \tau_2 \geq 0; \Delta_1, \Delta_2 \geq 0); V_1(t) = V_{L_{calc-1}} = V_{L_{calc-1}}(t); \\ &V_2(t) = V_{L_{calc-2}} = V_{L_{calc-2}}(t) \\ &V_{L_{calc-1}}(t) \to V_{L_{calc-1}}(t - \tau_1); V_{L_{calc-2}}(t) \to V_{L_{calc-2}}(t - \tau_2); \\ &X_1(t) \to X_1(t - \Delta_1); X_2(t) \to X_2(t - \Delta_2) \end{aligned}$$

We consider no delay effect on $\frac{dX_1}{dt}, \frac{dV_{L_{calc-1}}}{dt}, \frac{dX_2}{dt}, \frac{dV_{L_{calc-2}}}{dt}$. The RFID TAGs with double rectangular spiral antenna system differential equations under electromagnetic interferences (delays terms) influence only RFID first and second rectangular spiral antenna voltages $V_{L_{calc-1}}(t), V_{L_{calc-2}}(t)$ and voltages derivatives $X_1(t)$ and $X_2(t)$ respect to time, there is no influence on $\frac{dX_1(t)}{dt}; \frac{dV_{L_{calc-1}}(t)}{dt}; \frac{dX_2(t)}{dt}; \frac{dV_{L_{calc-2}}(t)}{dt}$.

$$\begin{pmatrix} \frac{dX_1}{dt} \\ \frac{dV_{L_{calc-1}}}{dt} \\ \frac{dX_2}{dt} \\ \frac{dV_{L_{calc-2}}}{dt} \end{pmatrix} = \begin{pmatrix} \Gamma_{11} & \cdots & \Gamma_{14} \\ \vdots & \ddots & \vdots \\ \Gamma_{41} & \cdots & \Gamma_{44} \end{pmatrix} \cdot \begin{pmatrix} X_1(t - \Delta_1) \\ V_{L_{calc-1}}(t - \tau_1) \\ X_2(t - \Delta_2) \\ V_{L_{calc-2}}(t - \tau_2) \end{pmatrix}$$

To find equilibrium points (fixed points) of the RFID TAGs with double rectangular spiral antenna system is by

$$\lim_{t\to\infty} V_{L_{calc-1}}(t-\tau_1) = V_{L_{calc-1}}(t); \lim_{t\to\infty} V_{L_{calc-2}}(t-\tau_2) = V_{L_{calc-2}}(t)$$
$$\lim_{t\to\infty} X_1(t-\Delta_1) = \lim_{t\to\infty} X_1(t); \lim_{t\to\infty} X_2(t-\Delta_2) = \lim_{t\to\infty} X_2(t)$$
$$\frac{dX_1(t)}{dt} = 0; \frac{dV_{L_{calc-1}}(t)}{dt} = 0; \frac{dX_2(t)}{dt} = 0; \frac{dV_{L_{calc-2}}(t)}{dt} = 0; \mathrm{t} \gg \tau_1; \mathrm{t} \gg \tau_2; \mathrm{t} \gg \Delta_1; \mathrm{t} \gg \Delta_2$$

We get four equations and the only fixed point is $E^{(0)}(X_1^{(0)}, V_{L_{calc-1}}^{(0)}, X_2^{(0)}, V_{L_{calc-2}}^{(0)})$ $E^{(0)}(X_1^{(0)}, V_{L_{calc-1}}^{(0)}, X_2^{(0)}, V_{L_{calc-2}}^{(0)}) = (0,0,0,0)$, and since

$$\eta_3 \neq 0 \,\&\, \eta_2 \neq 0 \Rightarrow \Gamma_{11} \neq 0; \xi_3 \neq 0 \,\&\, \xi_1 \neq 0 \Rightarrow \Gamma_{34} \neq 0; \eta_3 \neq 0 \,\&\, \eta_1 \neq 0 \Rightarrow \Gamma_{12} \neq 0; \xi_3 \neq 0 \,\&\, \xi_2 \neq 0 \Rightarrow \Gamma_{33} \neq 0$$

Stability analysis: The standard local stability analysis about any one of the equilibrium points of RFID TAGs with double rectangular spiral antenna system consists in adding to coordinates $[X_1\, V_{L_{calc-1}}\, X_2\, V_{L_{calc-2}}]$ arbitrarily small increments of exponential $[x_1\, v_{L_{calc-1}}\, x_2\, v_{L_{calc-2}}] \cdot e^{\lambda \cdot t}$, and retaining the first order terms in $X_1\, V_{L_{calc-1}}\, X_2\, V_{L_{calc-2}}$. The system of four homogeneous equations leads to a polynomial characteristics equation in the eigenvalues λ. The polynomial characteristics equations accept by set the below voltages and voltages derivative respect to time into two RFID TAGs with double rectangular spiral antenna system equations. RFID TAGs with double rectangular spiral antenna system fixed values with arbitrarily small increments of exponential form $[x_1\, v_{L_{calc-1}}\, x_2\, v_{L_{calc-2}}] \cdot e^{\lambda \cdot t}$ are: i = 0 (first fixed point), i = 1 (second fixed point), i = 2 (third fixed point), etc.,

$$X_1(t) = X_1^{(i)} + x_1 \cdot e^{\lambda \cdot t}; V_{L_{calc-1}}(t) = V_{L_{calc-1}}^{(i)} + v_{L_{calc-1}} \cdot e^{\lambda \cdot t}$$
$$X_2(t) = X_2^{(i)} + x_2 \cdot e^{\lambda \cdot t}; V_{L_{calc-2}}(t) = V_{L_{calc-2}}^{(i)} + v_{L_{calc-2}} \cdot e^{\lambda \cdot t}$$

We choose the above expressions for our $X_1(t), V_{L_{calc-1}}(t)$ and $X_2(t), V_{L_{calc-2}}(t)$ as small displacement $[x_1\, v_{L_{calc-1}}\, x_2\, v_{L_{calc-2}}]$ from the system fixed points at time t = 0.

$$X_1(t=0) = X_1^{(i)} + x_1; V_{L_{calc-1}}(t=0) = V_{L_{calc-1}}^{(i)} + v_{L_{calc-1}}$$
$$X_2(t=0) = X_2^{(i)} + x_2; V_{L_{calc-2}}(t=0) = V_{L_{calc-2}}^{(i)} + v_{L_{calc-2}}$$

For $\lambda < 0$, $t > 0$ the selected fixed point is stable otherwise $\lambda > 0$, $t > 0$ is Unstable. Our system tends to the selected fixed point exponentially for $\lambda < 0$, $t > 0$ otherwise go away from the selected fixed point exponentially. λ is the eigenvalue parameter which establish if the fixed point is stable or Unstable, additionally his absolute value ($|\lambda|$) establish the speed of flow toward or away from the selected fixed point (Table 8.2).

Table 8.2 RFID TAGs with double rectangular spiral antennas variables for different λ and t values

	λ < 0	λ > 0
t = 0	$X_1(t=0) = X_1^{(i)} + x_1$ $V_{L_{calc-1}}(t=0) = V_{L_{calc-1}}^{(i)} + v_{L_{calc-1}}$ $X_2(t=0) = X_2^{(i)} + x_2$ $V_{L_{calc-2}}(t=0) = V_{L_{calc-2}}^{(i)} + v_{L_{calc-2}}$	$X_1(t=0) = X_1^{(i)} + x_1$ $V_{L_{calc-1}}(t=0) = V_{L_{calc-1}}^{(i)} + v_{L_{calc-1}}$ $X_2(t=0) = X_2^{(i)} + x_2$ $V_{L_{calc-2}}(t=0) = V_{L_{calc-2}}^{(i)} + v_{L_{calc-2}}$
t > 0	$X_1(t) = X_1^{(i)} + x_1 \cdot e^{-\lvert\lambda\rvert \cdot t}$ $V_{L_{calc-1}}(t) = V_{L_{calc-1}}^{(i)} + v_{L_{calc-1}} \cdot e^{-\lvert\lambda\rvert \cdot t}$ $X_2(t) = X_2^{(i)} + x_1 \cdot e^{-\lvert\lambda\rvert \cdot t}$ $V_{L_{calc-2}}(t) = V_{L_{calc-2}}^{(i)} + v_{L_{calc-2}} \cdot e^{-\lvert\lambda\rvert \cdot t}$	$X_1(t) = X_1^{(i)} + x_1 \cdot e^{\lvert\lambda\rvert \cdot t}$ $V_{L_{calc-1}}(t) = V_{L_{calc-1}}^{(i)} + v_{L_{calc-1}} \cdot e^{\lvert\lambda\rvert \cdot t}$ $X_2(t) = X_2^{(i)} + x_1 \cdot e^{\lvert\lambda\rvert \cdot t}$ $V_{L_{calc-2}}(t) = V_{L_{calc-2}}^{(i)} + v_{L_{calc-2}} \cdot e^{\lvert\lambda\rvert \cdot t}$
t > 0 t → ∞	$X_1(t \to \infty) = X_1^{(i)}$ $V_{L_{calc-1}}(t \to \infty) = V_{L_{calc-1}}^{(i)}$ $X_2(t \to \infty) = X_2^{(i)}$ $V_{L_{calc-2}}(t \to \infty) = V_{L_{calc-2}}^{(i)}$	$X_1(t \to \infty, \lambda > 0) \approx x_1 \cdot e^{\lvert\lambda\rvert \cdot t}$ $V_{L_{calc-1}}(t \to \infty, \lambda > 0) \approx v_{L_{calc-1}} \cdot e^{\lvert\lambda\rvert \cdot t}$ $X_2(t \to \infty, \lambda > 0) \approx x_2 \cdot e^{\lvert\lambda\rvert \cdot t}$ $V_{L_{calc-2}}(t \to \infty, \lambda > 0) \approx v_{L_{calc-2}} \cdot e^{\lvert\lambda\rvert \cdot t}$

The speeds of flow toward or away from the selected fixed point for RFID TAGs with double rectangular spiral antenna system voltages and voltages derivatives respect to time are as follow:

$$\frac{dX_1(t)}{dt} = \lim_{\Delta t \to 0} \frac{X_1(t+\Delta t) - X_1(t)}{\Delta t} = \lim_{\Delta t \to 0} \frac{X_1^{(i)} + x_1 \cdot e^{\lambda \cdot (t+\Delta t)} - \left[X_1^{(i)} + x_1 \cdot e^{\lambda \cdot t}\right]}{\Delta t}$$

$$= \lim_{\Delta t \to 0} \frac{x_1 \cdot e^{\lambda t} \cdot \left[e^{\lambda \cdot \Delta t} - 1\right]}{\Delta t} \overset{e^{\lambda \cdot \Delta t} \approx 1 + \lambda \cdot \Delta t}{\longrightarrow} \lambda \cdot x_1 \cdot e^{\lambda \cdot t}$$

$$\frac{dV_{L_{calc-1}}(t)}{dt} = \lambda \cdot v_{L_{calc-1}} \cdot e^{\lambda \cdot t}; \frac{dV_{L_{calc-2}}(t)}{dt} = \lambda \cdot v_{L_{calc-2}} \cdot e^{\lambda \cdot t}; \frac{dX_2(t)}{dt} = \lambda \cdot x_2 \cdot e^{\lambda \cdot t}$$

First we take the RFID TAGs with double rectangular spiral antenna system voltages $V_{L_{calc-1}}, V_{L_{calc-2}}$ differential equations: $X_1 = \frac{dV_{L_{calc-1}}}{dt}; X_2 = \frac{dV_{L_{calc-2}}}{dt}$ and adding coordinates $[X_1 V_{L_{calc-1}} X_2 V_{L_{calc-2}}]$ arbitrarily small increments of exponential terms $[x_1 v_{L_{calc-1}} x_2 v_{L_{calc-2}}] \cdot e^{\lambda \cdot t}$ and retaining the first order terms in $x_1 v_{L_{calc-1}} x_2 v_{L_{calc-2}}$.

$$\lambda \cdot v_{L_{calc-1}} \cdot e^{\lambda \cdot t} = X_1^{(i)} + x_1 \cdot e^{\lambda \cdot t}; X_1^{(i=0)} = 0 \Rightarrow -\lambda \cdot v_{L_{calc-1}} + x_1 = 0$$

$$\lambda \cdot v_{L_{calc-2}} \cdot e^{\lambda \cdot t} = X_2^{(i)} + x_2 \cdot e^{\lambda \cdot t}; X_2^{(i=0)} = 0 \Rightarrow -\lambda \cdot v_{L_{calc-2}} + x_2 = 0$$

Second we take the RFID TAGs with double rectangular spiral antenna system's voltages derivatives X_1, X_2 differential equations:

$$\frac{dX_1}{dt} = \Gamma_{12} \cdot V_{L_{calc-1}} + \Gamma_{11} \cdot X_1; \frac{dX_2}{dt} = \Gamma_{34} \cdot V_{L_{calc-2}} + \Gamma_{33} \cdot X_2$$

Adding coordinates $[X_1 V_{L_{calc-1}} X_2 V_{L_{calc-2}}]$ arbitrarily small increments of exponential terms $[x_1 v_{L_{calc-1}} x_2 v_{L_{calc-2}}] \cdot e^{\lambda \cdot t}$ and retaining the first order terms in $x_1 v_{L_{calc-1}} x_2 v_{L_{calc-2}}$.

$$\lambda \cdot x_1 \cdot e^{\lambda \cdot t} = \Gamma_{12} \cdot [V^{(i)}_{L_{calc-1}} + v_{L_{calc-1}} \cdot e^{\lambda \cdot t}] + \Gamma_{11} \cdot [X^{(i)}_1 + x_1 \cdot e^{\lambda \cdot t}]$$
$$\lambda \cdot x_1 \cdot e^{\lambda \cdot t} = \Gamma_{12} \cdot V^{(i)}_{L_{calc-1}} + \Gamma_{11} \cdot X^{(i)}_1 + \Gamma_{12} \cdot v_{L_{calc-1}} \cdot e^{\lambda \cdot t} + \Gamma_{11} \cdot x_1 \cdot e^{\lambda \cdot t}$$

At fixed points $\Gamma_{12} \cdot V^{(i)}_{L_{calc-1}} + \Gamma_{11} \cdot X^{(i)}_1 = 0$

$$\Gamma_{12} \cdot V^{(i)}_{L_{calc-1}} + \Gamma_{11} \cdot X^{(i)}_1 = 0 \Rightarrow -\lambda \cdot x_1 + \Gamma_{11} \cdot x_1 + \Gamma_{12} \cdot v_{L_{calc-1}} = 0$$
$$\lambda \cdot x_2 \cdot e^{\lambda \cdot t} = \Gamma_{34} \cdot V^{(i)}_{L_{calc-2}} + v_{L_{calc-2}} \cdot e^{\lambda \cdot t} + \Gamma_{33} \cdot [X^{(i)}_2 + x_2 \cdot e^{\lambda \cdot t}]$$
$$\lambda \cdot x_2 \cdot e^{\lambda \cdot t} = \Gamma_{34} \cdot V^{(i)}_{L_{calc-2}} + \Gamma_{33} \cdot X^{(i)}_2 + \Gamma_{34} \cdot v_{L_{calc-2}} \cdot e^{\lambda \cdot t} + \Gamma_{33} \cdot x_2 \cdot e^{\lambda \cdot t}$$

At fixed points $\Gamma_{34} \cdot V^{(i)}_{L_{calc-2}} + \Gamma_{33} \cdot X^{(i)}_2 = 0$

$$\Gamma_{34} \cdot V^{(i)}_{L_{calc-2}} + \Gamma_{33} \cdot X^{(i)}_2 = 0 \Rightarrow -\lambda \cdot x_2 + \Gamma_{33} \cdot x_2 + \Gamma_{34} \cdot v_{L_{calc-2}} = 0$$

Double rectangular spiral coils system arbitrarily small increments equations:

$$-\lambda \cdot v_{L_{calc-1}} + x_1 = 0; -\lambda \cdot v_{L_{calc-2}} + x_2 = 0; -\lambda \cdot x_1 + \Gamma_{11} \cdot x_1 + \Gamma_{12} \cdot v_{L_{calc-1}} = 0$$
$$-\lambda \cdot x_2 + \Gamma_{33} \cdot x_2 + \Gamma_{34} \cdot v_{L_{calc-2}} = 0$$

We define the following expressions:

$$X_1(t - \Delta_1) = X^{(i)}_1 + x_1 \cdot e^{\lambda \cdot (t - \Delta_1)}; V_{L_{calc-1}}(t - \tau_1) = V^{(i)}_{L_{calc-1}} + v_{L_{calc-1}} \cdot e^{\lambda \cdot (t - \tau_1)}$$
$$X_2(t - \Delta_2) = X^{(i)}_2 + x_2 \cdot e^{\lambda \cdot (t - \Delta_2)}; V_{L_{calc-2}}(t - \tau_2) = V^{(i)}_{L_{calc-2}} + v_{L_{calc-2}} \cdot e^{\lambda \cdot (t - \tau_2)}$$

Then we get four delayed differential equations respect to coordinates $[X_1 V_{L_{calc-1}} X_2 V_{L_{calc-2}}]$ arbitrarily small increments of exponential $[x_1 v_{L_{calc-1}} x_2 v_{L_{calc-2}}] \cdot e^{\lambda \cdot t}$.

$$\lambda \cdot e^{\lambda \cdot t} \cdot x_1 = \Gamma_{11} \cdot e^{\lambda \cdot (t - \Delta_1)} \cdot x_1 + \Gamma_{12} \cdot e^{\lambda \cdot (t - \tau_1)} \cdot v_{L_{calc-1}}; \lambda \cdot e^{\lambda \cdot t} \cdot v_{L_{calc-1}} = e^{\lambda \cdot (t - \Delta_1)} \cdot x_1$$
$$\lambda \cdot e^{\lambda \cdot t} \cdot x_2 = \Gamma_{33} \cdot e^{\lambda \cdot (t - \Delta_2)} \cdot x_2 + \Gamma_{34} \cdot e^{\lambda \cdot (t - \tau_2)} \cdot v_{L_{calc-2}}; \lambda \cdot e^{\lambda \cdot t} \cdot v_{L_{calc-2}} = e^{\lambda \cdot (t - \Delta_2)} \cdot x_2$$

In the equilibrium fixed point $X^{(i=0)}_1 = 0, V^{(i=0)}_{L_{calc-1}} = 0; X^{(i=0)}_2 = 0, V^{(i=0)}_{L_{calc-2}} = 0$.

The small increments Jacobian of our RFID TAGs with double rectangular spiral antenna system is as bellow:

$$\Upsilon_{11} = -\lambda + \Gamma_{11} \cdot e^{-\lambda \cdot \Delta_1}; \Upsilon_{12} = \Gamma_{12} \cdot e^{-\lambda \cdot \tau_1}; \Upsilon_{13} = 0; \Upsilon_{14} = 0; \Upsilon_{21} = e^{-\lambda \cdot \Delta_1};$$
$$\Upsilon_{22} = -\lambda; \Upsilon_{23} = 0; \Upsilon_{24} = 0$$
$$\Upsilon_{31} = 0; \Upsilon_{32} = 0; \Upsilon_{33} = -\lambda + \Gamma_{33} \cdot e^{-\lambda \cdot \Delta_2}; \Upsilon_{34} = \Gamma_{34} \cdot e^{-\lambda \cdot \tau_2};$$
$$\Upsilon_{41} = 0; \Upsilon_{42} = 0; \Upsilon_{43} = e^{-\lambda \cdot \Delta_2}; \Upsilon_{44} = -\lambda$$

$$\begin{pmatrix} \Upsilon_{11} & \cdots & \Upsilon_{14} \\ \vdots & \ddots & \vdots \\ \Upsilon_{41} & \cdots & \Upsilon_{44} \end{pmatrix} \cdot \begin{pmatrix} x_1 \\ v_{L_{calc-1}} \\ x_1 \\ v_{L_{calc-2}} \end{pmatrix} = 0; A - \lambda \cdot I = \begin{pmatrix} \Upsilon_{11} & \cdots & \Upsilon_{14} \\ \vdots & \ddots & \vdots \\ \Upsilon_{41} & \cdots & \Upsilon_{44} \end{pmatrix}; \det |A - \lambda \cdot I| = 0$$

$$D(\lambda, \tau_1, \tau_2, \Delta_1, \Delta_2) = \lambda^4 + \Gamma_{12} \cdot \Gamma_{34} \cdot e^{-\lambda \cdot [\sum_{i=1}^{2} \tau_i + \sum_{j=1}^{2} \Delta_j]}$$
$$+ \lambda \cdot \{\Gamma_{11} \cdot \Gamma_{34} \cdot e^{-\lambda \cdot [\tau_2 + \sum_{j=1}^{2} \Delta_j]} + \Gamma_{33} \cdot \Gamma_{12} \cdot e^{-\lambda \cdot [\tau_1 + \sum_{j=1}^{2} \Delta_j]}\}$$
$$+ \lambda^2 \cdot \{-\Gamma_{34} \cdot e^{-\lambda \cdot (\Delta_2 + \tau_2)} - \Gamma_{12} \cdot e^{-\lambda \cdot (\Delta_1 + \tau_1)} + \Gamma_{11} \cdot \Gamma_{33} \cdot e^{-\lambda \cdot \sum_{j=1}^{2} \Delta_j}\}$$
$$- \lambda^3 \cdot \{\Gamma_{33} \cdot e^{-\lambda \cdot \Delta_2} + \Gamma_{11} \cdot e^{-\lambda \cdot \Delta_1}\}$$

We have three stability cases: $\tau_1 = \tau_2 = \tau \,\&\, \Delta_1 = \Delta_2 = 0$ Or $\tau_1 = \tau_2 = 0 \,\&\, \Delta_1 = \Delta_2 = \Delta$ or $\tau_1 = \tau_2 = \Delta_1 = \Delta_2 = \tau_\Delta$ otherwise $\tau_1 \neq \tau_2 \,\&\, \Delta_1 \neq \Delta_2$ and they are positive parameters. There are other possible simple stability cases:

$$\tau_1 = \tau; \tau_2 = 0; \Delta_1 = \Delta_2 = 0 \text{ or } \tau_1 = 0; \tau_2 = \tau; \Delta_1 = \Delta_2 = 0$$
$$\tau_1 = \tau_2 = 0; \Delta_1 = \Delta; \Delta_2 = 0 \text{ or } \tau_1 = \tau_2 = 0; \Delta_1 = 0; \Delta_2 = \Delta$$

We need to get characteristics equations for all above stability analysis cases. We study the occurrence of any possible stability switching resulting from the increase of value of the time delays $\tau, \Delta, \tau_\Delta$ for the general characteristic equation $D(\lambda, \tau/\Delta/\tau_\Delta)$. If we choose τ parameter then $D(\lambda, \tau) = P_n(\lambda, \tau) + Q_m(\lambda, \tau) \cdot e^{-\lambda\tau}$. The expression for $P_n(\lambda, \tau)$:$P_n(\lambda, \tau) = \sum_{k=0}^{n} P_k(\tau) \cdot \lambda^k = P_0(\tau) + P_1(\tau) \cdot \lambda + P_2(\tau) \cdot \lambda^2 + P_3(\tau) \cdot \lambda^3 + \ldots\ldots$

The expression for $Q_m(\lambda, \tau)$ is $Q_m(\lambda, \tau) = \sum_{k=0}^{m} q_k(\tau) \cdot \lambda^k = q_0(\tau) + q_1(\tau) \cdot \lambda + q_2(\tau) \cdot \lambda^2 + \ldots\ldots$

The case we analyze is when there is delay in RFID TAGs first and second rectangular spiral antennas voltages ($\tau_1 = \tau_2 = \tau$) and no delay in RFID TAGs first and second rectangular spiral antennas voltages derivatives. The general characteristic equation $D(\lambda, \tau)$ is as follow:

$$D(\lambda,\tau) = \lambda^4 - \lambda^3\cdot(\Gamma_{33}+\Gamma_{11}) + \lambda^2\cdot\Gamma_{11}\cdot\Gamma_{33} + \{\Gamma_{12}\cdot\Gamma_{34}\cdot e^{-\lambda\cdot\tau} + \lambda\cdot(\Gamma_{11}\cdot\Gamma_{34}+\Gamma_{12}\cdot\Gamma_{33}) - \lambda^2\cdot(\Gamma_{34}+\Gamma_{12})\}\cdot e^{-\lambda\cdot\tau}$$

Under Taylor series approximation: $e^{-\lambda\cdot\tau} \approx 1 - \lambda\cdot\tau + \frac{1}{2}\cdot\lambda^2\cdot\tau^2$

The Maclaurin series is a Taylor series expansion of a $e^{-\lambda\cdot\tau}$ function about zero (0). We get the following general characteristic equation D(λ, τ) under Taylor series approximation: $e^{-\lambda\cdot\tau} \approx 1 - \lambda\cdot\tau + \frac{1}{2}\cdot\lambda^2\cdot\tau^2$.

$$D(\lambda,\tau) = \lambda^4 - \lambda^3\cdot[\Gamma_{33}+\Gamma_{11}] + \lambda^2\cdot\Gamma_{11}\cdot\Gamma_{33} + \{\Gamma_{12}\cdot\Gamma_{34} + \lambda\cdot[\Gamma_{11}\cdot\Gamma_{34} + \Gamma_{12}\cdot\Gamma_{33} - \Gamma_{12}\cdot\Gamma_{34}\cdot\tau] + \lambda^2\cdot[\frac{1}{2}\cdot\Gamma_{12}\cdot\Gamma_{34}\cdot\tau^2 - \Gamma_{34} - \Gamma_{12}]\}\cdot e^{-\lambda\cdot\tau}$$

$$D(\lambda,\tau) = P_n(\lambda,\tau) + Q_m(\lambda,\tau)\cdot e^{-\lambda\tau}; n = 4; m = 2; n > m$$

The expression for $P_n(\lambda,\tau)$: $P_n(\lambda,\tau) = \sum_{k=0}^{n} P_k(\tau)\cdot\lambda^k$

$$P_n(\lambda,\tau) = \sum_{k=0}^{n} P_k(\tau)\cdot\lambda^k = P_0(\tau) + P_1(\tau)\cdot\lambda + P_2(\tau)\cdot\lambda^2 + P_3(\tau)\cdot\lambda^3 + P_4(\tau)\cdot\lambda^4 = \lambda^4 - \lambda^3\cdot[\Gamma_{33}+\Gamma_{11}] + \lambda^2\cdot\Gamma_{11}\cdot\Gamma_{33}$$

$$P_0(\tau) = 0; P_1(\tau) = 0; P_2(\tau) = \Gamma_{11}\cdot\Gamma_{33}; P_3(\tau) = -[\Gamma_{33}+\Gamma_{11}]; P_4(\tau) = 1$$

The expression for $Q_m(\lambda,\tau)$: $Q_m(\lambda,\tau) = \sum_{k=0}^{m} q_k(\tau)\cdot\lambda^k = q_0(\tau) + q_1(\tau)\cdot\lambda + q_2(\tau)\cdot\lambda^2$

$$Q_m(\lambda,\tau) = \sum_{k=0}^{m} q_k(\tau)\cdot\lambda^k = \Gamma_{12}\cdot\Gamma_{34} + \lambda\cdot[\Gamma_{11}\cdot\Gamma_{34} + \Gamma_{12}\cdot\Gamma_{33} - \Gamma_{12}\cdot\Gamma_{34}\cdot\tau] + \lambda^2\cdot[\frac{1}{2}\cdot\Gamma_{12}\cdot\Gamma_{34}\cdot\tau^2 - \Gamma_{34} - \Gamma_{12}]; q_0(\tau) = \Gamma_{12}\cdot\Gamma_{34}$$

$$q_1(\tau) = \Gamma_{11}\cdot\Gamma_{34} + \Gamma_{12}\cdot\Gamma_{33} - \Gamma_{12}\cdot\Gamma_{34}\cdot\tau; q_2(\tau) = \frac{1}{2}\cdot\Gamma_{12}\cdot\Gamma_{34}\cdot\tau^2 - \Gamma_{34} - \Gamma_{12}$$

The homogeneous system for $X_1 V_{L_{calc-1}} X_2 V_{L_{calc-2}}$ leads to a characteristic equation for the eigenvalue λ having the form $P(\lambda,\tau) + Q(\lambda,\tau)\cdot e^{-\lambda\cdot\tau} = 0$; $P(\lambda) = \sum_{j=0}^{4} a_j\cdot\lambda^j; Q(\lambda) = \sum_{j=0}^{2} c_j\cdot\lambda^j$ and the coefficients $\{a_j(q_i,q_k,\tau), c_j(q_i,q_k,\tau)\} \in \mathbb{R}$ depend on q_i, q_k and delay τ, q_i, q_k are any double rectangular spiral coils antennas system's parameters, other parameters keep as a constant [5, 6].

$$a_0 = 0; a_1 = 0; a_2 = \Gamma_{11} \cdot \Gamma_{33}; a_3 = -[\Gamma_{33} + \Gamma_{11}]; a_4 = 1$$
$$c_0 = \Gamma_{12} \cdot \Gamma_{34}; c_1 = \Gamma_{11} \cdot \Gamma_{34} + \Gamma_{12} \cdot \Gamma_{33} - \Gamma_{12} \cdot \Gamma_{34} \cdot \tau;$$
$$c_2 = \frac{1}{2} \cdot \Gamma_{12} \cdot \Gamma_{34} \cdot \tau^2 - \Gamma_{34} - \Gamma_{12}$$

The designation of the variation arguments (q_i, q_k) will subsequently be omitted from P, Q, a_j, c_j. The coefficients a_j, c_j are continuous, and differentiable functions of their arguments, and direct substitution shows that $a_0 + c_0 \neq 0$ for $\forall q_i, q_k \in \mathbb{R}_+$, i.e. $\lambda = 0$ is not a $P(\lambda, \tau) + Q(\lambda, \tau) \cdot e^{-\lambda \cdot \tau} = 0$. We assume that $P_n(\lambda, \tau)$ and $Q_m(\lambda, \tau)$ can't have common imaginary roots. That is for any real number ω:

$$\begin{aligned} p_n(\lambda = i \cdot \omega, \tau) + Q_m(\lambda = i \cdot \omega, \tau) \neq 0; p_n(\lambda = i \cdot \omega, \tau) \\ = \omega^4 + i \cdot \omega^3 \cdot (\Gamma_{33} + \Gamma_{11}) - \omega^2 \cdot \Gamma_{11} \cdot \Gamma_{33} \\ Q_m(\lambda = i \cdot \omega, \tau) = \Gamma_{12} \cdot \Gamma_{34} + i \cdot \omega \cdot [\Gamma_{11} \cdot \Gamma_{34} + \Gamma_{12} \cdot \Gamma_{33} - \Gamma_{12} \cdot \Gamma_{34} \cdot \tau] \\ - \omega^2 \cdot [\frac{1}{2} \cdot \Gamma_{12} \cdot \Gamma_{34} \cdot \tau^2 - \Gamma_{34} - \Gamma_{12}] \\ p_n(\lambda = i \cdot \omega, \tau) + Q_m(\lambda = i \cdot \omega, \tau) \\ = \omega^4 - \omega^2 \cdot [\frac{1}{2} \cdot \Gamma_{12} \cdot \Gamma_{34} \cdot \tau^2 - \Gamma_{34} - \Gamma_{12} + \Gamma_{11} \cdot \Gamma_{33}] \\ + \Gamma_{12} \cdot \Gamma_{34} + i \cdot \omega^3 \cdot (\Gamma_{33} + \Gamma_{11}) \\ + i \cdot \omega \cdot [\Gamma_{11} \cdot \Gamma_{34} + \Gamma_{12} \cdot \Gamma_{33} - \Gamma_{12} \cdot \Gamma_{34} \cdot \tau] \neq 0 \end{aligned}$$

$$\begin{aligned} |P(i \cdot \omega, \tau)|^2 &= \omega^8 + \omega^6 \cdot \{(\Gamma_{33} + \Gamma_{11})^2 - 2 \cdot \Gamma_{11} \cdot \Gamma_{33}\} + \omega^4 \cdot \Gamma_{11}^2 \cdot \Gamma_{33}^2 \\ |Q(i \cdot \omega, \tau)|^2 &= \Gamma_{12}^2 \cdot \Gamma_{34}^2 + \omega^2 \cdot \{[\Gamma_{11} \cdot \Gamma_{34} + \Gamma_{12} \cdot \Gamma_{33} - \Gamma_{12} \cdot \Gamma_{34} \cdot \tau]^2 \\ &\quad - 2 \cdot \Gamma_{12} \cdot \Gamma_{34} \cdot [\frac{1}{2} \cdot \Gamma_{12} \cdot \Gamma_{34} \cdot \tau^2 - \Gamma_{34} - \Gamma_{12}]\} \\ &\quad + \omega^4 \cdot [\frac{1}{2} \cdot \Gamma_{12} \cdot \Gamma_{34} \cdot \tau^2 - \Gamma_{34} - \Gamma_{12}]^2 \end{aligned}$$

We need to find the expression for $F(\omega, \tau) = |P(i \cdot \omega, \tau)|^2 - |Q(i \cdot \omega, \tau)|^2$

$$\begin{aligned} F(\omega, \tau) = |P(i \cdot \omega, \tau)|^2 - |Q(i \cdot \omega, \tau)|^2 = \omega^8 + \omega^6 \cdot \{(\Gamma_{33} + \Gamma_{11})^2 - 2 \cdot \Gamma_{11} \cdot \Gamma_{33}\} \\ + \omega^4 \cdot \{\Gamma_{11}^2 \cdot \Gamma_{33}^2 - [\frac{1}{2} \cdot \Gamma_{12} \cdot \Gamma_{34} \cdot \tau^2 - \Gamma_{34} - \Gamma_{12}]^2\} \\ - \omega^2 \cdot \{[\Gamma_{11} \cdot \Gamma_{34} + \Gamma_{12} \cdot \Gamma_{33} - \Gamma_{12} \cdot \Gamma_{34} \cdot \tau]^2 \\ - 2 \cdot \Gamma_{12} \cdot \Gamma_{34} \cdot [\frac{1}{2} \cdot \Gamma_{12} \cdot \Gamma_{34} \cdot \tau^2 - \Gamma_{34} - \Gamma_{12}]\} - \Gamma_{12}^2 \cdot \Gamma_{34}^2 \end{aligned}$$

We define the following parameters for simplicity:

$$\Xi_0 = -\Gamma_{12}^2 \cdot \Gamma_{34}^2; \Xi_2 = -[\Gamma_{11} \cdot \Gamma_{34} + \Gamma_{12} \cdot \Gamma_{33} - \Gamma_{12} \cdot \Gamma_{34} \cdot \tau]^2 + 2 \cdot \Gamma_{12} \cdot \Gamma_{34} \cdot [\frac{1}{2} \cdot \Gamma_{12} \cdot \Gamma_{34} \cdot \tau^2 - \Gamma_{34} - \Gamma_{12}]$$

$$\Xi_4 = \Gamma_{11}^2 \cdot \Gamma_{33}^2 - [\frac{1}{2} \cdot \Gamma_{12} \cdot \Gamma_{34} \cdot \tau^2 - \Gamma_{34} - \Gamma_{12}]^2;$$

$$\Xi_6 = (\Gamma_{33} + \Gamma_{11})^2 - 2 \cdot \Gamma_{11} \cdot \Gamma_{33}; \Xi_8 = 1$$

$$F(\omega, \tau) = |P(i \cdot \omega, \tau)|^2 - |Q(i \cdot \omega, \tau)|^2 = \Xi_0 + \Xi_2 \cdot \omega^2 + \Xi_4 \cdot \omega^4 + \Xi_6 \cdot \omega^6 + \Xi_8 \cdot \omega^8 = \sum_{k=0}^{4} \Xi_{2 \cdot k} \cdot \omega^{2 \cdot k}$$

Hence $F(\omega, \tau) = 0$ implies $\sum_{k=0}^{4} \Xi_{2 \cdot k} \cdot \omega^{2 \cdot k} = 0$ and its roots are given by solving the above polynomial. Furthermore $P_R(i \cdot \omega, \tau) = \omega^4 - \omega^2 \cdot \Gamma_{11} \cdot \Gamma_{33}$

$$P_I(i \cdot \omega, \tau) = \omega^3 \cdot (\Gamma_{33} + \Gamma_{11}); Q_R(i \cdot \omega, \tau) = \Gamma_{12} \cdot \Gamma_{34} - \omega^2 \cdot [\frac{1}{2} \cdot \Gamma_{12} \cdot \Gamma_{34} \cdot \tau^2 - \Gamma_{34} - \Gamma_{12}]$$

$$Q_I(i \cdot \omega, \tau) = \omega \cdot [\Gamma_{11} \cdot \Gamma_{34} + \Gamma_{12} \cdot \Gamma_{33} - \Gamma_{12} \cdot \Gamma_{34} \cdot \tau]$$

Hence

$$\sin \theta(\tau) = \frac{-P_R(i \cdot \omega, \tau) \cdot Q_I(i \cdot \omega, \tau) + P_I(i \cdot \omega, \tau) \cdot Q_R(i \cdot \omega, \tau)}{|Q(i \cdot \omega, \tau)|^2}$$

$$\cos \theta(\tau) = -\frac{P_R(i \cdot \omega, \tau) \cdot Q_R(i \cdot \omega, \tau) + P_I(i \cdot \omega, \tau) \cdot Q_I(i \cdot \omega, \tau)}{|Q(i \cdot \omega, \tau)|^2}$$

$$\sin \theta(\tau) = \frac{\begin{array}{c} -\{\omega^4 - \omega^2 \cdot \Gamma_{11} \cdot \Gamma_{33}\} \cdot \omega \cdot [\Gamma_{11} \cdot \Gamma_{34} + \Gamma_{12} \cdot \Gamma_{33} - \Gamma_{12} \cdot \Gamma_{34} \cdot \tau] \\ + \omega^3 \cdot (\Gamma_{33} + \Gamma_{11}) \cdot \{\Gamma_{12} \cdot \Gamma_{34} - \omega^2 \cdot [\frac{1}{2} \cdot \Gamma_{12} \cdot \Gamma_{34} \cdot \tau^2 - \Gamma_{34} - \Gamma_{12}]\} \end{array}}{\begin{array}{c} \Gamma_{12}^2 \cdot \Gamma_{34}^2 + \omega^2 \cdot \{[\Gamma_{11} \cdot \Gamma_{34} + \Gamma_{12} \cdot \Gamma_{33} - \Gamma_{12} \cdot \Gamma_{34} \cdot \tau]^2 \\ -2 \cdot \Gamma_{12} \cdot \Gamma_{34} \cdot [\frac{1}{2} \cdot \Gamma_{12} \cdot \Gamma_{34} \cdot \tau^2 - \Gamma_{34} - \Gamma_{12}]\} + \omega^4 \cdot [\frac{1}{2} \cdot \Gamma_{12} \cdot \Gamma_{34} \cdot \tau^2 - \Gamma_{34} - \Gamma_{12}]^2 \end{array}}$$

$$\cos \theta(\tau) = -\frac{\begin{array}{c} \{\omega^4 - \omega^2 \cdot \Gamma_{11} \cdot \Gamma_{33}\} \cdot \{\Gamma_{12} \cdot \Gamma_{34} - \omega^2 \cdot [\frac{1}{2} \cdot \Gamma_{12} \cdot \Gamma_{34} \cdot \tau^2 - \Gamma_{34} - \Gamma_{12}]\} \\ + \omega^4 \cdot (\Gamma_{33} + \Gamma_{11}) \cdot [\Gamma_{11} \cdot \Gamma_{34} + \Gamma_{12} \cdot \Gamma_{33} - \Gamma_{12} \cdot \Gamma_{34} \cdot \tau] \end{array}}{\begin{array}{c} \Gamma_{12}^2 \cdot \Gamma_{34}^2 + \omega^2 \cdot \{[\Gamma_{11} \cdot \Gamma_{34} + \Gamma_{12} \cdot \Gamma_{33} - \Gamma_{12} \cdot \Gamma_{34} \cdot \tau]^2 - 2 \cdot \Gamma_{12} \cdot \Gamma_{34} \\ \cdot [\frac{1}{2} \cdot \Gamma_{12} \cdot \Gamma_{34} \cdot \tau^2 - \Gamma_{34} - \Gamma_{12}]\} + \omega^4 \cdot [\frac{1}{2} \cdot \Gamma_{12} \cdot \Gamma_{34} \cdot \tau^2 - \Gamma_{34} - \Gamma_{12}]^2 \end{array}}$$

These are continuous and differentiable in τ based on Lemma 1.1. Hence we use Theorem 1.2 and this prove the Theorem 1.3.

Our RFID TAGs with double rectangular spiral system for $x_1 v_{L_{calc-1}} x_2 v_{L_{calc-2}}$ leads to a characteristic equation for the eigenvalue λ having the form $P(\lambda) + Q(\lambda) \cdot e^{-\lambda \cdot \tau} = 0$; The case $\tau_1 = \tau_2 = \tau$; $\Delta_1 = \Delta_2 = 0$.

$$D(\lambda, \tau_1 = \tau_2 = \tau, \Delta_1 = \Delta_2 = 0) = \lambda^4 - \lambda^3 \cdot (\Gamma_{33} + \Gamma_{11}) + \lambda^2 \cdot \Gamma_{11} \cdot \Gamma_{33} + \{\Gamma_{12} \cdot \Gamma_{34} \cdot e^{-\lambda \cdot \tau} + \lambda \cdot (\Gamma_{11} \cdot \Gamma_{34} + \Gamma_{12} \cdot \Gamma_{33}) - \lambda^2 \cdot (\Gamma_{34} + \Gamma_{12})\} \cdot e^{-\lambda \cdot \tau}$$

Under Taylor series approximation: $e^{-\lambda \cdot \tau} \approx 1 - \lambda \cdot \tau + \frac{1}{2} \cdot \lambda^2 \cdot \tau^2$. The Maclaurin series is a Taylor series expansion of a $e^{-\lambda \cdot \tau}$ function about zero (0). We get the following general characteristic equation D(λ, τ) under Taylor series approximation: $e^{-\lambda \cdot \tau} \approx 1 - \lambda \cdot \tau + \frac{1}{2} \cdot \lambda^2 \cdot \tau^2$.

$$D(\lambda, \tau) = \lambda^4 - \lambda^3 \cdot [\Gamma_{33} + \Gamma_{11}] + \lambda^2 \cdot \Gamma_{11} \cdot \Gamma_{33} + \{\Gamma_{12} \cdot \Gamma_{34} + \lambda \cdot [\Gamma_{11} \cdot \Gamma_{34} + \Gamma_{12} \cdot \Gamma_{33} - \Gamma_{12} \cdot \Gamma_{34} \cdot \tau] + \lambda^2 \cdot [\frac{1}{2} \cdot \Gamma_{12} \cdot \Gamma_{34} \cdot \tau^2 - \Gamma_{34} - \Gamma_{12}]\} \cdot e^{-\lambda \cdot \tau}$$

We use different parameters terminology from our last characteristics parameters definition: $k \to j; p_k(\tau) \to a_j; q_k(\tau) \to c_j; n = 4; m = 2; n > m$

Additionally $P_n(\lambda, \tau) \to P(\lambda); Q_m(\lambda, \tau) \to Q(\lambda)$ then $P(\lambda) = \sum_{j=0}^{4} a_j \cdot \lambda^j; Q(\lambda) = \sum_{j=0}^{2} c_j \cdot \lambda^j$

$$P_\lambda = \lambda^4 - \lambda^3 \cdot [\Gamma_{33} + \Gamma_{11}] + \lambda^2 \cdot \Gamma_{11} \cdot \Gamma_{33}$$

$$Q_\lambda = \Gamma_{12} \cdot \Gamma_{34} + \lambda \cdot [\Gamma_{11} \cdot \Gamma_{34} + \Gamma_{12} \cdot \Gamma_{33} - \Gamma_{12} \cdot \Gamma_{34} \cdot \tau] + \lambda^2 \cdot [\frac{1}{2} \cdot \Gamma_{12} \cdot \Gamma_{34} \cdot \tau^2 - \Gamma_{34} - \Gamma_{12}]$$

$n, m \in \mathbb{N}_0, n > m; a_j, c_j : \mathrm{R}_{+0} \to R$. They are continuous and differentiable function of τ such that $a_0 + c_0 \neq 0$. In the following "_" denotes complex and conjugate. Functions $P(\lambda), Q(\lambda)$ are analytic functions in λ and differentiable in τ. The coefficients: $\{a_j(C_1, R_1, \text{double rectangular spiral antennas parametrs})$ and $c_j(C_1, R_1, \tau, \text{double rectangular spiral antennas parametrs})\} \in \mathbb{R}$ depend on RFID TAGs with double rectangular spiral antennas system's C_1, R_1,τ values and antennas parameters.

$$a_0 = 0; a_1 = 0; a_2 = \Gamma_{11} \cdot \Gamma_{33}; a_3 = -[\Gamma_{33} + \Gamma_{11}]; a_4 = 1; c_0 = \Gamma_{12} \cdot \Gamma_{34};$$
$$c_1 = \Gamma_{11} \cdot \Gamma_{34} + \Gamma_{12} \cdot \Gamma_{33} - \Gamma_{12} \cdot \Gamma_{34} \cdot \tau$$
$$c_2 = \frac{1}{2} \cdot \Gamma_{12} \cdot \Gamma_{34} \cdot \tau^2 - \Gamma_{34} - \Gamma_{12}$$

Unless strictly necessary, the designation of the variation arguments. (R_1, C_1, τ, double rectangular spiral antennas parametrs) will subsequently be omitted from P, Q, a_j, c_j. The coefficients a_j, c_j are continuous, differentiable functions of their arguments and direct substitution shows that $a_0 + c_0 \neq 0; \Gamma_{12} \cdot \Gamma_{34} \neq 0$.

$$\frac{\eta_1 \cdot \xi_1}{\eta_3 \cdot \xi_3} = \frac{[\frac{1}{L_{calc-1}} + \frac{(R_{dc-1} + R_{dc-2})}{R_1 \cdot L_{calc-1}}] \cdot [\frac{1}{L_{calc-2}} + \frac{(R_{dc-1} + R_{dc-2})}{R_1 \cdot L_{calc-2}}]}{C_1^2 \cdot [1 + \frac{L_{calc-2}}{L_{calc-1}} + 2 \cdot K \cdot \sqrt{\frac{L_{calc-2}}{L_{calc-1}}}] \cdot [1 + \frac{L_{calc-1}}{L_{calc-2}} + 2 \cdot K \cdot \sqrt{\frac{L_{calc-1}}{L_{calc-2}}}]} \neq 0$$

$\forall C_1$, double rectangular spiral antennas parametrs $\in \mathbb{R}_+$ i.e. $\lambda = 0$ is not a root of characteristic equation. Furthermore $P(\lambda), Q(\lambda)$ are analytic function of λ for which the following requirements of the analysis (see kuang 1993, Sect. 3.4) can also be verified in the present case.

(a) If $\lambda = i \cdot \omega$, $\omega \in \mathbb{R}$ then $P(i \cdot \omega) + Q(i \cdot \omega) \neq 0$, i.e. P and Q have no common imaginary roots. This condition was verified numerically in the entire (R_1, C_1, double rectangular spiral antennas parameters) domain of interest.

(b) $|Q(\lambda)/P(\lambda)|$ is bounded for $|\lambda| \to \infty$, $\text{Re}\lambda \geq 0$. No roots bifurcation from ∞. Indeed, in the limit

$$|\frac{Q(\lambda)}{P(\lambda)}| = |\frac{\{\Gamma_{12} \cdot \Gamma_{34} + \lambda \cdot [\Gamma_{11} \cdot \Gamma_{34} + \Gamma_{12} \cdot \Gamma_{33} - \Gamma_{12} \cdot \Gamma_{34} \cdot \tau] + \lambda^2 \cdot \frac{1}{2} \cdot \Gamma_{12} \cdot \Gamma_{34} \cdot \tau^2 - \Gamma_{34} - \Gamma_{12}]\}}{\lambda^4 - \lambda^3 \cdot [\Gamma_{33} + \Gamma_{11}] + \lambda^2 \cdot \Gamma_{11} \cdot \Gamma_{33}}|$$

(c) $F(\omega) = |P(i \cdot \omega)|^2 - |Q(i \cdot \omega)|^2$

$$\begin{aligned} F(\omega, \tau) &= |P(i \cdot \omega, \tau)|^2 - |Q(i \cdot \omega, \tau)|^2 = \omega^8 + \omega^6 \cdot \{(\Gamma_{33} + \Gamma_{11})^2 - 2 \cdot \Gamma_{11} \cdot \Gamma_{33}\} \\ &+ \omega^4 \cdot \{\Gamma_{11}^2 \cdot \Gamma_{33}^2 - [\frac{1}{2} \cdot \Gamma_{12} \cdot \Gamma_{34} \cdot \tau^2 - \Gamma_{34} - \Gamma_{12}]^2\} \\ &- \omega^2 \cdot \{[\Gamma_{11} \cdot \Gamma_{34} + \Gamma_{12} \cdot \Gamma_{33} - \Gamma_{12} \cdot \Gamma_{34} \cdot \tau]^2 \\ &- 2 \cdot \Gamma_{12} \cdot \Gamma_{34} \cdot [\frac{1}{2} \cdot \Gamma_{12} \cdot \Gamma_{34} \cdot \tau^2 - \Gamma_{34} - \Gamma_{12}]\} - \Gamma_{12}^2 \cdot \Gamma_{34}^2 \end{aligned}$$

Has at most a finite number of zeros. Indeed, this is a polynomial in ω (degree in ω^8).

(d) Each positive root $\omega(R_1, C_1, \tau$, double rectangular spiral antennas parameters) of $F(\omega) = 0$ is continuous and differentiable with respect to R_1, C_1, τ, double rectangular spiral antennas parameters. This condition can only be assessed numerically.

In addition, since the coefficients in P and Q are real, we have $\overline{P(-i \cdot \omega)} = P(i \cdot \omega)$, and $\overline{Q(-i \cdot \omega)} = Q(i \cdot \omega)$ thus $\lambda = i \cdot \omega$, $\omega > 0$ may be on eigenvalue of characteristic equation. The analysis consists in identifying the roots of characteristic equation situated on the imaginary axis of the complex λ-plane, where by increasing the parameters R_1, C_1, τ, double rectangular spiral antennas parameters, Reλ may, at the crossing Change its sign from (−) to (+), i.e. from a stable focus $E^{(0)}(X_1^{(0)}, V_{L_{calc-1}}^{(0)}, X_2^{(0)}, V_{L_{calc-2}}^{(0)}) = (0,0,0,0)$ to an unstable one, or vice versa. This feature may be further assessed by examining the sign of the partial derivatives with respect to C_1, R_1, τ and double rectangular spiral coils antennas parameters [2, 3].

$$\wedge^{-1}(C_1) = \left(\frac{\partial \mathrm{Re}\lambda}{\partial C_1}\right)_{\lambda=i\cdot\omega}, \ \mathrm{R}_1, \tau, \text{double rectangular spiral antennas parametrs} = const$$

$$\wedge^{-1}(\mathrm{R}_1) = \left(\frac{\partial \mathrm{Re}\lambda}{\partial \mathrm{R}_1}\right)_{\lambda=i\cdot\omega}, C_1, \tau, \text{double rectangular spiral antennas parametrs} = const$$

$$\wedge^{-1}(L_{calc-1}) = \left(\frac{\partial \mathrm{Re}\lambda}{\partial L_{calc-1}}\right)_{\lambda=i\cdot\omega}, C_1, \mathrm{R}_1, \tau = const$$

$$\wedge^{-1}(L_{calc-2}) = \left(\frac{\partial \mathrm{Re}\lambda}{\partial L_{calc-2}}\right)_{\lambda=i\cdot\omega}, C_1, \mathrm{R}_1, \tau = const$$

$$\wedge^{-1}(\tau) = \left(\frac{\partial \mathrm{Re}\lambda}{\partial \tau}\right)_{\lambda=i\cdot\omega}, C_1, \mathrm{R}_1, \text{double rectangular spiral antennas parametrs} = const \text{ where } \omega \in \mathbb{R}_+.$$

For the case $\tau_1 = \tau_2 = \tau \,\&\, \Delta_1 = \Delta_2 = 0$ we get the following results:

$$P_R(i\cdot\omega,\tau) = \omega^4 - \omega^2\cdot\Gamma_{11}\cdot\Gamma_{33}; P_I(i\cdot\omega,\tau) = \omega^3\cdot(\Gamma_{33}+\Gamma_{11})$$

$$Q_R(i\cdot\omega,\tau) = \Gamma_{12}\cdot\Gamma_{34} - \omega^2\cdot\left[\frac{1}{2}\cdot\Gamma_{12}\cdot\Gamma_{34}\cdot\tau^2 - \Gamma_{34} - \Gamma_{12}\right];$$

$$Q_I(i\cdot\omega,\tau) = \omega\cdot[\Gamma_{11}\cdot\Gamma_{34} + \Gamma_{12}\cdot\Gamma_{33} - \Gamma_{12}\cdot\Gamma_{34}\cdot\tau]$$

$$\Xi_0 = -\Gamma_{12}^2\cdot\Gamma_{34}^2; \Xi_2 = -[\Gamma_{11}\cdot\Gamma_{34} + \Gamma_{12}\cdot\Gamma_{33} - \Gamma_{12}\cdot\Gamma_{34}\cdot\tau]^2 + 2\cdot\Gamma_{12}\cdot\Gamma_{34}\cdot\left[\frac{1}{2}\cdot\Gamma_{12}\cdot\Gamma_{34}\cdot\tau^2 - \Gamma_{34} - \Gamma_{12}\right]$$

$$\Xi_4 = \Gamma_{11}^2\cdot\Gamma_{33}^2 - \left[\frac{1}{2}\cdot\Gamma_{12}\cdot\Gamma_{34}\cdot\tau^2 - \Gamma_{34} - \Gamma_{12}\right]^2;$$

$$\Xi_6 = (\Gamma_{33}+\Gamma_{11})^2 - 2\cdot\Gamma_{11}\cdot\Gamma_{33}; \Xi_8 = 1$$

$$F(\omega,\tau) = |P(i\cdot\omega,\tau)|^2 - |Q(i\cdot\omega,\tau)|^2 = \Xi_0 + \Xi_2\cdot\omega^2 + \Xi_4\cdot\omega^4 + \Xi_6\cdot\omega^6 + \Xi_8\cdot\omega^8 = \sum_{k=0}^{4}\Xi_{2\cdot k}\cdot\omega^{2\cdot k}$$

Hence $F(\omega,\tau) = 0$ implies $\sum_{k=0}^{4}\Xi_{2\cdot k}\cdot\omega^{2\cdot k} = 0$ when writing $P(\lambda) = P_R(\lambda) + i\cdot P_I(\lambda)$ and $Q(\lambda) = Q_R(\lambda) + i\cdot Q_I(\lambda)$, and inserting $\lambda = i\cdot\omega$ into double rectangular spiral coils antennas system's characteristic equation, ω must satisfy the following:

$$\sin\omega\cdot\tau = g(\omega) = \frac{-P_R(i\cdot\omega)\cdot Q_I(i\cdot\omega) + P_I(i\cdot\omega)\cdot Q_R(i\cdot\omega)}{|Q(i\cdot\omega)|^2}$$
$$\cos\omega\cdot\tau = h(\omega) = -\frac{P_R(i\cdot\omega)\cdot Q_R(i\cdot\omega) + P_I(i\cdot\omega)\cdot Q_I(i\cdot\omega)}{|Q(i\cdot\omega)|^2}$$

Where $|Q(i\cdot\omega)|^2 \neq 0$ in view of requirement (a) above, $(g,h)\in R$. Furthermore, it follows above $\sin\omega\cdot\tau$ and $\cos\omega\cdot\tau$ equations that, by squaring and adding the sides, ω must be a positive root of $F(\omega) = |P(i\cdot\omega)|^2 - |Q(i\cdot\omega)|^2 = 0$. Note: $F(\omega)$ is dependent of τ. Now it is important to notice that if $\tau \notin I$(assume that $I \subseteq R_{+0}$ is the set where $\omega(\tau)$ is a positive root of $F(\omega)$ and for $\tau \notin I$, $\omega(\tau)$ is not define. Then for all τ in I $\omega(\tau)$ is satisfies that $F(\omega,\tau) = 0$). Then there are no positive $\omega(\tau)$ solutions for $F(\omega,\tau) = 0$, and we cannot have stability switches. For any $\tau \in I$, where $\omega(\tau)$ is a positive solution of $F(\omega,\tau) = 0$, we can define the angle $\theta(\tau) \in [0, 2\cdot\pi]$ as the solution of the below equations:

$$\sin\theta(\tau) = \frac{-P_R(i\cdot\omega)\cdot Q_I(i\cdot\omega) + P_I(i\cdot\omega)\cdot Q_R(i\cdot\omega)}{|Q(i\cdot\omega)|^2}$$
$$\cos\theta(\tau) = -\frac{P_R(i\cdot\omega)\cdot Q_R(i\cdot\omega) + P_I(i\cdot\omega)\cdot Q_I(i\cdot\omega)}{|Q(i\cdot\omega)|^2}$$

And the relation between the argument $\theta(\tau)$ and $\omega(\tau)\cdot\tau$ for $\tau \in I$ must be $\omega(\tau)\cdot\tau = \theta(\tau) + n\cdot 2\cdot\pi \; \forall\, n \in \mathbb{N}_0$. Hence we can define the maps $\tau_n : I \to R_{+0}$ given by $\tau_n(\tau) = \frac{\theta(\tau)+n\cdot 2\cdot\pi}{\omega(\tau)}; n \in \mathbb{N}_0, \tau \in I$. Let us introduce the functions $I \to R$; $S_n(\tau) = \tau - \tau_n(\tau), \tau \in I, n \in \mathbb{N}_0$ that are continuous and differentiable in τ. In the following, the subscripts $\lambda, \omega, R_1, C_1$ and RFID TAGs with double rectangular spiral antennas parameters $(A_{avg1}, B_{avg2}, A_{01}, B_{02}, N_{c1}, N_{c2}, g_1, g_2, \ldots)$ indicate the corresponding partial derivatives. Let us first concentrate on, $\wedge(x)$ remember in $\lambda(A_{avg1}, B_{avg2}, A_{01}, B_{02}, N_{c1}, N_{c2}, g_1, g_2, \ldots)$; $\omega(A_{avg1}, B_{avg2}, A_{01}, B_{02}, N_{c1}, N_{c2}, g_1, g_2, \ldots)$, and keeping all parameters except one (x) and τ. The derivation closely follows that in reference [BK]. Differentiating RFID TAGs with double rectangular spiral antennas characteristic equation $P(\lambda) + Q(\lambda)\cdot e^{-\lambda\cdot\tau} = 0$ with respect to specific parameter (x), and inverting the derivative, for convenience, one calculates:

Remark: $x = A_{avg1}, B_{avg2}, A_{01}, B_{02}, N_{c1}, N_{c2}, g_1, g_2, \ldots$

$$\left(\frac{\partial\lambda}{\partial x}\right)^{-1} = \frac{-P_\lambda(\lambda,x)\cdot Q(\lambda,x) + Q_\lambda(\lambda,x)\cdot P(\lambda,x) - \tau\cdot P(\lambda,x)\cdot Q(\lambda,x)}{P_x(\lambda,x)\cdot Q(\lambda,x) - Q_x(\lambda,x)\cdot P(\lambda,x)}$$

Where $P_\lambda = \frac{\partial P}{\partial\lambda}, \ldots$ etc., Substituting $\lambda = i\cdot\omega$, and bearing $\overline{P(-i\cdot\omega)} = P(i\cdot\omega)$, $\overline{Q(-i\cdot\omega)} = Q(i\cdot\omega)$ then $i\cdot P_\lambda(i\cdot\omega) = P_\omega(i\cdot\omega)$ and $i\cdot Q_\lambda(i\cdot\omega) = Q_\omega(i\cdot\omega)$ that on the surface $|P(i\cdot\omega)|^2 = |Q(i\cdot\omega)|^2$, one obtains

$$(\frac{\partial \lambda}{\partial x})|_{\lambda=i\cdot\omega}^{-1} = (\frac{i \cdot P_{\omega}(i \cdot \omega, x) \cdot \overline{P(i \cdot \omega, x)} + i \cdot Q_{\lambda}(i \cdot \omega, x) \cdot \overline{Q(\lambda, x)} - \tau \cdot |P(i \cdot \omega, x)|^2}{P_x(i \cdot \omega, x) \cdot \overline{P(i \cdot \omega, x)} - Q_x(i \cdot \omega, x) \cdot \overline{Q(i \cdot \omega, x)}})$$

Upon separating into real and imaginary parts, with $P = P_R + i \cdot P_I$; $Q = Q_R + i \cdot Q_I$

$$P_{\omega} = P_{R\omega} + i \cdot P_{I\omega}; Q_{\omega} = Q_{R\omega} + i \cdot Q_{I\omega}; P_x = P_{Rx} + i \cdot P_{Ix}; Q_x = Q_{Rx} + i \cdot Q_{Ix}$$

$P^2 = P_R^2 + P_I^2$. When (x) can be any RFID TAGs with double rectangular spiral antennas parameters R_1, C_1, And time delay τ etc. Where for convenience, we have dropped the arguments $(i \cdot \omega, x)$, and where

$$F_{\omega} = 2 \cdot [(P_{R\omega} \cdot P_R + P_{I\omega} \cdot P_I) - (Q_{R\omega} \cdot Q_R + Q_{I\omega} \cdot Q_I)]; F_x = 2 \cdot [(P_{Rx} \cdot P_R + P_{Ix} \cdot P_I) - (Q_{Rx} \cdot Q_R + Q_{Ix} \cdot Q_I)]$$

$\omega_x = -F_x/F_{\omega}$. We define U and V:

$$U = (P_R \cdot P_{I\omega} - P_I \cdot P_{R\omega}) - (Q_R \cdot Q_{I\omega} - Q_I \cdot Q_{R\omega});$$
$$V = (P_R \cdot P_{Ix} - P_I \cdot P_{Rx}) - (Q_R \cdot Q_{Ix} - Q_I \cdot Q_{Rx})$$

We choose our specific parameter as time delay x = τ.

$$P_{R\omega} = 2 \cdot \omega \cdot [2 \cdot \omega^2 - \Gamma_{11} \cdot \Gamma_{33}]; P_{I\omega} = 3 \cdot \omega^2 \cdot (\Gamma_{33} + \Gamma_{11});$$
$$P_{R\tau} = 0; P_{I\tau} = 0;$$
$$Q_{R\tau} = -\omega^2 \cdot \Gamma_{12} \cdot \Gamma_{34} \cdot \tau; Q_{I\tau} = -\omega \cdot \Gamma_{12} \cdot \Gamma_{34}$$
$$P_{R\omega} \cdot P_R = 2 \cdot \omega^3 \cdot [2 \cdot \omega^4 - 3 \cdot \omega^2 \cdot \Gamma_{11} \cdot \Gamma_{33} + \Gamma_{11}^2 \cdot \Gamma_{33}^2];$$
$$P_{I\omega} \cdot P_I = 3 \cdot \omega^5 \cdot (\Gamma_{33} + \Gamma_{11})^2; \omega_{\tau} = -F_{\tau}/F_{\omega}$$
$$P_{I\omega} \cdot P_I = 3 \cdot \omega^5 \cdot (\Gamma_{33} + \Gamma_{11})^2; \omega_{\tau} = -F_{\tau}/F_{\omega};$$
$$Q_{R\omega} = -2 \cdot \omega \cdot [\frac{1}{2} \cdot \Gamma_{12} \cdot \Gamma_{34} \cdot \tau^2 - \Gamma_{34} - \Gamma_{12}]$$
$$Q_{I\omega} = \Gamma_{11} \cdot \Gamma_{34} + \Gamma_{12} \cdot \Gamma_{33} - \Gamma_{12} \cdot \Gamma_{34} \cdot \tau;$$
$$Q_{I\omega} \cdot Q_I = \omega \cdot [\Gamma_{11} \cdot \Gamma_{34} + \Gamma_{12} \cdot \Gamma_{33} - \Gamma_{12} \cdot \Gamma_{34} \cdot \tau]^2$$
$$Q_{R\omega} \cdot Q_R = -2 \cdot \omega \cdot [\frac{1}{2} \cdot \Gamma_{12} \cdot \Gamma_{34} \cdot \tau^2 - \Gamma_{34} - \Gamma_{12}] \cdot [\Gamma_{12} \cdot \Gamma_{34} - \omega^2 \cdot (\frac{1}{2} \cdot \Gamma_{12} \cdot \Gamma_{34} \cdot \tau^2 - \Gamma_{34} - \Gamma_{12})]$$

$$F_{\tau} = 2 \cdot [(P_{R\tau} \cdot P_R + P_{I\tau} \cdot P_I) - (Q_{R\tau} \cdot Q_R + Q_{I\tau} \cdot Q_I)];$$
$$P_R \cdot P_{I\omega} = 3 \cdot \omega^4 \cdot (\omega^2 - \Gamma_{11} \cdot \Gamma_{33}) \cdot (\Gamma_{33} + \Gamma_{11})$$
$$F_{\tau} = 2 \cdot \omega^2 \cdot \Gamma_{12} \cdot \Gamma_{34} \cdot [\Gamma_{11} \cdot \Gamma_{34} + \Gamma_{12} \cdot \Gamma_{33} - \tau \cdot \omega^2 \cdot (\frac{1}{2} \cdot \Gamma_{12} \cdot \Gamma_{34} \cdot \tau^2 - \Gamma_{34} - \Gamma_{12})]$$

$$P_I \cdot P_{R\omega} = 2 \cdot \omega^4 \cdot (\Gamma_{33} + \Gamma_{11}) \cdot (2 \cdot \omega^2 - \Gamma_{11} \cdot \Gamma_{33}); V = (P_R \cdot P_{I\tau} - P_I \cdot P_{R\tau}) - (Q_R \cdot Q_{I\tau} - Q_I \cdot Q_{R\tau})$$
$$Q_R \cdot Q_{I\omega} = [\Gamma_{12} \cdot \Gamma_{34} - \omega^2 \cdot (\frac{1}{2} \cdot \Gamma_{12} \cdot \Gamma_{34} \cdot \tau^2 - \Gamma_{34} - \Gamma_{12})] \cdot [\Gamma_{11} \cdot \Gamma_{34} + \Gamma_{12} \cdot \Gamma_{33} - \Gamma_{12} \cdot \Gamma_{34} \cdot \tau]$$
$$Q_I \cdot Q_{R\omega} = -2 \cdot \omega^2 \cdot (\Gamma_{11} \cdot \Gamma_{34} + \Gamma_{12} \cdot \Gamma_{33} - \Gamma_{12} \cdot \Gamma_{34} \cdot \tau) \cdot (\frac{1}{2} \cdot \Gamma_{12} \cdot \Gamma_{34} \cdot \tau^2 - \Gamma_{34} - \Gamma_{12})$$
$$P_R \cdot P_{I\tau} = 0; P_I \cdot P_{R\tau} = 0; Q_R \cdot Q_{I\tau} = -\omega \cdot \Gamma_{12} \cdot \Gamma_{34} \cdot [\Gamma_{12} \cdot \Gamma_{34} - \omega^2 \cdot (\frac{1}{2} \cdot \Gamma_{12} \cdot \Gamma_{34} \cdot \tau^2 - \Gamma_{34} - \Gamma_{12})]$$
$$Q_I \cdot Q_{R\tau} = -\omega^3 \cdot \Gamma_{12} \cdot \Gamma_{34} \cdot \tau \cdot [\Gamma_{11} \cdot \Gamma_{34} + \Gamma_{12} \cdot \Gamma_{33} - \Gamma_{12} \cdot \Gamma_{34} \cdot \tau]; F(\omega, \tau) = 0$$

Differentiating with respect to τ and we get $F_{\omega} \cdot \frac{\partial \omega}{\partial \tau} + F_{\tau} = 0; \tau \in I \Rightarrow \frac{\partial \omega}{\partial \tau} = -\frac{F_{\tau}}{F_{\omega}}$

$$\wedge^{-1}(\tau) = (\frac{\partial \mathrm{Re}\lambda}{\partial \tau})_{\lambda = i \cdot \omega}; \wedge^{-1}(\tau) = \mathrm{Re}\{\frac{-2 \cdot [U + \tau \cdot |P|^2] + i \cdot F_{\omega}}{F_{\tau} + i \cdot 2 \cdot [V + \omega \cdot |P|^2]}\}; \frac{\partial \omega}{\partial \tau} = \omega_{\tau} = -\frac{F_{\tau}}{F_{\omega}}$$

$$sign\{\wedge^{-1}(\tau)\} = sign\{(\frac{\partial \mathrm{Re}\lambda}{\partial \tau})_{\lambda = i \cdot \omega}\}; sign\{\wedge^{-1}(\tau)\} = sign\{F_{\omega}\} \cdot sign\{\tau \cdot \frac{\partial \omega}{\partial \tau} + \omega + \frac{U \cdot \frac{\partial \omega}{\partial \tau} + V}{|P|^2}\}$$

We shall presently examine the possibility of stability transitions (bifurcations) in a RFID TAGs with double rectangular spiral antennas system, about the equilibrium point $E^{(0)}(X_1^{(0)}, V_{L_{calc-1}}^{(0)}, X_2^{(0)}, V_{L_{calc-2}}^{(0)})$ as a result of a variation of delay parameter τ. The analysis consists in identifying the roots of our system characteristic equation situated on the imaginary axis of the complex λ-plane where by increasing the delay parameter τ, Re λ may at the crossing, change its sign from − to +, i.e. from a stable focus $E^{(*)}$ to an unstable one, or vice versa. This feature may be further assessed by examining the sign of the partial derivatives with respect to τ, $\wedge^{-1}(\tau) = (\frac{\partial \mathrm{Re}\lambda}{\partial \tau})_{\lambda = i \cdot \omega}$

$\wedge^{-1}(\tau) = (\frac{\partial \mathrm{Re}\lambda}{\partial \tau})_{\lambda = i \cdot \omega}, C_1, R_1,$ RFID TAGs with double rectangular spiral antennas parameters $= const$ where $\omega \in \mathbb{R}_+$.

For our stability switching analysis we choose each of our system's rectangular spiral antenna on a substrate, width is 300 μm and $\mu_r = 450$. The antenna is constructed from silver ointment which his resistance is bigger than pure silver by 50 %. Track width is 20 μm, gap between tracks 20 μm, track depth is 20 μm up 10 100 μm. The requested rectangular spiral antenna inductance is 2.66 mH and parasitic resistance less than 10 Ω. We need to find the possible number of rectangular spiral antenna's turns (N_{c1} and N_{c2} for the first and second rectangular spiral antenna respectively). Since each RFID antenna substrate permeability is 450 ($\mu_r = 450$), our RFID antennas permeability is an average value between air permeability and magnet. We consider that the TAGs permeability is 100–300 (μ_r) and possible TAGs dimension: 5 mm × 5 mm, 6 mm × 6 mm, 7 mm × 7 mm, 8 mm × 8 mm and permeability 100, 200, 300 (μ_r). Tables 8.3, 8.4, and 8.5 describes the analysis for $\mu = \mu_r \cdot \mu_0$ (μ_r = 100, 200, 300). Table 8.3 is for $\mu = \mu_r \cdot \mu_0$; $\mu_r = 100$; $\mu = 125.66 \cdot (1e{-6})$ H/m. Table 8.4 is for $\mu = \mu_r \cdot \mu_0$; $\mu_r = 200$; $\mu = 251.32 \cdot (1e{-6})$ H/m and Table 8.5 is for $\mu = \mu_r \cdot \mu_0$; $\mu_r = 300$; $\mu = 376.98 \cdot (1e{-6})$ H/m.

Results: The most close inductance analysis to 2.66 mH is 2.7 mH and subcases.

Result Table 8.3: Lcalc = 2.7 mH, Nc = 60, (A_0 = 7 mm) × (B_0 = 7 mm); $\mu = \mu_r \cdot \mu_0$; $\mu_r = 100$; $\mu = 125.66 \cdot (1e{-6})$ H/m. The DC resistance of rectangular spiral RFID antenna: $R_{DC} = \frac{L_T}{\sigma \cdot S} = \frac{L_T}{\sigma \cdot \pi \cdot a^2}$. L_T—total length of the wire. σ—Conductivity of the wire (℧/m). S—Cross section area $\pi \cdot a^2$. a—radius of the wire.

$$R_{DC} = \frac{L_T}{\sigma \cdot S} = \frac{L_T}{\sigma \cdot \pi \cdot a^2}; R_{DC} = \frac{L_T}{\sigma \cdot \pi \cdot a^2} = \frac{2 \cdot \{(A_0 + B_0) \cdot (1 + N_C) - (w + g) \cdot [4 \cdot N_C - 3]\}}{\sigma \cdot \pi \cdot a^2}$$

Table 8.3 Rectangular spiral antenna L_{calc} as a function of different number of turns for (N_c) for $\mu = \mu_r \cdot \mu_0$; $\mu_r = 100$; $\mu = 125.66 \cdot (1e{-6})$ H/m

Nc	Lcalc (5 mm × 5 mm)—[H]	Lcalc (6 mm × 6 mm)—[H]	Lcalc (7 mm × 7 mm)—[H]	Lcalc (8 mm × 8 mm)—[H]
10	1.08×10^{-4}	1.37×10^{-4}	1.67×10^{-4}	1.98×10^{-4}
20	3.383×10^{-4}	4.38×10^{-4}	5.419×10^{-4}	6.48×10^{-4}
30	6.212×10^{-4}	8.25×10^{-4}	0.001	0.0013
60	0.0014	0.002	**0.0027 = 2.7 mH**	0.0035
100	0.001	0.0024	0.0041	0.0058
150	−0.0014	NaN	0.0021	0.0051
120	1.3598×10^{-4}	0.0018	0.0038	0.0061
170	−0.0040	−0.0012	2.5454×10^{-4}	0.0033
200	−0.0106	−0.0062	−0.0023	NaN
220	−0.0171	−0.0115	−0.0063	−0.0020
250	−0.0306	−0.023	−0.0158	−0.0092

Table 8.4 Rectangular spiral antenna L_{calc} as a function of different number of turns for (N_c) for $\mu = \mu_r \cdot \mu_0$; $\mu_r = 200$; $\mu = 251.32 \cdot (1e{-}6)$H/m

Nc	Lcalc (5 mm × 5 mm)—[H]	Lcalc (6 mm × 6 mm)—[H]	Lcalc (7 mm × 7 mm)—[H]	Lcalc (8 mm × 8 mm)—[H]
10	2.17×10^{-4}	2.75×10^{-4}	3.35×10^{-4}	3.97×10^{-4}
20	6.767×10^{-4}	8.76×10^{-4}	0.0011	0.0013
30	0.0012	0.001	0.0021	0.0025
60	**0.0027 = 2.7 mH**	0.004	0.0055	0.0069
100	0.0020	0.0049	0.0081	0.0116
150	−0.0028	NaN	0.0042	0.0102
120	2.719×10^{-4}	0.0035	0.0077	0.0122
170	−0.008	−0.0025	5.09×10^{-4}	0.0066
200	−0.0212	−0.0123	−0.0046	NaN
220	−0.0342	−0.0229	−0.0127	−0.0039
250	−0.0612	−0.046	−0.0316	−0.0184

Table 8.5 Rectangular spiral antenna L_{calc} as a function of different number of turns for (N_c) for $\mu = \mu_r \cdot \mu_0$; $\mu_r = 300$; $\mu = 376.98 \cdot (1e{-}6)$H/m

Nc	Lcalc (5 mm × 5 mm)—[H]	Lcalc (6 mm × 6 mm)—[H]	Lcalc (7 mm × 7 mm)—[H]	Lcalc (8 mm × 8 mm)—[H]
10	3.25×10^{-4}	4.13×10^{-4}	5.03×10^{-4}	5.95×10^{-4}
20	0.001	0.0013	0.0016	0.0019
30	0.0019	0.0025	0.0031	0.0038
60	0.0041	0.0061	0.0082	0.0104
100	0.003	0.0073	0.0122	0.0173
150	−0.004	NaN	0.0062	0.0153
120	4.079×10^{-4}	0.0053	0.0115	0.0183
170	−0.0119	−0.0037	7.63×10^{-4}	0.0099
200	−0.0317	−0.0185	−0.0069	NaN
220	−0.0513	−0.0344	−0.0190	−0.0059
250	−0.0918	−0.069	−0.0474	−0.0276

$A_0 + B_0 = 0.014\,\text{m}\,; N_c = 60; w + g = 40 \times 10^{-6} \Rightarrow L_T = 1.689\,\text{m}$. Cross section area $S = 20\,\mu\text{m} \cdot 20\,\mu\text{m} = 400 \times 10^{-12} \cdot \text{m}^2$. Conductivity of silver $\sigma = 6.1 \times 10^7\,(\mho/\text{m})$. Conductivity has SI units of Siemens per meter (S/m).$\sigma_{silver\,@\,20\,°\text{C}} = 6.3 \times 10^7$ (S/m).

The track depth (x) does not influence our total inductance, and then we can take it as a variable and find his minimum value for $R_{DC} < 10$ ▯. $S = 20\,\mu\text{m} \cdot x$.

$$\frac{1.689}{6.3 \times 10^7 \cdot 20 \times 10^{-6} \cdot x} < 10 \Rightarrow x > 1.3405 \times 10^{-4}\,\text{m} = 134.05\,\mu\text{m};$$

$$R_{DC@t=20\,\mu\text{m}} = 67\,\Omega$$

Actually the track is a mixture of silver then the conductivity is half of silver conductivity.

$$\frac{\sigma_{silver@20\,^\circ\mathrm{C}}}{2} = \frac{6.3 \times 10^7\ (\mathrm{S/m})}{2} = 3.15 \times 10^7\ (\mathrm{S/m}).$$

$$\frac{1.6890}{3.15 \times 10^7 \cdot 20 \times 10^{-6} \cdot x} < 10 \Rightarrow x > 2.681 \times 10^{-4}\,\mathrm{m} = 268.1\,\mu\mathrm{m};$$

$$R_{DC@t=20\,\mu\mathrm{m}} = 134\,\Omega$$

Result Table 8.4: Lcalc = 2.7 mH, Nc = 60, (A_0 = 5 mm) × (B_0 = 5 mm); $\mu = \mu_r \cdot \mu_0$; $\mu_r = 200$; $\mu = 251.32 \cdot (1e{-}6)$ H/m. The DC resistance of rectangular spiral RFID antenna: $R_{DC} = \frac{L_T}{\sigma \cdot S} = \frac{L_T}{\sigma \cdot \pi \cdot a^2}$. L_T—total length of the wire. σ—conductivity of the wire (℧/m). S—Cross section area $\pi \cdot a^2$. a—radius of the wire.

$$R_{DC} = \frac{L_T}{\sigma \cdot S} = \frac{L_T}{\sigma \cdot \pi \cdot a^2}; R_{DC} = \frac{L_T}{\sigma \cdot \pi \cdot a^2} = \frac{2 \cdot \{(A_0 + B_0) \cdot (1 + N_C) - (w + g) \cdot [4 \cdot N_C - 3]\}}{\sigma \cdot \pi \cdot a^2}$$

$$A_0 + B_0 = 0.01\,\mathrm{m}; N_c = 60; w + g = 40 \times 10^{-6} \Rightarrow L_T = 1.201\,\mathrm{m}$$

Cross section area $S = 20\,\mu\mathrm{m} \cdot 20\,\mu\mathrm{m} = 400 \times 10^{-12}\,\mathrm{m}^2$. Conductivity of Silver $\sigma = 6.1 \times 10^7\ (\mho/\mathrm{m})$. Conductivity has SI units of siemens per meter (S/m). $\sigma_{silver@20\,^\circ\mathrm{C}} = 6.3 \times 10^7\ (\mathrm{S/m})$. The track depth (x) does not influence our total inductance, and then we can take it as a variable and find his minimum value for $R_{DC} < 10\ \Omega$. $S = 20\,\mu\mathrm{m} \cdot x$.

$$\frac{1.201}{6.3 \times 10^7 \cdot 20 \times 10^{-6} \cdot x} < 10 \Rightarrow x > 9.5317 \times 10^{-5}\,\mathrm{m} = 95.317\,\mu\mathrm{m};$$

$$R_{DC@t=20\,\mu\mathrm{m}} = 47.65\,\Omega$$

Actually the track is a mixture of silver then the conductivity is half of silver conductivity.

$$\frac{\sigma_{silver@20\,^\circ\mathrm{C}}}{2} = \frac{6.3 \times 10^7\ (\mathrm{S/m})}{2} = 3.15 \times 10^7\ (\mathrm{S/m}).$$

$$\frac{1.201}{3.15 \times 10^7 \cdot 20 \times 10^{-6} \cdot x} < 10 \Rightarrow x > 19.06 \times 10^{-5}\,\mathrm{m} = 190.6\,\mu\mathrm{m};$$

$$R_{DC@t=20\,\mu\mathrm{m}} = 95.30\,\Omega$$

Follows 8.3 and 8.4 tables results we choose for Lcalc-1 = 2.7 mH

$R_{dc-1} = 134\,\Omega$; Lcalc-2 = 2.7 mH; $R_{dc-2} = 95.30\,\Omega$. Typical other values for our system R_1 = 100 kΩ, C_1 = 23 pF, K = 0.6 (RFID IC is represented as parallel circuit of capacitor C_1 and R_1). $\eta_1 = 371.21, \eta_2 = 3.395 \times 10^{-5}, \eta_3 = 7.36 \times 10^{-11}$

$$\xi_1 = 371.21, \xi_2 = 3.395 \times 10^{-5}, \xi_3 = 7.36 \times 10^{-11}; \Gamma_{21} = \Gamma_{43} = 1$$
$$\Gamma_{11} = -\frac{3.395 \times 10^{-5}}{7.36 \times 10^{-11}} = -4.6128 \times 10^5; \Gamma_{12} = -\frac{371.21}{7.36 \times 10^{-11}} = -5.0436 \times 10^{12}$$
$$\Gamma_{33} = -\frac{3.395 \times 10^{-5}}{7.36 \times 10^{-11}} = -4.6128 \times 10^5; \Gamma_{34} = -\frac{371.21}{7.36 \times 10^{-11}} = -5.0436 \times 10^{12}$$
$$\Gamma_{13} = \Gamma_{14} = \Gamma_{22} = \Gamma_{23} = \Gamma_{24} = \Gamma_{31} = \Gamma_{32} = \Gamma_{41} = \Gamma_{42} = \Gamma_{44} = 0$$

Then we get the expression for $F(\omega, \tau)$ typical RFID TAGs with double rectangular spiral antenna parameters values.

$$\begin{aligned} F(\omega,\tau) &= |P(i\cdot\omega,\tau)|^2 - |Q(i\cdot\omega,\tau)|^2 = \omega^8 + \omega^6 \cdot \{(\Gamma_{33}+\Gamma_{11})^2 - 2\cdot\Gamma_{11}\cdot\Gamma_{33}\} \\ &+ \omega^4 \cdot \{\Gamma_{11}^2 \cdot \Gamma_{33}^2 - [\frac{1}{2}\cdot\Gamma_{12}\cdot\Gamma_{34}\cdot\tau^2 - \Gamma_{34} - \Gamma_{12}]^2\} \\ &- \omega^2 \cdot \{[\Gamma_{11}\cdot\Gamma_{34} + \Gamma_{12}\cdot\Gamma_{33} - \Gamma_{12}\cdot\Gamma_{34}\cdot\tau]^2 \\ &- 2\cdot\Gamma_{12}\cdot\Gamma_{34}\cdot[\frac{1}{2}\cdot\Gamma_{12}\cdot\Gamma_{34}\cdot\tau^2 - \Gamma_{34} - \Gamma_{12}]\} - \Gamma_{12}^2\cdot\Gamma_{34}^2 \end{aligned}$$

$$\begin{aligned} F(\omega,\tau) &= |P(i\cdot\omega,\tau)|^2 - |Q(i\cdot\omega,\tau)|^2 \\ &= \omega^8 + \omega^6 \cdot \{8.5112 \times 10^{11} - 2\cdot 2.1278 \times 10^{11}\} \\ &+ \omega^4 \cdot \{4.5275 \times 10^{22} - [\frac{1}{2}\cdot 2.5438 \times 10^{25}\cdot\tau^2 \\ &+ 5.0436 \times 10^{12} + 5.0436 \times 10^{12}]^2\} \\ &- \omega^2 \cdot \{[2.3265 \times 10^{18} + 2.3265 \times 10^{18} - 2.5438 \times 10^{25}\cdot\tau]^2 \\ &- 2\cdot 2.5438 \times 10^{25}\cdot[\frac{1}{2}\cdot 2.5438 \times 10^{25}\cdot\tau^2 \\ &+ 5.0436 \times 10^{12} + 5.0436 \times 10^{12}]\} - 6.4709 \times 10^{50} \end{aligned}$$

We find those ω, τ values which fulfill $F(\omega, \tau) = 0$. We ignore negative, complex, and imaginary values of ω for specific τ values. $\tau \in [0.001\ldots10](\mathrm{s})$ and it can be express by 3D function $F(\omega, \tau) = 0$.

$$\begin{aligned} F(\omega,\tau) &= |P(i\cdot\omega,\tau)|^2 - |Q(i\cdot\omega,\tau)|^2 = \omega^8 + \omega^6\cdot 4.2556 \times 10^{11} \\ &+ \omega^4 \cdot \{4.5275 \times 10^{22} - [1.2719 \times 10^{25}\cdot\tau^2 + 1.0087 \times 10^{13}]^2\} \\ &- \omega^2 \cdot \{[4.653 \times 10^{18} - 2.5438 \times 10^{25}\cdot\tau]^2 \\ &-5.0876 \times 10^{25}\cdot[1.2719 \times 10^{25}\cdot\tau^2 + 10.0872 \times 10^{12}]\} - 6.4709 \times 10^{50} \end{aligned}$$

Hence $F(\omega, \tau) = 0$ implies $\sum_{k=0}^{4} \Xi_{2\cdot k} \cdot \omega^{2\cdot k} = 0$

$\Xi_j \rightarrow$ Phij (j = 0, 2, 4, 6, 8). Running MATLAB script for τ values (τ ∈ [0.001...10]). Phij(j = 1,3,5,7) = 0

MATLAB script: Tau = 0.001; Phi0 = -6.4709e50; Phi1 = 0; Phi2 = (4.653e18-2.5438e25 * Tau).^2-5.0876e25 * (1.2719e25 * Tau * Tau + 10.0872e12); Phi3 = 0; Phi4 = 4.5275e22-(1.2719e25 * Tau * Tau + 1.0087e13).^2; Phi5 = 0; Phi6 = 4.2556e11; Phi8 = 1; Phi7 = 0; p = [Phi8 Phi7 Phi6 Phi5 Phi4 Phi3 Phi2 Phi1 Phi0]; r = roots(p) (Tables 8.6, and 8.7).

We can summary our $\omega_i(\tau)$ results for $\omega_i(\tau) > 0$ and real number (ignore complex, negative and imaginary values). We exclude from our table (Table 8.12) the high and real $\omega_i(\tau)$ values (1.0e+009*, 1.0e+010*, 1.0e+011*, ...) and add results for τ = 15 s and τ = 20 s (Table 8.12). Next figure describes the RFID TAGs with double loop rectangular spiral antennas system, ω as a function of delay parameter τ (case $\tau_1 = \tau_2 = \tau; \Delta_1 = \Delta_2 = 0$) (Tables 8.8, 8.9, 8.10, 8.11, 8.14).

RFID TAGs with double loop rectangular spiral antennas system, ω as a function of delay parameter τ (Case $\tau_1 = \tau_2 = \tau; \Delta_1 = \Delta_2 = 0$) (Fig. 8.5).

Table 8.6 RFID TAGs with double rectangular spiral antennas systemroots $\omega_i(\tau)$

τ	τ = 0.001 s	τ = 0.01 s	τ = 0.1 s
ω_1	1.0e+009*	1.0e+010*	1.0e+011*
ω_2	−3.5664	−3.5664	−3.5664
ω_3	−0.0000 + 3.5664i	0.0000 + 3.5664i	−0.0000 + 3.5664i
ω_4	−0.0000 − 3.5664i	0.0000 − 3.5664i	−0.0000 − 3.5664i
ω_5	3.5664	3.5664	3.5664
ω_6	−0.0000 + 0.0000i	0.0000 + 0.0000i	−0.0000 + 0.0000i
ω_7	−0.0000 − 0.0000i	0.0000 − 0.0000i	−0.0000 − 0.0000i
ω_8	0.0000 + 0.0000i	-0.0000 + 0.0000i	0.0000 + 0.0000i
ω_9	0.0000 − 0.0000i	−0.0000 − 0.0000i	0.0000 − 0.0000i

Table 8.7 RFID TAGs with double rectangular spiral antennas system roots $\omega_i(\tau)$

τ	τ = 1 s	τ = 2 s	τ = 3 s
ω_1	1.0e+012*	1.0e+012*	1.0e+013*
ω_2	−3.5664	−7.1327	−1.0699
ω_3	0 + 3.5664i	−0.0000 + 7.1327i	−0.0000 + 1.0699i
ω_4	0 − 3.5664i	−0.0000 − 7.1327i	−0.0000 − 1.0699i
ω_5	3.5664	7.1327	1.0699
ω_6	−0.0000 + 0.0000i	0.0000 + 0.0000i	−0.0000 + 0.0000i
ω_7	-0.0000-0.0000i	0.0000-0.0000i	−0.0000 − 0.0000i
ω_8	0.0000 + 0.0000i	−0.0000 + 0.0000i	0.0000 + 0.0000i
ω_9	0.0000 − 0.0000i	−0.0000 − 0.0000i	0.0000 − 0.0000i

Table 8.8 RFID TAGs with double rectangular spiral antennas system roots $\omega_i(\tau)$

τ	$\tau = 4$ s	$\tau = 5$ s	$\tau = 6$ s
ω_1	1.0e+013*	1.0e+013*	1.0e+013*
ω_2	−1.4265	−1.7832	−2.1398
ω_3	0 + 1.4265i	0.0000 + 1.7832i	0 + 2.1398i
ω_4	0 − 1.4265i	0.0000 − 1.7832i	0 − 2.1398i
ω_5	1.4265	1.7832	2.1398
ω_6	−0.0000 + 0.0000i	0.0000 + 0.0000i	−0.0000 + 0.0000i
ω_7	−0.0000− 0.0000i	0.0000 − 0.0000i	−0.0000 − 0.0000i
ω_8	0.0000 + 0.0000i	−0.0000 + 0.0000i	0.0000 + 0.0000i
ω_9	0.0000 − 0.0000i	−0.0000 − 0.0000i	0.0000 − 0.0000i

Table 8.9 RFID TAGs with double rectangular spiral antennas system roots $\omega_i(\tau)$

τ	$\tau = 7$ s	$\tau = 8$ s	$\tau = 9$ s
ω_1	1.0e+013*	1.0e+013*	1.0e+013*
ω_2	−2.4965	−2.8531	−3.2097
ω_3	0.0000 + 2.4965i	−0.0000 + 2.8531i	−0.0000 + 3.2097i
ω_4	0.0000 − 2.4965i	−0.0000 − 2.8531i	−0.0000 − 3.2097i
ω_5	2.4965	2.8531	3.2097
ω_6	−0.0000 + 0.0000i	0.0000 + 0.0000i	−0.0000 + 0.0000i
ω_7	−0.0000 − 0.0000i	0.0000 − 0.0000i	−0.0000 − 0.0000i
ω_8	0.0000 + 0.0000i	−0.0000 + 0.0000i	0.0000 + 0.0000i
ω_9	0.0000 − 0.0000i	−0.0000 − 0.0000i	0.0000 − 0.0000i

Table 8.10 RFID TAGs with double rectangular spiral antennas system roots $\omega_i(\tau)$

τ	$\tau = 0$ s	$\tau = 10$ s
ω_1	1.0e+006*	1.0e+013*
ω_2	−3.4542	−3.5664
ω_3	3.4542	−0.0000 + 3.5664i
ω_4	0.0000 + 2.6095i	−0.0000 − 3.5664i
ω_5	0.0000−2.6095i	3.5664
ω_6	−0.1553 + 1.6727i	−0.0000 + 0.0000i
ω_7	−0.1553 − 1.6727i	−0.0000 − 0.0000i
ω_8	0.1553 + 1.6727i	0.0000 + 0.0000i
ω_9	0.1553 - 1.6727i	0.0000 − 0.0000i

Table 8.11 RFID TAGs with double rectangular spiral antennas system roots $\omega_i(\tau)$

τ	τ = 15 s	τ = 20 s
ω_1	1.0e+013*	1.0e+013*
ω_2	−5.3496	−7.1327
ω_3	0 + 5.3496i	0.0000 + 7.1327i
ω_4	0 − 5.3496i	0.0000 − 7.1327i
ω_5	5.3496	7.1327
ω_6	0.0000 + 0.0000i	−0.0000 + 0.0000i
ω_7	0.0000 − 0.0000i	−0.0000 − 0.0000i
ω_8	−0.0000 + 0.0000i	0.0000 + 0.0000i
ω_9	−0.0000 − 0.0000i	0.0000 − 0.0000i

Table 8.12 RFID TAGs with double rectangular spiral antennas system positive and real roots $\omega_i(\tau)$ values and sin ($\omega \cdot \tau$), cos($\omega \cdot \tau$) values

τ(s)	ω	sin($\omega \cdot \tau$)	cos($\omega \cdot \tau$)
0	3.4542	−1.43e−018	−9.98e−014
0.001…1	3.5664		
2	7.1327	−5.86e−016	4.14e−015
3	1.0699	−1.11e−015	1.44e−015
4	1.4265	−3.65e−016	9.77e−016
5	1.7832	−1.50e−016	6.51e−016
6	2.1398	−7.23e−017	4.59e−016
7	2.4965	−3.90e−017	3.39e−016
8	2.8531	−2.29e−017	2.60e−016
9	3.2097	−1.42e−017	2.06e−016
10	3.5664	−9.37e−018	1.67e−016
15	5.3496	−1.85e−018	7.43e−017
20	7.1327	−5.85e−019	4.18e−017

Hint: e − x = $\times 10^{-x}$

MATLAB script: plot([0 0.001 0.01 0.1 1 2 3 4 5 6 7 8 9 10 15 20], [3.4542 3.5664 3.5664 3.5664 3.5664 7.1327 1.0699 1.4265 1.7832 2.1398 2.4965 2.8531 3.2097 3.5664 5.3496 7.1327], '-or').

RFID TAGs with double loop rectangular spiral antennas system F(ω,τ) function (Fig. 8.6).

MATLAB script:

[w,t] = meshgrid(1:0.1:8,0:0.1:20); f = w.^8 + w.^6 * 4.2556e11 + w.^4. * (4.5275e22 − (1.2719e25. * t.^2 + 1.0087e13).^2) − w.^2. * ((4.653e18 − 2.5438e25. * t).^2 − 5.0876e25. * (1.2719e25. * t.^2 + 10.0872e12)) − 6.4709e50; set(gcf, 'renderer', 'painters'); meshc(f);%ω → w,τ → t.

Table 8.13 Single, two, and three turn rectangular planar coils L_0, M_+, M_-, and $\sum M$

	Single turn rectangular planar coil (N = 4)	Two turn rectangular planar coil (N = 8)	Three turn rectangular planar coil (N = 12)
$L_0 = \sum_{i=1}^{N} L_i$, $i = 1, 2, \ldots, N$	$\sum_{i=1}^{4} L_i = L_1 + L_2 + L_3 + L_4$	$\sum_{i=1}^{8} L_i$	$\sum_{i=1}^{12} L_i$
M_+	0	$2 \cdot (M_{1,5} + M_{2,6} + M_{3,7} + M_{4,8})$	$2 \cdot (M_{1,9} + M_{1,5} + M_{5,9} + M_{2,6} + M_{2,10} + M_{6,10} + M_{3,7} + M_{3,11} + M_{11,7} + M_{4,8} + M_{4,12} + M_{8,12})$
M_-	$2 \cdot (M_{1,3} + M_{2,4})$	$2 \cdot (M_{1,7} + M_{1,3} + M_{5,7} + M_{5,3} + M_{2,8} + M_{2,4} + M_{6,8} + M_{6,4})$	$2 \cdot (M_{1,11} + M_{1,7} + M_{1,3} + M_{5,11} + M_{5,7} + M_{5,3} + M_{9,11} + M_{9,7} + M_{9,3} + M_{2,12} + M_{2,8} + M_{2,4} + M_{6,12} + M_{6,8} + M_{6,4} + M_{10,12} + M_{10,8} + M_{10,4})$
$\sum M$	$\sum M = -M_- = -2 \cdot (M_{1,3} + M_{2,4})$	$2 \cdot (M_{1,5} + M_{2,6} + M_{3,7} + M_{4,8}) - 2 \cdot (M_{1,7} + M_{1,3} + M_{5,7} + M_{5,3} + M_{2,8} + M_{2,4} + M_{6,8} + M_{6,4})$	$2 \cdot (M_{1,9} + M_{1,5} + M_{5,9} + M_{2,6} + M_{2,10} + M_{6,10} + M_{3,7} + M_{3,11} + M_{11,7} + M_{4,8} + M_{4,12} + M_{8,12}) - 2 \cdot (M_{1,11} + M_{1,7} + M_{1,3} + M_{5,11} + M_{5,7} + M_{5,3} + M_{9,11} + M_{9,7} + M_{9,3} + M_{2,12} + M_{2,8} + M_{2,4} + M_{6,12} + M_{6,8} + M_{6,4} + M_{10,12} + M_{10,8} + M_{10,4})$

Table 8.14 Single, two turn, and three turn number of term contributing to M_+ and M_-

Rectangular planar coil type	N_+: number of term contributing to M_+ (number of positive mutual inductance terms)	N_-: number of terms contributing to M_-
Single turn rectangular planar coil (n = 1, Z_s = 4)	0	4
Two turn rectangular planar coil (n = 2, Z_s = 8)	8	16
Three turn rectangular planar coil (n = 3, Z_s = 12)	24	36

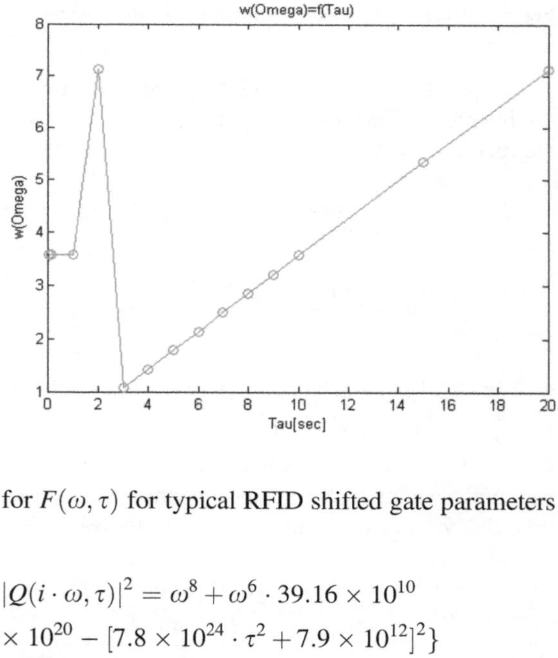

Fig. 8.5 RFID TAGs with double loop rectangular spiral antenna system ω as a function of delay parameter τ

Then we get the expression for $F(\omega, \tau)$ for typical RFID shifted gate parameters values.

$$\begin{aligned} F(\omega, \tau) &= |P(i \cdot \omega, \tau)|^2 - |Q(i \cdot \omega, \tau)|^2 = \omega^8 + \omega^6 \cdot 39.16 \times 10^{10} \\ &+ \omega^4 \cdot \{383.17 \times 10^{20} - [7.8 \times 10^{24} \cdot \tau^2 + 7.9 \times 10^{12}]^2\} \\ &- \omega^2 \cdot \{[34.94 \times 10^{17} - 15.6 \times 10^{24} \cdot \tau]^2 \\ &- 31.2 \times 10^{24} \cdot [7.8 \times 10^{24} \cdot \tau^2 + 7.9 \times 10^{12}]\} - 243.39 \times 10^{48} \end{aligned}$$

We find those ω, τ values which fulfill $F(\omega, \tau) = 0$. We ignore negative, complex, and imaginary values of ω for specific τ values. $\tau \in [0.001\ldots10]$ and we can be express by 3D function $F(\omega, \tau) = 0$. Since it is a very complex function

We recommend to solve it numerically rather than analytic.

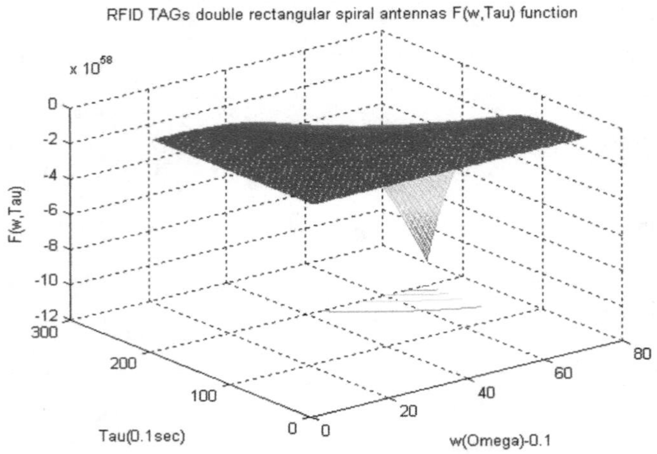

Fig. 8.6 RFID TAGs double rectangular spiral antennas F(ω,τ) function

We plot the stability switch diagram based on different delay values of our RFID double rectangular spiral coils antennas system. Since it is a very complex function we recommend to solve it numerically rather than analytic.

$$\wedge^{-1}(\tau) = (\frac{\partial \text{Re}\lambda}{\partial \tau})_{\lambda=i\cdot\omega} = \text{Re}\{\frac{-2\cdot[U+\tau\cdot|P|^2]+i\cdot F_\omega}{F_\tau+i\cdot 2\cdot[V+\omega\cdot|P|^2]}\}$$

$$\wedge^{-1}(\tau) = (\frac{\partial \text{Re}\lambda}{\partial \tau})_{\lambda=i\cdot\omega} = \frac{2\cdot\{F_\omega\cdot(V+\omega\cdot P^2)-F_\tau\cdot(U+\tau\cdot P^2)\}}{F_\tau^2+4\cdot(V+\omega\cdot P^2)^2}$$

The stability switch occurs only on those delay values (τ) which fit the equation: $\tau = \frac{\theta_+(\tau)}{\omega_+(\tau)}$ and $\theta_+(\tau)$ is the solution of $\sin\theta(\tau) = \ldots$; $\cos\theta(\tau) = \ldots$ when $\omega = \omega_+(\tau)$ if only ω_+ is feasible. Additionally when all double rectangular spiral coils antennas system's parameters are known and the stability switch due to various time delay values τ is describe in the following expression:

$$sign\{\wedge^{-1}(\tau)\} = sign\{F_\omega(\omega(\tau),\tau)\}\cdot sign\{\tau\cdot\omega_\tau(\omega(\tau))+\omega(\tau) + \frac{U(\omega(\tau))\cdot\omega_\tau(\omega(\tau))+V(\omega(\tau))}{|P(\omega(\tau))|^2}\}$$

Remark: we know $F(\omega,\tau)=0$ implies it roots $\omega_i(\tau)$ and finding those delays values τ which ω_i is feasible. There are τ values which ω_i are complex or imaginary number, then unable to analyse stability [5, 6].

8.3 Single-Turn Square Planar Straight Thin Film Inductors Antenna System Stability Optimization Under Microstrip Delayed in Time

We have a system of single turn square planar straight thin film inductors antenna (four segments). The system is constructed from four straight thin film inductors which are connected in a single turn square structure. The straight thin film inductors are connected by microstrip lines (A, B, and C). The single turn square planar straight thin film inductors antenna system is connected to transceiver module through two microstrip lines (D and E) [85]. Index (i) stands for the first (i = 1), second (i = 2), third (i = 3), and fourth (i = 4) straight thin film inductors. w_i is the width of straight thin film inductor (i) in cm, z_i is the thickness of straight thin film inductor (i) in cm, and l_i is the length of straight thin film conductor (inductor) in cm. The calculated inductance of straight thin film inductor (i) is as follow (L_i is the segment inductance in μH):

$$L_i = 0.002 \cdot l_i \cdot \{\ln[\frac{2 \cdot l_i}{w_i + z_i}] + 0.50049 + \frac{w_i + z_i}{3 \cdot l_i}\}[\mu\text{H}]; i = 1, 2, 3, 4$$

Remark: we assume that the magnetic permeability of the conductor material is 1 and the four straight thin film inductors are not identical $w_i \neq w_j; l_i \neq l_j\ z_i \neq z_j$ $(j \neq i; j = 1, 2, 3, 4; i = 1, 2, 3, 4)$ (Fig. 8.7).

N-turn planar rectangular coil structure: If we have N-turn planar rectangular coil structure the total inductance of this coil is equal to the sum of the self-inductance of each of the straight segment $(\sum_{i=1}^{N} L_i; i = 1, 2, \ldots, N)$ plus all the mutual inductances between the segments. The mutual inductance between segment (k) and (j) has a component $M_{k,j}$ caused by the current flowing in segment (k), and a component $M_{j,k}$ caused by the current flowing in segment (j). Since the frequency and phase in both segments are identical, the total mutual inductance linking them equals $M_{k,j} + M_{j,k}$. An analogous relationship exists between segment pairs 2-6, 3-7, 4-8, etc., in each of these pairs, current flow is in the same direction in both segments and all mutual inductances are positive. The mutual inductance between segment 1 and 7, on the other hand, has a component $M_{1,7}$ caused by the current in segment 1, and a component $M_{7,1}$ caused by the current in segment 7. It can be implemented in the same manner to other segments. The total mutual inductance linking these two segments equals $M_{1,7} + M_{7,1}$ but is negative because current flow in segment 1 is opposite in direction to current flow in segment 7 $(k = 1, 2, \ldots, N; j = 1, 2, \ldots, N; k \neq j; k, j \in \mathbb{N}_+)$. An analogous relationship exists between segment pairs 1-3, 5-7, 5-3, 2-8, 2-4, 6-8, and 6-4 (case of two-turn rectangular planar coil). Current magnitude is identical in all segments, with the result that $M_{j,k} = M_{k,j}$. The total inductance L_T for the case of two turn coil is as follow:

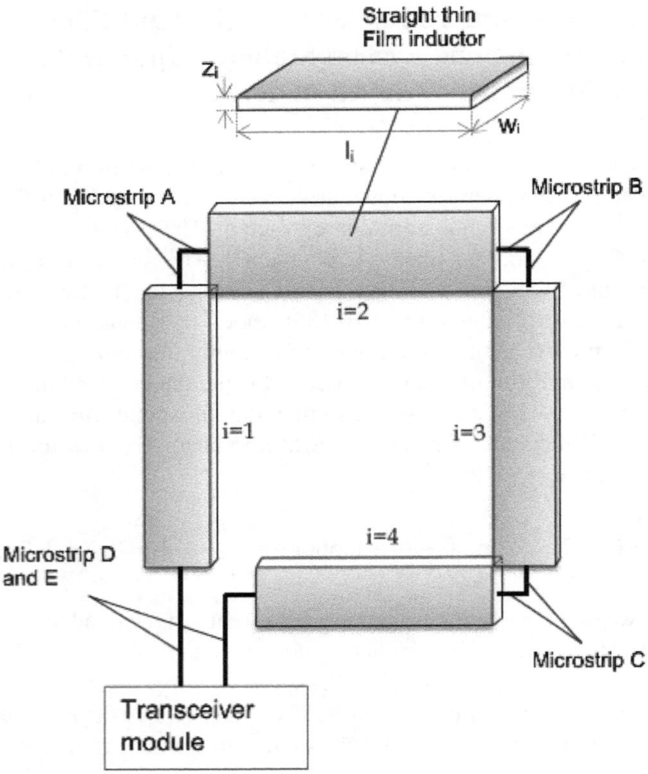

Fig. 8.7 Single turn square planar straight thin film inductors antenna system

$$L_T = \sum_{i=1}^{N=8} L_i + 2 \cdot (M_{1,5} + M_{2,6} + M_{3,7} + M_{4,8}) - 2 \cdot (M_{1,7} + M_{1,3} + M_{5,7} + M_{5,3} + M_{2,8} + M_{2,4} + M_{6,8} + M_{6,4})$$

We define the mutual inductances term as $\sum M$, $(L_0; L_0 = \sum_{i=1}^{N=8} L_i)$.

$$\sum M = 2 \cdot (M_{1,5} + M_{2,6} + M_{3,7} + M_{4,8}) - 2 \cdot (M_{1,7} + M_{1,3} + M_{5,7} + M_{5,3} + M_{2,8} + M_{2,4} + M_{6,8} + M_{6,4})$$

Where L_T is the total inductances, L_0 is the sum of the self-inductances of all the straight segments, and $\sum M$ is the sum of all the mutual inductances, both positive and negative.

$$M_{+} = 2 \cdot (M_{1,5} + M_{2,6} + M_{3,7} + M_{4,8}); M_{-}$$
$$= 2 \cdot (M_{1,7} + M_{1,3} + M_{5,7} + M_{5,3} + M_{2,8} + M_{2,4} + M_{6,8} + M_{6,4})$$

General case, N-turn planar rectangular coil structure $L_T = \sum_{i=1}^{N} L_i + M_{+} - M_{-}$ where M_+ is the sum of the positive mutual inductances and M_- is the sum of the negative mutual inductances. The mutual inductance between two parallel conductors is a function of the length of the conductors and of the geometric mean distance between them. The figure and Table 8.13 describe single-turn rectangular planar coil (N = 4), two-turn rectangular planar coil (n = 8), and three-turn rectangular planar coil (n = 12), sum of the self-inductances (L_O), sum of the positive mutual inductances (M+), sum of the negative mutual inductances (M−) and the sum of all the mutual inductances $\sum M$ (Table 8.15, Fig. 8.8).

We define (n) as the number of full turns and (Z_s) as the total number of segments. The number of term contributing to M_+ is N_+ and the number of terms contributing to M_- is N_-.

$$N_{+} = 4 \cdot n \cdot (n-1) + 2 \cdot n \cdot (Z_s - 4 \cdot n)$$
$$N_{-} = 4 \cdot n^4 + 2 \cdot n \cdot (Z_s - 4 \cdot n) + (Z_s - 4 \cdot n - 2) \cdot (Z_s - 4 \cdot n - 1) \cdot (Z_s - 4 \cdot n)/3$$

Table 8.15 Single turn square planar straight thin film inductors antenna equivalent circuit $V_{A_i}\ \forall\ i = 1, 2, \ldots, 14$ expressions

Node voltage	Expression
V_{A_1}	$I_{R_a} \cdot R_a$
V_{A_2}	$I_{R_a} \cdot R_a - L_a \cdot \frac{dI_{L'_1}}{dt}$
V_{A_3}	$I_{R_a} \cdot R_a - (L_a + L'_1) \cdot \frac{dI_{L'_1}}{dt}$
$V_{A_4} (V_{A_4} \approx V_{A_5})$	$I_{R_a} \cdot R_a - (L_a + L'_1) \cdot \frac{dI_{L'_1}}{dt} - I_{L'_1} \cdot R_1$
V_{A_5}	$I_{R_a} \cdot R_a - (L_a + L'_1) \cdot \frac{dI_{L'_1}}{dt} - I_{L'_1} \cdot R_1$
V_{A_6}	$I_{R_a} \cdot R_a - (L_a + L'_1) \cdot \frac{dI_{L'_1}}{dt} - I_{L'_1} \cdot R_1 - L'_2 \cdot \frac{dI_{L'_1}(t-\tau_1)}{dt}$
$V_{A_7} (V_{A_7} \approx V_{A_8})$	$I_{R_a} \cdot R_a - (L_a + L'_1) \cdot \frac{dI_{L'_1}}{dt} - I_{L'_1} \cdot R_1 - L'_2 \cdot \frac{dI_{L'_1}(t-\tau_1)}{dt} - I_{L'_1}(t-\tau_1) \cdot R_2$
V_{A_8}	$I_{R_a} \cdot R_a - (L_a + L'_1) \cdot \frac{dI_{L'_1}}{dt} - I_{L'_1} \cdot R_1 - L'_2 \cdot \frac{dI_{L'_1}(t-\tau_1)}{dt} - I_{L'_1}(t-\tau_1) \cdot R_2$
V_{A_9}	$I_{R_b} \cdot R_b + (L_b + L'_4) \cdot \frac{dI_{L'_1}(t-\sum_{i=1}^{3}\tau_i)}{dt} + I_{L'_1}(t - \sum_{i=1}^{3}\tau_i) \cdot R_4 + I_{L'_1}(t - \sum_{i=1}^{2}\tau_i) \cdot R_3$
$V_{A_{10}} (V_{A_{10}} \approx V_{A_{11}})$	$I_{R_b} \cdot R_b + (L_b + L'_4) \cdot \frac{dI_{L'_1}(t-\sum_{i=1}^{3}\tau_i)}{dt} + I_{L'_1}(t - \sum_{i=1}^{3}\tau_i) \cdot R_4$
$V_{A_{11}}$	$I_{R_b} \cdot R_b + (L_b + L'_4) \cdot \frac{dI_{L'_1}(t-\sum_{i=1}^{3}\tau_i)}{dt} + I_{L'_1}(t - \sum_{i=1}^{3}\tau_i) \cdot R_4$
$V_{A_{12}}$	$I_{R_b} \cdot R_b + (L_b + L'_4) \cdot \frac{dI_{L'_1}(t-\sum_{i=1}^{3}\tau_i)}{dt}$
$V_{A_{13}}$	$I_{R_b} \cdot R_b + L_b \cdot \frac{dI_{L'_1}(t-\sum_{i=1}^{3}\tau_i)}{dt}$
$V_{A_{14}}$	$I_{R_b} \cdot R_b$

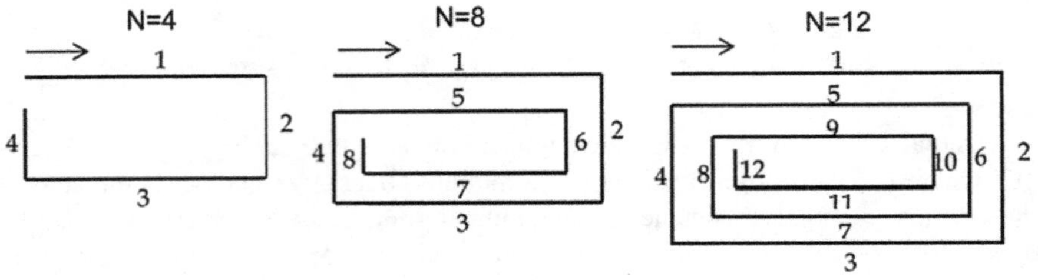

Fig. 8.8 Square planar straight thin film inductor antenna N=4, N=8, and N=12

Table 8.16 Single turn square planar straight thin film inductors antenna equivalent circuit's remarks 1.0–1.6 expressions

Remark no.	Expression
1.0	$\frac{d^2 I_{L'_1}(t-\sum_{i=1}^{3}\tau_i)}{dt^2} = \frac{d^2 I_{L'_1}(t)}{dt^2}\cdot e^{-\lambda\cdot\sum_{i=1}^{3}\tau_i}$
1.1	$\sum_{i=2}^{4} L'_1 \cdot \frac{d^2 I_{L'_1}(t-\sum_{k=1}^{i-1}\tau_k)}{dt^2} = \frac{d^2 I_{L'_1}(t)}{dt^2}\cdot \sum_{j=2}^{4} L'_j \cdot e^{-\lambda\cdot\sum_{k=1}^{j-1}\tau_k}$
1.2	$\sum_{i=2}^{4} \frac{\rho_i\cdot l_i}{w_i\cdot z_i} \cdot \frac{dI_{L'_1}(t-\sum_{k=1}^{i-1}\tau_k)}{dt} = \frac{dI_{L'_1}(t)}{dt}\cdot \sum_{j=2}^{4} \frac{\rho_j\cdot l_j}{w_j\cdot z_j} \cdot e^{-\lambda\cdot\sum_{k=1}^{j-1}\tau_k}$
1.3	$\frac{dI_{L'_1}(t-\sum_{i=1}^{3}\tau_i)}{dt} = \frac{dI_{L'_1}(t)}{dt}\cdot e^{-\lambda\cdot\sum_{i=1}^{3}\tau_i}$
1.4	$\frac{d^3 I_{L'_1}(t-\sum_{i=1}^{3}\tau_i)}{dt^3} = \frac{d^3 I_{L'_1}(t)}{dt^3}\cdot e^{-\lambda\cdot\sum_{i=1}^{3}\tau_i}$
1.5	$\sum_{i=2}^{4} L'_1 \cdot \frac{d^3 I_{L'_1}(t-\sum_{k=1}^{i-1}\tau_k)}{dt^3} = \frac{d^3 I_{L'_1}(t)}{dt^3}\cdot \sum_{j=2}^{4} L'_j \cdot e^{-\lambda\cdot\sum_{k=1}^{j-1}\tau_k}$
1.6	$\sum_{i=2}^{4} \frac{\rho_i\cdot l_i}{w_i\cdot z_i} \cdot \frac{d^2 I_{L'_1}(t-\sum_{k=1}^{i-1}\tau_k)}{dt^2} = \frac{d^2 I_{L'_1}(t)}{dt^2}\cdot \sum_{j=2}^{4} \frac{\rho_j\cdot l_j}{w_j\cdot z_j} \cdot e^{-\lambda\cdot\sum_{k=1}^{j-1}\tau_k}$

Mutual inductance between two parallel conductors (same length):

The mutual inductance $M_{j,k}$ between two parallel conductors (j and k segments $l_j, l_k; j \neq k; l_k = l_j = l$) is a function of the length of the conductors and of the Geometric Mean Distance (GMD) between them ($M_{j,k} = 2\cdot l\cdot Q$). $M_{j,k}$ is the mutual inductance in (nH), l is the conductor length in (cm), and Q is the mutual inductance parameter, calculated from the equation (Fig. 8.9):

$$Q = \ln\{\frac{l}{GMD} + \sqrt{1+\frac{l^2}{(GMD)^2}}\} - \sqrt{1+\frac{(GMD)^2}{l^2}} + \frac{GMD}{l}$$

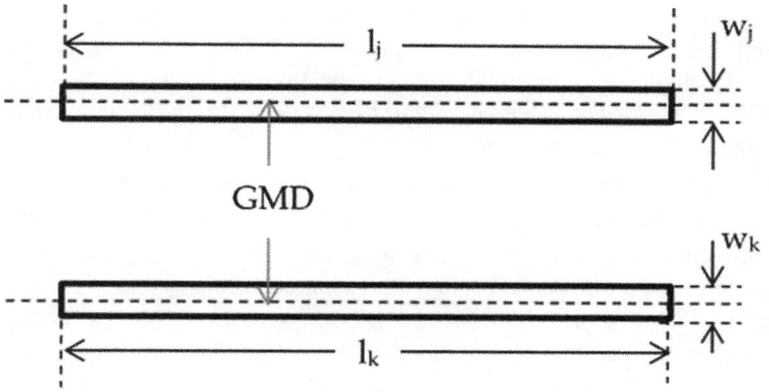

Fig. 8.9 Two parallel conductors (same length)

In this equation, l is the length corresponding to the subscript of Q, and GMD is the Geometric Mean Distance (GMD) between the two conductors (l_j, l_k), which is approximately equal to the distance $d_{j,k}(GMD \simeq d_{j,k} = d)$ between the track centers. The exact value of the GMD may be calculated from the equation:

$$\ln GMD = \ln d_{j,k} - \{\frac{1}{12 \cdot (\frac{d_{j,k}}{w})^2} + \frac{1}{60 \cdot (\frac{d_{j,k}}{w})^4} + \frac{1}{168 \cdot (\frac{d_{j,k}}{w})^6} + \frac{1}{360 \cdot (\frac{d_{j,k}}{w})^8} + \frac{1}{660 \cdot (\frac{d_{j,k}}{w})^{10}} + \ldots\}$$

where (w) is the track width $w_j = w_k = w$.

$$GMD = \exp[\ln d_{j,k} - \{\frac{1}{12 \cdot (\frac{d_{j,k}}{w})^2} + \frac{1}{60 \cdot (\frac{d_{j,k}}{w})^4} + \frac{1}{168 \cdot (\frac{d_{j,k}}{w})^6} + \frac{1}{360 \cdot (\frac{d_{j,k}}{w})^8} + \frac{1}{660 \cdot (\frac{d_{j,k}}{w})^{10}} + \ldots\}]$$

We can represent our single turn square planar straight thin film inductors antenna system equivalent circuit. Microstrip lines (A, B, and C) are represented as delay lines τ_1, τ_2, and τ_3 respectively. We neglect microstrip lines D and E parasitic effects (no delays). The sum of all the mutual

Inductances, both positive and negative is marked as $(SigmaM = \sum M)$ inductance. Transceiver module is represented as an equivalent circuit of mixer, with input and output impedances of the mixer board. The mixer itself has a common gate input. The input is dominated by parasitic impedances of the package and PCB stray, such as shunt capacitance of the device gates along with the bond wire inductances and resistance which is inversely proportional to the mixer current setting. In similar way as for the inputs, the impedance at the outputs can be modeled by a resistance with a shunt PCB stray capacitance and bond wires inductances. The resistance $R_i = \frac{\rho_i \cdot l_i}{A_i}$ of a straight thin film (strip) number $(i; i = 1, 2, 3, 4)$ is expected to depend on the DC resistivity ρ_i of the straight thin film strip material and the strip cross section $A_i = w_i \cdot z_i; R_i = \frac{\rho_i \cdot l_i}{w_i \cdot z_i}$. The dimension of the straight thin film strip that affects the inductance most strongly is the length l_i. The width w_i has much weaker influence, and straight thin film strip thickness can be neglected completely for the limit $(z_i \ll w_i)$. This is in contrast to the resistance $R_i = \frac{\rho_i \cdot l_i}{w_i \cdot z_i}$, which is inversely proportional to the straight thin film strip cross section $w_i \cdot z_i$ and depends on the material properties via its resistivity ρ_i [85] (Fig. 8.10).

Inductance of a single turn square planar coil (straight thin film inductors system) calculation:

$$L_i = 0.002 \cdot l_i \cdot \{\ln[\frac{2 \cdot l_i}{w_i + z_i}] + 0.50049 + \frac{w_i + z_i}{3 \cdot l_i}\}[\mu\text{H}]; \quad i = 1, 2, 3, 4$$

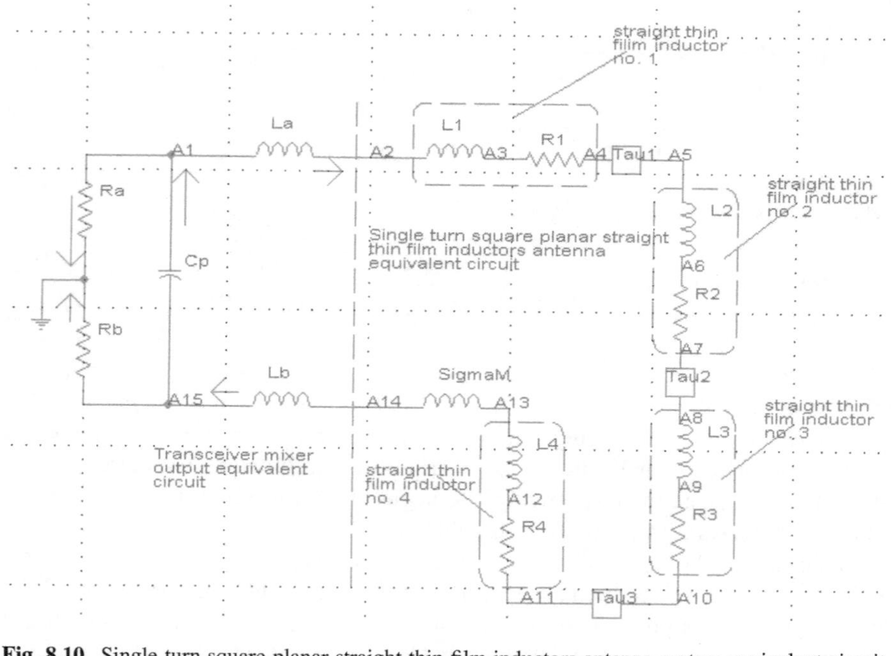

Fig. 8.10 Single turn square planar straight thin film inductors antenna system equivalent circuit

$$L_1 = 0.002 \cdot l_1 \cdot \{\ln[\frac{2 \cdot l_1}{w_1 + z_1}] + 0.50049 + \frac{w_1 + z_1}{3 \cdot l_1}\};$$
$$L_2 = 0.002 \cdot l_2 \cdot \{\ln[\frac{2 \cdot l_2}{w_2 + z_2}] + 0.50049 + \frac{w_2 + z_2}{3 \cdot l_2}\}$$
$$L_3 = 0.002 \cdot l_3 \cdot \{\ln[\frac{2 \cdot l_3}{w_3 + z_3}] + 0.50049 + \frac{w_3 + z_3}{3 \cdot l_3}\};$$
$$L_4 = 0.002 \cdot l_4 \cdot \{\ln[\frac{2 \cdot l_4}{w_4 + z_4}] + 0.50049 + \frac{w_4 + z_4}{3 \cdot l_4}\}$$

The derivations are producing the following calculation results (Grover):

$$L_T = \sum_{i=1}^{N=4} L_i + \sum M = \sum_{i=1}^{N=4} L_i + M_+ - M_-$$

Since the currents in parallel straight thin film segments flow in opposite directions, there is no positive mutual inductance in this segment coil; that is $M_+ = 0$. The negative mutual inductance is equal to the sum of $M_{1,3}, M_{3,1}$ and $M_{2,4}, M_{4,2}$ Or, since $M_{1,3}$ equals $M_{3,1}$ and $M_{2,4}$ equals $M_{4,2}$, $M_- = 2 \cdot (M_{1,3} + M_{2,4})$.

$$d_{j,k}; d_{1,3} = d_{3,1}; w_1 = w_3 = w_{1-3}$$

$$GMD_{1,3} = \exp[\ln d_{1,3} - \{\frac{1}{12 \cdot (\frac{d_{1,3}}{w_{1-3}})^2} + \frac{1}{60 \cdot (\frac{d_{1,3}}{w_{1-3}})^4} + \frac{1}{168 \cdot (\frac{d_{1,3}}{w_{1-3}})^6} + \frac{1}{360 \cdot (\frac{d_{1,3}}{w_{1-3}})^8} + \frac{1}{660 \cdot (\frac{d_{1,3}}{w_{1-3}})^{10}} + \ldots\}]$$

This expression and that for l_1, when substitute into $Q = \ldots$, yield a mutual inductance parameter Q_1 :

$$Q_1 = \ln\{\frac{l_1}{GMD_{1,3}} + \sqrt{1 + \frac{l_1^2}{(GMD_{1,3})^2}}\} - \sqrt{1 + \frac{(GMD_{1,3})^2}{l_1^2}} + \frac{GMD_{1,3}}{l_1}$$

Now, using $M_{j,k} = 2 \cdot l \cdot Q$ and the fact that l_1 equals l_3 we can write $M_{1,3} = 2 \cdot l_1 \cdot Q_1$

$$M_{1,3} = 2 \cdot l_1 \cdot Q_1$$
$$= 2 \cdot l_1 \cdot [\ln\{\frac{l_1}{GMD_{1,3}} + \sqrt{1 + \frac{l_1^2}{(GMD_{1,3})^2}}\} - \sqrt{1 + \frac{(GMD_{1,3})^2}{l_1^2}} + \frac{GMD_{1,3}}{l_1}]$$

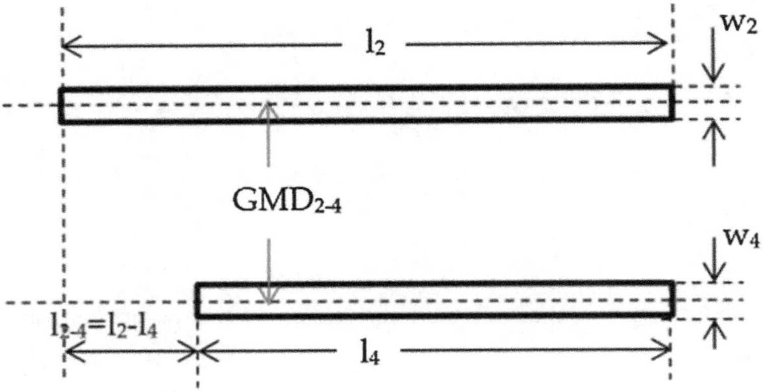

Fig. 8.11 Two parallel conductors (not the same length)

However, because l_2 does not equal $l_4(l_2 > l_4; l_{2-4} = l_2 - l_4)$, we use two parallel (no equal length) filament geometry calculation: Two segments of lengths l_2 and l_4, respectively, are separated by a Geometric Mean Distance ($GMD_{2\text{-}4}$). In this case, $2 \cdot M_{2,4} = (M_2 + M_4) - M_{2-4}$ and the individual M terms are calculated using equation $M = 2 \cdot l \cdot Q$ and the length corresponding to the

Subscript; that is, $M_2 = 2 \cdot l_2 \cdot Q_2$ where Q_2 is the mutual inductance parameter Q for $\frac{GMD_{2,4}}{l_2}$. We consider $w_2 = w_4 = w_{2-4}$ (Fig. 8.11).

$$2 \cdot M_{2,4} = (M_2 + M_4) - M_{2-4}; M_2 = 2 \cdot l_2 \cdot Q_2; M_4 = 2 \cdot l_4 \cdot Q_4; M_{2-4} = 2 \cdot (l_2 - l_4) \cdot Q_{2-4}$$

$$Q_2 = \ln\{\frac{l_2}{GMD_{2,4}} + \sqrt{1 + \frac{l_2^2}{(GMD_{2,4})^2}}\} - \sqrt{1 + \frac{(GMD_{2,4})^2}{l_2^2}} + \frac{GMD_{2,4}}{l_2}$$

$$Q_4 = \ln\{\frac{l_4}{GMD_{2,4}} + \sqrt{1 + \frac{l_4^2}{(GMD_{2,4})^2}}\} - \sqrt{1 + \frac{(GMD_{2,4})^2}{l_4^2}} + \frac{GMD_{2,4}}{l_4}$$

$$GMD_{2,4} = \exp[\ln d_{2,4} - \{\frac{1}{12 \cdot (\frac{d_{2,4}}{w_{2-4}})^2} + \frac{1}{60 \cdot (\frac{d_{2,4}}{w_{2-4}})^4} + \frac{1}{168 \cdot (\frac{d_{2,4}}{w_{2-4}})^6} + \frac{1}{360 \cdot (\frac{d_{2,4}}{w_{2-4}})^8} + \frac{1}{660 \cdot (\frac{d_{2,4}}{w_{2-4}})^{10}} + \ldots\}]$$

Since l_2 equals l_1 and the $GMD_{2\text{-}4}$ remains constant, Q_2 must equal Q_1 ($Q_2 = Q_1$) as calculated from $Q_2 = \ln\{\frac{l_2}{GMD_{2,4}} + \sqrt{1 + \frac{l_2^2}{(GMD_{2,4})^2}}\} - \ldots$. It follows that $M_2 = M_{1,3}$.

To obtain Q_4 and $Q_{2\text{-}4}$, however, $Q_4 = \ln\{\frac{l_4}{GMD_{2,4}} + \sqrt{1 + \frac{l_4^2}{(GMD_{2,4})^2}}\} - \ldots$ must be solved for a $GMD_{2,4}$.

$$\sum M = -M_- = -\{2 \cdot (M_{1,3} + M_{2,4})\} = -2 \cdot M_{1,3} - 2 \cdot M_{2,4} = -2 \cdot M_{1,3} - [(M_2 + M_4) - M_{2-4}]$$

$$\sum M = -4 \cdot l_1 \cdot Q_1 - 2 \cdot (l_2 \cdot Q_2 + l_4 \cdot Q_4) + 2 \cdot (l_2 - l_4) \cdot Q_{2-4}; M_+ = 0 \Rightarrow \sum M < 0$$

$$\begin{aligned}\sum M = {} & -4 \cdot l_1 \cdot \{\ln\{\frac{l_1}{GMD_{1,3}} + \sqrt{1 + \frac{l_1^2}{(GMD_{1,3})^2}}\} - \sqrt{1 + \frac{(GMD_{1,3})^2}{l_1^2}} + \frac{GMD_{1,3}}{l_1}\} \\ & - 2 \cdot (l_2 \cdot \{\ln\{\frac{l_2}{GMD_{2,4}} + \sqrt{1 + \frac{l_2^2}{(GMD_{2,4})^2}}\} - \sqrt{1 + \frac{(GMD_{2,4})^2}{l_2^2}} + \frac{GMD_{2,4}}{l_2}\} \\ & + l_4 \cdot \{\ln\{\frac{l_4}{GMD_{2,4}} + \sqrt{1 + \frac{l_4^2}{(GMD_{2,4})^2}}\} - \sqrt{1 + \frac{(GMD_{2,4})^2}{l_4^2}} + \frac{GMD_{2,4}}{l_4}\}) \\ & + 2 \cdot (l_2 - l_4) \cdot \{\ln\{\frac{l_4}{GMD_{2,4}} + \sqrt{1 + \frac{l_4^2}{(GMD_{2,4})^2}}\} - \ldots\}\end{aligned}$$

Result discussion: Negative mutual inductance (M_-) results from coupling between two conductors having current vectors in opposite directions. In electronic circuits, negative mutual inductance is usually much smaller in magnitude than overall inductance that it can be neglected with little effect. It is not so true in microelectronic circuits.

The sum of all the mutual inductances both positive and negative ($\sum M$) is a negative value and must be deduce from the sum of the self-inductances in our system for getting the total inductance of single turn square planar straight thin film inductors antenna system. Each straight thin film inductor segment is connected to the other by microstrip line. To analyze our system stability, we represent our equivalent circuit as a four inductors in series (with series resistance $R_i = \frac{\rho_i \cdot l_i}{A_i}$ of a straight thin film (strip) number $i; i = 1, 2, 3, 4$) and deduce from each segment inductance expression the quarter value of the sum of all mutual inductances ($\sum M$) (Fig. 8.12). .

$$L_i' = L_i + \frac{1}{4} \cdot \sum M; L_i \to L_i + \frac{1}{4} \cdot \sum M \; \forall \; i = 1, 2, 3, 4; \sum M < 0; L_i > \frac{1}{4} \cdot |\sum M|$$

$$L_1' = L_1 + \frac{1}{4} \cdot \sum M; L_2' = L_2 + \frac{1}{4} \cdot \sum M; L_3' = L_3 + \frac{1}{4} \cdot \sum M; L_4' = L_4 + \frac{1}{4} \cdot \sum M$$

$$Tau1 = \tau_1; Tau2 = \tau_2; Tau3 = \tau_3; V_{\tau_i} \to \varepsilon \; \forall \; i = 1, 2, 3; V_{\tau_1} \to \varepsilon; V_{\tau_2} \to \varepsilon; V_{\tau_3} \to \varepsilon$$

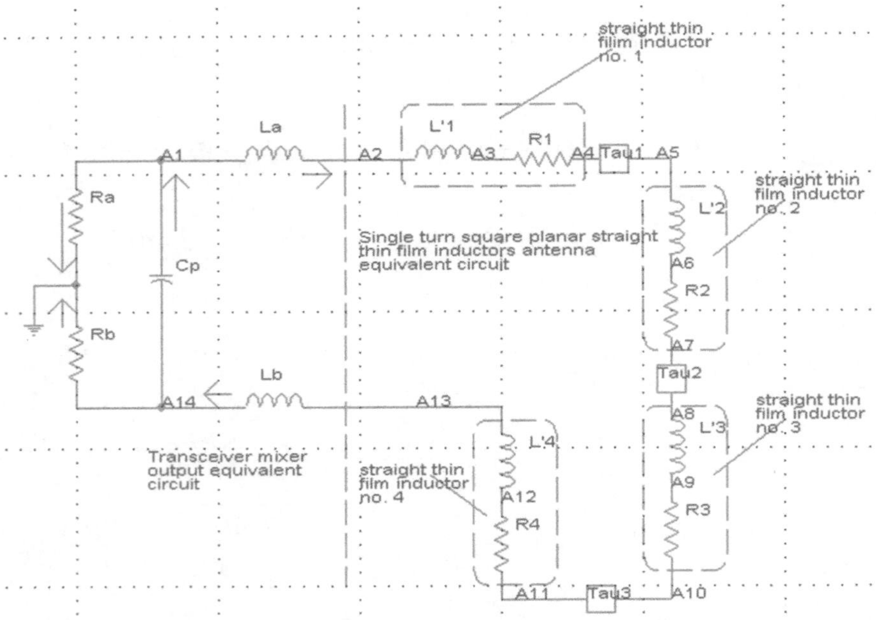

Fig. 8.12 Single turn square planar straight thin film inductors antenna system equivalent circuit with delay lines

$$I_{L'_1} = I_{R_1}; I_{L'_2} = I_{R_1}; I_{L'_3} = I_{R_3}; I_{L'_4} = I_{R_4}; I_{L'_2}(t) = I_{L'_1}(t - \tau_1); I_{L'_3}(t) = I_{L'_2}(t - \tau_2) = I_{L'_1}(t - \sum_{i=1}^{2} \tau_i)$$

$$I_{L'_4}(t) = I_{L'_3}(t - \tau_3) = I_{L'_2}(t - \sum_{i=2}^{3} \tau_i) = I_{L'_1}(t - \sum_{i=1}^{3} \tau_i); I_{L'_4} = I_{L_b}; I_{L'_1} = I_{L_a}; I_{R_a} = \frac{V_{A_1}}{R_a}; I_{R_b} = \frac{V_{A_{14}}}{R_b}$$

$$I_{C_p} = C_p \cdot \frac{d}{dt}(V_{A_{14}} - V_{A_1}); I_{L_b} = I_{C_p} + I_{R_b}; I_{C_p} = I_{L_a} + I_{R_a}; V_{A_4} \approx V_{A_5}; V_{A_7} \approx V_{A_8}; V_{A_{10}} \approx V_{A_{11}}$$

$$V_{A_1} - V_{A_2} = L_a \cdot \frac{dI_{L_a}}{dt}; V_{A_{13}} - V_{A_{14}} = L_b \cdot \frac{dI_{L_b}}{dt}; V_{A_2} - V_{A_3} = L'_1 \cdot \frac{dI_{L'_1}}{dt};$$

$$V_{A_5} - V_{A_6} = L'_2 \cdot \frac{dI_{L'_2}}{dt}$$

$$V_{A_8} - V_{A_9} = L'_3 \cdot \frac{dI_{L'_3}}{dt}; V_{A_{12}} - V_{A_{13}} = L'_4 \cdot \frac{dI_{L'_4}}{dt};$$

$$I_{R_1} = \frac{V_{A_3} - V_{A_4}}{R_1}; I_{R_2} = \frac{V_{A_6} - V_{A_7}}{R_2}; I_{R_3} = \frac{V_{A_9} - V_{A_{10}}}{R_3}$$

$$I_{R_4} = \frac{V_{A_{11}} - V_{A_{12}}}{R_4}; V_{C_p} + V_{L_a} + V_{L_b} + \sum_{i=1}^{4} V_{L'_i} + \sum_{i=1}^{4} V_{R_i} = 0$$

$$V_{A_4} \approx V_{A_5}; V_{A_4} \to V_{A_5}; I_{R_1} = \frac{V_{A_3} - V_{A_5}}{R_1}; V_{A_7} \approx V_{A_8}; V_{A_7} \to V_{A_8}; I_{R_2} = \frac{V_{A_6} - V_{A_8}}{R_2}$$

$$V_{A_{10}} \approx V_{A_{11}}; V_{A_{10}} \to V_{A_{11}}; I_{R_3} = \frac{V_{A_9} - V_{A_{11}}}{R_3}; I_{R_1} = I_{L'_1}; I_{R_1} \to I_{L'_1}; I_{L'_1} = \frac{V_{A_3} - V_{A_5}}{R_1}$$

$$I_{R_2} = I_{L'_2}; I_{R_2} \to I_{L'_2}; I_{L'_2} = \frac{V_{A_6} - V_{A_8}}{R_2}; I_{R_3} = I_{L'_3}; I_{R_3} \to I_{L'_3}; I_{L'_3} = \frac{V_{A_9} - V_{A_{11}}}{R_3}$$

$$I_{R_4} = I_{L'_4}; I_{R_4} \to I_{L'_4}; I_{L'_4} = \frac{V_{A_{11}} - V_{A_{12}}}{R_4}; I_{L'_2}(t) \to I_{L'_1}(t - \tau_1); I_{L'_3}(t) \to I_{L'_1}(t - \sum_{i=1}^{2} \tau_i)$$

$$I_{L'_4}(t) \to I_{L'_1}(t - \sum_{i=1}^{3} \tau_i)$$

System equation Version 1.0:

$$I_{L_a} = I_{L'_1}; I_{L_b} = I_{L'_4} = I_{L'_1}(t - \sum_{i=1}^{3} \tau_i); I_{R_a} = \frac{V_{A_1}}{R_a}; I_{R_b} = \frac{V_{A_{14}}}{R_b}; I_{L'_1} = \frac{V_{A_3} - V_{A_5}}{R_1};$$

$$I_{L'_2} = \frac{V_{A_6} - V_{A_8}}{R_2}; I_{L'_3} = \frac{V_{A_9} - V_{A_{11}}}{R_3}; I_{L'_4} = \frac{V_{A_{11}} - V_{A_{12}}}{R_4}; I_{C_p} = C_p \cdot \frac{d}{dt}(V_{A_{14}} - V_{A_1});$$

$$I_{L'_1}(t - \sum_{i=1}^{3} \tau_i) = I_{C_p} + I_{R_b}; I_{C_p} = I_{L'_1} + I_{R_a}; V_{A_1} - V_{A_2} = L_a \cdot \frac{dI_{L'_1}}{dt};$$

$$V_{A_{13}} - V_{A_{14}} = L_b \cdot \frac{dI_{L'_1}(t - \sum_{i=1}^{3} \tau_i)}{dt}; V_{A_2} - V_{A_3} = L'_1 \cdot \frac{dI_{L'_1}}{dt}$$

$$V_{A_5} - V_{A_6} = L'_2 \cdot \frac{dI_{L'_1}(t - \tau_1)}{dt}; V_{A_8} - V_{A_9} = L'_3 \cdot \frac{dI_{L'_1}(t - \sum_{i=1}^{2} \tau_i)}{dt};$$

$$V_{A_{12}} - V_{A_{13}} = L'_4 \cdot \frac{dI_{L'_1}(t - \sum_{i=1}^{3} \tau_i)}{dt}; V_{C_p} + V_{L_a} + V_{L_b} + \sum_{i=1}^{4} V_{L'_i} + \sum_{i=1}^{4} V_{R_i} = 0$$

&&&

$$V_{A_1} = I_{R_a} \cdot R_a; V_{A_{14}} = I_{R_b} \cdot R_b; V_{A_3} - V_{A_5} = I_{L'_1} \cdot R_1; V_{A_6} - V_{A_8} = I_{L'_2} \cdot R_2;$$

$$V_{A_9} - V_{A_{11}} = I_{L'_3} \cdot R_3$$

$$V_{A_{11}} - V_{A_{12}} = I_{L'_4} \cdot R_4; I_{C_p} = C_p \cdot \frac{d}{dt}(V_{A_{14}} - V_{A_1}) = C_p \cdot \frac{d}{dt}(I_{R_b} \cdot R_b - I_{R_a} \cdot R_a)$$

$$I_{C_p} = C_p \cdot R_b \cdot \frac{dI_{R_b}}{dt} - C_p \cdot R_a \cdot \frac{dI_{R_a}}{dt}; V_{A_1} - V_{A_2} = L_a \cdot \frac{dI_{L'_1}}{dt} \Rightarrow V_{A_2} = V_{A_1} - L_a \cdot \frac{dI_{L'_1}}{dt}$$

$$V_{A_2} = I_{R_a} \cdot R_a - L_a \cdot \frac{dI_{L'_1}}{dt}; V_{A_{13}} - V_{A_{14}} = L_b \cdot \frac{dI_{L_b}}{dt} \Rightarrow V_{A_{13}} = V_{A_{14}} + L_b \cdot \frac{dI_{L_b}}{dt} = V_{A_{14}} + L_b \cdot \frac{dI_{L'_4}}{dt}$$

$$V_{A_{13}} = I_{R_b} \cdot R_b + L_b \cdot \frac{dI_{L'_1}(t - \sum_{i=1}^{3} \tau_i)}{dt}; V_{A_2} - V_{A_3} = L'_1 \cdot \frac{dI_{L'_1}}{dt} \Rightarrow V_{A_3} = V_{A_2} - L'_1 \cdot \frac{dI_{L'_1}}{dt}$$

$$V_{A_3} = I_{R_a} \cdot R_a - L_a \cdot \frac{dI_{L'_1}}{dt} - L'_1 \cdot \frac{dI_{L'_1}}{dt} = I_{R_a} \cdot R_a - (L_a + L'_1) \cdot \frac{dI_{L'_1}}{dt}$$

$$V_{A_{12}} - V_{A_{13}} = L'_4 \cdot \frac{dI_{L'_1}(t - \sum_{i=1}^{3} \tau_i)}{dt} \Rightarrow V_{A_{12}} = V_{A_{13}} + L'_4 \cdot \frac{dI_{L'_1}(t - \sum_{i=1}^{3} \tau_i)}{dt}$$

$$V_{A_{12}} = I_{R_b} \cdot R_b + L_b \cdot \frac{dI_{L'_1}(t - \sum_{i=1}^{3} \tau_i)}{dt} + L'_4 \cdot \frac{dI_{L'_1}(t - \sum_{i=1}^{3} \tau_i)}{dt} = I_{R_b} \cdot R_b + (L_b + L'_4) \cdot \frac{dI_{L'_1}(t - \sum_{i=1}^{3} \tau_i)}{dt}$$

$$I_{L'_1} = \frac{V_{A_3} - V_{A_5}}{R_1} \Rightarrow V_{A_3} - V_{A_5} = I_{L'_1} \cdot R_1;$$

$$V_{A_5} = V_{A_3} - I_{L'_1} \cdot R_1 = I_{R_a} \cdot R_a - (L_a + L'_1) \cdot \frac{dI_{L'_1}}{dt} - I_{L'_1} \cdot R_1$$

$$V_{A_5} - V_{A_6} = L'_2 \cdot \frac{dI_{L'_2}}{dt} = L'_2 \cdot \frac{dI_{L'_1}(t - \tau_1)}{dt}; V_{A_6} = V_{A_5} - L'_2 \cdot \frac{dI_{L'_1}}{dt}$$

$$V_{A_6} = I_{R_a} \cdot R_a - (L_a + L'_1) \cdot \frac{dI_{L'_1}}{dt} - I_{L'_1} \cdot R_1 - L'_2 \cdot \frac{dI_{L'_1}(t - \tau_1)}{dt}$$

$$I_{L'_2} = \frac{V_{A_6} - V_{A_8}}{R_2} \Rightarrow V_{A_6} - V_{A_8} = I_{L'_2} \cdot R_2;$$

$$V_{A_8} = V_{A_6} - I_{L'_2} \cdot R_2 = V_{A_6} - I_{L'_1}(t - \tau_1) \cdot R_2$$

$$V_{A_8} = I_{R_a} \cdot R_a - (L_a + L'_1) \cdot \frac{dI_{L'_1}}{dt} - I_{L'_1} \cdot R_1 - L'_2 \cdot \frac{dI_{L'_1}(t - \tau_1)}{dt} - I_{L'_1}(t - \tau_1) \cdot R_2$$

$$I_{L'_4} = \frac{V_{A_{11}} - V_{A_{12}}}{R_4} \Rightarrow V_{A_{11}} - V_{A_{12}} = I_{L'_4} \cdot R_4 = I_{L'_1}(t - \sum_{i=1}^{3} \tau_i) \cdot R_4;$$

$$V_{A_{11}} = V_{A_{12}} + I_{L'_1}(t - \sum_{i=1}^{3} \tau_i) \cdot R_4$$

$$V_{A_{11}} = I_{R_b} \cdot R_b + (L_b + L'_4) \cdot \frac{dI_{L'_1}(t - \sum_{i=1}^{3} \tau_i)}{dt} + I_{L'_1}(t - \sum_{i=1}^{3} \tau_i) \cdot R_4$$

$$I_{L'_3} = \frac{V_{A_9} - V_{A_{11}}}{R_3} \Rightarrow V_{A_9} - V_{A_{11}} = I_{L'_3} \cdot R_3;$$

$$V_{A_9} = V_{A_{11}} + I_{L'_3} \cdot R_3 = V_{A_{11}} + I_{L'_1}(t - \sum_{i=1}^{2} \tau_i) \cdot R_3$$

$$V_{A_9} = I_{R_b} \cdot R_b + (L_b + L'_4) \cdot \frac{dI_{L'_1}(t - \sum_{i=1}^{3} \tau_i)}{dt} + I_{L'_1}(t - \sum_{i=1}^{3} \tau_i) \cdot R_4 + I_{L'_1}(t - \sum_{i=1}^{2} \tau_i) \cdot R_3$$

$$V_{C_p} + V_{L_a} + V_{L_b} + \sum_{i=1}^{4} V_{L'_1} + \sum_{i=1}^{4} V_{R_i} = 0; V_{C_p} = V_{A_{14}} - V_{A_1} = \frac{1}{C_p} \cdot \int I_{C_p} \cdot dt;$$

$$V_{L_a} = V_{A_1} - V_{A_2} = L_a \cdot \frac{dI_{L_a}}{dt}$$

$$V_{L_b} = V_{A_{13}} - V_{A_{14}} = L_b \cdot \frac{dI_{L_b}}{dt}; \sum_{i=1}^{4} V_{L'_i} = V_{L'_1} + V_{L'_2} + V_{L'_3} + V_{L'_4};$$

$$V_{L'_1} = V_{A_2} - V_{A_3} = L'_1 \cdot \frac{dI_{L'_1}}{dt}$$

$$V_{L'_2} = V_{A_5} - V_{A_6} = L'_2 \cdot \frac{dI_{L'_1}(t - \tau_1)}{dt}; V_{L'_3} = V_{A_8} - V_{A_9} = L'_3 \cdot \frac{dI_{L'_1}(t - \sum_{i=1}^{2} \tau_i)}{dt}$$

$$V_{L'_4} = V_{A_{12}} - V_{A_{13}} = L'_4 \cdot \frac{dI_{L'_1}(t - \sum_{i=1}^{3} \tau_i)}{dt}; \sum_{i=1}^{4} V_{R_i} = V_{R_1} + V_{R_2} + V_{R_3} + V_{R_4};$$

$$V_{R_1} = V_{A_3} - V_{A_4} = I_{L'_1} \cdot R_1$$

$$V_{R_1} = I_{L'_1} \cdot \frac{\rho_1 \cdot l_1}{w_1 \cdot z_1}; V_{R_2} = V_{A_6} - V_{A_7} = R_2 \cdot I_{L'_1}(t - \tau_1) = I_{L'_1}(t - \tau_1) \cdot \frac{\rho_2 \cdot l_2}{w_2 \cdot z_2}$$

$$V_{R_3} = V_{A_9} - V_{A_{10}} = R_3 \cdot I_{L'_3} = R_3 \cdot I_{L'_1}(t - \sum_{i=1}^{2} \tau_i) = I_{L'_1}(t - \sum_{i=1}^{2} \tau_i) \cdot \frac{\rho_3 \cdot l_3}{w_3 \cdot z_3}$$

$$V_{R_4} = V_{A_{11}} - V_{A_{12}} = R_4 \cdot I_{L'_4} = R_4 \cdot I_{L'_1}(t - \sum_{i=1}^{3} \tau_i) = I_{L'_1}(t - \sum_{i=1}^{3} \tau_i) \cdot \frac{\rho_4 \cdot l_4}{w_4 \cdot z_4}$$

$$\sum_{i=1}^{4} V_{R_i} = I_{L'_1} \cdot \frac{\rho_1 \cdot l_1}{w_1 \cdot z_1} + I_{L'_1}(t - \tau_1) \cdot \frac{\rho_2 \cdot l_2}{w_2 \cdot z_2} + I_{L'_1}(t - \sum_{i=1}^{2} \tau_i) \cdot \frac{\rho_3 \cdot l_3}{w_3 \cdot z_3}$$
$$+ I_{L'_1}(t - \sum_{i=1}^{3} \tau_i) \cdot \frac{\rho_4 \cdot l_4}{w_4 \cdot z_4}$$

$$\sum_{i=1}^{4} V_{R_i} = \frac{\rho_1 \cdot l_1}{w_1 \cdot z_1} \cdot I_{L'_1} + \sum_{i=2}^{4} \frac{\rho_i \cdot l_i}{w_i \cdot z_i} \cdot I_{L'_1}(t - \sum_{k=1}^{i-1} \tau_k); \sum_{i=1}^{4} V_{L'_1} = L'_1 \cdot \frac{dI_{L'_1}}{dt}$$
$$+ \sum_{i=2}^{4} L'_i \cdot \frac{dI_{L'_1}(t - \sum_{k=1}^{i-1} \tau_k)}{dt}$$

$$\sum_{i=1}^{4} V_{L'_1} = L'_1 \cdot \frac{dI_{L'_1}}{dt} + L'_2 \cdot \frac{dI_{L'_1}(t - \tau_1)}{dt} + L'_3 \cdot \frac{dI_{L'_1}(t - \sum_{i=1}^{2} \tau_i)}{dt} + L'_4 \cdot \frac{dI_{L'_1}(t - \sum_{i=1}^{3} \tau_i)}{dt}$$

$$I_{L_a} = I_{L'_1}; V_{L_a} = L_a \cdot \frac{dI_{L'_1}}{dt}; I_{L_b} = I_{L'_4} \Rightarrow V_{L_b} = L_b \cdot \frac{dI_{L'_4}}{dt} = L_b \cdot \frac{dI_{L'_1}(t - \sum_{i=1}^{3} \tau_i)}{dt}$$

$$V_{C_p} + V_{L_a} + V_{L_b} + \sum_{i=1}^{4} V_{L'_i} + \sum_{i=1}^{4} V_{R_i} = 0$$

$$\frac{1}{C_p} \cdot \int I_{C_p} \cdot dt + L_a \cdot \frac{dI_{L'_1}}{dt} + L_b \cdot \frac{dI_{L'_1}(t - \sum_{i=1}^{3} \tau_i)}{dt} + L'_1 \cdot \frac{dI_{L'_1}}{dt} + \sum_{i=2}^{4} L'_i \cdot \frac{dI_{L'_1}(t - \sum_{k=1}^{i-1} \tau_k)}{dt}$$
$$+ \frac{\rho_1 \cdot l_1}{w_1 \cdot z_1} \cdot I_{L'_1} + \sum_{i=2}^{4} \frac{\rho_i \cdot l_i}{w_i \cdot z_i} \cdot I_{L'_1}(t - \sum_{k=1}^{i-1} \tau_k) = 0$$

We derivate the above equation and get the following expression: (Table 8.17)

$$\frac{1}{C_p} \cdot I_{C_p} + L_a \cdot \frac{d^2 I_{L'_1}}{dt^2} + L_b \cdot \frac{d^2 I_{L'_1}(t - \sum_{i=1}^{3} \tau_i)}{dt^2} + L'_1 \cdot \frac{d^2 I_{L'_1}}{dt^2} + \sum_{i=2}^{4} L'_1 \cdot \frac{d^2 I_{L'_1}(t - \sum_{k=1}^{i-1} \tau_k)}{dt^2}$$
$$+ \frac{\rho_1 \cdot l_1}{w_1 \cdot z_1} \cdot \frac{dI_{L'_1}}{dt} + \sum_{i=2}^{4} \frac{\rho_i \cdot l_i}{w_i \cdot z_i} \cdot \frac{dI_{L'_1}(t - \sum_{k=1}^{i-1} \tau_k)}{dt} = 0$$

$$(*) \quad I_{C_p} = -C_p \cdot \{L_a \cdot \frac{d^2 I_{L'_1}}{dt^2} + L_b \cdot \frac{d^2 I_{L'_1}(t - \sum_{i=1}^{3} \tau_i)}{dt^2} + L'_1 \cdot \frac{d^2 I_{L'_1}}{dt^2} + \sum_{i=2}^{4} L'_i \cdot \frac{d^2 I_{L'_1}(t - \sum_{k=1}^{i-1} \tau_k)}{dt^2}$$
$$+ \frac{\rho_1 \cdot l_1}{w_1 \cdot z_1} \cdot \frac{dI_{L'_1}}{dt} + \sum_{i=2}^{4} \frac{\rho_i \cdot l_i}{w_i \cdot z_i} \cdot \frac{dI_{L'_1}(t - \sum_{k=1}^{i-1} \tau_k)}{dt}\}$$

Table 8.17 Single turn square planar straight thin film inductors antenna τ_1, τ_2, and τ_3 options

Case no.	τ_1	τ_2	τ_3
1	τ	0	0
2	0	τ	0
3	0	0	τ
4	τ	τ	0
5	0	τ	τ
6	τ	0	τ
7	τ	τ	τ

$$I_{L'_1}(t-\sum_{i=1}^{3}\tau_i) = I_{C_p} + I_{R_b} \Rightarrow I_{R_b} = I_{L'_1}(t-\sum_{i=1}^{3}\tau_i) - I_{C_p}; I_{C_p} = I_{L'_1} + I_{R_a} \Rightarrow I_{R_a} = I_{C_p} - I_{L'_1}$$

$$I_{C_p} = C_p \cdot \frac{d}{dt}(I_{R_b}\cdot R_b - I_{R_a}\cdot R_a) = C_p \cdot \frac{d}{dt}([I_{L'_1}(t-\sum_{i=1}^{3}\tau_i) - I_{C_p}]\cdot R_b - [I_{C_p} - I_{L'_1}]\cdot R_a)$$

$$(**)\, I_{C_p} = C_p \cdot (R_b \cdot \frac{dI_{L'_1}(t-\sum_{i=1}^{3}\tau_i)}{dt} + \frac{dI_{L'_1}}{dt}\cdot R_a - \frac{dI_{C_p}}{dt}\cdot [R_b + R_a])$$

Derivative equation (*):

$$\frac{dI_{C_p}}{dt} = -C_p \cdot \{L_a \cdot \frac{d^3 I_{L'_1}}{dt^3} + L_b \cdot \frac{d^3 I_{L'_1}(t-\sum_{i=1}^{3}\tau_i)}{dt^3} + L'_1 \cdot \frac{d^3 I_{L'_1}}{dt^3} + \sum_{i=2}^{4} L'_i \cdot \frac{d^3 I_{L'_1}(t-\sum_{k=1}^{i-1}\tau_k)}{dt^3}$$
$$+ \frac{\rho_1 \cdot l_1}{w_1 \cdot z_1}\cdot \frac{d^2 I_{L'_1}}{dt^2} + \sum_{i=2}^{4}\frac{\rho_i \cdot l_i}{w_i \cdot z_i}\cdot \frac{d^2 I_{L'_1}(t-\sum_{k=1}^{i-1}\tau_k)}{dt^2}\}$$

Expression (**):

$$- C_p \cdot \{L_a \cdot \frac{d^2 I_{L'_1}}{dt^2} + L_b \cdot \frac{d^2 I_{L'_1}(t-\sum_{i=1}^{3}\tau_i)}{dt^2} + L'_1 \cdot \frac{d^2 I_{L'_1}}{dt^2} + \sum_{i=2}^{4} L'_i \cdot \frac{d^2 I_{L'_1}(t-\sum_{k=1}^{i-1}\tau_k)}{dt^2}$$
$$+ \frac{\rho_1 \cdot l_1}{w_1 \cdot z_1}\cdot \frac{dI_{L'_1}}{dt} + \sum_{i=2}^{4}\frac{\rho_i \cdot l_i}{w_i \cdot z_i}\cdot \frac{dI_{L'_1}(t-\sum_{k=1}^{i-1}\tau_k)}{dt}\} = C_p \cdot (R_b \cdot \frac{dI_{L'_1}(t-\sum_{i=1}^{3}\tau_i)}{dt} + \frac{dI_{L'_1}}{dt}\cdot R_a$$
$$+ C_p \cdot \{L_a \cdot \frac{d^3 I_{L'_1}}{dt^3} + L_b \cdot \frac{d^3 I_{L'_1}(t-\sum_{i=1}^{3}\tau_i)}{dt^3} + L'_1 \cdot \frac{d^3 I_{L'_1}}{dt^3} + \sum_{i=2}^{4} L'_1 \cdot \frac{d^3 I_{L'_1}(t-\sum_{k=1}^{i-1}\tau_k)}{dt^3}$$
$$+ \frac{\rho_1 \cdot l_1}{w_1 \cdot z_1}\cdot \frac{d^2 I_{L'_1}}{dt^2} + \sum_{i=2}^{4}\frac{\rho_i \cdot l_i}{w_i \cdot z_i}\cdot \frac{d^2 I_{L'_1}(t-\sum_{k=1}^{i-1}\tau_k)}{dt^2}\}\cdot [R_b + R_a])$$

Remark 1.0

$$\frac{d^2 I_{L'_1}(t-\sum_{i=1}^{3}\tau_i)}{dt^2}; I_{L'_1}(t-\sum_{i=1}^{3}\tau_i) = I^*_{L'_1} + i_{L'_1}\cdot e^{\lambda\cdot(t-\sum_{i=1}^{3}\tau_i)}; I_{L'_1}(t) = I^*_{L'_1} + i_{L'_1}\cdot e^{\lambda\cdot t}$$

$$\frac{d^2I_{L_1'}(t)}{dt^2}=i_{L_1'}\cdot\lambda^2\cdot e^{\lambda\cdot t};\frac{d^2I_{L_1'}(t-\sum_{i=1}^{3}\tau_i)}{dt^2}=i_{L_1'}\cdot\lambda^2\cdot e^{\lambda\cdot(t-\sum_{i=1}^{3}\tau_i)};\frac{d^2I_{L_1'}(t-\sum_{i=1}^{3}\tau_i)}{dt^2}$$
$$=\frac{d^2I_{L_1'}(t)}{dt^2}\cdot e^{-\lambda\cdot\sum_{i=1}^{3}\tau_i}$$

Remark 1.1

$$\sum_{i=2}^{4}L_i'\cdot\frac{d^2I_{L_i'}(t-\sum_{k=1}^{i-1}\tau_k)}{dt^2}=L_2'\cdot\frac{d^2I_{L_1'}(t-\tau_1)}{dt^2}+L_3'\cdot\frac{d^2I_{L_1'}(t-\sum_{k=1}^{2}\tau_k)}{dt^2}+L_4'\cdot\frac{d^2I_{L_1'}(t-\sum_{k=1}^{3}\tau_k)}{dt^2}$$

$$I_{L_1'}(t-\tau_1)=I_{L_1'}^*+i_{L_1'}\cdot e^{\lambda\cdot(t-\tau_1)};I_{L_1'}(t-\sum_{k=1}^{2}\tau_k)=I_{L_1'}^*+i_{L_1'}\cdot e^{\lambda\cdot(t-\sum_{k=1}^{2}\tau_k)};$$

$$I_{L_1'}(t-\sum_{k=1}^{3}\tau_k)=I_{L_1'}^*+i_{L_1'}\cdot e^{\lambda\cdot(t-\sum_{k=1}^{3}\tau_k)}$$

$$I_{L_1'}(t)=I_{L_1'}^*+i_{L_1'}\cdot e^{\lambda\cdot t};\frac{d^2I_{L_1'}(t)}{dt^2}=i_{L_1'}\cdot\lambda^2\cdot e^{\lambda\cdot t};\frac{d^2I_{L_1'}(t-\tau_1)}{dt^2}=i_{L_1'}\cdot\lambda^2\cdot e^{\lambda\cdot t}\cdot e^{-\lambda\cdot\tau_1}$$

$$\frac{d^2I_{L_1'}(t-\tau_1)}{dt^2}=\frac{d^2I_{L_1'}(t)}{dt^2}\cdot e^{-\lambda\cdot\tau_1};\frac{d^2I_{L_1'}(t-\sum_{k=1}^{2}\tau_k)}{dt^2}=i_{L_1'}\cdot\lambda^2\cdot e^{\lambda\cdot t}\cdot e^{-\lambda\cdot\sum_{k=1}^{2}\tau_k}$$
$$\frac{d^2I_{L_1'}(t-\sum_{k=1}^{2}\tau_k)}{dt^2}=\frac{d^2I_{L_1'}(t)}{dt^2}\cdot e^{-\lambda\cdot\sum_{k=1}^{2}\tau_k};\frac{d^2I_{L_1'}(t-\sum_{k=1}^{3}\tau_k)}{dt^2}=\frac{d^2I_{L_1'}(t)}{dt^2}\cdot e^{-\lambda\cdot\sum_{k=1}^{3}\tau_k}$$

$$\sum_{i=2}^{4}L_i'\cdot\frac{d^2I_{L_i'}(t-\sum_{k=1}^{i-1}\tau_k)}{dt^2}=L_2'\cdot\frac{d^2I_{L_1'}(t)}{dt^2}\cdot e^{-\lambda\cdot\tau_1}+L_3'\cdot\frac{d^2I_{L_1'}(t)}{dt^2}\cdot e^{-\lambda\cdot\sum_{k=1}^{2}\tau_k}+L_4'\cdot\frac{d^2I_{L_1'}(t)}{dt^2}\cdot e^{-\lambda\cdot\sum_{k=1}^{3}\tau_k}$$
$$\sum_{i=2}^{4}L_i'\cdot\frac{d^2I_{L_i'}(t-\sum_{k=1}^{i-1}\tau_k)}{dt^2}=\frac{d^2I_{L_1'}(t)}{dt^2}\cdot[L_2'\cdot e^{-\lambda\cdot\tau_1}+L_3'\cdot e^{-\lambda\cdot\sum_{k=1}^{2}\tau_k}+L_4'\cdot e^{-\lambda\cdot\sum_{k=1}^{3}\tau_k}]$$
$$\sum_{i=2}^{4}L_i'\cdot\frac{d^2I_{L_i'}(t-\sum_{k=1}^{i-1}\tau_k)}{dt^2}=\frac{d^2I_{L_1'}(t)}{dt^2}\cdot\sum_{j=2}^{4}L_j'\cdot e^{-\lambda\cdot\sum_{k=1}^{j-1}\tau_k}$$

Remark 1.2

$$\sum_{i=2}^{4}\frac{\rho_i\cdot l_i}{w_i\cdot z_i}\cdot\frac{dI_{L_1'}(t-\sum_{k=1}^{i-1}\tau_k)}{dt};\frac{dI_{L_1'}(t-\tau_1)}{dt}=\frac{dI_{L_1'}(t)}{dt}\cdot e^{-\lambda\cdot\tau_1}$$

$$\begin{aligned}\sum_{i=2}^{4}\frac{\rho_i\cdot l_i}{w_i\cdot z_i}\cdot\frac{dI_{L_1'}(t-\sum_{k=1}^{i-1}\tau_k)}{dt}&=\frac{\rho_2\cdot l_2}{w_2\cdot z_2}\cdot\frac{dI_{L_1'}(t-\tau_1)}{dt}+\frac{\rho_3\cdot l_3}{w_3\cdot z_3}\\&\quad\cdot\frac{dI_{L_1'}(t-\sum_{k=1}^{2}\tau_k)}{dt}+\frac{\rho_4\cdot l_4}{w_4\cdot z_4}\cdot\frac{dI_{L_1'}(t-\sum_{k=1}^{3}\tau_k)}{dt}\end{aligned}$$

$$\begin{aligned}I_{L_1'}(t-\tau_1)&=I_{L_1'}^*+i_{L_1'}\cdot e^{\lambda\cdot(t-\tau_1)};\frac{dI_{L_1'}(t-\tau_1)}{dt}=i_{L_1'}\cdot\lambda\cdot e^{\lambda\cdot t}\cdot e^{-\lambda\cdot\tau_1};I_{L_1'}(t)\\&=I_{L_1'}^*+i_{L_1'}\cdot e^{\lambda\cdot t};\frac{dI_{L_1'}(t)}{dt}=i_{L_1'}\cdot\lambda\cdot e^{\lambda\cdot t}\end{aligned}$$

$$\frac{dI_{L_1'}(t-\sum_{k=1}^{2}\tau_k)}{dt}=\frac{dI_{L_1'}(t)}{dt}\cdot e^{-\lambda\cdot\sum_{k=1}^{2}\tau_k};\frac{dI_{L_1'}(t-\sum_{k=1}^{3}\tau_k)}{dt}=\frac{dI_{L_1'}(t)}{dt}\cdot e^{-\lambda\cdot\sum_{k=1}^{3}\tau_k}$$

$$\begin{aligned}\sum_{i=2}^{4}\frac{\rho_i\cdot l_i}{w_i\cdot z_i}\cdot\frac{dI_{L_1'}(t-\sum_{k=1}^{i-1}\tau_k)}{dt}&=\frac{\rho_2\cdot l_2}{w_2\cdot z_2}\cdot\frac{dI_{L_1'}(t)}{dt}\cdot e^{-\lambda\cdot\tau_1}\\&\quad+\frac{\rho_3\cdot l_3}{w_3\cdot z_3}\cdot\frac{dI_{L_1'}(t)}{dt}\cdot e^{-\lambda\cdot\sum_{k=1}^{2}\tau_k}+\frac{\rho_4\cdot l_4}{w_4\cdot z_4}\cdot\frac{dI_{L_1'}(t)}{dt}\cdot e^{-\lambda\cdot\sum_{k=1}^{3}\tau_k}\end{aligned}$$

$$\begin{aligned}\sum_{i=2}^{4}\frac{\rho_i\cdot l_i}{w_i\cdot z_i}\cdot\frac{dI_{L_1'}(t-\sum_{k=1}^{i-1}\tau_k)}{dt}&=\frac{dI_{L_1'}(t)}{dt}\cdot[\frac{\rho_2\cdot l_2}{w_2\cdot z_2}\cdot e^{-\lambda\cdot\tau_1}\\&\quad+\frac{\rho_3\cdot l_3}{w_3\cdot z_3}\cdot e^{-\lambda\cdot\sum_{k=1}^{2}\tau_k}+\frac{\rho_4\cdot l_4}{w_4\cdot z_4}\cdot e^{-\lambda\cdot\sum_{k=1}^{3}\tau_k}]\end{aligned}$$

$$\sum_{i=2}^{4}\frac{\rho_i\cdot l_i}{w_i\cdot z_i}\cdot\frac{dI_{L_1'}(t-\sum_{k=1}^{i-1}\tau_k)}{dt}=\frac{dI_{L_1'}(t)}{dt}\cdot\sum_{j=2}^{4}\frac{\rho_j\cdot l_j}{w_j\cdot z_j}\cdot e^{-\lambda\cdot\sum_{k=1}^{j-1}\tau_k}$$

Remark 1.3

$$\frac{dI_{L_1'}(t-\sum_{i=1}^{3}\tau_i)}{dt}=\frac{dI_{L_1'}(t)}{dt}\cdot e^{-\lambda\cdot\sum_{i=1}^{3}\tau_i}$$

Remark 1.4

$$\frac{d^3 I_{L_1'}(t - \sum_{i=1}^{3} \tau_i)}{dt^3} = \frac{d^3 I_{L_1'}(t)}{dt^3} \cdot e^{-\lambda \cdot \sum_{i=1}^{3} \tau_i}$$

Remark 1.5

$$\sum_{i=2}^{4} L_i' \cdot \frac{d^3 I_{L_1'}(t - \sum_{k=1}^{i-1} \tau_k)}{dt^3} = \frac{d^3 I_{L_1'}(t)}{dt^3} \cdot \sum_{j=2}^{4} L_j' \cdot e^{-\lambda \cdot \sum_{k=1}^{j-1} \tau_k}$$

$$\sum_{i=2}^{4} L_i' \cdot \frac{d^3 I_{L_1'}(t - \sum_{k=1}^{i-1} \tau_k)}{dt^3} = L_2' \cdot \frac{d^3 I_{L_1'}(t - \tau_1)}{dt^3} + L_3' \cdot \frac{d^3 I_{L_1'}(t - \sum_{k=1}^{2} \tau_k)}{dt^3} + L_4' \cdot \frac{d^3 I_{L_1'}(t - \sum_{k=1}^{3} \tau_k)}{dt^3}$$

$$\frac{d^3 I_{L_1'}(t - \tau_1)}{dt^3} = \frac{d^3 I_{L_1'}(t)}{dt^3} \cdot e^{-\lambda \cdot \tau_1}; \frac{d^3 I_{L_1'}(t - \sum_{k=1}^{2} \tau_k)}{dt^3} = \frac{d^3 I_{L_1'}(t)}{dt^3} \cdot e^{-\lambda \cdot \sum_{k=1}^{2} \tau_k}$$

$$\frac{d^3 I_{L_1'}(t - \sum_{k=1}^{3} \tau_k)}{dt^3} = \frac{d^3 I_{L_1'}(t)}{dt^3} \cdot e^{-\lambda \cdot \sum_{k=1}^{3} \tau_k}$$

$$\sum_{i=2}^{4} L_i' \cdot \frac{d^3 I_{L_1'}(t - \sum_{k=1}^{i-1} \tau_k)}{dt^3} = L_2' \cdot \frac{d^3 I_{L_1'}(t)}{dt^3} \cdot e^{-\lambda \cdot \tau_1} + L_3' \cdot \frac{d^3 I_{L_1'}(t)}{dt^3} \cdot e^{-\lambda \cdot \sum_{k=1}^{2} \tau_k} + L_4' \cdot \frac{d^3 I_{L_1'}(t)}{dt^3} \cdot e^{-\lambda \cdot \sum_{k=1}^{3} \tau_k}$$

$$\sum_{i=2}^{4} L_i' \cdot \frac{d^3 I_{L_1'}(t - \sum_{k=1}^{i-1} \tau_k)}{dt^3} = [L_2' \cdot e^{-\lambda \cdot \tau_1} + L_3' \cdot e^{-\lambda \cdot \sum_{k=1}^{2} \tau_k} + L_4' \cdot e^{-\lambda \cdot \sum_{k=1}^{3} \tau_k}] \cdot \frac{d^3 I_{L_1'}(t)}{dt^3}$$

Remark 1.6

$$\sum_{i=2}^{4} \frac{\rho_i \cdot l_i}{w_i \cdot z_i} \cdot \frac{d^2 I_{L_1'}(t - \sum_{k=1}^{i-1} \tau_k)}{dt^2} = \frac{d^2 I_{L_1'}(t)}{dt^2} \cdot \sum_{j=2}^{4} \frac{\rho_j \cdot l_j}{w_j \cdot z_j} \cdot e^{-\lambda \cdot \sum_{k=1}^{j-1} \tau_k}$$

$$\sum_{i=2}^{4}\frac{\rho_i\cdot l_i}{w_i\cdot z_i}\cdot\frac{d^2I_{L_1'}(t-\sum_{k=1}^{i-1}\tau_k)}{dt^2}=\frac{\rho_2\cdot l_2}{w_2\cdot z_2}\cdot\frac{d^2I_{L_1'}(t-\tau_1)}{dt^2}+\frac{\rho_3\cdot l_3}{w_3\cdot z_3}\cdot\frac{d^2I_{L_1'}(t-\sum_{k=1}^{2}\tau_k)}{dt^2}+\frac{\rho_4\cdot l_4}{w_4\cdot z_4}\cdot\frac{d^2I_{L_1'}(t-\sum_{k=1}^{3}\tau_k)}{dt^2}$$

$$\frac{d^2I_{L_1'}(t-\tau_1)}{dt^2}=\frac{d^2I_{L_1'}(t)}{dt^2}\cdot e^{-\lambda\cdot\tau_1};\frac{d^2I_{L_1'}(t-\sum_{k=1}^{2}\tau_k)}{dt^2}=\frac{d^2I_{L_1'}(t)}{dt^2}\cdot e^{-\lambda\cdot\sum_{k=1}^{2}\tau_k};\frac{d^2I_{L_1'}(t-\sum_{k=1}^{3}\tau_k)}{dt^2}=\frac{d^2I_{L_1'}(t)}{dt^2}\cdot e^{-\lambda\cdot\sum_{k=1}^{3}\tau_k}$$

$$\sum_{i=2}^{4}\frac{\rho_i\cdot l_i}{w_i\cdot z_i}\cdot\frac{d^2I_{L_1'}(t-\sum_{k=1}^{i-1}\tau_k)}{dt^2}=\frac{\rho_2\cdot l_2}{w_2\cdot z_2}\cdot\frac{d^2I_{L_1'}(t)}{dt^2}\cdot e^{-\lambda\cdot\tau_1}+\frac{\rho_3\cdot l_3}{w_3\cdot z_3}\cdot\frac{d^2I_{L_1'}(t)}{dt^2}\cdot e^{-\lambda\cdot\sum_{k=1}^{2}\tau_k}+\frac{\rho_4\cdot l_4}{w_4\cdot z_4}\cdot\frac{d^2I_{L_1'}(t)}{dt^2}\cdot e^{-\lambda\cdot\sum_{k=1}^{3}\tau_k}$$

$$\sum_{i=2}^{4}\frac{\rho_i\cdot l_i}{w_i\cdot z_i}\cdot\frac{d^2I_{L_1'}(t-\sum_{k=1}^{i-1}\tau_k)}{dt^2}=\frac{d^2I_{L_1'}(t)}{dt^2}\cdot[\frac{\rho_2\cdot l_2}{w_2\cdot z_2}\cdot e^{-\lambda\cdot\tau_1}+\frac{\rho_3\cdot l_3}{w_3\cdot z_3}\cdot e^{-\lambda\cdot\sum_{k=1}^{2}\tau_k}+\frac{\rho_4\cdot l_4}{w_4\cdot z_4}\cdot e^{-\lambda\cdot\sum_{k=1}^{3}\tau_k}]$$

We can summary remarks 1.0–1.6 in Table 8.16.
Expression (**):

$$-C_p\cdot\{L_a\cdot\frac{d^2I_{L_1'}}{dt^2}+L_b\cdot\frac{d^2I_{L_1'}(t)}{dt^2}\cdot e^{-\lambda\cdot\sum_{i=1}^{3}\tau_i}+L_1'\cdot\frac{d^2I_{L_1'}}{dt^2}+\frac{d^2I_{L_1'}(t)}{dt^2}\cdot\sum_{j=2}^{4}L_j'\cdot e^{-\lambda\cdot\sum_{k=1}^{j-1}\tau_k}$$
$$+\frac{\rho_1\cdot l_1}{w_1\cdot z_1}\cdot\frac{dI_{L_1'}}{dt}+\frac{dI_{L_1'}(t)}{dt}\cdot\sum_{j=2}^{4}\frac{\rho_j\cdot l_j}{w_j\cdot z_j}\cdot e^{-\lambda\cdot\sum_{k=1}^{j-1}\tau_k}\}=C_p\cdot(R_b\cdot\frac{dI_{L_1'}(t)}{dt}\cdot e^{-\lambda\cdot\sum_{i=1}^{3}\tau_i}+\frac{dI_{L_1'}}{dt}\cdot R_a$$
$$+C_p\cdot\{L_a\cdot\frac{d^3I_{L_1'}}{dt^3}+L_b\cdot\frac{d^3I_{L_1'}(t)}{dt^3}\cdot e^{-\lambda\cdot\sum_{i=1}^{3}\tau_i}+L_1'\cdot\frac{d^3I_{L_1'}}{dt^3}+\frac{d^3I_{L_1'}(t)}{dt^3}\cdot\sum_{j=2}^{4}L_j'\cdot e^{-\lambda\cdot\sum_{k=1}^{j-1}\tau_k}$$
$$+\frac{\rho_1\cdot l_1}{w_1\cdot z_1}\cdot\frac{d^2I_{L_1'}}{dt^2}+\frac{d^2I_{L_1'}(t)}{dt^2}\cdot\sum_{j=2}^{4}\frac{\rho_j\cdot l_j}{w_j\cdot z_j}\cdot e^{-\lambda\cdot\sum_{k=1}^{j-1}\tau_k}\}\cdot[R_b+R_a])$$

We define for simplicity the following functions:

$$\xi_1 = \xi_1(\lambda;\tau_i;i=1,2,3) = e^{-\lambda\cdot\sum_{i=1}^{3}\tau_i};\ \xi_2 = \xi_2(\lambda;L'_j;j=2,3,4;\tau_k;k=1,2,3) = \sum_{j=2}^{4} L'_j \cdot e^{-\lambda\cdot\sum_{k=1}^{j-1}\tau_k}$$

$$\xi_3 = \xi_3(\lambda;\tau_k;k=1,2,3;\rho_j,l_j,w_j,z_j;j=1,2,3,4) = \sum_{j=2}^{4}\frac{\rho_j\cdot l_j}{w_j\cdot z_j}\cdot e^{-\lambda\cdot\sum_{k=1}^{j-1}\tau_k}$$

$$\begin{aligned}
&-C_p\cdot\{L_a\cdot\frac{d^2I_{L'_1}(t)}{dt^2}+L_b\cdot\frac{d^2I_{L'_1}(t)}{dt^2}\cdot\xi_1+L'_1\cdot\frac{d^2I_{L'_1}(t)}{dt^2}+\frac{d^2I_{L'_1}(t)}{dt^2}\cdot\xi_2+\frac{\rho_1\cdot l_1}{w_1\cdot z_1}\cdot\frac{dI_{L'_1}(t)}{dt}+\frac{dI_{L'_1}(t)}{dt}\cdot\xi_3\}\\
&=C_p\cdot(R_b\cdot\frac{dI_{L'_1}(t)}{dt}\cdot\xi_1+\frac{dI_{L'_1}(t)}{dt}\cdot R_a+C_p\cdot\{L_a\cdot\frac{d^3I_{L'_1}(t)}{dt^3}+L_b\cdot\frac{d^3I_{L'_1}(t)}{dt^3}\cdot\xi_1+L'_1\cdot\frac{d^3I_{L'_1}(t)}{dt^3}\\
&+\frac{d^3I_{L'_1}(t)}{dt^3}\cdot\xi_2+\frac{\rho_1\cdot l_1}{w_1\cdot z_1}\cdot\frac{d^2I_{L'_1}(t)}{dt^2}+\frac{d^2I_{L'_1}(t)}{dt^2}\cdot\xi_3\}\cdot[R_b+R_a])
\end{aligned}$$

We define new variables: $X_1 = X_1(t); X_2 = X_2(t)$

$$X_1(t)=\frac{dI_{L'_1}(t)}{dt};X_2(t)=\frac{dX_1(t)}{dt}=\frac{d^2I_{L'_1}(t)}{dt^2};\frac{dX_2(t)}{dt}=\frac{d^3I_{L'_1}(t)}{dt^3};I_{L'_1}=I_{L'_1}(t)$$

$$\begin{aligned}
&-C_p\cdot\{L_a\cdot X_2+L_b\cdot X_2\cdot\xi_1+L'_1\cdot X_2+X_2\cdot\xi_2+\frac{\rho_1\cdot l_1}{w_1\cdot z_1}\cdot X_1+X_1\cdot\xi_3\}\\
&\quad=C_p\cdot(R_b\cdot X_1\cdot\xi_1+X_1\cdot R_a\\
&+C_p\cdot\{L_a\cdot\frac{dX_2}{dt}+L_b\cdot\frac{dX_2}{dt}\cdot\xi_1+L'_1\cdot\frac{dX_2}{dt}+\frac{dX_2}{dt}\cdot\xi_2+\frac{\rho_1\cdot l_1}{w_1\cdot z_1}\cdot X_2+X_2\cdot\xi_3\}\cdot[R_b+R_a])
\end{aligned}$$

$$\begin{aligned}
\frac{dX_2}{dt}=&-\frac{[\frac{\rho_1\cdot l_1}{w_1\cdot z_1}+\xi_3+R_b\cdot\xi_1+R_a]}{\{L_a+L_b\cdot\xi_1+L'_1+\xi_2\}\cdot[R_b+R_a]\cdot C_p}\cdot X_1\\
&-\frac{[L_a+L_b\cdot\xi_1+L'_1+\xi_2+C_p\cdot[R_b+R_a]\cdot(\frac{\rho_1\cdot l_1}{w_1\cdot z_1}+\xi_3)]}{\{L_a+L_b\cdot\xi_1+L'_1+\xi_2\}\cdot[R_b+R_a]\cdot C_p}\cdot X_2
\end{aligned}$$

We define new functions: $\psi_1 = \frac{[\frac{\rho_1\cdot l_1}{w_1\cdot z_1}+\xi_3+R_b\cdot\xi_1+R_a]}{\{L_a+L_b\cdot\xi_1+L'_1+\xi_2\}\cdot[R_b+R_a]\cdot C_p}$

$$\psi_2=\frac{[L_a+L_b\cdot\xi_1+L'_1+\xi_2+C_p\cdot[R_b+R_a]\cdot(\frac{\rho_1\cdot l_1}{w_1\cdot z_1}+\xi_3)]}{\{L_a+L_b\cdot\xi_1+L'_1+\xi_2\}\cdot[R_b+R_a]\cdot C_p}$$

We can summary our system differential equations:

$$\frac{dI_{L'_1}}{dt}=X_1;\frac{dX_1}{dt}=X_2;\frac{dX_2}{dt}=-\psi_1\cdot X_1-\psi_2\cdot X_2$$

Case 1.0: No delays, $\tau_1 = \tau_2 = \tau_3 = 0$

$$\xi_1(\tau_i = 0) = \xi_1(\lambda; \tau_i = 0; i = 1,2,3) = 1; \xi_2(\tau_k = 0) = \xi_2(\lambda; L'_j; j = 2,3,4; \tau_k = 0; k = 1,2,3) = \sum_{j=2}^{4} L'_j$$

$$\xi_3(\tau_k = 0) = \xi_3(\lambda; \tau_k = 0; k = 1,2,3; \rho_j, l_j, w_j, z_j; j = 1,2,3,4) = \sum_{j=2}^{4} \frac{\rho_j \cdot l_j}{w_j \cdot z_j}$$

$$\psi_1(\tau_{i,k} = 0) = \frac{[\sum_{j=1}^{4} \frac{\rho_j \cdot l_j}{w_j \cdot z_j} + R_b + R_a]}{\{L_a + L_b + \sum_{j=1}^{4} L'_j\} \cdot [R_b + R_a] \cdot C_p}; \psi_2(\tau_{i,k} = 0) = \frac{[L_a + L_b + \sum_{j=1}^{4} L'_j + C_p \cdot [R_b + R_a] \cdot \sum_{j=1}^{4} \frac{\rho_j \cdot l_j}{w_j \cdot z_j}]}{\{L_a + L_b + \sum_{j=1}^{4} L'_j\} \cdot [R_b + R_a] \cdot C_p}$$

At fixed point: $\frac{dI_{L'_1}}{dt} = 0; \frac{dX_1}{dt} = 0; \frac{dX_2}{dt} = 0; X_1^* = 0; X_2^* = 0$

Stability analysis: The standard local stability analysis about any one of the equilibrium points of Single turn square planar straight thin film inductors antenna system consists in adding to its coordinated $[X_1 X_2 I_{L'_1}]$ arbitrarily small increments of exponential terms $[x_1 x_2 i_{L'_1}] \cdot e^{\lambda \cdot t}$, and retaining the first order terms in $x_1 x_2 i_{L'_1}$. The system of eight homogeneous equations leads to a polynomial characteristic equation in the eigenvalueλ. The polynomial characteristic equation accepts by set the Single turn square planar straight thin film inductors antenna system equations. The Single turn square planar straight thin film inductors antenna system fixed values with arbitrarily small increments of exponential form $[x_1 x_2 i_{L'_1}] \cdot e^{\lambda \cdot t}$ are; i = 0 (first fixed point), i = 1 (second fixed point), i = 2 (third fixed point), etc., [2–4].

$$X_1(t) = X_1^{(i)} + x_1 \cdot e^{\lambda \cdot t}; X_2(t) = X_2^{(i)} + x_2 \cdot e^{\lambda \cdot t}; I_{L'_1}(t) = I_{L'_1}^{(i)} + i_{L'_1} \cdot e^{\lambda \cdot t}$$

$$\frac{dX_1(t)}{dt} = x_1 \cdot \lambda \cdot e^{\lambda \cdot t}; \frac{dX_2(t)}{dt} = x_2 \cdot \lambda \cdot e^{\lambda \cdot t}; \frac{dI_{L'_1}(t)}{dt} = i_{L'_1} \cdot \lambda \cdot e^{\lambda \cdot t}$$

$$\frac{dI_{L'_1}}{dt} = X_1 \Rightarrow i_{L'_1} \cdot \lambda \cdot e^{\lambda \cdot t} = X_1^{(i)} + x_1 \cdot e^{\lambda \cdot t}; X_1^{(i)} = 0; -i_{L'_1} \cdot \lambda + x_1 = 0$$

$$\frac{dX_1}{dt} = X_2 \Rightarrow x_1 \cdot \lambda \cdot e^{\lambda \cdot t} = X_2^{(i)} + x_2 \cdot e^{\lambda \cdot t}; X_2^{(i)} = 0; -x_1 \cdot \lambda + x_2 = 0$$

$$\frac{dX_2}{dt} = -\psi_1(\tau_{i,k} = 0) \cdot X_1 - \psi_2(\tau_{i,k} = 0) \cdot X_2$$

$$x_2 \cdot \lambda \cdot e^{\lambda \cdot t} = -\psi_1(\tau_{i,k} = 0) \cdot [X_1^{(i)} + x_1 \cdot e^{\lambda \cdot t}] - \psi_2(\tau_{i,k} = 0) \cdot [X_2^{(i)} + x_2 \cdot e^{\lambda \cdot t}]$$

$$x_2 \cdot \lambda \cdot e^{\lambda \cdot t} = -\psi_1(\tau_{i,k} = 0) \cdot X_1^{(i)} - \psi_2(\tau_{i,k} = 0) \cdot X_2^{(i)} - \psi_1(\tau_{i,k} = 0) \cdot x_1 \cdot e^{\lambda \cdot t} - \psi_2(\tau_{i,k} = 0) \cdot x_2 \cdot e^{\lambda \cdot t}$$

At fixed points: $-\psi_1(\tau_{i,k} = 0) \cdot X_1^{(i)} - \psi_2(\tau_{i,k} = 0) \cdot X_2^{(i)} = 0$

$$-x_2 \cdot \lambda - \psi_1(\tau_{i,k} = 0) \cdot x_1 - \psi_2(\tau_{i,k} = 0) \cdot x_2 = 0$$

We can summary our single turn square planar straight thin film inductors antenna system arbitrarily small increments equations:

$$-x_1 \cdot \lambda + x_2 = 0; -x_2 \cdot \lambda - \psi_1(\tau_{i,k} = 0) \cdot x_1 - \psi_2(\tau_{i,k} = 0) \cdot x_2 = 0; -i_{L'_1} \cdot \lambda + x_1 = 0$$

$$\begin{pmatrix} -\lambda & 1 & 0 \\ -\psi_1(\tau_{i,k} = 0) & -\lambda - \psi_2(\tau_{i,k} = 0) & 0 \\ 1 & 0 & -\lambda \end{pmatrix} \begin{pmatrix} x_1 \\ x_2 \\ i_{L'_1} \end{pmatrix} = 0; A - \lambda \cdot I = \begin{pmatrix} -\lambda & 1 & 0 \\ -\psi_1(\tau_{i,k} = 0) & -\lambda - \psi_2(\tau_{i,k} = 0) & 0 \\ 1 & 0 & -\lambda \end{pmatrix}$$

$$\det(A - \lambda \cdot I) = 0; \det \begin{pmatrix} -\lambda & 1 & 0 \\ -\psi_1(\tau_{i,k} = 0) & -\lambda - \psi_2(\tau_{i,k} = 0) & 0 \\ 1 & 0 & -\lambda \end{pmatrix} = 0$$

$$\det(A - \lambda \cdot I) = -\lambda \cdot \det\begin{pmatrix} -\lambda - \psi_2(\tau_{i,k} = 0) & 0 \\ 0 & -\lambda \end{pmatrix} - \det\begin{pmatrix} -\psi_1(\tau_{i,k} = 0) & 0 \\ 1 & -\lambda \end{pmatrix}$$

$$\det(A - \lambda \cdot I) = -\lambda \cdot (\lambda + \psi_2(\tau_{i,k} = 0)) \cdot \lambda - \psi_1(\tau_{i,k} = 0) \cdot \lambda = -\lambda \cdot [(\lambda + \psi_2(\tau_{i,k} = 0)) \cdot \lambda + \psi_1(\tau_{i,k} = 0)]$$

$$\det(A - \lambda \cdot I) = 0 \Rightarrow \lambda_1 = 0; \lambda^2 + \psi_2(\tau_{i,k} = 0) \cdot \lambda + \psi_1(\tau_{i,k} = 0) = 0$$

$$\lambda_{2,3} = \frac{-\psi_2(\tau_{i,k} = 0) \pm \sqrt{[\psi_2(\tau_{i,k} = 0)]^2 - 4 \cdot \psi_1(\tau_{i,k} = 0)}}{2};$$

$$\psi_1(\tau_{i,k} = 0) > 0; \psi_2(\tau_{i,k} = 0) > 0$$

$$\lambda_{2,3} = \frac{-\psi_2(\tau_{i,k} = 0) \pm \sqrt{[\psi_2(\tau_{i,k} = 0)]^2 - 4 \cdot \frac{\psi_1(\tau_{i,k}=0)}{[\psi_2(\tau_{i,k}=0)]^2} \cdot [\psi_2(\tau_{i,k} = 0)]^2}}{2}$$

$$\lambda_{2,3} = \frac{-\psi_2(\tau_{i,k} = 0) \pm \psi_2(\tau_{i,k} = 0) \cdot \sqrt{1 - 4 \cdot \frac{\psi_1(\tau_{i,k}=0)}{[\psi_2(\tau_{i,k}=0)]^2}}}{2};$$

$$\lambda_{2,3} = \frac{1}{2} \cdot \psi_2(\tau_{i,k} = 0) \cdot \{-1 \pm \sqrt{1 - 4 \cdot \frac{\psi_1(\tau_{i,k} = 0)}{[\psi_2(\tau_{i,k} = 0)]^2}}\}$$

$$\varsigma(\psi_1, \psi_2) = -1 \pm \sqrt{1 - 4 \cdot \frac{\psi_1(\tau_{i,k} = 0)}{[\psi_2(\tau_{i,k} = 0)]^2}}; \lambda_{2,3} = \frac{1}{2} \cdot \psi_2(\tau_{i,k} = 0) \cdot \varsigma(\psi_1, \psi_2)$$

The sign of $\varsigma(\psi_1, \psi_2)$ establish the sign of eigenvalues λ_2 and λ_3 $(\psi_2(\tau_{i,k} = 0) > 0)$

Case a: $1 - 4 \cdot \frac{\psi_1(\tau_{i,k}=0)}{[\psi_2(\tau_{i,k}=0)]^2} > 0 \Rightarrow 0 < \frac{\psi_1(\tau_{i,k}=0)}{[\psi_2(\tau_{i,k}=0)]^2} < \frac{1}{4}$

$$0 < \{\sqrt{1 - 4 \cdot \frac{\psi_1(\tau_{i,k} = 0)}{[\psi_2(\tau_{i,k} = 0)]^2}}\} < 1; \frac{\psi_1(\tau_{i,k} = 0)}{[\psi_2(\tau_{i,k} = 0)]^2} > 0$$
$$\Rightarrow \varsigma(\psi_1, \psi_2) < 0; \lambda_2 < 0; \lambda_3 < 0$$

Case b: $1 - 4 \cdot \frac{\psi_1(\tau_{i,k} = 0)}{[\psi_2(\tau_{i,k} = 0)]^2} = 0 \Rightarrow \frac{\psi_1(\tau_{i,k} = 0)}{[\psi_2(\tau_{i,k} = 0)]^2} = \frac{1}{4}$

$$\varsigma(\psi_1, \psi_2) = -1; \lambda_{2,3} = -\frac{1}{2} \cdot \psi_2(\tau_{i,k} = 0); \lambda_2 < 0; \lambda_3 < 0$$

Case c: $1 - 4 \cdot \frac{\psi_1(\tau_{i,k} = 0)}{[\psi_2(\tau_{i,k} = 0)]^2} < 0 \Rightarrow \frac{\psi_1(\tau_{i,k} = 0)}{[\psi_2(\tau_{i,k} = 0)]^2} > \frac{1}{4}$

$$\varsigma(\psi_1, \psi_2) = \alpha \pm \beta \cdot j; \varsigma(\psi_1, \psi_2) = -1 \pm \sqrt{1 - 4 \cdot \frac{\psi_1(\tau_{i,k} = 0)}{[\psi_2(\tau_{i,k} = 0)]^2}}$$

$$\imath(\psi_1, \psi_2) = 1 - 4 \cdot \frac{\psi_1(\tau_{i,k} = 0)}{[\psi_2(\tau_{i,k} = 0)]^2}; \imath(\psi_1, \psi_2) < 0; \sqrt{\imath(\psi_1, \psi_2)} = j \cdot \sqrt{|\imath(\psi_1, \psi_2)|}$$

$$\varsigma(\psi_1, \psi_2) = -1 \pm j \cdot \sqrt{|\imath(\psi_1, \psi_2)|}; \alpha = -1; \beta = \sqrt{|\imath(\psi_1, \psi_2)|}$$

$$\lambda_{2,3} = \frac{1}{2} \cdot \psi_2(\tau_{i,k} = 0) \cdot \varsigma(\psi_1, \psi_2) = \frac{1}{2} \cdot \psi_2(\tau_{i,k} = 0) \cdot (\alpha \pm \beta \cdot j); \psi_2(\tau_{i,k} = 0) > 0$$

We get three eigenvalues for our single turn square planar straight thin film inductors antenna system. The first eigenvalue (λ_1) is equal to zero and other

eigenvalues (λ_2, λ_3) can be real numbers ($\lambda_2, \lambda_3 \in \mathbb{R}$) and negative (cases a and b) or complex conjugate numbers ($\lambda_{2,3} = \frac{1}{2} \cdot \psi_2(\tau_{i,k} = 0) \cdot (\alpha \pm \beta \cdot j)$; $\psi_2(\tau_{i,k} = 0) > 0$).

If the first eigenvalue ($\lambda_1 = 0$) and two other eigenvalues ($\lambda_2, \lambda_3 \in \mathbb{R}$) are negative and real numbers then our single turn square planar straight thin film inductors antenna system fixed point is attracting line. If the first eigenvalue ($\lambda_1 = 0$) and two other eigenvalues ($\lambda_2, \lambda_3 \in \mathbb{C}$) are complex conjugate numbers with negative real part ($\alpha = -1; \alpha < 0$) then our single turn square planar straight thin film inductors antenna system fixed point is attracting stable spiral node line [2–4].

The next cases we analyze are when at least one of the delay parameters is real and positive value. We analyze our single turn square planar straight thin film inductors antenna system for the following cases:

Remark: It is reader exercise to analyze single turn square planar straight thin film inductors antenna system stability for cases 1 to 7 (at least one delay parameter τ_1, τ_2, τ_3 is positive and real number). The stability analysis is done by using geometric stability switch criteria in delay differential systems (E. Beretta and Y. Kuang). It is a practical guideline that combines graphical information with analytical work to effectively study the local stability of single turn square planar straight thin film inductors antenna system model involving delay dependent parameters. The stability of single turn square planar straight thin film inductors antenna system steady state is determined by the graphs of some functions of τ which can be expressed explicitly [5, 6].

The general geometric criterion: The occurrence of any possible stability switching resulting from the increase of value of the time delay τ for our single turn square planar straight thin film inductors antenna system characteristic equation.

$$D(\lambda, \tau_1, \tau_2, \tau_3) = \det(A - \lambda \cdot I); D(\lambda, \tau_1, \tau_2, \tau_3) = 0; D(\lambda, \tau) = P_n(\lambda, \tau) + Q_m(\lambda, \tau) \cdot e^{-\lambda \cdot \tau}$$

$$P_n(\lambda, \tau) = \sum_{k=0}^{n} p_k(\tau) \cdot \lambda^k; Q_m(\lambda, \tau) = \sum_{k=0}^{m} q_k(\tau) \cdot \lambda^k; n, m \in \mathbb{N}_0; n > m$$

$p_k(\bullet), q_k(\bullet) : \mathbb{R}_{+0} \to \mathbb{R}$ are continuous and differentiable functions of τ.

$$P_n(\lambda = 0, \tau) + Q_m(\lambda = 0, \tau) = p_0(\tau) + q_0(\tau) \neq 0 \; \forall \; \tau \in \mathbb{R}_{+0}$$

$P_n(\lambda, \tau), Q_m(\lambda, \tau)$ are analytic functions in λ and differentiable in τ for which we assume:

(I) If $\lambda = i \cdot \omega; \omega \in \mathbb{R}$ then $P_n(i \cdot \omega, \tau) + Q_m(i \cdot \omega, \tau) \neq 0; \tau \in \mathbb{R}$.
(II) $\limsup\{|Q_m(\lambda, \tau)/P_n(\lambda, \tau)| : |\lambda| \to \infty, \mathrm{Re}\lambda \geq 0\} < 1$ for any τ.
(III) $F(\omega, \tau) = |P_n(i \cdot \omega, \tau)|^2 - |Q_m(i \cdot \omega, \tau)|^2$ for each τ has at most a finite number of real zeros.
(IV) Each positive root $\omega(\tau)$ of $F(\omega, \tau) = 0$ is continuous and differentiable in τ whenever it exists.

8.4 Helix Antennas System Stability Analysis Under Parameters Variation

The helix antenna is a type of antenna which uses curved segments. The main issues regarding helix antenna are helix geometry and parameters, wire connection concepts, and the Perfect Electric Conductor (PEC) ground plane. A helical antenna is an antenna consisting of a conducting wire wound in the form of a helix. Helical antennas are mounted over a ground plane. The feed line is connected between the bottom of the helix and the ground plane.

Helical antennas can operate in one of two principal modes—normal mode or axial mode. In the normal mode or broad side helix, the dimensions of the helix are small compared with the wavelength. The antenna acts similarly to an electrically short dipole or monopole, and the radiation pattern has a maximum radiation at right angles to the helix axis. Manly used for compact antennas for portable and mobile two way radios, and for UHF broadcasting antennas. Additionally the normal mode helical antenna (NMHA) is used for applications such as mobile and satellite communication, RFID and medical devices. In the axial mode or end-fire helix, the dimensions of the helix are comparable to a wavelength. The antenna functions as a directional antenna radiating a beam off the ends of the helix, along the antenna's axis. It radiates circularly polarized radio waves and used for satellite communications. The normal mode helical antenna (NMHA) lumped-element equivalent circuit is represented the input impedance of a normal mode helical antenna. The normal mode helical antenna (NMHA) is modeled as the combination of a wire dipole antenna and distributed inductors. The equivalent circuit is divided into two parts: a five element circuit for the equivalent wire antenna with a different radius, and a three element sub-circuit representing the effects of the loops. The geometry of a typical helix antenna is shown in the below figure. There are some important parameters which characterize our helix antenna. The helix wire of radius, a, and uniformly wound with a constant pitch, S. The diameter of the imagined cylinder over which the axis of the helical conductor is wrapped is D and the radius is R (D = 2 • R). The number of turns is N and the half axial length of the antenna is h = N • S. We can modeled the helix antenna as a series of loops and linear conductors when the physical dimensions of the helix are much smaller than the wavelength [130–132]. We can represent each helix antenna turn as two radiating components, one an axial wire segment of length S, and the other a loop of diameter D. The wire and loop model can be representing as a short wire segments connected by lumped elements representing the inductance of the loop, where the loops are functioned as inductors (each one inductance L). Actually the helix structure model is a wire segment with a length of S and one inductive element per turn. We define new radius for the helix antenna wire segment $(a'; a \rightarrow a')$ since we need to keep the correct capacitance of the helix turn $(a' = S \cdot (\frac{a}{l})^{[\frac{S}{l}]})$. The length of helix antenna one turns, l where $l = \sqrt{(\pi \cdot D)^2 + S^2}$. The induced magnetic field of a single turn can be represented by an inductance L_{seg}, which includes the

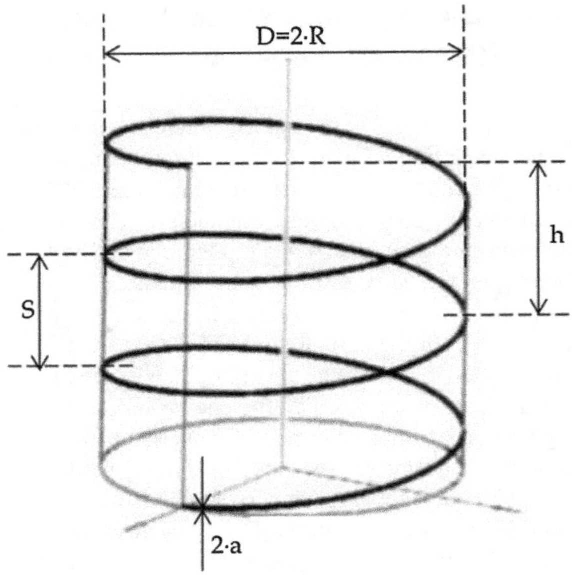

Fig. 8.13 Geometries of Helical antenna

self-inductance L_{self} of one turn and the mutual inductance M_{seg} coupled from two adjacent turns $L_{seg} = L_{self} + 2 \cdot M_{seg}$ (Fig. 8.13).

The L_{self} expression integrates the effects of pitch angle (α), μ_0 is the of free space.

$$L_{self} = \mu_0 \cdot R \cdot [\ln(\frac{8R}{a}) - 2] \cdot \cos(\alpha); M_{seg} = \frac{\pi \cdot \mu_0 \cdot R^4}{\sqrt{2} \cdot (R^2 + S^2)^{\frac{3}{2}}}$$

$$L_{seg} = \mu_0 \cdot R \cdot [\ln((\frac{8R}{a}) - 2] \cdot \cos(\alpha) + \frac{\pi \cdot \mu_0 \cdot R^4 \cdot \sqrt{2}}{(R^2 + S^2)^{\frac{3}{2}}}$$

The normal mode helical antenna (NMHA) equivalent circuit is divided to two parts: one modeling the equivalent wire antennas (five elements circuit) and the other modeling the inductive loops (three elements circuit) (Fig. 8.14).

We define two resonance frequencies for our helix antenna equivalent circuit (eight elements) $f_{01}(\omega_{01}); f_{02}(\omega_{02})$. Capacitor C_0 is the antenna capacitance at frequency below the first resonance frequency $f_{01}(\omega_{01})$. We select L_0 to resonate with C_0 at the second resonant frequency $f_{02}(\omega_{02})$. Inductor L_1 is chosen to resonate with C_1 at frequency $f_{02}(\omega_{02})$. We estimate C_1, L_1 and R_1 at the first resonant frequency $f_{01}(\omega_{01})$, the reactance of the antenna vanishes and the resistance is R_0.

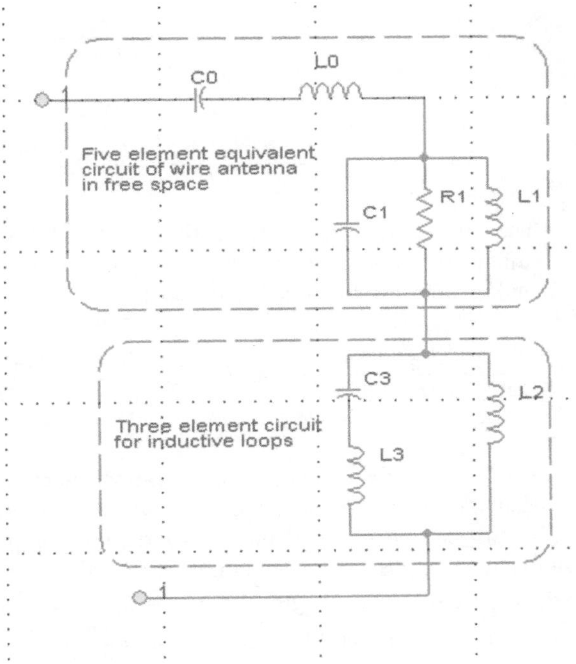

Fig. 8.14 Helix antenna equivalent circuit with eight frequency independent elements

$$C_0 = \frac{\pi \cdot \varepsilon_0 \cdot h}{\ln(\frac{h}{a'}) - 1}; L_0 = \frac{1}{\omega_{02}^2 \cdot C_0} = \frac{\ln(\frac{h}{a'}) - 1}{\omega_{02}^2 \cdot \pi \cdot \varepsilon_0 \cdot h}; A = \omega_{01} \cdot L_0 - \frac{1}{\omega_{01} \cdot C_0} = \frac{1}{C_0} \cdot [\frac{\omega_{01}}{\omega_{02}^2} - \frac{1}{\omega_{01}}]$$

$$A = \frac{\ln(\frac{h}{a'}) - 1}{\pi \cdot \varepsilon_0 \cdot h} \cdot [\frac{\omega_{01}}{\omega_{02}^2} - \frac{1}{\omega_{01}}]; C_1 = \frac{A}{R_0^2 + A^2} \cdot (\frac{\omega_{01}}{\omega_{01}^2 - \omega_{02}^2}); L_1 = \frac{R_0^2 + A^2}{A} \cdot (\frac{\omega_{01}}{\omega_{02}^2} - \frac{1}{\omega_{01}})$$

$$R_1 = \frac{R_0^2 + A^2}{R_0}; L_{unit} = \frac{L_{seg}}{S} = \mu_0 \cdot \frac{R}{S} \cdot [\ln(\frac{8R}{a}) - 2] \cdot \cos(\alpha) + \frac{\pi \cdot \mu_0 \cdot R^4 \cdot \sqrt{2}}{(R^2 + S^2)^{\frac{3}{2}}}$$

The axial wave number for the helical structure, k is $k = \frac{\omega}{v_p} = \sqrt{k_0^2 + \gamma^2}$, where v_p is the axial velocity of the wave in the helical structure, which is less than the light velocity, k_0 is a free space wave number, and γ is the radial wave number. We define factor F, which integrate k, h parameters [130, 131].

$$F = 1 + \frac{1}{(1 - \frac{2 \cdot k \cdot h}{\pi^2})} \cdot (\frac{\sin^2(k \cdot h)}{1 - \frac{\sin(2 \cdot k \cdot h)}{2 \cdot k \cdot h}} - 1); L_2 = \frac{h}{F} \cdot L_{unit} = \frac{h}{F} \cdot \frac{L_{seg}}{S}$$

Another factor is H which is frequency dependent. The expression for L_3 is dependent on operational wavelength and function of H. A wavelength of $4 \cdot l \cdot N$ is considered for our helix antenna ($k_0 = \frac{2 \cdot \pi}{4 \cdot l \cdot N}$).

$$H = H(\gamma, k_0, l, N_0) = \frac{\left(\frac{\pi^2}{4\cdot(1+\frac{\gamma^2}{k_0^2})} - \frac{l^2\cdot N^2\cdot k_0^2}{1+\frac{4\cdot\gamma_{def}^2\cdot l^2\cdot N^2}{\pi^2}}\right)}{\left(\frac{\pi^2}{4} - l^2\cdot N^2\cdot k_0^2\right)};$$

$$C_3 = \frac{2\cdot h/[L_{unit}\cdot\pi^2\cdot c^2]}{H} = \frac{2\cdot h/[\frac{L_{seg}}{S}\cdot\pi^2\cdot c^2]}{H}$$

The value of H varies significantly at low frequencies and changes little when k_0 is sufficiently large near the resonance. A relationship between k_0 and γ can be obtaining from the following expression:

$$\left(\frac{\gamma\cdot D}{k_0\cdot D}\right)^2\cdot\frac{I_0(\frac{\gamma\cdot D}{2})\cdot K_0(\frac{\gamma\cdot D}{2})}{I_1(\frac{\gamma\cdot D}{2})\cdot K_1(\frac{\gamma\cdot D}{2})}\cdot\tan^2(\alpha) = 1; L_3 = \frac{2\cdot L_{unit}\cdot l^2\cdot N^2}{h}\cdot\left(H - \frac{1}{1+\frac{4\cdot l^2\cdot N^2\cdot\gamma_{def}^2}{\pi^2}}\right)$$

where I_0, K_0, I_1, and K_1 are modified Bessel functions.

Remark: Helix antenna is modeled as a series of loops and linear conductors and each turn can be resolved into two radiating components: first is the axial wire segment of length S and second it a loop of diameter D. We define two parameters which related to wavelength, $S_\lambda(\lambda) = \frac{S}{\lambda}; C_\lambda(\lambda) = \frac{\pi\cdot 2\cdot R}{\lambda}$. We define the axial ratio of the equivalent wire and loop model $AR = \frac{2S_\lambda(\lambda)}{[C_\lambda(\lambda)]^2}$.

$AR = \frac{S\cdot\lambda}{2\cdot R^2\cdot\pi^2}$. The typical normal mode helical antennas, the axial ratio is much greater than one ($AR \gg 1 \Rightarrow \frac{S\cdot\lambda}{2\cdot R^2\cdot\pi^2} \gg 1 \Rightarrow S\cdot\lambda \gg 2\cdot R^2\cdot\pi^2$). The helical antenna wire and inductor model is presented in the below figure (Fig. 8.15).

One application of Helix antenna is RFID antennas. We use it for identification and tracking of objects using radio waves. RFID tags employ helical antennas embedded in a dielectric material. The antenna is designed to resonate at around

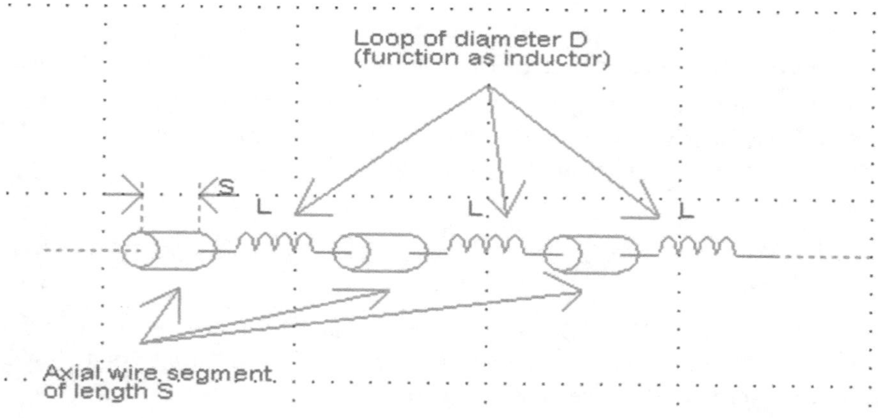

Fig. 8.15 Helical antenna wire and inductor model

specific RFID frequency. RFID IC can be represented as a parallel equivalent circuit of capacitor and resistor in parallel. The complete RFID tag with helical antenna equivalent circuit is describe in the below figure [85].

Remark: Helical antenna is connected to RFID IC through microstrip lines. We neglect in our analysis the parasitic effects of microstrip lines (delay in time) (Fig. 8.16).

$$I_{C_{rfic}} + I_{R_{rfic}} + I_{C_0} = 0\,(\text{KCL}\,@\,\text{A}_2),\ I_{C_0} = I_{L_0};\, I_{C_1} + I_{R_1} + I_{L_1} = I_{L_0}\,(\text{KCL}\,@\,\text{A}_4)$$

$$I_{C_1} + I_{R_1} + I_{L_1} = I_{C_3} + I_{L_2}\,(\text{KCL}\,@\,\text{A}_5),\ I_{L_3} = I_{C_3};\, I_{R_{rfid}} = \frac{V_{A_1} - V_{A_2}}{R_{rfid}};\, I_{C_{rfid}} = C_{rfid} \cdot \frac{d(V_{A_1} - V_{A_2})}{dt}$$

$$V_{L_2} = V_{A_1} - V_{A_5} = L_2 \cdot \frac{dI_{L_2}}{dt};\, V_{L_3} = V_{A_1} - V_{A_6} = L_3 \cdot \frac{dI_{L_3}}{dt};\, I_{C_3} = C_3 \cdot \frac{d(V_{A_6} - V_{A_5})}{dt}$$

$$I_{R_1} = \frac{V_{A_5} - V_{A_4}}{R_1};\, V_{L_1} = V_{A_5} - V_{A_4} = L_1 \cdot \frac{dI_{L_1}}{dt};\, I_{C_1} = C_1 \cdot \frac{d(V_{A_5} - V_{A_4})}{dt}$$

$$V_{L_0} = V_{A_4} - V_{A_3} = L_0 \cdot \frac{dI_{L_0}}{dt};\, I_{C_0} = C_0 \cdot \frac{d(V_{A_3} - V_{A_2})}{dt}$$

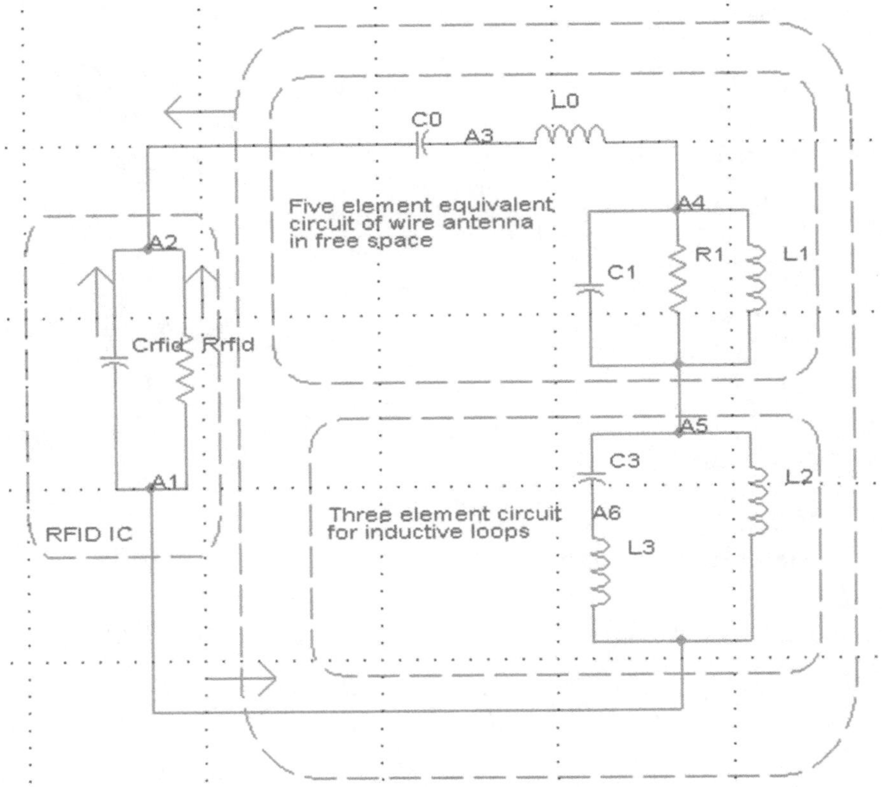

Fig. 8.16 Helix antenna equivalent circuit

$$I_{R_{rfid}} = \frac{V_{A_1} - V_{A_2}}{R_{rfid}} \Rightarrow V_{A_1} - V_{A_2} = R_{rfid} \cdot I_{R_{rfid}}; I_{C_{rfid}} = C_{rfid} \cdot \frac{d(V_{A_1} - V_{A_2})}{dt} = C_{rfid} \cdot R_{rfid} \cdot \frac{dI_{R_{rfid}}}{dt}$$
$$I_{R_1} = \frac{V_{A_5} - V_{A_4}}{R_1} \Rightarrow V_{A_5} - V_{A_4} = I_{R_1} \cdot R_1; L_1 \cdot \frac{dI_{L_1}}{dt} = V_{A_5} - V_{A_4} = I_{R_1} \cdot R_1; I_{C_1} = C_1 \cdot R_1 \cdot \frac{dI_{R_1}}{dt}$$
$$I_{C_1} = C_1 \cdot \frac{d(V_{A_5} - V_{A_4})}{dt} = C_1 \cdot \frac{d}{dt}(L_1 \cdot \frac{dI_{L_1}}{dt}) = C_1 \cdot L_1 \cdot \frac{d^2 I_{L_1}}{dt^2}; I_{C_1} = C_1 \cdot L_1 \cdot \frac{d^2 I_{L_1}}{dt^2}$$
$$R_1 \cdot \frac{dI_{R_1}}{dt} = L_1 \cdot \frac{d^2 I_{L_1}}{dt^2} \Rightarrow \frac{d^2 I_{L_1}}{dt^2} = \frac{R_1}{L_1} \cdot \frac{dI_{R_1}}{dt}.$$

$$\text{KVL}: V_{A_{kl}} = V_{A_k} - V_{A_l}; \sum V_{A_{kl}} = 0; k \neq l; k = 1, 2, \ldots, 6; l = 1, 2, \ldots, 6$$

$$V_{A_{12}} + V_{A_{23}} + V_{A_{34}} + V_{A_{45}} + V_{A_{51}} = 0; V_{A_{51}} = V_{A_{56}} - V_{A_{61}};$$
$$V_{A_{12}} = V_{A_1} - V_{A_2} = \frac{1}{C_{rfid}} \cdot \int I_{C_{rfid}} \cdot dt$$

$$V_{A_{23}} = V_{A_2} - V_{A_3} = -\frac{1}{C_0} \cdot \int I_{C_0} \cdot dt; V_{A_{34}} = V_{A_3} - V_{A_4} = -L_0 \cdot \frac{dI_{L_0}}{dt};$$
$$V_{A_{45}} = V_{A_4} - V_{A_5} = -\frac{1}{C_1} \cdot \int I_{C_1} \cdot dt$$
$$V_{A_{45}} = V_{A_4} - V_{A_5} = -L_1 \cdot \frac{dI_{L_1}}{dt} = -R_1 \cdot I_{R_1}; V_{A_{51}} = V_{A_5} - V_{A_1} = -L_2 \cdot \frac{dI_{L_2}}{dt}$$

$$\sum V_{A_{kl}} = 0 \Rightarrow \frac{1}{C_{rfid}} \cdot \int I_{C_{rfid}} \cdot dt - \frac{1}{C_0} \cdot \int I_{C_0} \cdot dt - L_0 \cdot \frac{dI_{L_0}}{dt}$$
$$- \frac{1}{C_1} \cdot \int I_{C_1} \cdot dt - L_2 \cdot \frac{dI_{L_2}}{dt} = 0$$

$$\frac{d}{dt}(\sum V_{A_{kl}}) = 0 \Rightarrow \frac{1}{C_{rfid}} \cdot I_{C_{rfid}} - \frac{1}{C_0} \cdot I_{C_0} - L_0 \cdot \frac{d^2 I_{L_0}}{dt^2} - \frac{1}{C_1} \cdot I_{C_1} - L_2 \cdot \frac{d^2 I_{L_2}}{dt^2} = 0$$

$$V_{A_{51}} = V_{A_{56}} - V_{A_{61}} \Rightarrow L_2 \cdot \frac{dI_{L_2}}{dt} = \frac{1}{C_3} \cdot \int I_{C_3} \cdot dt + L_3 \cdot \frac{dI_{L_3}}{dt};$$
$$L_2 \cdot \frac{d^2 I_{L_2}}{dt^2} = \frac{1}{C_3} \cdot I_{C_3} + L_3 \cdot \frac{d^2 I_{L_3}}{dt^2}$$

System differential equations (Ver.1):

$$I_{C_1} = C_1 \cdot R_1 \cdot \frac{dI_{R_1}}{dt}; I_{C_1} = C_1 \cdot L_1 \cdot \frac{d^2 I_{L_1}}{dt^2}; I_{C_{rfid}} = C_{rfid} \cdot R_{rfid} \cdot \frac{dI_{R_{rfid}}}{dt}$$
$$L_2 \cdot \frac{d^2 I_{L_2}}{dt^2} = \frac{1}{C_3} \cdot I_{C_3} + L_3 \cdot \frac{d^2 I_{L_3}}{dt^2}; \frac{d^2 I_{L_1}}{dt^2} = \frac{R_1}{L_1} \cdot \frac{dI_{R_1}}{dt}; \frac{dI_{L_1}}{dt} = \frac{R_1}{L_1} \cdot I_{R_1}$$
$$\frac{1}{C_{rfid}} \cdot I_{C_{rfid}} - \frac{1}{C_0} \cdot I_{C_0} - L_0 \cdot \frac{d^2 I_{L_0}}{dt^2} - \frac{1}{C_1} \cdot I_{C_1} - L_2 \cdot \frac{d^2 I_{L_2}}{dt^2} = 0$$

System differential equations (Ver.2): $I_{C_0} = I_{L_0}; I_{L_3} = I_{C_3}; I_{L_3} \to I_{C_3}$

$$I_{C_1} = I_{L_0} - I_{R_1} - I_{L_1}; I_{C_{rfid}} = -I_{R_{rfid}} - I_{C_0} = -I_{R_{rfid}} - I_{L_0}; I_{C_0} \to I_{L_0}$$

$$I_{L_0} - I_{R_1} - I_{L_1} = C_1 \cdot R_1 \cdot \frac{dI_{R_1}}{dt}; I_{L_0} - I_{R_1} - I_{L_1} = C_1 \cdot L_1 \cdot \frac{d^2 I_{L_1}}{dt^2}; I_{C_{rfid}} = C_{rfid} \cdot R_{rfid} \cdot \frac{dI_{R_{rfid}}}{dt}$$

$$L_2 \cdot \frac{d^2 I_{L_2}}{dt^2} = \frac{1}{C_3} \cdot I_{C_3} + L_3 \cdot \frac{d^2 I_{C_3}}{dt^2}; \frac{dI_{L_1}}{dt} = \frac{R_1}{L_1} \cdot I_{R_1}$$

$$\frac{1}{C_{rfid}} \cdot (-I_{R_{rfid}} - I_{L_0}) - \frac{1}{C_0} \cdot I_{L_0} - L_0 \cdot \frac{d^2 I_{L_0}}{dt^2} - \frac{1}{C_1} \cdot (I_{L_0} - I_{R_1} - I_{L_1}) - L_2 \cdot \frac{d^2 I_{L_2}}{dt^2} = 0$$

System differential equations (Ver.3):

$$\frac{dI_{R_1}}{dt} = \frac{1}{C_1 \cdot R_1} \cdot I_{L_0} - \frac{1}{C_1 \cdot R_1} \cdot I_{R_1} - \frac{1}{C_1 \cdot R_1} \cdot I_{L_1}; \frac{d^2 I_{L_1}}{dt^2} = \frac{1}{C_1 \cdot L_1} \cdot I_{L_0} - \frac{1}{C_1 \cdot L_1} \cdot I_{R_1} - \frac{1}{C_1 \cdot L_1} \cdot I_{L_1}$$

$$L_2 \cdot \frac{d^2 I_{L_2}}{dt^2} = \frac{1}{C_3} \cdot I_{C_3} + L_3 \cdot \frac{d^2 I_{C_3}}{dt^2}; \frac{dI_{L_1}}{dt} = \frac{R_1}{L_1} \cdot I_{R_1}; \frac{dI_{R_{rfid}}}{dt} = \frac{1}{C_{rfid} \cdot R_{rfid}} \cdot I_{C_{rfid}}$$

$$-\frac{1}{C_{rfid}} \cdot I_{R_{rfid}} - \left(\frac{1}{C_{rfid}} + \frac{1}{C_1} + \frac{1}{C_0}\right) \cdot I_{L_0} - L_0 \cdot \frac{d^2 I_{L_0}}{dt^2} + \frac{1}{C_1} \cdot I_{R_1} + \frac{1}{C_1} \cdot I_{L_1} - L_2 \cdot \frac{d^2 I_{L_2}}{dt^2} = 0$$

We define new variables: $\frac{dY_1}{dt} = \frac{d^2 I_{L_1}}{dt^2}; Y_1 = \frac{dI_{L_1}}{dt}; \frac{dY_2}{dt} = \frac{d^2 I_{L_2}}{dt^2}; Y_2 = \frac{dI_{L_2}}{dt}$

$$\frac{dY_3}{dt} = \frac{d^2 I_{C_3}}{dt^2}; Y_3 = \frac{dI_{C_3}}{dt}; \frac{dY_4}{dt} = \frac{d^2 I_{L_0}}{dt^2}; Y_4 = \frac{dI_{L_0}}{dt}$$

System differential equations (Ver.4):

$$\frac{dI_{R_1}}{dt} = \frac{1}{C_1 \cdot R_1} \cdot I_{L_0} - \frac{1}{C_1 \cdot R_1} \cdot I_{R_1} - \frac{1}{C_1 \cdot R_1} \cdot I_{L_1}; \frac{dY_1}{dt} = \frac{1}{C_1 \cdot L_1} \cdot I_{L_0} - \frac{1}{C_1 \cdot L_1} \cdot I_{R_1} - \frac{1}{C_1 \cdot L_1} \cdot I_{L_1}$$

$$\frac{dI_{R_{rfid}}}{dt} = \frac{1}{C_{rfid} \cdot R_{rfid}} \cdot I_{C_{rfid}}; \frac{dI_{C_3}}{dt} = Y_3$$

$$L_2 \cdot \frac{dY_2}{dt} = \frac{1}{C_3} \cdot I_{C_3} + L_3 \cdot \frac{dY_3}{dt}; \frac{dI_{L_1}}{dt} = \frac{R_1}{L_1} \cdot I_{R_1}; \frac{dI_{L_1}}{dt} = Y_1; \frac{dI_{L_2}}{dt} = Y_2; \frac{dI_{L_0}}{dt} = Y_4$$

$$-\frac{1}{C_{rfid}} \cdot I_{R_{rfid}} - \left(\frac{1}{C_{rfid}} + \frac{1}{C_1} + \frac{1}{C_0}\right) \cdot I_{L_0} - L_0 \cdot \frac{dY_4}{dt} + \frac{1}{C_1} \cdot I_{R_1} + \frac{1}{C_1} \cdot I_{L_1} - L_2 \cdot \frac{dY_2}{dt} = 0$$

At fixed points: $\frac{dI_{R_1}}{dt} = 0; \frac{dY_1}{dt} = 0; \frac{dI_{R_{rfid}}}{dt} = 0; \frac{dI_{C_3}}{dt} = 0; \frac{dY_2}{dt} = 0; \frac{dY_3}{dt} = 0$

$$\frac{dI_{L_1}}{dt} = 0; \frac{dI_{L_2}}{dt} = 0; \frac{dI_{L_0}}{dt} = 0; \frac{dY_4}{dt} = 0; I^*_{L_0} = I^*_{L_1}$$
$$\frac{dI_{R_1}}{dt} = 0 \Rightarrow I^*_{C_1} = 0; \frac{dY_1}{dt} = 0 \Rightarrow I^*_{C_1} = 0; \frac{dI_{R_{rfid}}}{dt} = 0 \Rightarrow I^*_{C_{rfid}} = 0; \frac{dI_{C_3}}{dt} = 0 \Rightarrow Y^*_3 = 0$$
$$\frac{dY_2}{dt} = 0; \frac{dY_3}{dt} = 0 \Rightarrow I^*_{C_3} = 0; \frac{dI_{L_1}}{dt} = 0 \Rightarrow I^*_{R_1} = 0; \frac{dI_{L_1}}{dt} = 0 \Rightarrow Y^*_1 = 0$$
$$\frac{dI_{L_2}}{dt} = 0 \Rightarrow Y^*_2 = 0; \frac{dI_{L_0}}{dt} = 0 \Rightarrow Y^*_4 = 0; I^*_{C_{rfid}} = 0 \Rightarrow I^*_{R_{rfid}} + I^*_{L_0} = 0$$
$$\frac{dY_4}{dt} = 0; \frac{dY_2}{dt} = 0 \Rightarrow -\frac{1}{C_{rfid}} \cdot I^*_{R_{rfid}} - (\frac{1}{C_{rfid}} + \frac{1}{C_1} + \frac{1}{C_0}) \cdot I^*_{L_0} + \frac{1}{C_1} \cdot I^*_{L_1} = 0; I^*_{L_0} = 0$$

$$I^*_{L_0} = I^*_{L_1} \Rightarrow I^*_{R_{rfid}} = -C_{rfid} \cdot (\frac{1}{C_{rfid}} + \frac{1}{C_0}) \cdot I^*_{L_0}; I^*_{R_{rfid}} + I^*_{L_0} = 0$$
$$\Rightarrow [1 - C_{rfid} \cdot (\frac{1}{C_{rfid}} + \frac{1}{C_0})] \cdot I^*_{L_0} = 0$$

Our circuit fixed point:

$$E^{(*)}(Y^*_1, Y^*_2, Y^*_3, Y^*_4, I^*_{L_0}, I^*_{L_1}, I^*_{L_2}, I^*_{C_1}, I^*_{C_3}, I^*_{R_1}, I^*_{C_{rfid}}, I^*_{R_{rfid}})$$
$$= (0, 0, 0, 0, 0, 0, 0, 0, 0, 0, 0, 0)$$

Stability analysis: The standard local stability analysis about any one of the equilibrium points of helix antenna system consists in adding to its coordinated $[Y_1 Y_2 Y_3 Y_4 I_{L_0} I_{L_1} I_{L_2} I_{C_1} I_{C_3} I_{R_1} I_{C_{rfid}} I_{R_{rfid}}]$ arbitrarily small increments of exponential terms $[y_1 y_2 y_3 y_4 i_{L_0} i_{L_1} i_{L_2} i_{C_1} i_{C_3} i_{R_1} i_{C_{rfid}} i_{R_{rfid}}] \cdot e^{\lambda \cdot t}$, and retaining the first order terms in $y_1 y_2 y_3 y_4 i_{L_0} i_{L_1} i_{L_2} i_{C_1} i_{C_3} i_{R_1} i_{C_{rfid}} i_{R_{rfid}}$. The system of nine homogeneous equations leads to a polynomial characteristic equation in the eigenvalue λ. The polynomial characteristic equation accepts by set the helix antenna system equations. The helix antenna system fixed *values with arbitrarily small increments* of exponential form $[y_1 y_2 y_3 y_4 i_{L_0} i_{L_1} i_{L_2} i_{C_1} i_{C_3} i_{R_1} i_{C_{rfid}} i_{R_{rfid}}] \cdot e^{\lambda \cdot t}$ are; i = 0 (first fixed point), i = 1 (second fixed point), i = 2 (third fixed point), etc., [2–4].

$$Y_1(t) = Y_1^{(i)} + y_1 \cdot e^{\lambda \cdot t}; Y_2(t) = Y_2^{(i)} + y_2 \cdot e^{\lambda \cdot t}; Y_3(t) = Y_3^{(i)} + y_3 \cdot e^{\lambda \cdot t}; Y_4(t) = Y_4^{(i)} + y_4 \cdot e^{\lambda \cdot t}$$
$$I_{L_0}(t) = I_{L_0}^{(i)} + i_{L_0} \cdot e^{\lambda \cdot t}; I_{L_1}(t) = I_{L_1}^{(i)} + i_{L_1} \cdot e^{\lambda \cdot t}; I_{C_1}(t) = I_{C_1}^{(i)} + i_{C_1} \cdot e^{\lambda \cdot t}; I_{C_3}(t) = I_{C_3}^{(i)} + i_{C_3} \cdot e^{\lambda \cdot t}$$
$$I_{R_1}(t) = I_{R_1}^{(i)} + i_{R_1} \cdot e^{\lambda \cdot t}; I_{C_{rfid}}(t) = I_{C_{rfid}}^{(i)} + i_{C_{rfid}} \cdot e^{\lambda \cdot t}; I_{R_{rfid}}(t) = I_{R_{rfid}}^{(i)} + i_{R_{rfid}} \cdot e^{\lambda \cdot t}$$

$$\frac{dY_1(t)}{dt} = y_1 \cdot \lambda \cdot e^{\lambda \cdot t}; \frac{dY_2(t)}{dt} = y_2 \cdot \lambda \cdot e^{\lambda \cdot t}; \frac{dY_3(t)}{dt} = y_3 \cdot \lambda \cdot e^{\lambda \cdot t}; \frac{dY_4(t)}{dt} = y_4 \cdot \lambda \cdot e^{\lambda \cdot t}$$
$$\frac{dI_{R_1}(t)}{dt} = i_{R_1} \cdot \lambda \cdot e^{\lambda \cdot t}; \frac{dI_{R_{rfid}}(t)}{dt} = i_{R_{rfid}} \cdot \lambda \cdot e^{\lambda \cdot t}; \frac{dI_{C_3}(t)}{dt} = i_{C_3} \cdot \lambda \cdot e^{\lambda \cdot t}; \frac{dI_{L_1}(t)}{dt} = i_{L_1} \cdot \lambda \cdot e^{\lambda \cdot t}$$
$$\frac{dI_{L_0}(t)}{dt} = i_{L_0} \cdot \lambda \cdot e^{\lambda \cdot t}; I_{L_2}(t) = I_{L_2}^{(i)} + i_{L_2} \cdot e^{\lambda \cdot t}; \frac{dI_{L_2}(t)}{dt} = i_{L_2} \cdot \lambda \cdot e^{\lambda \cdot t}$$

&&&

$$\frac{dI_{R_1}}{dt} = \frac{1}{C_1 \cdot R_1} \cdot I_{L_0} - \frac{1}{C_1 \cdot R_1} \cdot I_{R_1} - \frac{1}{C_1 \cdot R_1} \cdot I_{L_1}$$

$$i_{R_1} \cdot \lambda \cdot e^{\lambda \cdot t} = \frac{1}{C_1 \cdot R_1} \cdot [I_{L_0}^{(i)} + i_{L_0} \cdot e^{\lambda \cdot t}] - \frac{1}{C_1 \cdot R_1} \cdot [I_{R_1}^{(i)} + i_{R_1} \cdot e^{\lambda \cdot t}] - \frac{1}{C_1 \cdot R_1} \cdot [I_{L_1}^{(i)} + i_{L_1} \cdot e^{\lambda \cdot t}]$$
$$i_{R_1} \cdot \lambda \cdot e^{\lambda \cdot t} = \frac{1}{C_1 \cdot R_1} \cdot I_{L_0}^{(i)} - \frac{1}{C_1 \cdot R_1} \cdot I_{R_1}^{(i)} - \frac{1}{C_1 \cdot R_1} \cdot I_{L_1}^{(i)} + \frac{1}{C_1 \cdot R_1} \cdot i_{L_0} \cdot e^{\lambda \cdot t}$$
$$- \frac{1}{C_1 \cdot R_1} \cdot i_{R_1} \cdot e^{\lambda \cdot t} - \frac{1}{C_1 \cdot R_1} \cdot i_{L_1} \cdot e^{\lambda \cdot t}$$

At fixed points: $\frac{1}{C_1 \cdot R_1} \cdot I_{L_0}^{(i)} - \frac{1}{C_1 \cdot R_1} \cdot I_{R_1}^{(i)} - \frac{1}{C_1 \cdot R_1} \cdot I_{L_1}^{(i)}$

$$-i_{R_1} \cdot \lambda - \frac{1}{C_1 \cdot R_1} \cdot i_{R_1} + \frac{1}{C_1 \cdot R_1} \cdot i_{L_0} - \frac{1}{C_1 \cdot R_1} \cdot i_{L_1} = 0$$

$$\frac{dY_1}{dt} = \frac{1}{C_1 \cdot L_1} \cdot I_{L_0} - \frac{1}{C_1 \cdot L_1} \cdot I_{R_1} - \frac{1}{C_1 \cdot L_1} \cdot I_{L_1}$$

$$y_1 \cdot \lambda \cdot e^{\lambda \cdot t} = \frac{1}{C_1 \cdot L_1} \cdot [I_{L_0}^{(i)} + i_{L_0} \cdot e^{\lambda \cdot t}] - \frac{1}{C_1 \cdot L_1} \cdot [I_{R_1}^{(i)} + i_{R_1} \cdot e^{\lambda \cdot t}] - \frac{1}{C_1 \cdot L_1} \cdot [I_{L_1}^{(i)} + i_{L_1} \cdot e^{\lambda \cdot t}]$$
$$y_1 \cdot \lambda \cdot e^{\lambda \cdot t} = \frac{1}{C_1 \cdot L_1} \cdot I_{L_0}^{(i)} - \frac{1}{C_1 \cdot L_1} \cdot I_{R_1}^{(i)} - \frac{1}{C_1 \cdot L_1} \cdot I_{L_1}^{(i)} + \frac{1}{C_1 \cdot L_1} \cdot i_{L_0} \cdot e^{\lambda \cdot t}$$
$$- \frac{1}{C_1 \cdot L_1} \cdot i_{R_1} \cdot e^{\lambda \cdot t} - \frac{1}{C_1 \cdot L_1} \cdot i_{L_1} \cdot e^{\lambda \cdot t}$$

At fixed points: $\frac{1}{C_1 \cdot L_1} \cdot I_{L_0}^{(i)} - \frac{1}{C_1 \cdot L_1} \cdot I_{R_1}^{(i)} - \frac{1}{C_1 \cdot L_1} \cdot I_{L_1}^{(i)} = 0$

$$-y_1 \cdot \lambda + \frac{1}{C_1 \cdot L_1} \cdot i_{L_0} - \frac{1}{C_1 \cdot L_1} \cdot i_{R_1} - \frac{1}{C_1 \cdot L_1} \cdot i_{L_1} = 0$$

$$\frac{dI_{R_{rfid}}}{dt} = \frac{1}{C_{rfid} \cdot R_{rfid}} \cdot I_{C_{rfid}}; i_{R_{rfid}} \cdot \lambda \cdot e^{\lambda \cdot t} = \frac{1}{C_{rfid} \cdot R_{rfid}} \cdot [I_{C_{rfid}}^{(i)} + i_{C_{rfid}} \cdot e^{\lambda \cdot t}]$$

$$i_{R_{rfid}} \cdot \lambda \cdot e^{\lambda \cdot t} = \frac{1}{C_{rfid} \cdot R_{rfid}} \cdot I_{C_{rfid}}^{(i)} + \frac{1}{C_{rfid} \cdot R_{rfid}} \cdot i_{C_{rfid}} \cdot e^{\lambda \cdot t}$$

At fixed points: $\frac{1}{C_{rfid} \cdot R_{rfid}} \cdot I_{C_{rfid}}^{(i)} = 0; -i_{R_{rfid}} \cdot \lambda + \frac{1}{C_{rfid} \cdot R_{rfid}} \cdot i_{C_{rfid}} = 0$

$$\frac{dI_{C_3}}{dt} = Y_3 \Rightarrow i_{C_3} \cdot \lambda \cdot e^{\lambda \cdot t} = Y_3^{(i)} + y_3 \cdot e^{\lambda \cdot t}; Y_3^{(i)} = 0; -i_{C_3} \cdot \lambda + y_3 = 0$$

$$L_2 \cdot \frac{dY_2}{dt} = \frac{1}{C_3} \cdot I_{C_3} + L_3 \cdot \frac{dY_3}{dt}; L_2 \cdot y_2 \cdot \lambda \cdot e^{\lambda \cdot t} = \frac{1}{C_3} \cdot [I_{C_3}^{(i)} + i_{C_3} \cdot e^{\lambda \cdot t}] + L_3 \cdot y_3 \cdot \lambda \cdot e^{\lambda \cdot t}$$

$$L_2 \cdot y_2 \cdot \lambda \cdot e^{\lambda \cdot t} = \frac{1}{C_3} \cdot I_{C_3}^{(i)} + \frac{1}{C_3} \cdot i_{C_3} \cdot e^{\lambda \cdot t} + L_3 \cdot y_3 \cdot \lambda \cdot e^{\lambda \cdot t}; \frac{1}{C_3} \cdot I_{C_3}^{(i)} = 0$$

$$L_2 \cdot y_2 \cdot \lambda - L_3 \cdot y_3 \cdot \lambda = \frac{1}{C_3} \cdot i_{C_3}; L_2 \approx L_3 = L_{2-3}; (y_2 - y_3) \cdot L_{2-3} \cdot \lambda = \frac{1}{C_3} \cdot i_{C_3}$$

$$y_2 - y_3 \to \varepsilon; i_{C_3} \to \varepsilon; \frac{dI_{L_1}}{dt} = \frac{R_1}{L_1} \cdot I_{R_1}; i_{L_1} \cdot \lambda \cdot e^{\lambda \cdot t} = \frac{R_1}{L_1} \cdot (I_{R_1}^{(i)} + i_{R_1} \cdot e^{\lambda \cdot t})$$

$$i_{L_1} \cdot \lambda \cdot e^{\lambda \cdot t} = \frac{R_1}{L_1} \cdot I_{R_1}^{(i)} + \frac{R_1}{L_1} \cdot i_{R_1} \cdot e^{\lambda \cdot t}; \frac{R_1}{L_1} \cdot I_{R_1}^{(i)} = 0 \Rightarrow -i_{L_1} \cdot \lambda + \frac{R_1}{L_1} \cdot i_{R_1} = 0$$

$$\frac{dI_{L_1}}{dt} = Y_1; i_{L_1} \cdot \lambda \cdot e^{\lambda \cdot t} = Y_1^{(i)} + y_1 \cdot e^{\lambda \cdot t}; Y_1^{(i)} = 0; -i_{L_1} \cdot \lambda + y_1 = 0$$

$$\frac{dI_{L_2}}{dt} = Y_2; i_{L_2} \cdot \lambda \cdot e^{\lambda \cdot t} = Y_2^{(i)} + y_2 \cdot e^{\lambda \cdot t}; Y_2^{(i)} = 0; -i_{L_2} \cdot \lambda + y_2 = 0$$

$$\frac{dI_{L_0}}{dt} = Y_4; i_{L_0} \cdot \lambda \cdot e^{\lambda \cdot t} = Y_4^{(i)} + y_4 \cdot e^{\lambda \cdot t}; Y_4^{(i)} = 0; -i_{L_0} \cdot \lambda + y_4 = 0$$

$$-\frac{1}{C_{rfid}} \cdot I_{R_{rfid}} - (\frac{1}{C_{rfid}} + \frac{1}{C_1} + \frac{1}{C_0}) \cdot I_{L_0} - L_0 \cdot \frac{dY_4}{dt} + \frac{1}{C_1} \cdot I_{R_1} + \frac{1}{C_1} \cdot I_{L_1} - L_2 \cdot \frac{dY_2}{dt} = 0$$

$$-\frac{1}{C_{rfid}} \cdot (I_{R_{rfid}}^{(i)} + i_{R_{rfid}} \cdot e^{\lambda \cdot t}) - (\frac{1}{C_{rfid}} + \frac{1}{C_1} + \frac{1}{C_0}) \cdot (I_{L_0}^{(i)} + i_{L_0} \cdot e^{\lambda \cdot t})$$

$$- L_0 \cdot y_4 \cdot \lambda \cdot e^{\lambda \cdot t} + \frac{1}{C_1} \cdot (I_{R_1}^{(i)} + i_{R_1} \cdot e^{\lambda \cdot t}) + \frac{1}{C_1} \cdot (I_{L_1}^{(i)} + i_{L_1} \cdot e^{\lambda \cdot t}) - L_2 \cdot y_2 \cdot \lambda \cdot e^{\lambda \cdot t} = 0$$

$$-(\frac{1}{C_{rfid}} + \frac{1}{C_1} + \frac{1}{C_0}) \cdot I_{L_0}^{(i)} + \frac{1}{C_1} \cdot I_{R_1}^{(i)} + \frac{1}{C_1} \cdot I_{L_1}^{(i)} - \frac{1}{C_{rfid}} \cdot I_{R_{rfid}}^{(i)} - \frac{1}{C_{rfid}} \cdot i_{R_{rfid}} \cdot e^{\lambda \cdot t}$$

$$-(\frac{1}{C_{rfid}} + \frac{1}{C_1} + \frac{1}{C_0}) \cdot i_{L_0} \cdot e^{\lambda \cdot t} - L_0 \cdot y_4 \cdot \lambda \cdot e^{\lambda \cdot t} + \frac{1}{C_1} \cdot i_{R_1} \cdot e^{\lambda \cdot t}$$

$$+ \frac{1}{C_1} \cdot i_{L_1} \cdot e^{\lambda \cdot t} - L_2 \cdot y_2 \cdot \lambda \cdot e^{\lambda \cdot t} = 0$$

At fixed points: $-(\frac{1}{C_{rfid}} + \frac{1}{C_1} + \frac{1}{C_0}) \cdot I_{L_0}^{(i)} + \frac{1}{C_1} \cdot I_{R_1}^{(i)} + \frac{1}{C_1} \cdot I_{L_1}^{(i)} - \frac{1}{C_{rfid}} \cdot I_{R_{rfid}}^{(i)} = 0$

$$-(L_2 \cdot y_2 + L_0 \cdot y_4) \cdot \lambda - \frac{1}{C_{rfid}} \cdot i_{R_{rfid}} - (\frac{1}{C_{rfid}} + \frac{1}{C_1} + \frac{1}{C_0}) \cdot i_{L_0} + \frac{1}{C_1} \cdot i_{R_1} + \frac{1}{C_1} \cdot i_{L_1} = 0$$

$$L_2 \approx L_0 = L_{2-0}$$
$$\Rightarrow -(y_2 + y_4) \cdot L_{2-0} \cdot \lambda - \frac{1}{C_{rfid}} \cdot i_{R_{rfid}} - (\frac{1}{C_{rfid}} + \frac{1}{C_1} + \frac{1}{C_0}) \cdot i_{L_0} + \frac{1}{C_1} \cdot i_{R_1} + \frac{1}{C_1} \cdot i_{L_1} = 0$$

$$y_2 \approx y_4 \Rightarrow y_2 + y_4 \approx 2 \cdot y_2 \Rightarrow -y_2 \cdot \lambda - \frac{1}{2 \cdot L_{2-0} \cdot C_{rfid}} \cdot i_{R_{rfid}} - \frac{1}{2 \cdot L_{2-0}} \cdot (\frac{1}{C_{rfid}} + \frac{1}{C_1} + \frac{1}{C_0}) \cdot i_{L_0} + \frac{1}{2 \cdot L_{2-0} \cdot C_1} \cdot i_{R_1} + \frac{1}{2 \cdot L_{2-0} \cdot C_1} \cdot i_{L_1} = 0$$

$$y_2 \approx y_4 \Rightarrow y_2 + y_4 \approx 2 \cdot y_4 \Rightarrow -y_4 \cdot \lambda - \frac{1}{2 \cdot C_{rfid} \cdot L_{2-0}} \cdot i_{R_{rfid}} - \frac{1}{2 \cdot L_{2-0}} \cdot (\frac{1}{C_{rfid}} + \frac{1}{C_1} + \frac{1}{C_0}) \cdot i_{L_0} + \frac{1}{2 \cdot C_1 \cdot L_{2-0}} \cdot i_{R_1} + \frac{1}{2 \cdot C_1 \cdot L_{2-0}} \cdot i_{L_1} = 0$$

We can summary our helix antenna system arbitrarily small increments equations:

$$-i_{R_1} \cdot \lambda - \frac{1}{C_1 \cdot R_1} \cdot i_{R_1} + \frac{1}{C_1 \cdot R_1} \cdot i_{L_0} - \frac{1}{C_1 \cdot R_1} \cdot i_{L_1} = 0; -y_1 \cdot \lambda + \frac{1}{C_1 \cdot L_1} \cdot i_{L_0} - \frac{1}{C_1 \cdot L_1} \cdot i_{R_1} - \frac{1}{C_1 \cdot L_1} \cdot i_{L_1} = 0$$
$$-i_{R_{rfid}} \cdot \lambda + \frac{1}{C_{rfid} \cdot R_{rfid}} \cdot i_{C_{rfid}} = 0; -i_{C_3} \cdot \lambda + y_3 = 0; -i_{L_1} \cdot \lambda + \frac{R_1}{L_1} \cdot i_{R_1} = 0$$
$$-i_{L_1} \cdot \lambda + y_1 = 0; -i_{L_2} \cdot \lambda + y_2 = 0; -i_{L_0} \cdot \lambda + y_4 = 0$$
$$-y_2 \cdot \lambda - \frac{1}{2 \cdot L_{2-0} \cdot C_{rfid}} \cdot i_{R_{rfid}} - \frac{1}{2 \cdot L_{2-0}} \cdot (\frac{1}{C_{rfid}} + \frac{1}{C_1} + \frac{1}{C_0}) \cdot i_{L_0} + \frac{1}{2 \cdot L_{2-0} \cdot C_1} \cdot i_{R_1} + \frac{1}{2 \cdot L_{2-0} \cdot C_1} \cdot i_{L_1} = 0$$
$$-y_4 \cdot \lambda - \frac{1}{2 \cdot C_{rfid} \cdot L_{2-0}} \cdot i_{R_{rfid}} - \frac{1}{2 \cdot L_{2-0}} \cdot (\frac{1}{C_{rfid}} + \frac{1}{C_1} + \frac{1}{C_0}) \cdot i_{L_0} + \frac{1}{2 \cdot C_1 \cdot L_{2-0}} \cdot i_{R_1} + \frac{1}{2 \cdot C_1 \cdot L_{2-0}} \cdot i_{L_1} = 0$$

$$\begin{pmatrix} \Xi_{11} & \cdots & \Xi_{19} \\ \vdots & \ddots & \vdots \\ \Xi_{91} & \cdots & \Xi_{99} \end{pmatrix} \cdot \begin{pmatrix} i_{R_1} \\ y_1 \\ i_{R_{rfid}} \\ i_{C_3} \\ i_{L_1} \\ i_{L_2} \\ i_{L_0} \\ y_2 \\ y_4 \end{pmatrix} + \begin{pmatrix} \iota_{11} & \iota_{12} \\ \vdots & \vdots \\ \vdots & \vdots \\ \iota_{91} & \iota_{92} \end{pmatrix} \cdot \begin{pmatrix} i_{C_{rfid}} \\ y_3 \end{pmatrix} = 0$$

$$\text{Assumption:} \begin{pmatrix} i_{C_{rfid}} \\ y_3 \end{pmatrix} \to \varepsilon$$

$$\Xi_{11} = -\lambda - \frac{1}{C_1 \cdot R_1}; \Xi_{12} = \Xi_{13} = \Xi_{14} = 0; \Xi_{15} = -\frac{1}{C_1 \cdot R_1}; \Xi_{16} = 0; \Xi_{17} = \frac{1}{C_1 \cdot R_1}; \Xi_{18} = \Xi_{19} = 0$$
$$\Xi_{21} = -\frac{1}{C_1 \cdot R_1}; \Xi_{22} = -\lambda; \Xi_{23} = \Xi_{24} = 0; \Xi_{25} = -\frac{1}{C_1 \cdot L_1}; \Xi_{26} = 0; \Xi_{27} = \frac{1}{C_1 \cdot L_1}; \Xi_{28} = \Xi_{29} = 0$$
$$\Xi_{31} = \Xi_{32} = 0; \Xi_{33} = -\lambda; \Xi_{34} = \ldots = \Xi_{39} = 0; \Xi_{41} = \Xi_{42} = \Xi_{43} = 0; \Xi_{44} = -\lambda$$
$$\Xi_{45} = \ldots = \Xi_{49} = 0; \Xi_{51} = \frac{R_1}{L_1}; \Xi_{52} = \Xi_{53} = \Xi_{54} = 0; \Xi_{55} = -\lambda; \Xi_{56} = \ldots = \Xi_{59} = 0$$
$$\Xi_{61} = \ldots = \Xi_{65} = 0; \Xi_{66} = -\lambda; \Xi_{67} = 0; \Xi_{68} = 1; \Xi_{69} = 0; \Xi_{71} = \ldots = \Xi_{76} = 0; \Xi_{77} = -\lambda$$
$$\Xi_{78} = 0; \Xi_{79} = 1; \Xi_{81} = \frac{1}{2 \cdot L_{2-0} \cdot C_1}; \Xi_{82} = 0; \Xi_{83} = -\frac{1}{2 \cdot L_{2-0} \cdot C_{rfid}}; \Xi_{84} = 0; \Xi_{85} = \frac{1}{2 \cdot L_{2-0} \cdot C_1}$$
$$\Xi_{86} = 0; \Xi_{87} = -\frac{1}{2 \cdot L_{2-0}} \cdot \left(\frac{1}{C_{rfid}} + \frac{1}{C_1} + \frac{1}{C_0}\right); \Xi_{88} = -\lambda; \Xi_{89} = 0; \Xi_{91} = \frac{1}{2 \cdot L_{2-0} \cdot C_1}$$
$$\Xi_{92} = 0; \Xi_{93} = -\frac{1}{2 \cdot C_{rfid} \cdot L_{2-0}}; \Xi_{94} = 0; \Xi_{95} = \frac{1}{2 \cdot C_1 \cdot L_{2-0}}; \Xi_{96} = 0$$
$$\Xi_{97} = -\frac{1}{2 \cdot L_{2-0}} \cdot \left(\frac{1}{C_{rfid}} + \frac{1}{C_1} + \frac{1}{C_0}\right); \Xi_{98} = 0; \Xi_{99} = -\lambda$$

$$\imath_{11} = \imath_{12} = \imath_{21} = \imath_{22} = 0; \imath_{31} = \frac{1}{C_{rfid} \cdot R_{rfid}}; \imath_{32} = 0; \imath_{41} = 0; \imath_{42} = 1;$$
$$\imath_{51} = \imath_{52} = 0; \imath_{61} = \imath_{62} = 0$$
$$\imath_{71} = \imath_{72} = 0; \imath_{81} = \imath_{82} = 0; \imath_{91} = \imath_{92} = 0$$

$$A - \lambda \cdot I = \begin{pmatrix} \Xi_{11} & \cdots & \Xi_{19} \\ \vdots & \ddots & \vdots \\ \Xi_{91} & \cdots & \Xi_{99} \end{pmatrix}; \det(A - \lambda \cdot I) = 0; \det \begin{pmatrix} \Xi_{11} & \cdots & \Xi_{19} \\ \vdots & \ddots & \vdots \\ \Xi_{91} & \cdots & \Xi_{99} \end{pmatrix} = 0$$

Eigenvalues stability discussion: Our helix antenna system involving N_a variables ($N_a > 2$, $N_a = 9$), the characteristic equation is of degree $N_a = 9$ ($\sum_{k=0}^{9} \lambda^k \cdot \Pi_k = 0$) and must often be solved numerically. Expect in some particular cases, such an equation has ($N_a = 9$) distinct roots that can be real or complex. These values are the eigenvalues of the 9 × 9 Jacobian matrix (A). The general rule is that the Steady State (SS) is stable if there is no eigenvalue with positive real part. It is sufficient that one eigenvalue is positive for the steady state to be unstable. Our 9-variables ($Y_1 Y_2 Y_4 I_{L_0} I_{L_1} I_{L_2} I_{C_3} I_{R_1} I_{R_{rfid}}$) helix antenna system has nine eigenvalues. The type of behavior can be characterized as a function of the position of these eigenvalues in the Re/Im plane. Five non-degenerated cases can be distinguished: (1) the nine eigenvalues are real and negative (stable steady state), (2) the nine eigenvalues are real, eight of them are negative (unstable steady state), (3) and (4) two eigenvalues are complex conjugates with a negative real part and the other eigenvalues are real and negative (stable steady state), two cases can be distinguished depending on the relative value of the real part of the complex eigenvalues and of the real one, (5) two eigenvalues are complex conjugates with a negative real part and at least one eigenvalue is positive (unstable steady state).

Exercises

1. We have a system of three N-turn multilayer circular antennas which are integrated with RFID IC for complete RFID transponder. The first (i = 1) and second (i = 2) N-turn multilayer circular antennas are connected in parallel through microstrip lines 1 and 2. They are connected to switch S_1 which can be connected to third N turn multilayer circular antenna (S_1 in position a) or to capacitor C_x and microstrip line 3 (S_1 in position b). An N-turn multilayer circular coil antennas system is influenced by electromagnetic interferences, which affect their stability behavior but we neglect it in our analysis. Additionally, microstrip lines in the system have a parasitic effects, a delay lines Δ_{μ_1}, Δ_{μ_2} and Δ_{μ_3} respectively. We need to inspect our system performance under microstrip lines (1, 2, and 3) parasitic affects.

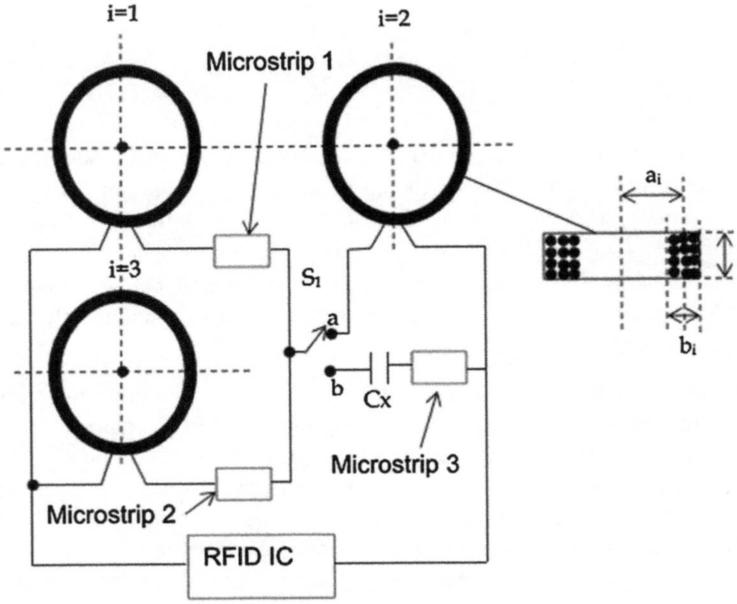

Every N-turn multilayer circular coil antenna has a parasitic DC resistance which need to be calculated. Index (i) indicated the first N-turn multilayer circular coil antenna (i = 1) or second N-turn multilayer circular coil antenna (i = 2) or third N-turn multilayer circular coil antenna (i = 3). We define RFID's

N-turn multilayer coil antenna parameters, a_i—Average radius of the coil in cm, N_i—number of turns, b_i—winding thickness in cm, S_i—wire cross section area, m_i—radius of the wire and h_i—winding height in cm. Integrating all those parameters gives the equation for N-turn multilayer circular coil antenna inductance calculation ($L_{calc-i} = \frac{0.31 \cdot (a_i \cdot N_i)^2}{6 \cdot a_i + 9 \cdot h_i + 10 \cdot b_i}$ [μH]). The length of N turn is $l_{N_i} = 2 \cdot \pi \cdot a_i \cdot N_i$ (Assumption: $a_i \gg b_i; a_i + b_i \approx a_i$). We consider system three N-turn multilayer circuit coil antennas and not identical. $a_k \neq a_l; N_k \neq N_l; h_k \neq h_l; b_k \neq b_l; k \neq l; k = 1, 2, 3; l = 1, 2, 3$

$a_k, N_k, h_k, b_k \in \mathbb{R}_+; a_l, N_l, h_l, b_l \in \mathbb{R}_+$

The DC resistance of the N-turn multilayer circular coil antenna: $R_{DC-i} = \frac{l_{N_i}}{\sigma_i \cdot S_i} = \frac{2 \cdot \pi \cdot a_i \cdot N_i}{\sigma_i \cdot S_i} = \frac{2 \cdot a_i \cdot N_i}{\sigma_i \cdot m_i^2}$. l_{N_i}—Total length of the wire, σ_i—Conductivity of the wire (S/m), S_i—wire cross section area ($\pi \cdot m_i^2$), m_i—Radius of the wire.

1.1 Write system differential equations for cases: S_1 in position (a) and S_1 in position (b).
1.2 Find system fixed points for the cases: S_1 in position (a) and S_1 in position (b).
1.3 Discuss stability and stability switching for the simple case $\Delta_{\mu_1} = 0$, $\Delta_{\mu_2} = 0$, $\Delta_{\mu_3} = 0$ Under variation of circuit parameters (switch S_1 in position (a).
1.4 Discuss stability and stability switching for the cases: $\Delta_\mu \in \mathbb{R}_+$
(1) $\Delta_{\mu_1} = \Delta_\mu; \Delta_{\mu_2} = 0; \Delta_{\mu_3} = 0$ (2) $\Delta_{\mu_1} = 0; \Delta_{\mu_2} = \Delta_\mu; \Delta_{\mu_3} = \Delta_\mu$.
(3) $\Delta_{\mu_1} = \Delta_{\mu_3} = \Delta_\mu; \Delta_{\mu_2} = 0$ (4) $\Delta_{\mu_1} = \Delta_\mu; \Delta_{\mu_2} = \Gamma \cdot \Delta_{\mu_1}; \Delta_{\mu_3} = 0$; $\Gamma \in \mathbb{R}_+$
Under variation of Δ_μ and Γ parameters (switch S_1 in position (a)).
1.5 Return (1.3) and (1.4) for the case, switch S_1 in position (b).
Remark: The delay is on the current that flow through microstrip line $I(t) \to I(t - \Delta_\mu)$, we consider that $V_{\Delta_\mu} \to \varepsilon$(neglect the voltage on microstrip line). Take care in your analysis and calculation the mutual inductances between each two N-turn multilayer circular antennas in our system ($\sum M = M_+ - M_-$).

2. We have a system of three N-turn multilayer circular antennas in series, which are integrated with RFID IC for complete RFID transponder system. Additionally, there is a capacitors bridge (C_{A1}, C_{A2}, C_{A3}, and C_{A4}) which is connected to our RFID transponder system and balance the circuit currents flow. An N-turn multilayer circular coil antennas system is influenced by electromagnetic interferences which affect their stability behavior but we neglect it in our analysis. Additionally, microstrip lines in the system have a parasitic effects, a delay lines Δ_{μ_1}, Δ_{μ_2} and Δ_{μ_3} respectively. We need to inspect our system performance under microstrip lines (1, 2, and 3) parasitic affects. Every N-turn multilayer circular coil antenna has a parasitic DC resistance which needs to be calculated. Index (i) indicated the first N-turn multilayer circular coil antenna (i = 1) or second N-turn multilayer circular coil antenna

(i = 2) or third N-turn multilayer circular coil antenna (i = 3). We define RFID's N-turn multilayer coil antenna parameters, a_i—Average radius of the coil in cm, N_i—number of turns, b_i—Winding thickness in cm, S_i—wire cross section area, m_i—radius of the wire and h_i—winding height in cm. Integrating all those parameters give the equation for N-turn multilayer circular coil antenna inductance calculation ($L_{calc-i} = \frac{0.31 \cdot (a_i \cdot N_i)^2}{6 \cdot a_i + 9 \cdot h_i + 10 \cdot b_i}$ [μH]). The length of N turn is $l_{N_i} = 2 \cdot \pi \cdot a_i \cdot N_i$ (Assumption: $a_i \gg b_i; a_i + b_i \approx a_i$). We consider System three N-turn multilayer circuit coil antennas and not identical

$$a_k \neq a_l; N_k \neq N_l; h_k \neq h_l; b_k \neq b_l; k \neq l; k = 1,2,3; l = 1,2,3$$
$$a_k, N_k, h_k, b_k \in \mathbb{R}_+; a_l, N_l, h_l, b_l \in \mathbb{R}_+$$

The DC resistance of the N-turn multilayer circular coil antenna:

$$R_{DC-i} = \frac{l_{N_i}}{\sigma_i \cdot S_i} = \frac{2 \cdot \pi \cdot a_i \cdot N_i}{\sigma_i \cdot S_i} = \frac{2 \cdot a_i \cdot N_i}{\sigma_i \cdot m_i^2}.$$

l_{N_i}—Total length of the wire, σ_i—Conductivity of the wire (S/m), S_i—wire cross section area ($\pi \cdot m_i^2$), m_i—Radius of the wire.

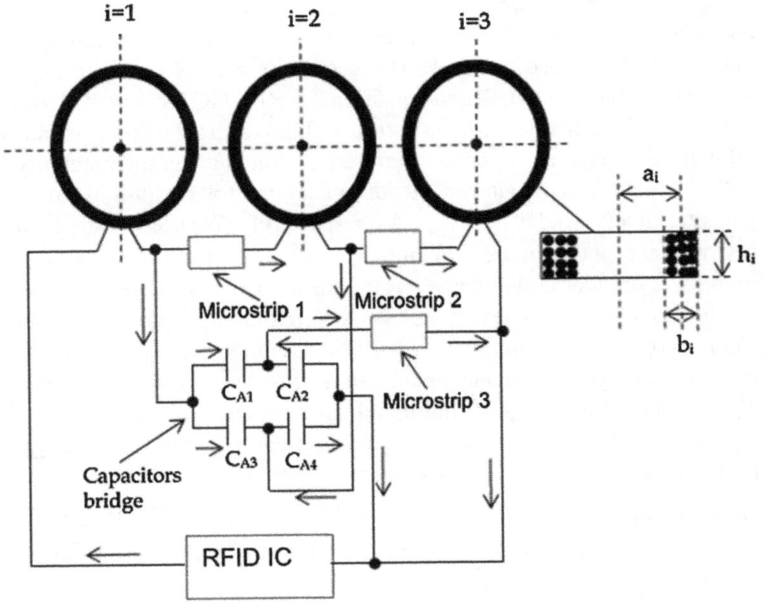

2.1 Write system differential equations and find fixed points.
2.2 Discuss system stability and stability switching for the simple case under circuit parameters variation ($\Delta_{\mu_1} = 0; \Delta_{\mu_2} = 0; \Delta_{\mu_3} = 0$).

2.3 Discuss stability and stability switching for the cases: $\Delta_{\mu} \in \mathbb{R}_{+}$
(1) $\Delta_{\mu_1} = \Delta_{\mu}; \Delta_{\mu_2} = 0; \Delta_{\mu_3} = 0$ (2) $\Delta_{\mu_1} = 0; \Delta_{\mu_2} = \Delta_{\mu}; \Delta_{\mu_3} = \Delta_{\mu}$.
(3) $\Delta_{\mu_1} = \Delta_{\mu_3} = \Delta_{\mu}; \Delta_{\mu_2} = 0$ (4) $\Delta_{\mu_1} = \Delta_{\mu}; \Delta_{\mu_2} = (1 + \sqrt[3]{\Gamma^2}) \cdot \Delta_{\mu_1}$; $\Delta_{\mu_3} = 0; \Gamma \in \mathbb{R}_{+}$
Under variation of Δ_{μ} and Γ parameters.

2.4 Bridge's capacitor C_{A3} is disconnected. How it influences circuit behavior? Find system differential equations, fixed points and discuss stability and stability switching under parameters variations.

2.5 Bridge's capacitor C_{A2} is shortened. How it influences circuit behavior? Find system differential equations, fixed points and discuss stability and stability switching under parameters variations.
Remark: The delay is on the current that flow through microstrip line $I(t) \to I(t - \Delta_{\mu})$, we consider that $V_{\Delta_{\mu}} \to \varepsilon$(neglect the voltage on microstrip line). Take care in your analysis and calculation the mutual inductances between each two N-turn multilayer circular antennas in our system ($\sum M = M_{+} - M_{-}$).

3. We have a system of two N-turn multilayer circular antennas which can be integrated with RFID IC for complete RFID transponder system. The first (i = 1) and second (i = 2) N-turn multilayer circular antennas are connected by two microstrip lines (1 and 2) and matching network. The matching network can be Pi-type or T-type. Switch S_1 has two positions: first position (a), RFID IC is connected to first (i = 1) N-turn multilayer circular antenna through matching network and microstrip line 1. The second (i = 2) N-turn multilayer circular antenna is disconnected. Second position (b), RFID IC is connected to two N-turn multilayer circular antennas. An N-turn multilayer circular coil antennas system is influenced by electromagnetic interferences which affect their stability behavior but we neglect it in our analysis. Additionally, microstrip lines in the system have parasitic effects, a delay lines $\Delta_{\mu_1}, \Delta_{\mu_2}$ respectively. We need to inspect our system performance under microstrip lines (1 and 2) parasitic affects. Every N-turn multilayer circular coil antenna has a parasitic DC resistance which needs to be calculated. Index (i) indicated the first N-turn multilayer circular coil antenna (i = 1) or second N-turn multilayer circular coil antenna (i = 2). We define RFID's N-turn multilayer coil antenna parameters, a_i—Average radius of the coil in cm, N_i—number of turns, b_i—Winding thickness in cm, S_i—wire cross section area, m_i—radius of the wire and h_i—winding height in cm. Integrating all those parameters give the equation for N-turn multilayer circular coil antenna inductance calculation L_{calc-i} ($L_{calc-i} = \frac{0.31 \cdot (a_i \cdot N_i)^2}{6 \cdot a_i + 9 \cdot h_i + 10 \cdot b_i}$ [μH]). The length of N turn is $l_{N_i} = 2 \cdot \pi \cdot a_i \cdot N_i$ (Assumption: $a_i \gg b_i; a_i + b_i \approx a_i$). Two N-turn multilayer circuit coil antennas are not identical.

$$a_1 \neq a_2; N_1 \neq N_2; h_1 \neq h_2; b_1 \neq b_2; i = 1, 2; a_i, N_i, h_i, b_i \in \mathbb{R}_{+}$$

The DC resistance of the N-turn multilayer circular coil antenna:

$$R_{DC-i} = \frac{l_{N_i}}{\sigma_i \cdot S_i} = \frac{2 \cdot \pi \cdot a_i \cdot N_i}{\sigma_i \cdot S_i} = \frac{2 \cdot a_i \cdot N_i}{\sigma_i \cdot m_i^2}.$$

l_{N_i}—Total length of the wire,
σ_i—Conductivity of the wire (S/m), S_i—wire cross section area $(\pi \cdot m_i^2)$,
m_i—Radius of the wire.

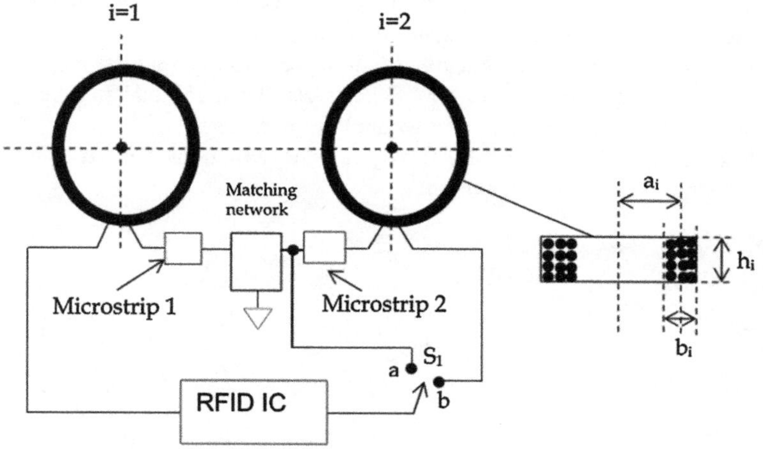

3.1 Write system differential equations for the cases: S_1 in position (a) and S_1 in position (b). Matching network is T-type.
3.2 Find system fixed points for the cases: S_1 in position (a) and S_1 in position (b). Matching network is T-type.
3.3 Discuss stability and stability switching for the simple case $\Delta_{\mu_1} = 0; \Delta_{\mu_2} = 0$ under variation of circuit parameters. Switch S_1 is in (a) position. Matching network is T-type.
3.4 Return (3.1), (3.2) and (3.3) for the case we use Pi-type matching network.
3.5 How the circuit dynamic is changed if switch S_1 is in (b) position? Return (3.3) for the case that switch S_1 is in position (b).
3.6 Return (3.3) for the case $\Delta_{\mu_1} = \Delta_\mu; \Delta_{\mu_2} = \Gamma \cdot \Delta_\mu; \Gamma, \Delta_\mu \in \mathbb{R}_+$. Discuss stability and stability switching for variation of Γ, Δ_μ parameters.
Remark: The delay is on the current that flow through microstrip line $I(t) \to I(t - \Delta_\mu)$, we consider that $V_{\Delta_\mu} \to \varepsilon$(neglect the voltage on microstrip line). Take care in your analysis and calculation the mutual inductances between two N-turn multilayer circular antennas in our system $(\sum M = M_+ - M_-)$.

4. We have a system of double rectangular spiral antennas which are connected in parallel through microstrip lines and discrete components. The antennas are

integrated with RFID IC to complete RFID transponder system. A double rectangular spiral antennas system influences by electromagnetic interferences which effect there stability behavior. Additionally, microstrip lines which are connect antennas in parallel and have parasitic effects, delays in time $\Delta_{\mu_1}, \Delta_{\mu_2}$ respectively. We inspect our system performances under electromagnetic interferences and microstrip lines parasitic effects. Every rectangular spiral antenna has a parasitic DC resistance which needs to be calculated. Index (i) indicates first rectangular spiral antenna (i = 1) or second rectangular spiral antenna (i = 2).
We define RFID's coil dimensional parameters. A_{0i}, B_{0i}—Overal dimensions of the coil. $Aavg_i$, $Bavg_i$—Average dimensions of the coil.
t_i—track thickness. w_i—track width. g_i—gaps between tracks. Nc_i—number of turns. d_i—equivalent diameter of the track. Average coil area $Ac_i = Aavg_i \cdot Bavg_i$. P_i—coil manufacturing technology parameter.
Integrating all those parameters gives the equations for rectangular spiral antenna inductance calculation:

$$L_{calc-i} = \frac{\mu_0}{\pi} \cdot \left(\sum_{k=1,k\neq 3}^{4} X_{ki} - X_3 \right) \cdot N_{ci}^{P_i};$$

$$X_{1i} = A_{avgi} \cdot \ln\left(\frac{2 \cdot A_{avgi} \cdot B_{avgi}}{d_i \cdot (A_{avgi} + \sqrt{A_{avgi}^2 + B_{avgi}^2})}\right)$$

$$X_{2i} = B_{avgi} \cdot \ln\left(\frac{2 \cdot A_{avgi} \cdot B_{avgi}}{d_i \cdot (B_{avgi} + \sqrt{A_{avgi}^2 + B_{avgi}^2})}\right);$$

$$X_{4i} = \frac{(A_{avgi} + B_{avgi})}{4}; d_i = \frac{2 \cdot (t_i + w_i)}{\pi}; A_{avgi} = A_{0i} - N_{ci} \cdot (g_i + w_i)$$
$$B_{avgi} = B_{0i} - N_{ci} \cdot (g_i + w_i)$$

The rectangular spiral antenna length is calculated as follows: l_{0i} is the length of the first turn $l_{0i} = 2 \cdot (A_{0i} + B_{0i}) - (w_i + g_i)$. l_k is the length of turn k + 1. We define the following:

$$l_{Ti} = l_{0i} + \sum_{k=1}^{N_{ci}-1} \{A_{0i} - [1 + (k-1) \cdot 2] \cdot (w_i + g_i) + B_{0i} - [2 + (k-1) \cdot 2] \cdot (w_i + g_i)$$
$$+ A_{0i} - [2 + (k-1) \cdot 2] \cdot (w_i + g_i) + B_{0i} - [3 + (k-1) \cdot 2] \cdot (w_i + g_i)\}$$

$$L_{Ti} = L_{0i} + 2 \cdot (A_{0i} + B_{0i}) \cdot (N_{Ci} - 1) - 8 \cdot (w_i + g_i) \cdot \sum_{k=1}^{N_C - 1} k; \sum_{k=1}^{N_C - 1} k = N_{Ci} - 1$$

$$L_{Ti} = 2 \cdot \{(A_{0i} + B_{0i}) \cdot (1 + N_{Ci}) - (w_i + g_i) \cdot [4 \cdot N_{Ci} - 3]\}$$

The DC resistance of rectangular spiral RFID antenna:

$$R_{DC-i} = \frac{l_{Ti}}{\sigma_i \cdot S_i} = \frac{l_{Ti}}{\sigma_i \cdot \pi \cdot a_i^2}.$$

l_{Ti}—total length of the wire. σ_i—conductivity of the wire (mΩ/m). S_i—Cross section area $\pi \cdot a_i^2$. a_i—radius of the wire.
Remark: $a_i^2 = w_i^2$.

$$R_{DC-i} = \frac{2 \cdot \{(A_{0i} + B_{0i}) \cdot (1 + N_{Ci}) - (w_i + g_i) \cdot [4 \cdot N_{Ci} - 3]\}}{\sigma_i \cdot \pi \cdot w_i^2}$$

Due to electromagnetic interferences there are different in time delays respect to first and second rectangular spiral antennas voltages and voltages derivatives. The delayed voltages are $V_1(t - \tau_1)$ and $V_2(t - \tau_2)$ respectively ($\tau_1 \neq \tau_2$) and delayed voltages derivatives are $dV_1(t - \Delta_1)/dt$, $dV_2(t - \Delta_2)/dt$ respectively $(\Delta_1 \neq \Delta_2; \tau_1 \geq 0); (\tau_2 \geq 0; \Delta_1, \Delta_2 \geq 0)$.

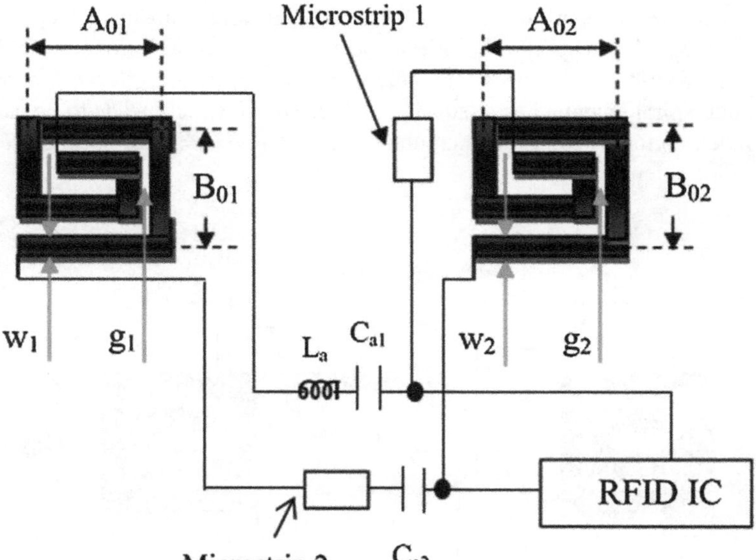

4.1 Write system differential equations and find fixed points for the following cases: $\tau, \Delta, \Delta_\mu \in \mathbb{R}_+$

(1) $\tau_1 = \tau; \tau_2 = 0; \Delta_1 = \Delta; \Delta_2 = 0; \Delta_{\mu_1} = \Delta_{\mu_2} = 0.$
(2) $\tau_1 = \tau_2 = \tau; \Delta_1 = \Delta_2 = 0; \Delta_{\mu_1} = \Delta_{\mu_2} = \Delta_\mu.$
(3) $\tau_1 = \tau_2 = 0; \Delta_1 = \Delta_2 = \Delta; \Delta_{\mu_1} = \Delta_\mu; \Delta_{\mu_2} = 0.$
(4) $\tau_1 = \tau_2 = \tau; \Delta_1 = \Delta_2 = \Delta; \Delta_{\mu_1} = 0; \Delta_{\mu_2} = \Delta_\mu.$

4.2 Discuss system stability and stability switching under variation of parameters $\tau, \Delta, \Delta_{\mu}$ for all cases in (4.1).
4.3 We short inductor L_a, How it influences our system dynamics and stability. Discuss stability and stability switching for different values of C_{a1} and C_{a2}.
4.4 We short capacitor C_{a1}, How it influences our system dynamic and stability? Discuss stability and stability switching for different values of L_a.
4.5 We short capacitor C_{a2}, How it influences our system dynamic and stability? Discuss stability and stability switching for different values of C_{a1}.
Remark: The delay is on the current that flow through microstrip line $I(t) \to I(t - \Delta_{\mu})$, we consider that $V_{\Delta_{\mu}} \to \varepsilon$ (neglect the voltage on microstrip line). Take care in your analysis and calculation the mutual inductances between double rectangular spiral antennas in our system $(\sum M = M_{+} - M_{-})$.

5. We have a system of double rectangular spiral antennas which are connected through switch S_1 to RFID IC. The antennas are integrated with RFID IC and discrete components $(C_{a1}, C_{a2}, L_a, L_b)$ to complete RFID transponder system. A double rectangular spiral antennas system influences by electromagnetic interferences which effect there stability behavior. Additionally, microstrip lines which are connect antennas to RFID IC and have parasitic effects, delays in time $\Delta_{\mu_1}, \Delta_{\mu_2}$ respectively. We inspect our system performances under electromagnetic interferences and microstrip lines parasitic effects. Every rectangular spiral antenna has a parasitic DC resistance which needs to be calculated. Index (i) indicates first rectangular spiral antenna (i = 1) or second rectangular spiral antenna (i = 2).

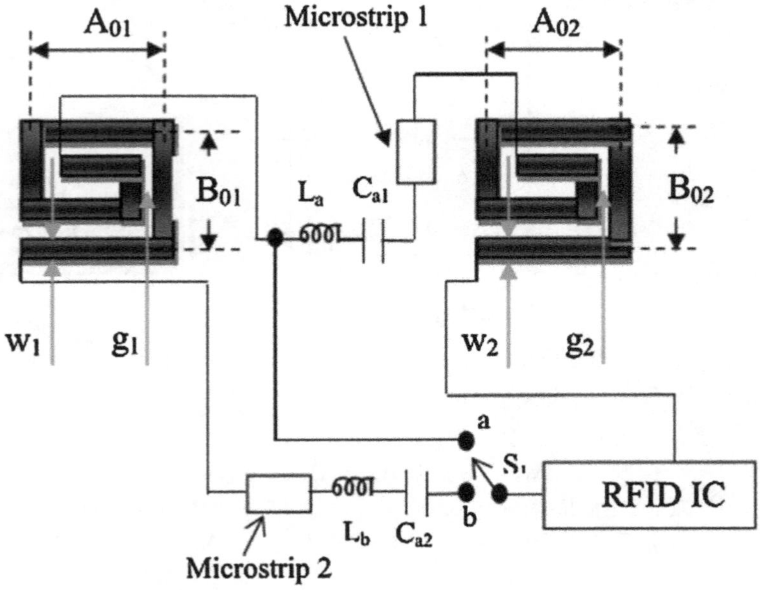

We define RFID's coil dimensional parameters. A_{0i}, B_{0i}—Overal dimensions of the coil. $Aavg_i$, $Bavg_i$—Average dimensions of the coil.
t_i—track thickness. w_i—track width. g_i—gaps between tracks. Nc_i—number of turns. d_i—equivalent diameter of the track. Average coil area $Ac_i = Aavg_i \cdot Bavg_i$. P_i—coil manufacturing technology parameter.
Integrating all those parameters gives the equations for rectangular Spiral antenna inductance calculation:

$$L_{calc-i} = \frac{\mu_0}{\pi} \cdot \left(\sum_{k=1, k \neq 3}^{4} X_{ki} - X_3 \right) \cdot N_{ci}^{P_i};$$

$$X_{2i} = B_{avgi} \cdot \ln\left(\frac{2 \cdot A_{avgi} \cdot B_{avgi}}{d_i \cdot (B_{avgi} + \sqrt{A_{avgi}^2 + B_{avgi}^2})}\right);$$

$$X_{4i} = \frac{(A_{avgi} + B_{avgi})}{4}; d_i = \frac{2 \cdot (t_i + w_i)}{\pi}; A_{avgi} = A_{0i} - N_{ci} \cdot (g_i + w_i)$$

$$B_{avgi} = B_{0i} - N_{ci} \cdot (g_i + w_i)$$

The rectangular spiral antenna length is calculated as follows: l_{0i} is the length of the first turn $l_{0i} = 2 \cdot (A_{0i} + B_{0i}) - (w_i + g_i)$. l_k is the length of turn k + 1. We define the following:

$$l_{Ti} = l_{0i} + \sum_{k=1}^{N_{ci}-1} \{A_{0i} - [1 + (k-1) \cdot 2] \cdot (w_i + g_i) + B_{0i} - [2 + (k-1) \cdot 2] \cdot (w_i + g_i)$$
$$+ A_{0i} - [2 + (k-1) \cdot 2] \cdot (w_i + g_i) + B_{0i} - [3 + (k-1) \cdot 2] \cdot (w_i + g_i)\}$$

$$L_{Ti} = L_{0i} + 2 \cdot (A_{0i} + B_{0i}) \cdot (N_{Ci} - 1) - 8 \cdot (w_i + g_i) \cdot \sum_{k=1}^{N_C - 1} k; \sum_{k=1}^{N_C - 1} k = N_{Ci} - 1$$

$$L_{Ti} = 2 \cdot \{(A_{0i} + B_{0i}) \cdot (1 + N_{Ci}) - (w_i + g_i) \cdot [4 \cdot N_{Ci} - 3]\}$$

The DC resistance of rectangular spiral RFID antenna:

$$R_{DC-i} = \frac{l_{Ti}}{\sigma_i \cdot S_i} = \frac{l_{Ti}}{\sigma_i \cdot \pi \cdot a_i^2}.$$

l_{Ti}—total length of the wire. σ_i—conductivity of the wire (mΩ/m). S_i—Cross section area $\pi \cdot a_i^2$. a_i—radius of the wire.
Remark: $a_i^2 = w_i^2$.

$$R_{DC-i} = \frac{2 \cdot \{(A_{0i} + B_{0i}) \cdot (1 + N_{Ci}) - (w_i + g_i) \cdot [4 \cdot N_{Ci} - 3]\}}{\sigma_i \cdot \pi \cdot w_i^2}$$

Due to electromagnetic interferences there are different in time delays respect to first and second rectangular spiral antennas voltages and voltages derivatives. The delayed voltages are $V_1(t - \tau_1)$ and $V_2(t - \tau_2)$ respectively ($\tau_1 \neq \tau_2$) and delayed voltages derivatives are $dV_1(t - \Delta_1)/dt$, $dV_2(t - \Delta_2)/dt$ respectively $(\Delta_1 \neq \Delta_2; \tau_1 \geq 0);(\tau_2 \geq 0; \Delta_1, \Delta_2 \geq 0)$.

5.1 Write system differential equations and find fixed points for the following cases: S_1 in position (a), S_1 in position (b) $\tau, \Delta, \Delta_\mu \in \mathbb{R}_+$ and for below subcases.

(1) $\tau_1 = \tau; \tau_2 = 0; \Delta_1 = \Delta; \Delta_2 = 0; \Delta_{\mu_1} = \Delta_{\mu_2} = 0.$
(2) $\tau_1 = \tau_2 = \tau; \Delta_1 = \Delta_2 = 0; \Delta_{\mu_1} = \Delta_{\mu_2} = \Delta_\mu.$
(3) $\tau_1 = \tau_2 = 0; \Delta_1 = \Delta_2 = \Delta; \Delta_{\mu_1} = \Delta_\mu; \Delta_{\mu_2} = 0.$
(4) $\tau_1 = \tau_2 = \tau; \Delta_1 = \Delta_2 = \Delta; \Delta_{\mu_1} = 0; \Delta_{\mu_2} = \Delta_\mu.$

5.2 Discuss system stability and stability switching under variation of parameters τ, Δ, Δ_μ for all cases in (5.1).

5.3 We short inductor L_a, How it influences our system dynamics and stability. Discuss stability and stability switching for different values of C_{a1} and C_{a2}.

5.4 We short capacitor C_{a1}, How it influences our system dynamic and stability? Discuss stability and stability switching for different values of L_a.

5.5 We short inductor L_b, How it influences our system dynamic and stability? Discuss stability and stability switching for different values of C_{a1}.

Remark: The delay is on the current that flow through microstrip line $I(t) \to I(t - \Delta_\mu)$, we consider that $V_{\Delta_\mu} \to \varepsilon$ (neglect the voltage on microstrip line). Take care in your analysis and calculation the mutual inductances between double rectangular spiral antennas in our system $(\sum M = M_+ - M_-)$.

6. We have system of almost two turn square planar straight thin film inductors antenna (seven segments). The system is constructed from seven straight thin film inductors which are connected in almost two turn square structure. The straight thin film inductors are connected by microstrip lines (A, B, C… F). The almost two turns square planar straight thin film inductors antenna system is connected to transceiver module (represent as a transceiver mixer output equivalent circuit) through two microstrip lines (G and H). Index (i) stands for straight thin film inductor in place (i). w_i is the width of straight thin film inductor (i) in cm, z_i is the thickness of straight thin film inductor (i) in cm, and l_i is the length of straight thin film conductor (inductor) in cm. The calculated inductance of straight thin film inductor (i) is as follow (L_i is the segment inductance in μH):

$$L_i = 0.002 \cdot l_i \cdot \{\ln[\frac{2 \cdot l_i}{w_i + z_i}] + 0.50049 + \frac{w_i + z_i}{3 \cdot l_i}\}[\mu\text{H}]; \quad i = 1, 2, 3, 4, \ldots, 7$$

Remark: we assume that the magnetic permeability of the conductor material is 1 and the seven straight thin film inductors are not identical

$$w_i \neq w_j; l_i \neq l_j \, z_i \neq z_j \quad (j \neq i; j = 1, 2, 3, 4, \ldots, 7; i = 1, 2, 3, 4, \ldots, 7).$$

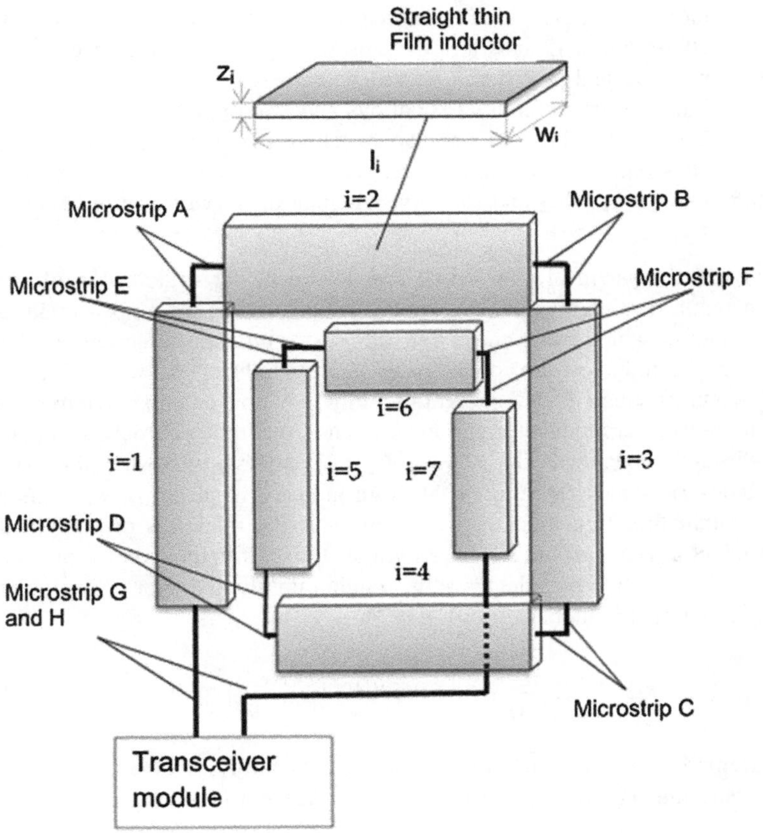

Consider in your analysis the DC resistivity ρ_i of the straight thin film strip material and the strip cross section $A_i = w_i \cdot z_i; R_i = \frac{\rho_i \cdot l_i}{w_i \cdot z_i}$. The dimension of the straight thin film strip that affects the inductance most strongly is the length l_i. The width w_i has much weaker influence, and straight thin film strip thickness can be neglected completely for the limit ($z_i \ll w_i$). This is in contrast to the resistance $R_i = \frac{\rho_i \cdot l_i}{w_i \cdot z_i}$, which is inversely proportional to the straight thin film strip cross section $w_i \cdot z_i$ and depends on the material properties via its resistivity ρ_i. The length of each straight thin film strip in our system is not the same to each other. Microstrip line in our system is represented as a delay line and the delay is on the current that flows through the microstrip line $(\tau_1, \tau_2, \ldots)$ respectively $(V_{\tau_i} \to \varepsilon; i = 1, 2, \ldots; A \to 1, B \to 2, \ldots)$.

6.1 Find the expression of system total inductance (L_T) which constructed from the sum of the self-inductances of all straight segments and the sum of all mutual inductances ($\sum M$), both negative and positive).
6.2 We short straight segment number 5, find the expression of system total inductance (L_T) which constructed from the sum of the self-inductances of all straight segments and the sum of all mutual inductances ($\sum M$), both negative and positive).
6.3 Find system differential equations and fixed points for (6.1) and (6.2).
6.4 Discuss stability and stability switching under variation of system parameters (simple case: no delays $\tau_1 = \tau_2 = \cdots = 0$).
6.5 Discuss stability and stability switching under variation of delay parameter τ $(\tau_1 = \tau_2 = \cdots = \tau)$.

7. We have system of almost two turn square planar straight thin film inductors antenna (six segments). The system is constructed from six straight thin film inductors which are connected in almost two turn square structure. The straight thin film inductors are connected by microstrip lines (A, B, C… E). The almost two turns square planar straight thin film inductors antenna system is connected to transceiver module (represent as a transceiver mixer output equivalent circuit) through capacitor bridge (C_a, C_b, C_c, C_d) and two microstrip lines (F and G). Index (i) stands for straight thin film inductor in place (i). w_i is the width of straight thin film inductor (i) in cm, z_i is the thickness of straight thin film inductor (i) in cm, and l_i is the length of straight thin film conductor (inductor) in cm. The calculated inductance of straight thin film inductor (i) is as follow (L_i is the segment inductance in μH):

$$L_i = 0.002 \cdot l_i \cdot \{\ln[\frac{2 \cdot l_i}{w_i + z_i}] + 0.50049 + \frac{w_i + z_i}{3 \cdot l_i}\} \,[\mu\text{H}]; \quad i = 1, 2, 3, 4, \ldots, 7$$

Remark: we assume that the magnetic permeability of the conductor material is 1 and the six straight thin film inductors are not identical

$$w_i \neq w_j; l_i \neq l_j \, z_i \neq z_j \quad (j \neq i; j = 1, 2, 3, 4, \ldots, 6; i = 1, 2, 3, 4, \ldots, 6).$$

Consider in your analysis the DC resistivity ρ_i of the straight thin film strip material and the strip cross section $A_i = w_i \cdot z_i; R_i = \frac{\rho_i \cdot l_i}{w_i \cdot z_i}$. The dimension of the straight thin film strip that affects the inductance most strongly is the length l_i. The width w_i has much weaker influence, and straight thin film strip thickness can be neglected completely for the limit $(z_i \ll w_i)$. This is in contrast to the resistance

$R_i = \frac{\rho_i \cdot l_i}{w_i \cdot z_i}$, which is inversely proportional to the straight thin film strip cross section $w_i \cdot z_i$ and depends on the material properties via its resistivity ρ_i. The length of each straight thin film strip in our system is not the same to each other. Microstrip line in our system is represented as a delay line and the delay is on the current that flows through the microstrip line $(\tau_1, \tau_2, \ldots)$ respectively $(V_{\tau_i} \to \varepsilon; i = 1, 2, \ldots; A \to 1, B \to 2, \ldots)$.

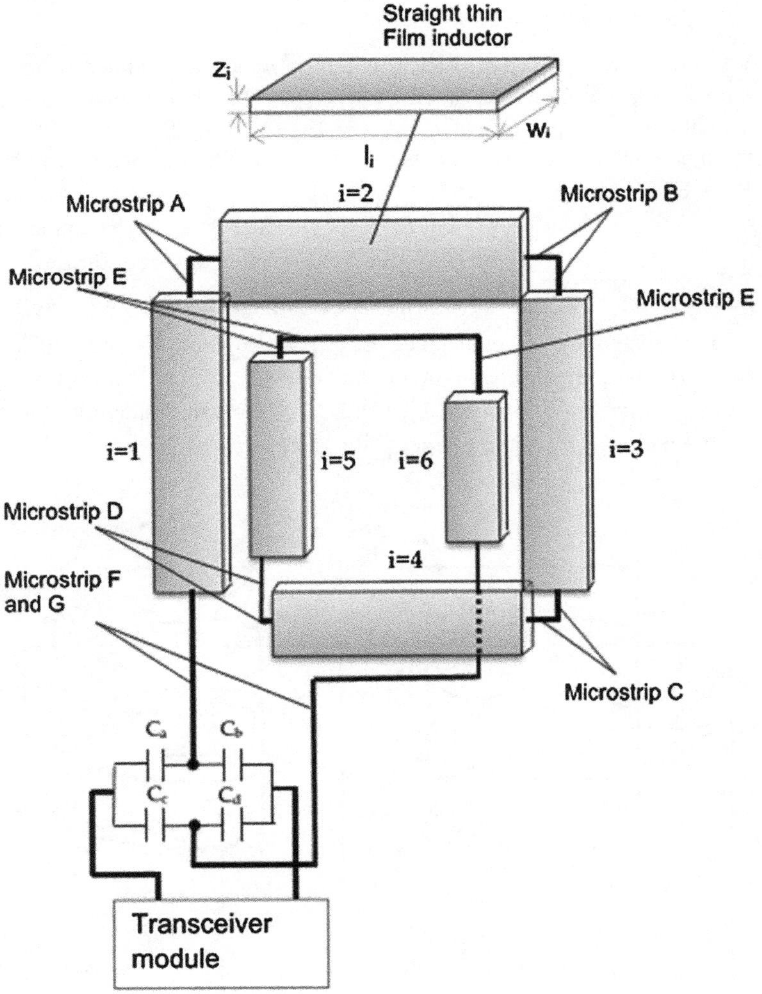

7.1 Find the expression of system total inductance (L_T) which constructed from the sum of the self-inductances of all straight segments and the sum of all mutual inductances ($\sum M$), both negative and positive).

7.2 We short straight segment number 5, find the expression of system total inductance (L_T) which constructed from the sum of the self-inductances of all straight segments and the sum of all mutual inductances ($\sum M$), both negative and positive).
7.3 Find system differential equations and fixed points for (7.1) and (7.2).
7.4 Discuss stability and stability switching under variation of system parameters (simple case: no delays $\tau_1 = \tau_2 = \cdots = 0$).
7.5 Discuss stability and stability switching under variation of delay parameter τ ($\tau_1 = \tau_2 = \cdots = \tau$).

8. We have a system of two helical antennas which are connected to RFID IC through capacitors network (C_a, C_{a1}, C_{a2}). The dimensional parameters of helical antennas are $h_i, S_i, a_i, R_i, D_i; i = 1, 2$ ($h_i, S_i, a_i, R_i, D_i \in \mathbb{R}_+$). We define the mathematical relationships between helical antennas parameters as follow: $R_2 = \Gamma_1 \cdot R_1; S_2 = S_1 \cdot \Gamma_2 + S_1 \cdot \Gamma_2^2$ $h_2 = \Gamma_3 \cdot \sqrt{h_1}; \frac{a_1}{a_2} = \sqrt{\Gamma_1 + \Gamma_2}; \Gamma_1, \Gamma_2, \Gamma_3 \in \mathbb{R}_+$. The two helical antennas are not identical. The length of helix antenna on turns, l_i where $l_i = \sqrt{(\pi \cdot D_i)^2 + S_i^2}; i = 1, 2$. The induced magnetic field of a single turn can be represented by an inductance $L_{seg(i)}$, which includes the self-inductance $L_{self(i)}$ of one turn and the mutual inductance $M_{seg(i)}$ coupled from two adjacent turns $L_{seg(i)} = L_{self(i)} + 2 \cdot M_{seg(i)}$. The $L_{self(i)}$ expression integrates the effects of pitch angel (α_i), μ_0 is the permeability of free space.

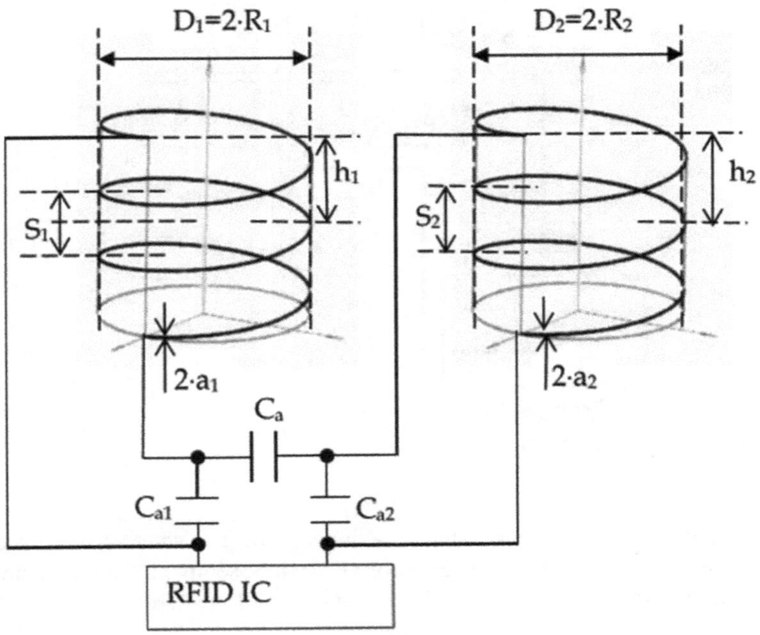

$$L_{self(i)} = \mu_0 \cdot R_i \cdot [\ln(\frac{8R_i}{a_i}) - 2] \cdot \cos(\alpha_i); M_{seg(i)} = \frac{\pi \cdot \mu_0 \cdot R_i^4}{\sqrt{2} \cdot (R_i^2 + S_i^2)^{\frac{3}{2}}}; i = 1, 2$$

$$L_{seg(i)} = \mu_0 \cdot R_i \cdot [\ln(\frac{8R_i}{a_i}) - 2] \cdot \cos(\alpha_i) + \frac{\pi \cdot \mu_0 \cdot R_i^4 \cdot \sqrt{2}}{(R_i^2 + S_i^2)^{\frac{3}{2}}}; i = 1, 2$$

The normal mode helical antenna (NMHA) equivalent circuit is divided to two parts: one modeling the equivalent wire antenna (five elements circuit) and the other modeling the inductive loops (three elements circuit). In your system analysis use the helical antenna equivalent circuit with eight frequency independent elements.

8.1 Find the ratio $\frac{L_{self(1)}}{L_{self(2)}}$ as a function of parameters Γ_1, Γ_2 and Γ_3. Draw 3D graph, Z-axis ($\frac{L_{self(1)}}{L_{self(2)}}$), Y-axis ($\Gamma_1$) and X-axis ($\Gamma_2$) for the case $\Gamma_2 = \Gamma_3$.

8.2 Find the ratio $\frac{L_{self(2)}}{L_{self(1)}}$ as a function of parameters Γ_1, Γ_2 and Γ_3. Draw 3D graph, Z-axis ($\frac{L_{self(2)}}{L_{self(1)}}$), Y-axis ($\Gamma_2$) and X-axis ($\Gamma_3$) for the case $\Gamma_1 = \Gamma_2$.

8.3 Write system differential equations and find fixed points.

8.4 Discuss stability and stability switching under variation of parameters Γ_1, Γ_2 and Γ_3 ($\Gamma_1, \Gamma_2, \Gamma_3 \in \mathbb{R}_+$).

8.5 We have a simple case $\Gamma = \Gamma_1 = \Gamma_2 = \Gamma_3$, discuss stability and stability switching under variation of Γ parameter $\Gamma \in \mathbb{R}_+$.

Remark: Take care in your analysis and calculation the mutual inductances between helical antennas in our system ($\sum M = M_+ - M_-$).

9. We have a system of two helical antennas which are connected to RFID IC through capacitors and inductor network (L_a, C_{a1}, C_{a2}) and selective switch S_1 (positions a, b). The dimensional parameters of helical antennas are $h_i, S_i, a_i, R_i, D_i; i = 1, 2$ ($h_i, S_i, a_i, R_i, D_i \in \mathbb{R}_+$). We define the mathematical relationships between helical antennas parameters as follow: $R_2 = \sqrt{\Gamma_1} \cdot R_1; S_2 = S_1 \cdot \sqrt{\Gamma_2} + S_1 \cdot \Gamma_2^2; h_2 = \Gamma_3 \cdot \sqrt{h_1}$
$\frac{a_1}{a_2} = \sqrt{\Gamma_1 + \sqrt{\Gamma_2}}; \Gamma_1, \Gamma_2, \Gamma_3 \in \mathbb{R}_+$. The two helical antennas are not identical.
The length of helix antenna on turns, l_i where $l_i = \sqrt{(\pi \cdot D_i)^2 + S_i^2}; i = 1, 2$. The induced magnetic field of a single turn can be represented by an inductance $L_{seg(i)}$, which includes the self-inductance $L_{self(i)}$ of one turn and the mutual inductance $M_{seg(i)}$ coupled from two adjacent turns $L_{seg(i)} = L_{self(i)} + 2 \cdot M_{seg(i)}$. The $L_{self(i)}$ expression integrates the effects of pitch angel (α_i), μ_0 is the permeability of free space.

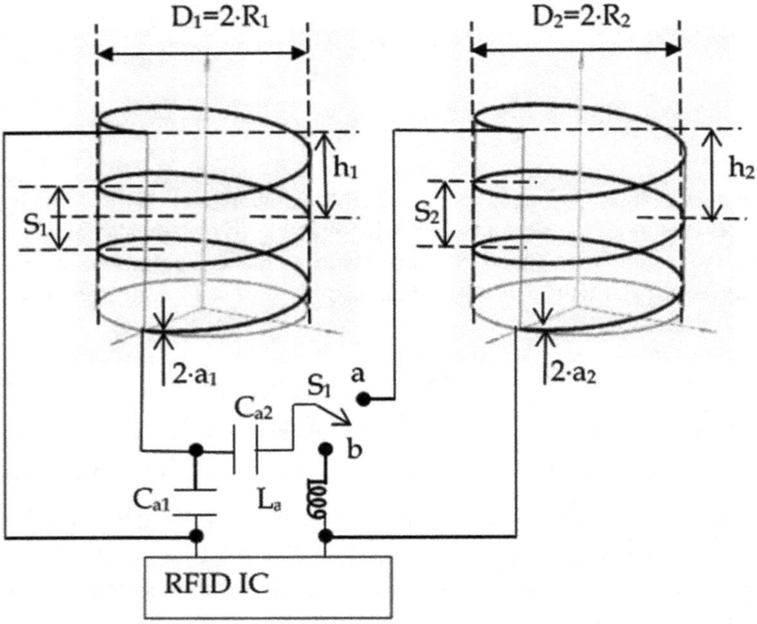

$$L_{self(i)} = \mu_0 \cdot R_i \cdot [\ln(\frac{8R_i}{a_i}) - 2] \cdot \cos(\alpha_i); M_{seg(i)} = \frac{\pi \cdot \mu_0 \cdot R_i^4}{\sqrt{2} \cdot (R_i^2 + S_i^2)^{\frac{3}{2}}}; i = 1,2$$

$$L_{seg(i)} = \mu_0 \cdot R_i \cdot [\ln(\frac{8R_i}{a_i}) - 2] \cdot \cos(\alpha_i) + \frac{\pi \cdot \mu_0 \cdot R_i^4 \cdot \sqrt{2}}{(R_i^2 + S_i^2)^{\frac{3}{2}}}; i = 1,2$$

The normal mode helical antenna (NMHA) equivalent circuit is divided to two parts: one modeling the equivalent wire antenna (five elements circuit) and the other modeling the inductive loops (three elements circuit). In your system analysis use the helical antenna equivalent circuit with eight frequency independent elements.

9.1 Write system differential equations and find fixed points, S_1 in a position.
9.2 Write system differential equations and find fixed points, S_1 in b position.
9.3 Discuss stability and stability switching under variation of parameters Γ_1, Γ_2 and Γ_3 ($\Gamma_1, \Gamma_2, \Gamma_3 \in \mathbb{R}_+$).
9.4 We disconnected capacitor C_{a1}, how it influences our system behavior? Discuss stability and stability switching under variation of Γ parameter ($\Gamma_1 = \Gamma, \Gamma_2 = \Gamma^2, \Gamma_3 = \sqrt{\Gamma} \in \mathbb{R}_+$).
9.5 We short capacitor C_{a2}, how it influences our system behavior? Discuss stability and stability switching for different values of Γ parameter

$(\Gamma_1 = \Gamma + 1, \Gamma_2 = \sqrt[3]{\Gamma^2}, \Gamma_3 = \sqrt{\Gamma} \in \mathbb{R}_+)$.
Remark: Take care in your analysis and calculation the mutual inductances between helical antennas in our system ($\sum M = M_+ - M_-$).

10. We have a system of two helical antennas which are connected to two RFID ICs through capacitors and inductor network ($L_a, C_{a1}, C_{a2}, C_{a3}$) and selective switch S_1 (positions a, b). The dimensional parameters of helical antennas are $h_i, S_i, a_i, R_i, D_i; i = 1, 2$ $(h_i, S_i, a_i, R_i, D_i \in \mathbb{R}_+)$. We define the mathematical relationships between helical antennas parameters as follow: $R_2 = \sqrt[3]{\sqrt{\Gamma_1}} \cdot R_1; S_2 = S_1 \cdot \sqrt{\Gamma_2} + S_1 \cdot \Gamma_2^2; h_2 = \Gamma_3 \cdot h_1$ $\frac{a_1}{a_2} = \sqrt{\Gamma_3 + \sqrt{\Gamma_2}}; \Gamma_1, \Gamma_2, \Gamma_3 \in \mathbb{R}_+$. The two helical antennas are not identical. The length of helix antenna on turns, l_i where $l_i = \sqrt{(\pi \cdot D_i)^2 + S_i^2}; i = 1, 2$. The induced magnetic field of a single turn can be represented by an inductance $L_{seg(i)}$, which includes the self-inductance $L_{self(i)}$ of one turn and the mutual inductance $M_{seg(i)}$ coupled from two adjacent turns $L_{seg(i)} = L_{self(i)} + 2 \cdot M_{seg(i)}$. The $L_{self(i)}$ expression integrates the effects of pitch angel (α_i), μ_0 is the permeability of free space.

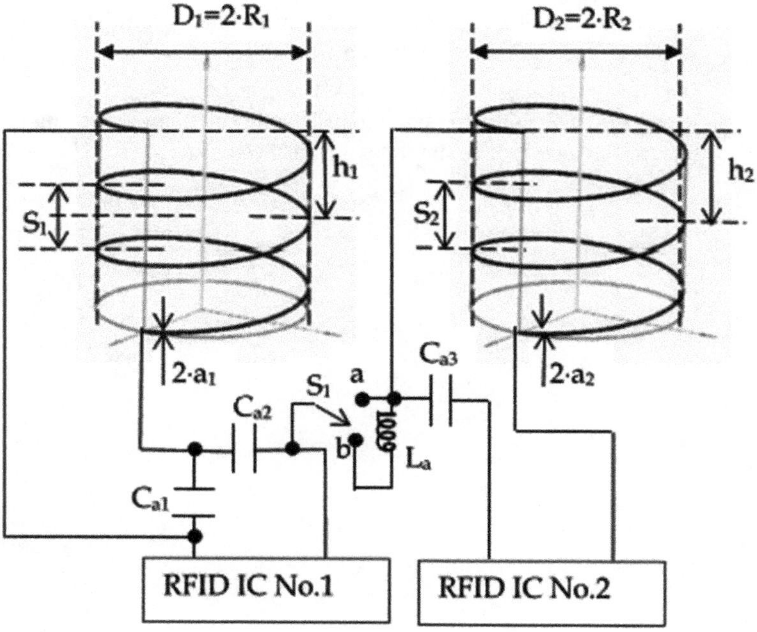

$$L_{self(i)} = \mu_0 \cdot R_i \cdot [\ln(\frac{8R_i}{a_i}) - 2] \cdot \cos(\alpha_i); M_{seg(i)} = \frac{\pi \cdot \mu_0 \cdot R_i^4}{\sqrt{2} \cdot (R_i^2 + S_i^2)^{\frac{3}{2}}}; i = 1,2$$

$$L_{seg(i)} = \mu_0 \cdot R_i \cdot [\ln(\frac{8R_i}{a_i}) - 2] \cdot \cos(\alpha_i) + \frac{\pi \cdot \mu_0 \cdot R_i^4 \cdot \sqrt{2}}{(R_i^2 + S_i^2)^{\frac{3}{2}}}; i = 1,2$$

The normal mode helical antenna (NMHA) equivalent circuit is divided to two parts: one modeling the equivalent wire antenna (five elements circuit) and the other modeling the inductive loops (three elements circuit). In your system analysis use the helical antenna equivalent circuit with eight frequency independent elements.

10.1 Write system differential equations and find fixed points, S_1 in a position.
10.2 Write system differential equations and find fixed points, S_1 in b position.
10.3 Discuss stability and stability switching under variation of parameters Γ_1, Γ_2 and Γ_3 ($\Gamma_1, \Gamma_2, \Gamma_3 \in \mathbb{R}_+$).
10.4 We disconnected capacitor C_{a3}, how it influences our system behavior? Discuss stability and stability switching under variation of Γ parameter ($\Gamma_1 = \Gamma^2, \Gamma_2 = \sqrt[3]{\Gamma^2}, \Gamma_3 = \sqrt{\Gamma} \in \mathbb{R}_+$).
10.5 We short capacitor C_{a2}, how it influences our system behavior? Discuss stability and stability switching for different values of Γ parameter ($\Gamma_1 = \sqrt{\Gamma + 1}, \Gamma_2 = \sqrt[3]{\Gamma^2}, \Gamma_3 = \sqrt{\Gamma} \in \mathbb{R}_+$).
Remark: Take care in your analysis and calculation the mutual inductances between helical antennas in our system ($\sum M = M_+ - M_-$).

Chapter 9
Microwave RF Antennas and Circuits Bifurcation Behavior, Investigation, Comparison and Conclusion

Microwave RF antennas are an integral part of every RF or microwave system. An antenna is an electrical device which converts electric power into radio waves, and vice versa. In many wireless applications antennas are required by radio receiver or transmitter to couple its electrical connection to the electromagnetic field. When we inspect system stability which includes radio waves, we inspect electromagnetic waves which carry signals through the space (or air) at the speed of light with almost no transmission loss. There are mainly two category antennas, the first is omnidirectional antenna which receives and/or radiate in all directions. The second is directional antenna which radiates in a particular direction or pattern. Antennas are characterized by a number of parameters, radiation pattern and the resulting gain. Antenna's gain is dependent on its power in the horizontal directions, and antenna's power gain takes into account the antenna's efficiency (figure of merit). The physical size of an antenna is a practical issue, particularly at lower frequencies. Resonant antennas mainly use a linear conductor or pair of such elements. When we implement RF or microwave antennas in higher frequency system (UHF, microwave), there is no essential need for a smaller physical size. Another important antenna parameter is the frequency range or bandwidth over which an antenna functions. The antenna bandwidth can be wide or narrow like in resonant antennas. In every RF system which includes antenna, we need to choose the suitable matching network between the transceiver and the antenna. Matching network is the practical circuit which is corresponding to maximize the power transfer or minimize signal reflection from the load RF antenna to the transceiver unit. In wireless application we can differentiate RF and microwave antennas. Radio spectrum antennas cover radio waves, microwaves and terahertz radiations. Optical spectrum covers infrared, visible, UV, X-rays and gamma radiations. Radio waves antennas range from 3 kHz to 300 GHz. Hence RF starts from much lower than the microwave starting range. Microwave antennas are mainly for EM waves above 1 GHz in frequency. RF and microwave antenna ranges are different in operation range and applications are concerned. Microwave range starts from 300 MHz to 300 GHz and most microwave applications range up to 100 GHz.

O. Aluf, *Microwave RF Antennas and Circuits*,
DOI 10.1007/978-3-319-45427-6_9

We can characterize microwave antenna as a high antenna gain and directivity, large bandwidth, travel by line of sight, microwaves penetrate ionosphere with less attenuation and less distortion, and in applications of 1–10 GHz range microwave noise level is very low and hence very low signal can be easily detected at receiver. The mainly application for RF antennas are mobile, AM/FM radio, television and the mainly applications for microwave antennas are radar, satellite and space communication. Antennas are used over much broader frequency ranges and are achieved using further techniques. The adjustment of an antenna matching network can allow for any antenna to be matched at any frequency. The loop antennas have a very narrow bandwidth and are tuned using a parallel capacitance which is adjusted according to the receiver tuning. The complex impedance of antennas is related to the electrical length of the antenna at the wavelength in use. The impedance of an antenna can be matched to the feed line and radio by adjusting the impedance of the feed line. The antenna feed line is as an impedance transformer and the impedance is adjusted at the load with an antenna tuner, a balun, a matching transformer, matching networks composed of inductors and capacitors (T-type, Pi-type, and L-type), or matching sections. We choose wide range of RFID antennas to fit an equally wide range of tags, readers, and systems. This includes UHF antennas, patch antennas, and linear or circular polarized antennas. Each RFID antenna has different strengths, and each fits specific types of RFID systems. In RFID system, tags are attached to all items that are to be tracked. These tags are made from tag chip (RFID IC), that is connected to an antenna that can be built into many and wide variety of industrial asset tags. The tag chip contains memory which stores the product EPC and other variable information so that it can be read and tracked by RFID reader anywhere. In our analysis we represent tag chip as a parallel resistor (R_{rfid}) and capacitor (C_{rfid}). An RFID reader is a network connected device (fixed or mobile) with an antenna that sends power as well as data and commands to the tags. The RFID reader acts like an access point for RFID tagged items. An RFID tags are comprised of an integrated circuit (RFID IC) attached to an antenna that has been printed, etched, stamped or vapor-deposited onto a mount which is often a paper substrate or PolyEthylene Terephthalate (PET). We inspect RF and microwave systems which involve, RF and microwave devices (RFID transponders, RF transistors, RF diodes, MMICs, Reflection Type Phase Shifter (RTPS), cylindrical RF network antennas, Tunnel Diode (TD), microwave field effect transistor (FETs), Impact Ionization Avalanche Transit Time (IMPATT), PIN diodes, Small Signal (SS) amplifiers, matching networks, Power Amplifiers (PAs), RF oscillators, RF filters, and antenna systems) as a dynamical system where a fixed rule describes the time dependent of specific RF circuit voltage in a geometrical space. Examples include the mathematical models that describe circuit with RF and microwave devices. At any time a dynamical system has a state given by a set of real numbers (a vector) which can be represented by a specific voltage in an appropriate state space (a geometrical manifold). Small changes in the state of the RF and microwave system create small changes in the numbers. The evolution rule of the RF and microwave dynamical system is a fixed rule that describes what future states follow from the current state. The rule is deterministic, for a given time interval only one future state follows from the current state. A RF and microwave dynamical system is a phase (or state) space endowed with a family of smooth evolution functions that for any element of time (t), map a point of the phase space back into the phase space. RF and microwave systems can be described by numbers. The state vector is a numerical description of the current configuration of a system. For example, RFID circuit with RFID IC and antenna (inductance element) can be described using some numbers: its voltages $(V_1, V_2, V_3, \ldots)$ and currents $(I_1, I_2, I_3, \ldots)$. Once we know these numbers $V_1, V_2, V_3, \ldots$ and $I_1, I_2, I_3, \ldots$ the voltages and currents trajectories are completely determined. The group of numbers $(V_1, V_2, V_3, \ldots, I_1, I_2, I_3, \ldots)$ is a vector which completely describes the state of our RF and microwave system and hence is called the state vector system. There are two main behaviors which are related to RF and microwave systems: (1) the system gravitates toward a fixed point, or (2) the system blowup. There are additional cases of oscillators related to RF and microwave system. We assume $f_i; i = 1, 2, 3, \ldots$ are differentiable with continuous derivatives. The vectors $\mathrm{V_i}$ are the state of the microwave and RF dynamical system, and the functions $f_i; i = 1, 2, 3, \ldots$ tell us how the system moves. In special circumstances, however, the system does not move. The system can be stuck (we will say fixed) in a special state; we call these states fixed points of the dynamical RF and microwave system. Not all fixed points are the same. We call some stable and others unstable. Is the specific fixed point V_i^*, I_i^* stable or unstable? The answer is, neither. To see that it is not stable, consider any points V_{i0}, I_{i0} near (but not equal to) V_i^*, I_i^* (i = 1, 2, 3...). At $t \to \infty$ the RF system never approaches (V_i^*, I_i^*). Further, V_i^*, I_i^* are not unstable. To be unstable, points near V_i^*, I_i^* must be sent "far" away from V_i^*, I_i^*. Clearly, if we start at certain distance from V_i^*, I_i^* the system does not get any farther away. Stable fixed points give excellent information about the fate of a dynamical system.

In our analysis we investigate RF and microwave circuits bifurcation and dynamical behavior. Bifurcation behavior in our RF system is the study of changes in the qualitative or topological structure of RF system, vector fields, and the solutions of a family of differential equations. A bifurcation occurs when a small smooth change made to the parameter values (the bifurcation parameters) of a RF and microwave system causes a sudden topological change in behavior. Bifurcations occur in both our continuous RF systems (ODEs, DDEs, and PDEs) and discrete systems (described by maps). We can inspect in our microwave and RF system two principal bifurcation classes: local bifurcation, which our RF system can be analyzed through changes in the local stability properties of equilibria, periodic orbits or other invariant sets as parameters cross through critical thresholds. Global bifurcations, which often occur in RF and microwave system happened when larger invariant sets of the system collide with each other, or with equilibria of the system. They cannot be detected only by stability analysis (fixed points). A local bifurcation occurs when a system parameter change causes the stability of an equilibrium (or fixed point) to change. In continuous system, this corresponds to the real part of an eigenvalue of an equilibrium passing through zero. Global bifurcations occur when larger invariant sets, such as periodic orbits, collide with equilibria. This causes changes in the topology of the trajectories in the phase space which cannot be confined to a small neighborhood (local bifurcation). The changes in system topology extend out to an arbitrarily large distance. In our analysis we pay attention to the co-dimension of a bifurcation which is the number of parameters which must be varied for the bifurcation to occur. The co-dimension of the parameter set for which the bifurcation occurs within the full space of RF system parameters. The simple case for stability analysis is when there are no time delay elements in our RF and microwave system ($\tau_1 = 0, \tau_2 = 0, \tau_3 = 0, \ldots; \Delta_1 = 0, \Delta_2 = 0, \Delta_3 = 0, \ldots; \Delta_{\mu_i} = 0; i = 1, 2.3, \ldots$).

If our RF and microwave system involving N_a variables ($N_a > 2$), the characteristic equation is of degree N_a ($\sum_{k=0}^{N_a} \lambda^k \cdot \Pi_k = 0$) and must often be solved numerically. Expect in some particular cases, such an equation has N_a distinct roots that can be real or complex. These values are the eigenvalues of the $Na \times Na$ Jacobian matrix (A). The general rule is that the Steady State (SS) is stable if there is no eigenvalue with positive real part. It is sufficient that one eigenvalue is positive for the steady state to be unstable. Our N_a-variables RF and microwave system has N_a eigenvalues. The type of behavior can be characterized as a function of the position of these eigenvalues in the Re/Im plane. Five non-degenerated cases can be distinguished: (1) the N_a eigenvalues are real and negative (stable steady state), (2) the N_a eigenvalues are real, $N_a - 1$ of them are negative (unstable steady state), (3) and (4) two eigenvalues are complex conjugates with a negative real part and the other eigenvalues are real and negative (stable steady state), two cases can be distinguished depending on the relative value of the real part of the complex eigenvalues and of the real one, (5) two eigenvalues are complex conjugates with a negative real part and at least one eigenvalue is positive (unstable steady state). The next case is when there are delay elements in our RF and microwave system and we can't neglect them. In that case, our RF and microwave system stability analysis is related to two main cases: first case, RF or microwave system is characterized by a set of voltages ($V_1, V_2, V_3, \ldots$) or/and currents ($I_1, I_2, I_3, \ldots$). Due to electromagnetic interferences there are differences in time delays with respect to system voltages and current variables

$$V_1(t) \to V_1(t-\tau_1); V_2(t) \to V_2(t-\tau_2); V_i(t) \to V_i(t-\tau_i); I_1(t) \to I_1(t-\tau_1)$$
$$I_2(t) \to I_2(t-\tau_2); I_i(t) \to I_i(t-\tau_i); i = 1, 2, 3, \ldots; \tau_k \neq \tau_l; k \neq l; \tau_i \in \mathbb{R}_+.$$

Sometimes the delay in time is related to circuit voltages derivatives:

$$\frac{dV_1(t)}{dt} \to \frac{dV_1(t-\Delta_1)}{dt}; \frac{dV_2(t)}{dt} \to \frac{dV_2(t-\Delta_2)}{dt};$$
$$\frac{dV_i(t)}{dt} \to \frac{dV_i(t-\Delta_i)}{dt}; \quad i = 1, 2, 3, \ldots$$

We assume $\Delta_k \neq \Delta_l; k \neq l; \Delta_i \in \mathbb{R}_+$. The stability of a given steady state is determined by the graphs of some function of $\tau_1, \tau_2, \tau_3, \ldots$ or/and $\Delta_1, \Delta_2, \Delta_3, \ldots$

which can be expressed explicitly and thus can be easily depicted by software. f need to look at one such function and locate the zeros. The stability switching is due to different values of delay parameters $\tau_1, \tau_2, \tau_3, \ldots$ or/and $\Delta_1, \Delta_2, \Delta_3, \ldots$. Second case, RF and microwave circuits include microstrip lines. Microstrip lines have parasitic effects, a delay in time $\Delta_{\mu_i}; i = 1, 2, 3, \ldots$. The delays are on the current that flow through microstrip lines $I(t) \to I(t - \Delta_{\mu_i})$; $i = 1, 2, 3, \ldots$. We consider $V_{\Delta_{\mu_i}} \to \varepsilon$ (neglect the voltage on microstrip lines). We inspect the stability behavior and stability switching under variation of delay parameters $\Delta_{\mu_i}; i = 1, 2, 3, \ldots$. In our RF and microwave systems which include antennas (inductances elements), we take care in our analysis and calculation the mutual inductances between antennas within the system ($\sum M = M_+ - M_-$).

If we minimize our RF and microwave system to specific case where there is only one time delay parameter τ then the general geometric criterion: The occurrence of any possible stability switching resulting from the increase of value of the time delay τ for our RF and microwave system characteristic equation.

$$D(\lambda, \tau_1, \tau_2, \tau_3, \ldots) = \det(A - \lambda \cdot I); D(\lambda, \tau_1, \tau_2, \tau_3, \ldots) = 0;$$
$$D(\lambda, \tau) = P_n(\lambda, \tau) + Q_m(\lambda, \tau) \cdot e^{-\lambda \cdot \tau}$$
$$P_n(\lambda, \tau) = \sum_{k=0}^{n} p_k(\tau) \cdot \lambda^k; Q_m(\lambda, \tau) = \sum_{k=0}^{m} q_k(\tau) \cdot \lambda^k; n, m \in \mathbb{N}_0; n > m$$

$p_k(\cdot), q_k(\cdot)$: $\mathbb{R}_{+0} \to \mathbb{R}$ are continuous and differentiable functions of τ.

$$P_n(\lambda = 0, \tau) + Q_m(\lambda = 0, \tau) = p_0(\tau) + q_0(\tau) \neq 0 \, \forall \, \tau \in \mathbb{R}_{+0}$$

$P_n(\lambda, \tau), Q_m(\lambda, \tau)$ are analytic functions in λ and differentiable in τ for which we assume ($P_n(\lambda, \tau) \to P, Q_m(\lambda, \tau) \to Q$):

(I) If $\lambda = i \cdot \omega; \omega \in \mathbb{R}$ then $P_n(i \cdot \omega, \tau) + Q_m(i \cdot \omega, \tau) \neq 0; \tau \in \mathbb{R}$.
(II) $\lim\sup\{|Q_m(\lambda, \tau)/P_n(\lambda, \tau)| : |\lambda| \to \infty, \mathrm{Re}\lambda \geq 0\} < 1$ for any τ.
(III) $F(\omega, \tau) = |P_n(i \cdot \omega, \tau)|^2 - |Q_m(i \cdot \omega, \tau)|^2$ for each τ has at most a finite number of real zeros.
(IV) Each positive root $\omega(\tau)$ of $F(\omega, \tau) = 0$ is continuous and differentiable in τ whenever it exists.

In addition, since the coefficients in P and Q are real, we have $\overline{P(-i \cdot \omega)} = P(i \cdot \omega)$, and $\overline{Q(-i \cdot \omega)} = Q(i \cdot \omega)$ thus $\lambda = i \cdot \omega$, $\omega > 0$ may be on eigenvalue of characteristic equation. The analysis consists in identifying the roots of characteristic equation situated on the imaginary axis of the complex λ—plane, where by increasing the parameters of RF and microwave system, Re λ may, at the crossing Change its sign from (−) to (+), i.e. from a stable focus $E^{(0)}$ to an unstable one, or vice versa. This feature may be further assessed by examining the sign of the partial derivatives with respect to τ and RF/Microwave system parameters [2, 3].

$$\wedge^{-1}(\tau) = \left(\frac{\partial \mathrm{Re}\lambda}{\partial \tau}\right)_{\lambda = i\cdot\omega},$$

Other RF and microwave parameters $= const$ where $\omega \in \mathbb{R}_+$.

Hence $F(\omega,\tau) = 0$ implies $\sum_{k=0}^{N} \Xi_{2\cdot k} \cdot \omega^{2\cdot k} = 0\,;\, N \in \mathbb{Z}_+$ when writing $P(\lambda) = P_R(\lambda) + i \cdot P_I(\lambda)$ and $Q(\lambda) = Q_R(\lambda) + i \cdot Q_I(\lambda)$, and inserting $\lambda = i \cdot \omega$ into RF system's characteristic equation, ω must satisfy the following:

$$\sin \omega \cdot \tau = g(\omega) = \frac{-P_R(i\cdot\omega)\cdot Q_I(i\cdot\omega) + P_I(i\cdot\omega)\cdot Q_R(i\cdot\omega)}{|Q(i\cdot\omega)|^2}$$

$$\cos \omega \cdot \tau = h(\omega) = -\frac{P_R(i\cdot\omega)\cdot Q_R(i\cdot\omega) + P_I(i\cdot\omega)\cdot Q_I(i\cdot\omega)}{|Q(i\cdot\omega)|^2}$$

where $|Q(i\cdot\omega)|^2 \neq 0$ in view of requirement (a) above, $(g,h) \in R$. Furthermore, it follows above $\sin \omega \cdot \tau$ and $\cos \omega \cdot \tau$ equations that, by squaring and adding the sides, ω must be a positive root of $F(\omega) = |P(i\cdot\omega)|^2 - |Q(i\cdot\omega)|^2 = 0$. Note: $F(\omega)$ is dependent of τ. Now it is important to notice that if $\tau \notin I$ (assume that $I \subseteq R_{+0}$ is the set where $\omega(\tau)$ is a positive root of $F(\omega)$ and for $\tau \notin I$, $\omega(\tau)$ is not define. Then for all τ in I, $\omega(\tau)$ is satisfies that $F(\omega,\tau) = 0$). Then there are no positive $\omega(\tau)$ solutions for $F(\omega,\tau) = 0$, and we cannot have stability switches. For any $\tau \in I$, where $\omega(\tau)$ is a positive solution of $F(\omega,\tau) = 0$, we can define the angle $\theta(\tau) \in [0, 2\cdot\pi]$ as the solution of the below equations:

$$\sin \theta(\tau) = \frac{-P_R(i\cdot\omega)\cdot Q_I(i\cdot\omega) + P_I(i\cdot\omega)\cdot Q_R(i\cdot\omega)}{|Q(i\cdot\omega)|^2}$$

$$\cos \theta(\tau) = -\frac{P_R(i\cdot\omega)\cdot Q_R(i\cdot\omega) + P_I(i\cdot\omega)\cdot Q_I(i\cdot\omega)}{|Q(i\cdot\omega)|^2}$$

And the relation between the argument $\theta(\tau)$ and $\omega(\tau)\cdot\tau$ for $\tau \in I$ must be $\omega(\tau)\cdot\tau = \theta(\tau) + n\cdot 2\cdot\pi\,\forall n \in \mathbb{N}_0$. Hence we can define the maps $\tau_n\colon I \to R_{+0}$ given by $\tau_n(\tau) = \frac{\theta(\tau)+n\cdot 2\cdot\pi}{\omega(\tau)}; n \in \mathbb{N}_0, \tau \in I$. Let us introduce the functions $I \to R$; $S_n(\tau) = \tau - \tau_n(\tau)$, $\tau \in I$, $n \in \mathbb{N}_0$ that are continuous and differentiable in τ. In the following, the subscripts λ, ω, R_1, C_1 and RF microwave parameters indicate the corresponding partial derivatives. Let us first concentrate on, $\wedge(x)$ remember in λ (parameters); ω (parameters), and keeping all parameters except one (x) and τ. The derivation closely follows that in reference [BK]. Differentiating RF system characteristic equation $P(\lambda) + Q(\lambda)\cdot e^{-\lambda\cdot\tau} = 0$ with respect to specific parameter (x), and inverting the derivative, for convenience, one calculates:

Remark: $x =$ RF system specific parameter

$$\left(\frac{\partial \lambda}{\partial x}\right)^{-1} = \frac{-P_{\lambda}(\lambda,x)\cdot Q(\lambda,x) + Q_{\lambda}(\lambda,x)\cdot P(\lambda,x) - \tau\cdot P(\lambda,x)\cdot Q(\lambda,x)}{P_x(\lambda,x)\cdot Q(\lambda,x) - Q_x(\lambda,x)\cdot P(\lambda,x)}$$

where $P_{\lambda} = \frac{\partial P}{\partial \lambda}, \ldots$ etc., Substituting $\lambda = i\cdot\omega$, and bearing $\overline{P(-i\cdot\omega)} = P(i\cdot\omega)$, $\overline{Q(-i\cdot\omega)} = Q(i\cdot\omega)$ then $i\cdot P_{\lambda}(i\cdot\omega) = P_{\omega}(i\cdot\omega)$ and $i\cdot Q_{\lambda}(i\cdot\omega) = Q_{\omega}(i\cdot\omega)$ that on the surface $|P(i\cdot\omega)|^2 = |Q(i\cdot\omega)|^2$, one obtains

$$\left(\frac{\partial \lambda}{\partial x}\right)^{-1}|_{\lambda=i\cdot\omega} = \left(\frac{i\cdot P_{\omega}(i\cdot\omega,x)\cdot\overline{P(i\cdot\omega,x)} + i\cdot Q_{\lambda}(i\cdot\omega,x)\cdot\overline{Q(\lambda,x)} - \tau\cdot|P(i\cdot\omega,x)|^2}{P_x(i\cdot\omega,x)\cdot\overline{P(i\cdot\omega,x)} - Q_x(i\cdot\omega,x)\cdot\overline{Q(i\cdot\omega,x)}}\right)$$

Upon separating into real and imaginary parts, with $P = P_R + i\cdot P_I; Q = Q_R + i\cdot Q_I$ $\quad P_{\omega} = P_{R\omega} + i\cdot P_{I\omega}; Q_{\omega} = Q_{R\omega} + i\cdot Q_{I\omega};$ $\quad P_x = P_{Rx} + i\cdot P_{Ix}; Q_x = Q_{Rx} + i\cdot Q_{Ix}$ $P^2 = P_R^2 + P_I^2$. When (x) can be any RF system parameters and time delay τ etc., where for convenience, we have dropped the arguments $(i\cdot\omega, x)$, and where $F_{\omega} = 2\cdot[(P_{R\omega}\cdot P_R + P_{I\omega}\cdot P_I) - (Q_{R\omega}\cdot Q_R + Q_{I\omega}\cdot Q_I)]$; $F_x = 2\cdot[(P_{Rx}\cdot P_R + P_{Ix}\cdot P_I) - (Q_{Rx}\cdot Q_R + Q_{Ix}\cdot Q_I)]$ $\omega_x = -F_x/F_{\omega}$. We define U and V: $U = (P_R\cdot P_{I\omega} - P_I\cdot P_{R\omega}) - (Q_R\cdot Q_{I\omega} - Q_I\cdot Q_{R\omega})$ $\quad V = (P_R\cdot P_{Ix} - P_I\cdot P_{Rx}) - (Q_R\cdot Q_{Ix} - Q_I\cdot Q_{Rx})$. We choose our specific parameter as time delay x = τ.

Differentiating with respect to τ and we get $F_{\omega}\cdot\frac{\partial\omega}{\partial\tau} + F_{\tau} = 0; \tau\in I \Rightarrow \frac{\partial\omega}{\partial\tau} = -\frac{F_{\tau}}{F_{\omega}}$

$$\wedge^{-1}(\tau) = \left(\frac{\partial \mathrm{Re}\lambda}{\partial\tau}\right)_{\lambda=i\cdot\omega};$$

$$\wedge^{-1}(\tau) = \mathrm{Re}\left\{\frac{-2\cdot[U + \tau\cdot|P|^2] + i\cdot F_{\omega}}{F_{\tau} + i\cdot 2\cdot[V + \omega\cdot|P|^2]}\right\};$$

$$\frac{\partial\omega}{\partial\tau} = \omega_{\tau} = -\frac{F_{\tau}}{F_{\omega}}$$

$$sign\{\wedge^{-1}(\tau)\} = sign\left\{\left(\frac{\partial \mathrm{Re}\lambda}{\partial\tau}\right)_{\lambda=i\cdot\omega}\right\};$$

$$sign\{\wedge^{-1}(\tau)\} = sign\{F_{\omega}\}\cdot sign\left\{\tau\cdot\frac{\partial\omega}{\partial\tau} + \omega + \frac{U\cdot\frac{\partial\omega}{\partial\tau} + V}{|P|^2}\right\}$$

We shall presently examine the possibility of stability transitions (bifurcations) in RF system, about the equilibrium point as a result of a variation of delay parameterτ. The analysis consists in identifying the roots of our system characteristic equation situated on the imaginary axis of the complex λ-plane where by increasing the delay parameter τ, Re λ may at the crossing, change its sign from

− to +, i.e. from a stable focus $E^{(*)}$ to an unstable one, or vice versa. This feature may be further assessed by examining the sign of the partial derivatives with respect toτ, $\wedge^{-1}(\tau) = (\frac{\partial \text{Re}\lambda}{\partial \tau})_{\lambda=i\cdot\omega}$.

Among microwave elements there are discrete circuit, packages diodes/transistors which are mounted in coax and waveguide assemblies, microwave integrated circuit and monolithic microwave integrated circuit. Stability analysis is also done for microwave transmission lines and full optimization. Many microwave circulators involve Reflection Type Phase Shifter (RTPS). Microstrip transmission lines are integral parts in RF circulator. A special type of antenna system is a cylindrical RF network antenna for coupled plasma sources which include copper legs. Many RF circuits include Tunnel Diode (TD) which is a p-n junction device that exhibits negative resistance. Microwave oscillators integrate Tunnel Diode (TD) elements and stability switching analysis is done. In many applications there are microwave semiconductor amplifiers. Bipolar transistor, FETs and IMPATT amplifiers are inspected when we integrate them in RF and microwave systems. Devices internal parameters influence the functionality of those circuit and RF modules. Internal parameters variation and circuit microstrip lines parasitic effects are inspected for best performance. There are many types of amplifiers, among them zero frequency amplifiers (DC amplifiers), audio amplifiers, RF amplifiers, and they come in three basic flavours: Common Base (CB) amplifiers, Common Collector (CC) amplifiers, and Common Emitter (CE) amplifiers. It is very important to design the right matching network which matches between the amplifier output and the load. Good matching avoids reflections and enhances the functionality of our RF system. Bias—T circuit is a very important element of any RF and microwave system which there is a need to combine RF and DC signals or to separate combined signal to RF and DC signals. Stability analysis of Bias—T circuit is done when integrating it in RF system to get the best performances. Power Amplifiers (PAs) are analysed for best performances and stability analysis is done. There are different types of amplifiers which classified according their circuit configurations and method of operation. There are two basic amplifier classes groups. The first are the classically controlled conduction angle amplifiers forming the more common amplifier classes (A, B, AB, and C). The second set amplifiers are the newer so-called "switching" amplifier classes (D, E, F, G, S, and T). We analyse the stability of these amplifiers by inspecting the equivalent circuit differential equations. BJT transistor is replaced by large signal model and more enhance model is Gummel-Poon model. We analyse the stability of wideband LNAs with negative feedback under circuit's parameters variation. A microwave oscillator is an active device to generate power and a resonator to control the frequency of the microwave signal. Important issues in oscillators are frequency stability, frequency tuning, and phase noise. Phase shift resonator circuit is widely used and stability analysis is done by considering BJT Small Signal (SS) equivalent circuit model. Closed loop functioning oscillator can be viewed as feedback system. There are three types of transistor LC oscillators, Colpitts, Hartley, and Clapp. Colpitt's oscillator circuit stability analysis is done by criterion of Liapunov. The target of analogy and RF filtering is to modify the magnitude and phase of signal frequency components. Many analogy or Radio Frequency (RF) circuits perform filtering on the signal passing through them. A diplexer is a passive device that implements frequency domain multiplexing. We analyse BPF diplexer circuit stability by using geometric stability switch criteria in delay differential systems. There are lattice crystal filter, half lattice and cascaded half lattice filters. A tunable BPF employing varactor diodes is ideal for many diverse wireless applications. BPF (varactor diodes) circuit involving N variables and stability behaviour is inspected. An antenna is a conductor or group of conductors used for radiating electromagnetic energy into space or collecting electromagnetic energy from space. There are many types of antennas and the operation at microwave frequencies is inspected for the best performances. N—turn multilayer circular coil antennas with RFID IC is investigated and stability optimization under delayed electromagnetic interference and parasitic effects is analysed. Double rectangular spiral antennas are constructed from two antennas and they are connected in series with microstrip lines and RFID IC. A system of single turn square planar straight thin film inductors antenna is constructed from four straight thin film inductors which are connected in a single turn square structure. A Helical antenna is an antenna consisting of a conducting wire wound in the form of a helix. Helix antenna system stability is inspected under parameters variation.

Appendix A
RFID LF TAG 125 kHz/134 kHz Design and Analysis

A.1 LF TAG 125 kHz/134 kHz Design and Analysis

We have RFID Antenna system on a substrate, width 300 μm and $\mu_r = 450$. The antenna is constructed from silver ointment which his resistance is bigger than pure silver by 50 %. Track width is 20 μm, gap between tracks 20 μm, track depth is 20 μm up to 100 μm. The requested antenna inductance is 2.66 mH and parasitic resistance less than 10 Ω. We need to find the number of rectangular spiral antenna's turn (Nc). Since the RFID antenna substrate permeability is 450 (μ_r = 450), RFID antenna permeability is average value between air permeability and the magnet. We consider that the TAG permeability is 100–300 (μ_r). Possible TAG dimensions: 5 mm × 5 mm, 6 mm × 6 mm, 7 mm × 7 mm, 8 mm × 8 mm and permeability 100, 200, 300 (μ_r) (Fig. A.1).

Part A: Analysis for $\mu = \mu_0$ and $\mu = \mu_r \cdot \mu_0$ ($\mu_r = 450$).

$$\mu_0 = 4 \cdot \pi \times 10^{-7} \frac{H}{m} \approx 1.2566 \times 10^{-6} \frac{H}{m};$$
$$\mu_r = \frac{\mu}{\mu_0} \Rightarrow \mu = \mu_r \cdot \mu_0 |_{\mu_r=450} = 450 \cdot 1.2566 \times 10^{-6} = 565.47 \times 10^{-6}$$

$$t = g = 20\,\mu\text{m} \Rightarrow d = 2 \cdot \frac{(t+w)}{\Pi} = 2 \cdot \frac{(20\,\mu\text{m} + 20\,\mu\text{m})}{3.14} = 25.47\,\mu\text{m};\ \text{t} = 20\,\mu\text{m}$$

d—Equivalent diameter of the track

$$A_{avg} = A_0 - N_c \cdot (g + w) = 0.005 - N_c \cdot (20\,\mu\text{m} + 20\,\mu\text{m}) = 0.005 - N_c \cdot 40\,\mu\text{m} = 0.005 - N_c \cdot 40 \times 10^{-6}$$
$$B_{avg} = B_0 - N_c \cdot (g + w) = 0.005 - N_c \cdot (20\,\mu\text{m} + 20\,\mu\text{m}) = 0.005 - N_c \cdot 40\,\mu\text{m} = 0.005 - N_c \cdot 40 \times 10^{-6}$$

O. Aluf, *Microwave RF Antennas and Circuits*,
DOI 10.1007/978-3-319-45427-6

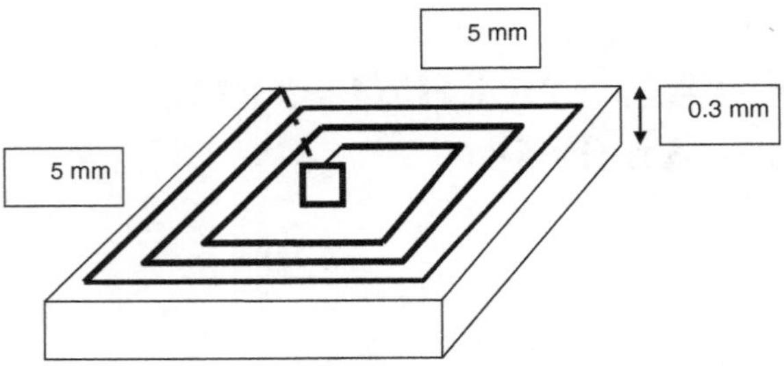

Fig. A.1 RFID antenna system constructed from silver ointment

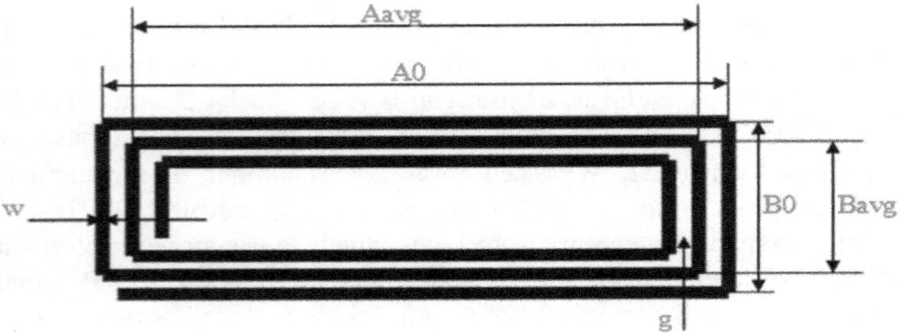

Fig. A.2 RFID rectangular spiral antenna overall parameters

$$d = 2 \cdot (t + w)/\pi; A_{avg} = a_0 - N_c \cdot (g + w); B_{avg} = b_0 - N_c \cdot (g + w)$$

A_0, B_0—Overall dimensions of the coil. Aavg, Bavg—Average dimensions of the coil. t—Track thickness, w—Track width, g—Gap between tracks. Nc—Number of turns, d—Equivalent diameter of the track. Average coil area; Ac = Aavg · Bavg. Integrating all those parameters give the equations for inductance calculation (Fig. A.2):

$$X_1 = A_{avg} \cdot \ln\left[\frac{2 \cdot A_{avg} \cdot B_{avg}}{d \cdot (A_{avg} + \sqrt{A_{avg}^2 + B_{avg}^2}}\right]; X_2 = B_{avg} \cdot \ln\left[\frac{2 \cdot A_{avg} \cdot B_{avg}}{d \cdot (B_{avg} + \sqrt{A_{avg}^2 + B_{avg}^2}}\right]$$

$$X_3 = 2 \cdot [A_{avg} + B_{avg} - \sqrt{A_{avg}^2 + B_{avg}^2}]; X_4 = \frac{(A_{avg} + B_{avg})}{4}$$

Table A.1 RFID coil manufacturing technology (wired, etched, printed)

Coil manufacturing technology	P
Wired	1.8–1.9
Etched	1.75–1.85
Printed	1.7–1.8

The RFID's coil calculation inductance expression is

$L_{calc} = \left[\frac{\mu_0}{\pi} \cdot (X_1 + X_2 - X_3 + X_4) \cdot N_c^p\right]$; $L_1 = L_{calc}$

Definition of limits, Estimations: Track thickness t, Al and Cu coils (t > 30 μm). The printed coils as high as possible. Estimation of turn exponent p is needed for inductance calculation (Table A.1).

We integrate the L_{calc} value inside the differential equations which characterize the RFID system with the Coil inductance. Nc → z; A_0 → x; B_0 → y

MATLAB Script:

Z = 10; x = 0.005 – z * 40 * (1e−6); y = 0.005 – z * 40 * (1e−6); x1 = x * log (2 * x * y/(25.47 * (1e−6) * (x + sqrt(x * x + y * y)))); x2 = y * log(2 * x * y/ (25.47 * (1e−6) * (y + sqrt(x * x + y * y)))); x3 = 2 * (x + y− sqrt(x * x + y * y)); x4 = (x + y)/4; l = ((1.2566 * (1e−6)/3.14) * (x1 + x2 − x3 + x4) * power(z,1.8))

The results in Table A.2:

Result: the most close inductance to our request 2.66 mH is 2.8 mH (Nc = 30, μ_r = 450).

Rectangular spiral RFID antenna length calculation & resistance

We have the following rectangular spiral RFID antenna and first we need to calculate the total length (Fig. A.3).

A_0, B_0—Overall dimensions of the coil. Aavg, Bavg—Average dimensions of the coil. w—Track width, g—Gap between tracks. Nc—Number of turns. L_0 is the length of the first turn $L_0 = 2 \cdot (A_0 + B_0) - (w + g)$. L_k is the length of turn k + 1.

$$k = 1 \Rightarrow L_1 = A_0 - (w+g) + B_0 - 2\cdot(w+g) + A_0 - 2\cdot(w+g) + B_0 - 3\cdot(w+g)$$
$$k = 2 \Rightarrow L_2 = A_0 - 3\cdot(w+g) + B_0 - 4\cdot(w+g) + A_0 - 4\cdot(w+g) + B_0 - 5\cdot(w+g)$$
$$k = 3 \Rightarrow L_3 = A_0 - 5\cdot(w+g) + B_0 - 6\cdot(w+g) + A_0 - 6\cdot(w+g) + B_0 - 7\cdot(w+g)$$

$$L_T = L_0 + \sum_{k=1}^{N_c-1} \{A_0 - [1 + (k-1)\cdot 2]\cdot(w+g) + B_0 - [2 + (k-1)\cdot 2]\cdot(w+g)$$
$$+ A_0 - [2 + (k-1)\cdot 2]\cdot(w+g) + B_0 - [3 + (k-1)\cdot 2]\cdot(w+g)\}$$

Table A.2 RFID rectangular spiral L_{calc} for different values of N_c and μ

Nc	$\mu = \mu_0$; $\mu_r = 1$	L_{calc} (5 mm × 5 mm) − [H]	L_{calc} (8 mm × 8 mm) − [H]
10	$\mu_0 = 1.25.66 \cdot (1e{-}6)$H/m	1.085 μH	1.98 μH
20	μ_0	3.38 μH	6.48 μH
30	μ_0	6.21 μH	12.5 μH
60	μ_0	1.35 μH	34.6 μH
100	μ_0	10.025 μH	55.7 μH
150	μ_0	−1.37 · 1e−5H (N/A)	50.7 μH
120	μ_0	1.35 μH	61 μH
170	μ_0	−3.97 · 1e−5H (N/A)	33 μH
200	μ_0	−1.0582 · 1e−4H (N/A)	NaN
220	μ_0	−1.7096 · 1e−4H (N/A)	−1.9 · 1e−5H (N/A)
250	μ_0	−3.059 · 1e−4H (N/A)	−9.19 · 1e−5H (N/A)
Nc	$\mu = \mu_r \cdot \mu_0$; $\mu_r = 450$	L_{calc} (5 mm × 5 mm) − [H]	L_{calc} (8 mm × 8) − [H]
10	565.47 · (1e−6)H/m	0.48 mH	0.89 mH
20	565.47 · (1e−6)H/m	1.5 mH	**2.9 mH**
30	565.47 · (1e−6)H/m	**2.8 mH**	5.7 mH
60	565.47 · (1e−6)H/m	6.1 mH	15.6 mH
100	565.47 · (1e−6)H/m	4.5 mH	26 mH
150	565.47 · (1e−6)H/m	−0.0062 N/A	22.8 mH
120	565.47 · (1e−6)H/m	0.611 mH	27.5 mH
170	565.47 · (1e−6)H/m	−0.0179 N/A	14.9 mH
200	565.47 · (1e−6)H/m	−0.0476 N/A	NaN
220	565.47 · (1e−6)H/m	−0.0769 N/A	−0.0088 N/A
250	565.47 · (1e−6)H/m	−0.1377 N/A	−0.0414 N/A

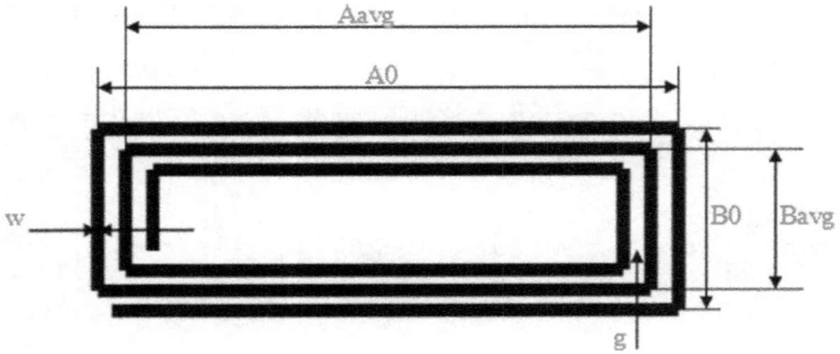

Fig. A.3 RFID rectangular spiral antenna overall parameters

$$\sum_{k=1}^{N_c-1} \{A_0 - [1+(k-1)\cdot 2]\cdot(w+g) + B_0 - [2+(k-1)\cdot 2]\cdot(w+g)$$
$$+A_0 - [2+(k-1)\cdot 2]\cdot(w+g) + B_0 - [3+(k-1)\cdot 2]\cdot(w+g)\}$$
$$= \sum_{k=1}^{N_C-1} \{2\cdot(A_0+B_0) - 8\cdot k\cdot(w+g)\}$$
$$= 2\cdot\sum_{k=1}^{N_C-1} \{(A_0+B_0) - 4\cdot k\cdot(w+g)\}$$
$$= 2\cdot(A_0+B_0)\cdot(N_C-1) - 2\cdot\sum_{k=1}^{N_C-1}[4\cdot k\cdot(w+g)]$$
$$= 2\cdot(A_0+B_0)\cdot(N_C-1) - 8\cdot(w+g)\cdot\sum_{k=1}^{N_C-1} k;\ \sum_{k=1}^{N_C-1} k = N_C - 1$$

$$\sum_{k=1}^{N_c-1} \{A_0 - [1+(k-1)\cdot 2]\cdot(w+g) + B_0 - [2+(k-1)\cdot 2]\cdot(w+g)$$
$$+A_0 - [2+(k-1)\cdot 2]\cdot(w+g) + B_0 - [3+(k-1)\cdot 2]\cdot(w+g)\}$$
$$= 2\cdot(A_0+B_0)\cdot(N_C-1) - 8\cdot(w+g)\cdot(N_C-1)$$
$$= 2\cdot(N_C-1)\cdot[A_0+B_0-4\cdot(w+g)]$$

$$L_T = L_0 + 2\cdot(N_C-1)\cdot[A_0+B_0-4\cdot(w+g)] = L_0 + 2\cdot(A_0+B_0)$$
$$-(w+g) + 2\cdot(N_C-1)$$
$$\cdot[A_0+B_0-4\cdot(w+g)] = 2\cdot(A_0+B_0) - (w+g) + 2\cdot(N_C-1)\cdot(A_0+B_0)$$
$$-8\cdot(N_C-1)\cdot(w+g)$$
$$= L_0 + 2\cdot(A_0+B_0)\cdot(1+N_C-1) - (w+g)\cdot[1+8\cdot(N_C-1)]$$
$$= L_0 + 2\cdot(A_0+B_0)\cdot N_C$$
$$-(w+g)\cdot(8\cdot N_C-7)$$

Final result:

$$L_T = L_0 + 2\cdot(A_0+B_0)\cdot N_C - (w+g)\cdot(8\cdot N_C-7)$$

$$L_T = 2\cdot(A_0+B_0) - (w+g) + 2\cdot(A_0+B_0)\cdot N_C - (w+g)\cdot(8\cdot N_C-7)$$
$$L_T = 2\cdot(A_0+B_0)\cdot(1+N_C) - (w+g)\cdot[1+8\cdot N_C-7]$$
$$L_T = 2\cdot(A_0+B_0)\cdot(1+N_C) - (w+g)\cdot[8\cdot N_C-6]$$
$$L_T = 2\cdot(A_0+B_0)\cdot(1+N_C) - 2\cdot(w+g)\cdot[4\cdot N_C-3]$$
$$L_T = 2\cdot\{(A_0+B_0)\cdot(1+N_C) - (w+g)\cdot[4\cdot N_C-3]\}$$

The DC resistance of rectangular spiral RFID antenna:

$$R_{DC} = \frac{L_T}{\sigma \cdot S} = \frac{L_T}{\sigma \cdot \pi \cdot a^2}$$

L_T—total length of the wire. σ—Conductivity of the wire (m Ω/m).
S—Cross section area $\pi \cdot a^2$. a—radius of the wire.

$$R_{DC} = \frac{L_T}{\sigma \cdot S} = \frac{L_T}{\sigma \cdot \pi \cdot a^2} = \frac{2 \cdot \{(A_0 + B_0) \cdot (1 + N_C) - (w + g) \cdot [4 \cdot N_C - 3]\}}{\sigma \cdot \pi \cdot a^2}$$

$$A_0 + B_0 = 0.01\,\text{m}; N_C = 30; w + g = 40 \times 10^{-6} \Rightarrow L_T = 0.6106\,\text{m} = 61.06\,\text{cm}$$

Cross section area

$$S = 20\,\mu\text{m} \times 20\,\mu\text{m} = 400 \times 10^{-12}\ \text{m}^2$$

Conductivity of Silver $\sigma = 6.1 \times 10^7 (\mho\,\text{m})$. Conductivity has SI units of Siemens per meter (S/m).

$$\sigma_{silver@20\,^\circ\text{C}} = 6.3 \times 10^7 (\text{S/m}).$$

The track depth (x) does not influence our total inductance, and then we can take it as a variable and find his minimum value for $R_{DC} < 10\ \Omega$. $S = 20\,\mu\text{m} \cdot x$.

$$\frac{0.6106}{6.3 \times 10^7 \cdot 20 \times 10^{-6} \cdot x} < 10 \Rightarrow x > 4.846 \times 10^{-5}\ \text{m} = 48.46\,\mu\text{m}; R_{DC@t=20\,\mu\text{m}} = 24.23\,\Omega$$

Actually the track is a mixture of silver then the conductivity is half of silver conductivity.

$$\frac{\sigma_{silver@20\,^\circ\text{C}}}{2} = \frac{6.3 \times 10^7\ (\text{S/m})}{2} = 3.15 \times 10^7\ (\text{S/m});$$

$$R_{DC@t=20\,\mu\text{m}} = 48.46\,\Omega.$$

$$\frac{0.6106}{3.15 \times 10^7 \cdot 20 \times 10^{-6} \cdot x} < 10 \Rightarrow x > 9.6921 \times 10^{-5}\ \text{m} = 96.92\ \mu\text{m}$$

Conclusion: In case of pure silver track. The track depth needs to be bigger than 48.46 μm to meet inductance resistance less than 10 Ω. In case of mixture of silver then track depth needs to be bigger than 96.92 μm.

Part B: Analysis for $\mu = \mu_r \cdot \mu_0$ (μ_r = 100, 200, 300) (Table A.3).

Results: The most close inductance analysis to 2.66 mH is 2.7 mH and subcases

Result B.1: L_{calc} = 2.7 mH, Nc = 60, (A_0 = 7 mm) × (B_0 = 7 mm); μ = μ_r μ_0; μ_r = 100

μ = 125.66 · (1e−6)H/m.

The DC resistance of rectangular spiral RFID antenna:

$$R_{DC} = \frac{L_T}{\sigma \cdot S} = \frac{L_T}{\sigma \cdot \pi \cdot a^2}$$

L_T—total length of the wire. σ—Conductivity of the wire (m Ω/m).

S—Cross section area $\pi \cdot a^2$. a—radius of the wire.

$$R_{DC} = \frac{L_T}{\sigma \cdot S} = \frac{L_T}{\sigma \cdot \pi \cdot a^2} = \frac{2 \cdot \{(A_0 + B_0) \cdot (1 + N_C) - (w + g) \cdot [4 \cdot N_C - 3]\}}{\sigma \cdot \pi \cdot a^2}$$

$$A_0 + B_0 = 0.014\,\text{m}; N_C = 60; w + g = 40 \times 10^{-6} \Rightarrow L_T = 1.689\,\text{m}$$

Cross section area $S = 20\,\mu\text{m} \cdot 20\,\mu\text{m} = 400 \times 10^{-12}\ \text{m}^2$

Conductivity of Silver $\sigma = 6.1 \times 10^7 (\mho/\text{m})$. Conductivity has SI units of Siemens per meter (S/m).

$$\sigma_{silver@20\,^\circ\text{C}} = 6.3 \times 10^7 (\text{S/m}).$$

The track depth (x) does not influence our total inductance, and then we can take it as a variable and find his minimum value for R_{DC} < 10 Ω. $S = 20\,\mu\text{m} \cdot x$.

$$\frac{1.689}{6.3 \times 10^7 \cdot 20 \times 10^{-6} \cdot x} < 10 \Rightarrow x > 1.3405 \times 10^{-4}\ \text{m} = 134.05\,\mu\text{m};$$

$$R_{DC@t=20\ \mu\text{m}} = 67\ \Omega$$

Actually the track is a mixture of silver then the conductivity is half of silver conductivity.

$$\frac{\sigma_{silver@20\,^\circ\text{C}}}{2} = \frac{6.3 \times 10^7 (\text{S/m})}{2} = 3.15 \times 10^7 (\text{S/m}). \quad \frac{1.6890}{3.15 \times 10^7 \cdot 20 \times 10^{-6} \cdot x} < 10$$

$$\Rightarrow x > 2.681 \times 10^{-4}\ \text{m} = 268.1\ \mu\text{m}; R_{DC@t=20\ \mu\text{m}} = 134\ \Omega$$

Result B.2: L_{calc}= 2.7 mH, Nc = 60, (A_0 = 5 mm) × (B_0 = 5 mm); μ = μ_r · μ_0; μ_r = 200

μ = 251.32 · (1e−6)H/m.

Table A.3 RFID rectangular spiral L_{calc} for different values of N_c and μ and tag overall dimension

Nc	$\mu = \mu_r \cdot \mu_0$; $\mu_r = 100$ $\mu = 125.66 \cdot (1e-6)$H/m	L_{calc} (5 mm × 5 mm) − [H]	L_{calc} (6 mm × 6 mm) − [H]	L_{calc} (7 mm × 7 mm) − [H]	L_{calc} (8 mm × 8 mm) − [H]
10	125.66 · (1e−6)H/m	1.085×10^{-4}	1.377×10^{-4}	1.677×10^{-4}	1.986×10^{-4}
20	125.66 · (1e−6)H/m	3.3837×10^{-4}	4.384×10^{-4}	5.4196×10^{-4}	6.4831×10^{-4}
30	125.66 · (1e−6)H/m	6.2129×10^{-4}	8.2571×10^{-4}	0.001	0.0013
60	125.66 · (1e−6)H/m	0.0014	0.002	**0.0027 = 2.7 mH**	0.0035
100	125.66 · (1e−6)H/m	0.001	0.0024	0.0041	0.0058
150	125.66 · (1e−6)H/m	−0.0014	NaN	0.0021	0.0051
120	125.66 · (1e−6)H/m	1.3598×10^{-4}	0.0018	0.0038	0.0061
170	125.66 · (1e−6)H/m	−0.0040	−0.0012	2.5454×10^{-4}	0.0033
200	125.66 · (1e−6)H/m	−0.0106	−0.0062	−0.0023	NaN
220	125.66 · (1e−6)H/m	−0.0171	−0.0115	−0.0063	−0.0020
250	125.66 · (1e−6)H/m	−0.0306	−0.023	−0.0158	−0.0092
Nc	$\mu = \mu_r \cdot \mu_0$; $\mu_r = 200$ $\mu = 251.32 \cdot (1e-6)$H/m	L_{calc} (5 mm × 5 mm) − [H]	L_{calc} (6 mm × 6 mm) − [H]	L_{calc} (7 × 7 mm) − [H]	L_{calc} (8 × 8 mm) − [H]
10	251.32 · (1e−6)H/m	2.17×10^{-4}	2.7539×10^{-4}	3.3553×10^{-4}	3.9719×10^{-4}
20	251.32 · (1e−6)H/m	6.7674×10^{-4}	8.7693×10^{-4}	0.0011	0.0013
30	251.32 · (1e−6)H/m	0.0012	0.0017	0.0021	0.0025
60	251.32 · (1e−6)H/m	**0.0027 = 2.7 mH**	0.004	0.0055	0.0069
100	251.32 · (1e−6)H/m	0.0020	0.0049	0.0081	0.0116
150	251.32 · (1e−6)H/m	−0.0028	NaN	0.0042	0.0102
120	251.32 · (1e−6)H/m	2.7196×10^{-4}	0.0035	0.0077	0.0122
170	251.32 · (1e−6)H/m	−0.008	−0.0025	5.09×10^{-4}	0.0066
200	251.32 · (1e−6)H/m	−0.0212	−0.0123	−0.0046	NaN
220	251.32 · (1e−6)H/m	−0.0342	−0.0229	−0.0127	−0.0039
250	251.32 · (1e−6)H/m	−0.0612	−0.046	−0.0316	−0.0184

Table A.3 (continued)

Nc	$\mu = \mu_r \cdot \mu_0$; $\mu_r = 300$ $\mu = 376.98 \cdot (1e-6)$ H/m	L_{calc} (5 mm × 5 mm) − [H]	L_{calc} (6 mm × 6 mm) − [H]	L_{calc} (7 mm × 7 mm) − [H]	L_{calc} (8 mm × 8 mm) − [H]
10	376.98 · (1e−6)H/m	3.2562×10^{-4}	4.1309×10^{-4}	5.0329×10^{-4}	5.9579×10^{-4}
20	376.98 · (1e−6)H/m	0.001	0.0013	0.0016	0.0019
30	376.98 · (1e−6)H/m	0.0019	0.0025	0.0031	0.0038
60	376.98 · (1e−6)H/m	0.0041	0.0061	0.0082	0.0104
100	376.98 · (1e−6)H/m	0.003	0.0073	0.0122	0.0173
150	376.98 · (1e−6)H/m	−0.0041	NaN	0.0062	0.0153
120	376.98 · (1e−6)H/m	4.0794×10^{-4}	0.0053	0.0115	0.0183
170	376.98 · (1e−6)H/m	−0.0119	−0.0037	7.6362×10^{-4}	0.0099
200	376.98 · (1e−6)H/m	−0.0317	−0.0185	−0.0069	NaN
220	376.98 · (1e−6)H/m	−0.0513	−0.0344	−0.0190	−0.0059
250	376.98 · (1e−6)H/m	−0.0918	−0.069	−0.0474	−0.0276

The DC resistance of rectangular spiral RFID antenna:

$$R_{DC} = \frac{L_T}{\sigma \cdot S} = \frac{L_T}{\sigma \cdot \pi \cdot a^2}$$

L_T—total length of the wire. σ—Conductivity of the wire (m Ω/m).
S—Cross section area $\pi \cdot a^2$. a—radius of the wire.

$$R_{DC} = \frac{L_T}{\sigma \cdot S} = \frac{L_T}{\sigma \cdot \pi \cdot a^2} = \frac{2 \cdot \{(A_0 + B_0) \cdot (1 + N_C) - (w + g) \cdot [4 \cdot N_C - 3]\}}{\sigma \cdot \pi \cdot a^2}$$

$$A_0 + B_0 = 0.01 \text{ m};\ N_C = 60;\ w + g = 40 \times 10^{-6} \Rightarrow L_T = 1.201 \text{ m}$$

Cross section area

$$S = 20\ \mu\text{m} \cdot 20\ \mu\text{m} = 400 \times 10^{-12}\ \text{m}^2$$

Conductivity of Silver $\sigma = 6.1 \times 10^7 (\mho/\text{m})$. Conductivity has SI units of Siemens per meter (S/m).

$$\sigma_{silver@20\,^\circ\text{C}} = 6.3 \times 10^7 (\text{S/m}).$$

The track depth (x) does not influence our total inductance, and then we can take it as a variable and find his minimum value for $R_{DC} < 10\ \Omega$. $S = 20\ \mu\text{m} \cdot x$.

$$\frac{1.201}{6.3 \times 10^7 \cdot 20 \times 10^{-6} \cdot x} < 10 \Rightarrow x > 9.5317 \times 10^{-5}\ \text{m} = 95.317\ \mu\text{m};\ R_{DC@t=20\ \mu\text{m}} = 47.65\ \Omega$$

Actually the track is a mixture of silver then the conductivity is half of silver conductivity.

$$\frac{\sigma_{silver@20\,^\circ\text{C}}}{2} = \frac{6.3 \times 10^7 (\text{S/m})}{2} = 3.15 \times 10^7 (\text{S/m}).$$

$$\frac{1.201}{3.15 \times 10^7 \cdot 20 \times 10^{-6} \cdot x} < 10 \Rightarrow x > 19.06 \times 10^{-5}\ \text{m} = 190.6\ \mu\text{m};$$

$$R_{DC@t=20\ \mu\text{m}} = 95.30\ \Omega$$

A.2 LF TAG 125 kHz/134 kHz Two Rectangular Spiral Antennas Design Analysis

We have RFID Antenna system on a substrate, width 300 μm and μ_r = 100… 450. The construction is made from two rectangular spiral antennas, the first rectangular spiral antenna is on the other rectangular spiral antenna (Symmetric or antisymmetric structure, mirror picture). We consider the symmetric structure for our calculation. Each antenna is constructed from silver ointment which his resistance is bigger than pure silver by 50 %. The track width is 20 μm, gap between tracks 20 μm, track depth is 20 μm up to 100 μm. The requested antennas total inductance is 2.66 mH and parasitic resistance less than 10 Ω. Both rectangular spiral antennas are in series. We need to find the number of rectangular spiral antenna's turn (Nc). We consider that the first and second spiral antennas have the same number of turns (Nc1 = Nc2 = Nc). Since the RFID antenna substrate permeability is between 100 and 450 (μ_r = 100… 450), RFID antenna permeability is an average value between air permeability and the magnet. We consider that the TAG permeability is 100–300 (μ_r). Possible Antennas dimensions: 2 mm × 2 mm, 3 mm × 3 mm, 4 mm × 4 mm, and permeability 100, 200, 300, 450 (μ_r). We neglect micro strip parasitic resistance (Fig. A.4).

Double rectangular spiral antennas can be represented as a two inductors in series (L_{calc-1} and L_{calc-2}), parasitic resistances (R_{DC-1} and R_{DC-2}) and micro strip (neglect parasitic resistance). The rectangular spiral antennas in series are connected in parallel to RFID TAG IC. The Equivalent Circuit of Passive RFID TAG with double rectangular antennas is Capacitor (C_1) and Resistor (R_1) in parallel with double rectangular antennas in the series (Fig. A.5).

L_{calc-1} and L_{calc-2} are mostly formed by traces on planar PCB. $2 \cdot L_m$ element represents the mutual inductance between L_{calc-1} and L_{calc-2}. Since two inductors (L_{calc-1}, L_{calc-2}) are in series and there is a mutual inductance between L_{calc-1} and L_{calc-2}, the total antenna inductance L_T: $L_T = L_{calc-1} + L_{calc-2} + 2 \cdot L_m$ and $L_m = K \cdot \sqrt{L_{calc-1} \cdot L_{calc-2}}$. L_m is the mutual inductance between L_{calc-1} and L_{calc-2}.

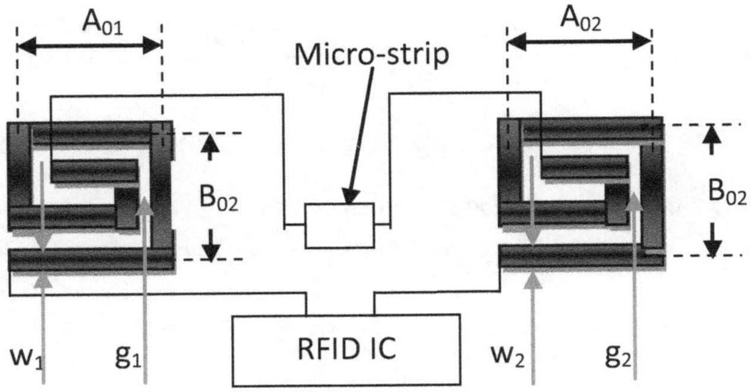

Fig. A.4 RFID rectangular spiral antennas system

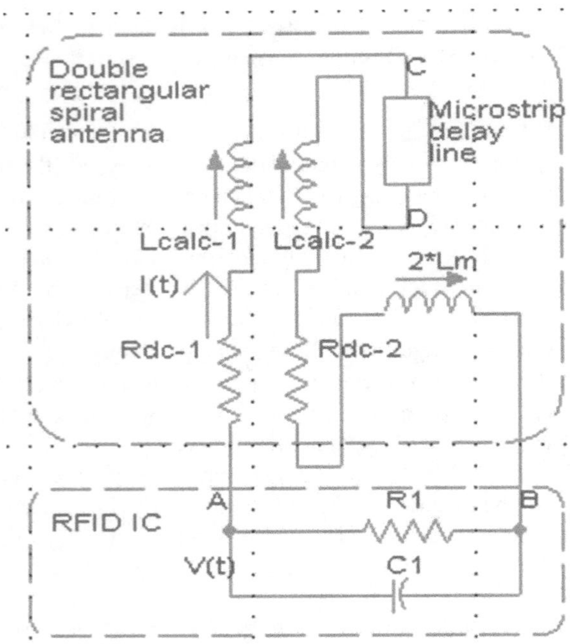

Fig. A.5 Double rectangular spiral antennas in series with RFID TAG IC equivalent circuit

K is the coupling coefficient of two inductors $0 \le K \le 1$. We do our analysis for K = 0.3, 0.5, 0.7. We consider for simplicity that two rectangular spiral antennas are identical (same parameter values).

Part A: Analysis for $\mu = \mu_0$ and $\mu = \mu_r \cdot \mu_0$ ($\mu_r = 450$).

$$\mu_0 = 4 \cdot \pi \times 10^{-7} \frac{H}{m} \approx 1.2566 \times 10^{-6} \frac{H}{m}; \mu_r = \frac{\mu}{\mu_0}$$
$$\Rightarrow \mu = \mu_r \cdot \mu_0|_{\mu_r=450} = 450 \cdot 1.2566 \times 10^{-6} = 565.47 \times 10^{-6}$$

$$t = g = 20\,\mu\text{m} \Rightarrow d = 2 \cdot \frac{(t+w)}{\Pi} = 2 \cdot \frac{(20\,\mu\text{m} + 20\,\mu\text{m})}{3.14} = 25.47\,\mu\text{m};\ \text{t} = 20\ \mu\text{m}$$

d—Equivalent diameter of the track (Fig. A.6)

$$A_{avg} = A_0 - N_c \cdot (g+w) = 0.005 - N_c \cdot (20\,\mu\text{m} + 20\,\mu\text{m}) = 0.005 - N_c \cdot 40\,\mu\text{m} = 0.005 - N_c \cdot 40 \times 10^{-6}$$
$$B_{avg} = A_0 - N_c \cdot (g+w) = 0.005 - N_c \cdot (20\,\mu\text{m} + 20\,\mu\text{m}) = 0.005 - N_c \cdot 40\,\mu\text{m} = 0.005 - N_c \cdot 40 \times 10^{-6}$$

$$d = 2 \cdot (t+w)/\pi; A_{avg} = a_0 - N_c \cdot (g+w); \quad B_{avg} = b_0 - N_c \cdot (g+w)$$

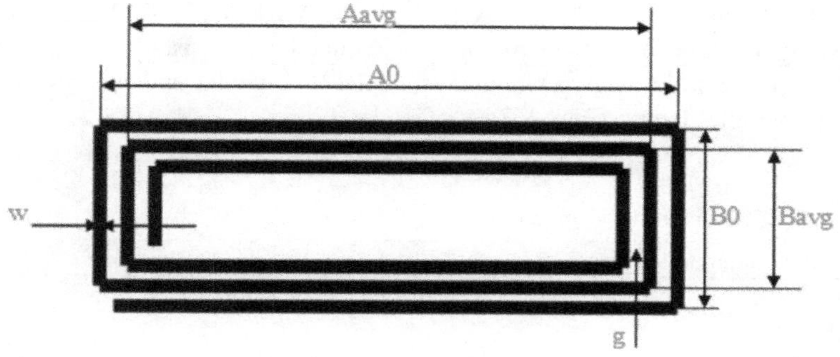

Fig. A.6 RFID rectangular spiral antenna overall parameters

A_0, B_0—Overall dimensions of the coil. Aavg, Bavg—Average dimensions of the coil. t—Track thickness, w—Track width, g—Gap between tracks. Nc—Number of turns, d—Equivalent diameter of the track. Average coil area; Ac = Aavg · Bavg. Integrating all those parameters give the equations for inductance calculation:

$$X_1 = A_{avg} \cdot \ln\left[\frac{2 \cdot A_{avg} \cdot B_{avg}}{d \cdot (A_{avg} + \sqrt{A_{avg}^2 + B_{avg}^2}}\right]; X_2 = B_{avg} \cdot \ln\left[\frac{2 \cdot A_{avg} \cdot B_{avg}}{d \cdot (B_{avg} + \sqrt{A_{avg}^2 + B_{avg}^2}}\right]$$

$$X_3 = 2 \cdot \left[A_{avg} + B_{avg} - \sqrt{A_{avg}^2 + B_{avg}^2}\right]; X_4 = \frac{(A_{avg} + B_{avg})}{4}$$

The RFID's coil calculation inductance expression is

$$L_{calc} = \left[\frac{\mu_0}{\pi} \cdot (X_1 + X_2 - X_3 + X_4) \cdot N_c^p\right]; L_1 = L_{calc}$$

Definition of limits, Estimations: Track thickness t, Al and Cu coils (t > 30 μm). The printed coils as high as possible. Estimation of turn exponent p is needed for inductance calculation (Table A.4).

We integrate the L_{calc} value inside the differential equations which characterize the RFID system with the Coil inductance. Nc → z; A_0 → x; B_0 → y

<u>MATLABScript</u>: z = 10; x = 0.005 − z * 40 * (1e−6); y = 0.005 − z * 40 * (1e−6); x1 = x * log(2 * x * y/(25.47 * (1e−6) * (x + sqrt(x * x + y * y)))); x2 = y * log(2 * x * y/(25.47 * (1e−6) * (y + sqrt(x * x + y * y)))); x3 = 2 * (x + y − sqrt(x * x + y * y)); x4 = (x + y)/4; l = ((1.2566 * (1e−6)/3.14) * (x1 + x2 − x3 + x4) * power(z,1.8))

The results in Tables A.5, A.6, A.7, A.8, A.9, A.10, A.11, A.12 and A.13).

Rectangular spiral RFID antenna length calculation & resistance

We have the following rectangular spiral RFID antenna and first we need to calculate the total length (Fig. A.7).

Table A.4 RFID coil manufacturing technology (wired, etched, printed)

Coil manufacturing technology	P
Wired	1.8–1.9
Etched	1.75–1.85
Printed	1.7–1.8

Table A.5 RFID rectangular spiral L_{calc} for different values of N_c and μ and tag overall dimension

Nc	$\mu = \mu_0;\ \mu_r = 1$	L_{calc} (2 mm × 2 mm) − [H]	L_{calc} (3 mm × 3 mm) − [H]	L_{calc} (4 mm × 4 mm) − [H]
10	$\mu_0 = 1.2566 \cdot (1e{-}6)$H/m	2.922×10^{-7}	5.3857×10^{-7}	8.048×10^{-7}
20	μ_0	7.0241×10^{-7}	1.5222×10^{-6}	2.425×10^{-6}
30	μ_0	8.532×10^{-7}	2.45×10^{-6}	4.266×10^{-6}
60	μ_0	-5.916×10^{-7}	2.009×10^{-6}	7.351×10^{-6}
100	μ_0	-1.767×10^{-7}	-6.6295×10^{-6}	NaN
150	μ_0	-9.167×10^{-5}	-6.3052×10^{-5}	-3.6674×10^{-5}
120	μ_0	-3.8527×10^{-5}	-2.1249×10^{-5}	-6.573×10^{-5}
170	μ_0	-1.45×10^{-4}	-1.074×10^{-4}	-7.2119×10^{-5}
200	μ_0	-2.577×10^{-4}	-2.047×10^{-4}	-1.538×10^{-4}
220	μ_0	-3.58×10^{-4}	-2.932×10^{-4}	-2.3072×10^{-4}
250	μ_0	-5.5179×10^{-4}	-4.673×10^{-4}	-3.8522×10^{-4}
Nc	$\mu = \mu_r \cdot \mu_0;\ \mu_r = 450$	L_{calc} (2 mm × 2 mm) − [H]	L_{calc} (3 mm × 3) − [H]	L_{calc} (4 mm × 4 mm) − [H]
10	$565.47 \cdot (1e{-}6)$H/m	1.3149×10^{-4}	2.4236×10^{-4}	3.621×10^{-4}
20	$565.47 \cdot (1e{-}6)$H/m	3.1608×10^{-4}	6.85×10^{-4}	0.0011
30	$565.47 \cdot (1e{-}6)$H/m	3.8394×10^{-4}	0.0011	0.0019
60	$565.47 \cdot (1e{-}6)$H/m	-2.6625×10^{-4}	9.04×10^{-4}	0.0033
100	$565.47 \cdot (1e{-}6)$H/m	−0.008	−0.003	NaN
150	$565.47 \cdot (1e{-}6)$H/m	−0.0413	−0.0284	−0.0165
120	$565.47 \cdot (1e{-}6)$H/m	−0.0173	−0.0096	−0.003
170	$565.47 \cdot (1e{-}6)$H/m	−0.0653	−0.0484	−0.0325
200	$565.47 \cdot (1e{-}6)$H/m	−0.116	−0.0921	−0.0692
220	$565.47 \cdot (1e{-}6)$H/m	−0.1611	−0.132	−0.1038
250	$565.47 \cdot (1e{-}6)$H/m	−0.2483	−0.2103	−0.1734

Table A.6 RFID rectangular spiral L_{calc}, L_m, and L_T for K = 0.3 (2 mm × 2 mm)

Nc	$L_{calc\text{-}1} = L_{calc\text{-}2} = L_{calc}$, $\mu = \mu_r \cdot \mu_0\ \mu_r = 450$ (2 mm × 2 mm) − [H]	$L_m = K \cdot \sqrt{L_{calc-1} \cdot L_{calc-2}}$ $L_m = K \cdot \sqrt{L_{calc} \cdot L_{calc}} = K \cdot L_{calc}$ (K = 0.3)	$L_T = L_{calc\text{-}1} + L_{calc\text{-}2} + 2 \cdot L_m = 2 \cdot L_{calc} + 2 \cdot L_m$ (K = 0.3)
10	1.3149×10^{-4}	3.9447×10^{-5}	3.4187×10^{-4}
20	3.1608×10^{-4}	9.4824×10^{-5}	8.218×10^{-4}
30	3.8394×10^{-4}	1.1518×10^{-4}	9.9824×10^{-4}

Table A.7 RFID rectangular spiral L_{calc}, L_m, and L_T for K = 0.5 (2 mm × 2 mm)

Nc	$L_{calc\text{-}1} = L_{calc\text{-}2} = L_{calc}$, $\mu = \mu_r \cdot \mu_0$ $\mu_r = 450$ (2 mm × 2 mm) − [H]	$L_m = K \cdot \sqrt{L_{calc-1} \cdot L_{calc-2}}$ $L_m = K \cdot \sqrt{L_{calc} \cdot L_{calc}} = K \cdot L_{calc}$ (K = 0.5)	$L_T = L_{calc\text{-}1} + L_{calc\text{-}2} + 2 \cdot L_m = 2 \cdot L_{calc} + 2 \cdot L_m$ (K = 0.5)
10	1.3149×10^{-4}	6.574×10^{-5}	3.9447×10^{-4}
20	$3.1608 \times 10 \times 10^{-4}$	1.58×10^{-4}	9.482×10^{-4}
30	3.8394×10^{-4}	1.9197×10^{-4}	0.0012

Table A.8 RFID rectangular spiral L_{calc}, L_m, and L_T for K = 0.7 (2 mm × 2 mm)

Nc	$L_{calc\text{-}1} = L_{calc\text{-}2} = L_{calc}$, $\mu = \mu_r \cdot \mu_0$ $\mu_r = 450$ (2 mm × 2 mm) − [H]	$L_m = K \cdot \sqrt{L_{calc-1} \cdot L_{calc-2}}$ $L_m = K \cdot \sqrt{L_{calc} \cdot L_{calc}} = K \cdot L_{calc}$ (K = 0.7)	$L_T = L_{calc\text{-}1} + L_{calc\text{-}2} + 2 \cdot L_m = 2 \cdot L_{calc} + 2 \cdot L_m$ (K = 0.7)
10	1.3149×10^{-4}	9.2043×10^{-5}	4.47×10^{-4}
20	3.1608×10^{-4}	2.212×10^{-4}	0.0011
30	3.8394×10^{-4}	2.6876×10^{-4}	0.0013

Table A.9 RFID rectangular spiral L_{calc}, L_m, and L_T for K = 0.3 (3 mm × 3 mm)

Nc	$L_{calc\text{-}1} = L_{calc\text{-}2} = L_{calc}$, $\mu = \mu_r \cdot \mu_0$ $\mu_r = 450$ (3 mm × 3 mm) − [H]	$L_m = K \cdot \sqrt{L_{calc-1} \cdot L_{calc-2}}$ $L_m = K \cdot \sqrt{L_{calc} \cdot L_{calc}} = K \cdot L_{calc}$ (K = 0.3)	$L_T = L_{calc\text{-}1} + L_{calc\text{-}2} + 2 \cdot L_m = 2 \cdot L_{calc} + 2 \cdot L_m$ (K = 0.3)
10	2.4236×10^{-4}	7.2708×10^{-5}	6.3014×10^{-4}
20	6.85×10^{-4}	2.055×10^{-4}	0.0018
30	0.0011	0.0033	**0.0029 = 2.9 mH**
60	9.04×10^{-4}	2.712×10^{-4}	**0.0024 = 2.4 mH**

Table A.10 RFID rectangular spiral L_{calc}, L_m, and L_T for K = 0.5 (3 mm × 3 mm)

Nc	$L_{calc\text{-}1} = L_{calc\text{-}2} = L_{calc}$, $\mu = \mu_r \cdot \mu_0$ $\mu_r = 450$ (3 mm × 3 mm) − [H]	$L_m = K \cdot \sqrt{L_{calc-1} \cdot L_{calc-2}}$ $L_m = K \cdot \sqrt{L_{calc} \cdot L_{calc}} = K \cdot L_{calc}$ (K = 0.5)	$L_T = L_{calc\text{-}1} + L_{calc\text{-}2} + 2 \cdot L_m = 2 \cdot L_{calc} + 2 \cdot L_m$ (K = 0.5)
10	2.4236×10^{-4}	1.2118×10^{-4}	7.27×10^{-4}
20	6.85×10^{-4}	3.425×10^{-4}	**0.0021 = 2.1 mH**
30	0.0011	5.5×10^{-4}	0.0033
60	9.04×10^{-4}	4.52×10^{-4}	**0.0027 = 2.7 mH**

A0, B0—Overall dimensions of the coil. Aavg, Bavg—Average dimensions of the coil. w—Track width, g—Gap between tracks. Nc—Number of turns. L_0 is the length of the first turn $L_0 = 2 \cdot (A_0 + B_0) - (w + g)$. L_k is the length of turn k + 1.

Table A.11 RFID rectangular spiral L_{calc}, L_m, and L_T for K = 0.7 (3 mm × 3 mm)

Nc	$L_{calc-1} = L_{calc-2} = L_{calc}$, $\mu = \mu_r \cdot \mu_0$ $\mu_r = 450$ (3 mm × 3 mm) − [H]	$L_m = K \cdot \sqrt{L_{calc-1} \cdot L_{calc-2}}$ $L_m = K \cdot \sqrt{L_{calc} \cdot L_{calc}} = K \cdot L_{calc}$ (K = 0.7)	$L_T = L_{calc-1} + L_{calc-2} + 2 \cdot L_m = 2 \cdot L_{calc} + 2 \cdot L_m$ (K = 0.7)
10	2.4236×10^{-4}	1.6965×10^{-4}	8.24×10^{-4}
20	6.85×10^{-4}	4.795×10^{-4}	**0.0023 = 2.3 mH**
30	0.0011	7.7×10^{-4}	0.0037
60	9.04×10^{-4}	6.328×10^{-4}	0.0031

Table A.12 RFID rectangular spiral L_{calc}, L_m, and L_T for K = 0.3, 0.5 (4 mm × 4 mm)

Nc	$L_{calc-1} = L_{calc-2} = L_{calc}$, $\mu = \mu_r \cdot \mu_0$ $\mu_r = 450$ (4 mm × 4 mm) − [H]	$L_m = K \cdot \sqrt{L_{calc-1} \cdot L_{calc-2}}$ $L_m = K \cdot \sqrt{L_{calc} \cdot L_{calc}} = K \cdot L_{calc}$ (K = 0.3)	$L_T = L_{calc-1} + L_{calc-2} + 2 \cdot L_m = 2 \cdot L_{calc} + 2 \cdot L_m$ (K = 0.3)
10	3.621×10^{-4}	1.0863×10^{-4}	9.4146×10^{-4}
20	0.0011	0.0033	**0.0029 = 2.9 mH**
30	0.0019	5.7×10^{-4}	0.0049
60	0.0033	9.9×10^{-4}	0.0086
Nc	$L_{calc-1} = L_{calc-2} = L_{calc}$, $\mu = \mu_r \cdot \mu_0$ $\mu_r = 450$ (4 mm × 4 mm) − [H]	$L_m = K \cdot \sqrt{L_{calc-1} \cdot L_{calc-2}}$ $L_m = K \cdot \sqrt{L_{calc} \cdot L_{calc}} = K \cdot L_{calc}$ (K = 0.5)	$L_T = L_{calc-1} + L_{calc-2} + 2 \cdot L_m = 2 \cdot L_{calc} + 2 \cdot L_m$ (K = 0.5)
10	3.621×10^{-4}	1.8105×10^{-4}	0.0011
20	0.0011	5.5×10^{-4}	0.0033
30	0.0019	9.5×10^{-4}	0.0057
60	0.0033	0.0017	0.0099

Table A.13 RFID rectangular spiral L_{calc}, L_m, and L_T for K = 0.7 (4 mm × 4 mm)

Nc	$L_{calc-1} = L_{calc-2} = L_{calc}$, $\mu = \mu_r \cdot \mu_0$ $\mu_r = 450$ (4 mm × 4 mm) − [H]	$L_m = K \cdot \sqrt{L_{calc-1} \cdot L_{calc-2}}$ $L_m = K \cdot \sqrt{L_{calc} \cdot L_{calc}} = K \cdot L_{calc}$ (K = 0.7)	$L_T = L_{calc-1} + L_{calc-2} + 2 \cdot L_m = 2 \cdot L_{calc} + 2 \cdot L_m$ (K = 0.7)
10	3.621×10^{-4}	2.5347×10^{-4}	0.0012
20	0.0011	7.7×10^{-4}	0.0037
30	0.0019	0.0013	0.0065
60	0.0033	0.0023	0.0112

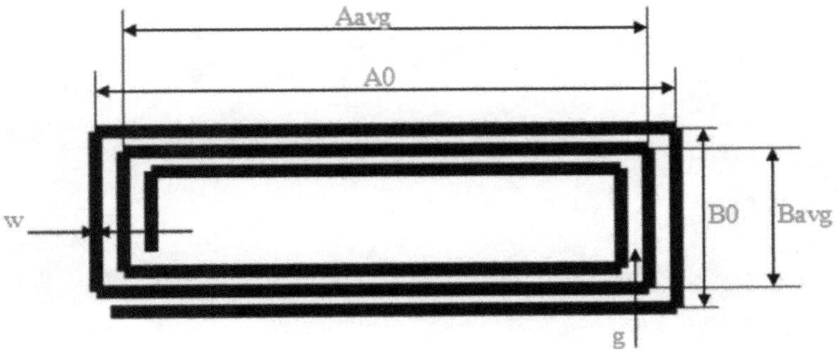

Fig. A.7 RFID rectangular spiral antenna overall parameters

$$k = 1 \Rightarrow L_1 = A_0 - (w+g) + B_0 - 2\cdot(w+g) + A_0 - 2\cdot(w+g) + B_0 - 3\cdot(w+g)$$
$$k = 2 \Rightarrow L_2 = A_0 - 3\cdot(w+g) + B_0 - 4\cdot(w+g) + A_0 - 4\cdot(w+g) + B_0 - 5\cdot(w+g)$$
$$k = 3 \Rightarrow L_3 = A_0 - 5\cdot(w+g) + B_0 - 6\cdot(w+g) + A_0 - 6\cdot(w+g) + B_0 - 7\cdot(w+g)$$

$$L_T = L_0 + \sum_{k=1}^{N_c-1} \{A_0 - [1+(k-1)\cdot 2]\cdot(w+g) + B_0 - [2+(k-1)\cdot 2]\cdot(w+g) + A_0 - [2+(k-1)\cdot 2]\cdot(w+g) + B_0 - [3+(k-1)\cdot 2]\cdot(w+g)\}$$

$$\begin{aligned}
&\sum_{k=1}^{N_c-1} \{A_0 - [1+(k-1)\cdot 2]\cdot(w+g) + B_0 - [2+(k-1)\cdot 2]\cdot(w+g) \\
&\quad + A_0 - [2+(k-1)\cdot 2]\cdot(w+g) + B_0 - [3+(k-1)\cdot 2]\cdot(w+g)\} \\
&= \sum_{k=1}^{N_C-1} \{2\cdot(A_0+B_0) - 8\cdot k\cdot(w+g)\} \\
&= 2\cdot\sum_{k=1}^{N_C-1} \{(A_0+B_0) - 4\cdot k\cdot(w+g)\} = 2\cdot(A_0+B_0)\cdot(N_C-1) \\
&\quad - 2\cdot\sum_{k=1}^{N_C-1} [4\cdot k\cdot(w+g)] \\
&= 2\cdot(A_0+B_0)\cdot(N_C-1) - 8\cdot(w+g)\cdot\sum_{k=1}^{N_C-1} k;\ \sum_{k=1}^{N_C-1} k = N_C - 1
\end{aligned}$$

$$\sum_{k=1}^{N_c-1} \{A_0 - [1+(k-1)\cdot 2]\cdot(w+g) + B_0 - [2+(k-1)\cdot 2]\cdot(w+g)$$
$$+A_0 - [2+(k-1)\cdot 2]\cdot(w+g) + B_0 - [3+(k-1)\cdot 2]\cdot(w+g)\}$$
$$= 2\cdot(A_0+B_0)\cdot(N_C-1) - 8\cdot(w+g)\cdot(N_C-1)$$
$$= 2\cdot(N_C-1)\cdot[A_0+B_0-4\cdot(w+g)]$$

$$L_T = L_0 + 2\cdot(N_C-1)\cdot[A_0+B_0-4\cdot(w+g)] = L_0 + 2\cdot(A_0+B_0)$$
$$-(w+g) + 2\cdot(N_C-1)$$
$$\cdot[A_0+B_0-4\cdot(w+g)] = 2\cdot(A_0+B_0)-(w+g)$$
$$+2\cdot(N_C-1)\cdot(A_0+B_0) - 8\cdot(N_C-1)\cdot(w+g)$$
$$= L_0 + 2\cdot(A_0+B_0)\cdot(1+N_C-1) - (w+g)\cdot[1+8\cdot(N_C-1)]$$
$$= L_0 + 2\cdot(A_0+B_0)\cdot N_C - (w+g)\cdot(8\cdot N_C - 7)$$

Final result:

$$L_T = L_0 + 2\cdot(A_0+B_0)\cdot N_C - (w+g)\cdot(8\cdot N_C - 7)$$

$$L_T = 2\cdot(A_0+B_0) - (w+g) + 2\cdot(A_0+B_0)\cdot N_C - (w+g)\cdot(8\cdot N_C-7)$$
$$L_T = 2\cdot(A_0+B_0)\cdot(1+N_C) - (w+g)\cdot[1+8\cdot N_C-7]$$
$$L_T = 2\cdot(A_0+B_0)\cdot(1+N_C) - (w+g)\cdot[8\cdot N_C-6]$$
$$L_T = 2\cdot(A_0+B_0)\cdot(1+N_C) - 2\cdot(w+g)\cdot[4\cdot N_C-3]$$
$$L_T = 2\cdot\{(A_0+B_0)\cdot(1+N_C) - (w+g)\cdot[4\cdot N_C-3]\}$$

The DC resistance of rectangular spiral RFID antenna:

$$R_{DC} = \frac{L_T}{\sigma\cdot S} = \frac{L_T}{\sigma\cdot\pi\cdot a^2}$$

L_T—total length of the wire. σ—Conductivity of the wire (m Ω/m).
S—Cross section area $\pi\cdot a^2$. a—radius of the wire.

$$R_{DC} = \frac{L_T}{\sigma\cdot S} = \frac{L_T}{\sigma\cdot\pi\cdot a^2} = \frac{2\cdot\{(A_0+B_0)\cdot(1+N_C) - (w+g)\cdot[4\cdot N_C-3]\}}{\sigma\cdot\pi\cdot a^2}$$

$$A_0 + B_0 = TBD; N_C = TBD; w + g = 40 \times 10^{-6} \Rightarrow L_T = TBD$$

To be define—value which need to be chosen according analysis results.
Cross section area

$$S = 20\,\mu\text{m} \cdot 20\,\mu\text{m} = 400 \times 10^{-12}\ \text{m}^2$$

Conductivity of Silver $\sigma = 6.1 \times 10^7 (\mho/\text{m})$. Conductivity has SI units of Siemens per meter (S/m).

$$\sigma_{silver@20^\circ\text{C}} = 6.3 \times 10^7 (\text{S/m}).$$

The track depth (x) does not influence our total inductance, and then we can take it as a variable and find his minimum value for $R_{DC} < 10\ \Omega$. $S = 20\,\mu\text{m} \cdot x$.

Part B: Analysis for $\mu = \mu_r \cdot \mu_0$ (μ_r = 100, 200, 300) (Tables A.14, A.15, A.16, A.17, A.18, A.19, A.20, A.21, A.22, A.23, A.24, A.25, A.26, A.27, A.28, A.29, A.30, A.31, A.32, A.33, A.34 and A.35).

The DC resistance of rectangular spiral RFID antenna:

$$R_{DC} = \frac{L_T}{\sigma \cdot S} = \frac{L_T}{\sigma \cdot \pi \cdot a^2}$$

L_T—total length of the wire. σ—Conductivity of the wire (m Ω/m).
S—Cross section area $\pi \cdot a^2$. a—radius of the wire.

$$R_{DC} = \frac{L_T}{\sigma \cdot S} = \frac{L_T}{\sigma \cdot \pi \cdot a^2} = \frac{2 \cdot \{(A_0 + B_0) \cdot (1 + N_C) - (w + g) \cdot [4 \cdot N_C - 3]\}}{\sigma \cdot \pi \cdot a^2}$$
$$A_0 + B_0 = TBD; N_C = TBD; w + g = 40 \times 10^{-6} \Rightarrow L_T = TBD$$

Cross section area

$$S = 20\,\mu\text{m} \cdot 20\,\mu\text{m} = 400 \times 10^{-12}\ \text{m}^2$$

Table A.14 RFID rectangular spiral L_{calc} for different values of N_c and μ and tag overall

Nc	$\mu = \mu_r \cdot \mu_0$; $\mu_r = 100$ $\mu = 125.66 \cdot (1e\text{-}6)$H/m	L_{calc} (2 mm × 2 mm) − [H]	L_{calc} (3 mm × 3 mm) − [H]	L_{calc} (4 mm × 4 mm) − [H]
10	125.66 · (1e−6)H/m	2.922×10^{-5}	5.3857×10^{-5}	8.0487×10^{-5}
20	125.66 · (1e−6)H/m	7.0241×10^{-5}	1.5222×10^{-4}	2.425×10^{-4}
30	125.66 · (1e−6)H/m	8.532×10^{-5}	2.4523×10^{-4}	4.266×10^{-4}
60	125.66 · (1e−6)H/m	-5.9168×10^{-5}	2.009×10^{-4}	7.351×10^{-4}
100	125.66 · (1e−6)H/m	−0.0018	-6.629×10^{-4}	NaN
150	125.66 · (1e−6)H/m	−0.0092	−0.0063	−0.0037
120	125.66 · (1e−6)H/m	−0.0039	−0.0021	-6.573×10^{-4}
170	125.66 · (1e−6)H/m	−0.0145	−0.0107	−0.0072
200	125.66 · (1e−6)H/m	−0.0258	−0.0205	−0.0154
220	125.66 · (1e−6)H/m	−0.0358	−0.0293	−0.0231
250	125.66 · (1e−6)H/m	−0.0552	−0.0467	−0.0385
Nc	$\mu = \mu_r \cdot \mu_0$; $\mu_r = 200$ $\mu = 251.32 \cdot (1e-6)$H/m	L_{calc} (2 mm × 2 mm) − [H]	L_{calc} (3 mm × 3 mm) − [H]	L_{calc} (4 mm × 4 mm) − [H]
10	251.32 · (1e−6)H/m	5.8439×10^{-5}	1.0771×10^{-4}	1.6097×10^{-4}
20	251.32 · (1e−6)H/m	1.4048×10^{-4}	3.0445×10^{-4}	4.8501×10^{-4}
30	251.32 · (1e−6)H/m	1.7064×10^{-4}	4.9045×10^{-4}	8.532×10^{-4}
60	251.32 · (1e−6)H/m	-1.1834×10^{-4}	4.0179×10^{-4}	0.0015
100	251.32 · (1e−6)H/m	−0.0035	−0.0013	NaN
150	251.32 · (1e−6)H/m	−0.0183	−0.0126	−0.0073
120	251.32 · (1e−6)H/m	−0.0077	−0.0042	−0.0013
170	251.32 · (1e−6)H/m	−0.029	−0.0215	−0.0144
200	251.32 · (1e−6)H/m	−0.0516	−0.0409	−0.0308
220	251.32 · (1e−6)H/m	−0.0716	−0.0587	−0.0461
250	251.32 · (1e−6)H/m	−0.1104	−0.0935	−0.077

(continued)

Table A.14 (continued)

Nc	$\mu = \mu_r \cdot \mu_0$; $\mu_r = 300$ $\mu = 376.98 \cdot (1e{-}6)$H/m	L_{calc} (2 mm × 2 mm) − [H]	L_{calc} (3 mm × 3 mm) − [H]	L_{calc} (4 mm × 4 mm) − [H]
10	376.98 · (1e−6)H/m	8.7659×10^{-5}	1.6157×10^{-4}	2.4146×10^{-4}
20	376.98 · (1e−6)H/m	2.1072×10^{-4}	4.5667×10^{-4}	7.2751×10^{-4}
30	376.98 · (1e−6)H/m	2.5596×10^{-4}	7.3568×10^{-4}	0.0013
60	376.98 · (1e−6)H/m	-1.775×10^{-4}	6.0269×10^{-4}	0.0022
100	376.98 · (1e−6)H/m	−0.0053	−0.002	NaN
150	376.98 · (1e−6)H/m	−0.0275	−0.0189	−0.011
120	376.98 · (1e−6)H/m	−0.0116	−0.0064	−0.002
170	376.98 · (1e−6)H/m	−0.0435	−0.0322	−0.0216
200	376.98 · (1e−6)H/m	−0.0773	−0.0614	−0.0462
220	376.98 · (1e−6)H/m	−0.1074	−0.088	−0.0692
250	376.98 · (1e−6)H/m	−0.1655	−0.1402	−0.1156

Table A.15 RFID rectangular spiral L_{calc}, L_m, and L_T for K = 0.3 (2 mm × 2 mm)

Nc	$L_{calc-1} = L_{calc-2} = L_{calc}$, $\mu = \mu_r \cdot \mu_0$ $\mu_r = 100$ (2 mm × 2 mm) − [H]	$L_m = K \cdot \sqrt{L_{calc-1} \cdot L_{calc-2}}$ $L_m = K \cdot \sqrt{L_{calc} \cdot L_{calc}} = K \cdot L_{calc}$ (K = 0.3)	$L_T = L_{calc-1} + L_{calc-2} + 2 \cdot L_m = 2 \cdot L_{calc} + 2 \cdot L_m$ (K = 0.3)
10	2.922×10^{-5}	8.766×10^{-6}	7.5972×10^{-5}
20	7.0241×10^{-5}	2.1072×10^{-5}	1.8263×10^{-4}
30	8.532×10^{-5}	2.5596×10^{-5}	2.2183×10^{-4}

Table A.16 RFID rectangular spiral L_{calc}, L_m, and L_T for K = 0.5 (2 mm × 2 mm)

Nc	$L_{calc-1} = L_{calc-2} = L_{calc}$, $\mu = \mu_r \cdot \mu_0$ $\mu_r = 100$ (2 mm × 2 mm) − [H]	$L_m = K \cdot \sqrt{L_{calc-1} \cdot L_{calc-2}}$ $L_m = K \cdot \sqrt{L_{calc} \cdot L_{calc}} = K \cdot L_{calc}$ (K = 0.5)	$L_T = L_{calc-1} + L_{calc-2} + 2 \cdot L_m = 2 \cdot L_{calc} + 2 \cdot L_m$ (K = 0.5)
10	2.922×10^{-5}	1.461×10^{-5}	8.766×10^{-5}
20	7.0241×10^{-5}	3.5121×10^{-5}	2.1072×10^{-4}
30	8.532×10^{-5}	4.266×10^{-5}	2.5596×10^{-4}

Table A.17 RFID rectangular spiral L_{calc}, L_m, and L_T for K = 0.7 (2 mm × 2 mm)

Nc	$L_{calc-1} = L_{calc-2} = L_{calc}$, $\mu = \mu_r \cdot \mu_0$ $\mu_r = 100$ (2 mm × 2 mm) − [H]	$L_m = K \cdot \sqrt{L_{calc-1} \cdot L_{calc-2}}$ $L_m = K \cdot \sqrt{L_{calc} \cdot L_{calc}} = K \cdot L_{calc}$ (K = 0.7)	$L_T = L_{calc-1} + L_{calc-2} + 2 \cdot L_m = 2 \cdot L_{calc} + 2 \cdot L_m$ (K = 0.7)
10	2.922×10^{-5}	2.0454×10^{-5}	9.9348×10^{-5}
20	7.0241×10^{-5}	4.9169×10^{-5}	2.3882×10^{-4}
30	8.532×10^{-5}	5.9724×10^{-5}	2.9009×10^{-4}

Table A.18 RFID rectangular spiral L_{calc}, L_m, and L_T for K = 0.3 (3 mm × 3 mm)

Nc	$L_{calc-1} = L_{calc-2} = L_{calc}$, $\mu = \mu_r \cdot \mu_0$ $\mu_r = 100$ (3 mm × 3 mm) − [H]	$L_m = K \cdot \sqrt{L_{calc-1} \cdot L_{calc-2}}$ $L_m = K \cdot \sqrt{L_{calc} \cdot L_{calc}} = K \cdot L_{calc}$ (K = 0.3)	$L_T = L_{calc-1} + L_{calc-2} + 2 \cdot L_m = 2 \cdot L_{calc} + 2 \cdot L_m$ (K = 0.3)
10	5.3857×10^{-5}	1.6157×10^{-5}	1.4003×10^{-4}
20	1.5222×10^{-4}	4.5666×10^{-5}	3.9577×10^{-4}
30	2.4523×10^{-4}	7.3569×10^{-5}	6.376×10^{-4}
60	2.009×10^{-4}	6.027×10^{-5}	5.2234×10^{-4}

Table A.19 RFID rectangular spiral L_{calc}, L_m, and L_T for K = 0.5 (3 mm × 3 mm)

Nc	$L_{calc-1} = L_{calc-2} = L_{calc}$, $\mu = \mu_r \cdot \mu_0$ $\mu_r = 100$ (3 mm × 3 mm) − [H]	$L_m = K \cdot \sqrt{L_{calc-1} \cdot L_{calc-2}}$ $L_m = K \cdot \sqrt{L_{calc} \cdot L_{calc}} = K \cdot L_{calc}$ (K = 0.5)	$L_T = L_{calc-1} + L_{calc-2} + 2 \cdot L_m$ $= 2 \cdot L_{calc} + 2 \cdot L_m$ (K = 0.5)
10	5.3857×10^{-5}	2.6929×10^{-5}	1.6157×10^{-4}
20	1.5222×10^{-4}	7.611×10^{-5}	4.5666×10^{-4}
30	2.4523×10^{-4}	1.2262×10^{-4}	7.35×10^{-4}
60	2.009×10^{-4}	1.0045	6.027×10^{-4}

Table A.20 RFID rectangular spiral L_{calc}, L_m, and L_T for K = 0.7 (3 mm × 3 mm) and K = 0.3, 0.5 (4 mm × 4 mm)

Nc	$L_{calc-1} = L_{calc-2} = L_{calc}$, $\mu = \mu_r \cdot \mu_0$ $\mu_r = 100$ (3 mm × 3 mm) − [H]	$L_m = K \cdot \sqrt{L_{calc-1} \cdot L_{calc-2}}$ $L_m = K \cdot \sqrt{L_{calc} \cdot L_{calc}} = K \cdot L_{calc}$ (K = 0.7)	$L_T = L_{calc-1} + L_{calc-2} + 2 \cdot L_m =$ $2 \cdot L_{calc} + 2 \cdot L_m$ (K = 0.7)
10	5.3857×10^{-5}	3.77×10^{-5}	1.8311×10^{-4}
20	1.5222×10^{-4}	1.0655×10^{-4}	5.1755×10^{-4}
30	2.4523×10^{-4}	1.7166×10^{-4}	8.3378×10^{-4}
60	2.009×10^{-4}	1.4063×10^{-4}	6.8306×10^{-4}
Nc	$L_{calc-1} = L_{calc-2} = L_{calc}$, $\mu = \mu_r \cdot \mu_0$ $\mu_r = 100$ (4 mm × 4 mm) − [H]	$L_m = K \cdot \sqrt{L_{calc-1} \cdot L_{calc-2}}$ $L_m = K \cdot \sqrt{L_{calc} \cdot L_{calc}} = K \cdot L_{calc}$ (K = 0.3)	$L_T = L_{calc-1} + L_{calc-2} + 2 \cdot L_m =$ $2 \cdot L_{calc} + 2 \cdot L_m$ (K = 0.3)
10	8.0487×10^{-5}	2.4146×10^{-5}	2.0927×10^{-4}
20	2.425×10^{-4}	7.275×10^{-5}	6.305×10^{-4}
30	4.266×10^{-4}	1.2798×10^{-4}	0.0011
60	7.351×10^{-4}	2.2053×10^{-4}	0.0019
Nc	$L_{calc-1} = L_{calc-2} = L_{calc}$, $\mu = \mu_r \cdot \mu_0$ $\mu_r = 100$ (4 mm × 4 mm) − [H] − [H]	$L_m = K \cdot \sqrt{L_{calc-1} \cdot L_{calc-2}}$ $L_m = K \cdot \sqrt{L_{calc} \cdot L_{calc}} = K \cdot L_{calc}$ (K = 0.5)	$L_T = L_{calc-1} + L_{calc-2} + 2 \cdot L_m =$ $2 \cdot L_{calc} + 2 \cdot L_m$ (K = 0.5)
10	8.0487×10^{-5}	4.0244×10^{-5}	2.4146×10^{-4}
20	2.425×10^{-4}	1.2125×10^{-4}	7.275×10^{-4}
30	4.266×10^{-4}	2.133×10^{-4}	0.0013
60	7.351×10^{-4}	3.6755×10^{-4}	**0.0022 = 2.2 mH**

Table A.21 RFID rectangular spiral L_{calc}, L_m, and L_T for K = 0.7 (4 mm × 4 mm) and K = 0.3, 0.5 (2 mm × 2 mm)

Nc	$L_{calc-1} = L_{calc-2} = L_{calc}$, $\mu = \mu_r \cdot \mu_0$ $\mu_r = 100$ (4 mm × 4 mm) − [H]	$L_m = K \cdot \sqrt{L_{calc-1} \cdot L_{calc-2}}$ $L_m = K \cdot \sqrt{L_{calc} \cdot L_{calc}} = K \cdot L_{calc}$ (K = 0.7)	$L_T = L_{calc-1} + L_{calc-2}$ $+ 2 \cdot L_m = 2 \cdot L_{calc} +$ $2 \cdot L_m$ (K = 0.7)
10	8.0487×10^{-5}	5.6341×10^{-5}	2.7366×10^{-4}
20	2.425×10^{-4}	1.6975×10^{-4}	8.245×10^{-4}
30	4.266×10^{-4}	2.9862×10^{-4}	0.0015
60	7.351×10^{-4}	5.1457×10^{-4}	**0.0025 = 2.5 mH**
Nc	$L_{calc-1} = L_{calc-2} = L_{calc}$, $\mu = \mu_r \cdot \mu_0$ $\mu_r = 200$ (2 mm × 2 mm) − [H]	$L_m = K \cdot \sqrt{L_{calc-1} \cdot L_{calc-2}}$ $L_m = K \cdot \sqrt{L_{calc} \cdot L_{calc}} = K \cdot L_{calc}$ (K = 0.3)	$L_T = L_{calc-1}$ $+ L_{calc-2} + 2 \cdot L_m =$ $2 \cdot L_{calc} + 2 \cdot L_m$ (K = 0.3)
10	5.8439×10^{-5}	1.7532×10^{-5}	1.5194×10^{-4}

(continued)

Table A.21 (continued)

Nc	$L_{calc-1} = L_{calc-2} = L_{calc}$, $\mu = \mu_r \cdot \mu_0$ $\mu_r = 200$ (2 mm × 2 mm) − [H]	$L_m = K \cdot \sqrt{L_{calc-1} \cdot L_{calc-2}}$ $L_m = K \cdot \sqrt{L_{calc} \cdot L_{calc}} = K \cdot L_{calc}$ (K = 0.3)	$L_T = L_{calc-1} + L_{calc-2} + 2 \cdot L_m = 2 \cdot L_{calc} + 2 \cdot L_m$ (K = 0.3)
20	1.4048×10^{-4}	4.2144×10^{-5}	3.6525×10^{-4}
30	1.7064×10^{-4}	5.1192×10^{-5}	4.4366×10^{-4}

Nc	$L_{calc-1} = L_{calc-2} = L_{calc}$, $\mu = \mu_r \cdot \mu_0$ $\mu_r = 200$ (2 mm × 2 mm) − [H]	$L_m = K \cdot \sqrt{L_{calc-1} \cdot L_{calc-2}}$ $L_m = K \cdot \sqrt{L_{calc} \cdot L_{calc}} = K \cdot L_{calc}$ (K = 0.5)	$L_T = L_{calc-1} + L_{calc-2} + 2 \cdot L_m = 2 \cdot L_{calc} + 2 \cdot L_m$ (K = 0.5)
10	5.8439×10^{-5}	2.922×10^{-5}	1.7532×10^{-4}
20	1.4048×10^{-4}	7.024×10^{-5}	4.2144×10^{-4}
30	1.7064×10^{-4}	8.532×10^{-5}	5.1192×10^{-4}

Table A.22 RFID rectangular spiral L_{calc}, L_m, and L_T for K = 0.7 (2 mm × 2 mm)

Nc	$L_{calc-1} = L_{calc-2} = L_{calc}$, $\mu = \mu_r \cdot \mu_0$ $\mu_r = 200$ (2 mm × 2 mm) − [H]	$L_m = K \cdot \sqrt{L_{calc-1} \cdot L_{calc-2}}$ $L_m = K \cdot \sqrt{L_{calc} \cdot L_{calc}} = K \cdot L_{calc}$ (K = 0.7)	$L_T = L_{calc-1} + L_{calc-2} + 2 \cdot L_m = 2 \cdot L_{calc} + 2 \cdot L_m$ (K = 0.7)
10	5.8439×10^{-5}	4.0907×10^{-5}	1.9869×10^{-4}
20	1.4048×10^{-4}	9.8336×10^{-5}	4.7763×10^{-4}
30	1.7064×10^{-4}	1.1945×10^{-5}	5.8018×10^{-4}

Table A.23 RFID rectangular spiral L_{calc}, L_m, and L_T for K = 0.3 (3 mm × 3 mm)

Nc	L_{calc}-1 = $L_{calc-2} = L_{calc}$, $\mu = \mu_r \cdot \mu_0$ $\mu_r = 200$ (3 mm × 3 mm) − [H]	$L_m = K \cdot \sqrt{L_{calc-1} \cdot L_{calc-2}}$ $L_m = K \cdot \sqrt{L_{calc} \cdot L_{calc}} = K \cdot L_{calc}$ (K = 0.3)	$L_T = L_{calc-1} + L_{calc-2} + 2 \cdot L_m = 2 \cdot L_{calc} + 2 \cdot L_m$ (K = 0.3)
10	1.0771×10^{-4}	3.2313×10^{-5}	2.8005×10^{-4}
20	3.0445×10^{-4}	9.1335×10^{-5}	7.9157×10^{-4}
30	4.9045×10^{-4}	1.4714×10^{-4}	0.0013
60	4.0179×10^{-4}	1.2054×10^{-4}	0.001

Table A.24 RFID rectangular spiral L_{calc}, L_m, and L_T for K = 0.5 (3 mm × 3 mm)

Nc	$L_{calc-1} = L_{calc-2} = L_{calc}$, $\mu = \mu_r \cdot \mu_0$ $\mu_r = 200$ (3 mm × 3 mm) − [H]	$L_m = K \cdot \sqrt{L_{calc-1} \cdot L_{calc-2}}$ $L_m = K \cdot \sqrt{L_{calc} \cdot L_{calc}} = K \cdot L_{calc}$ (K = 0.5)	$L_T = L_{calc-1} + L_{calc-2} + 2 \cdot L_m = 2 \cdot L_{calc} + 2 \cdot L_m$ (K = 0.5)
10	1.0771×10^{-4}	5.3855×10^{-5}	3.2313×10^{-4}
20	3.0445×10^{-4}	1.5223×10^{-4}	9.1335×10^{-4}
30	4.9045×10^{-4}	2.4523×10^{-4}	0.0015
60	4.0179×10^{-4}	2.009×10^{-4}	0.0012

Table A.25 RFID rectangular spiral L_{calc}, L_m, and L_T for K = 0.7 (3 mm × 3 mm), K = 0.3, 0.5 (4 mm × 4 mm)

Nc	$L_{calc-1} = L_{calc-2} = L_{calc}$, $\mu = \mu_r \cdot \mu_0$ $\mu_r = 200$ (3 mm × 3 mm) − [H]	$L_m = K \cdot \sqrt{L_{calc-1} \cdot L_{calc-2}}$ $L_m = K \cdot \sqrt{L_{calc} \cdot L_{calc}} = K \cdot L_{calc}$ (K = 0.7)	$L_T = L_{calc-1} + L_{calc-2} + 2 \cdot L_m = 2 \cdot L_{calc} + 2 \cdot L_m$ (K = 0.7)
10	1.0771×10^{-4}	7.5397×10^{-5}	3.6621×10^{-4}
20	3.0445×10^{-4}	2.1312×10^{-4}	0.001
30	4.9045×10^{-4}	3.4332×10^{-4}	0.0017
60	4.0179×10^{-4}	2.8125×10^{-4}	0.0014
Nc	$L_{calc-1} = L_{calc-2} = L_{calc}$, $\mu = \mu_r \cdot \mu_0$ $\mu_r = 200$ (4 mm × 4 mm) − [H]	$L_m = K \cdot \sqrt{L_{calc-1} \cdot L_{calc-2}}$ $L_m = K \cdot \sqrt{L_{calc} \cdot L_{calc}} = K \cdot L_{calc}$ (K = 0.3)	$L_T = L_{calc-1} + L_{calc-2} + 2 \cdot L_m = 2 \cdot L_{calc} + 2 \cdot L_m$ (K = 0.3)
10	1.6097×10^{-4}	4.8291×10^{-5}	4.1852×10^{-4}
20	4.8501×10^{-4}	1.455×10^{-4}	0.0013
30	8.532×10^{-4}	2.5596×10^{-4}	**0.0022 = 2.2 mH**
60	0.0015	4.5×10^{-4}	0.0039
Nc	$L_{calc-1} = L_{calc-2} = L_{calc}$, $\mu = \mu_r \cdot \mu_0$ $\mu_r = 200$ (4 mm × 4 mm) − [H]	$L_m = K \cdot \sqrt{L_{calc-1} \cdot L_{calc-2}}$ $L_m = K \cdot \sqrt{L_{calc} \cdot L_{calc}} = K \cdot L_{calc}$ (K = 0.5)	$L_T = L_{calc-1} + L_{calc-2} + 2 \cdot L_m = 2 \cdot L_{calc} + 2 \cdot L_m$ (K = 0.5)
10	1.6097×10^{-4}	8.0485×10^{-5}	4.8291×10^{-4}
20	4.8501×10^{-4}	2.425×10^{-4}	0.0015
30	8.532×10^{-4}	4.266×10^{-4}	**0.0026 = 2.6 mH**
60	0.0015	7.5×10^{-4}	0.0045

Table A.26 RFID rectangular spiral L_{calc}, L_m, and L_T for K = 0.7 (4 mm × 4 mm)

Nc	$L_{calc-1} = L_{calc-2} = L_{calc}$, $\mu = \mu_r \cdot \mu_0$ $\mu_r = 200$ (4 mm × 4 mm) − [H]	$L_m = K \cdot \sqrt{L_{calc-1} \cdot L_{calc-2}}$ $L_m = K \cdot \sqrt{L_{calc} \cdot L_{calc}} = K \cdot L_{calc}$ (K = 0.7)	$L_T = L_{calc-1} + L_{calc-2} + 2 \cdot L_m = 2 \cdot L_{calc} + 2 \cdot L_m$ (K = 0.7)
10	1.6097×10^{-4}	1.1268×10^{-4}	5.473×10^{-4}
20	4.8501×10^{-4}	3.3951×10^{-4}	0.0016
30	8.532×10^{-4}	5.9724×10^{-4}	**0.0029 = 2.9 mH**
60	0.0015	0.001	0.0051

Table A.27 RFID rectangular spiral L_{calc}, L_m, and L_T for K = 0.3 (2 mm × 2 mm)

Nc	$L_{calc-1} = L_{calc-2} = L_{calc}$, $\mu = \mu_r \cdot \mu_0$ $\mu_r = 300$ (2 mm × 2 mm) − [H]	$L_m = K \cdot \sqrt{L_{calc-1} \cdot L_{calc-2}}$ $L_m = K \cdot \sqrt{L_{calc} \cdot L_{calc}} = K \cdot L_{calc}$ (K = 0.3)	$L_T = L_{calc-1} + L_{calc-2} + 2 \cdot L_m = 2 \cdot L_{calc} + 2 \cdot L_m$ (K = 0.3)
10	8.7659×10^{-5}	2.6298×10^{-5}	2.2791×10^{-4}
20	2.1072×10^{-4}	6.3216×10^{-5}	5.4787×10^{-4}
30	2.5596×10^{-4}	7.6788×10^{-5}	6.655×10^{-4}

Table A.28 RFID rectangular spiral L_{calc}, L_m, and L_T for K = 0.5 (2 mm × 2 mm)

Nc	$L_{calc-1} = L_{calc-2} = L_{calc}$, $\mu = \mu_r \cdot \mu_0$ $\mu_r = 300$ (2 mm × 2 mm) − [H]	$L_m = K \cdot \sqrt{L_{calc-1} \cdot L_{calc-2}}$ $L_m = K \cdot \sqrt{L_{calc} \cdot L_{calc}} = K \cdot L_{calc}$ (K = 0.5)	$L_T = L_{calc-1} + L_{calc-2} + 2 \cdot L_m = 2 \cdot L_{calc} + 2 \cdot L_m$ (K = 0.5)
10	8.7659×10^{-5}	4.383×10^{-5}	2.6298×10^{-4}
20	2.1072×10^{-4}	1.0536×10^{-4}	6.3216×10^{-4}
30	2.5596×10^{-4}	1.2798×10^{-4}	7.6788×10^{-4}

Table A.29 RFID rectangular spiral L_{calc}, L_m, and L_T for K = 0.7 (2 mm × 2 mm)

Nc	$L_{calc-1} = L_{calc-2} = L_{calc}$, $\mu = \mu_r \cdot \mu_0$ $\mu_r = 300$ (2 mm × 2 mm) − [H]	$L_m = K \cdot \sqrt{L_{calc-1} \cdot L_{calc-2}}$ $L_m = K \cdot \sqrt{L_{calc} \cdot L_{calc}} = K \cdot L_{calc}$ (K = 0.7)	$L_T = L_{calc-1} + L_{calc-2} + 2 \cdot L_m = 2 \cdot L_{calc} + 2 \cdot L_m$ (K = 0.7)
10	8.7659×10^{-5}	6.1361×10^{-5}	2.9804×10^{-4}
20	2.1072×10^{-4}	1.475×10^{-4}	7.1645×10^{-4}
30	2.5596×10^{-4}	1.7917×10^{-4}	8.7026×10^{-4}

Table A.30 RFID rectangular spiral L_{calc}, L_m, and L_T for K = 0.3 (3 mm × 3 mm)

Nc	$L_{calc-1} = L_{calc-2} = L_{calc}$, $\mu = \mu_r \cdot \mu_0$ $\mu_r = 300$ (3 mm × 3 mm) − [H]	$L_m = K \cdot \sqrt{L_{calc-1} \cdot L_{calc-2}}$ $L_m = K \cdot \sqrt{L_{calc} \cdot L_{calc}} = K \cdot L_{calc}$ (K = 0.3)	$L_T = L_{calc-1} + L_{calc-2} + 2 \cdot L_m = 2 \cdot L_{calc} + 2 \cdot L_m$ (K = 0.3)
10	1.6157×10^{-4}	4.84×10^{-5}	4.2008×10^{-4}
20	4.5667×10^{-4}	1.37×10^{-4}	0.0012
30	7.3568×10^{-4}	2.207×10^{-4}	0.0019
60	6.0269×10^{-4}	1.8081×10^{-4}	0.0016

Table A.31 RFID rectangular spiral L_{calc}, L_m, and L_T for K = 0.5 (3 mm × 3 mm)

Nc	$L_{calc-1} = L_{calc-2} = L_{calc}$, $\mu = \mu_r \cdot \mu_0$ $\mu_r = 300$ (3 mm × 3 mm) − [H]	$L_m = K \cdot \sqrt{L_{calc-1} \cdot L_{calc-2}}$ $L_m = K \cdot \sqrt{L_{calc} \cdot L_{calc}} = K \cdot L_{calc}$ (K = 0.5)	$L_T = L_{calc-1} + L_{calc-2} + 2 \cdot L_m = 2 \cdot L_{calc} + 2 \cdot L_m$ (K = 0.5)
10	1.6157×10^{-4}	8.0785×10^{-5}	4.847×10^{-4}
20	4.5667×10^{-4}	2.2834×10^{-4}	0.0014
30	7.3568×10^{-4}	3.6784×10^{-4}	**0.0022 = 2.2 mH**
60	6.0269×10^{-4}	3.0135×10^{-4}	0.0018

Table A.32 RFID rectangular spiral L_{calc}, L_m, and L_T for K = 0.7 (3 mm × 3 mm)

Nc	$L_{calc\text{-}1} = L_{calc\text{-}2} = L_{calc}$, $\mu = \mu_r \cdot \mu_0$ $\mu_r = 300$ (3 mm × 3 mm) − [H]	$L_m = K \cdot \sqrt{L_{calc-1} \cdot L_{calc-2}}$ $L_m = K \cdot \sqrt{L_{calc} \cdot L_{calc}} = K \cdot L_{calc}$ (K = 0.7)	$L_T = L_{calc\text{-}1} + L_{calc\text{-}2} + 2 \cdot L_m = 2 \cdot L_{calc} + 2 \cdot L_m$ (K = 0.7)
10	1.6157×10^{-4}	1.131×10^{-4}	5.4934×10^{-4}
20	4.5667×10^{-4}	3.1967×10^{-4}	0.0016
30	7.3568×10^{-4}	5.1498×10^{-4}	**0.0025 = 2.5 mH**
60	6.0269×10^{-4}	4.2188×10^{-4}	0.002

Table A.33 RFID rectangular spiral L_{calc}, L_m, and L_T for K = 0.3 (4 mm × 4 mm)

Nc	$L_{calc\text{-}1} = L_{calc\text{-}2} = L_{calc}$, $\mu = \mu_r \cdot \mu_0$ $\mu_r = 300$ (4 mm × 4 mm) − [H]	$L_m = K \cdot \sqrt{L_{calc-1} \cdot L_{calc-2}}$ $L_m = K \cdot \sqrt{L_{calc} \cdot L_{calc}} = K \cdot L_{calc}$ (K = 0.3)	$L_T = L_{calc\text{-}1} + L_{calc\text{-}2} + 2 \cdot L_m = 2 \cdot L_{calc} + 2 \cdot L_m$ (K = 0.3)
10	2.4146×10^{-4}	7.2438×10^{-5}	6.278×10^{-4}
20	7.2751×10^{-4}	2.1825×10^{-4}	0.0019
30	0.0013	3.9×10^{-4}	0.0034
60	0.0022	6.6×10^{-4}	0.0057

Table A.34 RFID rectangular spiral L_{calc}, L_m, and L_T for K = 0.5 (4 mm × 4 mm)

Nc	$L_{calc\text{-}1} = L_{calc\text{-}2} = L_{calc}$, $\mu = \mu_r \cdot \mu_0$ $\mu_r = 300$ (4 mm × 4 mm) − [H]	$L_m = K \cdot \sqrt{L_{calc-1} \cdot L_{calc-2}}$ $L_m = K \cdot \sqrt{L_{calc} \cdot L_{calc}} = K \cdot L_{calc}$ (K = 0.5)	$L_T = L_{calc\text{-}1} + L_{calc\text{-}2} + 2 \cdot L_m = 2 \cdot L_{calc} + 2 \cdot L_m$ (K = 0.5)
10	2.4146×10^{-4}	1.2073×10^{-4}	7.2438×10^{-4}
20	7.2751×10^{-4}	3.6376×10^{-4}	**0.0022 = 2.2 mH**
30	0.0013	6.5×10^{-4}	0.0039
60	0.0022	0.0011	0.0066

Table A.35 RFID rectangular spiral L_{calc}, L_m, and L_T for K = 0.7 (4 mm × 4 mm)

Nc	$L_{calc\text{-}1} = L_{calc\text{-}2} = L_{calc}$, $\mu = \mu_r \cdot \mu_0$ $\mu_r = 300$ (4 mm × 4 mm) − [H]	$L_m = K \cdot \sqrt{L_{calc-1} \cdot L_{calc-2}}$ $L_m = K \cdot \sqrt{L_{calc} \cdot L_{calc}} = K \cdot L_{calc}$ (K = 0.7)	$L_T = L_{calc\text{-}1} + L_{calc\text{-}2} + 2 \cdot L_m = 2 \cdot L_{calc} + 2 \cdot L_m$ (K = 0.7)
10	2.4146×10^{-4}	1.6902×10^{-4}	8.2096×10^{-4}
20	7.2751×10^{-4}	5.0926×10^{-4}	**0.0025 = 2.5 mH**
30	0.0013	9.1×10^{-4}	0.0044
60	0.0022	0.0015	0.0075

Conductivity of Silver $\sigma = 6.1 \times 10^7 (\mho/\text{m})$. Conductivity has SI units of Siemens per meter (S/m).

$$\sigma_{silver@20\,^\circ\text{C}} = 6.3 \times 10^7 (\text{S/m}).$$

The track depth (x) does not influence our total inductance, and then we can take it as a variable and find his minimum value for $R_{DC} < 10\ \Omega$. $S = 20\,\mu\text{m} \cdot x$.

Appendix B
RF Amplifiers Basic and Advance Topics and Design Methods

B.1 Amplifier Design Concepts and Matching Guidelines

An amplifier is an active device that has the ability to amplify voltage, current and amplify both voltage and current. There are some types of amplifiers. Amplifiers types: zero frequency amplifiers (DC amplifiers), low frequency amplifiers (Audio amplifiers), and high frequency amplifiers (RF amplifiers). Power is $P = V \cdot I$, when current (I) or voltage (V) is raising and will create amplification. The main desirable specification when designing amplifier: High P1dB (high input power that cause the UUT gain to drop by 1dB from small signal value). P1dB (IP1dB, OP1dB), low noise (example LNA), and maximum efficiency (Efficiency (P.E %) = [output signal power]/[power supply power]). High Gain (S21), b_2 is output signal power, a_1 is input signal power $S21 = b_2/a_1$ for $a_2 = 0$ (no input signal at amplifier output). Good return loss (RL). S11 (IRL) $\rightarrow \varepsilon$ and (ORL) $\rightarrow 1$. Return Loss = Reflection Loss. Amplifiers come in three flavors: Common Base (CB) amplifiers, Common Collector (CC) amplifiers, and Common Emitter (CE) amplifiers. It depends whether the base, collector or emitter is common to both the input and output of the amplifier. Common Base (CB) amplifier, input signal inserted at emitter (E) and output signal taken from the collector (C). The CB amplifier can operate as a voltage amplifier for low input impedance circuits.

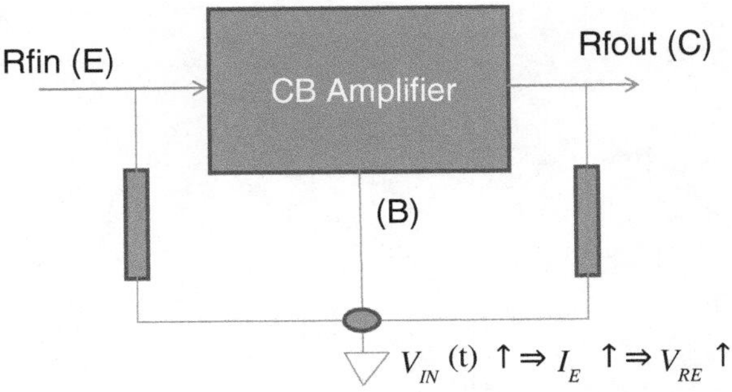

$V_{IN}(t) \uparrow \Rightarrow I_E \uparrow \Rightarrow V_{RE} \uparrow$
$\Rightarrow 0$ *degree phase shift*

Fig. B.1 Common base (CB) amplifier

O. Aluf, *Microwave RF Antennas and Circuits*,
DOI 10.1007/978-3-319-45427-6

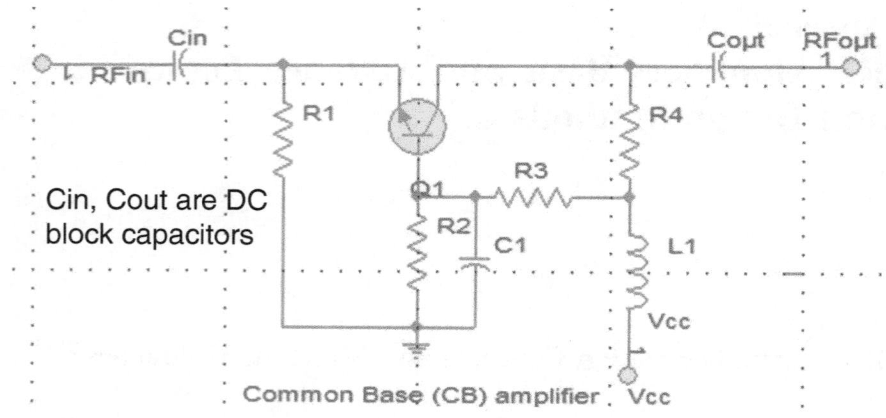

Fig. B.2 Common base (CB) amplifier schematic

CB amplifier can be found at the 50 Ω antenna input of the radio receiver (Figs. B.1 and B.2).

There is a JFET's CB amplifier circuit which can be used in receiver's IF unit. C_2, C_3, R_2 and RFC (RF choke) are for decoupling. C_4, C_6 are RF decoupling. C_5 is for flatter frequency response throughout its pass band. T_1 is for impedance matching (Fig. B.3).

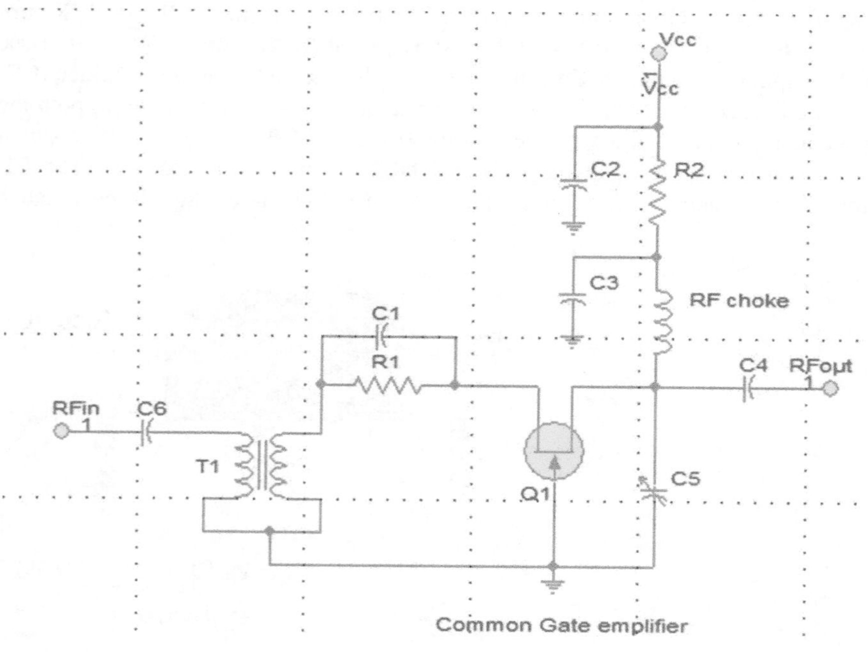

Fig. B.3 jfet common gate amplifier

There is a JFET's CB amplifier circuit which can be used in receiver's IF unit.

C_2, C_3, R_2 and RFC (RF choke) are for decoupling. C_5 is for flatter frequency response throughout its pass band. T_1 is for impedance matching. The most popular amplifier circuit is Common Emitter (CE). The CE amplifier has a greater current gain and voltage gain combination than any other type. CE amplifier make excellent

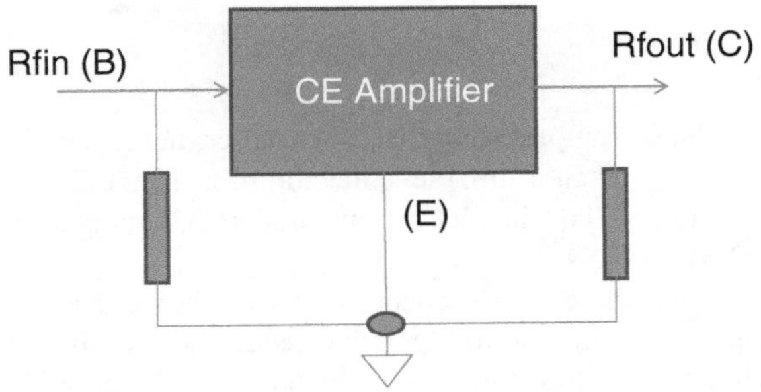

Fig. B.4 Common Emitter (CE) amplifier

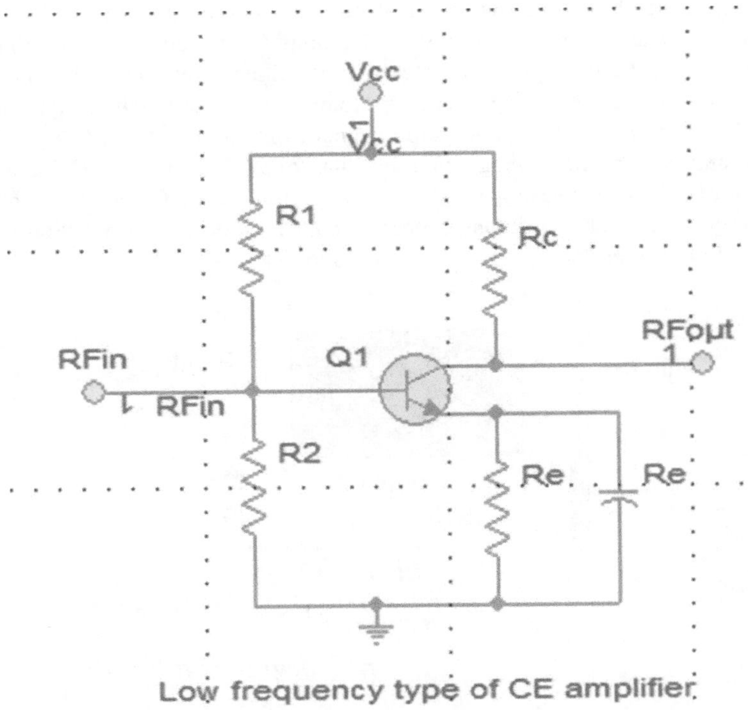

Fig. B.5 Common emitter (CE) amplifier schematic

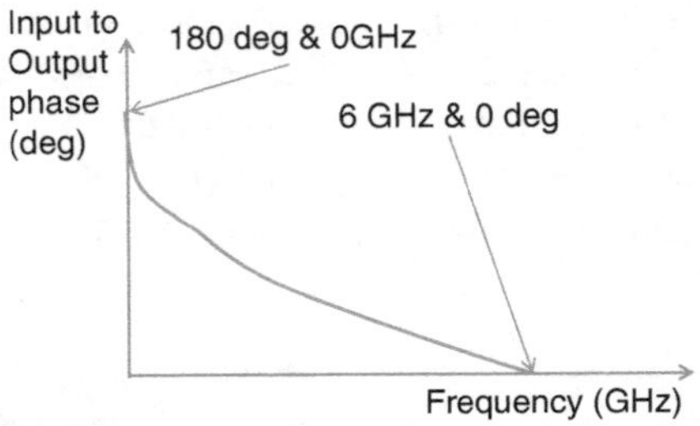

Fig. B.6 Input to output phase (deg) versus frequency (GHz)

power amplifier. The Common Emitter (CE) amplifier, input signal inserted at base (B) and output signal taken from the collector (C). The CE amplifier's output voltage is shifted by ~180° in phase compared to CE amplifier's input signal (Figs. B.4 and B.5).

At RF frequencies there is an effect of "positive feedback", which creates amplifier instability and oscillation. "positive feedback" is when there is internal feedback capacitance between transistor's collector and its base. It cause to undesired CE oscillations. Inter feedback capacitance can be as high as 25 pF or higher. At a specific frequency this capacitance will send an in phase signal back into the base input from the collector's output. This back in phase signal creates "oscillator". Transistor's internal resistance and capacitance along with other phase delays yield a powerful phase shift to normally out of phase 180° feedback signal. Only phase delays that are at a total 360° (0°) will bring amplifier instability and oscillations.

A phase = g(frequency) for a typical CE amplifier unmatched (Fig. B.6).

The Common Collector (CC) amplifier (emitter follower amplifier) has the input signal inserted into the base, and output signal from emitter. The CC amplifier has current and power gain, voltage gain less than one ($G_V < 1$). The CC amplifier's used as a buffer or active impedance matching circuit. The CC amplifier has high input impedance and low output impedance. There is no phase inversion between CC amplifier's input and output (Figs. B.7 and B.8).

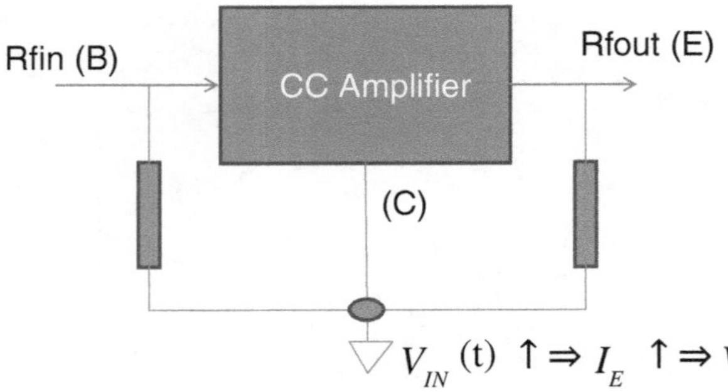

$$V_{IN}(t)\uparrow \Rightarrow I_E \uparrow \Rightarrow V_{RE}\uparrow$$
$$\Rightarrow 0\ \textit{degree phase shift}$$

Fig. B.7 Common collector (CC) amplifier

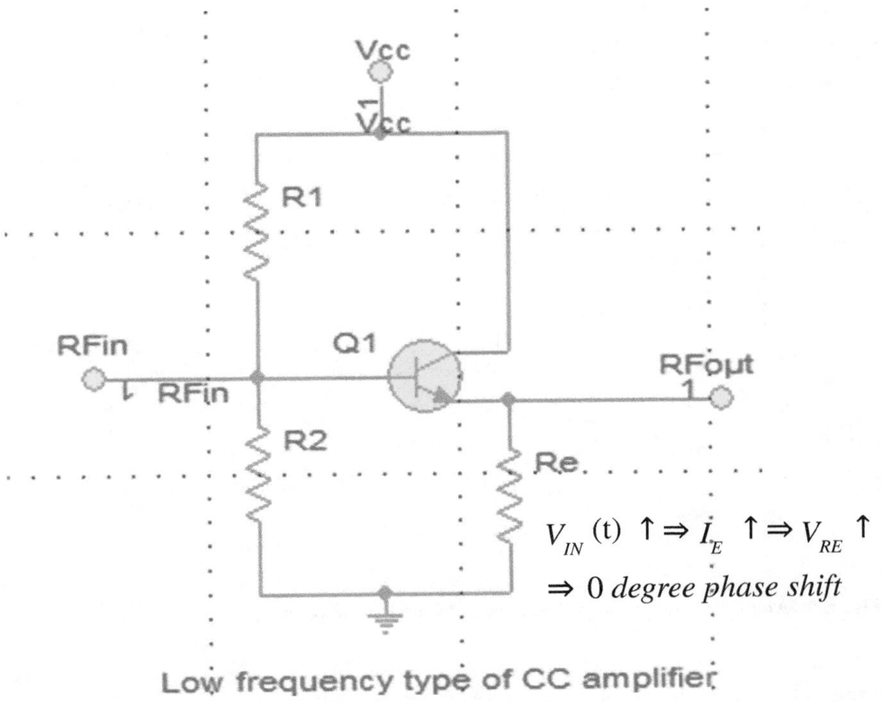

Fig. B.8 Common Collector (CC) amplifier schematic

When an amplifier's output impedance matches the load impedance, maximum power is transferred to the load ($R_S = R_L$). When an amplifier's output impedance matches the load impedance ($R_S = R_L$) all reflections are eliminated. When an amplifier's output impedance (R_S) unmatched the load impedance (R_L), there are reflections and less than maximum power is transferred to the load R_L. The amplifier's efficiency in its DC case is defined by (Fig. B.9 and Table B.1).

$$EFF(\%) = \left[\frac{R_{load}}{R_{load} + R_{source}}\right] \times 100\,\%.$$

$$\begin{bmatrix} \text{Point of max} \\ \text{Power transfer} \\ R_L = R_S \end{bmatrix} \neq \begin{bmatrix} \text{Point of max (DC)} \\ \text{amplifier efficiency} \\ EFF(\%) \rightarrow 100\,\% \\ R_L \gg R_S(R_S \rightarrow \varepsilon) \\ R_L \rightarrow \infty \end{bmatrix}$$

In order to develop maximum power the Z_{out} of the amplifier must be complex conjugate of the Z_{in} of the load. The low efficiency (EFF%) level can be increased if the load has a higher input resistance, thus dropping more power across the load

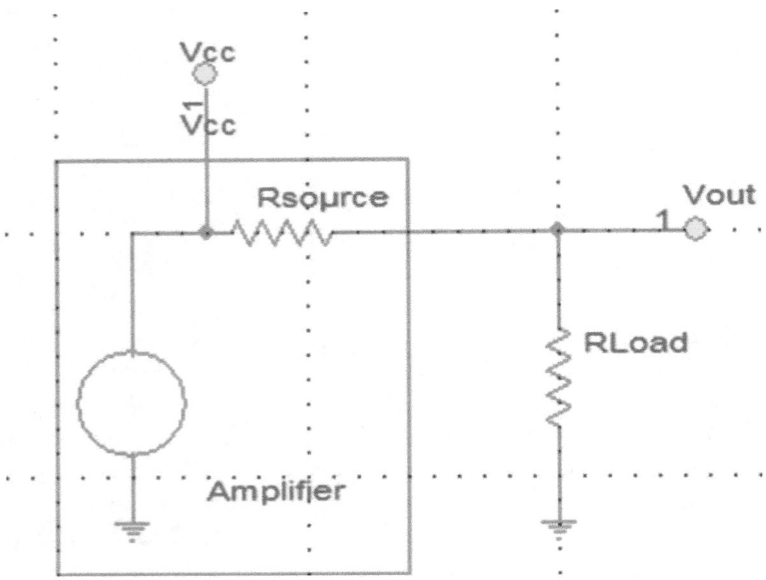

Fig. B.9 Amplifier's output impedance matches the load impedance

Table B.1 Amplifier's conditions and power across the load

Condition	Voltage or current on the load	Power across the load	Current or voltage drawn from Amplifier's power supply
$R_L \to \varepsilon$ $R_L = 0$ (Short load)	$V_{OUT} = V_L \to \varepsilon$ $V_L = 0$	$P_L = 0$ $P_{OUT} = P_L \to \varepsilon$	Load DC current = Vcc/Rs (maximum DC power)
$R_L \to \infty$ (No load)	$I_L \to \varepsilon$ $I_L = 0$	$P_L = 0$	Load DC voltage = Vcc but Iload = 0
$R_L = R_S$ (max power transfer)	$I_L = \frac{V_{cc}}{R_S + R_L}\Big\vert_{R_S = R_L} = \frac{V_{cc}}{2 \cdot R_S} = \frac{V_{cc}}{2 \cdot R_L}$	$P_{OUT} = P_L \to \varepsilon$ $P_L = I_L^2 \cdot R_L = \frac{V_{cc}^2}{4 \cdot R_L}$	Load DC voltage = Vcc/2 Load DC current = Vcc/(Rsource + Rload)

($Z_{load} \gg Z_{out}$). The total output power across load will be less in this condition ($Z_{load} \gg Z_{out}$) than if $Z_{load} = Z_{out}$ (pure resistive). The transfer of maximum power from the source to the load will not maximize efficiency (EEF%). Maximum power transfer only occurs when the source impedance equals the load impedance $Z_{load} = Z_{out}$ (pure resistive). Any impedance mismatches will end in a loss of power, Mismatch Loss (ML). ML(dB) = Mismatch Loss. VSWR = Voltage Stand

Fig. B.10 Amplifier's Z_{out} and load

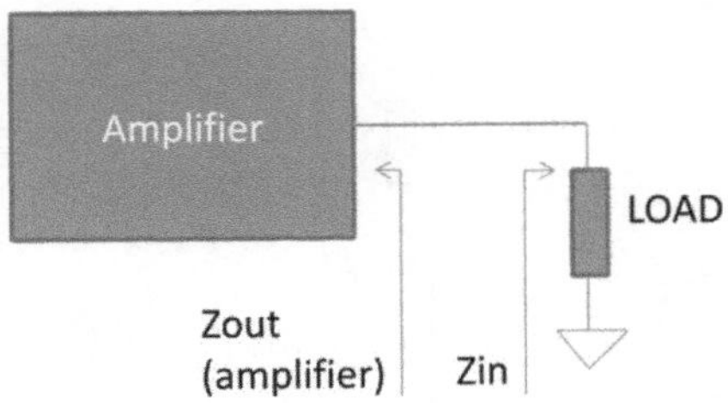

Wave Ratio (dimensionless units). $ML = 10 \cdot \log_{10}^{[1-(\frac{VSWR-1}{VSWR+1})]^2}$; Mismatch Loss (ML) (Fig. B.10).

$$ML = 10 \cdot \log_{10}^{[1-(\frac{VSWR-1}{VSWR+1})]^2}\Bigg|_{VSWR=\frac{1+|\Gamma|}{1-|\Gamma|}} = 10 \cdot \log_{10}^{\left[1-\left(\frac{\left\{\frac{1+|\Gamma|}{1-|\Gamma|}\right\}-1}{\left\{\frac{1+|\Gamma|}{1-|\Gamma|}\right\}+1}\right)\right]^2}$$

$$ML = 10 \cdot \log_{10}^{\left[1-\left(\frac{\left\{\frac{2|\Gamma|}{1-|\Gamma|}\right\}}{\left\{\frac{2}{1-|\Gamma|}\right\}}\right)\right]^2} = ML = 10 \cdot \log_{10}^{[1-|\Gamma|]^2}$$

Good amplifier match ($\Gamma = 0$), no reflections then VSWR = 1; ML = 10 · log (1) = 0. Bad amplifier match ($\Gamma = 1$), all power is reflected then VSWR $\rightarrow \infty$; 10 · log(0) = $-\infty$.

Amplifier matching allows: amplifier maximum power transfer and attenuation of harmonics to be achieved between stages. RF matching network able and facilitate impedance matching and filtering of signal, coupling between RF stages. The amplifier matching networks types are L matching network, T matching network, and PI matching network. In design of microwave matching network the device parasitic and the effect of length on RF circuit matching are very important.

L matching network: LC matching topology which especially for narrowband impedance matching. The L network name is due to its L shape and can furnish low pass filtering to decrease harmonic output. Two stages (amplifier, and load) with no matching network. Two stages (amplifier, load) with simple low pass L network between stages. Low pass L network can matches a higher output impedance source (Z_{high}) to a lower input impedance load (Z_{low}). Low pass L network can match also a lower output impedance source (Z_{low}) to a higher input impedance load (Z_{high}) (Fig. B.11).

Fig. B.11 Amplifier (source) and load poor match

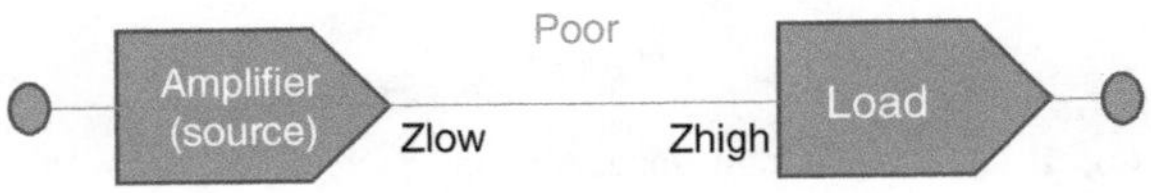

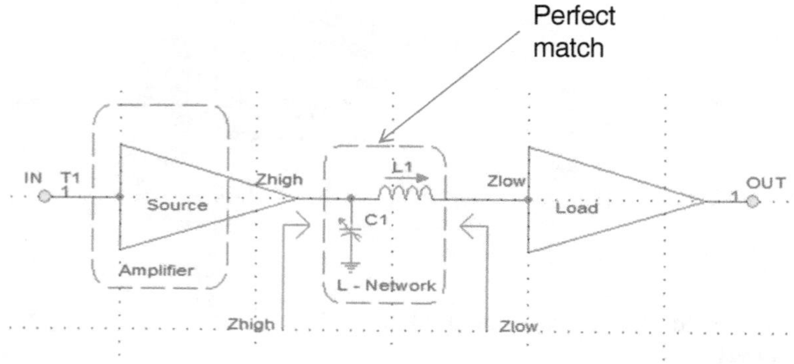

Fig. B.12 High to low impedance matching L network

Figure B.12 is a high to low ($Z_{high} \rightarrow Z_{low}$) impedance matching L network between two amplifiers.

Figure B.13 is a low to high ($Z_{low} \rightarrow Z_{high}$) impedance matching L network between two amplifiers.

A T matching network is a popular impedance matching network circuit. It can furnish almost any impedance matching level between two stages and we can selectable loaded Q (Fig. B.14).

A PI matching network can be applicable to many matching applications all types. We can alter the ratio between capacitors C_1 and C_2 in the next figure so the output impedance of the load can be matched to the source impedance. PI network also decreasing the harmonic output (Figs. B.15 and B.16).

PI network's topology is a low pass filter. It has a small resonant band pass like S21 gain pick and excellent return loss S11 at specific frequency (Fig. B.17).

It is very important in any microwave matching circuit first that device parasitic are part of any active or passive component and second the effect that length has a

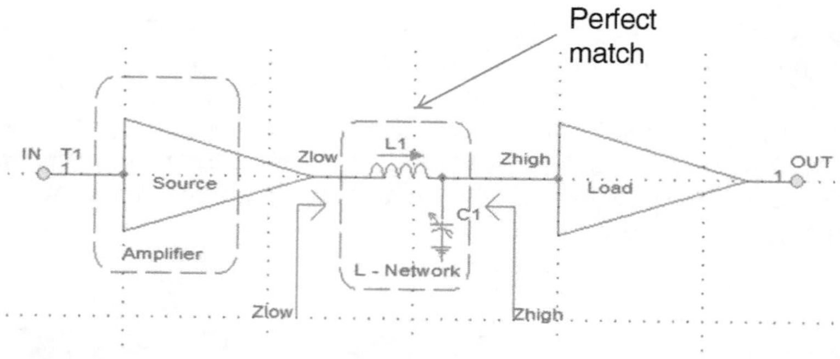

Fig. B.13 Low to high impedance matching L network

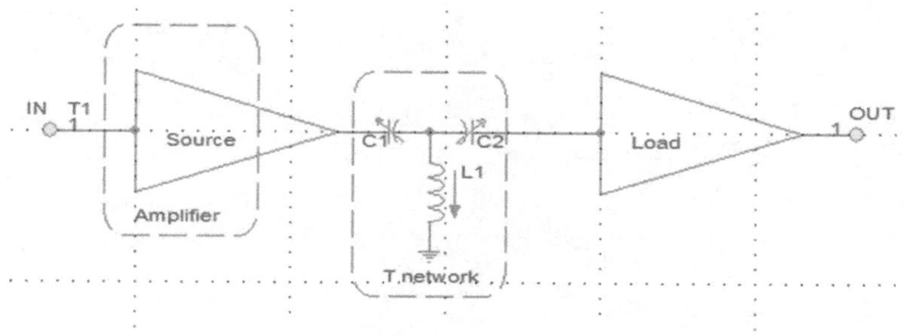

Fig. B.14 Impedance matching T network

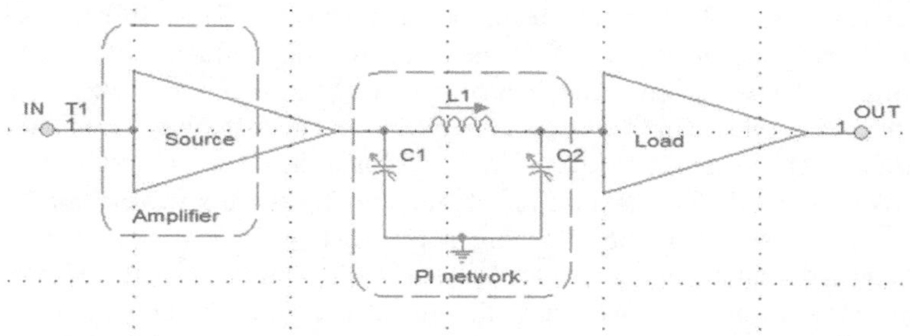

Fig. B.15 PI impedance matching network

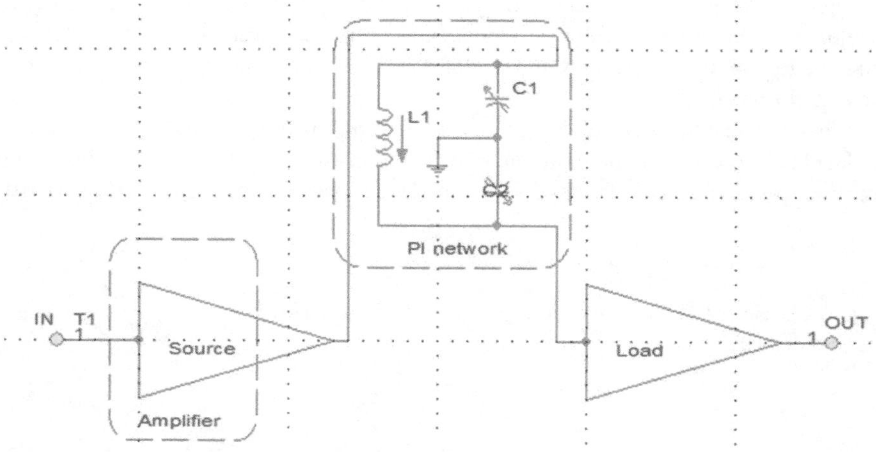

Fig. B.16 PI impedance matching network

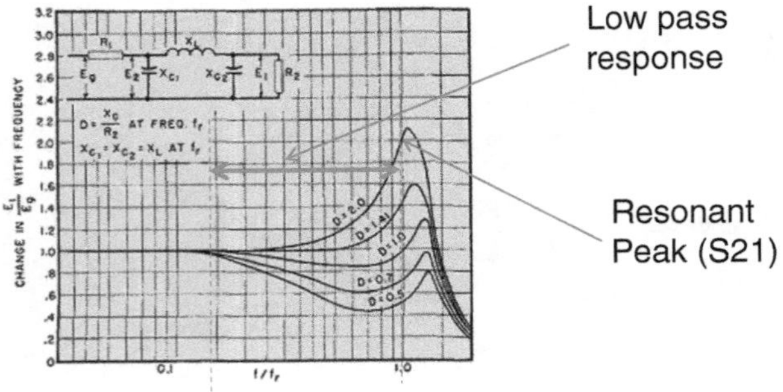

Fig. B.17 PI network's topology as a low pass filter

RF circuit matching. There are vital components parasitic effects. Effect that length: PCB trace and pad reactance (j · X elements), and strong influence of circuit length. In low frequency we can ignore effect of length. In low frequency we not consider PCB parasitic effects and distance between each matching component and source/load. At higher RF frequencies the distance between components and circuits is very critical. The distance affects the moving wave's wavelengths and the expected performance of the matching network. It is very important where the circuit and individual components are located on PCB. Any mismatched line that is a significant portion of a wavelength, the impedance will vary along that line. The variations on a mismatched line are due to the standing wave (SWR). The reflected RF wave is bouncing off of the mismatch load and interacting with forward wave. The interaction between reflected and forward waves, creating fixed peaks and valley of voltage and current. The fixed peaks and valleys of voltage and current are created at every half wavelength (λ/2) along the trace. The distance between each circuit element is varied; this distance will completely destroy any predicted RF match. Figure B.18 describes the voltage and current standing waves on a mismatched transmission line.

First the match is calculated with zero micro-strip length and second the calculated match will degrade with micro-strip length. We must take into consideration the micro-strips length effects in our impedance matching calculation (Fig. B.19).

Fig. B.18 Current and voltage signal amplitudes versus wavelength

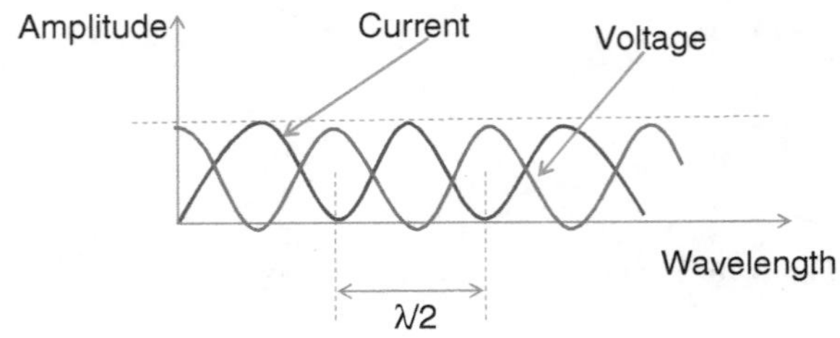

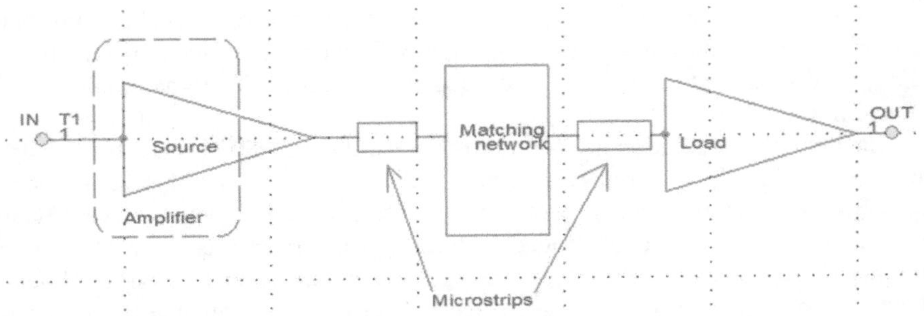

Fig. B.19 Amplifier and load, matching network with microstrips

B.2 Amplifier Distortion and Noise Products

There are two unavoidable and undesirable elements of any electronic circuits: distortion and noise. Distortion can deform the carrier and its sidebands at the transmitter and receiver. Distortion causes to spectral regrowth and adjacent channel interface. Additionally distortion causes faulty, distorted replica of the original baseband signal, and increasing the Bit Error Rate (BER). Noise degrades all important BER of the entire system. Distortion forms frequencies inter modulation products. Distortion frequency inter-modulation is cause by internal nonlinear mixing of any signal with one or more other signals, mixing with other signals, and modulated or unmodulated waveform that is altered is shape or amplitude from the original signal (improper circuit response). The distortion types are frequency distortion, amplitude and phase distortion, inter-modulation distortion, second order inter modulation distortion, harmonic distortion and noise. Frequency distortion happened when passive or active circuit increases or decreases the amplitude of particular frequencies differently than the other frequencies. Frequency distortion is a common problem wide band IF or RF amplifiers. The frequency limitations of amplifiers have many causes. The reasons for frequency limitation of amplitudes are active device's transit time, negative

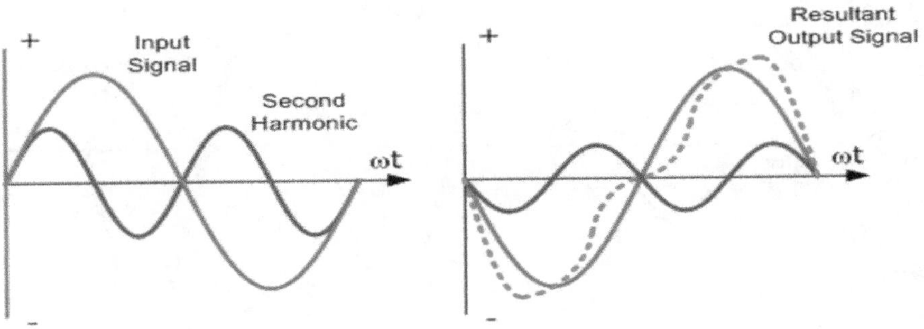

Fig. B.20 Input signal and resultant output signal

effects of junction capacitance, reactive nature of the transistor's matching, filtering and coupling, and decoupling networks (highly frequency dependent and act like a filter, band-pass, high-pass, etc.,). The frequency distortion due to harmonic is presented in Fig. B.20.

The amplitude distortion is a form of nonlinear distortion (nonlinear behaviour). It is produced by the incorrect biasing of an amplifier. It cause to either saturation or cut-off of the transistor and generates harmonics and IMD products. Overdriving the input of the amplifier (overload distortion) will create this same effect (flat topping). Overdriving the input of the amplifier is not depending on amplifier's bias. The harmonics and IMDs generated will produce interference to other services. The harmonics and IMDs generated will produce interference to adjacent channels. It will increase the system BER (Bit Error Rate) in a digital data radio. The voice band device will have an output signal with a harsh, coarse output. The amplitude distortion due to incorrect biasing is described in Fig. B.21.

The amplitude distortion due to clipping is described in Fig. B.22.

In CE amplifier during the amplification process of the signal waveform, some form of amplifier distortion has occurred. CE amplifier's amplification may not be

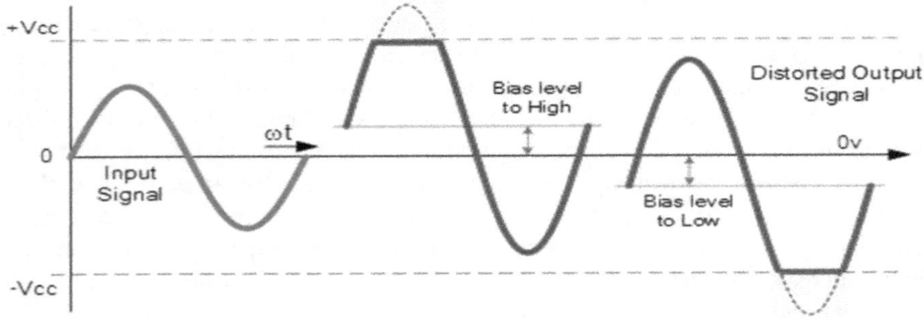

Fig. B.21 The amplitude distortion due to incorrect biasing

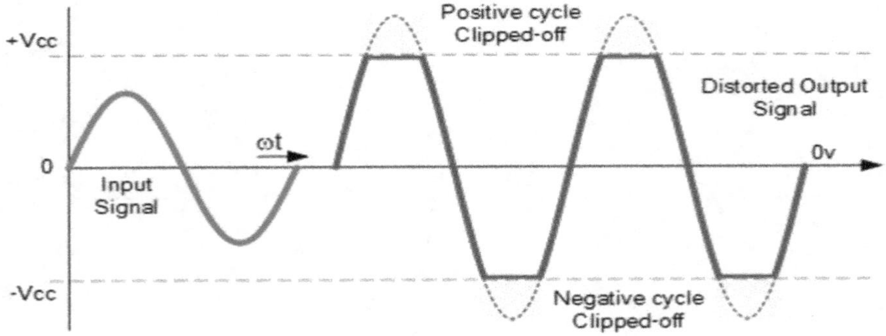

Fig. B.22 The amplitude distortion due to clipping

taking place over the whole signal cycle due to incorrect biasing. If CE amplifier's input signal is too large, it causes the amplifier to be limited by the supply voltage (clipping). CE amplifier's amplification may not be linear over the entire frequency range of inputs. CE amplifier's multiplication factor is called the Beta (β) value of the transistor. Common emitter or even common source type transistor circuits work fine for small AC input signals. CE amplifiers suffer from one major disadvantages: The bias Q-point of a bipolar amplifier depends on the same (β) value which may vary from transistors of the same type, Q-point for one transistor is not necessarily the same as the Q-point for another transistor of the same type due to the inherent manufacturing tolerances. If the CE amplifiers suffer from one major disadvantage the amplifier may not be linear, amplitude distortion will result, and a carful choice of the transistor and biasing components can minimize the effect of amplifier distortion. The CE amplifier's amplitude distortion is presented in Fig. B.23.

Phase distortion or delay distortion occurs in a nonlinear transistor amplifier when there is a time delay between the input signal and its appearance at the output (Fig. B.24).

The phase change between the input and the output is zero at the fundamental frequency. The resultant phase angle delay will be the different between the

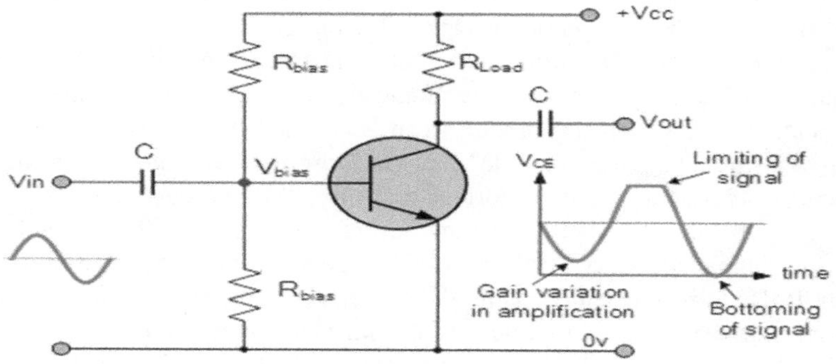

Fig. B.23 CE amplifier's amplitude distortion

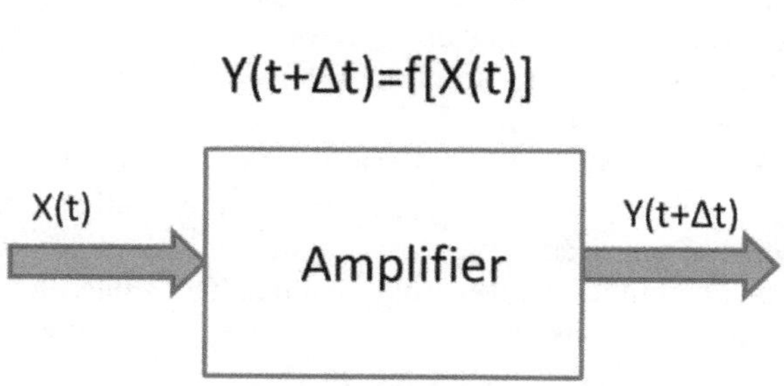

Fig. B.24 Amplifier's input signal X(t) and output signal Y(t + Δt)

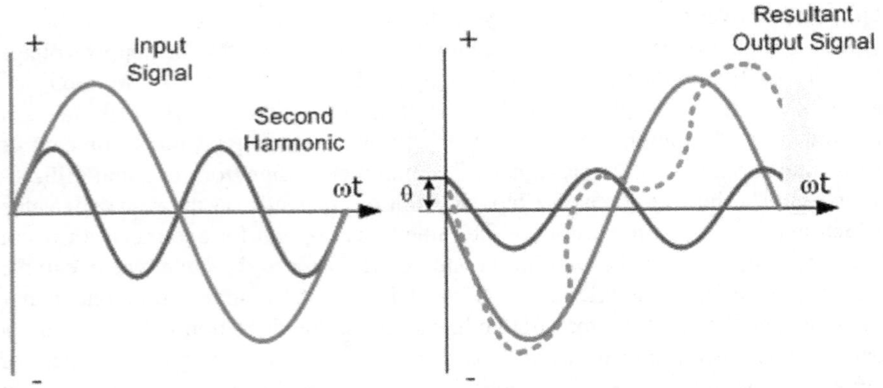

Fig. B.25 Amplifier's phase distortion due to delay

harmonic and fundamental. The time delay (Δ) will depend on the construction of the amplifier and will increase progressively with frequency within the bandwidth of the amplifier. Any practical amplifier will have a combination of both "frequency" and "phase" distortion together with amplitude distortion. Most applications such as in audio amplifiers or power amplifiers, unless the distortion is excessive or severe it will not generally affect the operation of the system. Figure B.25 describes the phase distortion due to delay.

The Intermodulation Distortion (ID), quite similar to the amplitude distortion. ID is produced when frequencies not harmonically related to the fundamental. Inter Modulation Distortion (IMD) products can be formed by mixing together of the carrier with interferers, harmonic, IMD products from other stages, other channels, or sideband, producing various spurious response. IMD products are in band and can swamp the desired signal, creating severs interference. When neighbouring transmitted signal arrives at a PA's stage, mix together with the transmitter's carrier, causing IMDs to be created. ID is produced when two or more frequencies mix in any nonlinear device. It causes numerous sum and different combinations of the original fundamental frequencies (second order products: $f_1 + f_2$, $f_1 - f_2$). It causes intermodulation products ($m \cdot f_1 + n$, $m \cdot f_1 \cdot f_2 - n \cdot f_2$), n and m are whole numbers. Third order ID products, which would be $2 \cdot f_1 + f_2$, $2 \cdot f_1 - f_2$, $2 \cdot f_2 + f_1$, $2 \cdot f_2 - f_1$ can be most damaging of the higher or lower IMDs. The second order IMD products would usually be too far from the receivers or transmitter's band pass to create many problems (Fig. B.26).

Third Order Intercept Point (TOIP, IP3): Third order spurious products will be created within nonlinearity of a device (linear amplifier, active filter, and mixer). Output IP3 point can never actually reached, since the amplifier will go into saturation before this amplitude is ever truly attained. The value of the IP3 must be measured only when it is in its linear operating range (DUT is not in compression). Typically amplifier's third order intercept point (IP3) is located approximately 10 to 15 dB above its P1 dB compression point. The output IP3 (OIP3) for a BJT

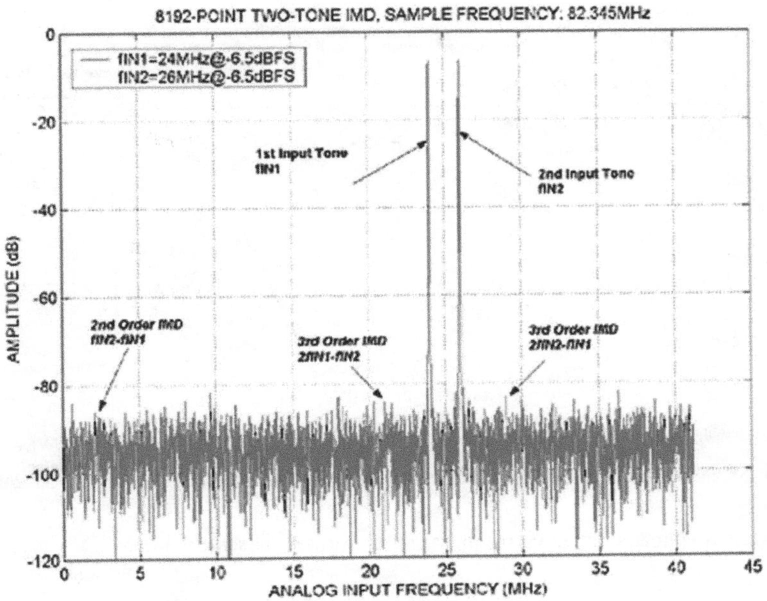

Fig. B.26 Amplifier's Inter modulation distortion (IMD)

amplifier approximated by $OIP3 = 10 \cdot \log[V_{CE} \cdot I_C \cdot 5]$, V_{CE} is transistor's collector to emitter voltage [V]. I_C is transistor's collector current [mA]. The higher the bias level, the higher will be the IP3 of the amplifier. I_C is the easiest bias parameter to increase for high IP3 (Figs. B.27 and B.28).

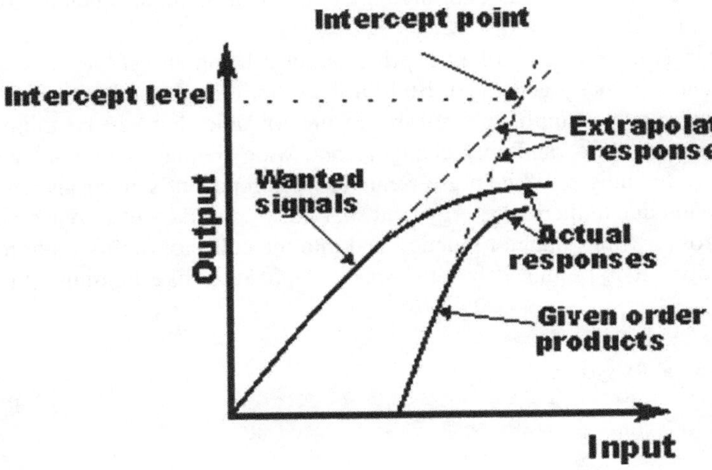

Fig. B.27 Amplifier's output versus input and intercept point

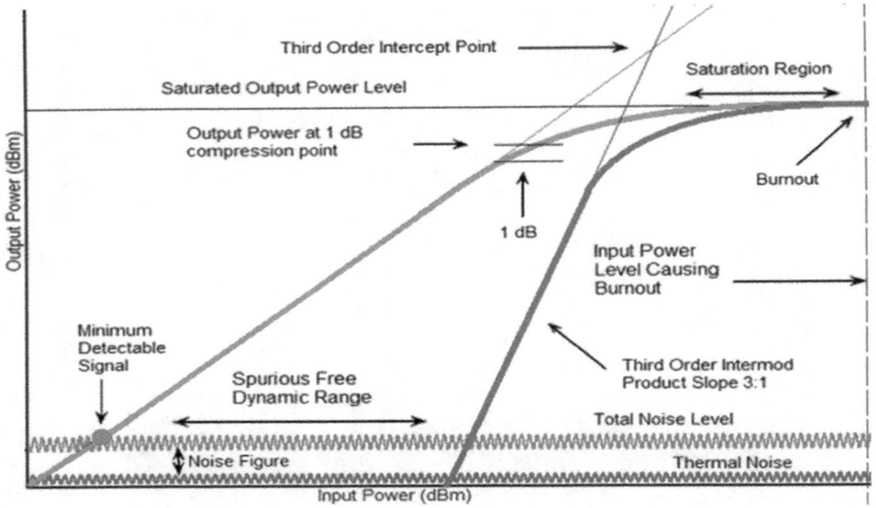

Fig. B.28 Amplifier's output versus input and intercept point saturation region

Harmonic Distortion (HD): HD occurs when an RF fundamental sine wave (f_r) is distorted due to nonlinearity within a circuit. It is generating harmonically related frequencies $(2 \cdot f_r, 3 \cdot f_r, \ldots, n \cdot f_r)$. Interference to receivers tuned to megahertz, or even gigahertz, away from the transmitter's output frequency (Fig. B.29).

The dominant cause of transmitted harmonics is overdriving a poorly filtered power amplifier. The cause to an extreme case of distortion resulting in the sine wave carrier actually is changing into rough square wave. These non- perfect square waves contain: fundamental frequency, odd harmonics, even harmonics. No active stage can be completely linear and there are number of harmonics being produced within all amplifiers.

Amplifier Noise: There are two principal classifications of noise, circuited generated and externally generated. Both of them limit the possible sensitivity and gain of the receiver. Amplifier's noises are unavoidable, but can be minimized. Circuit noise creates a randomly changing and wide frequency ranging voltage. Circuit generated noises: White noise created by a component's electrons randomly moving around due to thermal energy, and shot noise caused by electrons randomly moving across a semiconductor junction and into the collector drain of a transistor. External noise: It is produced by atmospheric upheavals like lighting and space

Fig. B.29 Tx and Rx system

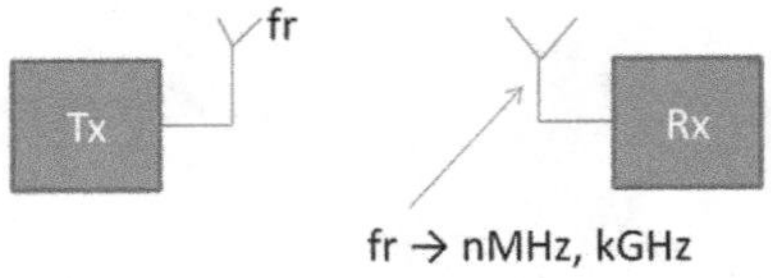

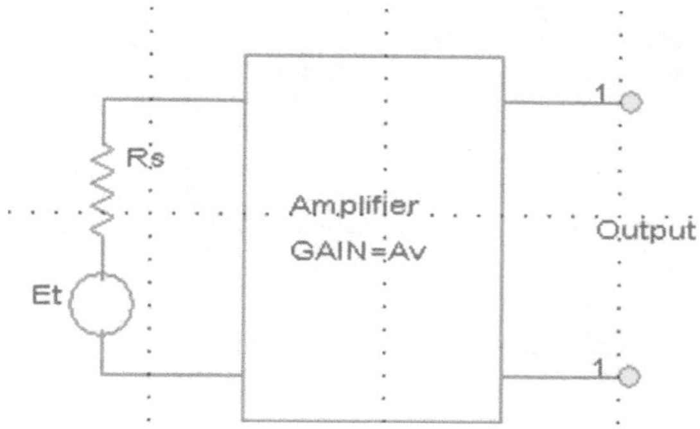

Fig. B.30 Amplifier's source resistance (R_S) which generate noise

noise caused by sunspots and solar flares. Cosmic noise is created by interfering signals from stars. Noise is generated in all stages in amplifier and radio receivers. The first stages, operating at the lowest signal levels, which are of main concern, particularly where low signals from aerials, microphones, etc., are not be amplified. Figure B.30 describes the source resistance (R_S) which generated noise.

Thermal noise is generated because there is no such a device as a perfect amplifier. Thermal noise is the resistance generates a noise voltage (E_t) as follow: $E_t = 2 \cdot \sqrt{K \cdot T \cdot B \cdot R_S}$; K—Boltzmann's constant, T—Absolute temperature, B—Bandwidth in hertz. At normal temperature ($\sim 17\,^{\circ}\mathrm{C}$), this is simplifies to $E_t = \sqrt{1.6 \times 10^{-20} \cdot B \cdot R_S}$ resistance's noise is the lowest noise which can be achieved at the amplifier input. The practical equivalent noise at the input is always higher than this resistance's noise. The noise whilst is generated by an amplifier system. The degree of noise is evaluated by referring it to the amplifier input. It is considered as equivalent noise at the input as though it were being generated at that point. The equivalent noise voltage (E_n) is calculated by dividing the noise measured at the amplifier output (E_{no}) by the gain of the amplifier (A_V), i.e, $E_n = E_{no}/A_V$. Noise in any system is dependent on bandwidth of the system and this must be specified when defining noise performance. It is common practice to define noise for a 1 Hz bandwidth. Noise voltage might be specified in Nano volts per square root of Hertz ($nV/\sqrt{Hz}$). Over a limited bandwidth, noise power can be considered to be proportional to bandwidth and the noise voltage is proportional to the square root of bandwidth. If noise voltage is defined for a 1 Hz bandwidth system, noise can be determined by multiplying by the square root of the system bandwidth. The level of noise generated by an amplifier system generally varies over a wide spectrum and for a wide bandwidth. Noise performance must be defined by plotting noise (say in $nV/\sqrt{Hz}$) against frequency. In solid state amplifiers, noise is often resolved into two components at the amplifier input. An equivalent noise voltage generator (V_n) and an equivalent noise current generator (I_n). The first component (V_n) is independent of the value of source resistance (R_s). The second component (I_n) develops a noise voltage across R_s and equal to $I_n \cdot R_s$. The noise voltage it develops is directly proportional to the value of R_s (Fig. B.31).

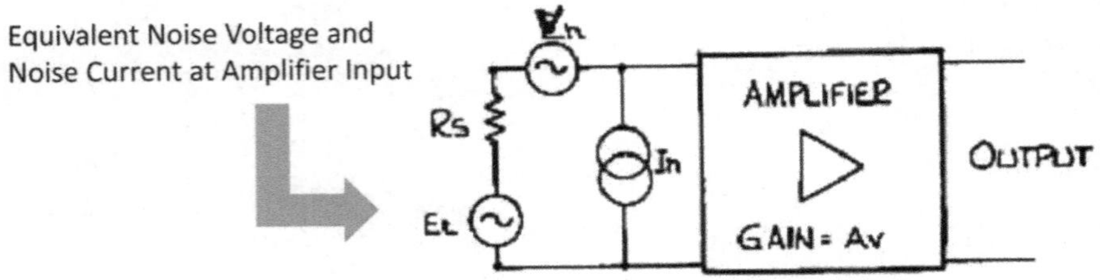

Fig. B.31 Equivalent noise voltage and noise current at amplifier input

The V_n can be separated from the voltage developed by I_n and also the thermal noise (E_t) by short circuiting the input terminals. One method of defining the noise performance of an amplifier is the noise figure (F). This can be defined as the ratio of equivalent noise power developed at the input to that generated by thermal noise in the source resistance (R_s). Noise figure is often expressed in decibel form and a perfect amplifier would have a Noise Figure (NF) of 0 dB, if such a device were possible. To establish Noise Figure (NF), the voltage gain (A_v) of the amplifier is measured and the noise voltage output (E_{no}) is measured at a known bandwidth (B). For the second measurement, the amplifier input must be terminated in a resistance (R_s) equal to the normal source resistance. Noise figure is calculated as follows:

$$F = 20 \cdot \log \left\{ \frac{E_{no}}{A_v \cdot \sqrt{1.6 \times 10^{-20} \cdot B \cdot R_S}} \right\} \text{dB}.$$

The noise figure formula assumes a high impedance input to the amplifier. The effective value of R_s as far as the calculation is concerned, is the parallel result of the source resistance and input resistance of the amplifier. If the source is a transmission line and it is terminated in its characteristic impedance (Z_0), then R_s should be substituted by a value Z_0 divided by 2 (Fig. B.32).

Small Signal (SS) amplifiers always bias in their linear region. Small Signal (SS) amplifier needed to increase tiny signal levels to proper levels required for a transmitter's final power amplifier (PA). A microwave receivers, first RF amplifier (class A SS), high gain type. The discreet RF amplifier design topics are choice of active device, input and output impedance matching network, bias circuit, and

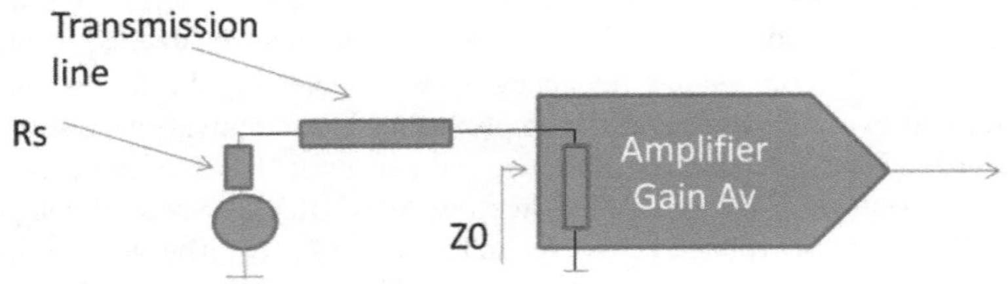

Fig. B.32 Source input, Rs and amplifier

physical layout. Typical transistor has not a 50 Ω resistive Z_{in} and Z_{out} and its reactance will vary over frequency, $A + i \cdot B \to B = g(f)$ then $A + i \cdot g(f)$. Matching network must be use to match the device. Using LC components, the match be perfect for narrow band of frequencies. SS amplifier has also inductive or capacitive parts when perfect match is $50 + i \cdot 0$.

The matching process is as follow: first to match the active device to the system's resistive impedance and second to cancel the innate reactive elements within the transistor, permit a perfect $50 + i \cdot 0$ match with no reactance (capacitive and inductive). This is calling conjugate matching. RF matching network: Take S-parameter two port file, which represent transistor and initially ignore any effects the added DC biasing network which may have on the active device in the final physical design. This assumption is valid only if small amounts of RF feedback are produced by the high values of R_f (R_B) (feedback resistor) in an amplifier's bias network. Low value resistor for R_f (R_B) employs heavy RF feedback. When we choose low value resistor for RF then device's S-parameter file calculation for the matching networks may no longer be completely valid for the transistor. It is accurate only when the bias network employs high resistance values within the bias network (Fig. B.33).

S-parameter files (*.S2P) contain only RF parameters for few frequencies (≤ 20). It is possible that interest frequency may falls between two published values. For accuracy we take mean value between two closest frequencies within

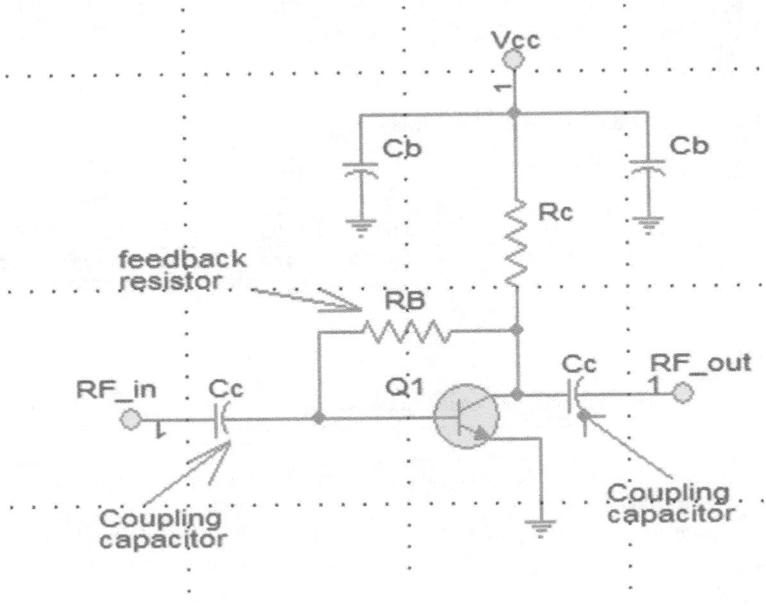

Fig. B.33 Collector feedback bias for BJT transistor circuit

the file. Example: S-parameters are given in a certain *.S2P file for 3 GHz and 4 GHz. Our design requires a centred frequency at 3.5 GHz. We take the mean value of each S-parameter at 3 and 4 GHz. To compute S_{12} at 3.6 GHz we use the following formula:

$$\frac{S_{12}MAG(@3\,\text{GHz}) + S_{12}MAG(@4\,\text{GHz})}{2} = S_{12}MAG(@3.5\,\text{GHz})$$

$$\frac{S_{12}\theta(@3\,\text{GHz}) + S_{12}\theta(@4\,\text{GHz})}{2} = S_{12}\theta(@3.5\,\text{GHz})$$

Filename: 21bfg425.001; BFG425WFieldC1; V1 = 7.884E001V, V2 = 2.000E+000V, I1 = 1.297E−005A, I2 = 1.000E−003A (Table B.2).

When we design a linear amplifier the amplifier need to be stable for our frequency and bias of interest, overall impedance variations, and very wide region of frequencies both low and high. We use the following K formula (Table B.3):

$$K = \frac{1 + (|D_s|^2 - |S_{11}|^2 - |S_{22}|^2)}{2 \cdot |S_{21}| \cdot |S_{12}|}; D_s = S_{11} \cdot S_{22} - S_{12}S_{21}$$

$$K = \frac{1 + (|D_s|^2 - |S_{11}|^2 - |S_{22}|^2)}{2 \cdot |S_{21}| \cdot |S_{12}|}; D_s = S_{11} \cdot S_{22} - S_{12}S_{21}$$

Example: We need to calculate whether transistor is stable at 1.5 GHz, with V_{CE} = 10 V and I_C = 6 mA. The S-parameters at that frequency and bias point are

Table B.2 S parameters for 3 and 4 GHz (Mag, Ang)

	S11		S21		S12		S22	
Freq (GHz)	Mag	Ang	Mag	Ang	Mag	Ang	Mag	Ang
3	0.63	−148	2.19	60.35	0.105	2.75	0.607	−77.07
4	0.591	167.91	1.76	30.28	0.104	−11.14	0.472	−97.41

Table B.3 Amplifier's K parameter values and outcome

K values	Outcome	Important!
K > 1	Active device stable for all Zins and Zouts presented its ports	Zin and Zout can be not exact match
K < 1	Device is potentially unstable	Zin and Zout must be very cautiously selected

found to be (transistor or S-parameter text file). The following calculation gives K > 1 @ 1.5 GHz with transistor bias conditions.

$$S_{11} = 0.195\angle 167.6°; S_{22} = 0.508\angle -32°; S_{12} = 0.139\angle 61.2°; S_{12} = 2.5\angle 62.4°$$

$$D_s = S_{11}\cdot S_{22} - S_{12}S_{21} = [0.195\angle 167.6°]\cdot[0.508\angle -32°] - [0.139\angle 61.2°]\cdot[2.5\angle 62.4°] = 0.25\angle -61.4°$$

$$K = \frac{1+(|D_s|^2-|S_{11}|^2-|S_{22}|^2)}{2\cdot|S_{21}|\cdot|S_{12}|} = \frac{1+(|0.25|^2-|0.195|^2-|0.508|^2)}{2\cdot|2.5|\cdot|0.139|} = 1.1$$

The maximum available gain (sometimes called MAG and sometimes called GMAX) of a device is only defined where K is greater than one. Algebraically, this is because the term under the square root becomes negative for values of K less than 1. If K < 1 then maximum available gain is infinite and infinite gain means oscillator. GMAX is calculated from stability factor K and the forward and reverse transmission coefficients (S21, S22).

$$G_{MAX} = (K - \sqrt{K^2-1})\cdot\left|\frac{S21}{S12}\right| \quad \text{for } K > 1.$$

If K = 1 then $G_{MAX} = \left|\frac{S21}{S12}\right|$ for $K = 1$ and available gain is undefined when K is less than one. That is when the square root of $(K^2 - 1)$ becomes imaginary.

$$G_{MAX} = (K - \sqrt{K^2-1})\cdot\frac{|S_{21}|}{|S_{12}|} = (K - \sqrt{K^2-1})\cdot\frac{(K+\sqrt{K^2-1})}{(K+\sqrt{K^2-1})}\cdot\frac{|S_{21}|}{|S_{12}|}$$

$$G_{MAX} = (K - \sqrt{K^2-1})\cdot\frac{|S_{21}|}{|S_{12}|} = \left[\frac{1}{(K+\sqrt{K^2-1})}\right]\cdot\frac{|S_{21}|}{|S_{12}|}$$

$$G_{MAX} = \left[\frac{1}{(K+\sqrt{K^2-1})}\right]\cdot\frac{|S_{21}|}{|S_{12}|}\bigg|_{K\gg 1\Rightarrow\sqrt{K^2-1}\approx K} = \frac{1}{2\cdot K}\cdot\frac{|S_{21}|}{|S_{12}|}$$

$$G_{MAX\ \log} = 10\cdot\log_{10}^{(K+\sqrt{K^2-1})\cdot\frac{|S_{21}|}{|S_{12}|}} = 10\cdot\log_{10}^{\left[\frac{|S_{21}|}{|S_{12}|}\right]} + 10\cdot\log_{10}^{(K+\sqrt{K^2-1})}$$

$$G_{MAX\ \log} = 10\cdot\log_{10}^{|S_{21}|} - 10\cdot\log_{10}^{|S_{12}|} + 10\cdot\log_{10}^{(K+\sqrt{K^2-1})}$$

$$S_{11} = 0.195\angle 167.6^\circ; S_{22} = 0.508\angle -32^\circ; S_{12} = 0.139\angle 61.2^\circ; S_{12} = 2.5\angle 62.4^\circ$$

$$D_s = S_{11}\cdot S_{22} - S_{12}S_{21} = [0.195\angle 167.6^\circ]\cdot[0.508\angle -32^\circ] - [0.139\angle 61.2^\circ]\cdot[2.5\angle 62.4^\circ] = 0.25\angle -61.4^\circ$$

$$K = \frac{1+(|D_s|^2-|S_{11}|^2-|S_{22}|^2)}{2\cdot|S_{21}|\cdot|S_{12}|} = \frac{1+(|0.25|^2-|0.195|^2-|0.508|^2)}{2\cdot|2.5|\cdot|0.139|} = 1.1$$

$$G_{MAX\ \log} = 10\cdot\log_{10}^{\left[\frac{|S_{21}|}{|S_{12}|}\right]} + 10\cdot\log_{10}^{(K+\sqrt{K^2-1})} = 10\cdot\log_{10}^{\left[\frac{|2.5|}{|0.139|}\right]} + 10\cdot\log_{10}^{(1.1\sqrt{1.1^2-1})} = 10.63\ \text{dB}$$

SS (Small Signal) amplifier stability: A typical amplifier must be unconditionally stable across all frequencies and input/output impedances. An amplifier may oscillate at his band when gain is higher than one. Unstable transistor causes to shift the bias point of the stage and it increases internal device dissipation and possibly causing its destruction. The display an oscillation is seen in the frequency domain on a spectrum analyser. It is distinguished by low voltage and low current spurs. Spurs which begin to shift frequency are viewing instabilities in amplifier. Instability spurs must be eliminated by stabilizing the circuit. Stability of an amplifier stage is dependent on transistor's temperature and bias, signal level, Hfe spread β(frequency), active device's positive internal feedback mechanism, excessively high gain outside of the desired bandwidth, external positive feedback caused by support components, PCB layout, and RF shield's box mode (RF shielding resonances). Additionally the stability of an amplifier stage is dependent on low frequency gain of a normal amplifier, transistor's possible instabilities when

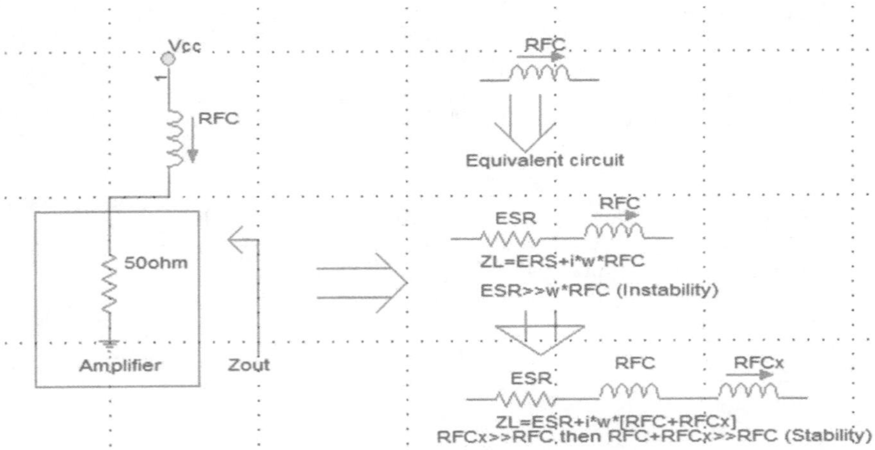

Fig. B.34 Amplifier possess a load that no longer appears as 50 Ω

presented with anything other than 50 Ω termination, RF coupling inductor which at low frequencies presents a true RF choke response over a higher band of limited frequencies. In low frequency the RF choke will begin to look more like a piece of straight, low impedance wire ($Z_{choke} = R_L + j \cdot \omega \cdot L$) then amplifier now possess a load that no longer appears as 50 Ω. This can cause oscillations in a conditionally stable amplifier. The solution is to add a high value low frequency choke in series with low value RF inductor (Fig. B.34).

Frequency decreases cause to lack of 50 Ω termination because the amplifier's matching circuit is good only over specific frequency band. The induction adopted for decoupling of the low impedance power supply becomes close to short circuit as frequency decreases ($f(\omega) \rightarrow \varepsilon$).

Another way to ensure no amplifier's low frequency oscillations is to employ a 50 Ω resistor at the DC end of the bias circuit. The 50 Ω termination resistance at low frequencies, the distributed RFC would have little effect. Capacitor CB helps to shunt low frequency RF to ground, further decreasing the disruptive low frequency RF gain. Another configuration is maintained strong decoupling from the power supply at low frequencies. The circuit uses both a low and high frequency choke to sustain high impedance into the power supply (Figs. B.35 and B.36).

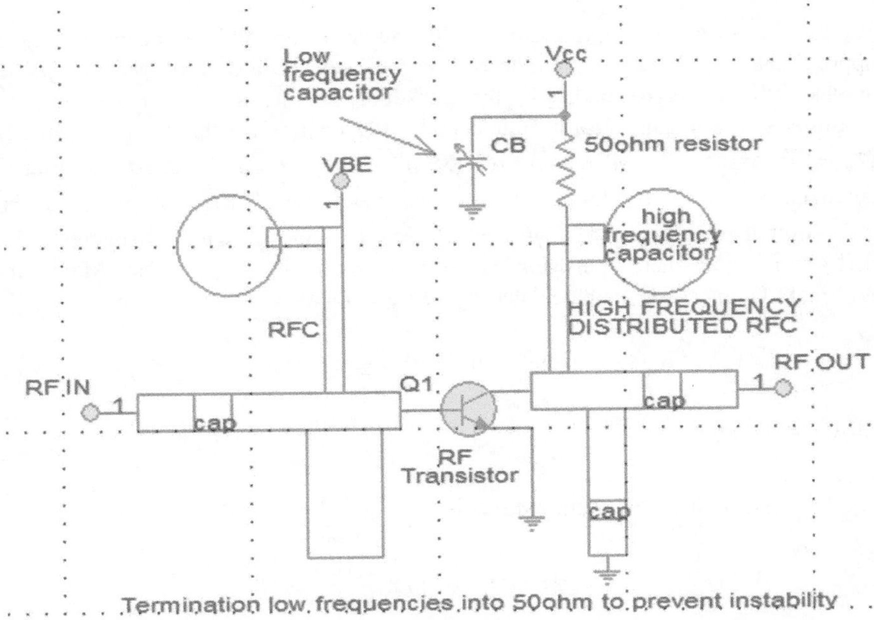

Fig. B.35 Amplifier circuit with termination low frequencies into 50 Ω to prevent instability

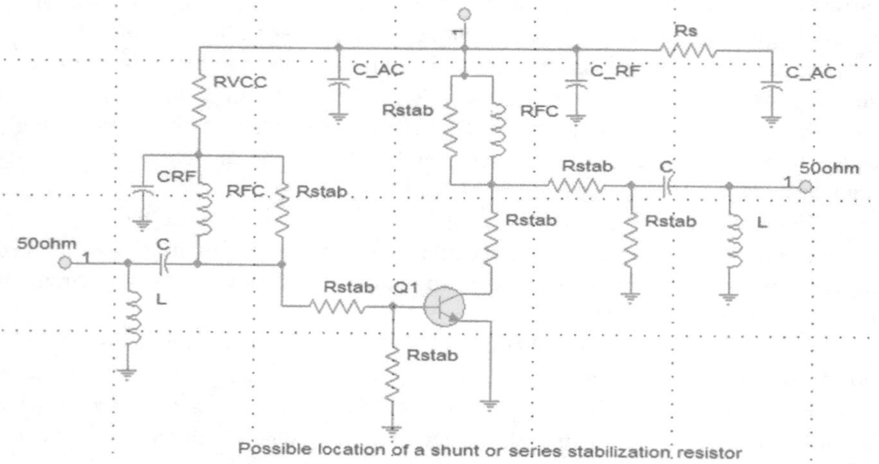

Fig. B.36 Amplifier circuit with possible location of a shunt or series stabilization resistor

B.3 Small Signal (SS) Amplifier Design & Matching Network

We get the SS amplifier gain and stability by scalar approximation. In the scalar approximation only the magnitude of the S-parameters are employed and phase angles are not employed. G_{tu} is the transducer unilateral gain in dB. It is an amplifier's power gain into an unmatched 50 Ω load, a worse case gain value is $G_{tu} = 10 \cdot \log[|S21|]^2$. Mismatch Loss (ML) is $(ML = \alpha \cdot p)$ at transistor's input in decibels: $\alpha \cdot p_{in} = -10 \cdot \log(1 - [S_{11}]^2)$. Mismatch Loss $(ML = \alpha \cdot p)$ at transistor's output in decibels: $\alpha \cdot p_{out} = -10 \cdot \log(1 - [S_{22}]^2)$. The Mismatch Loss (ML) for unmatched transistor is $\alpha \cdot P_{total} = \alpha \cdot p_{in} + \alpha \cdot p_{out}$. The Maximum Available Gain (MAG) is calculated by $MAG = G_{tu} + \alpha \cdot P_{total}$.

$$MAG = G_{tu} + \alpha \cdot P_{total} = 10 \cdot \log[|S21|]^2 - 10 \cdot \log(1 - [S_{11}]^2) - 10 \cdot \log(1 - [S_{22}]^2).$$

$$MAG = 10 \cdot \log_{10}^{\left\{\frac{|S_{21}|^2}{(1-S_{11}^2)\cdot(1-S_{22}^2)}\right\}}.$$

The Maximum Stable Gain (MSG) is

$$MSG = 10 \cdot \log(\frac{|S_{21}|}{|S_{12}|}).$$

If MAG<MSG then the transistor is stable and if MAG>MSG then the transistor is unstable. Example: we have transistor with the following S-parameters:

$$S_{11} = 0.195\angle 167.6^\circ \Rightarrow S_{11} = 0.195; S_{22} = 0.508\angle -32^\circ \Rightarrow S_{22} = 0.508$$
$$S_{12} = 0.139\angle 61.2^\circ \Rightarrow S_{12} = 0.139; S_{21} = 2.5\angle 62.4^\circ \Rightarrow S_{21} = 2.5.$$

$$G_{tu} = 10 \cdot \log_{10}^{(|2.5|^2)} = 7.96\,\text{dB}; \alpha Pin = -10 \cdot \log_{10}^{(1-0.195^2)} = 0.168\,\text{dB}$$
$$\alpha Pout = -10 \cdot \log_{10}^{(1-0.508)^2} = 1.29\,\text{dB}$$
$$\alpha Ptotal = \alpha Pin + \alpha Pout = 0.168\,\text{dB} + 1.29\,\text{dB} = 1.46\,\text{dB}$$
$$MAG = G_{tu} + \alpha Ptotal = G_{tu} + \alpha Pin + \alpha Pout = 7.96\,\text{dB} + 1.46\,\text{dB} = 9.42dB$$
$$MSG = 10 \cdot \log_{10}^{(|2.5|/|0.139|)} = 12.55\,\text{dB}; \quad MAG < MSG \Rightarrow transistor\ stable$$

The most amplifiers matching networks are type L (LC), type T, and type PI. We need to design the amplifier stage's matching network. Our target is that the amplifier's impedance is exactly matching the independences of the circuit. If $Z_S = Z_L^*$ than maximum power is transferred from the source to load (no power reflections) (Fig. B.37).

The amplifier and load $Z_S = R_s + j \cdot X_s; Z_L = R_L - j \cdot X_L$, only if $R_s = R_L$ and $X_s = X_L$ then there is a perfect match. There are three popular matching networks L, PI, T (Fig. B.38).

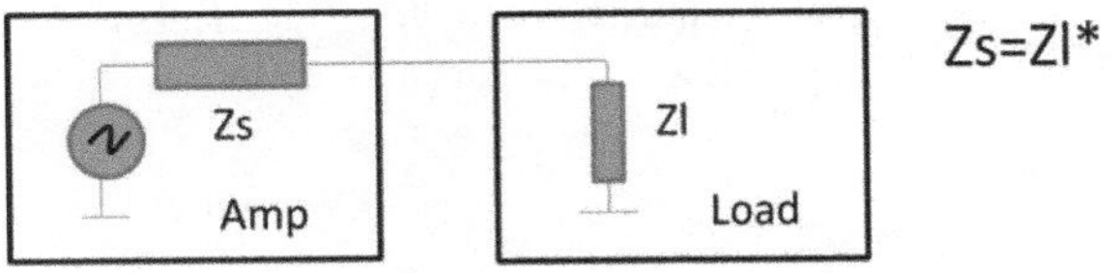

Fig. B.37 Amplifier and load system

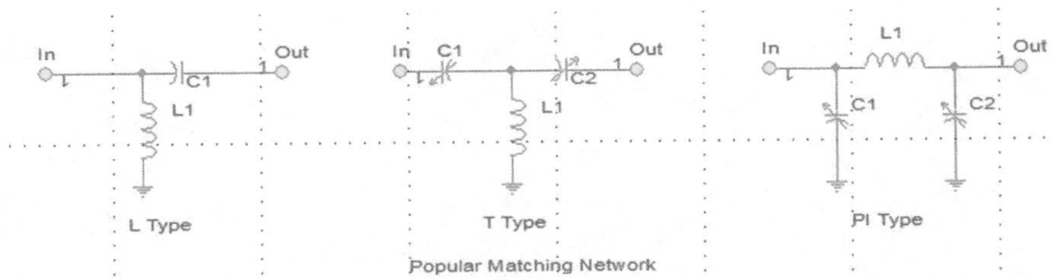

Fig. B.38 Popular matching network (L type, T type, and PI type)

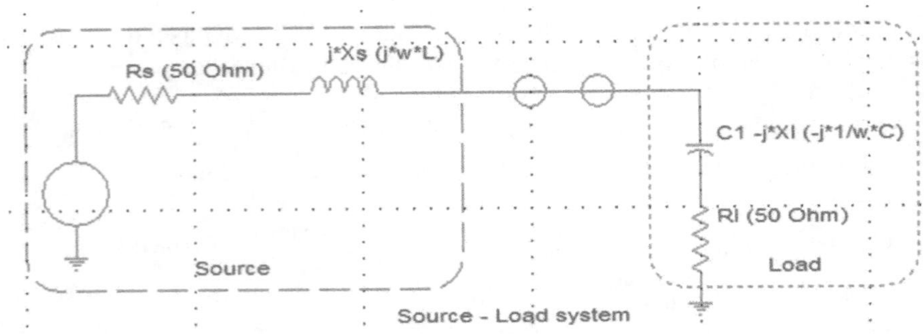

Fig. B.39 Source load system—matched

We get perfect match when

$$X_s = X_L \Rightarrow \omega \cdot L = \frac{1}{\omega \cdot C}; \omega^2 = \frac{1}{L \cdot C} \Rightarrow \omega = \frac{1}{\sqrt{L \cdot C}}$$

$$f = \frac{1}{2 \cdot \pi \cdot \sqrt{L \cdot C}}.$$

There is only one frequency which will be perfectly matched from source to load (Fig. B.39).

And unmatched system: (Fig. B.40).

Example A: $R_S < R_L$; $f = 1.5\,\text{GHz}$. If $Z_{in} = R_s$ then there is a perfect match otherwise $Z_{in} \neq R_s$ and there is unperfected match (Fig. B.41).

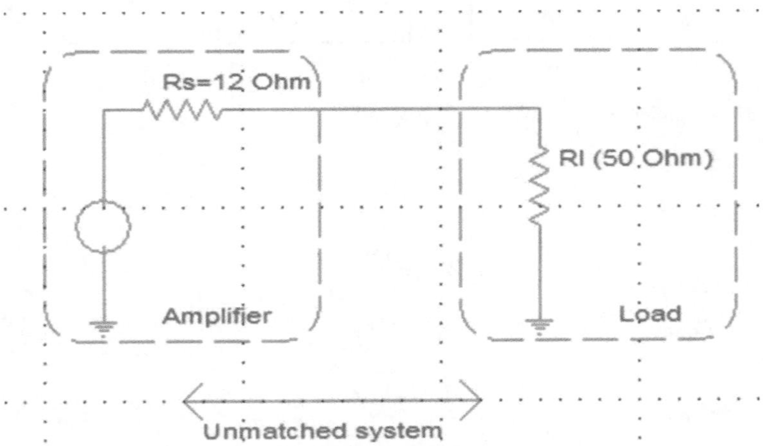

Fig. B.40 Source load system—unmatched

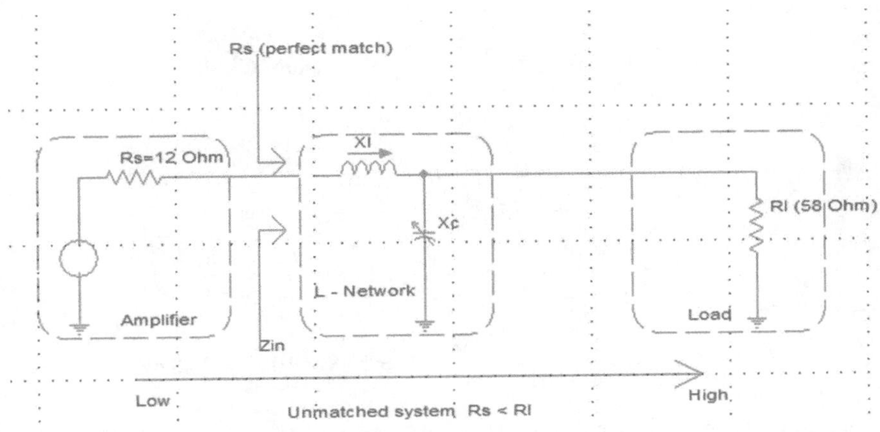

Fig. B.41 Source load system with L network, unmatched

$$X_L = j \cdot \omega \cdot L; X_C = \frac{1}{j \cdot \omega \cdot C}; Z_{in} = Z_L + Z_C \parallel Z_{Load} = Z_L + \frac{Z_C \cdot Z_{Load}}{Z_C + Z_{Load}}$$

$$Z_{in} = j \cdot \omega \cdot L + \frac{\frac{1}{j \cdot \omega \cdot C} \cdot R_L}{\frac{1}{j \cdot \omega \cdot C} + R_L} = j \cdot \omega \cdot L + \frac{R_L}{1 + j \cdot \omega \cdot C \cdot R_L}$$

$$Z_{in} = j \cdot \omega \cdot L + \frac{R_L}{1 + j \cdot \omega \cdot C \cdot R_L} \cdot \frac{(1 - j \cdot \omega \cdot C \cdot R_L)}{(1 - j \cdot \omega \cdot C \cdot R_L)} = j \cdot \omega \cdot L + \frac{R_L \cdot (1 - j \cdot \omega \cdot C \cdot R_L)}{1 + \omega^2 \cdot C^2 \cdot R_L^2}$$

$$Z_{in} = j \cdot \omega \cdot L - \frac{j \cdot \omega \cdot C \cdot R_L^2}{1 + \omega^2 \cdot C^2 \cdot R_L^2} + \frac{R_L}{1 + \omega^2 \cdot C^2 \cdot R_L^2} = j \cdot \omega \cdot \left[L - \frac{C \cdot R_L^2}{1 + \omega^2 \cdot C^2 \cdot R_L^2} \right] + \frac{R_L}{1 + \omega^2 \cdot C^2 \cdot R_L^2}$$

$$(1) \Rightarrow \frac{R_L}{1 + \omega^2 \cdot C^2 \cdot R_L^2} = R_S; \; (2) \Rightarrow L - \frac{C \cdot R_L^2}{1 + \omega^2 \cdot C^2 \cdot R_L^2} = 0;$$

$$(1) \Rightarrow \frac{R_L}{R_S} = 1 + \omega^2 \cdot C^2 \cdot R_L^2$$

$$(1) \Rightarrow \omega^2 \cdot C^2 \cdot R_L^2 = \frac{R_L}{R_S} - 1 \Rightarrow C^2 = \frac{1}{R_L \cdot R_S \cdot \omega^2} - \frac{1}{\omega^2 \cdot R_L^2} = \frac{1}{\omega^2 \cdot R_L^2} \cdot \left[\frac{R_L}{R_S} - 1 \right]$$

$$(1) \Rightarrow C = \frac{1}{\omega \cdot R_L} \cdot \sqrt{\left[\frac{R_L}{R_S} - 1 \right]} = \frac{1}{2 \cdot \pi \cdot 1.5\,\text{GHz} \cdot 58} \cdot \sqrt{\frac{58}{12} - 1} = 3.56 \times 10^{-12} = 3.56\,\text{pF}.$$

$$(2) \Rightarrow L - \frac{C \cdot R_L^2}{1 + \omega^2 \cdot C^2 \cdot R_L^2} = 0 \Rightarrow L = \frac{C \cdot R_L^2}{1 + \omega^2 \cdot C^2 \cdot R_L^2}$$

$$= \frac{\frac{1}{\omega \cdot R_L} \cdot \sqrt{\left[\frac{R_L}{R_S} - 1\right]} \cdot R_L^2}{1 + \omega^2 \cdot \frac{1}{\omega \cdot R_L} \cdot \left[\frac{R_L}{R_S} - 1\right] \cdot R_L^2}$$

$$(2) \Rightarrow L = \frac{\frac{R_L}{\omega} \sqrt{\left[\frac{R_L}{R_S} - 1\right]}}{1 + \left[\frac{R_L}{R_S} - 1\right]} = \frac{\frac{R_L}{\omega} \sqrt{\left[\frac{R_L}{R_S} - 1\right]}}{\frac{R_L}{R_S}} = \frac{R_S}{\omega} \cdot \sqrt{\left[\frac{R_L}{R_S} - 1\right]}$$

$$= \frac{12}{2 \cdot \pi \cdot 1.5\,\text{GHz}} \cdot \sqrt{\frac{58}{12} - 1} = 2.48 \times 10^{-9} = 2.48\,\text{nH}$$

Example B: $R_S > R_L$; $f = 1.5$ GHz and for match $Z_{in} = R_s$ (Fig. B.42).

$$Z_{in} = Z_C \Big\| (Z_L + R_L) = \frac{1}{j \cdot \omega \cdot C} \Big\| (j \cdot \omega \cdot L + R_L) = \frac{\frac{1}{j \cdot \omega \cdot C} \cdot (j \cdot \omega \cdot L + R_L)}{\frac{1}{j \cdot \omega \cdot C} + (j \cdot \omega \cdot L + R_L)}$$

$$= \frac{j \cdot \omega \cdot L + R_L}{j \cdot \omega \cdot C + R_L - \omega^2 \cdot C \cdot L + 1}$$

$$Z_{in} = \frac{R_L + j \cdot \omega \cdot L}{[1 - \omega^2 \cdot C \cdot L] + j \cdot \omega \cdot C \cdot R_L} = \frac{R_L + j \cdot \omega \cdot L}{[1 - \omega^2 \cdot C \cdot L] + j \cdot \omega \cdot C \cdot R_L}$$

$$\cdot \frac{[1 - \omega^2 \cdot C \cdot L] - j \cdot \omega \cdot C \cdot R_L}{[1 - \omega^2 \cdot C \cdot L] - j \cdot \omega \cdot C \cdot R_L}$$

$$Z_{in} = \frac{(R_L + j \cdot \omega \cdot L) \cdot \{[1 - \omega^2 \cdot C \cdot L] - j \cdot \omega \cdot C \cdot R_L\}}{[1 - \omega^2 \cdot C \cdot L]^2 + \omega^2 \cdot C^2 \cdot R_L^2}$$

$$= \frac{R_L \cdot [1 - \omega^2 \cdot C \cdot L]}{[1 - \omega^2 \cdot C \cdot L] - \omega^2 \cdot C^2 \cdot R_L^2}$$

$$- j \frac{\omega \cdot C \cdot R_L^2}{[1 - \omega^2 \cdot C \cdot L] - \omega^2 \cdot C^2 \cdot R_L^2} + j \cdot \frac{\omega \cdot L \cdot [1 - \omega^2 \cdot C \cdot L]}{[1 - \omega^2 \cdot C \cdot L] + \omega^2 \cdot C^2 \cdot R_L^2}$$

$$+ \frac{\omega^2 \cdot L \cdot C \cdot R_L}{[1 - \omega^2 \cdot C \cdot L]^2 + \omega^2 \cdot C^2 \cdot R_L^2}$$

$$Z_{in} = \frac{R_L}{[1 - \omega^2 \cdot C \cdot L]^2 - \omega^2 \cdot C^2 \cdot R_L^2} + j \cdot \frac{\{\omega \cdot L - \omega^3 \cdot C \cdot L^2 - \omega \cdot C \cdot R_L^2\}}{[1 - \omega^2 \cdot C \cdot L]^2 - \omega^2 \cdot C^2 \cdot R_L^2}$$

$$Z_{in} = \frac{R_L}{[1 - \omega^2 \cdot C \cdot L]^2 + \omega^2 \cdot C^2 \cdot R_L^2} + j \cdot \frac{\omega \cdot \{L - \omega^2 \cdot C \cdot L^2 - C \cdot R_L^2\}}{[1 - \omega^2 \cdot C \cdot L]^2 + \omega^2 \cdot C^2 \cdot R_L^2}$$

$$Z_{in} = R_S \Rightarrow (1) \frac{R_L}{[1 - \omega^2 \cdot C \cdot L]^2 + \omega^2 \cdot C^2 \cdot R_L^2} = R_S; \; (2) L - \omega^2 \cdot C \cdot L^2 - C \cdot R_L^2 = 0$$

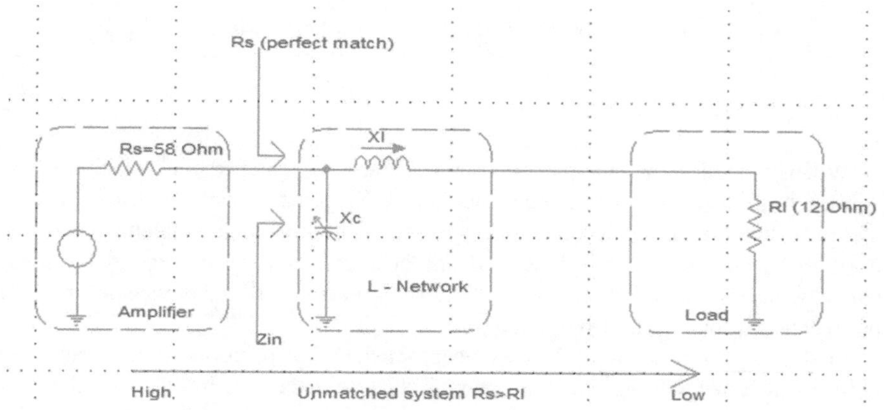

Fig. B.42 Source load system with L network unmatched

$$(2) L - \omega^2 \cdot C \cdot L^2 - C \cdot R_L^2 = 0 \Rightarrow C = \frac{L}{\omega^2 \cdot L^2 - R_L^2};\ (2) \Rightarrow (1) \Rightarrow$$

$$(1) \frac{R_L}{\left[1 - \omega^2 \cdot \frac{L}{\omega^2 \cdot L^2 - R_L^2} \cdot L^2\right] + \omega^2 \cdot \left[\frac{L}{\omega^2 \cdot L^2 - R_L^2}\right]^2 \cdot R_S} = R_S$$

$$(2) \frac{R_L}{R_S} = \left[1 - \frac{\omega^2 \cdot L^2}{\omega^2 \cdot L^2 - R^2}\right]^2 + \frac{\omega^2 \cdot L^2 \cdot R^2}{[\omega^2 \cdot L^2 - R^2]^2} \Rightarrow$$

$$(2)\, 0.2 = \frac{12}{58} = \left[1 - \frac{(2 \cdot \pi \cdot 1.5 \times 10^9)^2 \cdot L^2}{(2 \cdot \pi \cdot 1.5 \times 10^9)^2 \cdot L^2 - 12^2}\right]^2 + \frac{(2 \cdot \pi \cdot 1.5 \times 10^9)^2 \cdot L^2 \cdot 12^2}{\left[(2 \cdot \pi \cdot 1.5 \times 10^9)^2 \cdot L^2 - 12^2\right]^2}$$

NSolve[{(1 − ((2 * 3.14)^2
 * (1.5 * 1,000,000,000)^2 * x * x)/((2 * 3.14)^2 * (1.5 * 10,000,000,000)^2 * x * x * + 144))
 ^2 + ((2 * 3.14)^2 * (1.5 * 1,000,000,000)^2 * x * x * 144)/((2 * 3.14)^2
 * (1.5 * 10,000,000,000)^2 * x * x + 144)^2}□12/58, {x}]
 {{x → 2.49413 × 10^−9}, {x → 2.49413 × 10^19}}

$$\textit{Two solutions}: \mathrm{L}^1 = 2.49 \times 10^{-9};\ L^2 = -2.49 \times 10^{-9} \Rightarrow L = 2.49\,\mathrm{nH}$$

$$(2)\, C = \frac{L}{\omega^2 \cdot L^2 - R_L^2} \Rightarrow C = \frac{2.49\,\mathrm{nH}}{(2 \cdot \pi \cdot 1.5 \times 10^9)^2 \cdot [2.49\,\mathrm{nH}]^2 - 12^2}$$

$$= 3.587 \times 10^{-12} = 3.587\,\mathrm{pF}$$

EDU ≫ (2.49 · 0.000000001)/((2 * 3.14 * 1.5 * 2.49).^2 + 144)

ans =

3.5870e−012

When two different, but pure resistances must be matched the L-technique is applied to perform the task. If reactance X_L or X_C must be cancelled, we use two methods: first the absorption uses to reactance of the impedance matching network itself to be absorb the undesired load and/or source reactance. Second the resonance, which is utilized to resonant out the stray reactance of the device or circuit to be matched at our desired frequency.

First absorption method: this is accomplished by positioning the matching inductor in series with any load or source inductive reactance. In this way, the load or source's X_L becomes a part of the matching inductor. The same outcome can be attained by positioning a matching capacitor in parallel with any load or source X_C. Thus we are combining the two values into one larger value. This allows the internal stray reactance of both devices to contribute the matching network. This internal reactance is being subtracted from the calculated values of the LC matching components. The transistors own stray reactance is now becoming an additive part of the matching network. This absorption method is only useful if the stray internal reactance of the device is less than the calculated reactance required for a proper match. Figure B.43 describes the circuit that requires the addition of components to absorb reactance.

$L_S + L_1 \rightarrow L_1'$ is a new inductance value for matching L network. $C_S + C_1 \rightarrow C_1'$ is a new capacitor value for matching L network. The absorption methods are by using Z match network. The below figure describes the absorption methods by using Z matching network (Figs. B.44, B.45 and B.46).

Second absorption (resonance) method: Resonance technique is utilized to resonate out the stray reactance of the device or circuit to be matched at our desired frequency (f). It is done with a reactance that is equal in value, but opposite sign and then continuing on as if the matching problem were a completely resistive one $(R + j \cdot 0)$. This will make the internal stray reactance of the two devices or circuits

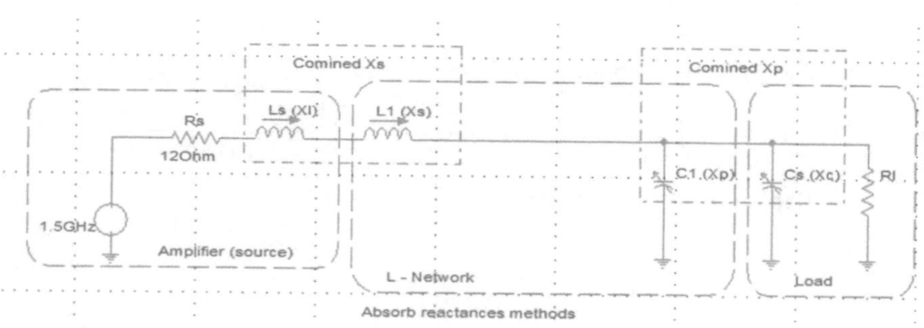

Fig. B.43 Amplifier (source) and load, absorb reactance methods

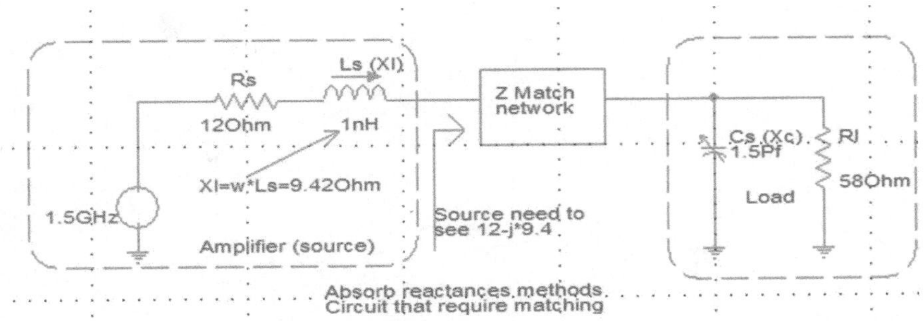

Fig. B.44 Amplifier (source) and load Z match network

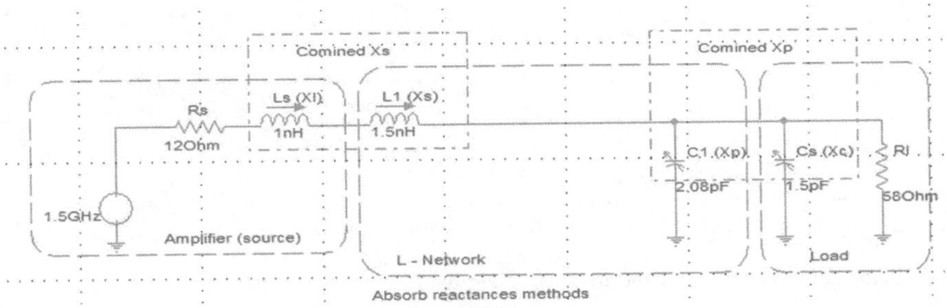

Fig. B.45 Amplifier (source) and load L – network

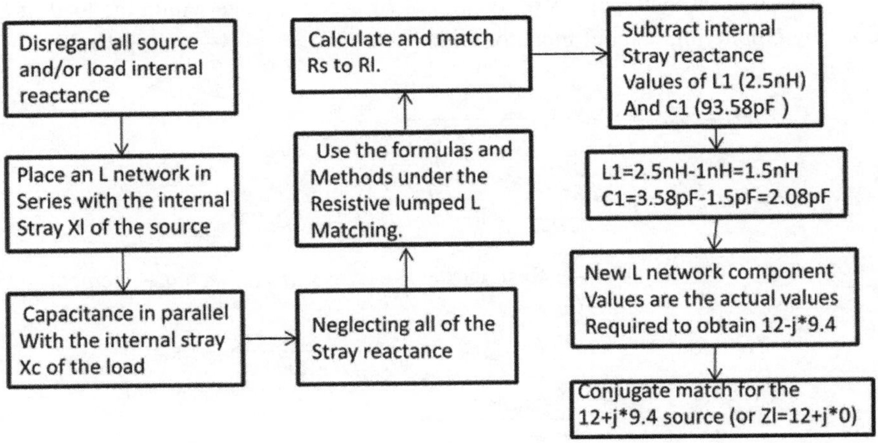

Fig. B.46 Amplifier (source) and load matching flow chart

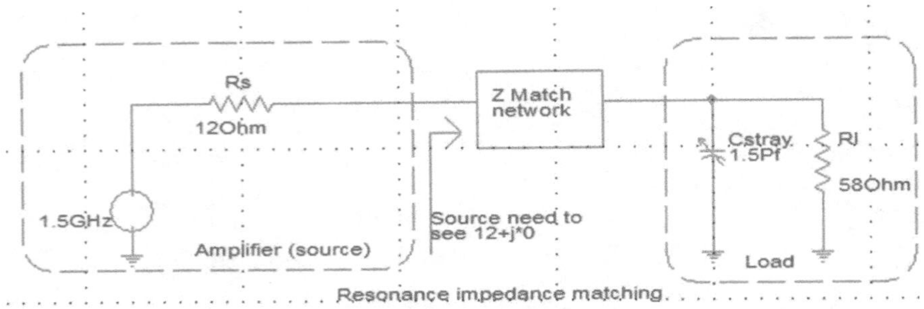

Fig. B.47 Amplifier (source) and load resonance impedance matching

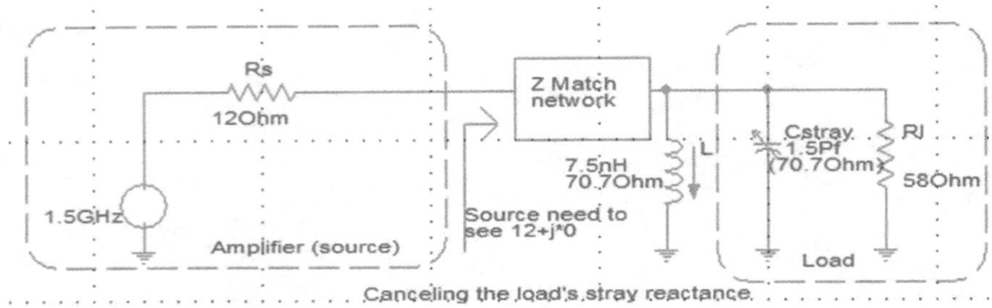

Fig. B.48 Amplifier (source) and load canceling the load's stray reactance

disappear and allowing only the pure resistances to be easily dealt with. We need to design a matching network which employing the second method. Figure B.47 describes the resonance impedance matching.

We need to resonate out 1.5 pF (C_{stray}) of stray capacitance within the load. It is done by employing a shunt inductor with value of $L = \frac{1}{|2\cdot\pi\cdot f|^2\cdot C_{stray}}$ (Fig. B.48).

$$C_{stray}||L \Rightarrow Z_T = j\cdot\omega\cdot L||\frac{1}{j\cdot\omega\cdot C_{stray}}$$

$$\Rightarrow Z_T = \frac{j\cdot\omega\cdot L\cdot\frac{1}{j\cdot\omega\cdot C_{stray}}}{j\cdot\omega\cdot L + \frac{1}{j\cdot\omega\cdot C_{stray}}} = \frac{j\cdot\omega\cdot L}{1-\omega^2\cdot L\cdot C_{stray}}$$

$$Z_T = j\cdot\frac{\omega\cdot L}{1-\omega^2\cdot L\cdot C_{stray}};\ \text{Resonate} \Rightarrow Z_T \to \infty(\textit{disconnected} - \text{element})$$

$$\Rightarrow Z_T \to \infty \Rightarrow 1 - \omega^2\cdot L\cdot C_{stray} \to \varepsilon \Rightarrow L \to \frac{1-\varepsilon}{\omega^2\cdot C_{stray}};\ \varepsilon = 0$$

$$\Rightarrow L = \frac{1}{\omega^2\cdot C_{stray}} = \frac{1}{|2\cdot\pi\cdot f|^2\cdot C_{stray}}$$

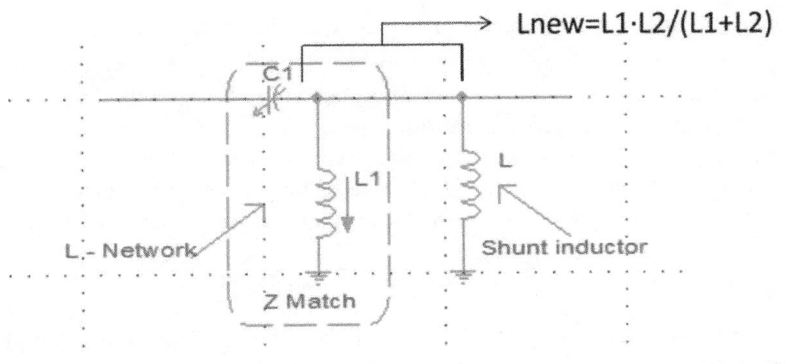

Fig. B.49 Combine shunt inductor with L_1 (L—network inductor)

Since $Z_T \to \infty$, the internal stray capacitance can be considered as no longer existing within the load. Since the source is purely resistive ($Z_s = R_s + j \cdot 0$) and the load is also pure resistive ($Z_L = R_L + j \cdot 0$) then we utilize the basic resistive lumped matching techniques. We need to design an L-network to match source to load. The Z match network is L-network, we can combining both of the inductors (L_1 & L), with a single inductor (Fig. B.49).

Three elements impedance matching (PI or T) networks are used in many narrow band applications. The narrow band is due to the higher loaded Q over what the L-network possesses. PI and T networks also permit any Q to be selected. We always consider that PI & T Q's $\geq$ L-network Q's. The Q desired for a particular application is calculated with the following formula,

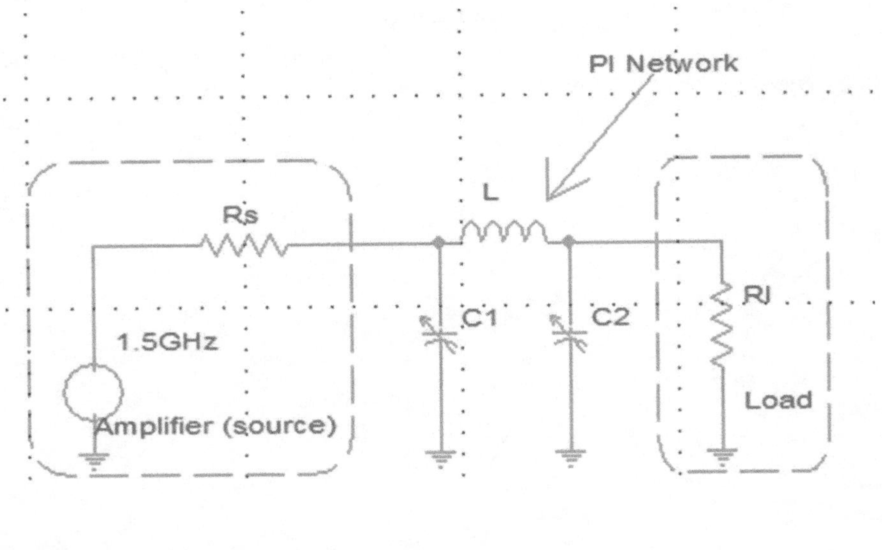

Fig. B.50 PI matching network between source and load

When utilizing high Q inductor $Q = \frac{f_c}{f_2 - f_1}$. Q is the loaded quality factor of the circuit. f_c is the center frequency of the circuit. f_2 is the upper frequency, need to pass with little loss. f_1 is the lower frequency, need to pass with little loss, we use PI network to match two different pure resistances (Fig. B.50).

We consider PI network as two L networks attached back to back. There is a virtual resistor "R" in the center, which is used only as an aid in designing these networks. The virtual "R" will not be in the final design. We choose (Fig. B.51)

$$X_{P_1} = 2.68\,\Omega, X_{S_1} = 2.55\,\Omega, X_{S_2} = 5.7\,\Omega, X_{P_2} = 5.8\,\Omega;\ Q = 10$$

R_L is pure resistance of the load. R_S is pure resistance of the source (amplifier). $R_h = \max(R_S, R_L)$.

$$\frac{''R''}{R_H} = \frac{1}{Q^2+1} \Rightarrow \frac{R_H}{''R''} = Q^2 + 1 \Rightarrow Q^2 = \frac{R_H}{''R''} - 1 \Rightarrow Q = \sqrt{\frac{R_H}{''R''} - 1}$$

$$R_H = \max(R_s, R_l);\ ''R'' = \frac{R_H}{Q^2+1} =_{\substack{Q=10\\R_{H=58}}} \frac{58}{10_2+1} = 0.57\,\Omega$$

$$R_L \| X_{P2} \Rightarrow Q = \frac{R_L}{X_{P2}} \Rightarrow X_{P2} = \frac{R_L}{Q} = \frac{58}{10} = 5.8\,\Omega$$

$$Q = \frac{X_{S2}}{''R''} \Rightarrow X_{S2} = Q \cdot ''R'' = 10 \cdot 0.57 = 5.7\,\Omega$$

$$X_{P1} = \frac{R_S}{Q_1};\ Q_1 = \sqrt{\frac{R_S}{''R''} - 1};\ X_{P1} = \frac{R_S}{Q_1} = \frac{R_S}{\sqrt{\frac{R_S}{''R''} - 1}} = 2.68\,\Omega$$

$$Q_1 = \sqrt{\frac{R_S}{''R''} - 1} = \sqrt{\frac{12}{5.7} - 1} = 4.48;\ Q_1 = \frac{X_{S1}}{''R''} \Rightarrow X_{S1}$$
$$= Q_1 \cdot ''R'' = 4.48 \cdot 0.57 = 2.55$$

We need to convert the reactance calculated to L and C values.

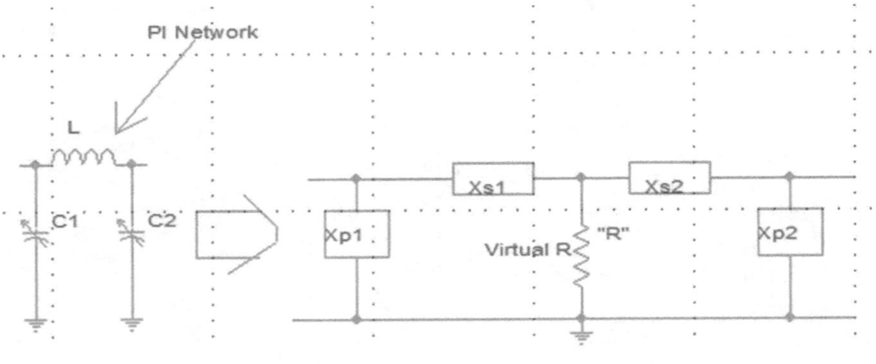

Fig. B.51 Virtual resistor and two L networks to design a PI network

$$X_{S1} + X_{S2} = \sum_{i=1}^{2} X_{Si};\ X_S = \omega \cdot L = 2 \cdot \pi \cdot f \cdot L \Rightarrow \frac{X_S}{2 \cdot \pi \cdot f}$$

$$X_P = \frac{1}{\omega \cdot C} = \frac{1}{2 \cdot \pi \cdot f \cdot C} \Rightarrow C = \frac{1}{2 \cdot \pi \cdot f \cdot X_P}$$

To match two stages with a PI network, while canceling reactance and matching resistances, first we convert the load/source to/from parallel or series equivalences and second we make it easier to absorb any reactance. Any PI network can be transformed to an equivalent T network. This is also known as the Wye-Delta transformation, which is the terminology used in power distribution and electrical engineering. The PI is equivalent to the Delta and the T is equivalent to the Wye (or star) form. The PI network and T network topologies are described in Fig. B.52.

The impedances of the PI network (Z_a, Z_b, Z_c) can be found from the impedances of the T-network with the following equations:

$$Z_a = [(Z_1 \cdot Z_2) + (Z_1 \cdot Z_3) + (Z_2 \cdot Z_3)]/Z_2;\ Z_b = [(Z_1 \cdot Z_2) + (Z_1 \cdot Z_3) + (Z_2 \cdot Z_3)]/Z_1$$
$$Z_c = [(Z_1 \cdot Z_2) + (Z_1 \cdot Z_3) + (Z_2 \cdot Z_3)]/Z_3.$$

The common numerator in all these expressions can prove useful in reducing the amount of computation necessary. The impedances of the T-network (Z_1, Z_2, Z_3) can be found from the impedances of the equivalent PI-network with the following equations. The next expression describes the PI network to T network transformation. There is a common denominator in these expressions.

$$Z_1 = (Z_a \cdot Z_c)/(Z_a + Z_b + Z_c);\ Z_2 = (Z_b \cdot Z_c)/(Z_a + Z_b + Z_c);$$
$$Z_3 = (Z_a \cdot Z_b)/(Z_a + Z_b + Z_c)$$

There is a case where all the impedances are creative (i.e. they are all in the form $j \cdot X$). In that case, the (−1) factors from squaring $j \cdot j$ on the top cancel the (−1)

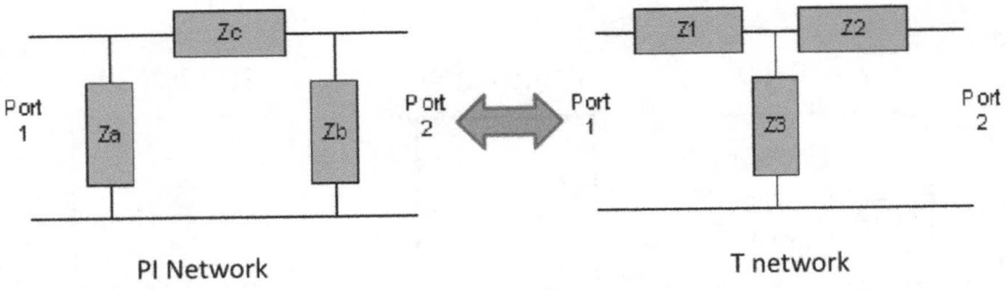

Fig. B.52 PI network and T network

from bringing the j in the denominator up top. In that case also all T-network impedances are reactive. The below equations describe the situation when all PI network impedances are reactive.

$$Z_1 = \frac{Z_a \cdot Z_c}{Z_a \cdot Z_b \cdot Z_c}; \; Z_2 = \frac{Z_b \cdot Z_c}{Z_a \cdot Z_b \cdot Z_c}; \; Z_3 = \frac{Z_a \cdot Z_b}{Z_a \cdot Z_b \cdot Z_c}$$

$$Z_a = j \cdot a; \; Z_b = j \cdot b; \; Z_c = j \cdot c$$

$$Z_1 = \frac{Z_a \cdot Z_c}{Z_a + Z_b + Z_c} = \frac{j \cdot a \cdot j \cdot c}{j \cdot a + j \cdot b + j \cdot c} = \frac{j^2 \cdot a \cdot c}{j \cdot (a+b+c)} = j \cdot \frac{a \cdot c}{(a+b+c)}$$

$$Z_2 = \frac{Z_b \cdot Z_c}{Z_a + Z_b + Z_c} = \frac{j \cdot a \cdot j \cdot c}{j \cdot a + j \cdot b + j \cdot c} = \frac{j^2 \cdot b \cdot c}{j \cdot (a+b+c)} = j \cdot \frac{b \cdot c}{(a+b+c)}$$

$$Z_1 = \frac{Z_a \cdot Z_c}{Z_a + Z_b + Z_c} = \frac{j \cdot a \cdot j \cdot b}{j \cdot a + j \cdot b + j \cdot c} = \frac{j^2 \cdot a \cdot b}{j \cdot (a+b+c)} = j \cdot \frac{a \cdot b}{(a+b+c)}$$

SS (Small Signal) amplifier synthesis of PI and T networks to transform resistances and create phase shifts. We assume that the desired port impedances are purely resistive (i.e. real). The design of T or PI network with purely reactive components is both to produce a desired phase shift (beta) and transform the impedances with the following equations. T and PI networks design (R_1, R_2, β).

$$Z_1 = \frac{-j \cdot \left[R_1 \cdot \cos\beta - \sqrt{R_1 \cdot R_2}\right]}{\sin\beta}; \; Z_2 = \frac{-j \cdot \left[R_2 \cdot \cos\beta - \sqrt{R_1 \cdot R_2}\right]}{\sin\beta}$$

$$Z_3 = \frac{-j \cdot \sqrt{R_1 \cdot R_2}}{\sin\beta}; \; Z_a = \frac{j \cdot R_1 \cdot R_2 \cdot \sin\beta}{\left[R_2 \cdot \cos\beta - \sqrt{R_1 \cdot R_2}\right]}$$

$$Z_b = \frac{j \cdot R_1 \cdot R_2 \cdot \sin\beta}{\left[R_1 \cdot \cos\beta - \sqrt{R_1 \cdot R_2}\right]}; \; Z_c = j \cdot \sqrt{R_1 \cdot R_2} \cdot \sin\beta$$

The beta (β) is the phase lag passing through the network from either port 1 to port 2 or vice versa. If beta (β) is 0 or π, these expressions break down, except if $R_1 = R_2$. To transform resistive impedances without any phase shift, we have to use

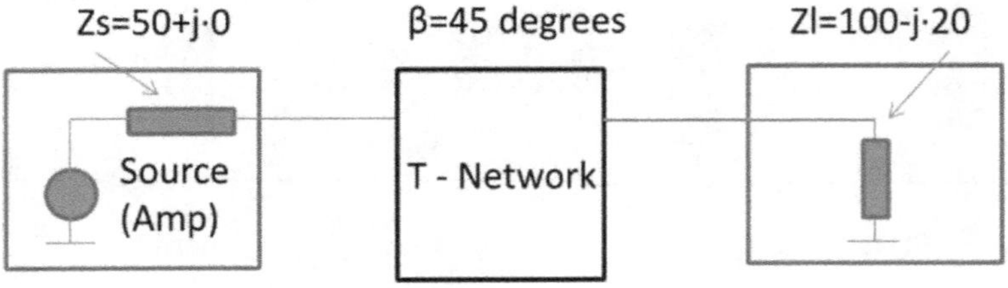

Fig. B.53 Source amplifier and load, T-network

a transformer. In many practical applications, the load or generator impedances may be reactive (i.e. Z (port 1) and Z (port 2) are some general R + j · X). This can be accommodated by absorbing the external reactive impedance into the network, reducing or increasing the series or shunt impedance as required. If T-network is required to connect between two impedances: 50 + j · 0 and 100 – j · 20 with 45 degrees of phase shift (β) (Fig. B.53).

$$Z_s = 50 + j \cdot 0;\ Z_l = 100 - j \cdot 20;\ R_1 = 50;\ R_2 = 100;\ \beta = 45$$

$$Cos\,\beta = \sin\,\beta = 0.707;\ Z_1 = \frac{-j \cdot \left[50 \cdot 0.707 - \sqrt{50 \cdot 100}\right]}{0.707} = j \cdot 50$$

$$Z_2 = \frac{-j \cdot \left[100 \cdot 0.707 - \sqrt{50 \cdot 100}\right]}{0.707} = 0;\ Z_3 = \frac{-j \cdot \sqrt{50 \cdot 100}}{0.707} = -j \cdot 100$$

What happened if $R_1 = R_2$? (Two equal resistive impedances), T-network.

$$Z_1 = \frac{-j \cdot \left[R_1 \cdot \cos\,\beta - \sqrt{R_1 \cdot R_2}\right]}{\sin\,\beta}\Big|_{R_1=R_2=R} = \frac{-j \cdot R \cdot [\cos\,\beta - 1]}{\sin\,\beta}$$

$$Z_2 = \frac{-j \cdot \left[R_2 \cdot \cos\,\beta - \sqrt{R_1 \cdot R_2}\right]}{\sin\,\beta}\Big|_{R_1=R_2=R} = \frac{-j \cdot R \cdot [\cos\,\beta - 1]}{\sin\,\beta}$$

$$Z_3 = \frac{-j \cdot \sqrt{R_1 \cdot R_2}}{\sin\,\beta} = \frac{-j \cdot R}{\sin\,\beta}$$

What happened if $R_1 = R_2$? (Two equal resistive impedances), PI-network.

$$Z_a = \frac{j \cdot R_1 \cdot R_2 \cdot \sin\,\beta}{\left[R_2 \cdot \cos\,\beta - \sqrt{R_1 \cdot R_2}\right]}\Big|_{R_1=R_2=R} = j \cdot \frac{R \cdot \sin\,\beta}{[\cos\,\beta - 1]}$$

$$Z_b = \frac{j \cdot R_1 \cdot R_2 \cdot \sin\,\beta}{\left[R_1 \cdot \cos\,\beta - \sqrt{R_1 \cdot R_2}\right]}\Big|_{R_1=R_2=R} = j \cdot \frac{R \cdot \sin\,\beta}{[\cos\,\beta - 1]}$$

$$Z_c = j \cdot \sqrt{R_1 \cdot R_2} \cdot \sin\,\beta$$

The certain high frequency microwave applications, we use distributed matching elements. It may be a lower cost and higher performance alternative to using lumped parts. The method limitation: inability to easily create series capacitors. We should employ a shunt distributed capacitors when matching impedances in our microwave designs. What happened when the series input impedance of the device is inductive? And we would like to tune it out. We use conjugate series capacitance to cancel the transistor's series input inductance (Fig. B.54).

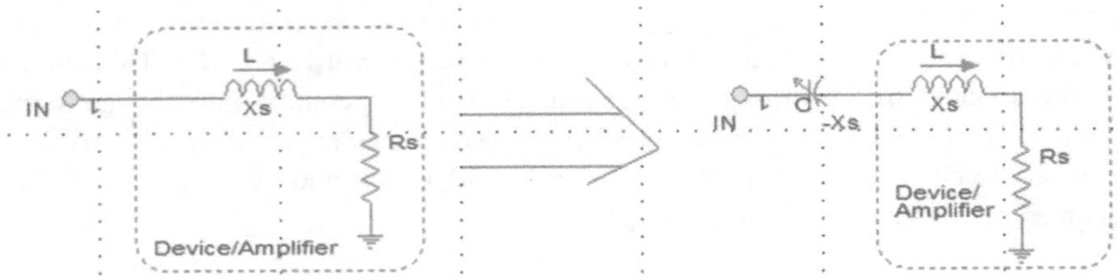

Fig. B.54 Conjugate series capacitance to cancel transistor's series input impedance

Hint: We would like to get away from using a lump series capacitor.

Solution: We convert the series input impedance of the device into equivalent parallel input impedance. The equivalent parallel input impedance circuit permits us to exploit a shunt distributed element. The shunt distributed element resonates out the input impedance of the device. Series input impedance: X_s series reactance (Ω) and R_s series resistance (Ω). Parallel input inductance: X_p equivalent parallel reactance (Ω) and X_p equivalent parallel resistance (Ω). We interest in amplifier's series input impedance circuit to parallel input impedance circuit conversion. We need to get the expressions for X_p and R_p as functions of X_s, R_s and developing the mathematical connections between those circuit's parameters (Fig. B.55).

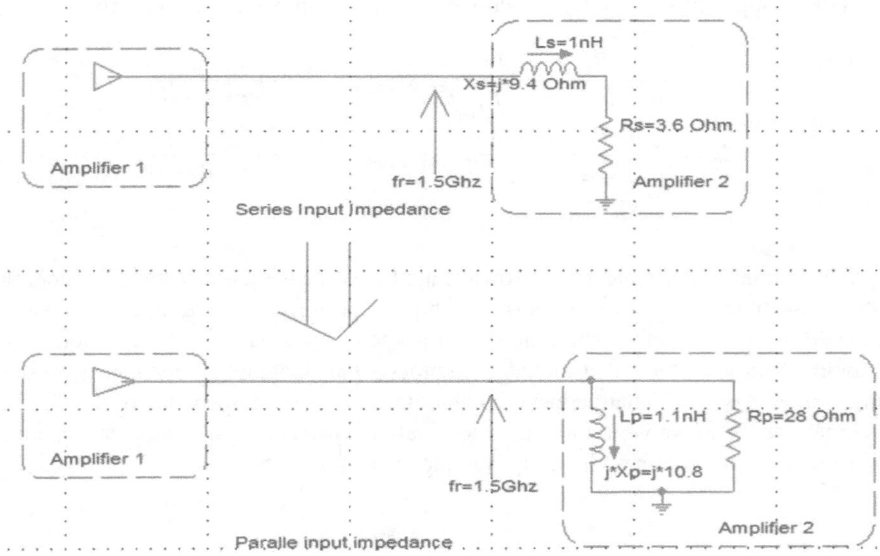

Fig. B.55 Series input impedance and parallel input impedance

$$j \cdot X_S + R_S = \frac{j \cdot X_P \cdot R_P}{j \cdot X_P + R_P} \Rightarrow j \cdot X_S + R_S = \frac{j \cdot X_P \cdot R_P}{j \cdot X_P + R_P} \cdot \left\{ \frac{-j \cdot X_P + R_P}{-j \cdot X_P + R_P} \right\}$$

$$\Rightarrow j \cdot X_S + R_S = \frac{X_P^2 \cdot R_P + j \cdot X_P \cdot R_P^2}{X_P^2 + R_P^2} \Rightarrow j \cdot X_S + R_S$$

$$= \frac{X_P^2 \cdot R_P}{X_P^2 + R_P^2} + j \cdot \frac{X_P \cdot R_P^2}{X_P^2 + R_P^2}$$

$$(1)\, X_S = \frac{X_P \cdot R_P^2}{X_P^2 + R_P^2} \Rightarrow X_P^2 + R_P^2 = \frac{X_P \cdot R_P^2}{X_S};$$

$$(2)\, R_S = \frac{X_P^2 \cdot R_P}{X_P^2 + R_P^2} \Rightarrow X_P^2 + R_P^2 = \frac{X_P^2 \cdot R_P}{R_S}$$

$$(1)\,\&\,(2) \Rightarrow \frac{X_P \cdot R_P^2}{X_S} = \frac{X_P^2 \cdot R_P}{R_S}$$

$$\Rightarrow \frac{R_P}{X_S} = \frac{X_P}{R_S} \Rightarrow X_P = \frac{R_P \cdot R_S}{X_S}; X_P^2 + R_P^2 = \frac{X_P^2 \cdot R_P}{R_S}$$

$$X_P^2 + R_P^2 = \frac{X_P^2 \cdot R_P}{R_S} \Rightarrow \left[\frac{R_P \cdot R_S}{X_S}\right] + R_P^2 = \frac{\left[\frac{R_P \cdot R_S}{X_S}\right]^2 \cdot R_P}{R_S} = \frac{R_P}{R_S} \cdot \left[\frac{R_P \cdot R_S}{X_S}\right]^2$$

$$\left[\frac{R_P \cdot R_S}{X_S}\right]^2 + R_P^2 = \frac{R_P}{R_S} \cdot \left[\frac{R_P \cdot R_S}{X_S}\right]^2$$

$$\Rightarrow R_P^2 + R_P^2 \cdot \frac{R_S^2}{X_S^2} = \frac{R_P}{R_S} \cdot R_P^2 \cdot \frac{R_S^2}{X_S^2} \Rightarrow R_P^2 \cdot \left\{1 + \frac{R_S^2}{X_S^2}\right\} = R_P^2 \cdot \frac{R_P}{R_S} \cdot \frac{R_S^2}{X_S^2}$$

$$R_P^2 \cdot \left\{1 + \frac{R_S^2}{X_S^2}\right\} = R_P \cdot \frac{R_P}{R_S} \cdot \frac{R_S^2}{X_S^2} \Rightarrow 1 + \frac{R_S^2}{X_S^2} = \frac{R_P}{R_S} \cdot \frac{R_S^2}{X_S^2} \Rightarrow 1 + \frac{R_S^2}{X_S^2} = R_P \cdot \frac{R_S}{X_S^2}$$

$$1 + \frac{R_S^2}{X_S^2} = R_P \cdot \frac{R_S}{X_S^2} \Rightarrow R_P \cdot R_S = X_S^2 + R_S^2 \Rightarrow R_P = \frac{X_S^2}{R_S} + R_S$$

Conclusion: We could design the distributed matching circuit as we would the lumped type. We simply substitute the equivalent distributed components (microwave as equivalent components).

$$(A) X_P = \frac{R_P \cdot R_S}{X_S}; \quad (\mathrm{B}) R_P = \frac{X_S^2}{R_S} + R_S$$

We substitute the equivalent distributed components by micro strip as equivalent components. For small (and large) signal devices, we implement quarter wave line transformer. We use a distributed transformer to match a 50 Ω resistive source to an

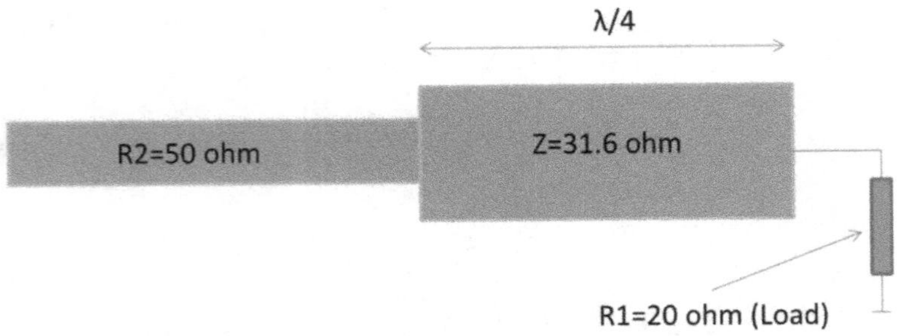

Fig. B.56 Distributed transformer to match a 50 Ω resistive source

unequal resistive load. The distributed transformer to match a 50 Ω resistive source to an unequal resistive load is as follow (Fig. B.56):

The matching can be accomplished as follow: calculate the input/output impedances of the device to be matched, it is series impedance, or $R + j \cdot X(R - j \cdot X)$. Otherwise we obtain these values from the data sheet. We convert series $R + j \cdot X(R - j \cdot X)$ to parallel as required. Whether we elect to utilize parallel or series will depend on the following. It would be easier, with micro strip, to resonate out the reactance in series or to resonate out the parallel circuit equivalent. If a distributed part must be used for this purpose, a shunt capacitor is always desire. We calculate the required micro strip width and length, at the frequency of interest. Simulate a lumped value that will cancel out the reactive component of the device being matched. We make the input or output R + j · 0. Lumped microwave capacitors and inductors can also be utilized if the micro strip part is unrealizable. We match the real (resistive) part of the transistor's input or output by employing the microstrip transformer. The microstrip transformer section is placed between the two mismatched impedances.

Example: 50 Ω for the system's transmission line impedance, and 20 Ω for the transistor's input resistance. The transformer segment will be $long = \frac{\lambda}{4} \cdot V_p$

V_p is the propagation velocity. As wide as a microstrip transmission line would be with an impedance of

$$Z = \sqrt{R_1 \cdot R_2}; R_1 = 20\Omega; R_2 = 50\,\Omega \Rightarrow Z = \sqrt{20 \cdot 50} = 31.6\,\Omega.$$

For other requirements we need to use equation to plug in different microstrip widths to obtain the desired impedance. The characteristic impedance of the microstrip is (Ω) and Z_0 is the microstrip's characteristic impedance. W is the width of the microstrip conductor. h is the thickness of the substrate between the ground plane and the microstrip conductor.

$$Z_0 = \frac{377}{\left[\frac{W}{h} + 1\right] \cdot \sqrt{E_r + \sqrt{E_r}}}$$

W and h use the same units and E_r is the dielectric constant of the board material. The dielectric constant of the medium does not have a unit because it is a ratio.

$$Z_0 = \frac{377}{\left[\frac{W}{h} + 1\right] \cdot \sqrt{E_r + \sqrt{E_r}}}$$

A 50 Ω microstrip is utilized in microwave circuits to prevent reflections and mismatch losses between physically separated components. A calculated nominal width will prevent the line from being either inductive or capacitive at any point along its length.

Our target: Source's output impedance matched to the microstrip. Microstrip matched to the input impedance of the load.

Results: No standing or reflected waves and no power dissipated at heat, except in the actual resistance of the copper and dielectric as $I^2 \cdot R$ loss. In microstrip, the dielectric constant (E_r) of the PCB's substrate material will not be the sole E_r. There is a flux leakage into the air above the PCB, combined with the flux penetrating into the dielectric. Actual effective dielectric constant E_{ff}, is true dielectric constant that the microstrip see. There is some value between the surrounding air and the true dielectric constant of the PCB. To avoid RF field leakage from microstrips, transmission lines should be isolated by at least two or more line widths and create spaces between traces and circuits for decreasing any mutual coupling effects.

We decrease any impedance bumps at high microwave frequencies by keeping microstrip always be run as short and as straight as possible. Microstrip's angle (turn) should using a mitered or slow turn bend (Fig. B.57).

We need to find the actual wavelength of the signal which it is being slowed down by the PCB's substrate material. We calculate the microstrip's velocity of propagation (V_p): First find the effective dielectric constant E_{ff} of the microstrip. The signal will be partly in the dielectric and partly in the air above the microstrip. This affects the propagation velocity through this combination of the two dielectric mediums. We need to find the actual wavelength of the signal in microstrip line. E_{ff} is the effective dielectric constant that the microstrip sees. E_r is the actual dielectric constant of the PCB's substrate material. h is the thickness of the substrate material between the top conductor and the bottom ground plane of the microstrip. W is the width of the top conductor of the microstrip (same units as h).

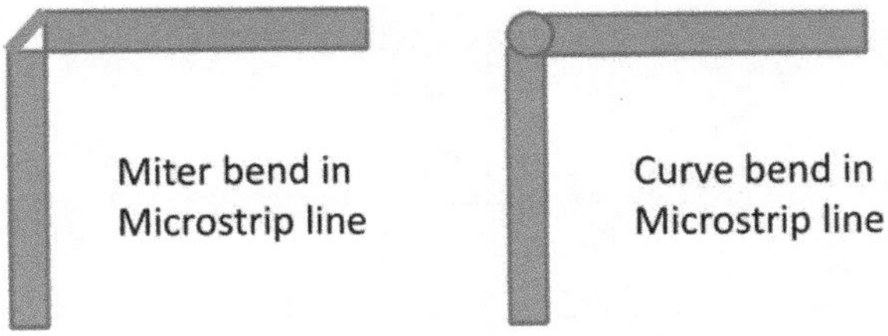

Fig. B.57 Miter bend and curve bend in microstrip line

V_p is the fraction of the speed of light as compared to light in a vacuum. λ_{vac} is the wavelength of the signal of interest in a perfect vacuum. f is the frequency of the signal of interest, GHz. The actual wavelength of the signal in microstrip line is as follows:

$$E_{eff} = \frac{E_r + 1}{2} + \left\{ \frac{(E_r - 1)}{2} \cdot \frac{1}{\sqrt{1 + (\frac{12 \cdot h}{W})}} \right\} \& V_P = \frac{1}{\sqrt{E_{eff}}}$$

$$\Rightarrow V_P = \frac{1}{\sqrt{\frac{E_r+1}{2} + \left\{ \frac{(E_r-1)}{2} \cdot \frac{1}{\sqrt{1+(\frac{12 \cdot h}{W})}} \right\}}}; \ \lambda_{vac} = 11{,}800/f$$

$$\lambda = V_P \cdot \lambda_{vac} = \frac{\lambda_{vac}}{\sqrt{\frac{E_r+1}{2} + \left\{ \frac{(E_r-1)}{2} \cdot \frac{1}{\sqrt{1+(\frac{12 \cdot h}{W})}} \right\}}} = \frac{11{,}800}{f \cdot \sqrt{\frac{E_r+1}{2} + \left\{ \frac{(E_r-1)}{2} \cdot \frac{1}{\sqrt{1+(\frac{12 \cdot h}{W})}} \right\}}}$$

The functional connection between microstrip characteristics impedance and PCB's parameters (λ,f,h,W) is $Z_0 = \xi(W, h, \lambda, f)$.

$$\lambda = \frac{11{,}800}{f \cdot \sqrt{\frac{E_r+1}{2} + \{\frac{(E_r-1)}{2} \cdot \frac{1}{\sqrt{1+(\frac{12 \cdot h}{W})}}\}}} \Rightarrow \sqrt{\frac{E_r + 1}{2} + \{\frac{(E_r - 1)}{2} \cdot \frac{1}{\sqrt{1 + (\frac{12 \cdot h}{W})}}\}}$$

$$= \frac{11{,}800}{\lambda \cdot f} \frac{E_r + 1}{2} + \{\frac{(E_r - 1)}{2} \cdot \frac{1}{\sqrt{1 + (\frac{12 \cdot h}{W})}}\} = \frac{11{,}800^2}{\lambda^2 \cdot f^2}$$

$$\Rightarrow \frac{E_r}{2} + \frac{1}{2} + \frac{E_r}{2} \cdot \frac{1}{\sqrt{1 + (\frac{12 \cdot h}{W})}} - \frac{1}{2} \cdot \frac{1}{\sqrt{1 + (\frac{12 \cdot h}{W})}} = \frac{11{,}800^2}{\lambda^2 \cdot f^2}$$

$$E_r \cdot \frac{1}{2} \cdot [1 + \frac{1}{\sqrt{1 + (\frac{12 \cdot h}{W})}}] + \frac{1}{2}[1 - \frac{1}{\sqrt{1 + (\frac{12 \cdot h}{W})}}] = \frac{11{,}800^2}{\lambda^2 \cdot f^2}$$

$$\Rightarrow E_r \cdot \frac{1}{2} \cdot [1 + \frac{1}{\sqrt{1 + (\frac{12 \cdot h}{W})}}] = \frac{11{,}800^2}{\lambda^2 \cdot f^2} - \frac{1}{2} \cdot [1 - \frac{1}{\sqrt{1 + (\frac{12 \cdot h}{W})}}]$$

$$E_r \cdot \frac{1}{2} \cdot [1 + \frac{1}{\sqrt{1 + (\frac{12 \cdot h}{W})}}] = \frac{11{,}800^2}{\lambda^2 \cdot f^2} - \frac{1}{2} \cdot [1 - \frac{1}{\sqrt{1 + (\frac{12 \cdot h}{W})}}]$$

$$\Rightarrow E_r = \frac{\frac{2 \cdot 11800^2}{\lambda^2 \cdot f^2} - [1 - \frac{1}{\sqrt{1+(\frac{12 \cdot h}{W})}}]}{[1 + \frac{1}{\sqrt{1+(\frac{12 \cdot h}{W})}}]}$$

$$
\begin{aligned}
Z_0 &= \frac{377}{\left[\frac{W}{h}+1\right]\cdot\sqrt{E_r+\sqrt{E_r}}} \\
&= \frac{377}{\left[\frac{W}{h}+1\right]\cdot\sqrt{\frac{\frac{2\cdot 11{,}800^2}{\lambda^2\cdot f^2}-\left[1-\frac{1}{\sqrt{1+(\frac{12\cdot h}{W})}}\right]}{\left[1+\frac{1}{\sqrt{1+(\frac{12\cdot h}{W})}}\right]}}+\sqrt{\frac{\frac{2\cdot 11{,}800^2}{\lambda^2\cdot f^2}-\left[1-\frac{1}{\sqrt{1+(\frac{12\cdot h}{W})}}\right]}{\left[1+\frac{1}{\sqrt{1+(\frac{12\cdot h}{W})}}\right]}}}
\end{aligned}
$$

Question: We have a system of amplifier and load. The source resistance (Rs) value is lower than load resistance (Rl) value. Rs = 15 Ω and our load resistance can't be lower than Γ0 [Ω] value and maximum load resistance variation can be Δ [Ω]. We use L—matching network which have fix inductor L_1 and fix capacitor C_1. We have trim capacitor in series to L—Network's C_1 capacitor, sign it as Cx. What is the variation gap of Cx trim capacitor, if we want to match our amplifier to load over all possible load values? Find Cx = ξ(Rs, Γ0, Δ) and possible maximum and minimum Cx values. f = 2 GHz (Fig. B.58).

Answer: First we need to get Zin expression. Capacitors C_1 and C_x are in series and the total Capacitance is C_{1-x}. $C_{1-x} = C_1 \cdot C_x/(C_1 + C_x)$; $Z_{c1-x} = 1/(j \cdot \omega C_{1-x})$.

$$
\begin{aligned}
C_{1-X} &= \frac{C_1\cdot C_X}{C_1+C_X};\, Z_{C_{1-X}} = \frac{1}{j\cdot\omega\cdot C_{1-X}} = \frac{1}{j\cdot\omega\cdot\left\{\frac{C_1\cdot C_X}{C_1+C_X}\right\}} \\
&= \frac{C_1+C_X}{j\cdot\omega\cdot C_1\cdot C_X} = -j\cdot\frac{C_1+C_X}{\omega\cdot C_1\cdot C_X}
\end{aligned}
$$

$$
\begin{aligned}
Z_{C_{1-X}} \parallel Z_{load} &= Z_{C_{1-X}} \parallel R_L;\, Z_{load} = R_L;\, Z_{C_{1-X}} \parallel R_L = \frac{Z_{C_{1-X}}\cdot R_L}{Z_{C_{1-X}}+R_L} \\
&= \frac{-j\cdot\frac{C_1+C_X}{\omega\cdot C_1\cdot C_X}\cdot R_L}{-j\cdot\frac{C_1+C_X}{\omega\cdot C_1\cdot C_X}+R_L}
\end{aligned}
$$

$$
\begin{aligned}
Z_{C_{1-X}} \parallel R_L &= \frac{-j\cdot(C_1+C_X)\cdot R_L}{-j\cdot(C_1+C_X)+R_L\cdot\omega\cdot C_1\cdot C_X} \\
&= \frac{-j\cdot(C_1+C_X)\cdot R_L}{R_L\cdot\omega\cdot C_1\cdot C_X-j\cdot(C_1+C_X)}
\end{aligned}
$$

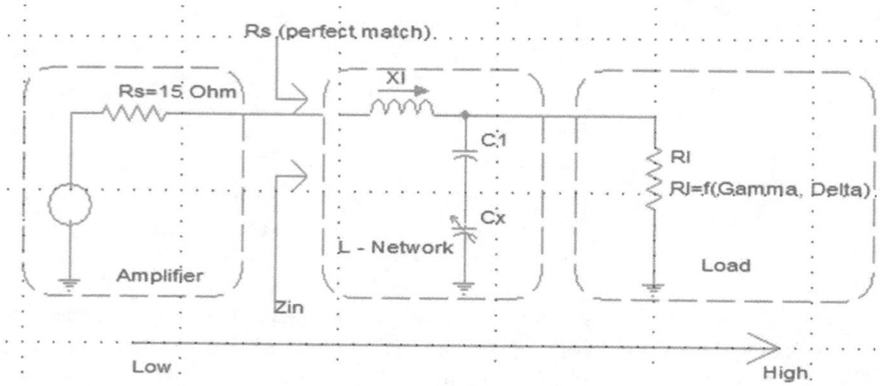

Fig. B.58 Unmatched system Rs < Rl

$$Z_{C_{1-X}} \parallel R_L = \frac{-j \cdot (C_1 + C_X) \cdot R_L}{R_L \cdot \omega \cdot C_1 \cdot C_X - j \cdot (C_1 + C_X)} \cdot \frac{R_L \cdot \omega \cdot C_1 \cdot C_X + j \cdot (C_1 + C_X)}{R_L \cdot \omega \cdot C_1 \cdot C_X + j \cdot (C_1 + C_X)}$$

$$Z_{C_{1-X}} \parallel R_L = \frac{(C_1 + C_X)^2 \cdot R_L - j \cdot (C_1 + C_X) \cdot R_L^2 \cdot \omega \cdot C_1 \cdot C_X}{R_L^2 \cdot \omega^2 \cdot C_1^2 \cdot C_X^2 + (C_1 + C_X)^2}$$

$$Z_{C_{1-X}} \parallel Z_{load} = Z_{C_{1-X}} \parallel R_L = \frac{(C_1 + C_X)^2 \cdot R_L}{R_L^2 \cdot \omega^2 \cdot C_1^2 \cdot C_X^2 + (C_1 + C_X)^2} - j \cdot \frac{(C_1 + C_X) \cdot R_L^2 \cdot \omega \cdot C_1 \cdot C_X}{R_L^2 \cdot \omega^2 \cdot C_1^2 \cdot C_X^2 + (C_1 + C_X)^2}$$

$$R_S[perfect\,match] = Z_{in} = R_S; Z_{in} = Z_L + Z_{C_{1-X}} \parallel Z_{load}; Z_L = j \cdot \omega \cdot L$$

$$Z_{in} = Z_L + Z_{C_{1-X}} \parallel Z_{load} = j \cdot \omega \cdot L + \frac{(C_1 + C_X)^2 \cdot R_L}{R_L^2 \cdot \omega^2 \cdot C_1^2 \cdot C_X^2 + (C_1 + C_X)^2} - j \cdot \frac{(C_1 + C_X) \cdot R_L^2 \cdot \omega \cdot C_1 \cdot C_X}{R_L^2 \cdot \omega^2 \cdot C_1^2 \cdot C_X^2 + (C_1 + C_X)^2}$$

$$Z_{in} = Z_L + Z_{C_{1-X}} \parallel Z_{load} = \frac{(C_1 + C_X)^2 \cdot R_L}{R_L^2 \cdot \omega^2 \cdot C_1^2 \cdot C_X^2 + (C_1 + C_X)^2} + j \cdot \omega \cdot \{L - \frac{(C_1 + C_X) \cdot R_L^2 \cdot C_1 \cdot C_X}{R_L^2 \cdot \omega^2 \cdot C_1^2 \cdot C_X^2 + (C_1 + C_X)^2}\}$$

For perfect match we need: $R_S = Z_{in}$; $Z_{in} = R_S$.

$$(1)\frac{(C_1+C_X)^2\cdot R_L}{R_L^2\cdot\omega^2\cdot C_1^2\cdot C_X^2+(C_1+C_X)^2}=R_S$$
$$(2)L-\frac{(C_1+C_X)\cdot R_L^2\cdot C_1\cdot C_X}{R_L^2\cdot\omega^2\cdot C_1^2\cdot C_X^2+(C_1+C_X)^2}=0$$

We know that

$$R_L\geq\Gamma_0\,\&\,R_L\leq\Gamma_0+\Delta$$

$$R_L\geq\Gamma_0\,\&\,R_L\leq\Gamma_0+\Delta\Rightarrow\Gamma_0\leq R_L\leq\Gamma_0+\Delta;R_L\in[\Gamma_0,\Gamma_0+\Delta]$$
$$(1)\ \mathrm{R}_S=\frac{(C_1+C_X)^2\cdot R_L}{R_L^2\cdot\omega^2\cdot C_1^2\cdot C_X^2+(C_1+C_X)^2}\Rightarrow R_L^2\cdot\omega^2\cdot C_1^2\cdot C_X^2\cdot\mathrm{R}_S$$
$$+(C_1+C_X)^2\cdot\mathrm{R}_S=(C_1+C_X)^2\cdot R_L$$
$$R_L^2\cdot\omega^2\cdot C_1^2\cdot C_X^2\cdot\mathrm{R}_S=(C_1+C_X)^2\cdot(R_L-\mathrm{R}_S)\Rightarrow(C_1+C_X)^2$$
$$=\frac{R_L^2\cdot\omega^2\cdot C_1^2\cdot C_X^2\cdot\mathrm{R}_S}{R_L-\mathrm{R}_S}$$

$$(1)\Rightarrow(2)\Rightarrow L-\frac{(C_1+C_X)\cdot R_L^2\cdot C_1\cdot C_X}{R_L^2\cdot\omega^2\cdot C_1^2\cdot C_X^2+(C_1+C_X)^2}=0$$
$$\Rightarrow L-\frac{(C_1+C_X)\cdot R_L^2\cdot C_1\cdot C_X}{R_L^2\cdot\omega^2\cdot C_1^2\cdot C_X^2+\frac{R_L^2\cdot\omega^2\cdot C_1^2\cdot C_X^2\cdot\mathrm{R}_S}{R_L-\mathrm{R}_S}}=0$$
$$L=\frac{(C_1+C_X)\cdot R_L^2\cdot C_1\cdot C_X}{R_L^2\cdot\omega^2\cdot C_1^2\cdot C_X^2+\frac{R_L^2\cdot\omega^2\cdot C_1^2\cdot C_X^2\cdot\mathrm{R}_S}{R_L-\mathrm{R}_S}}$$
$$\Rightarrow L=\frac{(C_1+C_X)\cdot R_L^2\cdot C_1\cdot C_X\cdot(R_L-\mathrm{R}_S)}{R_L^2\cdot\omega^2\cdot C_1^2\cdot C_X^2\cdot(R_L-\mathrm{R}_S)+R_L^2\cdot\omega^2\cdot C_1^2\cdot C_X^2\cdot\mathrm{R}_S}$$
$$R_L^2\cdot\omega^2\cdot C_1^2\cdot C_X^2\cdot(R_L-\mathrm{R}_S)\cdot L+R_L^2\cdot\omega^2\cdot C_1^2\cdot C_X^2\cdot\mathrm{R}_S\cdot L$$
$$=(C_1+C_X)\cdot R_L^2\cdot C_1\cdot C_X\cdot(R_L-\mathrm{R}_S)$$

$$R_L^2\cdot\omega^2\cdot C_1^2\cdot C_X^2\cdot(R_L-\mathrm{R}_S)\cdot L+R_L^2\cdot\omega^2\cdot C_1^2\cdot C_X^2\cdot\mathrm{R}_S\cdot L$$
$$=C_1\cdot R_L^2\cdot C_1\cdot C_X\cdot(R_L-\mathrm{R}_S)+C_X\cdot R_L^2\cdot C_1\cdot C_X\cdot(R_L-\mathrm{R}_S)$$
$$R_L^2\cdot\omega^2\cdot C_1^2\cdot C_X^2\cdot(R_L-\mathrm{R}_S)\cdot L+R_L^2\cdot\omega^2\cdot C_1^2\cdot C_X^2\cdot\mathrm{R}_S\cdot L$$
$$=C_1^2\cdot R_L^2\cdot C_X\cdot(R_L-\mathrm{R}_S)+C_X^2\cdot R_L^2\cdot C_1\cdot(R_L-\mathrm{R}_S)$$
$$C_X^2\cdot R_L^2\cdot C_1\cdot(R_L-\mathrm{R}_S)-R_L^2\cdot\omega^2\cdot C_1^2\cdot C_X^2\cdot(R_L-\mathrm{R}_S)\cdot L$$
$$-R_L^2\cdot\omega^2\cdot C_1^2\cdot C_X^2\cdot\mathrm{R}_S\cdot L+C_1^2\cdot R_L^2\cdot C_X\cdot(R_L-\mathrm{R}_S)=0$$
$$C_X^2\cdot\{R_L^2\cdot C_1\cdot(R_L-\mathrm{R}_S)-R_L^2\cdot\omega^2\cdot C_1^2\cdot(R_L-\mathrm{R}_S)\cdot L$$
$$-R_L^2\cdot\omega^2\cdot C_1^2\cdot\mathrm{R}_S\cdot L\}+C_1^2\cdot R_L^2\cdot C_X\cdot(R_L-\mathrm{R}_S)=0$$

$$C_X^2 \cdot \{R_L^2 \cdot C_1 \cdot R_L - R_L^2 \cdot C_1 \cdot \mathrm{R}_S - R_L^2 \cdot \omega^2 \cdot C_1^2 \cdot R_L \cdot L + R_L^2 \cdot \omega^2 \cdot C_1^2 \cdot \mathrm{R}_S \cdot L$$
$$- R_L^2 \cdot \omega^2 \cdot C_1^2 \cdot \mathrm{R}_S \cdot L\} + C_1^2 \cdot R_L^2 \cdot C_X \cdot (R_L - \mathrm{R}_S) = 0$$
$$C_X^2 \cdot \{R_L^2 \cdot C_1 \cdot R_L - R_L^2 \cdot C_1 \cdot \mathrm{R}_S - R_L^2 \cdot \omega^2 \cdot C_1^2 \cdot R_L \cdot L\} + C_1^2 \cdot R_L^2 \cdot C_X \cdot (R_L - \mathrm{R}_S) = 0$$
$$C_X^2 \cdot R_L^2 \cdot C_1 \cdot \{R_L - \mathrm{R}_S - \omega^2 \cdot C_1 \cdot R_L \cdot L\} + C_1^2 \cdot R_L^2 \cdot C_X \cdot (R_L - \mathrm{R}_S) = 0$$
$$C_X \cdot [C_X \cdot R_L^2 \cdot C_1 \cdot \{R_L - \mathrm{R}_S - \omega^2 \cdot C_1 \cdot R_L \cdot L\} + C_1^2 \cdot R_L^2 \cdot (R_L - \mathrm{R}_S)]] = 0$$
$$(1)\, C_X = 0; (2) C_X \cdot R_L^2 \cdot C_1 \cdot \{R_L - \mathrm{R}_S - \omega^2 \cdot C_1 \cdot R_L \cdot L\} + C_1^2 \cdot R_L^2 \cdot (R_L - \mathrm{R}_S)] = 0$$
$$(2) C_X \cdot R_L^2 \cdot C_1 \cdot \{R_L - \mathrm{R}_S - \omega^2 \cdot C_1 \cdot R_L \cdot L\} + C_1^2 \cdot R_L^2 \cdot (R_L - \mathrm{R}_S)] = 0 \Rightarrow$$
$$C_X = \frac{C_1^2 \cdot R_L^2 \cdot (R_S - \mathrm{R}_L)}{R_L^2 \cdot C_1 \cdot \{R_L - \mathrm{R}_S - \omega^2 \cdot C_1 \cdot R_L \cdot L\}} \Rightarrow C_X = \frac{C_1 \cdot (R_S - \mathrm{R}_L)}{R_L - \mathrm{R}_S - \omega^2 \cdot C_1 \cdot R_L \cdot L}$$

We need to get R_L as a function of Cx and additional components.

$$C_X = \frac{C_1 \cdot (R_S - \mathrm{R}_L)}{R_L - \mathrm{R}_S - \omega^2 \cdot C_1 \cdot R_L \cdot L} \Rightarrow C_X \cdot R_L - C_X \cdot \mathrm{R}_S - C_X \cdot \omega^2 \cdot C_1 \cdot R_L \cdot L$$
$$= C_1 \cdot R_S - C_1 \cdot \mathrm{R}_L$$
$$C_X \cdot R_L - C_X \cdot \omega^2 \cdot C_1 \cdot R_L \cdot L + C_1 \cdot \mathrm{R}_L = C_1 \cdot R_S + C_X \cdot \mathrm{R}_S$$
$$R_L \cdot \{C_X - C_X \cdot \omega^2 \cdot C_1 \cdot L + C_1\} = R_S \cdot (C_1 + C_X)$$
$$\Rightarrow R_L = \frac{R_S \cdot (C_1 + C_X)}{C_X - C_X \cdot \omega^2 \cdot C_1 \cdot L + C_1}$$
$$R_L \geq \Gamma_0 \Rightarrow R_L = \frac{R_S \cdot (C_1 + C_X)}{C_X - C_X \cdot \omega^2 \cdot C_1 \cdot L + C_1} \geq \Gamma_0 \Rightarrow \frac{R_S \cdot (C_1 + C_X)}{C_X \cdot \{1 - \omega^2 \cdot C_1 \cdot L\} + C_1} \geq \Gamma_0$$
$$\frac{R_S \cdot (C_1 + C_X)}{C_X \cdot \{1 - \omega^2 \cdot C_1 \cdot L\} + C_1} \geq \Gamma_0 \Rightarrow \frac{R_S \cdot (C_1 + C_X)}{C_X \cdot \{1 - \omega^2 \cdot C_1 \cdot L\} + C_1} - \Gamma_0 \geq 0$$

In case (=) exist we get

$$\frac{R_S \cdot (C_1 + C_X)}{C_X \cdot \{1 - \omega^2 \cdot C_1 \cdot L\} + C_1} - \Gamma_0 = 0$$

$$\frac{R_S \cdot (C_1 + C_X)}{C_X \cdot \{1 - \omega^2 \cdot C_1 \cdot L\} + C_1} - \Gamma_0 = 0 \Rightarrow \frac{R_S \cdot (C_1 + C_X)}{C_X \cdot \{1 - \omega^2 \cdot C_1 \cdot L\} + C_1} = \Gamma_0$$
$$R_S \cdot (C_1 + C_X) = C_X \cdot \{1 - \omega^2 \cdot C_1 \cdot L\} \cdot \Gamma_0 + C_1 \cdot \Gamma_0$$
$$R_S \cdot C_1 + R_S \cdot C_X = C_X \cdot \{1 - \omega^2 \cdot C_1 \cdot L\} \cdot \Gamma_0 + C_1 \cdot \Gamma_0$$
$$R_S \cdot C_1 - C_1 \cdot \Gamma_0 = C_X \cdot \{1 - \omega^2 \cdot C_1 \cdot L\} \cdot \Gamma_0 - R_S \cdot C_X$$
$$C_X \cdot \{\Gamma_0 - \omega^2 \cdot C_1 \cdot L \cdot \Gamma_0 - R_S\} = R_S \cdot C_1 - C_1 \cdot \Gamma_0 \Rightarrow C_X^{(1)}$$
$$= \frac{C_1 \cdot (R_S - \Gamma_0)}{\Gamma_0 \cdot [1 - \omega^2 \cdot C_1 \cdot L] - R_S}$$

The other option for case (=) is when upper limit is $\Gamma_0 + \Delta$, $\Gamma_0 \rightarrow \Gamma_0 + \Delta$ then

$$C_X^{(1)} = \frac{C_1 \cdot (R_S - \Gamma_0)}{\Gamma_0 \cdot [1 - \omega^2 \cdot C_1 \cdot L] - R_S}|_{\Gamma_0 \rightarrow \Gamma_0 + \Delta} \rightarrow \frac{C_1 \cdot (R_S - \Gamma_0 - \Delta)}{(\Gamma_0 + \Delta) \cdot [1 - \omega^2 \cdot C_1 \cdot L] - R_S}$$

$$C_X^{(2)} = \frac{C_1 \cdot (R_S - \Gamma_0 - \Delta)}{(\Gamma_0 + \Delta) \cdot [1 - \omega^2 \cdot C_1 \cdot L] - R_S}$$

Since $\Gamma 0 > 0$ & $\Delta > 0 \rightarrow C_X^{(2)} < C_X^{(1)}$

Additionally there is no way that $C_X^{(2)} \rightarrow \infty$ or $C_X^{(1)} \rightarrow \infty$

$$C_X^{(1)} = \frac{C_1 \cdot (R_S - \Gamma_0)}{\Gamma_0 \cdot [1 - \omega^2 \cdot C_1 \cdot L] - R_S} \Rightarrow \Gamma_0 \cdot [1 - \omega^2 \cdot C_1 \cdot L] - R_S \neq 0$$

$$\Rightarrow 1 - \omega^2 \cdot C_1 \cdot L \neq \frac{R_S}{\Gamma_0}$$

$$\omega^2 \cdot C_1 \cdot L \neq 1 - \frac{R_S}{\Gamma_0}$$

$$\Rightarrow \omega^2 \neq \frac{1}{C_1 \cdot L} \cdot (1 - \frac{R_S}{\Gamma_0}) \Rightarrow \omega \neq \frac{1}{\sqrt{C_1 \cdot L}} \cdot \sqrt{(1 - \frac{R_S}{\Gamma_0})}$$

$$\Rightarrow 2 \cdot \pi \cdot f \neq \frac{1}{\sqrt{C_1 \cdot L}} \cdot \sqrt{(1 - \frac{R_S}{\Gamma_0})} \Rightarrow f \neq \frac{1}{2 \cdot \pi \cdot \sqrt{C_1 \cdot L}} \cdot \sqrt{(1 - \frac{R_S}{\Gamma_0})};$$

$$1 - \frac{R_S}{\Gamma_0} > 0 \Rightarrow 1 > \frac{R_S}{\Gamma_0} \Rightarrow \Gamma_0 > R_S$$

$$C_X^{(2)} = \frac{C_1 \cdot (R_S - \Gamma_0 - \Delta)}{(\Gamma_0 + \Delta) \cdot [1 - \omega^2 \cdot C_1 \cdot L] - R_S} \Rightarrow (\Gamma_0 + \Delta) \cdot [1 - \omega^2 \cdot C_1 \cdot L] - R_S \neq 0$$

$$1 - \omega^2 \cdot C_1 \cdot L \neq \frac{R_S}{\Gamma_0 + \Delta}$$

$$\Rightarrow \omega^2 \cdot C_1 \cdot L \neq 1 - \frac{R_S}{\Gamma_0 + \Delta} \Rightarrow \omega^2 \neq \frac{1}{C_1 \cdot L} \cdot (1 - \frac{R_S}{\Gamma_0 + \Delta})$$

$$\omega \neq \frac{1}{\sqrt{C_1 \cdot L}} \cdot \sqrt{1 - \frac{R_S}{\Gamma_0 + \Delta}} \Rightarrow 2 \cdot \pi \cdot f \neq \frac{1}{\sqrt{C_1 \cdot L}} \cdot \sqrt{1 - \frac{R_S}{\Gamma_0 + \Delta}}$$

$$f \neq \frac{1}{2 \cdot \pi \cdot \sqrt{C_1 \cdot L}} \cdot \sqrt{1 - \frac{R_S}{\Gamma_0 + \Delta}} \Rightarrow 1 - \frac{R_S}{\Gamma_0 + \Delta} > 0$$

$$\Rightarrow 1 > \frac{R_S}{\Gamma_0 + \Delta} \Rightarrow \Gamma_0 + \Delta > R_S$$

Discussion No.1: check the values interval for lower limit Γ_0.

$$\frac{R_S \cdot (C_1 + C_X)}{C_X \cdot \{1 - \omega^2 \cdot C_1 \cdot L\} + C_1} - \Gamma_0 \geq 0 \Rightarrow \frac{R_S \cdot (C_1 + C_X)}{C_X \cdot \{1 - \omega^2 \cdot C_1 \cdot L\} + C_1} - \Gamma_0 > 0$$

$$\frac{R_S \cdot (C_1 + C_X) - C_X \cdot \{1 - \omega^2 \cdot C_1 \cdot L\} \cdot \Gamma_0 - C_1 \cdot \Gamma_0}{C_X \cdot \{1 - \omega^2 \cdot C_1 \cdot L\} + C_1} > 0$$

Case No.1.A:

$$R_S \cdot (C_1 + C_X) - C_X \cdot \{1 - \omega^2 \cdot C_1 \cdot L\} \cdot \Gamma_0 - C_1 \cdot \Gamma_0 > 0$$
$$\& \, C_X \cdot \{1 - \omega^2 \cdot C_1 \cdot L\} + C_1 > 0$$
$$R_S \cdot C_1 + R_S \cdot C_X - C_X \cdot \{1 - \omega^2 \cdot C_1 \cdot L\} \cdot \Gamma_0 - C_1 \cdot \Gamma_0 > 0$$
$$\& \, C_1 > C_X \cdot \{\omega^2 \cdot C_1 \cdot L - 1\}$$
$$C_X \cdot [R_S - \{1 - \omega^2 \cdot C_1 \cdot L\} \cdot \Gamma_0] > C_1 \cdot \Gamma_0 - R_S \cdot C_1$$
$$\& \, C_1 > C_X \cdot \{\omega^2 \cdot C_1 \cdot L - 1\}$$
$$C_X \cdot [R_S - \{1 - \omega^2 \cdot C_1 \cdot L\} \cdot \Gamma_0] > C_1 \cdot [\Gamma_0 - R_S]$$
$$\& \, C_1 > C_X \cdot \{\omega^2 \cdot C_1 \cdot L - 1\}$$

We consider $R_S - \{1 - \omega^2 \cdot C_1 \cdot L\} \cdot \Gamma_0 > 0$ and $\omega^2 \cdot C_1 \cdot L - 1 > 0$
Then

$$C_X > \frac{C_1 \cdot [\Gamma_0 - R_S]}{R_S - \{1 - \omega^2 \cdot C_1 \cdot L\} \cdot \Gamma_0} \; \& \; C_X < \frac{C_1}{\omega^2 \cdot C_1 \cdot L - 1}$$

$$\frac{C_1}{\omega^2 \cdot C_1 \cdot L - 1} > C_X > \frac{C_1 \cdot [\Gamma_0 - R_S]}{R_S - \{1 - \omega^2 \cdot C_1 \cdot L\} \cdot \Gamma_0}$$
$$\Rightarrow \frac{C_1}{\omega^2 \cdot C_1 \cdot L - 1} > C_X > \frac{C_1 \cdot [1 - \frac{R_S}{\Gamma_0}]}{\frac{R_S}{\Gamma_0} - \{1 - \omega^2 \cdot C_1 \cdot L\}}$$

We already know that $\Gamma_0 > R_S$ then Cx always has positive value.

$$\frac{R_S}{\Gamma_0} < 1 \Rightarrow 1 > [1 - \frac{R_S}{\Gamma_0}] > 0; \frac{C_1}{\omega^2 \cdot C_1 \cdot L - 1} > 0 \; \& \; \frac{C_1 \cdot [1 - \frac{R_S}{\Gamma_0}]}{\frac{R_S}{\Gamma_0} - \{1 - \omega^2 \cdot C_1 \cdot L\}} > 0$$

$$\omega^2 \cdot C_1 \cdot L - 1 > 0 \Rightarrow (\omega \cdot \sqrt{C_1 \cdot L} - 1) \cdot (\omega \cdot \sqrt{C_1 \cdot L} + 1) > 0$$
$$\Rightarrow \omega \cdot \sqrt{C_1 \cdot L} - 1 > 0$$

$$\omega \cdot \sqrt{C_1 \cdot L} - 1 > 0 \Rightarrow \omega \cdot \sqrt{C_1 \cdot L} > 1$$
$$\Rightarrow \omega > \frac{1}{\sqrt{C_1 \cdot L}} \Rightarrow 2 \cdot \pi \cdot f > \frac{1}{\sqrt{C_1 \cdot L}} \Rightarrow f > \frac{1}{2 \cdot \pi \cdot \sqrt{C_1 \cdot L}}$$

$$\frac{R_S}{\Gamma_0} - \{1 - \omega^2 \cdot C_1 \cdot L\} > 0$$

$$\Rightarrow (\frac{R_S}{\Gamma_0} - 1) + \omega^2 \cdot C_1 \cdot L > 0 \Rightarrow \frac{1}{C_1 \cdot L} \cdot (\frac{R_S}{\Gamma_0} - 1) + \omega^2 > 0$$

$$\frac{1}{C_1 \cdot L} \cdot (\frac{R_S}{\Gamma_0} - 1) + \omega^2 > 0 \Rightarrow \omega^2 - \frac{1}{C_1 \cdot L} \cdot (1 - \frac{R_S}{\Gamma_0}) > 0$$

$$\omega^2 - \frac{1}{C_1 \cdot L} \cdot (1 - \frac{R_S}{\Gamma_0}) > 0 \Rightarrow [\omega - \frac{1}{\sqrt{C_1 \cdot L}} \cdot \sqrt{(1 - \frac{R_S}{\Gamma_0})}]$$

$$\cdot [\omega + \frac{1}{\sqrt{C_1 \cdot L}} \cdot \sqrt{(1 - \frac{R_S}{\Gamma_0})}] > 0$$

$$\omega - \frac{1}{\sqrt{C_1 \cdot L}} \cdot \sqrt{(1 - \frac{R_S}{\Gamma_0})} > 0 \Rightarrow \omega > \frac{1}{\sqrt{C_1 \cdot L}} \cdot \sqrt{(1 - \frac{R_S}{\Gamma_0})}$$

$$\Rightarrow 2 \cdot \pi \cdot f > \frac{1}{\sqrt{C_1 \cdot L}} \cdot \sqrt{(1 - \frac{R_S}{\Gamma_0})}$$

$$2 \cdot \pi \cdot f > \frac{1}{\sqrt{C_1 \cdot L}} \cdot \sqrt{(1 - \frac{R_S}{\Gamma_0})} \Rightarrow f > \frac{1}{2 \cdot \pi \cdot \sqrt{C_1 \cdot L}} \cdot \sqrt{(1 - \frac{R_S}{\Gamma_0})}$$

We have two conditions:

$$\{f > \frac{1}{2 \cdot \pi \cdot \sqrt{C_1 \cdot L}}\} \cup \{f > \frac{1}{2 \cdot \pi \cdot \sqrt{C_1 \cdot L}} \cdot \sqrt{(1 - \frac{R_S}{\Gamma_0})}\}|_{\sqrt{(1-\frac{R_S}{\Gamma_0})} < 0}$$

$$= \{f > \frac{1}{2 \cdot \pi \cdot \sqrt{C_1 \cdot L}}\}$$

<u>Case No.1.B</u>:

$$\frac{R_S \cdot (C_1 + C_X)}{C_X \cdot \{1 - \omega^2 \cdot C_1 \cdot L\} + C_1} - \Gamma_0 \geq 0 \Rightarrow \frac{R_S \cdot (C_1 + C_X)}{C_X \cdot \{1 - \omega^2 \cdot C_1 \cdot L\} + C_1} - \Gamma_0 > 0$$

$$\frac{R_S \cdot (C_1 + C_X) - C_X \cdot \{1 - \omega^2 \cdot C_1 \cdot L\} \cdot \Gamma_0 - C_1 \cdot \Gamma_0}{C_X \cdot \{1 - \omega^2 \cdot C_1 \cdot L\} + C_1} > 0$$

$$R_S \cdot (C_1 + C_X) - C_X \cdot \{1 - \omega^2 \cdot C_1 \cdot L\} \cdot \Gamma_0 - C_1 \cdot \Gamma_0 < 0 \ \&$$

$$C_X \cdot \{1 - \omega^2 \cdot C_1 \cdot L\} + C_1 < 0$$

$$R_S \cdot C_1 + R_S \cdot C_X - C_X \cdot \{1 - \omega^2 \cdot C_1 \cdot L\} \cdot \Gamma_0 - C_1 \cdot \Gamma_0 < 0 \ \&$$

$$C_1 < C_X \cdot \{\omega^2 \cdot C_1 \cdot L - 1\}$$

$$C_X \cdot [R_S - \{1 - \omega^2 \cdot C_1 \cdot L\} \cdot \Gamma_0] < C_1 \cdot \Gamma_0 - R_S \cdot C_1 \ \& \ C_1 < C_X \cdot \{\omega^2 \cdot C_1 \cdot L - 1\}$$

$$C_X \cdot [R_S - \{1 - \omega^2 \cdot C_1 \cdot L\} \cdot \Gamma_0] < C_1 \cdot [\Gamma_0 - R_S] \ \& \ C_1 < C_X \cdot \{\omega^2 \cdot C_1 \cdot L - 1\}$$

We consider $R_S - \{1 - \omega^2 \cdot C_1 \cdot L\} \cdot \Gamma_0 > 0$ and $\omega^2 \cdot C_1 \cdot L - 1 > 0$
Then

$$C_X < \frac{C_1 \cdot [\Gamma_0 - R_S]}{R_S - \{1 - \omega^2 \cdot C_1 \cdot L\} \cdot \Gamma_0} \ \& \ C_X > \frac{C_1}{\omega^2 \cdot C_1 \cdot L - 1}$$

Then

$$C_X < \frac{C_1 \cdot [1 - \frac{R_S}{\Gamma_0}]}{\frac{R_S}{\Gamma_0} - \{1 - \omega^2 \cdot C_1 \cdot L\}} \ \& \ C_X > \frac{C_1}{\omega^2 \cdot C_1 \cdot L - 1}$$

$$\frac{C_1}{\omega^2 \cdot C_1 \cdot L - 1} < C_X < \frac{C_1 \cdot [1 - \frac{R_S}{\Gamma_0}]}{\frac{R_S}{\Gamma_0} - \{1 - \omega^2 \cdot C_1 \cdot L\}}$$

Finally for Case No.1 we have possible two options:

$$(1) \ \frac{C_1}{\omega^2 \cdot C_1 \cdot L - 1} > C_X > \frac{C_1 \cdot [1 - \frac{R_S}{\Gamma_0}]}{\frac{R_S}{\Gamma_0} - \{1 - \omega^2 \cdot C_1 \cdot L\}}$$

$$(2) \ \frac{C_1}{\omega^2 \cdot C_1 \cdot L - 1} < C_X < \frac{C_1 \cdot [1 - \frac{R_S}{\Gamma_0}]}{\frac{R_S}{\Gamma_0} - \{1 - \omega^2 \cdot C_1 \cdot L\}}$$

We need to decide which option is feasible for us. We define Ω as the difference between out limits bands (UL/LL).

$$\Omega = \frac{C_1}{\omega^2 \cdot C_1 \cdot L - 1} - \frac{C_1 \cdot [1 - \frac{R_S}{\Gamma_0}]}{\frac{R_S}{\Gamma_0} - \{1 - \omega^2 \cdot C_1 \cdot L\}}$$

and need to find if $\Omega > 0$ or $\Omega < 0$ or $\Omega = 0$.

$$\Omega = \frac{C_1}{\omega^2 \cdot C_1 \cdot L - 1} - \frac{C_1 \cdot [1 - \frac{R_S}{\Gamma_0}]}{\frac{R_S}{\Gamma_0} - \{1 - \omega^2 \cdot C_1 \cdot L\}}$$
$$= \frac{C_1 \cdot \left[\frac{R_S}{\Gamma_0} - \{1 - \omega^2 \cdot C_1 \cdot L\}\right] - C_1 \cdot [1 - \frac{R_S}{\Gamma_0}] \cdot [\omega^2 \cdot C_1 \cdot L - 1]}{[\omega^2 \cdot C_1 \cdot L - 1] \cdot \left[\frac{R_S}{\Gamma_0} - \{1 - \omega^2 \cdot C_1 \cdot L\}\right]}$$
$$\Omega = \frac{C_1 \cdot \frac{R_S}{\Gamma_0} - C_1 \cdot \{1 - \omega^2 \cdot C_1 \cdot L\} - [C_1 - C_1 \cdot \frac{R_S}{\Gamma_0}] \cdot [\omega^2 \cdot C_1 \cdot L - 1]}{[\omega^2 \cdot C_1 \cdot L - 1] \cdot \left[\frac{R_S}{\Gamma_0} - \{1 - \omega^2 \cdot C_1 \cdot L\}\right]}$$

$$\Omega = \frac{C_1 \cdot \frac{R_S}{\Gamma_0} - C_1 \cdot \{1 - \omega^2 \cdot C_1 \cdot L\} - \{C_1 \cdot [\omega^2 \cdot C_1 \cdot L - 1] - C_1 \cdot \frac{R_S}{\Gamma_0} \cdot [\omega^2 \cdot C_1 \cdot L - 1]\}}{[\omega^2 \cdot C_1 \cdot L - 1] \cdot \left[\frac{R_S}{\Gamma_0} - \{1 - \omega^2 \cdot C_1 \cdot L\}\right]}$$

$$\Omega = \frac{C_1 \cdot \frac{R_S}{\Gamma_0} - C_1 \cdot \{1 - \omega^2 \cdot C_1 \cdot L\} - C_1 \cdot [\omega^2 \cdot C_1 \cdot L - 1] + C_1 \cdot \frac{R_S}{\Gamma_0} \cdot [\omega^2 \cdot C_1 \cdot L - 1]}{[\omega^2 \cdot C_1 \cdot L - 1] \cdot \left[\frac{R_S}{\Gamma_0} - \{1 - \omega^2 \cdot C_1 \cdot L\}\right]}$$

$$\Omega = \frac{C_1 \cdot \frac{R_S}{\Gamma_0} - C_1 \cdot \{1 - \omega^2 \cdot C_1 \cdot L\} + C_1 \cdot [1 - \omega^2 \cdot C_1 \cdot L] + C_1 \cdot \frac{R_S}{\Gamma_0} \cdot \omega^2 \cdot C_1 \cdot L - C_1 \cdot \frac{R_S}{\Gamma_0}}{[\omega^2 \cdot C_1 \cdot L - 1] \cdot \left[\frac{R_S}{\Gamma_0} - \{1 - \omega^2 \cdot C_1 \cdot L\}\right]}$$

$$\Omega = \frac{C_1 \cdot \frac{R_S}{\Gamma_0} \cdot \omega^2 \cdot C_1 \cdot L}{[\omega^2 \cdot C_1 \cdot L - 1] \cdot \left[\frac{R_S}{\Gamma_0} - \{1 - \omega^2 \cdot C_1 \cdot L\}\right]} ; C_1 \cdot \frac{R_S}{\Gamma_0} \cdot \omega^2 \cdot C_1 \cdot L > 0$$

If $[\omega^2 \cdot C_1 \cdot L - 1] \cdot \left[\frac{R_S}{\Gamma_0} - \{1 - \omega^2 \cdot C_1 \cdot L\}\right] > 0$ then $\Omega > 0$

If $[\omega^2 \cdot C_1 \cdot L - 1] \cdot \left[\frac{R_S}{\Gamma_0} - \{1 - \omega^2 \cdot C_1 \cdot L\}\right] < 0$ then $\Omega < 0$

If $\Omega > 0$ then

$$\frac{C_1}{\omega^2 \cdot C_1 \cdot L - 1} > C_X > \frac{C_1 \cdot [1 - \frac{R_S}{\Gamma_0}]}{\frac{R_S}{\Gamma_0} - \{1 - \omega^2 \cdot C_1 \cdot L\}}$$

If $\Omega < 0$ then

$$\frac{C_1}{\omega^2 \cdot C_1 \cdot L - 1} < C_X < \frac{C_1 \cdot [1 - \frac{R_S}{\Gamma_0}]}{\frac{R_S}{\Gamma_0} - \{1 - \omega^2 \cdot C_1 \cdot L\}}$$

Discussion No.2: check the values interval for upper limit $\Gamma_0 + \Delta$.

$$\frac{R_S \cdot (C_1 + C_X)}{C_X \cdot \{1 - \omega^2 \cdot C_1 \cdot L\} + C_1} \leq \Gamma_0 + \Delta; \frac{R_S \cdot (C_1 + C_X)}{C_X \cdot \{1 - \omega^2 \cdot C_1 \cdot L\} + C_1} - \Gamma_0 - \Delta \leq 0$$

$$\Rightarrow \frac{R_S \cdot (C_1 + C_X)}{C_X \cdot \{1 - \omega^2 \cdot C_1 \cdot L\} + C_1} - (\Gamma_0 + \Delta) \leq 0$$

$$\frac{R_S \cdot (C_1 + C_X) - C_X \cdot \{1 - \omega^2 \cdot C_1 \cdot L\} \cdot (\Gamma_0 + \Delta) - C_1 \cdot (\Gamma_0 + \Delta)}{C_X \cdot \{1 - \omega^2 \cdot C_1 \cdot L\} + C_1} \leq 0$$

Appendix C
BJT Transistor Ebers-Moll Model and MOSFET Model

A bipolar junction transistor (BJT or bipolar transistor) is a type of transistor that relies on the contact of two types of semiconductor for its operation. BJTs can be used as amplifiers, optoisolation circuits, switches, or in oscillators in many industrial and commercial applications. BJTs can be found either as individual discrete components, or in large numbers as parts of integrated circuits. The operation of bipolar transistor involves both electron and holes. There are two kinds of charge carriers which characteristic of the two kinds of doped semiconductor material. Electrons are majority charge carriers in n-type semiconductors, whereas holes are majority charge carriers in p-type semiconductors. Unipolar transistors such as the field-effect transistors have only one kind of charge carrier. Charge flow in a BJT is due to diffusion of charge carriers across a junction between two regions of different charge concentrations. The regions of a BJT are called emitter, collector, and base. A discrete transistor has three leads for connection to these regions. Typically, the emitter region is heavily doped compared to the other two layers, whereas the majority charge carrier concentrations in base and collector layers are about the same. By design, most of the BJT collector current is due to the flow of charges injected from a high-concentration emitter into the base where there are minority carriers that diffuse toward the collector, and so BJTs are classified as minority-carrier devices. There are two types of BJT transistors, PNP and NPN based on the doping types of the three main terminal regions. An NPN transistor comprises two semiconductor junctions that share a thin p-doped anode region, and a PNP transistor comprises two semiconductor junctions that share a thin n-doped cathode region. In an NPN transistor, when positive bias is applied to the base–emitter junction, the equilibrium is disturbed between the thermally generated carriers and the repelling electric field of the n-doped emitter depletion region. This allows thermally excited electrons to inject from the emitter into the base region. These electrons diffuse through the base from the region of high concentration near the emitter towards the region of low concentration near the collector. The electrons in the base are called minority carriers because the base is doped p-type, which makes holes the majority carrier in the base.

O. Aluf, *Microwave RF Antennas and Circuits*,
DOI 10.1007/978-3-319-45427-6

The collector–emitter current can be viewed as being controlled by the base–emitter current (current control), or by the base–emitter voltage (voltage control). These views are related by the current–voltage relation of the base–emitter junction, which is just the usual exponential current–voltage curve of a p-n junction (diode). The Bipolar transistor exhibits a few delay characteristics when turning on and off. Most transistors and especially power transistors, exhibit long base-storage times that limit maximum frequency of operation in switching applications. One method for reducing this storage time is by using a Baker clamp. The proportion of electrons able to cross the base and reach the collector is a measure of the BJT efficiency. The heavy doping of the emitter region and light doping of the base region causes many more electrons to be injected from the emitter into the base than holes to be injected from the base into the emitter. The common-emitter current gain is represented by $\beta_F(\beta_f)$ or the h-parameter h_{FE}. It is approximately the ratio of the DC collector current to the DC base current in forward-active region. It is typically greater than 50 for small-signal transistors but can be smaller in transistors designed for high-power applications. Another important parameter is the common-base current gain $\alpha_F(\alpha_f)$. The common-base current gain is approximately the gain of current from emitter to collector in the forward-active region. This ratio usually has a value close to unity; between 0.98 and 0.998. It is less than unity due to recombination of charge carriers as they cross the base region.

$$\alpha_F = \frac{I_C}{I_E}; \beta_F = \frac{I_C}{I_B}; \beta_F = \frac{I_C}{I_E - I_C} = \frac{I_C/I_E}{1 - I_C/I_E} = \frac{\alpha_F}{1-\alpha_F}; \alpha_F = \frac{\beta_F}{\beta_F + 1}.$$

Transistors can be thought of as two diodes (P–N junctions) sharing a common region that minority carriers can move through. A PNP BJT will function like two diodes that share an N-type cathode region, and the NPN like two diodes sharing a P-type anode region. Connecting two diodes with wires will not make a transistor, since minority carriers will not be able to get from one P–N junction to the other through the wire. Both types of BJT function by letting a small current input to the base control an amplified output from the collector. The result is that the transistor makes a good switch that is controlled by its base input. The BJT also makes a good amplifier, since it can multiply a weak input signal to about 100 times its original strength. Networks of transistors are used to make powerful amplifiers with many different applications. In the discussion below, focus is on the NPN bipolar transistor. In the NPN transistor in what is called active mode, the base–emitter voltage V_{BE} and collector–base voltage V_{CB} are positive, forward biasing the emitter–base junction and reverse-biasing the collector–base junction. In the active mode of operation, electrons are injected from the forward biased n-type emitter region into the p-type base where they diffuse as minority carriers to the reverse-biased n-type

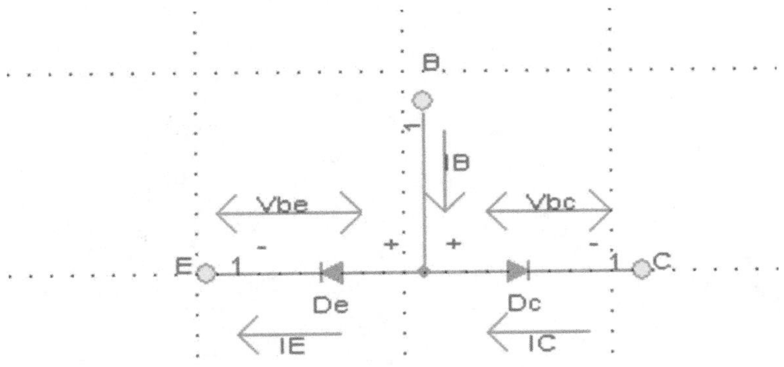

Fig. C.1 Bipolar transistor shown as two back to back p-n junction

collector and are swept away by the electric field in the reverse-biased collector–base junction. For a figure describing forward and reverse bias, see semiconductor diodes. The bipolar junction transistor can be considered essentially as two p-n junctions placed back to back, with the base p-type region being common to both diodes. This can be viewed as two diodes having a common third terminal as shown in the below figure. The two diodes are not in isolation, but are interdependent. This means that the total current flowing in each diode is influenced by the conditions prevailing in the other. In isolation, the two junctions would be characterized by the normal diode equation with a suitable notation used to differentiate between the two junctions as can be seen. When the two junctions are combined, to form a transistor, the base region is shared internally by both diodes even though there is an external connection to it (Fig. C.1).

In the forward active mode, α_F of the emitter current reaches the collector. This means that α_F of the diode current passing through the base-emitter junction contributes to the current flowing through the base-collector junction. Typically, α_F has a value of between 0.98 and 0.99. This is shown as the forward component of current as it applies to the normal forward active mode of operation of the device. This current is shown as a conventional current. It is equally possible to reverse the biases on the junctions to operate the transistor in the "reverse active mode". In this case, $\alpha_R(\alpha_r)$ times the collector current will contribute to the emitter current. For the doping ratios normally used the transistor will be much less efficient in the reverse mode and α_R would typically be in the range 0.1–0.5. The Ebers-Moll transistor model is an attempt to create an electrical model of the device as two diodes whose currents are determined by the normal diode law but with additional transfer ratios to quantify the interdependency of the junctions. Two dependent current sources are used to indicate the interaction of the junctions. Figure C.2 describes NPN Bipolar transistor Ebers Moll model.

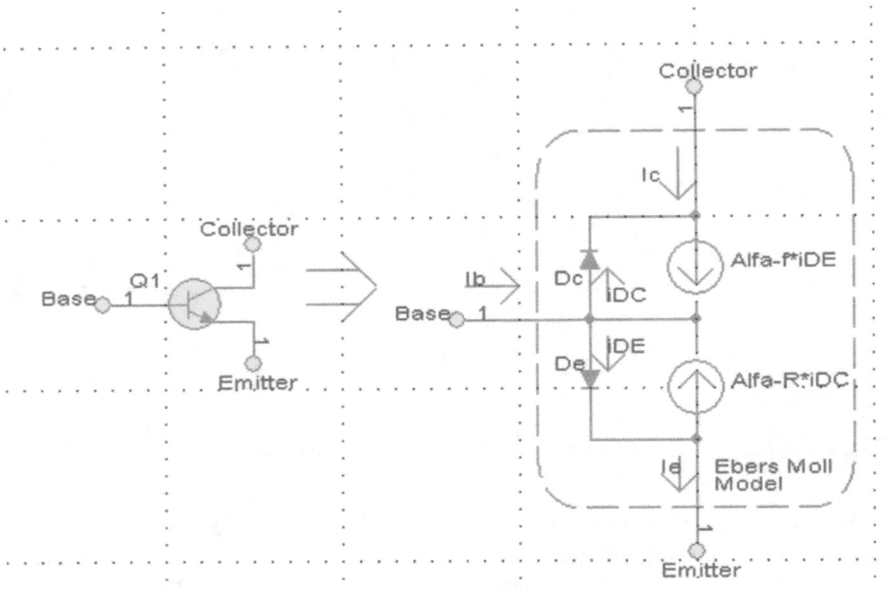

Fig. C.2 NPN bipolar transistor Ebers moll model

Applying Kirchoff's laws to the model gives the terminal current as:

$$I_{DE} = I_E + \alpha_r \cdot I_{DC}; I_C + I_{DC} = \alpha_f \cdot I_{DE}; \alpha_f \cdot I_{se} = \alpha_r \cdot I_{sc} = I_s; I_E = I_C + I_B$$

$\alpha_f = 0.98 - 0.99$ typically. $\alpha_r = 0.1 - 0.5$ typically. I_{se}: reverse saturation current of the base emitter diode. I_{sc}: reverse saturation current of the base collector diode. $I_{DC} = I_{sc} \cdot (e^{\frac{V_{BC}}{V_T}} - 1); I_{DE} = I_{se} \cdot (e^{\frac{V_{BE}}{V_T}} - 1)$. V_T – the thermal voltage $V_T \simeq \frac{k \cdot T}{q}$ (approximately 26 mV at 300 K (~room temperature). I_E is the transistor's emitter current. I_C is the transistor's collector current. I_B is the transistor's base current. The base internal current is mainly by diffusion (see Fick's law) and $J_{n(base)} = \frac{q \cdot D_n \cdot n_{bo}}{W} \cdot e^{\frac{V_{BE}}{V_T}}$. W is the base width. D_n is the diffusion constant for electron in the p type base.

$$I_{DE} = I_E + \alpha_r \cdot I_{DC} \Rightarrow I_{se} \cdot (e^{\frac{V_{BE}}{V_T}} - 1) = I_E + \alpha_r \cdot I_{sc} \cdot (e^{\frac{V_{BC}}{V_T}} - 1)$$
$$\Rightarrow I_E = I_{se} \cdot (e^{\frac{V_{BE}}{V_T}} - 1) - \alpha_r \cdot I_{sc} \cdot (e^{\frac{V_{BC}}{V_T}} - 1)$$
$$I_C + I_{DC} = \alpha_f \cdot I_{DE} \Rightarrow I_C + I_{sc} \cdot (e^{\frac{V_{BC}}{V_T}} - 1) = \alpha_f \cdot I_{se} \cdot (e^{\frac{V_{BE}}{V_T}} - 1)$$
$$\Rightarrow I_C = \alpha_f \cdot I_{se} \cdot (e^{\frac{V_{BE}}{V_T}} - 1) - I_{sc} \cdot (e^{\frac{V_{BC}}{V_T}} - 1)$$
$$I_B = I_E - I_C = (1 - \alpha_f) \cdot I_{se} \cdot (e^{\frac{V_{BE}}{V_T}} - 1) + (1 - \alpha_r) \cdot I_{sc} \cdot (e^{\frac{V_{BC}}{V_T}} - 1).$$

If we use the notation $\alpha_f \cdot I_{se} = \alpha_r \cdot I_{sc} = I_s; I_{sc} = \frac{I_s}{\alpha_r}; I_{se} = \frac{I_s}{\alpha_f}$ the following Ebers Moll equations:

$$I_E = \frac{I_s}{\alpha_f} \cdot (e^{\frac{V_{BE}}{V_T}} - 1) - \alpha_r \cdot \frac{I_s}{\alpha_r} \cdot (e^{\frac{V_{BC}}{V_T}} - 1) \Rightarrow I_E = \frac{I_s}{\alpha_f} \cdot (e^{\frac{V_{BE}}{V_T}} - 1) - I_s \cdot (e^{\frac{V_{BC}}{V_T}} - 1)$$

$$I_C = \alpha_f \cdot \frac{I_s}{\alpha_f} \cdot (e^{\frac{V_{BE}}{V_T}} - 1) - \frac{I_s}{\alpha_r} \cdot (e^{\frac{V_{BC}}{V_T}} - 1) \Rightarrow I_C = I_s \cdot (e^{\frac{V_{BE}}{V_T}} - 1) - \frac{I_s}{\alpha_r} \cdot (e^{\frac{V_{BC}}{V_T}} - 1)$$

$$I_B = (1 - \alpha_f) \cdot \frac{I_s}{\alpha_f} \cdot (e^{\frac{V_{BE}}{V_T}} - 1) + (1 - \alpha_r) \cdot \frac{I_s}{\alpha_r} \cdot (e^{\frac{V_{BC}}{V_T}} - 1); \frac{1}{\beta_f} = \frac{1 - \alpha_f}{\alpha_f}; \frac{1}{\beta_r} = \frac{1 - \alpha_r}{\alpha_r}$$

$$I_B = (1 - \alpha_f) \cdot \frac{I_s}{\alpha_f} \cdot (e^{\frac{V_{BE}}{V_T}} - 1) + (1 - \alpha_r) \cdot \frac{I_s}{\alpha_r} \cdot (e^{\frac{V_{BC}}{V_T}} - 1)$$

$$\Rightarrow I_B = \frac{I_s}{\beta_f} \cdot (e^{\frac{V_{BE}}{V_T}} - 1) + \frac{I_s}{\beta_r} \cdot (e^{\frac{V_{BC}}{V_T}} - 1)$$

The expressions for V_{BE}, V_{BC}, and V_{CE} are as follow:

$$I_E = I_{se} \cdot (e^{\frac{V_{BE}}{V_T}} - 1) - \alpha_r \cdot I_{sc} \cdot (e^{\frac{V_{BC}}{V_T}} - 1) \Rightarrow (e^{\frac{V_{BE}}{V_T}} - 1) = \frac{I_E + \alpha_r \cdot I_{sc} \cdot (e^{\frac{V_{BC}}{V_T}} - 1)}{I_{se}}$$

$$I_C = \alpha_f \cdot I_{se} \cdot (\frac{I_E + \alpha_r \cdot I_{sc} \cdot (e^{\frac{V_{BC}}{V_T}} - 1)}{I_{se}}) - I_{sc} \cdot (e^{\frac{V_{BC}}{V_T}} - 1) \Rightarrow I_C$$

$$= \alpha_f \cdot I_E + \alpha_f \cdot \alpha_r \cdot I_{sc} \cdot (e^{\frac{V_{BC}}{V_T}} - 1) - I_{sc} \cdot (e^{\frac{V_{BC}}{V_T}} - 1)$$

$$I_C = \alpha_f \cdot I_E + (\alpha_f \cdot \alpha_r - 1) \cdot I_{sc} \cdot (e^{\frac{V_{BC}}{V_T}} - 1) \Rightarrow e^{\frac{V_{BC}}{V_T}} = \frac{I_C - \alpha_f \cdot I_E}{(\alpha_f \cdot \alpha_r - 1) \cdot I_{sc}} + 1$$

$$e^{\frac{V_{BC}}{V_T}} = \frac{I_C - \alpha_f \cdot I_E}{(\alpha_f \cdot \alpha_r - 1) \cdot I_{sc}} + 1 \Rightarrow V_{BC} = V_T \cdot \ln\left[\frac{I_C - \alpha_f \cdot I_E}{(\alpha_f \cdot \alpha_r - 1) \cdot I_{sc}} + 1\right]$$

$$I_C = \alpha_f \cdot I_{se} \cdot (e^{\frac{V_{BE}}{V_T}} - 1) - I_{sc} \cdot (e^{\frac{V_{BC}}{V_T}} - 1) \Rightarrow (e^{\frac{V_{BC}}{V_T}} - 1) = \frac{\alpha_f \cdot I_{se} \cdot (e^{\frac{V_{BE}}{V_T}} - 1) - I_C}{I_{sc}}$$

$$I_E = I_{se} \cdot (e^{\frac{V_{BE}}{V_T}} - 1) - \alpha_r \cdot I_{sc} \cdot (\frac{\alpha_f \cdot I_{se} \cdot (e^{\frac{V_{BE}}{V_T}} - 1) - I_C}{I_{sc}}) \Rightarrow I_E$$

$$= I_{se} \cdot (e^{\frac{V_{BE}}{V_T}} - 1) - \alpha_r \cdot \alpha_f \cdot I_{se} \cdot (e^{\frac{V_{BE}}{V_T}} - 1) + \alpha_r \cdot I_C$$

$$I_E = (1 - \alpha_r \cdot \alpha_f) \cdot I_{se} \cdot (e^{\frac{V_{BE}}{V_T}} - 1) + \alpha_r \cdot I_C \Rightarrow e^{\frac{V_{BE}}{V_T}} = \frac{I_E - \alpha_r \cdot I_C}{(1 - \alpha_r \cdot \alpha_f) \cdot I_{se}} + 1$$

$$e^{\frac{V_{BE}}{V_T}} = \frac{I_E - \alpha_r \cdot I_C}{(1 - \alpha_r \cdot \alpha_f) \cdot I_{se}} + 1 \Rightarrow V_{BE} = V_T \cdot \ln\left[\frac{I_E - \alpha_r \cdot I_C}{(1 - \alpha_r \cdot \alpha_f) \cdot I_{se}} + 1\right];$$

$$V_{BE} = V_T \cdot \ln\left[\frac{\alpha_r \cdot I_C - I_E}{(\alpha_r \cdot \alpha_f - 1) \cdot I_{se}} + 1\right]$$

We can summery our intermediate results:

$$V_{BC} = V_T \cdot \ln\left[\left(\frac{I_C - \alpha_f \cdot I_E}{(\alpha_f \cdot \alpha_r - 1) \cdot I_{sc}}\right) + 1\right]; V_{BE} = V_T \cdot \ln\left[\left(\frac{\alpha_r \cdot I_C - I_E}{(\alpha_r \cdot \alpha_f - 1) \cdot I_{se}}\right) + 1\right]$$

$V_{CE} = V_{CB} + V_{BE}$, but $V_{CB} = -V_{BC}$. Then $V_{CE} = V_{BE} - V_{BC}$.

Remark: there is a use with capital and small letters in the Appendix compares to book chapter 1, consider the terminology is the same.

$$I_e = I_E; I_c = I_C; I_b = I_B; V_t = V_T\,; V_{be} = V_{BE}; V_{cb} = V_{CB}; V_{ce} = V_{CE}$$

$$V_{CB} = -V_{BC} = -V_T \cdot \ln\left[\left(\frac{I_C - \alpha_f \cdot I_E}{(\alpha_f \cdot \alpha_r - 1) \cdot I_{sc}}\right) + 1\right]; V_{CE} = V_{BE} - V_{BC}$$

$$V_{CE} = V_{BE} - V_{BC} = V_T \cdot \ln\left[\left(\frac{\alpha_r \cdot I_C - I_E}{(\alpha_r \cdot \alpha_f - 1) \cdot I_{se}}\right) + 1\right] - V_T \cdot \ln\left[\left(\frac{I_C - \alpha_f \cdot I_E}{(\alpha_f \cdot \alpha_r - 1) \cdot I_{sc}}\right) + 1\right]$$

$$V_{CE} = V_T \cdot \ln\left[\frac{\alpha_r \cdot I_C - I_E + (\alpha_r \cdot \alpha_f - 1) \cdot I_{se}}{(\alpha_r \cdot \alpha_f - 1) \cdot I_{se}}\right] - V_T \cdot \ln\left[\frac{I_C - \alpha_f \cdot I_E + (\alpha_f \cdot \alpha_r - 1) \cdot I_{sc}}{(\alpha_f \cdot \alpha_r - 1) \cdot I_{sc}}\right]$$

$$V_{CE} = V_T \cdot \ln\left\{\frac{[\alpha_r \cdot I_C - I_E + (\alpha_r \cdot \alpha_f - 1) \cdot I_{se}]}{(\alpha_r \cdot \alpha_f - 1) \cdot I_{se}} \cdot \frac{(\alpha_f \cdot \alpha_r - 1) \cdot I_{sc}}{[I_C - \alpha_f \cdot I_E + (\alpha_f \cdot \alpha_r - 1) \cdot I_{sc}]}\right\}$$

$$V_{CE} = V_T \cdot \ln\left\{\frac{[\alpha_r \cdot I_C - I_E + (\alpha_r \cdot \alpha_f - 1) \cdot I_{se}]}{[I_C - \alpha_f \cdot I_E + (\alpha_f \cdot \alpha_r - 1) \cdot I_{sc}]} \cdot \frac{I_{sc}}{I_{se}}\right\}$$

$$V_{CE} = V_T \cdot \ln\left\{\frac{[\alpha_r \cdot I_C - I_E + (\alpha_r \cdot \alpha_f - 1) \cdot I_{se}]}{[I_C - \alpha_f \cdot I_E + (\alpha_f \cdot \alpha_r - 1) \cdot I_{sc}]}\right\} + V_T \cdot \ln\left(\frac{I_{sc}}{I_{se}}\right); \frac{I_{sc}}{I_{se}} \approx 1 \Rightarrow \ln\left(\frac{I_{sc}}{I_{se}}\right) \to \varepsilon$$

$$V_{CE} \approx V_T \cdot \ln\left\{\frac{[\alpha_r \cdot I_C - I_E + (\alpha_r \cdot \alpha_f - 1) \cdot I_{se}]}{[I_C - \alpha_f \cdot I_E + (\alpha_f \cdot \alpha_r - 1) \cdot I_{sc}]}\right\}$$

Figure C.3 describes PNP Bipolar transistor Ebers Moll model.

$$I_{DE} = I_E + \alpha_r \cdot I_{DC} \Rightarrow I_{se} \cdot (e^{\frac{V_{EB}}{V_T}} - 1) = I_E + \alpha_r \cdot I_{sc} \cdot (e^{\frac{V_{CB}}{V_T}} - 1)$$
$$\Rightarrow I_E = I_{se} \cdot (e^{\frac{V_{EB}}{V_T}} - 1) - \alpha_r \cdot I_{sc} \cdot (e^{\frac{V_{CB}}{V_T}} - 1)$$

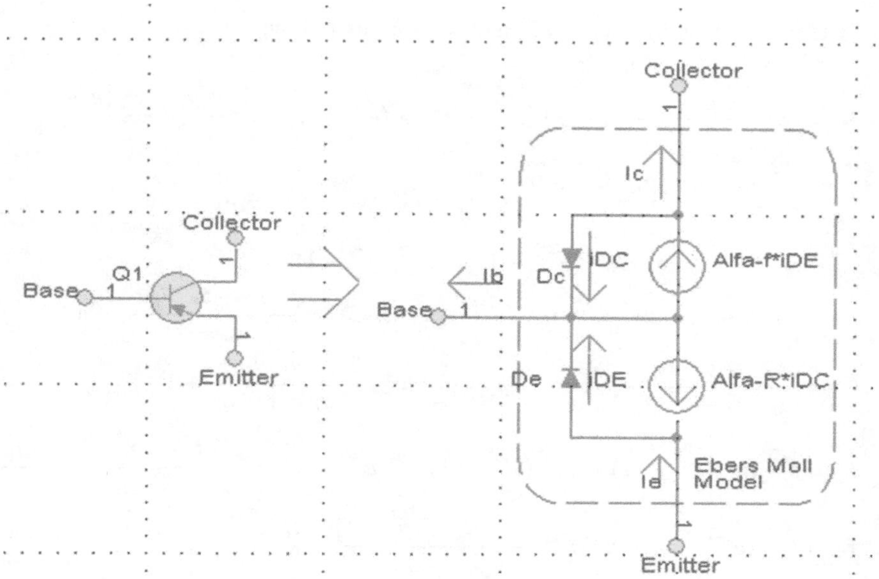

Fig. C.3 PNP Bipolar transistor Ebers Moll model

$$I_C + I_{DC} = \alpha_f \cdot I_{DE} \Rightarrow I_C + I_{sc} \cdot (e^{\frac{V_{CB}}{V_T}} - 1) = \alpha_f \cdot I_{se} \cdot (e^{\frac{V_{EB}}{V_T}} - 1)$$
$$\Rightarrow I_C = \alpha_f \cdot I_{se} \cdot (e^{\frac{V_{EB}}{V_T}} - 1) - I_{sc} \cdot (e^{\frac{V_{CB}}{V_T}} - 1)$$
$$I_B = I_E - I_C = (1 - \alpha_f) \cdot I_{se} \cdot (e^{\frac{V_{EB}}{V_T}} - 1) + (1 - \alpha_r) \cdot I_{sc} \cdot (e^{\frac{V_{CB}}{V_T}} - 1).$$

If we use the notation $\alpha_f \cdot I_{se} = \alpha_r \cdot I_{sc} = I_s; I_{sc} = \frac{I_s}{\alpha_r}; I_{se} = \frac{I_s}{\alpha_f}$ the following Ebers Moll equations:

$$I_E = \frac{I_s}{\alpha_f} \cdot (e^{\frac{V_{EB}}{V_T}} - 1) - \alpha_r \cdot \frac{I_s}{\alpha_r} \cdot (e^{\frac{V_{CB}}{V_T}} - 1) \Rightarrow I_E = \frac{I_s}{\alpha_f} \cdot (e^{\frac{V_{EB}}{V_T}} - 1) - I_s \cdot (e^{\frac{V_{CB}}{V_T}} - 1)$$
$$I_C = \alpha_f \cdot \frac{I_s}{\alpha_f} \cdot (e^{\frac{V_{EB}}{V_T}} - 1) - \frac{I_s}{\alpha_r} \cdot (e^{\frac{V_{CB}}{V_T}} - 1) \Rightarrow I_C = I_s \cdot (e^{\frac{V_{EB}}{V_T}} - 1) - \frac{I_s}{\alpha_r} \cdot (e^{\frac{V_{CB}}{V_T}} - 1)$$
$$I_B = (1 - \alpha_f) \cdot \frac{I_s}{\alpha_f} \cdot (e^{\frac{V_{EB}}{V_T}} - 1) + (1 - \alpha_r) \cdot \frac{I_s}{\alpha_r} \cdot (e^{\frac{V_{CB}}{V_T}} - 1); \frac{1}{\beta_f} = \frac{1 - \alpha_f}{\alpha_f}; \frac{1}{\beta_r} = \frac{1 - \alpha_r}{\alpha_r}$$
$$I_B = (1 - \alpha_f) \cdot \frac{I_s}{\alpha_f} \cdot (e^{\frac{V_{EB}}{V_T}} - 1) + (1 - \alpha_r) \cdot \frac{I_s}{\alpha_r} \cdot (e^{\frac{V_{CB}}{V_T}} - 1)$$
$$\Rightarrow I_B = \frac{I_s}{\beta_f} \cdot (e^{\frac{V_{EB}}{V_T}} - 1) + \frac{I_s}{\beta_r} \cdot (e^{\frac{V_{CB}}{V_T}} - 1)$$

The expressions for V_{EB}, V_{CB}, and V_{EC} are as follow:

$$I_E = I_{se} \cdot (e^{\frac{V_{EB}}{V_T}} - 1) - \alpha_r \cdot I_{sc} \cdot (e^{\frac{V_{CB}}{V_T}} - 1) \Rightarrow (e^{\frac{V_{EB}}{V_T}} - 1) = \frac{I_E + \alpha_r \cdot I_{sc} \cdot (e^{\frac{V_{CB}}{V_T}} - 1)}{I_{se}}$$

$$I_C = \alpha_f \cdot I_{se} \cdot (\frac{I_E + \alpha_r \cdot I_{sc} \cdot (e^{\frac{V_{CB}}{V_T}} - 1)}{I_{se}}) - I_{sc} \cdot (e^{\frac{V_{CB}}{V_T}} - 1) \Rightarrow I_C$$

$$= \alpha_f \cdot I_E + \alpha_f \cdot \alpha_r \cdot I_{sc} \cdot (e^{\frac{V_{CB}}{V_T}} - 1) - I_{sc} \cdot (e^{\frac{V_{CB}}{V_T}} - 1)$$

$$I_C = \alpha_f \cdot I_E + (\alpha_f \cdot \alpha_r - 1) \cdot I_{sc} \cdot (e^{\frac{V_{CB}}{V_T}} - 1) \Rightarrow e^{\frac{V_{CB}}{V_T}} = \frac{I_C - \alpha_f \cdot I_E}{(\alpha_f \cdot \alpha_r - 1) \cdot I_{sc}} + 1$$

$$e^{\frac{V_{CB}}{V_T}} = \frac{I_C - \alpha_f \cdot I_E}{(\alpha_f \cdot \alpha_r - 1) \cdot I_{sc}} + 1 \Rightarrow V_{CB} = V_T \cdot \ln[\frac{I_C - \alpha_f \cdot I_E}{(\alpha_f \cdot \alpha_r - 1) \cdot I_{sc}} + 1]$$

$$I_C = \alpha_f \cdot I_{se} \cdot (e^{\frac{V_{EB}}{V_T}} - 1) - I_{sc} \cdot (e^{\frac{V_{CB}}{V_T}} - 1) \Rightarrow (e^{\frac{V_{CB}}{V_T}} - 1) = \frac{\alpha_f \cdot I_{se} \cdot (e^{\frac{V_{EB}}{V_T}} - 1) - I_C}{I_{sc}}$$

$$I_E = I_{se} \cdot (e^{\frac{V_{EB}}{V_T}} - 1) - \alpha_r \cdot I_{sc} \cdot (\frac{\alpha_f \cdot I_{se} \cdot (e^{\frac{V_{EB}}{V_T}} - 1) - I_C}{I_{sc}}) \Rightarrow I_E$$

$$= I_{se} \cdot (e^{\frac{V_{EB}}{V_T}} - 1) - \alpha_r \cdot \alpha_f \cdot I_{se} \cdot (e^{\frac{V_{EB}}{V_T}} - 1) + \alpha_r \cdot I_C$$

$$I_E = (1 - \alpha_r \cdot \alpha_f) \cdot I_{se} \cdot (e^{\frac{V_{EB}}{V_T}} - 1) + \alpha_r \cdot I_C \Rightarrow e^{\frac{V_{EB}}{V_T}} = \frac{I_E - \alpha_r \cdot I_C}{(1 - \alpha_r \cdot \alpha_f) \cdot I_{se}} + 1$$

$$e^{\frac{V_{EB}}{V_T}} = \frac{I_E - \alpha_r \cdot I_C}{(1 - \alpha_r \cdot \alpha_f) \cdot I_{se}} + 1 \Rightarrow V_{EB} = V_T \cdot \ln[\frac{I_E - \alpha_r \cdot I_C}{(1 - \alpha_r \cdot \alpha_f) \cdot I_{se}} + 1];$$

$$V_{EB} = V_T \cdot \ln[\frac{\alpha_r \cdot I_C - I_E}{(\alpha_r \cdot \alpha_f - 1) \cdot I_{se}} + 1]$$

We can summery our result regarding I_C and I_E:

$$I_C = \alpha_f \cdot I_{se} \cdot (e^{\frac{V_{EB}}{V_T}} - 1) - I_{sc} \cdot (e^{\frac{V_{CB}}{V_T}} - 1); I_E = I_{se} \cdot (e^{\frac{V_{EB}}{V_T}} - 1) - \alpha_r \cdot I_{sc} \cdot (e^{\frac{V_{CB}}{V_T}} - 1)$$

$$I_B = I_E - I_C = I_{se} \cdot (e^{\frac{V_{EB}}{V_T}} - 1) - \alpha_r \cdot I_{sc} \cdot (e^{\frac{V_{CB}}{V_T}} - 1)$$

$$- [\alpha_f \cdot I_{se} \cdot (e^{\frac{V_{EB}}{V_T}} - 1) - I_{sc} \cdot (e^{\frac{V_{CB}}{V_T}} - 1)]$$

$$I_B = I_E - I_C = (1 - \alpha_f) I_{se} \cdot (e^{\frac{V_{EB}}{V_T}} - 1) + (1 - \alpha_r) \cdot I_{sc} \cdot (e^{\frac{V_{CB}}{V_T}} - 1);$$

$$V_{BE} = -V_{EB}; V_{BC} = -V_{CB}$$

$$I_C = \alpha_f \cdot I_{se} \cdot (e^{\frac{-V_{BE}}{V_T}} - 1) - I_{sc} \cdot (e^{\frac{-V_{BC}}{V_T}} - 1);$$

$$I_E = I_{se} \cdot (e^{\frac{-V_{BE}}{V_T}} - 1) - \alpha_r \cdot I_{sc} \cdot (e^{\frac{-V_{BC}}{V_T}} - 1)$$

$$I_B = I_E - I_C = (1 - \alpha_f) I_{se} \cdot (e^{\frac{-V_{BE}}{V_T}} - 1) + (1 - \alpha_r) \cdot I_{sc} \cdot (e^{\frac{-V_{BC}}{V_T}} - 1)$$

$V_{CE} = V_{CB} + V_{BE}$, but $V_{CB} = -V_{BC}$. Then $V_{CE} = V_{BE} - V_{BC}$.

$$V_{CB} = V_T \cdot \ln\left[\frac{I_C - \alpha_f \cdot I_E}{(\alpha_f \cdot \alpha_r - 1) \cdot I_{sc}} + 1\right]; V_{EB} = V_T \cdot \ln\left[\frac{\alpha_r \cdot I_C - I_E}{(\alpha_r \cdot \alpha_f - 1) \cdot I_{se}} + 1\right]$$

$$V_{BC} = -V_T \cdot \ln\left[\frac{I_C - \alpha_f \cdot I_E}{(\alpha_f \cdot \alpha_r - 1) \cdot I_{sc}} + 1\right]; V_{BE} = -V_T \cdot \ln\left[\frac{\alpha_r \cdot I_C - I_E}{(\alpha_r \cdot \alpha_f - 1) \cdot I_{se}} + 1\right]$$

$$V_{CE} = V_{CB} + V_{BE} = V_T \cdot \ln\left[\frac{I_C - \alpha_f \cdot I_E}{(\alpha_f \cdot \alpha_r - 1) \cdot I_{sc}} + 1\right] - V_T \cdot \ln\left[\frac{\alpha_r \cdot I_C - I_E}{(\alpha_r \cdot \alpha_f - 1) \cdot I_{se}} + 1\right]$$

$$V_{CE} = V_{CB} + V_{BE} = V_T \cdot \ln\left[\frac{I_C - \alpha_f \cdot I_E + (\alpha_f \cdot \alpha_r - 1) \cdot I_{sc}}{(\alpha_f \cdot \alpha_r - 1) \cdot I_{sc}}\right]$$
$$- V_T \cdot \ln\left[\frac{\alpha_r \cdot I_C - I_E + (\alpha_r \cdot \alpha_f - 1) \cdot I_{se}}{(\alpha_r \cdot \alpha_f - 1) \cdot I_{se}}\right]$$

$$V_{CE} = V_{CB} + V_{BE} = V_T \cdot \ln\left[\left\{\frac{I_C - \alpha_f \cdot I_E + (\alpha_f \cdot \alpha_r - 1) \cdot I_{sc}}{\alpha_r \cdot I_C - I_E + (\alpha_r \cdot \alpha_f - 1) \cdot I_{se}}\right\} \cdot \left\{\frac{I_{se}}{I_{sc}}\right\}\right]$$

$$V_{CE} = V_{CB} + V_{BE} = V_T \cdot \ln\left\{\frac{I_C - \alpha_f \cdot I_E + (\alpha_f \cdot \alpha_r - 1) \cdot I_{sc}}{\alpha_r \cdot I_C - I_E + (\alpha_r \cdot \alpha_f - 1) \cdot I_{se}}\right\}$$
$$+ V_T \cdot \ln\left\{\frac{I_{se}}{I_{sc}}\right\}; I_{se} \approx I_{sc}; \ln\left\{\frac{I_{se}}{I_{sc}}\right\} \to \varepsilon$$

$$V_{CE} = V_{CB} + V_{BE} \approx V_T \cdot \ln\left\{\frac{I_C - \alpha_f \cdot I_E + (\alpha_f \cdot \alpha_r - 1) \cdot I_{sc}}{\alpha_r \cdot I_C - I_E + (\alpha_r \cdot \alpha_f - 1) \cdot I_{se}}\right\}$$

$$V_{CE-NPN} \approx V_T \cdot \ln\left\{\frac{[\alpha_r \cdot I_C - I_E + (\alpha_r \cdot \alpha_f - 1) \cdot I_{se}]}{[I_C - \alpha_f \cdot I_E + (\alpha_f \cdot \alpha_r - 1) \cdot I_{sc}]}\right\}; V_{CE-PNP}$$
$$\approx V_T \cdot \ln\left\{\frac{I_C - \alpha_f \cdot I_E + (\alpha_f \cdot \alpha_r - 1) \cdot I_{sc}}{\alpha_r \cdot I_C - I_E + (\alpha_r \cdot \alpha_f - 1) \cdot I_{se}}\right\}$$

Summary of our BJT NPN and PNP transistors Ebers-Moll equations (Table C.1):

There are three basic circuit configurations to connect bipolar junction transistor. First: Common Base (CB), both the input and output share the base "in common". Second: Common Emitter (CE), both the input and output share the emitter "in common". Third: Common Collector (CC), both the input and output share the collector "in common". There are four bipolar junction transistor biasing modes. Active biasing is useful for amplifiers (most common mode). Saturation biasing mode is equivalent to an on state when transistor is used as a switch. Cutoff biasing mode is equivalent to an off state when transistor is used as a switch. Inverted biasing mode is rarely if ever used (Table C.2a, b).

Table C.1 Summary of our BJT NPN and PNP transistors Ebers-Moll equations

	BJT NPN transistor	BJT PNP transistor
I_C	$I_C = \alpha_f \cdot I_{se} \cdot (e^{\frac{V_{BE}}{V_T}} - 1) - I_{sc} \cdot (e^{\frac{V_{BC}}{V_T}} - 1)$	$I_C = \alpha_f \cdot I_{se} \cdot (e^{\frac{-V_{BE}}{V_T}} - 1) - I_{sc} \cdot (e^{\frac{-V_{BC}}{V_T}} - 1)$
I_E	$I_E = I_{se} \cdot (e^{\frac{V_{BE}}{V_T}} - 1) - \alpha_r \cdot I_{sc} \cdot (e^{\frac{V_{BC}}{V_T}} - 1)$	$I_E = I_{se} \cdot (e^{\frac{-V_{BE}}{V_T}} - 1) - \alpha_r \cdot I_{sc} \cdot (e^{\frac{-V_{BC}}{V_T}} - 1)$
I_B	$I_B = (1 - \alpha_f) \cdot \frac{I_s}{\alpha_f} \cdot (e^{\frac{V_{BE}}{V_T}} - 1) + (1 - \alpha_r) \cdot \frac{I_s}{\alpha_r} \cdot (e^{\frac{V_{BC}}{V_T}} - 1)$	$I_B = (1 - \alpha_f) I_{se} \cdot (e^{\frac{-V_{BE}}{V_T}} - 1) + (1 - \alpha_r) \cdot I_{sc} \cdot (e^{\frac{-V_{BC}}{V_T}} - 1)$
V_{CE}	$V_{CE} \approx V_T \cdot \ln\left\{\frac{[\alpha_r \cdot I_C - I_E + (\alpha_r \cdot \alpha_f - 1) \cdot I_{se}]}{[I_C - \alpha_f \cdot I_E + (\alpha_f \cdot \alpha_r - 1) \cdot I_{sc}]}\right\}$	$V_{CE} \approx V_T \cdot \ln\left\{\frac{I_C - \alpha_f \cdot I_E + (\alpha_f \cdot \alpha_r - 1) \cdot I_{sc}}{\alpha_r \cdot I_C - I_E + (\alpha_r \cdot \alpha_f - 1) \cdot I_{se}}\right\}$

Table C.2a Summary of NPN BJT transistor biasing mode

Biasing mode (NPN)	E-B junction bias (NPN)	C-B junction bias (NPN)	Applied voltages (NPN)
Saturation	Forward	Forward	$V_E < V_B > V_C$
Active (forward active)	Forward	Reverse	$V_E < V_B < V_C$
Inverted (reverse active)	Reverse	Forward	$V_E > V_B > V_C$
Cutoff	Reverse	Reverse	$V_E > V_B < V_C$

Table C.2b Summary of PNP BJT transistor biasing mode

Biasing mode (PNP)	E-B junction bias (PNP)	C-B junction bias (PNP)	Applied voltages (PNP)
Saturation	Forward	Forward	$V_E > V_B < V_C$
Active (forward active)	Forward	Reverse	$V_E > V_B > V_C$
Inverted (reverse active)	Reverse	Forward	$V_E < V_B < V_C$
Cutoff	Reverse	Reverse	$V_E < V_B > V_C$

The BJT transistor base current is much smaller that the emitter and collector currents in forward active mode. If the collector of an NPN transistor was open circuit, it would look like a diode. When forward biased, the circuit in the base-emitter junction would consist of holes injected into the emitter from the base and electrons injected into the base from the emitter. But since there are many more electrons in the emitter than holes in the base, the vast majority of the current will be due to electrons. When the reverse biased collector is added, It "sucks" the electrons out of the base. Thus, the base-emitter current is due predominantly to hole current (the smaller current component) while the collector-emitter current is due to electrons (larger current component due to more electrons from the n+ emitter doping). We define two BJT transistor performance parameters: emitter efficiency (γ) and base transport factor (α_T). Emitter efficiency parameter characterizes how effective the large hole current is controlled by the small electron current. Unity is best, zero is worst. Base transport factor characterizes how much of the injected hole current is lost to recombination in the base. Unity is best, zero is worst.

$$\gamma = \frac{I_{Ep}}{I_E} = \frac{I_{Ep}}{I_{Ep} + I_{En}}$$

$$\alpha_T = \frac{I_{Cp}}{I_{Ep}}.$$

We define some equations in active mode, common base characteristics. I_{CBo} is defined as the collector current when the emitter is open circuit. It is the collector base junction saturation current. I_C is the fraction of emitter current making it across the base + leakage current.

$$I_C = \alpha_{dc} \cdot I_E + I_{CBo},$$

where α_{dc} is the common base DC current gain.

$$I_{Cp} = \alpha_T \cdot I_{Ep} = \gamma \cdot \alpha_T \cdot I_E; I_C = I_{Cp} + I_{Cn} = \alpha_T \cdot I_{Ep} + I_{Cn} = \gamma \cdot \alpha_T \cdot I_E + I_{Cn}; \alpha_{dc} = \gamma \cdot \alpha_T$$

and $I_{CBo} = I_{Cn}$. We define some equations in active mode, common emitter characteristics. I_{CEo} is defined as the collector current when the base is open circuit. I_C is multiple of the base current making it across the base + leakage current. $I_C = \beta_{dc} \cdot I_B + I_{CEo}$; Where β_{dc} is the common emitter DC current gain. I_{CEo} is defines as the collector current when the base is open circuit.

$\alpha_F = \alpha_{dc}$ is common base current gain. $I_E = \alpha_R \cdot I_C; \alpha_R \neq \alpha_{DC}$. In inverse mode, the emitter current is the fraction of the collector current "collected".

$$I_E = I_C + I_B; I_C = \alpha_{dc} \cdot (I_C + I_B) + I_{CBo}; I_C = \frac{\alpha_{dc}}{1 - \alpha_{dc}} \cdot I_B + \frac{I_{CBo}}{1 - \alpha_{dc}}$$

$$\beta_{dc} = \frac{\alpha_{dc}}{1 - \alpha_{dc}}; I_{CEo} = \frac{I_{CBo}}{1 - \alpha_{dc}}; \beta_{dc} = \frac{I_C}{I_B}$$

We can break the BJT transistor up into a large signal analysis and a small signal analysis and "linearize" the non-linear behavior of the Ebers-Moll model. Small signal models are only useful for forward active mode and thus, are derived under this condition. Saturation and cutoff are used for switches which involve very large voltage/current swings from on to off states.

Small signal models are used to determine amplifier characteristics ("Gain" = increase in the magnitude of a signal at the output of a circuit relative to its magnitude at the input of the circuit). Just like when a diode voltage exceeds a certain value, the non-linear behavior of the diode leads to distortion of the current/voltage curves, if the inputs/outputs exceed certain limits, the full Ebers-Moll model must be used. There are physical meanings of β_f (β_F) and β_r (β_R). β_F is the current gain (I_C/I_B) of the device when it is operating with the emitter as the emitter and the collector as the collector in the active mode. β_R is the current gain of the device when it is operating with the emitter as a collector and the collector as an emitter in the reverse mode. The BJT device is made to have higher forward current gain than reverse current gain. The terminals for emitter and collector are not completely interchangeable due to different doping of the collector and emitter.

BJTtransistor modes of operation:

The Ebers-Moll BJT model is a good large signal, steady-state model of the transistor and allows the state of conduction of the device to be easily determined for different modes of operation of the device. The different modes of operation are determined by the manner in which the junctions are biased. BJT NPN transistor Ebers-Moll BJT model:

$$I_C = \alpha_f \cdot I_{se} \cdot (e^{\frac{V_{BE}}{V_T}} - 1) - I_{sc} \cdot (e^{\frac{V_{BC}}{V_T}} - 1); I_E = I_{se} \cdot (e^{\frac{V_{BE}}{V_T}} - 1) - \alpha_r \cdot I_{sc} \cdot (e^{\frac{V_{BC}}{V_T}} - 1)$$

$$I_B = (1 - \alpha_f) \cdot \frac{I_s}{\alpha_f} \cdot (e^{\frac{V_{BE}}{V_T}} - 1) + (1 - \alpha_r) \cdot \frac{I_s}{\alpha_r} \cdot (e^{\frac{V_{BC}}{V_T}} - 1);$$

$$\alpha_f \cdot I_{se} = \alpha_r \cdot I_{sc} = I_s; I_{sc} = \frac{I_s}{\alpha_r}; I_{se} = \frac{I_s}{\alpha_f}$$

$$I_B = (1 - \alpha_f) \cdot I_{se} \cdot (e^{\frac{V_{BE}}{V_T}} - 1) + (1 - \alpha_r) \cdot I_{sc} \cdot (e^{\frac{V_{BC}}{V_T}} - 1)$$

(A) Forward Active mode:

B-E forward biased, V_{BE} positive $e^{\frac{V_{BE}}{V_T}} \gg 1; (e^{\frac{V_{BE}}{V_T}} - 1) \approx e^{\frac{V_{BE}}{V_T}}$. B-C reverse biased, V_{BC} negative $e^{\frac{V_{BC}}{V_T}} \ll 1; (e^{\frac{V_{BC}}{V_T}} - 1) \approx -1$. Then from the Ebers-Moll model equations we get the following results:

$$I_E \simeq I_{se} \cdot e^{\frac{V_{BE}}{V_T}} + \alpha_r \cdot I_{sc} \approx I_{se} \cdot e^{\frac{V_{BE}}{V_T}}; I_{se} \cdot e^{\frac{V_{BE}}{V_T}} \gg \alpha_r \cdot I_{sc};$$

Relatively large.

$$I_C \simeq \alpha_f \cdot I_{se} \cdot e^{\frac{V_{BE}}{V_T}} + I_{sc} \approx \alpha_f \cdot I_{se} \cdot e^{\frac{V_{BE}}{V_T}} = \alpha_f \cdot I_E; \alpha_f \cdot I_{se} \cdot e^{\frac{V_{BE}}{V_T}} \gg I_{sc};$$

Relatively large.

$$I_B \simeq (1 - \alpha_f) \cdot I_{se} \cdot e^{\frac{V_{BE}}{V_T}} - (1 - \alpha_r) \cdot I_{sc} \approx (1 - \alpha_f) \cdot I_{se} \cdot e^{\frac{V_{BE}}{V_T}} = (1 - \alpha_f) \cdot I_E; (1 - \alpha_f) \cdot I_{se} \cdot e^{\frac{V_{BE}}{V_T}} \gg (1 - \alpha_r) \cdot I_{sc}$$

(B) Reverse active mode:
B-E reverse biased, V_{BE} negative $e^{\frac{V_{BE}}{V_T}} \ll 1; (e^{\frac{V_{BE}}{V_T}} - 1) \approx -1$. B-C forward biased, V_{BC} positive $e^{\frac{V_{BC}}{V_T}} \gg 1; (e^{\frac{V_{BC}}{V_T}} - 1) \approx e^{\frac{V_{BC}}{V_T}}$. The transistor conducts in the opposite direction. Then from the Ebers-Moll model equations we get the following results:

$$I_E \simeq -I_{se} - \alpha_r \cdot I_{sc} \cdot e^{\frac{V_{BC}}{V_T}} \approx -\alpha_r \cdot I_{sc} \cdot e^{\frac{V_{BC}}{V_T}}; \alpha_r \cdot I_{sc} \cdot e^{\frac{V_{BC}}{V_T}} \gg I_{se};$$

Moderately high.

$$I_C \simeq -\alpha_f \cdot I_{se} - I_{sc} \cdot e^{\frac{V_{BC}}{V_T}} \approx -I_{sc} \cdot e^{\frac{V_{BC}}{V_T}}; I_{sc} \cdot e^{\frac{V_{BC}}{V_T}} \gg \alpha_f \cdot I_{se}; \text{ Moderate.}$$

$$I_B \simeq -(1 - \alpha_f) \cdot I_{se} + (1 - \alpha_r) \cdot I_{sc} \cdot e^{\frac{V_{BC}}{V_T}} \approx (1 - \alpha_r) \cdot I_{sc} \cdot e^{\frac{V_{BC}}{V_T}}; (1 - \alpha_r) \cdot I_{sc} \cdot e^{\frac{V_{BC}}{V_T}} \gg (1 - \alpha_f) \cdot I_{se}$$

It is as high as $0.5 \cdot |I_C|$. This mode does not provide useful amplification but is used, mainly, for current steering in switching circuits, e.g. TTL.

(C) Cut-off mode:
B-E is unbiased, $V_{BE} = 0$ v. B-C is reverse biased, V_{BC} negative.

$$e^{\frac{V_{BE}}{V_T}} = 1; (e^{\frac{V_{BE}}{V_T}} - 1) \to \varepsilon = 0; e^{\frac{V_{BC}}{V_T}} \ll 1; (e^{\frac{V_{BC}}{V_T}} - 1) \approx -1$$

$I_E \simeq \alpha_r \cdot I_{sc}$; Leakage current nA. $I_C \simeq I_{sc}$; Leakage current nA. $I_B \simeq -(1 - \alpha_r) \cdot I_{sc}$.

This is equivalent to a very low conductance between collector and emitter, i.e. open switch (Fig. C.4).

(D) Saturation mode:
B-E is forward biased, V_{BE} is positive $e^{\frac{V_{BE}}{V_T}} \gg 1; (e^{\frac{V_{BE}}{V_T}} - 1) \approx e^{\frac{V_{BE}}{V_T}}$ and both junctions are forward biased. B-C is forward biased, $V_{BC} e^{\frac{V_{BC}}{V_T}} \gg 1; (e^{\frac{V_{BC}}{V_T}} - 1) \approx e^{\frac{V_{BC}}{V_T}}$. We get the following currents expressions:

$$I_C \approx \alpha_f \cdot I_{se} \cdot e^{\frac{V_{BE}}{V_T}} - I_{sc} \cdot e^{\frac{V_{BC}}{V_T}}; I_E \approx I_{se} \cdot e^{\frac{V_{BE}}{V_T}} - \alpha_r \cdot I_{sc} \cdot e^{\frac{V_{BC}}{V_T}}$$

$$I_B \approx (1 - \alpha_f) \cdot I_{se} \cdot e^{\frac{V_{BE}}{V_T}} + (1 - \alpha_r) \cdot I_{sc} \cdot e^{\frac{V_{BC}}{V_T}}$$

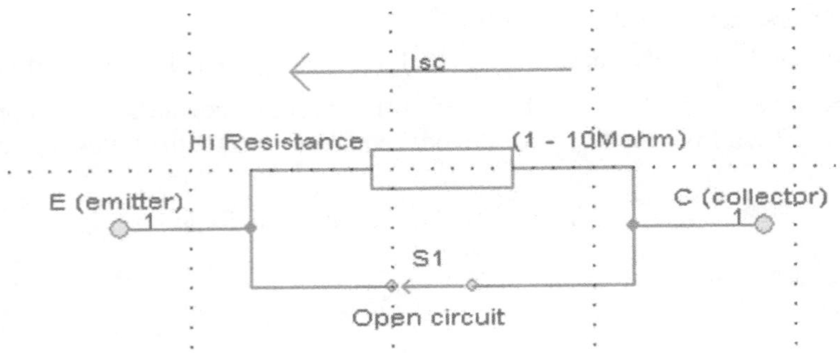

Fig. C.4 The cutoff mode of operation as equivalent to a leaky switch

In this case, with both junctions forward biased.

$$V_{BE} \approx 0.8\,\text{V};\ V_{BC} \approx 0.7\,\text{V};\ V_{CE} = V_{CB} + V_{BE};\ V_{CB} = -V_{BC};\ V_{CE} = V_{BE} - V_{BC} = 0.1\,\text{V}$$

There is a 0.1 V drop across the transistor from collector to emitter which is quite low while a substantial current flows through the device. In this mode it can be considered as having a very high conductivity and acts as a closed switch with a finite resistance and conductivity (Fig. C.5).

<u>BJT transistor avalanche breakdown region of operation</u>:

An avalanche transistor is a bipolar junction transistor designed for operation in the region of its collector-current/collector-to-emitter voltage characteristics beyond the collector-to-emitter breakdown voltage, called avalanche breakdown region. This region characterized by avalanche breakdown, a phenomenon similar to Negative Differential Resistance (NDR). Operation in the avalanche breakdown region is called avalanche-mode operation. It gives avalanche transistors the ability

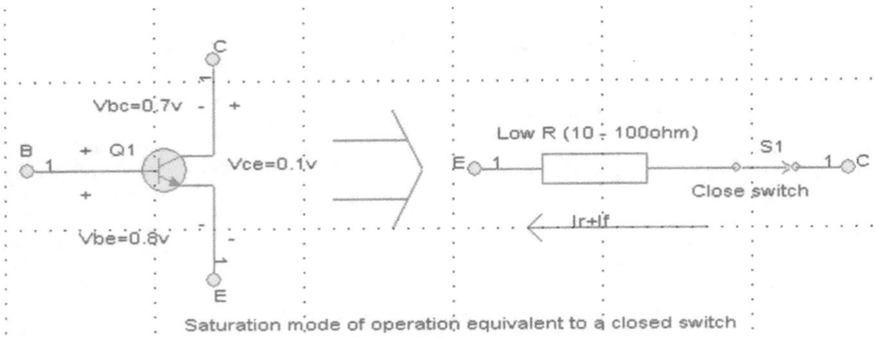

Fig. C.5 Saturation mode of operation equivalent to a closed switch

to switch very high currents with less than nanosecond rise and fall times (transition times). Transistors not specifically designed for the purpose can have reasonably consistent avalanche properties.

Static avalanche regioncharacteristics: The static characteristic of an avalanche transistor is I_C-V_{CE}. The static characteristic of an avalanche NPN transistor is the same as PNP devices only changing sign to voltages and currents accordingly. The avalanche breakdown multiplication is present only across the collector-base junction. The first step of the calculation is to determine collector current as a sum of various component currents through the collector since only those fluxes of charge are subject to this phenomenon. Applying Kirchhoff's current law (KCL) to a bipolar junction transistor, implies the following relation which satisfied by the collector current I_C ($I_C = I_E - I_B$) while for the same device working in the active region. $\alpha = \alpha_f; \beta = \beta_f; I_C = \beta \cdot I_B + (\beta + 1) \cdot I_{CBo}$, I_B is the base current. I_{CBo} is the collector-base reverse leakage current. I_E is the emitter current. B is the common emitter current gain of the transistor. Equating the two formulas for I_C gives the following result

$I_E = (\beta + 1) \cdot I_B + (\beta + 1) \cdot I_{CBo}$ and since $\alpha = \frac{\beta}{\beta + 1}$; α is the common base current gain of the transistor, then

$$\alpha \cdot I_E = \beta \cdot I_B + \beta \cdot I_{CBo} = I_C - I_{CBo} \Rightarrow I_C = \alpha \cdot I_E + I_{CBo}.$$

When the avalanche effects in a transistor collector are considered, the collector current I_C is given by $I_C = M \cdot (\alpha \cdot I_E + I_{CBo})$. M is miller's avalanche multiplication coefficient. It is the most important parameter in avalanche mode operation $M = \frac{1}{1 - (\frac{V_{CB}}{BV_{CBo}})^n}$. BV_{CBo} is the collector-base breakdown voltage. n is a constant depending on the semiconductor used for the construction of the transistor and doping profile of the collector-base junction. V_{CB} is the collector-base voltage. Using Kirchhoff's current law (KCL) for the bipolar junction transistor and the expression for M, the resulting expression for I_C is the following:

$$I_C = \frac{M}{(1 - \alpha \cdot M)} \cdot (\alpha \cdot I_B + I_{CBo}) \Rightarrow I_C = \frac{\alpha \cdot I_B + I_{CBo}}{1 - \alpha - (\frac{V_{CB}}{BV_{CBo}})^n}$$

$$V_{CB} = V_{CE} - V_{BE}; V_{BE} = V_{BE}(I_B)$$

where V_{BE} is the base-emitter voltage.

$$I_C = \frac{\alpha \cdot I_B + I_{CBo}}{1 - \alpha - (\frac{V_{CE} - V_{BE}(I_B)}{BV_{CBo}})^n} \simeq \frac{\alpha \cdot I_B + I_{CBo}}{1 - \alpha - (\frac{V_{CE}}{BV_{CBo}})^n}$$

Since $V_{CE} \gg V_{BE}$. This is the expression of the parametric family of the collector characteristics I_C–V_{CE} with parameter I_B (I_C) increases without limit if

$$(1-\alpha) = (\frac{V_{CE}}{BV_{CBo}})^n \Rightarrow V_{CE} = BV_{CEo} = BV_{CBo} \cdot \sqrt[n]{(1-\alpha)} = \frac{BV_{CBo}}{\sqrt[n]{\beta+1}};$$

$$1-\alpha = 1 - (\frac{\beta}{\beta+1}) = \frac{1}{\beta+1}$$

$$\beta \gg 1 \Rightarrow V_{CE} = BV_{CEo} = BV_{CBo} \cdot \sqrt[n]{(1-\alpha)}|_{\beta\gg 1} = \frac{BV_{CBo}}{\sqrt[n]{\beta}}.$$

where BV_{CEo} is the collector-emitter breakdown voltage.

Avalanche Multiplication: The maximum reverse biasing voltage which may be applied before breakdown between the collector and base terminals of the transistor, under the condition that the emitter lead be open circuited, is represented by the symbol BV_{CBo}. This breakdown voltage is a characteristic of the transistor alone. The breakdown may occur because of avalanche multiplication of the current I_{CO} that crosses the collector junction. As a result of this multiplication, the current becomes $M \cdot I_{CO}$, in which M is the factor by which the original I_{CO} is multiplies by the avalanche effect. It is possible to neglect leakage current, which does not flow through the junction and is therefore not subject to avalanche multiplication. At a high enough BV_{CBo}, the multiplication factor M becomes nominally infinite and the region of breakdown is then attained. The current rises abruptly, and large changes in current accompany small changes in applied voltage. The avalanche multiplication factor depends on the voltage V_{CB} between transistor's collector and base. If a current I_E is caused to flow across the emitter junction, then, neglecting the avalanche effect, a fraction $\alpha \cdot I_E$, where α is the common-base current gain, reaches the collector junction. If we take multiplication into account, I_C has the magnitude $M \cdot \alpha \cdot I_E$. In presence of avalanche multiplication, the transistor behaves as though its common base current gain where $M \cdot \alpha$. The maximum allowable collector to emitter voltage depends not only upon the transistor, but also upon the circuit in which it is used.

BJT transistor second breakdown avalanche mode: When the collector current rises above the data sheet limit I_{Cmax} a new breakdown mechanism happened, the second breakdown. This phenomenon is caused by excessive heating of some points (hot spots) in the base-emitter region of the bipolar junction transistor, which give rise to an exponentially increasing current through these points. This exponential rise of current in turn gives rise to even more overheating, originating a positive thermal feedback mechanism. While analyzing the I_C-V_{CE} static characteristic, the presence of this phenomenon is seen as a sharp collector voltage drop and a corresponding almost vertical rise of the collector current. While this phenomenon is destructive for bipolar junction transistors working in the usual way, it can be used to push up further the current and voltage limits of a device working in avalanche mode by limiting its time duration. The switching speed of the device is not negatively affected.

Small signal model of the BJT, base charging capacitance (diffusion capacitance). In active mode when the emitter-base is forward biased, the capacitance of the emitter-base junction is dominated by the diffusion capacitance (not depletion capacitance). Recall for a diode we define the following: $C_{Diffusion} = \frac{dQ_D}{dv'_D} = \frac{dQ_D}{dt} \cdot \frac{dt}{dv'_D}$. The sum up all minority carrier charges on either side of the junction.

$$Q_D = q \cdot A \cdot \int_0^{\infty} p_{no} \cdot (e^{\frac{v'_D}{V_T}} - 1) \cdot e^{-\frac{X}{L_P}} \cdot dX + q \cdot A \cdot \int_0^{\infty} n_{po} \cdot (e^{\frac{v'_D}{V_T}} - 1) \cdot e^{-\frac{X}{L_n}} \cdot dX$$

If we neglect charge injected from the base into the emitter due to p+ emitter in PNP then $Q_D = q \cdot A \cdot \int_0^{\infty} p_{no} \cdot (e^{\frac{v'_D}{V_T}} - 1) \cdot e^{-\frac{X}{L_P}} \cdot dX$. Excess charge stored is due almost entirely to the charge injected from the emitter. The BJT acts like a very efficient "siphon"; As majority carriers from the emitter are injected into the base and become "excess minority carriers", the collector "siphons them" out of the base. We can view the collector current as the amount of excess charge in the base collected by the collector per unit time and we can express the charge due to the excess hole concentration in the base as: $Q_B = i_c \cdot \tau_F$ or the excess charge in the base depends on the magnitude of current flowing and the "forward" base transport time, τ_F, the average time the carriers spend in the base. $\tau_F = \frac{W^2}{2 \cdot D_B}$, W is the base quasi-neutral region width. $D_B$is the minority carrier diffusion coefficient. Thus, the diffusion capacitance is

$$C_B = \frac{\partial Q_B}{\partial v_{BE}}|_{Q-point} = (\frac{W^2}{2 \cdot D_B}) \cdot \frac{\partial i_c}{\partial v_{BE}}|_{Q-point}; \quad C_B = \tau_F \cdot \frac{I_C}{V_T} = \tau_F \cdot g_m.$$

The upper operational frequencyof the transistor is limited by the forward base transport time $f \leq \frac{1}{2 \cdot \pi \cdot \tau_F}$. It is the similarity to the diode diffusion capacitance.

$$C_{Diffusion} = g_d \cdot \tau_t; \tau_t = \frac{|p_{no} \cdot L_p + n_{po} \cdot L_n| \cdot q \cdot A}{I_S};$$

$$C_{Diffusion} = g_d \cdot \frac{|p_{no} \cdot L_p + n_{po} \cdot L_n| \cdot q \cdot A}{I_S}$$

τ_t is the transit time. In active mode for small forward biases the depletion capacitance of the base-emitter junction can contribute to the total capacitance.

$$C_{jE} = \frac{C_{jEo}}{\sqrt{1 + \frac{V_{EB}}{V_{bi\ for\ emitter-base}}}}$$

C_{jE} ≡ zero bias depletion capacitance.

$V_{bi\ for\ emitter\text{-}base}$ ≡ built in voltage for E-B junction. Thus, the emitter-base capacitance is $C_\pi = C_B + C_{jE}$. In active mode when the collector-base is reverse biased, the capacitance of the collector-base junction is dominated by the depletion capacitance (not diffusion capacitance).

$$C_{\mu} = \frac{C_{\mu o}}{\sqrt{1 + \frac{V_{CB}}{V_{bi\ for\ collector-base}}}}.$$

$C_{\mu o}$ ≡ zero bias depletion capacitance.

$V_{bi\ for\ collector\text{-}base}$ ≡ built in voltage for the B-C junction. In some integrated BJTs (lateral BJTs in particular) the device has a capacitance to the substrate wafer it is fabricated in. This results from a "buried" reverse biased junction. Thus, the collector-substrate junction is reverse biased and the capacitance of the collector-substrate junction is dominated by the depletion capacitance (not diffusion capacitance).

$$C_{cs} = \frac{C_{cso}}{\sqrt{1 + \frac{V_{cs}}{V_{bi\ for\ collector-substrate}}}}.$$

C_{cs} ≡ zero bias depletion capacitance.

$V_{bi\ for\ collector\text{-}substrate}$ ≡ built in voltage for the C substrate junction.

Small signal model of the BJT, parasitic resistances:

r_b base resistance between metal inter connect and B-E junction.
r_c parasitic collector resistance.
r_{ex} emitter resistance due to polysilicon contact.

Complete BJT small signal model: (Fig. C.6).

What set the maximum limits of operation of the BJT circuit? Forward active mode lies between saturation and cutoff. Thus, the maximum voltage extremes that one can operate an amplifier over can easily be found by examining the boundaries between forward active and cutoff and the boundaries between forward active and saturation. Output signals that exceed the voltage range that would keep the transistor within its forward active mode will result in "clipping" of the signal leading to distortion. The maximum voltage swing allowed without clipping depends on the DC bias points.

MOSFET transistor model:

The basic static model of MOSFET transistor (Shichman and Hodges) is as follow (Fig. C.7):

$$I_{DS} = \mu_0 \cdot C_{ox} \cdot \frac{W}{L_{eff}} \cdot \left[(V_{GS} - V_{TH}) \cdot V_{DS} - \frac{V_{DS}^2}{2} \right]$$

$$I_{DSsat} = \frac{1}{2} \cdot \mu_0 \cdot C_{ox} \cdot \frac{W}{L_{eff}} \cdot (V_{GS} - V_{TH})^2$$

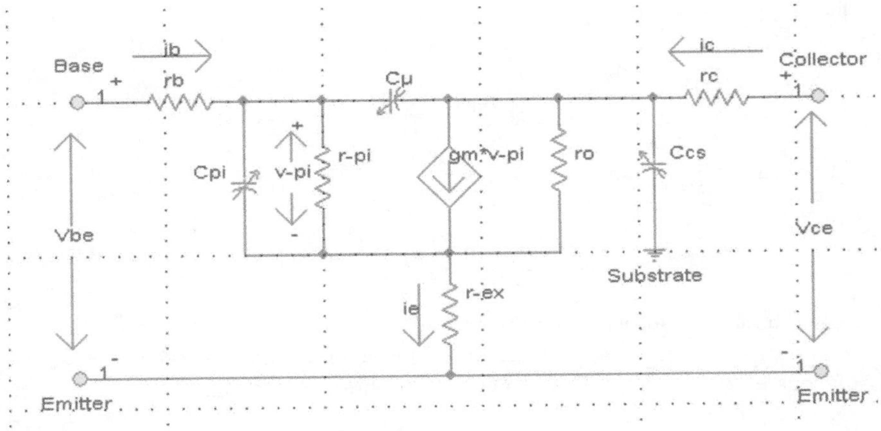

Fig. C.6 Complete BJT small signal model

Fig. C.7 MOSFET transistor model graph

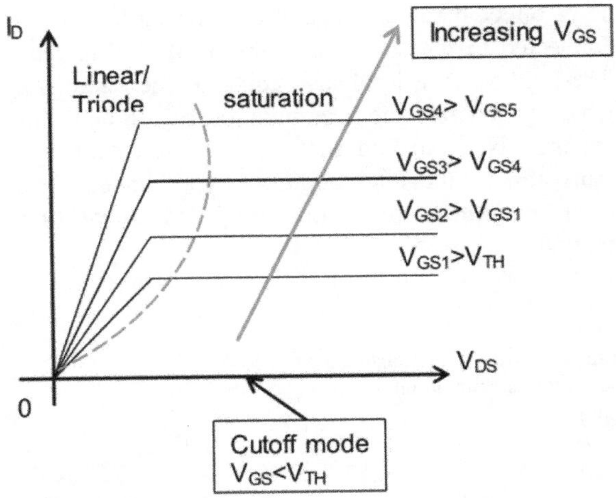

There is an empirical correction to these equations to account for the channel length modulation:

$$I_{DS} = \mu_0 \cdot C_{ox} \cdot \frac{W}{L_{eff}} \cdot \left[(V_{GS} - V_{TH}) \cdot V_{DS} - \frac{V_{DS}^2}{2}\right] \cdot [1 + \lambda \cdot V_{DS}]$$

$$I_{DSsat} = \frac{1}{2} \cdot \mu_0 \cdot C_{ox} \cdot \frac{W}{L_{eff}} \cdot (V_{GS} - V_{TH})^2 \cdot [1 + \lambda \cdot V_{DS}]$$

In the linear region:

$$I_{DS} = KP \cdot \frac{W}{(L - 2 \cdot X_{jl})} \cdot \left[(V_{GS} - V_{TH}) \cdot V_{DS} - \frac{V_{DS}^2}{2} \right] \cdot [1 + \lambda \cdot V_{DS}]$$

In the saturation region:

$$I_{Dsat} = \frac{KP}{2} \cdot \frac{W}{(L - 2 \cdot X_{jl})} \cdot (V_{GS} - V_{TH})^2 \cdot [1 + \lambda \cdot V_{DS}]$$

X_{jl} is the lateral diffusion parameter (Fig. C.8).

Threshold voltage (V_{TH}): The threshold voltage changes with changes in body-source voltage, V_{BS}. The expression for threshold voltage

$$V_{TH} = V_{TO} + \gamma \cdot (\sqrt{2 \cdot \phi_p - V_{BS}} - \sqrt{2 \cdot \phi_p})$$

where V_{TO} is the threshold voltage when the body-source voltage is zero, γ is the body effect parameter and Φ_p is the surface inversion potential. If the bulk is connected to the source (i.e. the MOSFET is acting as a 3 terminal device, the threshold voltage is always equal to the value V_{TO}). There is a depletion layer which grows into the accumulation region and thus for a given V_{GS}, cuts off the channel. Need to add more V_{GS} to re-establish the channel when we stacked transistors in integrated circuits. If you connected bulk to source on each transistor in an integrated circuit you would end up shorting many points in the circuit to ground.

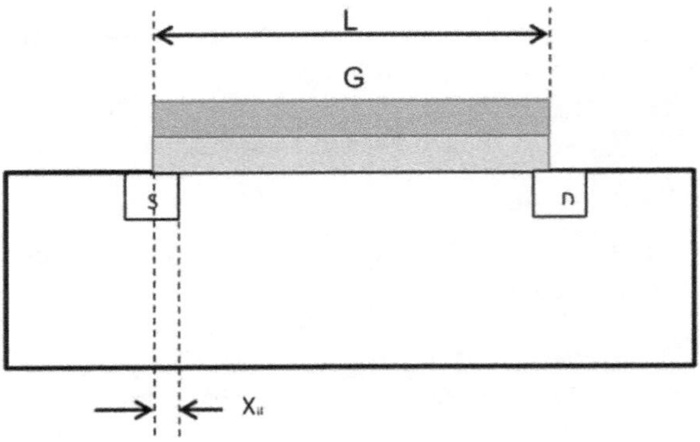

Fig. C.8 MOSFET transistor structure and important parameters

Complete DC model:

The model includes body-source and body-drain diodes. Equations used for the diode model, for forward bias on the body-source/body-drain diodes (Fig. C.9):

$$I_{BS} = I_{SS} \cdot [e^{\frac{V_{BS}}{V_t}} - 1] + GMIN \cdot V_{BS}; I_{BD} = I_{SD} \cdot \left[e^{\frac{V_{BD}}{V_t}} - 1\right] + GMIN \cdot V_{BD}$$

For the negative reverse bias on those diodes:

$$I_{BS} = I_{SS} \cdot \frac{V_{BS}}{V_t} + GMIN \cdot V_{BS}; \quad I_{BD} = I_{SD} \cdot \frac{V_{BD}}{V_t} + GMIN \cdot V_{BD}$$

MOSFET body diodes: The reverse bias terms are simply the first terms in a power series expansion of the exponential term. The GMIN convergence resistance. I_{SS} and I_{SD} are taken to be one constant in simulation.

DC MOSFET parameters: L = channel length, W = channel width, KP (kp) = The trans-conductance parameter, V_{TO} = Threshold voltage under zero bias conditions, GAMMA (γ) = Body effect parameter, PHI(Φ_p) = surface inversion potential, RS(R_S) = source contact resistance, RD(R_D) = Drain contact resistance, LAMBDA(λ) = channel length modulation parameter, XJ(X_{jl}) = lateral diffusion parameter. IS(I_{SS}, I_{SD}) = reverse saturation current of body-drain/source diodes.

Large signal transient model: We add some capacitances to the DC model to create the transient model to form the final transient model, as shown in Fig. C.10).

Capacitances: Static overlap capacitances between gate and drain (C_{GBO}), gate and source (C_{GSO}), and gate and bulk (C_{GBO}). These are fixed values, and are specified per unit width. In saturation,

$$C_{GS} = \frac{2}{3} \cdot C_{Ox} + C_{GSO} \cdot W; C_{GD} = C_{GDO} \cdot W$$

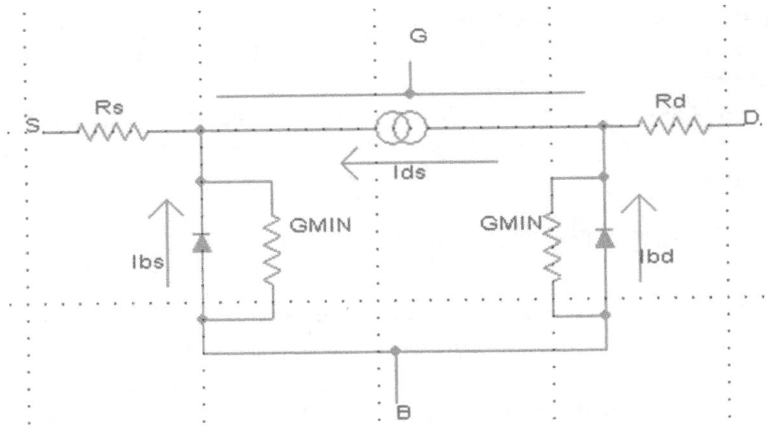

Fig. C.9 MOSFET complete DC model

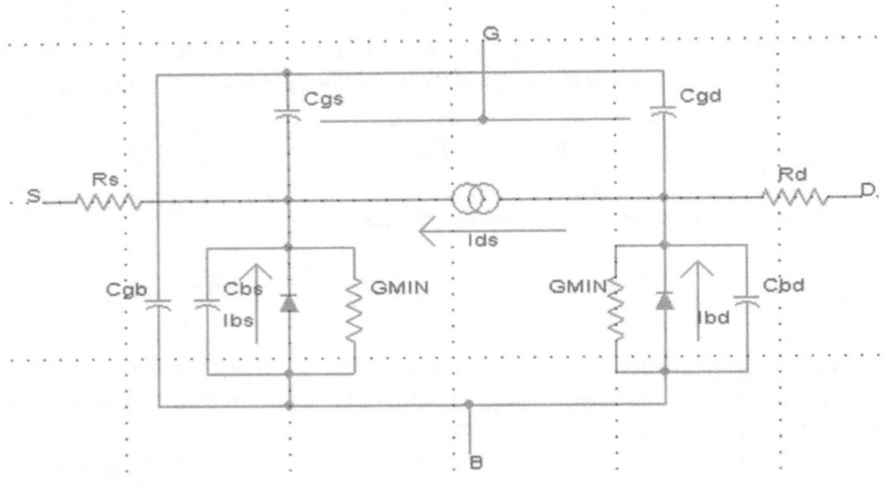

Fig. C.10 MOSFET large signal transient model

In saturation after pinch-off, it is assumed that altering the drain voltage does not have any effect on stored charge in the channel and thus the only capacitance between gate and drain is the overlap capacitance.

In the linear/triode region, in this region the following equations are used:

$$C_{GS} = C_{0x} \cdot \left\{ 1 - \left[\frac{V_{GS} - V_{DS} - V_{TH}}{2 \cdot (V_{GS} - V_{TH}) - V_{DS}} \right]^2 \right\} + C_{GS0} \cdot W$$

$$C_{GD} = C_{0x} \cdot \left\{ 1 - \left[\frac{V_{GS} - V_{TH}}{2 \cdot (V_{GS} - V_{TH}) - V_{DS}} \right]^2 \right\} + C_{GD0} \cdot W$$

As the device is moved further into the linear region, V_{GS} becomes large compared to (V_{DS}–V_{TH}) then the values of C_{GS} and C_{GD} become close to $C_{0x}/2$ (plus the relevant overlap capacitance).

The body diode capacitances: The capacitances of the body diodes are given by slightly modified expressions for junction capacitances of the diode model: The expression for a PN diode capacitance: $C_j = \frac{C_j(0)}{\sqrt{1-\frac{V}{V_0}}}$. The MOSFET equation is based on the following slightly modified equation:

$$C_j = \frac{C_j(0)}{\sqrt{1 - \frac{V}{V_0}}} + \frac{C_{jSW}(0)}{\sqrt{1 - \frac{V}{V_0}}}.$$

The junction capacitance is made up of two components. The main component, due to $C_j(0)$ is the normal junction capacitance. The second parameter is the perimeter junction capacitance of the diffused source. The diffusion capacitance is zero in reverse bias and the MOSFET must be operated with the bulk-drain and bulk-sourceBulk-source diodes in reverse bias to stop large bulk currents flowing. The additional parameters required for specifying the transient model in addition to those required by the DC model are thus:

CGD0(C_{GD0}) = Gate drain overlap capacitance per unit width of device.
CGS0(C_{GS0}) = Gate source overlap capacitance per unit width of device.
CJ(C_j) = Zero bias depletion capacitance for body diodes.
CJSW(C_{jsw}) = Zero bias depletion perimeter capacitance for body diodes.
TOX(t_{ox}) = Oxide thickness (used for calculating C_{ox}).

Bipolar transistor metrology and theory:

The interest topics regarding bipolar junction transistor (BJT) are operation, I-V characteristics, current gain and output conductance. High level injection and heavy doping induced band narrowing. SiGe transistor, transit time, and cutoff frequency are important parameters. There are several bipolar transistor models which are used (Ebers-Moll model, Small signal model, and charge control model). Each model has its own areas of applications. The metal-oxide-semiconductor (MOS) ICs have high density and low power advantages. The BJTs are preferred in some high frequency and analog applications because of their high speed, low noise, and high output power advantages such as in some cell phone amplifier circuits. A small number of BJTs are integrated into a high density complementary MOS (CMOS) chip integration of BJT and CMOS is known as the BiCMOS technology. The term bipolar refers to the fact that both electrons and holes are involved in the operation of a BJT. Minority carrier diffusion plays the leading role as in the PN diode junction diode. A BJT is made of a heavily doped emitter, a P-type base, and an N-type collector. This device is an NPN BJT, a PNP BJT would have a P^+ emitter, N-type base, and P-type collector. NPN transistor exhibit higher trans conductance and speed than PNP transistors because the electron mobility is larger than the hole mobility, BJTs are almost exclusively of the NPN type since high performance is BJT's competitive edge over MOSFETs (Fig. C.11).

When the base-emitter junction is forward biased, electrons are injected into the more lightly doped base. They diffuse across the base to the reverse biased base-collector junction which is the edge of the depletion layer and get swept into the collector. This produces a collector current, I_C. I_C is independent of V_{CB} as long as V_{CB} is a reverse bias or a small forward bias. I_C is determined by the rate of electron injection from the emitter into the base, determined by V_{BE}. The rate of electron injection is proportional to $e^{\frac{q \cdot V_{BE}}{k \cdot T}}$. The emitter is often connected to ground. The emitter and collector are the equivalents of source and drain of a MOSFET when the base is the equivalent of the gate. The $I_C$curve is usually plotted against V_{CE}. $V_{CE} = V_{CB} + V_{BE}$, below $V_{CE} = 0.3$ V the base-collector junction is strongly forward biased and I_C decreases. Because of the parasitic IR drops, it is difficult to accurately ascertain the true base-emitter junction voltage. The easily measurable

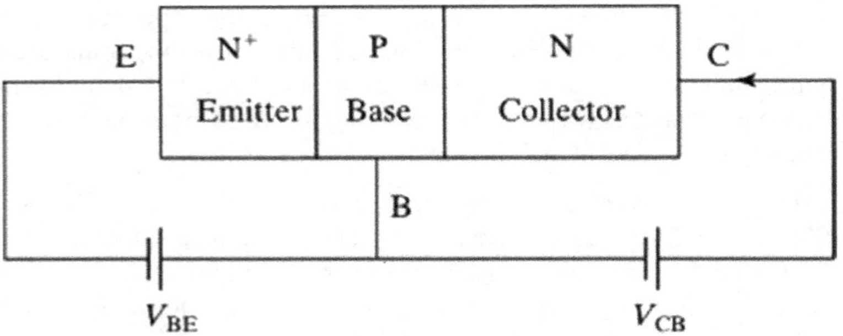

Fig. C.11 NPN BJT transistor voltages connection

base current I_B is commonly used as the variable parameter in lieu of V_{BE}, I_C is proportional to I_B (Fig. C.12).

Collector current: The collector current is the output current of a BJT transistor. Applying the electron diffusion equation to the base region gives in Fig. C.13

$$\frac{d^2n'}{dx^2} = \frac{n'}{L_B^2}; L_B = \sqrt{\tau_B \cdot D_B}; \frac{d^2n'}{dx^2} = \frac{n'}{\tau_B \cdot D_B}.$$

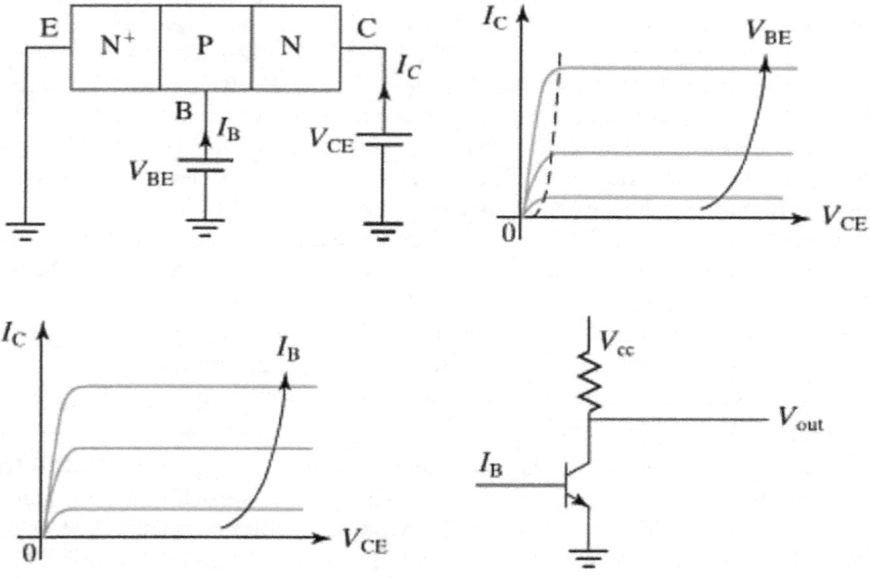

Fig. C.12 NPN transistor structure, connections and graphs

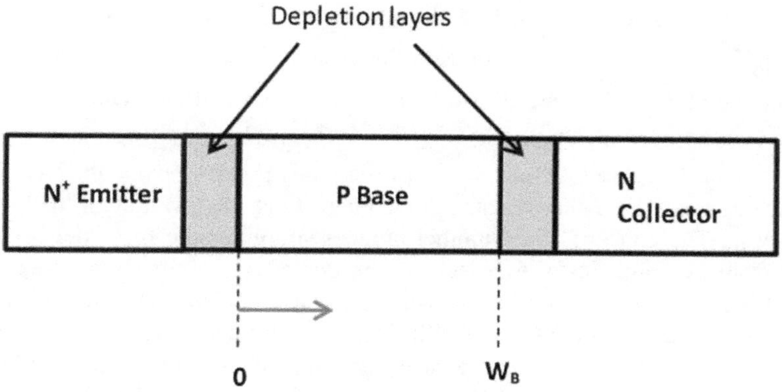

Fig. C.13 NPN transistor structure with depletion layers

τ_B and D_B are the recombination lifetime and the minority carrier (electron) diffusion constant in the base, respectively. The boundary conditions are as follow:

$$n'(0) = n_{B0} \cdot (e^{\frac{q \cdot V_{BE}}{k \cdot T}} - 1); n'(W_B) = n_{B0} \cdot (e^{\frac{q \cdot V_{BC}}{k \cdot T}} - 1) \approx -n_{B0} \approx 0.$$

where $n_{B0} = \frac{n_i^2}{N_B}$ and N_B is the base doping concentration. V_{BE} is normally a forward bias (positive value) and V_{BC} is a reverse bias (negative value).

We get the following expression for $n'(x) = n_{B0} \cdot (e^{\frac{q \cdot V_{BE}}{k \cdot T}} - 1) \cdot \frac{\sinh(\frac{W_B - x}{L_B})}{\sinh(\frac{W_B}{L_B})}$. Modern BJTs have base widths of about 0.1 μm. This is much smaller than the typical diffusion length of tens of microns. In the case of $W_B \ll L_B$ we get the $n'(x)$ expression: $n'(x) = n'(0) \cdot (1 - \frac{x}{W_B}) = \frac{n_{iB}^2}{N_B} \cdot (e^{\frac{q \cdot V_{BE}}{k \cdot T}} - 1) \cdot (1 - \frac{x}{W_B})$. n_{iB} is the intrinsic carrier concentration of the base material. The subscript B, is added to n_i because the base may be made of a different semiconductor such as SiGe alloy, which has a smaller band gap and therefore a larger n_i than the emitter and collector material. The minority carrier current is dominated by the diffusion current. The sign of I_C is positive and defined in the expression:

$$I_C = \left| A_E \cdot q \cdot D_B \cdot \frac{dn}{dx} \right| = A_E \cdot q \cdot D_B \cdot \frac{n'(0)}{W_B} = A_E \cdot q \cdot \frac{D_B}{W_B} \cdot \frac{n_{iB}^2}{N_B} \cdot (e^{\frac{q \cdot V_{BE}}{k \cdot T}} - 1).$$

A_E is the area of the BJT specifically the emitter area. There is a similarity between BJT transistor I_C current and the PN diode IV relation. Both are proportional to $(e^{\frac{q \cdot V}{k \cdot T}} - 1)$ and to $\frac{D \cdot n_i^2}{N}$. The only difference is that $\frac{dn'}{dx}$ has produced the $\frac{1}{W_B}$ term due to the linear n′ profile. We can condense the expression of I_C to $I_C = I_s \cdot (e^{\frac{q \cdot V_{BE}}{k \cdot T}} - 1)$, where I_s is the saturation current. $I_C = A_E \cdot \frac{q \cdot n_i^2}{G_B} \cdot (e^{\frac{q \cdot V_{BE}}{k \cdot T}} - 1)$ and $G_B = \frac{n_i^2}{n_{iB}^2} \cdot \frac{N_B}{D_B} \cdot W_B = \frac{n_i^2}{n_{iB}^2} \cdot \frac{p}{D_B} \cdot W_B$, where p is the majority carrier concentration in the base. It is valid even for no uniform base and high level injection condition if G_b is generalized to 1. $G_B = \int_0^{W_B} \frac{n_i^2}{n_{iB}^2} \cdot \frac{p}{D_B} \cdot dx$, G_B has the unusual dimension of s/cm^4 and is known as the base Gummel number. In the special case of $n_{iB} = n_i$, $D_B$is a constant, and $p(x) = N_B(x)$ which is low level injection. $G_B = \frac{1}{D_B} \cdot \int_0^{W_B} N_B(x) \cdot dx = \frac{1}{D_B} \times$ base dopant atoms per unit area. The base Gummel number is basically proportional to the base dopant density per area. The higher the base dopant density is, the lower the I_C will be for a given V_{BE}. The concept of a Gummel number simplifies the I_C model because it contains all the subtleties of transistor design that affect I_C; changing base material through $n_{iB}(x)$, non-constant D_B, non-uniform base dopant concentration through $p(x) = N_B(x)$ and even the high level injection condition, where $p > N_B$. Although many factors affect G_B, G_B can be easily determined from the Gummel plot. The inverse slope of the straight line can be described as 60 mV per decade. The extrapolated intercept of the straight line and $V_{BE} = 0$ yields I_s. G_B is equal to $A_E \cdot q \cdot n_i^2$ divided by the intercept (Fig. C.14).

The decrease in the slope of the curve at high I_C is called the high level injection effect. At large V_{BE}, n' can become larger than the base doping concentration N_B, $n' = p' \gg N_B$. The condition of $n' = p' \gg N_B$ is called high level injection. A consequence is that in the base

$$n \approx p \approx n_i \cdot e^{\frac{q \cdot V_{BE}}{2 \cdot k \cdot T}}; \quad G_B \propto n_i \cdot e^{\frac{q \cdot V_{BE}}{2 \cdot k \cdot T}}$$

Yield to $I_c \propto n_i \cdot e^{\frac{q \cdot V_{BE}}{2 \cdot k \cdot T}}$. Therefore, at high V_{BE} or high I_C, $I_c \propto e^{\frac{q \cdot V_{BE}}{2 \cdot k \cdot T}}$ and the inverse slope becomes 120 mV/decade. I_{KF}, the knee current, is the current at which the slope changes. It is a useful parameter in the BJT model for circuit simulation. The IR drop in the parasitic resistance significantly increases V_{BE} at very high I_C and further flattens the curve.

Base current: Whenever the base-emitter junction is forward biased, some holes are injected from the P-type into the N$^+$ emitter. These holes are provided by the base current I_B, I_B is an undesirable but inevitable side effect of producing I_C by

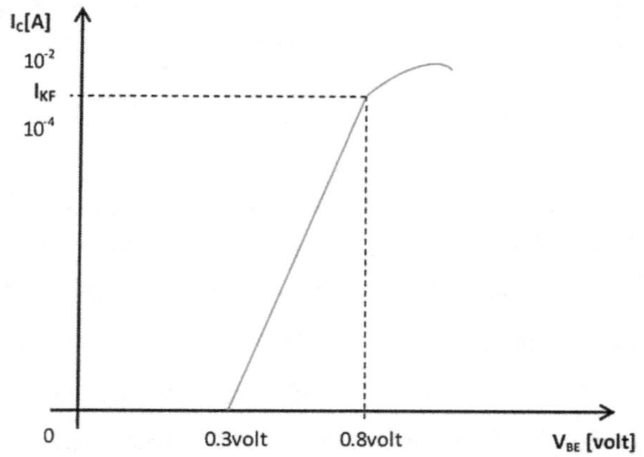

Fig. C.14 NPN transistor I_c [A] versus V_{BE} [volt]

forward biasing the BE junction. The analysis of I_B, the base to emitter injection current, is a perfect parallel of the I_C analysis. The base current can be expressed as

$$I_B = A_E \cdot \frac{q \cdot n_i^2}{G_E} \cdot (e^{\frac{q \cdot V_{BE}}{k \cdot T}} - 1); G_E = \int_0^{W_E} \frac{n_i^2}{n_{iE}^2} \cdot \frac{p}{D_E} \cdot dx.$$

G_E is the emitter Gummel number. In case of uniform emitter, where n_{iE}, N_E (emitter doping concentration) and D_E are not functions of x (Fig. C.15).

$$I_B = A_E \cdot q \cdot \frac{D_E}{W_E} \cdot \frac{n_{iE}^2}{N_E} \cdot (e^{\frac{q \cdot V_{BE}}{k \cdot T}} - 1).$$

Current gain: The most important DC parameter of a BJT is its common emitter current gain β_F. Another current ratio, the common base current gain, is defined by α_F. $\beta_F \equiv \frac{I_C}{I_B}; I_C = \alpha_F \cdot I_E;\ \alpha_F = \frac{I_C}{I_E} = \frac{I_C}{I_C + I_B} = \frac{I_C/I_B}{I_C/I_B + 1} = \frac{\beta_F}{1 + \beta_F}$. α_F is typically very close to unity, such as 0.99, because β_F is large. $\alpha_F = \frac{\beta_F}{1 + \beta_F}$;

$\beta_F = \frac{\alpha_F}{1 - \alpha_F}$. I_B is a load on the input signal source, an undesirable side effect of forward biasing the BE junction. I_B should be minimized (β_F should be maximized). $\beta_F = \frac{G_E}{G_B} = \frac{D_B \cdot W_E \cdot N_E \cdot n_{iB}^2}{D_E \cdot W_B \cdot N_B \cdot n_{iE}^2}$. A typical good β_F is 100. D and W cannot be changed very much. The most obvious way to achieve a high β_F, is to use a large N_E and a small N_B. A small N_B, would introduce too large a base resistance, which degrades the BJT's ability to operate at high current and high frequencies. Typically N_B is around 10^{18} cm^{-3}. An emitter is said to be efficient if the emitter current is mostly the useful electron current injected into the base with little useless hole current (the base current). The emitter efficiency is defined as $\gamma_E = \frac{I_E - I_B}{I_E} = \frac{I_C}{I_C + I_B} = \frac{1}{1 + G_B/G_E}$. To raise β_F, N_E is typically made larger than 10^{20} cm^{-3}. When N_E is very large, n_{iE}^2 becomes larger than n_i^2. This is called the heavy doping effect. $n_i^2 = N_C \cdot N_V \cdot e^{-\frac{E_g}{k \cdot T}}$, heavy doping can modify the Si crystal sufficient to reduce E_g and cause n_i^2 to increase significantly. Therefore, the heavy doping effect is also known as band gap narrowing. $n_{iE}^2 = n_i^2 \cdot e^{\frac{\Delta E_{gE}}{k \cdot T}}$, ΔE_{gE} is the

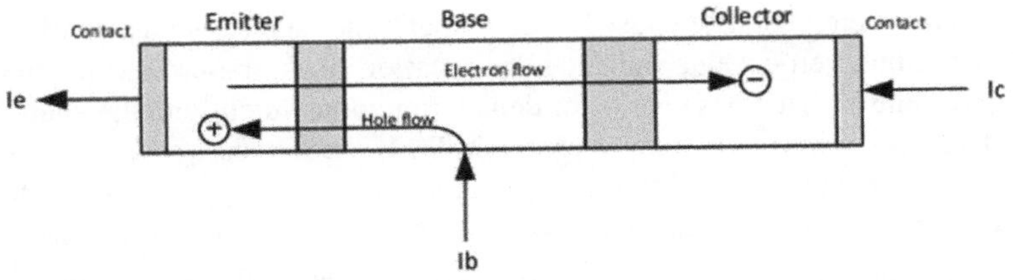

Fig. C.15 NPN transistor structure electron flow and hole flow

narrowing of the emitter band gap relative to lightly doped Si and is negligible for $N_E < 10^{18}$ cm^{-3}, 50 meV at 10^{19} cm^{-3}, 95 meV cm^{-3} at 10^{20} cm^{-3}, and 140 meV at 10^{21} cm^{-3}. To further elevate β_F, we can raise n_{iB} by using a base material that has a smaller band gap than the emitter material. $Si_{1-\eta}Ge_{\eta}$ is an excellent base material candidate for an Si emitter. With $\eta = 0.2$, E_{gB} is reduced by 0.1 eV. In a SiGe BJT, the base is made of high quality P-type epitaxial SiGe. In practice, η is graded such that $\eta = 0$ at the emitter end of the base and 0.2 at the drain end to create a built in field that improves the speed of the BJT. Because the emitter and base junction is made of two different semiconductors, the device is known as a heterojunction bipolar transistor or HBT. HBTs made of InP emitter ($E_g = 1.35$ eV) and InGaAs base ($E_g = 0.68$ eV) and GaAlAs emitter with GaAs base are other examples of well-studied HBTs. The ternary semiconductors are used to achieve lattice constant matching at the heterojunction. Whether the base material is SiGe or plain Si, a high performance BJT would have a relatively thick (>100 nm) layer of As doped N^+ poly-Si film in the emitter. Arsenic is thermally driven into the "base" by $\sim$20 nm and converts that single crystalline layer into a part of the N^+ emitter. This way, β_F is larger due to the large W_E, mostly made of the N^+ poly-Si. This is the poly-Silicon emitter technology. The simpler alternative, a deeper implanted or diffused N^+ emitter without the poly-Si film, is known to produce a higher density of crystal defects in the thin base causing excessive emitters to collector leakage current or even shorts in a small number of the BJTs. High speed circuits operate at high I_C, and low power circuits may operate at low I_C. Current gain β, drops at both high I_C and at low I_C. In Gummel plot the I_C flattens at high V_{BE} due to the high level injection effect in the base. That I_C curve arising from hole injection into the emitter, does not flatten due to this effect because the emitter is very heavily doped, and it is practically impossible to inject a higher density of holes than N_E. Over a wide mid-range of I_C, I_C and I_B are parallel, indicating that the ratio I_C/I_B, i.e., β_F is a constant. Above 1mA, the slope of I_C drops due to high level injection. Consequently, the I_C/I_B ratio or β_F decreases rapidly. This fall-off of current gain unfortunately degrades the performance of BJTs at high current where the BJTs speed is the highest. I_B is the base emitter junction forward bias current. The forward bias current slope decreases at low V_{BE} or very low current due to the Space Charge Region (SCR) current. As a result, the I_C/I_B ratio or β_F decreases at very low I_C.

As in MOSFETs, a large output conductance, $\frac{\partial I_C}{\partial V_{CE}}$, of BJTs is deleterious to the voltage gain of circuits. The cause of the output conductance is base-width modulation. The thick vertical line indicates the location of the base-collector junction. With increasing V_{ce}, the base-collector depletion region widens and the neutral base width decreases. This leads to an increase in I_C. If the curves I_C–V_{CE} are extrapolated, they intercept the $I_C = 0$ axis at approximately the same point. V_A is defined as early voltage. V_A is a parameter that describes the flatness of the I_C curves. Specifically, the output resistance can be expressed as V_A/I_C: $r_0 \equiv \left(\frac{\partial I_C}{\partial V_{CE}}\right)^{-1} = \frac{V_A}{I_C}$. A large V_A (large r_0) is desirable for high voltage gains. A typical V_A is 50 V. V_A is sensitive to the transistor design. We can except V_A and r_0 to increase, expect the base width modulation to be a smaller fraction of the base width, if we increase the base width, increase the base doping concentration N_B or decrease the collector doping concentration N_C. Increasing the base width would reduce the sensitivity to any given ΔW_B. Increasing the base doping concentration N_B would reduce the depletion region thickness on the base side because the depletion region penetrates less into the more heavily doped side of a PN junction. Decreasing the collector doping concentration N_C would tend to move the depletion region into the collector and thus reduce the depletion region thickness on the base side, too. Both increasing the base width and the base doping concentration N_B would depress β_F. Decreasing the collector doping concentration N_C is the most acceptable course of action. It is also reduces the base-collector junction capacitance, which is a good thing. Therefore, the collector doping is typically ten times lighter than the base doping. The larger slopes at $V_{CE} > 3$v are caused by impact ionization. The rise of I_C due to base-width modulation is known as the early effect. Model the collector current as a function of the collector voltage: $I_C = \beta_F \cdot I_B$ and differentiating with respect to V_C while I_B was held constant gave, $\frac{\partial I_C}{\partial V_C} = I_B \cdot \frac{\partial \beta_F}{\partial V_C}$. The question is how can β_F change with V_C, the collector depletion layer thickens as collector voltage is raised. The base gets thinner and current gain raises. Bipolar transistor transit time and charge storage: Static IV characteristics are only one part of the BJT theory. Another part is its dynamic behavior or its speed. When the BE junction is forward biased, excess holes are stored in the emitter, the base, and even the depletion layers. The sum of all excess hole charges everywhere Q_F. Q_F is the stored excess carrier charge. If $Q_F = 1$pC (Pico coulomb), there is +1 pC of excess hole charge and −1 pC of excess electron charge stored in the BJT. The ratio of Q_F to I_C is called the forward transit time τ_F ($\tau_F \equiv \frac{Q_F}{I_C}$). Ic and Q_F are related by a constant ratio τ_F. Q_F and therefore τ_F are very difficult to predict accurately for a complex device structure. τ_F can be measured experimentally and once τ_F is determined for a given BJT, equation $\tau_F \equiv \frac{Q_F}{I_C}$ becomes a powerful conceptual and mathematical tool giving Q_F as a function of I_C, and vice versa. τ_F sets a high frequency limit of BJT operation. The excess hole charge in the base Q_{FB}: $Q_{FB} = q \cdot A_E \cdot n'(0) \cdot W_B/2; \frac{Q_{FB}}{I_C} \equiv \tau_{FB} = \frac{W_B^2}{2 \cdot D_B}$. The base transit time can be further reduced by building into the base a drift field that aids the flow of electrons from the emitter to the collector. There are two ways of accomplishing this. The classical method is to use graded base doping (a large N_B near the EB junction), which gradually decreases toward the CB junction. Such a doping gradient is automatically achieved if the base is produced by dopant diffusion. The changing N_B creates a dE_v/dx and a dE_c/dx. This means that there is a drift field. Any electron injected into the base would drift toward the collector with a base transit time shorter than the diffusion transit time, $\frac{W_B^2}{2 \cdot D_B}$. In a SiGe BJT, P-type epitaxial $Si_{1-\eta}Ge_\eta$ is grown over the Si collector with a constant N_B and η linearly varying from about 0.2 at the collector end to 0 at the emitter end. A large dE_c/dx can be produced by the grading of E_{gB}. These high speed BJTs are used in high frequency communication circuits. Drift transistors can have a base transit time several times less than $\frac{W_B^2}{2 \cdot D_B}$, as short as 1psec. The total forward transit time, τ_F is known as the emitter to collector transit time. τ_{FB} is only one portion of τ_F. The base transit time typically contributes about half of τ_F. To reduce the transit (or storage) time in the emitter and collector, the emitter and the depletion layers must be kept thin. τ_F can be measured. τ_F starts to increase at a current density where the electron density corresponding to the dopant density in the collector (n = N_C) is insufficient to support the collector current even if the dopant induced electrons move at the saturation velocity. This intriguing condition of too few dopant atoms and too much current is lead to a reversal of the sign of the charge density in the depletion region.

$$I_C = A_E \cdot q \cdot n \cdot v_{sat}; \quad \rho = q \cdot N_C - q \cdot n = q \cdot N_C - \frac{I_C}{A_E \cdot v_{sat}}; \frac{d\wp(x)}{dx} = \frac{\rho}{\varepsilon_s}.$$

when I_C is small then $\rho(\rho = q \cdot N_C)$ as expected from the PN junction analysis, and the electric field in the depletion layer. The N^+ collector is always present to reduce the series resistance. No depletion layer is shown in the base for simplicity because the base is much more heavily doped than the collector. As I_C increases, ρ decreases and $\frac{d\wp(x)}{dx}$ decreases. The electric field drops to zero in the very heavily doped N^+ collector as expected. Because of the base widening, τ_F increases as a consequence. This is called the Kirk effect. Base widening can be reduced by increasing N_C and V_{CE}. The Kirk effect limits the peak BJT operating speed.

Bipolar transistor small signal model: The equivalent circuit for the behavior of a BJT in response to a small input signal (10 mV sinusoidal signal, superimposed on the DC bias) is presented in Fig. C.16. BJTs are often operated in this manner in analog circuits.

If V_{BE} is not close to zero, the "1" in $I_C = I_s \cdot (e^{\frac{q \cdot V_{BE}}{k \cdot T}} - 1)$ is negligible; in that case

$$I_C = I_s \cdot (e^{\frac{q \cdot V_{BE}}{k \cdot T}} - 1) \approx I_s \cdot e^{\frac{q \cdot V_{BE}}{k \cdot T}}.$$

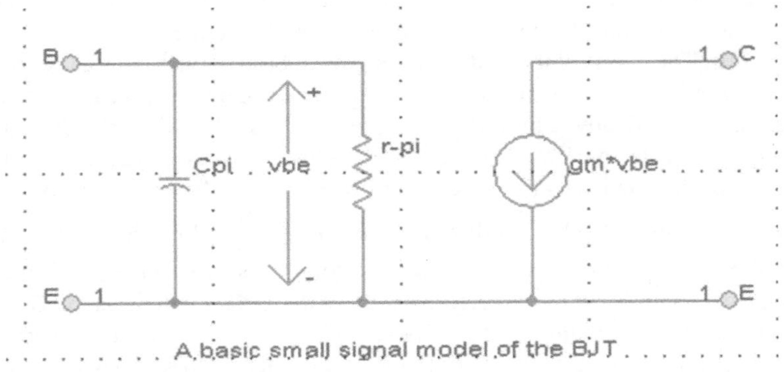

Fig. C.16 Bipolar transistor small signal model

When a signal v_{BE} is applied to the BE junction, a collector current $g_m \cdot v_{BE}$ is produced. g_m, the trans-conductance, is

$$g_m \equiv \frac{dI_C}{dV_{BE}} = \frac{d}{dV_{BE}}\left(I_s \cdot e^{\frac{q \cdot V_{BE}}{k \cdot T}}\right) = \frac{q}{k \cdot T} \cdot I_s \cdot e^{\frac{q \cdot V_{BE}}{k \cdot T}} = I_C / \frac{k \cdot T}{q}; \quad g_m = I_C / \frac{k \cdot T}{q}$$

At room temperature, $g_m = I_C/26\,\text{mV}$. The trans-conductance is determined by the collector bias current, I_C. The input node, the base, appears to the input drive circuit as a parallel RC circuit. $\frac{1}{r_\pi} = \frac{dI_B}{dV_{BE}} = \frac{1}{\beta_F} \cdot \frac{dI_C}{dV_{BE}} = \frac{g_m}{\beta_F}; r_\pi = \frac{\beta_F}{g_m}$. Q_F is the excess carrier charge stored in the BJT. If Q_F = 1 pC, there is +1pC of excess holes and −1pC of excess electrons in the BJT. All the excess hole charge, Q_F, is supplied by the base current, I_B. Therefore, the base presents this capacitance to the input drive circuit: $C_\pi = \frac{dQ_F}{dV_{BE}} = \frac{d}{dV_{BE}}[\tau_F \cdot I_C] = \tau_F \cdot g_m$. The capacitance C_Π may be called the charge storage capacitance, known as the diffusion capacitance. There is one charge component that is not proportional to I_C and therefore cannot be included in Q_F. That is the junction depletion layer charge. Therefore, a complete model of C_Π should include the BE junction depletion layer capacitance, C_{dBE}, $C_\pi = \tau_F \cdot g_m + C_{dBE}$. Once the parameters in the basic small signal model of the BJT have been determined, one can use the small signal model to analyze circuits with arbitrary signal source impedance network which composing resistors, capacitors, and inductors, and additionally load impedance network. r_O is the intrinsic output resistance, V_A/I_C. C_μ also arises from base width modulation; when V_{BC} varies, the base width varies; therefore, the base stored charge varies, thus giving rise to $C_\mu = \frac{dQ_{FB}}{dV_{CB}}$. C_{dBC} is the CB junction depletion layer capacitance. Model parameters are difficult to predict from theory with the accuracy required for commercial circuit design. Therefore, the parameters are routinely determined through comprehensive measurement of the BJT AC and DC characteristics.

Figure C.17 describes the small signal model which can be used to analyze a BJT circuit by hand.

Cutoff frequency: We consider small signal model when the load is a short circuit. The signal source is a current source i_b, at a frequency f. The question is at what frequency the AC current gain does $\beta \equiv i_c/i_b$ fall to unity?

$$v_{be} = \frac{i_b}{\text{input admittance}} = \frac{i_b}{1/r_\pi + j \cdot \omega \cdot C_\pi}; i_c = g_m \cdot v_{be}$$

$$\beta(\omega) \equiv \left|\frac{i_c}{i_b}\right| = \frac{g_m}{|1/r_\pi + j \cdot \omega \cdot C_\pi|} = \frac{1}{|1/g_m \cdot r_\pi + j \cdot \omega \cdot \tau_F + j \cdot \omega \cdot C_{dBE}/g_m|}$$

$$= \frac{1}{|1/\beta_F + j \cdot \omega \cdot \tau_F + j \cdot \omega \cdot C_{dBE} \cdot k \cdot T/q \cdot I_C|}$$

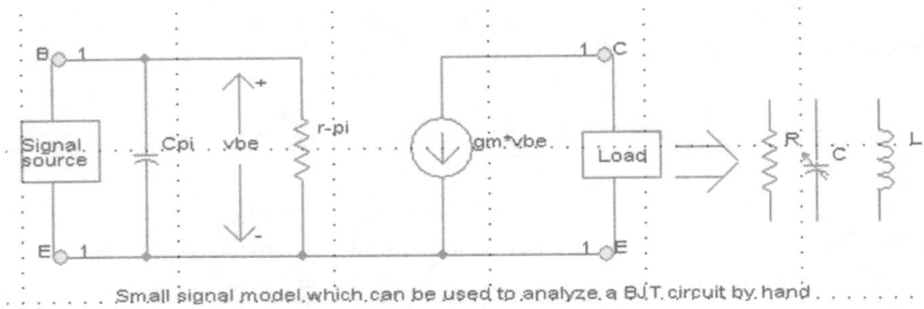

Fig. C.17 Bipolar transistor small signal model which can be used to analyze a BJT circuit by hand

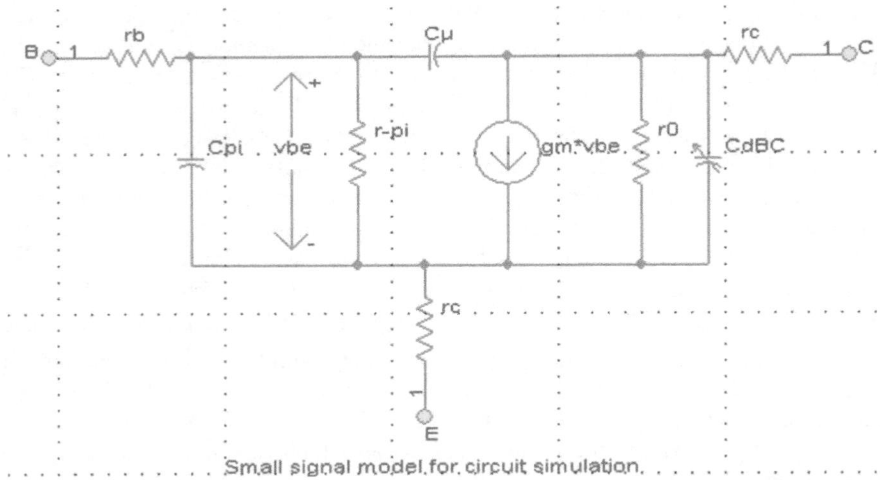

Fig. C.18 Bipolar transistor small signal model for circuit simulation

Figure C.18 describes the small signal model for circuit simulation by computer. At $\omega = 0$, i.e., DC $\beta(\omega) = \ldots$ equation reduces to β_F as expected. As ω increases, β drops. By carefully analyzing the $\beta(\omega)$ data, one can determine τ_F. If $\beta_F \gg 1$ so that $1/\beta_F$ is negligible, $\beta(\omega) \propto \frac{1}{\omega}$ and $\beta = 1$ at f_T, $f_T = \frac{1}{2\cdot\pi\cdot(\tau_F + C_{dBE}\cdot k\cdot T/q\cdot I_C)}$.

If we use a more complete small signal model, it can be shown that $f_T = \frac{1}{2\cdot\pi\cdot[\tau_F + (C_{dBE} + C_{dBC})\cdot k\cdot T/(q\cdot I_C) + C_{dBC}\cdot(r_e + r_c)]}$. f_T is the cutoff frequency and is commonly used to compare the speed of transistors. The above equations predict that f_T rises with increasing I_C due to increasing g_m, in agreement with the measured f_T. At very high I_C, τ_F increases due to base widening (Kirk effect), and therefore, f_T falls. BJTs are often biased near the I_C where f_T peaks in order to obtain the best high frequency performance. F_T is the frequency of unity power gain. The frequency of unity power gain, called the maximum oscillation frequency. $f_{\max} = \sqrt{\left(\frac{f_T}{8 \cdot \pi \cdot r_b \cdot C_{dBC}}\right)}$, it is therefore important to reduce the base resistance, r_b. While MOSFET scaling is motivated by the need for high packing density and large I_{dsat}, BJT scaling is often motivated by the need for high f_T and f_{max}. This involves the reduction of τ_F (thin base, etc,) and the reduction of parasitic $(C_{dBE}, C_{dBC}, r_b, r_e, r_c)$. We interested in BJT with poly-Si emitter, self-aligned base, and deep trench isolation. The base is contacted through two small P^+ regions created by boron diffusion from a P^+ poly-Si film. The film also provides a low resistance electrical connection to the base without introducing a large P^+ junction area and junction capacitance. To minimizing the base series resistance, the emitter opening is made very narrow. The lightly doped epitaxial N-type collector is contacted through a heavily doped sub-collector in order to minimize the collector series resistance. The substrate is lightly doped to minimize the collector capacitance. Both the shallow trench and the deep trench are filled with dielectrics (SiO_2) and serve the function of electrical isolation. The deep trench forms a rectangular moat that completely surrounds the BJT. It isolates the collector of this transistor from the collectors of neighboring transistors. The structure incorporates many improvements that have been developed over the past decades and have greatly reduced the device size from older BJT design. BJT is a larger transistor than a MOSFET.

Bipolar transistor charge control model: The small signal model is ideal for analyzing circuit response to small sinusoidal signals. If the signal is large, input is step function I_B switching from zero to 20 μA or by any $I_B(t)$ and then $I_C(t)$ is produced. The response is analyzed with the charge control model which is a simple extension of the charge storage concept. $I_C = \frac{Q_F}{\tau_F} \Rightarrow I_C(t) = \frac{Q_F(t)}{\tau_F}$, $I_C(t)$ becomes known if we solve for $Q_F(t)$. τ_F has to be characterized beforehand for the BJT being used. I_C is controlled by Q_F (charge control model). At DC condition $I_B = \frac{I_C}{\beta_F} = \frac{Q_F}{\tau_F \cdot \beta_F}$, the equation has a straightforward physical meaning. In order to sustain a constant excess hole charge in the transistor, holes must be supplied to the transistor through I_B to replenish the holes that are lost to recombination. Therefore, DC I_B is proportional to Q_F. When holes are supplied by I_B at the rate of $Q_F/\tau_F \cdot \beta_F$, the rate of hole supply is exactly equal to the rate of hole loss to recombination and Q_F remains at a constant value. In the case that I_B is larger than $Q_F/\tau_F \cdot \beta_F$.

$(I_B > Q_F/\tau_F \cdot \beta_F)$, holes flow into the BJT at a higher rate than the rate of hole loss and the stored hole charge Q_F increases with time $(\frac{dQ_F}{dt} = I_B(t) - \frac{Q_F}{\tau_F \cdot \beta_F})$.

The presented equations together constitute the basic charge control model.

For any given $I_B(t)$, equation $\frac{dQ_F}{dt} = I_B(t) - \frac{Q_F}{\tau_F \cdot \beta_F}$ can be solved for $Q_F(t)$ analytically or by numerical integration. Once $Q_F(t)$ is found, $I_C(t)$ becomes known from equation $I_C(t) = \frac{Q_F(t)}{\tau_F}$. Figure C.19 describes the charge control model. Excess hole charge Q_F rises or falls at the rate of supply current I_B minus loss ($\propto Q_F$).

Q_F is the amount of charges in the vessel, and $\frac{Q_F}{\tau_F \cdot \beta_F}$ is the rate of charge leakage. I_B is the rate of charges flowing into the vessel. The above figure is a basic version of the charge control model. We can introduce the junction depletion layer

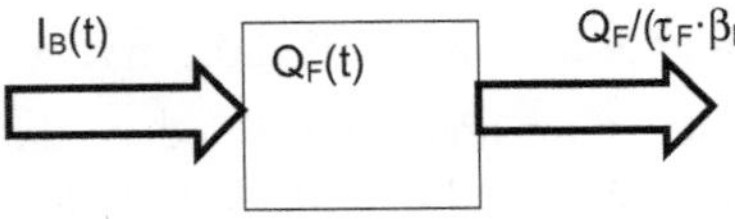

Fig. C.19 Bipolar transistor charge controlmodel

capacitances into equation $\frac{dQ_F}{dt} = I_B(t) - \frac{Q_F}{\tau_F \cdot \beta_F}$. Diverting part of I_B to charge the junction capacitances would produce an additional delay in I_C(t).

Bipolar transistor model for large signal circuit simulation: The BJT model used in circuit simulation can accurately represent the DC and dynamic currents of the transistor in response to V_{BE}(t) and V_{CE}(t). A typical circuit simulation model or compact model is made of the Ebers-Moll model when V_{BE} and V_{BC} are two driving forces for I_C and I_B, plus additional enhancements for high level injection, voltage dependent capacitances that accurately represent the charge storage in the transistor, and parasitic resistances as shown. This BJT model is known as the Gummel-Poon model. The two diodes represent the two I_B terms due to V_{BE} and V_{BC}. The capacitor labeled Q_F is voltage dependent such that the charge stored in it is equal to the Q_F described in the bipolar transistor transit time and charge storage discussion. Q_R is the counterpart of Q_F produced by a forward bias at the BC junction. Inclusion of Q_R makes the dynamic response of the model accurate even when V_{BC} is sometimes forward biased. C_{BE} and C_{BC} are the junction depletion layer capacitances. C_{CS} is the collector to substrate capacitance (Fig. C.20).

$$I_C = I_s' \cdot (e^{\frac{q \cdot V_{BE}}{k \cdot T}} - e^{\frac{q \cdot V_{BC}}{k \cdot T}}) \cdot (1 + \frac{V_{CB}}{V_A}) - \frac{I_s}{\beta_R} \cdot (e^{\frac{q \cdot V_{BC}}{k \cdot T}} - 1).$$

The $1 + \frac{V_{CB}}{V_A}$ factor is added to represent the early effect—I_C increasing with increasing V_{CB}. I_s' differs from I_s in that I_s' decreases at high V_{BE} due to the high level injection effect in accordance with equation

$$G_B \equiv \int_0^{W_B} \frac{n_i^2}{n_{iB}^2} \cdot \frac{p}{D_B} \cdot dx.$$

$$I_B = \frac{I_s}{\beta_F} \cdot (e^{\frac{q \cdot V_{BE}}{k \cdot T}} - 1) + \frac{I_s}{\beta_R} \cdot (e^{\frac{q \cdot V_{BC}}{k \cdot T}} - 1) + I_{SE} \cdot (e^{\frac{q \cdot V_{BE}}{n_E \cdot k \cdot T}} - 1).$$

I_{SE} and n_E parameters are determined from the measured BJT data as are all of the several dozens of model parameters. We can summery the current appendix discussion, that base emitter junction is usually forward biased while the base-collector junction is reverse biased. V_{BE} determines the rate of electron

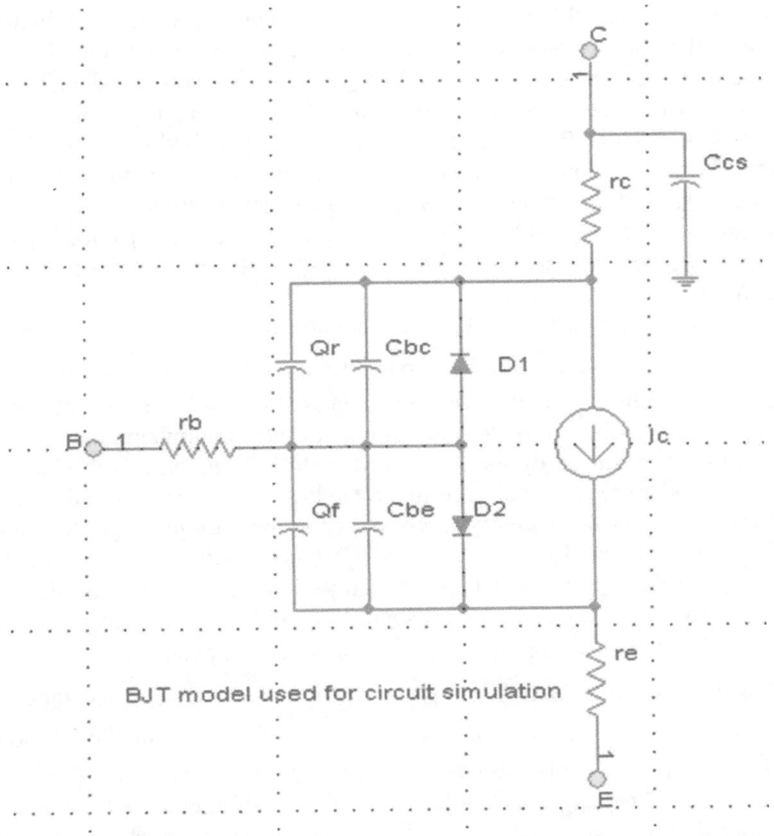

Fig. C.20 Bipolar transistor BJT model used for circuit simulation

injection from the emitter into the base, and thus uniquely determines the collector current, I_C regardless of the reverse bias V_{CB}.

$$I_C = A_E \cdot \frac{q \cdot n_i^2}{G_B} \cdot \left(e^{\frac{q \cdot V_{BE}}{k \cdot T}} - 1\right); \quad G_B \equiv \int_0^{W_B} \frac{n_i^2}{n_{iB}^2} \cdot \frac{p}{D_B} \cdot dx$$

G_B is the base Gummel number, which represents all the subtleties of BJT design that affects I_C; base material, non-uniform base doping, non-uniform material composition, and the high level injection effect. An undesirable but unavoidable side effect of the application on V_{BE} is a hole current flowing from the base, mostly into the emitter. This base input current, I_B, is related to I_C by the common emitter current gain β_F ($\beta_F = \frac{I_C}{I_B} \approx \frac{G_E}{G_B}$) where G_E is the emitter Gummel number. The common base current gain is $\alpha_F \equiv \frac{I_C}{I_E} = \frac{\beta_F}{1+\beta_F}$. The Gummel plot indicates that β_F falls off in the high I_C region due to high level injection in the base and also in the low I_C region due to excess base current. Base width modulation by V_{CB} results in a significant slope of the I_C-V_{CE}curve in the active region. This is the early effect. The slope, called the output conductance, limits the voltage gain that can be produced with a BJT. The early effect can be suppressed with a lightly doped collector. A heavily doped sub-collector is routinely used to reduce the collector resistance. Due to the forward bias, V_{BE}, a BJT stores a certain amount of excess hole charge, which is equal but of opposite sign to the excess electron charge. Its magnitude is called the excess carrier charge, Q_F. Q_F is linearly proportional to I_C $(Q_F \equiv I_C \cdot \tau_F)$.

τ_F is the forward transit time. If there were no excess carriers stored outside the base $\tau_F = \tau_{FB} = \frac{W_B^2}{2 \cdot D_B}$. τ_{FB} is the base transit time, $\tau_F > \tau_{FB}$ because excess carrier storage in the emitter and in the depletion layer is also significant. All these regions should be made small in order to minimize τ_F. Besides minimizing the base width, W_B, τ_{FB} may be reduced by building a drift field into the base with graded base doping (also with graded Ge content in a SiGe base). τ_{FB} is significantly increased at large I_C due to base widening which known at the Kirk effect. In the Gummel Poon model, both the DC and the dynamic (charge storage) currents are well modeled. The early effect and high level injection effect are included. Simpler models consisting of R, C, and current source are used for hand analysis of circuits. The small signal models employ parameters such as trans-conductance $g_m = \frac{dI_C}{dV_{BE}} = I_C / \frac{k \cdot T}{q}$ and input capacitance $C_\pi = \frac{dQ_F}{dV_{BE}} = \frac{d}{dV_{BE}}[\tau_F \cdot I_C] = \tau_F \cdot g_m$ and input resistance $r_\pi = \frac{dV_{BE}}{dI_B} = \frac{\beta_F}{g_m}$. The BJT's unity gain cutoff frequency at which β falls to unity is f_T. In order to raise device speed, device density, or current gain, a modern high performance BJT usually employs poly-Si emitter, self-aligned poly-Si base contacts, graded Si-Ge base, shallow oxide trench, and deep trench isolation. High performance BJTs excel over MOSFETs in circuits requiring the highest device g_m and speed.

Gummel-Pooncharge control model: The Gummel-Poon model is a detailed charge-controlled model of BJT dynamics, which has been adopted and elaborated by others to explain transistor dynamics in greater detail than the terminal-based models typically do. This model also includes the dependence of transistor β values upon the DC current levels in the transistor, which are assumed current independent in the Ebers-Moll model. A significant effect included in the Gummel-Poon model is the DC current variation of the transistor β_F and β_R. When certain parameters are omitted, the Gummel-Poon model reverts to the simpler Ebers-Moll model. The basic circuit which describes the Gummel-Poon model is the large signal schematic. It represents the physical transistor, a current-controlled output current sink, and two diode structures including their capacitors (Fig. C.21).

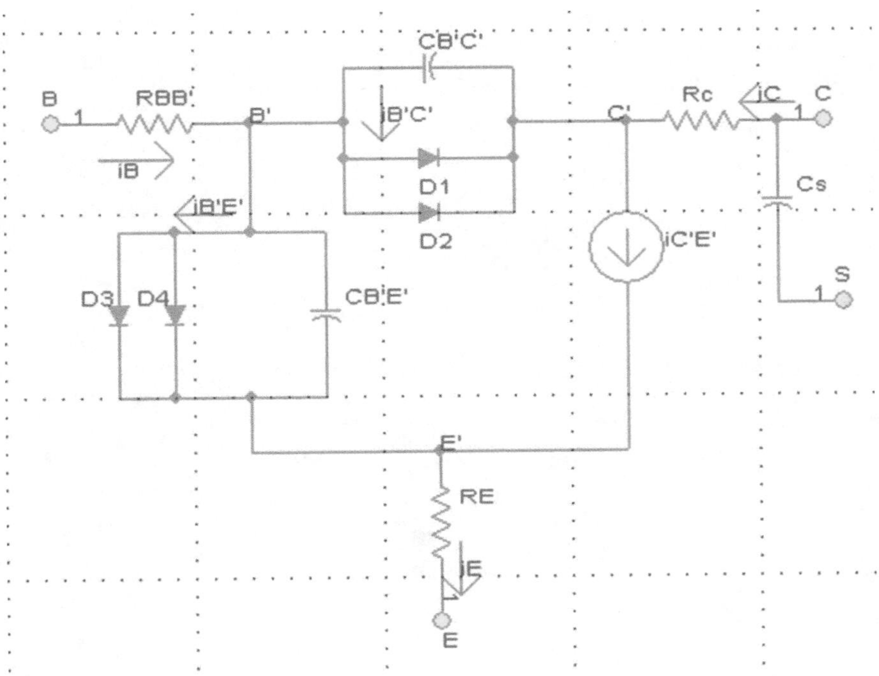

Fig. C.21 Gummel-Poon large signal schematic of the BJT

We can derive from the Gummel-Poon large signal schematics of the bipolar transistor the Small Signal (SS) schematics for high frequency simulations. This mean, for a given operating point, the DC currents are calculated and the model is linearized in this point. The schematic is a pure linear model. The AC Small Signal (SS) schematic of bipolar transistor is described in Fig. C.22.

For simplicity we assume no voltage drops at $R_{BB'}$, R_c, R_E then $V_{B'E'} = V_{BE}$, $V_{B'C'} = V_{BC}$, $V_{C'E'} = V_{CE}$.

The Gummel-Poon BJT full model analysis: The Gummel-Poon model is a compact model for bipolar junction transistors (BJT) which also takes into account effects of low currents and at high level injection (Fig. C.23).

Remark: we consider no voltage drop on $R_{BB'}$, R_c, R_E ($V_{EE'} \rightarrow \varepsilon$; $V_{CC'} \rightarrow \varepsilon$; $V_{BB'} \rightarrow \varepsilon$).

G_{min} is the minimum conductance which is automatically switched in parallel to each PN junction.

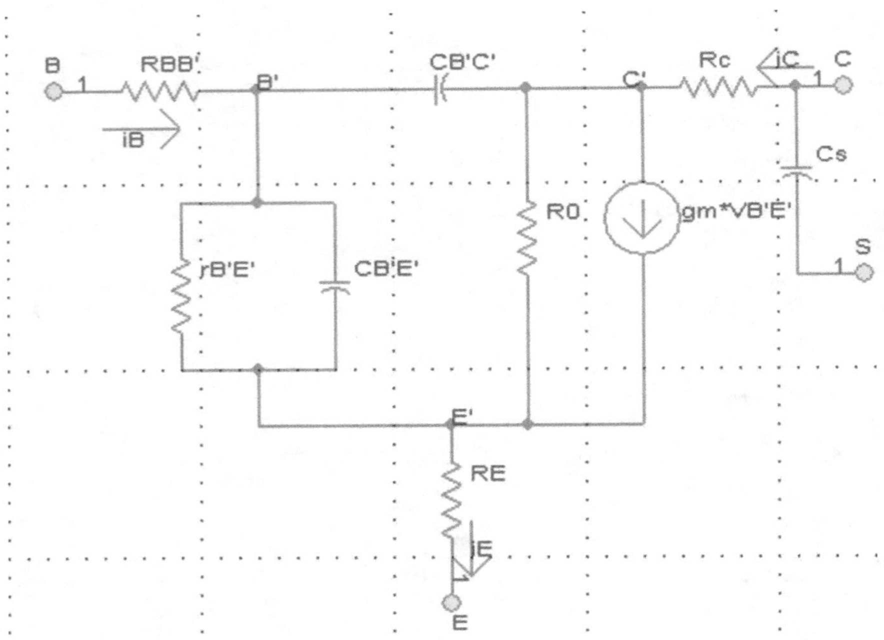

Fig. C.22 AC small signal schematic of the bipolar transistor

$$I_{cc} = \frac{I_{ss}}{q_b} \cdot (e^{\frac{q \cdot V_{be}}{n_f \cdot k_B \cdot T}} - 1); I_{ec} = \frac{I_{ss}}{q_b} \cdot (e^{\frac{q \cdot V_{bc}}{n_r \cdot k_B \cdot T}} - 1); I_{ct} = I_{cc} - I_{ec} = \frac{I_{ss}}{q_b} \cdot (e^{\frac{q \cdot V_{be}}{n_f \cdot k_B \cdot T}} - e^{\frac{q \cdot V_{bc}}{n_r \cdot k_B \cdot T}})$$

$$I_{D_1} = \frac{I_{ec}}{\beta_r} = \frac{I_{ss}}{q_b \cdot \beta_r} \cdot (e^{\frac{q \cdot V_{bc}}{n_r \cdot k_B \cdot T}} - 1); I_{D_3} = \frac{I_{cc}}{\beta_f} = \frac{I_{ss}}{q_b \cdot \beta_f} \cdot (e^{\frac{q \cdot V_{be}}{n_f \cdot k_B \cdot T}} - 1)$$

$$I_{D_2} = C_4 \cdot I_{ss} \cdot (e^{\frac{q \cdot V_{bc}}{n_{cl} \cdot k_B \cdot T}} - 1); I_{D_4} = C_2 \cdot I_{ss} \cdot (e^{\frac{q \cdot V_{be}}{n_{el} \cdot k_B \cdot T}} - 1)$$

The model distinguishes four operating region: normal active region, inverse region, saturated region, and off region.

Normal active region:

$$V_{be} > -\frac{5 \cdot n_f \cdot k_B \cdot T}{q}; \quad V_{bc} \leq -\frac{5 \cdot n_r \cdot k_B \cdot T}{q}$$

$$I_c = \frac{I_s}{q_b} \cdot [\exp(\frac{q \cdot V_{be}}{n_f \cdot k_B \cdot T}) + \frac{q_b}{\beta_r}] + C_4 \cdot I_s + [\frac{V_{be}}{q_b} - (\frac{1}{q_b} + \frac{1}{\beta_r}) \cdot V_{bc}] \cdot G_{\min}$$

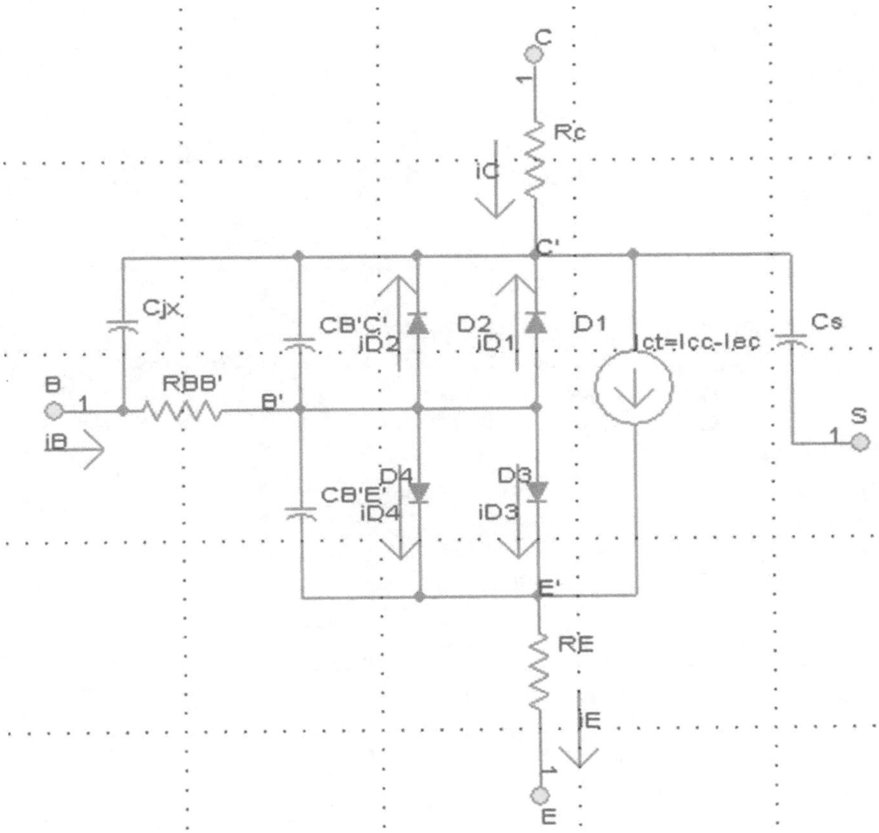

Fig. C.23 Gummel-Poon BJT full model

$$I_b = I_s \cdot [\frac{1}{\beta_f} \cdot [\exp(\frac{q \cdot V_{be}}{n_f \cdot k_B \cdot T}) - 1] - \frac{1}{\beta_r}] + C_2 \cdot I_s \cdot [\exp(\frac{q \cdot V_{be}}{n_{el} \cdot k_B \cdot T}) - 1]$$
$$- C_4 \cdot I_s + (\frac{V_{be}}{\beta_f} + \frac{V_{bc}}{\beta_r}) \cdot G_{\min}$$

<u>Inverse region</u>: $V_{be} \leq -\frac{5 \cdot n_f \cdot k_B \cdot T}{q}$; $V_{bc} > -\frac{5 \cdot n_r \cdot k_B \cdot T}{q}$

$$I_c = -\frac{I_s}{q_b} \cdot \{\exp(\frac{q \cdot V_{bc}}{n_r \cdot k_B \cdot T}) + \frac{q_b}{\beta_r} \cdot [\exp(\frac{q \cdot V_{bc}}{n_r \cdot k_B \cdot T}) - 1]\}$$
$$- C_4 \cdot I_s \cdot [\exp(\frac{q \cdot V_{bc}}{n_{cl} \cdot k_B}) - 1] + [\frac{V_{be}}{q_b} - (\frac{1}{q_b} + \frac{1}{\beta_r}) \cdot V_{bc}] \cdot G_{\min}$$

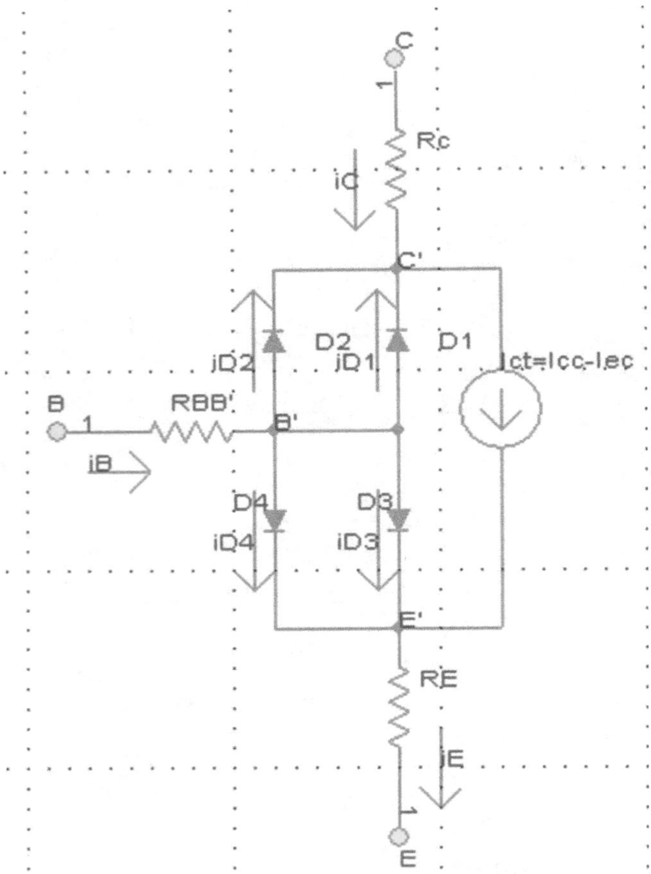

Fig. C.24 BJT NPN Gummel-Poon static model

$$I_b = I_s \cdot [\frac{1}{\beta_f} \cdot [\exp(\frac{q \cdot V_{be}}{n_f \cdot k_B \cdot T}) - 1] - \frac{1}{\beta_r}] + C_2 \cdot I_s \cdot [\exp(\frac{q \cdot V_{be}}{n_{el} \cdot k_B \cdot T}) - 1]$$
$$- C_4 \cdot I_s + (\frac{V_{be}}{\beta_f} + \frac{V_{bc}}{\beta_r}) \cdot G_{\min}$$

Saturated region: $V_{be} > -\frac{5 \cdot n_f \cdot k_B \cdot T}{q}$; $V_{bc} > -\frac{5 \cdot n_r \cdot k_B \cdot T}{q}$

$$I_c = \frac{I_s}{q_b} \cdot \{[\exp(\frac{q \cdot V_{be}}{n_f \cdot k_B \cdot T}) - \exp(\frac{q \cdot V_{bc}}{n_r \cdot k_B \cdot T})] - \frac{q_b}{\beta_r} \cdot [\exp(\frac{q \cdot V_{bc}}{n_r \cdot k_B \cdot T}) - 1]\}$$
$$- C_4 \cdot I_s \cdot [\exp(\frac{q \cdot V_{bc}}{n_{cl} \cdot k_B}) - 1] + [\frac{V_{be}}{q_b} - (\frac{1}{q_b} + \frac{1}{\beta_r}) \cdot V_{bc}] \cdot G_{\min}$$

$$I_b = I_s \cdot \{\frac{1}{\beta_f} \cdot [\exp(\frac{q \cdot V_{be}}{n_f \cdot k_B \cdot T}) - 1] + \frac{1}{\beta_r} \cdot [\exp(\frac{q \cdot V_{bc}}{n_r \cdot k_B \cdot T}) - 1]\}$$

$$+ C_2 \cdot I_s \cdot [\exp(\frac{q \cdot V_{be}}{n_{el} \cdot k_B \cdot t}) - 1] + C_4 \cdot I_s \cdot [\exp(\frac{q \cdot V_{bc}}{n_{cl} \cdot k_B \cdot T}) - 1] + (\frac{V_{be}}{\beta_f} + \frac{V_{bc}}{\beta_r}) \cdot G_{\min}$$

Off region: $V_{be} \leq -\frac{5 \cdot n_f \cdot k_B \cdot T}{q}$; $V_{bc} \leq -\frac{5 \cdot n_r \cdot k_B \cdot T}{q}$

$$I_c = \frac{I_s}{\beta_r} + C_4 \cdot I_s + [\frac{V_{be}}{q_b} - (\frac{1}{q_b} + \frac{1}{\beta_r}) \cdot V_{bc}] \cdot G_{\min};$$

$$I_b = -I_s \cdot [\frac{\beta_f + \beta_r}{\beta_f \cdot \beta_r}] - (C_2 + C_4) \cdot I_s + (\frac{V_{be}}{\beta_f} + \frac{V_{bc}}{\beta_r}) \cdot G_{\min}$$

By disconnecting all capacitor in Gummel-Poon BJT full model we get the NPN Gummel-Poon static model (Fig. C.24).

References

1. Kawdungta, S., Phongcharoenpanich, C., & Torrungrueng, D. (2008). *Novel design of double loop antennas by using a shifted Gate for the LF–RFID system*. Asian University, Thailand: Faculty of engineering and Technology.
2. Kuznetsov, Y. A. (1995). Elements of applied bifurcation theory. *Applied Mathematical Sciences*.
3. Hale, J. K. (2012). Dynamics and bifurcations. *Texts in Applied Mathematics, 3*.
4. Strogatz, S. H. (2014). *Nonlinear dynamics and chaos*. Boulder: Westview press.
5. Kuang, Y. (1993). *Delay differential equations with applications in population dynamics*. Boston: Academic Press.
6. Beretta, E., & Kuang, Y. (2002). Geometric stability switch criteria in delay differential systems with delay dependent parameters. *SIAM Journal on Mathematical Analysis, 33*, 1144–1165.
7. Aluf, O. (2008, May). RFID TAGs COIL's dimensional parameters optimization as excitable linear bifurcation systems. In *IEEE COMCAS2008 Conference*.
8. Aluf, O. (2011, November). RFID TAGs coil's system stability optimization under delayed electromagnetic interferences. In *IEEE COMCAS2011 Conference*.
9. Guckenheimer, J. (2002). Nonlinear oscillations, dynamical systems, and bifurcations of vector fields. *Applied Mathematical Sciences, 42*.
10. Wiggins, S. (2003). Introduction to applied nonlinear dynamical systems and chaos. *Text in Applied Mathematics* (Hardcover).
11. Haug, H., & Koch, S. W. (2008). *Quantum theory of the optical and electronic properties of semiconductors* (5th ed.). Singapore: World Scientific.
12. Kuang, J., & Cong, Y. (2007). *Stability of numerical methods for delay differential equations*. Amsterdam: Elsevier Science.
13. Balachandran, B., & Kalmár-Nagy, T., & Gilsinn, D. E. (2009, March 5). *Delay differential equations: Recent advances and new directions* (1st ed.) (Hardcover). Heidelberg: Springer.
14. Sze, S. M., & Ng, K. K. (2006). *Physics of semiconductor devices*. New York: Wiley Interscience (Hardcover—Oct 27).
15. Sah, C. T. *Fundamentals of solid state electronics*. Singapore: World Scientific.
16. Yang, W. Y., Cao, W., Chung, T.-S., & Morris, J. (2005). *Applied numerical methods using MATLAB*. New York: Wiley.
17. Band, A., & Unguris, J. *Optically isolated current-to-voltage converter for an electron optics system*. Electron physics group National Institute of Standards and technology Gaithersburg, Maryland 20899.
18. Beretta, E., & Kuang, Y. (2002). Geometric stability switch criteria in delay differential systems with delay dependent parameters. *SIAM Journal on Mathematical Analysis, 33*(5), 1144–1165 (Published electronically February 14, 2002).
19. Kuang, Y. (1993). Delay differential equations with applications in population dynamics. *Mathematics in Science and Engineering, 191*.

O. Aluf, *Microwave RF Antennas and Circuits*,
DOI 10.1007/978-3-319-45427-6

20. Flower, A. C., & Mcguinnless, M. J. (1982). A description of the Lorenz attractor at high prandtl number. *Physica 5D*, 149–182.
21. Dullin, H. R., Schmidt, S., Richter, P. H., & Grossmann, S. K. (2005, April 12). *Extended phase diagram of the Lorenz model*.
22. Sprott, J. C. Simplifications of the Lorenz attractor. *Nonlinear Dynamics Psychology and Life Sciences, 13*(3), 271–278.
23. van der Schrier, G., & Mass, L. R. M. (2000). The diffusionless Lorenz equations; Shil'nikov bifurcations and reduction to an explicit map. *Physica D, 141*, 19–36.
24. Marton, L. (1974). *Advanced electronics and electron physics* (Vol. 35). USA: Academic Press Inc. Elsevier.
25. Cripps, S. C. (2006). *RF power amplifiers for wireless communication*. London: Artech House microwave library.
26. Sayre, C. W. *Complete wireless design*. New York: McGraw-Hill companies.
27. Gopalsamy, K. *Stability and oscillation in delay differential equations of population dynamics*. Berlin: Kluwer Academic publishers.
28. Kuznetsov, Y. (2010). *Elements of applied bifurcation theory [Paperback]*. NY, LLC: Springer.
29. Butcher, E. A., Ma, H., Bueler, E., Averina, V., & Szabo, Z. (2004). Stability of linear time periodic delay differential equations via Chebyshev polynomials. *International Journal of Numerical Methods Engineering, 59*, 895–922.
30. Stepan, G., Insperger, T., & Szalai, R. (2005). Delay, parametric excitation, and the nonlinear dynamics of cutting process. *International Journal of Bifurcation and Chaos, 15* (9), 2783–2798.
31. Insperger, T., & Stepan, G. (2003). Stability of the damped Mathieu equation with time delay. *Journal of Dynamic Systems, Measurement, and Control, 125*.
32. Garg, N. K., Mann, B. P., Kim, N. H., & Kurdi, M. H. (2007, March). Stability of a time delayed system with parametric excitation. *Journal of Dynamic Systems, Measurement, and Control, 129*.
33. Allan, W. S. (2005). *Understanding microwaves*. New York: Wiley.
34. Rogers, J. W. M., Plett, C., & Marsland, I. (2014). *Radio frequency system architecture and design*. London: Artech house.
35. Linkhart, D. K. (2014). *Microwave circulator design* (2nd ed.). London: Artech house.
36. Tanaka, S., Shimomura, N., & Ohtake, K. (1965, March). Active circulators—The realization of circulators using transistors. *Proceedings of the IEEE*, 260–267.
37. Dougherty, R. (1989, June). Circulate signals with active devices on monolithic chips. *Microwave and RF*, 85–86, 89.
38. Looss, G., & Joseph, D. D. (1980). *Elementary stability and bifurcation theory*. Berlin: Springer.
39. Guckenheimer, J., & Holmes, P. (1983). Nonlinear oscillations, dynamical systems, and bifurcation of vector fields. *Applied Mathematical Sciences, 42*.
40. Perko, L. (1991). Differential equations and dynamics systems. *Texts in Applied Mathematics, 7*.
41. Kuang, J., & Cong, Y. (2005). *Stability of numerical methods for delay differential equations*. USA: Since press USA Inc.
42. Hollenstein, C., Guittienne, p., & Howling, A. A. Resonant RF network antennas for large-area and large-volume inductively coupled plasma sources.
43. Granas, A., & Dugundji, J. (2003, July 24). *Fixed point theory*. Berlin: Springer.
44. Border, K. C. (1989, July 28). *Fixed point theorems with application to economics and game theory*. Cambridge: Cambridge University Press (Rep sub edition).
45. Agarwal, R. P., Meehan, M., &O'Regan, D. (2009, March 19). *Fixed point theory and applications* (1st ed.). Cambridge: Cambridge University Press.
46. Istratescu, V. I. (2001, November 30). *Fixed point theory: An introduction*. Berlin: Springer.

47. Geller, S. B., & Mantek, P. A. (1962, January–March). Tunnel diode large-signal equivalent circuit study and the solutions of its nonlinear differential equations. *Journal of research of the National Bureau of Standards—C. Engineering and Instrumentation, 66C* (1).
48. Hines, M. E. (1960). High-frequency negative-resistance circuit principles for Esaki diode applications. *Bell System Technical Journal, 39*, 477.
49. Cohen, S. (1960). Tunnel diode characterization. *Electric Equipment & Engineering, 8*, 102.
50. Lowry, H. R., Giorgis, J., Gottlieb, E., & Weischedel, R. C. (1961). Tunnel diode manual. *General Electric*, 33–42.
51. Lefshetz, S. (1957). *Differential equations: Geometric theory* (p. 261). Geneva: Interscience Publishers.
52. Hsia, P. S. (1952). A graphical analysis for nonlinear systems. *Proceeding of IEEE, 99*, 125.
53. Crisson, G. (1931). Negative impedances in the twin 21 type repeater. *Bell System Technical Journal, 10*, 485.
54. Gautam, A. K., & Vishvakarma, B. R. (2006). Frequency agile microstrip antenna using symmetrically loaded tunnel diodes. *Indian Journal of Radio & Space physics, 35*, 212–216.
55. Srivastava, S., & Vishvakarma, B. R. (1999). Tunnel diode integrated rectangular patch antenna. In *Proceeding of the Radar Symposium India-99 (IRSI-99), Bangalore.*
56. Srivastava, S., & Vishvakarma, B. R. (2003). Tunnel diode integrated rectangular microstrip antenna for millimeter range. *IEEE Transactions on Antennas Propagation (USA), 51*, 750.
57. Woo, C. F. (1964). *Principles of tunnel diode circuits*. New York: Wiley.
58. Sylvesten, G. P. (1962). *Basic theory and applications of tunnel diode.* Princeton, New Jersey, USA: Van nostrand.
59. Soliman, F. A. S, & Kamh, S. A. (1993). Computer analysis for designing narrow band tunnel diode amplifier circuit. *Communication Faculty of Science University of Ankara Series A_2, A_3, 42*, 33–49.
60. Boric-Lubecke, O., Pan, D. S., & Itoh, T. (1995). Design and triggering of oscillators with a series connection of tunneling diodes. *Electronics and Engineering, 8*(2), 271–286.
61. Brown, E. R., Soderstrom, J. R., Paker, C. D., Mahoney, L. J., Molvar, K. M., & McGill, T. C. (1991). Oscillations up to 712 GHz in InAs/AISb resonant tunneling diodes. *Applied Physics Letters, 58*(20), 20.
62. Menozzi, R., Piazzi, A., & Contini, F. (1996). Small-signal modeling for microwave FET linear circuits based on a genetic algorithm. *IEEE Transaction on Circuits and Systems—I: Fundamental Theory and Applications, 43*(10).
63. Jerinic, G., Fines, J., Cobb, M., & Schindler, M. (1985). Ka/Q band GaAs IMPATT amplifier technology. *International Journal of Infrared and Millimeter Waves, 6*(2), 79–130.
64. Mishra, L. P., & Mitra, M. (2015). Design and characterization of Ka-Band reflection-type IMPATT amplifier. *Intelligent Computing, Communication and Devices Advanced Intelligent Systems and Computing, 308*, 487–492.
65. Al-Attar, T., & Lee, T. H. (2005). Monolithic integrated millimeter wave IMPATT transmitter in standard CMOS technology. *IEEE Transaction on Microwave Theory and Techniques, 53*(11).
66. Gupta, M. S., & Lomax, R. J. (1971). A self-consistent large-signal analysis of read-type IMPATT diode oscillator. *IEEE Transactions on Electron Devices, ED-18*(8), 544–550.

67. Gupta, M. S., & Lomax, R. J. (1971). Injection locking in IMPATT diode oscillators. In *Proceedings of the Third Biennial Cornell Electrical Engineering Conference* (pp. 215–223). Ithaca, New York: School of Electrical Engineering, Cornell University. Reprinted in: Avalanche Transit-Time Devices, G.I. Haddad, editor, Artech House, Dedham, Mass. 1973, pp. 457–460.
68. Gupta, M. S., & Lomax, R. J. (1973). A current-excited large-signal analysis of IMPATT devices and its circuit implications. *IEEE Transactions on Electron Devices, ED-20*(4), 395–399.
69. Gupta, M. S. (1973). Computer-aided characterization of IMPATT diodes. In *Proceedings of the Fourth Biennial Cornell Electrical Engineering Conference* (pp. 349–358). Ithaca, New York: School of Electrical Engineering, Cornell University.
70. Gupta, M. S. (1973). A small-signal and noise equivalent circuit for IMPATT diodes. *IEEE Transactions on Microwave Theory and Techniques, MTT-21*(9), 591–594.
71. Gupta, M. S. (1973). Large-signal equivalent circuit for IMPATT diode characterization and its application to amplifiers. *IEEE Transactions on Microwave Theory and Techniques, MTT-21*(11), 689–694.
72. Gupta, M. S. Lomax, R. J., & Haddad, G. I. (1974). Noise considerations in self-mixing IMPATT-diode oscillators for short-range doppler radar applications. *IEEE Transactions on Microwave Theory and Techniques, MTT-22*(1), 37–43.
73. Gupta, M. S. (1975). A simple approximate method of estimating the effect of carrier diffusion in IMPATT diodes. *Solid-State Electronics, 18*(4), 327–330.
74. Gupta, M. S. (1976). A nonlinear equivalent circuit for IMPATT diodes. *Solid-State Electronics, 19*(1), 23–26.
75. Atherton, D. P. (1981). *Stability of nonlinear system (Electronic & Electrical engineering research studies)*. New York: Wiley.
76. Xue, D., Chen, Y. Q., & Atherton, D. P. (2009). Linear feedback control: Analysis and design with MATLAB (advanced in design and control). *Society for industrial and applied mathematics* (1st ed.).
77. Atherton, D. P. (1982). *Nonlinear control engineering*. London: Chapman & Hall (stu sub edition).
78. Bar-yam, Y. (1997). *Dynamics of complex systems (studies in nonlinearity)* (1st ed.). Boulder: Westview press.
79. Steeb, W. H. (2014). *The nonlinear workbook: Chaos, fractals, cellular automata, generic algorithms, gene expression programming, support vector machine, wavelets, hidden... Java and symbolic C++ programs* (6th ed.). Singapore: World scientific.
80. Sternberg, S. (2012). *Curvature in mathematics and physics*. New York: Dover publications.
81. Sternberg, S. (2010). *Dynamical systems*. New York: Dover publications.
82. Scheinerman, E. R. (2012, January 18). *Invitation to dynamical systems*. New York: Dover publications (Reprint edition).
83. Abraha, R. (1983). *Dynamics the geometry of behavior part 2: Chaotic behavior (Visual mathematics library)*. Aerial Press.
84. Abraham, R. (1982). *Dynamics the geometry of behavior: Periodic behavior (Visual mathematics library)*. Aerial Press.
85. Ahson, S., & Ilyas, M. (2008). *RFID handbook, applications, technology, security, and privacy*. Boca Raton: CRC press, Taylor & Francis Group.
86. Aluf, O. (2014). Cylindrical RF network antennas for coupled plasma sources copper legs delayed in time system stability. *Transaction on networks and communications*, *2*(5), 116–146.
87. Cheban, D. N. (2015). *Global attractors of non-autonomous dynamical and control systems* (2nd ed.). Singapore: World scientific.
88. Li, C., Wu, Y., & Ye, R. (2013). *Recent advanced in applied nonlinear dynamics with numerical analysis*. Singapore: World scientific.

89. Sidorov, D. (2014). *Integral dynamical models*. Singapore: World scientific.
90. Pozar, D. M. (2005). *Microwave engineering* (3rd ed.). New York: Wiley.
91. Collin, R. E. (1992). *Foundation for microwave engineering* (2nd ed.). New York: McGraw-Hill.
92. Ludwig, R., & Bretchko, P. (2000). *RF circuit design—Theory and applications*. Upper Saddle River: Prentice-Hall.
93. Gonzalez, G. (1997). *Microwave transistor amplifiers—Analysis and design* (2nd ed.). Upper Saddle River: Prentice-Hall.
94. Vendelin, G. D., Pavio, A. M., & Rhode, U. L. (1990). *Microwave circuit design—using linear and nonlinear techniques*. New York: Wiley.
95. Gilmore, R., & Besser, L. (2003). *Practical RF circuit design for modem wireless System* (Vols. 1–2). Norwood: Artech House.
96. Iu, H. H. C., & Fitch, A. L. (2013). *Development of memristor based circuit*. Singapore: World Scientific.
97. Ling, B. W. K., & Iu, H. H. C. (2008). *Control of chaos in nonlinear circuits and systems*. Singapore: World Scientific.
98. Abarbanel, H. D. I., Rabinovich, M. I., & Sushchik, M. M. (1993). *Introduction to nonlinear dynamics for physicists*. Singapore: World Scientific.
99. Kilic, R. (2010). *A practical guide for studying chua's circuits*. Singapore: World Scientific.
100. Qiying, W. (2015). *Limit theorems for nonlinear cointegrating regression*. Singapore: World Scientific.
101. Tatarinova, T., & Schumitzky, A. (2015). *Nonlinear mixture models*. Singapore: World Sientific.
102. Blake, L. M., & Long, M. W. (2009). *Antennas: Fundametals, design, measurement*. USA: SciTech publishing Inc.
103. Milligon, T. A. (2005). *Modern antenna design*. New York: Wiley.
104. Hansen, R. C., & Collin, R. E. (2011). *Small antenna handbook*. New York: Wiley.
105. Volakis, J. L., Chen, C. C., & Fujimoto, K. (2010). *Small antennas: Miniaturization techniques & applications*. New York: Mc Graw Hill education.
106. Kumar, N., & Grebennikov, A. (2015). *Distributed power amplifiers for RF and microwave communications*. Norwood: Artech House.
107. Campos, R. S., & Lovisolo, L. (2015). RF positioning: Fundamentals, applications, and tools. Norwood: Artech House.
108. Wallace, R., & Andreasson, K. (2015). *Introduction to RF and microwave passive components*. Norwood: Artech House.
109. Gildenblat, G. (2010). *Compact modeling: Principles, techniques and applications*. Berlin: Springer.
110. Diks, C. (1999). *Nonlinear time series analysis*. Singapore: World Scientific.
111. Letellier, C., & Gilmore, R. (2013). *Topology and dynamics of chaos*. Singapore: World Scientific.
112. Wang, Q. (2015). *Limit theorems for nonlinear cointegrating regression*. Singapore: World Scientific.
113. Sibani, P., & Jensen, H. J. (2013). *Stochastic dynamics of complex systems*. Singapore: World Scientific.
114. Sprott, J. C. (2010). *Elegant chaos*. Singapore: World Scientific.
115. Elhadj, Z., & Sprott, J. C. (2011a). *Robust chaos and its applications*. Singapore: World Scientific.
116. Elhadj, Z., & Sprott, J. C. (2011b). *Frontiers in the study of chaotic dynamical systems with open problems*. Singapore: World Scientific.
117. Hoover, W. G. & Hoover, C. G. (2015). *Simulation and control of chaotic nonequilibrium systems*. Singapore: World Scientific.

118. Ivancevic, V. G., & Reid, D. J. (2014). *Complexity and control*. Singapore: World Scientific.
119. Nicolis, G., & Nicolis, C. (2012). *Foundations of complex systems*. Singapore: World Scientific.
120. Cencini, M., Cecconi, F., & Vulpiani, A. (2009). *Chaos*. Singapore: World Scientific.
121. Salleh, A., Abd aziz, M. Z. A., Misran, M. H., Othman, M. A., & Mohamad, N. R. (2013). *Design of wideband low noise amplifier using negative feedback topology for Motorola application* (Vol. 5, No. 1). ISSN: 2180-1843.
122. Duan, L., Huang, W., Ma, C., He, X., Jin, Y., & Ye, T. (2012). A single to differential low noise amplifier with low differential output imbalance. *Journal of Semiconductors, 33*(3), 035002.
123. Kuang, Y., & Cushing, J. M. (1996). Global stability in a nonlinear difference delay equation model of flour beetle population growth. *Journal of Difference Equations and Applications*, *2*, 31–37.
124. Elsayed, E. M., El-Dessoky, M. M., & Alotaibi, A. (2012). On the solutions of a general system of difference equations, Hindawi Publishing Corporation. *Discrete Dynamics in Nature and Society, 2012*(892571), 12.
125. Desoer, C. A. (1969). *Basic circuit theory* (1st ed.). New York: Mcgraw-Hill College.
126. Grover, F. W. (2004). Inductance calculations: Working formulas and tables. New York: Dover publications.
127. Greenhouse, H. M. (1974). Design of planar rectangular microelectronic inductors. *IEEE Transactions on Parts, Hybrids, and Packaging*, *PHP-10*(2).
128. Rodriguez, E. G. (2015). *Reconfigurable transceiver architecture for multiband RF frontends*. Berlin: Springer.
129. Steinberg, K., Scheffler, M., & Dressel, M. (2010). Microwave inductance of thin metal strips. *Journal of Applied Physics, 108*, 096102.
130. Liao, Y., Cai, K., Hubing, T. H., & Wang, X. (2014). Equivalent circuit of normal mode helical antennas using frequency—Independent lumped elements. *IEEE Transactions on Antennas and Propagation*, *62*, 5885–5888.
131. Su, C., Ke, H., & Hubing, T. H. (2010). A simplified model for normal mode helical antennas. *Applied Computational Electromagnetics Society Journal*, *25*(1), 32–40.
132. Su, C., Ke, H., & Hubing, T. H. (2010). Corrections to a simplified model for normal mode helical antennas. *Applied Computational Electromagnetics Society Journal*, *25*(8), 722.

Index

O. Aluf, *Microwave RF Antennas and Circuits*,
DOI 10.1007/978-3-319-45427-6

CPSIA information can be obtained
at www.ICGtesting.com
Printed in the USA
LVHW060201081118
596388LV00001B/32/P

9 783319 832913